Fundamental Skills for Biosciences

PEARSON

Custom Publishing

We work with leading authors to develop the strongest educational materials bringing cutting-edge thinking and best learning practice to a global market.

Under a range of well-known imprints, including Financial Times/Prentice Hall, Addison Wesley and Longman, we craft high quality print and electronic publications which help readers to understand and apply their content, whether studying or at work.

Pearson Custom Publishing enables our customers to access a wide and expanding range of market-leading content from world-renowned authors and develop their own tailor-made book. You choose the content that meets your needs and Pearson Custom Publishing produces a high-quality printed book.

To find out more about custom publishing, visit www.pearsoncustom.co.uk

A Pearson Custom Publication

Fundamental Skills for Biosciences

Compiled from:

Practical Skills in Biomolecular Sciences Third Edition
by Rob Reed, David Holmes, Jonathan Weyers
and Allan Jones

Practical Skills in Biology Fourth Edition
by Allan Jones, Rob Reed and Jonathan Weyers

PEARSON
Custom
Publishing

Pearson Education Limited
Edinburgh Gate
Harlow
Essex CM20 2JE

And associated companies throughout the world

Visit us on the World Wide Web at:
www.pearsoned.co.uk

First published 2009

Compiled from:

Practical Skills in Biomolecular Sciences Third Edition
by Rob Reed, David Holmes, Jonathan Weyers and Allan Jones
ISBN 978 0 13 239115 3

Practical Skills in Biology Fourth Edition
by Allan Jones, Rob Reed and Jonathan Weyers
ISBN 978 0 13 175509 3

ISBN 978 1 84776 434 8

Printed and bound in Great Britain by Ashford Colour Press, Gosport, Hampshire

Contents

Selected chapters from
Practical Skills in Biomolecular Sciences
Third Edition
by Rob Reed, David Holmes, Jonathan Weyers
and Allan Jones

Chapters from *Practical Skills in Biology* Fourth Edition by Allan Jones, Rob Reed and Jonathan Weyers

Study and examination skills

1 The importance of transferable skills

Skills terminology – *different phrases may be used to describe transferable skills, depending on place or context. These include: 'personal transferable skills' (PTS), 'key skills', 'core skills' and 'competences'.*

This chapter outlines the range of transferable skills and their significance to biomolecular scientists. It also indicates where practical skills fit into this scheme. Having a good understanding of this topic will help you place your work at university in a wider context. You will also gain an insight into the qualities that employers expect you to have developed by the time you graduate. Awareness of these matters will be useful when carrying out personal development planning (PDP) as part of your studies.

The range of transferable skills

Table 1.1 provides a comprehensive listing of university-level transferable skills under six skill categories. There are many possible classifications – and a different one may be used in your institution or field of study. Note particularly that 'study skills', while important, and rightly emphasised at the start of many courses, constitute only a subset of the skills acquired by most university students.

The phrase '*Practical Skills*' in the title of this book indicates that there is a special subset of transferable skills related to work in the laboratory. However, although this text deals primarily with skills and techniques required for laboratory practicals and associated studies, a broader range of material is included. This is because the skills concerned are important, not only in the biosciences but also in the wider world. Examples include time management, evaluating information and communicating effectively.

Using course materials – *study your course handbook and the schedules for each practical session to find out what skills you are expected to develop at each point in the curriculum. Usually the learning objectives/outcomes (p. 24) will describe the skills involved.*

KEY POINT ***Biomolecular sciences are essentially practical subjects, and therefore involve highly developed laboratory skills. The importance that your lecturers place on practical skills will probably be evident from the large proportion of curriculum time you will spend on practical work in your course.***

The word 'skill' implies much more than the robotic learning of, for example, a laboratory routine. Of course, some of the tasks you will be asked to carry out in practical classes *will* be repetitive. Certain techniques require manual dexterity and attention to detail if accuracy and precision are to be attained, and the necessary competence often requires practice to make perfect. However, a deeper understanding of the context of a technique is important if the skill is to be appreciated fully and then transferred to a new situation. That is why this text is not simply a 'recipe book' of methods and protocols and why it includes background information, tips and worked examples, as well as study exercises to test your understanding.

Example The skills involved in teamwork cannot be developed without a deeper understanding of the interrelationships involved in successful groups. The context will be different for every group and a flexible approach will always be required, according to the individuals involved and the nature of the task.

Transferability of skills

Transferable skills are those which allow someone with knowledge, understanding or ability gained in one situation to adapt or extend this for application in a different context. In some cases, the transfer of a skill is immediately obvious. Take, for example, the ability to use a spreadsheet to summarise biological data and create a graph to

Table 1.1 Transferable skills identified as important in the biosciences. The list has been compiled from several sources, including the UK Quality Assurance Agency for Higher Education Subject Benchmark Statement for the Biosciences and for Biomedical Sciences. Particularly relevant chapters are shown for the skills covered by this book (numbers in **bold coloured** text indicate a deeper, or more extensive, treatment)

Skill category	Examples of skills and competences	Relevant chapters in this textbook
Generic skills for bioscientists	Having an appreciation of the complexity and diversity of life and life processes	11 28 29 33 36
	Reading and evaluating biological literature with a full and critical understanding	4 8 9
	Ability to communicate a clear and accurate account of a biological topic, both verbally and in writing	14 15 16 17–19
	Applying critical and analytical skills to evaluate evidence regarding theories and hypotheses	9 30
	Using a variety of methods for studying the biosciences	32–62
	Having the ability to think independently, set personal tasks and solve problems	30 31 65
Intellectual skills	Recognising and applying biological theories, concepts and principles	9 30
	Analysing, synthesising and summarising information critically	9 19 63–67
	Obtaining evidence to formulate and test hypotheses; applying knowledge to address familiar and unfamiliar problems	28 29 30 31 67
	Recognising and explaining moral, ethical and legal issues in biological research	20 21 32 33 34
Experimental and observational skills	Carrying out basic laboratory techniques and understanding the principles that underlie them	20 21–30 32 41 43 60
	Working in the laboratory safely, responsibly and legally, with due attention to ethical aspects	20 21 31–34
	Designing, planning, conducting and reporting on biological investigations and data arising from them	14 15 18 30 31
	Obtaining, recording, collating and analysing biological data	28–34 35 36 37 63–67
	Carrying out basic techniques relevant to core subjects in biomedical science (biochemistry, molecular genetics, immunology, microbiology)	20–26 27 28–31 32 33 34 35 36 37 38 39 41 42 43 44–48 49 60 61 62–67
Numeracy, communication and IT skills	Understanding and using data in several forms (e.g. numerical, textual, verbal and graphical)	4 9 63–66
	Communicating in written, verbal, graphical and visual forms	14 15 16 17–19 63 64 65
	Citing and referencing the work of others in an appropriate manner	8 9 19
	Obtaining data, including the concepts behind sampling and sampling errors, calibration and types of error	27 28 30 31 37 65–67
	Processing, interpreting and presenting data, and applying appropriate statistical methods for summarising and analysing data	11 63–65 66 67
	Solving problems with calculators and computers, including the use of tools such as spreadsheets	10 11 12 20 65
	Using computer technology to communicate and as a source of biological information	10 11 12 13
Interpersonal and teamwork skills	Working individually or in teams as appropriate; identifying individual and group goals and acting responsibly and appropriately to achieve them	3
	Recognising and respecting the views and opinions of others	3
	Evaluating your own performance and that of others	3 7
	Appreciating the interdisciplinary nature of contemporary biosciences	1 19
Self-management and professional development skills	Working independently, managing time and organising activities	2 30 31
	Identifying and working towards targets for personal, academic and career development	1 7
	Developing an adaptable and effective approach to study and work (including revision and exam technique)	2 4 5 6

illustrate results. Once the key concepts and commands are learned (Chapter 12), they can be applied to many instances outside the biosciences where this type of output is used. This is not only true for similar data sets, but also in unrelated situations, such as making up a financial balance sheet and creating a pie chart to show sources of expenditure. Similarly, knowing the requirements for good graph drawing and tabulation (Chapters 63 and 64), perhaps practised by hand in earlier work, might help you use spreadsheet commands to make the output suit your needs.

Other cases may be less clear but equally valid. For example, towards the end of your undergraduate studies you may be involved in designing experiments as part of your project work. This task will draw on several skills gained at earlier stages in your course, such as preparing solutions (Chapters 22–25), deciding about numbers of replicates and experimental layout (Chapters 30 and 31) and perhaps carrying out some particular method of observation, measurement or analysis. How and when might you transfer this complex set of skills? In the workplace, it is unlikely that you would be asked to repeat the same process, but in critically evaluating a problem or in planning a complex project for a new employer, you will need to use many of the time management, organisational and analytical skills developed when designing and carrying out experiments. The same applies to information retrieval and evaluation and writing essays and dissertations, when transferred to the task of analysing or writing a business report.

Opportunities to develop and practise skills in your private or social life *– you could, for example, practise spreadsheet skills by organising personal or club finances using Microsoft Excel, or teamwork skills within any university clubs or societies you may join (see Chapter 7).*

Personal development planning

Many universities have schemes for personal development planning (PDP), which may go under slightly different names such as progress file or professional development plan. You will usually be expected to create a portfolio of evidence on your progress, then reflect on this, and subsequently set yourself plans for the future, including targets and action points. Analysis of your transferable skills profile will probably form part of your PDP. Other aspects commonly included are:

- your aspirations, goals, interests and motivations;
- your learning style or preference (see p. 21);
- your assessment transcript or academic profile information (e.g. record of grades in your modules);
- your developing CV (see p. 38).

Taking part in PDP can help focus your thoughts about your university studies and future career. This is important, as many biosciences degrees do not lead only to a single, specific occupation. The PDP process will introduce you to some new terms and will help you to describe your personality and abilities. This will be useful when constructing your CV and when applying for jobs.

Types of PDP portfolio and their benefits *– some PDP schemes are centred on academic and learning skills, while others are more focused on career planning. Some are carried out independently and others in tandem with a personal tutor or advisory system. Some PDP schemes involve creating an online portfolio, while others are primarily paper-based. Each method has specific goals and advantages, but whichever way your scheme operates, maximum benefit will be gained from being fully involved with the process.*

Definition

Employability – the 'combination of in-depth subject knowledge, work awareness, subject-specific, generic and career management skills, and personal attributes and attitudes that enable a student to secure suitable employment and perform excellently throughout a career spanning a range of employers and occupations.' *(Higher Education Academy Centre for Bioscience* definition of employability for bioscientists)

What your future employer will be looking for

At the end of your course, which may seem some time away, you will aim to get a job and start on your chosen career path. You will need to sell yourself to your future employer, firstly in your application form and curriculum vitae (Chapter 7), and perhaps later at interview. Companies rarely employ bioscience graduates simply because they know how to carry out a particular lab routine or because they can remember specific facts about their chosen degree subject. Instead, employers tend to look for a range of qualities and transferable skills that together define an attribute known as 'graduateness'. This encompasses, for example, the ability to work in a team, to speak effectively and write clearly about your work, to understand complex data and to manage a project to completion. All of these skills can be developed at different stages during your University studies.

KEY POINT Factual knowledge is important in degrees with a strong vocational element, but understanding how to find and evaluate information is usually rated more highly by employers than the ability to memorise facts.

Most likely, your future employer(s) will seek someone with a organised yet flexible mind, capable of demonstrating a logical approach to problems – someone who has a range of skills and who can transfer these skills to new situations. Many competing applicants will probably have similar qualifications. If you want the job, you will have to show that your additional skills place you above the other candidates.

Sources for further study

Drew, S. and Bingham, R. (2004) *The Student Skills Guide,* 2nd edn. Gower Publishing Ltd, Aldershot.

McMillan, K. and Weyers, J.D.B. (2006) *The Smarter Student: Skills and Strategies for Success at University*. Pearson Education, Harlow.

Race, P. (1999) *How to Get a Good Degree: Making the Most of Your Time at University*. Open University Press, Buckingham.

Study exercises

1.1 Evaluate your skills. Examine the list of skill topics shown in Table 1.1 (p. 4). Now create a new table with two columns, like the one on next page. The first half of this table should indicate *five* skills you feel confident about and show where you demonstrated this skill (for example, 'working in a team' and 'in a first year group project in molecular biology'). The second half of the table should show *five* skills you do not feel confident about, or you recognise need development (e.g. 'communicating in verbal form'). List these and then list ways in which you think the course material for your current modules will provide opportunities to develop these skills, or what activities you might take to improve them (e.g. 'forming a study group with colleagues').

1.2 Find skills resources. For at least one of the skills in the second half of Table 1.1, check your university's library database to see if there are any texts on that subject. Alternatively, carry out a search for relevant websites (there are many); decide which are useful and 'bookmark' them for future use (Chapter 10).

1.3 Analyse your goals and aspirations. Spend a little time thinking what you hope to gain from university. See if your friends have the same aspirations. Think about and/or discuss how these goals can be achieved, while keeping the necessary balance between university work, paid employment and your social life.

Study exercises (continued)

Skills I feel confident about	Where demonstrated
1.	
2.	
3.	
4.	
5.	
Skills that I could develop	**Opportunities for development**
6.	
7.	
8.	
9.	
10.	

2 Managing your time

Definition

Time management – a system for controlling and using time as efficiently and as effectively as possible.

Advantages of time management – these include:

- *a feeling of much greater control over your activities;*
- *avoidance of stress;*
- *improved productivity – achieve more in a shorter period;*
- *improved performance – work to higher standards because you are in charge;*
- *increase in time available for non-work matters – work hard, but play hard too.*

One of the most important activities that you can do is to organise your personal and working time effectively. There is a lot to do at university and a common complaint is that there isn't enough time to accomplish everything. In fact, research shows that most people use up a lot of their time without realising it through ineffective study or activities such as extended coffee breaks. Developing your time management skills will help you achieve more in work, rest and play, but it is important to remember that putting time management techniques into practice is an individual matter, requiring a level of self-discipline not unlike that required for dieting. A new system won't always work perfectly straight away, but through time you can develop a system that is effective for you. An inability to organise your time effectively, of course, results in feelings of failure, frustration, guilt and being out of control in your life.

Setting your goals

The first step is to identify clearly what you want to achieve, both in work and in your personal life. We all have a general idea of what we are aiming for, but to be effective, your goals must be clearly identified and priorities allocated. Clear, concise objectives can provide you with a framework in which to make these choices. Try using the 'SMART' approach, in which objectives should be:

Example The objective "to spend an extra hour each week on directed study in microbiology next term" fulfils the SMART criteria, in contrast to a general intention "to study more".

- Specific – clear and unambiguous, including what, when, where, how and why.
- Measurable – having quantified targets and benefits to provide an understanding of progress.
- Achievable – being attainable within your resources.
- Realistic – being within your abilities and expectations.
- Timed – stating the time period for completion.

Having identified your goals, you can now move on to answer four very important questions:

1. Where does your time go?
2. Where should your time go?
3. What are your time-wasting activities?
4. What strategies can help you?

Analysing your current activities

The key to successful development of time management is a realistic knowledge of how you currently spend your time. Start by keeping a detailed time log for a typical week (Fig. 2.1), but you will need to be truthful in this process. Once you have completed the log, consider the following questions:

- How many hours do I work in total and how many hours do I use for relaxation?
- What range of activities do I do?
- How long do I spend on each activity?
- What do I spend most of my time doing?

Time slots	Activity										Notes
7.00–7.15											
7.15–7.30											
7.30–7.45											
7.45–8.00											
8.00–8.15											
8.15–8.30											
8.30–8.45											
8.45–9.00											
9.00–9.15											

Fig. 2.1 Example of how to lay out a time log. Write activities along the top of the page, and divide the day into 15-minute segments as shown. Think beforehand how you will categorise the different things you do, from the mundane (laundry, having a shower, drinking coffee, etc.) to the well-timetabled (tutorial meeting, sports club meeting) and add supplementary notes if required. At the end of each day, place a dot in the relevant column for each activity and sum the dots to give a total at the bottom of the page. You will need to keep a diary like this for at least a week before you see patterns emerging.

- What do I spend the least amount of my time doing?
- Are my allocations of time in proportion to the importance of my activities?
- How much of my time is ineffectively used, e.g. for uncontrolled socialising or interruptions?

If you wish, you could use a spreadsheet (Chapter 12) to produce graphical summaries of time allocations in different categories as an aid to analysis and management. Divide your time into:

- Committed time – timetabled activities involving your main objectives/goals.
- Maintenance time – time spent supporting your general life activities (shopping, cleaning, laundry, etc.).
- Discretionary time – time for you to use as you wish, e.g. recreation, sport, hobbies, socialising.

Avoiding time-wasting activities

***Quality in time management** – avoid spending a lot of time doing unproductive studying, e.g. reading a textbook without specific objectives for that reading. Make sure you test your recall of the material, if you are working towards an examination (p. 20).*

Look carefully at those tasks that could be identified as time-wasting activities. They include gossiping, over-long breaks, uninvited interruptions and even ineffective study periods. Try to reduce these to a minimum, but do not count on eliminating them entirely. Remember also that some relaxation *should* be programmed into your daily schedule.

Organising your tasks

***Being assertive** – if friends and colleagues continually interrupt you, find a way of controlling them, before they control you. Indicate clearly on your door that you do not wish to be disturbed and explain why. Otherwise, try to work away from disturbance.*

Having analysed your time usage, you can now use this information, together with your objectives and prioritised goals, to organise your activities, both on a short-term and a long-term basis. Consider using a diary-based system (such as those produced by Filofax, TMI and Day-Timer) that will help you plan ahead and analyse your progress.

Divide your tasks into several categories, such as:

- Urgent – must be done as a top priority and at short notice (e.g. doctor's appointment).

WEEKLY DIARY

Week beginning:

	Sunday	Monday	Tuesday	Wednesday	Thursday	Friday	Saturday
DATE							

7–8 am		Breakfast	Breakfast	Breakfast	Breakfast	Breakfast	
8–9		Preparation	Preparation	Preparation	Preparation	Preparation	Breakfast
9–10	Breakfast	PE112(L)	PE112(L)	PE112(L)	PE112(L)	BIOL(P)	Travel
10–11	FREE	CHEM(L)	CHEM(L)	CHEM(L)	CHEM(L)	BIOL(P)	WORK
11–12	STUDY	STUDY	STUDY	STUDY	STUDY	BIOL(P)	WORK
12–1 pm	STUDY	BIOL(L)	BIOL(L)	BIOL(L)	BIOL(L)	TUTORIAL	WORK
1–2	Lunch	Lunch	Lunch	Lunch	Lunch	Lunch	Lunch
2–3	(VOLLEY-	CHEM(P)	STUDY	SPORT	PE112(P)	STUDY	WORK
3–4	BALL	CHEM(P)	STUDY	(VOLLEY-	PE112(P)	STUDY	WORK
4–5	MATCH)	CHEM(P)	STUDY	BALL	PE112(P)	SHOPPING	WORK
5–6	FREE	STUDY	STUDY	CLUB)	STUDY	TEA ROTA	WORK
6–7	Tea	Tea	Tea	Tea	Tea	Tea	Tea
7–8	FREE*	STUDY	STUDY	FREE*	STUDY	FREE*	FREE
8–9	FREE*	STUDY	STUDY	FREE*	STUDY	FREE*	FREE
9–10	FREE*	FREE*	STUDY	FREE*	STUDY	FREE*	FREE

Study (h)	2	10	11	4	11	6	0
Other (h)	13	5	4	11	4	9	15

Total study time = 44 h

Fig. 2.2 A weekly diary with example of entries for a first year science student with a Saturday job and active membership of a volleyball club. Note that 'free time' changes to 'study time', e.g. for periods when assessed work is to be produced or during revision for exams. Study time (including attendance at lectures, practicals and tutorials) thus represents between 42 and 50 per cent of the total time.

- Routine – predictable and regular and therefore easily scheduled (e.g. attending lectures or playing sport).
- One-off activities – usually with rather shorter deadlines and which may be of high priority (e.g. a tutorial assignment or seeking advice on a specific issue).
- Long-term tasks – sometimes referred to as 'elephant tasks' that are too large to 'eat' in one go (e.g. learning a language). These are best managed by scheduling frequent small 'bites' to achieve the task over a longer timescale.

You should make a weekly plan (Fig. 2.2) for the routine activities, with gaps for less predictable tasks. This should be supplemented by individual daily checklists, preferably written at the end of the previous working day. Such plans and checklists should be flexible, forming the basis for most of your activities except when exceptional circumstances intervene. The planning must be kept brief, however, and should be scheduled into your activities. Box 2.1 provides tips for effective time management during your studies.

***Use checklists as often as possible** – post your lists in places where they are easily and frequently visible, such as in front of your desk. Ticking things off as they are completed gives you a feeling of accomplishment and progress, increasing motivation.*

***Matching your work to your body's rhythm** – everyone has times of day when they feel more alert and able to work. Decide when these times are for you and programme your work accordingly. Plan relaxation events for periods when you tend to be less alert.*

***KEY POINT** Review each day's plan at the end of the previous day, making such modifications as are required by circumstances, e.g. adding an uncompleted task from the previous day or a new and urgent task.*

Box 2.1 Tips for effective planning and working

- Set guidelines and review expectations regularly.
- Don't procrastinate: don't keep putting off doing things you know are important – they will not go away but they will increase to crisis point.
- Don't be a perfectionist – perfection is paralysing.
- Learn from past experience – review your management system regularly.
- Don't set yourself unrealistic goals and objectives – this will lead to procrastination and feelings of failure.
- Avoid recurring crises – they are telling you that something is not working properly and needs to be changed.
- Learn to concentrate effectively and don't let yourself be distracted by casual interruptions.
- Learn to say 'no' firmly but graciously when appropriate.
- Know your own body rhythms: e.g. are you a morning person or an evening person?
- Learn to recognise the benefits of rest and relaxation at appropriate times.
- Take short but complete breaks from your tasks – come back feeling refreshed in mind and body.
- Work in suitable study areas and keep your own workspace organised.
- Avoid clutter (physical and mental).
- Learn to access and use information effectively (Chapter 9).
- Learn to read and write accurately and quickly (Chapters 4 and 16).

Sources for further study

Anon. *Day-Timer*. Available: http://www.daytimer.co.uk Last accessed: 01/04/07.
[Website for products of Day-Timers Europe Ltd., Chene Court, Poundwell Street, Modbury, Devon PL21 0QJ]

Anon. *Filofax*. Available: http://www.filofax.co.uk Last accessed: 01/04/07.
[Website for products of Filofax UK, Unit 3, Victoria Gardens, Burgess Hill, West Sussex, RH15 9NB]

Anon. *TMI Website*. Available: http://www.tmi.co.uk Last accessed: 01/04/07.
[Website for products of TMI (Time Manager International A/S), 50 High Street, Henley-in-Arden, Solihull, West Midlands B95 5AN]

Mayer, J.L. (1999) *Time Management for Dummies*, 2nd edn. IDG Books Worldwide, Inc., Foster City.

Study exercises

2.1 Evaluate your time usage. Compile a spreadsheet to keep a record of your daily activities in 15-minute segments for a week. Analyse this graphically and identify areas for improvement.

2.2 List your short-, medium- and long-term tasks and allocate priorities. Produce several lists, one for each of the three timescales, and prioritise each item. Use this list to plan your time management, by scheduling high priority tasks and leave low priority activities to 'fill in' the spare time that you may identify. This task should be done on a regular (monthly) basis to allow for changing situations.

2.3 Plan an 'elephant' task. Spend some time planning how to carry out a large or difficult task (learning a language or learning to use a complex computer program) by breaking it down into achievable segments ('bites').

3 Working with others

> **Definitions**
>
> **Team** – a team is not a bunch of people with job titles, but a congregation of individuals, each of whom has a role which is understood by other members. Members of a team seek out certain roles and they perform most effectively in the ones that are most natural to them.
>
> **Team role** – a tendency to behave, contribute and interrelate with others in a particular way.
>
> (both after Belbin, 1993)

It is highly likely that you will be expected to work with fellow students during practicals and study exercises. This might take the form of sharing tasks or casual collaboration through discussion, or it might be formally directed teamwork such as problem-based learning (Box 6.1) or preparing a poster (Chapter 14). Interacting with others can be extremely rewarding and realistically represents the professional world, where teamworking is common. The advantages of working with others include:

- Teamworking is usually synergistic in effect – it often results in better ideas, produced by the interchange of views, and better output, due to the complementary skills of team members.
- Working in teams can provide support for individuals within the team.
- Levels of personal commitment can be enhanced through concern about letting the team down.
- Responsibilities for tasks can be shared.

However, you can also feel both threatened and exposed if teamwork is not managed properly. Some of the main reasons for negative feelings towards working in groups include:

> ***Peer assessment*** *– this term applies to marking schemes in which all or a proportion of the marks for a teamwork exercise are allocated by the team members themselves. Read the instructions carefully before embarking on the exercise, so you know which aspects of your work your fellow team members will be assessing. When deciding what marks to allocate yourself, try to be as fair as possible with your marking.*

- Reservations about working with strangers – not knowing whether you will be able to form a friendly and productive relationship.
- Worries over rejection – a perception of being unpopular or being chosen last by the group.
- Concerns over levels of personal commitment – these can be enhanced through a desire to perform well, so the team as a whole achieves its target.
- Fear of being held back by others – especially for those who have been successful in individual work already.
- Lack of previous experience – worries about the kinds of personal interactions likely to occur and the team role likely to suit you best.
- Concerns about the outcomes of peer assessment – in particular, whether others will give you a fair mark for your efforts.

Teamwork skills

Some of the key skills you will need to develop to maximise the success of your teamworking activities include:

> ***Gaining confidence through experience*** *– the more you take part in teamwork, the more you know how teams operate and how to make teamwork effective for you.*

- Interpersonal skills. How do you react to new people? Are you able to both listen and communicate easily with them? How do you deal with conflicts and disagreements?
- Delegation/sharing of tasks. The primary advantage of teamwork is the sharing of effort and responsibility. Are you willing/able to do this? It involves trusting your team members. How will you deal with those group members who don't contribute fully?
- Effective listening. Successful listening is a skill that usually needs developing, e.g. during the exchange of ideas within a group.
- Speaking clearly and concisely. Effective communication is a vital part of teamwork, both between team members and when presenting team outcomes to others. Try to develop your communication skills through learning and practice (see Chapter 15).

- Providing constructive criticism. It is all too easy to be negative but only constructive criticism will have a positive effect on interactions with others.

Studying with others *– teaming up with someone else on your course for revision (a 'study buddy') is a potentially valuable activity and may especially suit some types of learners (p. 20). It can help keep your morale high when things get tough. You might consider:*

- *sharing notes, textbooks and other information;*
- *going through past papers together, dissecting the questions and planning answers;*
- *talking to each other about a topic (good for aural learners. p. 21);*
- *giving tutorials to each other about parts of the course that have not been fully grasped.*

Collaboration for learning

Much collaboration is informal and consists of pairs or groups of individuals getting together to exchange materials and ideas while studying. It may consist of a 'brainstorming' session for a topic or piece of work, or sharing efforts to research a topic. This has much to commend it and is generally encouraged. However, it is vital that this collaborative learning is distinguished from the collaborative writing of assessed documents: the latter is not usually acceptable and, in its most extreme form, is plagiarism, usually with a heavy potential punishment in university assessment systems. Make sure you know what plagiarism is, what unacceptable collaboration is, and how they are treated within your institution.

KEY POINT Collaboration is inappropriate during the final phase of an assessed piece of work unless you have been directed to produce a group report. Collaboration is encouraged during research and learning activities but the final write-up must normally be your own work. The extreme of producing copycat write-ups is regarded as plagiarism (p. 51) and will be punished accordingly.

Web-based resources and support for brainstorming *– websites such as http://www.brainstorming.co.uk give further information and practical advice for teamworking.*

The dynamics of teamworking

It is important that team activities are properly structured so that each member knows what is expected of them. Allocation of responsibilities usually requires the clear identification of a leader. Several studies of groups have identified different team roles that derive from differences in personality. You should be aware of such categorisations, both in terms of your own predispositions and those of your fellow team members, as it will help the group to interact more productively. Belbin (1993) identified eight such roles, recently extended to nine, as shown in Table 3.1. Several of the categories shown in Table 3.1 are suitable for a leader, including 'co-ordinator' and 'shaper'.

In formal team situations, your course organiser may deal with these issues; even if they do not, it is important that you are aware of these roles and their potential impact on the success or failure of teamwork. You should try to identify your own 'natural' role: if asked to form a team, bear these different roles in mind during your selection of colleagues and your interactions with them. The ideal team should contain members capable of adopting most of these roles. However, you should also note the following points:

- People will probably best fit one of these roles naturally as a function of their personality and skills.
- Group members may be suited to more than one role.
- In some circumstances, team members may be required to adapt and take a different role from the one that they feel suits them.
- No one role is 'better' than any other. For good teamwork, the group should have a balance of personality types present.
- People may have to adopt multiple roles, especially if the team size is small.

Recording group discussions *– make sure you structure meetings (including writing agendas) and note their outcomes (taking minutes and noting action points).*

Table 3.1 A summary of the team roles described by Belbin (1993). A good team requires members who are able to undertake appropriate roles at different times. Each role provides important strengths to a team, and its compensatory weaknesses should be accepted within the group framework

Team role	Personality characteristics	Typical function in a team	Strengths	Allowable weaknesses
Co-ordinator	Self-confident, calm and controlled	Leading: causing others to work towards shared goals	Good at spotting others' talents and delegating activities	Often less creative or intellectual than others in the group
Shaper	Strong need for achievement; outgoing; dynamic; highly strung	Leading: generating action within team; imposing shape and pattern to work	Providing drive and realism to group activities	Can be headstrong, emotional and less patient than others
Innovator	Individualistic, serious-minded; often unorthodox	Generating action; imposing shape and pattern to work activities	Creative, innovative and knowledgeable	Tendency to work in isolation; ideas may not always be practical
Monitor–evaluator	Sober, unemotional and prudent	Analysing problems and evaluating ideas	Shrewd judgement	May work slowly; not usually a good motivator
Implementer	Well-organised and self-disciplined, with practical common sense	Doing what needs to be done	Organising abilities and common sense	Lack of flexibility and tendency to resist new ideas
Teamworker	Sociable, mild and sensitive	Being supportive, perceptive and diplomatic; keeping the team going	Good listener; reliable and flexible; promotes team spirit	Not comfortable when leading; may be indecisive
Resource investigator	Extrovert, enthusiastic, curious and communicative	Exploiting opportunities; finding resources; external relations	Quick thinking; good at developing others' ideas	May lose interest rapidly
Completer–finisher	Introvert and anxious, painstaking, orderly and conscientious	Ensuring completion of activity to high standard	Good focus on fulfilling objectives and goals	Obsessive about details; may wish to do all the work to control quality
Specialist	Professional, self-motivated and dedicated	Providing essential skills	Commitment and technical knowledge	Contribute on a narrow aspect of project; tend to be single-minded

***KEY POINT** In formal teamwork situations, be clear as to how individual contributions are to be identified and recognised. This might require discussion with the course organiser. Make sure that recognition, including assessment, is truly reflective of effort. Failure to ensure that this is the case can lead to disputes and feelings of unfairness.*

Your lab partner(s)

Many laboratory sessions in the life sciences involve working in pairs or small groups. In some cases, you may work with the same partner(s) for series of practicals or for a complete module. The relationship you develop as a team is important to your progress, and can enhance your understanding of the material and the grades you obtain. Tips for building a constructive partnership include:

- Introduce yourselves at the first session and take a continuing interest in each other's interests and progress at university.
- At appropriate points, discuss the practical (both theory and tasks) and your understanding of what is expected of you.

- Work jointly to complete the practical effectively, avoiding the situation where either partner dominates the activities and gains most from the practical experience.
- Share tasks according to your strengths, but do this in such a way that one partner can learn new skills and knowledge from the other.
- Make sure you ask questions of each other and communicate any doubts about what you have to do.
- Discuss other aspects of your course, e.g. by comparing notes from lectures or ideas about in-course assessments.
- Consider meeting up outside the practical sessions to study, revise and discuss exams.

Text reference

Belbin, R.M. (1993) *Team Roles at Work*. Butterworth-Heinemann, Oxford.

Source for further study

Belbin, R.M. *The Belbin Website*. Available: http://www.belbin.com/
Last accessed: 01/04/07.

Study exercises

3.1 Evaluate your 'natural' team role(s). Using Table 3.1 as a source, decide which team role best fits your personality.

3.2 Keep a journal during a group activity. Record your feelings and observations about experiences of working with other students. After the event, review the journal, then draw up a strategy for developing aspects where you feel you might have done better.

3.3 Reflect upon your teamwork abilities. Draw up a list of your reactions to previous efforts at collaboration or teamwork and analyse your strengths and weaknesses. How could these interactions have been improved or supported more effectively?

4 Taking notes from lectures and texts

> ***Choose note-taking methods appropriately*** *– the method you choose to take notes might depend on the subject; the lecturer and their style of delivery; and your own preference.*

Note-taking is an essential skill that you will require in many different situations, such as:

- listening to lectures and seminars;
- attending meetings and tutorials;
- reading texts and research papers;
- finding information on the World Wide Web.

KEY POINT Good performance in assessments and exams is built on effective learning and revision (Chapters 5 and 6). However, both ultimately depend on the quality of your notes.

> ***Compare lecture notes with a colleague*** *– looking at different sets of notes for the same lecture may reveal interesting differences in approach, depth and detail.*

> ***Adjusting to the styles of your lecturers*** *– recognise that different approaches to lecture delivery demand different approaches to note-taking. For example, if a lecturer seems to tell lots of anecdotes or spend much of the time on examples during a lecture, do not switch off – you still need to be listening carefully to recognise the key take-home messages. Similarly, if a lecture includes a section consisting mainly of images, you should still try to take notes – names of organisms, locations, key features, even quick sketches. These will help prompt your memory when revising. Do not be deterred by lecturers' idiosyncrasies; in every case you still need to focus and take useful notes.*

Taking notes from lectures

Taking legible and meaningful lecture notes is essential if you are to make sense of them later, but many students find it difficult when starting their university studies. Begin by noting the date, course, topic and lecturer on the first page of each day's notes. Number every page in case they get mixed up later. The most popular way of taking notes is to write in a linear sequence down the page, emphasising the underlying structure via headings, as in Fig. 16.2. However, the 'pattern' and 'Mind Map' methods (Figs 4.1 and 4.2) have their advocates: experiment, to see which method you prefer.

Whatever technique you use, don't try to take down all of the lecturer's words, except when an important definition or example is being given, or when the lecturer has made it clear that he/she is dictating. Listen first, then write. Your goal should be to take down the structure and reasoning behind the lecturer's approach in as few words and phrases as possible. At this stage, follow the lecturer's sequence of delivery. Use headings and leave plenty of space, but don't worry too much about being tidy – it is more important that you get down the appropriate information in a readable form. Use abbreviations to save time. Recognise that you may need to alter your note-taking technique to suit different lecturers' styles.

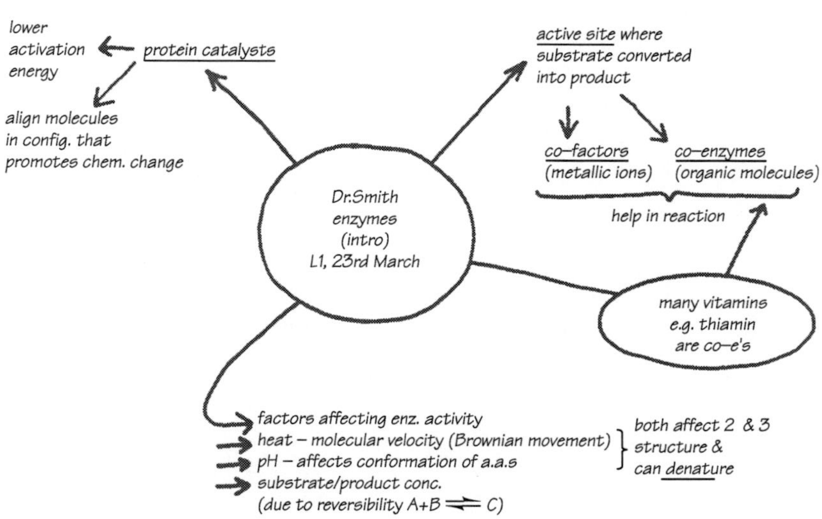

Fig. 4.1 An example of 'pattern' notes, an alternative to the more commonly used 'linear' format. Note the similarity to the 'spider diagram' method of brainstorming ideas (Fig. 16.2).

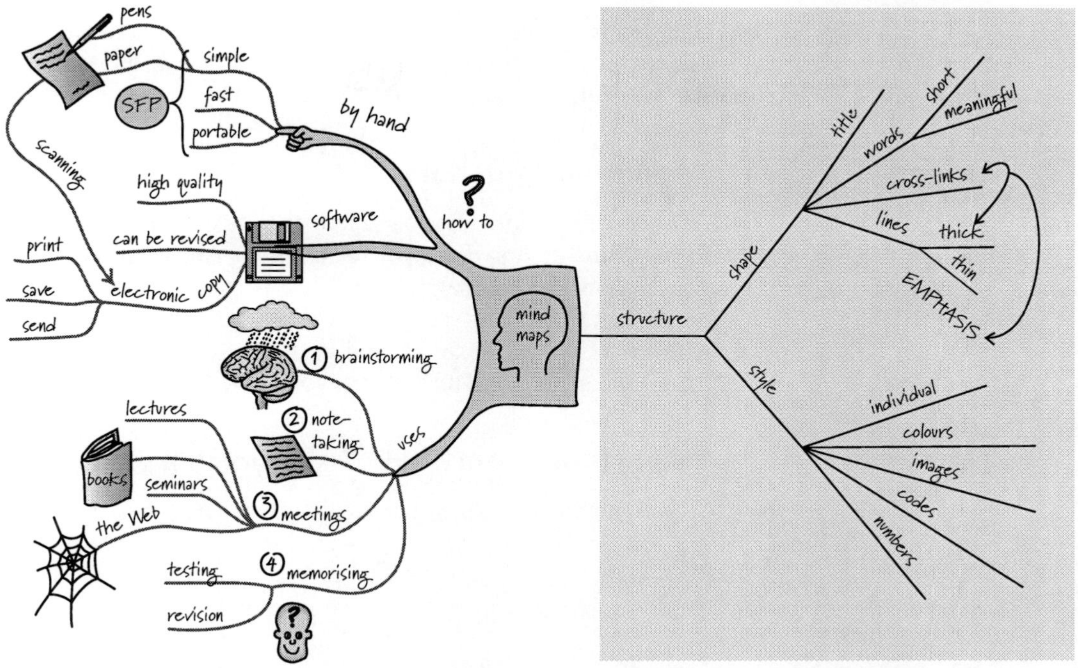

Fig. 4.2 Example of the 'Mind Map' approach to note-taking and 'brainstorming'. Start at the centre with the overall topic title, adding branches and sub-branches for themes and subsidiary topics. 'Basic' maps consist of a branched hierarchy overwritten with key words (e.g. shaded portion). Connections should be indicated with arrows; numbering and abbreviations are encouraged. To aid recall and creativity, Buzan (2006) recommends use of colour, different fonts, 3-dimensional doodles and other forms of emphasis (e.g. non-shaded portion).

Example Commonly used abbreviations include:

$\exists$	there are, there exist(s)
$\therefore$	therefore
$\because$	because
$\propto$	is proportional to
$\rightarrow$	leads to, into
$\leftarrow$	comes from, from
$\rightarrow\rightarrow$	involves several processes in a sequence
1°, 2°	primary, secondary (etc.)
$\approx$, $\cong$	approximately, roughly equal to
$=$, $\neq$	equals, not equal to
$\equiv$, $\not\equiv$	equivalent, not equivalent to
$<$, $>$	smaller than, bigger than
$\gg$	much bigger than
[X]	concentration of X
$\sum$	sum
Δ	change
f	function
#	number
∞	infinity, infinite

You should also make up your own abbreviations relevant to the context, e.g. if a lecturer is talking about photosynthesis, you could write 'PS' instead, etc.

Make sure you note down references to texts and take special care to ensure accuracy of definitions and numerical examples. If the lecturer repeats or otherwise emphasises a point, highlight (e.g. by underlining) or make a margin note of this – it could come in useful when revising. If there is something you don't understand, ask at the end of the lecture, or make an appointment to discuss the matter if there isn't time to deal with it then. Tutorials may provide an additional forum for discussing course topics.

Lectures delivered by PowerPoint or similar presentation programs

Some students make the mistake of thinking that lectures delivered as computer-based presentations with an accompanying handout or Web-resource require little or no effort by way of note-taking. While it is true that you may be freed from the need to copy out large diagrams and the basic text may provide structure, you will still need to adapt and add to the lecturer's points. Much of the important detail and crucial emphasis will still be delivered verbally. Furthermore, if you simply listen passively to the lecture, or worse, try to work from the handout alone, it will be far more difficult to understand and remember the content.

If you are not supplied with handouts, you may be given access to the electronic file, so that you can print out the presentation beforehand, perhaps in the '3 slides per page' format that allows space for notes alongside each slide (Fig. 4.3). Scan through this before the lecture if you can; then, during the presentation, focus on listening to what the lecturer has to say. Note down any extra details, points of emphasis and examples. After lectures, you could also add notes from supplementary reading. The text in presentations can be converted to word processor format if you have access to the electronic file. In PowerPoint, this can be achieved from the *Outline View* option on the *View* menu. You can copy and paste text between programs in the normal fashion, then modify font size and colour as appropriate.

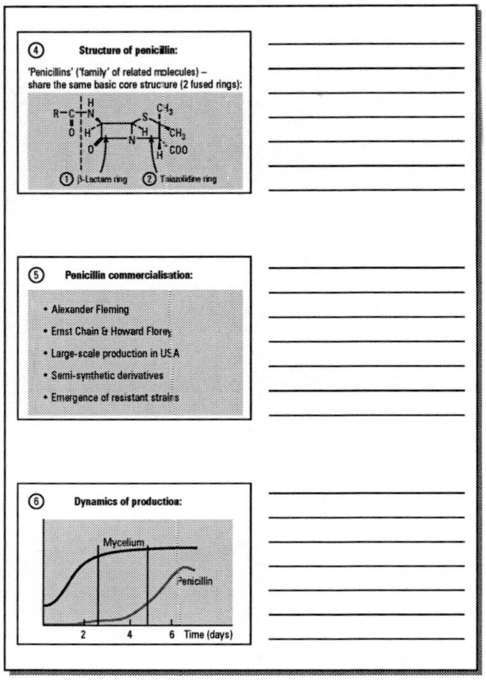

Fig. 4.3 An example of a printout from PowerPoint in 'Handouts (3 slides per page)' format.

'Making up' your notes

As soon as possible after each lecture, work through your notes, tidying them up and adding detail where necessary. Add emphasis to any headings you have made, so that the structure is clearer. If you feel it would be more logical for your purposes, change the order. Compare your notes with material in a textbook and correct any inconsistencies. Make notes from, or copy, any useful material you see in textbooks, ready for revision.

Taking notes from books and journal papers

Scanning and skimming are useful techniques.

Scanning

This involves searching for relevant content. Useful techniques are to:

- decide on key words relevant to your search;
- check that the book or journal title indicates relevance;
- look through the contents page (either paper titles in a journal volume, or chapter titles in a textbook);
- look at the index, if present.

Skimming

This is a valuable way to gain the maximum amount of information in the minimum amount of time, by reading as little of a text as is required. Essentially, the technique (also termed 'surveying') requires you to look at the structure of the text, rather than the detail. In a sense, you are trying to see the writer's original plan and the purpose behind each part of the text. Look through the whole of the piece first, to gain an overview of its scope and structure. Headings provide an obvious clue to structure, if present. Next, look for the 'topic sentence' in each paragraph (p. 103), which is often the first. You might then decide that the paragraph contains a definition that is important to note, or it may contain examples, so may not be worth reading for your purpose.

When you have found relevant material, note-taking fulfils the vital purpose of helping you understand and remember the information. If you simply read it, either directly or from a photocopy, you risk accomplishing neither. The act of paraphrasing (using different words to give the same meaning) makes you think about the meaning and forces you to express this for yourself. It is an important active learning technique. A popular method of skimming and note-taking is called the SQ3R technique (Box. 4.1).

***KEY POINT* Obtaining information and then understanding it are distinct, sequential parts of the process of learning. As discussed in Chapter 5 (Table 5.1), you must be able to do more than recall facts to succeed.**

Methods for finding and evaluating texts and articles are discussed further in Chapters 8 and 9.

Printing PowerPoint slides *– use the 'Black and White' option on the Print menu to avoid wasting ink on printing of coloured backgrounds. If you wish to use colour, remember that slides can be difficult to read if printed in small format. Always print a sample page before printing the whole lecture.*

Scanning effectively *– you need to stay focused on your key words, otherwise you may be distracted by apparently interesting but peripheral material.*

Spotting sequences *– writers often number their points (firstly, secondly, thirdly, etc.) and looking for these words in the text can help you skim it quickly.*

Making sure you have all the details of a source *– when taking notes from a text or journal paper: (a) always take full details (Chapter 8); (b) if copying word-for-word make sure you indicate this using quotes and take special care to ensure you do not alter the original wording.*

Box 4.1 The SQ3R technique for skimming texts

Survey Get a quick overview of the contents of the book or chapter, perhaps by rapidly reading the contents page or headings.

Question Ask yourself what the material covers and how precisely it relates to your study objectives.

Read Now read the text, paying attention to the ways it addresses your key questions.

Recall Recite to yourself what has been stated every few paragraphs. Write notes of this if appropriate, paraphrasing the text rather than copying it.

Review Think about what you have read and/or review your notes as a whole. Consider where it all fits in.

Sources for further study

Anon. *Mind Map*. Available: http://en.wikipedia.org/wiki/Mind_mapping. Last accessed: 01/04/07. [An independent review of mind-mapping and its history]

Anon. *FreeMind – Free Mind Mapping Software*. Available: http://freemind.sourceforge.net/wiki/index.php/Main_Page. Last accessed: 01/04/07. [An alternative to pen-and-paper mind mapping]

Buzan, A. (2006) *The Ultimate Book of Mind Maps*. Harper Thorsons, New York.

Morris, S. and Smith, J. (1998) *Understanding Mind Maps . . . in a Week*. [Foreword by Buzan, A.] Hodder and Stoughton, London.

Study exercises

4.1 Experiment with a new note-taking technique. If you haven't tried the pattern or mind-mapping methods (Figs 4.1 and 4.2), carry out a trial to see how they work for you. Research the methods first by consulting appropriate books or websites.

4.2 Carry out a 'spring clean' of your desk area and notes. Make a concerted effort to organise your notes and handouts, investing if necessary in files and folders. This will be especially valuable at the start of a revision period.

4.3 Try out the SQ3R technique. The next time you need to obtain information from a text, compare this method (Box 4.1) with others you may have adopted in the past. Is it faster, and does it aid your ability to recall the information?

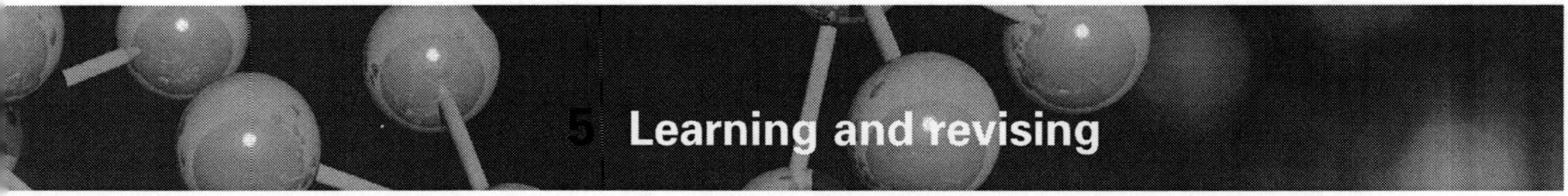

5 Learning and revising

There are many different ways of learning and at university you have the freedom to choose which approach to study suits you best. You should tackle this responsibility with an open mind, and be prepared to consider new options. Understanding how you learn best and how you are expected to think about your discipline will help you to improve your approach to study and to understand your branch of bioscience at a deeper level. Adopting active methods of studying and revision that are suited to your personality can make a significant difference to your performance. Your department will publish material that can help too. Taking account of learning outcomes and marking/assessment criteria, for example, can help you focus your revision.

***KEY POINT** At university, you are expected to set your own agenda for learning. There will be timetabled activities, assessments and exam deadlines, but it is your responsibility to decide how you will study and learn, how you will manage your time, and, ultimately, what you will gain from the experience. You should be willing to challenge yourself academically to discover your full potential.*

Learning styles

***Significance of learning styles** – no one learning style is 'better' than the others; each has its own strengths and weaknesses. However, since many university exams are conducted using 'reading and writing' modes of communication, you may need to find ways of expressing yourself appropriately using the written word (see Box 5.1).*

We don't all learn in the same way. Your preferred learning style is simply the one that suits you best for receiving, communicating and understanding information. It therefore involves approaches that will help you learn and perform most effectively. There are many different ways of describing learning styles, and you may be introduced to specific schemes during your studies. While methods and terminology may differ among these approaches, it is important to realise that the important thing is the *process* of analysing your learning style, together with the way you use the information to modify your approach to studying, rather than the specific type of learner that you may identify yourself to be.

***Learning styles and teaching styles** – there may be a mismatch between your preferred learning style and the corresponding 'teaching style' used by your lecturers, in which case you will need to adapt appropriately (see Box 5.1).*

A useful scheme for describing learning styles is the VARK system devised by Fleming (2001). By answering a short online questionnaire, you can 'diagnose' yourself as one of the types shown in Box 5.1, which also summarises important outcomes relating to how information and concepts can be assimilated, learned and expressed. People show different degrees of alignment with these categories; and research indicates that the majority of students are multi-modal learners – that is, falling into more than one category – rather than being only in one grouping. By carrying out an analysis like this, you can become more aware of your personal characteristics and think about whether the methods of studying you currently use are those that are best suited to your needs.

***KEY POINT** Having a particular learning preference or style does not mean that you are automatically skilled in using methods generally suited to that type of learner. You must work at developing your ability to take in information, study and cope with assessment.*

Box 5.1 How to diagnose your learning preferences using the VARK learning styles scheme

Visit www.vark-learn.com to carry out the online diagnostic test, reflect on whether it is a fair description of your preferences, and think about whether you might change the way you study to improve your performance. None of the outcomes should be regarded as prescriptive – you should mix techniques as you see fit and only use methods that you feel comfortable adopting. Adapted with permission from material produced by Fleming (2001).

	Outcomes for your learning, studying and exam technique		
Learning style ***Description of learning preferences***	**Advice for taking in information and understanding it**	**Best methods of studying for effective learning**	**Ways to cope with exams so you perform better**
Visual: *You are interested in colour, layout and design. You probably prefer to learn from visual media or books with diagrams and charts. You tend to add doodles and use highlighters on lecture and revision notes and express ideas and concepts as images.*	Use media incorporating images, diagrams, flow-charts etc. When constructing notes, employ underlining, different colours and highlighters. Use symbols as much as you can, rather than words. Leave plenty of white space in your notes. Experiment with the 'mind map' style of note-taking (p. 16).	Use similar methods to those described in column two. Reduce lecture notes to pictures. Try to construct your own images to aid understanding, then test your learning by redrawing these from memory.	Plan answers diagrammatically. Recall the images and doodles you used in your notes. Use diagrams in your answers (making sure they are numbered and fully labelled). As part of your revision, turn images into words.
Aural: *You prefer discussing subjects and probably like to attend tutorials and listen to lecturers, rather than read textbooks. Your lecture notes may be poor because you would rather listen than take notes.*	Make sure you attend classes, discussions and tutorials. Note and remember the interesting examples, stories, jokes. Leave spaces in your notes for later recall and 'filling'. Discuss topics with a 'study buddy'. Record lectures (with lecturer's permission).	Expand your notes by talking with others and making additional notes from the textbook. Ask others to 'hear' you talk about topics. Read your summarised notes aloud to yourself. Record your vocalised notes and listen to them later.	When writing answers, imagine you are talking to an unseen examiner. Speak your answers inside your head. Listen to your voices and write them down. Practise writing answers to old exam questions.
Read–Write: *You prefer using text in all formats. Your lecture notes are probably good. You tend to like lecturers who use words well and have lots of information in sentences and notes. In note-taking, you may convert diagrams to text and text to bullet points.*	Focus on note-taking. You may prefer the 'linear' style of note-taking (p. 101). Use the following in your notes: lists; headings; glossaries and lists of definitions. Expand you notes by adding further information from handouts, textbooks and library readings	Reduce your notes to lists or headings. Write out and read the lists again and again (silently). Turn actions, diagrams, charts and flowcharts into words. Rewrite the ideas and principles into other words. Organise diagrams and graphs into statements, e.g. 'The trend is . . .'.	Plan and write out exam answers using remembered lists. Arrange your words into hierarchies and points.

(continued)

Box 5.1 (continued)

Kinesthetic: *You tend to recall by remembering real events and lecturers' 'stories'. You probably prefer field excursions and lab work to theory and like lecturers who give real-life examples. Your lecture notes may be weak because the topics did not seem 'concrete' or 'relevant'.*	Focus on examples that illustrate principles. Concentrate on applied aspects and hands-on approaches, but try to understand the theoretical principles that underpin them. When taking in information, use all your senses – sight, touch, taste, smell, hearing.	Put plenty of examples, pictures and photographs into your notes. Use case studies and applications to help with principles and concepts. Talk through your notes with others. Recall your experience of lectures, tutorials, experiments or field trips.	Write practice answers and paragraphs. Recall examples and things you did in the lab or field trip. Role-play the exam situation in your own room.
Multi-modal: *Your preferences fall into two or more of the above categories. You are able to use these different modes as appropriate.*	If you are diagnosed as having two dominant preferences or several equally dominant preferences, read the study strategies above that apply to each of these. You may find it necessary to use more than one strategy for learning and communicating, feeling less secure with only one.		

Example A set of learning objectives taken from an introductory lecture on bacterial cell structure.
After this lecture, you should be able to:
- Define the following terms:
  - ✓ prokaryote
  - ✓ eukaryote
  - ✓ envelope
  - ✓ fimbriae
  - ✓ F pilus
  - ✓ plasmid
- Draw a labelled diagram to illustrate the principal components of a bacterial cell.
- Explain the functions of the major cellular components.
- Demonstrate knowledge of the relative magnitude of bacteria and eukaryotic cells, in terms of typical linear dimensions and volumes.
- Describe the basic process of cell division and give examples of typical timescales for different bacteria, e.g. *Escherichia coli, Clostridium perfringens, Mycobacterium tuberculosis.*

Thinking about thinking

The thinking processes that students are expected to carry out can be presented in a sequence, starting with shallower thought processes and ending with deeper processes, each of which builds on the previous level (see Table 5.1). The first two categories in this ladder apply to gaining basic knowledge and understanding, important when you first encounter a topic. Processes three to six are those additionally carried out by high-performing university students, with the latter two being especially relevant to final-year students, researchers and professionals. Naturally, the tutors assessing you will want to reward the deepest thinking appropriate for your level of study. This is often signified by the words they use in assessment tasks and marking criteria (column four, Table 5.1, and p. 108), and while this is not an exact process, being more aware of this agenda can help you to gain more from your studies and appreciate what is being demanded of you.

***KEY POINT** When considering assessment questions, look carefully at words used in the instructions. These cues can help you identify what depth is expected in your answer. Take special care in multi-part questions, because the first part may require lower-level thinking, while in later parts, marks may be awarded for evidence of deeper thinking.*

The role of assessment and feedback in your learning

Your starting point for assessment should be the learning outcomes or objectives for each module, topic or learning activity. You will usually find them in your module handbook. They state in clear terms what your tutors expect you to be able to accomplish after participating in each part and reading around the topic. Also of value will be marking/assessment

Table 5.1 A ladder of thinking processes, moving from shallower thought processes (top of table) to deeper levels of thinking (bottom of table). This table is derived from research by Benjamin Bloom *et al.* (1956). When considering the cue words in typical question instructions, bear in mind that the precise meaning will always depend on the context. For example, while 'describe' is often associated with relatively simple processes of recall, an instruction like 'describe how the human brain works' demands higher-level understanding. Note also that while a 'cue word' is often given at the start of a question/instruction, this is not universally so.

Thinking processes and description (in approximate order of increasing 'depth')	Example in life sciences	Example of typical question structure, with *cue word* highlighted	Other cue words used in question instructions
1. **Knowledge (knowing facts).** If you know information, you can *remember* or *recognise* it. This does not always mean you understand it at a higher level.	You might know the order of bases in a piece of DNA but not understand what this means.	*Describe* the main components of a biological membrane.	• define • list • state • identify
2. **Comprehension.** If you comprehend a fact, you *understand* what it means.	You might know the order of bases in a piece of DNA and understand that they code in triplets for specific amino acids.	*Explain* how membrane components are involved in the accumulation of solutes within living cells.	• contrast • compare • distinguish • interpret
3. **Application.** To apply a fact means that you can *put it to use* in a particular context.	You might be able to take the DNA base sequence and work out the amino acid sequence of the protein for which they code.	Using the Nernst equation, and realistic values for the membrane potential and solute concentrations, *demonstrate* how Na^+ ions must be actively transported out of the cells of marine organisms.	• calculate • illustrate • solve • show
4. **Analysis.** To analyse information means that you are able to *break it down into parts* and show how these components *fit together.*	You might be able to construct a three-dimensional model of a protein derived from the base sequence.	Drawing on information about membrane structure, *defend* the endosymbiotic theory of eukaryote evolution.	• compare • explain • consider • infer
5. **Synthesis.** To synthesise, you need to be able to *extract relevant facts* from a body of knowledge and use these to *address an issue in a novel way* or *create something new.*	You might be able to work out the function of a protein for which you know the sequence of bases, based on a comparison with other like proteins.	*Devise* an experiment to test the hypothesis that a specific membrane fraction contains a functional ATPase involved in glucose transport.	• design • integrate • test • create
6. **Evaluation.** If you evaluate information, you *arrive at a judgement* based on its importance relative to the topic being addressed.	You might be able to comment on theories about how a protein has evolved, by considering the structure of related proteins and relating this to the taxonomic position of their source species.	*Evaluate* the relative importance of passive and active transport in the accumulation of heavy metal salts by the main groups of soil fungi.	• review • assess • consider • justify

criteria or grade descriptors, which state in general terms what level of attainment is required for your work to reach specific grades. These are more likely to be defined at faculty/college/school/department level and consequently published in appropriate handbooks or websites. Reading learning outcomes and grade descriptors will give you a good idea of what to expect and the level of performance required to reach your personal goals. Relate them to both the material covered (e.g. in lectures and practicals, or online) and past exam papers. Doing this as you study and revise will indicate whether further reading and independent studying is required, and of what type. You will also have a much clearer picture of how you are likely to be assessed.

Definitions

Learning objectives/outcomes – statements of the knowledge, understanding or skills that a learner will be able to demonstrate on successful completion of a module, topic or learning activity.

Formative assessments – these may be mid-term or mid-semester tests and are often in the same format as later exams. They are intended to give you feedback on your performance. You should use the results to measure your performance against the work you put in, and to find out, either from grades or tutor's comments, how you could do better in future. If you don't understand the reason for your grade, talk to your tutor.

Summative assessments – these include end-of-year or end-of-module exams. They inform others about the standard of your work. In continuous or 'in-course' assessment, the summative elements are spread out over the course. Sometimes these exams may involve a formative aspect, if feedback is given.

***KEY POINT** Use the learning objectives for your course (normally published in the handbook) as a fundamental part of your revision planning. These indicate what you will be expected to be able to do after taking part in the course, so exam questions are often based on them. Check this by reference to past papers.*

There are essentially two types of assessment – formative and summative, although the distinction may not always be clear-cut (see margin). The first way you can learn from formative assessment is to consider the grade you obtained in relation to the work you put in. If this is a disappointment to you, then there must be a mismatch between your understanding of the topic and the marking scheme and that of the marker, or a problem in the writing or presentation of your assignment. This element of feedback is also present in summative assessment.

The second way to learn from formative assessment is through the written feedback and notes on your work. These comments may be cryptic, or scribbled hastily, so if you don't understand or can't read them, ask the tutor who marked the work. Most tutors will be pleased to explain how you could have improved your mark. If you find that the same comments appear frequently, it may be a good idea to seek help from your university's academic support unit. Take along examples of your work and feedback comments, so they can give you the best possible advice. Another suggestion is to ask to see the work of another student who obtained a good mark, and compare it with your own. This will help you judge the standard you should be aiming for.

Preparing for revision and examinations

Before you start revising, find out as much as you can about each exam, including:

- format and duration;
- date and location;
- types of question;
- whether any questions/sections are compulsory;
- whether the questions are internally or externally set or assessed;
- whether the exam is 'open book', and if so, which texts or notes are allowed.

Your course tutor is likely to give you details of exam structure and timing well beforehand, so that you can plan your revision; the course handbook and past papers (if available) can provide further useful details. Always check that the nature of the exam has not changed before you consult past papers.

__Time management when revising__ – this is vital to success and is best achieved by creating a revision timetable (Box 5.2).

Organising and using lecture notes, assignments and practical reports

__Filing lecture notes__ – make sure your notes are kept neatly and in sequence by using a ring binder system. File the notes in lecture or practical sequence, adding any supplementary notes or photocopies alongside.

Given their importance as a source of material for revision, you should have sorted out any deficiencies or omissions in your lecture notes and practical reports at an early stage. For example, you may have missed a lecture or practical due to illness, etc., but the exam is likely to assume attendance throughout the year. Make sure you attend classes whenever possible and keep your notes up to date. Your practical reports and any assignment work will contain specific comments from the teaching staff, indicating where

Box 5.2 How to prepare and use a revision timetable

1. **Make up a grid showing the number of days until your exams are finished.** Divide each day into several sections. If you like revising in large blocks of time, use am, pm and evening slots, but if you prefer shorter periods, divide each of these in two, or use hourly divisions (see also the table in study exercise 5.1).
2. **Write in your non-revision commitments,** including any time off you plan to allocate and physical activity at frequent intervals. Try to have about one-third or a quarter of the time off in any one day. Plan this in relation to your best times for useful work – for example, some people work best in the mornings, while others prefer evenings. If you wish, use a system where your relaxation time is a bonus to be worked for; this may help you motivate yourself.
3. **Decide on how you wish to subdivide your subjects** for revision purposes. This might be among subjects, according to difficulty (with the hardest getting the most time), or within subjects, according to topics. Make sure there is an adequate balance of time among topics and especially that you do not avoid working on the subject(s) you find least interesting or most difficult.
4. **Allocate the work to the different slots available on your timetable.** You should work backwards from the exams, making sure that you cover every exam topic adequately in the period just before each exam. You may wish to colour-code the subjects.
5. **As you revise, mark off the slots completed** – this has a positive psychological effect and will boost your self-confidence.
6. **After the exams, revisit your timetable** and decide whether you would do anything differently next time.

***Using feedback from tutors** – it is always worth reading any comments on your work as soon as it is returned. If you don't understand the comments, or are unsure about why you might have lost marks in an assignment, ask for an explanation.*

marks were lost, corrections, mistakes, inadequacies, etc. Most lecturers are quite happy to discuss such details with students on a one-to-one basis and this information may provide you with clues to the expectations of individual lecturers that may be useful in exams set by the same members of staff. However, you should never 'fish' for specific information on possible exam questions, as this is likely to be counter-productive.

Revision

Begin early, to avoid last-minute panic. Start in earnest several weeks beforehand, and plan your work carefully:

- Prepare a revision timetable – an 'action plan' that gives details of specific topics to be covered (Box 5.2). Plan your revision around when (and where) your examinations are to be held. Try to keep to your timetable. Time management during this period is as important as keeping to time during the exam itself.
- Study the learning objectives/outcomes for each topic (usually published in the course handbook) to get an idea of what lecturers expect from you.
- Use past papers as a guide to the form of exam and the type of question likely to be asked (Box 5.3).
- Remember to have several short (5 minute) breaks during each hour of revision and a longer break every few hours. In any day, try to work for a maximum of three-quarters of the time.
- Include recreation within your schedule: there is little point in tiring yourself with too much revision, as this is unlikely to be profitable.
- Make your revision as active and interesting as possible (see below): the least productive approach is simply to read and reread your notes.
- Ease back on the revision near the exam: plan your revision to avoid last-minute cramming and overload fatigue.

***Recognise when your concentration powers are dwindling** – take a short break when this happens and return to work refreshed and ready to learn. Remember that 20 minutes is often quoted as a typical limit to full concentration effort.*

Box 5.3 How to use past exam papers in your revision

Past exam papers are a valuable resource for targeting your revision.

1. **Find out where the past exam papers are kept.** Copies may be lodged in your department or the library; or they may be accessible online.
2. **Locate and copy relevant papers for your module(s).** Check with your tutor or course handbook that the style of paper will not change for the next set of examinations.
3. **Analyse the design of the exam paper.** Taking into account the length in weeks of your module, and the different lecturers and/or topics for those weeks, note any patterns that emerge. For example, can you translate weeks of lectures/practicals into numbers of questions or sections of the paper? Consider how this might affect your revision plans and exam tactics, taking into account (a) any choices or restrictions offered in the paper, and (b) the different types of questions asked (i.e. multiple choice, short-answer or essay).
4. **Examine carefully the style of questions.** Can you identify the expectations of your lecturers? Can you relate the questions to the learning objectives? How much extra reading do they seem to expect? Are the questions fact-based? Do they require a synthesis based on other knowledge? Can you identify different styles for different lecturers? Consider how the answers to these questions might affect your revision effort and exam strategy.
5. **Practise answering questions.** Perhaps with friends, set up your own mock exam once you have done a fair amount of revision, but not too close to the exams. Use a relevant past exam paper and don't study it beforehand. You need not attempt all of the paper at one sitting. You'll need a quiet room in a place where you will not be interrupted (e.g. a library). Keep close track of time during the mock exam and try to do each question in the length of time you would normally assign to it (see p. 31) – this gives you a feel for the speed of thought and writing required and the scope of answer possible. Mark each other's papers and discuss how each of you interpreted the question and laid out your answers and your individual marking schemes.
6. **Practise writing answer plans and starting answers.** This can save time compared to the 'mock exam' approach. Practice in starting answers can help you get over stalling at the start and wasting valuable time. Writing essay plans gets you used to organising your thoughts quickly and putting your thoughts into a logical sequence.

Aiding recall through effective note-taking *– the Mind Map technique (p. 16), when used to organise ideas, is claimed to enhance recall by connecting the material to visual images or linking it to the physical senses.*

Active revision

The following techniques may prove useful in devising an active revision strategy:

- 'Distil' your lecture notes to show the main headings and examples. Prepare revision sheets with details for a particular topic on a single sheet of paper, arranged as a numbered checklist. Wall posters are another useful revision aid.
- Confirm that you know the material by testing yourself – take a blank sheet of paper and write down all you know. Check your full notes to see if you missed anything out. If you did, go back immediately to a fresh blank sheet and redo the example. Repeat, as required.
- Memorise definitions and key phrases: definitions can be a useful starting point for many exam answers. Make up lists of relevant facts or definitions associated with particular topics. Test yourself repeatedly on these, or get a friend to do this. Try to remember *how many* facts or definitions you need to know in each case – this will help you recall them all during the exam.
- Use mnemonics and acronyms to commit specific factual information to memory. Sometimes, the dafter they are, the better they seem to work.

Question-spotting *– avoid this risky strategy. Lecturers are aware that this approach may be taken and try to ask questions in an unpredictable manner. You may find that you are unable to answer on unexpected topics which you failed to revise. Moreover, if you have a preconceived idea about what will be asked, you may also fail to grasp the nuances of the exact question set, and thereby fail to provide a focused answer.*

Revision checks *– it is important to test yourself frequently during revision, to ensure that you have retained the information you are revising.*

Final preparations *– try to get a good night's sleep before an exam. Last-minute cramming will be counter-productive if you are too tired during the exam.*

- Use pattern diagrams or mind maps as a means of testing your powers of recall on a particular topic (p. 16).
- Draw diagrams from memory: make sure you can label them fully.
- Try recitation as an alternative to written recall. Talk about your topic to another person, preferably someone in your class. Talk to yourself if necessary. Explaining something out loud is an excellent test of your understanding.
- Associate facts with images or journeys if you find this method works.
- Use a wide variety of approaches to avoid boredom during revision (e.g. record information on audio tape, use cartoons, or any other method, as long as it's not just reading).
- Form a revision group to share ideas and discuss topics with other students.
- Prepare answers to past papers, e.g. write essays or, if time is limited, write essay plans (see Box 5.3).
- If your subject involves numerical calculations, work through representative problems.
- Make up your own questions: the act of putting yourself in the examiner's mind-set by inventing questions can help revision. However, you should not rely on 'question spotting': this is risky!

The evening before your exam should be spent in consolidating material, and checking through summary lists and plans. Avoid introducing new material at this late stage: your aim should be to boost your confidence, putting yourself in the right frame of mind for the exam itself.

Text references

Bloom, B., Englehart, M. Furst, E., Hill, W. and Krathwohl, D. (1956) *Taxonomy of Educational Objectives: The Classification of Educational Goals. Handbook I: Cognitive Domain.* New York, Toronto: Longmans, Green.

Fleming, N.D. (2001) *Teaching and Learning Styles: VARK Strategies.* Neil Fleming, Christchurch.

Fleming, N.D. *VARK: A Guide to Learning Styles.* Available: http://www.vark-learn.com/ Last Accessed 01/04/07.

Sources for further study

Burns, R. (1997) *The Student's Guide to Passing Exams.* Kogan Page, London.

Hamilton, D. (1999) *Passing Exams: A Guide for Maximum Stress and Minimum Stress.* Cassell, London.

Many universities host study skills websites; these can be found using 'study skills', 'revision' or 'exams' as key words in a search engine.

Study exercises

5.1 **Draw up a revision timetable.** Use the techniques discussed in Box 5.2 to create a revision timetable for your forthcoming exams. You may wish to use or adapt the following arrangement, either on paper or within a spreadsheet.

A revision timetable planner

Date	Morning		Lunch	Afternoon		Tea/Dinner	Evening	
	Session 1	Session 2		Session 1	Session 2		Session 1	Session 2

5.2 **Make use of past exam papers.** Use the techniques discussed in Box 5.2 to improve your revision strategy: assess their effectiveness in a particular exam, or series of exams.

5.3 **Try out new active revision techniques.** Try any or all of the methods mentioned on pages 26–8 when revising. Compare notes with a colleague – which seems to be the most successful technique for you and for the topic you are revising?

6 Curriculum options, assessments and exams

> **Definition**
>
> **Transcript** – this is your record of achievement at university. Normally it will consist of details of each module or course you have taken, and an indication of the grade or mark achieved. It will also show your final (honours) classification, that is: first class, upper second class (2.1), lower second class (2.2), third class or unclassified (note: some UK universities do not differentiate second class degrees).

Many universities operate a modular system for their biosciences degree courses. This allows greater flexibility in subject choice and accommodates students studying on different degree paths. Modules also break a subject into discrete, easily assimilated elements. They have the advantage of spreading assessment over the academic year, but they can also tempt you to avoid certain difficult subjects or to feel that you can forget about a topic once the module is finished.

***KEY POINT** You should select your modules with care, mindful of potential degree options and how your transcript and CV will appear to a prospective employer. If you feel you need advice, consult your personal tutor or study advisor.*

> ***Aiming high** – your goal should be to perform at your highest possible level and not simply to fulfil the minimum criteria for progression. This will lay sound foundations for your later studies. Remember too that a future employer might ask to see your academic transcript, which will detail all your module grades including any fails/resits, and will not just state your final degree classification.*

As you move between levels of the university system, you will be expected to have passed a certain number of modules, as detailed in the progression criteria. These may be expressed using a credit point system. Students are normally allowed two attempts at each module exam and the resits often take place at the end of the summer vacation. If a student does not pass at the second attempt, they may be asked to 'carry' the subject in a subsequent year, and in severe cases of multiple failure, they may be asked to re-take the whole year or even leave the course. Consequently, it is worth finding out about these aspects of your degree. They are usually published in relevant handbooks.

You are unlikely to have reached this stage in your education without being exposed to the examination process. You may not enjoy being assessed, but you probably want to do well in your course. It is therefore important to understand why and how you are being tested. Identifying and improving the skills required for exam success will allow you to perform to the best of your ability.

Assessed coursework

> ***Avoiding plagiarism** – this is a key issue for assessed coursework - see p. 51 for a definition and Chapter 8 for appropriate methods of referring to the ideas and results of others using citation.*

There is a component of assessed coursework in many modules. This often tests specific skills, and may require you to demonstrate thinking at deeper levels (Table 5.1). The common types of coursework assessment are covered at various points in this book:

- practical exercises (Chapters 20–67);
- essays (Chapters 16 and 17);
- numerical problems (Chapter 65);
- data analysis (Chapters 65–67);
- poster and spoken presentations (Chapters 14 and 15);
- literature surveys and reviews (Chapter 19);
- project work (Chapters 18 and 31);
- problem-based learning (Box 6.1).

At the start of each year or module, read the course handbook or module guide carefully to find out when any assessed work needs to be submitted. Note relevant dates in your diary, and use this information to plan your work. Take special note if deadlines for different modules clash, or if they coincide with social or sporting commitments.

Box 6.1 Problem-based learning (PBL)

In this relatively new teaching method, you are likely to be presented with a 'real world' problem or issue, often working within a team. As you tackle the problem, you will gain factual knowledge, develop skills and exercise critical thinking (Chapter 9). Because there is a direct and relevant context for your work, and because you have to employ active learning techniques, the knowledge and skills you gain are likely to be more readily assimilated and remembered. This approach also more closely mimics workplace practices. PBL usually proceeds as follows:

1. **You are presented with a problem** (e.g. a case study, a hypothetical patient, a topical issue).
2. **You consider what issues and topics you need to research,** by discussion with others if necessary. You may need to identify where relevant resources can be found (Chapters 8–11).
3. **You then need to rank the issues and topics in importance,** allocating tasks to group members, if appropriate. A structured approach is an important aspect of PBL.
4. **Having carried out the necessary research, you should review what information has been obtained.** As a result, new issues may need to be explored and, where appropriate, allocated to group members.
5. **You will be asked to produce an outcome, such as a report, diagnosis, seminar presentation or poster.** An outline structure will be required, and for groups, further allocation of tasks will be needed to accomplish this goal.

If asked to carry out PBL as part of your course, it is important to get off to a good start. At first, the problem may seem unfamiliar. However, once you become involved in the work, you will quickly gain confidence. If working as part of a group, make sure that your group meets as early as possible, that you attend all sessions and that you do the necessary background reading. When working in a team, a degree of self-awareness is necessary regarding your 'natural' role in group situations (Table 3.1). Various methods are used for assessing PBL, including written, oral and poster presentations, and the assessment may involve peer marking.

***KEY POINT** If, for some valid reason (e.g. illness), you will be late with an assessment, speak to your tutors as soon as possible. They may be able to take extenuating circumstances into account by not applying a marking penalty. They will let you know what paperwork you may be required to submit to support your claim.*

Summative exams – general points

Summative exams (p. 24) normally involve you answering questions without being able to consult other students or your notes. Invigilators are present to ensure appropriate conduct, but departmental representatives may be present for some of the exam. Their role is to sort out any subject-related problems, so if you think something is wrong, ask at the earliest opportunity. It is not unknown for parts of questions to be omitted in error, or for double meanings to arise, for example.

Planning

When preparing for an exam, make a checklist of the items you'll need (see p. 35). On the day of the exam, give yourself sufficient time to arrive at the correct room, without the risk of being late. Double-check the times and places of your exams, both well before the exam, and also on arrival. If you arrive at the exam venue early, you can always rectify a mistake if you find you've gone to the wrong place.

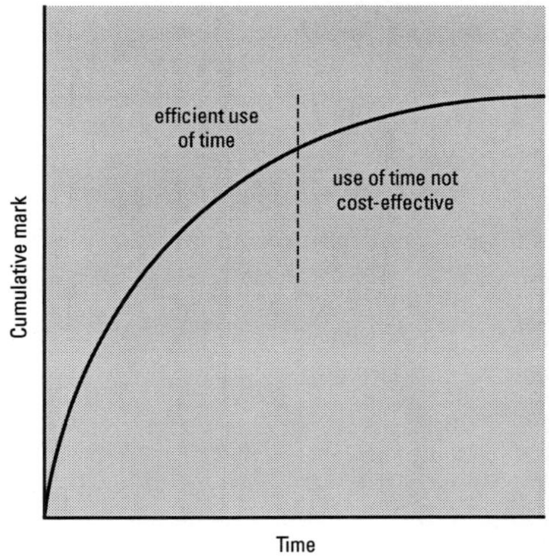

Fig. 6.1 Exam marks as a function of time. The marks awarded in a single answer will follow the law of diminishing returns – it will be far more difficult to achieve the final 25% of the available marks than the initial 25%. Do not spend too long on any one question.

***Using the question paper** – unless this is specifically forbidden, you should write on the question paper to plan your strategy, keep to time and organise answers.*

***Checking exam answers** – look for:*

- *errors of fact;*
- *missing information*
- *grammatical and spelling errors;*
- *errors of scale and units;*
- *errors in calculations.*

***Adopting different tactics according to the exam** – you should adjust your exam strategy (and revision methods) to allow for the differences in question types used in each exam paper.*

Tackling the paper

Begin by reading the instructions at the top of the exam paper carefully, so that you do not make any errors based on lack of understanding of the exam structure. Make sure that you know:

- how many questions are set;
- how many must be answered;
- whether the paper is divided into sections;
- whether any parts are compulsory;
- what each question/section is worth, as a proportion of the total mark;
- whether different questions should be answered in different books.

Do not be tempted to spend too long on any one question or section: the return in terms of marks will not justify the loss of time from other questions (see Fig. 6.1). Take the first 10 minutes or so to read the paper and plan your strategy, before you begin writing. Do not be put off by those who begin immediately; it is almost certain they are producing unplanned work of a poor standard.

Underline the key phrases in the instructions, to reinforce their message. Next, read through the set of questions. If there is a choice, decide on those questions to be answered and decide on the order in which you will tackle them. Prepare a timetable which takes into account the amount of time required to complete each section and which reflects the allocation of marks – there is little point in spending one-quarter of the exam period on a question worth only 5 per cent of the total marks. Use the exam paper to mark the sequence in which the questions will be answered and write the finishing times alongside; refer to this timetable during the exam to keep yourself on course.

Reviewing your answers

At the end of the exam, you should allow at least 10 minutes to check through your script. Make sure your name and/or ID number is on each exam book as required and on all other sheets of paper, including graph paper, even if securely attached to your script, as it is in your interest to ensure that your work does not go astray.

KEY POINT** **Never leave any exam early. Most exams assess work carried out over several months within a time period of 2–3 hours and there is always something constructive you can do with the remaining time to improve your script.

Special considerations for different types of exam question

Essay questions

Essay questions let examiners test the depth of your comprehension and understanding as well as your recall of facts. Essay questions give you plenty of scope to show what you know. They suit those with a good grasp of principles but who perhaps have less ability to recall individual details.

Before you tackle a particular question, you must be sure of what is required in your answer. Ask yourself 'What is the examiner looking for in this particular question?' and then set about providing a *relevant* answer. Consider each word in the question and highlight, underline or circle the key words. Make sure you know the meaning of the terms

Box 6.2 Writing under exam conditions

Make sure you go into an exam with a strategy for managing the available time.

- **Allocate some time (say 5 per cent of the total) to consider which questions to answer and in which order.**
- **Share the rest of the time among the questions, according to the marks available.** Aim to optimise the marks obtained. A potentially good answer should be allocated slightly more time than one you don't feel so happy about. However, don't concentrate on any one answer (see Fig. 6.1).
- **For each question divide the time into planning, writing and revision phases** (see p. 107).

Employ time-saving techniques as much as possible:

- **Use spider diagrams** (Fig. 16.2) **or mind maps** (Fig. 4.2) to organise and plan your answer.
- **Use diagrams and tables** to save time in making difficult and lengthy explanations, but make sure you refer to each one in the text.
- **Use standard abbreviations** to save time repeating text but always explain them at the first point of use (e.g. PCR, polymerase chain reaction).
- **Consider speed of writing and neatness** – especially when selecting the type of pen to use – ballpoint pens are fastest, but they can tend to smudge. You can only gain marks if the examiner can read your script.
- **Keep your answer simple and to the point**, with clear explanations of your reasoning.

Make sure your answer is relevant.

- **Don't include irrelevant facts** just because you memorised them during revision, as this may do you more harm than good. You must answer the specific question that has been set.
- **Time taken to write irrelevant material is time lost from another question.**

given in Table 17.1 so that you can provide the appropriate information, where necessary. Spend some time planning your writing (see Chapter 16). Refer back to the question frequently as you write, to confirm that you are keeping to the subject matter. Box 6.2 gives advice on writing essays under exam conditions.

It is usually a good idea to begin with the question that you are most confident about. This will reassure you before tackling more difficult parts of the paper. If you run out of time, write in note form. Examiners are usually understanding, as long as the main components of the question have been addressed and the intended structure of the answer is clear. Common reasons for poor exam answers in essay-style questions are listed in Box 6.3.

Penalties for guessing – *if there is a penalty for incorrect answers in a multiple choice test, the best strategy is not to answer questions when you know your answer is a complete guess. Depending on the penalty, it may be beneficial to guess if you can narrow the choice down to two options (but beware false or irrelevant alternatives). However, if there are no such penalties, then you should provide an answer to all questions.*

Multiple-choice and short-answer questions

Multiple-choice questions (MCQs) and short-answer questions (SAQs) are generally used to test the breadth and detail of your knowledge. The various styles that can be encompassed within the SAQ format allow for more demanding questions than MCQs, which may emphasise memory work and specific factual information.

A good approach for MCQ papers is as follows:

1. First trawl: read through the questions fairly rapidly, noting the 'correct' answer in those you can attempt immediately, perhaps in pencil.
2. Second trawl: go through the paper again, checking your original answers and this time marking up the answer sheet properly.
3. Third trawl: now tackle the difficult questions and those that require longer to answer (e.g. those based on numerical problems).

Box 6.3 Reasons for poor exam answers to essay-style questions

The following are reasons that lecturers often cite when they give low marks for essay answers:

- **Not answering the exact question set.** Either failing to recognise the specialist terms used in the question, failing to demonstrate an understanding of the terms by not providing definitions, failing to carry out the precise instruction in a question, or failing to address all aspects of the question.
- **Running out of time.** Not matching the time allocated to the extent of the answer. Frequently, this results in spending too long on one question and not enough on the others, or even failing to complete the paper.
- **Not answering all parts** of a multiple part question, or not recognising that one part (perhaps involving more complex ideas) may carry more marks than another.
- **Failing to provide evidence** to support an answer. Forgetting to state the 'obvious' – either basic facts or definitions.
- **Failing to illustrate an answer appropriately**, either by not including a relevant diagram, or by providing a diagram that does not aid communication, or by not including examples.
- **Incomplete answer(s).** Failing to answer appropriately due to lack of knowledge.
- **Providing irrelevant evidence** to support an answer. 'Waffling' to fill space.
- **Illegible handwriting.**
- **Poor English**, such that facts and ideas are not expressed clearly.
- **Lack of logic** or structure to the answer.
- **Factual errors**, indicating poor note-taking, poor revision or poor recall.
- **Failing to correct obvious mistakes** by re-reading an answer before submitting the script.

At higher levels, the following aspects are especially important:

- **Not providing enough in-depth information.**
- **Providing a descriptive rather than an analytical answer** – focusing on facts, rather than deeper aspects of a topic.
- **Not setting a problem in context,** or not demonstrating a wider understanding of the topic. However, make sure you don't overdo this, or you may risk not answering the question set.
- **Not giving enough evidence of reading around the subject.** Wider reading can be demonstrated by quoting relevant papers and reviews and by giving author names and dates of publication.
- **Not considering both sides of a topic/debate, or not arriving at a conclusion.**

One reason for adopting this three-phase approach is that you may be prompted to recall facts relevant to questions looked at earlier. You can also spend more time per question on the difficult ones.

When unsure of an answer, the first stage is to rule out options that are clearly absurd or have obviously been placed there to distract you. Next, looking at the remaining options, can you judge between contrasting pairs with alternative answers? Logically, both cannot be correct, so you should see if you can rule one of the pair out. Watch out, however, in case *both* may be irrelevant to the answer. If the question involves a calculation, try to do this independently from the answers, so you are not influenced by them.

Answer the question as requested *– this is true for all questions, but especially important for SAQs. If the question asks for a diagram, make sure you provide one; if it asks for a specified number of aspects of a topic, try to list this number of points; if there are two or more parts, provide appropriate answers to all aspects. This may seem obvious, but many marks are lost for not following instructions.*

In SAQ papers, there may be a choice of questions. Choose your options carefully – it may be better to gain half marks for a correct answer to half a question, than to provide a largely irrelevant answer that apparently covers the whole question but lacks the necessary detail. For the SAQ form of question, few if any marks are given for writing style. Think in 'bullet points' and list the crucial points only. The time for answering SAQ questions may be tight, so get down to work fast, starting with answers that demand remembered facts. Stick to your timetable by moving on to the next question as soon as possible.

Strategically, it is probably better to get part-marks for the full number of questions than good marks for only a few.

Practical and information-processing exams

The prospect of a practical or information-processing exam may cause you more concern than a theory exam. This may be due to a limited experience of practical examinations, or to the fact that practical and observational skills are tested, as well as recall, description and analysis of factual information. Your first thoughts may be that it is not possible to prepare for such exams but, in fact, you can improve your performance by mastering the various practical techniques described in this book.

You may be allowed to take your laboratory reports and other texts into the practical exam. Don't assume that this is a soft option, or that revision is unnecessary: you will not have time to read large sections of your reports or to familiarise yourself with basic principles, etc. The main advantage of 'open book' exams is that you can check specific details of methodology, reducing your reliance on memory, provided you know your way around your practical manual and notes. In all other respects, your revision and preparation for such exams should be similar to theory exams. Make sure you are familiar with all of the practical exercises, including any work carried out in class by your partner (since exams are assessed on individual performance). If necessary, check with the teaching staff to see whether you can be given access to the laboratory, to complete any exercises that you have missed.

At the outset of the practical exam, determine or decide on the order in which you will tackle the questions. A question in the latter half of the paper may need to be started early on in the exam period (e.g. an enzyme assay requiring 2-hour incubation in a 3-hour exam). Such questions are included to test your forward-planning and time-management skills. You may need to make additional decisions on the allocation of material; e.g. if you are given 30 sterile test tubes, there is little value in designing an experiment that uses 25 of these to answer question 1, only to find that you need at least 15 tubes for subsequent questions.

Make sure you explain your choice of apparatus and experimental design. Calculations should be set out in a stepwise manner, so that credit can be given, even if the final answer is incorrect. If there are any questions that rely on recall of factual information and you are unable to remember specific details, make sure that you describe the item fully, so that you gain credit for observational skills. Alternatively, leave a gap and return to the question at a later stage.

Examples These are principal types of question you are likely to encounter in a practical or information-processing exam:

Manipulative exercises Often based on work carried out during your practical course. Tests dexterity, specific techniques (e.g. sterile technique, p. 203).

'Spot' tests Short questions requiring an identification, or brief descriptive notes on a specific item (e.g. a prepared slide). Tests knowledge of seen material or ability to transfer this to a new example.

Calculations May include the preparation of aqueous solutions at particular concentrations and statistical exercises. Tests numeracy.

Data analyses May include the preparation and interpretation of graphs (p. 465) and numerical information, from data either obtained during the exam or provided by the examiner. Tests problem-solving skills.

Preparing specimens for examination with a microscope Tests staining technique and light microscopy technique (p. 163).

Interpreting images Sometimes used when it is not possible to provide living specimens, e.g. in relation to electron microscopy. Can test a variety of skills.

Oral exams and interviews

An oral interview is sometimes a part of final degree exams, representing a chance for the external examiner(s) to meet individual students and to test their abilities directly and interactively. In some departments, orals are used to validate the exam standard, or to test students on the borderline between exam grades. Sometimes an interview may form part of an assessment, as with project work or posters. This type of exam is often intimidating – many students say they don't know how to revise for an oral – and many candidates

***Terminology** – the oral exams are sometimes known simply as 'orals' or, borrowing Latin, as 'viva voce' (by the living voice) exams or 'vivas'.*

worry that they will be so nervous they won't be able to do themselves justice.

Preparation is just as important for oral exams as it is for written exams:

- Think about your earlier performances – if the oral follows written papers, it may be that you will be asked about questions you did not do so well on. These topics should be revised thoroughly. Be prepared to say how you would approach the questions if given a second chance.
- Read up a little about the examiner – he or she may focus their questions in their area of expertise.
- Get used to giving spoken answers – it is often difficult to transfer between written and spoken modes. Write down a few questions and get a friend to ask you them, possibly with unscripted follow-up queries.
- Research and think about topical issues in your subject area – some examiners will feel this reflects how interested you are in your subject.

Allow yourself to relax in oral exams *– external examiners are experienced at putting students at ease. They will start by asking 'simple-to-answer' questions, such as what modules you did, how your project research went, and what your career aspirations are. Look on the external as a friend rather than a foe.*

Your conduct during the oral exam is important, too:

- Arrive promptly and wear reasonably smart clothing. Not to do either might be considered disrespectful by the examiner.
- Take your time before answering questions. Even if you think you know the answer immediately, take a little while to check mentally whether you have considered all angles. A considered, logical approach will be more impressive than a quick but ill-considered response.
- Start answers with the basics, then develop into deeper aspects. There may be both surface and deeper aspects to a topic and more credit will be given to a student who mentions the latter.
- When your answer is finished, stop speaking. A short, crisp answer is better than a rambling one.
- If you don't know the answer, say so. To waffle and talk about irrelevant material is more damaging than admitting that you don't know.
- Make sure your answer is balanced. Talk about the evidence and opinions on both sides of a contentious issue.
- Don't disagree violently with the examiner. Politely put your point of view, detailing the evidence behind it. Examiners will be impressed by students who know their own mind and subject area. However, they will expect you to support a position at odds with the conventional viewpoint.
- Finally, be positive and enthusiastic about your topic.

Creating an exam action list *– knowing that you have prepared well, checked everything on your list and gathered together all you need for an exam will improve your confidence and reduce anxiety. Your list might include:*

- *Verify time, date and place of the exam.*
- *Confirm travel arrangements to exam hall.*
- *Double-check module handbooks and past papers for exam structure.*
- *Think through use of time and exam strategy.*
- *Identify a quiet place near the exam hall to carry out a last-minute check on key knowledge (e.g. formulae, definitions, diagram labels).*
- *Ensure you have all the items you wish to take to the exam, e.g.*
  - *pens, pencils (with sharpener and eraser);*
  - *ruler;*
  - *correction fluid;*
  - *calculator (allowable type), if required;*
  - *sweets and drink, if allowed;*
  - *tissues*
  - *watch or clock;*
  - *ID card*
  - *texts and/or notes, if an open book exam;*
  - *lucky charm/mascot.*
- *Lay out clothes (if exam is early in the morning).*
- *Set alarm and/or ask a friend or family member to check you are awake on time.*

Counteracting anxiety before and during exams

Adverse effects of anxiety need to be overcome by anticipation and preparation well in advance (Box 6.4). Exams, with their tight time limits, are especially stressful for perfectionists. To counteract this tendency, focus on the following points during the exam:

- Don't expect to produce a perfect essay – this won't be possible in the time available.

Box 6.4 Strategies for combating the symptoms of exam anxiety

Sleeplessness – this is common and does little harm in the short term. Get up, have a snack, do some light reading or other activity, then return to bed. Avoid caffeine (e.g. tea, coffee and cola) for several hours before going to bed.

Lack of appetite – again commonplace. Eat what you can, and take sugary sweets into the exam to keep energy levels up in case you become tired.

Fear of the unknown – it can be a good idea to visit the exam room beforehand, so you can become familiar with the location. By working through the points given in the exam action list on p. 35, you will be confident that nothing has been left out.

Worries about timekeeping – get a *reliable* alarm clock or a new battery for an old one. Arrange for an alarm phone call; ask a friend or relative to make sure you are awake on time. Make reliable travel arrangements, to arrive on time. If your exam is early in the morning, it may be a good idea to get up early for a few days beforehand.

Blind panic during an exam – explain how you feel to an invigilator. Ask to go for a supervised walk outside. Do some relaxation exercises (see below), then return to your work. If you are having problems with a specific question, it may be appropriate to speak to the departmental representative at the exam, to check that you are not misinterpreting the question.

Feeling tense – shut your eyes, take several slow, deep breaths, do some stretching and relaxing muscle movements. During exams, it can be a good idea to do this between questions, and possibly to have a complete rest for a minute or so. Before exams, try some exercise activity or escape temporarily from your worries by watching TV or a movie.

Running out of time – don't panic when the invigilator says 'five minutes left'. It is surprising how much you can write in this time. Write note-style answers or state the areas you would have covered: you may get some credit.

- Don't spend too long planning your answer – once you have an outline plan, get started.
- Don't spend too much time on the initial parts of an answer, at the expense of the main message.
- Concentrate on getting all of the basic points across – markers are looking for the main points first, before allocating extra marks for the detail.
- Don't be obsessed with neatness, either in handwriting, or in the diagrams you draw, but make sure your answers are legible.
- Don't worry if you forget something. You can't be expected to know everything. Most marking schemes give a first class grade to work that misses out on up to 30 per cent of the marks available.

***KEY POINT** Everyone worries about exams. Anxiety is a perfectly natural feeling. It works to your advantage, as it helps provide motivation and the adrenaline that can help you 'raise your game' on the day.*

***After the exam** – try to avoid becoming involved in prolonged analyses with other students over the 'ideal' answers to the questions; after all, it is too late to change anything at this stage. Go for a walk, watch TV for a while, or do something else that helps you relax, so that you are ready to face the next exam with confidence.*

There is a lot to be said for treating exams as a game. After all, they are artificial situations contrived to ensure that large numbers of candidates can be assessed together, with little risk of cheating. They have conventions and rules, just like games. If you understand the rationale behind them and follow the rules, this will aid your performance.

Sources for further study

Acres, D. (1998) *Passing Exams Without Anxiety: How to Get Organised, be Prepared and Feel Confident of Success*. How to Books, London.

Burns, R. (1997) *The Student's Guide to Passing Exams*. Kogan Page, London.

Hamilton, D. (1999) *Passing Exams: A Guide for Maximum Success and Minimum Stress*. Cassell, London.

Many universities host study skills websites; these can be found using 'study skills', 'revision' or 'exams' as key words in a search engine.

Study exercises

6.1 **Analyse your past performances.** Think back to past exams and any feedback you received from them. How might you improve your performance? Consider ways in which you might approach the forthcoming exam differently. If you have kept past papers and answers to continuous assessment exercises, look at any specific comments your lecturers may have made.

6.2 **Share revision notes with other students.** Make a revision plan (see pp. 10 and 26) and then allocate some time to discussing your revision notes with a colleague. Try to learn from his or her approach. Discuss any issues you do not agree upon.

6.3 **Plan your exam tactics.** Find out from your module handbook or past papers what the format of each paper will be. Confirm this if necessary with staff. Decide how you will tackle each paper, allocating time to each section and to each question within the sections (see p. 107). Write a personal checklist of requirements for the exam (see p. 35).

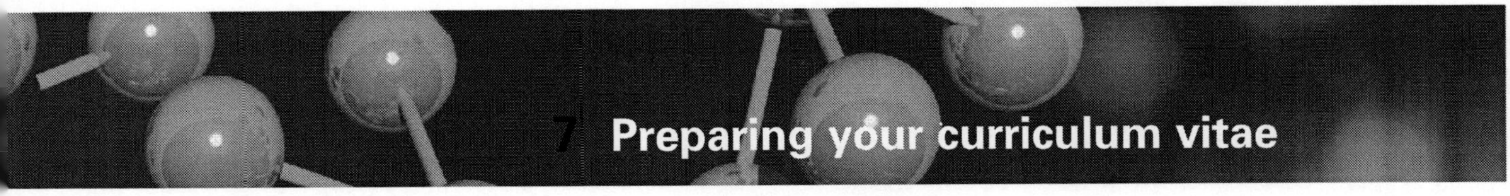

7 Preparing your curriculum vitae

Definition

Curriculum vitae (or CV for short) – a Latin phrase that means 'the course your life has taken'.

Many students only think about their curriculum vitae immediately before applying for a job. Reflecting this, chapters on preparing a CV are usually placed near the end of texts of this type. Putting the chapter near the beginning of this book emphasises the importance of focusing your thoughts on your CV at an early stage in your studies. There are four main reasons why this can be valuable:

1. Considering your CV and how it will look to a future employer will help you think more deeply about the direction and value of your academic studies.
2. Creating a draft CV will prompt you to assess your skills and personal qualities and how these fit into your career aspirations.
3. Your CV can be used as a record of all the relevant things you have done at university and then, later, will help you communicate these to a potential employer.
4. Your developing CV can be used when you apply for vacation or part-time employment.

Personal development planning (PDP) and your CV *– many PDP schemes (p. 5) also include an element of career planning that may involve creating a draft or generic CV. The PDP process can help you improve the structure and content of your CV, and the language you use within it.*

KEY POINT Developing your skills and qualities needs to be treated as a long-term project. It makes sense to think early about your career aspirations so that you can make the most of opportunities to build up relevant experience. A good focus for such thoughts is your developing curriculum vitae, so it is useful to work on this from a very early stage.

Skills and personal qualities

Understanding skills and qualities *– it may be helpful to think about how the skills and qualities in Tables 1.1 and 7.1 apply to particular activities during your studies, since this will give them a greater relevance.*

Skills (sometimes called competences) are generally what you have learned to do and have improved with practice. Table 1.1 summarises some important skills for bioscientists. This list might seem quite daunting, but your tutors will have designed your courses to give you plenty of opportunities to develop your expertise. Personal qualities, on the other hand, are predominantly innate. Examples include honesty, determination and thoroughness (Table 7.1). These qualities need not remain static, however, and can be developed or changed according to your experiences. By consciously deciding to take on new challenges and responsibilities, not only can you develop your personal qualities, but you can also provide supporting evidence for your CV.

Focusing on evidence *– it is important to be able to provide specific information that will back up the claims you make under the 'skills and personal qualities' and other sections of your CV. A potential employer will be interested in your level of competence (what you can actually do) and in situations where you have used a skill or demonstrated a particular quality. These aspects can also be mentioned in your covering letter or at interview.*

Personal qualities and skills are interrelated because your personal qualities can influence the skills you gain. For example, you may become highly proficient at a skill requiring manual dexterity if you are particularly adept with your hands. Being able to transfer your skills is highly important (Chapter 1) – many employers take a long-term view and look for evidence of the adaptability that will allow you to be a flexible employee and one who will continue to develop skills.

Developing your curriculum vitae

The initial stage involves making an audit of the skills and qualities you already have, and thinking about those you might need to develop. Tables 7.1 and 1.1 could form a basis of this self-appraisal. Assessing your skills may be easier than critically analysing your

Table 7.1 Some positive personal qualities

Adaptability
Conscientiousness
Curiosity
Determination
Drive
Energy
Enthusiasm
Fitness and health
Flexible approach
Honesty
Innovation
Integrity
Leadership
Logical approach
Motivation
Patience
Performance under stress
Perseverance
Prudence
Quickness of thought
Seeing others' viewpoints
Self-confidence
Self-discipline
Sense of purpose
Shrewd judgement
Social skills (sociability)
Taking initiative
Tenacity
Tidiness
Thoroughness
Tolerance
Unemotional approach
Willingness to take on challenges

Seeing yourself as others see you *– you may not recognise all of your personal qualities and you may need someone else to give you an honest appraisal. This could be anyone whose opinion you value: a friend, a member of your family, a tutor, or a careers adviser.*

Setting your own agenda *– you have the capability to widen your experience and to demonstrate relevant personal qualities through both curricular* and *extracurricular activities.*

Paying attention to the quality of your CV *– your potential employer will regard your CV as an example of your very best work and will not be impressed if it is full of mistakes or badly presented, especially if you claim 'good written communication' as a skill!*

personal characteristics. In judging your qualities, try to take a positive view and avoid being overly modest. It is important to think of your qualities in a specific context, e.g. 'I have shown that I am trustworthy, by acting as treasurer for the Biochemistry Society', as this evidence will form a vital part of your CV and job applications.

If you can identify gaps in your skills, or qualities that you would like to develop, especially in relation to the needs of your intended career, the next step is to think about ways of improving them. This will be reasonably easy in some cases, but may require some creative thinking in others. A relatively simple example would be if you decided to learn a new language or to keep up with one you learned at school. There are likely to be many local college and university courses dealing with foreign languages at many different levels, so it would be a straightforward matter to join one of these. A rather more difficult case might be if you wished to demonstrate 'responsibility', because there are no courses available on this. One route to demonstrate this quality might be to put yourself up for election as an officer in a student society or club; another could be to take a leading role in a relevant activity within your community (e.g. voluntary work such as hospital radio). If you already take part in activities like these, your CV should relate them to this context.

Basic CV structures and their presentation

Box 7.1 illustrates the typical parts of a CV and explains the purpose of each part. Employers are more likely to take notice of a well-organised and well-presented CV, in contrast to one that is difficult to read and assimilate. They will expect it to be concise, complete and accurate. There are many ways of presenting information in a CV, and you will be assessed partly on your choices.

- **Order**. There is some flexibility as to the order in which you can present the different parts (see Box 7.1). A chronological approach within sections helps employers gain a picture of your experience.
- **Personality and 'colour'**. Make your CV different by avoiding standard or dull phrasing. Try not to focus solely on academic aspects: you will probably work in a team, and the social aspects of teamwork will be enhanced by your outside interests. However, make sure that the reader does not get the impression that these interests dominate your life.
- **Style**. Your CV should reflect *your* personality, but not in such a way that it indicates too idiosyncratic an approach. It is probably better to be formal in both language and presentation, as flippant or chatty expressions will not be well received.
- **Neatness**. Producing a well-presented, word-processed CV is very important. Use a laser-quality printer and good-quality paper; avoid poor-quality photocopying at all costs.
- **Layout**. Use headings for different aspects, such as personal details, education, etc. Emphasise words (e.g. with capitals, bold, italics or underlining) sparingly and with the primary aim of making the structure clearer. Remember that careful use of white space is important in design.
- **Grammar and proofreading**. Look at your CV carefully before you submit it, as sloppy errors give a very poor impression. Even if you use a spell-checker, some errors may creep in. Ask someone whom you regard as a reliable proofreader to comment on it (many tutors will do this, if asked in advance).

Box 7.1 The structure and components of a typical CV and covering letter

There is no right or wrong way to write a CV, and no single format applies. It is probably best to avoid software templates and CV 'wizards' as they can create a bland, standardised result, rather than something that demonstrates your individuality.

You should include the following, with appropriate sub-headings, generally in the order given below:

1. **Personal details.** This section *must* include your full name and date of birth, your address (both home and term-time, with dates, if appropriate) and a contact telephone number at each address. If you have an email account, you might also include this. You need only mention sex if your name could be either male or female.
2. **Education.** Choose either chronological order, or reverse chronological order and make sure you take the same approach in all other sections. Give educational institutions and dates (month, year) and provide more detail for your degree course than for your previous education. Remember to mention any prizes, scholarships or other academic achievements. Include your overall mark for the most recent year of your course, if it seems appropriate. Make sure you explain any gap years.
3. **Work experience.** Include all temporary, part-time, full-time or voluntary jobs. Details include dates, employer, job title and major duties involved.
4. **Skills and personal qualities.** Tables 1.1 and 7.1 give examples of the aspects you might include. Emphasise your strengths, and tailor this section to the specific requirements of the post (the 'job description'): for example, the practical skills you have gained during your degree studies if the post is a biological one, but concentrate on generic transferable skills and personal qualities for other jobs. Provide supporting evidence for your statements in all cases.
5. **Interests and activities.** This is an opportunity to bring out the positive aspects of your personality, and explain their relevance to the post you are applying for. Aim to keep this section short, or it may seem that your social life is more important than your education and work experience. Include up to four separate items, and provide sufficient detail to highlight the *positive* aspects of your interests (e.g. positions of responsibility, working with others, communication, etc.). Use sections 4 and 5 to demonstrate that you have the necessary attributes to fulfil the major requirements of the post.
6. **Referees.** Include the names (and titles), job descriptions, full postal addresses, contact telephone numbers and email addresses of two referees (rarely, some employers may ask for three). It is usual to include your personal tutor or course leader at university (who among other things will verify your marks), plus another person – perhaps a current or former employer, or someone who runs a club or society and who knows your personal interests and activities. Unless you have kept in touch with a particular teacher since starting university, it is probably best to choose current contacts, rather than those from your previous education.

Some other points to consider:

- Try to avoid jargon and over-complicated phrases in your CV: aim for direct, active words and phrases (see Box 16.1).
- Most employers will expect your CV to be word-processed (and spell-checked). Errors in style, grammar and presentation will count against you, so be sure to check through your final version (and ask a reliable person to second-check it for you).
- Aim for a *maximum* length of two pages, printed single-sided on A4 paper, using a 'formal' font (e.g. Times Roman or Arial) of no less than 12 point for the main text. Always print onto good quality white paper. Avoid fussy use of colour, borders or fonts.
- Don't try to cram in too much detail. Use a clear and succinct approach with short sentences and lists to improve 'readability' and create structure. Remember that your aim is to catch the eye of your potential employer, who may have many applications to work through.
- It is polite to check that people are willing to act as a referee for you and to provide them with an up-to-date copy of your CV.

Your single-page covering letter should have four major components:

1. **Letterhead.** Include your contact details, the recipient's name and title (if known) and address, plus any job reference number.
2. **Introductory paragraph.** Explain who you are and state the post you are applying for.
3. **Main message.** This is your opportunity to sell yourself to a potential employer, highlighting particular attributes and experience. Keep it to three or four sentences at the most and relate it to the particular skills and qualities demanded in the job or person specification.
4. **Concluding paragraph.** A brief statement that you look forward to hearing the outcome of your application is sufficient.

Finally, add either 'Yours sincerely' (where the recipient's name is known) or 'Yours faithfully' (in a letter beginning 'Dear Sir or Madam') and then end with your signature.

- **Relevance**. If you can, slant your CV towards the job description and the qualifications required (see below). Make sure you provide evidence to back up your assertions about skills, qualities and experience.
- **Accuracy and completeness**. Check that all your dates tally; otherwise, you will seem careless. It is better to be honest about your grades and (say) a period of unemployment, than to cover this up or omit details that an employer will want to know. They may be suspicious if you leave things out.

Adjusting your CV

Creating a generic CV *– as you may apply for several jobs, it is useful to construct a CV in electronic format (e.g. as a Word file) which includes* all *information of potential relevance. This can then be modified to fit each post. Having a prepared CV on file will reduce the work each time you apply, while modifying this will help you focus on relevant skills and attributes for the particular job.*

You should fine-tune your CV for each post. Employers frequently use a 'person specification' to define the skills and qualities demanded in a job, often under headings such as 'essential' and 'desirable'. This will help you decide whether to apply for a position and it assists the selection panel to filter the applicants. Highlight relevant qualifications as early in your CV as possible. Be selective – don't include every detail about yourself. Emphasise relevant parts and leave out irrelevant details, according to the job. Similarly, your letter of application is not merely a formal document but is also an opportunity for persuasion (Box 7.1). You can use it to state your ambitions and highlight particular qualifications and experience. However, don't go over the top – always keep the letter to a single page.

KEY POINT A well-constructed and relevant CV won't necessarily guarantee you a job, but it may well get you onto the short list for interview. A poor-quality CV is a sure route to failure.

Sources for further study

Anon. (2000) *How to Write a Curriculum Vitae*. University of London Careers Service, London.

Anon. *Doctorjob.com Website*. Available: http://doctorjob. com
Last accessed: 01/04/07.

Anon. *Graduate Prospects Website*. Available: http://www.prospects.ac.uk/cms/ShowPage/Home_page/p!eLaXi
Last accessed: 01/04/07.
[Higher Education Careers Services Unit, containing good examples of CVs]

Anon. *Applying for Jobs*. Available: http://www.hero.ac.uk/uk/studying/careers_and_life long_learning/applying_for_jobs285.cfm
Last accessed: 01/04/07.

Study exercises

7.1 Evaluate your personal attributes. Using Table 7.1, list *five* qualities that you would use to best describe yourself, and cite the evidence you might give to a potential employer to convince them that this was the case. List *five* attributes you could develop, then indicate how you might do this.

7.2 Create a generic CV. Drawing on your school record of achievement, or any CV already prepared, e.g. for a part-time job, create a word-processed generic CV. Save the file in an appropriate (computer) folder and make a back-up copy. Print out a copy for filing. Periodically update the word-processed version. If appropriate, save different versions to be used in different contexts (e.g. when applying for a vacation job).

7.3 Think about your future career and ask for advice. Make an appointment with one of the advisors in your university's careers service. Ask about career options for graduates with your intended degree, or determine what qualifications or module options might be appropriate for occupations that interest you.

Information technology and library resources

8 Finding and citing published information

The ability to find scientific information is a skill required for many exercises in your degree programme. You will need to research facts and published findings as part of writing essays, literature reviews and project introductions, and when amplifying your lecture notes (p. 18) and revising for exams (p. 25). You must also learn how to follow scientific convention in citing source material as the authority for the statements you have made.

Browsing in a library *– this may turn up interesting material, but remember the books on the shelves are those not currently out on loan. Almost by definition, the latter may be more up-to-date and useful. To find out a library's full holding of books in any subject area, you need to search its catalogue (normally available as an online database).*

Sources of information

For essays and revision

You are unlikely to delve into the primary literature (p. 53) for these purposes – books and reviews are much more readable. If a lecturer or tutor specifies a particular book, then it should not be difficult to find out where it is shelved in your library, using the computerised index system. Library staff will generally be happy to assist with any queries. If you want to find out which books your library holds on a specified topic, use the system's subject index. You will also be able to search by author or by key words.

There are two main systems used by libraries to classify books: the Dewey Decimal system and the Library of Congress system. Libraries differ in the way they employ these systems, especially by adding further numbers and letters after the standard classification marks, e.g. to signify shelving position or edition number. Enquire at your library for a full explanation of local usage.

Example The book *The Selfish Gene* by Richard Dawkins (1976; Oxford University Press) is likely to be classified as follows:

Dewey Decimal system: 591.51
where 591 refers to zoology
591.5 refers to ecology of animals
591.51 refers to habits and behaviour patterns

Library of Congress system: QL751
where Q refers to science
QL refers to zoology
QL75 refers to animal behaviour
QL751 refers to general works and treatises

The World Wide Web is an ever-expanding resource for gathering both general and specific information (see Chapters 10 and 11). Sites fall into analogous categories to those in the printed literature: there are sites with original information, sites that review information and bibliographic sites. One considerable problem is that websites may be frequently updated, so information present when you first looked may be altered or even absent when the site is next consulted. Further, very little of the information on the Web has been monitored or refereed. Another disadvantage is that the site information may not state the origin of the material, who wrote it or when it was written. This is considered in more detail in Box 10.5.

Web resources *– your university library will provide you with access to a range of web-based databases and information systems. The library web pages will list these and provide links, which may be worth bookmarking on your web browser. Resources especially useful to bioscientists include:*

- ISI Web of Knowledge, *including the* Science Citation Index
- IngentaConnect *(previously known as BIDS), including* Ingenta Medline
- CSA Illumina
- Dialog
- PubMed
- ScienceDirect
- Scopus
- Ovid, *including* Cinahl *and* Medline

Most of these electronic resources operate on a subscription basis and may require an 'Athens' username and password – for details of how to obtain these consult library staff or your library's website.

For literature surveys and project work

You will probably need to consult the primary literature. If you are starting a new research project or writing a report from scratch, you can build up a core of relevant papers by using the following methods:

- Asking around: supervisors or their postgraduate students will almost certainly be able to supply you with a reference or two that will start you off.
- Searching an online database, either via the Internet (see Chapter 10) or on CD-ROM: these cover very wide areas and are a convenient way to start a reference collection, although a charge is often made for access and sending out a listing of the papers selected (your library may or may not pass this on to you).
- Consulting the bibliography of other papers in your collection – an important way of finding the key papers in your field. In effect, you

Definitions

Journal/periodical/serial – any publication issued at regular intervals. In biosciences, usually containing papers (articles) describing original research findings and reviews of literature.

eJournal – a journal published online, consisting of articles structured in the same way as a paper-based journal. A valid username and password may be required for access (arranged via your library, if it subscribes to the eJournal).

The primary literature – this comprises original research papers, published in specialist scientific periodicals. Certain prestigious general journals (e.g. *Nature*) contain important new advances from a wide subject area.

Monograph – a specialised book covering a single topic.

ebook – a book published online in downloadable form.

ebrary – a commercial service offering ebooks and other online resources.

HERON (Higher Education Resources ON-demand) – a national service for UK higher education offering copyright clearance, digitisation and delivery of book extracts and journal articles.

Review – an article in which recent advances in a specific area are outlined and discussed.

Proceedings – volume compiling written versions of papers read at a scientific meeting on a specific topic.

Abstracts – shortened versions of papers, often those read at scientific meetings. These may later appear in the literature as full papers.

Bibliography – a summary of the published work in a defined subject area.

Alternative methods of receiving information – *RSS (really simple syndication) feeds and email updates from publishers are increasingly used to provide automated information services to academic clients – for example, by supplying links to relevant contents of new editions of online journals.*

are taking advantage of the fact that another researcher has already done all the hard work.

- Referring to 'current awareness' online databases: these are useful for keeping you up to date with current research; they usually provide a monthly listing of article details (title, authors, source, author address) arranged by subject and cross-referenced by subject and author. Current awareness databases cover a wider range of primary literature than could ever be available in any one library. Examples include: *Current Contents Connect* (ISI), the *Current Advances* series (Elsevier), *Biological Abstracts* and *Zoological Record Archive* (Thompson Scientific). Some online databases also offer a service whereby they will email registered users with updates based on saved search criteria. Consult library staff or your library website to see which of these databases and services are available to you.
- Using the *Science Citation Index* (SCI): this is a valuable means of exploring the published literature in a given field, because it lets you see who has cited a given paper; in effect, SCI allows you to move forward through the literature from an existing reference. The Index is available on-line *via* ISI *Web of Science*.

For specialised information

You may need to consult reference works, such as encyclopaedias, maps and books providing specialised information. Some of this is now available on the Web, or online (consult your library's information service web pages). Three sources worth noting are:

- The *Handbook of Chemistry and Physics* (Lide, 2005; online via CHEMnetBASE): the Chemical Rubber Company's publication (affectionately known as the 'Rubber Bible') giving all manner of physical constants, radioisotope half-lives, etc.
- *The Merck Index* (O'Neil *et al.*, 2006), which gives useful information about organic chemicals, e.g. solubility, whether poisonous, etc., now also available online.
- The *Geigy Scientific Tables* (8th edition), a series of six volumes (Lentner, 1981; 1984; 1986; 1990; 1992; Lentner *et al.*, 1982) provides a wide range of information centred on biochemistry, e.g. buffer formulae, properties of constituents of living matter.

Obtaining and organising research papers

Obtaining a copy

It is usually more convenient to have personal copies of key research articles for direct consultation when working in a laboratory or writing. The simplest way of obtaining these is to photocopy the originals or download and/or print off copies online (e.g. as '.pdf' files). For academic purposes, this is normally acceptable within copyright law. If your library does not subscribe to the journal, it may be possible for them to borrow it from a nearby institute or obtain a copy via a national borrowing centre (an 'inter-library loan'). If the latter, you will have to fill in a form giving full bibliographic details of the paper and where it was cited, as well as signing a copyright clearance statement concerning your use of the copy. Alternatively, you might try emailing the communicating author and requesting an electronic copy ('.pdf' file) of the article.

Organising papers

Although the number of papers you accumulate may be small to start with, it is worth putting some thought into their storage and indexing before your collection becomes disorganised and unmanageable. Few things are more frustrating than not being able to lay your hands on a vital piece of information, and this can seriously disrupt your flow when writing or revising.

Indexing your references

Whether you have obtained a printed copy, have stored downloaded files electronically, or have simply noted the bibliographic details of a reference, you will need to index each resource. This is valuable for the following reasons:

- You will probably need the bibliographic information for creating a reference list for an assignment or report.
- If the index also has database features, this can be useful, allowing you to search for key words or authors.
- If you include an 'accession number' and if you then file printed material sequentially according to this number, then it will help you to find the hard copy.
- Depending on the indexing system used, you can add comments about the reference that may be useful at a later time, e.g. when writing an introduction or conclusion.

The simplest way to create an index system is to put the details on reference cards, but database software can be more convenient and faster to sort, once the bibliographic information has been entered. If you do not feel that commercial software is appropriate for your needs, consider using a word-processor or spreadsheet; their rudimentary database sorting functions (see Chapters 11 and 12) may be all that you require.

If you are likely to store lots of references and other electronic resources digitally, then you should consider carefully how this information is kept, for example by choosing file names that indicate what the file contains and that will facilitate sorting.

Copyright law – *In Europe, copyright regulations were harmonised in 1993 (Directive 93/98/EEC) to allow literary copyright for 70 years after the death of an author and typographical copyright for 25 years after publication. This was implemented in the UK in 1996, where, in addition, the Copyright, Designs and Patents Act (1988) allows the Copyright Licensing Agency to license institutions so that lecturers, students and researchers may take copies for teaching and personal research purposes – no more than a single article per journal issue, one chapter of a book, or extracts to a total of 5 per cent of a book.*

Storing research papers – *these can easily be kept in alphabetical order within filing boxes or drawers, but if your collection is likely to grow large, it will need to be refiled as it outgrows the storage space. You may therefore wish to add an 'accession number' to the record you keep in your database, and file the papers in sequence according to this, as they accumulate. New filing space is only required at the end and you can use the accession numbers to form the basis of a simple cross-referencing system.*

Using commercial bibliographic database software to organise your references – *for those with large numbers of references in their collection, and who may wish to produce lists of selected references in particular format, e.g. for inclusion in a project report or journal paper, systems like* EndNote, Reference Manager *or* ProCite *can reward the investment of time and money required to create a personal reference catalogue. Appropriate bibliographic data must first be entered into fields within a database (some versions assist you to search online databases and upload data from these). The database can then be searched and used to create customised lists of selected references in appropriate citation styles.*

Making citations in text

There are two main ways of citing articles and creating a bibliography (also referred to as 'references' or 'literature cited').

The Harvard system

For each citation, the author name(s) and the date of publication are given at the relevant point in the text. The bibliography is organised alphabetically, and by date of publication for papers with the same authors. Formats normally adopted are, for example, 'Smith and Jones (1983) stated that . . .' or 'it has been shown that . . . (Smith and Jones, 1983)'. Lists of references within parentheses are separated by semicolons, e.g. '(Smith and Jones, 1983; Jones and Smith, 1985)', normally in order of date of publication. To avoid repetition within the same

Examples Incorporating references in text – this sample shows how you might embed citations in text using the Harvard approach:

'... Brookes *et al.* (2001) proposed that protein A216 was involved in the degradation process. However, others have disputed this notion (Scott and Davis, 1997; Harley, 1998, 2000). Patel (1999a; 1999b) found that A216 is inactivated at pH values less than 5; while several authors (e.g. Hamilton, 1995; Drummond and Stewart, 2002) also report that its activity is strongly dependent on Ca^{2+} concentration ...'

Examples Incorporating references in text – this sample shows how you might embed citations in text using the Vancouver approach:

'... Brookes *et al.* proposed that protein A216 was involved in the degradation process.[1] However, others have disputed this notion.[2–4] Patel[5,6] found that A216 is inactivated at pH values less than 5; while several authors[7,8] also report that its activity is strongly dependent on Ca^{2+} concentration ...'

paragraph, an approach such as 'the investigations of Smith and Jones indicated that' could be used following an initial citation of the paper. Where there are more than two authors it is usual to write 'et al.' (or *et al.* if an italic font is available); this stands for the Latin *et alia* meaning 'and others'. If citing more than one paper with the same authors, put, for example, 'Smith and Jones (1987; 1990)' and if papers by a given set of authors appeared in the same year, letter them (e.g. Smith and Jones, 1989a; 1989b).

The numerical or Vancouver system

Papers are cited via a superscript or bracketed reference number inserted at the appropriate point. Normal format would be, for example: 'DNA sequences[4,5] have shown that ...' or 'Jones [55,82] has claimed that ...'. Repeated citations use the number from the first citation. In the true numerical method (e.g. as in *Nature*), numbers are allocated by order of citation in the text, but in the alpha-numerical method (e.g. the *Annual Review* series), the references are first ordered alphabetically in the bibliography, then numbered, and it is this number which is used in the text. Note that with this latter method, adding or removing references is tedious, so the numbering should be done only when the text has been finalised.

KEY POINT The main advantages of the Harvard system are that the reader might recognise the paper being referred to and that it is easily expanded if extra references are added. The main advantages of the Vancouver system are that it aids text flow and reduces length.

How to list your citations in a bibliography

Examples

Paper in journal:
Smith, A. B., Jones, C.D. and Professor, A. (2004). Innovative results concerning our research interest. *Journal of New Results,* **11**, 234–5.

Book:
Smith, A. B. (2005). *Summary of my Life's Work.* Megadosh Publishing Corp., Bigcity. ISBN 0-123-45678-9.

Chapter in edited book:
Jones, C. D. and Smith, A. B. (2003). Earth-shattering research from our laboratory. In: *Research Compendium 1998* (ed. A. Professor), pp. 123–456. Bigbucks Press, Booktown.

Thesis:
Smith, A. B. (2006). *Investigations on my Favourite Topic.* PhD thesis, University of Life, Fulchester.

Note: if your references are handwritten, you should indicate italics by underlining text or numerals.

Whichever citation method is used in the text, comprehensive details are required for the bibliography so that the reader has enough information to find the reference easily. Citations should be listed in alphabetical order with the priority: first author, subsequent author(s), date. Unfortunately, in terms of punctuation and layout, there are almost as many ways of citing papers as there are journals. Your department may specify an exact format for project work; if not, decide on a style and be consistent – if you do not pay attention to the details of citation you may lose marks. Take special care with the following aspects:

- Authors and editors: give details of *all* authors and editors in your bibliography, even if given as *et al.* in the text.
- Abbreviations for journals: while there are standard abbreviations for the titles of journals (consult library staff), it is a good idea to give the whole title, if possible.
- Books: the edition should always be specified as contents may change between editions. Add, for example, '(5th edition)' after the title of the book. You may be asked to give the International Standard Book Number (ISBN), a unique reference number for each book published.
- Unsigned articles, e.g. unattributed newspaper articles and instruction manuals – refer to the author(s) in text and bibliography as 'Anon.'.
- Websites: there is no widely accepted format at present. You should follow departmental guidelines if these are provided, but if these are not available, we suggest providing author name(s) and date in the

text when using the Harvard system (e.g. Hacker, 2006), while in the bibliography giving the URL details in the following format: Hacker, A. (2006) *University of Anytown Homepage on Aardvarks*. Available: http://www.myserver.ac.uk/homepage. Last accessed: 23/02/07. In this example, the webpage was constructed in 2006, but accessed in February, 2007. If no author is identifiable, cite the sponsoring body (e.g. University of Anytown, 2006), and if there is no author or sponsoring body, write 'Anon.' for 'anonymous', e.g. Anon. (2006), and use Anon. as the 'author' in the bibliography. If the web pages are undated, *either* use the 'last accessed' date for citation and put no date after the author name(s) in the reference list, *or* cite as 'no date' (e.g. Hacker, no date) and leave out a date after the author name(s) in the reference list – you should be consistent whichever option you choose.

- Unread articles: you may be forced to refer to a paper *via* another without having seen it. If possible, refer to another authority who has cited the paper, e.g. '... Jones (1980), cited in Smith (1990), claimed that ...'. Alternatively, you could denote such references in the bibliography by an asterisk and add a short note to explain at the start of the reference list.
- Personal communications: information received in a letter, seminar or conversation can be referred to in the text as, for example, '... (Smith, pers. comm.)'. These citations are not generally listed in the bibliography of papers, though in a thesis you could give a list of personal communicants and their addresses.
- Online material – some papers and articles are published solely online and others are available online ahead of publication in printed form. The item may be given a digital object identifier (DOI), allowing it to be cited and potentially tracked before and after it is allocated to a printed issue (see http://www.doi.org/). DOIs also allow for web page redirection by a central agency, and *CrossRef* (http://www.crossref.org/) is the official DOI registration organisation for scholarly and professional publications. DOIs can be used as 'live' hyperlinks in online articles, or cited in place of the volume and page numbers for the article, with the remainder of the details cited in the usual fashion, e.g. 'Smith. A. and Jones B. (2007) Our latest important research in the form of a web-published article. *Online Biosciences* 8/2006 (p. 781). Published Online: 26 March 2007. DOI: 10.1083/mabi.200680019'.

Finding dates on websites quickly – *when visiting a particular page, you can find occurrences of year dates beginning '200' by pressing the* Control *and* F *keys together, entering 200 in the* Find *window that appears, then carrying out a search using the* Find next *command.*

Adding web references in other systems – *if you are using a different referencing system than Harvard, consult Pears and Shields (2005) or McMillan and Weyers (2006) for further advice on how to cite websites in these systems.*

Text references and sources for further study

Anon. *CHEMnetBASE*. Available: http://www.chemnetbase.com Last accessed: 01/04/07. [Access to the *Handbook of Chemistry and Physics*]

Anon. *ISI Web of Science*. Available: http://wos.mimas.ac.uk Last accessed: 01/04/07. [Requires Athens password]

Lentner, C. (ed.) (1981) *Geigy Scientific Tables*, Vol. 1: *Units of Measurement, Body Fluids, Composition of the Body. Nutrition*, 8th edn. Ciba-Geigy, Basel.

Lentner, C. (ed.) (1984) *Geigy Scientific Tables*, Vol. 3: *Physical Chemistry, Composition of Blood, Hematology, Somatometric Data*, 8th edn. Ciba-Geigy, Basel.

Lentner, C. (ed.) (1986) *Geigy Scientific Tables*, Vol. 4: *Biochemistry, Metabolism of Xenobiotics, Inborn Errors of Metabolism, Pharmacogenetics, Ecogenetics*, 8th edn. Ciba-Geigy, Basel.

Lentner, C. (ed.) (1990) *Geigy Scientific Tables*, Vol. 5: *Heart and Circulation*, 8th edn. Ciba-Geigy, Basel.

Lentner, C. (ed.) (1992) *Geigy Scientific Tables*, Vol. 6: *Bacteria, Fungi, Protozoa, Helminths*, 8th edn. Ciba-Geigy, Basel.

Lentner, C., Diem, K. and Seldrup, J. (eds) (1982) *Geigy Scientific Tables*, Vol. 2: *Introduction to Statistics, Statistical Tables, Mathematical Formulae*, 8th edn. Ciba-Geigy, Basel.

Lide, D.R. (ed.) (2005) *CRC Handbook of Chemistry and Physics*, 86th edn. CRC Press, Boca Raton.

McMillan, K.M. and Weyers, J.D.B. (2006) *The Smarter Student: Skills and Strategies for Success at University*. Pearson Education, Harlow.

O'Neil, M.J., Heckelman, P.E., Koch, C.B. and Roman, K.J. (2006) *Merck Index: An Encyclopedia of Chemicals, Drugs, and Biologicals*, 14th edn. Merck & Co. Inc., Whitehouse Station.

Pears, R. and Shields, G. (2005) *Cite them Right: the Essential Guide to Referencing and Plagiarism*. Pear Tree Books, Newcastle upon Tyne.

Study exercises

8.1 Test your library skills. This exercise relies on the fact that most university-level libraries serving bioscience departments will take the scientific journal *Nature*. To help you answer these questions, it may be beneficial to attend a library induction session, if you haven't already. Alternatively, the library's help or enquiry desks may be able to assist you if you are having problems.

(a) First, find out and provide the name of the classification system that your University uses for cataloguing its books and periodicals.
(b) Using your library's cataloguing system (online, preferably) find out the appropriate local classification number for the journal *Nature*.
(c) Where is *Nature* shelved in your library? (Your answer need refer only to most recent issues if some have been archived.)
(d) What is the exact title of the landmark papers in the following two volumes? (i) *Nature* **171**, 737–738 (1953); (ii) *Nature* **408**, 796–815 (2000).

8.2 Explore different methods of citing references. Go to your library and seek out the journal area for biology or life sciences. Choose *three* different journals in your subject area and from a recent edition write down how they would print a typical citation for a multi-author journal paper in the 'references' or 'literature cited' section. Where used, indicate italicised text with normal underline and bold text with wavy underline. Pay attention to punctuation. Compare these methods with each other, with the methods recommended on pp. 47–8 of this book and with the recommendations your department or your course handbook makes. Are they all the same?

8.3 Make website citations. Use a search engine (pp. 63–4) to find an informative website that covers each of the following:

(a) The use of SI units.
(b) Information about the bacterial culture collection of the Pasteur Institute (l'Institut Pasteur).
(c) The different types of blood cells.

Indicate how you would cite each website at the end of an essay (follow your department's guidelines or use those in this chapter).

8.4 Compare the Harvard and Vancouver methods of citation. Pair up with a partner in your class. Each person should then pick one of the two main methods of citation and consider its pros and cons independently. Meet together and compare your lists. Given the choice, which method would you choose for (a) a handwritten essay; (b) a word-processed review; (c) an article in an academic journal, and why?

9 Evaluating information

Example A web search for the letters 'DNA' (e.g. using the search engine Google, p. 57), will reveal that this acronym appears in several hundred million web-sites. Not all of these deal with deoxyribonucleic acid – the listed websites include: *Doctors' Net Access*, a web resource for American physicians; the website of the *Dermatology Nurses Association*; and that of *Delta Nu Alpha*, an international organisation for those working in the professions. These examples are easy to identify as irrelevant to research on 'DNA, the molecule', but considering the many websites that do actually cover this topic, which might contain the exact information you seek? How valid is the information? Is it biased towards one viewpoint or hypothesis? Does it represent current mainstream thinking on the topic? These are some of the issues that an evaluation of the information sources might deal with.

Checking the reliability of information, assessing the relative value of different ideas and thinking critically are skills essential to the scientific approach (p. 188). You will need to develop your abilities to evaluate information in this way because:

- you will be faced with many sources of information, from which you will need to select the most appropriate material;
- you may come across conflicting sources of evidence and may have to decide which is the more reliable;
- the accuracy and validity of a specific fact may be vital to your work;
- you may doubt the quality of the information from a particular source;
- you may wish to check the original source because you are not sure whether someone else is quoting it correctly.

***KEY POINT** Evaluating information and thinking critically are regarded as higher order academic skills. The ability to think deeply in this way is greatly valued in the biosciences and will consequently be assessed in coursework and exam questions (see Chapter 5).*

The process of evaluating and using information can be broken down into four stages:

1. Selecting and obtaining material. How to find sources is covered in Chapter 8. Printed books and journals are important, but if you identify a source of this kind there may be delays in borrowing it or obtaining a photocopy. If the book or journal is available online, then downloading or printing sections or papers will be more convenient and faster. The Internet is often a first port of call if you wish to find something out quickly. However, for many websites, it can be difficult to verify the authenticity of the information given (see Box 10.5).
2. Assessing the content. You will need to understand fully what has been written, including any technical terms and jargon used. Establish the relevance of the information to your needs and assure yourself that the data or conclusions have been presented in an unbiased way.
3. Modifying the information. In order to use the information, you may need to alter it to suit your needs. This may require you to make comparisons, interpret or summarise. Some sources may require translation. Some data may require mathematical transformation before they are useful. There is a chance of error in any of these processes and also a risk of plagiarism (see Box 9.1).
4. Analysis. This may be your own interpretation of the information presented, or an examination of the way the original author has used the information.

Definition

Plagiarism – the unacknowledged use of another's work as if it were one's own. In this definition, the concept of 'work' includes ideas, writing, data or inventions and not simply words; and the notion of 'use' does not only mean copy 'word-for-word', but also 'in substance' (i.e. a copy of the ideas involved). Use of another's work is acceptable, if you acknowledge the source (Box 9.1).

***KEY POINT** Advances in communications and information technology mean that we can now access almost limitless knowledge. Consequently, the ability to evaluate information has become an extremely important skill.*

Box 9.1 How to avoid plagiarism and copyright infringement

Plagiarism is defined on page 51. Examples of plagiarism include:

- Copying the work of a fellow student (past or present) and passing it off as your own.
- Using 'essay-writing services', such as those on offer on certain websites.
- Copying text or images from a source (book, journal article or website, for instance) and using this within your own work without acknowledgement.
- Quoting others' words without indicating who wrote or said them.
- Copying ideas and concepts from a source without acknowledgement, even if you paraphrase them.

Most students would accept that some of the above can only be described as cheating. However, many students, especially at the start of their studies, are unaware of the academic rule that they must *always* acknowledge the originators of information, ideas and concepts, and that not doing so is regarded a form of academic dishonesty. If you adopt the appropriate conventions that avoid such accusations, you will achieve higher marks for your work as it will fulfil the marker's expectations for academic writing.

Universities have a range of mechanisms for identifying plagiarism, from employing experienced and vigilant coursework markers and external examiners to analysing students' work using sophisticated software programs. Plagiarism is always punished severely when detected. Penalties may include awarding a mark of zero to all involved – both the copier(s) and the person whose work has been copied (who is regarded as complicit in the crime). Further disciplinary measures may be taken in some instances. In severe cases, such as copying substantive parts of another's work within a thesis, a student may be dismissed from the university.

If you wish to avoid being accused of plagiarism, the remedies are relatively simple:

1. **Make sure the work you present is always your own.** If you have been studying alongside a colleague, or have been discussing how to tackle a particular problem with your peers, make sure you write on your own when working on your assignments.

2. **Never, ever be tempted to 'cut and paste'** from websites or online sources such as word-processed handouts. Read these carefully, decide what the important points are, express these *in your own words* and *provide literature citations to the original sources* (see Chapter 8). In some cases, further investigations may be required to find out details of the original sources. The lecturer's reading list or a book's references may help you here.

3. **Take care when note-taking.** If you decide to quote word-for-word, make sure you show this clearly in your notes with quotation marks. If you decide to make your own notes based on a source, make sure these are original and do not copy phrases from the text. In both cases, write down full details of the source at the appropriate point in your notes.

4. **Place appropriate citations throughout your text, where required.** If you are unsure about when to do this, study reviews and articles in your subject area (see also Chapter 8).

5. **Show clearly where you are quoting directly from a source.** For short quotes, this may involve using quotation marks and identifying the source afterwards, as in the example '... as Samuel Butler (1877) wrote: "a hen is only an egg's way of making another egg"'. For longer quotes (say 40 words or more), you should create a separate paragraph of quoted text, usually identified by inverted commas, indentation, italicisation or a combination of these. A citation must *always* be included, normally at the end. Your course handbook may specify a layout. Try not to rely too much on quotes in your work. If a large proportion of your work is made up from quotes, this will almost certainly be regarded as lacking in originality, scoring a poor mark.

Copyright issues are often associated with plagiarism, and refer to the right to publish (and hence copy) original material, such as text, images and music. Copyright material is indicated by a symbol © and a date (see, for example, page xiii of this book). Literary copyright is the aspect most relevant to students in their academic studies. UK Copyright Law protects authors' rights for life and gives their estates rights for a further 70 years. Publishers have 'typographical copyright' that lasts for 25 years. This means that it is illegal to photocopy, scan or print out copyright material unless you have permission, or unless your copying is limited to an extent that could be considered 'fair dealing'. For educational purposes – private study or research – in a scientific context, this generally means:

- no more than 5% in total of a work;
- one chapter of a book;
- one article per volume of an academic journal;
- 20% of a short book;
- one separate illustration or map.

You may only take one copy within the above limits, may not copy for others, and may not exceed these amounts *even if you own a copy of the original.* These rules also apply to web-based materials, but sometimes you will find sites where the copyright is waived. Some copying may be licensed; you should consult your library's website or helpdesk to see whether it has access to licensed material. Up-to-date copyright information is generally posted close to library and departmental photocopiers.

Evaluating sources of information

One way of assessing the reliability of a piece of scientific information is to think about how it was obtained in the first place. Essentially, facts and ideas originate from someone's research or scholarship, whether they are numerical data, descriptions, concepts or interpretations. Sources are divided into two main types:

Distinguishing between primary and secondary sources – *try the 'IMRaD test'. Many primary sources contain information in the order:* **I***ntroduction,* **M***aterials* **a***nd Methods,* **R***esults* **a***nd* **D***iscussion. If you see this format, and particularly if data from an experiment, study or observation are presented, then you are probably reading a primary source.*

1. Primary sources – those in which ideas and data are first communicated. The primary literature is generally published in the form of 'papers' (articles) in journals, whether printed or online. These are usually refereed by experts in the academic peer group of the author, and they will check the accuracy and originality of the work and report their opinions back to the editors of the journal. This peer review system helps to maintain reliability, but it is not perfect. Books and, more rarely, websites and articles in magazines and newspapers, can also be primary sources but this depends on the nature of the information published rather than the medium. These sources are not formally refereed, although they may be read by editors and lawyers, to check for errors and unsubstantiated or libellous allegations.
2. Secondary sources – those which quote, adapt, interpret, translate, develop or otherwise use information drawn from primary sources. It is the act of quoting or paraphrasing that makes the source secondary, rather than the medium. Reviews are examples of secondary scientific sources, and books and magazine articles are often of this type.

Example If a journalist wrote an article about a new 'flesh-eating bug' for the *New York Times* that was based on an article in the *British Medical Journal*, the *New York Times* article would be the secondary source, while the *British Medical Journal* article would be the primary source.

When information is modified for use in a secondary source, alterations are likely to occur, whether intentional or unintentional. Most authors do not deliberately set out to change the meaning of the primary source, but they may unwittingly do so, e.g. in changing text to avoid plagiarism or by oversimplification. Others may consciously or unconsciously exert bias in their reporting, for example, by quoting evidence that supports only one side of a debate. Therefore, the closer you can get to the primary source, the more reliable the information is likely to be. On the other hand, modification while creating a secondary source could involve correcting errors, or synthesising ideas and content from multiple sources.

Taking account of the changing nature of websites and wikis – *by their very nature, these sources may change. This means that it is important to quote accurately from them and to give a 'last accessed' date when citing (see p. 49).*

Authorship and provenance

Clearly, much depends on who is writing the source and on what basis (e.g. who paid them?). Consequently, an important way of assessing sources is to investigate the ownership and provenance of the work (who and where it originated from, and why).

Can you identify who wrote the information? If it is signed or there is a 'by-line' showing who wrote it, you might be able to make a judgement on the quality of what you are reading. This may be a simple decision, if you know or can assume that the writer is an authority in the area; otherwise a little research might help (for example, by putting the name into a web search engine). Of course, just because Professor X thinks something does not make it true. However, if you know that this opinion is backed up by years of research and experience, then you might take it a little more seriously than the thoughts of a school pupil. If an author is not cited, effectively nobody is taking responsibility for the content. Could there be a reason for this?

Finding out about authors and provenance – *these pieces of information are easy to find in most printed sources and may even be presented just below the title, for convenience. In the case of the Web, it may not be so easy to find what you want. Relevant clues can be obtained from 'home page' links and the header, body and footer information. For example, the domain (p. 60) may be useful, while the use of the tilde symbol (~) in a URL usually indicates a personal, rather than an institutional, website.*

***Assessing substance over presentation** – just because the information is presented well (e.g. in a glossy magazine or a particularly well-constructed website), this does not necessarily tell you much about its quality. Try to look below the surface, using the methods mentioned in this chapter.*

Is the author's place of work cited? This might tell you whether the facts or opinions given are based on an academic study. Is there a company with a vested interest behind the content? If the author works for a public body, there may be publication rules to follow and they may even have to submit their work to a publications committee before it is disseminated. They are certainly more likely to get into trouble if they include controversial material.

Evaluating facts and ideas

However reliable the source of a piece of information seems to be, it is probably a good idea to retain a slight degree of scepticism about the facts or ideas involved. Even information from impeccable primary sources may not be perfect – different approaches can give different outcomes, and interpretations can change with time and with further advances in knowledge. Table 9.1 provides a checklist for evaluating sources.

Critically examining facts and ideas is a complex task depending on the particular issues involved, and a number of different general approaches can be applied. You will need to decide which of the following general tips are useful in your specific case:

- Make cross-referencing checks – look at more than one source and compare what is said in each. The cross-referenced sources should be as independent as possible (for example, do not compare a primary source together with a secondary review based on it). If you find that all the sources give a similar picture, then you can be more confident about the reliability of the information.
- Look at the extent and quality of citations (Chapter 8) – if references are quoted, these indicate that a certain amount of research has been carried out beforehand, and that the ideas or results are based on genuine scholarship. If you are doubtful about the quality of the work, these references might be worth looking at. How up to date are they? Do they cite independent work, or is the author exclusively quoting their own work, or solely the work of one person?
- Consider the age of the source – the fact that a source is old is not necessarily a barrier to truth, but ideas and facts may have altered since the date of publication, and methods may have improved. Can you trace changes through time in the sources available to you? What key events or publications have forced any changes in the conclusions?
- Try to distinguish fact from opinion – to what extent has the author supported a given viewpoint? Have relevant facts been quoted, *via* literature citations or the author's own researches? Are numerical data used to substantiate the points used? Are these reliable and can you verify the information, for example, by looking at the sources cited? Might the author have a reason for putting forward biased evidence to support a personal opinion?
- Analyse the language used – words and their use can be very revealing. Subjective wording might indicate a personal opinion rather than an objective conclusion. Propaganda and personal bias might be indicated by absolute terms, such as 'everyone knows . . .'; 'It can be guaranteed that ...' or, a seemingly one-sided consideration of the evidence. How carefully has the author considered the topic? A less studious approach might be indicated by exaggeration, ambiguity or

Table 9.1 Checklist for assessing information in science. How reliable is the information you have been reading? The more 'yes' answers you can give below, the more trustworthy you can assume it to be

Assessing sources
❑ Can you identify the author's name?
❑ Can you determine what relevant qualifications he/she holds?
❑ Can you say who employs the author?
❑ Do you know who paid for the work to be done?
❑ Is this a primary or secondary source?
❑ Is the content original or derived from another source?
Evaluating information
❑ Have you checked a range of sources?
❑ Is the information supported by relevant literature citation?
❑ Is the age of the source likely to be important regarding the accuracy of the information?
❑ Have you focused on the substance of the information presented rather than its packaging?
❑ Is the information fact or opinion?
❑ Have you checked for any logical fallacies in the arguments?
❑ Does the language used indicate anything about the status of the information?
❑ Have the errors associated with any numbers been taken into account?
❑ Have the data been analysed using appropriate statistics?
❑ Are any graphs constructed fairly?

the use of 'journalese' and slang. Always remember, however, that content should be judged above presentation.

- Look closely at any numbers – if the information you are looking at is numerical in form, have statistical errors been taken into consideration, and, where appropriate, quantified? If so, does this help you arrive at a conclusion about how genuine the differences are between important values?
- Think carefully about any hypothesis-testing statistics used – are the methods appropriate? Are the underlying hypotheses the right ones? Have the results of any tests been interpreted correctly in arriving at the conclusion? To deal with these matters, you will need at least a basic understanding of the 'statistical approach' and of commonly used techniques (see Chapters 66 and 67).

Critical thinking

Learning from examples *– as your lecturers introduce you to case studies, you will see how life scientists have applied critical thinking to understand the nature of biomolecules, cells and organisms. Some of your laboratory sessions may mimic the processes involved – observation, hypothesis, experimental design, data gathering, analysis and formulating a conclusion (see Chapter 30). These skills and approaches can be applied in your degree programme, e.g. when writing about a biological issue or carrying out a research project.*

Critical thinking involves the application of logic to a problem, or case study issue. It requires a wide range of skills. Key processes involved include: acquiring and processing information; creating appropriate hypotheses and formulating conclusions; and acting on the conclusions towards a specific objective.

KEY POINT Critical thinking needs reliable knowledge, but it requires you to use this appropriately to analyse a problem. It can be contrasted with rote learning – where you might memorise facts without an explicit purpose other than building your knowledge base.

Critical thinking is particularly important in the biosciences, because the subject deals with complex and dynamic systems. These can be difficult to understand for several reasons:

- they are often multi-faceted, involving many interactions;
- it can be difficult to alter one variable in an experiment without producing confounding variables (see p. 190);
- many variables may be unmeasured or unmeasurable;
- heterogeneity (variability) is encountered at all scales from the molecular scale to the whole organism;
- perturbation of the system can lead to unexpected ('counter-intuitive') results.

'You can prove anything with statistics' *– leaving aside the issue that statistical methods deal with probability, not certainty (Chapter 67), it is often possible to analyse and present data in such a way that they support one chosen argument or hypothesis rather than another. Detecting a bias of this kind can be difficult, but the critical thinking skills involved are essential for all life scientists (see e.g. Box 63.3).*

As a result, conclusions in biological research are seldom clear-cut. Critical thinking allows you to arrive at the most probable conclusion from the results at hand; however, it also involves acknowledging that other conclusions might be possible. It allows you to weigh up these possibilities and find a working hypothesis or explanation, but also to understand that your conclusions are essentially dynamic and might alter when new facts are known. Hypothesis-testing with statistics (Chapter 67) is an important adjunct to critical thinking because it demands the formulation of simple hypotheses and provides rational reasons for making conclusions.

Recognising fallacies in arguments is an important aspect of critical thinking. Philosophers and logicians recognise different forms of argument and many different fallacies in each form. Damer (2004) provides an overview of this wide-ranging and complex topic.

***Analysing a graph** – this process can be split into six phases:*

1. *Considering the context and purpose of the graph.*
2. *Recognising the type of presentation and examining the axes.*
3. *Looking closely at the scale on each axis.*
4. *Examining the data presented (e.g. data points, symbols, curves).*
5. *Considering errors and statistics associated with the graph.*
6. *Reaching conclusions based on the above.*

These processes are amplified in Chapter 63.

***Analysing a table** – as with analysing a graph, this process can be split into six phases:*

1. *Considering the context and purpose.*
2. *Examining the subheadings to see what information is contained in the rows and columns.*
3. *Considering the units used and checking any footnotes.*
4. *Comparing the data values across rows and/or down columns, looking for patterns, trends and unusual values.*
5. *Taking into account any statistics presented.*
6. *Reaching conclusions based on the above.*

Interpreting data

Numerical data

Information presented in public, whether as a written publication or a spoken presentation, is rarely in the same form as it was when first obtained. Chapters 28 and 37 deal with processes in which data are recorded and manipulated – take particular care over percentages and proportions, while Chapter 66 describes the standard descriptive statistics used to 'encapsulate' large data sets. Chapter 65 covers some relevant mathematical techniques. Sampling (essentially, obtaining representative measurements) is at the heart of many observational and experimental approaches in the biosciences (see Chapters 28 and 30), and analysis of samples is a key component of hypothesis-testing statistics (Chapter 67). Understanding these topics and carrying out the associated study exercises will help you improve your ability to interpret numerical data.

Graphs

Frequently, understanding and analysis in science depend on your ability to interpret data presented in graphical form. Sometimes, graphs may mislead. This may be unwitting, as in an unconscious effort to favour a 'pet' hypothesis of the author. Graphs may be used to 'sell' a product, e.g. in advertising, or to favour a viewpoint as, perhaps, in politics. Experience in drawing and interpreting graphs will help you spot these flawed presentations, and understanding how graphs can be erroneously presented (Box 63.3) will help you avoid the same pitfalls.

Tables

Tables, especially large ones, can appear as a mass of numbers and thus be more daunting at first sight than graphs. In essence, however, most tables are simpler than most graphs. The construction of tables is dealt with in Chapter 64.

Text reference

Damer, T.E. (2004) *Attacking Faulty Reasoning: A Practical Guide to Fallacy-Free Arguments*, 5th edn. Wadsworth, Belmont, CA.

Sources for further study

Anon. *Critical Thinking.* Available: http://en.wikipedia.org/wiki/Critical_thinking
Last accessed: 01/04/07.
[A frequently updated review of terminology and techniques, with additional links and references]

Barnard, C.J., Gilbert, F.S. and MacGregor, P.K. (2001) *Asking Questions in Biology: Key Skills for Practical Assessments and Project Work*, 2nd edn. Prentice Hall, Harlow.

Smith, A. *Evaluation of Information Sources.* Available: http://www.vuw.ac.nz/staff/alastair_smith/evaln/evaln.htm
Last accessed: 01/04/07.
[Part of the web-based Information Quality Virtual Library]

Van Gelder, T. *Critical Thinking on the Web.* Available: http://www.austhink.org/critical/
Last accessed: 01/04/07.
[Includes a useful directory of web resources]

Study exercises

9.1 **Distinguish between primary and secondary literature.** Based on their titles and any research you can do in your library, determine whether the following journals are primary or secondary sources:

(a) *British Journal of Haematology*
(b) *Proceedings of the National Academy of Sciences*
(c) *Annual Review of Molecular Biology*
(d) *Essays in Biochemistry*
(e) *Microbiological Research*
(f) *Trends in Biotechnology*
(g) *The Lancet*
(h) *Biochemical Journal*
(i) *Scientific American*
(j) *Food Safety News*

9.2 **Consider a controversial issue from both sides.** Select a current biological topic being discussed in the newspapers or other media. Relatively controversial issues such as 'genetically modified crops' or 'stem cell research' would be good examples. Next, write out a statement that you might use for a motion in a debate, such as 'Genetic modification of crops is a good thing' or 'Creating chimeric organisms is acceptable if it results in medical advances'. Then write at least five points in support of either side of the argument, which you should organise in tabular form. If you can find more than five points, add these to your table, but for each point that you add to one side, you should add one to the other side.

9.3 **Analyse graphic presentations in the media.** Many newspapers provide graphic presentations related to current issues, and graphs are frequently used in television news reports. Practise critical thinking skills by determining whether the graphs presented are a fair representation of the facts.

10 Using online resources

> **Definitions**
>
> **Browser** – a program to display web pages and other Internet resources.
>
> **FAQ** – Frequently Asked Question; a file or web page giving information on common queries, sometimes used as a file extension (.faq).
>
> **FTP** – File Transfer Protocol; a mechanism of downloading files.
>
> **URL** – Uniform Resource Locator; the 'address' for web resources.

Information and communication technology (ICT) is vital in the modern academic world and 'IT literacy' is a core skill for all bioscientists. This involves a wide range of computer-based skills, including:

- Accessing web pages using a web browser such as Microsoft Internet Explorer, Mozilla Firefox or Opera.
- Searching the Web for useful information and resources using a search engine such as Google, or a meta-search engine such as Dogpile.
- Finding what you need within online databases, such as library catalogues or complex websites, such as your university's homepage.
- Downloading, storing and manipulating files.
- Communicating *via* the Internet.
- Using e-learning facilities effectively.
- Working with 'Office'-type programs and other software (dealt with in detail in Chapters 11 and 12).

You will probably receive an introduction to your university's networked IT systems and you will be required to follow rules and regulations that are important for the operation of these systems. Whatever your level of experience with PCs and the Internet, you should also follow the basic guidelines shown in Box 10.1. Reminding yourself of these from time to time will reduce your chances of losing data.

> ***Understanding the technology*** *– you do not need to understand the workings of the Internet to use it – most of it is invisible to the user. To ensure you obtain the right facilities, you may need to know some jargon, such as terms for the speed of data transfer (megabits) and the nature of internet addresses. Setting up a modem and/or local wireless network can be complex, but instructions are usually provided with the hardware. White and Downs (2005) and Gralla (2003) are useful texts if you wish to learn more about computing and the Internet.*

The Internet as a global resource

The Internet is a complex network of computer networks; it is loosely organised and no one group organises it or owns it. Instead, many private organisations, universities and government organisations fund and operate discrete parts of it.

The Web is the most popular application of the Internet. It allows easy links to information and files which may be located on networked computers across the world. The Web enables you to access millions of 'home pages' or 'websites' – the initial point of reference with many individuals, institutions and companies. Besides text and images, these sites may contain 'hypertext links', highlighted words or phrases that take you to another internet location *via* a single mouse click.

You can gain access to the Internet either through a network at your university, at most public libraries, at a commercial 'Internet cafe', or from home *via* a modem connected to a broadband or dial-up internet service provider (e.g. Zen, BT or AOL).

KEY POINT Most material on the Internet has not been subject to peer review or vetting. Information obtained from the Web or posted on newsgroups may be inaccurate, biased or spoof; do not assume that everything you read is true, or even legal.

Online communication

You will be allocated an email account by your university and should use this routinely for communicating with staff and fellow students, rather than using a personal account. You may be asked to use email to submit work as an attachment, or you may be asked to use a 'digital drop-box'

Box 10.1 Important guidelines for using PCs and networks

Hardware

- Don't drink or smoke around the computer.
- Try not to turn the computer off more than is necessary.
- Never turn off the electricity supply to the machine while in use.
- Switch off the computer and monitor when not in use (saves energy and avoids dangers of 'hijacking').
- Rest your eyes at frequent intervals if working for extended periods at a computer monitor. Consult Health and Safety Executive publications for up-to-date advice on working with display screens (http://www.hse.gov.uk/pubns/)
- Never try to reformat the hard disk without the help of an expert.

CDs and USB drives

- Protect CDs when not in use by keeping them in holders or boxes.
- Label USB (Universal Serial Bus) drives with your name and return details and consider adding these to a file stored on the drive.
- Try not to touch the surface of CDs, and if they need cleaning, do so carefully with a clean cloth, avoiding scratching. If floppy disks are used, keep these away from sources of magnetism (e.g. speakers).
- Keep disks and USB drives away from moisture, excess heat or cold.
- Don't use disks from others, unless you first check them for viruses.
- Don't insert or remove a disk or USB drive when it is operating (drive light on). Close all files before removing a USB drive and use the *Safely Remove Hardware* feature.
- Try not to leave a disk or USB drive in the computer when you switch off.

File management

- Organise your files in an appropriate set of folders.
- Always use virus-checking programs on copied or imported files before running them.
- Make backups of all important files at frequent intervals (say, every 10 minutes), e.g. when using a word processor or spreadsheet.
- Periodically clear out redundant files.

Network rules

- Never attempt to 'hack' into other people's files.
- Do not give out any of your passwords to others. Change your password from time to time. Make sure it is not a common word, is longer than 8 characters, and includes numerical characters and punctuation symbols, as well as upper and lower case letters.
- Never use network computers to access or provide financial or other personal information: spyware and Trojan programs may intercept your information.
- Never open email attachments without knowing where they came from; always virus-check attachments before opening.
- Remember to log out of the network when finished; others can access your files if you forget to log out.
- Be polite when sending email messages.
- Periodically reorganise your email folder(s). These rapidly become filled with acknowledgements and redundant messages that reduce server efficiency and take up your allocated filespace.
- Do not play games without approval – they can affect the operation of the system.
- If you are setting up your own network, e.g. in your flat, always install up-to-date firewall software, anti-spyware and anti-virus programs.

The Golden Rule – always make backup copies of important files and store them well away from your working copies. Ensure that the same accident cannot happen to both copies.

within the university's e-learning system (Box 10.2). When using email at university, follow conventions, including etiquette, carefully:

- Check your email account regularly (daily). Your tutors may wish to send urgent messages to you in this way.
- Respond promptly to emails. Even if you are just acknowledging receipt, it is polite to indicate that you have received and understood a message.
- Be polite. Email messages can seem to be abrupt and impersonal. Take care to read your messages through before sending and if you

Box 10.2 Getting to grips with e-learning

Some key aspects of tackling e-learning are outlined below.

1. **Develop your basic IT skills, if required.** E-learning requires only basic IT skills, such as: use of keyboard and mouse; word-processing; file management; browsing and searching. If you feel weak on any of these, seek out additional courses offered by the IT administration or your department.
2. **Visit your e-learning modules regularly.** You should try to get into a routine of doing this on a daily basis at a time that suits you. Staff will present up-to-date information (e.g. lecture room changes) via the 'announcements' section, may post information about assessments, or links to the assessments themselves, and you may wish to provide feedback or look at discussion boards and their threads.
3. **Participate.** E-learning requires an active approach.
   - At the start of each new course, spend some time getting to know what's been provided online to support your learning. As well as valuable resources, this may include crucial information such as learning objectives (p. 22), dates of submission for coursework and weighting of marks for different elements of the course.
   - If you are allowed to download lecture notes (e.g. in the form of PowerPoint presentations), do not think that simply reading through these will be an adequate substitute for attending lectures and making further notes (see page 16).
   - Do not be tempted to 'lurk' on discussion boards: take part. Ask questions; start new threads; answer points raised by others if you can.
   - Try to gain as much as you can from formative online assessments (p. 24). If these include feedback on your answers, make sure you learn from this and if you do not understand it, consult your tutors.
   - Learn from the critical descriptions that your lecturers provide of linked websites. These pointers may help you to evaluate such resources for yourself in future.
   - Don't think that you will automatically assimilate information and concepts, just because you are viewing them online. The same principles apply as with printed media: you must apply active learning methods (p. 20).
   - Help your lecturers by providing constructive feedback when they ask for it. You may find this easier to do when using the computer interface, and it may be more convenient than hurriedly filling out a feedback sheet at the end of a session.
4. **Organise files and web links.** Take the time to create a meaningful folder and file naming system for downloaded material in tandem with your own coursework files and set up folders on your browser for bookmarked websites (*Favorites* in Internet Explorer).
5. **Take care when submitting coursework.** Make sure you keep a back-up of any file you email or submit online and check the version you are sending carefully. Follow instructions carefully, for example regarding file type, or how to use your system's 'digital drop-box'.

Spam, junk mail and phishing *– these should be relatively easy to identify, and should never be responded to or forwarded. Some may look 'official' and request personal or financial details (for example, they may pretend to come from your bank, and ask for account details). Never send these details by email or your identity may be used illegally.*

are at all in doubt, do not send your message right away: re-read at a later time and consider how others might view what you say.

- Consider content carefully. Only send what you would be happy to hear being read out loud to classmates or family.
- When communicating with tutors, take care with language and names. Slang phrases and text message shorthand are unlikely to be understood. Over-familiarity does not go down well.
- Use email for academic purposes only – this includes discussing coursework with classmates, but not forwarding off-colour jokes, potentially offensive images, links to offensive websites, etc. In fact, doing so may break regulations and result in disciplinary action.
- Beware of spam, junk and 'phishing' *via* email.

Similar rules apply to discussion boards.

The Usenet Newsgroup service is an electronic discussion facility, and there are thousands of newsgroups representing different interests and

Using newsgroups – *these can be useful for getting answers to a specific problem: just post a query to the appropriate group and wait for someone to reply. Bear in mind that this may be the view of an individual person.*

topics. Any user can contribute to the discussion within a topic by posting their own message; it is like email, but without privacy, since your message becomes available to all other subscribers. To access a newsgroup, your system must be running, or have access to, a newsgroup server which has subscribed to the newsgroup of interest.

With mailing lists, messages sent to the list address are distributed automatically to all members of the mailing list, *via* their personal emailbox, keeping them up to date on the particular topic of the mailing list. To receive such messages, you will need to join the mailing list. Relevant mailing lists for biological sciences can be found at http://www.jiscmail.ac.uk/. Take care not to join too many lists, as you will receive a large number of messages, and many are likely to be of only marginal interest. A number of mailing lists also have archived files, offering a more selective means of locating relevant material.

Internet tools

The specific programs you will use for accessing the Internet will depend on what has been installed locally, on the network you are using, and on your internet service provider. The best way to learn the features of the programs is to try them out, making full use of whatever help services are available.

e-learning systems

Most university departments present their courses through a mixture of face-to-face sessions (e.g. lectures, tutorials, practicals) and online resources (e.g. lecture notes, websites, discussion boards, computerised tests and assessments). This constitutes 'blended learning' on your part, with the online component also being known as e-learning.

The e-learning element is usually delivered through an online module within a virtual learning environment (e.g. Blackboard, WebCT, Moodle). It is important not to neglect the e-learning aspects of your course just because it may not be as rigidly timetabled as your face-to-face sessions. This flexibility is to your advantage, as you can work when it suits you, but it requires discipline on your part. Box 10.2 provides tips for making the most of the e-learning components of your courses.

Internet browsers

These are software programs that interact with remote server computers around the world to carry out the tasks of requesting, retrieving and displaying the information you require. Many different browsers exist, but the most popular are Internet Explorer, Mozilla Firefox and Opera. These three browsers dominate the market and have plug-ins and add-on programs available that allow, for example, video sequences to be seen online. Many browsers incorporate email and newsgroup functions. The standard functions of browsers include:

- accessing web documents;
- following links to other documents;
- printing the current document;
- maintaining a history of visited URLs (including 'bookmarks' for key sites);
- searching for a term in a document;
- viewing images and image maps.

Definition

Bookmark – a feature of browsers that allows you to save details of websites you have visited. This is termed 'add to favorites' in Internet Explorer. Bookmarks save you the trouble of remembering complex URL names and of typing them into the browser's address window.

Examples Useful web portals
INTUTE – internet resources for Health and Life Sciences at: http://www.intute.ac.uk/healthandlifesciences/
OMIM – online Mendelian inheritance in man – http://www.ncbi.nlm.nih.gov/entrez./query.fcgi?db=OMIM

Examples Common domains and sub-domains include:

.ac	academic
.com	commercial
.co	commercial
.edu	education (USA mainly)
.gov	government (USA and UK)
.mil	military (USA only)
.net	Internet-based companies
.org	organisation
.uk	United Kingdom

***'Dissecting' a Web address** – you can often find out more about a particular site by progressively deleting sections of the URL from the right-hand side. This will often take you to the home page of the organisation or company involved.*

Browsers provide access to millions of websites. Certain sites specialise in providing catalogued links to other sites; these are known as portals and can be of enormous help when searching within a particular area of interest. Your university's library website will almost certainly provide a useful portal to catalogues and search services, often arranged by subject area, and this is often the first port of call for electronic resources; get to know your way around this part of the website as early as possible during your course.

When using a browser program to get to a particular page of information on the Web, all you require is the location of that page, i.e. the URL (Uniform Resource Locator). Most web page URLs take the form http://, or https://, followed by the various terms (domains and sub-domains) that direct the system to the appropriate site. If you don't have a specific URL in mind but wish to explore appropriate sites, you will need to use a search tool with the browser.

Search tools

With the proliferation of information on the Web, one of the main problems is finding the exact information you require. There are a variety of information services that you can use to filter the material on the network. These include:

- search engines (Boxes 10.3 and 10.4);
- meta-search engines;
- subject directories;
- subject gateways (portals).

Search engines such as Google (http://www.google.com), AltaVista (http://uk.altavista.com) and Lycos (http://www.lycos.com) are tools designed to search, gather, index and classify web-based information. Searching is usually by keyword(s), although specific phrases can be defined. Many search engines offer advanced searching tools such as the use of Boolean operators to specify combinations of keywords to more precisely filter the sites. Box 10.3 provides tips for refining keyword searches, while Box 10.4 provides tips for enhanced searching with Google.

It is important to realise that each search engine will cover at most about 40 per cent of the available sites and if you want to carry out an exhaustive search it is necessary to use several to cover as much of the Web as possible. Meta-search engines make this easier. These operate by combining collections of search engines. Examples include Mamma (http://www.mamma.com/), Dogpile (http://www.dogpile.com/index.gsp/) and Metacrawler (http://www.metacrawler.com/index.html/).

Some useful approaches to searching include the following:

- For a comprehensive search, use a variety of tools including search engines, meta-search engines and portals or directories.
- For a complex, finely specified search, employ Boolean operators and other tools to refine your keywords as fully as possible (Box 10.2). Some search engines allow you to include and exclude terms or restrict by date.
- Use 'cascading' searching when available – this is searching within the results of a previous search.
- Use advanced search facilities to limit your search, where possible, to the type of medium you are looking for (e.g. graphics, video), language, sites in a specific country (e.g. UK) or to a subject area (e.g. news only).

Box 10.3 Useful tips for using search engines

- **Keywords should be chosen with care.** Try to make them as specific as possible, e.g. search for 'archaebacteria' 'or Archaea', rather than 'bacteria' or 'micro-organisms'.
- **Most search engines are case-insensitive.** Thus 'Nobel Prize' will return the same number of hits as 'nobel prize'. If in doubt, use lower case throughout.
- **Putting keyword phrases in double quotes (e.g. "capillary electrophoresis") will result in a search for sites with the phrase as a whole** rather than sites with both (all) parts of the phrase as separate words (i.e. 'capillary' and 'electrophoresis' at different places within a site). This feature allows you to include common words normally excluded in the search, such as 'the'.
- **Use multiple words/phrases plus similar words to improve your search,** for example 'terrestrial crustacean "land crab" Australia'. If you can, use scientific terms, as you are likely to find more relevant sites, e.g. search for the name of a particular species such as *Coenobita compressus.*
- **Adding words preceded with + or − will add or exclude sites with that word present** (e.g. 'blood diseases − leukaemia' will search for all blood diseases excluding those mentioning leukaemia). This feature can also be used to include common words normally excluded by the search engine.
- **Check that your search terms have the correct spelling,** otherwise you may only find sites with the same misspelled word. In some cases, the search engine may prompt you with an alternative (correct) spelling. Remember that US and UK spellings may differ (e.g. hemolysis/haemolysis) and a search will only find hits from sites that use the spelling you specify.
- **Boolean operators (AND, OR, NOT) can be used with some search engines to specify combinations of keywords** to more precisely filter the sites identified (e.g. 'plant NOT engineering' will avoid sites about engineering plant and focus on biological material – though this combination might omit genetic engineering of plants).
- **Some search engines allow 'wildcards' to be introduced with the symbol*.** For example, this will allow you to specify the root of a word and include all possible endings, as with anthropomorph*, which would find anthropomorphic, anthropomorphism, etc. If the search engine does not allow wildcards (e.g. Google), then you will need to be especially careful with the keywords used, including all possible words of relevance.
- **Numbers can be surprisingly useful in search engines.** For example, typing in EC 1.1.1.1 will find sites concerned with alcohol dehydrogenase, as this is its Enzyme Commission code number (p. 381). If you know the phone number for a person, institute or company or the ISBN of the book, this can often help you find relevant pages quickly.
- **If you arrive at a large site and cannot find the point at which your searched word or phrase appears, press Control and F together** and a 'local' search window will appear, allowing you to find the point(s) where it is mentioned.

Downloading files from the Internet and emails *– read-only files are often available as 'pdf' files that can be viewed by Adobe reader software (available free from http://www.adobe.com), while other files may be presented as attachments to emails or as links from web pages that can be opened by suitable software (e.g. Microsoft Word or 'paint' programs like Paint Shop Pro). Take great care in the latter cases as the transfer of files can result in the transfer of associated viruses. Always check new files for viruses (especially .exe files) before running them, and make sure your virus-detecting software is kept up to date.*

However well defined your search is, you will still need to evaluate the information obtained. Chapter 9 covers general aspects of this topic while Box 10.5 provides specific advice on assessing the quality of information provided on websites.

Directories

A directory is a list of web resources organised by subject. It can usually be browsed and may or may not have a search facility. Directories often contain better quality information than the lists produced by search engines, as they have been evaluated, often by subject specialists or librarians. The BUBL information service directory of links, at: http://bubl.ac.uk, is a good example.

Box 10.4 Getting the most from Google searches

Google (http://www.google.com) has become the search engine of choice for millions of people, due to its simplicity and effectiveness. However, you may be able to improve your searches by understanding its default settings and how they can be changed.

- **Download the Google toolbar to your browser.** This is available from the Google homepage and will give you quick access to the Google search facility.
- **Understand how standard operators are used.** For combinations of keywords Google uses the 'minus' operator '−' instead of NOT (exclude) and the 'plus' operator '+' instead of AND (include). Since Google usually ignores small words ('stop words' such as *in* or *the*), use '+' to include them in a search. Where no operator is specified, Google assumes that you are looking for both terms (i.e. '+' is default). If you want to search for alternative words, you can use 'OR' (e.g. *sulphur OR sulfur*). Google does not allow brackets and also ignores most punctuation marks.
- **While wildcard truncation of words using '*' is not allowed, you can use '*' to replace a whole word (or number).** For example, if you type the phrase *"a virus is approximately * nanometres"* your results will give you results for web pages where the wildcard is replaced by a number.
- **Search for exact wording.** By placing text in double inverted commas (""), you can ensure that only websites with this exact phrasing will appear at the head of your search results.
- **Search within your results to improve the outcome.** If your first search has produced a large number of results, use the *Search within results* option near the bottom of each page to type in a further word or phrase.
- **Search for words within the title of a web page.** Use the command *intitle:* to find a web page, for example *intitle: "tissue culture"* returns web pages with this phrase in the title (note that phrases must always be in double speech marks, not single quotes).
- **Search within a website.** Use the *site:* command to locate words/phrases on a specific website, for example *site:unicef.org dysentery* returns only those results for this disease on the UNICEF website (unicef.org). Pressing *Control* and *F* when visiting a web page will give you a pop-up search window.
- **Locate definitions, synonyms and spellings.** The operator *define:* enables you to find the meaning of a word. If you are unsure as to the spelling of a word, try each possibility. Google will usually return more results for the correct spelling and will often also prompt you with the correct spelling (*Did you mean...?*).
- **Find similar web pages.** Simply click the *Similar pages* option at the end of a Google search result to list other sites (note that these sites will not necessarily include the term(s) searched for).
- **If a web link is unavailable, try the cached (stored) page.** Clicking on *Cached* at the end of a particular result should take you to the stored page, with the additional useful feature that the search term(s) will be highlighted.
- **Use the calculator function.** Simply enter a calculation and press *Enter* to display the result, for example '10+(2×4)' returns 18. The calculator function can also carry out simple interconversion of units, e.g. '2 feet 6 inches in metres' returns 0.762 (see Box 29.1 for interconversion factors between SI and non-SI units).
- **Try out the advanced search features.** In addition to the standard operators these include the ability to specify the number of results per page (e.g. 50) to reduce the use of the *next* button, language (e.g. English), file format (e.g. for PDF files), recently updated web pages (e.g. past 3 months), usage (e.g. free to use/share).
- **Find non-text material.** These include images, video and maps – always check that any material you use is not subject to copyright limitations (p. 47).
- **Use Google alerts to keep up to date.** This function (http://www.google.co.uk/alerts) enables you to receive regular updated searches by email.
- **Use Google Scholar to find articles and papers.** Go to http://scholar.google.com/ and type in either the general topic or specific details for a particular article, e.g. author names or words from the title. Results show titles/authors of articles, with links to either the full article, abstract or citation. A useful feature is the *Cited by* ... link, taking you to those papers that have cited the article in their bibliography and enabling you to carry out forward citation searching to locate more recent papers. Also try out the advanced scholar search features to limit your search to a particular author, journal, date or subject area. However, you should note that Google Scholar provides only a basic search facility to easily accessible articles and should not be viewed as a replacement for your library's electronic journal holdings and searching software. For example, if you find the title of a paper *via* Google Scholar you may then be able to locate the electronic version through your own library's databases, or request it *via* inter-library loan (p. 46). Another significant limitation is that older (more cited) references are typically listed first.
- **Use Google Earth to explore locations.** Download from http://earth.google.com/. This allows you to zoom in on satellite images to find locations.

Box 10.5 How to evaluate information on the World Wide Web

It is often said that 'you can find anything on the Web'. The two main disadvantages of this are, firstly, that you may need to sift through many sources before you find what you are looking for and, secondly, that the sources you find will vary in their quality and validity. *It is important to realise that evaluating sources is a key aspect of using the Internet for academic purposes, and one that you will need to develop during the course of your studies* (see also Chapter 9). The ease with which you can 'point and click' to reach various sources should not make you complacent about evaluating their information content. The following questions can help you to assess the quality of a website – the more times you can answer 'yes', the more credible the source is likely to be, and *vice versa:*

Authority

- Is the author identified?
- Are the author's qualifications or credentials given?
- Is the owner, publisher or sponsoring organisation identified?
- Is an address given (postal and/or email)?

It is sometimes possible to get information on authority from the site's metadata (try the 'View' 'Source' option in Internet Explorer, or look at the URL to see if it gives any clues as to the organisation, e.g. whether the domain name ends in .ac, .edu, .gov or .org, rather than .co or .com).

Content

- Is there any evidence that the information has been peer-reviewed (p. 113), edited or otherwise validated, or is it based on such sources?
- Is the information factual or based on personal opinion?
- Is the factual data original (primary) or derived from other sources (secondary)?
- Are the sources of specific factual information detailed in full (Chapter 8)?
- Is there any indication that the information is up to date, or that the site has been recently updated?
- What is the purpose of the site and who is it aimed at?
- Is the content relevant to the question you are trying to answer?
- Is there any evidence of a potential conflict of interest, or bias? (Is the information comprehensive and balanced, or narrowly focused?)
- Did you find the information *via* a subject-specific website (e.g. a bioscience gateway such as BIOME), or through a more general source, such as a search engine (e.g. Google)?

The above questions are similar to those that you would use in assessing the value of a printed resource (p. 53), and similar criteria should be applied to web-based information. You should be especially wary of sites containing unattributed factual information or data whose primary source is not given.

Presentation

- What is your overall impression of how well the site has been put together?
- Are there any grammatical or spelling mistakes?
- Are there links to other websites, to support statements and factual information?

The care with which a site has been constructed can give you an indication of the credibility of the author/organisation. However, while a poorly-presented site may cause you to question the credibility of the information, the reverse is not always necessarily true: don't be taken in by a slick, well-presented website – authority and content are *always* more important than presentation.

Using the Internet as a resource

A common way of finding information on the Web is by browsing or 'surfing'. However, this can be time-consuming; try to restrict yourself to sites known to be relevant to the topic of interest. Some of the most useful sites are those that provide hypertext links to other locations.

***Understanding* the impermanence of the Web** – *the temporary nature of much of the material on the Web is a disadvantage for academic purposes because it may change or even disappear after you have cited it. You may also find it difficult or impossible to find out who authored the material. A case in point are wikis, such as Wikipedia (www.wikipedia.org). This online encyclopaedia has many potential authors and the content may change rapidly as a result of new submissions or edits; nevertheless, it can be a useful resource for up-to-date general information about a wide range of topics, though it is not necessarily regarded as the best approach for researching scientific assignments.*

Using traditional sources – *remember that using the Internet to find information is not a substitute for visiting your university library. Internet resources complement rather than replace more traditional printed sources.*

Remembering useful websites – *create a 'bookmark' (=add a 'favorite') for the ones you find of value, to make revisiting easy. This can be done from the menu of your browser program. Make a copy of your bookmark file occasionally, to avoid loss of this information.*

Note that URLs may change – *make a keyword search using a search engine to find a particular site if the URL information you have does not lead you to an active page.*

Examples Selected websites of biological interest:
Biochemical Society:
http://www.biochemistry.org/
Genetics Society:
http://www.genetics.org.uk
Human Genome Organisation (HUGO):
http://www.hugo-international.org/
UK university library catalogues:
http://hero.ac.uk/sites/hero/uk/reference_and_subject_resources/institution_facilities/online_catalogues_alphabetic3793.cfm
WWW Virtual Library Biosciences:
http://vlib.org/Biosciences/

Some other resources you can use on the Web are:

- Libraries, publishers and commercial organisations. Your university library is likely to subscribe to one or more databases providing access to scientific articles; these include ISI Web of Science (http://wos.mimas.ac.uk), and Science Direct (http://www.sciencedirect.com). A password is usually required, especially for off-campus use; consult your library staff for further details. Some scientific database sites give free access, without subscription or password; these include the PubMed website of the National Center for Biotechnology Information (http://www.ncbi.nlm.nih.gov) and the HighWire Press (http://highwire.stanford.edu). Others allow free searching, but require payment for certain articles, e.g. the Scientific World (http://www.thescientificworld.com/) and Infotrieve (http://www4.infotrieve.com). Publishers such as Pearson and booksellers such as Amazon.com provide online catalogues and ecommerce sites that can be useful sources of information (see http://vig.pearsoned.co.uk and http://www.amazon.com).
- Online journals and e-books. A number of traditional journals have websites. You can keep up to date by visiting the websites of *Nature* (http://www.nature.com), *New Scientist* (http://www.newscientist.com) and *Scientific American* (http://www.sciam.com), or ScienceDirect journals (at http://www.sciencedirect.com/science/journals). Some scientific societies make their journals and other publications available *via* their websites, e.g. the American Society for Microbiology at http://www.asm.org. Journals solely published in electronic format are also available (e.g. *Molecular Vision*, http://www.molvis.org/molvis) but some require a subscription password for access; check whether your institute is a subscriber.
- Data and images. Archives of text material, video clips and photographs can be accessed, and much of the material is readily available. The HEA Centre for Bioscience ImageBank (http://heacademy.ac.uk/imagebank/) is a good example. When downloading such material, you should check that you are not breaching copyright and avoid potential plagiarism by giving a full citation of the source, if you use such images in an assignment (see p. 51).
- Biological institutions. Many culture collections, scientific societies and other biological institutions around the world are now online. Use their sites to obtain specific information about collections, resources, etc. They frequently provide lists of other relevant sites or topics, e.g. the Society for General Microbiology (http://www.sgm.ac.uk/links/).
- Databases. In addition to those covering the scientific literature (e.g. ISI Web of Science, or the PubMed service of the US National Library of Medicine), others focus on specific topics (e.g. biodiversity on the Internet, http://www.biodiversity.org.uk/ibs/; or academic employment, http://www.jobs.ac.uk). Chapter 11 provides further guidance on using online databases for the analysis of biomolecular structure (bioinformatics).

Text references

Gralla, P. (2003) *How the Internet Works*, 7th edn. Pearson, Harlow.

White, R. and Downs, T. (2005) *How Computers Work*, 8th edn. Pearson, Harlow.

Sources for further study

Anon. *Search Basics*. Available:
http://www.frame-uk.demon.co.uk/guide/search_basics.htm
Last accessed: 01/04/07.
[Demon Internet pages on browser searching]

Anon. *The Essentials of Google Search*. Available:
http://www.google.co.uk/intl/en/help/basics.html
Last accessed: 01/04/07.
[Advice on using the Google search engine]

Brandt, D.S. *Why We Need to Evaluate What We Find on the Internet*. Available: http://www.lib.purdue.edu/research/techman/eval.html
Last accessed: 01/04/07.

Cabibbo, A., Grant, R.P. and Citterich, M.H. (2004) *The Internet for Cell and Molecular Biologists: Current Applications and Future Potential*. Horizon Scientific Press, Wymondham.

Dussart, G. (2002) *Biosciences on the Internet*. Wiley, Chichester.

Grassian, E. *Thinking Critically about World Wide Web Resources*. Available:
http://www.library.ucla.edu/libraries/college/help/critical/index.htm
Last accessed: 01/04/07.

Study exercises

10.1 Explore the resources of the Web using a search engine. Using a search engine, find the answers to the following questions:

(a) Who is Barbara McClintock, and what is she famous for? What prize did she win and when?

(b) What is the common name for *Arabidopsis thaliana*?

(c) What is the postal address of the UK Biochemical Society?

10.2 Compare results from a variety of search engines. First, think of an appropriate biological keyword or phrase (e.g. a species name) and enter this into several search engines. Make sure that you include meta-search engines such as Dogpile. Compare the outcomes to reveal the strengths and weaknesses of the individual search engines. Work with a colleague to compare different searches on a quantitative (i.e. number of hits) and qualitative (quality of hits) basis.

10.3 Organise your bookmarks. Enter the *'Organize favorites'* menu for your preferred browser and create folders with appropriate headings. Move existing bookmarks to these folders and save any new ones appropriately. Doing this will help you find bookmarks more easily, rather than searching through long lists.

11 Bioinformatics – Internet resources

Definitions

Note that terms ending in 'ome' can be used to describe disciplines by changing the ending to 'omics'.

Genome – the entire complement of genetic information (coding and non-coding DNA) of an organism.

Transcriptome – the complement of mRNAs transcribed from the genome, weighted by the expression level of each RNA.

Proteome – the expressed protein complement of the genome.

Secretome – all of the secreted proteins from an organism.

Metabolome – all of the metabolites (low molecular mass biomolecules) of an organism.

Definition

Chemoinformatics – the application of information technology to chemical data and information, especially in the field of drug design/discovery.

Bioinformatics is a term used to describe the application of computer-based techniques in the biological sciences and, in particular, to the analysis of nucleic acid sequences (genomics, transcriptomics) and to the protein sequences predicted from nucleic acid sequence data (proteomics). These complex molecules contain a large amount of information within their structures: analysis of this information by visual examination is impractical for all but the smallest polynucleotide or polypeptide fragments.

***KEY POINT** The only practical approach to understanding the information in complex macromolecules is to use a computer – typically, by comparing sequence data for all or part of a specific biomolecule with that of others in a database.*

Table 11.1 gives details of some of the most important databases accessible *via* the Internet.

A range of programs and packages can be used to analyse nucleic acid and protein sequences. Many programs can be used online, *via* the Internet, while others can be downloaded, for local operation. In most instances, the information and software are freely available. The large number of different programs available is a reflection of the various algorithms used for analysis, and such programs are likely to be updated in parallel with advances in computer hardware/software. When using a new program, make use of the help facilities to familiarise yourself with the operating principles.

Primary sequence databases

Understanding the scale of primary sequence databases – in December 2006, the EBI-NCBI-DDBJ held over 80 million nucleic acid sequence records.

The major databases holding primary sequence information for nucleic acids and proteins are operated by the European Bioinformatics Institute (EBI) of the European Molecular Biology Laboratory, EMBL (at http://www.ebi.ac.uk), the National Center for Biotechnology Information (NCBI) GenBank database (at http://www.ncbi.nlm.nih.gov/Genbank/) and the DNA Data Bank of Japan, DDBJ (at http://www.ddbj.nig.ac.jp). These databases hold comprehensive information submitted by researchers and sequencing groups across the world, with collaborative data-sharing on a daily basis, as part of the International Nucleotide Sequence Database Collaboration.

Nucleic acid analysis

A typical application would be to investigate whether a nucleic acid sequence obtained, for example, as part of a cloning project, has been reported previously. Table 11.1 gives details of URLs for the nucleic acid databases of EMBL-EBI and NCBI GenBank. These databases allow you to:

Example Y10510 is the accession number for a human mRNA sequence for the CD67S protein in the EMBL-EBI database.

- enter a particular nucleic acid sequence and edit the sequence, if required;
- find and retrieve a nucleotide sequence from the database using a keyword, author name, organism name or an accession number, where known;

Table 11.1 Some useful databases and other internet locations relevant to bioinformatics

Major nucleic acid sequence databases	
EMBL-EBI nucleotide sequence database	http://www.ebi.ac.uk/embl/index.html
NCBL GenBank	http://www.ncbi.nlm.nih.gov/Genbank
Major protein sequence databases	
SWISS-PROT (EBI protein sequence database)	http://ca.expasy.org/
TrEMBL (EBI translated sequence database)	http://www.ebi.ac.uk/trembl
UniProt (universal protein resource)	http://www.ebi.ac.uk/uniprot/
PIR (protein information resource)	http://pir.georgetown.edu
Other bioinformatics databases	
Sanger Institute (links to various databases)	http://www.sanger.ac.uk/
PROSITE (protein motifs)	http://ca.expasy.org/prosite/
PRINTS (protein fingerprints)	http://bioinf.man.ac.uk/dbbrowser/PRINTS/
ExPaSy proteomics resource	http://ca.expasy.org/
PDB (Protein Data Bank)	http://www.rcsb.org/pdb
CAZy (carbohydrate-active enzymes)	http://afmb.cnrs-mrs.fr/CAZY/
MEROPS (protease/peptidase database)	http://merops.sanger.ac.uk
Transport protein database	http://tcdb.org/
DOLOP (database of bacterial lipoproteins)	http://www.mrc-lmb.cam.ac.uk/genomes/dolop/
CBS prediction servers (various databases)	http://www.cbs.dtu.dk/services/

Searching primary databases *– a search of the EMBL-EBI nucleic acid database can be carried out using BLAST (Fig. 11.1) to locate related sequences.*

- translate a nucleotide sequence into an amino acid sequence and *vice versa*;
- search for any resemblance between a particular sequence and sequences within the database, e.g. by comparing and aligning the sequence for several nucleic acids or proteins, to identify regions of sequence similarity (see Fig. 11.1);
- carry out phylogenetic analysis, constructing 'ancestry trees' to show the most likely evolutionary relationships between sequences from various organisms.

Other programs allow you to analyse nucleic acid sequence information in more detail, for example:

- to assemble/align sequence fragments into contiguous sequences, also known as 'contigs'. This type of alignment is particularly useful for sets of nucleic acid fragments from a clone library;
- to search for structural and functional 'motifs' within the nucleic acid sequence, including palindromes, restriction sites, likely promoter or control sites and regions that might form secondary structures, such as hairpin loops;
- to simulate *in vitro* manipulation and cloning procedures, including the preparation of restriction digests, designing oligonucleotide probes and primers, and simulating PCR.
- to identify likely coding regions (e.g. open reading frames, exons) and non-coding regions (e.g. introns, repeat sequences) in eukaryotic DNA, 'start' and 'stop' codons, etc. The starting point for this analysis is the creation of a 'six frame translation' of the target sequence (see Fig. 11.2).

Definition

Palindrome – a sequence reading the same in both directions, e.g. GACCAG. These are often sites at which restriction enzymes cleave DNA.

Remembering which is which *– INTrons (non-coding sequences) INTerrupt coding sequences while EXons are EXpressed (translated into an amino acid sequence).*

```
H. sapiens     atgtatggcattgagaatgaagtcttcctgagccttccatgtatcctcaatgcccgggg
               ||||| ||||| ||||| ||||||||||| || || || || ||||| ||||| |||||
R. norvegicus  atgtacggcatcgagaacgaagtcttcctcagtctcccgtgcatccttaatgctcgggg
```

Fig. 11.1 Representative output from a BLAST (basic local alignment search tool) alignment search for DNA sequences from part of the lactate dehydrogenase genes of *Homo sapiens* and *Rattus norvegicus*. Here, identical bases are shown in black, and non-identical bases are shown in blue. This region shows 49 identical bases out of 59, i.e. $49 \div 59 \times 100 = 83\%$ similarity (to the nearest integer).

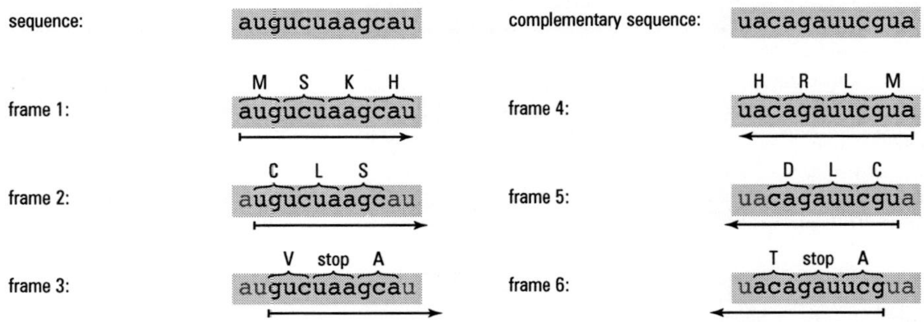

Fig. 11.2 'Six frame translation' of a DNA sequence. The triplet codons corresponding to each amino acid are given in Table 60.1 (p. 429) and the single letter amino acid code is explained in Table 50.1. Note that the complementary sequence is translated in the reverse direction to the original sequence.

Analysis of proteins and polypeptides

The principal protein sequence databases are SWISS-PROT/TrEMBL in Europe, together with PIR in the USA: Table 11.1 gives their current Internet locations. These databases can be searched using a range of programs equivalent to those described above for nucleic acid analysis, allowing the operator to:

- locate and retrieve a particular protein or amino acid sequence from the database by keyword, author name, organism name or accession number;
- enter and edit (cut and paste) a particular amino acid sequence, or convert a nucleotide sequence into an amino acid sequence, and *vice versa*;
- search the database for sequences with a perfect match, or for similarities in primary or secondary structure at a specified level of homology; such searches may be more informative than those based on DNA homology, since protein sequence information (based on 20 different amino acid residues) is more complex than nucleic acid sequence information (based on four different nucleotides);
- compare and align the amino acid sequences of several proteins, to identify particular regions of similarity and dissimilarity.

Using bioinformatics databases *– many of the database websites are rather com- plex and newcomers may find them a little difficult to navigate, especially if you are just browsing. The best approach is to try out the online tutorials, to get a feel for how the website operates. If the URL changes, use a search engine to locate the new website.*

Other bioinformatics databases

Pattern recognition databases for proteins

These have been created using primary sequence data, to provide information on the patterns identified within proteins. The representative examples given in Table 11.1 include the PROSITE and PRINTS databases, which identify protein families by their diagnostic 'signature' motifs, and the Expert Protein Analysis System (ExPaSy) website, which provides a gateway to the general analysis of protein structure, including three-dimensional imaging, while the CBS Prediction Servers can be used to identify particular motifs, as well as other aspects of genome analysis.

Many of these 'second-level' databases provide tools and software that can be used for a range of purposes, including:

- assembling overlapping peptide fragments, based on their amino acid sequences, e.g. during sequence analysis of an individual protein;

Understanding the terminology of protein structure:

- *primary structure – sequence of amino acids along a polypeptide;*
- *secondary structure – regular repeating structural motifs (α-helix and β-sheet), stabilised by hydrogen bonds;*
- *tertiary structure – overall three-dimensional form of the polypeptide;*
- *quaternary structure – where more than one polypeptide subunits form a functional complex.*

- predicting secondary structure from a primary sequence, identifying regions most likely to exist as α-helices or β-sheets, etc.;
- searching for motifs and patterns within a sequence that might provide information on the origins and/or possible function(s) of the protein;
- predicting potential sites of post-translational modification;
- modelling the likely three-dimensional structure of the protein, identifying likely membrane-spanning regions, active sites and other tertiary and quaternary features;
- investigating properties such as hydropathy, solvent accessibility and antigenicity;
- predicting and modelling the effects of changes in primary or secondary structure on protein characteristics and three-dimensional organisation.

Hydropathy *– hydrophobic interactions within a biomolecule; often important in determining folding and three-dimensional structure.*

Protein-specific databases

Several databases have been created to collect together information on a particular class of proteins – many are devoted to particular enzymes or groups of enzymes. An example of a second-level protein database is that devoted to the analysis of enzymes that degrade, modify or create glycosidic bonds, the carbohydrate-active enzymes (CAZy database, Table 11.1). This website brings together information on the various families of enzymes that interact with carbohydrates, enabling researchers to compare sequences against known catalytic and carbohydrate-binding modules, arranged by enzyme class and by organism. Links to related websites are also included. Another example is the MEROPS protease database, which provides a structure-based catalogue of proteases and peptidases, giving amino acid and nucleotide sequences, classification and nomenclature, plus links to relevant literature through the NCBI Entrez/PubMed search engine. If you are looking for information on a particular enzyme or other protein, it may be worth using a web search engine to see whether a specific database has been created (additional examples are given in Table 11.1).

Bioinformatics in drug design *– protein databases provide chemoinformatic information that can be used to locate new therapeutic targets, e.g. when designing novel antimicrobial agents.*

Example PyMOL is a molecular graphics programme that enables the user to visualise the structure of biomolecules – it can be obtained from the following website: http://pymol.sourceforge.net/

Protein structure

The Protein Data Bank (PDB) is the worldwide repository for three-dimensional structures of proteins (http://www.pdb.org/), also serving as a general portal for structural information on biological macromolecules. The database can be searched by identity number (e.g. 1ALB is the code for an adipocyte lipid-binding protein) or by name (note that this is a US resource, so using the US spelling 'hemoglobin' locates more structures than using the UK spelling). Structures can be viewed in a variety of formats, including those compatible with molecular graphics software such as PyMOL, RasMol and Chime. Additional information includes secondary structure (helices, sheets, hairpins, etc.), crystallographic data, geometry, etc., plus links to related sites. Figure 11.3 shows an example of a typical output – in this instance, for the 3D structure of the enzyme pectate lyase, a virulence factor in some plant pathogenic bacteria.

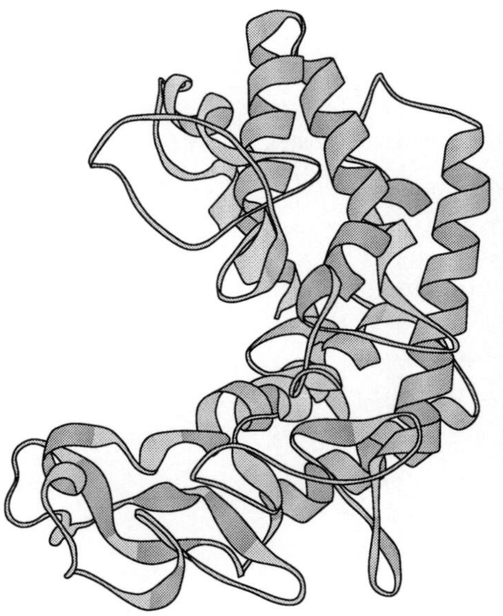

Fig. 11.3 Representation of the three-dimensional structure of a pectate lyase from *Cellvibrio japonicus*. This is a solid ribbon display, with helices and sheets shown in blue. The cleft, in which the plant polysaccharide is bound, is clearly visible on the left of the structure. Image courtesy of Prof. G. Black.

Other Internet resources

Several websites have been constructed to bring together bioinformatic data and related information for a single organism – many of these sites have been produced as a result of a genome-sequencing project for a particular organism, e.g. the human genome mapping project. In

Table 11.2 Other Internet locations relevant to bioinformatics

Databases and resources for particular organisms	
Arabidopsis information resource	http://www.arabidopsis.org
Drosophila database (flybase)	http://flybase.bio.indiana.edu
Caenorhabditis genetics centre	http://www.wormbase.org/
E. coli genome project	http://www.genome.wisc.edu
Human genome project (US site)	http://www.ornl.gov/sci/techresources/Human_Genome/home.shtml
HIV sequence database	http://www.hiv.lanl.gov/content/hiv-db/mainpage.html
Plant genome data and information centre	http://www.plantgdb.org/
Saccharomyces genome database	http://www.yeastgenome.org/
Mosquito genomics server	mosquito.colostate.edu/
Other resources	
Bioinformatics courses/lectures	http://www.bioinformaticscourses.com/
Directory of bioinformatics software and resources	http://www.bioinformatics.vg/
KEGG metabolic pathway database	http://www.genome.jp/kegg/pathway.html
MUSC DNA microarray database	http://proteogenomics.musc.edu/ma/

addition to the nucleic acid and/or protein sequence data available in the primary databases, the organism-specific websites provide a broader range of information and resources, often including links to other web locations. As such, they can be a useful means of accessing relevant internet sites: Table 11.2 gives examples, along with the locations of several other relevant websites that may be useful when researching a topic, e.g. as part of an assignment, or when planning a bioinformatics project.

Sources for further study

(Note also the websites given in Tables 11.1 and 11.2.)

Anon. *PubMed Homepage*. Available: http://www.ncbi.nlm.nih.gov/entrez/query.fcgi?db=PubMed
Last accessed: 01/04/07.
[Online access to details of bioinformatics papers and other publications]

Baxevanis, A.D. and Ouellette, B.F.F. (2005) *Bioinformatics: a Practical Guide to the Analysis of Genes and Proteins*, 3rd edn. Wiley, New York.

Brown, T.A. (2006) *Genomes 3*. Garland Science, New York.
[This book focuses on genomics, in contrast to more traditional molecular biology textbooks, which emphasise individual genes]

Campbell, A. and Heyer, L. (2006) *Discovering Genomics, Proteomics and Bioinformatics*. Benjamin Cummings, New York.
[An introductory text that provides hands-on exercises using web-based software]

Dardel, F., Kepes, F. and Hardy, N. (2006) *Bioinformatics: Genomics and Post-genomics*. Wiley, Chichester.

Dennis, C. and Gallagher, R. (2002) *The Human Genome*. Palgrave Macmillan, New York.

Lesk, A.M. (2005) *An Introduction to Bioinformatics*, 2nd edn. Oxford University Press, Oxford.

Mount, D.W. (2004) *Bioinformatics: Sequence and Genome Analysis*, 2nd edn. Cold Spring Harbor Laboratory Press, New York.
[Provides a good introduction and overview of bioinformatics]

Rashidi, H. and Buehler, L.K. (2000) *Bioinformatics Basics: Applications in Biological Science and Medicine*. CRC Press, Boca Raton.

Stekel, D. (2004) *Microarray Bioinformatics*. Cambridge University Press, Cambridge.
[Covers practical aspects of the use of DNA microarrays in bioinformatics]

Xoing, J. (2006) *Essential Bioinformatics*. Cambridge University Press, Cambridge.

Study exercises

11.1 Investigate the potential of the Internet for locating information relevant to bioinformatics. Using the Web, answer the following questions:
(a) How many nucleotide sequences are currently held by GenBank?
(b) What and where is Hinxton Hall?
(c) Where is the European Molecular Biology Laboratory (EMBL) Mouse Biology Program located?
(d) What is the total number of protein structures currently held by the Protein Data Bank?
(e) What is the full postal address of the US National Center for Biotechnology Information?

11.2 Search GenBank nucleotide database using a DNA sequence. Carry out a BLASTN search of GenBank (at: http://www.ncbi.nlm.nih.gov/BLAST) for the following nucleotide sequence: actgactcatagagtacccaccgtggtgct. How many 'hits' do you find? What is the length of the sequence showing complete identity?

11.3 Search the NCBI database using a protein sequence. Carry out a BLASTP search (at: http://www.ncbi.nlm.nih.gov/BLAST) for the following amino acid sequence: NVKAAWGKVGAHAGEYGAE
(a) What protein is this sequence most likely to correspond to? (b) Which sequence gives the highest alignment score? (c) What is the highest score for a sequence from *Homo sapiens*?

11.4 Convert a DNA base sequence to an amino acid sequence. Using the 'DNA→protein translate tool' of the ExPaSy server (at: http://ca.expasy.org/tools/dna.html), carry out a six-frame translation of the following nucleotide sequence: tcagagagaacccaccat. Which of the six frames contains both a start and stop codon and what is the predicted amino acid sequence for this frame?

11.5 Use the Protein Data Bank to investigate the structure of an enzyme. Search the PDB archive (at: http://www.rcsb.org/pdb/) using 'beta-galactosidase' as the keyword in the SearchLite search tool, and locate the entry for the β-galactosidase ($=\beta$-glycosidase) enzyme of *Sulfolobus solfataricus*, deposited on 19 September 1996. Use the 'explore' command to find out: (a) the number of sub-units in the holoenzyme; (b) the number of amino acids residues in each chain; (c) the number of helices in each chain. Next, use the 'view structure' option to look at the 3D configuration (either in quick PDB format, or using Rasmol, if this program is available on your PC, or network): (d) save an image as a file and then incorporate it into a Word document, as you might in an assignment.

12 Using spreadsheets

Definitions

Spreadsheet – a display of a grid of cells into which numbers, text or formulae can be typed to form a worksheet. Each cell is uniquely identifiable by its column and row number combination (i.e. its 2-D coordinates) and can contain a formula which makes it possible for an entry to one cell to alter the contents of one or more other cells.

Template – a pre-designed spreadsheet without data but including all formulae necessary for (repeated) data analysis.

Macro – a sequence of user-defined instructions carried out by the spreadsheet, allowing complex repeated tasks to be 'automated'.

Office suite – a package of complementary and integrated programs such as a word processor, a spreadsheet and a database.

The spreadsheet is one of the most powerful and flexible computer applications. It can be described as the electronic equivalent of a paper-based longhand calculation, where the sums are carried out automatically. Spreadsheets provide a dynamic method of storing, manipulating and analysing data sets. Advantages of spreadsheets include:

- Ease and convenience – especially when complex calculations are repeated on different sets of data.
- Accuracy – providing the entry data and cell formulae are correct, the result will be free of calculation errors.
- Improved presentation – data can be produced in graphical or tabular form to a very high quality.
- Integration with other programs – graphs and tables can be exported to other compatible programs, such as a word processor in the same office suite.
- Useful tools – advanced features include hypothesis-testing statistics, database features and macros.

Spreadsheets can be used to:

- store and manipulate raw data by removing the drudgery of repeated calculations, allowing easy transformation of data and calculation of statistics;
- graph out your data rapidly to get an instant evaluation of results. Printouts can be used in practical and project reports;
- carry out statistical analysis by built-in procedures or by allowing construction of formulae for specific tasks;
- model 'what if' situations where the consequences of changes in data can be seen and evaluated.

The spreadsheet (Fig. 12.1) is divided into rows (identified by numbers) and columns (identified by alphabetic characters). Each individual

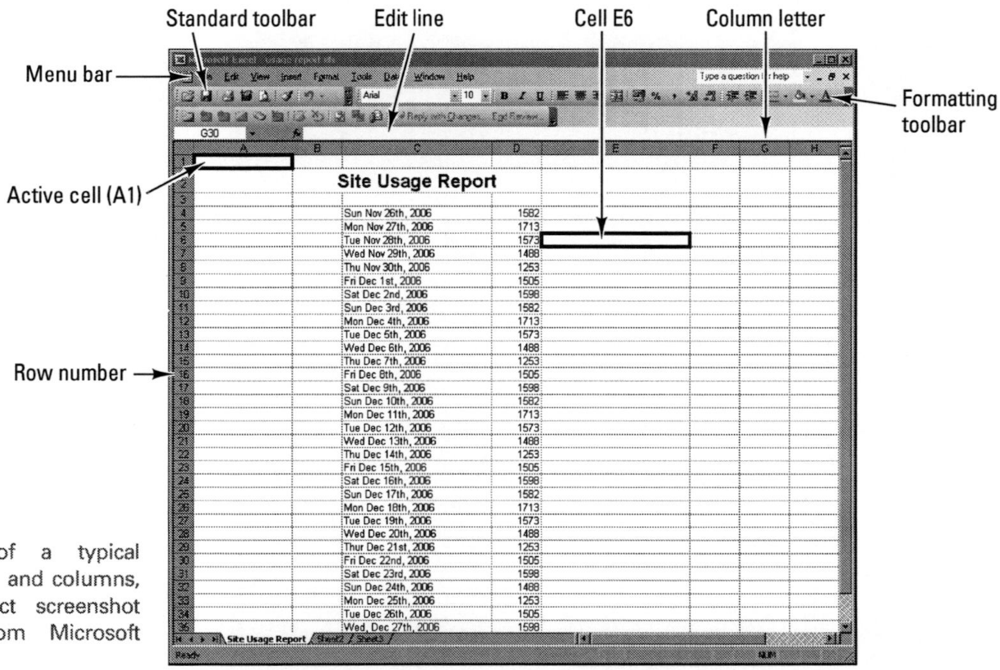

Fig. 12.1 The appearance of a typical spreadsheet, showing cells, rows and columns, toolbars, etc. Microsoft product screenshot reprinted with permission from Microsoft Corporation.

combination of column and row forms a cell which can contain either a data item, a formula, or a piece of text. Formulae can include scientific and/or statistical functions and/or a reference to other cells or groups of cells (often called a range). Complex systems of data input and analysis can be constructed. The analysis, in part or complete, can be printed out. New data can be added at any time and the sheet will recalculate automatically. The power a spreadsheet offers is directly related to your ability to create arrays of formulae (models) that are accurate and templates that are easy to use.

Data entry

***Data output from analytical instruments** – many devices provide output in spreadsheet-compatible form (e.g. a 'comma delimited' file). Once you have uploaded the information into a spreadsheet, you can manipulate, analyse and present it according to your needs. Consult instrument manuals and the spreadsheet help function for details.*

Spreadsheets have built-in commands which allow you to control the layout of data in the cells (see Fig. 12.2). These include number format, the number of decimal places to be shown (the spreadsheet always calculates using eight or more places), the cell width and the location of the entry within the cell (left, right or centre). An auto-entry facility assists greatly in entering large amounts of data by moving the entry cursor either vertically or horizontally as data are entered. Recalculation default is usually automatic so that when a new data value is entered the entire sheet is recalculated immediately.

The parts of a spreadsheet

***Using hidden (or zero-width) columns** – these are useful for storing intermediate calculations which you do not wish to be displayed on the screen or printout.*

Labels

These should be used to identify parts of the spreadsheet, for example, stating what data are contained in a particular column or indicating that a cell's contents represent the end point of a calculation. It may be useful to use the *Format* > *Cells* > *Border* and *Format* > *Cells* > *Patterns* functions to delimit numerical sections of your spreadsheet. Note that spreadsheet programs have been designed to make assumptions about the nature of data entry being made. If the first character is a number, then the

(a) (b)

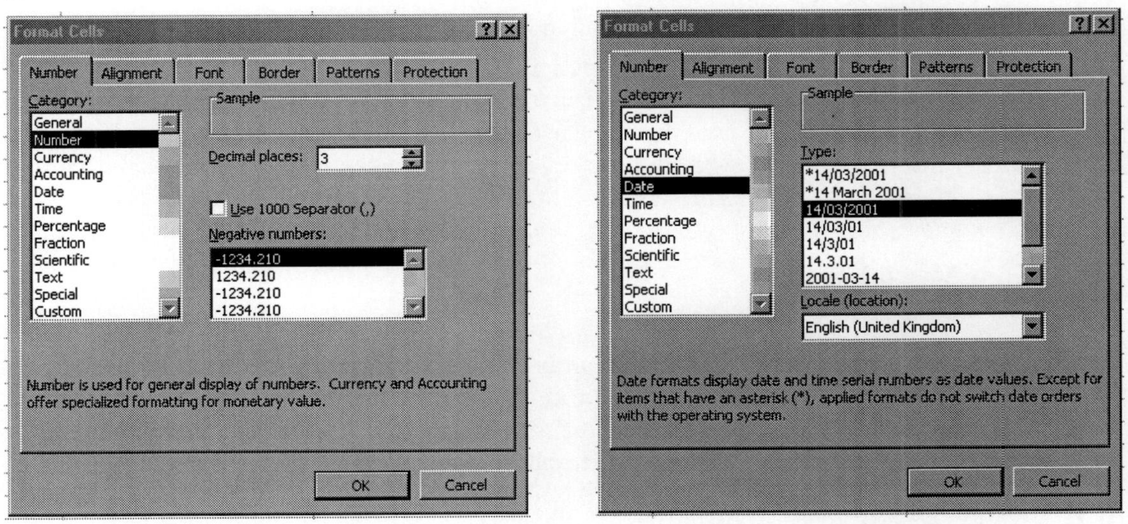

Fig. 12.2 Example of cell formatting options within a Microsoft Excel spreadsheet. These menus are accessed via the *Format* > *Cell* option and would apply to all of a range of selected cells. (a) Use of the number formatting option to specify that data will be presented to three decimal places (the underlying data will be held to greater accuracy). (b) Use of the date formatting option to specify that dates will be presented in day/month/year format (spreadsheet dates are stored numerically and converted to appropriate formats. This allows a period between two dates to be calculated more easily). Microsoft Product screenshots reprinted with permission from Microsoft Corporation.

entry is treated as numerical data; if it is a letter, then it is treated as a text entry; and if it is a specific symbol ('=' in Microsoft Excel), then what follows is a formula. If you wish to enter text that starts with a number, then you must type a designated character to show this (a single quote mark in Microsoft Excel).

Numbers
You can also enter numbers (values) in cells for use in calculations. Many programs let you enter numbers in more than one way and you must decide which method you prefer. The way you enter the number does not affect the way it is displayed on the screen as this is controlled by the cell format at the point of entry. There are usually special ways to enter data for percentages, currency and scientific notation for very large and small numbers.

Operators and brackets in spreadsheets *– the standard mathematical operators ÷ and × are usually replaced by / and * respectively, while ^ signifies 'to the power' (e.g. 10 ^ 4 = 10^4). In complex formulae, brackets should be used to separate the elements, otherwise the results may not be what you expect. For example, Excel will calculate =A1*B1/C1–D1 differently from (A1*B1)/(C1–D1).*

Formulae
These are the 'power tools' of the spreadsheet because they do the calculations. A cell can be referred to by its alphanumeric code, e.g. A5 (column A, row 5) and the value contained in that cell manipulated within a formula, e.g. =(A5 + 10) or =(A5 + B22) in another cell. Formulae can include various pre-programmed functions which can refer to a cell, so that if the value of that cell is changed, so is the result of the formula calculation. They may also include branching options through the use of logical operators (IF, TRUE, FALSE, OR, etc.).

Definition

Function – a pre-programmed code for the transformation of values (mathematical or statistical functions) or selection of text characters (string functions).

Functions

A variety of functions is usually offered, but only mathematical and statistical functions will be considered here.

Example =sin(A5) is an example of a function in Excel. If you write this in a cell, the spreadsheet will calculate the sine of the number in cell A5 (assuming it to be an angle in radians) and write it in the cell. Different programs may use a slightly different syntax.

Mathematical functions
Spreadsheets offer a range of functions, including trigonometrical functions, angle functions, logarithms and random number functions. Functions are invaluable for transforming sets of data rapidly and can be used in formulae required for more complex analyses. Spreadsheets work with an order of preference of the mathematical operators in much the same way as a standard calculator and this must always be taken into account when operators are used in formulae. They also require a very precise syntax – the program should warn you if you break this.

Working with empty cells *– note that these may be given the value 0 by the spreadsheet for certain functions. This may cause errors, e.g. by rendering a minimum value inappropriate. Also, an 'error return' may result for certain functions if the cell content is zero.*

Statistical functions
Modern spreadsheets incorporate many sophisticated statistical functions, and if these are not appropriate the spreadsheet can be used to carry out the calculations required for most of the statistical tests found in textbooks. The descriptive statistics normally available include:

- the sum of all data present in a column, row or block;
- the minimum and maximum of a defined range of cells;
- counts of cells – a useful operation if you have an unknown or variable number of data values;
- averages and other statistics describing location;
- standard deviations and other statistics describing dispersion.

A useful function where you have large numbers of data allows you to create frequency distributions using pre-defined class intervals.

The hypothesis-testing statistical functions are usually reasonably powerful (e.g. *t*-test, ANOVA, regression) and they often return the *probability* (*P*) of obtaining the test statistic when the null hypothesis (p. 492) is true (where $0 < P < 1$), so there may be no need to refer to statistical tables. Again, check on the effects of including empty cells within the statistical calculations.

Statistical calculations *– make sure you understand whether the functions you employ apply to populations or samples.*

Database functions

Many spreadsheets can be used as simple databases and offer a range of functions to support this, including filtering and sorting options. The rows and columns of the spreadsheet are used as the fields and records of the database (see Chapter 12). For many biological purposes, this form of database is perfectly adequate and should be seriously considered before using a full-feature database product.

Using text functions *– these allow you to manipulate text within your spreadsheet and include functions such as 'find and replace' and alphabetical or numerical 'sort'.*

Copying

All programs provide a means of copying (replicating) formulae or cell contents when required and this is a very useful feature. This is usually accomplished by 'dragging' a cell's contents to a new range using the mouse. When copying, references to cells may be either relative, changing with the row/column as they are copied, or absolute, remaining a fixed cell reference and not changing as the formulae are copied (Fig. 12.3).

KEY POINT The distinction between relative and absolute cell references is very important and must be understood; it provides one of the most common forms of error when copying formulae.

In Excel, copying is normally *relative* and if you wish a cell reference to be *absolute* when copied, this is done by putting a dollar ($) sign before and after the column reference letter e.g. $C$56.

(a)

	Cell	Formula	
Original → cell	A1	=B1+C1	← Original formula
Copied cells ↓	A2	=B2+C2	Copied formulae (relative)
	A3	=B3+C3	
	A4	=B4+C4	

(b)

	Cell	Formula	
Original → cell	A1	=B1/$C$1	← Original formula
Copied cells ↓	A2	=B2/$C$1	Copied formulae (mixed relative and absolute)
	A3	=B3/$C$1	
	A4	=B4/$C$1	

Fig. 12.3 Illustration of relative (a) and absolute (b) copying. In Excel, the $ sign before and after the column letter makes the cell reference absolute, as shown in (b).

Naming blocks

When a group of cells (a block) is carrying out a particular function, it is often easier to give the block a name which can then be used in all formulae referring to that block. This powerful feature also allows the spreadsheet to be more readable.

Spreadsheet templates

A template is a pre-constructed spreadsheet containing the formulae required for repeated data analysis. Data are added when they become available, and results are available as soon as the last item is entered. To create a template, the sequence of operations is:

1. Determine what information/statistics you want to produce.
2. Identify the variables you will need to use, both for original data that will be entered and for any intermediate calculations that might be required.

***Using templates** – these should contain:*

- *a data input section;*
- *data transformation and/or calculation sections;*
- *a results section, which can include graphics;*
- *text in the form of headings and annotations;*
- *a summary section.*

***Constructing a spreadsheet** – start with a simple model and extend it gradually, checking for correct operation as you go.*

3. Set up areas of the spreadsheet for data entry, calculation of intermediate values (statistical values such as sums of squares, etc.), calculation of final statistics and, if necessary, a summary area.
4. Establish the format of the numeric data if this is different from the default values. This can be done globally (affecting the entire spreadsheet) or locally (affecting only a specified part of the spreadsheet).
5. Establish the column widths required for the various activities.
6. Add text (labels) to identify input, intermediate formulae and output cells. This is valuable in error-tracking and when carrying out further development work. Text can be entered in designated cells, or cells can be annotated using the 'comments' feature (*Insert* > *Comment*).
7. Enter a test set of values to use during formula entry: use a fully worked example to check that formulae are working correctly.
8. Enter the formulae required to make all the calculations, both intermediate and final. Check that results are correct using the test data.

The spreadsheet is then ready for use. Delete all of the test data values and you have created your template. Save the template to a disk and it is then available for repeated operations.

Graphics display

Most spreadsheets now offer a wide range of graphics facilities which are easy to use, and this represents an ideal way to examine your data sets rapidly and comprehensively. The quality of the final graphics output (to a printer) is variable but is usually perfectly sufficient for data exploration and analysis. Many of the options are business graphics styles but there are usually histogram, bar chart, X-Y plotting, line and area graphics options available. Note that some spreadsheet graphics may not come up to the standards expected for the formal presentation of scientific data, unless you manipulate the initial output appropriately (see Box 63.2).

Printing spreadsheets

This is usually a straightforward, menu-controlled procedure, made difficult only by the fact that your spreadsheet may be too big to fit on one piece of paper. Try to develop an area of the sheet which contains only the data that you will be printing, i.e. perhaps a summary area. Remember that columns can usually be hidden for printing purposes and you can control whether the printout is in portrait or landscape mode, and for continuous paper or single sheets (depending on printer capabilities). Use a screen preview option, if available, to check your layout before printing. A 'print to fit' option is also available in some programs, making the output fit the page dimensions.

Sources for further study

Hart-Davies, G. (2003) *How to do Everything with Microsoft Office Excel 2003*. McGraw-Hill, Berkeley.

Harvey, G. (2003) *Excel 2003 for Dummies*. Wiley, New York. (And similar texts for other release versions.)

Study exercises

The instructions and tips for these problems assume that you have Excel (2003 or later) available. If not, they should be readily modified for most advanced spreadsheet programs. If you have problems with any of the tasks, consult Box 63.2 or try using the program's *Help* facility.

12.1 Create a spreadsheet and graph (introductory).

(i) Copy the information in the table below into a spreadsheet. Name and save the spreadsheet file appropriately.

(ii) From the copied information, create a pie chart using the *Chart Wizard function.*

(iii) Adjust the colours selected so the chart will print out in black and white. Save the final version of your spreadsheet. Print the chart out directly from Excel.

Relative percentages of the major types of leukocytes in a blood sample

Cell type	Percentage
Neutrophils	60
Eosinophils	4
Basophils	1
Monocytes	7
Lymphocytes	28
Total	100

12.2 Create a spreadsheet and graph (advanced).

(i) Copy the data in the table at the top of the next column into a spreadsheet. Name and save the file appropriately.

Time course of cell population growth

Time (hours)	Cell count (per unit volume)
1	2.00
2	6.00
6	50.0
12	6.00×10^3
18	2.20×10^5
24	1.65×10^7

(ii) Use the spreadsheet and chart-making facilities to explore by eye which of the following transformations would result in the best linear fit for these data: reciprocal, square root, cube root or log.

(iii) Add a linear trend-line to the chart for the most appropriate transformation.

(iv) Copy the graph to a file in Word and print out.

12.3 Use a spreadsheet as a simple database. Copy the data in the table below into cells within a spreadsheet. Modify the column widths so you can see all of the text on a single screen. Now sort the data in the following ways:

(a) by subject, in alphabetical order;

(b) by date and then by time of day;

(c) by topic, in reverse alphabetical order.

Use the *Column Hide* function (*Format* menu in Microsoft Excel) so that the information in columns 4 and 6 is not displayed. Find out how to undo this operation.

My exam timetable

Subject	Date	Time	Paper	Location	Question style
Biochemistry	3 Jun	Morning	1	Great Hall	Multiple choice
Biochemistry	17 Jun	Morning	2	Exam Hall 5	Essay paper
Biochemistry	2 Jun	Afternoon	3	Main Laboratory	Information processing
Genetics	3 Jun	Afternoon	A	Small Hall	Short answer questions
Genetics	14 Jun	Afternoon	B	Exam Hall 5	Essay paper
Cell Biology A	4 Jun	Morning	1	Small Hall	Short answer questions
Cell Biology B	1 Jul	Afternoon	1	Exam Hall 3	Short answer questions
Cell Biology A	13 Jun	Afternoon	2	Exam Hall 5	Essay paper
Cell Biology B	2 Jun	Morning	2	Main Laboratory	Practical exam

13 Word processors, databases and other packages

Word processors

Word processing is a transferable skill valuable beyond the immediate requirements of your course. The word processor has improved the process of writing because of the ease of revising text. Using a word processor should enhance your writing skills and speed because you can create, check and change your text on the screen before printing it as 'hard copy' on paper. Once entered and saved, multiple uses can be made of a piece of text with little effort.

When using a word processor you can:

- refine material many times before submission;
- insert material easily, allowing writing to take place in any sequence;
- use a spellchecker to check your text;
- use a thesaurus when composing your text;
- carry out ongoing checks of the word count;
- produce high quality final copies;
- reuse part or all of the text in other documents.

The potential disadvantages of using a word processor include:

- lack of ready access to a computer, software and/or a printer;
- time taken to learn the operational details of the program;
- the temptation to make 'trivial' revisions;
- loss of files due to computer breakdown, or disk loss, or failure.

> ***The computerised office*** *– many word processors are sold as part of an integrated office suite, e.g. Corel WordPerfect Office and Microsoft Office, with the advantage that they share a common interface in the different components (word processor, spreadsheet, database, etc.) and allow ready exchange of information (e.g. text, graphics) between component programs.*

> ***Using textbooks, manuals and tutorials*** *– most programs no longer come with paper-based manuals and support information is usually provided in one or more of the following ways: as a help facility within the program; as a help file on the program CD; or as an online help support site. It is still often worthwhile investing in one of the commercial textbooks that support specific programs.*

Word processors come as 'packages' comprising the program and a manual, often with a tutorial program. Examples are Microsoft Word and WordPerfect. Most word processors have similar general features but differ in operational detail; it is best to pick one and stick to it as far as possible so that you become familiar with it. Learning to use the package is like learning to drive a car – you need only to know how to drive the computer and its program, not to understand how the engine (program) and transmission (data transfer) work, although a little background knowledge is often helpful and will allow you to get the most from the program.

In most word processors, the appearance of the screen realistically represents what the printout on paper will look like. Because of variation in operational details, only general and strategic information is provided in this chapter: you must learn the details of your word processor through use of the appropriate manual and 'help' facilities.

Before starting you will need:

- the program (usually installed on a hard disk or available *via* a network);
- a medium for storage, retrieval and back-up of your own files when created;
- a draft page layout design: in particular you should have decided on page size, page margins, typeface (font) and size, type of text justification, and format of page numbering;
- an outline of the text content;
- access to a suitable printer: this need not be attached to the computer you are using since your file can be taken or sent to an office where a printer is available, providing it has the same word processing program.

> ***Using a word processor*** *– take full advantage of the differences between word processing and 'normal' writing (which necessarily follows a linear sequence and requires more planning):*
>
> - *Simply jot down your initial ideas for a plan, preferably at paragraph topic level. The order can be altered easily and if a paragraph grows too much it can easily be split.*
> - *Start writing wherever you wish and fill in the rest later.*
> - *Just put down your ideas as you think, confident in the knowledge that it is the concepts that are important to note; their order and the way you express them can be adjusted later.*
> - *Don't worry about spelling and use of synonyms – these can (and should) be checked during a separate revision run through your text, using the spell checker first to correct obvious mistakes, then the thesaurus to change words for style or to find the* mot juste.
> - *Don't forget that a draft printout may be required to check (a) for pace and spacing – difficult to correct for on-screen; and (b) to ensure that words checked for spelling fit the required sense.*

Laying out (formatting) your document

Although you can format your text at any time, it is good practice to enter the basic commands at the start of your document: entering them later can lead to considerable problems due to reorganisation of the text layout. If you use a particular set of layout criteria regularly, e.g. an A4 page with space for a letterhead, then make a template containing the appropriate codes that can be called up whenever you start a new document. Note that various printers may respond differently to particular codes, resulting in a different spacing and layout.

Typing the text

If new to word-processing, think of the screen as a piece of typing paper. The cursor marks the position where your text/data will be entered and can be moved around the screen by use of the cursor-control keys. When you type, don't worry about running out of space on the line because the text will wrap around to the next line automatically. Do not use a carriage return (usually the [ENTER] or [↵] key) unless you wish to force a new line, e.g. when a new paragraph is wanted. If you make a mistake when typing, correction is easy. You can usually delete characters or words or lines and the space is closed automatically. You can also insert new text in the middle of a line or word. You can insert special codes to carry out a variety of tasks, including changing text appearance such as <u>underlining</u>, **emboldening** and *italics*. Paragraph indentations can be automated using [TAB] or [⇆] as on a typewriter but you can also indent or bullet whole blocks of text using special menu options. The function keys are usually pre-programmed to assist in many of these operations.

> ***Deleting and restoring text*** *– because deletion can sometimes be made in error, there is usually an 'undelete' or 'restore' feature which allows the last deletion to be recovered, usually available through the program's* 'Edit' *menu or as an icon in the toolbar.*

Editing features

Word processors usually have an array of features designed to make editing documents easy. In addition to the simple editing procedures described above, the program usually offers facilities to allow blocks of text to be moved ('cut and paste'), copied or deleted.

An extremely valuable editing facility is the find or search procedure: this can rapidly scan through a document looking for a specified word, phrase or punctuation. This is particularly valuable when combined with a replace facility so that, for example, you could find the word 'test' and replace it with 'trial' throughout your document simply and rapidly.

Most word processors have a command which reveals the normally hidden codes controlling the layout and appearance of the printed text. When editing, this can be a very useful feature, since some changes to your text will cause difficulties if these hidden codes are not taken into account; in particular, make sure that the cursor is at the correct point before making changes to text containing hidden code, otherwise your text will sometimes change in apparently mystifying ways.

Fonts and line spacing

Most word processors offer a variety of fonts depending upon the printer being used. Fonts come in a wide variety of types and sizes, but they are defined in particular ways as follows:

- Typeface: the term for a family of characters of a particular design, each of which is given a particular name. The most commonly used for normal text is Times Roman (as used here for the main text) but

> ***Presenting your documents** – it is good practice not to mix typefaces too much in a formal document; also the font size should not differ greatly for different headings, subheadings and the text.*

> ***Preparing draft documents** – use double line spacing to allow room for your editing comments on the printed page.*

> ***Using fixed space fonts** – proportional fonts can create problems of base letter misalignment when representing DNA sequences (p. 69), so you should use a fixed space font, e.g. Courier.*

> ***Preparing final documents** – for most work, use a 12 point proportional serif typeface with spacing dependent upon the specifications for the work.*

many others are widely available, particularly for the better quality printers. They fall into three broad groups: serif fonts with curves and flourishes at the ends of the characters (e.g. Times Roman); sans serif fonts without such flourishes, providing a clean, modern appearance (e.g. Helvetica or Arial); and decorative fonts used for special purposes only, such as the production of newsletters and notices.

- Size: measured in points. A point is the smallest typographical unit of measurement, there being 72 points to the inch (about 28 points per cm). The standard sizes for text are 10, 11 and 12 point, but typefaces are often available up to 72 point or more.
- Appearance: many typefaces are available in a variety of styles and weights. Many of these are not designed for use in scientific literature but for desktop publishing.
- Spacing: can be either fixed, where every character is the same width, or proportional, where the width of every character, including spaces, is varied. Typewriter fonts such as Elite and Prestige use fixed spacing and are useful for filling in forms or tables, but proportional fonts make the overall appearance of text more pleasing and readable.
- Pitch: specifies the number of characters per horizontal inch of text. Typewriter fonts are usually 10 or 12 pitch, but proportional fonts are never given a pitch value since it is inherently variable.
- Justification: the term describing the way in which text is aligned vertically. Left justification is normal, but for some documents both left and right justification may be used (as here).

You should also consider the vertical spacing of lines in your document. Drafts and manuscripts are frequently double-spaced. If your document has unusual font sizes, this may well affect line spacing, although most word processors will cope with this automatically.

Table construction

Tables can be produced by a variety of methods:

- Using the tab key as on a typewriter: this moves the cursor to predetermined positions on the page, equivalent to the start of each tabular column. You can define the positions of these tabs as required at the start of each table.
- Using special table-constructing procedures (see Box 64.2). Here the table construction is largely done for you and it is much easier than using tabs, providing you enter the correct information when you set up the table.
- Using a spreadsheet to construct the table and then copying it to the word processor (see Box 64.2). This procedure requires considerably more manipulation than using the word processor directly and is best reserved for special circumstances, such as the presentation of a very large or complex table of data, especially if the data are already stored as a spreadsheet.

Graphics and special characters

Many word processors can incorporate graphics from other programs into the text of a document. Files must be compatible (see your manual) but if this is so, it is a relatively straightforward procedure. For professional documents this is a valuable facility, but for most undergraduate work it is probably better to produce and use graphics as a separate operation, e.g. using a spreadsheet (see Box 63.2).

Inserting special characters – *Greek letters and other characters are available using the 'Insert' and 'Symbol' features in Word.*

You can draw lines and other graphical features directly within most word processors and special characters (e.g. Greek characters) may be available, dependent upon your printer's capabilities.

Tools

Many word processors also offer you special tools, the most important of which are:

- Macros: special sets of files you can create when you have a frequently repeated set of keystrokes to make. You can record these keystrokes as a 'macro' so that it can provide a short-cut for repeated operations.
- Thesaurus: used to look up alternative words of similar or opposite meaning while composing text at the keyboard.
- Spellchecker: a very useful facility which will check your spellings against a dictionary provided by the program. This dictionary is often expandable to include specialist words which you use in your work. The danger lies in becoming too dependent upon such facilities as they all have limitations: in particular, they will not pick up incorrect words which happen to be correct in a different context (i.e. 'was' typed as 'saw' or 'see' rather than 'sea'). Be aware of American spellings in programs from the USA, e.g. 'color' instead of 'colour'. The rule, therefore, is to use the spell-check first and then carefully read the text for errors which have slipped through.
- Word count: useful when you are writing to a prescribed limit.

Using a spellcheck facility – *do not rely on this to spot all errors. Remember that spell-check programs do not correct grammatical errors.*

Printing from your program

If more than one printer is attached to your PC or network, you will need to specify which one to use from the word processor's print menu. Most printers offer choices as to text and graphics quality, so choose draft (low) quality for all but your final copy since this will save both time and materials.

Use a print preview option to show the page layout, if this is available. Assuming that you have entered appropriate layout and font commands, printing is a straightforward operation carried out by the word processor at your command. Problems usually arise because of some incompatibility between the criteria you have entered and the printer's own capabilities. Make sure that you know what your printer offers before starting to type: although specified details are modifiable at any time, changing the page size, margin size, font size, etc. all cause your text to be rearranged, and this can be frustrating if you have spent time carefully laying out the pages.

Using the print preview mode – *this can reveal errors of several kinds, e.g. spacing between pages, that can prevent you wasting paper and printer ink unnecessarily.*

***KEY POINT** It is vital to save your work frequently to a memory stick, hard drive or network drive. This should be done every 10 minutes or so. If you do not save regularly, you may lose hours or days of work. You should also save back-up copies on a regular basis. Many programs can be set to 'auto-save' every few minutes.*

Databases

Choosing between a database and a spreadsheet *– use a database only after careful consideration. Can the task be done better within a spreadsheet? A database program can be complex to set up and usually needs to be updated regularly.*

A database is an electronic filing system whose structure is similar to a manual record card collection. Its collection of records is termed a file. The individual items of information on each record are termed fields. Once the database is constructed, search criteria can be used to view files through various filters according to your requirements. The computerised catalogues in your library are just such a system; you enter the filter requirements in the form of author or subject keywords.

You can use a database to catalogue, search, sort and relate collections of information. The benefits of a computerised database over a manual card-file system are:

- The information content is easily amended/updated.
- Printout of relevant items can be obtained.
- It is quick and easy to organise through sorting and searching/selection criteria, to produce subgroups of relevant records.
- Record displays can easily be redesigned, allowing flexible methods of presenting records according to interest.
- Relational databases can be combined, giving the whole system immense flexibility. The older 'flat-file' databases store information in files which can be searched and sorted, but cannot be linked to other databases.

Using databases in bioformatics *– this is covered in Chapter 11.*

Relatively simple database files can be constructed within spreadsheets using the columns and rows as fields and records respectively. These are capable of reasonably advanced sorting and searching operations and are probably sufficient for most types of databases you are likely to require as an undergraduate. You may also make use of a bibliographic database specially constructed for that purpose (p. 41).

Statistical analysis packages

Using spreadsheet statistics functions *– before using a specific statistics package, check whether your spreadsheet is capable of carrying out the form of analysis you require, as this can often be the simpler option.*

Statistical packages vary from small programs designed to carry out very specific statistical tasks to larger packages (SYSTAT SigmaStat, Minitab, etc.) intended to provide statistical assistance, from experimental design to the analysis of results. Consider the following features when selecting a package:

- The data entry and editing section should be user-friendly, with options for transforming data.
- Data exploration options should include descriptive statistics and exploratory data analysis techniques.
- Hypothesis-testing techniques should include ANOVA, regression analysis, multivariate techniques, parametric and non-parametric statistics (see Chapter 67).
- The program may provide assistance with experimental design.
- Output facilities should be suitable for graphical and tabular formats.

Some programs have very complex data entry systems, limiting the ease of using data in different tests. The data entry and storage system should be based on a spreadsheet system, so that subsequent editing and transformation operations are straightforward.

Presentation using computer packages *– while many computer programs enhance presentational aspects of your work, there are occasions when they can make your presentation worse. Take care to avoid the following common pitfalls:*

- *Default or 'chart wizard' settings for graphs may result in output that is non-standard for the sciences (see Box 63.3).*
- *Fonts in labels and legends may not be consistent with other parts of your presentation.*
- *Some programs cannot produce Greek symbols (e.g. μ); do not use 'u' as a substitute. The same applies to scientific notation and superscripts: do not use 14C for ^{14}C, and replace e.g. 1.4E+09 with 1.4×10^9. First try cutting and pasting symbols from Word, or if this fails draw correct symbols by hand.*

KEY POINT Make sure that you understand the statistical basis for your test and the computational techniques involved before using a particular program.

Graphics/presentation packages

Microsoft Office programs can be used to achieve most coursework tasks, e.g. PowerPoint is useful for creating posters (Box 14.1) and for oral presentations (Box 15.1). Should you need more advanced features, additional software may be available on your network, for example:

- SigmaPlot can produce graphs with floating axes;
- Macromedia Freehand is useful for designing complex graphics;
- DreamWeaver enables you to produce high-quality web pages;
- MindGenius can be used to produce mind maps.

Important points regarding the use of such packages are:

- the learning time required for some of the more complex operations can be considerable;
- the quality of your printer will limit the quality of your output;
- not all files will readily import into a word processor such as Microsoft Word – you may need to save your work in a particular format. The different types of file are distinguished by the three-character filename extension, e.g. .jpg and .bmp refer to specific image file formats.

KEY POINT Computer graphics are not always satisfactory for scientific presentation. You should not accept the default versions produced – make appropriate changes to suit scientific standards and style. Box 63.1 gives a checklist for graph drawing and Box 63.2 provides guidelines for adapting Microsoft Excel output.

Image storage and manipulation

With the widespread use of digital images (p. 179), programs that facilitate the storage and manipulation of electronic image files have become increasingly important. These programs create a library of your stored images and provide a variety of methods for organising and selecting images.

The industry-standard program, Adobe Photoshop, is one many programs for image manipulation that vary widely in capability, cost and associated learning time. Many are highly sophisticated programs intended for graphic artists. However, for most scientific purposes, relatively limited functions are required.

Sources for further study

Gookin, D. (2003) *Word 2003 for Dummies*. Wiley, New York.

Kaufield, J. (2003) *Access 2003 for Dummies*. Wiley, New York.

Wang, W. (2003) *Microsoft Office 2003 for Dummies*. Wiley, New York.

(And similar texts for other packages and release versions.)

Study exercises

13.1 Investigate intermediate/advanced Word features. The tasks in the following list are likely to useful in preparing assignments and report writing within the life sciences. Can you carry out all of the tasks? If not, use either a manual or the online *Help* feature to find out how to accomplish them. Tips are given in the answer section.

(a) Sort information in a list into alphabetical order.

(b) Replace a text string word or phrase with a new text string throughout your document.

(c) Replace a text string in normal font with the same text string in italics throughout your document.

(d) Add a 'header' and 'footer' to your document, the former showing the document's title and the latter containing page numbers in the bottom centre of the page.

(e) Adjust the margins of the page to give a 5 cm margin on the left and a 2 cm margin on the right.

(f) Change the type of bullets used in a list from standard (• or ■) to a different form (e.g. –, ◆ or ☑).

(g) Use the 'thesaurus' option to find a different or more suitable word to express your meaning. Try, for example, to find alternatives to the word 'alternative'.

(h) Carry out a spell-check on your document.

(i) Carry out a word count on your document and on a selected part of it.

(j) Open two documents and switch between them.

13.2 Make precise copies of tables. Copy the following tables using a word processor such as Word.

	Results of analysis (units)		
Test organism	August	September	October
X			
Y			
Z			

	Results of analysis (units)		
Test organism	August	September	October
X			
Y			
Z			

13.3 Investigate what programs and packages are available to you as a student. Test each program with appropriate data, images, etc.

Communicating information

14 Organising a poster display

A scientific poster is a visual display of the results of an investigation, usually mounted on a rectangular board. Posters are used in undergraduate courses to display project results or assignment work, and at scientific meetings to communicate research findings.

__Learning from others__ – look at the various types of posters around your university and elsewhere; the best examples will be visual, not textual, with a clear structure that helps get the key messages across.

In a written report you can include a reasonable amount of specific detail and the reader can go back and reread difficult passages. However, if a poster is long-winded or contains too much detail, your reader is likely to lose interest.

***KEY POINT** A poster session is like a competition – you are competing for the attention of people in a room. Because you need to attract and hold the attention of your audience, make your poster as interesting as possible. Think of it as an advertisement for your work and you will not go far wrong.*

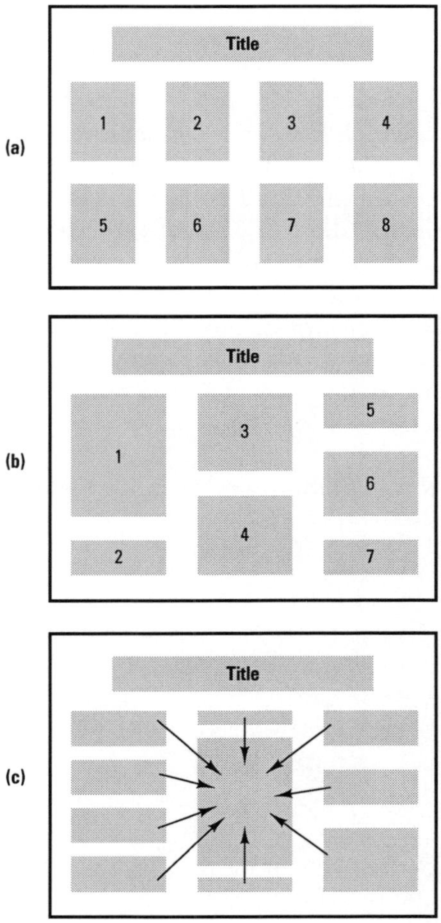

Fig. 14.1 Poster design. (a) An uninspiring design: sub-units of equal area, reading left to right, are not recommended. (b) This design is more interesting and the text will be easier to read (column format). (c) An alternative approach, with a central focus and arrows/tapes to guide the reader.

Preliminaries

Before considering the content of your poster, you should find out:

- the linear dimensions of your poster area, typically up to 1.5 m wide by 1.0 m high;
- the composition of the poster board and the method of attachment, whether drawing pins, Velcro tape, or some other form of adhesive; and whether these will be provided – in any case, it is safer to bring your own;
- the time(s) when the poster should be set up and when you should attend;
- the room where the poster session will be held.

Design

Plan your poster with your audience in mind, as this will dictate the appropriate level for your presentation. Aim to make your poster accessible to a broad audience. Since a poster is a *visual* display, you must pay particular attention to the presentation of information: work that may have taken hours to prepare can be ruined in a few minutes by the ill-considered arrangement of items (Fig. 14.1). Begin by making a draft sketch of the major elements of your poster. It is worth discussing your intended design with someone else, as constructive advice at the draft stage will save a lot of time and effort when you prepare the final version (or consult Simmonds and Reynolds, 1994).

Layout

One approach is to divide the poster into several smaller areas, perhaps six or eight in all, and prepare each as a separate item on a piece of card. Alternatively, you can produce a single large poster on one sheet of paper or card and store it inside a protective cardboard tube. However, a single large poster may bend and crease, making it difficult to flatten out. In addition, photographs and text attached to the backing sheet may work loose; a large printed poster with embedded images is an alternative approach.

Subdividing your poster means that each smaller area can be prepared on a separate piece of paper or card, of A4 size or slightly larger, making transport and storage easier. It also breaks the reading matter up into smaller pieces, looking less formidable to a potential reader. By using pieces of card of different colours you can provide emphasis for key aspects, or link text with figures or photographs.

You will need to guide your reader through the poster and headings/ sub-headings will help with this aspect. It may be appropriate to use either a numbering system, with large, clear numbers at the top of each piece of card, or a system of arrows (or thin tapes) to link sections within the poster (see Fig. 14.1). Make sure that the relationship is clear and that the arrows or tapes do not cross.

Presenting a poster at a formal conference *– it can be useful to include your photograph for identification purposes, e.g. in the top right-hand corner of the poster.*

Title

Your chosen title should be concise (no more than eight words), specific and interesting, to encourage people to read the poster. Make the title large and bold – it should run across the top of your poster, in letters at least 4 cm high, so that it can be read from the other side of the room. Coloured spirit-based marker and block capitals drawn with a ruler work well, as long as your writing is readable and neat (the colour can be used to add emphasis). Alternatively, you can print out each word in large font, using a word processor. Details of authors, together with their addresses (if appropriate), should be given, usually across the top of the poster in somewhat smaller lettering than the title.

Making up your poster *– text and graphics printed on good-quality paper can be glued directly onto a contrasting mounting card: use photographic spray mountant or Pritt rather than liquid glue. Trim carefully using a guillotine to give equal margins, parallel with the paper. Photographs should be scanned into an electronic format, or placed in a window mount to avoid the tendency for their corners to curl. Another approach is to trim pages or photographs to their correct size, then encapsulate in plastic film: this gives a highly professional finish and is easy to transport.*

Text

Write in short sentences and avoid verbosity. Keep your poster as visual as possible and make effective use of the spaces between the blocks of text. Your final text should be double-spaced and should have a minimum capital letter height of 8 mm (minimum font size 36 point), preferably greater, so that the poster can be read at a distance of 1 m. One method of obtaining text of the required size is to photo-enlarge standard typescript (using a good-quality photocopier), or use a high-quality (laser) printer. It is best to avoid continuous use of text in capitals, since it slows reading and makes the text less interesting to the reader. Also avoid italic, 'balloon' or decorative styles of lettering.

KEY POINT Keep text to a minimum – aim to have a maximum of 500 words in your poster.

Subtitles and headings

These should have a capital letter height of 15–20 mm, and should be restricted to two or three words. They can be produced by word processor, photo-enlargement, by stencilling or by hand, using pencilled guidelines (but make sure that no pencil marks are visible on your finished poster).

Producing composite material for posters *– PowerPoint is generally more useful than Word when you wish to include text, graphics and/or images on the same page. It is possible to use PowerPoint to produce a complete poster (Box 14.1), although it can be expensive to have this printed out commercially to A1 or A0 size.*

Colour

Consider the overall visual effect of your chosen display, including the relationship between your text, diagrams and the backing board. Colour can be used to highlight key aspects of your poster. However, it is very easy to ruin a poster by the inappropriate choice and application of colour. Careful use of two, or at most three, complementary colours and shades will be easier on the eye and should aid comprehension. Colour can be used to link the text with the visual images (e.g. by picking out a colour in a photograph and using the same colour on the mounting board for the accompanying text). For PowerPoint posters, careful choice of colours for the various elements will enhance the final product (Box 14.1). Use coloured inks or water-based paints to provide colour in diagrams and figures, as felt pens rarely give satisfactory results.

Presenting at a scientific meeting – *never be tempted to spend the minimum amount of time converting a piece of scientific writing into poster format – the least interesting posters are those where the author simply displays pages from a written communication (e.g. a journal article) on a poster board.*

Content

The typical format is that of a scientific report (see Box 18.1), i.e. with the same headings, but with a considerably reduced content. Keep references within the text to a minimum – interested parties can always ask you for further information. Also note that most posters have a summary/ conclusions section at the end, rather than an abstract.

Introduction

This should give the reader background information on the broad field of study and the aims of your own work. It is vital that this section is as interesting as possible, to capture the interest of your audience. It is often worth listing your objectives as a series of numbered points.

Designing the materials and methods section – *photographs or diagrams of apparatus can help to break up the text of this section and provide visual interest. It is sometimes worth preparing this section in a smaller typeface.*

Materials and methods

Keep this short, and describe only the principal techniques used. You might mention any special techniques, or problems of general interest.

Keeping graphs and diagrams simple – *avoid composite graphs with different scales for the same axis, or with several trend lines (use a maximum of three trend lines per graph).*

Results

Don't present your raw data: use data reduction wherever possible, i.e. figures and simple statistical comparisons. Graphs, diagrams, histograms and pie charts give clear visual images of trends and relationships and should be used in place of data tables. Final copies of all figures should be produced so that the numbers can be read from a distance of 1 m. Each should have a concise title and legend, so that it is self-contained: if appropriate, a series of numbered points can be used to link a diagram with the accompanying text. Where symbols are used, provide a key on each graph (symbol size should be at least 5 mm). Avoid using graphs straight from a written version, e.g. a project report, textbook or a paper, without considering whether they need modification to meet your requirements.

Listing your conclusions – *a series of numbered points is a useful approach, if your findings fit this pattern.*

Conclusions

This is where many readers will begin, and they may go no further unless you make this section sufficiently interesting. This part needs to be the strongest section of your poster, summarising the main points. Refer to your figures here to draw the reader into the main part of your poster. A slightly larger or bolder typeface may add emphasis, though too many different typefaces can look messy. For the reference list, a smaller font can be used.

Consider providing a handout – *this is a useful way to summarise the main points of your poster, so that your readers have a permanent record of the information you have presented.*

The poster session

A poster display session may be organised as part of the assessment of your coursework, and this usually means those held at scientific meetings and

Box 14.1 How to create a poster using PowerPoint

Software such as PowerPoint can be used to produce a high-quality poster, providing you have access to a good colour printer. However, you should avoid the standard templates available on the Web as they encourage unnecessary uniformity and stifle creativity, leading to a less satisfying end result. The following steps give practical advice on creating a poster as a single PowerPoint slide:

1. **Sketch out your plans.** Decide on the main poster elements (images, graphs, tables and text sections) and their relationship with each other and draw out a one-page 'storyboard' (see Fig. 14.1). Think about colours for background, text and graphics (use two or three complementary colours): dark text on a light background is clearer (high contrast), and uses less ink when printing. Also consider how you will link the elements in sequence, to guide readers through your 'story'.
2. **Get your material ready.** Collect together individual files for pictures, figures and tables. Make any required adjustments to images, graphs or tables before you import them into your poster.
3. **Create a new/blank slide.** Open PowerPoint and from *File* > *New* select *Blank presentation.* Next, choose the *Blank Presentation* option for *New Presentation* > *New option.* Then use the *File* > *Page setup* menu to select either *Landscape* or *Portrait* orientation and to set the correct page size (use *Width* and *Height* commands, or select a standard size such as A4, A3, A2, etc.). Right-click on the slide and select *Ruler* and *Guides* (to help position elements within the slide – the horizontal and vertical guidelines can be dragged to different positions at later stages, as required) and also select an appropriate *Background* colour. In general, avoid setting a picture as your background as this tends to detract from the content of the poster. Before going further, save your work. Repeat this frequently and in more than one location (e.g. hard drive and USB memory stick).
4. **Add graphics.** For images, use the drop-down *Insert* menu, select *Picture, From File* and browse to *Insert* the correct file. The *Insert, Object* command performs a similar function for Excel charts (graphs). Alternatively, use the copy-and-paste functions of complementary software. Once inserted, resize using the *sizing handles* in one of the corners (for photographs, take care not to alter one dimension relative to the other, or the image will be distorted). To reposition, put the mouse pointer over the image, left-click and hold, then drag to new location. While the *Drawing* toolbar offers standard shapes and other useful features, you should avoid clipart (jaded and over-used) and poor-quality images from the Web (always use the highest resolution possible) – if you do not have your final images, use blank text boxes to show their position within the draft poster.
5. **Add text.** Use either the *Drawing* toolbar to select a *Text box* and place this on your slide, then either type in your text (use the *Enter* key to provide line spacing within the box) or copy-and-paste text from a word-processed file. You will need to consider the font size for the printed poster (e.g. for an A0 poster (size 1189 × 841 mm), a printed font size of 24 point is appropriate for the main text, with larger fonts for headings and titles. If you find things difficult to read on-screen, use the *Zoom* function (either select a larger percentage in the *Zoom box* on the standard toolbar, or hold down the *Ctrl* key and use the mouse wheel to scroll up (*zoom*) or down (*reduce*). Use a separate text box for each element of your poster and don't be tempted to type too much text into each box – write in succinct phrases, using bullet points and numbered lists to keep text concise (aim for no more than 50 words per text box). Select appropriate font styles and colours using the *Format* > *Font* menu. For a background colour or surrounding line, right-click and use the *Format Text Box* command (line thickness and colour can then be altered using the *Drawing* toolbar). Present supplementary text elements in a smaller font. For example, details of methodology, references cited.
6. **Add boxes, lines and/or arrows** to link elements of the poster and guide the reader (see Fig. 14.1). These features are available from the *Drawing* toolbar. Note that new inserts are overlaid over older inserts – if this proves to be a problem, select the relevant item and use the *Draw* > *Order* functions to change its relative position.
7. **Review your poster.** Get feedback from another student, or your tutor, e.g. on a small printed version, or use a projector to view your poster without printing (adjust the distance between projector and screen to give the correct size).
8. **Revise and edit your poster.** Revisit your work and remove as much unnecessary text as possible. Delete any component that is not essential to the message of the poster. Keep graphs simple and clear. 'White space' is important in providing structure.
9. **Print the final version.** Use a high-resolution colour printer (this may be costly, so you should wait until you are sure that no further changes are needed).

***Coping with questions in assessed poster sessions** – you should expect to be asked questions about your poster, and to explain details of figures, methods, etc.*

conferences. Staff and fellow students (delegates at conferences) will mill around, looking at the posters and chatting to their authors, who are usually expected to be in attendance. If you stand at the side of your poster throughout the session you are likely to discourage some readers, who may not wish to become involved in a detailed conversation about the poster. Stand nearby. Find something to do – talk to someone else, or browse among the other posters, but remain aware of people reading your poster and be ready to answer any queries they may raise. Do not be too discouraged if you aren't asked lots of questions: remember, the poster is meant to be a self-contained, visual story, without need for further explanation.

A poster display will never feel like an oral presentation, where the nervousness beforehand is replaced by a combination of satisfaction and relief as you unwind after the event. However, it can be a very satisfying means of communication, particularly if you follow these guidelines.

Text reference

Simmonds, D. and Reynolds, L. (1994) *Data Presentation and Visual Literacy in Medicine and Science.* Butterworth-Heinemann, London.

Sources for further study

Alley, M. (2003) *The Craft of Scientific Presentations: Critical Steps to Succeed and Critical Errors to Avoid.* Springer-Verlag, New York.

Briscoe, M.H. (2000) *Preparing Scientific Illustrations: A Guide to Better Posters, Presentations and Publications,* 2nd edn. Springer-Verlag, New York.

Davis, M.F. (2005) *Scientific Papers and Presentations,* 2nd edn. Academic Press, New York.

Gosling, P.J. (1999) *Scientist's Guide to Poster Presentations.* Kluwer, New York.

Hess, G. and Liegel, L. *Creating Effective Poster Presentations.* Available: http://www.ncsu.edu/project/posters.
Last accessed 01/04/07.

Study exercises

14.1 Design a poster. Working with one or more partners from your year group, decide on a suitable poster topic (perhaps something linked to your current teaching programme). Working individually, make an outline plan of the major elements of the poster, with appropriate sub-headings and a brief indication of the content and relative size of each element (including figures, diagrams and images). Exchange draft plans with your partners and arrange a session where you can discuss their merits and disadvantages.

14.2 Prepare a checklist for assessing the quality of a poster presentation. After reading through this chapter, prepare a 10-point checklist of assessment criteria under the heading 'What makes a good poster presentation?' Compare your list with the one that we have provided– do you agree with our criteria, or do you prefer your own list (and can you justify your preferences)?

14.3 Evaluate the posters in your university. Most universities have a wide range of academic posters on display. Some may cover general topics (e.g. course structures), while others may deal with specific research topics (e.g. poster presentations from past conferences). Consider their good and bad features (if you wish to make this a group exercise, you might compare your evaluation with that of other students, in a group discussion session).

15 Giving a spoken presentation

Opportunities for practising speaking skills *– these include:*

- *answering lecturers' questions;*
- *contributing in tutorials;*
- *talking to informal groups;*
- *giving your views at formal (committee) meetings;*
- *demonstrating or explaining to other students, e.g. during a practical class;*
- *asking questions in lectures/ seminars;*
- *answering an examiner's questions in an oral exam.*

Most students feel very nervous about giving talks. This is natural, since very few people are sufficiently confident and outgoing that they look forward to speaking in public. Additionally, the technical nature of the subject matter may give you cause for concern, especially if you feel that some members of the audience have a greater knowledge than you have, e.g. your tutors. However, this is a fundamental method of scientific communication and an important transferable skill, therefore it forms an important component of many courses.

The comments in this chapter apply equally to informal talks, e.g. those based on assignments and project work, and to more formal conference presentations. It is hoped that the advice and guidance given below will encourage you to make the most of your opportunities for public speaking, but there is no substitute for practice. Do not expect to find all of the answers from this, or any other, book. Rehearse, and learn from your own experience.

***KEY POINT** The three 'Rs' of successful public speaking are: reflect – give sufficient thought to all aspects of your presentation, particularly at the planning stage; rehearse – to improve your delivery; revise – modify the content and style of your material in response to your own ideas and to the comments of others.*

Preparation

Preliminary information

Learning from experience *– use your own experience of good and bad lecturers to shape your performance. Some of the more common errors include:*

- *speaking too quickly;*
- *reading from notes or from slides and ignoring the audience;*
- *inexpressive, impersonal or indistinct speech;*
- *distracting mannerisms;*
- *poorly structured material with little emphasis on key information;*
- *factual information too complex and detailed;*
- *too few or too many visual aids.*

Begin by marshalling the details needed to plan your presentation, including:

- the duration of the talk;
- whether time for questions is included;
- the size and location of the room;
- the projection/lighting facilities provided, and whether pointers or similar aids are available.

It is especially important to find out whether the room has the necessary equipment for digital projection (e.g. PC, projector and screen, black-out curtains or blinds, appropriate lighting) or overhead projection before you prepare your audio-visual aids. If you concentrate only on the spoken part of your presentation at this stage, you are inviting trouble later on. Have a look around the room and try out the equipment at the earliest opportunity, so that you are able to use the lights, projector, etc. with confidence. For digital projection systems, check that you can load/present your material. Box 15.1 gives practical advice on the use of PowerPoint.

Testing the room *– if possible, try to rehearse your talk in the room in which it will be presented. This will help you to make allowance for layout of equipment, lighting, acoustics and sight lines that might affect the way you deliver your talk. It will also put you more at ease on the day, because of the familiarity of the surroundings.*

Audio-visual aids

If you plan to use overhead transparencies, find out whether your department has facilities for their preparation, whether these facilities are available for your use, and the cost of materials. Adopt the following guidelines:

- Keep text to a minimum: present only the key points, with up to 20 words per slide/transparency.

Box 15.1 Tips on preparing and using PowerPoint slides in a spoken presentation

Microsoft PowerPoint can be used to produce high-quality visual aids, assuming a computer and digital projector are available in the room where you intend to speak. The presentation is produced as a series of electronic 'slides' onto which you can insert images, diagrams and text. When creating your slides, bear the following points in mind:

- **Plan the structure of your presentation.** Decide on the main topic areas and sketch out your ideas on paper. Think about what material you will need (e.g. pictures, graphs) and what colours to use for background and text.
- **Choose slide layouts according to purpose.** Once PowerPoint is running, from the *Insert* menu select *New Slide > Choose an Autolayout*. You can then add material to each new slide to suit your requirements.
- **Select your background with care.** Many of the pre-set background templates available within the *Format* menu (*Apply design template* option) are best avoided, since they are over-used and fussy, diverting attention from the content of the slides. Conversely, flat, dull backgrounds may seem uninteresting, while brightly coloured backgrounds can be garish and distracting. Choose whether to present your text as a light-coloured font on a dark background (more restful but perhaps less engaging if the room is dark) or a dark-coloured font on a light background (more lively).
- **Use visual images throughout.** Remember the familiar maxim 'a picture is worth ten thousand words'. A presentation composed entirely of text-based slides will be uninteresting: adding images and diagrams will brighten up your talk considerably (use the *Insert* menu, *Picture* option). Images can be taken with a digital camera, scanned in from a printed version or copied and pasted from the Web, but you should take care not to break copyright regulations. 'Clipart' is copyright-free, but should be used sparingly as most people will have seen the images before and they are rarely wholly relevant. Diagrams can be made from components created using the *Drawing* toolbar while graphs and tables can be imported from other programs, e.g. Excel (Box 14.1 gives further specific practical advice on adding graphics, saving files, etc.).
- **Keep text to a minimum.** Aim for no more than 20 words on a single slide (e.g. four/five lines containing a few words per line). Use headings and sub-headings to structure your talk: write only key words or phrases as 'prompts' to remind you to cover a particular point during your talk – never be tempted to type whole sentences as you will then be reduced to reading these from the screen during your presentation, which is boring.
- **Use a large, clear font style.** Use the *Slide Master* option within the *View* menu to set the default font to a non-serif style such as Arial, or Comic Sans MS, and an appropriate colour. Default fonts for headings and bullet points are intentionally large, for clarity. Do not reduce these to anything less than 28 point font size (preferably larger) to cram in more words: if you have too much material, create a new slide and divide up the information.
- **Animate your material.** The *Slide Show* menu provides a *Custom Animation* function that enables you to introduce the various elements within a slide, e.g. text can be made to *Appear* one line at a time, to prevent the audience from reading ahead and help maintain their attention.
- **Don't overdo the special effects.** PowerPoint has a wide range of features that allow complex slide transitions and animations, additional sounds, etc. but these quickly become irritating to an audience unless they have a specific purpose within your presentation.
- **Always edit your slides before use.** Check through your slides and cut out any unnecessary words, adjust the layout and animation. Remember the maxim 'less is more' – avoid too much text; too many bullet points; too many distracting visual effects or sounds.

When presenting your talk:

- **Work out the basic procedures beforehand.** Practise, to make sure that you know how to move forwards and backwards in your slideshow, turn the screen on and off, hide the mouse pointer, etc.
- **Don't forget to engage your audience.** Despite the technical gadgetry, *you* need to play an active role in the presentation, as explained elsewhere in this chapter.
- **Don't go too fast.** Sometimes, new users tend to deliver their material too quickly: try to speak at a normal pace and practise beforehand.
- **Consider whether to provide a handout.** PowerPoint has several options, including some that provide space for notes (e.g. Fig. 4.3). However, a handout should not be your default option, as there is a cost involved.

Using audio-visual aids – *don't let the equipment and computer gadgetry distract you from the essential rules of good speaking (Box 15.2): remember that you are the presenter.*

- Make sure the text is readable: try out your material beforehand.
- Use several simpler figures rather than a single complex graph.
- Avoid too much colour on overhead transparencies: blue and black are easier to read than red or green.
- Don't mix slides and transparencies as this is often distracting.
- Use spirit-based pens for transparencies: use alcohol for corrections.
- Transparencies can be produced from typewritten or printed text using a photocopier, often giving a better product than pens. Note that you must use special heat-resistant acetate sheets for photocopying.

Electronic presentation software (e.g. PowerPoint) can replace these specialist requirements, as long as the necessary facilities are available for your talk (see below).

Pitching your talk at the right level – *the general rule should be: 'do not over-estimate the background knowledge of your audience'. This sometimes happens in student presentations, where fears about the presence of 'experts' can encourage the speaker to include too much detail, overloading the audience with facts.*

Audience

You should consider your audience at the earliest stage, since they will determine the appropriate level for your presentation. If you are talking to fellow students you may be able to assume a common level of background knowledge. In contrast, a research lecture given to your department, or a paper at a meeting of a scientific society, will be presented to an audience from a broader range of backgrounds. An oral presentation is not the place for a complex discussion of specialised information: build up your talk from a low level. The speed at which this can be done will vary according to your audience. As long as you are not boring or patronizing, you can cover basic information without losing the attention of the more knowledgeable members in your audience.

Content

While the specific details in your talk will be for you to decide, most spoken presentations share some common features of structure, as described below.

Getting the introduction right – *a good idea is to have an initial slide giving your details and the title of your talk, and a second slide telling the audience how your presentation will be structured. Make eye contact with all sections of the audience during the introduction.*

What to cover in your introduction – *You should:*

- *explain the structure of your talk;*
- *set out your aims and objectives;*
- *explain your approach to the topic.*

Introductory remarks

It is vital to capture the attention of your audience at the outset. Consequently, you must make sure your opening comments are strong, otherwise your audience will lose interest before you reach the main message. Remember it takes a sentence or two for an audience to establish a relationship with a new speaker. Your opening sentence should be some form of preamble and should not contain any key information. For a formal lecture, you might begin with 'Thank you for that introduction. My talk today is about ...' then restate the title and acknowledge other contributors, etc. You might show a transparency or slide with the title printed on it, or an introductory photograph, if appropriate. This should provide the necessary settling-in period.

After these preliminaries, you should introduce your topic. Begin your story on a strong note – avoid timid or apologetic phrases.

Opening remarks are unlikely to occupy more than 10% of the talk. However, because of their significance, you might reasonably spend up to 25% of your preparation time on them.

***KEY POINT** Make sure you have practised your opening remarks, so that you can deliver the material in a flowing style, with less chance of mistakes.*

The main message

This section should include the bulk of your experimental results or literature findings, depending on the type of presentation. Keep details of methods to the minimum needed to explain your data. This is *not* the place for a detailed description of equipment and experimental protocol (unless it is a talk about methodology). Results should be presented in an easily digested format.

***Allowing time for slides** – as a rough guide you should allow at least two minutes per illustration, although some diagrams may need longer, depending on content. Make a note of the halfway point, to help you check timing/pace.*

***KEY POINT** Do not expect your audience to cope with large amounts of data; use a maximum of six numbers per slide. Remember that graphs and diagrams are usually better than tables of raw data, since the audience will be able to see the visual trends and relationships in your data.*

Present summary statistics (Chapter 66) rather than individual results. Show the final results of any analyses in terms of the statistics calculated, and their significance, rather than dwelling on details of the procedures used. Figures should not be crowded with unnecessary detail. Every diagram should have a concise title and the symbols and trend lines should be clearly labelled, with an explanatory key where necessary. When presenting graphical data (Chapter 63) always 'introduce' each graph by stating the units for each axis and describing the relationship for each trend line or data set.

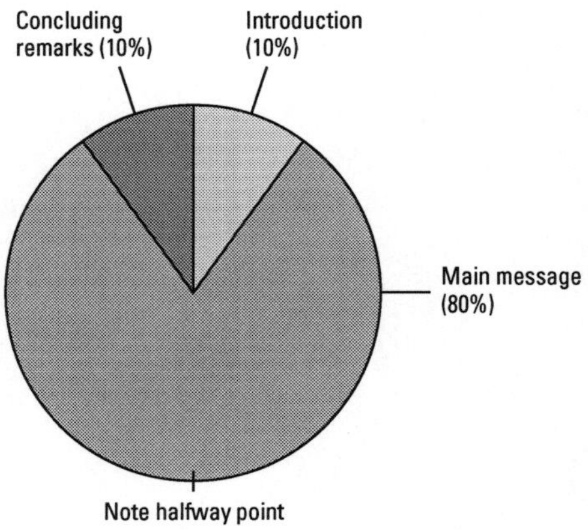

Fig. 15.1 Pie chart showing time allocation for a typical presentation.

***KEY POINT** Use summary slides at regular intervals, to maintain the flow of the presentation and to emphasise the main points.*

Take the audience through your story step-by-step at a reasonable pace. Try not to rush the delivery of your main message due to nervousness. Avoid complex, convoluted story-lines – one of the most distracting things you can do is to fumble backwards through PowerPoint slides or overhead transparencies. If you need to use the same diagram or graph more than once then you should make two (or more) copies. In a presentation of experimental results, you should discuss each point as it is raised, in contrast to written text where the results and discussion may be in separate sections. The main message typically occupies approximately 80% of the time allocated to an oral presentation (Fig. 15.1).

Concluding remarks

***Final remarks** – make sure you give the audience sufficient time to assimilate your final slide: some of them may wish to write down the key points. Alternatively, you might provide a handout, with a brief outline of the aims of your study and the major conclusions.*

Having captured the interest of your audience in the introduction and given them the details of your story in the middle section, you must now bring your talk to a conclusion. Do not end weakly, e.g. by running out of steam on the last slide. Provide your audience with a clear 'take-home message', by returning to the key points in your presentation. It is often appropriate to prepare a slide or overhead transparency listing your main conclusions as a numbered series.

Box 15.2 Hints on spoken presentations

In planning the delivery of your talk, bear the following aspects in mind:

- **Using notes.** Many accomplished speakers use abbreviated notes for guidance, rather than reading word-for-word from a prepared script. When writing your talk:
  - (a) **Consider preparing your first draft as a full script:** write in spoken English and keep the text simple, to avoid a formal, impersonal style. Your aim should be to *talk* to your audience, not to *read* to them.
  - (b) **If necessary, use note-cards with key words and phrases:** it is best to avoid using a full script in the final presentation. As you rehearse and your confidence improves, a set of note-cards may be an appropriate format. Mark the position of slides/key points, etc.: each note-card should contain details of structure as well as content. Your notes should be written/printed in text large enough to be read easily during the presentation (also check that the lecture room has a lectern light or you may have problems reading your notes if the lights are dimmed). Each note-card or sheet should be clearly numbered, so that you do not lose your place.
  - (c) **Decide on the layout of your talk:** give each sub-division a heading in your notes, so that your audience is made aware of the structure.
  - (d) **Memorise your introductory/closing remarks:** you may prefer to rely on a full written version for these sections, in case your memory fails, or if you suffer 'stage fright'.
  - (e) **Using PowerPoint** (Box 15.1): here, you can either use the 'notes' option (*View> Notes Page*), or you may even prefer to dispense with notes entirely, since the slides will help structure your talk, acting as an *aide-memoire* for your material.
- **Work on your timing.** It is essential that your talk is the right length and the correct pace:
  - (a) **Rehearse your presentation:** ask a friend to listen and to comment constructively on those parts that were difficult to follow, to improve your performance.
  - (b) **Use 'split times' to pace yourself:** following an initial run-through, add the times at which you should arrive at the key points of your talk to your notes. These timing marks will help you keep to time during the final presentation.
  - (c) **Avoid looking at your wristwatch when speaking;** this sends a negative signal to the audience. Use a wall clock (where available), or take off your watch and put it beside your notes so that you can glance at it without distracting the audience.
- **Consider your image.** Make sure that the image you project is appropriate for the occasion:
  - (a) **Think about what to wear:** aim to be respectable without 'dressing up', otherwise your message may be diminished.
  - (b) **Maintain a good posture:** it will help your voice projection if you stand upright, rather than slouching, or leaning over a lectern.
  - (c) **Deliver your material with expression:** project your voice towards the audience at the back of the room and make sure you look round to make eye contact with all sections of the audience. Arm movements and subdued body language will help maintain the interest of your audience. However, you should avoid extreme gestures (it may work for some TV personalities but it is not recommended for the beginner).
  - (d) **Try to identify and control any repetitive mannerisms:** repeated 'empty' words/phrases, fidgeting with pens, keys, etc. will distract your audience. Note-cards held in your hand give you something to focus on, while laser pointers will show up any nervous hand tremors. Practising in front of a mirror may help.
- **Think about questions.** Once again, the best approach is to prepare beforehand:
  - (a) **Consider what questions are likely to come up, and prepare brief answers.** However, do not be afraid to say 'I don't know': your audience will appreciate honesty rather than vacillation if you don't have an answer for a particular question.
  - (b) **If no questions are asked, you might pose a question yourself** and then ask for opinions from the audience: if you use this approach, you should be prepared to comment briefly if your audience has no suggestions, to avoid the presentation ending in an embarrassing silence.

Signal the end of your talk by saying 'finally ...', 'in conclusion ...', or a similar comment and then finish speaking after that sentence. Your audience will lose interest if you extend your closing remarks beyond this point. You may add a simple end phrase (for example, 'thank you') as you put your notes into your folder, but do not say 'that's all folks!', or make any similar offhand remark. Finish as strongly and as clearly as you started. Box 15.2 gives further advice on presentation.

Sources for further study

Alley, M. (2003) *The Craft of Scientific Presentations: Critical Steps to Succeed and Critical Errors to Avoid.* Springer-Verlag, New York.

Capp, C.C. and Capp, G.R. (1989) *Basic Oral Communication*, 5th edn. Prentice Hall, Harlow.

Matthews, C. and Marino, J. (1999) *Professional Interactions: Oral Communication Skills of Science, Technology and Medicine.* Pearson, Harlow.

Radel, J. *Oral Presentations.* Available: http://www.biology.eku.edu/RITCHISO/oralpres.html Last accessed: 01/04/07.

Study exercises

15.1 Prepare a checklist for assessing the quality of an oral presentation. After reading through this chapter, prepare a 10-point checklist of assessment criteria under the heading 'What makes a good oral presentation?'. Compare your list with the one that we have provided – do you agree with our criteria, or do you prefer your checklist? (Can you justify your preferences?)

15.2 Evaluate the presentation styles of other speakers. There are many opportunities to assess the streng-ths and weaknesses of academic 'public speakers', including your lecturers, seminar speakers, presenters of TV documentaries, etc. Decide in advance how you are going to tackle the evalua-tion (e.g. with a quantitative marking scheme, or a less formal procedure).

15.3 Rehearse a talk and get feedback on your performance. There are a number of approaches you might take, including: (i) recording and reviewing your presentation using a digital movie camera; or (ii) giving your talk to a small group of fellow students and asking them to provide constructive feedback.

16 General aspects of scientific writing

Written communication is an essential component of all sciences. Most courses include writing exercises in which you will learn to describe ideas and results accurately, succinctly and in an appropriate style and format. The following features/aspects are common to all forms of scientific writing.

Time management *– practical advice is given in Chapter 2.*

Organising your time

Making a timetable at the outset helps ensure that you give each stage adequate attention and complete the work on time (e.g. Fig. 16.1). To create and use a timetable:

1. Break down the task into stages.
2. Decide on the proportion of the total time each stage should take.
3. Set realistic deadlines for completing each stage, allowing some time for slippage.
4. Refer to your timetable frequently as you work: if you fail to meet one of your deadlines, make a serious effort to catch up as soon as possible.

Monday:	morning	Lectures (University)
	afternoon	Practical (University)
	evening	Initial analysis and brainstorming (Home)
Tuesday:	morning	Lectures (University)
	afternoon	Locate sources (Library)
	evening	Background reading (Library)
Wednesday:	morning	Background reading (Library)
	afternoon	Squash (Sports hall)
	evening	Planning (Home)
Thursday:	morning	Lectures (University)
	afternoon	Additional reading (Library)
	evening	Prepare outline (Library)
Friday:	morning	Lab class (University)
	afternoon	Write first draft (Home)
	evening	Write first draft (Home)
Saturday:	morning	Shopping (Town)
	afternoon	Review first draft (Home)
	evening	Revise first draft (Home)
Sunday:	morning	Free
	afternoon	Produce final copy (Home)
	evening	Proof read and print essay (Home)
Monday:	morning	Final read-through and check Submit essay (deadline midday)

Fig. 16.1 Example timetable for writing a short essay.

KEY POINT The appropriate allocation of your time to reading, planning, writing and revising will differ according to the task in hand (see Chapters 17–19).

Organising your information and ideas

Before you write, you need to gather and/or think about relevant material (Chapters 8 and 9). You must then decide:

- what needs to be included and what doesn't;
- in what order it should appear.

Start by jotting down headings for everything of potential relevance to the topic (this is sometimes called 'brainstorming'). A spider diagram (Fig. 16.2) or a Mind Map (Fig. 4.2) will help you organise these ideas. The next stage is to create an outline of your text (Fig. 16.3). Outlines are valuable because they:

- force you to think about and plan the structure;
- provide a checklist so nothing is missed out;
- ensure the material is balanced in content and length;
- help you organise figures and tables by showing where they will be used.

KEY POINT A suitable structure is essential to the narrative of your writing, and should be carefully considered at the outset.

Creating an outline *– an informal outline can be made simply by indicating the order of sections on a spider diagram (as in Fig. 16.2).*

In an essay or review, the structure of your writing should help the reader to assimilate and understand your main points. Subdivisions of the topic could simply be related to the physical nature of the subject matter (e.g. levels of organisation of a protein) and should proceed logically (e.g. primary structure then secondary, etc.). A chronological approach is good for evaluation of past work (e.g. the development of

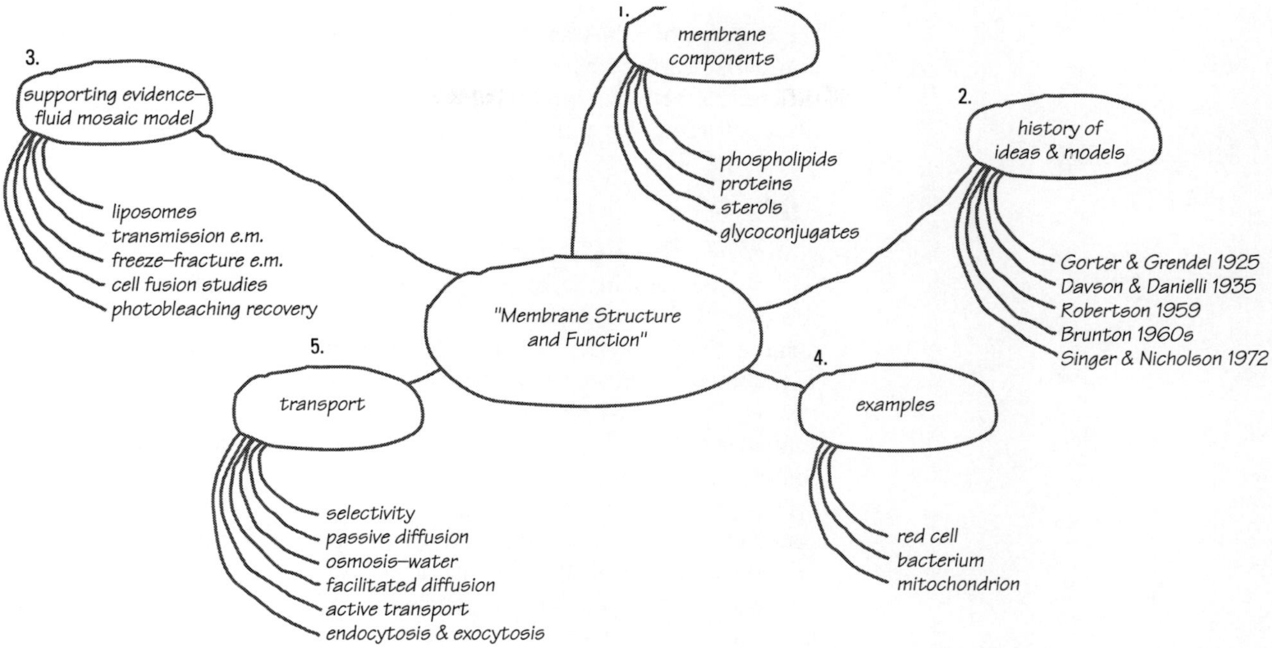

Fig. 16.2 Spider diagram showing how you might 'brainstorm' an essay with the title 'Membrane Structure and Function'. Write out the essay title in full to form the spider's body, and as you think of possible content place headings around this to form its legs. Decide which headings are relevant and which are not and use arrows to note connections between subjects. This may influence your choice of order and may help to make your writing flow because the links between paragraphs will be natural. You can make an informal outline directly on a spider diagram by adding numbers indicating a sequence of paragraphs (as shown). This method is best when you must work quickly, as with an essay written under exam conditions.

Membrane structure and function

Introduction
- membrane components
  - phospholipids
  - proteins
  - sterols
  - glycoconjugates
- history of ideas & models
  - Gorter & Grendel 1925
  - Davson & Danielli 1935
  - Robertson 1959
  - Brunton 1960s
  - Singer & Nicholson 1972

Supporting evidence – fluid mosaic model
- liposomes
- transmission e.m.
- freeze-fracture e.m.
- cell fusion studies
- photobleaching recovery

Transport
- selectivity
- passive diffusion
- osmosis – water
- facilitated diffusion
- active transport
- endocytosis & exocytosis

Examples
- red cell
- bacterium
- mitochondrion

(a)

Membrane structure and function

1. INTRODUCTION
   - 1.1 membrane components
     - 1.1.1 phospholipids
     - 1.1.2 proteins
     - 1.1.3 sterols
     - 1.1.4 glycoconjugates
   - 1.2 history of ideas & models
     - 1.2.1 Gorter & Grendel 1925
     - 1.2.2 Davson & Danielli 1935
     - 1.2.3 Robertson 1959
     - 1.2.4 Brunton 1960s
     - 1.2.5 Singer & Nicholson 1972
2. SUPPORTING EVIDENCE – FLUID MOSAIC MODEL
   - 2.1 liposomes
   - 2.2 transmission e.m.
   - 2.3 freeze-fracture e.m.
   - 2.4 cell fusion studies
   - 2.5 photobleaching recovery
3. TRANSPORT
   - 3.1 selectivity
   - 3.2 passive diffusion
   - 3.3 osmosis – water
   - 3.4 facilitated diffusion
   - 3.5 active transport
   - 3.5 endocytosis & exocytosis
4. EXAMPLES
   - 4.1 red cell
   - 4.2 bacterium
   - 4.3 mitochondrion

(b)

Fig. 16.3 Formal outlines. These are useful for a long piece of work where you or the reader might otherwise lose track of the structure. The headings for sections and paragraphs are simply written in sequence with the type of lettering and level of indentation indicating their hierarchy. Two different forms of formal outline are shown, a minimal form (a) and a numbered form (b). Note that the headings used in an outline are often repeated within the essay to emphasise its structure. The content of an outline will depend on the time you have available and the nature of the work, but the most detailed hierarchy you should reasonably include is the subject of each paragraph.

the concept of DNA as the hereditary material), whereas a step-by-step comparison might be best for certain exam questions (e.g. 'Discuss the differences between prokaryotes and eukaryotes'). There is little choice about structure for practical and project reports (see p. 111).

Writing

Adopting a scientific style

Your main aim in developing a scientific style should be to get your message across directly and unambiguously. While you can try to achieve this through a set of 'rules' (see Box 16.1), you may find other requirements driving your writing in a contradictory direction. For instance, the need to be accurate and complete may result in text littered with technical terms, and the flow may be continually interrupted by references to the literature. The need to be succinct also affects style and readability through the use of, for example, stacked noun-adjectives (e.g. 'restriction fragment length polymorphism') and acronyms (e.g. 'RFLP'). Finally, style is very much a matter of taste and each tutor, examiner, supervisor or editor will have pet loves and hates which you may have to accommodate. Different assignments will need different styles; Box 16.2 gives further details.

Developing technique

Improving your writing skills – *you need to take a long-term view if you wish to improve this aspect of your work. An essential preliminary is to invest in and make full use of a personal reference library (see Box 16.3).*

Writing is a skill that can be improved, but not instantly. You should analyse your deficiencies with the help of feedback from your tutors, be prepared to change work habits (e.g. start planning your work more carefully), and willing to learn from some of the excellent texts that are available on scientific writing (p. 106).

Getting started

A common problem is 'writer's block' – inactivity or stalling brought on by a variety of causes. If blocked, ask yourself these questions:

Talking about your work – *discussing your topic with a friend or colleague might bring out ideas or reveal deficiencies in your knowledge.*

Writing with a word processor – *use the dynamic/interactive features of the word processor (Chapter 13) to help you get started: first make notes on structure and content, then expand these to form a first draft and finally revise/improve the text.*

- Are you comfortable with your surroundings? Make sure you are seated comfortably at a reasonably clear desk and have minimised the possibility of interruptions and distractions.
- Are you trying to write too soon? Have you clarified your thoughts on the subject? Have you done enough preliminary reading?
- Are you happy with the underlying structure of your work? If you haven't made an outline, try this. If you are unhappy because you can't think of a particular detail at the planning stage, just start writing – it is more likely to come to you while you are thinking of something else.
- Are you trying to be too clever? Your first sentence doesn't have to be earth-shattering in content or particularly smart in style. A short statement of fact or a definition is fine. If there will be time for revision, first get your ideas down on paper and then revise grammar, content and order later.
- Do you really need to start writing at the beginning? Try writing the opening remarks after a more straightforward part. For example, with reports of laboratory work, the Materials and Methods section may be the easiest place to start.
- Are you too tired to work? Don't try to 'sweat it out' by writing for long periods at a stretch: stop frequently for a rest.

Box 16.1 How to achieve a clear, readable style

Words and phrases

- Choose short, clear words and phrases rather than long ones: e.g. use 'build' rather than 'fabricate'; 'now' rather than 'at the present time'. At certain times, technical terms must be used for precision, but don't use jargon if you don't have to.
- Don't worry too much about repeating words, especially when introducing an alternative might subtly alter your meaning.
- Where appropriate, use the first person to describe your actions ('We decided to'; 'I conclude that'), but not if this is specifically discouraged by your supervisor or department.
- Favour active forms of writing ('the observer completed the survey in ten minutes') rather than a passive style ('the survey was completed by the observer in ten minutes').
- Use tenses consistently. Past tense is always used for materials and methods ('samples were taken from...') and for reviewing past work ('Smith (1990) concluded that...'). The present tense is used when describing data ('Fig. 1 shows...'), for generalisations ('Most authorities agree that...') and conclusions ('To conclude, ...').
- Use statements in parentheses sparingly – they disrupt the reader's attention to your central theme.
- Avoid clichés and colloquialisms – they are usually inappropriate in a scientific context.

Punctuation

- Try to use a variety of types of punctuation, to make the text more interesting to read.
- Decide whether you wish to use 'closed' punctuation (frequent commas at the end of clauses) or 'open' punctuation (less frequent punctuation) and be consistent.
- Don't link two sentences with a comma. Use a full stop, this is an example of what *not* to do.
- Pay special attention to apostrophes, using the following rules:
  - To indicate possession, use an apostrophe before an 's' for a singular word (e.g. the rat's temperature was ...') and after the s for a plural word ending in s (e.g. the rats' temperatures were = the temperatures of the rats were). If the word has a special plural (e.g. woman ⟶ women) then use the apostrophe before the s (the women's temperatures were ...).
  - When contracting words, use an apostrophe (e.g. do not = don't; it's = it is), but remember that contractions are generally *not* used in formal scientific writing.
  - Do *not* use an apostrophe for 'its' as the possessive form of 'it' (e.g. 'the university and its surroundings') Note that 'it's' is reserved for 'it is'. This is an exception to the general rule and a very common mistake.
  - Never use an apostrophe to indicate plurals. Even for abbreviations, the accepted style is now to omit the apostrophe for the plural (e.g. write 'the ELISAs were').

Sentences

- Don't make them overlong or complicated.
- Introduce variety in structure and length.
- If unhappy with the structure of a sentence, try chopping it into a series of shorter sentences.

Paragraphs

- Get the paragraph length right – five sentences or so. Do *not* submit an essay that consists of a single paragraph, nor one that contains single sentence paragraphs.
- Make sure each paragraph is logical, dealing with a single topic or theme.
- Take care with the first sentence in a paragraph (the 'topic' sentence); this introduces the theme of the paragraph. Further sentences should then develop this theme, e.g. by providing supporting information, examples or contrasting cases.
- Use 'linking' words or phrases to maintain the flow of the text within a paragraph (e.g. 'for example'; 'in contrast'; 'however'; 'on the other hand').
- Make your text more readable by adopting modern layout style. The first paragraph in any section of text is usually *not* indented, but following paragraphs may be (by the equivalent of three character spaces). In addition, the space between paragraphs should be slightly larger than the space between lines. Follow departmental guidelines if these specify a format.
- Group paragraphs in sections under appropriate headings and sub-headings to reinforce the structure underlying your writing.
- Think carefully about the first and last paragraphs in any piece of writing: these are often the most important as they respectively set the aims and report the conclusions.

Note: If you're not sure what is meant by any of the terms used here, consult a guide on writing (see sources for further study).

Box 16.2 Using appropriate writing styles for different purposes (with examples)

Note that courses tend to move from assignments that are predominantly descriptive in the early years to a more analytical approach towards the final year (see Chapter 5). Also, different styles may be required in different sections of a write-up, e.g. descriptive for introductory historical aspects, becoming more analytical in later sections.

Descriptive writing

This is the most straightforward style, providing factual information on a particular subject and is most appropriate:

- in essays where you are asked to 'describe' or 'explain' (p. 108)
- when describing the results of a practical exercise, e.g.: 'The experiment shown in Figure 1 confirmed that enzyme activity was strongly influenced by temperature, as the rate observed at 37°C was more than double that seen at 20°C.'

However, in literature reviews and essays where you are asked to 'discuss' (p. 108) a particular topic, the descriptive approach is mostly inappropriate, as in the following example, where a large amount of specific information from a single scientific paper has been used, without any attempt to highlight the most important points:

'In a study carried out between July and October 2002, a total of 225 sputum samples from patients attending 25 different clinics in England and Wales were screened. Bacteria were isolated from 67.6% of these samples, with 47.42% of the samples giving *Pseudomonas aeruginosa*, 11.76% *Burkholderia cepacia* and 8.59% *Stentrophomonas maltophilia* (Grey and Gray, 2003).'

In the most extreme examples, whole paragraphs or pages of essays may be based on descriptive factual detail from a single source, often with a single citation at the end of the material, as above: such essays often score low marks in essays where evidence of deeper thinking is required (Chapter 5).

Comparative writing

This technique is an important component of academic writing, and it will be important to develop your comparative writing skills as you progress through your course. Its applications include:

- answering essay questions and assignments of the 'compare and contrast' type (p. 108)
- comparing your results with previously published work in the Discussion section of a practical report.

To use this style, first decide on those aspects you wish to compare and then consider the material (e.g. different literature sources) from these aspects – in what ways do they agree or disagree with each other? One approach is to compare/contrast a different aspect in each paragraph. At a practical level, you can use 'linking' words and phrases to help orientate your reader, as you move between aspects where there is agreement and disagreement. These include, for agreement: 'in both cases'; 'in agreement with'; 'is also shown by the study of'; 'similarly'; 'in the same way', and for disagreement: 'however'; 'although'; 'in contrast to'; 'on the other hand'; 'which differs from'. The comparative style is fairly straightforward, once you have decided on the aspects to be compared. The following brief example compares two different studies using this style:

'While Grey and Gray (2003) reported that *Pseudomonas aeruginosa* was present in 47.4% of 225 UK sputum samples, Black and White (2006) showed that 89.1% of sputum samples from 2592 patients were positive for this bacterium.'

Comparative text typically makes use of two or more references per paragraph.

Analytical writing

Typically, this is the most appropriate form of writing for:

- a review of scientific literature on a particular topic;
- an essay where you are asked to 'discuss' (p. 108) different aspects of a particular topic;
- evaluating a number of different published sources within the Discussion section of a final-year project dissertation.

By considering the significance of the information provided in the various sources you have read, you will be able to take a more critical approach. Your writing should evaluate the importance of the material in the context of your topic (see also Chapter 9). In analytical writing, you need to demonstrate critical thinking (p. 55) and personal input about the topic in a well-structured text that provides clear messages, presented in a logical order and demonstrating synthesis from a number of sources by appropriate use of citations (p. 47). Detailed information and relevant examples are used only to explain or develop a particular aspect, and not simply as 'padding' to bulk up the essay, as in the following example:

'*Pseudomonas aeruginosa* is often isolated from sputum samples of cystic fibrosis patients: a short-term UK study with a relatively small sample size (225 patients) isolated this bacterium from around half of all samples (Grey and Gray, 2003), while a longer-term study with a far larger sample size (2592 patients) gave an isolation rate of almost 90% (Black and White, 2006).'

Analytical writing is based on a broad range of sources, typically with several citations per paragraph.

Box 16.3 How to improve your writing ability by consulting a personal reference library

Using dictionaries

We all know that a dictionary helps with spelling and definitions, but how many of us use one effectively? You should:

- Keep a dictionary beside you when writing and always use it if in any doubt about spelling or definitions.
- Use it to prepare a list of words which you have difficulty in spelling: apart from speeding up the checking process, the act of writing out the words helps commit them to memory.
- Use it to write out a personal glossary of terms. This can help you memorise definitions. From time to time, test yourself.

Not all dictionaries are the same! Ask your tutor or supervisor whether he/she has a preference and why. Try out the *Oxford Advanced Learner's Dictionary*, which is particularly useful because it gives examples of use of all words and helps with grammar, e.g. by indicating which prepositions to use with verbs. Dictionaries of biology tend to be variable in quality, possibly because the subject is so wide and new terms are continually being coined. *Henderson's Dictionary of Biological Terms* is a useful example.

Using a thesaurus

A thesaurus contains lists of words of similar meaning grouped thematically; words of opposite meaning always appear nearby.

- Use a thesaurus to find a more precise and appropriate word to fit your meaning, but check definitions of unfamiliar words with a dictionary.
- Use it to find a word or phrase 'on the tip of your tongue' by looking up a word of similar meaning.
- Use it to increase your vocabulary.

Roget's Thesaurus is the standard. Collins also publishes a combined dictionary and thesaurus.

Using guides for written English

These provide help with the use of words.

- Use guides to solve grammatical problems such as when to use 'shall' or 'will', 'which' or 'that', 'effect' or 'affect', 'can' or 'may', etc.
- Use them for help with the paragraph concept and the correct use of punctuation.
- Use them to learn how to structure writing for different tasks.

Revising your text

Wholesale revision of your first draft is strongly advised for all writing, apart from in exams. When using a word processor, this can be a simple process. Where possible, schedule your writing so you can leave the first draft to 'settle' for at least a couple of days. When you return to it fresh, you will see more easily where improvements can be made. Try the following structured revision process, each stage being covered in a separate scan of your text:

1. Examine content. Have you included everything you need to? Is all the material relevant?
2. Check the grammar and spelling. Can you spot any 'howlers'?
3. Focus on clarity. Is the text clear and unambiguous? Does each sentence really say what you want it to say?
4. Be succinct. What could be missed out without spoiling the essence of your work? It might help to imagine an editor has set you the target of reducing the text by 15%.
5. Improve style. Could the text read better? Consider the sentence and paragraph structure and the way your text develops to its conclusion.

***Revising your text** – to improve clarity and shorten your text, 'distil' each sentence by taking away unnecessary words and 'condense' words or phrases by choosing a shorter alternative.*

***Learning from others** – ask a colleague to read through your draft and comment on its content and overall structure.*

Common errors

These include:

- Problems over singular and plural words ("a bacteria is"; "the results shows")
- Verbose text ("One definition that can be employed in this situation is given in the following sentence.")
- Misconstructed sentences ("Health and safety regulations should be made aware of . . .")
- Misuse of punctuation, especially commas and apostrophes (for examples, see Box 16.1).
- Poorly-constructed paragraphs (for advice, see Box 16.1).

Sources for further study

Burchfield, R.W. (ed.) (2004) *Fowler's Modern English Usage*, revised 3rd edn. Oxford University Press, New York.

Clark, R. *The English Style Book. A Guide to the Writing of Scholarly English*. Available: http://www.litency.com/stylebook/stylebook.php Last accessed: 01/04/07.

Kane, T.S. (1994) *The New Oxford Guide to Writing*. Oxford University Press, New York.
[This is excellent for the basics of English – it covers grammar, usage and the construction of sentences and paragraphs]

Lindsay, D. (1995) *A Guide to Scientific Writing*, 2nd edn. Longman, Harlow.

McMillan, K.M. and Weyers, J.D.B. (2006) *The Smarter Student: Skills and Strategies for Success at University*. Pearson, Harlow.

Partridge, E. (1978) *You Have a Point There*. Routledge, London.
[This covers punctuation in a very readable manner]

Pechnick, J.A. and Lamb, B.C. (1994) *How to Write about Biology*. Harper Collins, London.

Tichy, H.J. (1988) *Effective Writing for Engineers, Managers and Scientists*, 2nd edn. Wiley, New York.
[This is strong on scientific style and clarity in writing]

Study exercises

16.1 'Brainstorm' an essay title. Pair up with a partner in your class. Together, pick a suitable essay title from a past exam paper. Using the spider diagram or another technique, individually 'brainstorm' the title. Meet afterwards, compare your ideas, and discuss their relative merits and disadvantages.

16.2 Improve your writing technique. From the following checklist, identify the *three* weakest aspects of your writing, either in your own opinion or from essay/assignment feedback:

- grammar;
- paragraph organisation;
- presentation of work;
- punctuation;
- scientific style;
- sentence structure/variety;
- spelling;
- structure and flow;
- vocabulary.

Now either borrow a book from a library or buy a book that deals with your weakest aspects of writing. Read the relevant chapters or sections and for each aspect write down some tips that should help you in future.

16.3 Improve your spelling and vocabulary with two lists. Create a pair of lists and pin these up beside your desk. One should be entitled *Spelling Mistakes* and the other *New Words*. Now, whenever you make a mistake in spelling or have to look up how to spell a word in a dictionary, add the problem word to your spelling list, showing where you made the mistake. Also, whenever you come across a word whose meaning is unclear to you, look it up in a dictionary and write the word and its meaning in the 'new words' vocabulary list.

17 Writing essays

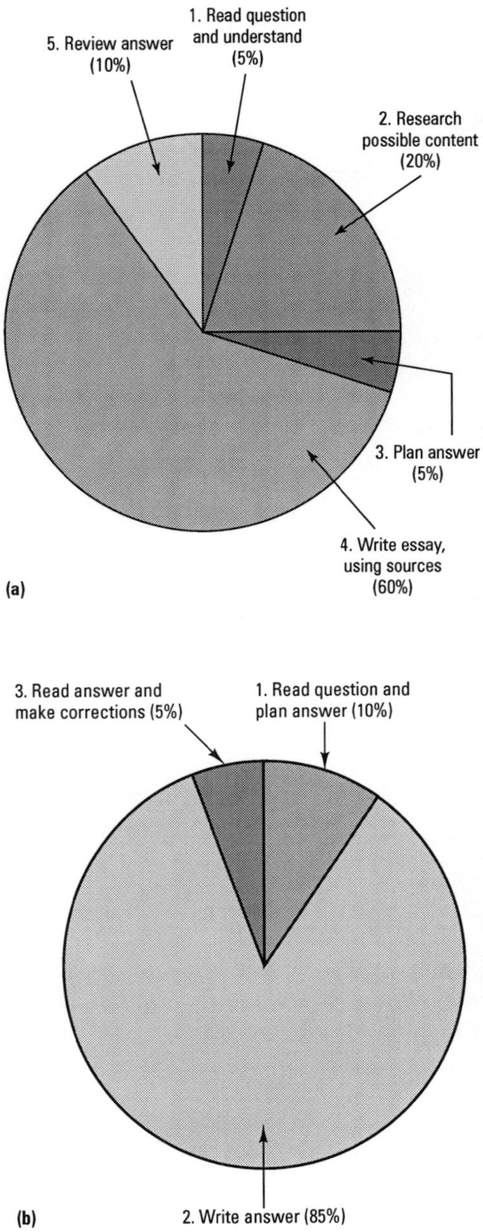

Fig. 17.1 Typical division of time for an essay written as part of an in-course assessment (a) or under exam conditions (b).

Considering essay content – *it is rarely enough simply to lay down facts for the reader – you must analyse them and comment on their significance (see p. 104).*

The function of an essay is to show how much you understand about a topic and how well you can organise and express your knowledge.

Organising your time

The way you should divide your time when producing an essay depends on whether you are writing it for in-course assessment or under exam conditions (Fig. 17.1). Essays written over a long period with access to books and other resources will probably involve a research element, not only before the planning phase but also when writing (Fig. 17.1a). For exams, it is assumed that you have revised appropriately (Chapter 5) and essentially have all the information at your fingertips. To keep things uncomplicated, the time allocated for each essay should be divided into three components – planning, writing and reviewing (Fig. 17.1b), and you should adopt time-saving techniques whenever possible (Box 6.2).

Making a plan for your essay

Dissect the meaning of the essay question or title

Read the title very carefully and think about the topic before starting to write. Consider the definitions of each of the important nouns (this can help in approaching the introductory section). Also think about the meaning of the verb(s) used and try to follow each instruction precisely (see Table 17.1). Don't get side-tracked because you know something about one word or phrase in the title: consider the whole title and all its ramifications. If there are two or more parts to the question, make sure you give adequate attention to each part.

Consider possible content and examples

Research content using the methods described in Chapters 8 and 9. If you have time to read several sources, consider their content in relation to the essay title. Can you spot different approaches to the same subject? Which do you prefer as a means of treating the topic in relation to your title? Which examples are most relevant to your case, and why?

Construct an outline

Every essay should have a structure related to its title.

***KEY POINT* Most marks for essays are lost because the written material is badly organised or is irrelevant. An essay plan, by definition, creates order and, if thought about carefully, should ensure relevance.**

Your plan should be written down (but scored through later if written in an exam book). Think about an essay's content in three parts:

1. The introductory section, in which you should include definitions and some background information on the context of the topic being considered. You should also tell your reader how you plan to approach the subject.
2. The middle of the essay, where you develop your answer and provide relevant examples. Decide whether a broad analytical approach is appropriate or whether the essay should contain more factual detail.

***Ten Golden Rules for essay writing** – these are framed for in-course assessments (p. 29), though many are also relevant to exams (see also Box 6.2).*

1. *Read the question carefully, and decide exactly what the assessor wants you to achieve in your answer.*
2. *Make sure you understand the question by considering all aspects – discuss your approach with colleagues or a tutor.*
3. *Carry out the necessary research (using books, journals, the Web), taking appropriate notes. Gain an overview of the topic before getting involved with the details.*
4. *Always plan your work in outline before you start writing. Check that your plan covers the main points and that it flows logically.*
5. *Introduce your essay by showing that you understand the topic and stating how you intend to approach it.*
6. *As you write the main content, ensure it is relevant by continually looking back at the question.*
7. *Use headings and sub-headings to organise and structure your essay.*
8. *Support your statements with relevant examples, diagrams and references where appropriate.*
9. *Conclude by summarising the key points of the topic, indicating the present state of knowledge, what we still need to find out and how this might be achieved.*
10. *Always review your essay before submitting it. Check grammar and spelling and confirm that you have answered* all *aspects of the question.*

Table 17.1 Instructions often used in essay questions and their meanings. When more than one instruction is given (e.g. compare and contrast; describe and explain), make sure you carry out *both* or you may lose a large proportion of the available marks (see also Table 5.1)

Account for:	give the reasons for
Analyse:	examine in depth and describe the main characteristics of
Assess:	weigh up the elements of and arrive at a conclusion about
Comment:	give an opinion on and provide evidence for your views
Compare:	bring out the similarities between
Contrast:	bring out dissimilarities between
Criticise:	judge the worth of (give both positive and negative aspects)
Define:	explain the exact meaning of
Describe:	use words and diagrams to illustrate
Discuss:	provide evidence or opinions about, arriving at a balanced conclusion
Enumerate:	list in outline form
Evaluate:	weigh up or appraise; find a numerical value for
Explain:	make the meaning of something clear
Illustrate:	use diagrams or examples to make clear
Interpret:	express in simple terms, providing a judgement
Justify:	show that an idea or statement is correct
List:	provide an itemised series of statements about
Outline:	describe the essential parts only, stressing the classification
Prove:	establish the truth of
Relate:	show the connection between
Review:	examine critically, perhaps concentrating on the stages in the development of an idea or method
State:	express clearly
Summarise:	without illustrations, provide a brief account of
Trace:	describe a sequence of events from a defined point of origin

3. The conclusion, which you can make quite short. You should use this part to summarise and draw together the components of the essay, without merely repeating previous phrases. You might mention such things as: the broader significance of the topic; its future; its relevance to other important areas of biology. Always try to mention both sides of any debate you have touched on, but beware of 'sitting on the fence'.

***KEY POINT** Use paragraphs to make the essay's structure obvious. Emphasise them with headings and sub-headings unless the material beneath the headings would be too short or trivial.*

***Using diagrams** – give a title and legend to each diagram so that it makes sense in isolation and point out in the text when the reader should consult it (e.g. 'as shown in Fig. 1 . . .' or 'as can be seen in the accompanying diagram, . . .').*

Start writing

- Never lose track of the importance of content and its relevance. Repeatedly ask yourself: 'Am I really answering this question?' Never waffle just to increase the length of an essay. Quality, rather than quantity, is important.
- Illustrate your answer appropriately. Use examples to make your points clear, but remember that too many similar examples can stifle the flow of an essay. Use diagrams where a written description would be difficult or take too long. Use tables to condense information.

- Take care with your handwriting. You can't get marks if your writing is illegible. Try to cultivate an open form of handwriting, making the individual letters large and distinct. If there is time, make out a rough draft from which a tidy version can be copied.

Reviewing your answer

***Learning from lecturers' and tutors' comments** – ask for further explanations if you don't understand a comment or why an essay was less successful than you thought it should have been.*

Make sure that you leave enough time to:

- re-read the question to check that you have answered all points;
- re-read your essay to check for errors in punctuation, spelling and content. Make any corrections obvious. In an exam, don't panic if you suddenly realise you've missed a large chunk out as the reader can be redirected to a supplementary paragraph if necessary.

Sources for further study

Anon. *Yahoo! Directory: Writing > Essays and Research Papers*. Available:
http://dir.yahoo.com/Social_Science/Communications/Writing/Essays_and_Research_Papers
Last accessed: 01/04/07.
[An extensive directory of web resources]

Anon. (2004) *Essay and Report Writing Skills*. Open University, Milton Keynes.

Good, S. and Jensen, B. (1995) *The Student's Only Survival Guide to Essay Writing*. Orca Book Publishers, Victoria, BC.

Study exercises

17.1 Practise dissecting essay titles. Use past exam papers, or make up questions based on learning objectives for your course and your lecture notes. Take each essay title and carefully 'dissect' the wording, working out exactly what you think the assessor expects you to do (see e.g. Table 17.1).

17.2 Write essay plans under self-imposed time limits. Continuing from study exercise 17.1, outline plans for essays from a past exam paper. Allow yourself a maximum of 5 minutes per outline. Within this time your main goal is to create an essay plan. To do this, you may need to 'brainstorm' the topic. Alternatively, if you allocate 10 minutes per essay, you may be able to provide more details, e.g. list the examples you could describe.

17.3 Practise reviewing your work carefully. For the next assignment you write, review it fully as part of the writing process. This will require you to finish the first draft about one week before the hand-in date, e.g. by setting yourself an earlier deadline than the submission date. This exercise is best done with a word-processed essay. Don't worry if it is a little over the word limit at this stage.

(a) Print out a copy of the essay. Do not look at it for at least two days after finishing this version.

(b) Review 1: spelling, grammar and sense. Read through the draft critically (try to imagine it had been written by someone else) and correct any obvious errors that strike you. Does the text make sense? Do sentences/paragraphs flow smoothly?

(c) Review 2: structure and relevance. Consider again the structure of the essay, asking yourself whether you have really answered the question that was set (see study exercise 17.1). Are all the parts in the right order? Is anything missed out? Have you followed precisely the instruction(s) in the title? Are the different parts of the essay linked together well?

(d) Review 3: shorten and improve style. Check the word count. Shorten the essay if required. Look critically at phrasing and, even if the essay is within the word limit, ask yourself whether any of the words are unnecessary or whether the text could be made more concise, more precise or more apt.

18 Reporting practical and project work

***Typical structure of scientific reports** – this usually follows the 'IMRaD' acronym: Introduction, Materials and Methods, Results and Discussion.*

Practical reports, project reports, theses and scientific papers differ greatly in depth, scope and size, but they all have the same basic structure. Some variation is permitted, however (see Box 18.1), and you should always follow the advice or rules provided by your department.

Additional parts may be specified: for project reports, dissertations and theses, a Title page is often required and a List of Figures and Tables as part of the Contents section. When work is submitted for certain degrees, you may need to include certain declarations and statements made by the student and supervisor. In scientific papers, a list of Key Words is often added following the Abstract: this information may be combined with words in the title for computer cross-referencing systems.

KEY POINT** **Department, school or faculty regulations may specify a precise format for producing your report or thesis. Obtain a copy of these rules at an early stage and follow them closely, to avoid losing marks.

Practical and project reports

These are exercises designed to make you think more deeply about your experiments and to practise and test the skills necessary for writing up research work. Special features are:

- Introductory material is generally short and, unless otherwise specified, should outline the aims of the experiment(s) with a minimum of background material.
- Materials and methods may be provided by your supervisor for practical reports. If you make changes to this, you should state clearly what you did. With project work, your lab notebook (see p. 176) should provide the basis for writing this section.
- Great attention in assessment will be paid to presentation and analysis of data. Take special care over graphs. Make sure your conclusions are justified by the evidence you present.

***Options for discussing data** – the main optional variants of the general structure include combining* Results and Discussion *into a single section and adding a separate* Conclusions *section.*

- *The main advantage of a joint* Results and Discussion *section is that you can link together different experiments, perhaps explaining why a particular result led to a new hypothesis and the next experiment. However, a combined Results and Discussion section may contravene your department's regulations, so you should check before using this approach.*
- *The main advantage of having a separate* Conclusions *section is to draw together and emphasise the chief points arising from your work, when these may have been 'buried' in an extensive* Discussion *section.*

Theses and dissertations

These are submitted as part of the examination for a degree following an extended period of research. They act to place on record full details about your experimental work and will normally only be read by those with a direct interest in it – your examiners or colleagues. Note the following:

- You are allowed scope to expand on your findings and to include detail that might otherwise be omitted in a scientific paper.
- You may have problems with the volume of information that has to be organised. One method of coping with this is to divide your thesis into chapters, each having the standard format (as in Box 18.1). A General Introduction can be given at the start and a General Discussion at the end. Discuss this with your supervisor.

***Oral assessments** – there may be an oral exam (viva voce) associated with the submission of a thesis or dissertation. The primary aim of the examiners will be to ensure that you understand what you did and why you did it.*

Box 18.1 The structure of reports of experimental work

Undergraduate practical and project reports are generally modelled on this arrangement or a close variant of it, because this is the structure used for nearly all research papers and theses. The more common variations include Results and Discussion combined into a single section and Conclusions appearing separately as a series of points arising from the work. In scientific papers, a list of Key Words (for computer cross-referencing systems) may be included following the Abstract. Acknowledgements may appear after the Contents section, rather than near the end. Department or faculty regulations for producing theses and reports may specify a precise format; they often require a Title page to be inserted at the start and a List of Figures and Tables as part of the Contents section, and may specify declarations and statements to be made by the student and supervisor.

Part (in order)	Contents/purpose	Checklist for reviewing content
Title	Explains what the project was about	Does it explain what the text is about succinctly?
Authors plus their institutions	Explains who did the work and where; also where they can be contacted now	Are all the details correct?
Abstract/Summary	Synopsis of methods, results and conclusion of work described. Allows the reader to grasp quickly the essence of the work	Does it explain why the work was done? Does it outline the whole of your work and your findings?
List of Contents	Shows the organisation of the text (not required for short papers)	Are all the sections covered? Are the page numbers correct?
Abbreviations	Lists all the abbreviations used (but not those of SI, chemical elements or standard biochemical terms)	Have they all been explained? Are they all in the accepted form? Are they in alphabetical order?
Introduction	Orientates the reader, explains why the work has been done and its context in the literature, why the methods used were chosen, why the experimental organisms were chosen. Indicates the central hypothesis behind the experiments	Does it provide enough background information and cite all the relevant references? Is it of the correct depth for the readership? Have all the technical terms been defined? Have you explained why you investigated the problem? Have you outlined your aims and objectives? Have you explained your methodological approach? Have you stated your hypothesis?
Materials and Methods	Explains how the work was done. Should contain sufficient detail to allow another competent worker to repeat the work	Is each experiment covered and have you avoided unnecessary duplication? Is there sufficient detail to allow repetition of the work? Are proper scientific names and authorities given for all organisms? Have you explained where you got them from? Are the correct names, sources and grades given for all chemicals?
Results	Displays and describes the data obtained. Should be presented in a form which is easily assimilated (graphs rather than tables, small tables rather than large ones)	Is the sequence of experiments logical? Are the parts adequately linked? Are the data presented in the clearest possible way? Have SI units been used properly throughout? Has adequate statistical analysis been carried out? Is all the material relevant? Are the figures and tables all numbered in the order of their appearance? Are their titles appropriate? Do the figure and table legends provide all the information necessary to interpret the data without reference to the text? Have you presented the same data more than once?
Discussion/ Conclusions	Discusses the results: their meaning, their importance; compares the results with those of others; suggests what to do next	Have you explained the significance of the results? Have you compared your data with other published work? Are your conclusions justified by the data presented?
Acknowledgements	Gives credit to those who helped carry out the work	Have you listed everyone that helped, including any grant-awarding bodies?
Literature Cited (Bibliography)	Lists all references cited in appropriate format: provides enough information to allow the reader to find the reference in a library	Do all the references in the text appear on the list? Do all the listed references appear in the text? Do the years of publications and authors match? Are the journal details complete and in the correct format? Is the list in alphabetical order, or correct numerical order?

Steps in the production of a practical report or thesis

Choosing between graphs and tables *– graphs are generally easier for the reader to assimilate, while tables can be used to condense a lot of data into a small space.*

Repeating your experiments *– remember, if you do an experiment twice, you have repeated it only once.*

Presenting your results *– remember that the order of results presented in a report need not correspond with the order in which you carried out the experiments: you are expected to rearrange them to provide a logical sequence of findings.*

Using the correct tense *– always use the past tense to describe the methodology used in your work, since it is now complete. Use the present tense only for generalisations and conclusions.*

Choose the experiments you wish to describe and decide how best to present them

Try to start this process before your lab work ends, because at the stage of reviewing your experiments, a gap may become apparent (e.g. a missing control) and you might still have time to rectify the deficiency. Irrelevant material should be ruthlessly eliminated, at the same time bearing in mind that negative results can be extremely important (see p. 188). Use as many different forms of data presentation as are appropriate, but avoid presenting the same data in more than one form. Relegate large tables of primary data to an appendix and summarise the important points within the main text (with a cross-reference to the appendix). Make sure that the experiments you describe are representative: always state the number of times they were repeated and how consistent your findings were.

Make up plans or outlines for the component parts

The overall structure of practical and project reports is well defined (see Box 18.1), but individual parts will need to be organised as with any other form of writing (see Chapter 16).

Write

The Materials and Methods section is often the easiest to write once you have decided what to report. Remember to use the past tense and do not allow results or discussion to creep in. The Results section is the next easiest as it should only involve description. At this stage, you may benefit from jotting down ideas for the Discussion – this may be the hardest part to compose as you need an overview both of your own work and of the relevant literature. It is also liable to become wordy, so try hard to make it succinct. The Introduction shouldn't be too difficult if you have fully understood the aims of the experiments. Write the Abstract and complete the list of references at the end. To assist with the latter, it is a good idea as you write to jot down the references you use or to pull out their cards from your index system.

Revise the text

Once your first draft is complete, try to answer all the questions given in Box 18.1. Show your work to your supervisors and learn from their comments. Let a friend or colleague who is unfamiliar with your subject read your text; they may be able to pinpoint obscure wording and show where information or explanation is missing. If writing a thesis, double-check that you are adhering to your institution's thesis regulations.

Prepare the final version

Markers appreciate neatly produced work but a well-presented document will not disguise poor science! If using a word processor, print the final version with the best printer available. Make sure figures are clear and in the correct size and format.

Submit your work

Your department will specify when to submit a thesis or project report, so plan your work carefully to meet this deadline or you may lose marks. Tell your supervisor early of any circumstances that may cause delay. And check to see whether any forms are required for late submission, or evidence of extenuating circumstances.

Producing a scientific paper

> **Definition**
>
> **Peer review** – the process of evaluation and review of a colleague's work. In scientific communication, a paper is reviewed by two or more expert reviewers for comments on quality and significance as a key component of the validation procedure.

Scientific papers are the means by which research findings are communicated to others. Peer-reviewed papers are published in journals; each covers a well-defined subject area and publishes details of the format they expect.

***KEY POINT** Peer review is an important component of the process of scientific publication; only those papers whose worth is confirmed by the peer-review process will be published.*

It would be very unusual for an undergraduate to submit a paper on his or her own – this would normally be done in collaboration with your project supervisor, and only then if your research has satisfied appropriate criteria. However, it is important to understand the process whereby a paper comes into being (Box 18.2), as this can help you understand and interpret the primary literature.

Box 18.2 Steps in producing a scientific paper

Scientific papers are the lifeblood of any science and it is a major landmark in your scientific career to publish your first paper. The main steps in doing this should include the following:

Assessing potential content

The work must be of an appropriate standard to be published and should be 'new, true and meaningful'. Therefore, before starting, the authors need to review their work critically under these headings. The material included in a scientific paper will generally be a subset of the total work done during a project, so it must be carefully selected for relevance to a clear central hypothesis – if the authors won't prune, the referees and editors of the journal certainly will.

Choosing a journal

There are thousands of journals covering biology and each covers a specific area (which may change through time). The main factors in deciding on an appropriate journal are the range of subjects it covers, the quality of its content and the number and geographical distribution of its readers. The choice of journal always dictates the format of a paper since authors must follow to the letter the journal's 'Instructions to Authors'.

Deciding on authorship

In multi-author papers, a contentious issue is often who should appear as an author and in what order they should be cited. Where authors make an equal contribution, an alphabetical order of names may be used. Otherwise, each author should have made a substantial contribution to the paper and should be prepared to defend it in public. Ideally, the order of appearance will reflect the amount of work done rather than seniority. This may not always happen in practice!

Writing

The paper's format will be similar to that shown in Box 18.1 and the process of writing will include outlining, reviewing, etc., as discussed elsewhere in this chapter. Figures must be finished to an appropriate standard and this may involve preparing photographs or digital images of them.

Submitting

When completed, copies of the paper are submitted to the editor of the chosen journal with a simple covering letter. A delay of one to two months usually follows while the manuscript is sent to two or more anonymous referees who will be asked by the editor to check that the paper is novel, scientifically correct and that its length is justified.

Responding to referees' comments

The editor will send on the referees' comments to the authors, who will then have a chance to respond. The editor will decide on the basis of the comments and replies to them whether the paper should be published. Sometimes quite heated correspondence can result if the authors and referees disagree.

Checking proofs and waiting for publication

If a paper is accepted, it will be sent off to the typesetters. The next the authors see of it is the proofs (first printed version in style of journal), which have to be corrected carefully for errors and returned. Eventually, the paper will appear in print, but a delay of six months following acceptance is not unusual. Nowadays, papers are often available electronically, *via* the Web, in PDF format - see p. 49 for advice on how to cite 'online early' papers using the DOI system.

Sources for further study

Berry, R. (2004) *The Research Project: How to Write it*, 5th edn. Routledge, London.

Davis, M. (2005) *Scientific Papers and Presentations*. Academic Press, London.

Day, R.A. and Gastel, B. (2006) *How to Write and Publish a Scientific Paper*, 6th edn. Cambridge University Press, Cambridge.

Lobban, C.S. and Schefter, M. (1992) *Successful Lab Reports: A Manual for Science Students*. Cambridge University Press, Cambridge.

Luck, M. (1999) *Your Student Research Project*. Gower, London.

Luey, B. (2002) *Handbook for Academic Authors*, 4th edn. Cambridge University Press, Cambridge.

Matthews, J.R., Bowen, J.M. and Matthews, R. (2000) *Successful Science Writing: a Step-by-Step Guide for the Biological and Medical Sciences*, 2nd edn. Cambridge University Press, Cambridge

Valiela, I. (2001) *Doing Science: Design, Analysis and Communication of Scientific Research*. Oxford University Press, Oxford.
[Covers scientific communication, graphical presentations and aspects of statistics]

Study exercises

18.1 Write a formal 'Materials and Methods' section. Adopting the style of a research paper (i.e. past tense, all relevant detail reported such that a competent colleague could repeat your work), write out the Materials and Methods for a practical you have recently carried out. Ask a colleague or tutor to comment on what you have written.

18.2 Describe a set of results in words. Again adopting the style of a research paper, write a paragraph describing the results contained in a particular table or graph. Ask a colleague or tutor to comment on your description, to identify what is missing or unclear.

18.3 Write an abstract for a paper. Pair up with a colleague. Each of you should independently choose a different research paper in a current journal. Copy the paper, but mask over the abstract section, having first counted the words used. Swap papers. Now, working to the same number of words as in the original, read the paper and provide an abstract of its contents. Then compare this with the real abstract. Compare your abstracts.

19 Writing literature surveys and reviews

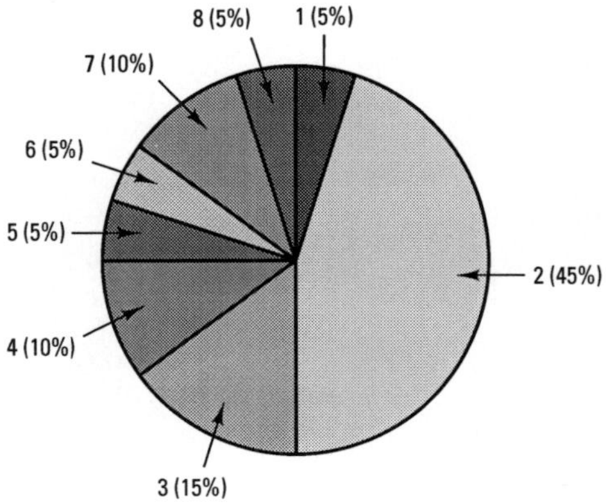

Fig. 19.1 Pie chart showing how you might allocate time for a literature survey:
1. select a topic;
2. scan the literature;
3. plan the review;
4. write first draft;
5. leave to settle;
6. prepare a structured review of text;
7. write final draft;
8. produce top copy.

The literature survey or review is a specialised form of essay which summarises and reviews the evidence and concepts concerning a particular area of research.

***KEY POINT** A literature review should **not** be a simple recitation of facts. The best reviews are those which analyse information rather than simply describe it.*

Making up a timetable

Figure 19.1 illustrates how you might divide up your time for writing a literature survey. There are many subdivisions in this chart because of the size of the task: in general, for lengthy tasks, it is best to divide up the work into manageable chunks. Note also that proportionately less time is allocated to writing itself than with an essay. In a literature survey, make sure that you spend adequate time on research and revision.

Selecting a topic

You may have no choice in the topic to be covered, but if you do, carry out your selection as a three-stage process:

1. Identify a broad subject area that interests you.
2. Find and read relevant literature in that area. Try to gain a broad impression of the field from books and general review articles. Discuss your ideas with your supervisor.
3. Select a relevant and concise title. The wording should be considered very carefully as it will define the content expected by the reader. A narrow subject area will cut down on the amount of literature you will be expected to review, but will also restrict the scope of the conclusions you can make (and *vice versa* for a wide subject area).

__Creating a glossary__ – one barrier to developing an understanding of a new topic is the jargon used. To overcome this, create your own glossary. You may wish to cross-reference a range of sources to ensure the definitions are reliable and context-specific. Remember to note your sources in case you wish to use the definition within your review.

Scanning the literature and organising your references

You will need to carry out a thorough investigation of the literature before you start to write. The key problems are as follows:

- Getting an initial toehold in the literature. Seek help from your supervisor, who may be willing to supply a few key papers to get you started. Hints on expanding your collection of references are given on p. 45.
- Assessing the relevance and value of each article. This is the essence of writing a review, but it is difficult unless you already have a good understanding of the field. Try reading earlier reviews in your area and discussing the topic with your supervisor or other academic staff.
- Clarifying your thoughts. Subdividing the main topic and assigning your references to these smaller subject areas may help you gain a better overview of the literature.

__Using index cards__ – these can help you organise large numbers of references. Write key points and author information on each card – this helps when considering where the reference fits into the literature. Arrange the cards in subject piles, eliminating irrelevant ones. Order the cards in the sequence you wish to write in.

Deciding on structure and content

The general structure and content of a literature survey are described below.

The *Annual Review* series (available in most university libraries) provides good examples of appropriate style for reviews of the biosciences.

Defining terms *– the introduction is a good place to explain the meaning of the key terms used in your survey or review.*

Introduction

The introduction should give the general background to the research area, concentrating on its development and importance. You should also make a statement about the scope of your survey; as well as defining the subject matter to be discussed, you may wish to restrict the period being considered.

Main body of text

The review itself should discuss the published work in the selected field and may be subdivided into appropriate sections. Within each portion of a review, the approach is usually chronological, with appropriate linking phrases (e.g. 'Following on from this, ...'; 'Meanwhile, Bloggs (2002) tackled the problem from a different angle ...'). However, a good review is much more than a chronological list of work done. It should:

Balancing opposing views *– even if you favour one side of a disagreement in the literature, your review should provide a balanced and fair description of all the published views of the topic. Having done this, if you do wish to state a preference, give reasons for your opinion.*

- allow the reader to obtain an overall view of the current state of the research area, identifying the key areas where knowledge is advancing;
- show how techniques are developing and discuss the benefits and disadvantages of using particular organisms or experimental systems;
- assess the relative worth of different types of evidence – this is the most important aspect (see Chapter 9). Do not be intimidated from taking a critical approach as the conclusions you may read in the primary literature aren't always correct;
- indicate where there is conflict in findings or theories, suggesting if possible which side has the stronger case;
- indicate gaps in current knowledge.

You do not need to wait until you have read all the sources available to you before starting to write the main body. Word processors allow you to modify and move pieces of text at any point and it will be useful to write paragraphs about key sources, or groups of related papers, as you read them. Try to create a general plan for your review as soon as possible. Place your draft sections of text under an appropriate set of subheadings that reflects your plan, but be prepared to rearrange these and re-title or re-order sections as you proceed. Not only will working in this way help to clarify your thoughts, but it may help you avoid a last-minute rush of writing near to the submission date.

Making citations *– a review of literature poses stylistic problems because of the need to cite large numbers of papers; in the* Annual Review *series this is overcome by using numbered references (see p. 48).*

Conclusions

The conclusions should draw together the threads of the preceding parts and point the way forward, perhaps listing areas of ignorance or where the application of new techniques may lead to advances.

References, etc.

The References or Literature Cited section should provide full details of all papers referred to in the text (see p. 48). The regulations for your department may also specify a format and position for the title page, list of contents, acknowledgements, etc.

Source for further study

Rudner, L.M. and Schafer, W.D. (1999) How to write a scholarly research report. *Practical Assessment, Research & Evaluation*, **6** (13).
Available: http://pareonline.net/getvn.asp?v = 6&n = 13
Last accessed: 01/04/07.

Study exercises

19.1 Summarise the main differences between a review and a scientific paper. From the many subject areas in the *Annual Review* series (find *via* your library's periodical indexing system), pick one that matches your subject interests, and within this find a review that seems relevant or interesting. Read the review and write down *five* ways in which the writing style and content differ from those seen in primary scientific papers.

19.2 Gather a collection of primary sources for a topic. From the journal section of the library, select an interesting scientific paper published about 5–10 years ago. First, work *back* from the references cited by that paper: can you identify from the text or the article titles which are the most important and relevant to the topic? List *five* of these, using the proper conventions for citing articles in a reference list (see Chapter 8). Note that each of these papers will also cite other articles, always going back in time. Now using the Science Citation Index or a similar system (e.g. the Web of Science website at http://wos.mimas.ac.uk or Google Scholar at http://scholar.google.com/), work *forward* and find out who has cited your selected article in the time since its publication. Again list the *five* most important articles found.

19.3 Write a synopsis of a review. Again using one of the *Annual Review* series as a source, allow yourself just *five* single-sentence bullet points to summarise the key points reported in a particular review.

Fundamental laboratory techniques

20 Your approach to practical work

Developing practical skills – *these will include:*

- *designing experiments*
- *observing and measuring*
- *recording data*
- *analysing and interpreting data*
- *reporting/presenting.*

All knowledge and theory in science have originated from practical observation and experimentation: this is equally true for disciplines as diverse as microscopy and molecular genetics. Practical work is an important part of most courses and often accounts for a significant proportion of the assessment marks. The abilities developed in practical classes will continue to be useful throughout your course and beyond, some within science and others in any career you choose (see Chapter 1).

Being prepared

KEY POINT ***You will get the most out of practicals if you prepare well in advance. Do not go into a practical session assuming that everything will be provided, without any input on your part.***

Using textbooks in the lab – *take this book along to the relevant classes, so that you can make full use of the information during the practical sessions.*

The main points to remember are:

- Read any handouts in advance: make sure you understand the purpose of the practical and the particular skills involved. Does the practical relate to, or expand upon, a current topic in your lectures? Is there any additional preparatory reading that will help?
- Take along appropriate textbooks, to explain aspects in the practical.
- Consider what safety hazards might be involved, and any precautions you might need to take, before you begin (p. 126).
- Listen carefully to any introductory guidance and note any important points: adjust your schedule/handout as necessary.
- During the practical session, organise your bench space – make sure your lab book is adjacent to, but not within, your working area. You will often find it easiest to keep clean items of glassware, etc. on one side of your working space, with used equipment on the other side.
- Write up your work as soon as possible, and submit it on time, or you may lose marks.
- Catch up on any work you have missed as soon as possible – preferably, before the next practical session.

SAFETY NOTE **Mobile phones** – *these should never be used in a lab class, as there is a risk of contamination from hazardous substances. Always switch off your mobile phone before entering a laboratory. Conversely, they are an extremely useful accessory for fieldwork.*

Ethical and legal aspects

You will need to consider the ethical and legal implications of biological work at several points during your studies:

- Safe working means following a code of safe practice, supported by legislation, alongside a moral obligation to avoid harm to yourself and others, as discussed in Chapter 21.
- Any laboratory work that involves working with animals must be carefully considered.
- Fieldwork can have a legal aspect, e.g. in relation to specimen collection, since legislation may protect particular habitats and/or species. Ethical and human aspects that must all be considered before any work is carried out include: possible damage to fauna and flora; safety of others within your group; implications of your work to others, e.g. landowners, or other users.

Getting to grips with bioethics – *in addition to any moral implications of your lab practicals and field work, you may have the opportunity to address broader issues within your course (see Box 19.1). Professional scientists should always consider the consequences of their work, and it is therefore important that you develop your appreciation of these issues alongside your academic studies.*

Box 20.1 Bioethics

Contemporary bioscience degree programmes place increasing emphasis on the ethical and social impacts of scientific advances, and on the need for scientists engaged in potentially controversial work to communicate their ideas to the general public. Bioscience research raises many moral and legal dilemmas, requiring difficult choices to be made (e.g. animal testing of medical products), and students are likely to be asked to reflect on bioethical topics, e.g. in group discussions and debates on current issues such as:

- environmental ethics (e.g. use of genetically modified plants, hunting of endangered species);
- animal ethics (e.g. factory farming, transgenic animals, xenotransplantation);
- medical and social ethics (e.g. human cloning; embryo research, genetic testing).

In discussing such topics (sometimes referred to as **e**thical, **l**egal and **s**ocial **i**ssues, ELSI), you will find that there is rarely a 'right' or 'wrong' answer, and it is important to be able to consider these issues in a logical manner, and to provide a reasoned argument in support of a particular viewpoint. Gaining experience in such debates should also help you understand some of the issues linked to the public understanding of science, and how these can be addressed (see: http://www.copus.org.uk/pubs_guides.html). While a full exposition is beyond the scope of this book, the following provides a framework of principles for considering particular topics:

- Beneficence – the obligation to do good (for example, if it is possible to prevent suffering by a particular course of action, then it should be carried out).
- Non-maleficence – the duty to cause no harm (contained within the Hippocratic Oath of medical practitioners).
- Justice – the obligation to treat all people fairly and impartially (for example, lack of discrimination between people on the grounds of race or sex).
- Autonomy – the duty to allow an individual to make their own choices, without constraints (this principle underlies the notion of informed consent in medical research).
- Respect – the need to show due regard for others (for example, by taking into account the rights and beliefs of all people equally).
- Rationality – the notion that a particular action or choice should be based on reason and logic (many scientists would argue that the scientific method is an example of rationality).
- Precautionary principle – the notion that it is better not to carry out an action if there is any risk of harm (for example, in deciding that the risks of building nuclear power stations outweigh their potential benefits).

Understanding the various theories of ethics may also help you formulate your ideas. These include:

- Utilitarianism – the notion that it is ethical to choose the action that produces the greatest good for the greatest number.
- Deontology – a theory that states than a particular action is either intrinsically good (right) or bad (wrong). According to deontological theory, decisions should be based on the actions themselves, rather than on their consequences.
- Virtue theory – the notion that making decisions according to established virtues (e.g. honesty, wisdom, justice) will lead to ethically valid choices.
- Objectivism – the theory that what is right and wrong is intrinsic and applies equally to all people, places and times (the alternative is that morality is subjective, being dependent on the views of each individual).

The principles and issues of bioethics are considered in a number of websites, including:

- introductory bioethics (e.g. http://www.accessexcellence.org/RC/AB/IE/);
- bioethics resources (e.g. http://bioethics.od.nih.gov/, http://www.ethicsweb.ca/resources/ and http://www.beep.ac.uk/content/index.php/);
- ethics discussion groups (e.g. http://www.beep.ac.uk/discuss/index.php?c=6 and http://www-hsc.usc.edu/~mbernste/).

The following introductory texts give further information and guidance:
Bryant *et al.* (2005),
Gert *et al.* (2006) and
Mepham (2005).

Basic requirements

Recording practical results

Presenting results – *while you don't need to be a graphic designer to produce work of a satisfactory standard, presentation and layout are important and you will lose marks for poorly presented work. Chapter 63 gives further practical advice.*

An A4 loose-leaf ring binder offers flexibility, since you can insert laboratory handouts or lined and graph paper at appropriate points. The danger of losing one or more pages from a loose-leaf system is the main drawback. Bound books avoid this problem, although those containing alternating lined/graph or lined/blank pages tend to be wasteful – it is often better to paste sheets of graph paper into a bound book, as required.

A good quality HB pencil or propelling pencil is recommended for recording your raw data, making diagrams, etc. as mistakes are easily corrected. Buy a black, spirit-based (permanent) marker for labelling experimental glassware, Petri plates, etc. Fibre-tipped fine line drawing/lettering pens are useful for preparing final versions of hand-drawn graphs and diagrams for assessment purposes. Use a see-through ruler (with an undamaged edge) for graph drawing, so that you can see data points and information below the ruler as you draw.

Calculators

Using calculators for numerical problems – Chapter 65 gives further advice.

These range from basic machines with no pre-programmed functions and only one memory, to sophisticated programmable portable computers with many memories. The following may be helpful when using a calculator:

- Power sources. Choose a battery-powered machine, rather than a mains-operated or solar-powered type. You will need one with basic mathematical/scientific operations, including powers, logarithms, roots and parentheses (brackets), together with statistical functions such as sample means and standard deviations (Chapter 66).
- Mode of operation. The older operating system used by e.g. Hewlett-Packard calculators is known as the reverse Polish notation: to calculate the sum of two numbers, the sequence is 2 [enter] 4 + and the answer 6 is displayed. The more usual method of calculating this equation is as $2 + 4 =$, which is the system used by the majority of modern calculators. Most newcomers find the latter approach to be more straightforward. Spend some time finding out how a calculator operates, e.g. does it have true algebraic logic ($\sqrt{}$ then number, rather than number then $\sqrt{}$)? How does it deal with (and display) scientific notation and logarithms?
- Display. Some calculators will display an entire mathematical operation (e.g. '$2 + 4 = 6$'), while others simply display the last number/operation. The former type may offer advantages in tracing errors.
- Complexity. In the early stages, it is usually better to avoid the more complex machines, full of impressive-looking, but often unused, pre-programmed functions – go for more memory, parentheses or statistical functions rather than engineering or mathematical constants. Programmable calculators may be worth considering for more advanced studies. However, it is important to note that such calculators are often unacceptable for exams.

Using inexpensive calculators – *many unsophisticated calculators have a restricted display for exponential numbers and do not show the 'power of 10', e.g. displaying 2.4×10^{-5} as 2.4^{-05}, or 2.4E–05, or even 2.4–05.*

Presenting more advanced practical work

Presenting graphs and diagrams – *ensure these are large enough to be easily read: a common error is to present graphs or diagrams that are too small, with poorly chosen scales.*

In some practical reports and in project work, you may need to use more sophisticated presentation equipment. Computer-based graphics packages can be useful – choose easily read fonts such as Arial or Helvetica for posters and consider the layout and content carefully. Alternatively, you

Printing on acetates *– standard overhead transparencies are not suitable for use in laser printers or photocopiers: you need to make sure that you use the correct type.*

could use fine line drawing pens and dry-transfer lettering/symbols, although this can be more time-consuming than computer-based systems, e.g. using Microsoft Excel.

To prepare overhead transparencies for spoken presentations, you can use spirit-based markers and acetate sheets. An alternative approach is to photocopy onto special acetates, or print directly from a computer-based package, using a laser printer and special acetates, or use a PC-based presentation package (e.g. PowerPoint) and a suitable projector. Further advice on content and presentation is given in Chapter 15.

Text references

Bryant, J., Baggott le Velle, L. and Searle, J. (2005) *Introduction to Bioethics*. Wiley, Chichester.

Gert, B., Culver, C.M. and Clouser, K.D. (2006) *Bioethics: a Systematic Approach,* 2nd edn. Oxford University Press, Oxford.

Mepham, B. (2005) *Bioethics: an Introduction for the Biosciences*. Oxford University Press, Oxford.

Sources for further study

Barnard, C.J., Gilbert, F.S. and MacGregor, P.K. (2001) *Asking Questions in Biology: Key Skills for Practical Assessments and Project Work,* 2nd edn. Prentice Hall, Harlow.

Howell, J.H., Sale, W.F. and Callahan, D. (2000) *Life Choices: A Hastings Center Introduction to Bioethics,* 2nd edn. Georgetown University Press, Washington, DC.

Mier-Jedrzejowicz, W.A.C. (1999) *A Guide to HP Handheld Calculators and Computers.* Wilson-Barnett, Tustin. [Provides further guidance on the use of Hewlett-Packard calculators (reverse Polish notation)]

Singer, P. and Kuhse, H. (2006) *Bioethics: an Anthology*. Blackwell, Oxford.

Study exercises

20.1 Consider the value of practical work. Spend a few minutes thinking about the purpose of practical work within a specific part of your course (e.g. a particular first year module) and then write a list of the six most important points. Compare your list with the generic list that we have provided, which is based on our experience as lecturers – does it differ much from your list, which is drawn up from a student perspective?

20.2 Make a list of items required for a particular practical exercise. This exercise is likely to be most useful if you can relate it to an appropriate practical session on your course. However, we have given a model list for a tissue dissection as an example.

20.3 Check your calculator skills. Carry out the following mathematical operations, using either a hand-held calculator or a PC with appropriate 'calculator' software.

(a) $5 \times (2 + 6)$

(b) $(8.3 \div [6.4 - 1.9]) \times 24$ (to four significant figures)

(c) $(1 \div 32) \times (5 \div 8)$ (to three significant figures)

(d) $1.2 \times 10^5 + 4.0 \times 10^4$ in scientific notation

(e) $3.4 \times 10^{-2} - 2.7 \times 10^{-3}$ in 'normal' notation (i.e. conventional notation, not scientific format) and to three decimal places.

(See also the numerical exercises in Chapter 65.)

21 Health and safety

Health and safety legislation – in the UK, the **Health & Safety at Work, etc. Act 1974** *provides the main legal framework for health and safety. The* **Control of Substances Hazardous to Health (COSHH) Regulations 2002** *impose specific legal requirements for risk assessment wherever hazardous chemicals or biological agents are used, with Approved Codes of Practice for the control of hazardous substances, carcinogens and biological agents, including pathogenic microbes.*

Health and safety law requires academic institutions to provide a working environment that is safe and without risk to health. Where appropriate, training and information on safe working practices must be provided. Students and staff must take reasonable care to ensure the health and safety of themselves and of others, and must not misuse any safety equipment.

***KEY POINT** All practical work must be carried out with safety in mind, to minimise the risk of harm to yourself and to others – safety is everyone's responsibility.*

Risk assessment

The most widespread approach to safe working practice involves the use of risk assessment, which aims to establish:

Definitions

Hazard – the ability of a substance or biological agent to cause harm.

Risk – the likelihood that a substance or biological agent might be harmful under specific circumstances.

1. The intrinsic chemical, biological and physical hazards, together with any maximum exposure limits (MELs) or occupational exposure standards (OESs), where appropriate. Chemical manufacturers provide data sheets listing the hazards associated with particular chemical compounds, while pathogenic (disease-causing) microbes are categorised according to their ability to cause illness (p. 205).
2. The risks involved, by taking into account the amount of substance to be used, the way in which it will be used and the possible routes of entry into the body (Fig. 21.1).

***Distinguishing between hazard and risk** – one of the hazards associated with water is drowning. However, the risk of drowning in a few drops of water is negligible.*

***KEY POINT** It is important to distinguish between the intrinsic hazards of a particular substance and the risks involved in its use in a particular exercise.*

3. The persons at risk, and the ways in which they might be exposed to hazardous substances, including accidental exposure (spillage).
4. The steps required to prevent or control exposure. Ideally, a non-hazardous or less hazardous alternative should be used. If this is not feasible, adequate control measures must be used, e.g. a fume cupboard or other containment system. Personal protective equipment (e.g. lab coats, safety glasses) must be used in addition to such containment measures. A safe means of disposal will be required.

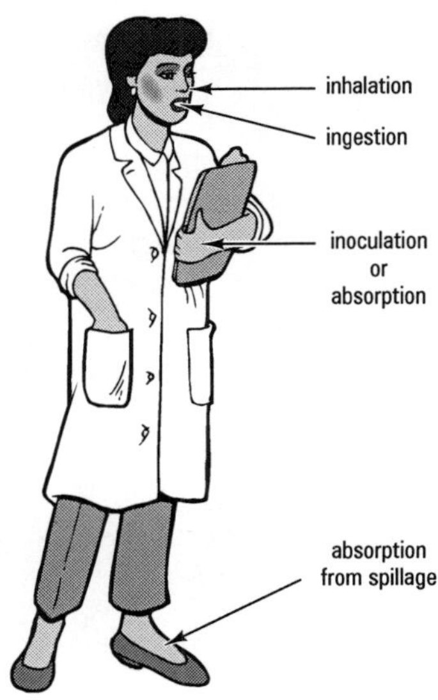

Fig. 21.1 Major routes of entry of harmful substances into the body.

The outcome of the risk assessment process must be recorded and appropriate safety information must be passed on to those at risk. For most practical classes, risk assessments will have been carried out in advance by the person in charge: the information necessary to minimise the risks to students may be given in the practical schedule. Make sure you know how your department provides such information and that you have read the appropriate material before you begin your practical work. You should also pay close attention to the person in charge at the beginning of the practical session, as they may emphasise the major hazards and risks. In project work, you will need to be involved in the risk assessment process along with your supervisor, before you carry out any practical work.

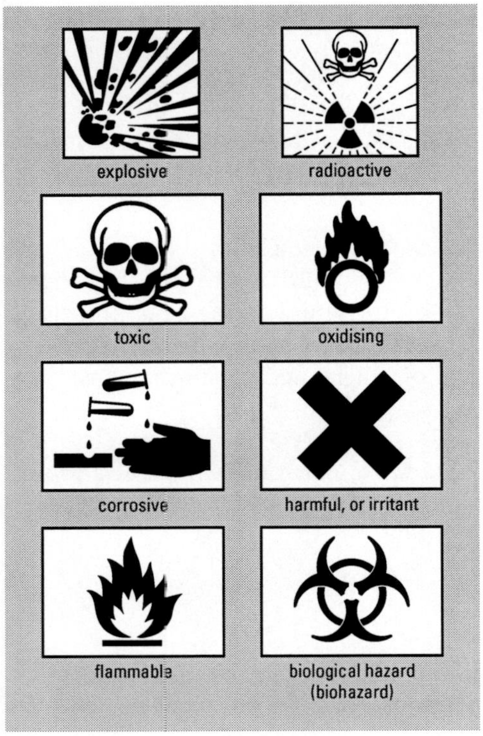

Fig. 21.2 Warning labels for specific chemical hazards. These appear on suppliers' containers and on tape used to label working vessels.

In addition to specific risk assessments, most institutions will have a safety handbook, giving general details of safe working practices, together with the names and telephone numbers of safety personnel, first aiders, hospitals, etc. Make sure you read this and follow any instructions given.

Basic rules for laboratory work

- Make sure you know what to do in case of fire, including exit routes, how to raise the alarm, and where to gather on leaving the building. Remember that the most important consideration at all times is human safety: do not attempt to fight a fire unless it is safe to do so.
- All laboratories display notices telling you where to find the first aid kit and who to contact in case of accident or emergency. Report all accidents, even those appearing insignificant – your department will have a formal recording procedure to comply with safety legislation.
- Wear appropriate protective clothing at all times – a clean lab coat (buttoned up), plus safety glasses if there is any risk to the eyes.
- Never smoke, eat or drink in any laboratory, because of the risks of contamination by inhalation or ingestion (Fig. 21.1).
- Never mouth pipette any liquid. Use a pipette filler (see p. 128) or, if appropriate, a pipettor (pp. 129–30).
- Take care when handling glassware – see p. 132 for details.
- Know the warning symbols for specific chemical hazards (see Fig. 21.2).
- Use a fume cupboard for hazardous chemicals. Make sure that it is working and then open the front only as far as is necessary: many fume cupboards are marked with a maximum opening.
- Always use the minimum quantity of any hazardous materials.
- Work in a logical, tidy manner and minimise risks by thinking ahead.
- Always clear up at the end of each session. This is an important aspect of safety, encouraging a responsible attitude towards laboratory work.
- Dispose of waste in appropriate containers. Most labs will have bins for sharps, glassware, hazardous solutions and radioactive waste.

Genetic engineering and molecular genetics

In the UK, the Genetically Modified Organisms (Contained Use) Regulations 2000 *define the legal requirements for risk assessment and for the notification of work involving genetic manipulation.*

The Advisory Committee on Genetic Manipulation provides guidance on the use of genetically manipulated organisms.

The Health and Safety Executive (HSE) has specific responsibility for the operation of these regulations and is the regulatory authority for genetic manipulation in the UK.

Additional legal constraints apply to practical work involving genetic manipulation. A specific risk assessment must be carried out for any experiment where a cell or organism is modified by genetic engineering techniques (Chapter 62) involving the insertion of DNA into a cell or organism in which it does not normally occur. Before any practical work can be carried out, it must be authorised by the establishment's genetic manipulation safety committee and notified to the relevant authority. Such work must be carried out with appropriate containment, to prevent the accidental release of genetically modified organisms into the environment.

Practicals in molecular genetics will involve some of the techniques of genetic manipulation. Typically, these will be examples of 'self-cloning' where recombinant DNA molecules are constructed from fragments of DNA which naturally occur in that organism, for example, the transformation of laboratory strains of *Escherichia coli* using a pUC plasmid.

Sources for further study

American Chemical Society (2003) *Safety in Academic Chemistry Laboratories*, 7th edn. Available: http://membership.acs.org/c/ccs/pub_3.htm SACL_Students.pdf
Last accessed: 01/04/07.

Anon. *BUBL Link – Biochemical Safety*. Available: http://bubl.ac.uk/link/b/biochemicalsafety.htm
Last accessed: 01/04/07.
[Provides a single-step link to a number of databases giving information on chemical and microbiological hazards]

Anon. *Howard Hughes Medical Institute On-line Safety Course*. Available: http://www.hhmi.org/about/labsafe/safescience.html
Last accessed: 01/04/07.
[This site, designed for laboratory workers, enables you to test your knowledge and understanding of laboratory safety]

Furr, A.K. (2000) *CRC Handbook of Laboratory Safety*. CRC Press, Boca Raton, FL.

Health and Safety Executive (2005) *Control of Substances Hazardous to Health Regulations*. HSE, London.

Study exercises

21.1 Test your knowledge of safe working procedures. After reading the appropriate sections of this book, can you remember the following:

(a) The four main steps involved in the process of risk assessment?

(b) The major routes of entry of harmful substances into the body?

(c) The warning labels for the major chemical hazard symbols (either describe them or draw them from memory)?

(d) The international symbol for a biohazard?

(e) The international symbol for radioactivity?

21.2 Locate relevant health and safety features in a laboratory. Find each of the following in one of the laboratories used as part of your course (draw a simple location map, if this seems appropriate):

(a) fire exit(s);

(b) fire-fighting equipment;

(c) first aid kit;

(d) 'sharps' container;

(e) container for disposal of broken glassware;

(f) eye wash station (where appropriate).

21.3 Investigate the health and safety procedures in operation at your university. Can you find out the following?

(a) your university's procedure in case of fire;
(b) the colour coding for fire extinguishers available in your department and their recommendations for use;
(c) the accident reporting procedure used in your department;
(d) your department's Code of Safe Practice relating to a specific aspect of bioscience, e.g. working with micro-organisms.

21.4 Carry out risk assessments for specific chemical hazards. Look up the hazards associated with the use of the following chemicals and list the appropriate protective measures required to minimise risk during use in a lab class:

(a) formaldehyde solution, used as a preservative for animal tissue, to be used for microscopic examination (Chapter 27);
(b) acetone, e.g. for use as a solvent for the quantitative analysis of plant pigments (Chapter 41);
(c) sodium hydroxide, used in solid form to prepare a dilute solution to be used for pH adjustment (Chapter 25).

22 Working with liquids

> ***SAFETY NOTE*** ***Using hazardous liquids and solutions*** *(flammable, corrosive, toxic, etc.) – make sure the liquid is properly contained and work out how to deal with any spillages before you begin.*

Measuring and dispensing liquids

The equipment you should choose to measure out liquids depends on the volumes being dispensed, the accuracy required and the number of times the job must be done (Table 22.1).

Table 22.1 Criteria for choosing a method for measuring out a liquid

Method	Best volume range	Accuracy	Usefulness for repetitive measurement
Pasteur pipette	30 μl to 2 ml	Low	Convenient
Conical flask/beaker	25–5000 ml	Very low	Convenient
Measuring cylinder	5–2000 ml	Medium	Convenient
Volumetric flask	5–2000 ml	High	Convenient
Burette	1–100 ml	High	Convenient
Glass pipette	1–100 ml	High	Convenient
Mechanical pipettor	5–1000 μl	High*	Convenient
Syringe	0.5–20μl	Medium**	Convenient
Microsyringe	0.5–50 μl	High	Convenient
Weighing	Any (depends on accuracy of balance)	Very high	Inconvenient

* If correctly calibrated and used properly (see p. 130).
** Accuracy depends on width of barrel: large volumes are less accurate.

> ***Reading any volumetric scale*** *– make sure your eye is level with the bottom of the liquid's meniscus and take the reading from this point.*

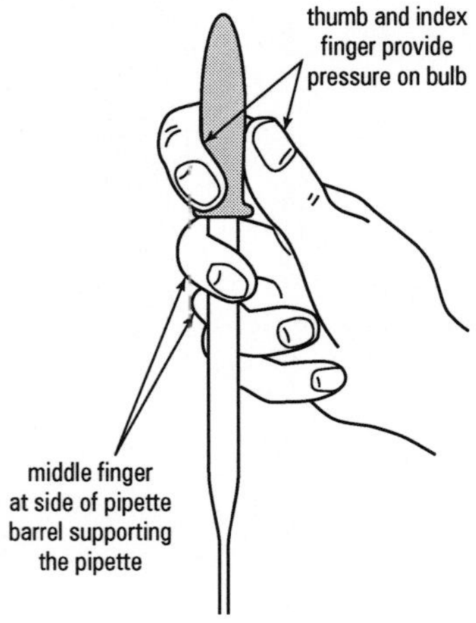

Fig. 22.1 How to hold a Pasteur pipette.

Certain liquids may cause problems:

- High viscosity liquids are difficult to dispense: allow time for all the liquid to transfer.
- Organic solvents may evaporate rapidly, making measurements inaccurate: work quickly; seal containers without delay.
- Solutions prone to frothing (e.g. protein and detergent solutions) are difficult to measure and dispense: avoid forming bubbles due to overagitation; do not transfer quickly.
- Suspensions (e.g. cell cultures) may sediment: thoroughly mix them before dispensing.

Pasteur pipettes

Hold correctly during use (Fig. 22.1) – keep the pipette vertical, with the middle fingers gripping the barrel while the thumb and index finger provide controlled pressure on the bulb. Squeeze gently to dispense individual drops. To avoid the risk of cross-contamination, take care not to draw up solution into the bulb or to lie the pipette on its side. Alternatively, use a plastic disposable 'Pastette'.

> ***SAFETY NOTE*** *Pasteur pipettes should be used with care for hazardous solutions: remove the tip from the solution before fully releasing pressure on the bulb – the air taken up helps prevent spillage.*

Measuring cylinders and volumetric flasks

These must be used on a level surface so that the scale is horizontal; you should first fill with solution until just below the desired mark; then fill slowly (e.g. using a Pasteur pipette) until the meniscus is level with the mark. Allow time for the solution to run down the walls of the vessel.

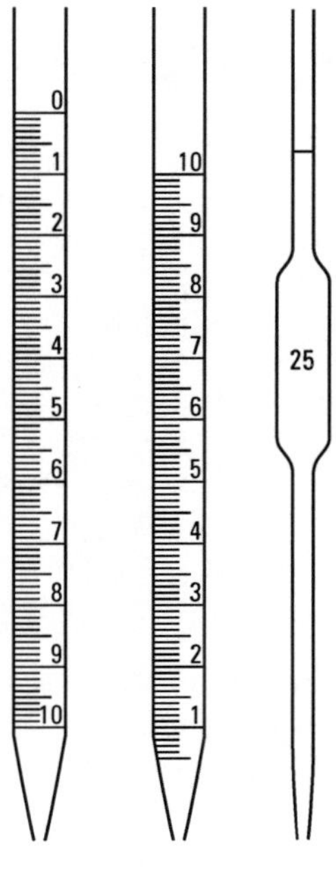

Fig. 22.2 Glass pipettes – graduated pipette, reading from zero to shoulder (a); graduated pipette, reading from maximum to tip, by gravity (b); bulb (volumetric) pipette, showing volume (calibration mark to tip, by gravity) above the bulb (c).

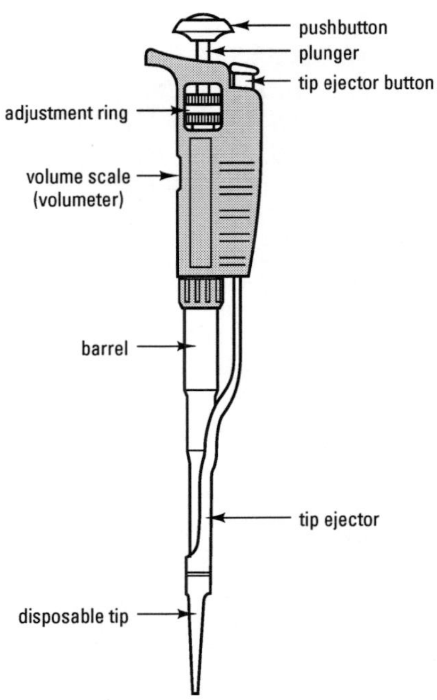

Fig. 22.3 A pipettor – the Gilson Pipetman.

Burettes

Burettes should be mounted vertically on a clamp stand – don't overtighten the clamp. First ensure the tap is closed and fill the body with solution using a funnel. Open the tap and allow some liquid to fill the tubing below the tap before first use. Take a meniscus reading, noting the value in your notebook. Dispense the solution *via* the tap and measure the new meniscus reading. The volume dispensed is the difference between the two readings. Titrations are usually performed on a magnetic stirrer.

Pipettes

These come in various designs, including graduated and bulb (volumetric) pipettes (Fig. 22.2). Take care to check the volume scale before use: some empty from full volume to zero, others from zero to full volume; some scales refer to the shoulder of the tip, others to the tip either by gravity or after blowing out.

KEY POINT For safety reasons, it is never permissible to mouth pipette – various aids are available such as the Pi-pump.

Pipettors (autopipettors)

These come in two basic types:

- Air displacement pipettors. For routine work with dilute aqueous solutions. One of the most widely used examples is the Gilson Pipetman (Fig. 22.3). Box 22.1 gives details on its use.
- Positive displacement pipettors. For non-standard applications, including dispensing viscous, dense or volatile liquids, or certain procedures in molecular genetics, e.g. the PCR, where an air displacement pipettor might create aerosols, leading to errors.

Air displacement and positive displacement pipettors may be:

- Fixed volume: capable of delivering a single factory-set volume.
- Adjustable: where the volume is determined by the operator across a particular range of values.
- Pre-set: movable between a limited number of values.
- Multi-channel: able to deliver several replicate volumes at the same time.

Whichever type you use, you must ensure that you understand the operating principles of the volume scale and the method for changing the volume delivered – some pipettors are easily misread.

A pipettor must be fitted with the correct disposable tip before use: each manufacturer produces different tips to fit particular models. Specialised tips are available for particular applications e.g. PCR.

Syringes

Syringes should be used by placing the tip of the needle in the solution and drawing the plunger up slowly to the required point on the scale. Check the barrel to make sure no air bubbles have been drawn up. Expel slowly and touch the syringe on the edge of the vessel to remove any liquid adhering to the end of the needle. Microsyringes should always be cleaned before and after use by repeatedly drawing up and expelling

Box 22.1 Using a pipettor to deliver accurate, reproducible volumes of liquid

A pipettor can be used to dispense volumes with accuracy and precision, by following this stepwise procedure:

1. **Select a pipettor that operates over the appropriate range.** Most adjustable pipettors are accurate only over a particular working range and should not be used to deliver volumes below the manufacturer's specifications (minimum volume is usually 10–20% of maximum value). Do not attempt to set the volume above the maximum limit, or the pipettor may be damaged.
2. **Set the volume to be delivered.** In some pipettors, you 'dial up' the required volume. Types like the Gilson Pipetman have a system where the scale (or 'volumeter') consists of three numbers, read from top to bottom of the barrel, and adjusted using the black knurled adjustment ring (Fig. 22.3). This number gives the first three digits of the volume scale and thus can only be understood by establishing the maximum volume of the Pipetman, as shown on the push-button on the end of the plunger (Fig. 22.3). The following examples illustrate the principle for two common sizes of Pipetman:

P1000 Pipetman (maximum volume 1000 μl) if you dial up

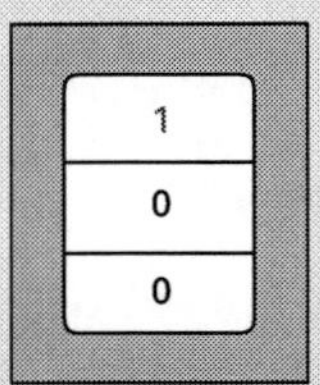

the volume is set at 1000 μl

P20 Pipetman (maximum volume 20 μl) if you dial up

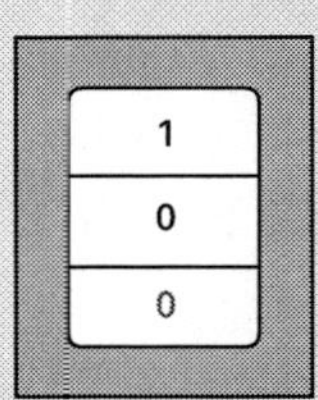

the volume is set at 10.0 μl

Note: The Pipetman scale is *not* a percentage one.

3. **Fit a new disposable tip to the end of the barrel.** Make sure that it is the appropriate type for your pipettor and that it is correctly fitted. Press the tip on firmly using a slight twisting motion – if not, you will take up less than the set volume and liquid will drip from the tip during use. Tips are often supplied in boxes, for ease of use: if sterility is important, make sure you use appropriate sterile technique at all times (p. 203). *Never, ever, try to use a pipettor without its disposable tip.*
4. **Check your delivery.** Confirm that the pipettor delivers the correct volume by dispensing volumes of distilled water and weighing on a balance, assuming 1 mg = 1 μl = 1 mm^3. The value should be within 1% of the selected volume. For small volumes, measure several 'squirts' together, e.g. 20 'squirts' of 5 μl = 100 mg. If the pipettor is inaccurate (p. 175) giving a biased result (e.g. delivering significantly more or less than the volume set), you can make a temporary correction by adjusting the volumeter scale down or up accordingly (the volume *delivered* is more important than the value *displayed* on the volumeter), or have the pipettor recalibrated. If the pipettor is imprecise (p. 175), delivering a variable amount of liquid each time, it may need to be serviced. After calibration, fit a clean (sterile) tip if necessary.
5. **Draw up the appropriate volume.** Holding the pipettor *vertically*, press down on the plunger/push-button until a resistance (spring-loaded stop) is met. Then place the end of the tip in the liquid. Keeping your thumb on the plunger/push-button, release the pressure slowly and evenly: watch the liquid being drawn up into the tip, to confirm that no air bubbles are present. Wait a second or so, to confirm that the liquid has been taken up, then withdraw the end of the tip from the liquid. Inexperienced users often have problems caused by drawing up the liquid too quickly/carelessly. If you accidentally draw liquid into the barrel, seek assistance from your demonstrator or supervisor as the barrel will need to be cleaned before further use.
6. **Make a quick visual check on the liquid in the tip.** Does the volume seem reasonable? (e.g. a 100 μl volume should occupy approximately half the volume of a P200 tip). The liquid will remain in the tip, without dripping, as long as the tip is fitted correctly and the pipettor is not tilted too far from a vertical position.
7. **Deliver the liquid.** Place the end of the tip against the wall of the vessel at a slight angle (10–15° from vertical) and press the plunger/push-button slowly and smoothly to the first (spring-loaded) stop. Wait a second or two, to allow any residual liquid to run down the inside of the tip, then press again to the final stop, dispensing any remaining liquid. Remove from the vessel with the plunger/push-button still depressed.
8. **Eject the tip.** Press the tip ejector button if present (Fig. 22.3). If the tip is contaminated, eject directly into an appropriate container, e.g. a beaker of disinfectant, for microbiological work, or a labelled container for hazardous solutions (p. 126). For repeat delivery, fit a new tip if necessary and begin again at step 5 above. Always make sure that the tip is ejected before putting a pipettor on the bench.

pure solvent. The dead space in the syringe needle can occupy up to 4% of the nominal syringe volume. A way of avoiding such problems is to fill the dead space with an inert substance (e.g. silicone oil) after sampling. Alternatively use a syringe where the plunger occupies the needle space (small volumes only).

Balances

These can be used to weigh accurately (p. 140) how much liquid you have dispensed. Convert mass to volume using the equation:

$$\text{mass/density} = \text{volume} \qquad [22.1]$$

Example Weighing 9 g of liquid with a density of $1.2\,\text{g ml}^{-1}$ will give a volume of $9 \div 1.2 = 7.5$ ml

Densities of common solvents can be found in Lide (2006). You will also need to know the liquid's temperature, as density is temperature dependent.

Holding and storing liquids

Test tubes

These are used for colour tests, small-scale reactions, holding cultures, etc. The tube can be sterilised by heating (p. 203) and maintained in this state with a cap or cotton wool plug.

Beakers

Beakers are used for general purposes, e.g. heating a solvent while the solute dissolves, carrying out a titration, etc.

Working with beakers and flasks *– remember that volume graduations, where present, are often inaccurate and should be used only where approximations will suffice.*

Conical (Erlenmeyer) flasks

These are used for storage of solutions: their wide base makes them stable, while their small mouth reduces evaporation and is easily sealed.

Bottles and vials

These are used when the solution needs to be sealed for safety, sterility or to prevent evaporation or oxidation. They usually have a screw top or ground glass stopper to prevent evaporation and contamination. Many types are available, including 'bijou', 'McCartney', 'universal' and 'Winkler'.

Storing light-sensitive chemicals *– use a coloured vessel or wrap aluminium foil around a clear vessel.*

You should clearly label all stored solutions (see p. 139), including relevant hazard information, preferably marking with hazard warning tape (p. 126). Seal vessels in an appropriate way, e.g. using a stopper or a sealing film such as Parafilm or Nescofilm to prevent evaporation. To avoid degradation store your solution in a fridge, but allow it to reach room temperature before use.

Storing an aqueous solution containing organic constituents *– unless this has been sterilised or is toxic, microbes will start growing, so store for short periods in a refrigerator: older solutions may not give reliable results.*

Creating specialised apparatus

Glassware systems incorporating ground glass connections such as Quickfit are useful for setting up combinations of standard glass components, e.g. for chemical reactions. In project work, you may need to adapt standard forms of glassware for a special need. A glassblowing service (often available in chemistry departments) can make special items to order.

SAFETY NOTE Special cleaning of glass – for an acid wash use dilute acid, e.g. $100\,\text{mmol}\,\text{l}^{-1}$ ($100\,\text{mol}\,\text{m}^{-3}$) HCl. Rinse thoroughly at least three times with distilled or deionised water. Glassware that must be exceptionally clean (e.g. for a micronutrient study) should be washed in a chromic acid bath, but this involves toxic and corrosive chemicals and should only be used under supervision.

Choosing between glass and plastic

Bear in mind the following points:

- Reactivity. Plastic vessels often distort at relatively low temperatures; they may be flammable, may dissolve in certain organic solvents and

Table 22.2 Spectral cutoff values for glass and plastics (λ_{50} = wavelength at which transmission of EMR is reduced to 50%)

Material	λ_{50} (nm)
Routine glassware	340
Pyrex glass	292
Polycarbonate	396
Acrylic	342
Polyester	318
Quartz	220

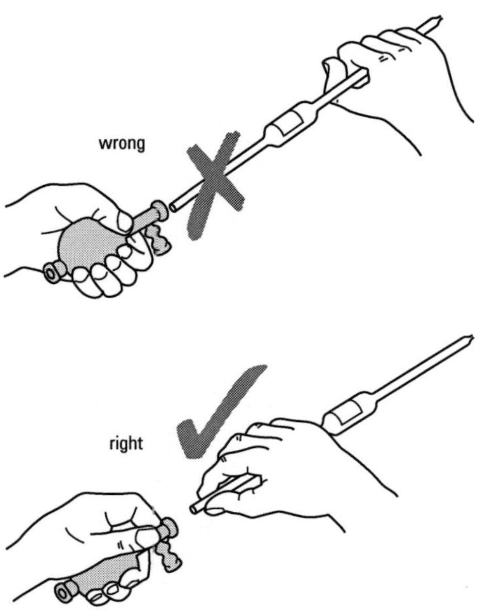

Fig. 22.4 Handling glass pipettes tubing.

may be affected by prolonged exposure to ultraviolet (UV) light. Some plasticisers may leach from vessels and have been shown to have biological activity. Glass may adsorb ions and other molecules and then leach them into solutions, especially in alkaline conditions. Pyrex glass is stronger than ordinary soda glass and can withstand temperatures up to 500 °C.

- Rigidity and resilience. Plastic vessels are not recommended where volume is critical as they may distort through time: use class A volumetric glassware for accurate work, e.g. preparing solutions (Chapter 24). Glass vessels are more easily broken than plastic, which is particularly important for centrifugation (see p. 301).
- Opacity. Both glass and plastic absorb light in the UV range of the EMR spectrum (Table 22.2). Quartz should be used where this is important, e.g. in cuvettes for UV spectrophotometry (see p. 280).
- Disposability. Plastic items may be cheap enough to make them disposable, an advantage where there is a risk of chemical or microbial contamination.

Cleaning glass and plastic

Take care to avoid the possibility of contamination arising from prior use of chemicals or inadequate rinsing following washing. A thorough rinse with distilled or deionised water immediately before use will remove dust and other deposits and is good practice in quantitative work, but ensure that the rinsing solution is not left in the vessel. 'Strong' basic detergents (e.g. Pyroneg) are good for solubilising acidic deposits. If there is a risk of basic deposits remaining, use an acid wash. If there is a risk of contamination from organic deposits, a rinse with Analar grade ethanol is recommended. Glassware can be disinfected by washing with a sodium hypochlorite bleach such as Chloros, with sodium metabisulphite or a blended commercial product such as Virkon – dilute as recommended before use and rinse thoroughly with sterile water after use. Alternatively, to sterilise glassware, heat to at least 121 °C for 15 min in an autoclave or 160 °C for 3 h in an oven.

Box 22.2 Safe working with glass

Many minor accidents in the laboratory are due to lack of care with glassware. You should follow these general precautions:

- **Always wear safety glasses when there is any risk of glass breakage** – e.g. when using low pressures, or heating solutions.
- **Take care when attaching tubing to glass tubes and when putting glass tubes into bungs** – always hold the tubing and glassware close together, as shown in Fig. 22.4 , and wear thick gloves when appropriate.
- **Use a 'soft' Bunsen flame when heating glassware** – this avoids creating a hot spot, where cracks may start: always use tongs or special heat-resistant gloves when handling hot glassware (never use a rolled-up paper towel).
- **Do not use chipped or cracked glassware** – it may break under very slight strain and should be disposed of in the broken glassware bin.
- **Never carry large glass bottles/flasks by their necks** – support them with a hand underneath or, better still, carry them in a basket.
- **Do not force bungs too firmly into bottles** – they can be extremely difficult to remove. If you need a tight seal, use a screw-top bottle with a rubber or plastic seal.
- **Dispose of broken glass thoroughly and carefully** – use disposable paper towels and wear thick gloves. Always put pieces of broken glass in the correct bin.

Text reference

Lide, D.R. (ed.) (2006) *CRC Handbook of Chemistry and Physics*, 87th edn. CRC Press, Boca Raton.

Sources for further study

Anon. *CHEMnet BASE*. Available http://www.chemnetbase.com Last accessed: 01/04/07.
[Online access to the *Handbook of Chemistry and Physics*]

Boyer, R. (2000) *Modern Experimental Biochemistry*, 3rd edn. Benjamin Cummings, San Francisco.

Henrikson, C., Byrd, L.C. and Hunter, N.W. (2005) *A Laboratory Manual for General, Organic and Biochemistry*, 5th edn. McGraw-Hill, New York.

Seidman, L.A. and Moore, C.J. (2000) *Basic Laboratory Methods for Biotechnology: Textbook and Laboratory Reference*. Prentice Hall, New Jersey.

Study exercises

22.1 Decide on the appropriate methods and equipment for the following procedures:

(a) Preparing one litre of ethanol at approximately 70% v/v in water for use as a general-purpose reagent.

(b) Adding 10 μl of a sample to the well of an agarose gel during a molecular biology procedure.

(c) Preparing a calibration standard of 100 ml of DNA, to contain 200.0 μg ml^{-1}, for spectrophotometry.

(d) Carrying out a titration curve for a buffer solution.

22.2 Write a protocol for calibrating and using a pipettor. After reading this chapter, prepare a detailed stepwise protocol explaining how to use a pipettor to deliver a specific volume, say of 500 μl (e.g. using a Gilson Pipetman, or an alternative if your department does not use this type). Ask another student to evaluate your protocol and provide you with written feedback – either simply by reading through your protocol, or by trying it out with a pipettor as part of a class exercise (check with a member of staff before you attempt this in a laboratory).

22.3 Determine the accuracy and precision of a pipette. Using the following data for three different models of pipettor, determine which pipettor is most *accurate* and which is most *precise* (check p. 175 if you are unsure of the definitions of these two terms). All three pipettors were set to deliver 1000 μl (1.000 ml) and 10 repetitive measurements of the weight of the volume of water in grams delivered were made using a three-place balance:

Model A pipettor: 0.986; 0.971; 0.993; 0.964; 0.983; 0.996; 0.977; 0.969; 0.982; 0.974

Model B pipettor: 1.013; 1.011; 1.010; 1.009; 1.011; 1.010; 1.011; 1.009; 1.011; 1.012

Model C pipettor: 0.985; 1.022; 1.051; 1.067; 0.973; 0.982; 0.894; 1.045; 1.062; 0.928

In your answer, you should support your conclusions with appropriate numerical (statistical) evidence (see Chapter 66 for appropriate measures of location and dispersion).

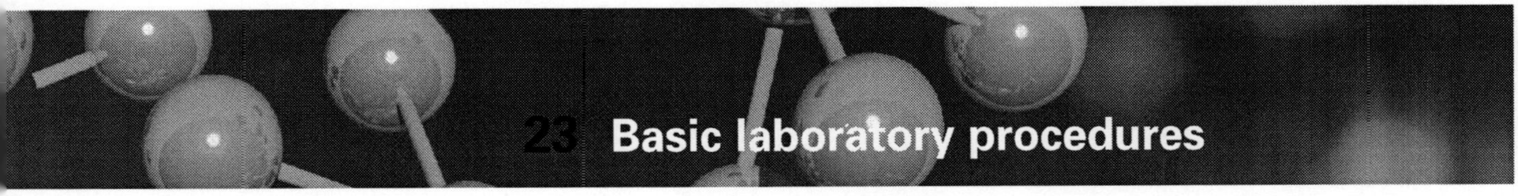

23 Basic laboratory procedures

> ***Using chemicals responsibly*** *– be considerate to others: always return store room chemicals promptly to the correct place. Report when supplies are getting low to the person who looks after storage/ordering. If you empty an aspirator or wash bottle, fill it up from the appropriate source.*

> ***Finding out about chemicals*** *– The* Merck Index *(O'Neil* et al.*, 2006: Fig. 23.1) and the* CRC Handbook of Chemistry and Physics *(Lide, 2006) are useful sources of information on the physical and biological properties of chemicals, including melting and boiling points, solubility, toxicity, etc.*

8671. Sodium Chloride. [7647-14-5] Salt; common salt. ClNa; mol wt 58.44. Cl 60.67%, Na 39.34%. NaCl. The article of commerce is also known as ***table salt, rock salt*** or ***sea salt***. Occurs in nature as the mineral ***halite***. Produced by mining (rock salt), by evaporation of brine from underground salt deposits and from sea water by solar evaporation: *Faith, Keyes & Clark's Industrial Chemicals*, F. A. Lowenheim, M. K. Moran, Eds. (Wiley-Interscience, New York, 4th ed., 1975) pp 722-730. Toxicity studies: E. M. Boyd, M. N. Shanas, *Arch. Int. Pharmacodyn.* **144,** 86 (1963). Comprehensive monograph: D. W. Kaufmann, *Sodium Chloride*, ACS Monograph Series no. **145** (Reinhold, New York, 1960) 743 pp.

Cubic, white crystals, granules, or powder; colorless and transparent or translucent when in large crystals. d 2.17. The salt of commerce usually contains some calcium and magnesium chlorides which absorb moisture and make it cake. mp 804° and begins to volatilize at a little above this temp. One gram dissolves in 2.8 ml water at 25°, in 2.6 ml boiling water, in 10 ml glycerol; very slightly sol in alcohol. Its soly in water is decreased by HCl. Almost insol in concd HCl. Its aq soln is neutral. pH: 6.7-7.3. d of satd aq soln at 25° is 1.202. A 23% aq soln of sodium chloride freezes at −20.5°C (5°F). LD_{50} orally in rats: 3.75 ±0.43 g/kg (Boyd, Shanas).

Note: ***Blusalt***, a brand of sodium chloride contg trace amounts of cobalt, iodine, iron, copper, manganese, zinc is used in farm animals.

Caution: Not generally considered poisonous. Accidental substitution of NaCl for lactose in baby formulas has caused fatal poisoning.

USE: Natural salt is the source of chlorine and of sodium as well as of all, or practically all, their compds, e.g., hydrochloric acid, chlorates, sodium carbonate, hydroxide, etc.; for preserving foods; manuf soap, dyes—to salt them out; in freezing mixtures; for dyeing and printing fabrics, glazing pottery, curing hides; metallurgy of tin and other metals.

THERAP CAT: Electrolyte replenisher; emetic; topical anti-inflammatory.

THERAP CAT (VET): Essential nutrient factor. May be given orally as emetic, stomachic, laxative or to stimulate thirst (prevention of calculi). Intravenously as isotonic solution to raise blood volume, to combat dehydration. Locally as wound irrigant, rectal douche.

Fig. 23.1 Example of typical *Merck Index* entry above showing type of information given for each chemical. From *The Merck Index: An Encyclopedia of Chemicals, Drugs and Biologicals*, Thirteenth edition, O'Neil, M.J., Smith, A., Heckelman, P.E. and Obenchain, J.R. Jr. (eds). Merck & Co, Inc., Whitehouse Station, NJ, USA, 2001. Reproduced with permission.

Using chemicals

Safety aspects

In practical classes, the person in charge has a responsibility to inform you of any hazards associated with the use of chemicals. For routine practical procedures, a risk assessment (p. 125) will have been carried out by a member of staff and relevant safety information will be included in the practical schedule: an example is shown in Table 23.1.

In project work, your first duty when using an unfamiliar chemical is to find out about its properties, especially those relating to safety. Your department must provide the relevant information to allow you to do this. If your supervisor has filled out the form, read it carefully before signing. Box 23.1 gives further advice.

> ***KEY POINT Before you use any chemical you must find out whether safety precautions need to be taken, and complete the appropriate forms confirming that you appreciate the risks involved.***

Selection

Chemicals are supplied in various degrees of purity and this is always stated on the manufacturer's containers. Suppliers differ in the names given to the grades and there is no conformity in purity standards. Very pure chemicals cost more, sometimes a lot more, and should only be used if the situation demands. If you need to order a chemical, your department will have a defined procedure for doing this.

Preparing solutions

Solutions are usually prepared with respect to their molar concentrations (e.g. mmol l^{-1}, or mol m^{-3}), or mass concentrations (e.g. g l^{-1}, or kg m^{-3}):

Table 23.1 Representative risk assessment information for a practical exercise in molecular biology, involving the isolation of DNA

Substance	Hazards	Comments
Sodium dodecyl sulphate (SDS)	Irritant Toxic	Wear gloves
Sodium hydroxide (NaOH)	Highly corrosive Severe irritant	Wear gloves
Isopropanol	Highly flammable Irritant/corrosive Potential carcinogen	No naked flames Wear gloves
Phenol	Highly toxic Causes skin burns Potential carcinogen	Use in fume hood Wear gloves
Chloroform	Volatile and toxic Irritant/corrosive Potential carcinogen	Use in fume hood Wear gloves

Box 23.1 Safe working with chemicals

You should always treat chemicals as potentially dangerous, following these general precautions:

- **Do not use any chemical until you have considered the risks involved** – for lab classes, you should carefully read all hazard and risk information provided *before* you start work. In project work, you may need to be involved in the risk assessment process with your supervisor.
- **Wear a laboratory coat at all times** – the coat should be fully fastened and cleaned appropriately, should any chemical compound be spilled on it. Closed-toe footwear will protect your feet should any spillages occur.
- **Make sure you know where the safety apparatus is kept before you begin working** – this apparatus includes eye bath, fire extinguishers and blanket, first aid kit.
- **Wear safety glasses and gloves when working with toxic, irritant or corrosive chemicals, and for any substances where the hazards are not yet fully characterised** – make sure you understand the hazard warning signs (p. 126) along with any specific hazard coding system used in your department. Carry out procedures with solid material in a fume cupboard.
- **Use aids such as pipette fillers to minimise the risk of contact with hazardous solutions** – these aids are further detailed on p. 129.
- **Never smoke, eat, drink or chew gum in a lab where chemicals are handled** – this will minimise the risk of ingestion.
- **Label all solutions appropriately** – use the appropriate hazard warning information (see p. 126/203/275).
- **Report all spillages of chemicals/solutions** – make sure that spillages are cleaned up properly.
- **Store hazardous chemicals only in the appropriate locations** – for example, a spark-proof fridge is required for flammable liquids; acids and solvents should not be stored together.
- **Dispose of chemicals in the correct manner** – if unsure, ask a member of staff (do not assume that it is safe to use the lab waste bin or the sink for disposal).
- **Wash hands after any direct contact with chemicals or biochemical material** – always wash your hands at the end of a lab session.

Examples
Using Eqn 23.1, 25 g of a substance dissolved in 400 ml of water would have a mass concentration of $25 \div 400 = 0.0625\ \mathrm{g\ ml^{-1}}$ ($\equiv 62.5\ \mathrm{mg\ ml^{-1}} \equiv 62.5\ \mathrm{g\ l^{-1}}$)

Using Eqn 23.1, 0.4 mol of a substance dissolved in 0.5 litres of water would have a molar concentration of $0.4 \div 0.5 = 0.8\ \mathrm{mol\ l^{-1}}$ ($\equiv 800\ \mathrm{mmol\ l^{-1}}$).

Solving solubility problems – *if your chemical does not dissolve after a reasonable time:*

- *check the limits of solubility for your compound (see* Merck Index, *O'Neil* et al. *2006);*
- *check the pH of the solution – solubility often changes with pH, e.g. you may be able to dissolve the compound in an acidic or an alkaline solution.*

both can be regarded as an amount of *substance* per unit volume of *solution*, in accordance with the relationship:

$$\text{Concentration} = \frac{\text{amount}}{\text{volume}} \qquad [23.1]$$

The most important aspect of Eqn [23.1] is to recognise clearly the units involved, and to prepare the solution accordingly: for molar concentrations, you will need the relative molecular mass of the compound, so that you can determine the mass of substance required. Further advice on concentrations and interconversion of units is given in Box 24.1.

Box 23.2 shows the steps involved in making up a solution. The concentration you require is likely to be defined by a protocol you are following, and the grade of chemical and supplier may also be specified. Success may depend on using the same source and quality, e.g. with enzyme work. To avoid waste, think carefully about the volume of solution you require, though it is always a good idea to err on the high side because you may spill some or make a mistake when dispensing it. Try to choose one of the standard volumes for vessels, as this will make measuring out easier.

Use distilled or deionised water to make up aqueous solutions and stir to make sure all the chemical is dissolved. Magnetic stirrers are the most convenient means of doing this: carefully drop a clean magnetic stirrer bar ('flea') in the beaker, avoiding splashing; place the beaker centrally on the stirrer plate, switch on the stirrer and gradually increase the speed of stirring. When the crystals or powder have completely dissolved, switch off and retrieve the flea with a magnet or another flea. Take care not to contaminate your solution when you do this, and rinse the flea with distilled water.

Box 23.2 How to make up an aqueous solution of known concentration from solid material

1. **Find out or decide the concentration of chemical required** and the degree of purity necessary.
2. **Decide on the volume of solution required.**
3. **Find out the relative molecular mass of the chemical (M_r).** This is the sum of the atomic (elemental) masses of the component elements and can be found on the container. If the chemical is hydrated, i.e. has water molecules associated with it, these must be included when calculating the mass required.
4. **Work out the mass of chemical that will give the concentration desired in the volume required.**

   Suppose your procedure requires you to prepare 250 ml of 0.1 mol l^{-1} NaCl.

   (a) Begin by expressing all volumes in the same units, either millilitres or litres (e.g. 250 ml as 0.25 litres).

   (b) Calculate the number of moles required from Eqn [23.1]: 0.1 = amount (mol) ÷ 0.25. By rearrangement, the required number of moles is thus 0.1 × 0.25 = 0.025 mol.

   (c) Convert from mol to g by multiplying by the relative molecular mass (M_r for NaCl = 58.44)

   (d) Therefore, you need to make up 0.025 × 58.44 = 1.461 g to 250 ml of solution, using distilled water.

   In some instances, it may be easier to work in SI units, though you must be careful when using exponential numbers (p. 475).

   Suppose your protocol states that you need 100 ml of 10 mmol l^{-1} KCl.

   (a) Start by converting this to 100 × 10^{-6} m^3 of 10 mol m^{-3} KCl.

   (b) The required number of mol is thus (100 × 10^{-6}) × (10) = 10^3.

   (c) Each mol of KCl weighs 72.56 g (M_r, the relative molecular mass).

   (d) Therefore you need to make up 72.56 × 10^{-3} g = 72.56 mg KCl to 100 × 10^{-6} m^3 (100 ml) with distilled water.

   See Box 24.1 for additional information.
5. **Weigh out the required mass of chemical to an appropriate accuracy.** If the mass is too small to weigh to the desired degree of accuracy, consider the following options:

   (a) Make up a greater volume of solution.

   (b) Make up a stock solution which can be diluted at a later stage (p. 134).

   (c) Weigh the mass first, and calculate what volume to make the solution up to afterwards using Eqn [23.1].
6. **Add the chemical to a beaker or conical flask then add a little less water than the final amount required.** If some of the chemical sticks to the paper, foil or weighing boat, use some of the water to wash it off.
7. **Stir and, if necessary, heat the solution to ensure all the chemical dissolves.** You can determine when this has happened visually by observing the disappearance of the crystals or powder.
8. **If required, check and adjust the pH of the solution when cool** (see p. 153).
9. **Make up the solution to the desired volume.** If the concentration needs to be accurate, use a class A volumetric flask; if a high degree of accuracy is not required, use a measuring cylinder (class B).

   (a) Pour the solution from the beaker into the measuring vessel using a funnel to avoid spillage.

   (b) Make up the volume so that the meniscus comes up to the appropriate measurement line (p. 128). For accurate work, rinse out the original vessel and use this liquid to make up the volume.
10. **Transfer the solution to a reagent bottle or a conical flask and label the vessel clearly.**

'Obstinate' solutions may require heating, but do this only if you know that the chemical will not be damaged at the temperature used. Use a stirrer-heater to keep the solution mixed as you heat it. Allow the solution to cool before you measure volume or pH as these change with temperature.

Stock solutions

Stock solutions are valuable when making up a range of solutions containing different concentrations of a reagent or if the solutions have some common ingredients. They also save work if the same solution is used over a prolonged period (e.g. a nutrient solution). The stock solution is more concentrated than the final requirement and is diluted as

Table 23.2 Use of stock solutions. Suppose you need a set of solutions 10 ml in volume containing differing concentrations of KCl, with and without reagent Q. You decide to make up a stock of KCl at twice the maximum required concentration (50 mmol l^{-1} = 50 mol m^{-3}) and a stock of reagent Q at twice its required concentration. The table shows how you might use these stocks to make up the media you require. Note that the total volumes of stock you require can be calculated from the table (end column)

Stock solutions	Volume of stock required to make required solutions (ml)						Total volume of stock required (ml)
	No KCl plus Q	No KCl minus Q	15 mmol l^{-1} KCl plus Q	15 mmol l^{-1} KCl minus Q	25 mmol l^{-1} KCl plus Q	25 mmol l^{-1} KCl minus Q	
50 mmol l^{-1} KCl	0	0	3	3	5	5	16
[reagent Q] × 2	5	0	5	0	5	0	15
Water	5	10	2	7	0	5	29
Total	10	10	10	10	10	10	60

appropriate when the final solutions are made up. The principle is best illustrated with an example (Table 23.2).

Preparing dilutions

Making a single dilution

You may need to dilute a stock solution to give a particular mass concentration, or molar concentration. Use the following procedure:

1. Transfer an accurate volume of stock solution to a volumetric flask, using appropriate equipment (Table 22.1).
2. Make up to the calibration mark with solvent – add the last few drops from a pipette or solvent bottle, until the meniscus is level with the calibration mark.
3. Mix thoroughly, either by repeated inversion (holding the stopper firmly) or by prolonged stirring, using a magnetic stirrer. Make sure you add the magnetic flea *after* the volume adjustment step.

Making a dilution *– use the relationship* $[C_1]V_1 = [C_2]V_2$ *to determine volume or concentration (see Box 24.1).*

Removing a magnetic flea from a volumetric flask *– use a strong magnet to bring the flea to the top of the flask, to avoid contamination during removal.*

Using the correct volumes for dilutions *– it is important to distinguish between the volumes of the various liquids: a one-in-ten dilution is obtained using 1 volume of stock solution plus 9 volumes of diluent (1+9 = 10). Note that when this is shown as a ratio, it may represent the initial and final volumes (e.g. 1:10) or, sometimes, the volumes of stock solution and diluent (e.g. 1:9).*

For routine work using dilute aqueous solutions where the highest degree of accuracy is not required, it may be acceptable to substitute test tubes or conical flasks for volumetric flasks. In such cases, you would calculate the volumes of stock solution and diluent required, with the assumption that the final volume is determined by the sum of the individual volumes of stock and diluent used (e.g. Table 23.2). Thus, a twofold dilution would be prepared using 1 volume of stock solution and 1 volume of diluent. The dilution factor is obtained from the ratio of the initial concentration of the stock solution and the final concentration of the diluted solution. The dilution factor can be used to determine the volumes of stock and diluent required in a particular instance. For example, suppose you wanted to prepare 100 ml of a solution of NaCl at 0.2 mol l^{-1}. Using a stock solution containing 4.0 mol l^{-1} NaCl, the dilution factor is $0.2 \div 4.0 = 0.05 = 1/20$ (a twentyfold dilution). Therefore, the amount of stock solution required is 1/20th of 100 ml = 5 ml and the amount of diluent needed is 19/20ths of 100 ml = 95 ml.

Preparing a dilution series

Dilution series are used in a wide range of procedures, including the preparation of standard curves for calibration of analytical instruments (Chapter 37), and in microbiology (Chapter 32) and immunoassay

Using diluents *– various liquids are used, including distilled or deionised water, salt solutions, buffers, Ringer's solution (p. 223), etc., according to the specific requirements of the procedure.*

(Chapter 38), where a range of dilutions of a particular sample is often required. A variety of different approaches can be used:

Linear dilution series Here, the concentrations are separated by an equal amount, e.g. a series containing protein at 0, 0.2, 0.4, 0.6, 0.8, 1.0 $\mu g\,ml^{-1}$. Such a dilution series might be used to prepare a calibration curve for spectrophotometric assay of protein concentration, or an enzyme assay. Use $[C_1]V_1 = [C_2]V_2$ (p. 145) to determine the amount of stock solution required for each member of the series, with the volume of diluent being determined by subtraction.

Logarithmic dilution series Here, the concentrations are separated by a constant proportion, often referred to as the step interval. This type of serial dilution is useful when a broad range of concentrations is required, e.g. for titration of biologically active substances (p. 238), making a plate count of a suspension of microbes (p. 206), or when a process is logarithmically related to concentration.

The most common examples are:

- Doubling dilutions – where each concentration is half that of the previous one (twofold step interval, log_2 dilution series). First, make up the most concentrated solution at twice the volume required. Measure out half of this volume into a vessel containing the same volume of diluent, mix thoroughly and repeat, for as many doubling dilutions as are required. The concentrations obtained will be 1/2, 1/4, 1/8, 1/16, etc. times the original (i.e. the dilutions will be two, four, eight and sixteenfold, etc.).
- Decimal dilutions – where each concentration is one-tenth that of the previous one (tenfold step interval, log_{10} dilution series). First, make up the most concentrated solution required, with at least a 10% excess. Measure out one-tenth of the volume required into a vessel containing nine times as much diluent, mix thoroughly and repeat. The concentrations obtained will be 1/10, 1/100, 1/1000, etc. times the original (i.e. dilutions of 10^{-1}, 10^{-2}, 10^{-3}, etc.). To calculate the actual concentration of solute, multiply by the appropriate dilution factor.

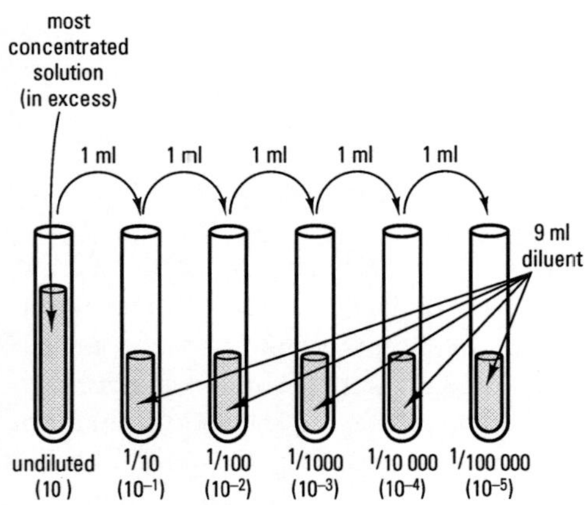

Fig. 23.2 Preparation of a dilution series. The example shown is a decimal dilution series, down to 1/100 000 (10^{-5}) of the solution in the first (left-hand) tube. Note that all solutions must be mixed thoroughly before transferring the volume to the next in the series. In microbiology and cell culture, sterile solutions and appropriate aseptic technique will be required (p. 203).

When preparing serial doubling or decimal dilutions, it is often easiest to add the appropriate amount of diluent to several vessels beforehand, as shown in the worked example in Figure 23.2. When preparing a dilution series, it is essential that all volumes are dispensed accurately, e.g. using calibrated pipettors (p. 130), otherwise any inaccuracies will be compounded, leading to gross errors in the most dilute solutions.

Harmonic dilution series Here, the concentrations in the series take the values of the reciprocals of successive whole numbers, e.g. 1, 1/2, 1/3, 1/4, 1/5, etc. The individual dilutions are simply achieved by a stepwise increase in the volume of diluent in successive vessels, e.g. by adding 0, 1, 2, 3, 4 and 5 times the volume of diluent to a set of test tubes, then adding a constant unit volume of stock solution to each vessel. Although there is no dilution transfer error between individual dilutions, the main disadvantage is that the series is non-linear, with a step interval that becomes progressively smaller as the series is extended.

***Preparing a dilution series using pipettes or pipettors** – use a fresh pipette or disposable tip for each dilution, to prevent carry-over of solutions.*

SAFETY NOTE Using a vortex mixer with open or capped test tubes – do not vortex too vigorously or liquid will spill from the top of the tube, creating a contamination risk.

SAFETY NOTE Cleaning up chemical spillages – you must always clean up any spillages of chemicals, as you are the only person who knows the risks from the spilled material.

Solutions must be thoroughly mixed before measuring out volumes for the next dilution. Use a fresh measuring vessel for each dilution to avoid contamination, or wash your vessel thoroughly between dilutions. Clearly label the vessel containing each dilution when it is made: it is easy to get confused! When deciding on the volumes required, allow for the aliquot removed when making up the next member in the series. Remember to discard any excess from the last in the series if volumes are critical.

Mixing solutions and suspensions

Various devices may be used, including:

- Magnetic stirrers and fleas. Magnetic fleas come in a range of shapes and sizes, and some stirrers have integral heaters. During use, stirrer speed may increase as the instrument warms up.
- Vortex mixers. For vigorous mixing of small volumes of solution, e.g. when preparing a dilution series in test tubes. Take care when adjusting the mixing speed – if the setting is too low, the test tube will vibrate rather than creating a vortex, giving inadequate mixing. If the setting is too high, the test tube may slip from your hand.
- Orbital shakers and shaking water baths. These are used to provide controlled mixing at a particular temperature, e.g. for long-term incubation and cell growth studies (p. 233).
- Bottle rollers. For cell culture work, ensuring gentle, continuous mixing.

Storing chemicals and solutions

Labile chemicals may be stored in a fridge or freezer. Take special care when using chemicals that have been stored at low temperature: the container and its contents must be warmed up to room temperature before use, otherwise water vapour will condense on the chemical. This may render any weighing you do meaningless and it could ruin the chemical. Other chemicals may need to be kept in a desiccator, especially if they are deliquescent (water-absorbing).

KEY POINT Label all stored chemicals clearly with the following information: the chemical name (if a solution, state solute(s), concentration(s) and pH if measured), plus any relevant hazard warning information, the date made up, and your name.

Separating components of mixtures and solutions

Particulate solids (e.g. soils) can be separated on the basis of size using sieves. These are available in stacking forms which fit on automatic shakers. Sieves with the largest pores are placed at the top and the assembly is shaken for a fixed time until the sample separates. Suspensions of solids in liquids may be separated out by centrifugation (see p. 297) or filtration. Various forms of filter paper are available having different porosities and purities. Vacuum-assisted filtration speeds up the process and is best carried out with a filter funnel attached to a filter flask. Filtration through pre-sterilised membranes with very small pores (e.g. the Millipore type) is an excellent method of sterilising small volumes of solution. Solvents can be removed from solutes by heating, using rotary film evaporation under low pressure and, for water, by freeze drying. The last two are especially useful for heat-labile solutes – refer to the manufacturers' specific instructions for use.

Using balances

Weighing – *never weigh anything directly onto a balance's pan: you may contaminate it for other users. Use a weighing boat or a slip of aluminium foil. Otherwise, choose a suitable vessel like a beaker, conical flask or aluminium tray.*

Electronic balances with digital readouts are now favoured over mechanical types: they are easy to read and their self-taring feature means the mass of the weighing boat or container can be subtracted automatically before weighing an object. The most common type offers accuracy down to 1 mg over the range 1 mg to 160 g, which is suitable for most biological applications.

To operate a standard self-taring balance:

1. Check that it is level, using the adjustable feet to centre the bubble in the spirit level (usually at the back of the machine). For accurate work, make sure a draught shield is on the balance.
2. Place an empty vessel in the middle of the balance pan and allow the reading to stabilise. *If the object is larger than the pan, take care that no part rests on the body of the balance or the draught shield as this will invalidate the reading.* Press the tare bar to bring the reading to zero.
3. Place the chemical or object carefully in the vessel (powdered chemicals should be dispensed with a suitably sized clean spatula). Take care to avoid spillages.
4. Allow the reading to stabilise and make a note of the value.
5. If you add excess chemical, take great care when removing it. Switch off if you need to clean any deposit accidentally left on or around the balance.

Deciding on which balance to use – *select a balance that weighs to an appropriate number of decimal places. For example, you should use a top-loading balance weighing to one decimal place for less accurate work. Note that a weight of 6.4 g on such a balance may represent a true value of between 6.350 g and 6.449 g (to three decimal places).*

Larger masses should be weighed on a top-loading balance to an appropriate degree of accuracy. Take care to note the limits for the balance: while most have devices to protect against overloading, you may damage the mechanism. In the field, spring or battery-operated balances may be preferred. Try to find a place out of the wind to use them. For extremely small masses, there are electrical balances that can weigh down to 1 μg, but these are very delicate and must be used under supervision.

Using a balance – *it is poor technique to use a large container to weigh out a small amount of a chemical: you are attempting to make accurate measurements of a small difference between two large numbers. Instead, use a small weighing container.*

Measuring and controlling temperature

Heating samples

Care is required when heating samples – there is a danger of fire whenever organic material is heated and a danger of scalding from heated liquids. Safety glasses should always be worn. Use a thermostatically controlled electric stirrer-heater if possible. If using a Bunsen burner, keep the flame well away from yourself and your clothing (tie back long hair). Use a non-flammable mat beneath a Bunsen to protect the bench. Switch off when no longer required. To light a Bunsen, close the air hole first, then apply a lit match or lighter. Open the air hole if you need a hotter, more concentrated flame: the hottest part of the flame is just above the apex of the blue cone in its centre.

Ovens and drying cabinets may be used to dry specimens or glassware. They are normally thermostatically controlled. If drying organic material for dry weight measurement, do so at about 80 °C to avoid caramelising the sample. Always state the actual temperature used as this affects results. Check that all water has been driven off by weighing until a constant mass is reached.

SAFETY NOTE Heating/cooling glass vessels – take care if heating or cooling glass vessels rapidly as they may break when heat stressed. Freezing aqueous solutions in thin-walled glass vessels is risky because ice expansion may break the glass.

Using thermometers *– some are calibrated for use in air, others require partial immersion in liquid and others total immersion – check before use.*

SAFETY NOTE If a mercury thermometer is broken, report the spillage, as mercury is a poison.

Example Water vapour can be removed by passing gas over dehydrated $CaCO_3$, and CO_2 may be removed by bubbling through KOH solution.

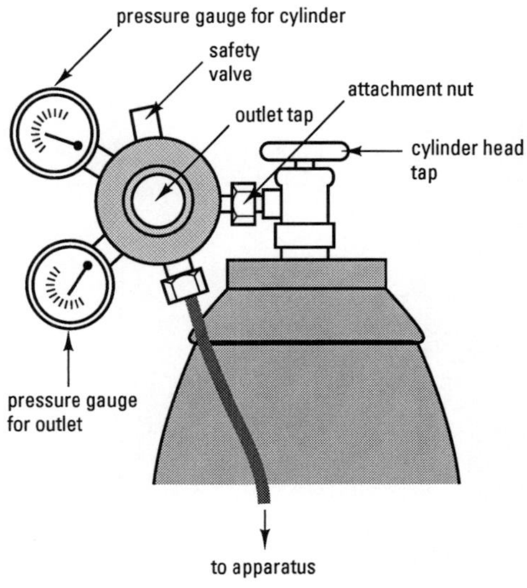

Fig. 23.3 Parts of a cylinder head regulator. The regulator is normally attached by tightening the attachment nut clockwise; the exception is with cylinders of hydrogen, where the special regulator is tightened *anticlockwise* to avoid the chance of this potentially explosive gas being incorrectly used.

Cooling samples and specimens

Fridges and freezers are used for storing stock solutions and chemicals that would either break down or become contaminated at room temperature. Normal fridge and freezer temperatures are about 4 °C and −15 °C respectively. Ice baths can be used when reactants must be kept close to 0 °C. Most bioscience departments will have a machine which provides flaked ice for use in these baths. If common salt is mixed with ice, temperatures below 0 °C can be achieved. A mixture of ethanol and solid CO_2 will provide a temperature of −72 °C if required. To freeze a specimen quickly, immerse in liquid nitrogen (−196 °C) using tongs and wearing an apron and thick gloves, as splashes will damage your skin. Always work in a well-ventilated room.

Maintaining constant temperature

Thermostatically controlled temperature rooms and incubators can be used to maintain temperature at a desired level. Always check with a thermometer or thermograph that the thermostat is accurate enough for your study. To achieve a controlled temperature on a smaller scale, e.g. for an oxygen electrode (pp. 339–341), use a water bath. These usually incorporate heating elements, a circulating mechanism and a thermostat. Baths for sub-ambient temperatures have a cooling element.

Controlling atmospheric conditions

Gas composition

The atmosphere may be 'scrubbed' of certain gases by passing through a U-tube or Dreschel bottle containing an appropriate chemical or solution.

For accurate control of gas concentrations, use cylinders of pure gas; the contents can be mixed to give specified concentrations by controlling individual flow rates. The cylinder head regulator (Fig. 23.3) allows you to control the pressure (and hence flow rate) of gas; adjust using the controls on the regulator or with spanners of appropriate size. Before use, ensure the regulator outlet tap is off (turn anticlockwise), then switch on at the cylinder (turn clockwise) – the cylinder dial will give you the pressure reading for the cylinder contents. Now switch on at the regulator outlet (turn clockwise) and adjust to desired pressure/flow setting. To switch off, carry out the above directions in reverse order.

To control dissolved gas composition in liquids, either 'degas' under vacuum or bubble with another gas, e.g. when preparing oxygen-free liquids for HPLC (p. 305) by bubbling with nitrogen.

Pressure

Many forms of pump are used to pressurise or provide partial vacuum, usually to force gas or liquid movement. Each has specific instructions for use. Many laboratories are supplied with 'vacuum' (suction) and pressurised air lines that are useful for procedures such as vacuum-assisted filtration. Make sure you switch off the taps after use. Take special care with glass items kept at very low or high pressures. These should be contained within a metal cage to minimise the risk of injury (e.g. a vacuum desiccator, as shown in Fig. 23.4).

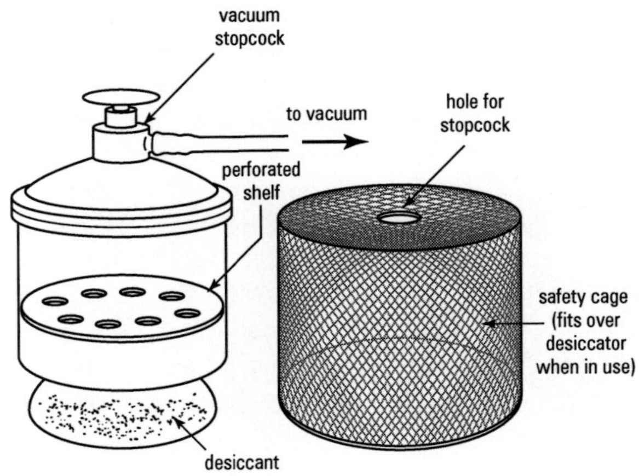

Fig. 23.4 Vacuum desiccator (with mesh safety cage).

Measuring time

Using a timer *– always set the alarm before the critical time, so that you have adequate time to react.*

Many experiments and observations need to be carefully timed. Large-faced stopclocks allow you to set and follow 'experimental time' and remove the potential difficulties in calculating this from 'real time' on a watch or clock. Some timers incorporate an alarm which you can set to warn when readings or operations must be carried out; 24-h timers are available for controlling light and temperature regimes.

Text references

Lide, D.R. (ed.) (2006) *CRC Handbook of Chemistry and Physics*, 87th edn. CRC Press, Boca Raton.

O'Neil, M.J., Smith, A. and Heckelman, P.E. (2006) *The Merck Index: An Encyclopedia of Chemicals, Drugs and Biologicals*, 14th edn. Merck & Co., Inc., Whitehouse Station.

Sources for further study

Anon. *CHEMnet BASE*. Available: http://www.chemnetbase.com
Last accessed: 01/04/07.
[Online access to the *Handbook of Chemistry and Physics*.]

Jack, C.R. (1995) *Basic Biochemical Laboratory Procedures and Computing*. Oxford University Press, Oxford.

Seidman, L.A. and Moore, C.J. (2000) *Basic Laboratory Methods for Biotechnology: Textbook and Laboratory Reference*. Prentice Hall, New Jersey.

Study exercises

23.1 Practise the calculations involved in preparing specific volumes of aqueous solutions (see also study exercise 24.1). What mass of substance would be required to prepare each of the following (answer in each case to three significant figures):

(a) 100 ml of NaCl at 50 mmol l^{-1} (M_r of NaCl = 58.44)?

(b) 250 ml of mannitol at 0.10 mol l^{-1} (M_r of mannitol = 182.17)?

(c) 200 ml of a bovine serum albumin solution at 800 μg ml^{-1}?

(d) 0.5 litres of $MgCl_2$ prepared from the hexahydrate salt at 22.5 mmol l^{-1} (M_r of $MgCl_2 \cdot 6H_2O$ = 203.30)?

(e) 400 ml of DNA at 20 ng μl^{-1}?

Study exercises (continued)

23.2 Practise the calculations involved in preparing dilutions (answer in each case to three significant figures).

(a) If you added 1.0 ml of an aqueous solution of NaCl at 0.4 mol l^{-1} to 9.0 ml of water, what would be the final concentration of NaCl in mmol l^{-1}?

(b) If you added 25 ml of an aqueous solution of DNA at 10 μg ml^{-1} to a 500 ml volumetric flask and made it up to the specified volume with water, what would be the final concentration of DNA, in ng ml^{-1}?

(c) If you added 10 μl of an aqueous solution of sucrose at 200 mmol l^{-1} to a 250 ml volumetric flask and made it up to the specified volume with water, what would be the final concentration of sucrose, in nmol ml^{-1}?

(d) How would you prepare 250 ml of KCl at a final concentration of 20.0 mmol l^{-1} from a solution containing KCl at 0.2 mol l^{-1}?

(e) How would you prepare 1×10^{-3} m^3 of glucose at a final concentration of 50 μmol m^{-3} from a stock solution containing glucose at 20.0 g m^{-3} (M_r of glucose = 180.16)?

23.3 Practise the calculations involved in using stock solutions. Suppose you had the following stock solutions: NaCl 100.0 mmol l^{-1}; KCl 200.0 mmol l^{-1}; $CaCl_2$ 160.0 mmol l^{-1}; glucose 5.0 mmol l^{-1}. Calculate the volumes of each stock solution and the volume of water required to prepare each of the following (answer in each case to three significant figures):

(a) 1.0 ml of a solution containing only KCl at 10.0 mmol l^{-1}

(b) 50 ml of a solution containing NaCl at 2.5 mmol l^{-1} and glucose at 0.5 mmol l^{-1}

(c) 100 ml of a solution containing NaCl at 5.0 mmol l^{-1}, KCl at 2.5 mmol l^{-1}, $CaCl_2$ at 40.0 mmol l^{-1} and glucose at 0.25 mmol l^{-1}

(d) 10 ml of a solution containing NaCl, $CaCl_2$ and KCl, all at 20.0 μmol ml^{-1}

(e) 25 ml of a solution containing $CaCl_2$ at 8.0 mmol l^{-1}, KCl at 50.0 mmol l^{-1} and glucose at 20.0 nmol ml^{-1}.

24 Principles of solution chemistry

> *SAFETY NOTE Working with solutions – many solutes and solvents used in biosciences are potentially toxic, corrosive, oxidising or flammable, and they may also be carcinogenic (see p. 126). Further, there is a risk of accident involving the vessels used when preparing, storing and dispensing solutions (see Box 22.2).*

A solution is a homogeneous liquid, formed by the addition of solutes to a solvent (usually water in biological systems). The behaviour of solutions is determined by the type of solutes involved and by their proportions, relative to the solvent. Many laboratory exercises involve calculation of concentrations, e.g. when preparing an experimental solution at a particular concentration (p. 136), or when expressing data in terms of solute concentration (p. 135). Make sure that you understand the basic principles set out in this chapter before you tackle such exercises.

Solutes can affect the properties of solutions in several ways, including:

> **Definitions**
>
> **Electrolyte** – a substance that dissociates, either fully or partially, in water to give two or more ions.
>
> **Relative atomic mass (A_r)** – the mass of an atom relative to $^{12}C = 12$.
>
> **Relative molecular mass (M_r)** – the mass of a compound's formula unit relative to $^{12}C = 12$.
>
> **Mole (of a substance)** – the equivalent in mass to relative molecular mass in grams.

Electrolytic dissociation

This occurs where individual molecules of an electrolyte dissociate to give charged particles (ions). For a strong electrolyte, e.g. NaCl, dissociation is essentially complete. In contrast, a weak electrolyte, e.g. acetic acid, will be only partly dissociated, depending upon the pH and temperature of the solution (pp. 152–3).

Osmotic effects

These are the result of solute particles lowering the effective concentration of the solvent (water). These effects are particularly relevant to biological systems since membranes are far more permeable to water than to most solutes. Water moves across biological membranes from the solution with the higher effective water concentration to that with the lower effective water concentration (osmosis).

Ideal/non-ideal behaviour

This occurs because solutions of real substances do not necessarily conform to the theoretical relationships predicted for dilute solutions of so-called ideal solutes. It is often necessary to take account of the non-ideal behaviour of real solutions, especially at high solute concentrations (see Lide, 2006, and Robinson and Stokes, 2002 for appropriate data).

Concentration

> ***Expressing solute concentrations** – you should use SI units wherever possible. However, you are likely to meet non-SI concentrations and you must be able to deal with these units too.*

In SI units (p. 183), the concentration of a solute in a solution is expressed in mol m^{-3}, which is convenient for most biological purposes. The concentration of a solute is usually symbolised by square brackets, e.g. [NaCl]. Details of how to prepare a solution using SI and non-SI units are given in Box 23.2.

A number of alternative ways of expressing the relative amounts of solute and solvent are in general use, and you may come across these terms in your practical work or in the literature:

Molarity

> **Example** A 1.0 molar solution of NaCl would contain 58.44 g NaCl (the relative molecular mass) per litre of solution.

This is the term used to denote molar concentration, $[C]$, expressed as moles of solute per litre volume of solution (mol l^{-1}). This non-SI term continues to find widespread usage, in part because of the familiarity of working scientists with the term, but also because laboratory glassware is calibrated in millilitres and litres, making the preparation of molar and

Box 24.1 Useful procedures for calculations involving molar concentrations

1. **Preparing a solution of defined molarity.** For a solute of known relative molecular mass, M_r, the following relationship can be applied:

$$[C] = \frac{\text{mass of solute/relative molecular mass}}{\text{volume of solution}} \quad [24.1]$$

So, if you wanted to make up 200 ml (0.2 l) of an aqueous solution of NaCl (M_r 58.44) at a concentration of 500 mmol l^{-1} (0.5 mol l^{-1}), you could calculate the amount of NaCl required by inserting these values into Eqn [24.1]:

$$0.5 = \frac{\text{mass of solute/58.44}}{0.2}$$

which can be rearranged to

$$\text{mass of solute} = 0.5 \times 0.2 \times 58.44 = 5.844\,\text{g}$$

The same relationship can be used to calculate the concentration of a solution containing a known amount of a solute, e.g. if 21.1 g of NaCl were made up to a volume of 100 ml (0.1 l), this would give

$$[\text{NaCl}] = \frac{21.1/58.44}{0.1} = 3.61\,\text{mol}\,l^{-1}$$

2. **Dilutions and concentrations.** The following relationship is very useful if you are diluting (or concentrating) a solution:

$$[C_1]V_1 = [C_2]V_2 \quad [24.2]$$

where $[C_1]$ and $[C_2]$ are the initial and final concentrations, while V_1 and V_2 are their respective volumes: each pair must be expressed in the same units. Thus, if you wanted to dilute 200 ml of 0.5 mol l^{-1} NaCl to give a final molarity of 0.1 mol l^{-1}, then, by substitution into Eqn [24.2]:

$$0.5 \times 200 = 0.1 \times V_2$$

Thus $V_2 = 1\,000$ ml (in other words, you would have to add water to 200 ml of 0.5 mol l^{-1} NaCl to give a final volume of 1 000 ml to obtain a 0.1 mol l^{-1} solution).

3. **Interconversion.** A simple way of interconverting amounts and volumes of any particular solution is to divide the amount and volume by a factor of 10^3: thus a molar solution of a substance contains 1 mol l^{-1}, which is equivalent to 1 mmol ml^{-1}, or 1 µmol µl^{-1}, or 1 nmol nl^{-1}, etc. You may find this technique useful when calculating the amount of substance present in a small volume of solution of known concentration, e.g. to calculate the amount of NaCl present in 50 µl of a solution with a concentration (molarity) of 0.5 mol l^{-1} NaCl:
   (a) this is equivalent to 0.5 µmol µl^{-1};
   (b) therefore 50 µl will contain 50×0.5 µmol = 25 µmol.

Alternatively, you may prefer to convert to primary SI units, for ease of calculation (see Box 23.1).

The 'unitary method' (p. 477) is an alternative approach to these calculations.

millimolar solutions relatively straightforward. However, the symbols in common use for molar (M) and millimolar (mM) solutions are at odds with the SI system and many people now prefer to use mol l^{-1} and mmol l^{-1} respectively, to avoid confusion. Box 24.1 gives details of some useful approaches to calculations involving molarities.

Molality

Example A 0.5 molal solution of NaCl would contain $58.44 \times 0.5 = 29.22$ g NaCl per kg of water.

This is used to express the concentration of solute relative to the *mass* of solvent, i.e. mol kg^{-1}. Molality is a temperature-independent means of expressing solute concentration, rarely used except when the osmotic properties of a solution are of interest (p. 148).

Per cent composition (% w/w)

Example A 5% w/w sucrose solution contains 5 g sucrose and 95 g water (= 95 ml water, assuming a density of 1 g ml^{-1}) to give 100 g of solution.

This is the solute mass (in g) per 100 g solution. The advantage of this expression is the ease with which a solution can be prepared, since it simply requires each component to be pre-weighed (for water, a volumetric measurement may be used, e.g. using a measuring cylinder) and then mixed together. Similar terms are parts per thousand (‰), e.g. mg g^{-1}, and parts per million (ppm), e.g. µg g^{-1} (see below).

Per cent concentration (% w/v and % v/v)

For solutes added in solid form, this is the number of grams of solute per 100 ml solution. This is more commonly used than per cent composition, since solutions can be accurately prepared by weighing out the required amount of solute and then making this up to a known volume using a volumetric flask. The equivalent expression for liquid solutes is % v/v.

The principal use of mass/mass or mass/volume terms (including $g\,l^{-1}$) is for solutes whose relative molecular mass is unknown (e.g. cellular proteins), or for mixtures of certain classes of substance (e.g. total salt in sea water). You should *never* use the per cent term without specifying how the solution was prepared, i.e. by using the qualifier w/w, w/v or v/v. For mass concentrations, it is simpler to use mass per unit volume, e.g. $mg\,l^{-1}$, $\mu g\,\mu l^{-1}$, etc.

Example A 5% w/v sucrose solution contains 5 g sucrose in 100 ml of solution. A 5% v/v glycerol solution would contain 5 ml glycerol in 100 ml of solution.

Note that when water is the solvent this is often not specified in the expression, e.g. a 20% v/v ethanol solution contains 20% ethanol made up to 100 ml of solution using water.

Alternative expressions *– parts per thousand (‰) w/v can be used as an alternative to $g\,l^{-1}$*

Parts per million (ppm) and parts per billion (ppb) concentration

Ppm is a non-SI weight per volume (w/v) concentration term commonly used in quantitative analysis such as flame photometry, atomic absorption spectroscopy and gas chromatography, where low concentrations of solutes are analysed. The term ppm is equivalent to the expression of concentration as $\mu g\,ml^{-1}$ ($10^{-6}\,g\,ml^{-1}$) and a 1.0 ppm solution of a substance will have a concentration of 1.0 $\mu g\,ml^{-1}$ ($1.0 \times 10^{-6}\,g\,ml^{-1}$).

Parts per billion (ppb) is an extension of this concentration term as $ng\,ml^{-1}$ ($10^{-9}\,g\,ml^{-1}$) and is commonly used to express concentrations of very dilute solutions. For example, the allowable concentration of arsenic in water is 0.05 ppm, more conveniently expressed as 50 ppb. Also note that ppm and ppb are sometimes used as a weight per weight (w/w) term, e.g. $\mu g\,g^{-1}$ and $ng\,g^{-1}$ respectively.

Example The concentration of a NaCl solution is stated as 3 ppm. This is equivalent to 3 $\mu g\,ml^{-1}$ (3 $mg\,l^{-1}$). The relative molecular mass of NaCl is 58.44 $g\,mol^{-1}$, so the solution has a concentration of $3 \times 10^{-6} \div 58.44\,mol\,ml^{-1} = 5.13 \times 10^{-8}\,mol\,ml^{-1} = 0.0513\,\mu mol\,ml^{-1} = 51.3\,\mu mol\,l^{-1}$.

Activity (a)

This is a term used to describe the *effective* concentration of a solute. In dilute solutions, solutes can be considered to behave according to ideal (thermodynamic) principles, i.e. they will have an effective concentration equivalent to the actual concentration. However, in concentrated solutions ($\geqslant 500\,mol\,m^{-3}$), the behaviour of solutes is often non-ideal, and their effective concentration (activity) will be less than the actual concentration $[C]$. The ratio between the effective concentration and the actual concentration is called the activity coefficient (γ) where

$$\gamma = \frac{a}{[C]} \qquad [24.3]$$

Equation [24.3] can be used for SI units ($mol\,m^{-3}$), molarity ($mol\,l^{-1}$) or molality ($mol\,kg^{-1}$). In all cases, γ is a dimensionless term, since a and $[C]$ are expressed in the same units. The activity coefficient of a solute is effectively unity in dilute solution, decreasing as the solute concentration increases (Table 24.1). At high concentrations of certain ionic solutes, γ may increase to become greater than unity.

Table 24.1 Activity coefficient of NaCl solutions as a function of molality. Data from Robinson and Stokes (2002)

Molality	Activity coefficient at 25 °C
0.1	0.778
0.5	0.681
1.0	0.657
2.0	0.668
4.0	0.783
6.0	0.986

KEY POINT Activity is often the correct expression for theoretical relationships involving solute concentration (e.g. where a property of the solution is dependent on concentration). However, for most practical purposes, it is possible to use the actual concentration of a solute rather than the activity, since the difference between the two terms can be ignored for dilute solutions.

Example A solution of NaCl with a molality of 0.5 mol kg^{-1} has an activity coefficient of 0.681 at 25 °C and a molal activity of $0.5 \times 0.681 = 0.340$ mol kg^{-1}.

The particular use of the term 'water activity' is considered below, since it is based on the mole fraction of solvent, rather than the effective concentration of solute.

Equivalent mass (equivalent weight)

Equivalence and normality are outdated terms, although you may come across them in older texts. They apply to certain solutes whose reactions involve the transfer of charged ions, e.g. acids and alkalis (which may be involved in H^+ or OH^- transfer), and electrolytes (which form cations and anions that may take part in further reactions). These two terms take into account the valency of the charged solutes. Thus the equivalent mass of an ion is its relative molecular mass divided by its valency (ignoring the sign), expressed in grams per equivalent (eq) according to the relationship:

Examples For carbonate ions (CO_3^{2-}), with a relative molecular mass of 60.00 and a valency of 2, the equivalent mass is $60.00/2 = 30.00$ g eq^{-1}.
For sulphuric acid (H_2SO_4, relative molecular mass 98.08), where 2 hydrogen ions are available, the equivalent mass is $98.08/2 = 49.04$ g eq^{-1}.

$$\text{equivalent mass} = \frac{\text{relative molecular mass}}{\text{valency}} \qquad [24.4]$$

For acids and alkalis, the equivalent mass is the mass of substance that will provide 1 mol of either H^+ or OH^- ions in a reaction, obtained by dividing the molecular mass by the number of available ions (n), using n instead of valency as the denominator in Eqn [24.4].

Example A 0.5 N solution of sulphuric acid would contain $0.5 \times 49.04 = 24.52$ g l^{-1}.

Normality

A 1 normal solution (1 N) is one that contains one equivalent mass of a substance per litre of solution. The general formula is:

$$\text{normality} = \frac{\text{mass of substance per litre}}{\text{equivalent mass}} \qquad [24.5]$$

Example Under ideal conditions, 1 mol of NaCl dissolved in water would give 1 mol of Na^+ ions and 1 mol of Cl^- ions, equivalent to a theoretical osmolarity of 2 osmol l^{-1}.

Osmolarity

This non-SI expression is used to describe the number of moles of osmotically active solute particles per litre of solution (osmol l^{-1}). The need for such a term arises because some molecules dissociate to give more than one osmotically active particle in aqueous solution.

Osmolality

This term describes the number of moles of osmotically active solute particles per unit mass of solvent (osmol kg^{-1}). For an ideal solute, the osmolality can be determined by multiplying the molality by n, the number of solute particles produced in solution (e.g. for NaCl, $n = 2$). However, for real (i.e. non-ideal) solutes, a correction factor (the osmotic coefficient, ϕ) is used:

Example A 1.0 mol kg^{-1} solution of NaCl has an osmotic coefficient of 0.936 at 25 °C and an osmolality of $1.0 \times 2 \times 0.936 = 1.872$ osmol kg^{-1}.

$$\text{osmolality} = \text{molality} \times n \times \phi \qquad [24.6]$$

If necessary, the osmotic coefficients of a particular solute can be obtained from tables (e.g. Table 24.2): non-ideal behaviour means that ϕ may have values >1 at high concentrations. Alternatively, the osmolality of a solution can be measured using an osmometer.

Table 24.2 Osmotic coefficients of NaCl solutions as a function of molality. Data from Robinson and Stokes (2002)

Molality	Osmotic coefficient at 25 °C
0.1	0.932
0.5	0.921
1.0	0.936
2.0	0.983
4.0	1.116
6.0	1.271

Colligative properties and their use in osmometry

Several properties vary in direct proportion to the effective number of osmotically active solute particles per unit mass of solvent and can be used

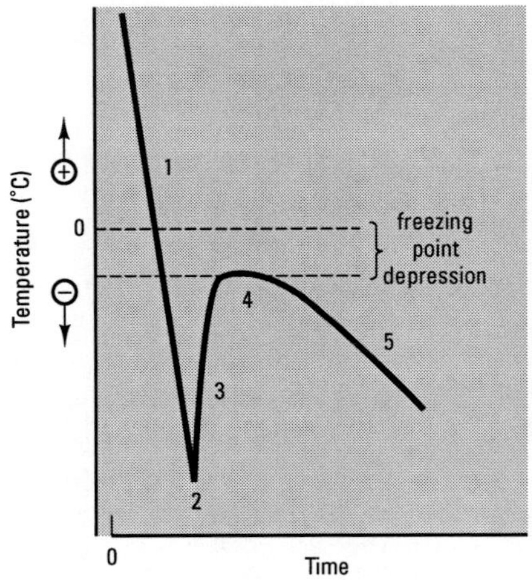

Fig. 24.1 Temperature responses of a cryoscopic osmometer. The response can be subdivided into:
1. initial supercooling
2. initiation of crystallisation
3. crystallisation/freezing
4. plateau, at the freezing point
5. slow temperature decrease.

to determine the osmolality of a solution. These colligative properties include freezing point, boiling point and vapour pressure.

An osmometer is an instrument which measures the osmolality of a solution, usually by determining the freezing point depression of the solution in relation to pure water, a technique known as cryoscopic osmometry. A small amount of sample is cooled rapidly and then brought to the freezing point (Fig. 24.1), which is measured by a temperature-sensitive thermistor probe calibrated in mosmol kg^{-1}. An alternative method is used in vapour pressure osmometry, which measures the relative decrease in the vapour pressure produced in the gas phase when a small sample of the solution is equilibrated within a chamber.

Using an osmometer *– it is vital that the sample holder and probe are clean, otherwise small droplets of the previous sample may be carried over, leading to inaccurate measurement.*

Osmotic properties of solutions

Several interrelated terms can be used to describe the osmotic status of a solution. In addition to osmolality, you may come across the following:

Osmotic pressure

This is based on the concept of a membrane permeable to water, but not to solute molecules. For example, if a sucrose solution is placed on one side and pure water on the other, then a passive driving force will be created and water will diffuse across the membrane into the sucrose solution, since the effective water concentration in the sucrose solution will be lower (see Fig. 24.2). The tendency for water to diffuse into the sucrose solution could be counteracted by applying a hydrostatic pressure equivalent to the passive driving force. Thus, the osmotic pressure of a solution is the excess hydrostatic pressure required to prevent the net flow of water into a vessel containing the solution. The SI unit of osmotic pressure is the pascal, Pa ($= kg\,m^{-1}\,s^{-2}$). Older sources may use atmospheres, or bars, and conversion factors are given in Table 29.1. Osmotic pressure and osmolality can be interconverted using the expression $1\ osmol\,kg^{-1} = 2.479\ MPa$ at 25 °C.

Example A 1.0 mol kg^{-1} solution of NaCl at 25 °C has an osmolalilty of 1.872 osmol kg^{-1} and an osmotic pressure of $1.872 \times 2.479 = 4.641$ MPa.

The use of osmotic pressure has been criticised as misleading, since a solution does not exhibit an ‘osmotic pressure’ unless it is placed on the other side of a selectively permeable membrane from pure water.

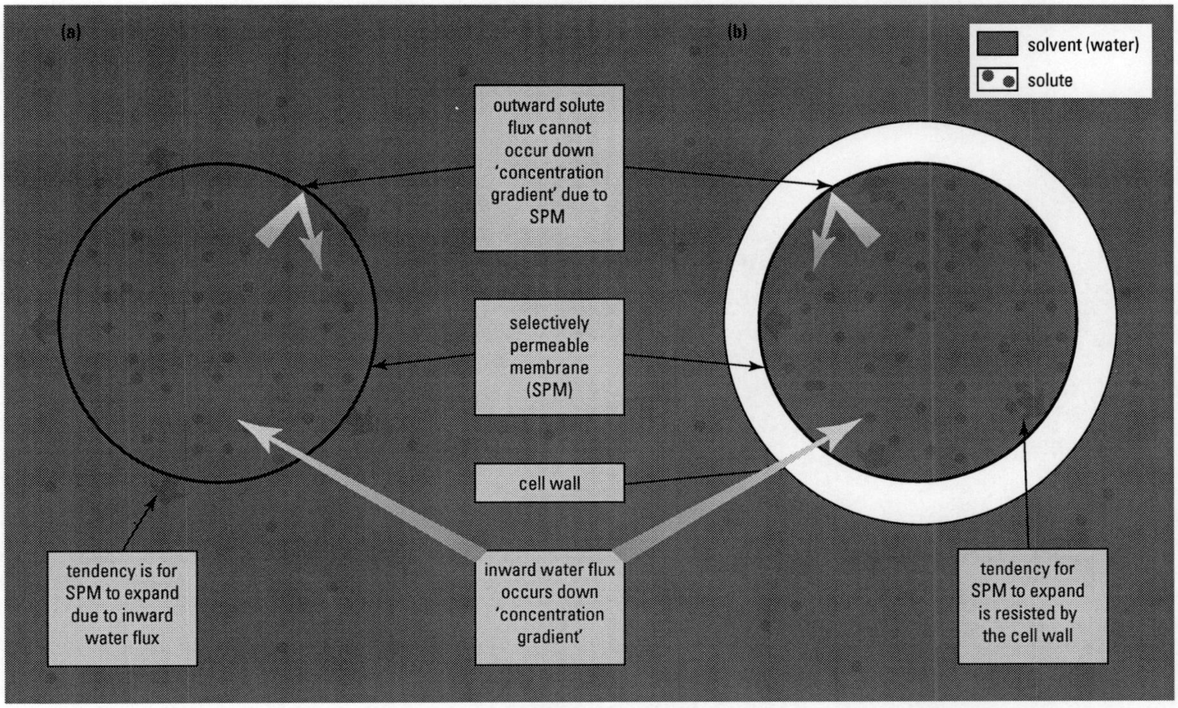

Fig. 24.2. Illustration of forces driving solvent (water) and solute movement across a selectively-permeable membrane (SPM). Energetically, both solutes and solvents tend to move down their respective 'concentration gradient' (strictly, down their chemical potential gradient). However, solute molecules cannot leave the model cells illustrated because they cannot pass through the SPM. In the situation illustrated in (a), water will tend to move from outside the cell to within because the solute molecules have effectively 'diluted' the water within the cell (illustrated by the density of point shading), creating a gradient in 'concentration' and because this molecule is able to pass through the SPM, the result will be an expansion of this model cell (short arrows). The osmotic pressure is the (theoretical) pressure that would need to be applied to prevent this. If the model cell were surrounded by a cell wall, as in (b), this would resist expansion, leading to internal pressurisation (turgor pressure, p. 150).

Water activity (a_w)

This is a term often used to describe the osmotic behaviour of microbial cells. It is a measure of the relative proportion of water in a solution, expressed in terms of its mole fraction, i.e. the ratio of the number of moles of water (n_w) to the total number of moles of all substances (i.e. water and solutes) in solution (n_t), taking into account the molal activity coefficient of the solvent, water (i.e. γ_w):

$$a_w = \gamma_w \frac{n_w}{n_t} \qquad [24.7]$$

The water activity of pure water is unity, decreasing as solutes are added. One disadvantage of a_w is the limited change which occurs in response to a change in solute concentration: a 1.0 mol kg^{-1} solution of NaCl has a water activity of 0.967 (Table 24.3).

Osmolality, osmotic pressure and water activity are measurements based solely on the osmotic properties of a solution, with no regard for any other driving forces, e.g. hydrostatic and gravitational forces. In circumstances where such other forces are important, you will need to measure a variable that takes into account these aspects of water status, namely water potential.

Table 24.3 Water activity (a_w) of NaCl solutions as a function of molality. Data from Robinson and Stokes (2002)

Molality	a_w
0.1	0.997
0.5	0.984
1.0	0.967
2.0	0.932
4.0	0.852
6.0	0.760

Water potential (hydraulic potential) and its applications

Water potential, Ψ_w, is the most appropriate measure of osmotic status in many areas of the biosciences. It is a term derived from the chemical potential of water. It expresses the difference between the chemical potential of water in the test system and that of pure water under standard conditions and has units of pressure (i.e. Pa). It is a more appropriate term than osmotic pressure because it is based on sound theoretical principles and because it can be used to predict the direction of passive movement of water, since water will flow down a gradient of chemical potential (i.e. osmosis occurs from a solution with a higher water potential to one with a lower water potential). Pure water at 20 °C and at 0.1 MPa pressure (i.e. ≈ atmospheric) has a water potential of zero. The addition of solutes will lower the water potential (i.e. make it negative), while the application of pressure, e.g. from hydrostatic or gravitational forces, will raise it (i.e. make it positive).

Examples A 1.0 mol kg^{-1} solution of NaCl has a (negative) water potential of −4.641 MPa.

Pure water at 0.2 MPa pressure (about 0.1 MPa above atmospheric pressure) has a (positive) water potential of 0.1 MPa.

Often, the two principal components of water potential are referred to as the solute potential, or osmotic potential (Ψ_s, sometimes symbolised as Ψ_π or π) and the hydrostatic pressure potential (Ψ_p) respectively. For a solution at atmospheric pressure, the water potential is due solely to the presence of osmotically active solute molecules (osmotic potential) and may be calculated from the measured osmolality (osmol kg^{-1}) at 25 °C, using the relationship:

$$\Psi_w\,(\text{MPa}) = \Psi_s\,(\text{MPa}) = -2.479 \times \text{osmolality} \qquad [24.8]$$

For aquatic microbial cells, e.g. algae, fungi and bacteria, equilibrated in their growth medium at atmospheric pressure, the water potential of the external medium will be equal to the cellular water potential ('isotonic') and the latter can be derived from the measured osmolality of the medium (Eqn [24.8]) by osmometry (p. 142). The water potential of such cells can be subdivided into two major parts, the cell solute potential (Ψ_s) and the cell turgor pressure (Ψ_p) as follows:

$$\Psi_w = \Psi_s + \Psi_p \qquad [24.9]$$

To calculate the relative contribution of the osmotic and pressure terms in Eqn [24.9], an estimate of the internal osmolality is required, e.g. by measuring the freezing point depression of expressed intracellular fluid. Once you have values for Ψ_w and Ψ_s, the turgor pressure can be calculated by substitution into Eqn [24.9].

Measuring water potential *– Eqn [24.9] ignores the effects of gravitational forces – for systems where gravitational effects are important an additional term is required (Nobel, 2005).*

For terrestrial plant cells, the water potential may be determined directly using a vapour pressure osmometer, by placing a sample of the material within the osmometer chamber and allowing it to equilibrate. If Ψ_s of expressed sap is then measured, Ψ_p can be determined from Eqn [24.9].

The van't Hoff relationship can be used to estimate Ψ_s, by summation of the osmotic potentials due to the major solutes, determined from their concentrations, as:

$$\Psi_s = -RTn\phi[C] \qquad [24.10]$$

where RT is the product of the universal gas constant and absolute temperature (2 479 J mol^{-1} at 25 °C), n and ϕ are as previously defined and $[C]$ is expressed in SI terms as mol m^{-3}.

Text references

Lide, D.R. (ed.) (2006) *CRC Handbook of Chemistry and Physics*, 87th edn. CRC Press, Boca Raton.

Nobel, P.S. (2005) *Physicochemical and Environmental Plant Physiology*, 2nd edn. Academic Press, New York.

Robinson, R.A. and Stokes, R.H. (2002) *Electrolyte Solutions*. Dover Publications, New York.

Sources for further study

Burtis, C.A. and Ashwood, E.R. (2001) *Fundamentals of Clinical Chemistry*, 5th edn. Saunders, Philadelphia.

Chapman, C. (1998) *Basic Chemistry for Biology*. McGraw-Hill, New York.

O'Neil, M.J., Smith, A. and Heckelman, P.E. (2006) *The Merck Index: An Encyclopedia of Chemicals, Drugs and Biologicals*, 14th edn. Merck & Co., Inc., Whitehouse Station.

Seidman, L.A. and Moore, C.J. (2000) *Basic Laboratory Methods for Biotechnology: Textbook and Laboratory Reference*. Prentice Hall, New Jersey.

Study exercises

24.1 Practise calculations involving molar concentrations (see also study exercises 23.1 and 23.2). What mass of substance would be required to prepare each of the following aqueous solutions (answer in grams, to three decimal places in each case):

(a) 1 litre of NaCl at a concentration of 1 molar? (M_r of NaCl = 58.44.)

(b) 250 ml of $CaCl_2$ at 100 mmol l^{-1}? (M_r of $CaCl_2$ = 110.99.)

(c) 2.5 l of mannitol at 10 nmol µl^{-1}? (M_r of mannitol = 182.17.)

(d) 400 ml of KCl at 5% w/v?

(e) 250 ml of glucose at 2.50 mol m^{-3}? (M_r of glucose = 180.16.)

24.2 Practise expressing concentrations in different ways. Express all answers to three significant figures:

(a) What is 5 g l^{-1} sucrose, expressed in terms of molarity? (M_r of sucrose = 342.3.)

(b) What is 1.0 mol m^{-3} NaCl, expressed in g l^{-1}? (M_r of NaCl = 58.44.)

(c) What is 5% v/v ethanol, expressed in terms of molarity? (M_r of ethanol = 46.06 and density of ethanol at 25°C = 0.789 g ml^{-1}.)

(d) What is 150 mmol l^{-1} glucose, expressed in terms of per cent concentration (% w/v)? (M_r of glucose = 180.16.)

(e) What is a 1.0 molal solution of KCl, expressed as per cent composition (% w/w)? (M_r of KCl = 74.55.)

24.3 Calculate osmolality and osmotic potentials. Answer to three significant figures in all cases.

(a) Assuming NaCl, KCl and $CaCl_2$ behave according to ideal thermodynamic principles, what would be the predicted osmolality of a solution containing:

(i) NaCl alone, at 50 mmol kg^{-1}?

(ii) KCl at 200 mmol kg^{-1} and $CaCl_2$ at 40 mmol kg^{-1}?

(iii) NaCl at 100 mmol kg^{-1}, KCl at 60 mmol kg^{-1} and $CaCl_2$ at 75 mmol kg^{-1}?

(b) What is the predicted osmotic pressure and osmotic potential of each of the solutions in (a) at 25°C?

25 pH and buffer solutions

Definitions

Acid – a compound that acts as a proton donor in aqueous solution.

Base – a compound that acts as a proton acceptor in aqueous solution.

Conjugate pair – an acid together with its corresponding base.

Alkali – a compound that liberates hydroxyl ions when it dissociates. Since hydroxyl ions are strongly basic, this will reduce the proton concentration.

Ampholyte – a compound that can act as both an acid and a base. Water is an ampholyte since it may dissociate to give a proton and a hydroxyl ion (amphoteric behaviour).

pH is a measure of the amount of hydrogen ions (H^+) in a solution: this affects the solubility of many substances and the activity of most biological systems from individual molecules to whole organisms. It is usual to think of aqueous solutions as containing H^+ ions (protons), though protons actually exist in their hydrated form as hydronium ions (H_3O^+). The proton concentration of an aqueous solution [H^+] is affected by several factors:

- Ionisation (dissociation) of water, which liberates protons and hydroxyl ions in equal quantities, according to the reversible relationship:

$$H_2O \rightleftharpoons H^+ + OH^- \qquad [25.1]$$

- Dissociation of acids, according to the equation:

$$H\text{–}A \rightleftharpoons H^+ + A^- \qquad [25.2]$$

  where H–A represents the acid and A^- is the corresponding conjugate base. The dissociation of an acid in water will increase the amount of protons, reducing the amount of hydroxyl ions as water molecules are formed (Eqn [25.1]). The addition of a base (usually, as its salt) to water will decrease the amount of H^+, due to the formation of the conjugate acid (Eqn [25.2]).
  Dissociation of alkalis, according to the relationship:

$$X\text{–}OH \rightleftharpoons X^+ + OH^- \qquad [25.3]$$

  where X–OH represents the undissociated alkali. Since the dissociation of water is reversible (Eqn [25.1]), in an aqueous solution the production of hydroxyl ions will effectively act to 'mop up' protons, lowering the proton concentration.

SAFETY NOTE Working with strong acids or alkalis – these can be highly corrosive; rinse with plenty of water, if spilled.

Many compounds act as acids, bases or alkalis: those which are almost completely ionised in solution are usually called strong acids or bases, while weak acids or bases are only slightly ionised in solution.

In an aqueous solution, most of the water molecules are not ionised. In fact, the extent of ionisation of pure water is constant at any given temperature and is usually expressed in terms of the ion product (or ionisation constant) of water, K_w:

$$K_w = [H^+][OH^-] \qquad [25.4]$$

Table 25.1 Effects of temperature on the ion product of water (K_w), H^+ ion concentration and pH at neutrality. Values calculated from Lide (2006)

Temp. (°C)	K_w (mol^2 l^{-2})	[H^+] at neutrality (nmol l^{-1})	pH at neutrality
0	0.11×10^{-14}	33.9	7.47
4	0.17×10^{-14}	40.7	7.39
10	0.29×10^{-14}	53.7	7.27
20	0.68×10^{-14}	83.2	7.08
25	1.01×10^{-14}	100.4	7.00
30	1.47×10^{-14}	120.2	6.92
37	2.39×10^{-14}	154.9	6.81
45	4.02×10^{-14}	199.5	6.70

where [H^+] and [OH^-] represent the molar concentration (strictly, the activity) of protons and hydroxyl ions in solution, expressed as mol l^{-1}. At 25 °C, the ion product of pure water (Table 25.1) is 10^{-14} mol^2 l^{-2} (i.e. 10^{-8} mol^2 m^{-6}). This means that the concentration of protons in solution will be 10^{-7} mol l^{-1} (10^{-4} mol m^{-3}), with an equivalent concentration of hydroxyl ions (Eqn [25.1]). Since these values are very low and involve negative powers of 10, it is customary to use the pH scale, where:

$$pH = -\log_{10}[H^+] \qquad [25.5]$$

and [H^+] is the proton activity in mol l^{-1} (see p. 146).

Example Human blood plasma has a typical H^+ concentration of approximately 0.4×10^{-7} mol l^{-1} (= $10^{-7.4}$ mol l^{-1}), giving a pH of 7.4.

***KEY POINT** While pH is strictly the negative logarithm (to the base 10) of H^+ activity, in practice H^+ concentration in mol l^{-1} (equivalent to kmol m^{-3} in SI terminology) is most often used in place of activity since the two are virtually the same, given the limited dissociation of H_2O. The pH scale is not SI: nevertheless, it continues to be used widely in biological science.*

***Measuring pH** – the pH of a neutral solution changes with temperature (Table 25.1), due to the enhanced dissociation of water with increasing temperature. This must be taken into account when measuring the pH of any solution and when interpreting your results.*

SAFETY NOTE Preparing a dilute acid solution using concentrated acid – always slowly add the concentrated acid to water, not the reverse, since the strongly exothermic process can trigger a violent reaction with water.

SAFETY NOTE Preparing an alkali solution – typically, the alkali will be in solid form (e.g. NaOH) and addition to water will rapidly raise the temperature of the solution: use only heat-resistant glassware, cooled with water, if necessary.

Table 25.2 Properties of some pH indicator dyes

Dye	Acid–base colour change	Useful pH range
Thymol blue (acid)	red–yellow	1.2–6.8
Bromophenol blue	yellow–blue	1.2–6.8
Congo red	blue–red	3.0–5.2
Bromocresol green	yellow–blue	3.8–5.4
Resazurin	orange–violet	3.8–6.5
Methyl red	red–yellow	4.3–6.1
Litmus	red–blue	4.5–8.3
Bromocresol purple	yellow–purple	5.8–6.8
Bromothymol blue	yellow–blue	6.0–7.6
Neutral red	red–yellow	6.8–8.0
Phenol red	yellow–red	6.8–8.2
Thymol blue (alkaline)	yellow–blue	8.0–9.6
Phenol-phthalein	none–red	8.3–10.0

The value where an equal amount of H^+ and OH^- ions are present is termed neutrality: at 25 °C the pH of pure water at neutrality is 7.0. At this temperature, pH values below 7.0 are acidic while values above 7.0 are alkaline.

Always remember that the pH scale is a logarithmic one, not a linear one: a solution with a pH of 3.0 is not twice as acidic as a solution of pH 6.0, but one thousand times as acidic (i.e. contains 1000 times the amount of H^+ ions). Therefore, you may need to convert pH values into proton concentrations before you carry out mathematical manipulations (see Box 66.2). For similar reasons, it is important that pH change is expressed in terms of the original and final pH values, rather than simply quoting the difference between the values: a pH change of 0.1 has little meaning unless the initial or final pH is known.

Measuring pH

pH electrodes

Accurate pH measurements can be made using a pH electrode, coupled to a pH meter. The pH electrode is usually a combination electrode, comprising two separate systems: an H^+-sensitive glass electrode and a reference electrode which is unaffected by H^+ ion concentration (Fig. 25.1). When this is immersed in a solution, a pH-dependent voltage between the two electrodes can be measured using a potentiometer. In most cases, the pH electrode assembly (containing the glass and reference electrodes) is connected to a separate pH meter by a cable, although some hand-held instruments (pH probes) have the electrodes and meter within the same assembly, often using an H^+-sensitive field effect transistor in place of a glass electrode, to improve durability and portability.

Box 25.1 gives details of the steps involved in making a pH measurement with a glass pH electrode and meter.

pH indicator dyes

These compounds (usually weak acids) change colour in a pH-dependent manner. They may be added in small amounts to a solution, or they can be used in paper strip form. Each indicator dye usually changes colour over a restricted pH range, typically 1–2 pH units (Table 25.2): universal indicator dyes/papers make use of a combination of individual dyes to measure a wider pH range. Dyes are not suitable for accurate pH measurement as they are affected by other components of the solution including oxidising and reducing agents and salts. However, they are useful for:

- estimating the approximate pH of a solution;
- determining a change in pH, for example at the end-point of a titration or the production of acids during bacterial metabolism (pp. 217–18);
- establishing the approximate pH of intracellular compartments, for example the use of neutral red as a 'vital' stain (p. 167).

Box 25.1 Using a glass pH electrode and meter to measure the pH of a solution

The following procedure should be used whenever you make a pH measurement: consult the manufacturer's handbook for specific information, where necessary. Do not be tempted to miss out any of the steps detailed below, particularly those relating to the effects of temperature, or your measurements are likely to be inaccurate.

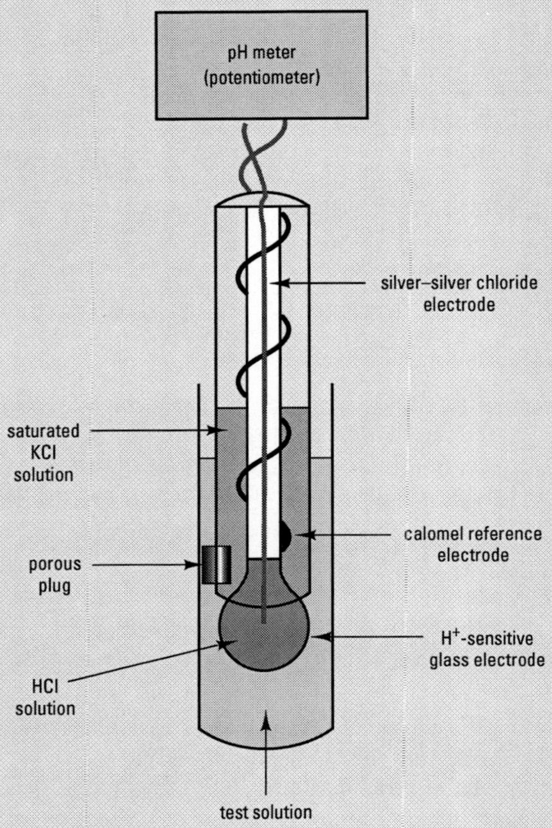

Fig. 25.1 Measurement of pH using a combination pH electrode and meter. The electrical potential difference recorded by the potentiometer is directly proportional to the pH of the test solution.

1. **Stir the test solution thoroughly before you make any measurement:** it is often best to use a magnetic stirrer. Leave the solution for sufficient time to allow equilibration at lab temperature.
2. **Record the temperature of every solution you use,** including all calibration standards and samples, since this will affect K_w, neutrality and pH.
3. **Set the temperature compensator on the meter to the appropriate value.** This control makes an allowance for the effect of temperature on the electrical potential difference recorded by the meter: it does *not* allow for the other temperature-dependent effects mentioned elsewhere. Basic instruments have no temperature compensator, and should only be used at a specified temperature, either 20 °C or 25 °C, otherwise they will not give an accurate measurement. More sophisticated systems have automatic temperature compensation.
4. **Rinse the electrode assembly with distilled water** and gently dab off the excess water onto a clean tissue: check for visible damage or contamination of the glass electrode (consult a member of staff if the glass is broken or dirty). Also check that the solution within the glass assembly is covering the metal electrode.
5. **Calibrate the instrument**: set the meter to 'pH' mode, if appropriate, and then place the electrode assembly in a standard solution of known pH, usually pH 7.00. This solution may be supplied as a liquid, or may be prepared by dissolving a measured amount of a calibration standard in water: calibration standards are often provided in tablet form, to be dissolved in water to give a particular volume of solution. Adjust the calibration control to give the correct reading. Remember that your calibration standards will only give the specified pH at a particular temperature, usually either 20 °C or 25 °C. If you are working at a different temperature, you must establish the actual pH of your calibration standards, either from the supplier, or from literature information.
6. **Remove the electrode assembly from the calibration solution and rinse again with distilled water:** dab off the excess water. Basic instruments have no further calibration steps (single-point calibration), while the more refined pH meters have additional calibration procedures.

   If you are using a basic instrument, you should check that your apparatus is accurate over the appropriate pH range by measuring the pH of another standard whose pH is close to that expected for the test solution. If the standard does not give the expected reading, the instrument is not functioning correctly: consult a member of staff.

   If you are using an instrument with a slope control function, this will allow you to correct for any deviation in electrical potential from that predicted by the theoretical relationship (at 25 °C, a change in pH of 1.00 unit should result in a change

Box 25.1 (continued)

in electrical potential of 59.16 mV) by performing a two-point calibration. Having calibrated the instrument at pH 7.00, immerse in a second standard at the same temperature as that of the first standard, usually buffered to either pH 4.00 or pH 9.00, depending upon the expected pH of your samples. Adjust the slope control until the exact value of the second standard is achieved (Fig. 25.2). A pH electrode and meter calibrated using the two-point method will give accurate readings over the pH range from 3 to 11: laboratory pH electrodes are not accurate outside this range, since the theoretical relationship between electrical potential and pH is no longer valid.

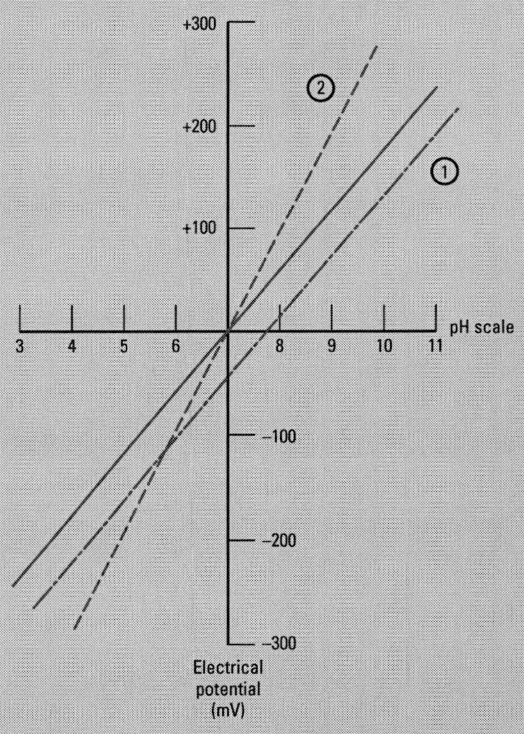

Fig. 25.2 The relationship between electrical potential and pH. The solid line shows the response of a calibrated electrode while the other plots are for instruments requiring calibration: 1 has the correct slope but incorrect isopotential point (calibration control adjustment is needed); 2 has the correct isopotential point but incorrect slope (slope control adjustment is needed).

7. **Once the instrument is calibrated, measure the pH of your solution(s)**, making sure that the electrode assembly is rinsed thoroughly between measurements. You should be particularly aware of this requirement if your solutions contain organic biological material, e.g. soil, tissue fluids, protein solutions, etc., since these may adhere to the glass electrode and affect the calibration of your instrument. If your electrode becomes contaminated during use, check with a member of staff before cleaning: avoid touching the surface of the glass electrode with abrasive material. Allow sufficient time for the pH reading to stabilise in each solution before taking a measurement: for unbuffered solutions, this may take several minutes, so do not take inaccurate pH readings due to impatience.

8. **After use, the electrode assembly must not be allowed to dry out.** Most pH electrodes should be stored in a neutral solution of KCl, either by suspending the assembly in a small beaker, or by using an electrode cap filled with the appropriate solution (typically 1.0 mol l^{-1} KCl buffered at pH 7.0). However, many labs simply use distilled water as a storage solution, leading to loss of ions from the interior of the electrode assembly. In practice, this means that pH electrodes stored in distilled water will take far longer to give a stable reading than those stored in KCl.

9. **Switch the meter to zero (where appropriate), but do not turn off the power**: pH meters give more stable readings if they are left on during normal working hours.

   Problems (and solutions) include: inaccurate and/or unstable pH readings caused by cross-contamination (rinse electrode assembly with distilled water and blot dry between measurements); development of a protein film on the surface of the electrode (soak in 1% w/v pepsin in 0.1 mol l^{-1} HCl for at least an hour); deposition of organic or inorganic contaminants on the glass bulb (use an organic solvent, such as acetone, or a solution of 0.1 mol l^{-1} disodium ethylenediamine-tetraacetic acid, respectively); drying out of the internal reference solutions (drain, flush and refill with fresh solution, then allow to equilibrate in 0.1 mol l^{-1} HCl for at least an hour); cracks or chips to the surface of the glass bulb (use a replacement electrode).

Buffers

Rather than simply measuring the pH of a solution, you may wish to *control* the pH, e.g. in metabolic experiments, or in a growth medium for cell culture (p. 233). In fact, you should consider whether you need to control pH in any experiment involving a biological system, whether whole organisms, isolated cells, subcellular components or biomolecules. One of the most effective ways to control pH is to use a buffer solution.

A buffer solution is usually a mixture of a weak acid and its conjugate base. Added protons will be neutralised by the anionic base while a reduction in protons, e.g. due to the addition of hydroxyl ions, will be counterbalanced by dissociation of the acid (Eqn [25.2]); thus the conjugate pair acts as a 'buffer' to pH change. The innate resistance of most biological fluids to pH change is due to the presence of cellular constituents that act as buffers, e.g. proteins, which have a large number of weakly acidic and basic groups in their amino acid side chains.

Definition

Buffer solution – one which resists a change in H^+ concentration (pH) on addition of acid or alkali.

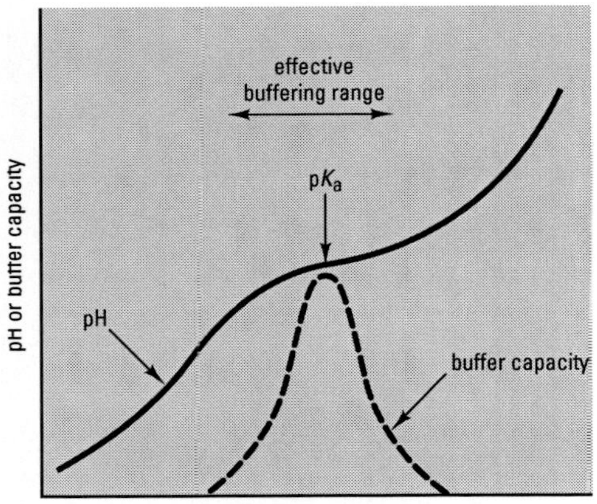

Fig. 25.3 Theoretical pH titration curve for a buffer solution. pH change is lowest and buffer capacity is greatest at the pK_a of the buffer solution.

Buffer capacity and the effects of pH

The extent of resistance to pH change is called the buffer capacity of a solution. The buffer capacity is measured experimentally at a particular pH by titration against a strong acid or alkali: the resultant curve will be strongly sigmoidal, with a plateau where the buffer capacity is greatest (Fig. 25.3). The mid-point of the plateau represents the pH where equal quantities of acid and conjugate base are present, and is given the symbol pK_a, which refers to the negative logarithm (to the base 10) of the acid dissociation constant, K_a, where

$$K_a = \frac{[H^+][A^-]}{[HA]} \quad [25.6]$$

By rearranging Eqn [25.6] and taking negative logarithms, we obtain:

$$pH = pK_a + \log_{10}\frac{[A^-]}{[HA]} \quad [25.7]$$

This relationship is known as the Henderson–Hasselbalch equation and it shows that the pH will be equal to the pK_a when the ratio of conjugate base to acid is unity, since the final term in Eqn [25.7] will be zero. Consequently, pK_a is an important factor in determining buffer capacity at a particular pH. In practical terms, a buffer solution will work most effectively at pH values about one unit either side of the pK_a.

Choosing a buffer *– an ideal buffer for biological purposes would possess the following characteristics:*

- *impermeability to biological membranes;*
- *biological stability and lack of interference with metabolic and biological processes;*
- *lack of significant absorption of ultraviolet or visible light;*
- *lack of formation of insoluble complexes with cations;*
- *minimal effect of ionic composition or salt concentration;*
- *limited pH change in response to temperature.*

Selecting an appropriate buffer

When selecting a buffer, you should be aware of certain limitations to their use. Citric acid and phosphate buffers readily form insoluble complexes with divalent cations, while phosphate can also act as a substrate, activator or inhibitor of certain enzymes. Both of these buffers contain biologically significant quantities of cations, e.g. Na^+ or K^+. TRIS (Table 25.3) is often toxic to biological systems: due to its high lipid solubility it can penetrate membranes, uncoupling electron transport reactions in whole cells and isolated organelles. In addition, it is markedly affected by temperature, with a tenfold increase in H^+ concentration from 4 °C to 37 °C. A number of zwitterionic molecules (having both positive

Table 25.3 pK_a values at 25 °C and M_r of some acids and bases (upper section) and some large organic zwitterions (lower section) commonly used in buffer solutions. For polyprotic acids, where more than one proton may dissociate, the pK_a values are given for each ionisation step. Only the trivial acronyms of the larger molecules are provided: their full names can be obtained from the catalogues of most chemical suppliers.

Acid or base	pK_a value(s)	M_r
Acetic acid	4.8	60.1
Boric acid	9.2	61.8
Citric acid	3.1, 4.8, 5.4	191.2
Glycylglycine	3.1, 8.2	132.1
Phosphoric acid	2.1, 7.1, 12.3	98.0
Phthalic acid	2.9, 5.5	166.1
Succinic acid	4.2, 5.6	118.1
TRIS (base)*	8.3	121.1
CAPS (free acid)	10.4	221.3
CHES (free acid)	9.3	207.3
HEPES (free acid)	7.5	238.3
MES (free acid)	6.1	213.2
MOPS (free acid)	7.2	209.3
PIPES (free acid)	6.8	302.4
TAPS (free acid)	8.4	243.3
TRICINE (free acid)	8.1	179.2

*Note that this compound is hygroscopic and should be stored in a desiccator; also see text regarding its biological toxicity (p. 156).

Table 25.4 Preparation of sodium phosphate buffer solutions for use at 25 °C. Prepare separate stock solutions of (a) disodium hydrogen phosphate (M_r = 141.96) and (b) sodium dihydrogen phosphate (M_r = 156.01), both at 200 mol m^{-3}. Buffer solutions (at 100 mol m^{-3}) are then prepared at the required pH by mixing together the volume of each stock solution shown in the table, then diluting to a final volume of 100 ml using distilled or deionised water

Required pH (at 25 °C)	Volume of stock (a) Na_2HPO_4 (ml)	Volume of stock (b) NaH_2PO_4 (ml)
6.0	6.2	43.8
6.2	9.3	40.7
6.4	13.3	36.7
6.6	18.8	31.2
6.8	24.5	25.5
7.0	30.5	19.5
7.2	36.0	14.0
7.4	40.5	9.5
7.6	43.5	6.5
7.8	45.8	4.2
8.0	47.4	2.6

and negative groups) have been introduced to overcome some of the disadvantages of traditional buffers. These newer compounds are often referred to as 'Good buffers', to acknowledge the work of Dr N.E. Good: HEPES is one of the most useful zwitterionic buffers, with a pK_a of 7.5 at 25 °C.

These zwitterionic substances are usually added to water as the free acid: the solution must then be adjusted to the correct pH with a strong alkali, usually NaOH or KOH. Alternatively, they may be used as their sodium or potassium salts, adjusted to the correct pH with a strong acid, e.g. HCl. Consequently, you may need to consider what effects such changes in ion concentration may have in a solution where zwitterions are used as buffers. In addition, zwitterionic buffers can interfere with protein determinations (e.g. Lowry method).

Figure 25.4 shows a number of traditional and zwitterionic buffers and their effective pH ranges. When selecting one of these buffers, aim for a pK_a which is in the direction of the expected pH change (Table 25.3). For example, HEPES buffer would be a better choice of buffer than PIPES for use at pH 7.2 for experimental systems where a pH increase is anticipated, while PIPES would be a better choice where acidification is expected.

Preparation of buffer solutions

Having selected an appropriate buffer, you will need to make up your solution to give the desired pH. You will need to consider two factors:

- The ratio of acid and conjugate base required to give the correct pH.
- The amount of buffering required; buffer capacity depends upon the absolute quantities of acid and base, as well as their relative proportions.

In most instances, buffer solutions are prepared to contain between 10 mmol l^{-1} and 200 mmol l^{-1} of the conjugate pair. While it is possible to calculate the quantities required from first principles using the Henderson–Hasselbalch equation (Eqn [25.7]), there are sources that tabulate the amount of substance required to give a particular volume of solution with a specific pH value for a range of buffers (e.g. Anon., 2007; Radiometer, 2006). For traditional buffers, it is customary to mix stock solutions of acidic and basic components in the correct proportions to give the required pH (Table 25.4). For zwitterionic acids, the usual procedure is to add the compound to

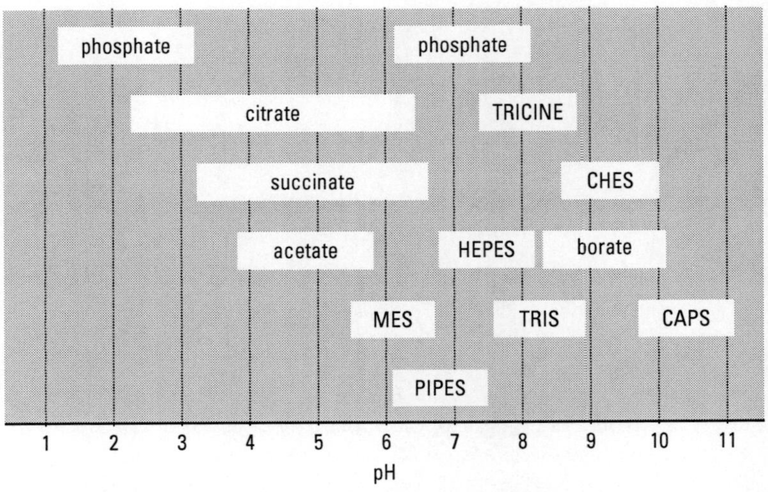

Fig. 25.4 Useful pH ranges of some commonly used buffers.

water, then bring the solution to the required pH by adding a specific amount of strong alkali or acid (obtained from tables). Alternatively, the required pH can be obtained by dropwise addition of alkali or acid, using a meter to check the pH, until the correct value is reached. When preparing solutions of zwitterionic buffers, the acid may be relatively insoluble. Do not wait for it to dissolve fully before adding alkali to change the pH – the addition of alkali will help bring the acid into solution (but make sure it has all dissolved before the desired pH is reached).

Remember that buffer solutions will only work effectively if they have sufficient buffering capacity to resist the change in pH expected during the course of the experiment. Thus a weak solution of HEPES (e.g. 10 mmol l^{-1}, adjusted to pH 7.0 with NaOH) will not be able to buffer the growth medium of a dense suspension of cells for more than a few minutes.

Finally, when preparing a buffer solution based on tabulated information, always confirm the pH with a pH meter before use.

Text references and sources for further study

Anon. (2007) *BioBasics Technical Library: Phosphate Buffer Table.* Available: http://www.sigmaaldrich.com/Area_of_Interest/Research_Essentials/Biochemicals/Key_Resources/Technical_Library.html Last accessed: 01/04/07.

Galster, H. (1991) *pH Measurement: Fundamentals, Methods, Applications, Instrumentation.* Wiley, New York.

Lide, D.R. (ed.) (2006) *CRC Handbook of Chemistry and Physics*, 87th edn. CRC Press, Boca Raton.

Radiometer (2006) *pH Theory and Practice: a Radiometer Analytical Guide.* Available: http://www.radiometer-analytical.com/all_resource_centre.asp?code=112 Last accessed: 01/04/07.

Rilbe, H. (1996) *pH and Buffer Theory – a New Approach.* Wiley, New York.

Study exercises

25.1 Practise interconverting pH values and proton concentrations. Express all answers to three significant figures.
(a) What is pH 7.4 expressed as [H^+] in mol l^{-1}?
(b) What is pH 4.1 expressed as [H^+] in mol m^{-3}?
(c) What is the pH of a solution containing H^+ at 2×10^{-5} mol l^{-1}?
(d) What is the pH of a solution containing H^+ at $10^{-12.5}$ mol l^{-1}?
(e) What is the pH of a solution containing H^+ at 2.8×10^{-5} mol m^{-3}?

25.2 Decide on a suitable buffer to use. In the following instances, choose a buffer that would be suitable:
(a) Maintaining the pH at 8.5 during an enzyme assay of a cell-free extract at 25 °C.
(b) Keeping a stable pH of 6.5 in an experiment to measure the uptake of radiolabelled glucose by a dense suspension of *E. coli.*
(c) Carrying out an assay of photosynthetic activity at pH 7.2 at temperatures of 10 °C, 20 °C and 30 °C.
(d) Stabilising pH at 5.5 during enzyme extraction, in a solution where you intend to measure total protein concentration at a later stage.

25.3 Practise using the Henderson–Hasselbalch equation. What are the relative proportions of deprotonated (A^-) and protonated (HA) forms of each substance at the following pH values:
(a) acetic acid ($pK_a = 4.8$) for use in an experiment at pH 3.8;
(b) boric acid ($pK_a = 9.2$) for use in an experiment at pH 9.5;
(c) HEPES ($pK_a = 7.5$) for use in an experiment at pH 8.1.

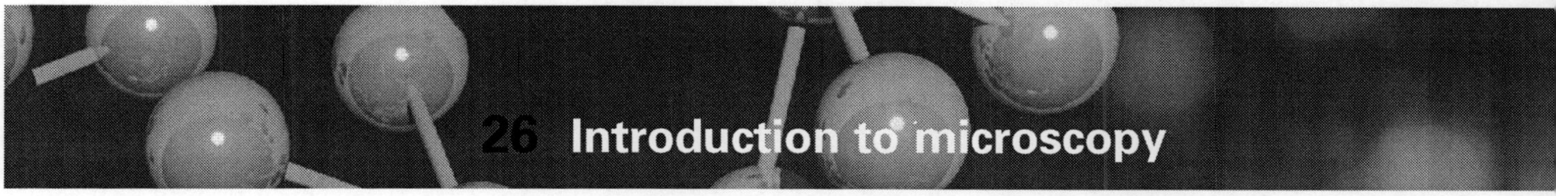

26 Introduction to microscopy

Many features of interest in biological systems are too small to be seen by the naked eye and can only be observed with a microscope. All microscopes consist of a coordinated system of lenses arranged so that a magnified image of a specimen is seen by the viewer (Figs 26.1 and 26.2). The main differences are the wavelengths of electromagnetic radiation used to produce the image, the nature and arrangement of the lens systems and the methods used to view the image.

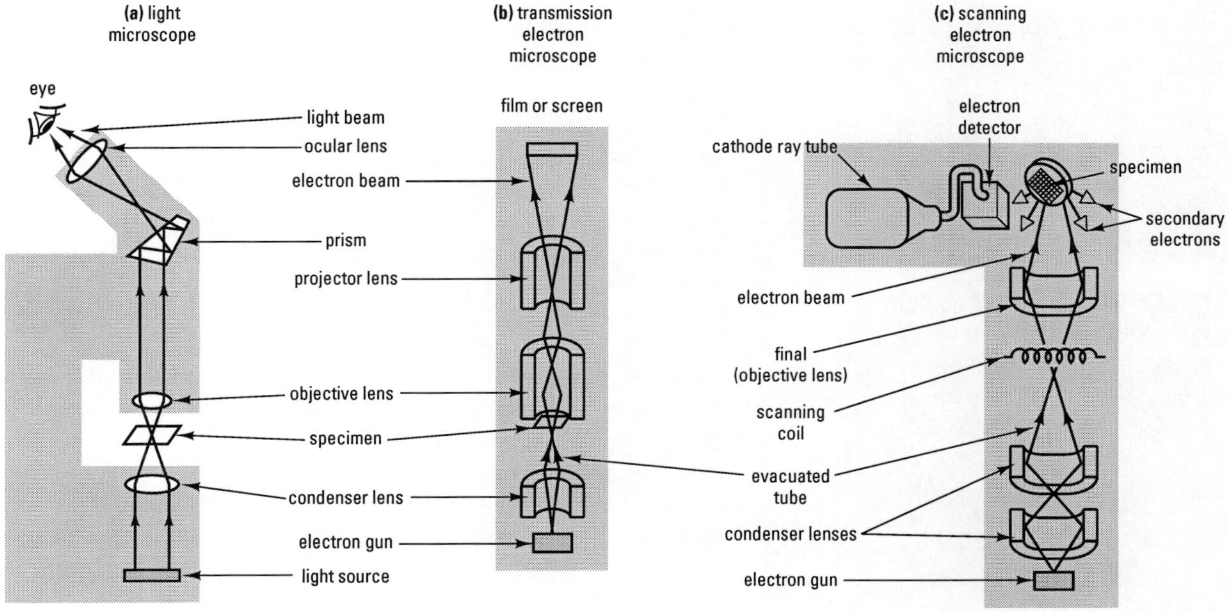

Fig. 26.1 Simplified diagrams of light and electron microscopes. Note that the electron microscopes are drawn upside down to aid comparison with the light microscope.

Table 26.1 Comparison of microscope types. Resolution is that obtained by a skilled user. LM, light microscope; SEM, scanning electron microscope; TEM, transmission electron microscope

	Type of microscope		
Property	LM	TEM	SEM
Resolution	200 nm	1 nm	10 nm
Depth of focus	Low	Medium	High
Field of view	Good	Limited	Good
Specimen preparation (ease)	Easy	Skilled	Easy
Specimen preparation (speed)	Rapid	Slow	Quite rapid
Relative cost of instrument	Low	High	High

Microscopes allow objects to be viewed with increased resolution and contrast. Resolution is the ability to distinguish between two points on the specimen – the better the resolution, the 'sharper' the image. Resolution is affected by lens design and inversely related to the wavelength of radiation used. Contrast is the difference in intensity perceived between different parts of an image. This can be enhanced (a) by the use of stains, and (b) by adjusting microscope settings, usually at the expense of resolution.

The three main forms of microscopy are light microscopy, transmission electron microscopy (TEM) and scanning electron microscopy (SEM). Their main properties are compared in Table 26.1 and their suitability for observing cells and organelles is shown in Table 26.2.

Light microscopy

Two forms of the standard light microscope are the binocular (compound) microscope described in detail in Chapter 27, and the dissecting microscope. These are the instruments most likely to be used in routine practical work. Figure 26.2(a) shows a typical image of cells

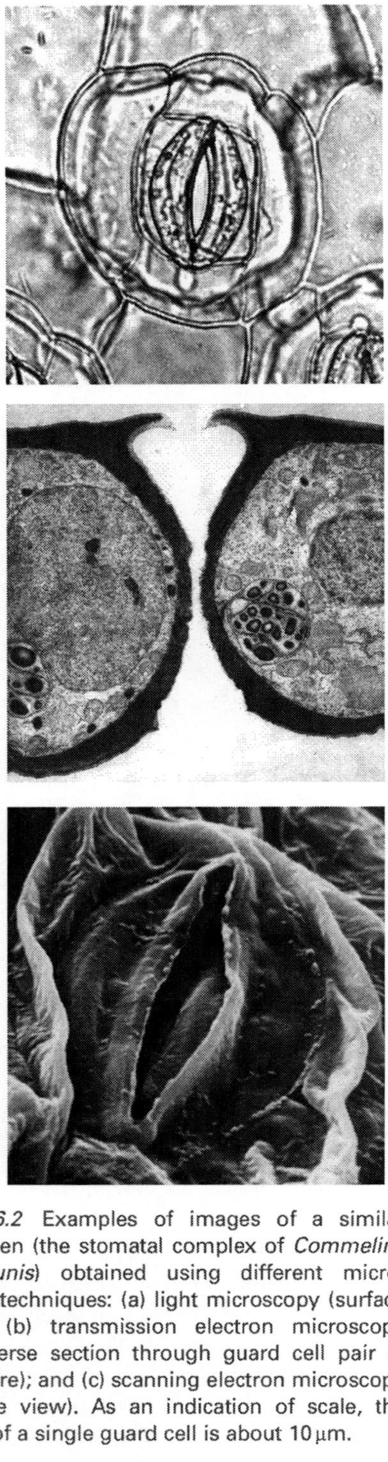

Fig. 26.2 Examples of images of a similar specimen (the stomatal complex of *Commelina communis*) obtained using different microscopic techniques: (a) light microscopy (surface view); (b) transmission electron microscopy (transverse section through guard cell pair at mid pore); and (c) scanning electron microscopy (surface view). As an indication of scale, the width of a single guard cell is about 10 μm.

viewed with a light microscope. In more advanced project work, you may use one or more of the following more sophisticated variants of light microscopy to improve image quality:

- Dark field illumination involves a special condenser which causes reflected and diffracted light from the specimen to be seen against a dark background. The method is particularly useful for near-transparent specimens and for delicate structures like flagella. Care must be taken with the thickness of slides used – air bubbles and dust must be avoided and immersion oil must be used between the dark field condenser and the underside of the slide.
- Ultraviolet microscopy uses short-wavelength UV light to increase resolution. Fluorescence microscopy uses radiation at UV wavelengths to make certain fluorescent substances (e.g. chlorophyll or fluorescent dyes which bind to specific cell components) emit light of visible wavelengths. Special light sources, lenses and mountants are required for UV and fluorescence microscopy and filters must be used to prevent damage to users' eyes.
- Phase contrast microscopy is useful for increasing contrast when viewing transparent specimens. It is superior to dark field microscopy because a better image of the interior of specimens is obtained. Phase contrast operates by causing constructive and destructive interference effects in the image, visible as increased contrast. Adjustments must be made, using a phase telescope in place of the eyepiece, for each objective lens and a matching phase condenser, and the microscope must be set up carefully to give optimal results.
- Nomarski or differential interference contrast (DIC) microscopy gives an image with a three-dimensional quality. However, the relief seen is optical rather than morphological and care should be taken in interpreting the result. One of the advantages of the technique is the extremely limited depth of focus which results: this allows 'optical sectioning' of a specimen.
- Polarised light microscopy can be used to reveal the presence and orientation of optically active components within specimens (e.g. starch grains, cellulose fibres), showing them brightly against a dark background.
- Confocal microscopy allows three-dimensional views of cells or thick sections. A finely focused laser is used to create electronic images of layered horizontal 'slices', usually after fluorescent staining. Images can be viewed individually or reconstructed to provide a 3D computer-generated image of the whole specimen.

Electron microscopes

Electron microscopes offer an image resolution up to 200 times better than light microscopes (Table 26.1) because they utilise radiation of shorter wavelength in the form of an electron beam. The electrons are produced by a tungsten filament operating in a vacuum and are focused by electromagnets. TEM and SEM differ in the way in which the electron beam interacts with the specimen: in TEM, the beam passes through the specimen (Fig. 26.1b), while in SEM the beam is scanned across the specimen and is reflected from the surface (Fig. 26.1c). In both cases, the beam must fall on a fluorescent screen before the image can be seen.

Table 26.2 Dimensions of some typical cells and organelles with an indication of suitable forms of microscopy for observing them. LM = light microscope; SEM = scanning electron microscope; TEM = transmission electron microscope. Column 2 data after Rubbi (1994).

Cell or organelle	Approximate diameter or width (μm)	Suitable form of microscopy
Prokaryote cell	0.15–5	LM, SEM, TEM
Eukaryote cell	10–100	LM, SEM, TEM
Fungal hypha	5–20	LM, SEM, TEM
Nucleus	5–25	LM, SEM, TEM
Mitochondrion	1–10	SEM, TEM
Chloroplast	2–8	LM, SEM, TEM
Golgi apparatus	1	SEM, TEM
Lysosome/ peroxisome	0.2–0.5	SEM, TEM
Plant cell wall	0.1–10	LM, SEM, TEM

Permanent images ('electron micrographs') are produced after focusing the beam on photographic film (Figs 26.2b and c).

You are unlikely to use either type of electron microscope as part of undergraduate practical work because of the time required for specimen preparation and the need for detailed training before these complex machines can be operated correctly. However, electron microscopy is extremely important in understanding cellular and sub-cellular structures and you may be shown electron micrographs, with one or more of the following objectives:

- to demonstrate cell ultrastructure (TEM);
- to show surface features of organisms, for example, when surface features are coated or when cells are 'freeze-fractured', then coated (SEM);
- to investigate changes in the number, size, shape and condition of cells and organelles (TEM);
- to carry out quantitative studies of cell and organelle disposition (TEM).

Preparative procedures

Without careful preparation of the material being studied, the biological structures viewed with any type of microscope can be rendered meaningless. Figure 26.3 summarises the processes involved for the main types of microscopy discussed above (see also Chapter 27 for light microscopy).

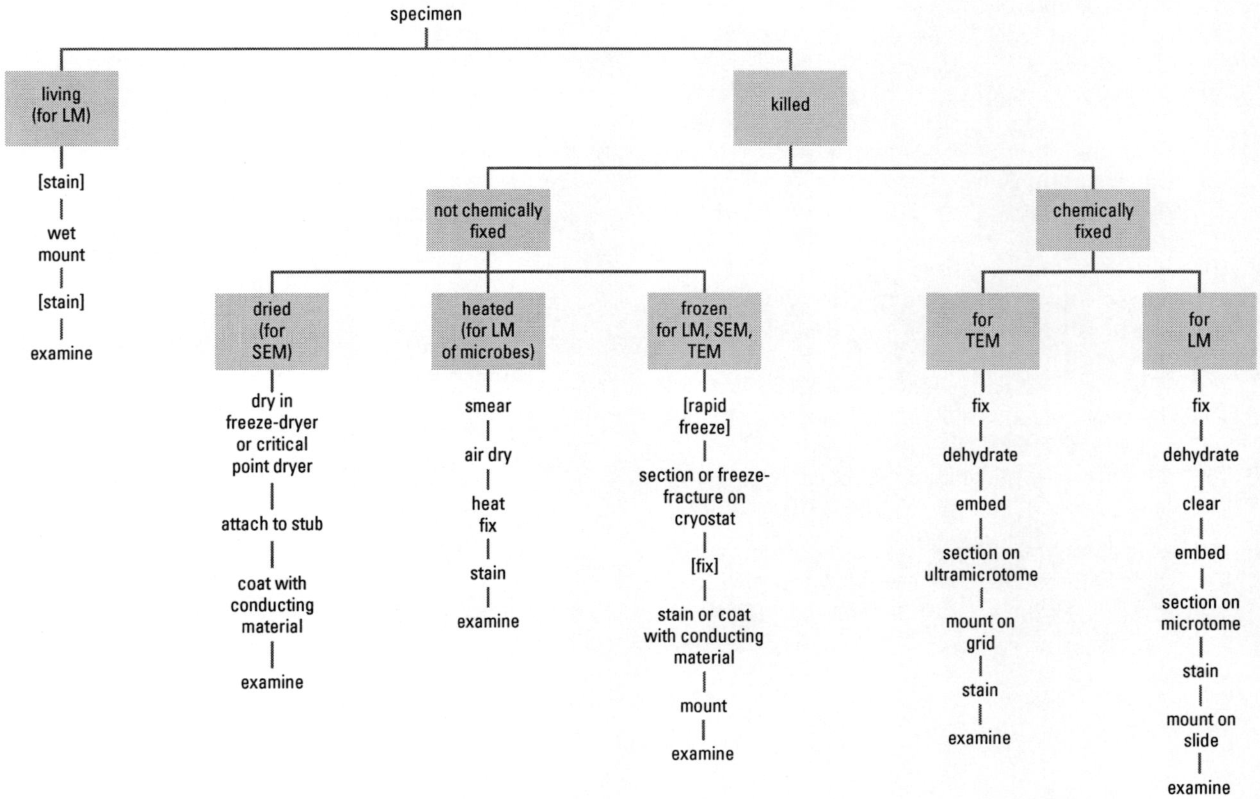

Fig. 26.3 Flowchart of procedures necessary to prepare specimens for different forms of microscopy. Steps enclosed in brackets are optional. LM, light microscope; SEM, scanning electron microscope; TEM, transmission electron microscope.

Text reference

Rubbi, C.P. (1994) *Light Microscopy Essential Data.* Wiley, Chichester.

Sources for further study

Bradbury, S. (1984) *An Introduction to the Optical Microscope.* Oxford University Press, Oxford.

Davidson, M.W. and Abramowitz, M. *Molecular Expressions. Exploring the World of Optics and Microscopy.* Available: http://micro.magnet.fsu.edu/index.html
Last accessed: 01/04/07.
[Covers many areas of basic knowledge underlying microscopy. Includes a microscopy primer]

Jeffries, C. *Microscopy Web Sites – by Organisation.* Available: http://www.ou.edu/research/electron/mirror/web-org .html
Last accessed: 01/04/07.
[Comprehensive set of links to microscopy websites]

Murphy, D.B. (2001) *Fundamentals of Light Microscopy and Electronic Imaging.* Wiley-Liss, New York.

Study exercises

26.1 Test your microscopy knowledge. Indicate whether the following statements about light microscopy, scanning electron microscopy (SEM) or transmission electron microscopy (TEM) are true or false.

(a) TEM allows you to see at finer resolution than light microscopy.
(b) TEM allows you to see surface features of specimens.
(c) SEM always requires staining of specimens.
(d) The resolution of TEM is about 200 times better than that of light microscopy.
(e) The resolution of a microscope is linked to the wavelength of electromagnetic radiation employed.
(f) The specimen in both TEM and SEM is viewed under near-vacuum conditions.
(g) Specimens for light microscopy can be living or dead.
(h) SEM provides better resolution than TEM.
(i) The depth of focus in light microscopy is greater than that in SEM.
(j) Light microscopy, SEM and TEM all involve the use of a condenser lens within the microscope.

26.2 Fill in the blanks in the following paragraph. Dark field microscopy involves shining reflected and _______ light on the specimen against a dark background. It is particularly useful for __________ specimens. UV microscopy uses short wavelength UV light in order to increase image _________. Phase contrast microscopy utilises constructive and destructive _______ effects to increase image ________. Nomarski microscopy provides a pseudo ____ image, with a very small depth of ________, allowing _____ _____ to be carried out. ______ light microscopy allows visualisation of optically active components in the specimen. Confocal microscopy involves the use of a ______ light source and can yield computer-generated 3D images.

26.3 Identify the missing preparative procedures. In each sequence below, one or two steps have been missed out. Using Fig. 26.3, identify the missing procedures.

(a) For light microscopy on a killed and fixed specimen: fix – dehydrate – clear – ________ – section – _________ – mount – examine.
(b) For light microscopy on a heat-fixed microbial specimen: smear – _______ – heat fix – _________ – examine.
(c) For TEM on a killed and fixed specimen: fix – _______ – embed – section – mount – stain – examine.

27 Setting up and using a light microscope

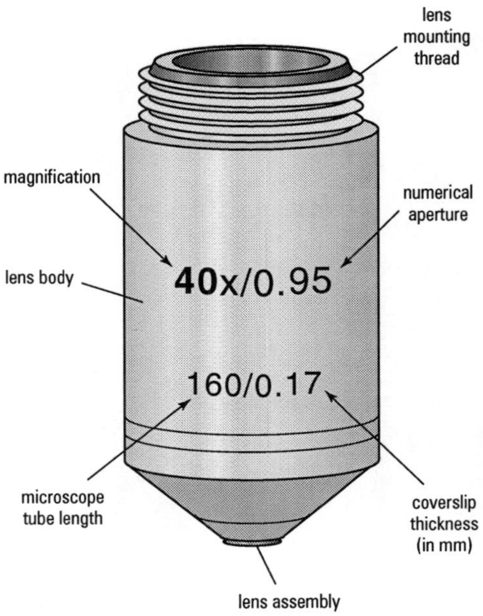

Fig. 27.1 Objective lens parameters. Most lenses are inscribed to show the details labelled above. The numerical aperture is a measure of the light-gathering power of the lens.

Correct use of the light microscope is one of the basic and essential laboratory skills. A standard undergraduate binocular microscope consists of three main types of optical unit: eyepiece, objective (Fig. 27.1) and condenser. These are attached to a stand which holds the specimen on a stage (Fig. 27.2). A monocular microscope is constructed similarly but has one eyepiece lens rather than two.

Setting up a binocular light microscope

Before using any microscope, familiarise yourself with its component parts.

***KEY POINT** Never assume that the previous person to use your microscope has left it set up correctly: apart from differences in users' eyes, the microscope needs to be properly set up for each lens combination used.*

The procedures outlined below are simplified to allow you to set up microscopes like those of the Olympus CX series (Fig. 27.2). For monocular microscopes, disregard instructions for adjusting eyepiece lenses in (5).

1. Place the microscope at a convenient position on the bench. Adjust your seating so that you are comfortable operating the focus and stage controls. Unwind the power cable, plug in and switch on after first ensuring that the lamp setting is at a minimum. Then adjust the lamp setting to about two-thirds of the maximum.
2. Select a low-power (e.g. ×10) objective. Make sure that the lens clicks home.
3. Set the eyepiece (ocular) lenses to your interpupillary distance; this can usually be read off a scale on the turret. You should now see a single circular field of vision. If you do not, try adjusting in either direction.
4. Put a prepared slide on the stage. Examine it first against a light source and note the position, colour and rough size of the specimen. Place the slide on the stage (coverslip up) and, viewing from the side, position it with the stage adjustment controls so that the specimen is illuminated.
5. Focus the image of the specimen using first the coarse and then the fine focusing controls (Fig. 27.3). The image will be reversed and upside down compared to that seen by viewing the slide directly.
   (a) If both eyepiece lenses are adjustable, set your interpupillary distance on the scale on each lens. Close your left eye, look through the right eyepiece with your right eye and focus the image with the normal controls. Now close your right eye, look through the left eyepiece with your left eye and focus the image by rotating the eyepiece holder. Take a note of the setting for future use.
   (b) If only the left eyepiece is adjustable, close your left eye, look with the right eye through the static right eyepiece and focus the image with the normal controls. Now close your right eye, look through the left eyepiece with your left eye and focus the image by rotating the eyepiece holder. Take a note of the setting for future use.

***Using binocular eyepieces** – if you do not know your interpupillary distance, ask someone to measure it with a ruler. You should stare at a fixed point in the distance while the measurement is taken. Take a note of the value for future use.*

***Issues for spectacle and contact lens wearers** – those who wear glasses can remove them for viewing, as microscope adjustments will accommodate most deficiencies in eyesight (except astigmatism). This is more comfortable and stops the spectacle lenses being scratched by the eyepiece holders. However, it may create difficulties in focusing when drawing diagrams. Those with contact lenses should simply wear them as normal for viewing.*

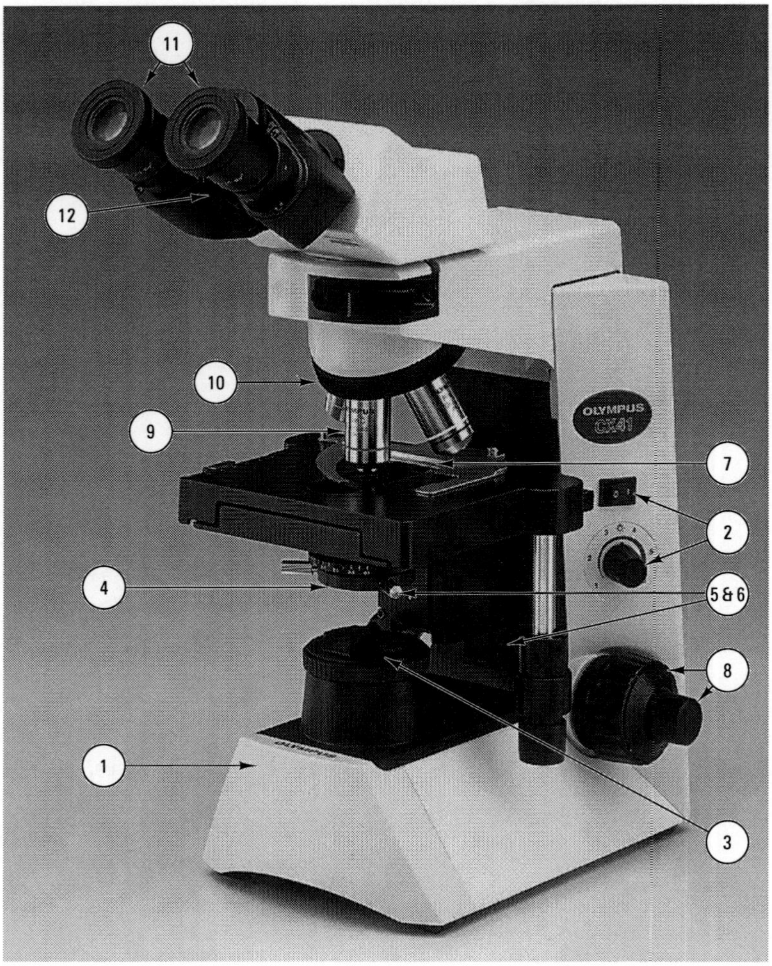

Fig. 27.2 Diagram of the Olympus binocular microscope model CX41.

- The lamp in the base of the stand (**1**) supplies light; its brightness is controlled by an on–off switch and voltage control (**2**). Never use maximum voltage or the life of the bulb will be reduced – a setting two-thirds to three-quarters of maximum should be adequate for most specimens. A field–iris diaphragm may be fitted close to the lamp to control the area of illumination (**3**).
- The condenser control focuses light from the condenser lens system (**4**) onto the specimen and projects the specimen's image onto the front lens of the objective. Correctly used, it ensures optimal resolution.
- The condenser–iris diaphragm (**5**) controls the amount of light entering and leaving the condenser; its aperture can be adjusted using the condenser–iris diaphragm lever (**6**). Use this to reduce glare and enhance image contrast by cutting down the amount of stray light reaching the objective lens.
- The specimen (normally mounted on a slide) is fitted to a mechanical stage or slide holder (**7**) using a spring mechanism. Two controls allow you to move the slide in the *x* and *y* planes. Vernier scales on the slide holder can be used to return to the same place on a slide. The fine and coarse focus controls (**8**) adjust the height of the stage relative to the lens systems. Take care when adjusting the focus controls to avoid hitting the lenses with the stage or slide.
- The objective lens (**9**) supplies the initial magnified image; it is the most important component of any microscope because its qualities determine resolution, depth of field and optical aberrations. The objective lenses are attached to a revolving nosepiece (**10**). Take care not to jam the longer lenses onto the stage or slide as you rotate the nosepiece. You should feel a distinct click as each lens is moved into position. The magnification of each objective is written on its side; a normal complement would be ×4, ×10, ×40 and ×100 (oil immersion).
- The eyepiece lens (**11**) is used to further magnify the image from the objective and to put it in a form and position suitable for viewing. Its magnification is written on the holder (normally ×10). By twisting the holder for one or both of the eyepiece lenses you can adjust their relative heights to take account of optical differences between your eyes. The interpupillary distance scale (**12**) and adjustment knob allow compensation to be made for differences in the distance between users' pupils.

From a photograph of an Olympus Model CX41 binocular microscope, supplied by Olympus microscopes, Olympus Optical Company (UK) Limited, published courtesy of Olympus Optical Co (UK) Ltd.

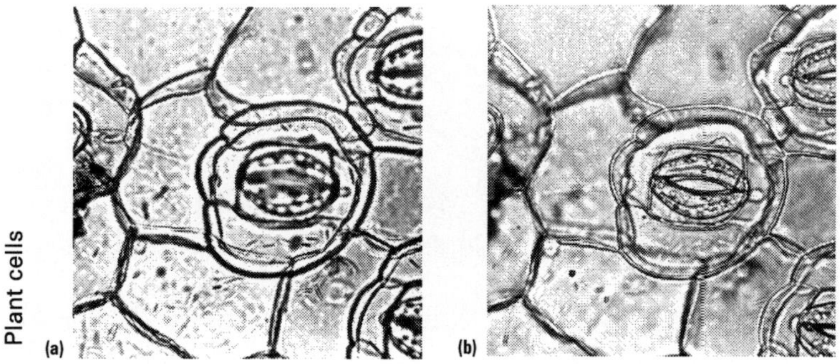

Fig. 27.3 Importance of correct focus in light microscopy. Stomatal complex of *Commelina communis*, a specimen that is a monolayer of cells approximately 30–50 μm thick. (a) Focal plane is on 'internal' walls of the cells; (b) focal plane is on the 'external' walls and stomatal pore. The two images are different and while it would not be possible to measure the stomatal pore in (a), it would not be possible to see the vacuolar crystals in (b). When looking at specimens, always use the fine focus control to view different focal planes.

SAFETY NOTE Take care when moving microscopes, not only because of the cost of replacement, but also because they weigh several kilograms and could cause injury if dropped. Always carry a microscope using two hands.

Adjusting a microscope with a field–iris diaphragm *– adjust this before the condenser–iris diaphragm: close it until its image appears in view as a circle of light, if necessary focusing on the edge of the circle with the condenser controls and centring it with the centring screws. Now open it so the whole field is just illuminated.*

Using high-power objectives *– never remove a slide while a high power objective lens (i.e. ×40 or ×100) is in position. Always turn back to the ×10 first. Having done this, lower the stage and remove the slide.*

6. Close the condenser–iris diaphragm (aperture–iris diaphragm), then open it to a position such that further opening has no effect on the brightness of the image (the 'threshold of darkening'). The edge of the diaphragm should not be in view. Turn down the lamp if it is too bright.
7. Focus the condenser. Place an opaque pointed object (the tip of a mounted needle or a sharp pencil point) on the centre of the light source. Adjust the condenser setting until both the specimen and needle tip/pencil point are in focus together. Check that the condenser–iris diaphragm is just outside the field of view.
8. For higher magnifications, swing in the relevant objective (e.g. ×40), carefully checking that there is space for it. Adjust the focus using the fine control only. If the object you wish to view is in the centre of the field with the ×10 objective, it should remain in view (magnified, of course) with the ×40. Adjust the condenser–iris diaphragm and condenser as before – the correct setting for each lens will be different.
9. When you have finished using the microscope, remove the last slide and clean the stage if necessary. Turn down the lamp setting to its minimum, then switch off. Clean the eyepiece lenses with lens tissue. Check that the objectives are clean. Unplug the microscope from the mains and wind the cable round the stand and under the stage. Replace the dust cover.

If you have problems in obtaining a satisfactory image, refer to Box 27.1; if this doesn't help, refer the problem to the class supervisor.

Procedure for observing transparent specimens

Some stained preparations and all colourless objects are difficult to see when the microscope is adjusted as above (Fig. 27.4). Contrast can be improved by closing down the condenser–iris diaphragm. Note that when you do this, diffraction haloes appear round the edges of objects. These obscure the image of the true structure of the specimen and may result in loss of resolution. Nevertheless, an image with increased contrast may be easier to interpret. Figure 27.5 shows several types of non-biological artefacts that you will come to recognise through practical experience.

Box 27.1 Problems in light microscopy and possible solutions

No image; very dark image; image dark and illuminated irregularly

- Microscope not switched on (check plug and base)
- Illumination control at low setting or off
- Objective nosepiece not clicked into place over a lens
- Diaphragm closed down too much or off-centre
- Lamp failure

Image visible and focused but pale and indistinct

- Diaphragm needs to be closed down further (see Fig. 27.4)
- Condenser requires adjustment

Image blurred and cannot be focused

- Dirty objective
- Dirty slide
- Slide upside down
- Slide not completely flat on stage
- Eyepiece lenses not set up properly for user's eyes
- Fine focus at end of travel
- Oil immersion objective in use, without oil

Dust and dirt in field of view

- Eyepiece lenses dirty
- Objective lens dirty
- Slide dirty
- Dirt on lamp glass or upper condenser lens

(a)
(b)
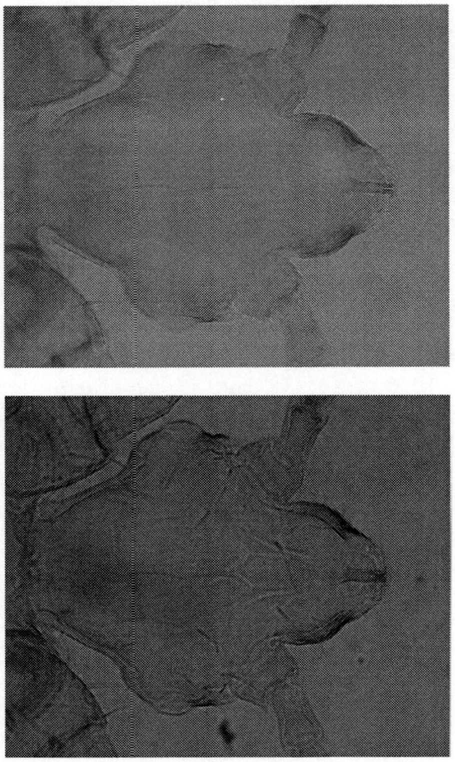

Fig. 27.4 Effect of closing the condenser–iris diaphragm on contrast. Head of human head louse, *Pediculus humanus capitis* (a) with condenser–iris diaphragm open; (b) with condenser–iris diaphragm closed (all other settings the same). Note the difference in detail that can be seen in (b), but also that image (b) is darker. When using the condenser–iris diaphragm in this way you may need to compensate by increasing the light setting.

Procedure for oil immersion objectives

These provide the highest resolution of which the light microscope is capable. They must be used with immersion oil filling the space between the objective lens and the top of the slide. The oil has the same refractive index as the glass lenses, so loss of light by reflection and refraction at the glass/air interface is reduced. This increases the resolution, brightness and clarity of the image and reduces aberration. Use oil immersion objective(s) as follows:

1. Check that the object of interest is in the field of view using e.g. the ×10 or ×40 objective.
2. Apply a single small droplet of immersion oil to the illuminated spot on the top of the slide, having first swung the ×40 objective away. Never use too much oil: it can run off the slide and mess up the microscope.
3. Move the high power (×100) oil immersion objective into position carefully, checking first that there is space for it. Focus on the specimen using the fine control only. You may need a higher brightness setting.
4. Perform condenser–iris diaphragm and condenser focusing adjustments as for the other lenses.
5. When finished, clean the oil immersion lens by gently wiping it with clean lens tissue. If the slide is a prepared one, wipe the oil off with lens tissue.

You should take great care when working with oil immersion lenses as they are the most expensive to replace. Because the working distance between the lens and coverslip is so short (less than 2 mm), it is easy to damage the lens surface by inadvertently hitting the slide or coverslip surface. You must also remember that they need oil to work properly. If working with an unfamiliar microscope, you can easily recognise oil immersion lenses. Look for a white or black ring on the lens barrel, near the lens, or for 'oil' clearly marked on the barrel.

Stains and staining procedures

Table 27.1 gives a number of widely used stains and their applications for animal, plant and microbial cells and tissues.

Establishing scale and measuring objects

Using an eyepiece graticule – *choose the eyepiece lens corresponding to your stronger eye and check that you have made the correct adjustments to the eyepiece lenses as detailed on* *p. 163.*

The magnification of a light microscope image is calculated by multiplying the objective magnification by the eyepiece magnification. However, the magnification of the image bears no certain relation to the magnification of any drawing of the image – you may equally well choose to draw the same image 10 mm or 10 cm long. For this reason, *it is essential to add a scale to all your diagrams.* You can provide either a bar giving the estimated size of an object of interest, or a bar of defined length (e.g. 100 μm.

The simplest method of estimating linear dimensions is to compare the size of the image to the diameter of the field of view. You can make a rough estimate of the field diameter by focusing on the millimetre scale of a transparent ruler using the lowest power objective. Estimate the diameter of this field directly, then use the information to work out the

Table 27.1 A selection of stains for light microscopy

	Stain	What it stains	Comments
Plant cells	Chlorazol black	Cell walls: black Nuclei: black, yellow or green Suberin: amber	The solvent used (70% ethanol in water or water alone) affects colours developed
	Neutral red	Living cells: pink (pH < 7)	A 'vital' stain used to determine cell viability or to visualise plant protoplasts in plasmolysis experiments; best used at neutral external pH
	Phloroglucinol/HCl	Lignified cell walls: red	Care is required because the acid may damage microscope lenses
	Ruthenium red	Pectins: red	Shows up the middle lamella
	Safranin + Fast green	Nuclei, chromosomes, cuticle and lignin: red Other components: green	Stain in safranin first, then counterstain with fast green (light green will substitute). A differentiation step is required
	Toluidine blue	Lignified cell walls: blue Cellulose cell walls: purple	Best to apply dilute and allow progressive staining to occur
Fungi and bacteria	Giemsa	Bacterial chromosome: purple Bacterial cytoplasm: colourless	Also used in zoology to stain protozoa
	Gram	Gram-positive bacteria: violet/purple Gram-negative bacteria: red/pink Yeasts: violet/purple	See p. 216 for procedure
	Gray	Bacterial flagella: red	Uses toxic chemicals: mercuric chloride and formaldehyde. Leifson's stain is an alternative
	Lactophenol cotton blue	Fungal cytoplasm: blue (hyphal wall unstained)	Shrinkage may occur
	Nigrosin or India ink	Background: grey–black	Negative stains for visualisation of capsules: requires a very thin film
	Shaeffer and Fulton	Bacterial endospores: green Vegetative cells: pink/red	Malachite green is primary stain, heated for 5 min. Counterstained with safranin
	Ziehl–Neelsen	Actinomycetes: red Bacterial endospores: red Other microbes: blue	Requires heat treatment of fuchsin primary stain, decolorisation with ethanol–HCl and a methylene blue counterstain (acid-fast structures remain red)
Animal cells	Azure A/eosin B	Nuclei, RNA: blue Basophilic cells: blue–violet Most other cells: pale blue Muscle cells: pink Necrosing cells: pink Cartilage matrix: red–violet Bone: pink Red blood cells: orange–red Mucins: green–blue/blue–violet	Used in pathology – shows up bacteria as blue; must be fresh; care required over pH: Mann's methyl blue/eosin gives similar results
	Chlorazol black	Chitin: greenish-black Nuclei: black, yellow or green Glycogen: pink or red	Solvent (70% v/v ethanol in water or water alone) affects colours formed
	Iron haematoxylin	Nuclei, chromosomes and red blood cells: black Other structures: grey or blue–black	Good for resolving fine detail; iron alum used as mordant before haematoxylin to differentiate
	Mallory	Nuclei: red Nucleoli: yellow Collagen, mucus: blue Red blood cells: yellow Cytoplasm: pink or yellow	Simple, one stage stain; fades within a year; not to be used with osmium-containing fixatives. Heidenhain's azan gives similar results but does not fade. Cason's one-step Mallory is a rapidly applied stain which is particularly good for connective tissue
	Masson's trichrome	Collagen, mucus: green Cytoplasm: orange or pink	Used as a counterstain after, for example, iron haematoxylin which will have stained nuclei black. Not to be used after osmium fixation
	Mayer's (haemalum and eosin; 'H&E')	Nuclei: blue/purple Cytoplasm: pink	Alum used as mordant for haematoxylin; eosin is the counterstain. To show up collagen, use van Gieson's stain as counterstain

field diameters at the higher powers *pro rata*. For example, if the field at an overall magnification of ×40 is 4 mm, at an overall magnification of ×100 it will be: 40/100×4 mm = 1.6 mm (1600 μm).

Greater accuracy can be obtained if an eyepiece micrometer (graticule) is used. This carries a fine scale and fits inside an eyepiece lens. The eyepiece micrometer is calibrated using a stage micrometer, basically a slide with a fine scale on it.

Figure 27.6 shows how to calibrate an eyepiece micrometer, along with a worked example. Once you have calibrated your eyepiece micrometer for each objective lens used, you can use it to measure objects: in the example shown in Fig. 27.6 the scale reading is multiplied by 2.65 μm to give the value in micrometres. So, if you measured the width of a human hair at 34 eyepiece micrometer units, then this will be equal to 34 × 2.65 = 90.1 μm. An alternative approach is to put a scale bar on a diagram, e.g. a 100 μm scale bar would be equivalent to the length of almost 38 eyepiece micrometer divisions. You should avoid putting too many significant figures in any estimates of dimensions: there may be quite large errors involved, which could make the implied accuracy misleading (spurious accuracy, p. 175).

Care and maintenance of your microscope

Microscopes are delicate precision instruments. Handle them with care and never force any of the controls. Never touch any of the glass surfaces with anything other than clean, dry lens tissue. Bear in mind that a replacement would be very expensive.

If moving a microscope, hold the stand above the stage with one hand and rest the base of the stand on your other hand. Always keep the microscope vertical (or the eyepieces may fall out). Put the microscope down gently.

(a) (b) (c) (d) (e) (f)

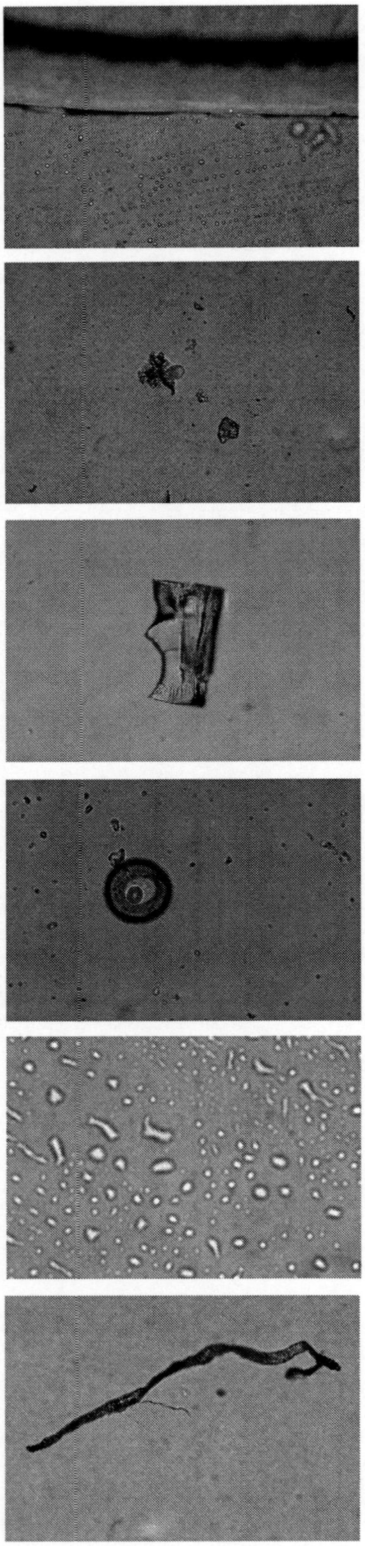

Fig. 27.5 Artefacts in light microscopy. Do not mistake any of the following for biological specimens: (a) edge of coverslip; (b) dust on slide; (c) shard of glass; (d) air bubble; (e) grease spots; (f) cellulose fibre.

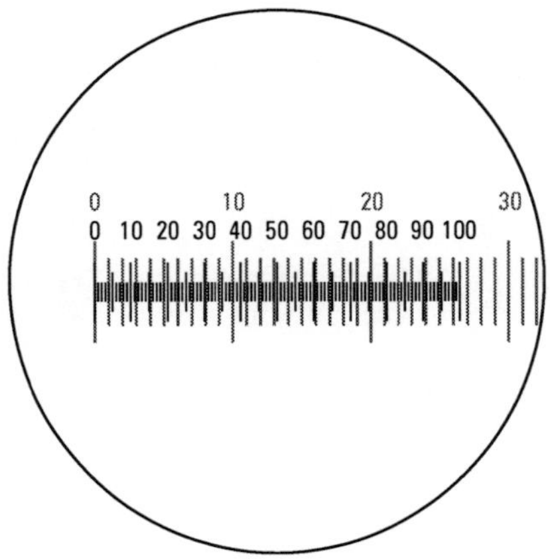

Fig. 27.6 Calibrating an eyepiece micrometer (graticule). Align two scales and read the number of stage micrometer divisions (shown in colour) for a particular number of eyepiece micrometer divisions (shown in black). In this case 26.5 scale divisions of 0.01 mm are equivalent to 100 eyepiece divisions, so each eyepiece division 0.265 ÷ 100 = 0.002 65 mm = 2.65 μm. This is a typical value for a total magnification of ×400 (e.g. ×40 objective and ×10 eyepiece).

Measuring specimens using a dissecting microscope *– because of the low magnification, sizes can generally be estimated by comparison with a ruler placed alongside the specimen. If accurate measurements are required, eyepiece graticules can be used.*

Clean lenses by gently wiping with clean, dry lens tissue. Use each piece of tissue once only. Try not to touch lenses with your fingers as oily fingerprints are difficult to clean off. Do not allow any solvent (including water) to come into contact with a lens; sea water is particularly damaging.

Sources for further study

Bradbury, S. and Bracegirdle, B. (1998) *Introduction to Light Microscopy*. Bios Scientific Publishers, Oxford.

Davidson, M.W. and Abramowitz, M. *Nikon Microscopy: Introduction to Microscope Objectives*. Available: http://www.microscopyu.com/articles/optics/objectiveintro.html
Last accessed: 01/04/07.

Olympus Microscopy Resource Centre. *Introduction to Optical Microscopy, Digital Imaging, and Photomicrography*. Available: http://www.olympusmicro.com/primer/index.html
Last accessed: 01/04/07.
[A primer on optical microscopy, including FAQs and links]

Study exercises

27.1 Test your knowledge of the parts of a binocular light microscope. Cover up the legend on the left of Fig. 27.2 with a piece of paper or card. Now identify the parts of the light microscope numbered on the diagram. Check your answers from the legend.

27.2 Identify roles of parts of a binocular light microscope. State briefly the *primary* role of each component of a standard binocular light microscope:
(a) condenser; (b) objective lens; (c) condenser–iris diaphragm; (d) interpupillary distance scale; (e) Vernier scales on the mechanical stage.

27.3 Identify the correct sequence of adjustments when setting up a light microscope.
(a) Focus specimen – set interpupillary distance – adjust condenser–iris diaphragm – make individual eyepiece adjustment – focus condenser – focus specimen.
(b) Make individual eyepiece adjustment – set interpupillary distance – focus condenser – focus specimen – adjust condenser–iris diaphragm – focus specimen.
(c) Set interpupillary distance – focus specimen – make individual eyepiece adjustment – adjust condenser–iris diaphragm – focus condenser – focus specimen.
(d) Make individual eyepiece adjustment – set interpupillary distance – focus specimen – adjust condenser–iris diaphragm – focus condenser – focus specimen.
(e) Focus specimen – make individual eyepiece adjustment – set interpupillary distance – focus condenser – adjust condenser–iris diaphragm.

27.4 Calculate the size of an object seen down a microscope using graticule readings. At 100× overall magnification, you find that 1 stage micrometer unit of 0.1 mm = 11 eyepiece micrometer units. The length of a parasitic worm on a slide viewed at this magnification appears to be 25 eyepiece units. What is its estimated true length to the nearest μm?

27.5 Select appropriate stains. From Table 27.1, identify stains you could use to help indicate the presence of the following:
(a) glycogen in a liver section; (b) woody (lignified) cells in a plant stem section; (c) a fungal pathogen in a leaf section; (d) living cells in an onion epidermal peel; (e) mucus in a lung section; (f) bacteria in a food sample.

The investigative approach

28 Making and recording measurements

> **Definitions**
>
> **Variable** – any characteristic or property which can take one of a range of values (contrast this definition with that for a **parameter,** which is a numerical constant in any particular instance).
>
> **Statistic** – an estimate of a parameter, obtained from a sample (e.g. the height of 18-year-old females based on those in your class).

The term data (singular = datum, or data value) refers to items of information, and you will use different types of data from a wide range of sources during your practical work. Consequently, it is important to appreciate the underlying features of data collection and measurement.

Variables

Biological variables (Fig. 28.1) can be classified as follows:

Quantitative variables

These are characteristics whose differing states can be described by means of a number. They are of two basic types:

- Continuous variables, such as length; these are usually measured against a numerical scale. Theoretically, they can take any value on the measurement scale. In practice, the number of significant figures of a measurement is directly related to the precision of your measuring system; for example, dimensions measured with Vernier callipers will provide readings of greater precision than a millimetre ruler. Many of the variables measured in biomolecular sciences are continuous and quantitative, e.g. mass, temperature, time, amount of product formed by an enzyme.
- Discontinuous (discrete) variables; these are always obtained by counting and therefore the data values must be whole numbers (integers), with no intermediate values.

> ***Working with discontinuous variables*** – *note that while the original data values must be integers, derived data and statistical values do not have to be whole numbers. Thus, it is perfectly acceptable to express the mean number of children per family as 2.4.*

Ranked variables

These provide data which can be listed in order of magnitude (i.e. ranked). A familiar example is the abundance of an organism in a sample, which is often expressed as a series of ranks, e.g. rare = 1, occasional = 2, frequent = 3, common = 4, and abundant = 5. When such data are given numerical ranks, rather than descriptive terms, they are sometimes called 'semi-quantitative data'. Note that the difference in magnitude between ranks need not be consistent. For example, regardless of whether there was a one-year or a five-year gap between offspring in a family; their ranks in order of birth would be the same.

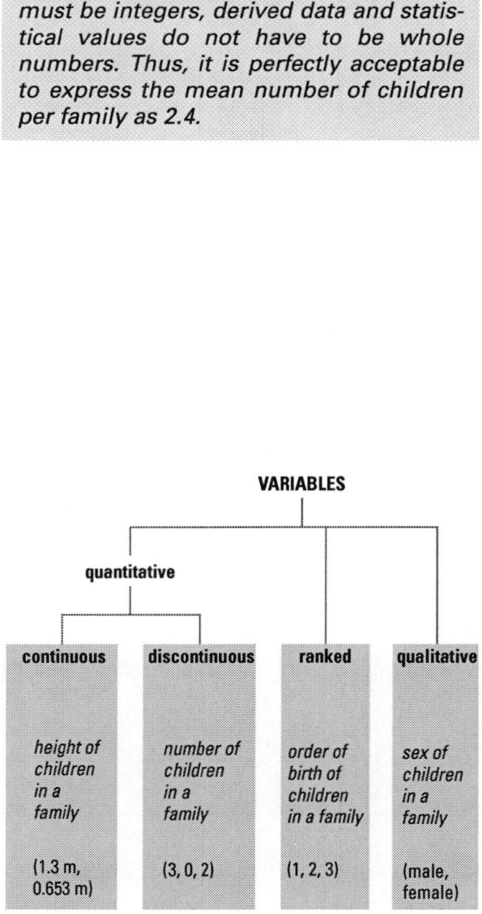

Fig. 28.1 Examples of the different types of variables as used to describe some characteristics of families.

Qualitative variables (attributes)

These are non-numerical and descriptive; they have no order of preference and, therefore, are not measured on a numerical scale nor ranked in order of magnitude, but are described in terms of categories. Examples include viability (i.e. dead or alive), shape (e.g. round, flat, elongated), presence/absence of a particular biomolecule in a sample.

Variables may be independent or dependent. Usually, the variable under the control of the experimenter (e.g. time) is the independent variable, while the variable being measured is the dependent variable (p. 189). Sometimes it is inappropriate to describe variables in this way and they are then referred to as interdependent variables (e.g. the length and breadth of an organism).

The majority of data values are recorded as direct measurements, readings or counts, but there is an important group, called derived (or computed), which result from calculations based on two or more data values, e.g. ratios, percentages, indices and rates.

Measurement scales

Variables may be measured on different types of scale:

- Nominal scale: this classifies objects into categories based on a descriptive characteristic. It is the only scale suitable for qualitative data.
- Ordinal scale: this classifies by rank. There is a logical order in any number scale used.
- Interval scale: this is used for quantitative variables. Numbers on an equal unit scale are related to an arbitrary zero point.
- Ratio scale: this is similar to the interval scale, except that the zero point now represents an absence of that character (i.e. it is an absolute zero). In contrast to the interval scale, the ratio of two values is meaningful (e.g. a temperature of 200 K is twice that of 100 K).

The measurement scale is important in determining the mathematical and statistical methods used to analyse your data. Table 28.1 presents a summary of the important properties of these scales. Note that you may be able to measure a characteristic in more than one way, or you may be able to convert data collected in one form to a different form. For instance, you might measure light in terms of the photon flux density

Examples A **nominal scale** for temperature is not feasible, since the relevant descriptive terms can be ranked in order of magnitude.

An **ordinal scale** for temperature measurement might use descriptive terms, ranked in ascending order, e.g. cold = 1, cool = 2, warm = 3, hot = 4.

The **Celsius scale** is an interval scale for temperature measurement, since the arbitrary zero corresponds to the freezing point of water (0 °C).

The **Kelvin scale** is a ratio scale for temperature measurement since 0 K represents a temperature of absolute zero (for information, the freezing point of water is 273.15 K on this scale).

Table 28.1 Some important features of scales of measurement

	Measurement scale			
	Nominal	Ordinal	Interval	Ratio
Type of variable	Qualitative (Ranked)* (Quantitative)*	Ranked (Quantitative)*	Quantitative	Quantitative
Examples	Species Sex Colour	Abundance scales Reproductive condition Optical assessment of colour development	Fahrenheit temperature scale Date (BC/AD)	Kelvin temperature scale Weight Length Response time Most physical measurements
Mathematical properties	Identity	Identity Magnitude	Identity Magnitude Equal intervals	Identity Magnitude Equal intervals True zero point
Mathematical operations possible on data	None	Rank	Rank Addition Subtraction	Rank Addition Subtraction Multiplication Division
Typical statistics used	Only those based on frequency of counts made: contingency tables, frequency distributions, etc. Chi-square test	Non-parametric methods, sign tests. Mann–Whitney *U*-test	Almost all types of test, *t*-test, analysis of variance (ANOVA), etc. (check distribution before using, Chapter 67)	Almost all types of test, *t*-test, ANOVA, etc. (check distribution before using, Chapter 67)

*In some instances (see text for examples).

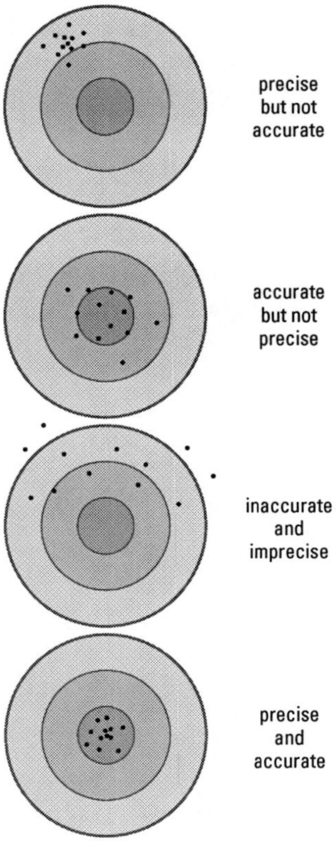

Fig. 28.2 'Target' diagrams illustrating precision and accuracy.

(p. 277) between particular wavelengths of the EMR spectrum (ratio scale), or simply as 'blue' or 'red' (nominal scale); you could find out the dates of birth of individuals (interval scale) but then use this information to rank them in order of birth (ordinal scale). Where there are no other constraints, you should use a ratio scale to measure a quantitative variable, since this will allow you to use the broadest range of mathematical and statistical procedures (Table 28.1).

Accuracy and precision

Accuracy is the closeness of a measured or derived data value to its true value, while precision is the closeness of repeated measurements to each other (Fig. 28.2). A balance with a fault in it (i.e. a bias, see below) could give precise (i.e. very repeatable) but inaccurate (i.e. untrue) results. Unless there is bias in a measuring system, precision will lead to accuracy and it is precision that is generally the most important practical consideration, if there is no reason to suspect bias. You can investigate the precision of any measuring system by repeated measurements of individual samples.

Absolute accuracy and precision are impossible to achieve, due to both the limitations of measuring systems for continuous quantitative data and the fact that you are usually working with incomplete data sets (samples). It is particularly important to avoid spurious accuracy in the presentation of results; include only those digits that the accuracy of the measuring system implies. This type of error is common when changing units (e.g. inches to metres) and in derived data, especially when calculators give results to a large number of decimal places.

Bias (systematic error) and consistency

Bias is a systematic or non-random distortion and is one of the most troublesome difficulties in using numerical data. Biases may be associated with incorrectly calibrated instruments, e.g. a faulty pipettor, or with experimental manipulations, e.g. shrinkage during the preservation of a specimen. Bias in measurement can also be subjective, or personal, e.g. an experimenter's preconceived ideas about an 'expected' result.

Bias can be minimised by using a carefully standardised procedure, with fully calibrated instruments. You can investigate bias in 'trial runs' by measuring a single variable in several different ways, to see whether the same result is obtained.

To avoid personal bias, 'blind' measurements should be made where the identity of individual samples is not known to the operator, e.g. using a coding system.

Minimising errors *– determine early in your study what the dominant errors are likely to be and concentrate your time and effort on reducing these.*

Measurement error

All measurements are subject to error, but the dangers of misinterpretation are reduced by recognising and understanding the likely sources of error and by adopting appropriate protocols and calculation procedures.

Working with derived data *– special effort should be made to reduce measurement errors because their effects can be magnified when differences, ratios, indices or rates are calculated.*

A common source of measurement error is carelessness, e.g. reading a scale in the wrong direction or parallax errors. This can be reduced greatly by careful recording and may be detected by repeating the measurement. Other errors arise from faulty or inaccurate equipment, but even a perfectly functioning machine has distinct limits to the accuracy and precision of its measurements. These limits are often quoted in manufacturers' specifications and are applicable when an instrument is new; however, you should allow for some deterioration with age. Further errors are introduced when the subject being studied is open to influences outside your control. Resolving such problems requires appropriate experimental design and representative sampling.

One major influence virtually impossible to eliminate is the effect of the investigation itself: even putting a thermometer in a liquid may change the temperature of the liquid. The very act of measurement may give rise to a confounding variable (p. 190), as discussed in Chapter 30. You should include descriptions of possible sources of errors and estimates of their likely importance in any report. However, do not use 'biological variability' as a catch-all excuse for poor technique or inadequacies in your experimental design.

Collecting and recording primary data

Understanding what's expected *– especially when taking notes for a lab-based practical, pay special attention to the aims and learning objectives (p. 22) of the session, as these will indicate the sorts of notes you should be taking, including content and diagrams, and the ways in which you should present these for assessment.*

Recording primary data *– never be tempted to jot down data on scraps of paper: you are likely to lose them, or to forget what individual values mean.*

When carrying out advanced lab work or research projects, you will need to master the important skill of managing data and observations and learn how to keep a record of your studies in a lab book. This is important for the following reasons:

- An accurate and neat record helps when using information later, perhaps for exam purposes or when writing a report.
- It allows you to practise important skills such as scientific writing, drawing diagrams, preparing graphs and tables and interpreting results.
- Analysing and writing up your data as you go along prevents a backlog at the end of your study time.
- You can show your work to a future employer to prove you have developed the skills necessary for writing up properly; in industry, this is vital so that others in your team can interpret and develop your work.

Recording numerical data *– write down only those figures that can be justified by the measurement technique (significant figures).*

KEY POINT A good set of lab notes should:

- ***outline the purpose of your experiment or observation;***
- ***set down all the information required to describe your materials and methods;***
- ***record all relevant information about your results or observations and provide a visual representation of the data;***
- ***note your immediate conclusions and suggestions for further experiments.***

Individual observations (e.g. laboratory temperature) can be recorded in the text of your notes, but tables are often the most convenient way to collect large amounts of information. When preparing a table for data collection, you should:

Designing a table for data collection *– use a spreadsheet or the table-creating facility in a word processor to create your table. This will allow you to reorganise it easily if required. Make sure there is sufficient space in each column for the values – if in doubt, err on the generous side.*

1. Use a concise title or a numbered code for cross referencing.
2. Decide on the number of variables to be measured and their relationship with each other, and lay out the table appropriately:

   (a) The first column of your table should show values of the independent (controlled) variable, with subsequent columns for the individual (measured) values for each replicate or sample.

(b) If several variables are measured for the same organism or sample, each should be given a row.
(c) In time-course studies, put the replicates as columns grouped according to treatment, with the rows relating to different times.

3. Make sure the arrangement reflects the order in which the values will be collected. Your table should be designed to make the recording process as straightforward as possible, to minimise the possibility of mistakes. For final presentation, a different arrangement may be best (Chapter 64).
4. Consider whether additional columns are required for subsequent calculations. Create a separate column for each mathematical manipulation, so the step-by-step calculations are clearly visible. Use a computer spreadsheet (Chapter 12) if you are manipulating lots of data.
5. Use a pencil to record data so that mistakes can be easily corrected.
6. Take sufficient time to record quantitative data unambiguously – use large clear numbers, making sure that individual numerals cannot be confused.
7. Record numerical data to an appropriate number of significant figures, reflecting the accuracy and precision of your measurement. Do not round off data values, as this might affect the subsequent analysis of your data.
8. Record discrete or grouped data as a tally chart, each row showing the possible values or classes of the variable. Provided that tally marks are of consistent size and spacing, this method has the advantage of providing an 'instant' frequency distribution chart.
9. Prepare duplicated recording tables if your experiments or observations will be repeated.
10. Explain any unusual data values or observations in a footnote. Don't rely on your memory.

Identifying your notes *– always put a date and time on each of your primary record sheets. You may also wish to add your name and details of the type of observation or experiment.*

Using photography and imaging to record information

These techniques are primarily for:

- documentation and validation of observations for project work (e.g. of agarose gels used for separation of nucleic acid molecules);
- producing 'hard copy' of microscope images (micrographs);
- preparation of visual material for talks and posters.

Photographs can be taken using two main systems: the digital camera and the film camera. While the basic operation of both types is similar, they differ in their light capturing and image recording systems.

Digital photography and imaging

This is the production of an electronically stored image, produced using a digital camera (still or video/movie). The image is held as a series of discrete picture elements called pixels (Fig. 28.3); the greater the number of pixels per unit surface area, the better the potential resolution of the image and the greater amount of detail captured.

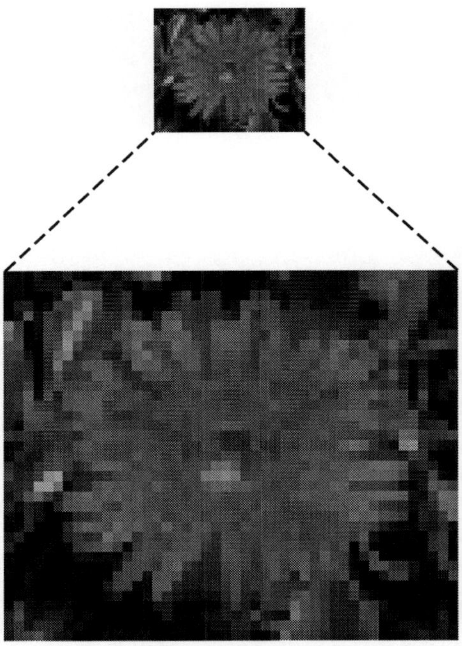

Fig. 28.3 Pixel structure in a digital image.

Film photography

Black and white or colour print photography involves capturing images on a film of plastic coated with light-sensitive silver halide salts. When exposed to light and subsequently chemically treated (developed), silver

Understanding the light sensitivity of photographic systems *– both digital and film media are rated in terms of their 'speed', measured in either ISO (=ASA) or DIN units. A good range for general purposes is 100 – 160 ISO (21–23 DIN).*

grains are produced to form a negative image, later converted to a positive image during printing. Transparency (slide or positive) films are based upon colour dyes and are effectively grainless. The choice of film is a compromise between speed and resolution: use slower film for fine detail. Faster film is more suitable for moving subjects or when light levels are low, but there is a loss of contrast and detail.

Comparing digital and film systems

Despite advances in digital image technology and reductions in the price of digital cameras and associated memory systems, even the best digital images are of somewhat lower quality (resolution) than high quality film images. However, digitised images have the advantages of speed and ease of production, since a developing stage is avoided, and are more readily manipulated. Because of the preview facility, they are also useful when it is important to confirm that a successful image has been taken, e.g. when an event is unrepeatable. On the other hand, images stored in magnetic form are not secure from computer failure, or damage from strong magnetic fields (see Box 10.1).

Camera and lens types

For both digital and film systems the primary difference is whether you preview the image through the lens (TTL), often referred to as single lens reflex (SLR), or whether a separate viewfinder is used. Digital cameras also incorporate a monitor which allows you to preview and review images. Lenses are either fixed, i.e. have a set focal length, or zoom, allowing a range of focal lengths, and hence magnification. Some digital systems have a digital zoom feature, often with loss of picture quality with increasing magnification. High quality images are usually taken with an SLR camera; such systems have better quality lenses and allow greater control over the final image.

Tips for better photography –

- *always use a tripod under low-light conditions;*
- *use electronic flash to provide uniformity of colour balance;*
- *shadowless lighting of small objects can be obtained using a ring-flash system.*

Measuring light for photography

The quantity of light can be measured by a photographic light meter, which may be external or built-in. The more light there is, the smaller the lens aperture you can use (larger 'f' number), and the greater will be the depth of field (= depth of focus). By using the camera on a tripod, you can use slow shutter speeds and allow larger 'f' numbers to be used to maximise depth of focus. The use of an electronic flash system to provide some or all of the lighting makes this even easier, since the effective shutter speed with electronic flash is extremely short (1 ms). Your choice of lighting arrangement will affect the quality of the picture: shadowless lighting is appropriate in some situations, but often shadows help to give three-dimensional form to subjects. In general, the use of more than one light source is advisable to prevent hard shadows.

***KEY POINT** To use an SLR camera properly, you need to understand the relationship between aperture (f-number), shutter speed and depth of field. For moving objects, give fast shutter speed priority by opening the lens aperture (small f-number). Where depth of field is required, close the lens aperture (high f-number); this will result in slower shutter speeds, so take care to avoid camera shake. Bracket your exposures (make extra exposures at plus and minus 1 f-number at same shutter speed) to ensure good results.*

Backing up digital image files – *since no physical copy exists, you must remember to download and back up digital image files to appropriate media (hard drive, network drive, CD or USB memory stick) as soon as possible.*

Image processing

Digital images are recorded as files on various forms of memory cards and are transferred to a computer for manipulation. For critical work, the rule is to capture images using the largest possible file size, as reductions (e.g. compression of file size) of digital information are possible, but you can never add information.

Digital image files are stored on disk in various formats (e.g. .bmp, .jpg) and can be manipulated using programs such as Adobe Photoshop or Paint Shop Pro. Aspects of image quality such as contrast and colour balance can be adjusted and images can be cropped and resized to suit your application (e.g. small file sizes are recommended for websites). Digital images are valuable for use with word processors and web-authoring software and can be transferred readily between such programs.

Film images require chemical development and are susceptible to irreversible damage by faulty development. The resulting image can only be modified to a limited extent during subsequent printing: this is one of the key reasons for selecting digital image systems.

Photomicrography

This requires the use of a camera system mounted on a microscope. The important operations for successful photomicrography are outlined below:

1. Carefully prepare the object: ensure thorough cleanliness of any slides or coverslips used.
2. Choose the correct film type and any filters required for alteration of colour balance, depending upon the type of light source available.
3. Decide on the magnification to be used: make sure that you know how this relates to the magnification of the negative/transparency. Always, include a photograph of a stage micrometer (p. 168) so the final magnification can be calculated and shown on the photograph.
4. Carefully focus the object onto the film plane: there will be a system-specific method for doing this (consult the manual, as this is a critical aspect).
5. Bracket extra exposures above and below the automatic settings, especially for colour photography.

Optimising print quality – *the quality of your print will depend upon the quality of your image, especially when producing colour prints from film.*

Printing images

The choice of print type will depend on its purpose. Glossy finish prints generally appear sharper than matt or other finishes. However, if preparing prints for display on a poster, matt/velvet finishes are often preferable, as there is less glare for the viewer. Printing from computers using ink-jet and similar printers requires care, since colours viewed on monitors do not always reproduce on paper as expected – experiment with settings and paper quality to obtain the appropriate results.

Using scales with images – *these are vital to the interpretation of biological images; always include information on the magnification used, as well as a scale bar.*

Adding scales and labels to images

Having acquired a suitable print, it is often necessary to add information to it:

- When digital images are copied into word processor or presentation software, additional information such as lettering, scales and labels can be added by overlaying a 'text box', or equivalent. You may need

Fig. 28.4 Use of white areas for labelling dark prints, illustrated by an SEM micrograph of a stellate hair on a leaf surface.

to make the background transparent, e.g. using the 'no fill' option in PowerPoint. Lettering can be added to printed media using transfer letters (e.g. Letraset).

- Choose a simple font type that will not distract and a size that is legible but not too dominant, e.g. 12 point in final reproduction. Use the same font type/size for related sets of images.
- Choose an appropriate part of the image for the lettering, i.e. a black area for a white letter. If the background is too variable, add a label on which to place the letters (see Fig. 28.4).

Recording details of project and research work

The recommended system is one where you make a dual record.

Primary record

The primary record is made at the bench or in the field. In this, you must concentrate on the detail of materials, methods and results. Include information that would not be used elsewhere, but which might prove useful in error tracing: for example, if you note how a solution was made up (exact volumes and weights used rather than concentration alone), this could reveal whether a miscalculation had been the cause of a rogue result. Note the origin, type and state of the chemicals and organism(s) used. Make rough diagrams to show the arrangement of replicates, equipment, etc. If you are forced to use loose paper to record data, make sure each sheet is dated and taped to your lab book, collected in a ring binder, or attached together with a treasury tag. The same applies to traces, printouts and graphs.

The basic order of the primary record should mirror that of a research report (see p. 111), including: the title and date; brief introduction; comprehensive materials and methods; the data and short conclusions.

Recording observations *– write down the actual observation, not your interpretation of it. For example, the colour of a biochemical test, rather than whether the test was positive or negative.*

Choosing a notebook for primary recording *– a spiral-bound notebook is good for making a primary record – it lies conveniently open on the bench and provides a simple method of dealing with major mistakes.*

Secondary record

You should make a secondary record concurrently or later in a bound book and it ought to be neater, in both organisation and presentation. This book will be used when discussing results with your supervisor, and when writing up a report or thesis, and may be part of your course assessment. While these notes should retain the essential features of the primary record, they should be more concise and the emphasis should move towards analysis of the experiment. Outline the aims more carefully at the start and link the experiment to others in a series (e.g. 'Following the results of Expt. D24, I decided to test whether . . .'). You should present data in an easily digested form, e.g. as tables of means or as summary graphs. Use appropriate statistical tests (Chapter 67) to support your analysis of the results. The choice of a bound book ensures that data are not easily lost.

Note that the dual method of recording deals with the inevitable untidiness of notes taken at the bench or in the field; these often have to be made rapidly, in awkward positions and in a generally complex environment. Writing a second, neater version forces you to consider again details that might have been overlooked in the primary record and provides a duplicate in case of loss or damage.

If you find it difficult to decide on the amount of detail required in Materials and Methods, the basic ground rule is to record enough information to allow a reasonably competent scientist to repeat your work

Choosing a book for secondary recording *– a hard-backed A4 size lined book is good because you will not lose pages. Graphs, printouts, etc. can be stuck in, as required.*

Formal aspects of keeping a record *– the diary aspect of the record can be used to establish precedence (e.g. for patentable research where it can be important to 'minute' where and when an idea arose and whose it was); for error tracing (e.g. you might be able to find patterns in the work affecting the results); or even for justifying your activities to a supervisor.*

Analyse your data as soon as possible – *always analyse and think about data immediately after collection as this may influence your subsequent activities.*

- *A graphical indication of what has happened can be particularly valuable.*
- *Carry out statistical analyses before moving on to the next experiment because apparent differences among treatments may not turn out to be statistically significant when tested.*
- *Write down any conclusions you make while analysing your data: sometimes those which seem obvious at the time of doing the work are forgotten when the time comes to write up a report or thesis.*
- *Note ideas for further studies as they occur to you – these may prove valuable later. Even if your experiment appears to be a failure, suggestions as to the likely causes might prove useful.*

SAFETY NOTE ***Maintaining and consulting communal lab records*** – *these activities may form a part of the safety requirements for working in a laboratory.*

exactly. You must tread a line between the extremes of pedantic, irrelevant detail and the omission of information essential for a proper interpretation of the data – better perhaps to err on the side of extra detail to begin with. An experienced worker can help you decide which subtle shifts in technique are important (e.g. batch numbers for an important chemical, or when a new stock solution is made up and used). Many important scientific advances have been made because of careful observation and record taking or because coincident data were recorded that did not seem of immediate value.

When creating a primary record, take care not to lose any of the information content of the data: for instance, if you only write down means and not individual values, this may affect your ability to carry out subsequent statistical analyses.

There are numerous ways to reduce the labour of keeping a record. Don't repeat Materials and Methods for a series of similar experiments; use devices such as 'method as for Expt. B4'. A photocopy might suffice if the method is derived from a text or article (check with supervisor). To save time, make up and copy a checklist in which details such as chemical batch numbers can be entered.

Using communal records

If working with a research team, you may need to use their communal databases. These avoid duplication of effort and ensure uniformity in techniques. You will be expected to use the databases carefully and to contribute to them properly. They might include:

- a shared notebook of common techniques (e.g. how to make up media or solutions);
- a set of simplified step-by-step instructions for use of equipment – manuals are often complex and poorly written and it may help to redraft them, incorporating any differences in procedure adopted by the group;
- an alphabetical list of suppliers of equipment and consumables (perhaps held on a card index system);
- a list of chemicals required by the group and where they are stored;
- the risk assessment sheets for dangerous procedures (p. 125);
- the record book detailing the use of radioisotopes and their disposal.

Sources for further study

Anon. *Measurement*. Available: http://wikipedia.org/wiki/Measurement
Last Accessed: 01/04/07.

Anon. *Suggestions for Keeping Laboratory Notebooks*. Available: http://otl.stanford.edu/inventors/resources/labnotebooks.html
Last accessed: 01/04/07.
[A Stanford University website which looks at the laboratory notebook from the patenting perspective]

Daney, C. *Photography and Digital Imaging Resources – a Collection of Links*. Available: http://www.mbay.net/~cgd/photo/pholinks.htm
Last accessed: 01/04/07.

Friedrich, G.W. *Basic Principles of Measurement. Methods of Inquiry*. Available: http://www.scils.rutgers.edu/~gusf/measurement.html
Last accessed: 01/04/07.
[Course notes covering valid aspects of measurement and inquiry]

National Instruments. *Measurement Encyclopedia*. Available: http://zone.ni.com/devzone/nidzgloss.nsf/glossary
Last accessed: 01/04/07.

Pechenik, J.A.A. (2001) *Short Guide to Writing about Biology*, 4th edn. Longman, Harlow.
[Chapter 8 is particularly relevant]

Study exercises

28.1 Classify variables. Decide on the type of variables used for the following measures, indicating whether they are quantitative or qualitative, continuous or discontinuous, and the type of scale that would be used.

(a) Number of organisms in a sample.
(b) Height of individuals in a sample.
(c) Eye colour of individuals in a sample.
(d) Species present in a sample.
(e) Date of a sample.
(f) Reproductive status of an animal.

28.2 Investigate types of error. A student weighed a set of standard masses on two electronic balances and obtained the readings shown in the table alongside. Explain these results in terms of the type of error involved in each case.

Comparison of weights of masses on two balances

	Standard mass (g)				
	10	25	50	100	250
Reading (balance A)	10.050	25.049	50.051	100.048	250.052
Reading (balance B)	10.004	25.011	50.021	100.039	250.102

28.3 Design a secondary record sheet for the collection and analysis of a set of count data for the number of bacterial colonies developed in each of 10 replicates for each of 5 different treatments. Assume that you need to calculate means, variable, etc. and then compare the results for each treatment.

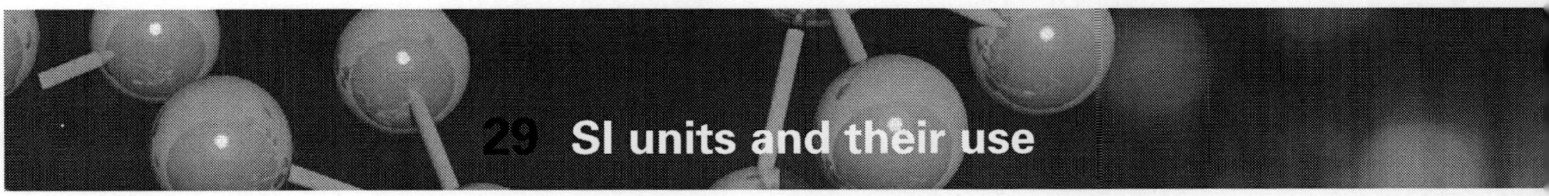

29 SI units and their use

Dimensionless measurements – *some quantities can be expressed as dimensionless ratios or logarithms (e.g. pH), and in these cases you do not need to use a qualifying unit.*

When describing a measurement, you normally state both a number and a unit (e.g. 'the length is 1.85 metres'). The number expresses the ratio of the measured quantity to a fixed standard, while the unit identifies that standard measure or dimension. Clearly, a single unified system of units is essential for efficient communication of such data within the scientific community. The Système International d'Unités (SI) is the internationally ratified form of the metre-kilogram-second system of measurement and represents the accepted scientific convention for measurements of physical quantities.

Another important reason for adopting consistent units is to simplify complex calculations where you may be dealing with several measured quantities. Although the rules of the SI are complex and the scale of the base units is sometimes inconvenient, to gain the full benefits of the system you should observe its conventions strictly.

The description of measurements in SI involves:

- seven base units and two supplementary units, each having a specified abbreviation or symbol (Table 29.1);
- derived units, obtained from combinations of base and supplementary units, which may also be given special symbols (Table 29.2);
- a set of prefixes to denote multiplication factors of 10^3, used for convenience to express multiples or fractions of units (Table 29.3).

Table 29.1 The base and supplementary SI units

Measured quantity	Name of SI unit	Symbol
Base units		
Length	metre	m
Mass	kilogram	kg
Amount of substance	mole	mol
Time	second	s
Electric current	ampere	A
Temperature	kelvin	K
Luminous intensity	candela	cd
Supplementary units		
Plane angle	radian	rad
Solid angle	steradian	sr

Table 29.2 Some important derived SI units

Measured quantity	Name of unit	Symbol	Definition in base units	Alternative in derived units
Energy	joule	J	$m^2\,kg\,s^{-2}$	N m
Force	newton	N	$m\,kg\,s^{-2}$	$J\,m^{-1}$
Pressure	pascal	Pa	$kg\,m^{-1}\,s^{-2}$	$N\,m^{-2}$
Power	watt	W	$m^2\,kg\,s^{-3}$	$J\,s^{-1}$
Electric charge	coulomb	C	A s	$J\,V^{-1}$
Electric potential difference	volt	V	$m^2\,kg\,A^{-1}\,s^{-3}$	$J\,C^{-1}$
Electric resistance	ohm	Ω	$m^2\,kg\,A^{-2}\,s^{-3}$	$V\,A^{-1}$
Electric conductance	siemens	S	$s^3\,A^2\,kg^{-1}\,m^{-2}$	$A\,V^{-1}$ or Ω^{-1}
Electric capacitance	farad	F	$s^4\,A^2\,kg^{-1}\,m^{-2}$	$C\,V^{-1}$
Luminous flux	lumen	lm	cd sr	
Illumination	lux	lx	$cd\,sr\,m^{-2}$	$lm\,m^{-2}$
Frequency	hertz	Hz	s^{-1}	
Radioactivity	becquerel	Bq	s^{-1}	
Enzyme activity	katal	kat	mol substrate s^{-1}	

Table 29.3 Prefixes used in the SI

Multiple	Prefix	Symbol	Multiple	Prefix	Symbol
10^{-3}	milli	m	10^3	kilo	k
10^{-6}	micro	μ	10^6	mega	M
10^{-9}	nano	n	10^9	giga	G
10^{-12}	pico	p	10^{12}	tera	T
10^{-15}	femto	f	10^{15}	peta	P
10^{-18}	atto	a	10^{18}	exa	E
10^{-21}	zepto	z	10^{21}	zetta	Z
10^{-24}	yocto	y	10^{24}	yotta	Y

Recommendations for describing measurements in SI units

Basic format

Example 10 μg is correct, while 10μg, 10 μg, and 10μ g are incorrect. 2.6 mol is right, but 2.6 mols is wrong.

- Express each measurement as a number separated from its units by a space. If a prefix is required, no space is left between the prefix and the unit it refers to. Symbols for units are only written in their singular form and do not require full stops to show that they are abbreviated or that they are being multiplied together.

- Give symbols and prefixes appropriate upper or lower case initial letters as this may define their meaning. Upper case symbols are named after persons but when written out in full they are not given initial capital letters.
- Show the decimal sign as a full point on the line. Some metric countries continue to use the comma for this purpose and you may come across this in the literature: commas should not therefore be used to separate groups of thousands. In numbers that contain many significant figures, you should separate multiples of 10^3 by spaces rather than commas.

Example n stands for nano and N for newtons.

Example 1 982 963.192 309 kg (perhaps better expressed as 1.982 963 192 309 Gg).

Compound expressions for derived units

- Take care to separate symbols in compound expressions by a space to avoid the potential for confusion with prefixes. Note, for example, that 200 m s (metre-seconds) is different from 200 ms (milliseconds).
- Express compound units using negative powers rather than a solidus (/): for example, write mol m^{-3} rather than mol/m^3. The solidus is reserved for separating a descriptive label from its units.
- Use parentheses to enclose expressions being raised to a power if this avoids confusion: for example, a photosynthetic rate might be given in mol CO_2 (mol photons)$^{-1}$ s^{-1}.
- Where there is a choice, select relevant (natural) combinations of derived and base units: e.g. you might choose units of Pa m^{-1} to describe a hydrostatic pressure gradient rather than kg m^{-2} s^{-2}, even though these units are equivalent and the measurements are numerically the same.

Use of prefixes

- Use prefixes to denote multiples of 10^3 (Table 29.3) so that numbers are kept between 0.1 and 1000.
- Treat a combination of a prefix and a symbol as a single symbol. Thus, when a modified unit is raised to a power, this refers to the whole unit including the prefix.
- Avoid the prefixes deci (d) for 10^{-1}, centi (c) for 10^{-2}, deca (da) for 10 and hecto (h) for 100 as they are not strictly SI.
- Express very large or small numbers as a number between 1 and 10 multiplied by a power of 10 if they are outside the range of prefixes shown in Table 29.3.
- Do not use prefixes in the middle of derived units: they should be attached only to a unit in the numerator (the exception is in the unit for mass, kg).

Examples

- 10 μm is preferred to 0.000 01 m or 0.010 mm.
- 1 mm^2 = 10^{-6} m^2 (not one-thousandth of a square metre).
- 1 dm^3 (1 litre) is more properly expressed in SI as 1×10^{-3} m^3.
- The mass of a neutrino is 10^{-36} kg.
- State as MW m^{-2} rather than W mm^{-2}.

***KEY POINT** For the foreseeable future, you will need to make conversions from other units to SI units, as much of the literature quotes data using imperial, c.g.s. or other systems. You will need to recognise these units and find the conversion factors required. Examples relevant to biology are given in Box 29.1. Table 29.4 provides values of some important physical constants in SI units.*

Box 29.1 Conversion factors between some redundant units and the SI

Quantity	SI unit/symbol	Old unit/symbol	Multiply number in old unit by this factor for equivalent in SI unit*	Multiply number in SI unit by this factor for equivalent in old unit*
Area	square metre/m^2	acre	$4.046\,86 \times 10^{3}$	$0.247\,105 \times 10^{-3}$
		hectare/ha	10×10^{3}	0.1×10^{-3}
		square foot/ft^2	0.092 903	10.763 9
		square inch/in^2	645.16×10^{-9}	$1.550\,00 \times 10^{6}$
		square yard/yd^2	0.836 127	1.195 99
Angle	radian/rad	degree/°	$17.453\,2 \times 10^{-3}$	57.295 8
Energy	joule/J	erg	0.1×10^{-6}	10×10^{6}
		kilowatt hour/kWh	3.6×10^{6}	$0.277\,778 \times 10^{-6}$
		calorie/cal	4.1868	0.2388
Length	metre/m	Ångstrom/Å	0.1×10^{-9}	10×10^{9}
		foot/ft	0.304 8	3.280 84
		inch/in	25.4×10^{-3}	39.370 1
		mile	$1.609\,34 \times 10^{3}$	$0.621\,373 \times 10^{-3}$
		yard/yd	0.914 4	1.093 61
Mass	kilogram/kg	ounce/oz	$28.349\,5 \times 10^{-3}$	35.274 0
		pound/lb	0.453 592	2.204 62
		stone	6.350 29	0.157 473
		hundredweight/cwt	50.802 4	$19.684\,1 \times 10^{-3}$
		ton (UK)	$1.016\,05 \times 10^{3}$	$0.984\,203 \times 10^{-3}$
Pressure	pascal/Pa	atmosphere/atm	101 325	$9.869\,23 \times 10^{-6}$
		bar/b	100 000	10×10^{-6}
		millimetre of mercury/mmHg	133.322	$7.500\,64 \times 10^{-3}$
		torr/Torr	133.322	$7.500\,64 \times 10^{-3}$
Radioactivity	becquerel/Bq	curie/Ci	37×10^{9}	$27.027\,0 \times 10^{-12}$
Temperature	kelvin/K	centigrade (Celsius) degree/°C	°C + 273.15	K − 273.15
		Fahrenheit degree/°F	(°F + 459.67) × 5/9	(K × 9/5) − 459.67
Volume	cubic metre/m^3	cubic foot/ft^3	0.028 316 8	35.314 7
		cubic inch/in^3	$16.387\,1 \times 10^{-6}$	$61.023\,6 \times 10^{3}$
		cubic yard/yd^3	0.764 555	1.307 95
		UK pint/pt	$0.568\,261 \times 10^{-3}$	1 759.75
		US pint/liq pt	$0.473\,176 \times 10^{-3}$	2 113.38
		UK gallon/gal	$4.546\,09 \times 10^{-3}$	219.969
		US gallon/gal	$3.785\,41 \times 10^{-3}$	264.172

*In the case of temperature measurements, use formulae shown.

Table 29.4 Some physical constants in SI terms

Physical constant	Symbol	Value and units
Avogadro's constant	N_A	$6.022\,174 \times 10^{23}$ mol^{-1}
Boltzmann's constant	k	$1.380\,626 \times 10^{-23}$ J K^{-1}
Charge of electron	e	$1.602\,192 \times 10^{-19}$ C
Gas constant	R	8.314 43 J K^{-1} mol^{-1}
Faraday's constant	F	$9.648\,675 \times 10^{4}$ C mol^{-1}
Molar volume of ideal gas at STP	V_0	0.022 414 m^3 mol^{-1}
Speed of light *in vacuo*	c	$2.997\,924 \times 10^{8}$ m s^{-1}
Planck's constant	h	$6.626\,205 \times 10^{-34}$ J s

Some implications of SI in bioscience

Using units for volume – *in this book, we use l and ml where you would normally find equipment calibrated in that way, but use SI units where this simplifies calculations. In formal scientific writing, constructions such as $1 \times 10^{-6}\,m^3$ (= 1 ml) and 1 mm^3 (= 1 μl) may be used.*

Volume

The SI unit of volume is the cubic metre, m^3, which is rather large for practical purposes. The litre (l) and the millilitre (ml) are technically obsolete, but are widely used and glassware is still calibrated using them. Note also that the US spelling is liter. In some instances you may find litre given the symbol L, rather than l to avoid confusion with the number 1 and with I, e.g. in certain font styles.

Mass

The SI unit for mass is the kilogram (kg) rather than the gram (g): this is unusual because the base unit has a prefix applied.

Expressing enzyme activity – *the derived SI unit is the katal (kat) which is the amount of enzyme that will transform 1 mol of substrate in 1 s.*

Amount of substance

You should use the mole (mol, i.e. Avogadro's constant, see Table 29.4) to express very large numbers. The mole gives the number of atoms in the atomic mass, a convenient constant. Always specify the elementary unit referred to in other situations (e.g. mol photons $m^{-2}\,s^{-1}$).

Converting between concentration units – *being able to express concentrations in different units (pp. 144–7) is important as this skill is frequently used when following instructions and interpreting data.*

Concentration

The SI unit of concentration, mol m^{-3}, is quite convenient for biological systems. It is equivalent to the non-SI term 'millimolar' (mM ≡ mmol l^{-1}) while 'molar' (M ≡ mol l^{-1}) becomes kmol m^{-3}. Note that the symbol M in the SI is reserved for mega and hence should not be used for concentrations. If the solvent is not specified, then it is assumed to be water (see Chapter 24).

Time

In general, use the second (s) when reporting physical quantities having a time element (e.g. give photosynthetic rates in mol $CO_2\,m^{-2}\,s^{-1}$). Hours (h), days (d) and years should be used if seconds are clearly absurd (e.g. samples were taken over a 5-year period). Note, however, that you may have to convert these units to seconds when doing calculations.

Definition

STP – standard temperature and pressure = 293.15 K and 0.101 325 MPa.

Temperature

The SI unit is the kelvin, K. The degree Celsius scale has units of the same magnitude, °C, but starts at 273.15 K, the melting point of ice at STP. Temperature is similar to time in that the Celsius scale is in widespread use, but note that conversions to K may be required for calculations. Note also that you *must not* use the degree sign (°) with K and that this symbol must be in upper case to avoid confusion with k for kilo; however, you *should* retain the degree sign with °C to avoid confusion with the coulomb, C.

Light

While the first six base units in Table 29.1 have standards of high precision, the SI base unit for luminous intensity, the candela (cd) and the derived units lm and lx (Table 29.2), are defined in 'human' terms. They are, in fact, based on the spectral responses of the eyes of 52 American GIs measured in

1923! Clearly, few organisms 'see' light in the same way as this sample of humans. Also, light sources differ in their spectral quality. For these reasons, it is better to use expressions based on energy or photon content (e.g. $W\,m^{-2}$ or mol photons $m^{-2}\,s^{-1}$) in studies other than those on human vision. Ideally you should specify the photon wavelength spectrum involved (see Chapter 40).

Sources for further study

Institute of Biology (1997) *Biological Nomenclature: Recommendations on Terms, Units and Symbols,* 2nd edn. Institute of Biology, London.

National Institute of Standards and Technology *The NIST Reference on Constants, Units and Uncertainty: International System of Units (SI)*. Available: http://physics.nist.gov/cuu/Units/
Last accessed: 01/04/07.

Pennycuick, C.J. (1988) *Conversion Factors: SI Units and Many Others: Over 2100 Conversion Factors for Biologists and Mechanical Engineers Arranged in 21 Quick-Reference Tables.* The University of Chicago Press, Chicago.

Rowlett, R. *How Many? A Dictionary of Units of Measurement.* Available: http://www.unc.edu/~rowlett/units
Last accessed: 01/04/07.

Tapson, F. *A Dictionary of Units.* Available:
http://www.ex.ac.uk/cimt/dictunit/dictunit.htm
Last accessed: 01/04/07.

Study exercises

29.1 Practise converting between units. Using Box 29.1 as a source, convert the following amounts into the units shown. Give your answers to three significant figures.

(a) 101 000 Pa into atmospheres
(b) One square yard into square millimetres
(c) One UK pint into millilitres
(d) 37 °C into kelvin
(e) 11 stone 6 pounds into kilograms.

29.2 Practise using prefixes appropriately. Simplify the following number/unit combinations using an appropriate prefix so that the number component lies between 0.1 and 1000.

(a) 10 000 mm
(b) 0.015 ml
(c) 5×10^9 J
(d) 65 000 m s^{-1}
(e) 0.000 000 000 1 g

29.3 Check units in an equation. The Hagan–Poiseuille equation describes water flux in smooth cylindrical pipes assuming laminar flow. This equation can be expressed as:

$$J_v = \frac{r^2 \times \delta P}{8\eta \times \delta x}$$

where

r = the radius of the cylindrical pipe (m);
η ('eta') = the viscosity of the liquid (Pa s);
δP = the pressure difference across the ends of the pipe (Pa);
δx = the length of the pipe (m).

Verify, by putting relevant units in place of the variables in this equation and simplifying the resulting relationship, that appropriate units for J_v are m s^{-1} (mean flow rate).

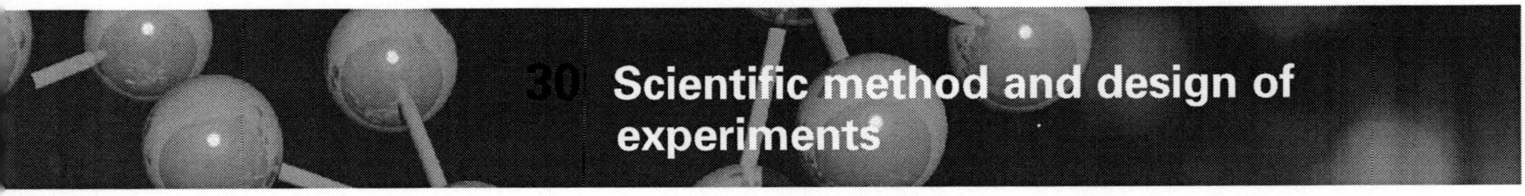

30 Scientific method and design of experiments

> **Definitions**
>
> There are many interpretations of the following terms. For the purposes of this chapter, the following will be used.
>
> **Paradigm** – theoretical framework so successful and well-confirmed that most research is carried out within its context and doesn't challenge it – even significant difficulties can be 'shelved' in favour of its retention.
>
> **Theory** – a collection of hypotheses that covers a range of natural phenomena – a 'larger-scale' idea than a hypothesis. Note that a theory may be 'hypothetical', in the sense that it is a tentative explanation.
>
> **Hypothesis** – an explanation tested in a specific experiment or by a set of observations. Tends to involve a 'small scale' idea.
>
> **(Scientific) Law** – this concept can be summarised as an equation (law) that provides a succinct encapsulation of a system, often in the form of a mathematical relationship. The term is often used in the physical sciences (e.g. 'Beer's Law', p. 280).

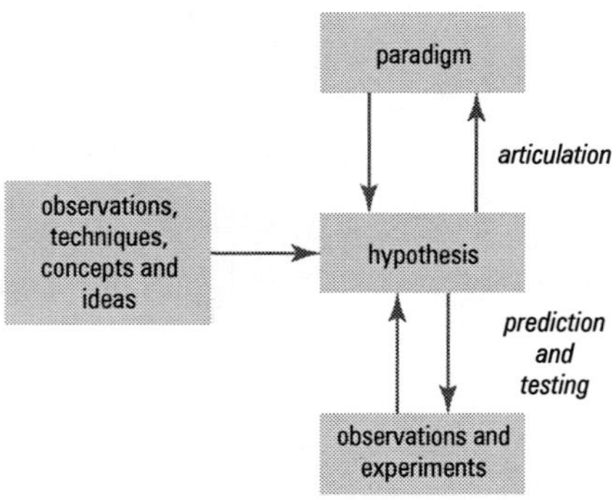

Fig. 30.1 A model of scientific method as used when testing hypotheses on a small scale. Hypotheses can arise as a result of various thought-processes on the part of the scientist, and are consistent with the overlying paradigm. Each hypothesis is testable by experiment or observation, leading to its confirmation or rejection. Confirmed hypotheses act to strengthen the status of the paradigm, but rejected ones do not immediately result in the paradigm's replacement.

Science is a body of knowledge based on observation and experiment. Biological scientists attempt to explain life in terms of theories and hypotheses. They make predictions from these hypotheses and test them by experiment or further observations. The philosophy and sociology that underlie this process are complex topics (see e.g. Chalmers, 1999) and any brief description must involve simplifications.

Figure 30.1 models the scientific process you are most likely to be involved in – testing hypotheses on a small scale. These represent the sorts of explanation that can give rise to predictions which can be tested by an experiment or a series of observations. For example, you might put forward the hypothesis that the rate of K^+ efflux from a particular cell type is dependent on the intracellular concentration of calcium ions. This might then lead to a prediction that the application of a substance known to decrease the intracellular concentration of calcium ions, would reduce K^+ efflux from the cells. An experiment could be set up to test this hypothesis and the results would either confirm or falsify the hypothesis.

If confirmed, a hypothesis is retained with greater confidence. If falsified, it is either rejected outright as false, or modified and retested. Alternatively, it might be decided that the experiment was not a valid test of the hypothesis (perhaps because it was later found that the applied substance could not penetrate the cell membrane to a presumed site of action).

Nearly all scientific research deals with the testing of small-scale hypotheses. These hypotheses operate within a theoretical framework that has proven to be successful (i.e. is confirmed by many experiments and is consistently predictive). This operating model or 'paradigm' is not changed readily, and even if a result appears that seems to challenge the conventional view, it would not be overturned immediately. The conflicting result would be 'shelved' until an explanation was found after further investigation. In the example used above, a relevant paradigm could be the notion that life processes are ultimately chemical in nature.

Although changes in paradigms are rare, they are important, and the scientists who recognise them become famous. For example, a 'paradigm shift' can be said to have occurred when Darwin's ideas about natural selection replaced special creation as an explanation for the origin of species. Generally, however, results from hypothesis-testing tend to support and develop ('articulate') the paradigm, enhancing its relevance and strengthening its status. Thus, research in the area of population genetics has developed and refined Darwinism.

Where do ideas for small-scale hypotheses come from? They arise from one or more thought processes on the part of a scientist:

- analogy with other systems;
- recognition of a pattern;
- recognition of departure from a pattern;
- invention of new analytical methods;
- development of a mathematical model;
- intuition;
- imagination.

Recently, it has been recognised that the process of science is not an entirely objective one. For instance, the choice of analogy which led to a new hypothesis might well be subjective, depending on past knowledge or understanding. Also, science is a social activity, where researchers put forward and defend viewpoints against those who hold an opposing view; where groups may work together towards a common goal; and where effort may depend on externally dictated financial opportunities and constraints. As with any other human activity, science is bound to involve an element of subjectivity.

Deciding whether to accept or reject a hypothesis *– this is sometimes clear-cut, as in some areas of genetics, where experiments can be set up to result in a binary outcome (Chapter 58). In many other cases, the existence of 'biological variation' means that statistical techniques need to be employed (Chapters 66 and 67; Box 58.2).*

No hypothesis can ever be rejected with certainty. Statistics allow us to quantify as vanishingly small the probability of an erroneous conclusion, but we are nevertheless left in the position of never being 100% certain that we have rejected all relevant alternative hypotheses, nor 100% certain that our decision to reject some alternative hypotheses was correct. However, despite these problems, experimental science has yielded and continues to yield many important findings.

KEY POINT ***The fallibility of scientific 'facts' is essential to grasp. No explanation can ever be 100% certain as it is always possible for a new alternative hypothesis to be generated. Our understanding of biology changes all the time as new observations and methods force old hypotheses to be retested.***

Definition

Mathematical model – an algebraic summary of the relationship between the variables in a system.

Quantitative hypotheses involve a mathematical description of the biological system. They can be formulated concisely by mathematical models. Formulating models is often useful because it forces deeper thought about mechanisms and encourages simplification of the system. A mathematical model:

- is inherently testable through experiment;
- identifies areas where information is lacking or uncertain;
- encapsulates many observations;
- allows you to predict the behaviour of the system.

Remember, however, that assumptions and simplifications required to create a model may result in it being unrealistic. Further, the results obtained from any model are only as good as the information put into it.

The terminology of experimentation

In many experiments, the aim is to provide evidence for causality. If x causes y, we expect, repeatably, to find that a change in x results in a change in y. Hence, the ideal experiment of this kind involves measurement of y, the dependent (measured) variable, at one or more values of x, the independent variable, and subsequent demonstration of some relationship between them. Experiments therefore involve comparisons of the results of treatments – changes in the independent variable as applied to an experimental subject. The change is engineered by the experimenter under controlled conditions.

Subjects given the same treatment are known as replicates (they may be called plots). A block is a grouping of replicates or plots. The blocks are contained in a field, i.e. the whole area (or time) available for the experiment (Fig. 30.2). These terms originated from the statistical analysis of agricultural experiments, but they are now used across all areas of bioscience.

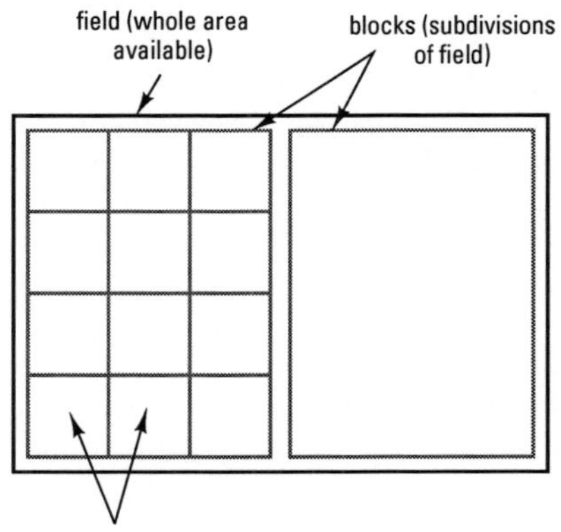

Fig. 30.2 Terminology and physical arrangement of elements in an experiment. Each block should contain the complete range of treatments (treatments may be replicated more than once in each block).

Why you need to control variables in experiments

Interpretation of experiments is seldom clear-cut because uncontrolled variables always change when treatments are given.

Confounding variables

Example Suppose you wish to investigate the effect of a metal ion on the growth of a culture. If you add the metal as a salt to the culture and then measure the growth, you will immediately introduce at least two confounding variables, compared with a control that has no salt added. Firstly, you will introduce an anion that may also affect growth in its own right, or in combination with the metal ion; secondly, you will alter the osmotic potential of the medium (see p. 150). Both of these effects could be tested using appropriate controls.

These increase or decrease systematically as the independent variable increases or decreases. Their effects are known as systematic variation. This form of variation can be disentangled from that caused directly by treatments by incorporating appropriate controls in the experiment. A control is really just another treatment where a potentially confounding variable is adjusted so that its effects, if any, can be taken into account. The results from a control may therefore allow an alternative hypothesis to be rejected. There are often many potential controls for any experiment.

The consequence of systematic variation is that you can never be certain that the treatment, and the treatment alone, has caused an observed result. By careful design, you can, however, 'minimise the uncertainty' involved in your conclusion. Methods available include:

- Ensuring, through experimental design, that the independent variable is the only major factor that changes in any treatment.
- Incorporating appropriate controls to show that potential confounding variables have little or no effect.
- Selecting experimental subjects randomly to cancel out systematic variation arising from biased selection.
- Matching or pairing individuals among treatments so that differences in response due to their initial status are eliminated.
- Arranging subjects and treatments randomly so that responses to systematic differences in conditions do not influence the results.
- Ensuring that experimental conditions are uniform so that responses to systematic differences in conditions are minimised. When attempting this, beware 'edge effects' where subjects on the periphery of the layout receive substantially different conditions from those in the centre.

Reducing edge effects *– one approach is to incorporate a 'buffer zone' of untreated subjects around the experiment proper.*

Nuisance variables

These are uncontrolled variables which cause differences in the value of y independently of the value of x, resulting in random variation. Experimental biology is characterised by the high number of nuisance variables that are found and their relatively great influence on results: biological data tend to have large errors. To reduce and assess the consequences of nuisance variables:

- incorporate replicates to allow random variation to be quantified;
- choose subjects that are as similar as possible;
- control random fluctuations in environmental conditions.

Constraints on experimental design

Evaluating design constraints *– a good way to do this is by processing an individual subject through the experimental procedures – a 'preliminary run' can help to identify potential difficulties.*

Box 30.1 outlines the important stages in designing an experiment. At an early stage, you should find out how resources may constrain the design. For example, limits may be set by availability of subjects, cost of treatment, availability of a chemical or bench space. Logistics may be a factor (e.g. time taken to record or analyse data).

Box 30.1 Checklist for designing and performing an experiment

1. **Preliminaries**
   (a) **Read background material** and decide on a subject area to investigate.
   (b) **Formulate a simple hypothesis to test.** It is preferable to have a clear answer to one question than to be uncertain about several questions.
   (c) **Decide which dependent variable you are going to measure and how.** Is it relevant to the problem? Can you measure it accurately, precisely and without bias?
   (d) **Think about and plan the statistical analysis of your results.** Will this affect your design?
2. **Designing**
   (a) **Find out the limitations on your resources.**
   (b) **Choose treatments that alter the minimum of confounding variables.**
   (c) **Incorporate as many effective controls as possible.**
   (d) **Keep the number of replicates as high as is feasible.**
   (e) **Ensure that the same number of replicates is present in each treatment.**
   (f) **Use effective randomisation and blocking arrangements.**
3. **Planning**
   (a) **List all the materials you will need.** Order any chemicals and make up solutions; grow, collect or breed the experimental subjects you require; check equipment is available.
   (b) **Organise space and/or time** in which to do the experiment.
   (c) **Account for the time taken to apply treatments and record results.** Make out a timesheet if things will be hectic.
4. **Carrying out the experiment**
   (a) **Record the results and make careful notes of everything you do** (see p. 176). Make additional observations to those planned if interesting things happen.
   (b) **Repeat the experiment** if time and resources allow.
5. **Analysing**
   (a) **Graph data as soon as possible** (during the experiment if you can). This will allow you to visualise what has happened and make adjustments to the design (e.g. timing of measurements).
   (b) **Carry out the planned statistical analysis.**
   (c) **Jot down conclusions and new hypotheses** arising from the experiment.

Your equipment or facilities may affect design because you cannot regulate conditions as well as you might desire. For example, you may be unable to ensure that temperature and lighting are equal over an experiment laid out in a glasshouse, or you may have to accept a great deal of initial variability if your subjects are collected from the wild. This problem is especially acute for experiments carried out under field conditions.

Using replicates

***Deciding the number of replicates in each treatment** – try to maximise the number of replicates in each treatment within the constraints of time and resources available.*

Replicate results show how variable the response is within treatments. They allow you to compare the differences among treatments in the context of the variability within treatments – you can do this *via* statistical tests such as analysis of variance (Chapter 67). Larger sample sizes tend to increase the precision of estimates of parameters and increase the chances of showing a significant difference between treatments if one exists. For statistical reasons (weighting, ease of calculation, fitting data to certain tests) it is often best to keep the number of replicates even. Remember that the degree of independence of replicates is important: sub-samples cannot act as replicate samples – they tell you about variability in the measurement method but not in the quantity being measured.

If the total number of replicates available for an experiment is limited by resources, you may need to compromise between the number of treatments and the number of replicates per treatment. Statistics can help here, for it is possible to work out the minimum number of replicates you would need to show a certain difference between pairs of means (say 10%) at a specified level of significance (say $P = 0.05$). For this, you need to obtain a prior estimate of variability within treatments (see Sokal and Rohlf, 1994).

Randomisation of treatments

The two aspects of randomisation you must consider are:

- positioning of treatments within experimental blocks;
- allocation of treatments to the experimental subjects.

For relatively simple experiments, you can adopt a completely randomised design; here, the position and treatment assigned to any subject are defined randomly. You can draw lots, use a random number generator on a calculator, or use the random number tables which can be found in most books of statistical tables (see Box 30.2).

Box 30.2 How to use random number tables to assign subjects to positions and treatments

This is one method of many that could be used. It requires two sets of n random numbers – where n is the total number of subjects used.

1. **Number the subjects in any arbitrary order** but in such a way that you know which is which (i.e. mark or tag them).
2. **Decide how treatments will be assigned,** e.g. first five subjects selected treatment A; second five – treatment B, etc.
3. **Use the first set of random numbers in the sequence obtained to identify subjects and allocate them to treatment groups** in order of selection as decided in (2).
4. **Map the positions for subjects in the block or field. Assign numbers to these positions using the second set of random numbers,** working through the positions in some arbitrary order, e.g. top left to bottom right.
5. **Match the original numbers given to subjects with the position numbers.**

To obtain a sequence of random numbers:

1. **Decide on the range of random numbers you need.**
2. **Decide how you wish to sample the random number tables** (e.g. row-by-row and top to bottom) and your starting point.
3. **Moving in the selected manner, read the sequence of numbers until you come to a group that fits your needs** (e.g. in the sequence 978186, 18 represents a number between 1 and 20). Write this down and continue sampling until you get a new number. If a number is repeated, ignore it. Small numbers need to have the appropriate number of zeros preceding (e.g. 5 = 05 for a range in the tens, 21 = 021 for a range in the hundreds).
4. **When you come to the last number required, you don't need to sample any more:** simply write it down.

Example: You find the following random number sequence in a table and wish to select numbers between 1 and 10 from it.

9059146823	4862925166	1063260345
1277423810	9948040676	6430247598
8357945137	2490145183	5946242208
6588812379	2325701558	3260726568

Working left to right and top to bottom, the order of numbers found is 5, 10, 3, 9, 4, 6, 2, 1, 8, 7 as indicated by coloured type. If the table is sampled by working row-by-row right to left from bottom to top, the order is 6, 10, 7, 2, 9, 3, 4, 8, 1, 5.

Example If you knew that soil type varied in a graded fashion across a field, you might arrange blocks to be long thin rectangles at right angles to the gradient to ensure conditions within the block were as even as possible.

3 × 3

A	C	B
B	A	C
C	B	A

4 × 4

A	C	B	D
D	B	C	A
C	D	A	B
B	A	D	C

Fig. 30.3 Examples of Latin square arrangements for 3 and 4 treatments. Letters indicate treatments; the number of possible arrangements for each size of square increases greatly as the size increases.

Example A Latin square format is used for agar bioassays of antibiotics and vitamins (p. 238). Replicates are arranged so that gradients across the plate are cancelled out.

A completely randomised layout has the advantage of simplicity but cannot show how confounding variables alter in space or time. This information can be obtained if you use a blocked design in which the degree of randomisation is restricted. Here, the experimental space or time is divided into blocks, each of which accommodates the complete set of treatments (Fig. 30.2). When analysed appropriately, the results for the blocks can be compared to test for differences in the confounding variables and these effects can be separated out from the effects of the treatments. The size and shape (or timing) of the block you choose is important: besides being able to accommodate the number of replicates desired, the suspected confounding variable should be relatively uniform within the block.

A Latin square is a method of placing treatments so that they appear in a balanced fashion within a square block or field. Treatments appear once in each column and row (see Fig. 30.3), so the effects of confounding variables can be 'cancelled out' in two directions at right angles to each other. This is effective if there is a smooth gradient in some confounding variable over the field. It is less useful if the variable has a patchy distribution, where a randomised block design might be better.

Latin square designs are useful in serial experiments where different treatments are given to the same subjects in a sequence (e.g. Fig. 30.4). A disadvantage of Latin squares is the fact that the number of plots is equal to the number of replicates, so increases in the number of replicates can only be made by the use of further Latin squares.

Pairing and matching subjects

The paired comparison is a special case of blocking used to reduce systematic variation when there are two treatments. Examples of its use are:

- 'Before and after' comparison. Here, the pairing removes variability arising from the initial state of the subjects, e.g. weight gain of mice on a diet, where the weight gain may depend on the initial weight.
- Application of a treatment and control to parts of the same subject or to closely related subjects. This allows comparison without complications arising from different origin of subjects, e.g. drug or placebo given to sibling rats, virus-containing or control solution swabbed on left or right halves of a leaf.
- Application of treatment and control under shared conditions. This allows comparison without complications arising from different environments of subjects, e.g. rats in a cage, plants in a pot.

Matched samples represent a restriction on randomisation where you make a balanced selection of subjects for treatments on the basis of some attribute or attributes that may influence results, e.g. age, sex, prior history. The effect of matching should be to 'cancel out' the unwanted source(s) of variation. Disadvantages include the subjective element in choice of character(s) to be balanced, inexact matching of quantitative characteristics, the time matching takes and possible wastage of unmatched subjects.

When analysed statistically, both paired comparisons and matched samples can show up differences between treatments that might otherwise be rejected on the basis of a fully randomised design, but note that the statistical analysis may be different.

month \ animal	1	2	3	4	5
1	A	C	D	B	E
2	D	B	E	A	C
3	E	D	B	C	A
4	C	E	A	D	B
5	B	A	C	E	D

Fig. 30.4 Example of how to use a Latin square design to arrange sequential treatments. The experimenter wishes to test the effect of drugs A–E on weight gain, but only has five animals available. Each animal is fed on a control diet for the first 3 weeks of each month, then on a control diet plus drug for the last week. Weights are taken at start and finish of each treatment. Each animal receives all treatments.

Definition

Interaction – where the effect of treatments given together is greater or less than the sum of their individual effects.

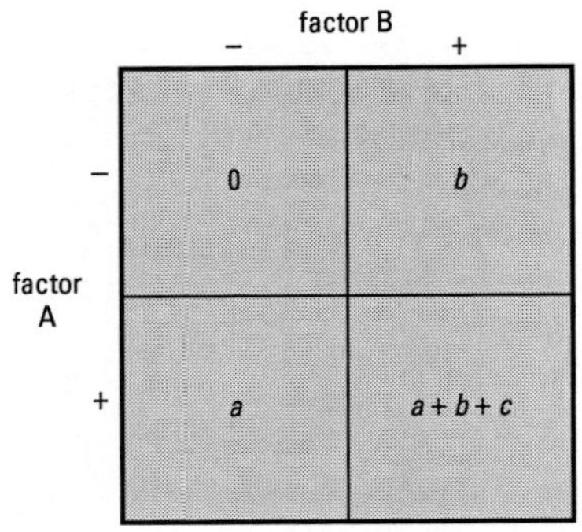

Fig. 30.5 Design of a simple multifactorial experiment. Factors A and B have effects *a* and *b* when applied alone. When both are applied together, the effect is denoted by $a + b + c$.

- If $c = 0$, there is no interaction (e.g. $2 + 2 + c = 4$).
- If *c* is positive, there is a positive interaction (synergism) between A and B (e.g. $2 + 2 + c = 5$).
- If *c* is negative, there is a negative interaction (antagonism) between A and B (e.g. $2 + 2 + c = 3$).

***Reporting results** – it is good practice to report how many times your experiments were repeated (in Materials and Methods); in the Results section, you should add a statement saying that the illustrated experiment is representative, or one explaining any differences between results obtained.*

Multifactorial experiments

The simplest experiments are those in which one treatment (factor) is applied at a time to the subjects. This approach is likely to give clear-cut answers, but it could be criticised for lacking realism. In particular, it cannot take account of interactions among two or more conditions that are likely to occur in real life. A multifactorial experiment (Fig. 30.5) is an attempt to do this; the interactions among treatments can be analysed by specialised statistics.

Multifactorial experiments are economical on resources because of 'hidden replication'. This arises when two or more treatments are given to a subject because the result acts statistically as a replicate for each treatment. Choice of relevant treatments to combine is important in multifactorial experiments; for instance, an interaction may be present at certain concentrations of a chemical but not at others (perhaps because the response is saturated). It is also important that the measurement scale for the response is consistent, otherwise spurious interactions may occur. Beware when planning a multifactorial experiment that the numbers of replicates do not get out of hand: you may have to restrict the treatments to 'plus' or 'minus' the factor of interest (as in Fig. 30.5).

Repetition of experiments

Even if your experiment is well designed and analysed, only limited conclusions can be made. Firstly, what you can say is valid for a particular place and time, with a particular investigator, experimental subject and method of applying treatments. Secondly, if your results were significant at the 5% level of probability, there is still an approximately one-in-twenty chance that the results did arise by chance. To guard against these possibilities, it is important that experiments are repeated. Ideally, this would be done by an independent scientist with independent materials. However, it makes sense to repeat work yourself so that you can have full confidence in your conclusions. Many scientists recommend that experiments are done three times in total, but this may not be possible in undergraduate practical classes or project work.

Text references

Chalmers, A.F. (1999) *What is This Thing Called Science?* 3rd edn. Open University Press, Buckingham.

Sokal, R.R. and Rohlf, F.J. (1994) *Biometry*, 3rd edn. W.H. Freeman, San Francisco.

Sources for further study

Heath, D. (1995) *An Introduction to Experimental Design and Statistics for Biology.* UCL Press, London.

Maber, J. (1999) *Data Analysis for Biomolecular Sciences.* Longman, Harlow.

Quinn, G.P. and Keough, M.J. (2002) *Experimental Design and Data Analysis for Biologists.* Cambridge University Press, Cambridge.

Study exercises

30.1 Consider the application of a Latin square design. Treatments W, X, Y and Z are to be applied to potted plants in a glasshouse where the researcher suspects there may be a slight gradation in temperature and light over an oblong bench that can hold a total of 48 pots. You decide to use an experimental design consisting of three 4×4 blocks of plants, each arranged in a different Latin square design (see below). Assign treatments to the locations in the diagram below. Explain why this design will help to eliminate the effects of the confounding variables.

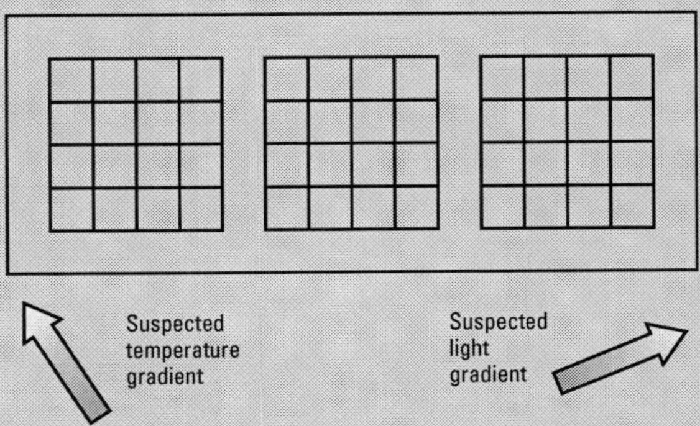

Layout for glasshouse experiment

30.2 Generate random numbers. Produce a list of 20 random whole numbers between 1 and 5 using a spreadsheet. If using MS Excel, investigate the RAND() and INT functions. The RAND() function produces a random number between 0 and 1, so you will need to multiply by a constant factor to scale your final output appropriately. Copy your test formula(e) to several cells to test empirically whether it works.

30.3 Investigate possible interactions. Treatments A, B and C involve tests of three different nutrients on proliferation of cells in a culture of carrot cells. Analyse the results in the table below to determine whether any interactions have occurred between the four possible combinations of treatments. No complex statistical analysis is required, simply a calculation of mean experimental effect (treatment minus controls) in each combined treatment and a comparison with the data for the relevant treatments on their own (in other words, assume that observed differences reflect true underlying differences). Classify the results as 'no interaction', 'antagonism' or 'synergism'.

Results of cell proliferation experiment

	Replicate (growth of callus in g)				
Treatment	1	2	3	4	5
Control	4.8	5.2	5.2	4.8	5.0
A	6.5	7.3	7.0	7.1	7.1
B	7.7	8.3	8.5	7.5	8.0
C	10.7	10.0	9.9	9.8	9.6
A+B	6.6	7.2	6.8	7.2	7.2
A+C	11.9	11.7	12.0	12.0	12.4
B+C	17.3	16.8	17.1	16.6	17.2
A+B+C	12.8	12.9	13.1	13.3	12.9

31 Project work

> ***Obtaining ethical approval*** *– if any aspect of your project involves work with human or animal subjects, then you must obtain the necessary ethical clearance before you begin; consult your department's ethical committee for details.*

Research projects are an important component of the final-year syllabus for most degree programmes in the biosciences, while shorter projects may also be carried out during courses in earlier years. Project work can be extremely rewarding, although it does present a number of challenges. The assessment of your project is likely to contribute significantly to your degree grade, so all aspects of this work should be approached in a thorough manner.

> ***Using the Internet as an information source for project work*** *– since many university departments and research groups have home pages on the World Wide Web, searches using relevant key words may indicate where research in your area is currently being carried out. Academics usually respond positively to emailed questions about their area of expertise.*

Deciding on a topic to study

Assuming you have a choice, this important decision should be researched carefully. Make appointments to visit possible supervisors and ask them for advice on topics that you find interesting. Use library texts and research papers to obtain further background information. Perhaps the most important criterion is whether the topic will sustain your interest over the whole period of the project. Other things to look for include:

- Opportunities to learn new skills. Ideally, you should attempt to gain experience and skills that you might be able to 'sell' to a potential employer.
- Ease of obtaining valid results. An ideal project provides a means to obtain 'guaranteed' data for your report, but also the chance to extend knowledge by doing genuinely novel research.
- Assistance. What help will be available to you during the project? A busy lab with many research students might provide a supportive environment should your potential supervisor be too busy to meet you often; on the other hand, a smaller lab may provide the opportunity for more personal interaction with your supervisor.
- Impact. Your project may result in publishable data: discuss this with your prospective supervisor.

> ***Asking around*** *– one of the best sources of information about supervisors, laboratories and projects is past students. Some of the postgraduates in your department may be products of your own system and they could provide an alternative source of advice.*

Planning your work

As with any lengthy exercise, planning is required to make the best use of the time allocated (p. 10). This is true on a daily basis as well as over the entire period of the project. It is especially important not to underestimate the time it will take to write and produce your thesis (see below). If you wish to benefit from feedback given by your supervisor, you should aim to have drafts in his/her hands in good time. Since a large proportion of marks will be allocated to the report, you should not rush its production.

If your department requires you to write an interim report, look on this as an opportunity to clarify your thoughts and get some of the time-consuming preparative work out of the way. If not, you should set your own deadlines for producing drafts of the introduction, materials and methods section, etc.

> ***Liaising with your supervisor(s)*** *– this is essential if your work is to proceed efficiently. Specific meetings may be timetabled, e.g. to discuss a term's progress, review your work plan or consider a draft introduction. Most supervisors also have an 'open-door' policy, allowing you to air current problems. Prepare well for all meetings: have a list of questions ready before the meeting; provide results in an easily digestible form (but take your lab notebook along); be clear about your future plans for work.*

KEY POINT ***Project work can be very time-consuming. Try not to neglect other aspects of your course – make sure your lecture notes are up-to-date and collect relevant supporting information as you go along.***

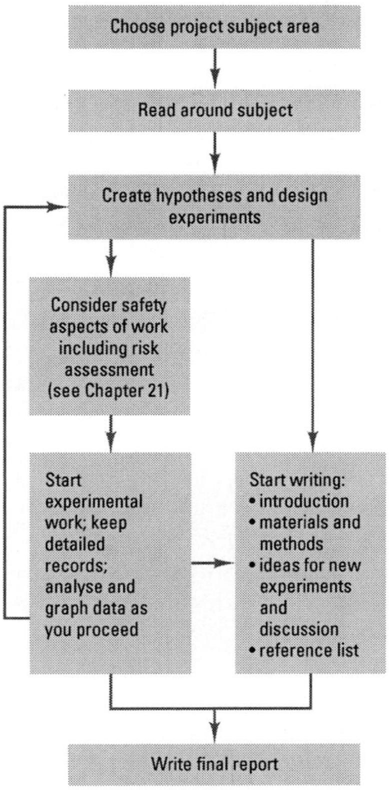

Fig. 31.1 Flowchart showing a recommended sequence of events in carrying out an undergraduate research project.

Getting started

Figure 31.1 is a flowchart illustrating how a project might proceed; at the start, don't spend too long reading the literature and working out a lengthy programme of research. Get stuck in and do an experiment. There's no substitute for 'getting your hands dirty' for stimulating new ideas:

- even a 'failed' experiment will provide some useful information which may allow you to create a new or modified hypothesis;
- pilot experiments may point out deficiencies in experimental technique that will need to be rectified;
- the experience will help you create a realistic plan of work.

Designing experiments or sampling procedures

Design of experiments and sampling procedure is covered in Chapter 30. Avoid being too ambitious at the start of your work. It is generally best to work with a simple hypothesis and design your experiments or sampling around this. A small pilot experiment or test sample will highlight potential stumbling blocks including resource limitations, whether in materials, or time, or both.

Working in a laboratory environment

During your time as a project student, you are effectively a guest in your supervisor's laboratory.

- Be considerate – keep your 'area' tidy and offer to do your share of lab duties such as calibrating the pH meter, replenishing stock solutions, distilled water, etc., maintaining cultures, tending plants or animals.
- Use instruments carefully – they could be worth more than you think. Careless use may invalidate calibration settings and ruin other people's work as well as your own.
- Do your homework on techniques you intend to use – there's less chance of making costly mistakes if you have a good background understanding of the methods you will be using.
- Always seek advice if you are unsure of what you are doing.

***KEY POINT** It is essential that you follow all the safety rules applying to the laboratory or field site. Make sure you know all relevant procedures – if in doubt, ask beforehand.*

Keeping notes and analysing your results

Tidy record keeping is often associated with good research, and you should follow the advice and hints given in Chapter 28 (p. 180). Try to keep copies of all files relating to your project. As you obtain results, you should always calculate, analyse and graph data as soon as you can (see Fig. 31.1). This can reveal aspects that may not be obvious in numerical or readout form. Don't be worried by negative results – these can sometimes be as useful as positive results if they allow you to eliminate hypotheses – and don't be too dispirited if things do not work first time. Thomas Edison's maxim 'Genius is one per cent inspiration and ninety-nine per cent perspiration' certainly applies to research work.

Sources of further information for project work:

- *experimental design checklist (Box 30.1);*
- *advice on recording results (p. 176);*
- *describing and analysing numerical data;*
- *checklist for presenting your final report (Box 18.1).*

Writing your project report

The structure of scientific reports is dealt with in Chapter 18. The following advice concerns methods of accumulating relevant information.

Brushing up on your IT skills *– word processors and spreadsheets are extremely useful when producing a thesis. Chapters 12 and 13 detail key features of these programs. You might benefit from attending courses on the relevant programs or studying manuals or texts so that you can use them more efficiently.*

Introduction This is a big piece of writing that can be very time-consuming. Therefore, the more work you can do on it early on the better. Allocate some time at the start for library work (without neglecting your lab work), to build up a database of references (Chapter 8). Photocopying can be expensive, but you will find it valuable to have copies of key reviews and references handy when writing away from the library. Discuss proposals for content and structure with your supervisor to ensure your effort is relevant. Leave space at the end for a section on aims and objectives. This is important to orientate readers (including assessors), but you may prefer to finalise the content after the results have been analysed.

Materials and Methods Note as many details as possible *when doing the experiment or making observations.* Don't rely on your memory or hope that the information will still be available when you come to write up. Even if it is, chasing these details can waste valuable time.

Using drawings and photographs *– these can provide valuable records of sampling sites or experimental set-ups and could be useful in your report. Plan ahead and do the relevant work at the time of carrying out your research rather than afterwards.*

Results Show your supervisor graphed and tabulated versions of your data promptly. These can easily be produced using a spreadsheet (p. 74), but you should seek your supervisor's advice on whether the design and print quality is appropriate to be included in your report. You may wish to use a specialist graphics program to produce publishable-quality graphs and charts: allow some time for learning its idiosyncrasies. If you are producing a poster for assessment (Chapter 14), be sure to mock up the design well in advance. Similarly, think ahead about your needs for any seminar or poster you will present.

Using a word processor to record your ideas *– remember that you can note down your thoughts and any other important information relevant to the Results and Discussion sections of your project in a file that can then form the basis of your first draft (p. 81); that way, you won't forget to include these points in your final report.*

Discussion Because this comes at the end of your report, and some parts can only be written after you have all the results in place, the temptation is to leave the discussion to last. This means that it might be rushed – not a good idea because of the weight attached by assessors to your analysis of data and thoughts about future experiments. It will help greatly if you keep notes of aims, conclusions and ideas for future work as you go along (Fig. 31.1). Another useful tip is to make notes of comparable data and conclusions from the literature as you read papers and reviews.

Acknowledgements Make a special place in your notebook for noting all those who have helped you carry out the work for use when writing this section of the report.

References Because of the complex formats involved (p. 48), these can be tricky to type. To save time, process them in batches as you go along – bibliographic software (e.g. Endnote) can help with organisation of references (p. 47).

KEY POINT ***Make sure you are absolutely certain about the deadline for submitting your report, and work towards submitting a day or so before it. If you leave things until the last moment, you may find access to printers, photocopiers and binding machines is difficult.***

Sources for further study

Barnard, C., Gilbert, F. and McGregor, P. (2001) *Asking Questions in Biology: Key Skills for Practical Assessments and Project Work,* 2nd edn. Prentice Hall, Harlow.

Ebel, H.F., Bliefert, C. and Russey, W.E. (2004) *The Art of Scientific Writing: from Student Reports to Professional Publications in Chemistry and Related Fields,* 2nd edn. Wiley-VCH, Weinheim, Germany.
[Covers record-keeping, notebooks, report writing and dissertations]

Jardine, F.H. (1994) *How to do your Student Project in Chemistry.* CRC Press, Boca Raton.
[Also contains general information relevant to bioscience projects]

Luck, M. (1999) *Your Student Research Project.* Gower, London.

Marshall, P. (1997) *Research Methods: How to Design and Conduct a Successful Project.* How To Books, Plymouth.

Ruxton, G. and Palgrave, N. (2006) *Experimental Design for the Life Sciences.* Oxford University Press, Oxford.

Sweetnam, D. (2004) *Writing your Dissertation: the Bestselling Guide to Planning, Preparing and Presenting First-Class Work,* 3rd edn. How To Books, Oxford.

Study exercises

Note: These exercises assume that you have started a research project, or are about to start one, as part of your studies.

31.1 Prepare a project plan. Make a formal plan for your research project, incorporating any milestones dictated by your department, such as interim reports and final submission dates. Discuss your plan with your supervisor and incorporate his or her comments. Refer back to the plan frequently during your project, to see how well you are meeting your deadlines.

31.2 Resolve to write up your work as you go along. Each time you complete an experiment or observation, write up the materials and methods, analyse the data and draw up the graphs as soon as you can. While you may reject some of your work at a later stage, you may wish to modify it; this will spread out the majority of the effort and allow time for critical thinking close to the final submission date.

31.3 Devise a computer database for keeping details of your references. Keeping these records up-to-date will save you a lot of time when writing up. You will need to decide on an appropriate referencing format, or find out about that followed by your department (see Chapter 8).

Handling cells and tissues

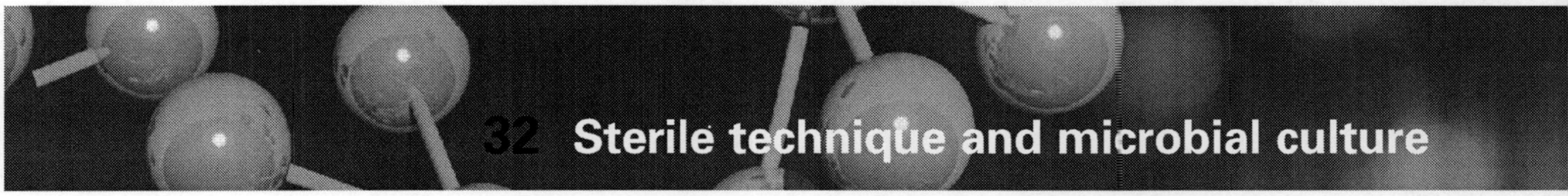

32 Sterile technique and microbial culture

Achieving a sterile state *– you should assume that all items of laboratory equipment have contaminating microbes on their surfaces, unless they have been destroyed by some form of sterilisation. Such items will only remain sterile if they do not come into contact with the non-sterile environment.*

Sterile technique (aseptic technique) is the name given to the procedures used in cell culture. While the same general principles apply to all cell types, you are most likely to learn the basic procedures using bacteria and most of the examples given in this section refer to bacterial culture.

Sterile technique serves two main purposes:

1. To prevent accidental contamination of laboratory cultures due to microbes from external sources, e.g. skin, clothing or the surrounding environment.
2. To prevent microbial contamination of laboratory workers, in this instance you and your fellow students.

KEY POINT ***All microbial cell cultures should be treated as if they contained potentially harmful organisms. Sterile technique forms an important part of safety procedures, and must be followed whenever cell cultures are handled in the laboratory.***

Care is required because:

- You may accidentally isolate a harmful microbe as a contaminant when culturing a relatively harmless strain.
- Some individuals are more susceptible to infection and disease than others – not everyone exposed to a particular microbe will become ill.
- Laboratory culture involves purifying and growing large numbers of microbial cells – this represents a greater risk than small numbers of the original microbe.
- A microbe may change its characteristics, perhaps as a result of gene exchange or mutation.

Fig. 32.1 International symbol for a biohazard. Usually red on a yellow background, or black on a red background.

The international biohazard symbol, shown in Figure 32.1, is used to indicate a significant risk due to a pathogenic microbe (p. 206).

Sterilisation procedures

Given the ubiquity of microbes, the only way to achieve a sterile state is by their destruction or removal. Several methods can be used to achieve this objective:

Heat treatment

This is the most widespread form of sterilisation and is used in several basic laboratory procedures including the following:

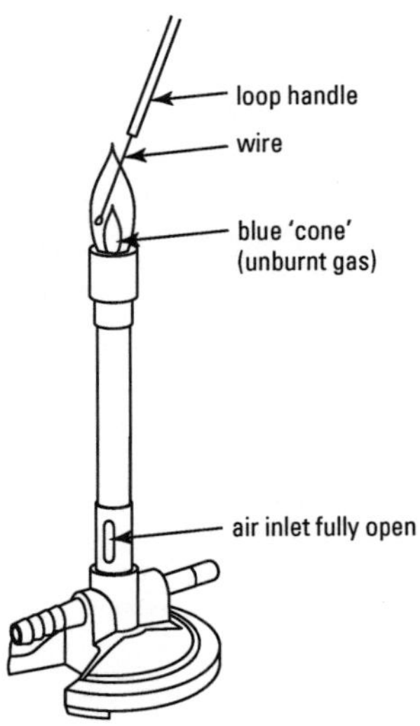

Fig. 32.2 'Flaming' a wire loop. Keep the loop in the hottest part of the Bunsen flame (just outside the blue 'cone') until the wire is red hot.

- Red heat sterilisation. Achieved by heating metal inoculating loops, forceps, needles, etc., in a Bunsen flame (Fig. 32.2). This is a simple and effective form of sterilisation as no microbe will survive even a brief exposure to a naked flame. Flame sterilisation using alcohol is used for glass rods and spreaders (see below).
- Dry heat sterilisation. Here, a hot air oven is used at a temperature of at least 160 °C for at least 2 h. This method is used for the routine sterilisation of laboratory glassware. Dry heat procedures are of little value for items requiring repeated sterilisation during use.

Fig. 32.3 Autoclave tape – the lower sample is untreated while the upper sample (with dark diagonal lines) has been autoclaved.

***Using a sterile filter** – most filters are supplied as pre-sterilised items. Make sure you follow a procedure that does not contaminate the filter on removal from its protective wrapping.*

SAFETY NOTE When working with biocides, take care to avoid skin contact or ingestion, as most are toxic and irritant. If contact does occur, rinse with plenty of water.

***Using molten agar** – a water bath (at 45–50 °C) can be used to keep an agar-based medium in its molten state after autoclaving. Always dry the outside of the container on removal from the water bath, to reduce the risk of contamination from microbes in the water, e.g. during pour plating (p. 207).*

- Moist heat sterilisation. This is the method of choice for many laboratory items, including most fluids, apart from heat-sensitive media. It is also used to decontaminate liquid media and glassware after use. The laboratory autoclave is used for these purposes. Typically, most items will be sterile after 15 min at 121 °C, although large items may require a longer period. The rapid killing action results from the latent heat of condensation of the pressurised steam, released on contact with cool materials in the autoclave. While special heat-sensitive tape (Fig. 32.3) is sometimes used to check that the autoclave is operating correctly, a better approach is to use spores of *Bacillus stearothermophilus*.

Radiation

Many disposable plastic items used in microbiology and cell biology are sterilised by exposure to UV or ionising radiation. They are supplied commercially in sterile packages, ready for use. Ultraviolet radiation has limited use in the laboratory, while ionising radiation (e.g. γ-rays) requires industrial facilities and cannot be operated on a laboratory scale.

Filtration

Heat-labile solutions (e.g. complex macromolecules, including proteins, antibiotics, serum) are particularly suited to this form of sterilisation. The filters come in a variety of shapes, sizes and materials, usually with a pore size of either 0.2 or 0.45 µm. The filtration apparatus and associated equipment are usually sterilised by autoclaving, or by dry heat. Passage of liquid through a sterile filter of pore size 0.2 µm into a sterile vessel is usually sufficient to remove bacteria but not viruses, so filtered liquids are not necessarily virus-free.

Chemical agents

These are known as disinfectants, or biocides, and are most often used for the disposal of contaminated items following laboratory use, e.g. glass slides and pipettes. They are also used to treat spillages. The term 'disinfection' implies destruction of disease-causing bacterial cells, although spores and viruses may not always be destroyed. Remember that disinfectants require time to exert their killing effect – any spillage should be covered with an appropriate disinfectant and left for at least 10 min before mopping up.

Use of laboratory equipment

Working area

One of the most important aspects of good sterile technique is to keep your working area as clean and tidy as possible. Start by clearing all items from your working surface, wipe the bench down with disinfectant and then arrange the items you need for a particular procedure so that they are close at hand, leaving a clear working space in the centre of your bench.

Media

Cells may be cultured in either a liquid medium (broth), or a solidified medium (p. 232). The gelling agent used in most solidified media is agar, a complex polysaccharide from red algae that produces a stiff transparent gel when used at 1–2% (w/v). Agar is used because it is relatively resistant to degradation by most bacteria and because of its rheological properties – an agar medium melts at 98 °C, remaining solid at all temperatures used for routine laboratory culture. However, once melted it does not solidify again until the temperature falls to about 44 °C. This means that heat-sensitive constituents (e.g. vitamins, blood, cells) can be added aseptically to the medium after autoclaving.

Working with plastic disposable loops – *these are used in many research laboratories: pre-sterilised and suitable for single use, they avoid the hazards of naked flames and the risk of aerosol formation during heating. Discard into disinfectant solution after use.*

Using a Bunsen burner to reduce airborne contamination – *working close to the updraught created by a Bunsen flame reduces the likelihood of particles falling from the air into an open vessel.*

Inoculating loops

The initial isolation and subsequent transfer of microbes between containers can be achieved using a sterile inoculating loop. Most teaching laboratories use nichrome wire loops in a metal handle. A wire loop can be repeatedly sterilised by heating the wire, loop downwards and almost vertical, in the hottest part of a Bunsen flame until the whole wire becomes red hot (Fig. 32.2). Then the loop is removed from the flame to minimise heat transfer to the handle. After cooling for 8–10 s (without touching any other object) it is ready for use.

When resterilising a contaminated wire loop in a Bunsen flame after use, do not heat the loop too rapidly, as the sample may spatter, creating an aerosol: it is better to soak the loop for a few minutes in disinfectant than to risk heating a fully charged (contaminated) inoculating loop.

Using glass pipettes – *these are plugged with cotton wool at the top before being autoclaved inside a metal can. Flame the open end of the can on removal of a pipette, to prevent contamination of the remaining pipettes. Autopipettors and sterile disposable tips (p. 129) offer an alternative approach.*

Containers

There is a risk of contamination whenever a sterile bottle, flask or test tube is opened. One method that reduces the chance of airborne contamination is quickly to pass the open mouth of the glass vessel through a flame. This destroys any microbes on the outer surfaces nearest to the mouth of the vessel. In addition, by heating the air within the neck of the vessel, an outwardly directed air flow is established, reducing the likelihood of microbial contamination.

It is general practice to flame the mouth of each vessel immediately after opening and then repeat the procedure just before replacing the top. Caps, lids and cotton wool plugs must not be placed on the bench during flaming and sampling: they should be removed and held using the smallest finger of one hand, to minimise the risk of contamination. This also leaves the remaining fingers free to carry out other manipulations. With practice, it is possible to remove the tops from two tubes, flame each tube and transfer material from one to the other while holding one top in each hand.

Table 32.1 Classification of microbes on the basis of hazard. The following categories are recommended by the UK Advisory Committee on Dangerous Pathogens (ACDP)

Hazard group	Comments
1	Unlikely to cause human disease
2	May cause disease: possible hazard to laboratory workers, minimal hazard to community
3	May cause severe disease: may be a serious hazard to laboratory workers, may spread to community
4	Causes severe disease: is a serious hazard to laboratory workers, high risk to community

Laminar flow cabinets

These are designed to prevent airborne contamination, e.g. when preparing media or subculturing microbes or tissue cultures. Sterile air is produced by passage through a high efficiency particulate air (HEPA) filter: this is then directed over the working area, either horizontally (towards the operator) or downwards. The operator handles specimens, media, etc. through an opening at the front of the cabinet. Note that standard laminar flow cabinets do *not* protect the worker from contamination and must not be used with pathogenic microbes: special safety cabinets and laboratories are used for work with ACDP hazard group 3 and 4 microbes (Table 32.1) and for samples that might contain such pathogens.

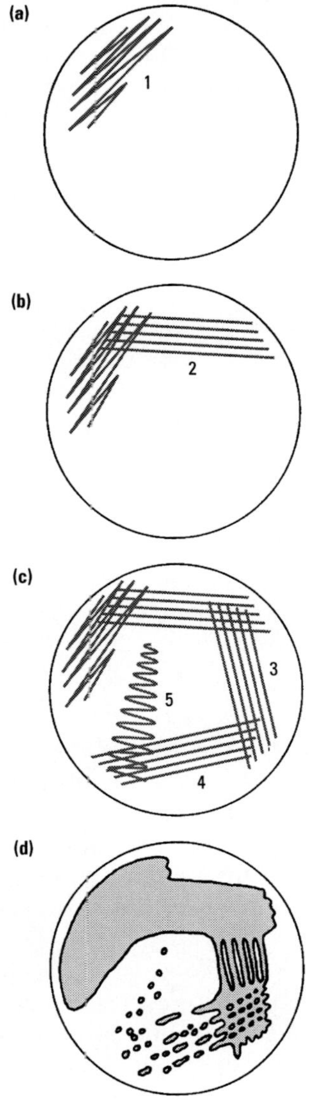

Fig. 32.4 Preparation of a streak plate for single colonies. (a) Using a sterile metal loop, take a small sample of the material to be streaked. Distribute the sample over a small sector of the plate (area 1), then flame the loop and allow to cool (approximately 8–10 s). (b) Make several small streaks from the initial sector into the adjacent sector (area 2), taking care not to allow the streaks to overlap. Flame the loop and allow to cool. (c) Repeat the procedure for areas 3 and 4, resterilising the loop between each step. Finally, make a single, long streak, as shown for area 5. (d) The expected result after incubation at the appropriate temperature (e.g. 37 °C for 24 h): each step should have diluted the inoculum, giving individual colonies within one or more sectors on the plate. Further subculture of an individual colony should give a pure (clonal) culture.

***KEY POINT** The most obvious risks when handling microbial cultures are those due to ingestion or entry via a cut in the skin – all cuts should be covered with a plaster or disposable plastic gloves. A less obvious source of hazard is the formation of aerosols of liquid droplets from microbial suspensions, with the risk of inhalation, or surface contamination of other objects.*

Microbiological hazards

The following steps will minimise the risk of aerosol formation:

- Use stoppered tubes when shaking, centrifuging or mixing microbial suspensions.
- Pour solutions gently, keeping the difference in height to a minimum.
- Discharge pipettes onto the side of the container.

Other general rules which apply in all laboratories include:

- Take care with sharp instruments, including needles and glass Pasteur pipettes.
- Do not pour waste cultures down the sink – they must be autoclaved.
- Put other contaminated items (e.g. slides, pipettes) into disinfectant after use.
- Wipe down your bench with disinfectant when practical work is complete.
- Always wash your hands before leaving the laboratory.

Plating methods

Many culture methods make use of a solidified medium within a Petri plate. A variety of techniques can be used to transfer and distribute the organisms prior to incubation. The three most important procedures are described below.

Streak dilution plate

Streaking a plate for single colonies is one of the most important basic skills in microbiology, since it is used in the initial isolation of a cell culture and in maintaining stock cultures, where a streak dilution plate with single colonies all of the same type confirms the purity of the strain. A sterile inoculating loop is used to streak the organisms over the surface of the medium, thereby diluting the sample (Fig. 32.4). The aim is to achieve single colonies at some point on the plate: ideally, such colonies are derived from single cells (e.g. in the case of unicellular bacteria, animal and plant cell lines) or from groups of cells of the same species (in filamentous or colonial forms). Single colonies, containing cells of a single species and derived from a single parental cell, form the basis of most pure culture methods (p. 211).

Note the following:

- Keep the lid of the Petri plate as close to the base as possible to reduce the risk of aerial contamination.
- Allow the loop to glide over the surface of the medium. Hold the handle at the balance point (near the centre) and use light, sweeping movements, as the agar surface is easily damaged and torn.
- Work quickly, but carefully. Do not breathe directly onto the exposed agar surface and replace the lid as soon as possible.

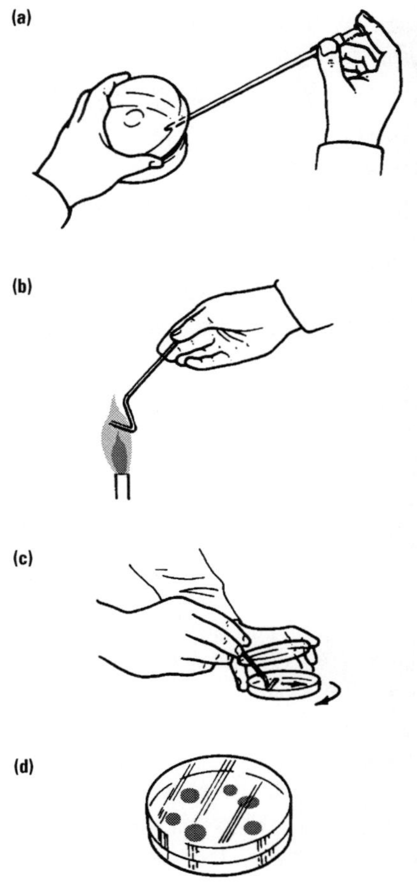

Fig. 32.5 Preparation of a spread plate. (a) Transfer a small volume of cell suspension (0.05–0.5 ml) to the surface of a solidified medium in a Petri plate. (b) Flame sterilise a glass spreader and allow to cool (8–10 s). (c) Distribute the liquid over the surface of the plate using the sterile spreader. Make sure of an even coverage by rotating the plate as you spread: allow the liquid to be absorbed into the agar medium. Incubate under suitable conditions. (d) After incubation, the microbial colonies should be distributed over the surface of the plate.

Spread plate

This method is used with cells in suspension, either in a liquid growth medium or in an appropriate sterile diluent. It is one method of quantifying the number of viable cells (or colony-forming units) in a sample, after appropriate dilution (p. 137).

An L-shaped glass spreader is sterilised by dipping the end of the spreader in a beaker containing a small amount of 70% v/v alcohol, allowing the excess to drain from the spreader and then igniting the remainder in a Bunsen flame. After cooling, the spreader is used to distribute a known volume of cell suspension across the plate (Fig. 32.5). *There is a significant fire risk associated with this technique*, so take care not to ignite the alcohol in the beaker, e.g. by returning an overheated glass rod to the beaker. The alcohol will burn with a pale blue flame that may be difficult to see, but will readily ignite other materials (e.g. a laboratory coat). Another source of risk comes from small droplets of flaming alcohol shed by an overloaded spreader onto the bench and this is why you *must* drain excess alcohol from the spreader *before* flaming. Some laboratories now provide plastic disposable spreaders for student use, to avoid the risk of fire.

Pour plate

This procedure also uses cells in suspension, but requires molten agar medium, usually in screw-capped bottles containing sufficient medium to prepare a single Petri plate (i.e. 15–20 ml), maintained in a water bath at 45–50 °C. A known volume of cell suspension is mixed with this molten agar, distributing the cells throughout the medium. This is then poured without delay into an empty sterile Petri plate and incubated, giving widely spaced colonies (Fig. 32.6). Furthermore, as most of the colonies are formed within the medium, they are far smaller than those of the surface streak method, allowing higher cell numbers to be counted (e.g. over 100 colonies per plate): some workers pour a thin layer of molten agar onto the surface of a pour plate after it has set, to ensure that no surface colonies are produced. Most bacteria and fungi are not killed by brief exposure to temperatures of 45–50 °C, though the procedure may be more damaging to microbes from low temperature conditions, e.g. psychrophilic bacteria.

One disadvantage of the pour plate method is that the typical colony morphology seen in surface-grown cultures will not be observed for those colonies that develop within the agar medium. A further disadvantage is that some of the suspension will be left behind in the screw-capped bottle. This can be avoided by transferring the suspension to the Petri plate, adding the molten agar, then swirling the plate to mix the two liquids. However, even when the plate is swirled repeatedly and in several directions, the liquids are not mixed as evenly as in the former procedure.

KEY POINT When working with molten agar, keep tubes and bottles of molten agar in a water bath until you are ready to use them, as they will begin to set within a couple of minutes at room temperature.

Working with phages

Bacterial viruses ('bacteriophages', or simply 'phages') are often used to illustrate the general principles involved in the detection and enumeration of viruses. They also have a role in genome mapping of bacteria. Individual

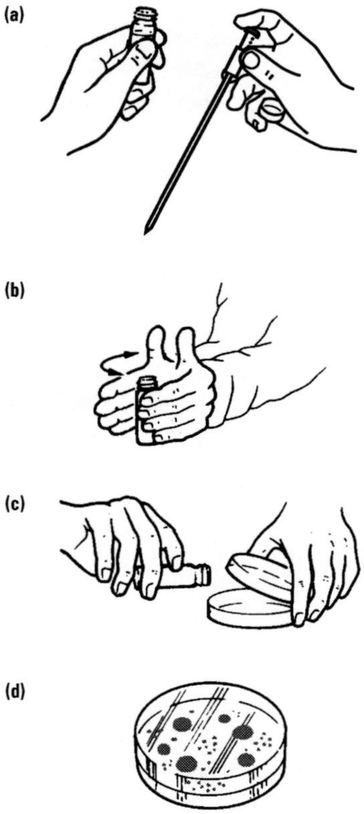

Fig. 32.6 Preparation of a pour plate. (a) Add a known volume of cell suspension (0.05–1.0 ml) to a small bottle of molten agar medium from a 45 °C water bath. (b) Mix thoroughly, by rotating between the palms of the hands: do not shake or this will cause frothing of the medium. (c) Pour the mixture into an empty, sterile Petri plate and allow to set. Incubate under suitable conditions. (d) After incubation, the microbial colonies will be distributed throughout the medium: any cells deposited at the surface will give larger, spreading colonies.

phage particles (virions) are too small to be seen by light microscopy, but are detected by their effects on susceptible hostcells:

- Virulent phages will infect and replicate within actively growing host cells, causing cell lysis and releasing new infective phages – this 'lytic cycle' takes $\approx$30 min for T-even phages of *E. coli*, e.g. T4.
- Temperate phages are a specialised group, capable either of lytic growth or an alternative cycle, termed lysogeny – the phage becomes latent within a host cell (lysogen), typically by insertion of its genetic information into the host cell genome, becoming a 'prophage'. At a later stage, termed induction, the prophage may enter the lytic cycle. A widely used example is λ phage of *E. coli.*

The lytic cycle can be used to detect and quantify the number of phages in a sample. A known volume of sample is mixed with susceptible bacterial cells in molten soft agar medium (45–50 °C), then poured on top of a plate of the same medium, creating a thin layer of 'top agar'. The upper layer contains only half the normal amount of agar, to allow phages to diffuse through the medium and attach to susceptible cells. On incubation, the bacteria will grow throughout the agar to produce a homogeneous 'lawn' of cells, except in those parts of the plate where a phage particle has infected and lysed the cells to create a clear area, termed a plaque (Fig. 32.7). Each plaque is due to a single functional phage (i.e. a plaque-forming unit, or PFU). A count of the number of plaques can be used to give the number of phages in a particular sample (e.g. as PFU ml^{-1}), with appropriate correction for dilution and the volume of sample counted in an analogous manner to a bacterial plate count (p. 237). When counting plaques in phage assays you should view them against a black background to make them easier to see: mark each plaque with a spirit-based marker to ensure an accurate count. Temperate phages often produce cloudy plaques, because many of the infected cells will be lysogenised rather than lysed, creating turbidity within the plaque. Samples of material from within the plaque can be used to subculture the phage for further study, perhaps in a broth culture where the phages will cause widespread cell lysis and a decrease in turbidity. Alternatively, phages can be stored by adding chloroform to aqueous suspensions – this will prevent contamination by cellular micro-organisms. A similar approach can be used to detect and count animal or human viruses, using a monolayer of a susceptible animal or human cells.

Electron microscopy (EM, p. 159) provides an alternative approach to the detection of viruses, avoiding the requirement for culture of infected

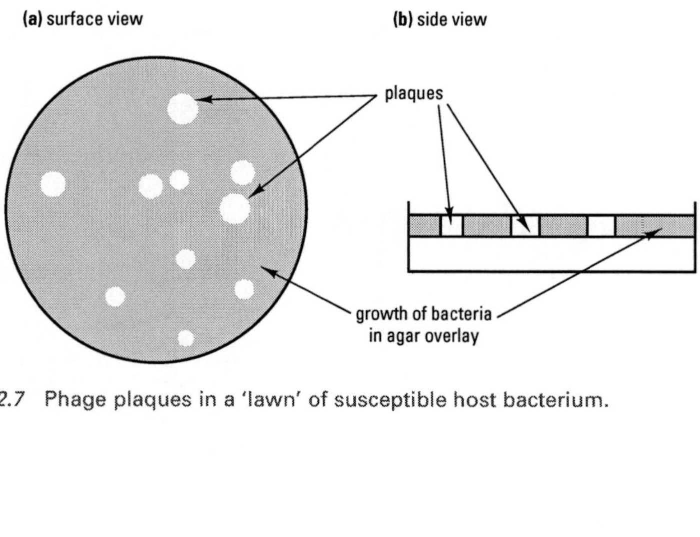

Fig. 32.7 Phage plaques in a 'lawn' of susceptible host bacterium.

host cells, and giving a faster result. However, it requires access to specialised equipment and expertise. EM counts are often higher than culture-based methods, for similar reasons to those described for bacteria (p. 237).

Labelling your plates and cultures

***Labelling Petri plates** – the following information should be recorded on the base:*

- *date*
- *growth medium*
- *your name or initials*
- *brief details of the organism, experimental treatment, etc.*

Petri plates should always be labelled on the *base*, rather than the lid, using a permanent marker. Restrict your labelling to the outermost region of the plate to avoid problems when counting colonies, assessing growth, etc. After labelling, Petri plates usually are incubated upside-down in a temperature-controlled incubator (often at 37 °C) for an appropriate period (usually 18–72 h). Plates are also usually kept upside-down on the lab bench – following incubation, the base (containing medium and microbes) can then be lifted from the lid and examined.

Sources for further study

Anon. (1995) *Advisory Committee on Dangerous Pathogens: Categorisation of Biological Agents According to Hazard and Categories of Containment*, 4th edn. HSE Books, London.

Anon. (2004) *Advisory Committee on Dangerous Pathogens: Approved List of Biological Agents*. Available: http://www.hse.gov.uk/pubns/misc208.pdf Last accessed: 01/04/07.

Bartelt, M.A. (1999) *Diagnostic Bacteriology: A Study Guide*. Davis, New York.

Collins, C.H., Lyne, P.M., Grange, J.M. and Falkingham, J. (2004) *Collins and Lyne's Microbiological Methods*, 8th edn. Hodder Arnold, London.

Hawkey. P. and Lewis, D. (2004) *Medical Bacteriology: a Practical Approach,* 2nd edn. Oxford University Press, Oxford.

Isenberg, H.D. (1998) *Essential Procedures for Clinical Microbiology*. American Society for Microbiology, Washington, DC.

Rhodes, P.M. and Stanbury, P.F. (1997) *Applied Microbial Physiology: a Practical Approach*. Oxford University Press, Oxford.

Study exercises

32.1 Check that you understand how to label a plate culture. List the key aspects and then compare your list against the one provided on p. 514.

32.2 Decide on the best method of sterilisation. What would be the most appropriate method of sterilisation for the following items?

(a) A box of 100 plastic tips to be used with a pipettor.
(b) A 50 ml batch of blood, for use in 5% v/v blood agar plates.
(c) A 1 litre batch of MacConkey agar.
(d) Ten 5 ml glass pipettes.
(e) A microbiological wire, used for 'stab' cultures.
(f) A 10 ml sample of a heat-sensitive solution of an antibiotic, to be used as a component of a selective isolation medium.

32.3 Find out the biohazard classification (UK ACDP categorisation – Table 32.1) for the following microbes:
(a) *Salmonella typhimurium* (b) *Leptospira interrogans* (c) *Shigella dysenteriae* (type 1) (d) *E. coli* K12 (e) *E. coli* O157 (f) Human immunodeficiency virus (HIV) (g) *Cryptococcus neoformans* var. *neoformans* (h) *Mycobacterium tuberculosis* (i) *Lactobacillus plantarum* (j) Marburg virus.

32.4 Consider the advantages and disadvantages of spread plating and pour plating methods. Having read through this chapter, list up to six pros/cons of each plating method and compare your answers either with the list on pp. 514–15, or with those of other students, as a group exercise.

33 Isolating, identifying and naming microbes

Micro-organisms have a broad range of applications in the life sciences:

- Microbes are widely used as model systems, as they are often easier to study under controlled laboratory conditions than 'higher' organisms.
- Bacteria and fungi are used as sources of particular biomolecules, e.g. for the characterisation of a specific enzyme *in vitro*.
- Pathogenic microbes are studied at the molecular level, to find new methods of identification using biochemical 'markers' and to investigate the molecular basis of their pathogenicity.
- Microbes are used in biotechnology and industrial microbiology as sources of particular biomolecules, e.g. production of blood-clotting factors using yeast cells.

Alternatives to traditional culture-based methods *– microbes can now be investigated by molecular methods, including the amplification of specific nucleic acid sequences by PCR, and by immunoassay, e.g. ELISA (p. 261).*

In your practical classes, you are likely to gain experience of a wide range of laboratory exercises involving microbes, particularly bacteria such as *Escherichia coli* (*E. coli*), using the procedures of sterile technique described in Chapter 32.

Sampling microbes

Microbes can be studied by taking samples for analysis, usually in one of the following ways:

- by direct examination of individual cells of a particular microbe, e.g. using fluorescence microscopy (p. 160);
- by isolating/purifying a particular species or related individuals of a taxonomic group, e.g. the faecal indicator bacterium *E. coli* in sea water;
- by studying microbial processes, rather than individual microbes, either *in situ* or in the laboratory.

Avoiding contamination during sampling *– always remember that you are the most important source of contamination of field samples: components of the oral or skin microflora are the most likely contaminants.*

Sampling techniques include the use of swabs, Sellotape strips and agar contact methods for sampling surfaces, bottles for aquatic habitats, plastic bags and corers for soils and sediments. A wide range of complex apparatus is available for accurately sampling water or soil at particular depths.

KEY POINT An important feature of all microbiological sampling protocols is that the sampling apparatus must be sterile; strict aseptic technique must be used throughout the sampling process (see Chapter 32).

SAFETY NOTE When working with newly isolated microbes, you should always treat them as potentially harmful (p. 203) until they have been identified.

The sampling method must minimise the chance of contamination with microbes from other sources, especially the exterior of the sampling apparatus and the operator. For example, if you are sampling an aquatic habitat, stand downstream of the sampling site. A portable Bunsen burner or spirit lamp can be used to assist sterile technique during field sampling, e.g. while flaming a loop (p. 203). Alternatively, use disposable single-use plastic loops.

Process the sample as quickly as possible to minimise any changes in microbiological status. As a general guideline, many procedures require samples to be analysed within 6 h of collection. Changes in aeration, pH and water content may occur after collection. Some microbes are more susceptible to such effects, e.g. anaerobic bacteria may not survive if the sample is exposed

Sub-sampling *– to minimise the effects of changes in temperature, aeration and water status during transportation, a primary sample may be returned to the laboratory, where the working sample (sub-sample) is then taken (e.g. from the centre of a large block of soil).*

to air. Sunlight can also inactivate bacteria; samples should be shielded from direct sunlight during collection and transport to the laboratory.

Soil and water samples are often kept cool (at 0–5 °C) during transport to the laboratory. In contrast, some microbes adapted to grow in association with warm-blooded animals may be damaged by low temperatures. An alternative approach is to keep the sample near the ambient sampling temperature using an insulated vessel (e.g. Thermos flask).

Isolating a particular microbe

Several different approaches may be used to obtain microbes in pure culture. The choice of method will depend upon the microbe to be isolated: some organisms are relatively easy to isolate, while others require more involved procedures.

Obtaining a pure culture *– if a single colony from a primary isolation medium is used to prepare a streak dilution plate and all the colonies on the second plate appear identical, then a pure culture has been established. Otherwise, you cannot assume that your culture is pure and you should repeat the subculture until you have a pure culture.*

Separation methods

Most microbial isolation procedures involve some form of separation to obtain individual microbial cells. The most common approach is to use an agar-based medium for primary isolation, with streak dilution, spread plating or pour plating to produce single colonies, each derived from a single type of microbe (pp. 206–7). It is often necessary to dilute samples before isolation, so that a small number of individual microbial cells are transferred to the growth medium. Strict serial dilution (pp. 137–8) of a known amount of sample is needed for quantitative work.

KEY POINT If your aim is to isolate a particular microbe, perhaps for further investigation, you will need to subculture individual colonies from the primary isolation plate to establish a pure culture, also known as an axenic culture.

Using a sonicator *– minimise heat damage with short treatment 'bursts' (typically up to 1 min), cooling the sample between bursts, e.g. using ice.*

Pure cultures of most microbes can be maintained indefinitely, using sterile technique and microbial culture methods (Chapter 32).

Other separation techniques include:

- Dilution to extinction. This involves diluting the sample to such an extent that only one or two microbes are present per millilitre: small volumes of this dilution are then transferred to a liquid growth medium (broth). After incubation, most of the tubes will show no growth, but some tubes will show growth, having been inoculated with a single viable microbe at the outset. This should give a pure culture, though it is wasteful of resources.
- Sonication/homogenisation. This is useful for separating individual microbial cells from each other and from inert particles, prior to isolation. However, some decrease in viability is likely.
- Filtration. This can be useful where the number of microbes is low. Samples can be passed through a sterile cellulose ester filter (pore size 0.2 μm), which is then incubated on the surface of an appropriate solidified medium. Sieving and filtration techniques (p. 139) are often used in soil microbiology to subdivide a sample on the basis of particle size.
- Micromanipulation. It may be possible to separate a microbe from contaminants using a micropipette and dissecting microscope. The microbe can then be transferred to an appropriate growth medium, to give a pure culture. However, this is rarely an easy task for the novice.

Definitions

Psychrophile – a microbe with an optimum temperature for growth of <20 °C (literally 'cold-loving').

Psychrotroph – a microbe with an optimum temperature for growth of ⩾ 20 °C, but capable of growing at lower temperature, typically 0–5 °C (literally 'cold-feeding').

Thermophile – a microbe with an optimum growth temperature of >45 °C (literally 'heat-loving').

Mesophile – a microbe with an optimum growth temperature of 20–45 °C (literally 'middle-loving').

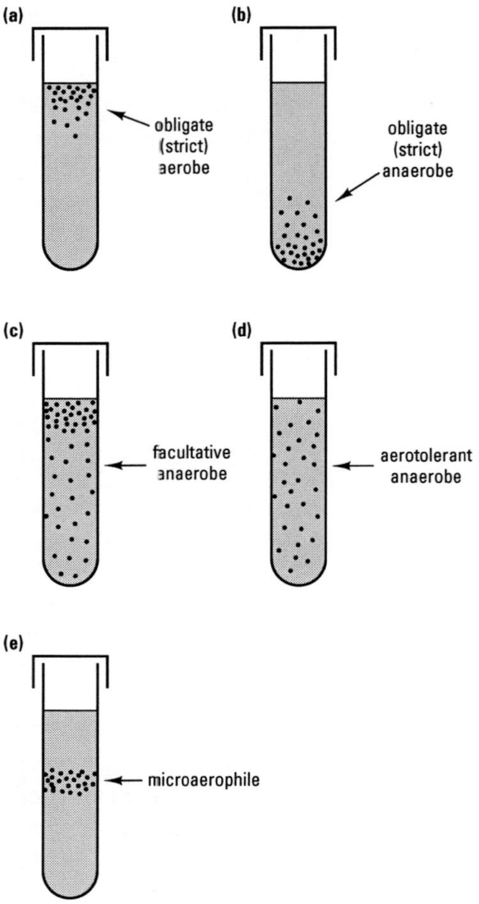

Fig. 33.1 Agar shake tubes. Bacteria are suspended in molten agar at 45–50 °C and allowed to cool. The growth pattern after incubation reflects the atmospheric (oxygen) requirements of the bacterium.

Definitions

Facultative anaerobe – a microbe that grows by aerobic respiration when oxygen is present, switching to fermentation under anaerobic conditions.

Aerotolerant anaerobe – a microbe that grows by fermentation, but which is insensitive to air/oxygen (in contrast to strict anaerobes, which are typically killed by exposure to air/oxygen).

Capnophile – a microbe that thrives in the presence of high levels of atmospheric carbon dioxide.

- Motility. Phototactic microbes (including photosynthetic flagellates and motile cyanobacteria) will move towards a light source; heterotrophic flagellate bacteria will move through a filter of appropriate pore size into a nutrient solution, or away from unfavourable conditions (chemotaxis).

Selective and enrichment methods

Laboratory incubation under selective/enrichment conditions will allow particular microbes to be isolated in pure culture. Selective and enrichment techniques can be considered together, since they both enhance the growth of a particular microbe when compared with its competitors and they are often combined in specific media (Box 33.1).

The fundamental difference between selective and enrichment techniques is that the former use growth conditions unfavourable for competitors while the latter provide improved growth conditions for the chosen microbes.

KEY POINT Selective methods are based on the use of physicochemical conditions that will permit the growth of a particular group of microbes while inhibiting others. Enrichment techniques encourage the growth of certain bacteria, usually by providing additional nutrients in the growth medium.

Methods based on specific physical conditions include:

- Temperature. Psychrophilic and psychrotrophic microbes can be isolated by incubating the growth medium at 4 °C, while thermophilic microbes require temperatures above 45 °C for isolation. Short-term heat treatment of samples can be used to select for endospore-forming bacteria, e.g. 70–80 °C for 5–15 min, prior to isolation.
- Atmosphere. Many eukaryotic microbes are obligate aerobes, requiring an adequate supply of oxygen to grow. Bacteria vary in their responses to oxygen: obligate anaerobes are the most demanding, growing only under anaerobic conditions (e.g. in an anaerobic cabinet or jar). Oxygen requirements can be determined using the agar shake tube method as part of the isolation procedure (Fig. 33.1). Some pathogenic bacteria grow best in an atmosphere with a reduced oxygen status and increased CO_2 concentration: such carboxyphilic bacteria (capnophiles) are grown in an incubator where the gas composition can be adjusted.
- Centrifugation. This can be used to separate buoyant microbes from their non-buoyant counterparts – on centrifugation, such organisms will collect at the surface while the remaining microbes will sediment. Alternatively, density gradient methods may be used (p. 298). Centrifugation can be combined with repeated washing, to separate microbes from contaminants.
- Ultraviolet irradiation. Some microbes are tolerant of UV treatment and can be selected by exposing samples to UV light. However, the survivors may show a greater rate of mutation.
- Illumination. Samples may be enriched for cyanobacteria and microalgae by incubation under a suitable light regime. For dilute samples, where the number of photosynthetic microbes is too low to give the sample any visible green coloration, there is a risk of photoinhibition and loss of viability if the irradiance is too high. Such samples need shading during initial growth.

Box 33.1 Differential media for bacterial isolation: an example

MacConkey agar is both a selective and a differential medium, useful for the isolation and identification of intestinal Gram-negative bacteria. Each component in the medium has a particular role:

- **Peptone:** (a meat digest) provides a rich source of complex organic nutrients, to support the growth of non-exacting bacteria.
- **Bile salts:** toxic to most microbes apart from those growing in the intestinal tract (selective agent).
- **Lactose:** present as an additional, specific carbon source (enrichment agent).
- **Neutral red:** a pH indicator dye, to show the decrease in pH which accompanies the breakdown of lactose (differential agent).
- **Crystal violet:** selectively inhibits the growth of Gram-positive bacteria. (This component is only present in certain formulations of MacConkey agar.)

Any intestinal Gram-negative bacterium capable of fermenting lactose will grow on MacConkey agar to produce large red–purple colonies, the red colouration being due to the neutral red indicator under acidic conditions while the purple colouration, often accompanied by a metallic sheen, is due to the precipitation of bile salts and crystal violet at low pH.

In contrast, enteric Gram-negative bacteria unable to metabolise lactose will give colonies with no obvious pigmentation. This differential medium has been particularly useful in medical microbiology, since many enteric bacteria are unable to ferment lactose (e.g. *Salmonella*, *Shigella*) while others metabolise this carbohydrate (e.g. *Escherichia coli*, *Klebsiella* spp.). Colonial morphology (Fig. 33.2) on such a medium can give an experienced bacteriologist important clues to the identity of an organism, e.g. capsulate *Klebsiella* spp. characteristically produce large, convex, mucoid colonies with a weak pink colouration, due to the fermentation of lactose, while *E. coli* produces smaller, flattened colonies with a stronger red colouration and a metallic sheen.

Chemical methods form the mainstay of bacteriological isolation techniques and various media have been developed for the isolation of specific groups of bacteria. The chemicals involved can be subdivided into the following groups:

- Nutrients that encourage the growth of certain microbes: including the addition of a particular carbon source, or specific inorganic nutrients.
- Selectively toxic substances: for example, salt-tolerant, Gram-positive cocci can be grown in a medium containing 7.5% w/v NaCl, which prevents the growth of most common heterotrophic bacteria. Several media include dyes as selective agents, particularly against Gram-positive bacteria.
- Antibiotics: for example, the use of antibacterial agents (e.g. penicillin, streptomycin, chloramphenicol) in media designed to isolate fungi, or the use of antifungal agents (e.g. cycloheximide, nystatin) in bacterial media. Some antibacterial agents show a narrow spectrum of toxicity and these can be incorporated into selective isolation media for resistant bacteria, e.g. metronidazole for anaerobic bacteria.
- Substances that affect the pH of the medium: for example, the use of alkaline peptone water at pH 8.6 for the isolation of *Vibrio* spp.

Example Cellulolytic bacteria can be isolated from soil or water using a growth medium that contains cellulose as the major source of carbon.

Example Slow-growing *Legionella pneumophila* can be isolated from water samples using media containing the antibiotics vancomycin, polymyxin and cycloheximide to suppress other microbes.

***KEY POINT* Note that subcultures from a primary isolation medium must be grown in a non-selective medium, to confirm the purity of the isolate.**

Many of the selective and enrichment media used in bacteriology are able to distinguish between different types of bacteria: such media are termed differential media or diagnostic media and they are often used in the preliminary stages of an identification procedure. Box 33.1 gives details of the constituents of MacConkey medium, a selective, differential

Table 33.1 Selective agents in bacteriological media

Substance	Selective for
Azide salts	*Enterococcus* spp.
Bile salts	Intestinal bacteria
Brilliant green	Gram-negative bacteria
Gentian violet	Gram-negative bacteria
Lauryl sulphate	Gram-negative bacteria
Methyl violet	*Vibrio* spp.
Malachite green	*Mycobacterium*
Polymyxin	*Bacillus* spp.
Sodium selenite	*Salmonella* spp.
Sodium chloride	Halotolerant bacteria *Staphylococcus aureus*
Sodium tetrathionate	*Salmonella* spp.
Tergitol/surfactant	Intestinal bacteria
Trypan blue	*Streptococcus* spp.

medium used in clinical microbiology (e.g. for the isolation of certain faecal bacteria), while Table 33.1 gives details of other selective agents.

Further details on methods can be found in Collins *et al.* (2004) or Hurst *et al.* (2002). Note that isolation procedures for a particular microbe often combine several of the techniques described above. For instance, a protocol for isolating a food-poisoning bacterium from a foodstuff might involve:

1. Homogenisation of a known amount of sample in a suitable diluent.
2. Serial decimal dilution.
3. Separation procedures using spread or pour plates to quantify the number of bacteria of a particular type present in the foodstuff and provide a plate count (p. 237).
4. Selective/enrichment procedures, e.g. specific media/temperatures/atmospheric conditions, depending on the bacteria to be isolated.
5. Confirmation of identity: any organism growing on a primary isolation medium would require subculture and further tests, to confirm the preliminary identification as discussed below.

Identifying a particular microbe

Most of the methods described in this chapter were developed for the identification of bacteria, and bacterial examples are used to illustrate the principles involved. While the basic techniques are equally applicable to other types of microbe, the identification systems for some protozoa, fungi and algae rely predominantly on microscopic appearance. Identification of viruses requires electron microscopy or immunological techniques.

KEY POINT Identification of bacteria is often based on a combination of a number of different features, including growth characteristics, microscopic examination, physiological or biochemical characterisation, and, where necessary, immunological tests.

Direct observation

Once a microbe has been isolated and cultured in the laboratory (Chapter 32), the visual appearance of individual colonies on the surface of a solidified medium may provide useful information. Bacteria typically produce smooth, glistening colonies, varying in diameter from <1 mm to >1 cm. Actinomycete colonies are often <1 cm, with a shrivelled, powdery surface. Filamentous fungi usually grow as large, spreading mycelia with a matt appearance and are identified by microscopy, using the morphological characteristics of their reproductive structures. Yeasts produce small, glistening colonies: identification usually involves microscopy, combined with physiological and biochemical tests similar to those used for bacteria.

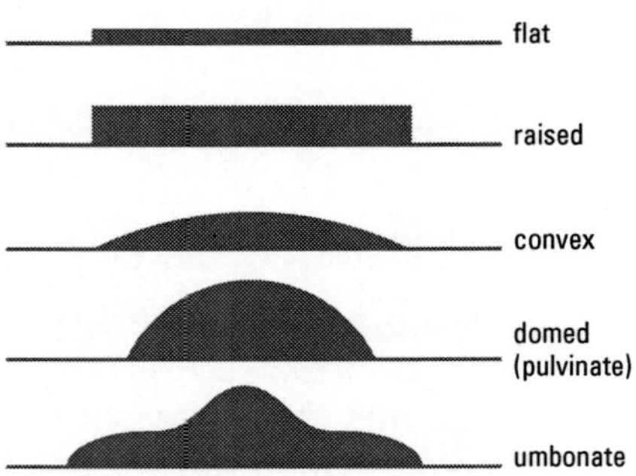

Fig. 33.2 Colony elevations (cross-sectional profile).

Colony characteristics

When measuring colony size, choose a typical colony, well spaced from any others as colony size is affected by competition for nutrients. The characteristics of a microbial colony on a particular medium include:

- Size: some bacteria produce punctiform colonies, with a diameter of less than 1 mm, while motile bacteria may spread over the entire plate.

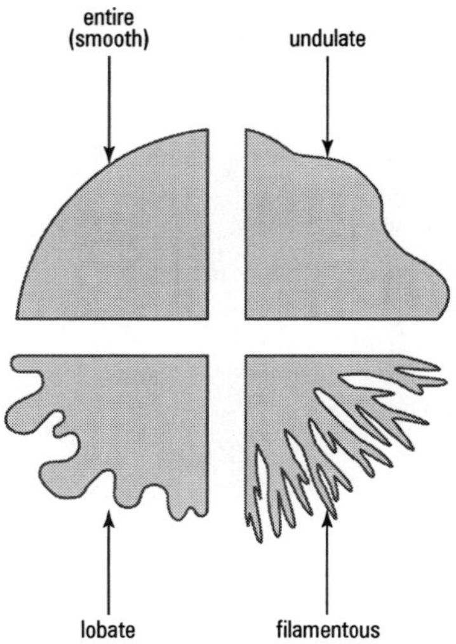

Fig. 33.3 Colony margins (surface view).

SAFETY NOTE Working with mould cultures – never attempt to smell mould cultures, because of the risk of inhaling large numbers of spores.

***Using cell shape in microbial identification** – many bacteria are pleomorphic, varying in size and shape according to the growth conditions and the age of the culture: thus, other characteristics are required for identification.*

SAFETY NOTE The Gram staining procedure involves toxic dyes and flammable solvents: avoid skin contact and extinguish any naked flames (e.g. nearby Bunsens).

***Assessing the Gram status of an unknown bacterium** – if a pure culture gives both Gram-positive and Gram-negative cells, identical in size and shape, it can be regarded as a Gram-positive organism that is demonstrating Gram-variability.*

- Form: colonies may be circular, irregular, lenticular (spindle-shaped) or filamentous.
- Elevation: colonies may be flat, raised, convex, etc. (see Fig. 33.2).
- Margin: the edge of a colony may be entire (smooth) or more distinctive, e.g. undulate, lobate or filamentous (Fig. 33.3).
- Consistency: colonies may be viscous (or mucoid), butyrous (of similar consistency to butter) or friable (dry and granular), etc.
- Colour: some bacteria produce characteristic pigments. A few pigments are fluorescent under UV light.
- Optical properties: colonies may be translucent or opaque.
- Haemolytic reactions on blood agar: many pathogenic bacteria produce characteristic zones of haemolysis. Alpha haemolysis is a partial breakdown of the haemoglobin from the erythrocytes, producing a green zone around the colony, while beta haemolysis is the complete destruction of haemoglobin, producing a clear zone.
- Odour: some actinomycetes and cyanobacteria produce earthy odours, while certain bacteria and yeasts produce fruity or 'off' odours. However, odour is not a reliable characteristic in bacterial identification and smelling plates creates a risk of inhaling microbes (p. 206).

Microscopic examination – cell shape

Bacteria are usually observed using an oil immersion objective at a total magnification of ×1000 (p. 166). Bacteria are subdivided into the following groups, based on their cell shape:

- Cocci (singular, coccus): spherical, or almost spherical, cells, sometimes growing in pairs (diplococci), chains or clumps.
- Rods: straight, cylindrical cells of variable length with flattened, tapered or rounded ends – sometimes termed bacilli. Short rods are sometimes called cocco-bacilli.
- Curved rods: the curvature varies according to the organism, from short curved rods, sometimes tapered at one end, to spiral shapes.
- Branched filaments: characteristic of actinomycete bacteria.

Gram staining

This is the most important differential staining technique in bacteriology (Box 33.2 gives details). It enables us to divide bacteria into two distinct groups, Gram-positive and Gram-negative, according to a particular staining procedure (the technique is given a capital letter, since it is named after its originator, H.C. Gram). The basis of the staining reaction is the different structure of the cell walls of Gram-positive and Gram-negative bacteria. Heat fixation of air-dried bacteria causes some shrinkage, but cells retain their shape: to measure cell dimensions use a chemical fixative.

Gram staining should be carried out using light smears of young, active cultures, since older cultures may give variable results. In particular, certain Gram-positive bacteria may stain Gram-negative if older cultures are used. This Gram-variability is due to autolytic changes in the cell wall of Gram-positive bacteria. Developing spores are often visible as unstained areas within older vegetative cells of *Bacillus* and *Clostridium*. Other stains are required to demonstrate spores, capsules or flagella (p. 167).

Box 33.2 Preparation of a heat-fixed, Gram-stained smear

Preparation of a heat-fixed smear
The following procedure will provide you with a thin film of bacteria on a microscope slide, for staining.

1. **Take a clean microscope slide and pass it through a Bunsen flame twice** to ensure it is free of grease. Allow to cool.
2. **Using a sterile inoculating loop, place a single drop of water in the centre of the slide and then mix in a small amount of sample** from a single bacterial colony with the drop, until the suspension is slightly turbid. Smear the suspension over the central area of the slide, to form a thin film. For liquid cultures, use a single drop of culture fluid, spread in a similar manner.
3. **Allow to air-dry at room temperature,** or high above a Bunsen flame: air-drying must proceed gently or the cells will shrink and become distorted.
4. **Fix the air-dried film by passage through a Bunsen flame.** Using a slide holder or forceps, pass the slide, film side up, rapidly through the hottest part of the flame (just above the blue cone). The temperature of the slide should be just too hot for comfort on the back of your hand: note that you must not overheat the slide or you may burn yourself (you will also ruin the preparation).
5. **Allow to cool:** the smear is now ready for staining.

Gram-staining procedure
The version given here is a modification of the Hucker method, since acetone is used to decolourise the smear. Note that some of the staining solutions used are flammable, especially the acetone decolourising solvent: you must make sure that all Bunsens are turned off during staining. The procedure should be carried out with the slides suspended over a sink, using a staining rack.

1. **Flood a heat-fixed smear with 2% w/v crystal violet in 20% v/v ethanol : water** and leave for 1 min.
2. **Pour off the crystal violet and rinse briefly with tap water. Flood with Gram's iodine** (2 g KI and 1 g I_2 in 300 ml water) for 1 min.
3. **Rinse briefly with tap water** and leave the tap running gently.
4. **Tilt the slide and decolourise with acetone** for 2–3 s: acetone should be added dropwise to the slide until no colour appears in the effluent. This step is critical, since acetone is a powerful decolourising solvent and must not be left in contact with the slide for too long.
5. **Immediately immerse the smear in a gentle stream of tap water** to remove the acetone.
6. **Pour off the water and counterstain for 10–15 s using 2.5% w/v safranin** in 95% v/v ethanol : water.
7. **Pour off the counterstain, rinse briefly with tap water, then dry the smear** by blotting gently with absorbent paper: all traces of water must be removed before the stained smear is examined microscopically.
8. **Place a small drop of immersion oil on the stained smear: examine directly** (without a coverslip) using an oil-immersion objective (p. 166).

Gram-positive bacteria retain the crystal violet (primary stain) and appear purple, while Gram-negative bacteria are decolourised by acetone and counterstained by the safranin, appearing pink or red when viewed microscopically.

Other decolourising solvents are sometimes used, including ethanol : water, ethyl ether : acetone and acetone : alcohol mixtures. The time of decolourisation must be adjusted, depending upon the strength of the solvents used, e.g. 95% v/v ethanol : water is less powerful than acetone, requiring around 30 s to decolourise a smear.

Motility

Wet mounts can be prepared by placing a small drop of bacterial suspension on a clean, degreased slide, adding a coverslip and examining the film by light microscopy without delay. For aerobes, areas near air bubbles or by the edge of the coverslip give best results, while anaerobes show greatest motility in the centre of the preparation, with rapid loss of motility due to oxygen toxicity.

__Using the hanging drop technique__ – place a drop of bacterial suspension on a coverslip and invert over a cavity slide so that the drop does not make contact with the slide: motile aerobes are best observed at the edge of the droplet, where oxygen is most abundant.

Prepare wet mounts using young cultures in exponential growth in a liquid medium (p. 233). It is best to work with cultures grown at 20 or 25 °C, since those grown at 37 °C may not be actively motile on cooling to room temperature. It is essential to distinguish between the following:

- True motility, due to the presence of flagella: bacteria dart around the field of view, changing direction in zigzag, tumbling movements.

Assessing motility *– if you have not seen bacterial motility before, it is worth comparing your unknown bacterium to a positive and a negative control.*

- Brownian motion: non-motile bacteria show a localised, vibratory, random motion, due to bombardment of bacterial cells by molecules in the solution.
- Passive motion, due to currents within the suspension: all cells will be swept in the same direction at a similar rate of movement.
- Gliding motility: a slower, intermittent movement, parallel to the longitudinal axis of the cell, requiring contact with a solid surface.

Basic laboratory tests

At least two simple biochemical tests are usually performed:

Performing the oxidase test *– never use a nichrome wire loop, as this will react with the oxidase reagent, giving a false positive result.*

1. Oxidase test This identifies cytochrome *c* oxidase, an enzyme found in obligate aerobic bacteria. Soak a small piece of filter paper in a fresh solution of 1% (w/v) *N-N-N′-N′*-tetramethyl-*p*-phenylenediamine dihydrochloride on a clean microscope slide. Rub a small amount from the surface of a young, active colony onto the filter paper using a glass rod, a *plastic* loop or a wooden applicator stick: a purple-blue colour within 10 s is a positive result.

SAFETY NOTE The catalase and oxidase reagents are irritants and could be harmful if swallowed. Avoid skin contact and ingestion.

2. Catalase test This identifies catalase, an enzyme found in obligate aerobes and in most facultative anaerobes, which catalyses the breakdown of hydrogen peroxide into water and oxygen ($2H_2O_2 \rightarrow 2H_2O + O_2$). Transfer a small sample of your unknown bacterium onto a coverslip using a disposable plastic loop or glass rod. Invert onto a drop of hydrogen peroxide: the appearance of bubbles within 30 s is a positive reaction. This method minimises the dangers from aerosols formed when gas bubbles burst.

The oxidase and catalase tests effectively allow us to subdivide bacteria on the basis of their oxygen requirements, without using agar shake cultures (p. 212) and overnight incubation, since, for the most part:

Avoiding false negatives *– ensure you use sufficient material during oxidase and catalase testing, otherwise you may obtain a false negative result: a clearly visible 'clump' of bacteria should be used.*

- obligate aerobes will be oxidase and catalase positive;
- facultative anaerobes are generally oxidase negative and catalase positive;
- microaerophilic bacteria, aerotolerant anaerobes and strict (obligate) anaerobes will be oxidase and catalase negative – the latter group will grow only under anaerobic conditions (p. 212).

Once you have reached this stage (colony characteristics, motility, shape, Gram reaction, oxidase and catalase status) it may be possible to make a tentative identification, at least for certain Gram-positive bacteria, at the generic level. To identify Gram-negative bacteria, particularly the oxidase-negative, catalase-positive rods, further tests are required.

Carbohydrate utilisation tests and isolation media *– many diagnostic agar-based media incorporate one or more specific carbohydrates and pH indicator dyes, thereby providing additional information as part of the isolation procedure (Box 33.1).*

Identification tables: further laboratory tests

Bacteria are asexual organisms and strains of the same species may give different results for individual biochemical/physiological tests. This variation is allowed for in identification tables, based on the results of a large number of tests. Identification tables are often used for particular subgroups of bacteria, after Gram staining and basic laboratory tests have been performed: an example is shown in Table 33.2.

Table 33.2 Identification table for selected Gram-negative rods

Bacterium	Biochemical test 1	2	3	4	5	6	7	8	9
Escherichia coli	v	+	−	+	−	v	v	+	−
Proteus mirabilis	−	−	v	−	+	−	+	−	+
Morganella morganii	−	−	−	−	+	−	+	+	−
Vibrio parahaemolyticus	−	+	v	−	−	+	+	+	−
Salmonella spp.	−	+	v	−	−	+	+	−	+

Key to biochemical tests and symbols
1. sucrose utilisation
2. mannitol utilisation
3. citrate utilisation
4. β-galactosidase activity
5. urease activity
6. lysine decarboxylase activity
7. ornithine decarboxylase activity
8. indole production
9. H_2S production

+, >90% of strains tested positive
−, <10% of strains tested positive
v, 10–90% of strains tested positive

A large number of specific biochemical and physiological tests are used in bacterial identification, including:

- Carbohydrate utilisation tests. Some bacteria can use a particular carbohydrate as a carbon and energy source. Acidic end-products can be identified using a pH indicator dye (p. 153) while CO_2 is detected in liquid culture using an inverted small test tube (Durham tube, Fig. 33.4). Aerobic breakdown (*via* respiration) is termed oxidation while anaerobic breakdown is known as fermentation. Identification tables usually incorporate tests for several different carbohydrates, e.g. Table 33.2.
- Enzyme tests. Most of these incorporate a substance which changes colour if the enzyme is present, e.g. a pH indicator, or a chromogenic or fluorogenic substrate.
- Tests for specific end-products of metabolism, e.g. the production of indole due to the metabolic breakdown of the amino acid tryptophan, or H_2S from sulphur-containing amino acids.

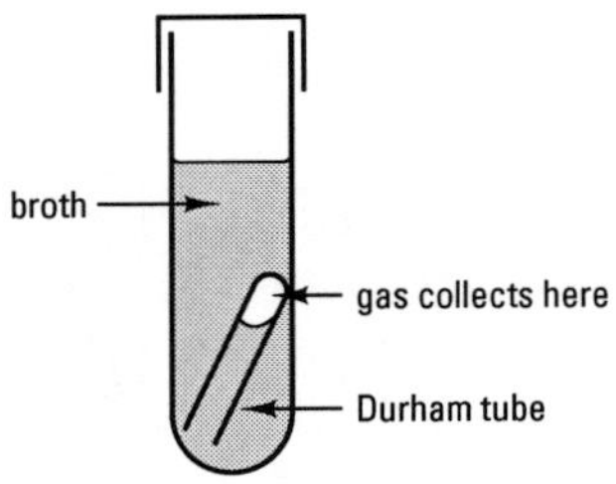

Fig. 33.4 Durham tube in carbohydrate utilisation broth. Air within the Durham tube is replaced by broth during the autoclaving procedure.

Identification kits

Some biochemical tests are now supplied in kit form, e.g. the API 20E system incorporates 20 tests within a sterile plastic strip (Fig. 33.5). After inoculation and overnight incubation, the results of the tests are converted into a seven-digit code, for comparison with known bacteria using either a reference book (the Analytical Profile Index), or a computer program. While kit identification systems save time and labour, they are more expensive and less flexible than conventional biochemical tests.

Molecular approaches to microbial identification *– several novel methods of detection and identification are based on nucleic acid techniques, including use of the PCR and Southern blotting to detect particular microbes.*

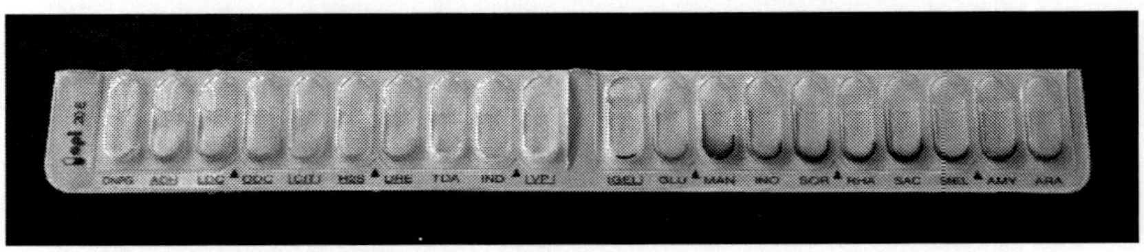

Fig. 33.5 Example of a bacterial identification kit (API 20E). Property of bioMérieux S.A./ photographer: Andrea Bannuscher.

Immunological tests

Tests used in diagnostic microbiology include:

- Agglutination tests: based on the reaction between specific antibodies and a particular bacterium (p. 258). These tests are particularly useful for subdividing biochemically similar bacteria.
- Fluorescent antibody tests: the reaction between a labelled antibody and a particular bacterium can be visualised using UV microscopy. The direct fluorescent antibody test uses fluorescein isothiocyanate as the label.
- Enzyme-linked immunoassay tests using antibodies labelled with a particular enzyme, e.g. the double antibody sandwich ELISA or competitive ELISA tests (p. 261).

While such tests can give specific and accurate confirmation of the identity of a bacterium under controlled laboratory conditions, they are often too expensive and time-consuming for routine identification purposes, especially when large numbers of tests are required.

Typing methods

The identification of bacteria at subspecies level is known as typing: this is usually done in a specialist laboratory, e.g. as part of an epidemiological study to establish the source of an infection. Various methods are used:

Practical applications of bacterial typing – E. coli *O157:H7 is a serotype of this bacterium that is capable of causing severe human disease: it can be identified on the basis of an agglutination reaction with an appropriate antiserum. Other applications include tracking the development and spread of particular types of antibiotic-resistant bacteria in hospitals and the community.*

- Antigen typing or serotyping is based on immunological tests.
- Phage typing is based on the susceptibility of different strains to certain bacterial viruses (phages).
- Biotyping is based on biochemical differences between different strains, e.g. enzyme profiles or antibiotic resistance screening ('antibiograms').
- Bacteriocin typing: bacteriocins are proteins released by bacteria which inhibit the growth of other members of the same species.

Naming microbes

The use of scientific names is fundamental to all aspects of biological science since it aims to provide a system of identification which is precise, fixed and of universal application. Without such a system, comparative studies would be impossible.

There are two possible bases for such classification:

- Phenetic taxonomy, which involves grouping on the basis of phenotypic similarity, frequently using complex statistical techniques to obtain objective measures of similarity. The characters used have been largely morphological and anatomical, but biochemical, cytological and other characters are increasingly used, especially for microbes, where structural characteristics are few.
- Phylogenetic (= phyletic) taxonomy, which involves grouping on the basis of presumed evolutionary, and therefore genetic, relationships.

These two systems are often broadly similar in outcome, since closely related organisms are usually fairly similar to each other and because judgements of evolutionary relationships are usually themselves based upon similarities. The situation is made more complex by phenomena such as convergent and divergent evolution; phylogenetic classifications are also liable to subjective bias.

Definitions

Systematics – the study of the diversity of living organisms and of the evolutionary relationships between them.

Classification (taxonomy) – the study of the theory and methods of organisation of taxa and therefore, a part of systematics.

Taxon (plural taxa) – an assemblage of organisms sharing some basic features.

Nomenclature – the allocation of names to taxa.

Identification – the placing of organisms into taxa.

The basis of classification

No single, simple definition of a species is possible, but there are two generally used definitions:

- A group of organisms capable of interbreeding and producing fertile offspring – this, however, excludes all asexual organisms, such as bacteria and some other microbes.
- A group of organisms showing a close similarity in phenotypic characteristics – this would include morphological, anatomical, biochemical, ecological and life history characters or, increasingly, similarities in nucleic acid sequence.

DNA-based definition of a species *– in bacteriology, members of a single species can be characterised by sharing at least 70% DNA, based on nucleic acid hybridisation studies.*

KEY POINT ***The basic unit of classification is the species, which represents a group of recognisably similar individuals, clearly distinct from other such groups.***

Writing taxonomic names *– always underline (handwritten text) or italicise (word-processed text) genus and species names to avoid confusion: thus bacillus is a descriptive term for a rod-shaped bacterium, while* <u>Bacillus</u> *is a generic name.*

Understanding microbiological terms *– the term* strain *is widely used, particularly in the context of the practice of lodging microbiological strains with culture collections, while the term* isolate *is often used for a pure culture derived from a natural (wild) population. Cell lines (Chapter 34) are often given code names and/or reference numbers.*

When species are compared, groups of species may show a number of features in common; they are then arranged into larger groupings known as genera (singular, genus). Cellular microbes (e.g. bacteria, archaea, fungi, protozoa and algae) are given two Latin terms to identify their genus and species (a Latin 'binomial'), for example *Escherichia coli*. All scientific names of organisms are either Latin or are treated as Latin, written in the Latin alphabet and subject to the rules of Latin grammar. Consequently, you must be very precise in your use of such names. When used in a formal scientific context, the specific name can be followed by the authority on which the name is based, i.e. the name of the person describing that species, and the year in which it was first described. For example, the full, formal name of baker's yeast is *Saccharomyces cerevisiae* Meyen ex Hansen 1883. Where specified, additional terms may follow the species name to indicate the type or subspecies, e.g. *Shigella dysenteriae* type 1. Many micro-organisms are now referred to by their generic and specific names followed by a culture collection reference number, e.g. *Bacillus subtilis* NCTC 10400, where NCTC stands for the National Collection of Type Cultures and 10400 is the reference number of that strain in the collection.

After first use in a text, the genus name may be abbreviated to a single letter, e.g. *E. coli*, as long as this does not cause confusion with other genera; for example, with *Escherichia coli* and *Enterococcus faecalis*, these could be abbreviated to *Esch. coli* and *Ent. faecalis*, to avoid confusion. Where the species name is unknown, the (non-italicised) abbreviation 'sp.' (singular) or 'spp.' (plural) should be used, e.g. *Enterococcus* sp. denotes a single unknown species of the genus *Enterococcus*.

Understanding the hierarchical taxonomic system *– the taxonomic groups, in decreasing level, are: kingdom, phylum or division, class, order, family, genus, species. An example of a full hierarchical classification would be: Bacteria (monera); Gracilicutes; Scotobacteria; Pseudomonadales; Pseudomonadaceae;* <u>Pseudomonas</u>*;* <u>Pseudomonas aeruginosa</u>*. Note that the names of orders normally end in '-ales' while family names usually end in '-aceae' and that all names of taxa apart from species begin with a capital letter.*

KEY POINT ***While the use of taxa above those of the genus are somewhat subjective and may vary between different authorities, the use of genus and species names are governed by strict internationally agreed conventions called Codes of Nomenclature (e.g. Lapage, 1992 for bacteria).***

The classification and nomenclature of viruses are less advanced than for cellular organisms and the current nomenclature has been arrived at on a piecemeal, *ad hoc* basis. The International Committee for Virus Taxonomy proposed a unified classification system, dividing viruses into 50 families on the basis of: host preference, nucleic acid type

Example The virus that causes tobacco mosaic disease belongs to the *Tobamovirus* group and can be referred to as tobacco mosaic tobamovirus, tobacco mosaic virus or TMV.

(i.e. DNA or RNA), whether the nucleic acid is single or double stranded and the presence or absence of a surrounding envelope. Many viruses are still referred to by trivial names or by code-names (sigla), e.g. the bacterial viruses ϕX174, T4, etc. Many of the names used reflect the diseases caused by the virus. Often, a three-letter abbreviation is used, e.g. HIV (for human immunodeficiency virus), TMV (for tobacco mosaic virus).

Text references

Collins, C.H., Lyne, P.M. and Grange, J.M. (2004) *Microbiological Methods*, 8th edn. Hodder-Arnold, London.

Hurst, C.J., Crawford, R.L. and Knudsen, G.R. (2002) *Manual of Environmental Microbiology*, 2nd edn. American Society for Microbiology, Washington, DC.

Lapage, S.P. (1992) *International Code of Nomenclature for Bacteria: 1990 revision.* American Society for Microbiology, Washington, DC.

Sources for further study

Atlas, R.M. (2004) *Handbook of Microbiological Media,* 3rd edn. CRC Press, Boca Raton.
[Gives details of culture media for a broad range of microbes]

Barrow, G.I. and Feltham, R.K.A. (1993) *Cowan and Steel's Manual for the Identification of Medical Bacteria,* 3rd edn. Cambridge University Press, Cambridge.
[A standard reference work for the identification of clinically important bacteria]

Boone, D.R. *Naming a New Procaryotic Taxon.* Available: http://methanogens.pdx.edu/naming.html
Last accessed: 01/04/07.
[Succinct indication of procedures for naming a new prokaryote species]

Buchen-Osmond, C. (2002) Universal Virus Database of the International Committee on Taxonomy of Viruses. Available: http://www.ncbi.nlm.nih.gov/ICTVdb/.
Last accessed: 01/04/07.

Cullimore, D.R. (2000) *Practical Atlas for Bacterial Identification.* CRC Press, Boca Raton.
[Detailed coverage of practical methods applicable to a wide range of bacteria]

Eaton, A.D., Clesceri, L.S., Rice, E.W. and Greenberg, A.E. (2005) *Standard Methods for Examination of Water and Wastewater,* 21st edn. American Public Health Association, Washington, DC.
[Gives standard US protocols for a range of indicator bacteria]

Fauquet, C.M., Mayo, M.A., Maniloff, J., Desselberger, U. and Ball, L.A. (2005) *Virus Taxonomy. Eighth Report of the International Committee on the Taxonomy of Viruses.* Elsevier, London.

Fisher, F., Cook, N.B., Fisher, F.W. and Kaszczuk, S. (1998) *Fundamentals of Diagnostic Mycology.* Saunders, Philadelphia.
[Covers identification techniques for medically important fungi]

Fox, A. (ed.) *Journal of Microbiological Methods.* Elsevier, London (available through Science Direct at: http://www.sciencedirect.com/)
[Provides information on novel developments in all aspects of microbiological methods, including microbial culture/isolation and molecular approaches]

Holt, J.G. (1994) *Bergey's Manual of Determinative Bacteriology,* 9th edn. Williams and Wilkins, Baltimore.

Institute of Biology (2000) *Biological Nomenclature: Recommendations on Terms, Units and Symbols.* Institute of Biology, London.
[Includes sections on taxonomy and classification of organisms]

Jeffrey, C. (1989) *Biological Nomenclature*, 3rd edn. Edward Arnold, London.

Levett, P.N. (ed.) (1991) *Anaerobic Microbiology: A Practical Approach.* IRL Press, Oxford.
[Provides details of the methods used to isolate and culture anaerobic micro-organisms]

Macfaddin, J.F. (2000) *Biochemical Tests for the Identification of Medical Bacteria*, 3rd edn. Williams and Wilkins, Philadelphia.
[Explains the operating principles underlying most of the biochemical tests in routine use in diagnostic bacteriology]

Pepper, I.L. and Gerba, C.P. (2005) *Environmental Microbiology: A Laboratory Manual*, 2nd edn. Academic Press, New York.

Rochelle, P.A. (2001) *Environmental Molecular Microbiology: Protocols and Applications*. Horizon Scientific Press, Norwich.
[Covers molecular approaches to the detection and identification of microbes in environmental samples]

Study exercises

33.1 Plan a collection/sampling strategy for a target group of microbes. What approaches might you take in the following instances?

(a) Collecting representatives of the normal skin microflora.
(b) Sampling psychrotrophic anaerobes from the subsurface mud of an estuary.
(c) Collecting a sample of sea water to enumerate faecal indicator bacteria.
(d) Mapping the microflora on the surface of a leaf.

33.2 Decide on an appropriate isolation procedure for a particular microbe. How might you isolate the following microbes from a sample?

(a) Photosynthetic flagellate algae in a sample of pond water.
(b) Bacteria growing as a biofilm on the surface of sand particles.
(c) Faecal streptococci (enterococci) present at low density (<1 per ml) in a sample of river water.
(d) A distinctively shaped bacterium present at very low numbers in a sample containing a high number of other microbes of different shape but with similar nutritional requirements.

33.3 Investigate the selective basis of microbiological media. Using textbooks on bacteriological methods or the Web, research how each of the following media operates, in terms of their selective and diagnostic (differential) features:

(a) Mannitol salt agar for *Staphylococcus aureus*.
(b) Membrane lauryl sulphate broth for coliforms and *E. coli*.
(c) Slanetz and Bartley medium for faecal streptococci (enterococci).
(d) Mannitol pyruvate egg yolk polymyxin (MPYP) medium for *Bacillus cereus*.

33.4 Describe the colonial characteristics of selected microbes. Either research the features of the following microbes (e.g. *via* the Web), or look at well-isolated individual colonies in the laboratory, following overnight growth on a suitable medium:

(a) *Escherichia coli*;
(b) *Pseudomonas aerogenes*;
(c) *Staphylococcus aureus*;
(d) *Streptococcus pneumoniae*;
(e) *Bacillus subtilis*;
(f) *Saccharomyces cerevisiae*.

33.5 Find out how some of the tests used in microbial identification work. Research the operating principles that underpin the following tests, using microbiological textbooks or the Web, and prepare brief notes explaining how each of the tests works:

(a) oxidase test;
(b) indole test;
(c) β-galactosidase test;
(d) urease test;
(e) lysine decarboxylase test;
(f) H_2S production.

33.6 Identify the following oxidase-negative, catalase-positive, Gram-negative rod-shaped bacteria, using Table 33.2.

(a) Positive for citrate, urease, ornithine decarboxylase and H_2S only.
(b) Positive for sucrose, mannitol, β-galactosidase, lysine decarboxylase and indole only.
(c) Positive for mannitol, citrate, lysine decarboxylase, ornithine decarboxylase and H_2S only.
(d) Positive for urease, ornithine decarboxylase and indole only.
(e) Positive for mannitol, β-galactosidase and indole only.

34 Working with animal and plant tissues and cells

> **Definitions**
>
> ***In vivo*** – occurring within a living organism (literally 'in life').
>
> ***In vitro*** – occurring outside a living organism, in an artificial environment (literally 'in glass').
>
> ***In silico*** – occurring within a computer system (literally 'in silicon').

While the aim of many studies is to isolate, quantify or characterise individual molecules from a biological system, e.g. the purification of a particular enzyme, this is not always the most appropriate course of action. Depending on the purpose of the investigation, it may be more relevant to study the functioning of biomolecules within more complex systems, in order to understand their role in a particular biological process. At one extreme this may be carried out *in vivo*, using whole multicellular organisms (e.g. individual animals or plants), while the other extreme is represented by *in vitro* studies, using subcellular 'cell-free' extracts. Between these two extremes, a range of tissue and cell culture techniques offers some of the biological complexity of the intact organism combined with a degree of experimental control that may not be obtainable *in vivo*. The ethics and costs of whole animal experimentation have provided an additional stimulus to the development of *in vitro* methods, e.g. toxicity tests using mammalian cell culture rather than laboratory animals, while developments in molecular genetics have led to further applications of cell and tissue culture.

Animal tissues and organs

> ***Working with vertebrate animals and their organs/tissues** – remember that procedures must be consistent with the law, i.e. in the UK, the* Animals Scientific Procedures Act 1986.

Physiological experiments are carried out using either whole organisms, or a range of animal organs and tissues, including heart, liver, muscle, etc.

> ***KEY POINT** A major practical consideration is that the tissue should be studied as soon as possible after the death of the animal, typically under laboratory conditions that mimic the* **in vivo** *environment as closely as possible.*

> **Using human tissues** – *in the UK, the* Human Tissue Act 2004 *applies and work is licensed and overseen by the Human Tissue Authority. Note that established cell lines (p. 225) are excluded from the regulations.*

In most instances, the experiments are relatively short term (<24 h) and the aim is to maintain the tissue in a physiological state similar to that within the living organism. For metabolic studies, the whole organ or a tissue slice (typically 1–10 mm thick) will be bathed in an appropriate perfusion fluid, supplied either by gravity or by peristaltic pump. Practical considerations include:

- Inorganic solute requirements – the chemical composition of the perfusion fluid is usually chosen to reflect the major inorganic ion requirements of the tissue. For short-term studies, a number of so-called 'physiological salt solutions' may be used, e.g. Ringer's solution, one formulation of which is given in Table 34.1.
- Oxygen requirements – it may be necessary to increase the O_2 content of the perfusion fluid by bubbling with air, in order to meet the oxygen demand of the innermost parts of the tissue. However, this can lead to oxygen toxicity in the outermost parts and an alternative approach is simply to increase the rate of perfusion.
- Physicochemical conditions – including temperature (usually controlled to ±1 °C of normal body temperature), water status (the perfusion fluid and the tissue should be isotonic, p. 150), pH and buffering capacity (e.g. some perfusion fluids have elevated an $NaHCO_3$ concentration, to mimic the buffering capacity of mammalian serum).

Table 34.1 Composition of Ringer's solution (simplified formulation, for amphibians, etc., pH 7.4–7.6)

Compound	Amount per litre (g)
NaCl	6.0
KCl	0.075
$CaCl_2$	0.1
$NaHCO_3$	0.1

Definitions

Apoplasm – that part of the plant body outside the symplasm.

Light compensation point – the amount of photosynthetically active radiation (PAR) where photosynthetic CO_2 uptake is balanced by CO_2 production due to respiration and photorespiration.

Plasmodesmata – transverse connections through the cell wall, linking the cytoplasm of adjacent plant cells and creating a symplasm.

Working with plant tissues and organs *– the ethical problems associated with animal and human tissues are avoided, and there is a decreased risk of infection to the laboratory worker; practical manipulations are often carried out in horizontal laminar-flow cabinets (p. 205).*

Table 34.2 Components of Long Ashton medium (nitrate version)

Stock solution: mass of component required per litre of solution (g)	Volume of stock solution to make 1 litre of medium (ml)
Major nutrients	
KNO_3: 50.60	8
$Ca(NO_3)_2$: 80.25	8
$MgSO_4 \cdot 7H_2O$: 46.00	8
$NaH_2PO_4 \cdot 2H_2O$: 52.00	4
Micronutrients	
FeKEDTA: 3.30	5
$MnSO_4 \cdot 4H_2O$: 2.23	1
$ZnSO_4 \cdot 7H_2O$: 0.29	1
$CuSO_4 \cdot 5H_2O$: 0.25	1
H_3BO_3: 3.10	1
$Na_2MoO_4 \cdot 2H_2O$: 0.12	1
NaCl: 5.85	1
$CoSO_4 \cdot 7H_2O$: 0.056	1

Effects of humans on plants *– remember that your exhaled breath will be nearly saturated with water vapour and will contain CO_2 at 3–4% v/v, some 100 times more concentrated than atmospheric CO_2: your breath can thus affect rates of transpiration and photosynthesis.*

- Organic nutrient requirements – for longer-term studies, suitable nutrients will be required: these may be chemically defined additives, e.g. vitamins, amino acids, proteins, etc., or biological fluids such as plasma or serum. Glucose is often added as a carbon and energy source (Freshney, 2005).

Plant tissues and organs

Individual plant components (e.g. leaves, leaf slices and epidermal strips) can be isolated from the main plant body for study under controlled conditions. Since photosynthetic plant parts are autotrophic, they may be maintained *in vitro* for longer than animal organs, given adequate light and CO_2. However, most plant cells are joined via plasmodesmata, and the separation of such connections when the component is removed from the plant often leads to death when these connections are broken.

KEY POINT Plants show wound responses that may affect the metabolic processes under study. For these reasons, the most suitable systems for longer-term studies are often whole organs, e.g. whole leaves or entire root systems.

Water culture (hydroponics), e.g. in Long Ashton medium (Table 34.2), is an alternative approach, offering greater control over the root environment (Dodds and Robert, 1995).

It is best to use vigorous, healthy stock plants and to follow a well-established procedure, taking account of the following:

- Sterility – strict attention to sterile technique can be essential to the success of many longer-term experiments. Decontamination of plant organs may be especially difficult where specimens are obtained from soil: to achieve this, use a surface wash with disinfectant (e.g. 10% w/v sodium hypochlorite), followed by several rinses with sterile water.
- Gaseous environment – in general, the experimental system should be well ventilated. Actively photosynthetic tissues will rapidly deplete the atmospheric CO_2 in a closed vessel: plant parts may also produce physiologically active gases, such as ethylene, especially at wound sites. Turgor loss may occur in isolated plant parts unless a high humidity is maintained.
- Nutrition – plant tissues may benefit from a supply of inorganic ions, including K^+, SO_4^{2-}, etc. and may require certain vitamins, micronutrients and plant hormones for prolonged studies.
- Physicochemical conditions – light is the most important environmental requirement for green plant parts. At atmospheric CO_2 concentrations, the light compensation point is about 5–10 μmol photons PAR $m^{-2} s^{-1}$ (p. 278) and photosynthesis is usually saturated between 500 and 2000 μmol photons PAR $m^{-2} s^{-1}$, depending on the plant type. Light quality and photoperiod (daylength) are also important: fluorescent tubes and incandescent bulbs that mimic the photosynthetic spectrum of sunlight are available (see p. 277), while the photoperiod can be controlled using a timer. The water potential of aqueous media can be adjusted using an impermeant osmoticum such as mannitol, and pH values are often kept close to those of the apoplasm (≈pH 6) using appropriate buffers, if necessary (p. 156).

Cell and tissue cultures

Many of the basic principles involved in culturing animal and plant cells are broadly similar to those described for microbial cell culture (p. 203). Definitions for several key terms are given in the margin.

Applications of cell and tissue culture

The main uses of animal and plant cell culture systems include:

- Experimental model systems in biochemistry, pharmacology and physiology: cell culture offers certain advantages over whole organism studies, with greater control over environmental conditions and biological variability. The use of genetically defined clones of cells may simplify the analysis of experimental data. Conversely, results obtained with specialised cell-based systems might be unrepresentative of a broader range of cell types and may be more difficult to interpret in terms of the whole organism.
- Studies of the growth requirements of particular cells: including studies of the positive effects of growth factors or growth-promoting substances, and the negative effects of xenobiotics or cytotoxic compounds. The use of cell culture in bioassays and mutagenicity testing is considered in Chapter 35.
- Studies of cell development and differentiation: including aspects of the cell cycle and gene expression. Cell cultures retaining their ability to differentiate *in vitro* are particularly interesting to researchers, while the lack of differentiation and unlimited growth of many animal cell lines makes them useful models of tumour development.
- Pathological studies: including culture of foetal cells for karyotyping, detection of genetic abnormalities, e.g. trisomy, etc.
- Genetic manipulation: cell culture techniques have played an essential role in the development of molecular biology, including the production of transgenic animals and plants by techniques such as transfection, etc.
- Biotechnology: including the industrial production of therapeutic proteins, vaccines and monoclonal antibodies using large-scale hybridoma culture techniques similar to those used in microbiology (pp. 233–5).
- Stem cell technology: increasingly important in biomedical research and development.

***KEY POINT** One of the main differences between organ and tissue incubation techniques and those used in cell/tissue culture is that the former aim simply to maintain metabolic and physiological activity for a limited period, while the latter provide conditions suitable for cell growth, division and development* in vitro *over an extended timescale, from a few days to several months.*

Animal cell culture systems

These may be established either from whole organisms (e.g. chick embryo), discrete organs (e.g. rat liver) or from blood (e.g. lymphocytes), typically using *mild* enzymic tissue disruption techniques, where necessary.

***KEY POINT** While, in theory, it is possible to culture nucleated cells from virtually any source, in practice the highest rates of success are most often achieved with young, actively growing tissues.*

Definitions

Cell line – a cell culture derived by passage of a primary culture.

Clone – a population of cells derived from a single original cell, i.e. sharing the same genotype.

Confluence/confluent growth – merging of individual cells to form a continuous layer.

Continuous cell line – a culture with the capacity for unlimited multiplication *in vitro*. Sometimes termed an established cell line.

Finite cell line – a culture with a limited capacity for growth *in vitro* (maximum number of cell doublings).

Hybridoma – a hybrid cell produced by fusion of a tumour (myeloma) cell and an antibody-producing B lymphocyte (p. 257). A cell line derived from a single hybridoma will produce a single antibody type (monoclonal).

Immortalisation – conversion of a finite cell line into a continuous cell line.

Passage – an alternative term for subculture.

Primary culture – a cell culture derived from tissue or organ fragments (explants). Primary culture ends on first subculture.

Senescence – the end-point in the limited lifespan of a finite cell line, characterised by the lack of proliferation. Conversion to an established (continuous) cell line requires 'escape' from senescence.

Stem cell – an undifferentiated animal cell that can generate new stem cells through mitosis, or can develop and differentiate into a particular specialised cell type.

Transformation – a permanent alteration in the growth characteristics of a finite cell line that may include (i) changes in morphology, (ii) an increased growth rate and/or (iii) the acquisition of an infinite lifespan, often termed immortalisation. Transformation may be spontaneous or may be induced by chemical agents or viruses, and often involves a change in chromosome number.

The principal considerations in animal cell culture are:

> *SAFETY NOTE Working with tissue cultures safely – do not confuse laminar flow hoods (p. 205) with biosafety cabinets, as they perform completely different functions. A horizontal laminar flow hood is designed to minimise contamination of the culture, rather than the worker.*

- Safety: it is important to be aware of the potential dangers of infection from cell cultures. Although avian and rodent cells present a reduced risk of disease transmission compared to human cells, all animal cell cultures must be regarded as a potential source of pathogenic microbes, and appropriate sterile technique (Box 34.1) must be used at all times (see also p. 203). Work involving human tissues and cell cultures must be carried out in a biosafety cabinet by trained, experienced personnel – you are unlikely to gain practical experience with such cultures in the early stages of your course. For other cell cultures, you will need to follow the code of safe practice of your department – consult your Departmental Safety Officer if you have any doubts about safe working procedures.

> *SAFETY NOTE When sterilising during tissue culture procedures, (Box 34.1), take care when using 70% v/v alcohol near a naked flame (e.g. Bunsen or spirit lamp) – it is easily ignited and burns with a weakly visible flame.*

- Whether to use a primary culture or a cell line: freshly isolated cells are more likely to reflect the biochemical activities of cells *in vivo*, though they will have a limited lifespan in culture, requiring repeated isolation for longer-term projects. Continuous cell lines are more easily cultured and offer the advantage that their growth requirements in culture may be known in some detail, especially for the more widely used cell lines (e.g. BHK, HeLa).
- The requirements for a solid substratum: some cells must be attached to a solid surface in order to grow. Anchorage dependence is a typical feature of primary cultures and finite cell lines – such cultures show density-dependent growth inhibition once the cells have formed a confluent monolayer on the surface of the substratum. An alternative approach is to grow such cells on a particulate support using 'microcarrier' beads. In contrast, many continuous cell lines can be maintained in suspension culture, as individual cells or aggregates.
- The physicochemical conditions, including pH (typically 7.2–7.5) and buffering capacity, osmolality (usually $300 \pm 20\,\text{mosmol}\,\text{kg}^{-1}$) and temperature (e.g. 35–37 °C for mammalian cells).

> **Example** Dulbecco's modified Eagle's medium contains Ca^{2+}, Fe^{3+}, Mg^{2+}, K^{+}, Na^{+}, Cl^{-}, SO_4^{2-}, PO_4^{2-}, glucose, 20 amino acids, 10 vitamins, inositol and glutathione. Foetal calf serum is usually added at up to 20% v/v.

- The requirements of the culture medium: these will include the provision of inorganic ions (as a balanced salt solution), a carbon/energy source plus other organic nutrients, and in some cases a supplement containing antimicrobial agents to counter the risks of contamination. For example, Dulbecco's modified Eagle's medium is used for many mammalian cell types that grow as adherent monolayers, while suspension cultures of continuous cell lines can be maintained using less stringent media. To support growth, the basal medium is usually supplemented with serum (usually foetal calf serum, at up to 20% v/v), or a chemically defined serum-like supplement containing a mixture of proteins, polypeptides, hormones, lipids and trace components. The *in vitro* level of CO_2 and O_2 must also be considered: many cell cultures are buffered using bicarbonate, and must be maintained in an atmosphere of elevated CO_2, either in a sealed culture vessel or in a CO_2 incubator, to maintain pH balance. In some cases, a pH-sensitive dye (e.g. phenol red, p. 153) may be incorporated into the growth medium, to provide a visual check on pH status during growth – a colour change indicates acidification of the medium and the growth medium should be renewed.

> ***Using antibiotics in cell culture*** *– a typical antimicrobial supplement might include antibacterial agents, e.g. penicillin and streptomycin, an antifungal agent, e.g. griseofulvin, and an antimycoplasmal agent, e.g. gentamicin. Use sparingly to treat contamination and discontinue once the contaminant has been eradicated.*

- The equipment required: this may include a laminar flow hood or biosafety cabinet to reduce the possibility of microbial contamination,

Box 34.1 Sterile technique and its application to animal and plant cell culture

While Chapter 32 gives general advice on the basic principles of sterile technique with microbial cultures, animal and plant cell cultures need additional precautions due to the complexity of some of the procedures and the likelihood of rapid overgrowth of any contaminating microbes. For routine work on the open bench, you should be aware of the following points:

1. **Consider personal clothing and hygiene:** long hair should be tied back, or retained by a net/cap. Avoid pendulous earrings and similar items. Wash your hands at the outset, to remove any loose skin flakes: surgical gloves may be worn, though there is loss of tactile sensitivity and comfort. Hands/gloves should be swabbed with 70% v/v alcohol to further reduce the risk of contamination. If you have a cold/cough, consider wearing a face mask.
2. **Work in a designated quiet area:** there should be no air currents (avoid open windows) and no 'through traffic' or other activity that might give rise to contamination (e.g. microbes should not be cultured in the same location). The term 'quiet area' also highlights that no talking should occur, to reduce the risk of aerosol contamination from the oral microflora.
3. **Organise you work surface:** clear everything from the bench, swab with 70% v/v alcohol, then position all items around a central area, so you can reach every item without reaching across anything. A well-organised workspace reduces the risk of accidental contact between sterile and non-sterile items. Swab bottles and flasks with 70% v/v alcohol before positioning them at the edge of your workspace. Also, swab your work surface between procedures and at the end of the session.
4. **Work close to a Bunsen flame,** positioned centrally within your workspace where convection currents create an upward airflow, reducing the likelihood of particles falling from the air into an open vessel. Always flame the tops of glass bottles (p. 205), but not plastic items, for 2-3 seconds both *before* and *after* opening and closing, rotating the bottles during exposure to the flame. Flame glass pipettes, as described on p. 205.
5. **Tilt flasks and bottles during use:** uncapped culture flasks and media bottles are best kept at a shallow angle, to minimise the risk of airborne contamination.
6. **Work without delay, but do not hurry:** always keep in mind that the air contains contaminant microbes and the longer you leave a vessel open to the air, the more likely it is that it will become contaminated.

When preparing to work in a laminar flow cabinet, the following additional aspects should be considered:

1. **Note whether the airflow is horizontal or vertical:** horizontal flow gives greatest protection to the work area and provides the most stable air flow, while vertical flow cabinets reduce exposure of the operator. A biosafety cabinet should be used for work with potentially hazardous cultures (e.g. human cells/tissues, or with any cells known to be infected with a virus).
2. **Prepare the cabinet:** switch on the cabinet and leave running for at least 10 minutes, then swab the work surface and other interior surfaces with 70% v/v alcohol. Swab the outsides of bottles, flasks, etc. before bringing them into the cabinet. Arrange items around a crescent-shaped central work area.
3. **Carry out your work with due regard for the airflow within the cabinet:** use of a Bunsen burner is often discouraged, as it disrupts the correct airflow, creates heat and can ignite flammable items within the restricted interior of the cabinet. Always try to keep your hands/arms further away from the sterile airstream than the items within the cabinet, e.g. avoid working *behind* an open vessel in a horizontal flow cabinet and *above* an open vessel in a vertical flow cabinet.
4. **Take particular care when using pipettes:** a laminar flow hood will have a restricted interior volume, and it is easy for the novice to touch a pipette tip on the hood, or on another item during use. When removing a pipette from its packaging and connecting to a pipette filler (p. 129) you should point the tip of the pipette away from the user and into the air flow, holding it well above the graduated scale markings to avoid contamination. The pipette and filler are then held horizontally until required: as this can be a tricky procedure for the novice, you should not take a pipette out of its container until you are ready to use it.
5. **Work carefully with open bottles and flasks:** caps can be either placed top-down on the work surface, or held in a crooked finger (Fig. 32.6a). Since the airflow is sterile, there is less need to work swiftly when compared to the open bench, but you should still re-cap all containers as soon as you have finished a particular procedure.

Definitions

Callus – an aggregation of undifferentiated plant cells in culture.

Embryogenic callus – tissue with the capacity to differentiate under defined laboratory conditions, typically in response to plant growth regulators in the medium.

Explant – a fragment of tissue used to initiate a culture (the term is also used in animal culture).

Protoplasts – cells lacking their cell walls.

Sphaeroplasts – cells with attached fragments of their cell walls: osmotically sensitive.

Totipotency – the ability (of any plant cell) to de-differentiate and redifferentiate into any of the cell types fo und in the mature plant.

suitable culture vessels (typically presterilised, disposable polystyrene dishes, bottles and flasks, treated to create a negatively charged, hydrophilic surface), a supply of high purity water (typically distilled, deionised and carbon filtered), a suitable incubator with temperature control of ±0.5 °C or better, often with CO_2 control and mechanical mixing, and an inverted microscope to examine adherent cell monolayers during growth.

The successive stages of isolation of animal components are shown in Fig. 34.1a. Box 34.1 gives advice on key aspects of sterile technique in cell culture; additional guidance on the principal practical procedures used in animal cell culture is given in Box 34.2.

Plant tissue and cell culture systems

Plant tissue cultures can be established by growing explants of sterilised tissue on the surface of an agar-based growth medium to give a callus of undifferentiated cells. Initial sterilisation is usually achieved by incubation for 15–20 min in 10% w/v sodium hypochlorite.

***KEY POINT** For most plants, cell cultures can be established from a broad range of tissue types, reflecting the totipotency of many plant cells.*

Embryogenic callus may be induced to differentiate, forming tissues and organs on a medium containing appropriate plant hormones: in many

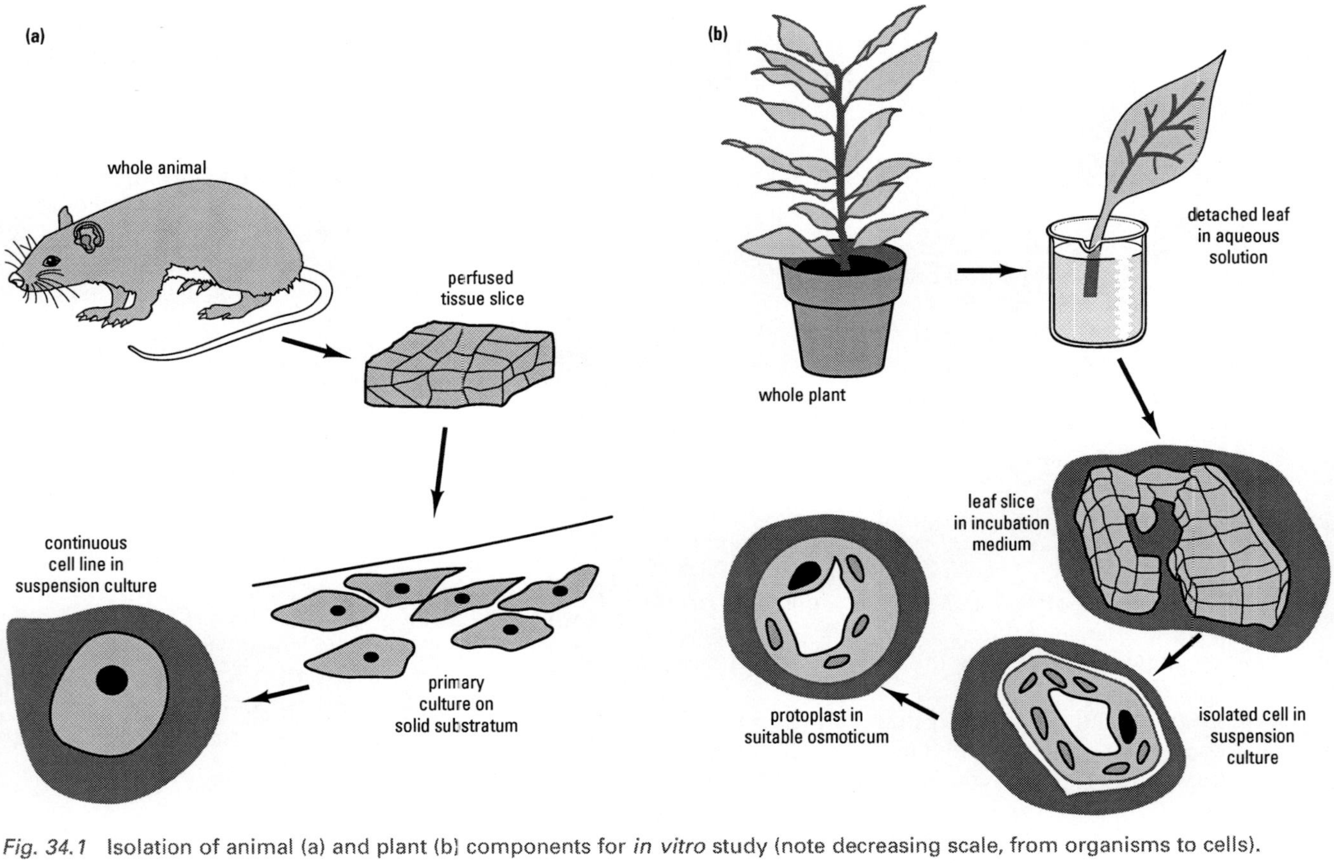

Fig. 34.1 Isolation of animal (a) and plant (b) components for *in vitro* study (note decreasing scale, from organisms to cells).

Box 34.2 Practical procedures in animal cell culture

The following activities are most likely to be encountered during the initial stages of your training:

- **Sterile technique (aseptic technique):** your first exercises are likely to involve transferring sterile liquids between vessels using pipettes, preparing solutions/media and making dilutions.
- **Preparing cells for initial culture:** if you are using a cryopreserved (frozen) vial containing a cell line, it will need to be thawed as rapidly as possible, to minimise the chance of intracellular ice crystal damage – thaw by immersion in water at 37°C (if the cells were initially stored in liquid nitrogen, thaw in a lidded container as there is a risk of explosion). The thawed culture should then be diluted slowly with new growth medium. If you are preparing a primary culture, you will first have to disrupt the tissue (Chapter 36 gives further details of methods available).
- **Checking for contamination:** this can be carried out by visual inspection, looking for: (i) a cloudy medium, sometimes with a film of surface growth; (ii) a change in pH; typically a sharp decrease due to bacterial contamination causes a colour change of the indicator dye (e.g. phenol red turns yellow). Alternatively, examine cultures microscopically, or subculture for specific microbes (Chapter 33).
- **Observing cells:** cells growing as an adherent monolayer can be observed by inverted microscope (x10 or x40 objective), looking for typical cell morphology, or cytopathic effects (e.g. rounding off and detachment, irregular shape, or internal vacuolation and granulation), as well as any evidence of microbial growth. Microscopy will also tell you when such cells are reaching confluent growth (p. 225), requiring subculture.
- **Counting cells:** while rough estimates of the growth stage and cell number may be made from direct microscopic examination of cultures, a haemocytometer is required for a fully quantitative count, as detailed in Box 35.1. Adherent cells would first be detached (e.g. using 0.25% w/v trypsin) and counted in suspension. Other quantitative approaches include electronic counting (p. 235), and flow cytometry where other cellular activities can be assessed simultaneously. Alternatively, cellular constituents e.g. DNA or protein (p. 353) can be assayed.
- **Checking viability:** this is usually carried out by assessing membrane integrity using either dye exclusion (1–2 min in 0.4% w/v trypan blue or 1.0% w/v naphthalene black) or, less frequently, dye uptake (10 min in 0.001% w/v diacetyl fluorescein or 0.5% w/v neutral red), with viability expressed in terms of the percentage of cells unstained or stained, respectively. When combined with haemocytometry (p. 236), the absolute numbers of live and dead cells can be estimated.
- **Feeding an adherent culture:** cells growing as a monolayer may need to have their growth medium replenished before they are ready for subculture, e.g. due to a drop in pH of the medium, or a deterioration in cell morphology. The original medium should be removed, usually by aspiration using a sterile pipette and suction line, and a replacement volume of fresh medium then added (the new medium should be pre-warmed to 37°C).
- **Subculturing cells ('passaging'):** typically this is carried out when adherent cells reach confluence, or when a suspension culture almost reaches the maximum population density in stationary phase (p. 233). Cells growing as a monolayer must first be detached from their substrate, typically using a small amount of 0.25% w/v trypsin for up to 15 min (it is important to dilute the trypsin following detachment, to prevent further digestion). Cells in suspension culture can simply be diluted in an appropriate volume of new medium. For both types of cell culture, the reseeding density is important, since over-diluted cultures will grow poorly. For primary cultures, low dilution factors (often termed 'split ratios') of between 1:2 and 1:8 are commonly used whereas some continuous cell lines can be diluted by up to 1:100, to give a cell density of $10^4 – 10^5$ per ml.
- **Harvesting cells:** gentle centrifugation (at 80–100 *g* for 10 min) is sufficient to pellet most animal cells, leaving the spent medium as a supernatant, which should be either aspirated, or removed using a pipettor (p. 129). Monolayer cultures would be centrifuged following trypsinisation (see above).
- **Freezing cells (cryopreservation):** cells are first suspended at high density (typically around 10^6 per ml) in a suitable freezing solution, usually growth medium or serum supplemented with a cryoprotectant such as 10% v/v DMSO or 20–30% v/v glycerol, dispensed into small vials ('ampoules'/'cryotubes'), and then frozen slowly (ideally, at around 1°C min^{-1}): one approach is to wrap the ampoules in cotton wool inside a small cardboard tube, then place them in an insulated container in a −70°C freezer for ⩾ 6 h, then transfer to a liquid nitrogen storage facility. An alternative approach is to use a proprietary system, e.g. 'Mr Frosty' (Nalgene), to provide a controlled rate of cooling.

cases, these cultures will develop to form plantlets that can be grown on to mature plants, or encapsulated to produce so-called 'artificial seeds'. This approach can be used to study the conditions necessary for differentiation and development, or to propagate rare plants and other valuable stock (e.g. virus-free stock, or genetically altered plants). Callus derived from anthers can be used to provide haploid cell cultures and haploid plants – these are often useful for experimental genetics and breeding purposes.

Using an orbital incubator *– continued rapid agitation (>100 r.p.m.) maintains a homogeneous suspension and encourages gas exchange between the culture medium and the atmosphere.*

Plant cell suspension cultures (Fig. 34.1b) are usually obtained by transferring fragments of actively growing callus to a 'shake' flask containing liquid medium in an orbital incubator: gentle agitation causes fragmentation of the callus tissue, to give a suspension culture that will contain individual plant cells and cell aggregates. This culture can be maintained by repeated subculturing of material from the upper layers of the liquid, encouraging the growth of small aggregates. However, in contrast to animal cell suspension cultures, it is rare for plant suspension cultures to be entirely unicellular, due to the presence of plasmodesmata and plant cell walls. Suspension cultures often require a minimum inoculum size on subculture – an inoculum volume of 10% v/v may be necessary to ensure successful subculture. It may also be helpful to add a small amount of 'conditioned' medium from a previous culture. Commercial-scale suspension cultures are used to produce certain plant pigments and secondary metabolites such as flavourings and high value pharmaceutical compounds.

Example Murashige and Skoog's medium contains a balanced mixture of the principal inorganic ions, plus 7 trace element compounds, 3 vitamins, inositol, glycine and sucrose (at 30 g l^{-1}) as the major carbon source.

The growth media used for callus and suspension cultures are more complex than those required for intact plant organs and tissues, with organic nutrient supplements in addition to a balanced salt solution. Such organic supplements often include a major carbon and energy source, e.g. sucrose, plus various vitamins and growth regulators (typically, at least one auxin and a cytokinin), together with undefined components such as yeast extract and hydrolysed casein in some instances. Otherwise, the techniques are broadly similar to those described for microbial systems (pp. 203–208).

Plant protoplasts

Working with protoplasts *– remember that all solutions must contain a suitable osmoticum, to prevent bursting. Mannitol is widely used at ≈400 mmol l^{-1}.*

Some experimental procedures using plant cells require the enzymatic removal of their cell walls, creating protoplasts that can be manipulated *in vitro*, then maintained under conditions that allow the regeneration of cell walls, with subsequent growth and differentiation to give genetically modified plants. Protoplast isolation often involves pretreatment in a concentrated osmoticum (e.g. sucrose or mannitol, at 300–500 mmol l^{-1}) to plasmolyse the cells, weakening the linkage between cell wall and plasma membrane. Enzymatic treatment is often prolonged, taking several hours in a suitable mixture of enzymes, e.g. cellulase and Macerozyme. The resulting material can be sieved through fine nylon mesh, centrifuged at low speed and then re-suspended, to remove subcellular debris and cell aggregates. The protoplast preparation can be checked for fragments of cell wall using a suitable stain, e.g. 0.1% w/v calcofluor white and a fluorescence microscope.

Assessing cell and protoplast viability *– cytochemical techniques are often used, e.g. exclusion of the mortal stain Trypan blue or Evan's blue. The fluorogenic vital stain fluorescein diacetate is an alternative: it is cleaved by esterases within living cells, liberating fluorescein and giving green–yellow fluorescent cells when viewed by UV fluorescence microscopy.*

The fusion of protoplasts from different plants can be used to produce a somatic hybrid: this process can be used to circumvent inter-species reproductive barriers, creating novel plants. Protoplast fusion can be induced by chemical 'fusogens', e.g. using polyethylene glycol (PEG) at high concentration, or by electrofusion (incubation under low alternating current to encourage aggregation, then brief exposure to a high voltage electrical field – typically $1000\,V\,cm^{-1}$ for 1–2 ms – causing protoplast fusion). Similar techniques can be used with animal cells. The successive stages of isolation of plant components are illustrated in Fig. 34.1b.

Text references

Dodds, J.H. and Robert, L.W. (1995) *Experiments in Plant Tissue Culture*, 3rd edn. Cambridge University Press, Cambridge.

Freshney, R.I. (2005) *Culture of Animal Cells: A Manual of Basic Technique*, 5th edn. Wiley, New York.

Sources for further study

Anon. (2006) *Fundamental Techniques in Cell Culture . . . a Laboratory Handbook*. Sigma-Aldrich/European Collection of Cell Cultures. Available at: http://www.sigmaaldrich.com/Area_of_Interest/Life_Science/Cell_Culture/Key_Resources/Cell_Culture_Manual.html. Last accessed: 01/04/07.

Butler, M. (2003) *Animal Cell Culture and Technology: The Basics,* 2nd edn. Garland Science Press, Oxford.

Doyle, A., Griffiths, J.B. and Newell, D.G. (1998) *Cell and Tissue Culture: Laboratory Procedures*. Wiley, New York.

Evans, D., Coleman, J. and Kearns, A. (2003) *Plant Cell Culture: the Basics*. Garland Science, Oxford.

Hall, R.D. (1999) *Plant Cell Culture Protocols. Methods in Molecular Biology Series*. Humana Press, New Jersey.

Helgason, C.D. (2004) *Basic Cell Culture Protocols,* 3rd edn. Humana Press, New Jersey.

Loyola-Vargas, V.M. and Vazques-Flota, F. (2005) *Plant Cell Culture Protocols,* 2nd edn. Humana Press, New Jersey.

Masters, J. (2000) *Animal Cell Culture: a Practical Approach,* 3rd edn. Oxford University Press, Oxford.

Razdan, M.K. (2003) *Introduction to Plant Tissue Culture*, 2nd edn. Intercept, Andover.

Study exercises

34.1 Calculate the osmolarity of Ringer's solution from its individual constituents. Use the data in Table 34.1 and the information in Chapter 24, and assume that each constituent behaves according to ideal thermodynamic principles. Note the following M_r values: NaCl 58.44; KCl 74.55; $CaCl_2$ 110.99; $NaHCO_3$ 84.01, required to convert the data in Table 34.1 to molar concentrations (see p. 136 if you are unsure about this conversion). Give your final answer in $mosmol\,kg^{-1}$, to one decimal place.

34.2 Find out about the availability of cell cultures using the Web. Research the culture collections of particular countries via their websites. How useful is each site in terms of features such as: ease of use; access to catalogues; information on individual strains, e.g. culture media; details of costs?

34.3 Research the origins of the code names used for individual cell lines. In many instances, the code names given to individual cell lines are derived from their original source. Use the Web to find out the origin of the following descriptors for cell lines:

(a) HeLa; (b) Vero; (c) BHK; (d) CHO; (e) TBY-2.

35 Culture systems and growth measurement

Definitions

Heterotroph – an organism that uses complex organic carbon compounds as a source of carbon and energy.

Photoautotroph – an organism that uses light as a source of energy and CO_2 as a carbon source (photosynthetic metabolism).

Chemoautotroph – an organism that acquires energy from the oxidation of simple inorganic compounds, fixing CO_2 as a source of carbon (chemosynthetic metabolism).

Microbial, animal and plant cell culture methods are based on the same general principles, requiring:

- a pure culture (also known as an axenic culture), perhaps isolated as part of an earlier procedure, or from a culture collection;
- a suitable nutrient medium to provide the necessary components for growth. This medium must be sterilised before use;
- satisfactory growth conditions including temperature, pH, atmospheric requirements, ionic and osmotic conditions;
- sterile technique (p. 203) to maintain the culture in pure form.

Heterotrophic animal cells, fungi and many bacteria require appropriate organic compounds as sources of carbon and energy. Non-exacting bacteria can utilise a wide range of compounds and they are often grown in media containing complex natural substances (including meat extract, yeast extract, soil, blood). Animal cells have more stringent growth requirements (p. 225).

Photoautotrophic bacteria, cyanobacteria and algae are grown in a mineral medium containing inorganic ions including chelated iron, with a light source and CO_2 supply. Plant cells may require additional vitamins and hormones (p. 226). For chemoautotrophic bacteria, the light source is replaced by a suitable inorganic energy source, e.g. H_2S for sulphur-oxidising bacteria, NH_3/NH_4^+ for nitrifying bacteria, etc.

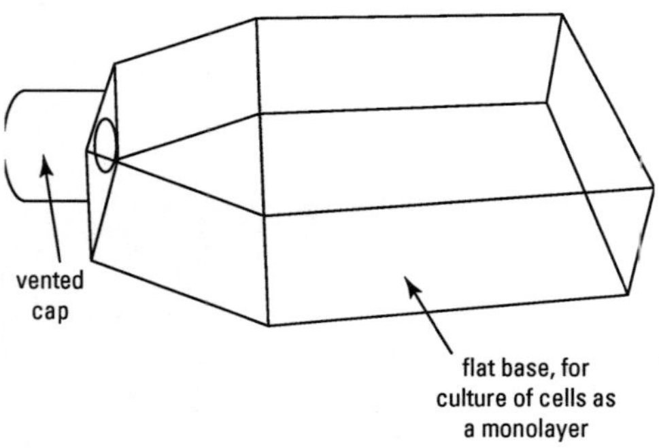

Fig. 35.1 Plastic flask for animal cell culture – this design provides a large surface area for growth of an adherent monolayer of cells.

Growth on solidified media

Many organisms can be cultured on an agar-based medium (p. 204).

***KEY POINT** An important benefit of agar-based culture systems is that an individual cell inoculated onto the surface can develop to form a visible colony: this is the basis of most microbial isolation and purification methods, including the streak dilution, spread plate and pour plate procedures (pp. 206–7).*

Animal cells are often grown as an adherent monolayer on the surface of a plastic or glass culture vessel (Fig. 35.1), rather than on an agar-based medium (p. 206).

Several types of culture vessel are used:

- Petri plates (Petri dishes): usually the presterilised, disposable plastic type, providing a large surface area for growth.
- Glass bottles or test tubes: these provide sufficient depth of agar medium for prolonged growth of bacterial and fungal cultures, avoiding problems of dehydration and salt crystallisation. Inoculate aerobes on the surface and anaerobes by stabbing down the centre, into the base (stab culture).
- Flat-sided bottles: these are used for animal cell culture, to provide an increased surface area for attachment and allow growth of cells as a surface monolayer. Usually plastic and disposable (Fig. 35.1).

__Harvesting bacteria from an agar plate__ – colonies can be harvested using a sterile loop, providing large numbers of cells without the need for centrifugation. The cells are relatively free from components of the growth medium; this is useful if the medium contains substances that interfere with subsequent procedures.

__Subculturing__ – when subculturing microbes from a colony on an agar medium, take your sample from the growing edge, so that viable cells are transferred.

The dynamics of growth are usually studied in liquid culture, apart from certain rapidly growing filamentous fungi, where increases in colony diameter can be measured accurately, e.g. using Vernier callipers.

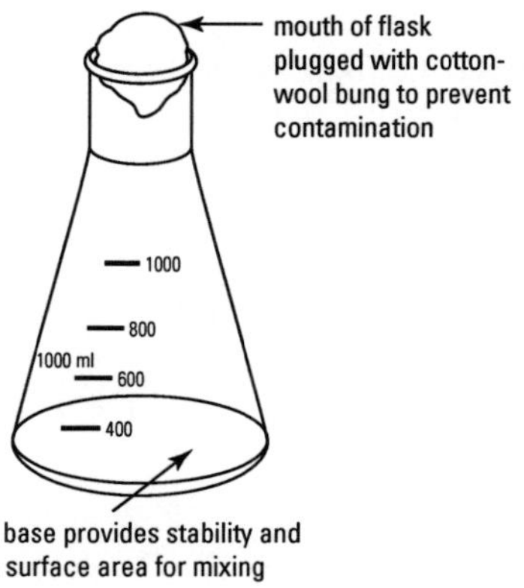

Fig. 35.2 Conical (Erlenmeyer) flask.

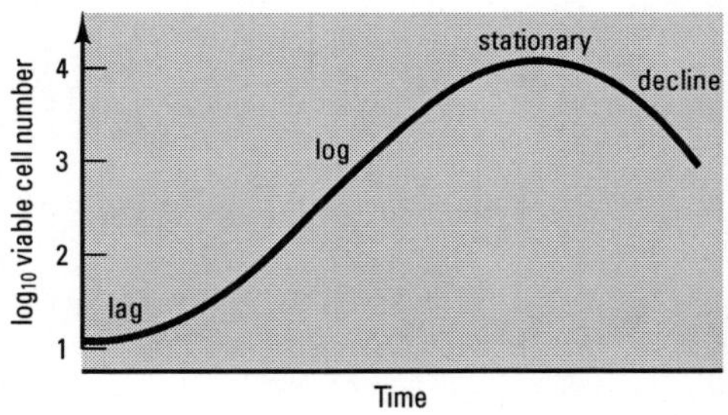

Fig. 35.3 Population growth curve for cells in batch culture (liquid medium).

Example Suppose you counted 2×10^3 cells ($\log_{10} = 3.30$) per unit volume at t_0 and 6.3×10^4 cells ($\log_{10} = 4.80$) after 2 h (t_x).

Substitution into Eqn [35.1] gives $[2.303(4.8 - 3.3)] \div 2 = 1.727\ \text{h}^{-1}$ (or $0.0288\ \text{min}^{-1}$).

Substituting the same values into Eqn [35.2] gives: $[0.301 \times 2] \div [4.8 - 3.3] = 0.40$ h (or 24 min).

Working with logarithms *– note that there is no log value for zero, so you cannot plot zero on a log–linear growth curve or on a death curve.*

Growth in liquid media

Many cells, apart from primary cultures of animal cells, can be grown as a homogeneous unicellular suspension in a suitable liquid medium, where growth is usually considered in terms of cell number (population growth) rather than cell size. Most liquid culture systems need agitation, to ensure adequate mixing and to keep the cells in suspension. An Erlenmeyer flask of 100–2000 ml capacity (Fig. 35.2) can be used to grow a batch culture on an orbital shaker, operating at 20–250 cycles per minute. For aerobic organisms, the surface area of such a culture should be as large as possible: restrict the volume of medium to not more than 20% of the flask volume. Larger cultures may need to be gassed with sterile air and mixed using a magnetic stirrer rather than an orbital shaker. The simplest method of air sterilisation is filtration, using glass wool, non-absorbent cotton wool or a commercial filter unit of appropriate pore size (usually 0.2 µm). Air is introduced via a sparger (a glass tube with many small holes, so that small bubbles are produced) near the bottom of the culture vessel to increase the surface area and enhance gas exchange. More complex systems have baffles and paddles to further improve mixing and gas exchange.

Liquid culture systems may be subdivided under two broad headings:

Batch culture

This is the most common approach for routine liquid culture. Cells are inoculated into a sterile vessel containing a fixed amount of growth medium. Your choice of vessel will depend upon the volume of culture required: larger-scale vessels (e.g. 1 litre and above) are often called 'fermenters' or 'bioreactors', particularly in biotechnology. Growth within the vessel usually follows a predictable S-shaped (sinusoidal) curve when plotted in log–linear format (Fig. 35.3), divided into four components:

1. Lag phase: the initial period when no increase in cell number is seen. The larger the inoculum of active cells the shorter the lag phase will be, provided the cells are transferred from similar growth conditions.
2. Log phase, or exponential phase: where cells are growing at their maximum rate. This may be quantified by the specific growth rate (μ), where:

$$\mu = \frac{2.303\,(\log N_x - \log N_0)}{(t_x - t_0)} \qquad [35.1]$$

where N_0 is the initial number of cells at time t_0 and N_x is the number of cells at time t_x. For times specified in hours, μ is expressed as h^{-1}.

Prokaryotes grow by binary fission while eukaryotes grow by mitotic cell division; in both cases each cell divides to give two identical offspring. Consequently, the doubling time or generation time (g, or T_2) is:

$$g = \frac{0.301\,(t_x - t_0)}{\log N_x - \log N_0} \qquad [35.2]$$

Cells grow at different rates, with doubling times ranging from under 20 min for some bacteria to 24 h or more for animal and plant cells. Exponential phase cells are often used in laboratory experiments, since growth and metabolism are nearly uniform.

3. Stationary phase: growth decreases as nutrients are depleted and waste products accumulate. Any increase in cell number is offset by death. This phase is usually termed the 'plateau' in animal cell culture.
4. Decline phase, or death phase: this is the result of prolonged starvation and toxicity, unless the cells are subcultured. Like growth, death often shows an exponential relationship with time, which can be characterised by a rate (specific death rate), equivalent to that used to express growth or, more often, as the decimal reduction time (d, or T_{90}), the time required to reduce the population by 90%:

$$d = \frac{t_x - t_0}{\log N_x - \log N_0} \qquad [35.3]$$

Some cells undergo rapid autolysis at the end of the stationary period while others show a slower decline.

Example Suppose you counted 5.2×10^5 cells (log10 = 5.716) per unit volume at t_0 and 3.7×10^3 cells (log10 = 3.568) after 60 min (t_x). Substitution into Eqn [35.3] gives 60 ÷ [2.148] = 27.9 min. To the nearest minute, this gives a value for d of 28 min.

Batch culture methods can be used to maintain stocks of particular organisms; cells are subcultured onto fresh medium before they enter the decline phase. However, primary cultures of animal cells have a finite life unless transformed to give a continuous cell line, capable of indefinite growth (p. 225).

Continuous culture

This is a method of maintaining cells in exponential growth for an extended period by continuously adding fresh growth medium to a culture vessel of fixed capacity. The new medium replaces nutrients and displaces some of the culture, diluting the remaining cells and allowing further growth.

After inoculating the vessel, the culture is allowed to grow for a short time as a batch culture, until a suitable population size is reached. Then medium is pumped into the vessel: the system is usually set up so that any increase in cell number due to growth will be offset by an equivalent loss due to dilution, i.e. the cell number within the vessel is maintained at a steady state. The cells will be growing at a particular rate (μ), counterbalanced by dilution at an equivalent rate (D):

$$D = \frac{\text{flow rate}}{\text{vessel volume}} \qquad [35.4]$$

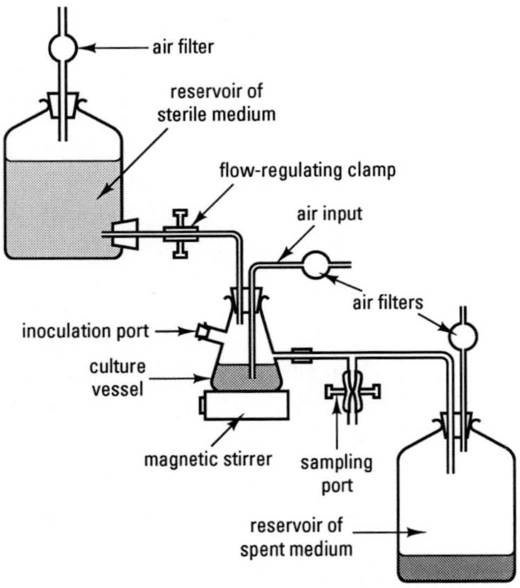

Fig. 35.4 Example of a two-dimensional lab equipment diagram of components of a chemostat.

where D is expressed per unit time (e.g. h^{-1}). In a chemostat, the growth rate is limited by the availability of some nutrient in the inflowing medium, usually either carbon or nitrogen (see Fig. 35.4). In a turbidostat, the input of medium is controlled by the turbidity of the culture, measured using a photocell. A turbidostat is more complex than a chemostat, with additional equipment and controls.

To determine the specific growth rate (μ) of a continuous culture:

1. Measure the flow of medium through the vessel over a known time interval (e.g. connect a sterile measuring cylinder or similar volumetric device to the outlet), to calculate the flow rate.
2. Divide the flow rate by the vessel volume (Eqn [35.4]) to give the dilution rate (D).
3. This equals the specific growth rate, since $D = \mu$ at steady state.

Example Suppose a continuous culture system of 2000 ml volume had a flow of 600 ml over a period of 40 min (flow rate 600 ÷ 40 = 15 ml min^{-1}. Substitution into Eqn [35.4] gives a dilution rate D of 15 ÷ 2000 = 0.00075 min^{-1} or 0.00075 × 60 = 0.45 h^{-1}.

Example Suppose you wanted to convert a doubling time of 20 min to a specific growth rate. Rearrangement of Eqn [35.5] gives $\mu = 0.693 \div 20 = 0.03465$ min^{-1} ($=2.08 h^{-1}$). For the example given on page 234, with a growth rate of $0.45 h^{-1}$, substitution into Eqn [35.5] gives $0.693/0.45 = 1.54$ h, (approx. 92 min).

4. If you want to know the doubling time (g), calculate using the relationship:

$$g = \frac{0.693}{\mu} \qquad [35.5]$$

(Note that Eqn [35.5] also applies to exponential phase cells in batch culture and is useful for interconverting g and μ.)

Continuous culture systems are more complex to set up than batch cultures. They are prone to contamination, having additional vessels for fresh medium and waste culture: strict aseptic technique is necessary when the medium reservoir is replaced, and during sampling and harvesting. However, they offer several advantages over batch cultures, including the following:

- The physiological state of the cells is more clearly defined, since actively growing cells at the same stage of growth are provided over an extended time period. This is useful for biochemical and physiological studies.
- Monitoring and control can be automated and computerised.
- Modelling can be carried out for biotechnology/fermentation technology.

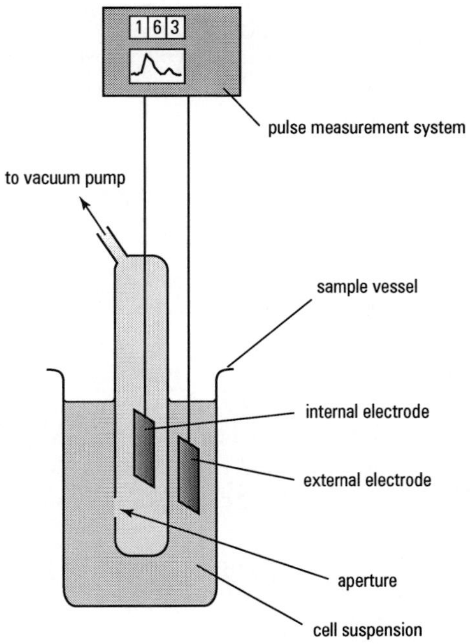

Fig. 35.5 Components of an electronic particle counter. During operation, the cell suspension is drawn through the aperture by the vacuum, creating a 'pulse' of resistance between the two electrodes as each cell passes through the aperture.

Measuring growth in cell cultures

The most widely used methods of measuring growth are based on cell number:

Direct microscopic counts

One of the simplest methods is to count the cells in a known volume of medium using a microscope and a counting chamber or haemocytometer (Box 35.1). While this gives a rapid assessment of the total cell number, it does not discriminate between living and dead cells. It is also time-consuming as a large number of cells must be counted for accurate measurement. It may be difficult to distinguish individual cells, e.g. for cells growing as clumps.

Electronic particle counters

These instruments can be used to give a direct (total) count of a suspension of microbial cells. The Coulter counter detects particles due to change in electrical resistance when they pass through a small aperture in a glass tube (Fig. 35.5). It gives a rapid count based on a larger number of cells than direct microscopy. It is well-suited for repeat measurements or large sample numbers and can be linked to a microcomputer for data processing. If correctly calibrated, the counter can also measure cell sizes. A major limitation of electronic counters is the lack of discrimination between living cells, dead cells, cell clumps and other particles (e.g. dust). In addition, the instrument must be set up and calibrated by trained personnel. Flow cytometry is a more specialised alternative, since particles can be sorted as well as counted.

Alternative approaches to measuring growth *– these include biomass, dry weight, turbidity, absorbance or any major cellular component, e.g. protein, nucleic acid, ATP, etc.*

Culture-based counting methods

A variety of culture-based techniques can be used to determine the number of microbes in a sample. A major assumption of such methods is

Box 35.1 How to use a counting chamber or haemocytometer

A counting chamber is a specially designed slide containing a chamber of known depth with a grid etched onto its lower surface. When a flat coverslip is placed over the chamber, the depth is uniform. Use as follows:

1. **Place the special coverslip over the chamber.** Press the edges firmly, to ensure that the coverslip makes contact with the surface of the slide, but take care that you do not break the slide or coverslip by using too much force. When correctly positioned, you should be able to see interference rings (Newton's rings) at the edge of the coverslip.
2. **Add a small amount of your cell suspension to fill the central space above the grid.** Place on the microscope stage and allow the cells to settle (2–3 min).
3. **Examine the grid microscopically,** using the ×10 objective lens first, since the counting chamber is far thicker than a standard microscope slide. Then switch to the ×40 objective: take care not to scratch the surface of the objective lens, as the special coverslip is thicker than a normal coverslip. For a dense culture, the small squares are used, while the larger squares are used for dilute suspensions. You may need to dilute your suspension if it contains more than 30 cells per small square.
4. **Count the number of cells in several squares:** at least 600 cells should be counted for accurate measurements. Include those cells that cross the upper and left-hand boundaries, but not those that cross the lower and right-hand rulings. A hand tally may be used to aid counting. Motile cells must be immobilised prior to counting (e.g. by killing with a suitable biocide).
5. **Divide the total number of cells (C) by the number of squares counted (S),** to give the mean cell count per square.
6. **Determine the volume (in ml) of liquid corresponding to a single square (V),** e.g. a Petroff–Hausser chamber has small squares of linear dimension 0.2 mm, giving an area of 0.04 mm^2; since the depth of the chamber is 0.02 mm, the volume is $0.04 \times 0.02 = 0.0008$ mm^3; as there are 1000 mm^3 in 1 ml, the volume of a small square is 8×10^{-7} ml; similarly, the volume of a large square (equal to 25 small squares) is 2×10^{-5} ml. Note that other types of counting chamber will have different volumes: check the manufacturer's instructions. For example, the improved Neubauer chamber (Fig. 35.6) has small squares of volume 0.000 25 mm^3 = 2.5×10^{-7} ml.

Fig. 35.6 Haemocytometer grid (Improved Neubauer rulings) viewed microscopically. The large square (delimited by triple etched lines) has a volume of 1/250 mm^3 (0.004 mm^3 = 4 µl) while each small square (16 contained within the large square) has a volume of 1/4000 mm^3 (0.00025 mm^3 = 0.25 µl). Note that the boundary line for squares delimited by triple-etched lines is the *middle* line, so this line must be used when counting (count cells straddling the top and left-hand gridlines and ignore those straddling the bottom and right-hand gridlines).

7. **Calculate the cell number per ml by dividing the mean cell count per square by the volume of a single square (in ml).**
8. **Remember to take account of any dilution of your original suspension** in your final calculation by multiplying by the reciprocal of the dilution (M), e.g. if you counted a 1 in 20 dilution of your sample, multiply by 20, or if you diluted to 10^{-5}, multiply by 10^5.

The complete equation for calculating the total microscopic count is:

$$\text{Total cell count (per ml)} = (C \div S \div V) \times M \quad [35.6]$$

e.g. if the mean cell count for a 100-fold dilution of a cell suspension, counted using a Petroff–Hausser chamber, was 12.4 cells in 10 small squares, the total count would be

$$(12.4 \div 10 \div 8 \times 10^{-7}) \times 10^2 = 1.55 \times 10^8 \text{ ml}^{-1}$$

A simpler, less accurate approach is to use a known volume of sample under a coverslip of known area on a standard glass slide, counting the number of cells per field of view using a calibrated microscope of known field diameter, then multiplying up to give the cell number per ml.

Box 35.2 How to make a plate count of bacteria using an agar-based medium

1. **Prepare serial decimal dilutions of the sample in a sterile diluent (pp. 137–8).** The most widely used diluents are 0.1% w/v peptone water or 0.9% w/v NaCl, buffered at pH 7.3. Take care that you mix each dilution before making the next one. For soil, food or other solid samples, make the initial decimal dilution by taking 1 g of sample and making this up to 10 ml using a suitable diluent. Gentle shaking or homogenisation may be required for organisms growing in clumps. The number of decimal dilutions required for a particular sample will be governed by your expected count: dilute until the expected number of viable cells is around 100–1000 ml^{-1}.
2. **Transfer an appropriate volume (e.g. 0.05–0.5 ml) of the lowest dilution to an agar plate** using either the spread plate method or the pour plate procedure (p. 207). At least two, and preferably more, replicate plates should be prepared for each sample. You may also wish to prepare plates for more than one dilution, if you are unsure of the expected number of viable cells.
3. **Incubate under suitable conditions for 18–72 h, then count the number of colonies on each replicate plate at the most appropriate dilution.** The most accurate results will be obtained for plates containing 30–300 colonies. Mark the base of the plate with a spirit-based pen each time you count a colony. Determine the mean colony count per plate at this dilution (*C*).
4. **Calculate the colony count per ml of that particular dilution** by dividing by the volume (in ml) of liquid transferred to each plate (*V*).
5. **Now calculate the count per ml of the original sample** by multiplying by the reciprocal of the dilution: this is the multiplication factor (*M*); e.g. for a dilution of 10^{-3}, the multiplication factor would be 10^3. For soil, food or other solid samples, the count should be expressed per g of sample.

The complete equation for calculating the viable count is:

$$\text{Count per ml (or per g)} = (C \div V) \times M \qquad [35.7]$$

e.g. for a sample with a mean colony count of 5.5 colonies per plate for a volume of 0.05 ml at a dilution of 10^{-7}, the count would be:

$$(5.5 \div 0.05) \times 10^7 = 1.1 \times 10^9 \text{ CFU ml}^{-1}$$

The count should be reported as colony-forming units (CFU) per ml, rather than as cells per ml, since a colony may be the product of more than one cell, particularly in filamentous microbes or in organisms with a tendency to aggregate. You should also be aware of the problems associated with counts of zero – these are best recorded as '<1', and you should then apply the appropriate correction factors for dilution and volume to obtain the detection limit. For example, a zero count (<1) of 100 μl of a five-fold dilution gives a detection limit of $(<1 \div 0.1) \times 5 = <50$ CFU ml^{-1}.

that, under suitable conditions, an individual viable microbial cell will be able to multiply and grow to give a visible change in the growth medium, i.e. a colony on an agar-based medium, or turbidity ('cloudiness') in a liquid medium. You are most likely to gain practical experience using bacterial cultures, counted by one or more of the following methods:

Definition

CFU – colony-forming unit: a cell or group of cells giving rise to a single colony on a solidified medium.

- Spread or pour plate methods ('plate counts', p. 207). The most widespread approach is to transfer a suitable amount of the sample to an agar medium, incubate under appropriate conditions and then count the resulting colonies (Box 35.2).
- Membrane filtration. For bacterial samples where the expected cell number is lower than 10 CFU ml^{-1}, pass the sample through a sterile filter (pore size 0.2 or 0.45 μm). The filter is then incubated on a suitable medium until colonies are produced, giving a count by dividing the mean colony count per filter by the volume of sample filtered.
- Multiple tube count, or most probable number (MPN). A bacteriological technique where the sample is diluted and known volumes are transferred to several tubes of liquid medium (typically,

Alternative approaches in plate counting *– when large numbers of samples have to be counted, a single agar plate can be divided into segments and a single droplet of each dilution placed into the appropriate segment ('Miles and Misra' droplet counting).*

Counting injured or stressed microbes *– a resuscitation stage may be required, to allow cells to grow under selective conditions, p. 212.*

five tubes at three volumes), chosen so that there is a low probability of the smallest volumes containing a viable cell. After incubation, the number of tubes showing growth (turbidity) is compared to tabulated values to give the most probable number (MPN per ml).

The principal advantage of culture-based counting procedures is that dead cells will not be counted. However, for such techniques the incubation conditions and media used may not allow growth of all cells, underestimating the true viable count. Further problems are caused by cell clumping and dilution errors. In addition, such methods require sterile apparatus and media and the incubation period is lengthy before results are obtained. An alternative approach is to use direct microscopy, combined with 'vital' or 'mortal' staining. For example, the direct epifluorescence technique (DEFT) uses acridine orange and UV epifluorescence microscopy to separate living and dead bacteria, while neutral red is a vital stain used for plant cells. Chapter 27 gives examples of vital/mortal stains for other cell types. A further approach is to use DNA-based methods, such as the polymerase chain reaction.

Bioassays and their applications

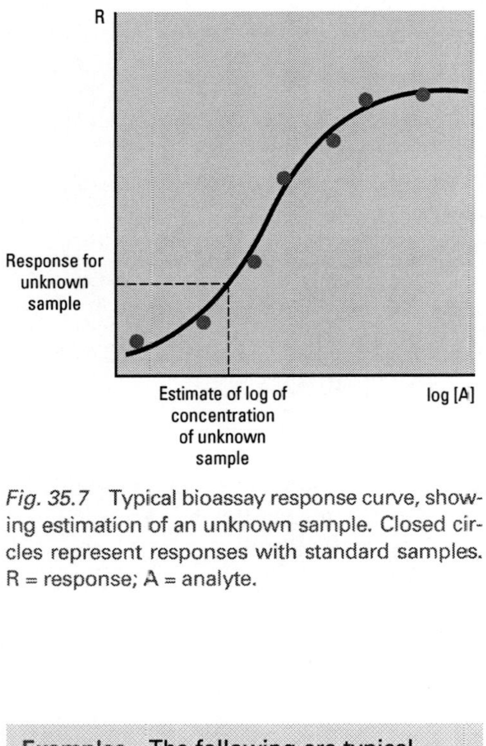

Fig. 35.7 Typical bioassay response curve, showing estimation of an unknown sample. Closed circles represent responses with standard samples. R = response; A = analyte.

A bioassay is a method of quantifying a chemical substance (analyte) by measuring its effect on a biological system under controlled conditions. The hypothetical underlying phenomena are summarised by the relationship:

$$A + Rec \rightleftharpoons ARec \rightarrow R \qquad [35.8]$$

where A is the analyte, Rec the receptor, ARec the analyte–receptor complex and R the response. This relation is analogous to the formation of product from an enzyme–substrate complex and, using similar mathematical arguments to those of enzyme kinetics, it can be shown that the expected relationship between [A] and rate of response is hyperbolic (sigmoidal in a log–linear plot like Fig. 35.7). This pattern of response is usually observed in practice if a wide enough range of [A] is tested.

To carry out the assay, the response elicited by the unknown sample is compared to the response obtained for differing concentrations of the substance, as shown in Fig. 35.7. When fitting a curve to standard points and estimating unknowns, the available methods, in order of increasing accuracy, are:

1. fitting by eye;
2. using linear regression on a restricted 'quasi-linear' portion of the assay curve;
3. linearisation followed by regression (e.g. by probit transformation);
4. non-linear regression (e.g. to the Morgan–Mercer–Flodin equation).

Examples The following are typical bioassays:

- measuring the amount of a plant growth substance in a plant extract using the coleoptile straight growth assay;
- measuring the effect of a drug on the dilation of a person's pupils;
- measuring the mutagenic properties of a chemical compound in the Ames test (see Box 35.3).

In general, bioassay techniques have more potential faults than physicochemical assay techniques. These may include the following:

- A greater level of variability: error in the estimate of the unknown compound will result because no two organisms will respond in exactly the same way. Assay curves vary through time, and because they are non-linear, a full standard curve is required each time the assay is carried out.

- Lack of chemical information: bioassays provide information about *biological* activity; they say little about the chemical structure of an unknown compound. The presence of a specific compound may need to be confirmed by a physicochemical method (e.g. mass spectrometry).
- Possibility of interference: while many bioassays are very specific, it is possible that different chemicals in the extract may influence the results.

Box 35.3 Mutagenicity testing using the Ames test – an example of a widely used bioassay

Chemical carcinogens can be identified by the formation of tumours in laboratory animals exposed to the compound under controlled conditions. However, such animal bioassays are time-consuming and expensive. Dr B.N. Ames and co-workers have shown that most carcinogens are also mutagens, i.e. they will induce mutational changes in DNA. The Ames test makes use of this correlation to provide a simple, rapid and inexpensive bioassay for the initial screening of potential carcinogens. The test makes use of particular strains of *Salmonella typhimurium* with the following characteristics:

- histidine auxotrophy – the tester strains are unable to grow on a minimal medium without added histidine: this characteristic is the result of specific mutational changes to the DNA of these strains, including base substitutions and frame shifts;
- increased cell envelope permeability, to permit access of the test compound to the cell interior;
- defects in excision repair systems and enhanced error-prone repair systems, to reduce the likelihood of DNA repair after treatment with a potential mutagen.

When grown in the presence of a chemical mutagen, the bacteria may revert to prototrophy as a result of back mutations that restore the wild-type phenotype: such revertants grow independently of external histidine and are able to form colonies on minimal medium, unlike the original test strains. The extent of reversion can be used to assess the mutagenic potential of a particular chemical compound. Since many chemicals must be activated *in vivo*, the test incorporates a rat liver homogenate (so-called 'S-9 activator', containing microsomal enzymes) to simulate the metabolic events within the liver. The tester strains are mixed with S-9 activator and a small amount of molten soft agar, then poured as a thin agar overlay (top agar) on a minimal medium plate. The top agar layer contains a very small amount of histidine, to allow the tester strains to divide a few times and express any mutational changes (i.e. prototrophy).

The test can be performed in one of two ways:

1. **Spot test:** a concentrated drop of the test compound is placed at the centre of the plate, either directly on the agar surface, or on a small filter paper disc. The test compound will diffuse into the agar and revertants appear as a ring of colonies around the site of inoculation, as shown in Fig. 35.8. The distance of the ring of colonies from this site provides a measure of the toxicity of the compound, while the number of colonies within the ring gives an indication of the relative mutagenicity of the test substance. The spot test is often carried out in a simplified form, without added S-9 activator, as a rapid preliminary test prior to quantitative analysis.

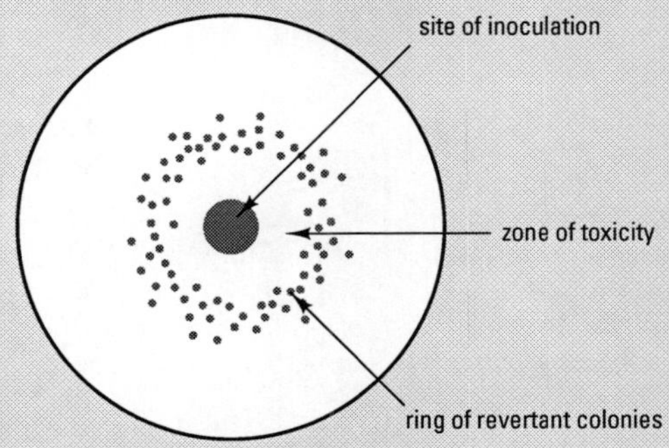

Fig. 35.8 Typical outcome of spot test (Ames test).

2. **Agar incorporation test:** known amounts of the test compound are mixed separately with molten top agar and the other constituents and poured onto separate minimal agar plates. After incubation for 48 h, revertants will appear as evenly dispersed colonies throughout the agar overlay. The number of colonies reflects the relative mutagenicity of the test compounds, with a direct relationship between colony count and the amount of mutagen. Agar incorporation tests can be used to generate dose–response curves similar to that shown in Fig. 35.7. The simplicity, sensitivity and reproducibility of the Ames test has resulted in its widespread use for screening potential carcinogens in many countries, though it is not an infallible test for carcinogenicity.

SAFETY NOTE *Correct handling procedures must be followed at all times, as the test substances may be carcinogenic – testers should wear gloves and avoid skin contact or ingestion.*

Despite these problems, bioassays are still much used. They are 'low-tech' and generally cheap to set up. They often allow detection at very low concentrations. Bioassays also provide the means to assess the biological activity of chemicals and to study changes in sensitivity to a chemical, which physicochemical techniques cannot do. Changes in sensitivity may be evident in the shape of the dose–response curve and its position on the concentration axis.

Bioassays can involve responses of whole organisms or parts of organisms. 'Isolated' responding systems (e.g. excised tissues or cells, Chapter 34) decrease the possibility of interference from other parts of the organism. Disadvantages include disruption of nutrition and wound damage during excision. Isolation can continue down to the molecular level, as in immunoassays (Chapter 38).

Bioassays are the basis for characterising the efficacy of drugs and the toxicity of chemicals. Here, response is often treated as a quantal (all-or-nothing) event. The E_{50} is defined as that concentration of a compound causing 50% of the organisms to respond. Where death is the observed response, the LD_{50} describes the concentration of a chemical that would cause 50% of the test organisms to die within a specified period under a specified set of conditions. Box 35.3 presents details of the Ames test, a widely used bioassay used to assess the mutagenicity of chemicals.

Legal use of bioassays *– in the UK, where bioassays involving 'higher' (vertebrate) animals are controlled by the* Animals Scientific Procedures Act (1986)*, they can only be carried out under the direct supervision of a scientist licensed by the Home Office.*

Considerations when setting up a bioassay

- The response should be easily measured and as metabolically 'close' to the initial binding event as possible.
- The experimental conditions should mimic the *in vivo* environment.
- The standards should be chemically identical to the compound being measured and spread over the expected concentration range being tested.
- The samples should be purified if interfering compounds are present and diluted so the response will be on the 'linear' portion of the assay curve.

To check for interference, the bioassay may be standardised against another method (preferably physicochemical). Related compounds known to be present in the analyte solution should be shown to have minimal activity in the bioassay. If an interfering compound is present, this may show up if a known amount of standard is added to sample vials – the result will not be the sum of independently determined results for the standard and sample.

Sources for further study

Ball, A.S. (1997) *Bacterial Cell Culture: Essential Data.* Wiley, New York.

Cann, A.J. (1999) *Virus Culture: a Practical Approach.* Oxford University Press, Oxford.

Cartledge, T.G. (ed.) (1992) *In Vitro Cultivation of Micro-organisms.* Butterworth-Heinemann, Oxford.
[Includes in-depth coverage of the mathematical principles underlying growth in batch and continuous culture]

Hewitt, W. and Vincent, S. (1989) *Theory and Application of Microbiological Assay.* Academic Press, New York.

Jennings, D.H. and Isaac, S. (1995) *Microbial Culture.* Bios, Oxford.

Rhodes, P.M. and Stanbury, P.F. (1997) *Applied Microbial Physiology: A Practical Approach.* Oxford University Press, Oxford.
[Provides details of growth requirements and culture methods applicable to a range of different microbes]

Salanki, J., Jeffrey, D. and Hughes, G.M. (1994) *Biological Monitoring of the Environment: A Manual of Methods.* CAB, Wallingford.

Study exercises

35.1 Calculate the specific growth rate and doubling time of cells in culture. What are the specific growth rates and doubling times of the following? (Give all answers to three significant figures.)

(a) A broth culture of the yeast *Saccharomyces cerevisiae* growing in exponential phase and containing 5.2×10^4 cells at 10 am and 3.4×10^6 cells at 7 pm.

(b) A log phase culture of *E. coli* containing 3.0×10^4 CFU ml^{-1} at the start of the experiment and 6.7×10^7 CFU ml^{-1} at the end of the experiment, 200 minutes later.

(c) An actively growing culture of *Bacillus subtilis* containing 32 bacteria in 25 squares of a haemocytometer chamber at 13.00 hours and 250 bacteria in 25 squares of the same haemocytometer chamber at 15.30 hours.

35.2 Practise the calculations involved in using a haemocytometer to make a direct microscopic count. Express your answers to three significant figures in all cases.

(a) The mean cell count of a yeast suspension per small square of an improved Neubauer counting chamber was 6.42. If the volume of a small square is 2.5×10^{-7} ml, what is the cell count per ml?

(b) The following counts were obtained for bacteria in 20 individual small squares of a Petroff–Hauser counting chamber: 26, 36, 42, 35, 27, 16, 29, 50, 24, 43, 41, 35, 18, 36, 33, 47, 25, 46, 32, 57. If the volume of each small square is 8×10^{-7} ml, what is the cell count per ml?

(c) A 10^{-2} dilution of a dense suspension of *E. coli* was examined microscopically using an improved Neubauer counting chamber, giving a total of 78 cells in a total of 25 small squares. If the volume of each small square is 2.5×10^{-7} ml what is the cell count per ml of the original (undiluted) suspension?

(d) A dilute suspension of yeast cells was concentrated tenfold by centrifuging 10 ml and resuspending the pellet in 1 ml. The mean cell count of the concentrated yeast suspension per large square of a Petroff–Hauser counting chamber was 6.8. If the volume of the large square is 2×10^{-5} ml, what is the cell count per ml of the original suspension?

35.3 Practise the calculations involved in making a plate count. Express your answers to three significant figures.

(a) The mean spread plate count for 100 µl of a 10^{-5} dilution of a culture of *E. coli* was 54.4 CFU. What is the mean plate count per ml of the original suspension?

(b) Three replicate plates of nutrient agar were each spread with 200 µl of a 10^{-3} dilution of a bacterial suspension, giving colony counts of 34, 40 and 37 after 24 h incubation. What is the mean plate count per ml of the original suspension?

(c) A 20-fold dilution of a yeast suspension was used to prepare four replicate pour plates, each containing 500 µl of this dilution and giving counts of 211, 186, 194 and 202 after incubation. What is the mean plate count per ml of the original suspension?

(d) A sample of 50 g of raw seafood was homogenised and diluted to 2% w/v in sterile saline solution. Three replicate pour plates were prepared using 1 ml of the diluted sample, giving counts for *E. coli* of 35, 41 and 32. What is the average count of *E. coli* per 100 g of raw seafood?

(e) Duplicate samples of 250 ml of river water were filtered through separate sterile membranes of pore size 0.2 µm. The membranes were then transferred to the surface of an agar-based medium and incubated for 48 h at 37 °C, giving colony counts of 45 and 32 respectively. What is the mean count per 100 ml of river water?

(f) A sample of bottled drinking water was processed by serial decimal dilution and 500 µl samples of each dilution were pour plated and incubated at 22 °C for 72 h. The colony counts obtained for the three replicate plates of the 10^{-1} dilution were 28, 32 and 39. Does this water meet the EU regulations for mineral water, which specify a maximum average plate count at 22 °C of 100 CFU ml^{-1}?

35.4 Use bioassay data to estimate the amount of analyte in unknown samples. Bioassay data obtained by exposing water fleas (*Daphnia* sp.) to a purified algal toxin are shown in the table on the next page.

(a) Graph the data and use the graph to estimate the toxin contents of two samples obtained from two lochs which gave (i) a heart rate of 130 beats per minute and (ii) a heart rate of 25 beats per minute.

(b) Which one of the two estimates would be the least reliable and why?

(c) How could you improve the reliability of this estimate, if you had more of the sample?

(continued)

Study exercises (continued)

(d) What unmeasured factors might limit the usefulness of this assay?

Responses of *Daphnia* to toxin

Toxin concentration ($mmol\,l^{-1}$)	Mean heart rate of *Daphnia* sp. (beats per minute)
1.00×10^{-3}	20
3.16×10^{-4}	23
1.00×10^{-4}	26
3.16×10^{-5}	50
1.00×10^{-5}	180
3.16×10^{-6}	192
1.00×10^{-6}	195

35.5 **Use agar diffusion bioassay data to quantify a growth inhibitor in a sample.** Agar diffusion bioassay is a simple means of estimating the amount of an inhibitory substance in a test sample, by adding the substance, in aqueous solution, to wells cut in a Petri plate of agar medium seeded with a microbe sensitive to the substance and then incubating the plate at an appropriate growth temperature. Diffusion of the inhibitory substance gives a zone of growth inhibition proportional to $\log_{10}$ of the concentration of substance. In the following example and table (below), the amount of the antibiotic nisin in a test sample can be determined by comparing the degree of growth inhibition of *Bacillus subtilis* for the test solution and a series of standards containing known concentrations of nisin. Prepare a calibration curve and determine the nisin concentration of solution 5 (unknown). Express your answer in $\mu g\,ml^{-1}$, to one decimal place.

Growth inhibition of *B. subtilis* in solutions containing different amounts of nisin

Solution	Nisin content ($\mu g\,ml^{-1}$)	Zone of growth inhibition (mm)
1	40	26.5
2	20	23
3	10	19.5
4	5	17
5	unknown (test solution)	22.5

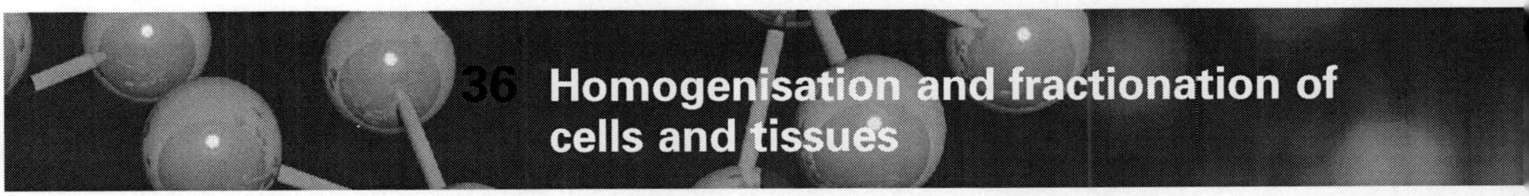

36 Homogenisation and fractionation of cells and tissues

Definitions

Disruption – a process involving structural damage to cells and tissues, to an extent where the biomolecule or complex of interest is released.

Homogenisation – a process where cells and tissues are broken into fragments small enough to create a uniform, stable emulsion (a homogenate).

Homogeniser – a general term for any equipment used to disrupt or homogenise cells and tissues.

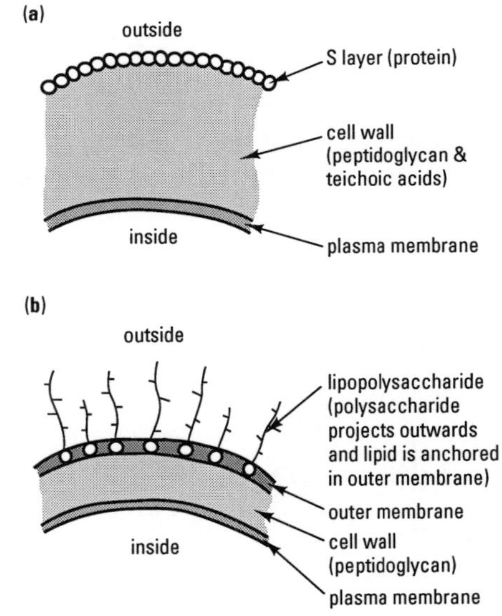

Fig. 36.1 Cell envelopes of (a) Gram-positive and (b) Gram-negative bacteria.

Most biological molecules and subcellular complexes must be isolated from their source material in order to be studied in detail. Unless the biomolecule is already a component of an aqueous medium (e.g. plasma, tissue exudate), the first step will be to disrupt the structure of the appropriate cells or tissues. Following disruption and/or homogenisation, the *in vitro* environment of the homogenate (disrupted tissue) will be very different from that of the intact cell, and it is important that the integrity of the biomolecule or subcellular complex is preserved as far as possible during the isolation procedure.

***KEY POINT** Disruption may be achieved by chemical, physical or mechanical procedures – the rigour of the technique(s) required will depend on the intracellular location of the molecule and the nature of the source material.*

Types of cell and their susceptibility to disruption

Cell walls are a major obstacle to disruption in many organisms and the technique used for a given application must take into account this aspect. With animal cells, there is no cell wall and the plasma membrane and cytoskeleton are relatively weak: unicellular suspensions of animal cells (e.g. blood cells) can be disrupted by gentle techniques. Animal cells within tissues are more difficult to disrupt, due to the presence of connective tissue. Muscle requires vigorous techniques, as a result of additional contractile proteins. For plant cells, the plasma membrane is surrounded by a cellulose cell wall, sometimes with additional components, e.g. lignin or waxes. As a result, large shear forces are often required to disrupt plant cells.

Most Gram-positive and Gram-negative bacteria are surrounded by a rigid, protective cell wall. Peptidoglycan is a major structural component and this may be degraded by the enzyme lysozyme, particularly in certain Gram-positive organisms. In contrast, the outer membrane of Gram-negative bacteria (Fig. 36.1b) protects against lysozyme: however, prior treatment of Gram-negative bacteria with EDTA destabilises the outer membrane by removing Ca^{2+}, making the cell sensitive to lysozyme. In an osmotically buffered solution, the cells will be converted to sphaeroplasts (p. 230), since components the Gram-negative envelope will remain after lysozyme treatment. Note also that some Gram-positive bacteria may be insensitive to lysozyme, due either to a modified peptidoglycan structure or to additional protective outer layers (proteinaceous S-layers: Fig. 36.1a).

The cell walls of filamentous fungi and yeasts are very robust, containing up to 90% polysaccharide (e.g. chitin, mannan), with embedded protein microfibrils – they are often difficult to disrupt, requiring enzymic digestion and/or mechanical homogenisation. Hopkins (1991) gives details of the various systems available for homogenisation and disruption of biological material.

Homogenising media

The solution used for homogenisation serves several purposes, since it acts as a solvent or suspension medium for the released components; it serves as a cooling medium (since many biological macromolecules are denatured by heat); and it contains various reagents that may help to preserve the biological integrity of components. A typical medium will contain:

- Buffer. This replaces the intracellular buffer systems and is needed to prevent pH changes that might denature proteins. TRIS and phosphate buffers are often used for 'physiological' pH values (7.0–8.0); alternatives include various zwitterionic buffers (p. 156).
- Inorganic salts. The intracellular ionic strength is quite high, so KCl and NaCl are often included to maintain the ionic strength of the homogenate. However, the total concentration of inorganic salts should be kept below 100 mmol l^{-1}, to avoid thickening of the homogenate due to solubilisation of structural proteins.
- EDTA. This chelates divalent cations and removes metal ions (e.g. Cu^{2+}, Pb^{2+}, Hg^{2+}) that inactivate proteins by binding to thiol groups. In addition, it removes Ca^{2+} which could activate certain proteases, nucleases and lipases in the homogenate.
- Sucrose. This can be used to prevent osmotic lysis of organelles (e.g. mitochondria, lysosomes), and stabilises proteins from hydrophobic intracellular environments by reducing the polarity of the aqueous medium.
- Mg^{2+}. This helps to preserve the integrity of membrane systems by counteracting the fixed negative charges of membrane phospholipids.
- Protease inhibitors (e.g. phenylmethanesulphonylfluoride, PMSF; L-*trans*-epoxysuccinylleucylamido-(4-guanodino)-butane, E-64; leupeptin). These protect solubilised proteins from digestion by intracellular proteases, mainly released from lysosomes on disruption of the cell. Lysosomal proteases have acid pH optima, another reason for maintaining the pH of the medium close to neutrality.
- Reducing agents (e.g. 2-mercaptoethanol, dithiothreitol, cysteine at ≈1 mmol l^{-1}). These reagents prevent oxidation of certain proteins, particularly those with free thiol groups that may be oxidised to disulphide bonds when released from the cell under aerobic conditions.
- Detergents (e.g. Triton X-100, SDS). These cause dissociation of proteins and lipoproteins from the cell membrane, aiding the release of membrane-bound and intracellular components.

__Using chelating agents__ – EDTA (ethylenediaminetetraacetic acid) and EGTA (ethylenebis(oxethylenenitrilo)-tetraacetic acid). If Mg^{2+} is an important component of the medium, use EGTA, which does not chelate Mg^{2+}.

__Extracting proteins from plants__ – carry out all post-homogenisation procedures as quickly as possible, because plant cell vacuoles may contain phenols that will inactivate proteins when released.

Methods of disrupting cells and tissues

Prior to disruption, animal tissues will need to be freed of any visible fat deposits and connective components, and then randomly sliced with fine scissors or a scalpel. Any fibrous and vascular tissue should be removed from plant material. Disruption can be achieved by mechanical and non-mechanical means: the principal applications of various methods are described in Table 36.1 for the major cell and tissue types.

Non-mechanical methods

- Osmotic shock: cells are first placed in a hypertonic solution of high osmolality (p. 147), e.g. 20% w/v sucrose, leading to loss of water. On dilution of this solution (e.g. by addition of water or transfer to a hypotonic solution), the cells will burst due to water influx. Osmotic shock treatment is effective for wall-less cells, sphaeroplasts (p. 230) or protoplasts.

Definitions

Isotonic – a medium with the same water potential (p. 150) as the cells or tissues.

Hypertonic – a medium with a more negative water potential, compared with the cells or medium.

Hypotonic – a medium with a less negative water potential, compared with the cells or medium.

Table 36.1 Summary of techniques for the disruption of tissues and cells – note that safety glasses should be worn for all procedures

Technique	Suitability	Comments
Non-mechanical methods		
Osmotic shock	Animal soft tissues Some plant cells	Small scale only
Freeze/thaw	Animal soft tissues Some bacteria	Time consuming; small scale; closed system – suitable for pathogens with appropriate safety measures; some enzymes are cold-labile
Lytic enzymes, e.g. lipases; proteases pectinase; cellulase	Animal cells Plant cells	Mild and selective; small scale; expensive; enzymes must be removed once lysis is complete
Lysozyme	Some bacteria	Gram-negative bacteria must be pretreated with EDTA. Suitable for some organisms resistant to mechanical disruption.
Mechanical methods		
Pestle and mortar + abrasives	Tough tissues	Not suitable for delicate tissues
Ball mills + glass beads	Bacteria and fungi	May cause organelle damage in eukaryotes
Blenders and rotor-stators	Plant and animal tissues	Ineffective for microbes
Homogenisers (glass and Teflon)	Soft, delicate tissues e.g. white blood cells, liver	Glass may shatter – wear safety glasses during use
Solid extrusion (Hughes press)	Tough plant material; bacteria; yeasts	Small scale
Liquid extrusion (French pressure cell)	Microbial cells	Small scale
Ultrasonication	Microbial cells	Cooling required; small scale; may cause damage to organelles, especially in eukaryotic cells.

- Freezing and thawing: causing leakage of intracellular material, following cell wall and membrane damage and internal disruption due to ice crystal formation.
- Lytic enzymes: damaging the cell wall and/or plasma membrane. Cells can then be disrupted by osmotic shock or gentle mechanical treatment.

Avoiding protein denaturation during homogenisation *– excessive frothing of the homogenate indicates denaturation of proteins (think of whipping egg whites for meringue).*

Mechanical methods

All mechanical procedures for cell disruption generate heat, and this may denature proteins. Therefore it is very important to cool the starting material, the homogenising medium, and, if possible, the homogeniser itself (to ≈4 °C). The homogenisation should be carried out in short bursts, and the homogenate should be cooled in an ice bath between each burst. Cooling will also reduce the activity of any degradative enzymes in the homogenate. Ideally, carry out homogenisation in a walk-in cooler (a 'cold room') typically at 4–10°C.

Equipment commonly used includes:

- Mixers and blenders. These are similar to domestic liquidisers, with a static vessel and rotating blades. The Waring blender is widely used: it has a stainless steel vessel that will stay cool if pre-chilled. The vessel and blades are designed to maximise turbulence, both disrupting and homogenising tissues and cells.

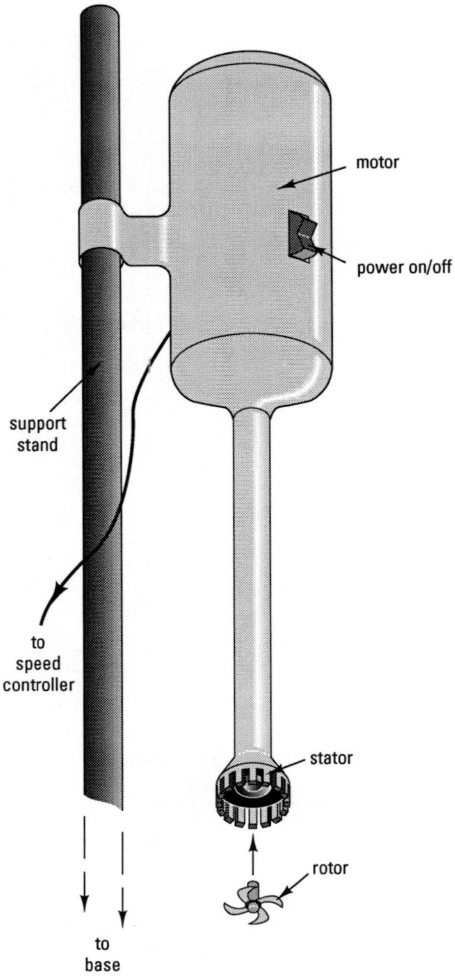

Fig. 36.2 Components of a rotor-stator homogeniser.

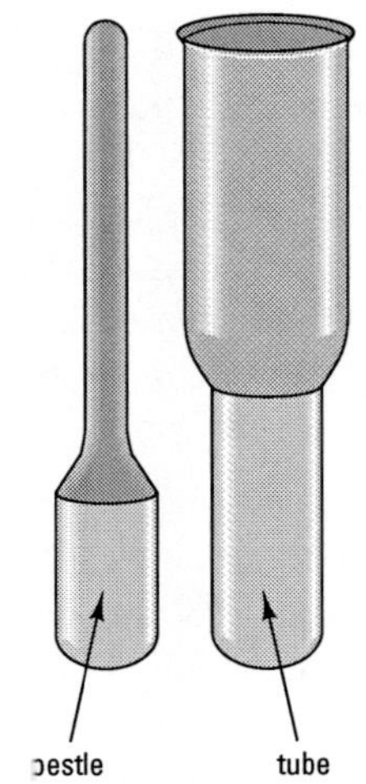

Fig. 36.3 Ground-glass homogeniser.

- Ball mills (e.g. Retch mixer mill, Mickle mill). These devices contain glass beads that vibrate and collide with each other and with tissues/cells, leading to disruption.
- Liquid extrusion devices (e.g. French pressure cell). Cells are forced from a vessel to the outside, through a very narrow orifice at high pressures ($\approx$100 MPa). The resulting pressure changes are a powerful means of disrupting cells.
- Solid extrusion (e.g. Hughes press). Here, a frozen cell paste is forced through a narrow orifice, where the shear forces and the abrasive properties of the ice crystals cause cell disruption.
- Rotor-stators (e.g. Ultra-turrax homogeniser). These have a rotor (a set of stainless steel blades) and a stator (a slotted stainless steel cylinder) at the tip of a stainless steel shaft, immersed in the homogenising medium: the arrangement is illustrated in Figure 36.2. The high speed of the rotor blades causes material in the homogenising fluid to be sucked into the dispersing head, where it is pressed radially through the slots in the stator. Along with the cutting action of the rotor blades, the material is subjected to very high shear and thrust and the resulting turbulence in the gap between rotor and stator gives effective mixing. The vigour of the homogenisation process can be altered by varying the rotor speed setting. Various sizes of rotor-stator are available, with typical diameters in the range 8–65 mm: the smaller sizes are particularly useful for small-scale preparations.
- Sonicators. Ultrasonic waves are transmitted to an aqueous suspension of cells via a metal probe. The ultrasound creates bubbles within the liquid and these produce shock waves when they collapse. Successful disruption depends on the correct choice of power and incubation time, together with pH, temperature and ionic strength of the suspension medium, often obtained by trial and error. You can reduce the effects of heating during ultrasonication by using short 'bursts' of power (10–30 s), with rests of 30–60 s in between, and by keeping your cell suspension on ice during disruption. An ultrasonic water bath provides a more gentle means of disrupting certain types of cells, e.g. some bacterial and animal cells.
- Homogenisers. These involve the reciprocating movement of a ground glass or Teflon pestle within a glass tube (Fig. 36.3). Cells are forced against the walls of the tube, releasing their contents. For glass pestles, the tubes also have ground glass homogenising surfaces and may have an overflow chamber. The homogeniser can either be hand operated (e.g. Dounce), or motorised (e.g. Potter-Elvejham). The clearance between the pestle and the tube (range 0.05–0.5 mm) must be chosen to suit the particular application.

Cell fractionation and the isolation of organelles

The fractionation and separation of organelles from a cell homogenate by differential centrifugation are described in Chapter 43. Particular organelles can be obtained by appropriate choice of source tissue and homogenisation method, as illustrated in Table 36.2 for the major types of organelle.

Table 36.2 Isolation and fractionation procedures for various organelles

Stage	Nuclei	Mitochondria	Microsomes	Chloroplasts*
Source	Thymus tissue, which has little cytoplasm, giving high yields.	Beef heart, with fat and connective tissue removed, then cubed and minced. Keep at pH 7.5 using TRIS buffer.	Rat liver, stored overnight to reduce glycogen content.	Spinach leaves, de-ribbed and cut into 1 cm strips.
Pretreatment	Rinse with buffered physiological saline. Suspend in homogenising medium.	Suspend in 2× volume of ice-cold homogenising medium. Squeeze through muslin.	Chop finely with scissors and wash in 2× volume of homogenising medium.	Rinse, then suspend in 3× volume of prechilled homogenising medium.
Homogenising medium	250 mmol l^{-1} sucrose; 10 mmol l^{-1} TRIS/HCl buffer, pH 7.6; 5 mmol l^{-1} $MgCl_2$; 0.2–0.5% v/v Triton X-100	250 mmol l^{-1} sucrose; 10 mmol l^{-1} TRIS/HCl buffer pH 7.7, containing 1 mmol l^{-1} succinic acid and 0.2 mmol l^{-1} EDTA.	250 mmol l^{-1} sucrose; 50 mmol l^{-1} TRIS/HCl buffer, pH 7.5; 25 mmol l^{-1} KCl; 5 mmol l^{-1} $MgCl_2$.	400 mmol l^{-1} sucrose; 25 mmol l^{-1} HEPES/NaOH buffer at pH 7.6; 2 mmol l^{-1} EDTA.
Homogenisation	Waring blender, low speed, 3 min.	Bring to pH 7.8 using 2 mol l^{-1} TRIS base: Waring blender, high speed, 15 s: check and adjust pH to 7.8 and repeat blending step, 5 s.	Potter-Elvejham glass homogeniser with a Teflon pestle – 3 × 5 min at 800 rpm.	Prechilled Waring blender or rotor-stator.
Filtration/ centrifugation	Filter through gauze. Spin at 2000 *g* for 10 min; discard supernatant. Repeat the homogenisation, filtration and centrifugation stages to improve purity of organelles.	Spin at 1200 *g* for 20 min: filter supernatant through muslin: centrifuge at 26 000 *g* for 15 min. Remove and discard upper (lighter) layer of pellet. Resuspend lower layer and rehomogenise (2 × 5 s). Centrifuge at 26 000 *g* for 15 min.	Centrifuge at 680 *g* for 10 min – discard pellet; centrifuge at 10 000 *g* for 10 min – discard pellet; centrifuge at 100 000 *g* for 60 min – retain pellet. Resuspend in buffer, pH 8.0 and re-centrifuge at 100 000 *g* for 60 min.	Pass through several layers of muslin (wear gloves); centrifuge at 2500 *g* for 60 s; resuspend pellet in buffer, pH 7.6; recentrifuge at 2500 *g* for 60 s. The colour of the supernatants gives a visual indication of chloroplast damage (e.g. if green).
Before use	Resuspend in homogenisation medium without Triton X-100.	Resuspend pellet in buffer, pH 7.8 and either use immediately, or store at −20 °C overnight.	Resuspend in buffer solution at pH 8.0.	Resuspend pellet in appropriate incubation medium, containing sucrose, e.g. for CO_2/O_2 studies, p. 400.

*An alternative approach is to use plant protoplasts (p. 230) as the starting material, releasing the chloroplasts by gentle lysis – diluting the medium with water.

Text reference

Hopkins, T. (1991) Physical and chemical cell disruption for the recovery of intracellular proteins. In: *Purification and Analysis of Recombinant Proteins* (ed R. Seetharam and S.K. Sharma). pp. 57–84. Marcel Dekker, New York.

Sources for further study

Anon. *Ultrasonic Homogenizers*. Available: http://www.biologics-inc.com/ultrasonic_homogenizers.htm Last accessed: 01/04/07.
[One of many websites dealing with commercial products]

Boyer, R. (2000) *Modern Experimental Biochemistry*, 3rd edn. Benjamin Cummings, San Francisco.
[Also covers centrifugation, spectroscopy, etc.]

Boyer, R. (2005) *Biochemistry Laboratory: Modern Theory and Techniques*. Benjamin Cummings, New York.

Graham, J.M. and Rickwood, D. (1997) *Subcellular Fractionation: A Practical Approach*. Oxford University Press, Oxford.

BioSpec Products Laboratory Cell Disrupters. Available: http://www.biospec.com/Lab%20Cell%20Disrupters%20Review.htm
Last accessed: 01/04/07.
[Provides a brief overview of methods]

Wilson, K. and Walker, J. (eds) (2005) *Principles and Techniques of Biochemistry and Molecular Biology*, 6th edn. Cambridge University Press, Cambridge.
[Also covers a range of other topics, including spectroscopy and chromatography]

Study exercises

36.1 Consider suitable disruption procedures for particular types of cell or tissue. After reading this chapter, identify the major problems associated with disrupting each of the following, and list suitable practical procedures to overcome these problems:
(a) Gram-negative bacteria;
(b) skeletal muscle;
(c) plant tissue.

36.2 Select a suitable disruption technique for the preparation of intracellular components. How might the procedure used to prepare a soluble cytoplasmic protein from fresh liver differ from that required for the isolation of intact organelles?

36.3 Test your understanding of the function of homogenisation medium components. The following list provides a number of possible components that might be used in a homogenisation medium:
(a) sucrose;
(b) low ionic strength buffer;
(c) dithiothreitol;
(d) detergent (e.g. Triton X-100).
Select one of these components to include in a medium for preparing:
(i) a membrane protein from liver tissue;
(ii) a mitochondrial suspension from liver tissue;
(iii) an oxygen-sensitive enzyme from pancreatic tissue;
(iv) a protein from the sarcolemma (cytoplasm) of skeletal muscle.
Give a brief explanation for your selection in each case.

Analytical techniques

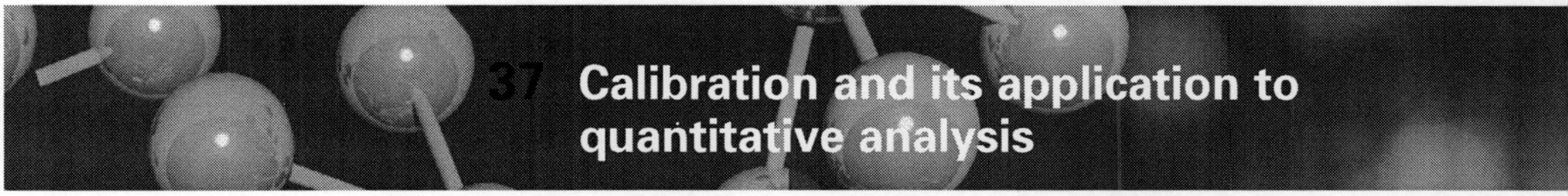

37 Calibration and its application to quantitative analysis

There are many instances where it is necessary to measure the quantity of a test substance using a calibrated procedure. You are most likely to encounter this approach in one or more of the following practical exercises:

- Quantitative spectrophotometric assay of biomolecules (Chapter 41).
- Flame or atomic absorption spectroscopic analysis of metal ions in biological solutions (p. 286).
- Using a chromogenic or fluorogenic substrate to determine the activity of an enzyme.
- Quantitative chromatographic analysis, e.g. GC or HPLC (Chapter 45).
- Using a bioassay system (p. 238) to quantify a test substance: examples include immunodiffusion (Chapter 38) and radioimmunoassay (p. 260).

***KEY POINT** In most instances, calibration involves the establishment of a relationship between the measured response (the 'signal') and one or more 'standards' containing a known amount of substance.*

***Calibrating laboratory apparatus** – this is important in relation to validation of equipment, e.g. when determining the accuracy and precision of a pipettor by the weighing method: see p. 130.*

In some instances, you can measure a signal due to an inherent property of the substance, e.g. the absorption of UV light by nucleic acids, whereas in other cases you will need to react it with another substance to see the result (e.g. molecular weight measurements of DNA fragments after electrophoresis, visualised using ethidium bromide, or to produce a measurable response (e.g. the reaction of copper(II) ions and peptide bonds in the biuret assay for proteins).

The different types of calibration curve

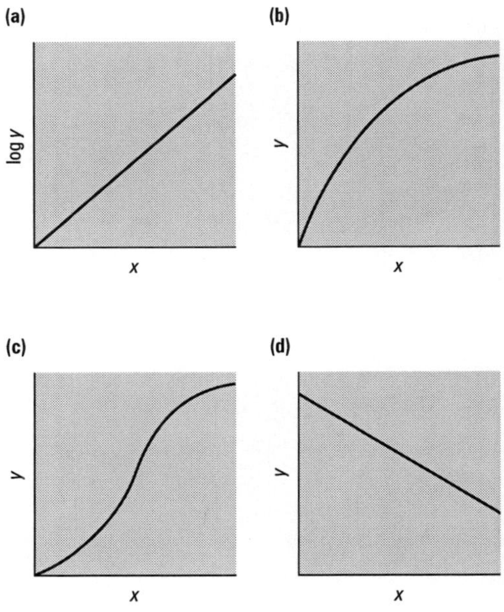

Fig. 37.1 Calibration curves: (a) log-linear; (b) curvilinear; (c) sigmoid, or S-shaped; (d) inverse.

By preparing a set of solutions (termed 'standards'), each containing either (i) a known *amount* or (ii) a specific *concentration* of the substance, and then measuring the response of each standard solution, the underlying relationship can be established in graphical form as a 'calibration curve', or 'standard curve'. This can then be used to determine either (i) the amount or (ii) the concentration of the substance in one or more test samples. Alternatively, the response can be expressed solely in mathematical terms: an example of this approach is the determination of chlorophyll pigments in plant extracts by measuring absorption at particular wavelengths, and then applying a formula based on previous (published) measurements for purified pigments (p. 282).

There are various types of standard curve: in the simplest cases, the relationship between signal and substance will be linear, or nearly so, and the calibration will be represented best by a straight-line graph (see Box 37.1). In some instances (Fig. 37.1a), you will need to transform either the x values or y values in order to produce a linear graph (e.g. in radial immunodiffusion bioassay, where the y values are squared, p. 259. In other instances, the straight-line relationship may only hold up to a certain value (the 'linear dynamic range') and beyond this point the graph may curve (e.g. in quantitative spectrophotometry, the Beer–Lambert relationship often becomes invalid at high absorbance, giving a curve, Fig. 37.1b). Some calibration curves are sigmoid (Fig. 37.1c). Finally, the signal may *decrease* in response to an increase in the substance (Fig. 37.1d), e.g. radioimmunoassay, where an inverse sigmoid calibration curve is obtained (p. 260). In some practical classes, you may be told that the relationship is expected to be linear, curvilinear, or

Box 37.1 The stages involved in preparing and using a calibration curve

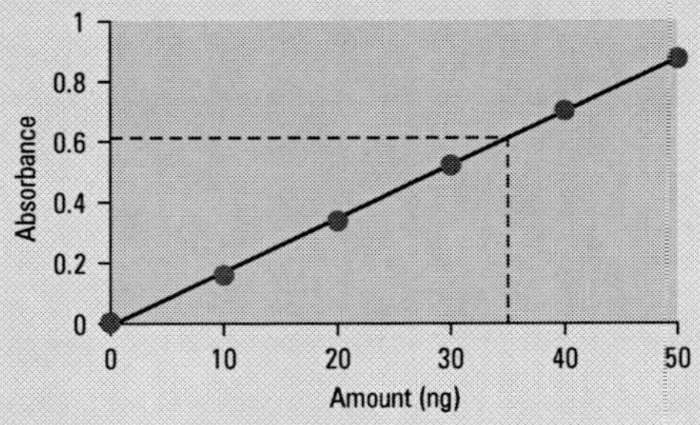

Fig. 37.2 Typical calibration curve for spectrophotometric analysis.

1. **Decide on an appropriate test method** – for example, in a project you may need to research the best approach to the analysis of a particular metabolite in your biological material.
2. **Select either (a) amount or (b) concentration, and an appropriate range and number of standards** – in practical classes, this may be given in your schedule, along with detailed instructions on how to make up the standard solutions. In other cases, you may be expected to work this out from first principles (Chapters 22–24 give worked examples) – aim to have evenly spaced values along the *x* axis.
3. **Prepare your standards very carefully** – due attention to detail is required: for example, you should ensure that you check the calibration of pipettors beforehand, using the weighing method (p. 130). Don't forget the 'zero standard' plus any other controls required, e.g. to test for interference due to other chemical substances. Your standards should cover the range of values expected in your test samples.
4. **Assay the standards and the unknown (test) samples** – preferably all at the same time, to avoid introducing error due to changes in the sensitivity or drift in the zero setting of the instrument with time. It is a good idea to measure all of your standard solutions at the outset, and then measure your test solutions, checking that the 'zero standard' and 'top standard' give the same values after, say, every six test measurements. If the re-measured standards do not fall within a reasonable margin of the previous value, then you will have to go back and recalibrate the instrument, and repeat the last six test measurements. If your test samples lie outside the range of your standards, you may need to repeat the assay using diluted test samples (extrapolation of your curve may not be valid).
5. **Draw the standard curve, or determine the underlying relationship** – Figure 37.2 gives an example of a typical linear calibration curve, where the spectrophotometric absorbance of a series of standard solutions is related to the amount of substance. When using a spreadsheet (Box 37.2) or a graphics package (Chapter 12), it is often appropriate to use Model I linear regression to produce a linear trend line (also termed the 'line of best fit') and you can then quote the value of r^2, which is a measure of the 'fit' of the measurements to the line. However, you should take care not to use a linear plot when the underlying relationship is clearly non-linear (Fig. 63.12) and you must consider whether the assumptions of the regression analysis are valid (e.g. for transformed data).
6. **Determine the amount or concentration in each unknown sample** – either by reading the appropriate value from the calibration curve, or by using the underlying mathematical relationship, i.e. $y = a + bx$. Make sure you draw any horizontal and vertical construction lines very carefully – students often lose marks unnecessarily by submitting poorly drawn construction lines within practical reports.
7. **Correct for dilution or concentration, where appropriate** – for example, if you diluted each test sample by 10-fold, then you would need to multiply by 10 to determine the value for the undiluted test sample. As another example, if you assayed 0.2 ml of test sample, you would need to multiply the value obtained from the calibration curve by 5, to give the value per ml.
8. **Quote your test results to an appropriate number of significant figures** – this should reflect the accuracy of the method used, not the size of your calculator's display.

whatever, while in others you may have to decide the form of the standard curve as part of the exercise.

Example A test tube containing 8 ml of water plus 2 ml of 1% w/v NaCl ($M_r = 58.44$) would have a *mass concentration* of 0.2% w/v NaCl (a five-fold dilution of the original NaCl solution), which can also be expressed as $2\,g\,l^{-1}$; in terms of *molar concentration* (p. 144), this would be equivalent to $2 \div 58.44 = 0.0342\,mol\,l^{-1}$ (to three significant figures). Expressed in terms of the *amount* of NaCl in the test tube, this would be 0.02 g in *mass*, or $0.02 \div 58.44 = 3.42 \times 10^{-4}$ mol (342 μmol) in *moles*.

Practical considerations

Amount or concentration?

This first step is often the most confusing for new students. It is vital that you understand the difference between *amount* of substance (e.g. mg, ng), and *concentration* (the amount of substance per unit volume, e.g. $mmol\,l^{-1}$, $mol\,m^{-3}$, % w/v) before you begin your practical work.

KEY POINT Essentially, you have to choose whether to work in terms of either (i) the total amount of substance in your assay vessel (e.g. test tube or cuvette) or (ii) the final concentration of the substance in your assay vessel, which is independent of the volume used.

The interconversion of amount and concentration is covered in more detail on p. 135. Either way, this is usually plotted on the x (horizontal) axis and the measured response on the y (vertical) axis.

Organisations providing national/ international standards *– these include:*

- *Laboratory of the Government Chemist (http://www.lgc.co.uk/) for UK standards and European Reference Materials.*
- *National Institute for Science and Technology (http://www.cstl.nist.gov/), for biochemical/chemical standards in the US.*
- *Institute for Reference Materials and Measurements (http://www.irmm.jrc.be/) for European standards, including BCR and ERM materials.*
- *OIE Biological Standards Commission (http://www.oie.int/bsc/eng/en_bsc.htm), for international animal standards.*

Choice of standards

In your early practical classes, you may be provided with a stock solution (pp. 136–7), from which you then have to prepare a specified number of standard solutions. In such cases, you will need to understand how to use dilutions to achieve the required amounts or concentrations (pp. 137–8). In later work and projects, you may need to prepare your standards from chemical reagents in solid form, where the important considerations are purity and solubility (p. 135). For professional analysis (e.g. in forensic science or clinical biochemistry), it is often important to be able to trace the original standard or stock solution back to national or international standards or to certified reference materials.

How many standards are required?

This may be given in your practical schedule, or you may have to decide what is appropriate (e.g. in project work and research). If the form and working range of the standard curve are known in advance, this may influence your choice – for example, linear calibration curves can be established with fewer standards than curvilinear relationships. In some instances, analytical instruments can be calibrated using a single standard solution, often termed a 'calibrator'. Replication of each standard solution is a good idea, since it will give you some information on the variability involved in preparing and assaying the standards. Consider whether you should plot mean values on your standard curve, or whether it is better to plot the individual values (if one value appears to be well off the line, you are likely to have made an error, and you may need to check and repeat).

Plotting a standard curve *– do not force your calibration line to pass through zero if it clearly does not. There is no reason to assume that the zero value is any more accurate than any other reading you have made.*

Preparing your standards

It is extremely important to take the greatest care to measure out all chemicals and liquids very accurately, to achieve the best possible standard curve. The grade of volumetric flask used and temperature of the solution also affect accuracy (grade A apparatus is best). You may also consider what other additives might be required in your standard solutions.

Box 37.2 How to use a spreadsheet (Microsoft Excel) to produce a linear regression plot

Two approaches are possible: either using the trendline feature or using the regression analysis function within the data analysis tool pack. Both make the assumption that the criteria for Model I linear regression are met. In the example shown below, the following simple data set has been used:

Amount (ng)	Absorbance
0	0.00
10	0.19
20	0.37
30	0.56
40	0.63
50	0.78

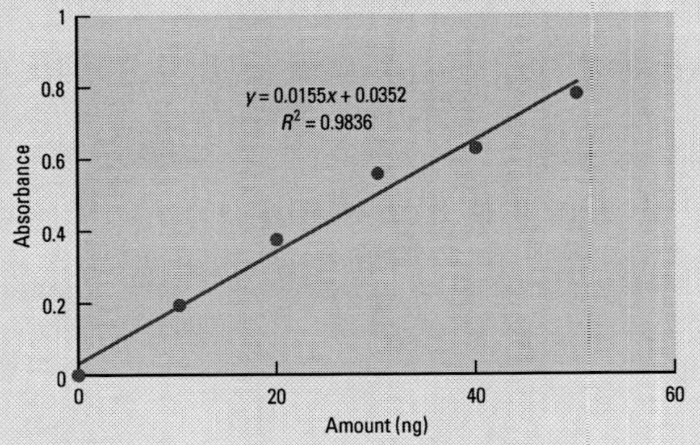

Fig. 37.3 Calibration curve showing line of best fit and details of linear regression equation.

Using the *Trendline* feature. This quick method provides a line of best fit on an Excel chart and can also provide set of equation values for predictive purposes.

1. **Create a graph (*chart*) of your data.** Enter the data in two columns within your spreadsheet, select the data array (highlight using left mouse button) and then, using the *Chart Wizard* icon, select *Chart type > XY (Scatter).*
2. **Add a trend line.** Right-click on any of the data points on your graph, and select the *Add Trendline* menu. Choose the *Linear* trend line option, but do not click *OK* at this stage. Rather, from the *Options* menu, select: (i) *Display equation on chart* and (ii) *Display R-squared value on chart.* Now click OK. The equation (shown in the form $y = bx + a$) gives the slope (b) and intercept (a) of the line of best fit, while the *R*-squared value (coefficient of determination) gives the proportional fit to the line (the closer this value is to 1, the better the fit of the data to the line).
3. **Modify the graph to improve its effectiveness.** For a graph that is to be used elsewhere (e.g. in a lab write-up or project report), adjust the display to remove the default background and gridlines and change the symbol shape (see Box 63.2 for more advice and examples). Right-click on the trend line and use the *Format Trendline > Custom* menu to adjust the *Weight* of the line to make it thinner. Drag and move the equation panel if you would like to alter its location on the chart, or delete it, having noted the values. Fig. 37.3 shows a calibration curve produced in this way for the data presented above.
4. **Use the regression equation to estimate unknown (test) samples.** By rearranging the equation for a straight line and substituting a particular *y*-value, you can predict the amount/concentration of substance (*x*-value) in a test sample. This is more precise than simply reading the values from the graph using construction lines (Box 37.1). If you are carrying out multiple calculations, the appropriate equation, $x = (y - a)/b$, can be entered into a spreadsheet, for convenience.

Using the *Regression Analysis* tool requires the 'data analysis tool pack' to be loaded beforehand (using *Tools > Add-Ins > Analysis ToolPak*) and provides summary output that contains details of slope, intercept) and coefficient of determination along with an analysis of variance (ANOVA) table (Chapter 67 gives further details of these aspects, including an example of output in study exercise 67.4).

***Dealing with interfering substances** – one approach is to use the method of 'standard additions', where the standards all contain a fixed additional amount of the sample (see e.g. Dean, 1997). Internal standards (p. 318) can also be used to detect such problems.*

For example, do your test samples have high levels of potentially interfering substances, and should these also be added to your standards? Also consider what controls and blank solutions to prepare.

***KEY POINT** The validity of your standard curve depends upon careful preparation of standards, especially in relation to accurate dispensing of the volumes of any stock solution and diluting liquid – the results for your test samples can only be as good as your standard curve.*

Preparing the calibration curve and determining the amount of the unknown (test) sample(s)

This is described in stepwise fashion in Box 37.1. Check you understand the requirements of graph drawing, especially in relation to plotted curves and the mathematics of straight-line graphs.

Spreadsheet programs such as Microsoft Excel can be used to produce a regression line for a straight line calibration plot (p. 252). Examples of how to do this are provided in Box 37.2.

Text reference

Dean, J.D. (1997) *Atomic Absorption and Plasma Spectroscopy*, 2nd edn. Wiley, Chichester.
[Chapter 1 deals with calibration, and covers the principle of standard additions]

Sources for further study

Brown, P.J. (1994) *Measurement, Regression, and Calibration.* Clarendon Press, Oxford.
[Covers advanced methods, including curve fitting and multivariate methods]

Mark, H. (1991) *Principles and Practice of Spectroscopic Calibration.* Wiley, New York.

Miller, J.N. and Miller, J.C. (2005) *Statistics and Chemometrics for Analytical Chemistry*, 5th edn. Prentice Hall, Harlow.
[Gives detailed coverage of calibration methods and the validity of analytical measurements]

Study exercises

37.1 Determine unknowns from a hand-drawn calibration curve. The following data are for a set of calibration standards for Zn, measured by atomic absorption spectrophotometry.

Absorbance measurements for a series of standard solutions containing different amounts of zinc

Zinc concentration ($\mu g\ ml^{-1}$)	Absorbance
0	0.000
1	0.082
2	0.174
3	0.257
4	0.340
5	0.408
6	0.463
7	0.511
8	0.543
9	0.561
10	0.575

Draw a calibration curve by hand using graph paper and estimate the concentration of zinc in the following water samples:

(a) an undiluted sample, giving an absorbance of 0.157;
(b) a 20-fold dilution, giving an absorbance of 0.304;
(c) a fivefold dilution, giving an absorbance of 0.550.

Give your answer to three significant figures in each case.

37.2 Determine unknowns from a calibration curve produced in Excel. The following data are for a set of calibration standards for protein content, measured by the Lowry (Folin–Ciocalteau) method.

Absorbance measurements for a series of protein standards

Protein (mg per tube)	Absorbance (A_{600})
0.00	0.000
0.02	0.161
0.04	0.284
0.06	0.438
0.08	0.572
0.10	0.762

(continued)

Study exercises (continued)

Using PC-based software (e.g. Excel), fit a trend line (linear regression) and determine the protein content of solutions with the following absorbances:
(a) 0.225 (b) 0.465 (c) 0.682
Give your answers to three significant figures in each case.

37.3 Identify the errors in a calibration curve. The figure shows a calibration curve of the type that might be submitted in a practical write-up.

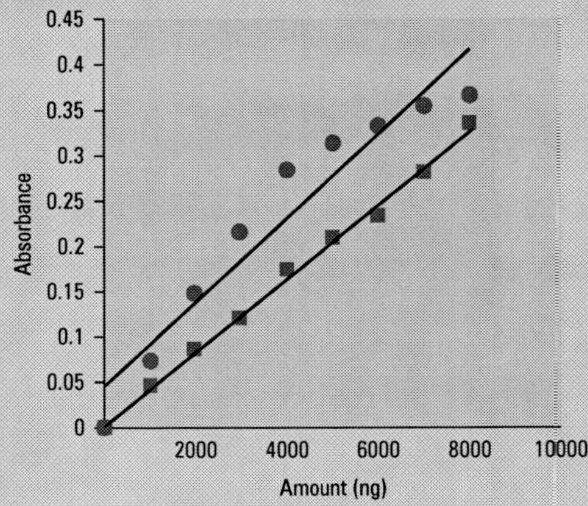

Calibration graph

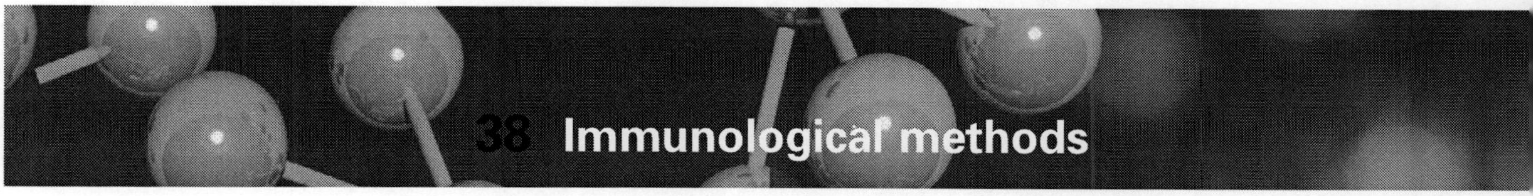

38 Immunological methods

Definitions

Antibody – a protein produced in response to an antigen (an *anti*body-*gen*erating foreign macromolecule).

Epitope – a site on the antigen that determines its interaction with a particular antibody.

Hapten – a substance that contains at least one epitope, but is too small to induce antibody formation unless it is linked to a macromolecule.

Ligand – a molecule or chemical group that binds to a particular site on another molecule.

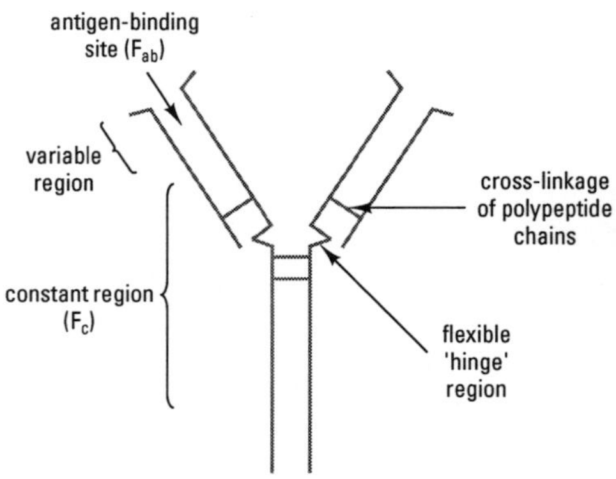

Fig. 38.1 Diagrammatic representation of IgG (antibody).

***Producing polyclonal antibodies** – in the UK, this is controlled by government regulations, since it involves vertebrate animals: personnel must be licensed by the Home Office and must operate in accordance with the **Animals Scientific Procedures Act (1986)** and with the **Code of Practice for the Housing and Care of Animals (2005)**.*

Antibodies are an important component of the immune system, which protects animals against certain diseases (see Delves *et al.,* 2006). They are produced by B lymphocytes in response to foreign macromolecules (antigens). A particular antibody will bind to a site on a specific antigen, forming an antigen–antibody complex (immune complex). Immunological bioassays use the specificity of this interaction for:

- identifying macromolecules, cellular components or whole cells;
- quantifying a particular substance.

Antibody structure

An antibody is a complex globular protein, or immunoglobulin (Ig). While there are several types, IgG is the major soluble antibody in vertebrates and is used in most immunological assays. Its main features are:

- Shape: IgG is a Y-shaped molecule (Fig. 38.1), with two antigen-binding sites.
- Specificity: variation in amino acid composition at the antigen-binding sites explains the specificity of the antigen–antibody interaction.
- Flexibility: each IgG molecule can interact with epitopes which are different distances apart, including those on different antigen molecules.
- Labelling: regions other than the antigen-binding sites can be labelled, e.g. using a radioisotope or an enzyme with fluorogenic or chromogenic detection.

KEY POINT The presence of two antigen-binding sites on a single flexible antibody molecule is relevant to many immunological assays, especially the agglutination and precipitation reactions.

Antibody production

Polyclonal antibodies

These are commonly used at undergraduate level. They are produced by repeated injection of antigen into a laboratory animal. After a suitable period (3–4 weeks) blood is removed and allowed to clot, leaving a liquid phase (polyclonal antiserum) containing many different IgG antibodies, resulting in:

- cross-reaction with other antigens or haptens;
- batch variation, as individual animals produce slightly different antibodies in response to the same antigen;
- non-specificity, as the antiserum will contain many other antibodies.

Standardisation of polyclonal antisera therefore is difficult. You may need to assess the amount of cross-reaction, interbatch variation or non-specific binding using appropriate controls, assayed at the same time as the test samples.

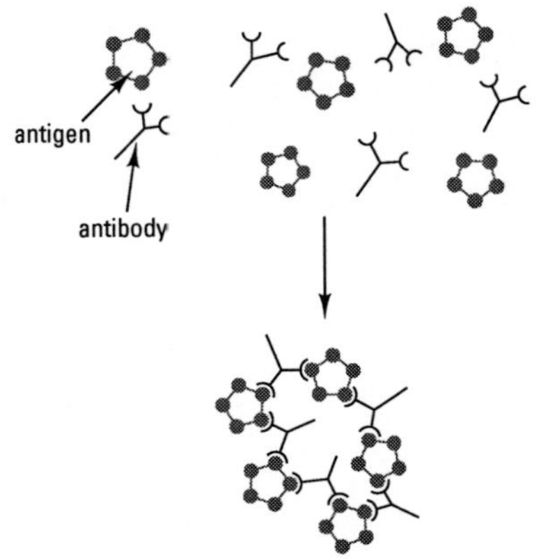

Fig. 38.2 Formation of an antigen–antibody complex (immune complex).

Monoclonal antibodies

These are specific to a single epitope and are produced from individual clones of cells (hybridomas, p. 225), grown using cell culture techniques (p. 232). Such cultures provide a stable source of antibodies of known, uniform specificity. While monoclonal antibodies are likely to be used increasingly in future years, polyclonal antisera are currently employed for many routine immunological assays.

Agglutination tests

When antibodies interact with a suspension of a particulate antigen, e.g. cells or latex particles, the formation of immune complexes (Fig. 38.2) causes visible clumping, termed agglutination. A positive haemagglutination reaction gives an even 'carpet' of red cells over the base of the tube, while a negative reaction gives a tightly packed 'button' of red cells at the bottom of the tube. Agglutination tests are used in several ways:

- Microbial identification: at the species or subspecies level (serotyping), e.g. mixing an unknown bacterium with the appropriate antiserum will cause the cells to agglutinate.
- Latex agglutination (bound antigens): by coating soluble (non-particulate) antigens onto microscopic latex spheres, their reaction with a particular antibody can be visualised.
- Latex agglutination (bound antibodies): antibodies can be bound to latex microspheres, leaving their antigen-binding sites free to react with soluble antigen.
- Haemagglutination: red blood cells can be used as agglutinating particles. However, in some instances, such reactions do not involve antibody interactions (e.g. some animal viruses may haemagglutinate unmodified red blood cells).

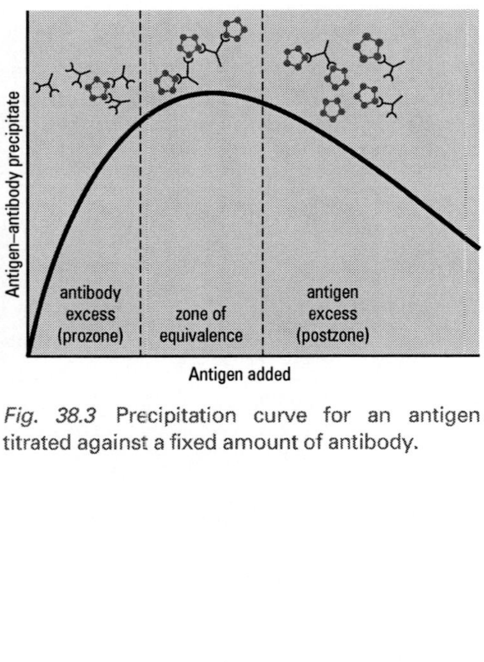

Fig. 38.3 Precipitation curve for an antigen titrated against a fixed amount of antibody.

Precipitin tests

Immune complexes of antibodies and soluble antigens (or haptens) usually settle out of solution as a visible precipitate: this is termed a precipitin test, or precipitation test. The formation of visible immune complexes in agglutination and precipitation reactions only occurs if antibody and antigen are present in an optimal ratio (Fig. 38.3). It is important to appreciate the shape of this curve: cross-linkage is maximal in the zone of equivalence, decreasing if either component is present in excess. The quantitative precipitin test can be used to measure the antibody content of a solution (Hay *et al.*, 2002 give details). Visual assessment of precipitation reactions forms the basis of several other techniques, described below.

Immunodiffusion assays

These techniques are easier to perform and interpret than the quantitative precipitin test. Precipitation of antibody and antigen occurs within an agarose gel, giving a visible line corresponding to the zone of equivalence (Fig. 38.3). Details of the main techniques are given in Box 38.1.

Box 38.1 How to carry out immunodiffusion assays

The two most widespread approaches are (i) single radial immunodiffusion (Mancini technique) and (ii) double diffusion immunoassay (Ouchterlony technique).

Single radial immunodiffusion (RID) (Mancini technique)
This is used to quantify the amount of antigen in a test solution, as follows:

1. **Prepare an agarose gel** (1.5% w/v), containing a fixed amount of antibody: allow to set on a glass slide or plate, on a level surface.
2. **Cut several circular wells in the gel.** These should be of a fixed size between 2 and 4 mm in diameter (see Fig. 38.4a). Cut your wells carefully – they should have straight sides and the agarose must not be torn or lifted from the glass plate. All wells should be filled to the top, with a flat meniscus, to ensure identical diffusion characteristics. Non-circular precipitin rings, resulting from poor technique, should not be included in your analysis.
3. **Add a known amount of the antigen or test solution to each well.**
4. **Incubate on a level surface at room temperature in a moist chamber:** diffusion of antigen into the gel produces a precipitin ring. This is usually measured after 2–7 days, depending on the molecular mass of the antigen.
5. **Examine the plates against a black background** (with side illumination), or stain using a protein dye (e.g. Coomassie blue).
6. **Measure the diameter of the precipitin ring,** e.g. using Vernier callipers.

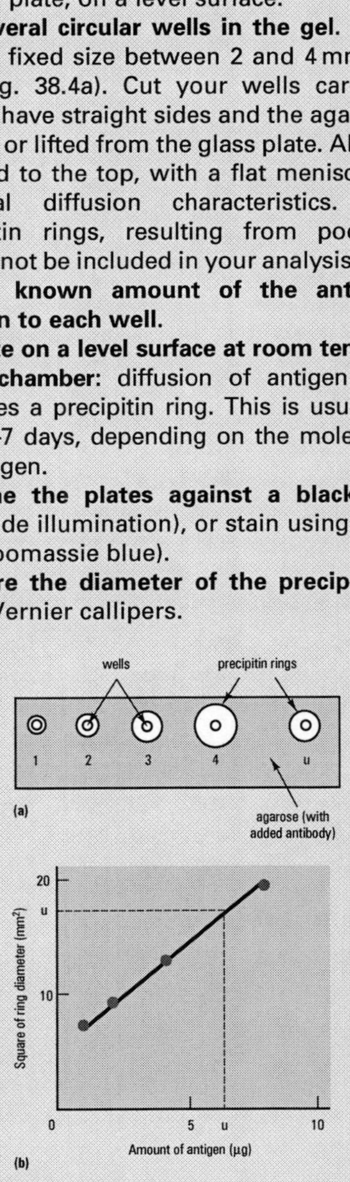

Fig. 38.4 Single radial immunodiffusion (RID). (a) Assay: four standards are shown (wells 1 to 4, each one double the strength of the previous standard), and an unknown (u), run at the same time. (b) Calibration curve. The unknown contains 6.25 µg of antigen. Note the non-zero intercept of the calibration curve, corresponding to the square of the well diameter: do not force such calibration lines through the origin.

7. **Prepare a calibration curve** (Chapter 37) from the samples containing known amounts of antigen (Fig. 38.4b): the squared diameter of the precipitin ring is directly proportional to the amount of antigen in the well.
8. **Quantify the amount of antigen in your test solutions** using a calibration curve prepared from standards assayed at the same time.

Double diffusion immunoassay (Ouchterlony technique)
This technique is widely used to detect particular antigens in a test solution, or to look for cross-reaction between different antigens.

1. **Prepare an agarose gel** (1.5% w/v) on a level glass slide or plate: allow to set.
2. **Cut several circular wells in the gel.**
3. **Add test solutions of antigen or polyclonal antiserum to adjacent wells.** Both solutions diffuse outwards, forming visible precipitin lines where antigen and corresponding antibody are present in optimal ratio (Fig. 38.5).

The various reactions between antigen and antiserum are:

- **Identity:** two wells containing the same antigen, or antigens with identical epitopes, will give a fused precipitin line (identical interaction between the antiserum and the test antigens, Fig. 38.5a).
- **Non-identity:** where the antiserum contains antibodies to two different antigens, each with its own distinct epitopes, giving two precipitin lines which intersect without any interaction (no cross-reaction, Fig. 38.5b).
- **Partial identity:** where two antigens have at least one epitope in common, but where other epitopes are present, giving a fused precipitin line with a spur (cross-reaction, Fig. 38.5c).

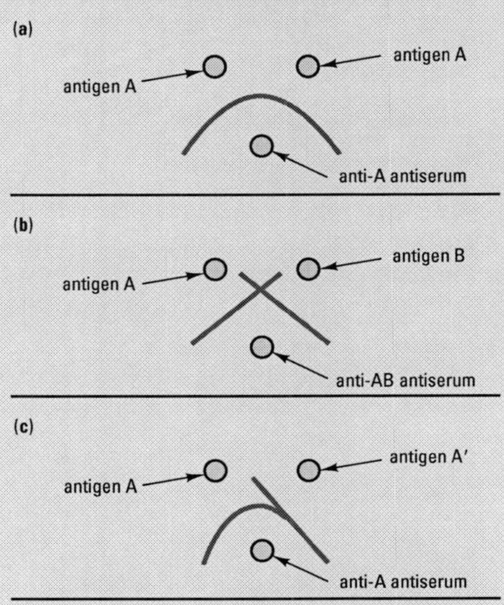

Fig. 38.5 Precipitin reactions in double diffusion immunoassay: (a) identity; (b) non-identity; (c) partial identity.

Immunoelectrophoretic assays

These methods combine the precipitin reaction with electrophoretic migration, providing sensitive, rapid assays with increased separation and resolution. The principal techniques are:

Cross-over electrophoresis (counter-current electrophoresis)

Similar to the Ouchterlony technique, since antigen and antibody are in separate wells. However, the movement of antigen and antibody towards each other is driven by a voltage gradient (p. 320): most antigens migrate towards the anode, while IgG migrates towards the cathode. This method is faster and more sensitive than double immunodiffusion, taking 15–20 min to reach completion.

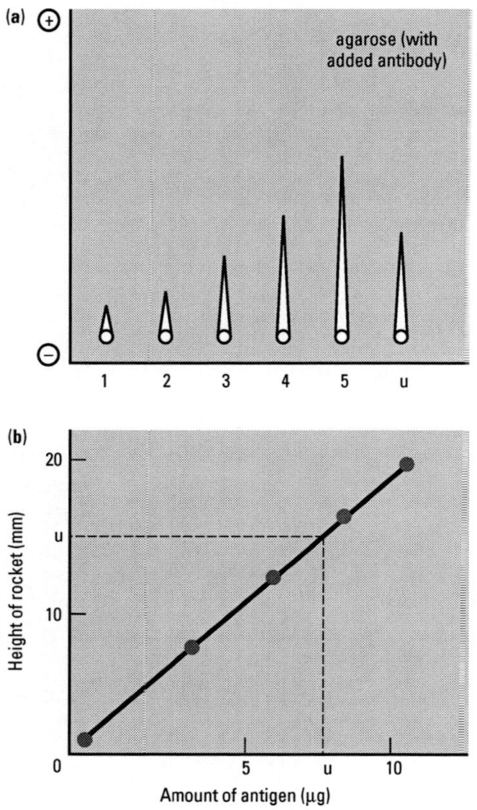

Fig. 38.6 Laurell rocket immunoelectrophoresis. (a) Assay: precipitin rockets are formed by electrophoresis of five standards of increasing concentration (wells 1 to 5) and an unknown (u). (b) Calibration curve: the unknown sample contains 7.7 μg of antigen.

Quantitative immunoelectrophoresis (Laurell rocket immunoelectrophoresis)

Similar to RID, as the antibody is incorporated into an agarose plate while the antigen is placed in a well. However, a voltage gradient moves the antigen into the gel, usually towards the anode, while the antibody moves towards the cathode, giving a sharply peaked, rocket-shaped precipitin line, once equivalence is reached (within 2–10 h). The height of each rocket shape at equivalence is directly proportional to the amount of antigen added to each well. A calibration curve for samples containing a known amount of antigen can be used to quantify the amount of antigen present in test samples (Fig. 38.6).

Radioimmunological methods

These methods use radioisotopes to detect and quantify the antigen–antibody interaction, giving improved sensitivity over agglutination and precipitation methods. The principal techniques are:

Radioimmunoassay (RIA)

This is based on competition between a radioactively labelled antigen (or hapten) and an unlabelled antigen for the binding sites on a limited amount of antibody. The quantity of antigen in a test solution can be determined using a known amount of radiolabelled antigen and a fixed amount of antibody (Fig. 38.7). As with other immunoassay methods, it is important to perform appropriate controls, to screen for potentially interfering compounds. The basic procedure for RIA is:

1. Add appropriate volumes of a sample to a series of small test tubes. Prepare a further set of tubes containing known quantities of the substance to be assayed to provide a standard curve.
2. Add a known amount of radiolabelled antigen (or hapten) to each tube (sample and standard).
3. Add a fixed amount of antibody to each tube (the antibody must be present in limited quantity).
4. Leave at constant temperature for a fixed time (usually 24 h), to allow antigen–antibody complexes to form.
5. Precipitate the antibody and bound antigen using saturated ammonium sulphate, followed by centrifugation.
6. Determine the radioactivity of the supernatant or the precipitate (pp. 269–71).

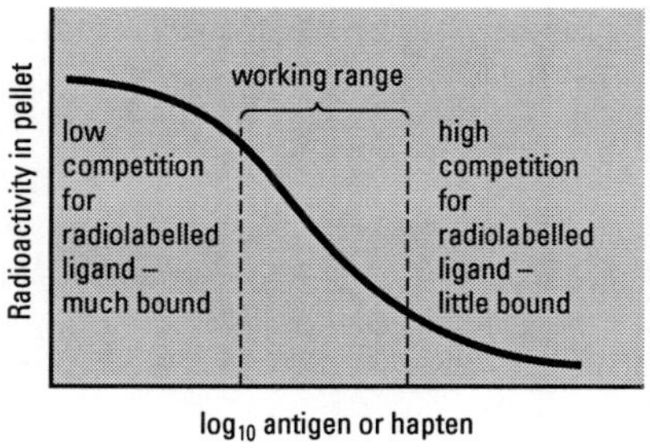

Fig. 38.7 Radioimmunoassay calibration curve. Note that the assay is insensitive at very low and very high antigen levels.

7. Prepare a calibration curve of radioactivity against $\log_{10}$ antigen (Fig. 38.7). The curve is most accurate in the central region, so adjust the amount of antigen in your test sample to fall within this range.

Note the following:

- You must be registered to work with radioactivity: check with the Departmental Safety Officer, if necessary (p. 273).
- Measure all volumes as accurately as possible as the end result depends on the volumetric quantities of unlabelled (sample) antigen, radiolabelled antigen and antibody: an error in any of these reagents will invalidate the assay.
- Incorporate replicates, so that errors can be quantified.
- Seek your supervisor's advice about fitting a curve to your data: this can be a complex process.

Immunoradiometric assay (IRMA)

This technique uses radiolabelled antibody, rather than antigen, for direct measurement of the amount of antigen (or hapten) in a sample. Most immunoradiometric assays are similar to the double antibody sandwich method described below, except that the second antibody is labelled using a radioisotope. The important advantages over RIA are:

- linear relationship between amount of radioactivity and test antigen;
- wider working range for test substance;
- improved stability/longer shelf-life.

Enzyme immunoassays (EIA)

These techniques are also known as enzyme-linked immunosorbent assays (ELISA). They combine the specificity of the antibody–antigen interaction with the sensitivity of enzyme assays using either an antibody or an antigen conjugated (linked) to an enzyme at a site which does not affect the activity of either component. The enzyme is measured by adding an appropriate chromogenic substrate, which yields a coloured product. Enzymes offer the following advantages over radioisotopic labels:

- Increased sensitivity: a single enzyme molecule can produce many product molecules, amplifying the signal.
- Simplified assay: enzyme assays are usually easier than radioisotope assays.
- Improved stability of reagents: components are generally more stable than their radiolabelled counterparts, giving them a longer shelf-life.
- No radiological hazard: no requirement for specialised containment/disposal facilities.
- Automation is straightforward: using disposable microtitre plates and an optical scanner.

The principal techniques are:

Double antibody sandwich ELISA

This is used to detect specific antigens, involving a three-component complex between a capture antibody linked to a solid support, the antigen, and a second, enzyme-linked antibody (Fig. 38.8). This can be used to detect a particular antigen, e.g. a virus in a clinical sample, or to quantify the amount of antigen.

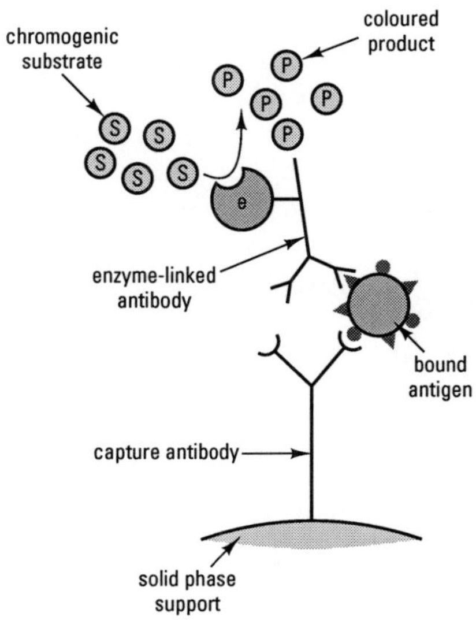

Fig. 38.8 Double antibody sandwich ELISA.

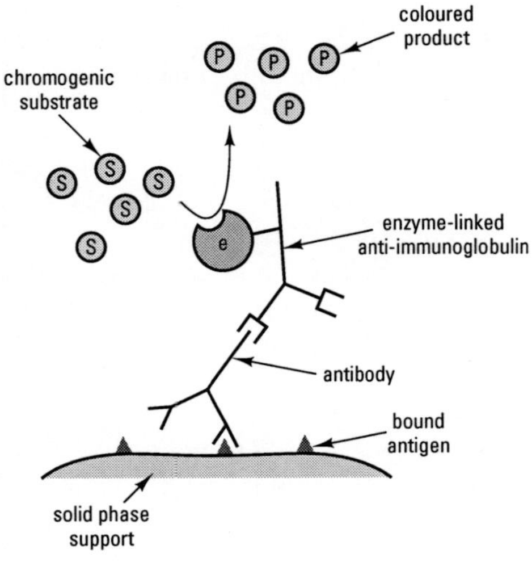

Fig. 38.9 Indirect ELISA.

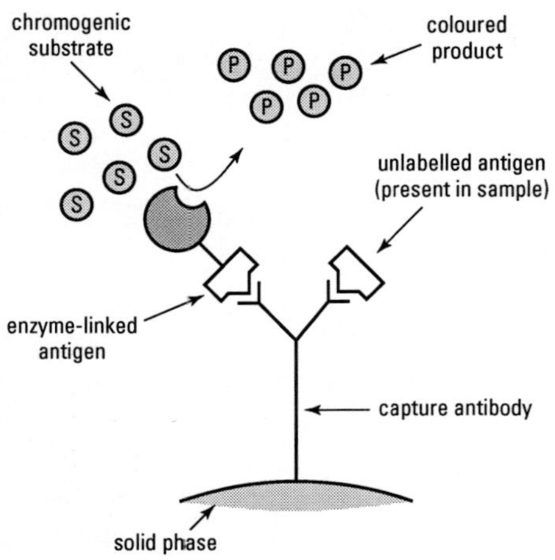

Fig. 38.10 Competitive ELISA

Indirect ELISA

This is used for antibody detection, with a specific antigen attached to a solid support. When the appropriate antibody is added, it binds to the antigen and will not be washed away during rinsing. Bound antibody is then detected using an enzyme-linked anti-immunoglobulin, e.g. a rabbit IgG antibody raised against human IgG (Fig. 38.9). One advantage of the indirect assay is that a single enzyme-linked anti-immunoglobulin can be used to detect several different antibodies, since the specificity is provided by the bound antigen.

Competitive ELISA

Here, any antigen present in a test sample competes with added enzyme-labelled antigen for a limited number of binding sites on the capture antibody (Fig. 38.10). Most commercial systems use 96-well microplates (12 columns by 8 rows), where each well is coated with the appropriate antibody. Following addition of known volumes of (i) sample and (ii) enzyme-labelled antigen, the plates are incubated (typically, up to 1 h), then washed thoroughly to remove all unbound material. Bound enzyme-labelled antigen is then detected using a suitable substrate. Quantitative results can be obtained by measuring the absorbance of each well using a spectrophotometric microplate reader: the absorbance at a particular wavelength is *inversely* proportional to the amount of antigen present in the test sample. Calibration standards (p. 252) are required to convert the readings to an amount or concentration. Alternatively, the test can be carried out in positive/negative format. Box 38.2 gives practical details for a sandwich-type ELISA.

Dip-stick (chromatographic) immunoassays

Improvements in technology have led to the development of a range of diagnostic kits that can be used either in a medical setting ('near-patient testing') or in the home, including immunoassays for cancer screening and drug testing. The first of these available to the general public was the home pregnancy test kit which measures the hormone human chorionic gonadotrophin (hCG) in urine. The level of this hormone is raised substantially during pregnancy. The principle of this test, which is similar to other home test kits, is that coloured plastic microbeads are conjugated to an anti-hCG monoclonal antibody and then coated onto an absorbent plastic or cellulose strip. When this strip is dipped into urine any hCG present will bind to the anti-hCG antibody and both hormone–antibody complexes and free antibody will then move up the strip by capillary action until they reach a second hCG-specific antibody that is coupled chemically to the strip so that it cannot move. Any hormone–antibody complex will bind to this 'fixed' antibody (through the hCG component); due to the presence of the coloured plastic microbeads, a distinct line will then appear at this location on the strip, signifying the presence of the hormone. If no hormone–antibody complex is present, the first antibody will simply continue up the strip until it reaches an antibody raised against the anti-hCG antibody itself, coupled to the strip at a higher point. This will bind any free anti-hCG antibody in the sample and a coloured line will appear at this point. Since there will always be an excess of the original anti-hCG antibody conjugated to the plastic beads, the appearance of

Box 38.2 How to perform an ELISA assay

While the following example is for a sandwich (capture) ELISA assay in microplate format, the same general principles apply to the other types of ELISA:

1. **Prepare the apparatus.** Switch on all equipment required:
   (i) The microplate reader – used to measure the absorbance of the solution in each well: set the reader to the required wavelength.
   (ii) The microplate washer (where used) – each well must be washed at various stages during the procedure. When using an automated washer, first check that the wash bottle contains sufficient diluent and then test using an old microplate, to check that all wells are being washed correctly. Where required, use a wire needle to clean any blocked wash delivery tubes and repeat. For manual washing, use a wash bottle or multichannel pipettor – make sure you fill each well and empty out all of the wash solution at the end of each wash stage.
   (iii) The computer – this will contain the software required to label the wells, draw the calibration curve and calculate the results for test samples: fill out the ELISA template with details of the assays to be carried out.
2. **Prepare the various solutions to be analysed.** These include:
   (i) Test samples – make sure that each sample is identified with a code that enables you to record what each test well contains.
   (ii) Calibrators/standards – including 'cutoff' calibrators and known positive standards.
   (iii) Controls – positive and negative controls and blanks.
3. **Coat the wells with the capture antibody.** Typically 100 μl of a solution of the appropriate monoclonal or polyclonal antibody is added to each well and microplates are then incubated overnight at 4°C to allow binding to the well.
4. **Wash the wells.** Transfer the microplate to the washer (or wash manually) – wash six times to remove excess coating (capture) antibody. The final rinse should be programmed so that the washer leaves the wells empty of diluent.
5. **Add blocking solution to each well.** Typically 100 μl of an inert protein solution (bovine serum albumin) is added to each well to block any free binding sites on the well. Microplates are incubated at room temperature for 30 min and then washed, as in step 4.
6. **Add test samples, calibrators and controls to wells.** Typically 100 μl of appropriately diluted test sample, control, etc. is added to each well. Microplates are incubated at room temperature for 90 min and then washed as step 4.
7. **Add the detection antibody to each well.** Add 100 μl of monoclonal or polyclonal detection antibody labelled with a suitable enzyme (e.g. horseradish peroxidise, HRP) to each well. The microplate is then incubated at room temperature for 30–60 min and then washed, as in step 4.
8. **Add chromogenic substrate to each well.** For example, with HRP-labelled antigen add 100 μl of a standard solution of tetramethylbenzidine (TMB) and hydrogen peroxide to each well and re-incubate in darkness for 30 min, to allow colour development. The TMB is oxidised in the presence of hydrogen peroxide to produce a blue colour.

   SAFETY NOTE TMB and hydrogen peroxide are harmful by inhalation – use a fume hood.
9. **Stop the reactions.** For example, by adding 100 μl of 2 mol l^{-1} sulphuric acid to denature the enzyme. The colour of the oxidised TMB will change from blue to yellow as a result of the pH shift. While the human eye can readily distinguish different shades of blue, it is more difficult to visually assess different shades of yellow, once the reaction has been stopped.
10. **Measure the absorbance of each sample/calibrator/control well.** Transfer the microplate to the reader and assay at an appropriate wavelength: for TMB, use 450 nm.
11. **Interpret the results.** Check that the absorbance values of calibrators and control are within the required range. Then, for each sample, either record the absolute value (convert to concentration or amount e.g. using a calibration curve, p. 252) or record as 'positive' or 'negative' (based on values for 'cutoff' calibrators), as appropriate.

two coloured lines means that the hormone is present and the test result is positive, while if only the second coloured line appears, it means that the test is negative (this also serves as a control to confirm that the test is working – lack of any coloured lines indicates that the test has failed). Fig. 38.11 shows the general operating principles.

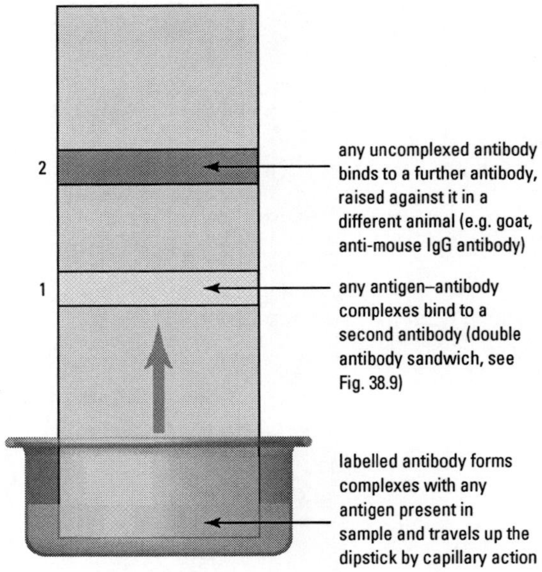

Fig. 38.11 Two-site dipstick immunoassay. For a positive test, coloured bands will appear at positions 1 and 2 while a negative test, as shown here, will give colour in band 2 only (any other colour options indicate a failed test).

Understanding complement-based assays *– the principal benefit from using the complement system is that it operates as an amplifying cascade of reactions, providing higher sensitivity than assays based on agglutination or precipitin reactions. For example, antibodies can be detected at concentrations of less than 1 μg ml^{-1}.*

Complement-based assays

Complement is a term applied to a group of around 20 different proteins which are involved in various aspects of immune function, including (i) control of inflammation, (ii) preparation of microbial cells for phagocytosis (opsonisation), (iii) lysis of target cells (e.g. antibody-coated cells) and (iv) activation of leucocytes (for details see Delves *et al.*, 2006). At a practical level, the complement system can be used to detect (i) antibodies, (ii) antigens or (iii) antibody–antigen complexes, often *via* its lytic action on antibody-coated ('sensitised') cells. The operating principles of a complement fixation assay are illustrated in Fig. 38.12 for antibody detection, with the following steps:

1. Prepare serial doubling dilutions (pp. 137–8) of a test sample containing an unknown amount of antibody.

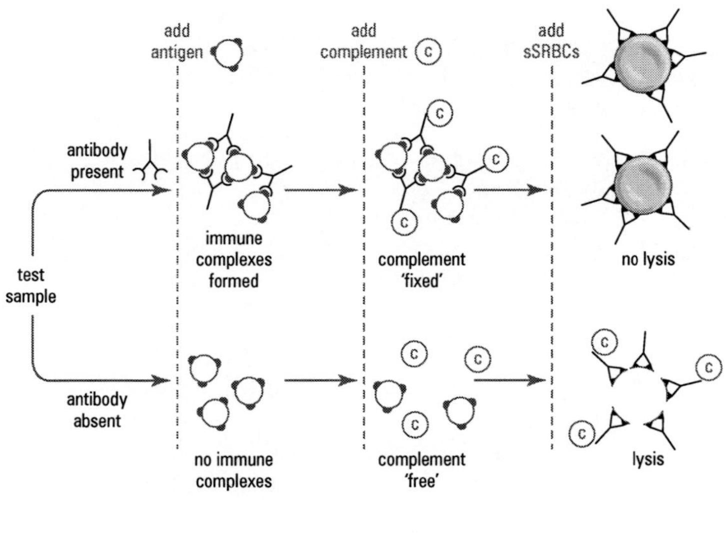

Fig. 38.12 Complement fixation test for antibody.

2. Add a standard amount of antigen: antigen–antibody complexes will be formed in direct proportion to the amount of antibody present in the sample.
3. Then, add a known amount of complement to the mixture: if immune complexes (Fig. 38.2) are present, they will 'fix' complement (i.e. the complement components will be consumed and no longer 'free' in solution), whereas if immune complexes are not present, or are at low concentration, complement will remain 'free' in solution, and available for subsequent reaction.
4. In the final step, add an indicator system for unbound complement – sensitised sheep red blood cells (sSRBCs) coated with rabbit anti-erythrocyte antibodies are often used. Unfixed complement is detected by lysis of the sensitised erythrocytes. The lowest dilution showing complement fixation (i.e. no erythrocyte lysis) gives the titre of antibody in the test sample.

Example Complement fixation tests are the basis of the Wasserman test to detect anti-treponemal antibodies in the serum of patients with syphilis.

The test can be modified to test for antigens in a sample, by titration against a fixed amount of added antibody. In all cases, suitable controls are required, to check for non-specific reactions and to confirm that immune complexes are not already present in the sample, since these will fix complement and invalidate the titration. Complement fixation tests are useful for many classes of IgG and for IgM, but not for most other antibody types.

Text references

Delves, P.J., Martin, S., Burton, D. and Roitt, I. (2006) *Roitt's Essential Immunology*, 10th edn. Blackwell, Oxford.

Hay, F.C., Westwood, O.M.R. and Nelson, P.N. (2002) *Practical Immunology*. Blackwell, Oxford.

Roitt, I.M., Broscoff, J. and Male, D.K. (eds) (2001) *Immunology*, 6th edn. Mosby, New York.

Sources for further study

Clausen, J. (1989) *Immunochemical Techniques for the Identification and Estimation of Macromolecules*, 3rd edn. Elsevier, Amsterdam.

Diamandis, E.P. and Christopoulos, T.K. (1997) *Immunoassay*. Academic Press, New York.

Gosling, J.R.G. (2000) *Immunoassays: a Practical Approach.* Oxford University Press, Oxford.

Lefkovits, I. (1997) *Immunology Methods Manual: The Comprehensive Sourcebook of Techniques.* Academic Press, New York.

Study exercises

38.1 Determine the amount of antigen in a test solution using Laurell rocket immunoelectrophoresis. The figure on the right represents the results from a rocket immunoelectrophoresis assay for a series of standards. Determine the amount of antigen in a test solution giving a rocket (precipitin line) height of 18.5 mm.

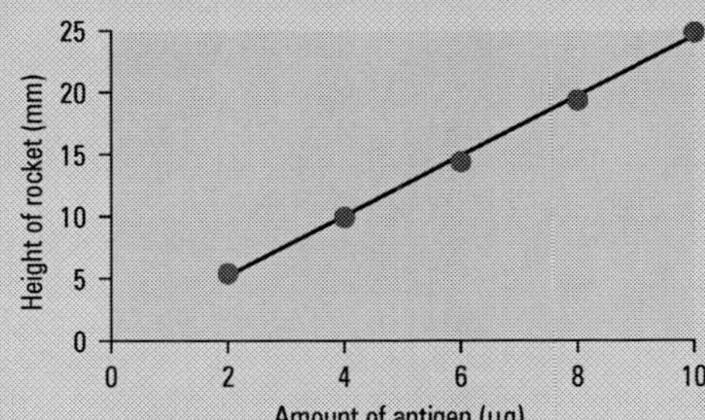

(continued)

Study exercises (continued)

38.2 Determine the amount of antigen in a test solution using single radial immunodiffusion (RID). Using the following data for a series of standards, determine the amount of antigen in a test solution giving a ring diameter of 6.5 mm.

Single radial immunodiffusion results for a series of standards containing 4–20 μg antigen

Amount of antigen (μg)	Ring diameter (mm)
4	3.5
8	5.5
12	7.0
16	8.0
20	9.0

38.3 Interpret data from a haemagglutination test. In some conditions, such as Hashimoto's disease, autoantibodies to the thyroid protein thyroglobulin are produced in large amounts. The level of these antibodies in serum can be measured by indirect haemagglutination: red blood cells coated with thyroglobulin are mixed with serial doubling dilutions of the test serum. The titre of antithyroglobulin autoantibodies is the lowest dilution at which haemagglutination occurs. The figure below represents the outcome of an antithyroglobulin haemagglutination test on a serum sample. The first well – numbered 1 – represents a 10-fold dilution of the serum sample, with serial doubling dilutions in subsequent wells.

(a) Write a brief explanation of the test results shown in the figure.

(b) Determine the autoantibody titre.

(c) Score the sample as 'positive' if the titre is more dilute than 1 in 5000 (i.e. 1:>5000) and 'negative' if it is less dilute than 1 in 5000 (i.e. 1:<5000).

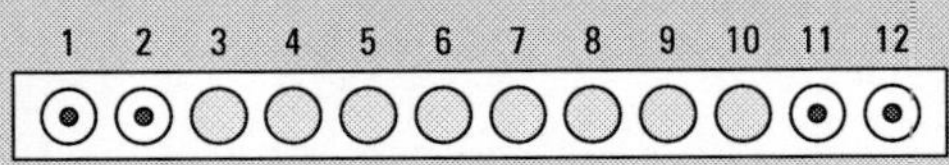

39 Radioactive isotopes and their uses

Examples $^{12}_{6}C$, $^{13}_{6}C$ and $^{14}_{6}C$ are three of the isotopes of carbon. About 98.9% of naturally occurring carbon is in the stable $^{12}_{6}C$ form. $^{13}_{6}C$ is also a stable isotope but it only occurs at 1.1% natural abundance. Trace amounts of radioactive $^{14}_{6}C$ are found naturally; this is a negatron-emitting radioisotope (see Table 39.2).

The isotopes of a particular element have the same number of protons in the nucleus but different numbers of neutrons, giving them the same proton number (atomic number) but different nucleon numbers (mass number, i.e. number of protons + number of neutrons). Isotopes may be stable or radioactive. Radioactive isotopes (radioisotopes) disintegrate spontaneously at random to yield radiation and a decay product.

Radioactive decay

There are three forms of radioactivity (Table 39.1) arising from three main types of nuclear decay:

Examples ^{226}Ra decays to ^{222}Rn by loss of an alpha particle, as follows:

$$^{226}_{88}Ra \rightarrow {}^{222}_{86}Rn + {}^{4}_{2}He^{2+}$$

^{14}C shows beta decay, as follows:

$$^{14}_{6}C \rightarrow {}^{14}_{7}N + \beta^{-}$$

^{22}Na decays by positron emission, as follows:

$$^{22}_{11}Na \rightarrow {}^{22}_{10}Ne + \beta^{+}$$

^{55}Fe decays by electron capture and the production of an X-ray, as follows:

$$^{55}_{26}Fe \rightarrow {}^{55}_{25}Mn + X$$

The decay of ^{22}Na by positron emission (β^{+}) leads to the production of a γ ray when the positron is annihilated on collision with an electron.

- Alpha decay involves the loss of a particle equivalent to a helium nucleus. Alpha (α) particles, being relatively large and positively charged, do not penetrate far in living tissue, but they do cause ionisation damage and this makes them generally unsuitable for tracer studies.
- Beta decay involves the loss or gain of an electron or its positive counterpart, the positron. There are three subtypes:
  (a) Negatron (β^{-}) emission: loss of an electron from the nucleus when a neutron transforms into a proton. This is the most important form of decay for radioactive tracers used in biology. Negatron-emitting isotopes of biological importance include ^{3}H, ^{14}C, ^{32}P and ^{35}S.
  (b) Positron (β^{+}) emission: loss of a positron when a proton transforms into a neutron. This only occurs when sufficient energy is available from the transition and may involve the production of gamma rays when the positron is later annihilated by collision with an electron.
  (c) Electron capture (EC): when a proton 'captures' an electron and transforms into a neutron. This may involve the production of X-rays as electrons 'shuffle' about in the atom (as with ^{125}I) and it frequently involves electron emission.
- Internal transition involves the emission of electromagnetic radiation in the form of gamma (γ) rays from a nucleus in a metastable state and always follows initial alpha or beta decay. Emission of gamma radiation leads to no further change in atomic number or mass.

Note from the above that more than one type of radiation may be emitted when a radioisotope decays. The main radioisotopes used in biology and their properties are listed in Table 39.2.

Table 39.1 Types of radioactivity and their properties

Radiation	Range of maximum energies (MeV*)	Penetration range in air (m)	Suitable shielding material
Alpha (α)	4–8	0.025–0.080	Unnecessary
Beta (β)	0.01–3	0.150–16	Plastic (e.g. Perspex)
Gamma (γ)	0.03–3	1.3–13†	Lead

*Note that 1 MeV = 1.6×10^{-13} J
†Distance at which radiation intensity is reduced to half

Table 39.2 Properties of some isotopes used commonly in the life sciences. Physical data obtained from Lide (2006)

Isotope	Emission(s)	Maximum energy (MeV)	Half-life	Main uses	Advantages	Disadvantages
^{3}H	β^-	0.01861	12.3 years	Suitable for labelling organic molecules in wide range of positions at high specific activity	Relatively safe	Low efficiency of detection, high isotope effect, high rate of exchange with environment
^{14}C	β^-	0.15648	5715 years	Suitable for labelling organic molecules in a wide range of positions	Relatively safe, low rate of exchange with environment	Low specific activity
^{22}Na	β^+ (90%) + γ, EC	2.842 (β^+)	2.60 years	Transport studies	High specific activity	Hazardous
^{32}P	β^-	1.710	14.3 days	Labelling proteins and nucleotides (e.g. DNA)	High specific activity, ease of detection	Short half-life, hazardous
^{33}P	β^-	0.248	25 days	Labelling nucleotides and proteins	Safer than ^{32}P	Moderate half-life
^{35}S	β^-	0.167	87.2 days	Labelling proteins and nucleotides	Low isotope effect	Low specific activity
^{36}Cl	β^-, β^+, EC	0.709 (β^-) 1.142 (β^+, EC)	300000 years	Transport studies	Low isotope effect	Low specific activity, hazardous
^{125}I	EC + γ	0.178 (EC)	59.9 days	Labelling proteins and nucleotides	High specific activity	Hazardous
^{131}I	β^- + γ	0.971 (β^-)	8.04 days	Labelling proteins and nucleotides	High specific activity	Hazardous

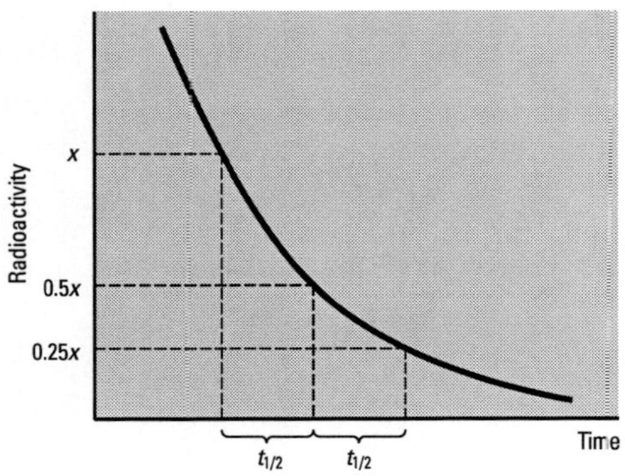

Fig. 39.1 Decay of a radioactive isotope with time. The time taken for the radioactivity to decline from x to $0.5x$ is the same as the time taken for the radioactivity to decline from $0.5x$ to $0.25x$, and so on. This time is the half-life ($t_{1/2}$) of the isotope.

Example For ^{35}S, with a half-life of 87.2 days (Table 39.2), the fraction of radioactivity remaining after 28 days would be worked out as follows: first, from Eqn [39.1] $x = (-0.693 \times 28) \div 87.2 = -0.222522936$, then using Eqn [39.1], $f = e^{-0.222522936} = 0.800496646$ (with appropriate rounding, 80.0% of original activity).

Each radioactive particle or ray carries energy, usually measured in electron volts (eV). The particles or rays emitted by a particular radioisotope exhibit a range of energies, termed an energy spectrum, characterised by the maximum energy of the radiation produced, E_{max} (Table 39.2). The energy spectrum of a particular radioisotope is relevant to the following:

- Safety: isotopes with the highest maximum energies will have the greatest penetrating power, requiring appropriate shielding (Table 39.1).
- Detection: different instruments vary in their ability to detect isotopes with different energies.
- Discrimination: some instruments can distinguish between isotopes, based on the energy spectrum of the radiation produced (p. 271).

The decay of an individual atom (a 'disintegration') occurs at random, but that of a population of atoms occurs in a predictable manner. The radioactivity decays exponentially, having a characteristic half-life ($t_{1/2}$). This is the time taken for the radioactivity to fall from a given value to half that value (Fig. 39.1). The $t_{1/2}$ values of different radioisotopes range from fractions of a second to more than 10^{19} years (see also Table 39.2). If $t_{1/2}$ is very short, as with ^{15}O ($t_{1/2} \approx 2\,\text{min}$), then it is generally impractical to use the isotope in experiments because you would need to account for the decay during the experiment and counting period.

To calculate the fraction (f) of the original radioactivity left after a particular time (t), use the following relationship:

$$f = e^x, \text{ where } x = -0.693t/t_{\frac{1}{2}} \qquad [39.1]$$

Note that the same units must be used for t and $t_{1/2}$ in this equation.

Measuring radioactivity

Table 39.3 Relationships between units of radioactivity. For abbreviations, see text

1 Bq = 1 d.p.s.	1 Sv = 100 rem
1 Bq = 60 d.p.m.	1 Gy = 100 rad
1 Bq = 27 pCi	1 Gy ≈ 100 roentgen
1 d.p.s. = 1 Bq	1 rem = 0.01 Sv
1 d.p.m. = 0.0167 Bq	1 rad = 0.01 Gy
1 Ci = 37 GBq	1 roentgen ≈ 0.01 Gy
1 mCi = 37 MBq	
1 μCi = 37 kBq	

The SI unit of radioactivity is the becquerel (Bq), equivalent to one disintegration per second (d.p.s.), but disintegrations per minute (d.p.m.) are also used. The curie (Ci) is a non-SI unit equivalent to the number of disintegrations produced by 1 g of radium (37 GBq). Table 39.3 shows the relationships between these units. In practice, most instruments are not able to detect all of the disintegrations from a particular sample, i.e. their efficiency is less than 100% and the rate of decay may be presented as counts min^{-1} (c.p.m.) or counts s^{-1} (c.p.s.). Most modern instruments correct for background radiation and inefficiencies in counting, converting count data to d.p.m. Alternatively, the results may be presented as the measured count rate, although this is only valid where the efficiency of counting does not vary greatly among samples.

Example If 0.4 ml of a ^{32}P-labelled DNA solution at a concentration of 50 μmol l^{-1} (amount = 0.4 × 50 ÷ 1000 = 0.02 μmol) gave a count of 2490 d.p.m. (= 41.5 Bq), using Eqn [39.2] this would correspond to a specific activity of 2490 ÷ 0.02 = 124500 dpm μmol^{-1} (or 2075 Bq μmol^{-1}).

KEY POINT The specific activity is a measure of the quantity of radioactivity present in a known amount of the substance:

$$\textbf{specific activity} = \frac{\textbf{radioactivity (Bq, Ci, d.p.m., etc.)}}{\textbf{amount of substance (mol, g, etc.)}} \qquad [39.2]$$

This is an important concept in practical work involving radioisotopes, since it allows interconversion of disintegrations (activity) and amount of substance (see Box 39.1).

Two SI units refer to doses of radioactivity and these are used when calculating exposure levels for a particular source. The sievert (Sv) is the amount of radioactivity giving a dose in humans equivalent to 1 gray (Gy) of X-rays: 1 Gy = an energy absorption of 1 J kg^{-1}. The dose received in most biological experiments is a negligible fraction of the maximum permitted exposure limit. Conversion factors from older units are given in Table 39.3.

The most important methods of measuring radioactivity for biological purposes are described below.

The Geiger–Müller (G–M) tube

This operates by detecting radiation when it ionises gas between a pair of electrodes across which a voltage has been applied. You should use a hand-held Geiger–Müller tube for routine checking for contamination (although it will not pick up ^{3}H activity).

The scintillation counter

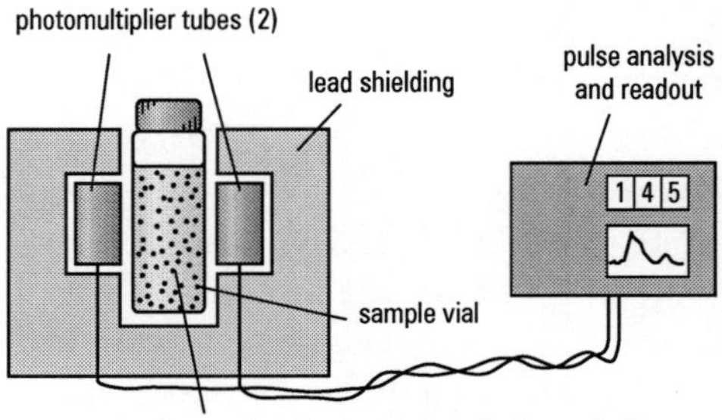

Fig. 39.2 Components of a scintillation counter. Note that in most modern instruments, all components are enclosed within a single cabinet.

This operates by detecting the scintillations (fluorescence 'flashes') produced when radiation interacts with certain chemicals called fluors (Fig. 39.2). In solid (or external) scintillation counters (often referred to as 'gamma counters') the radioactivity causes scintillations in a crystal of fluorescent material held close to the sample. This method is only suitable for radioisotopes producing penetrating radiation.

Liquid scintillation counters are mainly used for detecting beta decay and they are especially useful in biology. The sample is dispersed or dissolved in a suitable solvent containing the fluor(s) – the 'scintillation cocktail'. The radiation first interacts with the solvent, and the energy from this interaction is passed to the fluors which produce detectable light. The scintillations are measured by photomultiplier tubes which turn the light pulses into electronic pulses, the magnitude of which is directly related to

Box 39.1 How to determine the specific activity of an experimental solution

Suppose you need to make up a certain volume of an experimental solution, to contain a particular amount of radioactivity. For example, 50 ml of a mannitol solution at a concentration of 25 mmol l^{-1}, to contain 5 Bq μl^{-1} – using a manufacturer's stock solution of ^{14}C-labelled mannitol (specific activity = 0.1 Ci $mmol^{-1}$).

1. **Calculate the total amount of radioactivity in the experimental solution,** in this example 5 × 1000 (to convert μl to ml) × 50 (50 ml required) = 2.5 × 10^5 Bq (i.e. 250 kBq).

2. **Establish the volume of stock radioisotope solution required:** for example, a manufacturer's stock solution of ^{14}C-labelled mannitol contains 50 μCi of radioisotope in 1 ml of 90% v/v ethanol: water. Using Table 39.3, this is equivalent to an activity of 50 × 37 = 1850 kBq. So, the volume of solution required is 250/1850 of the stock volume, i.e. 0.135 1 ml (135 μl).

3. **Calculate the amount of non-radioactive substance required** as for any calculation involving concentration (see pp. 136, 145), e.g. 50 ml (0.05 l) of a 25 mmol l^{-1} (0.025 mol l^{-1}) mannitol (relative molecular mass 182.17) will contain 0.05 × 0.025 × 182.17 = 0.2277 g.

4. **Check the amount of radioactive isotope to be added.** In most cases, this represents a negligible amount of substance, e.g. in this instance, 250 kBq of stock solution at a specific activity of 14.8 × 10^6 kBq $mmol^{-1}$ (converted from 0.4 Ci $mmol^{-1}$ using Table 39.3) is equal to 250/14 800 000 = 16.89 nmol, equivalent to approximately 3 μg mannitol. This can be ignored in calculating the mannitol concentration of the experimental solution.

5. **Make up the experimental solution** by adding the appropriate amount of non-radioactive substance and the correct volume of stock solution.

6. **Measure the radioactivity in a known volume of the experimental solution.** If you are using an instrument with automatic correction to Bq, your sample should contain the predicted amount of radioactivity, e.g. an accurately dispensed volume of 100 μl of the mannitol solution should give a corrected count of 100 × 5 = 500 Bq (or 500 × 60 = 30 000 d.p.m.).

7. **Note the specific activity of the experimental solution:** in this case, 100 μl (1 × 10^{-4} l) of the mannitol solution at a concentration of 0.025 mol l^{-1} will contain 25 × 10^{-7} mol (2.5 μmol) mannitol. Dividing the radioactivity in this volume (30 000 d.p.m.) by the amount of substance (Eqn [39.2]) gives a specific activity of 30 000/2.5 = 12 000 d.p.m. μmol^{-1}, or 12 d.p.m. $nmol^{-1}$. This value can be used:

   (a) To assess the accuracy of your protocol for preparing the experimental solution: if the measured activity is substantially different from the predicted value, you may have made an error in making up the solution.

   (b) To determine the counting efficiency of an instrument; by comparing the measured count rate with the value predicted by your calculations.

   (c) To interconvert activity and amount of substance: the most important practical application of specific activity is the conversion of experimental data from counts (activity) into amounts of substance. This is only possible where the substance has not been metabolised or otherwise converted into another form; e.g. a tissue sample incubated in the experimental solution described above with a measured activity of 245 d.p.m. can be converted to nmol mannitol by dividing by the specific activity, expressed in the correct form. Thus 245/12 = 20.417 nmol mannitol.

***Correcting for quenching** – find out how your instrument corrects for quenching and check the quench indication parameter (QIP) on the printout, which measures the extent of quenching of each sample. Large differences in the QIP would indicate that quenching is variable among samples and might give you cause for concern.*

the energy of the original radioactive event. The spectrum of electronic pulses is thus related to the energy spectrum of the radioisotope.

Modern liquid scintillation counters use a series of electronic 'windows' to split the pulse spectrum into two or three components. This may allow more than one isotope to be detected in a single sample, provided their energy spectra are sufficiently different (Fig. 39.3). A complication of this approach is that the energy spectrum can be altered by pigments and chemicals in the sample, which absorb scintillations or interfere with the transfer of energy to the fluor; this is known as quenching (Fig. 39.3).

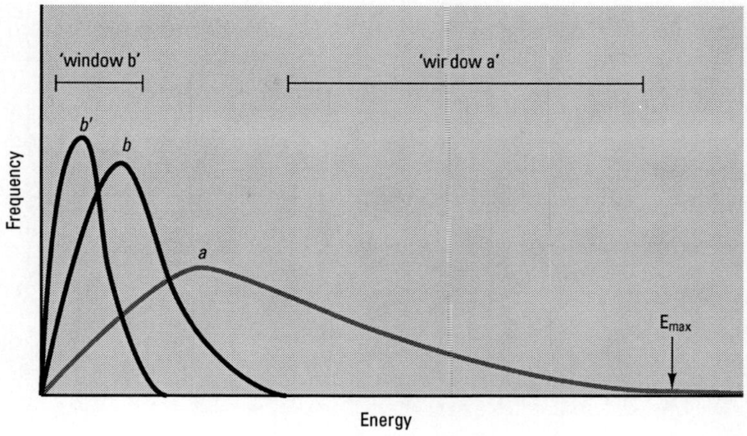

Fig. 39.3 Energy spectra for three radioactive samples, detected using a scintillation counter. Sample *a* is a high energy β-emitter while *b* contains a low energy β-emitter, giving a lower spectral range. Sample *b'* contains the same amount of low energy β-emitter, but with quenching, shifting the spectral distribution to a lower energy band. The counter can be set up to record disintegrations within a selected range (a 'window'). Here, 'window a' could be used to count isotope *a* while 'window b' could give a value of isotope *b*, by applying a correction for the counts due to isotope *a*, based on the results from 'window a'. Dual counting allows experiments to be carried out using two isotopes (double labelling).

Most instruments have computer-operated quench correction facilities (based on measurements of standards of known activity and energy spectrum) which correct for such changes in counting efficiency.

Many liquid scintillation counters treat the first sample as a 'background', subtracting whatever value is obtained from the subsequent measurements as part of the procedure for converting to d.p.m. If not, you will need to subtract the background count from all other samples. Make sure that you use an appropriate background sample, identical in all respects to your radioactive sample but with no added radioisotope, in the correct position within the machine. Check that the background reading is reasonable (15–30 c.p.m. is a reasonable background for most radioisotopes). Tips for preparing samples for liquid scintillation counting are given in Box 39.2.

Liquid scintillation counting of high energy β-emitters *– β-particles with energies greater than 1 MeV can be counted in water (Čerencov radiation), with no requirement for additional fluors (e.g. ^{32}P).*

Gamma-ray (γ-ray) spectrometry

This is a method by which a mixture of γ-ray-emitting radionuclides can be resolved quantitatively by pulse-height analysis. It is based on the fact that pulse heights (voltages) produced by a photomultiplier tube are proportional to the amounts of γ-ray energy arriving at the scintillant or a lithium-drifted germanium detector. The lithium-drifted germanium detector, which is abbreviated to Ge(Li) – pronounced 'jelly' – provides high resolution (narrow peaks), essential in the analysis of complex mixtures, such as in biological material.

Autoradiography

This is a method where photographic film is exposed to the isotope. It is used mainly to locate radioactive tracers in thin sections of an organism or on chromatography papers and gels, but quantitative work is possible. The radiation interacts with the film in a similar way to light, silver grains being formed in the developed film where the particles or rays have passed through. The radiation must have enough energy to penetrate into the film, but if it has too much energy the grain formation may be too distant from the point where the isotope was located to identify precisely the point of origin (e.g. high energy β-emitters). Autoradiography is a relatively specialised method and individual lab protocols should be followed for particular isotopes/applications.

Box 39.2 Tips for preparing samples for liquid scintillation counting

Modern scintillation counters are very simple to operate; problems are more likely to be due to inadequate sample preparation than to incorrect operation of the machine. Common pitfalls are the following:

- **Incomplete dispersal of the radioactive compound in the scintillation cocktail.** This may lead to underestimation of the true amount of radioactivity present:
  (a) Water-based samples may not mix with the scintillation cocktail – change to an emulsifier-based cocktail. Take care to observe the recommended limits, upper and lower, for amounts of water to be added or the cocktail may not emulsify properly.
  (b) Solid specimens may absorb disintegrations or scintillations: extract radiochemicals using an intermediate solvent like ethanol (ideally within the scintillation vial) and then add the cocktail. Tissue-solubilising compounds such as Soluene are effective, particularly for animal material, but extremely toxic, so the manufacturer's instructions must be followed closely. Radioactive compounds on slices of agarose or polyacrylamide gels may be extracted using a product such as Protosol. Agarose gels can be dissolved in a small volume of boiling water.
  (c) Particulate samples may sediment to the bottom of the scintillation vial – suspend them by forming a gel. This can be done with certain emulsifier-based cocktails by adding a specific amount of water.
- **Chemiluminescence.** This is where a chemical reacts with the fluors in the scintillation cocktail causing spurious scintillations, a particular risk with solutions containing strong bases or oxidising agents. Symptoms include very high initial counts which decrease through time. Possible remedies are:
  (a) Leave the vials at room temperature for a time before counting. Check with a suitable blank that counts have dropped to an acceptable level.
  (b) Neutralise basic samples with acid (e.g. acetic acid or HCl).
  (c) Use a scintillation cocktail that resists chemiluminescence, such as Hiconicfluor.
  (d) Raise the energy of the lower counts detected to about 8 keV – most chemiluminescence pulses are weak (0–7 keV). This approach is not suitable for weak emitters, e.g. ^{3}H.

Biological applications for radioactive isotopes

The main advantages of using radioactive isotopes in biological experiments are:

- Radioactivity is readily detected. Methods of detection are sufficiently sensitive to measure extremely small amounts of radioactive substances.
- Studies can be carried out on intact, living organisms. If care is taken, minimal disruption of normal conditions will occur when radiolabelled compounds are introduced.
- Protocols are simple compared to equivalent methods for chemical analysis.

The main disadvantages are:

- The 'isotope effect'. Molecules containing different isotopes of the same atom may react at slightly different rates and behave in slightly different ways from the natural isotope. The isotope effect is more extreme the smaller the atom, and is most important for ^{3}H-labelled compounds of low molecular mass.
- The possibility of mistaken identity. The presence of radioactivity does not tell you anything about the compound in which the radioactivity is present: it could be different from the one in which it was applied, due to metabolism or spontaneous breakdown of a ^{14}C-containing organic compound.

***Investigating the metabolic fate of radiolabelled compounds** – you may need to separate individual metabolites before counting, e.g. using chromatography (Chapter 44), or electrophoresis (Chapter 46).*

The main types of experiments are:

- Investigations of metabolic pathways: a radioactively labelled substrate is added (often to an *in vitro* experimental 'system' rather than a whole organism) and samples taken at different time intervals. By identifying the labelled compounds and plotting their appearance through time, an indication of the pathway of metabolism can be obtained.
- Translocation studies: radioisotopes are used to follow the fate of molecules within an organism. Uptake and translocation rates can be determined with relative ease.
- Ecological studies: radioisotopic tracers provide a convenient method for determining food web interrelationships and for investigating behaviour patterns, while environmental monitoring may involve following the 'spectral signature' of isotopes deliberately or accidentally released.
- Radio-dating: the age of plant or mineral samples can be determined by measuring the amount of a radioisotope in the sample. The age of the specimen can be found using the half-life by assuming how much was originally incorporated.
- Mutagenesis and sterilisation: radioactive sources can be used to induce mutations, particularly in micro-organisms. Gamma emitters of high energy will kill microbes and are used to sterilise equipment such as disposable Petri dishes.
- Assays: radioisotopes are used in several quantitative detection methods of value to biologists. Radioimmunoassay is described on p. 260. Isotope dilution analysis works on the assumption that introduced radiolabelled molecules will equilibrate with unlabelled molecules present in the specimen. The amount of substance initially present can be worked out from the change in specific activity of the radioisotope when it is diluted by the 'cold' material. A method is required whereby the substance can be purified from the sample and sufficient substance must be present for its mass to be measured accurately.

Example Carbon dating – living organisms have essentially the same ratio of ^{14}C to ^{12}C as the atmosphere; however, when an organism dies, its $^{14}C/^{12}C$ falls because the radioactive ^{14}C isotope decays. Since we know the half-life of ^{14}C (5715 years), a sample's $^{14}C/^{12}C$ ratio will allow us to estimate its age; e.g. if the ratio were exactly 1/8 of that in the atmosphere, the sample is three half-lives old and was formed 17 145 years before present. Such estimates carry an error of the order of 10% and are unreliable for samples older than 50 000 years, for which longer-lived isotopes can be used.

Billington *et al.* (1992) give further details and practical advice on using radioisotopes in biological experiments.

Working practices when using radioactive isotopes

By law, undergraduate work with radioactive isotopes must be very closely supervised. In practical classes, the protocols will be clearly outlined, but in project work you may have the opportunity to plan and carry out your own experiments, albeit under supervision. Some of the factors that you should take into account, based on the assumption that your department and laboratory are registered for radioisotope use, are discussed below:

Registering for radioisotope work *– in the UK, institutions must be registered for work with specific radioisotopes under the* Radioactive Substances Act (1993).

1. Must you use radioactivity? If not, it may be legal requirement that you use an alternative method.
2. Have you registered for radioactive work? Normal practice is for all users to register with a local Radiation Protection Supervisor. Details of the project may have to be approved by the appropriate administrator(s). You may have to have a short medical examination before you can start work.

Supervision of work with radioisotopes *– in the UK, the* Ionising Radiations Act (1985) *provides details of local arrangements for the supervision of radioisotope work.*

3. What labelled compound will you use? Radioactive isotopes must be ordered well in advance through your department's Radiation Protection Supervisor. Aspects that need to be considered include:
   (a) The radionuclide. With many organic compounds this will be confined to ^{3}H and ^{14}C (but see Table 39.2). The involvement of a significant 'isotope effect' may influence this decision (see p. 272).
   (b) The labelling position. This may be a crucial part of a metabolic study. Specifically labelled compounds are normally more expensive than those that are uniformly ('generally') labelled.
   (c) The specific activity. The upper limit for this is defined by the isotope's half-life, but below this, the higher the specific activity, the more expensive the compound.
4. Are suitable facilities available? You'll need a suitable work area, preferably out of the way of general lab traffic and within a fume cupboard for those cases where volatile radioactive substances are used or may be produced.

SAFETY NOTE Planning radioisotope work – each new experiment should be planned carefully and experimental protocols laid down in advance so you work as safely as possible and do not waste expensive radioactively labelled compounds.

In conjunction with your supervisor, decide whether your method of application will introduce enough radioactivity into the system, how you will account for any loss of radioactivity during recovery of the isotope and whether there will be enough activity to count at the end. You should be able to predict approximately the amount of radioactivity in your samples, based on the specific activity of the isotope used, the expected rate of uptake/exchange and the amount of sample to be counted. Use the isotope's specific activity to estimate whether the non-radioactive ('cold') compound introduced with the radiolabelled ('hot') compound may lead to excessive concentrations being administered. Advice for handling data is given in Box 39.1.

***Carrying out a 'dry run'** – consider doing this before working with radioactive compounds, perhaps using a dye to show the movement or dilution of introduced liquids, as this will lessen the risks of accident and improve your technique.*

Safety and procedural aspects

Make sure the bench surface is one that can be easily decontaminated by washing (e.g. Formica) and always use a disposable surfacing material such as Benchkote. It is good practice to carry out as many operations as possible within a Benchkote-lined plastic tray so that any spillages are contained. You will need a lab coat to be used exclusively for work with radioactivity, safety spectacles and a supply of thin latex or vinyl disposable gloves. Suitable vessels for liquid waste disposal will be required and special plastic bags for solids – make sure you know beforehand the disposal procedures for liquid and solid wastes. Wash your hands after handling a vessel containing a radioactive solution and again before removing your gloves. Gloves should be placed in the appropriate disposal bag as soon as your experimental procedures are complete.

SAFETY NOTE Using Benchkote – the correct way to use Benchkote and similar products is with the waxed surface down (to protect the bench or tray surface) and the absorbent surface up (to absorb any spillage). Write the date in the corner when you put down a new piece. Monitor using a G–M tube and replace regularly under normal circumstances. If you are aware of spillage, replace immediately and dispose of correctly.

It is important to comply with the following guidelines:

- Read and obey the local rules for safe usage of radiochemicals.
- Maximise the distance between you and the source as much as possible.
- Minimise the duration of exposure.
- Wear protective clothing (properly fastened lab coat, safety glasses, gloves) at all times.
- Use appropriate shielding at all times (Table 39.1).
- Monitor your working area for contamination frequently.
- Mark all glassware, trays, bench work areas, etc., with tape incorporating the international symbol for radioactivity (Fig. 39.4).

Fig. 39.4 Tape showing the international symbol for radioactivity.

- Keep adequate records of what you have done with a radioisotope – the stock remaining and that disposed of in waste form must agree.
- Store radiolabelled compounds appropriately and return them to storage areas immediately after use.
- Dispose of waste promptly and with due regard for local rules.
- Make the necessary reports about waste disposal, etc., to your departmental Radiation Protection Supervisor.
- Clear up after you have finished each experiment.
- Wash thoroughly after using radioactivity.
- Monitor the work area and your body when finished.

Text references and sources for further study

Billington, D., Jayson, G.G. and Maltby, P.J. (1992) *Radioisotopes*. Bios, Oxford.

L'Annunziata, M.F. (ed.) (2003) *Handbook of Radioactivity Analysis*, 2nd edn. Academic Press, San Diego.

Lide, D.R. (ed.) (2006) *CRC Handbook of Chemistry and Physics*, 87th edn. CRC Press, Boca Raton.

Slater, J. (ed.) (2002) *Radioisotopes in Biology: a Practical Approach,* 2nd edn. Oxford University Press: Oxford.

Study exercises

39.1 Carry out a half-life calculation. A rat dropping found in a pyramid in Egypt had a $^{14}C:^{12}C$ ratio that was 57.25% of a modern-day standard. Use this value to estimate the approximate date when the rat visited the pyramid, to the nearest century.

39.2 Practise radioactivity interconversions. Express the following values in the alternative units indicated, with appropriate prefixes as necessary. Answer to three significant figures.

(a) 72 000 d.p.m. as Bq;
(b) 20 μCi as d.p.m.;
(c) 44 400 Bq as μCi;
(d) 6.3×10^5 d.p.m. mol^{-1} as Bq g^{-1}, for a compound with a relative molecular mass of 350;
(e) 3108 d.p.m. as pmol, for a sample of a standard where the specific activity is stated as 50 Ci mol^{-1}.

39.3 Use the concept of specific activity in calculations. A researcher wishes to estimate the rate of uptake of the sugar galactose by carrot cells in a suspension culture. She prepares 250 ml of the cell culture medium containing 10^7 cells per ml and unlabelled galactose at a concentration of 5 mmol l^{-1}. She then 'spikes' this with 5 μl (regard this as an insignificant volume) of radioactive standard containing 55 MBq of ^{14}C-labelled galactose (regard as an insignificant concentration). Answer to two significant figures.

(a) Calculate the specific activity of the galactose in the culture solution in Bq mol^{-1}.
(b) If the total cell sample takes up 79.2×10^5 Bq in a 2-hour period, calculate the galactose uptake rate in mol s^{-1} $cell^{-1}$.

40 Light measurement

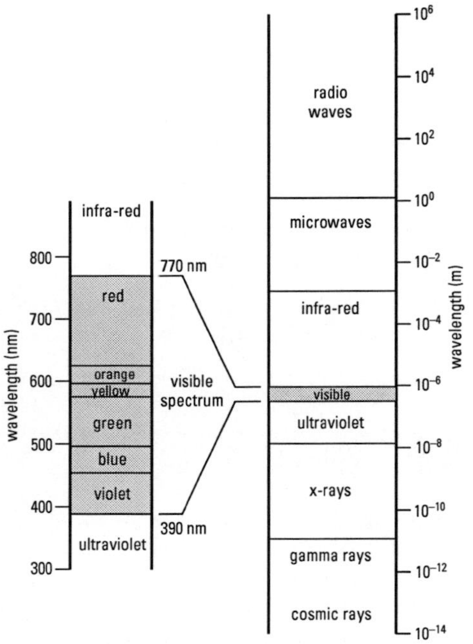

Fig. 40.1 The electromagnetic spectrum, *Methods for Physical and Chemical Analysis of Fresh Waters*, 2nd edn. International Biological Programme, No. 8, Blackwell Scientific (Golterman, H.L., Clymo, R.S. and Ohnstad, M.A.M., 1978).

Table 40.1 Wavelength ranges for UV, visible and IR radiation, compiled from several sources

Type	Wavelength range (nm)
UVC	100–290*
UVB	290–320
UVA	320–400
Violet	390–450
Blue	450–500
Green	500–560
Yellow	560–600
Orange	600–620
Red	620–770
IR	>770

* Note that UVC is absorbed by the Earth's ozone layer.

The measurement of light is directly relevant to several aspects of biology, including: photosynthesis, photomorphogenesis and photoperiodism in plants; perception and thermoregulation in animals; and aspects of aquatic biology.

The nature of light

Light is most strictly defined as that part of the spectrum of electromagnetic radiation detected by the human eye. However, the term is also applied to radiation just outside that visible range (e.g. UV and infra-red 'light'). Electromagnetic radiation is emitted by the Sun and by other sources (e.g. an incandescent lamp) and the electromagnetic spectrum is a broad band of radiation, ranging from cosmic rays to radio waves (Fig. 40.1). Most biological experiments involve measurements within the UV, visible and infra-red (IR) regions (generally, within the wavelength range 200–1000 nm; see Table 40.1).

Radiation has the characteristics of a particle and of a vibrating wave, travelling in discrete particulate units, or 'packets', termed photons. A quantum is the amount of energy contained within a single photon (it is important not to confuse these two terms, although they are sometimes used interchangeably in the literature). In some circumstances, it is appropriate to measure light in terms of the number of photons, usually expressed directly in moles (6.02×10^{23} photons = 1 mol); older textbooks may use the redundant term Einstein as the unit of measurement, where 1 Einstein = 1 mol photons. Alternatively, the energy content (power) may be measured (e.g. in $W\,m^{-2}$). Radiation also behaves as a vibrating electrical and magnetic field moving in a particular direction, with the magnetic and electrical components vibrating perpendicular to one another and perpendicular to the direction of travel. The wave nature of radiation gives rise to the concepts of wavelength (λ, usually measured in nm), frequency (v, measured in s^{-1}), speed (c, the speed of electromagnetic radiation, which is $3 \times 10^8\ m\ s^{-1}$ in a vacuum), and direction. In other words, radiation is a vector quantity, where

$$c = \lambda v \qquad [40.1]$$

Photometric and radiometric measurements

Photometric measurements

These are based on the energy perceived by a 'standard' human eye, with maximum sensitivity in the yellow–green region, around 555 nm (Fig. 40.2). The unit of measurement is the candela, a base unit in the SI system, defined in terms of the visual appearance of a specific quantity of platinum at its freezing point. Derived units are used for the luminous flow (lumen) and luminous flow per unit area (lux). These units were once used in photobiology and you may come across them in older literature. However, it is now recognised that such measurements are of little direct relevance to biologists, including even those who may wish to study visual responses, because they are not based on fundamental physical principles.

KEY POINT The human eye rapidly compensates for changes in light climate by varying the size of the pupil and is a very poor source of information on light quantity. It is important to make light measurements using reliable instruments and to express these measurements in appropriate units.

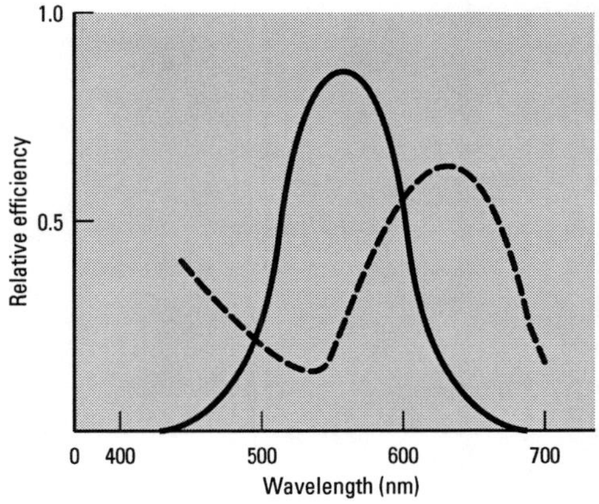

Fig. 40.2 Relative efficiency of vision (solid line) and photosynthesis (dotted line) as a function of wavelength.

Radiometric measurements

The radiometric system is based on physical properties of the electromagnetic radiation itself, expressed either as the number of photons, or their energy content. The following terms are used (units of measurement in parentheses):

- Photon flux (mol photons s^{-1}) is the number of photons arriving at an object within the specified time interval.
- Photon exposure (mol photons m^{-2}) is the total number of photons received by an object, usually expressed per unit surface area.
- Photon flux density (mol photons $m^{-2}\,s^{-1}$), or PFD, is the most commonly used term to describe the number of photons arriving at a particular surface, expressed per unit surface area and per unit time interval.
- Photosynthetically active radiation, or PAR, is radiation within the waveband 400–700 nm, since the photosynthetic pigments (chlorophylls, carotenoids, etc.) show maximum absorption within this band.
- Photosynthetic photon flux density (mol photons $m^{-2}\,s^{-1}$), or PPFD, is the number of photons within the waveband 400–700 nm arriving at a particular surface, expressed per unit surface area and per unit time interval. Often this term is used interchangeably with PFD.
- Irradiance ($J\,m^{-2}\,s^{-1} = W\,m^{-2}$) is the amount of energy arriving at a surface, expressed per unit surface area and per unit time interval.
- Photosynthetic irradiance ($W\,m^{-2}$), or PI, is the energy of radiation within the waveband 400–700 nm arriving at a surface, expressed per unit surface area and per unit time interval.

Choice of measurement scale

Photon flux density

This is often the most appropriate unit of measurement for biological systems where individual photons are involved in the underlying process, e.g. in photosynthetic studies, where PPFD is measured, since each photochemical reaction involves the absorption of a single photon by a pigment molecule. Most modern light-measuring instruments (radiometers) can measure this quantity, giving a reading in μmol photons $m^{-2}\,s^{-1}$.

Irradiance

This is appropriate if you are interested in the energy content of the light, e.g. if you are studying energy balance or thermal effects. Many radiometers measure photosynthetic irradiance within the waveband 400–700 nm, giving a reading in $W\,m^{-2}$. It is possible to make an approximate conversion between PPFD and PI measurements, providing the spectral properties of the light source are known (see Table 40.2).

Table 40.2 Approximate conversion factors for a photosynthetic irradiance (PI) of 1 $W\,m^{-2}$ to photosynthetic photon flux density (PPFD)

Source	PPFD (μmol photons $m^{-2}\,s^{-1}$)
Sunlight	4.6
'Cool white' fluorescent tube	4.6
Osram 'daylight' fluorescent tube	4.6
Quartz-iodine lamp	5.0
Tungsten bulb	5.0

(*Source:* Lüning, 1981. Reproduced by kind permission of Blackwell Publishing Ltd.)

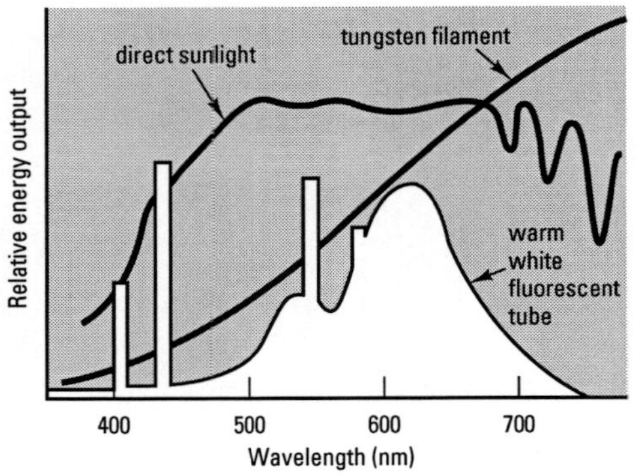

Fig. 40.3 Spectral distribution of energy output from various sources. (Adapted from Golterman *et al.*, 1978.

Spectral distribution

This can be determined using a spectroradiometer, e.g. to compare different light sources (Fig. 40.3). A spectroradiometer measures irradiance or photon flux density in specific wavebands. This instrument consists of a monochromator (p. 281) to allow separate narrow wavebands (5–25 nm bandwidth) to be measured by a detector; some instruments provide a plot of the spectral characteristics of the source.

Using a radiometer ('light meter')

The main components of a radiometer are:

- Receiver: either flat-plate, hemispherical or spherical, depending upon requirements. Most incorporate a protective diffuser, to reduce reflection.
- Detector: either thermoelectric or photoelectric. Some photoelectric detectors suffer from fatigue, with a decreasing response on prolonged exposure: check the manufacturer's handbook for exposure times.
- Processor and readout device to convert the output from the detector into a visible reading, in analogue or digital form.

Box 40.1 gives practical details of the steps involved in using a radiometer. In your write-up, give full details of how the measurement was made, e.g. the type of light source, instrument used, where the sensor was placed, whether an average was calculated, etc.

Box 40.1 Measuring photon flux density or irradiance using a battery-powered radiometer

1. **Check the battery.** Most instruments have a setting that gives a direct readout of battery voltage. Recharge if necessary before use.
2. **Select the appropriate type of measurement** (e.g. photon flux density or irradiance over the PAR waveband, or an alternative range): the simpler instruments have a selection dial for this purpose.
3. **Place the sensor in the correct location and position** for the measurement: it may be appropriate to make several measurements at different positions and take an average.
4. **Choose the most appropriate scale** for the readout device: for needle-type meters, the choice of maximum reading is usually selected by a dial, within the range 0.3 to 30 000. Start at a high range and work down until the reading is on the scale. Your final scale should be chosen to provide the most accurate reading, e.g. a reading of 15 μmol photons m^{-2} s^{-1} should be made using the 0–30 scale, rather than a higher range.
5. **Read the value from the meter.** For needle-type instruments there may be two scales, the upper one marked from 0 to 10 and the lower one from 0 to 3: make sure you use the correct one, e.g. a half-scale deflection on the 0–30 scale is 1.5.
6. **Check that the answer is realistic,** e.g. full sunlight has a PPFD of up to 2 000 μmol photons m^{-2} s^{-1} (PI ≡ 400 W m^{-2}), though the value will be far lower on a dull or cloudy day, while the PPFD at a distance of 1 m from a mercury lamp is around 150 μmol photons m^{-2} s^{-1}, and 50 μmol photons m^{-2} s^{-1} at the same distance from a fluorescent lamp.

Text references

Golterman, H.L., Clymo, R.S. and Ohnstad, M.A.M. (1978) *Methods for Physical and Chemical Analysis of Fresh Waters*. Blackwell Scientific Publications, Oxford.

Lüning, K.J. (1981) 'Light', in C.S. Lobban and M.J. Wynne (eds), *The Biology of Seaweeds*, pp. 326–55, Blackwell Scientific, Oxford.

Sources for further study

Anon. *International Light Homepage*.
Available: http://www.intl-lighttech.com
Last accessed: 01/04/07.
[Information on products including several types of detector and meter. Also includes access to a light measurement handbook in .pdf format]

Boyd, R.W. (1983) *Radiometry and the Detection of Optical Radiation*. Wiley, New York.

National Physics Laboratory *Beginner's Guide to Measurement – Light*.
Available: http://www.npl.co. uk/publications/light/
Last accessed: 01/04/07.
[Gives some background on the development and use of the candela, the SI unit of luminous intensity, and its relationship with human vision]

Study exercises

40.1 Carry out a calculation involving interconversion of photosynthetic photon flux and photosynthetic irradiance. For sunlight with a photosynthetic photon flux density (PPFD) of 1610 mol m^{-2}, what is the total amount of photosynthetic energy (in joules) falling on a leaf of area 45 cm^2 over a 30-minute experimental period?

40.2 Compare the photosynthetic photon flux densities (PPFD) of different locations. The following values represent typical light levels for a range of different situations, expressed either as photosynthetic irradiance (PI) or PPFD:

(a) outside, on a sunny day: PI = 300 W m^{-2};
(b) in a room lit by 'cool white' fluorescent tubes: PPFD = 6.50 nmol photons cm^{-2} s^{-1};
(c) under the leaf canopy in a forest: PPFD = 275 µmol photons m^{-2} s^{-1};
(d) in a growth cabinet lit by a bank of 'daylight' fluorescent lights: PI = 35.2 W m^{-2};
(e) in a room lit by a tungsten bulb: PI = 1.15 mW cm^{-2}.

Convert all to PPFD in the same units (expressed to three significant figures) and then rank the locations in order of decreasing PPFD.

40.3 Research the various types of commercial radiometers (light meters) available on the Web. Make a list of the features that you might want to find out about and see how much information you can find.

41 Basic spectroscopy

Using spectroscopy *– this technique is valuable for:*

- *tentatively identifying compounds, by determining their absorption or emission spectra;*
- *quantifying substances, either singly or in the presence of other compounds, by measuring the signal strength at an appropriate wavelength;*
- *determining molecular structure;*
- *following reactions, by measuring the disappearance of a substance, or the appearance of a product as a function of time.*

The absorption and emission of electromagnetic radiation of specific energy (wavelength) are characteristic features of many molecules, involving the movement of electrons between different energy states, in accordance with the laws of quantum mechanics and spectroscopic techniques are used to measure and interpret such interactions between molecules and radiation. Electrons in atoms or molecules are distributed at various energy levels, but are mainly at the lowest energy level, usually termed the ground state. When exposed to energy (e.g. from electromagnetic radiation), electrons may be excited to higher energy levels (excited states), with the associated absorption of energy at specific wavelengths giving rise to an absorption spectrum. One quantum of energy is absorbed for a single electron transition from the ground state to an excited state. On the other hand, when an electron returns to its ground state, one quantum of energy is released; this may be dissipated to the surrounding molecules (as heat) or may give rise to an emission spectrum. The energy change (ΔE) for an electron moving between two energy states, E_1 and E_2, is given by the equation:

$$\Delta E = E_1 - E_2 = h\nu \qquad [41.1]$$

where h is the Planck constant (p. 185) and ν is the frequency of the electromagnetic radiation expressed in Hz or s^{-1}). Frequency is related to wavelength (λ, usually expressed in nm) and the speed of electromagnetic radiation, c (p. 185) by the expression:

$$\nu = c \div \lambda \qquad [41.2]$$

UV/visible spectrophotometry

This is a widely used technique for measuring the absorption of radiation in the visible and UV regions of the spectrum. A spectrophotometer is an instrument designed to allow precise measurement at a particular wavelength, while a colorimeter is a simpler instrument, using filters to measure broader wavebands, e.g. light in the green, red or blue regions of the visible spectrum; Jones *et al.* (2007) gives details of how to use a colorimeter.

Definition

Absorbance (*A*) – this is given by:

$A = \log_{10}(I_0/I)$.

Usually shown as A_x where '*x*' is the wavelength, in nanometres. As an example, for incident light (I_0) = 1.00 and emergent light (I) = 0.16 (expressed in relative terms), $A = \log_{10}(1.00 \div 0.16) = \log_{10} 6.25 = 0.796$ (to three significant figures).

Principles of light absorption

Two fundamental principles govern the absorption of light passing through a solution:

- The absorption of light is exponentially related to the number of molecules of the absorbing solute that are encountered, i.e. the solute concentration $[C]$.
- The absorption of light is exponentially related to the length of the light path through the absorbing solution, l.

These two principles are combined in the Beer–Lambert relationship (sometimes referred to simply as 'Beer's Law'), which is usually expressed in terms of absorbance (A) – the logarithm of the ratio of the incident light (I_0) to the emergent light (I):

$$A = \varepsilon l[C] \qquad [41.3]$$

where A is absorbance, ε is a constant for the absorbing substance and the wavelength, termed the absorption coefficient or absorptivity, and $[C]$ is expressed either as mol l^{-1} or g l^{-1} (see p. 135) and l is given in cm.

***KEY POINT* The Beer–Lambert relationship, expressed in mathematical form in Eqn [41.3], states that there is a direct linear relationship between the concentration of a substance in a solution, [C], and the absorbance of that solution, A.**

Definition

Transmittance (T) – this is usually expressed as a percentage, at a particular wavelength, T_x, where

$$T_x = (I/I_0) \times 100\,(\%).$$

As an example, for incident light (I_0) = 1.00 and emergent light (I) = 0.275 (expressed in relative terms) then transmittance, $T = (0.275 \div 1.00) \times 100 = 27.5\%$.

This relationship is extremely useful, since most spectrophotometers are constructed to give a direct measurement of absorbance (A), sometimes also termed extinction (E), of a solution (older texts may use the outdated term optical density, OD). Note that for substances obeying the Beer–Lambert relationship, A is linearly related to $[C]$. Absorbance at a particular wavelength is often shown as a subscript, e.g. A_{550} represents the absorbance at 550 nm. The proportion of light passing through the solution is known as the transmittance (T), and is calculated as the ratio of the emergent and incident light intensities.

Some instruments have two scales:

- an exponential scale from zero to infinity, measuring absorbance;
- a linear scale from 0 to 100, measuring (per cent) transmittance.

For most practical purposes, the Beer–Lambert relationship applies and you should use the absorbance scale.

UV/visible spectrophotometer

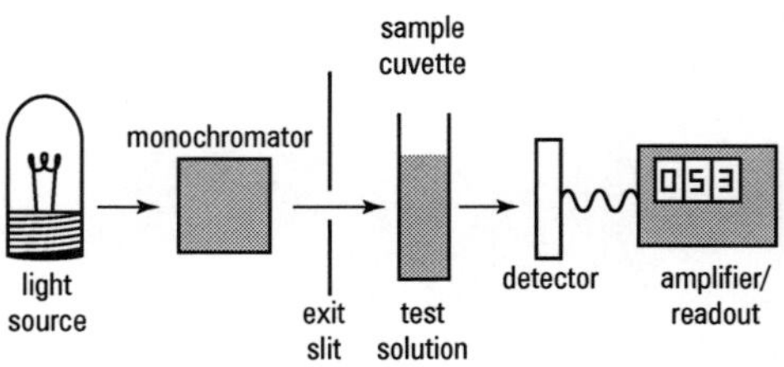

Fig. 41.1 Components of a UV/visible spectrophotometer.

The principal components of a UV/visible spectrophotometer are shown in Figure 41.1. High intensity tungsten bulbs are used as the light source in basic instruments, capable of operating in the visible region (i.e. 400–700 nm). Deuterium lamps are used for UV spectrophotometry (200–400 nm); these lamps are fitted with quartz envelopes, since glass does not transmit UV radiation.

A major improvement over the simple colorimeter is the use of a diffraction grating to produce a parallel beam of monochromatic light from the (polychromatic) light source. In practice the light emerging from such a monochromator is not of a single wavelength, but is a narrow band of wavelengths. This bandwidth is an important characteristic, since it determines the wavelengths used in absorption measurements – the bandwidth of basic spectrophotometers is around 5–10 nm while research instruments have bandwidths of less than 1 nm.

SAFETY NOTE *Working with spectrophotometers – take care not to spill water into the inside of the instrument, due to the risk of electric shock during use (switch off at mains and seek assistance if this should happen).*

Bandwidth is affected by the width of the exit slit (the slit width), since the bandwidth will be reduced by decreasing the slit width. To obtain accurate data at a particular wavelength setting, the narrowest possible slit width should be used. However, decreasing the slit width also reduces the amount of light reaching the detector, decreasing the signal-to-noise ratio. The extent to which the slit width can be reduced depends upon the sensitivity and stability of the detection/amplification system and the presence of stray light.

Using plastic disposable cuvettes – these are adequate for work in the near-UV region, e.g. for enzyme studies using nicotinamide coenzymes, at 340 nm (p. 282), as well as the visible range.

Most UV/visible spectrophotometers are designed to take cuvettes with an optical path length of 10 mm. Disposable plastic cuvettes are suitable for routine work in the visible range using aqueous and alcohol-based solvents, while glass cuvettes are useful for other organic solvents. Glass cuvettes are manufactured to more exacting standards, so use optically matched glass cuvettes for accurate work, especially at low absorbances

Examples The molar absorptivity of NADH is 6.22×10^3 l mol^{-1} cm^{-1} at 340 nm. For a test solution giving an absorbance of 0.21 in a cuvette with a light path of 5 mm, using Eqn [41.3] this is equal to a concentration of:

$0.21 = 6.22 \times 10^3 \times 0.5 \times [C]$
$[C] = 0.000\,067\,5$ mol l^{-1}
(or 67.5 μmol l^{-1}).

The specific absorptivity (10 g l^{-1}) of double-stranded DNA is 200 at 260 nm, therefore a solution containing 1 g l^{-1} will have an absorbance of 200/10 = 20. For a DNA solution, giving an absorbance of 0.35 in a cuvette with a light path of 1.0 cm, using Eqn [41.3] this is equal to a concentration of:

$0.35 = 20 \times 1.0 [C]$
$[C] = 0.0175$ g l^{-1}
(equivalent to 17.5 μg ml^{-1}).

Chlorophylls *a* and *b* of vascular plants and green algae can be extracted in 90% v/v acetone/water, assayed by measuring the absorbance of the mixed solution at two wavelengths, according to the formulae:

Chlorophyll *a* (mg l^{-1}) = $11.93 A_{664} - 1.93 A_{647}$

Chlorophyll *b* (mg l^{-1}) = $20.36 A_{647} - 5.5 A_{664}$

Note: different equations are required for other solvents.

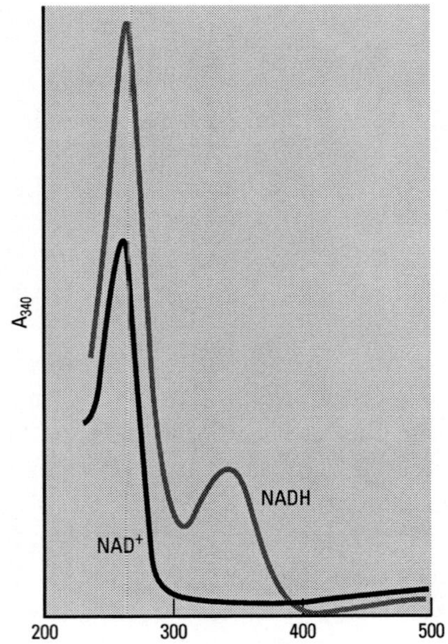

Fig. 41.2 Absorption spectra of nicotinamide adenine dinucleotide in oxidised (NAD^+) and reduced (NADH) form. Note the 340 nm absorption peak (A_{340}), used for quantitative work.

(< 0.1), where any differences in the optical properties of cuvettes for reference and test samples will be pronounced. Glass and plastic absorb UV light and quartz cuvettes must be used at wavelengths below 300 nm.

KEY POINT Before taking a measurement, make sure that cuvettes are clean, unscratched, dry on the outside, filled to the correct level and located in the correct position in their sample holders.

Proteins and nucleic acids in biological samples can accumulate on the inside faces of glass/quartz cuvettes, so remove any deposits using acetone on a cotton bud, or soak overnight in 1 mol l^{-1} nitric acid. Corrosive and hazardous solutions must be used in cuvettes with tightly fitting lids, to prevent damage to the instrument and to reduce the risk of accidental spillage.

Basic instruments use photocells similar to those used in simple colorimeters or photodiode detectors. In many cases, a different photocell must be used at wavelengths above and below 550–600 nm, due to differences in the sensitivity of such detectors over the visible waveband. The detectors used in more sophisticated instruments give increased sensitivity and stability when compared to photocells.

Digital displays are increasingly used in preference to traditional needle-type meters, as they are not prone to parallax errors and misreading of the absorbance scale. Some digital instruments can be calibrated to give a direct readout of the concentration of the test substance.

KEY POINT Basic spectrophotometers are most accurate within the absorbance range from 0.00 to 1.00 and your calibration standards and test solutions should be prepared to give readings within this range.

Types of UV/visible spectrophotometer

Basic instruments are single beam spectrophotometers in which there is only one light path. The instrument is set to zero absorbance using a blank solution, which is then replaced by the test solution, to obtain an absorbance reading. An alternative approach is used in double beam spectrophotometers, where the light beam from the monochromator is split into two separate beams, one beam passing through the test solution and the other through a reference blank. Absorbance is then measured by an electronic circuit which compares the output from the reference (blank) and sample cuvettes. Double beam spectrophotometry reduces measurement errors caused by fluctuations in output from the light source or changes in the sensitivity of the detection system, since reference and test solutions are measured at the same time (Box 41.1). Recording spectrophotometers are double beam instruments, designed for use with a chart recorder, either by recording the difference in absorbance between reference and test solutions across a predetermined waveband to give an absorption spectrum (Fig. 41.2), or by recording the change in absorbance at a particular wavelength as a function of time (e.g. in an enzyme assay).

Quantitative spectrophotometric analysis

A single (purified) substance in solution can be quantified using the Beer–Lambert relationship (Eqn [41.3]), provided its absorptivity is known at a particular wavelength (usually at the absorption maximum for the

Box 41.1 How to use a spectrophotometer

1. **Switch on and select the correct lamp** for your measurements (e.g. deuterium for UV, tungsten for visible light).
2. **Allow up to 15 min for the lamp to warm up** and for the instrument to stabilise before use.
3. **Select the appropriate wavelength:** on older instruments a dial is used to adjust the monochromator, while newer machines have microprocessor-controlled wavelength selection.
4. **Select the appropriate detector:** some instruments choose the correct detector automatically (on the basis of the specified wavelength), while others have manual selection.
5. **Choose the correct slit width** (if available): this may be specified in the protocol you are following, or may be chosen on the manufacturer's recommendations.
6. **Insert appropriate reference blank(s):** single beam instruments use a single cuvette, while double beam instruments use two cuvettes (a matched pair for accurate work). The reference blanks should match the test solution in all respects apart from the substance under test, i.e. they should contain all reagents apart from this substance. *Make sure that the cuvettes are positioned correctly, with their polished (transparent) faces in the light path, and that they are accurately located in the cuvette holder(s).*
7. **Check/adjust the 0% transmittance:** most instruments have a control which allows you to zero the detector output in the absence of any light (termed 'dark current' correction). Some microprocessor-controlled instruments carry out this step automatically.
8. **Set the absorbance reading to zero:** usually *via* a dial, or digital readout.
9. **Analyse your samples:** replace the appropriate reference blank with a test sample, allow the absorbance reading to stabilise (5–10 s) and read the absorbance value from the meter/readout device. For absorbance readings greater than 1 (i.e. $<10\%$ transmittance), the signal-to-noise ratio is too low for accurate results. Your analysis may require a calibration curve or you may be able to use the Beer–Lambert relationship (Eqn [41.3]) to determine the concentration of test substance in your samples.
10. **Check the scale zero at regular intervals** using a reference blank, e.g. after every 10 samples.
11. **Check the reproducibility of the instrument:** measure the absorbance of a single solution several times during your analysis. It should give the same value.

Problems (and solutions): inaccurate/unstable readings are most often due to incorrect use of cuvettes, e.g. dirt, fingerprints or test solution on outside of cuvette (wipe the clear faces using a soft tissue before insertion into the cuvette holder and handle only by the opaque faces), condensation (if cold solutions aren't allowed to reach room temperature before use), air bubbles (which scatter light and increase the absorbance; tap gently to remove), insufficient solution (causing refraction of light at the meniscus), particulate material in the solution (check for 'cloudiness' in the solution and centrifuge before use, where necessary) or incorrect positioning in light path (locate in correct position).

Measuring high absorbances in colorimetric analysis – *if any final solution has an absorbance that is too high to be read with accuracy on your spectrophotometer (i.e. $A > 2$), it is bad practice to dilute the solution so that it can be measured. This dilutes both the sample molecules and the colour reagents to an equal extent. Instead, you should dilute the original sample and reassay.*

substance, since this will give the greatest sensitivity). The molar absorptivity is the absorbance given by a solution with a concentration of $1\ \text{mol}\,\text{l}^{-1}$ ($= 1\ \text{kmol}\,\text{m}^{-3}$) of the compound in a light path of 1 cm. The appropriate value may be available from tabulated spectral data (e.g. Anon., 1963), or it can be determined experimentally by measuring the absorbance of known concentrations of the substance (Box 41.1) and plotting a standard curve (see Chapter 37). This should confirm that the relationship is linear over the desired concentration range and the slope of the line will give the molar absorptivity.

The specific absorptivity is the absorbance given by a solution containing $10\ \text{g}\,\text{l}^{-1}$ (i.e. 1% w/v) of the compound in a light path of 1 cm. This is useful for substances of unknown molecular weight, e.g. proteins or nucleic acids, where the amount of substance in solution is expressed in terms of its mass, rather than as a molar concentration. For

Measuring low absorbances in colorimetric analysis *– for accurate readings with highly dilute solutions, use a cuvette with a long optical path length (e.g. 5 cm, rather than 1 cm).*

use in Eqn [41.3], the specific absorptivity should be divided by 10 to give the solute concentration in $g\,l^{-1}$.

This simple approach cannot be used for mixed samples where several substances have a significant absorption at a particular wavelength. In such cases, it may be possible to estimate the amount of each substance by measuring the absorbance at several wavelengths, e.g. protein estimation in the presence of nucleic acids.

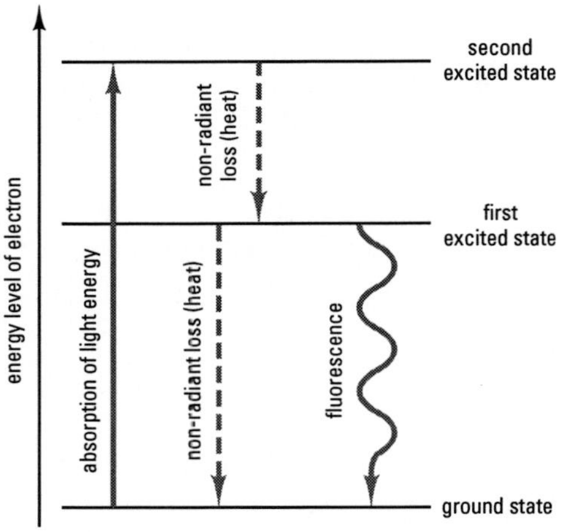

Fig. 41.3 Energy levels and energy transitions in fluorescence.

Fluorescence

With most molecules, after electrons are raised to a higher energy level by absorption of electromagnetic radiation, they soon fall back to the ground state by radiationless transfer of energy (heat) to the solvent. However, with some molecules, the events shown in Figure 41.3 may occur, i.e. electrons may lose only part of their energy by non-radiant routes and the rest may be emitted as electromagnetic radiation, a phenomenon known as fluorescence. Since not all of the energy that was absorbed is emitted (due to non-radiant loss), the wavelength of the fluorescent light is longer than the absorbed light (longer wavelength = lower energy). Thus, a fluorescent molecule has both an absorption spectrum and an emission spectrum. The difference between the excitation wavelength (λ_{ex}) and the emission wavelength (λ_{em}), measured in nm, is known as the Stokes shift, and is fundamental to the sensitivity of fluorescence techniques. The existence of a Stokes shift means that emitted light can be detected against a low background, independently of the excitation wavelength.

Most fluorescent molecules have the following features:

- A highly conjugated system (alternating double and single bonds), involving aromatic or heterocyclic rings, usually containing O or N (as heteroatoms).
- A condensed system of fused rings, with one or more heteroatoms.
- Electron-donating groups such as $-OH$, $-OCH_3$, $-NH_2$ and $-NR_2$, together with electron-attracting groups elsewhere in the molecule, in conjugation with the electron-donating groups.
- A rigid, planar structure.

Figure 41.4 illustrates many of these features for fluorescein, used in a range of biological applications including nucleic acid detection, fluorescent antibody tests, etc.

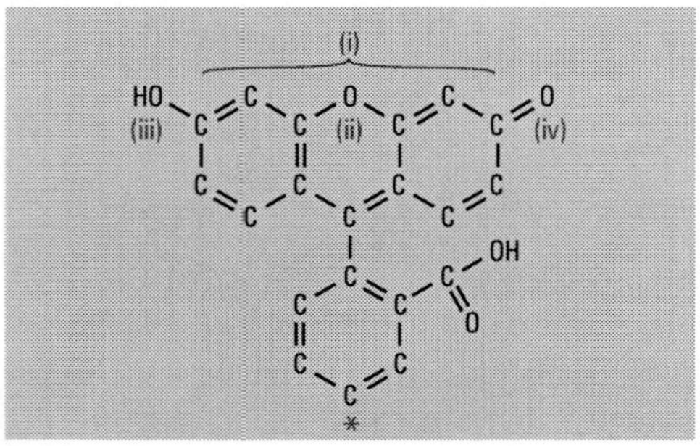

Fig. 41.4 Fluorescein – a widely used fluorescent label, showing (i) a planar conjugated system of fused rings; (ii) heteroatoms within the conjugated structures; (iii) an electron-donating group; and (iv) an electron-attracting group. In fluorogenic enzyme substrates (p. 382), linkage is usually *via* one of the two hydroxyl groups, while in fluoroscein-labelled proteins, linkage is *via* an isothiocyanate group (N=C=S) on the lowermost ring (*)

Fluorescence spectrophotometry

The principal components of a fluorescence spectrophotometer (fluorimeter) are shown in Figure 41.5. The instrument contains two monochromators, one to select the excitation wavelength and the other to monitor the light emitted, usually at 90° to the incident beam (though light is actually emitted in all directions). As an example, the wavelengths used to measure the highly fluorescent compound aminomethylcoumarin are 388 nm (excitation) and 440 nm (emission).

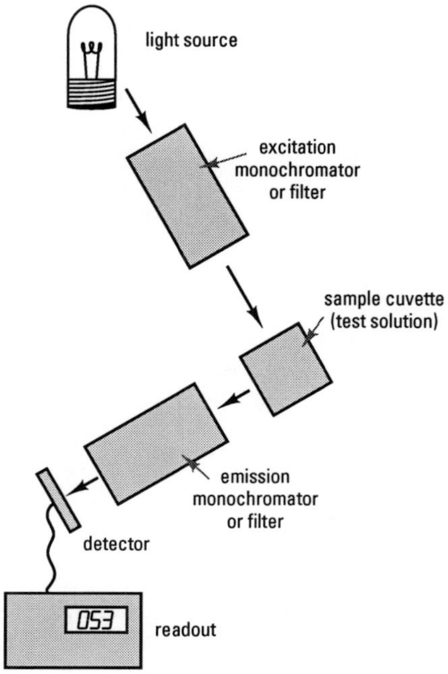

Fig. 41.5 Components of a fluorimeter (fluorescence spectrophotometer). Note that sample cuvettes for fluorimetry must have clear sides all round.

Table 41.1 Examples of compounds with intrinsic fluorescence

Drugs Aspirin, morphine, barbiturates, propranolol, ampicillin, tetracyclines
Vitamins Riboflavin, vitamins A, B6 and E, nicotinamide
Pollutants Naphthalene, anthracene, benzopyrene

Compared with UV/visible spectrophotometry, fluorescence spectroscopy has certain advantages, including:

- Enhanced sensitivity (up to 1000-fold), since the emitted light is detected against a background of zero, in contrast to spectrophotometry where small changes in signal are measured against a large 'background' (see Eqn [41.3]).
- Increased specificity, because not one but two specific wavelengths are required for a particular compound.

However, there are also certain drawbacks:

- Not all compounds show intrinsic fluorescence, limiting its application. However, some non-fluorescent compounds may be coupled to fluorescent dyes, or fluorophores (e.g. proteins may be coupled to fluorescamine).
- The light emitted can be less than expected due to quenching, i.e. when substances in the sample (e.g. oxygen) either interfere with energy transfer, or absorb the emitted light (in some instances, the sample molecules may self-quench if they are present at high concentration).

The sensitivity of fluorescence has made it invaluable in techniques in which specific antibodies are linked to a fluorescent dye, including:

- fluorescence immunoassay (FIA);
- immunohistochemistry, which requires the use of a fluorescence microscope, e.g. using fluorescent antibodies, or fluorescent *in situ* hybridisation (FISH) for nucleic acid detection;
- flow cytometry – a fluorescence-activated cell sorter (FACS) uses cell surface protein-specific monoclonal antibodies labelled with fluorescent dyes to separate and enumerate cells such as lymphocytes.

Phosphorescence and luminescence

A phenomenon related to fluorescence is phosphorescence, which is the emission of light following intersystem crossing between electron orbitals (e.g. between excited singlet and triplet states). Light emission in phosphorescence usually continues after the exciting energy is no longer applied and, since more energy is lost in intersystem crossing, the emission wavelengths are generally longer than with fluorescence. Phosphorescence has limited applications in biomolecular sciences.

Luminescence (or chemiluminescence) is another phenomenon in which light is emitted, but here the energy for the initial excitation of electrons is provided by a chemical reaction rather than by electromagnetic radiation. An example is the action of the enzyme luciferase, extracted from fireflies, which catalyses the following reaction:

$$\text{luciferin} + \text{ATP} + O_2 \Rightarrow \text{oxyluciferin} + \text{AMP} + PP_i + CO_2 + \text{light} \quad [41.4]$$

The light produced is either yellow–green (560 nm) or red (620 nm). This system can be used in biomolecular analysis of ATP, either to determine ATP concentration in a biological sample, or to follow a coupled reaction (p. 382). Measurement can be performed using the

Practical advantages of atomic absorption spectrophotometry over flame photometry *– these include:*

- *improved sensitivity;*
- *increased precision;*
- *decreased interference.*

focused through the flame and onto the detector. When the sample is introduced into the flame, it will decrease the light detected in direct proportion to the amount of metal present. Newer variants of this method include flameless atomic absorption spectrophotometry and atomic fluorescence spectrophotometry, both of which are more sensitive than the flame absorption technique.

Text references

Anon. (1963) *Tables of Spectrophotometric Absorption Data for Compounds used for the Colorimetric Detection of Elements (International Union of Pure and Applied Chemistry)*. Butterworth-Heinemann, London.

Jones, A.M., Reed, R.H. and Weyers, J.D.B. (2007) *Practical Skills in Biology*, 4th edn. Prentice Hall, Harlow.

Sources for further study

Anon. *Spectrophotometry Protocols*. Available: http://www.protocol-online.org/prot/General_Laboratory_Techniques/Spectrophotometry/
Last accessed: 01/04/07.
[Gives examples of spectrophotometric analyses of selected biomolecules]

Burgess, C. and Jones, D.G. (1995) *Spectrophotometry, Luminescence and Colour*. Elsevier, Amsterdam.

Gore, M.G. (ed.) (2000) *Spectrophotometry and Spectrofluorimetry: A Practical Approach*, 2nd edn. Oxford University Press, Oxford.

Van Dyke, K., van Dyke, C. and Woodfork, K. (2002) *Luminescence Biotechnology: Instruments and Applications*. CRC Press, Boca Raton.

Wilson, K. and Walker, J. (eds.) (2005) *Principles and Techniques of Practical Biochemistry and Molecular Biology*, 6th edn. Cambridge University Press, Cambridge.
[Also covers a range of other topics, including chromatography and centrifugation]

Study exercises

41.1 Write a protocol for using a spectrophotometer. After reading this chapter, prepare a detailed stepwise protocol explaining how to use one of the spectrophotometers in your department. Ask another student or a tutor to evaluate your protocol and provide you with feedback.

41.2 Use the Beer–Lambert relationship in quantitative spectrophotometric analysis. Calculate the following (express your answer to three significant figures):

(a) The concentration of NADH (μmol l^{-1}) in a test solution giving an absorbance at 340 nm (A_{340}) of 0.53 in a cuvette with a path length of 1 cm, based on a molar absorptivity for NADH of 6220 l mol^{-1} cm^{-1} at this wavelength.

(b) The amount of NADH (nmol) in 20 ml of a test solution where $A_{340} = 0.62$ in a cuvette with a path length of 5 cm, based on a molar absorptivity for NADH of 6220 l mol^{-1} cm^{-1} at this wavelength.

(c) The mass concentration (μg ml) of double-stranded DNA in a test solution giving an absorbance at 260 nm (A_{260}) of 0.57 in a cuvette of path length 5 mm, based on an absorptivity of 20 l g^{-1} cm^{-1}.

(d) The amount (ng) of double stranded DNA in a 50 μl sub-sample from a test solution where $A_{260} = 0.31$ in a cuvette of path length 1 cm, based on an absorptivity of 20 l g^{-1} cm^{-1}.

(e) The chlorophyll *a* content, in μg (g fresh weight)$^{-1}$, of a plant leaf of fresh weight 1.56 g extracted in 50 ml of 90% acetone, where $A_{664} = 0.182$ at 664 nm and $A_{647} = 0.035$, based on the following equation:

$$\text{Chlorophyll } a \text{ (mg l}^{-1}) = 11.93\, A_{664} - 1.93\, A_{647}$$

41.3 Determine the molar absorptivity of a substance in aqueous solution. A solution of *p*-nitrophenol containing 8.8 μg ml^{-1} gave an absorbance of 0.535 at 404 nm in a cuvette of path length 1 cm. What is the molar absorptivity of *p*-nitrophenol at 404 nm, expressed to three significant figures? (Note: M_r of *p*-nitrophenol is 291.27.)

(continued)

Practical advantages of atomic absorption spectrophotometry over flame photometry – these include:

- improved sensitivity;
- increased precision;
- decreased interference.

focused through the flame and onto the detector. When the sample is introduced into the flame, it will decrease the light detected in direct proportion to the amount of metal present. Newer variants of this method include flameless atomic absorption spectrophotometry and atomic fluorescence spectrophotometry, both of which are more sensitive than the flame absorption technique.

Text references

Anon. (1963) *Tables of Spectrophotometric Absorption Data for Compounds used for the Colorimetric Detection of Elements (International Union of Pure and Applied Chemistry)*. Butterworth-Heinemann, London.

Jones, A.M., Reed, R.H. and Weyers, J.D.B. (2007) *Practical Skills in Biology,* 4th edn. Prentice Hall, Harlow.

Sources for further study

Anon. *Spectrophotometry Protocols.* Available: http://www.protocol-online.org/prot/General_Laboratory_Techniques/Spectrophotometry/
Last accessed: 01/04/07.
[Gives examples of spectrophotometric analyses of selected biomolecules]

Burgess, C. and Jones, D.G. (1995) *Spectrophotometry, Luminescence and Colour*. Elsevier, Amsterdam.

Gore, M.G. (ed.) (2000) *Spectrophotometry and Spectrofluorimetry: A Practical Approach*, 2nd edn. Oxford University Press, Oxford.

Van Dyke, K., van Dyke, C. and Woodfork, K. (2002) *Luminescence Biotechnology: Instruments and Applications*. CRC Press, Boca Raton.

Wilson, K. and Walker, J. (eds.) (2005) *Principles and Techniques of Practical Biochemistry and Molecular Biology,* 6th edn. Cambridge University Press, Cambridge. [Also covers a range of other topics, including chromatography and centrifugation]

Study exercises

41.1 Write a protocol for using a spectrophotometer. After reading this chapter, prepare a detailed stepwise protocol explaining how to use one of the spectrophotometers in your department. Ask another student or a tutor to evaluate your protocol and provide you with feedback.

41.2 Use the Beer–Lambert relationship in quantitative spectrophotometric analysis. Calculate the following (express your answer to three significant figures):

(a) The concentration of NADH (μmol l^{-1}) in a test solution giving an absorbance at 340 nm (A_{340}) of 0.53 in a cuvette with a path length of 1 cm, based on a molar absorptivity for NADH of 6220 l mol^{-1} cm^{-1} at this wavelength.

(b) The amount of NADH (nmol) in 20 ml of a test solution where $A_{340} = 0.62$ in a cuvette with a path length of 5 cm, based on a molar absorptivity for NADH of 6220 l mol^{-1} cm^{-1} at this wavelength.

(c) The mass concentration (μg ml) of double-stranded DNA in a test solution giving an absorbance at 260 nm (A_{260}) of 0.57 in a cuvette of path length 5 mm, based on an absorptivity of 20 l g^{-1} cm^{-1}.

(d) The amount (ng) of double stranded DNA in a 50 μl sub-sample from a test solution where $A_{260} = 0.31$ in a cuvette of path length 1 cm, based on an absorptivity of 20 l g^{-1} cm^{-1}.

(e) The chlorophyll *a* content, in μg (g fresh weight)$^{-1}$, of a plant leaf of fresh weight 1.56 g extracted in 50 ml of 90% acetone, where $A_{664} = 0.182$ at 664 nm and $A_{647} = 0.035$, based on the following equation:

$$\text{Chlorophyll } a \text{ (mg l}^{-1}) = 11.93\, A_{664} - 1.93\, A_{647}$$

41.3 Determine the molar absorptivity of a substance in aqueous solution. A solution of *p*-nitrophenol containing 8.8 μg ml^{-1} gave an absorbance of 0.535 at 404 nm in a cuvette of path length 1 cm. What is the molar absorptivity of *p*-nitrophenol at 404 nm, expressed to three significant figures? (Note: M_r of *p*-nitrophenol is 291.27.)

(continued)

Study exercises (continued)

41.4 Determine the concentration of metal ions based on atomic spectroscopy of test and standard solutions. The following data represent a set of calibration standards for K^+ in aqueous solution, measured by flame photometry:

Absorbance of standard solutions containing K^+ at up to 0.5 mmol l^{-1}

K^+ concentration (mmol l^{-1})	Absorbance
0	0.000
0.1	0.155
0.2	0.279
0.3	0.391
0.4	0.537
0.5	0.683

Draw a calibration curve using the above data and use this to estimate the amount of K^+ in a test sample prepared by digestion of 0.482 g of tissue in a final volume of 25 ml of solution, giving an absorbance of 0.429 when measured at the same time as the standards shown above. Express your answer in μmol K^+ (g tissue)$^{-1}$, to three significant figures. See also study exercise 37.1 for a similar exercise, based on atomic absorption spectrophotometry of Zn.

41.5 Explain the basis of fluorescence. Why is the wavelength of the emitted light longer than that used to excite a fluorescent molecule?

41.6 Account for the difference in design between spectrophotometer and fluorimeter cuvettes. Why does a spectrophotometer cuvette have only two clear sides while a fluorimeter cuvette has four clear sides?

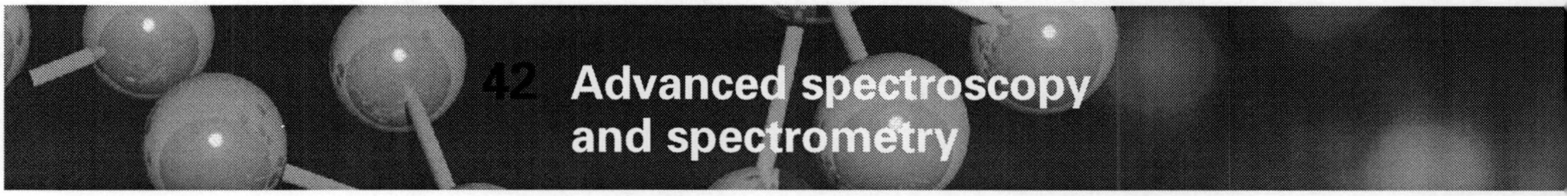

42 Advanced spectroscopy and spectrometry

***Identifying compounds** – the techniques described in this chapter can often provide sufficient information to identify a compound with a low probability of error.*

The following techniques are unlikely to be encountered in the early stages of your course, though they may be presented as a demonstration. Alternatively, you may be shown spectra for interpretation and analysis. The various types of spectroscopy, and their relationship with the electromagnetic spectrum, are shown in Table 42.1.

Definitions

Spectroscopy – any technique involving the production and subsequent recording of a spectrum of electromagnetic radiation, usually in terms of wavelength or energy.

Spectrometry – any technique involving the measurement of a spectrum, e.g. of electromagnetic radiation, molecular masses.

Wavenumber – the reciprocal of the wavelength (expressed as cm^{-1}): a term used widely in IR spectroscopy, but rarely in other types of analysis.

Infra-red (IR) and Raman spectroscopy

Both of these techniques involve the measurement of frequencies produced by the vibration of chemical bonds (bending and stretching). The IR/Raman region is generally considered to be from 800 to 2500 nm (for near-IR) and up to 16 000 nm (for mid-IR). Near-IR spectroscopy involves recording the spectrum in that region in a manner analogous to UV/visible spectroscopy, and quantitative analysis is possible. However, the most widely used technique is mid-IR spectroscopy, which allows identification of groups or atoms in a sample compound, but is inappropriate for quantitative measurement. A peak at a particular frequency can be identified by reference to libraries or computer databases of IR spectra, e.g. a peak at a wavenumber of 1730–1750 cm^{-1} corresponds to a carbonyl group – ($\sim$C=O), which is present in fatty acids and proteins. The 1400–600 cm^{-1} region is known as the 'fingerprint' region because no two compounds give identical spectra.

The value of IR spectroscopy is greatly enhanced by Fourier transformation (FT), named after the mathematician J.B. Fourier. FT is a procedure for interconverting frequency functions and time or distance functions. In FT-IR, information is obtained from an interferometer, which splits the incident beam so that it passes through both the sample and a reference. When the beam is recombined, interference patterns arise because the two path lengths are different. The interference pattern has the same relationship to a normal spectrum as a hologram has to a picture, and integral computers use FT to convert the pattern into a spectrum in under a minute. The overall result is a greatly enhanced signal-to-noise ratio.

***Understanding the origins of IR and Raman spectra** – the IR spectrum is due to changes in charge displacement in bonds. The Raman spectrum is due to changes in polarisability in bonds.*

Table 42.1 The electromagnetic spectrum and types of spectroscopy

Type of radiation	Origin	Wavelength	Type of spectroscopy
γ-rays	Atomic nuclei	<0.1 nm	γ-ray spectroscopy
X-rays	Inner shell electrons	0.1–1.0 nm	X-ray fluorescence (XRF)
Ultraviolet (UV)	Ionisation	10–200 nm	UV spectroscopy
UV/visible	Valency electrons	200–800 nm	UV/visible spectroscopy
Infra-red	Molecular vibrations	0.8–25 μm	Infrared spectroscopy (IR) and Raman
Microwaves	Electron spin alignment	400 μm–30 cm	Electron spin resonance (ESR)
Radiowaves	Nuclear spin resonance	>100 cm	Nuclear magnetic resonance (NMR)

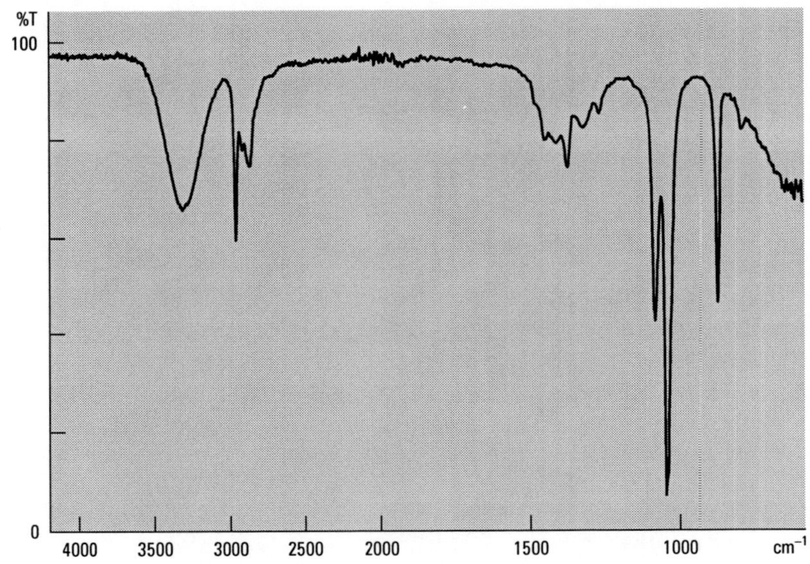

Fig. 42.1 IR spectrum for ethanol.

The use of wavenumber – *this is a well-established convention, since high wavenumber = high frequency = high energy = short wavelength. Expression of the IR range, 4000 cm^{-1} to 650 cm^{-1}, is in 'easy' numbers and the high energy is found on the left-hand side of the spectrum. Note that IR spectroscopists often refer to wavenumbers as 'frequencies', e.g. 'the peak of the C=O stretching 'frequency' is at 1720 cm^{-1}'.*

Using IR – *the rapid analysis possible with gaseous samples makes IR ideal for studying CO_2 metabolism in photosynthesis and respiration.*

IR spectra

A typical IR spectrum is shown in Figure 42.1 and you should note the following points:

- The *x*-axis, the wavelength of the radiation, is given in wavenumbers ($\bar{\nu}$) and expressed in reciprocal centimetres (cm^{-1}). You may still see some spectra from old instruments using microns (μ, equivalent to the SI unit 'micrometres', μm, at 1×10^{-6} m) for wavelength.
- The *y*-axis, expressing the amount of radiation absorbed by the molecule, is usually shown as % transmittance. When no radiation is absorbed (all is transmitted through the sample) there is 100% transmittance while 0% transmittance implies all radiation is absorbed at a particular wavenumber. Since the *y*-axis scale goes from 0 to 100% transmittance, the absorption peaks are displayed *down* from the 100% line; this is *opposite* to most other common spectra.
- The cells holding the sample usually display imperfections and are not completely transparent to IR radiation, even when empty. Therefore the base line of the spectrum is rarely set on 100% transmittance and quantitative applications of IR spectroscopy are more complex than for UV/visible spectrophotometry (p. 282).

Applications of IR and Raman spectroscopy

The principal use of IR and Raman spectroscopy is in the identification of drugs (e.g. penicillin), small peptides, pollutants and food contaminants. When an IR spectrometer is coupled to a gas–liquid chromatograph, it can be used for the analysis of drug metabolites.

One possible future use of IR is for the non-invasive analysis of organic compounds that are of diagnostic significance (e.g. glucose in diabetics). This involves illuminating a small area of skin with IR radiation and using the resultant IR spectrum to quantify a specific compound: an example of such a system is the non-invasive blood glucose monitor described by Malchoff *et al.*, 2002.

SAFETY NOTE Working with NMR spectrometers – these instruments can generate very high magnetic fields and can exert strong magnetic forces on materials such as steel items brought close to the magnet (<1 m). NMR magnets can also destroy data stored in magnetic form – keep items such as memory sticks, laptops and credit cards well away.

Example For an external magnetic field of 2.5 T (Tesla), ΔE for ^1H is 6.6×10^{-26} J, and, since $\Delta E = h\nu$, the corresponding frequency (ν) is 100 MHz; for ^{13}C in the same field, ΔE is 1.7×10^{-26} J, and ν is 25 MHz.

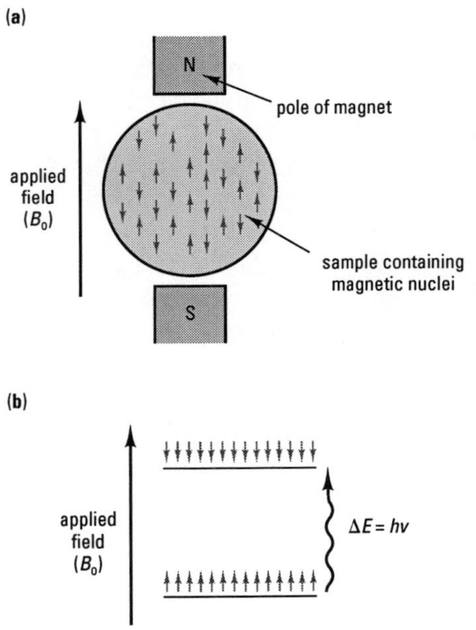

Fig. 42.2 Effect of an applied magnetic field, B_0, on magnetic nuclei. (a) Nuclei in magnetic field have one of two orientations – either with the field or against the field (in the absence of an applied field, the nuclei would have random orientation). (b) Energy diagram for magnetic nuclei in applied magnetic field.

Nuclear magnetic resonance (NMR)

Electromagnetic radiation (at radiofrequencies of 1–500 MHz) is used to identify and monitor compounds. This is possible because of differences in the magnetic states of atomic nuclei, involving very small transitions in energy levels. The atomic nuclei of isotopes of many elements are magnetic because they are charged and have spin. Typical magnetic nuclei are ^1H, ^{13}C, ^{14}N, ^{15}N, ^{19}F and ^{31}P. When these nuclei interact with a uniform external magnetic field, they behave like tiny compass needles and align themselves in a direction either parallel or antiparallel to the field. The two orientations have different energies, with the parallel direction having a lower energy than the antiparallel (Fig. 42.2). The energy difference between the two levels (ΔE) corresponds to a precise electromagnetic frequency (ν), according to similar quantum principles to those for the excitation of electrons (p. 280). When a sample containing an isotope with a magnetic nucleus is placed in a magnetic field and exposed to an appropriate radiofrequency, transitions between the energy levels of magnetic nuclei will occur when the energy gap and the applied frequency are in *resonance* (i.e. when they are matched exactly). Differences in energy levels, and hence resonance frequencies (ν_0), depend on the magnitude of the applied magnetic field (B_0) and the magnetogyric ratio (λ), according to the equation:

$$\nu_0 = \lambda B_0/2\pi \qquad [42.1]$$

The magnetogyric ratio varies from one isotope to another, so NMR is performed at different frequencies for different nuclei at any given value of B_0. The principal components of an NMR spectrometer are shown in Figure 42.3.

For magnetic nuclei in a given molecule, an NMR spectrum is generated because, in the presence of the applied field, different nuclei experience different local magnetic fields depending on the arrangement of electrons in their vicinity. The effective field (B) at the nucleus can be expressed as:

$$B = B_0(1 - \sigma) \qquad [42.2]$$

where σ (the shielding constant) expresses the contribution of the small secondary field generated by nearby electrons. The magnitude of σ is dependent on the electronic environment of a nucleus, so nuclei in different environments give rise to different resonance frequencies, according to the equation:

$$\nu_0 = \lambda B_0(1 - \sigma)/2\pi \qquad [42.3]$$

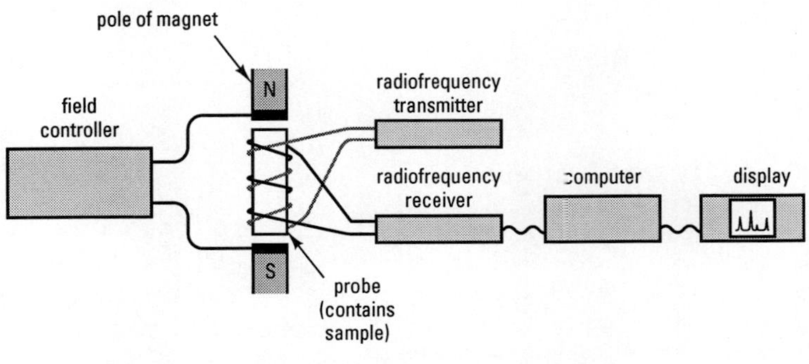

Fig. 42.3 Components of an NMR spectrometer.

***Measuring chemical shifts** – ppm is not a concentration term in NMR, but is used to reflect the very small frequency changes that occur relative to the reference standard, measured in proportional terms.*

The separation of resonance frequencies from a reference value is termed the *chemical shift*, and is expressed in dimensionless terms, as parts per million (ppm). By convention, the chemical shift is positive if the sample nucleus is less shielded than the reference and negative if it is more shielded.

Understanding NMR spectra of biomolecules

Figure 42.4 illustrates the principle underlying this technique – the different chemical environments of the hydrogen atoms in a simple biomolecule such as acetic acid result in two different ^1H resonances or 'signals', one corresponding to those of protons in $\sim CH_3$, and the other from that in $\sim COOH$. Furthermore, the relative intensities of the NMR signals, as measured by their areas, are proportional to the number of contributing nuclei, so the relative areas of the peaks due to the protons in $\sim CH_3$ and (undissociated) $\sim COOH$ would be 3 : 1. Similarly, with ^{31}P NMR of a biologically important molecule such as ATP, there are signals corresponding to the α, β and γ phosphates, the P nuclei of which are in different chemical environments (Fig. 42.5). Thus every molecule that contains one or more magnetic nuclei has its own characteristic NMR fingerprint that may be used for identification and analysis. Spectra such as those shown in Figs 42.4 and 42.5 can be obtained using FT of a large number of individual responses to radiowave pulses. Other factors such as the spin-lattice relaxation time (Gadian, 1996) can affect signal intensities (peak sizes), and resonances may be split into several lines due to spin–spin coupling (interactions between neighbouring nuclei).

In terms of resolution, narrow signals are obtained only from molecules that are fairly mobile, so most high resolution studies are carried out using solutions. However, since many metabolites in biological samples (cells, tissues, etc.) are in aqueous solution and freely mobile, they can give rise to high resolution spectra, as shown in Figure 42.5 for the ^{31}P-containing compounds, ATP, phosphocreatine and inorganic phosphate. On the other hand ^{31}P signals from immobile molecules such as DNA or phospholipids are very broad, and could be as wide as the whole scale shown in Fig. 42.5.

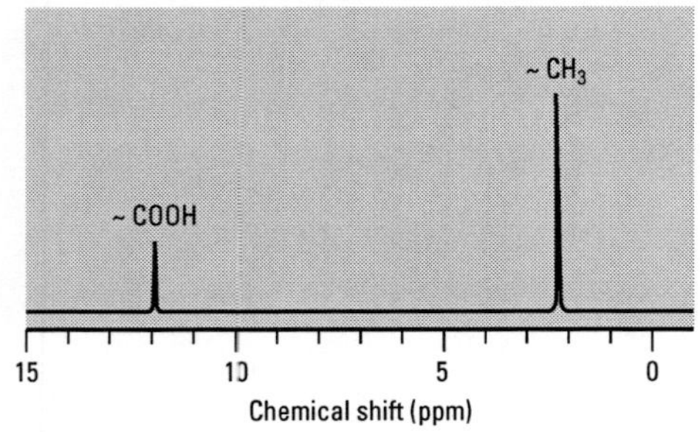

Fig. 42.4 ^1H NMR spectrum of acetic acid (CH_3COOH). The relative areas of the two signals are 1 : 3 and the frequencies (chemical shifts) are expressed in terms of ppm, relative to the reference signal (tetramethyl silane, TMS).

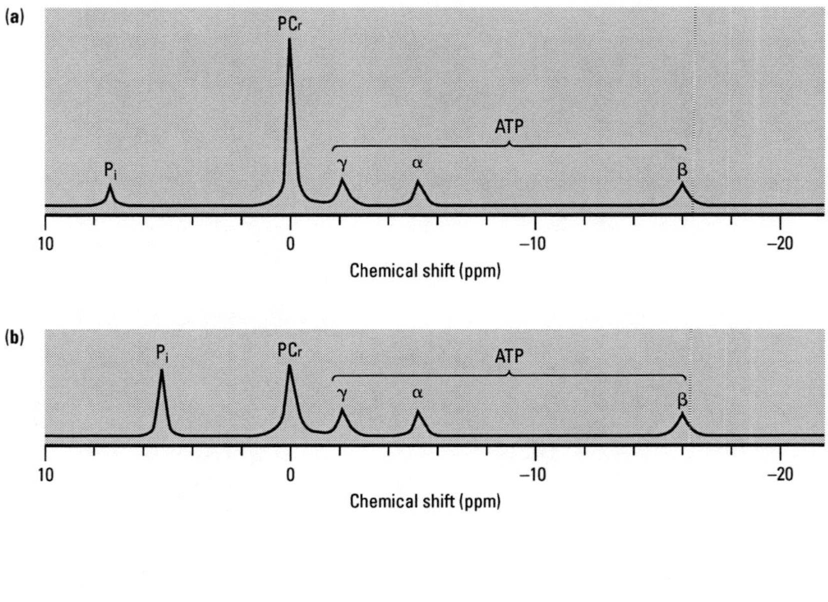

Fig. 42.5 Typical ^{31}P NMR spectrum from intact skeletal muscle (a) at rest and (b) after a fatiguing series of contractions. In (b) ATP levels are preserved at the expense of phosphocreatinine (PCr). Also, hydrolysis of ATP during contraction results in a large phosphate (P_i) peak which is shifted to the right compared to (a), reflecting the decrease in intracellular pH that accompanies glycolysis.

Table 42.2 The relative merits and disadvantages of various magnetic nuclei in biomolecular studies

Nucleus	Relative sensitivity	Natural abundance	Comments
^{1}H	100	99.98%	Multiple, but specific, spectral lines are obtained for individual biomolecules. For mixtures, the ubiquitous occurrence of ^{1}H gives complex, overlapping signals that are often difficult to interpret. Gives a large solvent peak with aqueous samples (can be avoided by using D_2O as solvent). Mainly used for structural studies of pure macromolecules. Essential for MRI.
^{31}P	6.6	100%	Very useful for studies on living systems, with narrow resonance peaks and a wide range of chemical shifts for different molecules. Spectra are simpler and easier to interpret than for ^{1}H, but are not as distinctive: different compounds may give similar ^{31}P spectra. Several important P-containing compounds (including ATP, ADP and inorganic phosphate) can be detected in intact cells – useful in bioenergetic studies.
^{13}C	0.016	1.1%	Gives narrow signals and a wide range of chemical shifts. Resolution is better than for ^{1}H, and a wide range of organic biomolecules can be detected. Low natural abundance gives low sensitivity, extending the time required to accumulate spectra. However, low natural abundance also means that specific metabolites can be selected for ^{13}C isotope enrichment, allowing particular metabolic pathways to be investigated, e.g. carbon assimilation.

Using NMR *– in contrast to most conventional metabolic studies, NMR is non-invasive and the time course of metabolic reactions can be followed using a single experimental subject or preparation, eliminating variation between samples.*

Understanding MRI *– body tissues have different water contents (typically 60–90% w/w). Signals arising from protons in water in different tissues can be used to differentiate between tissues (e.g. 'grey' and 'white' matter in the brain), and between normal tissue and tumours. Fat deposits can be detected, due to the difference between the ^{1}H signal from fatty acids and that of the water in the surrounding tissue.*

Biomolecular applications of NMR

The sensitivity of NMR has improved dramatically with the development of more powerful magnets. Details of the major uses of various magnetic nuclei are given in Table 42.2. The principal applications include:

- studies of the structure and function of macromolecules and biological systems, such as membranes;
- metabolic investigations on living organisms, including humans, since NMR can be used to obtain a 'fingerprint' of a particular molecule and changes in the intensity of spectra can be used for kinetic studies (e.g. Fig. 42.5). This involves use of 'surface coils' as sources of radiation, or placing the organism within the core of an electromagnet;
- measurement of intracellular pH by determination of the chemical shift of the phosphate peak, as this changes with pH in a predictable manner;
- magnetic resonance imaging (MRI), which is a form of proton NMR that uses a field gradient (as opposed to a uniform field) to produce signals that are translated by computers into anatomical images (Gadian, 1996).

Electron spin resonance (ESR)

This technique is based on energy transitions of spinning electrons in a magnetic field. As with NMR, the low energy state occurs when the electromagnetic field generated by the spinning electron is parallel to the externally generated field, while the high energy state occurs when the electron-generated field is antiparallel. ESR is very useful for studying metalloproteins and can be used to monitor the activity of such proteins (e.g. cytochrome oxidase) in intact mitochondria or chloroplasts. It can also be used to detect free radicals, for example in irradiated foodstuffs.

Mass spectrometry (MS)

***Understanding mass spectrometry** – since this technique does not involve the production and measurement of electromagnetic spectra and is not based on quantum principles, it should not be referred to as a spectroscopic technique.*

This technique involves the disintegration of organic compounds into fragment ions in a gas phase. These ions are accelerated to specific velocities using an electric field and then separated on the basis of their different masses. Each fragment of a particular mass is detected sequentially with time. The most widely used method in MS is electron impact ionisation (EI), where the electron source is a heated metal, such as a tungsten filament, subjected to an appropriate potential gradient. The stream of emitted electrons may then interact with biomolecules (M) in the sample by either:

1. electron removal – where an electron in a bond within the sample molecule is 'knocked out' by bombarding electrons, leaving the bond with only one unpaired electron and resulting in the production of a cationic free radical, i.e. $M + e^- \rightarrow M^{\dot{+}} + 2e^-$;
2. electron capture – where addition of an extra electron results in the production of an anionic free radical, i.e. $M + e^- \rightarrow M^{\dot{-}}$.

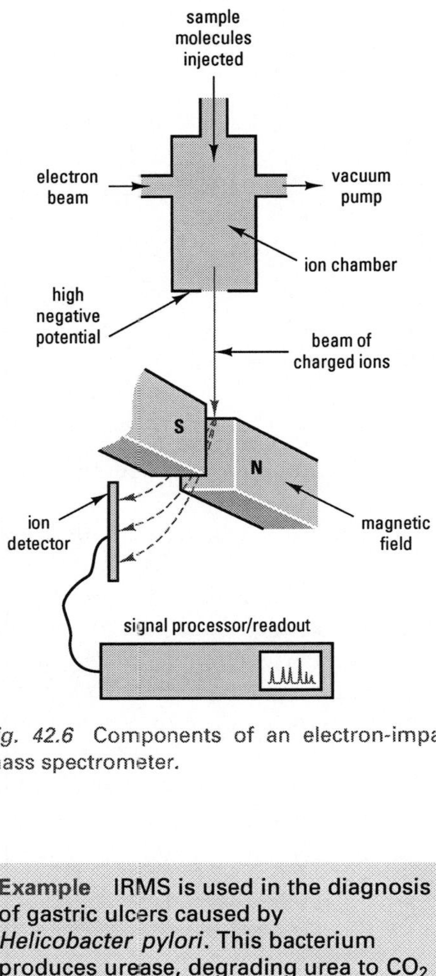

Fig. 42.6 Components of an electron-impact mass spectrometer.

Where it is not known whether cations or anions are formed, the radical is given the symbol ⌝, i.e. as $M^{\dot{\urcorner}}$. The first ions formed (parent ions) are unstable, and rapidly undergo further disintegration to give smaller fragments and daughter ions, subsequently separated in the mass spectrometer (Fig. 42.6), to give a mass spectrum such as that shown in Figure 42.7.

For *in vivo* metabolic studies, the technique of isotope ratio mass spectroscopy (IRMS) is useful since it removes the need to use radioactive isotopes. This technique exploits the ability of MS to distinguish between isotopes such as ^{13}C and ^{12}C. For example, compounds containing ^{13}C have a greater mass than the same compound containing ^{12}C and can be differentiated in the mass spectrum. By selectively labelling a key metabolite with a non-radioactive isotope, the fate of this metabolite can be followed by MS analysis of sequential samples. Normally, the materials used are combusted in oxygen to give gases such as $^{13}CO_2$ and $^{15}NO_2$, followed by exposure of such gases to an EI source; useful metabolic data can be obtained by determining the isotope ratio, e.g. of $^{13}CO_2^{\dot{\urcorner}}$ and $^{12}CO_2^{\dot{\urcorner}}$.

Example IRMS is used in the diagnosis of gastric ulcers caused by *Helicobacter pylori*. This bacterium produces urease, degrading urea to CO_2 and NH_3. A patient drinks a solution of ^{13}C-labelled urea and, if the organism is present in the stomach, $^{13}CO_2$ will appear at up to 5% v/v in the patient's breath, compared with the expected natural abundance value of 1.1% v/v.

Pyrolysis-mass spectrometry (PY-MS)

This is another variant of mass spectroscopy that is useful for biomolecular applications. The molecules in a sample are volatilised by heating to a precisely controlled high temperature for a specific time period. The volatile material is then removed by a vacuum, ionised by EI, and subjected to MS. Complex mixtures of pyrolysis products are produced, but interpretation is made easier by computer-based multivariate analysis. A useful application of PY-MS is in microbial identification, especially with small samples of organisms that are difficult to culture, e.g. mycobacteria.

Fast atom bombardment–mass spectrometry (FAB-MS)

This is particularly useful for biomolecular applications because it can be used with aqueous solutions. The solution containing the analyte is mixed with glycerol and applied to a probe which is inserted into a

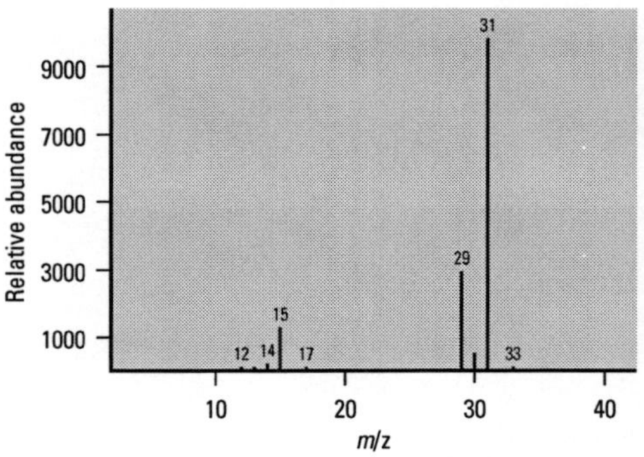

Fig. 42.7 Mass spectrum for methanol. *m/z* = charge ratio.

vacuum chamber. The mixture is bombarded with a high velocity stream of atoms (usually argon or xenon) rather than electrons, which induces fragmentation of the biomolecules and allows production of the mass spectrum. One problem with FAB-MS is the suppression effect phenomenon – with a mixed sample, not all components may be equally accessible to the atomic bombardment depending on the way they are distributed in the sample–glycerol mixture. This can be overcome by coupling FAB-MS to techniques that can give high resolution separation of components, e.g. gas–liquid chromatography (GC, p. 306), high performance liquid chromatography (HPLC, p. 305), or capillary electrophoresis (p. 331). Johnstone and Rose (1996) give further information on biomolecular aspects of MS.

Text references

Malchoff, C.D., Shoukri, K., Landau, J.I. and Buchert, J.M. (2002) Novel non-invasive blood glucose monitor. *Diabetes Care* **25:** 2268–2275.
Available: care.diabetesjournals.org/cgi/content/full/25/12/2268
Last accessed: 01/04/07.

Gadian, D.G. (1996) *NMR and its Applications to Living Systems*, 2nd edn. Oxford University Press, Oxford.

Johnstone, R.A.W. and Rose, M.E. (1996) *Mass Spectrometry for Chemists and Biochemists*, 2nd edn. Cambridge University Press, Cambridge.

Sources for further study

Guntzler, H. and Gremlich, H.U. (2002) *IR Spectroscopy: an Introduction.* Wiley-VCH, Mannheim.

Hore, P.J., Jones, J.A. and Wimperis, S. (2000) *NMR: the Toolkit.* Oxford University Press, Oxford.
[A 'primer' text – succinct, and includes relevant biological applications]

Keeler, J. (2005) *Understanding NMR Spectroscopy.* Wiley, New York.

Larkin, P.J. (2004) *An Introduction to Infrared and Raman Spectroscopy.* Jones and Bartlett, Sudbury.

Macomber, R.H. (2000) *A Complete Introduction to Modern NMR Spectrometry.* Wiley, New York.
[An accessible introductory text, with relevant biomedical information and examples]

Siudzak, G. (1996) *Mass Spectrometry for Biotechnology.* Academic Press, New York.
[Written for bioresearchers, it covers principles and applications without extensive use of maths]

Smith, R.M. and Butsch, K.L. (2004) *Understanding Mass Spectra: a Basic Approach.* Wiley, New York.

Westbrook, C. and Roth, C. (2005) *MRI in Practice*, 3rd edn. Blackwell, Oxford.
[Covers the basics of magnetic resonance imaging in medicine]

Study exercises

42.1 Interpret IR signals. The figure shows the IR spectra of acetic acid (ethanoic acid). Using the information provided below, identify the resonance corresponding to (a) the O–H and (b) C=O of the carboxylic acid group in the acetic acid spectrum.

Typical IR absorption ranges for different functional groups

Bond	Location	Wavenumber (cm^{-1})
C–O	esters, alcohols	1000–1300
C=O	ketones, aldehydes, carboxylic acids, esters, amides	1680–1750
O–H	carboxylic acids (H-bonded)	2500–3300
O–H	alcohols (H-bonded)	3230–3350
O–H	free	3580–3670
N–H	amines	3100–3500

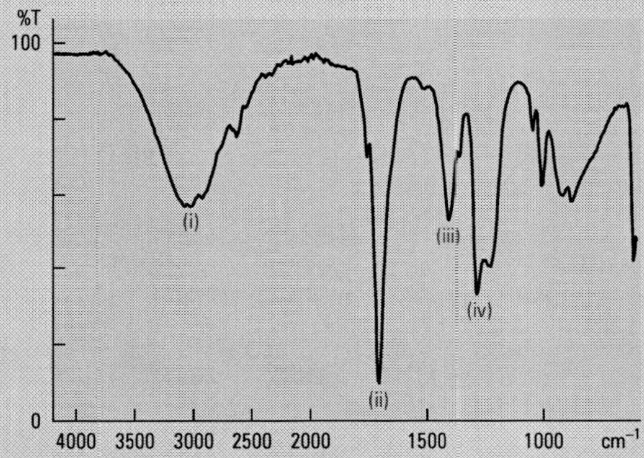

42.2 Investigate appropriate reference compounds for biological applications of NMR. Why are reference compounds important when measuring chemical shifts in NMR? Which compounds are suitable references for (a) ^{1}H, (b) ^{13}C and (c) ^{31}P NMR?

42.3 Test your understanding of the NMR terminology. Distinguish between the term ppm as used in NMR and as a concentration term (p. 146).

42.4 Interpret NMR signals from biomolecules in body fluids. A ^{1}H NMR spectrum (range 1–4.5 ppm) of a human urine sample obtained from a healthy adult is shown in the first figure below. The spectrum reveals a large number of components/peaks. The two most intense of these peaks (at approximately 3.05 and 4.05 ppm) correspond to creatinine (C), which is abundant in healthy urine. The molecular structure of creatinine is shown in the second figure below. Explain why the the ^{1}H NMR 'fingerprint' of creatinine in urine consists of just two discrete peaks.

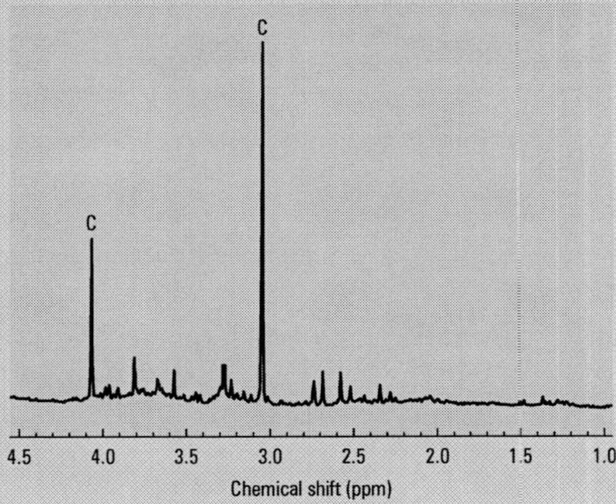

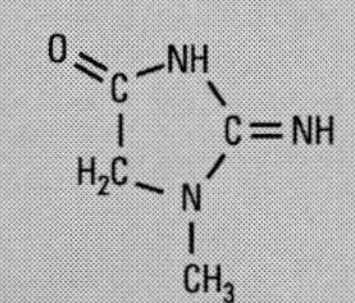

43 Centrifugation

> **SAFETY NOTE** *Working with centrifuges – these instruments generate very high centrifugal forces and should only be used after suitable instruction and training – if in doubt, ask.*

Particles suspended in a liquid will move at a rate that depends on:

- the applied force – particles in a liquid within a gravitational field, e.g. a stationary test tube, will move in response to the earth's gravity;
- the density difference between the particles and the liquid – particles less dense than the liquid will float upwards while particles denser than the liquid will sink;
- the size and shape of the particles;
- the viscosity of the medium.

For most biological particles (cells, organelles or molecules) the rate of flotation or sedimentation in response to the earth's gravity is too slow to be of practical use in separation.

> ***KEY POINT A centrifuge is an instrument designed to produce a centrifugal force far greater than the earth's gravity, by spinning the sample about a central axis (Fig. 43.1). Particles of different size, shape or density will thereby sediment at different rates, depending on the speed of rotation and their distance from the central axis.***

Table 43.1 Relationship between speed (r.p.m.) and acceleration (relative centrifugal field, RCF) for a typical bench centrifuge with an average radius of rotation, r_{av} = 115 mm

r.p.m.	RCF*
500	30
1000	130
1500	290
2000	510
2500	800
3000	1160
3500	1570
4000	2060
4500	2600
5000	3210
5500	3890
6000	4630

*RCF values rounded to nearest 10.

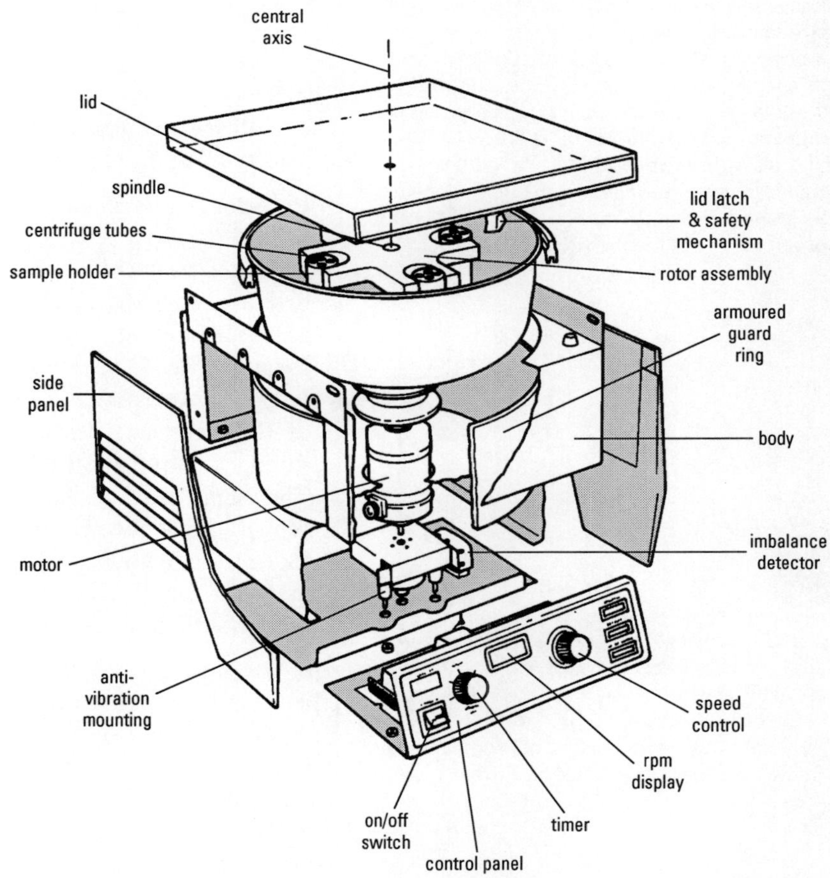

Fig. 43.1 Principal components of a low-speed bench centrifuge. Diagram of Fisher Scientific Model MSE Centaur 2, low-speed bench centrifuge, reproduced by kind permission of Fisher Scientific UK Ltd.

Working in SI units *– to convert RCF to acceleration in SI units, multiply by 9.80 m s^{-2}, e.g. an RCF of 290 is equivalent to an acceleration of 290 × 9.80 = 2842 m s^{-2}.*

Examples Suppose you wanted to calculate the RCF of a bench centrifuge with a rotor of r_{av} = 95 mm running at a speed of 3000 r.p.m. Using Eqn [43.1] the RCF would be: 1.118 × 95 × (3)^2 = 956 *g*.

You might wish to calculate the speed (r.p.m.) required to produce a relative centrifugal field of 2000 *g* using a rotor of r_{av} = 85 mm. Using Eqn [43.2] the speed would be: 945.7√(2000 ÷ 85) = 4587 r.p.m.

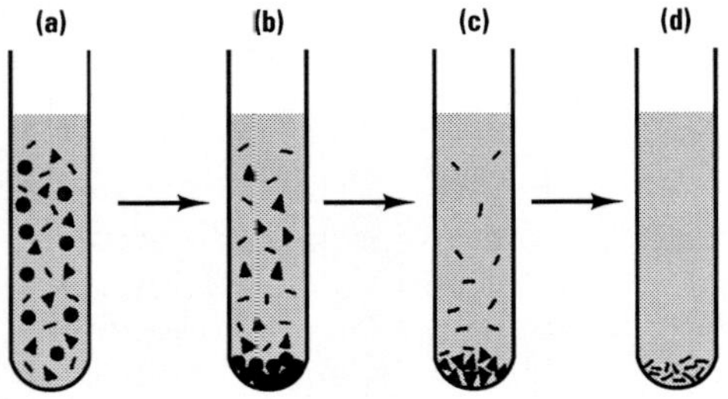

Fig. 43.2 Differential sedimentation. (a) Before centrifugation, the tube contains a mixed suspension of large, medium and small particles of similar density. (b) After low-speed centrifugation, the pellet is predominantly composed of the largest particles. (c) Further high-speed centrifugation of the supernatant will give a second pellet, predominantly composed of medium-sized particles. (d) A final ultracentrifugation step pellets the remaining small particles. Note that all of the pellets apart from the final one will have some degree of cross-contamination.

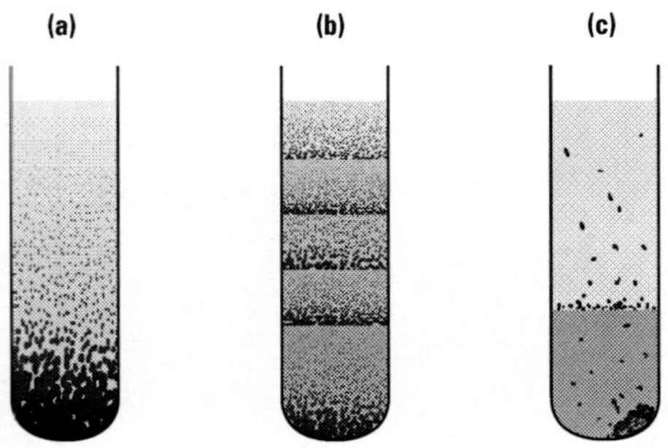

Fig. 43.3 Density gradients. (a) A continuous (linear) density gradient. (b) A discontinuous (stepwise) density gradient, formed by layering solutions of decreasing density on top of each other. (c) A single-step density barrier, designed to allow selective sedimentation of one type of particle.

How to calculate centrifugal acceleration

The acceleration of a centrifuge is usually expressed as a multiple of the acceleration due to gravity ($g = 9.80\,\mathrm{m\,s^{-2}}$), termed the relative centrifugal field (RCF, or '*g* value'). The RCF depends on the speed of the rotor (*n*, in revolutions per minute, r.p.m.) and the radius of rotation (*r*, in mm) where:

$$\mathrm{RCF} = 1.118\,r\left(\frac{n}{1000}\right)^2 \qquad [43.1]$$

This relationship can be rearranged, to calculate the speed (r.p.m.) for specific values of *r* and RCF:

$$n = 945.7\sqrt{\left(\frac{\mathrm{RCF}}{r}\right)} \qquad [43.2]$$

However, you should note that RCF is not uniform within a centrifuge tube: it is highest near the outside of the rotor (r_{max}) and lowest near the central axis (r_{min}). In practice, it is customary to report the RCF calculated from the average radius of rotation (r_{av}), as shown in Fig. 43.5. It is also worth noting that RCF varies relative to the *square* of the speed: thus the RCF will be doubled by an increase in speed of approximately 41% (Table 43.1).

Centrifugal separation methods

Differential sedimentation (pelleting)

By centrifuging a mixed suspension of particles at a specific RCF for a particular time, the mixture will be separated into a pellet and a supernatant (Fig. 43.2). The successive pelleting of a suspension by spinning for a fixed time at increasing RCF is widely used to separate organelles from cell homogenates. The same principle applies when cells are harvested from a liquid medium.

Density gradient centrifugation

The following techniques use a density gradient, a solution which increases in density from the top to the bottom of a centrifuge tube (Fig. 43.3).

- Rate-zonal centrifugation. By layering a sample onto a shallow preformed density gradient, followed by centrifugation, the larger particles will move faster through the gradient than the smaller ones, forming several distinct zones (bands). This method is time-dependent, and centrifugation *must* be stopped before any band reaches the bottom of the tube (Fig. 43.4).
- Isopycnic centrifugation. This technique separates particles on the basis of their buoyant density. Several substances form density gradients during centrifugation (e.g. sucrose, CsCl, Ficoll, Percoll, Nycodenz). The sample is mixed with the appropriate substance and then centrifuged – particles form bands where their density corresponds to that of the medium (Fig. 43.4). This method requires a steep gradient and sufficient time to allow gradient formation and particle redistribution, but is unaffected by further centrifugation.

Bands within a density gradient can be sampled using a fine Pasteur pipette, or a syringe with a long, fine needle. Alternatively, the tube may be punctured and the contents (fractions) collected dropwise in several tubes. For accurate work, an upward displacement technique can be used (see Ford and Graham, 1991, or Graham, 2001).

> ***Working with silicone oil*** *– the density of silicone oil is temperature-sensitive so work in a location with a known, stable temperature or the technique may fail.*

Density barrier centrifugation

A single step density barrier (Fig. 43.3c) can be used to separate cells from their surrounding fluid, e.g. using a layer of silicone oil adjusted to the correct density using dinonyl phthalate. Blood cell types can be separated using a density barrier of e.g. Ficoll.

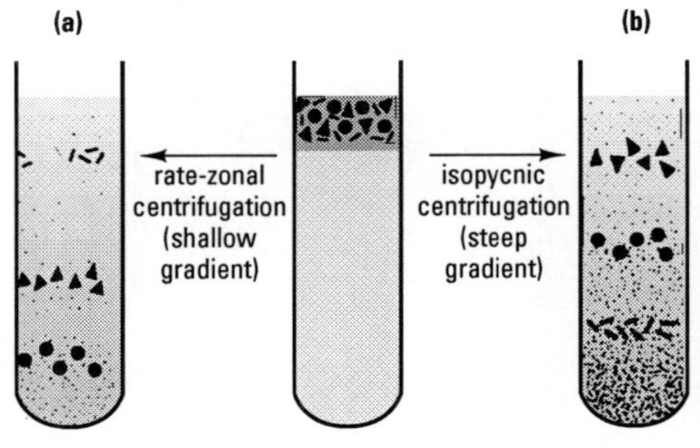

Fig. 43.4 Density gradient centrifugation. The central tube shows the position of the sample prior to centrifugation, as a layer on top of the density gradient medium. Note that particles sediment on the basis of size during rate-zonal centrifugation (a), but form bands in order of their densities during isopycnic centrifugation (b). ●, large particles, intermediate density; ▲, medium-sized particles, low density; ▬ small particles, high density.

Types of centrifuge and their uses

Low-speed centrifuges

These are bench-top instruments for routine use, with a maximum speed of 3000–6000 r.p.m. and RCF up to 6 000 *g* (Fig. 43.1). They are used to harvest cells, larger organelles (e.g. nuclei, chloroplasts) and coarse precipitates (e.g. antibody–antigen complexes, p. 258). Most modern machines also have a sensor that detects any imbalance when the rotor is spinning and cuts off the power supply (Fig. 43.1). However, some of the older models do not, and must be switched off as soon as any vibration is noticed, to prevent damage to the rotor or harm to the operator. Box 43.1 gives details of operation for a low-speed centrifuge.

Microcentrifuges (microfuges)

These are bench-top machines, capable of rapid acceleration up to 12 000 r.p.m. and 10 000 *g*. They are used to sediment small sample volumes (up to 1.5 ml) of larger particles (e.g. cells, precipitates) over short timescales (typically, 0.5–15 min). They are particularly useful for the rapid separation of cells from a liquid medium, e.g. silicone oil microcentrifugation.

Continuous flow centrifuges

Useful for harvesting large volumes of cells from their growth medium. During centrifugation, the particles are sedimented as the liquid flows through the rotor.

> ***Recording usage of high-speed centrifuges and ultracentrifuges*** *– most departments have a log book (for samples/speeds/times): make sure you record these details, as the information is important for servicing and replacement of rotors.*

High-speed centrifuges

These are usually larger, free-standing instruments with a maximum speed of up to 25 000 r.p.m. and RCF up to 60 000 *g*. They are used for microbial cells, many organelles (e.g. mitochondria, lysosomes) and protein precipitates. They often have a refrigeration system to keep the rotor cool at high speed. You would normally use such instruments only under direct supervision.

Ultracentrifuges

These are the most powerful machines, having maximum speeds in excess of 30 000 r.p.m. and RCF up to 600 000 *g*, with sophisticated refrigeration and vacuum systems. They are used for smaller organelles (e.g. ribosomes, membrane vesicles) and biological macromolecules. You would not normally use an ultracentrifuge, though your samples may be run by a member of staff.

> *SAFETY NOTE Changing a rotor – if you ever have to change a rotor, make sure that you carry it properly (don't knock/drop it), that you fit it correctly (don't cross-thread it, and tighten to the correct setting using a torque wrench) and that you store it correctly (clean it after use and don't leave it lying around).*

Rotors

Many centrifuges can be used with tubes of different size and capacity, either by changing the rotor, or by using a single rotor with different buckets/adaptors.

- Swing-out rotors: sample tubes are placed in buckets which pivot as the rotor accelerates (Fig. 43.5a). Swing-out rotors are used on many low-speed centrifuges: their major drawback is their extended path

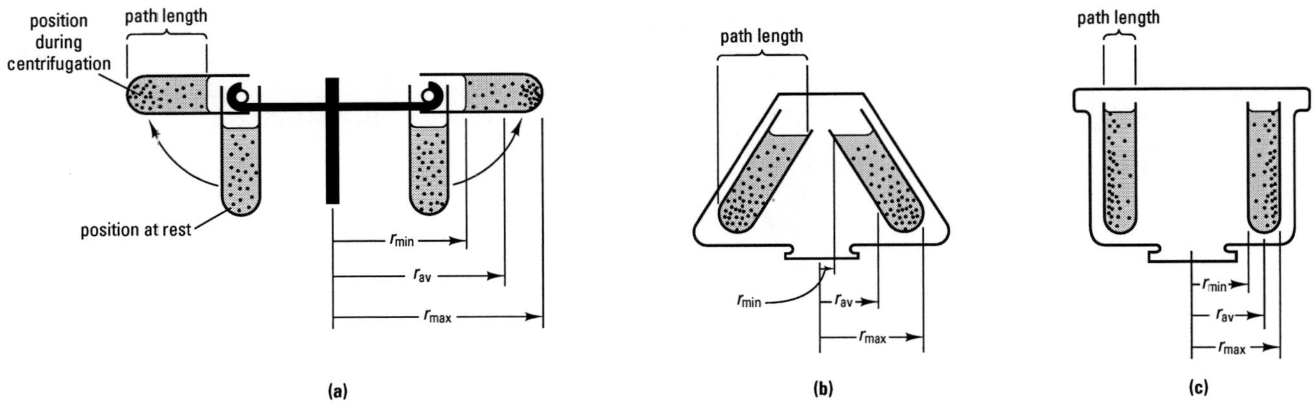

Fig. 43.5 Rotors: (a) swing-out rotor; (b) fixed angle rotor; (c) vertical tube rotor.

length and the resuspension of pellets due to currents created during deceleration.

- Fixed-angle rotors: used in many high-speed centrifuges and micro-centrifuges (Fig. 43.5b). With their shorter path length, fixed rotors are more effective at pelleting particles than swing-out rotors.

Box 43.1 How to use a low-speed bench centrifuge

1. **Choose the appropriate tube size and material for your application,** with caps where necessary. Most low-speed machines have four-place or six-place rotors – use the correct number of samples to *fill* the rotor assembly whenever possible.
2. **Fill the containers to the appropriate level:** do not overfill, or the sample may spill during centrifugation.
3. ⚠ **It is vital that the rotor is balanced during use.** Therefore, *identical* tubes must be prepared, to be placed opposite each other in the rotor assembly. This is particularly important for density gradient samples, or for samples containing materials of widely differing densities, e.g. soil samples, since the density profile of the tube will change during a run. However, for low-speed work using small amounts of particulate matter in aqueous solution, it is sufficient to counter-balance a sample with a second tube filled with water, or a saline solution of similar density to the sample.
4. ⚠ **Balance each pair of sample tubes** (plus the corresponding caps, where necessary) to within 0.1 g using a top-pan balance; add liquid dropwise to the lighter tube, until the desired weight is reached. Alternatively, use a set of scales. For small sample volumes (up to 10 ml) added to disposable, lightweight plastic tubes, accurate pipetting of your solution may be sufficient for low-speed use.
5. **For centrifuges with swing-out rotors,** check that each holder/bucket is correctly positioned in its locating slots on the rotor and that it is able to swing freely. All buckets must be in position on a swing-out rotor, even if they do not contain sample tubes – buckets are an integral part of the rotor assembly.
6. **Load the sample tubes into the centrifuge.** Make sure that the outside of the centrifuge tubes, the sample holders and sample chambers are dry: any liquid present will cause an imbalance during centrifugation, in addition to the corrosive damage it may cause to the rotor. For sample holders where rubber cushions are provided, make sure that these are correctly located. Balanced tubes must be placed opposite each other – use a simple code if necessary, to prevent mix-ups.
7. **Bring the centrifuge up to operating speed** by gentle acceleration. Do not exceed the maximum speed for the rotor and tubes used.
8. ⚠ **If the centrifuge vibrates at any time during use, switch off** and find the source of the problem.
9. **Once the rotor has stopped spinning, release the lid and remove all tubes.** If any sample has spilled, make sure you clean it up thoroughly using a non-corrosive disinfectant, e.g. Virkon, so that it is ready for the next user.
10. **Close the lid (to prevent the entry of dust) and return all controls to zero.**

- Vertical tube rotors: used for isopycnic density gradient centrifugation in high-speed centrifuges and ultracentrifuges (Fig. 43.5c). They cannot be used to harvest particles in suspension as a pellet is not formed.

Centrifuge tubes

SAFETY NOTE Working with centrifuge tubes – never be tempted to use a tube or bottle which was not designed to fit the machine you are using (e.g. a general-purpose glass test tube, or a screw-capped bottle), or you may damage the centrifuge and cause an accident.

These are manufactured in a range of sizes (from 1.5 ml up to 1000 ml) and materials. The following aspects may influence your choice:

- Capacity. This is obviously governed by the volume of your sample. Note that centrifuge tubes must be completely full for certain applications, e.g. for high-speed work.
- Shape. Conical-bottomed centrifuge tubes retain pellets more effectively than round-bottomed tubes, while the latter may be more useful for density gradient work.
- Maximum centrifugal force. Detailed information is supplied by the manufacturers. Standard Pyrex glass tubes can only be used at low centrifugal force (up to 2000 *g*).
- Caps. Most fixed-angle and vertical tube rotors require tubes to be capped, to prevent leakage during use and to provide support to the tube during centrifugation. For low speed centrifugation, caps must be used for any hazardous samples. Make sure you use the correct caps for your tubes.
- Solvent resistance. Glass tubes are inert, polycarbonate tubes are particularly sensitive to organic solvents (e.g. ethanol, acetone), while polypropylene tubes are more resistant. See manufacturer's guidelines for detailed information.
- Sterilisation. Disposable plastic centrifuge tubes are often supplied in sterile form. Glass and polypropylene tubes can be repeatedly sterilised. Cellulose ester tubes should *not* be autoclaved. Repeated autoclaving of polycarbonate tubes may lead to cracking/stress damage.
- Opacity. Glass and polycarbonate tubes are clear, while polypropylene tubes are more opaque.
- Ability to be pierced. If you intend to harvest your sample by puncturing the tube wall, cellulose acetate and polypropylene tubes are readily punctured using a syringe needle.

***Using microcentrifuge tubes** – the integral push-on caps of microcentrifuge tubes must be correctly pushed home before use or they can come off during centrifugation.*

Balancing the rotor

SAFETY NOTE Balancing tubes – never balance centrifuge tubes 'by eye' – use a balance. Note that a 35 ml tube full of liquid at an RCF of 3 000 g has an effective weight greater than a large adult man.

For the safe use of centrifuges, the rotor must be balanced during use, or the spindle and rotor assembly may be damaged permanently; in severe cases, the rotor may fail and cause a serious accident.

***KEY POINT** It is vital that you balance your loaded centrifuge tubes before use. As a general rule, balance all sample tubes to within 1% or better, using a top-pan balance or scales. Place balanced tubes opposite each other.*

Safe practice

Given their speed of rotation and the extremely high forces generated, centrifuges have the potential to be extremely dangerous, if used incorrectly. You should only use a particular centrifuge if you fully understand the operating principles: if unsure, check with a member of staff. For safety reasons, all centrifuges are manufactured with an armoured casing that should contain any fragments in cases of rotor failure. Machines usually have a safety lock to prevent the motor from being switched on unless the lid is closed and to stop the lid from being opened while the rotor is moving. Don't be tempted to use older machines without a safety lock, or centrifuges where the locking mechanism is damaged/inoperative. Be particularly careful to make sure that hair and clothing are kept well away from moving parts.

Text references

Ford, T.C. and Graham, J.M. (1991) *An Introduction to Centrifugation.* Bios, Oxford.

Graham, J. (2001) *Biological Centrifugation.* Bios, Oxford.

Sources for further study

Boyer, R.F. (2005) *Biochemistry Laboratory: Modern Theory and Techniques.* Benjamin Cummings, San Francisco.

Rickwood, D., Ford, T. and Steensgard, J. (1994) *Centrifugation: Essential Data.* Wiley, New York.

Study exercises

43.1 Decide on the type of centrifuge required for a particular application. What centrifuge would you use for each of the following?

(a) Separating mitochondria from a cell homogenate.
(b) Separating yeast cells from their surrounding medium in an experiment to study the uptake of a radiolabelled amino acid as a function of time.
(c) Harvesting cells from a bioreactor containing 25 litres of growth medium.

43.2 Determine centrifugal acceleration for a specific centrifuge. Calculate the centrifugal acceleration (RCF, i.e. the *g* value) of the following (express your answers to three significant figures):

(a) A bench centrifuge with an average radius of rotation of 125 mm operating at 4000 r.p.m.
(b) A bench-top microcentrifuge with an average radius of rotation of 60 mm, operating at 12 000 r.p.m.
(c) An ultracentrifuge with an average radius of rotation of 186 mm, operating at 30 000 r.p.m.

43.3 Determine the speed required to give a particular centrifugal acceleration (*g* value). Calculate the speed required (r.p.m. value) for each of the following (express your answers to three significant figures):

(a) RCF of 1500 *g*, using a bench centrifuge with an average radius of rotation of 95 mm.
(b) RCF of 50 000 *g*, using a high-speed centrifuge with an average radius of rotation of 135 mm.
(c) RCF of 13 000 *g*, using a bench-top microcentrifuge with an average radius of rotation of 5.5 cm.

43.4 Calculate the difference in centrifugal acceleration (*g* value) between the top, middle and bottom of a centrifuge tube. Assuming that the minimum, average and maximum radial distances of a centrifuge tube in a swing-out rotor of a bench centrifuge operating at 5000 r.p.m. are 45 mm, 70 mm and 95 mm respectively, what are the corresponding *g* values at the top, middle and bottom of the tube when the centrifuge is operating? (Express your answer to three significant figures.)

44 Chromatography – separation methods

Making compromises in chromatography – *the process is often a three-way compromise between:*

- *separation of analytes*
- *time of analysis*
- *volume of eluent.*

Thus, if you have a large sample volume and you want to achieve a good separation of a mixture of analytes, the time taken for chromatography will be lengthy.

SAFETY NOTE The solvents used as the mobile phases of chromatographic systems are often toxic and may produce noxious fumes – where necessary, work in a fume hood.

Chromatography is used to separate the individual constituents within a sample on the basis of differences in their physical characteristics, e.g. molecular size, shape, charge, volatility, solubility and/or adsorptivity. The essential components of a chromatographic system are:

- A stationary phase, either a solid, a gel or an immobilised liquid, held by a support matrix.
- A chromatographic bed: the stationary phase may be packed into a glass or metal column, spread as a thin layer on a sheet of glass or plastic, or adsorbed on cellulose fibres (paper).
- A mobile phase, either a liquid or a gas which acts as a solvent, carrying the sample through the stationary phase and eluting from the chromatographic bed.
- A delivery system to pass the mobile phase through the chromatographic bed.
- A detection system to monitor the test substances (Chapter 45).

Individual substances interact with the stationary phase to different extents as they are carried through the system, enabling separation to be achieved.

***KEY POINT** In a chromatographic system, those substances which interact strongly with the stationary phase will be retarded to the greatest extent while those which show little interaction will pass through with minimal delay, leading to differences in distances travelled or elution times.*

Types of chromatographic system

Chromatographic systems can be categorised according to the form of the chromatographic bed, the nature of the mobile and stationary phases and the method of separation.

Thin-layer chromatography (TLC) and paper chromatography

Here, you apply the sample as a single spot near one end of the sheet, by microsyringe or microcapillary. This sheet is allowed to dry fully, then it is transferred to a glass tank containing a shallow layer of solvent (Fig. 44.1). Remove the sheet when the solvent front has travelled across 80–90% of its length.

You can express movement of an individual substance in terms of its relative frontal mobility, or R_F value, where:

$$R_F = \frac{\text{distance moved by substance}}{\text{distance moved by solvent}} \quad [44.1]$$

Alternatively, you may express movement with respect to a standard of known mobility, as R_X, where:

$$R_X = \frac{\text{distance moved by test substance}}{\text{distance moved by standard}} \quad [44.2]$$

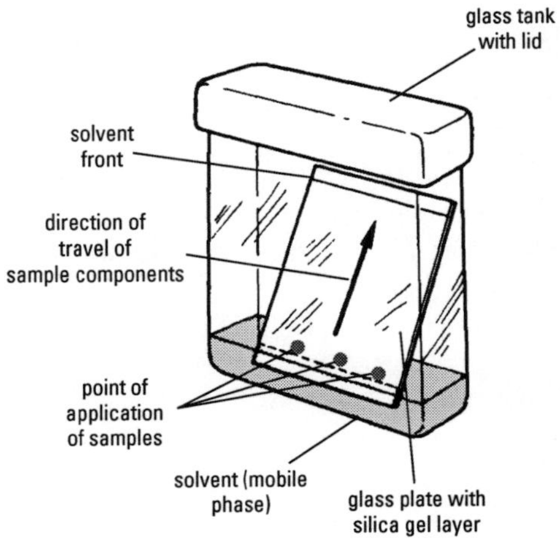

Fig. 44.1 Components of a TLC system.

Using a TLC system *– it is essential that you allow the solvent to pre-equilibrate in the chromatography tank for at least 2 h before use, to saturate the atmosphere with vapour. Deliver drops of sample with a blunt-ended microsyringe. Make sure you know exactly where each sample is applied, so that R_F values can be calculated.*

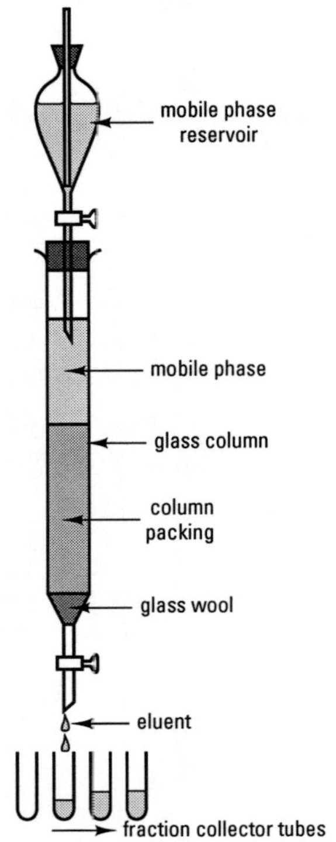

Fig. 44.2 Equipment for column chromatography (gravity feed system).

The R_F (or R_X) value is a constant for a particular substance and solvent system (under standard conditions) and closely reflects the partitioning of the substance between the stationary and mobile phases. Tabulated values are available for a range of biological molecules and solvents (e.g. Stahl, 1969, or Touchstone, 1980). However, you should analyse one or more reference compounds on the same sheet as your unknown sample, to check their R_F values.

Column chromatography

Here, you pack a glass column with the appropriate stationary phase and equilibrate the mobile phase by passage through the column, either by gravity (Fig. 44.2), or using a low pressure peristaltic pump. You can then introduce the sample to the top of the column, to form a discrete band of material. This is then flushed through the column by the mobile phase. If the individual substances have different rates of migration, they will separate within the column, eluting at different times as the mobile phase travels through the column.

You can detect eluted substances by collecting the mobile phase as it elutes from the column in a series of tubes (discontinuous monitoring), either manually or with an automatic fraction collector. Fractions of 2–5% of the bed volume are usually collected and analysed, e.g. by chemical assay. You can now construct an elution profile (or chromatogram) by plotting the amount of substance against either time, elution volume or fraction number, which should give a symmetrical peak for each substance (Fig. 44.3).

You can express the migration of a particular substance at a given flow rate in terms of its retention time (t), or elution volume (V_e). The separation efficiency of a column is measured by its ability to distinguish between two similar substances, assessed in terms of:

- selectivity (α), measured using the following equation, which takes into account the retention times of the two peaks (i.e. t_a and t_b), plus the column dead time (t_0):

$$\alpha = \frac{t_b - t_0}{t_a - t_0} \qquad [44.3]$$

The column dead time is the time it takes for an unretained compound to pass through the column without any interaction with the stationary phase.

- resolution (R), quantified in terms of the retention time and the base width (W) of each peak:

$$R = \frac{2(t_a - t_b)}{W_a + W_b} \qquad [44.4]$$

where the subscripts a and b refer to substances a and b respectively (Fig. 44.3). For most practical purposes, R values of 1 or more are satisfactory, corresponding to 98% peak separation for symmetrical peaks.

Non-symmetrical peaks may result from column overloading, co-elution of solutes, poor packing of the stationary phase, or interactions between the substances and the support material.

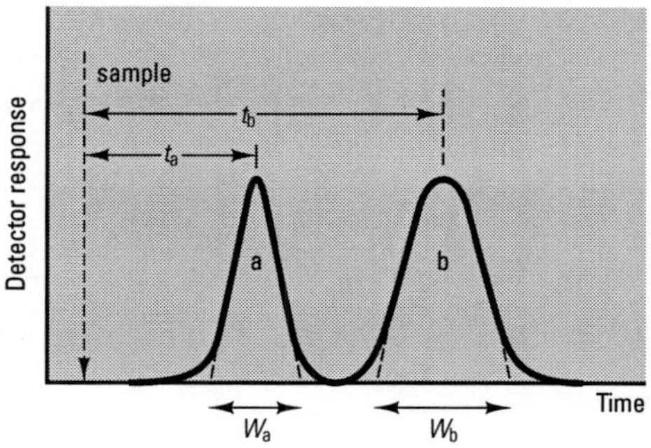

Fig. 44.3 Peak characteristics in a chromatographic separation, i.e. a chromatogram. For symbols, see Eqn [44.4].

***Using HPLC** – this is a versatile form of chromatography, used with a wide variety of stationary and mobile phases, to separate individual compounds of a particular class of molecules on the basis of size, polarity, solubility or adsorption characteristics.*

High performance liquid chromatography (HPLC)

Column chromatography originally used large 'soft' stationary phases that required low pressure flow of the mobile phase to avoid compression; separations were usually time-consuming and of low resolution ('low performance'). Subsequently, the production of small, incompressible, homogeneous particulate support materials and high pressure pumps with reliable, steady flow rates have enabled high performance systems to be developed. These systems operate at pressures up to 10 MPa, forcing the mobile phase through the column at a high flow rate to give rapid separation with reduced band broadening, due to smaller particle size.

HPLC columns are usually made of stainless steel, and all components, valves, etc., are manufactured from materials which can withstand the high pressures involved. The two main systems are:

- Isocratic separation: a single solvent (or solvent mixture) is used throughout the analysis.
- Gradient elution separation: the composition of the mobile phase is altered using a microprocessor-controlled gradient programmer, which mixes appropriate amounts of two different substances to produce the required gradient.

Most HPLC systems are linked to a continuous monitoring detector of high sensitivity, e.g. proteins may be detected spectrophotometrically by monitoring the absorbance of the eluent at 280 nm as it passes through a flow cell (cuvette). Other detectors can be used to measure changes in fluorescence, current or potential (Chapter 45). Most detection systems are non-destructive, which means that you can collect eluent with an automatic fraction collector for further study (Fig. 44.4).

The speed and sensitivity of HPLC make this the method of choice for the separation of many small molecules of biological interest, normally using reverse phase partition chromatography (p. 306). Separation of macromolecules (especially proteins and nucleic acids) usually requires 'biocompatible' systems in which stainless steel components are replaced by titanium, glass or fluoroplastics, using lower pressures to avoid denaturation, e.g. the Pharmacia FPLC system. Such separations are carried out using ion-exchange, gel permeation and/or hydrophobic interaction chromatography (p. 308).

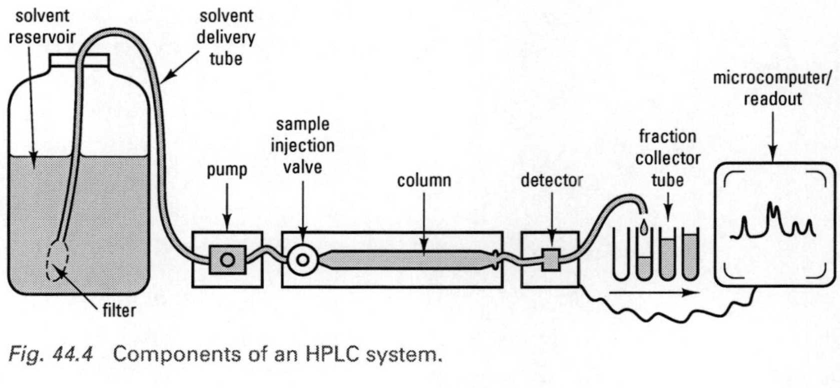

Fig. 44.4 Components of an HPLC system.

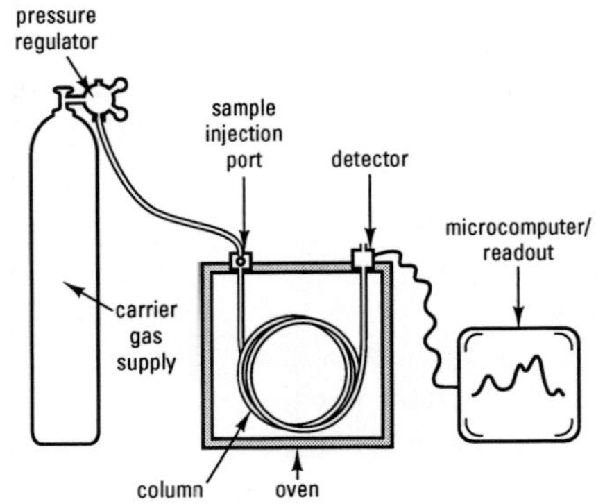

Fig. 44.5 Components of a GC system.

***Applications of gas chromatography** – GC is used to separate volatile, non-polar compounds: substances with polar groups must be converted to less polar derivatives prior to analysis, in order to prevent adsorption on the column, resulting in poor resolution and peak tailing.*

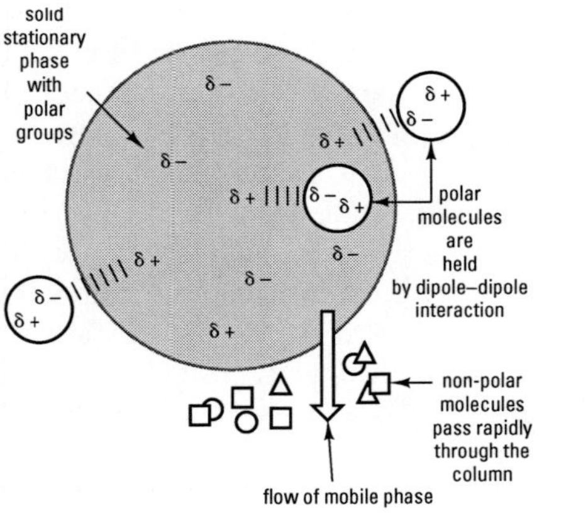

Fig. 44.6 Adsorption chromatography (polar stationary phase).

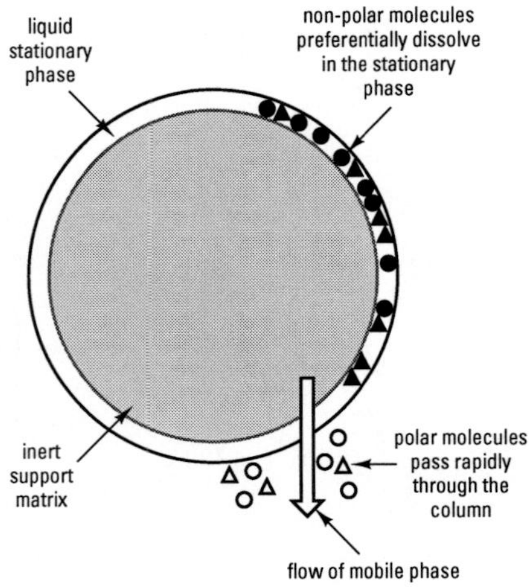

Fig. 44.7 Liquid–liquid partition chromatography, e.g. reverse-phase HPLC.

Gas chromatography (GC)

Modern GC uses capillary chromatography columns (internal diameter 0.1–0.5 mm) up to 50 m in length (Fig. 44.5). The stationary phase is generally a cross-linked silicone polymer, coated as a thin film on the inner wall of the capillary: at normal operating temperatures, this behaves in a similar manner to a liquid film but is far more robust. The mobile phase ('carrier gas') is usually nitrogen or helium. Selective separation is achieved as a result of the differential partitioning of individual compounds between the carrier gas and silicone polymer phases. The separation of most biomolecules is influenced by the temperature of the column, which may be constant during the analysis ('isothermal' – usually 50–250 °C) or, more commonly, may increase in a pre-programmed manner (e.g. from 50 °C to 250 °C at 10 °C per min). Samples are injected onto the 'top' of the column, through a sample injection port containing a gas-tight septum. The output from the column can be monitored by flame ionisation, electron capture or thermal conductivity (Chapter 45).

Spectrometric detection systems include mass spectrometry (GC-MS) and infra-red spectroscopy (GC-IR). GC can only be used with samples capable of volatilisation at the operating temperature of the column, e.g. short chain fatty acids. Other substances may need to be chemically modified to produce more volatile compounds, e.g. long chain saturated fatty acids are usually analysed as methyl esters while monosaccharides are converted to their trimethylsilyl derivatives.

Separation methods

Adsorption chromatography

This is a form of solid–liquid chromatography. The stationary phase is a porous, finely divided solid which adsorbs molecules of the test substance on its surface due to dipole–dipole interactions, hydrogen bonding and/or van der Waals interactions (Fig. 44.6). The range of adsorbents is limited, e.g. polystyrene-based resins (for non-polar molecules), silica, aluminium oxide and calcium phosphate (for polar molecules). Most adsorbents must be activated by heating to 110–120 °C before use, since their adsorptive capacity is significantly decreased in the presence of bound water. Adsorption chromatography can be carried out in column or thin-layer form, using a wide range of organic solvents.

Partition chromatography

This is based on the partitioning of a substance between two liquid phases, in this instance the stationary and mobile phases. Substances which are more soluble in the mobile phase will pass rapidly through the system while those which favour the stationary phase will be retarded (Fig. 44.7). In normal phase partition chromatography the stationary phase is a polar solvent, usually water, supported by a solid matrix (e.g. cellulose fibres in paper chromatography) and the mobile phase is an immiscible, non-polar organic solvent. For reverse-phase partition chromatography the stationary phase is a non-polar solvent (e.g. a C_{18} hydrocarbon, such as octadecylsilane) which is chemically bonded to a porous support matrix (e.g. silica), while the mobile phase can be chosen from a wide range of polar solvents, usually water or an aqueous buffered solution containing one or more organic solvents, e.g. acetonitrile. Solutes interact with the stationary phase through non-polar interactions and so the *least* polar solutes elute last from the column. Solute retention and separation are controlled by changing the composition of the mobile phase (e.g. % v/v acetonitrile). Reverse-phase high performance liquid chromatography (RP–HPLC) is used to separate a broad range of non-polar, polar and ionic biomolecules, including peptides, proteins, oligosaccharides and vitamins for ionic and ionisable solutes.

Selecting a separation method *– it is often best to select a technique that involves direct interaction between the substance(s) and the stationary phase, (e.g. ion exchange, affinity or hydrophobic interaction chromatography), due to their increased capacity and resolution compared to other methods, (e.g. partition or gel permeation chromatography).*

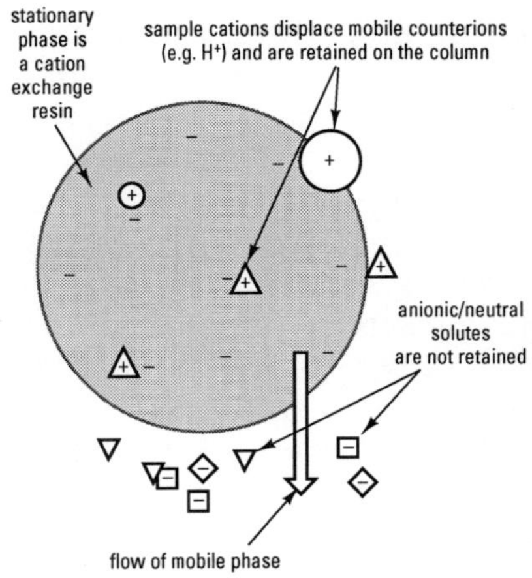

Fig. 44.8 Ion-exchange chromatography (cation exchanger).

Using a gel permeation system *– keep your sample volume as small as possible, in order to minimise band broadening due to dilution of the sample during passage through the column.*

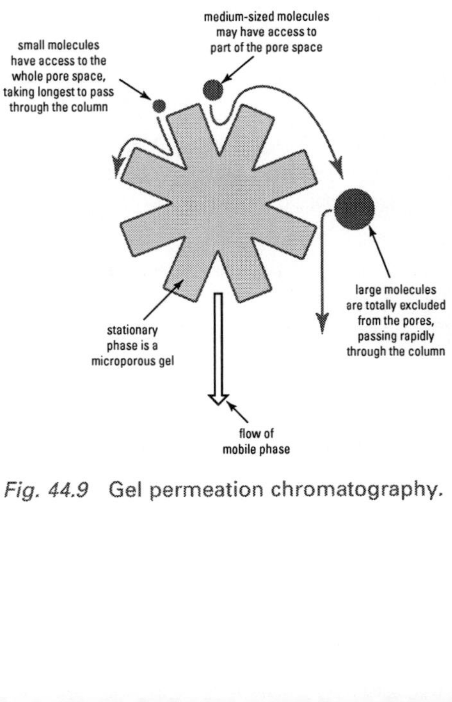

Fig. 44.9 Gel permeation chromatography.

Ion-exchange chromatography

Here, separations are carried out using a column packed with a porous matrix which has a large number of ionised groups on its surfaces, i.e. the stationary phase is an ion-exchange resin. The groups may be cation or anion exchangers, depending upon their affinity for positive or negative ions. The net charge on a particular resin depends on the pK_a of the ionisable groups and the pH of the solution, in accordance with the Henderson–Hasselbalch equation (p. 156).

For most practical applications, you should select the ion-exchange resin and buffer pH so that the test substances are strongly bound by electrostatic attraction to the ion-exchange resin on passage through the system, while the other components of the sample are rapidly eluted (Fig. 44.8). You can then elute the bound components by raising the salt concentration of the mobile phase, either stepwise or as a continuous gradient, so that exchange of ions of the same charge occurs at oppositely charged sites on the stationary phase. Weakly bound sample molecules will elute first, while more strongly bound molecules will elute at a higher concentration. Computer-controlled gradient formers are available: if two or more components cannot be resolved using a linear salt gradient, an adapted gradient can be used in which the rate of change in salt concentration is decreased over the range where these components are expected to elute. To maximise resolution in IEC (and HIC) keep your columns as short as possible. Once the sample components have been separated, they should be eluted as quickly as possible from the column in order to avoid band broadening resulting from diffusion of sample ions in the mobile phase.

Ion-exchange chromatography can be used to separate mixtures of a wide range of ionic biomolecules, including amino acids, peptides, proteins and nucleotides. Electrophoresis (Chapters 46 and 47) is an alternative means of separating charged biomolecules.

Gel permeation chromatography (GPC) or gel filtration

Here, the stationary phase is in the form of beads of a cross-linked gel containing pores of a discrete size (Fig. 44.9). The size of the pores is controlled so that at the molecular level, the pores act as 'gates' that will exclude large molecules and admit smaller ones. However, this gating effect is not an all or nothing phenomenon: molecules of intermediate size partly enter the pores. A column packed with such beads will have within it two effective volumes that are potentially available to sample molecules in the mobile phase, i.e. V_i, the volume surrounding the beads, and V_{ii}, the volume within the pores. If a sample is placed at the top of such a column, the mobile phase will carry the sample components down the column, but at different rates according to their molecular size. A very large molecule will have access to all of V_i but to none of V_{ii}, and will therefore elute in the minimum possible volume (the 'void volume', or V_o, equivalent to V_i). A very small molecule will have access to all of V_i and all of V_{ii}, and therefore

Table 44.1 Fractionation ranges of selected gel permeation chromatography media

M_r	Medium
50–1000	Sephadex G15 Biogel P-2
1000–5000	Sephadex G-25
1500–30 000	Sephadex G-50 Biogel P-10
4000–150 000	Sephadex G-100
5000–250 000	Sephadex G-200
20 000–1 500 000	Sephacryl S 300
60 000–20 000 000	Sepharose 4B

it has to pass through the total liquid volume of the column (V_t, equivalent to $V_i + V_{ii}$) before it emerges. Molecules of intermediate size have access to all of V_i but only part of V_{ii}, and will elute at a volume between V_o and V_t, in order of decreasing size depending on their access to V_{ii}.

Cross-linked dextrans (e.g. Sephadex), agarose (e.g. Sepharose) and polyacrylamide (e.g. Bio-gel) can be used to separate mixtures of macromolecules, particularly enzymes, antibodies and other globular proteins (see Table 44.1 for details). Selectivity in GPC is solely dependent on the stationary phase, with the mobile phase being used solely to transport the sample components through the column. Thus, it is possible to estimate the molecular mass of a sample component by calibrating a given column using molecules of known molecular mass and similar shape. A plot of elution volume (V_e) against $\log_{10}$ molecular mass is approximately linear. A further application of GPC is the general separation of low molecular mass and high molecular mass components, e.g. 'desalting' a protein extract by passage through a Sephadex G-25 column is faster and more efficient than dialysis.

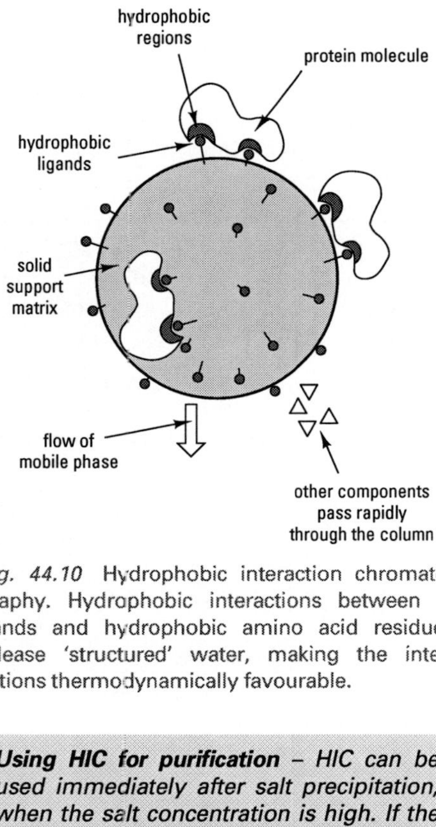

Fig. 44.10 Hydrophobic interaction chromatography. Hydrophobic interactions between ligands and hydrophobic amino acid residues release 'structured' water, making the interactions thermodynamically favourable.

***Using HIC for purification** – HIC can be used immediately after salt precipitation, when the salt concentration is high. If the desired biomolecule is then eluted using a gradient of reducing salt concentration, it may be possible to follow HIC directly by IEC (which requires an initial buffer of low ionic strength) without changing the buffer.*

Hydrophobic interaction chromatography (HIC)

This technique is used to separate proteins, and exploits the fact that many proteins have hydrophobic sites, with hydrophobic amino acid residues (e.g. leucine, isoleucine, valine, phenylalanine) on their surfaces. Proteins differ in the nature and extent of these hydrophobic regions. The underlying principle is similar to that of RP-HPLC in that it involves hydrophobic interactions between the sample components and a non-polar stationary phase. However, RP-HPLC is only useful for analytical separations of proteins where retention of biological activity is not required, and is unsuitable for separation of native proteins for several reasons:

- the stationary phase in RP-HPLC columns tends to be densely packed with hydrophobic groups, leading to tight protein binding, possibly through multi-site attachment;
- the use of polar organic solvents may lead to protein denaturation;
- the high pressures used to obtain rapid flow rates in HPLC may also denature proteins.

The groups used on HIC stationary phases are both less densely packed and less hydrophobic than those used in RP-HPLC, and this results in milder adsorption of proteins (octyl or phenyl groups are commonly used in HIC, rather than octadecyl groups). Furthermore, retention and elution can be achieved using aqueous solutions so that an individual protein can be isolated with its 3-D conformation intact.

Separations are based on interactions between the three components of the system, i.e. the hydrophobic stationary phase, the hydrophobic sample molecules, and the aqueous stationary phase. In an aqueous environment, hydrophobic groups tend to associate, and this results in certain proteins binding to the stationary phase, where the strength of binding is related to the degree of hydrophobicity of the protein (Fig. 44.10). This tendency is promoted by the presence of certain salts, most commonly ammonium sulphate, that produce 'salting out' effects. Elution of components can be achieved by a variety of means:

- reducing the ammonium sulphate concentration of the mobile phase, to decrease the 'salting out' effect;

Avoiding protein precipitation in HIC *– certain protein components may precipitate in high ionic strength buffers. If this occurs, dilute your sample and inject larger volumes.*

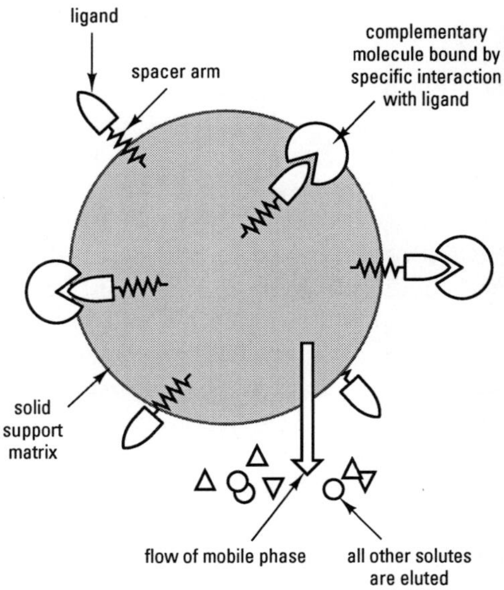

Fig. 44.11 Affinity chromatography.

Examples Biospecific molecules used in affinity chromatography include:

- enzymes and inhibitors/cofactors/substrates;
- hormones and receptor proteins;
- antibodies and antigens;
- complementary base sequences in DNA and RNA.

Elution of substances from an affinity system *– make sure that your elution conditions do not affect the interaction between the ligand and the stationary phase, or you may elute the ligand from the column.*

- changing the salt in the mobile phase to one that does not promote salting out;
- including non-ionic detergents (e.g. Triton X-100) to reduce hydrophobic interactions;
- including aliphatic alcohols, reducing the polarity of the mobile phase;
- changing the pH;
- reducing the temperature.

Affinity chromatography (AC)

Affinity chromatography enables biomolecules to be purified on the basis of their biological specificity rather than by differences in physicochemical properties, and a high degree of purification (>1000-fold) can be expected. It is especially useful for isolating small quantities of material from large amounts of contaminating substances. The technique involves the immobilisation of a complementary binding substance (the ligand) onto a solid matrix in such a way that the specific binding affinity of the ligand is preserved. When a biological sample is applied to a column packed with this affinity support matrix, the molecule of interest will bind specifically to the ligand, while contaminating substances will be washed through with buffer (Fig. 44.11). Elution of the desired molecule can be achieved by changing the pH or ionic strength of the buffer, to weaken the non-covalent interactions between the molecule and the ligand, or by the addition of other substances that have greater affinity for the ligand.

In AC, the ligand must show specific, but reversible, interaction with the molecule to be purified. It must also contain a reactive functional group, independent of the biospecific site, that will allow covalent attachment to the matrix. The support matrix must be free of non-specific adsorption effects and have sufficient reactive functional groups for the attachment of ligands. Agarose is an ideal support matrix for use in AC. If the ligand is small and the molecule to be purified is large, binding may be restricted due to the proximity of the matrix surface. This problem may be overcome by the introduction of a 'spacer arm' (e.g. a hexane group) between the ligand and the matrix (Fig. 44.11).

A potential disadvantage of the specificity of AC is that a new ligand may have to be sought for each individual separation – a potentially time-consuming and expensive process. It may be more practical to use 'group-specific adsorbents' which contain ligands that have an affinity for a class of biochemically related substances. Examples of group-specific adsorbents include:

- Lectins, which are a group of proteins produced from moulds, plants and animals, that bind reversibly with specific sugar residues. They are very useful for the purification of sugar-containing macromolecules such as glycoproteins, serum lipoproteins, and membrane proteins such as receptors. Different lectins show different specificities, e.g. concanavalin A and lentil lectin bind to sugars having –OH groups at C-3, C-4 and C-5 (i.e. mannose and glucose), while wheatgerm lectin binds to *N*-acetyl glucosamine residues. Once bound, substances can be resolved by eluting the column with a salt gradient, or by including free sugars to act as competitive binding agents. Lectins are sometimes called agglutinins because of their ability to agglutinate different types of erythrocyte (e.g. A, B and O) by binding to the specific receptors on their surfaces.

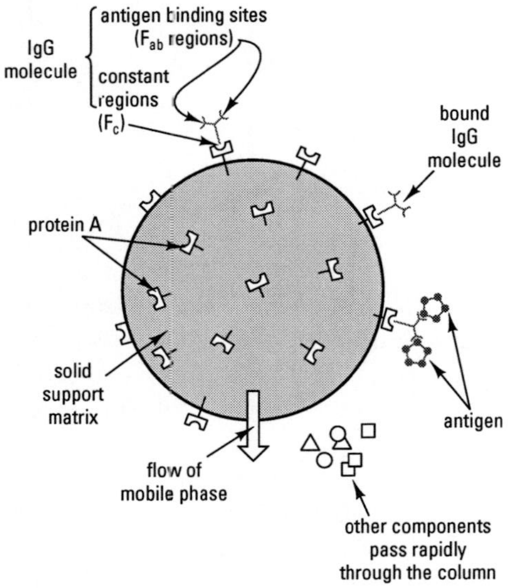

Fig. 44.12 Affinity chromatography using protein A as a ligand to bind IgG antibodies via the F_c region.

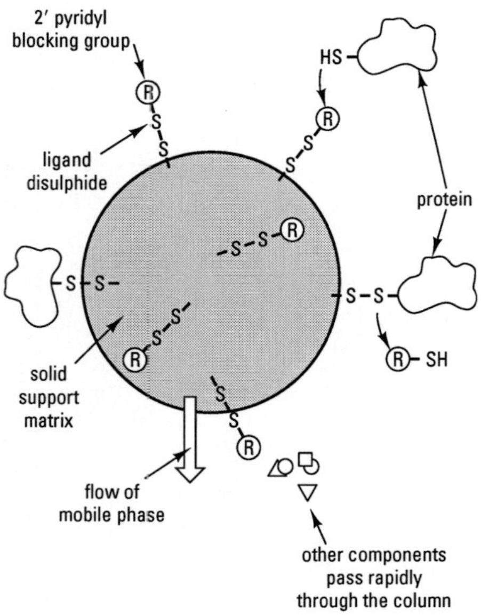

Fig. 44.13 Covalent chromatography. A protein with a free surface thiol group reduces the ligand disulphide, releasing the blocking group and forming a new disulphide.

- Protein A, which is a commercially available surface protein from the bacterium *Staphylococcus aureus* that has a specific binding capacity for the constant (F_c) region of IgG-type antibodies from most mammalian species (Fig. 44.12). As well as being useful in the purification of IgG, protein A can also be used for the isolation of other molecules, as long as an antibody can be raised against them. This is possible because interaction of protein A with IgG occurs via the F_c segment of immunoglobulins of this class, irrespective of their F_{ab} components (see Fig. 38.1). Immobilisation of a specific IgG antibody onto a protein A-agarose column produces an affinity support that can be used to purify the desired antigen.
- Immobilised dyes, which can be used to purify a wide range of enzymes and proteins. One of the most widely used dyes is Cibacron Blue F3G-A, available as an affinity support as Blue Sepharose. This dye has some structural similarities to nucleotide cofactors such as NAD^+ and $NADP^+$, and is useful for the purification of enzymes that require such cofactors. Chemical modification of such dyes increases their specificity, and it is likely that computer modelling will play a role in the future development of new dye-based affinity supports. Dyes are also used in the related non-chromatographic technique of affinity precipitation. This makes use of two dye molecules linked *via* a spacer molecule; when this 'bis-ligand' is added to a solution containing the molecule to be purified, the ligand specifically binds two desired molecules, forming an insoluble complex. Alternatively, a heterobifunctional ligand can be used – a combination of a specific ligand and a group that can be used to initiate precipitation once the desired molecule has bound to the ligand. Such interactions are equivalent to the precipitin reaction in immunology (p. 258).
- Poly(U)-agarose is an affinity support that can be used for the isolation of mRNA because of the biospecific hybridisation of poly(U) with the poly(A) 'tail' sequence characteristic of mRNAs. It can also be used to isolate proteins and enzymes that bind to RNA, such as reverse transcriptases.

Covalent chromatography

This is a variant of affinity chromatography which involves the formation of fairly strong, but reversible, covalent bonds between the affinity support and the molecule to be purified. One type of covalent chromatography is used for the purification of proteins containing thiol (–SH) groups (Fig. 44.13). If the sample is applied to a column containing matrix-attached disulphide 2′-pyridyl groups, thiol-containing proteins displace the 2′-pyridyl group on the support and become immobilised *via* disulphide bridges. Elution of bound components can be achieved by including thiol reagents such as cysteine, glutathione, 2-mercaptoethanol or dithiothreitol (DTT) in the mobile phase. Another form of covalent chromatography uses immobilised boronic acid, which binds certain carbohydrate groups (e.g. in glycoproteins).

Immobilised metal affinity chromatography (IMAC)

This exploits the ability of certain metal ions, especially Ni^{2+}, Cu^{2+}, Zn^{2+}, Hg^{2+} and Cd^{2+}, to bind to proteins by forming coordination complexes with imidazole groups of histidine, indole groups of tryptophan, or thiol groups of cysteine residues. The metal ion may be immobilised by chelation with iminodiacetic acid covalently bound to agarose. Proteins that bind to the metal ions can be eluted using free metal-binding ligands (e.g. amino acids) in the mobile phase.

A very useful application of IMAC is in the purification of 'histidine-tagged' recombinant proteins, e.g. using pET plasmid vectors.
The technique exploits the affinity of histidine-containing proteins for a chelating gel, typically Ni^{2+} immobilised on agarose. The histidine tag present at the *N*- or *C*-terminus of a recombinant protein forms complexes with the transition metal and is therefore bound to the gel. The proteins can be desorbed (eluted) from the gel with increasing concentration of a substrate with an affinity for the chelated metal ion, e.g. imidazole, using either a gradient or a stepwise approach. Alternatively, chelating agents (e.g. EDTA) or low pH can be used for protein elution.

Optimising chromatographic separations

In any chromatographic technique, sample components leave the column at different elution volumes/different times, and are then monitored by a suitable detector (Chapter 45). The responses of the detector are recorded on a chart or a screen in the form of a *chromatogram*. Ideally the sample biomolecules will be completely separated and detection of components will result in a series of discrete individual peaks corresponding to each type of biomolecule (e.g. Fig. 44.3). However, to minimise the possibility of overlapping peaks, or of peaks composed of more than one component, it is important to maximise the separation efficiency of the technique, which depends on:

__Assaying analytes__ – remember that you cannot quantify *a particular analyte without first* identifying *it: the presence of a single peak on a chromatogram does not prove conclusively that a single analyte is present.*

- the selectivity, as measured by the relative retention times of the two components, or by the volume of mobile phase between the peak maxima of the two components after they have passed through the column; this depends on the ability of the chromatographic method to separate two components with similar properties;
- the band-broadening properties of the chromatographic system, which influence the width of the peaks; these are mainly due to the effects of diffusion.

__Separating small biomolecules__ – these will diffuse faster than large molecules, so they should be separated using faster flow rates.

The separation efficiency, or resolution of two adjacent components, can be defined in terms of selectivity and peak broadening, using Eqn [44.4]. In practical terms, good resolution is achieved when there is a large 'distance' (either time or volume) between peak maxima, and the peaks are as narrow as possible. The resolution of components is also affected by the relative amount of each substance: for systems showing low resolution, it can be difficult to resolve small amounts of a particular component in the presence of larger amounts of a second component. If you cannot obtain the desired results from a poorly resolved chromatogram, other chromatographic conditions, or even different methods, should be tried in an attempt to improve resolution. For liquid chromatography, changes in the following factors may improve resolution:

- Stationary phase particle size – the smaller the particle, the greater the area available for partitioning between the mobile phase and the stationary phase. This partly accounts for the high resolution observed with HPLC and FPLC compared with low pressure methods.

- The slope of the salt gradient in eluting IEC or HIC columns, e.g. using computer-controlled adapted gradients.
- In low-pressure liquid chromatography, the flow rate of the mobile phase must be optimised because this influences two band-broadening effects which are dependent on diffusion of sample molecules, i.e. (i) the flow rate must be slow enough to allow effective partitioning between the mobile phase and the stationary phase, and (ii) it must be fast enough to ensure that there is minimal diffusion along the column once the molecules have been separated. To allow for these opposing influences, a compromise flow rate must be used.
- If you prepare your own columns, they must be packed correctly, with no channels present that might result in uneven flow and eddy diffusion.

Learning from experience *– if you are unable to separate your biomolecule using a particular method, do not regard this as a failure, but instead, think about what this tells you about either the substance(s) or the sample.*

Text references

Stahl, E. (1969) *Thin Layer Chromatography – a Laboratory Handbook*, 2nd edn. Springer-Verlag, Berlin.

Touchstone, J.C. (1980) *Thin Layer Chromatography: Quantitative Environmental and Clinical Applications.* Wiley, New York.

Sources for further study

Bliesner, D.M. (2006) *Validating Chromatographic Methods: a Practical Guide.* Wiley, New York.

Boyer, R.F. (2005) *Biochemistry Laboratory: Modern Theory and Techniques.* Benjamin Cummings, San Francisco.
[Also covers other topics, including centrifugation and spectroscopy]

Braithwaite, A. and Smith, F.J. (1996) *Chromatographic Methods*, 5th edn. Blackie Academic and Professional, London.

Cazes, J. (2005) *Encyclopedia of Chromatography.* CRC Press, Boca Raton.

Gooding, K.M. and Regnier, F.E. (2002) *HPLC of Biological Macromolecules Revised and Expanded.* Marcel Dekker, New York.

McMaster, M. (2005) *LC-MS – a Practical User's Guide.* Wiley, New York.
[Covers principles and practice of HPLC systems and MS detectors]

Miller, J.M. (2005) *Chromatography: Concepts and Contrasts,* 2nd edn. Wiley, New York.

Vijayalashmi, M.A. (2002) *Biochromatography.* Taylor & Francis, London.

Wellings, D.A. (2006) *A Practical Handbook of Preparative HPLC.* Elsevier, Amsterdam.
[Gives advice on column preparation and process optimisation]

Wilson, K. and Walker, J. (eds) (2005) *Principles and Techniques of Practical Biochemistry and Molecular Biology,* 6th edn. Cambridge University Press, Cambridge.
[Also covers a range of other topics, including spectroscopy and centrifugation]

Study exercises

44.1 Calculate R_F and R_X values from a chromatogram. The figure represents the separation of three pigments by thin-layer chromatography.

(a) What is the R_F value of each pigment?
(b) What is the mobility of pigments B and C, relative to pigment A (R_A)?

Express both answers to three significant figures.

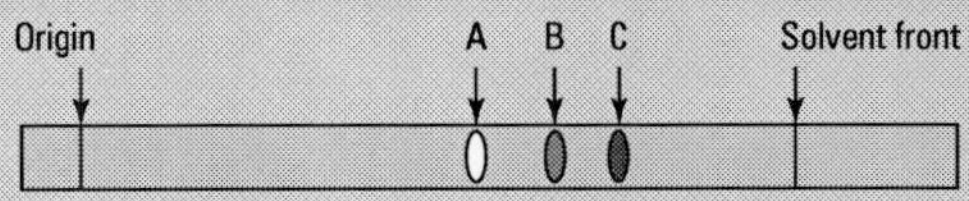

Thin-layer chromatographic separation of a mixture of three pigments (A, B and C).

44.2 Calculate the resolution and selectivity of two components from a chromatogram. Two compounds were separated by column chromatography, giving retention times of 4 min 30 s for A and 6 min 12 s for B, while a compound that was completely excluded from the stationary phase was eluted in 1 min 35 s. The base width of peak A was 40 s and the base width of peak B was 44 s. Calculate (a) the selectivity and (b) the resolution for these two compounds (express all answers to three significant figures).

44.3 Predict chromatographic behaviour on ion-exchange columns using protein titration curves. The variation of net charge with pH for two proteins, A and B, is shown in the figure.

(a) What is the isoelectric point (pI) of each protein?
(b) If a mixture of these proteins was passed down an anion exchange column (e.g. Q-Sepharose) equilibrated with buffer at pH 7.0, how would each protein behave in terms of elution and retention?
(c) If the proteins were bound to a cation exchanger (e.g. S-Sepharose) at pH 5.5, predict the order in which the proteins would elute using a salt gradient.

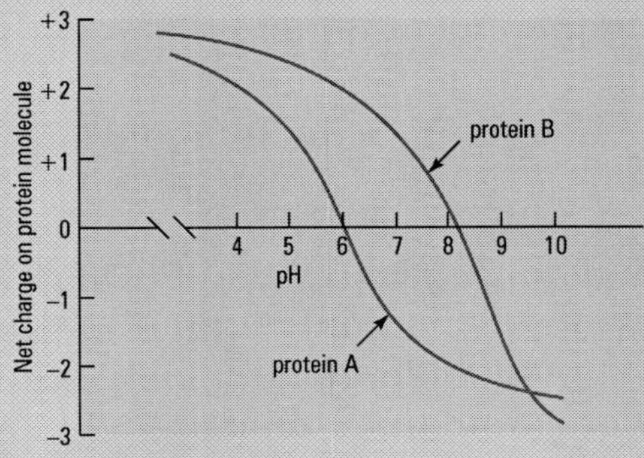

44.4 Test your understanding of the action of ammonium sulphate on proteins. Explain why mobile phases containing a concentration of ammonium sulphate promote the binding of many proteins to HIC columns.

45 Chromatography – detection and analysis

Once the individual components in a sample mixture have been separated by a given chromatographic technique (Chapter 44), a suitable detection system is required to monitor and record the elution of the components of the mixture as they pass through a 'flow cell'.

***KEY POINT** The most appropriate detector depends on the type of chromatography and the application: ideally, the detector should show high sensitivity, a low detection limit and minimal noise or drift.*

Liquid chromatography detectors

UV/visible detectors

***Understanding UV detection of proteins** – note that the absorbance at 280 nm is mostly due to tryptophan and tyrosine residues; if a protein contains low amounts of these amino acids, its absorbance at this wavelength will be low.*

These are widely used and have the advantages of versatility, sensitivity and stability. Such detectors are of two types: fixed wavelength and variable wavelength. Fixed-wavelength detectors are simple to use, with low operating costs. They usually contain a mercury lamp as a light source, emitting at several wavelengths between 254 and 578 nm; a particular wavelength is selected using suitable cut-off filters. The most frequently used wavelengths for analysis of biomolecules are 254 nm (for nucleic acids) and 280 nm (for proteins). Variable-wavelength detectors use a deuterium lamp and a continuously adjustable monochromator for wavelengths of 190–600 nm. For both types of detector, sensitivity is in the absorbance range 0.001–1.0 (down to $\cong$ 1 ng), with noise levels as low as 4×10^{-5}. Note that sensitivity is partly influenced by the path length of the flow cell (typically 10 mm). Monitoring at short wavelength UV (e.g. below 240 nm) may give increased sensitivity but decreased specificity, since many biological molecules absorb in this range. Additional problems with short wavelength UV detection include instrument instability, giving a variable baseline, and absorption by components of the eluting buffer (e.g. TRIS, which absorbs at 206 nm).

An important development in chromatographic monitoring is diode array detection (DAD). The incident light comprises the whole spectrum of light from the source, which is passed through a diffraction grating and the diffracted light detected by an array of photodiodes. A typical DAD can measure the absorbance of each sample component at 1–10 nm intervals over the range 190–600 nm. This gives an absorbance spectrum for each eluting substance which may be used to identify the compound and give some indication as to its purity. An example of a three-dimensional diode array spectrum is shown in Figure 45.1.

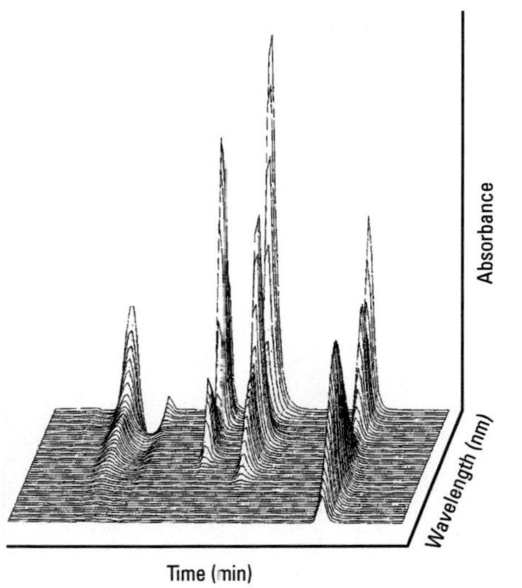

Fig. 45.1 Diode array detector absorption spectra of the eluent from an HPLC separation of a mixture of four steroids, taken every 15 seconds.

Fluorescence detectors

***Overcoming interference with fluorescence detectors** – use a dual flow cell to offset background fluorescence due to components of the mobile phase.*

Many biomolecules, including some vitamins, nucleotides and porphyrins show natural fluorescence (Table 41.1), or can be made to fluoresce by pre-column or post-column derivatisation with a fluorophore. Fluorescence detection is more sensitive than UV/visible detection (p. 382), and may allow analysis in the pg range. A fluorescence detector consists of a light source (e.g. a xenon lamp), a diffraction grating to supply light at the excitation wavelength, and a photomultiplier to monitor the emitted light (usually arranged to be at right angles to the

Maximising sensitivity with fluorescence detectors – *the concentration of other sample components, e.g. pigments, must not be so high that they cause quenching of fluorescence.*

excitation beam). The use of instruments with a laser light source can give an extremely narrow excitation waveband, and increased sensitivity and specificity.

Electrochemical detectors

These offer very high sensitivity and specificity, with the possibility of detection of fg amounts of electroactive compounds such as catecholamines, vitamins, thiols, purines, ascorbate and uric acid. The two main types of detector, amperometric and coulometric, operate on similar principles, i.e. by measuring the change in current or potential as sample components pass between two electrodes within the flow cell. One of these electrodes acts as a reference (or counter) electrode (e.g. calomel electrode), while the other – the working electrode – is held at a voltage that is high enough to cause either oxidation or reduction of sample molecules. In the oxidative mode, the working electrode is usually glassy carbon, while in reductive mode a mercury electrode is used. In either case, a current flow between the electrodes is induced and detected.

Optimising electrochemical detection – *the mobile phase must be free of any compounds that might give a response; all constituents must be of the highest purity.*

Gas chromatography detectors

The most commonly used detectors for GC analysis of biomolecules are:

Flame ionisation detector (FID)

This is a widely used detector, being particularly useful for the analysis of a broad range of organic biomolecules. It involves passing the exit gas stream from the column through a hydrogen flame that has a potential of >100 V applied across it (Fig. 45.2). Most organic compounds, on passage through this flame, produce ions and electrons that create a small current across the electrodes, and this is amplified for measurement purposes. The FID is very sensitive (down to ≈0.1 pg), with a linear response over a wide concentration range. One drawback is that the sample is destroyed during analysis.

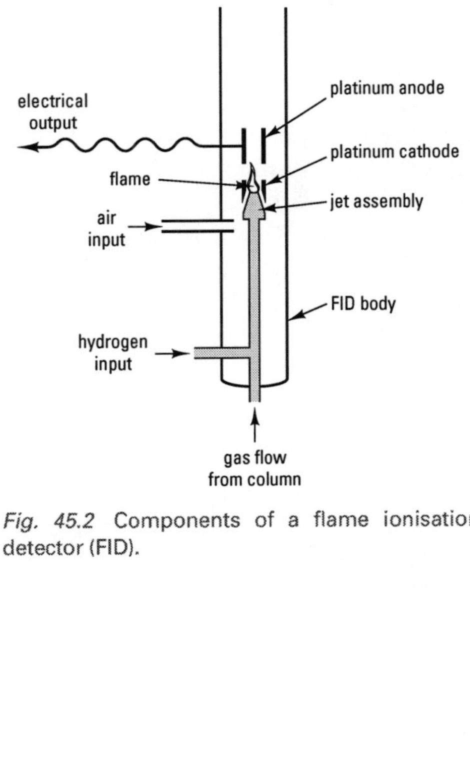

Fig. 45.2 Components of a flame ionisation detector (FID).

Thermal conductivity detector (TCD)

This simple detector is based on changes in the thermal conductivity of the gas stream brought about by the presence of separated sample molecules. The detector elements are two electrically heated platinum wires, one in a chamber through which only the carrier gas flows (the reference detector cell), and the other in a chamber that takes the gas flow from the column (the sample detector cell). In the presence of a constant gas flow, the temperature of the wires (and therefore their electrical resistance) is dependent on the thermal conductivity of the gas. Analytes in the gas stream are detected by temperature-dependent changes in resistance dependent on the thermal conductivity of each separated molecule; the size of the signal is directly related to concentration of the analyte.

The advantages of TCD include its applicability to a wide range of organic and inorganic biomolecules and its non-destructive nature, since the sample can be collected for further study. Its major limitation is its low sensitivity (down to ≈10 ng), compared with other systems.

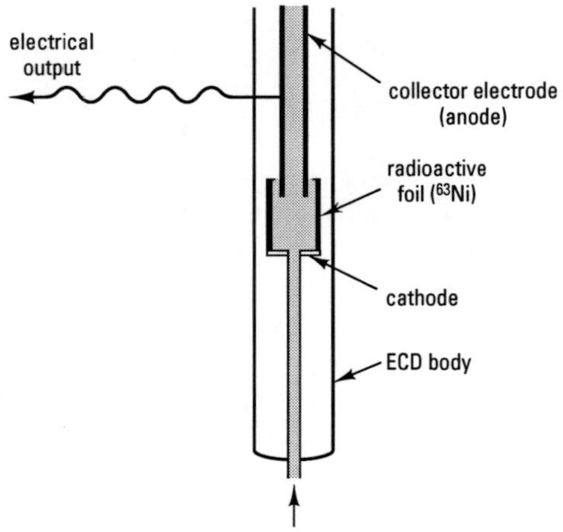

Fig. 45.3 Components of an electron capture detector (ECD).

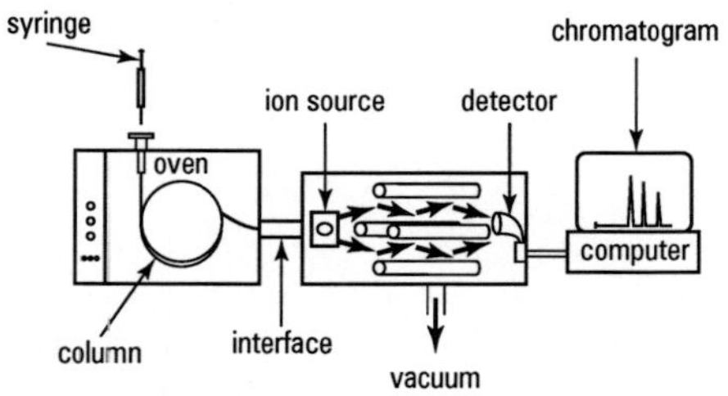

Fig. 45.4 Schematic diagram of a GCMS instrument.

Electron capture detector (ECD)

This highly sensitive detector (Fig. 45.3) is useful for the detection of certain compounds with electronegative functional groups, e.g. halogens, peroxides and quinones. The gas stream from the column passes over a β-emitter (p. 267) such as ^{63}Ni, which provides electrons that cause ionisation of the carrier gas (e.g. nitrogen). When carrier gas alone is passing the β-emitter, its ionisation results in a constant current flowing between two electrodes placed in the gas flow. However, when electron-capturing sample molecules are present in the gas flow, a decrease in current is detected. An example of the application of the ECD is in detecting and quantifying chlorinated pesticides.

Interfacing GC or HPLC with mass spectrometry

Mass spectrometry (p. 294) used in conjunction with chromatographic methods can provide a powerful tool for identifying the components of complex mixtures, e.g. aqueous pollutants. The procedure requires computer control of instruments and data storage/analysis (Fig. 45.4). One drawback is the limited capacity of the mass spectrometer – due to its vacuum requirements – compared with the volume of material leaving the chromatography column. For capillary GC, the relatively small output can be fed directly into the ionisation chamber of the mass spectrometer. For packed column GC, a 'jet separator' is used to remove most of the carrier gas from the molecules to be analysed before they enter the spectrometer. Similarly, in HPLC, devices have been developed for solving the problem of large solvent volumes, e.g. by splitting the eluent from the column so only a small fraction reaches the mass spectrometer.

The computer-generated outputs from the mass spectrometer are similar to chromatograms obtained from other methods, and show peaks corresponding to the elution of particular components. However, it is then possible to select an individual peak and obtain a mass spectrum for the component in that peak to aid in its identification (p. 294). This has helped to identify hundreds of components present in biological systems, including flavour molecules in food, drug metabolites and water pollutants.

Coupling capillary GC columns with FT-IR spectrometers (pp. 289–90) provides another powerful means of separating and identifying compounds in complex biological mixtures.

Recording and interpreting chromatograms

Recording detector output

For analytical purposes, the detector output is usually connected to a computer-based data acquisition and analysis system. This consists of a personal computer (PC) with data acquisition hardware to convert the analogue detector signal to digital format, plus software to control the data acquisition process, store the signal information and display the resulting chromatogram. The software will also detect peaks and calculate their retention times and sizes (areas) for quantitative analysis. The software often incorporates functions to control the chromatographic equipment, enabling automatic operation. In sophisticated systems, the detector output may be compared with that from a 'library' of chromatograms for known compounds, to suggest possible identities of unknown sample peaks.

Interpreting chromatograms – never *assume that a single peak is a guarantee of purity: there may be more than one compound with the same chromatographic characteristics.*

Using a fraction collector – *make sure you can relate individual fractions to the position of peaks on the chromatogram. Most fraction collectors send a signal to the recorder each time a fraction is changed.*

In simpler chromatographic systems, you may need to use a chart recorder for detector output. Two important settings must be considered before using a chart recorder:

1. The baseline reading – this should be set only after a suitable quantity of mobile phase has passed through the column (prior to injection of the sample) and stability is established. The chart recorder is usually set a little above the edge of the chart paper grid, to allow for any baseline drift.
2. The detector range – this must be set to ensure that the largest peaks do not go off the top of the chart. Adjustment may be based on the expected quantity of analyte, or by trial-and-error process. Use the maximum sensitivity that gives intact peaks. If peaks are still too large, you may need to reduce the amount of sample used, or prepare and analyse a diluted sample.

Interpretation of chromatograms

Make sure you know the direction of the horizontal axis (usually, either volume or time) – it may run from right to left or *vice versa*, and make a note of the detector sensitivity on the vertical axis.

Ideally, the baseline should be 'flat' between peaks, but it may drift up or down due to a number of factors including:

Avoiding problems with air bubbles in liquid chromatography – *always ensure that buffers are effectively degassed by vacuum treatment before use, and regularly clean the flow cell of the detector.*

- air bubbles (in liquid chromatography); if the buffers used in the mobile phase are not effectively degassed, air bubbles may build up in the flow cell of the detector, leading to a gradual upward drift of the baseline, followed by a sharp fall when the accumulated air is released. Small air bubbles that do not become trapped may give spurious small peaks.
- changes in the composition of the mobile phase (e.g. in gradient elution, p. 307);
- tailing of material from previous peaks;
- carry-over of material from previous samples; this can be avoided by efficient cleaning of columns between runs – allow sufficient time for the previous sample to pass through the column before you introduce the next sample;
- loss of stationary phase from the column (column 'bleed'), caused by extreme elution conditions;

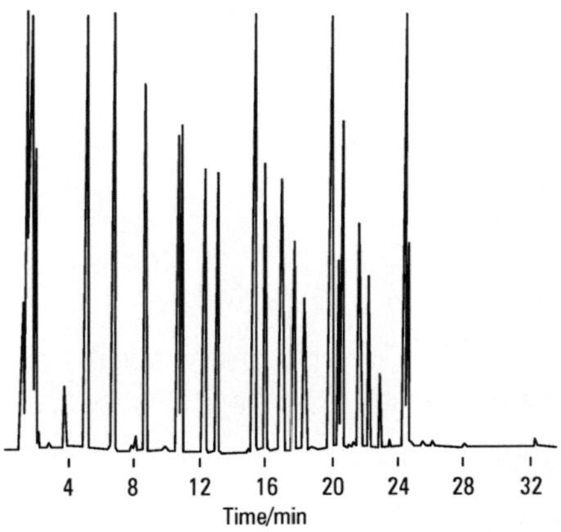

Fig. 45.5 A multicomponent chromatogram. Separation of many compounds, some that are well resolved, e.g. peaks at 12–13 min and others that are not, e.g. peaks at 24–25 min.

A peak close to the origin is likely to be due to non-retained sample molecules, flowing at the same rate as the mobile phase, or to artefacts, e.g. air (GC) or solvent (HPLC) in the sample. Whatever its origin, this peak can be used to measure the column dead time or column void volume (p. 304). No peaks from genuine sample components should appear before this peak.

Peaks can be denoted on the basis of their elution volume (used mainly in liquid chromatography) or their retention times, as in Fig. 45.5 (mainly in GC). If the peaks are not narrow and symmetrical, they may contain more than one component. Where peaks are more curved on the trailing side compared with the leading side, this may indicate too great an association between the component and the stationary phase, or may result from overloading of the column.

***Quantifying biomolecules** – note that quantitative analysis often requires assumptions about the identity of separated components (p. 349) and that further techniques may be required to provide information about the nature of the biomolecules present, e.g. mass spectrometry (see Chapter 42).*

***Using an external standard** – samples and standards should be analysed more than once, to confirm the reproducibility of the technique.*

***Using an internal standard** – you should add the internal standard to the sample at the first stage in the extraction procedure, so that any loss or degradation of test substance during purification is accompanied by an equivalent change in the internal standard, as long as the extraction characteristics of the internal standard and the test substance are very similar.*

Quantitative analysis

Most detectors and chemical assay systems give a linear response with increasing amounts of the test substance over a given 'working range' of concentration. Alternative ways of converting the measured response to an amount of substance are:

- External standardisation: this is applicable where the sample volume is sufficiently precise to give reproducible results (e.g. HPLC, column chromatography). You measure the peak areas (or heights) of known amounts of the substance to give a calibration factor or calibration curve which can be used to calculate the amount of test substance in the sample.
- Internal standardisation: where you add a known amount of a reference substance (not originally present in the sample) to the sample, to give an additional peak in the elution profile. You determine the response of the detector to the test and reference substances by analysing a standard containing known amounts of both substances, to provide a response factor (r), where

$$r = \frac{\text{peak area (or height) of test substance}}{\text{peak area (or height) of reference substance}} \qquad [45.1]$$

Use this response factor to quantify the amount of test substance (Q_t) in a sample containing a known amount of the reference substance (Q_r), from the relationship:

$$Q_t = \frac{\text{peak area (or height) of test substance}}{\text{peak area (or height) of reference substance}} \times (Q_r/r) \qquad [45.2]$$

Internal standardisation should be the method of choice wherever possible, since it is unaffected by small variations in sample volume (e.g. for GC microsyringe injection). The internal standard should be chemically similar to the test substance(s) and must give a peak that is distinct from all other substances in the sample. An additional advantage of an internal standard which is chemically related to the test substance is that it may show up problems due to changes in detector response, incomplete derivatisation, etc. A disadvantage is that it may be difficult to fit an internal standard peak into a complex chromatogram.

Sources for further study

Baugh, P. (1994) *Gas Chromatography: a Practical Approach.* Oxford University Press, Oxford.

Braithwaite, A. and Smith, F.J. (1996) *Chromatographic Methods*, 5th edn. Blackie Academic and Professional, London.

Cazes, J. (2005) *Encyclopedia of Chromatography*. CRC Press, Boca Raton.

Gooding, K.M. and Regnier, F.E. (2002) *HPLC of Biological Macromolecules Revised and Expanded.* Marcel Dekker, New York.

Grob, R.L. and Barry, E.F. (2004) *Modern Practice of Gas Chromatography,* 4th edn. Wiley, New York.

McKay, P. *An Introduction to Chromatography*. Available: http://www.accessexcellence.org/LC/SS/chromatography_background.html
Last accessed: 01/04/07.

Miller, J.M. (2005) *Chromatography: Concepts and Contrasts*, 2nd edn. Wiley, New York.

Scott, R.P.W. (1996) *Chromatography Detectors: Design, Function and Operation.* Marcel Dekker, New York.

Study exercises

45.1 Test your knowledge of detector terminology. Explain what the following acronyms stand for: (a) FID; (b) TCD; (c) ECD; (d) DAD.

45.2 Check your knowledge of liquid chromatography detectors. Make a list of the various major types of liquid chromatography detectors in order, from highest to lowest sensitivity. Which of these methods is most versatile and why?

45.3 Consider how fluorescence detection can be applied to non-fluorescent molecules. Fluoresence detection of natural molecules is limited to molecules such as porphyrins, nucleotides and some vitamins, e.g. riboflavin. What procedures can convert a non-fluorescent molecule into a form suitable for sensitive fluorescence detection? How does this increase the selectivity of the analysis?

45.4 Calculate the amount of substance in a chromatographic separation using an internal standard. The chromatogram shown in the figure on the right represents the separation of three carbohydrates by gas chromatography. Peak A corresponds to mannitol, peak B to sucrose and peak C to the internal standard, trehalose, at 1.50 mg. The results shown in the table were obtained for the retention times and areas of the three peaks.

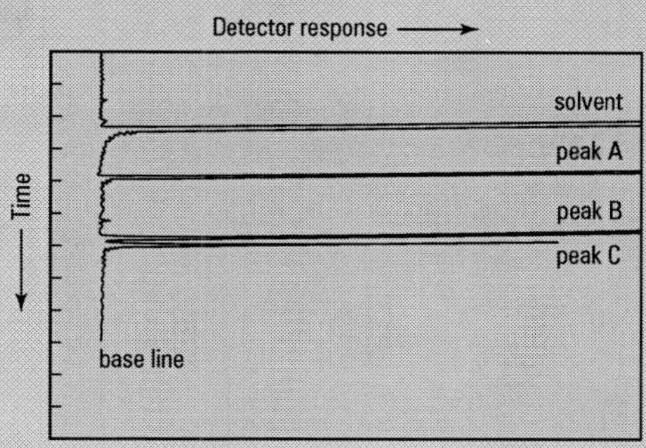

Gas-liquid chromatographic separation of three carbohydrates (A, B and C). Note that the plotter shows peaks A and B to be off-scale: however, the instrument still gives a valid area measurement for these two peaks (see table).

Retention times and peak areas for carbohydrates A, B and C

Peak	Retention time (min)	Area (relative)
A	3.92	2060
B	5.82	1898
C	6.03	604

Given a response factor (r) of 1.26 for mannitol and 0.92 for sucrose (relative to the internal standard), determine the amount of A and B in the sample (express your answer to three significant figures).

46 Principles and practice of electrophoresis

***KEY POINT** Electrophoresis is a separation technique based on the movement of charged molecules in an electric field. Dissimilar molecules move at different rates and the components of a mixture will be separated when an electric field is applied. It is a widely used technique, particularly for the analysis of complex mixtures or for the verification of purity (homogeneity) of isolated biomolecules.*

***Understanding electrophoresis** – this is, in essence, an incomplete form of electrolysis (p. 339), since the applied electrical field is switched off well before sample molecules reach the electrodes.*

While electrophoresis is mostly used for the separation of charged macromolecules, techniques are available for high resolution separations of small molecules such as amino acids (for example, by capillary electrophoresis, p. 331). This chapter deals mainly with the electrophoretic separation of proteins, which has many applications including clinical diagnosis. However, the principles apply equally to other molecules (separation of nucleic acids is considered in Chapter 60).

The electrophoretic mobility of a charged molecule depends on:

- Net charge – negatively charged molecules (anions) migrate towards the anode (+), while positively charged molecules (cations) migrate towards the cathode (–); highly charged molecules move faster towards the electrode of opposite charge than those with lesser charge.
- Size – frictional resistance exerted on molecules moving in a solution means that smaller molecules migrate faster than large molecules.
- Shape – the effect of friction also means that the shape of the molecule will affect mobility, e.g. globular proteins compared with fibrous proteins, linear DNA compared with circular DNA.
- Electrical field strength – mobility increases with increasing electrical potential (voltage), but there are practical limitations to using high voltages, especially due to heating effects.

Definitions

Electrophoretic mobility – the rate of migration of a particular type of molecule in response to an applied electrical field.

Electrical field strength – a measure of the magnitude of an electrical field, usually expressed in terms of electrical potential difference per unit length, as V m^{-1}.

The combined influence of net charge and size means that mobility (μ) is determined by the charge : density ratio or the charge : mass ratio, according to the formula:

$$\mu = \frac{qE}{r} \qquad [46.1]$$

where q is the net charge on the molecule, r is the molecular radius and E is the field strength.

Table 46.1 pK_a values of ionisable groups in selected amino acid residues of proteins

Group/residue	pK_a^*
Terminal carboxyl	3.1
Aspartic acid	4.4
Glutamic acid	4.4
Histidine	6.5
Terminal amino	8.0
Cysteine	8.5
Tyrosine	10.0
Lysine	10.0
Arginine	12.0

*Note that these are typical values: the pK_a will change with temperature and ionic strength. Acidic residues will tend to be negatively charged at pH values above their pK_a, while basic residues will tend to be positively charged below their pK_a.

Electrophoresis and the separation of proteins

The net charge of a sample molecule determines its direction of movement and significantly affects its mobility. The net charge of a protein molecule is pH dependent, and is determined by the relative numbers of positively and negatively charged amino acid side chains at a given pH (Table 46.1). The degree of ionisation of each amino acid side chain is pH dependent, resulting in a variation of net charge on the protein at different pH values (Fig. 46.1). Since an individual protein will have a unique content of ionisable amino acids, each protein will have a characteristic 'titration curve' when net charge is plotted against pH. Thus, electrophoresis is always carried out at *constant* pH and a suitable buffer must be present along with the sample in order to maintain that pH (Chapter 25). If the

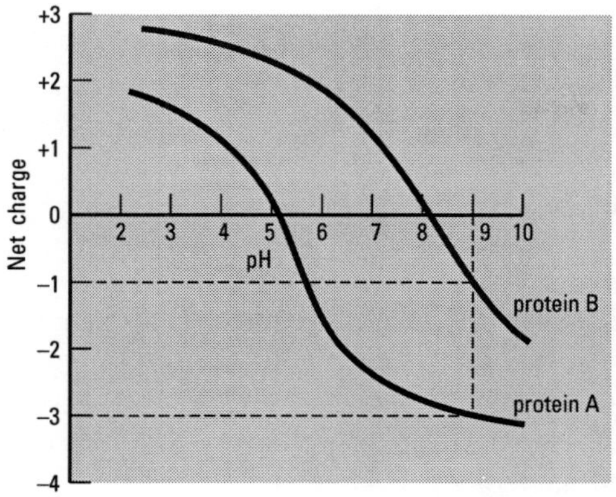

Fig. 46.1 Titration curves for two proteins, A and B containing different proportions of acidic and basic amino acid residues.

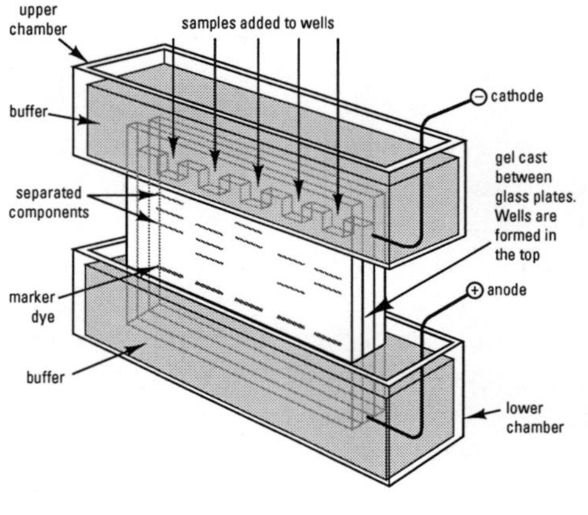

Fig. 46.2 Apparatus for vertical slab electrophoresis (components move downwards from wells, through the gel matrix).

***Minimising diffusion** – make the sample zone as narrow as possible, and fix and stain protein bands as soon as possible after the run.*

Definition

Ohm's law $V = IR$, where V = voltage, I = current and R = resistance.

***Optimising electrophoresis** – attempting to minimise heat production using very low currents is not practical, since it leads to long separation times, and therefore to increased diffusion.*

proteins shown in Figure 46.1 were subjected to electrophoresis at pH 9.0, and if the proteins were of similar size and shape, then the rate at which protein A (net charge, −3) migrates towards the anode would be faster than that for protein B (net charge, −1). Separation of proteins is usually carried out at alkaline pH, where most proteins carry a net negative charge.

Basic apparatus

Most types of electrophoresis using supporting media (described below) are simple to carry out and the apparatus can be easily constructed, although inexpensive equipment is commercially available. High resolution techniques such as 2D-electrophoresis and capillary electrophoresis require more sophisticated equipment, both for separation and analysis (p. 331).

Simple electrophoretic separations can be performed either vertically (Fig. 46.2) or horizontally (Fig. 60.2). The electrodes are normally made of platinum wire, each in its own buffer compartment. In vertical electrophoresis, the buffer solution forms the electrical contact between the electrodes and the supporting medium in which the sample separation takes place. In horizontal electrophoresis, electrical contact can be made by buffer-soaked paper 'wicks' dipping in the buffer reservoir and laid upon the supporting medium. The buffer reservoir normally contains a divider acting as a barrier to diffusion (but not to electrical current), so that localised pH changes which occur in the region of the electrodes (as a result of electrolysis, p. 339) are not transmitted to the supporting medium or the sample. Individual samples are spotted onto a solid supporting medium containing buffer or are applied to 'wells' formed in the supporting medium. The power pack used for most types of electrophoresis should be capable of delivering ≈500 V and ≈100 mA.

Using a supporting medium

The effects of convection currents (resulting from the heating effect of the applied field) and the diffusion of molecules within the buffer solution can be minimised by carrying out the electrophoresis in a porous supporting medium. This contains buffer electrolytes and the sample is added in a discrete location or zone. When the electrical field is applied, individual sample molecules remain in sharp zones as they migrate at different rates. After separation, post-electrophoretic diffusion of selected biomolecules (e.g. proteins) can be avoided by 'fixing' them in position on the supporting medium, e.g. using trichloracetic acid (TCA).

The heat generated during electrophoresis is proportional to the square of the applied current and to the electrical resistance of the medium: even when a supporting medium is used, heat production will lead to zone broadening by increasing the rate of diffusion of sample components and buffer ions. Heat denaturation of sample proteins may also occur, resulting in loss of biological activity e.g. with enzymes. Another problem is that heat will reduce buffer viscosity, leading to a decrease in resistance. If the electrophoresis is run at constant voltage, Ohm's law dictates that as resistance falls, the current will increase, leading to further heat production. This can be avoided by using a power pack that provides constant power. In practice, most electrophoresis equipment incorporates a cooling device;

even so, distortions of an electrophoretic zone from the ideal 'sharp, linear band' can often be explained by inefficient heat dissipation.

Types of supporting media

These can be subdivided into:

- Inert media – these provide physical support and minimise convection: separation is based on charge density only (e.g. cellulose acetate).
- Porous media – these introduce molecular sieving as an additional effect: their pore size is of the same order as the size of molecules being separated, restricting the movement of larger molecules relative to smaller ones. Thus, separation depends on both the charge density and the size of the molecule.

> **Definition**
>
> **Electro-osmotic flow** – the osmotically driven mass flow of water resulting from the movement of ions in an electrophoretic system.

With some supporting media, e.g. cellulose acetate, a phenomenon called electro-endosmosis or electro-osmotic flow (EOF) occurs. This is due to the presence of negatively charged groups on the surface of the supporting medium, attracting cations in the electrophoresis buffer solution and creating an electrical double layer. The cations are hydrated (surrounded by water molecules) and when the electric field is applied, they are attracted towards the cathode, creating a flow of solvent that opposes the direction of migration of anionic biomolecules towards the anode. The EOF can be so great that weakly anionic biomolecules (e.g. antibodies) may be carried towards the cathode.

Where necessary, EOF can be avoided by using supporting media such as agarose or polyacrylamide (p. 323), but it is not always a hindrance to electrophoretic separation. Indeed, the phenomenon of EOF is used in the high resolution technique of capillary electrophoresis (p. 331).

> ***Handling cellulose acetate*** *– the fragile strips must be carefully handled, avoiding touching the flat surfaces with your fingers.*

Cellulose acetate

Acetylation of the hydroxyl groups of cellulose produces a less hydrophilic structure than cellulose in the form of paper: as a result it holds less water and diffusion is reduced, with a corresponding increase in resolution. Cellulose acetate is often used in the electrophoretic separation of plasma proteins in clinical diagnosis – it can be carried out quickly (≈45 min) and its resolution is adequate to detect gross differences in various types of protein (e.g. paraproteins in myeloma). Cellulose acetate has a fairly uniform pore structure and the pores are large enough to allow unrestricted passage of all but the largest of molecules as they migrate through the medium.

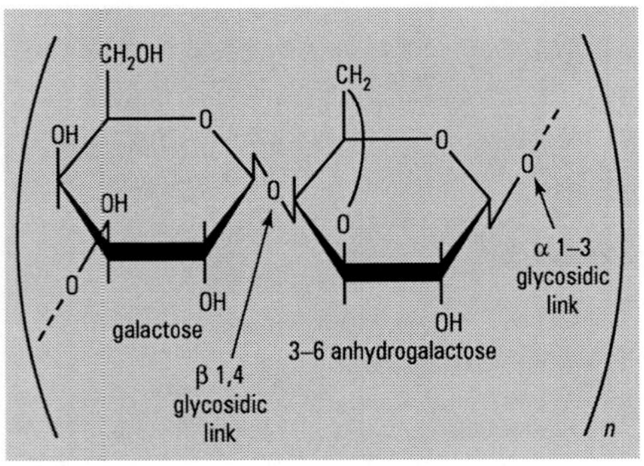

Fig. 46.3 Structure of agarose. Additional sulphate and pyruvyl groups are attached at selected hydroxyls in the polymer.

Agarose

Agarose is the neutral, linear polysaccharide component of agar (from seaweed), consisting of repeating galactose and 3,6-anhydrogalactose subunits (Fig. 46.3). Powdered agarose is mixed with electrophoresis buffer at concentrations of 0.5–3.0% w/v, boiled until the mixture becomes clear, poured onto a glass plate, then allowed to cool until it forms a gel. Gelation is due to the formation of hydrogen bonds both between and within the agarose polymers, resulting in the formation of pores. The pore size depends on the agarose concentration. Low concentrations produce gels with large pores relative to the size of proteins, allowing them to migrate relatively unhindered through the gel, as determined by their individual charge densities. Low concentrations of agarose gel are suitable for techniques such as immunoelectrophoresis (p. 260) and isoelectric focusing (p. 330), where charge is the main basis of separation. The smaller pores produced by higher concentrations of agarose may result in molecular sieving.

When agarose gels are used for the separation of DNA, the large fragment size means that molecular sieving is observed, even with low concentration gels. This is the basis of the electrophoretic separation of nucleic acids (see Chapter 60).

Advantages of polyacrylamide gels *– in addition to their versatility in terms of pore size, these gels are chemically inert, stable over a wide range of pH, ionic strength and temperature, and transparent.*

Polyacrylamide

Polyacrylamide gel electrophoresis (PAGE) has a major role in protein analysis, both for one-dimensional and two-dimensional separations. The gel is formed by polymerising acrylamide monomer into long chains and cross-linking these chains using N,N'-methylene bisacrylamide (often abbreviated to 'bis'). The process is shown in Figure 46.4. In most protocols, polymerisation is initiated by free radicals produced by ammonium persulphate in the presence of N,N,N',N'-tetramethylethylenediamine (TEMED). The photodecomposition of riboflavin can also be used as a source of free radicals.

SAFETY NOTE Preparing gels for PAGE – both acrylamide and bisacrylamide are extremely potent neurotoxins, so you must wear plastic gloves when handling solutions containing these reagents. Although the polymerised gel is non-toxic, it is still advisable to wear gloves when handling the gel, because some monomer may still be present.

The formation of polyacrylamide from its acrylamide monomers is extremely reproducible under standard conditions, and electrophoretic separations are correspondingly precise. The pore size, and hence the extent of molecular sieving, depends on the total concentration of monomer (% T), i.e. acrylamide plus bisacrylamide in a fixed ratio. This means that pores in the gel can be 'tailored' to suit the size of biomolecule to be separated: gels containing 3% acrylamide have large pores and are used in methods where molecular sieving should be avoided (e.g. in isoelectric focusing, p. 330), while higher concentrations of acrylamide (5–30% T) introduce molecular sieving to various degrees depending on the size of the sample components, i.e. with 30% acrylamide gels, molecules as small as M_r 2000 may be subject to molecular sieving. Gels of <2.5% are necessary for molecular sieving of molecules of $M_r > 10^6$, but such gels are almost fluid and require 0.5% agarose to make them solid. Note that a gel of 3% will separate DNA by molecular sieving, due to the large size of the nucleic acid molecules.

Preparing polyacrylamide gels *– most solutions used for gel preparation can be made in advance, but the ammonium persulphate solution must be prepared immediately before use.*

Before you embark on a particular PAGE protein separation, you will need to think about the general strategy for that separation and make certain choices, including whether to use:

Selecting and using gels in 2-D electrophoresis *– a rod gel may be used for the first dimension and a slab gel for the second dimension.*

- Rod or slab gels – flat slab gels are formed between glass plates, using plastic spacers 0.75–1.5 mm thick: rod gels are made in narrow bore tubes. For most separations using several samples, a slab gel saves time because up to 25 samples can be separated under identical conditions in a single gel, while rod gels can only be used for individual samples. Rectangular slab gels are also easier to read, by densitometry, and photograph. However, rod gels are useful in preliminary separations, for determining a suitable pH and gel concentration, and for applications where the gel is sliced in order to extract and assay proteins of interest.
- Dissociating or non-dissociating conditions – the most widely used PAGE protein separation technique uses an ionic detergent, usually sodium dodecyl sulphate (SDS), which dissociates proteins into their

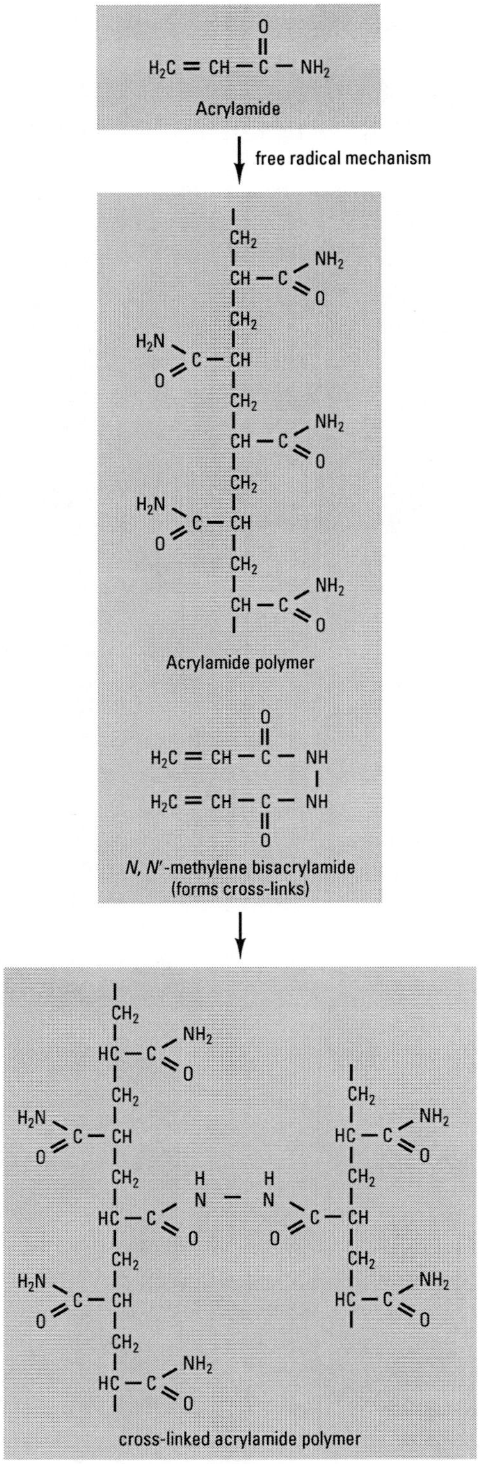

Fig. 46.4 Reactions involved in the formation of polyacrylamide gels.

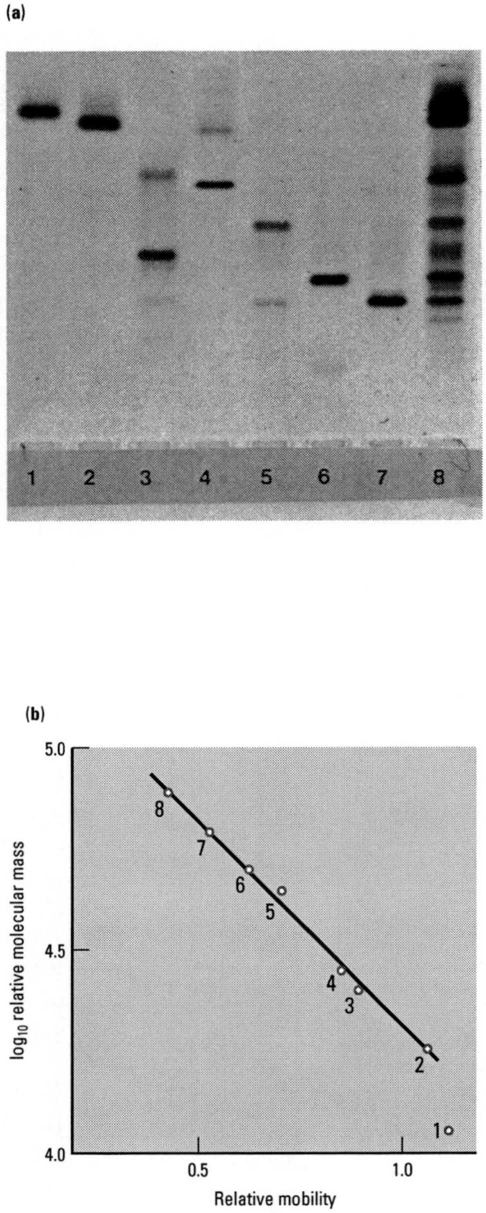

Fig. 46.5 Determination of relative molecular mass (M_r) of proteins by SDS-PAGE: (a) gel samples: 1, cytochrome c; 2, myoglobin; 3, γ-globulin; 4, carbonic anhydrase; 5, ovalbumin; 6, albumin; 7, transferrin; 8, mixture of samples 1–7 (photo courtesy of Pharmacia Biotech). (b) plot of log M_r against distance travelled through gel.

Table 46.2 Molecular masses of standard proteins used in electrophoresis

Protein	M_r	$\log_{10} M_r$
Cytochrome *c*	11 700	4.068
Myoglobin	17 200	4.236
γ-globulin (light chain)	23 500	4.371
Carbonic anhydrase	29 000	4.462
Ovalbumin	43 000	4.634
γ-globulin (heavy chain)	50 000	4.699
Human albumin	68 000	4.832
Transferrin	77 000	4.886
Myosin (heavy chain)	212 000	5.326

Following the progress of PAGE *– add bromophenol blue solution (0.002% w/v) to the sample in the ratio 1:25 (dye: sample). This highly ionic, small M_r dye migrates with the electrophoretic front.*

Choosing a buffer system for PAGE *– discontinuous systems are more time-consuming to prepare, but have the advantage over continuous systems in that relatively large volumes of dilute sample can be used and good resolution is still obtained.*

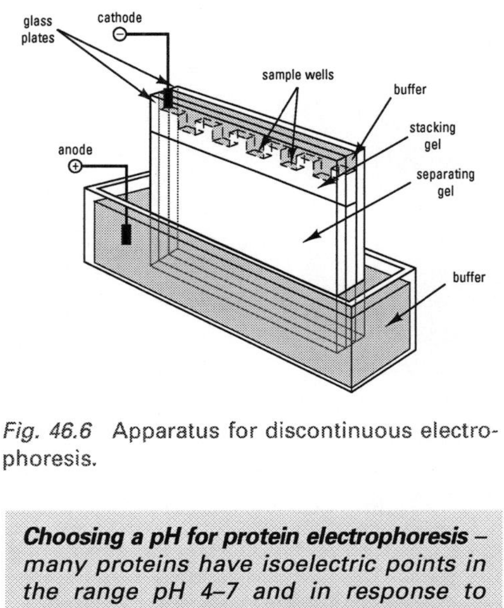

Fig. 46.6 Apparatus for discontinuous electrophoresis.

Choosing a pH for protein electrophoresis *– many proteins have isoelectric points in the range pH 4–7 and in response to electrophoresis with buffers in the region pH 8.0–9.5, most proteins will migrate towards the anode.*

individual polypeptide subunits and gives a uniform net charge along each denatured polypeptide. This technique, known as SDS-PAGE, requires only µg amounts of sample and is quick and easy to carry out. On the other hand, if it is necessary to preserve the native protein conformation and biological activity, non-dissociating conditions are used, i.e. no SDS is added. In SDS-PAGE the sample protein is normally heated to 100 °C for 2 min, in buffer containing 1% (w/v) SDS and 1% (w/v) 2-mercaptoethanol, the latter to cleave any disulphide bonds. The resultant polypeptides bind to SDS in a constant weight ratio, with 1.4 g of SDS per g of protein. As a result, the intrinsic net charge of each polypeptide is 'swamped' by the negative charge imposed by SDS, and there is a uniform negative charge per unit length of polypeptide. Since the polypeptides now have identical charge densities, when they are subject to PAGE (with SDS present) using a gel of appropriate pore size, molecular sieving will occur and they will migrate strictly according to polypeptide size. This not only gives effective separation, but the molecular mass of a given polypeptide can be determined by comparing its mobility to polypeptides of known molecular mass run under the same conditions (Fig. 46.5). Several manufacturers (e.g. Pharmacia, Sigma) supply molecular mass standard kits which may include polypeptides of M_r 11 700 to 212 000 (Table 46.2), together with details of their preparation and use. Where necessary, the treated sample can be concentrated by ultrafiltration and the buffer composition can be altered by diafiltration.

- Continuous or discontinuous buffer systems – a continuous system is where the same buffer ions are present in the sample, gel and buffer reservoirs, all at the same pH. The sample is loaded directly onto a gel (the 'separating gel' or 'resolving gel') that has pores small enough to introduce molecular sieving. In contrast, discontinuous systems have different buffers in the gel compared to the reservoirs, both in terms of buffer ions and pH. The sample is loaded onto a large-pore 'stacking gel', previously polymerised on top of a small-pore separating gel (Fig. 46.6). The individual proteins in the sample concentrate into very narrow zones during their migration through the large-pore gel and stack up according to their charge densities, prior to separation in the small-pore gel, giving enhanced results compared with continuous systems.
- pH and buffer for the separation – PAGE can be carried out between pH 3 and 10. The pH is not critical for continuous SDS-PAGE, since SDS-treated polypeptides are negatively charged over a wide pH range. However, when using non-dissociating systems, the pH can be critical, particularly if the biological activity of the molecules is to be retained.
- Gel concentration – at one extreme, a gel with a very high percentage T might totally exclude the sample components, while a gel with a very low percentage T might lead to all SDS-treated proteins migrating at the same rate, i.e. with the electrophoretic front. A sensible approach is to set up a series of rod gels in the range 5–15% T and observe the separation and resolution obtained. Alternatively, a *gradient gel* can be used, in which the percentage T increases, and hence pore size decreases, in the direction of protein migration. A useful gradient for a preliminary experiment would be 5–20% T. Such

Table 46.3 Preparation of gels for PAGE and SDS-PAGE. The gel solutions are made by mixing the components in the proportions and in the order shown. Figures are ml of each solution required to give the stated % gel strength

Solution (added in order shown)	PAGE			SDS-PAGE		
	3.5% gel (T = 3.6%)	5% gel (T = 5.1%)	7.5% gel (T = 7.7%)	5% gel (T = 5.1%)	7.5% gel (T = 7.7%)	10% gel (T = 10.2%)
1. Distilled water	19.3	14.9	7.5	14.9	7.5	—
2. TRIS-glycine buffer, pH 8.9, 0.1 mol $^{-1}$	33.0	33.0	33.0	—	—	—
3. Imidazole buffer, pH 7.0, 0.1 mol l^{-1} plus 0.2% w/v SDS	—	—	—	33.0	33.0	33.0
4. Acrylamide solution 22.2% w/v and 0.6% w/v bis	10.4	14.8	22.2	14.8	22.2	29.7
5. Ammonium persulphate solution, 0.15% w/v	3.2	3.2	3.2	3.2	3.2	3.2
6. TEMED	0.1	0.1	0.1	0.1	0.1	0.1
Final volume (ml)	66.0	66.0	66.0	66.0	66.0	66.0

Separating protein mixtures *– for high resolution, a combination of dissociating and discontinuous PAGE in slab gels is the system of choice.*

What to do if your polyacrylamide gels fail to set *– polymerisation is inhibited by oxygen, so solutions should be degassed, and the surfaces of the polymerisation mixture exposed to air should be overlaid with water; if your gels still do not polymerise, the most common cause is the use of 'old' ammonium persulphate stock solution. If low pH buffers are used, polymerisation may be delayed because TEMED is required in the free base form.*

gels are able to resolve protein mixtures with a wide range of molecular masses. Furthermore, as proteins migrate into regions of ever-decreasing pore size, the movement of the leading edge of a zone will become increasingly restricted. This allows the trailing edge to catch up, resulting in considerable zone sharpening. Gradient formers are fairly simple to make, and are commercially available.

Practical details of the preparation of PAGE and SDS-PAGE gels are given in Table 46.3 (see Westermeier, 2004, or Gersten, 1996 for further details).

Post-electrophoretic procedures – handling of the supporting medium, staining and analysis

For protein electrophoresis, the following stages are appropriate.

Handling

All types of supporting medium should be handled carefully: wearing gloves is advisable, for safety and to avoid transfer of proteins from the skin. Cellulose acetate strips should be immediately transferred to a fixing and staining solution. Agarose gels should be dried quickly before staining. Polyacrylamide gels in vertical slabs must be freed carefully from one of the glass plates in which they are formed, taking care to lever the glass at a point well away from the part of the gel containing the wells: once free, the gels should be immediately transferred to fixing or staining solution.

Rod gels are recovered by a process called 'rimming', a technique that takes a little practice to master. To remove the gel from the tube, hold the tube in one hand and, using a syringe with a long blunt needle, squirt water between the gel and the inner wall of the tube: while squirting, move the needle gently up and down and rotate the tube. If the gel does not become free, perform the same procedure at the other end of the tube, until the gel is released. If necessary, a Pasteur pipette bulb applied to one end can be used to push the gel out of the tube.

Fixing, staining and destaining

Handling gels *– avoid touching gels with paper as it sticks readily and is difficult to remove without tearing.*

To prevent the separated proteins from diffusing, they are usually fixed in position. Proteins on cellulose acetate strips can be fixed and stained using Ponceau S (0.2% w/v in a solution of 3% v/v aqueous trichloracetic acid, TCA).

Avoiding overloaded gels and band distortion *– determine the protein concentration of the sample beforehand. Around 100 µg of a complex mixture, or 1–10 µg of an individual component, will be sufficient, but bear in mind that underloading may result in bands being too faint to be detected.*

Fixing is also required for most types of gel electrophoresis: 3% v/v TCA is often used. The most widely used stain for protein separations in gels is Coomassie Blue R-250 (where R = reddish hue): the detection limit is ≈0.2 µg and staining is quantitative up to 20 µg for some proteins. It is normal for background staining of the gel to occur, and removal of background colour ('destaining') can be achieved either by diffusion or electrophoresis. To destain by diffusion, transfer the gel to isopropanol : acetic acid : water (12.5 : 10 : 77.5 v/v/v) and allow to stand for 48 h, or change the solution several times to speed up the staining process. Electrophoretic destaining can remove Coomassie Blue, which is anionic: stained gels are placed between porous plastic sheets with electrodes on each side, and the tank is filled with 7% acetic acid. Passing a current of up to 1.0 A through the gel for around 30–40 min. should result in substantial destaining.

If you need greater sensitivity (e.g. for ng to fg amounts), or when using high resolution techniques such as 2D-electrophoresis (p. 330), silver staining can be used. Depending on the protocol chosen and the proteins being stained, the silver technique can be 5 to 200-fold more sensitive than Coomassie Blue. The method involves a fixation step (e.g. with TCA), followed by exposure to silver nitrate solution and development of the stain. The silver ions are thought to react with basic and thiol groups in proteins, and subsequent reduction (e.g. by formaldehyde at alkaline pH, or by photodevelopment) leads to deposits of silver in the protein bands. Most proteins stain brown or black, but lipoproteins may stain blue, and some glycoproteins stain yellow or red. Some proteins lacking in amino acids with reducing groups (e.g. those lacking cysteine residues) may stain negatively, i.e. the bands are more transparent than the background staining of the gel. Although many protocols have been published, silver stain kits are commercially available, e.g. from Bio-Rad.

Optimising resolution *– keep the sample volume as small as possible; methods for concentrating protein solutions are considered. For vertical slab gels and for rod gels, include 10% w/v sucrose or glycerol to increase density and allow buffer solution to be overlaid on the sample without dilution.*

Although silver staining has clear advantages in terms of sensitivity, for routine work it is more laborious and expensive to carry out than the Coomassie Blue method. It also requires high purity water, otherwise significant background staining occurs. Another feature is that the staining can be non-specific, since DNA and polysaccharides may stain on the same gel as proteins.

Understanding the terminology of high purity water *– this is usually produced by reverse osmosis and its purity is often expressed with respect to its resistivity: ultrapure water has a resistivity of 18 MΩ cm^{-1} (informally referred to as 'eighteen megaohm water').*

Other methods for the detection of separated components include:

- autoradiography for proteins labelled with ^{32}P or ^{125}I (p. 271);
- fluorescence for proteins pre-labelled with fluorescent dyes;
- periodic acid-Schiff (PAS) stain using dansyl hydrazine for glycoproteins.

***Origins of 'Western' blotting** – following the description of 'Southern' blotting of DNA by Dr. Ed Southern, other points of the compass have been used to describe other forms of blotting, with 'Western' blotting for proteins.*

Blotting

The term 'blotting' refers to the transfer of separated proteins from the gel matrix to a thin sheet such as nitrocellulose membrane (commercially available from e.g. Millipore, Amersham). The proteins bind to this membrane, and are immobilised. Blotting of proteins is usually achieved by electrophoretic transfer, and this process is normally referred to as Western blotting (see also Southern blotting and Northern blotting for DNA and RNA respectively). Its major advantage is that the immobilised proteins on the surface of the membrane are readily accessible to detection reagents, and staining and destaining can be achieved in less than 5 min. Use of labelled antibodies to detect specific proteins (immunoblotting) can take less than 6 h. In addition, it is easy to dry and store Western blots for long periods, for further analysis.

Example For detection of lactate dehydrogenase (LDH): when a solution containing lactate, NAD^+, phenazine methasulphate (PMS) and methyl thiazolyl tetrazolium (MTT) is added to a gel containing LDH, a series of redox reactions occurs in the enzyme-containing regions, starting with the oxidation of lactate to pyruvate, and proceeding via NAD^+, PMS and MTT to the eventual reduction of MTT to formazan dye. After incubation at 37 °C in the dark (since MTT is light-sensitive), LDH is detected by the appearance of blue–black bands on the gel.

Detection of enzymes

If you need to detect enzyme activity you should use a non-denaturing gel. The gel matrix will hinder diffusion of the enzyme, but will allow access to the small molecular weight substrates, co-factors and dyes necessary to localise enzyme activity *in situ*. Most methods for detecting enzymes on gels are modifications of protocols originally developed by histochemists, e.g.:

- NAD^+-requiring oxidoreductases can be detected by incubating the gels with substrate, NAD^+ and a solution of a tetrazolium salt which, when reduced, forms an insoluble coloured formazan dye.
- Transferases and isomerases can be detected by coupling their reactions to an oxidoreductase-requiring reaction, visualised as described above.
- Hydrolases can be detected using appropriate chromogenic or fluorogenic substrates.
- Phosphatases can be detected by precipitating any phosphate released from the substrate with Ca^{2+}.

***Detecting enzymes in gels** – minimise zone spreading by incorporating substrates, etc. in a thin agarose indicator gel poured over the separating gel.*

For more details, see Manchenko (2002).

Recording and quantification of results

A number of expensive, dedicated instruments are available for the analysis of gels, e.g. laser densitometers. Alternatively, gel scanning attachments can be purchased for standard spectrophotometers, allowing measurement and recording of the absorbance of the Coomassie Blue stained bands at 560–575 nm: for instruments connected to a computer, quantification of individual components can be achieved by integrating the areas under the peaks.

***Measuring peak areas** – a valid 'low-tech' alternative to computer-based systems is to cut out and weigh peaks from a recorder chart.*

You can photograph gels using a conventional camera (fine grain film), digital camera, or using a photocopier. Alternatively, a dedicated image capture and analysis system may be used, e.g. GelDoc. The gel should be placed on a white glass transilluminator. A red filter will increase contrast with bands stained with Coomassie Blue. If the gels themselves need to be retained, they can be preserved in 7% acetic acid. Alternatively, they can be dried using a commercially available gel dryer.

Text references

Gersten, D. (1996) *Gel Electrophoresis: Proteins* (Essential Techniques Series). Wiley, New York.

Manchenko, G.P. (2002) *Handbook of Detection of Enzymes on Electrophoretic Gels,* 2nd edn. CRC Press, Boca Raton.

Westermeier, R. (2004) *Electrophoresis in Practice: A Guide to Methods and Applications of DNA and Protein Separations*, 4th edn. Wiley VCH, Berlin.

Sources for further study

Anon. *The American Electrophoresis Society Homepage.* Available: http://www.aesociety.org
Last accessed: 01/04/07.
[Includes details on a variety of electrophoretic techniques]

Anon. *British Society for Proteomics Research Homepage.* Available: http://www.bspr.org/
Last accessed: 01/04/07.
[Formerly the British Electrophoresis Society, includes links to other relevant websites and databases]

Hames, B.D. and Rickwood, D. (1998) *Gel Electrophoresis of Proteins: A Practical Approach*, 3rd edn. Oxford University Press, Oxford.

Martin, R.M. (1996) *Gel Electrophoresis: Nucleic Acids* (Introduction to Biotechniques Series). Bios, Oxford. [An introductory text, starting from first principles, with practical examples]

Rabilloud, T. (2004) *Proteome Research: Two-dimensional Gel Electrophoresis and Identification Methods.* Springer, Heidelberg.

Tietz, D. (1998) *Nucleic Acid Electrophoresis.* Springer-Verlag, Berlin.

Study exercises

46.1 Find out why the net charge on a protein molecule varies with pH. Identify the amino acids primarily responsible for determining the net charge on a protein molecule and draw simple diagrams to represent the ionisation of their side chains, indicating how you would expect these side chains to be charged at acid, neutral and alkaline pH values.

46.2 Explain the function of the various reagents used in SDS-PAGE analysis of proteins. What is the function of each of the following SDS-PAGE reagents? (a) acrylamide; (b) *N,N′*-methylene bisacrylamide; (c) ammonium persulphate; (d) *N,N,N′,N′*-tetramethylethylenediamine; (e) sodium dodecyl sulphate; (f) 2-mercaptoethanol; (g) buffer; (h) Coomassie Blue R-250; (i) bromophenol blue.

46.3 Test your knowledge of 'blotting' terminology. What is Western blotting, and how does it differ from Northern and Southern blotting?

46.4 Determine the M_r of subunits of a protein by SDS-PAGE. A number of proteins of known molecular weight were analysed by SDS-PAGE and the results are shown in the table. In the same experiment, a sample of purified lactate dehydrogenase (LDH), was treated in identical fashion and run, producing a band with a relative mobility of 0.77. (a) What is the M_r of the polypeptide produced from the LDH sample?

Relative mobilities of molecular weight standards by SDS-PAGE

Protein	M_r	Relative mobility
Myoglobin	17 200	1.00
Carbonic anhydrase	29 000	0.84
Ovalbumin	43 000	0.67
Human albumin	68 000	0.51
Transferrin	77 000	0.46

(b) If the M_r of native LDH is 136 000, what do you conclude about the subunit structure of this enzyme?

46.5 Find out about 'troubleshooting' problems with gels. Investigate the pitfalls of working with and polymerising gels.

47 Advanced electrophoretic techniques

Using electrophoresis in proteomics – *note that many of the techniques described in this chapter are employed to generate the information used in bio-informatic analysis (Chapter 11).*

Although the resolution obtained using the basic electrophoretic techniques described in Chapter 46 is adequate for many biomolecular applications, a number of advanced techniques are available that give very high resolution and which can be used with very small amounts of sample material.

Isoelectric focusing (IEF)

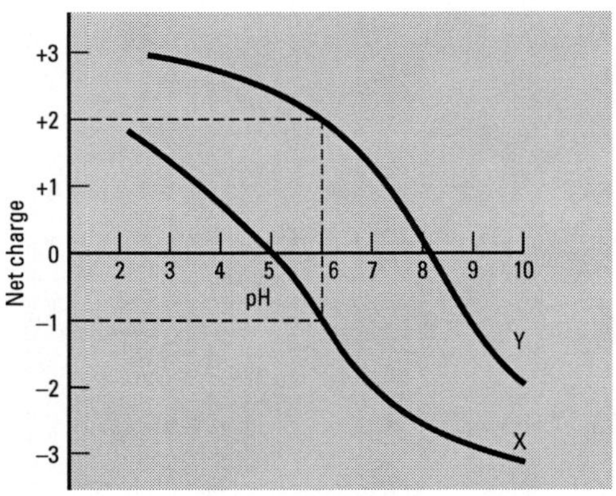

Fig. 47.1 Titration curves for two proteins, X and Y.

In contrast to electrophoresis, which is carried out at constant pH, IEF is carried out using a pH gradient. The gradient is formed using small molecular mass ampholytes, which are analogues and homologues of polyamino-, polycarboxylic acids that collectively have a range of isoelectric points (pI values) between pH 3 and 10. The mixture of ampholytes (p. 152), either in a gel or in free solution, is placed between the anode in acid solution (e.g. H_3PO_4), and the cathode in alkaline solution (e.g. NaOH). When an electric field is applied, each ampholyte migrates to its own pI and forms a stable pH gradient which will persist for as long as the field is applied. When a protein sample is applied to this gradient separation is achieved, since individual proteins will migrate to their isoelectric points. The net charge on the protein when first applied will depend on the specific 'titration curve' for that protein (Fig. 47.1). As an example, consider two proteins, X and Y, having pI values of pH 5 and pH 8 respectively, which are placed together on the gradient at pH 6 (Fig. 47.2). At that pH, protein X will have a net negative charge, and will migrate towards the anode, progressively losing charge until it reaches its pI (pH 5) and stops migrating. Protein Y will have a net positive charge at pH 6, and so will migrate towards the cathode until it reaches its pI (pH 8).

Using a polyacrylamide gel as a supporting medium and a narrow pH gradient, proteins differing in pI by 0.01 units can be separated. Even greater resolution is possible in free solution (e.g. in capillary electrophoresis, p. 331). Such resolution is possible because protein molecules that diffuse away from the pI will acquire a net charge (negative at increased pH, positive at decreased pH) and immediately be focused back to their pI. This focusing effect will continue for as long as the electric field is applied.

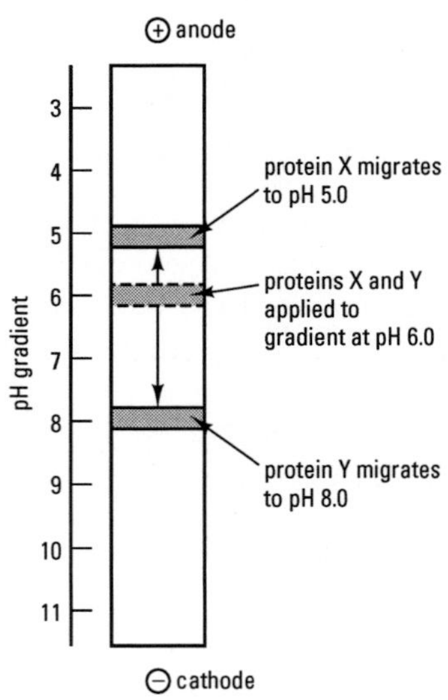

Fig. 47.2 The migration of two proteins, X and Y, in response to a pH gradient.

A useful variant of IEF is in obtaining *titration curves* for proteins. A pH gradient is set up, and the sample applied in a line at a right angle to the gradient. The net charge on a given protein will vary according to its position on the gradient – when electrophoresis is carried out at right angles to the pH gradient, the protein will migrate at a velocity and direction governed by that charge. When stained, each protein will appear as a continuous curved line, corresponding to its titration curve (Fig. 47.3). This technique can be usefully performed during protein purification, prior to ion-exchange chromatography (p. 307): by obtaining the titration curve for a protein of interest and those of major contaminants, the mobile phase pH that gives optimal separation can be selected.

In IEF, it is important that electro-osmotic flow (EOF, p. 322) is avoided, as this would affect the ability of the proteins to remain stationary at their pIs. For gel IEF, polyacrylamide minimises EOF, while capillary IEF uses narrow bore tubing with an internal polymer coating (p. 334).

Two-dimensional electrophoresis

Avoiding streaking in 2D electrophoresis – *ensure that the sample contains no particulate material (e.g. from protein aggregation); filter or centrifuge before use.*

Maximising resolution in 2D electrophoresis of proteins – *try to minimise nucleic acid contamination of your sample, as they may interact with polypeptides/proteins, affecting their movement in the gel.*

Freezing gel strips – *be sure to mark the identity and orientation of each gel strip before freezing, e.g. by inserting a fine wire into one end of the strip. Note that if urea is used in the gel strip, it will form crystals on freezing.*

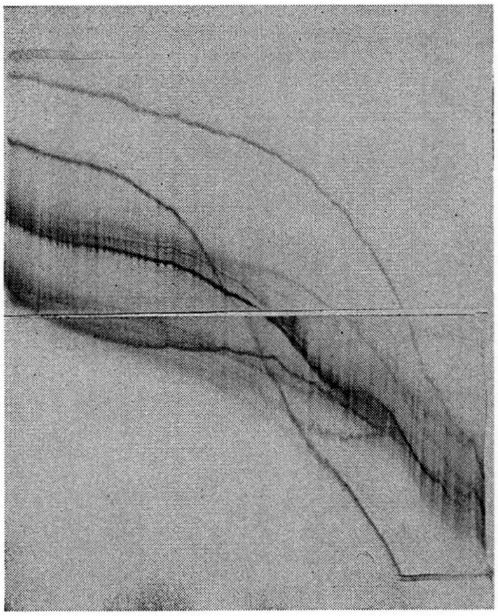

Fig. 47.3 Titration curves of bovine muscle proteins produced by electrofocusing-electrophoresis (photo courtesy of Pharmacia Biotech).

The most commonly used version of this high resolution technique involves separating proteins by charge in one dimension using IEF in polyacrylamide gel, followed by separation by molecular mass in the second dimension using denaturing SDS-PAGE (p. 325). The technique allows up to 1000 proteins to be separated from a single sample. Typically, the first dimension IEF run (pH 3–10) is carried out on gel strips of length 7–24 cm. Strips are run at a voltage of 500–3500 V for 1.5 h, then at 3500 V for a further 4 h. Gel strips can then be used immediately, or frozen until required.

It is common for the second-dimension SDS-PAGE separation to be carried out on a discontinuous slab gel 0.5–1.5 mm thick, which includes a low percentage T stacking gel and a separating gel with an exponential gradient of 10–16% T. The separating gel can be prepared in advance, but the stacking gel must be formed shortly before addition of the rod gel from the one dimensional run.

After equilibration with the buffer used in SDS-PAGE, the IEF gel strip is loaded onto the 2D gel (still between the glass plates in which it was formed) and sealed in position using acrylamide or agarose. Before the sealing gel sets, a well should be formed in it at one end to allow addition of molecular mass markers. The second-dimension is run at 100–200 V until the dye front is $\approx$1 cm from the bottom edge of the slab. After running, the gel is processed for the detection of polypeptides, e.g. using Coomassie Blue or silver stain. Analysis of the complex patterns that result from 2D electrophoresis requires computer-aided gel scanners to acquire, store and process data from a gel, such as that shown in Figure 47.4. These systems can compare, adjust and match up patterns from several gels, allowing both accurate identification of spots and quantification of individual proteins. Allowance is made for the slight variations in patterns found in different runs, using internal references ('landmarks'), which are either added standard proteins or particular spots known to be present in all samples.

Capillary electrophoresis

The technique of capillary electrophoresis (CE) combines the high resolving power of electrophoresis with the speed and versatility of HPLC (p. 305). The technique largely overcomes the major problem of carrying out electrophoresis without a supporting medium, i.e. poor resolution due to convection currents and diffusion. A capillary tube has a high surface area : volume ratio, and consequently the heat generated as a result of the applied electric current is rapidly dissipated. A further advantage is that very small samples (5–10 nl) can be analysed. The versatility of CE is demonstrated by its use in the separation of a range of biomolecules, e.g. amino acids, proteins, nucleic acids, drugs, vitamins, organic acids and inorganic ions; CE can even separate neutral species, e.g. steroids, aromatic hydrocarbons (see Weinberger, 2000).

The components of a typical CE apparatus are shown in Figure 47.5. The capillary is made of fused silica and externally coated with a polymer for mechanical strength. The internal diameter is usually 25–50 μm, a compromise between efficient heat dissipation and the need for a

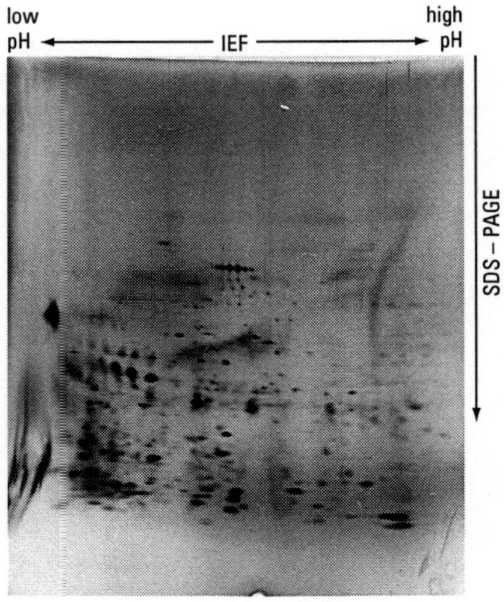

Fig. 47.4 Two-dimensional separation of proteins from 100x concentrated urine (2.5 μg total protein; silver stain. Courtesy of T. Marshall and K.M. Williams).

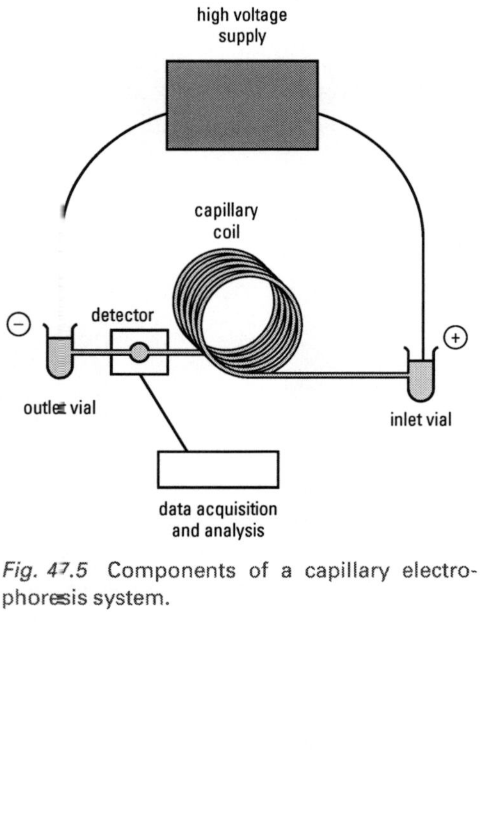

Fig. 47.5 Components of a capillary electrophoresis system.

light path that is not too short for detection using UV/visible spectrophotometry. A gap in the polymer coating provides a window for detection purposes. Samples are injected into the capillary by a variety of means, e.g. electrophoretic loading or displacement. In the former, the inlet end of the capillary is immersed in the sample and a pulse of high voltage is applied. The displacement method involves forcing the sample into the capillary, either by applying pressure in the sample vial using an inert gas, or by introducing a vacuum at the outlet. The detectors used in CE are similar to those used in chromatography (p. 314), e.g. UV/visible spectrophotometric systems. Fluorescence detection is more sensitive, but this may require sample derivitisation. Electrochemical and conductivity detection is also used in some applications, e.g. conductivity detection of inorganic cations such as Na^+ and K^+.

Electro-osmotic flow (EOF), described on page 322, is essential to the most commonly used types of CE. The existence of EOF in the capillary is the result of the net negative charge on the fused silica surface at pH values over 3.0. The resulting solvent flow towards the cathode is greater than the attraction of anions towards the anode, so they will flow towards the cathode (note that the detector is situated at the cathodic end of the capillary). The greater the net negative charge on an anion, the greater is its resistance to the EOF and the lower its mobility. Separated components migrate towards the cathode in the order: (1) cations, (2) neutral species, (3) anions.

Capillary zone electrophoresis (CZE)

This is the most widely used form of CE, and is based on electrophoresis in free solution and EOF, as discussed above. Separations are due to the charge : mass ratio of the sample components, and the technique can be used for almost any type of charged molecule, and is especially useful for peptide separation and confirmation of purity.

Micellar electrokinetic chromatography (MEKC)

This technique involves the principles of both electrophoresis and chromatography. Its main strength is that it can be used for the separation of neutral molecules as well as charged ones. This is achieved by including surfactants (e.g. SDS, Triton X-100) in the electrophoresis buffer at concentrations that promote the formation of spherical micelles, with a hydrophobic interior and a charged, hydrophilic surface. When an electric field is applied, these micelles will tend to migrate with or against the EOF depending on their surface charge. Anionic surfactants like SDS are attracted by the anode, but if the pH of the buffer is high enough to ensure that the EOF is faster than the migration velocity of the micelles, the net migration is in the direction of the EOF, i.e. towards the cathode. During this migration, sample components partition between the buffer and the micelles (acting as a pseudo-stationary phase); this may involve both hydrophobic and electrostatic interactions. For neutral species it is only the partitioning effect that is involved in separation; the more hydrophobic a sample molecule, the more it will interact with the micelle, and the longer will be its migration time, since the micelle resists the EOF. The versatility of MEKC enables it to be used for separations of biomolecules as diverse as amino acids and polycyclic hydrocarbons. MEKC is also known as micellar electrokinetic capillary chromatography (MECC).

Distinguishing between stereoisomers – *the* R *and* S *convention involves prioritising atoms or groups bonded to an asymmetric carbon atom in order of their atomic number. With the smallest atom or group pointing away from you, note the size of the remaining three groups or atoms. If the configuration in order of increasing size is clockwise, this is termed the* R*-configuration (L.* rectus*); if the order is anticlockwise, this is the* S*-configuration (L.* sinister*). Note that the older, alternative terminology of '*D*' (L.* dextro*) and '*L*' (L.* laevo*) isomers is widely used in the life sciences (p. 352), but the two terminologies do not always coincide (i.e.* R *is not always* D*, and* S *is not always* L*).*

Chiral capillary electrophoresis (CCE)

Resolution of a pair of chiral enantiomers (optical isomers) represents one of the biggest challenges for separation science, because each member of the pair will have identical physicochemical properties. CE offers an effective method of separating enantiomers by inducing a 'chiral selector' in the electrophoresis medium. The most commonly used chiral selectors are cyclodextrins such as the highly sulphated cyclodextrins (HSCDs) (Fig. 47.6). As the enantiomers migrate along the capillary, one will tend to interact more strongly than the other and its mobility will be reduced relative to the other. Figure 47.7 shows separation of the R and S forms of amphetamine using 5% HSCD in the electrophoresis buffer. The R-form has greater affinity for the HSCD used, so its retention time on the capillary is longer than that of the S-form.

Note that HSDCs can also be used in HPLC, but CCE is more effective, with shorter development times and lower reagent costs.

Capillary gel electrophoresis (CGE)

The underlying principle of this technique is directly comparable with that of conventional PAGE, i.e. the capillary contains a polymer that acts as a molecular sieve. As charged sample molecules migrate through the polymer network, larger molecules are hindered to a greater extent than smaller ones and will tend to move more slowly. CGE differs from CZE

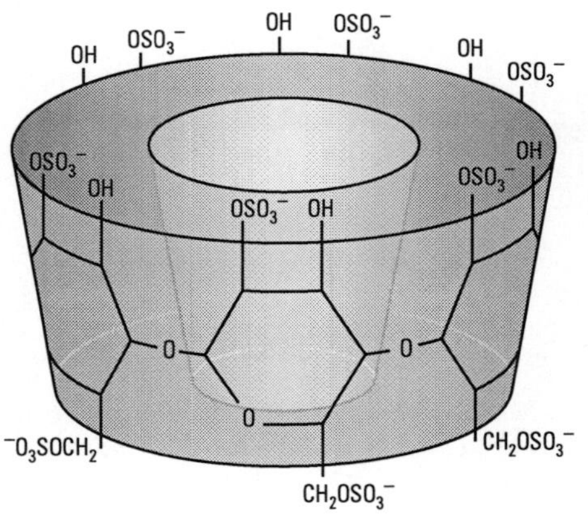

Fig. 47.6 General structure of highly sulphated cyclodextrins (HSCDs). The central cavity of the cyclodextrin (shaded blue) can interact differently with the R and S isomers of a biomolecule, enabling separation in CE. Copyright © 1999–2002 Beckman Coulter, Inc. http://www.beckmancoulter.com/products/splashpage/chiral38/default.asp/

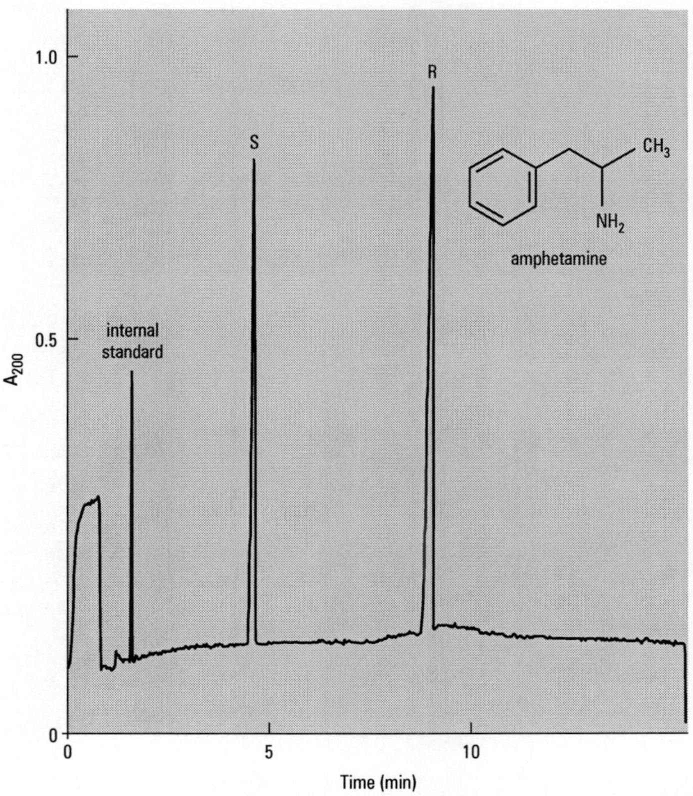

Fig. 47.7 CE separation of R and S enantiomers of amphetamine using Beckman Coulter highly sulphated gamma cyclodextrin. Copyright © 1999–2002 Beckman Coulter, Inc. http://www.beckmancoulter.com/products/splashpage/chiral38/default.asp/

and MEKC in that the inner surface of the capillary is polymer-coated in order to prevent EOF; this means that for most applications (e.g. polypeptide or oligonucleotide separations) sample components will migrate towards the anode at a rate determined by their size. The technique also differs from conventional PAGE in that a 'polymer network' is used rather than a gel: the polymer network may be polyacrylamide or agarose.

CGE offers the following advantages over conventional electrophoresis:

- efficient heat dissipation means that a high electrical field can be applied, giving shorter separation times;
- detection of the separated components as they move towards the anodic end of the capillary (e.g. using a UV/visible detector) means that staining is unnecessary;
- automation is feasible.

Choosing a detector for capillary electrophoresis – *most types of HPLC detector are suitable for CE and related applications (see Chapter 45).*

Capillary isoelectric focusing (CIEF)

This is used mainly for protein separation. Here, the principles of IEF are valid as long as EOF is prevented by using capillaries that are polymer-coated on their inner surface. Sample components migrate to their isoelectric points and become stationary. Once separated (<10 min), the components must be mobilised so that they flow past the detector. This is achieved by changing the NaOH solution in the cathodic reservoir with an NaOH/NaCl solution. When the electric field is reapplied, Cl^- enters the capillary, causing a decrease in pH at the cathodic end and the subsequent migration of sample components.

Text reference

Weinberger, R. (2000) *Practical Capillary Electrophoresis.* Academic Press, New York.

Sources for further study

Anon. *The American Electrophoresis Society Homepage.* Available: http://www.aesociety.org
Last accessed: 01/04/07.
[Includes details on electrophoretic techniques including IEF and CE]

Anon. *Cyclodextrin Resource.* Available: http://www.cyclodex.com/index.html
Last accessed: 01/04/07.
[Information on cyclodextrin structure, applications, etc.]

Cunico, R.L., Gooding, K.M. and Wehr, T. (1998) *Basic HPLC and CE of Biomolecules.* Bay Bioanalytical, Hercules, CA.

Khaledi, M.G. (1998) *High Performance Capillary Electrophoresis: Theory, Techniques and Applications.* Wiley, New York.
[Detailed manual of CE techniques and procedures]

Palfrey, S.M. (1999) *Clinical Applications of Capillary Electrophoresis.* Humana Press, New Jersey.

Strenge, M.A. and Lagu, A.L. (2004) *Capillary Electrophoresis of Proteins and Peptides.* Humana Press, New Jersey.

Westermeier, R. (2004) *Electrophoresis in Practice: A Guide to Methods and Applications of DNA and Protein Separations*, 3rd edn. Wiley-VCH, Berlin.
[Covers the principles underlying IEF]

Weston, A. and Brown, P.R. (1997) *HPLC and CE: Principles and Practice.* Academic Press, New York.
[Covers practical aspects including applications, optimisation and troubleshooting]

Study exercises

47.1 Consider the requirements for sample application in PAGE and IEF. Explain why in PAGE the sample is applied in a discrete narrow band, usually at the cathodic end of the gel, while in IEF the sample can be applied at any point along the length of the gel without concern about location or narrowness of the sample zone.

47.2 Choose a suitable CE technique for a range of applications. Which variant of CE would you consider most suitable for the following applications? (a) Separation of small peptides of similar size; (b) separation of oligonucleotides of different lengths; (c) separation of proteins with different isoelectric points; (d) a sample of urine tested for several anabolic steroids.

47.3 Check the similarities and differences between electrophoresis and isoelectric focusing. Complete the following table, indicating (i) whether each of the factors listed is relevant to each technique and (ii) briefly explain why, in each case.

Potential factors influencing electrophoresis (E) and isoelectric focusing (IEF)

Factor	E	IEF
Dependence upon ionisable groups in proteins		
Requirement for the application of an electric field		
Dependence upon differences in net charge between sample molecules		
Dependence upon differences in pI values between sample molecules		
Requirement for buffer at constant pH		
Requirement for a pH gradient		
Requirement for ampholytes to be included in solution		
Use of supporting medium with molecular sieving		
Treatment of samples with SDS is possible		
Diffusion of separated proteins is significant		

47.4 Identify proteins by 2D electrophoresis. The figure represents the separation of a mixture of 10 proteins by 2D electrophoresis. Using the information in the table, identify the individual proteins A–J.

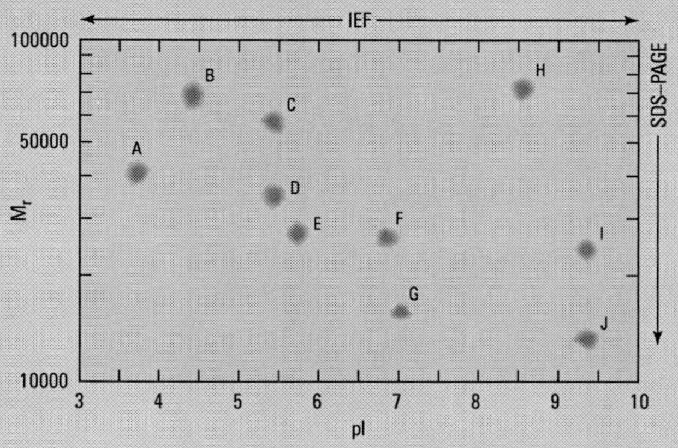

M_r and pI values of selected protein subunits

Protein	Subunit M_r	Subunit pI
Nerve growth factor	13 300	9.3
Ribonuclease	13 700	7.8
Haemoglobin	16 000	7.0
β-lactoglobulin	17 500	5.2
Ceramide trihexosidase	22 000	3.0
Trypsinogen	24 500	9.3
Triose phosphate isomerase	27 000	5.7
Galactokinase	27 000	5.7
Arginase	30 000	9.2
Glycerol-3-phosphate dehydrogenase	34 000	6.4
Alcohol dehydrogenase	35 000	5.4
Deoxyribonuclease II	38 000	10.2
Aldolase	40 000	5.2
Pepsinogen	41 000	3.7
Hexokinase	51 000	5.3
Lipoxidase	54 000	5.7
Catalase	57 500	5.4
Alkaline phosphatase	69 000	4.4
Acetylcholine esterase	70 000	4.5
Glyceraldehyde-3-phosphate dehydrogenase	72 000	8.5

48 Electroanalytical techniques

Electrochemical methods are used to quantify a broad range of different biomolecules, including ions, gases, metabolites, drugs and hormones.

Definitions

Oxidation – loss of electrons by an atom or molecule (or gain of O atoms, loss of H atoms, increase in positive charge, or decrease in negative charge).

Reduction – gain of electrons by an atom or molecule (or loss of O atoms, gain of H atoms, decrease in positive charge or increase in negative charge).

***KEY POINT** The basis of all electrochemical analysis is the transfer of electrons from one atom or molecule to another atom or molecule in an obligately coupled oxidation–reduction reaction (a redox reaction).*

It is convenient to separate such redox reactions into two half-reactions and, by convention, each is written as:

$$\text{oxidised form} + \text{electron(s)}\ (ne^-) \underset{oxidation}{\overset{reduction}{\rightleftharpoons}} \text{reduced form} \qquad [48.1]$$

You should note that the half-reaction is reversible: by applying suitable conditions, reduction *or* oxidation can take place. As an example, a simple redox reaction occurs when metallic zinc (Zn) is placed in a solution containing copper ions (Cu^{2+}), as follows:

$$Cu^{2+} + Zn \rightarrow Cu + Zn^{2+} \qquad [48.2]$$

The half-reactions are (i) $Cu^{2+} + 2e^- \rightarrow Cu$ and (ii) $Zn^{2+} + 2e^- \rightarrow Zn$. The oxidising power of (i) is greater than that of (ii), so in a coupled system the latter half-reaction proceeds in the opposite direction to that shown above, i.e. as $Zn - 2e^- \rightarrow Zn^{2+}$. When Zn and Cu electrodes are placed in separate solutions containing their ions, and connected electrically (Fig. 48.1), electrons will flow from the Zn electrode to Cu^{2+} via the Cu electrode due to the difference in oxidising power of the two half-reactions.

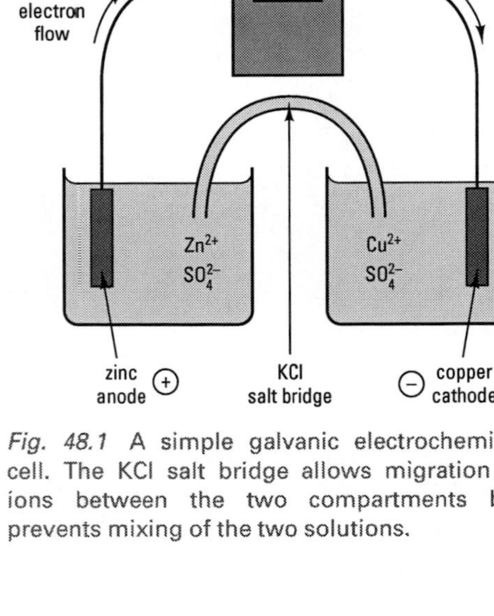

Fig. 48.1 A simple galvanic electrochemical cell. The KCl salt bridge allows migration of ions between the two compartments but prevents mixing of the two solutions.

By convention, the electrode potential of any half-reaction is expressed relative to that of a standard hydrogen electrode (half-reaction $2H^+ + 2e^- \rightarrow H_2$) and is called the standard electrode potential, E^o. Table 48.1 shows the values of E^o for selected half-reactions. With any pair of half-reactions from this series, electrons will flow from that having the lowest electrode potential to that of the highest. E^o is determined at $pH = 0$. It is often more appropriate to express standard electrode potentials at pH 7 for biological systems, and the symbol $E^{o\prime}$ is used: in all circumstances, it is important that the pH is clearly stated.

The arrangement shown in Figure 48.1 represents a simple galvanic cell where two electrodes serve as the interfaces between a chemical system and an electrical system. For analytical purposes, the magnitude of the potential (voltage) or the current produced by an electrochemical cell is related to the concentration (strictly the activity, *a*, p. 146) of a particular chemical species. Electrochemical methods offer the following advantages:

Definition

Galvanic cell – an electrochemical cell in which reactions occur spontaneously at the electrodes when they are connected externally by a conductor, producing electrical energy.

- excellent detection limits, and wide operating range (10^{-1}–10^{-8} mol l^{-1});
- measurements may be made on very small volumes (μl) allowing small amounts (pmol) of sample to be measured in some cases;
- miniature electrochemical sensors can be used for certain *in vivo* measurements, e.g. pH, glucose, oxygen content.

Table 48.1 Standard electrode potentials* (E^0) for selected half-reactions

Half-reaction	E^0 at 25 °C (V)
$Cl_2 + 2e^- \rightleftharpoons 2Cl^-$	+1.36
$O_2 + 4H^+ + 4e^- \rightleftharpoons 2H_2O$	+1.23
$Br_2 + 2e^- \rightleftharpoons 2Br^-$	+1.09
$Ag^+ + e^- \rightleftharpoons Ag$	+0.80
$Fe^{3+} + e^- \rightleftharpoons Fe^{2+}$	+0.77
$I_3^- + 2e^- \rightleftharpoons 3I^-$	+0.54
$Cu^{2+} + 2e^- \rightleftharpoons Cu$	+0.34
$Hg_2Cl_2 + 2e^- \rightleftharpoons 2Hg + 2Cl^-$	+0.27
$AgCl + e^- \rightleftharpoons Ag + Cl^-$	+0.22
$Ag(S_2O_3)_2^{3-} + e^- \rightleftharpoons Ag + 2S_2O_3^{2-}$	+0.01
$2H^+ + 2e^- \rightleftharpoons H_2$	+0.00
$AgI + e^- \rightleftharpoons Ag + I$	−0.15
$PbSO_4 + 2e^- \rightleftharpoons Pb + SO_4^{2-}$	−0.35
$Cd^{2+} + 2e^- \rightleftharpoons Cd$	−0.40
$Zn^{2+} + 2e^- \rightleftharpoons Zn$	−0.76

*From Milazzo *et al.* (1978).

***Using a calomel electrode** – always ensure that the KCl solution is saturated by checking that KCl crystals are present.*

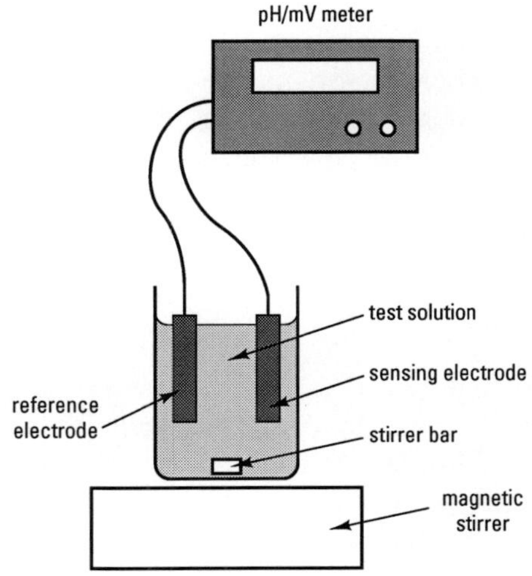

Fig. 48.2 Components of a potentiometric cell.

Potentiometry and ion-selective electrodes

Operating principles

These systems involve galvanic cells (p. 336) and are based on measurement of the potential (voltage) difference between two electrodes in solution when no net current flows between them: no net electrochemical reaction occurs and measurements are made under equilibrium conditions. These systems include methods for measuring pH, ions and gases such as CO_2 and NH_3. A typical potentiometric cell is shown in Figure 48.2. It contains two electrodes:

1. A ‘sensing’ electrode, the half-cell potential of which responds to changes in the activity (concentration) of the substance to be measured; the most common type of indicator electrodes are ion-selective electrodes (ISE).
2. A ‘reference’ electrode, the potential of which does not change, forming the second half of the cell.

To assay a particular analyte, the potential difference between these electrodes is measured by a millivolt meter (e.g. a standard pH meter).

Reference electrodes for potentiometry are of three main types:

1. The standard hydrogen electrode, which is the reference half-cell electrode, defined as 0.0 V at all temperatures, against which values of E^0 are expressed. H_2 gas at 1 atmosphere pressure is bubbled over a platinum electrode immersed in an acid solution with an activity of unity. This electrode is rarely used for analytical work, since it is unstable and other reference electrodes are easier to construct and use.
2. The calomel electrode (Fig. 48.3), which consists of a paste of mercury covered by a coat of calomel (Hg_2Cl_2), immersed in a saturated solution of KCl. The half-reaction: $Hg_2Cl_2 + 2e^- \rightarrow 2Hg + 2Cl^-$ gives a stable standard electrode potential of +0.24 V.
3. The silver/silver chloride electrode. This is a silver wire coated with AgCl and immersed in a solution of constant chloride concentration. The half-reaction: $AgCl + e^- \rightarrow Ag + Cl^-$ gives a stable, standard electrode potential of +0.20 V.

***KEY POINT** Ion-selective electrodes (ISEs) are based on measurement of a potential across a membrane which is selective for a particular analyte.*

An ISE consists of a membrane, an internal reference electrode, and an internal reference electrolyte of fixed activity. The ISE is immersed in a sample solution that contains the analyte of interest, along with a reference electrode. The membrane is chosen to have a specific affinity for a particular ion, and if the activity of this ion in the sample differs from that in the reference electrolyte, a potential develops across the membrane that is dependent on the ratio of these activities. Since the potentials of the two reference electrodes (internal and external) are fixed, and the internal electrolyte is of constant activity, the measured potential, E, is dependent on the membrane potential and is given by the Nernst equation (Robinson and Stokes, 2002) as:

$$E = K + 2.303\frac{RT}{zF}\log[a] \qquad [48.3]$$

Using ISEs, including pH electrodes – *standards and samples must be measured at the same temperature, as the Nernst equation shows that the measured potential is temperature-dependent.*

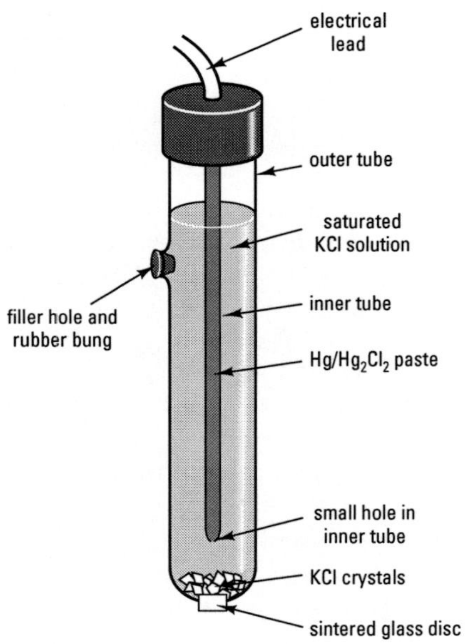

Fig. 48.3 A calomel reference electrode.

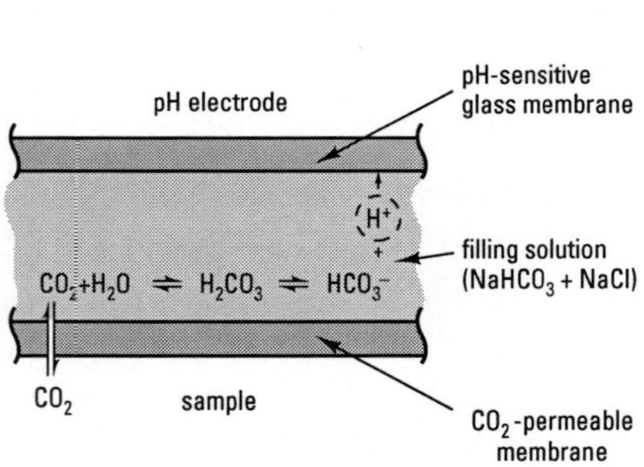

Fig. 48.4 Underlying principles of a gas-sensing electrode.

Using CO_2 electrodes – *applications include measurement of blood pCO_2 and in enzyme studies where CO_2 is utilised or released: calibration of the electrode is accomplished using 5% v/v and 10% v/v mixtures of CO_2 in an inert gas equilibrated against the measuring solution.*

where K represents a constant potential which is dependent on the reference electrode, z represents the net charge on the analyte, $[a]$ represents the activity of analyte in the sample and all other symbols and constants have their usual meaning (p. 185). For a series of standards of known activity, a plot of E against log $[a]$ should be linear over the working range of the electrode, with a slope of $2.303\,RT/zF$ (0.059 V at 25 °C). Although ISEs strictly measure *activity*, the potential differences can be approximated to concentration as long as (i) the analyte is in dilute solution (p. 144), (ii) the ionic strength of the calibration standards matches that of the sample, e.g. by adding appropriate amounts of a high ionic strength solution to the standards, and (iii) the effect of binding to sample macromolecules (e.g. proteins, nucleic acids) is minimal. Potentiometric measurements on undiluted biological fluids, e.g. K^+ and Na^+ levels in plasma, tissue fluids or urine, are likely to give lower values than flame emission spectrophotometry, since the latter procedure measures total ion levels, rather than just those in aqueous solution.

All of the various types of membrane used in ISEs operate by incorporating the ion to be analysed into the membrane, with the accompanying establishment of a membrane potential. The scope of electrochemical analysis has been extended to measuring gases and non-ionic compounds by combining ISEs with gas-permeable membranes, enzymes, and even immobilised bacteria or tissues.

Glass membrane electrodes

The most widely used ISE is the glass membrane electrode for pH measurement (p. 154). The membrane is thin glass (50 μm wall thickness) made of silica which contains some Na^+. When the membrane is soaked in water, a thin hydrated layer is formed on the surface in which negative oxide groups (Si-O^-) in the glass act as ion exchange sites. If the electrode is placed in an acid solution, H^+ exchanges with Na^+ in the hydrated layer, producing an external surface potential: in alkaline solution, H^+ moves out of the membrane in exchange for Na^+. Since the inner surface potential is kept constant by exposure to a fixed activity of H^+, a consistent, accurate potentiometric response is observed over a wide pH range. Glass electrodes for other cations (e.g. Na^+, NH_4^+) have been developed by changing the composition of the glass, so that it is predominantly sensitive to the particular analyte, though the specificity of such electrodes is not absolute. The operating principles and maintenance of such electrodes are broadly similar to those for pH electrodes (p. 154).

Gas-sensing glass electrodes

Here, an ISE in contact with a thin external layer of aqueous electrolyte (the 'filling solution') is kept close to the glass membrane by an additional, outer membrane that is selectively permeable to the gas of interest. The arrangement for a CO_2 electrode is shown in Figure 48.4: in this case the outer membrane is made of CO_2-permeable silicone rubber. When CO_2 gas in the sample selectively diffuses across the membrane and dissolves in the filling solution (in this case an aqueous $NaHCO_3$/NaCl mixture), a change in pH occurs due to the shift in the equilibrium:

$$CO_2 + H_2O \rightleftharpoons H_2CO_3 \rightleftharpoons H^+ + HCO_3^- \qquad [48.4]$$

The pH change is 'sensed' by the internal ion-selective pH electrode, and its response is proportional to the partial pressure of CO_2 of the solution

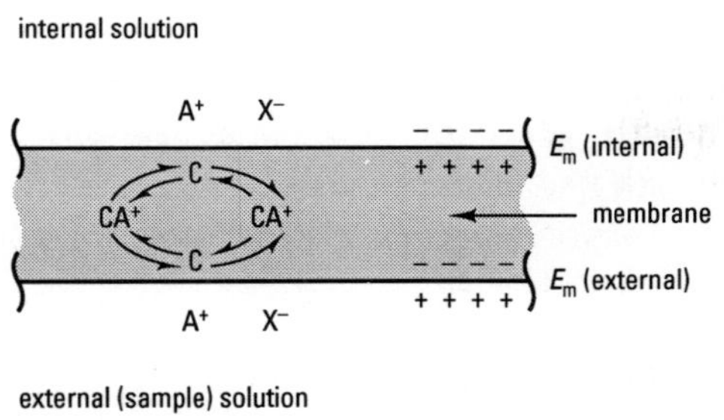

Fig. 48.5 Underlying principles of a liquid membrane ion-selective electrode. A^+ = analyte; C = neutral carrier ionophore; E_m = surface potential; membrane potential = E_m(internal) − E_m(external).

Definition

Ionophore – a compound that enhances membrane permeability to a specific ion: an ionophore may be incorporated into an ISE to detect that ion.

(pCO_2). A similar principle operates in the NH_3 electrode, where a Teflon membrane is used, and the filling solution is NH_4Cl.

Liquid and polymer membrane electrodes

In this type of ISE, the liquid is a water-insoluble viscous solvent containing a soluble ionophore, i.e. an organic ion-exchanger, or a neutral carrier molecule, that is specific for the analyte of interest. When this liquid is soaked into a thin membrane such as cellulose acetate, it becomes effectively immobilised. The arrangement of analyte (A^+) and ionophore in relation to this membrane is shown in Figure 48.5. The potential on the inner surface of the membrane is kept constant by maintaining a constant activity of A^+ in the internal solution, so the potential change measured is that which results from A^+ in the sample interacting with the ionophore in the outer surface of the membrane.

A relevant example of a suitable ionophore is the antibiotic valinomycin, which specifically binds K^+. Other ionophores have been developed for measurement of e.g. NH_4^+, Ca^{2+}, Cl^-. In addition, electrodes have been developed for organic species by using specific ion-pairing reagents in the membrane that interact with ionic forms of the organic compound, e.g. with drugs such as 5,5-diphenylhydantoin.

Solid-state membrane electrodes

These contain membranes made from single crystals or pressed pellets of salts of the analyte. The membrane material must show some permeability to ions and must be virtually insoluble in water. Examples include:

- the fluoride electrode, which uses LaF_3 impregnated with Eu^{2+} (the latter to increase permeability to F^-). A membrane potential is set up when F^- in the sample solution enters spaces in the crystal lattice;
- the chloride electrode, which uses a pressed-pellet membrane of Ag_2S and AgCl.

Voltammetric methods

Definitions

Electrolytic cell – an electrochemical cell in which reactions are driven by the application of an external voltage greater than the spontaneous potential of the cell.

Electrolysis – a non-spontaneous chemical reaction resulting from the application of a potential to an electrode.

Voltammetric methods are based on measurements made using an electrochemical cell in which electrolysis is occurring. Voltammetry, sometimes also called amperometry, involves the use of a potential applied between two electrodes (the working electrode and the reference electrode) to cause oxidation or reduction of an electroactive analyte. The loss or gain of electrons at an electrode surface causes current to flow, and the size of the current (usually measured in mA or μA) is directly proportional to the concentration of the electroactive analyte. The materials used for the working electrode must be good conductors and electrochemically inert, so that they simply transfer electrons to and from species in solution. Suitable materials include Pt, Au, Hg and glassy carbon.

Two widely used devices that operate on the voltammetric principle are the oxygen electrode and the glucose electrode. These are sometimes referred to as amperometric sensors.

Oxygen electrodes

The Clark (Rank) oxygen electrode

These instruments measure oxygen in solution using the polarographic principle, i.e. by monitoring the current flowing between two electrodes

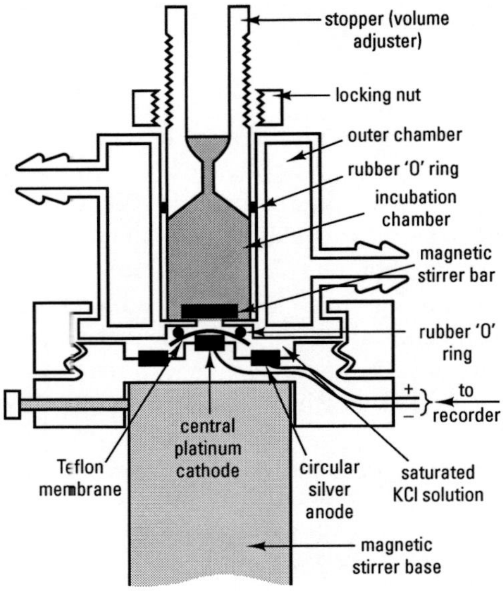

Fig. 48.6 Transverse section through a Clark (Rank) oxygen electrode.

when a voltage is applied. The most widespread electrode is the Clark type (Fig. 48.6), manufactured by Rank Bros, Cambridge, UK, which is suitable for measuring O_2 concentrations in cell, organelle and enzyme suspensions. Pt and Au electrodes are in contact with a solution of electrolyte (normally saturated KCl). The electrodes are separated from the medium by a Teflon membrane, permeable to O_2. When a potential is applied across the electrodes, this generates a current proportional to the O_2 concentration (Wilson and Walker, 2005). The reactions can be summed up as:

$4Ag \rightarrow 4Ag^{+} + 4e^{-}$ (at silver anode)

$O_2 + 2e^{-} + 2H^{+} \rightleftharpoons H_2O_2$ (in electrolyte solution; O_2 replenished by diffusion from test solution)

$H_2O_2 + 2e^{-} + 2H^{+} \rightleftharpoons 2H_2O$ (at platinum cathode)

Setting up and using a Clark (Rank) oxygen electrode

Box 48.1 describes the steps involved: if you are setting up from scratch, perform steps 1–13; if a satisfactory membrane is already in place, start at step 7. Box 48.2 gives advice on how to use a chart recorder with an oxygen electrode, including how to calculate rates of oxygen consumption/production, using data for oxygen saturation of air-equilibrated water shown in Table 48.2.

The temperature of the incubation vessel should be controlled by passing water (e.g. from a water bath) through the outer chamber (Fig. 48.6). Cells or organelles may be present in the solution added to the incubation chamber or can be added via the hole in the stopper using a syringe, as can chemicals such as inhibitors (Box 48.2). Take care not to introduce air bubbles, and remove any that appear by gently raising and lowering the stopper. The electrode can be used repeatedly, providing the membrane is

Table 48.2 Oxygen saturation values for distilled water and sea water at standard atmospheric pressure and a range of temperatures (derived from Green and Carritt, 1967)

	Distilled water			Sea water		
Temperature (°C)	mol m^{-3} (mmol l^{-1})	mg l^{-1} (p.p.m.)	ml l^{-1}	mol m^{-3} (mmol l^{-1})	mg l^{-1} (p.p.m.)	ml l^{-1}
0	0.460	14.7	10.3	0.359	11.5	8.04
2	0.435	13.9	9.75	0.342	10.9	7.65
4	0.413	13.2	9.24	0.326	10.4	7.30
6	0.392	12.5	8.78	0.311	9.95	6.97
8	0.373	11.9	8.35	0.298	9.54	6.66
10	0.355	11.4	7.95	0.285	9.12	6.37
12	0.339	10.8	7.59	0.273	8.74	6.11
14	0.324	10.4	7.25	0.261	8.35	5.85
16	0.310	9.92	6.94	0.251	8.03	5.62
18	0.297	9.50	6.65	0.241	7.71	5.40
20	0.285	9.12	6.38	0.232	7.42	5.19
22	0.274	8.77	6.12	0.224	7.17	5.00
24	0.263	8.42	5.89	0.215	6.88	4.82
26	0.253	8.1	5.67	0.208	6.66	4.64
28	0.244	7.81	5.46	0.200	6.40	4.48
30	0.235	7.52	5.27	0.193	6.18	4.32
37	0.211	6.75	4.71	0.174	5.57	3.88

Notes: mol m^{-3} ≡ mmol l^{-1} ≡ μmol ml^{-1} ≡ nmol μl^{-1}. Tabulated values assume atmospheric pressure = 101.3 kPa (≈760 mm Hg); for more accurate work, a correction for any deviation can be made by multiplying the appropriate figures from the table by the ratio of the real pressure to the assumed pressure.

satisfactory: remove solutions carefully (e.g. using a pipette, or vacuum line and trap). Keep water in the chamber when not in use. Replace the membrane if:

- the reading becomes noisy;
- the electrode will not zero after adding sodium dithionite;
- the response becomes too slow (check by switching off stirrer – oxygen concentration should drop rapidly as the available oxygen is consumed).

KEY POINT Successful operation of a Clark (Rank) oxygen electrode requires an intact Teflon membrane and clean electrodes. When replacing the Teflon membrane, you should always check that the electrode surfaces are clean (shiny): if necessary, use a mild abrasive paste to remove any oxidised material.

Box 48.1 How to set up a Clark (Rank) oxygen electrode

1. **Detach the base of the incubation vessel** (see Fig. 48.6) by unscrewing the locking ring.
2. **Add enough saturated KCl to cover the electrodes.**
3. **Cut a 1 mm square hole in the centre of a 10 × 10 mm square of lens tissue** and place this on the KCl solution so that the hole is over the central platinum cathode.
4. **Cut a 10 × 10 mm square of Teflon membrane and place over the lens tissue;** seal by gently lowering the incubation vessel and tightening the locking ring, making sure that the rubber O-ring is correctly positioned over the membrane.
   (a) Do not overtighten the locking ring.
   (b) Take care not to trap air bubbles beneath the membrane.
   (c) Make sure that the membrane does not become twisted.
5. **Clamp the electrode over the magnetic stirrer base using the clamping screw.**
6. **Connect the electrode leads to the polarising unit/recording device** (silver anode to positive, platinum cathode to negative). You can either use a digital readout system on the control unit or the output can be passed to a chart recorder, giving a readout of oxygen status as a function of time (see Box 48.2). Check that the polarising voltage is set to 0.60 V and adjust, if necessary, using the 'polarising voltage' control (typically, this requires a small screwdriver to adjust).
7. **Add air-saturated experimental solution and a small Teflon-coated magnetic stirrer bar to the chamber.** The volume of the incubation chamber can be adjusted by moving the locking nut on the stopper. To adjust, add the appropriate amount of liquid to the chamber using a pipette, insert the stopper and screw the locking nut until the solution just fills the incubation chamber.
8. **Gently push the stopper (volume adjuster) into position**, making sure that no air bubbles are trapped in the chamber, and switch on the stirrer. Adjust the rate of stirring using the 'stirrer' control, if required.
9. **Set the zero**: remove the stopper and add a few crystals of sodium dithionite – the dithionite ions ($S_2O_4^{2-}$) will be oxidised to sulphite (SO_3^{2-}) and sulphate ions (SO_4^{2-}), thereby consuming all of the oxygen in the solution within 3–5 min (an alternative is to bubble N_2 through the solution for > 10 min). Once the reading has stabilised, if this is not at zero, then adjust using the 'set zero' control (in practice, the zero is usually quite stable and often needs no adjustment).
10. **Adjust the sensitivity** – rinse out the dithionite solution thoroughly/repeatedly, then replace with air-equilibrated water and allow the reading to stabilise (this may take 5–10 min.). Then adjust the 'sensitivity' control of the electrode system to set the oxygen saturated value at an appropriate point (for a controller with a digital readout, this is best set to 100, and then the values represent percentage oxygen saturation). To check that the instrument is working correctly, switch off the stirrer for a few moments – the reading should drop quickly as oxygen is consumed at the electrode surface and should then rise rapidly when the stirrer is switched back on.
11. **Rinse the incubation chamber thoroughly and add fresh experimental solution.** Make sure that all traces of sodium dithionite are removed.
12. **Carry out your experiment.**
13. **Remove the solution and check the calibration.** If the reading for air – equilibrated water is different, the electode's sensitivity or the temperature may have changed and you may need to recalibrate and repeat the measurement.

Box 48.2 How to convert a chart recorder trace to a rate of O_2 consumption or production

Having set up your oxygen electrode (Box 48.1) or probe, the most common experiments are those in which you measure the rate of oxygen consumption (respiration) or production (photosynthesis) in a fixed volume of solution (e.g. in the chamber of a Rank (Clark) oxygen electrode. Typically, this involves attaching the electrode to a chart recorder and following the change in oxygen concentration with time. The principal steps are as follows:

1. **Calculate the amount of oxygen in the electrode chamber** – multiply the appropriate oxygen-saturation concentration, from Table 48.2, by the volume of the chamber. For example, if you are working in mmol l^{-1} (= μmol ml^{-1}) then at 20°C, air-equilibrated (oxygen-saturated) distilled water contains 0.285 μmol ml^{-1}. For an electrode chamber of volume 5 ml, there will be $0.285 \times 5 = 1.425$ μmol oxygen in the water within the chamber at saturation.
2. **Set the zero on the recorder** – following the addition of a small amount of sodium dithionite to the chamber, to remove oxygen (Box 48.1), the trace is monitored until it has stabilised. Then adjust the 'zero' or 'back off' control of the recorder until the chart trace is set to zero on the chart paper: this is marked as 'A' on Fig. 48.8.
3. **Set the oxygen-saturated reading on the recorder** – rinse out the dithionite solution thoroughly/repeatedly, replace with air-equilibrated water and allow the reading to stabilise (this may take 3–5 min). Then adjust the 'sensitivity' control of the chart recorder to set the oxygen-saturated value at an appropriate point on the chart – for respiration measurements, this can be close to the full width of the chart paper, whereas for photosynthesis measurements it is more usual to set it closer to the mid-point of the chart paper, to allow measurements over 100% to be recorded. In the example shown in Fig. 48.8, where the chart paper is calibrated in mm divisions, the oxygen saturated reading is set at 80 divisions (80 mm), marked as 'B'. Note that the recorder pen can be lifted between readings, and during rinsing of the chamber, to reduce the amount of 'noise' on the chart.
4. **Calculate the amount of oxygen equivalent to a single division of the chart paper** – divide the amount of oxygen in the chamber by the number of divisions (mm) to give the amount per division (e.g. $1.425 \div 80 = 0.0178125$ μmol (= 17.8125 nmol per division).
5. **Carry out your experiment** – the example shown in Fig. 48.8 represents the consumption of oxygen during respiration of a suspension of yeast cells, marked as 'C'. Continue the recording until you are satisfied that the rate has remained stable for at least 5 min. In many experiments, you will be adding substances that change the rate (e.g. metabolic inhibitors, or substrates) – this can be done by injecting a solution containing the substance through the small hole in the stopper of the electrode (keep the volume of this solution to < 1% of the total volume of the chamber, to minimise the effect of the added solution).
6. **Convert the gradient of the trace into a rate of oxygen consumed or produced** – using a transparent ruler, draw the line of best fit through that part of the trace showing a stable relationship (see Fig. 48.8). Use this line of best fit to work out the gradient of the trace, by converting from linear dimensions to an amount per unit time: e.g. the gradient shown in Fig. 48.8 is equivalent to a decrease of 24 chart divisions (mm) in 3 cm. Using the value calculated in step 4, the amount of oxygen consumed is thus $17.8125 \times 24 = 427.5$ nmol. For a chart speed of 0.2 cm min^{-1}, this represents the change occurring over a time interval of $3 \div 0.2 = 15$ min. Thus the rate of oxygen consumed is $427.5 \div 15 = 28.5$ nmol min^{-1}.
7. **Express the rate on an appropriate basis** – in most instances, this will be either per cell (e.g. determined using a haemocytometer, p. 236) or per unit mass of protein (p. 353), chlorophyll (p. 282), or an equivalent marker. For the example shown in Fig. 48.8, if there were 2.4×10^7 yeast cells per ml of suspension, then in the 5 ml chamber of the electrode there would be a total of $2.4 \times 10^7 \times 5 = 1.2 \times 10^8$ yeast cells, giving a rate of $28.5 \div (1.2 \times 10^8) = 2.375 \times 10^{-7}$ nmol $min^{-1} cell^{-1}$, probably better expressed as 237.5 amol min^{-1} $cell^{-1}$. Note that when inhibitors/substrates are used, you should always wait until a new stable rate has been achieved (for 5 min) before using the trace to determine the new rate of oxygen consumption/production.

Definition

Biosensor – a device for measuring a substance, combining the selectivity of a biological reaction with that of a sensing electrode.

Oxygen probes

Clark-type oxygen electrodes are also available in probe form for immersion in the test solution (Fig. 48.7), e.g. for field studies, allowing direct measurement of oxygen status *in situ*, in contrast to chemical assays. The main point to note is that the solution must be stirred during

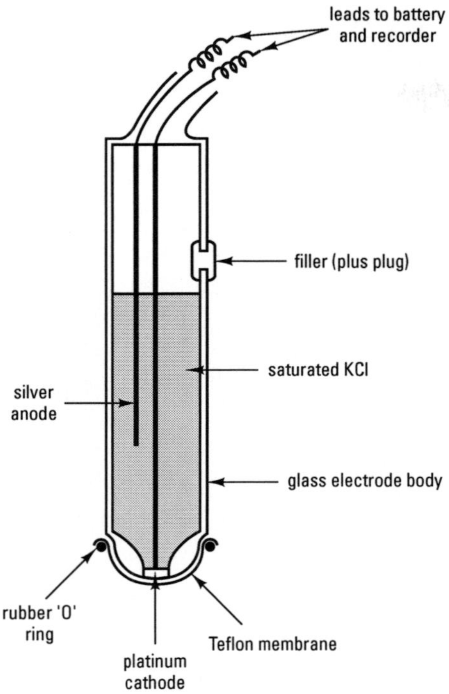

Fig. 48.7 A Clark-type oxygen probe.

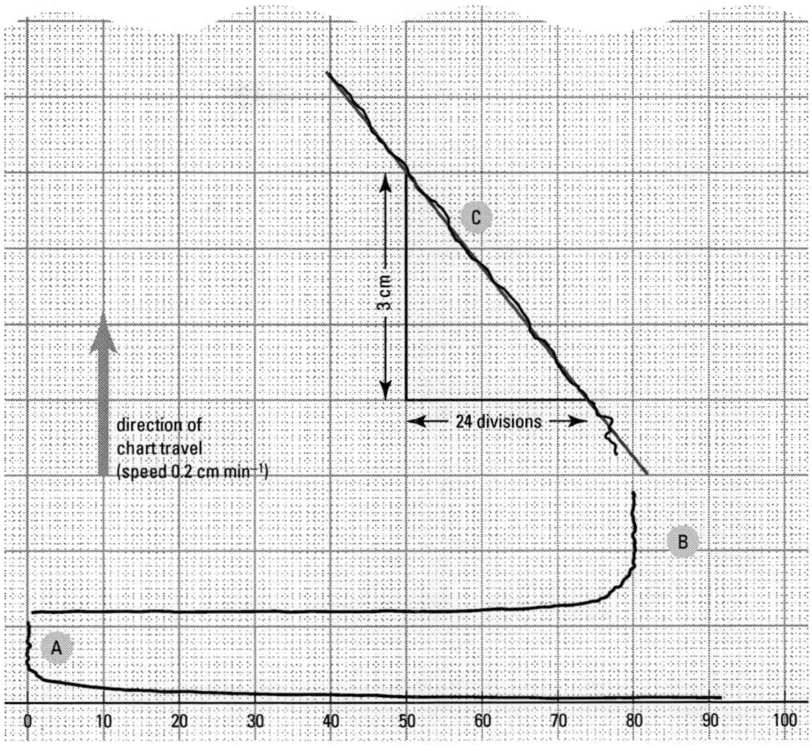

Fig. 48.8 Chart recorder output for a Clark (Rank) oxygen electrode. A = zero oxygen; B = oxygen saturation; C = respiration of yeast.

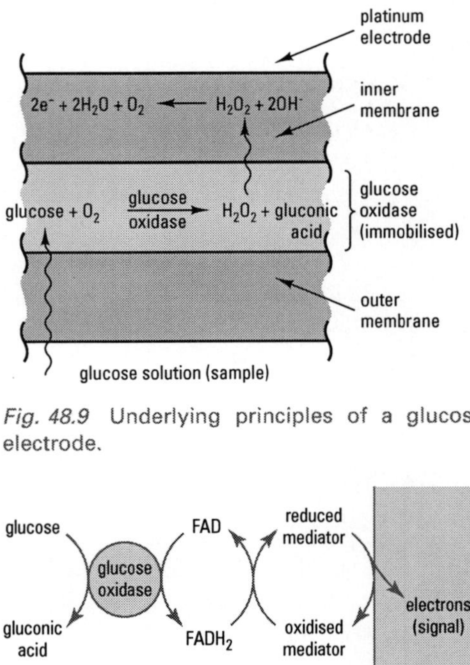

Fig. 48.9 Underlying principles of a glucose electrode.

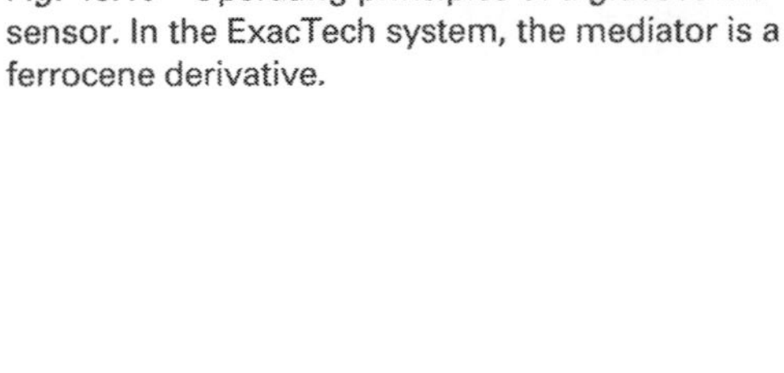

Fig. 48.10 Operating principles of a glucose biosensor. In the ExacTech system, the mediator is a ferrocene derivative.

measurement, to replenish the oxygen consumed by the electrode ('boundary layer' effect).

Glucose electrodes

These are simple types of biosensor; a simple design is shown in Figure 48.9. It consists of a Pt electrode, overlaid by two membranes. Sandwiched between these membranes is a layer of the immobilised enzyme glucose oxidase. The outer membrane is glucose-permeable and allows glucose in the sample to diffuse through to the glucose oxidase layer, where it is converted to gluconic acid and H_2O_2. The inner membrane is selectively permeable to H_2O_2, which is oxidised to O_2 at the surface of the Pt electrode. The current arising from this release of electrons is proportional to the glucose concentration in the sample within the range 10^{-7}–10^{-3} mol l^{-1}. The efficiency of a biosensor can be increased by including an electrode surface that is capable of capturing electrons directly. This is the basis of recently developed blood glucose meters for diabetics such as the ExacTech meter (Medisense-Abbott Laboratories). Here a pinprick blood sample is applied to a test strip containing glucose oxidase: if glucose is present a series of reactions occurs as shown in Figure 48.10.

The electrons produced are donated to the electrode surface, producing an initial current that is proportional to the initial rate at which glucose is oxidised which, in turn, is proportional to the glucose concentration in the blood sample. The future design of biosensors is likely to involve microchip technology in which the biocatalyst and the transducer are even more intimately linked (for more details, see Eggins, 2002).

Electrochemical detectors used in chromatography operate by voltammetric principles and currents are produced as the mobile phase flows over electrodes set at a fixed potential: to achieve maximum sensitivity, this potential must be set at a level that allows electrochemical reactions to occur in all analytes of interest.

Coulometric methods

Here, the charge required to completely electrolyse a sample is measured: the time required to titrate an analyte is measured at constant current and related to the amount of analyte using Faraday's law. There are few biological applications of this technique, though it is sometimes used for determination of Cl^- in serum and body fluids.

Text references

Eggins, B.R. (2002) *Chemical Sensors and Biosensors.* Wiley, New York.

Green, E.J. and Carritt, D.E. (1967) New Tables for Oxygen Saturation of Seawater, *Journal of Marine Research*, **25**, 140–7.

Milazzo, G., Caroli, S. and Sharma, V.K. (1978) *Tables of Standard Electrode Potentials.* Wiley, London.

Robinson, R.A. and Stokes, R.H. (2002) *Electrolyte Solutions.* Dover Publications, New York.

Wilson, K. and Walker, J. (eds) (2005) *Principles and Techniques of Practical Biochemistry and Molecular Biology*, 6th edn. Cambridge University Press, Cambridge.
[Also covers a range of other topics, including chromatography, spectroscopy and centrifugation]

Sources for further study

Anon *Rank Dissolved Oxygen Electrode*. Available: http://www.rankbrothers.co.uk/prod1.htm
Last accessed: 01/04/07.
[Homepage of the manufacturing company – contains technical details and contact information]

Fisher, A.C. (1996) *Electrode Dynamics*. Oxford University Press, Oxford.

Golterman, H.L., Clymo, R.S. and Ohnstad, M.A.M. (1978) *Methods for Physical and Chemical Analysis of Fresh Waters*. Blackwell Scientific Publications, Oxford.

Knopf, G.K. and Bassi, A.S. (2006) *Smart Biosensor Technology*. Taylor & Francis, New York.

Linek, V. (1988) *Measurement of Oxygen by Membrane-covered Probes*. Ellis-Horwood, Chichester.

Sanders, G.H.W., Compton, R.G. and Compton, S. (1996) *Electrode Potentials*. Oxford University Press, Oxford.
[Covers the underlying theory, plus applications relating to ion-selective electrodes]

Study exercises

48.1 Test your understanding of the circumstances under which ion-selective electrodes can be used. List the three main assumptions underlying the measurement of ion concentration using an ion-selective electrode.

48.2 Test your knowledge of biosensor design. List the two main functional components of biosensor such as the glucose electrode. Research a specific application – note that it should be different from the ones used in this chapter (e.g. using the Web).

(continued)

Study exercises (continued)

48.3 Calculate the oxygen content of specified volumes of water at defined temperatures. Using the information in Table 48.2 and assuming air equilibration at 101.3 kPa, what is the amount of oxygen in each of the following (give all answers to three significant figures):

(a) 4 ml of distilled water at 20 °C (express your answer in μmol);
(b) 20 ml of sea water at 12 °C (express your answer in μmol);
(c) 10 ml of distilled water at 15 °C (express your answer in μmol);
(d) 250 ml of distilled water at 37 °C (express your answer in mg);
(e) 200 ml of sea water at 25 °C (express your answer in ml).

48.4 Calculate respiration using data from an oxygen electrode readout. An oxygen electrode was set up to contain 5.0 ml of air-equilibrated distilled water at 30 °C, giving a full-scale reading on a chart recorder (200 divisions on the chart recorder paper), while a zero reading gave no deflection (0 divisions on the chart recorder paper). A yeast cell suspension (containing 4.1×10^9 cells in total) was added to the same electrode system along with a suitable growth medium, giving an approximately linear trace on the chart recorder with a slope of −25 units per cm, at a chart speed of 0.5 cm min^{-1}. What is the respiratory oxygen consumption in nmol min^{-1} $(10^9 \text{ cells})^{-1}$? (Give your answer to three significant figures.)

48.5 Determine net photosynthetic rate from oxygen electrode readings. The data below represent oxygen electrode readings taken every minute from 0 min for an illuminated suspension of a cyanobacterium in a 5.0 ml chamber of an oxygen electrode at 20 °C: 70.4, 73.6, 75.3, 78.5, 83.6, 85.9, 88.2, 91.4, 94.1. The system was calibrated to read 100.0 for air-equilibrated distilled water and 0.0 for anaerobic water, in effect making the scale read in terms of percentage oxygen saturation. The chlorophyll *a* content of an equivalent amount of cyanobacterial suspension to that used in the electrode chamber was measured at 2.13 μg ml^{-1}. What is the net photosynthetic rate of the cyanobacterial suspension, in μmol min^{-1} (mg chlorophyll *a*)$^{-1}$? (Express your answer to three significant figures.)

Genetics

58 Mendelian genetics

> **Definitions**
>
> **Phenotype** – the observable characteristics of an individual organism; the consequence of the underlying genotype and its interaction with the environment.
>
> **Genotype** – an individual's genetic make-up, i.e. the organism's genes.

Gregor Mendel, an Austrian monk, made pioneering studies of the genetics of eukaryotic organisms in the middle of the nineteenth century. He made crosses between different forms of flowering plants. Through careful examination and numerical analysis of the observable characteristics, or phenotype, of the parents and their progeny, Mendel was able to deduce much about their genetic characteristics, or genotype. The principles derived from these experiments explain the basis of heredity, and hence underpin our understanding of sexual reproduction, biodiversity and evolution. Mendelian genetics is concerned primarily with the transmission of genetic information, as opposed to molecular genetics which deals with the molecular details of the genome and techniques for altering genes (see Chapters 60–62) and genome analysis (genomics) in Chapter 11.

> ***Essential vocabulary for Mendelian genetics*** – *make sure you know what the following terms and symbols mean:*
> - *chromosome, sex chromosome, autosome;*
> - *gene, allele, locus;*
> - *dominant, recessive, lethal;*
> - *haploid, diploid, gamete, zygote;*
> - *heterozygous, homozygous;*
> - *P, F1, F2, ♀, ♂.*

KEY POINT A common initial stumbling block in genetics is terminology. In many cases the definitions are interdependent, so your success in this subject depends on your grasp of all of the definitions and underlying ideas explained below.

Important terms and concepts

> ***Genomes and chromosome numbers*** – *remember that mitochondria and chloroplasts contain DNA molecules, but these are not included when calculating the chromosome number.*

Each character in the phenotype is controlled by the organism's genes, the basic units of inheritance. Each gene includes the 'genetic blueprint' (DNA) which usually defines the amino acid sequence for a specific polypeptide or protein – often an enzyme or a structural protein. The protein gives rise to the phenotype through its activity in metabolism or its contribution to the organism's structure. The full complement of genes in an individual is known as its genome. Individual genes can exist in different forms, each of which generally leads to a different form of the protein it codes for. These different gene forms are known as alleles.

> **Definitions**
>
> **Meiosis** – division of a diploid cell which results in haploid daughter cells carrying half the original number of chromosomes. Occurs during gamete (sperm and egg) formation.
>
> **Mitosis** – division of a cell into two new cells, each with the same chromosome number. Occurs in somatic cells, e.g. during growth, development, repair, replacement.

In eukaryotes, the genes are located in a particular sequence on chromosomes within the nucleus. The number of chromosomes per cell is characteristic for each organism (its chromosome number, n). For example, the chromosome number for man is 23. In cells of most 'higher' organisms, there are two of each of the chromosomes ($2n$). This is known as the diploid state. As a result of the process of meiosis which precedes reproduction, special haploid cells are formed (gametes) which contain only one of each chromosome ($1n$). In sexual reproduction, haploid gametes from two individuals fuse to form a zygote, a diploid cell with a new genome, which gives rise to a new individual through the process of mitosis. Cell numbers are increased by this process, producing genetically identical cells.

Organisms vary in the span of the diploid and haploid phases. In some 'lower' organisms the haploid phase is the longer lasting form; in most 'higher' organisms the diploid phase is dominant. Since each diploid individual carries two of each chromosome, it has two copies of each gene in every cell. The number of alleles of each character present depends on whether the relevant genes are the same – the homozygous state – or whether they are different – the heterozygous state. Hence, while there may be many alleles for any given gene, an individual could have two at most, and might only have one if it were homozygous. This will depend on the alleles present in the parental gametes that fused when the zygote was formed. Offspring that inherit a different combination of alleles at the two loci compared with their parents are known as recombinants.

Example In the garden pea, *Pisum sativum*, studied by Mendel, the yellow seed allele Y was found to be dominant over the green seed allele y. A cross of YY x yy genotypes would give rise to Yy in all of the F_1 generation, all of which would thus have the yellow seed phenotype. If the F_1 generation were interbred, this Yy x Yy cross would lead to progeny in the next, F_2, generation with the genotypes YY : Yy : yy in the expected ratio 1 : 2 : 1. The expected phenotype ratio would be 3 : 1 for yellow : green seed.

The basis of Mendel's experiments and of many exercises in genetics are crosses, where individuals showing particular phenotypes are mated and the phenotypes of the offspring, or F_1 generation, are studied (see Box 58.1). If homozygous individuals carrying alternative alleles for a character are crossed, one of the alleles may be dominant, and all the F_1 generation will show that character in their phenotype. The character not evident is said to be recessive.

In describing crosses, geneticists denote each character with a letter of the alphabet, a capital letter being used for the dominant allele and lower case for the recessive. Taking, for example, a gene with dominant and recessive forms 'A' and 'a' respectively, there are three possibilities for each individual: it can be (a) homozygous recessive aa; (b) homozygous dominant AA; or (c) heterozygous Aa.

The reasons for dominance might relate to the activity of an enzyme coded by the relevant gene; for example, Mendel's yellow pea allele is dominant because the gene involved codes for the breakdown of chlorophyll. In the homozygous recessive case, none of the functional enzyme will be present, chlorophyll breakdown cannot occur and the seeds remain green. Not all alleles exhibit dominance in this form. In some cases, the heterozygous state results in a third phenotype (incomplete or partial dominance); in others, the heterozygous individual expresses both genotypes (codominance). Another possible situation is epistasis, in which one gene affects the expression of another gene that is independently inherited. Note also that many genes, such as those coding for human blood groups, have multiple alleles.

KEY POINT* *Genetics problems may well involve one of the 'standard' crosses shown in Box 58.1. Before tackling the problem, try to analyse the information provided to see if it fits one of these types of cross. Figure 58.1 is a flowchart detailing the steps you should take in answering genetics problems.

Unless otherwise stated or obvious from the evidence, you should assume that the genes being considered in any given case are on separate chromosomes. This is important because it means that they will assort independently during meiosis. Thus, the fact that an allele of gene A is present in any individual will not influence the possibility of an allele of gene B being present. This allows you to apply simple probability in predicting the genetic make-up of the offspring of any cross.

Where genes are present on the same chromosome, they are said to be linked genes, and thus it would appear that they would not be able to assort independently during metaphase 1 of meiosis, when homologous chromosomes are independently orientated. However, although physically attached to each other, they may become separated when crossing over occurs between the homologous chromosomes at an early stage of meiosis. Exchange of genetic information between homologous chromosomes is called recombination. Linkage can be detected from a cross between

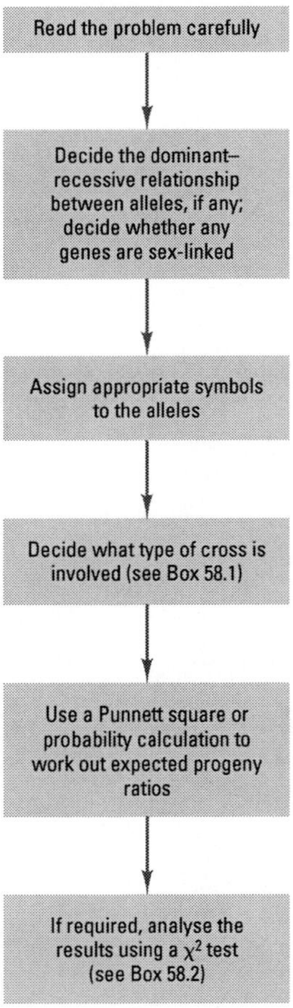

Fig. 58.1 Flowchart for tackling problems in Mendelian genetics.

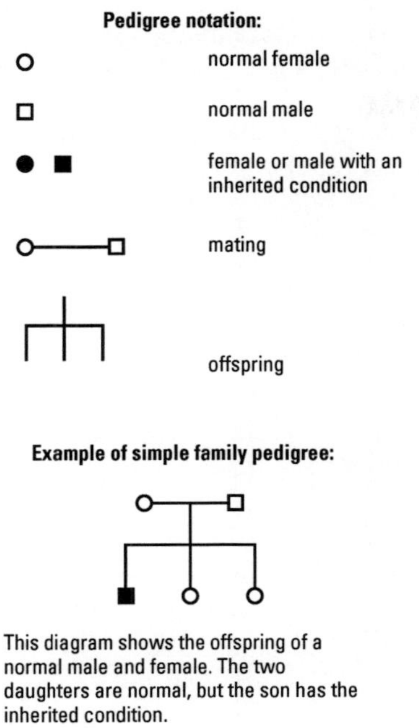

Fig. 58.2 Pedigree notation and family trees.

Denoting linked genes *– these are often shown diagrammatically, with a double line indicating the chromosome pair. For example, the two possible linkages for the genotype AaBb would be shown as:*

$$\frac{A \quad b}{a \quad B} \quad or \quad \frac{A \quad B}{a \quad b}$$

Using probability calculations *– this can be simpler and faster than Punnett squares when two or more genes are considered.*

individuals heterozygous and homozygous recessive for the relevant genes, e.g. AaBb × aabb. If the genes A and B are on different chromosomes, we expect the ratio of AaBb, aabb, Aabb and aaBb to be 1 : 1 : 1 : 1 in the F_1. However, if the dominant alleles of both genes occur on the same chromosome, the last two combinations will occur, but rarely. Just how rarely depends on how far apart they lie on the relevant chromosome – the further apart, the more likely it is that crossing over will occur. This is the basis of chromosome mapping (see Box 58.1).

Another complication you will come across is sex-linked genes. These occur on one of the X or Y chromosomes that control sex. Because one or other of the sexes – depending on the organism – is determined as XX and the other as XY (see Box 58.1), this means that rare recessive genes carried on the X chromosome may be expressed in XY individuals. Sex-linked genes are sometimes obvious from differences in the frequencies of phenotypes in male and female offspring. Pedigree charts (Fig. 58.2) are codified family trees that are often used to show the inheritance and expression of sex-linked characteristics through various generations.

Analysis of crosses

There are two basic ways of working out the results of crosses from known or assumed genotypes:

- The Punnett square method provides a good visual indication of potential combinations of gametes for a given cross. Lay out your Punnett squares consistently as shown in Figure 58.3. Then group together the like genotypes to work out the genotype ratio and proceed to work out the corresponding phenotype ratio if required.
- Probability calculations are based on the fact that the chance of a number of independent events occurring is equal to the probabilities of each event occurring, multiplied together. Thus if the probability P of a child being a boy is 0.5 and the probability of the child of particular parents being blue-eyed is 0.5, then the probability of that couple having a blue-eyed son is $0.5 \times 0.5 = 0.25$, and that of having two blue-eyed boys is:

$$P = (0.5 \times 0.5) \times (0.5 \times 0.5) = 0.0625$$

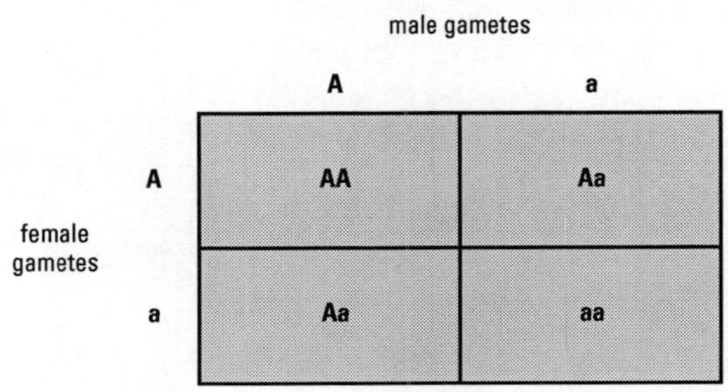

Fig. 58.3 Layout for a simple Punnett square for the cross Aa × Aa. The genotypic ratios for this cross are 1 : 2 : 1 for AA : Aa : aa, and the phenotypic ratio would be 3 : 1 for characteristic A to characteristic a. In this simple Punnett square, the allele frequencies are treated as equal (f = 0.5); if different from this, the probability of genotypes in each combination will be the relevant frequencies multiplied together.

Box 58.1 Types of cross and what you can (and can't) learn from them

Monohybrid cross – the simplest form of cross, considering two alleles of a single gene.

Example: AA × aa

If only the parental *phenotypes* are known, you can't always deduce the parental genotypes from the phenotype ratio in the F_1. An individual of dominant phenotype in the F_1 could arise from a homozygous dominant or heterozygous genotype. However, crossing the F_1 generation with themselves to give an F_2 generation may provide useful information from the phenotype ratios that are found.

Dihybrid cross – a cross involving two genes, each with two alleles.

Example: AaBB × AaBb

As with a monohybrid cross, you can't always deduce the parental genotype from the phenotype ratio in the F_1 generation alone.

Test cross – a cross of an unknown genotype with a homozygous recessive.

Example: AABb × aabb

A test cross is one between an individual dominant for A and B with one recessive for both genes. The progeny will all be dominant for A, revealing the homozygous nature of the parent for this gene, but the progeny phenotypes will be split approximately 1:1 dominant to recessive for gene B, revealing the heterozygous nature of the parent for this gene. This type of cross reveals the unknown parental genotype in the proportions of phenotypes in the F_1.

Sex-linked cross – a cross involving a gene carried on the X chromosome; this can be designated as dominant or recessive using appropriate superscripts (e.g. X^A and X^a).

Example: $X^AX^a \times X^AY$

In sex-linked crosses you need to know the basis of sex determination in the species concerned – which of XX or XY is male (for example, the former in birds, butterflies and moths, the latter in mammals and *Drosophila*). The expected ratios of phenotypes in the offspring will depend on this. Note that the recessive genotype a will be expressed in the X^aY case.

Crosses with linked genes – genes are linked if they are on the same chromosome. This is revealed from a cross between individuals heterozygous and homozygous recessive for the relevant genes.

Example 1 (genes on separate chromosomes): expected offspring frequency from the cross AaBb × aabb is AaBb, aabb, Aabb and aaBb in the ratio 1:1:1:1.

Example 2 (linked genes): the frequencies of the last two combinations in example 1 might be skewed according to the direction of parental linkage.

Chromosome mapping uses the frequency of crossing-over of linked genes to estimate their distance apart on the chromosome on the basis that crossing over is more likely, the further apart are the genes. So-called 'map units' are calculated on the following basis:

$$\frac{\text{No. of recombinant progeny}}{\text{Total no. of progeny}} \times 100 = \%\ \text{crossing over} \quad [58.1]$$

By convention, 1% crossing over = 1 map unit (centimorgan, cM). The order of a number of genes can be worked out from their relative distances from each other. Thus, if genes A and B are 12 map units apart, while A and B are respectively 5 and 7 map units from C, the assumed order on the chromosome is ACB (Fig. 58.4).

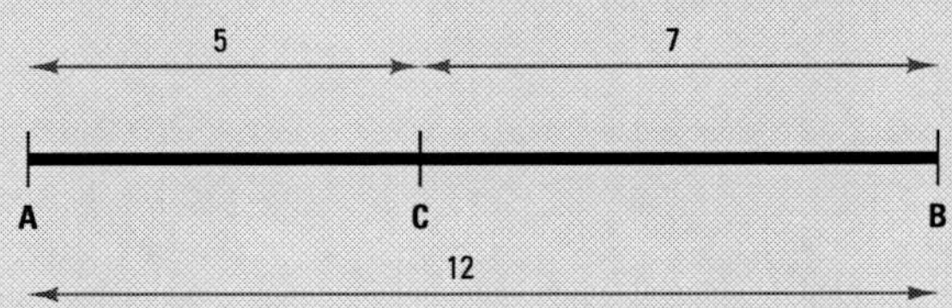

Fig. 58.4 Genetic map showing relative positions of genes A, B and C.

Carrying out genetic crosses *– typically, these are performed using organisms that have a large number of offspring, to even out random variation. Other useful attributes include: short generation time; ease of maintenance; readily observed mutant phenotypes.*

How do you decide whether the results of an experimental cross fit your expectation from theory as calculated above? This isn't easy, because of the element of chance in fertilisation. Thus, while you might expect to see a 3 : 1 phenotype ratio of progeny for a given cross, in 500 offspring you might actually observe a ratio of 379 : 121, which is a ratio of just over 3.13 : 1. Can you conclude that this is significantly different from 3 : 1 in the context of random error? The answer to this problem comes from statistics. However, the answer isn't certain, and your conclusion will be based on a balance of probabilities (see Box 58.2 and Table 58.1).

Population genetics

Population genetics is largely concerned with the frequencies of alleles in a population and how these may change in time. The Hardy–Weinberg Principle states that the frequency of alleles f remains the same between generations, unless influenced by some outside factor(s).

Limitation of the chi² test *– note that the formula cited in Box 58.2 is valid only if expected numbers are greater than 5.*

Box 58.2 Example of a chi² (χ^2) test

This test allows you to assess the difference between observed (O) and expected (E) values and is extremely useful in biology, particularly in determining whether progeny phenotype ratios fit your assumptions about their genotypes. The test is best illustrated by the use of an example. Assume that your null hypothesis is that the phenotypic ratio is 3 : 1 and you observe that in 500 offspring the phenotype ratio is 379 : 121 whereas the expected ratio is 375 : 125.

Start the test by calculating the test statistic χ^2. The general formula for calculating χ^2 is:

$$\chi^2 = \sum \frac{(O - E)^2}{E} \qquad [58.2]$$

In this example, this works out as:

$$\chi^2 = \frac{(379 - 375)^2}{375} + \frac{(121 - 125)^2}{125} = \frac{16}{375} + \frac{16}{125} = 0.171$$

The probability associated with this value can be obtained from χ^2 tables for $(n - 1)$ degrees of freedom (d.f.), where n = the number of categories = number of phenotypes considered. Here the d.f. value is $2 - 1 = 1$. Since the χ^2 value of 0.171 is lower than the tabulated value for 1 d.f. (3.84, Table 58.1), we therefore accept the null hypothesis and conclude that the difference between observed and expected results is not significant (since $P > 0.05$). Had χ^2 been $\geqslant 3.84$, then $P \leqslant 0.05$ and we would have rejected the null hypothesis and concluded that the difference was significant, i.e. that the progeny phenotype did not fit the expected ratio.

You can carry out χ^2 calculations using the CHITEST, CHIINV and CHIDIST functions in Microsoft Excel: the *Help* function within that program contains useful guidance.

Table 58.1 Values of chi² (χ^2) for which P = 0.05. The value for $(n - 1)$ degrees of freedom (d.f.) should be used, where n = the number of categories (= phenotypes) considered (normally fewer than 4 in genetics problems). If χ^2 is less than or equal to this value, accept the null hypothesis that the observed values arose by chance; if χ^2 is greater than this value, reject the null hypothesis and conclude that the difference between the observed and expected values is statistically significant

Degrees of freedom	χ^2 value for which $P = 0.05$
1	3.84
2	5.99
3	7.82
4	9.49

Definition

Hardy–Weinberg Principle – for a population in genetic equilibrium, the genotype frequencies at an autosomal locus will remain at particular equilibrium values, which can be expressed in terms of the allele frequencies at that locus.

To understand why this is the case, consider alleles H and h for a particular gene, which exist in the breeding population at frequencies p and q respectively. If the individuals carrying these alleles interbreed randomly, then the expected genotype and allele ratios in the F_1 generation can be calculated simply as:

$$f(\mathrm{HH}) = p^2;$$
$$f(\mathrm{Hh}) = 2pq; \quad \text{and}$$
$$f(\mathrm{hh}) = q^2$$

If you wish to confirm this, lay out a Punnett square with appropriate frequencies for each allele. Now, by summation, the frequency of H in the $F_1 = p^2 + pq$ (a similar calculation can be made for allele h); and since in this example there are only two alleles, $p + q = 1$ and so $q = (1 - p)$. Substituting $(1 - p)$ for q, the frequency of H in the F_1 is thus:

$$p^2 + p(1 - p) = p^2 + p - p^2 = p$$

i.e. the frequency of the allele is unchanged between generations. A similar relationship exists for the other alleles.

Example Cystic fibrosis occurs in 0.04% of Caucasian babies. If this condition results from a double recessive allele aa, then following the Hardy–Weinberg Principle, $q^2 = 0.0004$ (0.04% expressed as a fraction of 1) and so $q = \sqrt{0.0004} = 0.02$, or 2%. Since $p = (1 - q)$, $p = 0.98$, or 98%. The frequency of carriers of cystic fibrosis in the Caucasian population (people having the alleles Aa) is given by $2pq$. From the above, $2pq = 2(0.02 \times 0.98) = 0.0392$. Hence 3.92% of the Caucasian population are carriers (roughly one in 25).

The Hardy–Weinberg Principle was named after its first, independent, protagonists. It holds so long as the following criteria are satisfied:

1. random mating – so that no factors influence each individual's choice of a mate;
2. large population size – so that the laws of probability will apply;
3. no mutation – so that no new alleles are formed;
4. no emigration, immigration or isolation – so that there is no interchange of genes with other populations nor isolation of genes within the population;
5. no natural selection – so that no alleles have a reproductive advantage over others.

Population geneticists use the Hardy–Weinberg Principle to gain an idea of the rate of evolution and the influences on evolution. By ensuring that criteria 1–4 hold, if there are any changes in allele frequency between generations, then the rate of change of allele frequencies indicates the rate of evolutionary change (natural selection).

Sources for further study

Anon. *Online Mendelian Inheritance in Man*. Available: http://www.ncbi.nlm.nih.gov/entrez/query.fcgi?db=OMIM
Last accessed: 01/04/07.

Blumberg, R.B. *MendelWeb*. Available: http://www.mendelweb.org/
Last accessed: 01/04/07.
[A resource for students interested in the origins of classical genetics]

Falconer, D.S. and MacKay, T.F.C. (1996) *Introduction to Quantitative Genetics*, 4th edn. Longman, Harlow.

Klug, W.S. and Cummings, M.R. (2005) *Essentials of Genetics*, 5th edn. Prentice Hall, Upper Saddle River.

Mange, E.J. and Mange, A.P. (1998) *Basic Human Genetics*, 2nd edn. Sinauer, Sunderland.

Roberts, D.B. (1998) *Drosophila: a Practical Approach*, 2nd edn. Oxford University Press, Oxford.

Thomas, A. (2002) *Introducing Genetics: from Mendel to Molecule*. Taylor & Francis, Abingdon.

Winter, P.C., Hickey, G.I. and Fletcher, H.L. (1998) *Instant Notes in Genetics*. Bios, Oxford.

Study exercises

58.1 **Use a Punnett square to predict the outcome of a cross.** Lay out a Punnett square for a cross between genotypes RrOO × RrOo, where R is a semi-dominant gene for flower colour such that RR = red, Rr = pink and rr = white; and O is a dominant gene for corolla shape such that OO = closed corolla, Oo = closed corolla and oo = open corolla. From the Punnett square, derive both the genotypic and phenotypic ratios for the cross.

58.2 **Use probability to predict the outcome of a cross.** Two hazel-eyed parents are heterozygous for the eye-colour gene B. When expressed as bb, the individual is blue-eyed. Mum's hair is (genuine) blonde but Dad's is mousy-brown. In this case she is double recessive (mm) for the hair-colour gene M and he is heterozygous (Mm). What is the probability that they will have a blue-eyed, blonde daughter?

58.3 **Work out a likely genetic scenario for a given set of results.** Five tail-less female mice were crossed with normal males (with tails). There were 31 normal mice and 28 tail-less mice in the F_1 progeny. When pairs of tail-less mice from the F_1 generation were crossed, their (F_2) progeny were as follows: normal, 27; tail-less, 55; dead on birth, but tail-less, 30. In each case the ratio of males to females was roughly 50:50. Provide a logical explanation for these results.

58.4 **Predict parental genotypes from the results of crosses involving sex-linked genes.** A red-eye gene is known to be sex-linked in *Drosophila*; that is, the alleles R (red-eyed) or r (white-eyed) are carried on the X chromosome, while the Y chromosome does not carry these eye-colour alleles. In the fruit fly, XX = female and XY = male. Predict the possible parental genotypes from the following F_1 progeny ratios:

(a) 35 red-eyed female, 17 red-eyed male, 19 white-eyed male;
(b) 27 red-eyed male, 29 red-eyed female;
(c) 19 red-eyed female, 18 white-eyed female, 22 red-eyed male, 21 white-eyed male.

58.5 **Carry out a χ^2 (chi^2) test.** A geneticist expects the results of a test cross to be in the phenotype ratio 1:2:1. He observes 548 progeny from his cross in the ratio 125:303:120. What should he conclude?

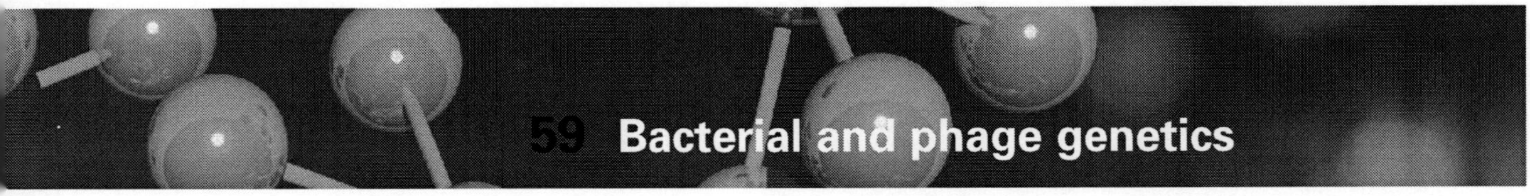

59 Bacterial and phage genetics

Definitions

Phage – a bacterial virus (bacteriophage).

Prophage – a bacterial virus genome integrated within the genome of a host bacterial cell.

Copy number – the average number of copies of a particular cellular molecule.

Transposon – a section of DNA coding for its own movement from one genomic location to another and carrying other genes in addition to those coding for transposition.

Merozygote – a cell containing two copies of a part of its genome, i.e. a partial diploid (sometimes also termed a merodiploid).

In eukaryotic organisms, genetic reassortment usually involves the fusion of two haploid gametes to form a zygote and a new generation.

KEY POINT ***Bacterial genetics is very different from eukaryotic genetics (Chapter 58), due to the nature of the bacterial genome, which typically consists of a single chromosome plus none, one or several types of plasmid and/or phage, depending on the particular strain of bacterium.***

In most bacteria, the chromosome is usually a covalently closed circular DNA molecule, carrying genes for essential metabolic functions and structural components. As a consequence, bacteria can be regarded as haploid organisms. Plasmids are additional 'mini-chromosomes', typically coding for non-essential features, e.g. antibiotic resistance, heavy metal tolerance. They are often present at a higher copy number than the chromosome and may also carry genes within mobile transposable elements (transposons). A single bacterium may contain more than one type of plasmid (though not if they are closely related plasmids, i.e. from the same incompatibility group). Plasmids can be introduced into a bacterial cell by conjugation, or by transformation. Bacteria can be 'cured' of their plasmids by chemical treatment, e.g. using acridine dyes that interfere with replication, or by growth under particular conditions, e.g. at high temperatures, where plasmid replication is unable to keep up with cell division.

A phage may replicate inside a bacterial cell or, in selected instances (temperate phages, p. 208), may exist within the cell in a non-replicating (latent) state, termed a prophage. As such, phages represent additional genetic elements that may be present within a bacterial cell, forming an important component of several aspects of bacterial genetics at the practical level.

Using and interpreting standard nomenclature in bacterial genetics *– different forms of three letter abbreviations are used for:*

- **phenotypic features** *– non-italic text, with superscripts where appropriate, e.g.* Lac^{+}*;*
- **genotypic features** *– lower case italic or underlined text, with individual letters to denote individual genes, e.g.* lacZ*;*
- **gene products** *(polypeptides and proteins) – the non-italicised equivalent of the abbreviation for the gene and with a capitalised first letter, e.g. the* LacZ *protein, which is a β-galactosidase and is the product of the* lacZ gene *in* Lac^{+} *cells.*
- **transposon** *– inactivated genes show the transposon after the individual gene, separated by two colons, e.g.* lacZ::Tn6.

Several abbreviations may be combined, e.g. a single strain might have the phenotype Amp^{r} Lac^{+} Trp^{-}.

The principal characteristics of experimental bacterial genetics ('crosses') are:

- the processes are completely distinct from sexual reproduction in eukaryotes;
- the processes are directional, from a donor cell (exogenote) to a recipient cell (endogenote);
- usually, only part of the donor cell's genome is transferred;
- in several instances, the recipient cell becomes a merozygote, with more than one copy of a gene, or genes. The merozygote may be a transient or a stable state, depending on circumstances;
- recombination (synapsis and 'crossing over') may or may not be involved, depending on the process and strains involved.

KEY POINT ***Bacterial crosses are best described as gene transfer rather than gene exchange, since the latter term suggests reciprocal DNA movement.***

Definitions

Minimal medium – a chemically defined medium, containing only sufficient nutrients to meet the requirements of wild-type cells, i.e. inorganic salts plus a particular carbon source.

Auxotrophs – mutants requiring an additional organic compound (e.g. an amino acid or growth factor) in order to grow in minimal medium.

Prototrophs – wild types (from which auxotrophs are derived) and all other strains capable of growth in minimal medium.

Differential medium – usually a complex medium with additional compounds that distinguish between two types of bacteria, e.g. wild-type and mutant strains, often using pH indicator dyes or chromogenic/fluorogenic substrates.

Visualising carbon source mutants – *chromogenic and fluorogenic enzyme substrates (p. 382) can be used to detect particular phenotypes; for example,* E. coli *mutant strains without a functional* LacZ *gene would give non-coloured colonies on a medium containing X-galactoside, in contrast to wild type strains, which would give blue-green colonies.*

Take care when distinguishing between carbon source and nutritional mutants – *note that a* Lac$^-$ *mutant is unable to grow if provided with lactose as the sole carbon source, while a* Trp$^-$ *mutant is unable to grow unless it is provided with tryptophan as a specific nutrient.*

Working with bacterial mutants

To study bacterial genetics, it is necessary to use mutant strains, which have phenotypic characteristics that allow them to be distinguished from wild-type strains. The principal types of bacterial mutant include:

- Morphological mutants, with different structural characteristics from the wild type, e.g. so-called 'rough' mutants of selected bacteria, such as *Streptococcus* and *Klebsiella*, are defective in their synthesis of capsular polysaccharides, giving small, dull colonies on agar-based media. This is in contrast to 'smooth' wild-type strains, where colonies are large and glistening due to the hydrophilic polysaccharide capsule. At a practical level, it is relatively straightforward to work with such mutants, since wild-type and mutant bacteria will grow on the same medium, the mutants having a feature that visibly distinguishes them from wild types. However, there are few examples in general use.
- Resistant mutants, which grow in the presence of an inhibitory substance such as an antibiotic, a toxic compound or a particular phage. The isolation of an ampicillin-resistant mutant (Amp^r phenotype) is made possible by including ampicillin in the growth medium, since the growth of sensitive wild-type cells (Amp^s phenotype) will be inhibited.
- Carbon source mutants, which are unable to use a particular substance as a source of carbon or energy, e.g. *E. coli* mutants unable to use lactose (Lac$^-$ phenotype), in contrast to Lac$^+$ wild types. A Lac$^-$ mutant would be unable to grow on a minimal medium containing lactose as the principal carbon source. In order to identify and characterise such mutants, a differential medium must be used, e.g. MacConkey agar (a complex medium containing lactose as a *supplementary* carbon source, plus a pH indicator dye – colonies of Lac$^+$ and Lac$^-$ strains are distinguished on the basis of size and pigmentation, p. 213).
- Nutritional mutants, which have an additional requirement for a particular nutrient, compared to the wild type. For example, a strain auxotrophic for the amino acid tryptophan (Trp$^-$) would grow only if the medium contained tryptophan (e.g. in a minimal medium plus tryptophan). Since it is not possible to devise a single medium which will allow an auxotroph to be distinguished from the corresponding prototrophic wild type, the selection and identification of such mutants require a different approach, involving replica plating from (a) nutrient-rich medium onto (b) minimal medium and (c) minimal medium supplemented with the particular nutrient (Fig. 59.1).
- Mutants created by transposon mutagenesis, where a gene has been inactivated by the insertion of a transposon, e.g. *lacZ* inactivated by Tn10.
- Strains with different forms of a particular gene on the chromosome and also on a plasmid (partial diploids): the genes carried on the plasmid can be shown as follows: *lacZ*::Tn10/F′*lacZ*$^+$ with the slash separating the transposon-inactivated chromosomal gene from the F′ plasmid with its functional *lacZ* gene.

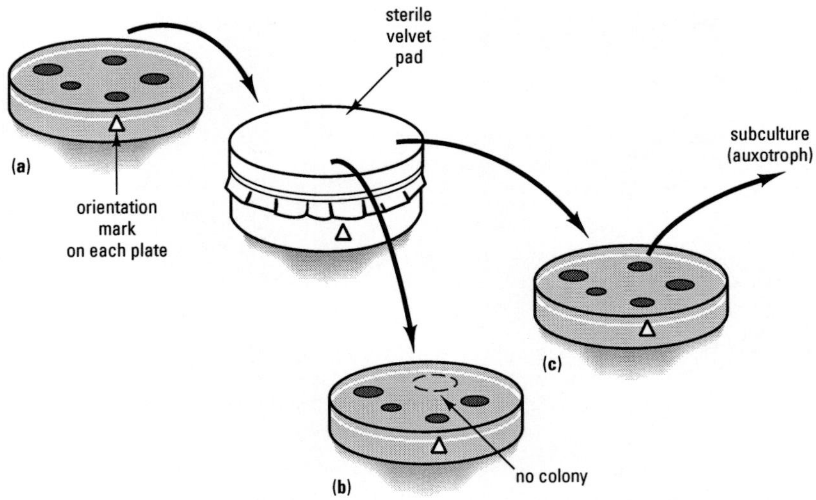

Fig. 59.1 Identification of auxotrophs using replica plating: a mixed suspension of wild-type and mutant cells is first cultured on the surface of a master plate containing nutrient-rich medium (a), then transferred to a sterile pad which is then used to inoculate minimal medium (b) and minimal medium plus a particular nutrient (c). Failure to grow on (b), coupled with growth on (c), implies auxotrophy for that particular nutrient and allows the auxotroph to be further subcultured and studied.

- Conditional lethal mutants, which have a defect that causes death under a specific set of circumstances (the 'restrictive condition'), e.g. in *E. coli* temperature-sensitive mutants that grow at 30 °C, but not 40 °C, often as a result of the temperature-dependent inactivation of mutant enzymes.

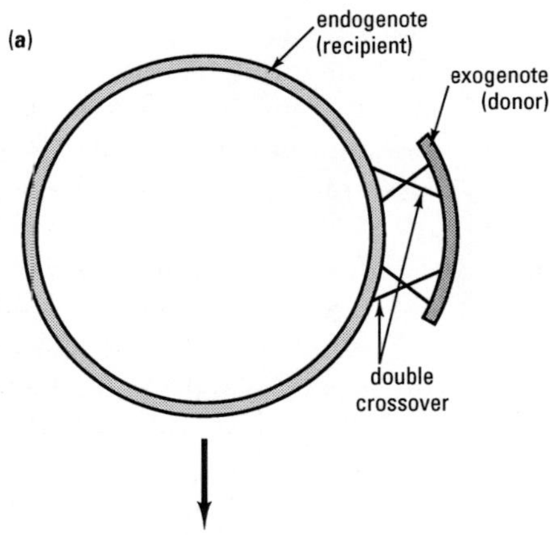

Fig. 59.2 Homologous recombination between donor DNA (from transformation, transduction or conjugation) and the recipient genome *via* a double crossover.

DNA transfer in bacteria

***KEY POINT** The traditional approach to genetic analysis in prokaryotes involves mapping the position of individual genes using information provided from 'crosses', based on gene transfer. Typically, the recipient cell will be a mutant and the donor DNA will carry wild-type genes, enabling the transfer of wild-type characteristics to be studied in the laboratory.*

The principal bacterial gene transfer processes are:

Natural transformation

In contrast to genetic engineering techniques, where DNA uptake is induced under specific laboratory conditions, natural transformation involves the release of DNA to the external medium (e.g. death and lysis of the donor cell) and its subsequent uptake and incorporation into the genome of the recipient cell. Natural transformation is restricted to a limited number of bacterial groups and only occurs if the recipient cells are in the correct physiological state, termed *competence*, often in the early exponential (log) growth phase (p. 233). Transformation is a relatively rare event, occurring at frequencies of $\leqslant 1$ transformant per 10^3 cells. In a competent cell, DNA will be taken up and, if homologous (i.e. from another strain of the same species), may then be incorporated into the genome of the recipient cell by homologous recombination *via* a double 'crossover' (two recombination events, as shown in Figure 59.2).

Performing co-transformation tests using auxotrophic recipients *– controls must be set up without added donor DNA to determine the number of revertants (spontaneous mutants), giving the actual rate of transformation by subtraction.*

The principal application of transformation has been in mapping the position of genes in those bacteria showing natural competence (e.g. *Bacillus subtilis*). The experiments are often easiest to perform with auxotrophic recipients and prototrophic (wild-type) donor DNA, since transformants can be selectively grown on media lacking one or more individual nutrients. The frequency of co-transformation of two genes is a measure of how close together they are likely to be on the donor DNA strand – a high co-transformation frequency implies that they are close together on the chromosome, reaching a recipient cell on the same fragment of DNA.

KEY POINT While co-transformation frequencies are inversely related to map distances, they are not directly equivalent to the recombination frequencies used in mapping eukaryotic genomes, because they are also influenced by the size distribution of the fragments of donor DNA and by the likelihood of homologous recombination.

Transformation mapping has several limitations, since it requires a fairly large number of complex replica plating experiments to produce a chromosomal map, and the relative position of genes that are very far apart cannot be determined directly – the 'jigsaw' requires a large number of available pieces, before the underlying structure can be seen. It is also insensitive for small map distances – two genes that are adjacent, or very close together, will give similar high co-transformation values.

Performing generalised transduction in **E. coli** *– P1 phage is often used, since it gives a high proportion of transducing particles (≈0.3% of the total), with a large fragment size (≈2.5% of the* E. coli *genome), making it easier to locate individual transductants.*

Transduction

Here, the DNA exchange is mediated by a phage (p. 208), in one of two processes:

1. Generalised transduction: occasionally, a fragment of chromosomal or plasmid DNA within an infected bacterial cell may be packaged within the protein coat of a phage, in place of the phage genome. This fragment might be derived from any part of the host cell genome (exogenote). After release, on lysis of the donor cell, the transducing particle may introduce the DNA fragment into a new recipient cell. The introduced DNA may then be incorporated into the host genome by a double crossover, in a homologous recombination event similar to that shown in Figure 59.2 for transformation. Generalised transduction can be used to establish gene order and for mapping purposes, using broadly similar principles to transformation, as only closely spaced genes will show co-transduction, with a co-transduction frequency that is inversely related to the distance between the two genes. Since a generalised transducing particle can carry a fairly small amount of DNA, the relative frequency of co-transduction can be used to provide finer detail of gene order over shorter distances than for transformation.
2. Specialised transduction: this is mediated only by a temperate phage, e.g. λ phage, which integrates at a specific site on the chromosome by a single crossover event. It involves a restricted number of genes – typically, a pair of genes on either side of the integration site for a particular temperate phage (Fig. 59.3). Specialised transduction will only occur if, on entry into the lytic cycle (phage replication, p. 208), there is an incorrect (abnormal) excision of the prophage and a part of the bacterial genome is incorporated, giving a specialised transducing

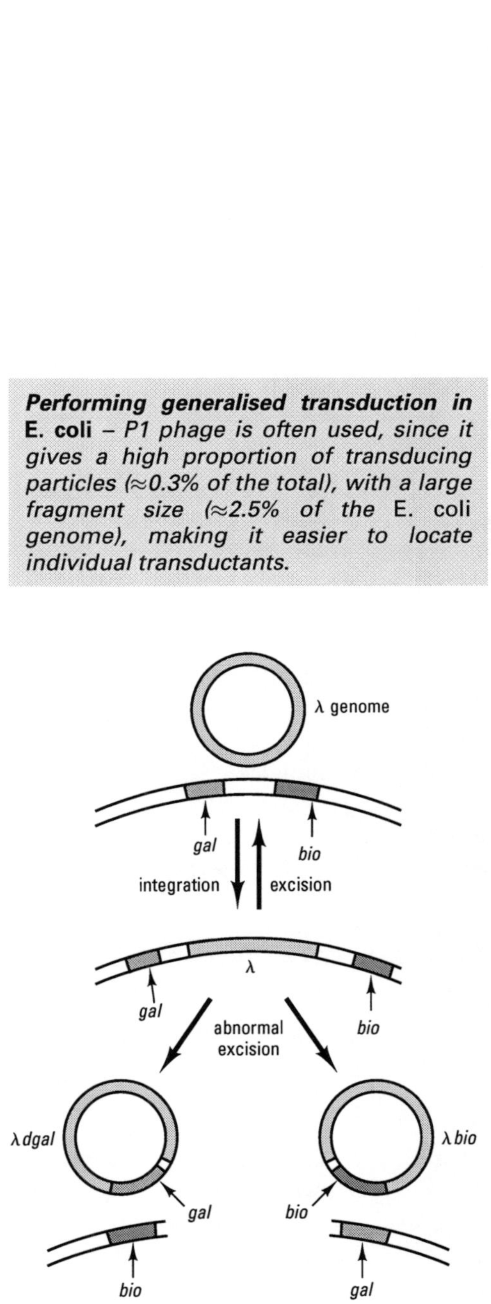

Fig. 59.3 Integration, excision and abnormal excision of λ phage, creating a modified phage genome (either λ*dgal* or λ*bio*), prior to specialised transduction.

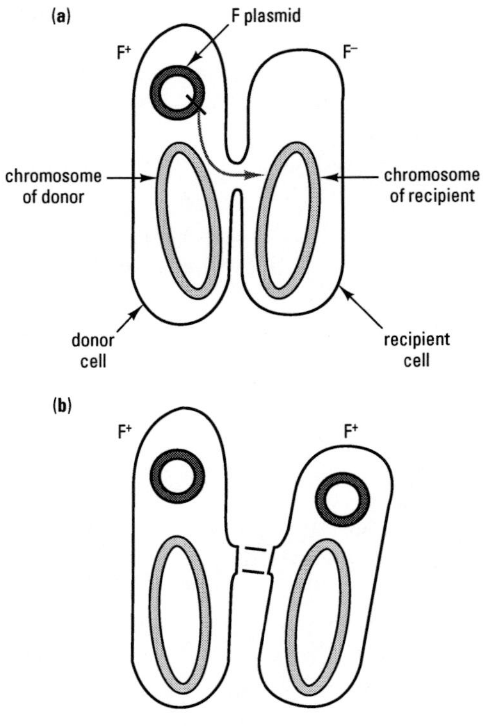

Fig. 59.4 Conjugation between F^+ donor and F^- recipient. A single strand of F plasmid DNA is cut and transferred in linear form to the recipient (a), followed by recirculatisation and complementary strand synthesis, converting the recipient to F^+ (b).

particle (a modified phage, carrying a bacterial gene). In the case of λ phage (Fig. 59.3), the transducing particle may carry either the biotin gene (λ*bio*), or the galactose gene (λ*gal*, also termed λ*dgal*, since the modified phage is defective due to the loss of an essential part of the phage genome, and therefore is incapable of replicating without the aid of a co-infecting, non-defective 'helper' phage). On entry into the recipient cell, the modified phage may become latent, creating a partial diploid. The practical applications of specialised transduction are limited to those genes flanking the integration sites of temperate phages. The most common investigations are of complementation in merozygotes containing two copies of a particular gene. Where donor and recipients of the same mutant phenotype have mutations in different genes, the creation of a partial diploid can produce a wild-type phenotype.

The integration and excision of phages is likely to be an important means of gene transfer in natural environments, playing a key role in the evolution and pathogenesis of bacteria, as has been demonstrated by recent genome sequencing projects that have demonstrated the role of prophages in toxigenic *Vibrio cholerae* and *E. coli.*

Conjugation

Here, the transfer occurs as a result of cell-to-cell contact, with direct transmission of DNA from donor to recipient cell. In *E. coli*, the donor cell carries specific surface pili (protein microtubules), allowing a donor cell to attach to receptors on the surface of a recipient cell and bringing the paired cells into close contact (Fig. 59.4). In the simplest instance, the donor cell carries an additional plasmid, the F plasmid (originally termed 'F factor'), that encodes the genes responsible for conjugation, including those for the protein subunits of the specialised pili. In conventional notation, the donor is termed F^+ and the recipient F^-. During conjugation between F^+ donor and F^- recipient, a single strand of the circular F plasmid is cut and transferred (in linear form) to the recipient cell, which becomes F^+ once the entire plasmid has been transferred and a complementary strand has been synthesised – this process takes a few minutes.

KEY POINT It is important to understand that the process of bacterial conjugation is completely different from sexual reproduction in eukaryotic organisms, since there are no gametes and no zygote or offspring is formed. Consequently, it is more appropriate to describe the participants as donor and recipient (of genetic information) rather than as male or female.

At a practical level, crosses involving F^+ donors give little useful information, apart from mapping the position of genes on the F plasmid. However, two other types of donor are more useful:

Example The *E. coli* strain HfrH has the F proplasmid integrated at a site close to genes for the synthesis of threonine, so these genes would be transferred early during HfrH × F^- conjugation.

1. Hfr strains, where the F plasmid DNA has become integrated into the chromosome, as a proplasmid, in an analogous manner to a temperate phage such as λ (Fig. 59.3). Such strains show a *high frequency of recombination* (hence Hfr), since chromosomal genes are transferred to the recipient cell at a far greater frequency than in crosses using F^+ donors. The Hfr × F^- cross is illustrated in Figure 59.5: a part of the F proplasmid is first to be transferred, followed by chromosomal DNA and, finally, the remaining fragment of the F proplasmid. After transfer, donor chromosomal DNA can be integrated into the

recipient's genome by homologous recombination (Fig. 59.2). The conjugating pair usually breaks apart before the process is complete and the recipient cell remains F$^-$, since only the leading fragment of the F plasmid reaches the recipient cell. A number of different Hfr strains of *E. coli* are available, with the F proplasmid inserted at different chromosomal locations.

2. F′ ('F prime') donors, with a modified F plasmid incorporating one or more chromosomal genes, formed as a result of defective excision of the F proplasmid from the chromosome in a similar manner to λ*bio* (Fig. 59.3). The F′ plasmid will transfer its chromosomal genes at very high frequency to a recipient, which will also become F′ once transfer of the F′ plasmid DNA is complete. This process is sometimes referred to as F-duction (or, less appropriately, as 'sexduction').

Example F′*lac* is an F′ plasmid that incorporates wild-type chromosomal DNA coding for lactose utilisation.

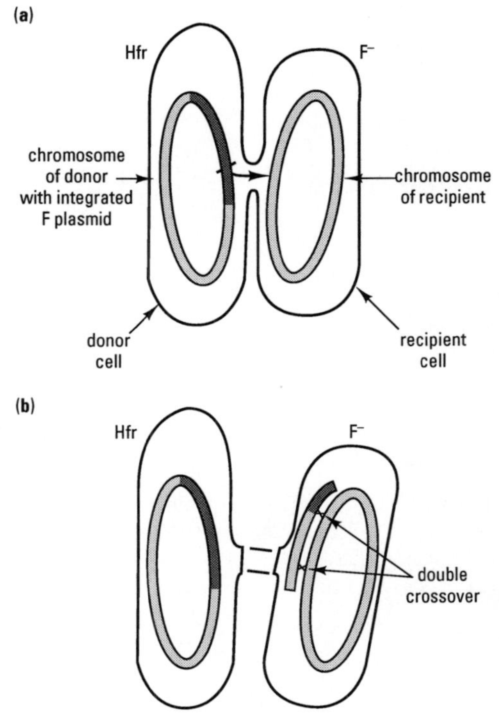

Fig. 59.5 Conjugation between Hfr donor and F$^-$ recipient. The F proplasmid is cut and leads the chromosome (in single stranded form) from the donor (a), followed by homologous recombination in the recipient (b).

Mapping with Hfr donors using interrupted conjugation

Mating an antibiotic-sensitive wild-type Hfr donor with an antibiotic-resistant mutant F$^-$ recipient provides a simple means of detecting recombinants, since the donor cells will be unable to grow on a medium containing the antibiotic while unmodified recipient cells will have a different phenotype from the recombinant recipient cells. The method is even simpler for the most common crosses, using nutritional mutants, where a prototrophic Hfr donor is crossed with an auxotrophic F$^-$ recipient and only the recombinants are able to grow on minimal medium with added antibiotic; this approach can be extended to multiple genes, using a multiple auxotroph F$^-$ recipient. By carrying out a series of experiments where the conjugation process is terminated at different times (e.g. using vigorous mixing in a vortex mixer, p. 139), the time required to transfer a particular characteristic can be determined as the earliest time at which this interruption no longer prevents recombinants from appearing. Typical results from a cross involving several genetic markers are shown in Figure 59.6a: by plotting the result graphically and extrapolating the curves to their intersect with the *x*-axis, the *time of entry* of a particular marker can be determined accurately.

The different times of entry of each characteristic reflect their relative positions on the chromosome, with genes near to the origin of transfer of

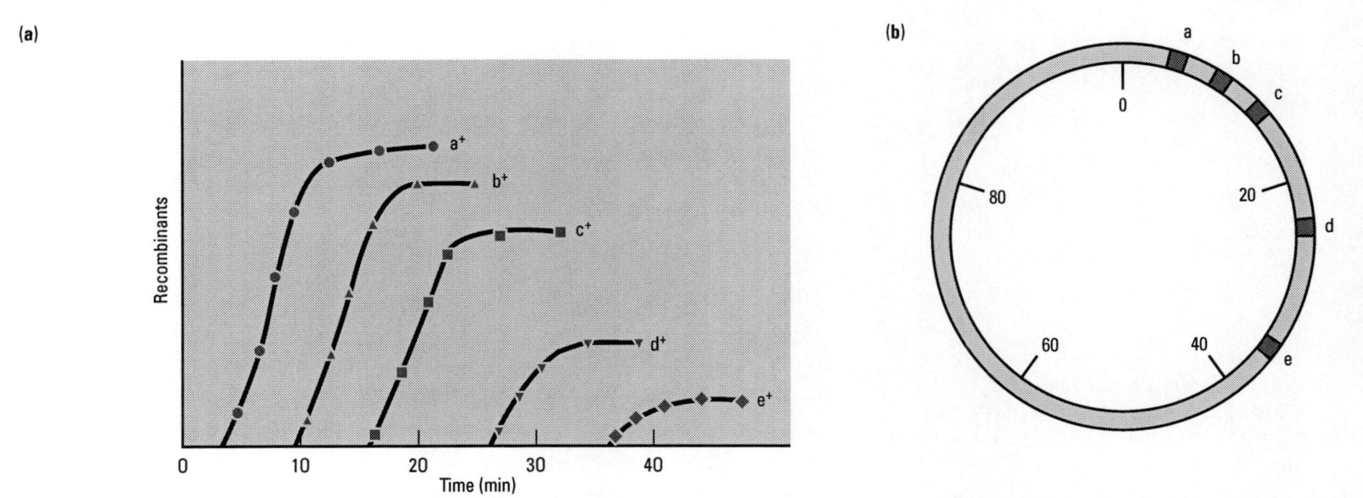

Fig. 59.6 Recombinants obtained from interrupted conjugation for a number of characters (a to e), plotted to show the time of entry of each character (a): the same data have been used to map the position of each character on the bacterial chromosome (b).

the F proplasmid having short times of entry and those further along the chromosome having later times of entry. The times of entry can be used to locate the genes on a chromosome map (Fig. 59.6b), representing map distances relative to the origin of transfer. This map differs from those produced from transformation and transduction analysis, since it is a transfer order map, rather than a linkage map. However, as transfer of the entire chromosome requires ≈100 minutes, it is difficult to map genes far from the origin of transfer, due to the decreasing number of recombinants with increasing time (Fig. 59.6a): this is overcome by combining mapping data from several Hfr strains, each with the F plasmid integrated at a different position. In fact, this approach was originally used to demonstrate the circularity of the *E. coli* chromosome and, subsequently, to determine the location of over 1500 individual genes in this organism.

Genome mapping using interrupted conjugation *– while this technique works well for widely spaced markers, it is less useful than transformation and transduction for closely linked genes.*

Genetic analysis using F′ plasmids

Since cells containing an F′ plasmid are stable merozygotes, their principal use is in studying the behaviour of genes under diploid conditions, for example:

- to determine whether a particular mutation (on the F′ plasmid) is dominant or recessive over another (on the chromosome);
- to test whether a particular gene complements another: if two mutations are in different genes, then they can complement each other in a stable merozygote, while two mutations in the same gene cannot complement each other in a stable merozygote;
- to examine whether a particular regulatory gene located on an F′ plasmid influences the expression of one or more chromosomal genes, i.e. whether it can operate in the *trans* position. The alternative test is to construct a merozygote with two genes in a chromosomal location (i.e. the *cis* position), for example, using specialised transduction. This type of analysis has been used to investigate the molecular basis of gene expression and the function of regulatory regions of the chromosome.

Definitions

Trans (Latin 'across') – on separate DNA molecules.

Cis (Latin 'here') – on the same DNA molecule.

KEY POINT Note that, while many of the above procedures have proved to be useful in the early stages of bacterial genome analysis, the techniques and methods of molecular biology (Chapters 60–62) can be used to provide more detailed genetic information, without the need to obtain certain types of mutant, or carry out specific crosses between particular donors and recipients.

Phage crosses

In practical classes, you may carry out experiments using phages, often using T-even phages of *E. coli* (e.g. T4 coliphage). The methods used to map the phage genome superficially resemble those used for eukaryotic organisms (Chapter 58) since the 'progeny' of two 'parental' phages with contrasting genotypes are analysed for recombinants and the frequency of recombination is used to measure the distance between linked genes.

Performing phage crosses *– large numbers of both types of phage must be used to ensure that every bacterial cell is infected with at least one of each type, i.e. there is a high multiplicity of infection.*

***KEY POINT** A phage cross (sometimes termed a 'mating') does not involve meiosis, gamete production or zygote formation, but is a result of a mixed infection, where a single bacterial cell is infected with phages of the two original genotypes.*

Following a single round of replication, the 'progeny' from this mixed infection are screened for the various possible genotypes, as plaques on a 'lawn' of a susceptible strain of *E. coli*, using conventional phage culture techniques (p. 208).

Phage mutants

The phage genes most widely used in crosses are those affecting:

- plaque morphology – for example, mutants may form large plaques as opposed to small plaques, or plaques with a different margin, e.g. a light turbid halo as opposed to a dark halo;
- host range – e.g. T-even mutants may infect different strains of *E. coli* from the parental types;
- conditional lethality, especially T4 phage mutants of the *rII* locus, unable to grow on *E. coli* K12(λ) in contrast to wild-type T4.

Understanding conditional lethal mutants *– an important group of phages are the so-called 'amber' mutants, with polypeptides containing a premature termination codon, UAG; such mutant strains are propagated in 'amber suppressor' mutant strains of* E. coli, *which contain a mutant tRNA that recognises UAG as coding for an amino acid.*

Phage mapping

By crossing a mutant for plaque morphology with a mutant for host range, the relative frequency of recombinants (i.e. progeny with double mutant and wild-type phenotypes) can be expressed in terms of the proportion of the total number of plaques, as a recombination frequency (*R*), where:

$$R = \frac{\text{number of double mutants and wild-type plaques}}{\text{total number of plaques}} \qquad [59.1]$$

This value is sometimes multiplied by 100 and expressed as a percentage, or as map units, equivalent to 'centimorgans' (1 cM = 1% crossing over) in conventional Mendelian crosses.

The pioneering work of Seymour Benzer established that the *rII* region of the T4 genome consisted of two distinct genes (*rIIA* and *rIIB*), each of which can give rise to the same mutant phenotype. T4 *rII* mutants are most often used for demonstrating *trans* complementation between these two genes and for demonstrating the principles of fine structure mapping of this region of the genome, since the conditional lethality of the rII phenotype simplifies the search for rare recombinant wild-type phages, as only these recombinants are able to form plaques on *E. coli* K12(λ).

Learning from Benzer's work *– detailed fine structure mapping of the rII region of T4 coliphage established that recombination can occur within a single gene, and even between adjacent nucleotides. Before these experiments, a gene was regarded as an indivisible unit – nowadays we appreciate that the smallest unit in genetics is the base pair.*

Sources for further study

Anon. *Information about bacteriophage lambda.* Available: http://www.asm.org/division/M/fax/LamFax.html
Last accessed: 01/04/07.
[Provides an overview of phage λ]

Birge, E.A. (2005) *Bacterial and Bacteriophage Genetics,* 4th edn. Springer, New York.

Dale, J.W. and Park, S.F. (2004) *Molecular Genetics of Bacteria*, 4th edn. Wiley, Chichester.

Eynard, N., and Teisse, J. (2000) *Electrotransformation of Bacteria*. Springer-Verlag, Berlin.

Hardy, K.G. (1994) *Plasmids: A Practical Approach.* Oxford University Press, Oxford.

Hartl, D.L. and Jones, E.W. *GeNETics on the Web*. Available: http://www.jbpub.com/genetics/essentials3e/geNETics_on_the_web_Links.cfm
Last accessed: 01/04/07.

Snyder, L. and Champness, W. (2002) *Molecular Genetics of Bacteria*, 2nd edn. American Society for Microbiology, Washington, DC.

Trun, N. and Trempy, J. (2003) *Fundamental Bacterial Genetics*. Blackwell, Oxford.

Study exercises

59.1 Test your understanding of bacterial and phage genetics. Distinguish between each of the following pairs of terms:
(a) zygote and merozygote;
(b) transformation and transposition;
(c) generalised and specialised transduction;
(d) prototroph and auxotroph;
(e) HFr and F′ strains.

59.2 Interpret the codes commonly used to describe bacterial mutant strains. Describe the *E. coli* phenotypes represented by the following:
(a) Ampr Lac^+ Trp^-
(b) His^- Lac^-
(c) *glnA*::Tn10
(d) *trpA*::Tn 6/F′*trpA*$^+$
(e) a Gal^+ strain lysogenised with λ*dgal*

59.3 Interpret results from an HFr mapping experiment. Three different HFr strains were used to map the genome of a bacterium in a series of interrupted conjugation experiments. The strains transferred genes to the F′ recipient cell in the following orders:
Strain (a): X W C Z A
Strain (b): Y D F X W
Strain (c): Z A E B Y
Assuming that all of these genes are present only once on a single circular chromosome, what is their order?

59.4 Construct a chromosomal map. The table below represents data for an interrupted conjugation study using an HFr donor strain containing four 'marker' genes. These were distinguished by plating onto four different media, with the following major components:
Medium A: glucose, histidine (trace amount), tryptophan (trace amount), ampicillin.
Medium B: glucose, histidine (trace amount).
Medium C: glucose, tryptophan (trace amount).
Medium D: lactose, histidine (trace amount), tryptophan (trace amount).
(a) What is tested for by each of these four different media?
(b) Based on the composition of these media and the data below, what is the phenotype of (i) the HFr donor and (ii) the F^- recipient?
(c) Given a chromosome map size of 60 min, where would each of these genes be located on the bacterial chromosome?

Colony numbers following interrupted mating, with plating on medium A, B, C or D

Time (min)	Medium A	Medium B	Medium C	Medium D
0	0	0	0	0
10	0	32	0	0
20	0	287	38	0
30	34	339	182	0
40	156	341	226	28
50	179	338	229	89
60	180	340	227	95

60 Molecular genetics I – fundamental principles

Deoxyribonucleic acid (DNA) is the genetic material of all cellular organisms.

***KEY POINT** The sequence of the bases A, G, T and C carries the genetic information of the organism. A section of DNA that encodes the information for a single polypeptide or protein is referred to as a gene, while the entire genetic information of an organism is called the genome.*

Definition

Units of nucleic acid size (length) –
Kilobase pair (kbp) = 10^3 base pairs
Megabase pair (Mbp) = 10^6 base pairs

The amount of DNA in the genome is usually expressed in terms of base pairs (bp), rather than M_r, and its size depends on the complexity of the organism: for example, the human papilloma virus has a genome of 8×10^3 base pairs (8 kbp), that of *Escherichia coli* is 4×10^6 base pairs (4 Mbp), while the human haploid genome is very large, comprising 3×10^9 base pairs (3000 Mbp). The human genome contains about 23 000 genes (distributed on 23 pairs of chromosomes), which represent only about 1 per cent of the total amount of genomic DNA (i.e. 99 per cent of human DNA is non-coding). Organisms with smaller genomes have smaller amounts of non-coding DNA: some viral genomes have 'overlapping' genes, where the same base sequences carry information for more than one protein.

The size of each individual gene varies considerably: the largest ones may exceed 10 Mbp. Chromosomes represent the largest organisational units of DNA: in eukaryotes, they are usually linear molecules, complexed with protein and RNA, varying in length from tens to hundreds of Mbp. The unit used to denote physical distance between genes (base pairs) differs from that used to describe genetic distance (centimorgan, cM), which is based on recombination frequency. In humans, 1 cM $\approx$ 1 Mbp, though this relationship varies widely depending on recombination frequency within particular regions of a chromosome.

Table 60.1 The Genetic Code – combinations of nucleotide bases coding for individual amino acids

1st Base	2nd Base *U*	*C*	*A*	*G*	3rd Base
U	F	S	Y	C	***U***
	F	S	Y	C	***C***
	L	S	*	*	***A***
	L	S	*	W	***G***
C	L	P	H	R	***U***
	L	P	H	R	***C***
	L	P	Q	R	***A***
	L	P	Q	R	***G***
A	I	T	N	S	***U***
	I	T	N	S	***C***
	I	T	K	R	***A***
	M	T	K	R	***G***
G	V	A	D	G	***U***
	V	A	D	G	***C***
	V	A	E	G	***A***
	V	A	E	G	***G***

* = termination codons.
Standard abbreviations for the above amino acids are as given in Table 50.1.
Note that AUG (=M) is the initiation codon.
The above codons are given for mRNA – the coding strand of DNA would have T in place of U, while the template strand would have complementary bases to those given above.

Each template for the synthesis of RNA (transcription) begins at a promoter site upstream of the coding sequence and terminates at a specific site at the end of the gene. The base sequence of this RNA is complementary to the 'template strand' and equivalent to the 'coding strand' of the DNA. In eukaryotic cells, transcription occurs in the nucleus, where the newly synthesised RNA, or primary transcript, is also subject to processing, or 'splicing', in which non-coding regions within the gene (introns) are excised, joining the coding regions (exons) together into a continuous sequence. Further processing results in the addition of a polyadenyl 'tail' at the 3′ end and a 7-methylguanosine 'cap' at the 5′ end of what is now mature eukaryotic messenger RNA (mRNA). The mRNA then migrates from the nucleus to the cytoplasm, where it acts as a template for protein synthesis (translation) at the ribosome: the translated portion of mRNA is read in coding units, termed codons, consisting of three bases.

Each codon corresponds to a specific amino acid, including a codon for the initiation of protein synthesis (Table 60.1). Individual amino acids are brought to the ribosome by specific transfer RNA (tRNA) molecules that recognise particular codons. The amino acids are incorporated into the growing polypeptide chain in the order dictated by the sequence of codons until a termination codon is recognised, after which the protein is released from the ribosome.

Good practice in molecular genetics *– this includes:*

- *accurate pipetting down to 1 µl, or less;*
- *steadiness of hand in sample loading;*
- *keeping enzyme solutions cold during use and frozen during storage;*
- *using sterile plasticware;*
- *using double-distilled ultrapure water;*
- *wearing disposable gloves to avoid contamination.*

The techniques described in this and subsequent chapters are used widely in many aspects of molecular biology, including the identification and characterisation of genes, gene cloning and genetic engineering, medical genetics and genetic fingerprinting.

KEY POINT An important characteristic of nucleic acids is their ability to hybridise: two single strands with complementary base pairs will hydrogen bond (anneal) to produce a duplex, as in conventional double stranded (ds)DNA. This duplex can be converted to single stranded (ss)DNA (i.e. 'melted') by conditions that disrupt hydrogen bonding, e.g. raising the temperature or addition of salt, and then reannealed by lowering the temperature or by removal of salt.

DNA can be easily purified and assayed, and it is relatively stable in its pure form. Most of the problems of working with such a large biomolecule have been overcome by the following:

Using restriction digests of viral genomes *– the controlled fragmentation pattern obtained when a phage or plasmid is cut with a specific restriction enzyme can be used to create a 'DNA ladder' – a set of DNA fragments of known length that can be used as reference standards in gel electrophoresis.*

- using restriction enzymes (see below) to cut DNA precisely and reproducibly: mammalian DNA may yield millions of fragments, with sizes ranging from a few hundred base pairs to tens of kbp, while small viral genomes may give only a few fragments;
- electrophoretic methods for the separation of DNA fragments on the basis of their sizes;
- segments of DNA can be detected by 'probes' that specifically hybridise with the DNA of interest;
- once separated, specific segments of DNA can be obtained in almost unlimited quantities by insertion into vectors and multiplication within suitable host cells;
- methods are now available for rapidly determining the base sequence of DNA segments, together with strategies for combining information from several adjacent segments to give contiguous sequences ('contigs') representing entire genomes in well-studied organisms;
- specific target sequences of DNA within a genome can be amplified by more than a billion-fold by the polymerase chain reaction.

Example ClustalW software can be used for multiple sequence alignment – it can be downloaded from the European Bioinformatics Institute website at: http://www.ebi.ac.uk/clustalw/

Producing DNA fragments

Fragments of DNA can be produced by mechanical shearing or ultrasonication, producing a random array of fragment sizes. Reproducible cleavage can be achieved using type II restriction endonucleases (commonly called restriction enzymes) which recognise and cleave at a specific palindromic sequence of double stranded DNA (usually four or six base pairs), known as the restriction site. Each enzyme is given a code name derived from the name of the organism from which it is isolated, e.g. *Hin* dIII was the third restriction enzyme to be obtained from *Haemophilus influenzae* strain Rd (Fig. 60.1). Most restriction enzymes will cut each DNA strand at a slightly different position within the restriction site to produce short, single stranded regions known as cohesive ends, or 'sticky ends', as shown in Figure 60.1. A few restriction enzymes cleave DNA to give blunt-ended fragments (e.g. *Hin* dII or *Sma* I).

Using restriction enzymes in genetic engineering *– those enzymes that cleave DNA to give single stranded 'sticky ends' are particularly useful in gene cloning. Two different molecules of DNA cut with the same restriction enzyme will have complementary single stranded regions, allowing them to anneal as a result of hydrogen bonding between individual bases within these regions.*

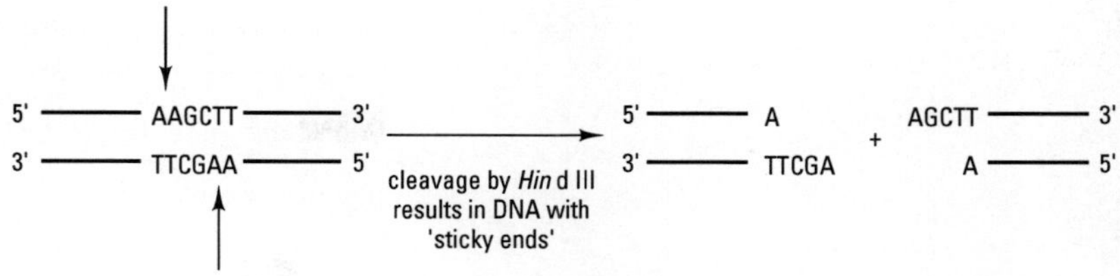

Fig. 60.1 Recognition base sequence and cleavage site for the restriction enzyme *Hin* dIII. This is the conventional representation of double-stranded DNA, showing the individual bases, where A is adenine, C cytosine, G guanine and T thymine. The cleavage site on each strand is shown by an arrow.

Maximising recovery of DNA *– these large molecules are easily damaged by mechanical forces, e.g. vigorous shaking or stirring during extraction. In addition, all glassware must be scrupulously cleaned and gloves must be worn, to prevent deoxyribonuclease contamination of solutions.*

Working with small volumes in molecular biology *– use a pipettor of appropriate size (e.g. P2 or P20 Gilson Pipetman) with a fine tip. For very small volumes, pre-wet the tip before delivering the required volume.*

SAFETY NOTE Using an electrophoresis tank – always ensure that the top cover is in position, to prevent evaporation of buffer solution and to reduce the possibility of electric shock.

Separation of nucleic acids using gel electrophoresis

Separation of DNA by agarose and polyacrylamide gel electrophoresis

Electrophoresis is the term used to describe the movement of ions in an applied electrical field. DNA molecules are negatively charged, migrating through an agarose gel towards the anode at a rate which is dependent upon molecular size – smaller, compact DNA molecules can pass through the sieve-like agarose matrix more easily than large, extended fragments. Electrophoresis of plasmid DNA is usually carried out using a submerged agarose gel (Fig. 60.2). The amount of agarose is adjusted, depending on the size of the DNA molecules to be separated, e.g. 0.3% w/v agarose is used for large fragments (>20 000 bp) while 0.8% is used for smaller fragments. Very small fragments are best separated using a polyacrylamide gel (Table 60.2). Note the following:

- Individual samples are added to pre-formed wells using a pipettor. The volume of sample added to each well is usually less than 25 µl so a steady hand and careful dispensing are needed to pipette each sample.
- The density of the samples is usually increased by adding a small amount of glycerol or sucrose, so that each sample is retained within the appropriate well.
- A water-soluble anionic tracking dye (e.g. bromophenol blue or xylene cyanol) is also added to each sample, so that migration can be followed visually.

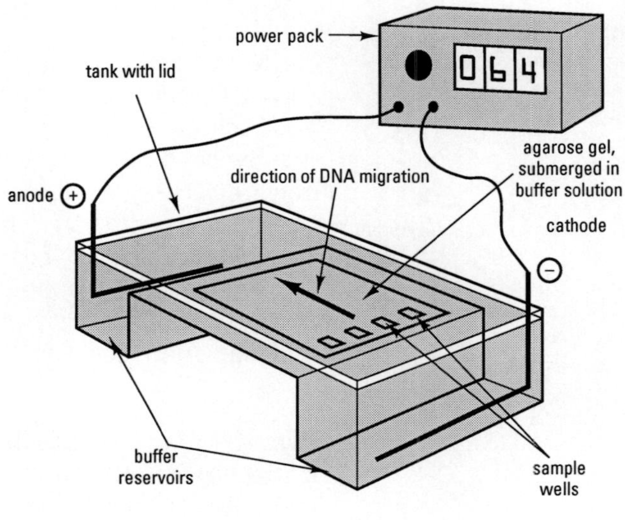

Fig. 60.2 Agarose gel electrophoresis of DNA.

Table 60.2 Gel concentrations for the separation of DNA of various sizes

Type of gel	% (w/v)	Range of resolution of DNA (bp)
Polyacrylamide	20.0	5–100
	15.0	20–150
	5.0	75–500
	3.5	100–1000
Agarose	2.0	100–5000
	1.2	200–8000
	0.8	400–20 000
	0.3	1000–70 000

Using DNA fragment size markers *– for accurate determination of fragment size (length) your standards must have the same conformation as the DNA in your sample, i.e. linear DNA standards for linear (restriction) fragments and closed circular standards for plasmid DNA.*

- DNA fragment size markers ('molecular weight' standards) are added to one or more wells. After electrophoresis, the relative positions of bands of known molecular weight can be used to prepare a calibration curve (usually by plotting $\log_{10}$ of the size (length) of each band against the distance travelled).
- The gel should be run until the bromophenol blue tracking dye has migrated across 80% of the gel (see manufacturer's instructions for appropriate voltages/times).
- After electrophoresis, the bands of DNA can be visualised by soaking the gel for around 5 min in 0.1–0.5 mg l^{-1} ethidium bromide, which binds to DNA by intercalation between the paired nucleotides of the double helix.
- Under UV light, bands of DNA are visible due to the intense fluorescence of the ethidium bromide. The limit of detection using this method is around 10 ng DNA per band. The migration of each band from the well can be measured using a ruler. Alternatively, a photograph can be taken, using a digital camera and adaptor, or a dedicated system, e.g. GelDoc from BioRad.
- If a particular band is required for further study (e.g. a plasmid), the piece of gel containing that band is cut from the gel using a clean scalpel. The DNA can be separated from the agarose by solubilising the gel slice and then binding the DNA to an anion exchange resin, followed by elution in water or buffer (e.g. TE buffer, containing 10 mmol l^{-1} TRIS/HCl at pH 8.0 plus 1 mmol l^{-1} EDTA). Various commercial kits based on this approach are also available, e.g. from Qiagen.

SAFETY NOTE Handling ethidium bromide – ethidium bromide is carcinogenic so always use gloves when handling stained gels and make sure you do not spill any staining solution, or use a safer alternative, such as SYBR Safe.

SAFETY NOTE Working with a UV radiation source – always wear suitable UV-filtering plastic safety glasses or goggles to protect your eyes.

Assaying nucleic acids in solution *– double stranded DNA at 50 µg ml^{-1} has an A_{260} of 1, and the same absorbance is obtained for single stranded DNA at 33 µg ml^{-1} and (single stranded) RNA at 40 µg ml^{-1}. These values can be used to convert the absorbance of a test solution to a concentration of nucleic acid.*

Electrophoretic separation of RNA

Total cellular RNA or purified mRNA can be separated on the basis of size by electrophoretic separations similar to those used for DNA fragments. However, under the conditions used to separate dsDNA, RNA molecules tend to develop a secondary structure, and this leads to anomalous mobilities. To eliminate RNA secondary structure, samples are pretreated by heating in dilute formamide or glyoxal, and electrophoresis is carried out in 'denaturing gels' which include buffers containing formaldehyde.

Pulsed field gel electrophoresis (PFGE)

If structural information is to be gained about large stretches of genomic DNA, then the order of the relatively short DNA segments (generated by

Definition

Chromosome walking – a method for analysing areas of interest in DNA, in which the end of a segment of DNA is used as a probe to locate other segments that overlap the first segment: long stretches of DNA can be analysed by subsequent use of probes made from the ends of successive overlapping segments.

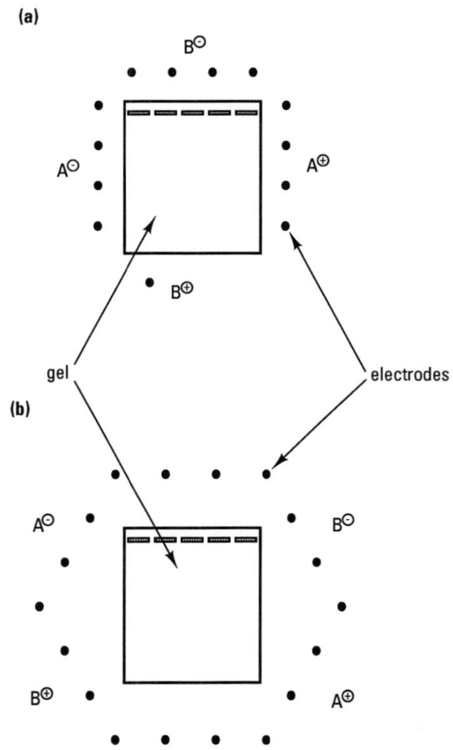

Fig. 60.3 The configuration of electrodes for conventional PFGE (a) and CHEF/PACE (b).

the restriction enzymes described above) needs to be established. This is technically possible by techniques such as 'chromosome walking', but is time-consuming and potentially difficult, especially when dealing with large chromosomes such as those from yeast (a few Mbp) and humans (50–100 Mb), in contrast to the smaller genomes of bacteria and viruses. The technique of PFGE allows separation of DNA fragments of up to ≈12 Mb.

Very large DNA fragments (>100 kbp) can be generated from chromosomal DNA by the use of certain restriction enzymes that recognise base sequences that are present at relatively low frequency, e.g. the enzyme *Not* I, which recognises a sequence of 8 bp rather than 4–6 bp. These enzymes are sometimes called 'rare cutters'. Genomic DNA prepared in the normal way is not suitable for digestion by these enzymes, as shearing during extraction fragments the DNA. Therefore, genomic DNA for analysis by PFGE is prepared as follows:

- cells are embedded in an agarose block;
- the block is incubated in solutions containing detergent, RNase and proteinase K, lysing the cells and hydrolysing RNA and proteins. The products of RNase and proteinase digestion diffuse away, leaving behind genomic DNA molecules exceeding several thousand kbp;
- the block is incubated *in situ* in a buffered solution containing an appropriate 'rare cutter': restriction fragments are produced, of up to ≈800 kbp.

PFGE differs from conventional electrophoresis in that it uses two or more alternating electric fields. An explanation for the effectivess of the technique is that large DNA fragments will be distorted by the voltage gradient, tending to elongate in the direction of the electric field and 'snaking' through pores in the gel. If the original electric field (Fig. 60.3a), is removed, and a second is applied at an angle to the first (Fig. 60.3b), the DNA must reorientate before it can migrate in the new direction. Larger (longer) DNA molecules will take more time to reorientate than smaller molecules, resulting in size-dependent separations.

The original configuration of electrodes used in PFGE is shown in Fig. 60.3a; this tends to produce 'bent' lanes that make lane-to-lane comparisons difficult. This can be overcome by using one of the many variants of the technique, one of which employs contour clamped homogeneous electric fields (CHEF). Here, multiple electrodes are arranged in a hexagonal array around the gel (Fig. 60.3b) and these are used to generate homogeneous electric fields with reorientation angles of up to 120°. A further development of CHEF involves programmable, autonomously controlled electrodes (PACE), which allows virtually unlimited variation of field and pulsing configurations, and can fractionate DNA molecules from 100 bp to 6 Mbp.

Identification of specific nucleic acid molecules using blotting and hybridisation techniques

Southern blotting

After separation by conventional agarose gel electrophoresis, the fragments of DNA can be denatured and immobilised on a filter membrane using a technique named after its inventor, E.M. Southern.

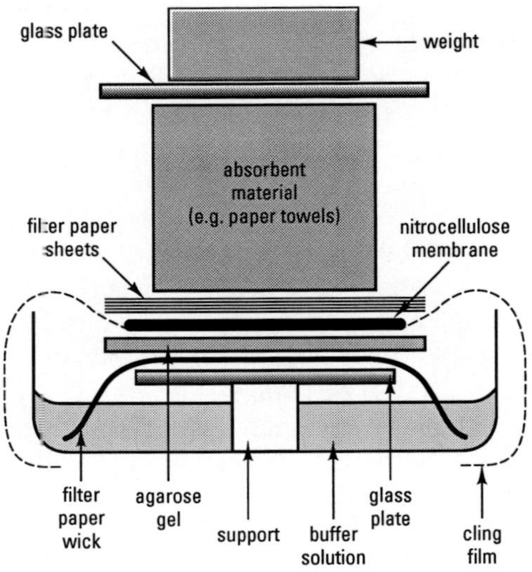

Fig. 60.4 Components of Southern blotting apparatus.

The main features of the conventional apparatus for Southern blotting are shown in Figure 60.4. The principal stages in the procedure are:

1. To ensure efficient transfer, the gel is first soaked in HCl, leading to random cleavage of the DNA into smaller fragments.
2. The gel is subsequently soaked in alkali to denature the dsDNA to ssDNA, then neutralised: this is necessary to allow hybridisation with probe DNA after blotting.
3. A nitrocellulose or nylon membrane is then placed directly on the gel, followed by several layers of absorbent paper. The DNA is 'blotted' onto the filter as the buffer solution soaks into the paper by capillary action.
4. The filter is baked in a vacuum oven at 80 °C for 3–5 h or exposed to UV light, in order to 'fix' the DNA.
5. Specific DNA fragments are identified by incubation (6–24 h) with complementary labelled probes of ssDNA, which will hybridise with a particular sequence, followed by visualisation, often using an enzyme-based system. If radiolabelled probes are used (see below), the desired fragments are located by autoradiography (p. 271).

Modifications of the method use either a vacuum apparatus, an electric field ('electroblotting') or positive pressure to transfer the DNA fragments from the gel to the membrane. These modifications reduce the time for transfer by around 10-fold.

Northern blotting

This process is virtually identical to Southern blotting, but RNA is the molecule that is separated and probed.

Dot blotting or slot blotting

Biomedical applications of dot blotting *– these include the detection of specific pathogenic microbes in samples and the identification of particular genetic markers, e.g. in inherited genetic disorders and forensic science.*

Here, samples containing denatured DNA or RNA samples are applied directly to the nitrocellulose membrane (*via* small individual round or slot-like templated holes) without prior digestion with restriction enzymes or electrophoretic separation. The 'blot' is then probed in a similar manner to that described for Southern blotting. This allows detection of a particular nucleic acid sequence in a sample.

Types of probe

Definition

Probe – a labelled DNA or RNA sequence used to detect the presence of a complementary sequence (by molecular hybridisation) in a mixture of DNA or RNA fragments.

The probes used in blotting and DNA hybridisation can be obtained from a variety of sources including:

- cDNA (complementary or copy DNA) which is produced from isolated mRNA using reverse transcriptase. This retroviral enzyme catalyses RNA-directed DNA synthesis (rather than the normal transcription of DNA to RNA). After the mRNA has been reverse transcribed, it is degraded by the addition of alkali or ribonuclease, leaving the ssDNA copy. This is then used as a template for a DNA polymerase, which directs the synthesis of the second complementary DNA strand to form dsDNA. This is denatured to ssDNA before use.
- Oligonucleotide probes (15–30 nucleotides) can be produced if the amino acid sequence of the gene product is known. Since the genetic code is degenerate, i.e. some amino acids are coded for by more than one codon (see Table 60.1) it may be necessary to synthesise a mixture of oligonucleotides to detect a particular DNA sequence: this mixture of oligonucleotides is termed a 'degenerate probe'.

Example For a simple tripeptide containing methionine, aspartate and phenylalanine, the synthetic oligonucleotide probes would include combinations of the following codons:
1st codon (met): ATG
2nd codon (asp): GAT, GAC
3rd codon (phe): TTT, TTC

- Specific genomic DNA sequences, where the gene has been characterised.
- PCR-generated fragments (Chapter 61).
- Heterologous probes, i.e. sequences for the same gene, or its equivalent, in another organism.

Labelling of probes

Detection of very low concentrations of target DNA sequences requires probes that can be detected with high sensitivity. This is achieved by radiolabelling, e.g. using ^{33}P, or by using enzyme-linked methods, which are often available in kit form:

Using commercial kits *– do not blindly follow the protocol given by the manufacturer without making sure you understand the principles of the method and the reasons for the procedure. This will help you to recognise when things go wrong and what you might be able to do about it.*

- Radiolabelled probes can be made by several methods, including the nick-translation technique. This uses DNA polymerase I from *E. coli*, which has (i) an exonuclease activity that 'nicks' dsDNA and removes a nucleotide, and (ii) a polymerase activity that can replace this with a labelled deoxyribonucleotide. After several cycles of nick-translation, the labelled DNA is denatured to ssDNA for use as probes. Newer approaches include random priming of single stranded template DNA, followed by the synthesis of radiolabelled DNA fragments complementary to the template using a DNA polymerase and a radiolabelled deoxynucleoside triphosphate (dNTP). Probe hybridisation to a target sequence is detected by autoradiography. Oligonucleotide probes cannot be labelled using the above methods and require 'end labelling', where a kinase is used to replace the 5′ terminal phosphate group with a radiolabelled group.
- Enzyme-linked methods involve incorporating a modified nucleotide precursor, such as biotinylated dTTP, into the DNA by nick-translation or random priming. When the probe hybridises with the target sequence, it can be detected by addition of an enzyme (e.g. horse radish peroxidase) coupled to streptavidin. The streptavidin binds specifically to the biotin attached to the probe and the addition of a suitable fluorogenic or chromogenic substrate for the enzyme allows the probe to be located.

Using labelled probes *– an alternative approach is the chemiluminescent system based on dioxygenin-labelled nucleotides.*

Hybridisation of probes

The stability of the duplex formed between the probe and its target is directly proportional to the number and type of complementary base pairs that can be formed between them: stability increases with the amount of G + C, since these bases form three hydrogen bonds per base pair, rather than two. Duplex stability is also influenced by temperature, ionic strength and pH of the hybridisation buffer, and these can be varied to suit the stringency of hybridisation required:

Example Low stringency hybridisation can be useful when using a heterologous probe, i.e. from another species.

Example High stringency conditions can be used for detecting a single base change in a mutant gene using a dot-blot hybridisation procedure with a specific oligonucleotide probe.

- In 'low stringency' hybridisation, duplex formation with less than perfect complementarity is promoted, either by lowering the temperature, or increasing ionic strength.
- 'Stringent' hybridisation conditions usually involve high temperatures, or decreased ionic strength, and will sustain only perfectly matched duplexes.

DNA sequencing

By fragmenting target DNA with several restriction enzymes and then sequencing the overlapping fragments, it is possible to determine the nucleotide sequences of very large stretches of DNA, including entire

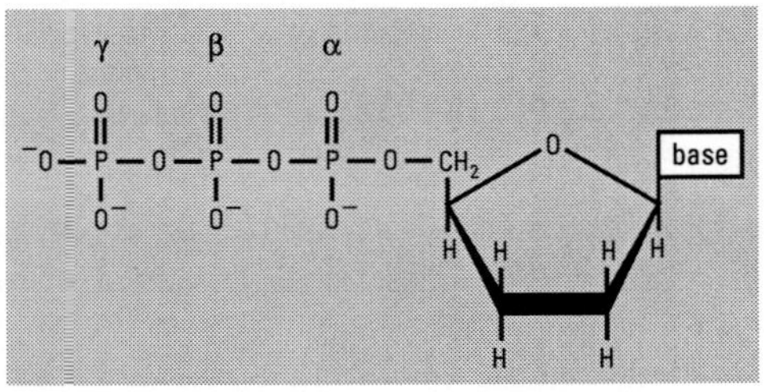

Fig. 60.5 A 2′,3′-dideoxynucleoside triphosphate.

genomes. Sequencing methods rely on polyacrylamide gel electrophoresis. The Sanger, or chain termination, method is the most widely used in DNA sequencing. This makes use of dideoxynucleotides (Fig. 60.5), which have no –OH group at either the C-2 or C-3 of ribose. A dideoxynucleoside triphosphate (ddNTP) can be added to a growing DNA chain, but since it lacks an –OH group at the C-3 position it cannot form a phosphodiester bond with the incoming dNTP of the growing chain. Therefore a dideoxynucleotide acts as a terminator at the site it occupies. Details of theSanger sequencing method are given in Box 60.1 and Figure 60.6.

Box 60.1 DNA sequencing using the chain termination (Sanger) method

The DNA to be sequenced must first be obtained as single-stranded fragments, typically around 200 bp, e.g. using a denatured plasmid. Sequencing is often performed using a commercial kit and the principal stages are as follows:

1. **Set up the strand synthesis reaction:** four separate tubes are required, each containing a small amount of one of the dideoxynucleoside triphosphates. Each tube contains all of the other components required for DNA synthesis, i.e. (i) the DNA selected for analysis (template DNA), e.g. denatured plasmid, (ii) a DNA polymerase (e.g. thermostable *Taq* polymerase), (iii) a [^{35}S] oligonucleotide primer of known sequence, to allow synthesis of the complementary DNA strand, (iv) dGTP, dATP, dCTP, dTTP (in excess) together with (v) one type of ddNTP (i.e. ddGTP, ddATP, ddCTP or ddTTP) in limited concentration. New DNA strands are synthesised by addition of dNTPs to the primer, guided by the template DNA until a ddNTP is added. As an example of the principle of the method, consider the ddTTP tube. When T is required to pair with A on the template DNA strand, the dTTP will be competing with the ddTTP, but because the dTTP is in excess this will normally be added to the chain at the appropriate position. However on occasions a ddTTP will be inserted at a given site and this will terminate DNA synthesis on the template strand. Thus, synthesis will be halted at all possible sites where ddT has substituted for dT and several strands of different length will be formed in the reaction mixture, each ending with ddT.

2. **Terminate the strand synthesis reaction:** after 5–10 min at 37 °C add a 'stop' solution containing formamide to disrupt hydrogen bonding between complementary bases and incubate at 80 °C for 15 min, to produce ssDNA.

3. **Separate the fragments using polyacrylamide gel electrophoresis:** 4–6% (w/v) denaturing acrylamide gels are used, containing urea (46% w/v) as the denaturing agent. Thin gels (0.35 mm thickness) are used, typically 20 cm wide by 50 cm long. The products in each of the four tubes are placed in four lanes side by side on the same gel and are separated by electrophoresis at 35–40 W (up to 32 mA, 1.5 kV) for 2.5 h. The high voltage raises the gel temperature to ≈50 °C, helping denaturation. A single gel can separate DNA fragments that differ in length by a single nucleotide.

4. **Locate the positions of individual bands by autoradiography:** the gel is fixed for 15 min with 10% (v/v) acetic acid, covered with Whatman 3MM paper and Saran wrap, then vacuum-dried, to avoid quenching the radioisotope signal (p. 270) and to prevent it from sticking to the X-ray film. The gel is unwrapped and placed next to an X-ray film for 24 h.

5. **Read the gel:** the nucleotide sequence can be read directly from the band positions that represent the newly made DNA segments of varying lengths (Fig. 60.6). The smallest segment is represented by the band at the bottom, since it travels furthest in the gel. The nucleotide sequence of the template strand can be deduced directly from that of the new strand, since the base on the template strand will be represented by an incomplete chain that terminates with the complementary dd nucleotide, i.e. template A with ddT, G with ddC, T with ddA, and C with ddG (Fig. 60.6).

Troubleshooting and other points to note: streaking can be due to damage to wells, air bubbles in wells/gel, or contamination by dirt/dust; faint and fuzzy gels may be due to insufficient template DNA, primer and/or dNTPs, due to poor annealing of primer and template, or to errors in preparing or running the gel; bands in more than one lane can indicate contamination of the template DNA, more than one primer site on the template, or secondary structure of DNA, giving 'ghost' banding.

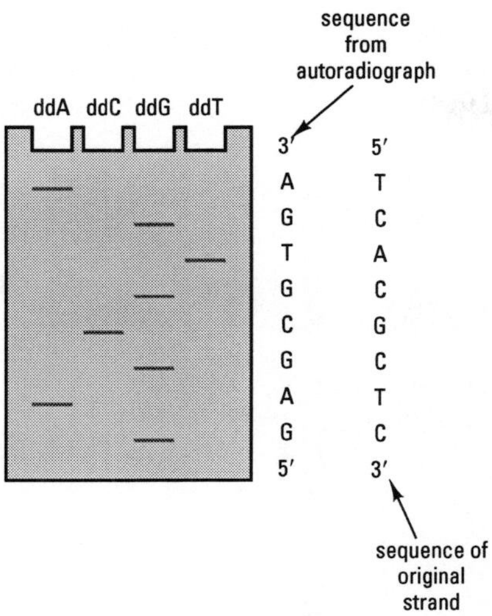

Fig. 60.6 Sanger sequencing gel, showing how the banding pattern is converted into a sequence of nucleotide bases.

Using PCR for sequencing *– with the four ddNTPs tagged using different fluorophores, the products can be run in the same lane and detected by laser excitation at the end of the gel, with the sequence of fluorophores giving the base sequence.*

DNA microarrays and bioinformatics *– interpretation of DNA microarray data is a growing area of bioinformatics, requiring dedicated software, e.g. the TM_4 microarray analysis freeware programs available from the Institute of Genomic Research, at: http://www.tm4.org/*

The alternative approach to sequencing is based on chemical degradation (the Maxam and Gilbert method), using different reagents to break the target DNA sequence into fragments which are then separated by polyacrylamide gel electrophoresis. The Maxam and Gilbert method is more involved than the Sanger technique, and is reserved for specialised applications, e.g. for studying the interaction between DNA and proteins.

Automated DNA sequencing

The development of automated DNA sequencing machines (so-called 'DNA sequenators') has enabled sequencing to be performed several orders of magnitude faster than with manual methods. Base-specificity is achieved by using primers labelled with fluorophores with different fluorescence characteristics in each of the four reaction tubes (Box 60.1). After the reactions are separately completed, the four sets of products can be pooled and fractionated by electrophoresis in a single lane. The fluorophore-labelled fragments are detected as they pass a scanning laser detector and the DNA sequence is determined by using both the specific wavelength emitted by the fragment (indicating the base) and by the migration time (indicating the fragment size, which corresponds to the base location in the DNA sequence). Nowadays, automated DNA sequencing tends to be performed by capillary electrophoresis rather than polyacrylamide gel electrophoresis.

DNA microarrays

This area has seen rapid, recent development. DNA microarrays ('gene chips') support the quantitative analysis of many different genes/sequences at the same time. This involves synthesising and fixing single-stranded probes for a large number (>1000) of different nucleotide sequences onto specific sites on a glass, quartz or silicon wafer. When single-stranded labelled nucleic acid is added, it will hybridise only with probes having a complementary sequence and these can be determined by the presence of an attached label, e.g. a fluorescent dye, detected using a scanning laser. One application involves the detection of labelled cDNA produced from the mRNA in an organism by using an 'expression' microchip, which contains probes for all of the expressed genes in an organism, enabling the relative expression of each gene to be determined from the fluorescence signal at each location on the microarray. Other applications include the use of single nucleotide polymorphism (SNP) microarrays to identify mutations and genetic variation, e.g. in screening for genetic diseases and in forensic DNA analysis. A commercial example is the Affymetrix GeneChip system.

Sources for further study

Anon. *NCBI Microarray Factsheet*. Available: http://www.ncbi.nlm.nih.gov/About/primer/microarrays.html Last accessed: 01/04/07.
[A basic guide to the principles of DNA microarrays]

Brown, T.A. (2002) *Essential Molecular Biology: A Practical Approach*, Vols 1 and 2, 2nd edn. Oxford University Press, Oxford.

Clark, M. (2003) *In Situ Hybridization: Laboratory Companion*. Wiley-VCH, Mannheim.

Darby, I.A. and Hewitson, T.D. (2003) *In Situ Hybridization Protocols*. Humana Press, New Jersey.

Hames, B.D. and Higgins, S.J. (1995) *Gene Probes: A Practical Approach*. Oxford University Press, Oxford.

Monaco, A.P. (1995) *Pulsed Field Gel Electrophoresis: A Practical Approach*. IRL Press, Oxford.

Nuber, U. (2005) *DNA Microarrays*. Bios (Advanced Methods Series). Taylor & Francis, Abingdon.

Sambrook, J. and Russell, D.W. (2001) *Molecular Cloning: A Laboratory Manual*, 3rd edn. Cold Spring Harbor Lab, Cold Spring Harbor.

Study exercises

60.1 Practise the calculations involved in working with nucleic acids. Imagine that you have been supplied with a freeze-dried sample of a single stranded DNA probe 24 bases in length, which you reconstitute in 1 ml of buffer (reconstituted probe). You take 5 μl, add it to 495 μl of water and determine the A_{260} as 0.14.

(a) What is the mass concentration of the probe in this solution, in μg ml^{-1}?

(b) Assuming the average M_r of a nucleotide is 325, what is the concentration of the probe in the solution, expressed in micromolar terms (i.e. as μmol l^{-1})?

(c) The concentration of probe required in a blotting procedure is 10 pmol ml^{-1} and a total of 10 ml of this solution is required. You have been asked to prepare 5 ml of a stock solution of the DNA probe at 100× the concentration required in the blotting procedure, so that 0.100 ml can be diluted to 10 ml for the final blotting procedure, while the rest will be stored at −20 °C, for subsequent experiments. How would you prepare the 100× stock solution?

Express all answers to three significant figures.

60.2 Select an electrophoretic technique for separating DNA fragments of different sizes. Which technique would you use to separate DNA fragments in the following ranges: (a) 1–70 kbp; (b) 75–500 bp; (c) 100–6 Mbp.

60.3 Test your understanding of the term 'stringency' in relation to nucleic acid probes. What does the term 'stringency' mean in the context of molecular biology? What factors cause increased stability between a nucleic acid probe and its target sequence? Give examples of when you might use 'low stringency' and 'high stringency' hybridisation.

60.4 Interpret the results of a Sanger sequencing gel. The figure represents a typical gel from a dideoxynucleotide sequencing procedure. Determine the sequence of the original DNA strand.

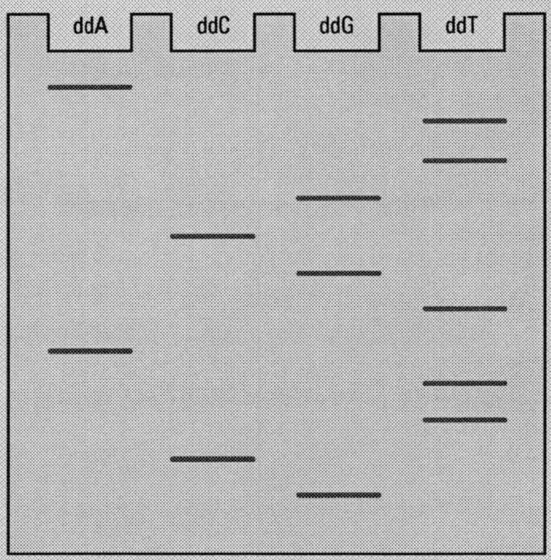

61 Molecular genetics II – PCR and related applications

***Applications of PCR** – these include:*

- *diagnosis and screening of genetic diseases and cancer;*
- *rapid detection of slowly growing microorganisms (e.g. mycobacteria) and viruses (e.g. HIV);*
- *HLA typing in transplantation;*
- *analysis of DNA in archival material;*
- *DNA fingerprinting in forensic science;*
- *preparation of nucleic acid probes;*
- *clone screening, mapping and sub-cloning.*

***Understanding the terminology of PCR** – cloned (amplified) DNA sequences produced using PCR techniques are often termed amplicons.*

The polymerase chain reaction (PCR) is a rapid, inexpensive and simple means of producing μg amounts of DNA from minute quantities of template.

***KEY POINT** PCR offers an alternative approach to gene cloning (Chapter 62) for the production of many copies of an identical sequence of DNA. The starting material may be genomic DNA (e.g. from a single cell), RNA, DNA from archival specimens, cloned DNA or forensic samples.*

The technique uses *in vitro* enzyme-catalysed DNA synthesis to create millions of identical copies of DNA. If the base sequence of the adjacent regions of the DNA to be amplified is known, this enables synthetic oligonucleotide primers to be constructed that are complementary to these so-called 'flanking regions'. Initiation of the PCR occurs when these primers are allowed to hybridise (anneal) to the component single strands of the target DNA, followed by enzymatic extension of the primers (from their 3′ ends) using a thermostable DNA polymerase. A single PCR cycle consists of three distinct steps, carried out at different temperatures (Fig. 61.1), as follows:

1. Denaturation of dsDNA by heating to 94–98 °C separating the individual strands of the target DNA.
2. Annealing of the primers, which occurs when the temperature is reduced to 37–65 °C.
3. Extension of the primers by a thermostable DNA polymerase (e.g. *Taq* polymerase, isolated from *Thermus aquaticus*) at 72 °C; this step should last long enough to generate the PCR product: approximately 1 min of reaction time is required per kbp of sequence.

In the first cycle, the product from one primer is extended beyond the region of complementarity of the other primer, so each newly synthesised strand can be used as a template for the primers in the second cycle (Fig. 61.1). Successive cycles will thus generate an exponentially increasing number of DNA fragments, the termini of which are bounded by the 5′ ends of the primers (length of each fragment = length of primers + length of target sequence). Since the amount of DNA produced doubles in each cycle, the amount of DNA produced = 2^n, where n is the number of cycles. Up to 1 μg of amplified target DNA can be produced in 25–30 cycles from a single-copy sequence within 50 ng of genomic DNA, assuming close to 100% efficiency during the cycling process. After electrophoresis, the PCR product is normally present in sufficient quantity to be visualised directly, for example with ethidium bromide or SYBR Safe.

The temperature changes in PCR are normally achieved using a thermal cycler, which is simply a purpose-built incubator block that can be programmed to vary temperatures, incubation times and cycle numbers.

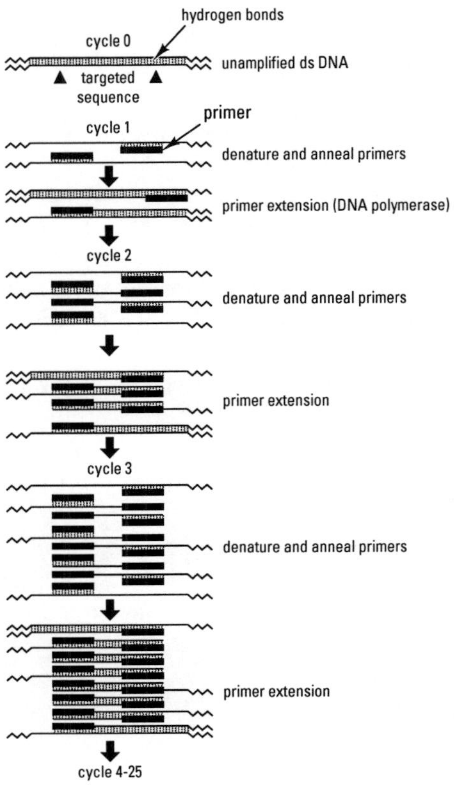

Fig. 61.1 The polymerase chain reaction (PCR).

***KEY POINT** PCR is so sensitive that one of the main problems associated with the technique is contamination with 'foreign' DNA. Great care is required to avoid sample contamination during* in vitro *amplification (Box 61.1).*

PCR components and conditions

These are readily available in kit form from commercial suppliers; they include a thermostable DNA polymerase, such as *Taq* polymerase, dNTPs, buffer, a detergent (e.g. Triton), KCl, $MgCl_2$ or $MgSO_4$, and primers. Protocols for PCR vary considerably for particular applications – a typical procedure is given in Box 61.1. However, if you are trying to develop your own PCR procedure, the following information may be useful:

Storing primers for PCR – *primers are best stored in ammonia solution, which remains liquid at −20°C, avoiding the need for repeated freezing and thawing when dispensed. Before use, aliquots of stock solution should be heated in a fume cupboard, to drive off the ammonia.*

- Primers need to be at least 18 nucleotides long (primers longer than 18 nucleotides should be unique, even in a large eukaryotic genome); both primers should have similar annealing temperatures, T_m, p. 372), with a minimal degree of self-complementarity (to avoid formation of secondary structures), and no complementarity to each other (so that primer dimers are not formed).
- For most applications, the final concentration of each primer should be 0.1–0.5 $\mu mol\, l^{-1}$, which gives an excess of primers of about 10^7 with respect to the template, e.g. target genomic DNA at a concentration of 50 ng per 10 µl reaction mixture.
- Annealing temperatures are based around T_m, the temperature at which 50% of the primers are annealed to their target sequence. For primers of <20 bases, T_m can be roughly calculated in °C from the equation:

$$T_m = 4(G + C) + 2(A + T) \quad [61.1]$$

Using dNTP solutions – *make up stock solutions, pool in small volumes (50–100 µl of each dNTP) and store separately at −20°C.*

where G, C, A and T are the number of bases in the primer. Using this as a starting point, the optimum annealing temperature can be determined by trial and error. The T_m values of the two primers should be within 5 °C of each other (and, therefore, be ideally identical in G + C content). The annealing temperature is then set 5 °C below the lowest T_m of the primer pair.
- The last two bases at the 3′ end (where elongation is initiated) should be either G or C (3 hydrogen bonds, giving strong annealing), rather than either A or T (only 2 hydrogen bonds).
- dNTPs should be used at equal concentrations of 200 $\mu mol\, l^{-1}$, which should provide the initial excess required for incorporation into DNA.
- One of the key variables in PCR is the Mg^{2+} concentration; Mg^{2+} is required as a cofactor for the thermostable DNA polymerase (Anon., 2002). Excess Mg^{2+} stabilises dsDNA and may prevent complete denaturation of product at each cycle; it also promotes spurious annealing of primers, leading to the formation of undesired products. However, very low Mg^{2+} concentration impairs polymerisation. The purpose of gelatin and Triton X-100 is to stabilise the DNA polymerase during thermal cycling.
- The most frequently used thermostable DNA polymerase is *Taq* polymerase, which extends primers at a rate of 2–4 kbp per min at 72 °C. It should be used at a concentration of $\geqslant$1 $nmol\, l^{-1}$ ($\geqslant$0.1 U per 5 µl reaction mixture). A disadvantage of *Taq* polymerase is that it has a relatively high rate of misincorporation of bases (one aberrant nucleotide per 100 000 nucleotides per cycle). Other polymerases are available (e.g. *Pfu* polymerase from *Pyrococcus furiosus*, KOD polymerase from *Thermococcus kodakarensis* or genetically modified forms of *Taq* polymerase), which have lower misincorporation rates.

Establishing the optimum concentration of Mg^{2+} – *this can be determined by trial and error, and can be a useful starting point for any new PCR; try a range of different Mg^{2+} concentrations and select the one giving the strongest intensity of the target band following gel electrophoresis.*

Understanding codes for primers and probes – *oligomeric nucleotides are often referred to by the number of bases, e.g. a 20-mer primer will contain 20 nucleotide bases.*

Box 61.1 How to carry out the polymerase chain reaction (PCR)

The protocol given below is typical for a standard PCR. Note that temperatures, incubation times and the number of cycles will vary with the particular application, as discussed in the text.

1. **Make sure you have the required apparatus and reagents to hand**, including: (i) a thermal cycler; (ii) template thermostable DNA ($\geqslant$50 ng/μl); (iii) stock solution of all dNTPs (5 mmol l^{-1} for each dNTP); (iv) a DNA polymerase (at 5 U/μl); (v) primers at e.g. 30 μmol l^{-1}; (vi) stock buffer solution, e.g. containing 100 mmol l^{-1} TRIS (pH 8.4), 500 mmol l^{-1} KCl, 15 mmol l^{-1} $MgCl_2$, 1% (w/v) gelatin, 1% (v/v) Triton X-100 (this stock is often termed '10x PCR buffer stock').

2. **Prepare a reaction mixture:** for example, a mixture containing: 1.0 μl target DNA; 2.5 μl stock buffer solution; 1.0 μl primer 1; 1.0 μl primer 2; 1.0 μl of each of the stock solutions of dNTPs, 0.1 μl *Taq* polymerase; 15.4 μl distilled deionised water, to give a total volume of 25.0 μl.

3. **Use appropriate positive and negative controls**: a positive control is a PCR template that is known to work under the conditions used in the laboratory, e.g. a plasmid, with appropriate primers, known to amplify at the annealing temperature to be used. A commonly used negative control is the PCR mixture minus the template DNA, though negative controls can be set up lacking any one of the reaction components.

4. **Cycle in the thermal cycler**: for example, an initial period of 5 min at 94 °C, followed by 30 cycles of 94 °C for 1 min (denaturation), then 50 °C for 1 min (primer annealing; temperature depends upon G + C and A + T content, Eqn [61.1] p. 440), then 72 °C for 1 min (chain extension).

5. **Assess the effectiveness of the PCR**: for example, by gel electrophoresis and ethidium bromide staining.

Troubleshooting

- If no PCR product is detected, repeat the procedure, checking carefully that all components are added to the reaction mixture. If there is still no product, check that the annealing temperature is not too high, or the denaturing temperature is not too low.
- If too many bands are present, this may indicate that (i) the primers may not be specific, (ii) the annealing temperature is too low, or (iii) there is an excess of Mg^{2+}, dNTPs, primers or enzyme.
- Bands corresponding to primer-dimers indicate that (i) the 3′ ends of the primers show partial complementarity, (ii) the annealing temperature is not high enough, or (iii) the concentration of primers is too high.

Avoiding contamination in diagnostic PCR

The sensitivity of PCR is also the major drawback, since the technique is susceptible to contamination, particularly from DNA from the skin and hair of the operator, from previous PCR products, from airborne microbes and from positive control plasmids. A number of routine precautions can be taken to avoid such contamination:

- use a laminar flow cabinet (p. 205) dedicated to PCR use and located in a separate lab from that used to store PCR products or prepare clones;
- keep separate supplies of pipettors, tips, microfuge tubes and reagents – these should be exclusive to the PCR, with separate sets for sample preparation, reagents and product analysis;
- autoclave all buffers, distilled deionised water, pipette tips and tubes;
- wear disposable gloves at all times and change them frequently: protective coverings for the face and hair are also advisable;
- avoid contamination due to carry-over by including dUTP in the PCR mixture instead of dTTP. Thus copies will contain U rather than T. Before the template denaturing step, treat the mixture with uracil-N-glycosylase (UNG, available commercially as AmpEase): this will destroy any strands containing U, i.e. any strands carried over from a previous reaction, or any contaminating material from another PCR. The target DNA will contain T, rather than U and will not be degraded by UNG. At the first heating step, the UNG will be denatured, so any newly synthesised U-containing copies will remain intact.
- for very sensitive work, use a strong UVC light (p. 276) inside the PCR workstation for 20–30 minutes before starting your work, to degrade any contaminating DNA.

SAFETY NOTE Note that you must not expose your skin or eyes to such a UVC source.

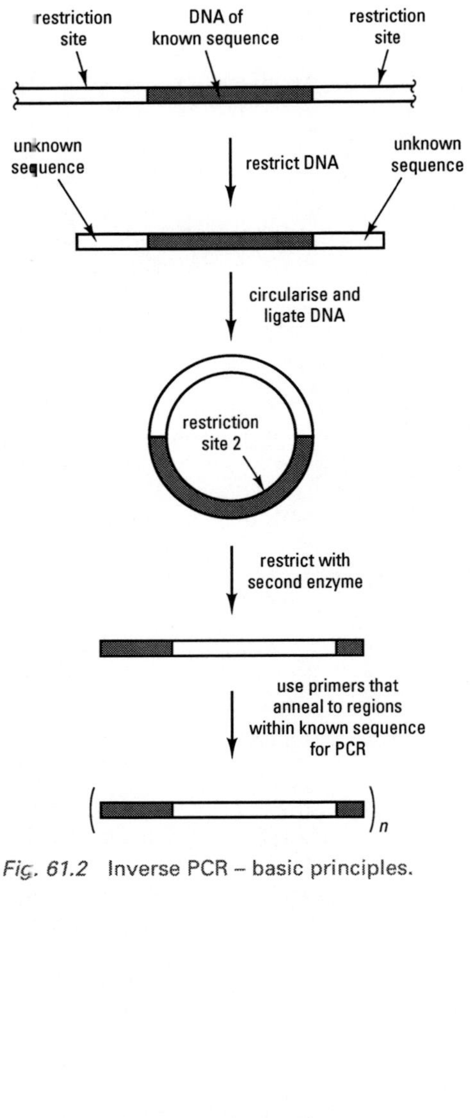

Fig. 61.2 Inverse PCR – basic principles.

PCR variations

Nested PCR

This can be used when the target sequence is known, but the number of DNA copies is very small (e.g. a single DNA molecule from a microbial genome), or if the sample is degraded (e.g. a forensic sample). The process involves two consecutive 'rounds' of PCR. The first PCR uses so-called 'external' primers, and the second PCR uses two 'internal' (or 'nested') primers that anneal to sequences within the product of the first PCR. This increases the likelihood of amplification of the target sequences by selecting for it using different primers during each round. Thus, nested PCR also increases the specificity of the reaction, since a single set of primers used in isolation may give a reasonable yield but several bands, while the use of a second set of primers ensures that a unique sequence is amplified, e.g. in microbial diagnostics.

Inverse PCR

This is a useful technique for amplifying a DNA sequence flanking a region of known base sequence (Fig. 61.2), e.g. to provide material for characterising an unknown region of DNA. The DNA is cut with a restriction enzyme so that both the region of known sequence and the flanking regions are included. This restriction fragment is then circularised and cut with a second restriction enzyme with specificity for a region in the known sequence. The now linear DNA will have part of the known sequence at each terminus, and by using primers that anneal to these parts of the known sequence, the unknown region can be amplified by conventional PCR. The product can then be sequenced and characterised (Chapter 60).

Reverse transcriptase-PCR (RT-PCR)

This technique is useful for detecting cell-specific gene expression (as evident by the presence of specific mRNA) when the amount of biological material is limited. Using either an oligo-dT primer to anneal to the 3′ polyadenyl 'tail' of the mRNA, or random hexamer primers, together with reverse transcriptase, cDNA is produced which is then amplified by PCR. RT-PCR is often a useful method of generating a probe, the identity of which can be confirmed by sequencing.

Amplification fragment length polymorphism (AFLP)

This term refers to several closely related techniques in which a single oligonucleotide primer of arbitrary sequence is used in a PCR reaction under conditions of low stringency, so that the primer is able to anneal to a large number of different sites within the target DNA. Some of the multiple amplification products will be polymorphic (e.g. the presence or absence of a particular annealing site will result in presence or absence of a particular band on the gel, after PCR and electrophoresis). Such polymorphisms can be used to detect differences between dissimilar DNA sequences.

Example Random amplification of polymorphic DNA (RAPD-PCR) can be used as a form of 'molecular typing' (p. 219), to identify a particular strain of an organism, e.g. in tracing the route of transmission of a pathogenic microbe.

PCR and DNA fingerprinting techniques

Genetic mapping and sequencing studies have led to the discovery of highly variable regions in the non-coding regions of DNA between different individuals. These hypervariable regions, often termed 'minisatellite DNA', are found at many sites throughout the genome.

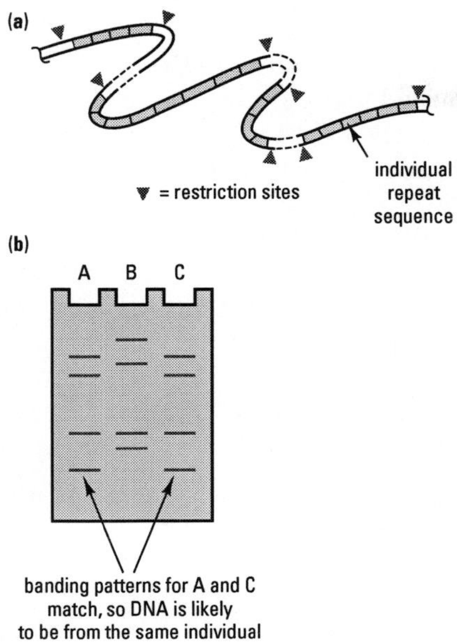

Fig. 61.3 Simplified representation of DNA fingerprinting: (a) tandem repeat sequences within DNA are cut using a restriction enzyme to yield fragments of different size; (b) these fragments are then separated by agarose gel electrophoresis on the basis of their size (M_r), giving a banding pattern that is a characteristic of the DNA used.

Each minisatellite contains a defined sequence of nucleotides which is repeated a number of times in a tandem fashion (Fig. 61.3); the greater the number of repeats, the longer the minisatellite. The number of tandem repeats in any particular minisatellite varies from one person to another (i.e. they have a variable number of tandem repeats or VNTRs). This is exploited in identifying individuals on the basis of their DNA profile, by carrying out the following steps:

1. Extraction of DNA from cells of the individual (e.g. leukocytes, buccal cells, spermatozoa).
2. Digestion of the DNA with a restriction enzyme that cuts at sites other than those within the minisatellite, to produce a series of fragments of different M_r.
3. Electrophoresis and Southern blotting of the restriction fragments using probes that are specific to the particular minisatellite.

The size of the fragments identified will depend on the number of minisatellites that each fragment contains, and the pattern obtained in the Southern blot is characteristic of the individual being profiled (Fig. 61.3).

PCR is widely used in DNA profiling in circumstances where there is a limited amount of starting material. By selecting suitable primers, highly variable regions can be amplified from very small amounts of DNA, and the information from several such regions is used to decide, with a very low chance of error, whether any two samples of DNA are from the same individual or not, e.g. in forensic science (see Kirby, 2002).

Real-time PCR

Conventional PCR techniques rely on end-point detection of amplified product, e.g. by electrophoretic separation and staining. However, such methods are time-consuming and are only semi-quantitative, since they are based predominantly on the detection of an amplified fragment (band) in a sample, rather than being designed to give exact information on its abundance (copy number). Quantitative analysis is only feasible during the early stages of PCR, where reagents are in excess and where the amount of amplified product is small, thereby avoiding the problems of product hybridisation, which would compete with primer binding. Real-time PCR enables the simultaneous amplification and quantification of template DNA in a sample by establishing the number of copies present by working in the exponential (early) phase of amplification. Currently, while several different formats are available for real-time PCR, all rely on the generation of a fluorescent signal.

Dye fluorescence

This is the simplest and least expensive approach. A fluorescent dye such as SYBR Green is included in the PCR mixture and the level of fluorescence is monitored as the reaction proceeds; since SYBR Green binds strongly to dsDNA, showing an enhancement in fluorescence of over 100-fold, any increase in fluorescence is directly proportional to the amount of dsDNA produced. Calibration is achieved by running a series of standards containing known amounts of dsDNA. This approach works well for optimised single PCR product reactions where non-specific reactions are minimised.

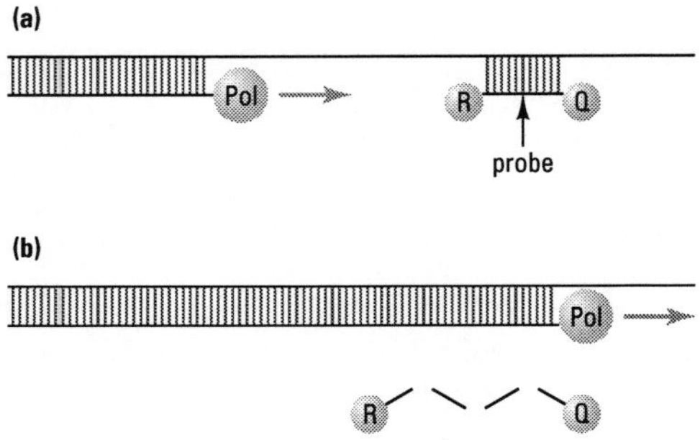

Fig. 61.4 Real-time PCR using a fluorescent reporter probe. In (a) the probe is bound to the target sequence while in (b) the 5′ nuclease activity of the DNA polymerase (Pol) has cleaved the reporter dye (R) and quencher (Q), resulting in fluorescence.

Fluorescent reporter probes

Here a fluorescent reporter dye is covalently bound to the 5′ end of an oligonucleotide probe while a 'quencher' group is attached to the 3′ end (Fig. 61.4). The probe is designed to hybridise to an internal region of the target sequence. During PCR, when the DNA polymerase molecule reaches the hybridised probe, the 5′ nuclease activity of the polymerase will cleave the reporter dye from the rest of the probe, causing an increase in fluorescence with each cycle that is in direct proportion to the amount of PCR product being formed, which is itself directly related to the original number of copies of the target DNA sequence. A commercial example is the TaqMan series of probes – while this approach is more accurate and reliable than dye fluorescence, it is also far more expensive, since a specific reporter probe must be synthesised for each target sequence. Other variants rely on changes in 3D conformation of the probe when it binds to the target sequence, causing an increase in fluorescence, e.g. Molecular Beacons and Scorpion probes.

***KEY POINT** You should note that the PCR technique is continually being updated, with subtle new variations being produced on a regular basis – novel approaches and novel acronyms are very likely to be reported within the lifetime of this book.*

Text references

Anon. *PCR Optimization*. Available: http://www.promega.com/paguide/chap1.htm Last accessed: 01/04/07. [Details of how to optimise Mg^{2+} and polymerase concentration, etc.]

Kirby, L.T. (2002) *DNA Fingerprinting: An Introduction. Breakthroughs in Molecular Biology*. Oxford University Press, Oxford.

Sources for further study

Altshuler, M.L. (2006) *PCR Troubleshooting: the Essential Guide*. Caister Academic Press, Wymondham.

Diffenback, C.W. and Dveksler, G.S. (2003) *PCR Primer: a Laboratory Manual*, 2nd edn. Cold Spring Harbor Laboratory, Cold Spring Harbor.

Dorak, M.T. (2006) *Real-Time PCR*. Taylor & Francis, Abingdon.

McPherson, M.J. and Moller, S.G. (2006) *PCR Basics*. Taylor & Francis, Abingdon.

Study exercises

61.1 Investigate the role of magnesium ions in PCR. Find out why optimising the Mg^{2+} concentration is important for successful amplification of target DNA in PCR.

61.2 Check your knowledge of primer design. List the principal factors that should be taken into account when designing primers for PCR.

61.3 Carry out a T_m calculation for a primer. For the following primer, what would be the temperature at which 50% of the molecules are annealed to their target sequence (i.e. T_m)?

`5′-TGCATGGCTGGATTAGCG-3′`

Study exercises (continued)

61.4 Consider how the risk of contamination can be reduced in PCR. List the most important practical steps that can be taken to reduce the possibility of contamination.

61.5 Select appropriate primers for a specific PCR amplification. The sequence in the box below represents a region of DNA that is to be amplified by PCR.

Which of the following three pairs of primers is most suited to amplify a PCR product from the above sequence?

(a) 5′-TCGCTGAAGGACATGTCG-3′ and 5′-CGATCTAGGCGACATGTCC-3′

(b) 5′-TCGCTGAAGGACATGTCG-3′ and 5′-TCCTTAGGTCTCGACCGTA-3′

(c) 5′-TCGCTGAAGGACATGTCG-3′ and 5′-GTCTCGACCGTAGCTAGCATC-3′

61.6 Find out what types of DNA analysis might be suitable for forensic examination of degraded DNA samples. In 1918 Tsar Nicholas of Russia, his wife Tsarina Alexandra, and their five children were executed and buried in a shallow grave in Yekaterinberg, Russia. Samples from the bones of the family were analysed by the UK Forensic Science Service (FSS). What types of DNA-based tests would you expect to have been carried out to confirm that a family group was present in the grave?

```
5′-TCGCTGAAGGACATGTCGATGCTAGCTACGGTCGAGACGTAAGGACATGTCGGCTAGATCGC-3′
3′-AGCGACTTCCTGTACAGCTACGATCGATGCCAGCTCTGCATTCCTGTACAGCGGATCTAGCG-5′
```

62 Molecular genetics III – genetic engineering techniques

Advances in the procedures used to manipulate nucleic acids *in vitro* have increased our understanding of the structure and function of genes at the molecular level. Additionally, these techniques can be used to alter the genome of an organism (genetic engineering or gene cloning), e.g. to create a bacterium capable of synthesising a foreign protein such as a potentially useful hormone or vaccine component, or a novel protein. The procedures are often termed 'recombinant DNA technology', since they involve the creation of novel combinations of DNA (i.e. recombinant DNA) under controlled laboratory conditions.

In the UK, the Genetically Modified Organisms (Contained Use) Regulations (2000) *provide the regulatory framework for all research procedures involving the genetic modification of organisms.*

KEY POINT ***While genetic manipulation must be carried out under strict containment, in accordance with appropriate legislation, the procedures involved in the isolation, amplification, recombination and cloning of DNA are often used at undergraduate level, to illustrate the general features of the techniques.***

Basic principles

Gene cloning involves several steps:

1. Isolation of the DNA sequence (gene) of interest from the genome of an organism, or from a gene library. This usually involves DNA purification followed by enzymic digestion or mechanical fragmentation, to liberate the target DNA sequence, or PCR amplification of the target DNA sequence.
2. Creation of an artificial recombinant DNA molecule (rDNA), by inserting the target sequence into a DNA molecule capable of replicating in a host cell, i.e. a 'cloning vector'. Suitable cloning vectors for bacterial cells include plasmids bacteriophages and cosmids.
3. Introduction of the recombinant DNA molecule into a suitable host, e.g. *E. coli*. The process is termed transformation when a plasmid is used, or transfection for viral nucleic acid.
4. Selection and growth of the transformed (or transfected) cell, using the techniques of cell culture. Since a single transformed host cell can be grown to give a clone of genetically identical cells, each carrying the target DNA of interest, the technique is often referred to as 'gene cloning', or molecular cloning.

Definitions

Bacteriophage (phage) – a bacterial virus. Useful as they possess a means of penetrating the bacterial cell.

Cosmids – hybrid plasmid vectors containing the *cos* (cohesive) sites from phage λ, enabling *in vitro* packaging into phage capsids. Useful for cloning large segments of DNA, typically up to 50 kb.

Plasmids – circular molecules of DNA capable of autonomous replication within a bacterial cell. Can be isolated, manipulated and reintroduced into bacterial cells.

Subcloning – procedures that isolate and characterise smaller portions of a particular DNA sequence, e.g. during the search for a specific gene.

Transfection – in bacteria: uptake of viral nucleic acid; in eukaryotes: uptake of any naked, foreign DNA.

Transformation – in bacteria: stable incorporation of external DNA, e.g. a plasmid; in eukaryotes: the conversion of a cell culture of finite life to a continuous (immortal) cell line (also occurs in cancer).

Extraction and purification of plasmid DNA

Specific details of the steps involved in the isolation of DNA vary, depending upon the source material. However, the following protocol shows the principal stages in the purification of plasmid DNA from bacterial cells:

__Preparing glassware__ – all glassware for DNA purification must be siliconised, to prevent adsorption of DNA. All glass and plastic items must be sterilised before use.

1. Cell wall digestion: incubation of bacteria in a lysozyme solution will hydrolyse the peptidoglycan cell wall. This is often carried out under isotonic conditions, to stop the cells from bursting open and releasing chromosomal DNA. Note that Gram-negative bacteria are relatively insensitive to lysozyme, requiring additional treatment to allow the

SAFETY NOTE Working with solvents in molecular biology – note that phenol is toxic and corrosive while chloroform is potentially carcinogenic (p. 134). Take appropriate safety precautions (e.g. wear gloves, extinguish all naked flames, use a fume hood, where available).

Using ultrapure sterile water in molecular biology *– to avoid contamination, prepare all solutions using pre-sterilised (autoclaved) deionised purified water (e.g. MilliQ water).*

Extracting DNA from organisms other than bacteria *– different homogenisation techniques will be required (see Chapter 36). DNA can be extracted from the resulting homogenate by (i) complexation with the detergent cetrylmethylammonium bromide (CTAB), e.g. for plant DNA, or (ii) anion-exchange column chromotography.*
Many laboratories now use small-scale chromatographic columns for routine extraction, e.g. Qiagen miniprep columns.

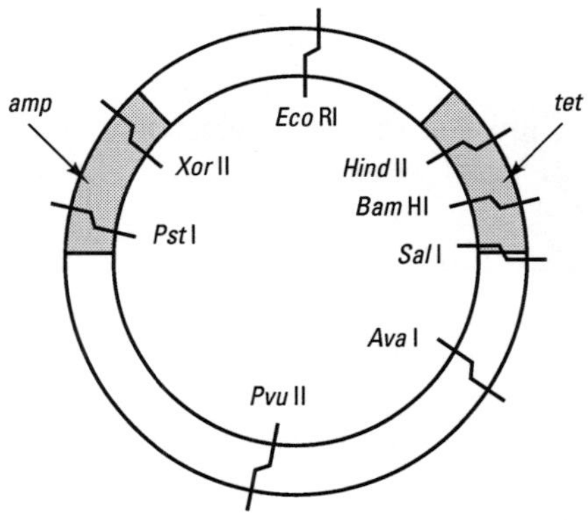

Fig. 62.1 Restriction map of the plasmid pBR322. The position of individual restriction sites is shown together with the genes for ampicillin resistance (*amp*) and tetracycline resistance (*tet*).

Definition

Type II restriction endonucleases – intracellular enzymes, produced by certain strains of bacteria: their function is to restrict the growth of phages within the cell by cutting phage DNA at particular sites (host DNA is protected by methylation at these sites).

enzyme to reach the peptidoglycan layer (p. 243), e.g. osmotic shock, or incubation with a chelating agent, e.g. ethylenediaminetetra-acetate (EDTA). The latter treatment will also inactivate any bacterial deoxyribonucleases (DNases) in the solution, preventing enzymic degradation of plasmid DNA during extraction.
2. Lysis using strong alkali (NaOH) and a detergent, e.g. sodium dodecyl sulphate, to solubilise the cellular membranes and partially denature the proteins. Neutralisation of this solution (e.g. using potassium acetate) causes the chromosomal DNA to aggregate as an insoluble mass, leaving the plasmid DNA in solution.
3. Removal of other macromolecules, particularly RNA and proteins, by, for example, enzymic digestion using ribonuclease and proteinase. Additional steps give further increases in purity: most laboratories now use small-scale chromatographic columns containing an anion exchange resin for routine extraction of nucleic acids. These are available in kit form, e.g. Wizard miniprep columns. This approach gives more reliable purification than traditional methods, such as phenol/chloroform extraction.
4. Precipitation of DNA by adding two volumes of absolute ethanol to one volume of aqueous extract, followed by centrifugation, to recover the DNA as a pellet. Further washing with 70% v/v ethanol will remove any residual salt from the previous stages. After removal of residual ethanol by evaporation, the purified DNA is dissolved in buffer and can be stored for months at −20 °C.

The simplest approach to quantifying the amount of nucleic acid in an aqueous solution is to measure the absorbance of the solution at 260 nm (A_{260}) using a spectrophotometer, as detailed on page 283. Any contaminating protein would also contribute to A_{260}; protein can be detected by measuring the absorbance of the solution at 280 nm. Purified nucleic acids have a value for A_{260}/A_{280} of 1.8–2.0 and contaminating protein will give a lower ratio. If your solution gives a ratio substantially lower than this, you should carry out further purification steps.

Enzymatic manipulation of DNA

Restriction enzymes

Type II restriction endonucleases can be used to produce linear fragments of DNA with either single stranded 'sticky' ends or 'blunt' ends (Fig. 60.1): most commercial preparations of restriction enzymes will completely cleave a sample of DNA, producing a 'restriction digest', within 1 h at 37 °C. The position of individual restriction sites can be used to create a diagnostic restriction map for a particular molecule, e.g. a plasmid cloning vector (Fig. 62.1). An important additional feature is that, under appropriate conditions, the sticky ends of any two restriction fragments cut with the same enzyme (Fig. 62.2) can anneal (base pair), due to the formation of hydrogen bonds between individual bases within this region, allowing them to be joined together (ligated), irrespective of the sources of the two restriction fragments.

DNA ligase

To construct a recombinant DNA molecule, a suitable vector must be ligated to the DNA fragment to be cloned: this is performed using another microbial enzyme, DNA ligase (usually obtained from T4-infected

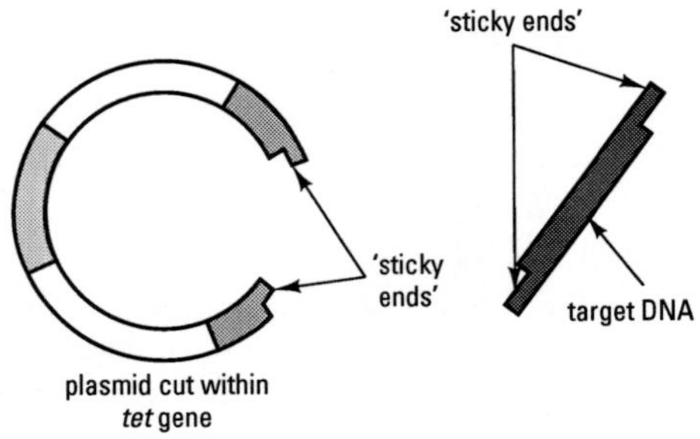

Fig. 62.2 Restriction of plasmid pBR322 and foreign DNA with *Hin* dIII.

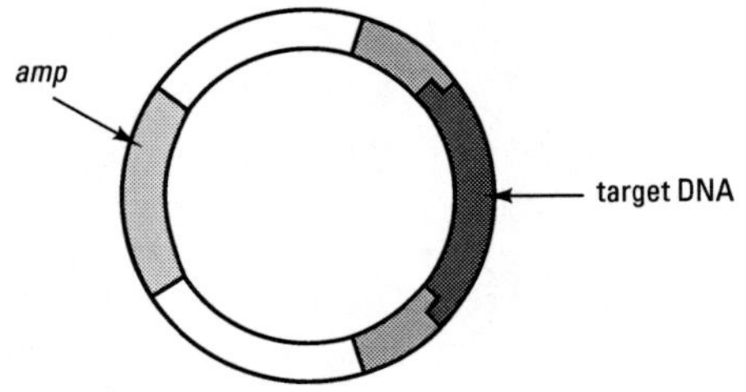

Fig. 62.3 Annealing and ligation of plasmid pBR 322 and target DNA to give a recombinant plasmid which confers resistance to ampicillin only. As target DNA has been inserted within the *tet* gene (Fig. 62.1), the gene is now discontinuous and inactive.

***Understanding the nomenclature for plasmid vectors** – pBR322 is named according to standard rules: 'p' indicates the vector is a plasmid, 'BR' identifies the researchers Bolivar and Rodriguez and '322' is the specific code given to this plasmid, to distinguish it from others developed by the same workers.*

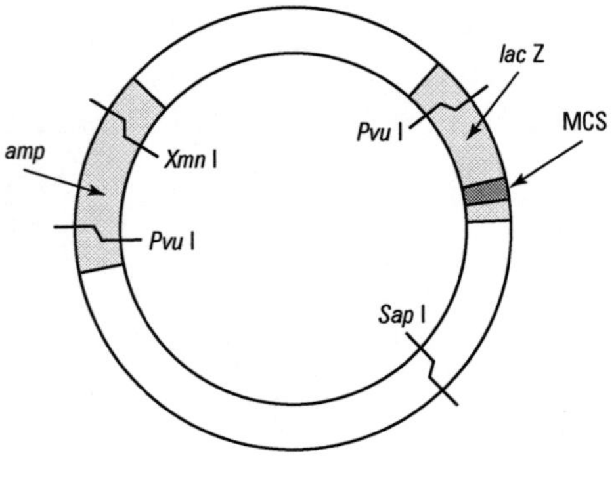

Fig. 62.4 Restriction map of the plasmid pUC19. The position of some individual restriction sites is shown together with the genes for ampicillin resistance (*amp*) and β-galactosidase (*lacZ*). MCS = multiple cloning site for 40 restriction enzymes within the *lacZ* gene.

E. coli). This ATP-dependent enzyme is capable of forming covalent phosphodiester bonds between annealed DNA molecules, thereby creating recombinant DNA (Fig. 62.3). When the two molecules involved are the cloning vector and the target DNA, the size of the recombinant molecule can be predicted (e.g. a plasmid of 4.5 kbp, plus a target DNA fragment of 2.5 kbp will give a recombinant molecule of 7 kbp), allowing separation and recovery by agarose gel electrophoresis. Ligation is usually carried out at lower temperatures (to encourage annealing), over an extended time period of several hours (to allow the enzyme to operate), e.g. an overnight incubation at ≃16 °C. The volume of the ligation mixture is kept as low as possible (typically, <10 µl), with approximately a 2 : 1 ratio of target DNA to vector DNA, to encourage annealing between the two different types of DNA molecule and to reduce the chance of circularisation of vector or target DNA fragments.

Choosing a suitable vector

The features to be considered when selecting a suitable vector are:

- the ease of purification – e.g. reliable procedures have been developed to allow plasmid DNA to be extracted and purified from bacterial cells;
- the efficiency of insertion of the recombinant vector into a new host cell;
- the presence of single copies of suitable restriction sites;
- the presence of selectable markers (e.g. 'reporter genes';
- the size of the DNA insert to be cloned;
- the copy number of the vector in the host cell.

Plasmids

The simplest bacterial cloning vectors are those based on small plasmids – circular DNA molecules, capable of autonomous replication within a bacterial cell. The general characteristics of plasmids are described in Chapter 59. One of the first cloning vectors to be developed for *E. coli* was pBR322, a genetically engineered plasmid of 4.4 kbp containing an origin of replication, two antibiotic resistance genes and single sites for a range of restriction enzymes (Fig. 62.1). Subsequently, other plasmid vectors have been developed, with additional features: for example, plasmids of the pUC series (Fig. 62.4) are now widely used for transformation of *E. coli*, having the following advantages over earlier types:

- high copy number: several hundred identical copies of the plasmid may be present in each bacterial cell, giving improved yield of plasmid DNA;
- single-step selection of recombinants using the *lacZ′* gene;
- clustering of restriction sites within a short region of the *lacZ′* gene – restriction using two enzymes (a 'double digest') cuts a small fragment from this 'multiple cloning site' or 'polylinker', producing a cleaved plasmid with two different 'sticky ends' (e.g. *Eco* RI at one end and *Hin* dIII at the other), allowing complementary fragments of target DNA to be ligated in a particular orientation ('directional cloning').

The pET series of plasmid vectors (Novagen; Fig. 62.5) are now widely used, since they offer several additional advantages, including:

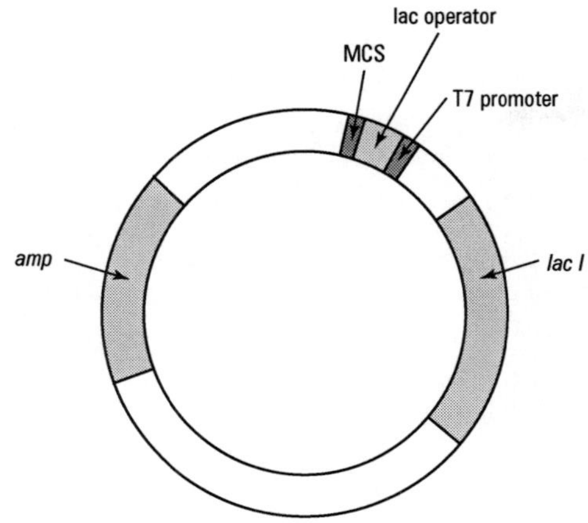

Fig. 62.5 Map of a pET plasmid vector (Novagen) showing the location of the genes for ampicillin resistance (*amp*) and lactose repressor protein (*lacI*) which controls expression of the T7 promoter *via* the lac operator (binding of the Lac repressor protein is countered by addition of isothiopropylgalactoside, IPTG). Expression of a gene cloned into the multiple cloning site (MCS) can thus be controlled, since DNA transcription only occurs when IPTG is added to the growth medium.

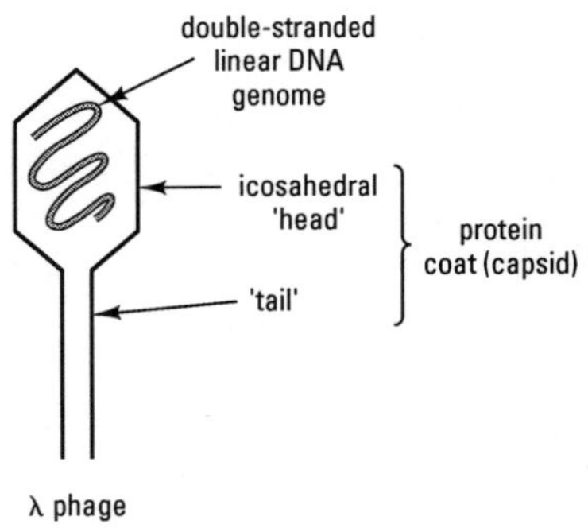

Fig. 62.6 Bacteriophage λ.

Definition

Phagemid – a recombinant entity created by fusion of a plasmid with the origin of replication of a phage: can be propagated and packaged into phage particles with the aid of a 'helper' phage.

- high levels of protein expression, as the cloned DNA is under the control of the T7 promoter and the lac operon system;
- specialised vectors and hosts for the production of soluble proteins, disulphide bond formation, protein export, etc;
- the attachment of a short histidine 'tag' on the end of an expressed protein, to enable the protein to be purified by immobilised metal affinity chromatography (IMAC).

Phages

Bacteriophage vectors can offer several advantages over plasmid vectors:

- some phages have been engineered to carry larger fragments of foreign DNA – plasmid vectors work best with relatively short inserts of 1–2 kbp, while some phage vectors can accept 10 kbp or more of foreign DNA;
- a recombinant phage has a ready-made mechanism for transferring rDNA to a new host cell: the process is broadly similar to phage transduction.

A number of vectors have been developed from *E. coli* phage λ (genome = 49 kbp of dsDNA in linear form, Fig. 62.6). For example, λZAPII, an insertion vector capable of accepting up to 10 kbp of additional (foreign) DNA and with the *lacZ′* gene and multiple cloning site for selection, or λEMBL4, a replacement vector, where part of the vector DNA is removed and replaced by foreign DNA (up to 23 kbp) during the cloning procedure.

Most vectors based on λ are virulent phages, released by lysis of the infected cell at the end of the replication phase: such vectors are cloned using standard techniques for phages, e.g. as plaques of lysis on a 'lawn' of bacteria growing on an agar-based medium. Note that suitable strains of *E. coli* must be used for each of the various cloning vectors, e.g. vectors containing *lacZ′* can only be used with those *E. coli* strains that are deficient in this part of their genome (*lacZ′* codes for the α-component of the enzyme β-galactosidase, present in wild-type *E. coli*). In some instances, a second phage (a 'helper phage') must be used along with the vector, to carry out some of the functions removed from the vector genome, e.g. phage packaging and assembly functions. In some instances, more specialised phage vectors may be required, e.g. M13mp19, a single stranded phage vector, sometimes used for sequencing studies.

Cosmids and other vectors

Cosmid vectors contain the single stranded cohesive terminal sequences from λ phage (so-called 'cos' sites) inserted into a plasmid cloning vector. They can be used to clone up to 40 kbp of foreign DNA, using the 'cos' sites to allow *in vitro* packaging into the λ phage coat: such large fragments are required for the production of gene libraries, as described below. Other hybrid vectors have been produced by the fusion of a phage and a plasmid vector – these phagemids can be manipulated in the laboratory as plasmids for ease of DNA uptake, or converted to phage-like particulate form for ease of storage.

Creating a gene library

For viruses with small genomes, an individual gene may be identified by hybridisation of a single fragment with a suitable probe, following

***Screening clones** – a particular gene can be detected using Southern blotting, or the expressed gene product (protein) may be detected by a suitable method, e.g. an immunoassay. While it is relatively simple to describe the screening process, finding a specific gene (= clone) in a clone library is likely to be a time-consuming process.*

***Using cloned cDNA** – information from the sequence of a particular cDNA clone may be useful in designing a suitable oligonucleotide probe to locate a particular gene in a genomic library.*

Table 62.1 Examples of genes used to detect transformants

Gene	Product/assay
lacZ/*lacZ'*	β-galactosidase/chromogenic substrate (e.g. X-GAL) or fluorogenic substrate (e.g. MU-GAL)
uidA	β-glucuronidase/chromogenic substrate (e.g. X-GLUC) or fluorogenic substrate (e.g. MU-GLAC)
lux	luciferase/bioluminescence in the presence of luciferin
bla/*amp*	β-lactamase/resistance to ampicillin
cat	chloramphenicol acetyltransferase/resistance to chloramphenicol
gfp	green fluorescent protein from *Aequorea victoria*, visualised under UV light

digestion with a particular restriction enzyme (Chapter 60). However, for most genomes, such a digest would contain a very complex mixture of fragments and a 'clone library' must be created. The isolation and identification of a particular gene are then carried out by screening individual clones from this library. Two types of clone library can be used:

1. A genomic library – prepared from the entire genome of the organism under study. The genome is fragmented to give overlapping fragments, e.g. by partial restriction (incomplete digestion, e.g. at low temperature), or by mechanical shearing. Individual fragments are then incorporated into a suitable vector, e.g. a cosmid, to create a vector library (e.g. a cosmid library). Each recombinant vector is then transferred to a separate host cell, which is cultured, giving a collection of transformants that represents the clone library. This approach is sometimes termed 'shotgun cloning'.
2. A cDNA library (cDNA = complementary, or copy DNA) – prepared by converting mRNA into DNA using the retroviral enzyme reverse transcriptase. Such a library consists only of genes expressed in those cells used to create the library, making individual genes easier to locate and identify, since the number of clones represented is usually smaller than for a genomic library. Another feature of cDNA is that it is complementary to the processed transcript (mRNA), so it contains no information on non-coding regions (introns) or transcriptional sequences (promoter regions, etc.).

Transferring rDNA to a suitable host cell

- Once a recombinant vector has been produced *in vitro*, it must be introduced into a suitable host cell. Packaged phage vectors have a built-in transfer mechanism, while naked phage DNA and plasmids must be introduced by treatments that cause a temporary increase in membrane permeability resulting in transformation. These treatments include:
- physicochemical shock treatment – Box 62.1 gives details of a typical procedure using $CaCl_2$ and heat shock treatment to transform *E. coli* with the plasmid pUC19;
- electroporation – cells or protoplasts are subjected to electric shock treatment (typically, $>10\,kV\,cm^{-1}$) for very short periods ($<10\,ms$);
- a range of techniques can be used for animal and plant cells, e.g. electroporation of protoplasts, or various micro-injection treatments, either using a microsyringe, DNA-coated microprojectiles ('biolistics') or cationic lipids (Fig. 62.7) which form lipid–DNA complexes that fuse with cell membranes and allow DNA entry. *Agrobacterium* can be used for rDNA transfer in certain plants.

Selection and detection of transformants

As the efficiency of transformation is very low, many of the plasmid vectors used in genetic engineering carry genes coding for antibiotic resistance, e.g. pBR322 carries separate genes for ampicillin resistance, *amp*, and tetracycline resistance, *tet* (Fig. 62.1). These genes act as 'markers' for the vector. One gene (e.g. *amp*) can be used to select for transformants, which would form colonies on an agar-based medium

***Recognising transformants** – after plating bacteria onto medium containing ampicillin, you may notice a few small 'satellite' colonies surrounding a single larger (transformant) colony. These satellite colonies are derived from non-transformed cells which survive due to the breakdown of antibiotic in the medium around the transformant colony, and should not be selected for subculture.*

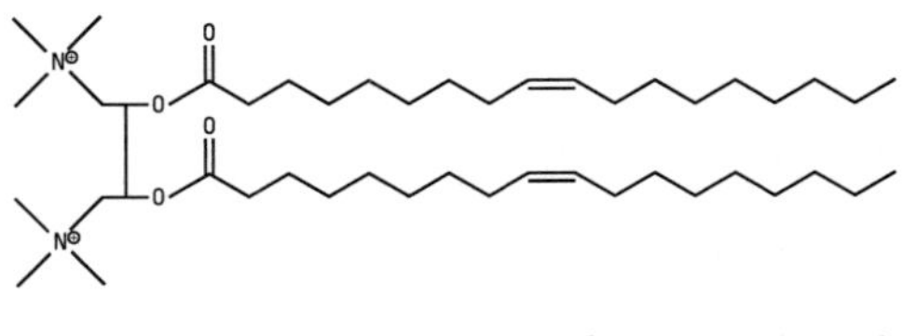

Fig. 62.7 General structure of a synthetic cationic lipid, e.g. Tfx.

containing the antibiotic, while non-transformed (ampicillin-sensitive) cells would be killed. The other gene (e.g. *tet*) can be used as a marker for the recombinant plasmid vector, since ligation of the target sequence into this gene causes insertional inactivation (Fig. 62.3). Thus, cells transformed with the recombinant plasmid will be resistant to ampicillin only, while cells transformed with the recircularised ('native') plasmid will be resistant to both antibiotics. These two types of transformant can be distinguished using replica plating.

For those genes where the product is an enzyme, the presence or absence of the functional gene can be assessed using a suitable substrate. For example, the insertional inactivation of *lacZ′* can be detected by including a suitable inducer of β-galactosidase (e.g. isopropylthiogalactoside, IPTG) and the chromogenic substrate 5-bromo-4-chloro-3-indolyl-β-D-galactoside (X-GAL) within the agar medium: a transformant colony derived from the native plasmid will be blue-green while a transformant containing the recombinant molecule (inactive *lacZ′*) will grow to produce a white colony. A number of other relevant examples are given in Table 62.1.

Box 62.1 Transformation of *E. coli* and selection of transformants

The following procedure illustrates the principal stages of the process: the efficiency of transformation will depend on the choice of *E. coli* strain and plasmid, and on the handling procedures prior to and during the process – for example, some strains do not require all of the stages listed below (for further details, see Hanahan *et al.*, 1995). Sterile equipment and appropriate technique (Chapter 32) are required at all times:

1. **Grow the cells under appropriate conditions:** the best results are often obtained using actively growing mid-log phase cells, rather than cells that have been grown to stationary phase. Actively growing cells can be stored under appropriate conditions, e.g. as a 'frozen stock' – in suspension in concentrated glycerol at −85 °C.
2. **Induce cell competence by transfer to a suitable sterile transformation buffer:** this would normally contain divalent and monovalent salts, e.g. a salt solution containing $CaCl_2$ at 50–100 mmol l^{-1}, plus KCl and $MnCl_2$ at 10–20 mmol l^{-1}. Cells are usually kept on ice in this solution for 10–15 min, to encourage the binding of plasmid DNA, added at a later stage.
3. **Add reagents to increase the permeability of cellular membranes:** typically dimethyl sulphoxide (DMSO, at up to 7% v/v) and dithiothreitol (DTT at up to 0.2 mmol l^{-1}), incubated on ice for a few minutes. DMSO is readily oxidised, reducing the effectiveness of the transformation procedure, and should be stored at −80 °C when not in use, to minimise oxidation.
4. **Add plasmid DNA:** typically at 10–1000 ng per transformation. Maintain on ice for at least 10 min in a minimal volume of solution, to allow the plasmid to become associated with the cell surface.
5. **Heat shock the cells:** briefly raise the temperature to 42–45 °C for 60–120 s, then return to ice for a few minutes. It is important that this treatment gives a *rapid* change in temperature – use a small, thin-walled (disposable) sterile plastic tube containing the minimum volume of solution to maximise the rate of temperature change.
6. **Allow the cells to recover from heat shock and to express any new genes** (e.g. to synthesise enzymes conferring antibiotic resistance in transformed cells): add sterile nutrient broth (e.g. 1 ml) and then incubate at 37 °C for up to 60 min.
7. **Plate onto an appropriate medium, to allow selection and detection of transformants:** for example, using pUC19, surface spread the suspension onto a medium containing ampicillin (at 25 mg l^{-1}), plus IPTG (at 15 mg l^{-1}) and X-GAL (at 25 mg l^{-1}), to detect the *lacZ′* gene product, as described in the text.

Text reference

Hanahan, D., Jessee, J. and Bloom, F.R. (1995) 'Techniques for transformation of *E. coli*' in Glover, D.M. and Hames, B.D. *DNA Cloning I*, Practical Approach Series. IRL Press, Oxford.

Sources for further study

Brown, T.A. (2002). *Essential Molecular Biology: A Practical Approach*, Vols 1 and 2, 2nd edn. Oxford University Press, Oxford.
[A practical manual on the basic principles involved in gene cloning and recombinant DNA technology]

Brown, T.A. (2006) *Gene Cloning and DNA Analysis*, 5th edn. Blackwell, Oxford.

Dale, J.W. and Schantz, M.V. (2002) *From Genes to Genomes: Concepts and Applications of DNA Technology*. Wiley, Chichester.

Hardin, C., Edwards, J., Riell, A., Presutti, D., Miller, W. and Robertson, D. (2001) *Cloning, Gene Expression, and Protein Purification: Experimental Procedures and Process Rationale*. Oxford University Press, Oxford.
[A practical handbook and laboratory manual, with basic experimental protocols together with underlying theoretical principles]

Jones, P. (1997) *Vectors: Expression Systems: Essential Techniques*. Wiley, New York.

Lodge, J., Lund, P. and Michin, S. (2006) *Gene Cloning: Principles and Applications*. Taylor & Francis, Abingdon.

Primrose, S.B. and Twyman, R.M. (2004) *Principles of Gene Manipulation and Genomics*, 7th edn. Blackwell, Oxford.

Russell, P. (2005) *iGenetics: a Molecular Approach*. Addison-Wesley, Harlow.

Study exercises

62.1 Explain the terminology of restriction enzymes. Why were *Hin* dIII and *Eco* RI given these names?

62.2 Explain the purpose of the various stages involved in the extraction and purification of DNA. Having read through this chapter, briefly describe the main purpose of each of the following reagents within a DNA extraction protocol, and rearrange them into their order of use within the procedure:

(a) anion exchange resin column chromatography;
(b) incubation in 70% v/v ethanol : water;
(c) ribonuclease and proteinase digestion;
(d) incubation with lysozyme/EDTA;
(e) sodium dodecyl sulphate incubation;
(f) NaOH treatment, followed by neutralisation with potassium acetate.

62.3 Construct a restriction map. The figure alongside represents an electrophoretic separation of plasmid pRD, digested with different restriction enzymes, both individually and in combination. All of the enzymes cut the plasmid only once. Use this information to construct a simple restriction map of the plasmid. Key to lanes: 1 = DNA standards (0.5–8 kbp); 2 = *Eco* RI; 3 = *Hin* dIII; 4 = *Pvu* II; 5 = *Eco* RI and *Hin* dIII; 6 = *Eco* RI and *Pvu* II; 7 = *Hin* dIII and *Pvu* II; 8 = *Eco* RI, *Hin* dIII and *Pvu* II; 9 = DNA standards (0.5–8 kbp).

62.4 Calculate transformation efficiency. A volume of 100 μl of *E. coli* cells was transformed using 20 μl of plasmid DNA (prepared at 0.1 ng μl^{-1}) and then made up to a final volume of 500 μl in buffer. A 10-fold dilution of this suspension was prepared and 200 μl of this dilution was surface spread (p. 206) onto a suitable medium containing an antibiotic to select for transformants, giving 180 colonies on the spread plate after overnight incubation. What is the transformation efficiency, expressed as the number of transformants (colony-forming units, CFU) per μg DNA? (Give your answer in exponential notation, to three significant figures.)

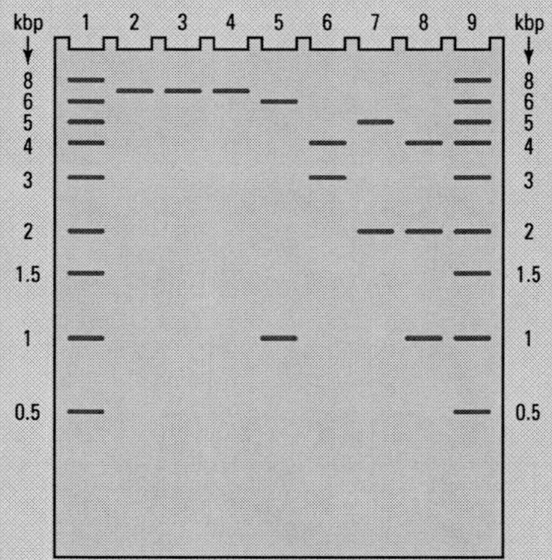

Analysis and presentation of data

63 Using graphs

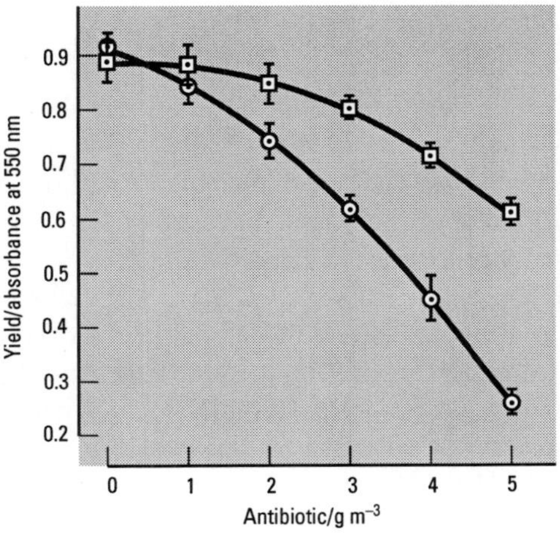

Fig. 63.1 Effect of antibiotic on yield of two bacterial isolates: ○, sensitive isolate; □, resistant isolate. Vertical bars show standard errors ($n = 6$).

Criteria for preparing graphs:

- *Clarity – show the relationships within the data.*
- *Economy – leave out superfluous information/text.*
- *Integrity – present the data without trying to mislead.*

Selecting a title *– it is a common fault to use titles that are grammatically incorrect. A widely applicable format is to state the relationship between the dependent and independent variables within the title, e.g. 'The relationship between enzyme activity and external pH'.*

Remembering which axis is which *– a way of remembering the orientation of the x-axis is that 'x' is 'a cross', and it runs 'across' the page (horizontal axis) while 'y' is the first letter of 'yacht', with a large vertical mast (vertical axis).*

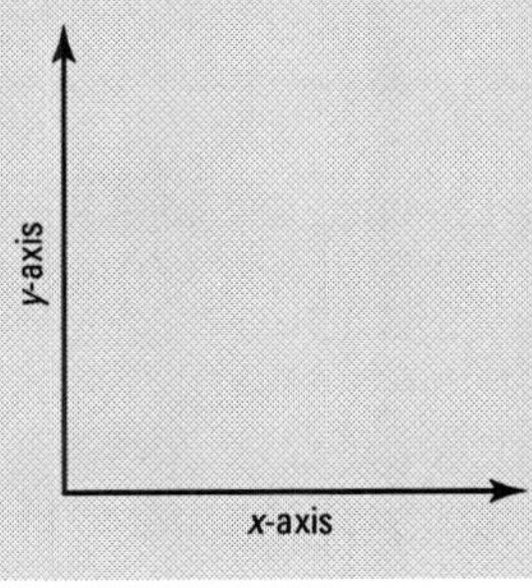

Graphs can be used to show detailed results in an abbreviated form, displaying the maximum amount of information in the minimum space. Graphs and tables present findings in different ways. A graph (figure) gives a visual impression of the content and meaning of your results, while a table provides an accurate numerical record of data values. You must decide whether a graph should be used, e.g. to illustrate a pronounced trend or relationship, or whether a table (Chapter 64) is more appropriate.

A well-constructed graph will combine simplicity, accuracy and clarity. Planning of graphs is needed at the earliest stage in any write-up as your accompanying text will need to be structured so that each graph delivers the appropriate message. Therefore, it is best to decide on the final form for each of your graphs before you write your text. The text, diagrams, graphs and tables in a laboratory write-up or project report should be complementary, each contributing to the overall message. In a formal scientific communication it is rarely necessary to repeat the same data in more than one place (e.g. as a table and as a graph). However, graphical representation of data collected earlier in tabular format may be applicable in laboratory practical reports.

Practical aspects of graph drawing

The following comments apply to graphs drawn for laboratory reports. Figures for publication, or similar formal presentation, are usually prepared according to specific guidelines, provided by the publisher/organiser.

KEY POINT ***Graphs should be self-contained – they should include all material necessary to convey the appropriate message without reference to the text. Every graph must have a concise explanatory title to establish the content. If several graphs are used, they should be numbered so they can be quoted in the text.***

- Consider the layout and scale of the axes carefully. Most graphs are used to illustrate the relationship between two variables (x and y) and have two axes at right angles (e.g. Fig. 63.1). The horizontal axis is known as the abscissa (x-axis) and the vertical axis as the ordinate (y-axis).
- The axis assigned to each variable must be chosen carefully. Usually the x-axis is used for the independent variable (e.g. treatment) while the dependent variable (e.g. biological response) is plotted on the y-axis (p. 189). When neither variable is determined by the other, or where the variables are interdependent, the axes may be plotted either way round.
- Each axis must have a descriptive label showing what is represented, together with the appropriate units of measurement, separated from the descriptive label by a solidus or 'slash' (/), as in Figure 63.1, or by brackets, as in Figure 63.2.
- Each axis must have a scale with reference marks ('tics') on the axis to show clearly the location of all numbers used.
- A figure legend should be used to provide explanatory detail, including a key to the symbols used for each data set.

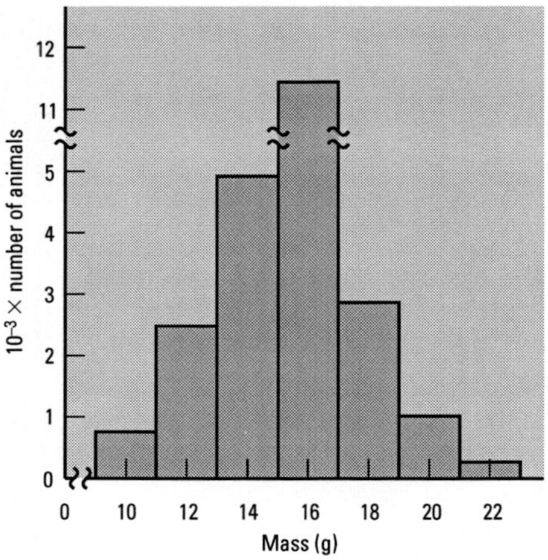

Fig. 63.2 Frequency distribution (histogram) of masses for a sample of animals (sample size 24 085); the size class interval is 2 g.

Handling very large or very small numbers

To simplify presentation when your experimental data consist of either very large or very small numbers, the plotted values may be the measured numbers multiplied by a power of 10: this multiplying power should be written immediately before the descriptive label on the appropriate axis (as in Fig. 63.2). However, it is often better to modify the primary unit with an appropriate prefix (p. 183) to avoid any confusion regarding negative powers of 10.

Size

Remember that the purpose of your graph is to communicate information. It must not be too small, so use at least half an A4 page and design your axes and labels to fill the available space without overcrowding any adjacent text. If using graph paper, remember that the white space around the grid is usually too small for effective labelling. The shape of a graph is determined by your choice of scale for the *x* and *y*-axes which, in turn, is governed by your experimental data. It may be inappropriate to start the axes at zero (e.g. Fig. 63.1). In such instances, it is particularly important to show the scale clearly, with scale breaks where necessary, so the graph does not mislead. Note that Fig. 63.1 is drawn with 'floating axes' (i.e. the *x* and *y*-axes do not meet in the lower left-hand corner), while Fig. 63.2 has clear scale breaks on both *x* and *y* axes.

Example For a data set where the smallest number on the log axis is 12 and the largest number is 9000, three-cycle log-linear paper would be used, covering the range 10–10 000.

Choosing between a histogram and a bar chart – *use a histogram for* continuous quantitative *variables and a bar chart for* discrete *variables (see Chapter 28 for details of these types of measurement scale).*

Graph paper

In addition to conventional linear (squared) graph paper, you may need the following:

- Probability graph paper. This is useful when one axis is a probability scale.
- Log-linear graph paper. This is appropriate when one of the scales shows a logarithmic progression, e.g. the exponential (log) growth of cells in liquid culture (p. 233). Log-linear paper is defined by the number of logarithmic divisions (usually termed 'cycles') covered (e.g. Fig. 63.3) so make sure you use a paper with the appropriate number of cycles for your data. An alternative approach is to plot the log-transformed values on 'normal' graph paper.
- Log-log graph paper. This is appropriate when both scales show a logarithmic progression.

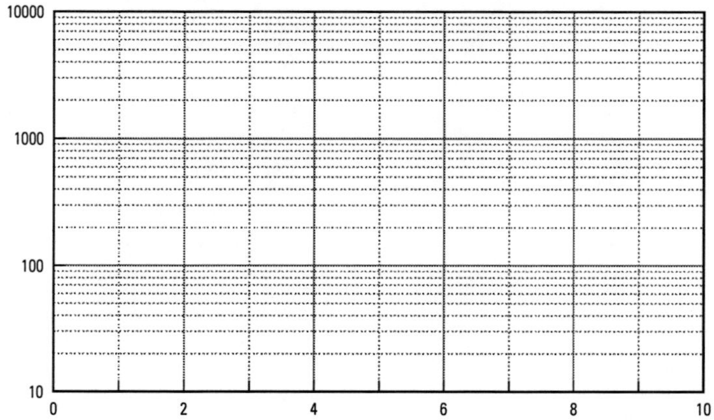

Fig. 63.3 Representation of three-cycle log-linear graph paper, marked up to show a *y*-axis (log) scale from 10 to 10000 and an *x*-axis (linear) scale from 0 to 10.

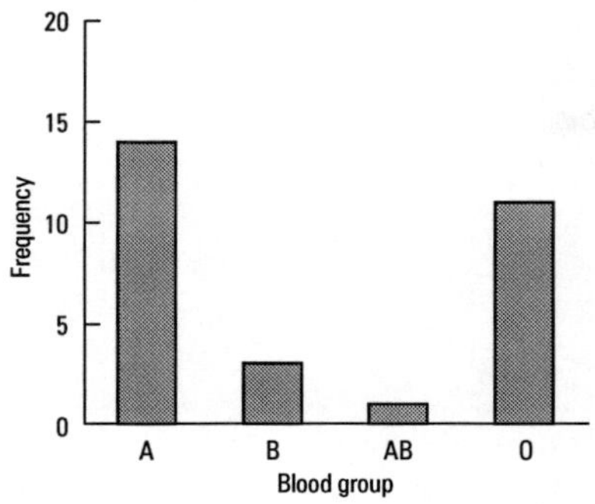

Fig. 63.4 Bar chart, showing the number of students belonging to each ABO blood group ($n = 29$).

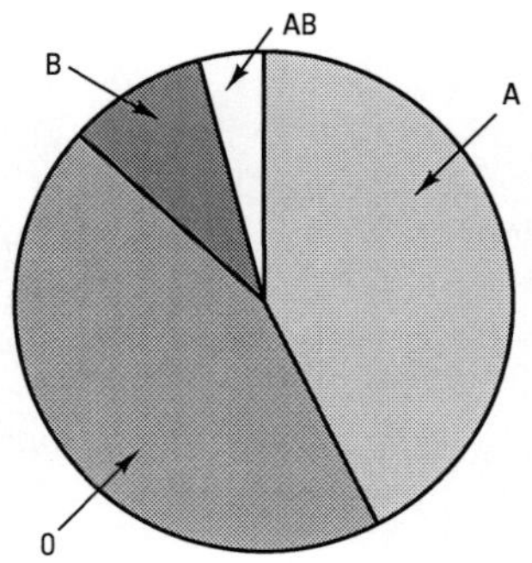

Fig. 63.5 Pie chart: relative abundance of human blood groups.

***Choosing graphical symbols** – plotted curves are usually drawn using a standard set of symbols: ●, ○, ■, □, ▲, △, ◆, ◇. By convention, paired symbols ('closed' and 'open') are often used to represent 'plus' (treatment) and 'minus' (control) treatments.*

Types of graph

Different graphical forms may be used for different purposes, including:

- Plotted curves – used for data where the relationship between two variables can be represented as a continuum (e.g. Fig. 63.1).
- Scatter diagrams – used to visualise the relationship between individual data values for two interdependent variables (e.g. Fig. 67.7) often as a preliminary part of a correlation analysis.
- Three-dimensional graphs show the interrelationships of three variables, often one dependent and two independent (e.g. Fig. 45.1). A contour diagram is an alternative method of representing such data.
- Histograms represent frequency distributions of continuous variables (e.g. Fig. 63.2).
- Frequency polygons emphasise the form of a frequency distribution by joining the coordinates with straight lines, in contrast to a histogram. This is particularly useful when plotting two or more frequency distributions on the same graph.
- Bar charts represent frequency distributions of a discrete qualitative or quantitative variable (e.g. Fig. 63.4). An alternative representation is the line chart (e.g. Fig. 67.3).
- Pie charts illustrate portions of a whole (e.g. Fig. 63.5).

The plotted curve

This is the commonest form of graphical representation used in biology. The key features are outlined below and in checklist form in Box 63.1, while Box 63.2 advises on using Microsoft Excel.

Data points

Each data point must be shown accurately, so that any reader can determine the exact values of x and y. The results of each treatment must be readily identifiable. A useful technique is to use a dot for each data point, surrounded by a hollow symbol for each treatment (see Fig. 63.1) or

Box 63.1 Checklist for the stages in drawing a graph

The following sequence can be used whenever you need to construct a plotted curve: it will need to be modified for other types of graph.

1. **Collect all of the data values and statistical values** (in tabular form, where appropriate).
2. **Decide on the most suitable form of presentation:** this may include transformation to convert data to linear form.
3. **Choose a concise descriptive title,** together with a reference (figure) number and date, where necessary.
4. **Determine which variable is to be plotted on the *x*-axis and which on the *y*-axis.**
5. **Select appropriate scales for both axes** and make sure that the numbers and their location (scale marks) are clearly shown, together with any scale breaks.
6. **Decide on appropriate descriptive labels for both axes,** with SI units of measurement, where appropriate.
7. **Choose the symbols for each set of data points** and decide on the best means of representation for statistical values.
8. **Plot the points** to show the coordinates of each value with appropriate symbols.
9. **Draw a trend line for each set of points.** Use a see-through ruler, so you can draw the line to have an equal number of points on either side of it.
10. **Write a figure legend,** to include a key which identifies all symbols and statistical values and any descriptive footnotes.

Box 63.2 How to create and amend graphs within a spreadsheet (Microsoft Excel) for use in coursework reports and dissertations

Excel can be used to create plotted curves, bar charts and histograms of reasonable quality, but only if you know how to amend the default settings to improve the overall effect, so that they meet the standard required for practical and project reports. As with a hand-drawn graph, the basic stages in graph drawing (Box 63.1) still apply.

Producing a plotted curve or bar chart

1. **Create the appropriate type of graph (chart) for your data.** Enter your data into the spreadsheet (e.g. in two columns), select the data array (highlight the appropriate cells by clicking and holding down the left mouse button and dragging) then use the 'Chart Wizard' icon (or the *Insert > Chart* option) to select a particular *Chart type*. For a plotted curve, use the *XY (Scatter)* option with unconnected points as the *Chart sub-type* – the line will be added at a later stage (note: never use the *Line plot* option because this will create a graph with evenly spaced *x*-axis entries, regardless of their actual values). For a bar chart, select the *Column* option. Work through each step of the Chart Wizard, selecting appropriate options for *Chart source, Chart options* (including labels for *x* and *y* axes) and *Location*. The graphs shown in Figure 63.6a and Figure 63.7a were produced using the default settings in Excel 2003.
2. **Change the default settings to improve the overall appearance.** Consider each element of the image in turn, including the overall size, height and width of your graph (click on the chart using the left-hand mouse button to show the 'sizing handles' and drag these to resize). The graphs shown in Fig. 63.6b and Fig. 63.7b were produced by altering the settings, typically by moving the cursor over the feature and clicking the right mouse button to reveal an additional menu of editing/formatting options.

The examples given below are for illustrative purposes only, and should not necessarily be regarded as a prescriptive list:

Example for a plotted curve (compare Fig. 63.6a with Fig. 63.6b):

- Grey background removed using *Clear* function, and border line around graph removed using *Format Chart Area > Patterns > Border > None.*
- Horizontal gridlines can be either removed using *Clear* function or, if desired (Fig. 63.6b), changed using *Chart Options > Gridlines* to show major and minor gridlines, then selecting *Format Gridlines* option to change the *Colour, Scale* and *Weight* of the gridlines, to make them more like those of conventional graph paper.
- Unnecessary legend box on right-hand side removed using the *Clear* option.
- Axes reformatted using the *Format Axis* menu *(Patterns, Scale* and *Number* options).
- Axis label changes achieved using the *Format Axis Title > Font* menu options (including superscript for selected parts – alternatively, cut-and-paste symbols/text from Word).
- Data point style changed using the *Format Data Series > Patterns > Marker > Custom* options for *Style, Foreground* (edge of symbol), *Background* (centre of symbol) and *Size* (note: *Line* function not

(a)

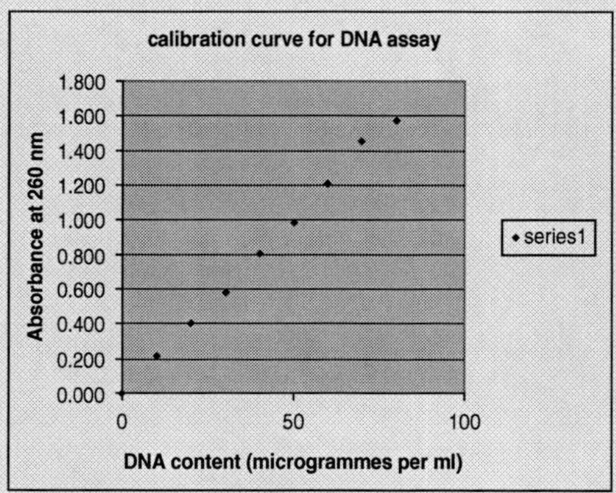

(b)

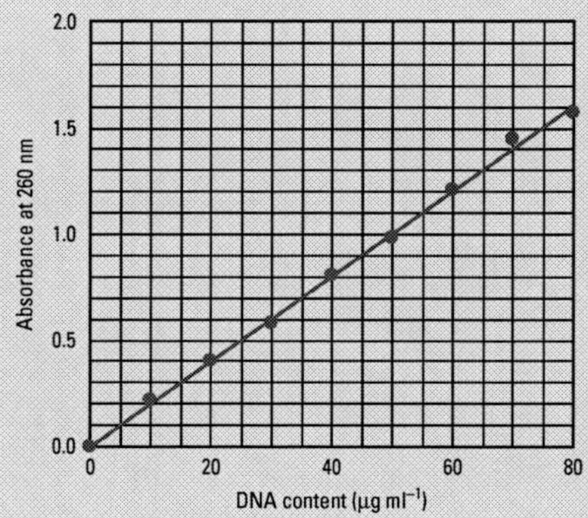

Fig. 1 Calibration curve for DNA assay. Performed using an A100X spectrophotometer (on 01.04.07). Values shown are averages of triplicate measurements.

Fig. 63.6 Examples of a plotted curve produced in Microsoft Excel using (a) default settings and (b) modified (improved) settings.

Box 63.2 (continued)

used, since it joins individual data points, rather than giving a linear plot).

- Straight line of best fit added using the *Format Data Series > Add Trendline > Type > Linear* option, with the line thickness then adjusted using the *Format Trendline > Patterns > Custom > Weight* option (see also Box 37.2).

Example for a bar chart (compare Fig. 63.7a with Fig. 63.7b):

- Grey background removed using *Clear* function, and border line around graph removed using *Format Chart Area > Patterns > Border > None;* colour of horizontal gridlines changed to light grey using *Format Gridlines > Patterns > Custom > Colour* option.
- Unnecessary legend box on right-hand side removed using the *Clear* option.
- Axes reformatted (tic marks, scales and font) using the *Format Axis* menu *(Patterns, Scale* and *Font* options).
- Axis labels changed to Times New Roman regular 12 point font size using *Format Axis Title > Font, > Font style* and *> Size* options.
- Bar colour removed using *Format Data Series > Patterns > Area,* selecting the white colour (not option *None*).
- Individual numbers shown using the *Format Data Series > Labels > Value* option; bar width altered using *Format Data Series > Options > Gap Width.*

Note that for both types of graph, it is probably better not to use the default *Chart Title* option within Excel, which places the title at the top of the chart (as in Fig. 63.6a and Fig. 63.7a), but to cut-and-paste your untitled graph into a word processor such as Microsoft Word and then type a formal figure legend below the graph itself (as in Fig. 63.6b and Fig. 63.7b). However, once your graph is embedded into Word, it is generally best not to make further amendments – go back to the original Excel file, make the required change(s) and re-insert the graph into Word.

Producing a histogram

The histogram function in Excel requires a little more effort to master, compared with the 'Chart Wizard' graph drawing function. Essentially, a histogram is a graphical display of frequencies (counts) for a continuous quantitative variable (p. 173), where the data values are grouped into discrete classes. It is possible to select the upper limit for each class into which you want your data to be grouped (termed 'bin range values' in Excel).

An alternative approach is to let the software select the group ranges for you: Excel selects evenly distributed bins between the minimum and maximum values (this is often less effective than selecting your own class boundaries).

The following steps outline the procedure used to create the histograms shown in Figure 63.8 for the table of data below (length, in mm, of 24 leaf petioles from a single plant).

7.2	6.5	7.1	8.5	6.6	7.2
7	7.3	8.6	9.1	7.5	8.3
7.1	5.7	7.3	7.6	6.9	7.1
8.3	7.6	5.4	8.6	7.9	8

(a)

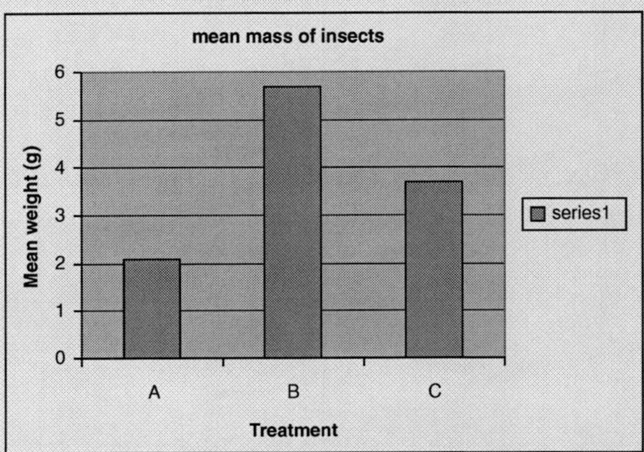

(b)

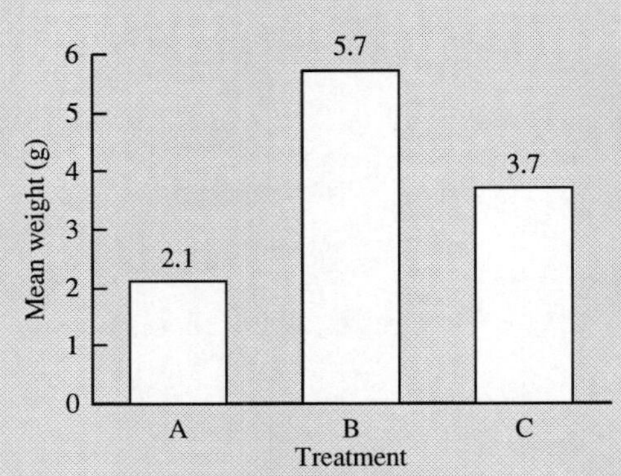

Fig. 2 Bar chart for mean mass of insects treated using three different methods. Key: A = chloroform, B = saline, C = formalin.

Fig. 63.7 Examples of a bar chart produced in Microsoft Excel using (a) default settings and (b) modified (improved) settings.

(continued)

Box 63.2 (continued)

1. **Enter the raw data values** e.g. as a single column of numbers, or as an array as above.
2. **Decide the class intervals to be used.** Base your choice on the number of data points and the maximum and minimum values (for a small data set, you may be able to do this by quickly examining the data set, whereas for a large data set, use the Excel *COUNT, MAX* and *MIN* functions) – a typical histogram would have 4–10 classes depending on the level of discrimination required. Enter the upper limit for each class (bin range values) in a separate array of cells close to the data, in ascending order (in the above example, 6, 7, 8 and 9 were chosen – the few data values above the final bin value will be shown on the histogram as a group labelled 'more').
3. **Select the histogram function, then input your data and class interval values.** From the dropdown menu under *Tools* select *Data Analysis > Histogram* (if the *Data Analysis* option is not already activated on the PC you are using, select using the *Add Ins* function within this menu (but note that this addition may require the Microsoft Office software CD) then *Histogram.* Input your data into the *Input Range* box (highlight the appropriate cells by clicking and dragging, while holding down the left mouse button) and similarly input the *Bin Range* values into the appropriate box (if this is left empty, Excel will select default *Bin Range* values) – most of the remaining boxes can be left empty, though you must click the last box to get a *Chart Output,* otherwise the software will give the numerical counts for each group, without a histogram. Click *OK* and entries in a new worksheet will be created showing the upper limits of each group (column labelled *Bin*) along with the counts for each group (column labelled *Frequency),* plus a poorly constructed chart based on default settings, as shown in Figure 63.8a (note that the default output is a bar chart rather than a histogram, since there are gaps between the groups).

Example for a histogram Fig. 63.8b shows the following modifications compared with Fig. 63.8a:

- Chart resized to increase height (use sizing handles); grey background removed using *Clear* function and border line around graph removed using *Format Chart Area > Patterns > Border > None.*
- Title and unnecessary legend box removed using *Clear* option; *y* scale amended using *Format Axis > Scale* command; *x* labels (classes) amended by typing class limits directly into the cells containing the bin values.
- Axes reformatted using the *Format Axis* menu, and new text typed directly into the axis label box (double-click on the chart to access).
- Bar colour changed to light grey using *Format Data Series > Patterns > Area* option.
- Bar chart converted to true histogram (no gaps between bars) using the *Format Data Series > Options menu,* setting *Gap Width* = 0.
- Figure legend added in Microsoft Word (below figure), following copying and pasting of the Excel histogram into a Word file.

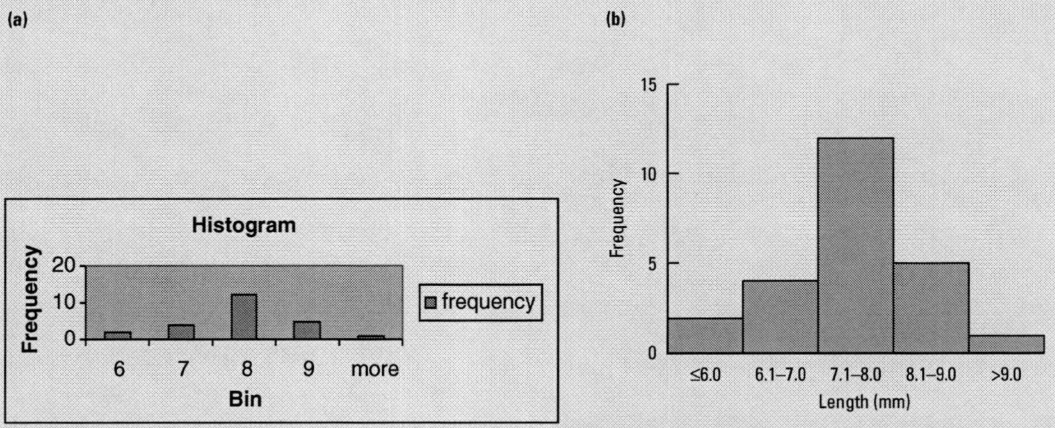

Fig. 3 Histogram of petiole lengths (mm) in a single plant (n = 24).

Fig. 63.8 Examples of histogram output from Microsoft Excel using (a) default settings and (b) modified (improved) settings.

to use symbols only (Fig. 59.6), though the coordinates of each point are defined less accurately. Use the same symbol for the same entity if it occurs in several graphs and provide a key to all symbols.

Statistical measures

If you are plotting average values for several replicates and if you have the necessary statistical knowledge, you can calculate the standard error, or the 95% confidence limits for each mean value and show these on your graph as a series of vertical bars (see Fig. 63.1). Make it clear in the legend whether the bars refer to standard errors or 95% confidence limits and quote the value of n (the number of replicates per data point). Another approach is to add a least significant difference bar to the graph.

Adding error bars to Microsoft Excel graphs *– you can do this as follows (for Excel 2003):*

1. *On your graph, right-click on one of the data points of the series to which you want to add error bars;*
2. *Select* Format Data Series *and use either the* Y Error Bars *tab (vertical error bars) or the* X Error Bars *tab (horizontal error bars):*
   - *select the type of error bar that you want under* Display *(e.g.* Both*);*
   - *select the method to be used to input the error values under* Error amount*; typically, you will want to select* Custom*, to input pre-calculated numerical values. Then, in the* Plus *and* Minus *boxes, either specify the range of cells in your worksheet where the error values are located, or enter the relevant numbers directly, separated by commas.*

Interpolation

Once you have plotted each point, you must decide whether to link them by straight lines or a smoothed curve. Each of these techniques conveys a different message to your reader. Joining the points by straight lines may seem the simplest option, but may give the impression that errors are very low or non-existent and that the relationship between the variables is complex. Joining points by straight lines is appropriate in certain graphs involving time sequences (e.g. the number of animals at a particular site each year), or for repeat measurements where measurement error can be assumed to be minimal (e.g. recording a patient's temperature in a hospital, to emphasise any variation from one time point to the next). However, in most plotted curves the best straight line or curved line should be drawn (according to appropriate mathematical or statistical models, or by eye) to highlight the relationship between the variables – after all, your choice of a plotted curve implies that such a relationship exists. Don't worry if some of your points do not lie on the line: this is caused by errors of measurement and by biological variation. Most curves drawn by eye should have an equal number of points lying on either side of the line. You may be guided by 95% confidence limits, in which case your curve should pass within these limits wherever possible.

Curved lines can be drawn using a flexible curve, a set of French curves or freehand. In the latter case, turn your paper so that you can draw the curve in a single, sweeping stroke by a pivoting movement at the elbow (for larger curves) or wrist (for smaller ones). Do not try to force your hand to make complex, unnatural movements as the resulting line will not be smooth.

Conveying the correct message *– the golden rule is: 'always draw the simplest line that fits the data reasonably well and is biologically reasonable'.*

Extrapolation

Be wary of extrapolation beyond the upper or lower limit of your measured values. This is rarely justifiable and may lead to serious errors. Whenever extrapolation is used, a dotted line ensures that the reader is aware of the uncertainty involved. Any assumptions behind an extrapolated curve should also be stated clearly in your text.

Extrapolating plotted curves *– try to avoid the need to extrapolate by better experimental design.*

The histogram

Drawing a histogram *– each datum is represented by a column with an area proportional to the magnitude of y: in most cases, you should use columns of equal width, so that the height of each column is then directly proportional to y. Shading or stippling may be used to identify individual columns, according to your needs.*

While a plotted curve assumes a continuous relationship between the variables by interpolating between individual data points, a histogram involves no such assumptions. Histograms are also used to represent frequency distributions, where the y-axis shows

Box 63.3 How graphs can misrepresent and mislead

1. **The 'volume' or 'area' deception** – this is mainly found in histogram or bar chart presentations where the size of a symbol is used to represent the measured variable. For example, the amount of hazardous waste produced in different years might be represented on a chart by different sizes of a chemical drum, with the *y*-axis (height of drum) representing the amount of waste. However, if the symbol retains its *shape* for all heights as in Figure 63.9a, its *volume* will increase as a cubic function of the height, rather than in direct proportion. To the casual observer, a twofold increase may look like an eightfold one, and so on. Strictly, the *height* of the symbol should be the measure used to represent the variable, with no change in symbol width, as in Fig. 63.9b.
2. **Effects of a non-zero axis** – A non-zero axis acts to emphasise the differences between measures by reducing the range of values covered by the axis. For example, in Figure 63.10a, it looks as if there are large differences in mass between males and females; but if the scale is adjusted to run from zero (Fig. 63.10b), then it can be seen that the differences are not large as a proportion of the overall mass. Always scrutinise the scale values carefully when interpreting any graph.
3. **Use of a relative rather than absolute scale** – this is similar to the above, in that data compared using relative scales (e.g. percentage or ratio) can give the wrong impression if the denominator is not the same in all cases. In Figure 63.11a, two treatments are shown as equal in *relative* effect, both resulting in 50% relative response compared (say) to the respective controls. However, if treatment A is 50% of a control value of 200 and treatment B is 50% of a control value of 500, then the actual difference in *absolute* response would have been masked, as shown by Fig. 63.11b.
4. **Effects of a non-linear scale** – when interpreting graphs with non-linear (e.g. logarithmic) scales, you may interpret any changes on an imagined linear scale. For example, the pH scale is logarithmic, and linear changes on this scale mean less in terms of absolute H^+ concentration at high (alkaline) pH than they do at low (acidic) pH. In Figure 63.12a, the cell density in two media is compared on a logarithmic scale, while in Fig. 63.12b, the same data are graphed on a linear scale. Note, also, that the log *y*-axis scale in Fig. 63.12a cannot be shown to zero, because there is no logarithm for 0.

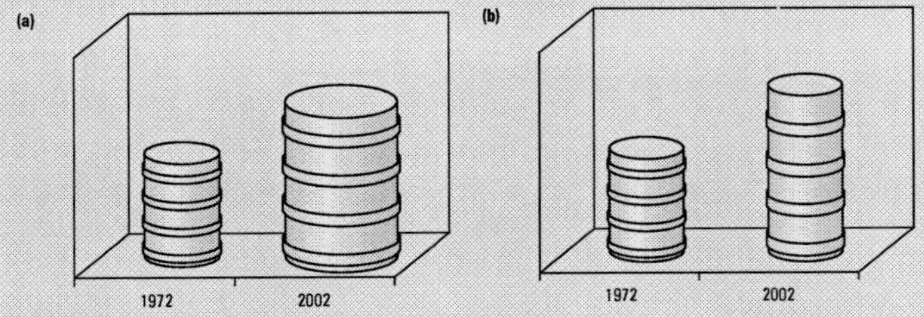

Fig. 63.9 Increase in pesticide use over a 30 year period.

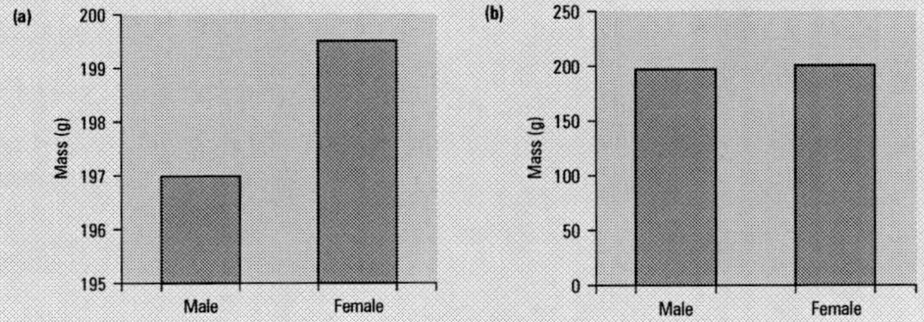

Fig. 63.10 Average mass of males and females in test group.

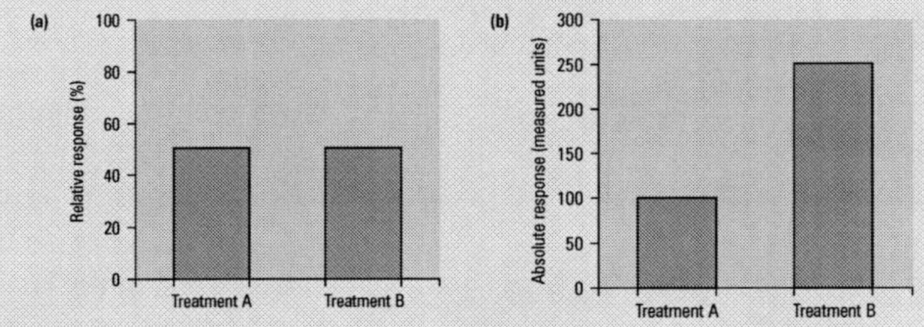

Fig. 63.11 Responses to treatments A and B.

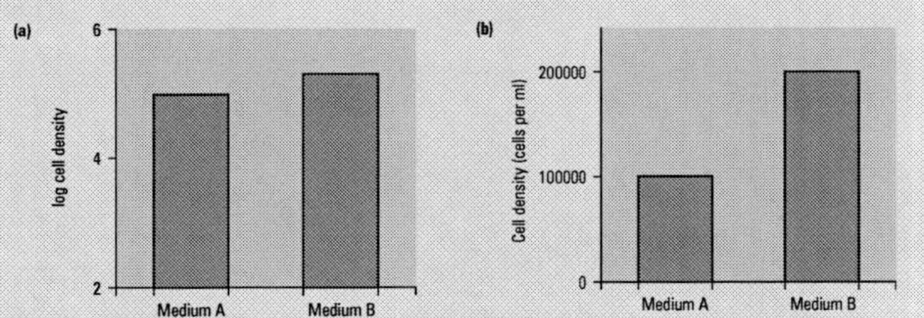

Fig. 63.12 Effect of different media on cell density.

Box 63.3 (continued)

5. **Unwarranted extrapolation** – a graph may be extrapolated to indicate what would happen if a trend continued, as in Figure 63.13a. However, this can only be done under certain assumptions (e.g. that certain factors will remain constant or that relationships will hold under new conditions). There may be no guarantee that this will actually be the case. Figure 63.13b illustrates other possible outcomes if the experiment were to be repeated with higher values for the *x*-axis.
6. **Failure to account for data point error** – this misrepresentation involves curves that are overly complex in relation to the scatter in the underlying data. When interpreting graphs with complex curves, consider the errors involved in the data values. It is probably unlikely that the curve would pass through all the data points unless the errors were very small. Figure 63.14a illustrates a curve that appears to assume zero error and is thus overly complex, while Fig. 63.14b shows a curve that takes possible errors of the points into account.
7. **Failure to reject outlying points** – this is a special case of the previous example. There may be many reasons for outlying data, from genuine mis-takes to statistical 'freaks'. If a curve is drawn through such points on a graph, it indicates that the point carries equal weight with the other points, when in fact, it should probably be ignored. To assess this, consider the accuracy of the measurement, the number and position of adjacent points, and any special factors that might be involved on a one-off basis. Figure 63.15a shows a curve where an outlier has perhaps been given undue weight when showing the presumed relationship. If there is good reason to think that the point should be ignored, then the curve shown in Fig. 63.15b would probably be more valid.
8. **Inappropriate fitted line** – here, the mathematical function chosen to represent a trend in the data might be inappropriate. A straight line might be fitted to the data, when a curve would be more correct, or *vice versa*. These cases can be difficult to assess. You need to consider the theoretical validity of the model used to generate the curve (this is not always stated clearly). For example, if a straight line is fitted to the points, the implicit underlying model states that one factor varies in direct relation to another, when the true situation may be more complex. In Figure 63.16a, the relationship has been shown as a linear relationship, whereas an exponential relationship, as shown in Fig. 63.16b, could be more correct.

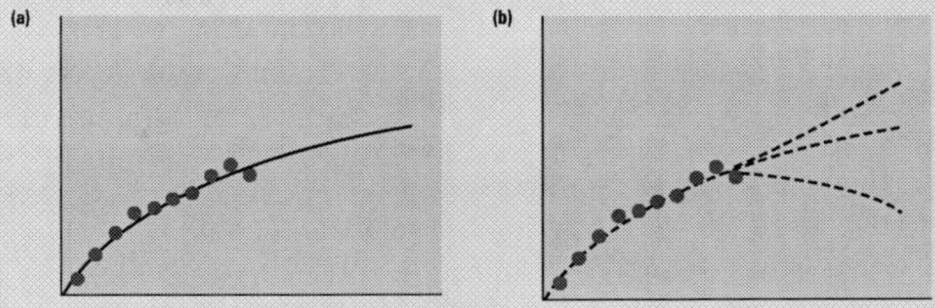

Fig. 63.13 Extrapolation of data under different assumptions.

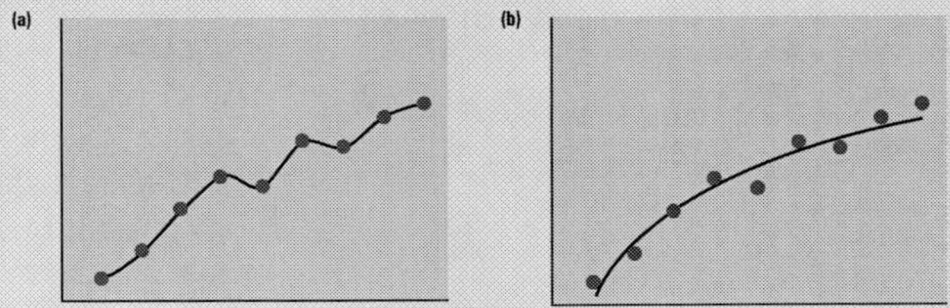

Fig. 63.14 Fitted curves under different assumptions of data error.

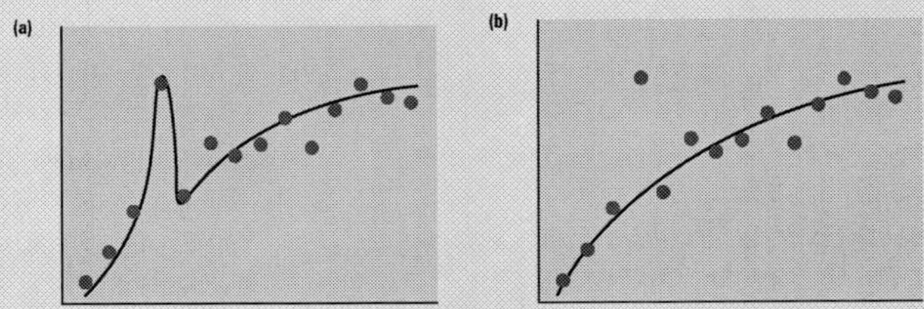

Fig. 63.15 Curves with and without outlier taken into account.

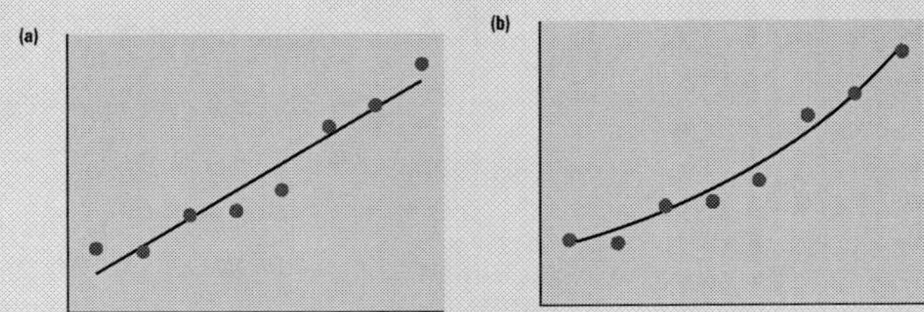

Fig. 63.16 Different mathematical model used to represent trends in data.

the number of times a particular value of x was obtained (e.g. Fig. 63.2). As in a plotted curve, the x-axis represents a continuous quantitative variable which can take any value within a given range (e.g. plant height), so the scale must be broken down into discrete classes and the scale marks on the x-axis should show either the mid-points (mid-values) of each class, or the boundaries between the classes.

The columns are contiguous (adjacent to each other) in a histogram, in contrast to a bar chart, where the columns are separate because the x-axis of a bar chart represents discrete values.

Using computers to produce graphs – *never allow a computer program to dictate size, shape and other aspects of a graph: find out how to alter scales, labels, axes, etc. and make appropriate selections (see Box 62.2 for Microsoft Excel). Draw curves freehand if the program only has the capacity to join the individual points by straight lines.*

Interpreting proportions and percentages – *you will need to establish how the values have been calculated (see p. 76).*

Examining graphs – *don't be tempted to look at the data displayed within a graph before you have considered its context, read the legend and the scale of each axis.*

Understanding graphs within scientific papers – *the legend should be a succinct summary of the key information required to interpret the figure without further reference to the main text. This is a useful approach when 'skimming' a paper for relevant information (p. 18).*

Interpreting graphs

The process of analysing a graph can be split into five phases:

1. Consider the context. Look at the graph in relation to the aims of the study in which it was reported. Why were the observations made? What hypothesis was the experiment set up to test? This information can usually be found in the Introduction or Results section of a report. Also relevant are the general methods used to obtain the results. This might be obvious from the figure title and legend, or from the Materials and Methods section.
2. Recognise the graph form and examine the axes. First, what kind of graph is presented (e.g. histogram, plotted curve)? You should be able to recognise the main types summarised on page 457 and their uses. Next, what do the axes measure? You should check what quantity has been measured in each case and what units are used.
3. Look closely at the scale of each axis. What is the starting point and what is the highest value measured? For the x-axis, this will let you know the scope of the treatments or observations (e.g. whether they lasted for 5 min or 20 years; whether a concentration span was two-fold or 50-fold). For each axis, it is especially important to note whether the values start at zero; if not, then the differences between any treatments shown may be magnified by the scale chosen (see Box 63.3).
4. Examine the symbols and curves. Information will be provided in the key or legend to allow you to determine what these refer to. If you have made your own photocopy of the figure, it may be appropriate to note this directly on it. You can now assess what appears to have happened. If, say, two conditions have been observed while a variable is altered, when exactly do they differ from each other; by how much; and for how long?
5. Evaluate errors and statistics. It is important to take account of variability in the data. For example, if mean values are presented, the underlying errors may be large, meaning that any difference between two treatments or observations at a given x-value could simply have arisen by chance. Thinking about the descriptive statistics used (Chapter 66) will allow you to determine whether apparent differences could be significant in both statistical and biological senses.

Sometimes graphs are used to mislead. This may be unwitting, as in an unconscious favouring of a 'pet' hypothesis of the author. Graphs may be used to 'sell' a product in the field of advertising or to favour a viewpoint as, perhaps, in politics. Experience in drawing and interpreting graphs will help you spot these flawed presentations, and understanding how graphs can be erroneously presented (Box 63.3) will help you avoid the same pitfalls.

Sources for further study

Anon. (2000) *Biological Nomenclature: Recommendations on Terms, Units and Symbols.* Institute of Biology, London.
[Includes a section on presentation of data]

Briscoe, M.H. (1996) *Preparing Scientific Illustrations: A Guide to Better Posters, Presentations and Publications.* Springer-Verlag, New York.

Carter, M., Wiebe, E.N. and Ferzli, M. *Graphing with Excel.* Available: http://www.ncsu.edu/labwrite/res/gt/gt-menu.html
Last accessed: 01/04/07.
[Online-tutorial from the US–NSF LabWrite 2000 Project]

Robbins, N.B. (2005) *Creating More Effective Graphs.* Wiley, New York.

Study exercises

63.1 Select appropriate graphical presentations (see also study exercise 37.3). Choose an appropriate graphical form for each of the following examples:

(a) Interaction between pH and cation concentration on enzyme activity.
(b) Proportion of different eye colours in a survey of undergraduates attending your university.
(c) Relationship between pulse rate and age in humans.
(d) Number of bacteria per field of view for several different samples.
(e) Effect of copper concentration on the activity of an enzyme.

63.2 Create a pie chart. Display the following information in the form of a hand-drawn pie chart. Do *not* use a spreadsheet for this exercise.

Number of bacterial strains isolated from a sample

Bacterium	Number of strains isolated
Escherichia coli	37
Aeromonas hydrophila	8
Klebsiella spp.	13
Salmonella enterica	2
Enterococcus spp.	25

63.3 Create a frequency distribution histogram. The table below gives data for the haemoglobin levels of 100 people. Plot a histogram showing the frequency distribution of the data. Write a brief description of the important features of the distribution.

Haemoglobin content ($g\,l^{-1}$) in blood

11.1	14.2	13.5	9.8	12.0	13.9	14.1	14.6	11.0	12.3
13.4	12.9	12.9	10.0	13.1	11.8	12.6	10.7	8.1	11.2
13.8	12.4	12.9	11.3	12.7	12.4	14.6	15.1	11.2	9.7
11.3	14.7	10.8	13.3	11.9	11.4	12.5	13.0	11.6	13.1
9.3	13.5	14.6	11.2	11.7	10.9	12.4	12.0	12.1	12.6
10.9	12.1	13.4	9.5	12.5	11.6	12.2	8.8	10.7	11.1
10.2	11.7	10.4	14.0	14.9	11.5	12.0	13.2	12.1	13.3
12.4	9.4	13.2	12.5	10.8	11.7	12.7	14.1	10.4	10.5
13.3	10.6	10.5	13.7	11.8	14.1	10.3	13.6	10.4	13.9
11.7	12.8	10.4	11.9	11.4	10.6	12.7	11.4	12.9	12.1

63.4 Find examples of misleading graphs. Create a portfolio of examples of misleading graphs taken from newspapers. For each graph, state what aspect is misleading (see Box 63.3) and, where possible, attempt to show the data correctly in a new graph.

64 Presenting data in tables

Alternatives to tables for presenting numerical data – *if you have only a few numbers, consider simply presenting these within the text; an alternative approach is to show the data values on a bar chart (e.g. Fig. 63.7b).*

A table is often the most appropriate way to present numerical data in a concise, accurate and structured form. Assignments and project reports should contain tables which have been designed to condense and display results in a meaningful way and to aid numerical comparison. The preparation of tables for recording primary data is discussed on page 176.

Decide whether you need a table, or whether a graph is more appropriate. Histograms and plotted curves can be used to give a visual impression of the relationships within your data. On the other hand, a table gives you the opportunity to make detailed numerical comparisons.

KEY POINT ***Always remember that the primary purpose of your table is to communicate information and allow appropriate comparison, not simply to put down the results on paper.***

Preparation of tables

Title

Constructing titles – *take care over titles as it is a common mistake in student practical reports to present tables without titles, or to misconstruct the title.*

Every table must have a brief descriptive title. If several tables are used, number them consecutively so they can be quoted in your text. The titles within a report should be compared with one another, making sure they are logical and consistent and that they describe accurately the numerical data contained within them.

Structure

Saving space in tables – *you may be able to omit a column of control data if your results can be expressed as percentages of the corresponding control values.*

Display the components of each table in a way that will help the reader understand your data and grasp the significance of your results. Organise the columns so that each category of like numbers or attributes is listed vertically, while each horizontal row shows a different experimental treatment, organism, sampling site, etc. (as in Table 64.1). Where appropriate, put control values near the beginning of the table. Columns that need to be compared should be set out alongside each other. Use rulings to subdivide your table appropriately, but avoid cluttering it up with too many lines.

Table 64.1 Characteristics of selected photoautotrophic microbes

			Intracellular carbohydrate	
Division	Species	Optimum [NaCl]* (mol m^{-3})	Identity	Quantity† (nmol (g dry wt)$^{-1}$)
Chlorophyta	*Scenedesmus quadruplicatum*	340	Sucrose	49.7
	Chlorella emersonii	780	Sucrose	102.3
	Dunaliella salina	4700	Glycerol	910.7
Cyanobacteria	*Microcystis aeruginosa*	<20‡	None	0.0
	Anabaena variabilis	320	Sucrose	64.2
	Rivularia atra	380	Trehalose	ND

* Determined after 28-day incubation in modified Von Stosch medium.
† Individual samples, analysed by gas-liquid chromatography.
‡ Poor growth in all media with added NaCl (minimum NaCl concentration 5 mol m^{-3}).
ND Sample lost: no quantitative data.

Headings and subheadings

These should identify each set of data and show the units of measurement, where necessary. Make sure that each column is wide enough for the headings and for the longest data value.

Numerical data

Within the table, do not quote values to more significant figures than necessary, as this will imply spurious accuracy (p. 175). By careful choice of appropriate units for each column you should aim to present numerical data within the range 0 to 1000. As with graphs, it is less ambiguous to use derived SI units, with the appropriate prefixes, in the headings of columns and rows, rather than quoting multiplying factors as powers of 10. Alternatively, include exponents in the main body of the table (see Table 25.1), to avoid any possible confusion regarding the use of negative powers of 10.

Examples If you measured the width of a fungal hypha to the nearest one-tenth of a micrometre, quote the value in the form '52.6 μm'.

Quote the width of a fungal hypha as 52.6 μm, rather than 0.000 052 6 m or 52.6 10^{-6} m.

Other notations

Avoid using dashes in numerical tables, as their meaning is unclear; enter a zero reading as '0' and use 'NT' not tested or 'ND' if no data value was obtained, with a footnote to explain each abbreviation. Other footnotes, identified by asterisks, superscripts or other symbols in the table, may be used to provide relevant experimental detail (if not given in the text) and an explanation of column headings and individual results, where appropriate. Footnotes should be as condensed as possible. Table 64.1 provides examples.

Saving further space in tables *– in some instances a footnote can be used to replace a whole column of repetitive data.*

Statistics

In tables where the dispersion of each data set is shown by an appropriate statistical parameter, you must state whether this is the (sample) standard deviation, the standard error (of the mean) or the 95% confidence limits and you must give the value of n (the number of replicates). Other descriptive statistics should be quoted with similar detail, and hypothesis-testing statistics should be quoted along with the value of P (the probability). Details of any test used should be given in the legend, or in a footnote.

Text

Sometimes a table can be a useful way of presenting textual information in a condensed form (see examples on pp. 268 and 293).

When you have finished compiling your tabulated data, carefully double-check each numerical entry against the original information, to ensure that the final version of your table is free from transcriptional errors. Box 64.1 gives a checklist for the major elements of constructing a table.

Using spreadsheets and word processing packages *– these can be used to prepare high quality versions of tables for project work (Box 64.2).*

Box 64.1 Checklist for preparing a table

Every table should have the following components:

1. **A title,** plus a reference number and date where necessary.
2. **Headings for each column and row,** with appropriate units of measurement.
3. **Data values,** quoted to the nearest significant figure and with statistical parameters, according to your requirements.
4. **Footnotes** to explain abbreviations, modifications and individual details.
5. **Rulings to emphasise groupings** and distinguish items from each other.

Box 64.2 How to use a word processor (Microsoft Word) or a spreadsheet (Microsoft Excel) to create a table for use in coursework reports and dissertations

Creating tables with Microsoft Word: word-processed tables are suitable for text-intensive or number-intensive tables, although in the second case entering data can be laborious. When working in this way, the natural way to proceed is to create the 'shell' of the table, add the data, then carry out final formatting on the table.

1. **Move the cursor to the desired position in your document.** This is where you expect the top left corner of your table to appear. Go to the *Table* command, then choose *Insert > Table.*
2. **Select the appropriate number of columns and rows.** Don't forget to add rows and columns for headings. As default, a full-width table will appear, with single rulings for all cell boundaries, with all columns of equal width and all rows of equal height.

   Example of a 4 x 3 table:

3. **Customise the columns.** By placing the cursor over the vertical rulings then 'dragging', you can adjust their width to suit your heading text entries, which should now be added.

Heading 1	Heading 2	Heading 3	Heading 4

4. **Work through the table adding the data.** Entries can be numbers or text.

Heading 1	Heading 2	Heading 3	Heading 4
xx	xx	xx	xx
xx	xx	xx	

5. **Make further adjustments to column and row widths to suit.** For example, if text fills several rows within a cell, consider increasing the column width and if a column contains only single or double digit numbers, consider shrinking its width. To combine cells, first highlight them, then use *Table > Merge Cells.* You may wish to reposition text within a cell using *Format > Paragraph > Spacing.*

Heading 1	Heading 2	Heading 3	Heading 4
xx	xx	xx	xx
	xx	xx	xx

6. **Remove selected borders to cells.** One way to do this is using the table borders options accessed from the *Table borders* button on the toolbar, so that your table looks like the examples shown in this chapter.
7. **Add a table title.** This should be positioned *above* the table (*cf* a figure title/legend), legend and footnotes.

Final version of the Table:

***Table xx.* A table of some data.**

Heading 1	Heading 2*	Heading 3	Heading 4
Aaa	xx	xx	xx
	xx	xx	xx

*An example of a footnote

Creating tables with Microsoft Excel: tables derived from spreadsheets are effective when you have lots of numerical data, especially when these are stored or created using the spreadsheet itself. When working in this way, you can design the table as part of an output or summary section of the spreadsheet, add explanatory headings, format, then possibly export to a word processor when complete.

1. **Design the output or summary section.** Plan this as if it were a table, including adding text headings within cells.

17				
18	Heading 1	Heading 2	Heading 3	Heading 4
19				
20				
21				
22				

2. **Insert appropriate formulae within cells to produce data.** If necessary, formulae should draw on the other parts of the spreadsheet.

17				
18	Heading 1	Heading 2	Heading 3	Heading 4
19	Aaa	=A1	=C3*5	=SDEV(A1:A12)
20	Bbb	=A2	=F45/G12	=SDEV(B1:B12)
21				
22				

Box 64.2 (continued)

3. **Format the cells.** This is important to control the number of decimal places presented (*Format > Cells > Number).*
4. **Adjust column width to suit.** You can do this *via* the column headings, by placing the cursor over the rulings between columns then 'dragging'.

17				
18	Heading 1	Heading 2	Heading 3	Heading 4
19	Aaa	=A1	=C3*5	=SDEV(A1:A12)
20	Bbb	=A2	=F45/G12	=SDEV(B1:B12)
21				
22				

5. **Add rulings as appropriate.** Use the borders menu on the toolbar as described above.

17				
18	Heading 1	Heading 2	Heading 3	Heading 4
19	Aaa	=A1	=C3*5	=SDEV(A1:A12)
20	Bbb	=A2	=F45/G12	=SDEV(B1:B12)
21				

6. **Add 'real' data values to the spreadsheet.** This should result in the summary values within the table being filled. Check that these are presented with the appropriate number of significant figures.
7. **The table can now be copied and pasted to a Word document.** If you wish to link the spreadsheet and the word-processed document so that the latter is updated whenever changes are made to the spreadsheet values, then click *Paste* on the *Formatting* toolbar, click *Paste Options* next to the data, and then select *Match Destination Table Style and Link to Excel* or *Keep Source Formatting and Link to Excel.*

Sources for further study

Kirkup, L. (1994) *Experimental Methods: An Introduction to the Analysis and Presentation of Data.* Wiley, New York.

Simmonds, D. and Reynolds, L. (1994) *Data Presentation and Visual Literacy in Medicine and Science.* Butterworth-Heinemann, London.

Study exercises

64.1 Redesign a table of data. Using the following example, redraft the table to improve layout and correct inconsistencies.

Concentrations of low molecular weight solutes in bacteria

Bacterium	Constituent	Concentration
Escherichia coli	Proline	21.0 mmol l^{-1}
Escherichia coli	Trehalose	1.547×10^2 kmol m^{-3}
Bacillus subtilis	Proline	39.7 mmol l^{-1}
Bacillus subtilis	Glutamate	0.0521 mmol cm^{-3}
*Staphylococcus aureus**	Glutamate	15 260 mmol m^{-3}
Escherichia coli	Glutamate	0.50% w/v
Bacillus subtilis	Trehalose	<0.001% w/v

* Proline and trehalose were not measured.

64.2 Devise a text-based table. After reading through this chapter, and working from memory, draw up a table listing the principal components of a typical table in the first column, and brief comments on the major features of each component in the second column.

64.3 Interpret data from a table. The table alongside shows representative data for energy, protein and niacin (vitamin B3) as a function of age for human males and females. The typical average mass ('weight') of each class is also shown.

(a) In what age class does the mass of females exceed that of males?

(b) In what age class is the energy requirement of males greatest in proportion to the energy requirement of females?

(c) Over what age range do males and females have the same protein requirement?

(d) In what age class do males have their highest niacin requirement, and how does this compare with females?

Selected age-related characteristics of human males and females

Age	Average mass (kg)		Energy requirement (MJ)		Protein requirement (g)		Niacin requirement (mg)	
(years)	Male	Female	Male	Female	Male	Female	Male	Female
<1	8	8	3.4	3.2	14	14	6	6
1–3	13	13	5.2	4.9	16	16	9	9
4–6	20	20	7.2	6.5	24	24	12	12
7–10	28	28	8.2	7.3	28	28	13	13
11–14	45	46	9.3	7.7	45	46	17	15
15–18	66	55	11.5	8.8	59	44	20	15
19–50	79	63	10.6	8.1	63	50	19	15
>50	77	65	9.5	7.7	63	50	15	13

65 Hints for solving numerical problems

Advice on dealing with numerical procedures *– see:*

- *Box 23.2: preparing solutions (p. 136);*
- *Box 24.1: molar concentrations (p. 145);*
- *Box 35.1: cell counting chambers (p. 236);*
- *Box 35.2: plate (colony) counts (p. 237);*
- *Box 37.1: calibration curves (p. 252);*
- *Box 39.1 radioactivity (p. 270).*

Biology often requires a numerical or statistical approach. Not only is mathematical modelling an important aid to understanding, but computations are often needed to turn raw data into meaningful information or to compare them with other data sets. Moreover, calculations are part of laboratory routine, perhaps required for making up solutions of known concentration (see p. 136 and below) or for the calibration of a microscope (see p. 163). In research, 'trial' calculations can reveal what input data are required and where errors in their measurement might be amplified in the final result (see p. 175).

KEY POINT If you find numerical work difficult, practice at problem-solving is especially important.

Practising at problem-solving:

- demystifies the procedures involved, which are normally just the elementary mathematical operations of addition, subtraction, multiplication and division (Table 65.1);
- allows you to gain confidence so that you don't become confused when confronted with an unfamiliar or apparently complex form of problem;
- helps you recognise the various forms a problem can take as, for instance, in crossing experiments in classical genetics (Chapter 58).

Table 65.1 Sets of numbers and operations

Sets of numbers	
Whole numbers:	0, 1, 2, 3, ...
Natural numbers:	1, 2, 3, ...
Integers:	... −3, −2, −1, 0, 1, 2, 3, ...
Real numbers:	integers and anything between (e.g. −5, 4.376, 3/16, π, √5)
Prime numbers:	subset of natural numbers divisible by 1 and themselves only (i.e. 2, 3, 5, 7, 11, 13, ...)
Rational numbers:	p/q where p (integer) and q (natural) have no common factor (e.g. 3/4)
Fractions:	p/q where p is an integer and q is natural (e.g. −6/8)
Irrational numbers:	real numbers with no exact value (e.g. π)
Infinity:	(symbol ∞) is larger than any number (technically not a number as it does not obey the laws of algebra)
Operations and symbols	
Basic operators:	+, −, × and ÷ will not need explanation; however, / may substitute for ÷, * may substitute for × or this operator may be omitted
Powers:	a^n, i.e. 'a to the power n', means a multiplied by itself n times (e.g. $a^2 = a \times a$ = 'a squared', $a^3 = a \times a \times a$ = 'a cubed'). n is said to be the index or exponent. Note $a^0 = 1$ and $a^1 = a$
Logarithms:	the common logarithm (log) of any number x is the power to which 10 would have to be raised to give x (i.e. the log of 100 is 2; $10^2 = 100$); the antilog of x is 10^x. Note that there is no log for 0, so take this into account when drawing log axes by breaking the axis. Natural or Napierian logarithms (ln) use the base e (= 2.718 28 ...) instead of 10
Reciprocals:	the reciprocal of a real number a is $1/a$ ($a \neq 0$)
Relational operators:	$a > b$ means 'a is greater (more positive) than b', < means less than, ⩽ means less-than-or-equal-to and ⩾ means greater-than-or-equal-to
Proportionality:	$a \propto b$ means 'a is proportional to b' (i.e. $a = kb$, where k is a constant). If $a \propto 1/b$, a is inversely proportional to b ($a = k/b$)
Sums:	Σx_i is shorthand for the sum of all x values from $i = 0$ to $i = n$ (more correctly the range of the sum is specified under the symbol)
Moduli:	$\|x\|$ signifies modulus of x, i.e. its absolute value (e.g. $\|4\| = \|-4\| = 4$)
Factorials:	$x!$ signifies factorial x, the product of all integers from 1 to x (e.g. $3! = 6$). Note $0! = 1! = 1$

Steps in tackling a numerical problem

Tracing errors in mathematical problems – *this is always easier when all the stages in a calculation are laid out clearly.*

The step-by-step approach outlined below may not be the fastest method of arriving at an answer, but most mistakes occur where steps are missing, combined or not made obvious, so a logical approach is often better. Error tracing is distinctly easier when all stages in a calculation are laid out.

Using a computer spreadsheet for numerical problems – *this may be very useful in repetitive work or for 'what if?' case studies (see Chapter 12).*

Have the right tools ready

Scientific calculators (p. 123) greatly simplify the numerical part of problem-solving. However, the seeming infallibility of the calculator may lead you to accept an absurd result which could have arisen because of faulty key-pressing or faulty logic. Make sure you know how to use all the features on your calculator: especially how the memory works; how to introduce a constant multiplier or divider; and how to obtain an exponent (note that the 'exp' button on most calculators gives you 10^x, not 1^x or y^x; so 1×10^6 would be entered as [1] [exp] [6], *not* [10] [exp] [6]).

Approach the problem thoughtfully

If the individual steps have been laid out on a worksheet, the 'tactics' will already have been decided. It is more difficult when you have to adopt a strategy on your own, especially if the problem is presented as a story and it isn't obvious which equations or rules need to be applied.

Table 65.2 Simple algebra – rules for manipulating

If $a = b + c$, then $b = a - c$ and $c = a - b$
If $a = b \times c$, then $b = a \div c$ and $c = a \div b$
If $a = b^c$, then $b = a^{1/c}$ and $c = \log a \div \log b$
$a^{1/n} = \sqrt[n]{a}$
$a^{-n} = 1 \div a^n$
$a^b \times a^c = a^{(b+c)}$ and $a^b \div a^c = a^{(b-c)}$
$(a^b)^c = a^{(b \times c)}$
$a \times b = \text{antilog}(\log a + \log b)$

- Read the problem carefully as the text may give clues as to how it should be tackled. Be certain of what is required as an answer before starting.
- Analyse what kind of problem it is, which effectively means deciding which equation(s) or approach will be applicable. If this is not obvious, consider the dimensions/units of the information available and think how they could be fitted to a relevant formula. In examinations, a favourite ploy of examiners is to present a problem such that the familiar form of an equation must be rearranged (see Table 65.2 and Box 65.1). Another is to make you use two or more equations in series (see Box 65.2). If you are unsure whether a recalled formula is correct, a dimensional analysis can help: write in all the units for the variables and make sure that they cancel out to give the expected answer.
- Check that you have, or can derive, all of the information required to use your chosen equation(s). It is unusual but not unknown for examiners to supply redundant information. So, if you decide not to use some of the information given, be sure why you do not require it.
- Decide on what format and units the answer should be presented in. This is sometimes suggested to you. If the problem requires many changes in the prefixes to units, it is a good idea to convert all data to base SI units (multiplied by a power of 10) at the outset.
- If a problem appears complex, break it down into component parts.

Present your answer clearly

Presenting calculations in assessed work – *always show the steps in your calculations, as most markers will only penalise a mistake once and part marks will be given if the remaining operations are performed correctly. This can only be done if those operations are visible.*

The way you present your answer obviously needs to fit the individual problem. The example shown in Box 65.2 has been chosen to illustrate several important points, but this format would not fit all situations. Guidelines for presenting an answer include:

(a) Make your assumptions explicit. Most mathematical models of biological phenomena require that certain criteria are met before they can be

Box 65.1 Example of using the algebraic rules of Table 65.2

Problem: if $a = (b - c) \div (d + e^n)$, find e.

1. Multiply both sides by $(d + e^n)$; formula becomes: $a(d + e^n) = (b - c)$
2. Divide both sides by a; formula becomes: $d + e^n = \frac{b - c}{a}$
3. Subtract d from both sides; formula becomes: $e^n = \frac{b - c}{a} - d$
4. Raise each side to the power $1/n$; formula becomes: $e = \left\{\frac{b - c}{a} - d\right\}^{1/n}$

Units – never write any answer without its unit(s) unless it is truly dimensionless.

Rounding off to a specific number of significant figures – do not round off numbers until you arrive at the final answer or you will introduce 'rounding' errors into the calculation.

legitimately applied (e.g. 'assuming the tissue is homogeneous ...'), while some approaches involve approximations which should be clearly stated (e.g. 'to estimate the mouse's skin area, its body was approximated to a cylinder with radius x and height y ...').

(b) Explain your strategy for answering, perhaps giving the applicable formula or definitions which suit the approach to be taken. Give details of what the symbols mean (and their units) at this point.

(c) Rearrange the formula to the required form with the desired unknown on the left-hand side (see Table 65.2).

(d) Substitute the relevant values into the right-hand side of the formula, using the units and prefixes as given (it may be convenient to convert values to SI beforehand). Convert prefixes to appropriate powers of 10 as soon as possible.

(e) Convert to the desired units step by step, i.e. taking each variable in turn.

(f) When you have the answer in the desired units, rewrite the left-hand side and underline the answer, for emphasis. Make sure that the result is presented to an appropriate number of significant figures (see below).

Check your answer

Having written out your answer, you should check it methodically, answering the following questions:

- Is the answer realistic? You should be alerted to an error if a number is absurdly large or small. In repeated calculations, a result standing out from others in the same series should be double-checked.
- Do the units make sense and match up with the answer required? Don't, for example, present a volume in units of m^2.
- Do you get the same answer if you recalculate in a different way? If you have time, recalculate the answer using a different 'route', entering the numbers into your calculator in a different form and/or carrying out the operations in a different order.

Rounding: decimal places and significant figures

In many instances, the answer you produce as a result of a calculation will include more figures than is justified by the accuracy and precision of the original data. Sometimes you will be asked to produce an answer to a specified number of decimal places or significant figures, and other times you will be expected to decide for yourself what would be appropriate.

Box 65.2 Model answer to a mathematical problem

Problem

Estimate the total length and surface area of the fibrous roots on a maize seedling from measurements of their total fresh weight and mean diameter. Give your answers in m and cm^2 respectively to four significant figures.

Measurements

Fresh weight[a] = 5.00 g, mean diameter[b] = 0.5 mm.

Answer

Assumptions: (1) the roots are cylinders with constant radius[c] and the 'ends' have negligible area; (2) the root system has a density of $1000\,kg\,m^{-3}$ (i.e. that of water[d]).

Strategy: from assumption (1), the applicable equations are those concerned with the volume and surface area of a cylinder, namely:

$$V = \pi r^2 h \qquad [65.1]$$
$$A = 2\pi rh \text{ (ignoring ends)} \qquad [65.2]$$

where V is volume (m^3), A is surface area (m^2), $\pi \approx 3.14159$, h is height (m) and r is radius (m). The total length of the root system is given by h and its surface area by A. We can find h by rearranging Eqn [65.1] and then substitute its value in Eqn [65.2] to get A.

To calculate total root length: rearranging Eqn [65.1], we have $h = V/\pi r^2$. From measurements[e], $r = 0.25\,mm = 0.25 \times 10^{-3}\,m$.

From density = weight/volume,

$$V = \text{fresh weight/density}$$
$$= 5\,g/1000\,kg\,m^{-3}$$
$$= 0.005\,kg/1000\,kg\,m^{-3}$$
$$= 5 \times 10^{-6}\,m^3$$

Total root length,

$$h = V/\pi r^2$$
$$5 \times 10^{-6}\,m^3/3.14159 \times (0.25 \times 10^{-3}\,m)^2$$

<u>Total root length = 25.46 m</u>

To calculate surface area of roots: substituting the value for h obtained above into Eqn [65.2], we have:

Root surface area

$$= 2 \times 3.14159 \times 0.25 \times 10^{-3}\,m \times 25.46\,m$$
$$= 0.04\,m^2$$
$$= 0.04 \times 10^4\,cm^2$$
(there being $100 \times 100 = 10^4\,cm^2$ per m^2)

<u>Root surface area = 400.0 cm^2</u>

Notes

(a) The fresh weight of roots would normally be obtained by washing the roots free of soil, blotting them dry and weighing.

(b) In a real answer you might show the replicate measurements giving rise to the mean diameter.

(c) In reality, the roots will differ considerably in diameter and each root will not have a constant diameter throughout its length.

(d) This will not be wildly inaccurate as about 95% of the fresh weight will be water, but the volume could also be estimated from water displacement measurements.

(e) Note conversion of measurements into base SI units at this stage and on line 3 of the root volume calculation. Forgetting to halve diameter measurements where radii are required is a common error.

***KEY POINT** Do not simply accept the numerical answer from a calculator or spreadsheet, without considering whether you need to modify this to give an appropriate number of significant figures or decimal places.*

Rounding to n *decimal places*

This is relatively easy to do.

1. Look at the number to the right of the *n*th decimal place.
2. If this is less than five, simply 'cut off' all numbers to the right of the *n*th decimal place to produce the answer (i.e. round down).
3. If the number is greater than five, 'cut off' all numbers to the right of the *n*th decimal place and add one to the *n*th decimal place to produce the answer (i.e. round up).

Examples
The number 4.123 correct to two decimal places is 4.12
The number 4.126 correct to two decimal places is 4.13
The number 4.1251 correct to two decimal places is 4.13
The number 4.1250 correct to two decimal places is 4.12
The number 4.1350 correct to two decimal places is 4.14
The number 99.99 correct to one decimal place is 100.0.

4. If the number is 5, then look at further numbers to the right to determine whether to round up or not.
5. If the number is *exactly* 5 and there are no further numbers to the right, then round the *n*th digit to the nearest even number. *Note:* When considering a large number of calculations, this procedure will not affect the overall mean value. Some rounding systems do the opposite to this (i.e. round to the nearest odd number), while others always round up where the number is exactly 5 (which *will* affect the mean). Take advice from your tutor and stick to one system throughout a series of calculations.

Whenever you see any numbers quoted, you should assume that the last digit has been rounded. For example, in the number 22.4, the '.4' is assumed to be rounded and the calculated value may have been between 22.35 and 22.45.

Examples
The number of significant figures in 194 is three
The number of significant figures in 2305 is four
The number of significant figures in 0.003 482 is four
The number of significant figures in 210×10^8 is 3 (21×10^9 would be two).

Quoting to n *significant figures*

The number of significant figures indicates the degree of approximation in the number. For most cases, it is given by counting all the figures except zeros that occur at the beginning or end of the number. Zeros *within* the number are always counted as significant. The number of significant figures in a number like 200 is ambiguous and could be one, two or three; if you wish to specify clearly, then quote as e.g. 2×10^2 (one significant figure), 2.0×10^2 (two significant figures), etc. to avoid spurious accuracy. When quoting a number to a specified number of significant figures, use the same rules as for rounding to a specified number of decimal places, but do not forget to keep zeros before or after the decimal point. The same principle is used if you are asked to quote a number to the 'nearest 10', 'nearest 100', etc.

When deciding for yourself how many significant figures to use, adopt the following rules of thumb:

- Always round *after* you have done a calculation. Use *all* significant figures available in the measured data during a calculation.
- If adding or subtracting with measured data, then quote the answer to the number of decimal places in the data value with the least number of decimal places (e.g. $32.1 - 45.67 + 35.6201 = 22.1$, because 32.1 has one decimal place).
- If multiplying or dividing with measured data, keep as many significant figures as are in the number with the least number of significant places (e.g. $34\,901 \div 3445 \times 1.341\,0344 = 13.59$, because 3445 has four significant figures).
- For the purposes of significant figures, assume 'constants' (e.g. number of mm in a m) have an infinite number of significant figures.

Examples
The number of significant figures in 3051.93 is six
To five significant figures, this number is 3051.9
To four significant figures, this number is 3052
To three significant figures, this number is 3050
To two significant figures, this number is 3100
To one significant figure, this number is 3000
3051.93 to the nearest 10 is 3050
3051.93 to the nearest 100 is 3100
Note that in this last case you must include the zeros before the decimal point to indicate the scale of the number (even if the decimal point is not shown). For a number less than 1, the same would apply to the zeros before the decimal point. For example, 0.003 051 93 to three significant figures is 0.003 05. Alternatively, use scientific notation (in this case, 3.05×10^{-3}).

Some reminders of basic mathematics

Errors in calculations sometimes appear because of faults in mathematics rather than computational errors. For reference purposes, Tables 65.1 and 65.2 give some basic mathematical principles that may be useful. Eason *et al.* (1992) or Stephenson (2003) should be consulted for more advanced/specific needs.

Examples
1/8 as a percentage is $1 \div 8 \times 100 = 100 \div 8 = 12.5\%$
0.602 as a percentage is $0.602 \times 100 = 60.2\%$.

Percentages and proportions

A percentage is just a fraction expressed in terms of hundredths, indicated by putting the percentage sign (%) after the number of hundredths. So 35% simply means 35 hundredths. To convert a fraction to a percentage, just multiply the fraction by 100. When the fraction is in decimal form, multiplying by 100 to obtain a percentage is easily achieved just by moving the decimal point two places to the right.

Examples
190% as a decimal fraction is $190 \div 100 = 1.9$
5/2 as a percentage is $5 \div 2 \times 100 = 250\%$.

To convert a percentage to a fraction, just remember that, since a percentage is a fraction multiplied by 100, the fraction is the percentage divided by 100. For example: $42\% = 42/100 = 0.42$. In this example, since we are dealing with a decimal fraction, the division by 100 is just a matter of moving the decimal point two places to the left (42% could be written as 42.0%). Percentages greater than 100% represent fractions greater than 1. Percentages less than 1 may cause confusion. For example, 0.5% means half of one per cent (0.005) and must not be confused with 50% (which is the decimal fraction 0.5).

Example A population falls from 4 million to 3.85 million. What is the percentage change? The decrease in numbers is $4 - 3.85 = -0.15$ million. The fractional decrease is $-0.15 \div 4 = -0.0375$ and we multiply by 100 to get the percentage change = minus 3.75%.

To find a percentage of a given number, just express the percentage as a decimal fraction and multiply the given number. For example: 35% of 500 is given by $0.35 \times 500 = 175$. To find the percentage change in a quantity, work out the difference (= value 'after' – value 'before'), and divide this difference by the original value to give the fractional change, then multiply by 100.

Exponents

Exponential notation is an alternative way of expressing numbers in the form a^n ('a to the power n'), where a is multiplied by itself n times. The number a is called the base and the number n the exponent (or power or index). The exponent need not be a whole number, and it can be negative if the number being expressed is less than 1. See Table 65.2 for other mathematical relationships involving exponents.

Example $2^3 = 2 \times 2 \times 2 = 8$.

Scientific notation

In scientific notation, also known as 'standard form', the base is 10 and the exponent a whole number. To express numbers that are not whole powers of 10, the form $c \times 10^n$ is used, where the coefficient c is normally between 1 and 10. Scientific notation is valuable when you are using very large numbers and wish to avoid suggesting spurious accuracy. Thus if you write 123 000, this may suggest that you know the number to ±0.5, whereas 1.23×10^5 might give a truer indication of measurement accuracy (i.e. implied to be ±500 in this case). Engineering notation is similar, but treats numbers as powers of 10 in groups of 3, i.e. $c \times 10^0$, 10^3, 10^6, 10^9, etc. This corresponds to the SI system of prefixes (p. 183).

Example Avogadro's number, ≈602 352 000 000 000 000 000 000, is more conveniently expressed as $6.023\,52 \times 10^{23}$.

A useful property of powers when expressed to the same base is that when multiplying two numbers together, you simply add the powers, while if dividing, you subtract the powers. Thus, suppose you counted 8 bacteria in a known value of a 10^{-7} dilution, there would be 8×10^7 in the same volume of undiluted solution; if you now dilute this 500-fold (5×10^2), then the number present in the same volume would be $8/5 \times 10^{(7-2)} = 1.6 \times 10^5 = 160\,000$.

Examples The logarithm to the base 10 ($\log_{10}$) of 1000 is 3, since $10^3 = 1000$. The logarithm to the base e ($\log_e$ or ln) of 1000 is 6.907 755 (to six decimal places).

Examples (use to check the correct use of your own calculator)
102 963 as a log (to base 10) = 5.012 681 (to six decimal places)
$10^{5.012681} = 102962.96$
(Note loss of accuracy due to loss of decimal places.) 102 963 as a natural logarithm (ln) = 11.542 125 (to six decimal places) thus $2.718282^{11.542125} = 102963$.

Logarithms

When a number is expressed as a logarithm, this refers to the power n that the base number a must be raised to give that number. Any base could be used, but the two most common are 10, when the power is referred to as $\log_{10}$ or simply log, and the constant e (2.718 282), used for mathematical convenience in certain situations, when the power is referred to as $\log_e$ or ln (natural logarithm). Note that (a) logs need not be whole numbers; (b) there is no log value for the number zero; and (c) $\log_{10} = 0$ for the number 1.

To obtain logs, you will need to use the log key on your calculator, or special log tables (now largely redundant). To convert back ('antilog') use

- the [10^x] key, with x = log value;
- the [inverse] then the [log] key; or
- the [y^x] key, with $y = 10$ and x = log value.

If you have used log tables, you will find complementary antilogarithm tables to do this.

There are many uses of logarithms in biology, including pH ($= -\log[H^+]$), where $[H^+]$ is expressed in mol l^{-1} (see p. 152); the exponential growth of micro-organisms, where if log(cell number) is plotted against time, a straight-line relationship is obtained (see p. 233); and allometric studies of growth and development, where if data are plotted on log axes, a series of straight-line relationships may be found.

Example Using Eqn [65.3], the predicted value for y for a linear function where $a = 2$ and $b = 0.5$, where $x = 8$ is: $y = 2 + (0.5 \times 8) = 6$.

Linear functions and straight lines

One of the most straightforward and widely used relationships between two variables x and y is that represented by a straight-line graph, where the corresponding mathematical function is known as the equation of a straight line, where:

$$y = a + bx \qquad [65.3]$$

In this relationship, a represents the intercept of the line on the y (vertical) axis, i.e. where $x = 0$, and therefore $bx = 0$, while b is equivalent to the slope (gradient) of the line, i.e. the change in y for a change in x of 1. The constants a and b are sometimes given alternative symbols, but the mathematics remains unchanged, e.g. in the equivalent expression for the slope of a straight line, $y = mx + c$. Figure 65.1 shows what happens when these two constants are changed, in terms of the resultant straight lines.

The two main applications of the straight-line relationship are:

1. Function fitting. Here, you determine the mathematical form of the function, i.e. you estimate the constants a and b from a data set for x and y, either by drawing a straight line by eye and then working out the slope and y intercept, or by using linear regression to obtain the most probable values for both constants. When putting a straight line of best fit by eye on a hand-drawn graph, note the following:

   - Always use a *transparent* ruler, so you can see data points on either side of the line.

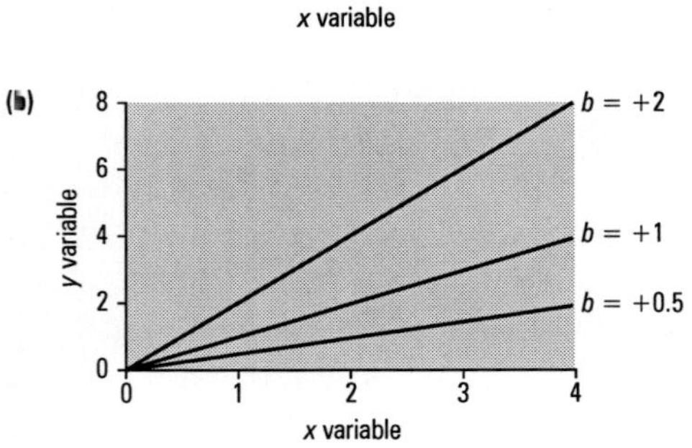

Fig. 65.1 Straight-line relationships ($y = a + bx$), showing the effects of (a) changing the intercept at constant slope, and (b) changing the slope at constant intercept.

- For a data series where the points do not fit a perfect straight line, try to have an equal number of points on either side of the line, as in Figure 65.2a, and try to minimise the average distance of these points from the line.
- Once you have drawn the line of best fit use this line, rather than your data values, in all subsequent procedures (e.g. in a calibration curve, Chapter 37).
- Tangents drawn to a curve give the slope (gradient) at a particular point, e.g. in an enzyme reaction progress curve. These are best drawn by bringing your ruler up to the curve at the exact point where you wish to estimate the slope and then trying to make the two angles immediately on either side of this point approximately the same, by eye (Fig. 65.2b).
- Once you have drawn the straight line or tangent, choose two points reasonably far apart at either end of your line and then draw construction lines to represent the change in y and the change in x between these two points: make sure that your construction lines are perpendicular to each other. Determine the slope as the change in y divided by the change in x (Fig. 65.2b).

Example Using Eqn [65.4], the predicted value for x for a linear function where $a=1.5$ and $b=2.5$, where $y=-8.5$ is: $x=(-8.5-1.5)\div 2.5=-4$.
Using Eqn [65.4] the predicted x intercept for a linear function where $a=0.8$ and $b=3.2$ is: $x=(0-0.8)\div 3.2=-0.25$.

2. Prediction. Where a and b are known, or have been estimated, you can use Eqn [65.3] to predict any value of y for a specified value of x, e.g. during exponential growth of a cell culture (p. 233), where $\log_{10}$ cell number (y) increases as a linear function of time (x): note that in this example the dependent variable has been transformed to give a linear relationship. You will need to rearrange Eqn [65.3] in cases where a prediction of x is required for a particular value of y (e.g. in calibration curves, p. 252, or bioassays, p. 238), as follows:

$$x = (y - a) \div b \qquad [65.4]$$

This equation can also be used to determine the intercept on the x (horizontal) axis, i.e. where $y=0$.

Example A lab schedule states that 5 g of a compound with a relative molecular mass of 220 are dissolved in 400 ml of solvent. For writing up your Materials and Methods, you wish to express this as mol l^{-1}.
1. If there are 5 g in 400 ml, then there are 5 ÷ 400 g in 1 ml.
2. Hence, 1000 ml will contain 5 ÷ 400 × 1000 g = 12.5 g.
3. 12.5 g = 12.5 ÷ 220 mol = 0.0568 mol, so [solution] = 56.8 mmol l^{-1} (= 56.8 mol m^{-3}).

Hints for some typical problems

Calculations involving proportions or ratios

The 'unitary method' is a useful way of approaching calculations involving proportions or ratios, such as those required when making up solutions from stocks (see also Chapter 24) or as a subsidiary part of longer calculations.

1. If given a value for a multiple, work out the corresponding value for a single item or unit.
2. Use this 'unitary value' to calculate the required new value.

Examples For a geometric dilution series involving ten-fold dilution steps, calculation of concentrations is straightforward, e.g. two serial decimal dilutions (= 100-fold dilution) of a solution of NaCl of 250 mmol l^{-1} will produce a dilute solution of 250 ÷ 100 = 2.5 mmol l^{-1}. Similarly, for an arithmetic dilution series, divide by the overall dilution to give the final concentration, e.g. a sixteen-fold dilution of a solution of NaCl of 200 mg ml^{-1} will produce a dilute solution of 200 ÷ 16 = 12.5 mg ml^{-1}.

Calculations involving series

Series (used in e.g. dilutions, see also p. 137) can be of three main forms:

1. Arithmetic, where the *difference* between two successive numbers in the series is a constant, e.g. 2, 4, 6, 8, 10, ...
2. Geometric, where the *ratio* between two successive numbers in the series is a constant, e.g. 1, 10, 100, 1000, 10 000, ...
3. Harmonic, where the values are reciprocals of successive whole numbers, e.g. 1, $\frac{1}{2}$, $\frac{1}{3}$, $\frac{1}{4}$, ...

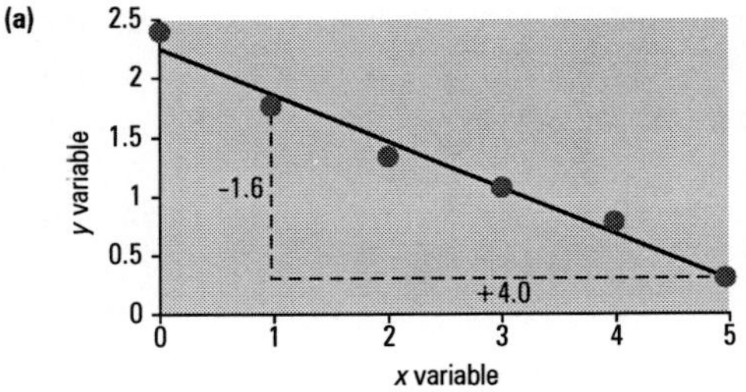

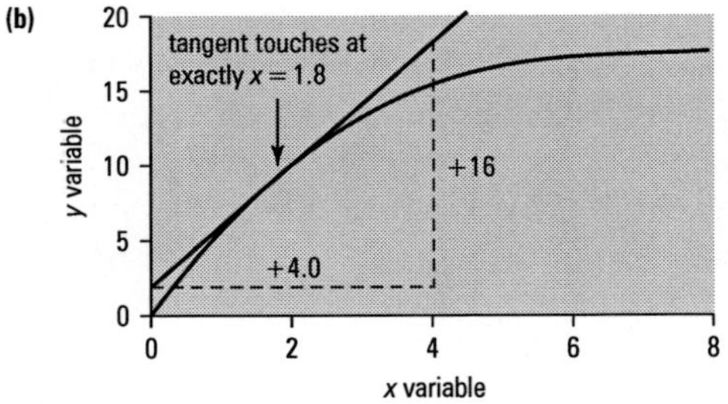

Fig. 65.2 Drawing straight lines. (a) Simple linear relationship, giving a straight line with an intercept of 2.3 and a slope of $-1.6 \div 4.0 = 0.4$. (b) Tangent drawn to a curve at $x = 1.8$, giving a slope of $16 \div 4 = 4$.

Note that the logs of the numbers in a geometric series will form an arithmetic series (e.g. 0, 1, 2, 3, 4,... in the above case). Thus, if a quantity y varies with a quantity x such that the rate of change in y is proportional to the value of y (i.e. it varies in an exponential manner), a semi-log plot of such data will form a straight line. This form of relationship is relevant for exponentially growing cell cultures (p. 233) and radioactive decay (p. 268).

Statistical calculations

The need for long, complex calculations in statistics has largely been removed because of the widespread use of spreadsheets with statistical functions (Chapter 12) and specialised programs such as SPSS and Minitab. It is, however, important to understand the principles behind what you are trying to do (see Chapters 66 and 67) and interpret the program's output correctly, either using the 'help' function or a reference manual.

Problems in Mendelian genetics

These cause difficulties for many students. The key is to recognise the different types of problem and to practise so you are familiar with the techniques for solving them. Chapter 58 deals with the different types of cross you will come across and methods of analysing them, including the use of the χ^2 (chi^2) test.

Text references

Eason, G., Coles, C.W. and Gettinby, G. (1992) *Mathematics and Statistics for the Bio-Sciences*. Ellis Horwood, Chichester.

Stephenson, F.H. (2003) *Calculations in Molecular Biology and Biotechnology*. Academic Press, London.

Sources for further study

Anon. *S.O.S. Mathematics*.
Available: http://www.sosmath.com
Last accessed: 01/04/07.
[A basic Web-based guide with very wide coverage]

Britton, N.F. (2003) *Essential Mathematical Biology*. Springer-Verlag. New York.
[Describes a range of mathematical applications in population dynamics, epidemiology, genetics, biochemistry, and medicine]

Cann, A.J. (2002) *Maths from Scratch for Biologists*. Wiley, Chichester.
[Deals with basic manipulations, formulae, units, molarity, logs and exponents as well as basic statistical procedures]

Causton, D.R. (1992) *A Biologist's Basic Mathematics*. Cambridge University Press, Cambridge.

Forster, P.C. (1999) *Easy Mathematics for Biologists*. Harwood, Amsterdam.
[Covers basic principles, units, logarithms, ratios and proportions, concentrations, equations, rates and graphs]

Harris, M., Taylor, G. and Taylor, J. (2005) *Catch Up Maths and Stats for the Life and Medical Sciences*. Scion, Bloxham.
[Covers a range of basic mathematical operations and statistical procedures]

Koehler, K.R. *College Physics for Students of Biology and Chemistry*. Available: http://www.rwc.uc.edu/koehler/biophys/text.html
Last accessed: 01/04/07.
[A 'hypertextbook' written for first-year undergraduates. Assumes that you have a working knowledge of algebra]

Lawler, G. (2006) *Understanding Maths. Basic Mathematics Explained. Studymates*, Abergele.

Study exercises

65.1 Rearrange a simple formula. The Beer–Lambert relationship, Eqn [41.3] is written in the form $A = \varepsilon l [C]$. Rearrange, in the form:

(a) $[C] =$
(b) $\varepsilon =$

65.2 Rearrange the following formulae:

(a) If $y = ax + b$, find b
(b) If $y = ax + b$, find x
(c) If $x = y^3$, find y
(d) If $x = 3^y$, find y
(e) If $x = (1 - y)(z^p + 3)$, find z
(f) If $x = (y - z)^{1/n}/pq$, find n

65.3 Work with decimal places or significant figures. Give the following numbers to the accuracy indicated:

(a) 214.51 to three significant figures
(b) 107 029 to three significant figures
(c) 0.0450 to one significant figure
(d) 99.817 to two decimal places
(e) 99.897 to two decimal places
(f) 99.997 to two decimal places
(g) 6255 to the nearest 10
(h) 134 903 to the nearest ten thousand

State the following:

(i) the number of significant figures in 3400
(j) the number of significant figures in 3400.3
(k) the number of significant figures in 0.001 67
(l) the number of significant figures in 1.001 67
(m) the number of decimal places in 34.46
(n) the number of decimal places in 0.001 67

65.4 Carry out calculations involving percentages. Answer the following questions, giving your answers to two decimal places:

(a) What is 6/35ths expressed as a percentage?
(b) What is 0.0755 expressed as a percentage?
(c) What is 4.35% of 322?
(d) A rat's weight increased from 55.23 to 75.02 g. What is the % increase in its weight?

65.5 Practise using exponents and logarithms.

(a) Write out (i) the charge on an electron and (ii) the speed of light *in vacuo* in longhand (i.e. without using powers of 10). See Table 29.4 (p. 185) for values given in scientific notation.
(b) Compute the following values: $10^{3.624}$; $\log(6.37)$; $e^{-2.32}$; $\ln(1123)$; $6^{3.2}$. A calculator should be used, but round the output to give five significant figures.

65.6 Practise working with linear functions (note also that Chapter 37 includes study exercises based on linear functions and plotting straight lines). Assuming a linear relationship between x and y, calculate the following (give your answers to three significant figures):

(a) x, where $y = 7.0$, $a = 4.5$ and $b = 0.02$;
(b) x, where $y = 15.2$, $a = -2.6$ and $b = -4.46$;
(c) y, where $x = 10.5$, $a = 0.2$ and $b = -0.63$;
(d) y, where $x = 4.5$, $a = -1.8$ and $b = 4.1$.

66 Descriptive Statistics

Use of symbols *– Y is used in Chapters 66 and 67 to signify the dependent variable in statistical calculations (following the example of Sokal and Rohlf, 1994, Heath, 1995, and Wardlaw, 2000). Note, however, that some authors use X or x in analogous formulae and many calculators refer to e.g. $\bar{x}$, Σx^2, etc. for their statistical functions.*

The purpose of most practical work is to observe and measure a particular characteristic of a biomolecule or biological system. However, it would be extremely rare if the same value was obtained every time the characteristic was measured, or with every experimental subject. More commonly, such measurements will show variability, due to measurement error, sampling variation and/or biological variability (p. 175). Such variability can be displayed as a frequency distribution (e.g. Fig. 63.2), where the y-axis shows the number of times (frequency, f) each particular value of the measured (dependent) variable (Y) has been obtained. Descriptive (or summary) statistics quantify aspects of the frequency distribution of a sample. You can use them to condense a large data set, for presentation in figures or tables. An additional application of descriptive statistics is to provide estimates of the true values of the underlying frequency distribution of the population being sampled, allowing the significance and precision of the observations to be assessed.

Summarising your results *– original data belong only in your primary record, either in laboratory books or as computer records. You should produce summary tables to condense and describe original data.*

KEY POINT The appropriate descriptive statistics to choose will depend on both the type of data, i.e. whether quantitative, ranked or qualitative (see p. 173) and the nature of the underlying frequency distribution.

Initial steps – organising numbers and displaying distributions

In order to organise and summarise data, you should:

- Display the data in graphical form; this provides an immediate visual summary which is relatively easy to interpret.
- Look for an overall pattern in the data – avoid getting lost in the details at this stage.
- Look for any striking exceptions to that pattern (outliers) – they often point to special cases of particular interest or to errors in the data produced through mistakes during the acquisition, recording or copying of data.
- Move from graphical interpretations to numerical summaries and/or verbal descriptions including, where applicable, an explanatory hypothesis.

Colour	Tally	Total
Green	III	3
Blue	~~IIII~~ III	8
Red	IIII	4
White	~~IIII~~ ~~IIII~~ II	12
		= 27

Fig. 66.1 An example of a tally chart.

After collecting data, the first step is often to count how often each value occurs and to produce a frequency table. The frequency is simply the number of times a value occurs in the data set, and is, therefore, a count. The raw data could be acquired using a tally chart system to provide a simple frequency table to construct a tally chart (e.g. Fig. 66.1):

Producing a histogram *– a neatly constructed tally chart doubles as a rough histo-gram or bar chart, with a horizontal layout of the 'columns' (Fig. 66.1)*

- enter only one tally at a time;
- if working from a data list, cross out each item on the list as you enter it onto the tally chart, to prevent double entries;
- check that all values are crossed out at the end and that the totals agree.

Table 66.1 An example of a frequency table

Size class	Frequency	Relative frequency (%)
0–4.9	7	2.6
5–9.9	23	8.6
10–14.9	56	20.9
15–19.9	98	36.7
20–24.9	50	18.7
25–29.9	30	11.2
30–34.9	3	1.1
Total	267	99.8*

* ≠ 100 due to rounding error

stem	leaves
8	233
8	45555
8	77
8	888899
9	0000111111
9	2333333
9	44555555555
9	66677777
9	88888999
10	00

Fig. 66.2 A simple 'stem and leaf' plot of a data set. The 'stem' shows the common component of each number, while the 'leaves' show the individual components, e.g. the top line in this example represents the numbers 82, 83 and 83.

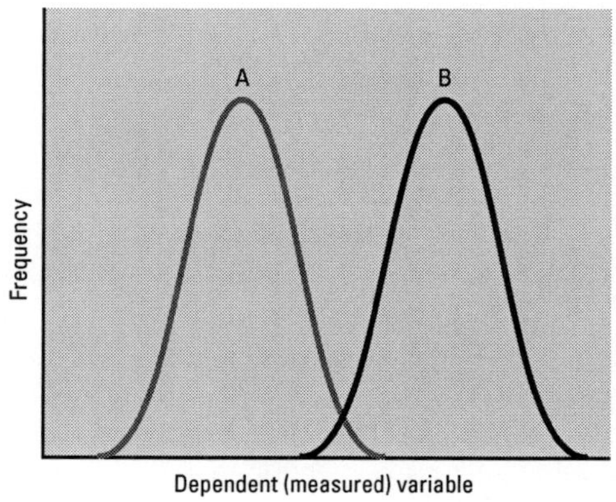

Fig. 66.3 Two distributions with different locations but the same dispersion. The data set labelled B could have been obtained by adding a constant to each datum in the data set labelled A.

Convert the data to a formal table when complete (e.g. Table 66.1). Because proportions are easier to compare than class totals, the table may contain a column to show the relative frequency of each class. Relative frequency can be expressed in decimal form (as a proportion of 1) or as a percentage (as a proportion of 100).

A visual display of a distribution of values is often useful for quantitative variables measured on an interval or ratio scale. The distribution of a variable can be displayed by a frequency table for each value (Table 66.1) or, if the possible values are numerous, groups (classes) of values of the variable. Graphically, there are two main ways of viewing such data:

- histograms, or frequency distributions, generally used for large samples;
- stem and leaf plots (e.g. Fig. 66.2), often used for samples of less than 100: these retain the actual values and are faster to draw by hand. The main drawback is the limitation imposed by the choice of stem values since the class boundaries may obscure some features of the distribution.

These displays allow you to look at the overall shape of a distribution and to observe any significant deviations from the idealised theoretical ones. Where necessary, you can use data transformations to investigate any departures from standard distribution patterns.

In many instances, the normal (Gaussian) distribution best describes the observed pattern for a quantitative variable, giving a symmetrical, bell-shaped frequency distribution, for example, measurements of a quantitative continuous variable in a number of individuals (e.g. blood plasma pH of several adult males), or replicate measurements of a particular characteristic (e.g. repeated measurements of the blood plasma pH of an individual adult male). If you have no clear theoretical grounds for assuming what the underlying frequency distribution is like, graph one or more sample frequency distributions, ideally with a sample size >100.

Three important features of a frequency distribution that can be summarised by descriptive statistics are:

- the sample's location, i.e. its position along a given dimension representing the dependent (measured) variable (Fig. 66.3);
- the dispersion of the data, i.e. how spread out the values are (Fig. 66.4);
- the shape of the distribution, i.e. whether symmetrical, skewed, U-shaped (Fig. 66.5).

Measuring location

Here, the objective is to pinpoint the 'centre' of the frequency distribution, i.e. the value about which most of the data are grouped. The chief measures of location are the mean, median and mode. Figure 66.6 shows how to choose among these for a given data set.

Mean

The mean (denoted $\bar{Y}$ and also referred to as the arithmetic mean) is the average value of the data. It is obtained from the sum of all the data values divided by the number of observations (in symbolic terms, $\Sigma Y/n$). The mean is a good measure of the centre of symmetrical frequency distributions. It

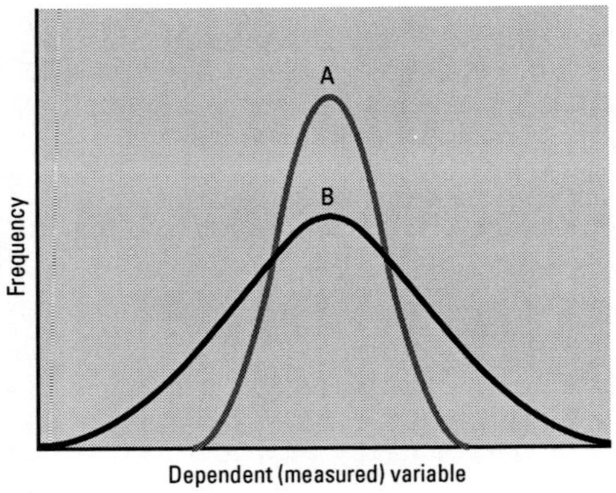

Fig. 66.4 Two distributions with different dispersions but the same location. The data set labelled A covers a relatively narrow range of values of the dependent (measured) variable while that labelled B covers a wider range.

Example Box 66.1 shows a set of data and the calculated values of the measures of location, dispersion and shape for which methods of calculation are outlined here. Check your understanding by calculating the statistics yourself and confirming that you arrive at the same answers.

Definitions

An outlier – any datum that has a value much smaller or bigger than most of the other data values.

Rank – the position of a data value when all the data are placed in order of ascending magnitude. If ties occur, an average rank of the tied variates is used. Thus, the rank of the datum 6 in the sequence 1,3,5,6,8,8,10 is 4; the rank of each datum with value 8 is 5.5.

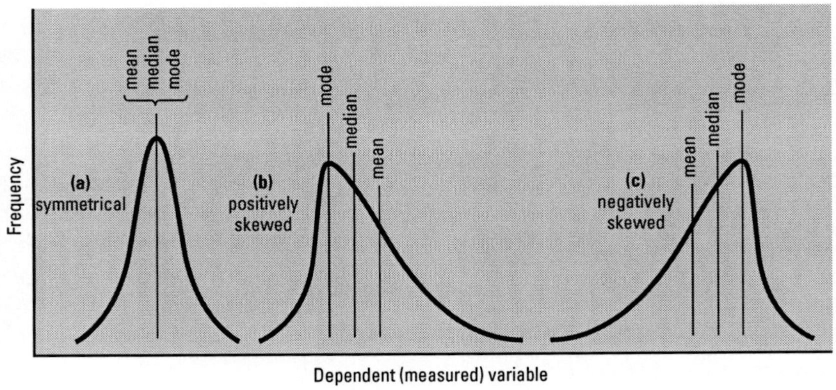

Fig. 66.5 Symmetrical and skewed frequency distributions, showing relative positions of mean, median and mode.

uses all of the numerical values of the sample and therefore incorporates all of the information content of the data. However, the value of a mean is greatly affected by the presence of extreme values (outliers). The arithmetic mean is a widely used statistic in the biosciences, but there are situations when you should be careful about using it (see Box 66.2 for examples).

Median

The median is the mid-point of the observations when ranked in increasing order. For odd-sized samples, the median is the middle observation; for even-sized samples it is the mean of the middle pair of observations. Where data are grouped into classes, the median can only be estimated. This is most simply done from a graph of the cumulative frequency distribution, but can also be worked out by assuming the data to be evenly spread within the class. The median may represent the

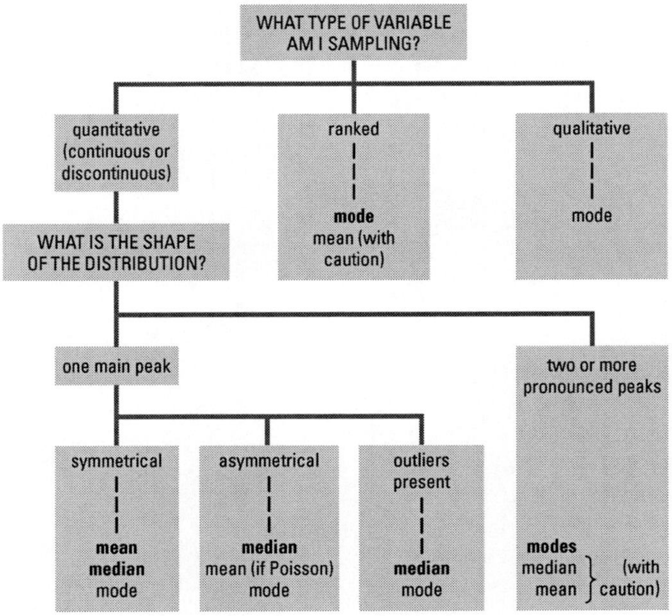

Fig. 66.6 Choosing a statistic for characterising a distribution's location. Statistics written in bold are the preferred option(s).

Box 66.1 Descriptive statistics for a sample of data – an example

Value (Y)	Frequency (f)	Cumulative frequency	fY	fY^2
1	0	0	0	0
2	1	1	2	4
3	2	3	6	18
4	3	6	12	48
5	8	14	40	200
6	5	19	30	180
7	2	21	14	98
8	0	21	0	0
Totals	$21 = \Sigma f\ (= n)$		$104 = \Sigma fY$	$548 = \Sigma fY^2$

In this example, for simplicity and ease of calculation, integer values of Y are used. In many practical exercises, where continuous variables are measured to several significant figures and where the number of data values is small, giving frequencies of 1 for most of the values of Y, it may be simpler to omit the column dealing with frequency and list all the individual values of Y and Y^2 in the appropriate columns. To gauge the underlying frequency distribution of such data sets, you would need to group individual data into broader classes (e.g. all values between 1.0 and 1.9, all values between 2.0 and 2.9, etc.) and then draw a histogram. Calculation of certain statistics for data sets that have been grouped in this way (e.g. median, quartiles, extremes) can be tricky and a statistical text should be consulted.

Statistic	Value*	How calculated
Mean	4.95	$\Sigma fY/n$, i.e. 104/21
Median	5	Value of the $(n+1)/2$ variate, i.e. the value ranked $(21+1)/2 = 11$th (obtained from the cumulative frequency column)
Mode	5	The most common value (Y value with highest frequency)
Upper quartile	6	The upper quartile is between the 16th and 17th values, i.e. the value exceeded by 25% of the data values
Lower quartile	4	The lower quartile is between the 5th and 6th values, i.e. the value exceeded by 75% of the data values
Semi-interquartile range	1.0	Half the difference between the upper and lower quartiles, i.e. $(6-4)/2$
Upper extreme	7	Highest Y value in data set
Lower extreme	2	Lowest Y value in data set
Range	5	Difference between upper and lower extremes
Variance (s^2)	1.65	$s^2 = \dfrac{\Sigma fY^2 - (\Sigma fY)^2/n}{n-1} = \dfrac{548 - (104)^2/21}{20}$
Standard deviation (s)	1.28	$\sqrt{s^2}$
Standard error (SE)	0.280	$s/\sqrt{n}$
95% confidence limits	4.36 – 5.54	$\bar{Y} \mp t_{0.05}[20] \times \text{SE}$, (where $t_{0.05}[20] = 2.09$, Table 67.2)
Coefficient of variation (cov)	25.9%	$100s/\bar{Y}$

*Rounded to three significant figures, except when it is an exact number.

location of the main body of data better than the mean when the distribution is asymmetric or when there are outliers in the sample.

Mode

The mode is the most common value in the sample. The mode is easily found from a tabulated frequency distribution as the most frequent value. If data have been grouped into classes then the term modal class is used for the class containing most values. The mode provides a rapidly and easily found estimate of sample location and is unaffected by outliers. However, the mode is affected by chance variation in the shape of a sample's distribution and it may lie distant from the obvious centre of the distribution.

Describing the location of qualitative data *– the mode is the only statistic that is suitable for this task. For example, 'the modal (most frequent) eye colour was blue'.*

The mean, median and mode have the same units as the variable under discussion. However, whether these statistics of location have the same or similar values for a given frequency distribution depends on the symmetry and shape of the distribution. If it is near-symmetrical with a single peak, all three will be very similar; if it is skewed or has more than one peak, their values will differ to a greater degree (see Fig. 66.5).

Measuring dispersion

Here, the objective is to quantify the spread of the data about the centre of the distribution. Figure 66.7 indicates how to decide which measure of dispersion to use.

Range

The range is the difference between the largest and smallest data values in the sample (the extremes) and has the same units as the measured variable. The range is easy to determine, but is greatly affected by outliers. Its value

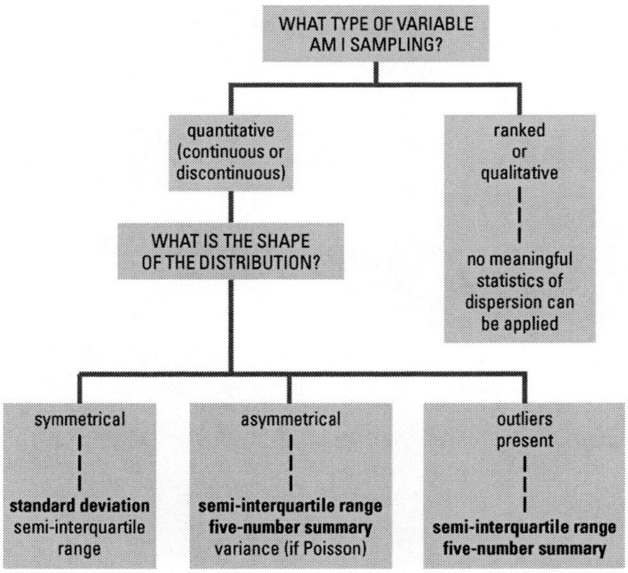

Fig. 66.7 Choosing a statistic for characterising a distribution's dispersion. Statistics written in bold are the preferred option(s). Note that you should match statistics describing dispersion with those you have used to describe location, i.e. standard deviation with mean, semi-interquartile range with median.

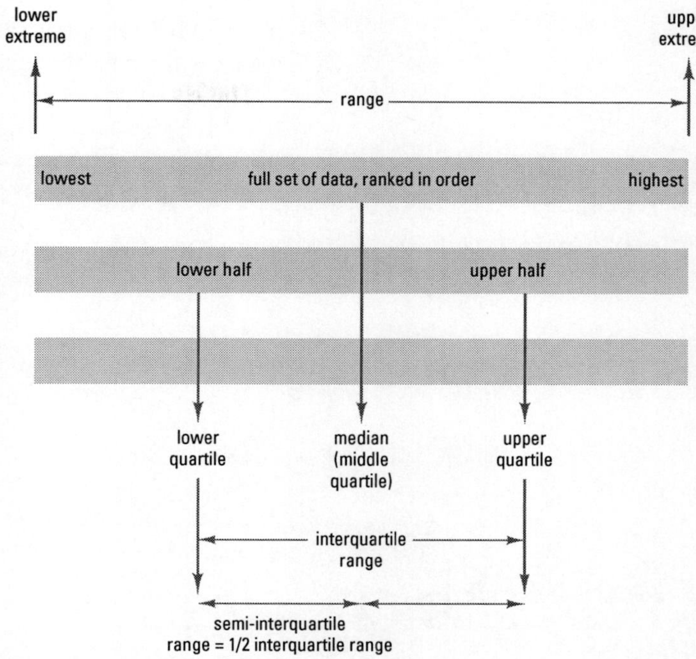

Fig. 66.8 Illustration of median, quartiles, range and semi-interquartile range.

Example In a sample of data with values 3, 7, 15, 8, 5, 10 and 4, the range is 12 (i.e. the difference between the highest value, 15, and the lowest value, 3).

may also depend on sample size: in general, the larger this is, the greater will be the range. These features make the range a poor measure of dispersion for many practical purposes.

Semi-interquartile range

The semi-interquartile range is an appropriate measure of dispersion when a median is the appropriate statistic to describe location. For this, you need to determine the first and third quartiles, i.e. the medians for those data values ranked below and above the median of the whole data set (see Fig. 66.8). To calculate a semi-interquartile range for a data set:

1. Rank the observations in ascending order.
2. Find the values of the first and third quartiles.
3. Subtract the value of the first quartile from the value of the third.
4. Halve this number.

For data grouped in classes, the semi-interquartile range can only be estimated. Another disadvantage is that it takes no account of the shape of the distribution at its edges. This objection can be countered by using the so-called 'five number summary' of a data set, which consists of the three quartiles and the two extreme values; this can be presented on graphs as a box and whisker plot (see Fig. 66.9) and is particularly useful for summarising skewed frequency distributions. The corresponding 'six number summary' includes the sample's size.

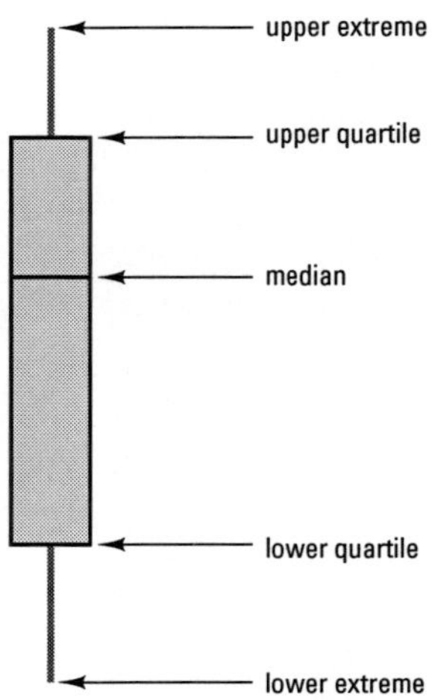

Fig. 66.9 A box and whisker plot, showing the 'five number summary' of a sample as it might be used on a graph.

Variance and standard deviation

For symmetrical frequency distributions, an ideal measure of dispersion would take into account each value's deviation from the mean and provide a measure of the average deviation from the mean. Two such statistics are the sample variance, which is the sum of squares ($\Sigma(Y - \bar{Y})^2$) divided by

$n - 1$ (where n is the sample size), and the sample standard deviation, which is the positive square root of the sample variance.

The variance (s^2) has units which are the square of the original units, while the standard deviation (s, or SD) is expressed in the original units, one reason s is often preferred as a measure of dispersion. Calculating s or s^2 longhand is a tedious job and is best done with the help of a calculator or computer. If you don't have a calculator that calculates s for you, an alternative formula that simplifies calculations is:

$$s = +\sqrt{\frac{\Sigma Y^2 - (\Sigma Y)^2/n}{n-1}} \qquad [66.1]$$

To calculate s using a calculator:

1. Obtain ΣY, square it, divide by n and store in memory.
2. Square Y values, obtain ΣY^2, subtract the memory value from this.
3. Divide this answer by $n - 1$.
4. Take the positive square root of this value.

Using a calculator for statistics *– make sure you understand how to enter individual data values and which keys will give the sample mean (usually shown as $\bar{X}$ or $\bar{x}$) and sample standard deviation (often shown as σ_{n-1}). In general, you should not use the population standard deviation (usually shown as σ_n).*

Take care to retain significant figures, or errors in the final value of s will result. If continuous data have been grouped into classes, the class mid-values or their squares must be multiplied by the appropriate frequencies before summation (see example in Box 66.1). When data values are large, longhand calculations can be simplified by coding the data, e.g. by subtracting a constant from each datum, and decoding when the simplified calculations are complete (see Sokal and Rohlf, 1994).

Coefficient of variation

The coefficient of variation (CoV) is a dimensionless measure of variability relative to location which expresses the sample standard deviation, usually as a percentage of the sample mean, i.e.

$$\text{CoV} = 100s/\bar{Y}\,(\%) \qquad [66.2]$$

Example Consider two methods of bioassay for a toxin in fresh water. For a given standard, Method A gives a mean result of = 50 'response units' with $s = 8$, while Method B gives a mean result of = 160 'response units' with $s = 18$. Which bioassay gives the more reproducible results? The answer can be found by calculating the CoV values, which are 16 and 11.25% respectively. Hence, Method B is the more precise (≡ reproducible), even though the absolute value of s is larger.

This statistic is useful when comparing the relative dispersion of data sets with widely differing means, or where different units have been used for the same or similar quantities.

A useful application of the CoV is to compare different analytical methods or procedures, so that you can decide which involves the least proportional error – create a standard stock solution, then base your comparison on the results from several subsamples analysed by each method. You may find it useful to use the CoV to compare the precision of your own results with those of a manufacturer, e.g. for a pipettor (p. 130). The smaller the CoV, the more precise (repeatable) is the apparatus or technique (note: this does not mean that it is necessarily more *accurate*, see p. 175).

Measuring the precision of the sample mean as an estimate of the true value using the standard error

Most practical exercises are based on a limited number of individual data values (a sample) which are used to make inferences about the population from which they were drawn. For example, the haemoglobin content might be measured in blood samples from 100 adult females and used as an estimate of the adult female haemoglobin content, with the sample mean ($\bar{Y}$) and sample standard deviation (s) providing estimates of the

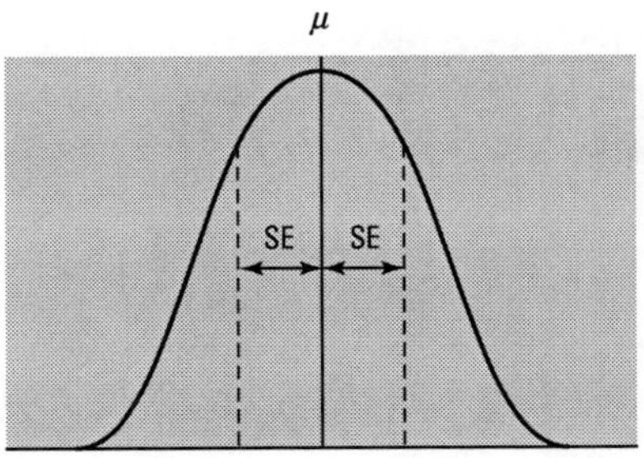

Fig. 66.10 Frequency distribution of sample means around the population mean (μ). Note that SE is equivalent to the standard deviation of the sample means, for sample size $= n$.

Example Summary statistics for the sample mean and standard error for the data shown in Box 66.1 would be quoted as 4.95 ± 0.280 ($n = 21$).

Calculating the extent of skew and kurtosis of a data set *– use the SKEW and KURT functions in Microsoft Excel.*

true values of the underlying population mean (μ) and the population standard deviation (σ). The reliability of the sample mean as an estimate of the true (population) mean can be assessed by calculating a statistic termed the standard error of the sample mean (often abbreviated to standard error or SE), from:

$$\text{SE} = s/\sqrt{n} \qquad [66.3]$$

Strictly, the standard error is an estimate of the dispersion of repeated sample means around the true (population) value: if several samples were taken, each with the same number of data values (n), then their means would cluster around the population mean (μ) with a standard deviation equal to SE, as shown in Figure 66.10. Therefore, the *smaller* the SE, the more reliable the sample mean is likely to be as an estimate of the true value, since the underlying frequency distribution would be more tightly clustered around μ. At a practical level, Eqn [66.3] shows that SE is directly affected by the dispersion of individual data values within the sample, as represented by the sample standard deviation (s). Perhaps more importantly, SE is inversely related to the *square root* of the number of data values (n). Therefore, if you wanted to increase the precision of a sample mean by a factor of 2 (i.e. to reduce SE by half), you would have to increase n by a factor of 2^2 (i.e. fourfold).

Summary descriptive statistics for the sample mean are often quoted as $\bar{Y} \pm \text{SE}\,(n)$, with the SE being given to one significant figure more than the mean. You can use such information to carry out a t-test between two sample means (Box 67.1); the SE is also useful because it allows calculation of confidence limits for the sample mean.

Describing the 'shape' of frequency distributions

Frequency distributions may differ in the following characteristics:

- number of peaks;
- skewness or asymmetry;
- kurtosis or pointedness.

The shape of a frequency distribution of a small sample is affected by chance variation and may not be a fair reflection of the underlying population frequency distribution: check this by comparing repeated samples from the same population or by increasing the sample size. If the original shape were due to random events, it should not appear consistently in repeated samples and should become less obvious as sample size increases.

Genuinely bimodal or polymodal distributions may result from the combination of two or more unimodal distributions, indicating that more than one underlying population is being sampled (Fig. 66.11). An example of a bimodal distribution is the height of adult humans (females and males combined).

A distribution is skewed if it is not symmetrical, a symptom being that the mean, median and mode are not equal (Fig. 66.5). Positive skewness is where the longer 'tail' of the distribution occurs for higher values of the measured variable; negative skewness where the longer tail occurs for lower values. Some biological examples of characteristics distributed in a skewed fashion are volumes of plant protoplasts, insulin levels in human plasma and bacterial colony counts.

Box 66.2 Three examples where simple arithmetic means are inappropriate

Mean	n
6	4
7	7
8	1

1. **If means of samples are themselves meaned, an error can arise if the samples are of different size.** For example, the arithmetic mean of the means in the table shown left is 7, but this does not take account of the different 'reliabilities' of each mean due to their sample sizes. The correct weighted mean is obtained by multiplying each mean by its sample size (n) (a 'weight') and dividing the sum of these values by the total number of observations, i.e. in the case shown, $(24 + 49 + 8)/12 = 6.75$.
2. **When making a mean of ratios (e.g. percentages) for several groups of different sizes, the ratio for the combined total of all the groups is not the mean of the proportions for the individual groups.** For example, if 20 rats from a batch of 50 are male, this implies 40% are male. If 60 rats from a batch of 120 are male, this implies 50% are male. The mean percentage of males $(50 + 40)/2 = 45\%$ is *not* the percentage of males in the two groups combined, because there are $20 + 60 = 80$ males in a total of 170 rats $= 47.1\%$ approx.
3. **If the measurement scale is not linear, arithmetic means may give a false value.** For example, if three media had pH values 6, 7 and 8, the appropriate mean pH is not 7 because the pH scale is logarithmic. The definition of pH is $-\log_{10}[H^+]$, where $[H^+]$ is expressed in mol l^{-1} ('molar'); therefore, to obtain the true mean, convert data into $[H^+]$ values (i.e. put them on a linear scale) by calculating $10^{(-\text{pH value})}$ as shown. Now calculate the mean of these values and convert the answer back into pH units. Thus, the appropriate answer is pH 6.43 rather than 7. Note that a similar procedure is necessary when calculating statistics of dispersion in such cases, so you will find these almost certainly asymmetric about the mean.

pH value	$[H^+]$ (mol l^{-1})
6	1×10^{-6}
7	1×10^{-7}
8	1×10^{-8}
mean	3.7×10^{-7}
$-\log_{10}$ mean	6.43

Mean values of log-transformed data are often termed geometric means – they are sometimes used in microbiology and in cell culture studies, where log-transformed values for cell density counts are averaged and plotted (p. 233), rather than using the raw data values. The use of geometric means in such circumstances serves to reduce the effects of outliers on the mean.

Kurtosis is the name given to the 'pointedness' of a frequency distribution. A platykurtic frequency distribution is one with a flattened peak, while a leptokurtic frequency distribution is one with a pointed peak (Fig. 66.12). While descriptive terms can be used, based on visual observation of the shape and direction of skew, the degree of skewness

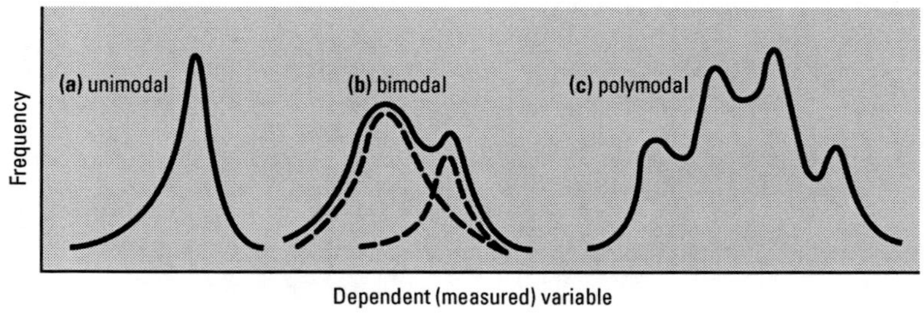

Fig. 66.11 Frequency distributions with different numbers of peaks. A unimodal distribution (a) may be symmetrical or asymmetrical. The dotted lines in (b) indicate how a bimodal distribution could arise from a combination of two underlying unimodal distributions. Note here how the term 'bimodal' is applied to any distribution with two major peaks – their frequencies do not have to be exactly the same.

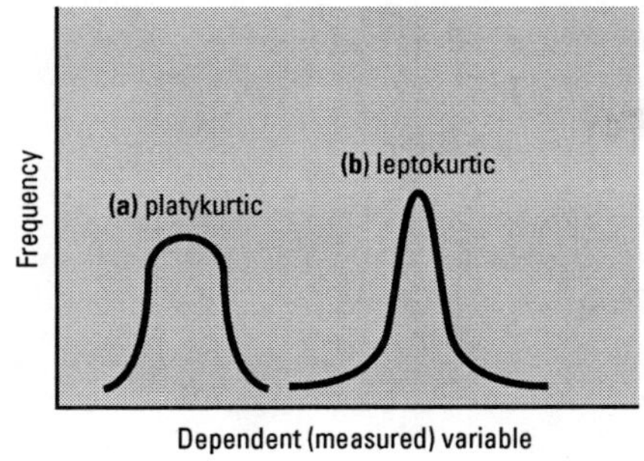

Fig. 66.12 Examples of the two types of kurtosis.

and kurtosis can be quantified and statistical tests exist to test the 'significance' of observed values (see Sokal and Rohlf, 1994), but the calculations required are complex and best done with the aid of a computer.

Using computers to calculate descriptive statistics

There are many specialist statistical packages (e.g. SPSS) that can be used to simplify the process of calculation of statistics. Note that correct interpretation of the output requires an understanding of the terminology used and the underlying process of calculation, and this may best be obtained by working through one or more examples by hand before using these tools. Spreadsheets offer increasingly sophisticated statistical analysis functions, some examples of which are provided in Box 66.3 for Microsoft Excel.

Box 66.3 How to use a spreadsheet (Microsoft Excel) to calculate descriptive statistics

Method 1: Using spreadsheet functions to generate the required statistics. Suppose you had obtained the following set of data, stored within an array (block of columns and rows) of cells (A2:L6) within a spreadsheet:

	A	B	C	D	E	F	G	H	I	J	K	L
1	**My data set**											
2	4	4	3	3	5	4	3	7	7	3	5	3
3	6	2	9	7	3	4	5	6	6	9	4	8
4	5	3	2	5	4	5	7	2	8	3	6	3
5	11	3	5	2	4	3	7	8	4	4	4	3
6	3	6	8	5	6	4	3	4	3	6	10	5

The following functions could be used to extract descriptive statistics from this data set:

Descriptive statistic	Example of use of function[a,b]	Result for the above data set
Sample size *n*	=COUNT((A2:L6)	60
Mean	=AVERAGE(A2:L6)[c]	4.9
Median	=MEDIAN(A2:L6)	4.0
Mode	=MODE(A2:L6)	3
Upper quartile	=QUARTILE(A2:L6,3)[d]	6.0
Lower quartile	=QUARTILE(A2:L6,1)	3.0
Semi-interquartile range	=QUARTILE(A2:L6,3)-QUARTILE(A2:L6,1)	3.0
Upper extreme	=QUARTILE(A2:L6,4) *or* =MAX(A2:L6)	11
Lower extreme	=QUARTILE(A2:L6,0) *or* =MIN(A2:L6)	2
Range	=MAX(A2:L6)- MIN(A2:L6)[e]	9.0
Variance	=VAR(A2:L6)	4.464
Standard deviation	=STDEV(A2:L6)	2.113
Standard Error	=STDEV(A2:L6)/(SQRT(COUNT(A2:L6)))[f]	0.273
Coefficient of variation	=100*STDEV(A2:L6)/AVERAGE(A2:L6)	43.12%

Notes:

[a] Typically, in an appropriate cell, you would *Insert* > *Function* > COUNT, then select the input range and press return.

[b] Other descriptive statistics can be calculated – these mirror those shown in Box 66.1, but for this specific data set.

[c] There is no 'MEAN' function in Microsoft Excel.

[d] The first argument within the brackets relates to the array of data, the second relates to the quartile required (consult *Help* feature for further information).

[e] There is no direct 'RANGE' function in Microsoft Excel.

[f] There is no direct 'STANDARD ERROR' function in Microsoft Excel. The SQRT function returns a square root and the COUNT function determines the number of filled data cells in the array.

(continued)

Box 66.3 (continued)

Method 2: using the *Tools > Data Analysis* option. This can automatically generate a table of descriptive statistics for the data array selected, although the data must be presented as a single row or column. This option might need to be installed for your network or personal computer before it is available to you (in the latter case use the *Add Ins > Analysis ToolPak* option from the *Tools* menu – consult the *Help* feature for details). Having entered or rearranged your data into a row or column, the steps involved are as follows:

1. Select *Tools > Data Analysis.*
2. From the *Data Analysis box,* select *Descriptive Statistics.*
3. Input your data location into the *Input Range* (left click and hold down to highlight the column of data).
4. From the menu options, select *Summary Statistics* and *Confidence Level for Mean: 95%.*
5. When you click *OK,* you should get a new worksheet, with descriptive statistics and confidence limits shown. Alternatively, at step 3, you can select an area of your current worksheet as a data output range (select an area away from any existing content as these cells would otherwise be overwritten by the descriptive statistics output table).
6. Change the format of the cells to show each number to an appropriate number of decimal places. You may also wish to make the columns wider so you can read their content.
7. For the data set shown above, the final output table should look as shown below:

Descriptive statistics for a data set

Column1[a,b]	
Mean	4.9
Standard error	0.27
Median	4.0
Mode	3
Standard deviation	2.113
Sample variance	4.464
Kurtosis	0.22
Skewness	0.86
Range	9.00
Minimum	2.0
Maximum	11.0
Sum	294
Count	60
Confidence level (95.0%)	0.55

Notes:
[a] These descriptive statistics are specified (and are automatically presented in this order) – any others required can be generated using Method 1.
[b] A more descriptive heading can be added if desired – this is the default.

Text references and sources for further study

Dytham, C. (2003) *Choosing and Using Statistics: A Biologists' Guide*, 2nd edn. Blackwell, Oxford.

Heath, D. (1995) *An Introduction to Experimental Design and Statistics for Biology*. UCL Press Ltd, London.

Maber, J. (1999) *Data Analysis for Biomolecular Sciences.* Longman, Harlow.

Schmuller, J. (2005) *Statistical Analysis with Excel for Dummies*. Wiley, Hoboken.

Sokal, R.R. and Rohlf, F.J. (1994) *Biometry*, 3rd edn. W.H. Freeman and Co., San Francisco.

Wardlaw, A.C. (2000) *Practical Statistics for Experimental Biologists*, 2nd edn. Wiley, New York.

Study exercises

66.1 Practise calculating descriptive statistics. Using the data set given in Box 66.3, calculate the following statistics:

(a) range
(b) variance
(c) standard deviation
(d) coefficient of variation
(e) standard error.

Answers (b) to (e) should be given to three significant figures.

66.2 Calculate and interpret standard errors. Two samples, A and B, gave the following descriptive statistics (measured in the same units): Sample A, mean = 16.2, standard deviation = 12.7, number of data values = 12; Sample B, mean = 13.2, standard deviation 14.4, number of data values = 20. Which has the lower standard error in absolute terms and in proportion to the sample mean? (Express answers to three significant figures.)

66.3 Compute a mean value correctly. A researcher finds that the mean size of larvae in three replicate samples designated A, B and C is 3.0, 2.5 and 2.0 mm respectively. He computes the mean larval size as 2.5 mm, but forgets that the sample sizes were 24, 37 and 6 respectively. What is the true mean size of the larvae? (Answer to three significant figures.)

67 Choosing and using statistical tests

This chapter outlines the philosophy of hypothesis-testing statistics, indicates the steps to be taken when choosing a test, and discusses features and assumptions of some important tests. For details of the mechanics of tests, consult appropriate texts (e.g. Sokal and Rohlf, 1994; Heath, 1995; Wardlaw, 2000). Most tests are now available in statistical packages for computers (see p. 84) and many in spreadsheets (Chapter 12).

To carry out a statistical test:

1. Decide what it is you wish to test (create a null hypothesis and its alternative).
2. Determine whether your data fit a standard distribution pattern.
3. Select a test and apply it to your data.

Setting up a null hypothesis

Hypothesis-testing statistics are used to compare the properties of samples either with other samples or with some theory about them. For instance, you may be interested in whether two samples can be regarded as having different means, whether the counts of an organism in different locations can be regarded as randomly distributed, or whether characteristic A of an organism is linearly related to characteristic B.

KEY POINT You can't use statistics to prove any hypothesis, but they can be used to assess how likely it is to be wrong.

Statistical testing operates in what at first seems a rather perverse manner. Suppose you think a treatment has an effect. The theory you actually test is that it has no effect; the test tells you how improbable your data would be if this theory were true. This 'no effect' theory is the null hypothesis (NH). If your data are very improbable under the NH, then you may suppose it to be wrong, and this would support your original idea (the 'alternative hypothesis'). The concept can be illustrated by an example. Suppose two groups of subjects were treated in different ways, and you observed a difference in the mean value of the measured variable for the two groups. Can this be regarded as a 'true' difference? As Figure 67.1 shows, it could have arisen in two ways:

- Because of the way the subjects were allocated to treatments, i.e. all the subjects liable to have high values might, by chance, have been assigned to one group and those with low values to the other (Fig. 67.1a).
- Because of a genuine effect of the treatments, i.e. each group came from a distinct frequency distribution (Fig. 67.1b).

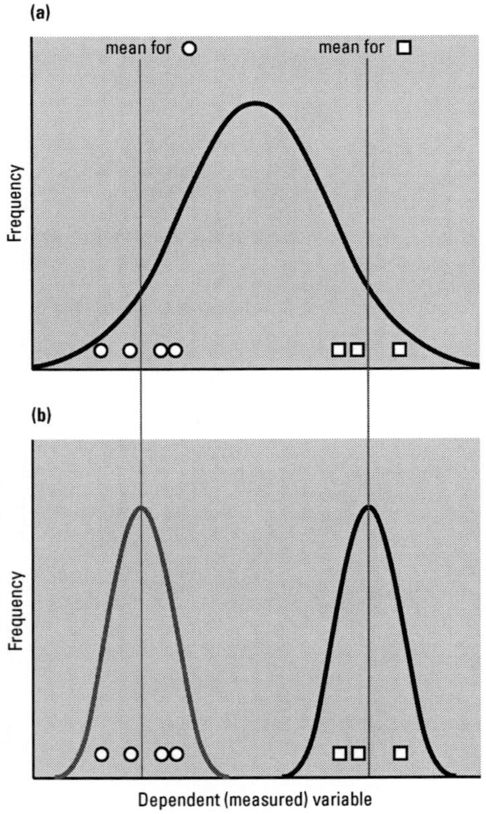

Fig. 67.1 Two explanations for the difference between two means. In case (a) the two samples happen by chance to have come from opposite ends of the same frequency distribution, i.e. there is no true difference between the samples. In case (b) the two samples come from different frequency distributions, i.e. there is a true difference between the samples. In both cases, the means of the two samples are the same.

A statistical test will indicate the probabilities of these options. The NH states that the two groups come from the same population (i.e. the treatment effects are negligible in the context of random variation). To test this, you calculate a test statistic from the data, and compare it with tabulated critical values giving the probability of obtaining the observed or a more extreme result by chance (see Boxes 67.1 and 67.2). This probability is sometimes called the significance of the test.

Note that you must take into account the degrees of freedom (d.f.) when looking up critical values of most test statistics. The d.f. is related to the size(s) of the samples studied; formulae for calculating it depend on the test being used. Biologists normally use two-tailed tests, i.e. we have no expectation beforehand that the treatment will have a positive or negative effect compared to the control (in a one-tailed test we expect one particular treatment to be bigger than the other). Be sure to use critical values for the correct type of test.

Definition

Modulus – the absolute value of a number, e.g. modulus −3.385 = 3.385.

By convention, the critical probability for rejecting the NH is 5% (i.e. $P = 0.05$). This means we reject the NH if the observed result would have come up by chance a maximum of one time in twenty. If the modulus of the test statistic is less than or equal to the tabulated critical value for $P = 0.05$, then we accept the NH and the result is said to be 'not significant' (NS for short). If the modulus of the test statistic is greater than the tabulated value for $P = 0.05$, then we reject the NH in favour of the alternative hypothesis that the treatments had different effects and the result is 'statistically significant'.

***Quoting significance** – the convention for quoting significance levels in text, tables and figures is as follows:*
$P > 0.05$ = 'not significant' (or NS)
*$P \leqslant 0.05$ = 'significant' (or *)*
*$P \leqslant 0.01$ = 'highly significant' (or **)*
*$P \leqslant 0.001$ = 'very highly significant' (or ***)*
Thus, you might refer to a difference in means as being 'highly significant ($P \leqslant 0.01$)'. For this reason, the word 'significant' in its everyday meaning of 'important' or 'notable' should be used with care in scientific writing.

Two types of error are possible when making a conclusion on the basis of a statistical test. The first occurs if you reject the NH when it is true and the second if you accept the NH when it is false. To limit the chance of the first type of error, choose a lower probability, e.g. $P = 0.01$, but note that the critical value of the test statistic increases when you do this and results in the probability of the second error increasing. The conventional significance levels given in statistical tables (usually 0.05, 0.01, 0.001) are arbitrary. Increasing use of statistical computer programs now allows the actual probability of obtaining the calculated value of the test statistic to be quoted (e.g. $P = 0.037$).

Note that if the NH is rejected, this does not tell you which of many alternative explanations are true. Also, it is important to distinguish between statistical significance and biological relevance: identifying a statistically significant difference between two samples doesn't mean that this will carry any biological importance.

Comparing data with parametric distributions

A parametric test is one that makes particular assumptions about the mathematical nature of the population distribution from which the samples were taken. If these assumptions are not true, then the test is obviously invalid, even though it might give the answer we expect. A non-parametric test does not assume that the data fit a particular pattern, but it may assume some things about the distributions. Used in appropriate circumstances, parametric tests are better able to distinguish between true but marginal differences between samples than their non-parametric equivalents (i.e. they have greater 'power').

***Choosing between parametric and non-parametric tests** – always plot your data graphically when determining whether they are suitable for parametric tests as this may save a lot of unnecessary effort later.*

The distribution pattern of a set of data values may be biologically relevant, but it is also of practical importance because it defines the type of statistical tests that can be used. The properties of the main distribution types found in biology are given below, with both rules-of-thumb and more rigorous tests for deciding whether data fit these distributions.

Binomial distributions

These apply to samples of any size from populations when data values occur independently in only two mutually exclusive classes (e.g. type A or type B). They describe the probability of finding the different possible

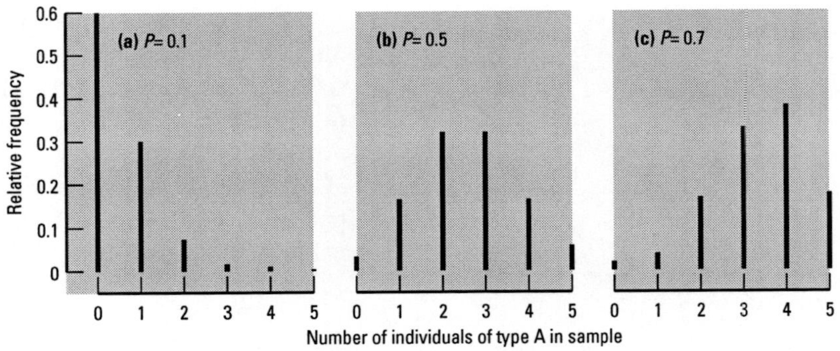

Fig. 67.2 Examples of binomial frequency distributions with different probabilities. The distributions show the expected frequency of obtaining *n* individuals of type A in a sample of 5. Here *P* is the probability of an individual being type A rather than type B.

***Quantifying skew** – the Microsoft Excel SKEW function can be used to assess the extent of skewness in a data set.*

combinations of the attribute for a specified sample size k (e.g. out of 10 specimens, what is the chance of 8 being type A). If p is the probability of the attribute being of type A and q the probability of it being type B, then the expected mean sample number of type A is kp and the standard deviation is $\sqrt{kpq}$. Expected frequencies can be calculated using mathematical expressions (see Sokal and Rohlf, 1994). Examples of the shapes of some binomial distributions are shown in Figure 67.2. Note that they are symmetrical in shape for the special case $p = q = 0.5$ and the greater the disparity between p and q, the more skewed the distribution.

Some biological examples of data likely to be distributed in binomial fashion are: possession of two alleles for seed coat morphology (e.g. smooth and wrinkly); whether an organism is infected with a microbe or not; whether an animal is male or female. Binomial distributions are particularly useful for predicting gene segregation in Mendelian genetics and can be used for testing whether combinations of events have occurred more frequently than predicted (e.g. more siblings being of the same sex than expected). To establish whether a set of data is distributed in binomial fashion: calculate expected frequencies from probability values obtained from theory or observation, then test against observed frequencies using a χ^2-test or a G-test (see Wardlaw, 2000).

Poisson distributions

***Tendency towards the normal distribution** – under certain conditions, binomial and Poisson distributions can be treated as normally distributed:*

- *where samples from a binomial distribution are large (i.e. > 15) and p and q are close to 0.5;*
- *for Poisson distributions, if the number of counts recorded in each outcome is greater than about 15.*

These apply to discrete characteristics which can assume low whole number values, such as counts of events occurring in area, volume or time. The events should be 'rare' in that the mean number observed should be a small proportion of the total that could possibly be found. Also, finding one count should not influence the probability of finding another. The shape of Poisson distributions is described by only one parameter, the mean number of events observed, and has the special characteristic that the variance is equal to the mean. The shape has a pronounced positive skewness at low mean counts, but becomes more and more symmetrical as the mean number of counts increases (Fig. 67.3).

Some examples of characteristics distributed in a Poisson fashion are: number of plants in a quadrat; number of microbes per unit volume of medium; number of animals parasitised per unit time; number of radioactive disintegrations per unit time. One of the main uses for the Poisson distribution is to quantify errors in count data such as estimates of

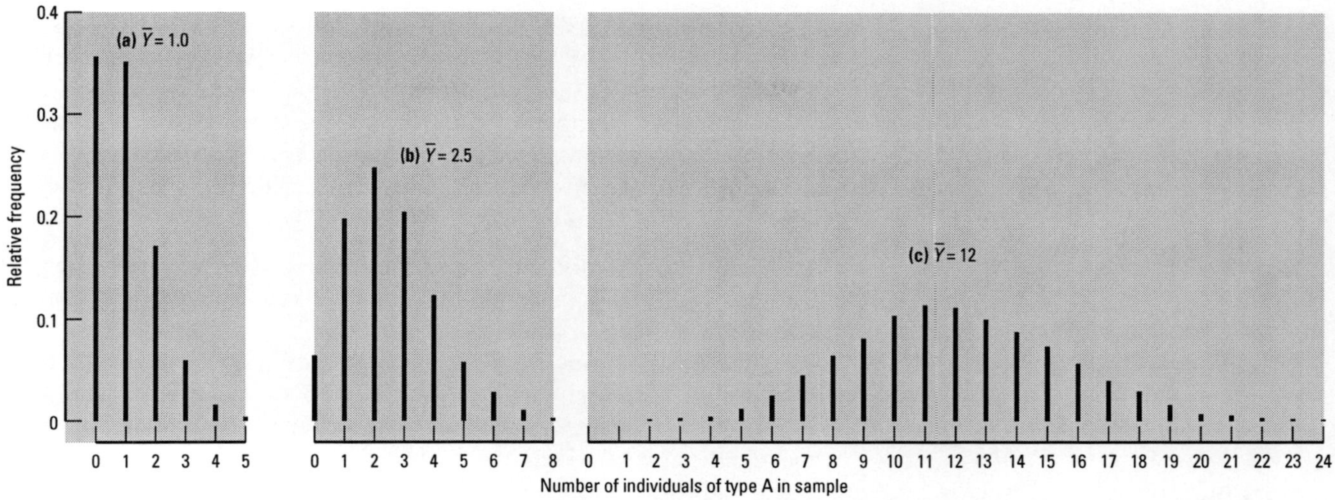

Fig. 67.3 Examples of Poisson frequency distributions differing in mean. The distributions are shown as line charts because the independent variable (events per sample) is discrete.

cell densities in dilute suspensions (see p. 236). To decide whether data are Poisson distributed:

> **Definition**
>
> **Coefficient of dispersion** $= s^2/\bar{Y}$. This is an alternative measure of dispersion to the coefficient of variation.

- Use the rule of thumb that if the coefficient of dispersion ≈ 1, the distribution is likely to be Poisson.
- Calculate 'expected' frequencies from the equation for the Poisson distribution and compare with actual values using a χ^2-test or a *G*-test.

It is sometimes of interest to show that data are *not* distributed in a Poisson fashion, e.g. the distribution of parasite larvae in hosts. If $s^2/\bar{Y} > 1$, the data are 'clumped' and occur together more than would be expected by chance; if $s^2/\bar{Y} < 1$, the data are 'repulsed' and occur together less frequently than would be expected by chance.

Normal distributions (Gaussian distributions)

These occur when random events act to produce variability in a continuous characteristic (quantitative variable). This situation occurs frequently in biology, so normal distributions are very useful and much used. The bell-like shape of normal distributions is specified by the population mean and standard deviation (Fig. 67.4): it is symmetrical and

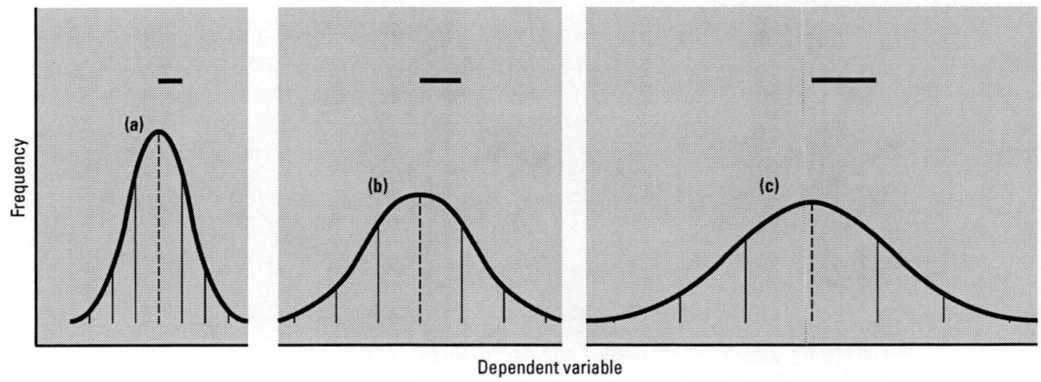

Fig. 67.4 Examples of normal frequency distributions differing in mean and standard deviation. The horizontal bars represent population standard deviations for the curves, increasing from (a) to (c). Vertical dashed lines are population means, while vertical solid lines show positions of values ± 1, 2 and 3 standard deviations from the means.

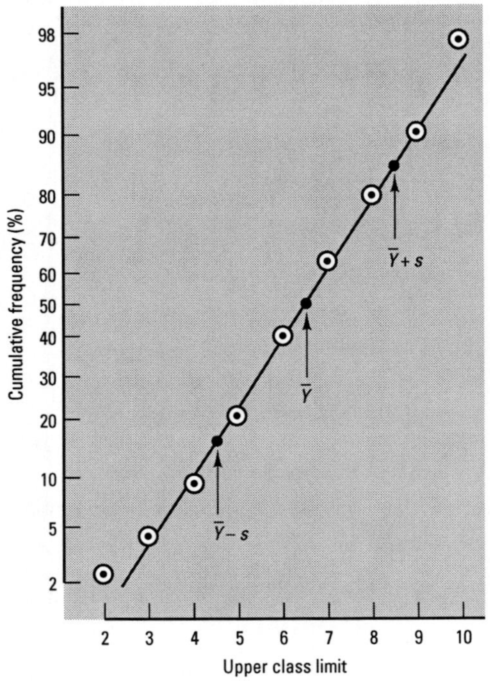

Fig. 67.5 Example of a normal probability plot. The plotted points are from a small data set where the mean $\bar{Y} = 6.93$ and the standard deviation $s = 1.895$. Note that values corresponding to 0% and 100% cumulative frequency cannot be used. The straight line is that predi-cted for a normal distribution with $\bar{Y} = 6.93$ and $s = 1.895$. This is plotted by calculating the expected positions of points for $\bar{Y} \pm s$. Since 68.3% of the distribution falls within these bounds, the relevant points on the cumulative frequency scale are $50 \pm 34.15\%$; thus this line was drawn using the points (4.495, 15.85) and (8.285, 84.15) as indicated on the plot.

configured such that 68.27% of the data will lie within ±1 standard deviation of the mean, 95.45% within ±2 standard deviations of the mean, and 99.73% within ±3 standard deviations of the mean.

Some biological examples of data likely to be distributed in a normal fashion are: fresh weight of plants of the same age; linear dimensions of bacterial cells; height of either adult female or male humans. To check whether data come from a normal distribution, you can:

- Use the rule of thumb that the distribution should be symmetrical and that nearly all the data should fall within $\pm 3s$ of the mean and about two-thirds within $\pm 1s$ of the mean.
- Plot the distribution on normal probability graph paper. If the distribution is normal, the data will tend to follow a straight line (see Fig. 67.5). Deviations from linearity reveal skewness and/or kurtosis, the significance of which can be tested statistically (see Sokal and Rohlf, 1994).
- Use a suitable statistical computer program to generate predicted normal curves from the $\bar{Y}$ and s values of your sample(s). These can be compared visually with the actual distribution of data and can be used to give 'expected' values for a χ^2-test or a G-test.

The wide availability of tests based on the normal distribution and their relative simplicity means you may wish to transform your data to make them more like a normal distribution. Table 67.1 provides transformations that can be applied. The transformed data should be tested for normality as described above before proceeding – don't forget that you may need to check that transformed variances are homogeneous for certain tests (see below).

A very important theorem in statistics, the Central Limit Theorem, states that as sample size increases, the distribution of a series of means from any frequency distribution will become normally distributed. This fact can be used to devise an experimental or sampling strategy that ensures that data are normally distributed, i.e. using means of samples as if they were primary data.

Choosing a suitable statistical test

Comparing location (e.g. means)

If you can assume that your data are normally distributed, the main test for comparing two means from independent samples is Student's t-test

Table 67.1 Suggested transformations altering different types of frequency distribution to the normal type. To use, modify data by the formula shown; then examine effects with the tests described.

Type of data; distribution suspected	Suggested transformation(s)
Proportions (including percentages); binomial	arcsine $\sqrt{x}$ (also called the angular transformation)
Scores; Poisson	$\sqrt{x}$ or $\sqrt{(x+1/2)}$ if zero values present
Measurements; negatively skewed	x^2, x^3, x^4, etc. (in order of increasing strength)
Measurements; positively skewed	$1/\sqrt{x}$, $\sqrt{x}$, $\ln x$, $1/x$ (in order of increasing strength)

Definition

Homogeneous variance – uniform (but not necessarily identical) variance of the dependent variable across the range of the independent variable. The term homoscedastic is also used in this sense. The opposite of homogeneous is heterogeneous (= heteroscedastic).

Understanding 'degrees of freedom' – *this depends on the number of values in the data set analysed, and the method of calculation depends on the statistical test being used. It relates to the number of observations that are free to vary before the remaining quantities for a data set can be determined.*

Checking the assumptions of a test – *always acquaint yourself with the assumptions of a test. If necessary, test them before using the test.*

(see Boxes 67.1 and 67.2, and Table 67.2). This assumes that the variances of the data sets are homogeneous. Tests based on the *t*-distribution are also available for comparing means of paired data or for comparing a sample mean with a chosen value.

When comparing means of two or more samples, analysis of variance (ANOVA) is a very useful technique. This method also assumes data are normally distributed and that the variances of the samples are homogeneous. The samples must also be independent (e.g. not subsamples). The test statistic calculated is denoted *F* and it has two different degrees of freedom related to the number of means tested and the pooled number of replicates per mean. The nested types of ANOVA are useful for letting you know the relative importance of different sources of variability in your data. Two-way and multi-way ANOVAs are useful for studying interactions between treatments.

For data satisfying the ANOVA requirements, the least significant difference (LSD) is useful for making planned comparisons among several means (see Sokal and Rohlf, 1994). Any two means that differ by more than the LSD will be significantly different. The LSD is useful for showing on graphs.

The chief non-parametric tests for comparing the locations of two samples are the Mann–Whitney *U*-test and the Kolmogorov–Smirnov test. The former assumes that the frequency distributions of the samples are similar, whereas the latter makes no such assumption. In both cases the sample's size must be $\geqslant 4$ and for the Kolmogorov–Smirnov test the samples must have equal sizes. In the Kolmogorov–Smirnov test, significant differences found with the test could be due to differences in location or shape of the distribution, or both.

Suitable non-parametric comparisons of location for paired data (sample size $\geqslant 6$) include Wilcoxon's signed rank test, which is used for

Table 67.2 Critical values of Student's *t* statistic (for two-tailed tests). Reject the null hypothesis at probability *P* if your calculated *t* value equals or exceeds the value shown for the appropriate degrees of freedom $= (n_1 - 1) + (n_2 - 1)$

Degrees of freedom	Critical values for $P = 0.05$	Critical values for $P = 0.01$	Critical values for $P = 0.001$
1	12.71	63.66	636.62
2	4.30	9.92	31.60
3	3.18	5.84	12.94
4	2.78	4.60	8.61
5	2.57	4.03	6.86
6	2.45	3.71	5.96
7	2.36	3.50	5.40
8	2.31	3.36	5.04
9	2.26	3.25	4.78
10	2.23	3.17	4.59
12	2.18	3.06	4.32
14	2.14	2.98	4.14
16	2.12	2.92	4.02
20	2.09	2.85	3.85
25	2.06	2.79	3.72
30	2.04	2.75	3.65
40	2.02	2.70	3.55
60	2.00	2.66	3.46
120	1.98	2.62	3.37
∞	1.96	2.58	3.29

Box 67.1 How to carry out a *t*-test

The *t*-test was devised by a statistician who used the pen-name 'Student', so you may see it referred to as Student's *t*-test. It is used when you wish to decide whether two samples come from the same population or from different ones (Fig. 67.1). The samples might have been obtained by observation, or by applying two different treatments to an originally homogeneous population (Chapter 30).

The null hypothesis (NH) is that the two groups can be represented as samples from the same overlying population (Fig. 67.1a). If, as a result of the test, you accept this hypothesis, you can say that there is no significant difference between the group means.

The alternative hypothesis is that the two groups come from different populations (Fig. 67.1b). By rejecting the NH as a result of the test, you can accept the alternative hypothesis and say that there is a significant difference between the sample means, or, if an experiment were carried out, that the two treatments affected the samples differently.

How can you decide between these two hypotheses? On the basis of certain assumptions (see below), and some relatively simple calculations, you can work out the probability that the samples came from the same population. If this probability is very low, then you can reasonably reject the NH in favour of the alternative hypothesis, and if it is high, you will accept the NH.

To find out the probability that the observed difference between sample means arose by chance, you must first calculate a '*t* value' for the two samples in question. Some computer programs (e.g. Minitab) provide this probability as part of the output, otherwise you can look up statistical tables (e.g. Table 67.2). These tables show 'critical values' – the borders between probability levels. If your value of *t* equals or exceeds the critical value for probability *P*, you can reject the NH at this probability ('level of significance').

Note that:

- for a given difference in the means of the two samples, the value of *t* will get larger the smaller the scatter within each data set; and
- for a given scatter of the data, the value of *t* will get larger the greater the difference between the means.

So, at what probability should you reject the NH? Normally, the threshold is arbitrarily set at 5% – you quite often see descriptions like 'the sample means were significantly different ($P < 0.05$)'. At this 'significance level' there is still up to a 5% chance of the *t* value arising by chance, so about 1 in 20 times, on average, the conclusion will be wrong. If *P* turns out to be lower, then this kind of error is much less likely.

Tabulated probability levels are generally given for 5%, 1% and 0.1% significance levels (see Table 67.2). Note that this table is designed for 'two-tailed' tests, i.e. where the treatment or sampling strategy could have resulted in either an increase or a decrease in the measured values. These are the most likely situations you will deal with in biology.

Examine Table 67.2 and note the following:

- The larger the size of the samples (i.e. the greater the 'degrees of freedom') the smaller *t* needs to be to exceed the critical value at a given significance level.
- The lower the probability the greater *t* needs to be to exceed the critical value.

The mechanics of the test

A calculator that can work out means and standard deviations is helpful.

1. **Work out the sample means $\bar{Y}_1$ and $\bar{Y}_2$ and calculate the difference between them.**
2. **Work out the sample standard deviations s_1 and s_2.** (NB if your calculator offers a choice, choose the '$n-1$' option for calculating *s*).
3. **Work out the sample standard errors $SE_1 = s_1/\sqrt{n_1}$ and $SE_2 = s_2/\sqrt{n_2}$; now square each, add the squares together, then take the positive square root of this** (n_1 and n_2 are the respective sample sizes, which may, or may not, not be equal).
4. **Calculate *t*** from the formula:

$$t = \frac{\bar{Y}_1 - \bar{Y}_2}{\sqrt{((SE_1)^2) + (SE_2)^2)}} \qquad [67.1]$$

The value of *t* can be negative or positive, depending on the values of the means; this does not matter and you should compare the modulus (absolute value) of *t* with the values in tables.

5. **Work out the degrees of freedom** $= (n_1 - 1) + (n_2 - 1)$.
6. **Compare the *t* value with the appropriate critical value (see e.g. Table 67.2) and decide on the significance of your finding.**

Box 67.2 provides a worked example – use this to check that you understand the above procedures.

Assumptions that must be met before using the test

The most important assumptions are:

- The two samples are independent and randomly drawn (or if not, drawn in a way that does not create bias). The test assumes that the samples are quite large.
- The underlying distribution of each sample is normal. This can be tested with a special statistical test, but a rule of thumb is that a frequency distribution of the data should be (a) symmetrical about the mean and (b) nearly all of the data should be within 3 standard deviations of the mean and about two-thirds within 1 standard deviation of the mean.
- The two samples should have uniform variances. This again can be tested (by an *F*-test), but may be obvious from inspection of the two standard deviations.

Box 67.2 Worked example of a *t*-test

Suppose the following data were obtained in an experiment (the units are not relevant):

Control: 6.6, 5.5, 6.8, 5.8, 6.1, 5.9
Treatment: 6.3, 7.2, 6.5, 7.1, 7.5, 7.3

Using the steps outlined in Box 67.1, the following values are obtained (denoting control with subscript 1, treatment with subscript 2):

1. $\bar{Y}_1 = 6.1167$; $\bar{Y}_2 = 6.9833$: difference between means $= \bar{Y}_1 - \bar{Y}_2 = -0.8666$
2. $s_1 = 0.495\,65$; $s_2 = 0.475\,04$
3. $SE_1 = 0.495\,65/2.449\,49 = 0.202\,348$
   $SE_2 = 0.475\,04/2.449\,49 = 0.193\,934$
4. $t = \dfrac{-0.8666}{\sqrt{(0.202\,348^2 + 0.193\,934^2)}} = \dfrac{-0.8666}{0.280\,277} = -3.09$
5. d.f. $= (5 + 5) = 10$
6. Looking at Table 67.2, we see that the modulus of this t value exceeds the tabulated value for $P = 0.05$ at 10 degrees of freedom ($= 2.23$). We therefore reject the NH, and conclude that the means are different at the 5% level of significance. If the modulus of t had been $\leqslant$ 2.23, we would have accepted the NH. If modulus of t had been > 3.17, we could have concluded that the means are different at the 1% level of significance.

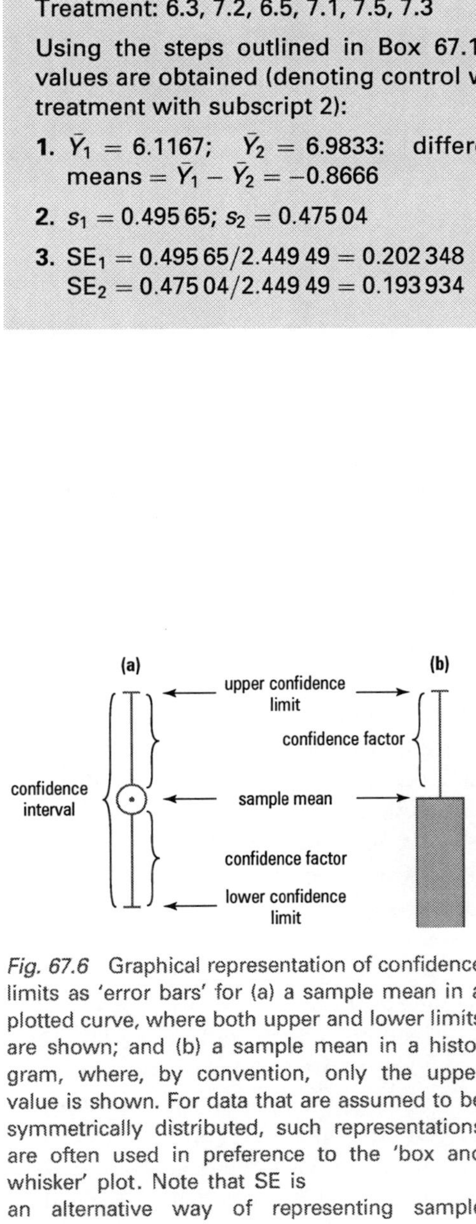

Fig. 67.6 Graphical representation of confidence limits as 'error bars' for (a) a sample mean in a plotted curve, where both upper and lower limits are shown; and (b) a sample mean in a histogram, where, by convention, only the upper value is shown. For data that are assumed to be symmetrically distributed, such representations are often used in preference to the 'box and whisker' plot. Note that SE is an alternative way of representing sample imprecision/error (e.g. Fig. 63.1).

***Confidence limits for statistics other than the mean** – consult an advanced statistical text (e.g. Sokal and Rohlf, 1994) if you wish to indicate the reliability of estimates of e.g. population variances.*

quantitative data and assumes that the distributions have similar shape. Dixon and Mood's sign test can be used for paired data scores where one variable is recorded as 'greater than' or 'better than' the other.

Non-parametric comparisons of location for three or more samples include the Kruskal–Wallis H-test. Here, the number of samples is without limit and they can be unequal in size, but again the underlying distributions are assumed to be similar. The Friedman S-test operates with a maximum of five samples and data must conform to a randomised block design. The underlying distributions of the samples are assumed to be similar.

Comparing dispersions (e.g. variances)

If you wish to compare the variances of two sets of data that are normally distributed, use the F-test. For comparing more than two samples, it may be sufficient to use the F_{max}-test, on the highest and lowest variances. The Scheffé–Box (log-ANOVA) test is recommended for testing the significance of differences between several variances. Non-parametric tests exist but are not widely available: you may need to transform the data and use a test based on the normal distribution.

Determining whether frequency observations fit theoretical expectation

The χ^2-test (Box 58.2) is useful for tests of 'goodness of fit', e.g. comparing expected and observed progeny frequencies in genetical experiments or comparing observed frequency distributions with some theoretical function. One limitation is that simple formulae for calculating χ^2 assume that no expected number is less than 5. The G-test ($2I$ test) is used in similar circumstances.

Comparing proportion data

When comparing proportions between two small groups (e.g. whether 3/10 is significantly different from 5/10), you can use probability tables such as those of Finney *et al.* (1963) or calculate probabilities from formulae; however, this can be tedious for large sample sizes. Certain proportions can be transformed so that their distribution becomes normal.

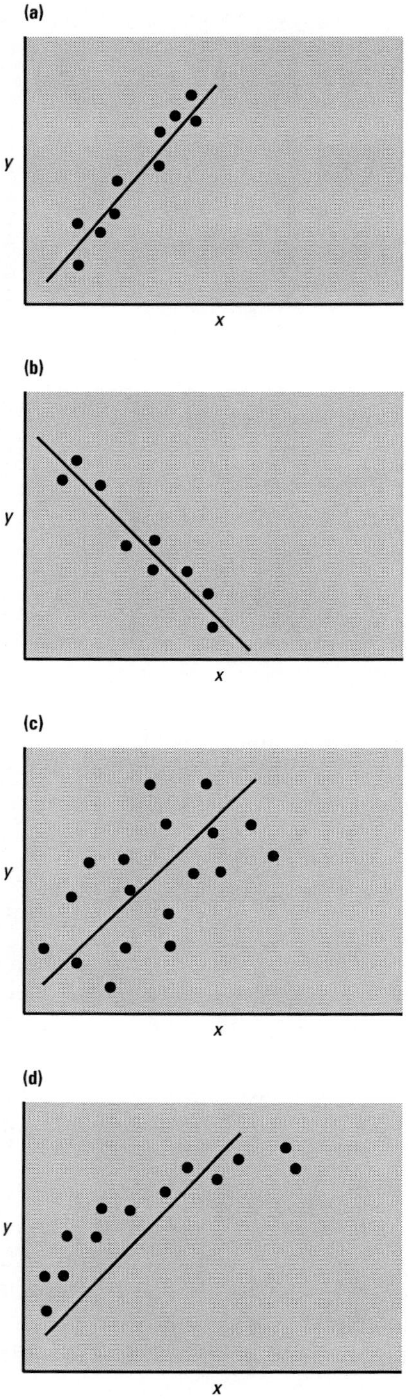

Fig. 67.7 Examples of correlation. The linear regression line is shown. In (a) and (b), the correlation between x and y is good: for (a) there is a positive correlation and the correlation coefficient, r, would be close to 1; for (b) there is a negative correlation and the correlation coefficient would be close to -1. In (c) there is a weak positive correlation and r would be close to 0. In (d) the correlation coefficient may be quite large, but the choice of linear regression is clearly inappropriate.

Placing confidence limits on an estimate of a population parameter

On many occasions, a sample statistic is used to provide an estimate of a population parameter, and it is often useful to indicate the reliability of such an estimate. This can be done by putting confidence limits on the sample statistic, i.e. by specifying an interval around the statistic within which you are confident that the true value (the population parameter) is likely to fall, at a specified level of probability. The most common application is to place confidence limits on the mean of a sample taken from a population of normally distributed data values. In practice, you determine a confidence factor for a particular level of probability which is added to and subtracted from the sample mean ($\bar{Y}$) to give the upper confidence limit and lower confidence limit respectively. These are calculated as:

$$\bar{Y} + (t_{P[n-1]} \times \text{SE}) \text{ for the upper limit and}$$
$$\bar{Y} - (t_{P[n-1]} \times \text{SE}) \text{ for the lower limit} \qquad [67.2]$$

where $t_{P[n-1]}$ is the tabulated critical value of Student's t-statistic for a two-tailed test with $n - 1$ degrees of freedom at a specified probability level (P) and SE is the standard error of the sample mean. The 95% confidence limits (i.e. $P = 0.05$) tells you that on average, 95 times out of 100, the interval between the upper and lower limits will contain the true (population) value. Confidence limits are often shown as 'error bars' for individual sample means plotted in graphical form. Figure 67.6 illustrates how this is applied to plotted curves and histograms (note that this can be carried out for data series within a Microsoft Excel graph (chart) using the *Format data series* and *Y error bars* commands).

Correlation and regression

These methods are used when testing the relationship between data values for two variables. Correlation is used to measure the extent to which changes in the two sets of data values occur together in a linear manner. If one variable can be assumed to be dependent upon the other (i.e. a change in X causes a particular change in Y), then regression techniques can be used to provide a mathematical description of the underlying relationship between the variables, e.g. to find a line of best fit for a data series. If there is no *a priori* reason to assume dependency, then correlation methods alone are appropriate.

A correlation coefficient measures the strength of the linear relationship between two variables, but does not describe the relationship. The coefficient is expressed as a number between -1 and $+1$: a positive coefficient indicates a direct relationship, where the two variables change in the same direction, while a negative coefficient indicates an inverse relationship, where one variable decreases as the other increases (Fig. 67.7). The nearer the coefficient is to -1 or $+1$, the stronger the linear relationship between the variables, i.e. the less 'scatter' there would be about a straight line of best fit (note that this does *not* imply that one variable is dependent upon the other). A coefficient of 0 implies that the two variables show no linear association and therefore the closer the correlation coefficient is to zero, the weaker the linear relationship. The importance of graphing data is shown by the case illustrated in Fig. 67.7d.

Pearson's product moment correlation coefficient (r) is the most commonly used statistic for testing correlations. The test is valid only if both variables are normally distributed. Statistical tests can be used to decide whether the correlation is significant, e.g. using a one-sample t-test to see whether r is significantly different from zero, based on the following equation:

$$t = r \div \sqrt{[(1 - r^2) \div (n - 2)]} \qquad [67.3]$$

at $n - 2$ degrees of freedom, where n is the number of paired observations. If one or both variables are not normally distributed, then you should calculate an alternative non-parametric coeffcient, e.g. Spearman's coefficient of rank correlation (r_S) or Kendall's coefficient of rank correlation (τ). These require the two sets of data to be ranked separately, and the calculation can be complex if there are tied (equal) ranks. Spearman's coefficient is said to be better if there is any uncertainty about the reliability of closely ranked data values.

Using more advanced types of regression *– these include:*

- *Model II linear regression, which applies to situations where a dependent variable Y varies with an independent variable X, and where both variables may have error terms associated with them.*
- *Multiple regression, which applies when there is a relationship between a dependent variable and two or more independent variables.*
- *Non-linear regression, which extends the principles of linear regression to a wide range of functions. Technically, this method is more appropriate than transforming data to allow linear regression.*

Advanced statistics books should be consulted for details of these methods, which also may be offered by some statistical computer programs.

If underlying theory or empirical graphical analysis indicates a linear relationship between a dependent and an independent variable, then linear regression can be used to estimate the mathematical equation that links the two variables. Model I linear regression is the standard approach, and is available within general-purpose software programs such as Microsoft Excel (Box 67.3), and on some scientific calculators. It is suitable for experiments where a dependent variable Y varies with an *error-free* independent variable X in accordance with the relationship $Y = a + bX + e_Y$, where e_Y represents the residual (error) variability in the Y variable. For example, this relationship might apply in a laboratory procedure where you have carefully controlled the independent variable and the X values can be assumed to have zero error (e.g. in a calibration curve, see Chapter 37, or in a time course experiment where measurements are made at exact time points). The regression analysis gives estimates for a and b (equivalent to the slope and intercept of the line of best fit): computer-based programs usually provide additional features, e.g. residual values for Y (e_Y), estimated errors for a and b, predicted values of Y along with graphical plots of the line of best fit (the trend line) and the residual values. In order for the model to be valid, the residual (error) values should be normally distributed around the trend line and their variance should be uniform (homogeneous), i.e. there should be a similar scatter of data points around the trend line along the x-axis (independent variable).

If the relationship is not linear, try a transformation. For example, this is commonly done in analysis of enzyme kinetics (see Fig. 55.4). However, you should be aware that the transformation of data to give a straight line can lead to errors when carrying out linear regression analysis: take care to ensure that (a) the assumptions listed in the previous paragraph are valid for the transformed data set, and (b) the data points are evenly distributed throughout the range of the independent variable. If these criteria cannot be met, non-linear regression may be a better approach, but for this you will require a suitable computer program, e.g. GraphPad Prism.

Example If a regression analysis gives a value for r^2 of 0.75 (i.e. $r = 0.84$), then 75% of the variance in Y can be explained by the trend line, with $1 - r^2 = 0.25$ (25%) remaining as unexplained (residual) variation.

The strength of the relationship between Y and X in model I linear regression is best estimated by the coefficient of determination (r^2 or R^2), which is equivalent to the square of the Pearson correlation coefficient. The coefficient of determination varies between 0 and +1 and provides a measure of the goodness of fit of the Y data to the regression line: the closer the value is to 1, the better the fit. In effect, r^2 represents the fraction of the variance in Y that can be accounted for by the regression equation. Conversely, if you subtract this value from 1, you will obtain the residual (error) component, i.e. the fraction of the variance in Y that cannot be explained by the line of best fit. Multiplying the values by 100 allows you to express these fractions in percentage terms.

Using computers to calculate hypothesis-testing statistics

As with the calculation of descriptive statistics, specialist statistical packages such as SPSS and MINITAB can be used to simplify the calculation of hypothesis-testing statistics. The correct use of the software and interpretation of the output requires an understanding of relevant terminology and of the fundamental principles governing the test, which is probably best obtained by working through one or more examples by hand before using these tools (e.g. Box 58.2; Box 67.2). Spreadsheets offer increasingly sophisticated statistical analysis functions, three examples of which are provided in Box 67.3.

Box 67.3 Using a spreadsheet (Microsoft Excel) to calculate hypothesis-testing statistics

Presented below are three examples of the use of Microsoft Excel to investigate hypotheses about specific data sets. In each case, there is a brief description of the problem; a table showing the data analysed; an outline of the Excel commands used to carry out the analysis; and an annotated table of results from the spreadsheet (you can use the same data to check your ability to carry out the procedures).

Example 1: a *t*-test. As part of a project, a student applied a chemical treatment to a series of flasks containing fungal cultures with nutrient solution. An otherwise similar set of control flasks received no chemical treatment. After 3 weeks' growth, the student measured the wet mass of the filtered cultures:

Wet mass of samples (g)

Replicate	1	2	3	4	5	6	7	8	Mean	Variance
Treated with ZH52	2.342	2.256	2.521	2.523	2.943	2.481	2.601	2.449	2.515	0.042
Control	2.658	2.791	2.731	2.402	3.041	2.668	2.823	2.509	2.703	0.038

The student proposed the null hypothesis that there was no difference between the two means and tested this using a *t*-test, and had evidence from other studies that the fungal masses of replicate flasks were normally distributed. The student had also established, by calculation, that the assumption that the populations had homogeneous variances was likely to be valid. Using the *Tools > Data Analysis > t-Test: Two-Sample Assuming Equal Variance* option, with *Hypothesized Mean Difference* = 0 and *Alpha* (= *P*) = 0.05, and adjusting the number of significant figures displayed, the following table was obtained:

***t*-Test: two-sample assuming equal variances**

	Variable 1	*Variable 2*
Mean	2.515	2.703
Variance	0.042	0.038
Observations	8	8
Pooled variance	0.040	
Hypothesized mean difference	0	
df	14	
t Stat	−1.881	
$P(T \leqslant t)$ one-tail	0.040	
t Critical one-tail	1.761	
$P(T \leqslant t)$ two-tail	0.081	
t Critical two-tail	2.145	

The value of t obtained was −1.881 (row 7 't Stat') and the probability of obtaining this value for a two-tailed test (row 10) was 0.081 (or 8.1%), so the student was able to accept the null hypothesis and conclude that ZH52 had no significant effect on fungal growth in these circumstances.

Example 2: an ANOVA test. A biochemist made six replicate measurements of four different batches (A–D) of alcohol dehydrogenase, obtaining the following data:

Box 67.3 (continued)

Alcohol dehydrogenase activity (U l^{-1})

Batch \ replicate	1	2	3	4	5	6	Mean	Variance
A	0.562	0.541	0.576	0.545	0.542	0.551	0.552833	0.000 189
B	0.531	0.557	0.537	0.521	0.559	0.538	0.540500	0.000 221
C	0.572	0.568	0.551	0.549	0.564	0.559	0.560500	0.000 085
D	0.532	0.548	0.541	0.538	0.547	0.536	0.540333	0.000 039

The biochemist wanted to know whether the observed differences were statistically significant, so he carried out an ANOVA test, assuming the samples were normally distributed and the variances in the three populations were homogeneous. Using the *Tools > Data Analysis > Anova: Single Factor* option, with *Alpha* (= *P*) = 0.05, and adjusting the number of significant figures displayed, the following table was obtained:

ANOVA: single factor

SUMMARY

Groups	*Count*	*Sum*	*Average*	*Variance*
A	6	3.317	0.552833	0.000 189
B	6	3.243	0.5405	0.000 221
C	6	3.363	0.5605	8.51E-05
D	6	3.242	0.540333	3.95E-05

ANOVA:

Source of variation	*SS*	*df*	*MS*	*F*	*P-value*	*F crit*
Between groups	0.001 761	3	0.000 587	4.397 856	0.015669	3.098 391
Within groups	0.002 669	20	0.000 133			
Total	0.004 43	23				

The *F* value calculated was 4.397 856. This comfortably exceeds the stated critical value (F_{crit}) of 3.098 391 and the probability of obtaining this result by chance (*P*-value) was calculated as 0.015669 (1.57% to three significant figures), hence the biochemist was able to reject the null hypothesis and conclude that there was a significant difference in average enzyme activity between the four batches, since $P < 0.05$. Such a finding might lead onto an investigation into why there was batch variation, e.g. had they been stored differently?

Example 3: testing the significance of a correlation. A researcher wanted to know whether there was any correlation between the levels of tar and nicotine in cigarettes. The researcher made measurements of both consitituents for 10 different brands and obtained the following results:

Tar and nicotine content (mg g^{-1}) of cigarettes

Brand	*Tar*	*Nicotine*
1	12.6	0.92
2	8.5	0.58
3	15.7	1.14
4	32.5	2.16
5	14.5	1.03
6	3.1	0.17
7	18.9	1.32
8	7.7	0.98
9	10.7	1.13
10	6.2	0.55

The researcher assumed that both variables were normally distributed and used the Microsoft Excel function *PEARSON (array1, array 2)* to obtain a value of +0.950 260 385 for the Pearson's product moment correlation coefficient *r*, specifying the tar content data as array1 and the nicotine content data as array2 (the Excel *CORREL* function can also be used to carry out the same task). The researcher then used a spreadsheet to calculate the *t* statistic for this *r* value, using the formula for Eqn [67.3]. The calculated value of *t* was 8.6296, with 8 degrees of freedom. The critical value from tables (e.g. Table 67.2) at $P = 0.001$ is 5.04, so the researcher concluded that there was a very highly significant positive correlation between the two constituents.

Text references and sources for further study

Dytham, C. (2003) *Choosing and Using Statistics: A Biologists' Guide*, 2nd edn. Blackwell, Oxford.

Finney, D.J., Latscha, R., Bennett, B.M. and Hsu, P. (1963) *Tables for Testing Significance in a 2 × 2 Table.* Cambridge University Press, Cambridge.

Heath, D. (1995) *An Introduction to Experimental Design and Statistics for Biology*. UCL Press Ltd, London.

Maber, J. (1999) *Data Analysis for Biomolecular Sciences.* Longman, Harlow.

Schmuller, J. (2005) *Statistical Analysis with Excel for Dummies.* Wiley, Hoboken.

Sokal, R.R. and Rohlf, F.J. (1994) *Biometry*, 3rd edn. W.H. Freeman and Co., San Francisco.

Wardlaw, A.C. (2000) *Practical Statistics for Experimental Biologists*, 2nd edn. Wiley, New York.

Study exercises

67.1 Calculate 95% confidence limits. What are the 95% confidence limits of a sample with a mean = 24.7, standard deviation = 6.8 and number of data values = 16? (Express your answer to three significant figures.)

67.2 Use the Poisson distribution. In a sample of 15 snails, a researcher finds the following number of parasite larvae per snail: 0, 0, 0, 0, 1, 1, 1, 3, 4, 5, 5, 7, 7, 9, 9. Using the rule of thumb decide whether the parasites are 'clumped' or 'repulsed' in distribution on their host. What might this mean in biological terms?

67.3 Practise using a *t*-test. A biology student examined the effect of adding a plant hormone to pea plants. She dissolved an appropriate amount of the compound in ethanol and applied 25 μl of this to the topmost parts of the treated plants. With the controls, she applied the same amount of pure ethanol. After 3 days, she measured the distance between the second and third internodes of the plants and obtained the results shown below. Carry out a *t*-test on the data and draw appropriate conclusions.

Internode distance in cm

Control	7.5	8.1	7.6	6.2	7.5	7.8	8.9
Treatment	5.6	7.5	8.2	6.7	3.5	6.5	5.9

67.4 Interpret the output from Microsoft Excel linear regression analysis. The following output represents a regression analysis for an experiment measuring the uptake of an amino acid by a cell suspension (in pmol cell^{-1}) against time (in minutes). Based on this output, what is the form and strength of the underlying linear relationship? (Express the coefficients to three significant figures.)

Output from Excel spreadsheet linear regression analysis

Summary output

Regression statistics	
Multiple R	0.985 335 951
R square	0.970 886 937
Adjusted R square	0.963 608 672
Standard error	2.133 876 419
Observations	6

ANOVA

	df	SS	MS	F	Significance F
Regression	1	607.406 2857	607.4063	133.3954	0.000 320 975
Residual	4	18.213 714 29	4.553 429		
Total	5	625.62			

	Coefficients	Standard error	*t* Stat	*P*-value
Intercept	1.171 428 571	1.544 386 367	0.758 507	0.490 383
X variable 1	2.945 714 286	0.255 047 014	11.549 69	0.000 321

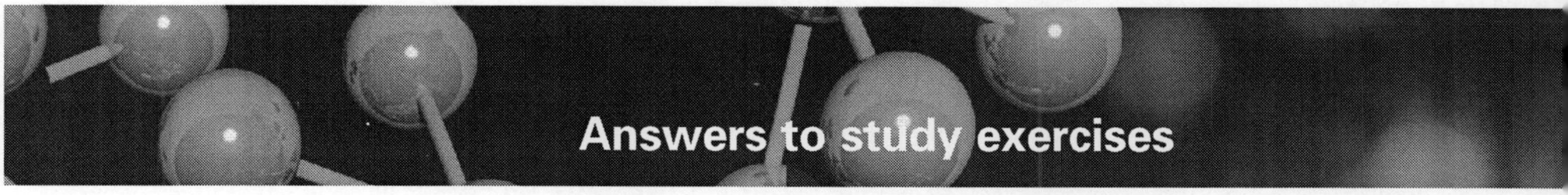

Answers to study exercises

We have attempted to provide an 'answer' to all of the study exercises. Where the question is open-ended and no 'correct' answer can be given, we have provided a ***Tip*** which should either help with your general approach, indicate which resources are worth consulting, or provide a pointer to relevant material within the book. Where a non-numerical question has a 'correct' answer, we have provided a model text-based answer. If a calculation is involved, we have shown the steps involved and have indicated the **correct answer underlined in bold**.

1 The importance of transferable skills

1.1 ***Tips:*** Where you feel confident about skills you have learned at school, take care to consider whether they might need to be upgraded for university/college use. When thinking about opportunities for developing skills, remember that these may occur outside university, perhaps in work or social contexts.

1.2 ***Tip:*** Possible keywords/phrases for a Web search include: biology skills, C&IT skills, numeracy skills, study skills, time management.

1.3 ***Tip:*** You could create a grid to record your thoughts, with column headings 'University'; 'Work' and 'Social' and rows for 'Short term'; 'Medium term' and 'Long term'. Short term and medium term could refer to your time at university, while long term could refer to after graduation.

2 Managing your time

2.1 ***Tips:*** Model your spreadsheet on Fig. 2.1. It isn't too difficult to fill in details at the end of each day either directly into the spreadsheet or on a printed version. You will need to collect information for at least a week and possibly longer before you see patterns emerging.

2.2 ***Tips:*** It is important to include both social and study-related activities on your lists. Short-term lists deal with issues of the day or week; medium-term for the term or semester and longer-term for one year onwards. Relate each of these lists to appropriate goals you may have over these timescales.

2.3 ***Tips:*** This task could be related to (a) your self-analysis of transferable skills (Chapter 1), or (b) to some large task such as carrying out a final-year project. In the latter case, for example, there are small jobs that can be done as you proceed with the larger task, such as writing parts of the Introduction or compiling a list of references, and these will reduce time pressure towards the end of the task.

3 Working with others

3.1 ***Tips:*** Think about your contributions to past group activities while doing this exercise, and, if necessary, get feedback from friends and colleagues. You may feel that you fit into more than one 'natural' role: this is quite often the case.

3.2 ***Tip:*** This exercise might best be associated with a specific teamwork activity, but this need not be restricted to study – group contributions with clubs, societies or in employment are equally valid. Strategies for improvement will depend on your identified weaknesses. For example, if you felt too shy to contribute effectively, you might want to practise public speaking (see Chapter 15) or attend an assertiveness workshop.

3.3 ***Tip:*** Columns 4 and 5 of Table 3.1 are relevant, particularly if you are able to identify a 'natural' role or roles for yourself in group situations (as in exercise 3.1). Can you relate your strengths and weaknesses to specific events that have occurred during teamwork, and might you handle the situation differently in future?

4 Taking notes from lectures and texts

4.1 ***Tips:*** If you are used to using a particular note-taking method that seems to work well, you may feel reluctant to experiment. Try out the new method by taking notes at a seminar, tutorial or meeting where it will not matter so much if the technique does not work well at the first attempt.

4.2 ***Tip:*** Make a list of any missing notes or handouts and ask a colleague if you can work from their notes to fill in these gaps.

4.3 ***Tip:*** To compare methods, test your recall some time afterwards. Take a blank sheet of paper and see what you can write about each topic.

5 Learning and revising

5.1 ***Tips:*** When constructing your revision timetable, remember the following important points:

- Break large topics into 'bite-sized' units.
- Mix up hard or less interesting topics with those you enjoy revising (don't ignore them).
- Keep an appropriate balance between work and relaxation.
- Offer yourself rewards if work targets are completed.

5.2 ***Tips:*** You may wish to keep one of the past papers (preferably a recent one) aside to use in a mock exam. Remember that learning objectives can be just as important as exam papers in helping you to revise.

5.3 ***Tips:*** Read through the list of active revision tips on pp. 26–7 and decide which might work best for you. Carry out a trial of the method with one of your topics where the material has seemed difficult to learn. Experiment until you arrive at a solution that works, remembering (a) that different techniques might work best in different subjects and (b) a little variety might make the revision process more interesting.

6 Curriculum options, assessments and exams

6.1 *Tip:* Some questions to ask yourself about past exam performances... Did you prepare well enough? Did you run out of time during the exam? Did you spend too long on a particular question or section, and have to rush another? Were your answers direct and at the appropriate depth? Did you misinterpret any questions, or miss out part of any answers? Were your writing skills, including planning, up to the standard required?

6.2 *Tip:* When revising, try to find a topic you find hard to understand and see whether your colleague has approached it in the same way as you. If not, you may be able to learn from this.

6.3 *Tip:* Discuss your tactics with a colleague sitting the same exam. What topics do you agree are likely to come up?

7 Preparing your curriculum vitae

7.1 *Tip:* This exercise might be attempted with a friend you can trust to give you a frank opinion. Remember that your personal qualities can be developed in both curricular and extra-curricular activities.

7.2 *Tip:* Working with a friend or group from your class, compare notes on how you have organised your CVs and then modify your own CV, taking on board good ideas from your peers. Create a physical file for your revised CV and add handwritten notes, updating it as and when required.

7.3 *Tip:* Use the 'Prospects' website noted in the sources for further study to gain ideas before your appointment with the careers service. The library or careers resource centre may have useful information, often accessible *via* the Web.

8 Finding and citing published information

8.1

(a) The Dewey Decimal Classification system; the US Library of Congress system; or other, as appropriate.
(b) *Nature*: Dewey: 505.0942; Library of Congress: Q1.N3; (or other, as appropriate). Note: the 'Per' or 'per' that may precede these numbers refers to the fact that this is a periodical, rather than a book.
(c) *Tip:* Answer as appropriate – this will depend on the layout of your particular library.
(d) (i) 'A structure for deoxyribose nucleic acid'; (ii) 'Analysis of the genome sequence of the flowering plant *Arabidopsis thaliana*'.

8.2 *Tips:* The main point here is that methods of writing down references are diverse. You should also find that the details recorded are the same, although they may be in a different order or font style. You must pay attention to the *precise* instructions given for your course, to ensure that you get this right when writing up a project report or dissertation.

8.3 *Tips:* Three possible sites are:
(a) SI units: http://www.ex.ac.uk/cimt/dictunit/dictunit.htm (last accessed 01/04/07)
(b) l'Institut Pasteur: http://www.pasteur.fr/externe (last accessed 01/04/07) is your starting point. Reaching the culture collection is a little tricky – try *via* '*Scientific Research*', '*Biological Resources*', '*Bacteria Strain Search*'.
(c) Blood cells: http://www.ultranet.com/~jkimball/Biology Pages/B/Blood.html (last accessed 01/04/07).

8.4 *Tips:* When writing a handwritten essay, the variant of the Vancouver method in which references are listed in the order cited might be the most convenient to use, because you could write down all the references cited on a separate sheet as you wrote out the essay. However, if you felt that it was important for the reader to see the author(s) and publication dates of the papers, then the Harvard method might be more appropriate – however, you would need to wait until the essay was fully written before organising the literature cited section in author and date order. In a word processed review, you might prefer the Vancouver method as this might allow the text to flow more smoothly without being interrupted by lists of author names and dates. On the other hand, the Harvard system is easier to use as you write with a word processor, because you can simply 'slot' each new reference into the correct place in the list. In an academic journal, it might be assumed that the readership would want to know the authors and dates of the references cited, so the Harvard system would probably be the most appropriate. In contrast, the journal *Nature* uses the Vancouver system in order to present important new scientific advances to a wide audience. *Nature* readers want to be able to assimilate each paper's findings rapidly without getting distracted by citation details.

9 Evaluating information

9.1 (a) P; (b) P; (c) S; (d) S; (e) P; (f) S; (g) P; (h) P; (i) S; (j) S; (where P = primary, S = secondary). Where journals contain both original papers and reviews, we have classed them as primary.

9.2

Some potential debating points on whether '*Genetic modification of crops is a good thing*'

For	Against
Large increases in yield could be engineered, helping to alleviate hunger in developing countries	Transfer of new genes to other crops might occur, with unknown and unpredictable effects
The range of possible modifications to crops is much greater than in conventional breeding – entirely new forms of crops are possible	'Super weeds' derived from genetically modified plants might spread uncontrollably and disrupt the ecological balance
It is possible to modify crops much faster than with conventional breeding	The long-term effects of e.g. modified proteins on human health are untested

(*continued*)

Table continued

For	Against
Very specific changes are possible, e.g. adding a particular vitamin or specific amino acid to a crop, effectively improving nutrition rather than calorific intake	Our effort (and money) would be better applied to birth control, basic medicine and better implementation of cropping practice
Genetic modification can be used to deter pests, reducing pesticide use and thereby benefiting the environment	The public are so concerned about the safety of foods derived from genetically modified crops that they will fail economically anyway, since consumers will avoid food labelled as containing genetically modified material

9.3 ***Tip:*** Use Box 63.3 as a source to indicate ways in which graphs may be used to misrepresent information. You may find this exercise easier to accomplish with printed graphs. Those on television may not be shown for long enough to allow you to carry out a detailed analysis. Nevertheless, they represent a rich source of material to criticise!

10 Using online resources

10.1

(a) Barbara McClintock is famous for her work on the discovery of mobile genetic elements (transposons): she won the 1983 Nobel Prize in Medicine. See: http://www.nobel.se/medicine/laureates/1983/

(b) *Arabidopsis thaliana* is also known as mouse-ear cress. See: http://plants.usda.gov/cgi_bin/plant_profile.cgi?symbol=ARTH

(c) The postal address of the UK Biochemical Society is The Biochemical Society, 59 Portland Place, London, W1B 1QW, United Kingdom. See: http://www.biochemistry.org/feedback.htm

10.2 ***Tips:*** Note the number of hits found for each search engine. Make a note or printout of the top 10 sites located by each search engine. Compare these site listings to reveal the quality of data obtained. Repeat this exercise for meta-search engines. Make lists of the component search engines used by the main meta-search engines to help you devise a strategy for the best coverage of the Internet.

10.3 ***Tips:*** Within the bookmark editor for your browser, create folders for each of your study courses and then create sub-folders for each module that you are doing. Now use some searches to put relevant bookmarks in each folder. Don't forget to include a folder for transferable skills sites; there are a lot of supportive sites available that are worth visiting.

11 Bioinformatics – Internet resources

11.1

(a) The Entrez nucleotides database of GenBank, available at: http://www.ncbi.nlm.nih.gov/entrez/query.fcgi?db=Nucleotide contained **65 369 091 950 (over 65 billion) bases as of August 2006** (note that this is likely to show a substantial increase over the life of this textbook).

(b) Hinxton Hall is the location of the UK campus of the European Bioinformatics Institute, located at Hinxton, Cambridge (see: http://www.ebi.ac.uk/Information/Travel/travel.html).

(c) Monterotondo, in the Lazio region of Italy, north of Rome (see: http://www.embl-monterotondo.it/ and: http://www.emma.rm.cnr.it/images/map1.html).

(d) PDB held **29 937 structures as of Jan 2007** (see: http://www.wwpdb.org/stats.html). Note that this number will increase over the life of this textbook.

(e) National Center for Biotechnology Information, National Library of Medicine, Building 38A, Bethesda, MD 20894, USA (see: http://www.ncbi.nlm.nih.gov/About/glance/contact_info.html).

11.2 The sequence gives a single 'hit' for part of the *Homo sapiens* chromosome 15 clone RP11-46M12 map 15q15 (accession code: AC012652), with a total of 17 consecutive identical bases (shown in colour): actgactcatagagtacccac cgtggtgct.

11.3

(a) The amino acid sequence gives a number of 'hits' for the alpha chain of haemoglobin.

(b) The haemoglobin alpha-3 chain from chimpanzee (accession code: HAG03C) gives **100% identity** (all 19 amino acids match).

(c) The highest score for a human sequence is for the alpha-1 globin chain of haemoglobin (AF349571), which gives 18 identical residues out of 19 (**≈95% identity**).

11.4 The results of the six-frame translation are: 5′ to 3′ frame 1: SERTHH; frame 2: QREPT; frame 3: RENPP; 3′ to 5′ frame 1: MetVGSLStop; frame 2: WWVLS; frame 3: GGFSL, so, **3′ to 5′ frame 1 contains a start (methionine) and a stop codon**. The predicted amino acid sequence for this frame is **MVGSL ≡ met-val-gly-ser-leu.** (Note that this sequence is far shorter than in a functional protein.)

11.5 ***Tip:*** Select *Summary of PDB structure* for key information on structure. The ID number of this enzyme in PDB is 1GOW.

(a) The enzyme is a **homotetramer**.

(b) There are two types of chain in the crystal structure: A and B, each containing **489 residues** with an identical amino acid sequence.

(c) There are **23 helices** in each chain. The structure is mainly parallel β-sheets (β-α-β units). Note that the crystal structure is a dimer whereas the functional enzyme is a tetramer (comprising 2 dimers).

(d) ***Tips:*** To save an image from a Web page, use the menu provided by clicking the right-hand mouse button ('*Save Picture As*' function in Internet Explorer); to insert an image into a Word document, use the '*Insert*' menu, then the '*Picture*' and '*From File*' options.

12 Using spreadsheets

12.1 See the figure below. ***Tips:*** Depending on the precise method you use, you may need to provide the *Name*, *Values* and *Category Labels* for the *Chart Wizard* or adjust those assumed. When creating the graph, select the 'as new sheet' option when prompted for a *Chart Location*. For printing in black and white, select black and white hatching/shading options after click-selecting chart segments. Ensure the segments of the chart are labelled (i.e. not the default legend option) using the *Chart/Chart Options* menu.

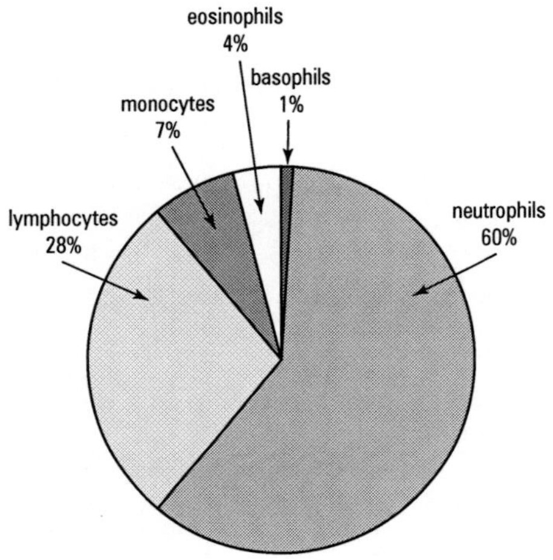

Relative abundance of the major types of leukocytes in adult blood.

12.2 See the figure below. ***Tips:*** When setting up the spreadsheet, create appropriate formulae in new columns for the transformed data; graph using the '*XY (Scatter)*' option in the Microsoft Excel *Chart Wizard* to make sure the data are correctly spaced on the *x*-axis. Plot time on the *x*-axis and the transformed cell count data on the *y*-axis. You will need to create a separate graph to examine the effects of each transformation, because the *y*-axis scales will be different. Select the *Line None* option for the relevant (log transformed) series on the *Data Series* menu, then use the *Add Trendline* option (linear model) within the *Chart* menu to add a linear regression line (note: this can also be done using the *Regression* command in the *Data Analysis* option on the *Tools* menu, but the procedures required are much more complex). Adjust the chart output to suit, e.g. by changing gridlines, background and legends, before copying and pasting to the MS Word document.

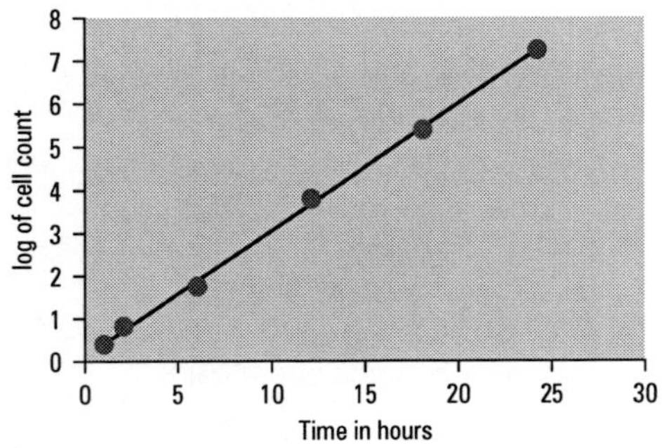

Log-linear graph of data of the table in study exercise 12.2.

12.3 ***Tips:*** Take care when entering the date data in the spreadsheet. You may need to specify the format of the cells for the date column (*Format* menu, *Cells* option). Highlight all parts of the spreadsheet table including column headers, then use the *Data* menu, *AZ↓ Sort* option to rearrange data. You do not need to fill in all the response boxes and can select 'none' if a box contains criteria from a previous sort.

13 Word processors, databases and other packages

13.1 ***Tips:***

(a) Ensure list components are in separate paragraphs; highlight list; from the *Table* menu select *AZ↓ Sort* function; ensure options are set to *paragraphs, text, ascending*.

(b) From the *Edit* menu, select *Replace* option, then fill in the relevant boxes.

(c) From *Edit* menu, select *Replace* option, then fill in the relevant boxes; select *More*▾ then *Format, Font, Italic*.

(d) From the *View* menu, select *Header and Footer* and use the small on-screen menu to add text and numbering as desired. Formatting commands (e.g. centre) will work in the header and footer boxes. As an alternative for adding page numbers, this can be done from the *Insert* menu, *Page Numbers*... option.

(e) Margins can be adjusted from *File* menu, *Page Setup, Margins* option.

(f) Select paragraphed text, then use the *Format* menu, *Bullets and Numbering* option, using *Customize* if the option you desire is not displayed within the panel.

(g) Type 'alternative' within a Word page. Ensure cursor is within word (or highlight word); from *Tools* menu select *Language, Thesaurus* options. Perhaps 'option' might be a useful synonym?

(h) From *Tools* menu, select *Spelling and Grammar* to start a spell check. Take care if the checker queries UK spellings (it may be set to accept US spellings). Note: this is *not* a substitute for reading your document – the spell checker will miss typing errors like 'form' instead of 'from'.

(i) For whole document, triple click in left margin to select all, then use *Tools* menu, *Word Count* option. For selected text, highlight (select) the part you wish to count, then carry out same instructions.

(j) While in first document, use *File* menu, *Open* option to open another. You can switch between them using the *Window* menu, where you will see all open documents listed. Note: remember to save changes to all documents when closing down your editing session.

13.2 ***Tips:*** Although there are several ways to modify tables, you may find it useful to try the following commands (think carefully about their order).

- *Table, Insert Table*
- *Table, Merge Cells*
- *Table, Distribute Columns Evenly*
- *Tables and borders*, modify borders commands (various)

Consult Box 64.2 or use the *Help* feature if you run into problems.

13.3 ***Tips:*** You might start your search on the networked 'desktop' and icons visible to you when you log on. Use 'help' menus and 'wizards' to investigate how to use the programs, or consult any manuals available in the computer suite.

14 Organising a poster display

14.1 ***Tip:*** use Figure 14.1 and the subheadings of Chapter 14 as a starting point for assessing the pros and cons of your designs.

14.2 Ten-point checklist – what makes a good poster presentation?

- ☐ *Title* – is this concise, specific and interesting?
- ☐ *Author(s)* – does the poster show details of the names and addresses of all contributors?
- ☐ *Structure* – is the overall structure clear to the observer, with good use of subheadings?
- ☐ *Flow* – is the layout of the poster clear, in terms of how the audience is expected to 'read' each section?
- ☐ *Content* – does the presentation have a clear introduction, main message and conclusion?
- ☐ *Text* – does the poster contain an appropriate amount of written material in a 'readable' font?
- ☐ *Graphics* – does the poster make good use of diagrams, figures and images?
- ☐ *Size and scale* – does the poster make effective use of the space available?
- ☐ *Colour* – does the presentation use colour effectively?
- ☐ *Conclusion* – is the 'take-home message' clear to the observer?

14.3 ***Tip:*** you might use the 10-point checklist given above, or your own version from exercise 14.2 to assess selected posters by giving a mark out of 10 for each item in the list. In a group exercise, you can see how your 'scores' compare with those of other students, and discuss items where scores are different.

15 Giving a spoken presentation

15.1 Ten-point checklist – what makes a good spoken presentation?

- ☐ *Introduction* – does the speaker explain the structure and rationale of the topic at the outset?
- ☐ *Audience* – is the material presented at an appropriate level?
- ☐ *Content* – does the presentation cover the topic effectively?
- ☐ *Speed and structure* – are the main elements of the presentation presented clearly, and at an appropriate pace?
- ☐ *Voice* – is the presentation delivered in an engaging, active vocal style?
- ☐ *Audio-visual aids* – does the speaker make good use of visual images (e.g. graphs, photographs, etc.)?
- ☐ *Image* – does the presenter make good use of body language?
- ☐ *Notes* – does the speaker use notes to structure the talk, rather than reading from a written text?
- ☐ *Conclusion* – is a clear 'take-home message' delivered in the final stages of the presentation?
- ☐ *Questions* – does the presenter deal with questions effectively?

15.2 ***Tips:*** You could try marking speakers out of 10 for each of the points listed in exercise 15.1 (either using the checklist provided above or your own version), or you could simply prepare a list of the good and bad points of the presentations you have evaluated.

15.3 ***Tips:*** When evaluating your own performance: (i) remember that most of us are overcritical when viewing ourselves on video, tending to focus on relatively minor points, e.g. facial expression, rather than broader aspects, e.g. clarity of delivery and content; (ii) when giving a talk to a group of fellow students you might ask them to use all or part of the checklist from exercise 15.1 as a way of providing feedback.

16 General aspects of scientific writing

16.1 ***Tips:*** When brainstorming, think of all aspects of the topic that could be relevant (even if you could not give precise details at this stage). When you meet together, discuss which of your combined ideas would be most important in crafting an answer and what order you might place them in.

16.2 ***Tip:*** Try to apply what you have learned in the next assignment you write – ask a tutor for specific feedback if you are unsure about whether your technique has improved.

16.3 ***Tips:*** Use the spelling list to help you when writing and make a conscious effort to memorise the spelling each time you use it. The act of writing the vocabulary list will assist you to memorise new words and their meanings. If you have trouble learning definitions, e.g. terms in molecular biology, you could extend these ideas to create a personal glossary of terms and acronyms.

17 Writing essays

17.1 ***Tip:*** Can you determine what lies behind each question? Some question titles are quite direct, but others ask for information in a more subtle way. Use your imagination to think beyond the immediate question, linking to learning objectives or to other relevant material covered in lectures.

17.2 ***Tips:*** Compare your ideas with those of a partner, or ask tutors or lecturers what they think of your efforts. A useful adjunct to this exercise is to write the first sentence of your answer to avoid writer's block (p. 102).

17.3 ***Tips:*** Note that each of the stages (b)–(d) requires a separate 'review' – it is difficult, if not impossible to make a good job of all aspects in a single 'pass' through the text. Ideally, you would reprint the essay between reviews, having made corrections.

18 Reporting practical and project work

18.1 ***Tips:*** To assimilate the required style for Materials and Methods, you may find it appropriate to read some research papers from the journal section of your library. Note particularly that Materials and Methods is a description of 'what was done', not a set of instructions.

18.2 ***Tips:*** Again, you may find it appropriate to read some research papers from the journal section of your library to assimilate the normal style for describing research results.

18.3 ***Tips:*** Abstracts are difficult to write: try to include a synopsis of all the main sections of the paper, including its aims, methods, key results and conclusions. Ignore the word length restriction until you have noted the main points, then try to reduce (or, less likely, expand) your effort to match the requirement.

19 Writing literature surveys and reviews

19.1 Five differences between reviews and papers are:

1. Reviews deal with material in general terms taking an overview, whereas papers tend to look at a focused, single aspect of a topic.
2. The subdivisions are generally not the same – papers usually follow the IMRaD format (p. 110), while reviews are structured according to the major components of the topic.
3. There is little or no reference to original data in reviews.
4. There is little *detail* of methods in reviews.
5. They tend to use different reference citation systems, with Harvard (p. 47) most commonly used in papers and Vancouver (p. 48) used in reviews.

19.2 ***Tip:*** It should be possible to do the second part of study exercise 19.2 online at your library – you may need to enlist the help of library staff if unsure of how to do this.

19.3 ***Tip:*** Obviously, the Abstract and Conclusions sections will provide information about the key points addressed. It might be helpful to use a highlighter on a photocopied version of the review to note important statements within the main body of the review.

20 Your approach to practical work

20.1 Possible reasons why practical work might be of value in a university course include:

- practicals reinforce lecture material, in agreement with the adage: 'I hear and I forget, I see and I remember, I do and I understand';
- practical procedures provide students with an opportunity to develop their manual skills and laboratory competences;
- investigative procedures and problem-solving practical exercises enable students to improve their skills in experimental design and the application of the scientific method;
- practicals enable students to develop their abilities to observe and measure biological systems, and to record the outcome;
- students are able to analyse and interpret the data from practical exercises and to consider the validity of their results, including sources of error, etc.;
- the results of practical procedures give students an opportunity to develop skills in reporting and presenting 'experiments' in written format.

20.2 ***Tip:*** Possible items for a snail dissection: scalpel; blunt seekers; coarse and fine scissors; wax dish and pins; pointed seeker; dissection microscope; paper tissues; water; Pasteur pipette; first aid kit near by – in case of accident.

20.3

(a) **40**
(b) 44.266 66°, expressed to 4 significant figures = **44.27**
(c) 0.019 531 25, expressed to 3 significant figures = **0.0195**
(d) $\mathbf{1.6 \times 10^5}$
(e) 0.0313, expressed to 3 decimal places = **0.031**

21 Health and safety

21.1

(a) p. 125.
(b) Fig. 21.1 (p. 125).
(c) Fig. 21.2 (p. 126).
(d) Fig. 32.1 (p. 203).
(e) Fig. 39.4 (p. 274).

21.2 ***Tip:*** The locations of each item will vary, according to the layout of your chosen laboratory. You should ask someone in charge (e.g. demonstrator, lecturer, technician or laboratory manager) if you are unable to find any of the items listed.

21.3

(a) ***Tip:*** The answer will vary, according to your university, but is likely to cover the main points listed on p. 126.
(b) In the UK, water-filled fire extinguishers are red – use for wood, paper and textiles; CO_2 extinguishers are black, or have a black panel – use for flammable liquids and electrical fires; dry powder extinguishers are blue, or have a blue panel – use for all types of fires (wood, paper, textiles, gaseous fires, flammable liquids and electrical fires); foam extinguishers are beige, or have a beige panel – use for flammable liquids (note that other colour conventions may apply outside the UK).
(c) ***Tip:*** This will vary, depending upon your department's procedure. However, it is likely to include a specific written record book for accidents.
(d) ***Tip:*** There will often be separate codes of practice for work involving non-pathogenic microbes (e.g. general microbiology classes) and pathogenic microbes (e.g. in medical microbiology courses). The

latter requires a higher level of safe working practice, and may include the use of disposable aprons, etc.

21.4 ***Tip:*** Use either your department's chemical hazard information system or a Web database (e.g. the National Institute for Occupational Safety & Health Databases at: http://www.cdc.gov/niosh/database.html; or the Agency for Toxic Substances and Disease Registry ToxFAQs database at: http://www.atsdr.cdc.gov/toxfaq.html. Alternatively, you could use a 'portal' site such as the ChemPortal site at: http://webnet3.oecd.org/echemportal/Contents.aspx?ContentName = PortalInfo; or the material safety data sheet (MSDS) site at http://www.ilpi.com/msds/index.html).

(a) Formaldehyde is toxic and potentially carcinogenic (and also highly flammable in pure form): protective gloves and eye/face protection should be used during preparation of stock solutions. Formaldehyde can also cause allergic skin reactions and breathing the vapour should be avoided – working solutions should be used in a well-ventilated place or fume hood.

(b) Acetone is highly flammable and can form explosive peroxides on contact with strong oxidants. It should be used in a well-ventilated room, away from naked flames or spark-prone electrical equipment. Disposable gloves can be used to prevent skin contact.

(c) Sodium hydroxide is highly corrosive and an irritant. Protective gloves and safety glasses should be worn when preparing stock solutions using solid NaOH. Contact with moisture generates considerable heat, so it should always be added in small amounts to a large volume of water.

22 Working with liquids

22.1

(a) Use measuring cylinders (e.g. measure out 700 ml of ethanol using a 1000 ml measuring cylinder and measure out 300 ml of water using a 500 ml measuring cylinder: mix together in a large conical flask (e.g. 1000 ml capacity) or beaker.

(b) Use a pipettor (e.g. P20 Gilson Pipetman) set to deliver 10.0 µl and sterile RNAse-free tips, manufactured for molecular biology (note also that some disposable tips are designed to have extended, fine tips, to aid gel loading).

(c) Weigh out 20.0 mg (0.0200 g) of DNA standard using a four-place balance, add this to a 100 ml volumetric flask and make up to 100 ml with distilled, deionised water.

(d) Use a burette (e.g. 50 ml burette) and beaker or flask (e.g. 250 ml capacity), plus a magnetic stirrer and 'flea' (p. 139) for mixing.

22.2 ***Tip:*** Box 22.1 (p. 132) gives stepwise details of how to use a pipettor.

22.3 ***Tip:*** You may find this type of calculation easier using a spreadsheet, rather than a calculator (Box 66.3 gives further details). The model A pipettor gave a mean delivery of 0.9795 ml, with a standard deviation of 0.010 40 ml (to 5 decimal places). The model B pipettor delivered, on average 1.0107 ml, with a standard deviation of 0.001 25 ml. The model C pipettor gave an average delivery of 1.0009 ml, with a standard deviation of 0.058 82 ml. Therefore, model C was the most accurate pipettor (with a mean value closest to the true value of 1.0000 ml) and model B was most precise (giving the most reproducible results, as shown by the lowest standard deviation). The model A pipettor was the least accurate (mean value furthest from the true value) while model C was the least precise (having the highest standard deviation).

23 Basic laboratory procedures

23.1 ***Tip:*** Box 23.2 and Box 24.1 give advice on preparing solutions.

(a) 50 ÷ 1000 (number of moles required for 1 litre) × 0.1 (volume required, expressed in litres) × 58.44 (relative molecular mass of NaCl) = 0.2922 = **0.292 g**

(b) 0.1 (number of moles required for 1 litre) × 0.25 (volume required, expressed in litres) × 182.17 (relative molecular mass of mannitol) = 4.554 25 = **4.55 g**

(c) 800 (number of µg in 1 ml) × 200 (volume, in ml) = 160 000 µg = **160 mg**

(d) 22.5 ÷ 1000 (number of moles required for 1 litre) × 0.5 (volume required, in litres) × 203.30 (relative molecular mass of $MgCl_2.6H_2O$) = 2.287 125 = **2.29 g**

(e) 20 (number of ng in 1 µl) × 1000 (number of µl in 1 ml) × 400 (volume required, in ml) = 8 000 000 ng = **8.00 mg**

23.2 ***Tip:*** All of these can be calculated using the relationship $[C_1]V_1 = [C_2]V_2$ (Eqn [24.2]). Do not forget to include the volumes of *both* solutions when calculating the final volume – e.g. in example (a) below, the final volume is 10 ml (1 + 9) and *not* 9 ml.

(a) $1 \times 0.4 = [C_2] \times 10$, therefore $[C_2] = (1 \times 0.4) \div 10 =$ 0.04 mol l^{-1} and therefore 0.04 × 1000 = **40.0 mmol l^{-1}**.

(b) $10 \times 25 = [C_2] \times 500$, therefore $[C_2] = (10 \times 25) \div 500 =$ 0.5 µg ml and therefore 0.5 × 1000 = **500 ng ml^{-1}**.

(c) $200 \times 0.01 = [C_2] \times 250$ (note that both volumes are expressed in ml), therefore $[C_2] = (200 \times 0.01) \div 250 =$ 0.008 mmol l^{-1} and therefore 0.008 × 1 000 000 = 8000 nmol l^{-1} and therefore 8000 ÷ 1000 = **8.00 nmol ml^{-1}**

(d) $0.2 \times V_1 = 0.02 \times 250$ (note that both concentrations are expressed in mol l^{-1}), therefore $V_1 = (0.02 \times 250) \div 0.2 =$ **25.0 ml** of KCl solution, made up to a final volume of 250 ml with water.

(e) First express the concentration of the glucose solution in mol m^{-3}, by dividing by M_r, therefore 20 ÷ 180.16 = 0.111 012 4 mol m^{-3}. Then use the relationship $[C_1]V_1 = [C_2]V_2$. Thus $0.111\,012\,4 \times V_1 = 50 \times 10^{-6} \times 1 \times 10^{-3}$, therefore $V_1 = (50 \times 10^{-6} \times 1 \times 10^{-3}) \div 0.111\,012\,4 = 4.504 \times 10^{-7}\ m^3$ = **450 µl**. Make up 450 µl of glucose solution to $1 \times 10^{-3}\ m^3$ (1000 ml) with water.

23.3 ***Tip:*** Either use the relationship $[C_1 1]V_1 = [C_2]V_2$ (Eqn [24.2]) to calculate the volumes required, or calculate the dilution factor from the ratio of the concentration in the stock solution to that required in the final solution.

(a) The dilution factor for KCl is 20-fold. To prepare the solution, use **0.050 ml (≡50.0 μl) of KCl stock solution, plus 0.950 ml (≡950 μl) of water**.

(b) The dilution factor is 40-fold for NaCl and 10-fold for glucose. To prepare the solution, use **1.25 ml of NaCl stock solution, 5.00 ml of glucose stock solution and 43.75 ml of water**.

(c) The dilution factor for NaCl is 20-fold, for KCl is 80-fold, for $CaCl_2$ is 4-fold and for glucose is 20-fold. To prepare the solution, use **5.00 ml of NaCl stock solution, 1.25 ml of KCl stock solution, 25.00 ml of $CaCl_2$ stock solution, 5.00 ml of glucose stock solution and 63.75 ml of water**.

(d) First, convert the required concentrations to the same units as those of the stock solutions. In this case, $\mu mol\,ml^{-1}$ and $mmol\,l^{-1}$ are directly equivalent (divide by 10^3 to convert from μmol to mmol, then multiply by 10^3 to convert from ml to l) so the numerical values remain the same. The dilution factor for NaCl is 5-fold, for $CaCl_2$ is 8-fold and for KCl is 10-fold. To prepare the solution, use **2.00 ml of NaCl stock solution, 1.25 ml of $CaCl_2$ stock solution, 1.00 ml of KCl stock solution and 5.75 ml of water**.

(e) First, convert the glucose concentration from $nmol\,ml^{-1}$ to $mmol\,l^{-1}$ (divide by 10^6 to convert nmol to mmol and multiply by 10^3 to convert ml to l), giving glucose at $0.2\,mmol\,l^{-1}$ (all other concentrations are in the same units as those of the stock solutions). The dilution factor for $CaCl_2$ is 20-fold, for KCl is 4-fold and for glucose is 25-fold. To prepare the solution, use **1.25 ml of $CaCl_2$ stock solution, 6.25 ml of KCl stock solution, 1.00 ml of glucose stock solution and 16.50 ml of water**.

24 Principles of solution chemistry

24.1 ***Tip:*** Box 23.2 and Box 24.1 give useful procedures and example calculations for molar concentrations.

(a) 1 molar $= 1\,mol\,l^{-1} = 58.44$ g in 1 litre of solution $=$ **58.440 g**.

(b) $110.99 \times 100 \div 1000$ (number of g required for $100\,mmol\,l^{-1}$) $\times 250 \div 1000$ (number of g in 250 ml) $= 2.774\,75 =$ **2.775 g**.

(c) $10\,nmol\,\mu l^{-1}$ is equivalent to $10\,mmol\,l^{-1} \div 1000$ (convert to $mol\,l^{-1}$) $\times 182.17$ (number of $g\,l^{-1}$) $\times 2.5$ (number of g in 2.5 l) $= 4.554\,25 =$ **4.554 g**.

(d) 5% w/v is equivalent to 5 g in 100 ml × 4 (number of g in 400 ml) $=$ **20.000 g**.

(e) 2.50×180.16 (convert from $mol\,m^{-3}$ to $g\,m^{-3}$) $\times 250 \div 1\,000\,000$ (number of g in 250 ml) $= 0.1126 =$ **0.113 g**.

24.2 ***Tip:*** Make stepwise changes and always show your working out in any written report or assessment including calculations involving the interconversion of concentrations.

(a) $5 \div 342.3$ (convert $g\,l^{-1}$ to $mol\,l^{-1}$) $= 0.014\,607\,1 =$ **0.0146 mol l^{-1}**. Note that this can also be expressed as $14.6\,mmol\,l^{-1}$.

(b) 1×58.44 (convert $mol\,m^{-3}$ to $g\,m^{-3}$) $\div 1000$ (convert $g\,m^{-3}$ to $g\,l^{-1}$) $= 0.058\,44 =$ **0.0584 g l^{-1}**. Note that this could also be expressed as $58.4\,mg\,l^{-1}$.

(c) 5×0.789 (convert % v/v to % w/v) $\times 10$ (convert % w/v to $g\,l^{-1}$) $\div 46.06$ (convert $g\,l^{-1}$ to $mol\,l^{-1}$) $= 0.856\,491\,5 =$ **0.856 mol l^{-1}**. Note that this could also be expressed as $856\,mmol\,l^{-1}$.

(d) $150 \div 1000$ (convert from $mmol\,l^{-1}$ to $mol\,l^{-1}$) $\times 180.16$ (convert from $mol\,l^{-1}$ to $g\,l^{-1}$) $\div 10$ (convert from $g\,l^{-1}$ to % w/v) $=$ **2.70% w/v**.

(e) 1×74.55 (convert from $mol\,kg^{-1}$ to $g\,kg^{-1}$) $\div 10$ (convert from $g\,kg^{-1}$ to g per 100 g, i.e. % w/w) $= 7.455 =$ **7.46% w/w**.

24.3

(a) ***Tip:*** Use Eqn [24.6] and set $\phi = 1$ (ideal behaviour) to calculate the osmolality of individual solutes in each solution. (i) NaCl $= 50 \times 2 =$ **100 mosmol kg^{-1}**; (ii) KCl $= 200 \times 2 = 400\,mosmol\,kg^{-1}$; $CaCl_2 = 40 \times 3 = 120\,mosmol\,kg^{-1}$, giving a total osmolality of $400 + 120 =$ **520 mosmol kg^{-1}**. (iii) NaCl $= 100 \times 2 = 200\,mosmol\,kg^{-1}$; KCl $= 60 \times 2 = 120\,mosmol\,kg^{-1}$; $CaCl_2 = 75 \times 3 = 225\,mosmol\,kg^{-1}$, giving a total osmolality of $200 + 120 + 225 =$ **545 mosmol kg^{-1}**.

(b) ***Tip:*** To calculate osmotic pressure, use the relationship $1\,osmol\,kg^{-1} = 2.479$ MPa at 25 °C (p. 150), as follows: $100 \div 1000$ (convert $mosmol\,kg^{-1}$ to $osmol\,kg^{-1}$) $\times 2.479 = 0.2479 =$ **0.248 MPa**. (i) $520 \div 1000$ (convert $mosmol\,kg^{-1}$ to $osmol\,kg^{-1}$) $\times 2.479 = 1.289\,08 =$ **1.29 MPa**. (ii) $545 \div 1000$ (convert mosmol kg^{-1} to $osmol\,kg^{-1}$) $\times 2.479 = 1.351\,055 =$ **1.35 MPa**.
Tip: To calculate osmotic potential you can use Eqn [24.8] to calculate the value for each solution. In each case, the numerical value stays the same as that for osmotic pressure, as shown above, but there is a negative sign for osmotic potential, compared to osmotic pressure: (i) **−0.248 MPa**. (ii) **−1.29 MPa**. (iii) **−1.35 MPa**.

25 pH and buffer solutions

25.1 ***Tip:*** Use Eqn [25.5] to interconvert between pH and $[H^+]$.

(a) $7.4 = -\log_{10}[H^+]$, therefore $[H^+] =$ **3.98×10^{-8} mol l^{-1}**.

(b) $4.1 = -\log_{10}[H^+]$, therefore $[H^+] = 7.94 \times 10^{-5}\,mol\,l^{-1}$ and therefore **7.94×10^{-2} mol m^{-3}**.

(c) $pH = -\log_{10}[2 \times 10^{-5}]$, therefore pH $=$ **4.70**.

(d) $pH = -\log_{10}[10^{-12.5}]$, therefore pH $=$ **12.5**.

(e) First convert $[H^+]$ from $mol\,m^{-3}$ to $mol\,l^{-1}$. $[H^+] = 2.8 \times 10^{-8}\,mol\,l^{-1}$. Then use Eqn [25.5] $pH = -\log_{10}[2.8 \times 10^{-8}]$, therefore pH $=$ **7.55**.

25.2

(a) Since the assay does not involve whole cells or organelles (where toxicity would be a factor to consider), TRIS buffer would be suitable. It has a pK_a of 8.3 (Table 25.3), which is close to the required pH. If the enzymic reaction involves no major production of acids or bases a buffer concentration of $50\,mmol\,l^{-1}$ would be sufficient, otherwise a higher concentration (e.g. of $200\,mmol\,l^{-1}$) might be more appropriate.

(b) Here, you could use one of the zwitterionic 'Good buffers' such as MES, which has a pK_a of 6.1 (Table 25.3), and this is in the direction of any expected pH change during the experiment (*E. coli* cells normally show a decrease in pH during growth, as substrates are metabolised to acidic end-products). A reasonably high concentration of buffer would be appropriate (e.g. 200 mmol l^{-1}) if the experiment is to run for any length of time.

(c) TRIS buffer would *not* be suitable, as it shows a large change in pH with temperature (p. 156) and can also inhibit photosynthetic activity – either MOPS, HEPES or PIPES (Table 25.3) would be suitable alternatives, based on their pK_a values.

(d) Zwitterionic 'Good buffers' would *not* be suitable, as they interfere with some protein assays (e.g. Lowry method, p. 354) – citrate or acetate buffer would be suitable alternatives.

25.3 ***Tip:*** Rearrange Eqn [25.7] to give $\log_{10}[A^-]/[HA] = pH - pK_a$.

(a) $\log_{10}[A^-]/[HA] = 3.8 - 4.8 = -1$, so $\underline{\mathbf{[A^-]/[HA] = 0.1}}$ (1 to 10);

(b) $\log_{10}[A^-]/[HA] = 9.5 - 9.2 = 0.3$ so $\underline{\mathbf{[A^-]/[HA] = 1.995}}$ (approximately 2 to 1);

(c) $\log_{10}[A^-]/[HA] = 8.1 - 7.5 = 0.6$ so $\underline{\mathbf{[A^-]/[HA] = 3.98}}$ (approximately 4 to 1).

26 Introduction to microscopy

26.1 (a) T; (b) F; (c) F; (d) T; (e) T; (f) T; (g) T; (h) F; (i) F; (j) T.

26.2 Missing words are: diffracted; near-transparent; resolution; interference; contrast; 3D; focus; optical sectioning; polarised; laser.

26.3

(a) Embed; stain

(b) Air dry; stain

(c) Dehydrate.

27 Setting up and using a light microscope

27.1 ***Tip:*** To memorise the parts, it may help to imagine yourself setting up a microscope.

27.2

(a) Provides optimal resolution by ensuring light source is focused on slide.

(b) Provides magnified image to the eyepiece lenses.

(c) Enhances image contrast by cutting down on stray light reaching the objective lens.

(d) Sets up the microscope for the distance between your pupils, so you can use both eyepieces.

(e) Allows you to return to same place on a slide.

27.3 Answer (c) is correct (see pp. 163–65).

27.4 One eyepiece micrometer unit = 100 µm/11 = 9.09 µm; 25 eyepiece units = 25 × 9.09 µm = **<u>227 µm</u>**

27.5

(a) Chlorazol black

(b) Phloroglucinol/HCL or toluidine blue

(c) Lactophenol cotton blue

(d) Neutral red

(e) Mallory

(f) Gram

28 Making and recording measurements

28.1

(a) quantitative, discontinuous, ratio

(b) quantitative, continuous, ratio

(c) qualitative, discontinuous, nominal

(d) quantitative, discontinuous, ratio

(e) qualitative, discontinuous, nominal

(f) qualitative, discontinuous, ordinal

28.2 The results indicate that balance A has a bias of +0.05 g across the weighing range, while balance B has a consistent bias of about +0.04%. Both balances are precise, but not accurate.

28.3 ***Tip:*** Use a spreadsheet to lay out the table once you have decided what it needs to contain and what statistics you might wish to calculate. You can use spreadsheet functions to help you calculate some of the statistics.

29 SI units and their use

29.1

(a) **<u>0.997 atmospheres</u>**

(b) **<u>836 000 mm²</u>**

(c) **<u>568 ml</u>**

(d) **<u>310 K</u>**

(e) **<u>72.6 kg</u>**

29.2

(a) **<u>10 m</u>**

(b) **<u>15 µl</u>**

(c) **<u>5 gJ</u>**

(d) **<u>65 km s⁻¹</u>**

(e) **<u>100 pg</u>**

29.3 ***Tip:*** The units can be put in place of the symbols and cancelled out as follows:

$$\frac{m^2 \times Pa}{Pa\,s \times m} = \frac{m^2}{m\,s} = m\,s^{-1}$$

30 Scientific method and design of experiments

30.1 A possible answer (one of many) is given in the figure:

W	Z	Y	X
X	W	Z	Y
Y	X	W	Z
Z	Y	X	W

Z	W	X	Y
W	Z	Y	X
Y	X	W	Z
X	Y	Z	W

W	Y	X	Z
Y	Z	W	X
Z	X	Y	W
X	W	Z	Y

A possible layout of treatments in the glasshouse experiment.

The Latin square design ensures that treatments appear with equal probability in each row and column; hence, if there are gradients of potentially confounding variables, such as temperature and light in this case, their effects are spread evenly among treatments.

30.2 *Tip:* For Microsoft Excel, try the formula = INT(5*(RAND()) + 1).

30.3 *Tip:* It might be easier to lay out your answer like the table below:

Interactions between treatments in the cell proliferation experiment

Treatment	Mean result	Observed effect of treatment(s)	Expected effect of treatments	Classification
Control (baseline)	5.0	0.0	–	–
A	7.0	2.0	–	–
B	8.0	3.0	–	–
C	10.0	5.0	–	–
A + B	7.0	2.0	2 + 3 = 5	Antagonism
A + C	12.0	7.0	2 + 5 = 7	No interaction
B + C	17.0	12.0	3 + 5 = 8	Synergism
A + B + C	13.0	8.0	2 + 3 + 5 = 10	Antagonism

Notes:

- The observed effect is the mean result minus the control.
- The expected result is the additive effect of individual treatments.
- The classification is based on the difference between the observed and expected effects, shown in columns 3 and 4 above.

31 Project work

31.1 *Tips:*

- Divide the time available for your project into relevant parts concerned with different aspects of your work (note some of these could run in parallel, such as experimental work and research for your Introduction).
- A Gantt chart might be an appropriate method of presenting this (see e.g. http://freegroups.net/gantt).
- Set yourself realistic target dates for completion of each section.
- Identify important potential 'sticking points' (e.g. time taken for delivery of reagents and other supplies) and work around these (e.g. by doing library work).
- Make sure you allow some flexibility (say 5%) to allow for slippage or unforeseen problems.

31.2 *Tip:* Even if you prefer to write up your work on paper first, create appropriate computer files soon afterwards. Devise a simple and logical system for naming the files.

31.3 *Tip:* Remember that a word processor or a spreadsheet may have sufficient functions for your database needs for storing reference details (see p. 74).

32 Sterile technique and microbial culture

32.1

- Use a permanent marker;
- Write only on the *base* of the plate;
- Record information around the outermost region of the base;
- Record (i) date, (ii) medium, (iii) your name/initials, (iv) details of organism/treatment, etc.

32.2 *Tip:* Consider the suitability of the various types of sterilisation procedures listed in Chapter 32, and whether they might be applicable for each item.

(a) Tips for pipettors are often sterilised by irradiation during the manufacturing process and may be supplied in sterile form. Autoclaving (p. 204) is the easiest approach to routine laboratory sterilisation of these items.

(b) Blood is collected and handled using sterile technique, to avoid contamination, and therefore it does not require sterilisation – it is added to sterilised agar medium after sterilisation, once the medium has cooled to below 50 °C.

(c) Autoclave, e.g. at 121 °C for 15 min.

(d) Either dry heat sterilisation in a metal pipette can within a hot air oven at 160 °C for ≥2 h, or autoclaving at 121 °C for 15 min in the same type of pipette can.

(e) Red heat sterilisation in a Bunsen flame (Fig. 32.2).

(f) Filter sterilisation through a pre-sterilised filter unit, typically with a pore size of 0.45 or 0.2 μm. The filter-sterilised antibiotic solution is mixed thoroughly with the molten agar medium at 50 °C, then poured immediately. The filter unit may be purchased ready sterilised, or prepared by autoclaving.

32.3 *Tip:* You will need to consult UK ACDP guidelines and categorisation (for details, see sources for further study in Chapter 32) – some information on categorisation and safe working with specific pathogens is available on the Web (note that other countries may use a slightly different categorisation, e.g. biosafety levels, BSL 1–4 in the USA):

(a) Category 2
(b) Category 2 (gloves should also be worn)
(c) Category 3
(d) Category 1
(e) Category 3
(f) Category 3
(g) Category 2 (a microbiological safety cabinet should be used)
(h) Category 3
(i) Category 1
(j) Category 4

32.4 The following list covers six important points, and is based on material in Chapter 32. However, you and your fellow students may think of other valid points:

- Spread plates show the typical colonial form of the microbe, whereas most of the colonies in a pour plate are contained within the agar and typical colony morphology cannot be observed.

- Pour plates are more suitable for counting higher numbers of microbes, since the colonies are smaller and are contained within the agar medium, in contrast to spread plates.
- Heat-sensitive microbial cells may be damaged by the temperature of the molten agar used in pour plating (45–50 °C) and may not grow on subsequent incubation.
- Some liquid will adhere to the spreader, and this may give a lowered count compared to a pour plate. Similarly, when the sample and molten agar are pre-mixed before pouring (see Fig. 32.5), some of the cells may not reach the pour plate, leading to a reduced count (note that this is one reason why some workers carry out the mixing of sample and molten agar within the Petri plate, rather than in the bottle, though this needs to be done thoroughly to ensure an even distribution of sample within the molten medium).
- Each type of plating procedure requires different apparatus – spread plating requires a sterile spreader and pre-poured plates, while pour plating requires molten agar at 50 °C (usually kept in a water bath until needed).
- There is a fire risk associated with using ethanol to flame-sterilise a glass spreader. This risk is absent in pour plates, and can be removed in spread plates by using a disposable plastic spreader.

33 Isolating, identifying and naming microbes

33.1

(a) A range of approaches is possible, including taking swabs, which are then surface spread onto an agar-based medium (e.g. a general-purpose medium such as nutrient agar, or blood agar), or direct 'imprints' from the skin surface onto agar-based 'contact' plates.

(b) Important factors to consider here are, firstly, limiting the exposure of the sample to oxygen – for example, but taking a core through the mud and putting the core into an anaerobic jar (p. 212) immediately after sampling and, secondly, keeping the sample cool, since the organisms are psychrotrophs – for example, by putting the anaerobic jar into an ice bucket during transportation to the laboratory.

(c) You will need to consider (i) where to sample; (ii) how much water to collect (typically, faecal bacteria are often counted in volumes of up to 100 ml of sea water); and (iii) how many samples to collect (typically, at least three separate volumes of 100 ml). The samples should not be exposed to sunlight, as this may inactivate the bacteria, should be kept at 1–4 °C during transit to the laboratory and processed within 6 h of collection.

(d) One approach is to take 'contact' plates by pressing the selected surface of the leaf onto a suitable agar-based medium (e.g. a non-selective medium such as place count agar for heterotrophic bacteria). This should be done as soon as possible after collecting the leaves, and is feasible under field conditions. Incubation at different temperatures would give counts for psychrophiles, mesophiles and thermophiles.

33.2

(a) Use positive phototaxis towards a unidirectional light source. 'Sloppy' agar (e.g. 0.5% w/v agar) can help reduce contamination due to other microorganisms.

(b) Rinse free of other microbes using sterile water, then use sonication to disrupt the biofilm and create a suspension for further investigation (e.g. culture). Alternatively, you might try an enzyme mixture formulated to hydrolyse the biofilm polymers, if their composition is known.

(c) Use membrane filtration (e.g. pore size 0.45 μm) to 'concentrate' the enterococci, then transfer the membrane (face up) to a suitable selective medium (e.g. glucose azide medium) for enumeration after incubation at an appropriate temperature, e.g. at 44 °C.

(d) Use a micromanipulator to separate the target bacterium from the other microbes. Alternatively, you might try dilution to extinction (p. 211), but you would be likely to look through a large number of cultures before finding the distinctively shaped bacterium.

33.3

(a) NaCl at 7.5% is selective for halotolerant bacteria, while mannitol is present as a specific carbohydrate nutrient source for *S. aureus*. Phenol red in the medium turns yellow when mannitol is metabolised to acidic end-products. *S. aureus* grows as larger colonies with a yellow 'halo' in an otherwise red medium, due to mannitol breakdown. Other *Staphylococcus* spp. grow as smaller colonies, with no yellow 'halo'.

(b) The surfactant sodium lauryl sulphate is selective for coliform bacteria. Lactose is present as a specific carbohydrate nutrient source, plus phenol red as a pH indicator. Coliforms, including *E. coli*, grow in the presence of sodium lauryl sulphate and metabolise lactose to acidic end-products, giving large yellow colonies. A selective temperature of 44 °C can be used to grow thermotolerant (faecal) coliforms, most of which are *E. coli*.

(c) Sodium azide is the selective ingredient, inhibiting the growth of most obligate aerobes and facultative anaerobes but leaving *Enterococcus* spp. unaffected, as they are aerotolerant anaerobes (see Fig. 33.1). The medium also contains a tetrazolium salt as a general indicator of metabolic activity, used to help visualise the small blue–purple colonies.

(d) Polymyxin is present as a selective agent for *Bacillus* spp. Mannitol and pyruvate are present as specific carbohydrate nutrient sources. *B. cereus*, in contrast to most other species of *Bacillus*, can utilise pyruvate but not mannitol, causing the pH to rise (most other *Bacillus* spp. use both mannitol and pyruvate, causing the pH to drop or remain stable). Bromophenol blue changes from green (neutral) to peacock blue (alkaline) for *B. cereus*. The egg yolk lecithin is broken down by a lecithinase produced by this bacterium, giving an opaque 'halo' of precipitated degradation products around the colonies.

33.4 Colony appearance is typically as follows (note that the growth medium can modify the appearance from that noted below):

(a) Shiny, off-white, mucoid, slightly raised circular colonies with an entire (smooth) margin, 2–3 mm in diameter.

(b) Mucoid, flat spreading colonies with an undulate margin, often with green pigmentation, 4–5 mm in diameter.

(c) Circular 'pinpoint' colonies, convex and often with yellow or orange pigmentation, 1–2 mm in diameter.

(d) Circular off-white colonies, with a depressed centre ('draughtsman' colonies), α-haemolytic on blood agar, 1–3 mm diameter.

(e) Flat irregular off-white colonies with a matt surface texture and lobed margin, 2–4 mm diameter.

(f) Raised circular white colonies with an entire (smooth) margin, 3–4 mm in diameter.

33.5

(a) *N′-N′-N′-N′*-tetramethyl-*p*-phenylenediamine acts as an electron donor to cytochrome oxidase and is converted from the colourless, reduced form to the oxidised form, which has a strong purple–blue colour (a positive result).

(b) Indole, produced as a result of the breakdown of tryptophan, can be detected by a variety of methods including Kovac's reagent (*para*-dimethylamino-benzaldehyde in alcohol).

(c) β-galactosidase can be detected using a suitable chromogenic substrate (a synthetic substrate that liberates a coloured product when cleaved by the target enzyme), such as *o*-nitrophenyl-β-D-galactoside (ONP-GAL), or a fluorogenic substrate (a synthetic substrate that liberates a fluorescent product when cleaved by the target enzyme), such as 4-methylumbelliferyl-β-D-galactoside (MU-GAL).

(d) Hydrolysis of urea to ammonia and CO_2 is most easily demonstrated by including a pH indicator such as phenol red, to visualise the alkalinisation that accompanies this reaction. In this instance, the medium turns from yellow to red as the pH rises.

(e) The decarboxylation of lysine under anaerobic conditions can be detected by an alkalinisation of the medium, usually visualised using bromophenol purple, which changes from yellow to purple as the pH rises due to the action of lysine decarboxylase.

(f) H_2S production from S-containing amino acids is most easily demonstrated by incorporating an iron-containing salt in the medium (e.g. ferrous ammonium sulphate). The H_2S produced by the growing bacteria combines with the ferrous ions to form a black deposit of iron sulphide. In tube tests, a filter paper strip containing lead acetate turns black on exposure to H_2S.

33.6

(a) *Proteus mirabilis*.
(b) *Escherichia coli*.
(c) *Salmonella* spp.
(d) *Morganella morganii*.
(e) *Escherichia coli*.

34 Working with animal and plant tissues and cells

34.1 ***Tip:*** See Chapter 24 for details of osmolarity. Assuming that each solute behaves in an ideal manner, then the osmolarity of each constituent will be equivalent to the amount in grams per litre divided by the relative molecular mass and multiplied by the number of osmotically active particles produced from each solute molecule when it dissolves in water. Thus NaCl $= 6 \div 58.44 \times 2 = 0.205\,338\,8$ osmol l^{-1}; KCl $= 0.075 \div 74.55 \times 2 = 0.002\,012\,1$ osmol l^{-1}; $CaCl_2 = 0.1 \div 110.99 \times 3 = 0.002\,702\,9$ osmol l^{-1}; $NaHCO_3 = 0.1 \div 84.01 \times 2$ (note that $NaHCO_3$ gives only two osmotically active entities in water) $= 0.002\,380\,7$ osmol l^{-1}. Added together, this gives an osmolarity of $0.212\,434\,5$ osmol l^{-1} = **212.4 mosmol l^{-1}**. Note that you should use the full numerical values until the last step, to avoid introducing 'rounding' errors.

34.2 ***Tip:*** The UK national culture collection website at http://www.ukncc.co.uk provides coordinated information for a number of individual UK collections. The UK National Collection of Type Cultures (NCTC) can be accessed at http://www.phls.co.uk/services/nctc and the American Collection of Type Cultures (ATCC) is at http://www.atcc.org while the World Federation of Culture Collections has a website at http://wdcm.nig.ac.jp/wfcc/index.xml. Box 10.3 gives further details on the evaluation of web-based resources.

34.3

(a) HeLa cells are named after their source, Ms Henrietta Lacks, of Baltimore, USA.

(b) Vero cells are derived from the kidney of an African green monkey (*Cercopithecus aethiops*), known as the vervet monkey, hence 'vero'.

(c) BHK stands for 'baby hamster kidney', the original source of this cell line.

(d) CHO stands for 'Chinese hamster ovary', the original source of this cell line.

(e) TBY-2 stands for tobacco bright yellow strain 2, a plant cell line.

35 Culture systems and growth measurement

35.1 ***Tip:*** Use Eqns [35.1] and [35.2] for these calculations.

(a) Substituting the appropriate values into Eqn [35.1], μ = [2.303 (6.531 478 9 − 4.716 003 3)] ÷ 9 = 0.464 56 = **0.465 h^{-1}**. Similarly, using Eqn [35.2], g = (0.301 × 9) ÷ (6.531 478 9 − 4.716 003 3) = 1.492 171 = **1.49 h** (approximately 1 h 29 min).

(b) Substituting the appropriate values into Eqn [36.1], μ = [2.303 (7.826 074 8 − 4.477 121 3) ÷ 200 = 0.038 563 2 = **0.0386 min^{-1}**. Similarly, using Eqn [35.2], g = (0.301 × 200) ÷ (7.826 074 8 − 4.477 121 3) = 17.975 765 = **18.0 min**.

(c) Substituting the appropriate values into Eqn [35.1], μ = [2.303 (2.397 94 − 1.505 15)] ÷ 2.5 = 0.822 438 2 = **0.822 h^{-1}**. Similarly, using Eqn [35.2], g = (0.301 × 2.5) ÷ (2.397 94 − 1.505 15) = 0.842 863 4 = **0.843 h** (approximately 51 min). Note that the time difference in this example is two and a half hours, not 2.3 hours!

35.2 ***Tip:*** Use Eqn [35.6] (Box 35.1) to carry out these calculations.

(a) 6.42 ÷ 1 ÷ (2.5 × 10^{-7}) × 1 = 2.568 × 10^{-7} = **2.57 × 10^{7} ml^{-1}**.

(b) Total cell count = 698, so 698 ÷ 20 ÷ (8 × 10^{-7}) × 1 = 4.3625 × 10^{7} = **4.36 × 10^{7} ml^{-1}**.

(c) 78 ÷ 25 ÷ (2.5 × 10^{-7}) × 10^{2} = 1.248 × 10^{9} = **1.25 × 10^{9} ml^{-1}** (note the correction factor of 10^{2}, to account for the dilution factor of 10^{-2}).

(d) 6.8 ÷ 1 ÷ (2 × 10^{-5}) × 0.1 = **3.40 × 10^{4} ml^{-1}** (note the correction factor of 0.1, because the suspension was concentrated by a factor of 10-fold).

35.3 ***Tip:*** Use Eqn [35.7] (Box 35.2) to carry out the calculations.

(a) (54.4 ÷ 0.1) × 10^{5} = **5.44 × 10^{7} CFU ml^{-1}**.

(b) The mean count per spread plate is (34 + 40 + 37) ÷ 3 = 37.0 CFU. So (37.0 ÷ 0.2) × 10^{3} = **1.85 × 10^{5} CFU ml^{-1}**.

(c) The mean count per pour plate is (211 + 186 + 194 + 202) ÷ 4 = 198.25 CFU. So (198.25 ÷ 0.5) × 20 = **7.93 × 10^{3} CFU ml^{-1}**.

(d) The mean count per plate is (35 + 41 + 32) ÷ 3 = 36 CFU. So (36 ÷ 1) × 50 = 1800 per g and therefore 1800 × 100 = **1.80 × 10^{5} CFU per 100 g**. Note that the multiplication factor to correct for dilution was 50 in this case, since the seafood had been diluted to 2% w/v.

(e) This example does not require Eqn [35.7] in full. The mean count for 250 ml was (45 + 32) ÷ 2 = 38.5 CFU. Therefore, the count per 100 ml is 38.5 ÷ 250 × 100 = **15.4 CFU per 100 ml**.

(f) The mean count per plate is (28 + 32 + 39) ÷ 3 = 33.0, so using Eqn [35.7], (33.0 ÷ 0.5) × 10^{1} = **660 CFU ml^{-1}**. This value is higher than the EU regulations (maximum 100 CFU ml^{-1}), so this water sample does not meet the specification required of mineral water.

35.4 ***Tip:*** You need to graph heart rate against log concentration values; this should give a sigmoidal curve (see below). Having obtained log concentration estimates graphically from the heart rate values for the two loch samples, convert back to mmol l^{-1} concentration by raising 10 to the power of the estimated log concentration (see p. 476).

(a) The estimate for (i) is approximately **1.58 × 10^{-5} mol m^{-3}**; and for (ii) **approximately 6.31 × 10^{-5} mol m^{-3}**.

(b) Estimate (ii) is the least reliable owing to the shallowness of the log concentration versus heart rate curve at this point: a small error in heart rate will lead to a large difference in estimated concentration.

(c) If sample (ii) were diluted by about fivefold or so, then tested again, this would shift the 'expected' heart rate downwards to the most sensitive part of the graph. The dilution factor would need to be accounted for when estimating the true concentration.

(d) There is no guarantee that the only substance affecting the *Daphnia* sp. heart rate in the loch samples is the toxin tested. Other substances present might increase or decrease heart rate.

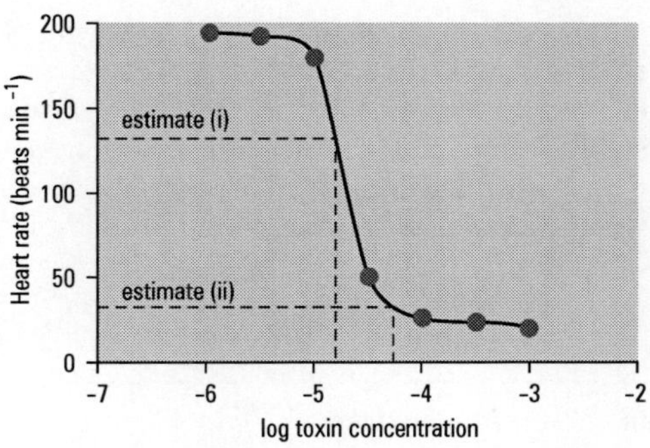

Responses of *Daphnia* to toxin.

35.5 ***Tip:*** You will first need to convert the nisin concentrations of the standard solutions to $\log_{10}$ values, before plotting zone diameter against $\log_{10}$ nisin concentration (see figure). Based on the calibration curve, solution 5 contains antilog$_{10}$ 1.24 = **17.4 µg ml^{-1}** (to 1 decimal place).

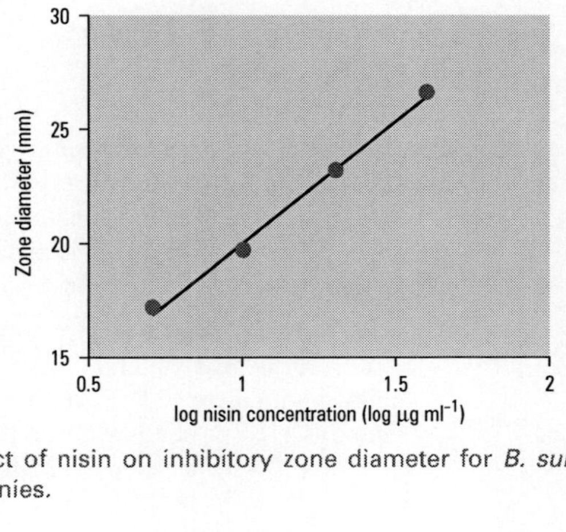

Effect of nisin on inhibitory zone diameter for *B. subtilis* colonies.

36 Homogenisation and fractionation of cells and tissues

36.1 ***Tip:*** Table 36.1 provides a summary of techniques.

(a) Gram-negative bacteria have strong, multi-layered cell envelope (Fig. 36.1), best disrupted by EDTA treatment (to destabilise the outer membrane) and lysozyme (to digest the peptidoglycan that provides most of the structural strength of the cell wall). The procedure should be carried out in a dilute medium, to encourage the cells to burst. Mechanical techniques such as ultrasonication and the French pressure cell (Table 36.1) can be used, but are often less effective.

(b) Skeletal muscle is difficult to disrupt because of its high content of fibrous contractile proteins and connective tissue components such as collagen: vigorous mechanical methods are best, e.g. a rotor-stator.

(c) Plant tissue is usually quite resilient, due to major structural components in the cell wall such as cellulose, lignins and waxes. Incubation in a lytic enzyme mixture, followed by mechanical disruption (e.g. using a blender) is often the best approach.

36.2 Vigorous methods such as use of a Waring blender or a rotor-stator (Fig. 36.2) would be suitable for isolating a soluble protein, as it is not necessary to maintain the structure of organelles. However, less vigorous disruption, e.g. using a ground-glass or Teflon homogeniser (Fig. 36.3), would ensure least disruption when isolating intact organelles.

36.3

Selection of components, with explanation

Selection	Explanation
(i): component (d)	A detergent such as Triton X-100 is required to disrupt the membrane, by interacting with the lipid component, and thereby allowing solubilisation of the membrane protein component (which will normally have extensive hydrophobic regions).
(ii): component (a)	When preparing mitochondria, sucrose is used at a concentration suitable to maintain the osmolarity of the medium at a level that will not result in osmotic lysis of the organelle. (Osmotic properties of solutions are further discussed in Chapter 24.)
(iii): component (c)	Dithiothreitol is a strong reducing agent that can prevent oxidation of thiol groups in proteins (p. 244) and thereby maintains biological activity of oxygen-sensitive enzymes and other proteins.
(iv): component (b)	Low ionic strength buffer is used because the contractile proteins actin and myosin are solubilised with increased salt concentration, leading to a highly viscous solution that is difficult to process further when high ionic strength buffers are used.

37 Calibration and its application to quantitative analysis

37.1 See the figure below – note that this calibration line is approximately linear up to $\approx 4.0\,\mu g\,ml^{-1}$ but it becomes increasingly curved above this point. Water sample (a) contains Zn at **<u>1.70 μg ml⁻¹</u>**; water sample (b) contains Zn at $3.5 \times 20 =$ **<u>70.0 μg ml⁻¹</u>**; water sample (c) contains $8.3 \times 5 =$ **<u>41.5 μg ml⁻¹</u>** (note that the graph curves at higher concentrations, making the estimate for water sample (c) less reliable than the other two – a better approach might have been to dilute this sample further and reassay).

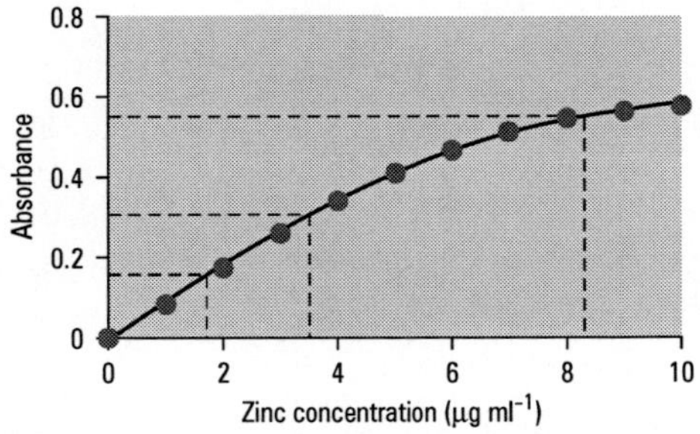

Calibration curve for a series of solutions containing zinc, based on data in the table in study exercise 37.1.

37.2 Note that this data set is a very good fit to a straight line – linear regression analysis carried out using Excel gives a value for r^2 of 0.996 829 051, with an intercept of 0.001 714 286 and a slope of 7.424 285 714. The protein content of each solution could be determined directly from a print-out of the graph (e.g. the figure below). However, the calculations below show how you can substitute values into the equation for a straight line, $y = a + bx$, to give the required answers.

(a) $0.225 = 0.001\,714\,286 + 7.424\,285\,714x$, so $x = (0.225 - 0.001\,714\,286) \div 7.424\,285\,714 = 0.030\,075\,043 =$ **<u>0.0301 mg protein</u>**.

(b) $0.465 = 0.001\,714\,286 + 7.424\,285\,714x$, so $x = (0.465 - 0.001\,714\,286) \div 7.424\,285\,714 = 0.062\,401\,385 =$ **<u>0.0624 mg protein</u>**.

(c) $0.682 = 0.001\,714\,286 + 7.424\,285\,714x$, so $x = (0.682 - 0.001\,714\,286) \div 7.424\,285\,714 = 0.091\,629\,786 =$ **<u>0.0916 mg protein</u>**.

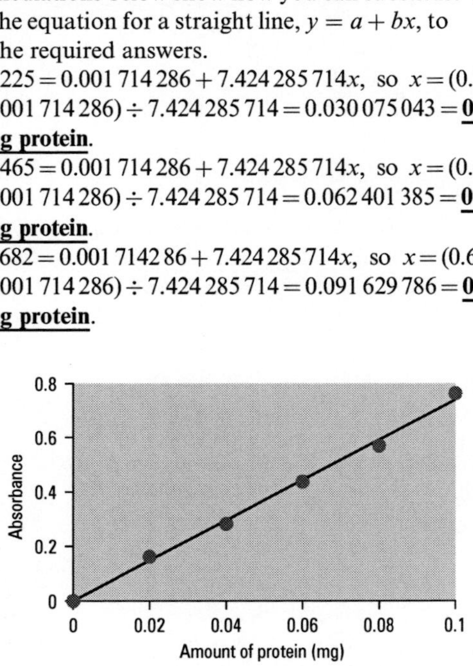

Calibration curve for a series of protein standards, based on data in the table in study exercise 37.2.

37.3 ***Tip:*** Check that you understand the general requirements for good graph drawing, detailed in Chapter 63.

(a) The title of the figure in study exercise 37.3 ('calibration graph') does not give sufficient information to enable the reader to understand the content.
(b) No details are given as to the nature of each set of data values – are these curves for two separate substances?
(c) The x-axis has a poor choice of units – it would be better to use µg, then all of the values would be 1000 times smaller.
(d) The x-axis runs beyond the highest value (8 µg), wasting around 20% of the horizontal space.
(e) The scale marks (tic marks) on the x-axis are missing.
(f) The x-axis has no zero shown.
(g) The y-axis has too many numbers, and too many minor scale marks – it is probably better to have a tic mark every 0.1 absorbance units. (Note that absorbance (A) is a dimensionless term, so no units are required – this is *not* an error.)
(h) The data values represented by the square symbols show a reasonable fit to the lower linear trend line, but the data values represented by the round symbols do not – the latter set of data shows a clear curvilinear relationship and therefore a linear trend line is not appropriate.

38 Immunological methods

38.1 ***Tip:*** Take care to make sure that your construction lines are carefully drawn by hand – they must be *parallel* to the appropriate axis. The test sample contains 7.5 µg of antigen.

38.2 ***Tip:*** Remember that you need to plot the *square* of the ring diameter versus the amount of antigen in each standard. The test sample has a squared ring diameter of 42.25 mm^2 and contains 11 µg of antigen (see the figure below).

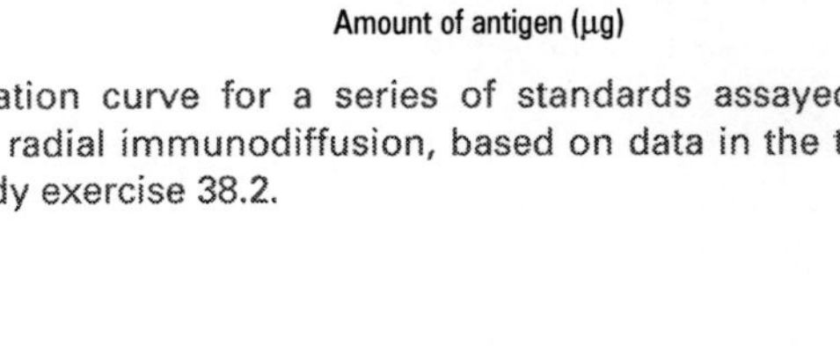

Calibration curve for a series of standards assayed by single radial immunodiffusion, based on data in the table in study exercise 38.2.

38.3

(a) Wells 1 and 2 in the figure in study exercise 38.3 show the 'prozone' phenomenon, with a negative haemagglutination test result (a small 'button' of cells), due to an excess of antibody. Wells 4 to 10 show strong positive haemagglutination, with an even 'carpet' of red blood cells rather than a 'button'. Wells 11 and 12 show a negative reaction, due to the further dilution of the test serum.
(b) The lowest dilution giving a positive result is well 10, which is nine doubling dilutions of the original 10-fold dilution, i.e. $10 \times 2^9 = 5120$-fold. Therefore the titre is 1 in 5120.
(c) The test serum is positive, since the titre is greater than 1 in 5000.

39 Radioactive isotopes and their uses

39.1 Assume a ^{14}C half-life of 5715 years (Table 39.2). Using Eqn [39.1], $e^x = 0.5725$, and so $x = \ln(0.5725) = -0.5578$. If $-0.5578 = -0.693(t/5715)$, then $t = (-0.5578/-0.693) \times 5715 = 4600$. **The rat visited about 4600 years ago.**

39.2

(a) **1200 Bq**.
(b) **4.44×10^7 d.p.m.**
(c) **1.20 µCi.**
(d) **30.0 Bq g^{-1}**.
(e) **28.0 pmol.**

39.3

(a) The solution contains $250 \times 10^{-6} \times 5\,\text{mol} = 1.25 \times 10^{-3}\,\text{mol}$. The specific activity is therefore $55 \times 10^6 / 1.25 \times 10^{-3}\,\text{Bq mol}^{-1}$ = **$4.4 \times 10^{10}\,\text{Bq mol}^{-1}$**.
(b) $79.2 \times 10^5 / 4.4 \times 10^{10} = 1.8 \times 10^{-4}$ mol per 2 h per 10^7 cells is equivalent to **2.5 fmol s^{-1} cell^{-1}**.

40 Light measurement

40.1 First convert PFFD to photosynthetic irradiance (PI) using Table 40.2: $1610 \div 4.6 = 350\,\text{W m}^{-2}$ ($\equiv \text{J m}^{-2}\,\text{s}^{-1}$). Then, calculate the PI for the leaf area of 45 cm^2 ($\equiv 0.0045\,\text{m}^2$): $350 \times 0.0045 = 1.575\,\text{J s}^{-1}$. Finally, calculate the total energy supplied over the 30 min (1800 s) period: $1.575 \times 1800 = 2835\,\text{J}$ = **2840 J**.

40.2 ***Tip:*** Convert values to PPFD expressed in µmol photons m^{-2} s^{-1} using the conversion factors given in Table 40.2, where necessary.

(a) 300×4.6 (to convert to PPFD) = **1380 µmol photons m^{-2} s^{-1}**;
(b) $6.5 \div 1000$ (to convert to µmol) $\times$ 10 000 (to convert to m^{-2}) = **65.0 µmol photons m^{-2} s^{-1}**;
(c) this value is already expressed in the appropriate units, **275 µmol photons m^{-2} s^{-1}**;
(d) 35.2×4.6 (to convert to PPFD) = 161.92 = **162 µmol photons m^{-2}s^{-1}**;
(e) $1.15 \div 1000$ (to convert to W) $\times$ 10 000 (to convert to m^{-2}) $\times$ 5.0 (to convert to PPFD) = **57.5 µmol photons m^{-1} s^{-1}**;

The locations ranked in order of decreasing PPFD are therefore as follows: (a); (c); (d); (b); (e).

40.3 ***Tip:*** If you simply use the term 'light meter' in a general-purpose search engine (e.g. Google), you will find out mostly about photographic light meters, and these are not suitable for scientific purposes – you'll need to add other terms, such as photosynthesis, quantum, irradiance, photon flux density, etc. Relevant features might include: measuring scale(s); range(s); detector/receiver type; cost; portability/weight; power supply; the type of sensor (e.g. 3D, flat or rod-shaped); whether the sensor can be used in water; whether the meter can be used with a data logger.

41 Basic spectroscopy

41.1 ***Tip:*** Box 41.1 gives stepwise instructions on using a spectrophotometer – check that you have covered all of the major points. However, the exact details will vary from instrument to instrument.

41.2 ***Tip:*** Substitute values into Eqn [41.3] and solve for [*C*] (margin examples are given on p. 273).

(a) Using Eqn [41.3]: 0.53 = 6220 × 1 × [*C*], therefore [*C*] = 0.000 085 209 mol l^{-1} = **85.2 μmol l^{-1}**.

(b) First calculate the concentration, using Eqn [41.3]: 0.62 = 6220 × 5 × [*C*] = 0.000 019 935 7 mol l^{-1} = 19.9357 μmol l^{-1}. Next, calculate the amount in 20 ml: 19.9357 ÷ 1000 (amount in 1 ml) × 20 (amount in 20 ml) = 0.398 714 μmol = **398 nmol**.

(c) Using Eqn [41.3] (note that you must convert the path length from mm to cm beforehand): 0.57 = 20 × 0.5 × [*C*] = 0.057 g l^{-1} = **57.0 μg ml^{-1}**.

(d) First calculate the concentration, using Eqn [41.3]: 0.31 = 20 × 1 × [*C*] = 0.0155 g l^{-1}. Next, express the concentration in the required units: 0.0155 × 10^9 (convert to ng) ÷ 10^6 (convert to μl) = 15.5 ng μl^{-1}. Then, calculate the amount in 50 μl: 15.5 × 50 = **775 ng**.

(e) Substitute the values into the equation given: (11.93 × 0.182) – (1.93 × 0.035) = 2.103 71 mg l^{-1}. Next, calculate the amount in 50 ml: 2.103 71 ÷ 1000 (amount in 1 ml) × 50 (amount in 50 ml) = 0.105 185 5 mg. Since this is the amount of chlorophyll in 1.56 g fresh weight of leaf, divide by the fresh weight to give the content: 0.105 185 5 ÷ 1.56 = 0.067 426 6 mg g fresh weight^{-1}. Finally, convert to the specified units: 0.067 426 6 × 1000 (convert from mg to μg) = 67.4266 = **67.4 μg (g fresh weight)$^{-1}$**.

41.3 First determine the concentration of *p*-nitrophenol, in mol l^{-1}. Thus 8.8 μg ml^{-1} ÷ 10^6 (convert to g ml^{-1}) × 10^3 (convert to g l^{-1}) ÷ 291.27 (convert to mol l^{-1}) = 0.000 030 212 5 mol l^{-1}. Next, substitute values into Eqn [41.3] and solve for ε: 0.535 = ε × 1 × 0.000 030 212 5, so ε = 17 707.902 36 = **17 700 l mol^{-1} cm^{-1}**. Note that you should use the full numerical value until the final stage, to avoid introducing 'rounding' errors.

41.4 The calibration graph is shown in the figure (next column). The test solution has a K$^+$ concentration of 0.32 mmol l^{-1}. This is equivalent to 0.32 ÷ 1000 × 25 = 0.008 mmol in 25 ml (the test sample volume). Next, divide by the weight of tissue used in grams: 0.008 ÷ 0.482 = 0.016 598 mmol (g tissue)$^{-1}$ = **16.6 μmol (g tissue)$^{-1}$**.

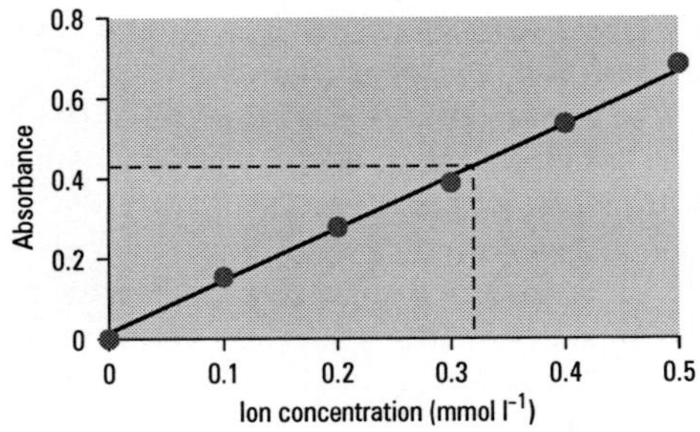

Calibration curve for a series of standards, assayed for K$^+$ and based on the data in the table in study exercise 41.4.

41.5 ***Tip:*** Fig. 44.3 shows the principles involved. Some of the excitation light energy absorbed by the molecule is dissipated as heat during the short time the molecule is in the excited state (typically ≈10^{-9} seconds), resulting in a decrease in energy from the second excited state to the first. Thus less energy is available for emission. Since the wavelength of light increases with decreasing energy (see Eqns [41.1] and [41.2]), the emitted light has a longer wavelength than the light used to excite the molecule. The difference between the emission wavelength (λ_m) and the excitation wavelength (λ_{ex}) is known as the Stokes shift.

41.6 ***Tip:*** A comparison of the light paths in Figs 41.1 and 41.5 should give you some clues to the answer. The spectrophotometer cuvette has two clear sides to allow transmitted light to pass directly through the solution from the source to the photocell detector. The opaque sides are for handling the cuvette (you should never handle spectrophotometer cuvettes by their clear faces). In contrast, a fluorimeter cuvette has four clear sides, since emitted light is detected at right angles to the excitation beam, resulting in low background readings and high sensitivity. In practical terms, this means that you must handle fluorimeter cuvettes close to the top, to avoid contaminating these surfaces (the excitation and emission beams are usually measured close to the centre of the cuvette).

42 Advanced spectroscopy and spectrometry

42.1

(a) Peak (i).

(b) Peak (ii).

42.2 ***Tip:*** the term 'chemical shift' refers to the position in a spectrum of a signal from an absorbing magnetic nucleus relative to the signal from a reference compound. Ideally, the reference compound will give a single, unambiguous, sharp resonance at a frequency relatively far from the resonances of the compounds under analysis.

(a) Tetramethylsilane (TMS) [$(CH_3)_4Si$] is a suitable reference for ^{1}H as it contains 12 protons in identical environments, giving a single peak well-separated from those of the protons of most other molecules.

(b) TMS is also suitable as a reference compound for ^{13}C NMR in most biological applications.

(c) Phosphocreatine is a useful reference compound for metabolic studies using ^{31}P, as it is often present in the tissue or extract being studied, and therefore does not need to be added (for more detail, see Macomber, 2000, p. 295).

42.3 In both instances, ppm is an abbrevation for 'parts per million', and the units must be established from the context in which the expression is used. In NMR, ppm refers to the small frequency changes that occur relative to the reference standard, expressed in proportional terms. Thus 1 ppm in NMR means a frequency difference of 1 Hz per MHz, e.g. 270 Hz in a 270 mHz spectrum. When used as a concentration term, ppm is equivalent to the expression of a concentration in parts per million, e.g. as $\mu g\,ml^{-1}$ ($10^{-6}\,g\,ml^{-1}$) in w/v terms, or $\mu g\,g^{-1}$ ($10^{-6}\,g\,g^{-1}$) in w/w terms.

42.4 ***Tip:*** to explain the ^1H NMR spectrum, you need to consider the presence of exchangeable, ionisable or dissociable hydrogen atoms in the molecule's structure. The dissociable hydrogens in a molecule will readily exchange with any others in the sample, particularly with those of water, and therefore do not yield a discrete ^1H NMR signal. The molecular structure of creatinine contains hydrogen atoms in four distinct environments (see the second figure in study exercise 42.4). Of these, two are attached to nitrogen atoms and are likely to be easily dissociated in aqueous systems: the majority of hydrogens bonded to atoms other than carbon tend to be readily dissociated, especially oxygen-bonded and nitrogen-bonded hydrogens. Those hydrogens bonded to the carbon skeleton are not readily exchanged and therefore give the observed signal (two distinct peaks) in the first figure in study exercise 42.4.

43 Centrifugation

43.1

(a) Use a high speed centrifuge and differential centrifugation.
(b) Use silicone oil microcentrifugation (e.g. using an Eppendorf-type microfuge).
(c) Use continuous flow centrifugation.

43.2 ***Tip:*** Substitute values into Eqn [43.1].
(a) RCF = 1.118 × 125 (4000 ÷ 1000)^2 = **2240 *g***;
(b) RCF = 1.118 × 60 (12 000 ÷ 1000)^2 = **9660 *g***;
(c) RCF = 1.118 × 186 (30 000 ÷ 1000)^2 = **187 000 *g***.

43.3 ***Tip:*** Substitute values into Eqn [43.2].
(a) 945.7 $\sqrt{(1500 \div 95)}$ = **3760 r.p.m.**
(b) 945.7 $\sqrt{(50\,000 \div 135)}$ = **18 200 r.p.m.**
(c) First convert the radius of rotation from cm to mm before substituting into Eqn [43.2]:
945.7 $\sqrt{(13\,000 \div 55)}$ = **14 500 r.p.m.**

43.4 ***Tip:*** Use Eqn [43.1] with each of the values in turn:
For r_{min} = 45 mm, 1.118 × 45 (5)^2 = **1260 *g***.
For r_{av} = 70 mm, 1.118 × 70 (5)^2 = **1960 *g***.
For r_{max} = 95 mm, 1.118 × 95 (5)^2 = **2660 *g***.
This calculation illustrates the range of *g* values possible within a single centrifuge tube, with the maximum value being more than double the minimum value in this instance.

44 Chromatography – separation methods

44.1 ***Tip:*** use Eqn [44.1] to determine R_F values and Eqn [44.2] for R_A values, based on the measured distances travelled by each component, relative to the solvent front.
(a) Using Eqn [44.1], the corresponding R_F values are:
pigment A = 63 ÷ 114 = **0.553**;
pigment B = 76 ÷ 114 = **0.667**;
pigment C = 86 ÷ 114 = **0.754**.
(b) Using Eqn [44.2], the corresponding R_A values are:
pigment B = 76 ÷ 63 = 1.206 349 2 = **1.21**.
pigment C = 86 ÷ 63 = 1.365 079 4 = **1.37**.

44.2

(a) ***Tip:*** Calculate the selectivity using Eqn [44.3], using the elution time for the compound excluded from the stationary phase as the column dead time (t_0), with all times converted to seconds. Thus α = (372 – 95) ÷ (270 – 95) = 1.582 857 1 = **1.58**.
(b) ***Tip:*** Calculate the resolution using Eqn [44.4]. Thus R = 2(372 – 270) ÷ (40 + 44) = 2.428 571 4 = **2.43**. This is a good level of resolution for quantitative work, giving clear separation of the two peaks.

44.3

(a) ***Tip:*** The isoelectric point (pI) is the intercept on the *x*-axis. Protein A has a pI of **6.0**. Protein B has a pI of **8.2**.
(b) ***Tip:*** Use the pI values to predict the net charge of each protein at the pH used. Protein A has a pI of 6.0 so at pH 7.0 it will have a net negative charge and bind to the anion exchanger; at pH 7.0, protein B (pI = 8.2) will have a net positive charge and will not be bound to the positively charged matrix and will be rapidly eluted.
(c) At pH 5.5, both proteins will have a net positive charge and will bind to the negatively charged cation exchanger. However, the net charge on protein A will be less than that on protein B and the order of elution on a salt gradient will be A, followed by B.

44.4 ***Tip:*** Chapter 54 discusses the interaction of ammonium sulphate and proteins. Ammonium sulphate ions are hydrated and limit the water required to keep many proteins solvated. In interactions similar to 'salting out' (p. 376), proteins with hydrophobic regions on their surface interact more strongly with hydrophobic ligands on the stationary phase in the presence of high concentrations of ammonium sulphate.

45 Chromatography – detection and analysis

45.1

(a) Flame ionisation detector;
(b) Thermal conductivity detector;
(c) Electron capture detector;
(d) Diode array detector.

45.2 Electrochemical detectors, then fluorescence detectors, then UV/visible detectors. Many biomolecules, either intrinsically or after reaction with a colour reagent, can be detected by UV/visible spectroscopy. However, fewer biomolecules show native fluorescence or the electrical properties necessary for electrochemical detection, making UV/visible detection the most versatile approach.

45.3 Some biomolecules that do not fluoresce can be detected by carrying out either pre-column or post-column derivatisation with a suitable fluorophore, e.g. dansyl chloride is used to detect very small amounts of protein by means of the intense fluorescence of the sulphonamide formed when it reacts with the terminal amino group of proteins or peptides. Other molecules that can be used as fluorescent labels are detailed on p. 285. The selectivity of the analysis is increased since only certain molecules in the sample will be capable of the chemical reaction required.

45.4 ***Tip:*** Substitute relative area values for each peak in Eqn [45.2], as follows:
For mannitol (peak A): $Q_A = [2060 \div 604] \times [1.50 \div 1.26] =$ 4.060 233 4 = **4.06 mg**.
For sucrose (peak B): $Q_B = [1898 \div 604] \times [1.50 \div 0.92] =$ 5.123 452 3 = **5.12 mg**.

46 Principles and practice of electrophoresis

46.1 ***Tip:*** Table 50.1 shows the ionisable side chains of amino acids in proteins. The dissociation of two important amino acids is shown below:

Glutamate	R-COOH *low pH*	⇔	$R\text{-}COO^- + H^+$ *neutral/high pH*
Lysine	$R\text{-}NH_3^+$ *low/neutral pH*	⇔	$R\text{-}NH_2 + H^+$ *high pH*

Note that the charge on a given amino acid side chain at a given pH will depend on the dissociation constant (pK_a, see p. 156) of the ionisable group.

46.2 ***Tip:*** the basic principles are explained in Chapter 46 – for further details, consult Hames and Rickwood (1998) – see p. 323. (a) Acrylamide is the monomer which is polymerised to give long chains of polyacrylamide within the gel; (b) *N,N′*-methylene bisacrylamide is used to form cross-links between the polyacrylamide strands within the gel; (c) ammonium persulphate produces the free radicals required for the polymerisation mechanism *via* a base-catalysed mechanism: (d) *N,N,N′,N′*-tetramethylethylene-diamine is the base used to catalyse the production of free radicals by ammonium persulphate; (e) sodium dodecyl sulphate is a detergent that binds to proteins, dissociating them into subunits and imparting a uniform net negative charge per unit length of the polypeptide; (f) 2-mercaptoethanol reduces disulphide bonds in proteins, causing dissociation of subunits and disruption of tertiary structure; (g) buffer is required to maintain a constant pH during electrophoresis and determines the net charge on each protein molecule; (h) Coomassie Blue R-250 is a useful stain for proteins; (i) bromophenol blue is a small, highly ionic dye that can be used to follow the progress of electrophoresis because it migrates with the electrophoretic front.

46.3 ***Tip:*** The different forms of blotting are listed on pp. 369–70. Transfer of proteins onto gels is termed Western blotting, while DNA transfer is known as Southern blotting and RNA transfer as Northern blotting (p. 328 explains the origins of the terms).

46.4 ***Tip:*** First convert M_r data to log format, then plot a graph of log M_r against relative mobility for the proteins of known M_r (see figure) and then use this to establish the M_r of the unknown protein (LDH).

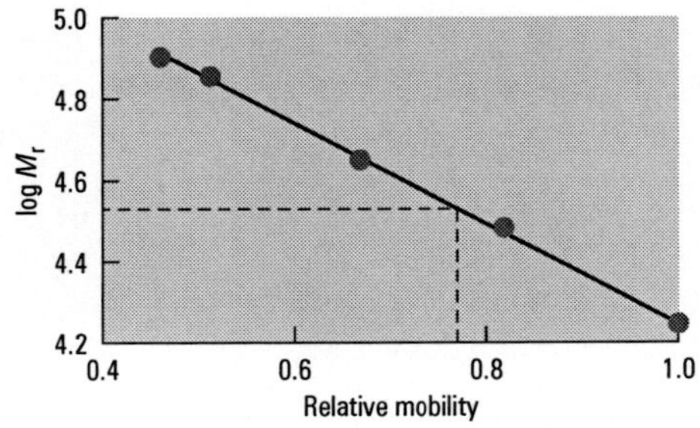

Log_{10} M_r versus relative mobility for SDS-PAGE separation of a series of molecular weight standards (based on data in exercise 46.4), and showing the relative mobility of LDH (dotted line).

(a) From the figure, log M_r of the polypeptide derived from SDS-treated LDH (relative mobility 0.77) is approximately 4.53, giving an antilog value of 33 884.416 = **34 900** (to 2 significant figures).
(b) Since the M_r of native LDH is 136 000, and only a single band of $M_r \approx 34\,000$ is produced on SDS-PAGE, native LDH must contain 136 000 ÷ 34 000 = 4 subunits of identical M_r.

46.5 The gel may not polymerise, or may polymerise unevenly: this is often due to the use of old ammonium persulphate (a fresh solution must be prepared daily, p. 323). Alternatively, the problem may be caused by a high level of dissolved oxygen in the gel (solutions can be degassed before use, if necessary, and the gel solution should be gently poured down the sides of the apparatus to avoid introducing air), or by the use of low pH buffers that slow down the polymerisation reaction (p. 319). Alternatively, the gel may polymerise too quickly, due to either (i) too much ammonium persulphate or TEMED (reduce the amounts used), or (ii) a high laboratory temperature (the gel mixture can be cooled before adding catalysts). For further details, consult Hames and Rickwood (1998) or Westermeier (2004), see p. 329.

47 Advanced electrophoretic techniques

47.1 In PAGE any diffusion of sample proteins that occurs reduces resolution, so it is best to start with as narrow a sample zone as possible. For most protein separations the buffer pH is chosen so that the proteins are negatively charged and will migrate from the cathodic end to the anodic end of the gel. In IEF, the separation is carried out on a pH gradient: because proteins will all migrate to their pI values irrespective of the starting point, the position of sample application is not important. The only exception would be if any of the proteins to be separated were unstable at extreme pH values, in which case it would be best to avoid application at either end of the pH gradient (e.g. pH 3.0 and pH 10.0). Once a given protein has reached the position corresponding to its pI, it will remain in a narrow band because any diffusion away from the pI will result in the protein acquiring a net charge and being focused back to its pI.

47.2

(a) CE in free solution would be appropriate, which would separate the peptides on the basis of differences in net charge at the pH chosen for the analysis. CGE would be inappropriate because the peptides are of similar size.

(b) CGE would be appropriate here because the oligonucleotides will have a uniform net negative charge per unit length, and using a gel in the capillary will introduce a molecular sieving effect.

(c) CIEF will successfully separate proteins that differ in pI.

(d) MEKC, which separates molecules partly on the basis of the principles of CE (electrophoresis and electroosmotic flow) and partly on partitioning of neutral molecules between micelles formed by detergents included in the electrophoresis medium, would be an appropriate technique for hydrophobic compounds such as steroids.

47.3

Potential factors influencing electrophoresis and isoelectric focusing

Factor	Electrophoresis	Isoelectric focusing
Dependence upon ionisable groups in proteins	Yes – this determines the net charge of each protein at a given pH	Yes – the number and types will determine the pI
Requirement for the application of an electric field	Yes – charged molecules migrate to the electrode of opposite charge	Yes – a pH gradient will be set up and proteins will migrate to their pI
Dependence upon differences in net charge between sample molecules	Yes – movement is dependent upon size, shape and net charge of molecules	No
Dependence upon differences in pI values between sample molecules	Only in that pI is related to the net charge at a given pH	Yes – molecules move to a position where pH = pI
Requirement for buffer at constant pH	Yes – proteins will have different net charges at a particular pH and will migrate towards the electrode of opposite charge	No
Requirement for a pH gradient	No	Yes – proteins will migrate to their pI and remain in place for as long as the electric field is applied
Requirement for ampholytes to be included in solution	No	Yes – these form the pH gradient (p. 330)
Use of a supporting medium with molecular sieving properties	Yes – this is one aspect of the separation process	No – it is important that proteins can migrate freely to their pI
Treatment of samples with SDS is possible	Yes – this gives protein subunits a uniform net charge per unit length and allows separation by molecular sieving, on the basis of M_r	No – SDS treatment eliminates differences in pI between native proteins
Diffusion of separated proteins is significant	Yes – though this is reduced by the presence of a supporting medium	No – proteins that move away from their pI acquire a net charge and are attracted back to their pI

47.4 ***Tip:*** Use a ruler across the scale marks, to help you to locate each component – but remember that the SDS-PAGE separation (y-axis) is based on log M_r.
A = pepsinogen. B = alkaline phosphatase; C = catalase; D = alcohol dehydrogenase; E = galactokinase; F = triose phosphate isomerase; G = haemoglobin; H = glyceraldehyde-3-phosphate dehydrogenase; I = trypsinogen; J = nerve growth factor.

48 Electroanalytical techniques

48.1 These assumptions are given on p. 338. The electrical potential measured at an ISE can be approximate to ion concentration assuming (i) the ion is in dilute solution (since the relationship is based on activity, Eqn [48.3]); (ii) the ionic strength of the calibration standards matches that of the test sample and (iii) there is no substantial binding of the ion to sample components (e.g. to proteins or nucleic acids).

48.2 The two functional components of a biosensor are (i) an immobilised enzyme (the biocatalyst) and (ii) a transducer, to convert the biochemical signal into the electrical signal that can be measured. A representative example (there are many others) would be the Biotrace luminescence-based system for monitoring microbial ATP in food, using luciferase (see: http://www.biotrace.com).

48.3

(a) 0.285 ÷ 1000 × 4 (to calculate mmol in 4 ml) × 1000 (to convert from mmol to μmol) = **1.14 μmol**.

(b) 0.273 ÷ 1000 × 20 (to calculate mmol in 20 ml) × 1000 (to convert from mmol to μmol) = **5.46 μmol**.

(c) 0.317 (estimated as the mid-point between the values at 14 and 16 °C) ÷ 1000 × 10 (to calculate mmol in 10 ml) × 1000 (to convert from mmol to μmol) = **3.17 μmol**.

(d) 6.75 ÷ 1000 × 250 (to calculate mg in 250 ml) = 1.6875 = **1.69 mg**.

(e) 4.73 (estimated as the mid-point between the values at 24 and 26 °C) ÷ 1000 × 200 (to calculate ml in 200 ml) = 0.946 ml.

48.4 First, calculate the oxygen content of the electrode chamber as 0.235 ÷ 1000 × 5 = 0.001 175 mmol (≡1.175 μmol). This is equal to 200 divisions on the chart recorder paper, so therefore each division will correspond to 1.175 ÷ 200 = 0.005 875 μmol (≡ 5.875 nmol). The rate of oxygen consumption was −25 units (−146.875 nmol) per 2 min, equivalent to −146.875 ÷ 2 = −73.44 nmol min^{-1}. Since this was the rate for 4.1 × 10^9 cells, the rate per 10^9 cells is −73.44 ÷ 4.1 = **−17.9 nmol min^{-1} (10^9 cells)$^{-1}$**.

48.5 ***Tip:*** Use an Excel spreadsheet (i) to convert the data to μmol and (ii) to plot the converted data and (iii) to estimate the slope of the line (Box 37.2). The calculations are as follows: first calculate the amount of oxygen in the chamber, as 0.285 (value in mmol l^{-1} from Table 48.2) ÷ 1000 (amount in 1 ml) × 5 (amount in 5 ml) × 1000 (convert from mmol to μmol) = 1.425 μmol. Next, convert the data from study exercise 48.5 (p. 345) from percentage to μmol by dividing by 100 and multiplying by 1.425, to give the values shown in the table in the next column.

These values are then plotted as a graph (see figure), with the slope giving the rate of net photosynthesis: in this instance, the slope (rate of net oxygen production) is 0.043 083 μmol min^{-1} (based on linear regression analysis, in Excel). Next, calculate the amount of chlorophyll *a* in the electrode chamber, as 2.13 (μg ml^{-1}) × 5 (volume of chamber, in ml) = 10.65 μg. Therefore the rate of net photosynthesis is 0.043 083 ÷ 10.65 = 0.004 045 352 1 mol min^{-1} (μg chlorophyll *a*)$^{-1}$ and therefore **4.05 μmol min^{-1} (mg chlorophyll *a*)$^{-1}$**.

Oxygen content of illuminated electrode chamber containing a cyanobacterial suspension, based on data in the table in study exercise 48.5

Time	Oxygen content of chamber (μmol)
0	1.003 2
1	1.048 8
2	1.073 025
3	1.118 625
4	1.191 3
5	1.224 075
6	1.256 85
7	1.302 45
8	1.340 925

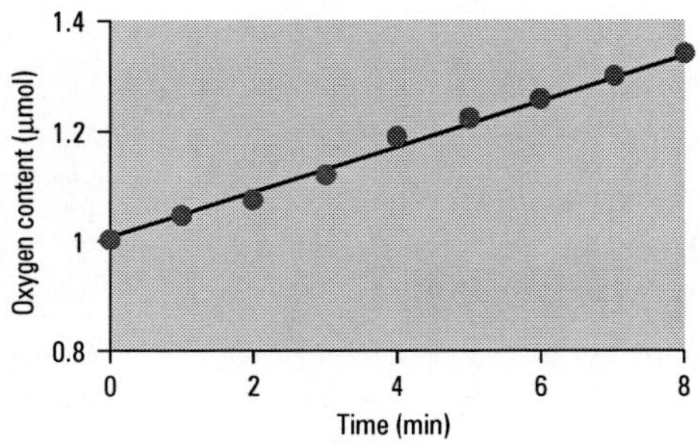

Oxygen content of an illuminated electrode chamber containing a cyanobacterial suspension, based on data in Table 48.1.

58 Mendelian genetics

58.1 The table shows the layout of the Punnett square.

Punnett square for cross RrOO × RrOo

Gametes		Parent 1 RO	Parent 1 rO
Parent 2	RO	RROO	RrOO
	Ro	RROo	RrOo
	rO	RrOO	rrOO
	ro	RrOo	rrOo

Genotype ratios: RROO : RROo : RrOo : RrOO : rrOO : rrOo : in ratio 1 : 1 : 2 : 2 : 1 : 1
Phenotype ratio: red closed : pink closed : white closed : in ratio 1 : 2 : 1

58.2 Probability of blue-eyed = 0.25
Probability of blonde = 0.5
Probability of girl = 0.5
Overall probability = 0.25 × 0.5 × 0.5 = **0.0625**

58.3 Assuming normal mice are homozygous for tail presence, The approximately 1 : 1 phenotype ratio in the F_1 indicates that the tail-less phenotype is heterozygous and possibly dominant. The approximately 1 : 2 : 1 ratio of phenotypes in the F_2 generation (counting 'dead on birth' as a phenotype) also indicates that tail-less might be dominant over normal. The results could be explained if the tail-less gene is fatal when double recessive. There is no sex linkage because the ratio of males to females is roughly equivalent in all progeny phenotypes.

58.4 Use notation X = female, Y = male, X^R = red allele, X^r = white allele.

(a) Assume progeny ratio is 2 : 1 : 1, then cross must be X^RX^r (red-eyed heterozygous female) × X^RY (red-eyed male), giving progeny genotypes X^RX^R, X^RX^r, X^RY and X^rY in the ratio 1 : 1 : 1 : 1 and phenotypes as observed.

(b) Assume progeny ratio is 1 : 1, then cross must be *either* X^RX^R (red-eyed homozygous female) × X^rY (white-eyed male), giving progeny genotypes X^RX^r, X^RY in the ratio 1 : 1 and the phenotypes as observed; *or* X^RX^R (red-eyed homozygous female) × X^RY (red-eyed male), giving progeny genotypes X^RX^R, X^RY in the ratio 1 : 1 and the phenotypes as observed.

(c) Assume progeny ratio is 1 : 1 : 1 : 1, then cross must be X^RX^r (red-eyed heterozygous female) × X^rY (white-eyed male), giving progeny genotypes X^RX^r, X^rX^r, X^RY and X^rY in the ratio 1 : 1 : 1 : 1 and the phenotypes as observed.

58.5 The expected values are 137 : 274 : 137, based on a 1 : 2 : 1 ratio. The χ^2 value is 1.05 + 3.07 + 2.11 = 6.23 (see Box 58.2 for method of working). Since this value is greater than the critical value for χ^2 for 2 degrees of freedom ($n - 1$ categories = 3 − 1) at $P = 0.05$ (which is 5.99, Table 58.1), the geneticist should reject the null hypothesis of a 1 : 2 : 1 outcome.

59 Bacterial and phage genetics

59.1

(a) A zygote, formed by the fusion of two gametes, contains two copies of each chromosome and is diploid, while a merozygote is a cell containing two copies of part of its genome (a partial diploid).

(b) Tranformation is the uptake and assimilation of external genetic material into the genome of a cell, while transposition is the movement of genetic material due to the action of a transposon (a small mobile DNA sequence, coding for its own movement within the genome).

(c) In generalised transduction, a suitably sized fragment of the bacterial genome is packaged inside the protein coat of a phage and then introduced into a new cell, where it may be incorporated by homologous recombination, while specialised transduction involves the inclusion of one or more bacterial genes during excision of a latent phage, followed by transfer and reintegration into a new cell.

(d) While a prototroph can grow in unsupplemented minimal medium, an auxotroph requires a specific additional organic compound in order to grow in minimal medium.

(e) HFr strains have F plasmid DNA integrated into the chromosome and show a high frequency of recombination for chromosomal genes during conjugation, while F′ strains have a modified F plasmid containing one or more chromosomal genes which are transferred at very high frequency during conjugation.

59.2

(a) Ampicillin-resistant, capable of growth on lactose, and a tryptophan auxotroph, requiring tryptophan in the growth medium.

(b) Histidine auxotroph, incapable of using lactose as a carbon source for growth.

(c) Glutamine auxotroph, due to the insertional inactivation of the *gln*A gene by transposon 10.

(d) Prototroph – capable of growing in the absence of tryptophan (note that the chromosomal *trpA* gene is inactivated by Tn6, but the F′ plasmid-encoded *trpA* gene will complement the inactivated chromosomal gene).

(e) The functional *gal* gene present in the *λdgal* genome would complement the chromosomal deficiency (Gal^-), giving a strain that would be capable of growth on lactose.

59.3 ***Tip:*** Look for overlapping sequences, and remember that the F plasmid DNA may integrate either way round. The sequences overlap as follows:

Strain (b) Y D F X W
Strain (a) X W C Z A
Strain (c) Z A E B Y

So, the complete (linear) sequence is **Y D F X W C Z A E B**, shown in circular form in the figure.

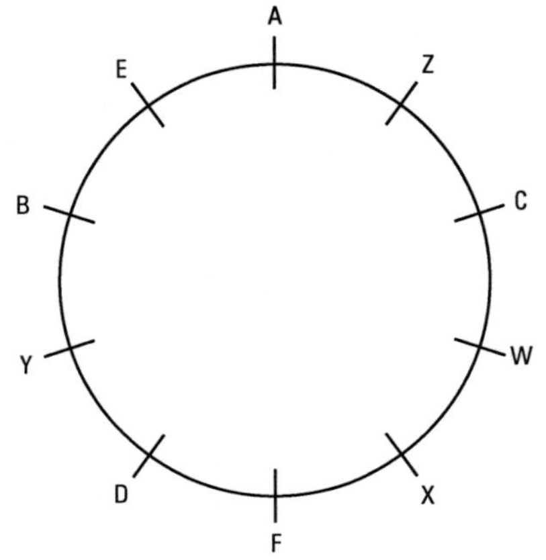

Chromosome map for study exercise 59.3 (note that the map shows only the position of each gene relative to the others, not the exact location).

59.4

(a) Medium A tests for ampicillin resistance; medium B tests for the ability to grow without added tryptophan (tryptophan prototrophy); medium C tests for the ability to grow without added histidine; medium D tests for the growth on lactose as the sole carbon and energy source (the small amounts of tryptophan and histidine would not support significant growth).

(b) For each character, the HFr strain must carry the wild-type gene and the recipient must carry a defective (mutant) form. Thus (i) the HFr donor has

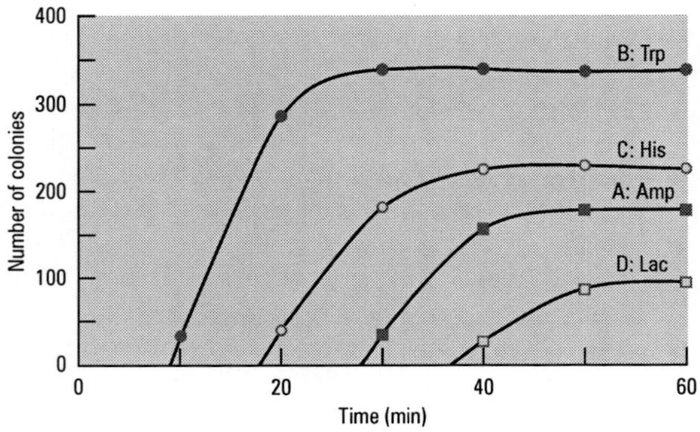

Number of recombinant colonies as a function of time for an interrupted conjugation experiment (data from table in study exercise 59.4).

the phenotype Amp^r, His^+, Trp^+, Lac^+ and (ii) the F^- recipient strain has the phenotype Amp^s His^- Trp^- Lac^-.

(c) ***Tip:*** Begin by plotting the number of recombinant colonies against time – extrapolation to the x-axis gives the time of entry for each gene – as in Fig. 59.6. This is shown in the figure here, giving the following times of entry, in minutes: **Trp = 8; His = 17; Amp = 27; Lac = 36**. This could also be represented in map form, as in the figure in exercise 59.3.

60 Molecular genetics I – fundamental principles

60.1

(a) ***Tip:*** Use the relationship given, where an absorbance value of 1 corresponds to a single stranded DNA concentration of 33 μg ml^{-1}. Thus, $0.14 \times 33 = 4.62$ μg ml^{-1} (DNA in the spectrophotometer cuvette) × 100 (to account for the 100-fold dilution step, i.e. 5 + 495 μl) = 462 μg ml^{-1}.

(b) ***Tip:*** It is essential that you understand the difference between *amount* of substance and *concentration*, as discussed, before you perform these calculations. Chapters 23 and 24 also give advice on calculations involving molar concentrations, while Chapter 29 covers units and prefixes. One approach is to divide by the M_r of the DNA probe (in this case, 24 bases × 325 = 7800) to covert μg ml^{-1} to μmol ml^{-1}, thus 462 ÷ 7800 = 0.059 230 7 μmol ml^{-1}, equivalent to **59.2 μmol l^{-1}** ('micromolar').

(c) The working solution of the probe is required at 100 × 10 pmol ml^{-1} = 1000 pmol ml^{-1} ≡ 1 nmol ml^{-1}. The reconstituted stock contains the DNA probe at a concentration of 0.059 230 7 μmol ml^{-1} (≡ 59.2307 nmol ml^{-1}). So, a dilution of 1 in 59.2 (3 significant figures) is required, for a volume of 5 ml, i.e. **take 84.4 μl of reconstituted stock and make up to a final volume of 5 ml with an appropriate buffer solution**.

60.2 ***Tip:*** Table 60.2 gives advice for standard polyacrylamide and agarose gels, while pulse field gel electrophoresis is covered on pp. 368–69.

(a) Use an agarose gel, e.g. at a concentration of 0.3% w/v.

(b) Use a polyacrylamide gel, e.g. at a concentration of 5% w/v.

(c) Use PACE, which is a variant of pulsed field gel electrophoresis that is able to fractionate DNA molecules of widely different sizes.

60.3 Stringency is a term used to describe the interaction between a nucleic acid probe and its target sequence – high stringency refers to conditions where the hybridisation requires exact complementarity between the two, while low stringency implies that a degree of base mismatching will be tolerated. Stability depends on the number and type of complementary base pairs that can be formed between the probe and the target sequence. The more G + C pairs (with three H-bonds, rather than the two between A-T), the more stable the binding. Stability also increases with probe size, but decreases with increasing incubation temperature, ionic strength and the pH of the hybridisation buffer. Low stringency hybridisation is useful if you are working with a probe obtained from a species different from the source of target DNA. Stringent conditions are required when attempting to detect a single base change, e.g. in hybridisation of mutant and wild-type DNA.

60.4 *Tip:* The gel is read from top (largest fragment at 3′ end) to bottom (smallest fragment at 5′ end) – note also that the original strand has the complementary sequence to that read from the gel. The sequence from the gel is ATTGCGTATTCG, while the (complementary) sequence of the original strand is therefore TAACGCATAAGC.

61 Molecular genetics II – PCR and related applications

61.1 Insufficient Mg^{2+} can impair the polymerisation reaction, since it is an essential co-factor for thermostable DNA polymerases. Excess Mg^{2+} stabilises double stranded DNA and may prevent complete denaturation of the PCR product after each cycle, thereby reducing yield. Excess Mg^{2+} can also promote spurious annealing of primers, and this can lead to production of non-specific products (for more details, see: http://www.promega.com/guides/pcr_guide/070_04/promega.html).

61.2

(i) They should be at least 18 nucleotides long (optimally, 18–30 nucleotides).
(ii) The 3′ ends of primers should end with a G or C.
(iii) They should have similar G + C content so that they have similar annealing temperatures, ideally within 5 °C of each other, based on Eqn [61.1].
(iv) The primers should have minimal self-complementarity, and minimal complementarity to each other, so that primer dimers are not formed.

61.3 *Tip:* Use Eqn [61.1] to calculate T_m values for the primer illustrated. The number of bases is: G = 7, C = 3; A = 3; T = 5, therefore: $T_m = 4(7+3) + 2(3+5) = 56\,°C$ (if the sequence is short enough, you can add up the individual 2s and 4s in your head as you work along the primer).

61.4 *Tip:* Box 61.1 covers the key aspects. Use a laminar flow hood to avoid sample contamination during processing (built-in UV lighting can be used to degrade contaminating DNA); use separate laboratories for (i) initial preparation (pre-PCR), (ii) PCR processing and (iii) storage and analysis of products (post-PCR); keep separate supplies of disposable plastic items (tips, tubes, etc.) for (i) sample preparation, (ii) PCR reagents and (iii) product analysis; autoclave all reagents, buffers and plastic items; wear disposable gloves throughout, changing them between procedures.

61.5 *Tip:* You'll need to rewrite the primer sequences in 3′ to 5′ order, to check alignment against the lower strand in the target sequence, since this runs from the 3′ to the 5′ end (in all three cases, the second of the two primers is directed towards the lower strand).

Primer pair (a) is not suitable: these oligonucleotides may lead to the formation of primer dimers because the two sequences partially complement one another, as follows:

```
5'-TCGCTGAAGGACATGTCG-3'
          3'-CCTGTACAGCGGATCTAGC-5'
```

The GC content of the two primers also differs, with the upper primer containing 10 G + C and the lower primer containing 11 G + C. Ideally, they should have the same G + C content, and therefore T_m, if possible.

Primer pair (b) is also unsuitable: although their GC contents are identical, the second primer contains a single mismatch error, and thus will not hybridise perfectly with the central section of the (upper) single stranded template, as follows:

```
5'-26basesTACGGTCGAGACGTAAGGA17bases-3'
     3'-ATGCCAGCTCTGGATTCCT-5'
```

Also, the 3′ end of the second primer (where elongation is initiated) ends with TA. For optimal hybridisation, it should end in either GC, CG, GG or CC.

Primer pair (c): these two primers have a similar GC content, 3′ ends comprising two G or C residues, no mismatched bases and they have a low degree of complimentarity. They are therefore the most suited pair for the amplification of the desired DNA sequence. Note that sometimes a 'perfect' pair of primers cannot be designed because of the constraints imposed by the target DNA sequence, and high AT contents, or 3′ ends that include A or T residues have to be used.

61.6 *Tip:* Use the Web to research forensic applications of PCR and DNA testing. In this instance, the tests carried out included PCR targeted at: (i) DNA-based sex testing, to establish which bones were from males and which from females, and whether the number of each was consistent with the Tsar's family, (ii) short tandem repeat (STR) analysis, a variation of VNTR, to establish that a single family group was involved, and (iii) mitochondrial DNA (mitDNA) analysis. Mitochondrial DNA is very useful in forensic studies as it is more stable than chromosomal DNA and can be amplified when samples are too small or too old for conventional DNA analysis. Furthermore, if no soft tissue is available, mitDNA can be extracted from solid material such as bone. Since mitDNA is inherited through the maternal line, it can be used to establish family relationships between generations, especially if it can be compared with known members of the family (in this case, the mitDNA of Prince Philip, the Duke of Edinburgh was used, as he is the grandnephew of Tsarina Alexandra). Using these techniques the FSS was able to confirm that the bones found in Yekaterinberg were from Tsar Nicholas and his family. More details, plus other examples of DNA analysis in forensic science, can be found at: http://www.forensic.gov.uk/forensic/entry.htm (look under 'news and casefiles').

62 Molecular genetics III – genetic engineering techniques

62.1 *Tip:* The origins of restriction enzyme names are explained in Chapter 60 (p. 366). *Hin* dIII was the third restriction enzyme to be isolated from *Haemophilus influenzae* strain Rd, while *Eco* RI was the first isolated from *Escherichia coli* strain R. Note that in most cases, the first letter of the genus and the first two letters of the species are used to construct the first part of the restriction enzyme code, e.g. *Ava* I from *Anabaena variabilis* (Fig. 62.1).

62.2

(a) Purification step for DNA, reducing the level of most proteins and non-anionic cellular constituents.
(b) Precipitates DNA and reduces salt contamination.
(c) Degrades contaminating RNA (ribonuclease) and proteins (protease).
(d) Lysozyme digests the peptidoglycan of the bacterial cell wall and EDTA aids its access through the Gram-negative bacterial outer membrane (p. 243) to reach the peptidoglycan layer.
(e) Sodium dodecyl sulphate at high concentration solubilises membranes and causes cell lysis.
(f) NaOH aids cell lysis and disrupts macromolecules. Neutralisation with potassium acetate allows plasmid DNA to return to a soluble form, while leaving chromosomal DNA and other macromolecules as an insoluble mass.

The order of use of these reagents is: (d); (e); (f); (c); (a); (b).

62.3 First, determine the fragment sizes (in kbp) for each digest:

Eco RI: 7 kbp
Hin dIII: 7 kbp
Pvu II: 7 kbp

The results for the individual enzymes confirm that the plasmid is cut once, creating a single linear molecule of 7 kbp (note that an intact, circular, plasmid would run further on the gel, since it would be more compact).

Eco RI and *Hin* dIII: 1 and 6 kbp
Eco RI and *Pvu* II: 3 and 4 kbp
Hin dIII and *Pvu* II: 2 and 5 kbp
Eco RI, *Hin* dIII and *Pvu* II: 1, 2 and 4 kbp

The approach to working out the relative locations of each restriction site is to match up the sizes of the band, or bands, that no longer appear on the gel when the enzymes are combined and then try to piece together the fragments, to see how they were combined in the intact molecule by sketching out possible configurations. The exercise is not unlike a jigsaw puzzle. The results for the paired enzyme combinations confirm that each cuts the plasmid once, since two fragments are obtained. The minimum distance between two enzymes is given by the smaller band in each case: 1 kbp between *Eco* RI and *Hin* dIII, 3 kbp between *Eco* RI and *Pvu* II and 2 kbp between *Hin* dIII and *Pvu* II. In this instance, the results when all three enzymes are combined can then be compared to those for any pair of enzymes (e.g. *Eco* RI and *Pvu* II), showing that the *Hin* dIII site must lie between *Eco* RI and *Pvu* II sites, since the 3 kbp fragment is cleaved into a 1 kbp and a 2 kbp fragment, giving the map shown below. (Note that the results cannot tell us the *absolute* orientation of the sites, but only their relative position: a mirror image of the map shown in the figure would give the same results. Complex restriction digests are best analysed using an appropriate software package: alternatively, restriction sites can be predicted on the basis of DNA sequence information.)

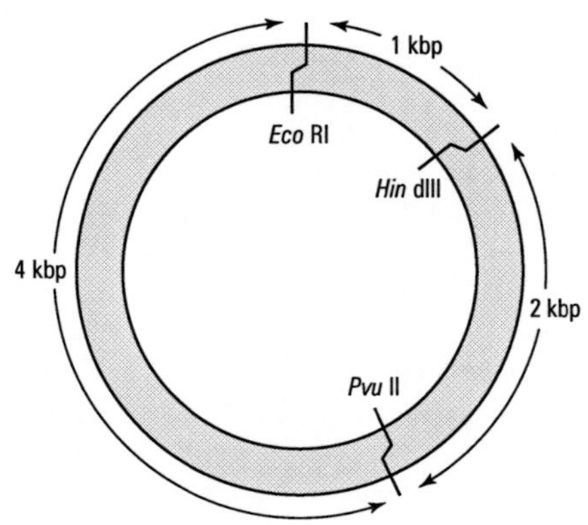

Restriction map of plasmid pRD.

62.4 *Tip:* Work through the calculation in small steps, taking into account the various volumes and dilutions used, to arrive at the correct answer. First, calculate the number of transformants in the entire sample: 180×10 (to correct for 10-fold dilution) $\div 200 \times 500$ (to convert from 200 μl to 500 μl) $= 4500$ CFU. Next, calculate the amount of plasmid DNA (in μg) used in the procedure: $20 \times 0.1 = 2.0$ ng $= 0.002$ μg. Finally, divide the number of transformants by the amount of DNA, to give the answer: $4500 \div 0.002 = 2\,250\,000 =$ **2.25×10^6 CFU (μg DNA)$^{-1}$**. An alternative approach is to calculate the amount of plasmid DNA (in μg) present in the subsample used on the spread plate, as 0.1×20 (ng added) $\div 10$ (dilution factor) $\times 200 \div 500$ (fractional volume spread) $= 0.08$ ng $=$ 0.000 08 μg. Next, divide the number of transformants on the spread plate by this value to give the answer: $180 \div 0.000\,08 =$ **2.25×10^6 CFU (μg DNA)$^{-1}$**.

63 Using graphs

63.1

(a) Three-dimensional graph.
(b) Pie chart.
(c) Scatter diagram, possibly with an added trend line.
(d) Bar chart.
(e) Plotted curve.

63.2 *Tip:* The data must first be calculated as a proportion of the total, and then the degrees of the relevant 'slice' worked out by multiplying by 360 (the number of degrees in a circle), as shown in the table overleaf. Use a compass to make a circle for your graph, then use a protractor to mark out the number of degrees required per slice. The resulting graph should look like the figure overleaf.

Number of strains of bacteria isolated from a water sample

Bacterium	Number of strains	Proportion of total	Degrees for 'slice'
Escherichia coli	37	0.435	156.6
Aeromonas hydrophila	8	0.294	105.8
Klebsiella spp.	13	0.153	55.1
Salmonella enterica	2	0.094	33.9
Enterococcus spp.	25	0.024	8.6
Total	85	1	360

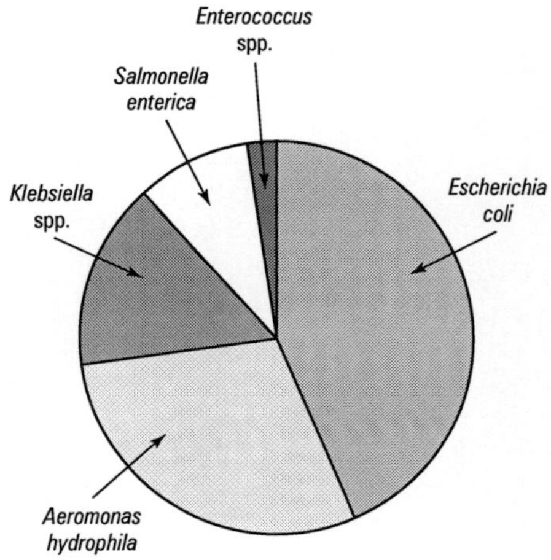

Pie chart showing the relative proportions of strains of five bacteria isolated from a water sample.

63.3 ***Tip:*** First decide on the class boundaries and number of classes – in this case, the values range from 8.1 to 15.1. The subdivisions shown in the table below provide eight classes, which gives a frequency distribution with a reasonable number of classes to aggregate the data sufficiently, but also provide enough detail to indicate the overall shape of the distribution. Note that Excel also has a useful histogram function, but you will need to learn how to set the class boundaries (termed 'bin values') to make effective use of this function (see Box 63.2).

A histogram for these data is shown in the figure below. Your brief description should include something about the location, dispersion and shape of the distribution (see Chapter 66), for example: 'The modal class of the distribution is 12.0–12.9 $g\,l^{-1}$, with a range of 7.0 $g\,l^{-1}$ (15.5–8.5 $g\,l^{-1}$). The data show a slight negative skew, with 28 values in the three classes above the mode and 46 in the four classes below it'.

Data of table in study exercise 63.6 presented as a frequency distribution

Class	Frequency
8.0–8.9	2
9.0–9.9	5
10.0–10.9	17
11.0–11.9	22
12.0–12.9	26
13.0–13.9	17
14.0–14.9	10
15.0–15.9	1

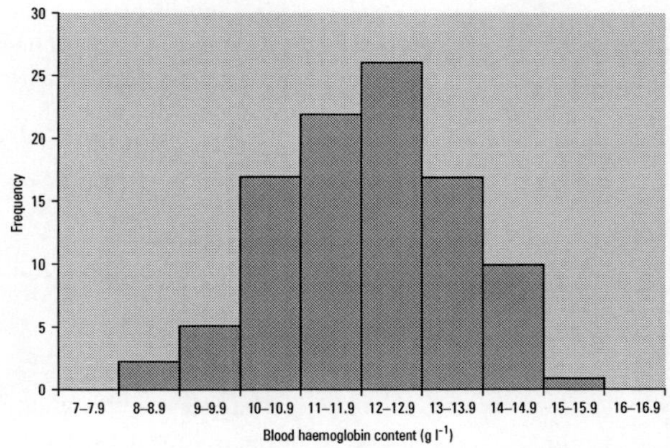

Frequency distribution of data of table in study exercise 63.3.

63.4 ***Tip:*** The Sunday newspaper supplements often analyse news stories with the aid of graphs, and might be a good source for material, as their articles are often slanted towards a particular viewpoint, and you may find that the graphical presentation has been chosen to support this.

64 Presenting data in tables

64.1 ***Tip:*** To convert all of the concentrations to the same units, you will need to know the relative molecular mass of each constituent (glutamate $M_r = 147.1$; proline $M_r = 115.1$; trehalose $M_r = 342.3$). We have used $mmol\,l^{-1}$ in the example below, but other units would be equally valid, as long as you are consistent. Data expressed as % (w/v) were first converted to $g\,l^{-1}$ (multiplied by 10), then to $mmol\,l^{-1}$.

Concentrations of low molecular weight solutes in bacteria.

	Solute concentration ($mmol\,l^{-1}$)*		
Bacterium	**Glutamate**	**Proline**	**Trehalose**
Bacillus subtilis	52.1	39.7	<0.1
Escherichia coli	34.0	21.0	154.7
Staphylococcus aureus	15.3	NT	NT

* All values expressed to 1 decimal place.
NT = not tested.

64.2 ***Tip:*** Box 64.1 gives the key aspects. The following table gives one possible layout.

The principal components of a typical table

Component	Comment:
Title	Concise, self-contained description of table contents, with numbering where appropriate
Column/row headings	Identifies the content of each column and row, showing units of measurement, where appropriate
Data values	Quoted to an appropriate number of significant figures/decimal places
Footnotes	All abbreviations and other individual details must be explained
Rulings	Used to emphasise any groupings within the table, and to separate individual items

64.3

(a) **11–14 years**.

(b) ***Tip:*** Calculate the ratio of male : female energy requirement, in order to establish which is greatest in proportional terms, as in the following table.

Male : female energy requirements, calculated from the table in study exercise 64.3

Age	Male : female energy requirement
<1	1.0625
1–3	1.0612
4–6	1.1077
7–10	1.1233
11–14	1.2078
15–18	1.3068
19–50	1.3086
>50	1.2338

This shows that the age class where the energy requirement of males is greatest in proportion to that of females is **19–50 years**.

(c) **Birth to 10 years**.

(d) **15–18 years for males**, **11–50 years for females**.

65 Hints for solving numerical problems

65.1

(a) $[C] = A/\varepsilon l$

(b) $\varepsilon = A/[C]l$

65.2

(a) Subtract ax from both sides and then swap sides, so: $\mathbf{b = y - ax}$

(b) 1. Subtract b from both sides and then swap sides, so: $ax = y - b$

2. Divide both sides by a, so: $\mathbf{x = (y - b)/a}$

(c) Raise each side to the power $\frac{1}{3}$, then swap sides, so: $\mathbf{y = x^{1/3}}$

(d) 1. Take logs of both sides, so: $\log x = \log(3^y)$

2. Noting $\log(3^y) = y \log 3$, substitute on right-hand side, so $\log x = y \log 3$

3. Divide both sides by log 3, then swap sides, so: $\mathbf{y = \log x/\log 3}$

(e) 1. Divide both sides by $1 - y$, so: $z^p + 3 = x/(1 - y)$

2. Subtract 3 from both sides, so: $z^p = x/(1 - y) - 3$

3. Raise both sides to the power $1/p$, so: $\mathbf{z = ((x \div (1 - y)) - 3)^{1/p}}$

(f) 1. Multiply both sides by pq, so: $zpq = (y - z)^{1/n}$

2. Raise each side to the power n, so: $zpq^n = (y - z)$

3. Take logs of both sides, so: $n\log(zpq) = \log(y - z)$

4. Hence $\mathbf{n = (\log(zpq)) \div (\log(y - z))}$

65.3

(a) **215**

(b) **107 000**

(c) **0.04**

(d) **99.82**

(e) **99.90**

(f) **100.00**

(g) **6260**

(h) **130 000**

(i) **2 3 or 4, depending on circumstances (see p. 474)**

(j) **5**

(k) **3**

(l) **6**

(m) **2**

(n) **5**

65.4

(a) **17.14%**

(b) **7.55%**

(c) **14.01**

(d) 100 × (75.02 – 55.23) ÷ 55.23 = 35.831 975 = **35.83%**

65.5

(a) (i) **0.000 000 000 000 000 000 160 219 2 C**;

(ii) **299 792 400 mws^{-1}**

(b) **4207.3**; **0.804 14**; **0.098 27**; **7.0238**; **309.09**

65.6 ***Tip:*** To calculate x, substitute values into Eqn [65.4].

(a) $x = (7.0 - 4.5) \div 0.02 =$ **125**;

(b) $x = (15.4 - -2.6) \div -4.46 = -4.035\,874\,439 =$ **−4.04**.

Tip: to calculate y, substitute values into Eqn [65.3].

(c) $y = 0.2 + (-0.63 \times 10.5) = -6.415 =$ **−6.42**;

(d) $y = -1.8 + (4.1 \times 4.5) = 16.65 =$ **16.6**.

66 Descriptive statistics

66.1

(a) **Range = 9**

(b) **Variance = 4.46**

(c) **Standard deviation = 2.11**

(d) **Coefficient of variation = 43.1%** (CoV = 100 × 2.113/4.9)

(e) **Standard error = 0.273**

66.2 ***Tip:*** Substitute values into Eqn [66.3] to calculate the SE of each sample.

For sample A: 12.7 ÷ 3.464 101 615 = 3.666 174 209 = **3.67**.
For sample B: 14.4 ÷ 4.472 135 955 = 3.219 937 888 = **3.22**.
Thus, in absolute terms, sample B has the lower standard error (note that the higher standard deviation of sample B is more than offset by the increased number of data values).
To calculate the proportional standard error, divide the SE by the mean, as follows:
For sample A: 3.666 174 209 ÷ 16.2 = 0.226 307 05 = **0.226**.
For sample B: 3.219 937 888 ÷ 13.2 = 0.243 934 688 = **0.244**.
Thus, in proportion to the mean, sample A has the lower standard error.

66.3

Larval size data (recalculated)

Replicate sample	Mean sample larval size (mm)	Sample number	Mean size × sample size
A	3.0	24	72.0
B	2.5	37	92.5
C	2.0	6	12.0
Total		67	176.5

True mean = 176.5/67 = **2.63 mm**

67 Choosing and using statistical tests

67.1 ***Tip:*** First calculate the standard error using Eqn [66.3], as 6.8 ÷ 4 = 1.7. Then, substitute values into Eqn [67.2] to calculate the 95% confidence limits (note that you also need the value for $t_{0.05[15]} = 2.131$). The upper limit is calculated as: 24.7 + (2.131 × 1.7) = 28.3227 = **28.3**. The lower limit is calculated as 24.7 − (2.131 × 1.7) = 21.0773 = **21.1**.

67.2 ***Tip:*** The variance of the data, s^2, computes as 10.543 and the mean as 3.4. Hence s^2 ÷ mean = **3.1**, hence the data appear clumped. In biological terms this might mean that if a snail is infected once, then it has a greater chance of multiple infection. Perhaps parasite eggs are clumped in the snails' environment.

67.3 ***Tip:*** The calculated t = **2.136** with 12 degrees of freedom; the critical value at P = 5% is 2.18 (Table 67.2); hence you would accept the null hypothesis that the samples came from the same population and hence conclude that the plant hormone had no significant effect.

67.4 ***Tip:*** Amino acid uptake will be the dependent (y) variable and time the independent (x) variable. From the Excel output, the coefficients are: intercept = **1.17** (to 3 significant figures) and slope (labelled as 'X variable 1') = **2.95** (to 3 significant figures), so the linear relationship has the following form: $y = 1.17 + 2.95x$. The strength of the relationship is measured by the coefficient of determination, termed 'R squared' in the Excel output. In this case, a value of **0.971** (to 3 significant figures) shows that there is a very strong fit of the y data to the trend line.

Manipulating and observing specimens

40 The purpose and practice of dissection

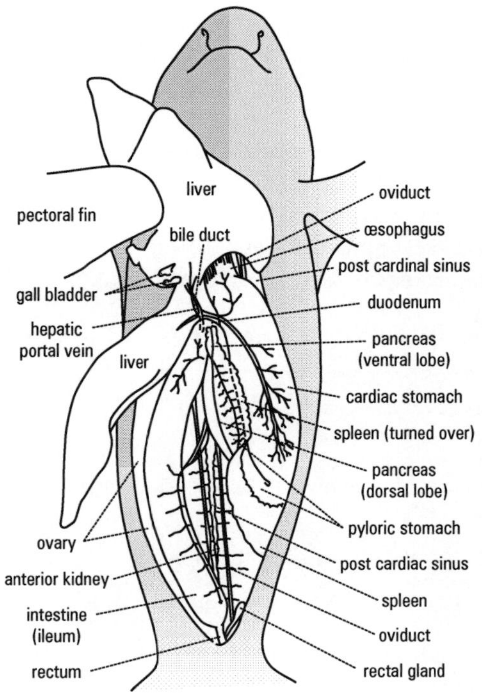

Fig. 40.1 Labelled diagram of abdomen of young female dogfish. Good drawings are essential to get the most out of dissections (see Chapter 28).

Dissection in zoology involves the display or removal of parts of any dead animal, whereas vivisection is an operation on a living animal. Dissection is usually associated with teaching animal structure, but the skills involved are used widely within the life sciences for:

- investigating anatomy and morphology;
- making physiological preparations of nerves, muscles and other organs;
- investigating parasites in various body organs;
- removing specific body organs/tissues for chemical analysis;
- investigating reproductive status;
- removing organs/tissues for histological/histochemical investigation;
- manipulating living material, as in grafting processes and preparing diagrams to show selected features (Fig. 40.1).

There is considerable debate about the ethics of using animals for dissection (p. 116) but this must be distinguished from that of vivisection. It is part of a complex and emotive issue, but one viewpoint is that dissection is required for effective zoology teaching: experience of dissection at an *appropriate* stage of the curriculum is both enlightening and teaches an essential technical skill. However, if dissection is definitely not for you, it is best to discover this as early as possible in your career.

KEY POINT The primary objectives of dissection are:
- ***personal exploration of animal structure and function;***
- ***development of manipulative skills;***
- ***production of reports based upon personal observation.***

The ground rules of dissection

There are some important rules to be considered if dissection is to be an acceptable procedure in zoological teaching.

Humane treatment

The use of any animal, whatever its level of organisation, for experimentation or dissection must be a considered act with due regard for humane treatment and killing (see Chapter 19). Remember there are very specific regulations for the use of vertebrate species and some higher invertebrates: check with your supervisor.

Maximum benefit

Any animal should be used for as many investigations as possible.

Preparation

Prepare for the exercise by ensuring that practical schedules and relevant texts are consulted before attempting a dissection. If you have to miss a practical, inform the organiser in advance to prevent animals being killed or prepared unnecessarily.

Your responsibility *– because of the ethical aspect to dissection, you have a particular responsibility to attend to and make careful and effective use of a dissection specimen.*

Types of dissections

Dissection can be carried out at three levels of sophistication, related directly to the size of the organism or structure being investigated. Gross dissection of large organisms requires equipment very different from that used for 'normal' dissection in that large knives replace scalpels, etc. Normal-scale dissection involves creatures from a dog down to an earthworm, and requires equipment typical of commercial dissection kits/ instruments. Fine-scale dissection for the removal or display of organs, glands, etc., from small animals usually requires only mounted needles, fractured glass edges for cutting and a dissecting microscope or magnifier. Fine-scale dissection requires extensive practice, and the preparation of material (relaxation and fixation, p. 206) is much more critical than for larger specimens.

SAFETY NOTE Taking care with dissection equipment – sharp blades should be used with extreme caution and disposed of immediately after use in a sharps container.

Equipment

Table 40.1 lists basic equipment required for normal dissection: some comments on use are given below (see also Fig. 40.2). Commercial dissection kits often contain inappropriate components and you should buy your instruments individually from specialist suppliers, if possible. Equipment for fine-scale dissections can be assembled easily to your own specifications.

Table 40.1 List of basic equipment recommended for dissection.

Quantity	Description
1	All-metal scalpel, stainless steel, 45 mm blade
1 each	Swann–Morton scalpel handles, sizes 3 and 4
1 each	Swann–Morton blades, packets of nos. 10, 11, 12, 15, 22 and 24
2	Dissecting needles, straight, stainless steel
2	Blunt seekers, stainless steel: metal handles
1	Fine forceps, blunt points, stainless steel, 112 mm length
1	Coarse forceps, blunt points, stainless steel, 112 mm length
1	Coarse scissors, open shanks, straight points, stainless steel, 150 mm length
1	Fine scissors, open shanks, straight fine points, stainless steel, 110 mm length
1	Section lifter

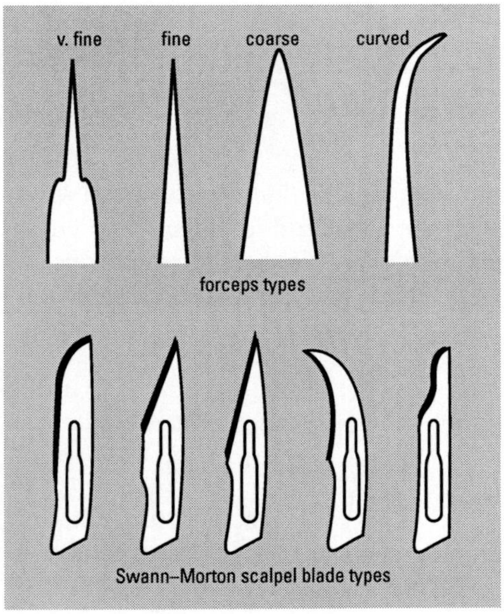

Fig. 40.2 Variations in design of dissection equipment

Scalpels

There are two basic types of scalpels, fixed blade and replaceable blade. Buy at least one 45-mm blade length solid forged scalpel, preferably made of stainless steel: use this for coarser cutting procedures as it can be resharpened using an oil-stone. The Swann–Morton scalpel comprises a handle and a disposable blade: blades come in a variety of shapes and sizes (Fig. 40.2), curved-edge ones being the most useful. Blades fit only specific handle sizes so make sure that they match.

Forceps

Buy at least one each of coarse and fine stainless steel forceps. The latter are very delicate and you should check that the points meet precisely before purchasing. Look after them very carefully as the points are easily damaged: use only on soft tissues. Use large and small forceps for general purposes but keep fine ones for delicate work only.

Dissecting scissors

Buy two pairs of stainless steel, pointed scissors, one medium-large for coarse work and cutting small bones and one fine pair for delicate work. Points and edges are easily damaged and require professional sharpening when repair is needed, so use carefully. A pair of bone cutters is optional but do *not* use scissors to cut large bones.

Dissecting (mounted) needles

Buy at least two with metal handles and protect the points; use them for dissecting membranes in areas where damage is acceptable. Needles have many non-dissection functions associated with other fine manipulative techniques.

Box 40.1 Basic stages of an animal dissection

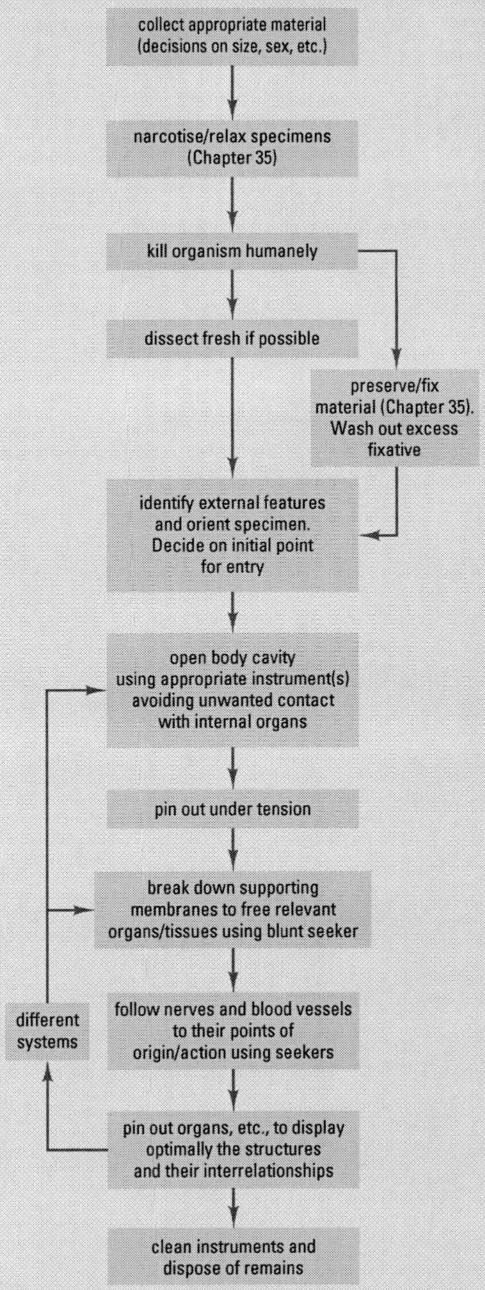

Fig. 40.3 Flowchart for dissection

The basic sequence of steps is outlined in Fig. 40.3.

1. **The animal should have been killed as humanely and as recently as possible**: this will probably be done for you by the class supervisor or technician according to the rules appropriate to the type of animal. The method of killing should be chosen carefully to keep the specimen relaxed. If preserved material is used, wash out excess fixative thoroughly before dissecting.
2. **Orient the specimen carefully**; determine the dorsal/ventral, anterior/posterior or other oral/aboral axes (see Fig. 44.4) and work out the correct orientation of the specimen for dissection. Invertebrates are usually dissected from the dorsal surface (the nerve cord being ventral in position) and vertebrates from the ventral surface (the nerve cord being dorsal in position): however, special objectives may require a different orientation.
3. **Open the body cavity carefully.** This is usually done using forceps to lift the skin away from underlying organs while using a fine pair of scissors to make an initial opening; the scissors are then used to extend this opening in anterior–posterior and lateral directions until the skin flaps can be pinned back. Pin out by placing tension on the skin flaps, breaking down any restricting membranes using a seeker.
4. **Subsequent dissection procedure depends upon your objectives.** To display the system being investigated as clearly and neatly as possible requires:
   - **(a)** identification of organs, blood vessels and nerves initially visible;
   - **(b)** separation of these structures from the membranes that hold them in place. This is best achieved using a blunt instrument such as a seeker to avoid damage;
   - **(c)** removal or displacement of organs that obscure parts of the system you wish to display. Displacement using pins to hold the organ in position is the preferred option when possible;
   - **(d)** if your objective is to display a blood system or a nervous system, you must follow individual vessels/nerves from their point of origin to their destination organ; do this using a blunt seeker to remove covering membranes by working carefully along the structure. Do not pull sideways during this procedure as this often results in breakages; such tissues are usually much stronger in the direction of their length than when pulled laterally.
   - **(e)** tidy up loose pieces and wash away residual blood, etc.; use more pins to finalise the display of the system and then make notes and drawings as necessary (see e.g. Figs 28.3 and 40.1).

Box 40.2 Tips for improving dissection technique

The following tips should help you develop your dissection technique:

- **Use a blunt seeker and controlled tension on the tissue to remove the connective tissue** that binds tissues and organs together: use scalpels sparingly.
- **Insert pins/awls obliquely** so that they do not interfere with further dissection. In segmented animals such as earthworms, fine pins can be used to mark the position of specific segments.
- **Dissect most invertebrates under water:** the water buoys up the tissues/organs and assists dissection. Change the water if it becomes clouded but do not allow flowing water to run directly on to the specimen as it will damage delicate structures.
- **Dissect vertebrates and larger invertebrates in air** on a dissection board, using cotton-wool swabs to remove excess blood and other body fluids.
- **Keep tissues under tension while dissecting,** but avoid damaging them with forceps/fingers.
- **Cut away from the organs when using scalpels or scissors** and keep scissor points away from structures you are attempting to free. This is essential when opening the body cavities of small animals.
- **Use appropriately sized equipment:** e.g. do not attempt to use large scissors to open an earthworm. Never use delicate instruments for coarse work.
- ***Never* remove anything until you know what you are removing.**
- **Dissect along structures and not across them,** particularly for tubular structures such as nerves and blood vessels.
- **Use fresh material whenever possible.** Fixatives make tissues more brittle and inelastic: alcohol storage tends to harden skin, muscle and connective tissue.
- **Keep your instruments clean and sharp:** you can't dissect with blunt or dirty equipment. Dry instruments after washing and wipe with an oily cloth. Cut nothing but tissues with scalpels and scissors. Do not sharpen pencils with scalpels or stick mounted needles into the bench or the dissecting board.
- **Be hygienic.** Wear rubber gloves if you have cuts or lesions on your hands. Wash hands and equipment thoroughly when finished. Dispose of animal remains carefully as instructed.

Dissecting seekers

These have blunt points and should have metal handles. Use for breaking down membranes holding delicate organs together. Again, they have multiple uses.

Recommended accessory equipment

A teat (Pasteur) pipette is valuable for washing delicate organs and tissues. A camel-hair brush is useful for removing material from delicate structures. Pins and awls are usually provided by your department and are used for fixing the dissection specimen to the board (awls) or wax dish (pins): small pins are invaluable for pinning organs aside for display purposes. For measurement, and as an aid to drawing, a pair of dividers is very useful, as is a small steel ruler.

Carrying out the dissection

The basic sequence of steps in a typical dissection is shown in Box 40.1 and Fig. 40.3, while Box 40.2 gives tips to help you improve your dissection technique.

Sources for further study

About Inc. *About Biology: Online dissections.* Available: http://biology.about.com/od/onlinedissections/Online_Dissections.htm
Last accessed 09/04/07
[A compendium of Web-based dissections.]

Fishbeck, D. and Sebastini, A. (2001) *Comparative Anatomy: Manual of Vertebrate Dissection.* Morton Publishing Co., Englewood, NJ.

Kardong, K.V. and Zalisko, E.J. (1998) *Comparative Vertebrate Anatomy: A Laboratory Dissection Guide.* McGraw-Hill, Boston.

Morgan, M. *Earthworm Dissection.* Available: http://www.microscopy-uk.org.uk/mag/articles/worm.html
Last accessed 09/04/07.
[A simple, descriptive account of an earthworm dissection.]

Walker, W.F. and Homberger, D. (1997) *Anatomy and Dissection of the Rat,* 3rd edn. W.H. Freeman, New York.

Whitehouse, R.H. and Grove, A.J. (1947) *The Dissection of the Crayfish.* University Tutorial Press, London.
[Part of a classic series including manuals on frog, rabbit, dogfish, earthworm and cockroach. Includes lots of information on how to carry out detailed dissections.]

Study exercises

40.1 Make a list of situations where dissection is used. There are many areas of zoology (and botany) where dissection technique is required. List up to ten situations where you would require to use some form of dissection technique to acquire material to work with, or to carry out manipulations before or after experiments.

40.2 Decide on appropriate dissection instruments. What instruments should you use for the following operations:

(a) following a nerve or blood vessel from source to ending?
(b) opening up the body cavity of the earthworm?
(c) removing or severing muscle tissue in the body wall of a vertebrate?

40.3 Research general rules of dissection. What is the general rule for deciding on which side a general dissection should be commenced: (a) for invertebrates; and (b) for vertebrates?

41 Introduction to microscopy

Many features of interest in biological systems are too small to be seen by the naked eye and can only be observed with a microscope. All microscopes consist of a coordinated system of lenses arranged so that a magnified image of a specimen is seen by the viewer (Figs. 41.1 and 41.2). The main differences are the wavelengths of electromagnetic radiation used to produce the image, the nature and arrangement of the lens systems and the methods used to view the image.

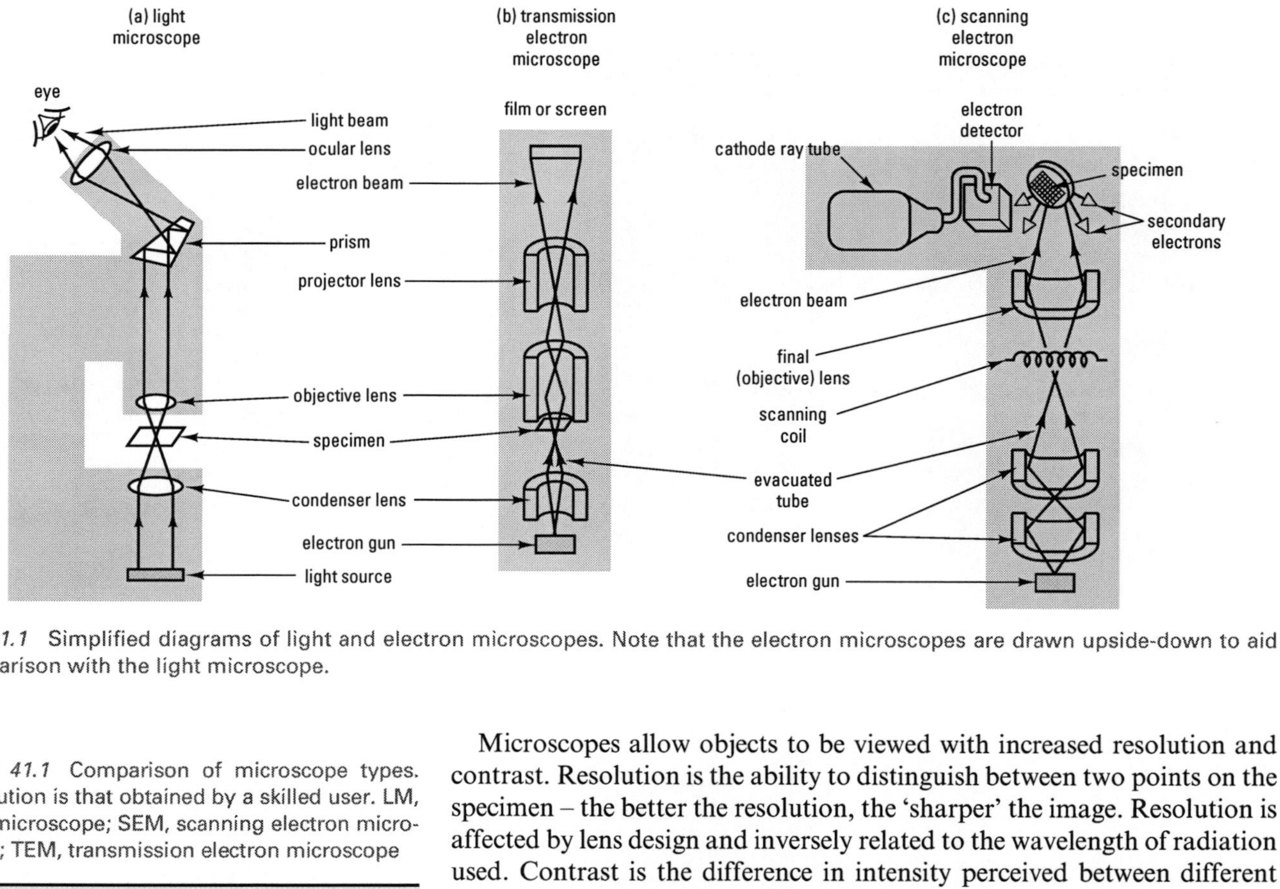

Fig. 41.1 Simplified diagrams of light and electron microscopes. Note that the electron microscopes are drawn upside-down to aid comparison with the light microscope.

Table 41.1 Comparison of microscope types. Resolution is that obtained by a skilled user. LM, light microscope; SEM, scanning electron microscope; TEM, transmission electron microscope

	Type of microscope		
Property	LM	TEM	SEM
Resolution	200 nm	1 nm	10 nm
Depth of focus	Low	Medium	High
Field of view	Good	Limited	Good
Specimen preparation (ease)	Easy	Skilled	Easy
Specimen preparation (speed)	Rapid	Slow	Quite rapid
Relative cost of instrument	Low	High	High

Microscopes allow objects to be viewed with increased resolution and contrast. Resolution is the ability to distinguish between two points on the specimen – the better the resolution, the 'sharper' the image. Resolution is affected by lens design and inversely related to the wavelength of radiation used. Contrast is the difference in intensity perceived between different parts of an image. This can be enhanced (a) by the use of stains, and (b) by adjusting microscope settings, usually at the expense of resolution.

The three main forms of microscopy are light microscopy, transmission electron microscopy (TEM) and scanning electron microscopy (SEM). Their main properties are compared in Table 41.1 and their suitability for observing cells and organelles is shown in Table 41.2.

Light microscopy

Two forms of the standard light microscope, the binocular (compound) microscope and the dissecting microscope, are described in detail in Chapter 43. These are the instruments most likely to be used in routine practical work. Figure 41.2(a) shows a typical image from a light microscope. In more advanced project work, you may use one or more

(a)

(b)

(c)

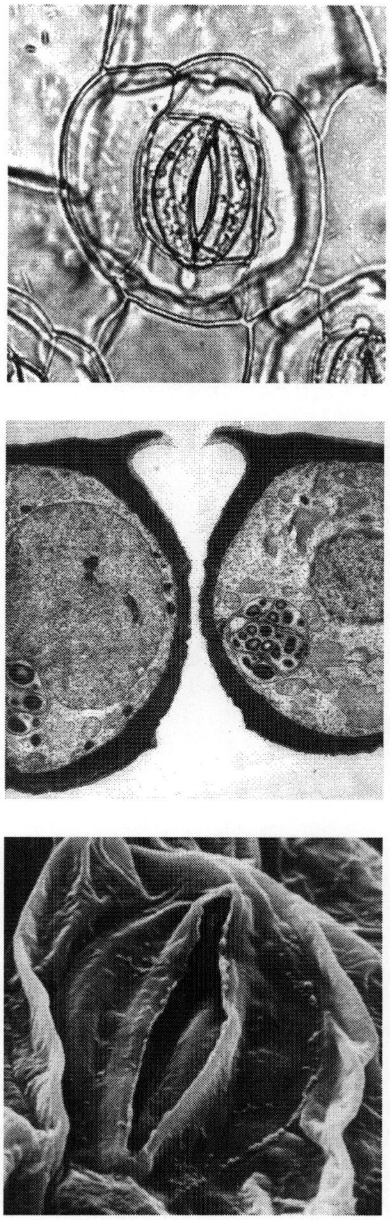

Fig. 41.2 Examples of images of a similar specimen (the stomatal complex of *Commelina communis* L.) obtained using different microscopic techniques: (a) light microscopy (surface view); (b) transmission electron microscopy (transverse section through guard cell pair at mid pore); and (c) scanning electron microscopy (surface view). As an indication of scale, the width of a single guard cell is about 10 μm.

of the following more sophisticated variants of light microscopy to improve image quality:

- Dark-field illumination involves a special condenser that causes reflected and diffracted light from the specimen to be seen against a dark background. The method is particularly useful for near-transparent specimens and for delicate structures like flagella. Care must be taken with the thickness of slides used – air bubbles and dust must be avoided and immersion oil must be used between the dark-field condenser and the underside of the slide.
- Ultraviolet microscopy uses short-wavelength UV light to increase resolution. Fluorescence microscopy uses radiation at UV wavelengths to make certain fluorescent substances (e.g. chlorophyll or fluorescent dyes that bind to specific cell components) emit light of visible wavelengths. Special light sources, lenses and mountants are required for UV and fluorescence microscopy, and filters must be used to prevent damage to users' eyes.
- Phase-contrast microscopy is useful for increasing contrast when viewing transparent specimens. It is superior to dark-field microscopy because a better image of the interior of specimens is obtained. Phase contrast operates by causing constructive and destructive interference effects in the image, visible as increased contrast. Adjustments must be made, using a phase telescope in place of the eyepiece, for each objective lens and a matching phase condenser, and the microscope must be set up carefully to give optimal results.
- Nomarski or Differential Interference Contrast (DIC) microscopy gives an image with a three-dimensional quality. However, the relief seen is optical rather than morphological, and care should be taken in interpreting the result. One of the advantages of the technique is the extremely limited depth of focus that results: this allows 'optical sectioning' of a specimen.
- Polarised-light microscopy can be used to reveal the presence and orientation of optically active components within specimens (e.g. starch grains, cellulose fibres), showing them brightly against a dark background.
- Confocal microscopy allows three-dimensional views of cells or thick sections. A finely focused laser is used to create electronic images of layered horizontal 'slices', usually after fluorescent staining. Images can be viewed individually or reconstructed to provide a 3-D computer-generated image of the whole specimen.

Electron microscopes

Electron microscopes offer an image resolution up to 200 times better than light microscopes (Table 41.1) because they utilise radiation of shorter wavelength in the form of an electron beam. The electrons are produced by a tungsten filament operating in a vacuum and are focused by electromagnets. TEM and SEM differ in the way in which the electron beam interacts with the specimen: in TEM, the beam passes through the specimen (Fig. 41.1(b)), while in SEM the beam is scanned across the specimen and is reflected from the surface (Fig. 41.1(c)). In both cases, the beam must fall on a fluorescent screen before the image can be seen. Permanent images ('electron micrographs') are produced after focusing the beam on photographic film (Figs. 41.2(b) and (c)).

Table 41.2 Dimensions of some typical cells and organelles with an indication of suitable forms of microscopy for observing them. LM = light microscope; SEM = scanning electron microscope; TEM = transmission electron microscope. Column 2 data after Rubbi (1994).

Cell or organelle	Approximate diameter or width (μm)	Suitable form of microscopy
Prokaryote cell	0.15–5	LM, SEM, TEM
Eukaryote cell	10–100	LM, SEM, TEM
Fungal hypha	5–20	LM, SEM, TEM
Nucleus	5–25	LM, SEM, TEM
Mitochondrion	1–10	SEM, TEM
Chloroplast	2–8	LM, SEM, TEM
Golgi apparatus	1	SEM, TEM
Lysosome/ peroxisome	0.2–0.5	SEM, TEM
Plant cell wall	0.1–10	LM, SEM, TEM

You are unlikely to use either type of electron microscope as part of undergraduate practical work because of the time required for specimen preparation and the need for detailed training before these complex machines can be operated correctly. However, electron microscopy is extremely important in understanding cellular and subcellular structures, and you may be shown electron micrographs with one or more of the following objectives:

- to demonstrate cell ultrastructure (TEM);
- to show surface features of organisms, e.g. when surface features are coated or when cells are 'freeze-fractured', then coated (SEM);
- to investigate changes in the number, size, shape and condition of cells and organelles (TEM);
- to carry out quantitative studies of cell and organelle disposition (TEM).

Aspects of the interpretation of electron micrographs are dealt with in Chapter 44.

Preparative procedures

Without careful preparation of the material being studied, the biological structures viewed with any microscope can be rendered meaningless. Figure 41.3 summarises the processes involved for the main types of microscopy discussed above. The processes involved in preparing material for light microscopy are outlined in Chapter 42.

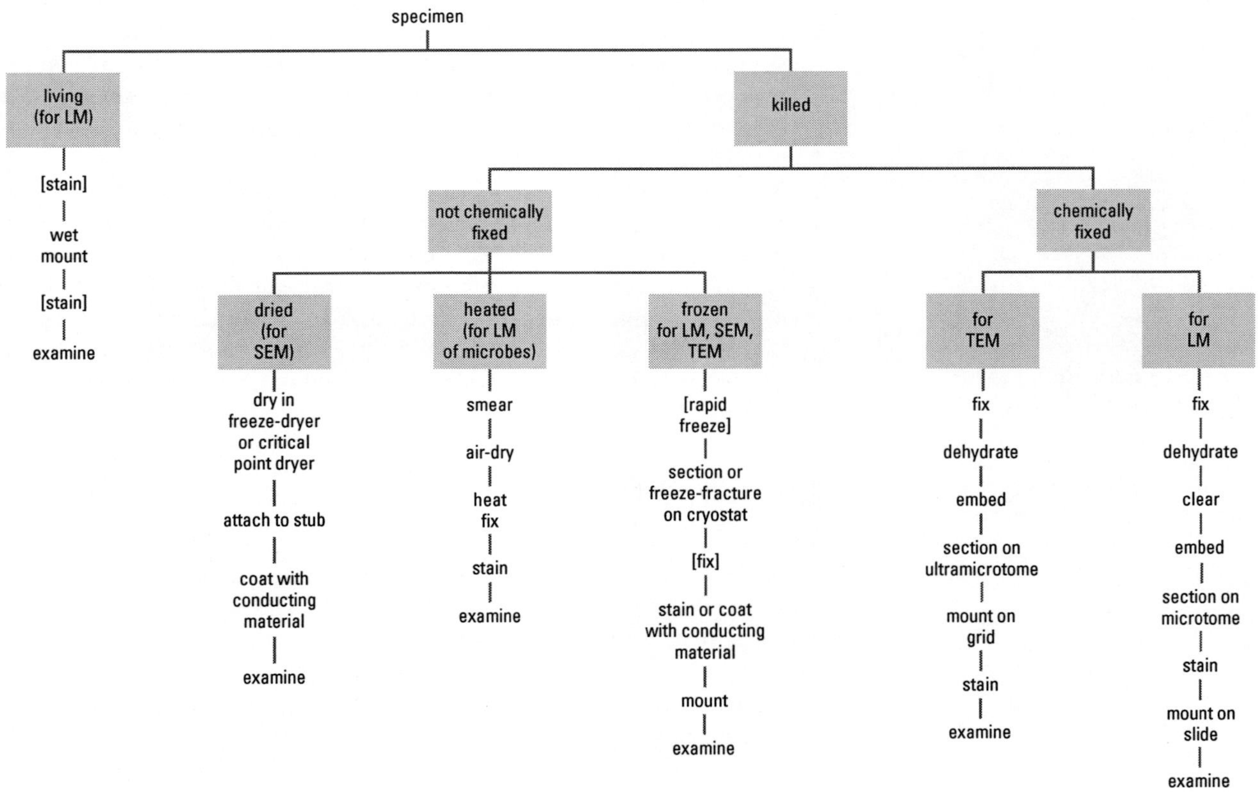

Fig. 41.3 Flowchart of procedures necessary to prepare specimens for different forms of microscopy. Steps enclosed in brackets are optional. LM, light microscope; SEM, scanning electron microscope; TEM, transmission electron microscope.

Text reference

Rubbi, C.P. (1994) *Light Microscopy Essential Data*. Wiley, Chichester.

Sources for further study

Bradbury, S. (1976) *The Optical Microscope in biology*. Edward Arnold, London.

Bradbury, S. (1984) *An Introduction to the Optical Microscope*. Oxford University Press, Oxford.

Davidson, M.W. and Abramowitz, M. *Molecular Expressions. Exploring the World of Optics and Microscopy*. Available: http://micro.magnet.fsu.edu/index.html
Last accessed 09/04/07.
[Covers many areas of basic knowledge underlying microscopy. Includes a microscopy primer.]

Jeffries, C. *Microscopy WWW Sites – by Organisation*. Available: http://www.ou.edu/research/electron/mirror/web-org.html
Last accessed 09/04/07.
[Comprehensive set of links to microscopy websites.]

Murphy, D.B. (2001) *Fundamentals of Light Microscopy and Electronic Imaging*. Wiley-Liss, New York.

van Egmond, W. and Walker, D. *A Virtual Pond-dip*. Available: http://www.microscopy-uk.org.uk/ponddip/x_index.html
Last accessed 09/04/07.
[Introduction to things you can see in pond water using a standard light microscope.]

Study exercises

41.1 Test your microscopy knowledge. Indicate whether the following statements about light microscopy, scanning electron microscopy (SEM) or transmission electron microscopy (TEM) are true or false.

(a) TEM allows you to see at finer resolution than light microscopy.
(b) TEM allows you to see surface features of specimens.
(c) SEM always requires staining of specimens.
(d) The resolution of TEM is about 200 times better than that of light microscopy.
(e) The resolution of a microscope is linked to the wavelength of electromagnetic radiation employed.
(f) The specimen in both TEM and SEM is viewed under near-vacuum conditions.
(g) Specimens for light microscopy can be living or dead.
(h) SEM provides better resolution than TEM.
(i) The depth of focus in light microscopy is greater than that in SEM.
(j) Light microscopy, SEM and TEM all involve the use of a condenser lens within the microscope.

41.2 Fill in the blanks in the following paragraph. Dark-field microscopy involves shining reflected and _______ light on the specimen against a dark background. It is particularly useful for __________ specimens. UV microscopy uses short-wavelength UV light in order to increase image _________. Phase-contrast microscopy utilises constructive and destructive _______ effects to increase image ________. Nomarski microscopy provides a pseudo ____ image, with a very small depth of ________, allowing _____ _____ to be carried out. _____ light microscopy allows visualisation of optically active components in the specimen. Confocal microscopy involves the use of a ______ light source and can yield computer-generated 3-D images.

41.3 Identify the missing preparative procedures. In each sequence below, one or two steps have been missed out. Using Fig. 41.3, identify the missing procedures.

(a) For light microscopy on a killed and fixed specimen: fix – dehydrate – clear – ________ – section – _________ – mount – examine.
(b) For light microscopy on a heat-fixed microbial specimen: smear – _______ – heat fix – _________ – examine.
(c) For TEM on a killed and fixed specimen: fix – _______ – embed – section – mount – stain – examine.

42 Preparing specimens for light microscopy

SAFETY NOTE The fixatives, solvents, embedding media and stains used in the preparation of specimens for light microscopy should all be assumed to be toxic, and treated accordingly. Take care to follow the safety advice provided by laboratory staff and tutors.

Preparative techniques are crucial to successful microscopical investigation because the chemical and physical processes involved have the potential for making the material difficult to work with and for producing artefacts. The basic steps (outlined in Fig. 41.3) are similar in most cases, but the exact details (e.g. timing, chemicals used and their concentrations) differ according to the material being examined and the purpose of the investigation. It is usually best to follow a recipe that has worked in the past for your material (see Grimstone and Skaer, 1972, Kiernan, 1999).

Chemical fixation

The main purpose of fixation is to preserve material in a lifelike manner. The process of fixation for microscopy is much more critical than for whole specimens and only small pieces of tissue should be used. The fixation solutions used for microscopy are intended to:

- penetrate rapidly to prevent post-mortem changes in the cells;
- coagulate the cell contents into insoluble substances;
- protect tissues against shrinkage and distortion during subsequent processing;
- allow cell parts to become selectively and clearly visible when stained.

__Fixing specimens__ – tissues must be fixed as soon as possible after death. If this is not possible, specimens should be stored at low temperature (4°C) and for as short a time as possible.

Fixative solutions are usually mixtures of chemicals selected for their combined properties. Your choice of fixative from the numerous recipes available in reference texts will depend upon both the type of investigation and the nature of the material.

***KEY POINT** Poor fixation can produce artefacts, particularly where coagulant fixatives are used.*

When using a fixative for microscopy, observe the following points:

- Use fresh solutions: some of the fluids are unstable and do not keep well. Do not reuse fixative.
- Always use plenty of fixing fluid compared with the volume of material to be fixed (not less than a 10 : 1 fixative : sample volume ratio).
- Avoid underfixation or overfixation: in general the optimum time will be a function of several factors including:
  - (a) The penetration capacity of the fixative.
  - (b) The size of the piece of tissue: this should always be small and have as large a surface : volume ratio as possible.
  - (c) The type of tissue to be fixed: uniform tissues fix more quickly than complex tissues, where one component may form a barrier to others. The presence of chitin usually means a slow rate of penetration. Tissues filled with air can be difficult to submerge and infiltrate; this can be overcome by fixation in a partial vacuum.
  - (d) The temperature: increased temperature results in increased penetration rate, but also tends to make tissue brittle.
- Wash the specimen thoroughly after fixation: residues of fixative can interfere with subsequent processes. The washing may be in water or another appropriate solution.

__Decalcifying specimens__ – this may be necessary if calcareous structures remain after fixation: this is usually done using a 5–10 per cent solution of EDTA followed by thorough washing.

Dehydration and clearing

Allowing time for dehydration – *never be tempted to rush, because incomplete dehydration will do more than anything else to ruin a preparation.*

A high water content in tissues will usually hinder subsequent processing so they must be dehydrated. This is done with an organic solvent using a series of solutions graded from pure water to pure solvent. Ethanol/water mixtures are often used in histology. Dehydration must be carried out carefully, using prescribed time schedules to avoid distortion and hardening. Protect delicate specimens by mixing solutions in the container rather than by transferring the specimen as this is when damage will occur.

The chemicals used for dehydration are not usually soluble in the embedding medium and tissues must therefore be infiltrated with an intermediate fluid, miscible with the waxes and resins normally used: such fluids make the tissues transparent and are termed clearing agents. The most widely used ones are hydrocarbons such as xylene; many are volatile, pose significant health risks, and tend to harden tissues rapidly. Clearing oils such as clove oil and cedarwood oil are safer, but slower in action. Terpineol is useful since it does not require such complete dehydration and can be used straight from 90% v/v ethanol. Note that every trace of oil must be removed during the infiltration (embedding) stage, or specimens will not embed properly.

Embedding and sectioning

SAFETY NOTE The sharp blades use in hand sectioning are an obvious hazard. Identify beforehand the location of the nearest first aid kit. Take care to carry out such procedures with care and away from others who may inadvertently brush against you.

Embedding involves infiltrating the specimen with a medium that will solidify and support it when sectioned. The specimen is either passed through a series of gradually increasing concentrations of the embedding material (e.g. wax or epoxy resin) dissolved in the dehydrating or clearing agent, or placed straight into pure embedding agent. Several changes of embedding material are made to remove the last traces of solvent, then the embedding agent is solidified by cooling or by a polymerisation treatment. Precise protocols can be found in specialist books (e.g. Kiernan, 1999). The specimen must be oriented carefully during the solidification process to aid section-cutting in known planes.

The aim of sectioning is to provide a thin slice through the tissues suitable for observing cellular details at maximum resolution. After trimming, blocks of embedded material are sectioned using a microtome. Section-cutting techniques are complex and require time for familiarisation. Sections are attached to slides after flattening the sections by a flotation procedure carried out either directly on the glass slide or in a water bath. The sections are attached to the slide by coating the latter with a very thin layer of albumen before drying them down on to this sticky surface.

For some procedures, specimens are frozen in isopentane, freon or dry-ice/ethanol mixtures and then maintained below −20 °C until sectioned. Sectioning is performed in a cold chamber (cryostat) where there is a microtome; sections are removed directly to glass slides. If storage is necessary, slides are placed over a desiccant in a refrigerator, but this is possible only for short periods.

Hand-cut sections can be made through certain relatively stiff materials, notably stems and roots of herbaceous plants:

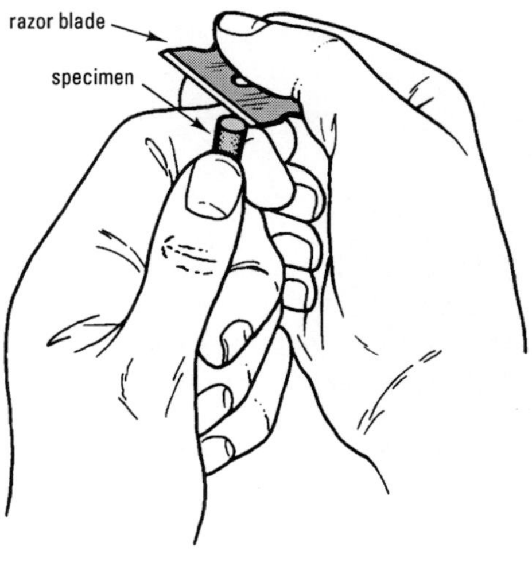

Fig. 42.1 Preparation of a hand-cut section

1. Grasp the object firmly between your index finger and thumb.
2. Brace your elbows against your ribs to steady your hands.
3. Rest the side of a fresh and sharp razor blade on the index finger holding the specimen (Fig. 42.1).

(a)

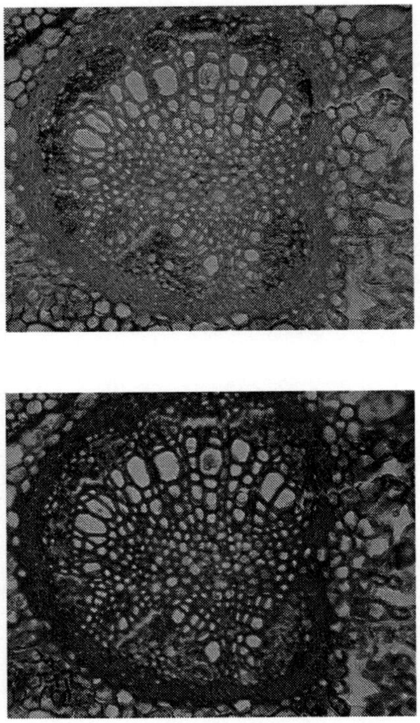

(b)

Fig. 42.2 Simulation of the introduction of contrast into a biological specimen by staining. Leaf midrib transverse section showing in greyscale the effects 'before' (a) and 'after' (b) staining xylem tissue red with phloroglucinol. In colour, the effects of staining are even more distinct.

4. Pull the blade towards your body so that you cut the specimen with a slicing action.
5. Repeat this action quickly, pushing the object slowly upwards with your thumb.
6. Float off the cut sections on to a watch glass of water.

This is a relatively tricky procedure, requiring practice. You may find it helps to lubricate the razor and object by wetting them before cutting. Thin material, such as a leaf, is best sectioned when supported between the halves of a longitudinally split cylinder of pith or fresh carrot. Always cut lots of sections, because only a small proportion will be thin enough to use – the best ones are often wedge-shaped, tapering off to thin edges.

Staining

The purpose of staining in microscopy is to:

- add contrast to the image;
- identify chemical components of interest;
- locate particular tissues, cells or organelles.

This is achieved in different ways for different types of microscopy. In standard light microscopy, contrast is achieved by staining the structure of interest with a coloured dye (e.g. Fig. 42.2); in UV microscopy, contrast is obtained using fluorescent stains. Physico-chemical properties of the stain cause it to attach to certain structures preferentially or be taken up across cell membranes.

Stains for light microscopy are categorised according to the charge on the dye molecule. Stains like haematoxylin, whose coloured part is a cation (i.e. basic dyes), stain acidic, anionic substances like nucleic acids: such structures are termed basophilic. Stains like eosin, whose coloured part is an anion (i.e. acid dyes), stain basic, cationic substances: such structures are termed acidophilic. Acid dyes tend to stain all tissue components, especially at low pH, and are much used as counterstains. Staining is progressive if it results in some structures taking up the dye preferentially. Staining is regressive if it involves initial over-staining followed by decolorisation (differentiation) of those structures that do not bind the dye tightly (e.g. Gram staining).

Certain 'vital' stains (e.g. neutral red) are used to determine cell viability or the pH of cell compartments such as plant vacuoles. 'Mortal' stains (e.g. Evans' blue) are excluded from living cells but diffuse into dead ones and are used to assay cell mortality.

Stains and staining procedures

There is a huge range of stains for light microscopy but the features of those used commonly for cytology in botany, microbiology and zoology are given in Table 42.1. Consult appropriate texts for full details of (a) how to make up stains and (b) the protocol to use. Results depend on technique: follow the recommended procedures carefully. There are many stains used in histochemistry for identifying various classes of macromolecules such as DNA, RNA, proteins, lipids, carbohydrates (chitin, cellulose, starch, callose, pectins, glycogen) and heteropolymers (e.g. lipopolysaccharides, peptido-glycans, proteoglycans). Consult specialist texts for methods (e.g. Grimstone and Skaer, 1972; Horobin and Kiernan, 2002).

Definitions

Metachromic stains – have the capacity to stain different structures different colours.

Orthochromic stains – never change colour whatever they stain.

Mordants – chemicals (salts and hydroxides of divalent and trivalent metals) that increase the efficiency of stains usually by forming complexes with the stain.

Counterstains – stains that apply a background colour to contrast with stained structures.

Negative staining – where the background is stained rather than the structure of interest.

Table 42.1 A selection of stains for light microscopy of sections

	Stain	What it stains	Comments
Plant cells	Chlorazol black	Cell walls: black Nuclei: black, yellow or green Suberin: amber	The solvent used (70 per cent ethanol in water or water alone) affects colours developed
	Neutral red	Living cells: pink (pH < 7)	A 'vital' stain used to determine cell viability or to visualise plant protoplasts in plasmolysis experiments; best used at neutral external pH
	Phloroglucinol/HCl	Lignified cell walls: red	Care is required because the acid may damage microscope lenses
	Ruthenium red	Pectins: red	Shows up the middle lamella
	Safranin + Fast green	Nuclei, chromosomes, cuticle and lignin: red Other components: green	Stain in safranin first, then counterstain with fast green (light green will substitute). A differentiation step is required
	Toluidine blue	Lignified cell walls: blue Cellulose cell walls: purple	Best to apply dilute and allow progressive staining to occur
Fungi and bacteria	Giemsa	Bacterial chromosome: purple Bacterial cytoplasm: colourless	Also used in zoology to stain protozoa
	Gram	Gram-positive bacteria: violet/purple Gram-negative bacteria: red/pink Yeasts: violet/purple	
	Gray	Bacterial flagella: red	Uses toxic chemicals: mercuric chloride and formaldehyde. Leifson's stain is an alternative
	Lactophenol cotton blue	Fungal cytoplasm: blue (hyphal wall unstained)	Shrinkage may occur
	Nigrosin or India ink	Background: grey–black	Negative stains for visualisation of capsules: requires a very thin film
	Shaeffer and Fulton	Bacterial endospores: green Vegetative cells: pink/red	Malachite green is primary stain, heated for 5 min. Counterstained with safranin
	Ziehl–Neelsen	Actinomycetes: red Bacterial endospores: red Other microbes: blue	Requires heat treatment of fuchsin primary stain, decolorisation with ethanol–HCl and a methylene blue counterstain (acid-fast structures remain red)
Animal cells	Azure A/eosin B	Nuclei, RNA: blue Basophilic cells: blue–violet Most other cells: pale blue Muscle cells: pink Necrosing cells: pink Cartilage matrix: red–violet Bone: pink Red blood cells: orange–red Mucins: green–blue/blue–violet	Used in pathology – shows up bacteria as blue; must be fresh; care required over pH: Mann's methyl blue/eosin gives similar results
	Chlorazol black	Chitin: greenish-black Nuclei: black, yellow or green Glycogen: pink or red	Solvent (70% v/v ethanol in water or water alone) affects colours formed
	Iron haematoxylin	Nuclei, chromosomes and red blood cells: black Other structures: grey or blue–black	Good for resolving fine detail; iron alum used as mordant before haematoxylin to differentiate
	Mallory	Nuclei: red Nucleoli: yellow Collagen, mucus: blue Red blood cells: yellow Cytoplasm: pink or yellow	Simple, one-stage stain; fades within a year; not to be used with osmium-containing fixatives. Heidenhain's azan gives similar results but does not fade. Cason's one-step Mallory is a rapidly applied stain that is particularly good for connective tissue
	Masson's trichrome	Collagen, mucus: green Cytoplasm: orange or pink	Used as a counterstain after, for example, iron haematoxylin, which will have stained nuclei black. Not to be used after osmium fixation
	Mayer's (haemalum and eosin; 'H&E')	Nuclei: blue/purple Cytoplasm: pink	Alum used as mordant for haematoxylin; eosin is the counterstain. To show up collagen, use van Gieson's stain as counterstain

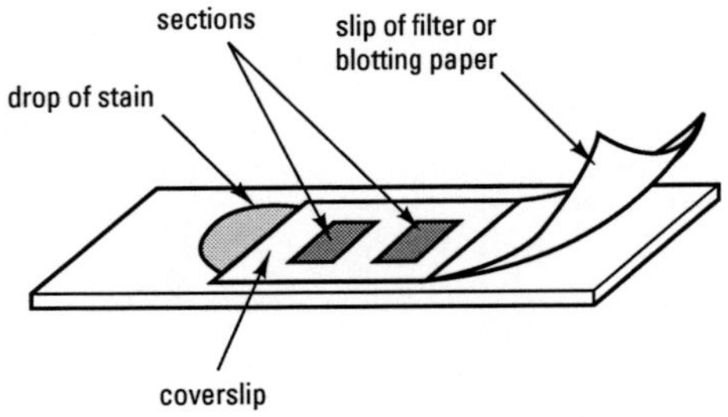

Fig. 42.3 How to irrigate a section with stain by drawing it through with a piece of filter paper.

***Transferring sections** – a camel-hair paintbrush is useful for transferring sections between staining solutions and on to a slide.*

Stains for light microscopy are normally applied by one of four methods:

- Floating the sections on the stain.
- Applying the stain to a smear fixed on to a slide, e.g. when staining bacteria and blood.
- Drawing the stain through under a coverslip as shown in Fig. 42.3.
- Immersing slides with sections attached into a staining trough. This is best for bulk staining.

Most stains need to act in an aqueous medium, so sections that have been embedded in wax must be rehydrated before staining. The wax is dissolved, e.g. in Histo-clear for 1–3 min, the Histo-clear replaced by 100 per cent ethanol (1 min) then 70% (v/v) ethanol : water for 1 min., finally transferring to distilled water. Fresh sections of plants or heat-fixed smears of microbes generally require no pretreatment before staining if they are not to be retained after examination.

Mounting sections

Wet mounts

These are used for observing fresh specimens. The following steps are involved:

1. Isolate the specimen.
2. Place the specimen in a small droplet of the relevant fluid (fresh water, sea water, etc.) on a microscope slide.
3. Gently lower a coverslip on to the droplet, using forceps or two needles and avoiding bubbles.
4. Remove any excess water on or around the coverslip with absorbent paper.

***Using coverslips** – all wet specimens for microscopic examination must be covered with a coverslip to protect the objective lens from water, oil, stains and dirt.*

Entire specimens can be examined under the light microscope providing they are small enough to be mounted on a glass slide. They may be mounted in cavity slides or by using ring mounts.

Temporary mounts

These essentially involve wet mounting in a mountant with a short useful life, e.g. for identification purposes. It may be desirable to clear the specimen first and a dual-purpose substance such as lactophenol, which will clear from 70% (v/v) ethanol, is recommended.

Permanent mounts

***Using mountants** – most mountants require that all water is removed from the section by transfer through increasing concentrations of ethanol until 100 per cent ethanol is reached.*

These protect sections during examination and allow storage without deterioration. A permanent mount involves sealing your section under a coverslip in a mountant. The mountants used are clear resins dissolved in a slowly evaporating solvent. A good mountant has a similar refractive index to the tissue being mounted, remains clear through time, is chemically inert and will harden quickly. Natural resins like Canada balsam take a long time to dry, are variable in quality and tend to colour up and crack in time. The newer synthetic resins and plastics such as DPX

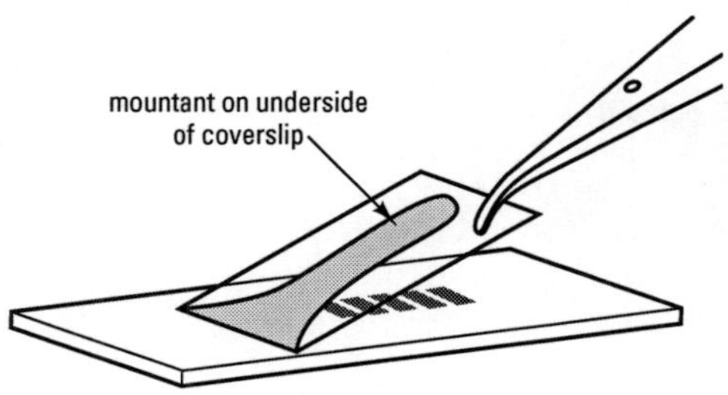

Fig. 42.4 How to lower a coverslip painted with mountant on its underside on to sections on a slide.

***Making squash preparations** – these may be required for any type of mount. The smallest specimens can be squashed after mounting by applying gentle pressure on the coverslip with your forceps. Larger specimens can be squashed between two slides before fixing and mounting – this ensures that higher pressures are applied evenly.*

mountant are superior: they dry quickly, are available in a range of refractive indices, and do not yellow with age. For tissue components soluble in organic solvents, aqueous mounting media based on, for example, gelatine or glycerol should be used. Find out which mountant is recommended for your particular sections.

The recommended procedure when mounting sections on a slide is:

1. Apply a little mountant to a coverslip of appropriate size.
2. Turn the coverslip over and place on its edge to one side of the sections as in Fig. 42.4.
3. Lower the coverslip slowly down on to the sections so as to displace all the air and sandwich the sections between the slide and the coverslip.
4. Press firmly from the centre outwards to distribute the mounting medium evenly.
5. Allow the solvent to evaporate – best results come from slow drying when time allows, but many synthetic mountants will tolerate brief heating when speed is essential.

Text references and sources for further study

Davidson, M.W. and Abramowitz, M. *Molecular Expressions. Exploring the World of Optics and Microscopy*. Available: http://micro.magnet.fsu.edu/index.html
Last accessed 09/04/07.
[Covers many areas of basic knowledge underlying microscopy, including preparative procedures.]

Grimstone, A.V. and Skaer, R.J. (1972) *A Guidebook to Microscopical Methods*. Cambridge University Press, Cambridge.

Horobin, R.W. and Kiernan, J.A. (eds) (2002) *Conn's Biological Stains. A Handbook of Dyes, Stains and Fluorochromes for use in biology and Medicine*, 10th edn. Bios Scientific Publishers, Oxford.

Kiernan, J.A. (1999) *Histological and Histochemical Methods: Theory and Practice*, 3rd edn. Scion Publishing, Bloxham.

Study exercises

42.1 Select appropriate stains. From Table 42.1, identify a stain you could use to help indicate the presence of the following: (a) glycogen in a liver section; (b) woody (lignified) cells in a plant stem section; (c) a fungal pathogen in a leaf section; (d) living cells in an onion epidermal peel; (e) mucus in a lung section; (f) bacteria in a food sample.

42.2 Explain the processes involved in preparing a section for light microscopy. Prepare one-sentence summaries of the reasons for carrying out the following procedures: (a) fixation; (b) dehydration; (c) embedding; (d) sectioning; (e) staining; (f) permanent mounting.

42.3 Investigate toxicity and safety aspects of some of the staining reagents used for light microscopy. Many chemicals used in staining are toxic or otherwise dangerous. As examples, investigate the specific hazards associated with the use of the following stains: (a) chlorazol black; (b) neutral red; (c) malachite green; (d) safranin.

43 Setting up and using a light microscope

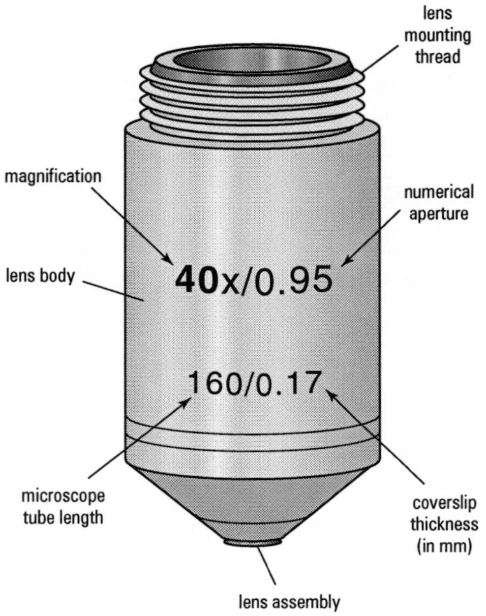

Fig. 43.1 Objective lens parameters. Most lenses are inscribed to show the details labelled above. The numerical aperture is a measure of the light-gathering power of the lens.

The light microscope is probably the most important instrument used in biology practicals and its correct use is one of the basic and essential skills of biology. A standard undergraduate binocular microscope consists of three main types of optical unit: eyepiece, objective (Fig. 43.1) and condenser. These are attached to a stand that holds the specimen on a stage (Fig. 43.2). A monocular microscope is constructed similarly but has one eyepiece lens rather than two.

Setting up a binocular light microscope

Before using any microscope, familiarise yourself with its component parts.

***KEY POINT** Never assume that the previous person to use your microscope has left it set up correctly: apart from differences in users' eyes, the microscope needs to be properly set up for each lens combination used.*

The procedures outlined below are simplified to allow you to set up microscopes like those of the Olympus CX series (Fig. 43.2). For monocular microscopes, disregard instructions for adjusting eyepiece lenses in (5).

Using binocular eyepieces *– if you do not know your interpupillary distance, ask someone to measure it with a ruler. You should stare at a fixed point in the distance while the measurement is taken. Take a note of the value for future use.*

1. Place the microscope at a convenient position on the bench. Adjust your seating so that you are comfortable operating the focus and stage controls. Unwind the power cable, plug in and switch on after first ensuring that the lamp setting is at a minimum. Then adjust the lamp setting to about two-thirds of the maximum.
2. Select a low-power (e.g. ×10) objective. Make sure that the lens clicks home.
3. Set the eyepiece (ocular) lenses to your interpupillary distance; this can usually be read off a scale on the turret. You should now see a single circular field of vision. If you do not, try adjusting in either direction.
4. Put a prepared slide on the stage. Examine it first against a light source and note the position, colour and rough size of the specimen. Place the slide on the stage (coverslip up!) and, viewing from the side, position it with the stage adjustment controls so that the specimen is illuminated.
5. Focus the image of the specimen using first the coarse and then the fine focusing controls (Fig. 43.3). The image will be reversed and upside down compared with that seen by viewing the slide directly.
   (a) If both eyepiece lenses are adjustable, set your interpupillary distance on the scale on each lens. Close your left eye, look through the right eyepiece with your right eye and focus the image with the normal controls. Now close your right eye, look through the left eyepiece with your left eye and focus the image by rotating the eyepiece holder. Take a note of the setting for future use.
   (b) If only the left eyepiece is adjustable, close your left eye, look with the right eye through the static right eyepiece and focus the image with the normal controls. Now close your right eye, look through the left eyepiece with your left eye and focus the image by rotating the eyepiece holder. Take a note of the setting for future use.

Issues for spectacle and contact lens wearers *– those who wear glasses can remove them for viewing, as microscope adjustments will accommodate most deficiencies in eyesight (except astigmatism). This is more comfortable and stops the spectacle lenses being scratched by the eyepiece holders. However, it may create difficulties in focusing when drawing diagrams. Those with contact lens should simply wear them as normal for viewing.*

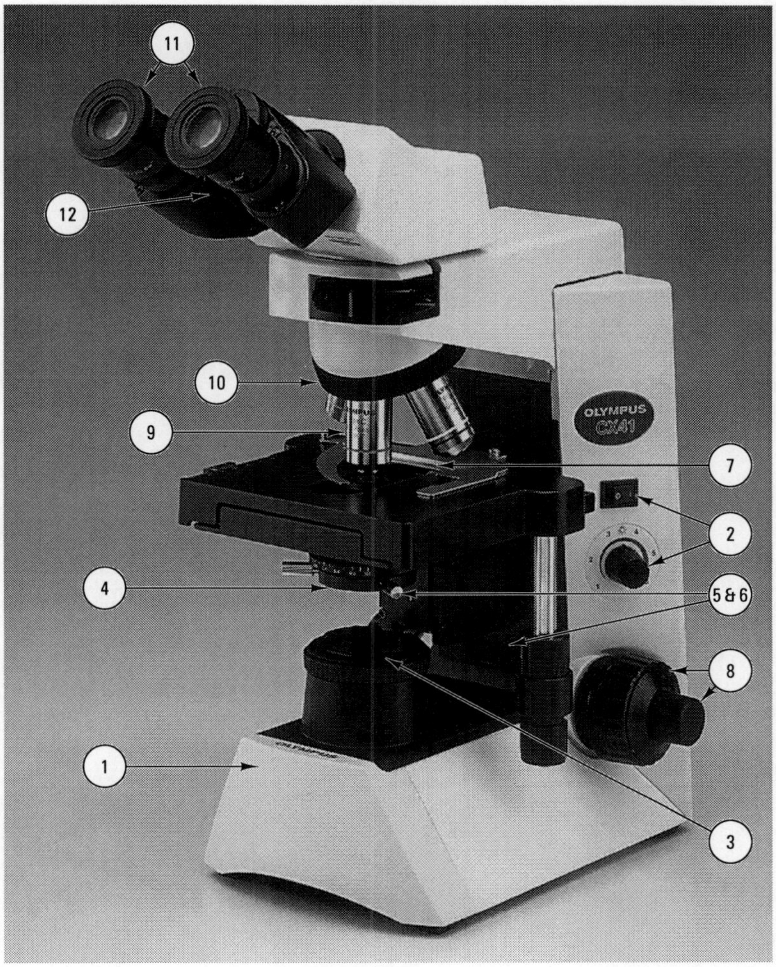

Fig. 43.2 Diagram of the Olympus binocular microscope model CX41.

- The lamp in the base of the stand (**1**) supplies light; its brightness is controlled by an on–off switch and voltage control (**2**). Never use maximum voltage or the life of the bulb will be reduced – a setting two-thirds to three-quarters of maximum should be adequate for most specimens. A field–iris diaphragm may be fitted close to the lamp to control the area of illumination (**3**).
- The condenser control focuses light from the condenser lens system (**4**) on to the specimen and projects the specimen's image on to the front lens of the objective. Correctly used, it ensures optimal resolution.
- The condenser–iris diaphragm (**5**) controls the amount of light entering and leaving the condenser; its aperture can be adjusted using the condenser–iris diaphragm lever (**6**). Use this to reduce glare and enhance image contrast by cutting down the amount of stray light reaching the objective lens.
- The specimen (normally mounted on a slide) is fitted to a mechanical stage or slide holder (**7**) using a spring mechanism. Two controls allow you to move the slide in *x* and *y* planes. Vernier scales (see p. 134) on the slide holder can be used to return to the same place on a slide. The fine and coarse focus controls (**8**) adjust the height of the stage relative to the lens systems. Take care when adjusting the focus controls to avoid hitting the lenses with the stage or slide.
- The objective lens (**9**) supplies the initial magnified image; it is the most important component of any microscope because its qualities determine resolution, depth of field and optical aberrations. The objective lenses are attached to a revolving nosepiece (**10**). Take care not to jam the longer lenses on the stage or slide as you rotate the nosepiece. You should feel a distinct click as each lens is moved into position. The magnification of each objective is written on its side; a normal complement would be ×4, ×10, ×40 and ×100 (oil immersion).
- The eyepiece lens (**11**) is used to further magnify the image from the objective and to put it in a form and position suitable for viewing. Its magnification is written on the holder (normally ×10). By twisting the holder for one or both of the eyepiece lenses you can adjust their relative heights to take account of optical differences between your eyes. The interpupillary distance scale (**12**) and adjustment knob allow compensation to be made for differences in the distance between users' pupils.

From a photograph of Olympus Model CX41 binocular microscope, supplied by Olympus Microscopes, Olympus Optical Company (UK) Ltd, published courtesy of Olympus Optical Company (UK) Ltd.

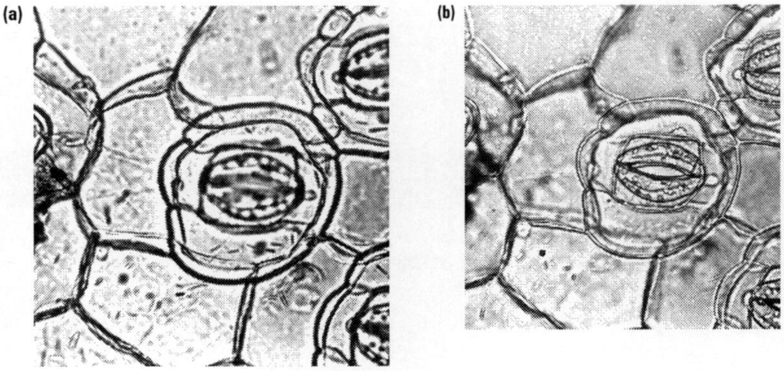

Fig. 43.3 Importance of correct focus in light microscopy. Stomatal complex of *Commelina communis* L., a specimen that is a monolayer of cells approximately 30–50 μm thick. (a) Focal plane is on 'internal' walls of the cells; (b) focal plane is on the 'external' walls and stomatal pore. The two images are different, and while it would not be possible to measure the stomatal pore in (a), it would not be possible to see the vacuolar crystals in (b). When looking at specimens, always use the fine-focus control to view different focal planes.

Adjusting a microscope with a field–iris diaphragm – *adjust this before the condenser–iris diaphragm: close it until its image appears in view as a circle of light, if necessary focusing on the edge of the circle with the condenser controls and centring it with the centring screws. Now open it so the whole field is just illuminated.*

High-power objectives – *never remove a slide while a high power objective lens (i.e. ×40 or ×100) is in position. Always turn back to the ×10 first. Having done this, lower the stage and remove the slide.*

6. Close the condenser–iris diaphragm (aperture–iris diaphragm), then open it to a position such that further opening has no effect on the brightness of the image (the 'threshold of darkening'). The edge of the diaphragm should not be in view. Turn down the lamp if it is too bright.
7. Focus the condenser. Place an opaque pointed object (the tip of a mounted needle or a sharp pencil point) on the centre of the light source. Adjust the condenser setting until both the specimen and needle tip/pencil point are in focus together. Check that the condenser–iris diaphragm is just outside the field of view.
8. For higher magnifications, swing in the relevant objective (e.g. ×40), carefully checking that there is space for it. Adjust the focus using the fine control only. If the object you wish to view is in the centre of the field with the ×10 objective, it should remain in view (magnified, of course) with the ×40. Adjust the condenser–iris diaphragm and condenser as before – the correct setting for each lens will be different.
9. When you have finished using the microscope, remove the last slide and clean the stage if necessary. Turn down the lamp setting to its minimum, then switch off. Clean the eyepiece lenses with lens tissue. Check that the objectives are clean. Unplug the microscope from the mains and wind the cable round the stand and under the stage. Replace the dust cover.

If you have problems in obtaining a satisfactory image, refer to Box 43.1; if this doesn't help, refer the problem to the class supervisor.

Procedure for observing transparent specimens

Some stained preparations and all colourless objects are difficult to see when the microscope is adjusted as above (Fig. 43.4). Contrast can be improved by closing down the condenser–iris diaphragm. Note that when you do this, diffraction haloes appear round the edges of objects. These obscure the image of the true structure of the specimen and may result in loss of resolution. Nevertheless, an image with increased contrast may be easier to interpret.

Box 43.1 Problems in light microscopy and possible solutions

No image; very dark image; image dark and illuminated irregularly
- Microscope not switched on (check plug and base)
- Illumination control at low setting or off
- Objective nosepiece not clicked into place over a lens
- Diaphragm closed down too much or off-centre
- Lamp failure

Image visible and focused but pale and indistinct
- Diaphragm needs to be closed down further (see Fig. 43.4)
- Condenser requires adjustment

Image blurred and cannot be focused
- Dirty objective
- Dirty slide
- Slide upside down
- Slide not completely flat on stage
- Eyepiece lenses not set up properly for user's eyes
- Fine focus at end of travel
- Oil-immersion objective in use, without oil

Dust and dirt in field of view
- Eyepiece lenses dirty
- Objective lens dirty
- Slide dirty
- Dirt on lamp glass or upper condenser lens

(a)

(b)

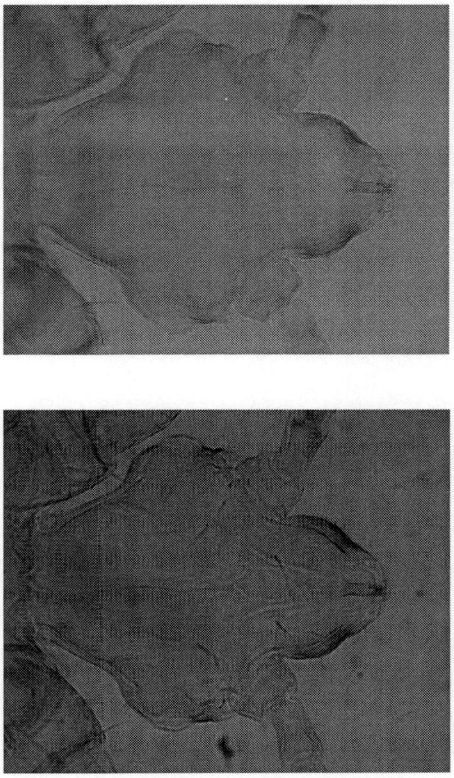

Fig. 43.4 Effect of closing the condenser–iris diaphragm on contrast. Head of human head louse, *Pediculus humanus capitus* DeGeer; (a) with condenser – iris diaphragm open; (b) with condenser – iris diaphragm closed (all other settings the same). Note difference in detail that can be seen in (b), but also that image (b) is darker – when using the condenser–iris diaphragm in this way you may need to compensate by increasing the light setting.

SAFETY NOTE Take care when moving microscopes, not only because of the cost of replacement, but also because they weigh several kilograms and could cause injury if dropped. Always carry a microscope using two hands.

***Measuring specimens using a dissecting microscope** – because of the low magnification, sizes can generally be estimated by comparison with a ruler placed alongside the specimen. If accurate measurements are required, eyepiece graticules can be used.*

Procedure for oil-immersion objectives

These provide the highest resolution of which the light microscope is capable. They must be used with immersion oil filling the space between the objective lens and the top of the slide. The oil has the same refractive index as the glass lenses, so loss of light by reflection and refraction at the glass/air interface is reduced. This increases the resolution, brightness and clarity of the image and reduces aberration. Use oil-immersion objective(s) as follows:

1. Check that the object of interest is in the field of view using e.g. the ×10 or ×40 objective.
2. Apply a single small droplet of immersion oil to the illuminated spot on the top of the slide, having first swung the ×40 objective away. Never use too much oil: it can run off the slide and mess up the microscope.
3. Move the high-power (×100) oil-immersion objective into position carefully, checking first that there is space for it. Focus on the specimen using the fine control only. You may need a higher brightness setting.
4. Perform condenser–iris diaphragm and condenser focusing adjustments as for the other lenses.
5. When finished, clean the oil immersion lens by gently wiping it with clean lens tissue. If the slide is a prepared one, wipe the oil off with lens tissue.

You should take great care when working with oil immersion lenses as they are the most expensive to replace. Because the working distance between the lens and coverslip is so short (less than 2 mm), it is easy to damage the lens surface by inadvertently hitting the slide or coverslip surface. You must also remember that they need oil to work properly. If working with an unfamiliar microscope, you can easily recognise oil-immersion lenses. Look for a white or black ring on the lens barrel, near the lens, or for 'oil' clearly marked on the barrel.

Care and maintenance of your microscope

Microscopes are delicate precision instruments. Handle them with care and never force any of the controls. Never touch any of the glass surfaces with anything other than clean, dry lens tissue. Bear in mind that a replacement would be very expensive.

If moving a microscope, hold the stand above the stage with one hand and rest the base of the stand on your other hand. Always keep the microscope vertical (or the eyepieces may fall out). Put the microscope down gently.

Clean lenses by gently wiping with clean, dry lens tissue. Use each piece of tissue once only. Try not to touch lenses with your fingers as oily fingerprints are difficult to clean off. Do not allow any solvent (including water) to come into contact with a lens; sea water is particularly damaging.

The dissecting (stereoscopic) microscope

The dissecting microscope (Fig. 43.5) is a form of stereoscopic microscope used for observations at low total magnification (×4 to ×50) where a large

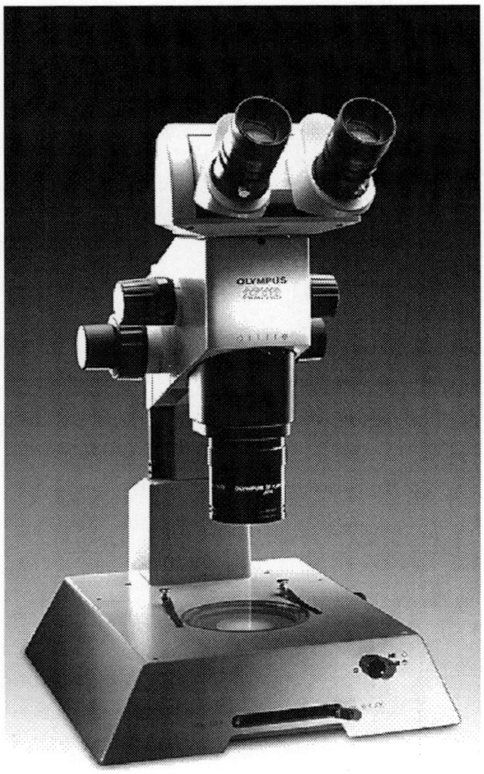

Fig. 43.5 From a photograph of Olympus Model SZX12 dissecting microscope, supplied by Olympus Microscopes, Olympus Optical Company (UK) Ltd, published courtesy of Olympus Optical Company (UK) Ltd.

working distance between objectives and stage is required, perhaps because the specimen is not flat or dissecting instruments are to be used. A stereoscopic microscope essentially consists of two separate lens systems, one for each eye. Some instruments incorporate zoom objectives. The eyepiece–objective combinations are inclined at about 15° to each other and the brain resolves the compound image in three dimensions as it does for normal vision. The image is right side up and not reversed, which is ideal for dissections. Specimens are often viewed in a fresh state and need not be placed on a slide – they might be in a Petri dish or on a white tile. Illumination can be from above or below the specimen, as desired.

Most of the instructions for the binocular microscope given above apply equally well to dissecting microscopes, although the latter do not normally have adjustable condensers or diaphragms. With stereoscopic microscopes, make specially sure to adjust the eyepiece lenses to suit your eyes so that you can take full advantage of the stereoscopic effect.

Sources for further study

Bradbury, S. and Bracegirdle, B. (1998) *Introduction to Light Microscopy*. Bios Scientific Publishers, Oxford.

Davidson, M.W. and Abramowitz, M. *Nikon MicroscopyU: Introduction to Microscope Objectives.* Available: http://www.microscopyu.com/articles/optics/objectiveintro.html.
Last accessed 09/04/07.
[Covers the optics of objective lenses.]

Olympus Microscopy Resource Centre *Introduction to Optical Microscopy, Digital Imaging, and Photomicrography* Available:
http://www.olympusmicro.com/primer/index.html
Last accessed 09/04/07.
[A primer on optical microscopy, including FAQs and links to further resources.]

Study exercises

43.1 Test your knowledge of the parts of a binocular light microscope. Cover up the legend on the left of Fig. 43.2 with a piece of paper or card. Now identify the parts of the light microscope numbered on the diagram. Check your answers from the legend.

43.2 Identify roles of parts of a binocular light microscope. State briefly the *primary* role of each component of a standard binocular light microscope:

(a) condenser
(b) objective lens
(c) condenser–iris diaphragm
(d) interpupillary distance scale
(e) Vernier scales on the mechanical stage.

43.3 Identify the correct sequence of adjustments when setting up a light microscope.

(a) Focus specimen – set interpupillary distance – adjust condenser–iris diaphragm – make individual eyepiece adjustment – focus condenser – focus specimen.
(b) Make individual eyepiece adjustment – set interpupillary distance – focus condenser – focus specimen – adjust condenser–iris diaphragm – focus specimen.
(c) Set interpupillary distance – focus specimen – make individual eyepiece adjustment – adjust condenser–iris diaphragm – focus condenser – focus specimen.
(d) Make individual eyepiece adjustment – set interpupillary distance – focus specimen – adjust condenser–iris diaphragm – focus condenser – focus specimen.
(e) Focus specimen – make individual eyepiece adjustment – set interpupillary distance – focus condenser – adjust condenser–iris diaphragm.

44 Interpreting microscope images

Microscope images, whether viewed directly or as photomicrographs, need to be interpreted with extreme care. This is not only because of the potentially damaging procedures involved in specimen preparation, but also because scale, section orientation and staining combine to create images that can be misinterpreted by the human mind.

Eliminating non-biological artefacts

There are several types of non-biological artefacts (e.g. those shown in Fig. 44.1). It is only through experience that you can learn to recognise these. Sometimes they are mistaken for specimens, because the specimen itself is hard to see. A useful tip is to look at your slide carefully before placing it on the stage; if you do this against a light background, you may be able to see the specimen and judge its location as you move the slide under the objective. For slides where specimens are tiny and spread out, such as blood or bacterial smears, it can be useful to focus on the edge of the coverslip (as in Fig. 44.1(a)) to determine the correct focal plane before searching. Correctly setting up the microscope is crucial in such cases.

Establishing scale and measuring objects

Working out the length of objects on photomicrographs

On a micrograph, the scale will usually be provided as a magnification factor (e.g. ×500) or in the form of a bar of defined length (e.g. 100 nm). If you need to estimate the dimensions of an object in the micrograph, follow the steps below:

1. Measure the object as it appears on the micrograph with a ruler or set of Vernier calipers. It may be difficult to decide exactly where the boundary of a structure lies; rather than a discrete boundary, you may be dealing with shades of grey. It is essential to be consistent!
2. If the scale is given as a bar, measure the bar too and find the object's size by proportion. For example, if the object measures 32 mm and a bar representing 100 nm is 20 mm long, then the size of the object is $32/20 \times 100\,\text{nm} = 160\,\text{nm}$.
3. If the scale is given as a magnification factor, divide your measurement by this number to obtain the object's size, taking care to enter the correct units in your calculator. For example, if the object measures 32 mm and the print magnification is stated as ×200 000, then its size is $32 \times 10^{-3}\,\text{m}/200\,000 = 1.6 \times 10^{-7}\,\text{m} = 160\,\text{nm}$.

Avoid putting too many significant figures in any estimates of dimensions: there may be quite large errors in estimating print magnifications that could make the implied accuracy meaningless.

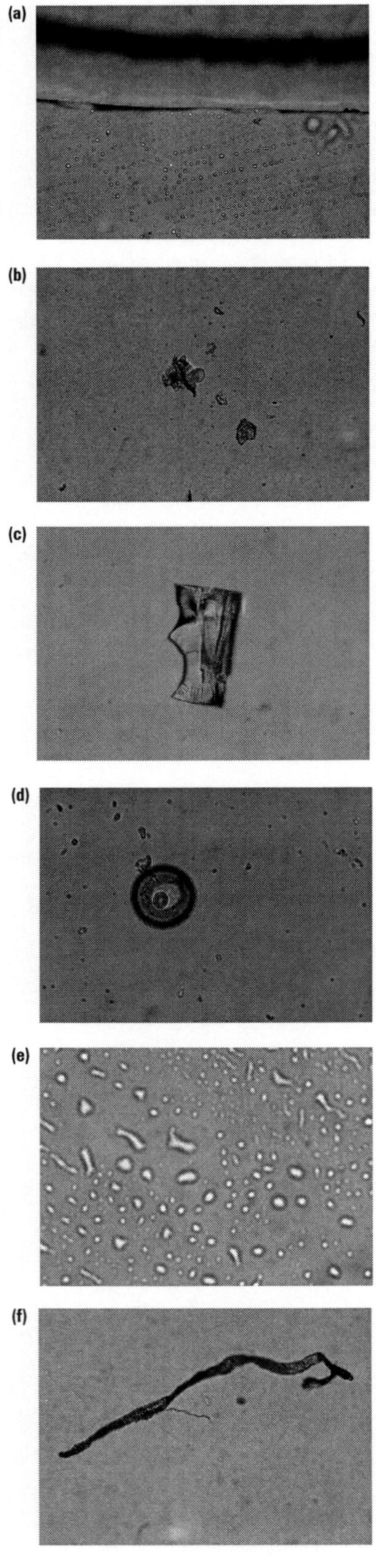

Fig. 44.1 Artefacts in microscopy. Do not mistake any of the following for biological specimens: (a) edge of coverslip; (b) dust on slide; (c) shard of glass; (d) air bubble; (e) grease spots; (f) cellulose fibre.

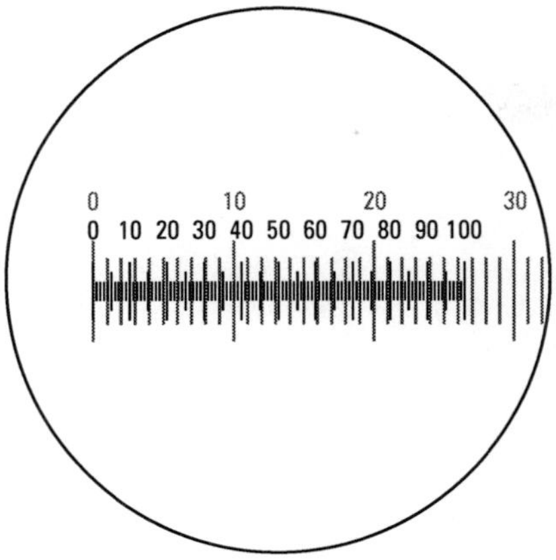

Fig. 44.2 Calibrating an eyepiece micrometer (graticule). Align the two scales and read the number of stage micrometer divisions (shown in colour) for a particular number of eyepiece micrometer divisions (shown in black). In this case 26.5 scale divisions of 0.01 mm (= 2.65 mm) are equivalent to 100 eyepiece divisions, so each eyepiece division 0.265 ÷ 100 = 0.00265 mm = 2.65 μm. This is a typical value for a total magnification of ×400 (e.g. ×40 objective and ×10 eyepiece).

Adding linear scales to drawings

The magnification of a light-microscope image is calculated by multiplying the objective magnification by the eyepiece magnification. However, the magnification of the image bears no certain relation to the magnification of any drawing of the image – you may equally well choose to draw the same image 10 mm or 10 cm long. For this reason, *it is essential to add a scale to all your diagrams*. You can provide either a bar giving the estimated size of an object of interest, or a bar of defined length (e.g. 100 μm).

The simplest method of estimating linear dimensions is to compare the size of the image to the diameter of the field of view. You can make a rough estimate of the field diameter by focusing on the millimetre scale of a transparent ruler using the lowest power objective. Estimate the diameter of this field directly, then use the information to work out the field diameters at the higher powers *pro rata*. For example, if the field at an overall magnification of ×40 is 4 mm, at an overall magnification of ×100 it will be: $40/100 \times 4\,\text{mm} = 1.6\,\text{mm}$ (1600 μm).

Greater accuracy can be obtained if an eyepiece micrometer (graticule) is used. This carries a fine scale and fits inside an eyepiece lens. The eyepiece micrometer is calibrated using a stage micrometer, basically a slide with a fine scale on it. Figure 44.2 shows how to calibrate an eyepiece micrometer, along with a worked example. Once you have calibrated your eyepiece micrometer for each objective lens used, you can use it to measure objects: in the example shown in Fig. 44.2, the scale reading is multiplied by 2.65 μm to give the value in micrometres. So, if you measured the width of a human hair at 34 eyepiece micrometer units, then this will be equal to 34 × 2.65 = 90.1 μm. An alternative approach is to put a scale bar on a diagram, e.g. a 100 μm scale bar would be equivalent to the length of almost 38 eyepiece micrometer divisions.

Using an eyepiece graticule *– choose the eyepiece lens corresponding to your stronger eye and check that you have made the correct adjustments to the eyepiece lenses as detailed on p. 252.*

Estimating the area and area density of objects

You can calculate the area of the field at each magnification using a field diameter estimate obtained as above (obtain the area from πr^2, but don't forget that radius r = diameter ÷ 2). The units you use to measure the diameter will be squared; thus, a field diameter of 0.45 mm converts to a field area of 0.159 mm^2. Use the field area to estimate the cross-sectional area of a specimen by proportion.

If several objects appear in a field, you can express their frequency on an area basis by dividing the number seen by the field area. It is nearly always best to take an average from several fields of view before calculating.

Examples An object occupying one-third of the area of a field of diameter 0.45 mm would have a cross-sectional area of 0.159/3 (= 0.053 mm^2).

Seven objects in a field of area 0.159 mm^2 would be present at a density of 7 ÷ 0.159 (= 44 objects mm^{-2}).

Volume–density estimates

The density of particles in a liquid suspension can be *roughly* estimated as follows: place a drop of well-mixed suspension on to a slide with a Pasteur pipette. You can estimate the drop volume by weighing drops of water and assuming 1 mg = 1 mm^3. Alternatively, use a pipettor. Quickly cover the drop with a coverslip, avoiding air bubbles – a 22 × 22 mm coverslip should allow a 30 mm^3 drop to spread out completely. Now count the numbers of the object of interest within each of several fields and calculate a mean value. The volume of each field will be:

$$\text{(drop volume} \times \text{field area)} \div \text{coverslip area} \qquad [44.1]$$

and the density will be:

$$\text{(mean no. of objects per field)} \div \text{field volume} \qquad [44.2]$$

Example For a field area of 0.159 mm^2, a droplet volume of 30 mm^3 and a coverslip of area 484 mm^2, the field volume would be 30 × (0.159 ÷ 484) = 0.01 mm^3, so if you saw 25 objects per field, their density would be 25 ÷ 0.01 = 2 500 mm^{-3}.

Where accurate estimates of volume–density are required, you should use a haemocytometer.

Determining the position and orientation of the section within the specimen

When interpreting microscope images, you need to determine and take into account what position and orientation the section was taken from.

Always remember that a section is a two-dimensional slice through a three-dimensional structure. Thus, a circular profile in two dimensions could arise from sectioning numerous three-dimensional shapes (Fig. 44.3); you cannot determine which shape without further information (e.g. from serial sections or sections taken in another plane). Likewise, the size of a profile in a section may reveal little about the size of the structure from which it arose. Also bear in mind that very thin sections (say 75 nm thick) may not provide a representative sample of all the organelles present in a cell (which may be 1000–100 000 nm in width).

Always relate what you see to what you know; at a minimum, you should take account of your basic knowledge of animal, microbial or plant cell structure. Note that, in the absence of a defining characteristic, you cannot be sure about the nature of a specimen, because it may simply be due to chance that the section does not include the feature.

The terminology used for sections taken in different planes is shown in Figs. 44.4 and 44.5.

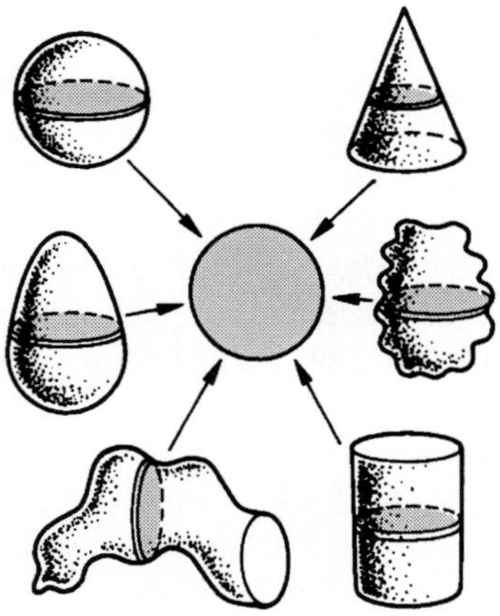

Fig. 44.3 A two-dimensional section may be representative of many different three-dimensional structures.

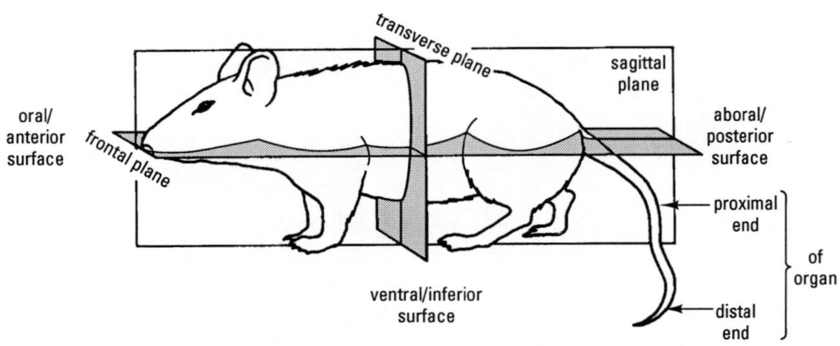

Fig. 44.4 The terms used for various parts and sections through the animal body.

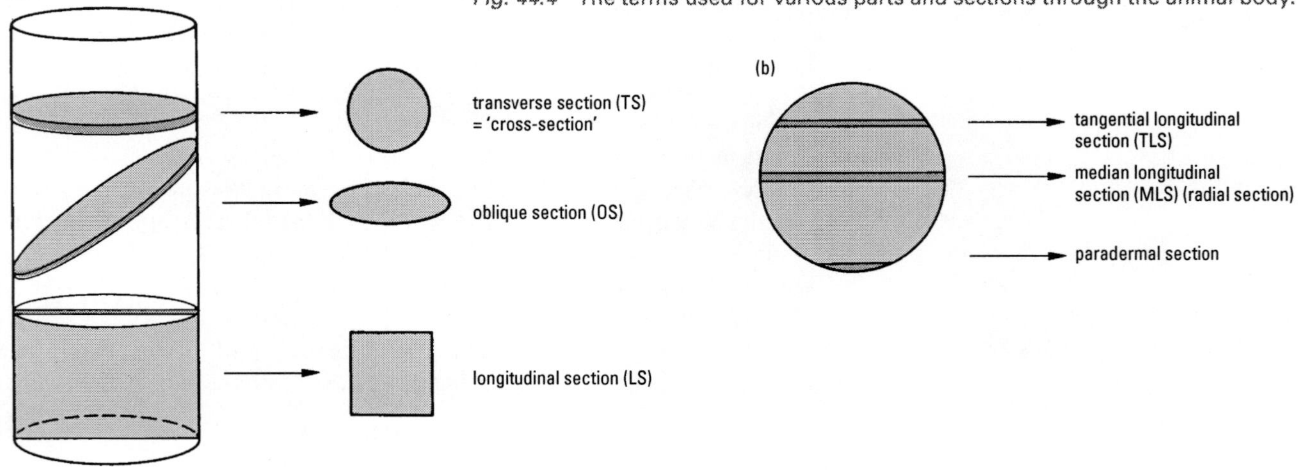

Fig. 44.5 The main section planes: (a) sections through a cylindrical object; (b) various types of longitudinal sections as seen in transverse orientation.

Table 44.1 The main stains used in electron microscopy

Stain	What it stains
Lead citrate	Lipoproteins Ribonucleoproteins Glycogen
Osmium tetroxide	Lipids and proteins
Phosphotungstic acid	Proteins (at pH >3) Polysaccharides Glycoproteins (at pH <3)
Uranyl acetate	Proteins Nucleic acids (at pH 3.5)

***Comparing methods** – it is informative to compare the same tissue prepared and viewed by different methods.*

Accounting for methodology

As discussed, a great deal of preparative work is required before specimens can be viewed under LM or TEM. Under normal circumstances, it is hoped that the stain has the greatest effect on the image. The colours produced in various cell types and structures by the main stains used in light microscopy are tabulated. You should bear in mind that colours are subjective and that dyes may exhibit different colours in different chemical environments (e.g. at different pH values).

Stains for TEM are generally compounds or ions containing elements of high atomic number, such as the heavy metals, because these scatter electrons strongly. On a positive micrograph from a positively stained specimen, dark regions show where electrons have been scattered, i.e. where the stain has attached. Light regions show where electrons have passed relatively unaffected through the specimen, i.e. where the stain has not been taken up. Table 44.1 shows the main stains used in TEM and the specific cell components to which they attach.

Images from SEM are relatively easy to interpret because the brain is used to interpreting surface features. However, it is possible for optical illusions to occur – for instance, it may be difficult to decide whether a shadowed feature projects out of a surface or represents a depression in it.

Artefacts

The following is a list of the main types of artefact occurring during the preparation process:

- Fixation: swelling or shrinkage may occur, often due to the use of hypotonic or hypertonic fixative solutions. Glutaraldehyde-fixed tissue is particularly likely to shrink if this fixative is used at high concentration.
- Dehydration: shrinkage may occur as water is withdrawn from the tissue. Lipids may dissolve in the solvent used and be lost from the tissue.
- Embedding: difficulties in sectioning may be caused by incorrect hardness of embedding resins.
- Sectioning: too thick a section can result in a blurred image. Compression effects, tearing and knife chattering may occur.
- Mounting: folds may be introduced into the section. These can be detected by the presence of discontinuities in features that cross the fold. In TEM, the pattern of the metal grid may sometimes obscure the image (as a solid black area).
- The stain(s) used on the section: if staining is not carried out correctly, precipitates of stain may appear, often in crystalline form.

KEY POINT No preparative procedure gives a 'perfect' image. Artefacts can be introduced at many stages.

Stereological studies

Stereology allows the study of the three-dimensional organisation of specimens based on two-dimensional information. The subject is a complex one, relying on geometric and statistical principles, and specialist texts should be consulted for detailed background. There are three main approaches:

- Cells or organelles are assumed to have a regular geometric shape, whose surface area or volume can be estimated from measurements of the principal axes. Cell and organelle shapes are generally so irregular as to make such estimates very approximate.
- A physical or computer-generated model of structures in three dimensions is built up from a set of serial sections. Considerable expertise is required to obtain the basic information required.
- Representative sections are sampled and information derived from the profiles seen. This method can be based on simple count data and is surprisingly informative. Three parameters of interest are:
  (a) Numerical density, N_V (the numbers of a component found per unit volume). For spherical components, this can be estimated from:

  $$N_V = N_A/\bar{D} \qquad [44.3]$$

  where N_A is the number of components per unit area and $\bar{D}$ is their mean diameter. N_A is simply obtained by counting component profiles in defined areas. The mean diameter (of spheres) will be approximately equal to the largest diameters seen in the sample, or may be calculated from $\bar{D} = 4/\pi\bar{d}$, where $\bar{d}$ is the mean component diameter of the sample. The situation for non-spherical components is more complex.
  (b) Surface density, S_V (the surface area of the component per unit containing volume). For any shape of object, this can be estimated from:

  $$S_V = 4/\pi B_A \qquad [44.4]$$

  where B_A is the length of profile boundaries per unit area of section. Profile boundary lengths can be measured by fitting a thread or by using a map-measuring wheel or a planimeter. An even simpler method is to count the number of intersections of the bounding membrane I_i with test lines of total length L_T on an overlay (Fig. 44.6). Then S_V is obtained from:

  $$S_V = 2I_i/L_T \qquad [44.5]$$

  Note that this method depends on the components being randomly oriented with respect to the lines on the overlay; this can be tested by comparing results with the lines placed in different orientations.
  (c) Volume density, V_V (the volume of the component as a proportion of the total volume). This is numerically equivalent to A_A, the component density on the test area. A_A can be most simply estimated:
  (i) Destructively, by the cut-and-weigh method. Here, the total area to be considered is cut out, weighed, then the component profiles are cut out and weighed: the ratio of weights equals A_A.
  (ii) Non-destructively, by point counting, where a patterned overlay or lattice (e.g. Fig. 44.6) is placed over the micrograph and counts made of the number of points overlaying the feature of interest and the total area. A_A is estimated by the ratio of these point counts.

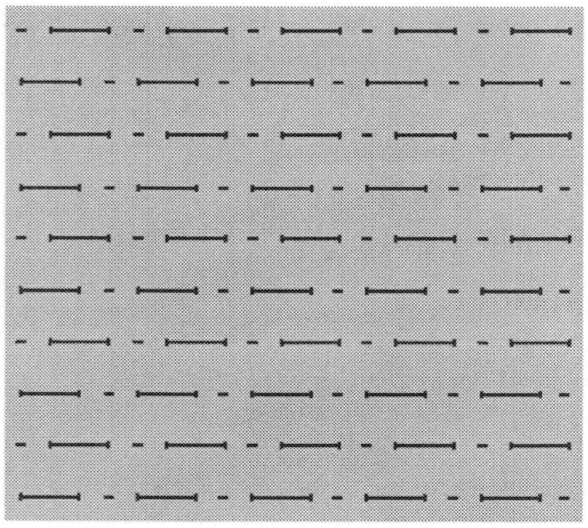

Fig. 44.6 Weibel's multipurpose test overlay for stereological studies. The number of intersections of a membrane with the solid lines can be used when estimating surface densities, while point counts for estimates of profile density on test areas can be made using the ends of the lines.

Both of these methods of estimating A_A require that a reasonably large representative area is sampled. Other methods for estimating A_A include planimetry and linear integration.

Sources for further study

Anon. *Size and Scale: on the Relative Size of Things in the Universe*. Available: http://invsee.asu.edu/nmodules/sizes/sizescalemod/microscopy.htm
Last accessed 09/04/07.
[Scale in relation to the history of microscopy.]

Crang, R.F.E. and Klomparens, K.L. (1988) *Artefacts in Biological Electron Microscopy*. Plenum Press, New York.

Russ, J.C. and Dehoff, R.T. (2000) *Practical Stereology*, 2nd edn. Plenum Publishing Corporation, New York.

Study exercises

44.1 Calculate the size of an object seen down a microscope using graticule readings. At ×100 overall magnification, you find that 1 stage micrometer unit of 0.1 mm = 11 eyepiece micrometer units. The length of a parasitic worm on a slide viewed at this magnification appears to be 25 eyepiece units. What is its estimated true length to the nearest μm?

44.2 Calculate the area density of objects seen down a microscope. A light microscope at ×400 overall magnification has a field diameter estimated as 0.425 mm. Observing a sample of leaf surface at this magnification, you count 25 hairs in a single field of view. What is your estimate of the area density (frequency) of hairs, expressed on a per mm^2 basis, to three significant figures?

44.3 Calculate the volume density of objects seen down a microscope. A diluted blood sample viewed at the same magnification as in Study exercise 43.2 gives a mean count of 32 white blood cells per field of view. If the thickness of the liquid layer below the coverslip is 0.1 mm, what is the volume density of white blood cells (express on a per mm^3 basis, to four significant figures)?

44.4 Use point count data to estimate relative volumes of cellular components. A research team fixed and embedded leaf mesophyll cells, sectioned them randomly, stained the sections appropriately and used a TEM to obtain photomicrographs. Then, using an overlay like the one illustrated in Fig. 44.6, they made point counts within each of 25 cell outlines and took a mean of the results (see table below). From their data, answer the following questions, to two decimal places:

(a) What percentage of the cell volume did the cell wall account for?
(b) What percentage of the cell volume was taken up by the cytoplasm?
(c) What is the estimated cyoplasm : vacuole ratio for this cell?
(d) What percentage of the cytoplasm was taken up by chloroplasts?

Point counts over random mesophyll sections

Organelle	Mean point count
Cell wall	34.0
Cytosol	35.0
Nucleus	16.9
Mitochondria	5.1
Chloroplasts	160.8
Other cytoplasmic bodies	2.6
Vacuole	264.7
Total	519.1

45 Sterile technique

***Achieving a sterile state** – you should assume that all items of laboratory equipment have contaminating microbes on their surfaces, unless they have been destroyed by some form of sterilisation. Such items will only remain sterile if they do not come into contact with the non-sterile environment.*

Fig. 45.1 International symbol for a biohazard. Usually red on a yellow background, or black on a red background.

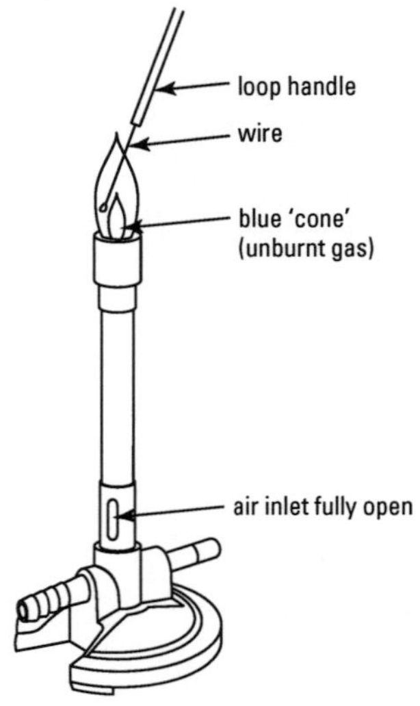

Fig. 45.2 'Flaming' a wire loop. Keep the loop in the hottest part of the Bunsen flame (just outside the blue 'cone') until the wire is red-hot.

Sterile technique (aseptic technique) is the name given to the procedures used in cell culture. Although the same general principles apply to all cell types, you are most likely to learn the basic procedures using bacteria and most of the examples given in this section refer to bacterial culture.

Sterile technique serves two main purposes:

1. To prevent accidental contamination of laboratory cultures due to microbes from external sources, e.g. skin, clothing or the surrounding environment.
2. To prevent microbial contamination of laboratory workers, in this instance you and your fellow students.

KEY POINT** All **microbial cell cultures should be treated as if they contained potentially harmful organisms. Sterile technique forms an important part of safety procedures, and must be followed whenever cell cultures are handled in the laboratory.

Care is required because:

- You may accidentally isolate a harmful microbe as a contaminant when culturing a relatively harmless strain.
- Some individuals are more susceptible to infection and disease than others – not everyone exposed to a particular microbe will become ill.
- Laboratory culture involves purifying and growing large numbers of microbial cells – this represents a greater risk than small numbers of the original microbe.
- A microbe may change its characteristics, perhaps as a result of gene exchange or mutation.

The international biohazard symbol, shown in Fig. 45.1, is used to indicate a significant risk due to a pathogenic microbe.

Sterilisation procedures

Given the ubiquity of microbes, the only way to achieve a sterile state is by their destruction or removal. Several methods can be used to achieve this objective:

Heat treatment

This is the most widespread form of sterilisation and is used in several basic laboratory procedures including the following:

- Red-heat sterilisation. Achieved by heating metal inoculating loops, forceps, needles, etc. in a Bunsen flame (Fig. 45.2). This is a simple and effective form of sterilisation as no microbe will survive even a brief exposure to a naked flame. Flame sterilisation using alcohol is used for glass rods and spreaders (see below).
- Dry-heat sterilisation. Here, a hot-air oven is used at a temperature of at least 160 °C for at least 2 hours. This method is used for the routine sterilisation of laboratory glassware. Dry-heat procedures are of little value for items requiring repeated sterilisation during use.

Fig. 45.3 Autoclave tape – the bottom sample is untreated while the upper sample (with dark diagonal lines) has been autoclaved.

- Moist-heat sterilisation. This is the method of choice for many laboratory items, including most fluids, apart from heat-sensitive media. It is also used to decontaminate liquid media and glassware after use. The laboratory autoclave is used for these purposes. Typically, most items will be sterile after 15 minutes at 121 °C, although large items may require a longer period. The rapid killing action results from the latent heat of condensation of the pressurised steam, released on contact with cool materials in the autoclave. Although special heat-sensitive tape (Fig. 45.3) is sometimes used to check that the autoclave is operating correctly, a better approach is to use spores of *Bacillus stearothermophilus*.

Radiation

Many disposable plastic items used in microbiology and cell biology are sterilised by exposure to UV or ionising radiation. They are supplied commercially in sterile packages, ready for use. Ultraviolet radiation has limited use in the laboratory, while ionising radiation (e.g. γ-rays) requires industrial facilities and cannot be operated on a laboratory scale.

Filtration

Using a sterile filter – *most filters are supplied as pre-sterilised items. Make sure you follow a procedure that does not contaminate the filter on removal from its protective wrapping.*

Heat-labile solutions (e.g. complex macromolecules, including proteins, antibiotics, serum) are particularly suited to this form of sterilisation. The filters come in a variety of shapes, sizes and materials, usually with a pore size of either 0.2 µm or 0.45 µm. The filtration apparatus and associated equipment is usually sterilised by autoclaving, or by dry heat. Passage of liquid through a sterile filter of pore size 0.2 µm into a sterile vessel is usually sufficient to remove bacteria but not viruses, so filtered liquids are not necessarily virus-free.

Chemical agents

SAFETY NOTE When working with biocides, take care to avoid skin contact or ingestion, as most are toxic and irritant. If contact does occur, rinse with plenty of water.

These are known as disinfectants, or biocides, and are most often used for the disposal of contaminated items following laboratory use, e.g. glass slides and pipettes. They are also used to treat spillages. The term 'disinfection' implies destruction of disease-causing bacterial cells, although spores and viruses may not always be destroyed. Remember that disinfectants require time to exert their killing effect – any spillage should be covered with an appropriate disinfectant and left for at least 10 minutes before mopping up.

Use of laboratory equipment

Working area

One of the most important aspects of good sterile technique is to keep your working area as clean and tidy as possible. Start by clearing all items from your working surface, wipe the bench down with disinfectant and then arrange the items you need for a particular procedure so that they are close at hand, leaving a clear working space in the centre of your bench.

Using molten agar – *a water bath (at 45–50 °C) can be used to keep an agar-based medium in its molten state after autoclaving. Always dry the outside of the container on removal from the water bath, to reduce the risk of contamination from microbes in the water, e.g. during pour plating.*

Media

Cells may be cultured in either a liquid medium (broth), or a solidified medium . The gelling agent used in most solidified media is agar, a complex polysaccharide from red algae that produces a stiff transparent gel when used at 1–2% (w/v). Agar is used because it is relatively resistant to degradation by most bacteria and because of its rheological properties – an agar medium melts at 98 °C, remaining solid at all temperatures used for routine laboratory culture. Once melted however, it does not solidify again until the temperature falls to about 44 °C. This means that heat-sensitive constituents (e.g. vitamins, blood, cells, etc.) can be added aseptically to the medium after autoclaving.

Plastic disposable loops *– these are used in many research laboratories: pre-sterilised and suitable for single use, they avoid the hazards of naked flames and the risk of aerosol formation during heating. Discard into a disinfectant solution after use.*

Using a Bunsen burner to reduce airborne contamination *– working close to the updraught created by a Bunsen flame reduces the likelihood of particles falling from the air into an open vessel.*

Using glass pipettes *– these are plugged with cotton wool at the top before being autoclaved inside a metal can. Flame the open end of the can on removal of a pipette, to prevent contamination of the remaining pipettes. Autopipettors and sterile disposable tips offer an alternative approach.*

Inoculating loops

The initial isolation and subsequent transfer of microbes between containers can be achieved by using a sterile inoculating loop. Most teaching laboratories use nichrome wire loops in a metal handle. A wire loop can be repeatedly sterilised by heating the wire, loop downwards and almost vertical, in the hottest part of a Bunsen flame until the whole wire becomes red-hot. Then the loop is removed from the flame to minimise heat transfer to the handle. After cooling for 8–10 seconds (without touching any other object), it is ready for use.

When re-sterilising a contaminated wire loop in a Bunsen flame after use, do not heat the loop too rapidly, as the sample may spatter, creating an aerosol: it is better to soak the loop for a few minutes in disinfectant than to risk heating a fully charged (contaminated) inoculating loop.

Containers

There is a risk of contamination whenever a sterile bottle, flask or test tube is opened. One method that reduces the chance of airborne contamination is quickly to pass the open mouth of the glass vessel through a flame. This destroys any microbes on the outer surfaces nearest to the mouth of the vessel. In addition, by heating the air within the neck of the vessel, an outwardly directed air flow is established, reducing the likelihood of microbial contamination.

It is general practice to flame the mouth of each vessel immediately after opening and then repeat the procedure just before replacing the top. Caps, lids and cotton wool plugs must not be placed on the bench during flaming and sampling: they should be removed and held using the smallest finger of one hand, to minimise the risk of contamination. This also leaves the remaining fingers free to carry out other manipulations. With practice, it is possible to remove the tops from two tubes, flame each tube and transfer material from one to the other while holding one top in each hand.

Table 45.1 Classification of microbes on the basis of hazard. The following categories are recommended by the UK Advisory Committee on Dangerous Pathogens (ACDP).

Hazard group	Comments
1	Unlikely to cause human disease
2	May cause disease: possible hazard to laboratory workers, minimal hazard to community
3	May cause severe disease: may be a serious hazard to laboratory workers, may spread to community
4	Causes severe disease: is a serious hazard to laboratory workers, high risk to community

Laminar-flow cabinets

These are designed to prevent airborne contamination, e.g. when preparing media or subculturing microbes or tissue cultures. Sterile air is produced by passage through a high-efficiency particulate air (HEPA) filter: this is then directed over the working area, either horizontally (towards the operator) or downwards. The operator handles specimens, media, etc., through an opening at the front of the cabinet. Note that standard laminar-flow cabinets do *not* protect the worker from contamination and must not be used with pathogenic microbes: special safety cabinets are used for work with ACDP hazard group 3 and 4 microbes (Table 45.1) and for samples that might contain such pathogens.

Microbiological hazards

***KEY POINT** The most obvious risks when handling microbial cultures are those due to ingestion or entry via a cut in the skin – all cuts should be covered with a plaster or disposable plastic gloves. A less obvious source of hazard is the formation of aerosols of liquid droplets from microbial suspensions, with the risk of inhalation, or surface contamination of other objects.*

The following steps will minimise the risk of aerosol formation:

- Use stoppered tubes when shaking, centrifuging or mixing microbial suspensions.
- Pour solutions gently, keeping the difference in height to a minimum.
- Discharge pipettes on to the side of the container.

Other general rules that apply in all laboratories include:

- Take care with sharp instruments, including needles and glass Pasteur pipettes.
- Do not pour waste cultures down the sink – they must be autoclaved.
- Put other contaminated items (e.g. slides, pipettes) into disinfectant after use.
- Wipe down your bench with disinfectant when practical work is complete.
- Always wash your hands before leaving the laboratory.

Plating methods

Many culture methods make use of a solidified medium within a Petri plate. A variety of techniques can be used to transfer and distribute the organisms prior to incubation. The three most important procedures are described below.

Streak-dilution plate

Streaking a plate for single colonies is one of the most important basic skills in microbiology, since it is used in the initial isolation of a cell culture and in maintaining stock cultures, where a streak-dilution plate with single colonies all of the same type confirms the purity of the strain. A sterile inoculating loop is used to streak the organisms over the surface of the medium, thereby diluting the sample (Fig. 45.4). The aim is to achieve single colonies at some point on the plate: ideally, such colonies are derived from single cells (e.g. in the case of unicellular bacteria, animal and plant cell lines) or from groups of cells of the same species (in filamentous or colonial forms). Single colonies, containing cells of a single species and derived from a single parental cell, form the basis of all pure culture methods.

Note the following:

- Keep the lid of the Petri plate as close to the base as possible to reduce the risk of aerial contamination.
- Allow the loop to glide over the surface of the medium. Hold the handle at the balance point (near the centre) and use light, sweeping movements, as the agar surface is easily damaged and torn.
- Work quickly, but carefully. Do not breathe directly on to the exposed agar surface and replace the lid as soon as possible.

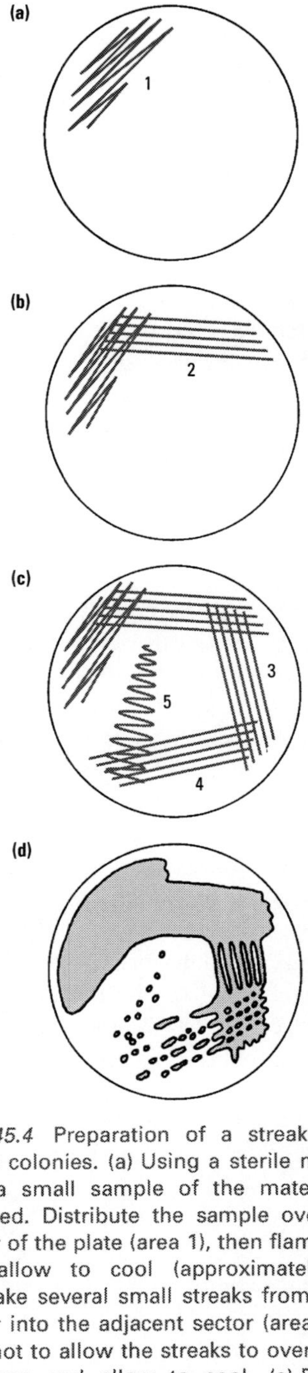

Fig. 45.4 Preparation of a streak plate for single colonies. (a) Using a sterile metal loop, take a small sample of the material to be streaked. Distribute the sample over a small sector of the plate (area 1), then flame the loop and allow to cool (approximately 8–10 s). (b) Make several small streaks from the initial sector into the adjacent sector (area 2), taking care not to allow the streaks to overlap. Flame the loop and allow to cool. (c) Repeat the procedure for areas 3 and 4, re-sterilising the loop between each step. Finally, make a single, long streak, as shown for area 5. (d) The expected result after incubation at the appropriate temperature (e.g. 37 °C for 24 h): each step should have diluted the inoculum, giving individual colonies within one or more sectors on the plate. Further subculture of an individual colony should give a pure (clonal) culture.

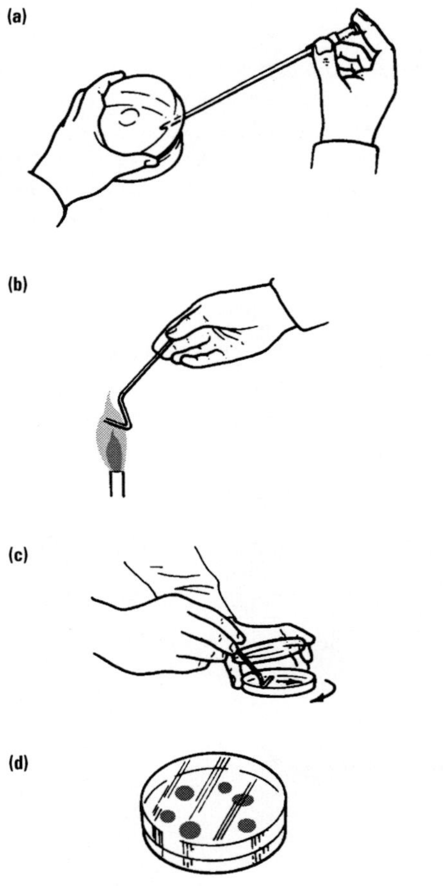

Fig. 45.5 Preparation of a spread plate. (a) Transfer a small volume of cell suspension (0.05–0.5 ml) to the surface of a solidified medium in a Petri plate. (b) Flame sterilise a glass spreader and allow to cool (8–10 s). (c) Distribute the liquid over the surface of the plate using the sterile spreader. Make sure of an even coverage by rotating the plate as you spread: allow the liquid to be absorbed into the agar medium. Incubate under suitable conditions. (d) After incubation, the microbial colonies should be distributed over the surface of the plate.

Spread plate

This method is used with cells in suspension, either in a liquid growth medium or in an appropriate sterile diluent. It is one method of quantifying the number of viable cells (or colony forming units) in a sample, after appropriate dilution.

An L-shaped glass spreader is sterilised by dipping the end of the spreader in a beaker containing a small amount of 70% v/v alcohol, allowing the excess to drain from the spreader and then igniting the remainder in a Bunsen flame. After cooling, the spreader is used to distribute a known volume of cell suspension across the plate (Fig. 45.5). *There is a significant fire risk associated with this technique*, so take care not to ignite the alcohol in the beaker, e.g. by returning an overheated glass rod to the beaker. The alcohol will burn with a pale blue flame that may be difficult to see, but will readily ignite other materials (e.g. a laboratory coat). Another source of risk comes from small droplets of flaming alcohol shed by an overloaded spreader on to the bench and this is why you *must* drain excess alcohol from the spreader *before* flaming. Some laboratories now provide plastic disposable spreaders for student use, to avoid the risk of fire.

Pour plate

This procedure also uses cells in suspension, but requires molten agar medium, usually in screw-capped bottles containing sufficient medium to prepare a single Petri plate (i.e. 15–20 ml), maintained in a water bath at 45–50 °C. A known volume of cell suspension is mixed with this molten agar, distributing the cells throughout the medium. This is then poured without delay into an empty sterile Petri plate and incubated, giving widely spaced colonies (Fig. 45.6). Furthermore, as most of the colonies are formed within the medium, they are far smaller than those of the surface streak method, allowing higher cell numbers to be counted (e.g. up to 1000 colonies per plate): some workers pour a thin layer of molten agar on to the surface of a pour plate after it has set, to ensure that no surface colonies are produced. Most bacteria and fungi are not killed by brief exposure to temperatures of 45–50 °C, though the procedure may be more damaging to microbes from low temperature conditions, e.g. psychrophilic bacteria.

One disadvantage of the pour-plate method is that the typical colony morphology seen in surface-grown cultures will not be observed for those colonies that develop within the agar medium. A further disadvantage is that some of the suspension will be left behind in the screw-capped bottle. This can be avoided by transferring the suspension to the Petri plate, adding the molten agar, then swirling the plate to mix the two liquids. However, even when the plate is swirled repeatedly and in several directions, the liquids are not mixed as evenly as in the former procedure.

Working with phages

Bacterial viruses ('bacteriophages', or simply 'phages') are often used to illustrate the general principles involved in the detection and enumeration of viruses. They also have a role in genome mapping of bacteria. Individual phage particles (virions) are too small to be seen by light microscopy, but are detected by their effects on susceptible host cells:

- Virulent phages will infect and replicate within actively growing host cells, causing cell lysis and releasing new infective phages – this 'lytic cycle' takes ≈30 min for T-even phages of *E. coli*, e.g. T4.

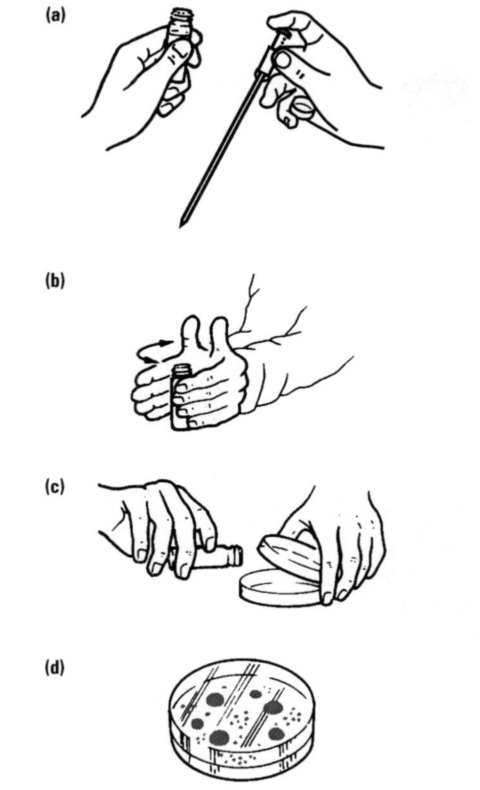

Fig. 45.6 Preparation of a pour plate. (a) Add a known volume of cell suspension (0.05–1.0 ml) to a small bottle of molten agar medium from a 45 °C water bath. (b) Mix thoroughly, by rotating between the palms of the hands: do not shake or this will cause frothing of the medium. (c) Pour the mixture into an empty, sterile Petri plate and allow to set. Incubate under suitable conditions. (d) After incubation, the microbial colonies will be distributed throughout the medium: any cells deposited at the surface will give larger, spreading colonies.

- Temperate phages are a specialised group, capable either of lytic growth or an alternative cycle, termed lysogeny – the phage becomes latent within a host cell (lysogen), typically by insertion of its genetic information into the host cell genome, becoming a 'prophage'. At a later stage, termed induction, the prophage may enter the lytic cycle. A widely used example is λ phage of *E. coli.*

The lytic cycle can be used to detect and quantify the number of phages in a sample. A known volume of sample is mixed with susceptible bacterial cells in molten soft agar medium (45–50 °C), then poured on top of a plate of the same medium, creating a thin layer of 'top agar'. The upper layer contains only half the normal amount of agar, to allow phages to diffuse through the medium and attach to susceptible cells. On incubation, the bacteria will grow throughout the agar to produce a homogeneous 'lawn' of cells, except in those parts of the plate where a phage particle has infected and lysed the cells to create a clear area, termed a plaque (Fig. 45.7). Each plaque is due to a single functional phage (i.e. a plaque-forming unit, or PFU). A count of the number of plaques can be used to give the number of phages in a particular sample (e.g. as PFU ml^{-1}), with appropriate correction for dilution and the volume of sample counted in an analogous manner to a bacterial plate count. When counting plaques in phage assays you should view them against a black background to make them easier to see: mark each plaque with a spirit-based marker to ensure an accurate count. Temperate phages often produce cloudy plaques, because many of the infected cells will be lysogenised rather than lysed, creating turbidity within the plaque. Samples of material from within the plaque can be used to subculture the phage for further study, perhaps in a broth culture where the phages will cause widespread cell lysis and a decrease in turbidity. Alternatively, phages can be stored by adding chloroform to aqueous suspensions – this will prevent contamination by cellular micro-organisms. A similar approach can be used to detect and count animal or human viruses, using a monolayer of susceptible host cells.

Electron microscopy (EM) provides an alternative approach to the detection of viruses, avoiding the requirement for culture of infected host cells, and giving a faster result. However, it requires specialised equipment and expertise. EM counts are often higher than culture-based methods, for similar reasons to those described for bacteria.

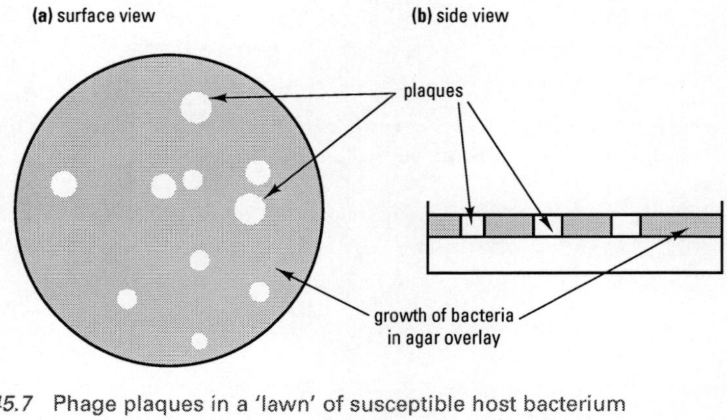

Fig. 45.7 Phage plaques in a 'lawn' of susceptible host bacterium

Labelling Petri plates – *the following information should be recorded on the base:*

- *date*
- *the growth medium*
- *your name or initials*
- *brief details of the experimental treatment.*

Labelling your plates and cultures

Petri plates should always be labelled on the *base*, rather than the lid. Restrict your labelling to the outermost region of the plate, to avoid problems when counting colonies, assessing growth, etc. After labelling, Petri plates usually are incubated upside down in a temperature-controlled incubator (often at 37 °C) for an appropriate period (usually 18–72 hours). Plates are also usually kept upside down on the lab bench – following incubation, the base (containing medium and microbes) can then be lifted from the lid and examined.

Sources for further study

Anon. (1995) *Advisory Committee on Dangerous Pathogens: Categorisation of Biological Agents According to Hazard and Categories of Containment*, 4th edn. HSE Books, London.

Anon. (2004) *Advisory Committee on Dangerous Pathogens: Approved List of Biological Agents.* Available: http://www.hse.gov.uk/pubns/misc208.pdf Last accessed 09/04/07.

Collins, C.H., Lyne, P.M., Grange, J.M. and Falkinham, J. (2004) *Collins and Lyne's Microbiological Methods*, 8th edn. Hodder-Arnold, London.

Hawkey. P. and Lewis, D. (2004) *Medical Bacteriology: A Practical Approach,* 2nd edn. Oxford University Press, Oxford.

Rhodes, P.M. and Stanbury, P.F. (1997) *Applied Microbial Physiology: a Practical Approach.* Oxford University Press, Oxford.

Study exercises

45.1 Decide on the best method of sterilisation. What would be the most appropriate method of sterilisation for the following items?

(a) A box of 100 plastic tips to be used with a pipettor.
(b) A 50 ml batch of blood, for use in 5% v/v blood agar plates.
(c) A 1 litre batch of MacConkey agar.
(d) Ten 5 ml glass pipettes.
(e) A microbiological wire, used for 'stab' cultures.
(f) A 10 ml sample of a heat-sensitive solution of an antibiotic, to be used as a component of a selective isolation medium.

45.2 Find out the biohazard classification (UK ACDP categorisation – Table 45.1) for the following microbes:

(a) *Salmonella typhimurium*
(b) *Leptospira interrogans*
(c) *Shigella dysenteriae* (type 1)
(d) *E. coli* K12
(e) *E. coli* O157
(f) Human immunodeficiency virus (HIV)
(g) *Cryptococcus neoformans* var. *neoformans*
(h) *Mycobacterium tuberculosis*
(i) *Lactobacillus plantarum*
(j) Marburg virus.

45.3 Consider the advantages and disadvantages of spread-plating and pour-plating methods. Having read through this chapter, list up to six pros/cons of each plating method and compare your answers either with the list we have provided or with those of other students, as a group exercise.

46 Cell culture

Definitions

Chemoautotroph – an organism that acquires energy from the oxidation of simple inorganic compounds, fixing CO_2 as a source of carbon (chemosynthetic metabolism).

Heterotroph – an organism that uses complex organic carbon compounds as a source of carbon and energy.

Photoautotroph – an organism that uses light as a source of energy and CO_2 as a carbon source (photosynthetic metabolism).

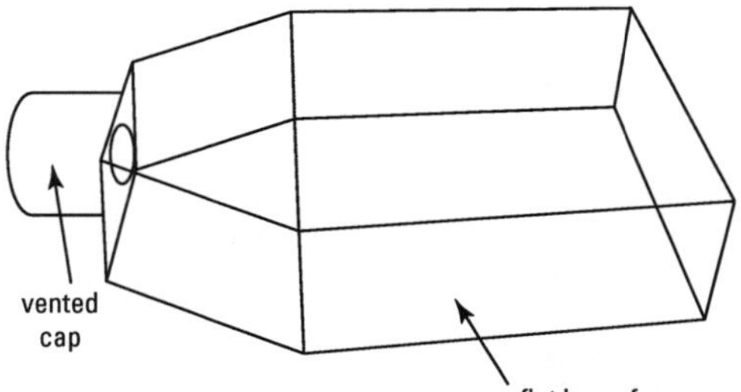

Fig. 46.1 Plastic flask for animal cell culture – this design provides a large surface area for growth of an adherent monolayer of cells.

Harvesting bacteria from an agar plate *– colonies can be harvested using a sterile loop, providing large numbers of cells without the need for centrifugation. The cells are relatively free from components of the growth medium; this is useful if the medium contains substances that interfere with subsequent procedures.*

Subculturing *– when subculturing microbes from a colony on an agar medium, take your sample from the growing edge, so that viable cells are transferred.*

Microbial, animal and plant cell culture methods are based on the same general principles, requiring:

- a pure culture (also known as an axenic culture), perhaps isolated as part of an earlier procedure, or from a culture collection;
- a suitable nutrient medium to provide the necessary components for growth. This medium must be sterilised before use;
- satisfactory growth conditions including temperature, pH, atmospheric requirements, ionic and osmotic conditions;
- sterile technique to maintain the culture in pure form.

Heterotrophic animal cells, fungi and many bacteria require appropriate organic compounds as sources of carbon and energy. Non-exacting bacteria can utilise a wide range of compounds and they are often grown in media containing complex natural substances (including meat extract, yeast extract, soil, blood). Animal cells have more stringent growth requirements.

Photoautotrophic bacteria, cyanobacteria and algae are grown in a mineral medium containing inorganic ions including chelated iron, with a light source and CO_2 supply. Plant cells may require additional vitamins and hormones. For chemoautotrophic bacteria, the light source is replaced by a suitable inorganic energy source, e.g. H_2S for sulphur-oxidising bacteria, NH_3/NH_4^+ for nitrifying bacteria, etc.

Growth on solidified media

Many organisms can be cultured on an agar-based medium.

KEY POINT An important benefit of agar-based culture systems is that an individual cell inoculated onto the surface can develop to form a visible colony: this is the basis of most microbial isolation and purification methods, including the streak dilution, spread plate and pour plate procedures.

Animal cells are often grown as an adherent monolayer on the surface of a plastic or glass culture vessel (Fig. 46.1), rather than on an agar-based medium.

Several types of culture vessel are used:

- Petri plates (Petri dishes): usually the pre-sterilised, disposable plastic type, providing a large surface area for growth.
- Glass bottles or test tubes: these provide sufficient depth of agar medium for prolonged growth of bacterial and fungal cultures, avoiding problems of dehydration and salt crystallisation. Inoculate aerobes on the surface and anaerobes by stabbing down the centre, into the base (stab culture).
- Flat-sided bottles: these are used for animal cell culture, to provide an increased surface area for attachment and allow growth of cells as a surface monolayer. Usually plastic and disposable.

The dynamics of growth are usually studied in liquid culture, apart from certain rapidly growing filamentous fungi, where increases in colony diameter can be measured accurately, e.g. using Vernier calipers.

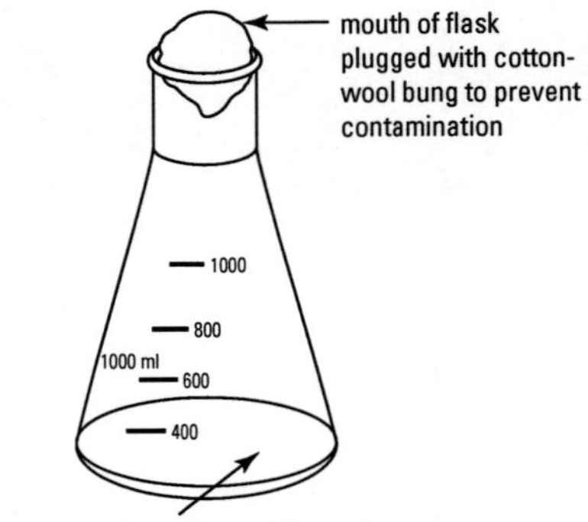

Fig. 46.2 Conical (Erlenmeyer) flask

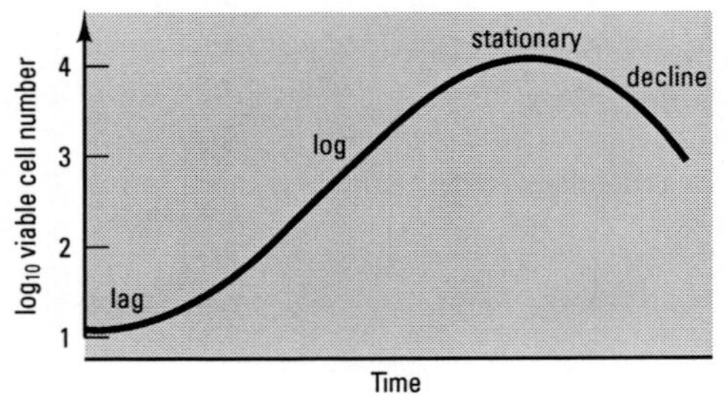

Fig. 46.3 Population growth curve for cells in batch culture (liquid medium).

Example Suppose you counted 2×10^3 cells ($\log_{10} = 3.30$) per unit volume at t_0 and 6.3×10^4 cells ($\log_{10} = 4.80$) after 2 h (t_x).

Substitution into eqn [46.1] gives $[2.303(4.8 - 3.3)] \div 2 = 1.727\ h^{-1}$ (or $0.0288\ min^{-1}$).

Substituting the same values into eqn [46.2] gives: $[0.301 \times 2] \div [4.8 - 3.3] = 0.40$ h (or 24 min).

***Working with logarithms** – note that there is no log value for zero, so you cannot plot zero on a log–linear growth curve or on a death curve.*

Growth in liquid media

Many cells, apart from primary cultures of animal cells, can be grown as a homogeneous unicellular suspension in a suitable liquid medium, where growth is usually considered in terms of cell number (population growth) rather than cell size. Most liquid culture systems need agitation, to ensure adequate mixing and to keep the cells in suspension. An Erlenmeyer flask of 100–2000 ml capacity (Fig. 46.2) can be used to grow a batch culture on an orbital shaker, operating at 20–250 cycles per minute. For aerobic organisms, the surface area of such a culture should be as large as possible: restrict the volume of medium to not more than 20 per cent of the flask volume. Larger cultures may need to be gassed with sterile air and mixed using a magnetic stirrer rather than an orbital shaker. The simplest method of air sterilisation is filtration, using glass wool, non-absorbent cotton wool or a commercial filter unit of appropriate pore size (usually 0.2 μm). Air is introduced via a sparger (a glass tube with many small holes, so that small bubbles are produced) near the bottom of the culture vessel to increase the surface area and enhance gas exchange. More complex systems have baffles and paddles to further improve mixing and gas exchange.

Liquid culture systems may be subdivided under two broad headings:

Batch culture

This is the most common approach for routine liquid culture. Cells are inoculated into a sterile vessel containing a fixed amount of growth medium. Your choice of vessel will depend upon the volume of culture required: larger-scale vessels (e.g. 1 litre and above) are often called 'fermenters' or 'bioreactors', particularly in biotechnology. Growth within the vessel usually follows a predictable S-shaped (sinusoidal) curve when plotted in log–linear format (Fig. 46.3), divided into four components:

1. Lag phase: the initial period when no increase in cell number is seen. The larger the inoculum of active cells the shorter the lag phase will be, provided the cells are transferred from similar growth conditions.
2. Log phase, or exponential phase: where cells are growing at their maximum rate. This may be quantified by the growth rate constant, or specific growth rate (μ), where:

$$\mu = \frac{2.303\,(\log N_x - \log N_0)}{(t_x - t_0)} \qquad [46.1]$$

where N_0 is the initial number of cells at time t_0 and N_x is the number of cells at time t_x. For times specified in hours, μ is expressed as h^{-1}.

Prokaryotes grow by binary fission, whereas eukaryotes grow by mitotic cell division; in both cases each cell divides to give two identical offspring. Consequently, the doubling time or generation time (g, or T_2) is:

$$g = \frac{0.301\,(t_x - t_0)}{\log N_x - \log N_0} \qquad [46.2]$$

Cells grow at different rates, with doubling times ranging from under 20 min for some bacteria to 24 h or more for animal and plant cells. Exponential phase cells are often used in laboratory experiments, since growth and metabolism are nearly uniform.

3. Stationary phase: growth decreases as nutrients are depleted and waste products accumulate. Any increase in cell number is offset by death. This phase is usually termed the 'plateau' in animal cell culture.
4. Decline phase, or death phase: this is the result of prolonged starvation and toxicity, unless the cells are subcultured. Like growth, death often shows an exponential relationship with time, which can be characterised by a rate constant (death rate constant), equivalent to that used to express growth or, more often, as the decimal reduction time (d, or T_{90}), the time required to reduce the population by 90 per cent:

$$d = \frac{t_x - t_0}{\log N_x - \log N_0} \qquad [46.3]$$

Some cells undergo rapid autolysis at the end of the stationary period while others show a slower decline.

Example Suppose you counted 5.2 × 10^5 cells ($\log_{10}$ = 5.716) per unit volume at t_o and 3.7 × 10^3 cells ($\log_{10}$ = 3.568) after 60 min (t_x). Substitution into eqn [46.3] gives 60 ÷ [2.148] = 27.9 min. To the nearest minute, this gives a value for d of 28 min.

Batch-culture methods can be used to maintain stocks of particular organisms; cells are subcultured on to fresh medium before they enter the decline phase. However, primary cultures of animal cells have a finite life unless transformed to give a continuous cell line, capable of indefinite growth.

Continuous culture

This is a method of maintaining cells in exponential growth for an extended period by continuously adding fresh growth medium to a culture vessel of fixed capacity. The new medium replaces nutrients and displaces some of the culture, diluting the remaining cells and allowing further growth.

After inoculating the vessel, the culture is allowed to grow for a short time as a batch culture, until a suitable population size is reached. Then medium is pumped into the vessel: the system is usually set up so that any increase in cell number due to growth will be offset by an equivalent loss due to dilution, i.e. the cell number within the vessel is maintained at a steady state. The cells will be growing at a particular rate (μ), counterbalanced by dilution at an equivalent rate (D):

$$D = \frac{\text{flow rate}}{\text{vessel volume}} \qquad [46.4]$$

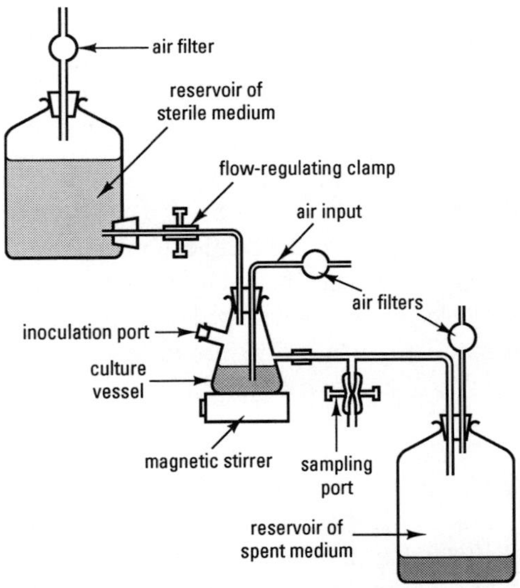

Fig. 46.4 Example of a two-dimensional lab equipment diagram of components of a chemostat

where D is expressed per unit time (e.g. h^{-1}). In a chemostat, the growth rate is limited by the availability of some nutrient in the inflowing medium, usually either carbon or nitrogen (see Fig. 46.4). In a turbidostat, the input of medium is controlled by the turbidity of the culture, measured using a photocell. A turbidostat is more complex than a chemostat, with additional equipment and controls.

To determine the specific growth rate (μ) of a continuous culture:

1. Measure the flow of medium through the vessel over a known time interval (e.g. connect a sterile measuring cylinder or similar volumetric device to the outlet), to calculate the flow rate.
2. Divide the flow rate by the vessel volume (eqn [46.4]) to give the dilution rate (D).
3. This equals the specific growth rate, since $D = \mu$ at steady state.

Example Suppose a continuous culture system of 2000 ml volume had a flow of 600 ml over a period of 40 min (flow rate 600 ÷ 40 = 15 ml min^{-1}. Substitution into eqn [46.4] gives a dilution rate D of 15 ÷ 2000 = 0.000 75 min^{-1} or 0.000 75 × 60 = 0.45 h^{-1}.

Example Suppose you wanted to convert a doubling time of 20 min to a specific growth rate. Rearrangement of eqn [46.5] gives $\mu = 0.693 \div 20 = 0.03465\ \text{min}^{-1}$ ($=2.08\ \text{h}^{-1}$). For the example given on page 273, with a growth rate of $0.45\ \text{h}^{-1}$, substitution into eqn [46.4] gives $0.693/0.45 = 1.54$ h, (approx. 92 min).

4. If you want to know the doubling time (g), calculate using the relationship:

$$g = \frac{0.693}{\mu} \qquad [46.5]$$

(Note that eqn [46.5] also applies to exponential phase cells in batch culture and is useful for interconverting g and μ.)

Continuous culture systems are more complex to set up than batch cultures. They are prone to contamination, having additional vessels for fresh medium and waste culture: strict aseptic technique is necessary when the medium reservoir is replaced, and during sampling and harvesting. However, they offer several advantages over batch cultures, including the following:

- The physiological state of the cells is more clearly defined, since actively growing cells at the same stage of growth are provided over an extended time period. This is useful for biochemical and physiological studies.
- Monitoring and control can be automated and computerised.
- Modelling can be carried out for biotechnology/fermentation technology.

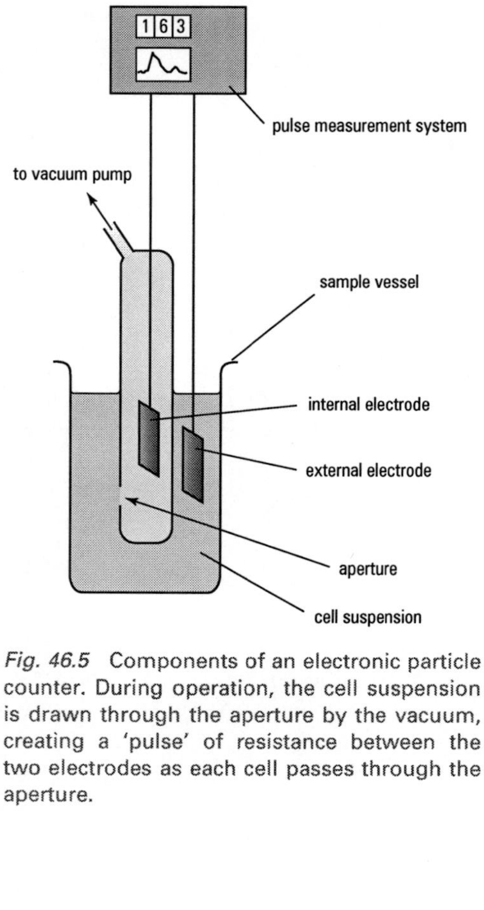

Fig. 46.5 Components of an electronic particle counter. During operation, the cell suspension is drawn through the aperture by the vacuum, creating a 'pulse' of resistance between the two electrodes as each cell passes through the aperture.

Measuring growth in cell cultures

The most widely used methods of measuring growth are based on cell number.

Direct microscopic counts

One of the simplest methods is to count the cells in a known volume of medium using a microscope and a counting chamber or haemocytometer (Box 46.1). While this gives a rapid assessment of the total cell number, it does not discriminate between living and dead cells. It is also time-consuming as a large number of cells must be counted for accurate measurement. It may be difficult to distinguish individual cells, e.g. for cells growing as clumps.

Electronic particle counters

These instruments can be used to give a direct (total) count of a suspension of microbial cells. The Coulter counter detects particles due to change in electrical resistance when they pass through a small aperture in a glass tube (Fig. 46.5). It gives a rapid count based on a larger number of cells than direct microscopy. It is well suited for repeat measurements or large sample numbers and can be linked to a microcomputer for data processing. If correctly calibrated, the counter can also measure cell sizes. A major limitation of electronic counters is the lack of discrimination between living cells, dead cells, cell clumps and other particles (e.g. dust). In addition, the instrument must be set up and calibrated by trained personnel. Flow cytometry is a more specialised alternative, since particles can be sorted as well as counted.

Alternative approaches to measuring growth *– these include biomass, dry weight, turbidity, absorbance or any major cellular component, e.g. protein, nucleic acid, ATP, etc.*

Culture-based counting methods

A variety of culture-based techniques can be used to determine the number of microbes in a sample. A major assumption of such methods is

Box 46.1 How to use a counting chamber or haemocytometer

A counting chamber is a specially designed slide containing a chamber of known depth with a grid etched on to its lower surface. When a flat coverslip is placed over the chamber, the depth is uniform. Use as follows:

1. **Place the special coverslip over the chamber.** Press the edges firmly, to ensure that the coverslip makes contact with the surface of the slide, but take care that you do not break the slide or coverslip by using too much force. When correctly positioned, you should be able to see interference rings (Newton's rings) at the edge of the coverslip.
2. **Add a small amount of your cell suspension to fill the central space above the grid.** Place on the microscope stage and allow the cells to settle (2–3 min).
3. **Examine the grid microscopically.** using the ×10 objective lens first, since the counting chamber is far thicker than a standard microscope slide. Then switch to the ×40 objective: take care not to scratch the surface of the objective lens, as the special coverslip is thicker than a normal coverslip. For a dense culture, the small squares are used, while the larger squares are used for dilute suspensions. You may need to dilute your suspension if it contains more than thirty cells per small square.
4. **Count the number of cells in several squares**: at least 600 cells should be counted for accurate measurements. Include those cells that cross the upper and left-hand boundaries, but not those that cross the lower or right-hand rulings (p. 181). A hand tally may be used to aid counting. Motile cells must be immobilised prior to counting (e.g. by killing with a suitable biocide).
5. **Divide the total number of cells (*C*) by the number of squares counted (*S*)**, to give the mean cell count per square.
6. **Determine the volume (in ml) of liquid corresponding to a single square (*V*)**, e.g. a Petroff–Hausser chamber has small squares of linear dimension 0.2 mm, giving an area of 0.04 mm^2; since the depth of the chamber is 0.02 mm, the volume is $0.04 \times 0.02 = 0.0008\ mm^3$; as there are 1 000 mm^3 in 1 ml, the volume of a small square is 8×10^{-7} ml; similarly, the volume of a large square (equal to 25 small squares) is 2×10^{-5} ml. Note that other types of counting chamber will have different volumes: check the manufacturer's instructions. For example, the improved Neubauer chamber (Fig. 46.6) has small squares of volume $0.000\,25\ mm^3 = 2.5 \times 10^{-7}$ ml.

Fig. 46.6 Haemocytometer grid (improved Neubauer rulings) viewed microscopically. The large square (delimited by triple etched lines) has a volume of 1/250th mm^3 ($0.004\ mm^3 = 4\ \mu l$) while each small square (16 contained within the large square) has a volume of 1/4000 mm^3 ($0.00025\ mm^3 = 0.25\ \mu l$). Note that the boundary line for squares delimited by triple-etched lines is the *middle* line, so this line must be used when counting (see Fig. 30.4, p. 181, for how to deal with objects straddling the gridlines).

7. **Calculate the cell number per ml by dividing the mean cell count per square by the volume of a single square (in ml).**
8. **Remember to take account of any dilution of your original suspension** in your final calculation by multiplying by the reciprocal of the dilution (*M*), e.g. if you counted a one in twenty dilution of your sample, multiply by twenty, or if you diluted to 10^{-5}, multiply by 10^5.

The complete equation for calculating the total microscopic count is:

$$\text{Total cell count (per ml)} = (C \div S \div V) \times M \quad [46.6]$$

For example, if the mean cell count for a hundred-fold dilution of a cell suspension, counted using a Petroff–Hausser chamber, was 12.4 cells in ten small squares, the total count would be

$$(12.4 \div 10 \div 8 \times 10^{-7}) \times 10^2 = 1.55 \times 10^8\ ml^{-1}.$$

A simpler, less accurate approach is to use a known volume of sample under a coverslip of known area on a standard glass slide, counting the number of cells per field of view using a calibrated microscope of known field diameter, then multiplying up to give the cell number per ml (see p. 259).

Box 46.2 How to make a plate count of bacteria using an agar-based medium

1. **Prepare serial decimal dilutions of the sample in a sterile diluent.** The most widely used diluents are 0.1% w/v peptone water or 0.9% w/v NaCl, buffered at pH 7.3. Take care that you mix each dilution before making the next one. For soil, food, or other solid samples, make the initial decimal dilution by taking 1 g of sample and making this up to 10 ml using a suitable diluent. Gentle shaking or homogenisation may be required for organisms growing in clumps. The number of decimal dilutions required for a particular sample will be governed by your expected count: dilute until the expected number of viable cells is around 100–1 000 ml^{-1}.

2. **Transfer an appropriate volume (e.g. 0.05–0.5 ml) of the lowest dilution to an agar plate** using either the spread-plate method or the pour-plate procedure. At least two, and preferably more, replicate plates should be prepared for each sample. You may also wish to prepare plates for more than one dilution, if you are unsure of the expected number of viable cells.

3. **Incubate under suitable conditions for 18–72 h, then count the number of colonies on each replicate plate at the most appropriate dilution.** The most accurate results will be obtained for plates containing 30–300 colonies. Mark the base of the plate with a spirit-based pen each time you count a colony. Determine the mean colony count per plate at this dilution (*C*).

4. **Calculate the colony count per ml of that particular dilution** by dividing by the volume (in ml) of liquid transferred to each plate (*V*).

5. **Now calculate the count per ml of the original sample** by multiplying by the reciprocal of the dilution: this is the multiplication factor (*M*); e.g. for a dilution of 10^{-3}, the multiplication factor would be 10^3. For soil, food or other solid samples, the count should be expressed per g of sample.

The complete equation for calculating the viable count is:

$$\text{Count per ml (or per g)} = (C \div V) \times M \quad [46.7]$$

For example, for a sample with a mean colony count of 5.5 colonies per plate for a volume of 0.05 ml at a dilution of 10^{-7}, the count would be:

$$(5.5 \div 0.05) \times 10^7 = 1.1 \times 10^9 \text{ CFU ml}^{-1}$$

The count should be reported as colony-forming units (CFU) per ml, rather than as cells per ml, since a colony may be the product of more than one cell, particularly in filamentous microbes or in organisms with a tendency to aggregate. You should also be aware of the problems associated with counts of zero – these are best recorded as '<1', and you should then apply the appropriate correction factors for dilution and volume to obtain the detection limit. For example, a zero count (<1) of 100 μl of a five-fold dilution gives a detection limit of $(<1 \div 0.1) \times 5 = <50$ CFU ml^{-1}.

that, under suitable conditions, an individual viable microbial cell will be able to multiply and grow to give a visible change in the growth medium, i.e. a colony on an agar-based medium, or turbidity ('cloudiness') in a liquid medium. You are most likely to gain practical experience using bacterial cultures, counted by one or more of the following methods:

Definition

CFU – colony-forming unit: a cell or group of cells giving rise to a single colony on a solidified medium.

- Spread- or pour-plate methods ('plate counts'. The most widespread approach is to transfer a suitable amount of the sample to an agar medium, incubate under appropriate conditions and then count the resulting colonies (Box 46.2).
- Membrane filtration. For bacterial samples where the expected cell number is lower than 10 CFU ml^{-1}, pass the sample through a sterile filter (pore size 0.2 μm or 0.45 μm). The filter is then incubated on a suitable medium until colonies are produced, giving a count by dividing the mean colony count per filter by the volume of sample filtered.
- Multiple tube count, or most probable number (MPN). A bacteriological technique where the sample is diluted and known volumes are transferred to several tubes of liquid medium (typically, five tubes at three volumes), chosen so that there is a low probability of the smallest volumes containing a viable cell. After incubation, the

Alternative approaches in plate counting *– when large numbers of samples have to be counted, a single agar plate can be divided into segments and a single droplet of each dilution placed into the appropriate segment ('Miles and Misra' droplet counting).*

number of tubes showing growth (turbidity) is compared to tabulated values to give the most probable number (MPN per ml).

Counting injured or stressed microbes *– a resuscitation stage may be required, to allow cells to grow under selective conditions.*

The principal advantage of culture-based counting procedures is that dead cells will not be counted. However, for such techniques, the incubation conditions and media used may not allow growth of all cells, underestimating the true viable count. Further problems are caused by cell clumping and dilution errors. In addition, such methods require sterile apparatus and media and the incubation period is lengthy before results are obtained. An alternative approach is to use direct microscopy, combined with 'vital' or 'mortal' staining. For example, the direct epifluorescence technique (DEFT) uses acridine orange and UV epifluorescence microscopy to separate living and dead bacteria, while neutral red is a vital stain used for plant cells. Chapter 42 gives examples of vital/mortal stains for other cell types. A further approach is to use DNA-based methods, such as the polymerase chain reaction (PCR).

Sources for further study

Ball, A.S. (1997) *Bacterial Cell Culture: Essential Data.* Wiley, New York.

Cann, A.J. (1999) *Virus Culture: A Practical Approach.* Oxford University Press, Oxford.

Cartledge, T.G. (ed.) (1992) *In Vitro Cultivation of Micro-organisms.* Butterworth-Heinemann, Oxford. [Includes in-depth coverage of the mathematical principles underlying growth in batch and continuous culture.]

Jennings, D.H. and Isaac, S. (1995) *Microbial Culture.* Bios, Oxford.

Rhodes, P.M. and Stanbury, P.F. (1997) *Applied Microbial Physiology: A Practical Approach.* Oxford University Press, Oxford. [Provides details of the growth requirements and culture methods applicable to a range of different microbes.]

Study exercises

46.1 Calculate the specific growth rate and doubling time of cells in culture. What are the specific growth rates and doubling times of the following (give all answers to three significant figures)?

(a) A broth culture of the yeast *Saccharomyces cerevisiae* growing in exponential phase and containing 5.2×10^4 cells at 10 a.m. and 3.4×10^6 cells at 7 p.m.

(b) A log phase culture of *E. coli* containing 3.0×10^4 CFU ml^{-1} at the start of the experiment and 6.7×10^7 CFU ml^{-1} at the end of the experiment, 200 minutes later.

(c) An actively growing culture of *Bacillus subtilis* containing 32 bacteria in 25 squares of a haemocytometer chamber at 13.00 hours and 250 bacteria in 25 squares of the same haemocytometer chamber at 15.30 hours.

46.2 Practise the calculations involved in using a haemocytometer to make a direct microscopic count. Express your answers to three significant figures in all cases.

(a) The mean cell count of a yeast suspension per small square of an improved Neubauer counting chamber was 6.42. If the volume of a small square is 2.5×10^{-7} ml, what is the cell count per ml?

(b) The following counts were obtained for bacteria in 20 individual small squares of a Petroff–Hauser counting chamber: 26, 36, 42, 35, 27, 16, 29, 50, 24, 43, 41, 35, 18, 36, 33, 47, 25, 46, 32, 57. If the volume of each small square is 8×10^{-7} ml, what is the cell count per ml?

(c) A 10^{-2} dilution of a dense suspension of *E. coli* was examined microscopically using an improved Neubauer counting chamber, giving a total of 78 cells in a total of 25 small squares. If the volume of each small square is 2.5×10^{-7} ml what is the cell count per ml of the original (undiluted) suspension.

(d) A dilute suspension of yeast cells was concentrated ten-fold by centrifuging 10 ml and

(continued)

Study exercises (continued)

resuspending the pellet in 1 ml. The mean cell count of the concentrated yeast suspension per large square of a Petroff–Hauser counting chamber was 6.8. If the volume of the large square is 2×10^{-5} ml, what is the cell count per ml of the original suspension?

46.3 Practise the calculations involved in making a plate count. Express your answers to three significant figures.

(a) The mean spread-plate count for 100 μl of a 10^{-5} dilution of a culture of *E. coli* was 54.4 CFU. What is the mean plate count per ml of the original suspension?

(b) Three replicate plates of nutrient agar were each spread with 200 μl of a 10^{-3} dilution of a bacterial suspension, giving colony counts of 34, 40 and 37 after 24 h incubation. What is the mean plate count per ml of the original suspension?

(c) A twenty-fold dilution of a yeast suspension was used to prepare four replicate pour plates, each containing 500 μl of this dilution and giving counts of 211, 186, 194 and 202 after incubation. What is the mean plate count per ml of the original suspension?

(d) A sample of 50 g of raw seafood was homogenised and diluted to 2% w/v in sterile saline solution. Three replicate pour plates were prepared using 1 ml of the diluted sample, giving counts for *E. coli* of 35, 41 and 32. What is the average count of *E. coli* per 100 g of raw seafood?

(e) Duplicate samples of 250 ml of river water were filtered through separate sterile membranes of pore size 0.2 μm. The membranes were then transferred to the surface of an agar-based medium and incubated for 48 h at 37°C, giving colony counts of 45 and 32 respectively. What is the mean count per 100 ml of river water?

(f) A sample of bottled drinking water was processed by serial decimal dilution and 500 μl samples of each dilution were pour plated and incubated at 22 °C for 72 h. The colony counts obtained for the three replicate plates of the 10^{-1} dilution were 28, 32 and 39. Does this water meet the EU regulations for mineral water, which specify a maximum average plate count at 22 °C of 100 CFU ml^{-1}?

47 Working with animal and plant tissues and cells

Definitions

In vivo – occurring within a living organism (literally 'in life').

In vitro – occurring outside a living organism, in an artificial environment (literally 'in glass').

Although the aim of many studies is to isolate, quantify or characterise individual molecules from a biological system, e.g. the purification of a particular enzyme, this is not always the most appropriate course of action. Depending on the purpose of the investigation, it may be more relevant to study the functioning of biomolecules within more complex systems, in order to understand their role in a particular biological process. At one extreme this may be carried out *in vivo*, using whole multicellular organisms (e.g. individual animals or plants), while the other extreme is represented by *in vitro* studies, using subcellular 'cell-free' extracts. Between these two extremes, a range of tissue and cell culture techniques offers some of the biological complexity of the intact organism combined with a degree of experimental control that may not be obtainable *in vivo*. The ethics and costs of whole animal experimentation have provided an additional stimulus to the development of *in vitro* methods, e.g. toxicity tests using mammalian cell culture, while developments in molecular genetics have led to further applications of cell and tissue culture.

Animal tissues and organs

Physiological experiments are carried out using either whole organisms, or a range of animal organs and tissues, including heart, liver, muscle, etc.

Working with vertebrate animals and their organs/tissues – *remember that procedures must be consistent with the law, i.e. in the UK, the* Animals Scientific Procedures Act 1986.

KEY POINT A major practical consideration is that the tissue should be studied as soon as possible after the death of the animal, typically under laboratory conditions that mimic the* in vivo *environment as closely as possible.

In most instances, the experiments are relatively short term (<24 h) and the aim is to maintain the tissue in a physiological state similar to that within the living organism. For metabolic studies, the whole organ or a tissue slice (typically 1–10 mm thick) will be bathed in an appropriate perfusion fluid, supplied either by gravity or by peristaltic pump. Practical considerations include:

Using human tissues – *in the UK, the* Human Tissue Act 2004 *applies and work is licensed and overseen by the Human Tissue Authority. Note that established cell lines are excluded from the regulations.*

- Inorganic solute requirements – the chemical composition of the perfusion fluid is usually chosen to reflect the major inorganic ion requirements of the tissue. For short-term studies, a number of so-called 'physiological salt solutions' may be used, e.g. Ringer's solution, one formulation of which is given in Table 47.1.
- Oxygen requirements – it may be necessary to increase the O_2 content of the perfusion fluid by bubbling with air, in order to meet the oxygen demand of the innermost parts of the tissue. However, this can lead to oxygen toxicity in the outermost parts and an alternative approach is simply to increase the rate of perfusion.
- Physico-chemical conditions – including temperature (usually controlled to $\pm 1\,°C$ of normal body temperature), water status (the perfusion fluid and the tissue should be isotonic), pH and buffering capacity (e.g. some perfusion fluids have elevated an $NaHCO_3$ concentration, to mimic the buffering capacity of mammalian serum).

Table 47.1 Composition of Ringer's solution (simplified formulation, for amphibians, etc., pH 7.4–7.6)

Compound	Amount per litre (g)
NaCl	6.0
KCl	0.075
$CaCl_2$	0.1
$NaHCO_3$	0.1

Definitions

Apoplasm – that part of the plant body outside the symplasm.

Light compensation point – the amount of photosynthetically active radiation (PAR) where photosynthetic CO_2 uptake is balanced by CO_2 production due to respiration and photorespiration.

Plasmodesmata – transverse connections through the cell wall, linking the cytoplasm of adjacent plant cells and creating a symplasm.

Working with plant tissues and organs – *the ethical problems associated with animal and human tissues are avoided, and there is a decreased risk of infection to the laboratory worker; practical manipulations are often carried out in horizontal laminar-flow cabinets.*

Table 47.2 Components of Long Ashton medium (nitrate version)

Stock solution: mass of component required per litre of solution (g)	Volume of stock solution to make 1 litre of medium (ml)
Major nutrients	
KNO_3: 50.60	8
$Ca(NO_3)_2$: 80.25	8
$MgSO_4{\cdot}7H_2O$: 46.00	8
$NaH_2PO_4{\cdot}2H_2O$: 52.00	4
Micronutrients	
FeKEDTA: 3.30	5
$MnSO_4{\cdot}4H_2O$: 2.23	1
$ZnSO_4{\cdot}7H_2O$: 0.29	1
$CuSO_4{\cdot}5H_2O$: 0.25	1
H_3BO_3: 3.10	1
$Na_2MoO_4{\cdot}2H_2O$: 0.12	1
NaCl: 5.85	1
$CoSO_4{\cdot}7H_2nO$: 0.056	1

Effects of humans on plants – *remember that your exhaled breath will be nearly saturated with water vapour and will contain CO_2 at 3–4% v/v, some 100 times more concentrated than atmospheric CO_2: your breath can thus affect rates of transpiration and photosynthesis.*

- Organic nutrient requirements – for longer-term studies, suitable nutrients will be required: these may be chemically defined additives, e.g. vitamins, amino acids, proteins, etc., or biological fluids such as plasma or serum. Glucose is often added as a carbon and energy source (Freshney, 2005).

Plant tissues and organs

Individual plant components (e.g. leaves, leaf slices and epidermal strips) can be isolated from the main plant body for study under controlled conditions. Since photosynthetic plant parts are autotrophic, they may be maintained *in vitro* for longer than animal organs, given adequate light and CO_2. However, most plant cells are joined via plasmodesmata, and the separation of such connections when the component is removed from the plant often leads to death when these connections are broken.

KEY POINT Plants show wound responses that may affect the metabolic processes under study. For these reasons, the most suitable systems for longer-term studies are often whole organs, e.g. whole leaves or entire root systems.

Water culture (hydroponics), e.g. in Long Ashton medium (Table 47.2), is an alternative approach, offering greater control over the root environment (Dodds and Robert, 1995).

It is best to use vigorous, healthy stock plants and to follow a well-established procedure, taking account of the following:

- Sterility – strict attention to sterile technique can be essential to the success of many longer-term experiments. Decontamination of plant organs may be especially difficult where specimens are obtained from soil: to achieve this, use a surface wash with disinfectant (e.g. 10% w/v sodium hypochlorite), followed by several rinses with sterile water.
- Gaseous environment – in general, the experimental system should be well ventilated. Actively photosynthetic tissues will rapidly deplete the atmospheric CO_2 in a closed vessel: plant parts may also produce physiologically active gases, such as ethylene, especially at wound sites. Turgor loss may occur in isolated plant parts unless a high humidity is maintained.
- Nutrition – plant tissues may benefit from a supply of inorganic ions, including K^+, SO_4^{2-}, etc., and may require certain vitamins, micronutrients and plant hormones for prolonged studies.
- Physico-chemical conditions – light is the most important environmental requirement for green plant parts. At atmospheric CO_2 concentrations, the light compensation point is about 5–10 μmol photons PAR $m^{-2}\,s^{-1}$ and photosynthesis is usually saturated between 500 and 2000 μmol photons PAR $m^{-2}\,s^{-1}$, depending on the plant type. Light quality and photoperiod (daylength) are also important: fluorescent tubes and incandescent bulbs that mimic the photosynthetic spectrum of sunlight are available, while photoperiod can be controlled using a timer. The water potential of aqueous media can be adjusted using an impermeant osmoticum such as mannitol, and pH values are often kept close to those of the apoplasm (≈pH 6) using appropriate buffers, if necessary.

Cell and tissue cultures

Many of the basic principles involved in culturing animal and plant cells are broadly similar to those described for microbial cell culture. Definitions for several key terms are given in the margin.

Definitions

Cell line – a cell culture derived by passage of a primary culture.

Clone – a population of cells derived from a single original cell, i.e. sharing the same genotype.

Confluence/confluent growth – merging of individual cells to form a continuous layer.

Continuous cell line – a culture with the capacity for unlimited multiplication *in vitro*. Sometimes termed an established cell line.

Finite cell line – a culture with a limited capacity for growth *in vitro* (maximum number of cell doublings).

Hybridoma – a hybrid cell produced by fusion of a tumour (myeloma) cell and an antibody-producing B lymphocyte. A cell line derived from a single hybridoma will produce a single antibody type (monoclonal).

Immortalisation – conversion of a finite cell line into a continuous cell line.

Passage – an alternative term for subculture.

Primary culture – a cell culture derived from tissue or organ fragments (explants). Primary culture ends on first subculture.

Senescence – the end point in the limited lifespan of a finite cell line, characterised by the lack of proliferation. Conversion to an established (continuous) cell line requires 'escape' from senescence.

Stem cell – an undifferentiated animal cell that can generate new stem cells through mitosis, or can develop and differentiate into a particular specialised cell type.

Transformation – a permanent alteration in the growth characteristics of a finite cell line that may include (i) changes in morphology, (ii) an increased growth rate and/or (iii) the acquisition of an infinite lifespan, often termed immortalisation. Transformation may be spontaneous or may be induced by chemical agents or viruses, and often involves a change in chromosome number.

Applications of cell and tissue culture

The main uses of animal and plant cell culture systems include:

- Experimental model systems in biochemistry, pharmacology and physiology: cell culture offers certain advantages over whole organism studies, with greater control over environmental conditions and biological variability. The use of genetically defined clones of cells may simplify the analysis of experimental data. Conversely, results obtained with specialised cell-based systems might be unrepresentative of a broader range of cell types and may be more difficult to interpret in terms of the whole organism.
- Studies of the growth requirements of particular cells: including studies of the positive effects of growth factors or growth-promoting substances, and the negative effects of xenobiotics or cytotoxic compounds. The use of cell culture in bioassays and mutagenicity testing is considered in Chapter 49.
- Studies of cell development and differentiation: including aspects of the cell cycle and gene expression. Cell cultures retaining their ability to differentiate *in vitro* are particularly interesting to researchers, while the lack of differentiation and unlimited growth of many animal cell lines makes them useful models of tumour development.
- Pathological studies: including culture of foetal cells for karyotyping and detection of genetic abnormalities, e.g. trisomy, etc.
- Genetic manipulation: cell culture techniques have played an essential role in the development of molecular biology, including the production of transgenic animals and plants by techniques such as transfection, etc.
- Biotechnology: including the industrial production of therapeutic proteins, vaccines and monoclonal antibodies using large-scale hybridoma culture techniques similar to those used in microbiology.
- Stem cell technology: increasingly important in biomedical research and development.

***KEY POINT** One of the main differences between organ and tissue incubation techniques and those used in cell/tissue culture is that the former aim simply to maintain metabolic and physiological activity for a limited period, while the latter provide conditions suitable for cell growth, division and development* in vitro *over an extended timescale, from a few days to several months.*

Animal cell culture systems

These may be established either from whole organisms (e.g. chick embryo), discrete organs (e.g. rat liver) or from blood (e.g. lymphocytes), typically using *mild* enzymic tissue disruption techniques, where necessary.

KEY POINT ***Although, in theory, it is possible to culture nucleated cells from virtually any source, in practice, the highest rates of success are most often achieved with young, actively growing tissues.***

SAFETY NOTE When working with tissue cultures safely, do not confuse laminar-flow hoods with biosafety cabinets.

SAFETY NOTE When sterilising during tissue culture (Box 47.1), take care when using 70% v/v alcohol near a naked flame (e.g. Bunsen or spirit lamp) – it is easily ignited and burns with a weakly visible flame.

The principal considerations in animal cell culture are:

- Safety: it is important to be aware of the potential dangers of infection from cell cultures. Although avian and rodent cells present a reduced risk of disease transmission compared with human cells, all cell cultures must be regarded as a potential source of pathogenic microbes, and appropriate sterile technique (Box 47.1) must be used at all times. Work involving human tissues and cell cultures must be carried out in a biosafety cabinet by trained, experienced personnel – you are unlikely to gain practical experience with such cultures in the early stages of your course. For other cell cultures, you will need to follow the code of safe practice of your department – consult your departmental safety officer if you have any doubts about safe working procedures.
- Whether to use a primary culture or a cell line: freshly isolated cells are more likely to reflect the biochemical activities of cells *in vivo*, though they will have a limited lifespan in culture, requiring repeated isolation for longer-term projects. Continuous cell lines are more easily cultured and offer the advantage that their growth requirements in culture may be known in some detail, especially for the more widely used cell lines (e.g. BHK, HeLa).
- The requirements for a solid substratum: some cells must be attached to a solid surface in order to grow. Anchorage dependence is a typical feature of primary cultures and finite cell lines – such cultures show density-dependent growth inhibition once the cells have formed a confluent monolayer on the surface of the substratum. An alternative approach is to grow such cells on a particulate support using 'microcarrier' beads. In contrast, many continuous cell lines can be maintained in suspension culture, as individual cells or aggregates.
- The physico-chemical conditions, including pH (typically 7.2–7.5) and buffering capacity, osmolality (usually $300 \pm 20\,\text{mosmol}\,\text{kg}^{-1}$) and temperature (e.g. 35–37 °C for mammalian cells).
- The requirements of the culture medium: these will include the provision of inorganic ions (as a balanced salt solution), a carbon/energy source plus other organic nutrients, and in some cases a supplement containing antimicrobial agents to counter the risks of contamination. For example, Dulbecco's modified Eagle's medium is used for many mammalian cell types that grow as adherent monolayers, while suspension cultures of continuous cell lines can be maintained using less stringent media. To support growth, the basal medium is usually supplemented with serum (usually foetal calf serum, at up to 20% v/v), or a chemically defined serum-like supplement containing a mixture of proteins, polypeptides, hormones, lipids and trace components. The *in vitro* level of CO_2 and O_2 must also be considered: many cell cultures are buffered using bicarbonate, and must be maintained in an atmosphere of elevated CO_2, either in a sealed culture vessel or in a CO_2 incubator, to maintain pH balance. In some cases, a pH-sensitive dye (e.g. phenol red) may be incorporated into the growth medium, to provide a visual check on pH

Example Dulbecco's modified Eagle's medium contains Ca^{2+}, Fe^{3+}, Mg^{2+}, K^+, Na^+, Cl^-, SO_4^{2-}, PO_4^{2-}, glucose, 20 amino acids, 10 vitamins, inositol and glutathione. Foetal calf serum is usually added at up to 20% v/v.

Using antibiotics in cell culture *– a typical antimicrobial supplement might include antibacterial agents, e.g. penicillin and streptomycin, an antifungal agent, e.g. griseofulvin, and an antimycoplasmal agent, e.g. gentamicin. Use sparingly to treat contamination and discontinue once the contaminant has been eradicated.*

Box 47.1 Sterile technique and its application to animal and plant cell culture

Although Chapter 45 gives general advice on the basic principles of sterile technique with microbial cultures, animal and plant cell cultures need additional precautions due to the complexity of some of the procedures and the likelihood of rapid overgrowth of any contaminating microbes. For routine work on the open bench, you should be aware of the following points:

1. **Consider personal clothing and hygiene**: long hair should be tied back, or retained by a net/cap. Avoid pendulous earrings and similar items. Wash your hands at the outset, to remove any loose skin flakes: surgical gloves may be worn, though there is loss of tactile sensitivity and comfort. Hands/gloves should be swabbed with 70% v/v alcohol to further reduce the risk of contamination. If you have a cold/cough, consider wearing a face mask.
2. **Work in a designated quiet area**: there should be no air currents (avoid open windows) and no 'through traffic' or other activity that might give rise to contamination (e.g. microbes should not be cultured in the same location). The term 'quiet area' also highlights that no talking should occur, to reduce the risk of aerosol contamination from the oral microflora.
3. **Organise your work surface**: clear everything from the bench, swab with 70% v/v alcohol, then position all items around a central area, so you can reach every item without reaching across anything. A well-organised workspace reduces the risk of accidental contact between sterile and non-sterile items. Swab bottles and flasks with 70% v/v alcohol before positioning them at the edge of your workspace. Also, swab your work surface between procedures and at the end of the session.
4. **Work close to a Bunsen flame**, positioned centrally within your workspace where convection currents create an upward airflow, reducing the likelihood of particles falling from the air into an open vessel. Always flame the tops of glass bottles, but not plastic items, for 2–3 seconds both *before* and *after* opening and closing, rotating the bottles during exposure to the flame. Flame glass pipettes, as described .
5. **Tilt flasks and bottles during use**: uncapped culture flasks and media bottles are best kept at a shallow angle, to minimise the risk of airborne contamination.
6. **Work without delay, but do not hurry**: always keep in mind that the air contains contaminant microbes and the longer you leave a vessel open to the air, the more likely it is that it will become contaminated.

When preparing to work in a laminar-flow cabinet, the following additional aspects should be considered:

1. **Note whether the airflow is horizontal or vertical**: horizontal flow gives greatest protection to the work area and provides the most stable air flow, while vertical-flow cabinets reduce exposure of the operator. A biosafety cabinet should be used for work with potentially hazardous cultures (e.g. human cells/tissues, or with any cells known to be infected with a virus).
2. **Prepare the cabinet**: switch on the cabinet and leave running for at least 10 minutes, then swab the work surface and other interior surfaces with 70% v/v alcohol. Swab the outsides of bottles, flasks, etc., before bringing them into the cabinet. Arrange items around a crescent-shaped central work area.
3. **Carry out your work with due regard for the airflow within the cabinet**: use of a Bunsen burner is often discouraged, as it disrupts the correct airflow, creates heat and can ignite flammable items within the restricted interior of the cabinet. Always try to keep your hands/arms further away from the sterile airstream than the items within the cabinet, e.g. avoid working *behind* an open vessel in a horizontal-flow cabinet and *above* an open vessel in a vertical-flow cabinet.
4. **Take particular care when using pipettes**: a laminar-flow hood will have a restricted interior volume, and it is easy for the novice to touch a pipette tip on the hood, or on another item during use. When removing a pipette from its packaging and connecting to a pipette filler you should point the tip of the pipette away from the user and into the air flow, holding it well above the graduated scale markings to avoid contamination. The pipette and filler are then held horizontally until required: as this can be a tricky procedure for the novice, you should not take a pipette out of its container until you are ready to use it.
5. **Work carefully with open bottles and flasks**: caps can be either placed top-down on the work surface, or held in a crooked finger. Since the airflow is sterile, there is less need to work swiftly when compared to the open bench, but you should still re-cap all containers as soon as you have finished a particular procedure.

***Using a horizontal laminar-flow hood** – note that this is designed to minimise contamination of the culture, rather than the worker.*

Definitions

Callus – an aggregation of undifferentiated plant cells in culture.

Embryogenic callus – tissue with the capacity to differentiate under defined laboratory conditions, typically in response to plant growth regulators in the medium.

Explant – a fragment of tissue used to initiate a culture (the term is also used in animal culture).

Protoplasts – cells lacking their cell walls.

Sphaeroplasts – cells with attached fragments of their cell walls: osmotically sensitive.

Totipotency – the ability (of any plant cell) to de-differentiate and re-differentiate into any of the cell types found in the mature plant.

status during growth – a colour change indicates acidification of the medium and the growth medium should be renewed.

- The equipment required: this may include a laminar-flow hood or biosafety cabinet to reduce the possibility of microbial contamination, suitable culture vessels (typically pre-sterilised, disposable polystyrene dishes, bottles and flasks, treated to create a negatively charged, hydrophilic surface), a supply of high-purity water (typically distilled, deionised and carbon filtered), a suitable incubator with temperature control of $\pm 0.5\,^{\circ}C$ or better, often with CO_2 control and mechanical mixing, and an inverted microscope to examine adherent cell monolayers during growth.

The successive stages of isolation of animal components are shown in Fig. 47.1(a). Box 47.1 gives advice on key aspects of sterile technique in cell culture; additional guidance on practical procedures in animal cell culture is given by Reed *et al.* (2007).

Plant tissue and cell culture systems

Plant tissue cultures can be established by growing explants of sterilised tissue on the surface of an agar-based growth medium to give a callus of undifferentiated cells. Initial sterilisation is usually achieved by incubation for 15–20 min in 10% w/v sodium hypochlorite.

***KEY POINT** For most plants, cell cultures can be established from a broad range of tissue types, reflecting the totipotency of many plant cells.*

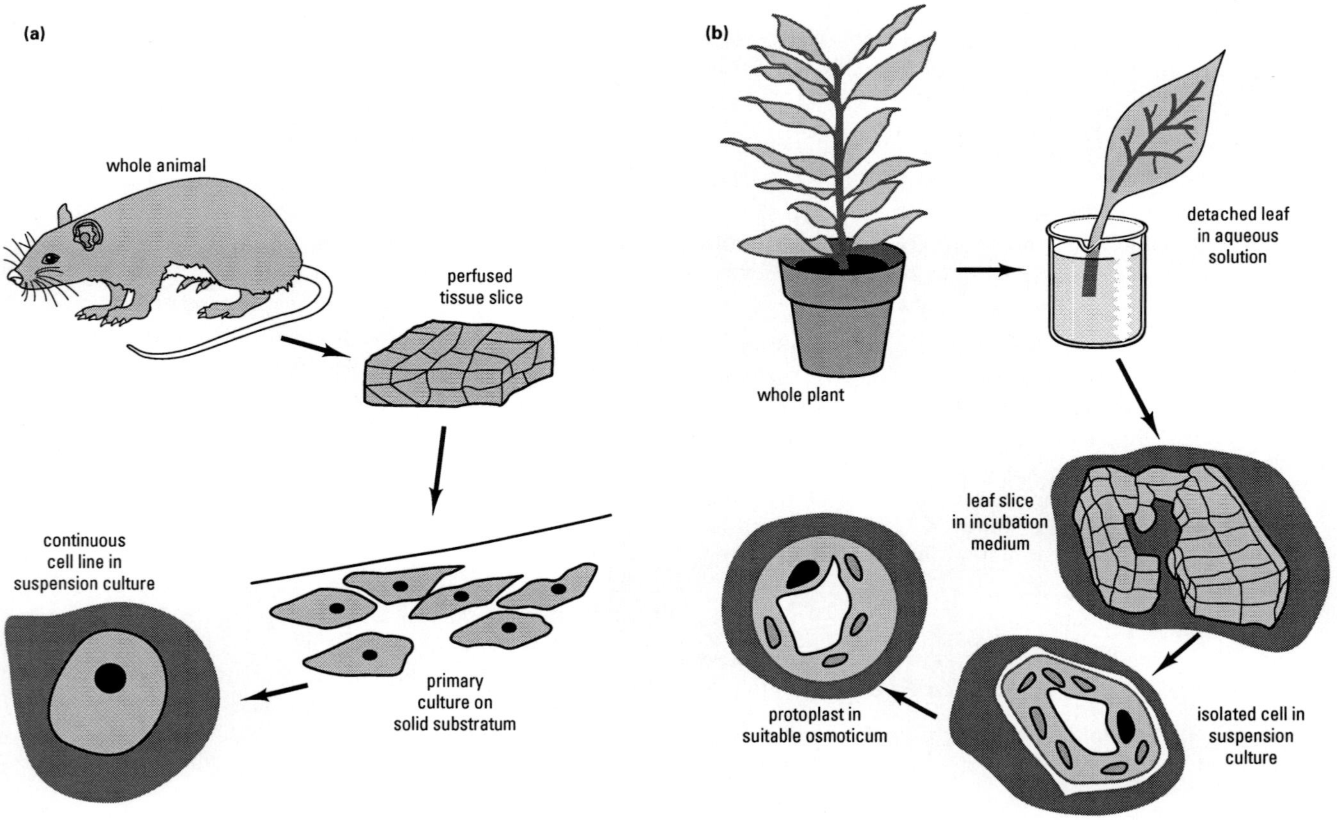

Fig. 47.1 Isolation of animal (a) and plant (b) components for *in vitro* study (note decreasing scale, from organisms to cells).

***Using an orbital incubator** – continued rapid agitation (>100 r.p.m.) maintains a homogeneous suspension and encourages gas exchange between the culture medium and the atmosphere.*

Embryogenic callus may be induced to differentiate, forming tissues and organs on a medium containing appropriate plant hormones: in many cases, these cultures will develop to form plantlets that can be grown on to mature plants, or encapsulated to produce so-called 'artificial seeds'. This approach can be used to study the conditions necessary for differentiation and development, or to propagate rare plants and other valuable stock (e.g. virus-free stock, or genetically altered plants). Callus derived from anthers can be used to provide haploid cell cultures and haploid plants – these are often useful for experimental genetics and breeding purposes.

Plant cell suspension cultures (Fig. 47.1(b)) are usually obtained by transferring fragments of actively growing callus to a 'shake' flask containing liquid medium in an orbital incubator: gentle agitation causes fragmentation of the callus tissue, to give a suspension culture that will contain individual plant cells and cell aggregates. This culture can be maintained by repeated subculturing of material from the upper layers of the liquid, encouraging the growth of small aggregates. However, in contrast to animal cell suspension cultures, it is rare for plant suspension cultures to be entirely unicellular, due to the presence of plasmodesmata and plant cell walls. Suspension cultures often require a minimum inoculum size on subculture – an inoculum volume of 10% v/v may be necessary to ensure successful subculture. It may also be helpful to add a small amount of 'conditioned' medium from a previous culture. Commercial-scale suspension cultures are used to produce certain plant pigments and secondary metabolites such as flavourings and high-value pharmaceutical compounds.

Example Murashige and Skoog's medium contains a balanced mixture of the principal inorganic ions, plus seven trace element compounds, three vitamins, inositol, glycine and sucrose (at 30 g/l) as the major carbon source.

The growth media used for callus and suspension cultures are more complex than those required for intact plant organs and tissues, with organic nutrient supplements in addition to a balanced salt solution. Such organic supplements often include a major carbon and energy source, e.g. sucrose, plus various vitamins and growth regulators (typically, at least one auxin and a cytokinin), together with undefined components such as yeast extract and hydrolysed casein in some instances. Otherwise, the techniques are broadly similar to those described for microbial systems.

Plant protoplasts

***Working with protoplasts** – remember that all solutions must contain a suitable osmoticum, to prevent bursting. Mannitol is widely used at ≈400 mmol l^{-1}.*

Some experimental procedures using plant cells require the enzymatic removal of their cell walls, creating protoplasts that can be manipulated *in vitro*, then maintained under conditions that allow the regeneration of cell walls, with subsequent growth and differentiation to give genetically modified plants. Protoplast isolation often involves pretreatment in a concentrated osmoticum (e.g. sucrose or mannitol, at 300–500 mmol l^{-1}) to plasmolyse the cells, weakening the linkage between cell wall and plasma membrane. Enzymatic treatment is often prolonged, taking several hours in a suitable mixture of enzymes, e.g. cellulase and Macerozyme. The resulting material can be sieved through fine nylon mesh, centrifuged at low speed and then resuspended, to remove sub-cellular debris and cell aggregates. The protoplast preparation can be checked for fragments of cell wall using a suitable stain, e.g. 0.1% w/v calcofluor white and a fluorescence microscope.

***Assessing cell and protoplast viability** – cytochemical techniques are often used, e.g. exclusion of the mortal stain Trypan blue or Evan's blue. The fluorogenic vital stain fluorescein diacetate is an alternative: it is cleaved by esterases within living cells, liberating fluorescein and giving green–yellow fluorescent cells when viewed by UV fluorescence microscopy.*

The fusion of protoplasts from different plants can be used to produce a somatic hybrid: this process can be used to circumvent interspecies reproductive barriers, creating novel plants. Protoplast fusion can be induced by chemical 'fusogens', e.g. using polyethylene glycol (PEG) at high concentration, or by electrofusion (incubation under low alternating current to encourage aggregation, then brief exposure to a high-voltage electrical field – typically 1 000 $V cm^{-1}$ for 1–2 ms – causing protoplast fusion). Similar techniques can be used with animal cells. The successive stages of isolation of plant components are illustrated in Fig. 47.1(b).

Text references

Dodds, J.H. and Robert, L.W. (1995) *Experiments in Plant Tissue Culture*, 3rd edn. Cambridge University Press, Cambridge.

Freshney, R.I. (2005) *Culture of Animal Cells: A Manual of Basic Technique*, 5th edn. Wiley, New York.

Reed, R.H., Holmes, D., Weyers, J.D.B. and Jones, A.M. (2007) *Practical Skills in Biomolecular Sciences*, 3rd edn. Prentice Hall, Harlow.

Sources for further study

Anon. (2006) *Fundamental Techniques in Cell Culture – A Laboratory Handbook*. Sigma-Aldrich/European Collection of Cell Cultures. Available at: http://www.sigmaaldrich.com/Area_of_Interest/Life_Science/Cell_Culture/ Key_Resources/ECACC_Handbook.html. Last accessed: 09/04/07.

Butler, M. (2003) *Animal Cell Culture and Technology: The Basics*, 2nd edn. Garland Science Press, Oxford.

Doyle, A., Griffiths, J.B. and Newell, D.G. (1998) *Cell and Tissue Culture: Laboratory Procedures*. Wiley, New York.

Evans, D., Coleman, J. and Kearns, A (2003) *Plant Cell Culture: The Basics*. Garland Science Press, Oxford.

Helgason, C.D. (2004) *Basic Cell Culture Protocols*, 3rd edn. Humana Press, New Jersey.

Loyola-Vargas, V. M. and Vazques-Flota, F. (2005) *Plant Cell Culture Protocols*, 2nd edn. Humana Press, New Jersey.

Masters, J. (2000) *Animal Cell Culture: A Practical Approach*, 3rd edn. Oxford University Press, Oxford.

Razdan, M.K. (2003) *Introduction to Plant Tissue Culture*, 2nd edn. Intercept, Andover.

Study exercises

47.1 Calculate the osmolarity of Ringer's solution from its individual constituents. Use the data in Table 47.1 and the information in Chapter 23, and assume that each constituent behaves according to ideal thermodynamic principles. Note the following M_r values: NaCl 58.44; KCl 74.55; $CaCl_2$ 110.99; $NaHCO_3$ 84.01, required to convert the data in Table 47.1 to molar concentrations. Give your final answer in mosmol kg^{-1}, to one decimal place.

47.2 Find out about the availability of cell cultures using the World Wide Web. Research the culture collections of particular countries via their websites. How useful is each site in terms of features such as: ease of use; access to catalogues; information on individual strains, e.g. culture media; details of costs?

47.3 Research the origins of the code names used for individual cell lines. In many instances, the code names given to individual cell lines are derived from their original source. Use the Web to find out the origin of the following descriptors for cell lines: (a) HeLa; (b) Vero; (c) BHK; (d) CHO; (e) TBY-2.

48 Photography and imaging

Definitions

The following terms are used when discussing the quality and features of images:

Colour balance – the way in which the perceived natural colours of a subject are reproduced in an image: this can be affected by image capture, lighting and printing.

Colour depth – in digital images, the amount of information used to define the colour of each pixel: at 1 bit per pixel, images are black and white; at 24 bits per pixel, images are displayed as over 16 million colours, beyond the colour-discerning ability of the human eye.

Contrast – the degree of gradation between colours or the number of grey shades in black-and-white film: the higher the contrast, the sharper is the gradation.

Depth of field – the range of shot that is in focus for a given camera setting: a low value may be required to highlight a specific subject or you may wish the whole field to be in focus, as in a panoramic view.

Focal length – the distance (in mm) at which a lens system will focus an image when set to 'infinity': this affects the magnification of the image and the angle of view, which are inversely related.

Grain – flaws in a film image, caused by the method of image capture.

Image stabilisation (antishake) – a feature available on digital cameras that compensates for camera shake on long exposures.

Noise – flaws, often white pixels, in a digital image caused by sensor errors.

Resolution – the sharpness and quality of an image, often expressed in 'dots per inch' (a non-SI unit).

Saturation – the term used to describe the intensity of a colour: a saturated colour is an intense colour.

Zoom – lens feature that magnifies an image by altering focal length. Optical zoom accomplishes this using the lens configuration, without great loss of image quality; however, camera shake may become a significant factor at high magnifications. Digital zooms achieve this electronically, and there is always loss of image quality.

Photography and imaging techniques are used primarily for:

- documentation and validation of observations for projects or coursework;
- recording visual information for later measurement, counting and/or analysis;
- preparation of material for illustrating talks;
- preparation of poster displays;
- producing 'hard copy' of microscope images.

Professional technical assistance may be available for this type of work but there is an important role for you in defining your requirements precisely. A basic understanding of the key factors that affect the final product will help you do this.

Types of photography

Photographs can be taken using two main systems: the digital camera and the film camera. Although the basic operation of both types is similar, they differ in their light-capturing and image-recording systems.

Digital photography and imaging

This is the production of an electronically stored image, using either a digital camera (still or video) or a scanning device from a pre-existing image. The image is held as a series of discrete picture elements called pixels (Fig. 48.1); the greater the number of pixels per unit surface area, the better the potential resolution of the image and the greater the amount of detail captured. The image may be in black/white, greyscale or colour. Image resolution is constrained by the nature of the sensor array and the amount of memory available for recording, while size and quality will also depend on the display or printer used.

Film photography

Black-and-white or colour-print photography involves capturing images on a film of plastic coated with light-sensitive silver halide salts. When exposed to light and subsequently chemically treated (developed), silver grains are produced to form a negative image, later converted to a positive image during printing. Transparency (slide or positive) films are based on colour dyes and are effectively grainless. Table 48.1 summarises the relationship between speed, grain and resolution for print films. The choice of film is a compromise between speed and resolution. Slower film is best for fine detail, but requires high light levels for effective results. Faster film is more suitable for moving subjects or when light levels are low, but there is a loss of contrast and detail.

Comparing digital and film systems

Despite advances in digital image technology and reductions in the price of digital cameras and associated memory systems, even the best digital images are of somewhat lower quality (resolution) than high-quality film images (see Table 48.2). However, digitised images have the advantages of

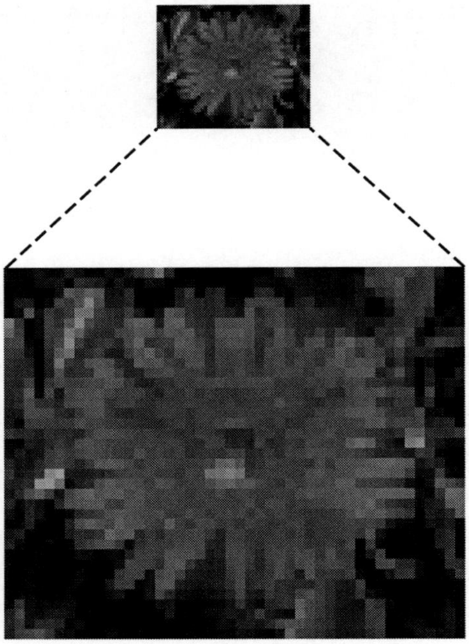

Fig. 48.1 Pixel structure in a digital image

Table 48.1 The speed and grain relationship in film

Speed of film	ISO number	DIN number	Grain	Resolution
Slow	25–64	15–19	Very fine	Very sharp
Medium	64–200	19–24	Fine	Sharp
Fast	200–400	24–27	Medium	Medium
Extra-fast	400–1600	27–32	Coarse	Poor to medium

speed and ease of production, since a developing stage is avoided, and they are also more readily manipulated. The preview facility on digital cameras is extremely useful when it is important to confirm that a successful image has been taken, e.g. when an event is unrepeatable. On the other hand, digital images are stored electronically and are not secure from computer failure, or damage from strong magnetic fields (see Box 10.1).

Other points of comparison include:

- the need for precision in exposure for digital systems is higher than for colour-print films: this means that the exposure timing is more critical to the quality achieved;
- digital systems are weak at recording highlights but tend to provide better shadow detail than film;
- digital systems tend to offer greater depth of field (the range of shot that is in focus), but this is not always an advantage as it can then be

Table 48.2 A comparison of digital and film photography. This table assumes you would be using a 6 megapixel SLR (single-lens reflex) digital camera and a 35 mm SLR film camera with 100 ASA film. Although it is not the only factor determining image quality, results would in general be poorer for a compact digital camera or one offering fewer megapixels.

Aspect	SLR digital camera	SLR film camera
Resolution	Primarily limited by number of pixels, i.e. 6 megapixels.	Limited by grain density of film and generally better. For scanned images, probably higher than 6 megapixels (up to 25 megapixels, depending on scanner).
Image quality	Better for shadow detail. Long-exposure images may be of poor quality. Delays in image capture may make it difficult to capture moving subjects. Depth of field longer with compact digital cameras.	Better for highlights. Grain is present, and will show if exposure time is long and/or image is magnified. Good-quality long exposures are possible. Easier to control depth of field.
Convenience	Easier to take repeated shots of same subject; easier to send images to others digitally; easier to incorporate into word-processed documents.	Results not known until film is developed – hence use of bracketing, which is wasteful of film. Unless scanned, less easy to send or incorporate into word-processed documents.
Costs	Costs of equipment, operation and printing similar. No film or processing costs, but memory card required.	Costs of equipment, operation and printing similar, but film must be bought and subsequently processed, with additional cost. Similar cost of printing.
Post-exposure image manipulation	Easy using commercially available software.	Possible at developing stage but only after scanning thereafter.
Storage and copying of images	Electronic. Digital images can be stored without loss of quality and occupy little physical space. Copies of files are exact facsimiles. Back-up easier.	Physical (hard copy). Colour-film images may fade though time and negatives and prints occupy significant space. There is always loss of quality when they are copied. Back-up less easy, unless scanned.

difficult to focus on critical features against a blurred background, in contrast to film;

- moving subjects can be easier to capture using film as there is often a significant delay between pressing the shutter and image capture in digital cameras.
- digital cameras frequently offer a 'movie' option, and most digital movie cameras provide a 'still' facility allowing you to capture an event at a critical point.

Capturing images

Camera and lens types

For both digital and film systems, the primary difference is whether you preview the image through the lens (TTL), often referred to as single-lens reflex (SLR), or whether a separate viewfinder is used. Digital cameras also incorporate a monitor that allows you to preview and review images. Lenses are either fixed, i.e. have a set focal length, or zoom, allowing a range of focal lengths, and hence magnification and angle of view. In general, higher-quality images can be taken with an SLR camera; these systems have better-quality lenses and allow greater control over the final image.

Setting up for photography

In the biosciences, this is usually carried out in one of three general situations:

1. laboratory or studio environments – where conditions are under your control and problems are relatively easy to overcome;
2. field situations – where many factors may be difficult to control, and you may need to use artificial means to enhance the shot (e.g. flash photography, or fast film speed);
3. working with specialised equipment – such as photo-microscopes or electron microscopes, where control of the photographic process is largely managed by the instrument's operating system.

The amount and type of light available to create an image is very important, especially for colour photography. Colour-image creation requires the appropriate colour temperature balance to be selected. Colour temperature is measured in Kelvin (K: see Fig. 48.2). In digital cameras this is usually automatic, but can be manually controlled via the 'white balance' settings. Filters are used with film cameras, to control overall colour balance, but the most important factor is the film type selected:

- daylight film is balanced for daylight conditions (5400 K) and electronic flash;
- artificial light types A (3400 K) or B (3200 K), are designed for studio lighting conditions.

Digital images are also modifiable after being taken, unlike film images where minimal adjustment is possible after taking the photograph.

The quantity of light can be measured by a photographic light meter, which may be external or built-in. The more light there is, the smaller the lens aperture you can select (larger f-number), and the greater will be the depth of field (= depth of focus). With the camera on a tripod, you can select slow shutter speeds and larger aperture settings (f-numbers), to

Understanding the light sensitivity of photographic systems *– both digital and film media are rated in terms of their 'speed', measured in either ISO (= ASA) or DIN units. A good range for general purposes is 100–160 ISO (21–23 DIN). Film of 1600 ISO is available, as are digital cameras with settings as high as 3200 ISO. These are useful for low light situations, but with some loss of image quality.*

Capturing moving images *– this may be valuable in biology, for example, to record animal behaviour, speed up plant movements or study fungal growth patterns.*

Focusing images correctly *– the autofocus feature on cameras may result in lack of focus on the parts of the image you are interested in because the algorithm used (generally spot or area) may not be suitable for your purpose. Better results may be obtained using manual focusing.*

Tips for better photography:

- *Use your camera on a tripod whenever possible, and use high shutter speed to minimise the effect of movements. Otherwise, adopt a posture with your elbows firmly locked against your body to reduce camera shake.*
- *Use electronic flash wherever possible as it provides uniformity of colour balance. Unless you want shadowless lighting, do not place the flash on the camera (hot-shoe) connection.*
- *Shadowless lighting of smaller objects is usefully obtained through a ring-flash system.*

Storing colour film *– colour film is highly sensitive to environmental factors such as heat and humidity, which cause changes in film speed and colour rendition. Make sure that your film has a sufficiently long expiry date. Store film for extended periods in a refrigerator or freezer: if stored in a freezer, allow a 24 h thawing period before use.*

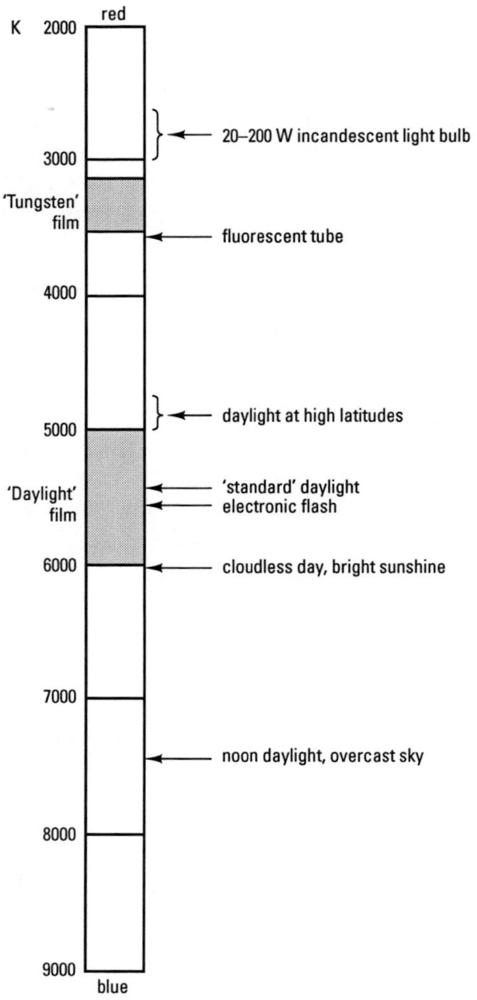

Fig. 48.2 Colour temperature and the Kelvin scale in relation to sensitivity of film type

Backing up digital image files *– since no physical copy exists, you must remember to download and back up digital image files to appropriate media (hard drive, network drive, CD or USB memory stick) as soon as possible.*

maximise depth of focus. Where an electronic flash system is employed to provide some or all of the lighting this is even easier, since the effective shutter speed with electronic flash is extremely short (1 ms). Remember that your choice of lighting arrangement will affect the quality of the picture: shadowless lighting is appropriate in some situations, but often shadows help to give three-dimensional form to subjects. In general, the use of more than one light source is advisable to prevent hard shadows.

KEY POINT ***To use an SLR camera properly, you need to understand the relationship between aperture (f-number), shutter speed and depth of field. For moving objects, give priority to fast shutter speed by opening the lens aperture (small f-number). Where depth of field is required, close the lens aperture (high f-number); this will result in slower shutter speeds, so take care to avoid camera shake. Bracket your exposures (make extra exposures at plus and minus 1 f-number at same shutter speed) to ensure good results.***

Scanning

Scanners can be useful to digitise a printed image. The quality of the scanned image is defined in terms of the number of dots per inch (dpi). Dedicated facilities may be available in your university; however, many modern printers also have a scan feature, though often at lower resolution. A scanner with a value of 600 dpi is adequate for most purposes.

Image processing

Images are recorded in digital cameras as files on various forms of memory cards and the data are later transferred either to a computer for manipulation or directly to a printer. An appropriate card reader or USB cable may be required. For critical work, the rule is to capture images using the largest possible file size, as reductions (e.g. compression of file size) of digital information are possible, but you can never add information.

Digital image files can be stored in various formats (e.g. .bmp, .jpg) and can be manipulated using programs such as Adobe Photoshop or Paint Shop Pro. Aspects of image quality such as contrast and colour balance can be adjusted and images can be cropped and resized to suit your application (e.g. small file sizes are recommended for websites). Digital images are valuable for use with word processors and Web-authoring software and can be transferred readily between such programs.

Film images require chemical development, usually a professional task. A film image may be impaired irreversibly by faulty development and the resulting image can only be modified to a limited extent during subsequent printing. This is a key reason for selecting digital image systems.

Photomicrography

This requires the use of a camera system mounted on a microscope. The important operations for successful photomicrography are outlined below:

1. Carefully prepare the specimen: ensure thorough cleanliness of any slides or coverslips used.
2. Choose the correct film type and any filters required for alteration of colour balance, depending upon the type of light source available.

3. Decide on the magnification to be used: make sure that you know how this relates to the magnification of the negative/transparency. Always include a photograph of a stage micrometer so the final magnification can be calculated and given in the legend to the photograph.
4. Carefully focus the specimen onto the film plane: there will be a system-specific method for doing this (consult the manual, as this is a critical aspect).
5. Bracket extra exposures above and below the automatic settings, especially for colour photography.

Printing images

Using scales with images – *these are vital to the interpretation of any biological image (see Chapter 44 for information on different types and how to calculate them).*

The most appropriate choice of print type will depend on your purpose. Glossy-finish prints generally appear sharper than matt or other finishes. Moreover, the addition of lettering and scales is often only possible on the smooth surface of a glossy print. However, when preparing prints for display on a poster, matt/velvet finishes are often preferable, as there is less glare for the viewer. Printing from computers using ink-jet and similar printers requires care, since colours viewed on monitors do not always reproduce on paper as expected. You may need to experiment with settings and paper quality to obtain the appropriate results.

Adding scales and labels to images

Having acquired a suitable print, it is often necessary to add information to it:

- When digital images are copied into word-processor or presentation software, additional information such as lettering, scales and labels can be added by overlaying a 'text box', or equivalent. You may need to make the background transparent, e.g. using the 'no fill' option in PowerPoint. Lettering can be added to printed media using transfer letters (e.g. Letraset).
- Choose a simple font type that will not distract and a size that is legible but not too dominant, e.g. 12 point in final reproduction. Always use the same font and type size for a related sets of images.
- Choose an appropriate part of the image for the lettering, i.e. a black area for a white letter. If the background is variable, create a coloured or white area and place the letters on this (see Fig. 48.3).

Fig. 48.3 Use of white areas for labelling dark prints, illustrated by an SEM micrograph of a stellate hair on a leaf surface.

Storing and mounting images

Digital images

Indexing software can be used to organise and catalogue your images (e.g. Picasa, available for free download at: http://picasa.google.com/). These programs allow you to add appropriate information and store this electronically along with the image, facilitating viewing and selection when your collection of images is large.

Prints and transparencies

Prints, negatives and transparencies are easily damaged and should be stored with care:

- Prints should be stored flat in boxes or in albums. When used for display, mount on stiff board either with modern photographic adhesives or dry-mounting tissue.

- Negative strips should be stored in transparent or translucent paper filing sheets. Obtain a set of contact prints for each film and store this with the negatives for easy reference.
- Transparencies (slides) should be mounted in plastic mounts but not between glass, as this often causes more problems than it solves. Labelling of transparencies is best done on the mount, using small, self-adhesive labels.

__Recording information about images__ – this should include subject, location, image-capture details and processing procedures. Bear in mind that these data may be of use when the image is used in a different context from the one you initially planned.

Beware of cheap plastic filing materials as some contain residual chemicals that can cause damage over time. Avoid dampness: silica gel desiccant can be used where necessary. Both physical copies and records should be well organised, including all potentially relevant information.

Sources for further study

Adobe Systems Inc. *Digital Imaging and Photo Editing Software for Digital Photography*. Available: http://www.adobe.com/digitalimag/
Last accessed: 09/04/07

Ang, T. (2002) *Digital Photographer's Handbook*. Dorling Kindersley, London.

Daney, C. *Photography and Digital Imaging Resources – a Collection of Links*. Available: http://www.mbay.net/~cgd/photo/pholinks.htm
Last accessed: 09/04/07

Gookin, D. (2000) *Digital Scanning and Photography*. Microsoft Press International, Redmond, Washington.

McCartney, S. (2001) *Mastering the Basics of Photography*. Allworth Press, New York.

Oldfield, R., Dickes, F.W. and Rost, F.W.D. (2000) *Photography with a Microscope*. Cambridge University Press, Cambridge.

Study exercises

48.1 Choose an appropriate image-recording method. You wish to record a stage in an experiment as an image. What are the relative merits of doing this using a film camera relative to using a digital camera?

48.2 Make a list of situations where an image is often required. When in biology project work might you require to use images?

48.3 Choose an appropriate combination of film/lighting/image for specific purposes. What film, lighting and image types would you select to record the following items *in situ*?
(a) a flower in a field
(b) a field site
(c) plants in a greenhouse
(d) mice in a laboratory experiment.

49 Measuring growth and responses

The ability of organisms to grow, develop and respond to stimuli is fundamental to life. The aims of studies concerned with these phenomena include:

- investigating developmental processes;
- comparing developmental patterns;
- understanding behavioural responses;
- characterising physiological processes;
- measuring the levels of a particular compound (bioassays);
- assessing sensitivity to a stimulus.

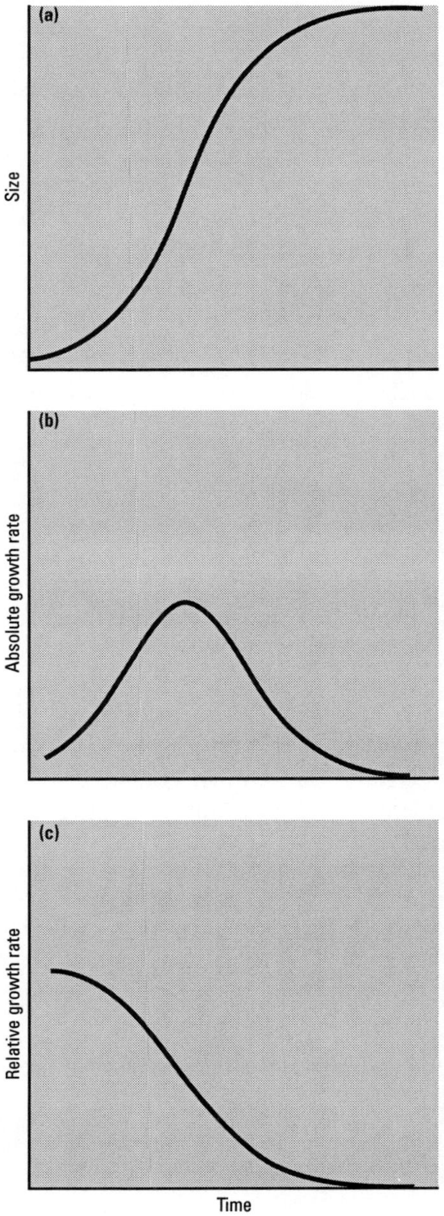

Fig. 49.1 Different ways of analysing growth: (a) simple plot of size against time; (b) plot of absolute growth rate for the same data; (c) plot of relative growth rate for the same data.

***KEY POINT** The growth and development of a multicellular organism is a complex phenomenon that cannot be easily measured by a single variable. An overall picture emerges from changes in size, weight, number, complexity and function evaluated at cell, tissue and organ levels.*

Growth and development

Length and area are relatively easy to measure (see techniques) though curved organisms may present difficulties. Volume measurements can be made from the displacement of water employing Archimedes' principle, or derived from fresh weight if the density of the organism can be estimated. Growth may not occur smoothly through time (e.g. plants increase in height mainly at night, when cell turgor is highest). For rapidly growing organisms it is desirable to have a continuous photographic or video record, though this may only be feasible with sedentary organisms. Care is required to ensure that focus is maintained on the points that will be measured, or your data will be inaccurate. If you are interested in growth of different parts of the organism, use reference points, e.g. recognisable morphological features such as hairs or marks made with ink or paint.

Fresh weight might at first sight appear a useful measure of growth: however, water content may vary according to availability and consumption. Dry weight does not have this disadvantage, but it may decrease when respiration dominates metabolism (e.g. during seed germination), despite the fact that growth is obviously occurring. Protein or nitrogen content may be better correlated with 'mass of protoplasm', but they are both complex to measure and are 'destructive' measurements.

Whichever of the above variables is measured, growth generally shows a sigmoidal increase in time (Fig. 49.1(a)) that is reminiscent of the population growth curve of cells in batch culture. The slope of this curve at any given time represents the absolute growth rate, with units of increment in size per unit time. Plotting this against time generally shows a bell-shaped curve (Fig. 49.1(b)). The relative growth rate or rate of increase in size per unit of size measures the 'efficiency' of growth (Fig. 49.1(c)).

Cell number is a common measure of the growth of unicellular suspensions. For multicellular organisms, this can be estimated from measurements of average cell size and total volume. Numbers of organs (e.g. leaves), specialised cells (e.g. hairs) or organelles (e.g. mitochondria) can indicate the progress of differentiation and development, but in general these are difficult processes to characterise quantitatively. Changes in numbers can also be expressed as absolute and relative rates, as above.

Display and analysis of growth data

Care should be taken when plotting a growth curve. When plotting change in size against time for a single organism, a smooth curve passing through all measurement points is appropriate (i.e. measurement error is assumed to be negligible). If dealing with a sample of organisms, the approach depends on whether (a) the same sample is measured each time (as with non-destructive harvesting, e.g. for linear measurement), when the curve should pass through each mean point, or (b) the samples are independent (as with destructive harvesting, e.g. for dry-weight measurement), when a smooth trend line should be drawn that need not pass through all of the points. Regarding statistical analysis, note that time-series measurements can only be treated as independent if the experimental design is as in (b) above; otherwise, in comparing changes in size with time, you should use tests for paired data. Note that variance may increase with size, so non-parametric tests are likely to be appropriate.

When presenting growth rates, use histograms if the data are obtained from differences in mean values divided by the time interval between measurements. A smooth curve can be used if it represents the slope (i.e. first derivative) of a function that has been fitted to the growth curve. Growth data are often log-normally distributed, so curves should be fitted to logarithmically transformed data. Fitting functions to growth curves is complex and specialist advice should be sought.

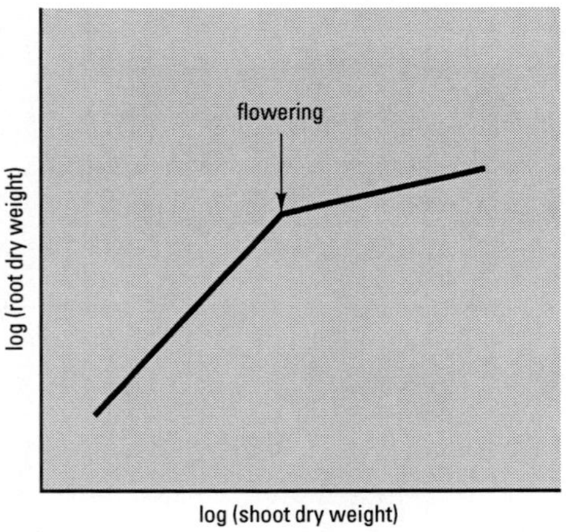

Fig. 49.2 Example showing changes in an allometric relationship based on a published analysis of the growth of a grass. Both root and shoot dry weight increase as the plant ages. The onset of flowering causes a clear change in their relationship.

Allometry

This is the study of relationships between the dimensions of organisms. It can be used for describing changes in the allocation of resources within a single species (e.g. root-to-shoot ratio in plants, Fig. 49.2) or among species (e.g. brain size in relation to body mass). The underlying relationship between dimensions x and y is generally:

$$y = ax^b \qquad [49.1]$$

where a is a positive constant. The coefficients a and b can be estimated from a log–log plot (e.g. Fig. 49.2), which may show one or more linear phases of the form:

$$\log y = \log a + b \log x \qquad [49.2]$$

Care should be taken in interpreting data of this type: remember that eqn [49.1] is an experimentally derived relationship rather than a theoretical one. Consult a specialist text (e.g. Causton and Venus, 1981) if you wish to fit a line or lines to such data.

Analysing allometric data *– when fitting eqn [49.2] to allometric data, you should use Model II regression as the variables are interdependent. Note that for monophasic log–log relationships,* **a** *can be determined from the intercept (log* **a***) and* **b** *from the slope (log* **b***).*

Responses to stimuli

Responses to stimuli are involved in many activities essential for life. For example, they are required to capture resources, escape or deter predators, move towards or away from a given set of conditions or interact with other individuals in a community. Understanding these responses requires answers to the following questions:

- What is the precise response and what causes it?
- How does the response develop?

- What is the function of the response?
- How might the response have evolved?
- How much of the response is inherited and how much learned?

The first stage when investigating a response is to describe it fully (preferably under conditions as near to those in nature as possible) and determine the precise stimuli that cause it. For instance, a study of plant phototropism might start with research into which cells perceive the light stimulus and which respond; effective wavelengths and doses of light might then be investigated. When working with animals, avoid anthropomorphism and describe objectively what happens. All circumstances of potential relevance (e.g. weather, time of day) should be recorded until a clear pattern emerges. A full description of most responses obviously requires physiological and biochemical techniques whose scope is too wide to be covered here.

Definition

Anthropomorphism – attributing human characteristics to animals.

You may wish to follow an individual animal in space or time and this will require either recognising its features or marking it in some way that does not interfere with the response (e.g. leg-ringing birds or painting symbols on insects). When recording the elements of a behavioural pattern, a coding system may be helpful (e.g. 1 = flaps wings, 2 = drops neck, etc.). As with growth, photography and video techniques are good aids for recording movements and analysing them later. Tape recorders can be used to record a description of responses so long as your voice does not interfere with the response. They can also record the sound component of a response and might be used later to elicit a response under controlled conditions (e.g. for a suspected alarm cry).

The second question listed above requires a study of the response throughout the developmental stages of the organism. The third and fourth questions require a certain amount of interpretive skills but, as with the fifth question, are amenable to experimentation. You may be presented with a natural experiment if conditions change in the right way in the course of your observations. Experiments in a laboratory give you control over conditions but this may well be at the expense of realism.

Distinguishing between innate and learned behaviour – *innate behaviour is genetically determined; it is always executed perfectly and in its entirety the first and every subsequent time the stimulus occurs. Learned behaviour is modified by experience.*

In designing experiments to investigate animal behaviour, you must take account of the fact that an animal's behaviour may change according to its experience. If not actually investigating learning *per se*, you need to ensure that the opportunity for a subject to learn is the same for each treatment, or bias will occur. Ways round this problem are either to use naïve animals for every replicate or to adopt a Latin square design so that experiences are shared evenly among treatments (fully randomised or randomised block designs may not achieve this).

Bioassays and their applications

A bioassay is a method of quantifying a chemical substance (analyte) by measuring its effect on a biological system under controlled conditions. The hypothetical underlying phenomena are summarised by the relationship:

$$\mathrm{A} + \mathrm{Rec} \rightleftharpoons \mathrm{ARec} \rightarrow\rightarrow \mathrm{R} \qquad [49.3]$$

where A is the analyte, Rec the receptor, ARec the analyte–receptor complex and R the response. This relation is analogous to the formation

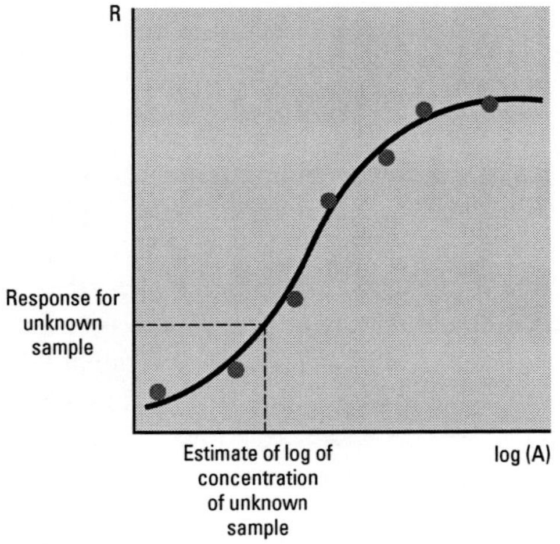

Fig. 49.3 Typical bioassay response curve, showing estimation of an unknown sample. Closed circles represent responses with standard samples. R = response; A = analyte.

of product from an enzyme–substrate complex and, using similar mathematical arguments to those of enzyme kinetics, it can be shown that the expected relationship between [A] and rate of response is hyperbolic (sigmoidal in a log–linear plot like Fig. 49.3). This pattern of response is usually observed in practice if a wide enough range of [A] is tested.

To carry out the assay, the response elicited by the unknown sample is compared with the response obtained for differing concentrations of the substance, as shown in Fig. 49.3. When fitting a curve to standard points and estimating unknowns, the available methods, in order of increasing accuracy, are:

1. fitting by eye;
2. using linear regression on a restricted 'quasi-linear' portion of the assay curve;
3. linearisation followed by regression (e.g. by probit transformation);
4. non-linear regression (e.g. to the Morgan–Mercer–Flodin equation).

In general, bioassay techniques have more potential faults than physico-chemical assay techniques. These may include the following:

- A greater level of variability: error in the estimate of the unknown compound will result because no two organisms will respond in exactly the same way. Assay curves vary through time, and because they are non-linear, a full standard curve is required each time the assay is carried out.
- Lack of chemical information: bioassays provide information about *biological* activity; they say little about the chemical structure of an unknown compound. The presence of a specific compound may need to be confirmed by a physico-chemical method (e.g. mass spectrometry).
- Possibility of interference: though many bioassays are very specific, it is possible that different chemicals in the extract may influence the results.

Examples The following are typical bioassays:

- measuring the amount of a plant growth substance in a plant extract using the coleoptile straight growth assay;
- measuring the effect of a drug on the dilation of a person's pupils;
- measuring the mutagenic properties of a chemical compound in the Ames test (see Box 49.1).

Despite these problems, bioassays are still much used. They are 'low tech' and generally cheap to set up. They often allow detection at very low concentrations. Bioassays also provide the means to assess the biological activity of chemicals and to study changes in sensitivity to a chemical, which physico-chemical techniques cannot do. Changes in sensitivity may be evident in the shape of the dose–response curve and its position on the concentration axis.

Bioassays can involve responses of whole organisms or parts of organisms. 'Isolated' responding systems (e.g. excised tissues or cells) decrease the possibility of interference from other parts of the organism. Disadvantages include disruption of nutrition and wound damage during excision. Isolation can continue down to the molecular level, as in immunoassays.

Legal use of bioassays *– in the UK, where bioassays involving 'higher' (vertebrate) animals are controlled by the* Animals Scientific Procedures Act (1986)*, they can only be carried out under the direct supervision of a scientist licensed by the Home Office.*

Bioassays are the basis for characterising the efficacy of drugs and the toxicity of chemicals. Here, response is often treated as a quantal (all-or-nothing) event. The E_{50} is defined as that concentration of a compound causing 50 per cent of the organisms to respond. Where death is the observed response, the LD_{50} describes the concentration of a chemical that would cause 50 per cent of the test organisms to die within a specified period under a specified set of conditions. Box 49.1 presents details of the Ames test, a widely used bioassay to assess the mutagenicity of chemicals.

Box 49.1 Mutagenicity testing using the Ames test – an example of a widely used bioassay

Chemical carcinogens can be identified by the formation of tumours in laboratory animals exposed to the compound under controlled conditions. However, such animal bioassays are time-consuming and expensive. Dr B.N. Ames and co-workers have shown that most carcinogens are also mutagens, i.e. they will induce mutational changes in DNA. The Ames test makes use of this correlation to provide a simple, rapid and inexpensive bioassay for the initial screening of potential carcinogens. The test makes use of particular strains of *Salmonella typhimurium* with the following characteristics:

- histidine auxotrophy – the tester strains are unable to grow on a minimal medium without added histidine: this characteristic is the result of specific mutational changes to the DNA of these strains, including base substitutions and frame shifts;
- increased cell envelope permeability, to permit access of the test compound to the cell interior;
- defects in excision repair systems and enhanced error-prone repair systems, to reduce the likelihood of DNA repair after treatment with a potential mutagen.

When grown in the presence of a chemical mutagen, the bacteria may revert to prototrophy as a result of back mutations that restore the wild-type phenotype: such revertants grow independently of external histidine and are able to form colonies on minimal medium, unlike the original test strains. The extent of reversion can be used to assess the mutagenic potential of a particular chemical compound. Since many chemicals must be activated *in vivo*, the test incorporates a rat liver homogenate (so-called 'S-9 activator', containing microsomal enzymes) to simulate the metabolic events within the liver. The tester strains are mixed with S-9 activator and a small amount of molten soft agar, then poured as a thin agar overlay (top agar) on a minimal medium plate. The top agar layer contains a very small amount of histidine, to allow the tester strains to divide a few times and express any mutational changes (i.e. prototrophy).

The test can be performed in one of two ways:

1. **Spot test:** a concentrated drop of the test compound is placed at the centre of the plate, either directly on the agar surface, or on a small filter paper disc. The test compound will diffuse into the agar and revertants appear as a ring of colonies around the site of inoculation, as shown in Fig. 49.4. The distance of the ring of colonies from this site provides a measure of the toxicity of the compound, while the number of colonies within the ring gives an indication of the relative mutagenicity of the test substance. The spot test is often carried out in a simplified form, without added S-9 activator, as a rapid preliminary test prior to quantitative analysis.

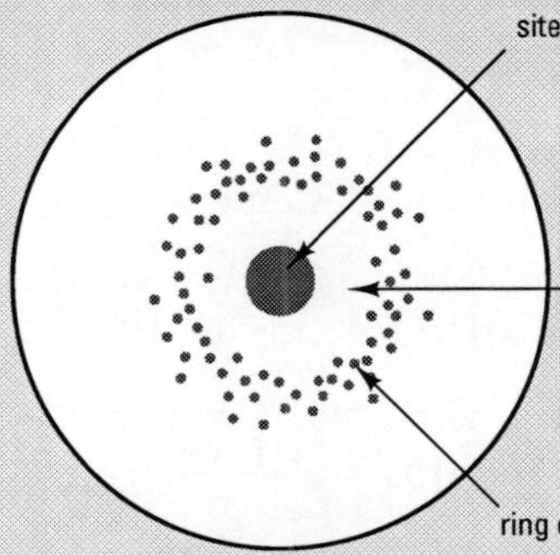

Fig. 49.4 Typical outcome of spot test (Ames test)

2. **Agar incorporation test:** known amounts of the test compound are mixed separately with molten top agar and the other constituents and poured on to separate minimal agar plates. After incubation for 48 h, revertants will appear as evenly dispersed colonies throughout the agar overlay. The number of colonies reflects the relative mutagenicity of the test compounds, with a direct relationship between colony count and the amount of mutagen. Agar incorporation tests can be used to generate dose–response curves similar to that shown in Fig. 49.3. The simplicity, sensitivity and reproducibility of the Ames test has resulted in its widespread use for screening potential carcinogens in many countries, though it is not an infallible test for carcinogenicity.

SAFETY NOTE **Correct handling procedures must be followed at all times, as the test substances may be carcinogenic – testers should wear gloves and avoid skin contact or ingestion.**

Considerations when setting up a bioassay

- The response should be easily measured and as metabolically 'close' to the initial binding event as possible.
- The experimental conditions should mimic the *in vivo* environment.
- The standards should be chemically identical to the compound being measured and spread over the expected concentration range being tested.

- The samples should be purified if interfering compounds are present and diluted so the response will be on the 'linear' portion of the assay curve.

To check for interference, the bioassay may be standardised against another method (preferably physico-chemical). Related compounds known to be present in the analyte solution should be shown to have minimal activity in the bioassay. If an interfering compound is present, this may show up if a known amount of standard is added to sample vials – the result will not be the sum of independently determined results for the standard and sample.

Text reference

Causton, D.R. and Venus, J.C. (1981) *The Biometry of Plant Growth*. Edward Arnold, London.

Sources for further study

Drickamer, L.C., Vessey, S.H. and Jakob, E.M. (2001) *Animal Behavior: Mechanisms, Ecology, Evolution*. McGraw-Hill, Dubuque, Iowa.

Hunt, R. (1990) *Basic Growth Analysis*. Unwin Hyman, London.

Martin, P. and Bateson, P. (1993) *Measuring Behaviour: An Introductory Guide*, 2nd edn. Cambridge University Press, Cambridge.

Price, E.O. and Stokes, A.W. (1975) *Animal Behavior in Laboratory and Field*, 2nd edn. W.H. Freeman, San Francisco.

Salanki, J., Jeffrey, D. and Hughes, G.M. (1994) *Biological Monitoring of the Environment. A Manual of Methods*. International Union of Biological Sciences Methodology Series, CAB International, Oxford.

Study exercises

49.1 Estimate relative growth rate. A baby hamster weighed at the start of a week has a mass of 35 g. By the end of one week it has a mass of 38.5 g.

(a) What is its absolute growth rate on a daily basis, expressed in mg day^{-1}?

(b) What is its relative growth rate on a weekly basis, expressed as a percentage?

49.2 Use bioassay data to estimate the amount of analyte in unknown samples. Bioassay data obtained by exposing water fleas (*Daphnia* sp.) to a purified algal toxin are shown in the table opposite.

(a) Graph the data and use the graph to estimate the toxin contents of two samples obtained from two lochs that gave (i) a heart rate of 130 beats per minute and (ii) a heart rate of 25 beats per minute.

(b) Which one of the two estimates would be the less reliable and why?

(c) How could you improve the reliability of this estimate, if you had more of the sample?

(d) What unmeasured factors might limit the usefulness of this assay?

Responses of *Daphnia* to toxin

Toxin concentration (mmol l^{-1})	Mean heart rate of *Daphnia* sp. (beats per minute)
1.00×10^{-3}	20
3.16×10^{-4}	23
1.00×10^{-4}	26
3.16×10^{-5}	50
1.00×10^{-5}	180
3.16×10^{-6}	192
1.00×10^{-6}	195

(continued)

Study exercises (continued)

49.3 Use agar diffusion bioassay data to quantify a growth inhibitor in a sample. Agar diffusion bioassay is a simple means of estimating the amount of an inhibitory substance in a test sample, by adding the substance, in aqueous solution, to wells cut in a Petri plate of agar medium seeded with a microbe sensitive to the substance and then incubating the plate at an appropriate growth temperature. Diffusion of the inhibitory substance gives a zone of growth inhibition proportional to $\log_{10}$ of the concentration of substance. In the following example, the amount of the antibiotic nisin in a test sample can be determined by comparing the degree of growth inhibition of *Bacillus subtilis* for the test solution and a series of standards containing known concentrations of nisin. Prepare a calibration curve and determine the nisin concentration of solution 5 (unknown). Express your answer in $\mu g\ ml^{-1}$, to one decimal place.

Growth inhibition of *B. subtilis* in solutions containing different amounts of nisin

Solution	Nisin content ($\mu g\ ml^{-1}$)	Zone of growth inhibition (mm)
1	40	26.5
2	20	23
3	10	19.5
4	5	17
5	unknown (test solution)	22.5

Advanced laboratory techniques

52 Enzyme studies

Enzymes are globular proteins that increase the rate of specific biochemical reactions. Each enzyme operates on a limited number of substrates of similar structure to generate products under well-defined conditions of concentration, pH, temperature, etc. In metabolism, groups of enzymes work together in sequential pathways to carry out a particular process, e.g. the multi-reaction conversion of glucose to lactate (glycolysis).

Example Enzyme EC 1.1.1.1 is usually known by its trivial name, alcohol dehydrogenase.

KEY POINT Enzymes are categorised according to the chemical reactions involved, leading to a four-figure Enzyme Commission code number and a systematic name for each enzyme. Most enzymes also have a recommended trivial name, often denoted by the suffix 'ase'.

Measuring enzyme reactions

Activity

This is measured in terms of the rate of enzyme reaction. Activity may be expressed directly as amount of substrate utilised per unit time (e.g. nmol min^{-1}, etc.), or in terms of the non-SI international unit (U, or sometimes IU), defined as the amount of enzyme that will convert 1 μmol of substrate to product(s) in 1 minute under specified conditions. However, the recommended (SI) unit of enzyme activity is the katal (kat), which is the amount of enzyme that will convert 1 mol of substrate to product(s) in 1 second under optimal conditions, determined from the following equation:

Example Using eqn [52.1], the conversion of 4.5 μmol (4.5 x 10^{-6} mol) of substrate in 5 min (300 s) would give an enzyme activity of 4.5 x 10^{-6} ÷ 300 = 1.5 x 10^{-8} kat = 15 nkat.

$$\text{enzyme activity (kat)} = \frac{\text{substrate converted (mol)}}{\text{time (s)}} \qquad [52.1]$$

This unit is relatively large (1 kat = 6 × 10^7 U) so SI prefixes are often used, e.g. nkat or pkat. Note that the units are *amount of substrate* (mol), not concentration (mol l^{-1} or mol m^{-3}).

For enzymes with macromolecular substrates of unknown molecular weight (e.g. deoxyribonuclease, amylase), activity can be expressed as the mass of substrate consumed (e.g. ng DNA min^{-1}), or amount of product formed (e.g. nmol glucose min^{-1}). You must ensure that your units clearly specify the substrate or product used, especially when the enzyme transformations involve different numbers of substrate or product molecules.

Example The hydrolysis of 1 molecule of maltose to give 2 glucose molecules by α-glucosidase means that enzyme activity specified in terms of substrate consumption (nmol maltose) would be half the value expressed with respect to product formation (nmol glucose).

The turnover number of an enzyme is the *amount of substrate* (mol) converted to product in 1 second by 1 mol of enzyme operating at maximum rate under optimum conditions. In practice, this requires information on the molecular weight of the enzyme, the amount of enzyme present and its maximum activity.

Purity

The specific activity of an enzyme preparation expresses the enzyme activity in terms of the quantity of protein present:

Example For an enzyme preparation with an activity of 1.5 x 10^{-8} kat in 24 mg (= 2.4 x 10^{-8} kg), eqn [52.2] gives a specific activity of 1.5 x 10^{-8} ÷ 2.4 x 10^{-8} = 0.625 kat kg^{-1}.

$$\text{specific activity} = \frac{\text{enzyme activity (kat)}}{\text{mass of protein (kg)}} \qquad [52.2]$$

Alternative units may be used, e.g. U mg^{-1}, $\text{ng min}^{-1}\text{ mg}^{-1}$. The specific activity is a useful way of comparing the purity of different enzyme preparations (purified enzyme preparations have high specific activities), comparing the stages in purification of an enzyme (specific activity increases as other proteins are eliminated), and assessing enzyme stability (an unstable enzyme will show a decrease in specific activity with time).

Enzyme activity and purity are often expressed in terms of the protein content of the enzyme preparation. Most assays for protein content do not give an absolute value, but require standard solutions, containing appropriate amounts of a particular protein, to be analysed at the same time, enabling a standard curve to be constructed. Bovine serum albumin (BSA) is commonly used as a general-purpose protein standard. However, you may need to use an alternative standard if your enzyme has an amino acid composition that is markedly different from BSA, depending on the specific method you use (Box 52.1).

Box 52.1 Methods of determining the amount of protein in an aqueous solution

For all of the following methods, the amounts are appropriate for semi-micro cuvettes (1.5 ml volume, path length 1 cm). Appropriate controls (blanks) must be analysed, to assess possible interference (e.g. due to buffers, etc.).

Biuret method

This is based on the specific reaction between cupric ions (Cu^{++}) in alkaline solution and two adjacent peptide bonds, as found in proteins. As such, it is not significantly affected by differences in amino acid composition between proteins.

1. **Prepare protein standards over an appropriate range** (typically, between 1–20 mg ml^{-1}).
2. **Add 1 ml of each standard solution to separate test tubes. Prepare a reagent blank, using 1 ml of distilled water, or an appropriate solution.**
3. **Add 1 ml of each unknown solution to separate test tubes.**
4. **Add 1 ml of Biuret reagent (1.5 g $CuSO_4 \cdot 5H_2O$, 6.0 g sodium potassium tartrate in 300 ml of 10 w/v NaOH) to all standard and unknown tubes and to the reagent blank.**
5. **Incubate at 37 °C for 15 min.**
6. **Read the absorbance of each solution at 520 nm against the reagent blank.** The colour is stable for several hours.

The main limitation of the Biuret method is its lack of sensitivity – it is unsuitable for solutions with a protein content of less than 1 mg ml^{-1}.

Direct measurement of UV absorbance (Warburg–Christian method)

Proteins absorb EMR maximally at 280 nm (due to the presence of aromatic amino acids) and this forms the basis of the method. The principal advantages of this approach are its simplicity and the fact that the assay is non-destructive. The most common interfering substances are nucleic acids, which can be assessed by measuring the absorbance at 260 nm: a pure solution of protein will have a ratio of absorption (A_{280}/A_{260}) of approximately 1.8, decreasing with increasing nucleic acid contamination. Note also that any free aromatic amino acids in your solution will absorb at 280 nm, leading to an overestimation of protein content. The simplest procedure, which includes a correction for small amounts of nucleic acid, is as follows (use quartz cuvettes throughout):

1. **Measure the absorbance of your solution at 280 nm (A_{280}): if A_{280} is greater than 1, dilute by an appropriate amount and re-measure.**
2. **Repeat at 260 nm (A_{260}).**
3. **Estimate the approximate protein content using the following relationship:**

$$\text{[protein] mg ml}^{-1} = 1.45\,A_{280} - 0.74\,A_{260} \qquad [52.3]$$

This equation is based on the work of Warburg and Christian (1942), for enolase. For other proteins, it should not be used for quantitative work, since it gives only a rough approximation of the amount of protein present, due to variations in aromatic amino acid composition.

(continued)

Box 52.1 (continued)

Lowry (Folin–Ciocalteau) method

This is a colorimetric assay, based on a combination of the Biuret method, described above, and the oxidation of tyrosine and tryptophan residues with Folin and Ciocalteau's reagent to give a blue–purple colour. The method is extremely sensitive (down to a protein content of 20 μg ml^{-1}), but is subject to interference from a wide range of non-protein substances, including many organic buffers (e.g. TRIS, HEPES, etc.), EDTA, urea and certain sugars. Choice of an appropriate standard is important, as the intensity of colour produced for a particular protein is dependent on the amount of aromatic amino acids.

1. **Prepare protein standards within an appropriate range for your samples** (the method can be used from 0.02–1.00 mg ml^{-1}).
2. **Add 1 ml of each standard solution to separate test tubes. Prepare a reagent blank, using 1 ml of distilled water, or an appropriate solution.**
3. **Add 1 ml of each of your unknown solutions to separate test tubes.**
4. **Then, add 5 ml of 'alkaline solution' (prepared by mixing 2% w/v Na_2CO_3 in 0.1 mol l^{-1} NaOH, 1% w/v aqueous NaKtartrate in the ratio 100 : 1 : 1).** Mix thoroughly and then allow to stand for 3 min.
5. **Add 0.5 ml of Folin–Ciocalteau reagent (commercial reagent, diluted 1 : 1 with distilled water on the day of use).** Mix rapidly and thoroughly and then allow to stand for 30 min.
6. **Read the absorbance of each sample at 600 nm.**

Dye-binding (Bradford) method

Coomassie brilliant blue combines with proteins to give a dye–protein complex with an absorption maximum of 595 nm. This provides a simple and sensitive means of measuring protein content, with few interferences. However, the formation of dye–protein complex is affected by the number of basic amino acids within a protein, so the choice of an appropriate standard is important. The method is sensitive down to a protein content of approximately 5 μg ml^{-1} but the relationship between absorbance and concentration is often non-linear, particularly at high protein content.

1. **Prepare protein standards over an appropriate range (between 5 and 100 μg ml^{-1}).**
2. **Add 100 μl of each standard solution to separate test tubes. Prepare a reagent blank, using 100 μl of distilled water, or an appropriate solution** (note that these small volumes must be accurately dispensed, e.g. using a calibrated pipettor).
3. **Add 100 μl of your unknown solutions to separate test tubes.**
4. **Add 5.0 ml of Coomassie brilliant blue G250 solution (0.1 g l^{-1}).**
5. **Mix and incubate for at least 5 min: read the absorbance of each solution at 595 nm.**

Other methods are less widely used. They include determination of the total amount of nitrogen in solution (e.g. using the Kjeldahl technique) and calculating the protein content, assuming a nitrogen content of 16 per cent. An alternative approach is to precipitate the protein (e.g. using trichloracetic acid, tannic acid, or salicylic acid) and then measure the turbidity of the resulting precipitate (using a nephelometer, or a spectrophotometer).

Types of assay

The rate of substrate utilisation or product formation must be measured under controlled conditions, using some characteristic that changes in direct proportion to the concentration of the test substance.

Spectrophotometric assays

Many substrates and products absorb visible or UV light, and the change in absorbance at a particular wavelength provides a convenient assay method. Artificial substrates are used where no suitable spectrophotometric assay is available for the natural substrates. In most cases, these are chromogenic analogues of the natural substrate, producing coloured products. In other cases, a product may be measured by a colorimetric chemical reaction.

Example α-Glucosidase activity can be measured using the artificial substrate *p*-nitrophenol-α-D-glucose, which liberates *p*-nitrophenol (yellow) with an absorption maximum at 404 nm.

Example Phosphoenolpyruvate carboxylase can be assayed by coupling to malate dehydrogenase: the first enzyme converts phosphoenolpyruvate to oxaloacetate which is then oxidised to malate by the second enzyme (present in excess) with a stoichiometric (1 : 1) reduction of NAD^+. The coupled assay is monitored spectrophotometrically as an increase in A_{340} against time.

Several assays are based on interconversion of the nicotinamide adenine dinucleotide coenzymes NAD^+ or $NADP^+$, which are reversibly reduced by many enzymes. The reduced form (either NADH or NADPH) can be detected at 340 nm, where the oxidised form has negligible absorbance. An alternative approach is to use a coupled-enzyme assay, where the product of the test enzyme is used as a substrate for a second enzyme reaction that involves oxidation/reduction of nucleotide coenzymes. Such assays are particularly useful for continuous monitoring of enzyme activity (see below) and for reactions where the product from the test substance is too low to detect by other methods, since coupled assays are more sensitive.

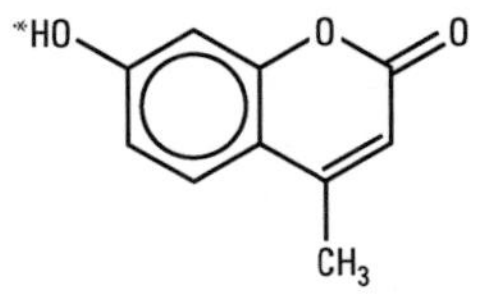

Fig. 52.1 Structure of 4-methylumbelliferone (4-MU). The free hydroxlyl group* is used to create fluorogenic substrates, e.g. coupled to galactose (4-MU-GAL) it can be used to asay β-galactosidase activity, since the galactoside derivative is non-fluorescent and hydrolysis can be detected by the fluorescence of liberated 4-MU.

Fluorimetric assays

Certain substrates liberate fluorescent products as a result of enzyme activity, providing a highly sensitive assay method. Fluorogenic substrates include fluorescein and methylumbelliferone derivatives (Fig. 52.1). Care is required, since impurities in the enzyme preparation may produce background fluorescence, or quenching (reduction) of the signal.

Radioisotopic assays

These are useful where the substrate and product can be easily separated, e.g. in decarboxylase assays using a ^{14}C-labelled substrate, where gaseous $^{14}CO_2$ is produced.

Electrochemical assays

Enzyme reactions involving acids and bases can be monitored using a pH electrode, though the change in pH will affect the activity of the test enzyme. An alternative approach is to measure the amount of acid or alkali required to maintain a constant pre-selected pH in a pH-stat.

An oxygen electrode can be used if O_2 is a substrate or a product. Other ion-specific electrodes can monitor ammonia, nitrate, etc.

Monitoring substrate utilisation/product formation

Continuous assays (kinetic assays)

The change in substrate or product is monitored as a function of time, to provide a progress curve for the reaction (Fig. 52.2). These curves start off in a near-linear manner, decreasing in slope as the reaction proceeds and substrate is used up. The initial velocity of the reaction (v_0) is obtained by drawing a tangent to the curve at zero time and measuring its slope. Continuous monitoring can be used when the test substance can be assayed rapidly and non-destructively, e.g. using a chromogenic substrate.

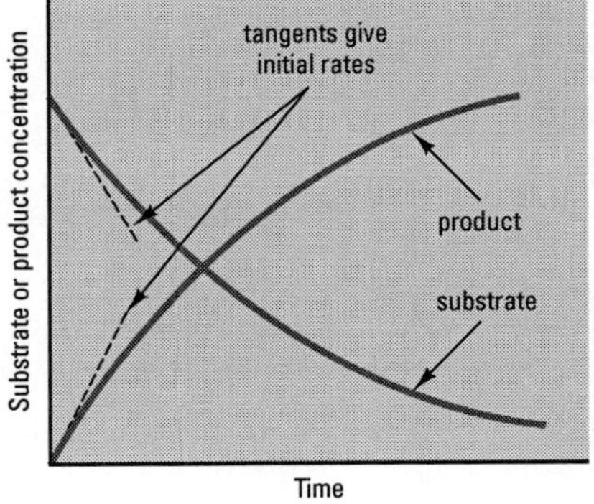

Fig. 52.2 Enzyme reaction progress curve: substrate utilisation/product formation as a function of time.

Discontinuous assays (fixed-time assays)

It is sometimes necessary to measure the amount of substrate consumed or product formed after a fixed time period, e.g. where the test substance is assayed by a (destructive) colorimetric chemical method. It is vital that the time period is kept as short as possible, with the change in substrate concentration limited to around 10 per cent, so that the assay is within the linear part of the progress curve (Fig. 52.2). This may need to be established in a preliminary experiment, e.g. using a continuous assay.

Enzyme purification

The purification of an enzyme usually involves several stages:

1. Tissue/cell disruption, often using mechanical/ultrasonic homogenisation.
2. Differential centrifugation, removing larger particulate components.
3. Ammonium sulphate fractionation: selective precipitation by the stepwise addition of ammonium sulphate at low temperature (<10 °C). The precipitated protein is then redissolved in a fresh buffer solution.
4. Chromatographic and/or electrophoretic separation, including ion exchange and gel filtration chromatography and polyacrylamide gel electrophoresis or isoelectric focusing. This stage may involve several individual steps and can lead to a fully purified enzyme (purification to homogeneity), with maximum specific activity.

At each step in the procedure the percentage yield of enzyme in an individual fraction can be determined:

$$\text{yield} = \frac{\text{enzyme in fraction (kat or U)}}{\text{enzyme in original preparation (kat or U)}} \times 100 \qquad [52.4]$$

Example If an original preparation of 50 ml had an activity of 100 pkat ml^{-1} (total 5 nkat) and a subsequent fraction of 5 ml volume had an activity of 200 pkat ml^{-1} (total 1 nkat), using eqn [52.4] this would represent a percentage yield of 20 per cent.

Note that the yield equation does not use a concentration-based measure of enzyme activity, as this would be affected by the volumes of the solutions involved. The relative purification of an enzyme is usually expressed as 'n-fold purification', where:

$$n = \frac{\text{specific activity of fraction}}{\text{specific activity of original preparation}} \qquad [52.5]$$

Example Using eqn [52.5] an increase in specific activity from 10 nkat (mg protein)$^{-1}$ to 120 nkat (mg protein)$^{-1}$ represents a twelve-fold purification.

Enzyme kinetics

For most enzymes, when the initial velocity (v_0) using a fixed amount of an enzyme is plotted as a function of the concentration of a single substrate [S] with all other substrates present in excess, a rectangular hyperbola is obtained (Fig. 52.3). At low substrate concentrations v_0 is directly proportional to [S], with a decreasing proportional response as substrate concentration is increased until saturation is achieved. The shape of this plot can be described by a mathematical relationship, known as the Michaelis–Menten equation:

$$v_0 = \frac{V_{max}[S]}{K_m + [S]} \qquad [52.6]$$

This equation makes use of two kinetic constants:

- V_{max}, the maximum velocity of the reaction (at infinite substrate concentration).
- K_m, the Michaelis constant, is the substrate concentration where $v_0 = \frac{1}{2}V_{max}$.

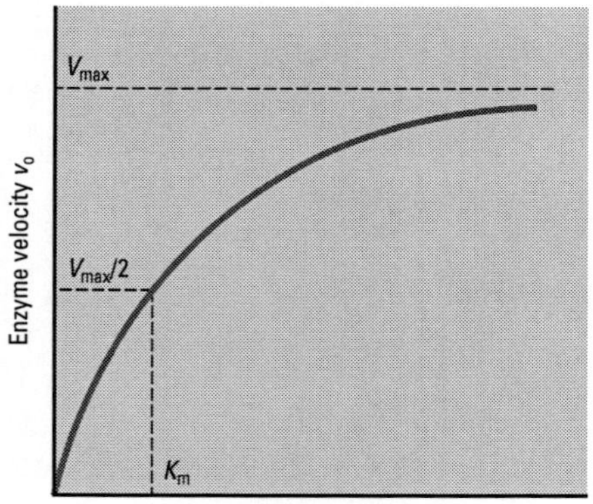

Fig. 52.3 Effect of substrate concentration on enzyme activity

V_{max} is a function of the amount of enzyme and is the appropriate rate to use when determining the specific activity of a purified enzyme.

The Michaelis constant can provide a measure of the substrate affinity of an enzyme and is an important characteristic of a particular enzyme. Thus, an enzyme with a large K_m usually has a low affinity for the substrate whereas an enzyme with a small K_m usually has a high affinity.

Your first step in determining the kinetic constants for a particular enzyme is to measure the rate of reaction at several substrate concentrations,

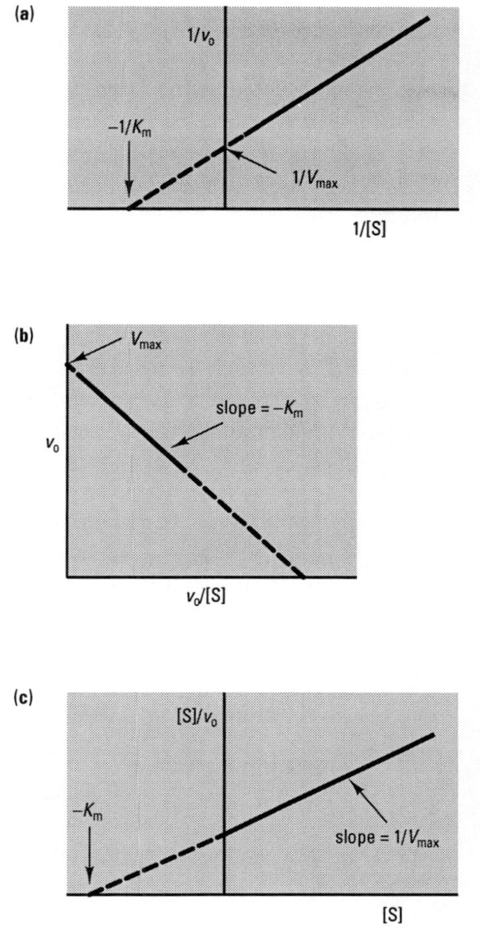

Fig. 52.4 Graphical transformations for determining the kinetic constants of an enzyme. (a) Lineweaver–Burk plot. (b) Eadie–Hofstee plot. (c) Hanes–Woolf plot.

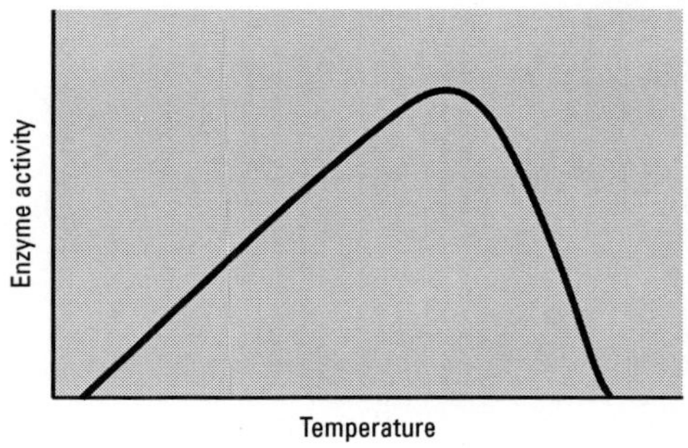

Fig. 52.5 Effect of temperature on enzyme activity

as in Fig. 52.3. There are various ways to obtain K_m and V_{max} from such data, mostly involving drawing a graph representing a linear transformation of eqn [52.6]:

- The Lineweaver–Burk plot: a graph of the reciprocal of the reaction rate ($1/v_0$) against the reciprocal of the substrate concentration (1/[S]) gives $-1/K_m$ as the intercept of the x axis and $1/V_{max}$ as the intercept of the y axis (Fig. 52.4(a)). Note that the slope of the plot is most affected by the least-accurate values, i.e. those measured at low substrate concentration.
- The Eadie–Hofstee plot: v_0 against v_0/[S], where the intercept on the y axis gives V_{max} and the slope equals $-K_m$ (Fig. 52.4(b)).
- The Hanes–Woolf plot: [S]/v_0 against [S], giving $-K_m$ as the intercept of the x axis and $1/V_{max}$ from the slope (Fig. 52.4(c)).

There are several computer packages that will plot the above relationships and calculate the kinetic constants from a given set of data using linear regression analysis. While the Eadie–Hofstee and Hanes–Woolf plots distribute the data points more evenly than the Lineweaver–Burk plot, the best approach to such data is to use non-linear regression on untransformed data. This is usually outside the scope of the simpler computer programs, though tailor-made commercial packages can carry out such analyses. Note also that some enzymes do not show Michaelis–Menten kinetics, e.g. those with more than one active site per molecule (allosteric enzymes), which often give sigmoid curves of v_0 against [S]. Such enzymes are usually involved in the control of metabolism.

Factors affecting enzyme activity

If you want to measure the maximum rate of a particular enzyme reaction, you will need to optimise the following:

Substrate concentration

The substrates must be present in excess, to ensure maximum reaction velocity. Equation [52.6] shows that a substrate concentration equivalent to $10 \times K_m$ will give 91% of V_{max}, while a concentration of $100 \times K_m$ will give 99 per cent of the maximum rate.

Cofactors

Many enzymes require appropriate concentrations of specific cofactors for maximum activity. These are subdivided into coenzymes (soluble, low molecular weight organic compounds that are actively involved in catalysis by accepting or donating specific chemical groups, i.e. they are co-substrates of the enzyme; examples include NAD^+ and ADP); and activators (inorganic metal ions, required for maximal activity, e.g. Mg^{2+}).

Temperature

Enzyme activity increases with temperature, until an optimum is reached. Above this point, activity decreases as a result of protein denaturation (Fig. 52.5). Note that the optimum temperature for enzyme *activity* may not be the same as that for maximum *stability* (enzymes are usually stored at temperatures near to or below 0 °C, to maximise stability).

Using stored enzymes *– When removing a bottle of freeze-dried enzyme from a freezer or fridge, do not open it until it has been warmed to room temperature or water may condense on the contents – this will make weighing inaccurate and may lead to loss of enzyme activity.*

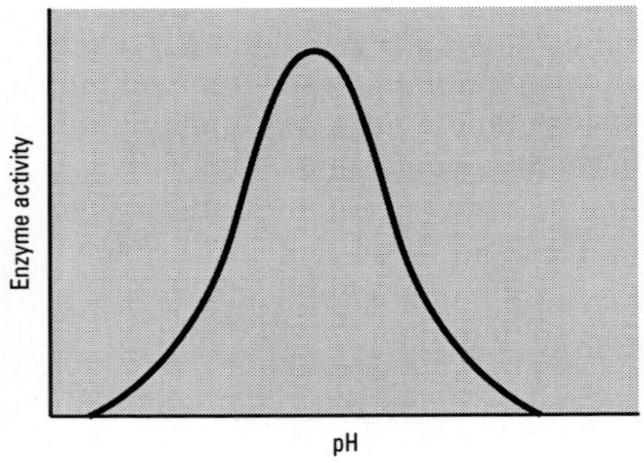

Fig. 52.6 Effect of pH on enzyme activity

pH

Enzymes work best at a particular pH, due to changes in ionisation of the substrates or of the amino acid residues within the enzyme (Fig. 52.6). Most enzyme assays are performed in buffer solutions, to prevent changes in pH during the assay.

KEY POINT Note that temperature and pH optima are dependent upon reaction conditions – you should therefore specify the experimental conditions under which such optima are determined.

Inhibitors

Many compounds can reduce the rate of an enzyme reaction, e.g. substances that compete for the active site of the enzyme due to similarities in chemical structure with the natural substrate (competitive inhibitors), or substances that bind to the enzyme at other sites, inhibiting normal function (non-competitive inhibitors). Most non-competitive inhibitors are not chemically related to the natural substrate, e.g. heavy-metal ions, such as Hg^{2+} and Cd^{+}, acting as irreversible, non-competitive inhibitors. Competitive inhibitors reduce K_m but have no effect on V_{max} whereas non-competitive inhibitors reduce V_{max} but have no effect on K_m.

Text reference

Warburg, O. and Christian, W. (1942) 'Isolierung und Kristallisation des Garungferments Enolase', *Biochemische Zeitschrift*, **310**, pp. 384–421.

Sources for further study

Copeland, R.A. (2000) *Enzymes: A Practical Introduction to Structure, Mechanism, and Data Analysis*. Wiley, New York.
[Covers basic principles, kinetics and experimental systems]

Purich, D.L. and Allison, R.D. (2000) *Handbook of Biochemical Kinetics*. Academic Press, New York.

Study exercises

52.1 Determine enzyme activity and purity (specific activity). An extract of α-glucosidase containing 2.5 mg of protein was incubated with *p*-nitrophenol-α-D-glucose at 1.0 mmol l^{-1} and assayed by following the liberation of *p*-nitrophenol at 404 nm. This gave a progress curve with an initial velocity equal to an absorbance change of 0.25 min^{-1}. Given that the conversion of 10 nmol of substrate to product results in an absorbance change of 0.122:

(a) Determine the amount of product formed in nmol min^{-1}.
(b) Calculate the enzyme activity in kat.
(c) Determine the specific activity, in terms of (i) nmol min^{-1} (mg protein)$^{-1}$ and (ii) kat (mg protein)$^{-1}$.

Quote all answers to three significant figures.

52.2 Calculate yield and relative purification of an enzyme. The table opposite shows representative data for an enzyme purification involving five stages. Calculate the yield and *n*-fold purification at each stage (give all answers to three significant figures).

(continued)

Study exercises (continued)

Volume, protein concentration and enzyme activity at several stages during a protein purification procedure

Stage	Total volume of extract (ml)	Enzyme activity ($U\,ml^{-1}$)	Protein concentration ($mg\,ml^{-1}$)
1	78	1.2	4.8
2	70	1.2	1.7
3	10	6.4	2.5
4	5	10.8	1.4
5	0.6	65.7	3.4

52.3 **Estimate K_m and V_{max} for an enzyme**. The table below shows representative data for enzyme activity as a function of substrate concentration. Use this to determine the kinetic constants K_m and V_{max} of the enzyme (give your answers to three significant figures).

Enzyme activity at various substrate concentrations

Substrate concentration ($\mu mol\,l^{-1}$)	Enzyme activity (U)
0	0
5	16.2
7	20.5
10	28.1
15	32.7
25	39.2
50	56.7
100	68.2

Answers to study exercises

We have attempted to provide an 'answer' to all of the study exercises. Where the question is open-ended and no 'correct' answer can be given, we have provided a ***Tip*** or ***Tips***, which should either help with your general approach, indicate which resources are worth consulting, or provide a pointer to relevant material within the book. Where a non-numerical question has a 'correct' answer, we have provided a model text-based answer. If a calculation is involved, we have shown the steps involved and have indicated the **correct answer underlined and in bold**.

40 The purpose and practice of dissection

40.1 ***Tip:*** This question requires considerable thought as even cell biology lines originally started from organs dissected out of animals. Chemical analysis of tissue samples will likewise require some form of dissection to remove the sample. Don't forget the dissection of plant organs – grafting is a form of dissection technique.

40.2

(a) Blunt seeker that allows separation of tissues with minimal risk of damaging the nerve or blood vessel.
(b) Fine scissors that allow incisions to be finely controlled so as not to damage the tissues within the cavity.
(c) Scalpel as the easiest and cleanest tool for removing tissues in relative bulk.

40.3 The usual rule is to dissect from the side opposite the central nervous system, i.e. from the dorsal surface of most invertebrates and the ventral surface of vertebrates.

41 Introduction to microscopy

41.1 (a) T; (b) F; (c) F; (d) T; (e) T; (f) T; (g) T; (h) F; (i) F; (j) T.

41.2 Missing words are: diffracted; near-transparent; resolution; interference; contrast; 3-D; focus; optical sectioning; polarised; laser.

41.3

(a) Embed; stain
(b) Air dry; stain
(c) Dehydrate

42 Preparing specimens for light microscopy

42.1

(a) Chlorazol black
(b) Phloroglucinol/HCl or toluidine blue
(c) Lactophenol cotton blue
(d) Neutral red
(e) Mallory
(f) Gram

42.2

(a) To preserve material and prevent degradation.
(b) To assist further processing of the specimen by removing water.
(c) To assist sectioning.
(d) To provide a thin slice of specimen suitable for viewing at best resolution.
(e) To enhance the contrast of the specimen and to identify chemical constituents.
(f) To protect specimens and allow them to be stored.

42.3 ***Tip:*** Use a Web database to find out about chemical hazards. The following answers were compiled using the Biochemical Safety website at: http://biochemlinks.com/bclinks/safety.cfm (accessed 09/04/07).
In general, all of the stains cause irritation to skin and eyes, and can be harmful if ingested in large amounts. Safety glasses and disposable gloves should be worn to reduce the risk of exposure during use. Specific additional comments for each stain include:

(a) Chlorazol black is toxic, is known to be carcinogenic and may cause foetal abnormalities.
(b) Neutral red is known to be mutagenic and to cause chromosomal abnormalities in animals.
(c) Malachite green is toxic and has caused reproductive and foetal abnormalities in laboratory animals.
(d) Safranin can cause permanent eye damage in humans and animals, and its toxicological properties remain to be established.

43 Setting up and using a light microscope

43.1 ***Tip:*** To memorise the parts, it may help to imagine yourself setting up a microscope.

43.2

(a) Provides optimal resolution by ensuring light source is focused on slide.
(b) Provides magnified image to the eyepiece lenses.
(c) Enhances image contrast by cutting down on stray light reaching the objective lens.
(d) Sets up the microscope for the distance between your pupils, so you can use both eyepieces.
(e) Allows you to return to same place on a slide.

43.3 Answer (c) is correct.

44 Interpreting microscope images

44.1 One eyepiece micrometer unit = 100 μm/11 = 9.09 μm; 25 eyepiece units = 25 × 9.09 μm = <u>**227 μm**</u>.

44.2 Area of field = $\pi r^2 = 3.14159(0.2125^2) = 0.1419$ mm^2. Area density = 25/0.1419 = <u>**176 mm^{-2}**</u>.

44.3 Volume of field = area × thickness = 0.1419 × 0.1 = 0.01419 mm^3. Volume density = 32/0.01419 = <u>**2255 mm^{-3}**</u>.

44.4

(a) $100 \times (34.0 \div 519.1) = $ **6.55%**

(b) Mean count for cytoplasm = counts for cytosol + nucleus + mitochondria + chloroplasts + other cytoplasmic bodies = $(35.0 + 16.9 + 5.1 + 160.8 + 2.6) =$ 220.4. Expressed as percentage = $100 \times (220.4 \div 519.1)$ = **42.46%**

(c) From (b) mean cytoplasmic count = 220.4, hence cytoplasm : vacuole ratio = 220.4 : 264.7 = **0.83**.

(d) $160.8 \div 220.4 = $ **72.96%**

45 Sterile technique

45.1 ***Tip:*** Consider the suitability of the various types of sterilisation procedures listed in Chapter 45, and whether they might be applicable for each item.

(a) Tips for pipettors are often sterilised by irradiation during the manufacturing process and may be supplied in sterile form. Autoclaving is the easiest approach to routine laboratory sterilisation of these items.

(b) Blood is collected and handled using sterile technique, to avoid contamination, and therefore it does not require sterilisation – it is added to sterilised agar medium after autoclaving, once the medium has cooled to below 50 °C.

(c) Autoclave, e.g. at 121 °C for 15 minutes.

(d) Either dry-heat sterilisation in a metal pipette can within a hot-air oven at 160 °C for ⩾2 h, or autoclaving at 121 °C for 15 min in the same type of pipette can.

(e) Red-heat sterilisation in a Bunsen flame (Fig. 45.2).

(f) Filter-sterilisation through a pre-sterilised filter unit, typically with a pore size of 0.45 μm or 0.2 μm. The filter-sterilised antibiotic solution is mixed thoroughly with the molten agar medium at 50 °C, then poured immediately. The filter unit may be purchased ready-sterilised, or prepared by autoclaving.

45.2 ***Tip:*** You will need to consult UK ACDP guidelines and categorisation (Anon., 1995, p. 270) – some information on categorisation and safe working with specific pathogens is available on the Web (note that other countries may use a slightly different categorisation, e.g. biosafety levels, BSL 1–4 in the USA):

(a) Category 2
(b) Category 2 (gloves should also be worn)
(c) Category 3
(d) Category 1
(e) Category 3
(f) Category 3
(g) Category 2 (a microbiological safety cabinet should be used)
(h) Category 3
(i) Category 1
(j) Category 4

45.3 The following list covers six important points, and is based on material in Chapter 45. However, you and your fellow students may think of other valid points:

- Spread plates show the typical colonial form of the microbe, whereas most of the colonies in a pour plate are contained within the agar and typical colony morphology cannot be observed.
- Pour plates are more suitable for counting higher numbers of microbes, since the colonies are smaller and are contained within the agar medium, in contrast to spread plates.
- Heat-sensitive microbial cells may be damaged by the temperature of the molten agar used in pour plating (45–50 °C) and may not grow on subsequent incubation.
- Some liquid will adhere to the spreader, and this may give a lowered count compared with a pour plate. Similarly, when the sample and molten agar are pre-mixed before pouring (see Fig. 45.6), some of the cells may not reach the pour plate, leading to a reduced count (note that this is one reason why some workers carry out the mixing of sample and molten agar within the Petri plate, rather than in the bottle).
- Each type of plating procedure requires different apparatus – spread plating requires a sterile spreader and pre-poured plates, while pour plating requires molten agar at 50 °C (usually kept in a water bath until needed).
- There is a fire risk associated with using ethanol to flame-sterilise a glass spreader. This risk is absent in pour plates, and can be removed in spread plates by using a disposable plastic spreader.

46 Cell culture

46.1 ***Tip:*** Use eqns [46.1] and [46.2] for these calculations.

(a) Substituting the appropriate values into eqn [46.1], $\mu = [2.303\ (6.5314789 - 4.7160033)] \div 9 = 0.46456 =$ **0.465 h^{-1}**. Similarly, using eqn [46.2], $g = (0.301 \times 9) \div (6.5314789 - 4.7160033) = 1.492\,171 =$ **1.49 h** (approximately 1 h 29 min, *not* 1 h 49 min).

(b) Substituting the appropriate values into eqn [46.1], $\mu = [2.303\ (7.8260748 - 4.4771213) \div 200 = 0.0385632 =$ **0.0386 min^{-1}**. Similarly, using eqn [46.2], $g = (0.301 \times 200) \div (7.8260748 - 4.4771213) = 17.975765 =$ **18.0 min**.

(c) Substituting the appropriate values into eqn [46.1], $\mu = [2.303\ (2.39794 - 1.50515)] \div 2.5 = 0.8224382 =$ **0.822 h^{-1}**. Similarly, using eqn [46.2], $g = (0.301 \times 2.5) \div (2.39794 - 1.50515) = 0.8428634 =$ **0.843 h** (approximately 51 min). Note that the time difference in this example is two and a half hours, not 2.3 hours.

46.2 ***Tip:*** Use eqn [46.6] (Box 46.1) to carry out these calculations.

(a) $6.42 \div 1 \div (2.5 \times 10^{-7}) \times 1 = 2.568 \times 10^{-7} =$ **2.57×10^{7} ml^{-1}**

(b) Total cell count = 698, so $698 \div 20 \div (8 \times 10^{-7}) \times 1 = 4.3625 \times 10^7 =$ **4.36×10^7 ml^{-1}**.

(c) $78 \div 25 \div (2.5 \times 10^{-7}) \times 10^2 = 1.248 \times 10^9 =$ **1.25×10^9 ml^{-1}** (note the correction factor of 10^2, to account for the dilution factor of 10^{-2}).

(d) $6.8 \div 1 \div (2 \times 10^{-5}) \times 0.1 =$ **3.40×10^4 ml^{-1}** (note the correction factor of 0.1, because the suspension was concentrated by a factor of 10).

46.3 ***Tip:*** Use eqn [46.7] (Box 46.2) to carry out the calculations.

(a) $(54.4 \div 0.1) \times 10^5 =$ **5.44×10^7 CFU ml^{-1}**.

(b) The mean count per spread plate is $(34 + 40 + 37) \div 3 = 37.0$ CFU. So $(37.0 \div 0.2) \times 10^3 =$ **1.85×10^5 CFU ml^{-1}**.

(c) The mean count per pour plate is $(211 + 186 + 194 + 202) \div 4 = 198.25$ CFU. So $(198.25 \div 0.5) \times 20 =$ **7.93×10^3 CFU ml^{-1}**.

(d) The mean count per plate is $(35 + 41 + 32) \div 3 = 36$ CFU. So $(36 \div 1) \times 50 = 1800$ per g and therefore $1800 \times 100 =$ **1.80×10^5 CFU per 100 g**. (Note that the multiplication factor to correct for dilution was 50 in this case, since the seafood had been diluted to 2% w/v.)

(e) This example does not require eqn [46.7] in full. The mean count for 250 ml was $(45 + 32) \div 2 = 38.5$ CFU. Therefore, the count per 100 ml is $38.5 \div 250 \times 100 =$ **15.4 CFU per 100 ml**.

(f) The mean count per plate is $(28 + 32 + 39) \div 3 = 33.0$, so using eqn [46.7], $(33.0 \div 0.5) \times 10^1 =$ **660 CFU ml^{-1}**. This value is higher than the EU regulations (maximum 100 CFU ml^{-1}), so this water sample does not meet the specification required of mineral water.

47 Working with animal and plant tissues and cells

47.1 Assuming that each solute behaves in an ideal manner, then the osmolarity of each constituent will be equivalent to the amount in grams per litre divided by the relative molecular mass and multiplied by the number of osmotically active particles produced from each solute molecule when it dissolves in water. Thus $NaCl = 6 \div 58.44 \times 2 = 0.2053388$ osmol l^{-1}; $KCl = 0.075 \div 74.55 \times 2 = 0.0020121$ osmol l^{-1}; $CaCl_2 = 0.1 \div 110.99 \times 3 = 0.0027029$ osmol l^{-1}; $NaHCO_3 = 0.1 \div 84.01 \times 2$ (note that $NaHCO_3$ gives only two osmotically active entities in water) $= 0.0023807$ osmol l^{-1}. Added together, this gives an osmolarity of 0.2124345 osmol l^{-1} = **212.4 mosmol l^{-1}**. Note that you should use the full numerical values until the last step, to avoid introducing 'rounding' errors.

47.2 ***Tip:*** The UK National Culture Collection website at http://www.ukncc.co.uk/ (accessed 09/04/07) provides coordinated information for a number of individual UK collections. The UK National Collection of Type Cultures (NCTC) can be found at http://www.hpa.org.uk/nctc/default.htm/ (accessed 09/04/07); the American Collection of Type Cultures (ATCC) is at http://www.atcc.org/; the World Federation of Culture Collections has a website at http://wdcm.nig.ac.jp/wfcc/index.xml (accessed 09/04/07). Box 10.5 gives further details on the evaluation of Web-based resources.

47.3

(a) HeLa cells are named after their source, Ms Henrietta Lacks, of Baltimore, USA.

(b) Vero cells are derived from the kidney of an African Green Monkey (*Cercopithecus aethiops*), known as the vervet monkey, hence 'vero'.

(c) BHK stands for 'baby hamster kidney', the original source of this cell line.

(d) CHO stands for 'Chinese hamster ovary', the original source of this cell line.

(e) TBY-2 stands for tobacco bright yellow strain 2.

48 Photography and imaging

48.1 ***Tip:*** The key issues here relate to:

- The ability to take a good image in the first place; film still provides the highest-quality image but you do not know what it is like until it has been processed – by this time you might not be able to repeat the photograph. Digital cameras provide instant information as to the success of the image creation, allowing repetition until it is adequate.
- The ability to subsequently manipulate the image to improve it. Digital images are easily manipulated but, of course, film material can also be digitised once it has been processed.

48.2 ***Tip:*** The key point about making an image is that it provides a record of an event or an action and this can be particularly important when things cannot be repeated. In addition, there are many modern microscopic methods that depend on image creation for their use, e.g. scanning electron microscopy.

48.3 ***Tip:*** Decide first on the available lighting options available for the imaging equipment. This will to a large extent determine what type of film will be required. Poor lighting requires fast films, but these generally produce poorer image quality, so maximum effort should be made to provide good lighting conditions. This might require the use of artificial aids such as electronic flashes.

49 Measuring growth and responses

49.1

(a) $(38.5 - 35)/7 =$ **500 mg day^{-1}**.

(b) $((38.5 - 35)/35) \times 100 =$ **10%**.

49.2 ***Tip:*** You need to graph heart rate against log concentration values; this should give a sigmoidal curve (see below). Having obtained log concentration estimates graphically from the heart rate values for the two loch samples, convert back to mmol l^{-1} concentration by raising 10 to the power of the estimated log concentration.

(a) The estimate for (i) is **approximately 1.58×10^{-5} mol m^{-3}**; and for (ii) **approximately 6.31×10^{-5} mol m^{-3}**.

(b) Estimate (ii) is the least reliable owing to the shallowness of the log concentration versus heart rate curve at this point: a small error in heart rate will lead to a large difference in estimated concentration.

(c) If sample (ii) were diluted approximately 4-fold, then tested again, this would shift the 'expected' heart rate downwards to the most sensitive part of the graph. The dilution factor would need to be accounted for when estimating the true concentration.

(d) There is no guarantee that the only substance affecting the *Daphnia* sp. heart rate in the loch samples is the toxin tested. Other substances present might increase or decrease heart rate.

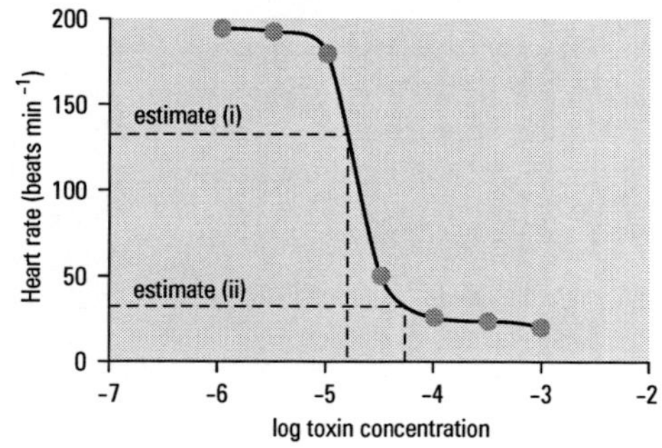

Responses of *Daphnia* to toxin

49.3 ***Tip:*** You will first need to convert the nisin concentrations of the standard solutions to $\log_{10}$ values, before plotting zone diameter against $\log_{10}$ nisin concentration (see figure below). Based on the calibration curve, solution 5 contains antilog$_{10}$ 1.24 = **17.4 μg ml^{-1}**.

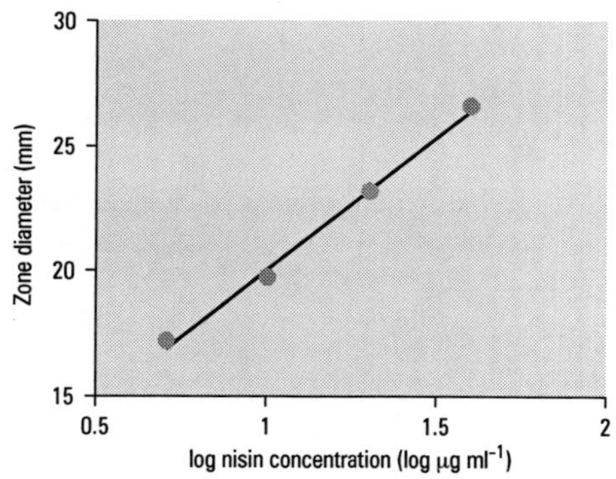

Effect of nisin on inhibitory zone diameter for *B. subtilis* colonies

52 Enzyme studies

52.1

(a) ***Tip:*** Tackle this problem by converting the absorbance change per minute to amount of product formed, in nmol: 0.25 ÷ 0.122 × 10 = 20.491803 nmol min^{-1} = **20.5 nmol min^{-1}**.

(b) ***Tip:*** Express the activity in the appropriate units (mol s^{-1}) to give the number of katals. From part (a), enzyme activity = 20.491803 nmol min^{-1} ÷ 60 = 0.3415301 nmol s^{-1} = 0.3415301 × 10^{-9} mol s^{-1} = 0.3415301 × 10^{-9} kat = **0.342 × 10^{-9} kat** (or 0.342 nkat). Note that you should use the full numerical values in these calculations and not the answers expressed to 3 significant figures, to avoid introducing 'rounding' errors.

(c) ***Tip:*** Divide the enzyme activity by the amount of protein present in the assay to give the specific activity. Thus, in terms of nmol: 20.491803 ÷ 2.5 = 8.1967212 = **8.20 nmol min^{-1} (mg protein)$^{-1}$0**. In terms of kat: 0.3415301 × 10^{-9} ÷ 2.5 = 0.136612 × 10^{-9} = **0.137 × 10^{-9} kat (mg protein)$^{-1}$**.

52.2 ***Tip:*** Prepare a table, with columns for each of the additional steps required, i.e. (i) total enzyme activity (activity per ml × total volume of extract), (ii) total protein content (protein concentration per ml × total volume of extract), and (iii) specific activity (total enzyme activity ÷ total protein content). Then, use eqns [52.4] and [52.5] to calculate yield and *n*-fold purification and add these to two further columns. The table below shows how this can be applied in this case (note that the values are not rounded off until the final stage).

52.3 ***Tip:*** If you do not have specialised software for determining kinetic constants, you can use Microsoft Excel to prepare a suitable plot, e.g. a Lineweaver–Burk plot. Enter the data from the table in the question into two columns, then prepare a further two columns to contain

Calculation of the yield and purification of an enzyme using data from table in Study exercise 52.2

Stage:	Total enzyme activity (U)	Total protein content (mg)	Specific activity (U mg^{-1})	Percentage yield (to 3 significant figures)	*n*-fold purification (to 3 significant figures)
1	93.6	374.4	0.25	100	1.00
2	84.0	119.0	0.705882352	89.7	2.82
3	64.0	25.0	2.56	68.4	10.3
4	54.0	7.0	7.714285714	57.7	30.9
5	39.42	2.04	19.32352941	42.1	77.3

the reciprocals of each value (note that you cannot plot the zero values, since there is no reciprocal). Then use the chart 'wizard' to prepare a double reciprocal plot: note that Microsoft Excel will only draw a trend line between the points – you will need either to extend this line by hand, as shown in the figure below, or use linear regression analysis to give the y intercept and slope (if you divide the y intercept by the slope, this will give you the x intercept, but without the negative sign). In the example shown in the figure below, the y intercept ($1/V_{max}$) is 0.013443707, so $V_{max} = 74.38424731 =$ **74.4 U**. Similarly, the x intercept ($-1/K_m$) is -0.055498669, so $K_m = 18.01845 =$ **18.0 μmol l⁻¹**.

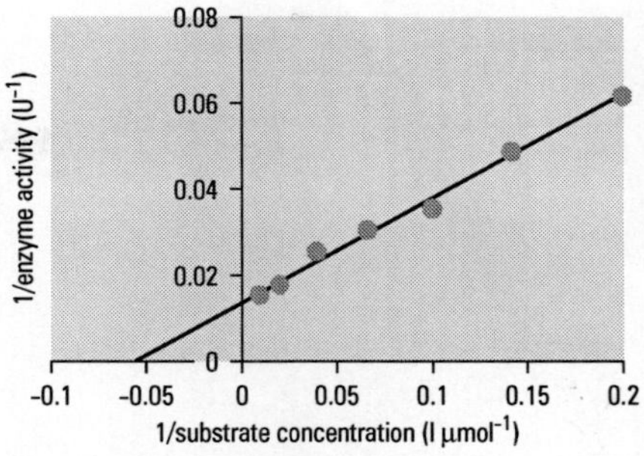

Lineweaver–Burk double reciprocal plot.

Les Sabres

portés par l'Armée Française

par

Jean Lhoste
Patrick Resek

Editions du Portail

Mais l'article II exclut de ces mesures draconiennes certaines catégories de militaires

II

Les militaires servant dans les troupes à cheval, les officiers généraux, les militaires attachés à leur état-major, les commandans et adjudans-majors des bataillons d'infanterie et les guides à cheval, auront seuls le droit d'avoir un sabre de la longueur susdite, tout le temps qu'ils seront en activité de service dans les armées de la République

Après la menace de sanctions, l'article IV devient tentateur

IV

En recevant les dits sabres, les commissaires-ordonnateurs et commissaires de guerres les feront estimer par des experts, et en fourniront à l'instant le récépissé, dont le montant sera payé par les payeurs généraux ou payeurs particuliers, sur les ordonnances des commissaires ordonnateurs.

Puis les articles suivants donnent des indications pour la remise des sabres aux différents corps d'armées, dont la liste n'est pas sans enseignement sur leur répartition géographique : armée du Nord, des Ardennes, de la Moselle, du Rhin, des Alpes, d'Italie, des Pyrénées orientales, des Pyrénées occidentales, de l'Est, de Brest, de Cherbourg.

Cet important décret est signé :

A Paris, le dix neuvième jour de Ventôse, an second de la République Française, une et indivisible. Signé GOHIER, président du Conseil exécutif provisoire. Contre-signé GOHIER. Et scellé du sceau de la République

Signé GUINOT, secrétaire général

On comprendra, après une telle collecte, pourquoi les armes blanches dont les lames mesurent plus de 30 pouces (67,5 cm), antérieures à 1792, soient devenues rares…

Le second document intéresse la Campagne d'Egypte de Bonaparte.

ORDRE DU JOUR DU 30 THERMIDOR AN 7

Avec cet ordre du jour, nous sommes sous le Directoire (août 1799). Depuis une année, Bonaparte commande le corps expéditionnaire français en Egypte. Entouré d'une cohorte de savants, le général en chef n'oublie pas ses valeureux soldats et délivre à certains d'entre eux des armes de récompense, dont les sabres énumérés dans la liste donnée ci-après.

Au quartier-général du Kaire, le 30 thermidor an 7 de la République Française, une et indivisible

ORDRE DU JOUR DU 30 THERMIDOR AN 7

ETAT

SABRES ACCORDES PAR LE GENERAL EN CHEF

1 *26 vendémiaire.....Au citoyen Pierre Laurent, chasseur à la 21ème légère... à la bataille de Sediman*

2 *2 nivôse..........Au chef de brigade du génie Samson... pour services rendus en Italie et en Egypte.*

3 *27 nivôse........Au citoyen Gonin, grenadier de la 13ème... pour la prise d'Alexandrie*

4 *3 ventôse........Au citoyen Pierre Germain, sergent major de la 85ème... au siège d'el-A'rich*

5 *23 ventôse......Au citoyen Nollet, sergent-major des grenadiers de la 69ème... à Jaffa.*

6 *29 floréal........ Au citoyen Molière, grenadier de la 25ème...au siège d'Acre.*

7 *19 messidor....Au capitaine Laplane, de la 32ème... pour ses services en Italie, en Egypte*

8 *Idem...............Au citoyen Fournier, lieutenant de la 52ème... pour le siège d'Acre.*

9 *Idem..............Au citoyen Gude, tambour-major de la 32ème... à Acre.*

10 *15 thermidor...Au chef d'escadron Lecaire, du 7ème Hussard... à Abou-Qyr.*

11 *29 thermidor.....Au citoyen Combelle, capitaine de grenadiers de la 18ème demi-brigade... à Acre.*

Cette liste évoque les principales batailles de la Campagne d'Egypte : Sediman, el A'rich, Jaffa, Acre, Alexandrie, About Qyr... de vieux noms qui, un instant, font encore rêver.

Origine du sabre

L'expression "armes blanches" dont le "Dictionnaire de l'Académie", dans son édition de 1776, détermine ainsi la définition : " les armes offensives, comme épées, hallebardes ", par opposition aux " armes à feu ", était en réalité employée à cette époque pour opposer l'état des lames finies aux lames en cours de fabrication. Il y avait ainsi à la Manufacture de Klingenthal un " contrôleur d'armes noires " qui examinait les sabres et les baïonnettes pendant qu'elles étaient dans les mains du forgeur, et un contrôleur d'armes blanches " qui les vérifiait après leur passage dans les mains du trempeur et de l'aiguiseur ". *(in* Le Tromblon n°6 par Charles Buttin - juin/ juillet 1960.)

En étudiant les épées, les fortes épées, les wallonnes, ... c'est avec grand intérêt que nous avons retrouvé le symbolisme, les légendes attachées à ces armes blanches (voir J. Lhoste et J.J. Buigné 1999). En ce qui concerne le sabre, la poésie, le rêve ne se mêlent pas à son histoire.

En effet, le sabre, dans sa forme moderne, n'a guère plus de cinq siècles d'existence. Sa caractéristique essentielle est qu'il peut frapper d'estoc et de taille, alors que l'épée sera affinée dans le but de ne frapper que d'estoc.

Bien sûr, on sait que les anciens Grecs utilisaient déjà des lames courbes pour leur *machaira* ou leur *kopis* et que la Légion romaine d'Ibérie utilisait la *falcata*, à large lame courbée d'origine ibérique. Plus tard, et apparemment sans rapport, les hommes de guerre Francs possédaient le *scramasaxe* à lame incurvée à un seul tranchant (voir J. Lhoste, 1997, p. 65). D. Venner (1986) écrit que *les paysans libres portaient souvent un coutelas courbe, le Bauernwehr, ou Hauswehr qu'en France on nommait badelaire ou malchus.* D'après cet auteur, ces coutelas seraient les ancêtres directs des couteaux de vénerie dont certains, aux XVIIème et XVIIIème siècles possédaient bien une lame courbe.... (fig 1. a et b)

Voici donc l'origine roturière du sabre : le couteau !

Au XVIème siècle apparaît le Düssack germanique. (fig 1.c) C'est un

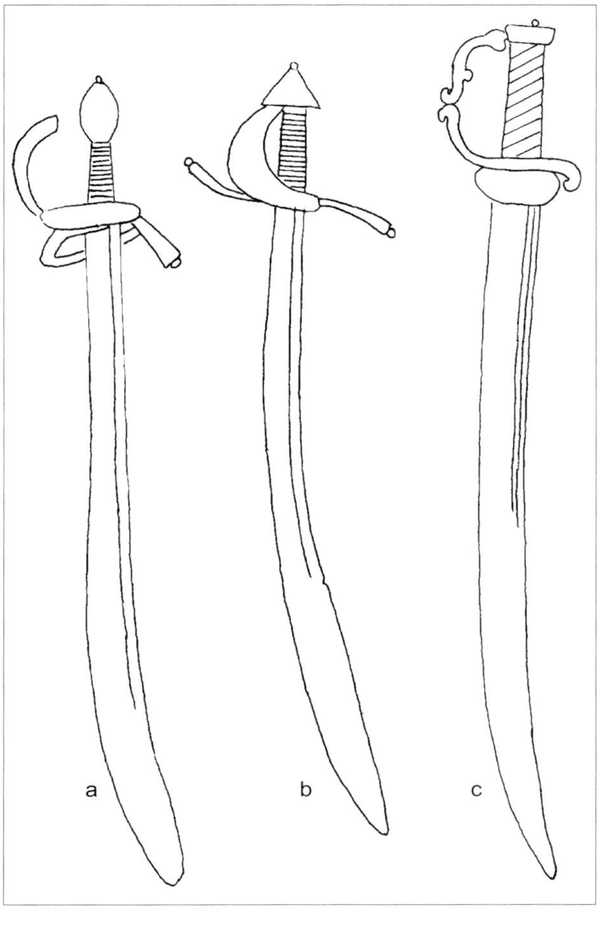

Figure 1 - Ancêtres des sabres

- **a** et **b** : Deux exemplaires de sabres germaniques, dits *Hauswehr.*
- **c** : Un exemplaire de *Düssack.*

Ces sabres du XVIème siècle montrent une évidente influence orientale relayée par la Hongrie.

véritable sabre à forte lame courbe, rainurée dorsalement. Ce sabre sera certainement assez largement répandu et fabriqué par les manufacturiers rhénans avec la collaboration des artisans d'Italie dont le nom peut figurer sur la lame, tel celui d'Andrea Ferara, actif durant la seconde moitié du XVIème siècle. Il ne semble pas que ce soit une invention purement germanique. Ce serait plutôt une copie inspirée des armes des pays de l'Est de l'Europe, eux-mêmes influencés par les Turcs.

Düssacks et couteaux de chasse furent utilisés comme armes de guerre comme le rapporte le comte de Rabutin en décrivant, vers le milieu du XVIème siècle, l'armée de Henri II rassemblée à Metz. Le comte décrit ainsi l'armement des cavaliers : *la lance, l'espée, l'estoc, le coutelas... Les archers* (portent) *la demi-lance... l'espée ou le coutelas... Près de deux mille chevau-légers... estoient armés... de la demi-lance ou le pistolet ou le coutelas...*

Parallèlement au développement des coutelas badelaires et des sabres courts, vers la fin du XVème siècle, certains forgeurs de lames, notamment en Helvétie, en Germanie, et en Italie, proposent des lames nettement plus longues et incurvées. Ces sabres portent le nom de Hiebmesseur (*) et sont passés dans la ceinture. Les mercenaires, les lansquenets, les cavaliers des XVIème et XVIIème siècles en font leur arme préférée.

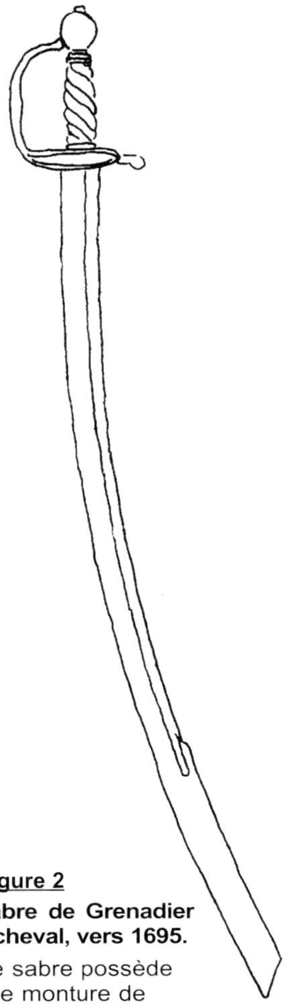

Figure 2

Sabre de Grenadier à cheval, vers 1695.

Ce sabre possède une monture de forte épée du style dit de 1695, mais la lame à gouttière, avec un contre-tranchant et surtout courbe, annonce les lames des sabres de Hussards.

Progressivement, on arrive aux premiers sabres qui deviendront réglementaires dans l'armée française, tel le sabre de Grenadier à cheval de la Maison du Roy. Ce sabre possède une monture en laiton moulé du type 1695, à une branche, plateau de garde et quillon, donc tout à fait comparable à la monture de la forte épée de la fin du XVIIème siècle. Mais cette arme possède une lame courbe, à rainure dorsale et pan creux dont l'extrémité est taillée en biseau et dont le dos se prolonge par un contre-tranchant. Ce montage exceptionnel sera éphémère. (fig. 2)

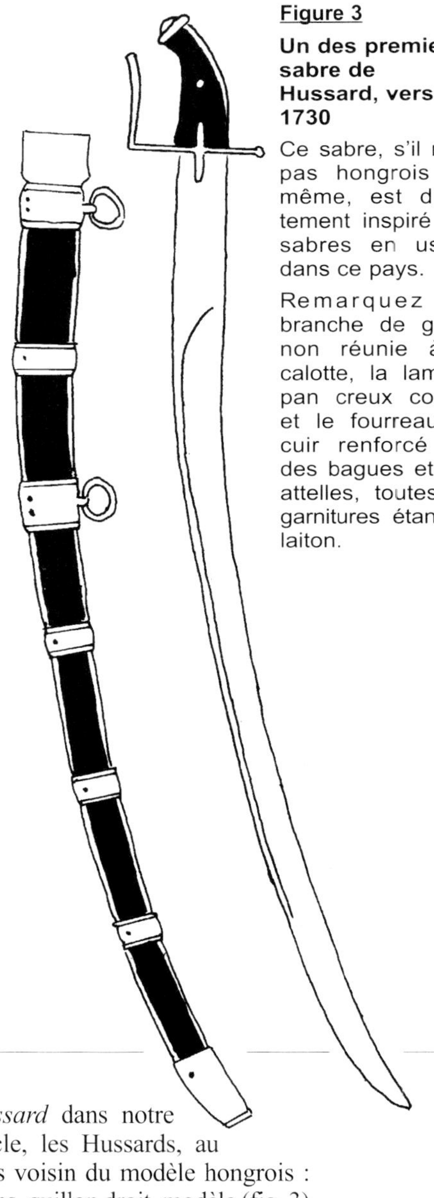

Figure 3

Un des premier sabre de Hussard, vers 1730

Ce sabre, s'il n'est pas hongrois lui-même, est directement inspiré des sabres en usage dans ce pays.

Remarquez la branche de garde non réunie à la calotte, la lame à pan creux courbe et le fourreau de cuir renforcé par des bagues et des attelles, toutes les garnitures étant de laiton.

En revanche, le sabre qui armait les cavaliers hongrois, les *Huszars*, devait faire une longue carrière dans l'armée française. Le mot hongrois devient *Houzard*, *Hussar* et *Hussard* dans notre langue, en 1605. Au milieu du XVIIIème siècle, les Hussards, au service de la France, furent armés d'un sabre très voisin du modèle hongrois : branche unique, fusée recouverte de basane, oreillons, quillon droit, modèle (fig. 3) qui préfigure le modèle typiquement français de 1786.

(*) Hieb = coup, entaille, balafre.
Messer = couteau.

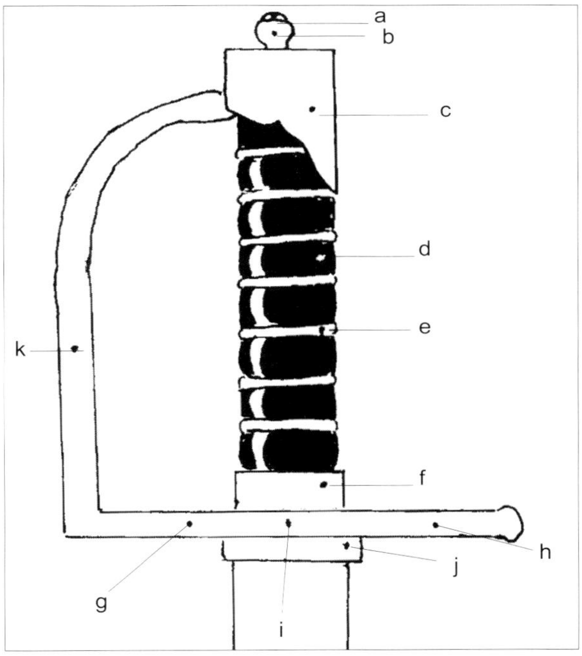

Revenons sur quelques parties importantes.

Figure 4

Composants d'une monture à calotte à queue courte
- **a**, rivure de la soie de la lame ;
- **b**, bouton de rivure ;
- **c**, calotte à courte queue ;
- **d**, poignée de bois recouverte de basane, galuchat, ... ou en corne ;
- **e**, filigrane de fil de laiton, de fer, d'argent ;
- **f**, virolle basale ;
- **g**, croisière qui peut être élargie en plateau et porter des oreillons (voir p. 20-21) ;
- **h**, partie postérieure de la croisière prenant le nom de quillon qui peut être droit ou infléchi, se terminant parfois par un bouton ciselé en bouton floral, en tête d'animal fabuleux ;
- **i**, noeuds de corps ;
- **j**, couvre-chape assurant l'étanchéité du fourreau ; cette étanchéité peut être assurée par une pièce de cuir ou par une cravate de laine ;
- **k**, arc de jointure s'insérant par un tenon dans la calotte et se prolongeant à sa partie inférieure par la croisière ; l'arc peut être partiellement droit ou former une courbe plus ou moins régulière ; on donne souvent le nom de branche de garde à cet arc de jointure.

Figure 5

Composants d'une monture à longue queue de calotte
- **a**, rivure de la soie ;
- **b**, bouton de rivure qui peut être absent ;
- **c**, calotte prolongée par une longue queue (**d**) ; ce type de calotte est parfois dit en cuillère, la calotte et la queue étant d'une seule pièce
- **e**, filigrane de fer, de laiton, d'argent ;
- **f**, poignée de bois recouvert de basane, de galuchat,... ou en corne, voire de matière plastique dans certains modèles récents ;
- **g**, arc de jointure s'insérant par un tenon dans la calotte, pouvant être partiellement droit ou plus ou moins régulièrement courbé ;
- **h**, branches secondaires, ici au nombre de deux mais pouvant être jusqu'à quatre et plus ; les branches secondaires peuvent arriver séparément sur l'arc de jointure ou bien converger ou fusionner en atteignant l'arc et partir soit de la croisière ou du plateau ou encore du médaillon servant de garde ;
- **i**, coquille d'où partent les branches secondaires ;
- **j**, quillon plus ou moins long, droit ou infléchi et se terminant par un bouton ou tout autre motif ;
- **k**, nœud de corps.

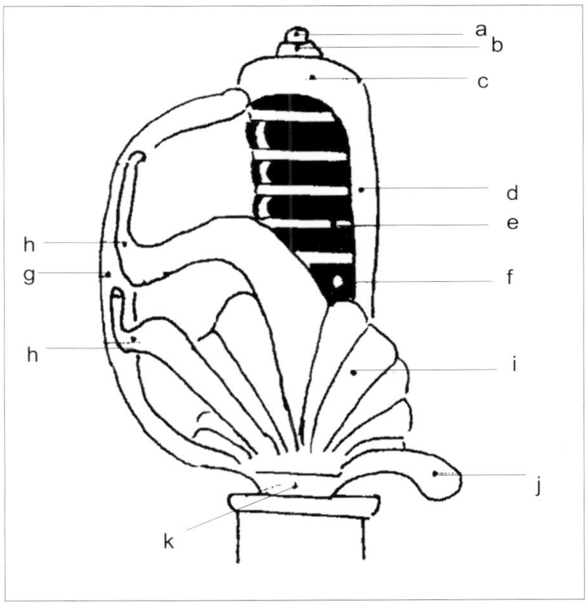

La calotte

La raison d'être de la *calotte* est de renforcer la partie supérieure de la poignée pour soutenir le bouton de rivure de la soie de la lame. La calotte est souvent prolongée d'une *courte queue* parfois cloutée à sa base ou d'une *longue queue* généralement munie d'un tenon qui entre dans un logement ménagé dans la croisière.

Les pommeaux, pièces sphéroïdes ayant la même fonction que la calotte, sont dits *figuratifs* quand ils sont travaillés en têtes de lion, de béliers, d'aigle, de coq, de bonnet... Dans le cas contraire on parle de pommeaux *non figuratifs*. Les pommeaux figuratifs sont liés, selon le symbole utilisé, à une légende mythologique gréco-latine, au corps d'armée, au régime...

La poignée

La *poignée* est une pièce permettant la prise en main de l'arme. Elle est percée de part en part pour recevoir la *soie* de la lame qui est rivée sur le dessus. Parfois, notamment sur les sabres orientaux ou à l'orientale, la *poignée* n'est pas percée totalement, ces lames n'étant pas fixée par un bouton de rivure mais par un ou plusieurs rivets, garnis de rosettes, traversant la poignée sur les côtés. Elle est le plus souvent à section carrée, ovale ou cylindrique et taillée dans du bois de hêtre. Elle est généralement encordée quant elle est recouverte de cuir, de basane ou de *galuchat* (*) encollés et garnie d'un filigrane métallique. Lorqu'il n'y a pas de filigrane, on trouve des clous qui peuvent être ronds, ovales ou en fleurettes. Il existe des poignées de bois nu plus précieux que le hêtre, tel l'ébène, qui sont agrémentées de quadrillages ou de cannelures. Sur certains sabres la poignée est entièrement métallique. A la partie inférieure et quelques fois supérieure, on trouve des bagues métalliques *les viroles*, qui masquent et maintiennent les extrémités du filigrane.

Le filigrane

Elément servant essentiellement à une meilleure tenue en main du sabre, le *filigrane* peut être en fer, en laiton, en cuivre et parfois même en argent. Il est plus ou moins important et élaboré. L'artisan, soucieux de présenter des armes agréables à l'œil, enjolive les fils métalliques en jouant sur leur grosseur, leur présentation, leur association… avec un sens artistique indiscutable comme le montrent les exemples donnés (fig. 6b à h). Parfois la poignée est totalement filigranée (fig. 6a). Il est à noter que, quelquefois, la disposition du filigrane fait partie intégrante de la description des modèles réglementaires.

Figure 6 - Quelques modèles de filigranes

La description des filigranes donnée ci-après est celle utilisée dans le texte.

Poignée non encordée
- **a**, entièrement filigranée de fils torsadés disposés en chaînette.

Poignée encordée
- **b**, poignée à filigrane de fils torsadés se terminant par quatre spires serrées ;
- **c**, poignée à deux fils torsadés jumeaux ;
- **d**, poignée à filigrane, à deux fils torsadés disposés en chaînette ;
- **e**, poignée à fils torsadés bordés de deux fils simples et virole basale ;
- **f**, poignée à filigrane complexe : au milieu de chaque spire se trouvent deux fils torsadés disposés en chaînette, bordés d'un fil simple très fin, lui-même accompagné de deux petits fils torsadés ;
- **g**, poignée dont le filigrane est constitué d'une bandelette de métal bordée de deux fils simples ;
- **h**, poignée à bandelette métallique bordée de deux fils torsadés.

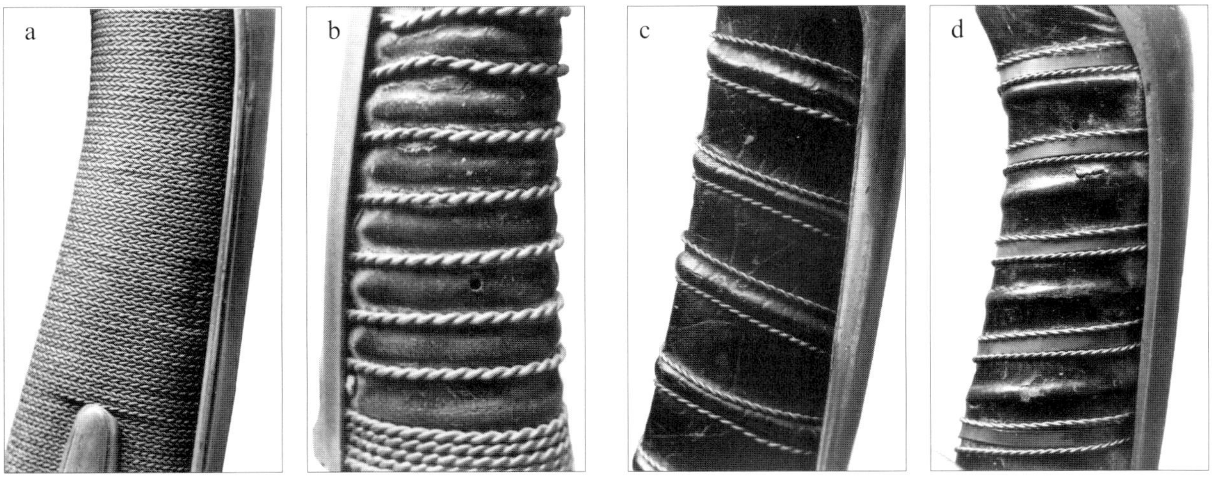

(*) Peau de raie ou de requin tannée particulièrement résistante.

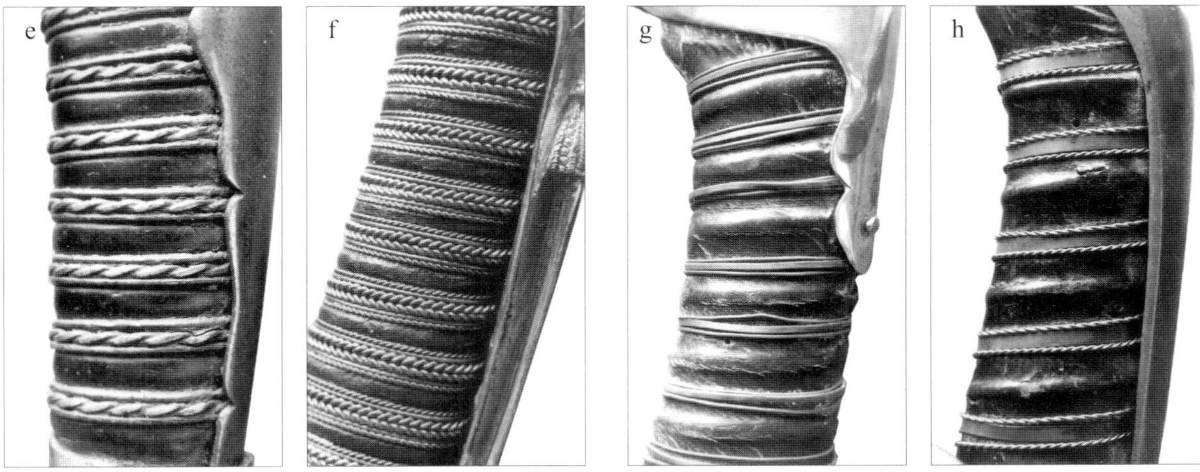

La croisière

La *croisière* est une pièce métallique horizontale faisant office de garde. Elle est fixée sous la poignée des armes blanches, épées, glaives, sabres et dont le centre se nomme *noeud de corps*. La croisière était appellée *barrette* pour les épées du Moyen Age (voir Lhoste 1997 p. 61). Cette croisière évoluera au cours des temps, s'infléchira symétriquement ou asymétriquement, les extrémités s'appelant des *quillons*. Ceux-ci peuvent se terminer en masque léonin, en boule, en fleur, en olive pour les plus courants, mais l'on trouve aussi d'autres figures ou motifs. La croisière, horizontale, peut être simple ou prolongée par l'arc de jointure, dit aussi branche de garde, et terminée à l'opposé par le quillon. Elle peut supporter les coquilles, les oreillons le ou les branches de garde secondaires. L'ensemble de ces pièces forme la *garde*.

La coquille

Sur certains sabres part de la croisière une coquille servant à une meilleure protection de la main. Cette coquille est frontale et peut être prolongée vers l'arrière d'une contre-coquille (contre-garde) plus ou moins réduite. Cette coquille peut être rapportée par brasure ou bien souvent coulée d'un seul tenant avec la croisière et la ou les branches de garde, dans ce cas la croisière n'est plus visible. Sur quelques sabres, la coquille peut être fixée par tenons (voir fig. 7).

Figure 7 – Système spécial de fixation de la coquille sur un sabre de chasseur du règne de Louis XVI.

- **a**, vue par dessous des tenons s'engageant dans des mortaises ménagées dans le plateau de garde ;
- **b**, vue par dessus d'un tenon soudé sur la coquille ;
- **c**, vue de la vis maintenant la branche secondaire supérieure sur l'arc de jointure.

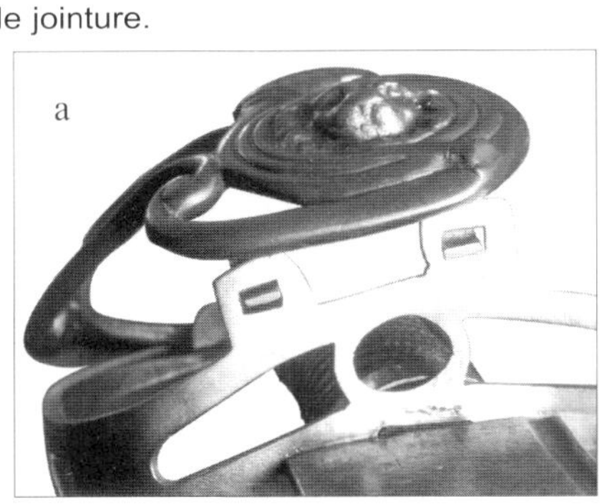

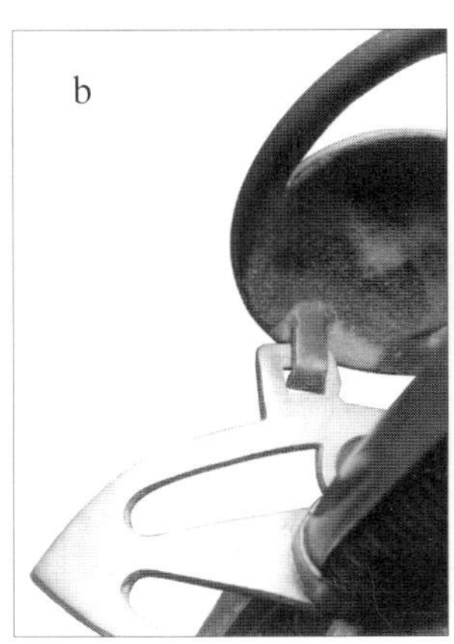

En effet sur cet exemplaire de sabre de chasseur à pied du règne de Louis XVI (fig. 7), on trouve un montage très particulier : en place des soudures, il existe des tenons sur la partie basale de la coquille. Ces tenons s'engagent dans des mortaises ménagées dans deux prolongements latéraux du plateau (fig. 7a). La branche secondaire supérieure est fixée par une vis sur l'arc de jointure (fig. 7a, b et c). Ce mode de fixation présente l'avantage d'être facilement démontable. Certainement une spécialité d'un fourbisseur ingénieux, ce mode de montage reste très rare. On le retrouve sur les sabres des XVIIème et début du XVIIIème mais il disparaît dès le milieu du XVIIIème.

Les oreillons

Les oreillons sont des pièces métalliques en général fondues avec la croisière. Nous attacherons beaucoup d'intérêt aux oreillons car leur présence – ou leur absence – est un caractère important dans la classification et l'identification des sabres.

Dans le cas où les oreillons sont présents, on fera les distinctions suivantes :

- demi-oreillons inférieurs, dont le rôle est de bien maintenir le sabre dans le fourreau ;

- demi-oreillons supérieurs, dont le rôle initial est d'empêcher la poignée de pivoter. Ces demi-oreillons peuvent être également décoratifs, notamment dans les fabrications tardives des briquets de Grenadiers du modèle 1767 à monture de laiton moulé.

- oreillons entiers, dont le rôle est double : celui des demi-oreillons inférieurs et supérieurs. Ces oreillons se trouvent notamment sur les sabres de Hussards, de Chasseurs de la Garde Impériale de Napoléon 1er….

On trouvera (fig. 8, a à i) les principaux types d'oreillons pouvant se trouver sur les croisières de sabres.

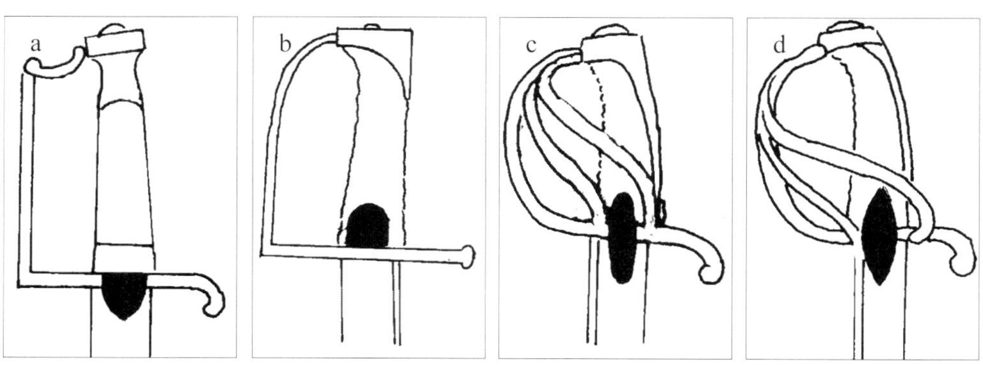

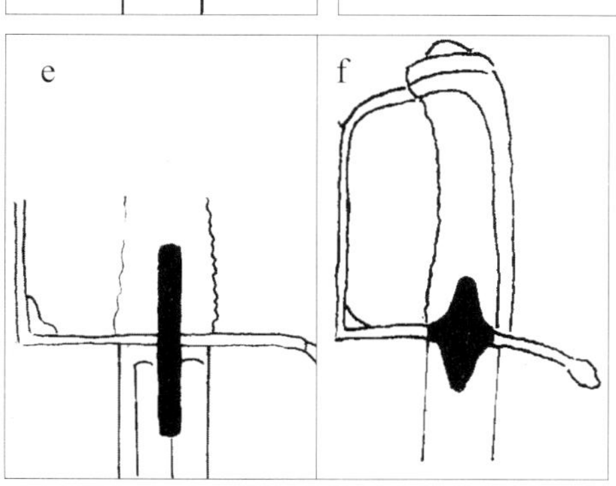

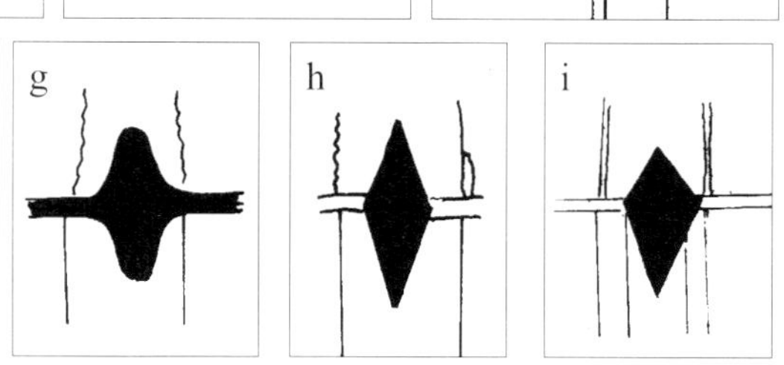

Figure 8 – Différents types d'oreillons

- **a**, demi-oreillon inférieur dit également basal ;
- **b**, demi-oreillon supérieur, partie intégrante de la croisière ;
- **c**, oreillon entier ovoïde, dit au aussi en olive ;
- **d**, oreillon en navette ;
- **e**, oreillon en baguette, celle-ci pouvant être arrondie à ses extrémités ou à coupure franche ;
- **f**, oreillon en losange curviligne ;
- **g**, idem mais l'oreillon fait partie intégrante de la croisière ;
- **h**, oreillon en losange régulier ;
- **i**, oreillon en losange dont la partie basale est plus grande que la partie supérieure.

Les plateaux de garde

Le *plateau de garde* des sabres peut être considéré comme un élargissement de la croisière dans certaines fabrications. Parfois, le plateau, l'arc de jointure et les branches secondaires éventuelles sont découpés dans une plaque métallique avant d'être mis en forme par estampage. Dans ce cas, on ne peut plus parler de croisière et la partie postérieure du plateau prend le nom de *bec*.

Dans les dernière années du règne de Louis XVI, apparaît une nouvelle catégorie de sabres : les *petits Montmorency* (*). Ces sabres d'officiers ont un plateau allégé grâce à un travail à jour de découpages plus ou moins artistiques. Ces plateaux sont souvent ignorés, les photographies de profil, les plus fréquentes, ne permettent pas d'en apprécier la qualité. C'est la raison qui nous a conduits à en donner les dessins des figures (fig. 9, a à i) et les photographies (fig. 10). Cette catégorie de plateaux disparaîtra pour faire place, entre autres, aux sabres d'Infanterie dits briquets, destinés à la troupe.

Figure 9 Différents motifs de découpe de garde des sabres à lame à la Montmorency.

- Type **a**, c'est le plus allégé. Le plateau est largement découpé longitudinalement avec, vers le milieu, deux anneaux de renfort ;
- Type **b**, c'est le même type de découpe, mais les anneaux sont remplacés par une courte languette. Souvent d'autres découpes se trouvent à la partie antérieure du plateau ;
- Type **c**, dans ce type, les découpes latérales sont associées à des cercles et à de nombreux trous circulaires.
- Type **d**, le plateau est découpé en soleil, les branches multiples partant du centre de la garde, les rayons sont simples ;
- Type **e**, le plateau est également découpé en soleil, mais les rayons sont plus finement travaillés ;
- Type **f**, dans ce cas, l'ornementation est très allégée et composée de rubans qui se chevauchent ;
- Types **g**, **h**, **i**, nous avons regroupé dans cette catégorie, très arbitrairement, des motifs rares figurant des grenades enflammées, des faisceaux de licteur, des ailes... motifs assurément faits sur commande et marquant bien leur époque.

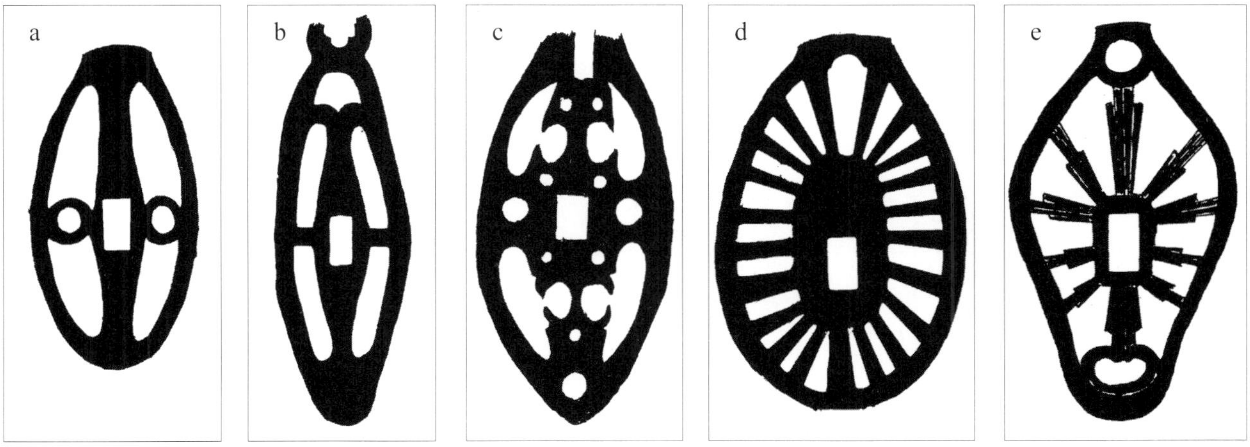

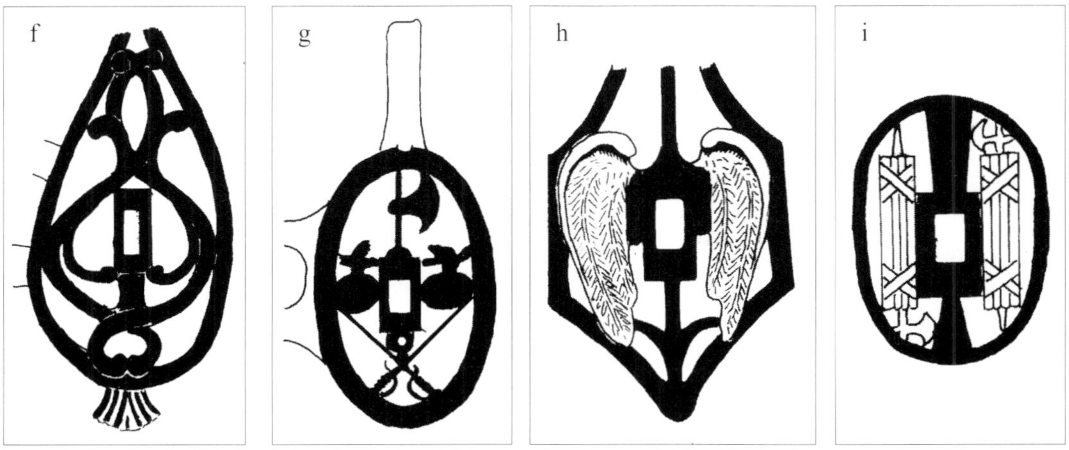

(*) Le nom de Montmorency provient du type de lames qui dès 1770 étaient montées sur les sabres de Dragons de Montmorency qui devinrent Chasseurs des Evêchés en 1788 et qui formèrent, sous la Révolution et l'Empire, le Deuxième Régiment de Chasseurs.

Figure 10 Autres exemples de plateaux

Différents types et variantes de plateaux sont présentés grâce à des photographies qui ont le mérite de bien les situer au sein de la monture. Ces plateaux sont les suivants :

- Type **a**, variante à plateau simple et allégé ;
- Type **b**, variante dont le plateau se termine par une palmette ;
- Type **c**, variante élaborée à branches découpées et palmette ;
- Type **d**, au soleil et terminé par une palmette ;
- Type **e**, découpes figurant deux épées, deux grenades et une hallebarde ;
- Type **f**, aux faisceaux de licteur.

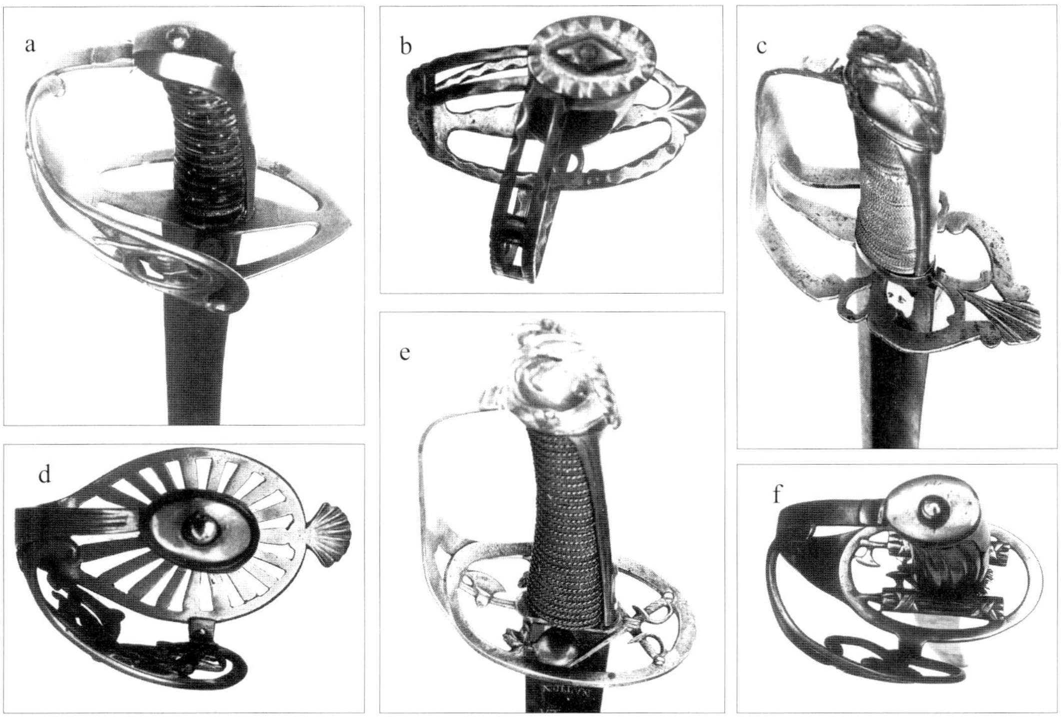

La décoration des sabres

Dans un ouvrage précédent, les auteurs (Lhoste et Buigné 1999) se sont surtout intéressés à l'ornementation des armes blanches directement liée à sa signification militaire. En poursuivant leurs investigations, les auteurs se sont rendu compte que nombre d'ornements étaient inspirés par l'art architectural de l'Antiquité et du Moyen Age. Bien entendu, ces ornements seront évoqués dans la description de chaque sabre mais, pour une question de terminologie et de meilleure compréhension, quelques exemples seront donnés ci-après (*).

Acrostale

L'*acrostale* (*acrostalium*) est l'ornement qui décorait la proue des navires de l'Antiquité (fig.11a). Cette pièce était détachée des navires vaincus, rapportée comme trophée et figurait dans les bas reliefs des monuments commémoratifs. Cette acrostale devenait progressivement symbole de victoire. Les poignées en crosse des nombreuses armes blanches et tout particulièrement des sabres trouvent-elles leur origine dans ce symbolisme ? On peut émettre cette hypothèse…

(*) Les dessins sont effectués à partir des gravures de l'ouvrage de E. Bose du XIXème siècle, réédition Hugues de Fleurville 1984.

Bagues et bracelets

Dans l'Antiquité on donnait aux *bracelets* et aux *bagues*, les noms d'*annelets* ou d'*armilles* qui ornaient les colonnes. Or, de telles bagues se rencontrent sur de nombreuses montures de sabres et sur les fourreaux où ils prennent aujourd'hui, le nom de bracelets de bélière. (fig. 11b et 11c).

Palmettes

On trouve des *palmettes* (fig. 11, d) dans les angles des corniches doriques vers le XIIème siècle avant J.C. aussi bien qu'au Moyen Age et qu'au temps de la Renaissance. Les palmettes ciselées ou découpées dans le métal ornent souvent les gardes des sabres, les oreillons, les fourreaux…

L'acanthe

L'*acanthe* est une plante aux feuilles découpées particulièrement décoratives et fut utilisée pour la décoration architecturale par les Gréco-Romains et sous Byzance. Cette ornementation a été reprise par les fourbisseurs qui en décorèrent les montures (fig. 11e), les oreillons de sabres d'officiers supérieurs, voire les bracelets de bélière. En fait, deux espèces d'acanthe furent utilisées : l'acanthe molle (*Acanthe mollis*) et l'acanthe épineuse (*A. spinosus*) mais la stylisation des feuilles ne permet guère de définir l'espèce utilisée.

Guirlandes et festons

Ces *guirlandes*, souvent nommées *festons*, sont apparues en architecture vers le XIIIème siècle pour orner les dessus des portes, les manteaux de cheminées (fig. 11, f seule la partie droite est figurée). Ces guirlandes sont constituées de feuilles et de fruits assemblés et dans celle figurée, il s'agit de feuilles de laurier semées de fruits. Un tel décor peut orner le plateau, le clavier et le fourreau de sabres et d'épées, notamment sous le Premier Empire et la Restauration.

Fleurs, fleurons et fleurettes

La fleur de lys (*Lilum candidum*) ne sera pas examinée ici et pour plus d'information on se reportera aux ouvrages de J. Lhoste et J.J. Buigné 1994 ou 1999. Sous ce titre et notamment sous celui de *fleuron*, on désigne en

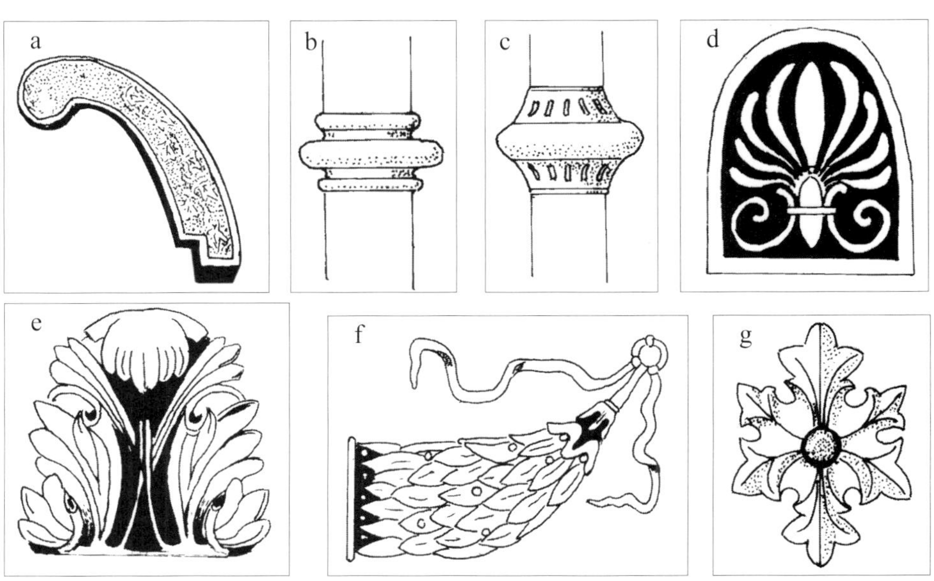

Figure 11 - Motifs de décoration

- **a**, acrostale ;
- **b** et **c**, bagues ;
- **d**, palmette ;
- **e**, feuille d'acanthe ;
- **f**, demi-guirlande ;
- **g**, fleuron.

architecture une rose épanouie, une fleur à quatre pétales qui prend alors le nom de fleuron crucifère, ou encore des motifs floraux hautement stylisés (fig. 11, g) . Ces fleurons sont largement utilisés à partir du XIIIème siècle. Sur les armes blanches, ils apparaissent au début du XVIIIème siècle et fleurissent surtout sous la Restauration, au sommet de l'arc de jointure, sur les parties latérales du plateau. Dans ce cas, le fleuron est, en général, inscrit dans un cercle.

Chapelets

Les *chapelets* sont des suites de petits motifs divers qui, le plus souvent, soulignent une ornementation plus importante. Il s'agit fréquemment de perles (fig. 12, a) et de graines que d'aucuns assimilent à des olives (fig. 12, b) ou à des olives séparées par des sortes de lentilles uniques ou doublées (fig. 12, c). Ce type de décoration date de l'Antiquité et se retrouve sur les armes blanches dès que les wallonnes disparurent pour être remplacées par des armes à monture en laiton qui se prêtaient mieux à la ciselure fine.

Câble

Ce n'est qu'une variante des frises. Mais ici le *câble* peut être serré (fig. 12, d) ou bien les spires peuvent être lâches et on aboutit à une représentation en ressort...

Oves

Les *oves* ont été utilisés depuis l'Antiquité et ce nom, en latin signifie œuf (*ovum*). En général, la partie supérieure étant masquée, on parle de quart de rond. Les oves, toujours convexes, sont jointifs ou bien séparés par un ornement : palmette, culot, dard ainsi que dans le modèle présenté (fig. 12, e). Les oves apparaissent sur les armes blanches dès le XVIIIème siècle et sur les calottes des épées jusqu'au Second Empire.

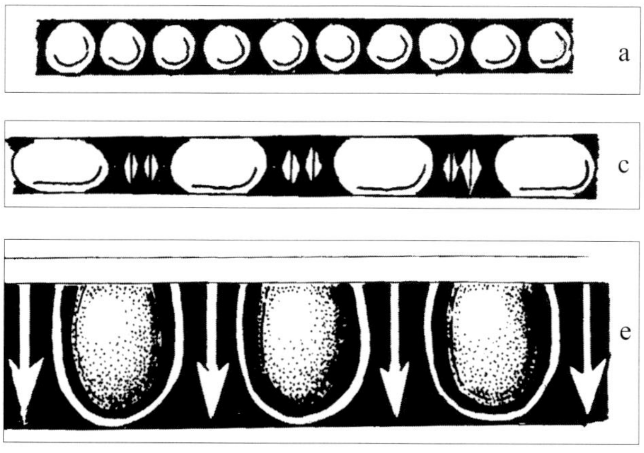

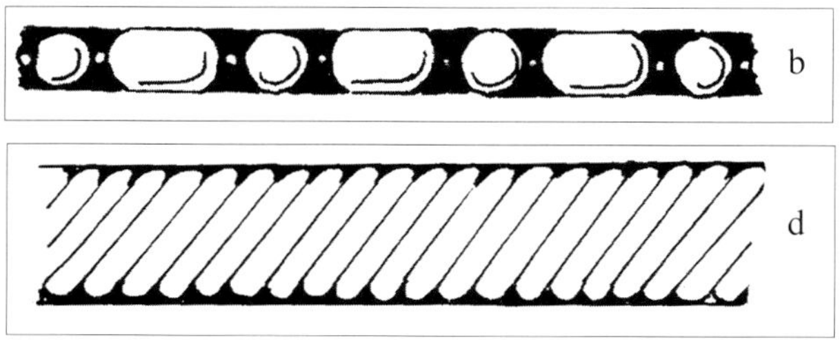

Figure 12

- **a**, frise de perles ;
- **b** et **c**, frises de graines ;
- **d**, frise de cordage ;
- **e**, frise d'oves.

Cet inventaire n'est pas exhaustif. Les ornementations plus exceptionnelles seront mentionnées à l'occasion des descriptions détaillées. De plus, le chêne, le laurier, la foudre (ou le foudre), la toile d'araignée ont été traités par J. Lhoste et J.J. Buigné dans leur ouvrage " Armes blanches, Symbolisme, Inscription, Marquages "... de 1994, réédité et complété en 1999.

Arc de jointure

Appelé souvent *branche de garde*. Notons en premier lieu, que certains sabres sont dépourvus d'arc de jointure et sont couramment nommés glaives, mais généralement, les ingénieurs militaires, ou les fourbisseurs, ont cherché à protéger la main du combattant. Pour ce faire, ils ont placé, autour de la poignée des tiges métalliques fixées d'une part sur la calotte ou sur l'arc de jointure et d'autre part, sur la croisière ou le plateau : ce sont les *branches de garde*. La branche principale joint la calotte à la partie antérieure de la croisière ou du plateau : c'est l'*arc de jointure*. Les autres branches, quand elles existent, partent généralement du plateau ou de la coquille, et rejoignent l'arc de jointure sur lequel elles sont fixées : ce sont les *branches secondaires*. On remarquera que le nombre de ces branches secondaires diffère suivant les modèles et que leurs parcours varient : tantôt elles rejoignent isolément l'arc, tantôt elles fusionnent en l'atteignant.

LES LAMES

Dans notre ouvrage sur les " Armes blanches " (1999), au sujet de la Manufacture de Klingenthal nous avons donné quelques précisions sur la fabrication des lames. On ajoutera ci-après, la liste des multiples opérations qui sont pratiquées pour les terminer et les rendre propres à être montées.

Fabrication des lames

Les opérations sont les suivantes selon *les procédés de fabrication des Armes blanches publiés par ordre du Comité de Salut public en l'An 2* :

Aiguiser à la grande meule, en travers,

Aiguiser en travers le chanfrein et le biseau,

Aiguiser le dos en travers,

Aiguiser en long le tranchant, pour mettre la lame dans sa largeur,

Aiguiser les pans creux en long,

Aiguisage des pans creux en travers,

Aiguisage en long pour ajuster le tranchant, et égaliser le dos,

Polir les pans creux en long,

Polir en long le plat et le biseau vers la pointe, ainsi que chanfrein d'un bout à l'autre,

Polir le long du dos,

Polir en travers les pans creux et faces plates qui doivent l'être,

Repolir avec de l'émeri plus fin,

Brunir les lames.

Pour les lames, l'opération n'est pas la plus simple. Les auteurs ajoutent

> *" Ce n'est encore que la série des quatre mêmes opérations ; on les fait avec les mêmes polissoires et de la même manière ; à cela près qu'elles sont alors desséchées, frottées de charbon, et polies avec l'agate. Le bruni que donnent les petites meules employées en travers, est toujours faible et peu brillant, parce que la vitesse au contact est moindre que lorsqu'on emploie les grandes polissoires. C'est ce qui fait que le bruni des deux premiers pouces de la base dans toutes les lames, excepté celles d'artillerie à pied et de grenadiers d'infanterie, tranche, comme moins noir, sur celui de tout le reste. Après cette opération, il ne reste plus qu' à saupoudrer les lames de cendres, à bien les essuyer, et à les porter chez le trempeur qui dresse encore celles qui en ont besoin. Les lames courtes ne sont jamais dans ce cas ; l'aiguiseur reprend les lames aussitôt que le trempeur les a dressées et il les reporte de suite au magasin pour être contrôlées. "*

On ne peut qu'être admiratif devant un tel art et une telle conscience professionnelle.

Principaux types de lames

Les types les plus simples sont à dos plat (plus rarement à dos rond dit en baguette ou à jonc), sans pan creux ni gouttière, donc à *section en triangle isocèle* très allongé (fig. 13, a), à quelque niveau que ce soit. Certaines lames possèdent un dos qui va s'amenuisant pour former un *contre-tranchant* (fig. 13, b). Il existe également des lames à une *gouttière dorsale*, soit tout au long de la lame soit seulement sur un partie, en général médiane (fig. 13, b et 13, c). Cette gouttière peut être accompagnée d'une rainure.

Les lames les plus fréquemment rencontrées sont à pan creux (*)avec ou sans gouttière dorsale (fig. 13, d) : ce sont les lames dites *à la Montmorency*. Toujours dans le souci d'alléger la lame sans lui enlever de sa robustesse, certaines lames sont à deux pans creux sur toute leur longueur (fig. 13e) ou seulement sur une portion de celle-ci. Il existe même, chez certaines lames de la fin du XIXème siècle des lames à trois pans creux sur une partie plus ou moins courte de leur longueur.

Les glaives ont des lames à *double tranchant*. Les unes sont à section losangique très aplatie (fig. 13, f), les autres, légèrement convexes ont un pan creux sur une longue partie de leur longueur, les autres encore ont deux pans creux près du talon et un autre plus long et médian (fig. 13, g).

Bien entendu, chaque description de sabres intègrera non seulement la monture mais aussi la forme de la lame.

Il nous semble bon de rappeler les inscriptions qui intéressent la manufacture, voire le modèle et la date de fabrication, les devises et professions de foi, le corps auquel l'arme est attribuée, le nom de son propriétaire... L'inventaire de nombre de ces inscriptions a été donné dans notre ouvrage intitulé *" Armes blanches "*. En conséquence nous renvoyons à cet ouvrage le lecteur intéressé.

(*) Nous mettons *pan creux* au singulier (sous-entendu sur les deux faces) pour éviter toute confusion avec les lames à *deux pans creux* sur les deux faces de la lame.

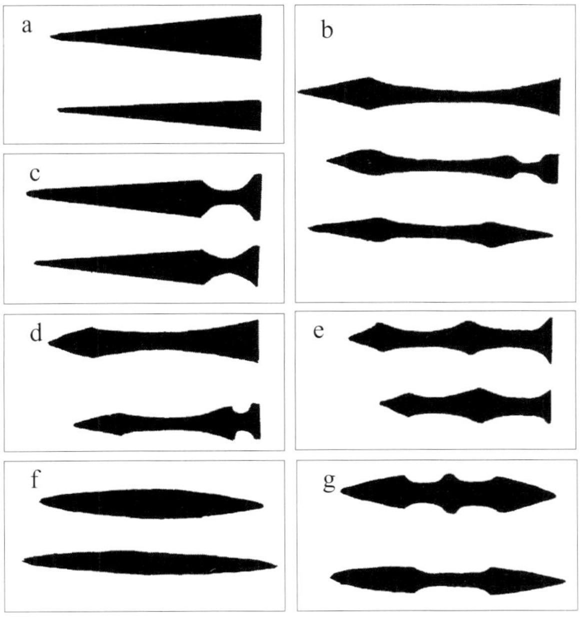

Figure 13 – Coupes de lames de différents modèles de sabres

- **a**, sabre de Grenadier de l'époque révolutionnaire. En haut, section à 5 cm du talon ; en bas, à demi-longueur ;
- **b**, sabre de Hussard du Premier Empire. En haut, section 5 cm du talon, à mi-longueur ; en bas, à 65 cm du talon, montrant à droite, le contre tranchant ;
- **c**, sabre de Grenadier de l'époque révolutionnaire. En haut, section à 5 cm du talon ; en bas, à demi-longueur ;
- **d**, sabre de Grenadier, vers 1790. En haut, à 5 cm du talon : en bas, à demi-longueur ;
- **e**, sabre à garde de bataille (1784) à 8 cm du talon ; en bas à mi-longueur ;
- **f**, glaive de Marine de la fin du XVIIIème siècle ; en haut, 2 cm du talon ; en bas à mi-longueur ;
- **g**, glaive de la Restauration (1816). En haut, à 5 cm du talon ; en bas deux tiers de la longueur.

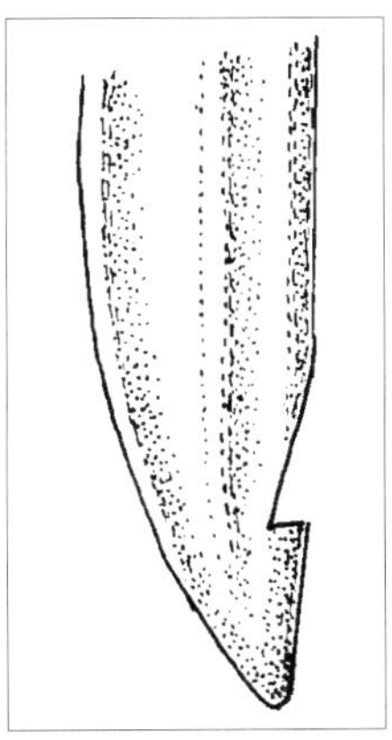

On observe parfois, près de la pointe et sur le dos de la lame, une petite encoche (fig. 15). Cette encoche volontaire est surtout présente sur les sabres de Hussards ou de Dragons dont le courage était joint à une certaine efficacité. En effet, l'encoche permet à la lame du sabre de se transformer en un petit harpon : en retirant le sabre de la plaie, le petit harpon aggravait les déchirures des tissus profonds... Ces cavaliers n'étaient pas des tendres ! Une autre hypothèse est la suivante : le petit harpon pouvait servir à accrocher le cavalier adverse et ainsi le désarçonner....

Figure 15 :

Encoche faite à l'extrémité de la lame à double pans creux d'un sabre de Dragon à garde de bataille pour rendre les blessures plus meurtrières ou pour accrocher et désarçonner l'adversaire.

La courbure des lames

Si l'on désire donner une idée exacte de l'importance de la courbure d'une lame, une seule solution : évaluer la longueur de la flèche. On commence par dresser une ligne qui s'étend de la partie postérieure du talon à la pointe de la lame. Pour évaluer la flèche a-b, on élève une perpendiculaire au niveau de la courbure maximale (a), l'arc étant constitué par la lame et la corde étant la droite tracée entre le talon et la pointe. Dans l'exemple qui est donné, la flèche est de 12 cm (fig. 16).

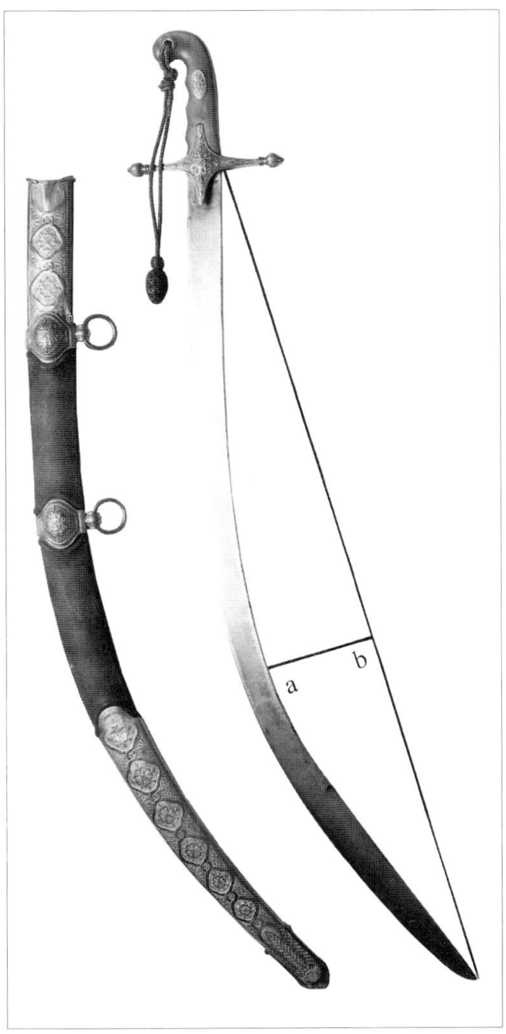

Figure 16 :

Flèche (a-b) d'un sabre de la Restauration à la mode orientale

Nota – Lorsque nous sommes dans l'impossibilité de mesurer la flèche ou si l'on juge qu'il est inutile de lui donner une valeur exacte, on a recours aux approximations détaillées dans le tableau ci-après :

Tableau d'évaluation de la courbure des lames

Courbure de la lame	Flèche
Très forte courbure	8 cm et plus
Forte courbure	Entre 4 et 8 cm
Courbée	Entre 2 et 4 cm
Faiblement courbée	Moins de 2 cm

Ces valeurs sont évidemment approximatives.

Production des lames à la Manufacture de Klingenthal en l'An II

Avant de confier le sabre nu au gainier et d'examiner son travail, donnons une idée sur la cadence de la production des ouvriers de Klingenthal selon l'ouvrage déjà cité publié par le Comité de Salut Public et dont nous donnons ci-après de longs extraits en ce qui concerne les lames.

Etat du travail
qu'exécute, par semaine, un ouvrier de chaque catégorie

	Nombre de lames
FORGEURS	
De lames de sabres du second régiment de Chasseurs, acier à trois marques, 1ère qualité	72
Grenadiers d'infanterie, à deux marques	180
Artillerie à pied, à 3 marques, 2ème qualité	108
Artillerie à cheval, idem	120
Carabiniers, idem	108
Cavalerie, Dragons à 3 marques, 1ère qualité	60
Hussards à 3 marques, 2ème qualité	72
Chasseurs à cheval, idem	72
Gendarmerie à cheval, idem	120
TREMPEURS	
De la lame du 2ème régiment de Chasseurs	400
Grenadiers d'infanterie	1200
Artillerie à pied	1200
Artillerie à cheval	1200
Carabiniers	600
Cavalerie, dragons	360
Hussards	900
Chasseurs à cheval	900
Gendarmerie à cheval	900

GRAVEURS	Nombre de lames
De lames de grenadiers d'infanterie	1200
De hussards	1200
De carabiniers	70
AIGUISEURS	
De lames de grenadiers du 2ème régiment de chasseurs	36
De lames de grenadiers	70
D'artillerie à pied	50
D'artillerie à cheval	50
De carabiniers	72
De cavalerie, de dragons	36
De hussards	40
De chasseurs à cheval	40
De gendarmerie à cheval	60

On comprend peut-être mieux, après avoir parcouru cette liste que les manufactures d'armes blanches étaient de véritables ruchers bruyants.

LES FOURREAUX

Pour des raisons de sécurité, mais aussi pour assurer la protection de la lame contre les intempéries, le forgeron, le gainier, le peaussier, le fourbisseur... inventèrent le fourreau.

Dans les temps protohistoriques (voir J. Lhoste p. 40 – 1997), le *fourreau* fut de bronze, de fer mais alourdissait considérablement l'arme. Dans la période historique, peut-être en premier lieu chez les Grecs, puis en Europe de l'Ouest des matériaux plus légers furent mis en œuvre : le bois, les tissus, la peau ou le cuir... Mais peu de ces fourreaux en matières organiques nous sont parvenus, attaqués et détruits par les bactéries, les moisissures, les insectes, les rongeurs....

Lorsque le sabre fut introduit en France, le fourreau fut de cuir jusqu'à la Révolution. La tôle d'acier n'apparut que vers la fin du XVIIIème siècle. Sous le Consulat, jusqu'au Second Empire, les fourreaux de laiton laminé furent à la mode chez les officiers.

Les sabres qui équipèrent les armées françaises possèdent des parties bien distinctes qui doivent retenir toute notre attention et qui seront étudiées ci-après : le matériel utilisé pour le fourreau proprement dit, la chape, les bracelets et les pitons supportant les anneaux de bélières, la bouterolle et le dard, les éventuelles attelles latérales de renforcement (fig. 17).

Le matériel utilisé

Ainsi que nous venons de l'écrire, le cuir est le matériel utilisé jusqu'à la fin du XVIIIème siècle. Il s'agit de *cuir de vache* noirci, plus rarement coloré en brun ou d'une autre couleur. La couture est au milieu de la face postérieure, exceptionnellement latérale pour les glaives de Second Empire de cuir bouilli et frontale pour certains sabres à l'orientale où la couture est remplacée par une sorte de chaînette en laiton ou en argent. Sur de nombreux fourreaux, le cuir est décoré d'une *rainure latérale* dans un but décoratif et pour rigidifier le cuir. Pour renforcer les fourreaux de cuir, il peut exister des *attelles de bois de hêtre* provenant généralement de la Forêt de Villers-Cotterêts.

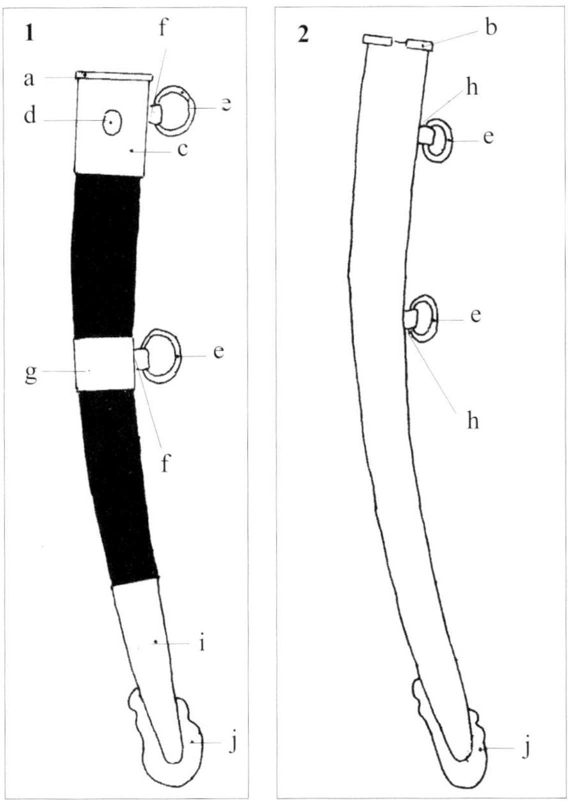

<u>Figure 17</u> – Schémas de trois fourreaux

1 - fourreau de cuir à garnitures de laiton ou de fer ;

2 - fourreau métallique (laiton ou tôle de fer) ;

3 - fourreau pour glaive en cuir et garnitures de laiton.

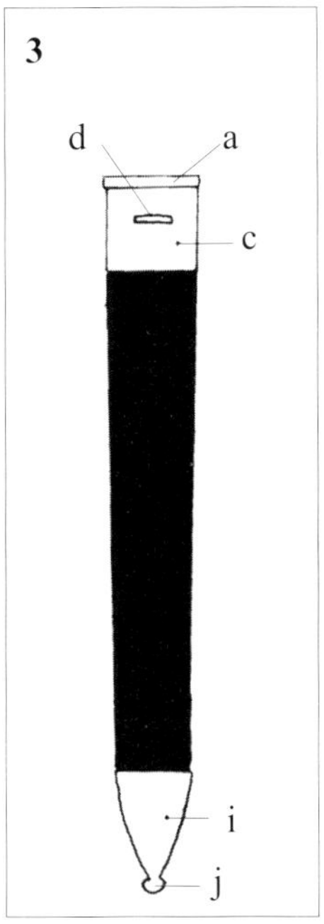

- **a**, cuvette ;
- **b**, cuvette échancrée pour le passage des oreillons ;
- **c**, chape ;
- **d**, bouton de chape ou pontet pour le port ;
- **e**, anneaux de bélière ;
- **f**, pitons ;
- **g**, bracelet de bélière qui peut faire défaut ;
- **h**, piton de bélière soudés directement sur le fourreau métallique ; parfois chape et bracelets sont figurés et non rapportés ;
- **i**, bouterolle plus ou moins longue, sur les modèles à très longue bouterolle, un anneau peut être soudé à son sommet ;
- **j**, dard asymétrique ou symétrique, en fer ou en laiton.

Il semble bien que les sabres de Cavalerie de l'An IX soient les premiers de la période moderne à recevoir un fourreau de *tôle d'acier* d'un modèle qui perdurera jusqu'à nos jours. Mais, au cours des ans il ne cessera de s'alléger. Le *laiton laminé* est un alliage qui flatte l'œil et, sous le Consulat, les fourbisseurs proposèrent aux officiers, de leur faire fabriquer des fourreaux de cette matière brillante : ils eurent le plus grand succès. De plus, cet alliage se prête bien à tous les travaux de ciselure et nombreux sont les fourreaux qui sont richement décorés de fines gravures et même de motifs en relief repoussés. Certains sabres de laiton sont dorés ou argentés.

On trouvera dans la description des sabres des indications quant à la nature de leur matière et de leur décoration.

La chape

La *chape* est la garniture métallique du fourreau de cuir qui protège son entrée : elle est le plus généralement de laiton, parfois de fer, surtout en période révolutionnaire du fait de manque de métaux cuivreux. La chape possède un autre rôle non moins important : permettre le port de l'arme. A cet effet on peut distinguer plusieurs systèmes qui seront présentés dans un ordre approximativement chronologique :

Languette de cuir

Dans ce système, la chape est pourvue d'une fente horizontale (fig. 18, a) par laquelle passe une languette de cuir pourvue à son extrémité d'une boutonnière (fig. 18, b) destinée à s'attacher sur un bouton de ceinturon.

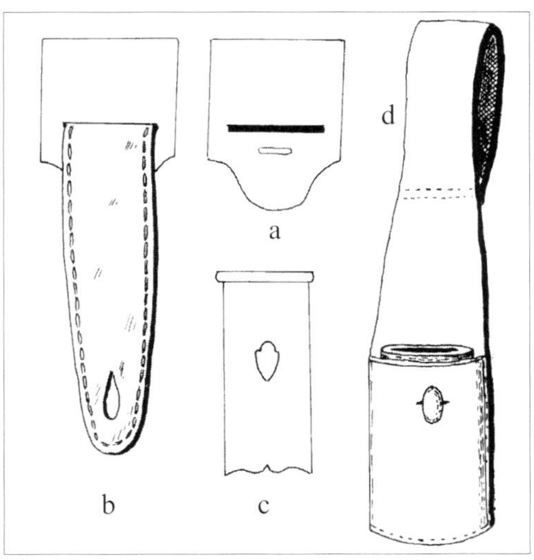

<u>Figure 18</u> – La chape et le port du sabre d'Infanterie

- **a**, chape agraffée et pourvue d'une fente horizontale ;
- **b**, languette de port qui se relève pour être boutonnée sur le ceinturon ;
- **c**, chape à bouton ;
- **d**, porte-sabre dit gousset avec le bouton de chape en place.L'ensemble sera passé dans le ceinturon.

Bouton

La chape a perdu sa fente qui est remplacée par un bouton dit de chape. Ce bouton succède aux crochets liés directement sur le haut des fourreaux qui se rencontrent fréquemment dans la première moitié du XVIIIème siècle, notamment pour le port des fortes épées. Le bouton est destiné à passer dans la boutonnière d'un étui de cuir enveloppant la chape : côté corps, l'étui se prolonge en une large bande de cuir terminée en boucle pour passer le ceinturon (fig. 18, d) ou est un étui fixé dans le prolongement d'une bretelle : le baudrier (fig. 19).

Figure 19 – Baudrier pour le port des sabres à chape à bouton.

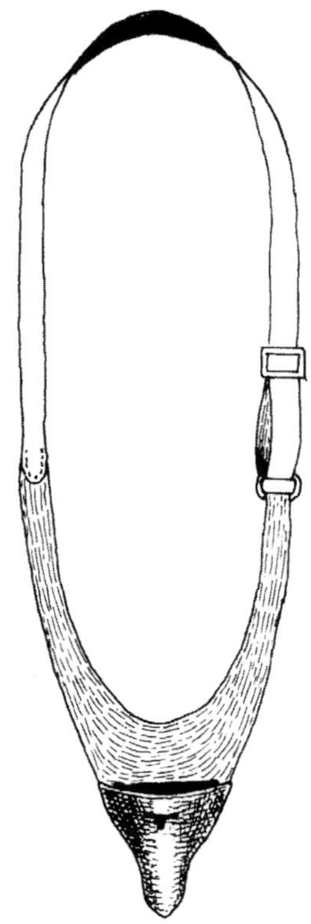

Pontet

Plus tardivement, le bouton de chape peut être remplacé par une tige métallique horizontale recourbée à ses deux extrémités soudées sur la chape. Ce dispositif forme un petit pont : c'est le pontet (fig. 20, a). Le porte-sabre, dans ce cas, est constitué d'une lanière de cuir pourvue d'un bouton et d'une boutonnière permettant, une fois fermée, de passer le ceinturon (fig. 20, b et c).

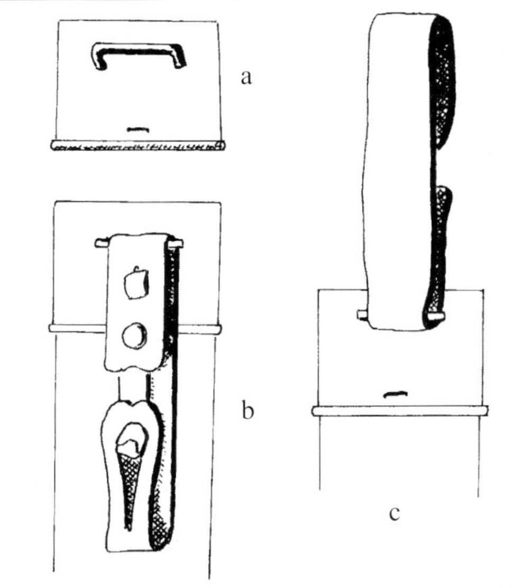

Figure 20
La chape à pontet et porte-glaive.

- **a**, chape à pontet ;
- **b**, détail du boutonnage du porte-glaive ;
- **c**, porte-glaive en place passant dans le ceinturon.

C'est vers la fin du XVIIIème siècle, qu'un dispositif va empêcher la lame de battre dans le fourreau : les lames ressorts et la cuvette.

Lame-ressorts et cuvette

Ce dispositif consiste en une ceinture de fer souple vissée, rivée ou soudée à l'intérieur du haut de la chape (fig. 21, a à e). Cette ceinture se prolonge par deux lamelles faisant ressort qui pincent la lame du sabre la maintenant bien en place.

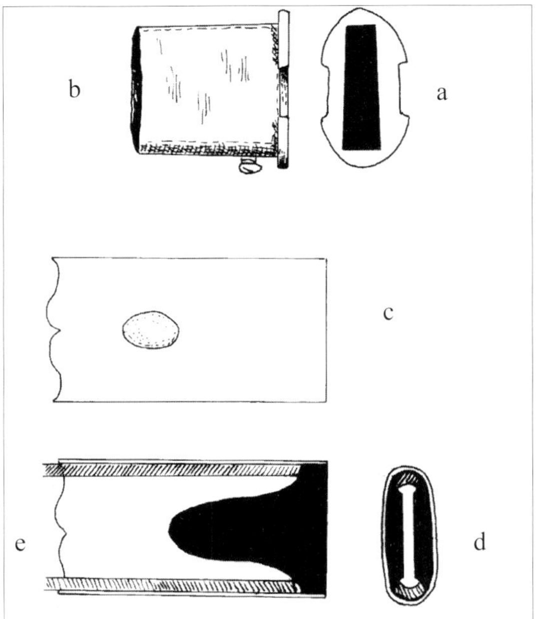

Figure 21 – Chape et cuvette

- **a** et **b**, chape d'un sabre dont la croisière est munie d'oreillons, en conséquence, la cuvette présente deux encoches latérales ;
- **c**, chape à bouton d'un sabre de Mineur ;
- **d**, la cuvette vue de haut, montre (en noir) le début des ressorts qui maintiennent la lame en place (en hachure, le cuir) ;
- **e**, coupe de la chape montrant les ressorts.

La cuvette, d'acier ou de laiton, est une pièce rapportée dont la partie supérieure réduit et protège l'ouverture du fourreau et épouse la forme de la lame (fig. 22). Cette cuvette est montée sur un support métallique qui s'emboîte dans le fourreau et qui est maintenu soit par une vis, soit par un rivet, soit par soudure.

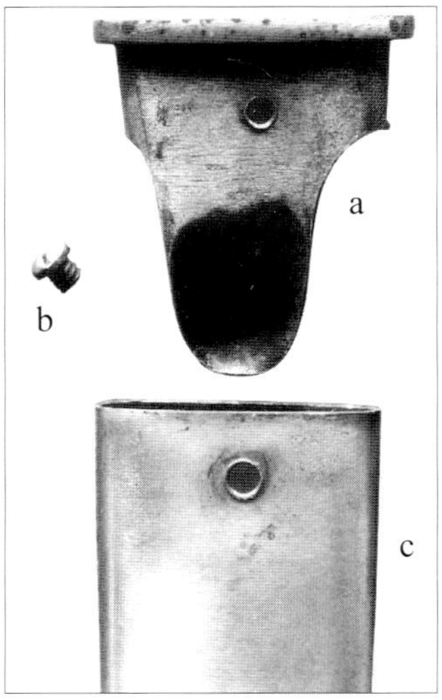

Figure 22 – Montage d'une cuvette

- **a**, cuvette à lame-ressort maintenant la lame du sabre dans le fourreau. Remarquer le trou ménagé pour le passage de la vis ;
- **b**, vis très courte passant par le trou de la chape et se vissant sur la cuvette ;
- **c**, trou de chape pour le passage de la vis maintenant la cuvette.

Nota - bien souvent, en place de la vis se trouve un rivet, mais le montage est identique.

Bracelets et anneaux de bélières

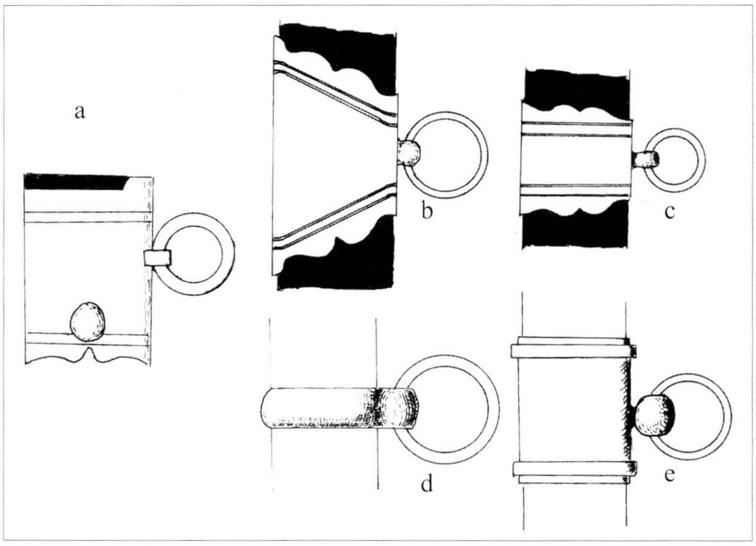

Figure 23 – Chape et bracelets à anneau de bélière.

- **a**, chape en laiton à bouton et anneau passant dans un piton d'un sabre à garde pivotante de la Révolution. Remarquer l'entaille frontale (en noir) ménagée pour la branche mobile ;
- **b**, bracelet en trapèze, de laiton découpé, sur le fourreau de cuir d'un sabre de Chasseur à pied de la fin du XVIIIème siècle ;
- **c**, bracelet subquadrangulaire, en laiton découpé, sur le fourreau de cuir d'un petit Montmorency d'Infanterie légère ;
- **d**, bracelet de fer réduit à une bague sur le fourreau de tôle d'un sabre de Cavalerie légère du modèle de l'An IX ;
- **e**, bracelet quadrangulaire, de laiton doré sur le fourreau en tôle d'un sabre de Chasseur du Premier Empire.

Les *anneaux de bélières* sont présents sur les sabres de Cavalerie et d'Officiers montés, trop longs pour être portés verticalement. Mais il existe des exceptions, pour certains sabres d'Infanterie, le premier anneau est alors fixé sur la chape et le second sur un bracelet (fig. 23). Lorsque la bouterolle est très longue, l'anneau peut être soudé à son extrémité supérieure. Les *bracelets* et les anneaux peuvent être d'acier ou de laiton. Chez certains sabres ils sont soudés directement sur les fourreaux métalliques.

Les fourreaux de laiton possèdent également des bracelets rapportés mais parfois ils sont simplement figurés par ciselure.

Tous ces sabres à deux anneaux sont portés grâce à des *bélières* de longueur variable, fixées sur le ceinturon (fig. 24 et 25).

Durant une grande partie du XIXème siècle, et même avant, la grande mode pour les cavaliers démontés était de régler les suspentes de manière que le dard de leur fourreau traîne sur le sol, d'où le sobriquet de *traîneurs de sabre* qui leur était attribué par l'Infanterie.

Dès lors on ne sera pas étonné de déceler une usure anormale de la face antérieure des dards... et aussi l'écrasement de la bouterolle qui restait coincée par la porte trop rapidement rabattue ou par les coups d'éperons.

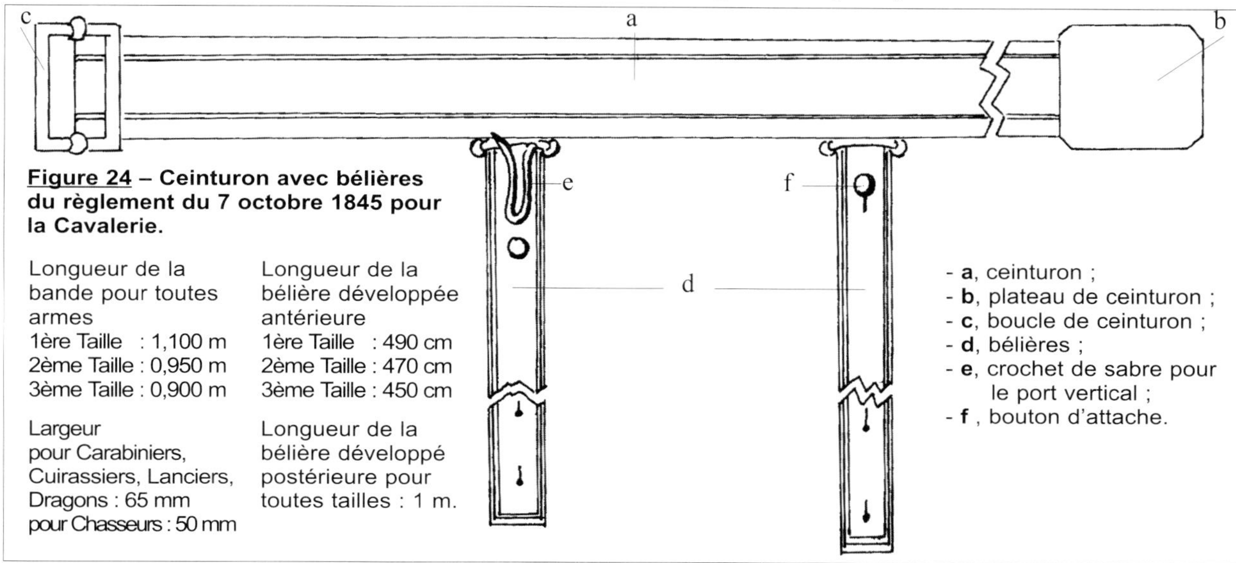

Figure 24 – Ceinturon avec bélières du règlement du 7 octobre 1845 pour la Cavalerie.

Longueur de la bande pour toutes armes
1ère Taille : 1,100 m
2ème Taille : 0,950 m
3ème Taille : 0,900 m

Largeur pour Carabiniers, Cuirassiers, Lanciers, Dragons : 65 mm
pour Chasseurs : 50 mm

Longueur de la bélière développée antérieure
1ère Taille : 490 cm
2ème Taille : 470 cm
3ème Taille : 450 cm

Longueur de la bélière développé postérieure pour toutes tailles : 1 m.

- **a**, ceinturon ;
- **b**, plateau de ceinturon ;
- **c**, boucle de ceinturon ;
- **d**, bélières ;
- **e**, crochet de sabre pour le port vertical ;
- **f** , bouton d'attache.

A partir de 1882, le sabre de Cavalerie étant porté par l'arçon de selle, un seul anneau se révéla nécessaire et suffisant. Le second bracelet et son anneau disparurent.

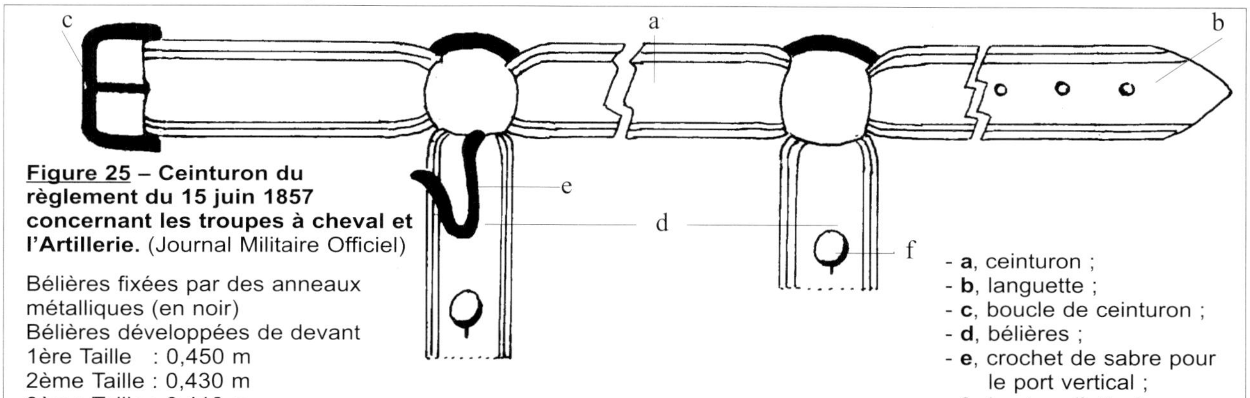

Figure 25 – Ceinturon du règlement du 15 juin 1857 concernant les troupes à cheval et l'Artillerie. (Journal Militaire Officiel)

Bélières fixées par des anneaux métalliques (en noir)
Bélières développées de devant
1ère Taille : 0,450 m
2ème Taille : 0,430 m
3ème Taille : 0,410 m

- **a**, ceinturon ;
- **b**, languette ;
- **c**, boucle de ceinturon ;
- **d**, bélières ;
- **e**, crochet de sabre pour le port vertical ;
- **f** , bouton d'attache.

Les pitons

Les *pitons*, soudés sur les bracelets de bélières ou directement sur le fourreau sont de laiton ou de fer. Leur forme est assez constante, mais leur examen permet de constater que leur bord inférieur présente parfois une usure évidente : cette usure est due à un port prolongé...

La bouterolle et le dard

La *bouterolle* est une pièce métallique qui protège (de l'usure) l'extrémité inférieure des fourreaux et ainsi maintient son étanchéité.

C'est une invention très ancienne. En effet, on trouve des bouterolles dès l'Âge du Fer (voir J. Lhoste 1997, p. 45) aussi bien que sur les épées gauloises.

Les bouterolles des sabres modernes peuvent ne pas dépasser quelques centimètres ou bien atteindre 30, voire 45 centimètres. Vers la fin du XVIème siècle mais surtout un peu plus tard, les fourbisseurs font ajouter une pièce de fer ou de laiton pour protéger la bouterolle : c'est le dard (*) . Ces dards sont classables en deux catégories : les dards *asymétriques* et les *dards symétriques*.

Dards symétriques

Les dards symétriques les plus anciens ne sont constitués que d'un bouton, aussi bien sur les bouterolles des épées que des sabres (fig. 26, a à d).

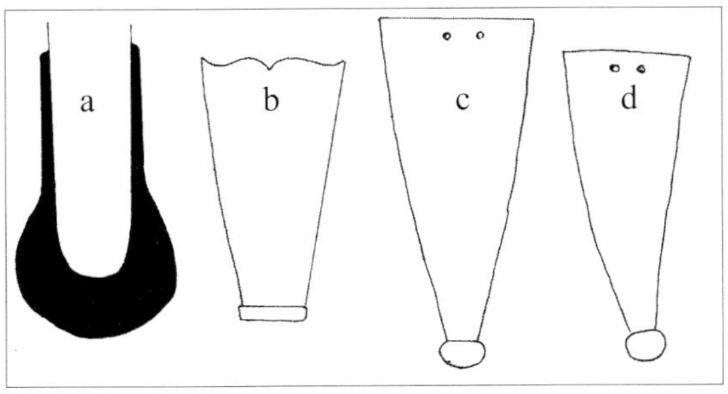

Figure 26 – Dards symétriques.

- **a**, dard en fer, symétrique, dit en lyre, pour les fourreaux des lames droites (lattes) de Cavalerie ;
- **b**, dard réduit à une plaquette de laiton d'un glaive de Marine de la fin du XVIIIème ;
- **c** et **d**, dards réduits à une petite sphère de laiton, ou bouton, sur la bouterolle en laiton de glaive, bouterolle symétrique ou légèrement infléchie.

Au cours des ans, ce bouton sera enjolivé et pourra devenir un véritable petit bulbe (fig. 27). Le dard plat et symétrique apparaît sur les sabres à lame droite de la Cavalerie du Consulat (fig. 26, a) et ordinairement qualifié de dard en lyre. La lyre perdurera sur ce type de sabres jusqu'à nos jours.

Figure 27 – Dard en bouton décoré.

- **a**, bouton ciselé de la bouterolle de laiton d'un briquet de la Garde Impériale ;
- **b**, bouton ciselé de la bouterolle de laiton d'un sabre d'Infanterie du modèle 1821 ;
- **c**, bouton constitué d'une demi-lentille de fer sur la bouterolle en laiton d'un sabre d'officier de la Maréchaussée du modèle 1787 ;
- **d**, bouton en demi-sphère, à gouge, ciselé, à l'extrémité d'un fourreau de tôle de Chasseur à cheval de la Restauration.

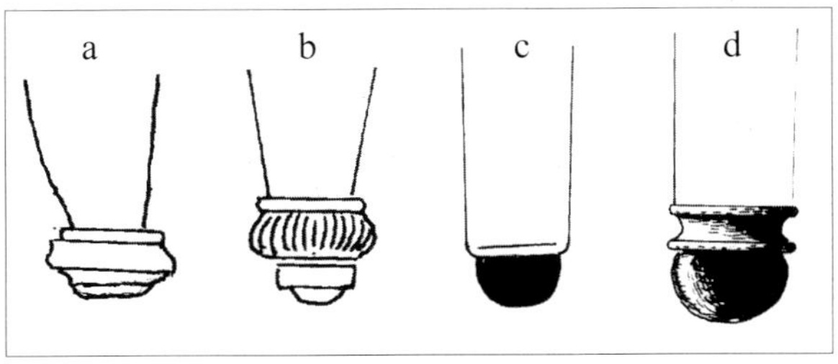

Dards asymétriques

Il semble que les dards les plus anciens des sabres aient été constitués d'une pièce de fer ou de laiton ne protégeant que le côté antérieur de la bouterolle et le plus souvent, sa base. Cette pièce est simple ou découpée uniquement par souci esthétique (fig. 28).

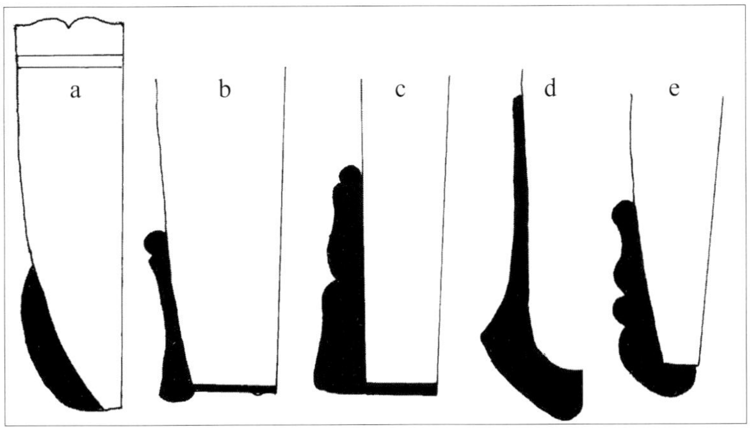

Figure 28 – Les dards frontaux de la bouterolle.

- **a**, dard en laiton, demi-lune, d'un sabre à la Montmorency de Chasseurs ;
- **b**, dard frontal en laiton prolongé à la base de la bouterolle par une plaque vissée d'un sabre de Volontaire National, vers 1791-1792 ;
- **c**, dard en fer découpé sur la partie frontale se prolongeant sous la base de la bouterolle en laiton d'un sabre à garde de bataille de Dragons, vers 1780 ;
- **d**, dard en fer découpé du même type que le précédent, mais plus simple, d'un sabre à fourreau de laiton, fin du Consulat ;
- **e**, dard en fer d'une bouterolle de laiton avec renforcement apical d'un sabre d'Infanterie du Premier Empire.

(*) dard : vient du latin *dardus*, arme de jet garnie de barbelures en forme de dard. Par analogie, la pièce protégeant la bouterolle dépassant celle-ci, peut être assimilée à une barbelure, donc à un dard (simple hypothèse).

D'autres modèles de dards enveloppent toute l'extrémité de la bouterolle, mais toujours la protection latéro-antérieure et basale est plus importante que la protection latéro-postérieure (fig. 29).

Figure 29 – Les dards frontaux de la bouterolle.

- **a**, dard en fer presque symétrique sur la bouterolle en laiton d'un sabre d'officier de Marine du Directoire ;
- **b**, dard en fer aux extrémités découpées sur la bouterolle en laiton d'un sabre de Cavalerie légère de la Révolution ;
- **c**, grand et fort dard en fer sans enjolivure d'un fourreau de tôle de Cavalerie légère de l'An IX ;
- **d**, dard en fer très renforcé d'un sabre de Cavalerie légère de la Révolution ;
- **e**, grand dard en laiton enjolivé à ses extrémités, d'un fourreau de laiton d'un sabre de Cavalerie légère de la Révolution.

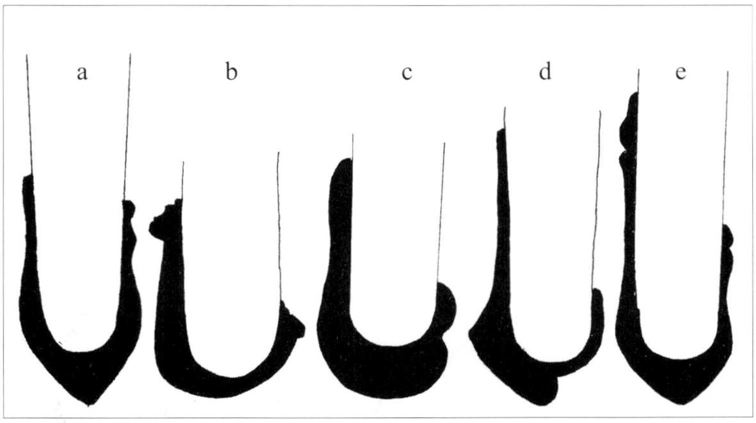

Les attelles extérieures

Les *attelles* sont des lames de laiton façonnées en gouttières et qui sont maintenues en place par introduction partielle sous les garnitures du fourreau de cuir. Elles ont pour mission de renforcer la rigidité du fourreau et de le protéger des pliures dûes aux accidents de selles.

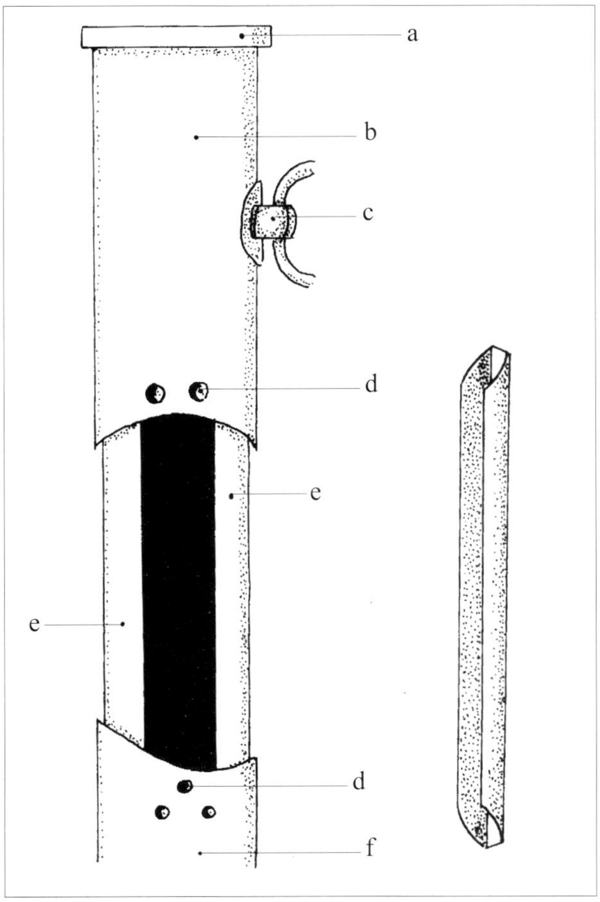

Figure 30 – Les attelles.

- **a**, cuvette ;
- **b**, chape ;
- **c**, anneau de bélière ;
- **d**, poinçonnage de fixation (ou par agrafage) ;
- **e**, attelles ;
- **f**, garniture inférieure.

A gauche :
attelles (e) en place, passant sous la chape (b) et sous la garniture inférieure (f).

A droite :
attelle isolée, en forme de gouttière, on remarque que l'attelle elle-même est plus longue qu'il n'apparaît lorsqu'elle est en place dans le fourreau (fig. gauche) afin d'assurer un bon maintien.

Maintenant, les pièces composant les sabres ayant été passées en revue et décrites avec précision, surtout pour les jeunes collectionneurs, on peut passer au choix des critères choisis pour établir la classification proposée.

Présentation illustrée des sabres

Nous avons donné, à propos de la description succincte des branches de garde, un premier aperçu de la classification que nous avons adoptée. Il va de soi, que dans chaque partie proposée, nous affinons cette classification pour faciliter l'identification des sabres dans un esprit très voisin de celui que nous avons mis en œuvre dans *" Les épées portées en France "*, à partir du Chapitre IV. En effet, à l'intérieur de ces *Parties*, en fonction de la nature du métal employé, de la disposition des branches secondaires, de la présence ou l'absence d'oreillons, de la longueur de la queue de calotte, de la forme de la calotte elle-même, de la présence ou l'absence de coquille… les sabres sont regroupés en *Chapitres*, en *Familles*, en *Sous-familles,* de la façon suivante, afin de faciliter les recherches du lecteur.

Première partie

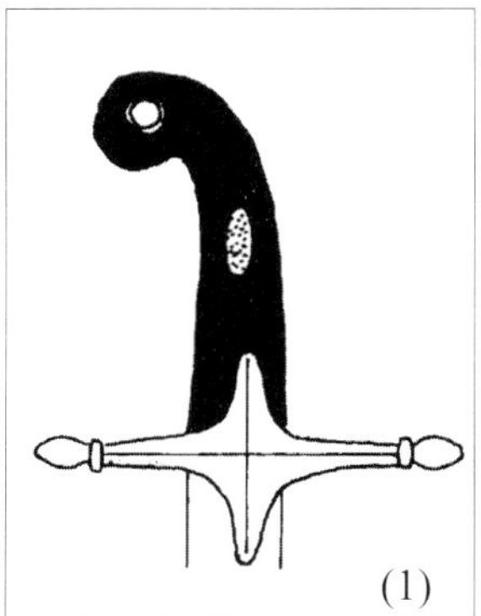
(1)

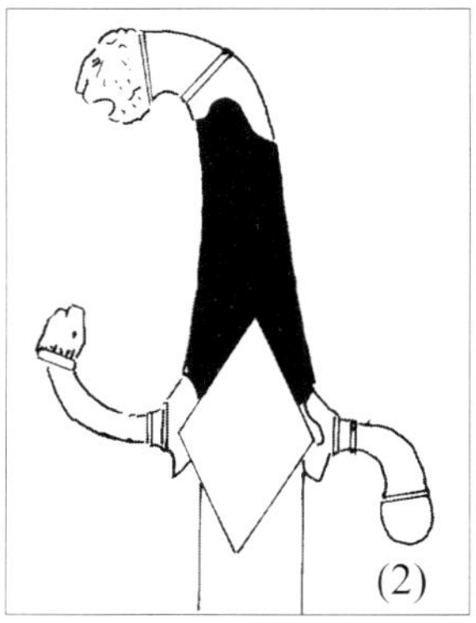
(2)

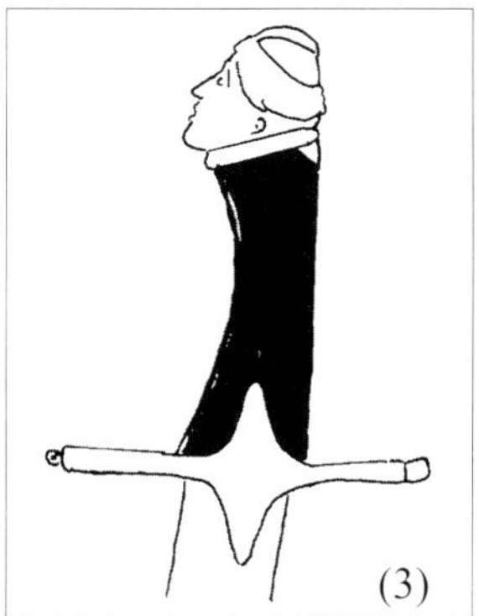
(3)

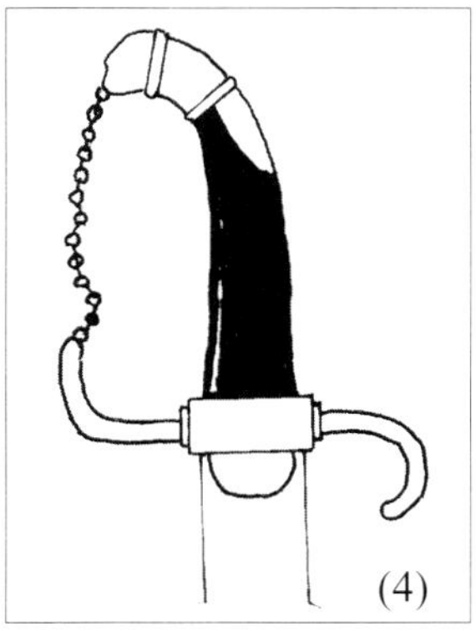
(4)

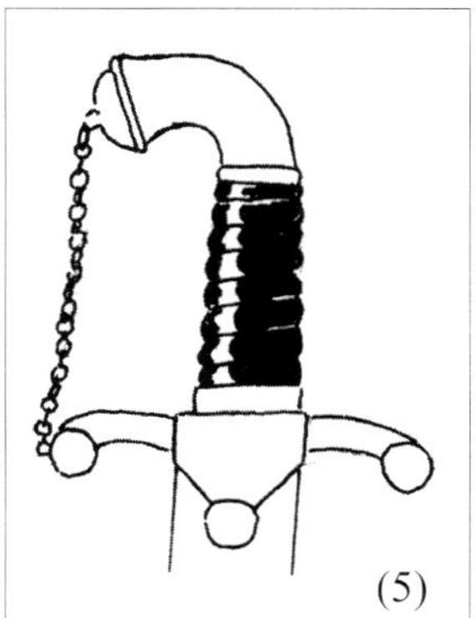
(5)

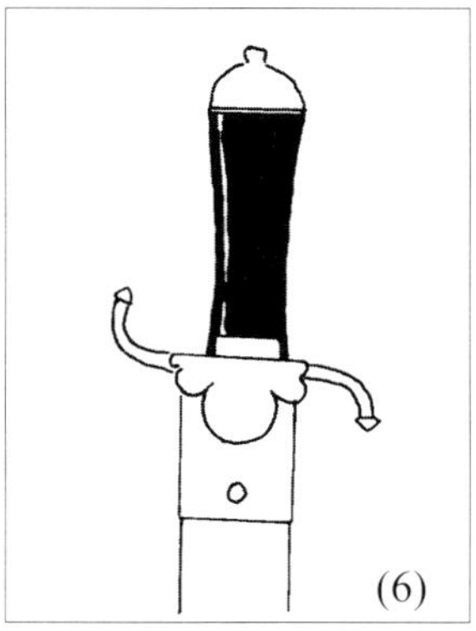
(6)

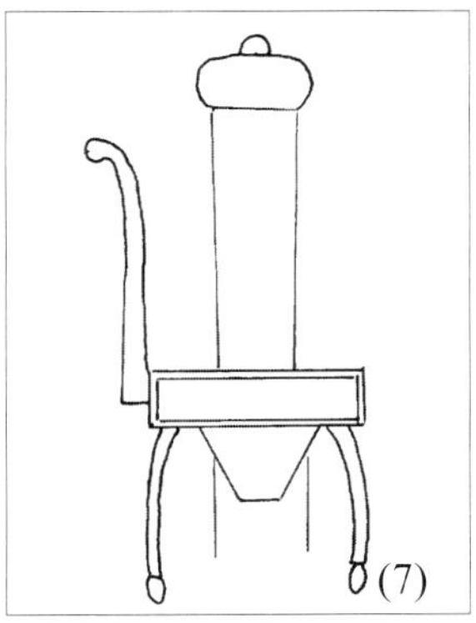
(7)

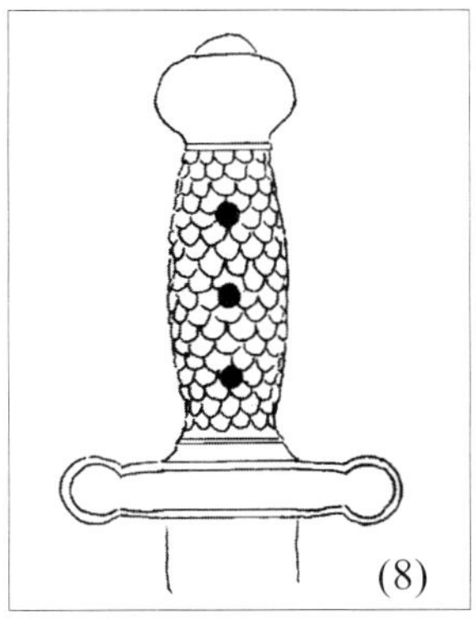
(8)

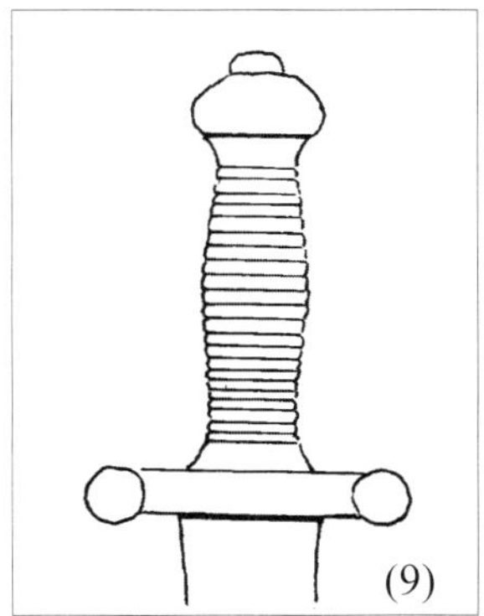
(9)

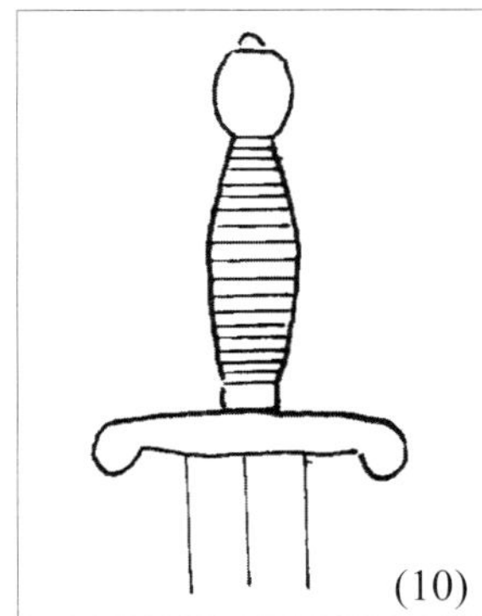
(10)

(11)

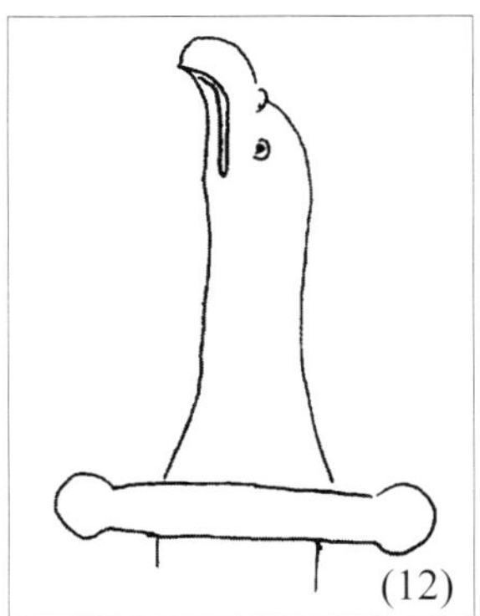
(12)

(13)

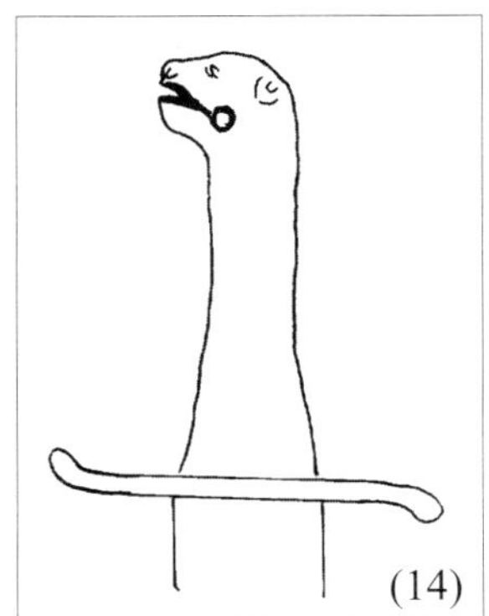
(14)

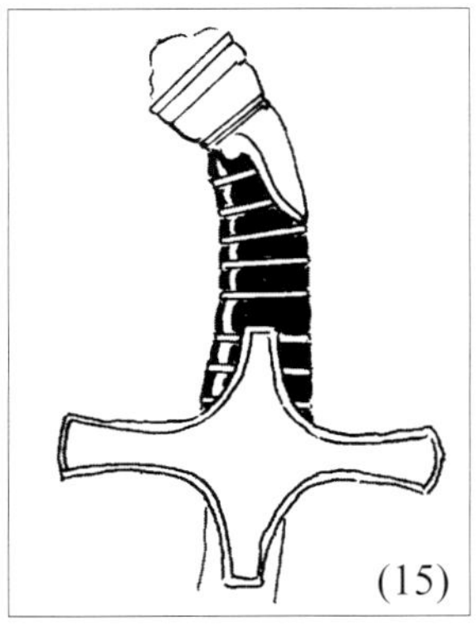
(15)

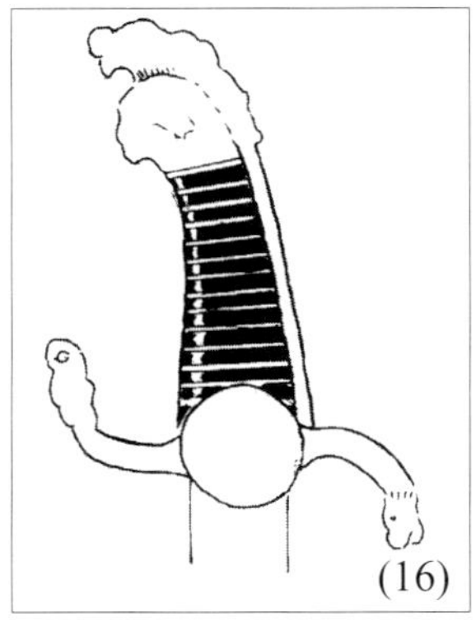
(16)

Seconde partie

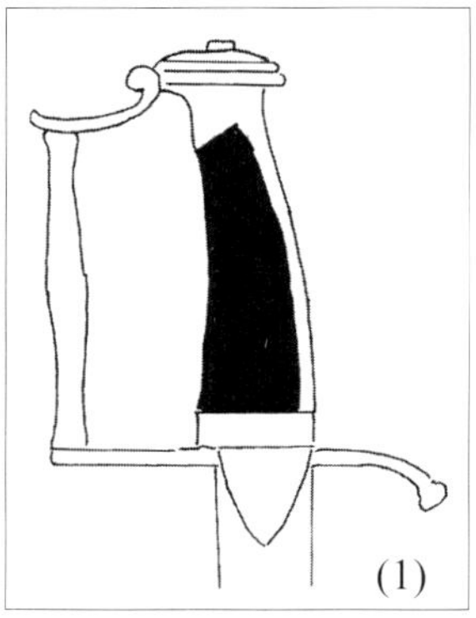
(1)

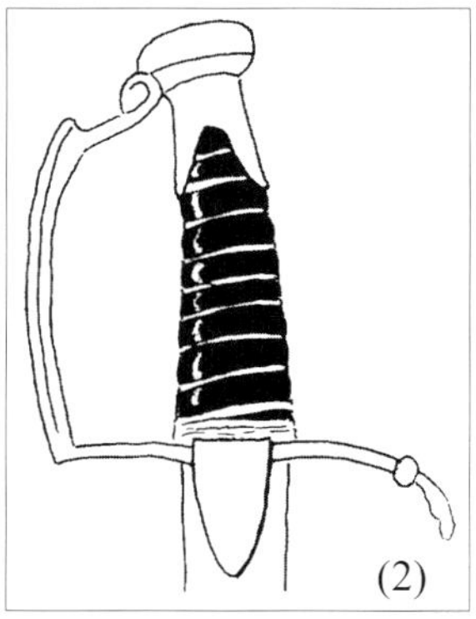
(2)

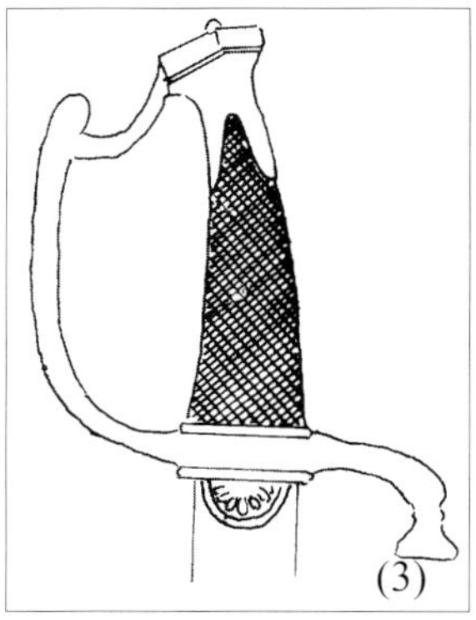
(3)

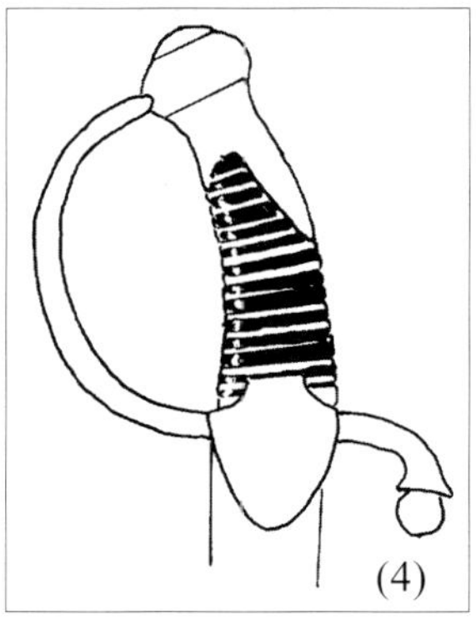
(4)

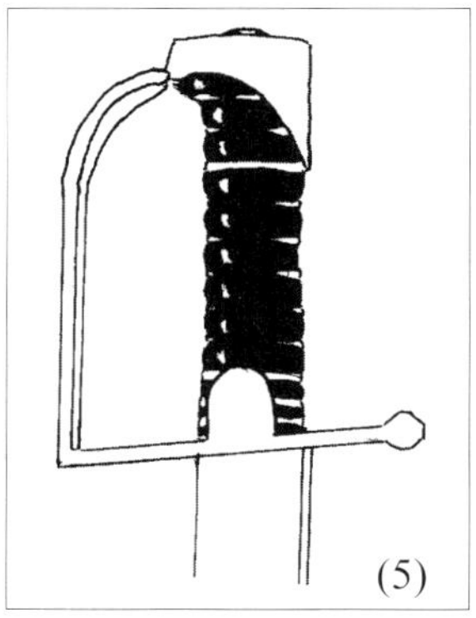
(5)

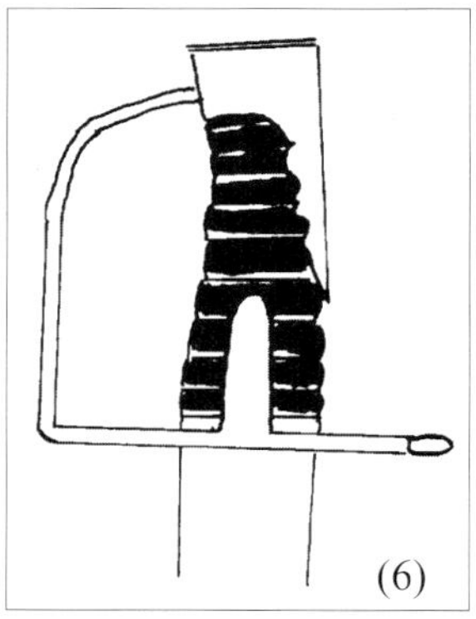
(6)

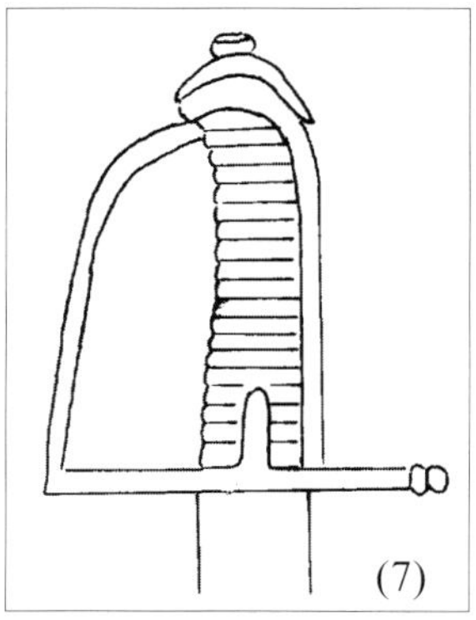
(7)

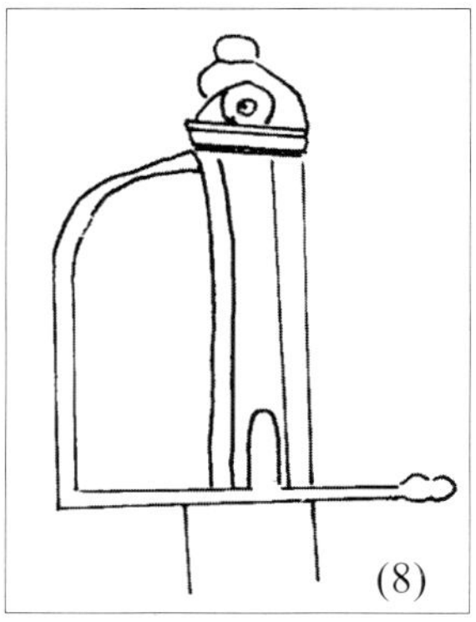
(8)

(1) On trouvera également quelques montures à oreillons aux chapitres suivants :
Quatrième Partie, Chapitre I, Famille III, p. 271.
Cinquième Partie, Chapitre II, Famille III, p. 305.

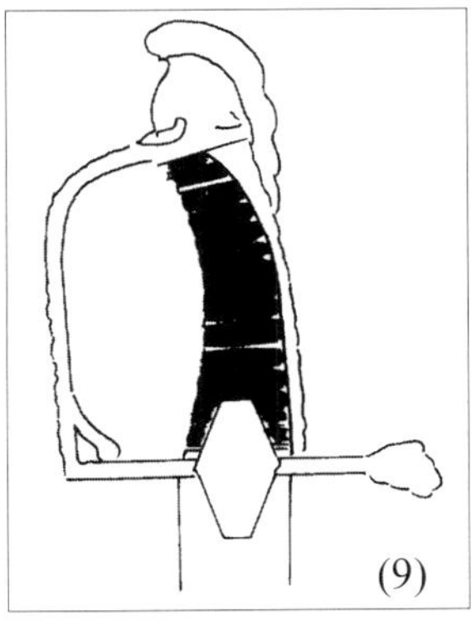
(9)

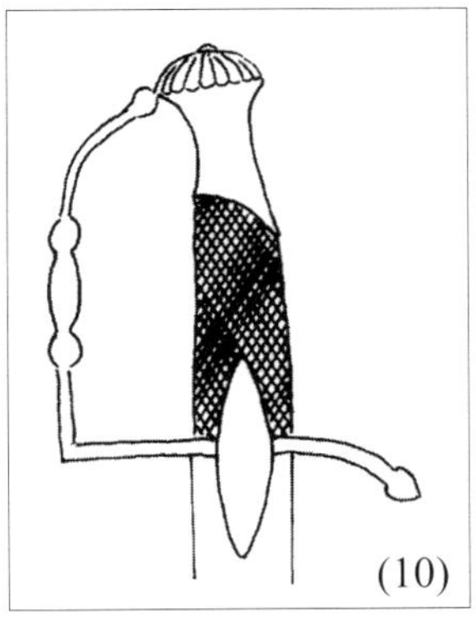
(10)

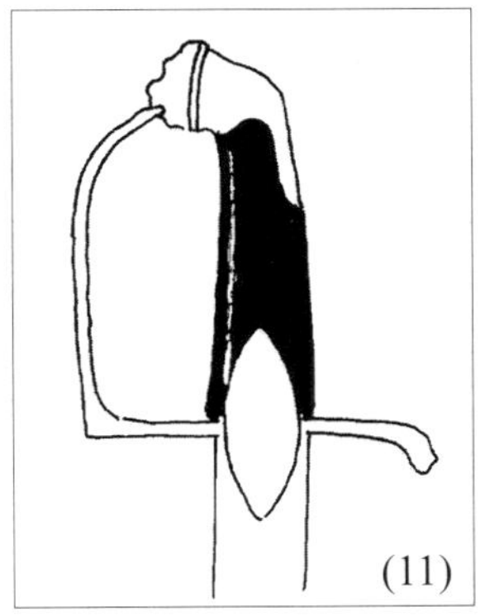
(11)

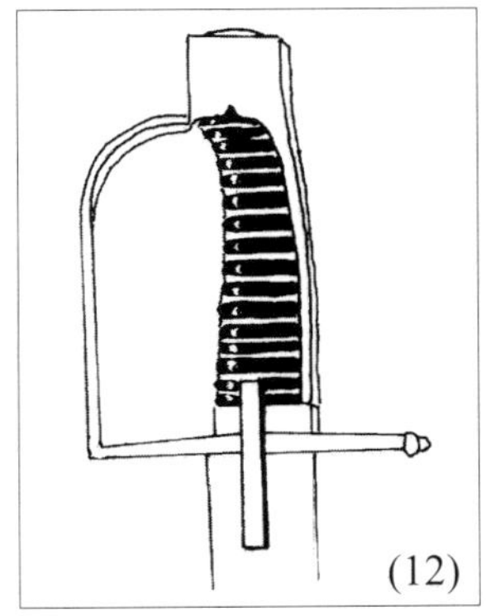
(12)

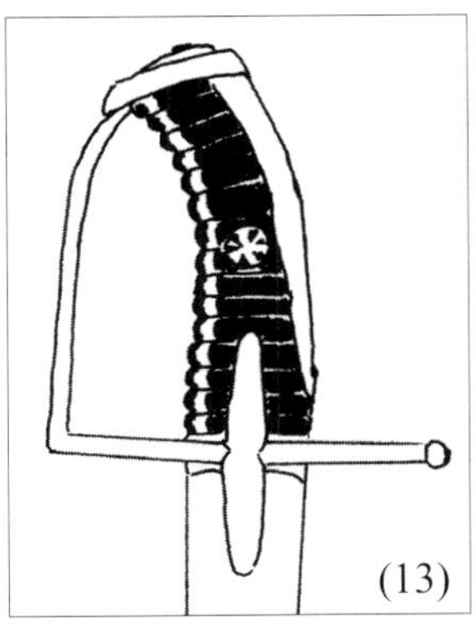
(13)

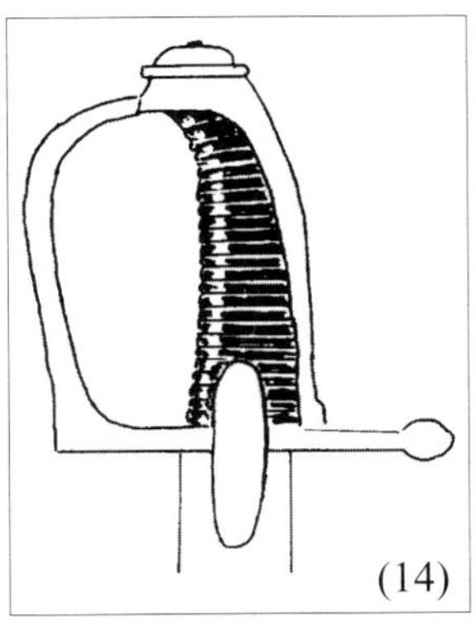
(14)

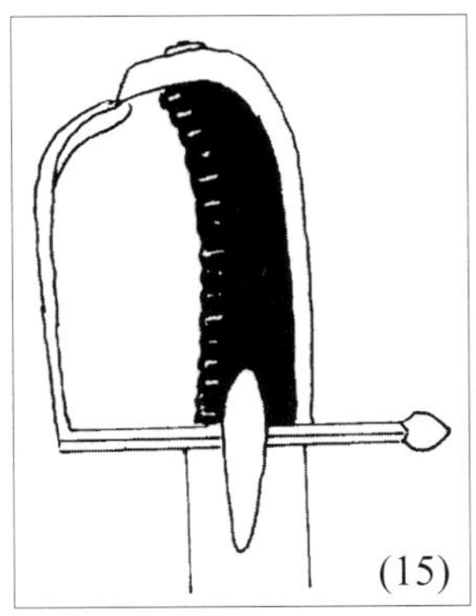
(15)

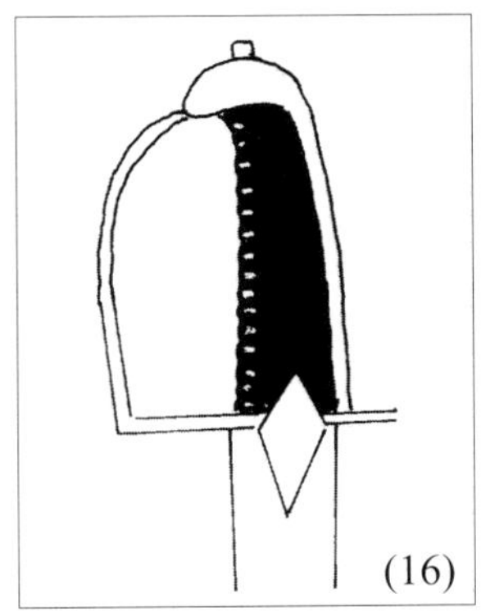
(16)

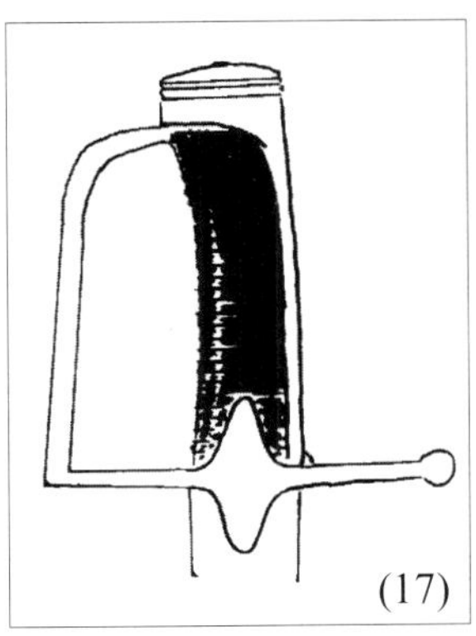
(17)

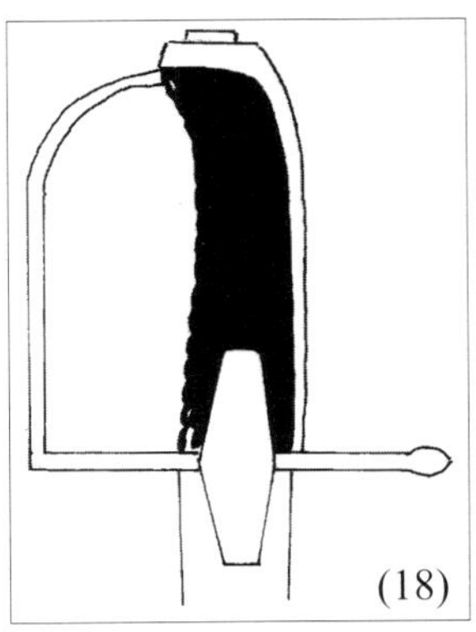
(18)

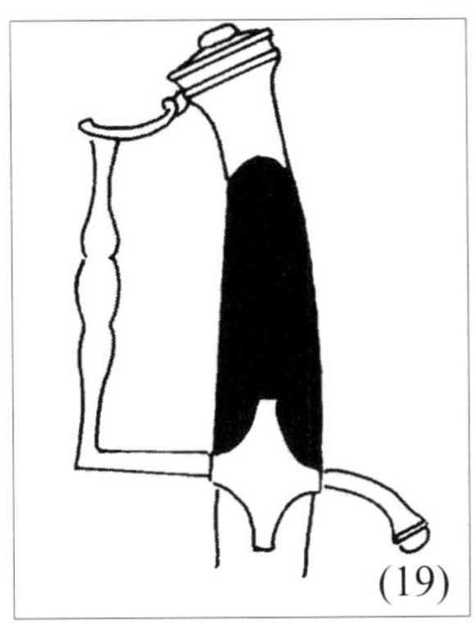
(19)

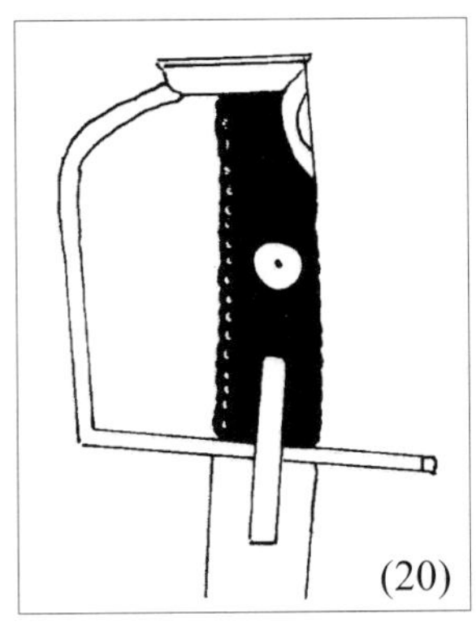
(20)

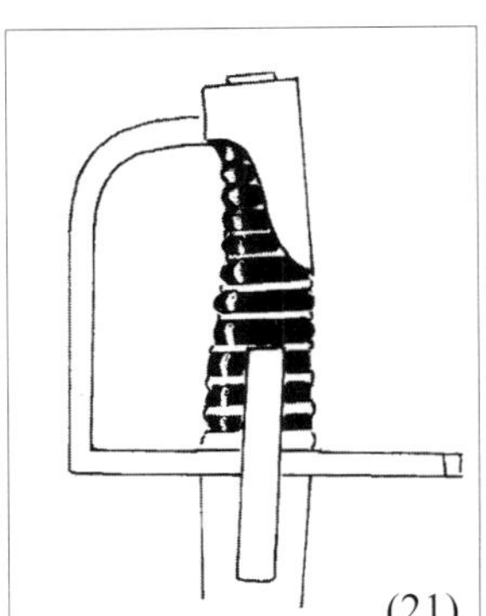
(21)

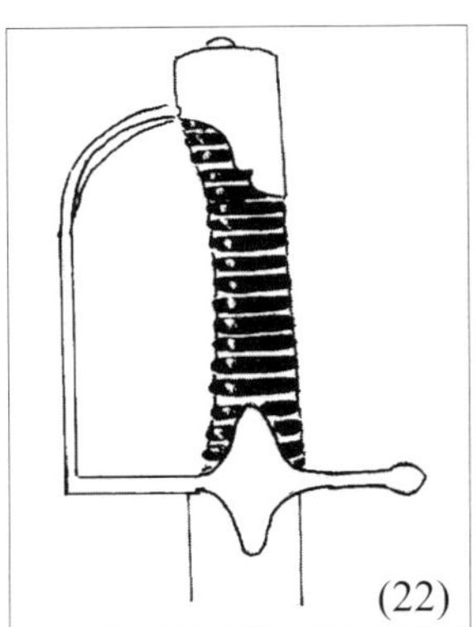
(22)

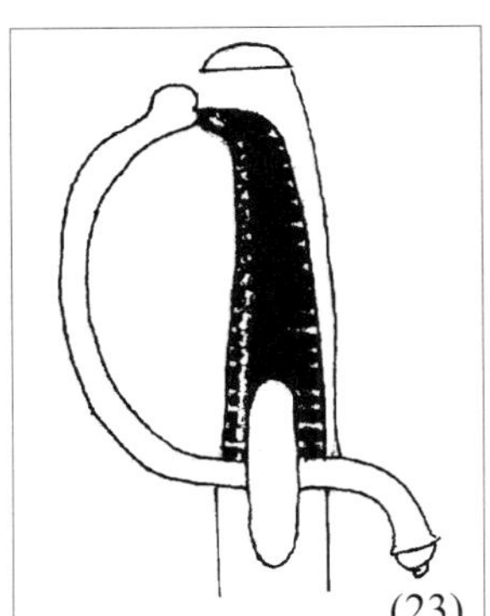
(23)

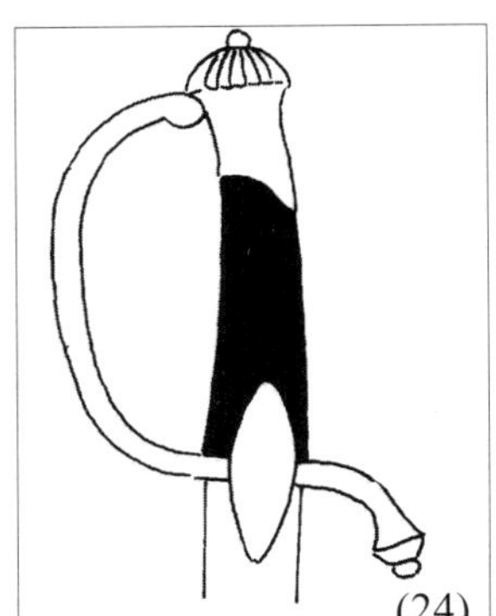
(24)

Troisième partie

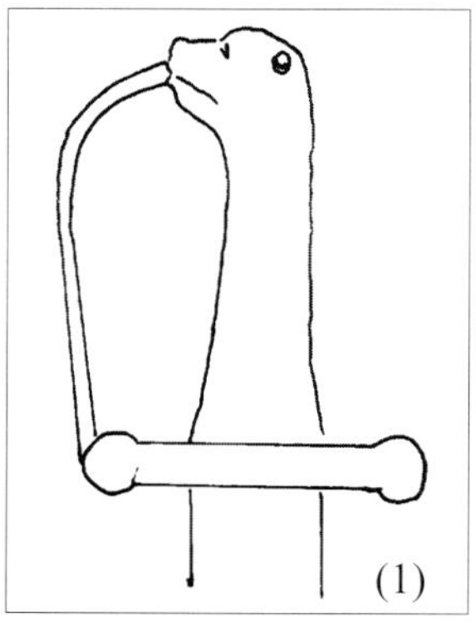
(1)

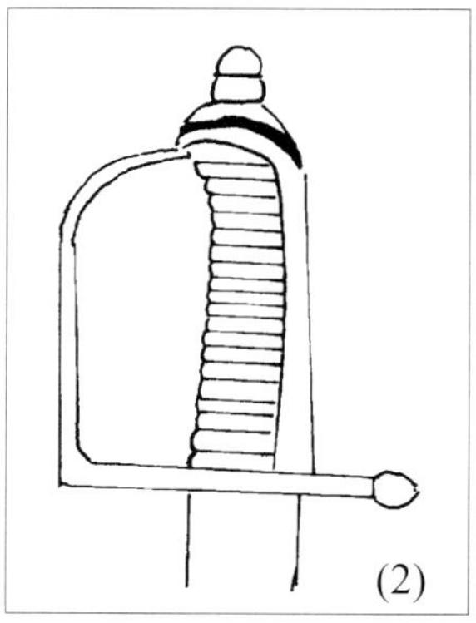
(2)

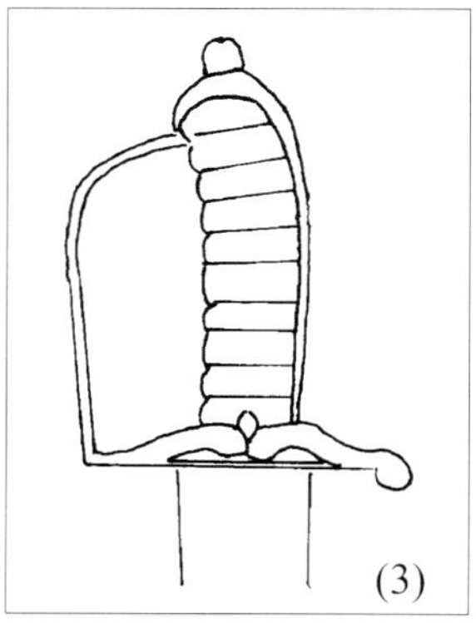
(3)

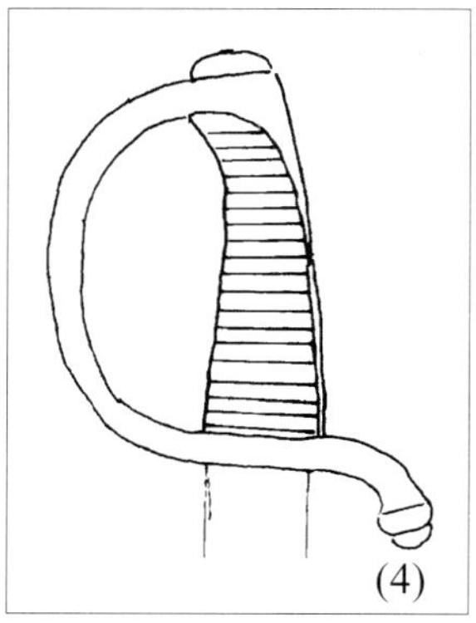
(4)

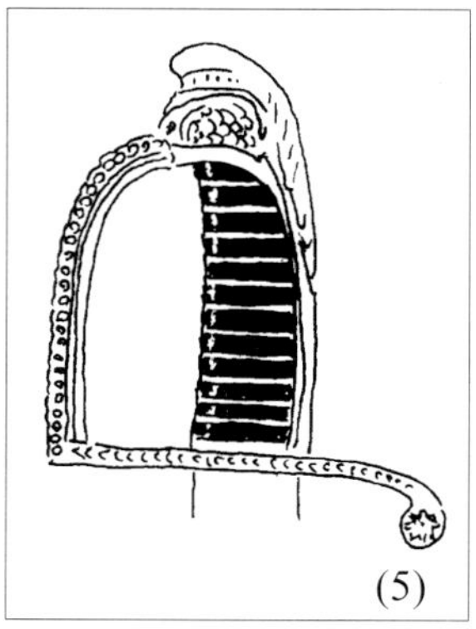
(5)

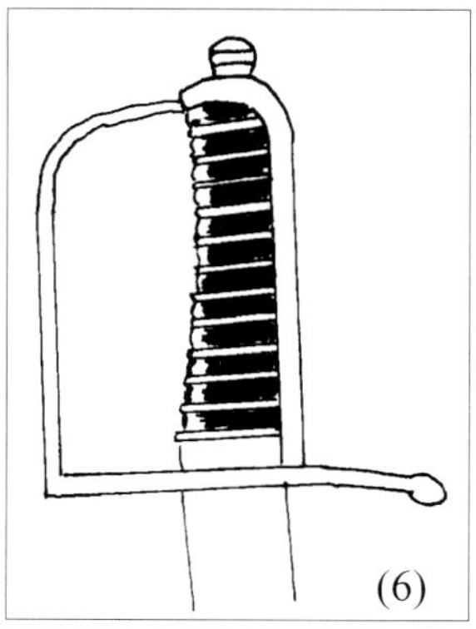
(6)

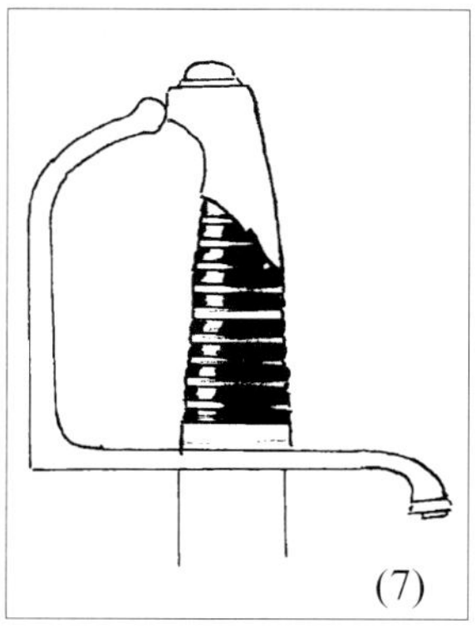
(7)

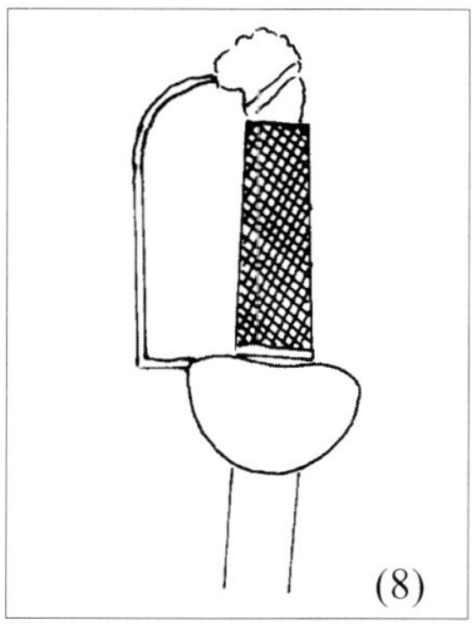
(8)

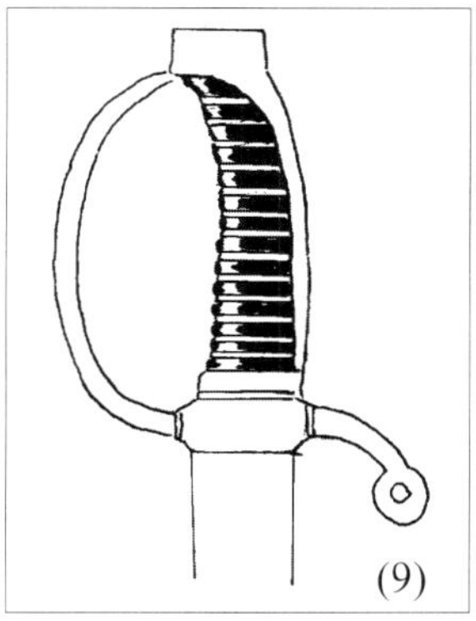
(9)

(10)
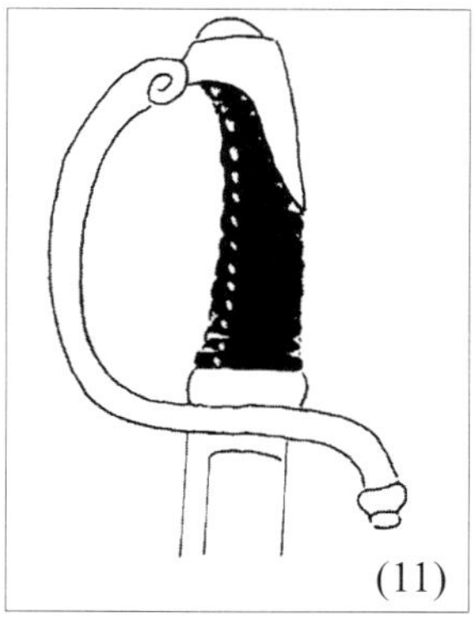
(11)
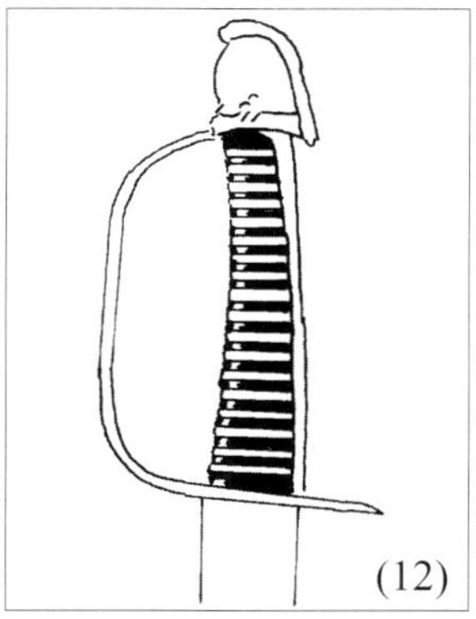
(12)
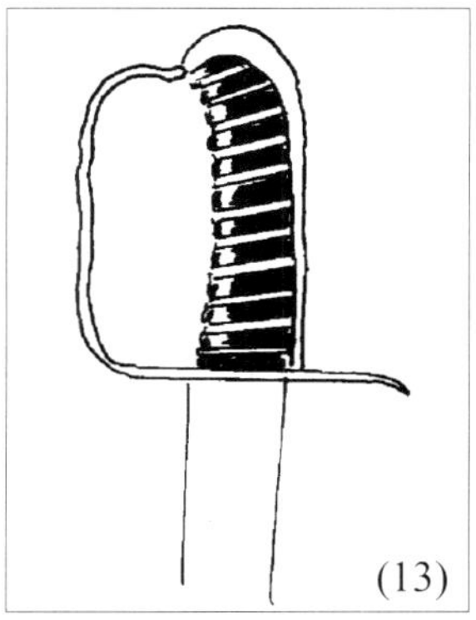
(13)
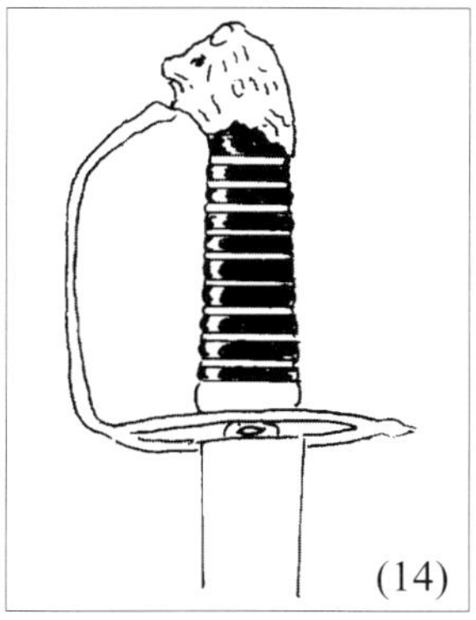
(14)
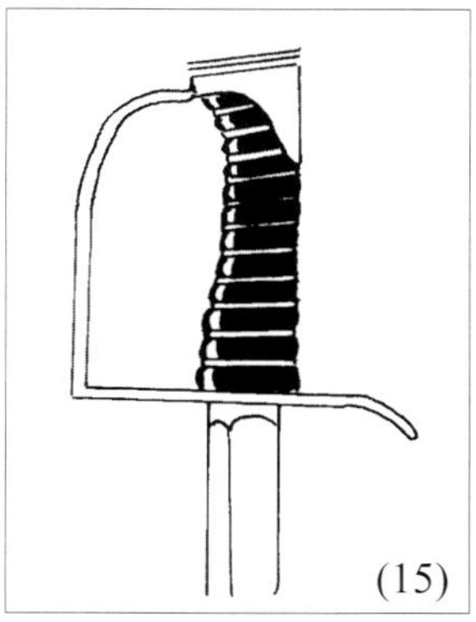
(15)
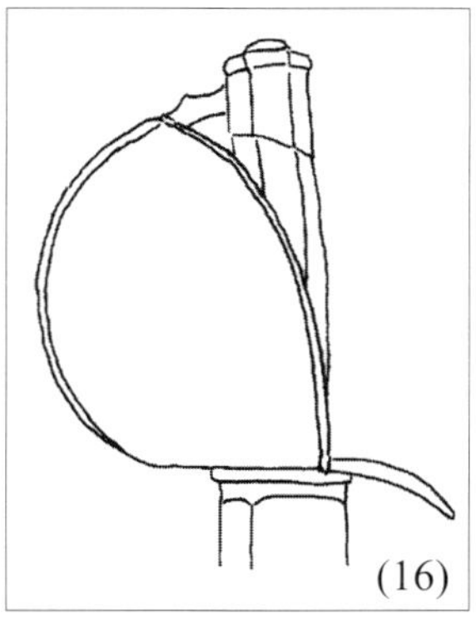
(16)

Quatrième partie

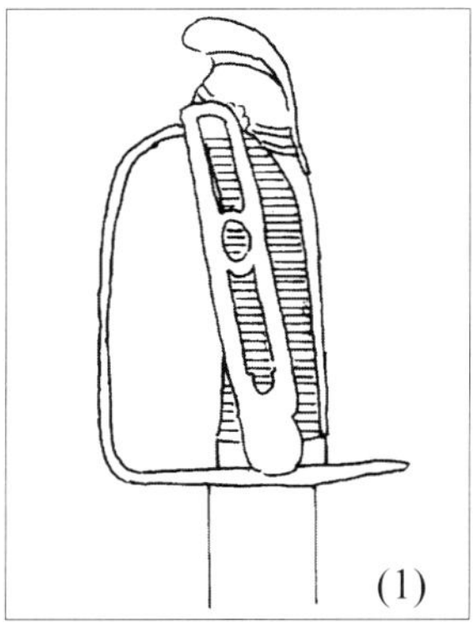
(1)
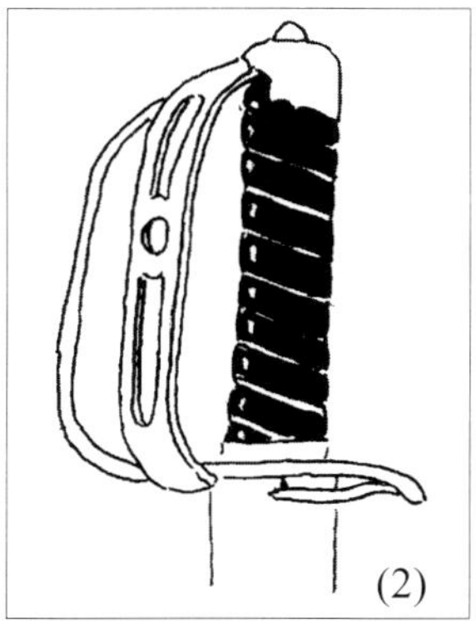
(2)
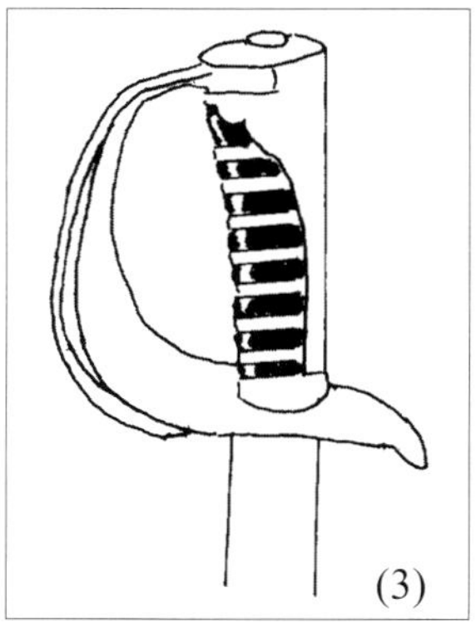
(3)
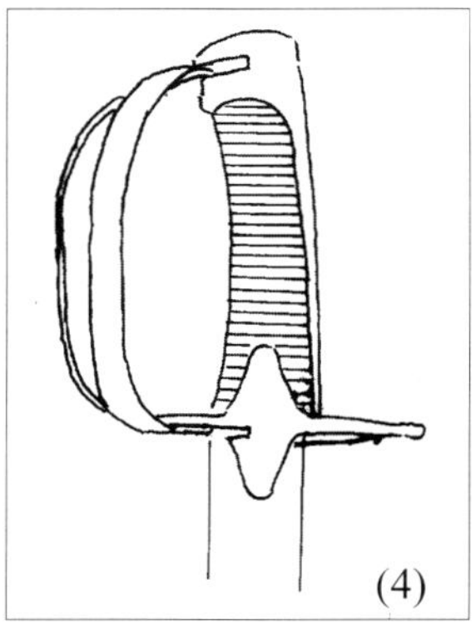
(4)

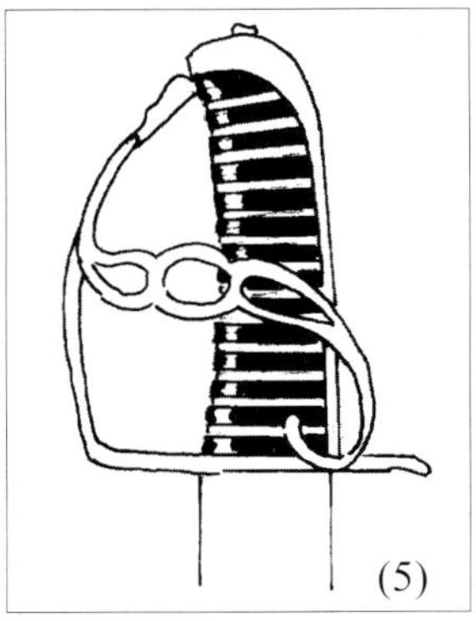
(5)

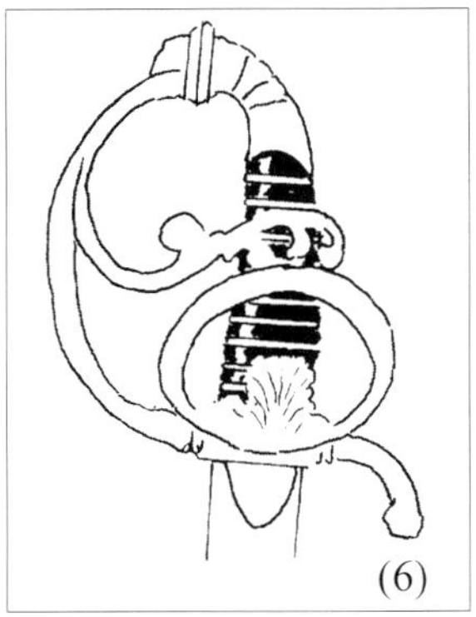
(6)

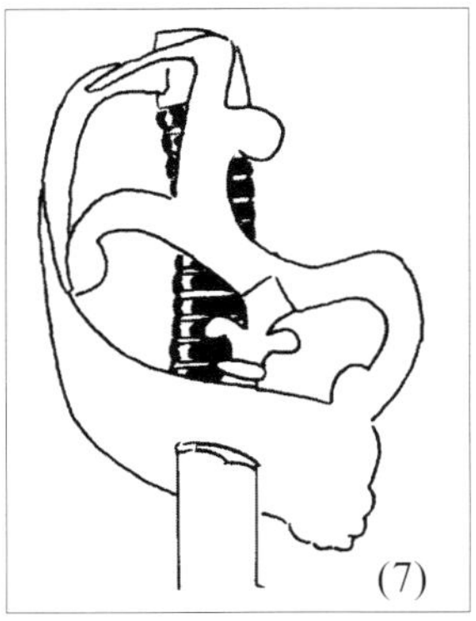
(7)

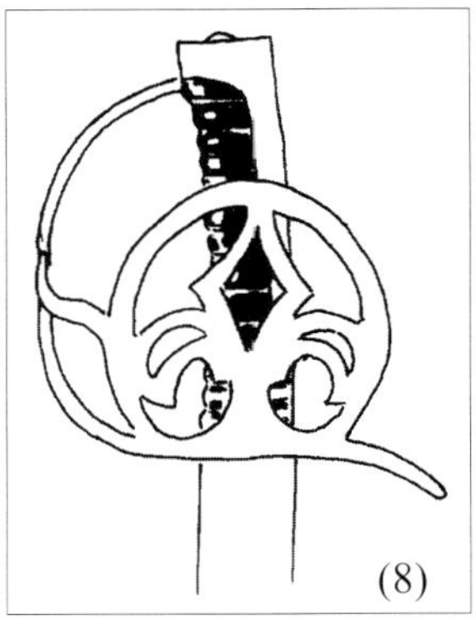
(8)

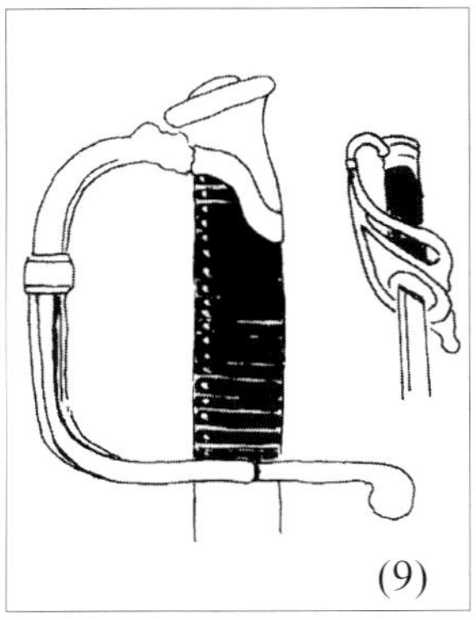
(9)

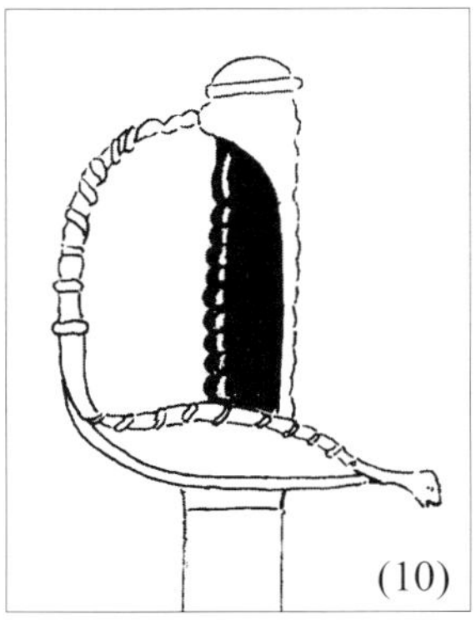
(10)

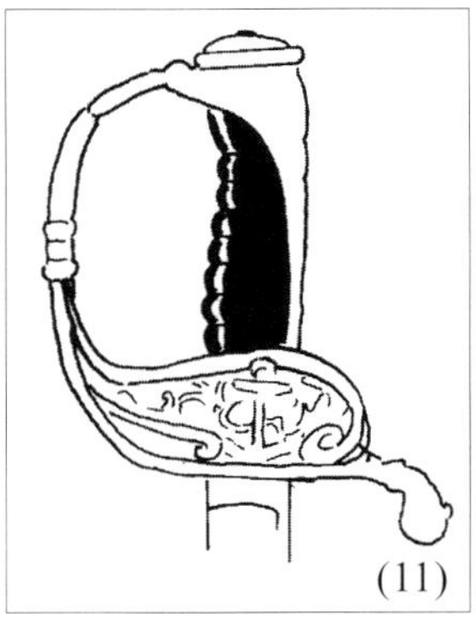
(11)

Cinquième partie

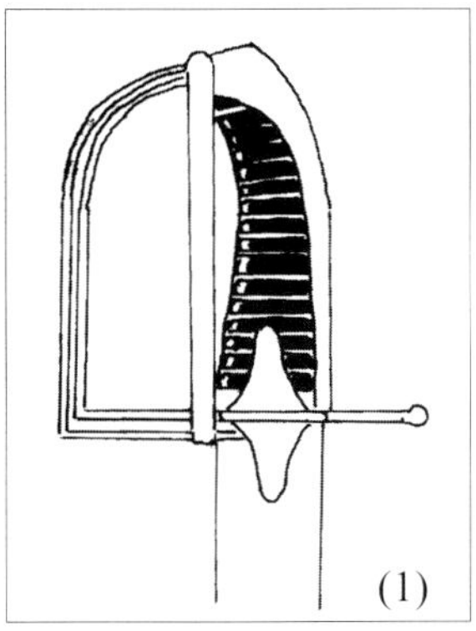
(1)

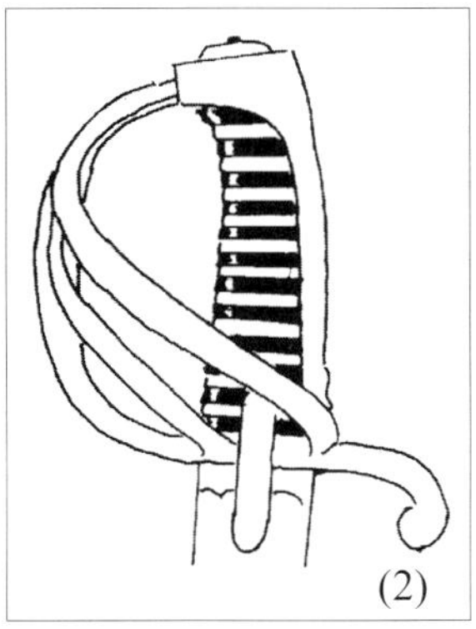
(2)

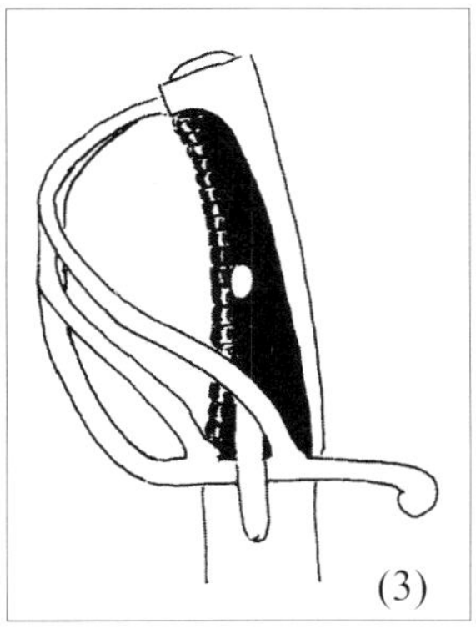
(3)

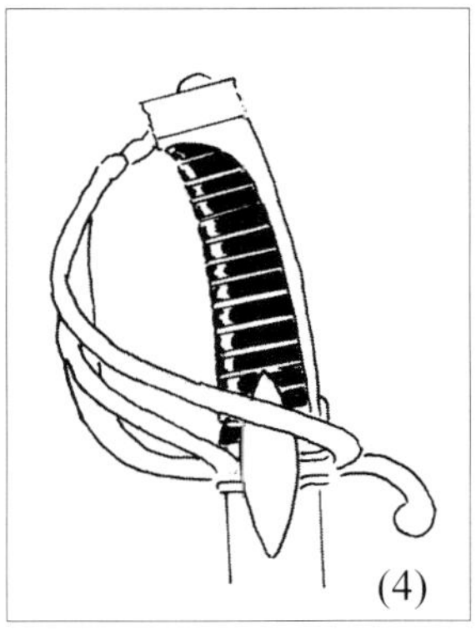
(4)

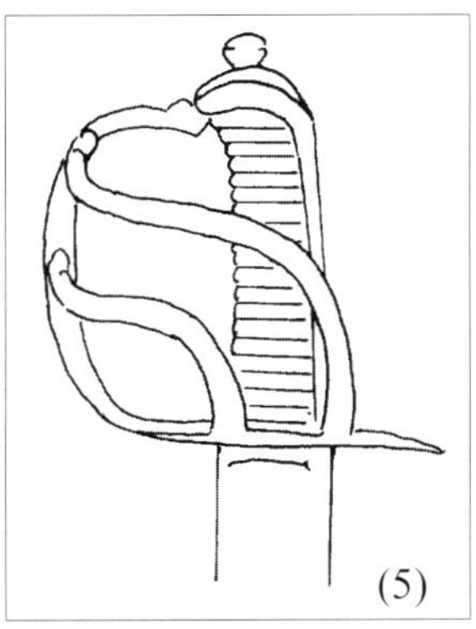
(5)

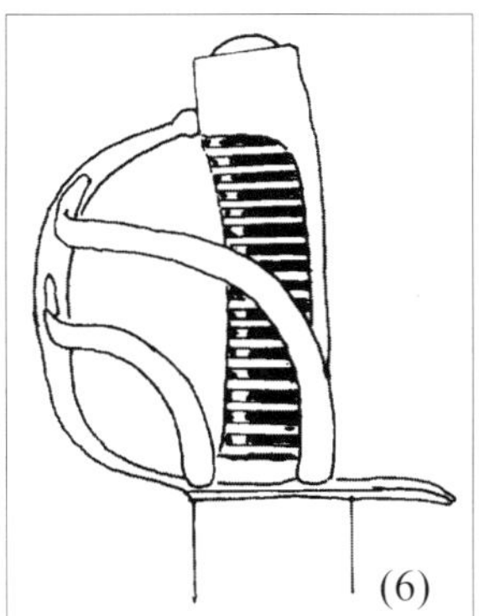
(6)

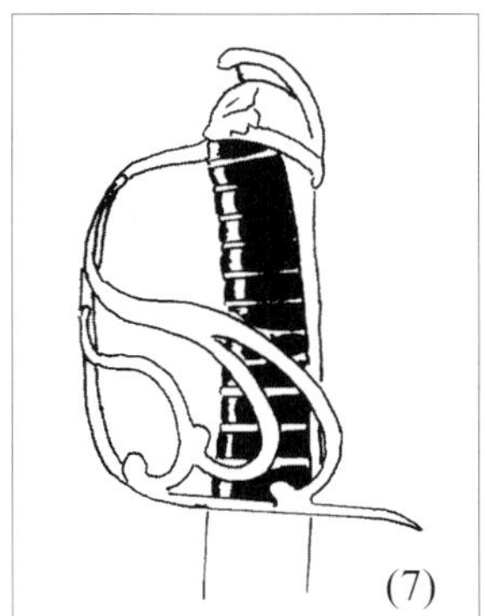
(7)

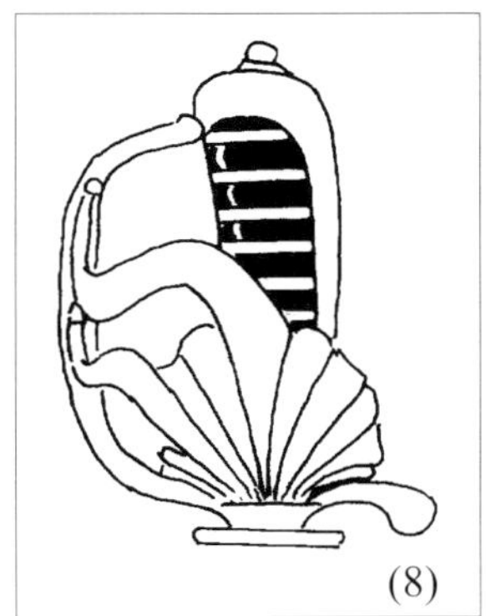
(8)

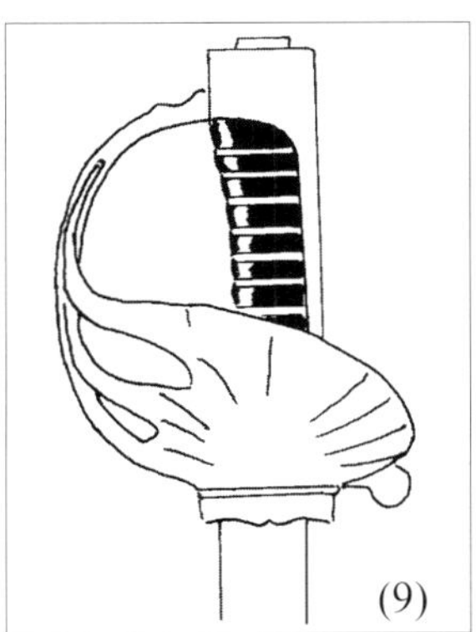
(9)

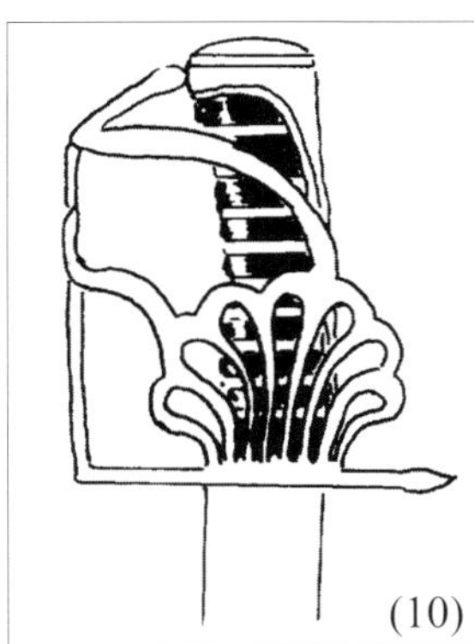
(10)

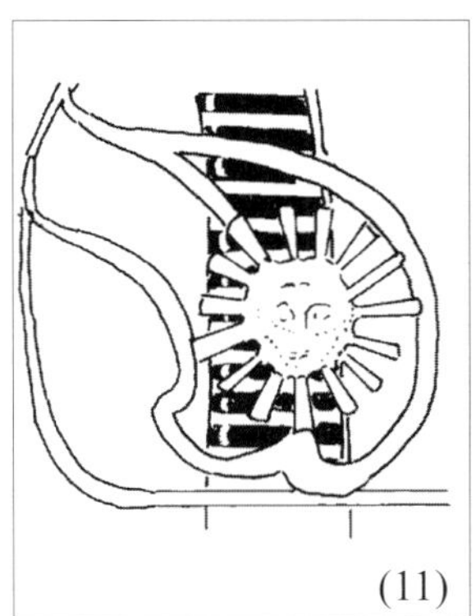
(11)

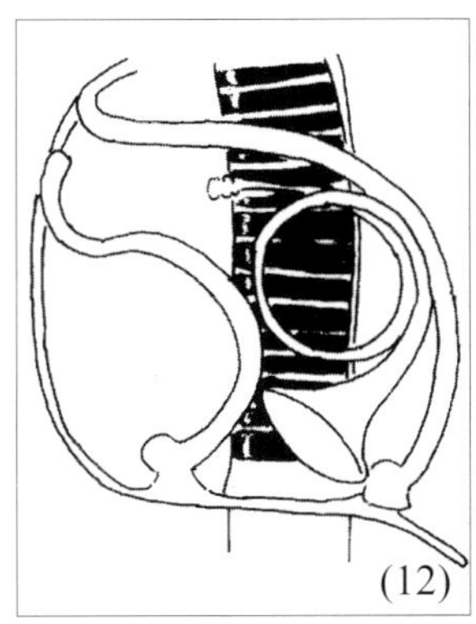
(12)

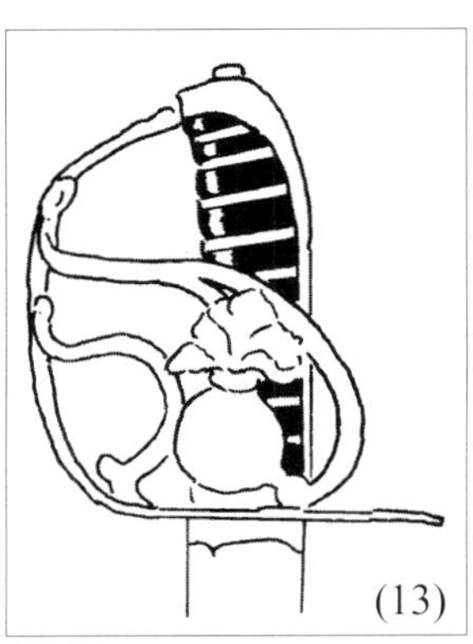
(13)

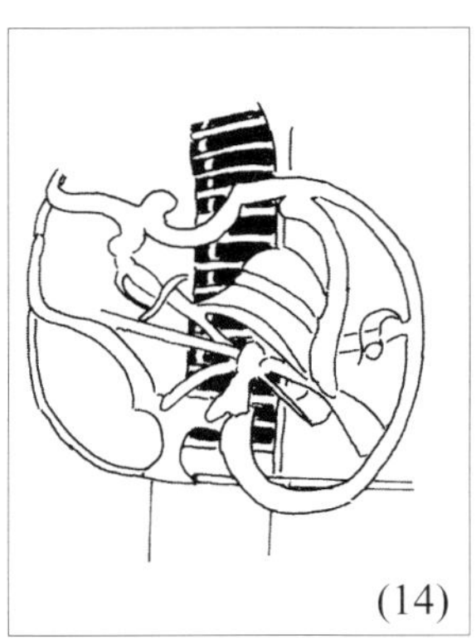
(14)

(15)

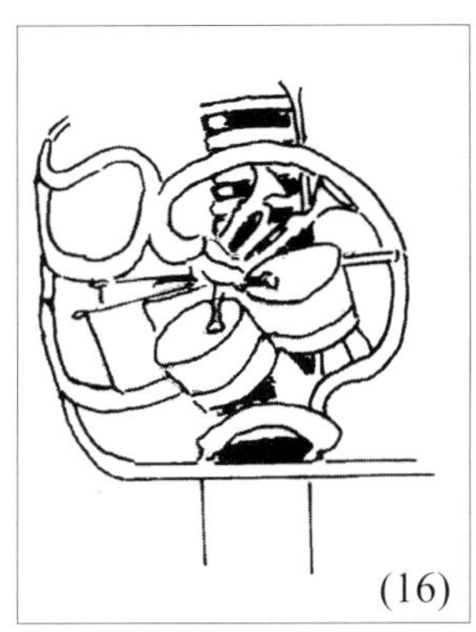
(16)

(17)

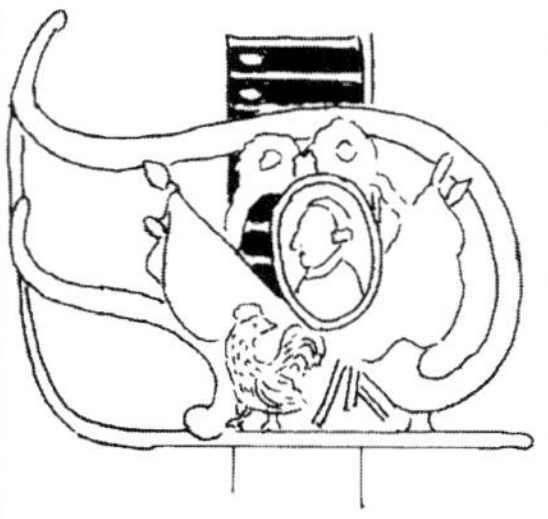
(18)

(19)

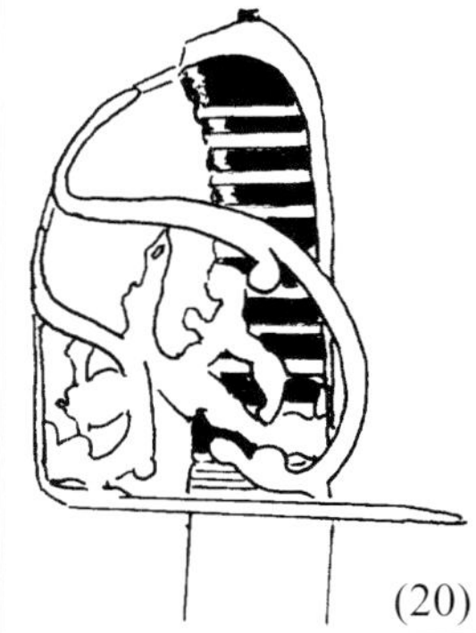
(20)

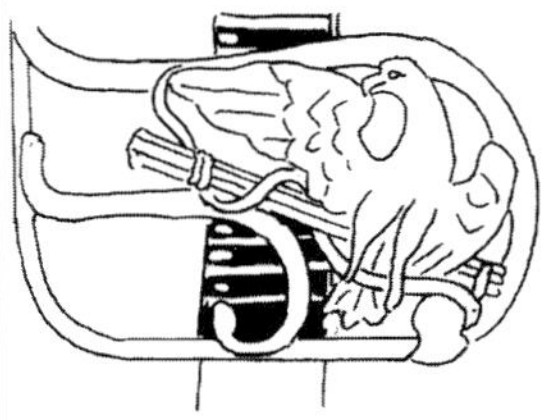
(21)

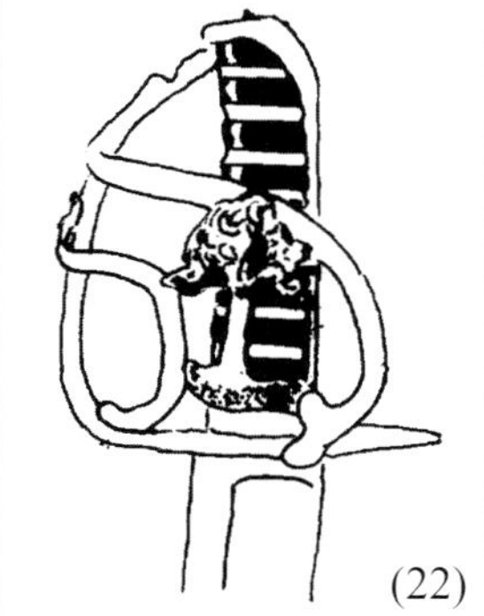
(22)

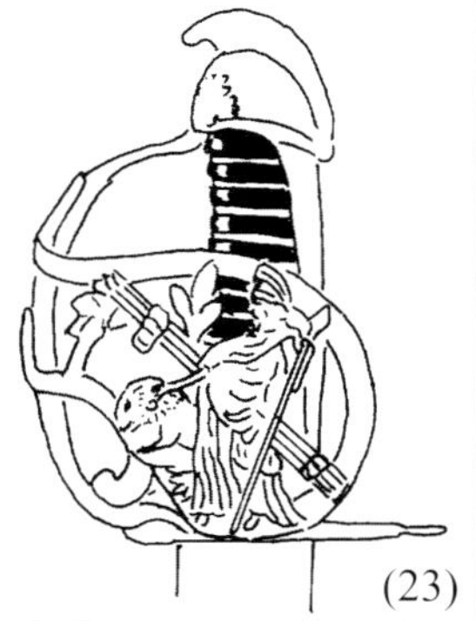
(23)

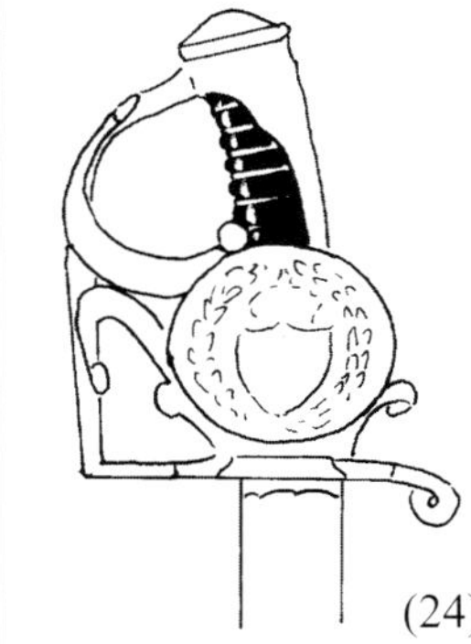
(24)

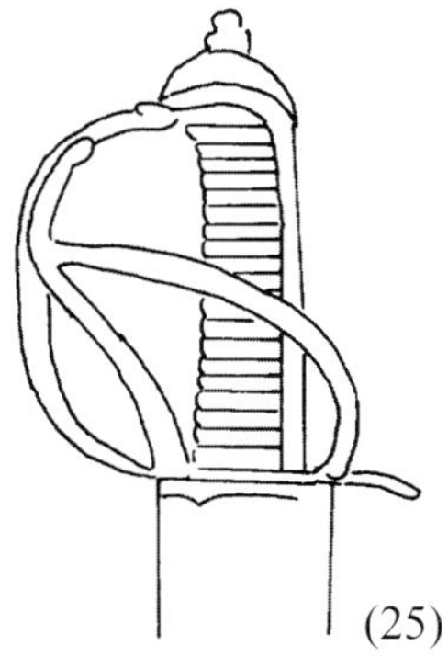
(25)

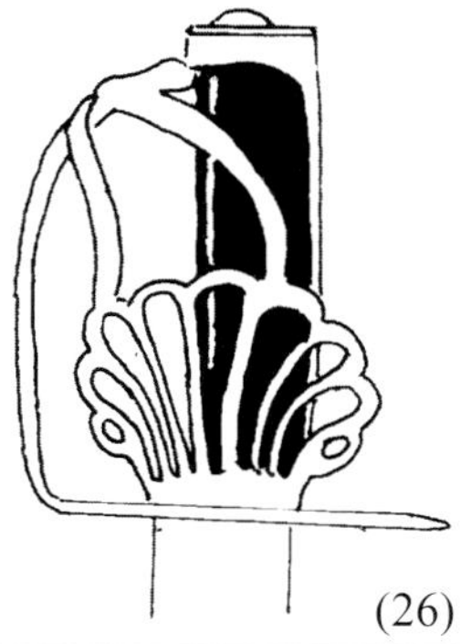
(26)

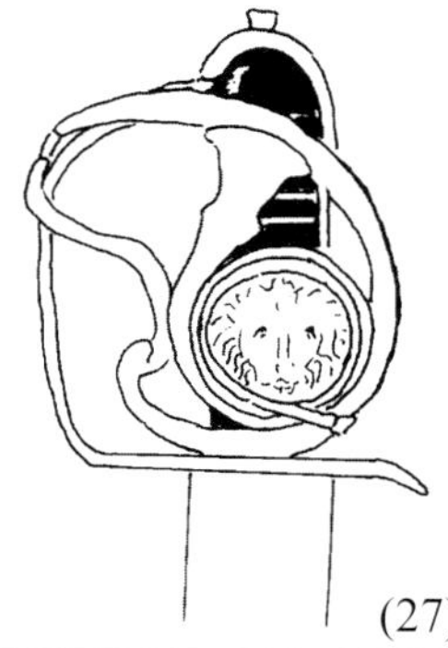
(27)

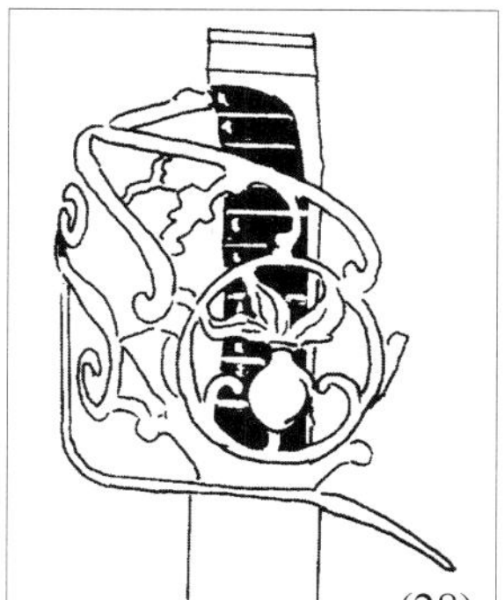
(28)

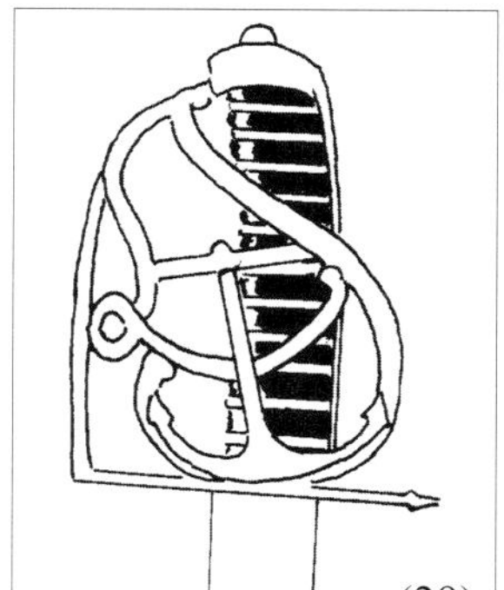
(29)

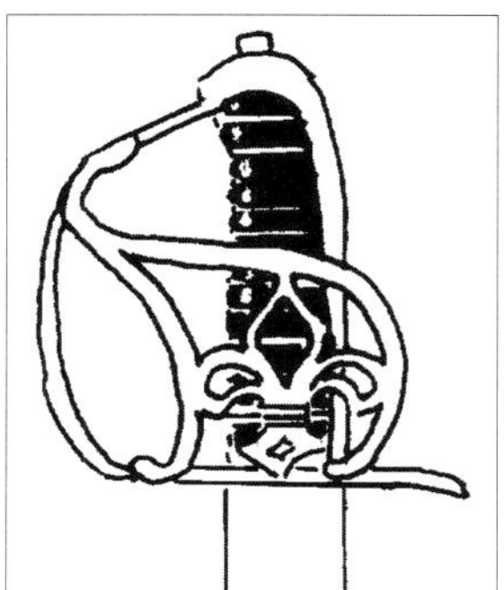
(30)

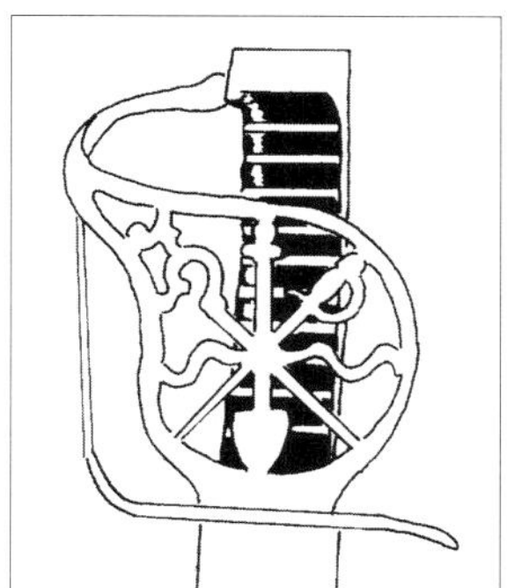
(31)

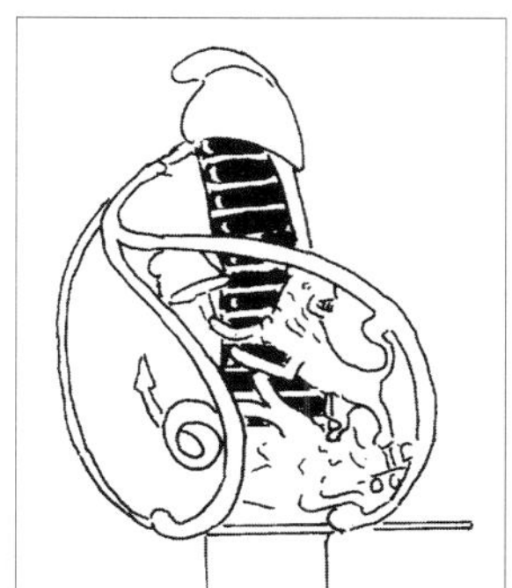
(32)

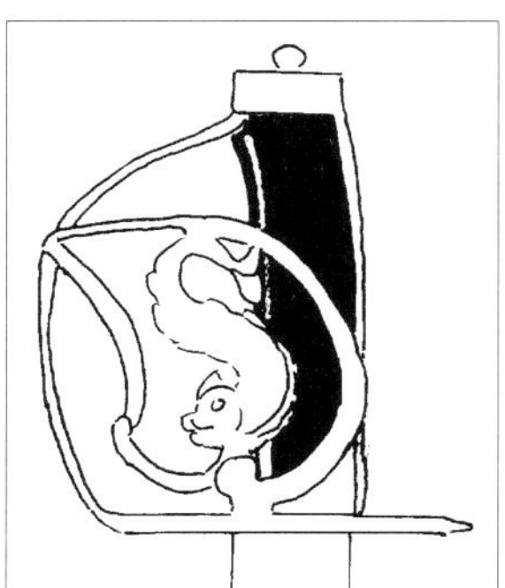
(33)

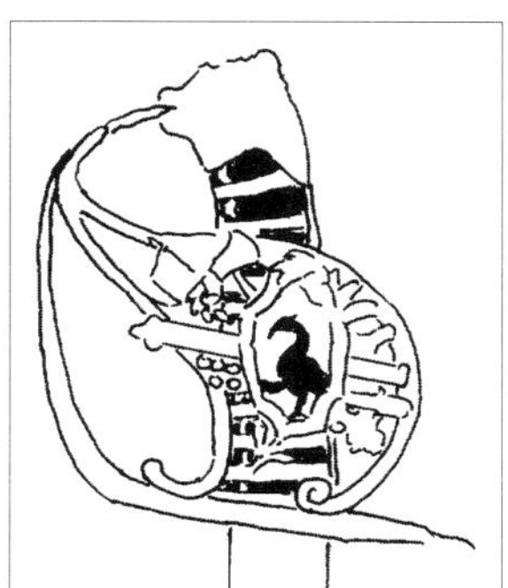
(34)

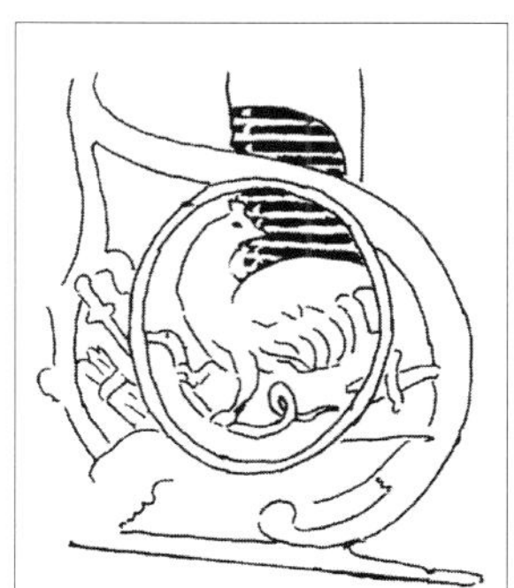
(35)

(36)

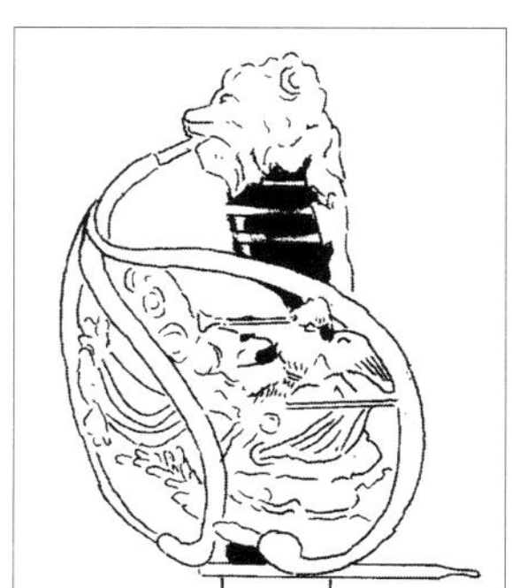
(37)

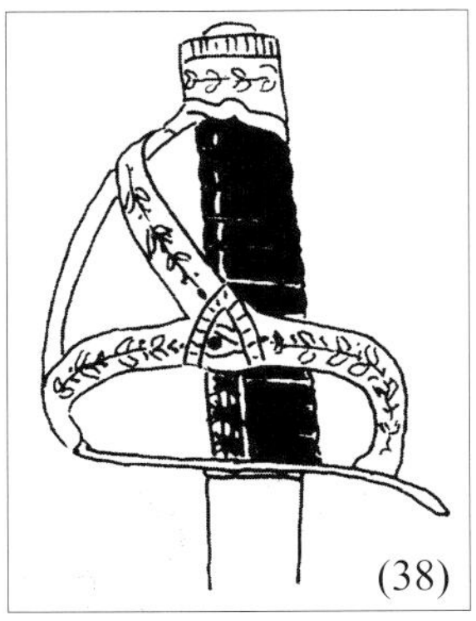
(38)

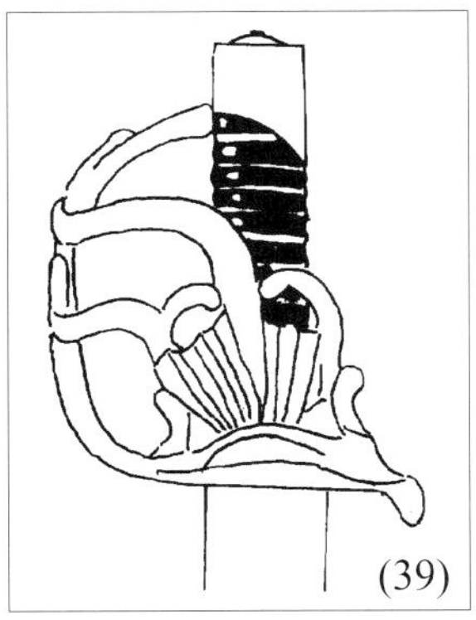
(39)

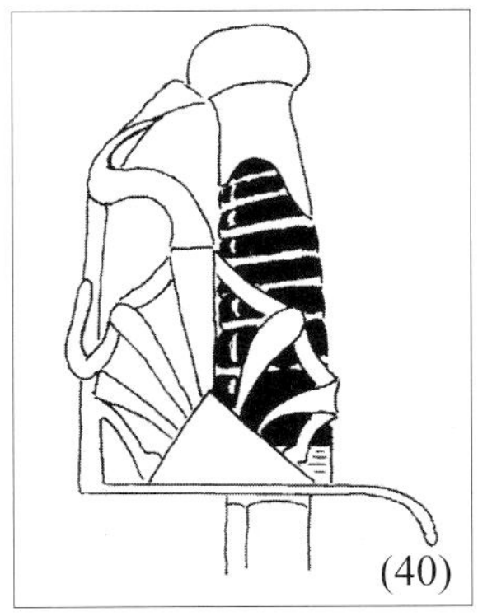
(40)

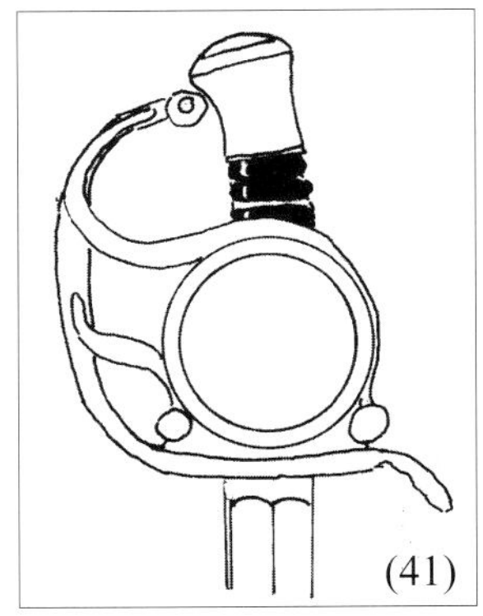
(41)

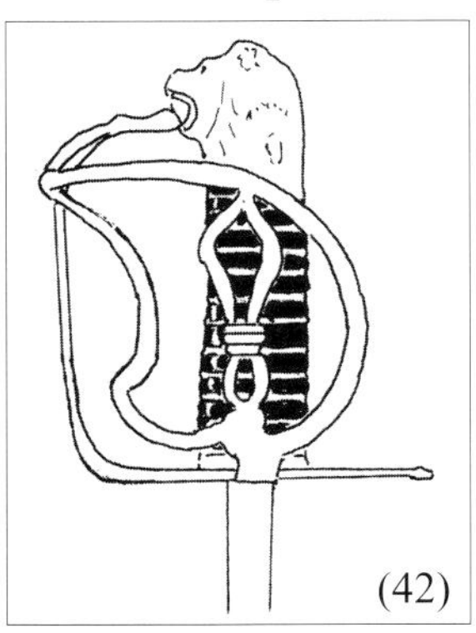
(42)

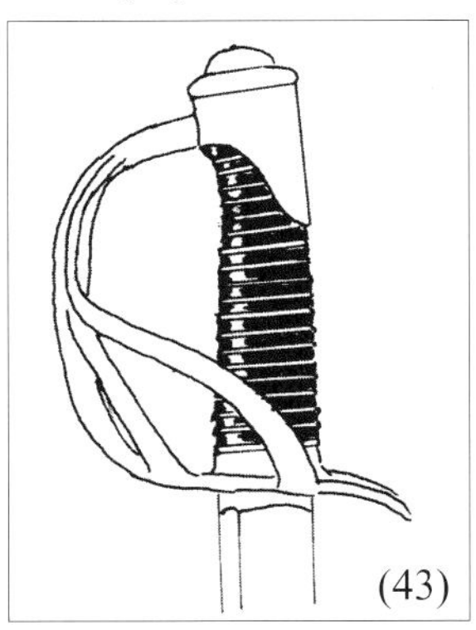
(43)

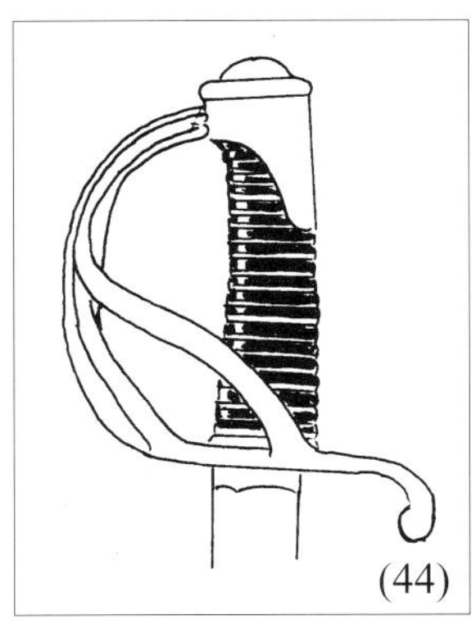
(44)

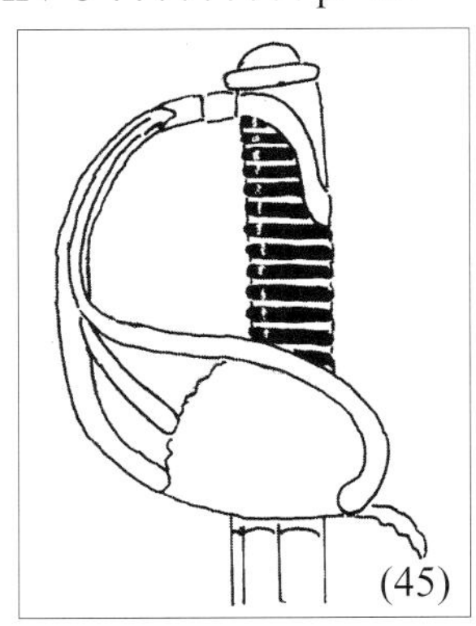
(45)

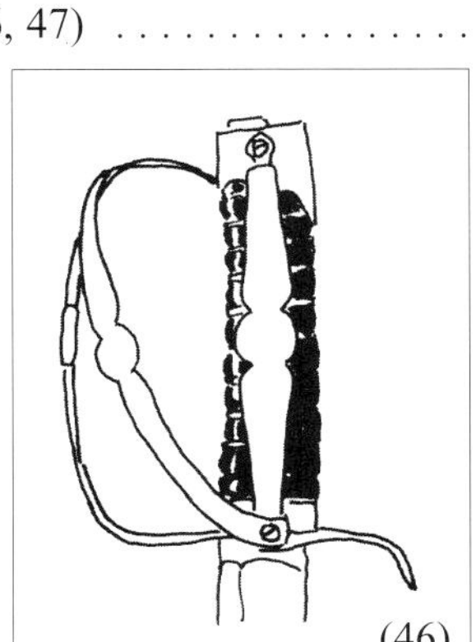
(46)

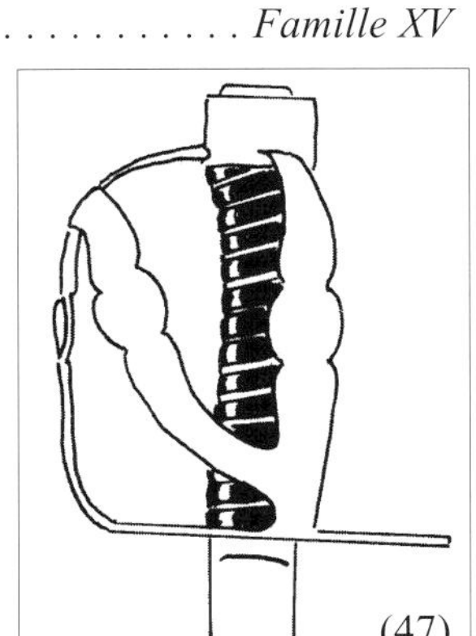
(47)

Sixième partie

Sabres à quatre branches : un arc de jointure et trois branches secondaires . .p. 417

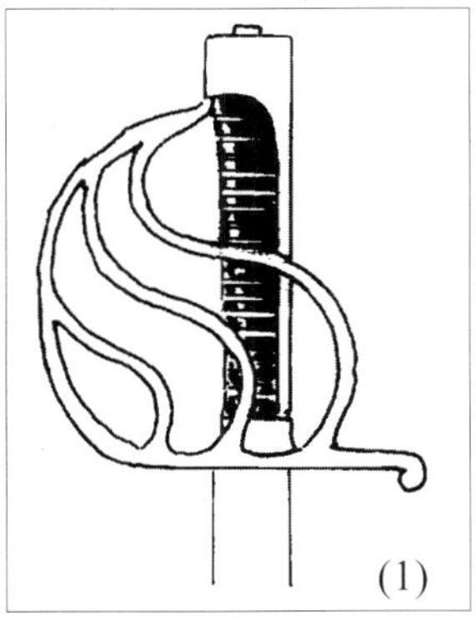
(1)

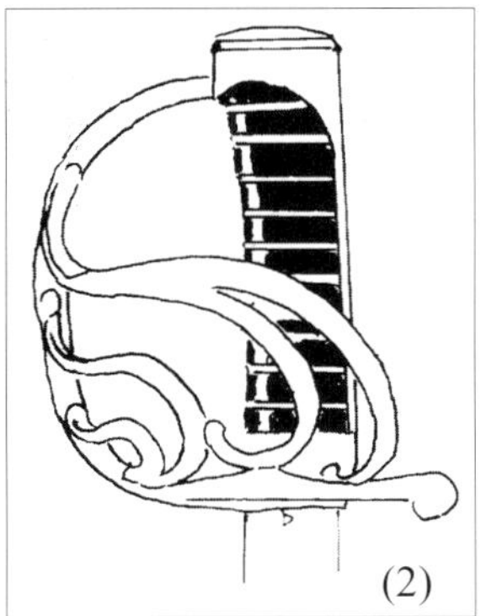
(2)

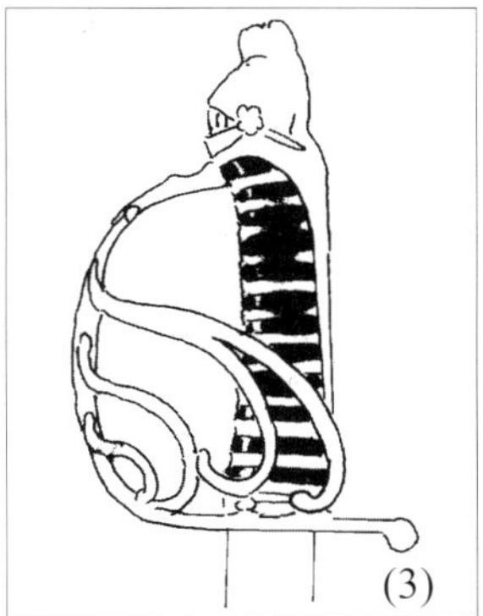
(3)

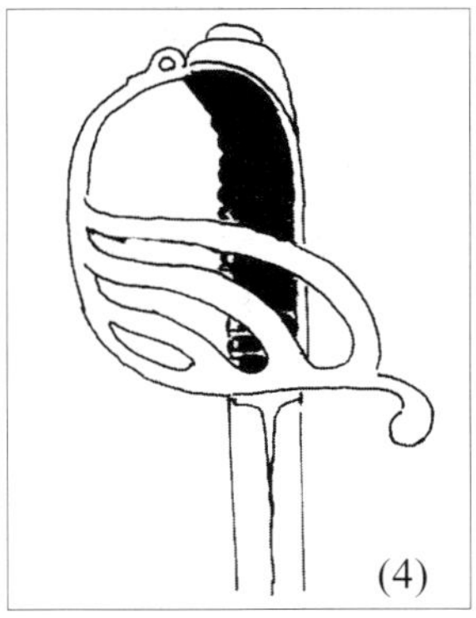
(4)

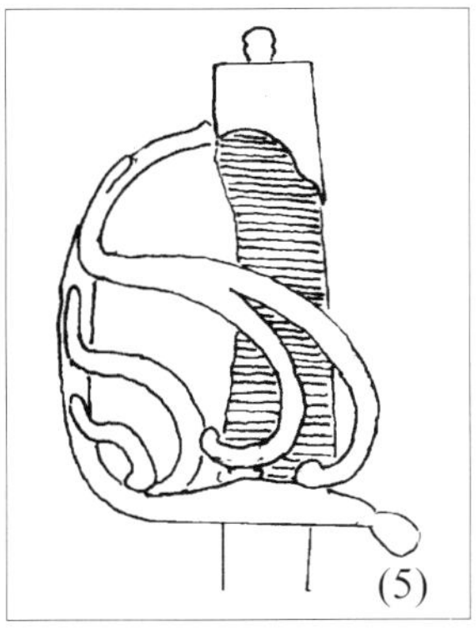
(5)

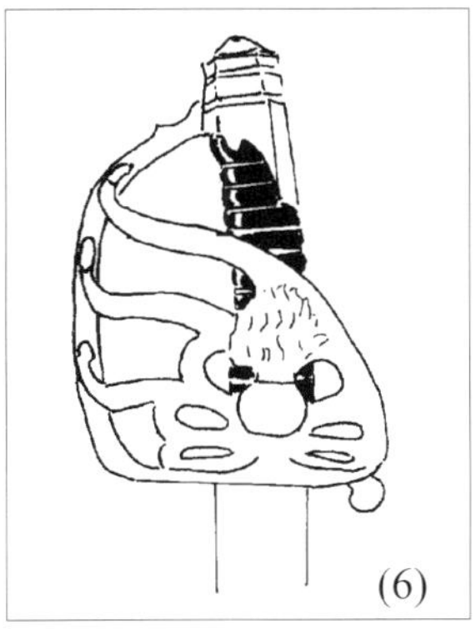
(6)

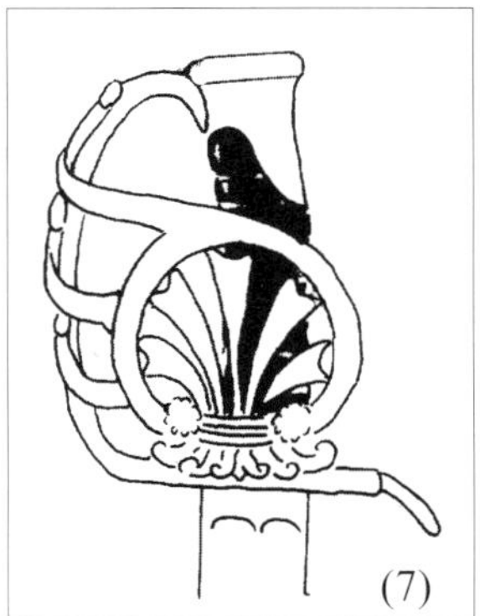
(7)

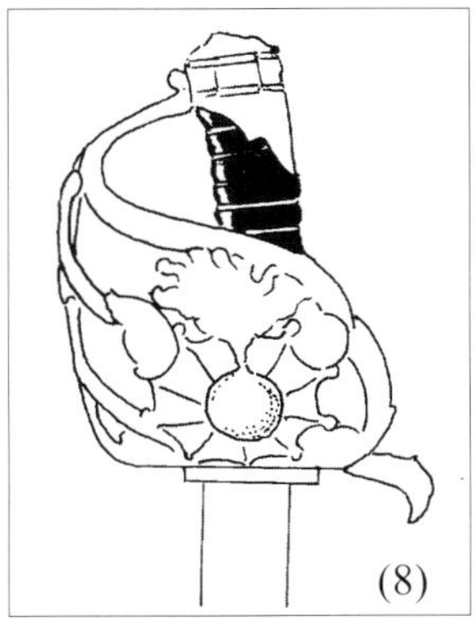
(8)

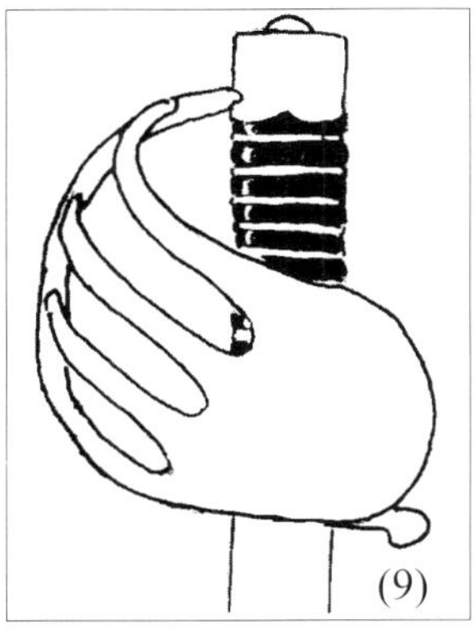
(9)

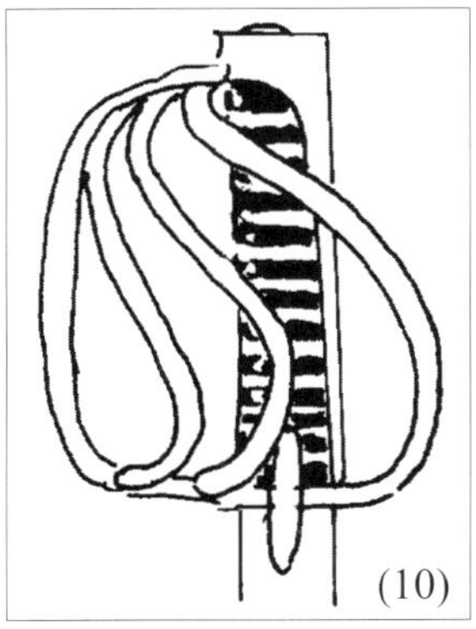
(10)

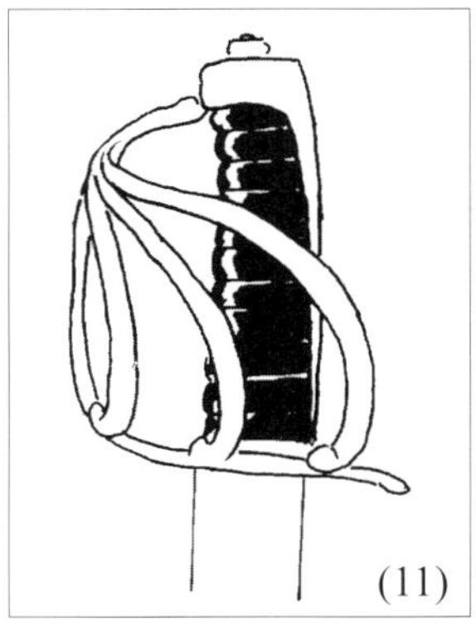
(11)

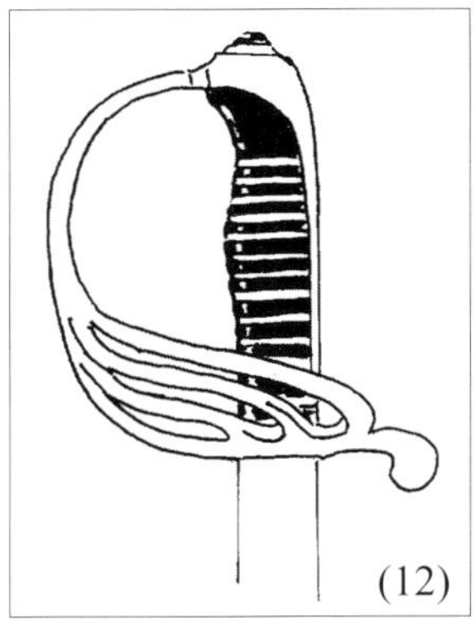
(12)

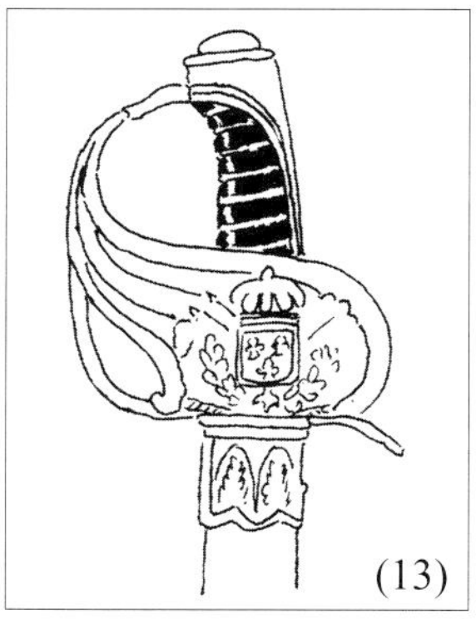
(13)

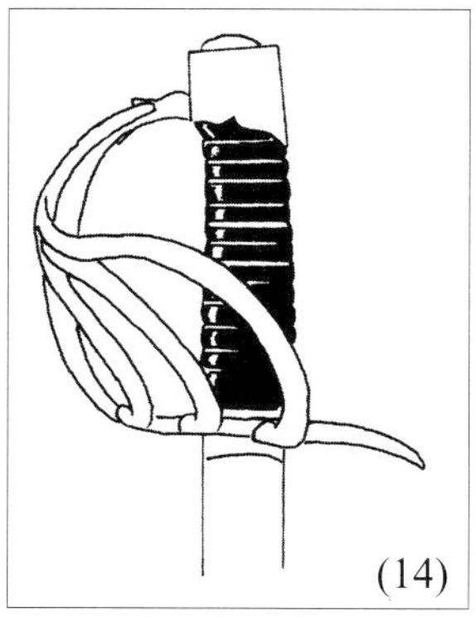
(14)

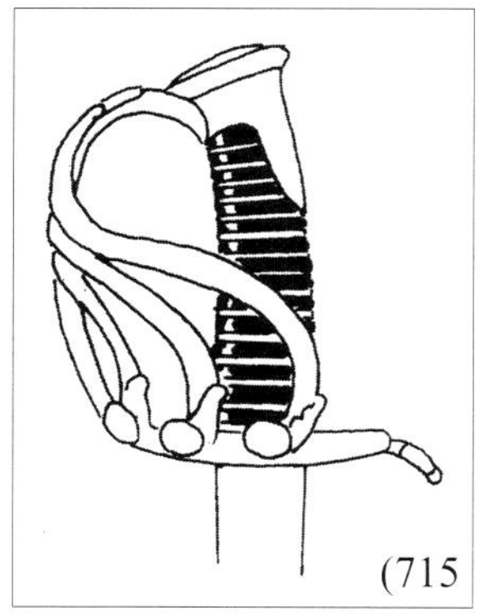
(715

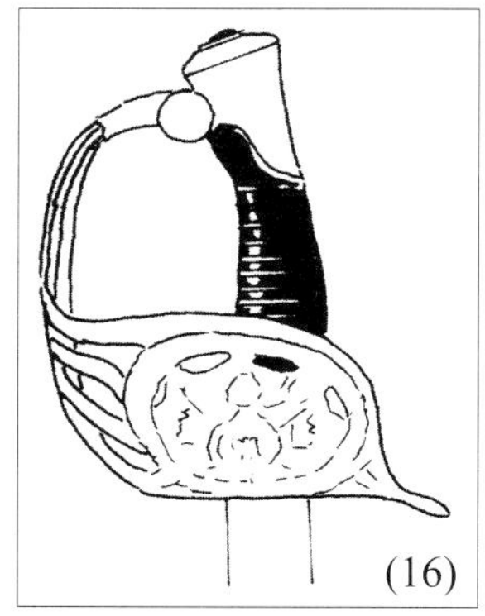
(16)

Septième partie

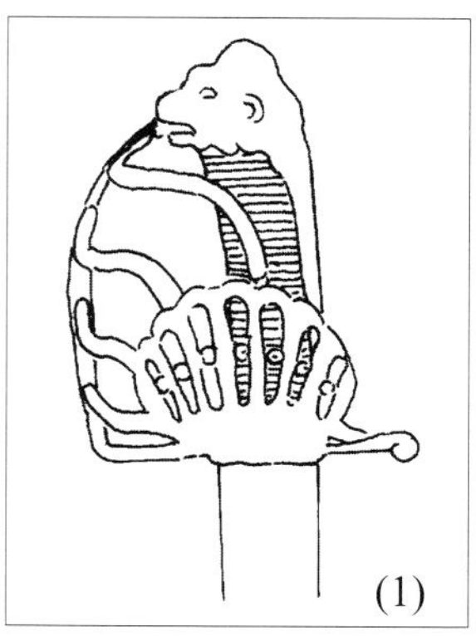
(1)

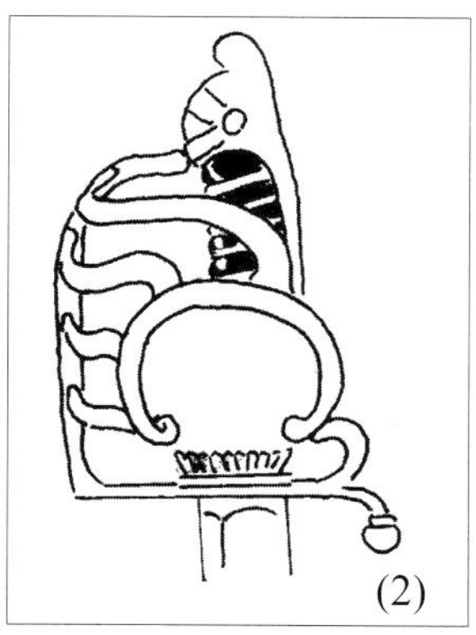
(2)

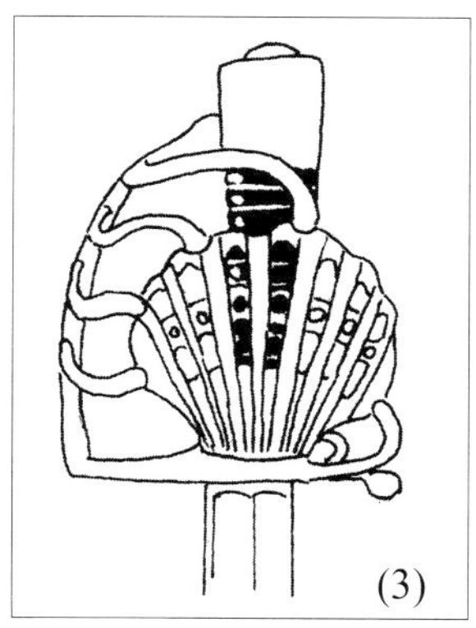
(3)

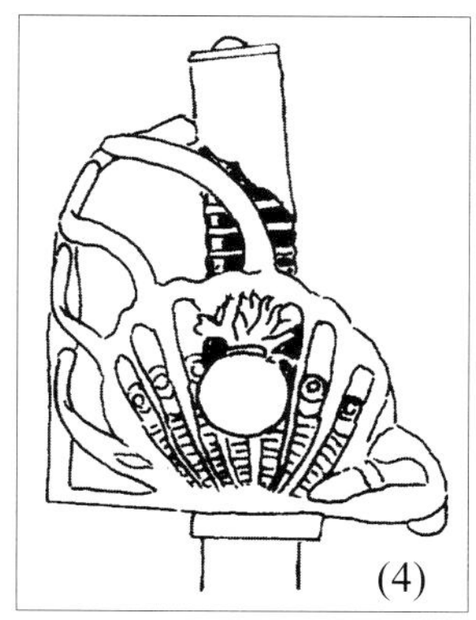
(4)

(5)

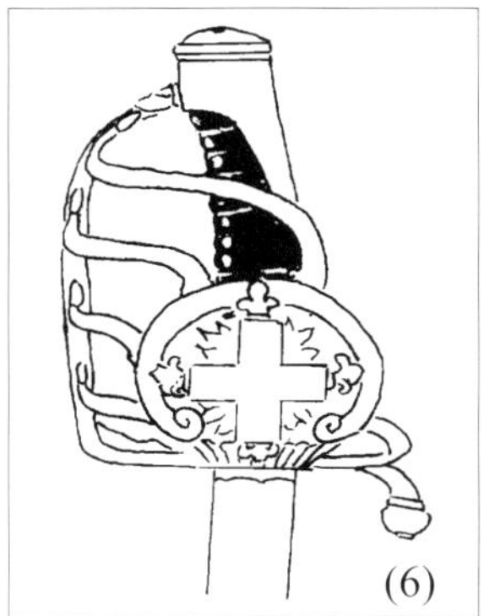
(6)

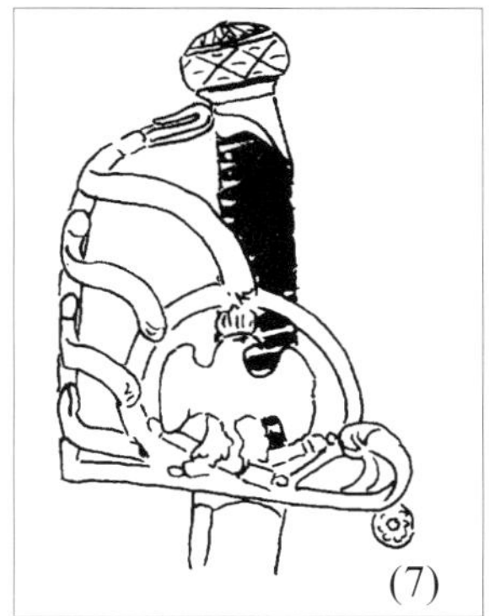
(7)

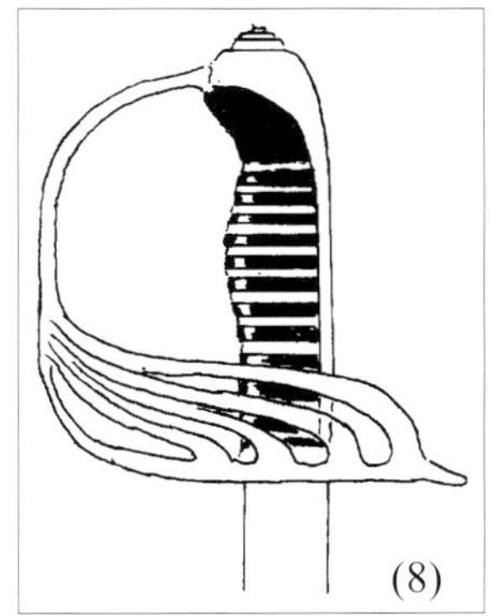
(8)

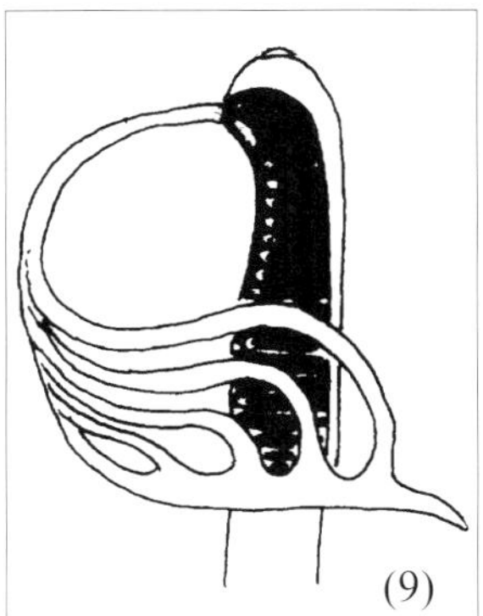
(9)

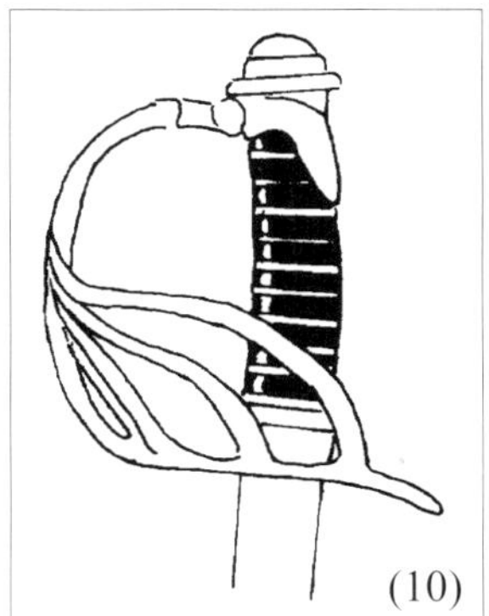
(10)

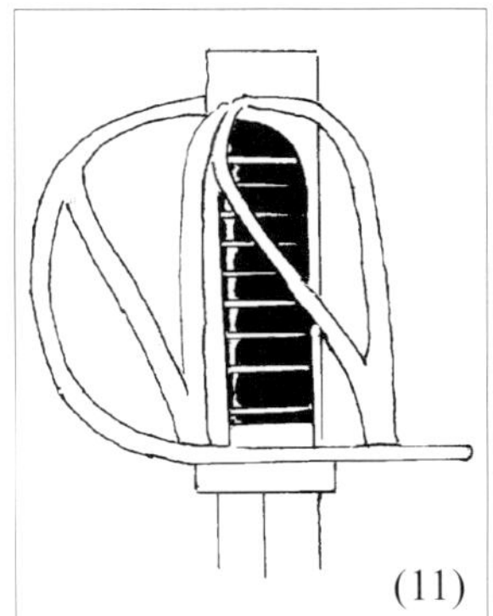
(11)

Huitième partie

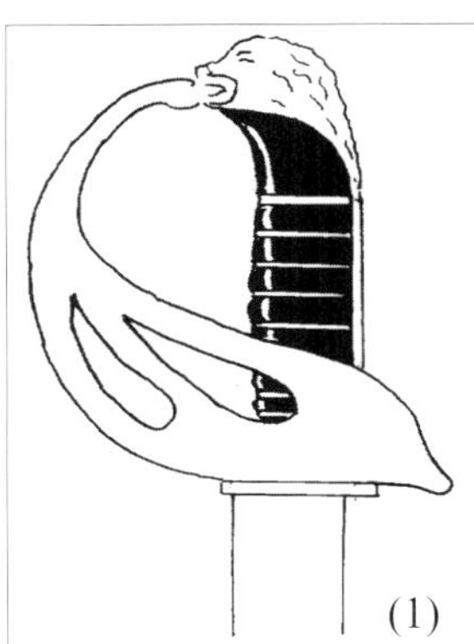
(1)

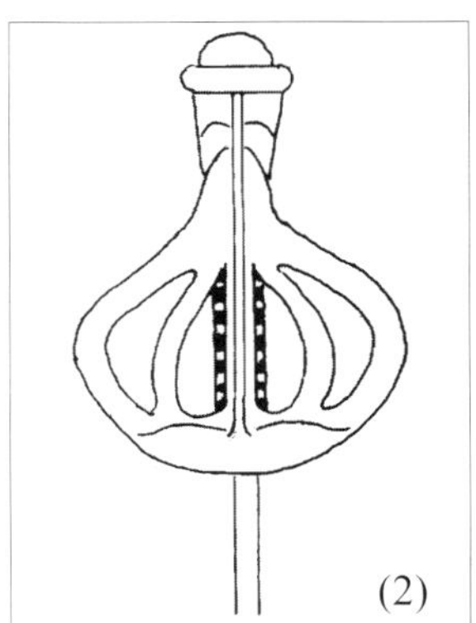
(2)

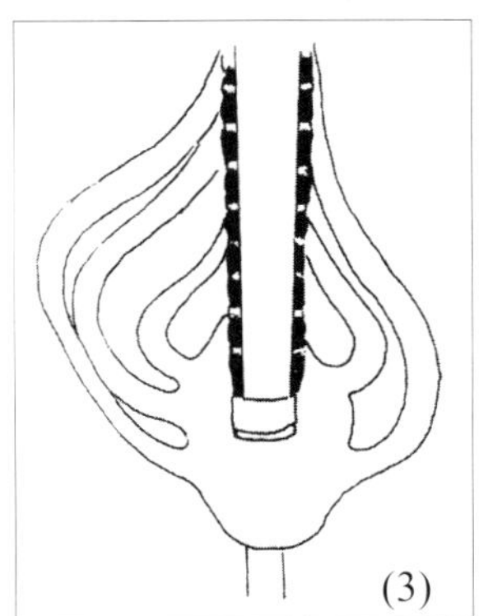
(3)

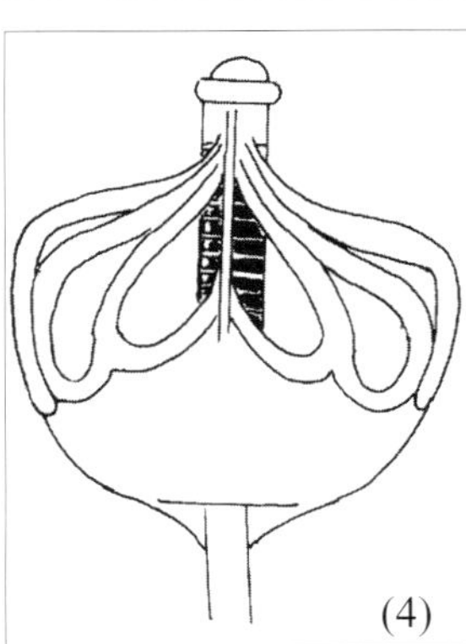
(4)

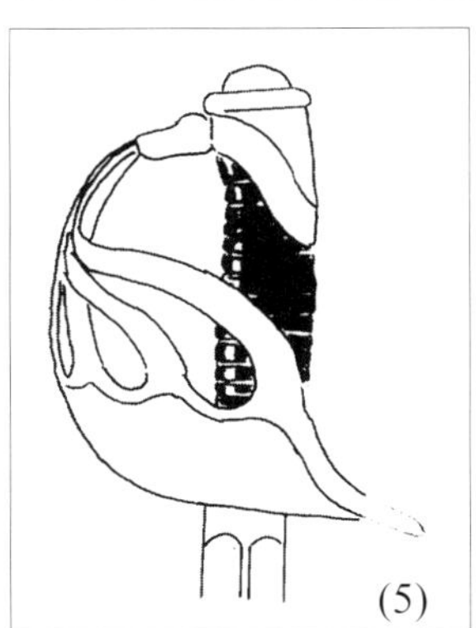
(5)

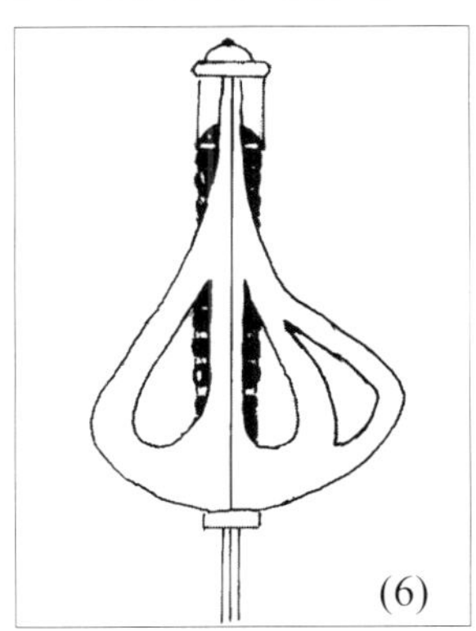
(6)

Première partie

Sabres sans arc de jointure

I - Poignée en crosse de divers matériaux organiques : bois, corne, ivoire ... non filigranée

II - Poignée droite en laiton moulé

III - Poignée de bois recouvert de basane, souvent filigranée

Tous les sabres groupés en cette Première Partie sont caractérisés par l'absence d'arc de jointure. Hors cette particularité, les modèles sont très variés : les uns ont une poignée sans filigrane, à croisière ornée d'oreillons en losange à de rares exceptions près. Ce sont essentiellement des sabres à lame à l'orientale.

Sous le Premier Empire des sabres de Marine sont d'un style comparable mais avec seulement des oreillons basaux.

Les autres sabres ont des montures plus diversifiées, tels les sabres de la Légion Corse à la calotte figurant une tête de Maure enturbannée ou les premiers sabres attribués aux Tambour-Majors à tête de coq, d'aigle, de lion. Puis les sabres de ce corps deviennent réglementaires en 1821 - 1822 et possèdent une poignée filigranée inclinée vers l'avant, terminée par une face léonine.

Un autre groupe de sabres, à poignée droite et de laiton moulé , les sabres de Marine à tête léonine dite à tête de chien de mer, les glaives de l'Ecole de Mars et ses dérivés, les glaives à calotte en olive placée horizontalement armant les troupes à pied surtout depuis 1831, les sabres de Sapeurs à calotte en tête de coq ou d'aigle forment un ensemble important.

Chapitre I

Poignée en crosse de divers matériaux organiques : bois, corne, ivoire ... non filigranée

Les troupes françaises rencontrèrent les *sabres orientaux* lors de la Campagne d'Egypte dirigée par le général Bonaparte. Ces troupes débarquèrent à Alexandrie le 1er juillet 1798. Mais la destruction de la flotte française au tout début du mois d'août immobilisa les armées du général qui purent encore repousser les troupes turques en avril 1799 lors de la bataille du Mont Thabor mais qui ne purent se rendre maître, en juillet 1799 de Saint-Jean d'Acre. Rappelé par le Directoire, Bonaparte quitte l'Egypte le 22 août 1799, rapportant dans ses bagages et dans ceux des officiers l'accompagnant, les premiers sabres islamiques. En conséquence, ces sabres, souvent désignés sous le nom de *sabres du Retour d'Egypte*, ont frappé les militaires par leur beauté mais aussi par leur redoutable efficacité. (*)

UNE LAME ORIENTALE DE LA FIN DU XVIIIÈME SIÈCLE

Cette lame fait partie, vraisemblablement, des lames non montées qui furent ramenées en France, suite à la malheureuse Campagne d'Egypte. La soie non perforée en est la preuve. Ses dimensions sont les suivantes : longueur 83 cm, largeur maximale 3,3 cm, flèche 10,3 cm. Cette lame est totalement gravée et certains décors sont damasquinés à l'or fin : signature du graveur (a, b), la partie

a

- **a**, *recto* de la partie basale de la lame,
de bas en haut : signature du graveur, fusils, étendards, croissant de lune et étoile. Sur le reste de la lame se succèdent les versets du Coran (voir verso, c et d).
- **b**, *verso* de la lame : signature du graveur et partie basale d'un soldat en armure, casqué, armé d'une lance et d'un bouclier.
- **c**, verset du Coran avec l'ornement étoilé le séparant du suivant.
- **d**, pointe du sabre gravé jusqu'à son extrémité de versets.

b

c

d

(*) On trouvera dans Tradition, n° 95, p. 6 à 11, un important inventaire des sabres à l'orientale de Jean-François Teulière et les photograhies de Jean-Louis Viau.

basale jusqu'à l'étoile, le croissant et tous les petits ornements qui séparent les versets du Coran (c, d). On comprendra aisément que de telles lames aient attisé la convoitise de Bonaparte et de ses soldats.

Le capitaine Devernois (*in* J.F. Teulière, 1995) décrit ainsi le sabre qu'il conquit sur un mamelouk :

" C'était un magnifique sabre ciselé en or au fourreau et à la garde. La monture en corne de rhinocéros, la lame en damas noir de la plus grande beauté, la dragonne en velours ".

Au point de vue efficacité, le général de Colbert écrit (*in* J.F. op. cit.) :

" Manié par des mains exercées, ce sabre produit des effets extraordinaires. On ne frappe pas avec lui, mais la lame, soit poussée par l'impulsion du cheval ou de la main, soit brusquement ramenée en arrière, glisse ou plutôt scie, et pénètre de manière à produire d'effroyables blessures ". et le général ajoute *: " Dans une charge, un Hussard du septième régiment, eut le corps coupé en deux ! Il ne tenait plus que par la colonne vertébrale ".*

De son côté, le capitaine François, du Régiment des Dromadaires, témoigne : *" Après le combat de Salaheih, j'ai vu plusieurs de nos cavaliers tués et j'ai remarqué, ainsi que mes camarades, que plusieurs avaient la tête coupée, d'autres à moitié... J'ai vu un Chasseur du 22ème Régiment, fendu depuis l'épaule gauche jusqu'au milieu des reins... "*

Les sabres à l'orientale qui nous sont parvenus sont de différentes origines ; on peut distinguer les catégories suivantes :

- les sabres recueillis sur le champ de bataille et ramenés en France comme trophées. Il sont de pure facture islamique.
- les sabres dont seule la lame a été recueillie et montée ou remontée par des fourbisseurs français.
- les sabres entièrement fabriqués en France, dits *sabres à l'orientale*, lames et montures.

Boutet en imagina de superbes qu'il fit fabriquer par la Manufacture de Versailles, mais d'autres fourbisseurs montèrent des sabres à l'orientale de qualité plus variable, tel Fatou, Le Page, Manceaux...

On distingue trois Familles :

Famille I – Poignée en crosse, à oreillons entiers losangéïformes

Les sabres à l'orientale furent très à la mode durant une grande partie du XIXème siècle. Ils furent portés par des officiers européens, des officiers de Marine, les officiers des Mameluks de la Garde Impériale. Sous la Restauration et la Monarchie de Juillet, ces sabres perdurèrent notamment grâce à la Campagne d'Algérie, où les sabres à l'orientale étaient tout indiqués. Sous le Second Empire de tels sabres seront portés par des officiers de Cavalerie légère et par des officiers de la Garde Impériale.

A noter que pour les Mameluks de la Garde Impériale, l'arme fut réglementaire. Pour toutes les autres formations, elle ne fut que de fantaisie et seuls les officiers pouvaient se permettre de les commander et de les porter.

Cette mode déborda nos frontières. On trouve des sabres à l'orientale portés par des officiers prussiens, (fig. 20) britanniques, belges et parfois encore au XXème siècle.

La totalité de ces sabres constitue la Famille I dans laquelle on a distingué deux types de montures de la façon suivante :

Sous-famille I a - Crosse simple

La *poignée* en forme de crosse est constituée de deux plaquettes de bois, de corne, d'ivoire... fixées sur la soie de la lame par des boutons de rivure. En général, au sommet de la crosse un orifice est ménagé pour le passage de la dragonne. La croisière peut être droite ou à quillons inversés. En son milieu, elle est élargie en un grand losange pour former les oreillons, plus ou moins décorés.

La *lame* est très courbe, ayant autour de 10 - 12 cm de flèche. La largeur est variable mais voisine les 3 cm pour les fabrications françaises, souvent à contre-tranchant, à pan creux, parfois à gouttières pouvant être plusieurs. La lame est souvent décorée par gravure, damasquinage sur fond bleui ou bruni.

Le *fourreau* est constitué d'une armature de bois recouverte de tissus ou de cuir dans les parties visibles. Il est garni d'une chape surmontant un bracelet de bélière, un second bracelet et une longue bouterolle.

Nota - Une autre caractéristique des sabres à l'orientale est la suivante : la chape est fendue sur le côté postérieur. Cette fente est indispensable pour rengainer ou dégainer le sabre en raison de sa forte courbure. Cette fente est souvent fermée par une pièce métallique mobile, maintenue par un système de ressort.

Les sabres à l'orientale variant à l'infini, on ne peut donner dans cet ouvrage la description complète de tous les types rencontrés. Aussi ne seront données ci-après que des représentations exemplaires agrémentées d'un court commentaire.

On remarquera que ne sont figurés que des sabres utilisés dans l'Armée française depuis le Consulat jusque sous la Monarchie de Juillet.

1

1- Sabre oriental d'officier.
Consulat.
Monture en laiton. Poignée à plaquette de corne blonde. Quillons droits, à oreillons losangiformes et terminés en olive. Lame partiellement damasquinée d'argent. Il s'agit d'une arme de prise.

1a

1a -Sabre oriental d'officier.
Consulat (suite).
Fort fourreau de bois recouvert de chagrin cousu de fils d'argent sur une face ; quatre garnitures en vermeil : longue chape, deux bracelets et longue bouterolle terminée en pointe. Ce sabre a conservé son rare cordon d'attache d'origine, en fils tressés bleu ciel et bleu foncé (J.C. D.).

2

2 - Sabre oriental d'officier.
Consulat.
Poignée à plaquettes de bois, croisière et oreillons losangéïformes en laiton finement ciselé. Lame à dos plat débordant, relativement peu courbée, simplement décorée d'une gouttière bordée d'un filet. Fourreau en bois recouvert de chagrin à trois garnitures en argent : chape ornée de motifs végétaux et terminée par un bracelet de bélière, bracelet isolé et longue bouterolle (J.C. D.).

3 - Sabre d'officier à l'orientale.
Première République.
Monture en laiton. Poignée et crosse à plaquettes de corne, largement perforée sur la partie supérieure pour le passage de la dragonne, à bouton de rivure ovalaire et partie antérieure de la fusée à quatre dépressions correspondant aux prises de doigts de la main, (le pouce excepté). Croisière classique. L'intérêt de cet exemplaire réside dans la lame de sabre de Hussard Mle An IV, *République* sur une face et *Françoise* sur l'autre. Fourreau de fer à longs bracelets de bélières.

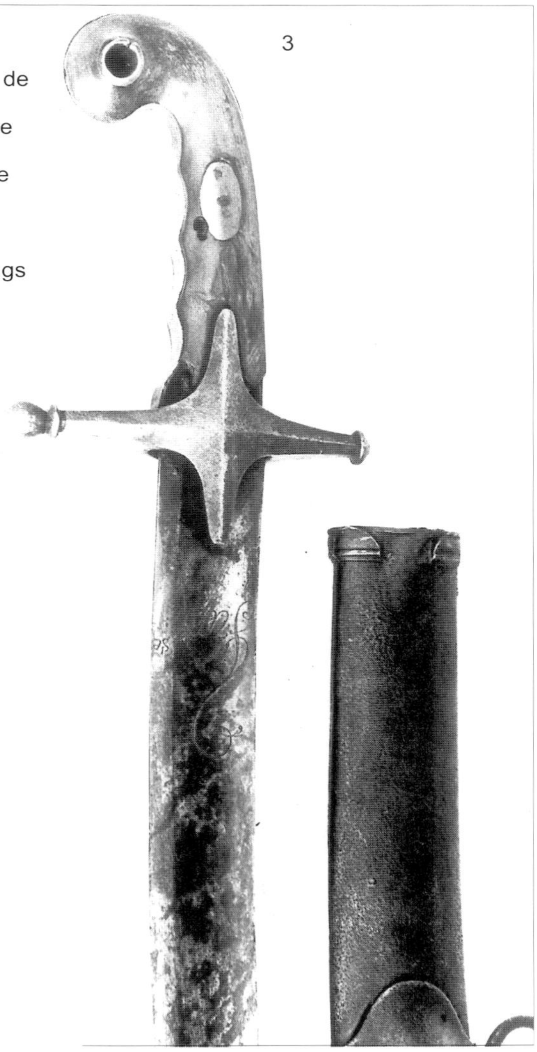

3

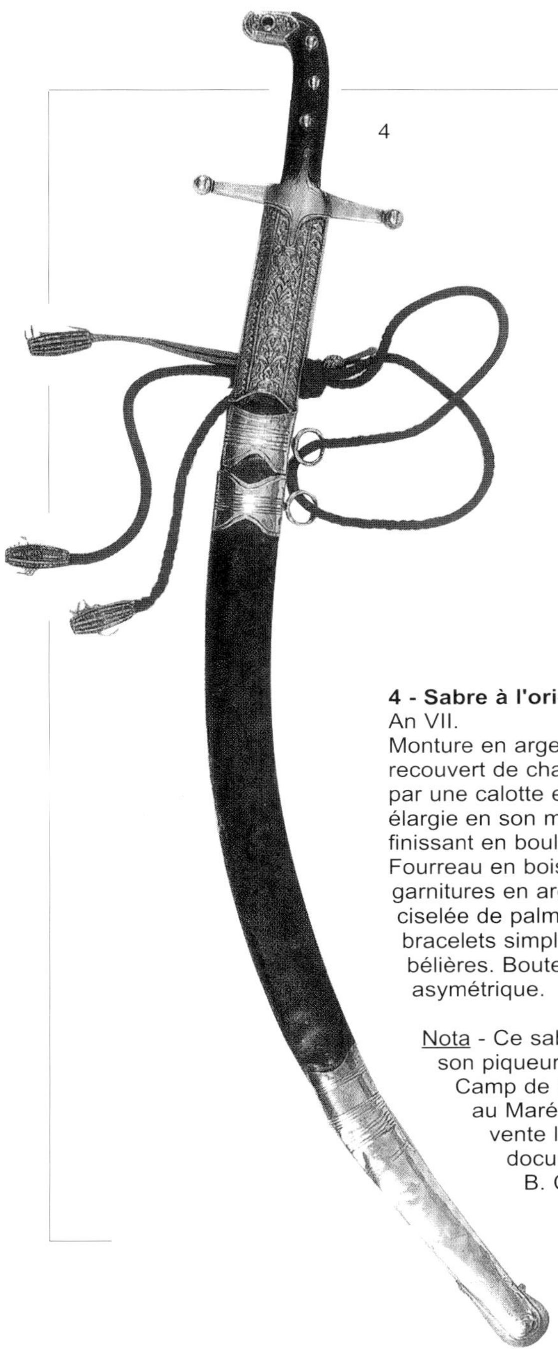

4

4 - Sabre à l'orientale.
An VII.
Monture en argent du type shamshir. Poignée en bois recouvert de chagrin, à trois boutons de rivure se terminant par une calotte en bec, percée d'un orifice. Croisière droite, élargie en son milieu pour former les oreillons. Quillons finissant en boule. Lame en damas, sans inscription. Fourreau en bois recouvert également de chagrin à quatre garnitures en argent, chape à décors repoussés, finement ciselée de palmettes, bordée de motifs géométriques. Deux bracelets simplement rainurés portant les anneaux de bélières. Bouterolle décorée en suite et dard en lyre asymétrique.

Nota - Ce sabre exceptionnel a été offert par Bonaparte à son piqueur Lavigne, pendant la Campagne d'Egypte au Camp de Saint-Jean d'Acre, en l'An VII. Il a été offert au Maréchal Foch en 1918. Ce sabre a été mis en vente le 6 novembre 1998, accompagné de documents justifiant son authenticité (Experts B. Croissy et J.C. Dey).

5 - Sabre d'officier à l'orientale.
Premier Empire.
Poignée en crosse, à deux plaquettes de corne fixées sur la soie de la lame par deux boutons de rivure et portant à son sommet l'orifice pour passer le cordon de la dragonne. Garde à deux longs quillons droits terminés par des boutons travaillés à facettes et supportant les deux oreillons losangiformes. Lame à flèche de 11 cm, longue de 85 cm, large au talon de 3 cm, dos en baguette se terminant par un contre-tranchant de 20 cm. Cette lame est finement gravée d'écailles, de la lune, du soleil, de grottesques... Fourreau de bois recouvert de cuir noirci, à couture frontale ; chape fendue latéralement sur toute la longueur du côté postérieur pour faciliter le maniement du sabre, deux bracelets à tête de lion (5a), bouterolle de 25,5 cm décorée de deux doubles traits avec un dard de laiton asymétrique.

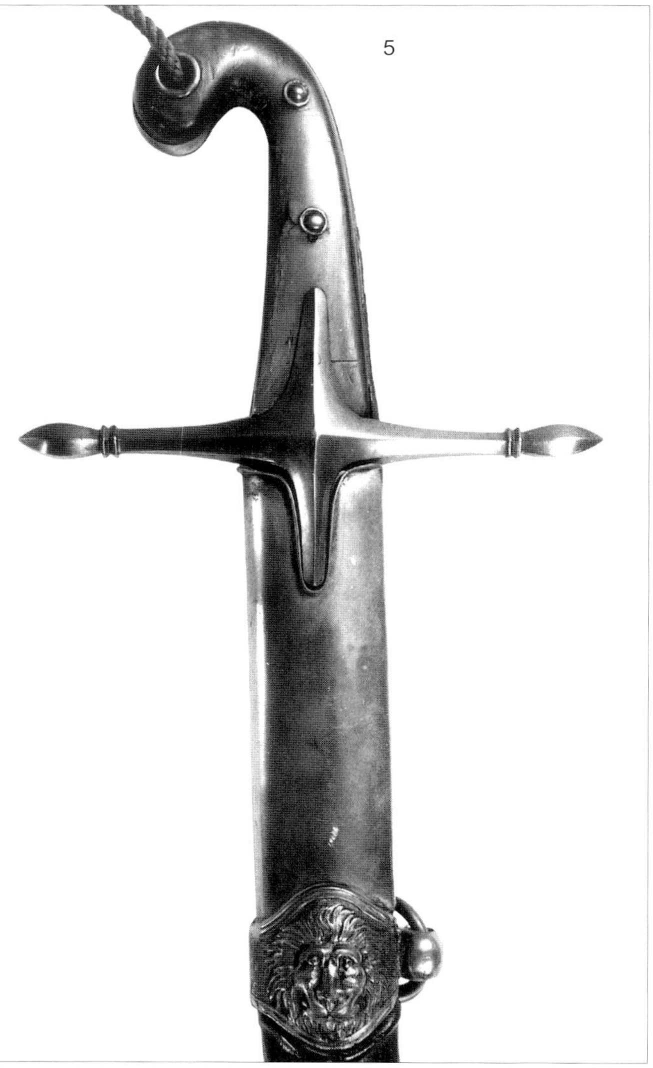
5

5a

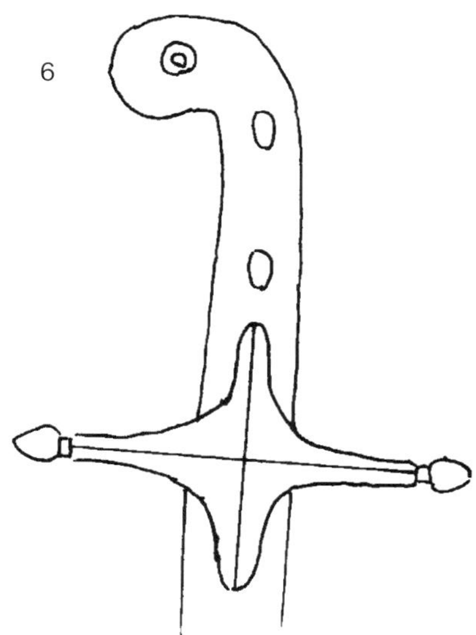
6

6 - Sabre d'officier de Mameluk de la Garde Impériale.
Les sabres d'officiers sont conçus sur le même schéma que les sabres de la troupe mais plus richement traités et avec de nombreuses variantes. La poignée, en crosse, est formée de la soie de la lame recouverte de deux plaquettes de corne brune ou blonde, d'ivoire... fixées par deux boutons de rivure de grandeur variable. Le sommet de la crosse est percé pour passer le cordon de dragonne soit directement, soit par l'intermédiaire d'un anneau. Une bande de laiton, plus ou moins ouvragée masque la tranche de la soie et des plaquettes. La lame de 80 cm, large au talon de 3 cm est très courbée, la flèche pouvant dépasser les 12 cm. Elle est souvent gravée de motifs divers, copies fantaisistes et maladroites des hiéroglyphes. Fourreau de bois recouvert de chagrin, de galuchat, de maroquin de couleurs variées. Chape et bouterolle à dard asymétrique, avec entre elles deux bracelets en bossettes plus ou moins ornées portant les anneaux de bélière.

Nota - On peut, parfois, trouver sur le dos de la lame bleuie et dorée, la signature : *Mfture Imple du Klingenthal Coulaux frères* (M.E. - S. de P.).

7

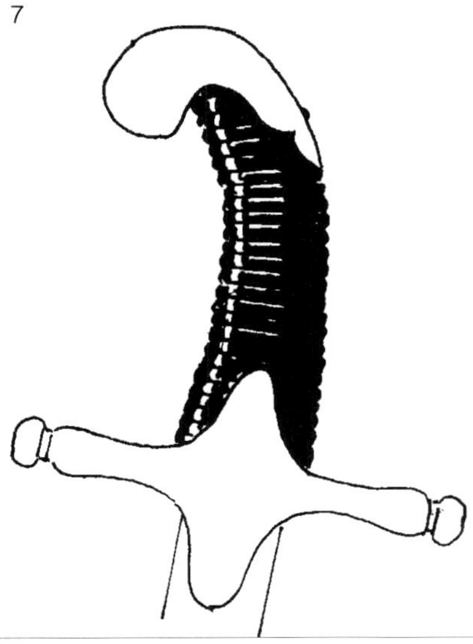

7 - Sabre de Mameluk de la Garde.
1802 - 1805, variante. (voir également le n°127, p. 106).
Monture de laiton. Calotte en boule, percée ou non, faisant crosse avec courte queue maintenue par un petit clou. Poignée de bois cannelé recouverte de basane noircie, sans filigrane. Croisière élargie en son milieu pour former deux grands oreillons et terminée par deux quillons droits, à boules... Lame de 80 cm, à flèche d'environ 12 cm, large au talon de 3 cm, sans décors. Fourreau de cuir noir, sans alèse à chape largement échancrée dorsalement pour permettre la sortie et la rentrée de la lame. Longue bouterolle à dard asymétrique, avec deux bracelets en bossettes unies portant chacune un anneau de bélière (M.E. - S. de P.).

8

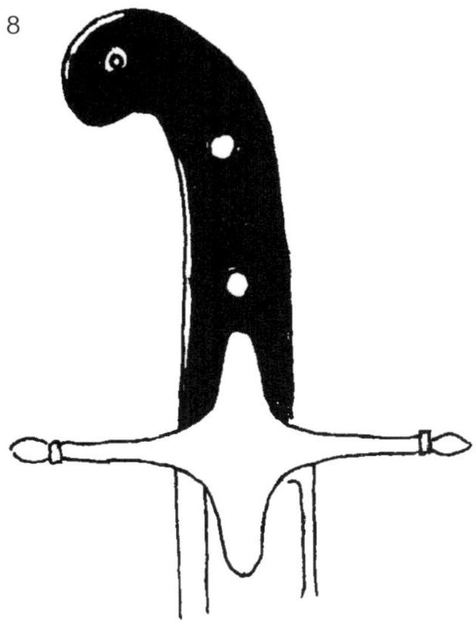

8 - Sabre fantaisie de Premier Maître des Marins de la Garde Impériale.
Premier Empire.
Monture en laiton et poignée en crosse à plaquettes de corne noire maintenue par deux boutons de rivure. Lame courbe à pans creux portant l'inscription *Garde Impériale* limitée par deux ancres, dans un cartouche décoré à ses extrémités de palmettes. Fourreau à trois bracelets unis et bouterolle terminée par un dard asymétrique. Sabre exceptionnel cité par C. Ariès, XX, 2ème fasc. 1971.

9

9 - Sabre d'officier à l'orientale de Marine.
Consulat - Premier Empire.
Monture en laiton. Poignée en corne. La calotte, à courte queue, est reliée à la croisière par une chaînette. Lame à l'orientale peu courbée, poinçonnée au talon d'une ancre de marine. Au tiers, un motif qui reste énigmatique constitué de deux étoiles (ou soleils) de part et d'autre d'un petit cercle portant deux ailes (ou deux plumes), probablement poinçon du fourbisseur.

10

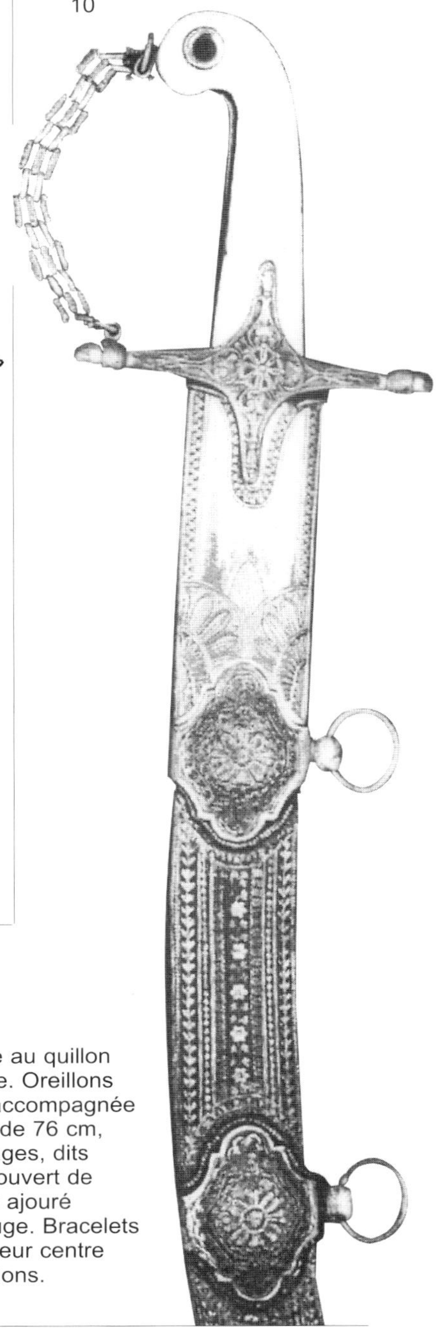

10 - Sabre d'officier.
Restauration.
Poignée à plaquettes d'ivoire, reliée au quillon antérieur par une chaînette fantaisie. Oreillons décorés en leur centre d'une fleur accompagnée de motifs végétaux. Lame courbée de 76 cm, gravée sur les deux faces de feuillages, dits Koftagari, dorés. Fourreau bois recouvert de laiton à décors repoussés, ciselé et ajouré laissant apparaître un maroquin rouge. Bracelets de bélière en bossette décorés en leur centre d'une fleur rappelant celle des oreillons. (Sotheby's).

11

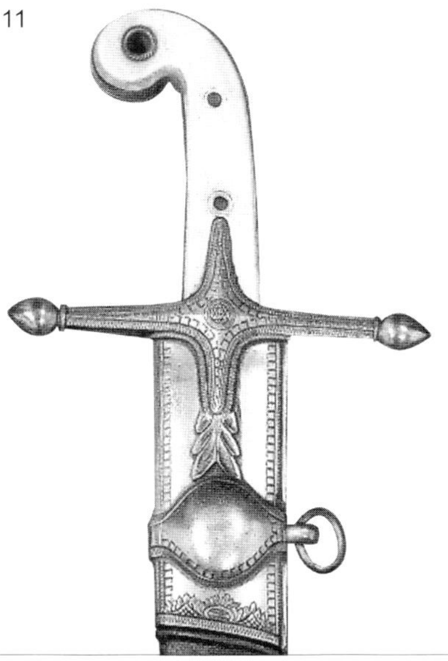

11 - Sabre à l'orientale.
Milieu du XIXème siècle. Monture en laiton doré. Poignée en crosse à plaquettes d'ivoire fixées par deux boutons de rivure. Croisière finement ciselée de motifs géométriques. Lame relativement peu courbée, en damas rehaussé d'un liseré poli glace. Fourreau en bois recouvert de chagrin, à quatre garnitures décorées en suite de motifs géométriques, deux bracelets en bossette.

13

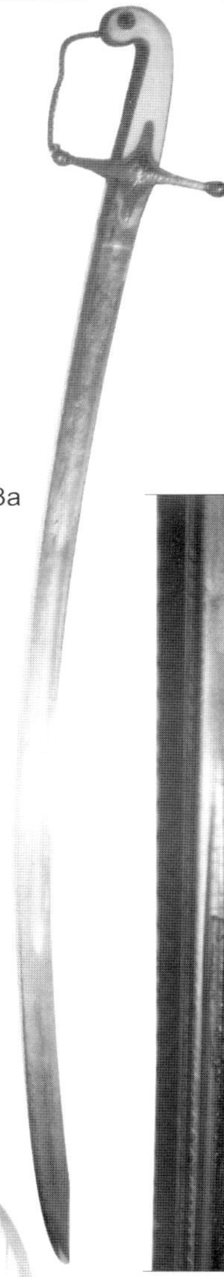

13a

13b

12 - Sabre à l'orientale de Louis-Antoine de Bourbon.
1816 - 1818.
Poignée en crosse à plaquettes de corne blonde fixées par deux boutons de rivure en rosette. Croisière ciselée de rinceaux, à quillons inversés se terminant par une face léonine. Lame courbée en damas, à deux gouttières, pointe en langue de carpe et marquée au talon *Fque d'armes Pirmet à Paris.* Fourreau en acier, à deux bossettes ciselées du profil de Henri IV et chape gravée en grandes lettres L A pour Louis Antoine sous couronne fleur-de-lysée des enfants de France.

12

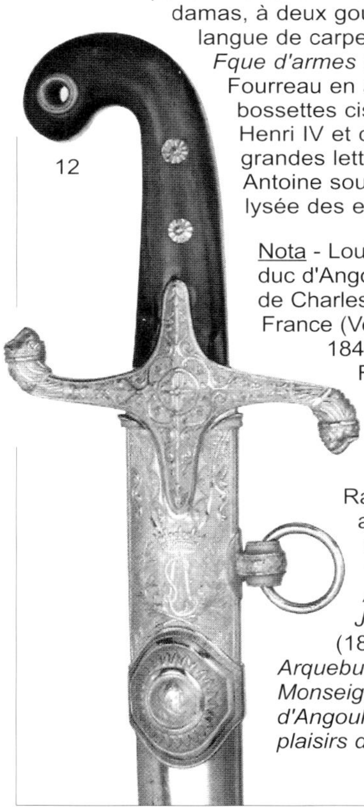

Nota - Louis Antoine de Bourbon, duc d'Angoulême est le fils aîné de Charles X. Dernier dauphin de France (Versailles 1775 - Goritz 1844) fut grand amiral de France. Il commanda l'Expédition d'Espagne de 1823 et abdiqua ainsi que Charles X en 1830.
Rappelons que Pirmet fut arquebusier fourbisseur à Paris de l'An IX à 1818, sous les titres de *Arquebusier du roi Jérôme de Westphalie* (1807-1813) puis de *Arquebusier fourbisseur de Monseigneur le duc d'Angoulême et des menus plaisirs du roi* (1816-1830).

13c

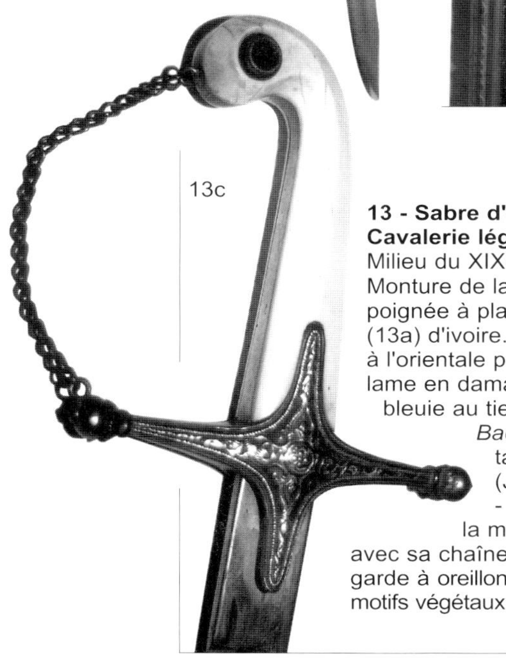

13 - Sabre d'officier de Cavalerie légère.
Milieu du XIXème siècle. Monture de laiton doré, poignée à plaquettes (13a) d'ivoire. Ce sabre à l'orientale possède une lame en damas doré et bleuie au tiers, signée *Backes* sur le talon (13b). (J.P. Aumo).
- **c**, détail de la monture avec sa chaînette et sa garde à oreillons ciselés de motifs végétaux.

14 - Sabre d'officier supérieur de Cavalerie légère.
Modèle à l'orientale, vers 1830.
Monture à bronze doré. Poignée en crosse recouverte de plaquettes de nacre ciselée de feuillage et d'attributs militaires. Chaînette reliant le sommet de la crosse au quillon antérieur. Croisière droite, élargie en son milieu pour former les grands oreillons ciselés de rinceaux. Lame en damas très courbée, de 76,7 cm, gravée aux deux tiers de sa longueur, de feuillages avec un cartouche médian dans lequel on distingue deux croissants de lune côté talon et un soleil, côté pointe, l'ensemble étant sur fonds doré. Sur le talon se trouve la signature du fourbisseur *P. KNECHT in Solingen*. Fourreau de laiton doré entièrement décoré sur le *recto*. Au niveau des deux cartouches supérieurs sont soudés les deux anneaux de bélières. Sur le *verso* décors gravés de toiles d'araignée et de feuillages. Dard symétrique à bordure festonnée (J.C. D.)

14

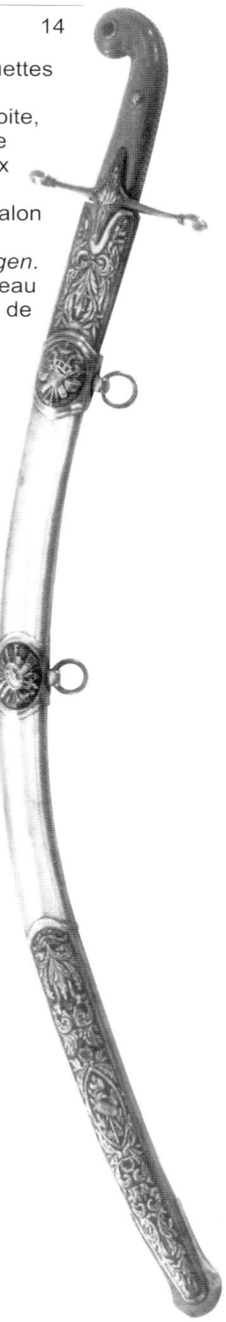

15

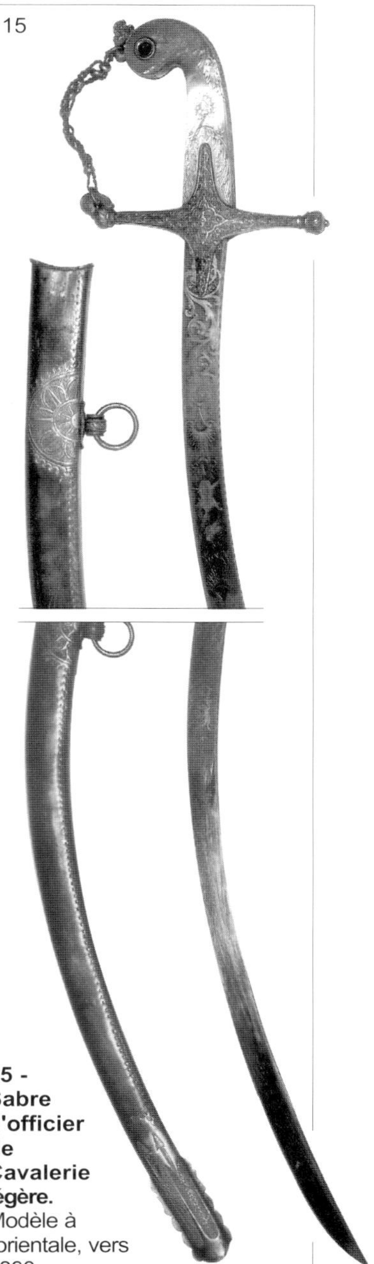

16 - Sabre d'officier de Cavalerie légère à l'orientale.
Restauration, vers 1830.
Poignée à plaquettes de nacre finement et richement ciselées de feuillages et d'attributs militaires. Monture en bronze ciselé et doré. Branche de garde en chaînette. Garde à deux quillons droits terminés en boules et deux oreillons finement ciselés de rinceaux. Lame courbe de 76,7 cm plate, à dos plat et contre-tranchant, gravée aux 2/3 de turqueries et de rinceaux feuillagés, sur fond de damas, en partie doré, signée au talon *P. KNECHT in Solingen*. Riche fourreau en laiton doré, à deux anneaux de suspente, enrichi de quatre médaillons, à décor d'attributs guerriers ciselés en ronde-bosse, gravé au dos de toiles d'araignée et de feuillages.

16

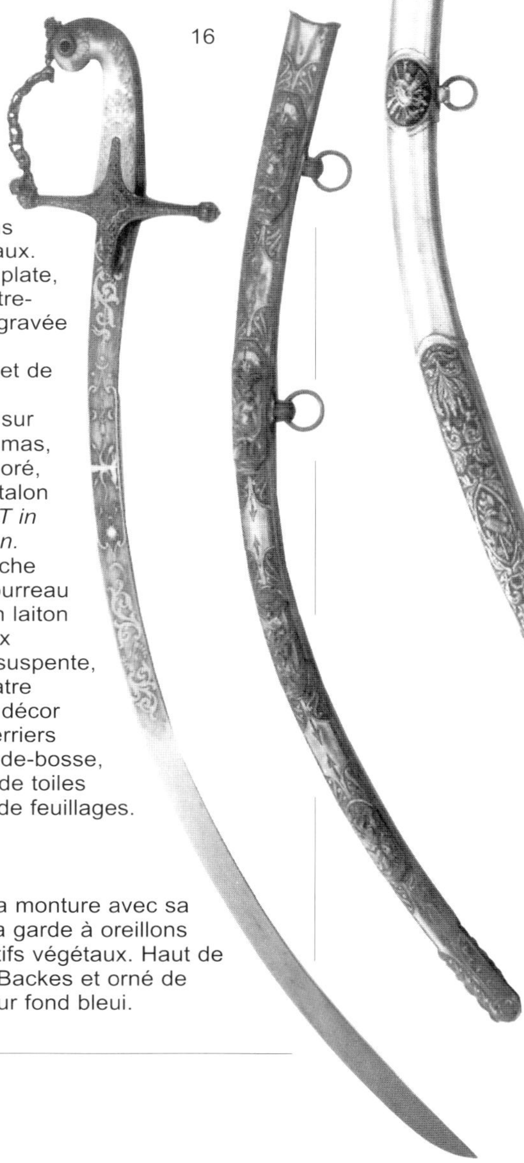

16a

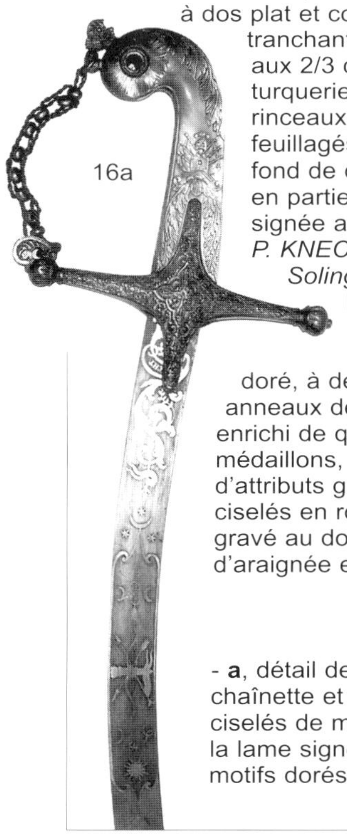

- **a**, détail de la monture avec sa chaînette et sa garde à oreillons ciselés de motifs végétaux. Haut de la lame signé Backes et orné de motifs dorés sur fond bleui.

15 - Sabre d'officier de Cavalerie légère.
Modèle à l'orientale, vers 1830.
Monture de fer. Poignée à plaquettes de corne blonde fixées par un bouton de rivure en rosette de fer. Croisière aux extrémités en olive et élargie au milieu pour former les longs oreillons décorés de ciselures en harmonie avec le décor du fourreau. Lame de 81 cm, très courbée, à dos plat. Fourreau de fer à trois garnitures à décors repoussé et finement ciselées de feuillages. Dard symétrique en fer. (J.C. D.).

17

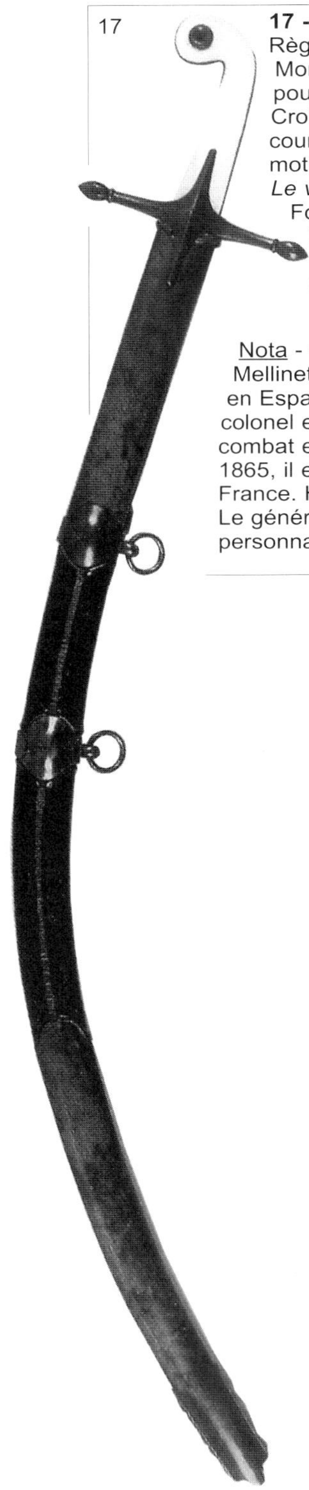

17 - Sabre à l'orientale du général Castelnau.
Règne de Louis-Philippe et de Napoléon III.
Monture en acier jaspé. Poignée en crosse, à plaquettes en ivoire, pourvue d'un œillet à son extrémité pour le passage de la dragonne. Croisière à oreillons en losange et quillons terminés en olive. Lame très courbée en fin damas, gravée sur une face d'un cartouche doré orné de motifs orientaux et sur l'autre, la dédicace :
Le vieux Mellinet à son ami le général Castelnau
Fourreau de bois recouvert de chagrin, à longue chape, à deux bracelets en bossettes d'acier jaspé et terminé par une longue bouterolle à dard symétrique à pointe médiane en acier jaspé (B. C.).

Nota - Grâce aux recherches de B. Croissy on connaît bien Henri Mellinet qui naquit à Nantes en 1798. Sous-lieutenant en 1815, blessé en Espagne, fait la conquête de l'Algérie à partir de 1841 et promu colonel en 1849, il fonde la ville de Sidi-bel-Abès. Rallié à Napoléon III, il combat en Orient en 1855 et devient général de brigade en 1859. En 1865, il est élu sénateur, devient grand maître de la Franc-Maçonnerie en France. H. Mellinet mourra en 1894, à l'âge de 96 ans !
Le général Castelnau (1814-1896) aide de camp de Napoléon III est une personnalité beaucoup mieux connue que le *vieux Mellinet.*

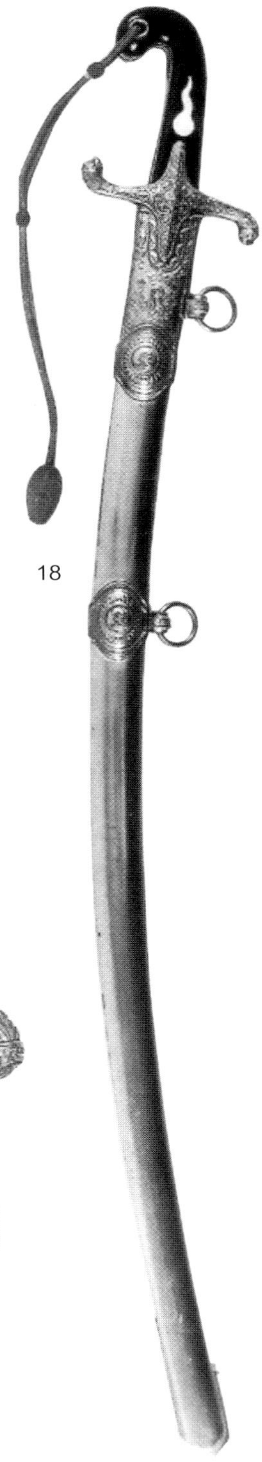

18

18 - Sabre à l'orientale de Louis-Antoine de Bourbon.
Restauration.
Monture en laiton doré. Poignée en crosse de pistolet à plaquettes d'ébène à liseré d'ivoire et décorée d'une larme flammée en ivoire, symbole de deuil (18a). Croisière et chape au même décor que le n° 12 frappée également des lettres " L.A. " pour Louis Antoine, et bracelets au profil de Henry IV. Lame modérément courbée, à double gouttière, à pointe en langue de carpe, gravée de motifs orientaux et signée au talon : *Fque d'armes Pirmet à Paris.*

Nota - B. Croissy émet l'hypothèse très vraisemblable que ce sabre fut porté en signe de deuil par Louis Antoine après l'assassinat par Louvel en 1820, du duc de Berry (Photo Sotheby's).

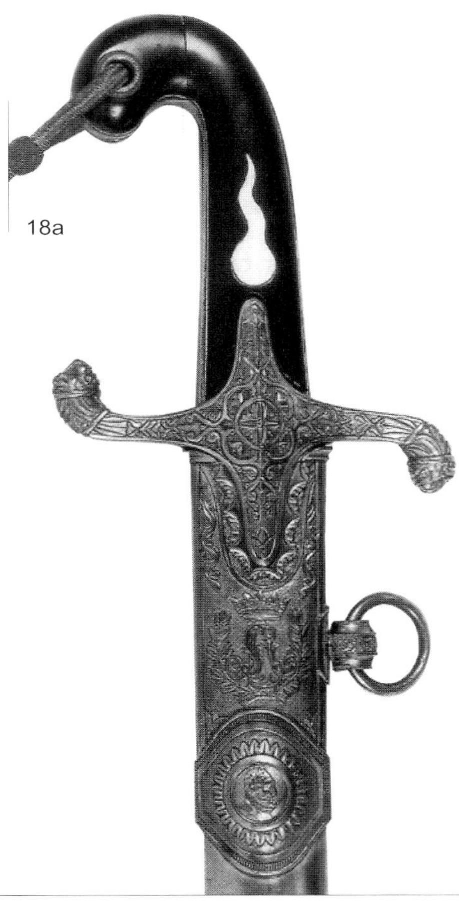

18a

19

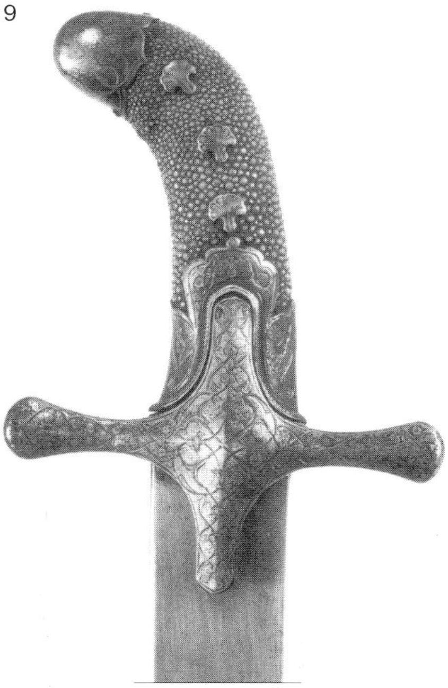

19 - Sabre à l'orientale du général Kleber.
Ce sabre dont nous donnons la silhouette est assez traditionnel dans sa forme générale. La poignée est recouverte de peau de requin et à trois boutons de rivure en forme de champignon, très caractéristique. Toutes les parties métalliques sont finement décorées au trait représentant des motifs floraux très stylisés, voire géométriques (M.A.P. J – Po 2712 – (A 12730)).

20

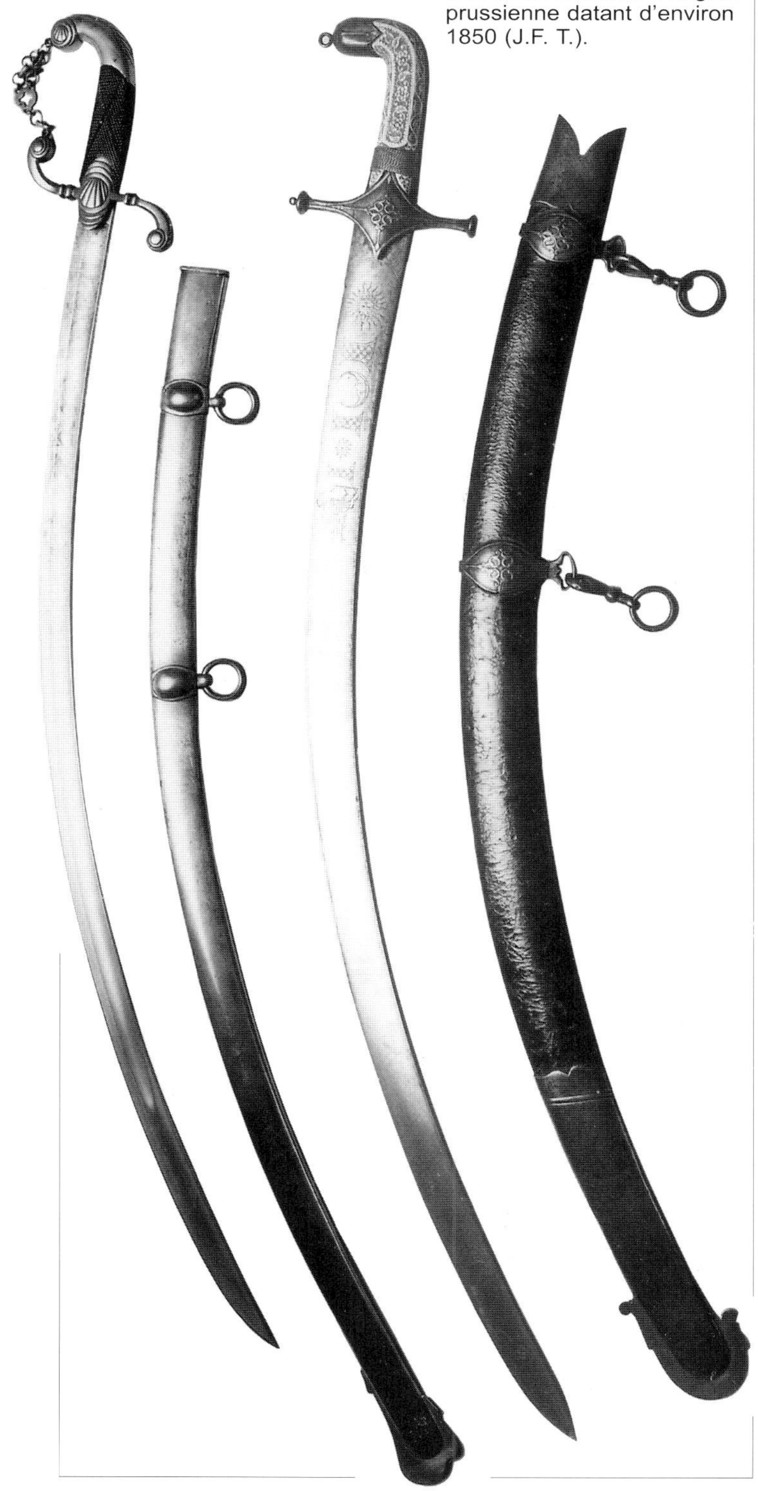

20 - A titre d'exemples. sont donnés ci-contre deux sabres à l'orientale d'origine prussienne datant d'environ 1850 (J.F. T.).

Sous famille I b – Crosse figurative

Les premiers de ces sabres ont été créées par N. Boutet en s'inspirant des sabres à l'orientale rencontrés lors de la Campagne d'Egypte. Leur principales caractéristiques sont les suivantes : calotte à courte queue, en crosse terminée le plus souvent par une tête de lion tenant dans sa gueule une chaînette qui rejoint le quillon antérieur relevé alors que le postérieur est infléchi vers la lame. La croisière est formée d'oreillons losangiques. Poignée en ébène quadrillée ou parfois sculptée en écailles. La lame est de forme variée, plus ou moins courbée. Fourreau recouvert de chagrin, ou métallique à trois ou quatre garnitures.

Lors de la bataille de Marengo, le 14 mai 1800, Bonaparte portait une arme de cette facture, d'où le nom générique *" à la Marengo "* qui est donné à ce type de sabre. Tous les sabres à la Marengo sont des sabres de luxe, en bronze doré et finement ciselés. L'un d'entre eux fut offert au général Victor avec l'inscription suivante, entre la cuvette du fourreau et le premier bracelet de bélière : *Bataille / de Marengo / le 25 Prairial an 8 / commandée / en personne / par le 1er Consul / donné par le Gouvt / de la Rép. Fran. / au Général / Victor* (Musée de l'Armée).

21

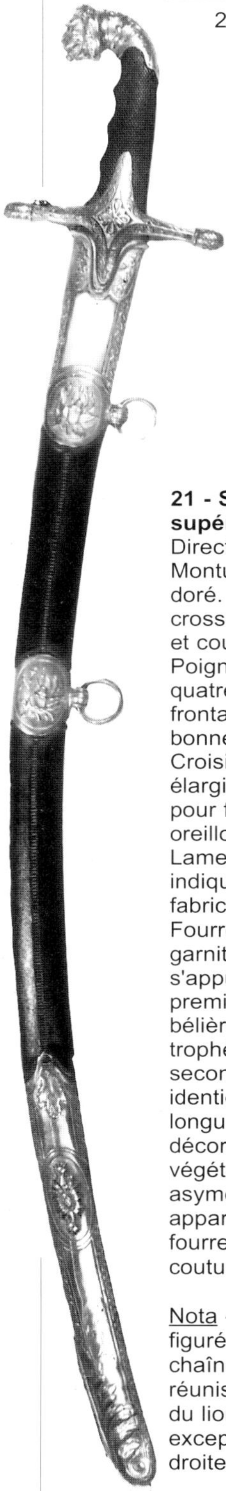

21 - Sabre d'officier supérieur.
Directoire, Consulat. Monture en laiton doré. Calotte en crosse, à tête de lion et courte queue. Poignée d'ébène à quatre dépressions frontales pour une bonne prise en main. Croisière droite, élargie en son milieu pour former les deux oreillons à l'orientale. Lame peu courbée indiquant une fabrication française. Fourreau à quatre garnitures : chape s'appuyant sur le premier bracelet de bélière orné d'un trophée d'armes, un second bracelet identique au premier, longue bouterolle décorée de motifs végétaux, à dard asymétrique. Le cuir apparent de ce fourreau est à couture frontale.

<u>Nota</u> - L'exemplaire figuré a perdu la chaînette qui réunissait la gueule du lion à la croisière, exceptionnellement droite.

22

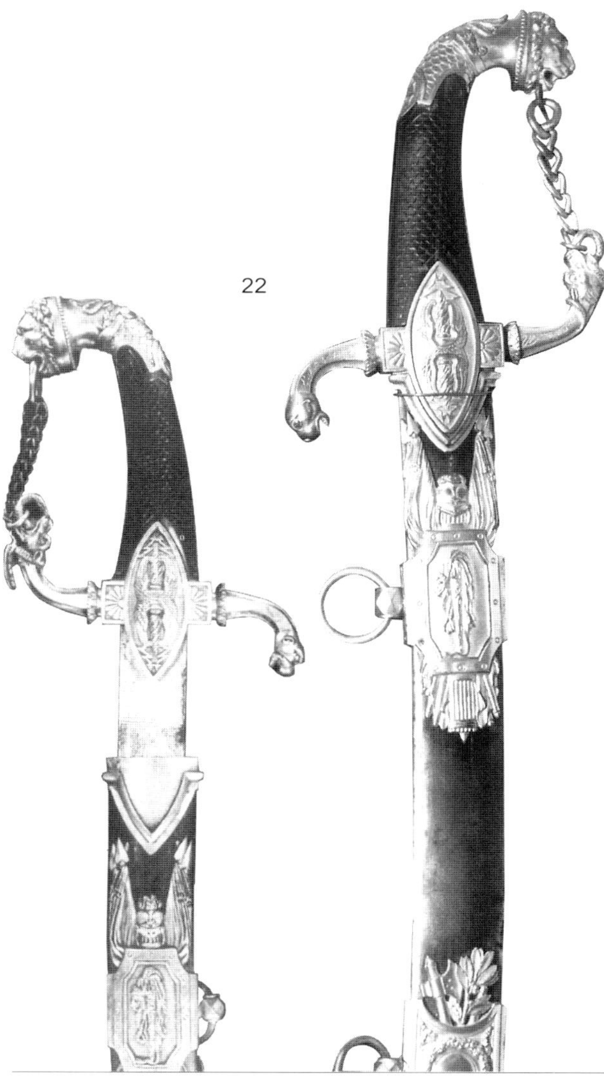

22 - Sabre de général dit " à la Marengo ".
Directoire Consulat. Monture en laiton ciselé et doré. Poignée en crosse, à calotte à courte queue et terminée par un mufle de lion. Fusée en ébène sculptée d'écailles. Quillons inversés, l'antérieur unissant la calotte par une chaînette. Oreillons en navette ciselés de foudres ailées. Lame légèrement courbée. Fourreau de fer à quatre garnitures en laiton doré ciselées et découpées : chape formée d'un faisceau de drapeaux surmonté d'un bracelet, second bracelet posé sur une hache et une branche de laurier ; troisième garniture représentant un soubassement à l'antique et bouterolle ornée d'un carquois et ses flèches.

23

23 - Sabre de Tambour-Major.
Vers 1810.
Monture en laiton doré. Calotte en crosse à tête de lion à courte queue. Poignée de corne noircie. Quillons inversés, terminés par une tête de bélier, l'antérieur étant réuni à la calotte par une chaînette. Oreillons losangéïformes ciselés de la tête du dieu Mars casquée. Lame courbe, longue, suivant les exemplaires, de 73 à 92 cm, à pan creux et gouttière au fort. Fourreau en laiton ciselé de feuillages sur toute sa longueur garni d'un médaillon festonné médian dans lequel est inscrit une rose. Le haut du fourreau porte deux crochets latéraux de suspension ; dard asymétrique terminé par deux rosaces.

25

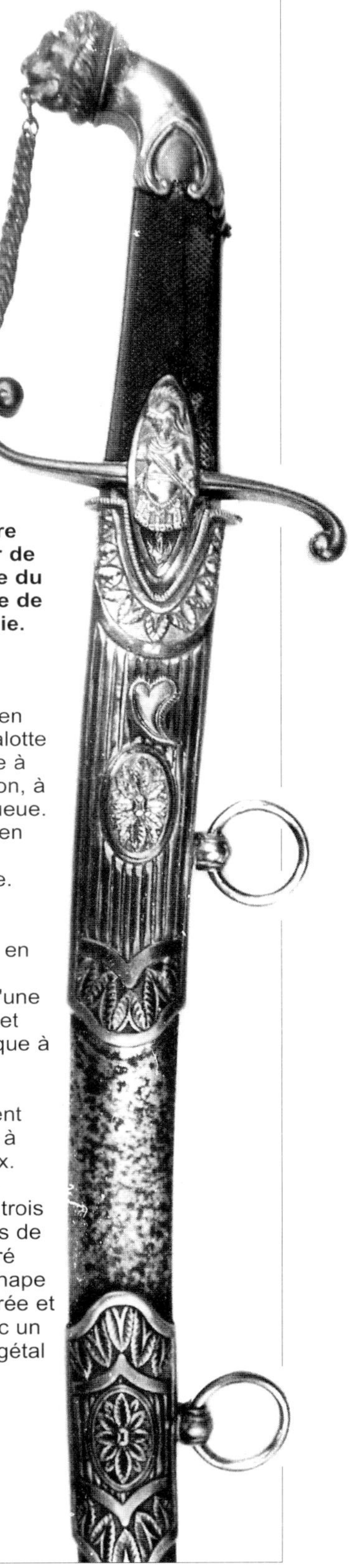

25 - Sabre d'officier de Cavalerie du Royaume de Wesphalie.
Epoque Premier Empire.
Monture en laiton. Calotte en crosse à tête de lion, à courte queue. Poignée en ébène quadrillée. Quillons inversés. Oreillons en navette ciselée d'une cuirasse et d'un casque à l'antique. Lame légèrement courbée, à pan creux. Fourreau d'acier à trois garnitures de laiton doré dont la chape est rainurée et orné avec un décor végétal central.

24

24 - Sabre d'officier supérieur ou de Tambour-Major.
Premier Empire.
Ce sabre est surtout caractérisé par ses oreillons en losanges et par sa poignée d'ébène finement quadrillée. D'autre part son fourreau est sobrement décoré.

26 - Sabre de Tambour-Major.
Début du Premier Empire.
Monture en laiton, dite à la Marengo. Calotte au masque léonin, à courte queue. Poignée d'ébène quadrillée. Croisière aux quillons inversés, terminés par des mufles de molosses. Oreillons en forme d'écu timbré d'une aigle posée sur des foudres. Lame à faible courbure, longue de 76 cm, à pan creux et gouttière, bleuie et gravée de motifs dorés classiques. Fourreau en laiton, ciselé d'une frise de palmettes sur les deux faces et sur le *recto* de trophées, de scènes mythologiques et de combat, avec sur le bas, un faisceau de licteur et une pointe de flèche stylisée. Bouton de suspension ovalaire frappé d'une aigle couronnée (non figuré).

27 - Sabre à l'orientale offert à Bonaparte.
Consulat.
Ce sabre unique en son genre a été conçu par A. Fillberg de Stockholm et répond à la description ci-dessous.
Monture en bronze doré dont la calotte est en tête de cheval ainsi que le montre le schéma ci-joint. Poignée d'ébène finement quadrillée et ornée de deux médaillons, figurant l'un Athéna et l'autre la République. Croisière décorée de six médaillons représentant le même visage moustachu. Oreillons en ovale ciselés du portrait d'une reine égyptienne (?) de grande beauté. Lame gravée en lettres arabes et dorées *" Tu anéantiras tes ennemis et tu protègeras les Musulmans "*. Chape du fourreau, sous le triangle et l'œil de la vigilance, décorée de deux médaillons, l'un représentant la République avec les lettres de fantaisie R.F. sur un autel et l'autre, le Premier Consul à cheval au milieu de ses généraux (M.A.P. Ca 12 – J 387/1 – 5027).

27

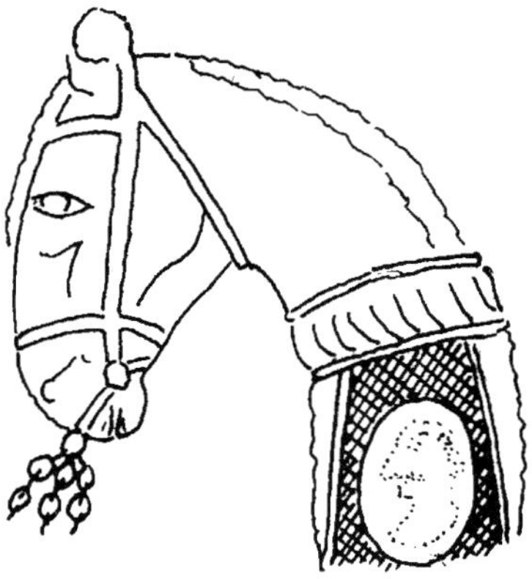

28

Famille II – Poignée en crosse, à demi-oreillons basaux

Les sabres de cette Famille sont également inspirés, dans leur conception, des sabres à l'orientale.

28 - Sabre de Tambour-Major.
Consulat - Premier Empire.
Monture en laiton doré dite *" à la Marengo "*. Calotte en crosse terminée par une tête de lion à courte queue. Poignée en ébène quadrillée avec quatre dépressions frontales pour une bonne prise en main. Croisière à quillons inversés, à tête de chien, l'antérieur étant relié à la crosse par une chaînette. Les oreillons basaux sont ornés d'une toile d'araignée. Lame courbe, gravée. Fourreau en bois recouvert de cuir à couture frontale en chaînette, à trois garnitures en laiton doré : chape ciselée d'un trophée de drapeaux, de tambour, de grenade... bracelet médian ciselé de drapeaux et de feuillage, longue bouterolle décorée en suite à dard asymétrique.

29 - Sabre de luxe d'officier général.
Consulat - Premier Empire.
Monture en laiton, dite *" A la Marengo "*. Calotte en crosse terminée par une face léonine à courte queue décorée de palmette et de volutes. Poignée à plaquettes de nacre gravée d'un motif énigmatique. Quillons inversés : l'antérieur en tête de molosse tenant dans sa gueule un serpent soutenant une chaînette qui aboutit dans la gueule du lion de la calotte ; le quillon postérieur est en tête de bélier. Oreillons en losange ciselé d'une tête à l'égyptienne. Lame de 80 cm, plate, à forte courbure. Fourreau entièrement ciselé de palmes, de volutes sur presque toute sa surface ; bracelet en bossette à l'orientale, moulurée sur les bordures et décorées de motifs végétaux en leur centre. Dard en laiton, symétrique, crénelé sur les bords. Pour faciliter les mouvements de la lame, la chape du fourreau est munie d'une lame d'acier mobile (non figuré).

Le 12 thermidor de l'An IX, le Premier Consul attribue à six compagnies la charge de l'exécution *des règlements relatifs à la surveillance, garde et police des dits ports et arsenaux ainsi que ceux qui concernent l'inscription maritime* rappellent D. Casanova et J.B. Bastié (1993). En 1813, neuf compagnies sont installées à Anvers, le Havre, Brest, Lorient, Rochefort, Toulon, Gêne, Amsterdam et Hambourg.

Un sabre particulier, très inspiré des sabres à l'orientale, est attribué aux officiers et aux hommes de troupe.

30 - Sabre d'officier de la Gendarmerie des ports et Arsenaux,
Début XIXème siècle.
Monture de laiton. Poignée en crosse de corne noire à six ou sept cordons, sans filigrane ; une chaînette part de la calotte et rejoint le quillon antérieur, identique au quillon postérieur, tous deux inclinés et ornés d'une fleurette. Entre ces deux quillons, large demi-oreillon frappé d'une ancre de marine flanquée de deux rameaux de laurier et d'un soleil à sa base (30a). Lame à flèche de 10 - 12 cm, à pans creux. Fourreau à trois garnitures simplement gravées au trait à deux anneaux de bélière, dard asymétrique (M. G. de Melun).

30

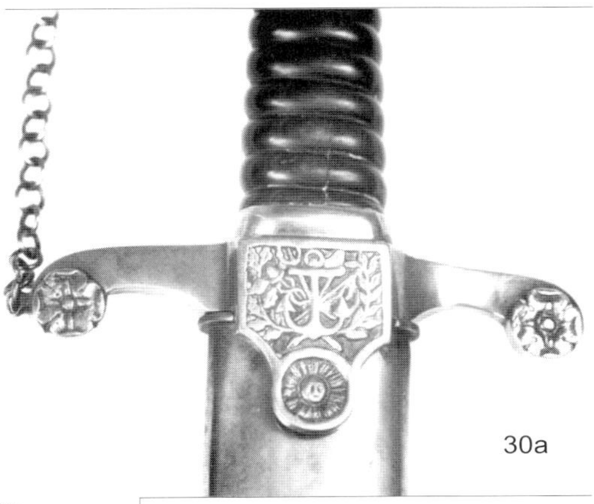

30a

31

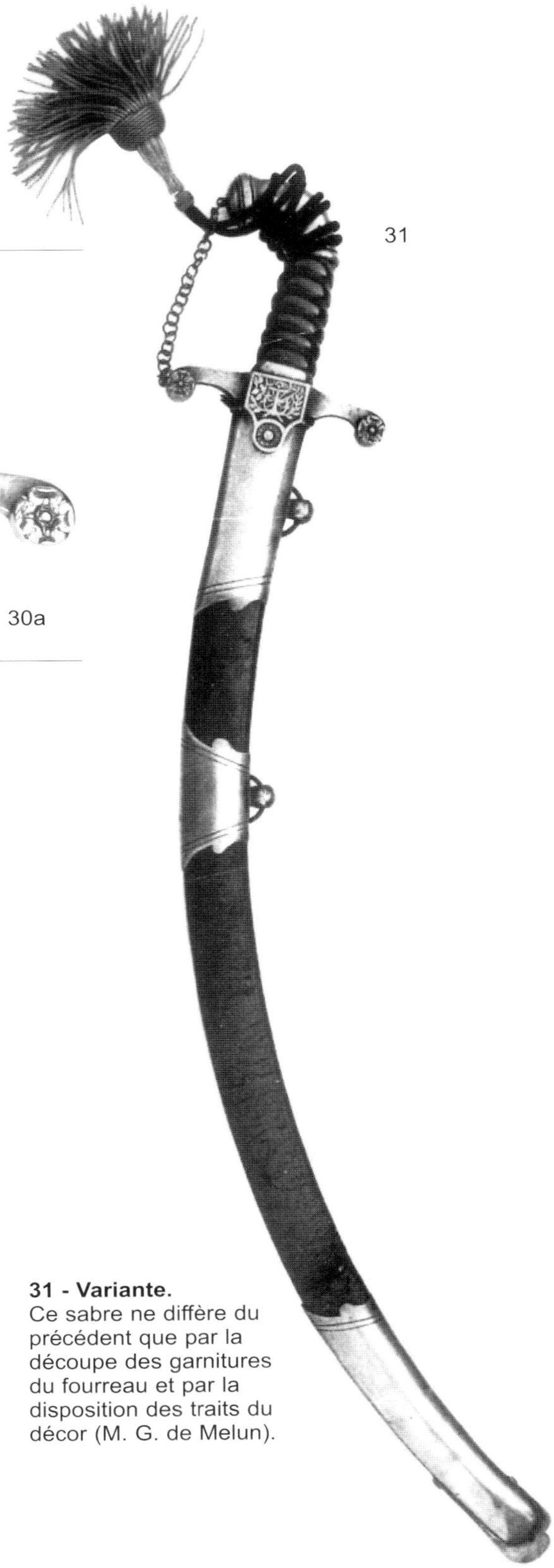

31 - Variante.
Ce sabre ne diffère du précédent que par la découpe des garnitures du fourreau et par la disposition des traits du décor (M. G. de Melun).

32

32a

32 - Sabre d'officier de Gendarmerie des Ports.
Début XIXème siècle.
Monture en laiton. Calotte en crosse. Poignée en corne à cinq cannelures dans le fond desquelles court un fin filigrane torsadé. Arc de jointure remplacé par une longue chaînette. Croisière à quillons inversés, élargie en sa partie médiane pour former un oreillon basal arrondi. L'ensemble est décoré d'étoiles à ses extrémités, l'oreillon d'une ancre non étalinguée, flanquée de deux rameaux feuillus s'étendant sur la croisière (32a). Lame de 81 cm plate, très courbée, sans ornementation. Fourreau de cuir, à trois garnitures de laiton, se terminant par un dard symétrique (M.R.).

33

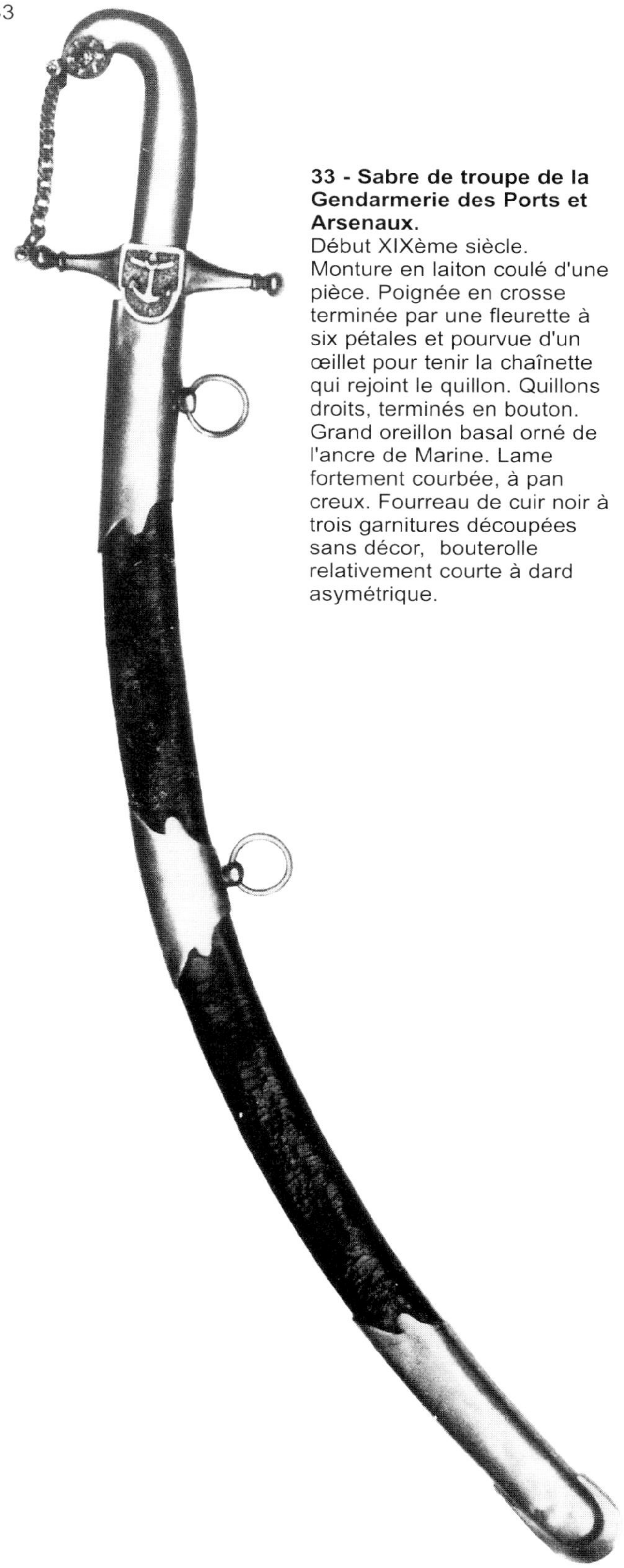

33 - Sabre de troupe de la Gendarmerie des Ports et Arsenaux.
Début XIXème siècle.
Monture en laiton coulé d'une pièce. Poignée en crosse terminée par une fleurette à six pétales et pourvue d'un œillet pour tenir la chaînette qui rejoint le quillon. Quillons droits, terminés en bouton. Grand oreillon basal orné de l'ancre de Marine. Lame fortement courbée, à pan creux. Fourreau de cuir noir à trois garnitures découpées sans décor, bouterolle relativement courte à dard asymétrique.

Famille III – Poignée droite terminée par une tête humaine, d'animal ou par une calotte simple

Dans la Famille III nous avons réunis quelques sabres présentant les caractères communs suivants : pas d'arc de jointure, poignée généralement d'ébène non filigranée... C'est dans cette famille que l'on trouvera les sabres de la Légion Corse à tête de Maure enturbannée, les premiers modèles non réglementaires destinés aux Tambour-Majors à calotte au coq ou à l'aigle... A noter que la calotte de ces sabres, ou glaives, peut ne pas être figurative et que la croisière peut porter des oreillons ou en être démunie. un glaive de *structure légère* dont des exemples ont été cités dans " Les Epées " de J. Lhoste (1997, p. 449 - 461) a été inclus dans cette Famille III.

34

34 - Sabre de la Légion Corse 1769 - .1795
C. Ariès considère ce sabre comme étant celui d'un officier subalterne de Grenadiers (XVIII, 4, 1970). Il s'agit d'un sabre à l'orientale conçu 30 années avant la grande mode consécutive à la Campagne d'Egypte. Il semble que ce sabre ait été inspiré par l'armement des pirates tunisiens qui étaient alors maîtres de l'île ...
Monture en laiton. Calotte figurant une tête de Maure moustachu. Poignée en ébène quadrillée. Croisière droite avec un petit anneau terminant le quillon antérieur, s'élargissant pour former des oreillons en losange curviligne. Lame longue de 54,1 cm, flèche de 10,4 cm gravée *LEGION CORSE N° 1* et d'une fleur de lys au talon sur le *verso*. Fourreau de cuir noir ou fauve sur alèse de bois, à trois garnitures décorées, au trait. Chape à anneau de laiton, bouterolle terminée par un bouton plat (catalogue " Le Hussard " n° 21 p. 34, figure 559).

Nota - Certains exemplaires ne portent pas l'inscription Légion Corse mais seulement une fleur de lys près du talon de la lame, sur les deux faces.

35

35 - Sabre de Tambour-Major.
Consulat.
Monture en laiton. Calotte en tête de coq à courte queue. Poignée en ébène cannelée verticalement. Quillons horizontaux terminés par des faces léonines. Oreillons ovoïdes, à bordure finement ciselée dans lequel est inscrit Hercule en pied avec sa massue. Lame unie légèrement courbée, à pan creux. Fourreau en laiton doré à deux bracelets ciselés, encadrés de feuilles stylisées ; le dernier tiers du fourreau, à l'emplacement de la bouterolle, est décoré de motifs végétaux, feuilles et fleurettes ; dard antérieur et basal.
Ce type de sabre possède une chape à deux crochets sur la chape ou un bouton.

36 - Sabre de Tambour-Major.
Début du Premier Empire.
Monture en laiton. Calotte en tête d'aigle, sans queue. Poignée en ébène quadrillée. Quillons inversés terminés par une tête vaguement léonine. Oreillons en navette, ciselés d'une aigle couronnée, posée sur un bâton de foudre. Longue lame à pan creux (M.E. - S. de P.).

36

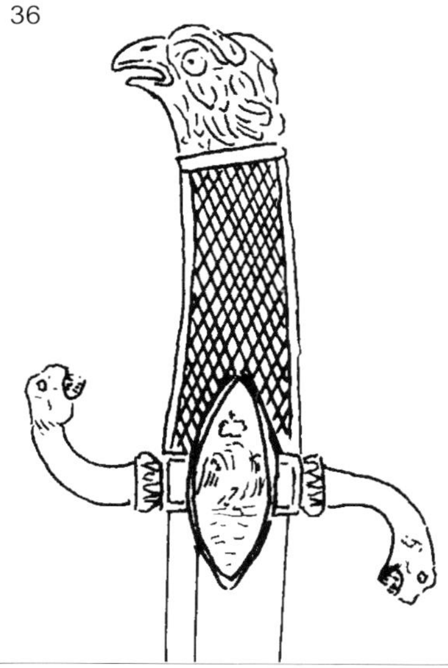

37 - Sabre à l'orientale.
Vers 1830.
Monture en argent. Poignée en courte crosse et croisière droite. Lame courbée, à large pan creux, signée *MANCEAU à PARIS*. Fourreau en argent à trois bracelets entre lesquels sont ciselés des décors très fins parmi lesquels se trouvent trois trophées ; à noter que seul le bracelet situé à la base de la chape est muni d'un anneau de bélière. Le fourreau est signé *MANCEAU à PARIS*, accompagné de J. M. pour préciser qu'il s'agit de Jules Manceau.

Nota - Il existe plusieurs sabres de ce modèle.
Un exemplaire daté de 1839 se trouvait dans la collection Grancsay. Un autre est conservé au Musée de l'Empéri à Salon de Provence et aurait été offert par le duc d'Aumale lors de la reddition de l'Emir Abd-El-Kader, le 24 décembre 1847, à Djemmaa-Ghazaouet. Mais Abd-El-Kader ne l'accepta qu'un instant et l'offrit au général La Moricière.

37

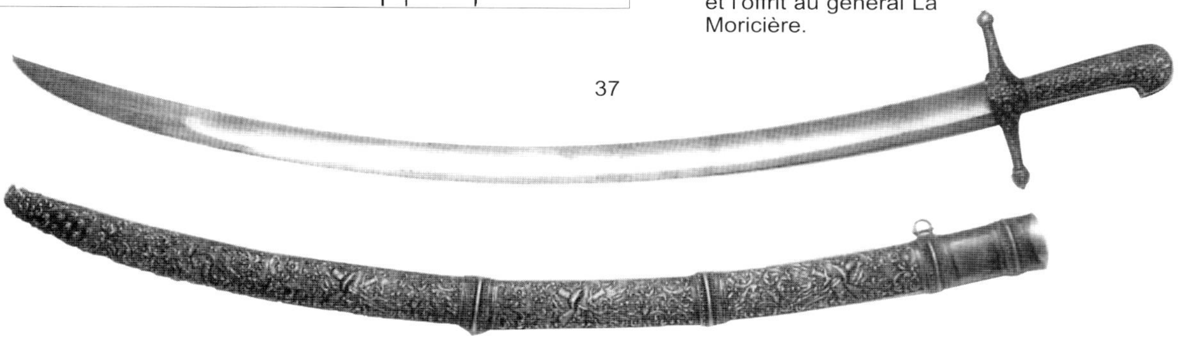

38

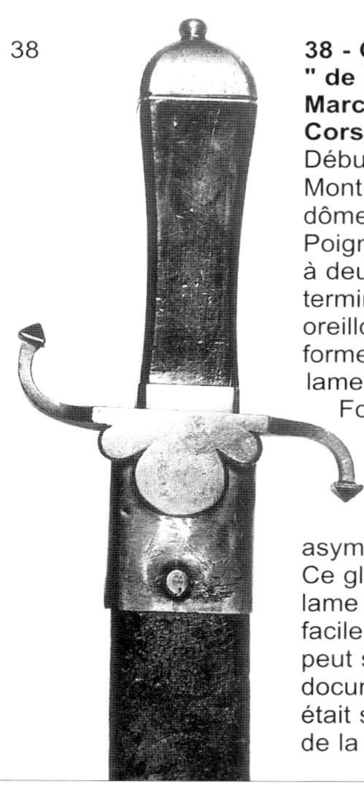

38 - Glaive dit " de Port " ou " de Bord " de la Marine Marchande voire de Corsaire.
Début du XIXème siècle.
Monture en laiton. Calotte en dôme à bouton de rivure. Poignée en corne noire. Garde à deux quillons inversés terminés par un bouton, deux oreillons basaux découpés en forme de trèfle. Large et courte lame légèrement courbée.
Fourreau de cuir noir, chape à bouton, bouterolle à bordure découpée en accolade et protégée par un petit dard asymétrique.
Ce glaive remarquable par sa lame légèrement incurvée est facilement reconnaissable. On peut supposer, en l'absence de documentation officielle, qu'il était surtout destiné aux marins de la Marine Marchande.

39 - Glaive de l'Administration.
Second Empire.
Monture en laiton. Poignée en corne, à pans et calotte hémisphérique en laiton. Croisière terminée par deux quillons terminés en boules. Lame à deux pans creux et deux tranchants, signée au talon *Klingenthal* sur une face et *Coulaux* sur l'autre. Fourreau de cuir, à deux garnitures agrafées, chape à pontet.

39

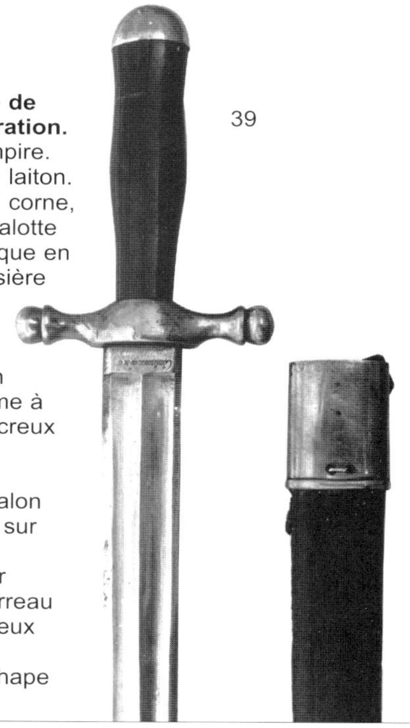

40

40 - Glaive des Services administratifs de la Marine.
Second Empire.
Poignée de bois noirci et verni, à calotte de laiton hémisphérique sans bouton de rivure. Croisière droite à quillons terminés en boules, avec sur le noeud de croisière un médaillon ovale frappé d'une ancre en relief sur fond martelé. Lame droite de 49,5 cm, large au talon de 3 cm, à deux pans creux sur toute sa longueur. Cette lame est signée sur la face frontale *Klingenthal*, avec un poinçon au " B " couronné d'un contrôleur d'environ 1840 (40a), et sur la face postérieure CoulauxAc et poinçon au " D ", probablement d'un maître-ouvrier (40b). Fourreau de cuir à couture médiane postérieure, à deux garnitures de chape à barette, bouterolle terminée par une boule striée verticalement.

40a

- **a** , médaillon à l'ancre. Sur la lame, signature de Klingenthal et poinçon au " B " couronné sur la face frontale.

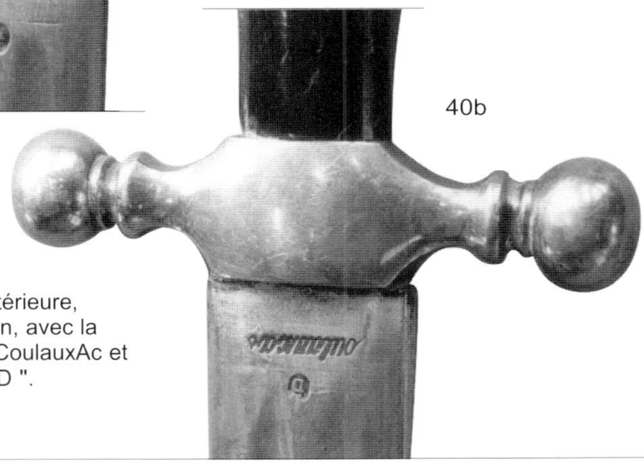

40b

- **b** , face postérieure, sans médaillon, avec la signature de CoulauxAc et poinçon au " D ".

41

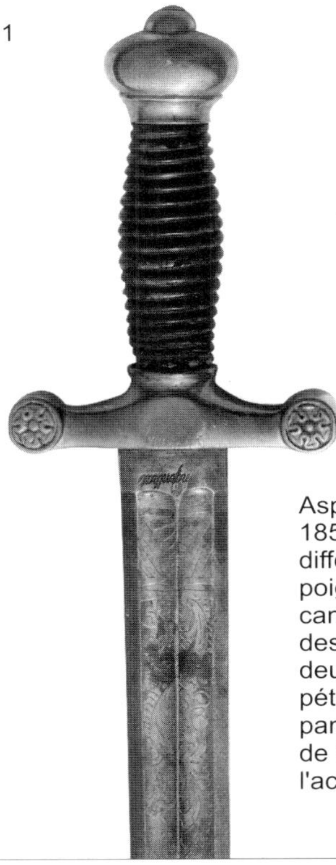

41 - Glaive fantaisie de la Garde Nationale .
Second Empire.
Monture en laiton.
Aspect général du modèle 1855, mais avec les différences suivantes : poignée en bois à seize cannelures aux extrémités des quillons ornés de deux fleurettes à six pétales, la lame à deux pans creux est décorée de feuillages gravés à l'acide (B. L.)

42 - Glaive de garde ou de surveillant de la voierie.
Second Empire.
Voirie / vicinale en lettres manuscrites gravées à l'acide sur la lame d'un glaive du Second Empire, dans la tradition des petits glaives civils. Pommeau orné de feuilles d'acanthe. Fusée en ébène. Croisière terminée par des têtes de chiens avec une petite aigle sur le nœud de corps sur la face antérieure et un écusson vierge sur la face postérieure.

43

43 - Epée-glaive de membre de Tribunat.
Consulat et Premier Empire (1800 - 1807).
Pommeau ovalaire, tronqué en haut et en bas, orné du profil d'Athéna et monté sur un col lisse reposant sur une bague ciseléede feuillage. Fusée en ébène cannelée verticalement. Quillons se terminant chacun par une tête de bélier avec l'oeil de vigilance, symbole maçonnique, sur l'écusson de la croisière. Plateau ovalaire, convexe, ajouré et portantles Tables de la Loi sur deux galives croisées (43a). Lame à section losangéïforme (losangée), bleuie et gravée de motifs dorés. Fourreau de cuir noirci à deux garniture, ornés de bagues et bouton de chape en palmette (B. C.)

43a

44

44 - Glaive de cérémonie de général.
Restauration.
Monture en bronze doré. Pommeau au buste de Minerve. Fusée à plaquette de nacre cannelée. Croisière droite à quillons en palmettes. Lame losangée, dorée et bleuie au tiers. Fourreau de bois recouvert de galuchat blanc. Longue chape à bouton en anneau ciselée de motifs floaux ainsi que la bouterolle. Le bracelet médian porte les Armes de France sur un trophée d'armes.

Chapitre II

Poignée droite en laiton moulé

Tous les sabres-glaives à monture en laiton moulé ont été réunis dans ce chapitre. Quatre familles bien distinctes ont été créées dont on trouvera les caractéristiques ci-après :

Famille IV – Fourreau plat à découpes ressortant sur fond de couleur

Le type le plus connu de ces sabres est le glaive de l'Ecole de Mars. Mais il existe d'autres modèles issus des fantasmes révolutionnaires de 1789 et qui trouveront leur prolongement au centenaire de la Révolution et dans les armes de théâtre. Ces sabres sont présentés ainsi :

Sous famille IV a – Monture à antennes

Le modèle le plus représentatif de cette famille est le glaive de l'Ecole de Mars dont on trouvera ci-après la description. Disons simplement que le fourreau est la partie du glaive la plus typique : attelles de laiton consolidées par des barrettes transversales laissant apparaître dans des *crevés* les alèses de bois recouvertes de drap coloré, avec son dard arrondi orné d'une palmette. Or, ce glaive n'est pas le seul à présenter un tel fourreau. Mais toutes les autres composantes des glaives apparentés, sont différentes ainsi que nous allons le voir. Que sont ces glaives ? A qui étaient-ils attribués ? Deux questions qui restent encore aujourd'hui sans réponses satisfaisantes, faute de documents d'époque, nous ne pouvons donc faire que des hypothèses.

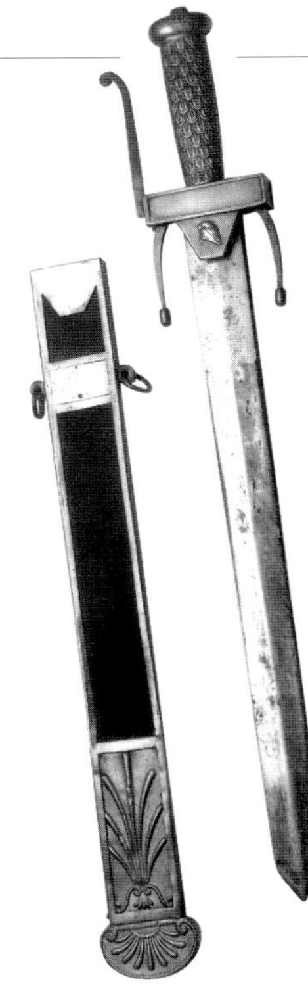

45

45 - Glaive de l'Ecole de Mars.
1er juin 1794 - fin septembre de la même année.
Monture en laiton. Pommeau ovoïde à bouton de rivure. Poignée à douze rangées d'écailles (45a). Croisière rectangulaire à oreillons trapézoïdaux ciselés d'un bonnet phrygien en relief (45b). De l'intérieur de la croisière partent vers le haut une branche de garde en crosse et vers le bas deux branches également en fer, terminées par une olive en laiton, les quillons. Lame droite de 48 à 55,5 cm de long, large au talon d'environ 4 cm, à arête médiane atténuée et pointe en triangle dans le prolongement de l'arête

45a

45b
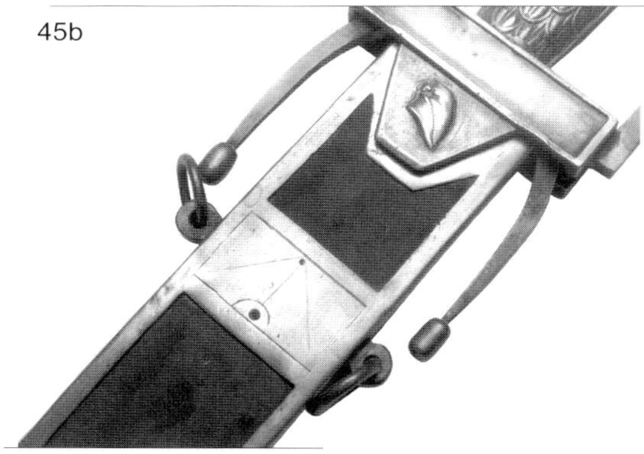

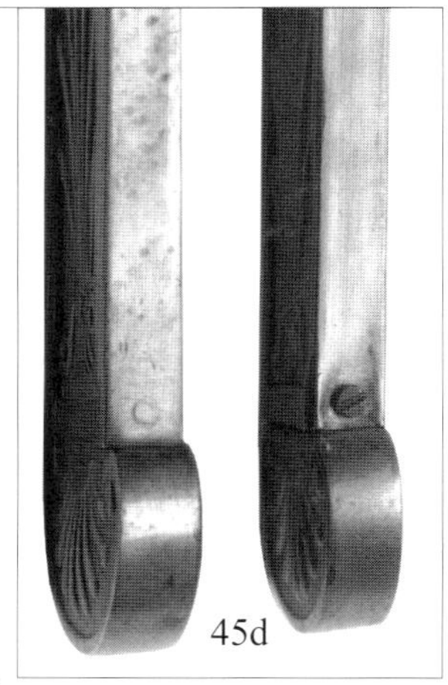
45d

médiane. Fourreau en laiton laminé, plat et découpé de deux crevés, ménageant une sorte de trapèze, un peu plus grand que les oreillons, en guise de chape, une barrette servant de bracelet pour y fixer les anneaux de bélière, placés symétriquement, pour le port au baudrier. Cette barrette est parfois gravée du niveau de l'égalité (45b). Le fourreau se termine par une bouterolle estampée d'une longue palmette (45c) et d'un dard en demi lune fixé par des rivets ou des vis (45d). Dans les crevés, entre les garnitures, apparaît un drap de couleur rouge, vert, jaune ou même bleu pour les Centuries (voir Ariès).

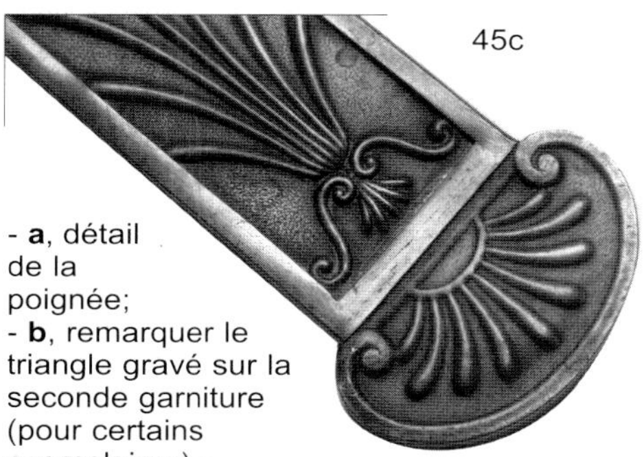
45c

- **a**, détail de la poignée;
- **b**, remarquer le triangle gravé sur la seconde garniture (pour certains exemplaires) ;
- **c**, détail de la bouterolle et du dard à palmette ;
- **d**, fixation du dard, à gauche par rivure, à droite par vis.

Marquage – La lame est souvent poinçonnée au faisceau de licteur, sous l'oreillon et de ce fait peu visible. Le poinçon au coq de Klingenthal se rencontre parfois, ainsi que la marque du fourbisseur Bru.

<u>Nota</u> - M. Pétard (1990) écrit : *" La fabrication puis la livraison de 3500 sabres ne peut être réalisée que par fractions échelonnées sur deux ou trois mois. Un telle quantité dût mobiliser plusieurs ateliers... d'où d'immanquables variations de détails dans la longueur de la lame voire dans la confection du fourreau "*.
En ce qui concerne l'Ecole de Mars et la vie de ses élèves, se reporter à l'étude de G. Cottereau (1905) ou à J. Lhoste (1996).

46
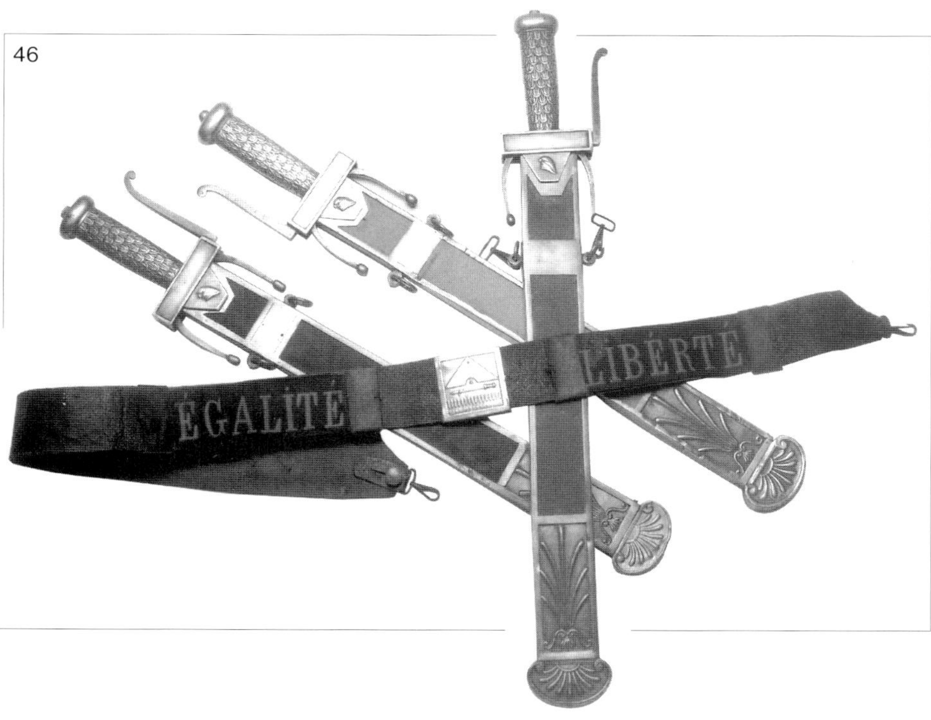

46 - Trois glaives de Mars. De gauche à droite, à drap rouge (le plus foncé), le jaune (le plus clair) et, placé verticalement, le vert. Remarquer le baudrier marqué *EGALITE - LIBERTE* avec boucle ornée du niveau, d'un glaive et d'une frise d'épis de blé. (B. C.)

47 - Sabre de Représentant du Peuple
Vers 1792 - 1794.
Ce sabre dessiné par le peintre David est en fait, inspiré du sabre marocain dit " Nimcha ". On remarquera les trois antennes pendantes servant de quillons et l'antenne dressée qui sert de branche de garde. Longue chape ornée d'un bâton de foudres ailé et la bague sur laquelle est inscrit *PEUPLE / FRANÇAIS /.* Ce sabre appartint à Billaud-Varenne et est conservé au Musée Carnavalet de Paris.

47

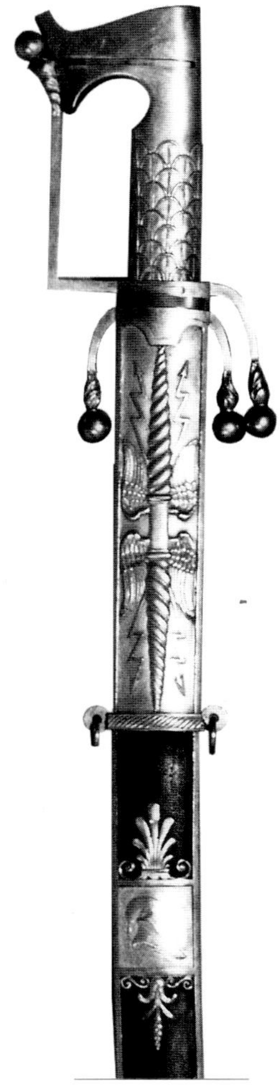

Sous-famille IV b – Monture dépourvue d'antenne

Les sabres de cette Sous-famille gardent une partie de leur mystère quant à leur attribution. Celles que nous donnons ne sont, bien souvent, que des hypothèses.

48 - Glaive dérivé du modèle de l'Ecole de Mars.
Monture en laiton moulé. Pommeau à côtes de melon (48a) surmontée d'un bouton de rivure. Poignée de 11 cm décorée de trois languettes frontales, croisière rectangulaire, unie, avec des oreillons trapézoïdaux ainsi que le sabre de Mars, mais sans décor (48b). Lame de 40 cm de longueur, à faible arête médiane, large au talon de 3 cm. Fourreau de laiton à crevés faisant apparaître le tissus rouge : chape sans décor, ainsi que le bracelet médian portant les anneaux de bélières symétriques, bouterolle ornée d'une palmette en relief, dard du même modèle que celui du sabre de l'Ecole de Mars (48c).

Remarque - Ainsi que cela vient d'être décrit, ce glaive ressemble beaucoup au glaive de l'Ecole de Mars mais en diffère nettement par l'absence d'antenne. Son attribution reste énigmatique.

48

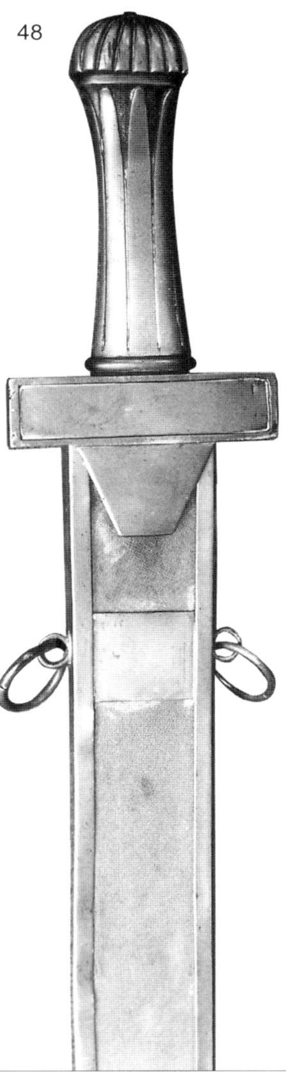

48a

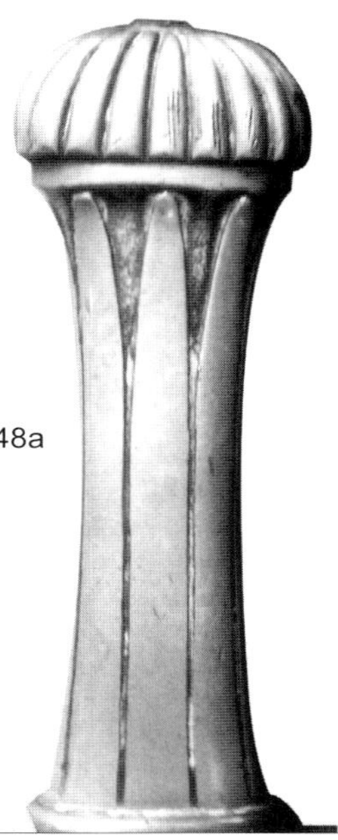

48b

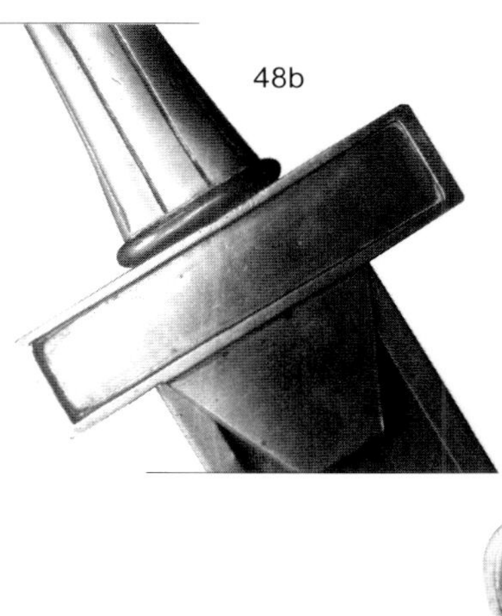

48c

49

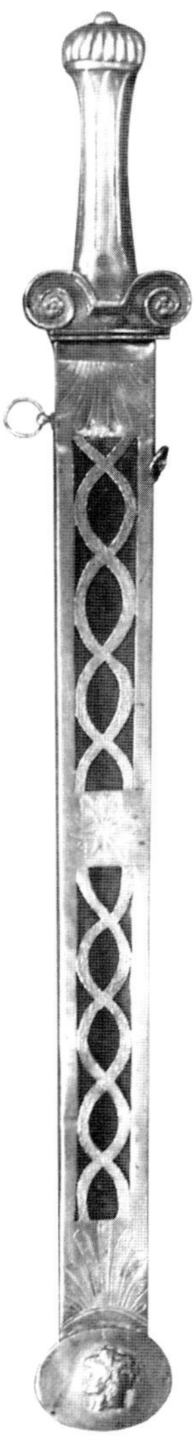

Ci-dessus : vue d'ensemble.
- **a**, détail de la poignée. Remarquer la position des anneaux de bélière ;
- **b**, détail des guirlandes découpées ornées de feuillages stylisés et de la fleur ornant la garniture médiane ;
- **c**, base du fourreau gravée d'une palmette rappelant celle du haut du fourreau (a) et dard ovoïde surmonté d'un profil à la Janus avec une tête d'homme (Mars ?) accolé à celui de la République.

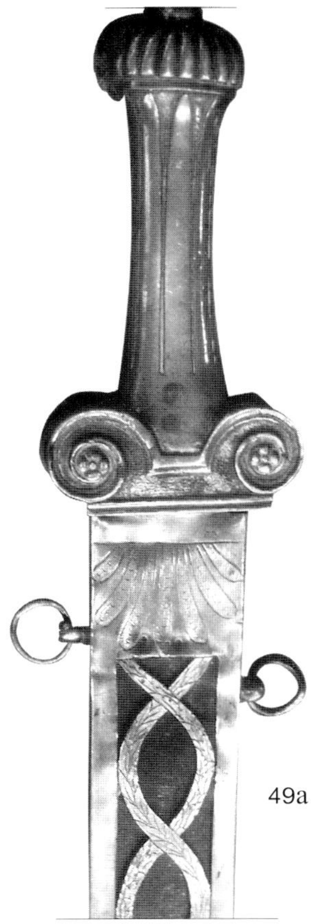

49a

49 - Glaive de cérémonie.
Révolution.
Monture en bronze doré. Poids total : 1,350 kg dont 0,500 Kg pour le fourreau. Pommeau ovoïde en côtes de melon, surmontée d'un bouton de rivure. Poignée décorée de trois languettes unies, partant à 2 cm de la croisière, se terminant en pointe sous la calotte et ménageant des intervalles sablés (49a). Croisière étroite, large de seulement 8,3 cm et constituée essentiellement de deux enroulements aboutissant chacun à une fleurette à quatre pétales.
Lame droite de 53,2 cm, large au talon de 4 cm et de 3,1 cm juste avant la pointe triangulaire. Une arête médiane parcourt toute la longueur de la lame. Or, cette lame est poinçonnée au talon du faisceau de licteur. Toutes ces caractéristiques laissent supposer que cette lame est de fabrication révolutionnaire. Fourreau long de 53,4 cm, large de 4,8 cm et de 1,5 cm d'épaisseur. En laiton ciselé et découpé, il présente les ornementations suivantes : sur la face frontale, un à plat long de 4,6 cm est ciselé de palmette (49a) puis un crevé d'environ 17 cm de long découpé de manière à ménager deux guirlandes de feuillage stylisé se croisant quatre fois, un aplat gravé d'une fleur (49b), un nouveau crevé identique au précédent et un aplat gravé de palmettes. Le fourreau se termine par un grand dard ovoïde (6 cm de long pour 3,7 cm de large). Sur ce dard, une pièce rapportée, en relief, figure le profil dun homme et le profil droit d'une femme qui ne peut être que celui de la République (49c). Le *verso* du fourreau est sans décor, avec les mêmes découpes que celui du glaive de Mars. A l'intérieur du fourreau, il existe des alèses de bois cousues sur un côté et maintenues par un chagrin bordeaux collé sur l'autre côté.

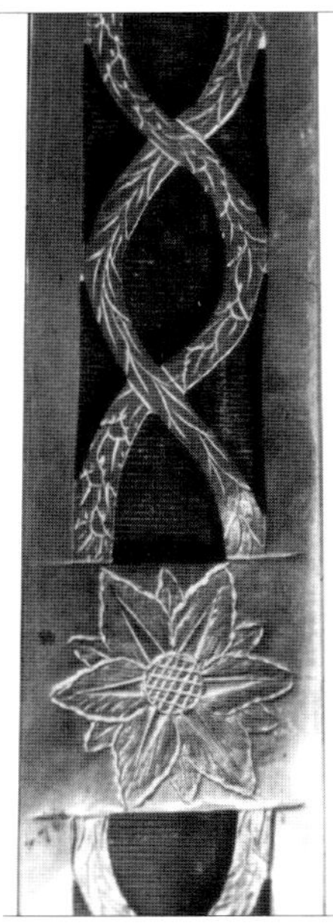

49b

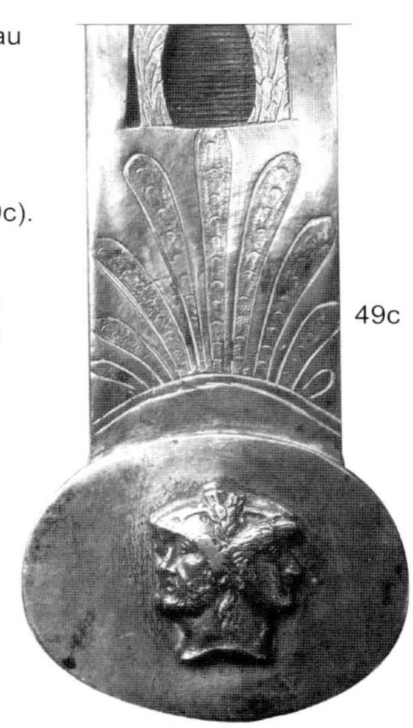

49c

50

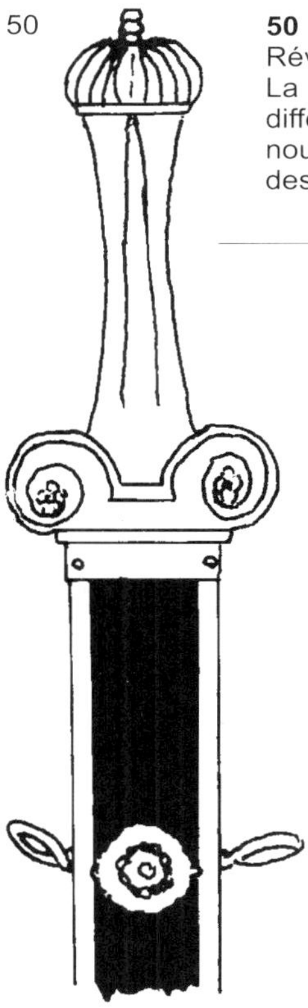

50 - Glaive des Cinq Cents (?).
Révolution.
La poignée de ce glaive présente beaucoup d'affinités avec le glaive n° 49. Seul le fourreau est différent en sa partie supérieure avec sa rosace située au niveau des anneaux. La partie inférieure nous reste inconnue. Selon F. Wilkinson, il s'agirait d'un glaive attribué aux membres du Conseil des Cinq Cents créé par la Constitution de l'An III.

51

51 - Glaive de notable ?
Révolution.
Ce glaive est à rattacher, par sa structure générale, au glaive de l'Ecole de Mars. Il en diffère par de nombreux détails : calotte unie à sa partie supérieure pour permettre la rivure, poignée de corne décorée de motifs végétaux très stylisés, se terminant à sa base par un fil torsadé bordé de petites perles, croisière rectangulaire, décorée d'une palme médiane flanquée de deux rameaux très fins. Lame droite unie, longue de 49 cm large au talon de 4,5 cm, plate.

Beaucoup plus spécifique est le fourreau. Il est recouvert de drap beige à trois garnitures : chape à deux anneaux soudés, ornée d'une palme dorée et prolongée d'une palme concave contenant une sorte de gland. Ce même motif se retrouve inversé, sur la courte bouterolle. Entre ces garnitures se trouve un long motif fixé par des griffes sur les bords du fourreau ; ce motif est constitué d'une couronne de feuillage flanquée de deux palmettes découpées, l'une vers la chape, l'autre vers la bouterolle.

En l'absence de tout document officiel, à notre connaissance, on peut être tenté de considérer ce glaive comme une arme de théâtre, alors que sa qualité de fabrication est bien supérieure à celle d'un accessoire de scène. Cette classification nous semble trop facile : bien des notables du Directoire ou même du Consulat ont bien pu se faire offrir une telle arme d'inspiration antique, toute allusion à l'ancien Empire Romain étant très à la mode en cette fin du XVIIIème siècle...

52

52 - Glaive fantaisie.
XIXème siècle.
Poignée en bronze moulé. Calotte en côtes de melon, Poignée ornée à sa partie supérieure d'une frise de motifs foliacés, croisière à deux quillons infléchis vers la lame, celle-ci mesurant 73,5 cm, à l'extrémité arrondie. Fourreau disparu.
A n'en point douter, il s'agit d'un montage composite.

53

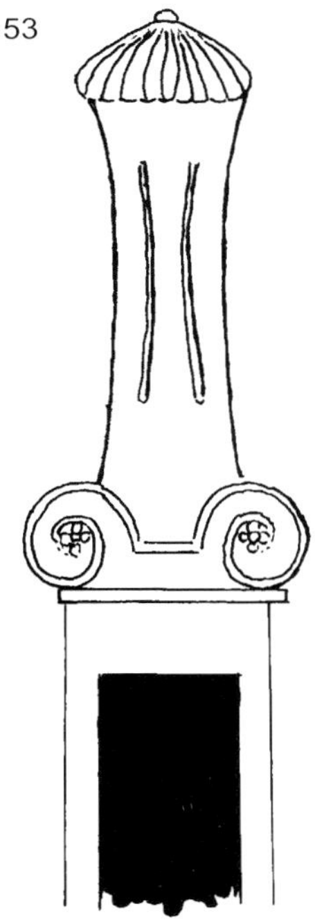

53 - Glaive fantaisie.
XIXème siècle.
Monture en acier fondu.
Pommeau en tranches de melon, à petit bouton de rivure. Poignée rétrécie en son milieu et ornée de deux longues côtes. Croisière étroite, à deux enroulements opposés terminés par une fleurette à quatre pétales. Lame à arête médiane. Fourreau construit sur le modèle de celui du glaive de l'Ecole de Mars, à deux anneaux de bélière, cuir rouge apparaissant dans les crevés et dard ovoïde à palmette.

54 - Glaive de cérémonie pour un haut dignitaire ou un officier supérieur.
Période révolutionnaire.
Ce glaive quoique bien différent du Glaive de l'Ecole de Mars, présente, avec ce dernier des affinités certaines : monture, en laiton doré à découpes sur un drap de couleur rouge sur la poignée. (le fourreau a disparu). Toutes les autres parties sont particulières à ce glaive : pommeau en tête d'aigle, fusée à médaillon ovale dans lequel est inscrit le buste de la République au front ceint de la couronne de laurier et placé sur deux armes d'hast en sautoir, et palme à sa base. Croisière droite avec en son milieu une étoile à cinq branches sur fond rayonnant, quillons en têtes de lion. Lame droite ; récupérée sur un couteau de vénerie, longue de 45 cm, large au talon de 4,5 cm, de l'Ancien Régime. En effet, celle-ci est à large pan creux et contre-tranchant, gravée au talon d'entrelacs inscrits dans un cartouche, puis d'un trophée d'armes et de végétaux, d'un ovale dans lequel est inscrit un dextrochère surmonté d'un croissant et de deux étendards, et pour finir un buste de personnage enturbanné.

Ce glaive est évidemment, du fait de la présence de la figuration de la République, d'au moins 1792. Que ce soit une arme d'apparat, cela ne fait guère de doute. Mais qui avait l'honneur de la porter ?

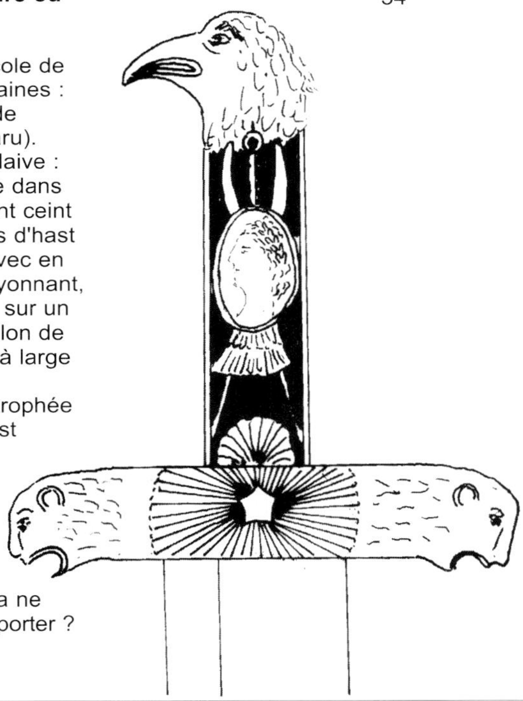

54

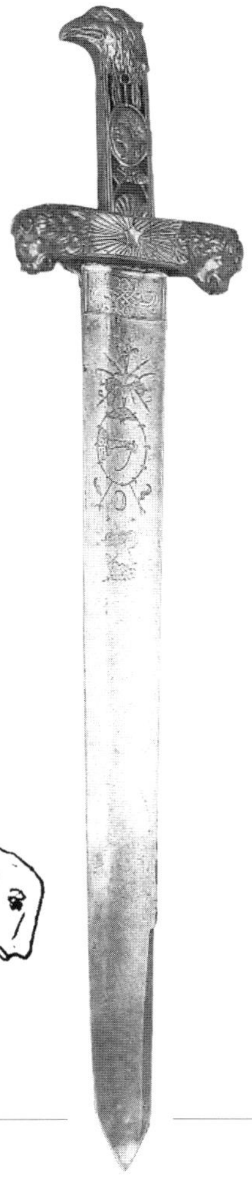

55

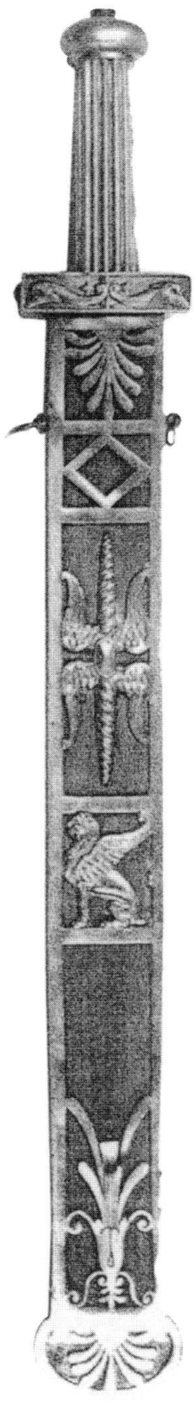

55 - Glaive de cérémonie ou de dignitaire.
Epoque révolutionnaire.
Monture en laiton doré. Pommeau ovoïde, à bouton de rivure. Poignée cannelée. Croisière rectangulaire ciselée de deux dauphins. Lame du type du glaive de l'Ecole de Mars, ainsi que le fourreau mais avec différents motifs : on reconnaît, de haut en bas, une palmette ciselée sur la chape, un carré ajouré à deux bâtons de foudre ailés, un sphinx et sur la bouterolle un motif végétal stylisé. Dard conforme à celui du glaive de Mars. Deux anneaux de bélière symétriques. Le fourreau de bois est recouvert de drap rouge qui apparaît sous les découpes.

Nota - Ce glaive surprenant est décrit à partir d'une photographie, sans mention de poinçons, et fut mis en vente le 25 janvier 1988 à l'Hôtel Drouot, à Paris, sous la mention " Glaive de cérémonie ". C'est une hypothèse plausible, mais la personnalité qui le porta reste énigmatique.

56

56 - Glaive de fantaisie de l'époque révolutionnaire (?).
Autre modèle de glaive énigmatique apparenté au glaive de l'Ecole de Mars. Monture en laiton dorée. Poignée terminée par deux cônes réunis par leur base, rainurés verticalement. Croisière à quillons enroulés. La lame est comparable à celle des glaives de l'Ecole de Mars mais le fourreau est tout à fait différent : sur un fond de velours cramoisi très fin, se détachent, de haut en bas, une palmette, une lyre à cols de cygnes surmontée d'un petit soleil, une femme dénudée portant au-dessus de sa tête un bouquet de palme. Le dard est orné d'une cuirasse sur fond de quatre drapeaux.

Ce dessin nous a été communiqué par M.G. Richard Collin qui nous a fourni les renseignements suivants :
" je vous adresse un dessin exécuté en 1957, relatif à un glaive que m'avait cédé Monsieur Raymond Naudin, frère du peintre militaire Bernard Naudin... " et M.G. Richard Collin ajoute : *" ce glaive que nous attribuions à un objet de théâtre a été vendu par moi, à l'expert de l'époque Pierre Fourry qui l'avait apprécié, mais n'étant guère captivé, je ne sais ce qu'il est à présent devenu... Ce glaive avait une étiquette portant manuscrit le nom de Monsieur Virille, à Châteauroux ".*

Tous nos remerciements à M.G. Rochard Collin pour l'aide qu'il apporte à percer le mystère des glaives de cette famille.

Famille V – Fourreau en cuir, pommeau ovoïde horizontal

Les glaives regroupés dans la Famille V sont évidemment une réminiscence du glaive romain (voir J. Lhoste, 1998, p. 57-59). Le premier modèle est mis en service en 1816 et il est destiné à remplacer les briquets de l'Artillerie. Un second modèle est créé pour l'Infanterie, en 1831 et son usage, moyennant quelques aménagements, perdurera jusque sous le Second Empire.

D'autres modèles, en général de plus petite taille, seront attribués aux Pompiers, aux officiers des Eaux et Forêts, après avoir été utilisés par les Gardes Nationaux... Nous en donnons quelques exemples à la fin de cette présentation.

Pour faciliter l'identification des glaives militaires, nous avons créé deux Sous-familles décrites ci-dessous :

Dans cette Sous-famille V b, a été inclus un glaive des Services administratifs à pommeau déporté vers la gauche (n° 85, p. 91).

Sous-famille V a – Poignée à écailles

Le modèle de 1816 est à écailles et ce décor se maintiendra jusque sous le Second Empire pour certains corps telles la Gendarmerie, la Garde Nationale... et souvent le pommeau porte un symbole militaire en relief, alors que pour le modèle initial, la fleur de lys est incrustée.

57

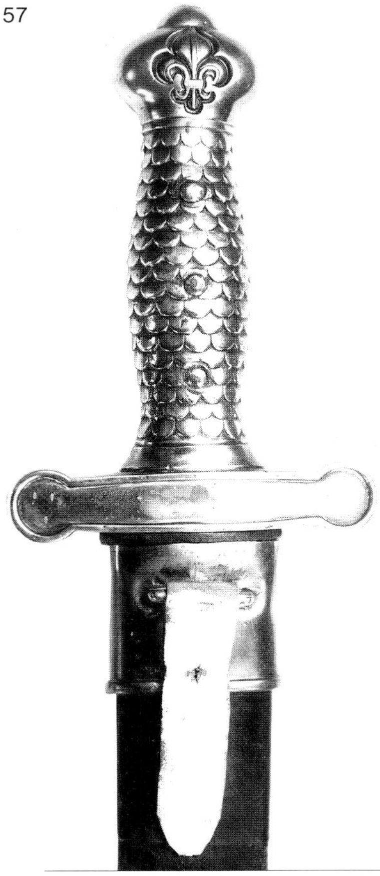

57 - Glaive d'artillerie à pied.
Modèle 1816.
Pommeau ovoïde placé horizontalement, profondément ciselé d'une fleur de lys (57a). Ce pommeau est surmonté d'un dôme recevant le bouton de rivure. Poignée en balustre décorée de 18 rangs d'écailles, avec trois rivets d'acier pour assujettir la lame. Croisière droite, unie, à quillons arrondis. Lame en langue de carpe, longue de 48,7 cm, large au talon de 4,2 cm, à deux gouttières de 9 cm de long et larges de 0,9 cm à sa base puis d'un pan creux médian de 21 cm de long et de 1,2 cm de large. Fourreau de cuir noirci et cousu sur le dos, chape à pontet, à bordure renforcée d'un bourrelet ainsi que la bouterolle terminée par un bouton ovale.

La lame est généralement signée de la *Manufacture Royale de Klingenthal* sur une face et sur l'autre la date de mise en service pour les premiers modèles.

57a

58 - Glaive d'Artillerie à pied, modèle 1816.
On distingue les deux gouttière de la lame.

58

59

59 - Glaive d'Artillerie à pied, modèle 1816, fabriqué sous la Monarchie de Juillet.
On remarquera que la fleur de lys de la calotte a disparu.

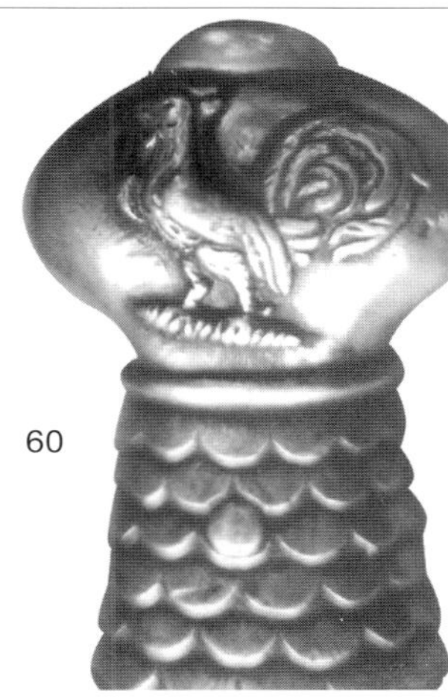

60

60 - Pommeau au coq.
Modèle aux trois rivets figurés.
Le coq est le symbole de la Monarchie de Juillet adoptée en 1830 par Louis-Philippe.

61

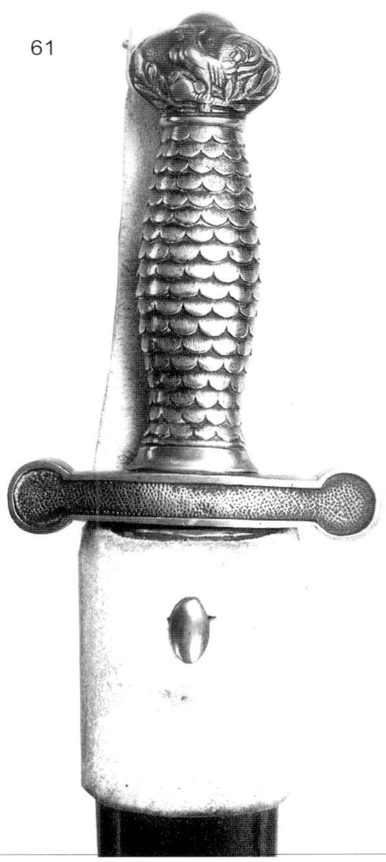

61 - Glaive d'Artillerie à pied, modèle de la Monarchie de Juillet.
Pommeau au coq, la patte droite posée sur un globe et flanqué de rameaux de laurier sur fond sablé (60a). Poignée dépourvue de rivets, à écailles. Fond de la croisière sablé. Lame du modèle 1816 signée au talon, en toutes lettres : *Manufacture de Klingenthal* sur une face, et sur l'autre *Coulaux Aîné et Compagnie*. Fourreau en cuir noir comprimé ; chape à bouton en oliveet bouterolle sans bordure en relief.

Nota - Ce glaive, par son fourreau en cuir comprimé, remplace le fourreau à couture médian (à partir de 1847), peut être daté des dernières années du règne de Louis Philippe.

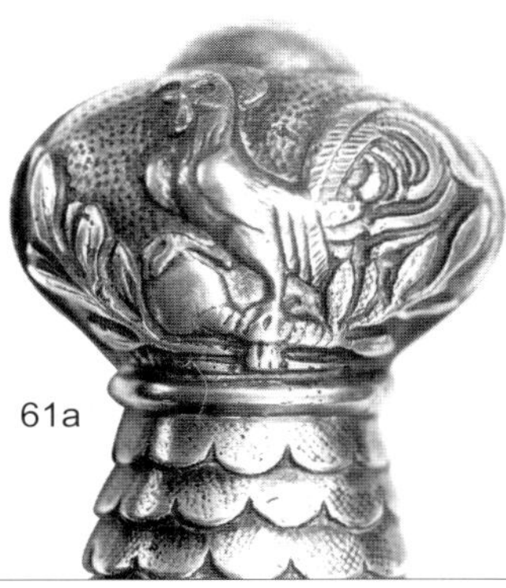

61a

61a - Pommeau au coq, à partir de 1830.
La patte droite est posée sur le globe terrestre, l'ensemble flanqué de rameau de laurier.

62

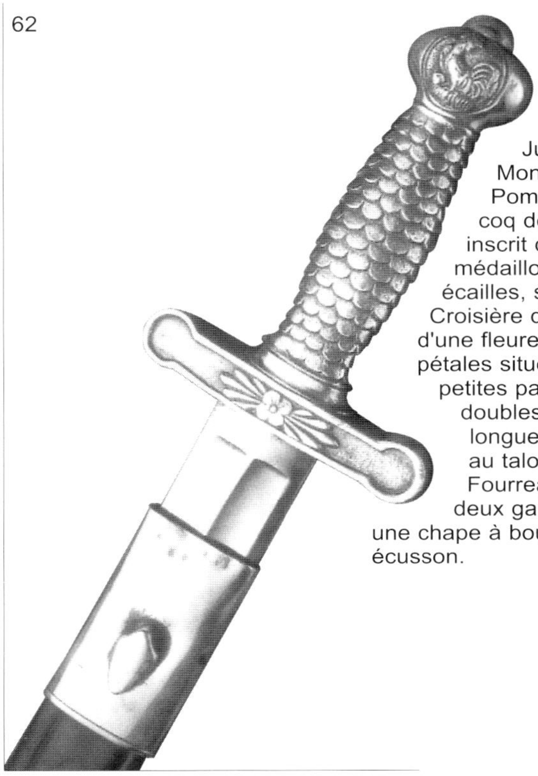

62 - Glaive d'officier de la Garde Nationale.
Monarchie de Juillet.
Monture de laiton. Pommeau orné du coq de Louis-Philippe inscrit dans le médaillon. Poignée à écailles, sans rivet. Croisière droite, ornée d'une fleurette à quatre pétales située entre deux petites palmettes. Lame à doubles pans creux, longue de 49 cm, large au talon de 4,3 cm. Fourreau de cuir à deux garnitures avec une chape à bouton en écusson.

63

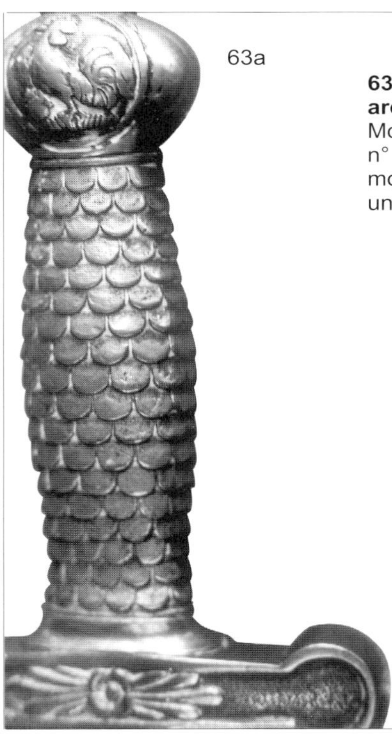

63a

63 - Variante à lame à arête médiane.
Modèle plus ancien que le n° 62, avec la même monture (63a) mais avec une lame à arête médiane.

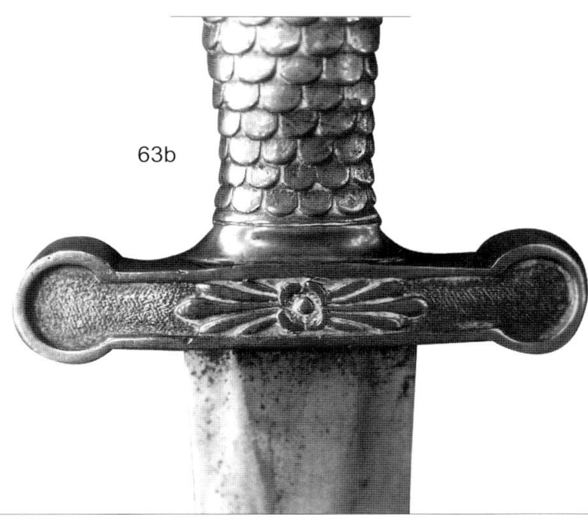

63b

64

64 - Variante.
Lame à deux gouttières creux, marquée au talon *IR.HSGLI* dont la signification nous reste inconnue. (B.L.).

65

65 - Variante.
Lame à arête médiane signée au talon *Manuf*^{re} *R*^{ale} */ de Châtellerault* (B.L.).

66

66b

66a

66 - Glaive d'artillerie de la Garde Nationale.
Monarchie de Juillet.
Pommeau différant de celui du modèle 1816 ; il est subglobuleux, et décoré d'une grenade enflammée (66a). Poignée, croisière et lame identiques au modèle 1816. Fourreau en cuir noirci, à couture dorsale médiane ; chape à pontet et bouterolle renforcée (66b).

67 - Glaive de la Garde Nationale.
Second Empire.
Monture générale identique au modèle 1816, mais à pommeau à l'aigle impériale en relief (67a). Poignée à 18 rangs d'écailles. Croisière à fond sablé. Lame de récupération en langue de carpe à deux gouttières près du talon et une gouttière médiane vers la pointe, provenant des réserves et signée en abrégé *Manufacture Royale de Klingenthal et datée du 8bre 1817*, par exemple. Fourreau de cuir noir comprimé, chape à pontet et bouterolle, sans bordure.

67

67a

Sous-famille V b – Poignée à cannelures

Le 19 avril 1831, le maréchal Soult, Ministre de la Guerre, annonce qu'*" un nouveau modèle de sabre pour l'Infanterie a été adopté par le Roi "*. Le 12 mai de la même année ; le Comité d'Infanterie en donne la description suivante :... *en adoptant un nouveau modèle de sabre paraît avoir été de se procurer une arme qui peut, en campagne faire office d'un outil tranchant ... et à l'occasion, faire office de hache de campement, voire de serpe pour couper le fourrage destiné aux chevaux*... Ce modèle, avec diverses modifications mineures, sera en service pendant plus de quarante ans !

Les glaives à cannelures ou dits aussi à cordons sont les glaives les plus fréquemment rencontrés ayant armé les troupes de la Monarchie de Juillet et du Second Empire. Leur pommeau ne porte pas d'emblèmes et les derniers modèles ont un fourreau en cuir comprimé, sans couture apparente.

C'est dans ce groupe que nous inclurons les petits glaives de différentes organisations paramilitaires.

68- Glaive des troupes à pied, modèle 1831.
Monture en laiton moulé, longue de 14,8 cm. Pommeau ovoïde surmonté d'un dôme servant de bouton de rivure. Poignée à 26 cannelures. Croisière droite de 10,5 cm, à quillons à rouelle. Lame de 47,5 - 48,7 cm, à deux tranchants, en langue de carpe, à méplat triangulaire au talon, une arête peu saillante se prolongeant jusqu'à la pointe. Fourreau en cuir noirci, d'abord à couture puis en cuir compressé à partir de 1847 ; chape à pontet, bouterolle se terminant par un petit bouton en olive.
La lame sera ultérieurement à un pan creux.

68a

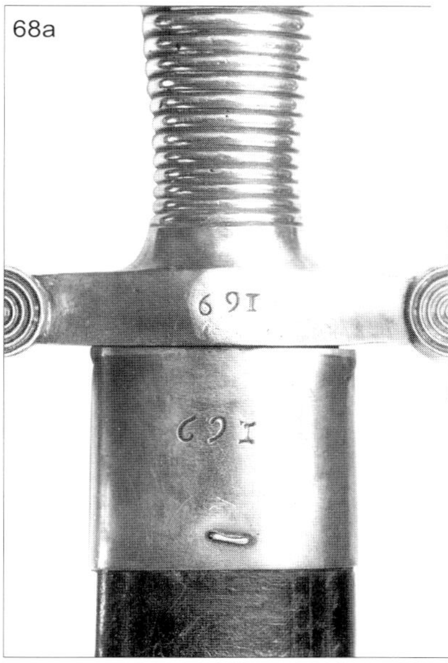

Nota - Ce glaive aux vertus guerrières discutables, fut nommé, par dérision coupe-choux

Marquages - En plus des poinçons de l'époque, on trouve sur le dos de la croisière et de la chape le numéro de la formation ou de matricule militaire (68a).

68

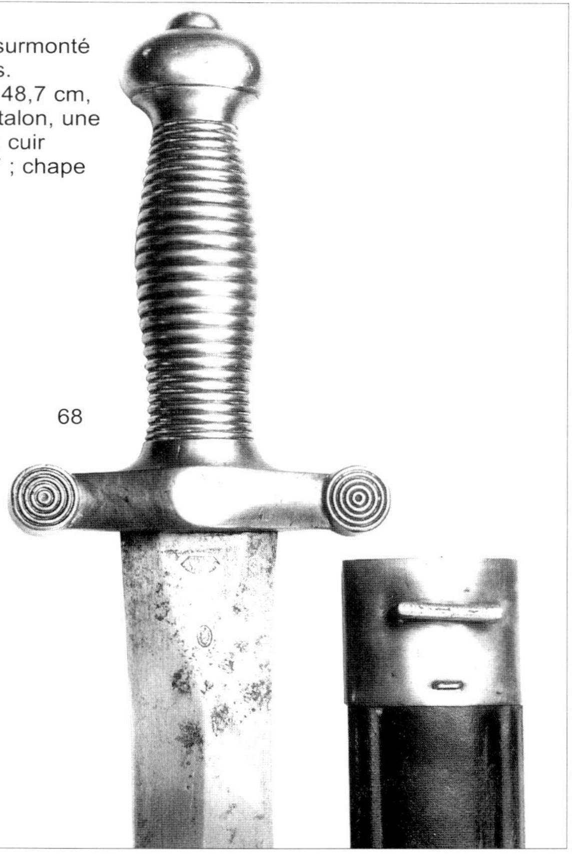

Variantes

Il est évident que ce sabre, au cours des ans, rappelons qu'il ne cessera d'être en usage qu'en 1871, va subir des aménagements, des perfectionnements plus ou moins heureux et que même des *essais* ne verront pas leur introduction dans l'armement. On ne retiendra que les modifications qui ont été adoptées, que l'on peut ainsi résumer :

6 septembre 1846 – Le fourreau " Manceaux " devient en cuir comprimé, à couture latérale et rentrée. La bouterolle est maintenue par un cône intérieur muni d'une tige qui s'engage dans le trou vertical du bouton, supprimant ainsi l'agrafe et le collage.

14 décembre 1852 – La lame est à pan creux médian.

19 octobre 1855 – Mention de l'indication du modèle en ces termes : Troupe à pied, Mle 1831

On peut lire dans l'Aide-Mémoire à l'Usage des Officiers d'Artillerie, daté de 1856, la description suivante : " *même nomenclature que le sabre du modèle 1831 dont il ne diffère que par les dimensions* ". Le petit modèle fut porté et acheté par les Gardes Nationaux et un glaive encore plus réduit fut attribué aux Cantinières des troupes à pied.

Les petits modèles, ressemblent aux glaives du modèle 1831. Un autre modèle à pommeau en olive verticale, à poignée cannelées, à croisière plane, et à deux boutons plats inclinés vers le bas fut également porté par les Gardes Nationaux. (voir n° 86 p. 92).

69

69 - Variante.
Croisière gravée en son milieu du mot *GOZ*, qui peut être la signature du nommé Goze, armurier de Verdun jusque vers 1855. Lame à pan creux (B.L.).

70

70 - Glaive de la Garde Nationale.
Modèle 1855.
Exemplaire dont la lame est marquée dans les pans creux : *GARDE NATIONALE / DE PARIS*, avec au-dessus, dans un petit cartouche 1848. (B.L.).

71

71 - Variante.
Présence d'un bouton sur le noeud de corps activant une lame-ressort pour fixation du fourreau. (B.L.).

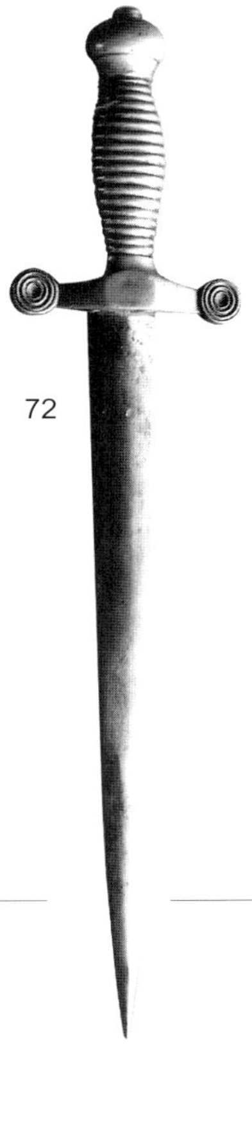

72

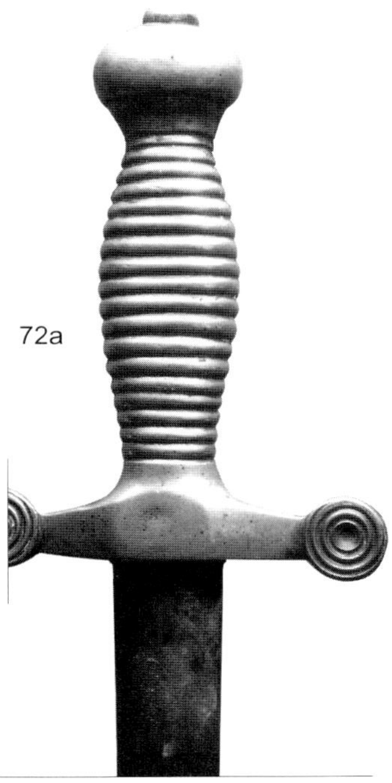

72a

72 - Glaive de cantinière.
Second Empire.
Monture en laiton, très comparable en sa forme générale au glaive du modèle 1831, mais de taille réduite. En effet, la longueur totale du poignard ne dépasse pas 30 cm. La poignée de 7,2 cm est à 18 cannelures, la croisière est longue de 6,8 cm, terminée par des rouelles de 1,3 cm de diamètre (72a). La lame de 21 cm, large au talon de 1,7 cm est unie, à section lenticulaire. Fourreau de cuir à garnitures de laiton, chape à bouton en écusson.

73

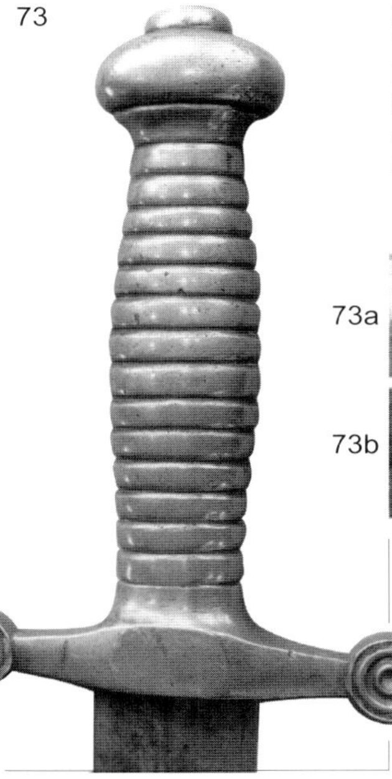

73 - Glaive de Garde Nationale de la Seine.
Troisième République.
Poignée en laiton moulé, à 14 cannelures légèrement aplatis. Pommeau plus large que haut, ovoïde, à bouton de rivure aplati. Croisière droite à quillons à rouelles. Lame de 42 cm, large au talon de 3 cm, gravée à l'acide, sur une face *VIVE LA REPUBLIQUE FRANÇAISE* (73a) et sur l'autre, *GARDE NATIONALE DE LA SEINE* (73b), (M.R.).

73a

73b

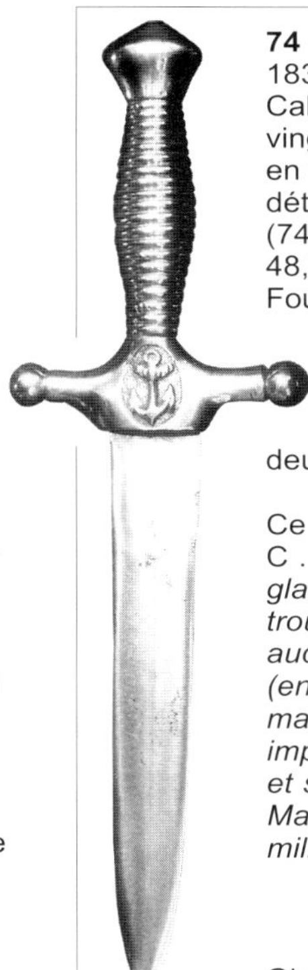

74

74 - Glaive des troupes de Marine.
1830 - 1848.
Calotte bi-conique et poignée à une vingtaine de cannelures. Croisière avec en son milieu, un ovale dans lequel se détache en relief, une ancre de marine (74a). Quillons à boules. Lame de 48,7 cm à arête médiane atténuée. Fourreau en cuir noirci, à deux garnitures : chape à bouton ; les premiers modèles possèdent une couture dorsale médiane et à partir de 1847, le fourreau est en cuir comprimé à deux rainures latérales (B. M.).

Ce glaive a été signé par M. Cottaz (*in* C . Ariès I, I, fig. 8) : " *Une variante du glaive d'Infanterie a été donné aux troupes de Marine. A notre connaissance, aucun texte officiel n'a encore été trouvé (en 1966) justifiant cette arme, mais son existence en nombre important à l'Arsenal de Toulon et sa présence au Musée de la Marine (A Paris) en font une arme militaire incontestable* ".

Glaive 1830 - 1840 retaillé en poignard.

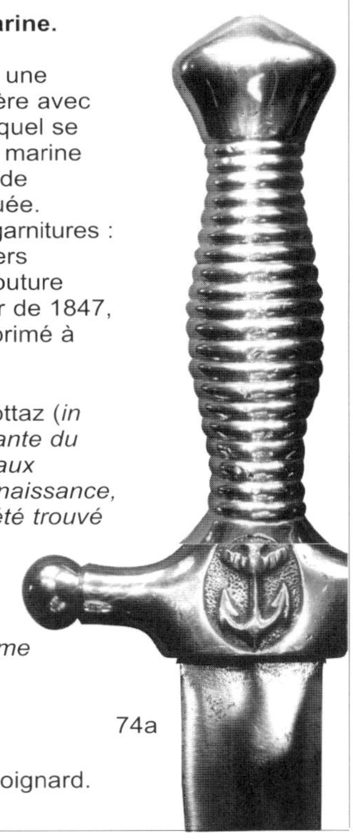

74a

75 - Glaive fantaisie de Sapeur de la Garde Nationale de Louis-Philippe.
Pommeau ovoïde, à dôme servant de bouton de rivure. Poignée à 26 cannelures. Croisière à deux quillons en têtes léonines (75a). Lame de 53,5 cm, large au talon de 3 cm et à deux longues gouttières se terminant à 6,5 cm de la pointe, signée au talon *DEBRYE*. Fourreau de cuir noir à couture dorsale et deux garnitures : chape à gros bouton ciselé d'un masque de lion (75b).

75

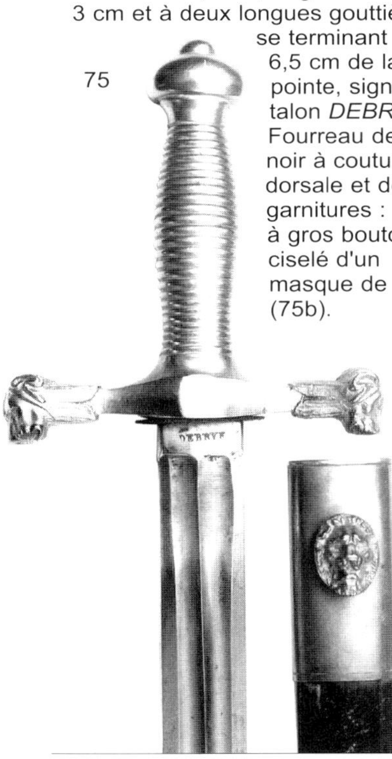

75a

75b

76

Les glaives des Sapeurs Pompiers

Sous le Second Empire, les Sapeurs Pompiers se voient attribuer le petit glaive de 1855. C'est le seul modèle qui sera examiné ici. L'on dira simplement que les officiers portaient également de petits glaives plus ou moins richement ornés, de bronze argenté ou non. Ces glaives sont généralement reconnaissables au fait qu'ils portent sur le pommeau ou la croisière deux haches en sautoir. Plus rarement ils en sont dépourvus et portent alors une grenade qui ne désigne pas un glaive de Grenadier militaire.

76 - Glaive de Pompier, modèle 1855.
Monture en laiton, voisine de celle du modèle 1831, mais de taille réduite : monture de 12,5 cm. Poignée à 26 cannelures. Croisière de 9 cm, terminée par des rouelles dont le centre est profondément évidé, voire percé. Lame de 45 - 48 cm, large au talon de 2,7 cm, à deux longues gorges. Fourreau de cuir, à couture dorsale : chape à bouton ovoïde ou en écusson.

Remarque - Certains glaives ont une lame gravée de feuillages à l'acide : ceux-ci peuvent être attribués aux officiers et officiers-supérieurs.

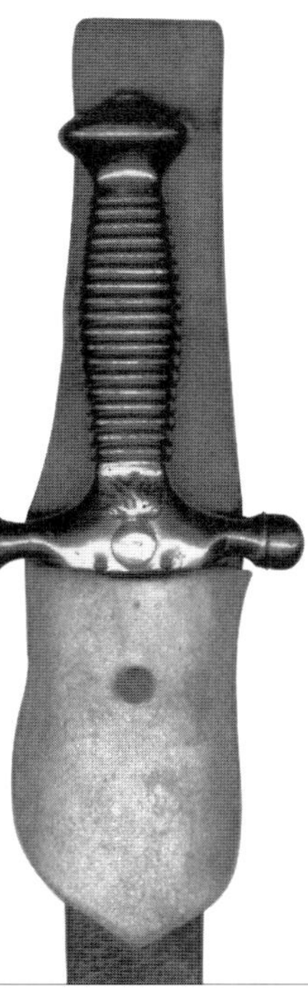

77

recto

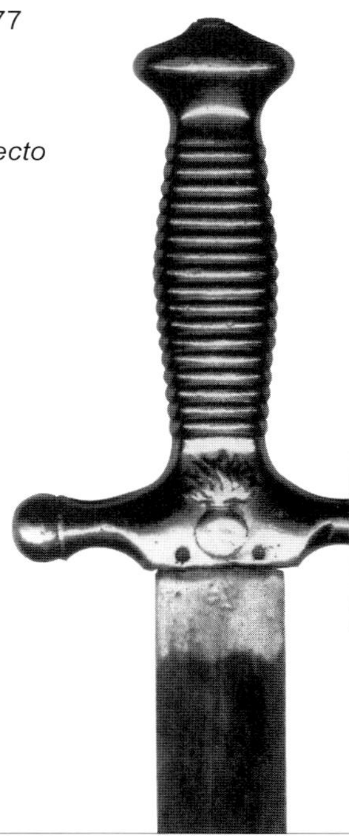

verso

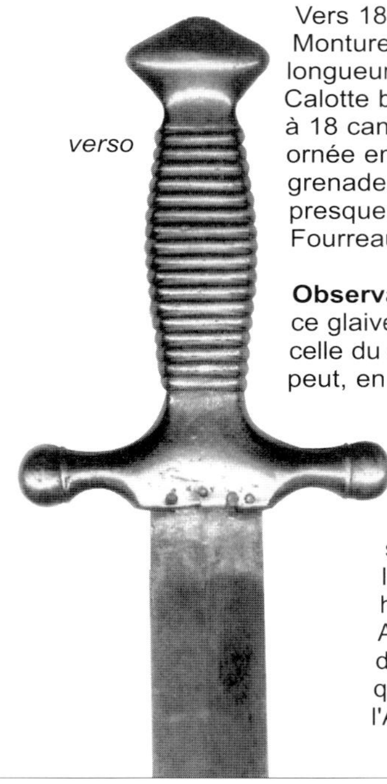

77 - Glaive de Pompier.
Vers 1855.
Monture de laiton moulé, longueur totale de 60,5 cm. Calotte bi-pyramidale. Poignée à 18 cannelures, croisière ornée en son centre d'une grenade enflammée. Lame presque plate de 47,8 cm. Fourreau à deux garnitures.

Observation - La monture de ce glaive est très voisine de celle du glaive n° 74, p. 87. On peut, en conséquence penser à son appartenance à la Marine, mais l'absence d'ancre même sous la forme de poinçon d'arsenal semble conduire à l'élimination de cette hypothèse. Pourtant, A. Riches, propriétaire de ce glaive, précise qu'il fut longtemps à l'Arsenal de Libourne.

78

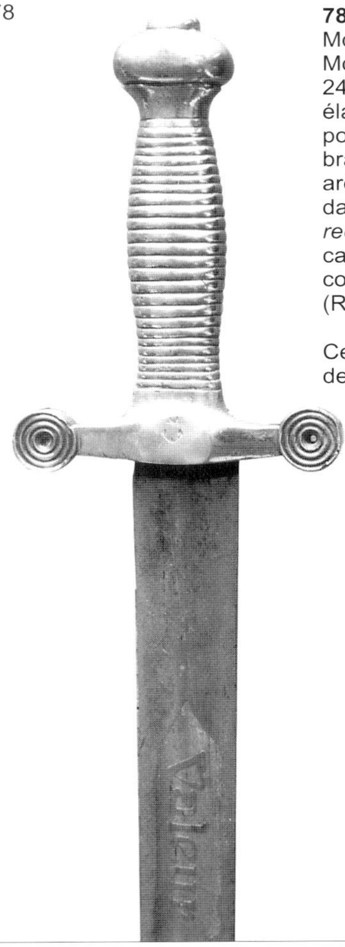

78 - Glaive de Pompier de la Marine.
Modèle 1855.
Monture en laiton doré. Poignée à 24 cannelures, longue de 12,5 cm, calotte élargie. Croisière à rouelles perforées, poinçonnée en son milieu d'une ancre aux bras arrondis (78a). Lame de 46,5 cm, à arête atténuée médiane, gravée à l'acide dans un cartouche à fond grisé. *Valeur* sur le *recto* (78b) et *Discipline* sur le *verso* (78c) en caractères droits. Fourreau de cuir noirci à couture médiane dorsale : chape à bouton (Remontage ?).

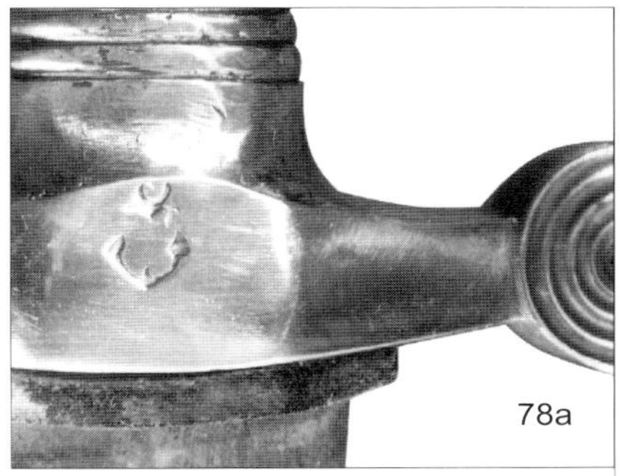

78a

Ce glaive peut être attribué aux sous-officiers ou aux officiers du Corps des Pompiers de la Marine.

78b

78c

79 - Variante.
Modèle dont le fourreau comporte une chape à bouton en écusson.

79

80 - Variante.
Modèle avec une chape de fourreau à bouton en olive.

80

81

81 - Glaive de garde des Eaux-et-Forêts.
Second Empire.
Monture en laiton moulé de 13,7 cm de hauteur. Poignée à 26 cannelures. Croisière de 9,5 cm de longueur, terminée par des rouelles sans autre ornement. Lame à arête médiane peu saillante, longue de 42 cm, large au talon de 3,4 cm. Fourreau de cuir noirci : chape à bouton en forme de gland, bouterolle terminée par un bouton précédé d'un renflement circulaire.

Nota - Ce glaive peut avoir été porté par un garde-champêtre, voire un gardien de propriété !

82

82 - Glaive d'officier supérieur de Sapeur Pompier.
Second Empire.
Monture en laiton. Calotte ovoïde, plus haute que large, ciselée d'un casque placé sur deux haches en sautoir. Poignée rainurée verticalement. Croisière à noeud de corps à la grenade placée sur deux haches en sautoir. Quillons symétriques, rabattus vers la lame. Couvre-chape. Lame à deux pans creux. (B.L.).

83

83 - Glaive de Pompier savoyard.
Type du modèle français de 1855. Monture en laiton. Croisière ornée en son milieu de la croix de la Maison de Savoie. Lame à pan creux médian.

84

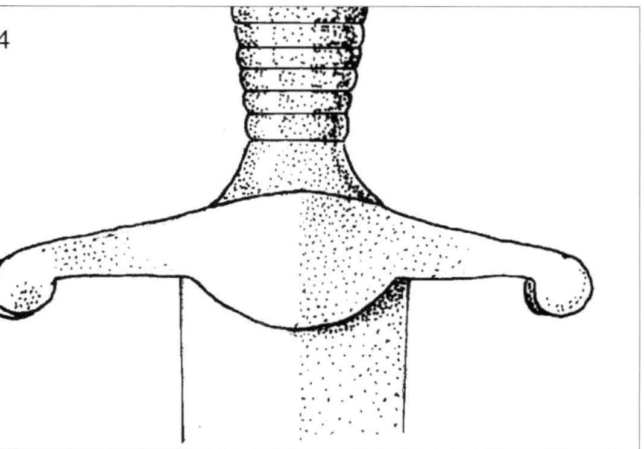

84 - Glaive fantaisie de Garde-Champêtre.
Fin XIXème siècle.
Monture en laiton, du type 1831, mais de dimensions réduites. Poignée à 18 cannelures. Remarquer la croisière : les quillons sont infléchis vers la lame et le milieu prend la forme d'un oreillon basal ébauché. Lame à arête médiane atténuée.

85

85a

85 - Glaive de l'administration.
Second Empire. Monture en laiton monobloc. Calotte asymétrique, inclinée vers la gauche. Poignée à 19 cannelures. Croisière terminée à ses deux extrémités de boutons rehaussés de deux cannelures en bague. Lame à deux pans creux et deux tranchants, marquée au talon *Klingenthal* sur le *recto* et *Coulaux* sur le *verso*. Fourreau de cuir à bouterolle et chape en laiton, celle-ci à bouton en écusson (85a).

Famille VI – Fourreau en cuir, pommeau de formes et de décors variés, plus haut que large

Bien que de structures variées, les sabres de la Famille VI, possèdent une calotte plus haute que large ce qui les différencie nettement des glaives des modèles 1816 et 1831 dont la calotte ovoïde est plus large que haute.

86

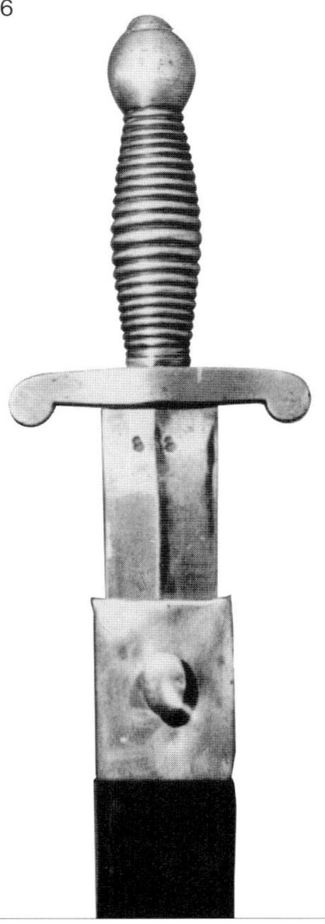

86 - Glaive de la Garde Nationale.
Règne de Louis-Philippe.
Monture en laiton fondu. Calotte ovoïde, placée verticalement et terminée par un bouton de rivure. Poignée à cordons dont les quatre médians sont plus forts que les autres. Croisière plane à deux quillons arrondis et infléchis vers la lame. Lame à arête médiane saillante, poinçonnée au talon de deux petites grenades enflammées. Fourreau de cuir noirci, chape à bouton ovoïde uni et courte bouterolle terminée par un bouton de renfort.

87 - Exemplaire de glaive du Second Empire, très comparable à celui de gauche. Il s'en distingue essentiellement par le pontet de la chape.

87

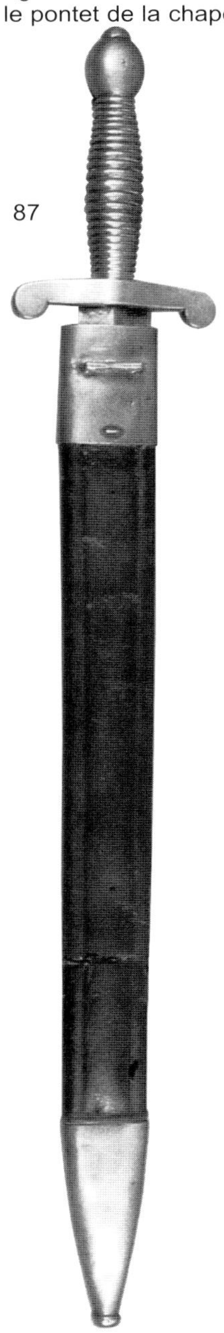

88

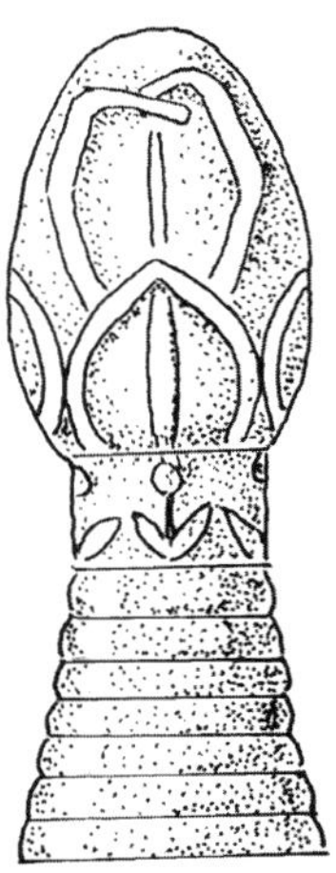

88 - Glaive d'Infanterie de la Garde Nationale.
Monarchie de Juillet.
Monture en bronze fondu et ciselé. Pommeau en bouton foliacé. Poignée à quinze cannelures. Croisière droite, du type 1816, mais décorée de motifs foliacés de part et d'autre d'une fleurette centrale. Lame à gorge médiane au talon de laquelle, sur une face on lit *Acier fondu* et sur l'autre *Charrière à Paris* qui est le fourbisseur. Fourreau de cuir noirci, chape à bouton. Ce glaive fantaisie a dû appartenir à un officier ou sous-officier dit *Tête de colonne*.

89

89 - Glaive fantaisie d'officier de Sapeur-Pompier.
Fin du Second Empire.
Monture en bronze doré. Pommeau subglobuleux surmonté d'un gros bouton de rivure ciselé d'une grenade sur le médaillon frontal. Poignée moulée, ornée d'une spirale de feuillage et d'un filigrane simulé. Croisière horizontale, frappée d'une grenade sur le nœud de corps, à deux quillons ciselés d'une fleurette à quatre pétales. Lame à section losangique, gravée de la dédicace : *Au lieutenant Rollet les sapeurs de St Ouen*. Fourreau de cuir noirci à chape décorée d'une frise de feuilles de laurier stylisées et d'un bouton ovale entouré de petites perles ; bouterolle décorée en suite. Ce glaive a été monté de divers éléments, ce qui était courant pour armer les pompiers de village. (C.B.). (Voir également p. 90, n° 82).

Famille VII – Fourreau en cuir, pommeau en tête d'animal

Le pommeau en tête de coq apparaît en 1775 puis sur les glaives attribués à la Garde Constitutionnelle et à la Garde Nationale après la proclamation de la Première République et perdurera jusque sous la Monarchie de Juillet et le Second Empire. Les glaives à tête d'aigle sont attribués, au XVIIIème siècle à l'Artillerie, puis adoptés sous le Premier Empire par les Sapeurs de la Garde Impériale ainsi que sous le Second Empire. Quant aux glaives à tête de lion, d'aucuns les nomment chiens de mer, tant leur figuration est fantaisiste. Ils apparurent avec les Gardes Suisses au service du Roi et furent ensuite attribués, vers 1790, à l'Artillerie de Marine.

Nous avons divisé la Famille VII en trois Sous-familles :

Sous-famille VII a – A tête de coq

La tête de coq est toujours tournée vers la gauche, placée horizontalement ou le bec levé quant il salue le soleil ou la victoire. Le cou est plus ou moins emplumé et, en général, soigneusement repris à partir du moulage par le ciseleur. La croisière est assez variable de forme et la lame peut être à dos de scie, surtout pour les pontonniers et les sapeurs.

90

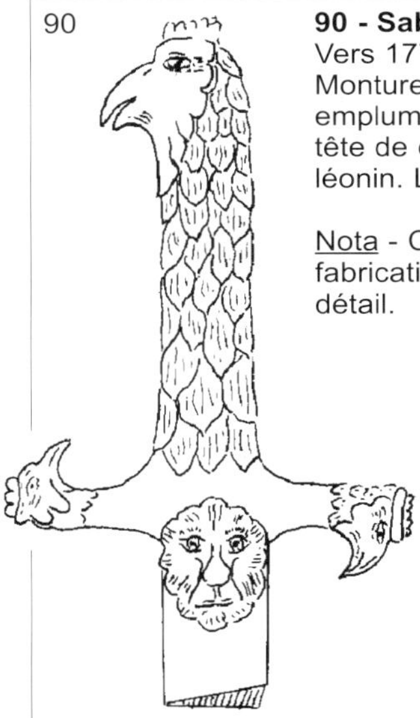

90 - Sabre de Sapeur d'Infanterie
Vers 1775.
Monture en laiton moulé. Calotte en tête de coq. Poignée emplumée jusqu'à la croisière terminée par deux quillons en tête de coq, inversés et pourvue d'un oreillon au masque léonin. Lame plate à dos plat.

Nota - Ce sabre qui annonce les sabres de Sapeurs de fabrication ultérieure présente de nombreuses variantes de détail.

91

91 - Sabre de Sapeurs de la Garde Constitutionnelle.
Ce sabre fut en service de 1789 à 1792. Monture en laiton doré. Pommeau à tête de coq chantant, la langue bien séparée du bec. Poignée entièrement emplumée. Croisière moulée séparément et terminée par deux quillons infléchis, décorés d'une rainure striée sur le pourtour. Lame longue de 68 cm, large au talon d'environ 5 cm, dos dentelé en scie, à contre-tranchant et signée au talon de *COULLIER / FOURBISSEUR / DE LA MAISON / DU ROY / RUE ST HONORE / N° 574 / PARIS* ou *COULLIER / RUE ST HONORE N° 574 / PARIS*
Fourreau à deux garnitures découpée.

92

92 - Sabre de Sapeur de la Première République.
Ce sabre en service dès l'avénement de la Première République, est identique à celui de la Garde Constitutionnelle. La lame peut ne pas dépasser les 48,3 cm. La signature de Coullier a disparu, vraisemblablement par prudence en ces temps difficiles. Sur le talon de la lame, un petit poinçon rectangulaire aux lettes RF rappelle la proclamation de la République. Sur l'autre face, un gros poinçon aux lettres " BC " ou " BG " indique peut-être le nom du fourbisseur.

92a, variante.
Sur cet exemplaire on notera que la croisière est unie, sans la bordure striée traditionnelle. (J.C. D).

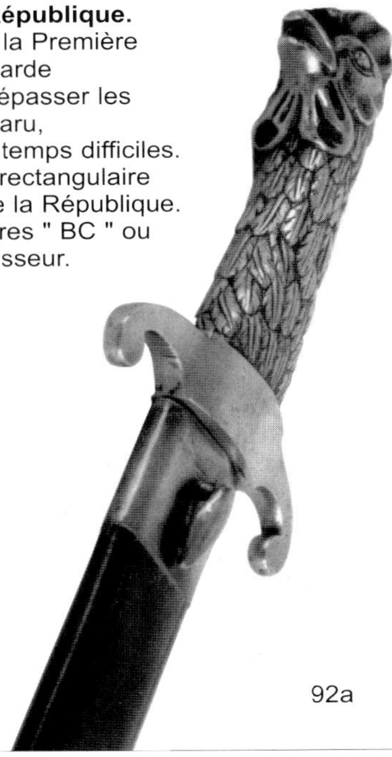

92a

93

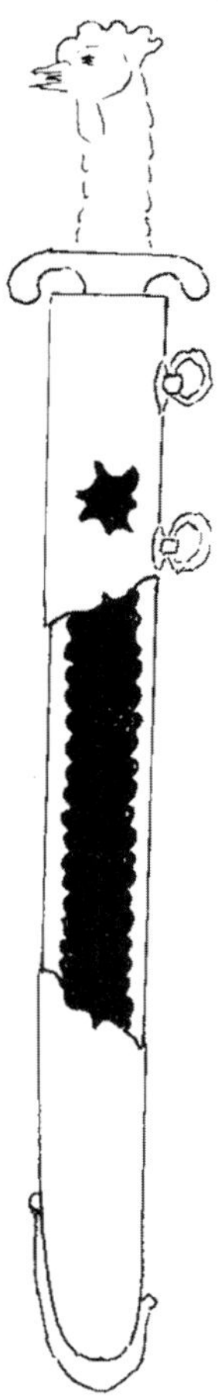

93 - Variante.
Parfois, les Sapeurs préfèrent mettre leurs sabres réglementaires dans des fourreaux plus ou moins extravagants, dont le schéma ci-contre donne une idée.

94 - Sabre de Sapeur.
Première République.
Ce sabre est probablement un montage car on décèle sur la lame, le mot *" PARIS "* seul vestige de la signature de Coullier ayant disparu en 1792.
Pommeau au coq chantant, à longue crête et cou emplumé seulement sur le tiers supérieur de la poignée, les deux tiers restant étant à pans lisses. Croisière à deux quillons infléchis, de la même facture que celle du modèle de la Garde Constitutionnelle. Lame relativement large, plate et droite. Fourreau de cuir noirci, à chape décorée au trait de part et d'autre du bouton en olive ; bordure inférieure en accolade, ménageant un orifice arrondi médian laissant voir le cuir. Il s'agit de toute évidence d'un fourreau du premier modèle. (Voir également p. 95, n° 97 et 98).

94

95

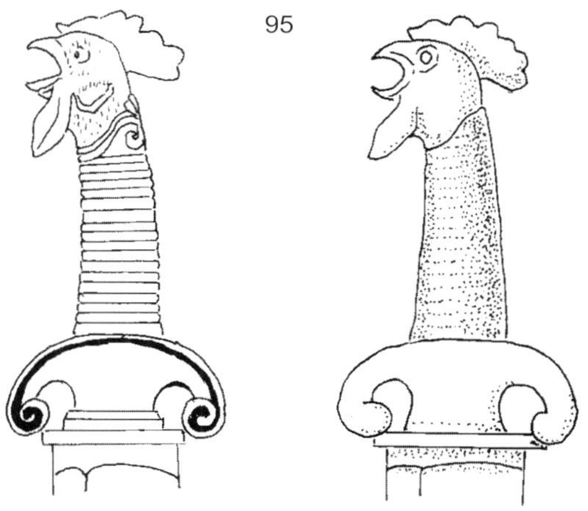

95 - Variantes - Sabres de Sapeur d'infanterie.
Révolution.
Ces modèles de la Révolution sont du premier modèle au coq chantant, Remarquer les fines cannelures de la poignée et la croisière étriquée, qui ne peut servir de garde, et terminée par deux quillons à enroulements.

96

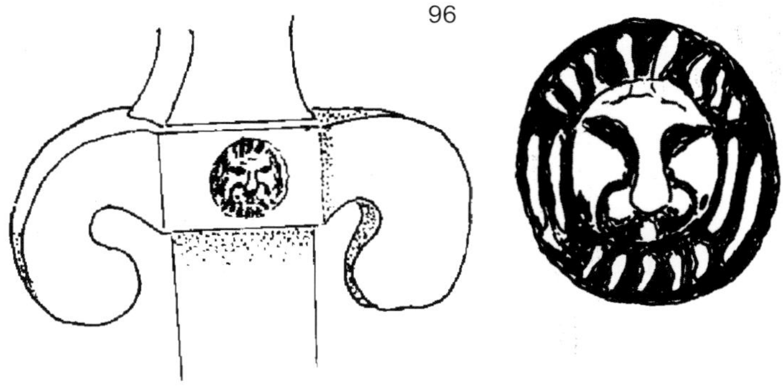

96 - Sabre de Sapeur du 14ème de ligne.
Premier Empire.
Dans son étude sur les Sapeurs du 14ème de ligne du Premier Empire, A. Pigeard (1997) montre une photographie du glaive les armant. Le pommeau au coq chantant et la poignée lisse à pans est comparable à celle du modèle de la Révolution. Mais le sabre du 14ème de ligne présente une croisière plus large, moins élégante et un nœud de corps quadrangulaire orné d'un masque léonin. Par ailleurs, la lame est courte, la chape du fourreau à bouton, le dard carré ce qui laisse supposer que cette lame est en scie.

Nota - *Depuis le 18 février 1808, les Sapeurs du 14ème de ligne sont au nombre de quatre dans chaque bataillon d'Infanterie - ligne ou légère* - (A. Pigeard *op. cit.*) et sont commandés par un caporal.

97

97 - Sabre de Sapeur.
1805 - 1810.
Monture en laiton, parfois dorée. Pommeau en tête de coq chantant, la langue collée sur la partie inférieure du bec, crête et barbillons grossièrement sablés. Poignée emplumée au tiers de sa longueur, à huit pans subégaux. Croisière fondue séparément, terminée par deux quillons arqués et décorés de la même rainure striée. Lame de 71 cm, large au talon de 3,5 cm et à flèche de 4,1 cm, à pan creux sur une cinquantaine de centimètres. Fourreau à deux garnitures de laiton, très variable notamment en ce qui concerne le bouton de chape qui peut être ovoïde et lisse ou ciselé d'une tête léonine ou de tout autre motif.

Nota - De nombreux exemplaires ont le dos de la lame dentelé en scie. (Voir n° 98.)

98

98 - Variante.
Autre exemplaire pourvu d'une lame à dos en scie, avant 1810.

99

99 - Sabre de Sapeur.
Premier Empire.
Remarquer le bouton de chape figurant une tête humaine d'inspiration égyptienne. Souvenir de la campagne d'Egypte ?
(G. Lavallay Op. cit. Pl. XV).

100

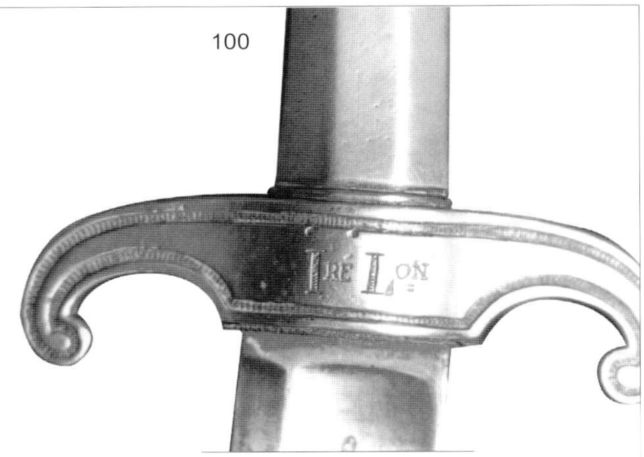

100a

100 - Sabre de Sapeur.
Premier Empire, avant 1810. Poignée de laiton à tête de coq chantant, emplumé sur le tiers supérieur, puis à huit pans unis. Croisière décorée d'une rainure finement striée, aux quillons infléchis vers le bas ; au centre de cette croisière, on lit " *Ire L^{ON}* " pour *Première Légion*, Forte lame, large au talon de 4 cm, longue de 73 cm, à flèche d'environ 1 cm ; dos de la lame en scie à partir de 5,8 cm du talon jusqu'à 19 cm de la pointe à contre-tranchant. On relève sur le talon de la lame la signature *M^{fre} Imp^{le} de Klingenthal* (100a) et dans le pan creux, les poinçons de Jean Pache et vraisemblablement celui de Joseph Ambroise Bisch. Fourreau de cuir noir encollé sur des alèses de bois, à deux garnitures simplement décorées d'un double trait : chape de 12,8 cm à bouton de port ovalaire uni, bouterolle de 19 cm à dard de laiton légèrement asymétrique.

101

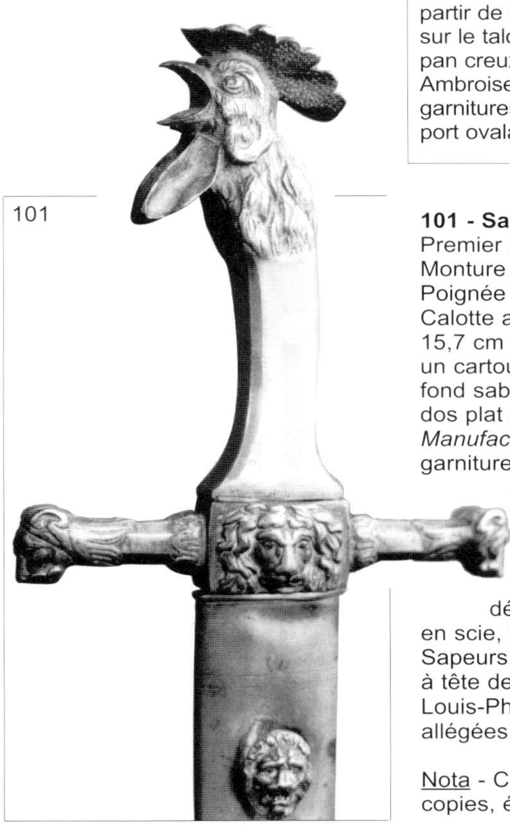

101 - Sabre de Sapeur.
Premier Empire, après 1810.
Monture en laiton doré ou non, allégé par rapport au sabre de la Garde. Poignée longue de 14,8 cm, large en son milieu de 2,5 cm à huit pans lisses. Calotte au coq chantant. Crête sablée sur sa face inférieure. Croisière de 15,7 cm terminée par des quillons en têtes léonines avec, au nœud de corps, un cartouche de 4,5 sur 3,4 cm ciselé d'une tête de lion à crinière frisée sur fond sablé. Lame de 75 cm, large au talon de 3,2 cm et flèche de 4,5 cm, à dos plat et pan creux, terminée en biseau et signée sur le dos de la lame *Manufacture Impériale de Klingenthal*. Fourreau de cuir noirci à deux garnitures décorées au trait ; chape à bouton figurant une tête léonine ; fort dard asymétrique. Poinçon de Levavasseur sur le talon de la lame.

Variantes
Les variantes sont nombreuses, ce sabre ayant été porté jusqu'en 1870. La monture peut être d'une seule pièce, la langue de coq bien dégagée, la lame plus ou moins forte, plus ou moins courbée et dentée en scie, bleuie et gravée de motifs dorés en particulier pour les caporaux-Sapeurs qui portaient leur sabre la lame nue selon M. Bottet (1968). Ce sabre à tête de coq fut également porté par les Sapeurs de la Garde Nationale de Louis-Philippe, et de Napoléon III, les dernières fabrications étant souvent très allégées.

Nota - C'est peut-être le sabre qui a fait l'objet du plus grand nombre de copies, étant très apprécié des collectionneurs. On reconnaît, parfois avec

difficulté, les copies des originaux. Il faut examiner avec soin la qualité du moulage, les traces de microfusion, la qualité des dents, certains faussaires reprenant des lames de briquets et taillant les dos en scie.
De plus, il ne faut pas oublier que des sabres à tête de coq ont été en usage dans divers pays limitrophes.

102

102 - Sabres de Sapeur de la Garde Impériale.
Après 1810.
Monture en bronze doré. Poignée de 15 cm et au milieu de 3 cm. Pommeau au coq chantant, entièrement sablé sauf les barbillons. Croisière de 16,5 cm, épaisse en ses parties lisses de 1,13 cm, terminée par les têtes léonines classiques avec au nœud de corps, dans un cartouche de 4,6 cm sur 3 cm, la tête de lion à crinière frisée. Lame de 72 cm, large au talon de 4 cm, en scie sur 47 cm et flèche de 1,5 cm. Au *recto*, à sa base, on lit : *à / Klingenthal et au verso Coulaux / Frères* (102a). Fourreau de cuir à deux garnitures : chape de 13 cm, à cuvette et ornée de quatre double traits, avec en son milieu, un bouton ciselé en tête de vieillard (Mars ? 102b). Bouterolle en suite, à deux doubles traits et dard en laiton asymétrique. Poids total 2,08 Kg.

102a

102b

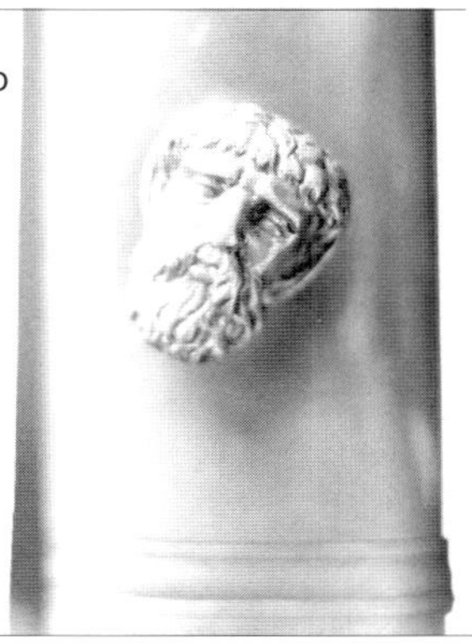

103

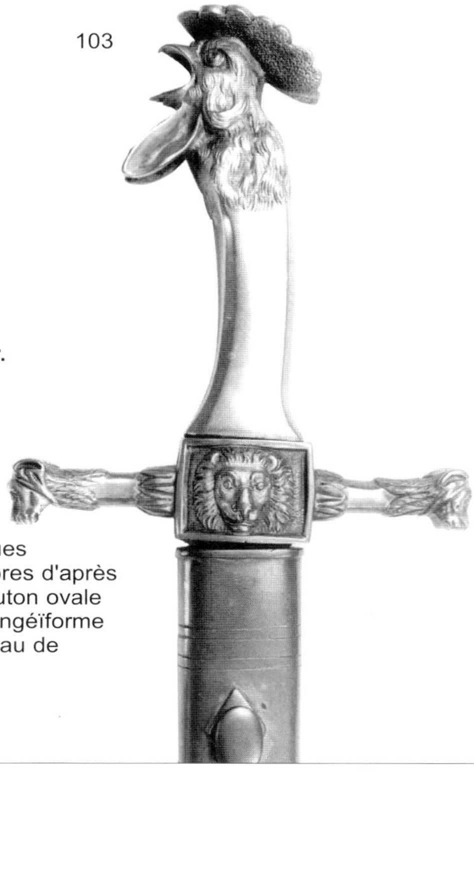

103 - Sabre de Sapeur.
Modèle tardif.
Remarquer la face léonine du nœud de corps : la crinière est dense, courte, faisant une auréole autour du mufle. Toutes les autres caractéristiques sont identiques aux sabres d'après 1810. Le fourreau à bouton ovale monté sur embase losangéïforme est, peut être, un fourreau de circonstance.

104

104 - Sabre de Sapeur-Pompier officier ou sous-officier.
Règne de Louis-Philippe et de Napoléon III.
Monture en laiton. Pommeau à tête de coq au bec à demi ouvert (104a), le cou emplumé sur les deux tiers de la longueur de la poignée, celle-ci à pans. Croisière avec les quillons inversés et nœud de corps ciselé de deux haches en sautoir, liées par le milieu de leur manche (104b). Lame blanche, plate, longue de 59 cm et flèche de 2,3 cm. Fourreau de cuir noir, à deux garnitures, la chape étant à pontet (M.R.).
Ce sabre est évidemment une fantaisie de fourbisseur ou de l'officier qui en passa la commande.

104a

- **a**, calotte de coq chantant, mais sur ce sabre, le bec est à demi ouvert.

104b

- **b**, noeud de corps aux deux haches en sautoir.

106

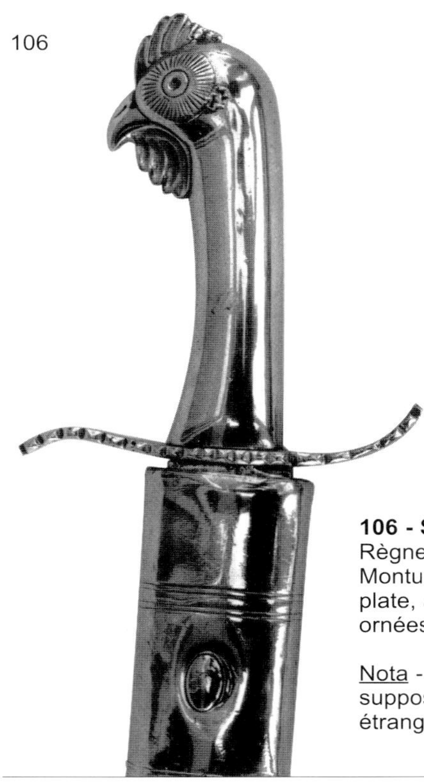

105 - Sabre de Sapeur de la Garde Nationale.
Règne de Louis-Philippe.
Même aspect général que les sabres de la page 96 mais à la lame demi-courbe, de 58 cm, est marquée dans un cartouche, *" Garde Nationale "* en lettres d'or sur fond bleu (non figuré).

106 - Sabre de Sapeur de la Garde Nationale.
Règne de Louis Philippe.
Monture en laiton. Pommeau en tête de coq, le bec clos. Poignée à pans. Croisière plate, à bordure crénelée, courbée en " S ". Fourreau de cuir à deux garnitures ornées de rainures jumelées et bouton ovoïde de chape.

<u>Nota</u> - On remarquera que la tête de coq est traitée d'une façon très naïve et laisse supposer que ce sabre appartint à la Garde Nationale ou qu'il soit d'origine étrangère...

Sous-famille VII b – A tête d'aigle

C'est vers le milieu du XVIIIème siècle que les glaives à tête d'aigle sont apparus. D'abord placée horizontalement, la tête à été relevée pour faciliter la prise en main. Le modèle le plus fréquemment rencontré est le modèle 1771 d'Artillerie et des exemplaires plus légers et plus tardifs qui ont été adoptés par la Garde Nationale. Sous le Consulat l'aigle réapparaît et persistera dans l'armement de la Garde Impériale.

107 - Glaive de Sapeur du Règlement de 1767.
Laiton moulé. Poignée terminée par une tête d'aigle placée horizontalement, au cou totalement emplumé. Croisière aux quillons rabattus vers la lame, celle-ci droite, longue de 73 cm, large au talon de 4 cm, à dos taillé en scie.

107

108

Type a

108 - Glaive d'Artillerie, modèle 1771.
Monture en laiton coulé et moulé. Poignée à tête d'aigle levée à la verticale, au cou emplumé. Croisière en barrette aux extrémités arrondies. Lame droite, longue de 48 - 49 cm, large au talon de 4,5 cm, à deux pans creux jointifs près du talon, puis à pan médian. La lame est assujettie par trois rivets et noter la soie apparente qui affleure le cou de l'aigle.
On distingue deux types de fabrication :

- Type a : En 1771, la monture est fondue en deux parties et les ailes de la soie apparaissent sur presque toute la longueur de la poignée (108a).

- **a**, l'aile de la soie apparaît sur presque toute la longueur du dos de la poignée. Croisière marquée au *verso*, 91.

108a

Marquages - Pour les lames les plus anciennes, on trouve le poinçon de Gau de Vaumarin. Plus tard ce sont les poinçons de J.F. Perrier, de F.A. Bisch, de Juchereau, de de Cabannes qui apparaissent.

Nota - Les officiers possèdent le même type d'arme, mais avec souvent des enjolivures. Aries et Baudriot (1979) signalent un exemplaire dont la croisière est ciselée d'écailles et se termine aux deux extrémités par un bouton. Le talon de la lame, signée *Coulaux / Frères* est gravé de trophée surmonté d'une aigle couronnée, laissant supposer que cette arme a appartenu à un militaire de l'Empire.

109b

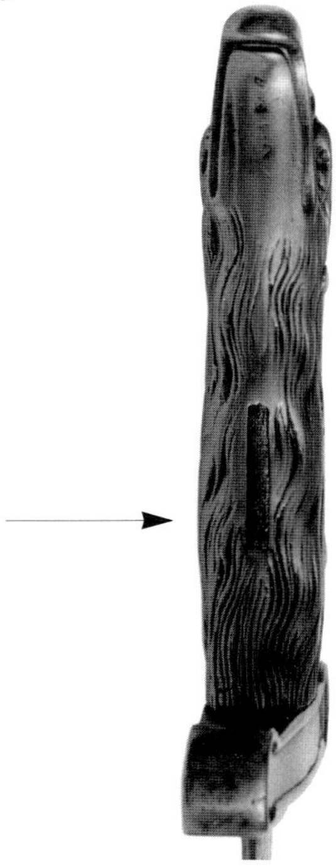

La flèche indique l'apparition de l'aile de la soie sur environ 25 mm, sur le devant de la poignée.

109
Type b

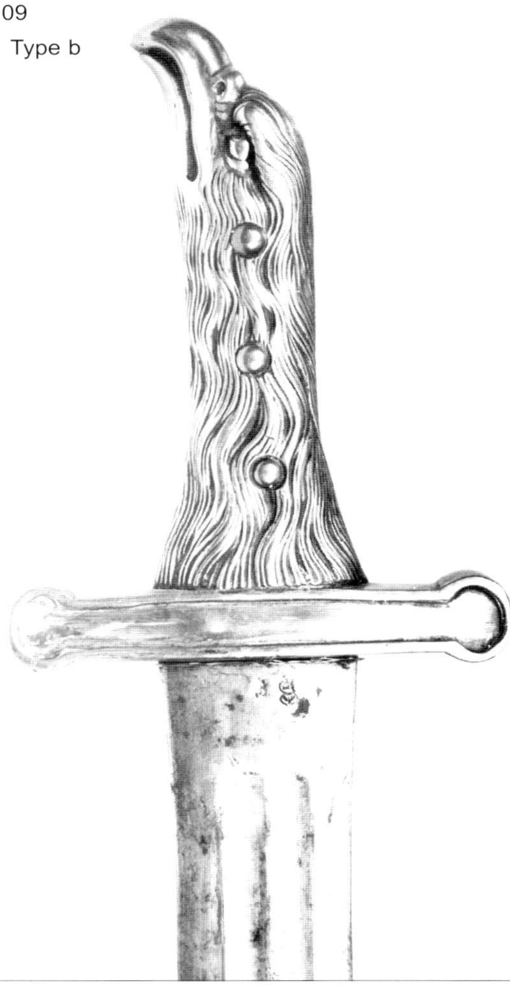

109 - Glaive d'Artillerie, modèle 1771.

Seconde fabrication (type b)

- Type b : dans les fabrications plus tardives, la monture est monobloc et les ailes de la scie n'apparaissent que sur 25 mm environ (109b). Fourreau à deux garnitures, chape à fente pour la languette de tirant en cuir, bouterolle à bouton.

110

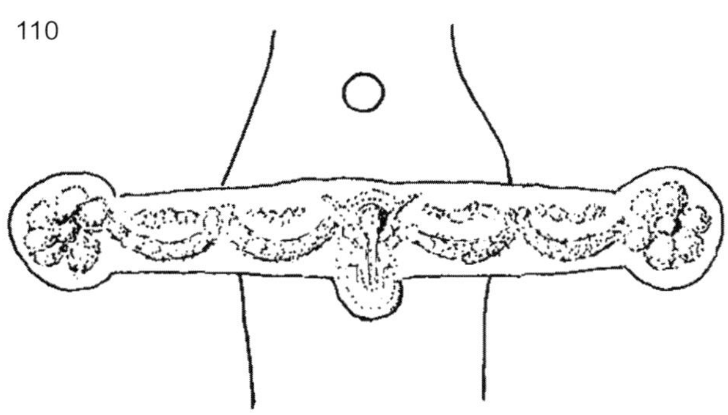

110 - Glaive d'Artillerie, officier, m**odèle 1771**.
Ce glaive est remarquable par sa croisière. M. Pétard écrit (1979) : *La monture très remarquablement ciselée et dorée se distingue surtout par le décor de la croisière ; elle comporte une frise typique empruntée à l'ornementation romaine.*
De plus, la lame est dorée, sur la bordure des deux pans creux avec un décor floral au talon et marquée *ALSACE*. Fourreau de cuir noir à deux garnitures dorées.

111

111 - Glaive de canonnier des Gardes Françaises.
Vers 1780.
Monture en laiton coulé. Poignée à tête d'aigle placée verticalement, au cou totalement emplumé et s'évasant vers la base pour épouser la forme de la croisière. A noter que sur la face antérieure du cou, la soie de la lame n'est pas apparente. Croisière à quillons infléchis vers la lame, avec nœud de corps rectangulaire. Lame d'environ 48 cm parfois aussi large que la base du cou, d'autres fois ne dépassant que de peu le nœud de corps, droite et à large pan creux et signée au talon *COULLIER / FOURBISSEUR / DE LA MAISON / DU ROY / RUE ST HONORE / N° 574 / PARIS*. Le plat de la lame peut être marqué : *Canonnier des Gardes françaises*. Fourreau de cuir noir, chape à bouton en olive, bouterolle terminée en demi-cercle, à dard asymétrique.
Ce glaive fut également attribué aux charpentiers du même corps.

112 - Glaive de canonnier des Gardes Françaises.
Vers 1780.
On notera que sur cet exemplaire l'inscription *DE LA MAISON / DU ROY* a été biffée ainsi que les fleurs de lys et la couronne royale, vraisemblablement en 1792, pour un service qui a continué sous la Première République.

Nota - La mutilation des lames aux fleurs de lys fut très courante, et moins onéreuse que le remplacement de l'arme.

112

113

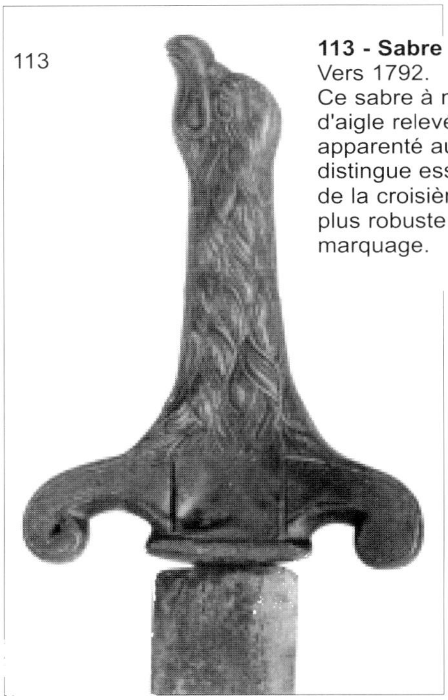

113 - Sabre d'Artillerie à cheval.
Vers 1792.
Ce sabre à monture de laiton en tête d'aigle relevée est évidemment apparenté au modèle précédent. Il s'en distingue essentiellement par la forme de la croisière ; celle-ci est plus épaisse, plus robuste et la lame est sans marquage.

114

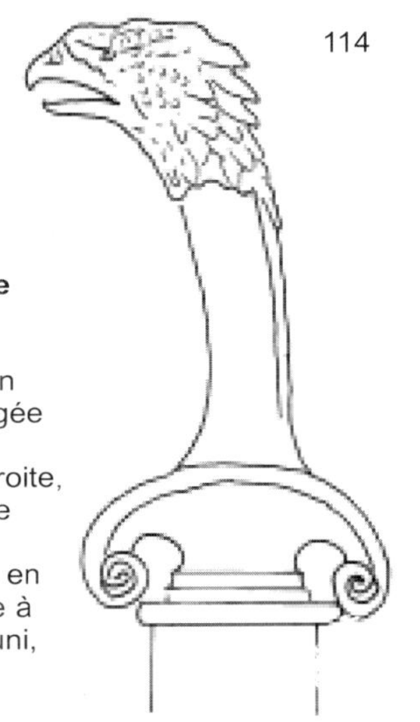

114 - Sabre de Sapeur de Garde Impériale.
Vers 1804 - 1806
Bronze fondu et moulé. Calotte en tête d'aigle, le bec ouvert, prolongée par une poignée à huit pans, légèrement courbée. Croisière étroite, arquée, à quillons enroulés. Lame légèrement courbée, signée *M^fre^ Imp^le^ de Klingenthal*, avec le dos en scie. Fourreau à chape découpée à sa base et pourvue d'un bouton uni, en olive, bouterolle à dard asymétrique.

115

115 - Sabre de Sapeur de la Garde Impériale.
Vers 1810 - 1812.
Monture en laiton. Poignée à tête d'aigle au cou emplumé, unie et à pans. Croisière à quillons infléchis vers la lame. Lame droite, large, longue de 75 cm, avec le dos en scie. Fourreau de cuir noir à deux garnitures de laiton décorées au trait : chape à bouton, bouterolle à dard en laiton légèrement asymétrique.

115 bis

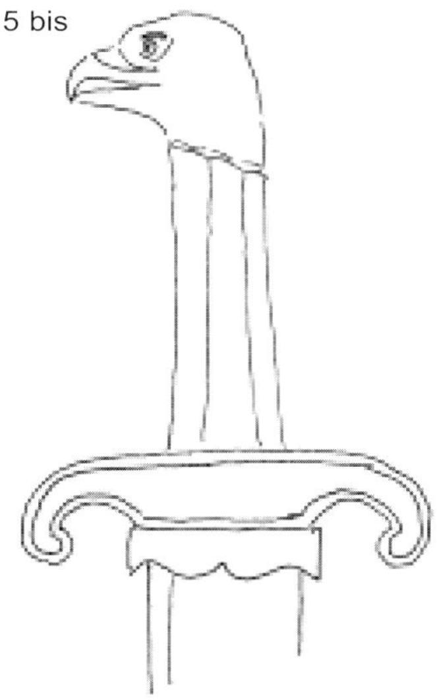

115bis - Variante Sapeur de la Garde Impériale.
1810 - 1812.
Cet exemplaire présente la particularité de posséder un couvre-chape.

115 ter

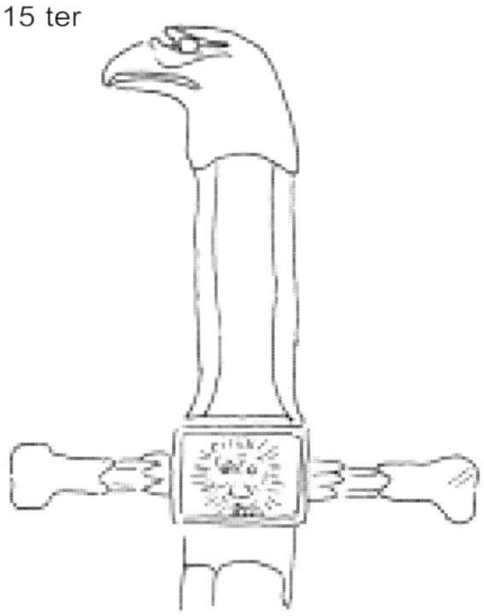

115ter - Sabre de Sapeur de la Garde Impériale - Musée de l'Empéri.
Second Empire.
Monture en laiton fondu. Poignée terminée par une tête d'aigle au bec fermé surmontant une poignée à huit pans. Croisière droite dont les quillons sont en tête de béliers avec sur le nœud de corps, dans un rectangle, une face léonine. Lame à pan creux et dos en scie, signée *Coulaux et Cie Klingenthal.* Fourreau de cuir, chape à bouton ciselé d'une tête de lion et bouterolle asymétrique (M.E. - S. de P.).

116

116a

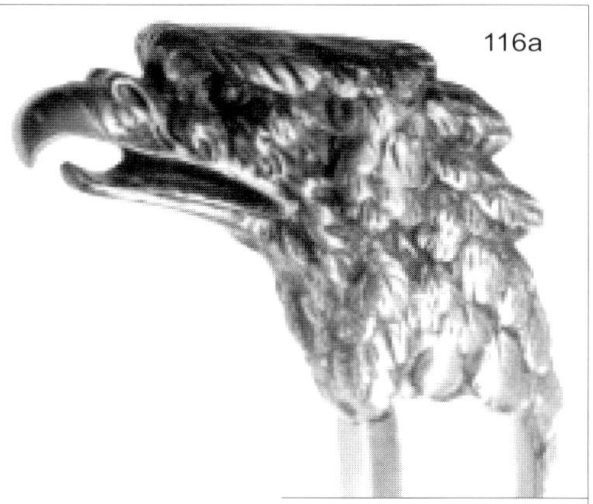

116 - Sabre de Sapeur de la Garde Impériale.
Second Empire.
Monture en laiton, dorée ou non. Pommeau en tête d'aigle (116a), le bec entrouvert, avec le haut de la poignée emplumé et à quatre pans s'élargissant sur une base pyramidale de 1,7 cm de haut et de 4,5 cm de large. Croisière fondue séparément à nœud de corps quadrangulaire de 4,5 cm de largeur et de 3,6 cm de hauteur, ciselé d'une tête de lion sur fond sablé. De part et d'autre du nœud de corps, deux quillons droits, à base ornée de couronnes de feuilles et se terminant par des têtes léonines. Lame longue de 73,7 cm, large au talon de 3,2 cm, presque droite, la flèche n'étant que de 0,6 cm, à long pan creux relativement étroit (1,4 cm) sur presque toute la longueur de la lame. Fourreau de cuir à deux garnitures de laiton : chape à bouton ovalaire, longue bouterolle terminée par un dard de laiton, asymétrique.

Marquage - La signature de Duc est poinçonnée sur le haut du côté antérieur du cartouche et en bas du nombre 38, désignant probablement un corps.

117

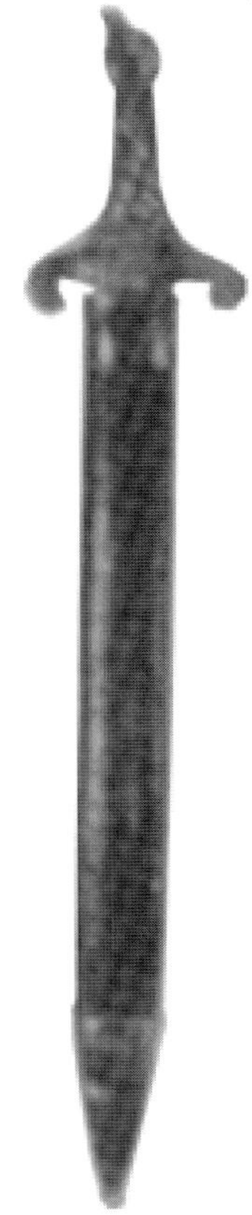

117 - Glaive d'Artillerie de la Garde Nationale.
Restauration.
Monture en laiton. Poignée en tête d'aigle rappelant le modèle 1771 d'Artillerie à pied mais avec le cou à plumes plus grandes et à croisière recourbée vers la lame. Lame et fourreau du modèle du glaive 1816.

Sous-famille VII c – A tête de lion

Les glaives à tête de lion semblent avoir été introduits dans les armées françaises par les Gardes Suisses, ce décor étant très ancien en Helvétie, ils voulurent lui rester fidèles. Moins figuratif, le roi des animaux fait très triste figure sur les glaives de l'Artillerie de Marine des modèles 1784 - 1792.

118

118 - Glaive de soldat charpentier des Gardes Suisses.
Règne de Louis XVI.
Laiton moulé. Poignée en tête de lion, à pans unis. Croisière importante, recourbée à ses extrémités. Lame à pans creux avec une arête médiane.

119 - Sabres de Sapeurs des Régiments Suisses.
Ancien Régime.
Les Suisses au service du Roy de France ont conservé leur armement d'origine. Ce sont des glaives en laiton fondu à tête de lion avec, le plus généralement, une croisière aux quillons infléchis vers la lame. Un des derniers modèles est probablement celui qui est figuré ci-contre et répondant à la description suivante : calotte en tête de lion terminant une poignée à section quadrangulaire. Croisière plus légère que celle des modèles antérieurs. Poignée et croisière bordées de rainures dont le fond est finement strié. Lame de forme variée ; plate ou à une ou deux gorges profondes. Fourreau de cuir noirci, à bouton ou à pontet. Courte bouterolle à dard symétrique en forme de lyre (M.E. - S. de P.).

119

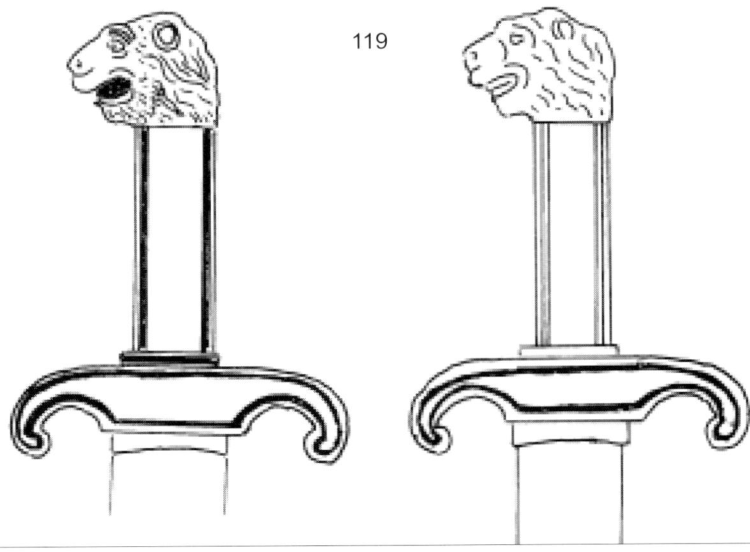

120

120 - Glaive de l'Artillerie des Colonies.
Modèle 1784.
Poignée de laiton moulé à tête léonine placée horizontalement. Pas de rivet sur la poignée, la lame étant matée sur la tête. Croisière unie, droite, aux extrémités arrondies. Lame à arête médiane poinçonnée de F.A. Bisch, de Juchereau sur le talon. Chape du fourreau de cuir, à fente pour la languette de cuir. Longueur totale de l'arme : 61,7 cm, dont 12,7 cm pour la monture. Poids 1,360 Kg.

Nota - Ce glaive fut attribué à l'Artillerie de Marine en 1790.

121 - Glaive d'Artillerie de Marine.
Modèle 1792.
Ce modèle diffère des sabres précédents sur les points suivants : tête léonine relevée à 45°, lame fixée sur la poignée par trois rivets, avec les ailes de la soie de 2,6 - 2,8 cm, apparaissant sur le dos de la poignée et sur le côté antérieur. Croisière constituée d'une lame de laiton, plate, à quillons inversés très allongés. Lame sans pan creux, longue de 48,5 cm, en langue de carpe. Fourreau de cuir collé sur des alèses de bois, à deux garnitures en laiton découpé : chape à bouton, bouterolle terminée par un aplat.

Marquage - Eventuellement, poinçon de de Cabannes sur le talon de la lame.

121

122 - Variante du modèle 1792.
La croisière est en laiton aux quillons relevés à ses deux extrémités.

122

123 - Variante.
On encontre le plus fréquemment des exemplaires à lame à trois pans creux.

123

124

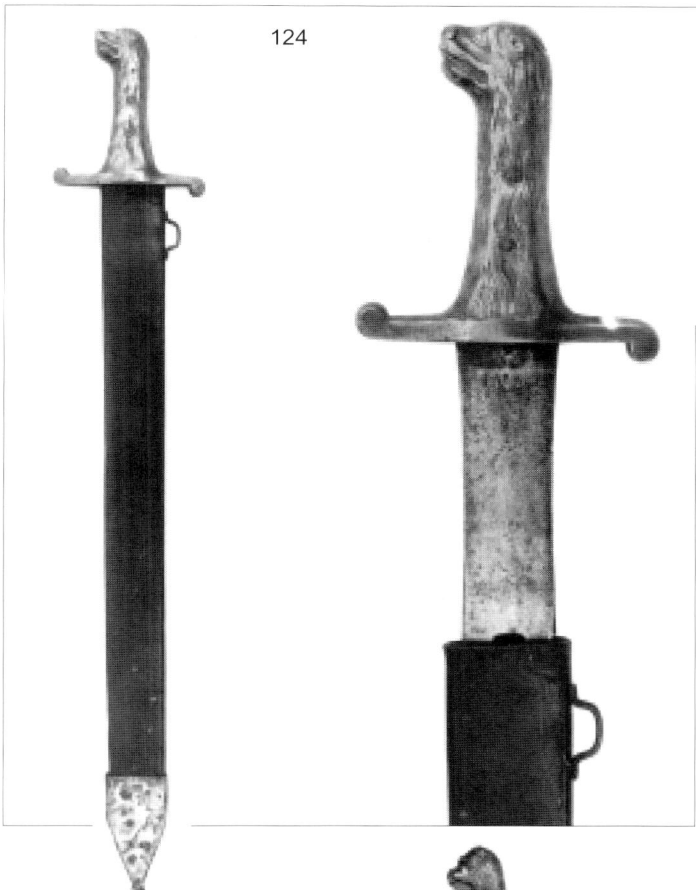

124 - Sabre allemand sur le modèle d'Artillerie de Marine français.
Ce glaive à tête de lion levée, ainsi que le modèle 1771, possède une croisière moulée à quillons inversés. On remarquera de plus l'anneau métallique fixé sur le haut du dos du fourreau.

Nota - Ce modèle de glaive a été fabriqué également pour des Marines voisines de notre pays et il est parfois difficile de reconnaître les fabrications françaises.

126

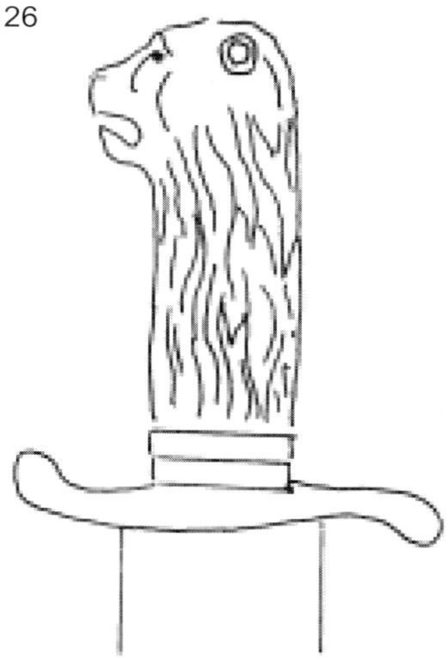

125

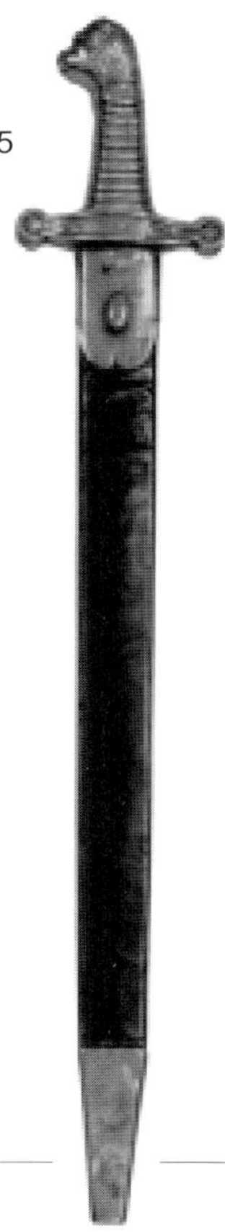

125 - Glaive d'Artillerie de marine.
Epoque révolutionnaire.
Poignée en laiton moulé. Calotte en tête de lion à crinières latérales, séparées par un aplat et terminée par une bague portant le nombre 24. Fusée aplatie, à 18 cannelures. Croisière oblongue, peut-être rapportée, sablée, terminée par deux arrondis grossièrement ciselés d'une spirale. Lame de 50 cm, à longue gouttière médiane, large de 11 - 12 mm. Fourreau de cuir , à couture postérieure, à deux garnitures : chape à bouton découpée en deux lobes, bouterolle symétrique, terminée par un aplat d'acier.

126 - Sabre de Charpentier de Régiment Suisse.
Fin XVIIIème siècle.
Monture en laiton moulé. Calotte en tête de lion à long cou pour former la poignée. Croisière en " S " très allongée. Forte lame légèrement courbée, à large pan creux et dos en scie. Ce sabre très volumineux est une exception dans l'armement comme le précise C. Ariès en ces termes (XVIII, 5) : ...*On est en présence d'une arme très extraordinaire... La taille et le poids sont très extraordinaires, le caractère est très archaïque...*

Mais cette arme est bien de fabrication française comme en témoignent sur le *recto* de la lame *KLINGENTHAL* et sur le *verso* la signature *COULAUX / FRERES*.

Chapitre III

Poignée en bois recouvert de basane, souvent filigranée

Les sabres à poignée en crosse de pistolet filigranée se distinguent souvent très nettement des sabres à l'orientale et couvrent un large ensemble s'étendant des sabres de Marine aux sabres de Cavalerie et sabres de Tambour-Major pour ne citer que les groupes les plus importants. L'unique exemplaire de sabre typiquement à l'orientale à fusée filigranée est ainsi décrit par Aries (XXVII, 5) : le filigrane est formé d'un faisceau de soie jaune enveloppé par un étroit ruban de tissu métallique doré ...

Tous ces sabres ont été réunis, assez arbitrairement, dans la Famille VIII elle-même subdivisée de la façon suivante :

Famille VIII – Poignée en crosse, oreillons entiers

Sous-famille VIII a - Croisière droite ou non, oreillons en losange, en navette

Cette Sous-Famille inclut essentiellement les sabres de Tambour-Major et un sabre de Mameluk de la Garde Impériale

127 - Sabre de Mameluk de la Garde Impériale, troupe.
Vers 1802 - 1815.
Monture en laiton. Courte calotte dont la partie terminale est perforée pour recevoir la dragonne. Poignée en crosse en bois cannelé, recouverte de basane maintenue par un filigrane de fil de laiton simple. Croisière droite, à deux oreillons losangéïformes et quillons terminés en boules sur un desquels pour l'exemplaire décrit, sont frappées les initiales AC et NB. Lame plate, longue de 77,2 cm à flèche de 12,2 cm sans décor, signée : *Mfture de Klingenthal Coulaux frères Entreprs* et poinçonnée par Mouton et Levavasseur avec sur le plat PA 29. Fourreau de bois recouvert de cuir : chape unie, échancrée dorsalement pour faciliter le passage de la lame, deux anneaux de bélière à bossette unie, bouterolle de 33 cm terminée par un dard asymétrique en fer (J.C. D.)

127

128 - Sabre fantaisie d'officier d'Etat-Major.
Premier Empire.
Monture en laiton doré. Calotte en crosse à face léonine et courte queue gravée de palme, d'écailles, d'une toile d'araignée. Poignée encordée, recouverte de basane et filigranée d'un fil doré torsadé disposé en chaînette (128a). Oreillons en navette ornés d'un soleil rayonnant (128b). Croisière en " S ", à quillons inversés, terminés par des têtes de dragon (ou de chien), l'antérieur relevé et tenant dans sa gueule une chaîne la reliant à celle du lion. Lame à pan creux, longue de 81 cm, large au talon de 3 cm. Fourreau de cuir à deux longues garnitures : chape gravée d'une aigle inscrite dans un ovale au niveau du bracelet flanqué de motifs foliacés, bouterolle décorée en suite et portant le bracelet de bélière (M.R.).

128

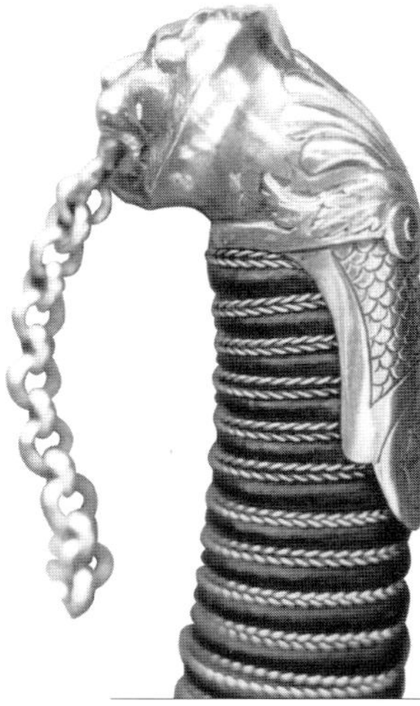

128a

- **a**, calotte à tête de lion d'où part la chaînette qui rejoint le quillon antérieur.

128b

- **b**, oreillon au soleil rayonnant.

Sabres de Tambour-Major

Avant la création du modèle 1822, aucun écrit officiel ne réglementait les sabres de Tambour-Major. Sous le Consulat il existait un sabre dont la poignée était terminée par une tête de coq d'où partait une chaînette rejoignant le quillon antérieur d'une croisière rectiligne portant un médaillon représentant Hercule. Sous le Premier Empire, la tête d'aigle succède à la tête de coq : la poignée est en ébène quadrillée, les quillons courbés sont inversés, l'un relevé, l'autre incliné vers la lame. Les oreillons apparaissent et sont en navette ciselée d'une aigle couronnée. Puis la crosse se dessine et la tête d'aigle est remplacée par une tête de lion avec une chaînette aboutissant au quillon antérieur. Le nœud de corps est masqué par un grand oreillon frontal losangéïforme orné d'une tête de guerrier coiffée d'un casque surmonté d'un volatile aux ailes déployées. Les lames sont de types variés ainsi que les fourreaux en général très ornés.

129 - Sabre de Tambour-Major.
Début du XIXème siècle, avant 1822.
Monture en laiton à la Marengo. Calotte au masque léonin, à courte queue. Poignée de corne cannelée et filigranée. Croisière aux deux quillons inversés, décorés de palmettes et terminés par des mufles de molosses. Oreillons en navette, ciselés d'un trophée. Lame légèrement courbée de 82 cm, bleuie et gravée de motifs dorés. Fourreau de laiton à deux crochets de suspension, décoré d'instruments de musique militaire, de palmes, de fleurons. Dard en laiton découpé, subtriangulaire dont la base est ornée de deux rosaces symétriques.
Ce modèle, non figuré, annonce nettement le modèle réglementaire de 1822.

130

Le sabre de Tambour-Major dit de 1822.

Le 4 mai 1821, le général Curial, Président du Comité des Inspecteurs généraux souhaite armer les Tambour-Majors d'un sabre un peu orné ayant une lame de 23 à 30 pouces. François Manceaux établit, en 1822, un modèle qui fut approuvé le 16 janvier 1823 et 104 exemplaires lui furent immédiatement commandés.

130 - Sabre de Tambour-Major.
Début XIXème siècle.
Ce modèle, assez variable dans ses détails, n'est pas réglementaire et c'est le colonel du régiment qui le choisit. Néanmoins, par sa décoration somptueuse, il annonce le modèle réglementaire dit de 1822.
Calotte en crosse terminée par une face de lion, à courte queue. Poignée en corne brune filigranée de fils torsadés. Quillons inversés, l'antérieur dirigé vers le haut, le postérieur vers le bas, tous deux ciselés de feuillages et terminés par une tête d'animal mythique. Entre les deux quillons, sur le nœud de corps, un cartouche rectangulaire est orné d'un casque flanqué à gauche d'un drapeau et à droite, d'une poignée de glaive. De part et d'autre de la croisière, deux larges oreillons décorés de palmettes à sept branches. Lame bleuie et gravée de motifs dorés. Fourreau de laiton entièrement ciselé de feuillages stylisés et d'attributs de musique. En haut du fourreau, deux crochets symétriques sont brasés pour le port de l'arme.

131

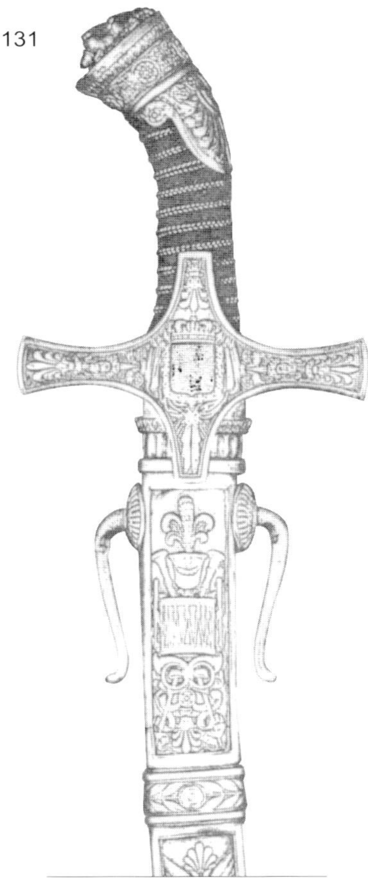

131 - Premier modèle.
Monture en bronze doré. Calotte à couvercle orné d'une tête de lion cachant l'écrou fixant la lame, à courte queue ciselée d'une rosace et d'une palmette sur sa partie postérieure. Poignée gainée de veau, à une dizaine de spires d'un filigrane en chaînette. Croisière dite à la turque élargie pour former les deux oreillons tronqués ainsi que les deux quillons élargis vers leurs extrémités ; au centre de cette croisière ciselée de feuillages stylisés, se trouvent les Armes de France couronnées, avec le collier du Saint-Esprit. Lame longue de 81,21 cm, à flèche de 5,19 cm, à pan creux du type Montmorency, à dos plat, signé *Manufacture Royale de Klingenthal mai 1823*. Fourreau de laiton divisé en quatre parties par des bracelets ornés de feuilles de laurier : la première partie, qui supporte les crochets de suspension, est ciselée d'une fleur de lys et d'instruments de musique ; la seconde et la quatrième, de feuilles de laurier, la troisième, d'arabesques et de rosaces. Bouterolle décorée d'un masque léonin, d'une bague ; dard trilobé à palmette et rosaces. Seuls les bracelets et le dard sont décorés du côté corps.
A propos de ce sabre, M. Pétard écrit avec juste raison en 1993 : *Tout amateur distingué d'armes blanches vous dira qu'il connaît le sabre de tambour-major de 1822. Il est vrai que sa physionomie impressionnante, invariable durant un demi-siècle, le rendit singulièrement repérable…*

132

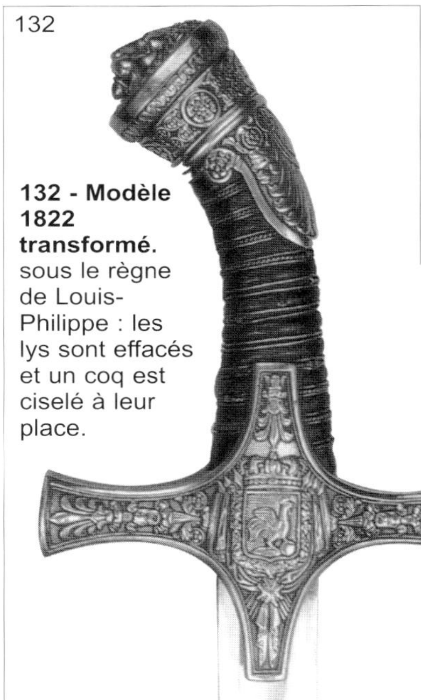

132 - Modèle 1822 transformé.
sous le règne de Louis-Philippe : les lys sont effacés et un coq est ciselé à leur place.

132 - Modèle 1822 transformé.
Ce sabre présente la particularité de posséder sur la croisière, un écusson au coq sous couronne royale sur la face frontale et sur la face postérieure, les lettre " CP " avec le nombre 22. Lame longue de 78,2 cm, marquée à l'envers *Manufare R*[ale] *de Klingenthal, Mai 1823*, utilisé sous Louis-Philippe après effacement des lys des Bourbon. Fourreau de laiton ciselé, marqué lui aussi du nombre 22 et signé *Manceaux / à / Paris.* (Sotheby's, Sussex).

A l'avènement de Louis-Philippe, certains des 104 sabres du modèle 1822 fabriqués par François Manceaux sont réactualisés dès le 11 Septembre 1830, en remplaçant les Armes de France par le coq du roi bourgeois (voir p. 108, n° 134). Il y aura de nouvelles commandes de l'Armée et de la Garde Nationale. Notons que le coq figure sur un blason couronné.

Marquage - Sur le haut du fourreau, on relève la signature de *Manceaux à Paris* en lettres cursives.

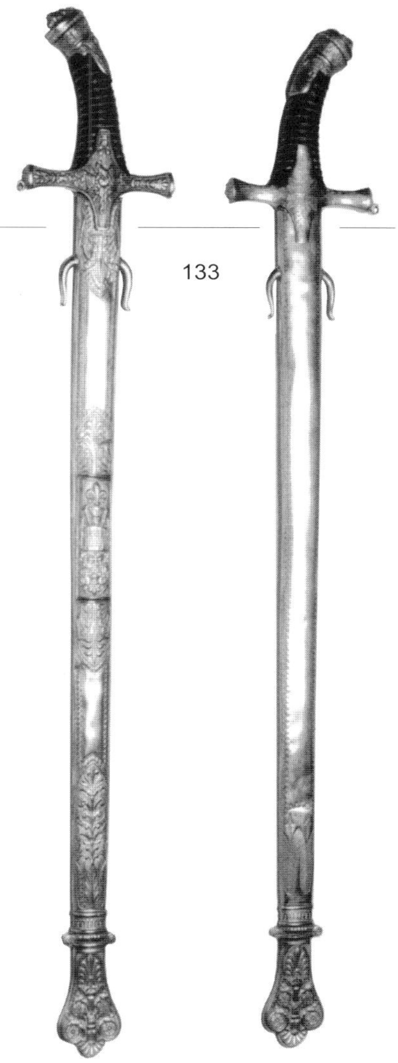

133

133 - Type 1830.
Le sabre de Tambour-Major du règne de Louis-Philippe conserve la forme générale du modèle 1822. La calotte est en crosse terminée par une tête de lion, à courte queue, la poignée est légèrement arquée et filigranée (133a). La particularité essentielle est la décoration de la croisière : le symbole du régime, le coq, est placé sur une couronne de laurier sur fond de quatre drapeaux, l'oreillon supérieur et les quillons sont ciselés de palmettes, (133b) avec une face léonine de profil (133c). Fourreau en laiton à deux crochets de suspente, longuement ciselé sur le *recto* de palmettes, d'un tambour, allusion à la fonction du militaire (133e) ; dard en trèfle orné de fleurettes et de palmettes (133d). La face postérieure du fourreau est beaucoup plus sobrement traitée, sauf en sa partie basale et son dard.

133a

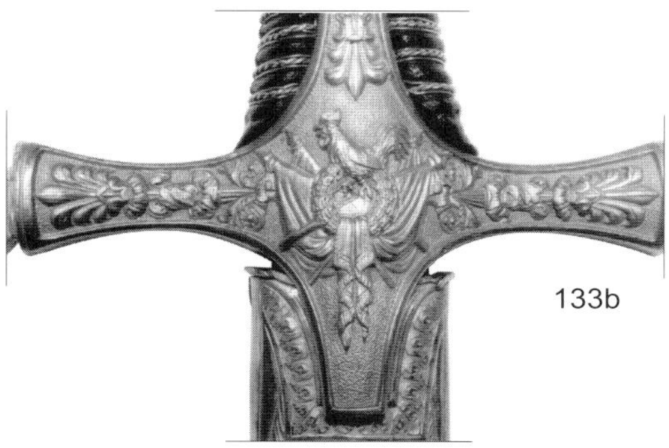

133b

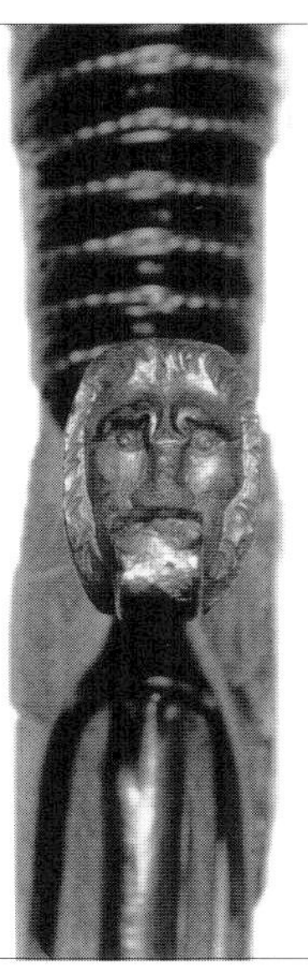

133c

133d

133e

- **a**, détail de la poignée côté *recto* ;
- **b**, croisière au coq ;
- **c**, face léonine terminant les quillons ;
- **d**, suite de la figure a ;
- **e**, détail du motif principal : le tambour.

135

134 - Type 1848.
Le terme de modèle 1848 n'est que partiellement justifié : il s'agit surtout de faire disparaître le coq de la Monarchie de Juillet. On hésite, puis, quelques sabres sont ornés du faisceau de licteur sur la croisière. Ce changement s'accompagne, selon M. Pétard (1993) de modifications mineures :
Poignée recouverte de chagrin ou en corne . Plaquettes ciselées du fourreau plus étroites, collées et non plus brasées, chape et bouterolle vissées... L'auteur ajoute : *Nous ignorons s'il s'agit de fourreaux de remplacement ou d'une production parallèle de moindre prix contractée près d'un atelier de second choix* (non figuré).

135 - Type 1848-1852.
L'exemplaire de la Présidence est déjà orné de l'aigle sur les oreillons, aigle non couronnée mais placée en gloire. Le fourreau conserve le shako et le tambour comme ornement principal.

136 - Type 1852.
A l'avènement de Napoléon III, le sabre de Tambour-Major perdure dans sa forme d'origine, à un détail près : la croisière, dorénavant, sera ornée d'un nouvel écusson ciselé à l'emplacement du coq de Louis-Philippe et figurant l'aigle impériale aux ailes déployées (136a). Puis une nouvelle croisière sera élaborée où l'aigle est inscrite dans une couronne de feuilles de chêne (136b, 136c). Ce modèle semble avoir été également attribué à la Garde Nationale.

Marquage - Le sabre est maintenant signé, en lettres cursives du fils de François Manceaux, J. (pour Jules) Manceaux à Paris et la lame est signée de la Manufacture d'Armes Châtellerault.

136a

136b

136c

Sous-famille VIII b – Croisière à oreillons en disque

Les oreillons en disque différencient bien ces armes des sabres traités dans les Chapitres et Familles précédents. Seulement trois exemplaires seront décrits.

137

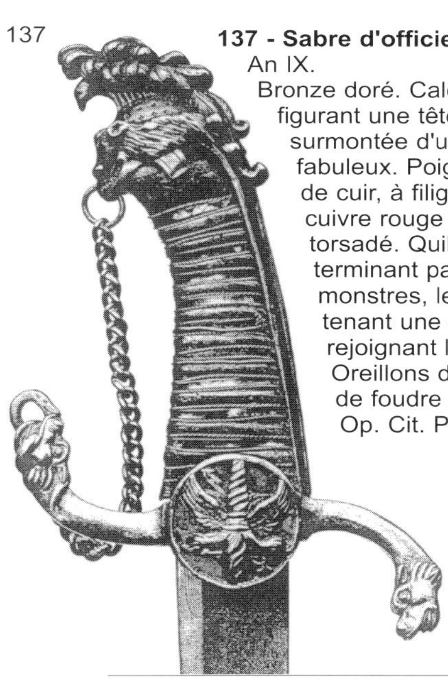

137 - Sabre d'officier d'Etat-Major
An IX.
Bronze doré. Calotte en casque figurant une tête humaine surmontée d'un oiseau fabuleux. Poignée recouverte de cuir, à filigrane en ruban de cuivre rouge bordé d'un fin fil torsadé. Quillons inversés se terminant par des têtes de monstres, le quillon antérieur tenant une chaînette rejoignant la calotte. Oreillons discoïdes ciselés de foudre ailé (G. Lavallay, Op. Cit. Pl. XIV).

138

138 - Sabre de Tambour-Major.
Monarchie de juillet.
Monture de laiton doré. Calotte en tête de coq, à courte queue. Poignée en bois noirci (ébène ?) à rainures. Croisière droite à nœud de corps en médaillon ciselé d'un Hercule sur fond sablé. Quillons en mufle de lion. Lame courbée, de récupération, étant marquée *MANUF*^{RE} *du Klingenthal 8*^{bre} *1814*.
Fourreau de laiton, crochet à palmette ; deux bracelets à bourrelets marginaux. Le corps du fourreau est décoré d'une toile d'araignée, de feuilles d'acanthe entre les bracelets et celle du niveau de la bouterolle est surmontée d'un coq.

139

139 - Sabre de Tambour-Major Type 1852 pour les Grenadiers de la Garde Impériale
Les oreillons de la croisière sont arrondis autour de l'aigle centrale, symbole du Second Empire, chaque quillon est orné d'une grenade, symbole du corps de troupe. Présenté ici avec son ceinturon, le porte-fourreau et les épaulettes du militaire.

Marquage -
Sabre signé *Manceaux à Paris*, sans prénom.

Famille IX – Poignée droite

Ont été regroupées dans cette Famille, quelques armes assez spéciales ne pouvant être incluses dans les catégories précédentes. Leurs seuls points communs est de posséder une poignée droite et d'être dépourvues d'arc de jointure…(*)

140

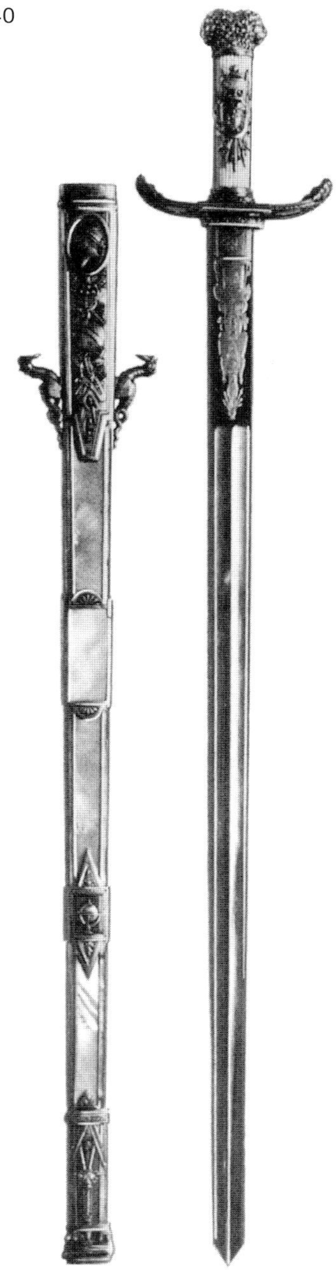

140 - Glaive de consul.
Vers 1800.
Monture en laiton doré. Calotte à deux têtes de lion accolées dos à dos. Poignée d'ivoire décorée d'un trophée d'armes. Croisière aux quillons relevés. Lame à deux pans creux, sauf au talon qui est plat, bleui et orné de motifs dorés. Fourreau en bois recouvert de nacre, à longue chape ciselée pourvue latéralement de deux êtres fabuleux à tête d'oiseau et double queue en tire-bouchon servant d'attache pour le port ; vers l'extrémité un petit encart est ciselé d'une sphère terrestre et courte bouterolle se terminant par un aplat.

Nota - Ce glaive, élaboré par N. Boutet, appartint à Bonaparte.

141

141 - Glaive de Commissaire des Guerres Directoire.
Monture en bronze argenté. Calotte en urne surmontée d'une pièce de rivure. Poignée en ébène cannelée garnie d'un écusson uni en ivoire. Croisière droite terminée par des têtes de bélier (141a). La partie centrale, élargie est gravée de frises de feuillage. Lame bleuie et gravée de motifs dorés, à deux tranchants et méplat médian. Fourreau de cuir à trois garnitures dont une longue chape décorée transversalement de lignes jumelées, bouton en olive (M.R.).

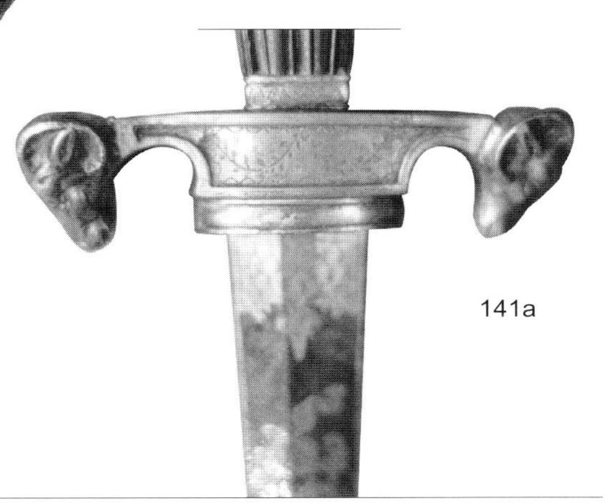

141a

(*) Etant donné qu'entre les épées et les sabres d'une part et les glaives d'autre part, on trouve de nombreuses formes intermédiaires, certains glaives ont été décrits dans " Les Epées ", p. 453 - 461 (J. Lhoste, 1997). On aura intérêt pour avoir une connaissance plus complète de ces armes, de se rapporter à cet ouvrage.

142

142 - Glaive de dignitaire Franc-Maçon.
Début XIXème siècle. Monture en bronze bruni. La partie supérieure de la poignée est gravée d'une aigle, la poignée elle-même étant constituée de feuilles d'acanthe. La croisière supporte un bâton de foudre terminé par des éclairs. Fourreau à chape également à feuilles d'acanthe avec un bracelet portant deux anneaux de bélière symétriquement placés. Sur le cuir on lit : F ∴ pour frère.

143

143 - Glaive de dignitaire.
Restauration.
Monture en laiton ciselé. Calotte octogonale à la fleur de lys, surmontée d'un panache. Poignée à plaquettes de nacre unie. Croisière terminée par des quillons en trèfle (143a) ornés d'une fleurette à huit pétales, avec sur le nœud de corps, une fleur de lys sur la face frontale (143a) et, sur la face postérieure, une fleurette à quatre grands pétales et quatre petits (143b). Lame à deux tranchants et méplat médian. Fourreau de cuir à trois garnitures de laiton (M.R.).

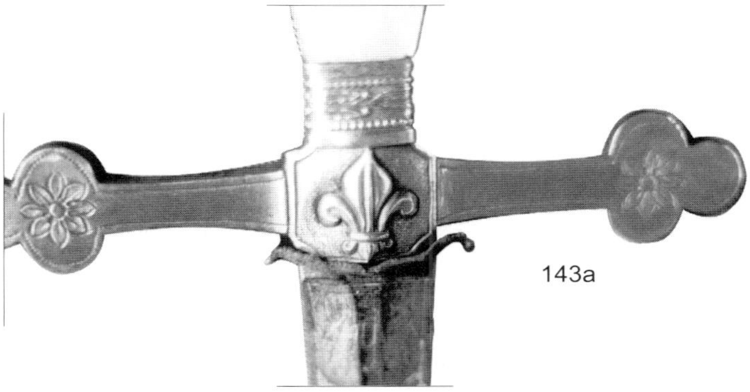

143a

143b

144 - Epée-glaive de Pair de France.
Restauration.
Pommeau réniforme au profil d'Henri IV. Fusée à plaquettes de nacre ornée des Armes de France couronnées (144a). Croisière à médaillons avec une fleur de lys sur fond rayonnant sur le noeud de corps (144b). Fourreau de cuir à longues garnitures dorées (M. R.)

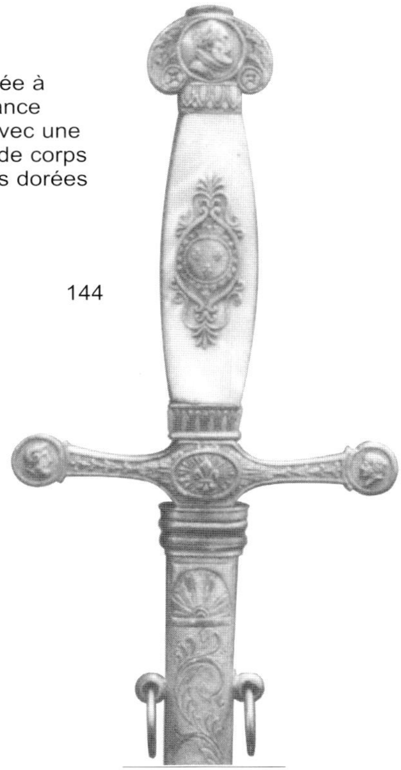
144

144a

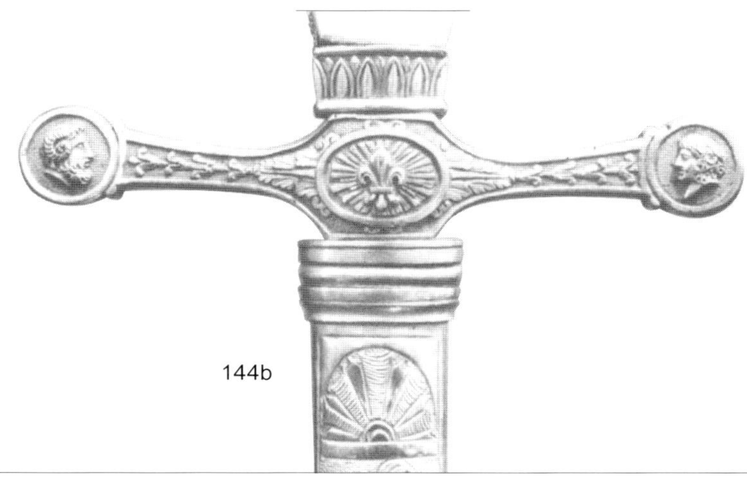
144b

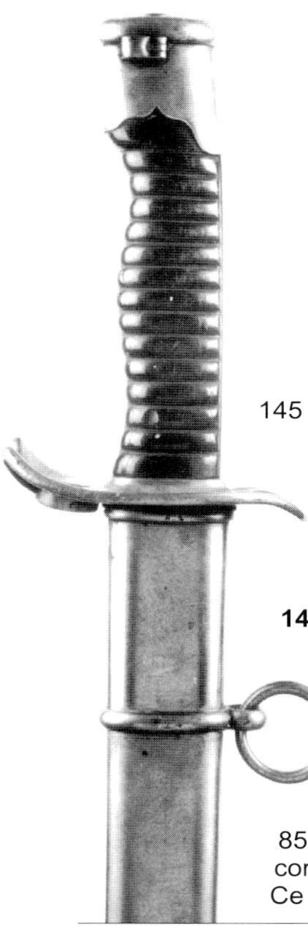
145

145 - Sabre-lance Treuille de Beaulieu.
Modèle 1854.
Ce sabre-lance d'essai était destiné à être fixé comme une baïonnette, au mousqueton du modèle 1854. Un fois monté, sabre et mousqueton faisaient une lance de plus de 2 m.
Le sabre lance a une monture en laiton, dont la calotte à courte queue est frappée du " N " et de la couronne impériale. Poignée de corne de buffle noire.
Primitivement la lame devait mesurer 100 cm. Peu maniable, elle fut réduite à 85,5 cm. L'exemplaire figuré possède une lame de 86,3 cm, avec un fourreau correspondant exactement à cette longueur.
Ce sabre-lance était destiné aux Cent Gardes. L'exemplaire décrit porte le numéro 58.

Seconde partie

Sabres à un seul arc de jointure, croisière à oreillons

I - Croisière à demi-oreillons basaux

II - Croisière à demi-oreillons supérieurs

III - Croisière à oreillons entiers, arc de jointure arrivant en angle droit sur la croisière

IV - Croisière à oreillons entiers, arc de jointure arrondi

Après la guerre de Succession d'Autriche, le duc de Choiseul devient ministre de la Guerre. Par ordonnance du 25 avril 1767, il entreprend de réglementer l'armement et les premiers sabres de Hussards de fabrication française apparaissent bientôt. Ces sabres possèdent une nouvelle pièce sur la croisière : les oreillons. Ces oreillons, quand ils sont basaux, évitent les vibrations de la lame dans le fourreau et quand ils sont supérieurs empêchent la poignée de tourner si la rivure ne joue plus complètement son rôle. Les oreillons sont de formes très variées , supérieurs ou inférieurs entiers, en baguettes, en losanges, hexagonaux... et certains sont ornés de symboles, de l'emblème du corps ou simplement de motifs décoratifs.

L'arc de jointure peut arriver en angle droit sur la croisière où au contraire être plus ou moins régulièrement arrondi. La queue de calotte peut se prolonger jusqu'à la croisière ou elle est fixée souvent par un tenon ou être très courte. Dans ce dernier cas, on parle de *jupe* ou de courte queue en fonction de sa forme.

Prenant en considération toutes ces caractéristiques nous avons établi la classification suivante :

Chapitre I

Croisière à demi-oreillons basaux

Les sabres de ce chapitre I sont assez hétérogènes : les uns sont à montures de fer, d'inspiration germanique, les autres à monture de laiton, parfois doré, sont plus soignés dans leur présentation. Les uns sont à longue queue de calotte, les autres à courte queue, l'arc de jointure arrive en angle droit sur la croisière ou bien est régulièrement arqué, ce qui nous a incité à créer les deux familles suivantes :

Famille I – Arc de jointure arrivant en angle droit sur la croisière

Les sabres de cette famille, à demi-oreillons basaux, possèdent soit une calotte à longue queue, soit une courte queue, d'où les deux Sous-familles proposées :

Sous-famille I a - Calotte à longue queue

Les sabres les plus remarquables de cette Sous-famille sont ceux qui furent en service sous le Directoire, notamment les sabres d'officiers de Marine.

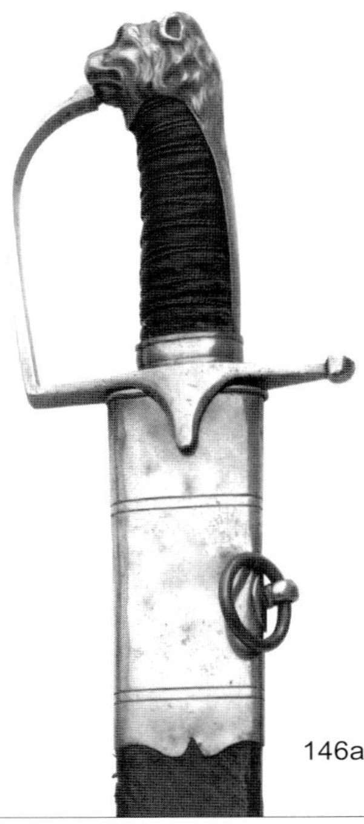

146a

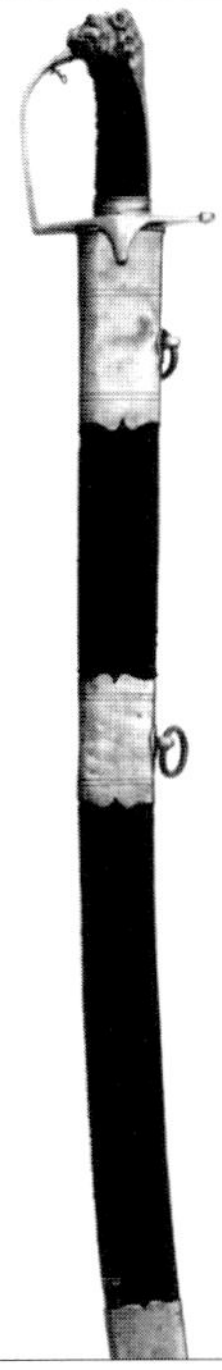

146

146 - Sabre d'officier de Cavalerie.
Règne de Louis XVI.
Monture en laiton doré. Calotte en tête de lion, à longue queue (146a). Poignée recouverte de basane, filigranée et à bague basale. Arc de jointure arqué puis arrivant en angle droit sur la croisière, à oreillons basaux et à quillon rectiligne terminé par une boule. Forte lame. Fourreau à trois garnitures décorées au trait, à deux anneaux de bélière.

147

147 - Sabre de récompense offert par le Directoire.
1800.
C'est un sabre très richement orné en bronze ciselé : calotte en casque à l'antique, prolongée d'une longue queue, fusée filigranée, branche principale s'ajustant à la croisière par un angle droit, quillon terminé par une tête de bélier. Lame bleuie, à motifs dorés, longue de 93 cm, légèrement courbe. Fourreau entièrement de bronze doré et ciselé d'oves au niveau des deux bracelets, de trophées d'armes et d'objets évoquant les arts et les sciences (147c). Sur une plaquette fixée au nœud de corps, on lit sur une face : *Le général Kléber au général Verdier* (147a) et sur l'autre face, *Au nom du Directoire Exécutif.* Au-dessous de la plaquette se trouve un demi-oreillon basal.Sur le haut du fourreau est inscrit l'acte héroïque *Bataille du Mont Tabor, Floréal An 7 et combat du Béga de Damiette, 10 Brumaire An 8* (147b). La lame est signée sur le dos : Dehèque Klingenthal. (Musée Paul Dupuy, Toulouse).

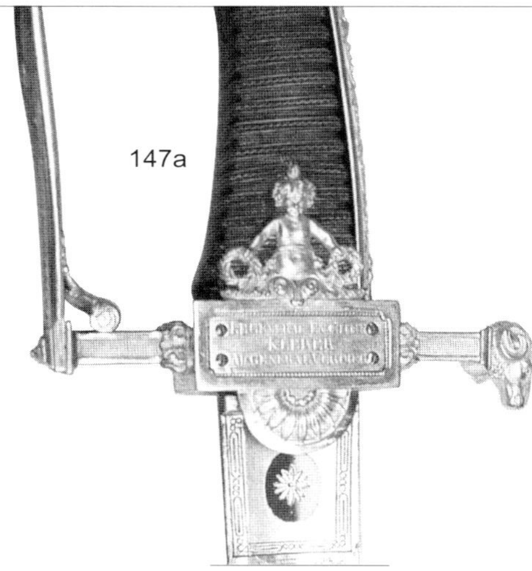

147a

147b

147c

148

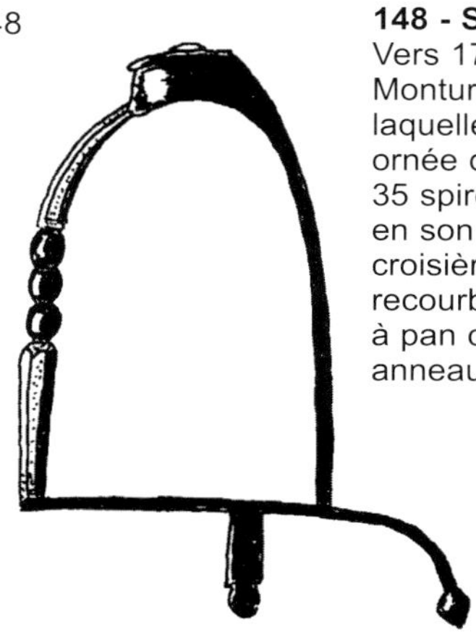

148 - Sabre de Cavalerie légère.
Vers 1793.
Monture en fer. Calotte recouverte d'une pièce soudée sur laquelle la lame est matée, et suivie d'une longue queue ornée d'un filet : fusée à filigrane serré, faisant quelque 35 spires d'un fil de fer torsadé. Arc de jointure à filet orné en son milieu de trois perles et arrivant en angle droit sur la croisière. Celle-ci porte deux petits oreillons basaux et est recourbée avant de se terminer par un bouton. Lame blanche à pan creux. Fourreau métallique, en général de fer, à deux anneaux de bélière brasés directement sur le métal.

149 - Variante de luxe.
Ariès décrit un sabre de ce type conservé au Musée des Hussards, dont la lame porte l'inscription en grandes lettres de fantaisie : *Toujours au chemin de la gloire.* Le fourreau est de laiton ciselé de trois trophées où l'on reconnaît

des piques, drapeaux, tambours, casques... séparés par des cartouches ornés de fleurettes et de soleil rayonnant (non figuré)

Nota - Ces sabres sont à rapprocher des sabres de Cavalerie légère dépourvus d'oreillons basaux.

150 - Sabre d'officier d'Artillerie
Régiment d'Hector, vers 1794 - 1795. Monture en laiton. Poignée recouverte de galuchat et filigranée. Calotte plate, à longue queue portant l'inscription *ART-FRANC-EMIG = A / 110* Croisière à oreillons subtriangulaires. Forte lame droite, à dos plat et contre-tranchant. Fourreau de cuir à trois garnitures de laiton, chape portant l'inscription *KNUBLEY LONDON A / 110.* Il est évident que ce sabre est de fabrication anglaise et que le comte d'Hector le reçut après avoir quitté la France. Ce sabre représente un épisode important de notre histoire ainsi que J-C. Dey s'est plu à le conter (1).

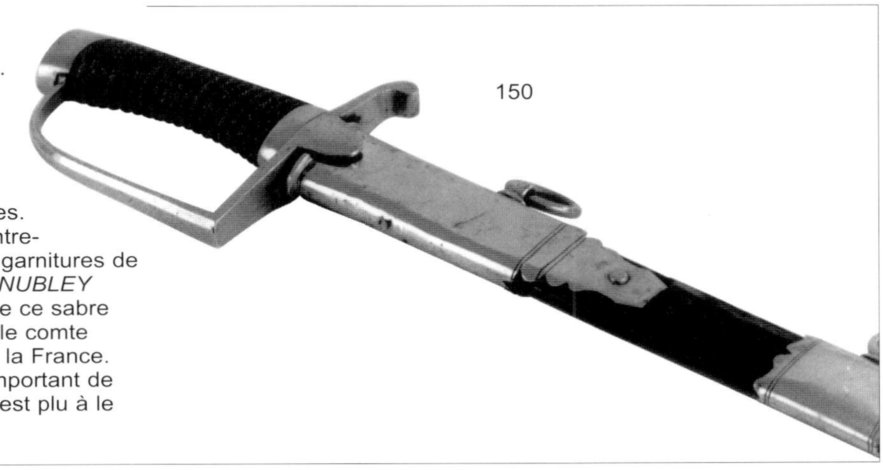
150

151a

151 - Sabre d'officier de Marine.
Directoire.
Calotte plate, à bordure moulée surmontée d'un bouton de rivure également à moulures, prolongée d'une longue queue creusée de cinq gouges sur la moitié basale. Fusée d'ébène quadrillée. Croisière ornée d'un demi-oreillon basal, en triangle curviligne, frappé d'une ancre de marine (151a). Quillon terminé par un bouton tronqué. Lame de 84,5 cm, large au talon de 3 cm, flèche de 6,5 cm, à pan creux et signé sur le dos : *Pierre Guillaume Knecht à Solingen.* Fourreau de laiton à deux bracelets rainurés et anneau de bélière de même métal ; dard asymétrique en fer.

Ce sabre, non réglementaire est, semble-t-il, le premier modèle conçu pour les officiers de Marine qui ne portaient jusqu'alors que l'épée. Parmi les autres exemplaires qui nous sont parvenus, certains ont une calotte à courte queue et gravée d'une ancre ainsi que sur les demi-oreillons. Le fourreau peut être de cuir à trois garnitures, avec un bouton de chape qui peut être à palmette.

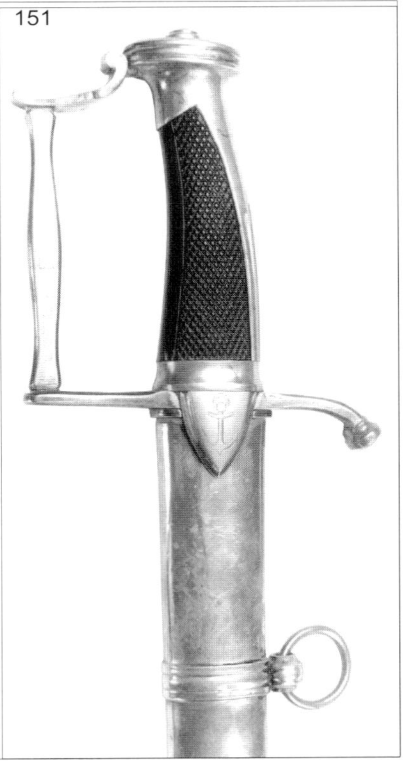
151

(1) En septembre 1794, Jean-Charles comte d'Hector, Lieutenant Général des Armées Navales, émigré à Londres, est chargé de lever un régiment dit *de la Marine Royale* dont les officiers devaient être choisis parmi ceux de l'ancienne Marine française. Les hommes sont recrutés en partie chez les marins républicains prisonniers. Ils sont équipés d'uniformes et d'armes spécifiquement anglaises. Le Régiment d'Hector, à effectif de 700 hommes environ, s'embarque à bord de vaisseaux anglais et le 27 juin 1795 a lieu le débarquement à Carnac. Après quelques succès en soutien des Chouans et l'occupation de la presqu'île de Quiberon, la situation se renverse. Plus de la moitié du régiment est hors de combat. Les officiers sont condamnés à mort par les commissions militaires et exécutés à Vannes et à Quiberon au nombre de 60 environ. Les sous-officiers et les soldats sont presque tous réincorporés dans les rangs républicains.

152

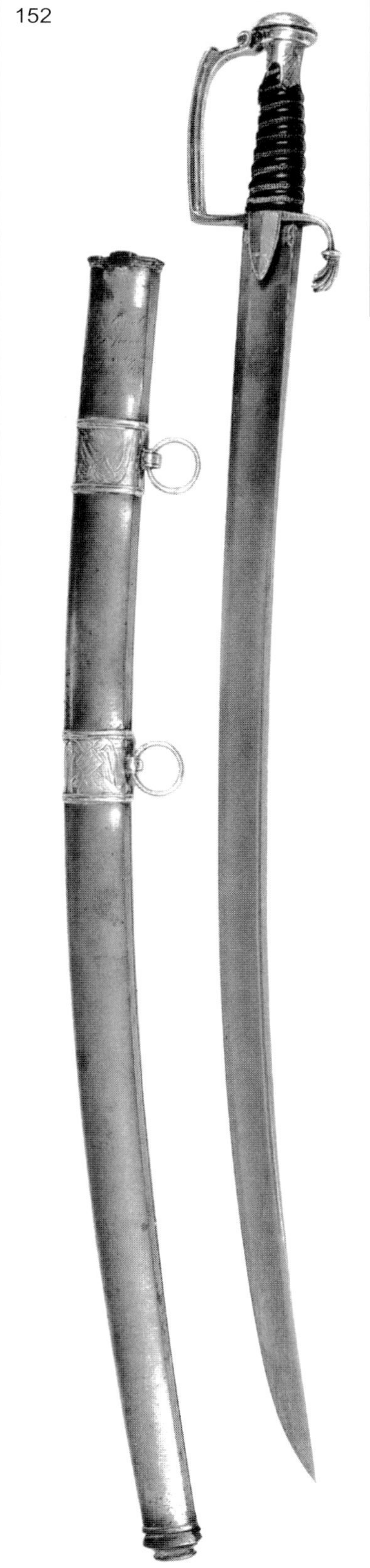

Sous-famille I b – Calotte à courte queue

Cette Sous-famille se distingue par une queue de calotte courte. Elle comprend de nombreux modèles plus ou moins dérivés des modèles réglementaires : c'est dans cette Sous-famille que l'on trouve les arcs de jointures à colonne, très fréquente sous le Directoire et le Consulat. Destinés aux officiers, ces sabres sont finement ciselés, leurs lames sont bleuies avec des gravures dorées et souvent les fourreaux sont de laiton laminé plus ou moins richement décorés. Quelques exemplaires perdureront sous le Premier Empire.

152 - Sabre d'honneur.
Modèle de l'An VIII pour la Cavalerie.
Monture argentée. Calotte bombée, à jupe symétrique profondément échancrée en son milieu. Poignée recouverte de basane à filigrane d'argent composé de deux fils torsadés disposés en chaînette. Arc de jointure pouvant être divisé en deux parties : enroulement près de la calotte, partie légèrement arquée arrivant sur la croisière en angle presque droit. Croisière pourvue d'un oreillon basal en écusson. Quillon recourbé terminé en palmette. Lame à pan creux, longueur variable en fonction du récipiendaire selon qu'il est fantassin ou cavalier. Fourreau de tôle mince, à deux bracelets argentés, ciselés de trophées et dard en deux boutons superposés.

Marquage - La dédicace est inscrite, en général, sur le haut du fourreau (152a). La lame est poinçonnée de Levavasseur.

Remarques - Certains modèles sont plus richement décorés que le modèle figuré. En particulier, la calotte peut être à bordure perlée. En revanche les bracelets de bélières peuvent être plus étroits et sans décor (152b).

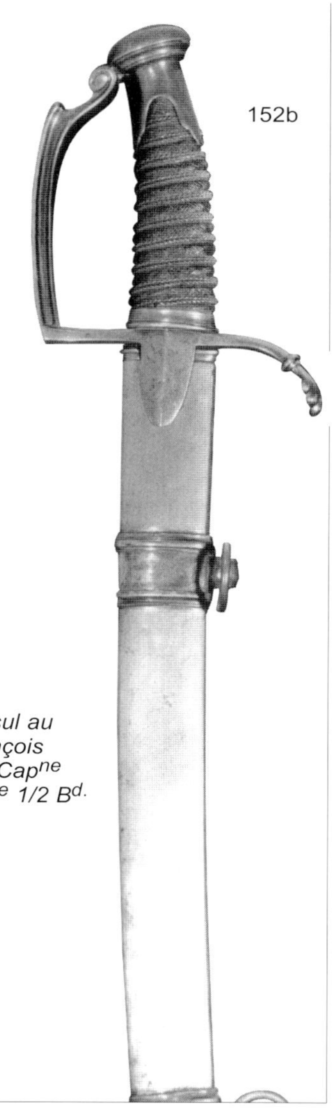

152b

Le 1er Consul au C^{en} 3^{n} François Hennequin Cap^{ne} dans la 108e 1/2 B$^{d.}$

152a

153 - Sabres des Quatre Cents Braves. (1)
Ces sabres ont été offerts par le ministre de la Guerre du Directoire. Monture dorée ou argentée. Calotte peu épaisse surmontée d'un dôme masquant la rivure de la soie. Courte queue de calotte mais dépassant la demi-longueur de la poignée et maintenue par un clou. Poignée recouverte de basane et filigranée d'un double fil torsadé. Branche en colonne plate reliée au pommeau par un petit brin arqué et arrivant sur la croisière en angle droit. Oreillons basaux en triangle à côtés arrondis, le frontal portant la dédicace. Quillon recourbé terminé par un bouton à facettes. Lame à pan creux, longue de 67,7 cm, légèrement courbe. Fourreau de cuir noirci à deux garnitures dorées ou argentées en harmonie avec la monture.

Les sabres des Quatre Cents Braves ne sont pas homogènes dans leur présentation. Parfois, sur les oreillons sont gravés *Manu^{fre} / A / Versailles* sur l'un et sur l'autre *Boutet / Directeur / Artiste.* Il semble que Bonaparte ait préféré faire graver la dédicace sur le haut du fourreau en raison d'une plus grande place disponible.

153

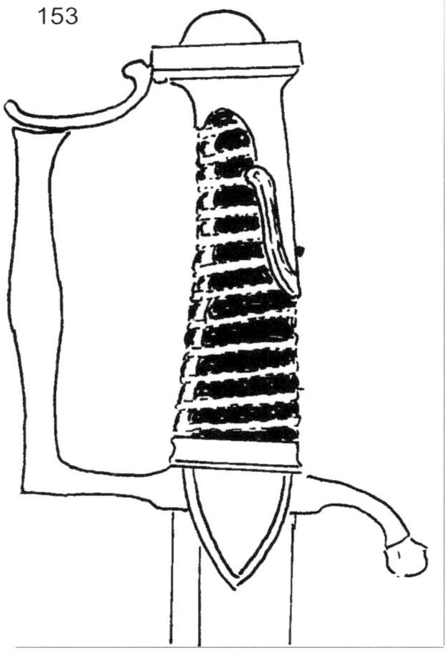

154 - Sabres de Récompense de la 3ème Compagnie de la 19ème demi-brigade d'Infanterie de Ligne.
1798 - 1799.
Gérard Jaeger (1998, Tradition 139, 12-15) rappelle que ...*le six floréal An VI, le général Vaubois demande un sabre de récompense pour tous les officiers, sous-officiers et grenadiers composant la fraction de la 19ème demi-brigade de Ligne qui se distinguèrent en Corse. Une centaine de sabres fut fabriquée ; dix neuf seulement auraient été distribués.*
Monture en laiton dorée du type à colonne et semblable à celle des sabres des Quatre Cents Braves. Calotte en disque surmonté d'un bouton masquant la rivure de la lame. Courte queue de calotte atteignant la première moitié de la poignée, fixée par un clou doré, et à filet marginal en sa demi-longueur. Poignée recouverte de galuchat maintenu par un double filigrane doré disposé en chaînette. Virole basale. Arc de jointure en colonne relié à la calotte par un brin arqué. Croisière portant des oreillons en écusson ; sur le frontal on lit : *LE MINISTRE / DE LA GUERRE / à la 3^{e}Comp^{e} / de la 19e ½ / Brig^{e}* et sur le *verso AB LD / MANUFACTURE / VERSAILLES / BOUTET*, Boutet étant en plus petits caractères et inscrit dans un cartouche.
Lame à pan creux, large au talon de 3,5 cm, longue de 64,5 cm et flèche de 2,2 cm. La signature *Klingenthal* figure sur le dos. Sur les deux faces, sont gravés, à partir du talon, les motifs suivants : arabesques fleuries, trophée, couronne de laurier, motif végétal. Fourreau de cuir à deux garnitures : chape à bouton, à base découpée, bouterolle découpée en suite et dard en bouton. Les deux garnitures sont ajourées d'un petit losange.

154

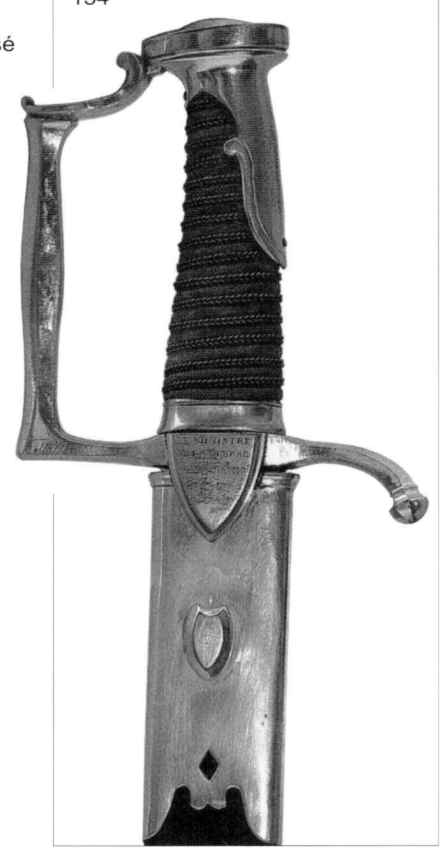

155 - Variante.
G. Jaeger (op. cit.) signale une variante surtout caractérisée par sa lame : celle-ci est large à talon de 3,7 cm, longue de 65,5 cm et bleuie sur 23 cm aux motifs dorés classiques, plus sur le *verso*, un faisceau de licteur sur feuillage. La signature de *Klingenthal* se trouve également sur le dos (non figuré).

La dédicace est ainsi libellée
LE MINISTRE / DE LA GUERRE / à la 3^{e}Comp^{e} / de la 21e ½ / Brig^{e}
sur l'oreillon frontal et sur le postérieur
AB LD / MANUFACTURE / VERSAILLES / BOUTET

Pour les officiers

Il existe un modèle pour les officiers qui se différencie du modèle de troupe par une calotte en côte de melon, une poignée en ébène quadrillée, une colonne de garde cannelée et l'oreillon frontal décoré d'écailles. L'oreillon postérieur est lisse pour recevoir la dédicace.

(1) Ce sabre est destiné aux 300 militaires commandés par le général Junot qui ont vaincu 3 000 cavaliers orientaux à Nazareth, Bataille qui fut livrée le 8 avril 1799. On parle de quatre cents sabres car la commande passée à Boutet en mentionnait ce nombre.

156

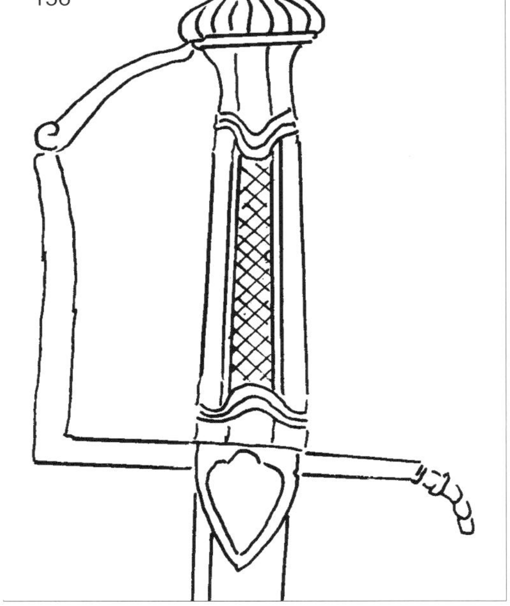

156 - Sabre commémoratif du 19 Brumaire de l'An VIII.
Pour commémorer la coup d'Etat de Bonaparte contre le Directoire, annonçant le Consulat, des sabres de modèles assez variés ont été offerts par le Premier Consul à ses fidèles. La chape de ces sabres porte l'inscription :
Journée de St Cloud / 19 Brumaire an 8
Et sur le demi-oreillon frontal *Manufre / à / Versailles* et sur le demi-oreillon postérieur *Boutet / Directeur / Artiste*

Nota : La silhouette du sabre figuré est empruntée à C. Aries, 3e fasc. 1970).
Il existe plusieurs modèles commémoratifs de cette journée.

157 - Sabre d'officier d'Etat-Major, variante du modèle Vendémiaire An XII.
Vers 1800.
Monture de bronze doré. Calotte à courte queue. Poignée d'ébène finement quadrillée. Branche de garde en colonne, reliée à la calotte par une petite branche évidée, ciselée de feuillage et terminée en crosse. La partie droite de la branche cannelée est coudée en angle droit pour former la croisière dont la partie jouxtant le nœud de corps est décorée de feuillage. Nœud de corps rectangulaire frappé d'un casque sur fond de drapeaux et d'épée, et prolongée par un oreillon basal à une palmette de sept feuilles. Quillon décoré de feuilles à sa base, infléchi et terminé par un mufle de lion. Lame à faible courbure, bleuie et gravée des motifs classiques dorés. Fourreau d'acier noirci, à trois garnitures de bronze doré ; bracelet supérieur lié à la chape, à bordure moulurée et décoré de deux couronnes de laurier ainsi que le bracelet inférieur ; longue bouterolle à bague ciselée supérieure, se terminant par un dard à deux enroulements avec une protection antérieure (description d'après B. Bruel, 1996.).
Ce sabre non figuré, possède donc beaucoup d'affinités avec le sabre suivant.

158

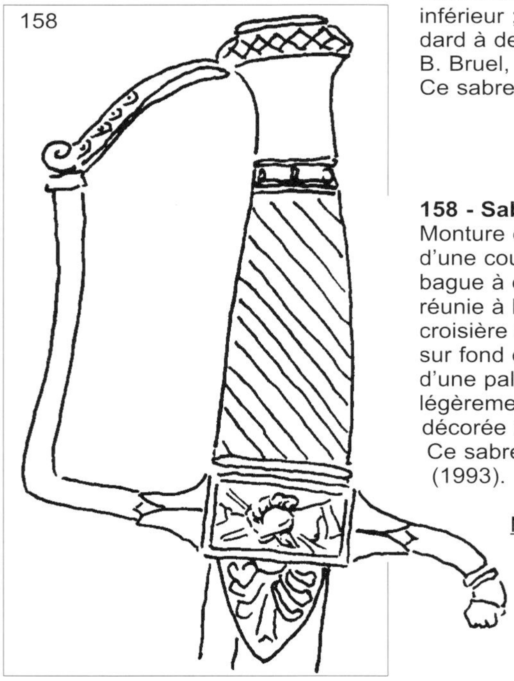

158 - Sabre d'Aide de Camp, modèle de l'An XII.
Monture en laiton doré. Calotte à bouton de rivure orné à sa partie supérieure d'une couronne de feuillage ; courte queue unie, terminée par une petite bague à décor perlée. Poignée en ébène striée obliquement. Branche droite, réunie à la calotte par un petit brin à enroulement distal et rejoignant la croisière en angle droit. Ecusson rectangulaire décoré d'un casque à l'antique sur fond de glaive, de drapeau, de trompette... Demi-oreillon triangulaire orné d'une palmette. Quillon incurvé terminés par un bouton ciselé. Lame légèrement courbe, à pan creux. Fourreau en tôle d'acier, à trois garnitures décorée de rainures épousant leur contour.
Ce sabre a appartenu à Cazimir Girod de Novillart et figuré par B. Malvaux (1993).

Nota - Rappelons que le règlement du 12 Vendémiaire An XII stipule que *Les adjudants-commandants, les adjoints et les aides de camp auront, à cheval, un sabre demi-courbe dont la poignée sera en ébène, les garnitures en métal doré et les fourreaux en fer bronzé.*

159 - Sabre d'officier d'Etat-Major.
Consulat - Type Vendémiaire l'An XII.
Monture en laiton doré. Calotte ornée de feuillages, à courte jupe bordée d'une moulure perlée. Poignée d'ébène sculptée de rainures obliques. Arc de jointure légèrement infléchi, lié à la calotte par un brin perforé pour placer la dragonne. Croisière à cartouche rectangulaire aux casque, fanion, trompette, prolongée par un oreillon basal orné d'une palme. Quillon courbé, terminé par un mufle léonin. Fourreau de laiton doré : chape et bracelet ornés de trophées. (J.C. D.)

159

160

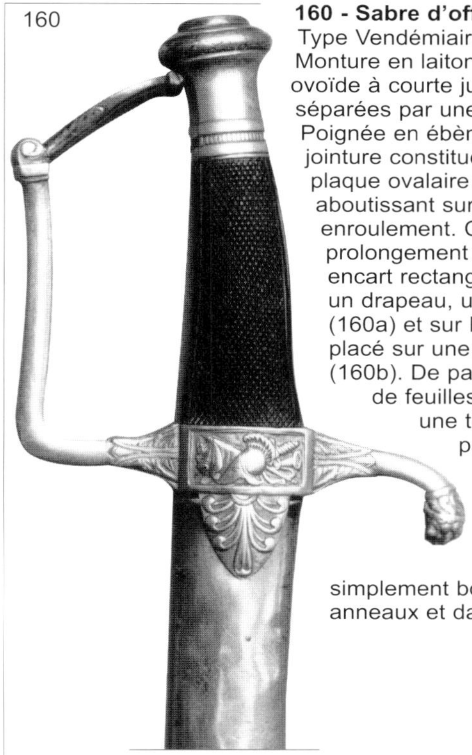

160 - Sabre d'officier d'Etat-Major.
Type Vendémiaire An XII.
Monture en laiton. Calotte à bouton de rivure ovoïde à courte jupe terminée par deux moulures séparées par une bague striée verticalement. Poignée en ébène finement quadrillée. Arc de jointure constitué, à partir de la calotte, d'une plaque ovalaire ciselée à lumière médiane et aboutissant sur la partie verticale par un enroulement. Croisière formée par le prolongement de l'arc, ornée en son milieu d'un encart rectangulaire : casque à crête posé sur un drapeau, une épée, une trompette au *recto* (160a) et sur le *verso* du même casque mais placé sur une pique, une hache, un drapeau (160b). De part et d'autre de cet encart, décor de feuilles d'acanthe. Quillon terminé par une tête léonine. Oreillon en écu à palmette ciselée. Lame de 81 cm, bleuie sur 21,5 cm aux gravures dorées classiques, pan creux, large au talon de 3 cm et flèche de 5 cm. Fourreau de laiton, à deux bracelets de bélière simplement bordés de deux moulures jumelles ; anneaux et dard asymétrique en laiton. (M.R.).

160a

160b

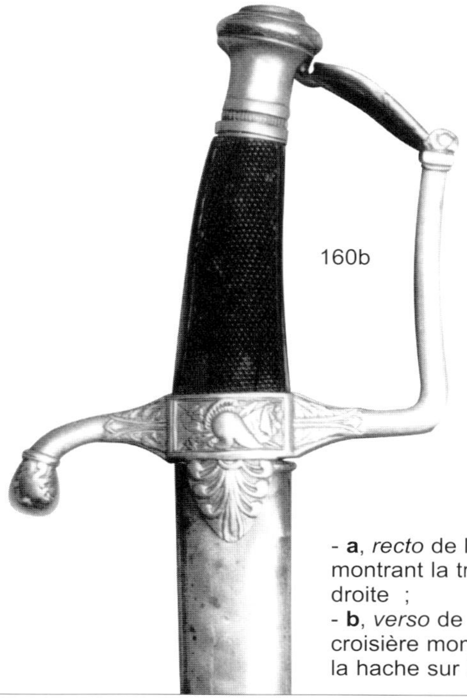

- **a**, *recto* de la croisière montrant la trompette à droite ;
- **b**, *verso* de la croisière montrant la hache sur la gauche.

161

161 - Sabre d'officier de Cavalerie d'Etat Major.
Type Vendémiaire An XII.
Variante du modèle classique avec une poignée en ébène striée en oblique. Fourreau de cuir. Bouterolle à décor *recto-verso*, rappelant celui de l'oreillon basal en acier ou fer (161a, 161b).

161a

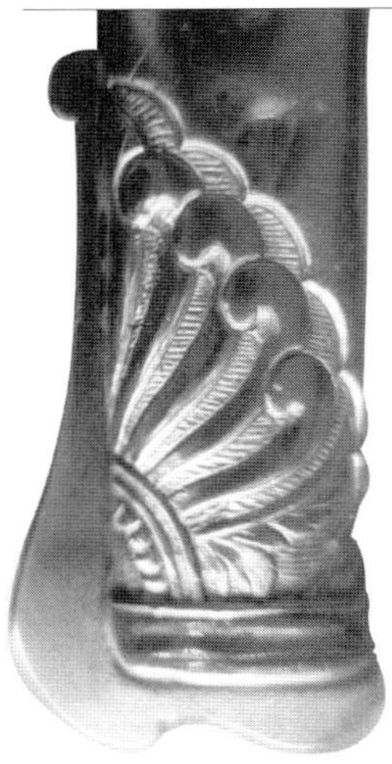

161b

162 - Sabre d'officier d'Etat-Major, variante du modèle Vendémiaire de l'An XII.
Fabrication vers 1810.
La ligne de ce sabre est très comparable à celle datée d'environ 1800. Il s'en distingue sur les points suivants : calotte ciselée sur le dessus, selon B. Bruel (1996) *d'une fleur d'héliotrope au symbole solaire* et bordure à frise de losange. Queue de calotte courte, subcylindrique. Poignée d'ébène à cannelures obliques. Lame de Solingen bleuie à la base et gravée d'astres, de trophées et d'un cartouche à calligraphie ésotérique. Fourreau en acier : chape découpée laissant apparaître le fer ; deux bracelets à bordure moulurée et décorés chacun d'un carré ; bouterolle longue, ornée d'une bague et terminée par un dard asymétrique (description d'après B. Bruel).
Ce sabre a appartenu à Marie, Jean-Baptiste Le Monnier Delfosse qui finit sa carrière comme lieutenant-colonel sous la Restauration (non figuré).

Nota - A partir de 1810, les officiers d'Etat-Major abandonnèrent progressivement les sabres du modèle Vendémiaire au profit d'armes de luxe à l'orientale, à la chasseur ou à la hussarde.

163 - Sabre d'officier de Hussard du Consulat.
Monture en bronze. Calotte plate, à bordures saillantes, surmontée d'un petit bouton de rivure, prolongée d'une queue antérieure moins longue que la queue postérieure. Arc de jointure en colonne, relié à la calotte par une petite branche arquée et fixée sur le dessus de la croisière. Demi-oreillons en écusson (163a).

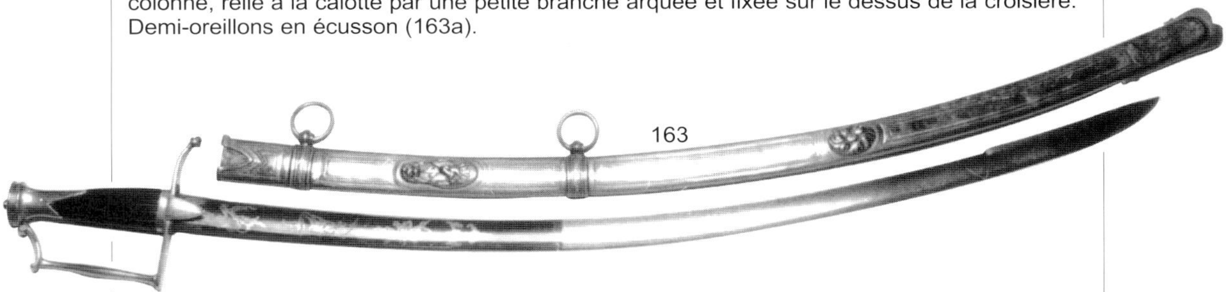

163

<u>Nota</u> - Il existe des différences notables entre les exemplaires de ce sabre : l'oreillon au soleil peut être remplacé par un oreillon à la toile d'araignée, la sirène peut être nue ou porter sur le bas-ventre, pour ménager la pudeur de certains, un médaillon à l'ancre ou même une jupette...

173

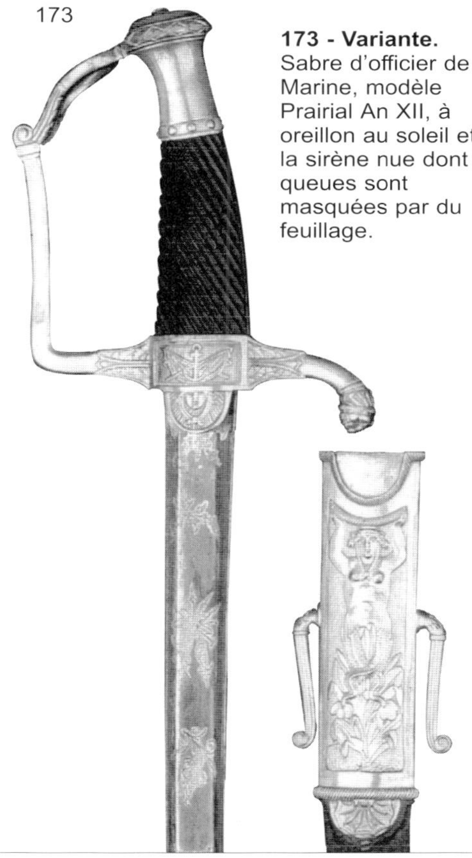

173 - Variante.
Sabre d'officier de la Marine, modèle Prairial An XII, à oreillon au soleil et à la sirène nue dont les queues sont masquées par du feuillage.

174

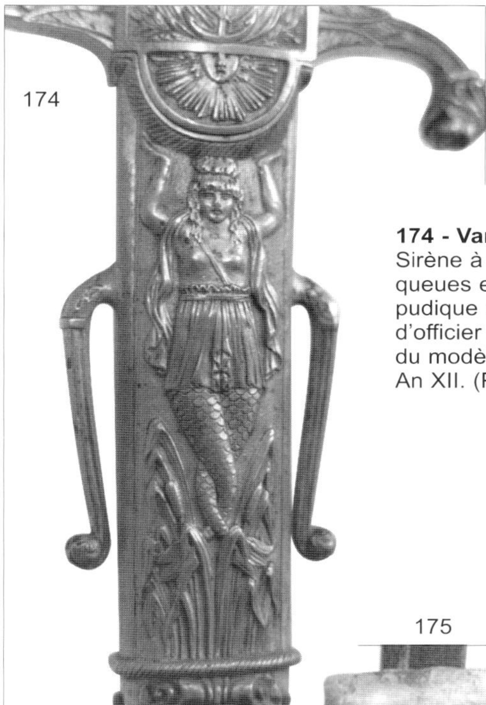

174 - Variante.
Sirène à deux queues et jupette pudique sur un sabre d'officier de Marine du modèle Prairial An XII. (Photo B. C.).

175

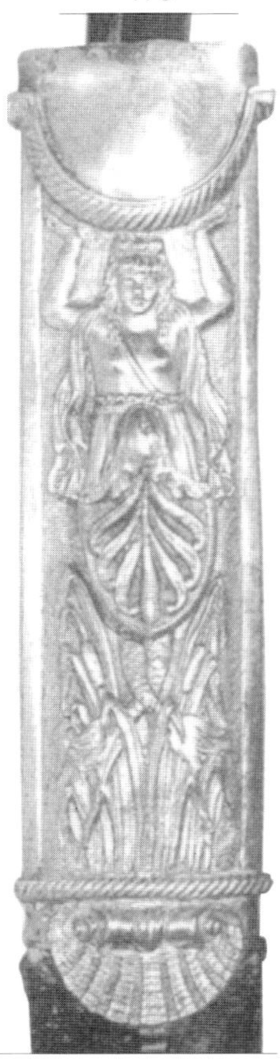

175 - Variante du sabre de Marine.
Prairial An XII.
Monture en bronze doré. Ce sabre figuré par G. Jaeger (2000) et dont nous donnons le détail de la chape montre bien la sirène dont les deux queues écaillées apparaissent sous la palmette. On remarque que son buste est probablement recouvert d'un voile léger et que sa jupette est maintenue par une sangle en bandoullière. Sous la sirène, est placé un motif foliacé constitué par des Typhas (que l'on retrouve sur la bouterolle). La chape se termine par une coquille sur laquelle se trouvent deux fûts de canons.

176

176 - Sabres d'Amiral.
Variante du modèle de Prairial An XII.
Monture en bronze doré. Calotte à couvercle à tête de lion et couronne de laurier. Sur la courte jupe, trois étoiles, puis poignée à écailles, l'ensemble en métal fondu et ciselé.
Arc de jointure en colonne légèrement incurvée, arrivant en angle droit sur la croisière et fixé sur la calotte par un petit brin à tête léonine. Nœud de corps avec, dans un cartouche, une sorte de bouclier à bordure perlée, à deux têtes de coqs, le tout sur fond de drapeaux. Oreillon basal en triangle curviligne, orné d'une tête de Gorgone. Lame de 70,4 cm, demi-courbe, à pan creux, bleuie et gravée de motifs dorés classiques. Fourreau de cuir, à couture métallique : chape à deux longs crochets avec dans un cartouche, Neptune et son trident. Au-dessous, dans un autre cartouche plus petit, la face d'Hercule sur deux massues en sautoir ; bouterolle, avec comme décoration principale, deux flèches qui rappellent probablement les dents du trident de Neptune.

<u>Nota</u> - Certains modèles peuvent être pourvus d'un fourreau entièrement en laiton.

177

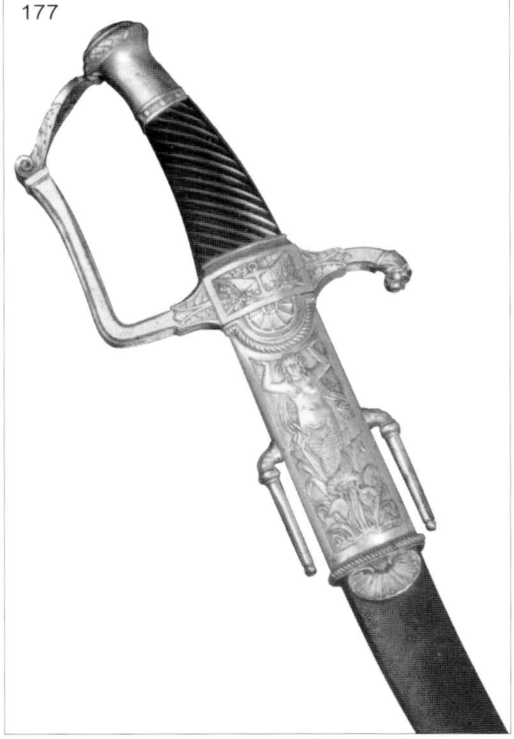

177 - Variante du modèle de Prairial An XII.
Remarquer le cartouche du noeud de corps à l'ancre posée sur des pavillons.

178 - Variante du modèle de Prairial An XII.
Remarquer l'ancre inscrite dans un ovale qui masque le bas-ventre de la sirène.

178

179

179 - Sabre d'officier général d'Etat Major.
Règlement du 1er Vendémiaire An XII.
Même monture que celle du sabre d'Amiral (voir p. 129, n° 176) avec les différences suivantes :
sur le nœud de corps, sont gravées les lettres " RF " pour République Française.
Fourreau de fer noirci à chape décorée du coq les ailes déployées, posé sur un bâton de foudre d'où partent quatre éclairs et surmontant deux casques à l'antique ; bracelets et anneaux de bélière ; bouterolle ciselée de motifs végétaux ; dard frontal à enroulement.

180 - Sabre d'officier des Légions vénètes.
Les Légions vénètes ont été levées par Bonaparte à Venise lors de la Campagne d'Italie. Des sabres aux armes de Venise ont été attribués aux Officiers :
calotte inclinée vers l'avant, à dessus plat et courte queue légèrement courbée et ornée d'une palmette. Poignée filigranée. Branche cannelée, interrompue en son milieu par une bague ciselée et formant un angle droit avec la croisière, celle-ci ornée d'une bague avant le nœud de corps.
Nœud de corps rectangulaire, orné de la face du lion ailé, emblème de Venise, prolongé par un oreillon frontal en trapèze décoré de deux glaives en sautoir passant à travers une couronne de feuillage et sur la partie opposée, oreillon à écailles. Quillon recourbé et terminé par un bouton ciselé. Lame demi-courbe, à dos plat et pan creux. Fourreau de cuir à deux longues garnitures : chape à bouton à bordure quadrillée, bouterolle ornée d'une bague semblable à celle de la chape et décorée de palmettes ; dard en boule striée (non figuré).
Ces sabres fabriqués à l'entreprise peuvent présenter quelques variations dans le décor, (voir le modèle de 1911, n° 181).

181

181 - Sabre d'essai pour officier de Cavalerie.
Modèle 1911.
Ce sabre est une véritable résurrection du sabre des officiers des Légions vénètes de Bonaparte. Il fut proposé par Maurice Berteaux et en lisant L'illustration de juin 1911, on voit que ce sabre était destiné à accompagner un projet proposant d'introduire les uniformes kakis pour l'Infanterie en place du bleu horizon.
Si le sabre d'essai 1911 est comparable, dans ses grandes lignes, au sabre de la Légion, il en diffère sur les points suivants : allègement de l'ensemble, soie fixée par une vis qui tient lieu de bouton de rivure. Poignée presque droite, chape à pontet. Aucun marquage extérieur. Sur la soie facilement accessible grâce à la vis de fixation, on trouve les lettre PDS suivies d'un X et sur le haut de la fusée de bois, on découvre deux X (XX) tracés au couteau.

181a

- **a**, bouterolle ciselée du sabre d'essai de 1911.

182

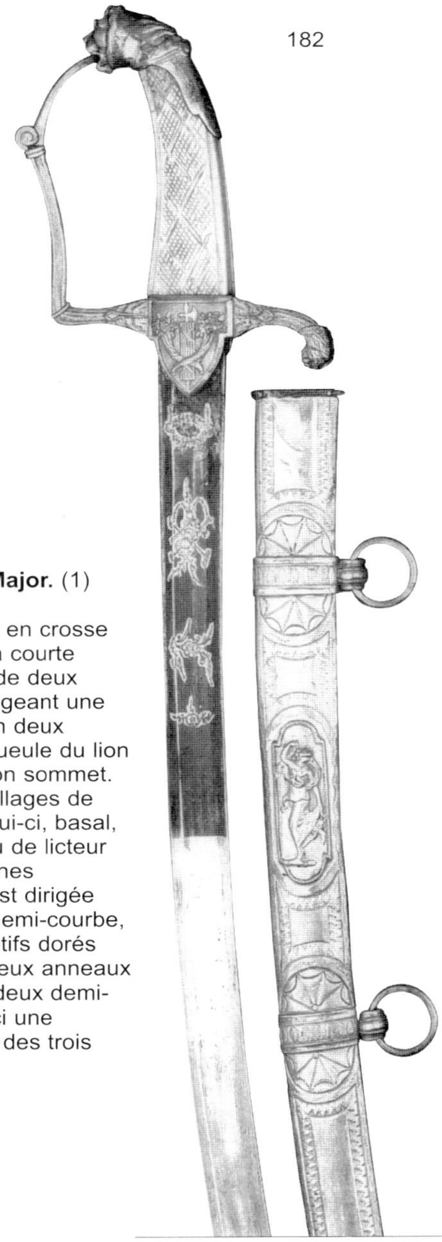

182 - Sabre d'officier d'Etat Major. (1)
Premier Empire.
Monture en laiton doré. Calotte en crosse terminée par une tête de lion, à courte queue. Poignée d'ivoire gravé de deux systèmes de losanges et ménageant une bordure unie. Arc de jointure en deux parties : un fin brin sort de la gueule du lion et rejoint la partie verticale à son sommet. Croisière droite, ciselée de feuillages de part et d'autre de l'oreillon ; celui-ci, basal, est en écu frappé d'un faisceau de licteur sur lequel se croisent deux cornes d'abondance dont l'ouverture est dirigée vers le haut. Lame de 84 cm, demi-courbe, bleuie au tiers et gravée de motifs dorés basaux. Fourreau de laiton à deux anneaux de bélière, chacun flanqué de deux demi-toiles d'araignées, entre ceux-ci une danseuse qui pourrait être une des trois Grâces. (Sotheby's, Sussex).

(1) Nota - Ce sabre est certainement celui du Général Jean Lemarois (1776-1836) qui fut successivement aide-de-camp de Bonaparte sous le Directoire, sous le Consulat et sous l'Empire.

Famille II – Arc de jointure arqué, à calotte à courte queue

Cette Famille se distingue par l'arc de jointure arrondi. Cette modification de la monture concerne notamment des sabres de Hussards et des sabres de Marine aux grands oreillons frappés d'une ancre.

Deux Sous-familles ont été créées :

Sous-famille II a – Calotte dans le prolongement de la poignée

182

On rencontrera, dans cette Sous-famille II a, des sabres de Cavalerie aussi bien que de Marine et quelques sabres de haut luxe.

182 - Sabre de fantaisie d'un officier de Cavalerie légère
Consulat.
Monture en laiton. Calotte en côtes de melon, à bordure ciselée d'une frise et jupe assez longue atteignant presque la moitié de la poignée (182a). Celle-ci est en ébène quadrillé. Arc de jointure très spécial : droit sur les 4/5èmes de sa longueur et rattaché à la calotte par un brin incurvé qui aboutit à un centimètre environ de la partie supérieure de l'arc. La partie basale est courbée pour aboutir à un quillon infléchi terminé par une boule. Au niveau du milieu de l'arc, dans un médaillon en navette est inscrit un faisceau de licteur à un fer (182b). Sur la croisière est soudé un demi-oreillon basal subtriangulaire ciselé d'une palmette. Lame bleuie gravée des motifs dorés classiques. Fourreau de fer, à trois longues garnitures. Dard asymétrique protégeant la face frontale et l'extrémité de la bouterolle (182c). M.R.

182a

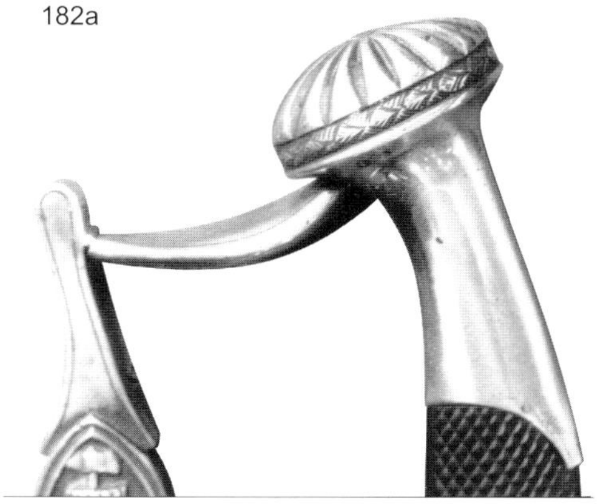

182c

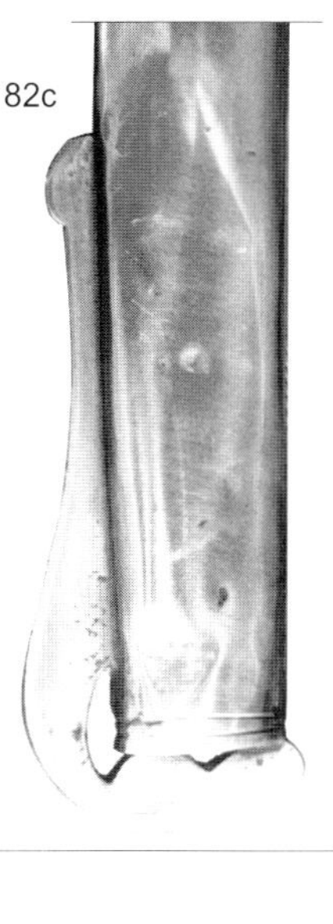

182b

- **a**, détail de la calotte à bordure d'une frise de petits losanges ciselés ;
- **b**, arc de jointure à médaillon au faisceau de licteur ;
- **c**, dard protégeant la base et la partie frontale de la bouterolle.

183

183 - Sabre d'officier de la Garde d'Honneur de Vienne.
Premier Empire.
Monture dite *" à la Marengo "*. Calotte hexagonale à gouge latérale et bouton de rivure. Branche à petite courbure concave près de la calotte puis à grande partie arrondie, à gorge médiane. Croisière lisse, ornée de deux bourrelets de bordure et munie, à sa base seulement d'un oreillon en demi-lune ciselé d'une palmette. Quillon également à gorge, terminé par un bouton évasé, à côtes, rappelant celles de la palmette de l'oreillon. Poignée en ébène quadrillée. Lame de 87 cm, large au talon de 3,3 cm, à flèche de 7 cm. Fourreau en laiton, marqué près de l'ouverture *" Garde d'Honneur de Vienne "* (183a) (voir Lhoste et Buigné 1999, p. 158) à trois pans frontaux, de deux bracelets de bélière et dard asymétrique en laiton.

183a

184 - Sabre d'officier de Cavalerie légère.
Premier Empire.
Monture à la Marengo en bronze doré. Grosse calotte ovoïde à côtes de melon, débordant largement la jupe (184a) Poignée en ébène quadrillée. Branche constituée d'une petite partie concave liée à la grande partie bien arquée. Croisière à bordures plates ; demi-oreillons en demi-lune décorés d'une palmette.
Quillon finissant par un bouton conique à côtes et se terminant par un méplat. Toute la garde est à large gorge. Lame de 82,5 cm, large au talon de 3 cm, flèche de 5,6 cm, contre-tranchant d'environ 22 cm ; bleuie au tiers et gravée de motifs floraux dorés et d'un trophée. Fourreau de laiton, à deux bracelets plats ; dard en fer asymétrique.

184a

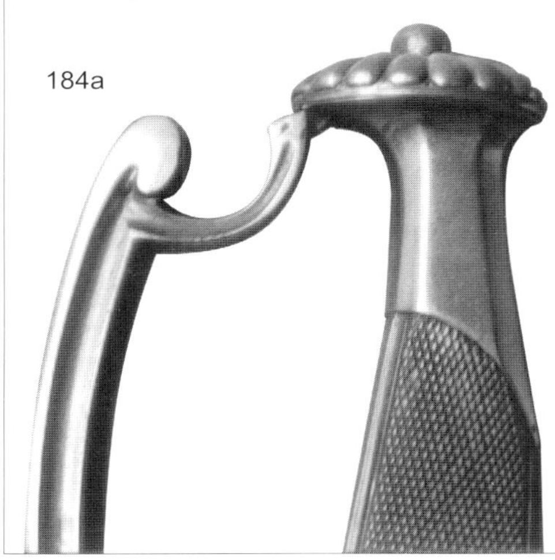

184

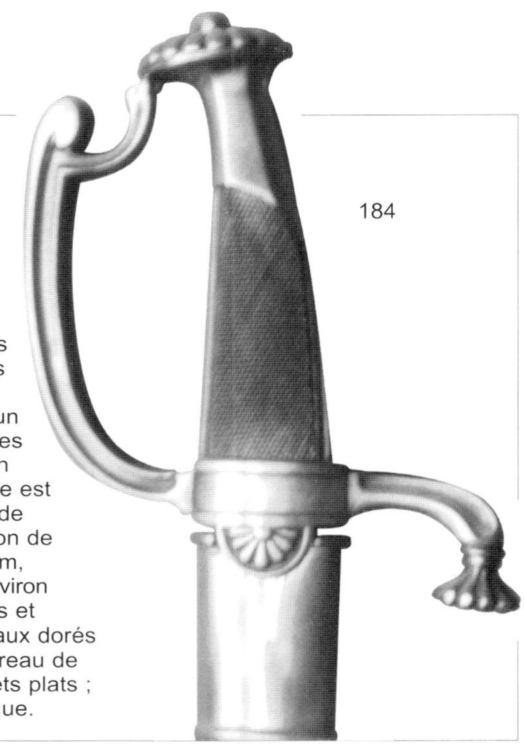

185

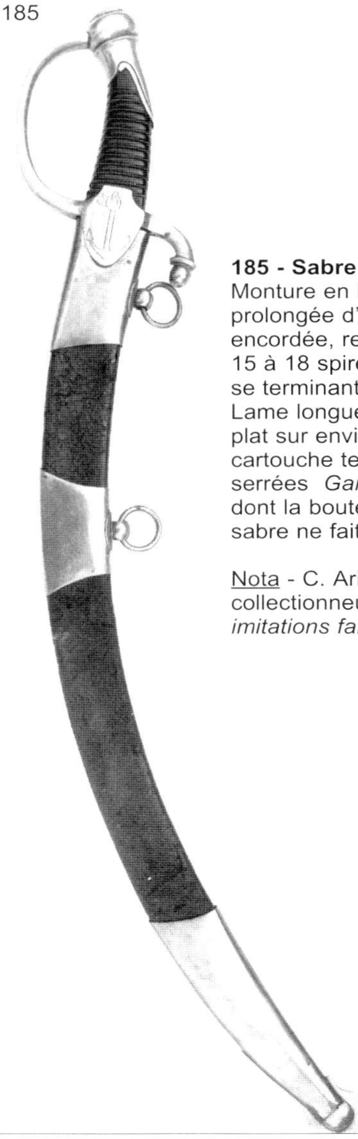

Les sabres des Marins de la Garde Impériale

Le corps des Marins de la Garde a été créé le 17 septembre 1803, lors du fameux Camp de Boulogne. A ce moment, on comptait 737 hommes qui bientôt reçurent leur sabre dont le style est très influencé par la mode orientale qui sévissait depuis la Campagne d'Egypte. Les Marins de la Garde furent engagés en Allemagne, en Espagne, au Portugal et en Russie.

185 - Sabre du modèle 1804.
Monture en laiton. Calotte arrondie surmontée d'un petit dôme servant de bouton de rivure et prolongée d'une queue courte atteignant la demi-longueur de la poignée. Celle-ci est encordée, recouverte de basane à filigrane constitué de fils simples ou torsadés faisant de 15 à 18 spires. Branche de garde unie, régulièrement arquée, se poursuivant par la croisière et se terminant par une bague et une boule. Oreillon frontal soudé, frappé d'une ancre de marine. Lame longue de 71 à 72 cm, large au talon de 4 à 4,3 cm, flèche de 7 à 7,5 cm ; le dos est plat sur environ 45 cm et se termine par un contre-tranchant ; sur le plat de la lame, dans un cartouche terminé à ses deux extrémités par une ancre, est gravé en cursives hautes et serrées *Garde Impériale* (185a). Fourreau en bois encollé de veau cousu, à trois garnitures dont la bouterolle qui se termine par un bulbe enveloppé d'un dard asymétrique en laiton. Ce sabre ne fait l'objet d'aucun règlement officiel et fut fourbi généralement par Duc.

Nota - C. Ariès écrit avec juste raison : " *Mais il importe de mettre en garde ceux-ci* (les collectionneurs) *car ces armes sont relativement peu nombreuses comparativement aux imitations faites dès la fin de l'Empire* ".

185a

185bis - Variante.
Modèle 1804.
Le modèle 1804 qui vient d'être décrit a subi des modifications de détail découvertes, en particulier, par M. Pétard (1994). Ces modifications intéressent :
- la poignée qui peut être à filigrane de fantaisie,
- la lame peut être à dos en jonc,
- la lame peut être très courbe et semble appartenir à des sabres faits pour la Marine marchande ou pour les Gardes-Côtes,
- la lame, sur un exemplaire conservé au Musée de l'Empéri, est gravée dans un écusson, de deux haches d'abordage en sautoir liées d'un ruban,
- des sabres ne portent plus l'inscription *Garde Impériale* sur la lame ; ces sabres auraient été fabriqués, ou modifiés, pour être attribués à la Garde Nationale après la chute de l'Empire ou pour être attribués à une milice Garde-Côtes comme celle de La Teste, en Gironde créée en 1849.
- On trouve également des lames signées *Manuf. De Klingenthal, Coulaux aîné et Cie*, donc postérieures au Premier Empire (non figuré).

186

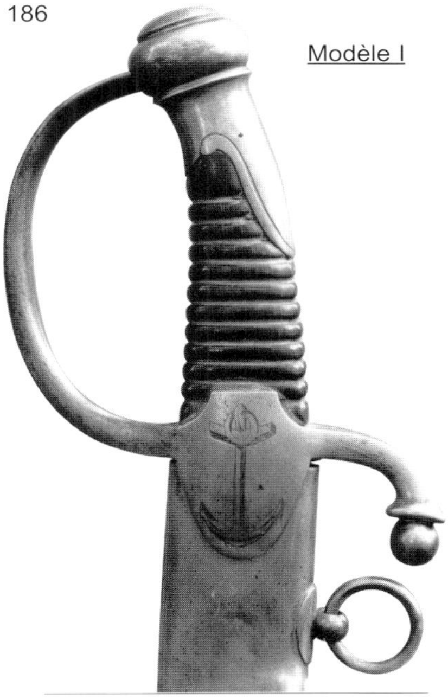

Modèle I

186 - Sabre des Gardes-Côtes.

Il semble bien, qu'en 1849, une milice des Garde-Côtes ait été créée à La Teste, en Gironde. M. Pétard précise (1994) : *Dans le texte concernant l'uniforme des officiers, il est noté – sabre modèle des anciens marins de la garde.*

Les sabres des Gardes -Côtes dits également de la Gendarmerie des Ports sont caractérisés par une fusée de corne, filigranée ou non et par l'oreillon fondu avec la croisière.

Modèle I – Ce modèle semble dater de la Seconde République. Il est, dans sa forme générale tout à fait conforme aux sabres des Marins de la Garde si ce n'est la poignée en corne cannelée et bien entendu, la gravure *" Garde Impériale "* qui a disparu. (M.R.).

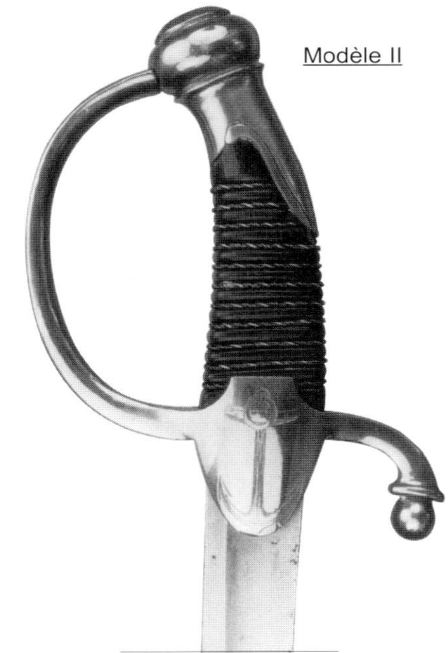

Modèle II

Modèle I – Ce modèle est décrit par M. Pétard (1994). La forme générale reste celle du modèle 1804, mais la monture est plus riche en cuivre, l'oreillon frontal est frappé à l'ancre,et il est fondu avec la croisière (186a). La poignée de corne cannelée est filigranée de cuivre et de laiton torsadés faisant 11 spires. Lame ayant environ 68 cm, étroite, plus courbée, sans gravure si ce n'est la signature *Manuf. de Klingenthal, Coulaux et Cie* sur le dos. Trois garnitures sur le fourreau avec la bouterolle terminée en bulbe, en supposant que cette garniture soit d'origine (186a).

Modèle II - poignée de corne cannelée dont le filigrane semble avoir disparu. Oreillon fondu avec la croisière. (M.R.) ;
- **a**, sabre de Garde-Côtes. Bouterolle tout à fait conforme au modèle 1804.

186a

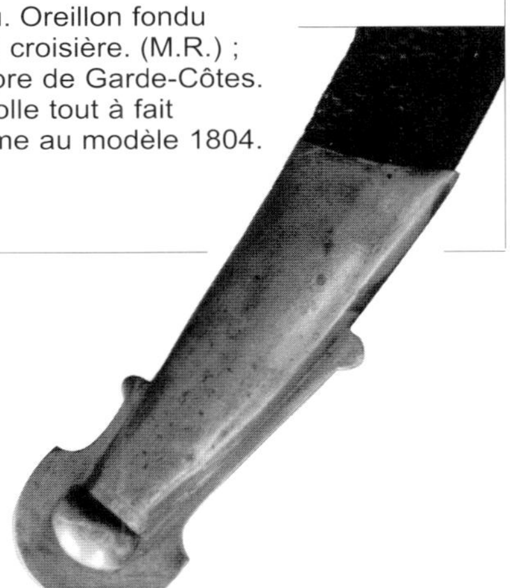

187

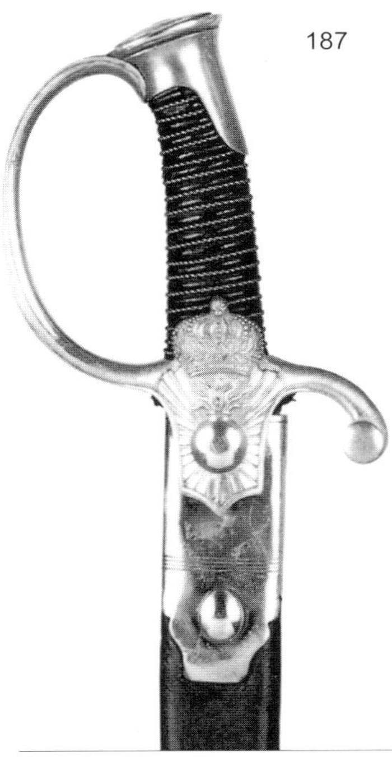

187 - Sabre de sous-officier des Cent-Suisses ?

Après 1822.

Monture en laiton. Calotte simple, à courte queue. Poignée recouverte de basane filigranée d'un fil torsadé. Arc de jointure arrondi terminé par un quillon infléchi. Oreillon du *recto* en écusson à la grenade sur fond rayonnant et surmonté de la couronne royale. Sur le *verso* l'écusson est nu et dépourvu de la couronne. Lame à pan creux, longue de 70 cm, signée *Coulaux à Klingenthal.* Fourreau de cuir à deux garnitures, chape à bouton en écusson (M.E. - S. de P.).

Sous-famille II b – Calotte en crosse

Les sabres de cette Sous-famille sont peu nombreux, mais ce sont des armes de prestige.

188 - Sabre d'officier d'Etat-Major.
Type Vendémiaire An XII. Monture en laiton doré. Calotte en crosse terminée par une tête de lion, à courte queue. Poignée d'ébène à stries obliques. Arc de jointure arrondi, ciselé de motifs végétaux et quillon recourbé. La croisière porte un demi-oreillon basal décoré d'un casque à l'antique sur fond de glaive, de haches et de piques (188a). Lame bleuie gravée de motifs dorés. Fourreau de laiton ciselé à sa partie supérieure d'un trophée d'armes (188b) et de deux scènes de bataille, l'une représentant la canonnade d'une forteresse (188c), l'autre un combat au sabre de cavaliers alors qu'un blessé vise de son pistolet l'un des combattants (188d). (M.R.).

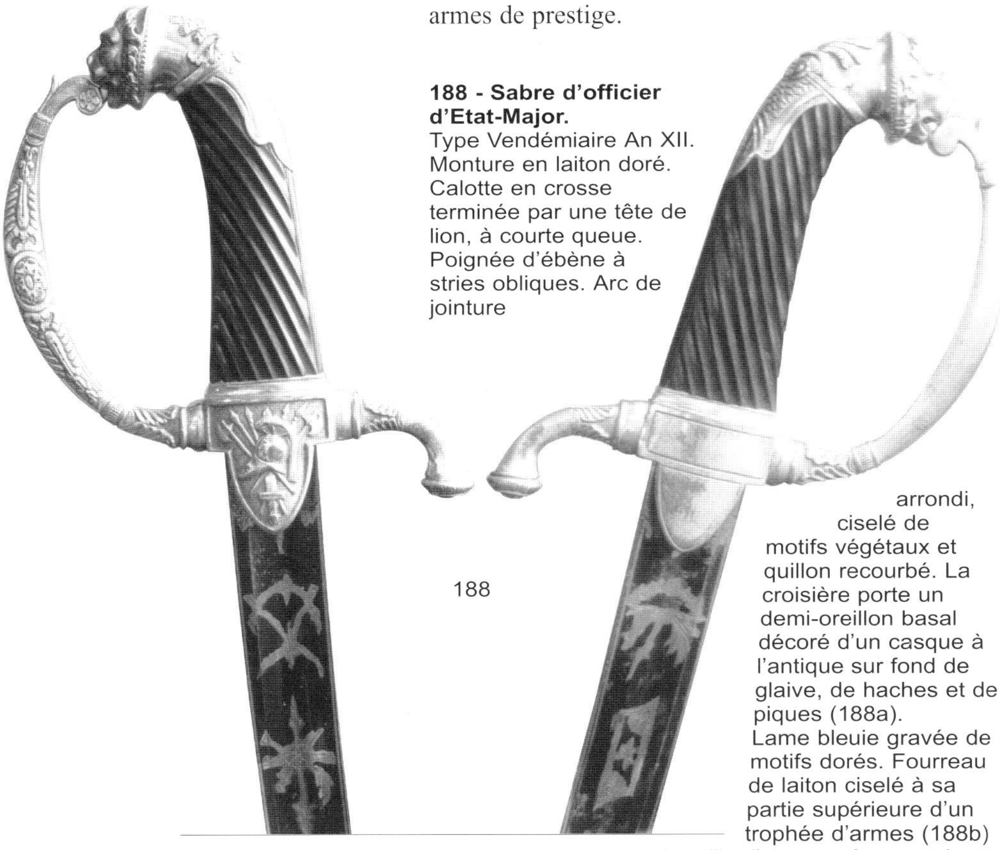

188

188b

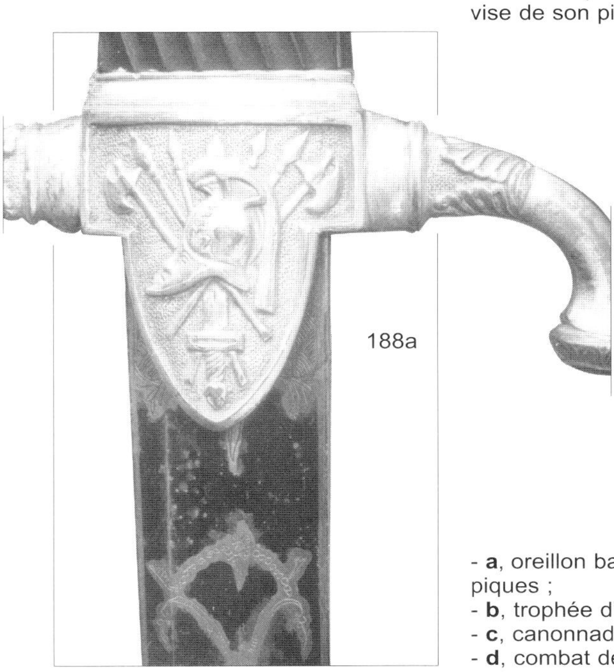

188a

188c

188d

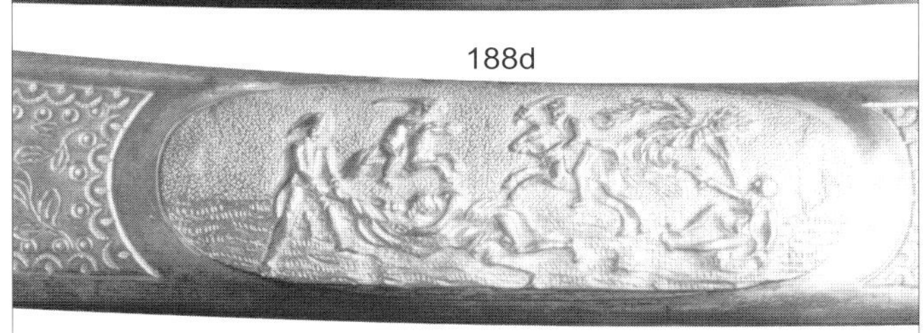

- **a**, oreillon basal au casque à l'antique sur fond de glaive, de haches, de piques ;
- **b**, trophée d'armes ;
- **c**, canonnade d'une forteresse ;
- **d**, combat de Cavaliers et tir d'un blessé.

189

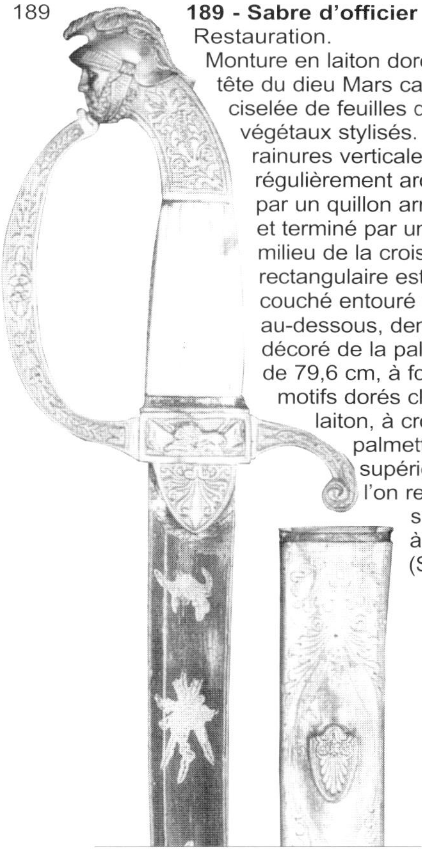

189 - Sabre d'officier d'Etat-Major.
Restauration.
Monture en laiton doré. Calotte en crosse à tête du dieu Mars casqué, à courte queue ciselée de feuilles d'acanthe et d'autres végétaux stylisés. Poignée de nacre à six rainures verticales. Arc de jointure régulièrement arquée se prolongeant par un quillon arrondi en sens contraire et terminé par un enroulement. Au milieu de la croisière, un encart rectangulaire est ciselé d'un lion couché entouré de drapeaux : au-dessous, demi-oreillon en écu décoré de la palmette classique. Lame de 79,6 cm, à fond bleui et gravée des motifs dorés classiques. Fourreau de laiton, à crochet orné d'une palmette et gravé au tiers supérieur d'un trophée où l'on reconnaît un tambour surmonté d'une coiffure à haut panache (Sotheby's, Sussex).

190 - Sabre d'officier de Cavalerie Légère.
Premier Empire.
Monture en laiton. Calotte en crosse finissant par un aplat, à bouton de rivure et courte queue. Poignée en ébène finement quadrillée. Arc de jointure arrondi, à profond sillon longitudinal se poursuivant par un quillon infléchi et terminé par un petit cône. Oreillon basal en palmette à neuf folioles (190a). Lame bleuie et gravée de trophées, de motifs végétaux dorés, longue de 75 cm, large au talon de 3 cm et flèche de 4,1 cm. Fourreau de cuir à deux longues garnitures : chape ornée de filets de part et d'autre du bouton de port, bouterolle en suite terminée par un dard asymétrique (190b). Chez certains exemplaires, le fourreau est de fer, voire de bronze.
D'après G. Le Diberder, (1989), ce sabre a été également offert comme armes d'honneur (R.J.).

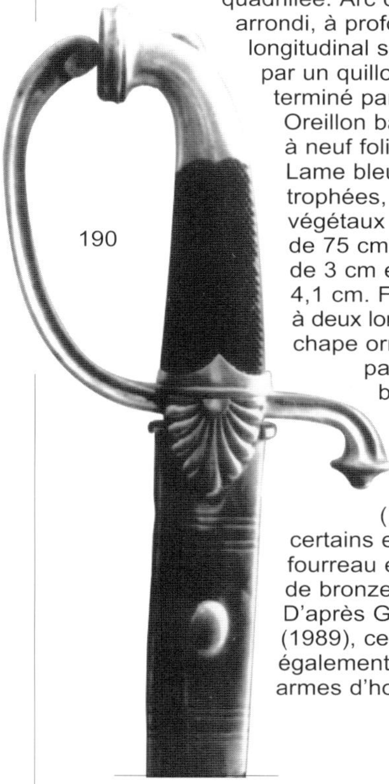

190

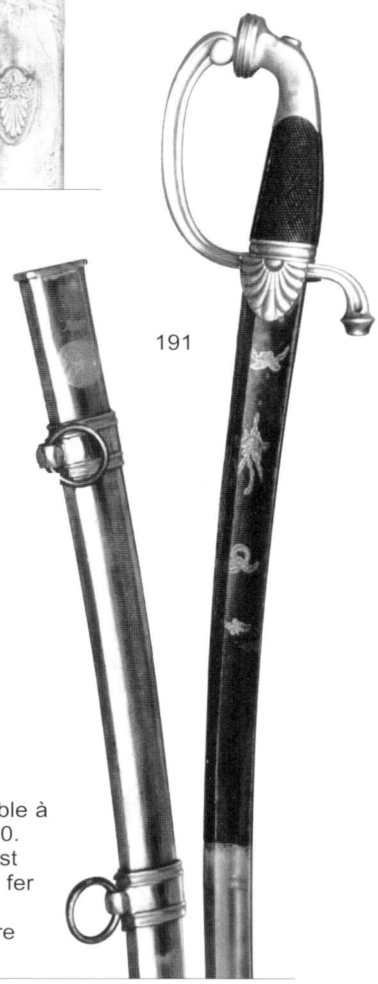

191

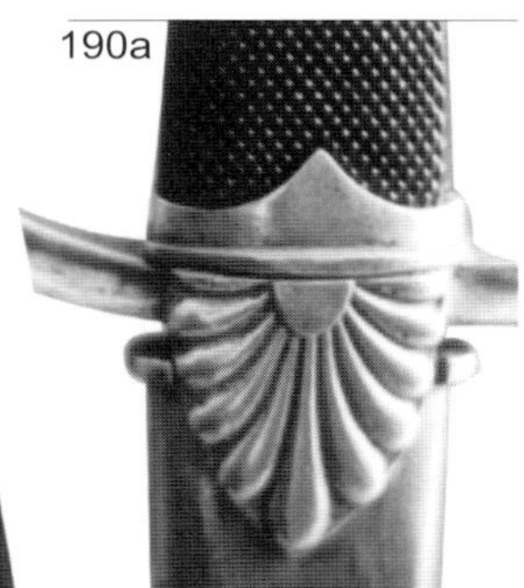

190a

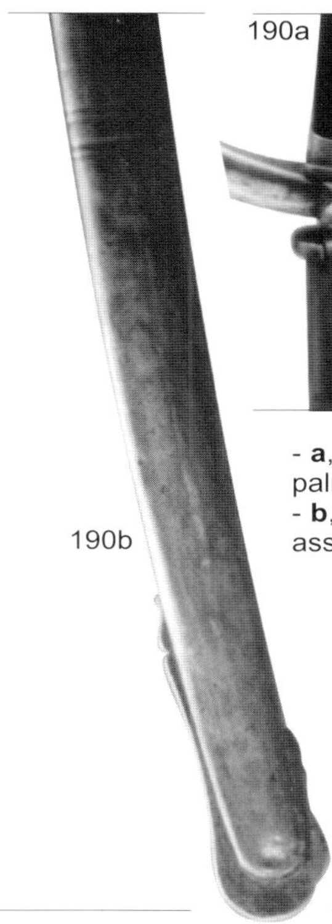

190b

- **a**, oreillon à palmette ;
- **b**, dard assymétrique.

191 - Variante.
Monture comparable à l'exemplaire n° 190. Seul le fourreau est différent : il est de fer bronzé à deux bracelets à bordure dorée (J.C. D.).

192

192 - Sabre du général Moncey.

Premier Inspecteur général de Gendarmerie, 1802 - 1814.

Monture en bronze doré. Calotte en crosse terminée par une tête de lion et courte queue (192a). Poignée encordée, recouverte de basane et filigranée. Arc de jointure arrondi avec un petit brin à demi-palmette rejoignant un cercle à sept étoiles finissant par des rosettes. Demi-oreillon basal également ciselé d'une palmette. Quillon infléchi vers la lame. Lame courbée. Fourreau de cuir noirci à couture en chaînette frontale apparaissant entre deux longues garnitures : chape ciselée dans un cartouche de l'aigle impériale posée sur un bâton de foudre placé sur un petit cartouche à une étoile, puis un faisceau de licteur à deux fers, flanqué de branches de laurier : la chape se termine par un bracelet décoré de motifs foliacés supportant l'anneau de bélière supérieur (192b). Bouterolle ornée de deux longs cartouches, le supérieur figurant un trophée d'armes enrubanné (192c et d), l'inférieur à motifs végétaux et se terminant par un dard asymétrique constitué de deux lames finissant chacune par un enroulement (192e) (M.G. de Melun).

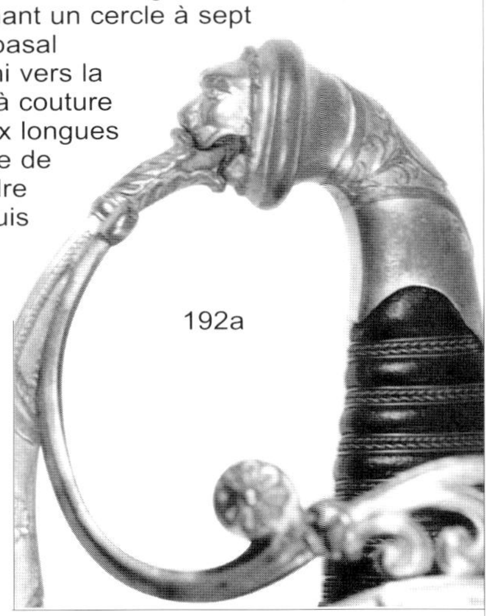

192a

- **a**, détail de la partie supérieure de la monture ;
- **b**, détail de la chape montrant le faisceau de licteur surmonté de l'étoile et de l'aigle ;
- **c, d** et **e**, décor de la longue bouterolle au trophée d'armes, de flèches, de feuillages en sa partie supérieure et de palmes en sa partie inférieure.

192c

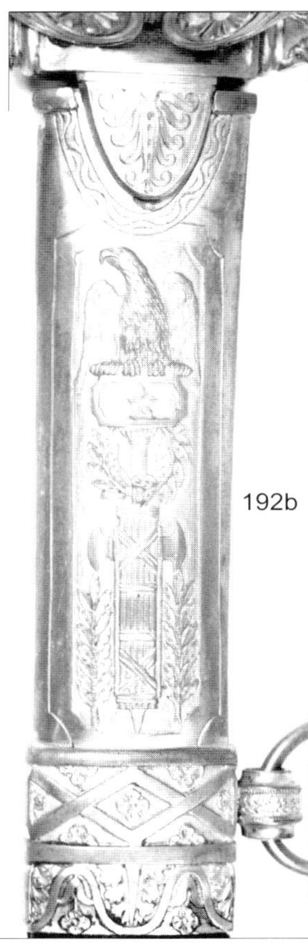

192b

192d

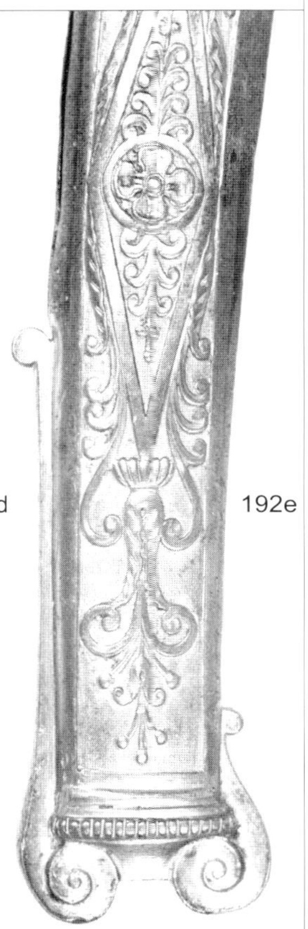

192e

206 - 207

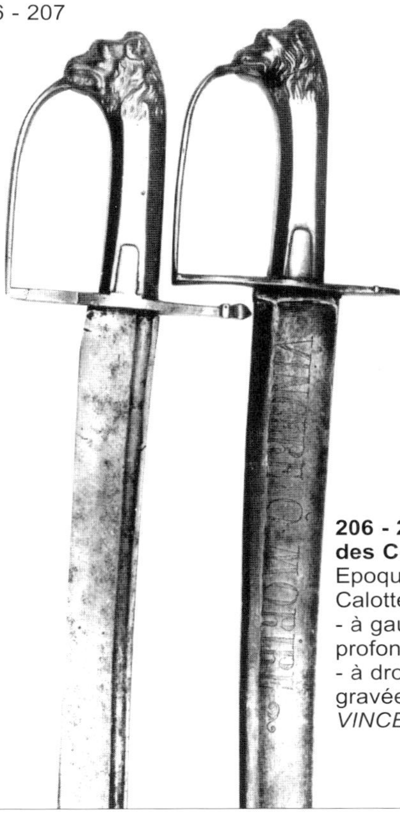

206 - 207 - Deux sabres des Compagnies d'Elite.
Epoque révolutionnaire.
Calotte en tête de lion.
- à gauche, lame unie à profonde gouttière dorsale ;
- à droite, lame à pan creux gravée de la devise *VINCERE O MORIR.*

208 - Sabre des Compagnies d'Elite.
Période révolutionnaire.
Calotte en tête léonine curieusement traitée ...

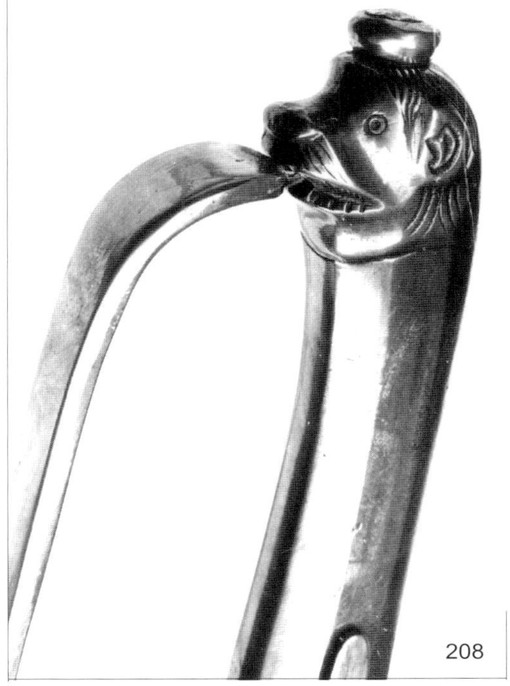

208

209a

209 - Exemplaire de sabre.
portant l'inscription sur le *recto, Vivre Libre Ou Mourir* (209a) et sur le *verso, Constitution Liberté* (209b) en lettres de fantaisie d'or sur fond bleui. Cette lame est signée *Dida / Marchand / Fourbisseur / Rue Ste Catrine* (sic) *à BorDeau* (sic) (M.R.).

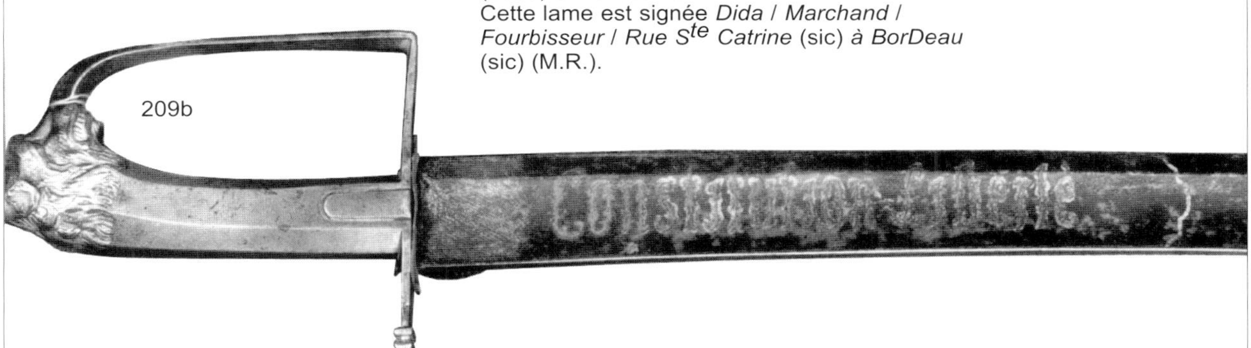

209b

210 - Sabre d'officier des Compagnies d'Elite.
Fin du XVIIIème siècle.
Monture de laiton fondu. Calotte à bouton de rivure en tête de lion prolongée par une longue queue. Poignée à une quinzaine de cannelures. Arc de jointure droit dans les deux tiers inférieurs et arrivant en angle droit sur la croisière portant des oreillons supérieurs. Quillon terminé par deux boules successives. Lame plate légèrement courbe. Fourreau de cuir à deux garnitures unies, chape à bouton, bouterolle à extrémité renforcée en guise de dard.

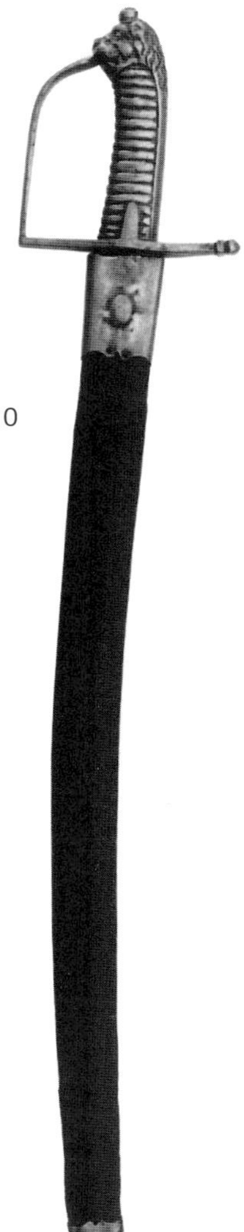

210

211a

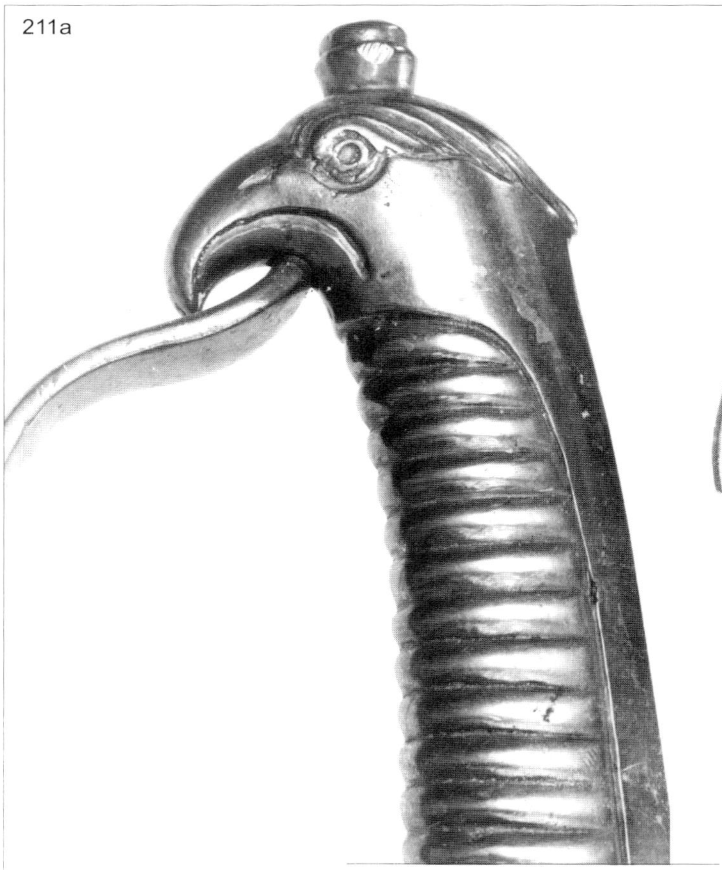

211

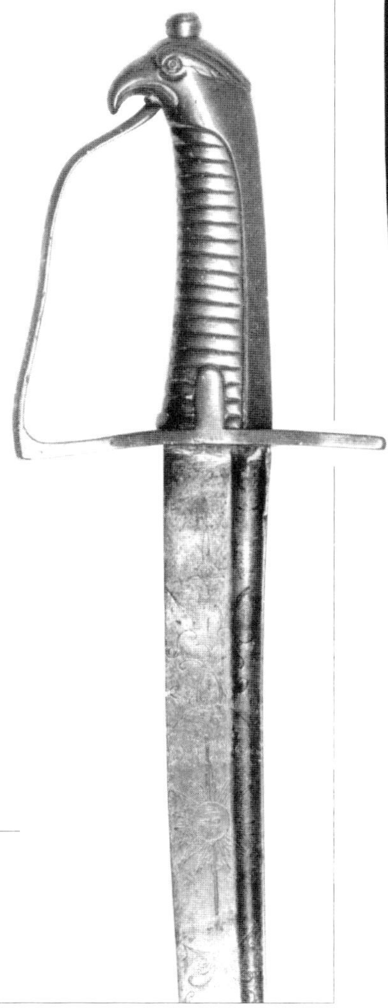

211 - Sabre des Compagnies d'Elite.
Epoque révolutionnaire.
Monture à la tête d'aigle (211a). Poignée à 19 cannelures plates. La garde est montée sur une lame de l'Ancien Régime gravée d'un soleil et de motifs végétaux stylisés.

212

212 - Variante.
Remarquer l'arête qui parcourt la lame à partir du talon et les gravures partiellement effacées où l'on discerne un soleil, un trophée d'armes et des grotesques.

213 - Sabre au casque empanaché et à grille.
Calotte au casque dit " à l'antique ", à grand panache ciselé, à couronne de laurier et à grille. La calotte se prolonge par une longue queue séparée de la poignée par une profonde rainure. Poignée à pans, à deux demi-oreillons supérieurs figurés. Quillon droit terminé par une barrette et une pièce subpyramidale. Lame de 65 cm, large au talon de 4 cm et flèche de 1 cm, gravée sur le *recto* de trophées et d'un bras armé et sur le *verso* de signes cabalistiques. Lame probablement de récupération (213a et 213b) ; (voir J. Lhoste et J.J. Buigné 1999 p.110-113).

213

Lame de récupération de la fin du XVIIIème siècle
- en haut, *verso* aux signes cabalistiques (213a) ;
- en bas, *recto* au trophée et au bras armé qui rappelle les lames signées Cassagnard (213b).

213a

213b

214

214 - Sabre au bonnet empanaché et à cocarde.
Période révolutionnaire
Modèle au bonnet à cocarde surmonté d'un gros bouton de rivure. Poignée à pans, le pan frontal étant orné d'une grenade enflammée. Lame de sabre plat, avec sa cravate.

Chapitre III

Croisière à oreillons entiers, arc de jointure arrivant en angle droit sur la croisière

Les termes oreillons entiers définissent les oreillons qui maintiennent à la fois la poignée et le fourreau. Les sabres de ce chapitre sont très nombreux et comprennent aussi bien des armes destinées à la Cavalerie qu'aux troupes à pied.

Nous basant sur la forme de l'arc de jointure, de celle de la calotte et des oreillons, trois Familles ont été définies, elles-mêmes éventuellement subdivisées en Sous-familles :

Famille V – Calotte à ornement figuratif

Les sabres de cette Famille sont présentés en trois Sous-familles de la façon suivante :

Sous-famille V a – Calotte à longue queue partant d'un casque à l'antique, d'une tête de lion ou d'aigle

Ce type de sabre se rencontre du règne de Louis XV jusqu'au début du Premier Empire. Il est en général l'attribut des Hussards.

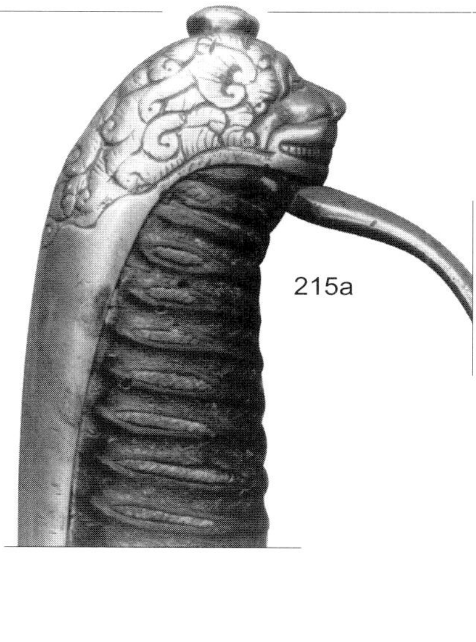

215a

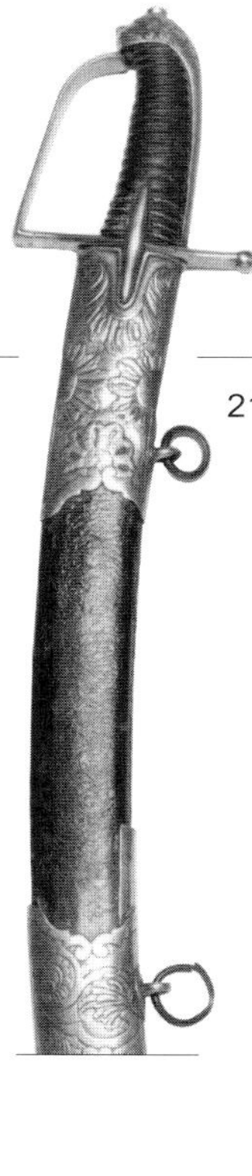

215

215 - Sabre d'officier de Hussards.
Règne de Louis XV.
Bronze doré. Calotte à face léonine (215a), à bouton de rivure et longue queue. Poignée encordée, recouverte de basane non filigranée. Arc de jointure droit sur les 2/3 de sa longueur inférieure, rejoignant en oblique la croisière en formant un angle d'environ 80°. Croisière rectiligne terminée par un bouton. Oreillons longs et étroits, en navette, à extrémités arrondies. Forte lame de 76 cm, large au talon de 4 cm, à flèche de 5 cm, gravée en partant du talon, d'entrelacs, du petit Hussard chargeant, d'un

- **a**, calotte en tête de lion. On remarquera que, la basane étant usée, on distingue nettement la corde sous-jascente.

215b

215c

bandeau avec *VIVAT* et d'entrelacs de nouveau (215b). Fourreau de cuir à trois garnitures gravées de palmettes et de fleurs genre tournesol (215c) ; attelles de protection latérales en laiton sur toute la bordure antérieure et entre le bracelet médian et la bouterolle sur la bordure postérieure (M.R.).

- **b**, lame au Hussard chargeant ;
- **c**, chape aux fleurs de tournesol autre emblême des Hussards.

216 - Sabre d'Infanterie de la Légion Corse.
Vers 1769.
Monture en laiton. Calotte en tête de lion incomplète suivie d'une longue queue. Poignée recouverte de basane collée, non cannelée ni filigranée. Arc de jointure droit sur les 2/3 de sa longueur et faisant un angle quelque peu supérieur à 90° avec la croisière rectiligne. Oreillons petits et ovales. Quillon terminé par un bouton. Lame de 59,6 cm, large au talon de 4,1 cm, peu courbée et terminée par un contre-tranchant saillant. Fourreau de cuir, à deux garnitures, chape à fente horizontale.

216

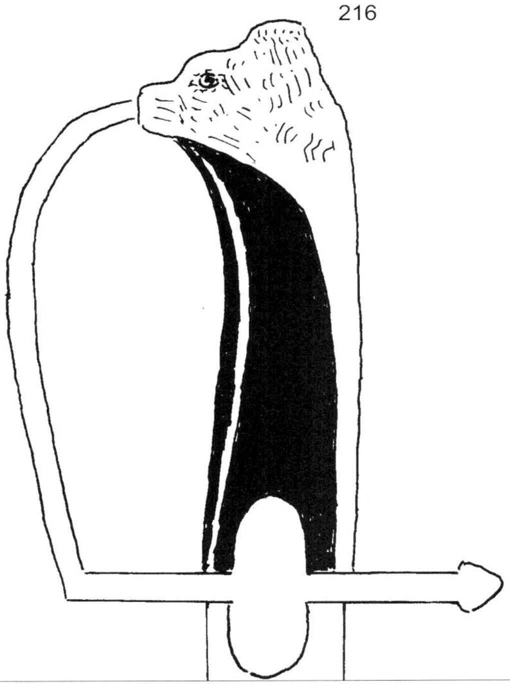

217

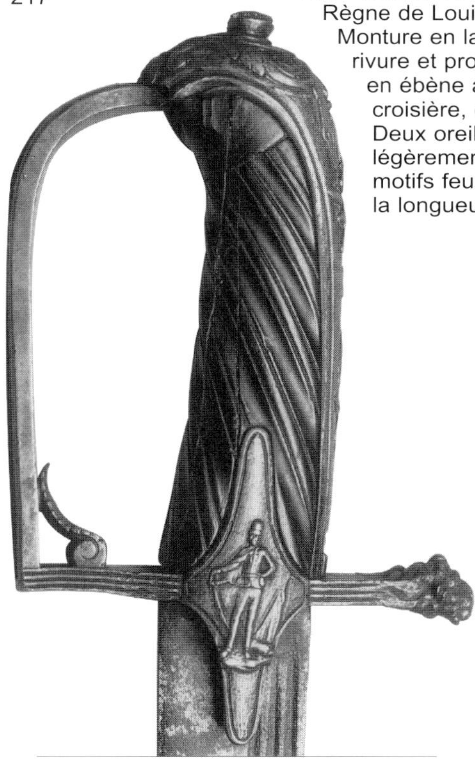

217 - Sabre d'officier de Hussards.
Règne de Louis XVI.
Monture en laiton anciennement dorée. Calotte en cuillère, surmontée d'un bouton de rivure et prolongée d'une longue queue ornée à son sommet d'un trophée. Poignée en ébène à cannelures obliques. Arc de jointure arrivant en angle droit sur la croisière, celle-ci droite, rainurée et se terminant par un quillon à mufle léonin. Deux oreillons, en losange curviligne, ciselés d'un Hussard à pied. Lame légèrement courbée, à trois gouttières dorsales. Fourreau à chape décorée de motifs feuillus, à bracelets de bélière. Bouterolle remontant à plus de la moitié de la longueur du fourreau, ciselée latéralement de frises et dard en lyre symétrique.

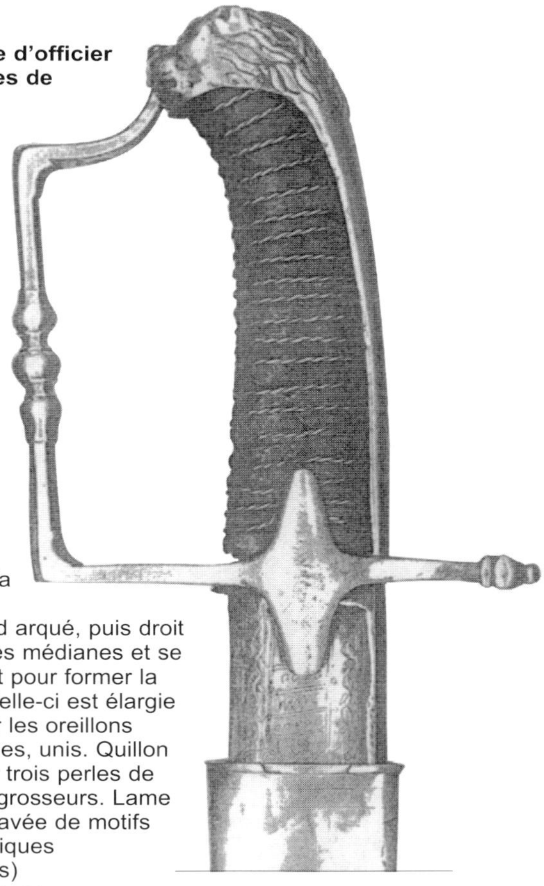

218 - Sabre d'officier des Troupes de Marine.
Règne de Louis XVI.
Monture en laiton. Calotte en tête de lion à longue queue. Poignée recouverte de basane filigranée de fils torsadés faisant 20 spires. Arc de jointure partant de la gueule du lion, d'abord arqué, puis droit à trois perles médianes et se prolongeant pour former la croisière. Celle-ci est élargie pour former les oreillons losangiformes, unis. Quillon terminé par trois perles de différentes grosseurs. Lame bleuie et gravée de motifs dorés classiques (grottesques) plus un poinçon ovalaire surmonté d'un noeud et dans lequel est inscrit une ancre (218a), indiquant son affectation aux Troupes embarquées.

218a

219 - Sabre d'officier de Chasseur à cheval.
Fin du XVIIIème siècle.
Monture en laiton. Calotte à demi-tête de lion, à bouton de rivure et longue queue. Poignée recouverte de basane à une trentaine de spires d'un fil torsadé. Arc de jointure arrivant en angle droit sur la croisière, celle-ci élargie en son milieu en oreillon en losange curviligne. Lame à gouttière dorsale, gravée d'un croissant de lune.

219

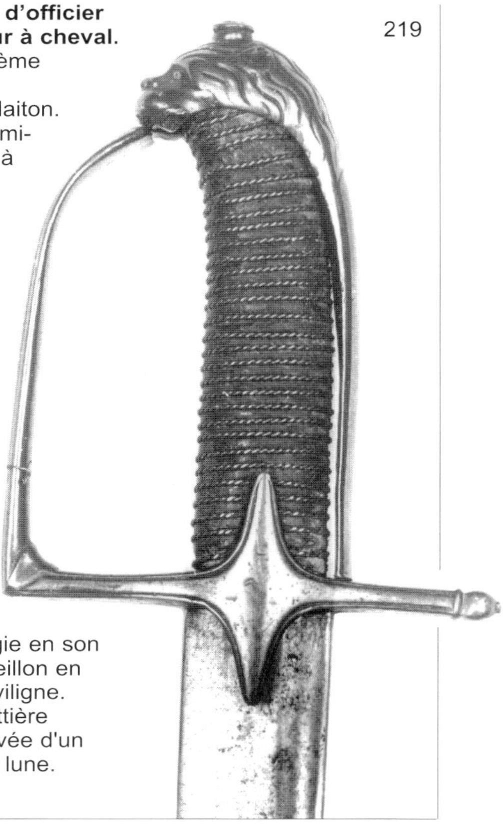

219 - Sabre d'officier de Cavalerie légère.
Fin du XVIIIème siècle.
Monture en laiton. Calotte à face léonine écrasée et longue queue. Poignée recouverte de basane, à une quinzaine de spires d'un filigrane de fils torsadés. Arc de jointure arrivant en angle droit sur la croisière. Oreillons ovalaires, le frontal orné d'une face humaine grimaçante, genre Gorgone. Lame à pan creux. Fourreau de cuir, à trois garnitures et attelles latérales de laiton.

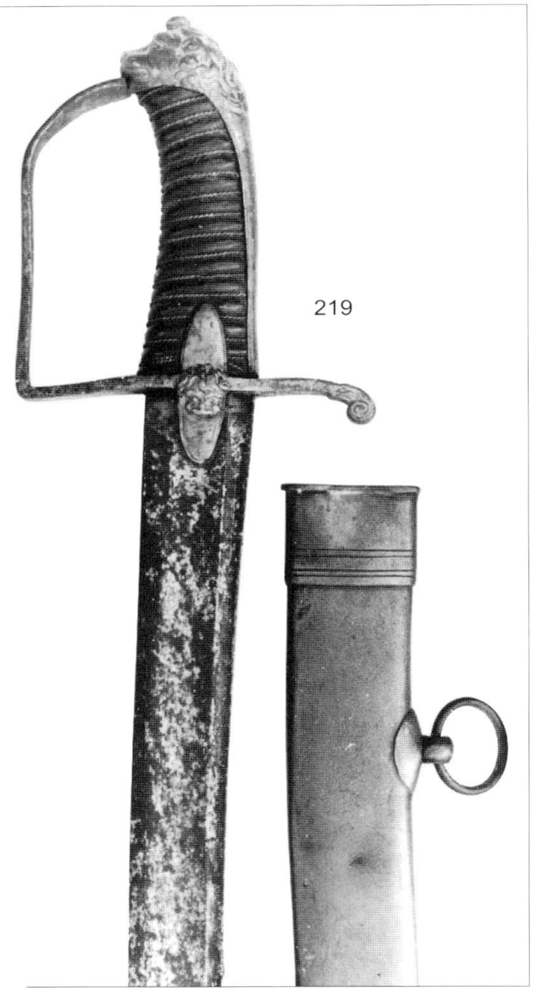

219

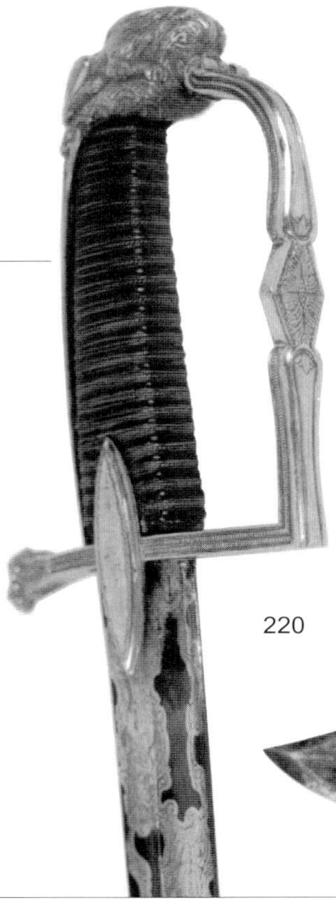

220 - Sabre d'officier supérieur.
Consulat.
Monture en laiton doré. Calotte en dépouille de lion, à longue queue. Poignée à section subtriangulaire, recouverte de basane à vingt et une spires d'un filigrane composite. Arc de jointure découpé d'un motif losangiforme, arrivant en angle droit sur la croisière, portant des oreillons en navettes.

Ce sabre est surtout remarquable par sa lame entièrement bleuie avec de nombreuses et fines gravures recouvertes de dorures (220a). Parmi les motifs dorés on distingue, çà et là, des trophées d'armes, un croissant de lune, un soleil, des étoiles, des fleurettes...(J.C. D.).

220

220a

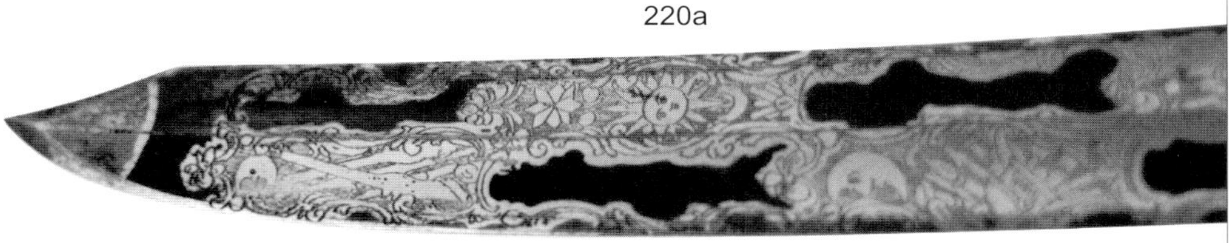

221

221 - Sabre du général divisionnaire Jean-Baptiste Kléber
Monture en bronze doré. Calotte en tête de lion et longue queue ciselée. Poignée recouverte de basane à 23 spires d'un filigrane à double fil torsadé. Arc de jointure, croisière et oreillons en baguettes aux extrémités arrondies, entièrement décorés de motifs végétaux. Quillon en tête de lion. Lame gravée à l'acide portant en caractères gothiques " *Don patriotique offert par la ville de Belfort / au général divisionnaire Kléber* " (M.N.N. de Juan-les-Pins).

Rappelons que J.B.Kléber est né à Strasbourg en 1753. Officier issu de l'Académie Militaire de Munich, il revient en France et s'établit comme architecte. En 1792 il prend la tête d'un contingent de Volontaires alsaciens, combat en Vendée et s'oppose aux sans-culottes. Kléber se distingue à Fleurus (1793), prend Maëstricht (1794) et démissionne en 1796. Il accompagne Bonaparte en Egypte et lorsque celui-ci revient en France, il donne à Kléber le commandement en chef des troupes françaises. Il prend le Caire où il est assassiné par un Turc le 14 juin 1800.

222

222 - Sabre de Chasseur à cheval.
Fin du XVIIIème début du XIXème siècle.
Monture en laiton. Calotte en tête de lion et longue queue. Poignée à 12 cannelures, recouverte de basane et filigranée. Arc de jointure arrivant en angle droit sur la croisière terminée par un quillon à bouton. Oreillons en losange curviligne à filet marginal. Lame très légèrement courbe. Chape à bracelet figuré et anneau de bélière en laiton ; longue bouterolle à anneau en suite avec la chape, les deux garnitures étant reliées par des attelles de laiton dentelées (M.R.).

223

223 - Variante.
Fin du XVIIIème début du XIXème siècle.

Sur la lame est gravé *L'UNITE DE LA NATION / DONNE DE BONNES LOIS* sur les 2 faces (J.C. D.).

224 - Sabre d'officier général de Cavalerie légère.
Vers 1792.
Monture en laiton.
Calotte en tête de lion suivie d'une longue queue (224a). Poignée totalement filigranée de cuivre. Arc de jointure courbé, puis droit et faisant un angle droit avec la croisière sur laquelle sont fixés deux oreillons en losanges, légèrement curvilignes. Ces oreillons sont ornés d'une fleurette et de quatre feuilles d'acanthe stylisées. Très forte lame de 87,9 cm, large, courbée, plate, bleuie et dorée, portant l'inscription sur le *recto* *VAINCRE OU MOURIR* (224b) et sur le *verso* *POUR LA PATRIE*. Fourreau de laiton ciselé : entre les deux bracelets figurés se trouve un faisceau de licteur au bonnet phrygien, flanqué de deux drapeaux (224c) ; entre le second bracelet et le dard asymétrique en fer sont gravés des drapeaux, des lances, des instruments de musique …

224

224a

- **a**, belle tête de lion formant la calotte. Remarquer la ciselure de l'arc de jointure à feuilles de chêne et glands.

224b

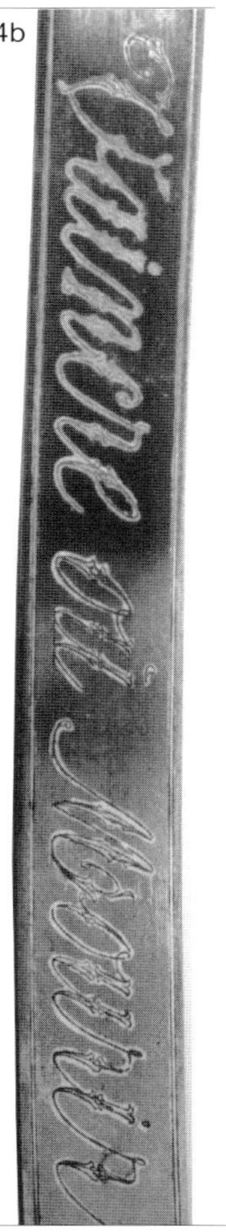

224c

- **b**, *Vaincre ou mourir* en lettres dorées sur le fond bleu de la lame ;
- **c**, faisceau typiquement révolutionnaire, à décor repoussé, gravé sur le fourreau.

225

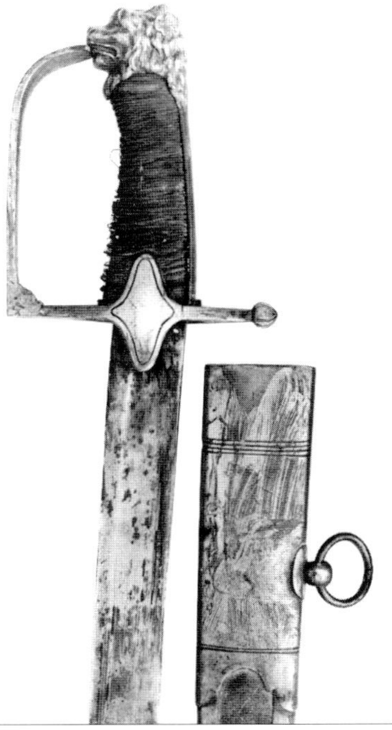

225 - Variante.
Fin du XVIIIème siècle.
Sur cet exemplaire, la poignée est totalement filigranée, la gueule du lion est fermée et l'oreillon est simplement orné d'un filet marginal, dit *décor au trait*.

226

226 - Variante.
Fin du XVIIIème siècle.
Par fantaisie, la bordure de l'arc de jointure, de la croisière, des oreillons et du quillon est guillochée.

227

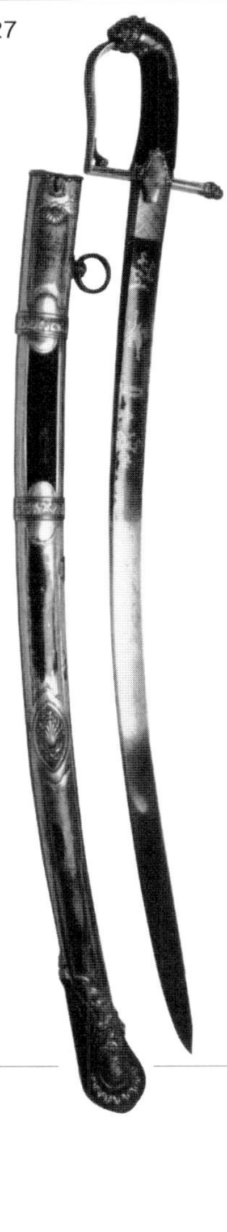

227a

227 - Sabre d'honneur du général Le Fèvre.
Directoire.
Monture en laiton doré.
Calotte en crosse à tête de lion, à courte queue.
Poignée d'ébène. Arc de jointure orné sur sa face antérieure d'une frise de feuilles de chêne (227a).
Croisière rectiligne portant deux oreillons hexagonaux unis, mais bordure ciselée de feuillages. Lame signée Klingenthal (227b) bleuie et gravée de motifs dorés

227b

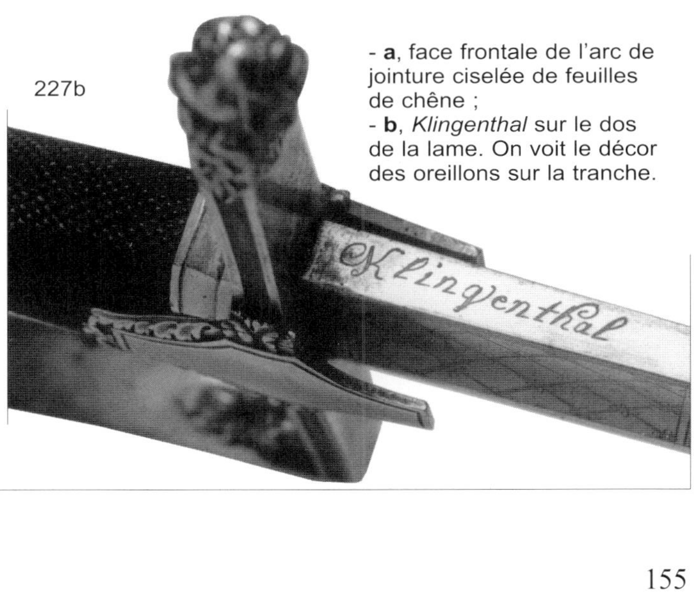

- **a**, face frontale de l'arc de jointure ciselée de feuilles de chêne ;
- **b**, *Klingenthal* sur le dos de la lame. On voit le décor des oreillons sur la tranche.

227c

227d

227e

- **c**, dédicace du Directoire au Général Le Fèvre ;
- **d**, signature de Boutet ;
- **e**, dard.

classiques. Chape à anneau de bélière, unique, gravée dans un cartouche : *Le Directoire / Exécutif / AU / GENERAL / LE FEVRE sur le recto* (227c) *et sur le verso* (227d) *MANFACTURE / à Versailles / BOUTET / Directeur / Artiste.* Le second bracelet ne porte pas d'anneau de bélière. Bouterolle ornée d'un enroulement avec un dard asymétrique en fer (227e). (J.C. D.)

228

228 - Sabre d'officier de Chasseur à cheval de la Garde Impériale.
Premier Empire.
Monture en laiton doré. Calotte en crosse à tête de lion suivie d'une longue queue à filet marginal. Fusée recouverte de basane à dix spires d'un filigrane de fil torsadé bordé de deux fils simples. Arc de jointure partant de la gueule de la tête de lion, courbé au tiers supérieur, puis droit et rejoignant la croisière par un angle à 90°. Oreillons en losanges curvilignes, le frontal est ciselé d'une tête d'Athéna. Quillon infléchi à bouton décoré de feuilles stylisées. Lame demi-courbe, à jonc, bleuie au tiers et gravée dans le pan creux de motifs traditionnels dorés, plus, dans un médaillon ovale, de l'aigle couronnée. Fourreau en laiton doré, à deux anneaux de bélière ciselés d'une frise de feuilles stylisées et flanqués de part et d'autre d'un motif rappelant un demi-soleil rayonnant ; entre les anneaux, dans un hexagone est inscrite une aigle couronnée. (Ch. B.).

240

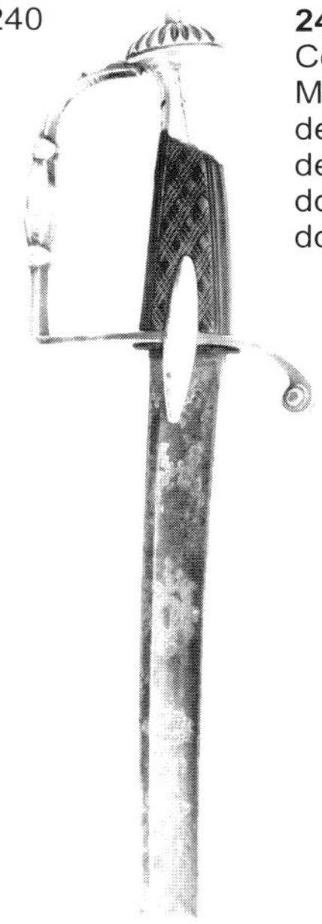

240 - Sabre d'officier de Cavalerie Légère.
Consulat ou Premier Empire.
Monture en laiton argenté. Arc de jointure à olive médiane limitée par deux perles. Poignée d'ébène quadrillée ménageant des losanges et des bordures à gouttière. Lame demi-courbe bleuie et ornée de motifs dorés. Fourreau de laiton argenté à deux bracelets de bélière à double moulure latérale.

241 - Sabre d'officier de Cavalerie légère.
Premier Empire.
Monture en bronze doré. Calotte à côtes de melon surmontée d'un bouton de rivure et à courte queue. Arc de jointure denté à sa partie supérieure et décoré de bagues et d'une perle allongée au milieu de la partie droite qui rejoint la croisière par un angle droit. Croisière et quillon rainurés. Oreillons en navette à bordure déprimée. Poignée en noyer noirci et quadrillé. Lame à dos plat, longue de 76 cm, large au talon de 2,8 cm et flèche de 4 cm, avec un cartouche doré au talon et gravée au tiers de motifs floraux. Fourreau de laiton uni, à deux larges bracelets plats dont les bordures sont marquées d'un trait (M.R.).

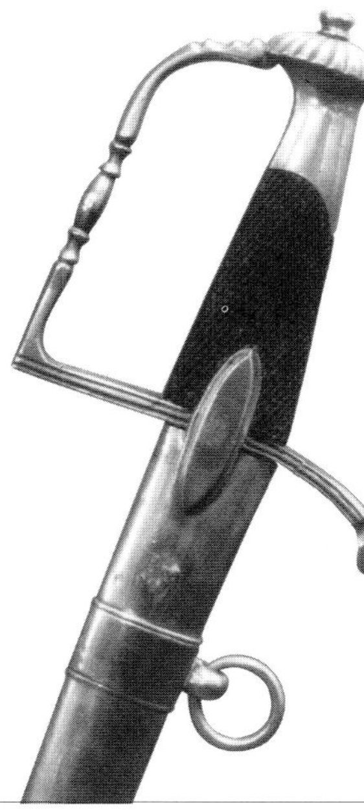

241

242

242 - Sabre de sous-officier d'Infanterie.
Consulat.
Monture en laiton. Arc de jointure légèrement découpé en sa partie médiane. Poignée d'ébène quadrillée. Oreillons en navette à filet marginal. Quillon terminé par un petit cône. Lame peu courbée, blanche et sans gouttière. Fourreau à deux garnitures, chape à bouton.

Remarque - Ce sabre a conservé, exceptionnellement, sa " cravate " protégeant l'entrée du fourreau. L'épaisseur de la cravate explique que bien souvent on constate que celle-ci disparue, le fourreau ne recouvre pas totalement la lame !

243

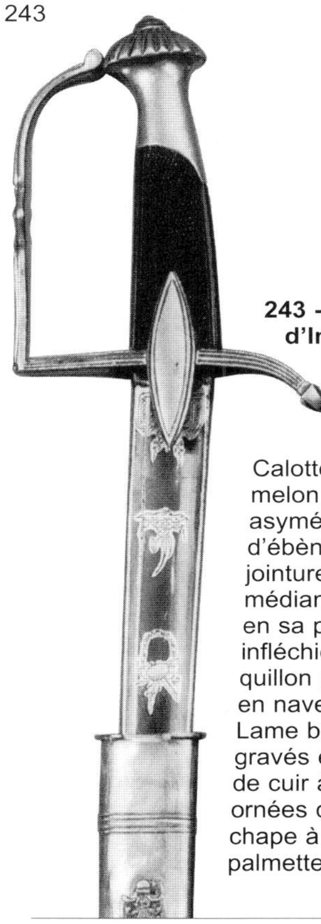

243 - Sabre officier d'Infanterie
Consulat ou Premier Empire.
Monture en bronze doré.
Calotte en côtes de melon, à courte jupe asymétrique. Poignée d'ébène quadrillée. Arc de jointure à petite décoration médiane. Croisière droite en sa partie antérieure et infléchie pour former un quillon postérieur. Oreillons en navette bordée d'un filet. Lame bleuie à décors gravés et dorés. Fourreau de cuir à deux garnitures ornées de doubles traits, chape à bouton ciselé d'une palmette.

244

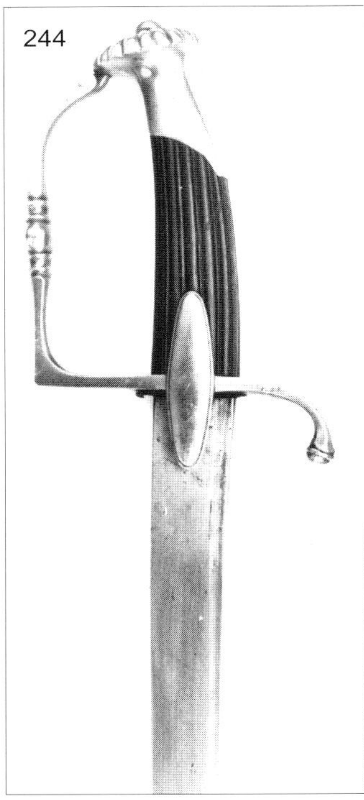

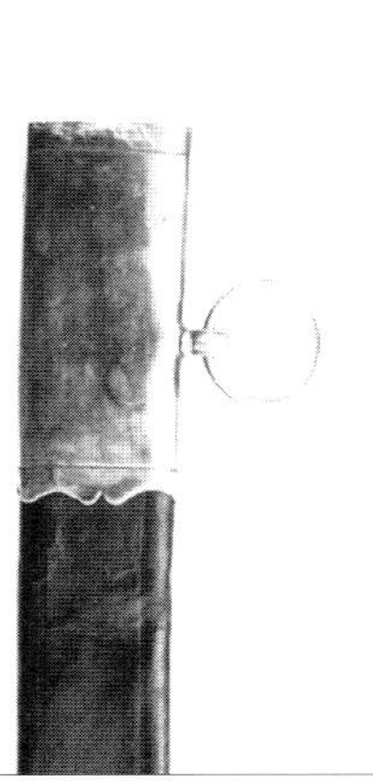

244 - Sabre d'officier de Cavalerie légère.
Consulat ou Premier Empire.
Monture en laiton. Calotte à côtes de melon aplati et décentré. Arc de jointure à bague médiane avec de part et d'autre deux petites perles placées entre deux disques. Poignée de corne à cannelures verticales. Oreillons ovalaires à liseré marginal. Lame à gouttière dorsale, blanche. Fourreau de cuir à trois garnitures de laiton, découpées en accolades, dard asymétrique.

245

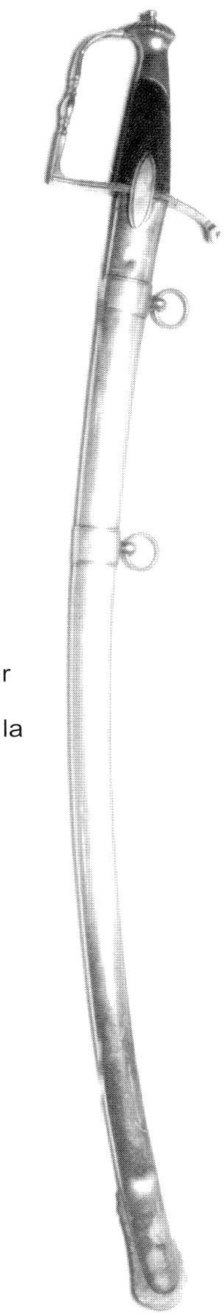

245 - Sabre d'officier de Cavalerie légère.
Premier Empire - Restauration.
Monture en laiton. Ce sabre est intéressant par ses oreillons nettement ovoïdes et surtout par son fourreau en tôle d'acier. C'est une fabrication tardive ou la lame a bénéficié d'un nouveau fourreau sous la Restauration.

246

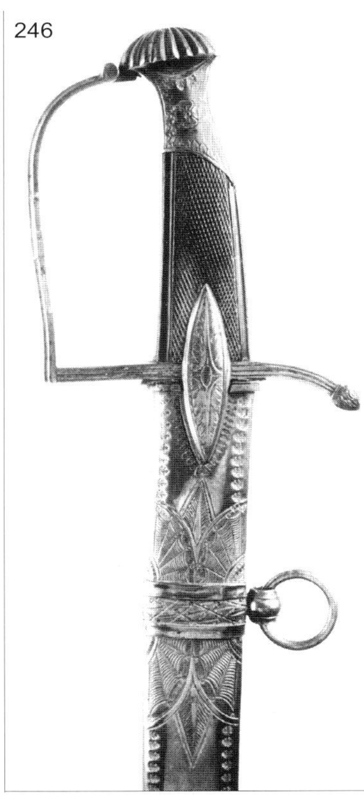

246 - Sabre d'officier d'Etat-Major.
Vers 1800.
Monture en laiton doré. Calotte en côtes de melon et courte queue (246a). Poignée en ébène quadrillée. Arc de jointure arrivant sur la croisière en angle droit avec deux oreillons en navette ciselée de feuillage. Quillon infléchi et ciselé. Lame courbe de 85,1 cm, à dos plat, pan creux, bleuie et gravée de motifs dorés classiques. Fourreau en laiton, à deux bracelets rainurés et flanqués de toile d'araignée et de palmette.

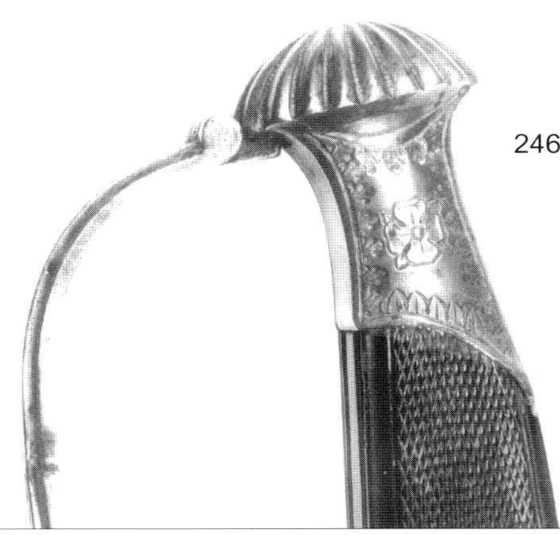

246a

- **a**, remarquer le décor floral de la jupe.

247

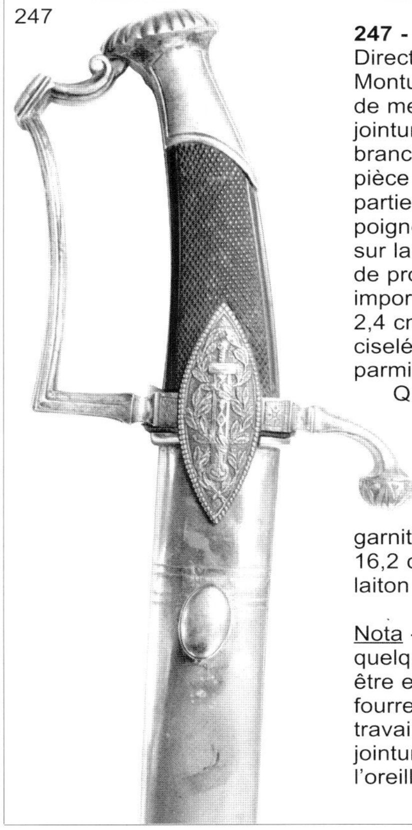

247 - Sabre d'officier supérieur.
Directoire.
Monture en laiton doré. Calotte en côtes de melon, à queue courte. Arc de jointure en trois parties : une petite branche arquée reliant la calotte et une pièce quadrangulaire d'où part la longue partie, légèrement incurvée vers la poignée et aboutissant en angle droit sur la croisière, l'ensemble étant décoré de profondes rainures. Oreillons importants, en navette de 7 cm sur 2,4 cm, bordés de petites perles et ciselés d'un glaive à l'antique placé parmi les rameaux de laurier (247a).
Quillon courbé à 90° et finissant en fleur. Lame de 61 cm, large au talon de 3 cm, avec une flèche de 3 cm, à pan creux, bleuie sur 22 cm et gravée de motifs floraux dorés. Fourreau à deux garnitures décorées au trait : chape de 16,2 cm, bouterolle de 20 cm et dard de laiton asymétrique.

Nota - Ce modèle peut présenter quelques modifications : la calotte peut être en crosse et les garnitures du fourreau peuvent être plus richement travaillées. Sont constants l'arc de jointure en trois parties, la poignée en ébène, le quillon orné d'une fleurette et surtout l'oreillon au glaive lauré.

247a

- **a**, remarquer le glaive flanqué de laurier.

248

248 - Sabre d'officier supérieur de Cavalerie légère.
Directoire.
Monture en laiton doré. Calotte en côtes de melon à pièce de rivure hémisphérique surmontant une jupe à pans, le frontal ciselé d'un rameau de

248a

248b

248c

laurier. Arc de jointure légèrement incurvé, relié à la calotte par un brin arqué, arrivant en angle presque droit sur la croisière à quillon infléchi. Oreillon en navette ciselée de feuilles d'acanthe disposées en croix, à bouton central (248a). Poignée en ébène quadrillée. Lame de 81 cm, large au talon de 2,7 cm, à pan creux, bleuie et gravée de motifs dorés classiques. Fourreau de laiton, décoré de feuilles d'acanthe et vers le milieu d'un bâton de foudre ailé (248b). Bracelets de bélière ciselés de feuilles (248c) et de fruits de laurier stylisés. M.R.

249 - Sabre d'officier supérieur de Cavalerie.

Consulat, Premier Empire.

Monture en laiton doré. Calotte en côtes de melon, à jupe à pans décorés en alternance, de cercles et de fleurettes. Arc de jointure à décor en relief. Poignée d'ébène quadrillée en losange. Oreillons en navette à bordure moulurée avec au centre sur fond rayonnant un visage de femme très chevelue (Méduse ?) (249a). Lame demi-courbe, bleuie et à motifs dorés sans originalité. Fourreau de laiton à frise marginale et deux bracelets de bélière ciselés avec, de part et d'autre, une demie toile d'araignée. (M.R.).

249

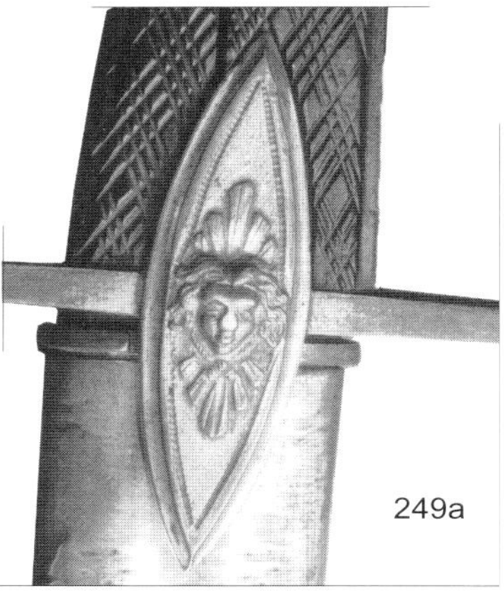
249a

250 - Sabre de Cavalerie légère.

Consulat - Premier Empire.

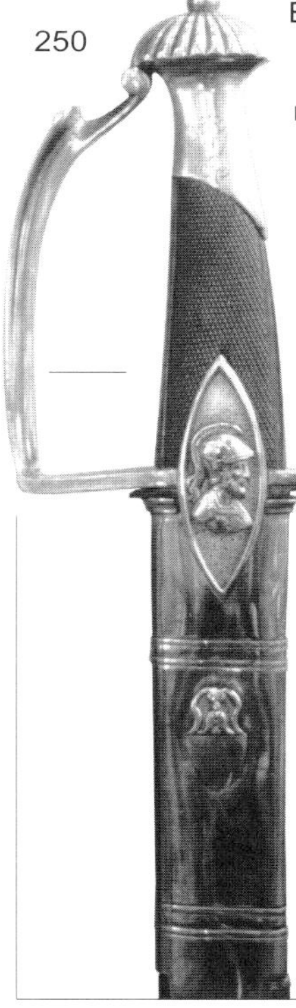
250

Monture en laiton. Calotte en côtes de melon à courte queue. Poignée en ébène finement quadrillée. Arc de jointure à décrochement supérieur, arqué et arrivant en angle droit sur la croisière terminée par un quillon en palmette. Deux oreillons en navette, le frontal décoré du buste du Dieu de Mars (250a), le postérieur à simple fleurette à six pétales (250b). Lame légèrement courbée, à pan creux. Fourreau de cuir à longues garnitures : chape à crochets enjolivés, bouterolle à dard asymétrique. M.R.

250a

250b

251

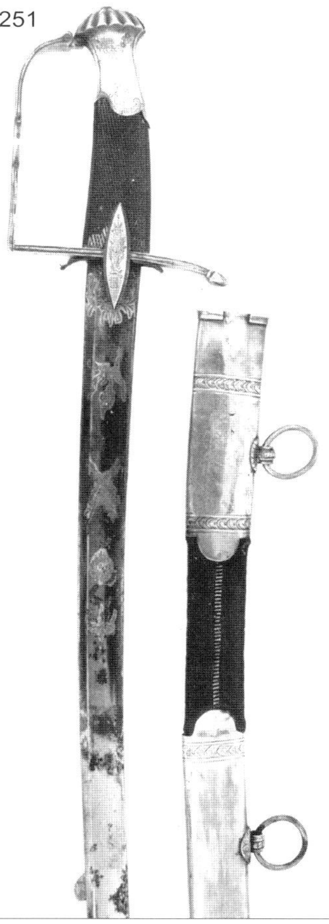

251 - Sabre de Commissaire de Guerre.
Premier Empire.
Monture en argent poinçonné à la tête de Vieillard (1798-1809) et d'un poinçon de maître (losange incluant les lettres " FAC " autour d'un " I "). Calotte à côtes de melon et courte jupe ornée de motifs végétaux stylisés. Poignée d'ébène quadrillée. Arc de jointure droit sur plus des 4/5èmes de sa longueur et décoré d'une gerbe de blé que l'on retrouve sur les oreillons en navettes. Lame demi-courbe de 77,5 cm, à pan creux, bleuie, gravée de motifs classiques dorés et signée *Pierre Guillaume Knecht fabricant à Solingen.* Fourreau argenté à chape et bracelets de bélière ornés de frises de disques partiellement superposés, le cuir étant maintenu par une couture métallique frontale rare sur les sabres français. (J.C. D.).

Les sabres des officiers des Régiments d'Eclaireurs sous le Premier Empire

Rappelons que les Régiments d'Eclaireurs ont été créés par l'Empereur en décembre 1813 qui les répartit en trois Régiments de chacun 250 hommes. Le premier accompagne les Grenadiers à cheval, le second les Dragons et le troisième le 1er Régiment de Chevaux légers-lanciers.

Les éclaireurs participèrent aux nombreux combats de la Campagne de France avec honneur et furent dissous en 1814, sous la Restauration.

Les sabres d'Eclaireurs sont assez variables de présentation, les difficultés d'alors en sont certainement pour quelque chose...

Le sabre suivant est probablement un des plus anciens modèles.

252

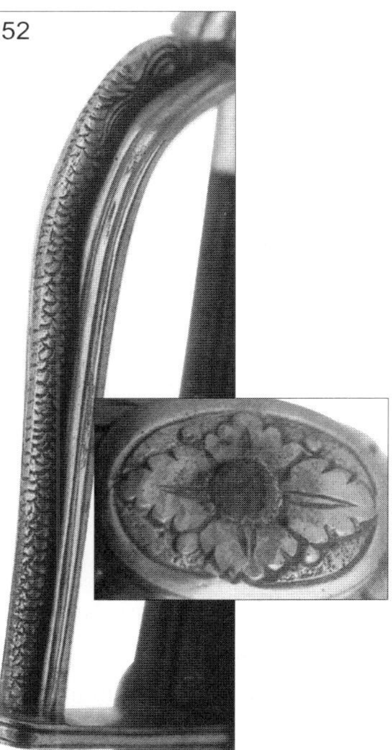

252 - Montures en laiton. Calotte en tranches de melon, mais avec un méplat supérieur orné de quatre folioles disposées en croix autour de la rivure de la lame. Courte jupe coiffant la fusée d'ébène rainurée verticalement. Arc de jointure courbé en sa partie supérieure mais arrivant en angle droit sur la croisière. Cet arc est remarquable par le serpent qui couvre toute sa longueur, la tête arrivant au niveau de la calotte. Ce serpent peut symboliser la protection, la défense... (voir Lhoste et J.J. Buigné, op. cit. 1999, p. 42). Croisière droite terminée en goutte d'eau et portant deux oreillons en navettes surchargés d'une étoile inscrite dans un cercle (voir p. 166, n° 253). Lame de 87,5 cm, large au talon de 3,1 cm, légèrement courbe, bleuie sur le tiers basal et gravée de motifs dorés classiques. Fourreau de bois recouvert de cuir noirci, à trois garnitures de laiton réunies par des attelles, ménageant ainsi deux faux crevés. Dard asymétrique.

253

253 - Variante.
Facture générale très comparable aux sabres précédents mais avec les différences suivantes : Monture en laiton argenté. Arc de jointure décoré du serpent, oreillons en losange avec l'étoile inscrite dans un cercle, mais avec leurs extrémités plus arrondies que pour le sabre précédemment cité. Lame bleuie et dorée au tiers. Fourreau à trois garnitures en laiton argenté, à pans et reliées entre elles par des attelles (B.C.).

Nota - Certains de ces sabres peuvent porter sur la lame l'inscription *Eclaireur : Garde Impériale*, en caractères de fantaisie. D'autre part, G. Jaeger (*op. cit.* 1998) émet l'hypothèse très vraisemblable suivante : *Ces sabres argentés seraient attribués au 1er et 3ème Eclaireurs dont les boutons et la passementerie sont argentés.*

254

254 - Variante.
Premier Empire. Monture en laiton doré. Calotte en côtes de melon légèrement décentrée. Arc de jointure dépourvu de serpent, à filet, arrivant en angle droit sur la croisière à deux filets et terminée par un bouton recourbé. Poignée d'ébène à stries latérales. Oreillons en navette sur laquelle est fixé le cercle dans lequel est inscrite l'étoile à six branches. Fourreau à section hexagonale, de cuir noir ; chape également à section hexagonale, unie, sur laquelle est soudé un anneau de bélière, bracelet à anneaux et dard asymétrique de fer.

255 - Variante.
G. Jaeger (1998) signale l'existence d'un modèle différent des précédents par les caractéristiques suivantes : Calotte, en côtes de melon, mais en son milieu une fleur d'héliotrope, jupe plus longue, à pans dont un sur deux est finement quadrillé ; poignée d'ébène, à pans quadrillés finement et surtout à oreillons en olive allongée, à filet marginal et ciselés de deux épées, la pointe vers le haut et nouées par un ruban par leur milieu. A ce sujet, G. Jaeger écrit : *... Cet ornement ressemble aux médaillons de vétérance de la fin du XVIIIème siècle et l'on peut imaginer que, parmi les officiers des trois régiments d'éclaireurs, les vétérans des guerres de la République et de l'Empire ont voulu se singulariser et bien marquer leur ancienneté.* Lame de 84 cm, large au talon de 3,3 cm, très comparable par ailleurs aux modèles précédents. Fourreau de laiton renforcé par des alèses de bois à deux bracelets de bélière, dard asymétrique en fer.

255

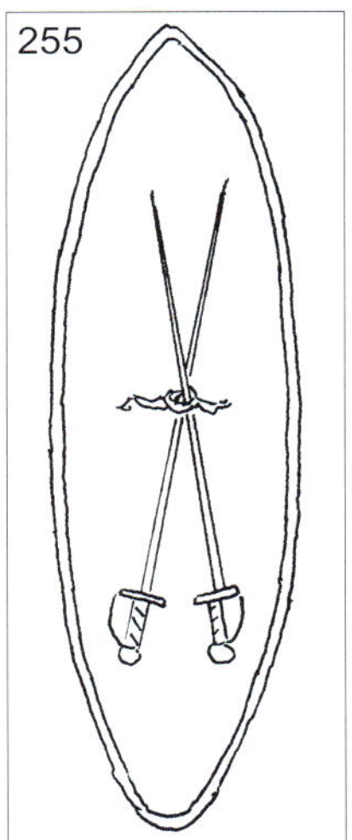

256

256 - Sabre d'officier d'Infanterie de la Restauration.
Monture en laiton argenté. Ce sabre par ses caractéristiques est évidemment du modèle du Consulat. On sait que ces sabres furent en usage sous la Restauration et celui-ci en apporte une preuve indiscutable : une fleur de lys est ciselée sur l'arc de jointure !

257

Sous-famille V c – Calotte à courte queue en crosse, souvent à tête léonine

La calotte en crosse, à courte queue, est la caractéristique principale de cette Sous-famille. Ces sabres sont particulièrement soignés dans leur exécution, les oreillons sont ciselés d'emblèmes ainsi que les garnitures du fourreau, ou en totalité lorsque ceux-ci sont en laiton laminé. Ce sont tous des sabres d'officiers supérieurs.

257 - Sabre d'officier supérieur.
Directoire.
Monture en laiton doré. Calotte à courte queue, en crosse à tête de lion (257a). Poignée en ébène quadrillée. Arc de jointure avec un petit arc entre la calotte et la branche de garde. Oreillons en navette, ciselés d'une tête de lion située entre deux palmettes (257b). Quillon postérieur recourbé. Lame à pan creux. Fourreau en laiton ciselé : chape à trophée de drapeaux et et d'instruments de musique (257c), trophée d'armes, rectangle uni bordé de palmette précédant un décor floral (257d). Dard en fer orné d'une palmette en laiton (257e). M.R.

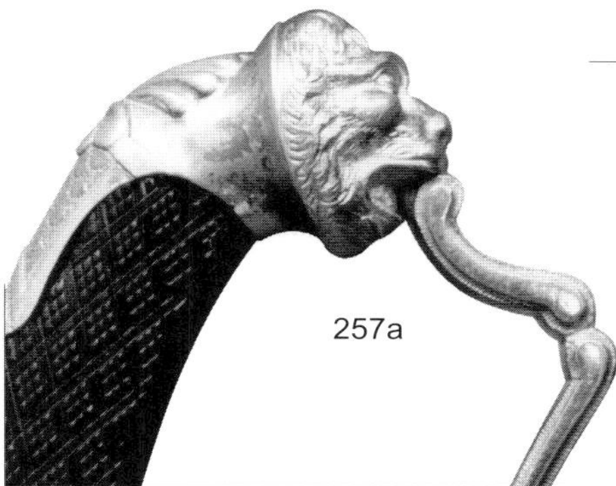

257a

257b

257c

257d

257e

258

258 - Sabre d'officier de Cavalerie légère.

Consulat - Premier Empire.

Monture en laiton doré. Calotte en crosse, à face léonine et courte queue (258a). Poignée d'ébène quadrillée. Arc de jointure décoré d'une cuirasse arrivant en angle droit sur la croisière, celle-ci portant deux oreillons en navette, de grande taille et ciselés d'un buste casqué (258b). Quillon infléchi à son extrémité et décoré de feuillage. Forte lame courbée de 81,3 cm, traitée, en damas, de turqueries... Fourreau de bois recouvert de chagrin à trois garnitures : chape à long médaillon au trophée d'armes, terminée par une demi-lune feuillue (258b), bracelet au trophée prolongé de la même demi-lune de part et d'autre (258c) ; bouterolle à décor en suite. Anneaux de bélière en laiton.

258a

- **a**, calotte à face de lion et courte queue.

258c

- **c**, bracelet au trophée où l'on distingue nettement deux sabres orientaux en sautoir.

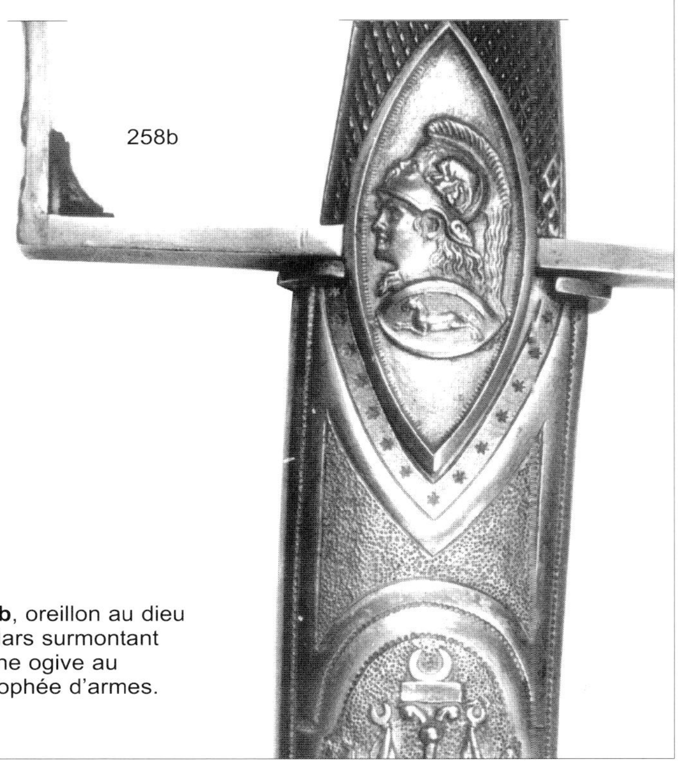

258b

- **b**, oreillon au dieu Mars surmontant une ogive au trophée d'armes.

259

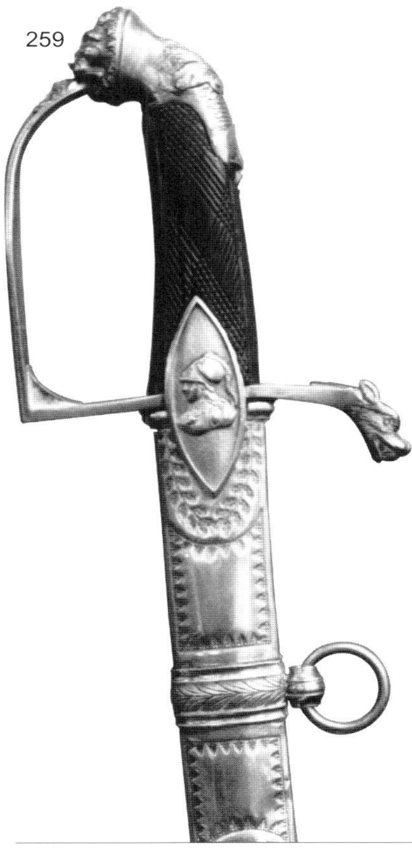

259 - Sabre d'officier supérieur de Cavalerie légère.
Premier Empire.
Monture en laiton doré. Calotte en crosse se terminant par une tête de lion et courte queue ciselée (259a). Poignée en ébène à stries contrariées formant des losanges. Arc de jointure se poursuivant par un quillon à tête de chien. Oreillons en navette à bordure moulurée dans lesquels est inscrit un buste d'Athéna (259b). Lame demi-courbe, bleuie et à motifs dorés. Fourreau de laiton doré à deux bracelets de bélière ciselés d'une frise de feuilles de laurier, avec en son milieu un cartouche (259c) où l'on reconnaît, entre autres, des feuillages, des tambours, des pavillons, des piques ... (M.R.).

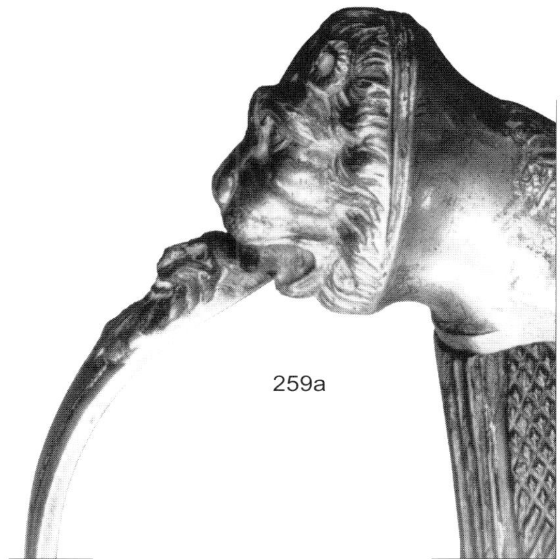

259a

259b

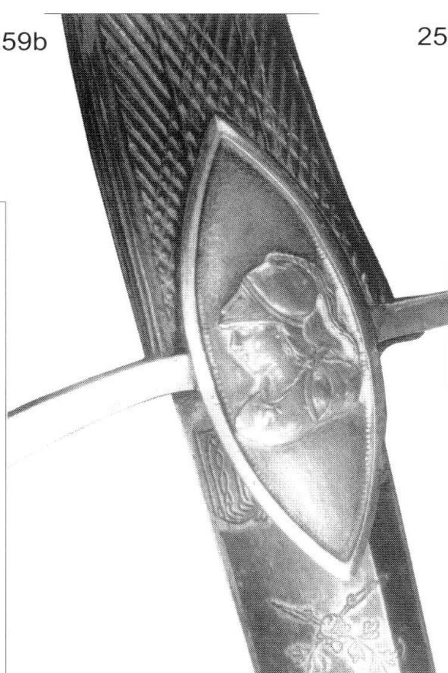

259c

- **a**, calotte à tête de lion et début de l'arc également ciselé d'une petite face léonine ;
- **b**, oreillon au buste d'Athéna ;
- **c**, cartouche médian du fourreau de laiton au trophée où l'on reconnait des instruments de musique, des boucliers ...

Nota - Il existe des sabres à monture identique mais à fourreau de chagrin rouge à trois garnitures dorées décorées d'une grande fleur à quatre pétales.

260

260 - Sabre d'officier supérieur de Cavalerie.
Consulat - Premier Empire.
Monture en laiton doré. Calotte en crosse terminée par une tête de lion à courte queue. Poignée en ébène quadrillée. Arc de jointure partant de la gueule du lion et faisant un angle droit avec la croisière. Grands oreillons en navettes ciselées d'attributs militaires. Quillon recourbé à tête de chien. Lame courbe de 77 cm, à dos plat, pan creux, bleuie au tiers et gravée de motifs dorés. Fourreau de laiton entièrement gravé de motifs végétaux sur toute sa longueur, à deux bracelets ciselés.

261

recto

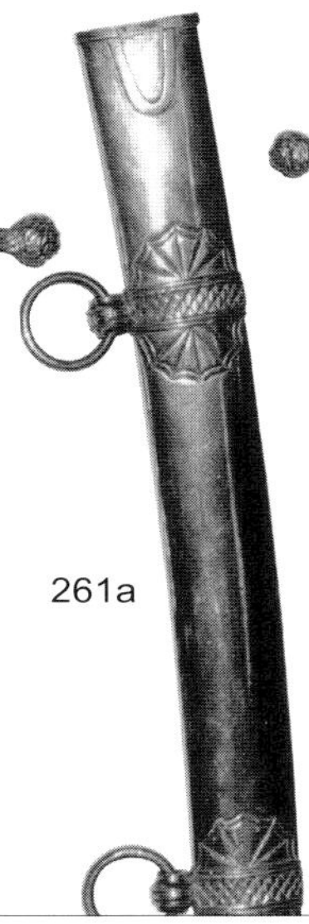

261a

verso

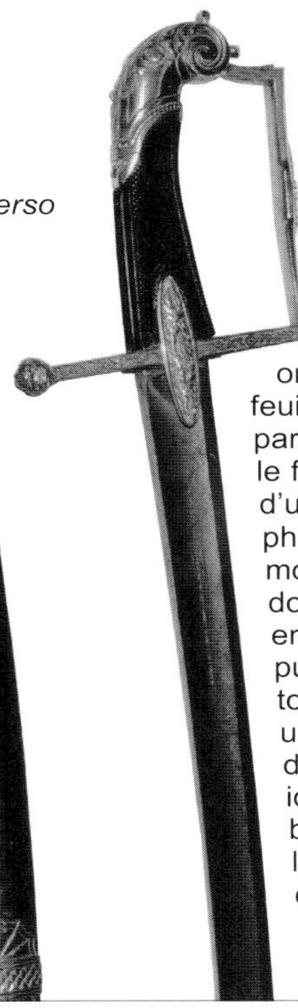

261 - Sabre d'officier de Cavalerie légère.
Consulat.
Monture en laiton doré. Calotte se terminant par un enroulement formant une petite crosse, à courte queue. Poignée d'ébène finement quadrillée. Arc de jointure à angle droit à ses deux extrémités, élargi en son milieu pour supporter un ornement végétal. Croisière décorée de feuillages, de petits losanges et terminée par une boule ciselée. Oreillons ovoïdes, le frontal étant orné d'un écu surmonté d'une coiffure qui paraît être un bonnet phrygien. Lame demi-courbe, gravée de motifs peu lisibles. Fourreau de laiton doré, entièrement décoré : près de son embouchure deux branches de laurier, puis un bracelet quadrillé placé sur une toile d'araignée, un trophée inscrit dans un hexagone constitué de drapeaux, d'un casque... et le second bracelet identique au premier. Bouterolle bordée de deux frises de laurier. Sur le *verso* du fourreau ne subsiste, comme décoration, que les bracelets de bélière (261a) (J.C. D.).

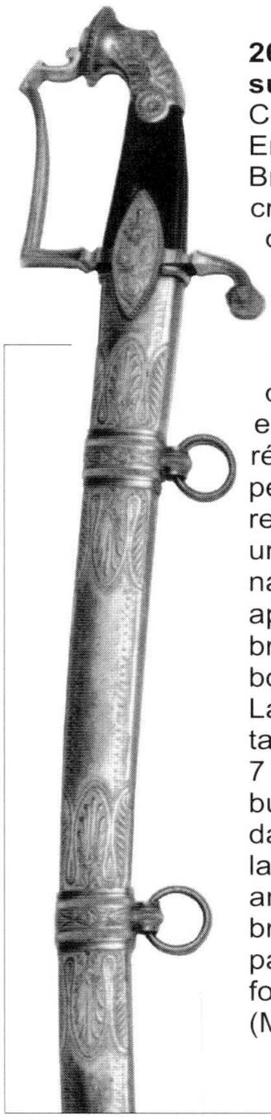

262 - Sabre d'officier supérieur de Cavalerie
Consulat - début du Premier Empire.
Bronze doré. Calotte en crosse à partie terminale décorée d'un trèfle, à queue courte ornée d'une feuille d'acanthe et d'une rosace basale (262a). Poignée d'ébène quadrillée. Arc de jointure en colonne infléchie et réunie à la calotte par une petite branche. Croisière recourbée et terminée par un gros bouton. Oreillon en navette ciselée d'un glaive apparaissant sous des branches de laurier et à bordure de perles (262b). Lame de 87 cm, large au talon de 3 cm, à flèche de 7 cm, à jonc décorée d'un buste de mameluk en damas doré. Fourreau de laiton à deux bracelets et anneaux de bélière, chaque bracelet étant flanqué de palmettes ciselées sur le fourreau lui-même (262a), (M.R.).

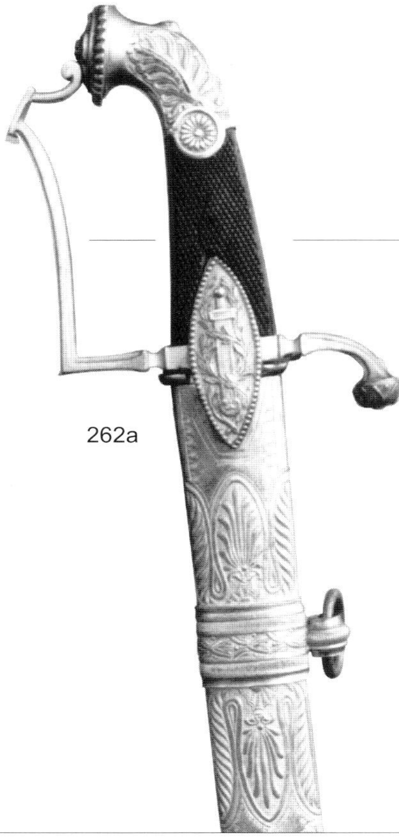

262a

262b

Famille VI – Calotte non figurative à longue queue

La forme des oreillons est assez variable. Tantôt ils sont en baguette, en navette allongée, tantôt la partie supérieure est plus courte que la partie inférieure, ou encore en losange, en losange curviligne, en hexagone... La forme de ces oreillons correspondant souvent au corps auquel le sabre est attribué, nous avons été amenés à distinguer les Sous-familles suivantes :

Sous-famille VI a – Oreillons en baguettes aux extrémités tronquées

Les sabres de cette Sous-famille sont essentiellement des sabres de Hussards. Les premières fabrications en Europe de l'Ouest sont vraisemblablement d'origine germanique. Dès 1752, la forme définitive est acquise.

De nombreux sabres de Hussards de fabrication française, sont signés de fourbisseurs renommés, tels Berger, Boutet, Juste, Manceaux, Duc, Coullier, Spol, Goze, Coulaux...

263 - Sabre de Hussard.
Vers 1750.
Monture en laiton. Calotte plus large que le diamètre de la poignée, à longue queue. Cette poignée est ficelée, recouverte de basane et avec un clou médian de fixation. Arc de jointure courbé sur presque la moitié de sa longueur, arrivant en angle droit sur la croisière rectiligne, se terminant par un quillon en boule. Oreillons en baguettes tronquées... Lame courbe. Fourreau de cuir, à deux garnitures.

263

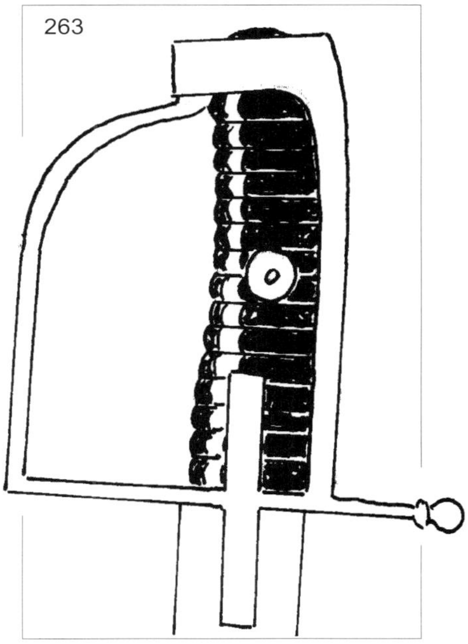

264 - Sabre d'officier de Hussards.
Règne de Louis XVI.
Monture à la hongroise en bronze doré. Calotte à pans sans bouton de rivure apparent, décorée au trait, à longue queue. L'arc de jointure, la croisière, les oreillons et le quillon droit également décorés au trait. Poignée recouverte de basane avec deux clous à tête en fleurette de quatre pétales. Petit carré au nœud de corps dont les angles sont orientés vers le milieu des oreillons en bâtonnets, et les angles latéraux vers la croisière. Lame courbe, à simple pan creux et gorge le long du dos, dorée et bleuie sur la moitié de sa longueur. Fourreau à quatre garnitures : chape cannelée, deux bracelets ornés d'un médaillon à trophée ciselé, longue bouterolle cannelée à médaillon ovale décoré de motifs militaires ; demi-dard en laiton sur la face antérieure (non figuré).

265

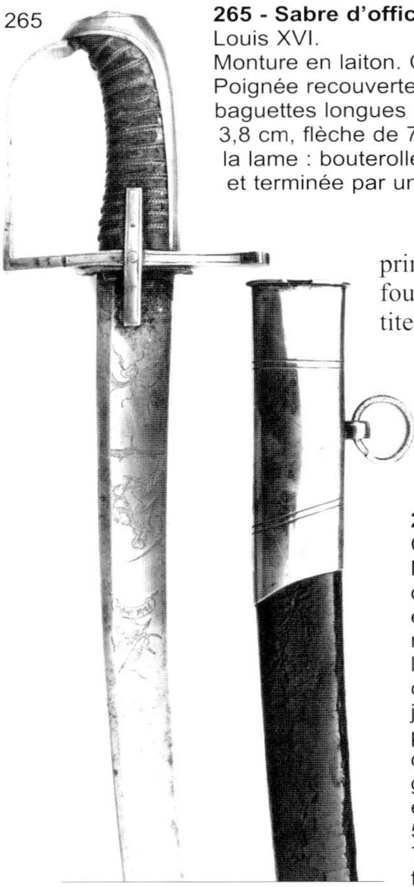

265 - Sabre d'officiers de Hussards.
Louis XVI.
Monture en laiton. Calotte arrondie se poursuivant par une longue queue à filets latéraux. Poignée recouverte de basane et filigranée d'un double fil de laiton torsadé. Oreillons en baguettes longues de 6 cm. Lame à pan creux, de 81,2 cm, épaisse au talon de 1 cm et large de 3,8 cm, flèche de 7,2 cm. Fourreau à chape de 15 cm côté tranchant et de 12,6 cm côté dos de la lame : bouterolle de 51,5 cm côté tranchant et de 46,7 cm de longueur du côté dos de la lame et terminée par un dard de fer asymétrique.

Les sabres de ce type sont sujets à des variations importantes intéressant principalement les dimensions, la flèche de la lame et la taille des garnitures du fourreau. Sur cet exemplaire, au dessus du Hussard chargeant se trouve une petite banderolle avec *Vivat Hussar*.

266

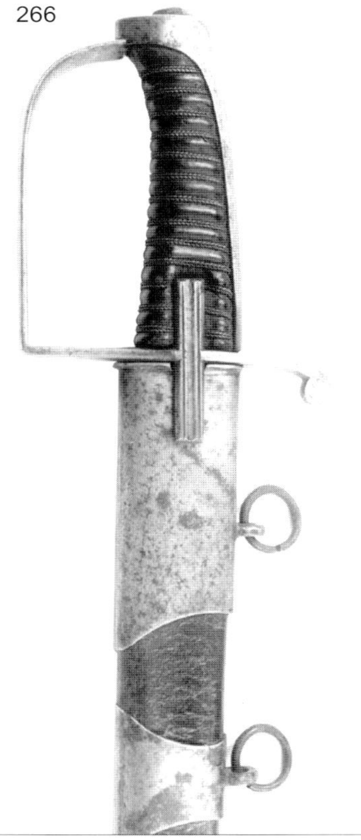

266 - Sabre d'officier de Hussards.
Consulat.
Monture en fer. Calotte légèrement décentrée, pièce de rivure en losange et longue queue unie. Poignée ficelée recouverte de basane à filigrane de laiton en double fil torsadé disposé en chaînette faisant 13 spires. Arc de jointure arrondi à sa partie supérieure puis rectiligne et aboutissant en angle droit sur la croisière qui se termine en goutte d'eau aplatie ; deux oreillons en baguette cannelée longue de 5,8 cm et large de 1 cm. Lame de 78 cm, large au talon de 3,5 cm et flèche de 2,3 cm, à pan creux, bleuie sur 21 cm et gravée de décors classiques dorés ; sur le dos plat, on lit :
Schaberg fils de françoise fabriquant à Solingen
Fourreau de cuir à trois garnitures en fer découpées en suite : chape longue de 12 cm sur la face frontale, avec un piton pour l'anneau de bélière en fer, bracelet trapézoïdal à longueur maximale de 5,5 cm et minimale de 2,8 cm, bouterolle de 24,5 cm terminée par un bouton prolongé sur la face frontale d'un demi-dard (M.R.).

267 - Sabre de Cavalerie légère.
Règne de Louis XVI.
Monture de laiton. Calotte moulurée à longue queue. Poignée recouverte de basane, filigranée. Arc de jointure perlé, arrivant en angle droit sur la croisière rectiligne, garnie de deux longs oreillons en baguettes tronquées. Lame de 80 cm de longueur, large au talon de 3 cm, à flèche de 2,5 cm, gravée de turqueries, de flèches, de motifs végétaux... dorés sur fond bleuie. Fourreau à deux garnitures ciselées de motifs végétaux et d'étoiles multibranches (M.R.), (non figuré).

268

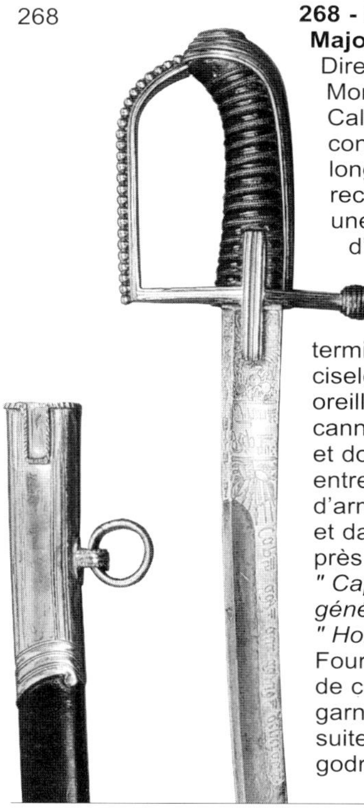

268 - Sabre d'officier d'Etat-Major.
Directoire - Consulat.
Monture en laiton doré. Calotte légèrement convexe prolongée d'une longue queue. Poignée recouverte de basane à une quinzaine de spires d'un triple filigrane de cuivre rouge. Arc de jointure perlé. Croisière droite terminée par un gros bouton ciselé et portant des oreillons en baguettes cannelées. Lame en damas et dorée gravée sur le *verso* entre deux trophées d'armes *" Rép^{que} / Fran^{se} "* et dans le pan creux doré près du dos de la lame *" Cap^{tn} adj^{t} aux adjud^{t} généraux "* et sur le *recto* *" Honneur et Courage "*
Fourreau de bois recouvert de cuir noirci, à deux garnitures décorées en suite de filets et de godrons.

269

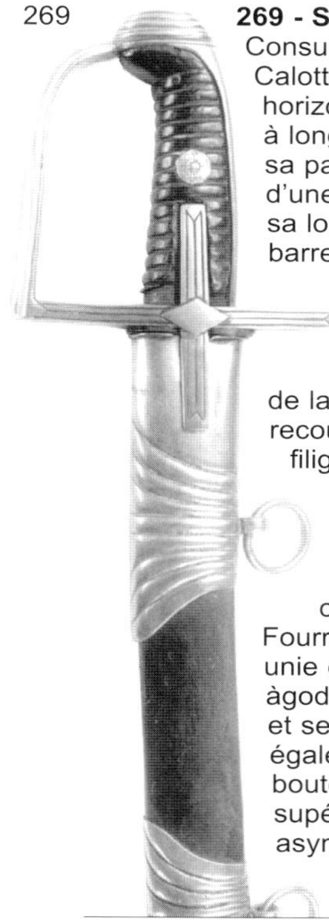

269 - Sabre d'officier de Hussards.
Consulat - Premier Empire.
Calotte à quatre gorges horizontales, sans bouton de rivure, à longue queue. Branche élargie en sa partie supérieure et parcourue d'une arête peu saillante sur toute sa longueur. Longs oreillons en barrettes, rectangulaires, gravés de trois traits se terminant en queue d'aronde ; un losange décoratif est placé à la jonction des oreillons et de la croisière. Poignée encordée, recouverte de basane, sans filigrane mais avec un clou à tête ciselée en fleurette. Lame de 72,8 cm de long, de 3 cm de largeur au talon et 3 cm de flèche ; large pan creux et contre-tranchant de 16 cm.
Fourreau à trois garnitures : chape unie en sa première moitié et àgodrons dans la seconde moitié et servant de bracelet, bracelet également à godrons ainsi que la bouterolle dans sa partie supérieure ; dard en fer, asymétrique.

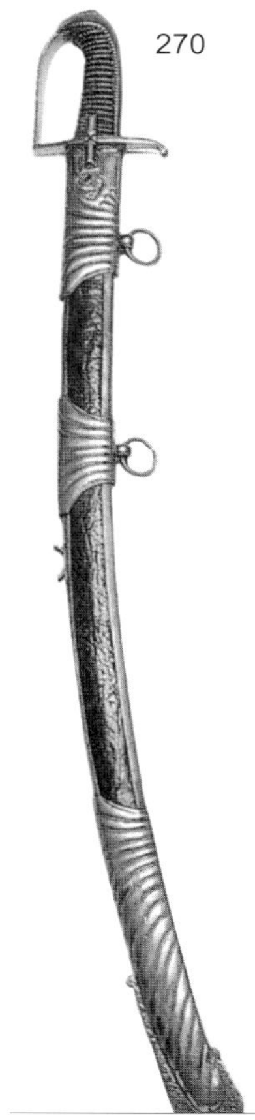

270

270 - Variante.
Monture en laiton. Calotte en cuillère prolongée d'une longue queue. Poignée recouverte de basane filigranée. Arc de jointure rejoignant la croisière par un angle droit. Lame de 75,7 cm, courbée, à dos plat et pan creux, gravée au tiers d'attributs militaires et du petit Hussard chargeant. Fourreau recouvert de chagrin noir entre les garnitures, toutes trois ornées de godrons et reliées entre elles par des attelles latérales. Fort dard de fer asymétrique. Cet exemplaire est très voisin du sabre n° 269. et ne s'en différencie que par l'absence de clou sur la poignée, par une lame de 2,9 cm plus longue...

271

271 - Sabre d'officier de Hussards.
Période révolutionnaire.
On notera que l'arc de jointure est perlé et que les oreillons sont en baguettes d'égale longueur de part et d'autre de la croisière et qu'ils sont tronqués à leurs extrémités. De plus, les deux longues garnitures sont gravées de deux médaillons rendus illisibles par suite de leur usure. Attelles de renfort pour le fourreau.

Sous-famille VI b – Oreillons en navette plus ou moins allongée aux extrémités arrondies

Dans ce groupe les sabres possèdent des oreillons en navette dont la longueur est au minimum égale à quatre fois la largeur au niveau de la croisière. Parfois ces oreillons peuvent être posés sur un carré et ornés d'un emblème de corps.

272

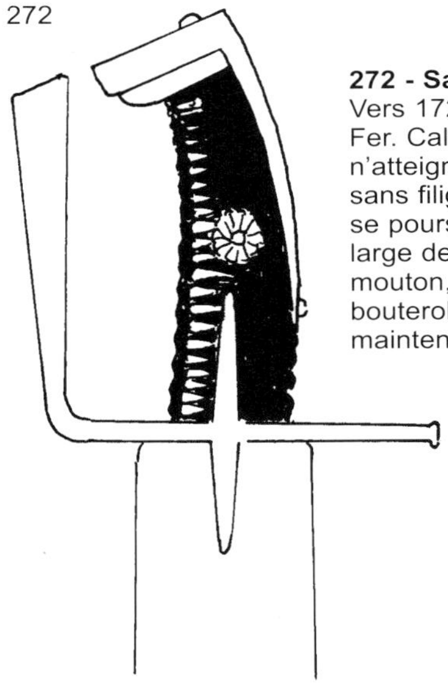

272 - Sabre de Bercheny.
Vers 1720.
Fer. Calotte plate, débordant largement la fusée sur l'avant, à longue queue rapportée n'atteignant pas la base de la poignée et cloutée sur le dos. Celle-ci est de bois cannelé, sans filigrane mais ornée d'une rosette. Branche de garde plate, non fixée à la calotte et se poursuivant par une croisière droite, portant deux longs oreillons. Lame de 75 à 86 cm, large de 3,4 à 3,8 cm, à flèche de 3,6 à 5,6 cm. Fourreau de bois recouvert de peau de mouton, souvent à deux paires de bracelets jumelées avec chacun un anneau de bélière ; bouterolle réduite à une cuvette à fond plat soutenue par deux attelles latérales maintenues à leur partie supérieure par une bague.

273

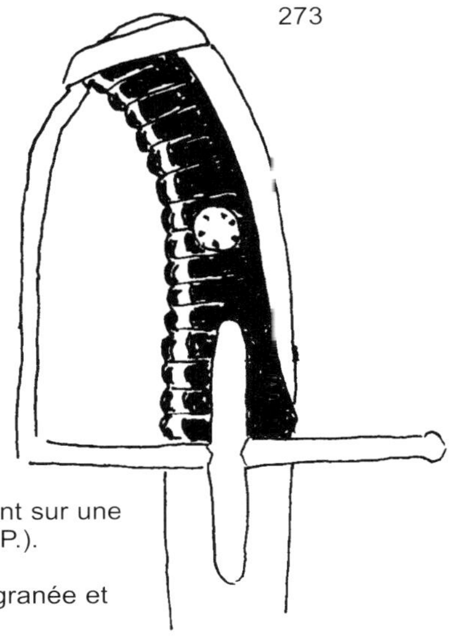

273 - Sabre d'officier de Hussard du Régiment de Ferrary.
Modèle 1752.
Monture de laiton. Calotte plane et mince, prolongée par une longue queue rapportée, fixée à son extrémité basale par un clou. Poignée inclinée vers l'avant, en bois sculpté de grosses cannelures et recouverte de cuir bouilli, collé et maintenu par un clou en rosette. Arc de jointure arrivant en angle droit sur la croisière celle-ci portant deux oreillons en baguettes arrondies à leurs extrémités. Lame de 95 cm environ, large au talon de 3,2 cm, très courbée, à pan creux et signée *" Berger, marchand fourbisseur à Strasbourg "*. De plus cette lame porte les inscriptions suivantes : *" Régiment / de / Ferrary "* avec les Armes de France couronnées et le Hussard chargeant sur une face et sur l'autre "*Vive le Roy*". (M.E. - S. de P.).

Nota - Des modèles voisins ont une fusée filigranée et sans clou en rosette.

274

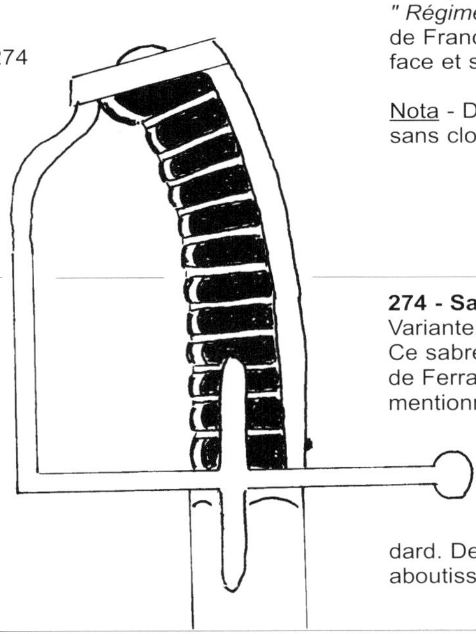

274 - Sabre de Hussard
Variante du modèle 1752.
Ce sabre est d'une conception très voisine de celle du modèle du Régiment de Ferrary mais la fusée n'est plus *couverte de cuir bouilli* crénelé comme le mentionne l'Ordonnance de 1752 ; elle est recouverte de basane filigranée d'un fil de laiton torsadé faisant onze spires. La queue de calotte est fixée à sa base par un clou sur la poignée et est forgée séparément. Lame à pan creux et gouttière dorsale au fort, gravée du Hussard chargeant. Fourreau à quatre garnitures : chape, deux bracelets à anneau, trois bagues de renfort et courte bouterolle sans dard. Deux attelles latérales, de laiton, partent du second bracelet et aboutissent à la bouterolle.

275 - Sabre de Hussard, variante du modèle 1752.
Vers 1764.
Monture en fer. Calotte plate. Queue rapportée, fixée sur la croisière. Poignée sans filigrane mais avec un clou à tête sphérique. Oreillons en longues baguettes à extrémités arrondies. Lame plate de 86,6 cm gravée du Hussard chargeant, des Armes de France couronnées, de *Vive le Roy*. L'exemplaire présenté par C. Ariès (4ème fasc. 1969, fig. 3) porte la mention *Régiment Esterhazy* (Musée de l'Emperi). Lame signée *Guillaume Mumm Négociant à Sohligen*. Fourreau à trois garnitures dont une longue bouterolle terminée par un dard asymétrique.

275

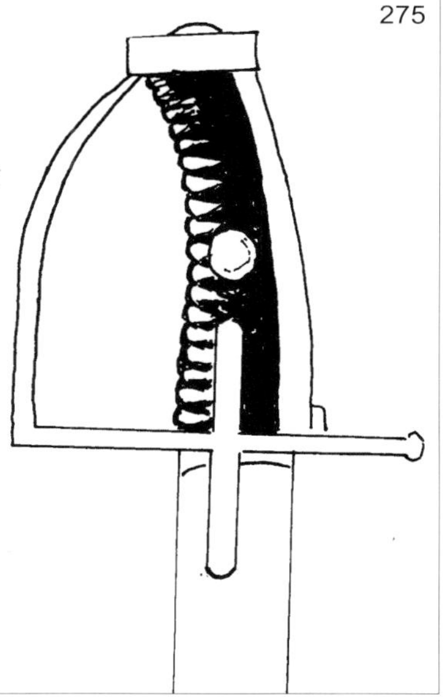

276 - Sabre d'officier de Hussard.
Louis XV.
Monture à la hongroise, en laiton. Arc de jointure gravé d'une palme et de volutes, arrivant en angle droit sur la croisière. Calotte à cache-rivure et longue queue décorée en suite. Oreillons en baguette à extrémités arrondies. Poignée encordée, recouverte de basane non filigranée. Longue lame de 80 cm, à faible courbure, à pan creux et dos plat. Fourreau à alèse de bois recouvert de cuir, à trois garnitures de laiton ciselé de volutes, de torons, jointes entre elles par des bandes de renfort rivées ménageant des crevés à extrémités concaves ; la chape est moulurée autour de l'emplacement des oreillons ; bouterolle se terminant par un dard de fer très étroit et lui même protégé par un bouton apical (non figuré).

277

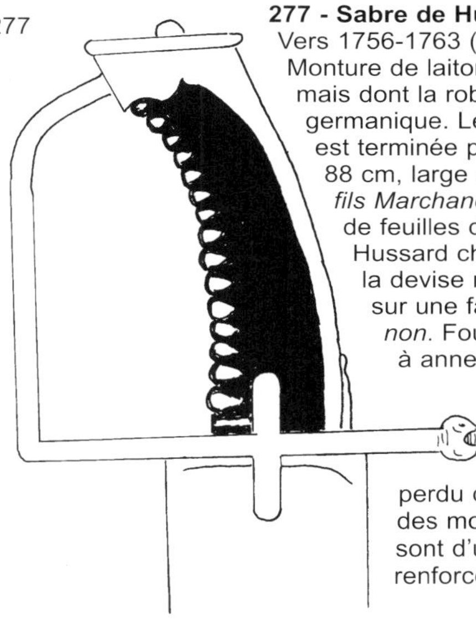

277 - Sabre de Hussard.
Vers 1756-1763 (Guerre de Sept Ans).
Monture de laiton, voisine de celle du modèle de 1752, mais dont la robustesse indique une fabrication germanique. Les oreillons sont très courts et la croisière est terminée par une tête d'animal fantastique. Lame de 88 cm, large de 6,3 cm signée *La Veuve Jean Knecht et fils Marchand à Sohlingen.* Cette lame plate est gravée de feuilles d'acanthe, des Armes de France, du Hussard chargeant, de *Vive le Roy*,... et au talon de la devise menaçante : *Fuyes le bras / qui me dirige* sur une face et sur l'autre *Soyes En garde / ou Si non.* Fourreau à trois garnitures : chape et bracelet à anneaux de bélière, bouterolle à petit dard asymétrique.

Nota - Malgré les dimensions extraordinaires de sa lame, ce sabre a perdu quelques-uns des caractères archaïques des modèles précédents : la calotte et sa queue sont d'une seule pièce, le fourreau n'est plus renforcé par des bagues (cf. Ariès, fasc. XI, fig. 3).

278

278 - Sabre d'officier de Cavalerie légère.
Règne de Louis XV.
Monture à la hongroise. Calotte ovalaire, à bouton de rivure en losange, unie et se poursuivant par une longue queue à bordures découpées. Branche élargie près de la calotte, rejoignant la croisière par un angle droit, la face externe de cette branche étant ornée d'une longue crête percée d'une série de vingt trous. Oreillons en ovale très allongé. Bouton du quillon en petite hémisphère. Poignée recouverte de basane à filigrane composé de deux fils de cuivre rouge, torsadés formant chaînette, celle-ci limitée par un fil également torsadé mais beaucoup plus fin ; on compte treize spires complètes. Lame de 77,5 cm, large au talon de 2,5 cm, à flèche de 3 cm, à dos plat, gouttière profonde partant du talon et finissant au début du contre- tranchant de 22 cm ; la lame est gravée de la lune, du soleil, d'étoiles et du bras armé. Fourreau à couture rabattue, à trois garnitures : chape à anneau, bracelet intermédiaire portant le second anneau, bouterolle à dard de laiton, asymétrique, toutes les garnitures sont découpées en deux lobes et ornées au trait.

279

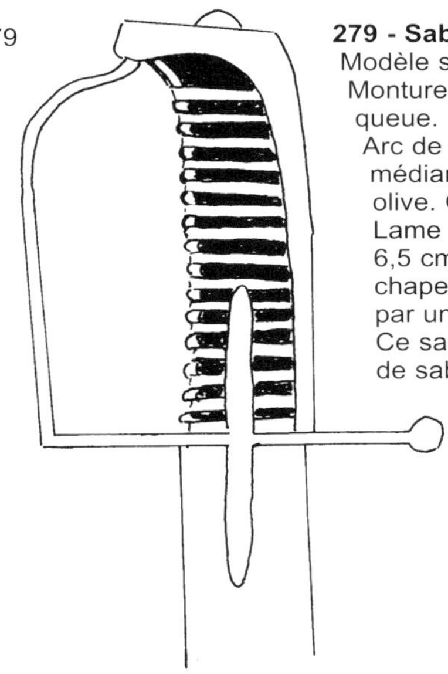

279 - Sabre de Hussard.
Modèle selon l'Ordonnance du 2 avril 1767.
Monture de laiton . Calotte plate, à petit bouton de rivure et à longue queue. Poignée recouverte de basane à 14 spires d'un gros fil torsadé. Arc de jointure partant du dessous de la calotte, droite dans sa partie médiane et arrivant en angle droit sur la croisière terminée par une olive. Oreillons en navettes de 6,5 cm de long à extrémités arrondies. Lame plate de 90 - 91 cm de long, large au talon de 3,6 cm, à flèche de 6,5 cm environ. Fourreau de bois recouvert de cuir à deux garnitures : chape d'une quinzaine de centimètres et bouterolle de 46 cm terminée par un petit dard asymétrique en fer.
Ce sabre probablement créé en 1767, est l'aboutissement d'une série de sabres dont l'origine remonte vers 1750 (Musée de Strasbourg).

280a

280c

280- Sabre d'officier de Cavalerie légère.
Règne de Louis XVI.
Monture en laiton. Calotte drapée se poursuivant par une longue queue. Poignée encordée, recouverte de basane à treize spires d'un filigrane torsadé bordé de fils plus fins également torsadés. Arc de jointure surmonté d'une crête perforée (280a). Croisière décorée au trait se terminant par un quillon à triple boutons. Oreillons en navette très allongée bordée d'un trait. Lame à pan creux, bleuie et gravée de motifs classiques plus des armes à l'orientale coiffées d'un turban (280b). Fourreau de cuir à trois garnitures de laiton, découpées en suite avec sur la chape deux drapeaux et deux lances gravés au trait dans un ovale, anneau de bélière et bouton de suspente (280c) (M.R.).

280b

280

281

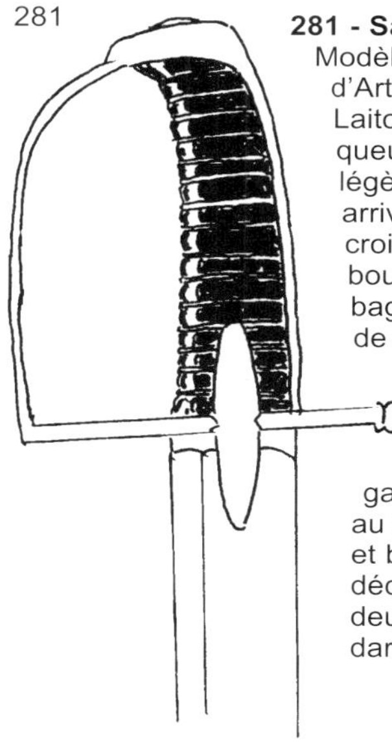

281 - Sabre de Hussard.
Modèle 1783 (Archives d'Artillerie 4 f48-3689 (13). Laiton. Calotte plate à longue queue. Arc de jointure légèrement courbé mais arrivant en angle droit sur la croisière terminée par un bouton. Oreillons en baguettes curvilignes. Lame de 81,2 à 89 cm, large d'environ 3,8 cm, à pan creux et flèche de 6 à 7 cm. Fourreau à deux garnitures : chape de 15 cm au maximum de sa longueur et bouterolle de 51,4 cm, à découpes en arrondis, à deux anneaux de bélière et dard asymétrique.

282

282 - Sabre d'officier de Hussards.
Règne de Louis XVI. Monture en laiton. Calotte subcylindrique, ornée de six traits jumelés et prolongée par une longue queue. Poignée recouverte de basane, à 23 spires d'un double filigrane torsadé disposé en chaînette. Arc de jointure droit sur les deux tiers basaux rejoignant en angle droit la croisière terminée par un quillon emboulé et portant deux oreillons en baguettes décorés au trait en leur partie médiane. Lame à pan creux et gouttière dorsale au fort et gravée à partir du talon d'un motif foliacé (grottesque), des Armes de France couronnées, de trois étoiles disposées en triangle.

283

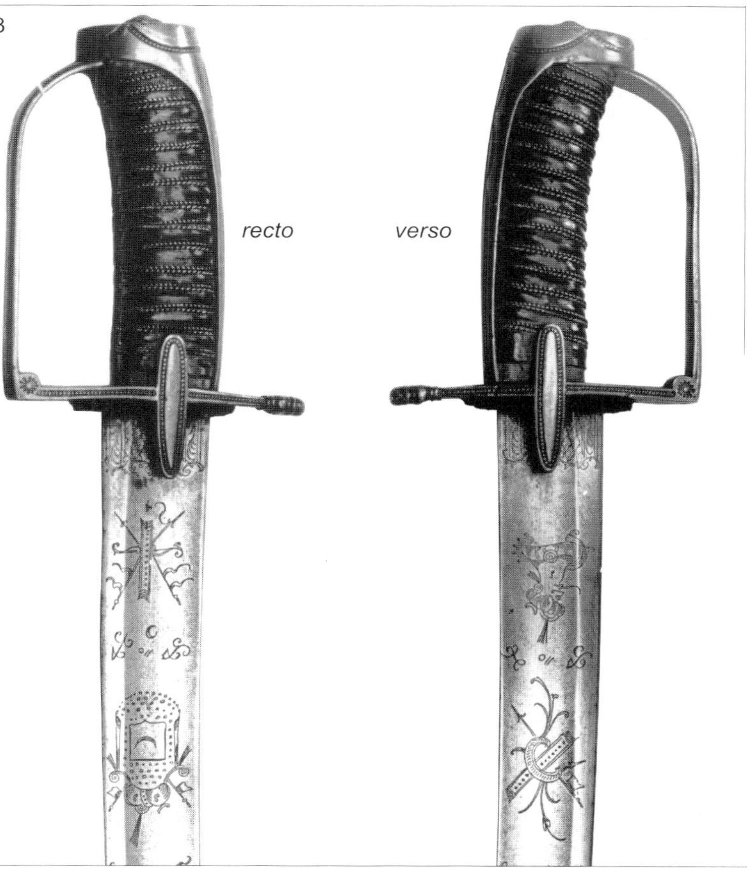

283 - Sabre d'officier de Hussards.
Règne de Louis XVI.
Monture en laiton. Remarquer la calotte décorée d'une frise de petites perles, motif que l'on retrouve sur les oreillons. Sur la lame des turqueries, très à la mode alors, sont gravées : au *recto* à partir de la garde, on identifie une sorte de trophée, un croissant, des signes cabalistiques, des armoiries au croissant, de nouveau des signes énigmatiques, des étoiles... sur le *verso*, un buste évidemment de Turc, des signes, une sorte de trophée, un turban (?). J. Lhoste et J.J. Buigné ont donné certaines significations des signes cabalistiques (op. cit. 1999, p.110-113).

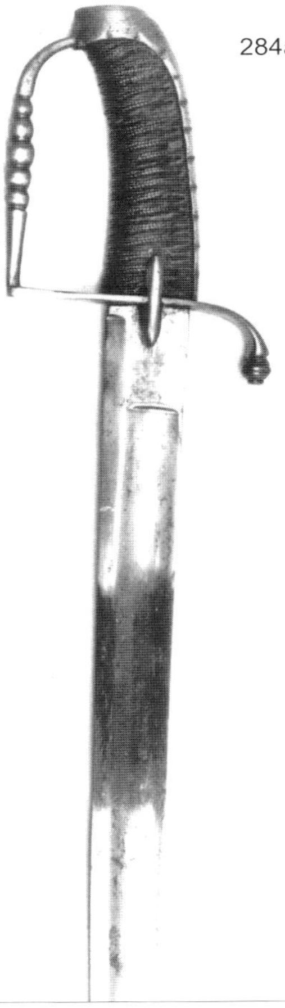

284a

284b

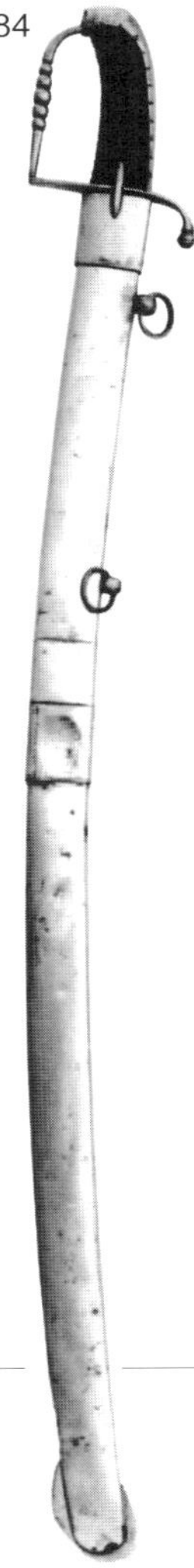

284

284 - Sabre d'officier de Hussard.
Période révolutionnaire.
Monture en fer. Calotte et sa longue queue régulièrement dentées marginalement. Poignée recouverte de basane, filigranée. Arc de jointure orné de cinq perles en sa partie droite (284a, 284b). Oreillons en olive étroite. Quillon recourbé terminé par un bouton. Lame demi-courbe, blanche, à fort pan creux (284a). Fourreaux de fer à deux anneaux de bélière, ayant été réparé par des bagues de renfort près de l'entrée et vers le milieu. Il est évident que le fourreau n'est pas conforme au modèle.

285

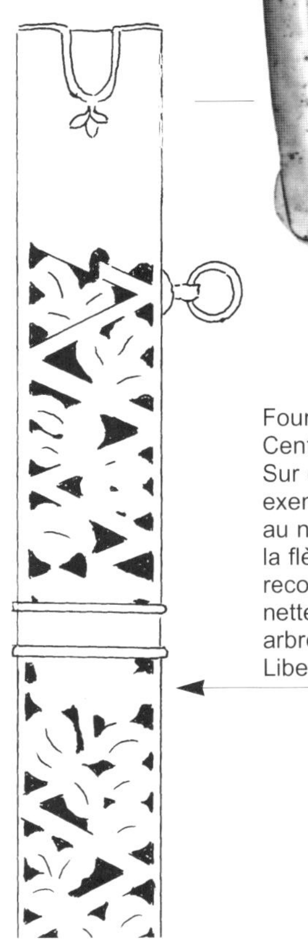

Fourreau des Cent Braves. Sur cet exemplaire, au niveau de la flèche, on reconnait nettement un arbre de la Liberté.

285 - Sabre de Récompense du Directoire, dit des Cent Sabres pour l'Armée d'Italie.
Modèle d'Infanterie.
Monture en laiton. Calotte fortement inclinée vers l'avant dite à la hongroise et prolongée par une longue queue unie. Poignée recouverte de basane et filigranée. Arc de jointure plan, ciselé à jour de motifs végétaux, et rejoignant la croisière en angle droit. Quillon droit, terminé par une petite olive. Oreillons assez larges, arrondis à leur extrémité. Lame large, à pan creux, à cartouche destiné à la gravure de l'attribution. Fourreau de bois recouvert de basane noire apparaissant sous un fourreau de laiton ajouré figurant une ligne brisée avec, dans les triangles ainsi ménagés des motifs végétaux représentant des feuilles trilobées, voire un Arbre de la Liberté (sous le bracelet de la figure) ou des bonnets phrygiens sur certains exemplaires. Extrémité du fourreau en bouterolle à dard en lyre.

286 - Sabre d'officier de Hussards.
Révolution.
Monture de laiton. Calotte plane prolongée d'une longue queue à bordure finement dentée. Arc de jointure orné d'une frise de perles. Poignée recouverte de basane, filigranée d'un fil de laiton torsadé faisant 19 spires. Longs oreillons à extrémités arrondies. Croisière droite terminée par un double bouton (286a). Lame demi-courbe, longue de 83 cm, large au talon de 3,5 cm, à deux pans creux au fort, longs de 53 cm ; le pan près du dos est à trois profondes rainures de 36 cm. Cette lame est gravée sur sa base, d'un trophée aux carquois en sautoir et d'un croissant de lune sur le *recto* et sur le *verso*, d'un trophée aux fanions, au canon, et d'un soleil. Fourreau de cuir noirci, à longue chape à bordure découpées en accolade, d'un bracelet trapézoïdal, d'une longue bouterolle à dard en fer, asymétrique ; les garnitures sont reliées entre elles par des cornières latérales de protection à bordures festonnées (M.R.).

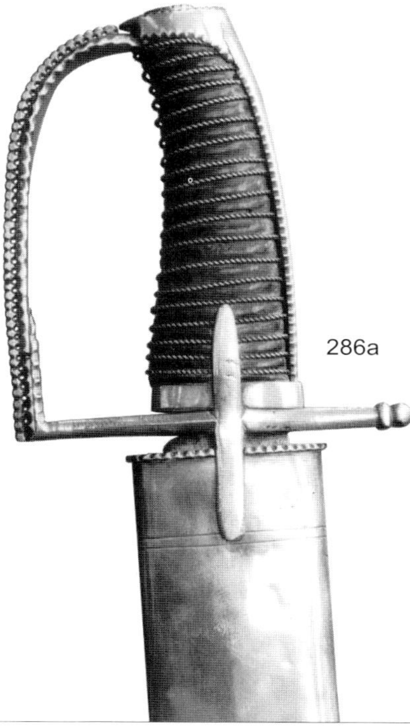
286a

286
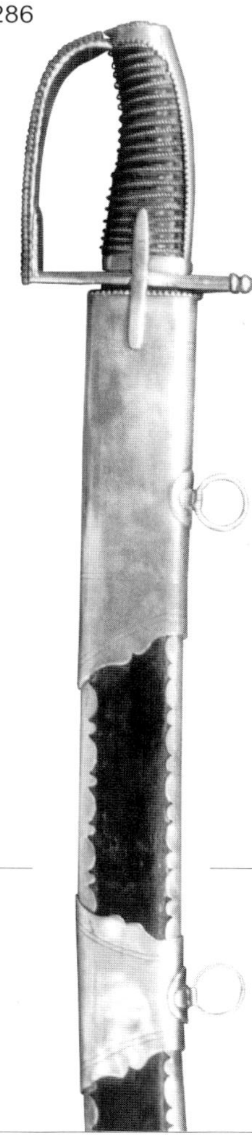

287
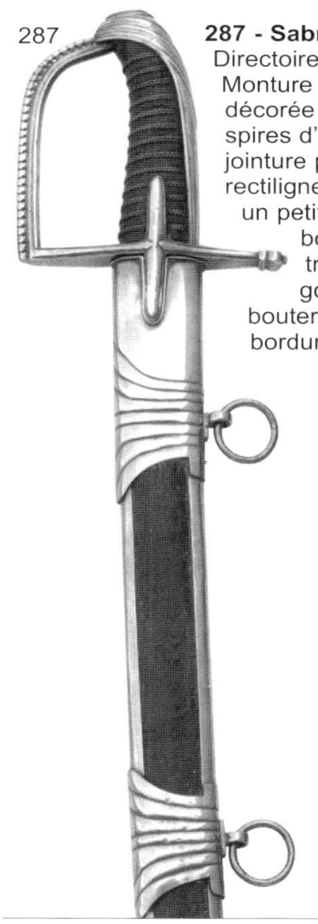

287 - Sabre d'officier de Hussards.
Directoire - Consulat.
Monture en laiton. Calotte en cuillère à longue queue décorée au trait. Poignée recouverte de basane à treize spires d'un filigrane torsadé disposé en chaînette. Arc de jointure perlé arrivant en angle droit sur la croisière rectiligne terminée par une bague, une sorte de cube et un petit bouton. Oreillons en navette allongée à fine bordure. Lame légèrement courbée. Fourreau à trois garnitures : chape à bracelet figuré, à godrons, bracelet médian en suite ainsi que la bouterolle. Le fourreau est renforcé par des attelles de bordure.

288
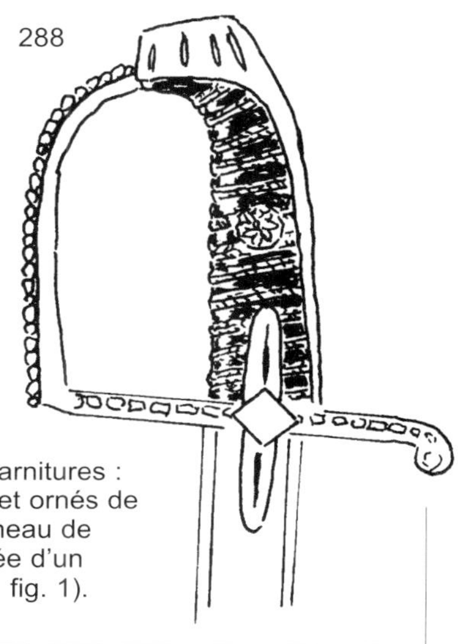

288 - Sabre d'officier de Hussards.
Vers 1790.
Laiton doré. Calotte plate et longue queue. Poignée à 13 spires d'un double filigrane disposé en chaînette, avec un clou en rosette. Arc de jointure bordé d'une rangée de perles. Croisière rectiligne également décorée de perles et terminée en crosse. Oreillons en navette arrondie à leurs extrémités. Longue lame demi-courbe, plus étroite que pour la troupe, gravée de motifs dorés sur fond bleui. Fourreau à quatre garnitures : chape rainurée, premier et second bracelet ornés de médaillon à trophées avec chacun un anneau de bélière, longue bouterolle rainurée et ornée d'un médaillon à motif floral (Voir Ariès, XIII, 2, fig. 1).

289

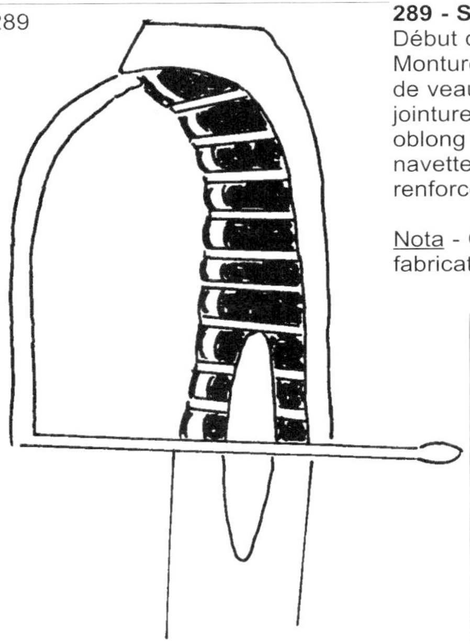

289 - Sabre de Hussard.
Début de la Révolution, vers 1790.
Monture en laiton. Calotte plate, en cuillère et longue queue. Poignée recouverte de veau brun, à filigrane de fils torsadés jumelés faisant neuf spires. Arc de jointure courbé en sa moitié supérieure et arrivant en angle droit sur un plateau oblong terminé par un bouton ; ce plateau est traversé par des oreillons en navette. Lame de 92 cm, gravée de motifs dorés. Fourreau à deux garnitures renforcées par des alèses de bois internes (Musée de l'Armée).

Nota - Ce sabre est évidemment inspiré du style germanique ou même de fabrication d'outre-Rhin.

290 - Sabre d'officier de Cavalerie légère.
Vers 1792.
Monture en fer. Calotte et queue en cuillère sans ornement. Poignée recouverte de basane filigranée faisant dix-sept spires et virole basale. Oreillons très étroits terminés en pointes. Lame demi-courbe, blanche. Deux anneaux de bélière brasées directement sur le fourreau uni.

290

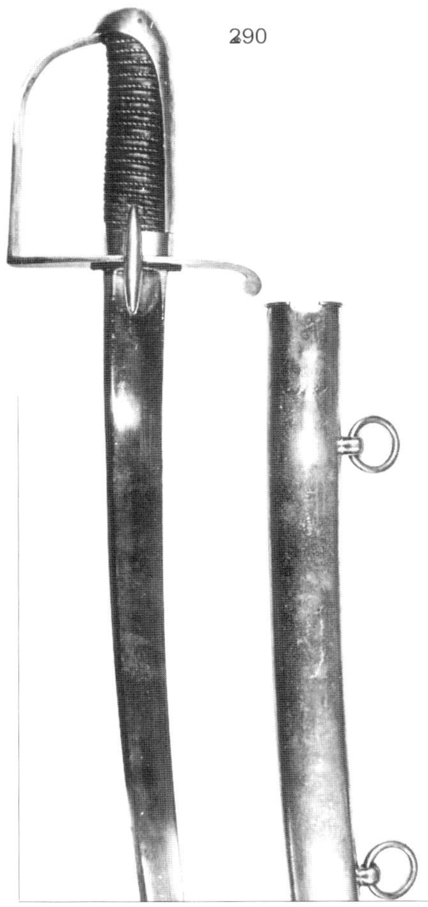

291

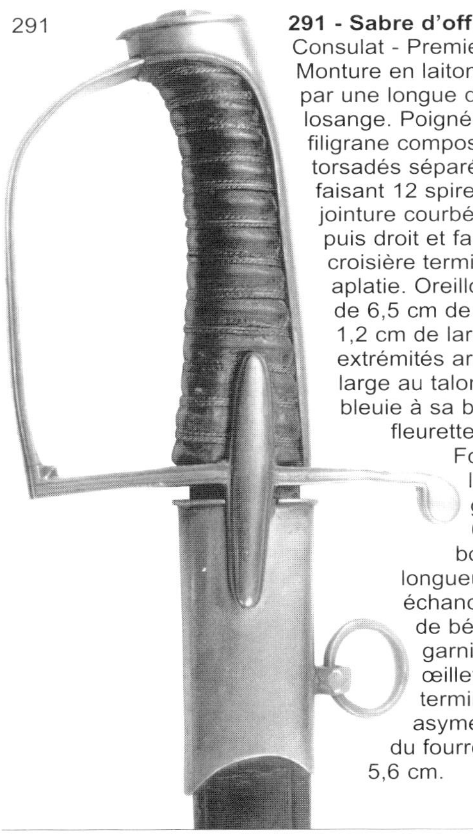

291 - Sabre d'officier de Cavalerie légère.
Consulat - Premier Empire.
Monture en laiton. Calotte simple prolongée par une longue queue. Bouton de rivure en losange. Poignée recouverte de cuir à filigrane composé de deux fils de laiton torsadés séparés par un fil simple et faisant 12 spires, sans virole. Arc de jointure courbé en son quart supérieur puis droit et faisant un angle droit avec la croisière terminée en goutte d'eau aplatie. Oreillons en olive très allongée, de 6,5 cm de hauteur et seulement de 1,2 cm de largeur maximale, à extrémités arrondies. Lame de 75 cm, large au talon de 2,8 cm, à pan creux, bleuie à sa base et gravée de fleurettes, de trophée dorés. Fourreau de cuir rainuré latéralement, à trois garnitures : chape de 6,8 cm, bracelet de 3,7 cm, bouterolle de 17,3 cm de longueur maximale, régulièrement échancrée en arrondi. Anneaux de bélière en laiton, fixés sur les garnitures supérieures par des œillets à facettes. Dard de fer terminant la bouterolle, asymétrique, protégeant le bout du fourreau et sa partie frontale sur 5,6 cm.

292

292 - Sabres de Guides du général Mortier.
1803 - 1804.
Monture en laiton. Calotte en cuillère et longue queue. Poignée déportée vers l'avant, recouverte de basane filigranée d'un fil de laiton simple. Arc de jointure arrivant en angle droit sur la croisière droite, à deux oreillons allongés aux extrémités arrondies. Lame à pan creux, longue de 83 cm, marquée :

GUIDE / DU / GENRAL / MORTIER

en lettres d'or sous un cavalier gravé au trait doré (292). Fourreau de cuir à longue chape à anneau de bélière soudé. Bouterolle recouvrant largement la seconde moitié du fourreau, à anneau soudé à sa partie supérieure. Dard arrondi, plus haut sur la partie antérieure que sur la partie postérieure.

<u>Nota</u> - Ce sabre ressemble beaucoup aux sabres des années 1786 à l'An IV destinés aux Hussards. Un exemplaire de ce modèle est figuré par Bernard Bruel (1996) *in* Tradition, 109, 16-17, sous le titre : *Un sabre des Guides du général Mortier, Formation de 1803.*

293 - Sabre d'officier de Hussards.
Consulat et Premier Empire.
Partie supérieure de la calotte plate, débordant la calotte proprement dite, à bouton de rivure en losange ; longue queue. Arc de jointure légèrement élargi près de la calotte et faisant un angle droit avec la croisière. Longs oreillons étroits (6,5 sur 1,2 cm), à extrémités arrondies. Toute la monture est ornée de traits. Poignée entièrement filigranée d'un fin fil torsadé disposé en chaînette. Lame de 77 cm, large au talon de 2,6 cm, à flèche de 4,3 cm, bleuie sur 29,5 cm et gravée, en plus de motifs végétaux, d'un petit Hussard chargeant doré. Fourreau à trois garnitures : chape de 14,8 cm au maximum, à anneau soudé, bracelet de 9 cm au maximum et bouterolle de 32,5 cm, à dard asymétrique de fer ; deux longues baguettes de renfort latérales de laiton, la basane apparaissant entre les garnitures.

293

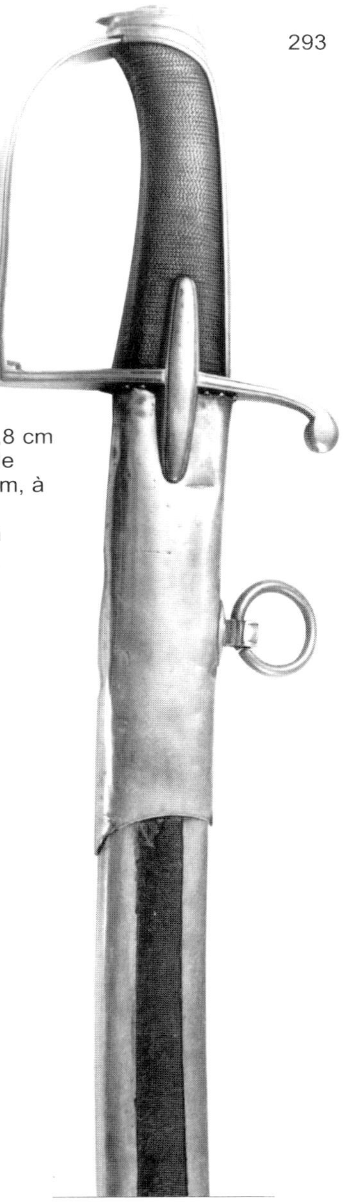

294

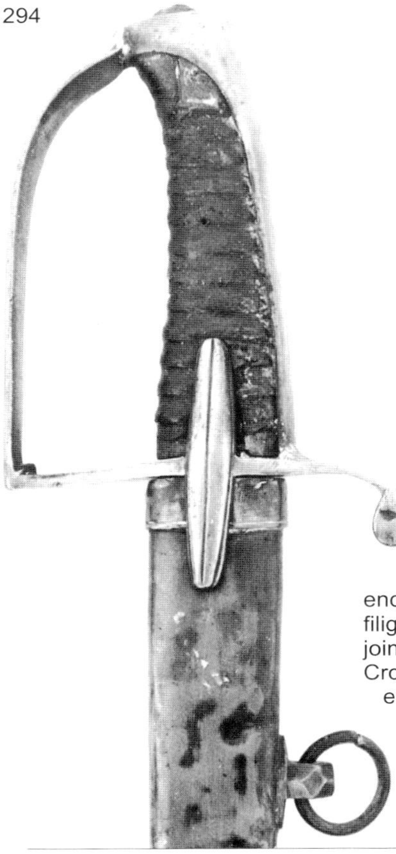

294 - Sabre de Cavalerie légère ou d'Artillerie montée.
Période révolutionnaire.
Monture en laiton. Calotte dite en cuillère, à longue queue unie. Poignée encordée, recouverte de basane, sans filigrane (peut-être disparu ?). Arc de jointure droit dans ses 2/3 inférieurs. Croisière terminée par un quillon infléchi, en goutte d'eau. Oreillons en olive très allongée à rainure médiane. Fourreau en laiton portant deux anneaux de bélière directement soudés sur le dos. Lame gravée de motifs classiques.

295 - Sabre d'officier de Hussards.
Vers 1790 - 1800.
Calotte à bouton de rivure en losange, débordant la poignée prolongée par une longue queue. Branche élargie près de la calotte, plate, rejoignant la croisière par un angle droit. Fins oreillons finissant en pointe. Quillon se terminant par une pièce quadrangulaire surmontée d'un élément rhomboïde à petit bouton. Fusée ficelée recouverte de basane à double filigrane formant chaînette et faisant 24 spires. Lame presque droite à flèche de 1,5 cm, longue de 78 cm, large au talon de 3,7 cm, à pan creux, gouttière et contre-tranchant de 10 cm. Fourreau à deux garnitures, la basane fixée sur des alèses entre les deux attelles latérales de laiton, la chape et la bouterolle ; chape longue au maximum de 13 cm, gravée au trait figurant un bracelet de bélière ; bouterolle de 48 cm avec le même décor que la chape à sa partie supérieure et finissant par un dard de fer asymétrique.

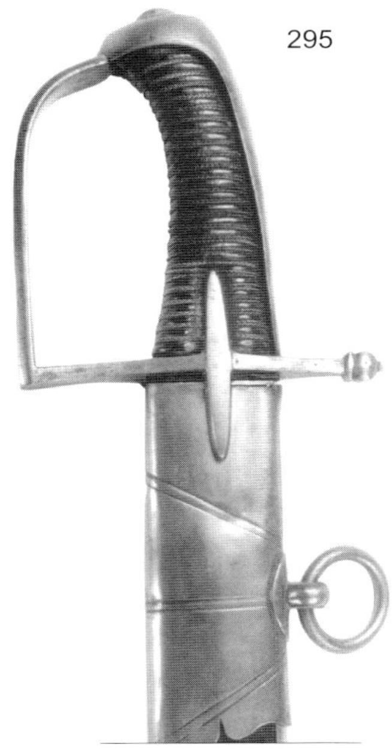
295

296
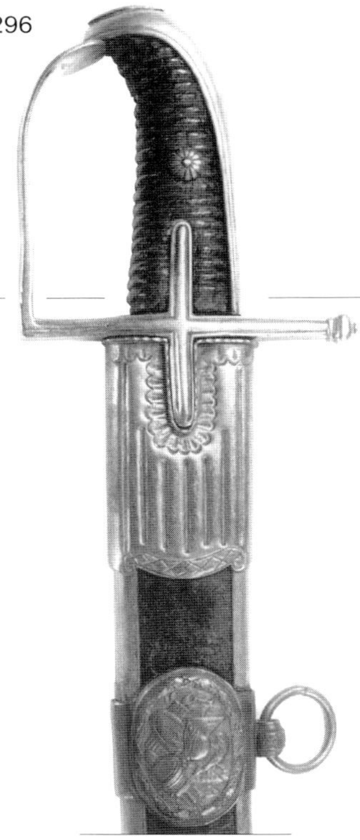

296 - Sabre d'officier supérieur de Hussards.
Règne de Louis XVI.
Calotte débordant la poignée en sa partie supérieure et se prolongeant en une longue queue ; une pièce en écusson masque la rivure de la soie. Arc de jointure nettement élargi à son départ de la calotte et s'amenuisant pour rejoindre, par un angle droit, la croisière. Fins oreillons rainurés ainsi que l'épaisseur de la croisière. Quillon droit, terminé par une pièce quadrangulaire à petit bouton, Poignée encordée, recouverte de basane avec un clou en tête de fleurette. Lame à flèche de 5,8 cm, longue de 88,5 cm et large au talon de 3,5 cm ; pan creux peu profond, à gouttière dorsale sur 39 cm environ et contre-tranchant de 24,5 cm. Fourreau à alèses recouvertes de basane, à quatre garnitures à décor estampé : chape à bordure ciselée de motifs foliacés autour de la gouge recevant l'oreillon basal, de motifs géométriques à sa base et de sept profondes rainures ; deux bracelets ovales bordés d'une frise de feuilles de laurier avec comme motif central un trophée composé de casque empanaché, carquois, bouclier,... sur fond de quatre drapeaux et de piques (296a, 296b,) ; longue bouterolle de 31 cm à frise identique à celle de la chape puis sept longues et profondes rainures (296c) . Longues attelles de laiton. Dard en fer, asymétrique.

296a

296b

296c

- **a**, bracelet supérieur au casque ;
- **b**, bracelet inférieur où le casque est remplacé par une timbale ;
- **c**, bouterolle cannelée et dard de fer asymétrique.

297

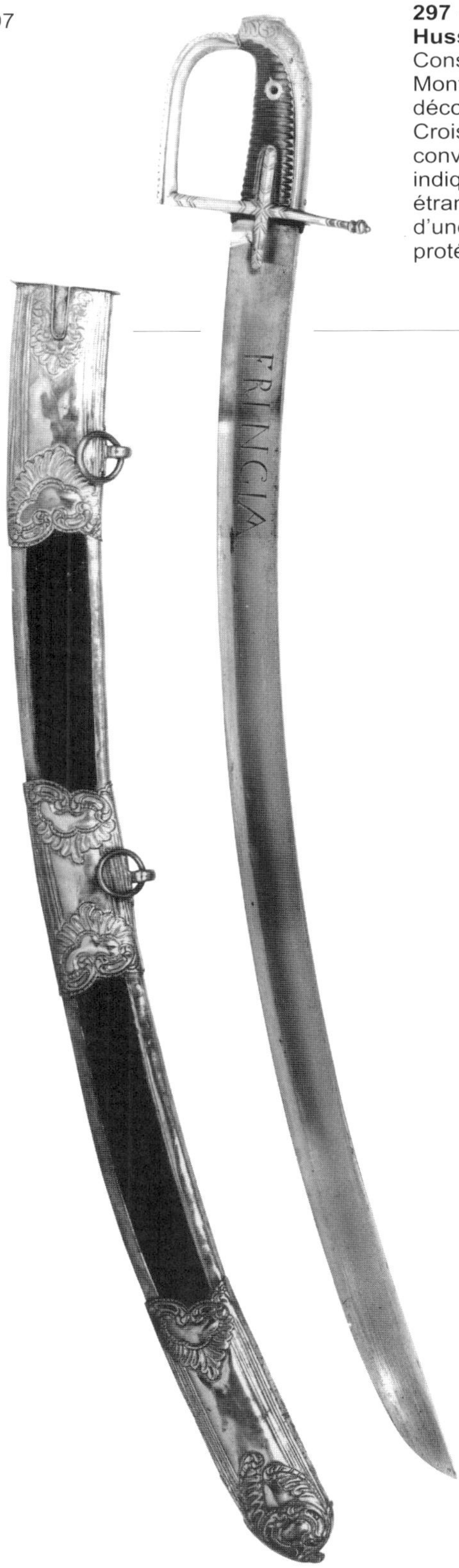

297 - Sabre d'officier supérieur de Hussards.
Consulat.
Monture en laiton. Calotte et longue queue décorées de ciselures. Arc de jointure perlé. Croisière et oreillons ornés de doubles lignes convergeantes. Lame marquée *FRINCIA*, (s c), indiquant peut-être que l'officier était d'origine étrangère. Fourreau à trois garnitures ornées d'une sorte de palmette ; attelles latérales protégeant le cuir ; petit dard frontal et basa .

298

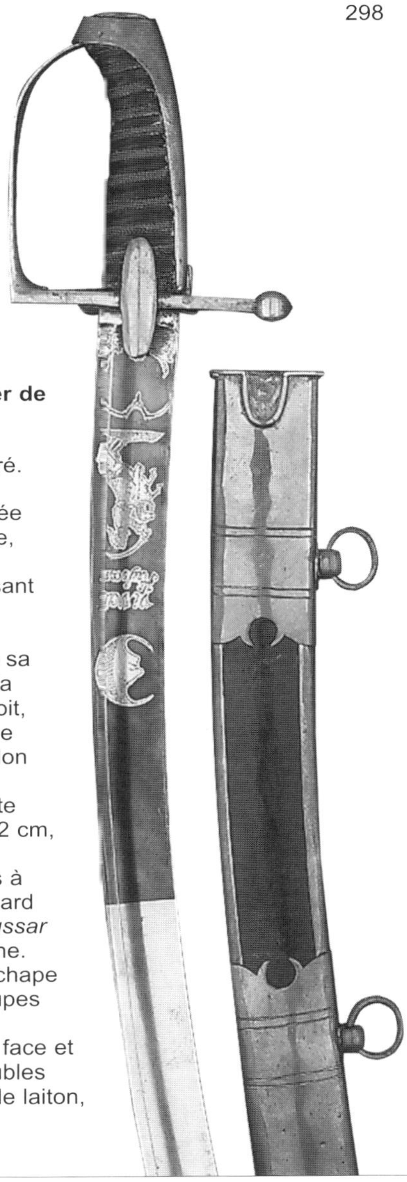

298 - Sabre d'officier de Hussards.
Fin XVIIIème siècle.
Monture en laiton doré. Calotte en cuillère à longue queue. Poignée recouverte de basane, filigranée de trois fils torsadés jumelés faisant 10 spires. Arc de jointure droit sur les deux tiers basaux de sa longueur, rejoignant la croisière en angle droit, celle-ci rectiligne et se terminant par un quillon en boule. Oreillons ovoïdes à légère arête médiane. Lame de 82 cm, bleuie et gravée des motifs dorés suivants à partir du talon : Hussard chargeant, *Vivat / Hussar* et un croissant de lune. Fourreau de laiton : chape et bouterolle à découpes assimilables à des croissants se faisant face et gravées de deux doubles traits. Deux attelles de laiton, (Sotheby's, Sussex).

299 - Sabre de Chasseur à cheval de la Garde Impériale
Monture en laiton. Calotte, à bouton de rivure, prolongée d'une longue queue. Poignée recouverte de cuir de veau, filigranée d'un fil de laiton torsadé faisant treize spires. Branche arquée, rejoignant la croisière par un angle droit. Oreillon en ovale très allongé, avec une arête médiane verticale. Quillon droit, emboulé. Lame de 85,5 à 86 cm, flèche de 6,2 à 6,4 cm à large pan creux. De 1803 à 1811, le crevé du fourreau apparaît entre les deux attelles de laiton réunissant la chape et la bouterolle, toutes deux portant un anneau de bélière (299a). A partir de 1811, le fourreau est d'une seule pièce avec un crevé de 17 à 18 cm (299b, 299c) ; dard en fer, petit au début et de plus grande taille au cours des ans pour atteindre sa grandeur maximum en 1813.

Marquage - Jusqu'en 1804, les lames sont marquées *M^{ture} N^{ale} du Klingenthal - Coulaux frères $Entrp^{rs}$*, parfois sans le Nationale. A partir de cette date, la signature devient *M^{ture} Imp^{le} du Klingenthal Coulaux frères.* Les sabres étant souvent montés à la Manufacture de Versailles on trouve souvent le poinçon Boutet ou VERSAILLES et le " V " pour vérification. On peut également déceler des poinçons de contrôleurs et de réviseurs, de Mouton, de Levavasseur,... Un marquage 2 RLG, sur le haut du fourreau peut signifier 2ème Régiment des Lanciers de la Garde, selon M. Pétard (1990). Bien qu'au total, environ 7000 sabres de ce modèle auraient été produits, il reste très rare.

- **a**, type 1803, à attelles indépendantes ;
- **b**, type 1811, le fourreau est d'une seule pièce.

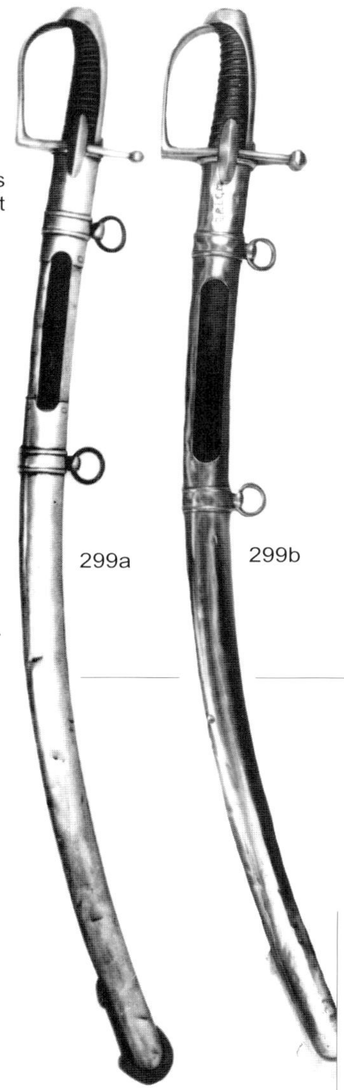
299a 299b

300

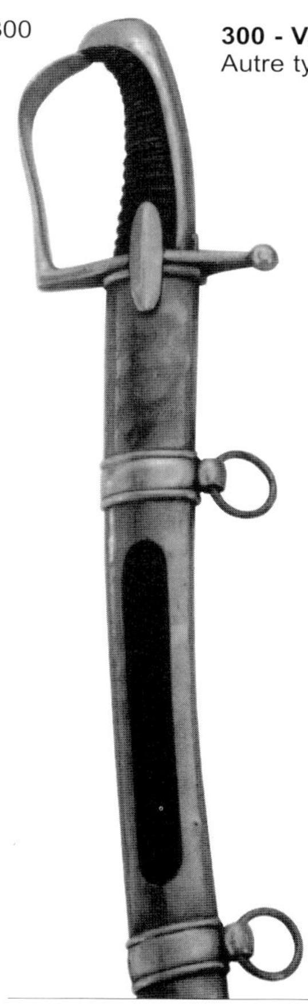

300 - Variante.
Autre type b, vers 1813.

301 - Sabre de chasseur de la Garde Consulaire.
Modèle 1800.
Monture en laiton. Calotte à dôme-bouton de rivure, à longue queue décorée d'un filet. Poignée recouverte de basane et filigranée de fil de laiton torsadé faisant 24 spires. Arc de jointure plat, courbé en sa partie supérieure et droite sur les 2/3 inférieurs. Croisière rectiligne, à filet. Oreillons ovoïdes, légèrement étranglés au niveau de la croisière. Quillon à bouton à six pans. Lame de 81,22 cm à pan creux, assez fortement courbée ; sur le dos se trouve la signature : *Mf^{ture} de Klingenthal.* Cette lame est montée à Versailles comme en témoigne une barrette au nom de Boutet et le " V " de la vérification. Ce sabre non figuré ne fut en service que pendant deux ans et remplacé par le modèle 1802.

302

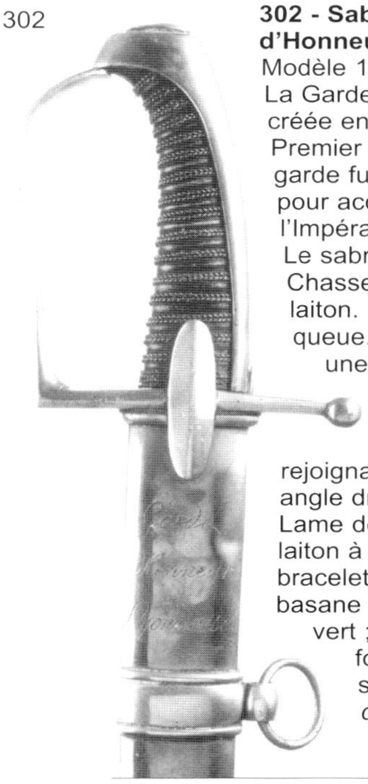

302 - Sabre de la Garde d'Honneur Lyonnaise.
Modèle 1811.
La Garde d'Honneur de Lyon fut créée en 1802 pour le séjour du Premier Consul Bonaparte. Cette garde fut réorganisée en 1805 pour accueillir l'Empereur et l'Impératrice Joséphine.
Le sabre figuré est du type Chasseur à cheval. Monture en laiton. Calotte simple et longue queue. Poignée filigranée sur une basane teintée en vert. Arc de jointure largement arrondi en sa partie supérieure et rejoignant la croisière par un angle droit. Oreillons ovoïdes. Lame de 85,9 cm. Fourreau en laiton à long crevé entre les bracelets laissant voir une basane teinte en rouge ou en vert ; entre la bouche du fourreau et le bracelet supérieur, on lit : *Garde / d'Honneur / Lyonnaise*.

304

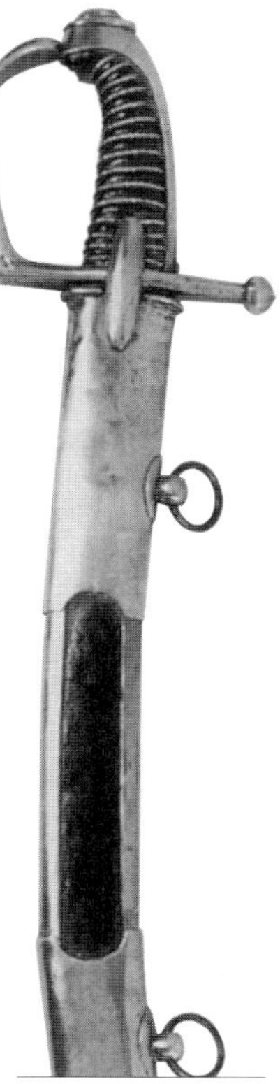

304 - Sabre de chasseur à cheval de la garde consulaire.
Epoque du Consulat.
Garde en laiton à une branche, poinçonnée de la barette : *" Boutet "* et d'un *" V "*. Lame poinçonnée du *" M "* de Mouton surmonté d'une roue et du *" LW "* de Levavasseur et marquée sur le dos : *" Mture du Klingenthal "*. Fourreau de bois recouvert de cuir, à deuxgarnitures et attelles en laiton.

305

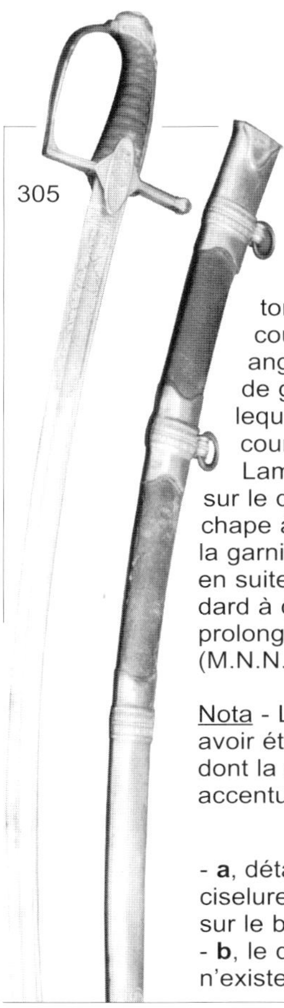

305 - Sabre de Trompette Major.
Premier Empire.
Monture en laiton, à la hussard. Calotte décorée de feuillage ainsi que la longue queue. Poignée encordée, recouverte de basane et filigranée d'un double fil de laiton torsadé disposé en chaînette. Arc de jointure courbé en sa partie supérieure et venant en angle droit sur la croisière. Celle-ci est pourvue de grands oreillons constitués d'un losange sur lequel est appliqué un ovale ciselé d'une couronne de laurier et d'une trompette (305a). Lame courbe gravée à l'acide d'un décor foliacé sur le quart basal. Fourreau à trois garnitures : une chape à bracelet ciselé d'une frise d'oves ainsi que la garniture médiane ; longue bouterolle à bracelet en suite, gravée de motifs foliacés à son extrémité et dard à deux enroulements ajourés dont l'antérieur est prolongé pour assurer la protection du fourreau (305b). (M.N.N. de Juan-les-Pins).

Nota - Le sabre de Trompette Major ne semble pas avoir été strictement réglementé. Il existe des modèles dont la courbure de la lame est plus ou moins accentuée et le fourreau peut être entièrement de laiton.

- **a**, détail de la monture. Noter la forme de l'oreillon et la ciselure du bracelet de la chape, motif que l'on retrouve sur le bracelet et la bouterolle ;
- **b**, le dard vu de la face postérieure et montrant qu'il n'existe aucune ciselure sur le dos du fourreau.

305a

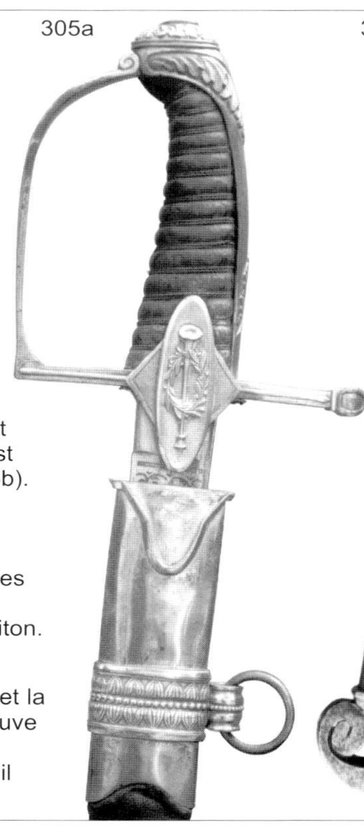

305b

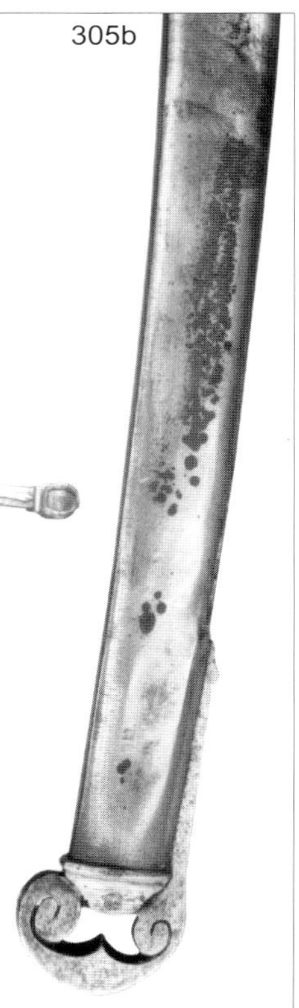

Sous-famille VI c – Oreillons d'inégale longueur

Dans cette Sous-famille on trouvera un sabre du Régiment des Volontaires Etrangers de Clermont Prince puis des sabres de Hussards dont les oreillons supérieurs sont d'inégales longueurs.

306

306 - Sabre d'officier d'Infanterie du Régiment des Volontaires Etrangers de Clermont-Prince.
1759 - 1766.
On doit à Gérard Jaeger (1998) une description très précise de ce sabre qui, selon l'auteur, a été fabriqué entre 15 à 40 exemplaires. Il répond à la description suivante : Monture en fer. Calotte en cuillère à longue queue. Fort bouton de rivure cylindrique. Poignée entièrement filigranée d'un fil de laiton torsadé disposé en chaînette. Arc de jointure courbé en sa partie supérieure puis droit sur les 2/3 de sa longueur. Croisière droite à oreillons passant sous la croisière, dont la partie supérieure est plus longue que la partie inférieure (306). Quillon se terminant par une petite pyramide. Lame de 68,2 cm, à pan creux à extrémité en langue de carpe et marquée, à partir du talon :
Régiment / Des volontaires / De Clermont / Prince en petites lettres, puis en grandes lettres manuscrites de fantaisie *Vive le* (Armes de France) *Roy.*
Fourreau de cuir à trois garnitures de laiton : courte chape à anneau de bélière, bracelet triangulaire avec le second anneau, petite bouterolle à dard en laiton asymétrique.

307 - Sabre de Hussard
1792 - 1793.
Monture de laiton. Poignée encordée recouverte de basane et filigranée d'un fil de laiton torsadé. Oreillons en olive allongée, la partie inférieure plus longue que la partie supérieure, sur une croisière rectiligne. Lame à pan creux, poinçonnée de J.G. Bick et de Cabannes avec l'abréviation classique, *Rep*^{e} sur le *verso* et *Fran*^{se} sur le *recto*, longue de 86 cm. Fourreau à garnitures découpées en demi- cercle plus ou moins ouvert.

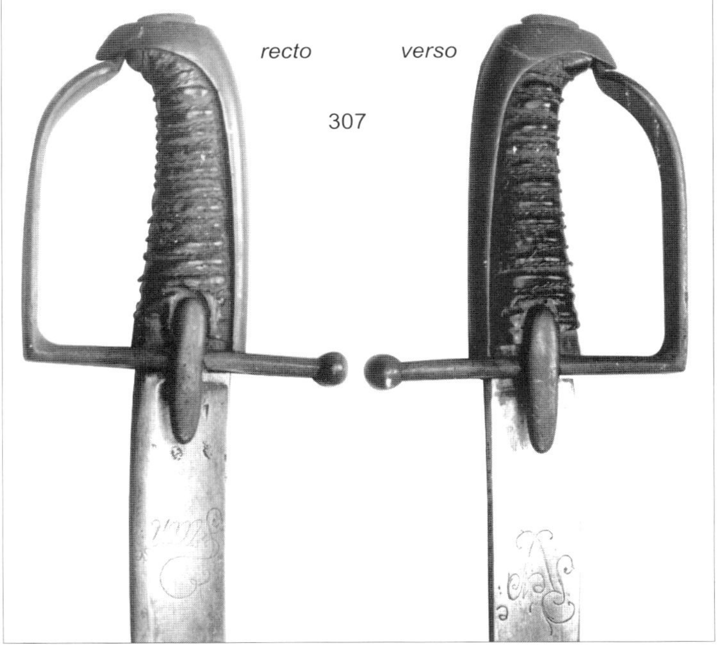

recto *verso*

307

308 - Sabre d'officier du 15ème chasseur
Période révolutionnaire.
Monture de laiton. Calotte à bouton de rivure et longue queue simple. Arc de jointure arrivant en angle droit sur la croisière portant des oreillons à extrémités arrondies, le supérieur plus court que l'inférieur. Quillon infléchi et à bouton. Lame demi-courbe gravée de feuillages, à médaillon constitué d'un cercle dans lequel est inscrit *15*, avec au-dessous, un faisceau de licteur puis la signature de Coulaux. Fourreau de fer à deux bracelets de bélière du même métal.

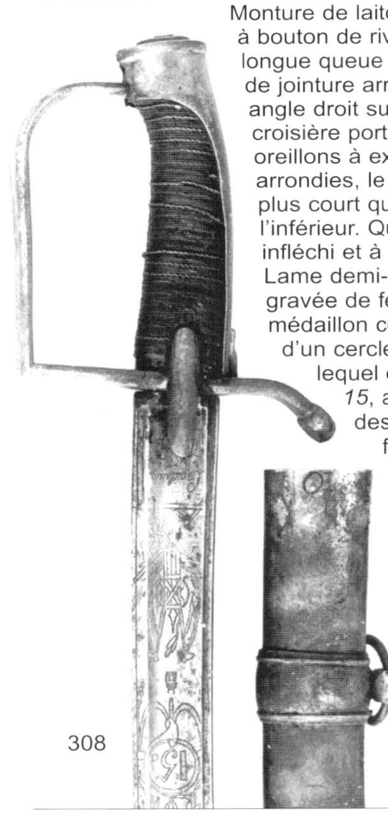

308

Les sabres de Hussards du comte Saint-Germain

Le modèle de 1786 qui devait, moyennant certains aménagements, être en service jusqu'à la fin du Premier Empire est le descendant du modèle du 19 août 1776 créé par le comte de Saint-Germain. Les petites modifications que ce modèle subit au cours des ans ne sont pas toujours facilement décelables comme le montre le tableau ci-dessous.

Modèles	Longueur de la lame	Largeur au talon	Flèche	Forme de la lame
1776	81,22 cm	4,06 cm	4,51-4,74 cm	plate
1778	81,22 cm	3,83 cm	7,22 cm	pan creux
1783	81,22 cm	3,83 cm	7,22 cm	pan creux
1786	81,22 cm	3,83 cm	6,99 cm	pan creux

A noter que certains sabres d'avant 1786 possèdent un dard ne protégeant que la face frontale de la bouterolle. Mais toutes les autres parties de la monture sont déjà bien voisines du modèle 1786.

309 - Sabre de Hussard.
Modèle 1776.
Monture en laiton. Calotte plate surmontée d'une pièce de rivure et à longue queue. Poignée de bois ficelé recouvert de basane encollée, sans filigrane, forée d'un œillet pour passer la dragonne dans les premières fabrications. Arc de jointure arqué en sa partie supérieure et droite avant de rejoindre perpendiculairement la croisière. Croisière droite terminée par un quillon à bouton. Oreillons plus courts en leur partie supérieure qu'en leur partie inférieure. Lame plate de 81,2 cm, à flèche de 4,51 cm, large au talon de 4 cm, ornée à sa base d'une fleur de lys (310a). Fourreau à alèse de bois recouvert de cuir noirci, à deux longues garnitures : chape mesurant 15,1cm du côté antérieur et seulement 12,6 cm du côté postérieur ; bouterolle longue de 51,5 cm du côté antérieur et de 46,70 du côté postérieur. Dard frontal en fer. Poids de 2 kg.

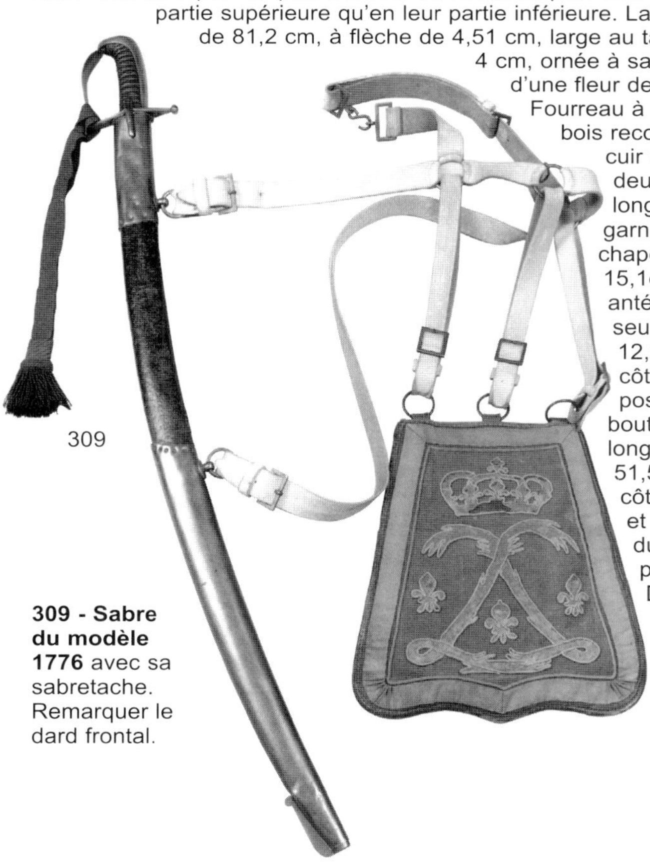

309

309 - Sabre du modèle 1776 avec sa sabretache. Remarquer le dard frontal.

310 - Sabre de Hussard.
Modèle 1778.
La lame est à pan creux.

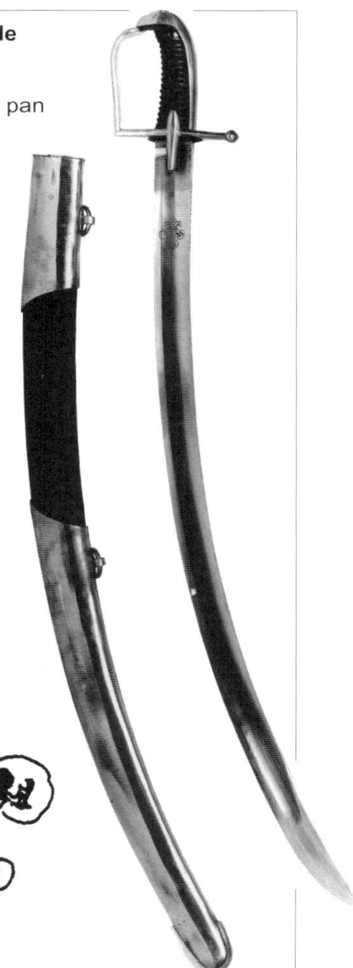

310a

- **a**, détail de la fleur de lys.

311

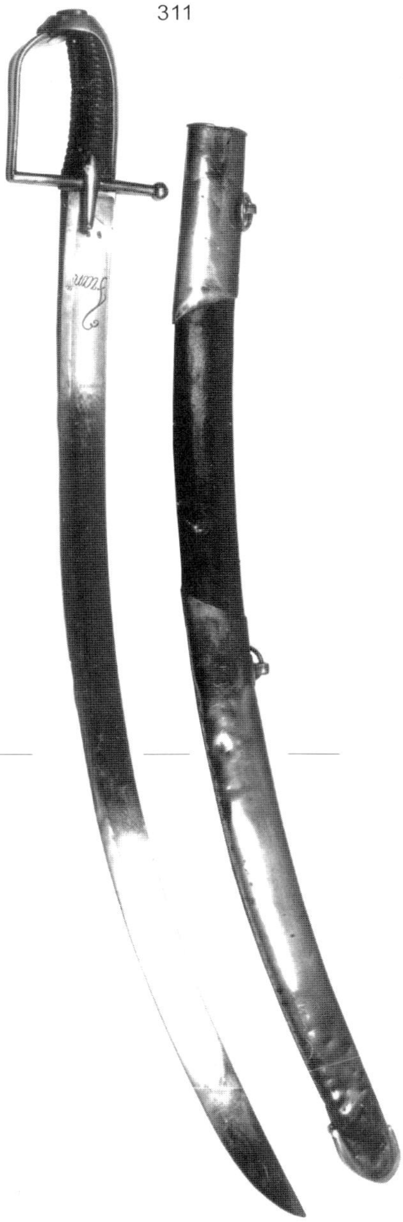

311 - Modèle de l'An IV (1795).
Monture de laiton très comparable à celle du modèle 1786. Le sabre de l'An IV se reconnaît à l'inscription République Française en lettres de fantaisie en abrégé : *Rep*^{e} / *Fran*^{se} en place de la fleur de lys du modèle 1786 (311a).

Nota - Ce sabre étant fabriqué dans une période révolutionnaire difficile et l'approvisionnement en cuivre faisant défaut, on rencontre ce modèle avec un fourreau à garnitures de fer (311b). Certains émettent l'hypothèse que c'est une caractéristique du modèle attribué à l'Artillerie montée. Nous aurons garde de prendre partie...

311a

- a, *Rep*^{e} *Fran*^{ce}
Abréviation pour République Française figurant sur le talon des sabres de Hussard de l'An IV. A noter que *REPe* se trouve sur le *verso* de la lame alors que *FRANse* se trouve sur le *recto*.

311b

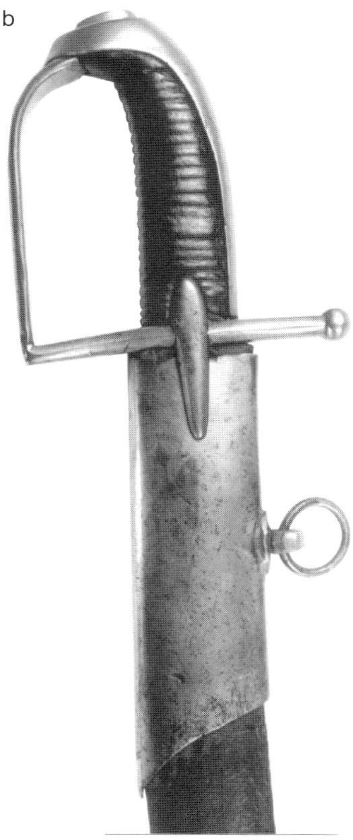

311b - Modèle de l'An IV dont les garnitures sont de fer.

312

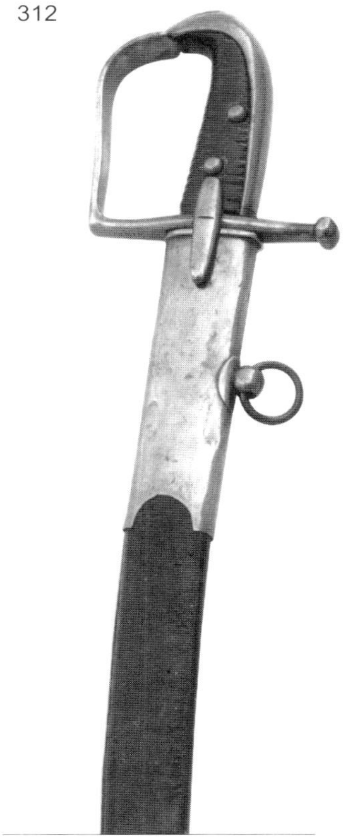

312 - Modèle de l'An IX.
Le sabre de l'An IX fut bien mal défini par une *Commission des Armes* tenue en août et septembre 1800. M. Pétard écrit en 1995 : " *Un nouveau modèle de sabre est arrêté pour les Hussards, mais le rapport ne nous indique que deux détails : la calotte est à pans et la lame mesure 84 cm. C'est tout !* "
Monture de laiton. Calotte en cuillère à longue queue, unie. Poignée ficelée recouverte de basane non filigranée à deux clous. Arc de jointure d'abord courbé, puis droit et faisant un angle de 90° avec la croisière également rectiligne se prolongeant par le quillon terminé par un bouton. Fixés sur la croisière, des oreillons dont la partie supérieure mesure 2 cm et la partie inférieure, plus fine, mesure 3,5 cm. Lame de 86 cm, à pan creux, courbée, à flèche de 7,2 cm, souvent marquée République Française et signée *Klingental*. Fourreau de bois recouvert de basane, à deux longues garnitures de laiton échancrées en suite ; anneaux de bélière soudés directement vers le milieu de la garniture supérieure et au sommet de la garniture inférieure, celle-ci terminée par un dard asymétrique de fer.
Ce modèle ne sera *réglementaire* que pendant deux ans. En effet, dès l'An XI, officiellement, les Hussards reçoivent le sabre de Cavalerie légère. Mais ce règlement n'est guère suivi et la troupe restera le plus longtemps possible fidèle à son sabre de l'An IX.

313 - Sabre de Hussard.
Période révolutionnaire. An IX. Variante.
Monture en fer. Calotte en cuillère à longue queue, unie. Poignée en bec recouverte de basane, non filigranée et ornée de deux clous. Arc de jointure élargi en sa partie courbe et s'amenuisant en rejoignant la croisière terminée par un quillon emboulé. Oreillons d'inégale longueur, le supérieur étant plus court que l'inférieur. Lame légèrement courbée. Fourreau à deux garnitures unies : chape découpée en demi-lune et anneau de bélière maintenu par un piton ; longue bouterolle portant à sa partie supérieure le second anneau de bélière. Attelles de laiton réunissant chape et bouterolle.

313a

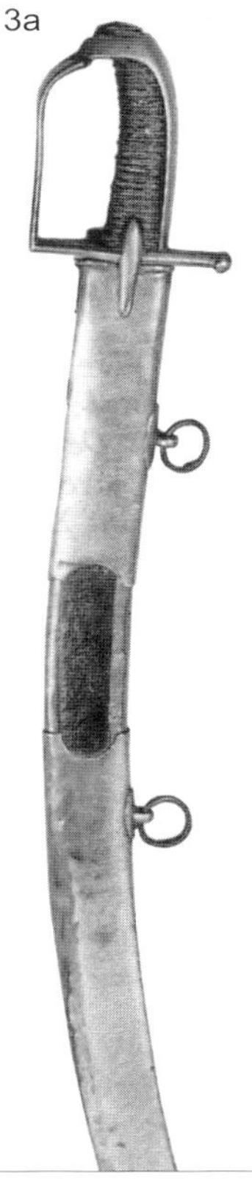

313b

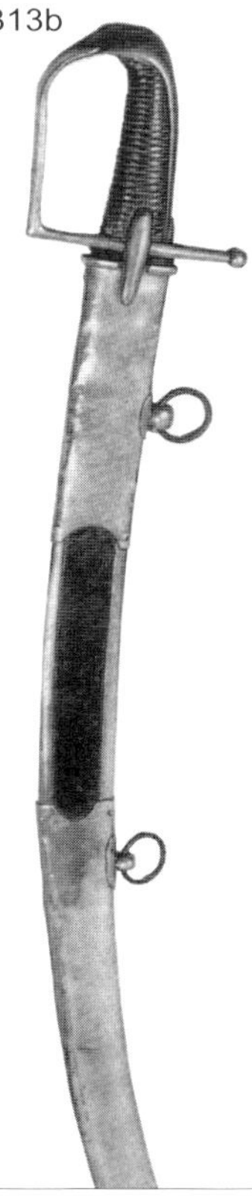

313c

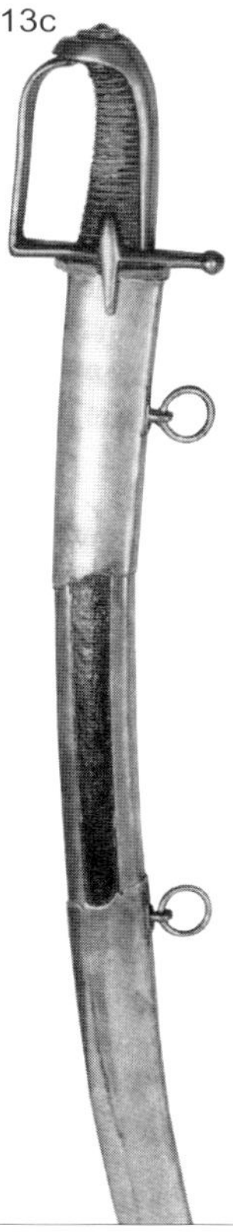

Différentes variations du fourreau du Modèle de l'An IX.
Remarquer la longueur variable des attelles latérales laissant apparaître le cuir sur des longueurs différentes.
- **a**, garde gravée SIMONET lame avec *Rep*^{e} *Fran*^{se} et sur le dos KLINGENTHAL ;
- **b**, lame signée sur le dos *M*^{ture} *de Klingenthal.* Garde poinçonnée de de Cabannes et du " B " surmonté du bonnet phrygien de J.G. Bick ;
- **c**, garde poinçonnée de F.A. et L.A. Gau. Sur le dos de la lame : T. Craven.

314

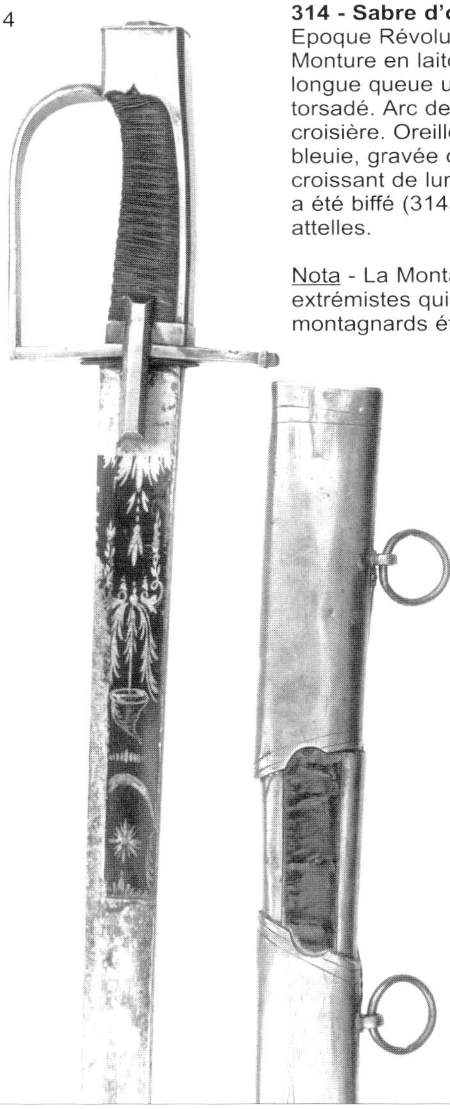

314 - Sabre d'officier de Hussards.
Epoque Révolution.
Monture en laiton. Calotte plate, quadrangulaire à bordure découpée en accolade, longue queue unie. Poignée recouverte de basane à une douzaine de spires d'un fil torsadé. Arc de jointure droit sur les 3/4 de sa longueur et arrivant en angle droit sur la croisière. Oreillons en baguette plus longue sur la partie placée sous la croisière. Lame bleuie, gravée de motifs dorés représentant des feuillages, un bonnet phrygien, un croissant de lune... et portant sur l'autre face : *Vive La Montagne* mais ce dernier mot a été biffé (314a). Fourreau à deux longues garnitures de laiton réunies par de courtes attelles.

<u>Nota</u> - La Montagne est le nom donné aux députés de la Convention les plus extrémistes qui siégeaient dans la partie haute de l'assemblée. Les chefs des montagnards étaient Robespierre, Saint-Just et Marat.

314a

- a, *verso* de la lame où le mot montagne a été biffé, ne laissant en clair que Vive / La ...

315

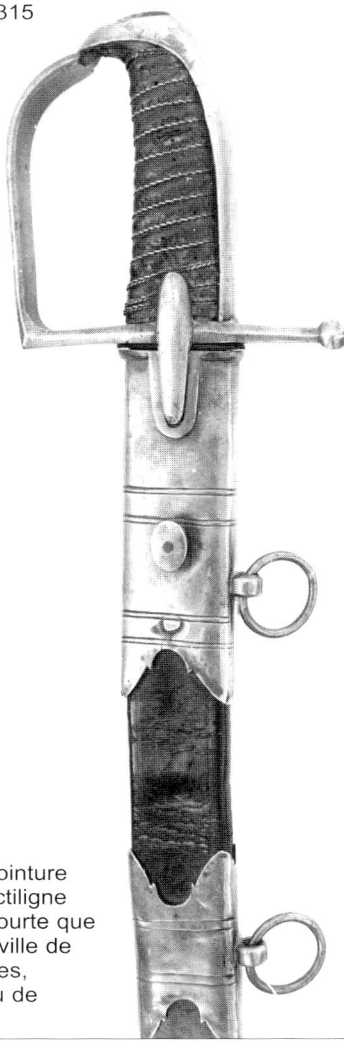

315 - Sabre d'officier de Hussard.
Epoque Révolution.
Monture en laiton. Calotte en cuillère à longue queue unie. Poignée recouverte de basane à 12 - 13 spires d'un filigrane torsadé. Arc de jointure élargi en sa partie courbée puis droit sur les 2/3 basaux. Croisière rectiligne terminée par un bouton. Oreillons dont la partie supérieure est plus courte que la partie basale. Lame courbée, de 73 cm, à pan creux, signée de la ville de Langres, célèbre pour sa coutellerie. Fourreau de cuir à trois garnitures, échancrées en symétrie, décorées au trait, chape à bouton et anneau de bélière.

Sous-famille VI d – Oreillons en losange plus ou moins curviligne

Les sabres de cette Sous-famille possèdent des oreillons en losange dont les côtés sont rectilignes ou nettement courbés.

316

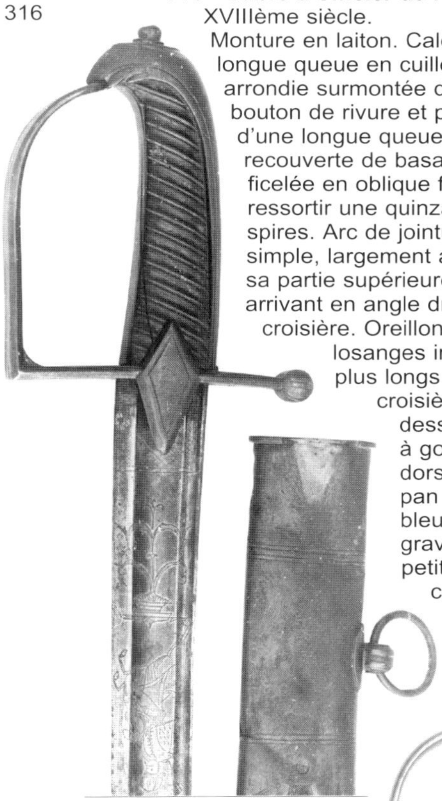

316 - Sabre d'officier de Hussards.
XVIIIème siècle.
Monture en laiton. Calotte à longue queue en cuillère arrondie surmontée d'un bouton de rivure et prolongée d'une longue queue. Poignée recouverte de basane, ficelée en oblique faisant ressortir une quinzaine de spires. Arc de jointure simple, largement arrondi en sa partie supérieure et arrivant en angle droit sur la croisière. Oreillons en losanges irréguliers, plus longs sous la croisière qu'au-dessus. Lame à gouttière dorsale et pan creux, bleuie et gravée du petit Hussard chargeant.

317

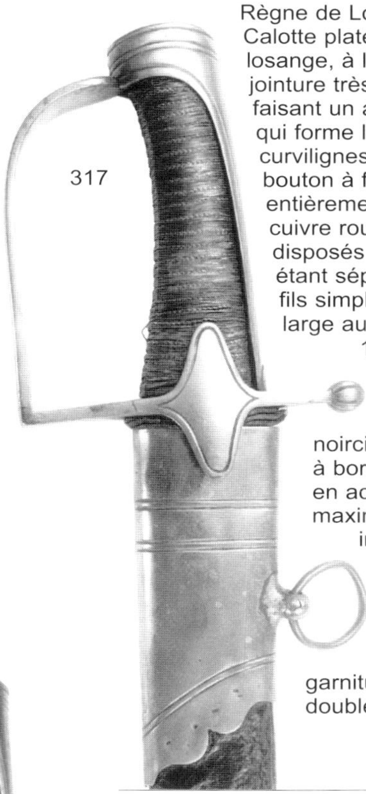

317 - Sabre d'officier de Chasseurs à cheval.
Règne de Louis XVI.
Calotte plate, à bouton de rivure en losange, à longue queue. Arc de jointure très élargi près de la calotte, faisant un angle droit avec la croisière qui forme les oreillons en losanges curvilignes. Quillon terminé par un bouton à facettes. Poignée entièrement filigranée de fils de cuivre rouge : deux fils torsadés disposés en chaînette, chaque spire étant séparée de la suivante par des fils simples. Lames de 74,5 cm, large au talon de 3,5 cm, flèche de 1,5 cm, contre-tranchant de 15 cm et gouttière de 30 cm. Fourreau à alèse de bois recouvert de cuir noirci, à trois garnitures : chape à bordure oblique et découpée en accolade, avec une longueur maximum de 12 cm, bracelet intermédiaire trapézoïdal découpé ainsi que la chape et la bouterolle longue de 23,5 cm ; dard uniquement placé sur le côté antérieur. Toutes les garnitures sont décorées de doubles traits.

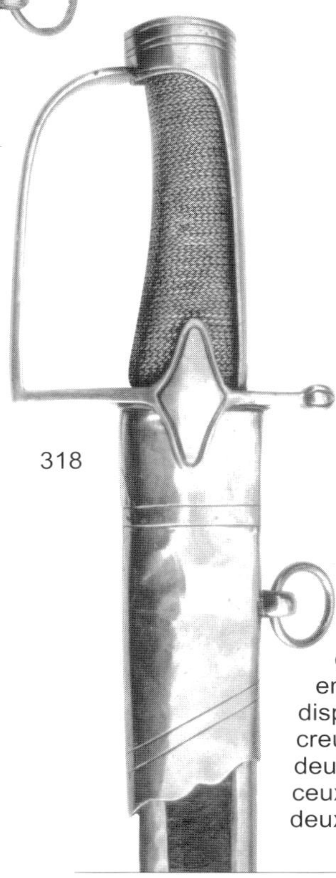

318

318 - Sabre d'officier de Chasseurs à cheval.
Règne de Louis XVI.
Monture en laiton. Remarquer la calotte à filets ainsi que la longue queue et surtout la poignée entièrement filigranée d'un fil torsadé disposé en chaînette. Lame à pan creux. Fourreau à trois garnitures et deux anneaux de bélière avec, entre ceux-ci, le cuir apparaissant entre les deux attelles latérales.

319 - Sabre d'officier de Chasseurs à cheval.
Règne de Louis XVI. Monture en laiton. Calotte à longue queue, en cuillère. Poignée recouverte de basane à filigrane simple faisant dix huit spires. Arc de jointure arrivant en angle droit sur la croisière, celle-ci élargie en un losange légèrement curviligne en guise d'oreillon. Quillon terminé par une petite bague suivie d'une boule à petit piton. Lame à pan creux avec gouttière au fort et ornée, à partir du talon, de motifs foliacés stylisés, d'un trophée où l'on reconnaît un clairon, un canon, des lances, d'un autre motif végétal et d'un soleil aux rayons droits et sinueux alternés.

319

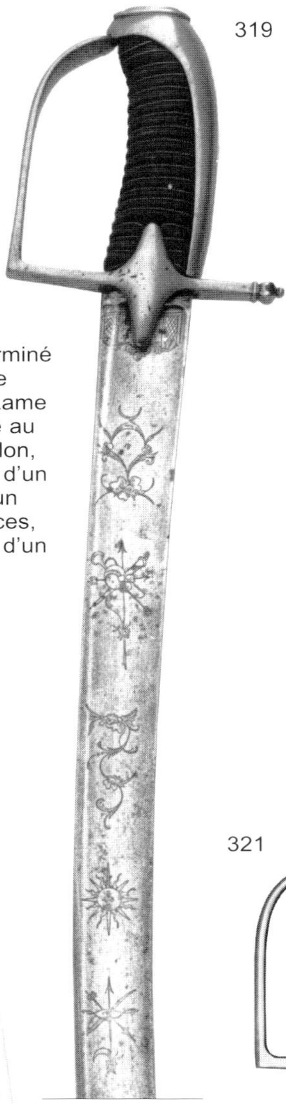

320 - Sabre de Chasseurs à cheval.
Modèle réglementaire de 1792, variante. Laiton. Calotte rainurée à pièce de rivure hexagonale, prolongée d'une longue queue décorée au trait. Poignée recouverte de basane, à fils torsadés faisant une quinzaine de spires. Arc de jointure droit sur les 3/4 de sa longueur. Croisière élargie pour former deux oreillons en losanges curvilignes. Quillon droit à bouton terminal. Lame plate, à dos poinçonnée d'une étoile. Fourreau de cuir à trois garnitures de laiton découpées en très petite accolade (M.R.).

320

321a

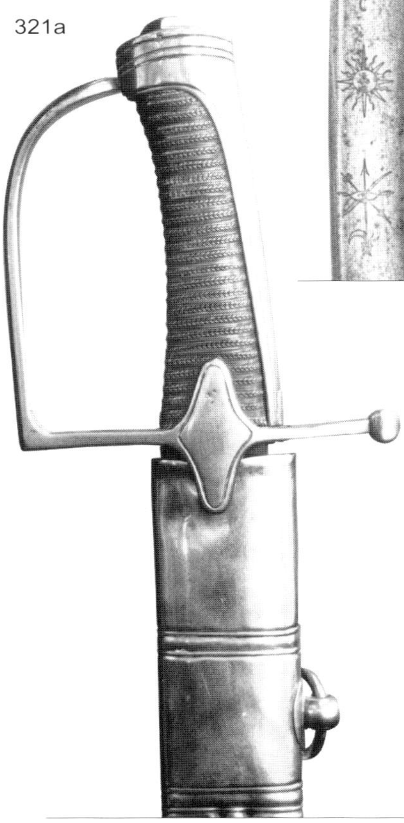

321

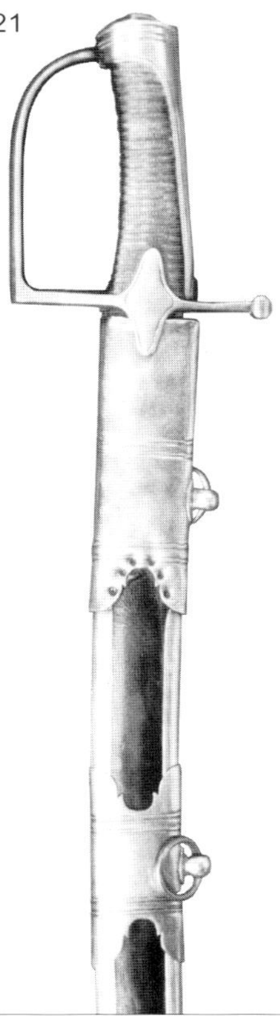

321 - Sabre d'officier de Chasseurs.
Modèle 1792.
Calotte plate, ornée de trois traits, surmontée d'un bouton de rivure en losange, et se prolongeant par une longue queue également bordée d'un trait partant de la base de la calotte. Arc de jointure élargi à son départ de la calotte et rejoignant la croisière en formant un angle droit, celle-ci à oreillons en losange curviligne souligné d'un trait. Quillon droit terminé par un bouton à quatre facettes. Poignée entièrement filigranée de fils de cuivre rouge torsadés en alternance avec un ruban (321a). Lame de 82 cm, large au talon de 3,5 cm, à flèche de 2,2 cm, à large pan creux, gouttière d'environ 43 cm et contre-tranchant de 20 cm environ. Fourreau à alèse de bois, à trois garnitures : chape longue de 15,5 cm, décorée aux traits à bord inférieur à six enfoncements pour agripper le cuir du fourreau en sa partie médiane et en sa partie basale ; pitons d'anneaux de bélière soudés ; bouterolle de 36 cm, décorée au trait ; dard en fer asymétrique dont la partie antérieure mesure 14,56 cm. Deux baguettes de laiton protégent les parties latérales du fourreau. Entre les garnitures apparaît la basane collée sur les alèses.

322

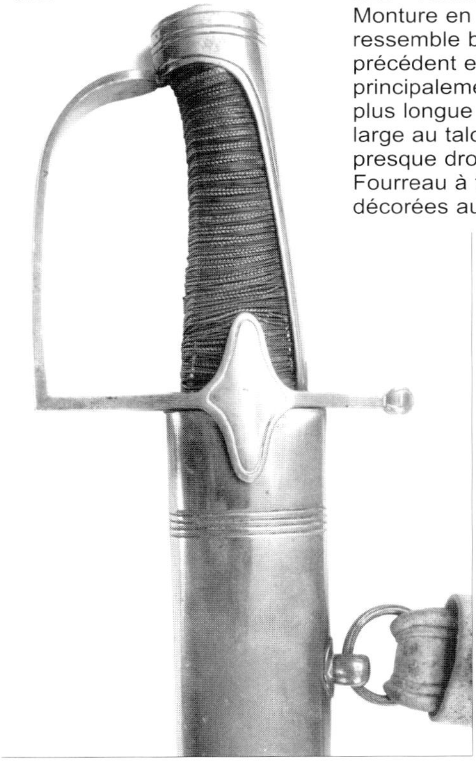

322 - Variante.
Monture en laiton. Ce modèle ressemble beaucoup au sabre précédent et s'en différencie principalement par une lame plus longue atteignant 84 cm, large au talon d'environ 4 cm, presque droite, à pan creux. Fourreau à trois garnitures décorées au trait.

323

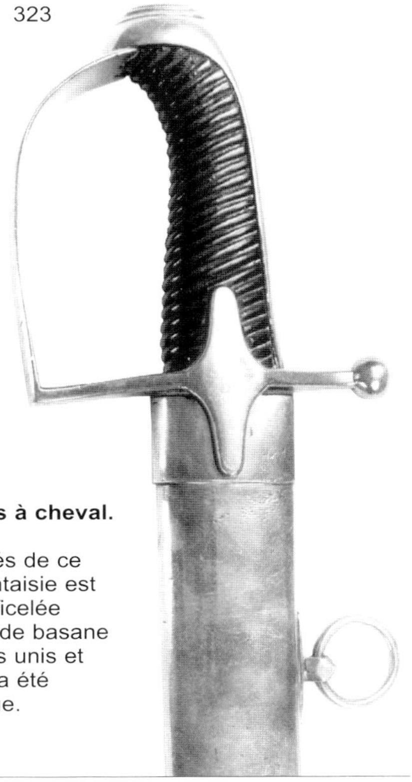

323 - Sabre de Chasseurs à cheval.
Variante du modèle 1792.
Les principales particularités de ce sabre apparemment de fantaisie est de posséder une poignée ficelée obliquement et recouverte de basane non filigranée, des oreillons unis et un fourreau dont la chape a été renforcée d'une large bague.

324 - Sabre d'officier de Chasseurs à cheval.
Vers 1792.
Monture en laiton. Calotte plate à double rainures et longue queue. Poignée filigranée. Arc de jointure droit sur les 3/4 de sa longueur et arrivant en angle droit sur la croisière. Oreillons en triangle curviligne. Lame à pan creux portant sur le *recto Vivent les Etats* et sur le *verso, Vivent les Patriotes* (324a), ces deux inscriptions se terminant par le triangle maçonnique. Fourreau à trois garnitures, à bordures découpées en accolade et à rainures obliques. Entre ces garnitures, le cuir noir apparaît bordé d'attelles de laiton. Dard en fer asymétrique. De part et d'autre de la chape un " R " majuscule et un " F " également en majuscule pour République Française.

324, *recto*

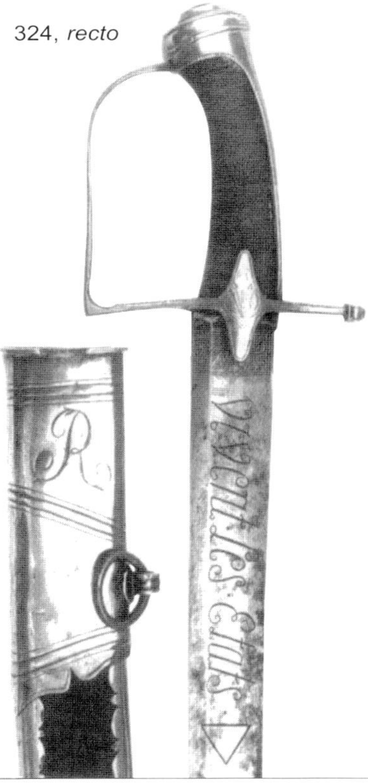

324a, *verso*

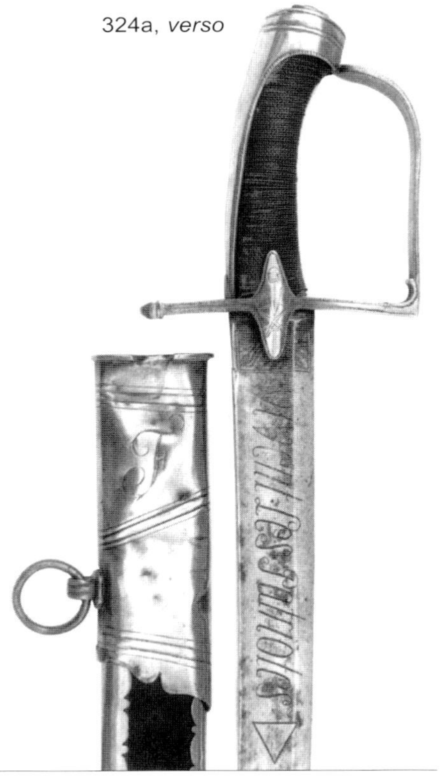

325

325 - Sabre d'officier de Chasseurs à cheval. Période révolutionnaire. Monture en laiton. Calotte en cuillère décorée d'une dépouille de lion et à longue queue. Poignée recouverte de basane à dix neuf spires d'un fil torsadé. Arc de jointure très élargi en sa partie courbée et s'amenuisant en rejoignant la croisière, celle-ci portant des oreillons en losanges curvilignes bordés d'un trait. Quillon terminé en boule. Lame légèrement courbe, à pan creux, bleuie, au tiers et gravée de motifs dorés figurant des végétaux stylisés, un trophée, des feuillages.

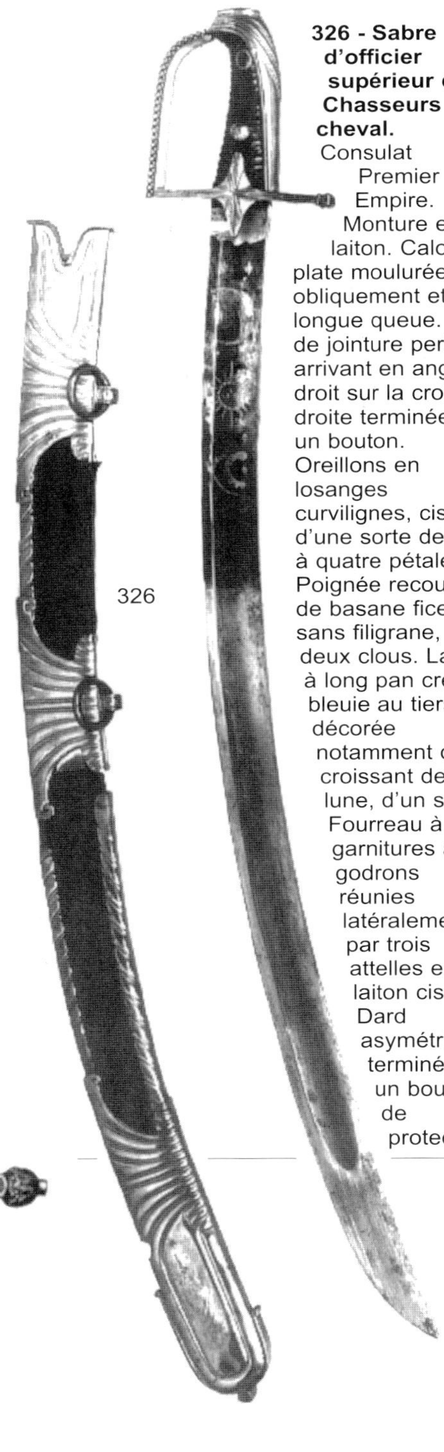

326

326 - Sabre d'officier supérieur de Chasseurs à cheval. Consulat Premier Empire. Monture en laiton. Calotte plate moulurée obliquement et longue queue. Arc de jointure perlé arrivant en angle droit sur la croisière droite terminée par un bouton. Oreillons en losanges curvilignes, ciselés d'une sorte de fleur à quatre pétales. Poignée recouverte de basane ficelée, sans filigrane, avec deux clous. Lame à long pan creux, bleuie au tiers et décorée notamment d'un croissant de lune, d'un soleil. Fourreau à trois garnitures à godrons réunies latéralement par trois attelles en laiton ciselé. Dard asymétrique terminé par un bouton de protection.

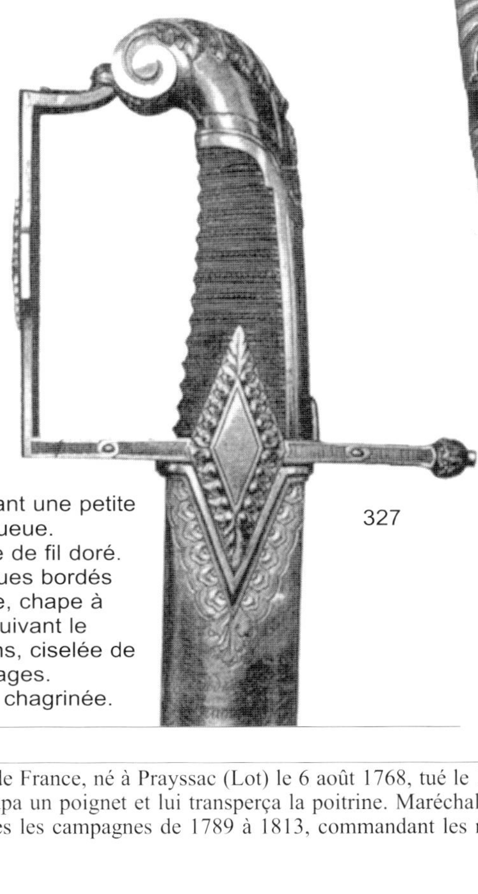

327

327 - Sabre du général Bessières (*). Consulat. Monture en laiton doré. Calotte se terminant par un enroulement formant une petite crosse à longue queue. Poignée à filigrane de fil doré. Oreillons losangiques bordés d'une frise foliacée, chape à entrée moulurée suivant le dessin des oreillons, ciselée de guirlande de feuillages. Fourreau en peau chagrinée.

(*) Bessières, maréchal de France, né à Prayssac (Lot) le 6 août 1768, tué le 1er mai 1813 par un boulet qui lui coupa un poignet et lui transperça la poitrine. Maréchal d'Empire le 19 mai 1804, a fait toutes les campagnes de 1789 à 1813, commandant les régiments les plus glorieux.

328 - Sabre d'officier des Chasseurs à pied de la Garde Impériale.

Premier Empire.

Monture de bronze doré. Calotte surmontée d'un bouton de rivure conique ciselée de feuilles, à bordure moulurée et longue queue ornée d'une frise verticale de feuilles de laurier. Poignée de bois recouvert de galuchat, filigranée souvent de fil d'argent. Arc de jointure arrivant en angle droit sur la croisière élargie en son milieu d'un losange curviligne orné d'un buste en argent de l'empereur sur la face frontale (328a) et d'un motif ciselé en feuilles d'acanthe sur la face postérieure. Lame à rainure et pan creux, longue de 65 cm, large au talon de 3,2 cm et à flèche de 2,7 cm. Sur le *verso* de la lame on lit, dans un cartouche en lettres d'or *GARDE IMPERIALE* et sur le *recto CHASSEURS A PIED* (328c). Le cartouche encadré, à partir du talon signé Duc, un motif foliacé avec le " N " dans une couronne, l'aigle et une fleurette (328b). Fourreau à deux garnitures à double rainures de part et d'autre du bouton ovalaire. Bouterolle ornée de feuillages se terminant par un dard symétrique à trois enroulements (M.R.).

<u>Nota</u> - Ce modèle aurait été dessiné vers 1800 par Duc qui bénéficiera plus tard du titre de *Fourbisseur de la Garde Impériale*. La Manufacture de Versailles produira également ce sabre, d'où quelques petites variantes de détail. Ces variantes concernent la longueur de la lame qui peut atteindre 76 cm, la forme de la branche de garde qui peut être arquée, le quillon qui peut se terminer par une embase pyramidale précédant un bouton.

On peut rencontrer ce modèle utilisé sous la Restauration voire sous le règne de Louis-Philippe. Dans ce cas, les aigles ont été surchargées d'un trophée et *GARDE IMPERIALE* remplacée par *GARDE ROYALE* sous la Restauration, et *GARDE NATIONALE* sous Louis-Philippe alors que le *CHASSEUR A PIED* subsiste.

328

- **a**, profil de Napoléon Ier, en argent, sur l'oreillon frontal d'un sabre de la Garde Impériale. Remarquer que le profil est placé sur quatre feuille d'acanthe ;
- **b**, ornement de la lame faisant suite à l'inscription Garde Impériale ;
- **c**, *recto* et *verso* de la lame.

328b

328c

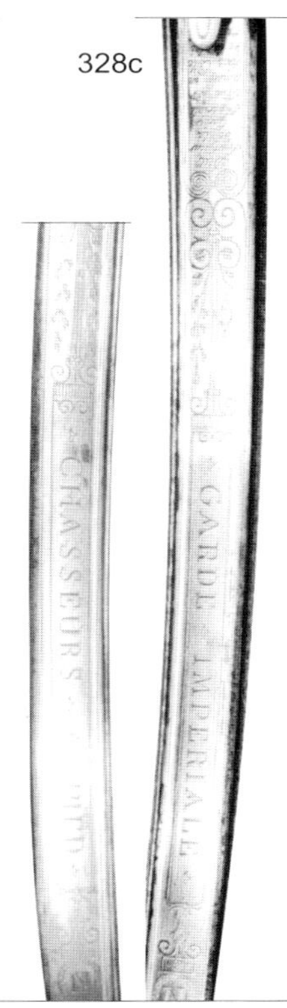

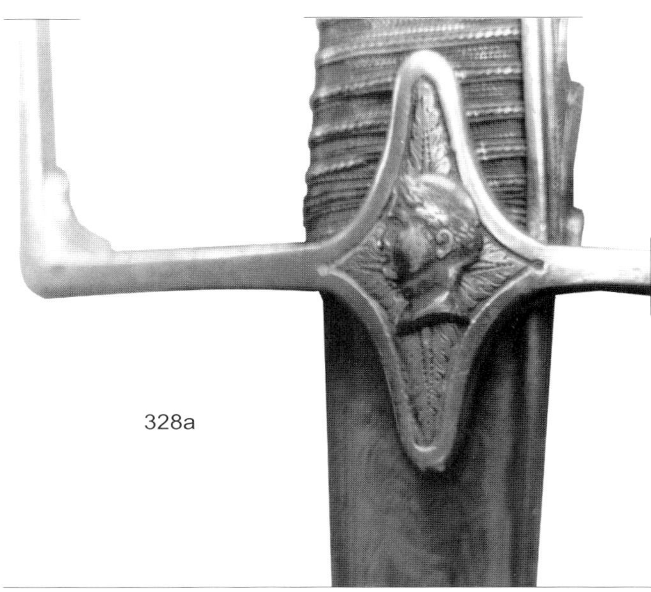

328a

329, *recto*

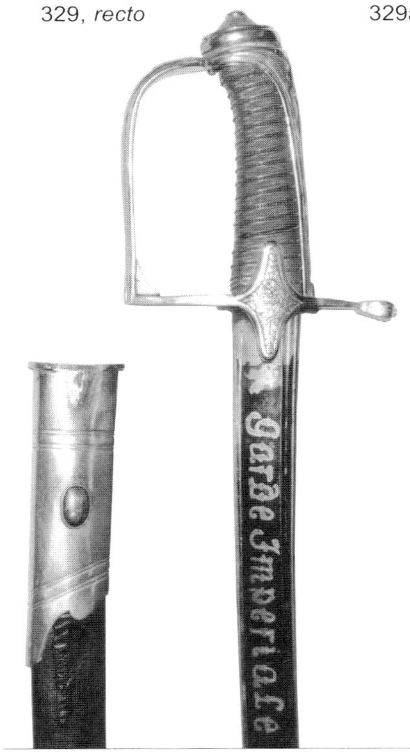

329a, *verso*

329 - Variante.
Monture classique des officiers de la Garde Impériale, mais la lame ne porte que *Garde Impériale* en caractères dorés sur fond bleui. (J.C. D.)

- **a**, *verso* : remarquer que l'oreillon postérieur est aux seules feuilles d'acanthe.

330

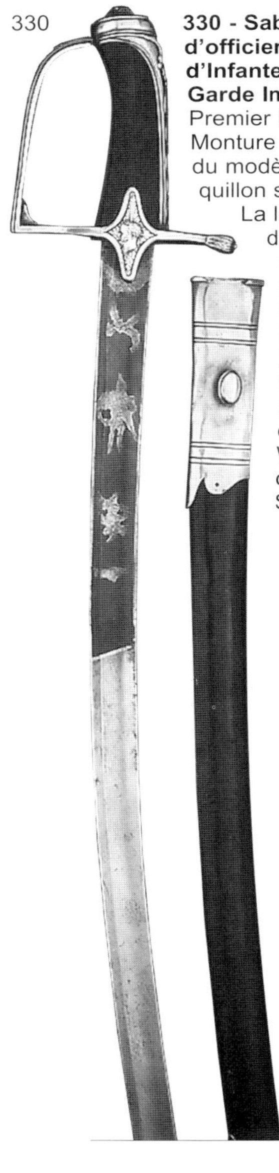

330 - Sabre d'officier d'Infanterie de la Garde Impériale.
Premier Empire.
Monture classique du modèle avec un quillon signé Duc.
La lame bleuie, de 71,3 cm ne porte aucune référence à la Garde Impériale et est poinçonnée du " W " de Weyersberg de Solingen.

331

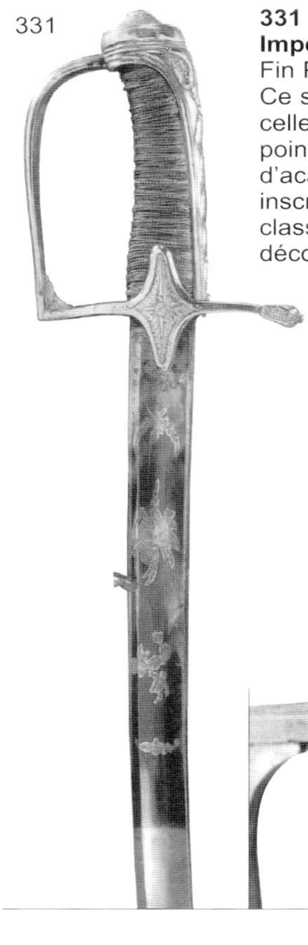

331 - Sabre d'officier d'Infanterie de la Garde Impériale dit souvent de la Jeune Garde.
Fin Premier Empire.
Ce sabre possède une monture très comparable à celle des premiers modèles, mais en diffère sur deux points : l'oreillon frontal est ciselé de quatre feuilles d'acanthe disposées en croix (331a) et la lame, sans inscription, est bleuie avec des motifs dorés classiques. Fourreau à deux garnitures de laiton décorées au trait.

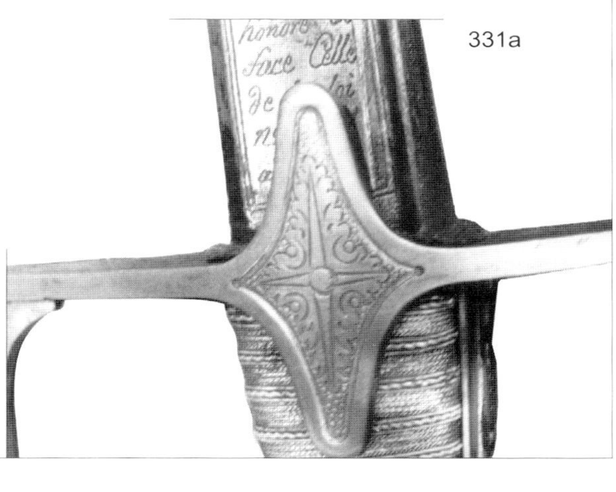

331a

332 - Sabre d'officier de Cavalerie légère.
Premier Empire.
Ce sabre est évidemment une prise de guerre sur les Coalisés germaniques lors des campagnes napoléoniennes. Il a été *naturalisé* en plaçant sur la croisière un oreillon au profil de l'Empereur généralement réservé à la Garde (J.C. D.).

332

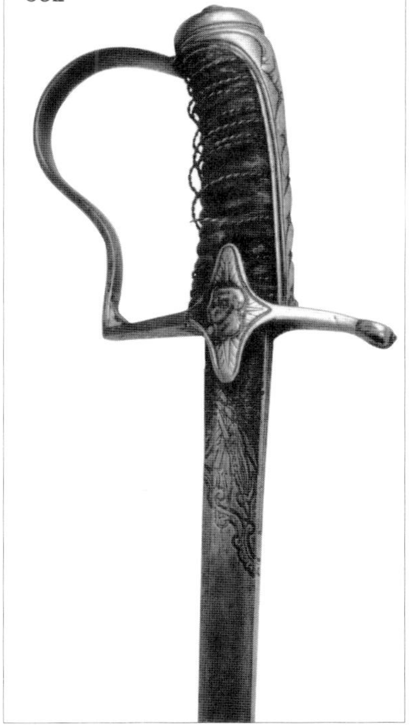

333

333 - Sabre d'Artillerie à pied de la Garde Impériale.
Premier Empire.
La monture de ce sabre en diffère sur deux points principaux : oreillons en losanges curvilignes et décorés d'une bombe enflammée ; la lame portant l'inscription *Garde Impériale*, l'adjectif *Impériale* a été effacé vraisemblablement sous la Restauration.

334 - Sabres d'officiers supérieurs des Chasseurs à cheval de la Garde Impériale.
Premier Empire.
Monture en bronze doré. Pommeau en crosse figurant une tête de lion, se prolongeant en une longue queue. Branche de garde courbée en sa partie supérieure, devenant droite pour faire un angle droit avec la croisière, quillon se terminant par une griffe d'aigle enserrant un globe. Au milieu de la croisière sont fixés des oreillons losangiformes frappés d'une tête de bacchante ou de Méduse. Poignée en ébène quadrillée. Lame bleuie et gravée de motifs dorés au 1/3, de 84 cm de longueur, à flèche de 5,9 cm à deux pans creux et se terminant en biseau ; dos en baguette marqué *" Chasseurs à cheval de la Garde Impériale "*. Fourreau en laiton doré, à deux bracelets de bélière placés au centre d'une toile d'araignée ; entre ceux-ci une aigle couronnée est ciselée ; entre le second bracelet et la bouterolle se trouve un motif foliacé inscrit dans un losange ; bouterolle très ornée avec sur le dard en acier, une étoile de bronze (M.E. S. de P.), (non figuré).

Nota - Ce sabre de très belle facture présente des variations notables : fusée filigranée, lame à un pan creux et M. Boutet écrit : *" demi-oreilles timbrées d'une tête de bacchante "*. L'étoile du dard peut être absente.
Ce sabre a été dessiné par N. Boutet et monté à Versailles et chez d'autres fourbisseurs, d'où quelques petites variations.

335

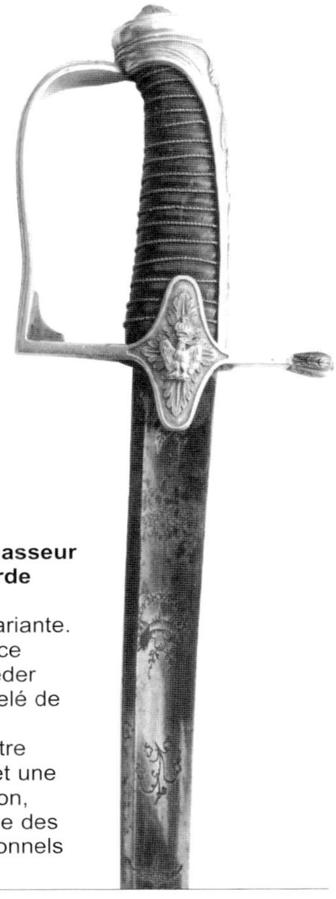

335 - Sabre de Chasseur à cheval de la Garde Impériale.
Premier Empire. Variante.
La particularité de ce sabre est de posséder l'oreillon frontal ciselé de l'aigle impériale couronnée sur quatre feuilles d'acanthe et une lame sans inscription, simplement décorée des motifs dorés traditionnels sur fond bleu.

336

336 - Sabre d'officier du 33ème Régiment d'Infanterie Légère
Sabre dérivé des modèles de la Garde Impériale. Il en diffère essentiellement par une poignée en ébène quadrillée et par l'inscription : *33ème Régiment d'Infanterie Légère* en grandes lettres manuscrites évidées. Fourreau de cuir, à chape décorée de deux doubles rainures, bracelet en suite, bouterolle à dard symétrique en laiton (M.E. - S. de P.).

337

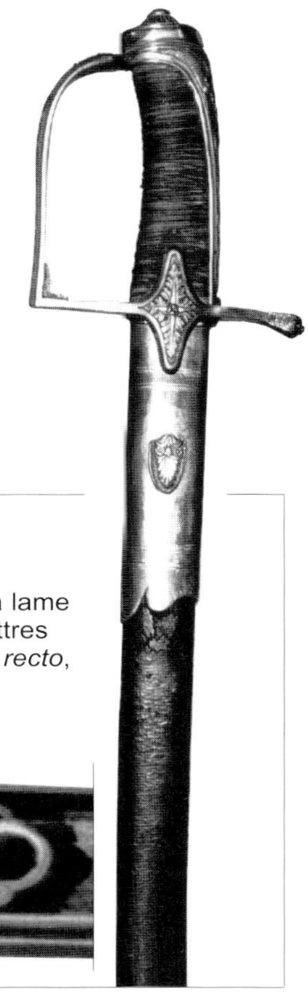

337 - Sabre de la Garde Nationale.
Louis-Philippe.
Sabre du type de la Jeune Garde Impériale. La lame porte la devise de l'époque : *ordre public* en lettres d'or sur fond bleui sur le *verso* (337a) et sur le *recto*, *garde nationale*. (J.C. D.).

337a

338 - Sabre d'officier de Cavalerie légère.
Premier Empire.
Monture en bronze argenté. Calotte en cuillère à longue queue. Arc de jointure droit sur la croisière terminée par un quillon infléchi. Oreillons en losanges curvilignes allongés, à deux stries verticales. Lame à pan creux, bleuie au tiers basal, gravée de motifs assez classiques parmi lesquels on reconnaît un soleil. Fourreau à trois garnitures en laiton argenté découpé, largement échancrées en suite, chape et bracelet munis d'un anneau de bélière (J.C. D.).

338

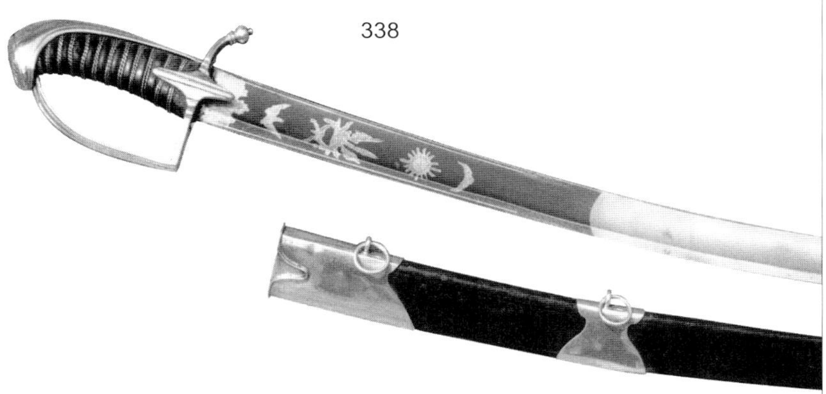

339 - Sabre d'officier de Marine

Directoire (1).

Monture en laiton doré. Calotte à bouton de rivure conique à bordure de feuilles cerclée de points ; longue queue ciselée de feuilles d'acanthe et de cinq groupes de trois feuilles (laurier) séparés par un point (fruit ?) (339a). Poignée en ébène à sept pans unis, sans filigrane. Arc de jointure courbé dans sa partie supérieure, rectiligne sur environ 9 cm et orné d'une crête qui pourrait être un cordage stylisé. Cet arc fait un angle droit avec la croisière horizontale, oreillons en losange curviligne gravé en relief, d'une ancre sur fond sablé (339b). Lame bleuie sur 24 cm avec des gravures banales dorées, longue de 73 cm, large au talon de 3,5 cm, légèrement courbée, la flèche n'étant que de 1,5 cm, à pan creux et gouttière dorsale. Fourreau de cuir noirci à deux garnitures : chape de 13,7 cm à bouton en olive, à fin liseré ; bouterolle de 16 cm à dard presque symétrique, terminé en pointe.

Marquages - Aucun marquage de manufacture ni de fourbisseur. Un exemplaire possède une lame bleuie et gravée sur le *recto* de la profession de foi : *Pour / La Loi / La / Patrie* en lettres d'or. Le *verso* est débleui et dédoré, sauf au niveau des initiales du propriétaire laissant supposer qu'il n'y a pas eu changement de main mais changement d'idée ! Sur cette face on lit avec difficulté tant la gravure est maladroite *LA / REPU / BLIQUE* avec au-dessus un bonnet phrygien (voir Casanova 1992. J. Lhoste et J.J. Buigné, 1999, p. 87 - 90).

339a

339

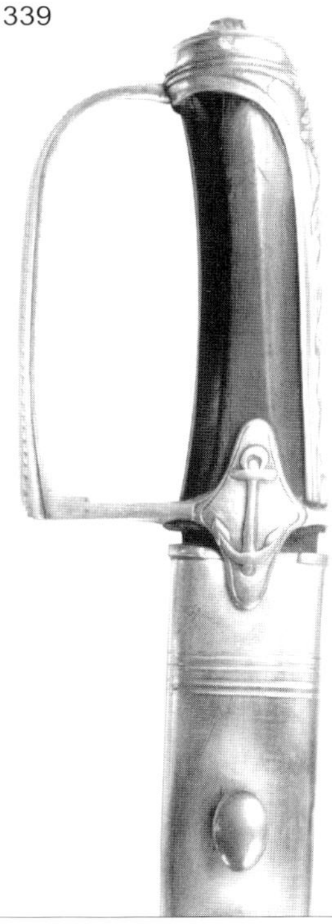

339b

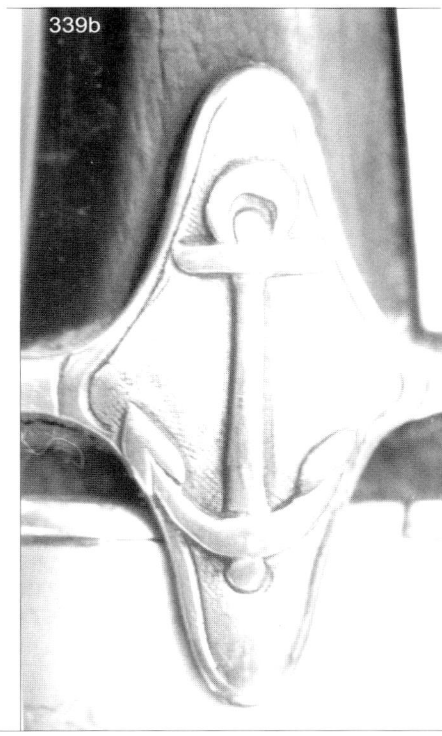

- **a**, détail de l'ornementation de calotte à longue queue représentant une frise de feuilles de laurier et de fruits ;
- **b**, détail de l'oreillon en triangle curviligne frappé de l'ancre de Marine non étalinguée.

(1) Le sabre de Marine a été officialisé le 7 Paririal An XII, et nommé Modèle Prairial An XII (voir p. 128).

340

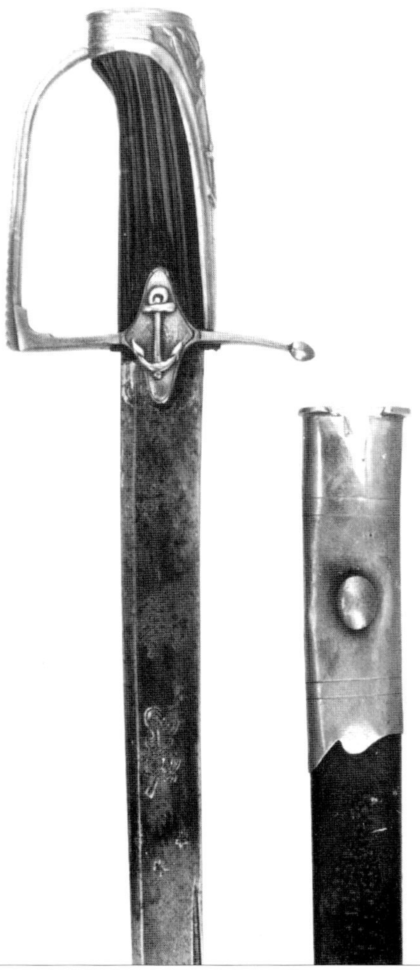

340a

340 - Variante.
Cet exemplaire possède également une poignée en ébène, mais ici, elle est cannelée. De plus, la queue de la calotte est ornée d'une ancre de marine suspendue par un orin sur lequel, à sa partie supérieure, est plaqué un trophée constitué d'un casque sur fond de deux pavillons (340a).

342

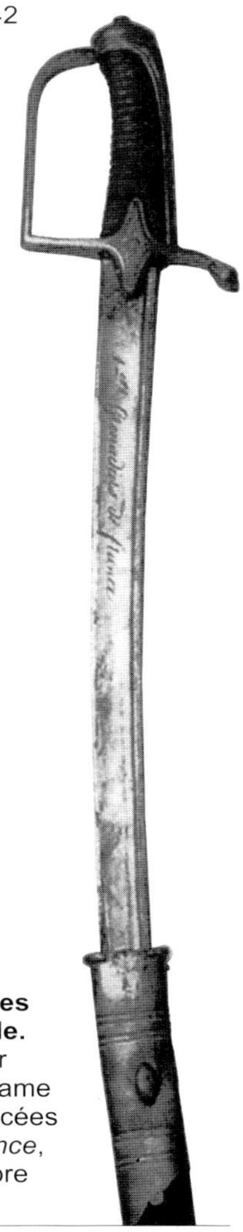

341 - Sabre du Corps des Marins de la Garde Impériale, modèle pour les officiers.
Premier Empire.
Monture en laiton (bronze) des sabres de la Garde de Napoléon Ier. Calotte assez mince surmontée d'un bouton de rivure légèrement conique et orné d'une rosace foliacée, à longue queue ciselée de feuilles de laurier. Poignée filigranée. Oreillons en losanges curvilignes à décor de feuilles d'acanthe disposées en croix. Arc de jointure arrivant en angle droit sur la croisière. Lame signée *Duc*, gravée en lettres capitales, sur le *recto CORPS DES MARINS* et sur le *verso GARDE IMPERIALE.* Fourreau de cuir noirci, à deux garnitures : chape à palmette, longue bouterolle abondamment décorée à bordure sinueuse et dard asymétrique en laiton (non figuré).

342 - Sabre d'officier du 1er Régiment des Grenadiers de France de la Garde Royale.
Ce sabre de la Garde Impériale du Premier Empire (voir p. 195, n° 328) possède une lame dont les inscriptions impériales ont été affacées et remplacées par *1ers Grenadiers de France*, en 1814. Exemple de la pérennité d'un sabre sous deux régimes antagonistes.

Sous-famille VI e – Oreillons hexagonaux

Les sabres à oreillons hexagonaux sont rares et sont souvent de facture germanique ou hongroise comme en témoignent les exemples que nous donnons ci-après.

343 - Sabre des Volontaires de Benyowsky.
Règne de Louis XV.
Monture en fer. Calotte prolongée d'une longue queue. Poignée recouverte de basane. Arc de jointure arrivant en angle droit sur la croisière rectiligne portant deux oreillons hexagonaux à filet marginal. Lame de 64,9 cm, à contre-tranchant à la pointe, large au talon de 5 cm. Cette lame est gravée sur le *recto* des deux " *L* " entrelacés de Louis XV et sur le *verso* : " *Régiment / Des / Volontaires / De / Benyowsky* ". Fourreau de cuir recouvert d'une feuille de tôle ajourée de cercles et demi-cercles en suite (342a) ; à noter que le fourreau est dépourvu de système de suspension. On peut supposer qu'il était cousu sur le pendant du ceinturon.
Le corps des Volontaires de Benyowsky a été levé pendant la Guerre de Sept ans pour servir à Madagascar où il fut anéanti (C. Blondieau, 1989).

343

343a, schéma des découpes du haut du fourreau.

344

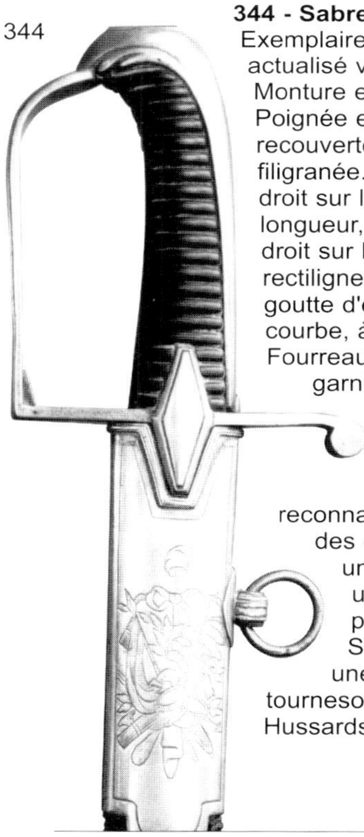

344 - Sabre de Hussard
Exemplaire du règne de Louis XVI actualisé vers 1789 - 1792.
Monture en laiton. Calotte en cuillère, à longue queue. Poignée encordée, recouverte de basane, non filigranée. Arc de jointure droit sur les 4/5èmes de sa longueur, arrivant en angle droit sur la croisière rectiligne finissant en goutte d'eau aplatie. Lame courbe, à pan creux. Fourreau en cuir à trois garnitures dont une longue chape ciselée sur le *recto* d'un trophée où l'on reconnaît des timbales, des drapeaux, ... et une lance portant un bonnet phrygien (343a). Sur le *verso* figure une fleur de tournesol, emblème des Hussards (343b). M.R.

344a

344b

- **a**, trophée d'armes surmonté d'un bonnet phrygien sur le *recto* de la chape ;
- **b**, fleur de tournesol sur le *verso* de la chape.

345

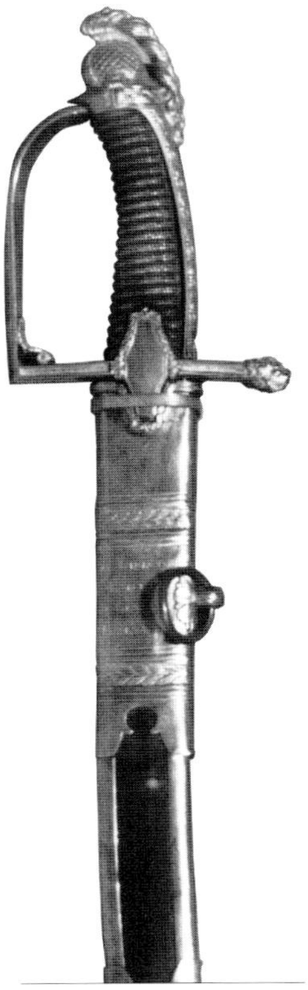

345 - Sabre donné au Général Masséna par le Directoire.
Monture en laiton doré. Calotte en casque empanaché, à longue queue. Poignée encordée, recouverte de basane. Oreillons hexagonaux, à bordure ciselée. Quillon terminé par une tête canine. Fourreau de laiton ciselé de bagues foliacées, deux attelles latérales laissant apparaître le cuir entre les deux bracelets de bélières ; longue bouterolle décorée de feuillages stylisés.

Nota - *André Masséna (Nice 1756 - Paris 1817) après des aventures diverses est nommé général en 1793, maréchal sous l' Empire, devient duc de Rivoli en 1808 et prince d'Essling en 1810. Il guerroie en Suisse, en Italie où il se rend maître du Royaume de Naples (1806), au Portugal qu'il occupe en 1810 mais qu'il doit quitter en 1811. En 1814, Masséna se rallie à Louis XVIII.*

346 - Sabres donnés par le Directoire.
Jusqu'à l'An XII.
Ces sabres sont souvent à calotte au casque empanaché et longue queue. La poignée est recouverte de basane et filigranée ou en ébène. L'ensemble est toujours luxueux et ces sabres sont facilement reconnaissables par l'inscription qu'ils portent soit sur les oreillons losangiques, soit sur le haut du fourreau. On lit :
DONNE / PAR LE / DIRECTOIRE / EXECUTIF
et souvent *Manufacture / Versailles*, accompagné de
BOUTET / DIRECTEUR / ARTISTE.

346

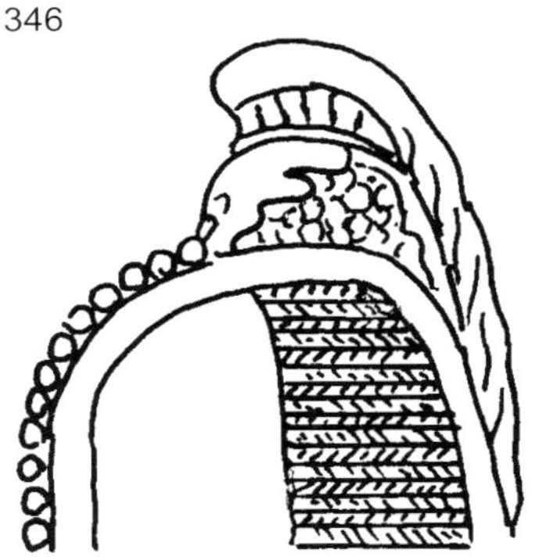

347

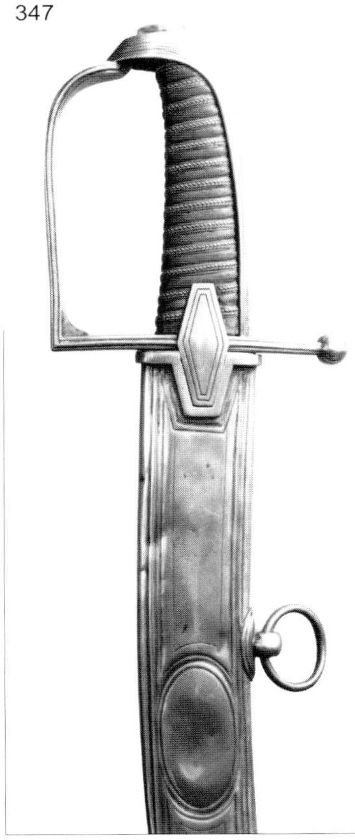

347 - Sabre d'officier de Hussards ou de Cavalerie légère
Premier Empire.
Monture en laiton. Calotte à longue queue à filet marginal et bouton de rivure en losange. Arc de jointure élargi à sa partie supérieure et arrivant en angle droit sur la croisière décorée, ainsi que l'arc, de deux filets jumelés, terminée par un bouton à facettes ébauchées. Poignée recouverte de basane à filigrane torsadé disposé en chaînette. Oreillons hexagonaux à deux filets de bordure (346a). Lame courbe, de 83 cm, large au talon de 3,5 cm, bleuie au quart de trophée d'armes doré. Fourreau de laiton à rainures marginales et ciselé de deux ovales et d'un losange, la partie inférieure étant cannelée. Dard en fer uniquement frontal (M.R.).

347a

- **a**, détail du filigrane, de la croisière, de l'oreillon frontal et du haut de la chape.

348 - Sabre du général Hoche.
Directoire.
Monture en laiton doré et ciselé. Calotte, ornée d'un casque avec cimier à crinière, et longue queue décorée de guirlande de feuilles de chêne ainsi que l'arc de jointure qui arrive en angle droit sur la croisière. Quillon en mufle de lion. Poignée à filigrane d'argent. Oreillons hexagonaux à tête de Gorgone. Lame à faible courbure, longue de 83 cm faite à Solingen, bleuie et gravée du Hussard chargeant, de trophées, de feuillages dorés. Fourreau de bois recouvert de maroquin noirci à deux importantes garnitures jointes par deux attelles de laiton sur les bords ; chape à entrée moulurée suivant le dessin des oreillons, ciselée de guirlandes de feuillage, multilobes et frises finement gravées ; bouterolle ciselée en suite.

348

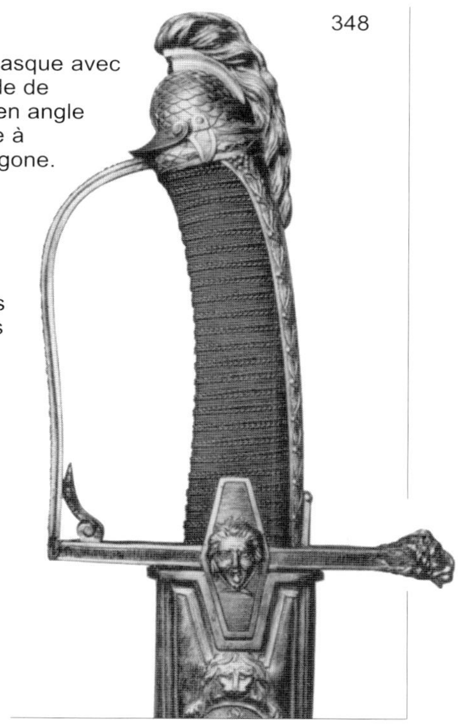

349

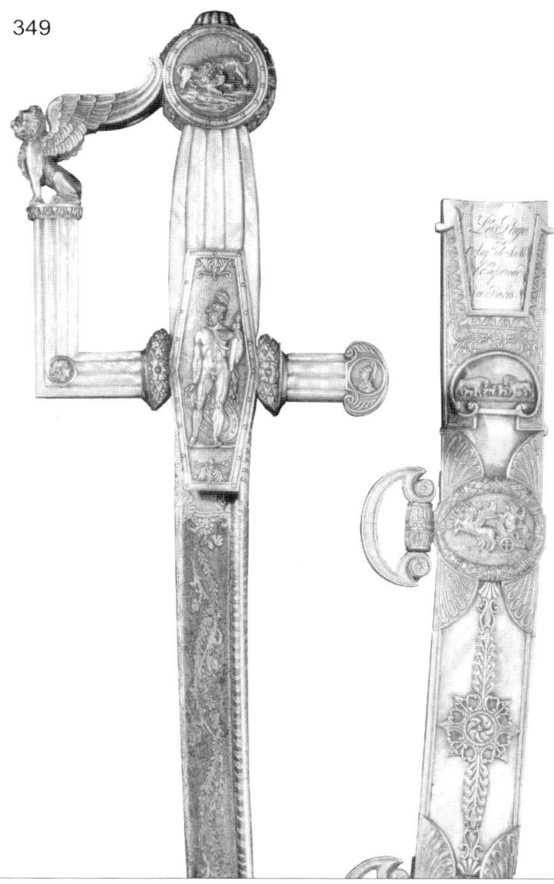

349 - Sabre offert au maréchal Bessière, duc d'Istrie.
Vers 1810.
Sabre de luxe à monture en bronze doré. Calotte en disque placé verticalement, ciselée d'un lion terrassant un autre félin. Poignée, arc de jointure et croisière en nacre striée, arrivant en angle droit sur la croisière et réunie au pommeau par une chimère. Quillon terminé par un médaillon à la figure d'Athéna. Longs oreillons hexagonaux, le frontal portant le dieu Mars en pied. Lame demi-courbe, décorée de fin damas polychrome de feuillages en guirlandes entrelacées ; dos à jonc et contre-tranchant. Fourreau de bois recouvert de plaques de nacre unies à quatre garnitures dorées : chape ciselée à jour d'une famille de lions, premier bracelet gravé du dieu Apollon conduisant un char tiré par deux chevaux ailés, le second bracelet représentant également Apollon ; on rencontre ensuite une frise verticale de cinq rosaces puis, la bouterolle ciselée sur toute sa longueur de motifs végétaux stylisés et d'un écusson à attributs guerriers ; anneaux en demi-lune. Dard en fer bleui et ajouré, à enroulement, décoré de petites étoiles et protégé par deux fines baguettes remontant sur les côtés de la bouterolle.

Marquage - Gravé au dos de la chape : *Le Page / Arqer de S.M. / l'Empereur / à Paris* en belle calligraphie (349).

Famille VII – Calotte non figurative à courte queue

Les sabres à calotte non figurative et courte queue de calotte ou à jupe, forment une famille assez homogène. Néanmoins, pour faciliter leur connaissance, cette Famille VII a été divisée en trois Sous-familles en tenant compte de la forme de la partie terminale de la poignée, de la façon suivante :

Sous-famille VII a – Calotte à jupe coiffant la poignée

Ces sabres sont à calotte plate, avec ou sans bouton de rivure, souvent bordée de deux cannelures séparées par une dépression. Certains exemplaires ont des oreillons ciselés d'un décor végétal ou d'un emblème de corps. Ce sont des sabres d'officiers de Cavalerie légère, de Chasseurs voire de Grenadiers à pied.

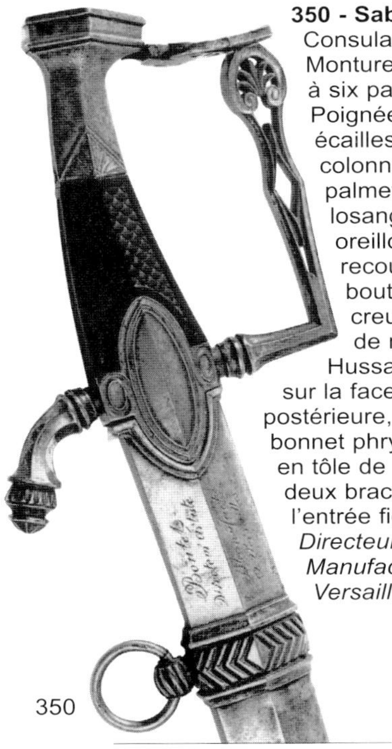

350

350 - Sabre d'officier de Hussards. Consulat. Monture en fer doré. Calotte plate à six pans, prolongée d'une jupe. Poignée en ébène sculptée en écailles. Arc de jointure en colonne, ciselé à jour d'une palmette et d'un évidé losangique. Croisière à deux oreillons ovalaires et quillon recourbé se terminant par un bouton décoré. Lame à pan creux, bleuie au tiers et gravée de motifs dorés dont : le Hussard chargeant, un soleil... sur la face frontale et sur la face postérieure, une colonne coiffée du bonnet phrygien, un soleil... Fourreau en tôle de fer, à six pans unis et à deux bracelets rainurés. Près de l'entrée figure la signature *Boutet / Directeur Artiste* avec au-dessous *Manufacture / à Versailles.*

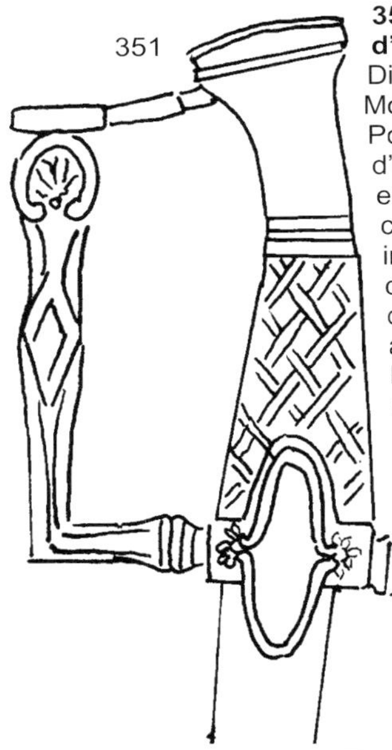

351

351 - Sabre d'officier d'Infanterie. Directoire - Consulat. Monture en acier bruni. Poignée sculptée d'entrelacs. Arc de jointure en colonne, rattaché à la calotte par un brin incurvé. Lame à pan creux, légèrement courbée. Fourreau en acier bruni, à deux bracelets de laiton ciselés d'entrelacs. On relève la signature : *Boutet / directeur artistique / Manuf^re^ à Versailles.* Un exemplaire de ce type est décrit par C. Ariès (XIX,5,).

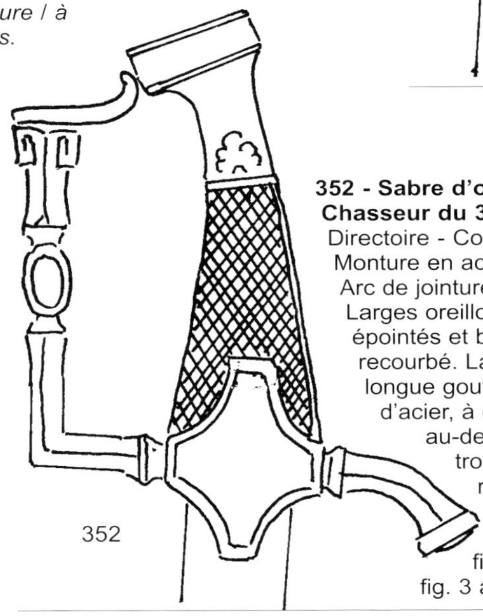

352

352 - Sabre d'officier de Hussards ou de Chasseur du 3ème Régiment. Directoire - Consulat. Monture en acier. Poignée en ébène quadrillée. Arc de jointure en colonne à ovale médian. Larges oreillons en losanges curvilignes épointés et bordés d'un liseré. Quillon recourbé. Lame courbe à pan creux, à longue gouttière dorsale. Fourreau de tôle d'acier, à deux bracelets finement ciselés : au-dessus du bracelet supérieur se trouve le chiffre 3 flanqué de deux rameaux de laurier et entre les bracelets est gravé un trophée. Un sabre de même facture est figuré par C. Ariès (XIX,5, 1971, fig. 3 a à 3 e).

353 - Sabre d'officier de Chasseurs à pied.
Consulat.
Calotte plate, à bouton de rivure, bordée de deux bourrelets, à courte queue. Arc de jointure en colonne découpée d'un ovale médian et deux renflements de part et d'autre. Cette branche est reliée à la calotte par un petit brin arqué et vient rejoindre la croisière en angle droit. Gros oreillons en losanges curvilignes, aux angles tronqués et ciselés d'un cornet dont le pavillon est frappé de trois points (franc-maçons ?). Poignée d'ébène sculptée en losanges. Quillon légèrement recourbé, évasé et terminé par un bouton. Lame de 71,5 cm de long, large au talon de 2,4 cm et flèche de 4,8 cm, gravée d'un petit soleil dans un cartouche, de motifs végétaux, de trophées. Fourreau de cuir noirci à deux garnitures : chape à bouton ovale, bouterolle à extrémité plate à dard de laiton protégeant seulement le côté antérieur.

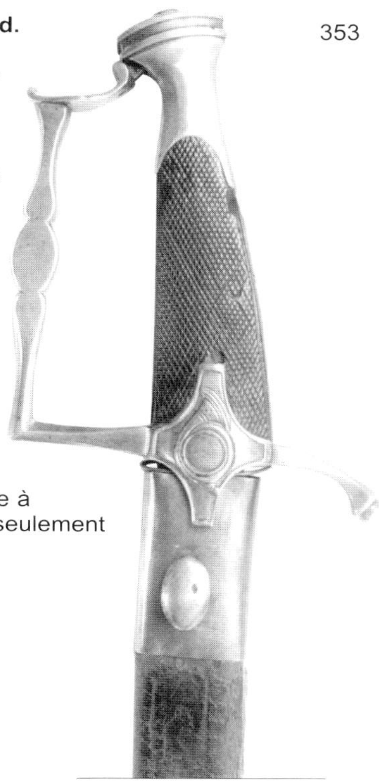

353

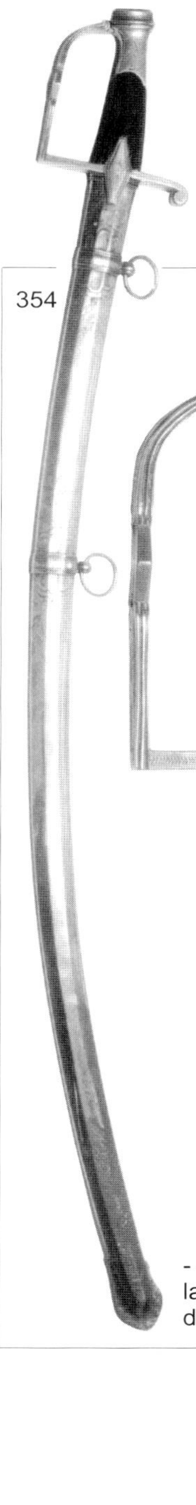

354

354 - Sabre d'officier de Cavalerie légère.
Consulat - Premier Empire.
Monture en laiton. Calotte cannelée et courte jupe presque symétrique. Poignée en ébène striée verticalement. Arc de jointure au trait latéral, courbé en sa partie supérieure avec un ornement quadrangulaire en son milieu. Croisière finement décorée et terminée par un bouton recourbé ciselé d'une fleurette. Oreillons subhexagonaux, à marge moulurée et ornés d'un motif floral stylisé (354a). Lame courbe, bleuie et dorée. Fourreau de laiton, à pans et deux bracelets de bélière. Ceux-ci sont placés au centre d'un groupe de motifs en navette. Dard en laiton, asymétrique.

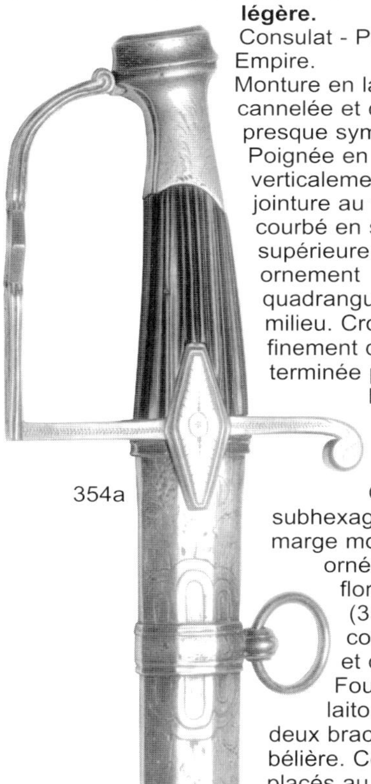

354a

- **a**, partie supérieure de la monture ; noter la poignée en bois strié verticalement et la décoration du haut du fourreau.

355 - Sabre d'officier de Cavalerie légère
Consulat ou début du Premier Empire.
Calotte plate, creusée d'un trait latéral, à jupe unie, découpée. Arc de jointure, orné dans sa partie droite d'un losange découpé, joignant la croisière par un angle droit ; oreillons hexagonaux décorés de motifs végétaux ; quillon se terminant par un petit enroulement. Poignée d'ébène striée longitudinalement. Lame longue de 84 cm, large au talon de 3 cm et flèche de 5 cm, à double pan creux et gravée au tiers de motifs floraux et de trophées. Fourreau de laiton à pans, ciselé à sa partie supérieure de motifs végétaux : deux bracelets de bélière à traits creusés marginaux (M.R.).

355

356

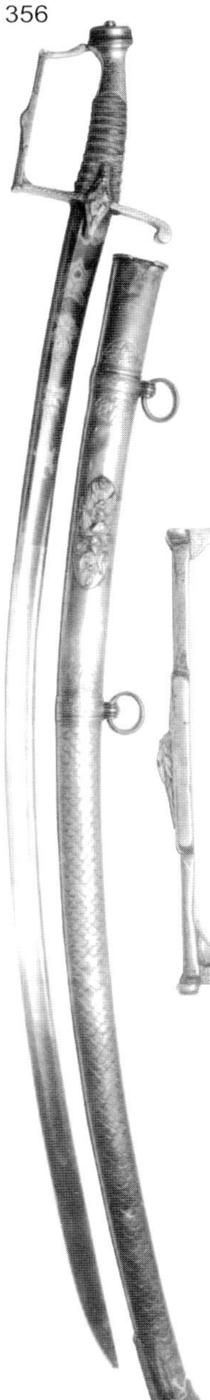

356 - Sabre d'officier de Cavalerie légère.
Consulat - Premier Empire.
Monture en laiton. Calotte à gorge et courte jupe ciselée de feuillage stylisé et d'une fleur à quatre pétales. Arc de jointure à colonne ornementée en son milieu d'un motif en relief. Croisière ciselée d'une frise de feuilles de laurier se terminant par un quillon en bouton recourbé. Oreillons hexagonaux à quatre côtés incurvés avec un motif végétal en relief (356a). Lame demi- courbe, bleuie et décorée de gravures dorées banales. Fourreau de laiton à deux anneaux de bélière fixés entre deux demies fleurs, la partie inférieure étant décorée d'écailles. Entre les deux bracelets un cartouche contient un trophée où l'on reconnaît une face léonine, des piques, un casque à l'antique, une cuirasse, un tambour,... (356b).

- **a**, partie supérieure du sabre. Remarquer les oreillons ;
- **b**, trophée d'armes ciselé entre les deux bracelets.

356a

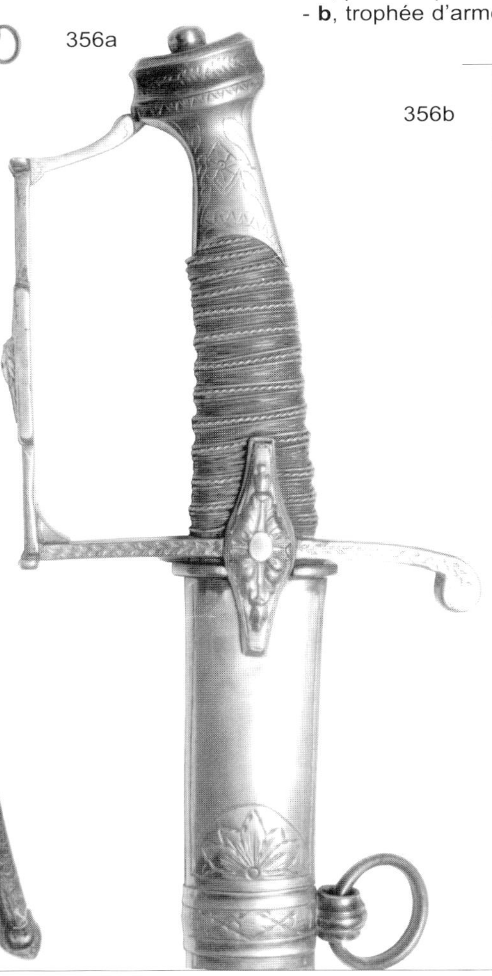

356b

357

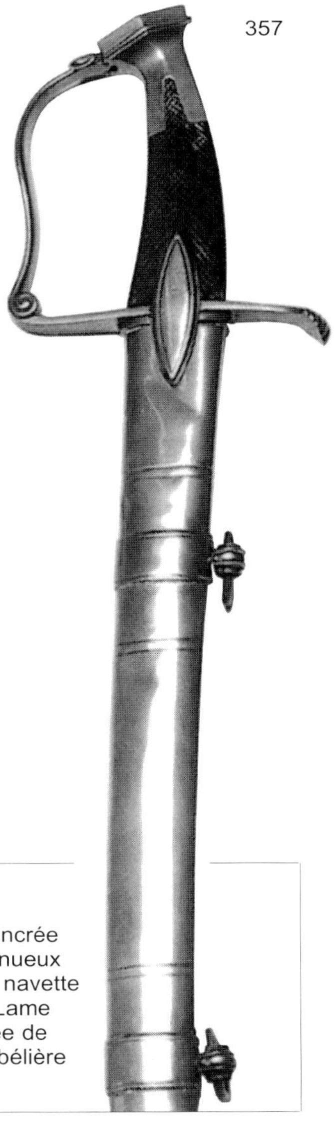

357 - Sabre de Cavalerie légère.
Consulat.
Monture en laiton. Calotte à pans, à jupe profondément échancrée latéralement. Poignée en ébène quadrillée. Arc de jointure sinueux aboutissant sur la croisière par un enroulement. Oreillons en navette bordée d'un filet. Quillons en palmette légèrement infléchie. Lame de 83,6 cm, courbe, à dos plat et pans creux, bleuie et gravée de motifs dorés. Fourreau de laiton, à deux larges bracelets de bélière encadrés de deux paires de traits.

358 - Sabre d'officier du 8ème Hussard.
Consulat.
Calotte à jupe ornée d'une dépression horizonale et surmontée d'un petit bouton de rivure. Branche en colonne composée d'une perle, d'une longue pyramide à facettes, d'une grosse perle, puis d'une pyramide inversée et d'une petite perle. Oreillon en navette (6 sur 1,5 cm) ; quillon à enroulement figuré. Poignée d'ébène quadrillée. Lame de 85 cm, large au talon de 3 cm et à flèche de 5,2 cm. Longue gouttière partant du talon, pan creux, pointe en biseau et gravée de petits motifs végétaux et d'un Hussard chargeant, surmonté d'une banderole portant l'inscription *8ème Hussard.* Fourreau de laiton à pans, deux bracelets de bélière, dard symétrique en laiton.

358

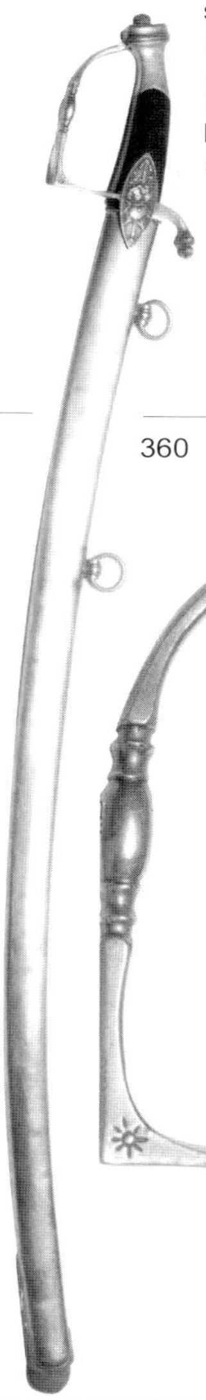

359 - Sabre d'officier de Cavalerie légère.
Consulat - Premier Empire.
Monture en laiton doré. Calotte à jupe légèrement inclinée vers l'avant, à bordure perlée. Poignée en ébène striée horizontalement. Arc de jointure incurvé en partant de la calotte, puis droit jusqu'à la croisière rectiligne se terminant par un quillon à tête de chien. Oreillons en navette ciselée d'une femme se mirant dans un miroir pouvant représenter Vénus. Lame de 81,3 cm, demi-courbe, à pan creux, bleuie au tiers et gravée de motifs dorés classiques. Cette lame est signée AW pour Weyersberg, fourbisseur à Solingen. Fourreau de laiton à deux bracelets figurés ciselés de feuilles de laurier flanqués chacun de quatre longues feuilles avec entre elles, des tigelles terminées par trois fruits qui font penser aux fruits de *Sparganium* (roseaux). Les bordures du fourreau sont ornées d'une frise de petites oves (359a) (Sotheby's, Sussex).

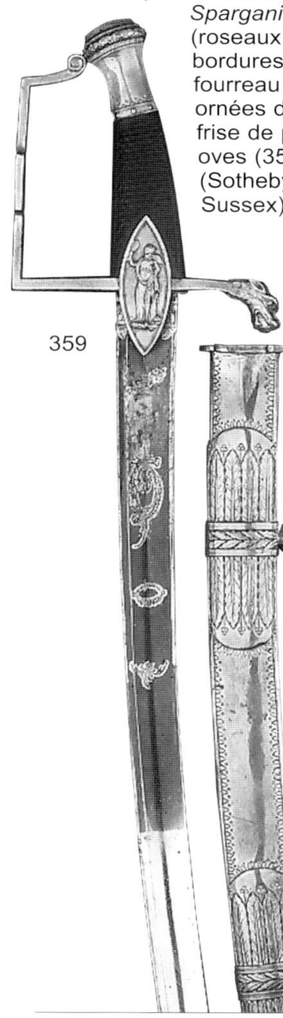

359

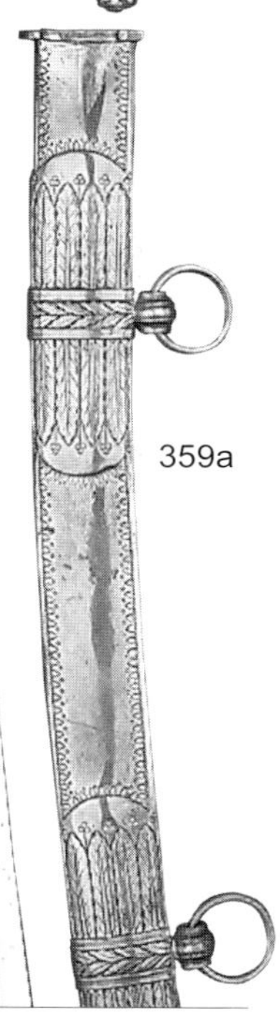

359a

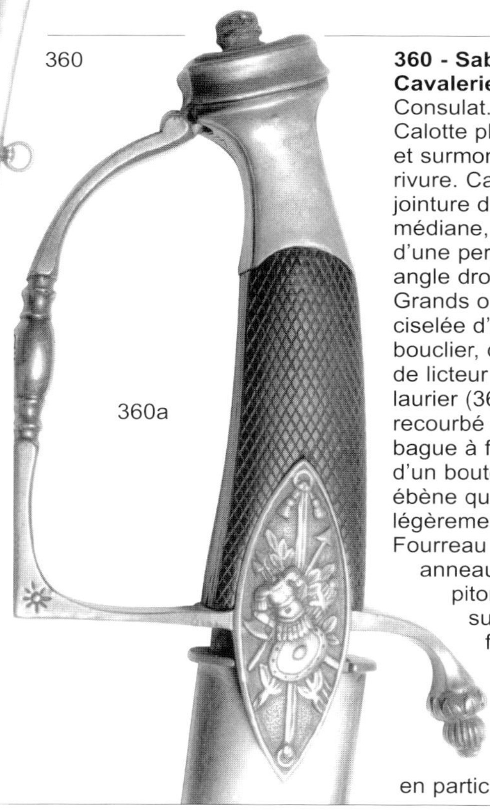

360

360a

360 - Sabre d'officier de Cavalerie légère d'Etat-Major.
Consulat.
Calotte plate, à gorge horizontale et surmontée d'un bouton de rivure. Calotte en jupe. Arc de jointure droit en sa partie médiane, décoré de bagues et d'une perle en olive et formant un angle droit avec la croisière. Grands oreillons en navette ciselée d'un trophée constitué de bouclier, de cuirasse, de faisceau de licteur sur fond de flèches, de laurier (360a). Quillon très recourbé se terminant par une bague à frise de feuilles puis, d'un bouton ciselé. Poignée en ébène quadrillée. Lame légèrement courbe, à pan creux. Fourreau de laiton uni à deux anneaux de bélière sur des pitons directement soudés sur le bord dorsal du fourreau (M.R.).

- **a**, détail de la monture et des oreillons en particulier.

361

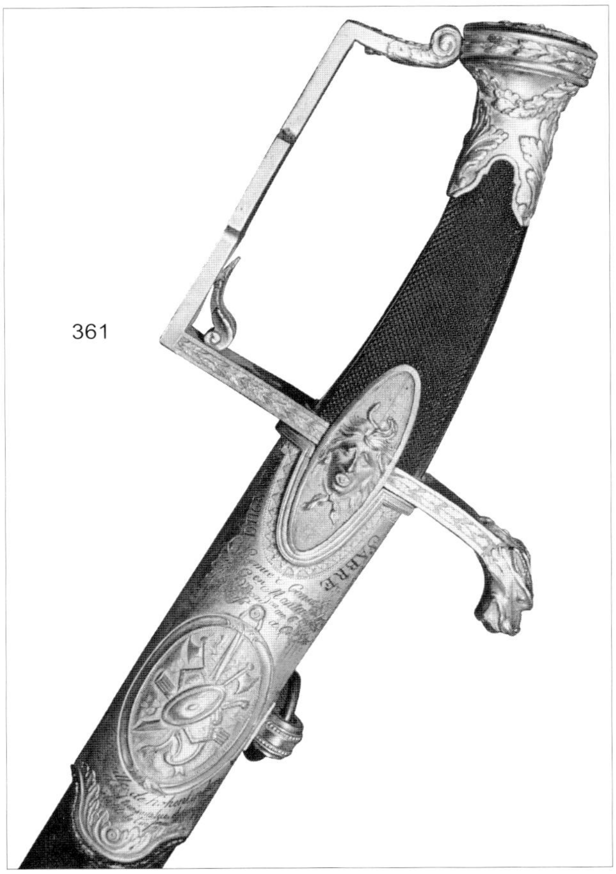

361 - Sabre d'Honneur.
Consulat.
Monture de laiton doré. Calotte plate ciselée d'une frise de feuilles de laurier, à courte jupe échancrée ornée de feuilles d'acanthe et d'une frise de feuilles de chêne. Poignée en ébène finement quadrillée. Arc de jointure droit relié à la calotte par un petit brin ciselé en enroulement. Croisière droite ciselée d'une frise de feuilles de chêne, portant des oreillons ovalaires, le frontal à tête de Gorgone, et se terminant par une tête d'animal mythique infléchie. Chape ornée d'un trophée d'armes inscrit dans un ovale. Au-dessus de cet ovale, on lit : *SABRE D HONNEUR* (sic), puis :
Le premier Consul / au C^{en} Montault / Chef d'Escadon au 6 Reg^{t} / de Chafs^{r} à cheval
et au-dessous :
à la bataille de hohenlinden / eut son cheval tué sous lui / à la tête de l'Inf^{erie}.

Sous-famille VII b – Calotte partant d'un disque aplati débordant la largeur de la poignée

Les sabres ayant une calotte en disque débordant la largeur de la poignée et d'où part une courte queue sont donc caractérisés par une monture dite à la hongroise. Cette Sous-famille VII b, en conséquence, regroupe essentiellement des sabres de Hussards des XVIIIème et XIXème siècles.

C'est sous le règne de Louis XIV que les sabres de Hussards sont introduits en France. Ils furent apportés dans notre pays par les troupes alliées telles celles de Kroneberg en 1692, de Montany en 1693... puis celles de Bercheny en 1720 qui entrent au service du roi. Ces sabres, ancêtres des fabrications françaises, sont d'une facture très spéciale comme en témoignent les figures rassemblées ci-après. On remarquera que les queues de calotte n'atteignent pas la croisière bien que chez certains exemplaires elle soit relativement longue, parfois fixée par un clou. La poignée est recouverte de basane, filigranée ou non. Dans ce dernier cas, le revêtement peut être renforcé, en plus d'un collage soigné par un ou plusieurs clous à tête enjolivée. La croisière est rectiligne et porte deux oreillons. La lame mesure de 75 à 90 cm, large au talon de 3,4 à 4 cm, à flèche de 3,5 à près de 6 cm. Le fourreau est de bois recouvert de cuir et généralement à trois garnitures parfois richement ciselées.

Lorsque les fabrications françaises commencèrent, des règlements furent pris, mais toujours assez imprécis comme les Ordonnances du 25 avril 1767, de 1783, du 17 mars 1786, de l' An IV, de l'An IX, ou de l'An XI. Si la Manufacture de Klingenthal respectait les décisions prises, de nombreux fourbisseurs n'étaient pas sans apporter des variantes fantaisistes, fantaisies éventuellement accentuées par les Hussards eux-mêmes, surtout s'ils étaient officiers.

Les sabres de Hussards furent en service jusqu'en 1812.

362

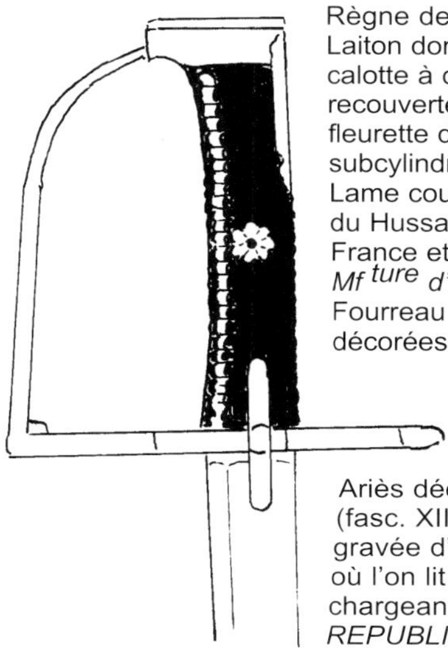

362 - Sabre d'officier de Hussards.
Règne de Louis XVI.
Laiton doré. Monture à la hongroise, calotte à courte queue. Poignée recouverte de basane et clou médian à fleurette de laiton. Oreillons subcylindriques à extrémités arrondies. Lame courbée, gravée d'un trophée, du Hussard chargeant, des Armes de France et signée *Mf*^{ture} *d'Alsace*.
Fourreau de cuir à trois garnitures décorées au trait.

Remarque - Ce type de sabre se retrouvera sous le Premier Empire.
Ariès décrit un sabre du même type (fasc. XIII, 2, fig. 3) dont la lame est gravée d'un trophée, d'une banderole où l'on lit *VIVAT*, d'un Hussard chargeant et de *HUSARD / DE LA / REPUBLIQUE / FRANCOISE.*

363

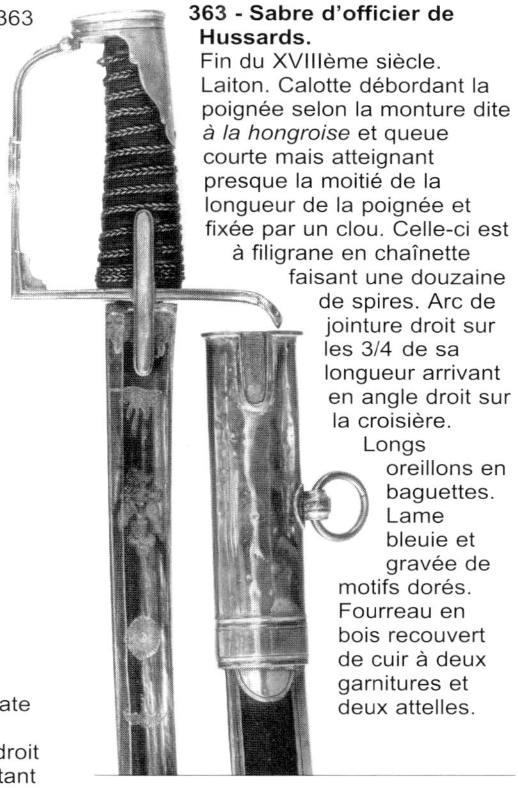

363 - Sabre d'officier de Hussards.
Fin du XVIIIème siècle.
Laiton. Calotte débordant la poignée selon la monture dite *à la hongroise* et queue courte mais atteignant presque la moitié de la longueur de la poignée et fixée par un clou. Celle-ci est à filigrane en chaînette faisant une douzaine de spires. Arc de jointure droit sur les 3/4 de sa longueur arrivant en angle droit sur la croisière.
Longs oreillons en baguettes. Lame bleuie et gravée de motifs dorés. Fourreau en bois recouvert de cuir à deux garnitures et deux attelles.

364 - Sabre d'officier supérieur de Hussards.
Règne de Louis XVI.
Monture en laiton doré décoré au trait. Calotte plate débordant la poignée, à courte queue. Arc de jointure arrivant en angle droit sur la croisière portant des oreillons en bâtonnets tronqués. Poignée recouverte de basane à deux clous en croix. Lame à pan creux, à gorge le long du dos, bleuie sur la moitié de sa longueur et gravée de grottesques dorés. Fourreau de cuir à quatre garnitures en laiton repoussé : chape rainurée verticalement, deux bracelets à trophées, longue bouterolle rainurée, en suite avec la chape, à trophée médian, dard en laiton asymétrique.

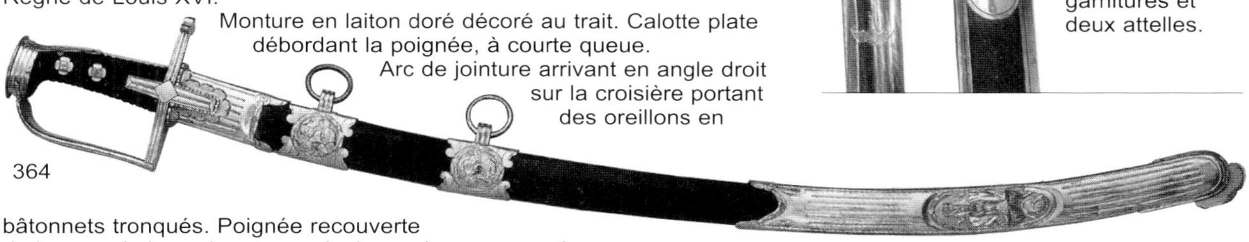

364

365

365 - Sabre d'officier de Hussard.
Règne de Louis XVI.
Monture en laiton doré. Calotte arrondie et plate, *à la hongroise*, et courte queue. Poignée recouverte de basane à filigrane de laiton torsadé faisant une vingtaine de spires. Arc de jointure et croisière rainurés, ornés de deux oreillons en baguette avec un petit carré médian. Lame en langue de carpe signée *" Berger fourbisseur à la tête noire à Strasbourg "* au talon avec, au-dessous, le petit Hussard chargeant et une banderole portant *" Vivat Hussard "*.
Fourreau à quatre garnitures dorées : courte chape, deux bracelets ciselés de trophées et bouterolle, celle-ci décorée en suite avec la chape.

366

366 - Sabre d'officier supérieur de Hussards.
Règne de Louis XVI.
Monture en laiton doré.
Calotte plate *à la hongroise* ciselée de motifs végétaux et gravée sur le dos de la courte queue cloutée, du chiffre *B* de la *M* (366a).
Poignée recouverte de basane maintenue par un clou. Oreillons en baguettes tronquées à leurs extrémités, décorés en leur milieu d'une petite rosette avec de part et d'autre une petite palmette.
Lame courbe, longue de 81,4 cm, à dos rond, contre-tranchant, pan creux et gouttière. Fourreau de cuir à quatre garnitures : chape ciselée de feuilles rayonnantes, bracelets à trophées de casque, de flèches, bordées d'une couronne de laurier (366b), bouterolle à godrons et dard asymétrique en fer (366c).

366a

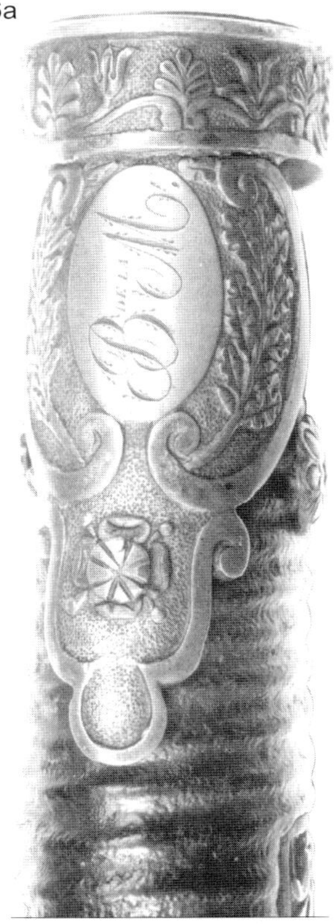

366c

366b

- **a**, queue de calotte signée des initiales " B " de la " M " ;
- **b**, décoration du bracelet de bélière médian ;
- **c**, extrémité de la bouterolle (*verso*) ciselée d'une palmette et dard asymétrique.

367

367 - Sabre d'officier de Hussards.
Règne de Louis XVI, Monarchie constitutionnelle.
Monture en laiton doré. Calotte plate, *à la hongroise*, à godrons et courte queue (367a). Cette courte queue de calotte est maintenue par un clou et décorée de faisceau d'armes et du chiffre *C R* sous couronne de laurier. Poignée recouverte de basane filigranée. Oreillons en baguettes tronquées décorées en leur milieu d'une fleurette à quatre pétales. Lame de 84 cm, à dos plat, contre-tranchant, pan creux, bleuie et gravée de motifs végétaux classiques plus un faisceau d'armes et un guerrier en arme. Fourreau à trois garnitures : chape à godrons, à base oblique, bracelets et bouterolle en suite, chape et garnitures étant ornées d'un macaron ovale à motifs d'attributs militaires et d'un faisceau de licteur coiffé d'un bonnet phrygien. Les deux anneaux de bélière et le dard asymétrique sont en fer.

367a

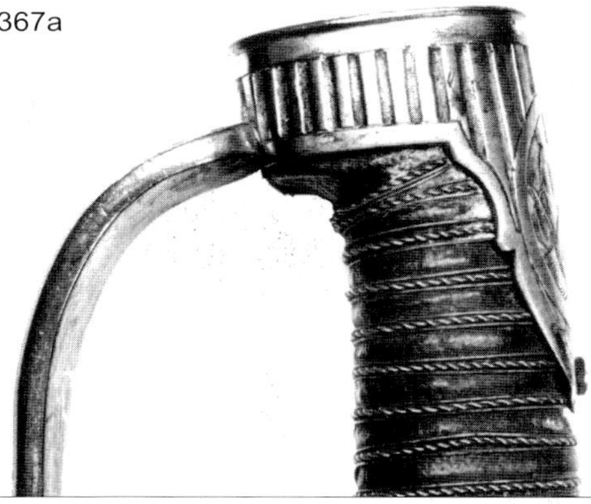

- **a**, détail de la calotte. Remarquer la courte queue fixée par un clou.

368

368 - Sabre d'officier de Hussards.
Consulat.
Monture en laiton doré. Calotte arrondie et plate *à la hongroise* et courte queue. Poignée encordée sous basane, sans filigrane. Croisière droite, décorée au trait, à deux longs oreillons à décor en suite, avec un petit carré au niveau de la jonction croisière-oreillon (ce petit carré se rencontre assez fréquemment sur les sabres de Hussards de cette époque). Lame à large pan creux, bleuie et gravée de motifs dorés classiques. Fourreau à couture métallique et à quatre garnitures : une courte chape rainurée, deux bracelets ornés de trophées et une bouterolle décorée en suite avec la chape.

369

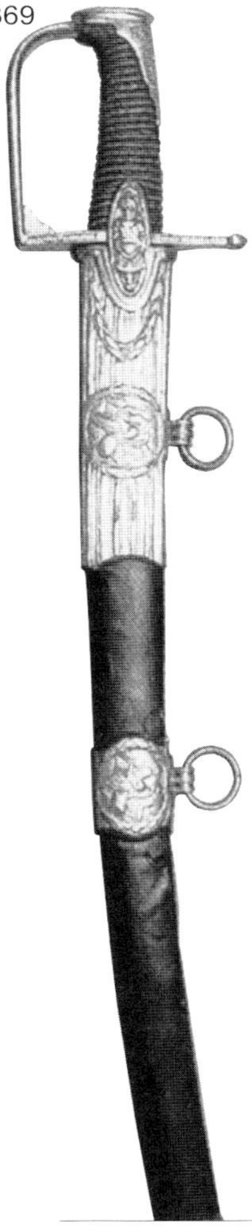

369 - Sabre d'officier supérieur de Hussards.
Premier Empire.
Monture de laiton doré. Calotte plate, à bouton de rivure masqué par une pièce ciselée, à courte queue (369a) décorée de feuilles d'acanthe. Arc de jointure arrivant en angle droit sur la croisière, celle-ci ornée de deux lignes sinueuses parallèles. Oreillons ovoïdes, le frontal ciselé d'un casque à l'antique et d'une armure, le postérieur d'une rosace à seize folioles (369b, 369c). Poignée recouverte de basane filigranée. Lame demi-courbe de 88 cm de longueur, bleuie sur le tiers basal et gravée de motifs dorés représentant une couronne, un trophée d'armes, un soleil. Fourreau de cuir à trois garnitures de laiton doré : longue chape cannelée, ornée à sa partie supérieure d'une guirlande de feuilles de laurier et d'un faux bracelet à macaron au trophée d'armes (369d). Le second anneau de bélière est porté par un bracelet qui est la reproduction du macaron de la chape, longue bouterolle cannelée et dard asymétrique en laiton (M.R.).

369a

369c

- **a**, ornement de la queue de calotte ;
- **b**, oreillon frontal ciselé d'une armure ;
- **c**, oreillon postérieur orné d'un motif végétal (tournesol) ;
- **d**, chape du fourreau ciselé d'un bracelet de bélière.

369b

369d

370

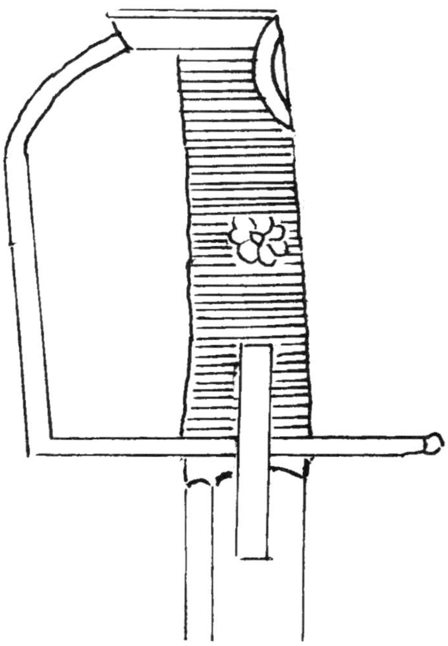

370 - Sabre d'officier de Hussards.
Vers 1805.
Monture en laiton doré. Calotte plate *à la hongroise* avec un médaillon en guise de courte queue. Poignée presque cylindrique finement ficelée et recouverte de basane, maintenue par un clou à tête de fleurette. Arc de jointure légèrement galbé en sa partie supérieure et arrivant en angle droit sur la croisière rectiligne. Oreillons en baguettes à extrémités tronquées. Lame bleuie et gravée de motifs dorés. Fourreau à quatre garnitures : chape rainurée, premier et second bracelets ornés d'un trophée et portant les anneaux de bélière, longue bouterolle rainurée à médaillon orné d'un casque, dard asymétrique (M.E. - S. de P.).

371

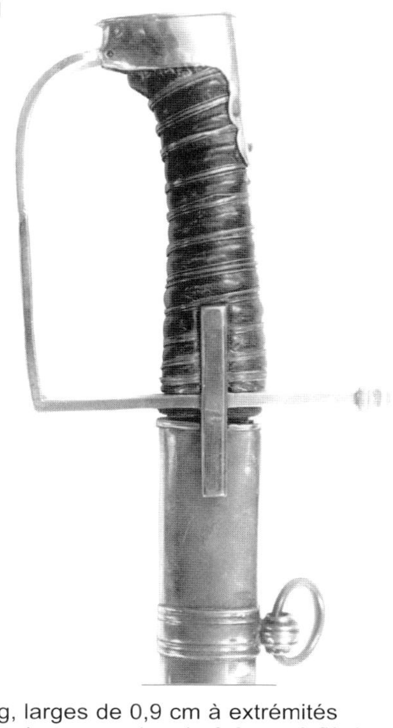

371 - Sabre d'officier de Hussards.
Consulat - Premier Empire.
Monture en bronze doré : calotte *à la hongroise*, plate, à couvercle soudé cachant le bouton de rivure, et prolongée d'une petite queue lobée, fixée par un clou. Poignée cordée, recouverte de basane et filigranée d'un mince ruban de cuivre rouge bordé de fils très fins. Arc de jointure plat, élargi au tiers supérieur et rainuré. Croisière droite terminée par un bouton subpyramidal fixé sur une embase quadrangulaire. Oreillons en baguettes de 6,4 cm de long, larges de 0,9 cm à extrémités tronquées. Lame de 72,7 cm, large au talon de 2,6 cm, à flèche de 5,5 cm, gravée en partant du talon, de motifs foliacés, du Hussard chargeant, d'une banderolle avec *VIVAT HUSSAR* et d'un trophée où l'on reconnaît une pique, un étendard, un tambour, une trompette... Fourreau de laiton laminé, uni, à deux bracelets et anneaux de bélière, dard en fer, asymétrique.

372

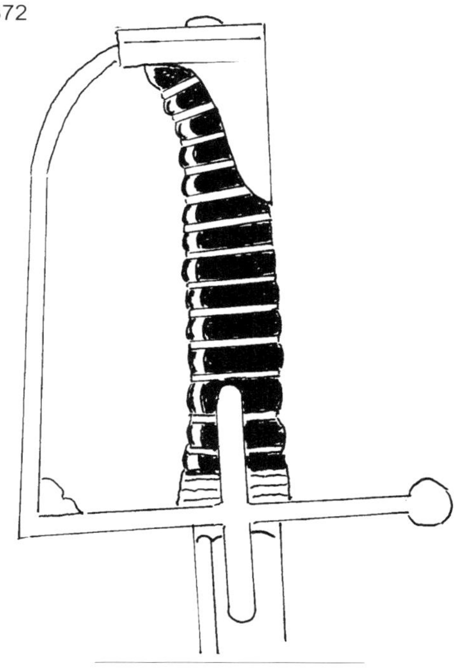

372 - Sabre d'officier de Chasseurs à pied.
Consulat, vers 1798.
Monture en fer. Calotte à couvercle plat, petit bouton de rivure, à courte queue. Poignée à double filigrane de fil de fer torsadé faisant une douzaine de spires, plus quelques spires serrées en place de virole basale. Arc de jointure droit sur les 2/3 inférieurs arrivant en angle de 80° sur la croisière rectiligne et terminée par une boule. Deux oreillons en baguettes à extrémités arrondies. Lame relativement courte et étroite. Fourreau d'acier, à deux bracelets à piton supportant les deux anneaux de bélière.

373

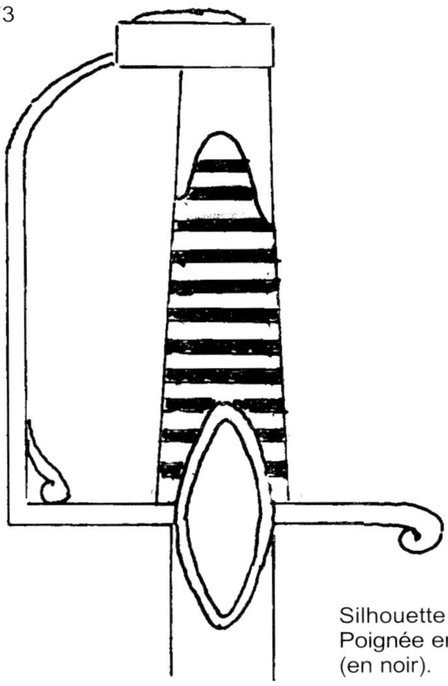

373 - Sabre de général de Hussards.
Premier Empire.
Sabre décrit et figuré par B. Bruel (1997 - Tradition n° 122, p. 48).

Monture en bronze doré. Calotte *à la hongroise* et courte jupe, ciselée d'une fleur d'héliotrope, rappelant le soleil, emblème des Hussards. Poignée d'ivoire filigranée de onze spires en chaînette. Croisière droite terminée par un rouleau pendant, et portant, deux oreillons en navette décorées de trophées. Lame légèrement courbe, longue de 86 cm, à gouttière et dos rond, contre-tranchant. Cette lame est bleuie et gravée de trophées, de symboles maçonniques... incrustés d'or. Fourreau de chagrin noir, à couture métallique frontale en chaînette et trois garnitures : longue chape ciselée de trophées, à bordure inférieure arrondie et à la toile d'araignée ainsi que le bracelet à motif floral ; bouterolle à faisceau de baguettes et de feuilles d'acanthe alternées.

B. Bruel signale la présence d'un sabre voisin au Musée de l'Empéri à Salon-de-Provence.

Silhouette d'après B. Bruel.
Poignée en ivoire, filigrane doré (en noir).

374

374 - Sabre d'officier de Grenadier à pied.
Consulat - Premier Empire.
Monture en laiton. Calotte *à la hongroise* en laiton, jupe découpée. Arc de jointure en colonne aplatie reliée à la calotte par une petite branche arquée. Fusée en ébène quadrillée. Croisière portant des oreillons en écu, le frontal orné d'une grenade enflammée (374a), le postérieur avec le numéro de la 1/2 brigade. Quillon recourbé en bouton. Lame de 80 cm environ. Fourreau de cuir à deux garnitures en laiton, à bordures découpées en suite et décorées au trait (M.E. - S. de P.).

374a

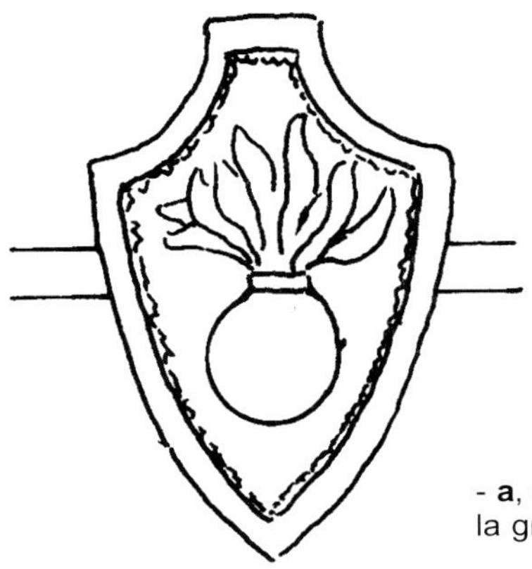

- **a**, oreillon frontal à la grenade

375

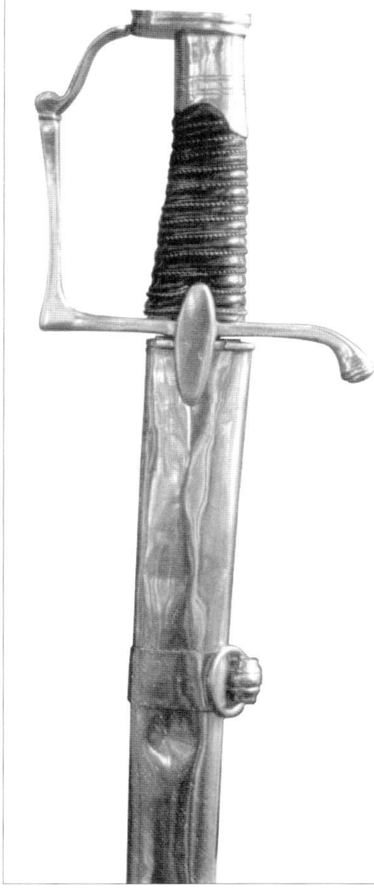

375 - Sabre d'officier de Cavalerie légère.
Consulat - Premier Empire.
Calotte *à la hongroise* débordant largement la fusée, à double cannelures latérales, jupe légèrement asymétrique maintenue postérieurement par un clou et décorée de deux doubles traits (375a). Le bouton de rivure est masqué par une pièce ovale ciselée d'un trophée où l'on reconnaît un casque, deux étendards, deux boucliers et un canon (375b). Poignée recouverte de basane et filigranée d'un fil de laiton torsadé faisant treize spires. Arc de jointure à colonne, celle-ci étant reliée à la calotte par un petit brin sinueux. Quillon infléchi, à bouton tronqué. Oreillons ovoïdes, unis. Lame demi-courbe, longue de 81,5 cm, blanche, à pan creux. Fourreau en laiton uni, à deux bracelets plats de 2 cm de hauteur. Dard asymétrique en fer (M.R.).

- **a**, détail de la calotte ;
- **b**, couvercle de la calotte.

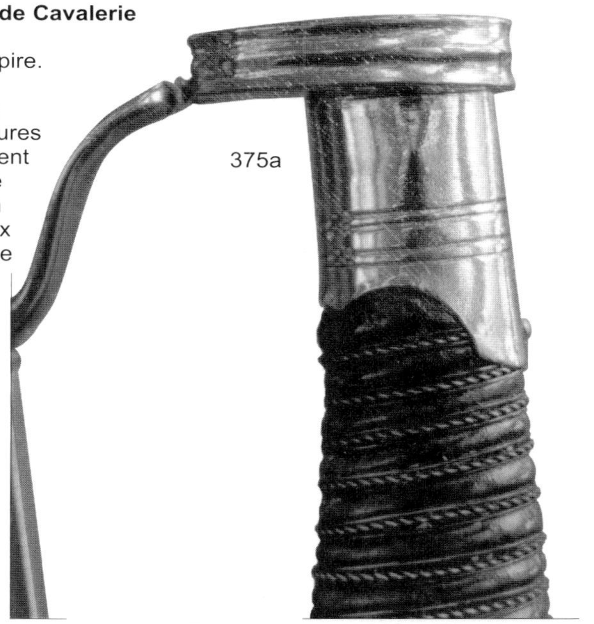

375a

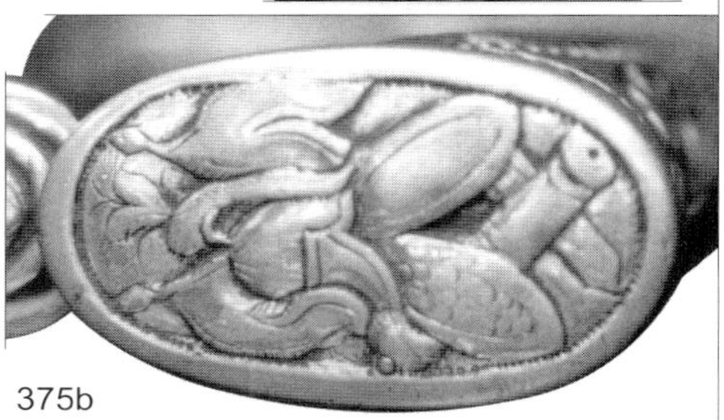

375b

376

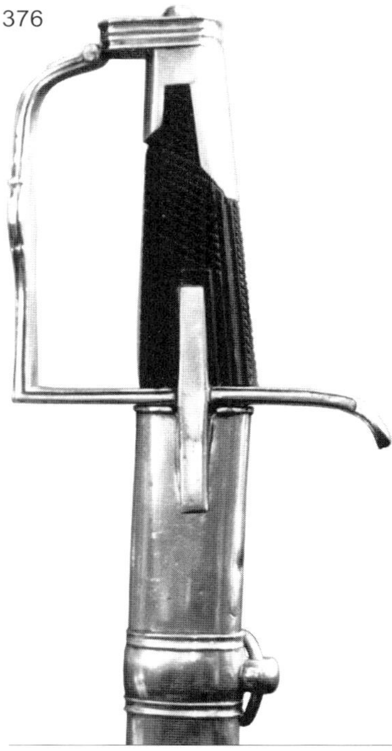

376 - Sabre d'officier de Hussards.
Premier Empire.
Monture en laiton. Calotte *à la hongroise*, à bouton de rivure, plate, débordant largement la jupe fortement échancrée, formant ainsi une courte queue postérieure. Poignée d'ébène à pans, quadrillée. Arc de jointure à décor médian, droit sur les 4/5èmes de sa longueur, arrivant en angle droit sur la croisière terminée par un petit fléchissement. Oreillons en baguettes tronquées. Lame de 83 cm, large au talon de 3 cm, à pan creux, gravée sur fond bleui d'attributs militaires dorés. Fourreau de laiton, uni, à deux bracelets de bélière légèrement bombés et bordés de moulures (M.R.).

377

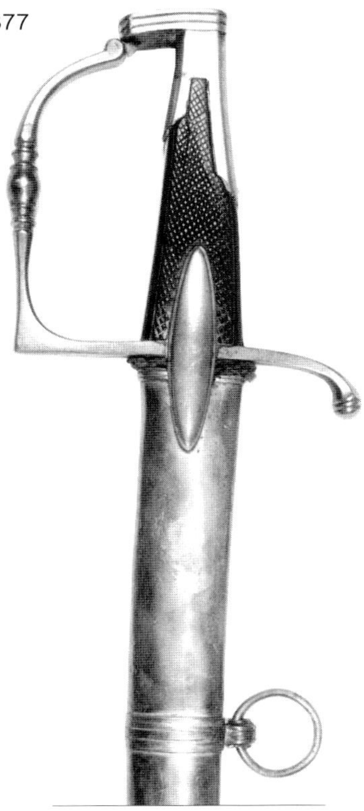

377 - Sabre d'officier de Cavalerie légère.
Premier Empire.
Monture de laiton, *à la hongroise* avec sa calotte plate et décentrée placée sur une jupe nettement divisée en deux parties, la frontale plus courte que la postérieure. Poignée d'ébène finement quadrillée. Arc de jointure avec en son milieu, une perle entourée de deux petits disques. Oreillons en navette dont les extrémités sont arrondies. Quillon infléchi, terminé par un bouton strié horizontalement rappelant la décoration latérale de la calotte. Lame demi-courbe, blanche. Fourreau de laiton, uni, à deux anneaux de bélière sobrement décorés.

378

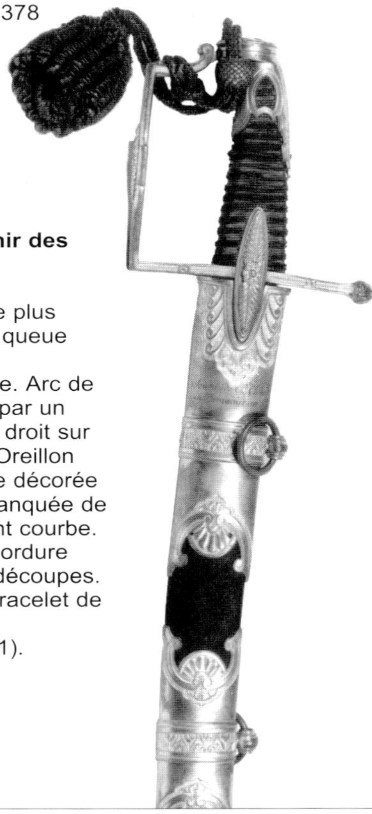

378 - Sabre d'officier Souvenir des Journées de Saint-Cloud.
19 Brumaire an 8.
Monture en laiton doré. Calotte plus large que la poignée et courte queue ajourée latéralement. Poignée recouverte de basane filigranée. Arc de jointure droit, relié à la calotte par un brin arqué et arrivant en angle droit sur la croisière. Quillon en boule. Oreillon frontal en navette très allongée décorée en son milieu d'une fleurette flanquée de deux palmes. Lame légèrement courbe. Fourreau à trois garnitures à bordure découpée et palmette en ces découpes. Chape gravée au-dessus du bracelet de la mention : *Souvenir de St Cloud / 19 Brumaire an 8.* (1).

(1) Sieyès fait transporter le Conseil des Anciens du Directoire à Saint-Cloud pour faciliter l'intervention de Bonaparte de retour d'Egypte. En effet, celui-ci renverse le Directoire et instaure un nouveau régime politique : le Consulat composé de trois consuls Cambacérès, Lebrun et Bonaparte qui se fait nommer Premier Consul.

379

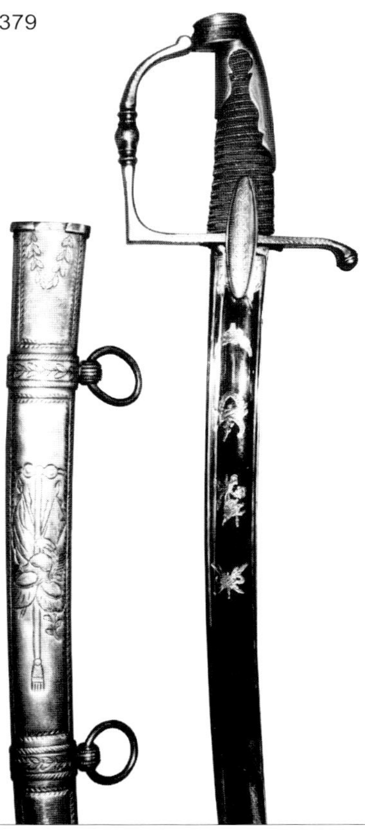

379 - Sabre d'officier de Cavalerie légère.
Consulat, Premier Empire.
Monture en laiton. Calotte *à la hongroise*, queue dépassant la moitié de la longueur de la poignée et profondément échancrée latéralement. Poignée recouverte de basane et filigranée de fils torsadés et formant une vingtaine de spires. Arc de jointure à perle médiane entourée de deux petites moulures et arrivant en angle droit sur la croisière. Oreillons ovoïdes décorés de feuillages. Quillon légèrement infléchi. Lame demi-courbe, bleuie et gravée de motifs classiques. Fourreau en laiton, avec entre les deux bracelets, un long trophée aux drapeaux et aux tambours (J.C. D.).

Famille IX – Calotte à courte queue

Cette Famille IX est surtout caractérisée par son arc de jointure faisant un " S " avec le quillon et une queue de calotte courte ou le plus souvent en jupe. Les oreillons sont de formes variées, parfois décorés ou portant les armes du corps.

389

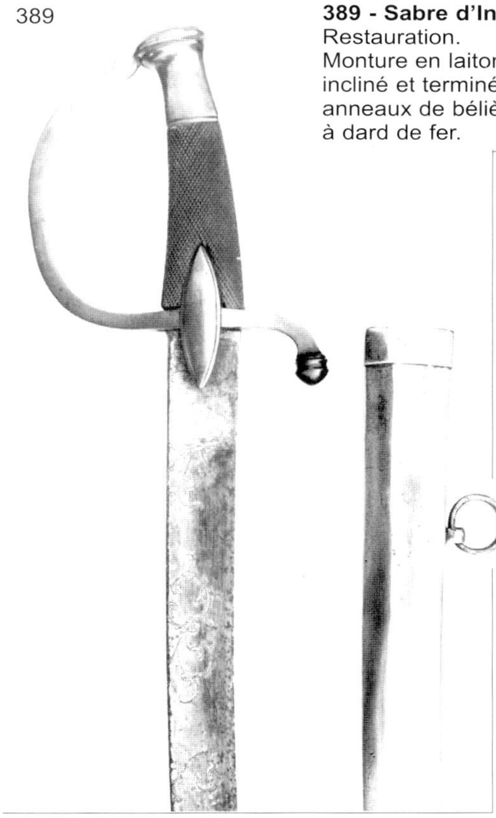

389 - Sabre d'Infanterie.
Restauration.
Monture en laiton. Calotte à jupe. Arc de jointure uni se poursuivant par un quillon incliné et terminé en boule. Oreillon en navette à liseré. Lame à pan creux. Deux anneaux de bélière dont les pitons sont soudés directement sur le fourreau de laiton, à dard de fer.

390

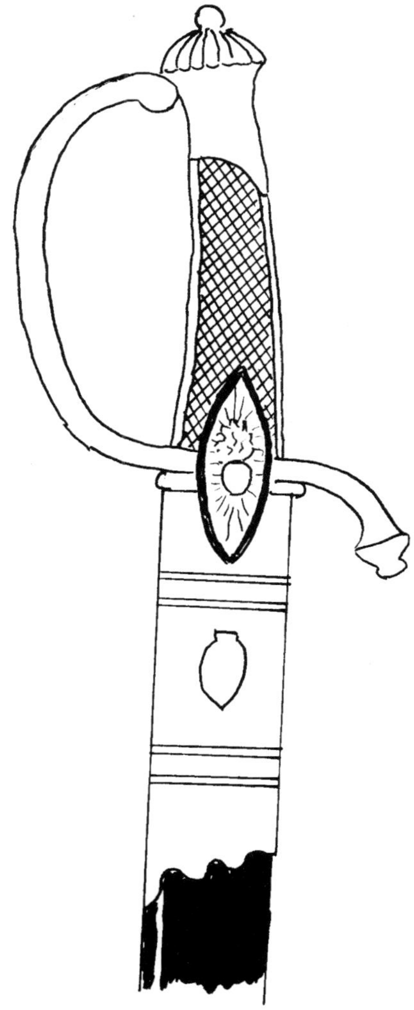

390 - Sabre des Cent Suisses de la Garde Royale.
Restauration, 1816, troupe.
Monture en laiton. Calotte en côtes de melon, à courte queue. Poignée en ébène quadrillée. Arc de jointure arrondi et se poursuivant par un quillon infléchi. Oreillons en navette gravés d'une grenade enflammée sur fond rayonnant (390a). Lame à pan creux, longue d'environ 70 cm. Fourreau en cuir de vache, à deux garnitures : chape à bouton en écussons, coupée en accolade oblique, en suite avec la bouterolle.

390a

- **a**, détail de l'oreillon.

391

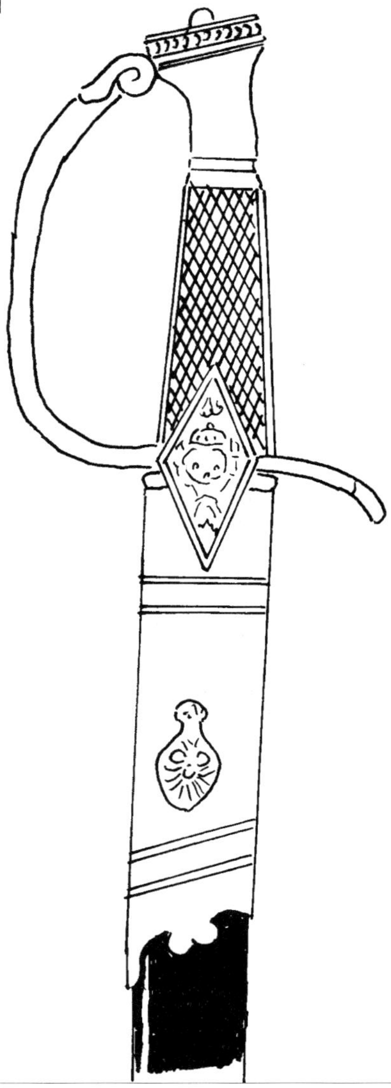

391 - Sabre de la Garde Royale.
1822, officier.
Monture en argent. Calotte aplatie, en jupe légèrement inclinée vers l'avant, à bordure en cordon. Poignée d'ébène quadrillée. Arc de jointure arrondi et terminé par un quillon infléchi. Oreillons en losanges, seul le frontal est frappé des Armes de France couronnées, flanquées de deux rameaux de laurier (391a). Lame à pan creux, longue d'environ 70 cm. Fourreau de cuir de vache, à deux garnitures : chape à bouton décoré d'une palmette, bouterolle à dard asymétrique se terminant par deux enroulements à son extrémité (M.E. - S. de P.).

391a

- a, détail de l'oreillon.

392

392 - Sabre de général ou d'amiral.
Directoire - Consulat.
Monture en laiton doré. Calotte en crosse à tête de lion et courte queue. Poignée d'ébène finement quadrillée. Arc de jointure courbé, à brin supérieur aboutissant à la croisière marquée par un visage humain (Neptune ?) en guise d'oreillon. Quillon recourbé terminé par une tête canine. Fourreau de cuir à trois longues garnitures : chape à médaillon ovale dans lequel est inscrit une tête léonine, un trophée d'armes, l'ensemble bordé d'une couronne de feuillage ; la chape se termine par une toile d'araignée. Bracelet à la même tête que sur la croisière avec de part et d'autre, des toiles d'araignée, bouterolle en suite.

393

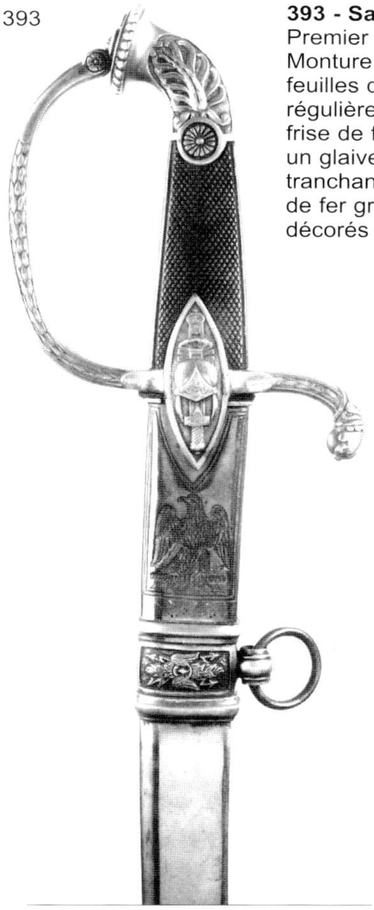

393 - Sabre d'officier d'Etat-Major.
Premier Empire.
Monture en laiton doré. Calotte en crosse et courte queue décorée d'une rosace et de feuilles d'acanthe stylisées (393a). Poignée en bois quadrillé et noirci. Arc de jointure régulièrement courbé, faisant un " S " avec le quillon, arc et quillons étant ciselés d'une frise de feuilles de laurier. Oreillon en navette décorée d'un casque vu de face, placé sur un glaive à l'antique (393b). Lame de 86,2 cm, courbe, dos à jonc, pan creux, contre-tranchant et gravée sur fond d'or de rinceaux, foudre, trophée d'armes (393b). Fourreau de fer gravé à sa partie supérieure de l'aigle, à deux bracelets de laiton à grosse bordure, décorés de foudres ailées (393c), bouterolle à feuillage et palmes.

393b

393c

393a

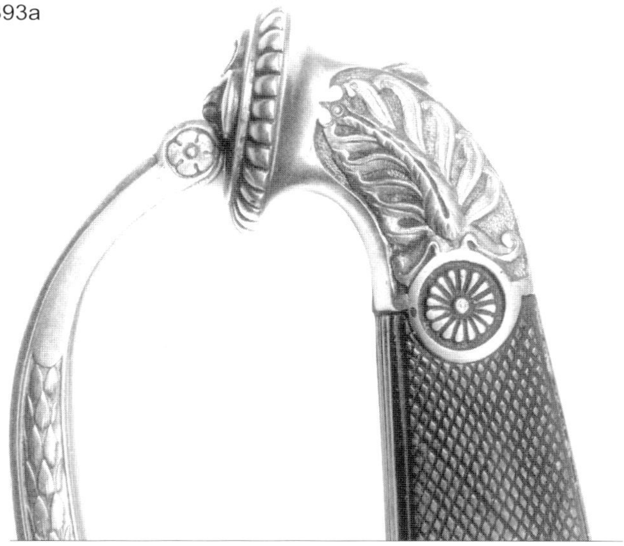

- **a**, gros plan sur la calotte en crosse montrant la palme et la rosette, la poignée finement quadrillée et une partie de la décoration de l'arc de jointure.
- **b**, oreillon et lame en damas très ornée, remarquer le bâton de foudre ;
- **c**, présence de l'aigle au-dessus du bracelet supérieur, lui-même frappé d'un bâton de foudre ailé.

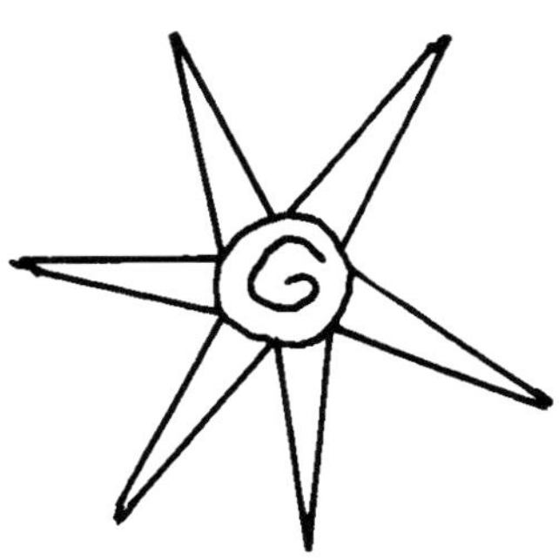

Troisième partie

Sabres à un arc de jointure, croisière dépourvue d'oreillon

I - Poignée en laiton moulé

II - Poignée en bois recouvert ou non de basane, arc de jointure se poursuivant par la croisière

III - Poignée en bois recouvert ou non de basane, arc de jointure se prolongeant par un plateau ovoïde découpé

IV - Monture en fer, à grosse coquille noircie

Les sabres décrits dans cette Troisième Partie sont groupés en quatre Chapitres de la façon suivante :

Prenant en considération toutes ces caractéristiques nous avons établi la classification suivante :

Chapitre I

Poignée en laiton moulé

Ce chapitre inclut des sabres obtenus par moulage de la poignée et pourvus d'un arc de jointure. Pas d'oreillons sur la croisière. Nous avons distingué deux cas, créant ainsi deux Familles ainsi caractérisées :

Famille I – Arc de jointure arrivant en angle droit sur la croisière

Occasionnellement, les Manufactures d'Armes ou les fourbisseurs, depuis le règne de Louis XVI jusque sous celui de Louis-Philippe, ont fabriqué par moulage des sabres de différents corps, aussi bien pour les troupes à pied que pour la Marine. La forme de leur calotte nous a conduit à créer deux Sous-familles ainsi définies :

Sous-famille I a - Calotte figurative

La poignée se termine par une tête de lion dite aussi de chien de mer, d'aigle ou de bonnet phrygien durant la période révolutionnaire.

394

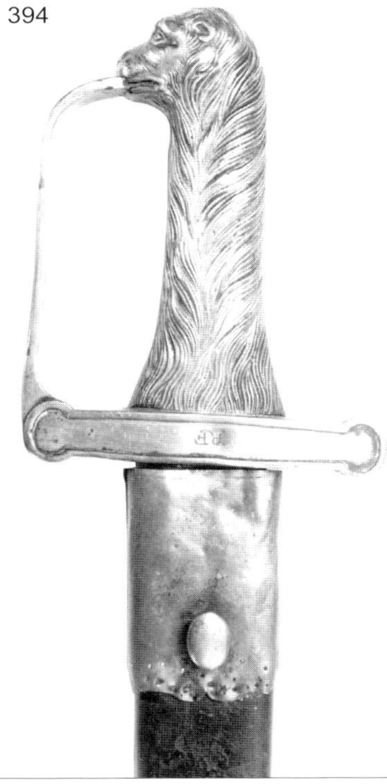

394 - Sabre de Grenadier.
Poignée de laiton fondu et moulé, à tête léonine placée horizontalement avec une crinière qui s'étend jusqu'à la croisière et pas de rivet visible. Arc de jointure fiché dans la gueule du lion, d'abord incurvé puis droit sur les trois quarts inférieurs. Croisière plate, à extrémités arrondies, l'ensemble étant à bourrelet ; chez les exemplaires en service après 1792, le milieu de la croisière peut être gravé de " RF ", pour République Française (394a). Lame souvent de récupération, pouvant mesurer 3,7 cm au talon, longue d'environ 60 cm, à large rainure sur 45 cm, contre-tranchant et flèche de 1,5 cm. Fourreau de cuir noirci, à chape à bouton ovalaire, fixée sur le cuir par martelage ainsi que la bouterolle légèrement asymétrique.

394a

- **a**, croisière marquée " RF " en son milieu pour République Française.

395

395 - Sabre de l'Artillerie montée.
Modèle 1792.
Monture en laiton moulé. Poignée à trois rivets, terminée par une tête léonine à 45° et entièrement velue. Branche simple, arrivant sur la croisière en angle de 90°. Croisière en barrette, aux extrémités arrondies. Lame d'environ 59 cm, large au talon de 3,8 cm, à flèche de 2,3 cm, à soie apparente (395a, 395b). Fourreau de cuir noirci à deux garnitures : chape à fente, bouterolle à bouton.

395a

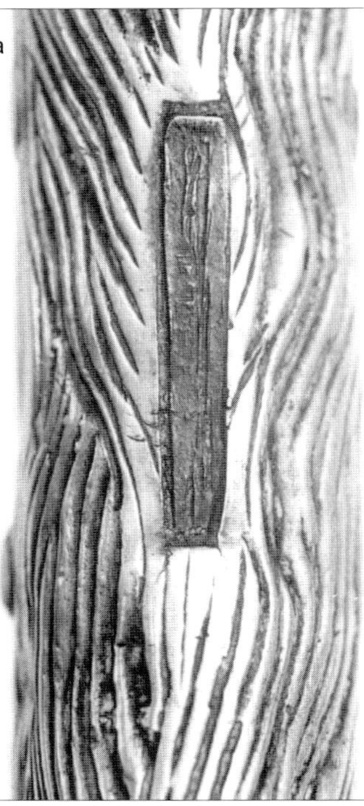

395b

Marques - Certains exemplaires peuvent être marqués du poinçon de Bick, du bonnet phrygien surmontant le " B ", de Jean Georges Bick et du faisceau de licteur.

Variantes - Il existe des dérivés de ce sabre dont la fusée moulée est à cordons ou même de bois entièrement filigrané (voir C. Ariès 2ème fasc. 1969).

- **a**, vue de l'aileron sur le dos de la poignée ;
- **b**, position de l'aileron sur la poignée.

396

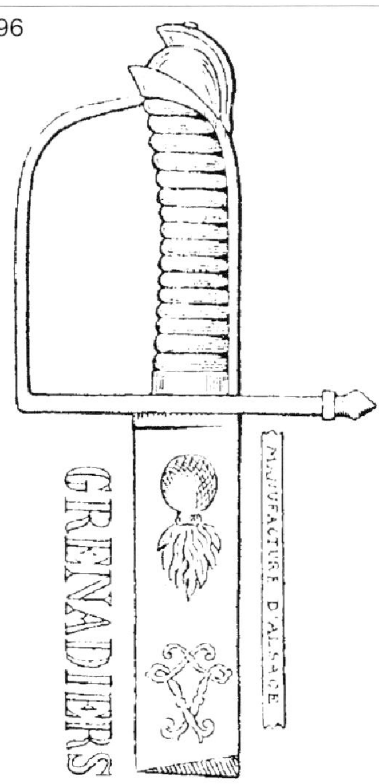

396 - Sabre de Grenadier.
Fin du XVIIIème siècle.
Monture en laiton. Calotte en casque à cimier et longue queue. Poignée à 16 torons.Lame plate marquée *GRENADIERS*, avec au talon une grenade enflammée et les deux " L " entrelacés, signée sur le dos *MANUFACTURE D'ALSACE*. Fourreau à deux garnitures, chape à fente (d'après M. Bottet).

397

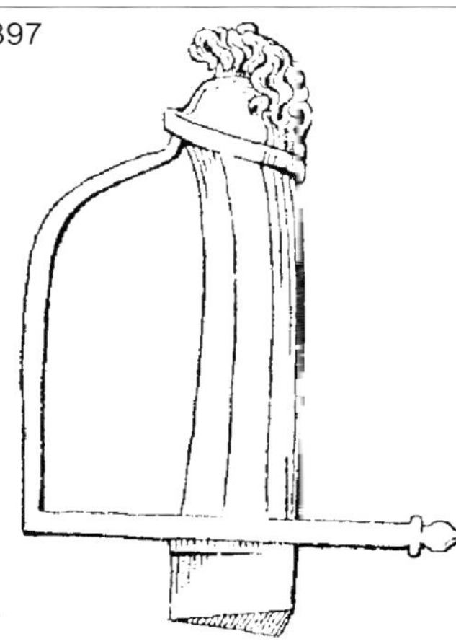

397 - Sabre d'Infanterie.
Fin du XVIIIème siècle.
Calotte au casque à panache.
Monture en laiton. Poignée à pans.
Lame plate. (d'après M. Bottet).

Les sabres des Compagnies d'Elite Epoque Révolution

Ces sabres d'Infanterie hérités des troupes de l'Ancien régime ont été plus ou moins modifiés avec l'avènement de la République. Le caractère le plus évident est la forme du pommeau. On en a distingué quatre modèles principaux : calotte en tête de lion, en tête d'aigle, en casque, en bonnet généralement à cocarde tricolore.

Les montures sont en laiton moulé, à l'arc de jointure arrivant en angle droit sur la croisière légèrement élargie au niveau de la poignée et terminée par un quillon droit à bouton.

On trouve des têtes de lion, de différentes factures, très détaillées, mais aussi très épurées tel l'exemple figuré ci-contre.

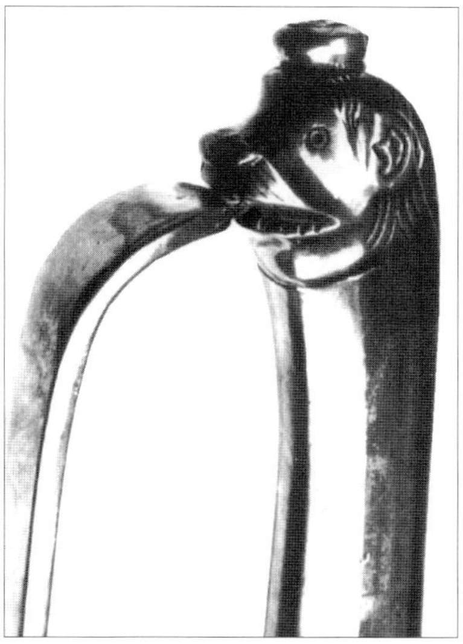

398

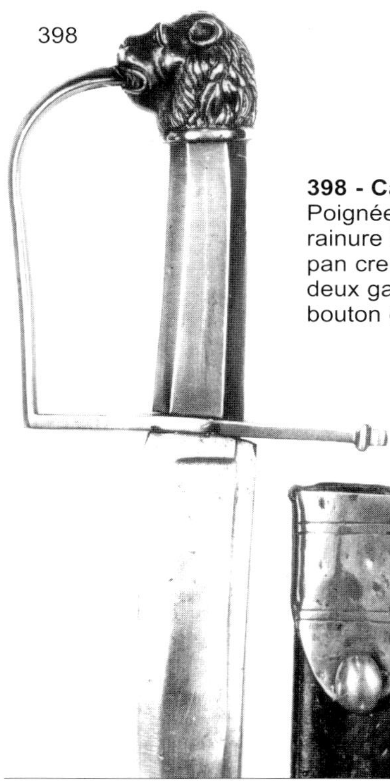

398 - Calotte à tête de lion.
Poignée à pans. Lame à rainure le long du dos et grand pan creux. Fourreau de cuir à deux garnitures, chape à bouton et anneaux (398a).

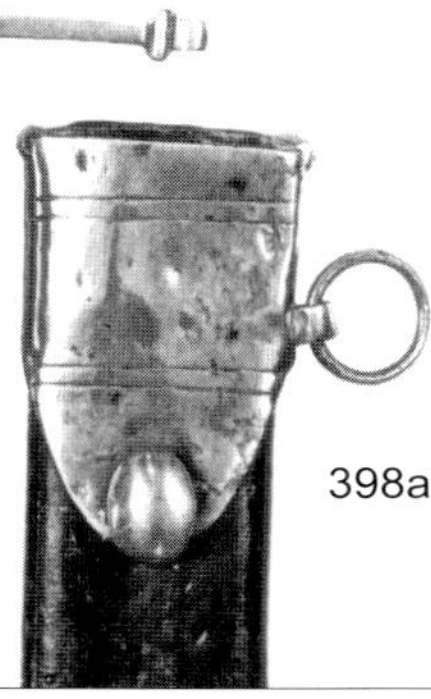

398a

399

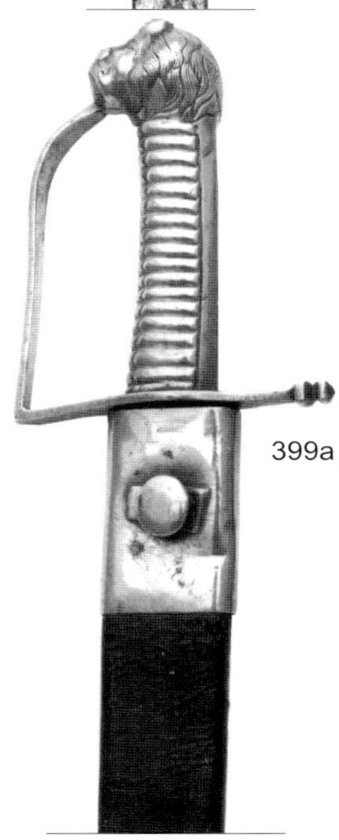

399a

399 - Calotte à tête de lion.
Poignée à cordon. Lame plate. Fourreau à bouton (399a).

400

400a

400 - Calotte à tête d'aigle.
Poignée à pans. Lame à long pan creux. Fourreau à deux garnitures : courte chape à bouton (400a).

401

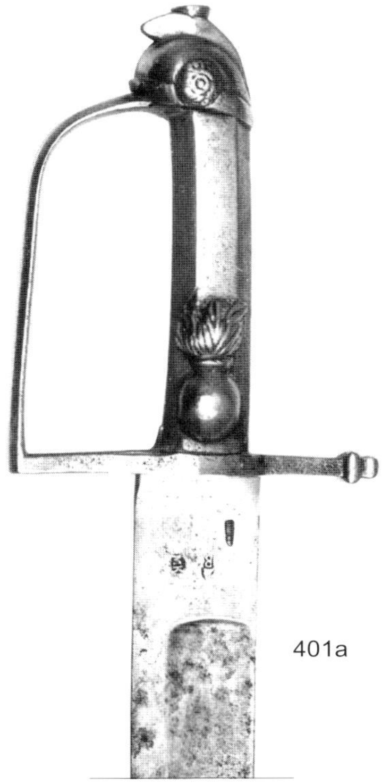

401a

401 - Calotte au casque à la cocarde.
Officier de Grenadiers. Après 1792. Poignée à pans. Lame à large pan creux, poinçonnée au faisceau de licteur, au " B " surmonté du bonnet phrygien de J.G. Bick et du " MK " de Cabannes (401a).

402

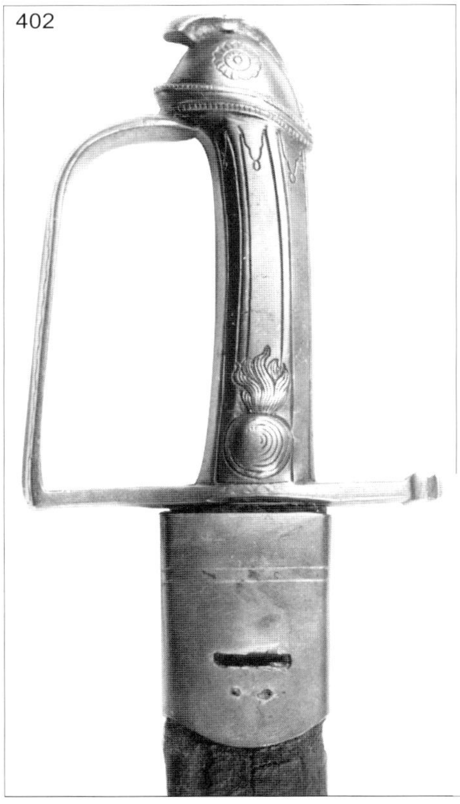

402 -

Officier de Grenadiers.
Après 1792.
Calotte au casque frappé de la cocarde. Arc de jointure ciselée de deux lignes marginales sinueuses. Poignée à huit pans peu marqués, séparés par des filets réunis, à leur partie supérieure, par un petit décor gravé ; sur la base du *recto*, une grenade en relief décorée de cercles concentriques dont la flamme est striée de lignes épousant sa forme ; sur le *verso*, oreillon supérieur figuré orné d'un filet. Lame plate, longue de 64,5 cm, large au talon de 4 cm et à flèche de 2,3 cm. Fourreau avec chape à fente, bouterolle terminée par un méplat, l'une et l'autre étant décorées de deux traits.

Sous-famille I b - Calotte non figurative

Nous avons inclus dans cette Sous-famille des sabres à croisière linéaire et des sabres ornés d'un petit motif latéral que l'on peut considérer comme une coquille embryonnaire.

403 - Sabre de Troupe de Garde Suisse, Régiment d'Erlach.
Vers 1765.
Monture en laiton. Monture fondue d'une seule pièce. Arc de jointure droit sur les 4/5èmes et faisant un angle droit avec la croisière rectiligne. Poignée à pans. Lame de Solingen, à gouttière dorsale, gravée à partir du talon d'entrelacs, de *REG*^{MT} */ D'HERLAG / SUISSE*, d'un trophée constitué d'un tambour, d'un canon sur fond de piques, de trompettes... (M.E. - S. de P.).

403

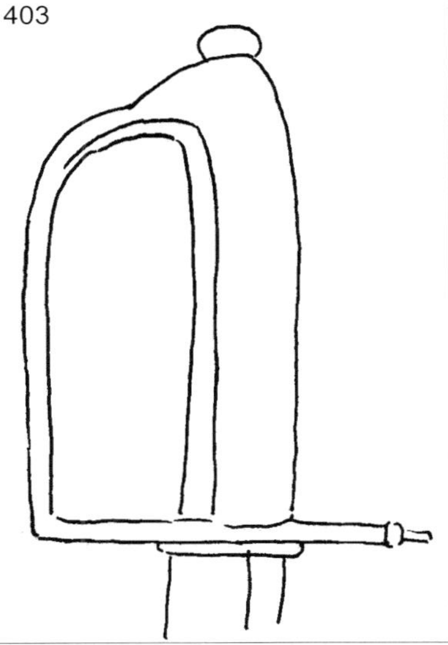

404 - Sabre d'Infanterie du Régiment de Nassau.
1767 - 1790.

G. Jaeger (1998) a examiné avec grand soin des sabres du Régiment de Nassau, régiment au service du roi de France.
Le sabre attribué à l'Infanterie est très voisin du briquet de 1767 mais est dépourvu d'oreillons, réels ou figurés. Ce sabre répond à la description suivante :

Monture en laiton moulé. Calotte simple, à bouton de rivure subconique inversé et longue queue. Poignée à 20 cordons. Arc de jointure arrivant en angle droit sur la croisière se terminant par un bouton. Lame plate, longue de 58 cm, large au talon de 3,35 cm et légère flèche de 1,5 cm, marquée sur le plat :

et sur le dos *M*^{ture} *de Solingen*... suivi du nom des fourbisseurs *Schnitzler et Kirschbaum*, un poinçon " S K " étant près du talon. Fourreau de cuir à deux garnitures de laiton du modèle 1767 (non figuré).

405

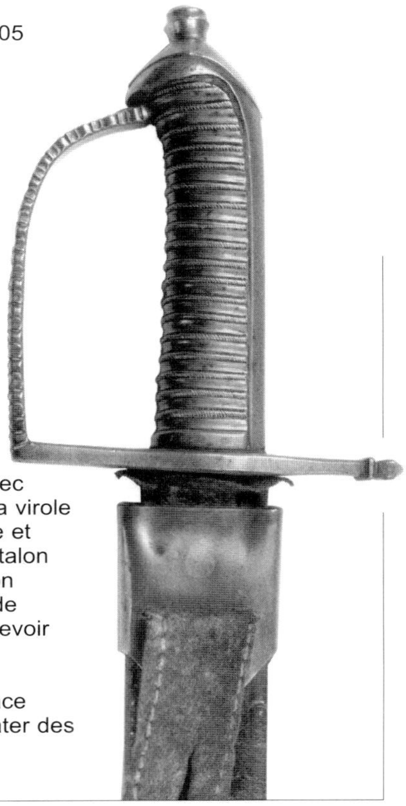

405 - Sabre d'officier d'Infanterie.
Vers 1774 - 1784, variante du modèle 1767.
Monture moulée en deux pièces. Calotte simple prolongée par une longue queue figurée. Poignée à 18 cannelures en creux, avec entre chacun d'eux, un filigrane venu lors du moulage ainsi que la virole basale. Arc de jointure latéralement cannelé. Croisière plate, unie et terminée par un bouton. Lame de 62,5 cm de longueur, large au talon de 3,5 cm, à flèche de 0,9 cm, à large gouttière finissant à environ 13 cm de la pointe, puis contre- tranchant. Fourreau de fort cuir de vache à deux garnitures : une chape à fente horizontale pour recevoir la languette de port, petite bouterolle à bouton.

Ce sabre très particulier n'est pas signé de la Manufacture d'Alsace mais la lame est poinçonnée de L.A. Gau, ce qui permet de la dater des années 1774 - 1784. Il s'agit selon toute évidence d'un sabre de fantaisie commandé par un officier.

406

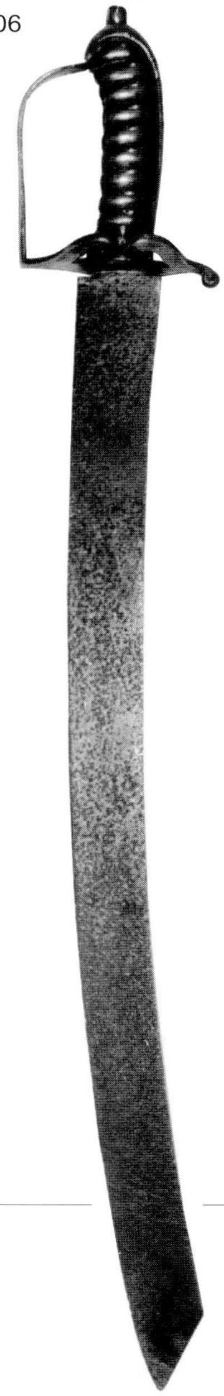

406 - Sabre des troupes de Marine.
antérieur à 1772.
Monture en laiton moulé, particulièrement riche en cuivre. Calotte à gros bouton de rivure et prolongée d'une longue queue. Poignée à dix gros cordons (406a). Branche partant du premier cordon et arrivant en angle droit sur la croisière s'élargissant pour former un étroit plateau orné d'une petite garde formée de deux petites branches courbées en " S " se rejoignant sur le plan médian (406b). Court quillon en bouton infléchi. Lame plate de 56,8 cm, large au talon de 4,1 cm et flèche de 1,8 cm, avec son extrémité taillée en biseau. Fourreau de cuir noir, chape à fente et tirant ou bouton, bouterolle courte terminée par un bouton (M.E. - S. de P.).

Ce sabre était attribué aux bas-officiers des Fusiliers et des Canonniers du Corps Royal de la Marine.

406a

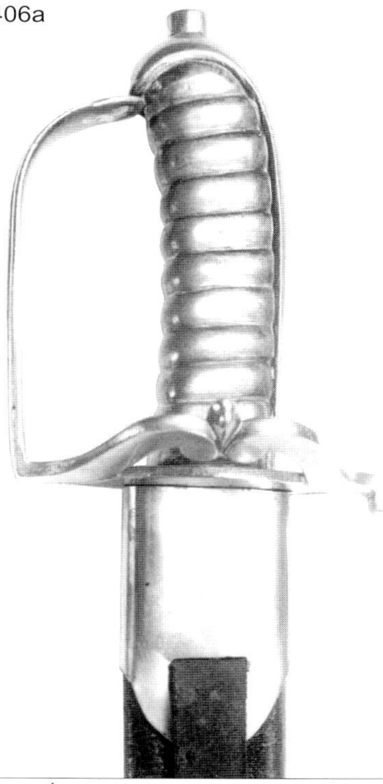

406b

- **a**, détail de la monture ;
- **b**, détail de la petite garde frontale précédent un plateau en navette.

407

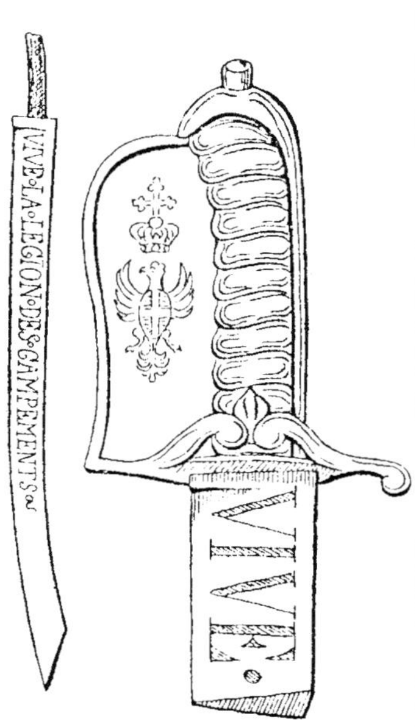

407 - Sabre de la Légion des Campements
Vers 1774 - 1776.
Ce sabre de Marine présente la particularité de posséder une lame sur laquelle on lit
VIVE . LA . LEGION . DES . CAMPEMENTS
avec l'aigle couronnée à la croix de Savoie, sur le *verso* et sur le *recto*
VIVE . LE . ROY . DE . SARDAIGNE

Nota - *Nous n'avons pas retenu, en général, les sabres sur lesquels existe l'inscription Vive le Roy de Sardaigne qui sont d'une facture étrangère à notre pays. Le sabre présent est donc une exception et nous ne l'avons retenu simplement que comme témoignage d'un fait historique : l'Alliance de la Maison de Savoie et du Royaume de Sardaigne lors de la Guerre de Succession au trône de Pologne.* (d'après M. Bottet).

408 - Sabre des Troupes embarquées.
Fin du XVIIIème siècle.
Monture de laiton moulé. Il diffère du sabre de Grenadier d'Infanterie sur les points suivants : la poignée compte 20 cannelures, l'oreillon figuré supérieur a disparu, toute la monture étant d'un seul bloc. La chape du fourreau est munie d'un bouton et d'un anneau de bélière et la lame est poinçonnée d'un ovale dans lequel est inscrit une ancre de marine.

Famille II – Arc de jointure régulièrement arqué

Les trois premiers sabres décrits ont une parenté évidente avec les sabres de la Famille I. Les autres modèles figurés apparaissent en l'An IX et annoncent toute une série de sabres qui prendront le surnom de " briquets ".

409 - Sabre de Grenadiers.
Règlement du 19 janvier 1747.
Monture en laiton moulé. Calotte en casque rudimentaire ou en poire. Poignée à grosses rainures en hélice, *afin que les poignées filées... ne rapent pas et détruisent les habits* (*in* J. Margerand, 1945, p. 54). Branche arrondie, se poursuivant par la croisière et finissant par le quillon. Lame primitivement de 81,2 - 83,9 cm, puis plate et ramenée à 59,5 cm de longueur, large au talon de 3,6 cm et flèche de 0,9 cm. Fourreau de cuir de vache avec éclisses, chape de 3 cm, bouterolle de 5,4 cm à bouton (d'après J. Margerand, 1945, p.53), (non figuré).

410 - Sabre des troupes de Marine.
Entre 1772 et 1783.
Monture en laiton moulé. Calotte à longue queue du modèle de 1772 mais plus droite. Poignée jusqu'à 19 cordons. Branche de garde bien arrondie à ses deux extrémités. Petite garde constituée d'une palmette à cinq éléments, soutenus par deux petites branches en " S ". Lame plate de 56,8 cm environ, large au talon de 4,1 cm avec une flèche de 1,8 - 2 cm et se terminant en pointe au niveau du dos. Fourreau de cuir noir, chape à fente et tirant de cuir, bouterolle à bouton terminal.

410

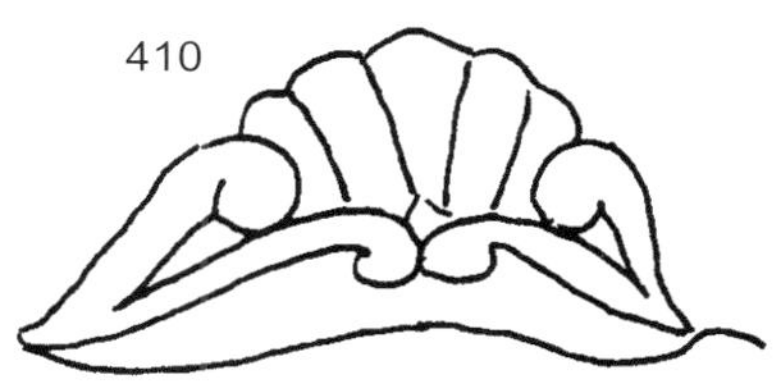

Nota - On observe quelques variations dans la forme de la lame qui peut être munie d'une gouttière.

Inscriptions - Rares inscriptions concernant le corps, telles sur la lame : *Corps Royal de la Marine* sur une face et sur l'autre *Grenadier du rgt de Brest* en lettres manuscrites. Il existait également des troupes de Marine de ce corps à Toulon, à Rochefort...

411a

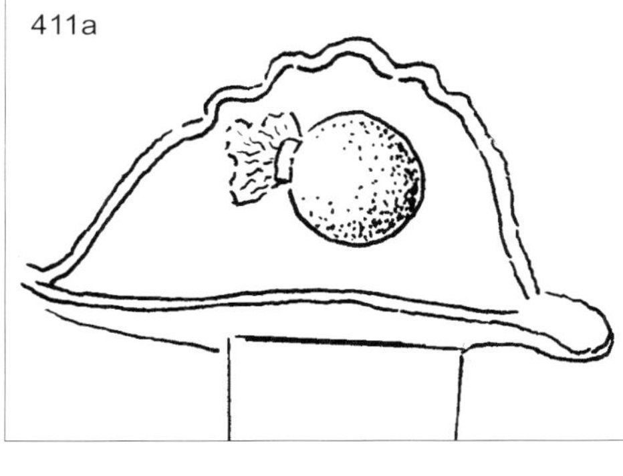

411 - Sabre de Bombardiers du Corps royal de la Marine.
Modèle 1772.
Ce sabre est au modèle 1772-1783 et s'en distingue par le présence d'une bombe enflammmée en relief sous la garde frontale (411a). (M. Pétard, 1991).

411

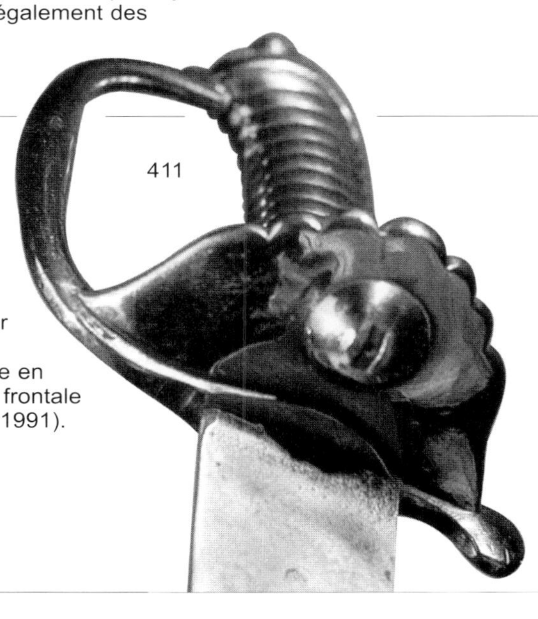

On peut rencontrer des sabres dont la lame est marquée :

(inscription se lisant Grenadiers Postiches)

Le mot *postiche* est explicité par M. Pétard (Gaz. Armes, 281, 23-26) en ces termes :

... dès le XVIIIème siècle, il était d'usage courant d'appeler postiche tout homme militaire ou domestique introduit discrètement dans le rang afin de combler les manques d'un effectif établi... Autrement dit, le *postiche* est une sorte de *suppléant* et par extension, dans le cas présent, un *aspirant grenadier,* selon l'ordonnance du 28 janvier 1746. L'autre face de la lame porte sa marque d'origine, soit *Manufacture R^{ale} d'Alsace,* également écrit en grandes lettres manuscrites évidées.

Les briquets (*)

Ce modèle de sabre a été créé en l'An IX. Originellement, la lame est courbe, sans pan creux ni biseau, longue de 59,6 cm et de 2,6 cm ,de flèche. Monture de laiton coulé et moulé d'une seule pièce. La branche de garde est arrondie et se termine par un quillon en bouton surmontant une embase en tronc de pyramide. La poignée comporte 36 cannelures. Le fourreau est à deux garnitures, en cuir noir, sans éclisse ; chape à pontet et tirant de buffle, bouterolle terminée par un bouton. Dès l'An XI, ce sabre subit deux modifications : le quillon ne se termine plus que par un bouton en goutte d'eau et les cannelures sont réduites à 28.

Les troupes embarquées seront armées de ces modèles, mais dépendant de la Marine, le milieu de la croisière est frappé d'une petite ancre de marine.

De 1800 à 1815, 418 639 briquets des années IX et XI ont été fabriqués par les Manufactures de Versailles, de Klingenthal et de Turin.

412

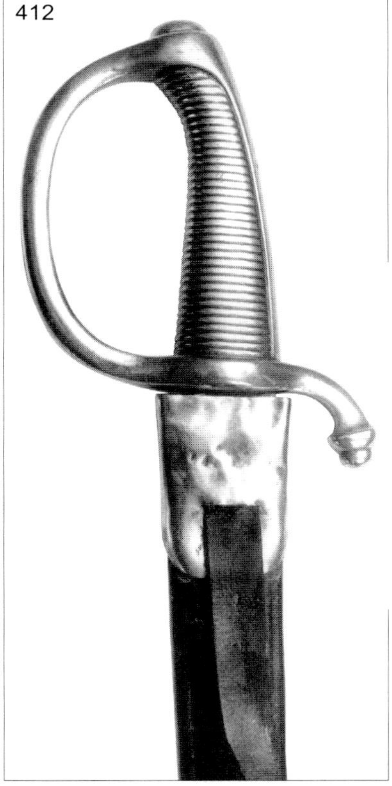

412 - Sabre d'Infanterie dit briquet.
Modèle de l'An IX.
Monture en laiton moulé, dessiné par N. Boutet, Entrepreneur à Versailles. Calotte à dôme et à longue queue figurée. Poignée à 36 cannelures aplatis. Arc de jointure arrondi pouvant porter au *verso*, le poinçon de Boutet et le " V " de vérification. L'arc se poursuit en " S " pour aboutir au quillon recourbé se terminant en *" trompette "*. Lame plate, longue de 59,5 - 62 cm, large au talon de 3,38 - 3,5 cm avec une flèche de 1,12 - 2 cm. En fonction de la date de fabrication, cette lame peut être signée de la *Manufacture de Klingenthal, Coulaux Entrepreneur* et plus tard de la *Manufacture Impériale de Klingenthal, Coulaux Frères...* et être poinçonnée par François Louis Lobstein, Jean-Georges Bick, Joseph Innocent Krantz, Simon Ferdinand Beaumaretz, Jean-Jacques Mouton... Fourreau de cuir noirci à deux garnitures : chape à fente horizontale pour assujettir la languette de buffle pour le port, bouterolle à bouton. Chape et bouterolle sont souvent ornées d'une moulure marginale.

(*) Le surnom péjoratif de briquet fut donné aux sabres d'Infanterie par les cavaliers, ceux-ci jugeaient que ce sabre à lame courte était impropre au combat et ne pouvait servir qu'à battre le briquet. Toutefois, il ne faut pas oublier qu'au XVIIIème siècle, le briquet était déjà un long couteau.

413

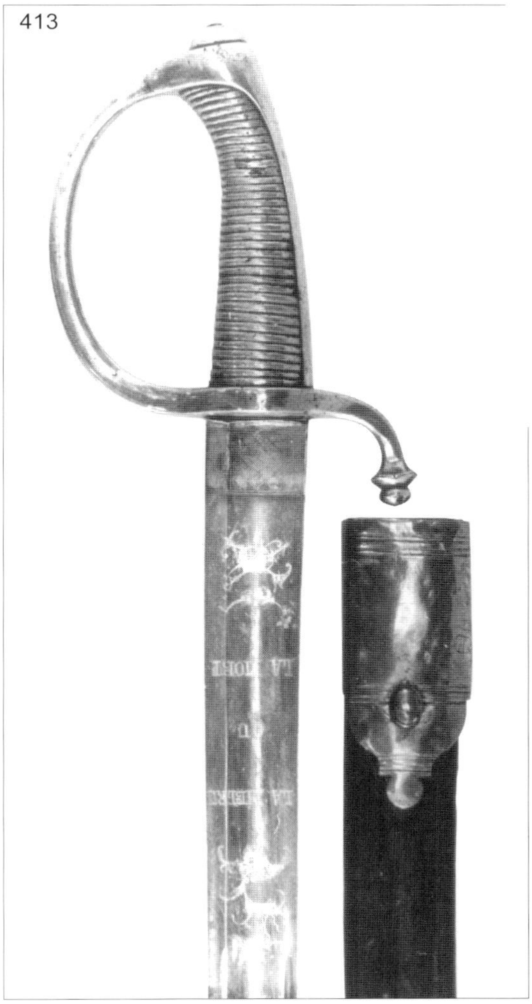

413 - Sabre de l'An IX modifié.
Ce modèle est nettement allégé, de fabrication très soignée et fut certainement commandé à un fourbisseur privé par un officier. On remarquera tout particulièrement que la lame est une lame de Cavalerie légère, bleuie et portant, sur le *recto*, en capitales dorées *LA LIBERTE / OU / LA MORT* et sur le *verso VIVE / LA / NATION* entre un trophée et des motifs végétaux. Chape à bouton.

414

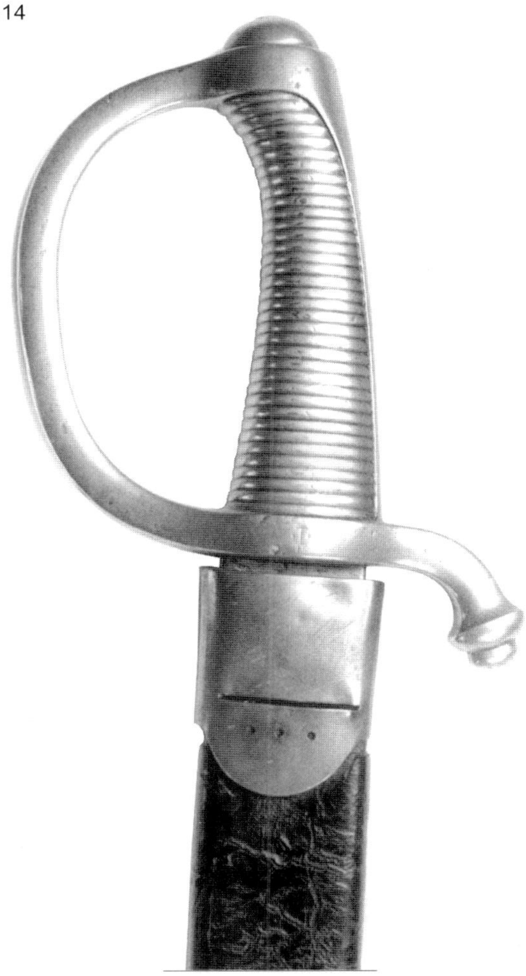

414 - Sabre-briquet des troupes de Marine, modèle de l'An IX.
Ce modèle est très semblable à celui de l'Infanterie : 36 cannelures sur la poignée, quillon en forme plus ou moins pyramidale surmonté d'un bouton. Il en diffère par la présence d'une petite ancre aux bras bien ouverts au milieu de la croisière (414a). La lame est un peu plus longue que celle de l'Infanterie et mesure 65 cm mais avec la même largeur au talon et la même flèche ; elle est gravée, sur chaque face, au trait, d'une grande ancre placée obliquement, les bras placés vers le talon de la lame. Fourreau de cuir noirci, avec une chape à bordure inférieure arrondie et pourvue d'une fente pour insérer la languette de port.

Marquage - Sur le dos de la lame, on peut lire, en abrégé, *" Manufacture Nationale du Klingenthal, Coulaux Frères Entrepreneurs "*. Sur le talon de la lame, on reconnaît le poinçon J.G. Bick et une petite ancre identique à celui de la croisière qui est un poinçon d'Arsenal.

- **a**, briquet des troupes de marine, modèle de l'An IX, avec le poinçon de réception de la Marine, au centre de la croisière.

414a

415

415 - Sabre briquet d'Infanterie.
Modèle de l'An XI.
Monture en laiton moulé. Chez ce modèle, on remarque que les cannelures ne sont plus que 28 et plus saillants que dans le modèle de l'An IX. Le quillon se termine en goutte d'eau. Lame de 59,5 cm de long à flèche de 2 cm environ. Même fourreau mais la chape est à pontet.

Marquage - Le dos de l'arc de jointure ou de la lame porte la signature de la Manufacture de Versailles (415a) avec les poinçons de " L.F. " Levavasseur, de " S.F. " Beaumaretz et de " J.I. " Krantz. Puis à partir de 1804, de la Manufacture Impériale de Klingenthal, le mois et la date de fabrication. Ce modèle fut également fourni sous la Restauration comme en témoigne le marquage de la lame, *Manufacture Royale de Klingenthal, Octobre 1814.*

415a

- **a**, sur l'envers de l'arc de jointure on peut découvrir chez certains exemplaires le poinçon de Versailles, celui de Boutet et un poinçon " L.F. ".

416 - Sabre-briquet des Troupes de Marine.
Modèle de l'An XI.
Ce modèle est très comparable au modèle de l'Infanterie :
28 cannelures sur la poignée, quillon recourbé et à goutte d'eau. Lame plate de 61 - 62 cm, donc un peu plus longue que celle de l'Infanterie, large au talon de 3,5 cm, légèrement courbe avec une flèche d'environ 2 cm et sur le dos on lit en abrégé, "*Manufacture Impériale de Châtellerault,* le mois et la date". Croisière poinçonnée d'une petite ancre de Marine, aux bras droits et relevés, formant un " V " (416a). Fourreau en cuir bouilli noirci, sans couture apparente ; chape à base rectiligne, à pontet de suspension. Encore en service sous le Second Empire.

416

416a

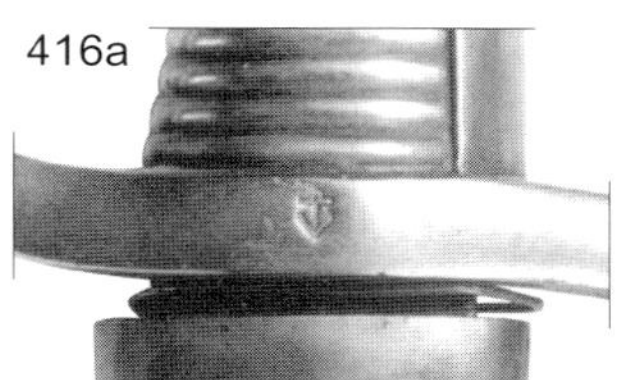

417 - Sabre-briquet des troupes de Marine.
Modèle de l'An XI.
Ce briquet, en plus de l'ancre gravée au milieu de la croisière est remarquable par la signature qui figure sur la lame. En effet, on lit clairement Marseille. Cette signature, rare, permet de faire plusieurs hypothèse : Une marque d'arsenal ? Une fabrication marseillaise ? ...

427 - Sabre de bas-officier de Cavalerie légère.
Vers 1790.
Monture en laiton. Calotte simple, à gros bouton de rivure et longue queue. Poignée recouverte de basane à filigrane de laiton torsadé faisant une douzaine de spires. Arc de jointure droit sur environ les 4/5èmes de sa longueur et arrivant en angle droit sur la croisière terminée par un bouton à facettes. Lame blanche d'environ 84 cm, à pans creux et gouttière au fort. Fourreau de cuir noirci, à trois garnitures de laiton.

427

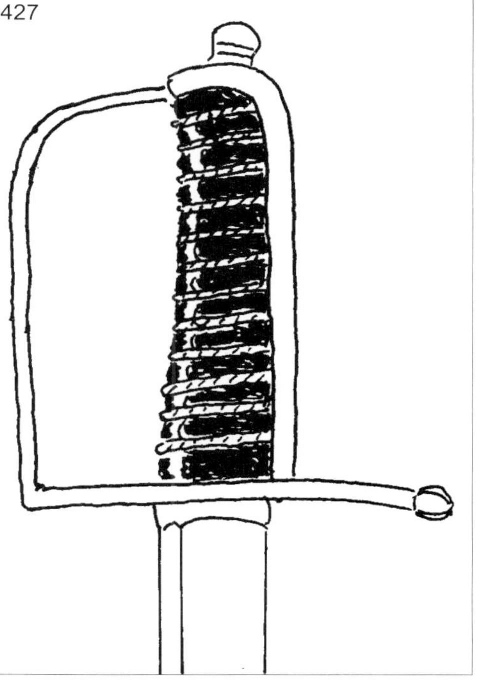

428

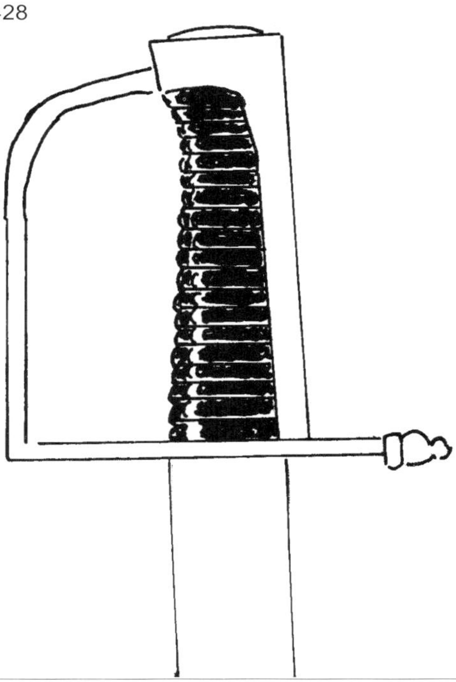

428 - Sabre de Cavalerie légère.
Milieu du XVIIIème siècle.
Monture en laiton. Calotte surmontée d'une pièce de rivure en dôme très aplati et longue queue. Poignée encordée recouverte de basane non filigranée. Arc de jointure arqué en son tiers supérieur et ensuite rectiligne, arrivant en angle droit sur la croisière également rectiligne et terminée par un bouton trilobé. Lame légèrement courbée.

429

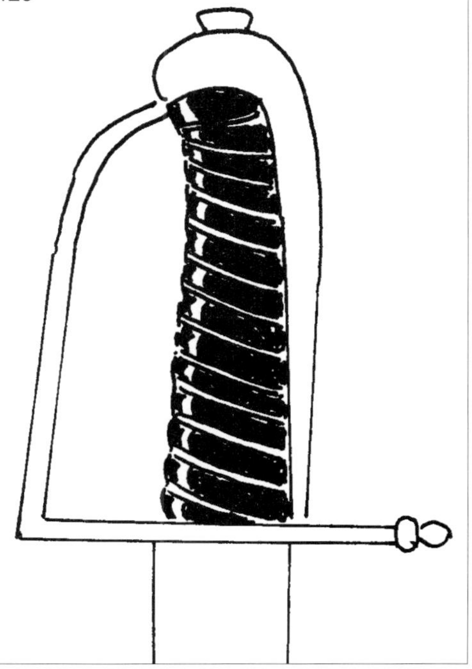

429 - Sabre de Troupe de Garde Suisse, Régiment de Lochmann.
1752 - 1777.
Monture en laiton. Calotte simple à longue queue. Poignée recouverte de basane, fixée par un fil de fer faisant 13 spires. Arc de jointure oblique, arrivant sur la croisière rectiligne, en angle de 80° environ. Lame marquée *Reg*^{t} *De Lochmann Suisse* (M.E. - S. de P.).

430

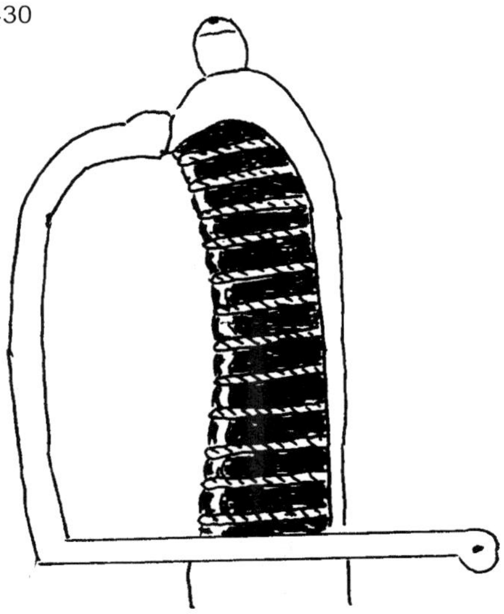

430 - Sabre de Troupe Suisse, Régiment de Salis - Samade.
1782 - 1792.
Monture en laiton. Calotte à fort bouton de rivure et longue queue. Poignée recouverte de basane filigranée d'un fil de laiton torsadé faisant une douzaine de spires. Arc de jointure arrivant sur la croisière en angle presque droit. Exceptionnellement ce modèle n'est pourvu que d'un seul oreillon supérieur sur la face postérieure. Lame plate de 62,3 cm à très faible flèche. L'exemplaire conservé au Musée de la Manufacture d'Armes de Saint-Etienne est marqué : *Modèle / de sabre / de / Salis / Samade Suisse.* Chape à bouton.

432

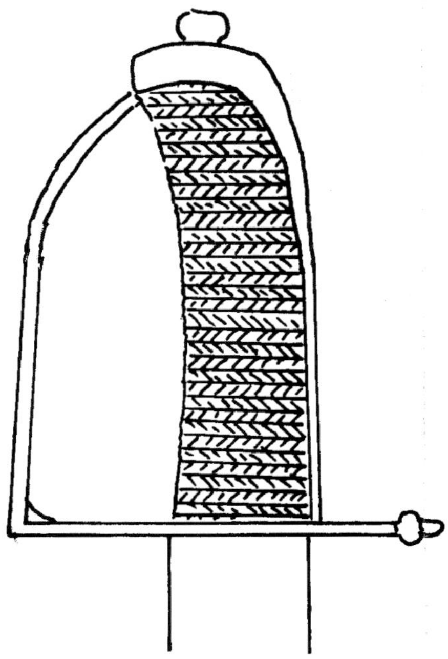

432 - Sabre de Grenadier.
Règne de Louis XVI
Monture en laiton. Calotte en cuillère à longue queue. Arc de jointure peu courbé, arrivant en angle presque droit sur la croisière, celle-ci rectiligne et terminée par un bouton. Poignée entièrement filigranée de fil de laiton. Lame droite marquée au talon de la signature du fourbisseur, par exemple : *Briet / m^{a} / fourbisseur / Pont / St Michel / à Paris.* Fourreau de cuir à deux garnitures de laiton, la chape étant à fente pour la languette de port. (d'après C. Ariès.IX.6).

431

431 - Sabre d'officier de Cavalerie légère.
Révolution - Directoire.
Monture en laiton doré. Calotte plate décorée au trait et longue queue. Poignée entièrement filigranée d'un fil de laiton torsadé. Arc de jointure arrivant en angle droit sur la croisière dont le quillon est fortement recourbé. Lame légèrement fléchée. Fourreau de cuir à deux garnitures : chape ornée de lignes sinueuses plates, avec dans les parties concaves un motif végétal. Bouterolle atteignant le milieu du fourreau, décorée en suite. Chape et bouterolle sont réunies par des attelles latérales.

433

433 - Sabre d'officier de Cavalerie légère.
Ce sabre, de facture austro-hongroise, armait des troupes étrangères au service de Louis XVI. Monture en fer. Calotte à bouton de rivure, unie, à longue queue. Poignée recouverte de basane filigranée de fils de cuivre rouge. Branche arrivant sur la croisière en angle droit. Quillon infléchi. Lame légèrement courbée, à dos plat. Fourreau de fer à deux longues garnitures à découpes de motifs géométriques constituées essentiellement de cercles (433a, 433b). Dard légèrement asymétrique.

- **a**, détail de la chape ;
- **b**, détail de la découpe de la chape d'un sabre de même conception.

433a

433b

434

434 - Sabre de Cavalerie légère.
Fin XVIIIème siècle.
Monture en fer.
Calotte à couvercle soudé et longue queue. Arc de jointure à cinq perles médianes. Poignée ficelée obliquement à une quinzaine de spires d'un fin fil de laiton torsadé. Croisière plate et étroite, terminée par un bouton replié en angle droit. Lame blanche de 83 cm de longueur, à gorge dorsale. Fourreau de cuir à trois garnitures de fer ; chape à bordure découpée en oblique avec anneau de bélière (434a), bracelet découpé en suite à anneau et longue bouterolle.

434a

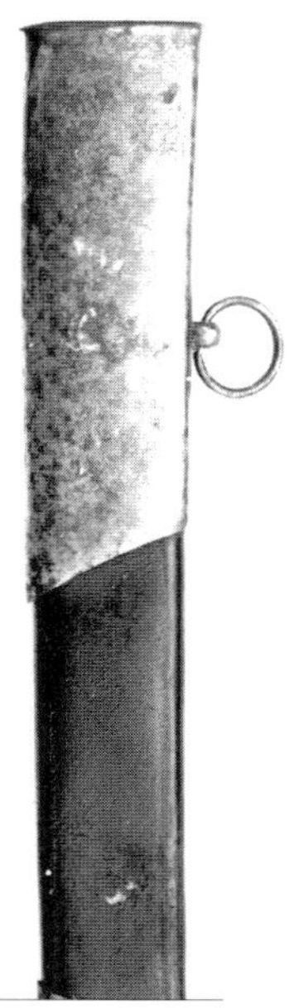

435 - Variante.
Fin du XVIIIème siècle.
Monture en fer. Calotte plate à longue queue. Arc de jointure, orné en sa partie médiane de cinq perles, arrivant en angle droit sur la croisière plate, à quillon infléchi terminé par un bouton allongé. Poignée recouverte de basane à filigrane faisant 22 spires, constitué d'un fil de laiton torsadé bordé d'un fil très fin également torsadé. Lame de 88 cm, raccourcie d'environ 4 cm, large au talon de 3,7 cm et à flèche de quelque 4 cm. Fourreau de fer uni, à deux anneaux de bélière fixés par des pitons. Dard asymétrique.

435

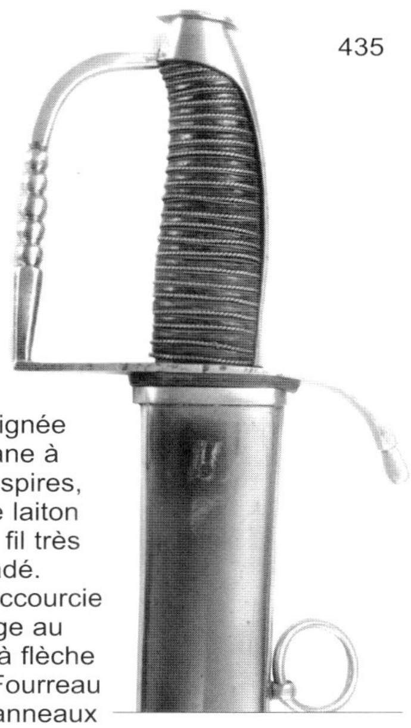

<u>Nota</u> - Ces deux derniers sabres sont à rapprocher du sabre décrit par Ariès (3ème fasc. 1970, fig. 3) : calotte à couvercle soudé, fusée à filigrane serré, anneaux de bélière directement brasés sur le métal du fourreau. Ariès émet l'hypothèse que ces sabres ont pu être portés par des Volontaires de la Révolution ou par les Cavaliers de la Légion germanique.

Famille IV – Arc de jointure arrivant en angle droit sur la croisière, calotte à courte queue

Le sabre le plus important de cette série est le sabre de la Garde des Consuls de l'An VIII. Un curieux sabre de Marine dont la croisière est pourvue d'un clavier nous a conduits à l'isoler dans une Sous-famille de la façon suivante :

Sous-famille IV a - Croisière dépourvue de clavier

Cette Sous-famille est composée de quelques sabres très différents de style dont les uns possèdent une fusée recouverte de basane filigranée ou de bois strié en quadrillage ou en oblique.

436

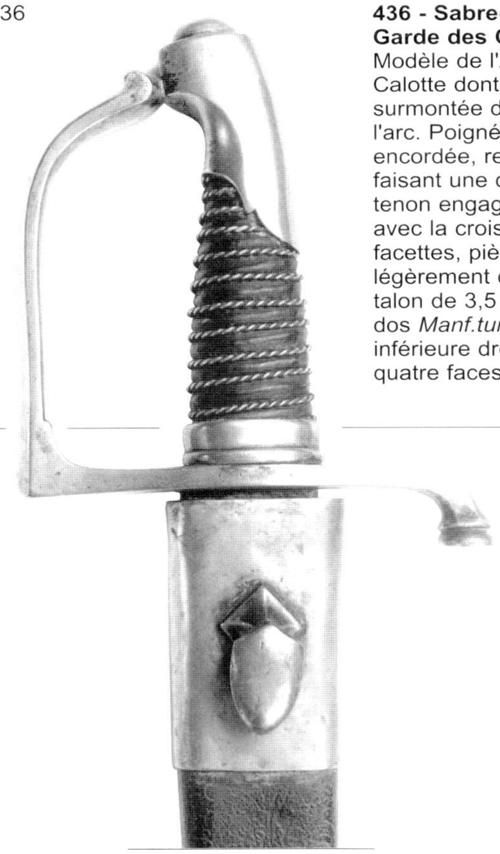

436 - Sabre-briquet de Grenadiers de la Garde du Directoire et de la Garde des Consuls. (1)
Modèle de l'An VIII. Monture en laiton.
Calotte dont la queue arrive à la moitié de la longueur de la poignée, surmontée d'un dôme servant de bouton de rivure, déprimée sous l'insertion de l'arc. Poignée plus large à la base (3,5 cm) qu'à la partie supérieure (2,3 cm), encordée, recouverte de basane et filigranée d'un gros fil de laiton torsadé, faisant une dizaine de spires ; grosse bague basale. Arc de jointure fixé par un tenon engagé sur la calotte, d'abord courbé, puis droit et faisant un angle droit avec la croisière. Croisière droite, se terminant par un quillon recourbé à huit facettes, pièces sur laquelle est plaquée une couronne ovalaire et plate. Lame légèrement courbée à flèche de 1 cm environ, longue de 63,8 cm, large au talon de 3,5 cm, à pan creux et double tranchant sur 11 cm et signée sur le dos *Manf.ture à Klingenthal*. Fourreau de cuir noirci, à chape à bordure inférieure droite, à gros bouton en écusson fixé sur une embase pyramidale à quatre faces ; bouterolle à bordure courbe, se terminant par un bouton ovoïde.

437 - Variante.
G. Jaëger (1999) signale un exemplaire quelque peu différent : la poignée est en ébène quadrillée et la monture est poinçonnée de BOUTET, " LD " et " AB " donc issue de la Manufacture de Versailles. La lame plate mesure 64 cm de longueur, large au talon de 3,1 cm et est gravée en grandes lettres verticales serrées les unes contre les autres *L'honneur est mon guide*. Le fourreau est à deux garnitures, la chape est à olive pour le port, sans embase pyramidale et la bouterolle se termine par un bouton. Ce sabre, selon C. Ariès (op. cit.) aurait été attribué aux officiers (non figuré).

(1) C'est le 17 ventôse de l'An X (2 mars 1802) que le premier Consul crée définitivement la Garde des Consuls. Cette garde se compose alors de quatre officiers généraux, d'un Inspecteur, d'un Capitaine du Génie, d'un Commissaire des Guerres, d'un Corps de Grenadiers et d'un corps de Chasseurs. Chaque Corps comprend un Etat-Major et deux Bataillons de huit Compagnies. A la Garde des Consuls, le 14 prairial de l'An XI seront adjoints la Légion d'Elite de la Gendarmerie et le 30 fructidor de la même année, un Corps de Marins.
Remarque : Pour être inclus dans ces Corps d'Elite, il fallait avoir montré son courage au Combat, être d'une morale exemplaire et mesurer, pour les Grenadiers 1,80 m et pour les Chasseurs 1,70 m.

438

438 - Sabre d'officier de Cavalerie légère.
Consulat.
Monture en laiton. Calotte à la hongroise avec une jupe, la queue postérieure étant un peu plus longue que l'antérieure. Arc de jointure en colonne ornée d'une perle médiane et relié à la calotte par une petite branche arquée. Poignée en ébène striée obliquement. Lame de prise espagnole aux armes de Navarre et portant l'inscription *" Viva Espana "*. Fourreau de laiton à deux bracelets aux bords moulurés, anneaux de bélière et dard du même métal.

440

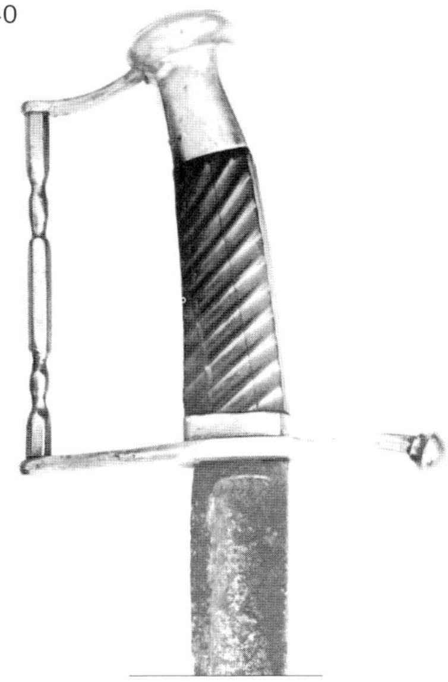

441

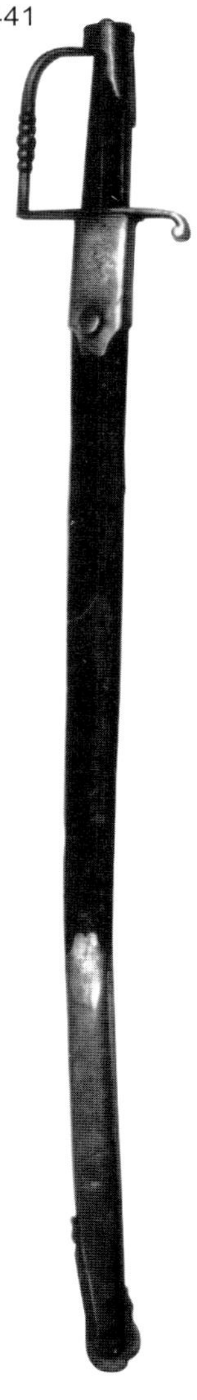

440 - Sabre d'officier d'Etat-Major de l'Infanterie.
Premier Empire.
Monture en laiton. Calotte plate, moulurée, à jupe subcylindrique, poignée d'ébène à cannelures obliques. Arc de jointure en colonne à pans et quatre étranglements limitant une longue *perle* entre deux plus petites. Croisière horizontale terminée par un bouton conique. Lame à large pan creux.

439

439 - Sabre d'officier d'Infanterie.
Consulat.
Monture en laiton. Calotte en tranches de melon et courte jupe. Poignée octogonale en ébène. Arc de jointure droit sur les 4/5èmes de sa longueur et décoré de trois perles placées entre deux petites pyramides. Quillon en goutte d'eau. Lame droite à pan creux, anciennement bleuie et dorée.

441 - Sabre d'officier d'Infanterie .
Premier Empire.
Monture en laiton. Calotte simple, à bouton de rivure saillant et courte queue. Poignée recouverte de basane et filigranée. Arc de jointure arqué près de la calotte puis droite et décorée de cinq perles, avant de rejoindre la croisière en angle droit. Quillon recourbé. Lame presque droite. Fourreau de cuir noirci, à deux garnitures dont la chape à bouton en olive et assez longue bouterolle terminée par un dard asymétrique.

Sous-famille IV b – Croisière à clavier

Les sabres exceptionnels constituant cette Sous-famille, préfigurent dans une certaine mesure, l'épée des Services administratifs, les présents modèles datant du Premier Empire. (voir J. Lhoste, Les Epées, 1997, p. 297, fig. 526).

442

442 - Sabre d'officier des Services administratifs de la Marine.
Premier Empire.
Monture en laiton argenté. Calotte en tête de lion au col à décor de toile d'araignée. Poignée en ébène finement quadrillée. Arc de jointure arqué en sa partie supérieure et arrivant en angle droit sur la croisière. Clavier en demi-lune à l'ancre à orin formant trois boucles, sans début ni fin. Lame à pan creux, bleuie et ornée de gravures banales dorées. Fourreau de cuir à deux garnitures argentées : chape à bouton ornée de quatre doubles traits et à bordure découpée (442a) ; bouterolle découpée et ornée en suite, avec un dard asymétrique.

442a

443

443 - Sabre du général Leclerc.
Premier Empire.
Monture en laiton doré. Calotte surmontée d'un gros bouton de rivure et à courte queue. Poignée en ébène cannelée verticalement à cartouche dans lequel est inscrit un personnage ailé et armé d'un arc (Eros ?). Arc de jointure en colonne genre tronc de palmier. Clavier à trophée d'armes. Fourreau à trois garnitures, décorées au trait, chape découpée et bouton de suspente.

Famille V - Arc de jointure régulièrement arqué et quillon infléchi en " S ", calotte à longue queue

444 - Sabre d'officier des Compagnies d'Elite.
Vers 1812.
Sabre tout à fait conforme au sabre d'officier de la Jeune Garde Impériale, sauf la calotte à longue queue à décor d'abeille et de feuilles d'acanthe. Lame bleuie et dorée, signée des Frères Weyersberg de Solingen, (non figuré).

445

445 - Sabre d'officier de Cavalerie légère de la Garde Royale
1815 - 1830.
Monture en laiton. Calotte à fleur de lys et longue queue. Poignée filigranée. Branche de garde régulièrement arquée, ciselée de feuilles de laurier stylisées avec une fleurette médiane ; nœud de corps décoré d'une fleur de lys placée entre deux rosaces. Quillon recourbé à bouton en fleurette. Lame à dos rond, contre-tranchant, pan creux et gouttière, portant les inscriptions suivantes : sur le *recto MANUFRE Royale du Kligenthal Coulaux Frères* au talon sur fond d'or, puis les Armes de France et *GARDE ROYALE* ; sur le *verso* : *Duc Fourbisseur rue St Honoré en face celle de Richelieu à Paris* N° 251 au talon, puis les Armes de France, *GARDE ROYALE*, fleurs de lys et le chiffre de Louis XVIII. Fourreau de cuir à deux garnitures décorées au trait, chape à bouton.

Famille VI - Arc de jointure irrégulièrement arqué, calotte à courte queue

Nous commencerons cette série de sabres par un rare modèle à calotte en tête de bélier de la période révolutionnaire. Nous isolons ce sabre dans la Sous-famille VI a. Tous les autres sabres entrent dans la Sous-famille VI b, qui ne sont pas sans affinité, quant à leur forme générale, avec les sabres en laiton moulé de la Famille II.

Ces deux Sous-familles sont donc ainsi caractérisées :

Sous-famille VI a - Calotte figurative

Il semble évident que le seul modèle répertorié est une fantaisie de fourbisseur mais n'oublions pas que le bélier est hautement symbolique (voir J. Lhoste et J.J. Buigné 1999, p. 39).

446

446 - Sabre d'officier de Volontaire National.
Période révolutionnaire, probablement entre 1789 et 1792.
Monture en laiton argenté. Courte queue de calotte en forme de tête de bélier. Arc de jointure, croisière et quillon en gros fil torsadé. Poignée en ébène. Lame gravée de motifs végétaux et de trophées d'armes. Fourreau à deux garnitures.

Nota - Le symbolisme de la tête de bélier est assez rare mais il en existe quelques exemples chez les sabres à grosse coquille des Volontaires Nationaux.

Sous-famille VI b - Calotte non figurative

Sont présentés dans cette Sous-famille VI b, des sabres briquets de la Garde Impériale et des sabres de cavalerie, notamment le sabre de Canonnier monté du modèle 1829.

G. Jaeger (1999) a apporté une utile contribution à la connaissance de plusieurs variantes du sabre-briquet de la Garde Impériale dont on a extrait les caractères essentiels suivants :

447

447 - Sabre-briquet de la Garde Impériale.
1803 et ultérieurement.
Il existe plusieurs variantes du sabre briquet de la Garde mais toutes ont en commun les caractéristiques suivantes : monture en laiton, calotte à dôme et courte queue, poignée recouverte de basane et filigranée, avec bague basale, arc de jointure à enroulement supérieur figuré, arrondi, se terminant par un quillon infléchi, à bague et bouton, l'ensemble formant un " S ". Lame à pans creux, longue d'environ 68 - 69 cm, légèrement courbée. Fourreau de cuir, à deux garnitures : chape à base coupée droit et bouton de suspente plus ou moins arrondi placé sur une embase pyramidale, bouterolle à bordure supérieure à découpe concave et terminée par un bouton plus ou moins ciselé.

448 - Variante 1. Vers 1803 - 1807. Poignée à 13 - 14 spires d'un double filigrane de cuivre rouge torsadé disposé en chaînette. Arc de jointure gravé *VERSAILLES* et poinçonné de Jacquesson et de J. Casamajou. Lame de 68 cm, large au talon de 3,4 cm et flèche d'environ 1,5 cm signée sur le dos :
Manf^{re} n^{ale} du Klingenthal Coulaux frères Entrep^{rs} avec les poinçons de J.J. Mouton et de S.F. Beaumaretz.

449 - Variante 2. Vers 1806 - 1807. Poignée à 18 spires d'un fil de cuivre torsadé disposé en chaînette. Arc de jointure poinçonné de J.J. Mouton, F.L. Lobstein, S.F. Beaumaretz. Lame longue de 68,5 cm, large au talon de 3,6 cm, flèche de 2,6 cm, signée sur le dos :
MF^{ture} Imp^{le} du Klingenthal Coulaux frères
avec poinçons de J.J. Mouton et de S.F. Beaumaretz

450 - Variante 3. Vers 1812 - 1813. Poignée à 13 spires d'un fil de cuivre torsadé disposé en chaînette. Arc de jointure marqué *VERSAILLES* avec les poinçons de Cl. Marion et J.J. Mouton. Signature sur le dos de la lame identique à la Variante 2.

451 - Variante 4. Vers 1814 ? Poignée à 12 spires d'un fil de cuivre torsadé disposé en chaînette. Monture et lame sans marquage, celle-ci mesurant 60 cm, large au talon de 3,3 cm avec une flèche de 1,3 cm. Chape tout à fait classique mais poinçonnée au *verso* de J.G. Bick et de F. Lobstein.
Cet exemplaire est certainement un remontage.

452 - Autres variantes. Certains exemplaires peuvent avoir des lames de 70 cm. Les marquages sur le dos des lames peuvent être
Manuf^{re} Imp^{le} de (ou *du*) *Klingenthal Coulaux frères.*
Il y eut encore les fabrications non signées que Louis-Philippe fit faire pour armer la Garde Nationale.

453 - Sabre d'Honneur.
Janvier 1800 - Mai 1802.
Monture argentée, la forme générale est celle du sabre briquet d'Infanterie qui sera adoptée en 1802. Lame de 70 - 73 cm, large au talon de 3,5 cm environ. Fourreau de cuir de vache à deux garnitures argentées : chape simple pour la troupe, à cuvette fendue juste pour le passage de la lame pour les sous-officiers, (non figuré).

460

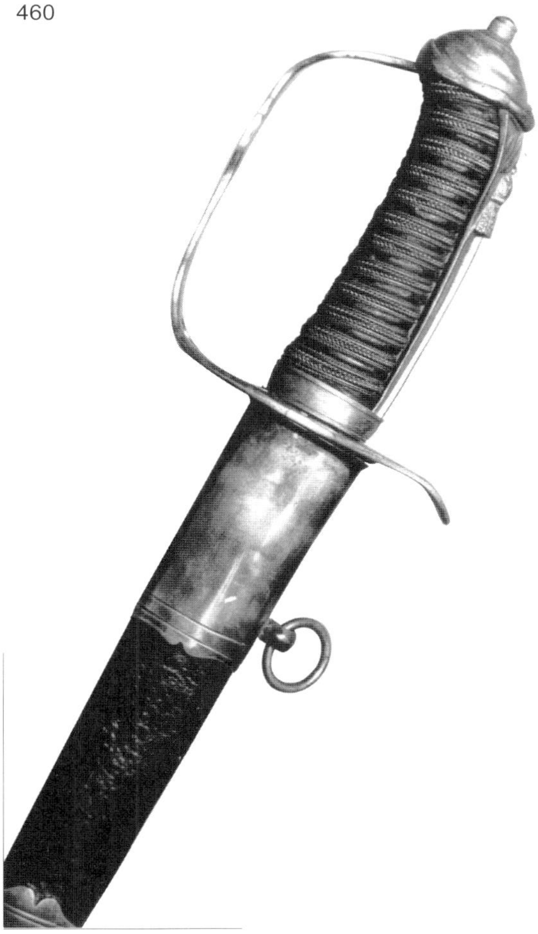

460 - Sabre d'officier de Volontaires Nationaux.
Révolution.
Laiton doré. Calotte en bonnet surmonté d'un bouton de rivure, longue queue ornée en relief, d'une pique dont la hampe part de la base pour se terminer par une pointe et un gland près du bonnet. Poignée recouverte de basane (460a), à 12 spires d'un filigrane composé de deux fils torsadés. Virole à deux rainures. Arc de jointure plat, à bords ondulés, se poursuivant par un plateau découpé du type e (voir p. 21). Lame de 75,5 cm, plate, large au talon de 3,4 cm, à flèche de 2 cm ; contre-tranchant saillant de 23 cm ; la lame est gravée au talon d'un motif floral doré, puis bleuie sur 22 cm et porte l'inscription suivante, en grandes lettres fleuries et dorées sur le *recto* : *P^r La Nation La Loi Le (Roi)* ; le mot Roi est effacé (460b) ce qui permet de conclure que ce sabre fut en service sous la Monarchie constitutionnelle avant de l'être sous la Première République ; sur le *verso*, avec les mêmes caractères, *Vaincre ou Mourir*. Fourreau renforcé par des éclisses de bois intérieures, à trois garnitures ; chape et bracelet pourvus chacun d'un anneau de bélière.

460a

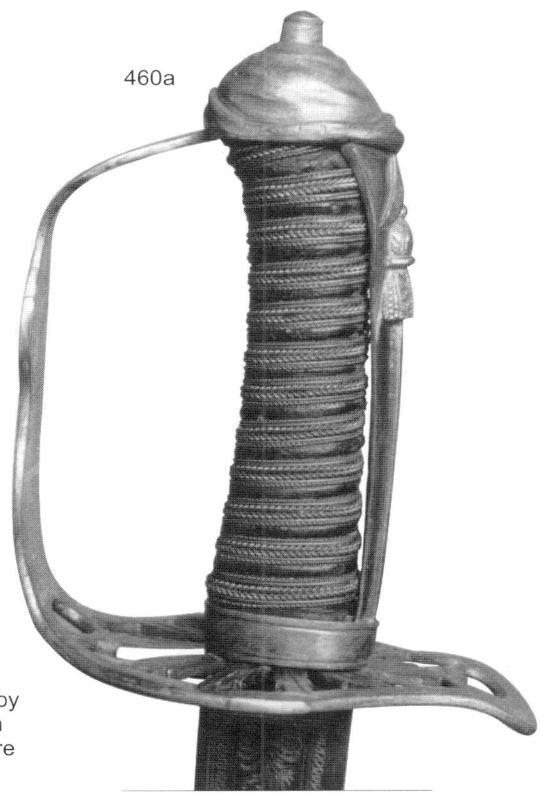

- **a**, calotte au bonnet, avec la queue ornée d'une pique. La prise de vue permet de voir le plateau en soleil du type e. ;
- **b**, remarquer, sur le *recto* de la lame que le mot Roy a été martelé. Ce sabre est donc une transformation d'un sabre de la Monarchie constitutionnelle en sabre révolutionnaire.

460b

Sous-famille VII b - Calotte non figurative

C'est un sabre simple, probablement de bas-officier. Nous n'en donnerons qu'un seul exemple.

461

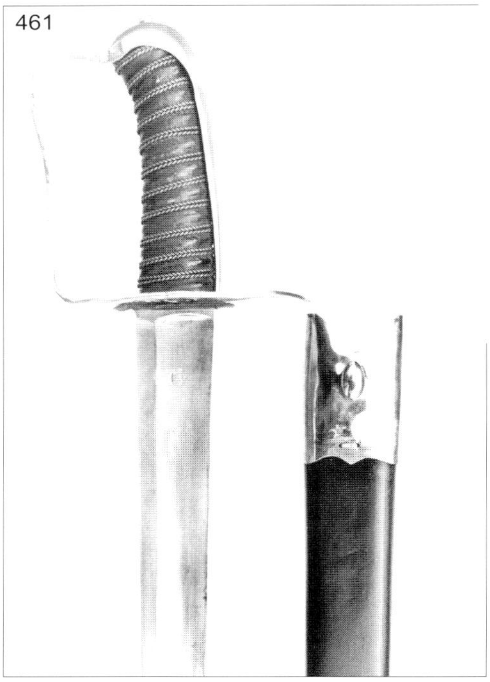

461 - Sabre d'Infanterie.
Période Révolutionnaire.
Monture en laiton. Calotte simple à petit bouton de rivure et prolongée par une longue queue unie. Arc de jointure plat, se poursuivant par un plateau de garde ajouré. Poignée recouverte de basane, filigranée d'un double fil de cuivre torsadé disposé en chaînette. Lame droite, à pan creux, poinçonnée de Cabannes. Fourreau de cuir à deux garnitures, chape à bouton.

Famille VIII - Calotte à courte queue

La monture est comparable à celle des sabres de la Famille VII, l'arc de jointure se prolongeant pour former le plateau découpé. Mais, cette famille possède une courte queue de calotte, celle-ci pouvant être figurative ou simple, ce qui nous a conduit à créer les deux Sous-familles suivantes :

462

Sous-famille VIII a - Calotte figurative

L'exemplaire que nous décrivons dans cette Sous-famille est à calotte à tête de lion.

462 - Sabre d'officier d'Infanterie.
XVIIIème siècle.
Monture en laiton. Calotte à tête de lion , dépourvue de queue. Poignée recouverte de basane et à une douzaine de spires d'un filigrane double disposé en chaînette. Arc de jointure simple découpé d'un triangle à sa base et se poursuivant par un plateau ajouré, du type a, (p. 22). Lame à pan creux, blanche.

Sous famille VIII b - Calotte non figurative

De tels sabres se rencontre à partir de l'Ancien Régime jusque sous le Second Empire.

463 - Sabre d'officier de Chasseurs à pied.
Fin de l'Ancien Régime.
Montur en fer. Calotte simple, plate et à courte queue. Poignée recouverte de basane à filigrane constitué de deux fils de laiton jumelés non torsadés faisant une douzaine de spires. Arc de jointure arrivant en angle droit sur la croisière, celle-ci formant le milieu du plateau ovalaire largement ajouré, à bordures légèrement galbées. Lame de 75,8 cm, à pan creux, étroite et à légère flèche, gravée de rinceaux et de feuillages. Fourreau de cuir noirci, à trois garnitures de fer dont les bordures sont découpées en accolades.

403

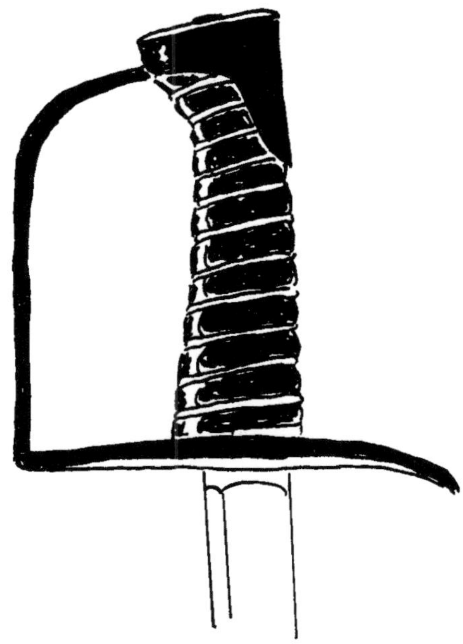

404

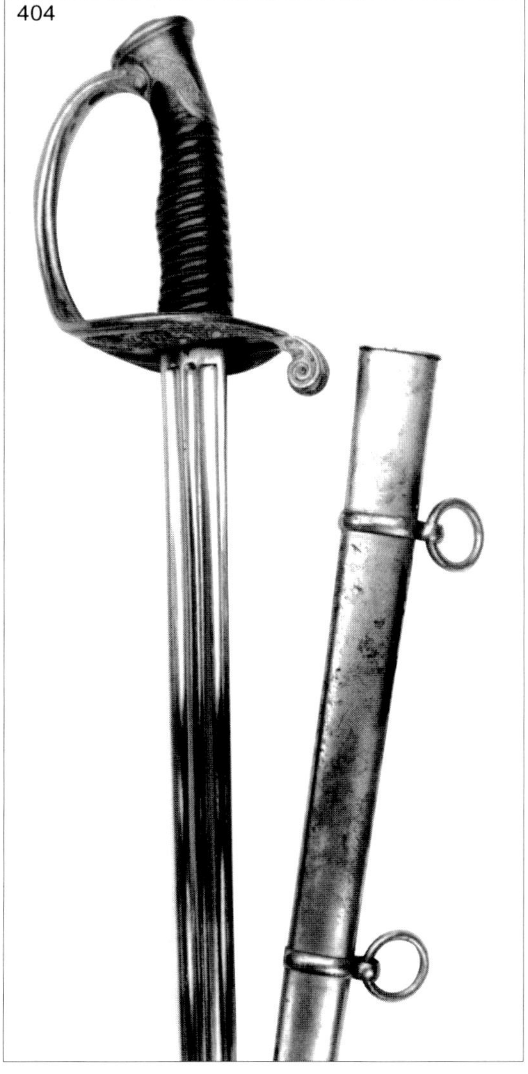

464 - Sabre d'officier subalterne d'Infanterie de Marine, modèle 1845.
Sabre à monture identique à celle des officiers mais avec une lame à légère flèche et un fourreau en tôle en acier gainé de cuir noir (voir p. 279).

404 - Sabre d'officiers d'Infanterie, modèle 1855
Ce sabre possède les mêmes caractéristiques que le modèle 1845, mais en diffère par la lame et le fourreau : lame plate et gouttière dorsale ou à section ovalaire à double tranchant, à gouge presque médiane moins longue que la lame. Fourreau en tôle d'acier poli à deux anneaux de bélière jusqu'au 22 décembre 1883, date où une note ministérielle impose la suppression du second anneau, le sabre étant porté par la selle pour des officier montés.

Chapitre IV
Monture en fer, à grosse coquille noircie

Ce chapitre ne comporte qu'une seule famille, la Famille IX réunissant des sabres à grosse coquille de fer. Les sabres répertoriés ci-après présentent des affinités certaines entre eux et sont bien connus sous les noms de sabres de bord, de sabres d'abordage ou encore, du fait de la forme de la garde, sous le sobriquet de cuillère à pot.

Remarques *- Les sabres de bord de la Marine de l'Ancien Régime étaient en laiton. Sous la Révolution, le cuivre venant à faire défaut, le sabre à monture de fer fut créé en l'An IX... et la tradition fit que ce matériau fut utilisé sous le Consulat et les régimes suivants, jusque sous le règne de Napoléon III. Au cours des ans, quelques modifications furent apportées ainsi que nous allons le voir en examinant les trois modèles plus ou moins réglementaires.*

Famille IX - Sabres d'abordage

466

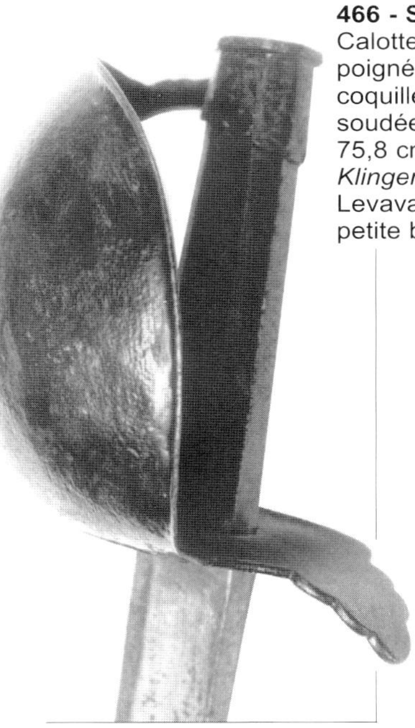

466 - Sabre de bord modèle de l'An IX.
Calotte à huit pans, courte et bouton de rivure, recouverte de tôle ainsi que la poignée également à pans. Branche régulièrement arrondie supportant une grande coquille très convexe, large au milieu de 15 cm. Cette coquille est également soudée sur le plateau terminé par une palmette. Lame à pan creux, longue de 75,8 cm large au talon de 3,8 cm à légère flèche signée sur le dos *M^fture^ de Klingenthal Coulaux frères entrep^r^* et poinçons de J.J. Mouton, de BPC Levavasseur. Fourreau en cuir, à garnitures de laiton : chape à bouton en olive et petite bouterolle.

467

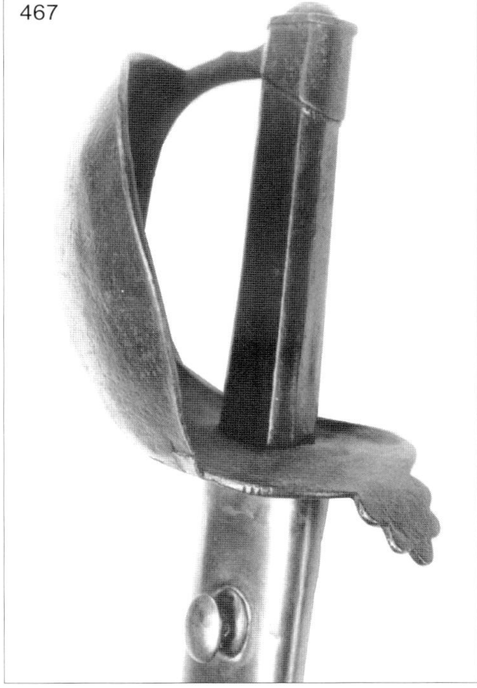

467 - Sabre de bord.
Modèle 1811 dit *" Sabre de bord n°1 "*.
Monture comparable au modèle de l'An IX, mais avec les caractéristiques suivantes : coquille dont la largeur en son milieu n'est que de 12 cm. Plateau terminé par une palmette. Lame de 67 - 68 cm ; large au talon de 3,7 cm et gravée d'une ancre de marine les bras dirigés vers le talon (voir 468a). Fourreau en cuir noir, chape à bouton en demi-olive, bouterolle terminée par un bouton. Sur les fabrications tardives, la chape peut être à pontet.

Marquage - Sur le dos de la lame on lit *M^facture^ Nationale* (ou *Impériale*) du *Klingenthal*, avec le mois et l'année de mise en service ; sur le talon, on peut trouver les poinçons de Krantz, Bick ou Lobstein.

468

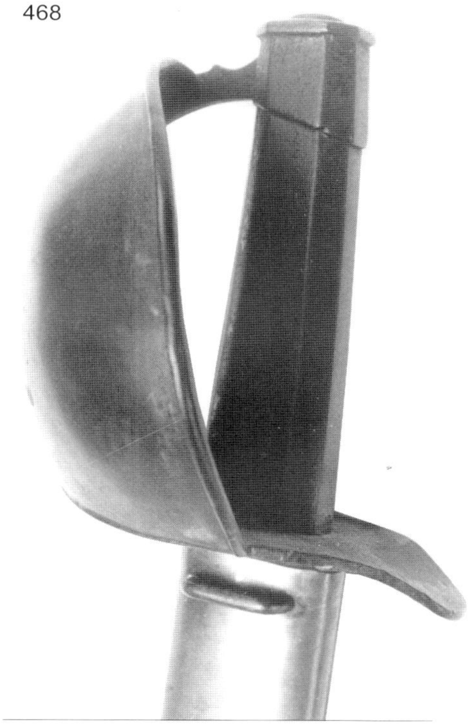

468a

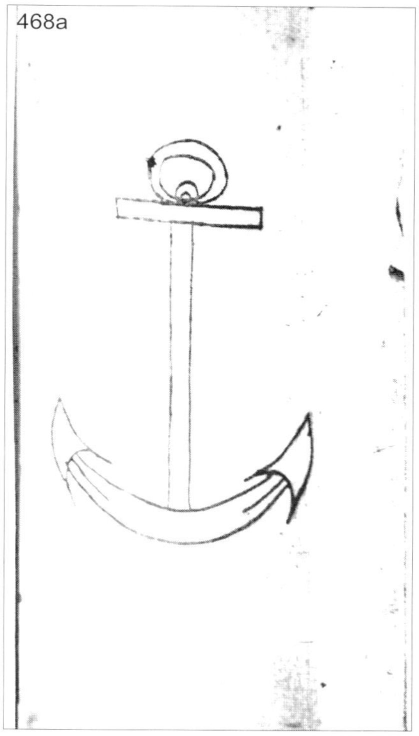

468 - Sabre de bord.
Modèle 1833.
Monture conçue dans le même style que le modèle précédent. Elle en diffère sur les points suivants : si la coquille mesure toujours en son milieu 12 cm, le plateau a perdu sa palmette qui est remplacée par un simple bec. Lame de 67,6 cm de long, large au talon de 3,65 cm, à pan creux et gravée d'une ancre placée verticalement, les bras dirigés vers la pointe (468a). Fourreau de cuir pressé sans couture, à deux garnitures de laiton, chape à pontet.

Marquage - Ce sabre est en service jusqu'à la fin du Second Empire. En conséquence on trouve sur le dos de la lame les signatures : *Manuf[re] R[le] de Châtellerault* ou *Manufacture Impériale*, ces deux signatures étant suivies du mois et de l'année de la fabrication.

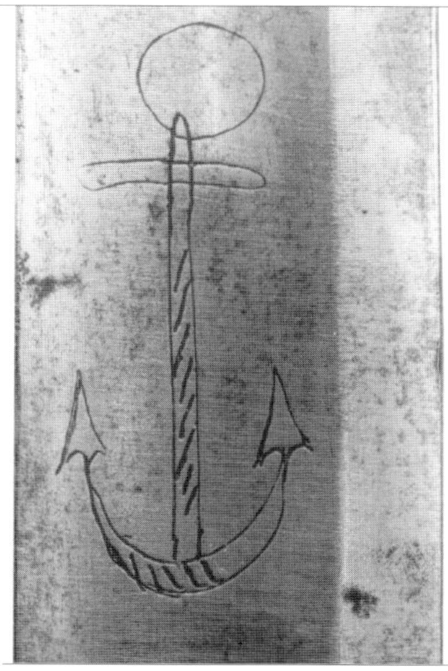

468bis - Sabre d'abordage modifié.
Modèle de 1833 présentant la particularité d'être gravé, *recto-verso*, sur la lame d'une ancre d'une facture exceptionnelle (468bis, a). Sa longueur est de 4,7 cm, l'espace entre les deux pointes des bras, de 2,7 cm. Sur le dos on remarque la signature de *Coulaux & Cie : Klingenthal* (468bis, b). Cette gravure de fantaisie est probablement celle d'un matelot dont la lame n'avait pas été marquée par la manufacture.

468bis, a

- **a**, ancre de fantaisie ;
- **b**, signature de Coulaux & Cie : Klingenthal

468bis, b

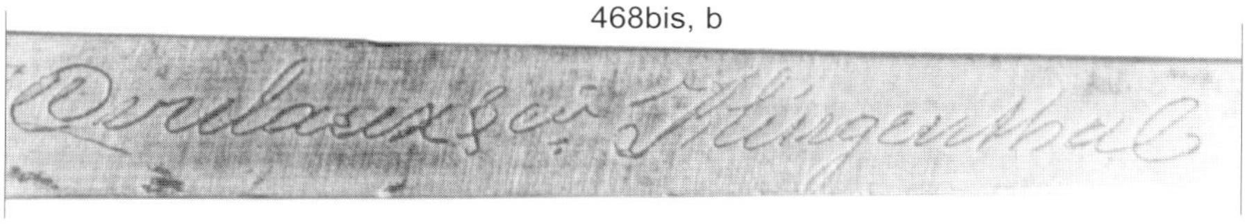

468ter - Sabre d'abordage modifié.
Modèles An IX et 1833.
On rencontre parfois des sabres d'abordage des modèles de l'An IX et surtout de 1833, dont la coquille en fer a été coupée au ras de l'arc de jointure et dont le plateau a été rétréci si bien qu'il ne déborde plus que de 1 cm de part et d'autre de la poignée (voir La Sabretache, n° 139, p. 12).

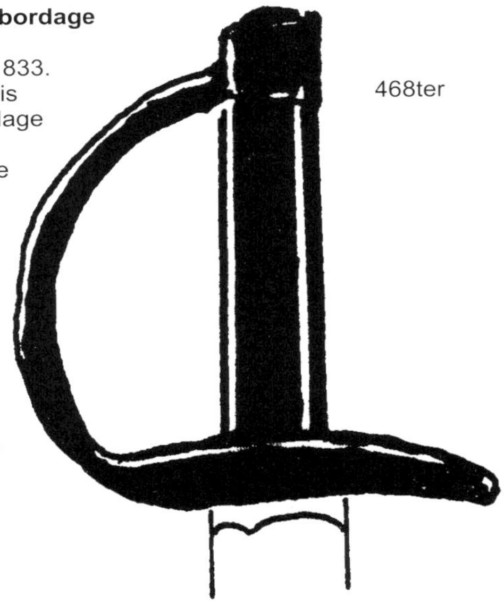

468ter

Il est certain que ces sabres ainsi modifiés sont d'un port beaucoup plus aisé et deviennent de véritables briquets. Etant donné que ces sabres sont connus à d'assez nombreux exemplaires, l'hypothèse d'une fantaisie individuelle semble pouvoir être rejetée. L'hypothèse la plus vraisemblable est une modification faite en arsenal pour équiper des troupes de débarquement, donc d'Infanterie de Marine, peut- être celle engagée au moment de la colonisation de l'Indochine.

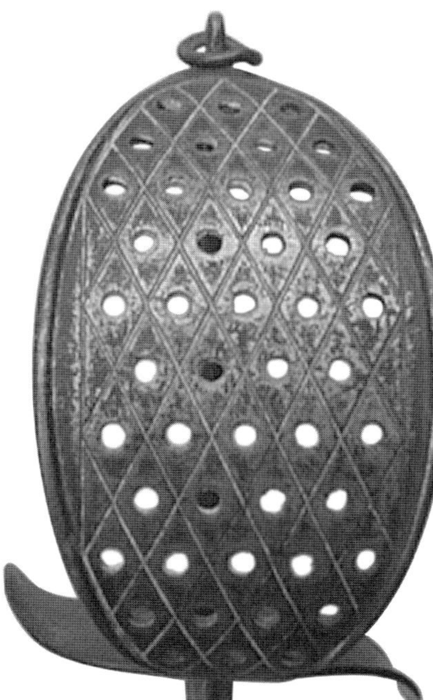

469

coquille vue de face.

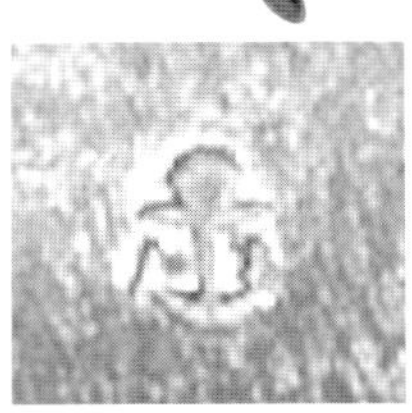

- **a**, poinçon d'Arsenal.

469 - Sabre de bord.
Modèle 1872.
Monture en fer noirci. Calotte inclinée, à courte queue. Poignée de bois dur, à 12 cannelures, sans filigrane. Arc de jointure à anneau près de la calotte, élargi symétriquement pour former une coquille. Celle-ci est percée de trous réguliers placés en losanges tracés par des lignes gravées. Garde infléchie. Lame plate, de 74 cm, à légère flèche. Fourreau de cuir comprimé, sans couture visible, à garniture de fer, bouterolle à bouton. Poinçon d'Arsenal (469a).

Remarque : Ce sabre est nettement inspiré par un modèle de sabre anglais (G.C. Neuman, et J. Baudurat, 1967). On n'en connait que quelques exemplaires, ce modèle ne semblant pas avoir été adopté par les Autorités Maritimes.

Quatrième partie

Sabres à un arc de jointure plus une branche secondaire

I - Monture à une branche secondaire tournante

II - Monture à une branche secondaire fixe

Ces sabres sont couramment désignés sous le nom de sabres à deux branches.

Nous les avons séparé en deux chapitres ainsi caractérisés : :

Chapitre I

Monture à une branche secondaire tournante

Les sabres présentés dans ce chapitre possèdent la particularité suivante : une branche peut pivoter sur un quart de tour grâce à un axe placé dans la calotte et à un autre petit axe riveté sous le plateau de garde, de telle sorte, que pour le port, la branche mobile puisse venir chevaucher l'arc de jointure fixe (a). Pour maintenir la branche tournante en position de combat et assurer la protection de la main, il existe sous la partie latérale du plateau de garde, une barrette ressort rivée près du quillon. A l'autre extrémité, une pointe vient s'engager à la base de la branche mobile pour la maintenir en position (b). A travers la bordure du plateau, un poussoir permet de libérer la branche tournante (c et d).

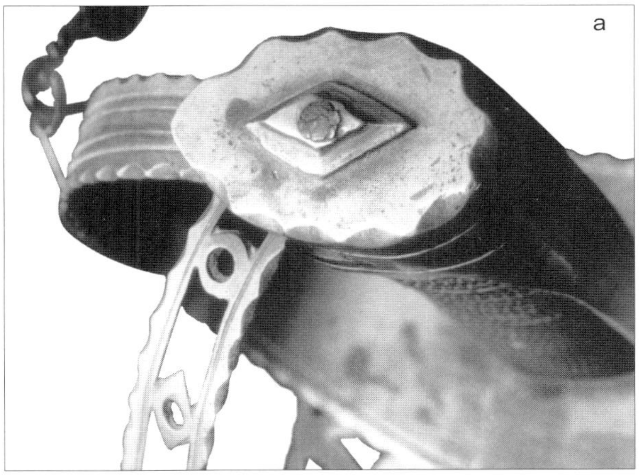

- a, calotte de départ de l'arc de jointure et de la branche pivotante ajourée.

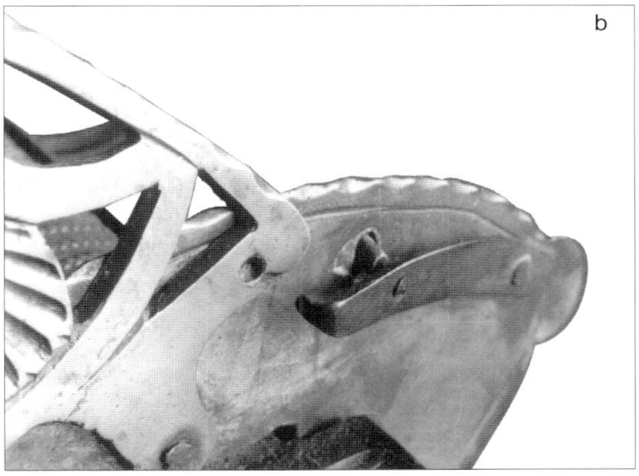

- b, barette-ressort pour maintenir en position déployée la branche mobile.

- c, poussoir remplacé ici par une vis pour libérer la branche mobile.

- d, bouton-poussoir pour libérer la branche mobile.

LES COMPOSANTES DES SABRES À UNE BRANCHE TOURNANTE

La poignée est évidemment la partie de la monture la plus caractéristique

La poignée - On rencontre des poignées recouvertes ou non de basane, avec ou sans filigrane, de bois noirci ou d'ébène, de corne. La calotte est plate ou figurative, à longue queue généralement.

Les branches de garde - Les branches de garde sont décorées, soit d'une découpe composée de cercles et de barrettes, ou pleines, soit de symboles politiques : fleur de lys pour les plus anciens sous l'Ancien Régime, d'un faisceau de licteur, de l'épée, de la crosse et de la bêche pour les exemplaires créés lors de la Monarchie constitutionnelle ou pendant la période révolutionnaire.

Le plateau de garde - Il est le plus souvent ovale et découpé plus ou moins simplement, parfois à décor rayonnant et, dans son ensemble tout à fait comparable à celui des sabres dits petits Montmorency (p. 21 et 22), et se termine souvent par un bec en forme de palmette. Nous avons vu que c'est sous le plateau qu'est fixé le ressort. D'autres fois, ce plateau est entier, qu'il soit de fer ou de laiton.

Les lames - En plus des lames commandées aux Manufactures ou aux forgeurs locaux, on rencontre fréquemment des lames récupérées sur des sabres d'Infanterie plus en service ou dont la garde était endommagée. Elles sont à gouttière près du dos, à pan creux et à contre-tranchant vers la pointe. Ces lames mesurent au talon entre 3,3 et 3,7 cm de large et sont longues de 74-75 cm, parfois plus courtes (71 cm) ou plus longues (78-84 cm), à légère flèche. Rarement signées, les lames peuvent être bleuies et gravées de motifs dorés, porter des inscriptions ou des emblèmes révolutionnaires.

Les fourreaux - Les fourreaux sont en cuir noirci, à deux ou trois garnitures, celles-ci pouvant être réunies par des attelles de laiton. On notera que la chape est pourvue d'une encoche permettant la rotation de la branche tournante.

Selon C. Ariès, l'origine de ces sabres serait liée à l'ordonnance du 8 août 1784.

M. Bottet écrit (ces sabres) *ont été portés sous la Révolution par les officiers de tous corps et principalement de la Marine. On les rencontre dans l'armement des troupes régulières et des corps francs.* En fait, il semble bien que cette catégorie de sabres ait été attribuée principalement aux officiers et sous-officiers de l'Infanterie et plus exceptionnellement pour les lames plus longues, aux officiers montés. Ces sabres resteront en service jusque vers 1798.

Si, comme nous venons de le dire, les lames sont rarement signées, on peut néanmoins en trouver portant leur origine telle *Manufacture de Klingenthal*, des poinçons d'inspecteur ou / et de contrôleur ou le nom du fourbisseur tel *Jacques Spol marchand fourbisseur à METZ.*

Les sabres de ce chapitre ont été divisés en trois Familles en prenant comme critères, la structure du plateau ou le système utilisé pour faire pivoter la branche secondaire :

Famille I – Plateau ajouré

Le plateau est ajouré selon les types figurés p 21-22. Les calottes sont différemment ornées et permettent de distinguer deux Sous-familles :

Sous-famille I a - Calotte figurative

La figuration de la calotte représente une tête de lion ou un casque à l'antique dans les exemples que nous donnons ci-après.

470 - Sabre d'officier de troupe à pied.
Fin XVIIIème siècle, pré-révolutionnaire.
Monture en laiton. Calotte en tête de lion (470a), prolongée d'une longue queue. Poignée entièrement filigranée. Les deux branches ont la même découpe, motif qui se retrouve sur la plateau (470b). Lame de 74 cm, large au talon de 3,5 cm, et flèche de 1,5 cm, blanche, à pan creux. Cet exemplaire est marqué du nom de son propriétaire : *VIBERT*.

470

470a

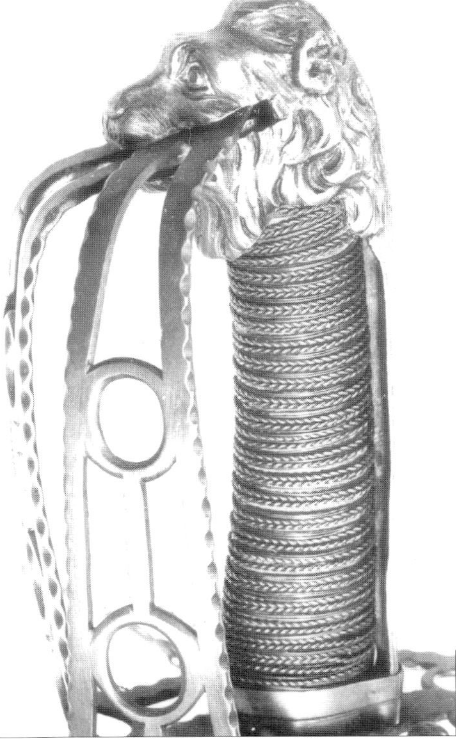

470b

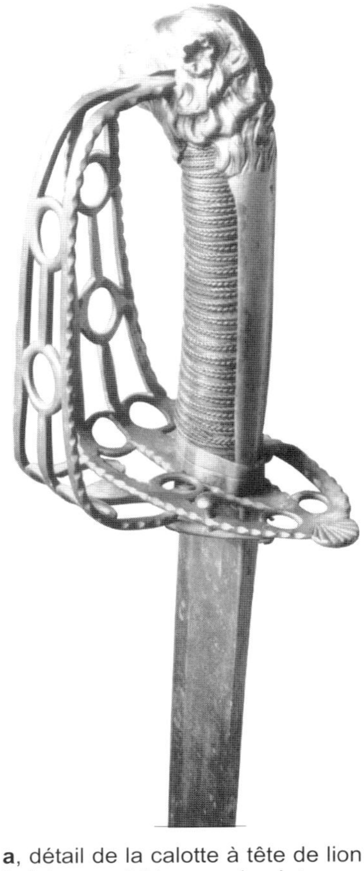

- **a**, détail de la calotte à tête de lion
- **b**, ici, on voit bien que le plateau est le prolongement de l'arc de jointure.

471

471 - Sabre d'officier supérieur de grosse Cavalerie.
Fin du XVIIIème siècle.
Monture en laiton doré. Calotte en tête de lion et longue queue. Poignée recouverte de basane et filigranée. Arc de jointure fixe, guilloché en bordure ainsi que la branche secondaire tournante. Plateau découpé. Longue lame plate, damasquinée d'or au talon et tout au long du dos. Fourreau de cuir noir, chape à anneau de bélière, décorée au trait et découpée en accolade ainsi que le bracelet à anneau et la bouterolle terminée par un dard de fantaisie constitué de deux baguettes à extrémités supérieures recourbées, l'antérieure étant deux fois plus longue que la postérieure et ne protégeant pas la base de la dite bouterolle.

472 - Sabre d'officier de grosse Cavalerie.
Fin XVIIIème siècle.
Monture en laiton. Calotte en tête de lion et longue queue sans décor. Poignée entièrement filigranée d'un fin fil torsadé en alternance avec un fil simple. Grosse virole. Branches sans décor et plateau ajouré. Lame presque droite, à pan creux. Fourreau de cuir à trois garnitures : chape à bouton et anneau de bélière (472a).

472

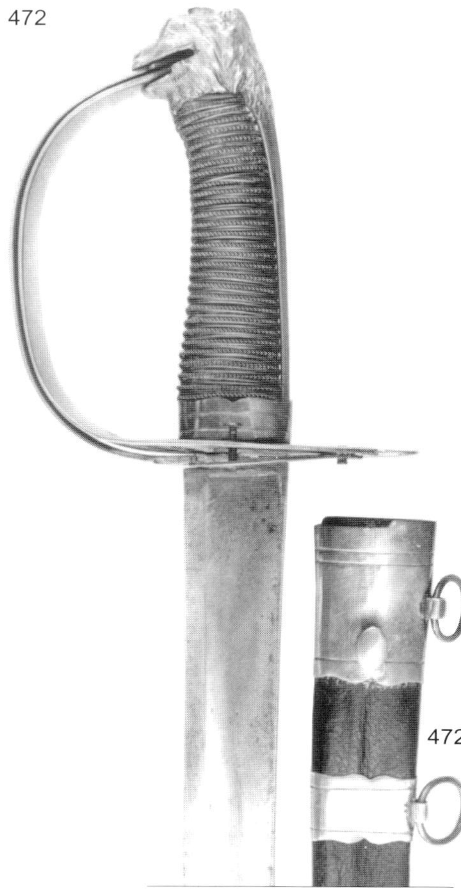

472a

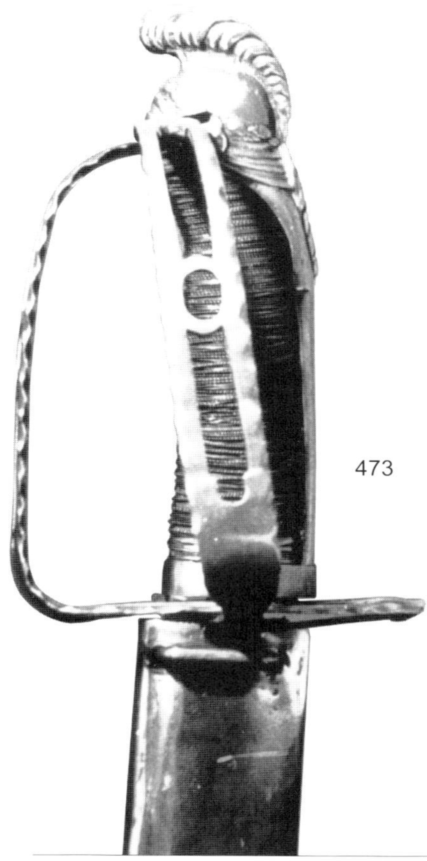

473

473 - Sabre d'officier de troupe montée.
Fin du XVIIIème siècle, Révolutionnaire.
Monture en laiton. Calotte au casque à l'antique et longue queue. Poignée recouverte de basane filigranée de fils torsadés en alternance avec un simple fil. Branches découpées ainsi que le plateau. Lame de 72,5 cm, large au talon de 3,5 cm à flèche de 2 cm. Fourreau de cuir à trois garnitures de laiton.

Sous-famille I b – Calotte non figurative

Ces modèles à calotte simple, sans ornement figuratif, sont les plus fréquemment rencontrés. En général la calotte est à longue queue mais nous présentons un exemplaire à courte queue assez exceptionnel.

474

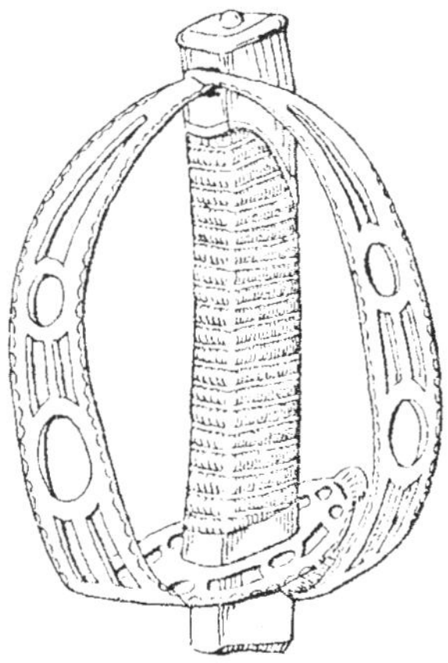

474 - Sabre de Cavalerie.
Règne de Louis XVI.
Monture en laiton. Poignée à section rectangulaire, recouverte de basane filigranée. Calotte à courte queue. Arc de jointure et branche pivotante au même décor : deux découpes ovales séparées par des découpes subrectangulaires. Plateau ajouré voisin du type b (voir p. 21) (d'après M. Bottet.).

- **a**, découpe du plateau.

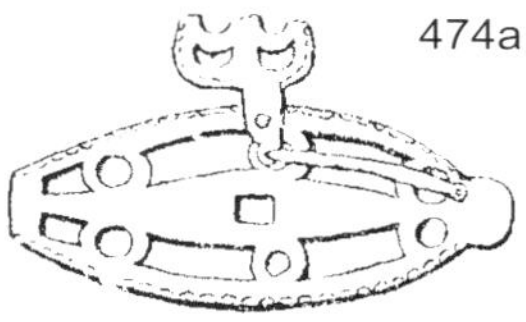

474a

475

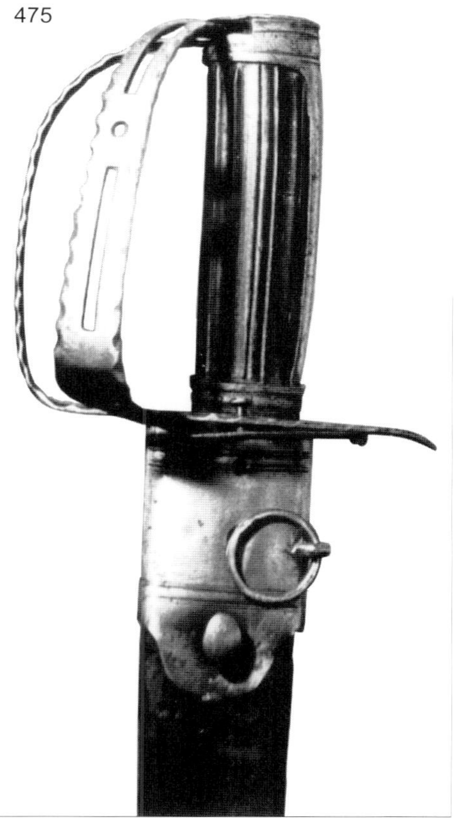

475 - Sabre de Marine.
Fin du XVIIIème siècle.
Monture en fer. Calotte plate, à longue queue. Poignée de bois noirci, rainuré verticalement. Branches à découpes classiques, guillochées sur les bords, l'arc de jointure se poursuivant pour former le plateau à mêmes découpes que les branches. Lame plate, longue de 77 cm, large au talon de 3,6 cm et flèche de 1,9 cm, blanche. Fourreau à deux garnitures de laiton fixées par martelage : chape à bouton et anneau de bélière, bouterolle à extrémité aplatie, dépourvue de dard.

476

476 - Sabre d'officier monté.
Période pré-révolutionnaire.
Monture en laiton. Calotte plate à bouton de rivure en losange. Poignée à section quadrangulaire, totalement filigranée. Les deux branches ont la même découpe que le plateau. Lame à pan creux et gouttière près du dos, longue de 75 cm, large au talon de 3,8 cm et flèche de 2 cm. Fourreau à trois garnitures : chape décorée de filets, à bouton et anneau de bélière, bracelet découpé en suite à anneau et bouterolle à dard basal et remontant sur le côté antérieur du fourreau.

477

477 - Sabre d'officier monté.
Fin XVIIIème siècle.
Monture en laiton. Calotte plate, rainurée, prolongée d'une longue queue et bouton de rivure en losange. Poignée recouverte de basane à treize spires constituées de trois fils torsadés. Arc de jointure plein, à bordure découpée, se prolongeant en plateau au soleil terminé par une petite palmette ; branche tournante à bordure identique à l'arc mais découpée (477a). Lame à pan creux et gouttière dorsale, longue de 75,5 cm, large au talon de 3,8 cm et flèche de 1,6 cm. Fourreau à alèses de bois, à trois garnitures décorées en suite, de trois filets jumelés : chape à anneau de bélière ainsi que le bracelet trapézoïdal, bouterolle de 33,5 cm à dard asymétrique.

477a

- **a**, sabre vu de face, montrant l'arc de jointure plein et la branche tournante découpée.

478

478 - Sabre d'officier d'Infanterie.
Fin XVIIIème siècle.
Monture en laiton. Calotte à longue queue et bouton de rivure en losange. Arc de jointure plein et branche tournante à bordures guillochées et à découpes. Poignée recouverte de basane et filigranée. Lame à pan creux. Fourreau à deux garnitures, courte chape à bouton et anneau de bélière et bracelet à anneau. (J.C. D.)

479

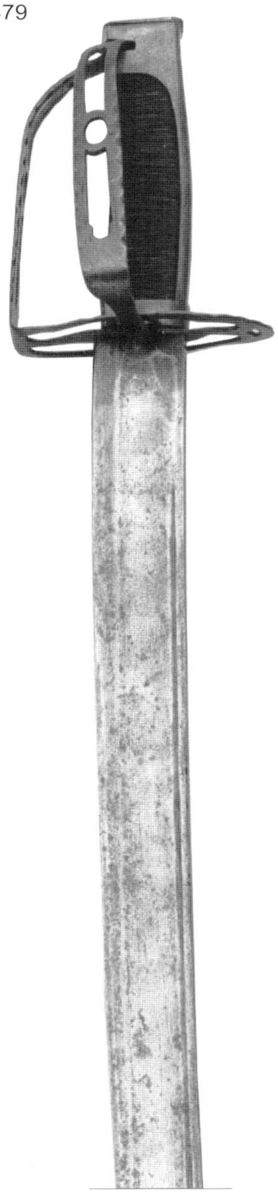

480

479 - Sabre de sous-officier d'Infanterie.
Fin XVIIIème siècle.
Monture en laiton. Calotte plate, carrée, à longue queue unie. Poignée recouverte de basane et filigranée. L'arc de jointure et la branche pivotante ont la même découpe. Remarquer la forte lame à pan creux et la longue gouttière dorsale.

480 - Sabre d'officier d'Infanterie.
Fin XVIIIème siècle.
Monture en laiton. Calotte plate, à longue queue. L'arc de jointure et la branche pivotante ne sont pas découpés mais simplement à bordures guillochées. Poignée recouverte de basane et filigranée. Ce sabre est probablement d'un officier car on décèle encore, sur la lame, malgré les dommages qu'elle a subi, des traces de gravures.

481 - Sabre d'officier monté.
Fin XVIIIème siècle.
Monture en laiton. Calotte plate à bouton de rivure en losange et longue queue. Poignée totalement filigranée d'un fil tenu torsadé. Arc de jointure plein et branche pivotante découpée, ainsi que le plateau. Lame de 71 cm, plate, blanche.

481

482 - Sabre d'officier de troupe à pied.
Fin XVIIIème siècle.
Monture en cuivre rouge. Poignée de corne noire cannelée et filigranée. Calotte légèrement conique à bouton de rivure et courte queue. Arc de jointure et branche pivotante à découpe identique et plateau découpé en suite. Lame à pan creux et longue gouttière dorsale.

482

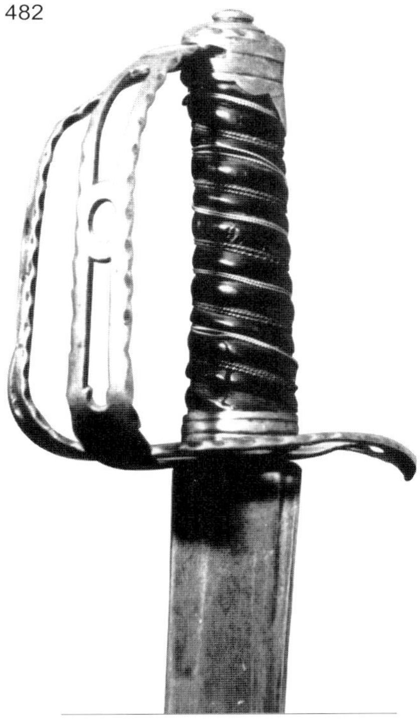

483 - Sabre d'officier monté.
Fin XVIIIème siècle - Révolution.
Monture en laiton. Calotte a bouton de rivure, plate et longue queue unie. Poignée totalement filigranée d'un fil de laiton torsadé (483a). Arc de jointure plein se prolongeant par un plateau ajouré (483b), branche pivotante au même décor ajouré que le plateau. Lame demi-courbe de 78 cm, à pan creux bleuie, gravée d'arabesques foliacées sur le *recto* et portant l'inscription *VAINCRE OU MOURIR* sur le *verso*. Fourreau à deux garnitures : chape à bouton avec de part et d'autre deux filets, bouterolle découpée et gravée en suite, à dard asymétrique (M.R.).

483

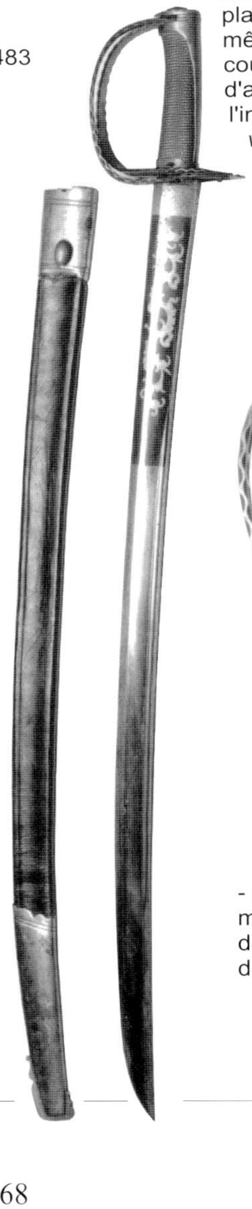

483a

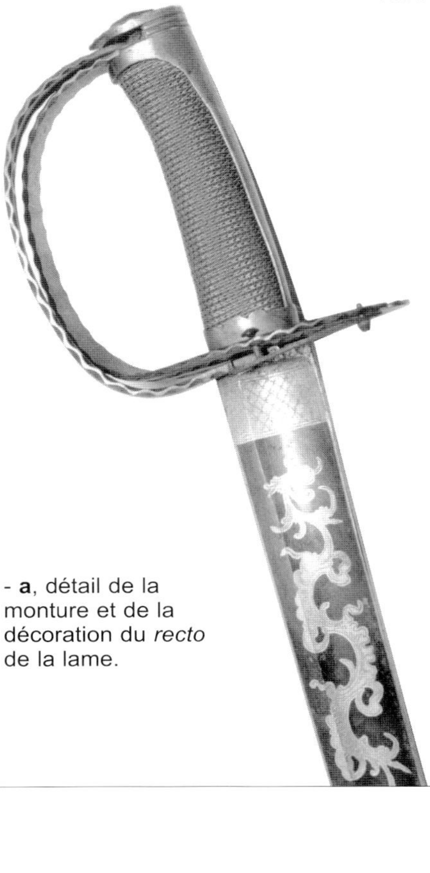

- **a**, détail de la monture et de la décoration du *recto* de la lame.

483b

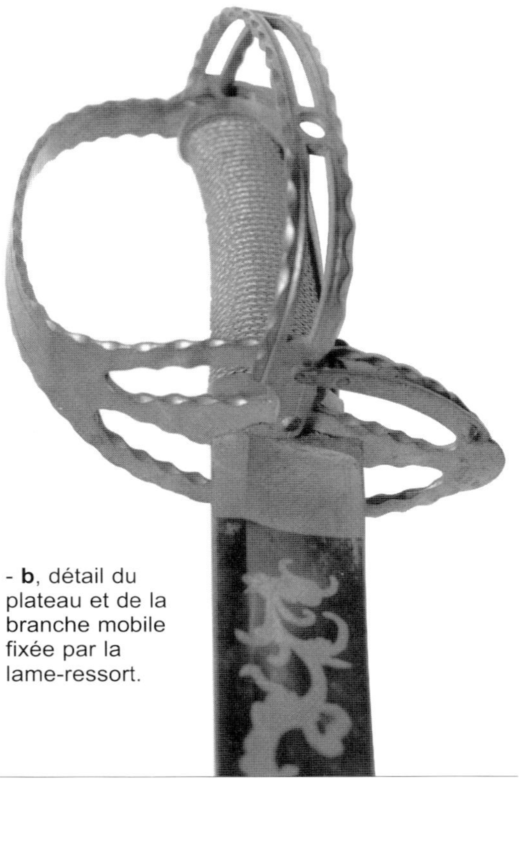

- **b**, détail du plateau et de la branche mobile fixée par la lame-ressort.

484

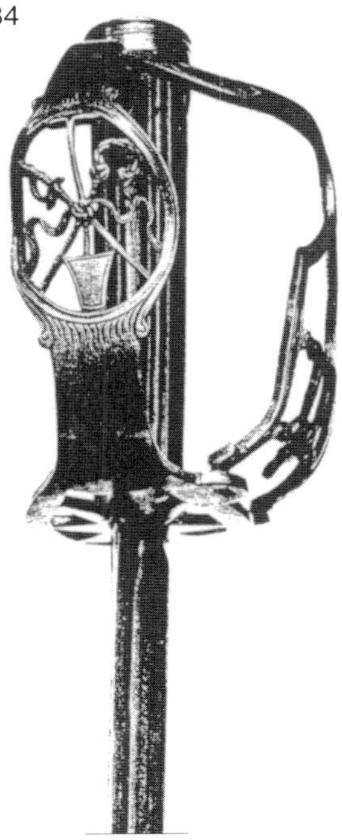

484 - Sabre d'officier dit aux Trois Ordres (1).
Vers 1789 - 1790.
Ce sabre, dont il existe encore plusieurs exemplaires, est remarquable par le décor aux Trois Ordres ciselé à jour dans l'arc de jointure. Poignée de bois cannelé verticalement. Plateau au soleil (Lavallay op. cit. 1945, pl. VI).

(1) La *crosse* pour le Clergé, l'*épée* pour la Noblesse, la *bêche* pour le Tiers Etat.

485

485 - Variante.
Poignée filigranée et longue queue de calotte. Remarquer que c'est la branche mobile qui porte, en découpe, l'emblème des Trois Ordres (d'après Lavallay).

Famille II – Plateau plein

Contrairement à la Famille I, ici, le plateau n'est pas ajouré. Ces fabrications semblent être réalisées par de petits fourbisseurs insuffisamment équipés pour préparer des plateaux découpés ou pour assurer une meilleure protection de la main.

486

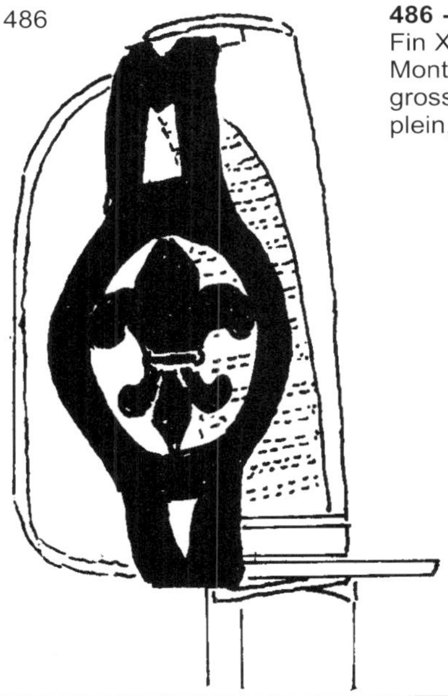

486 - Sabre d'officier subalterne des bataillons provinciaux.
Fin XVIIIème siècle.
Monture en laiton. Calotte plate à longue queue. Branche pivotante ornée d'une grosse fleur de lys découpée. Poignée recouverte de basane et filigranée. Plateau plein, avec poucier sur la partie postérieure (d'après c. Ariès, XIII, 1975).

487

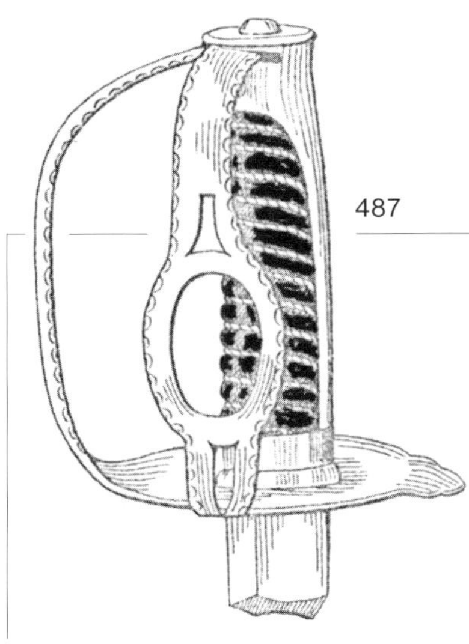

487 - Sabre de Cavalerie.
Règne de Louis XVI.
Monture en laiton. Calotte plate à longue queue. Poignée recouverte de basane filigranée. Arc de jointure plat et étroit se prolongeant par un plateau plein. Branche tournante à grand ovale découpé, avec à ses pôles, deux petites découpes secondaires. Les deux branches possèdent des bordures guillochées. Lame à pan creux (d'après M. Bottet).

488

488a

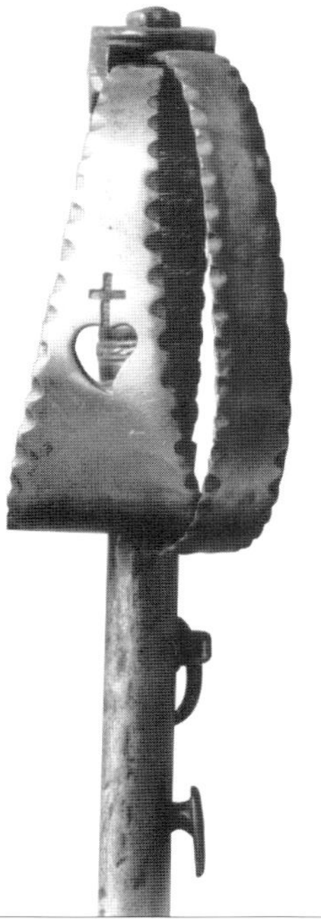

488 - Sabre des troupes vendéennes.
Période révolutionnaire.
Monture en fer. Calotte plate à bouton de rivure ciselé latéralement ; longue queue. Poignée recouverte de basane à huit rangs d'un triple filigrane. C'est la branche fixe qui est ornée de l'emblème découpé des troupes chouannes ; la branche mobile est pleine, ce qui permet de masquer, hors des combats, le cœur surmonté de la croix. Ruse de royalistes ? Plateau plein dans le prolongement de la branche fixe. Toutes les bordures de la monture sont guillochées. Lame blanche, de 70 cm, large au talon de 3,8 cm flèche de 1,5 cm, à pan creux peu marqué. Cette lame de mauvaise qualité, avec de nombreuses pailles, est évidemment de fabrication artisanale. Fourreau à deux garnitures fixées par martelage ; chape en fer à bouton et anneau de bélière en laiton, bouterolle en laiton à dard asymétrique

- **a**, arc de jointure découpée du coeur et de la croix. La branche mobile, en se repliant masque l'emblème adapté par les royalistes de l'ouest.

489 - Sabre d'officier monté.
Période révolutionnaire.
Monture en laiton doré. Calotte à bouton de rivure en losange et longue queue. Poignée totalement filigranée d'un double fil torsadé disposé en chaînette en alternance avec un fil simple, l'ensemble étant de cuivre rouge. Arc de jointure rainuré longitudinalement et branche tournante ornée d'un faisceau de licteur en découpe avec à sa base une palmette (489a, 489b). Lame gravée, en lettres dorées *vive La nation* (489c) dans un bandeau et sur l'autre face la figuration d'une femme, la Nation, comme en témoigne l'inscription sur son bouclier : *LA NATION* (489d). Cette lame mesure 80 cm de long, 3,7 cm de large au talon avec une flèche de 2 cm.

489

489a

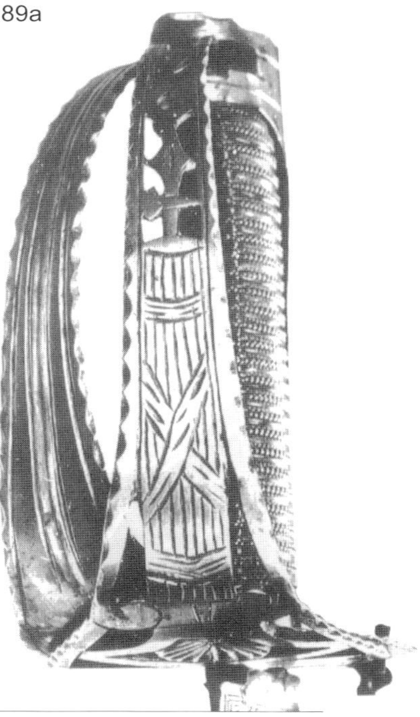

- **a**, branche mobile décorée d'un faisceau de licteur.

489b

489c

489d

- **b**, base de la branche mobile ornée d'une palmette ;
- **c**, *Vive la nation* dans un bandeau sur le *recto* de la lame ;
- **d**, sur le *verso* de la lame, Athéna ou la figuration de la Nation.

Famille III – Système spécial de mobilité de la branche tournante et oreillons

De fait, cette famille n'est composée que d'un sabre probablement d'origine germanique.

490

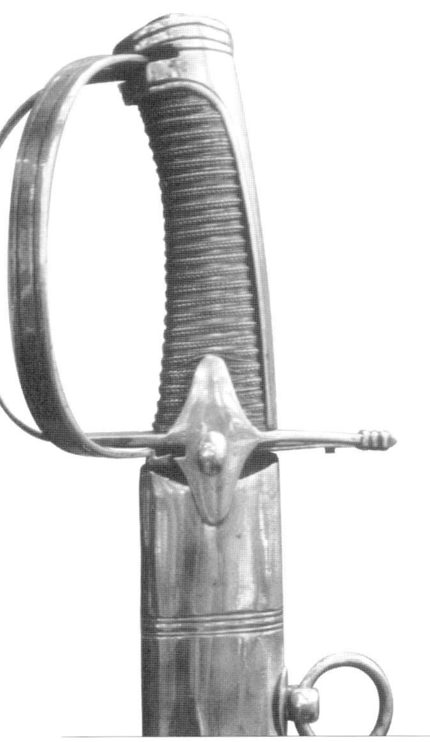

490 - Sabre d'officier de Chasseurs à cheval.
Fin XVIIIème siècle.
Monture de bronze. Calotte rainurée horizontalement, à longue queue. Poignée entièrement filigranée de fils de cuivre rouge torsadé et de ruban de même métal, en alternance. Arc de jointure se poursuivant par une croisière terminée par un triple bouton. La particularité de ce sabre est le système qui permet de faire pivoter la branche mobile : celle-ci est maintenue par deux axes placés l'un dans la calotte, l'autre à la base de la fusée, en conséquence la calotte est échancrée ainsi que l'oreillon frontal en losange curviligne ; un système de blocage pour maintenir la branche dépliée est commandé par un bouton placé au centre de cet oreillon. Large lame à dos plat, à pan creux et dorée au talon. Fourreau de laiton, à crevé médian garni de cuir noir ; longue chape à bracelet figuré à anneau de bélière, longue bouterolle également à bracelet figuré et anneau ; dard asymétrique (M.R.).

Chapitre II

Monture à une branche secondaire fixe

Dans ce chapitre sont présentés les sabres à deux branches fixes. Chez les uns, la branche secondaire aboutit au sommet de l'arc de jointure, chez d'autres cette branche s'arrête au niveau de la bague médiane de l'arc. Enfin, chez les sabres cités en dernier la branche secondaire borde une petite coquille ciselée à jour. En tenant compte de ces détails de la monture, on propose le classement suivant :

Famille IV – Branche secondaire atteignant la partie supérieure de l'arc de jointure

Trois Sous-familles ont été créées pour faciliter le classement et l'identification des sabres :

Sous-famille IV a – Branche secondaire simple, ou ajourée ou à l'ancre

La monture est en fer découpé ou en laiton.

491 - Sabre de représentant du Peuple.
1792 - 1793.
Monture en laiton. Calotte rainurée à longue queue. Poignée filigranée, soit sur bois, soit sur basane. Arc de jointure arrivant en angle droit sur la croisière, couvre-chape et quillon à bouton (491a). Branche secondaire atteignant presque le sommet de l'arc et s'élargissant pour couvrir une grande partie de la croisière. Lame et fourreau assez variables ; lame parfois bleuie et gravée de motifs dorés.

<u>Nota</u> - Cette arme civile présente de grandes variations dans la monture. Ariès en donne six variantes (3ème fasc. 1970).

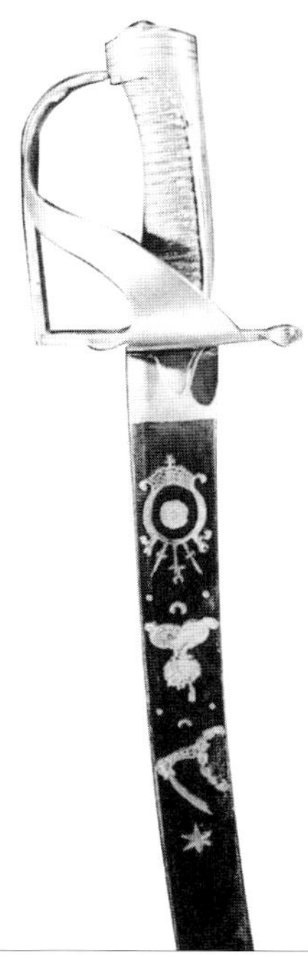

491

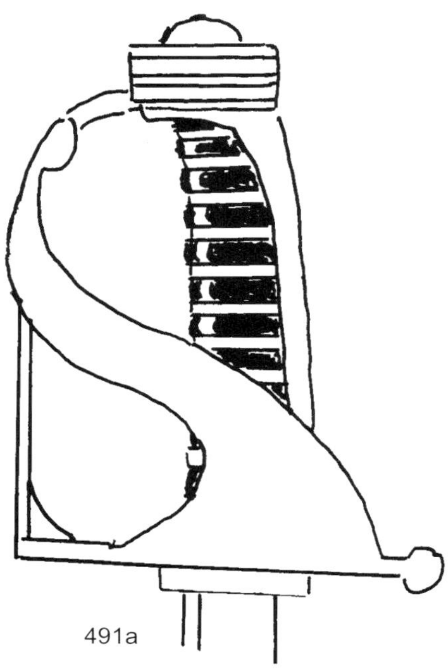

491a

492

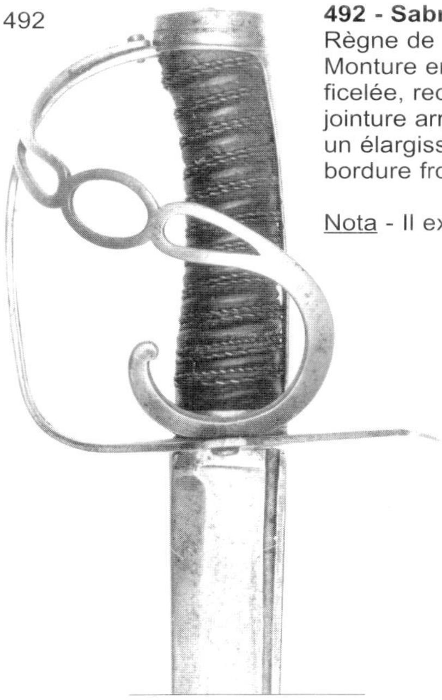

492 - Sabre d'officier subalterne d'Infanterie.
Règne de Louis XVI.
Monture en fer. Calotte plate, moulurée et prolongée d'une longue queue. Poignée ficelée, recouverte de basane, à une douzaine de spires d'un double filigrane. Arc de jointure arrondi suivi d'un plateau. Branche secondaire en " S " avec en sa partie médiane un élargissement ajouré ; cette branche est rivée sur le sommet de l'arc et soudée sur la bordure frontale du plateau. Lame à la Montmorency à profonde gouttière dorsale.

Nota - Il existe des sabres identiques à monture de laiton.

493 - Sabre d'Infanterie de Marine.
Vers 1774 et en usage sous la Révolution.
R.D. Stiot décrit un sabre très particulier dont le schéma ci-contre donne la forme générale. Il s'agit d'un ancien sabre dont la branche mobile a été remplacée par une ancre de Marine munie d'un orin. La calotte est à dôme et à longue queue. Poignée cannelée et à filigrane doré. Lame de 74,1 cm, à légère flèche de 0,17 cm, large de 6,5 cm. Cette lame porte sur le *recto*, les Armes de France (martelées) *Vive / Le Roy /* (martelé) et trois fleurs de lys également martelées. Sur le *verso* on trouve la signature du fourbisseur.
Dida / Marchd (&) Fourbisseur / dans La Rue / Ste Catrine / à / Bordeaux
Fourreau dont la bouterolle est munie d'une roulette pour faciliter le port, l'arme pesant 2,6 kg.

493

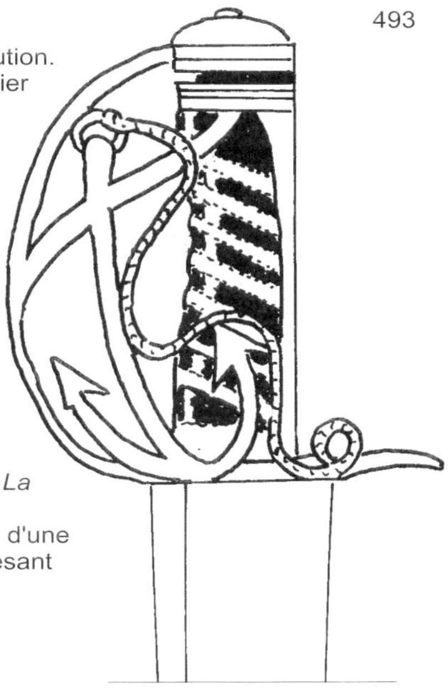

Sous-famille IV b - Branche secondaire à fleurons

On donne le nom de *fleurons* à de petites digitations décorant la branche secondaire. Un motif variable relie la base de cette branche au plateau. Il s'agit d'une fleur de lys pour les plus anciens, d'un faisceau de licteur pour les sabres fabriqués en période révolutionnaire, ou d'un motif sans signification précise à notre connaissance. Ces sabres de Cavalerie lourde seront en service de 1779 jusqu'à la fin de la Révolution, pour le moins et armeront les Dragons en particulier.

494

494 - Sabre de Cavalerie.
Modèle 1779.
Monture en fer. Calotte ovale, plate, surmontée d'un petit bouton de rivure et percée d'un trou pour le passage de la dragonne. Courte queue à bordure découpée en accolade. Poignée de bois, liée d'une ficelle et recouverte de basane maintenue par un fil de fer torsadé. A sa base, une sorte de poucier constitué d'une lanière de cuir. Arc de jointure régulièrement courbé, sur lequel une branche secondaire développant quatre fleurons, est soudée en trois points. Une fleur de lys est brasée entre le plateau et la base de la branche secondaire. Lame droite à deux pans creux, longue de 97,5 cm, large au talon de 3,4 cm. Fourreau de cuir noirci, sans alèse, à deux garnitures dont une bouterolle à dard asymétrique protégeant sa face antérieure et son extrémité.
Ce modèle de 1779 donnera naissance aux modèles 1784 et de l'An IV. Poinçons de Bisch, de Juchereau et du " K " pour Klingenthal.

495

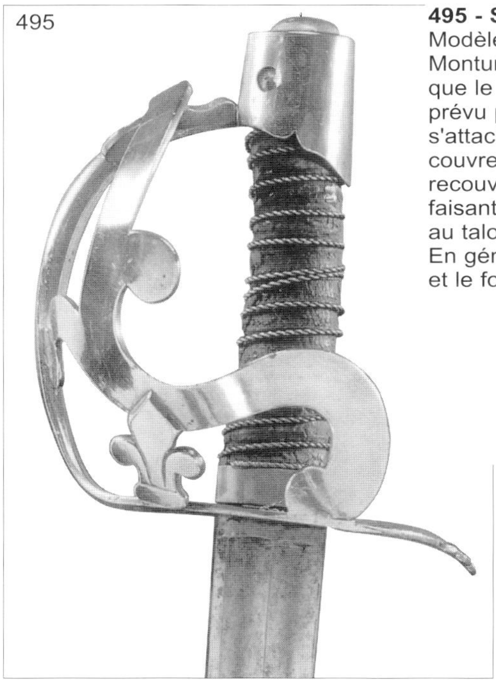

495 - Sabre de Cavalerie .
Modèle 1784.
Monture en " arco " (1). Ce sabre conserve la même facture que le modèle 1779. A noter qu'un second fourreau était prévu pour protéger le fourreau proprement dit : il s'attachait par une languette au bouton de chape. Pas de couvre-chape. Fusée de bois de hêtre, encordée, recouverte de basane et filigranée d'un fil de laiton torsadé faisant une douzaine de spires. Lame de 97, 45 cm, large au talon de 3,6 cm. Poids total du sabre : 1,77 Kg.
En général, la lame est poinçonnée de Juchereau, de Bisch et le fourreau peut être marqué du K de Klingenthal.

496 - Modèle 1784, variante pour officier
Cette variante est caractérisée par une grosse fleur de lys en ronde bosse (496a), couvre-chape. Lame à pan creux, longue de 98, 4 cm, à contre-tranchant, pointe dans le prolongement du dos et le tiers basal ciselé d'arabesques dorées. Fourreau à chape à bouton (J.C. D.).

496

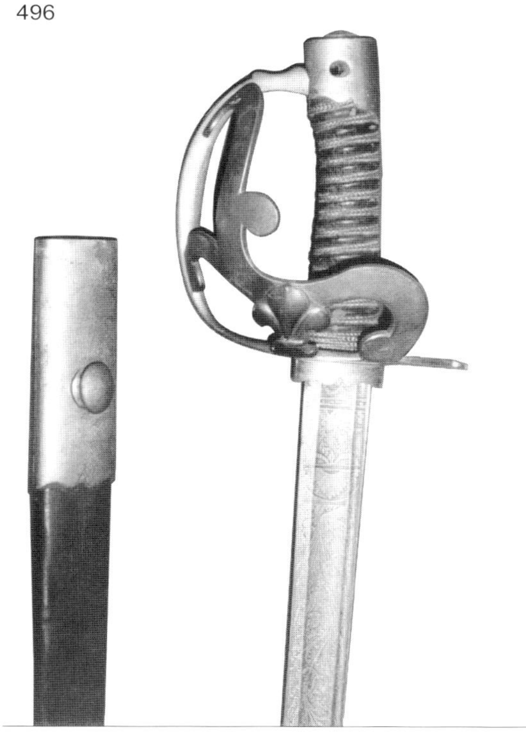

496a

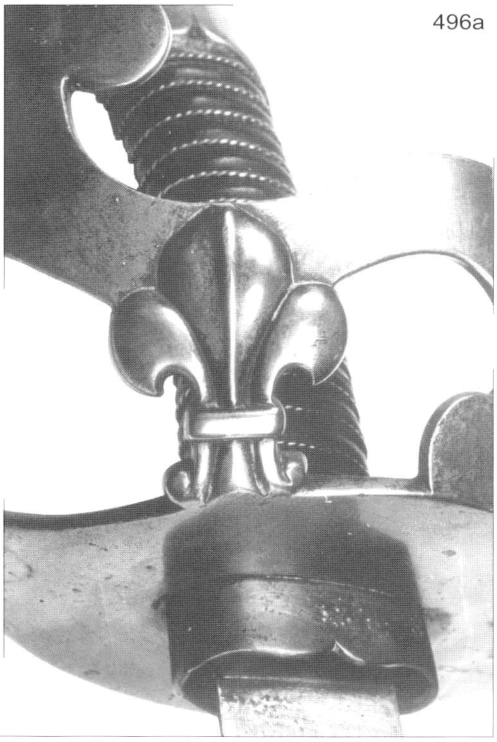

- **a**, détail de la fleur de lys et du couvre-chape.

(1) Arco est un alliage qui est riche en cuivre.

497

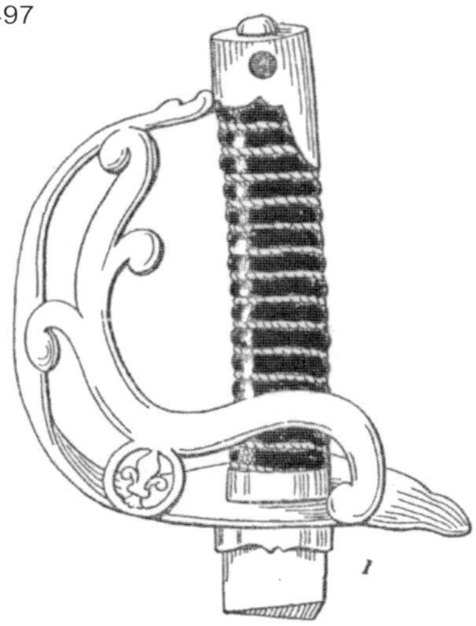

498

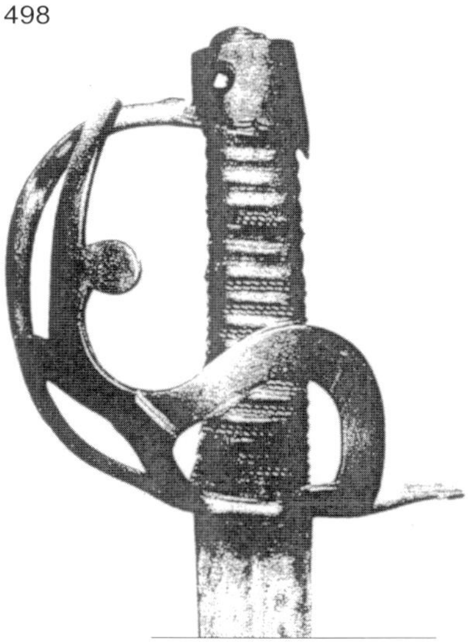

497 - Variante.
La variante ci-contre est signalée par M. Bottet : la fleur de lys découpée est inscrite dans un cercle.

498 - Variante, vers 1792.
Sabre de Dragons du modèle 1784 à monture en laiton. Les pétales latéraux de la fleur de lys ont été sectionnés, la transformant en un faisceau de licteur rudimentaire (Lavallay 1945, pl. XIII).

499

499a

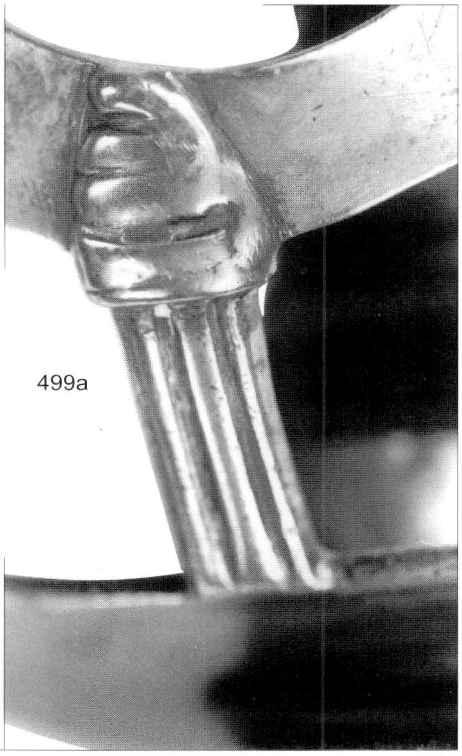

499 - Variante de la période révolutionnaire.
Sabre de Dragons du modèle 1784 dont la fleur de lys a été remplacée sous la Révolution par le faisceau de licteur coiffé du bonnet phrygien (499a).

500

500a

500 - Variante du sabre modèle 1784.
Fabrication révolutionnaire. Monture en laiton. La branche secondaire est reliée au plateau par un faisceau de licteur à bonnet phrygien. Lame à double pans creux, fourreau cuir à trois garnitures.

501

501 - Sabre de Cavalerie.
Modèle de l'An IV.
Monture en laiton. Ce modèle est nettement inspiré du modèle 1784. Le trou de la dragonne a disparu de la calotte, celle-ci étant surmontée d'un petit bouton de rivure en dôme. La branche secondaire est reliée au plateau par un faisceau de licteur à bonnet phrygien (500a). La languette de cuir servant de poucier a parfois disparu. Lame droite, de 97,5 cm, à un pan creux et large talon (onze centimètres chez certains exemplaires). Fourreau de cuir noirci, garnitures en fer, dard protégeant la partie antérieure et l'extrémité de la bouterolle.

Nota - Il existe des fourreaux à trois garnitures de laiton.

Marquages - Lame signée sur le dos *Klingenthal* et sur le plat, poinçons de J.G. Bick, de de Cabannes, du coq , du faisceau de licteur....

502

502 - Variante.
Monture en laiton. Branche à fleurons entièrement ornée à feuillages gravés. Faisceau de licteur liant la branche secondaire au plateau également gravé au trait. Ce sabre est évidemment une fantaisie du début de la période révolutionnaire, vers 1792. (J.C. D.).

503

503 - Variante, après 1792.
Sabre de Dragon du modèle de l'An IV.
Monture en laiton, signée de la Manufacture de Klingenthal et poinçonné de Levavasseur (?) sur la lame et sur la garde. La fleur de lys a été complètement enlevée et aucun motif ornemental ou symbolique n'a été mis en sa place. (J.C. D.).

504

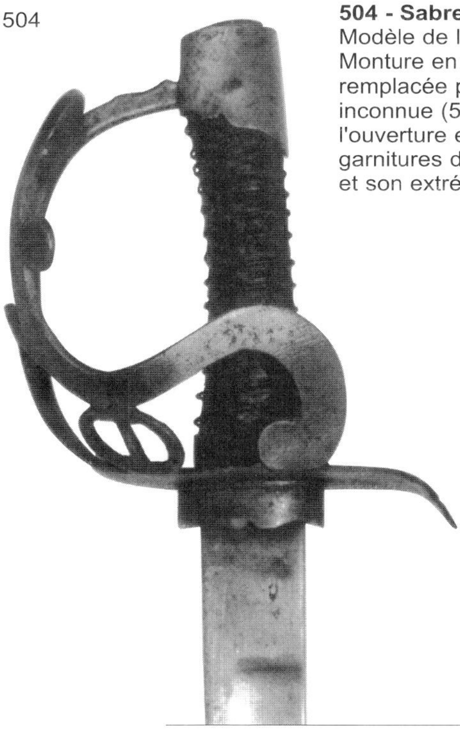

504 - Sabre de Dragons.
Modèle de l'An IV.
Monture en fer à calotte non percée du trou pour la dragonne. La fleur de lys est remplacée par un cercle traversé d'une barrette dont la signification nous reste inconnue (504a).Filigrane simple. Lame de 97,3 cm à pan creux. Fourreau dont l'ouverture est protégée par un couvre-chape fixé sous le plateau de garde, à trois garnitures de fer, bouterolle à dard asymétrique ne protégeant que sa face antérieure et son extrémité (voir G. Jaeger, 1997).

504a

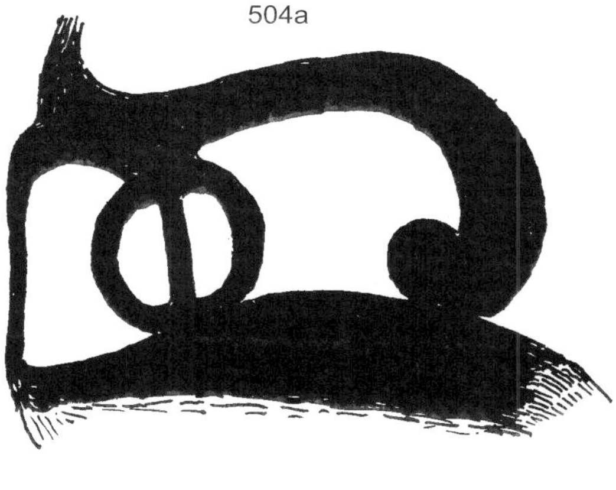

Marquages - Sur le dos de la lame *Klingenthal* et sur la monture : poinçons de J.G. Bick, de de Cabannes.

505 - Diverses autres ornementations du sabre de Cavalerie des modèles 1784 et de l'An IV.
Ces ornementations se rencontrent en période révolutionnaire, avec l'objectif d'effacer tout symbole monarchique :
- **a**, la fleur de lys a été découpée en laissant subsister uniquement le pétale médian ;
- **b**, l'arbre de la Liberté remplace la fleur de lys ;
- **c**, dans un cercle, une colonne est inscrite, allusion à la République romaine ?
- **d**, fût de colonne, même allusion à l'antiquité ?

505a

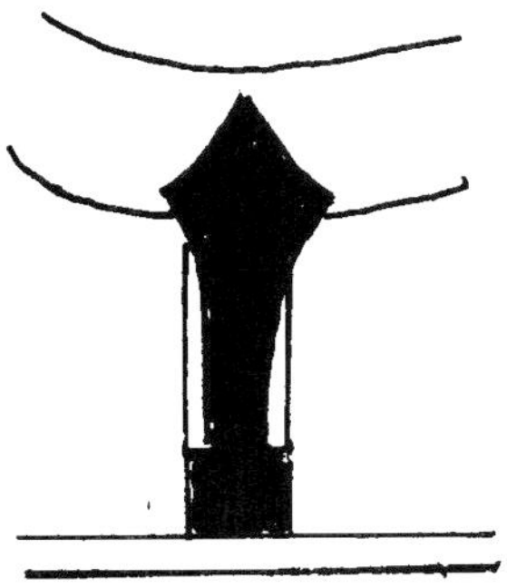

505b

505c

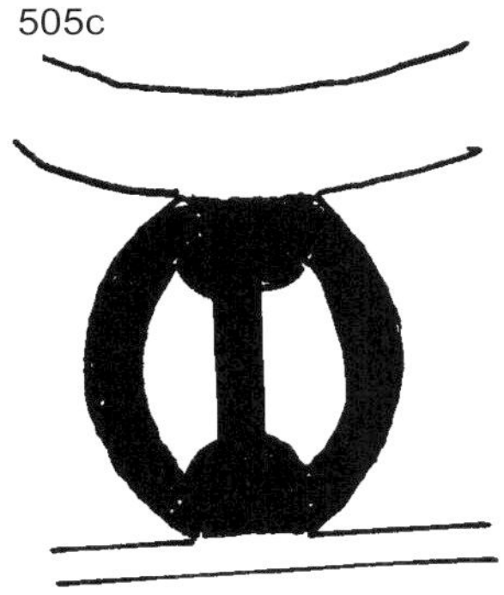

505d

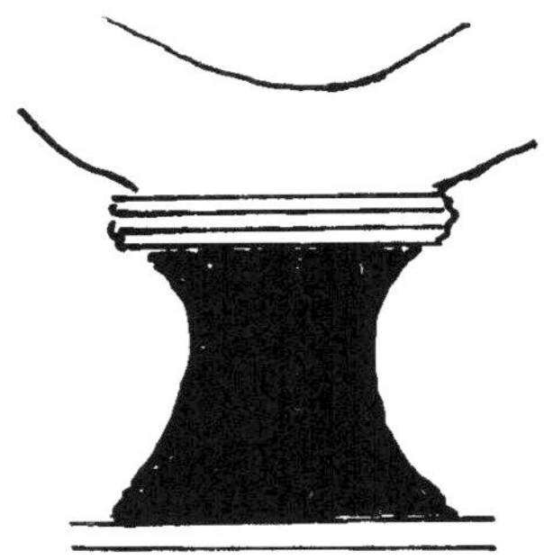

Sous-famille IV c – Branche secondaire reliant une coquille découpée à jour à l'arc de jointure

La coquille limitée par la branche secondaire est ciselée de différents motifs. A noter que cette coquille peut être traversée par l'embryon d'une troisième branche. Les sabres répertoriés sont essentiellement attribués aux bas-officiers.

506

506 - Sabre de Milice.
Vers 1790.
Monture en laiton. Calotte plate, cylindrique, à longue queue. Poignée recouverte de basane à huit spires d'un simple fil doublé. Arc de jointure légèrement arrondi se poursuivant par un plateau ajouré. Branche secondaire transformée en médaillon dans lequel est inscrit un " L " majuscule pouvant signifier Louis, ou le nom de la ville dans laquelle officiait la Milice. Lame plate, sans décor.

507 - Sabre du Régiment de Salis-Samade Suisse.
Modèle de 1791.
Monture en laiton. Calotte plate simple, à bouton de rivure et longue queue. Poignée recouverte de basane et filigranée de fils torsadés. Branche arrondie, se poursuivant par un plateau et se terminant par un quillon plat en palmette. Une branche secondaire relie la principale au médaillon découpé d'une grande fleur de lys stylisée. Lame à dos carré, longue de 62,2 cm, large au talon de 3,6 cm et flèche de 1,1 cm. Fourreau de cuir noir à deux garnitures de laiton, la chape étant à bouton (d'après Margerand, 1945).

507

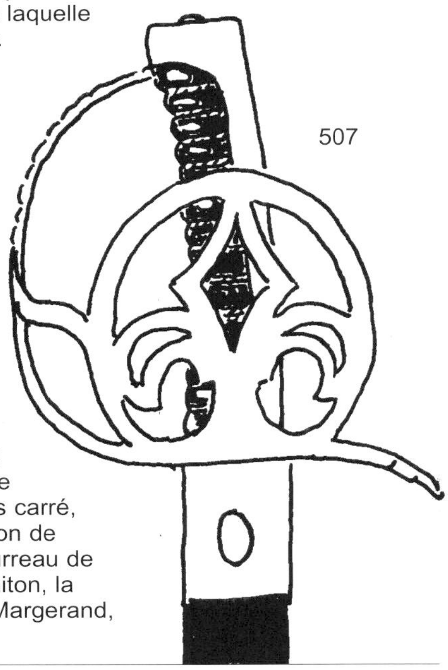

508

508 - Sabres des officiers de Zouaves.
Second Empire.
Ce sabre ne semble pas avoir été réglementé. Monture de métal blanc (nickelé).
Calotte à longue queue, inclinée vers l'avant. Poignée de corne noire, cannelée à filigrane blanc. Les branches et la coquille sont estampées dans la même feuille de métal. L'arc de jointure est plat. La branche secondaire contourne une coquille découpée en médaillon dans lequel est inscrit le " Z " initiale de Zouave. Petite coquille postérieure, également découpée. Lame droite ou courbe, à pan creux. Fourreau métallique à deux anneaux de bélière.

509 - Sabres d'officiers supérieurs, d'officiers subalternes et d'adjudants d'Infanterie.
Modèle 1845.
Monture en laiton doré pour les officiers et simplement poli pour les grades inférieurs. Calotte à bordure de feuilles de laurier le long de la courte queue. Poignée en corne de buffle plus ou moins noircie, rainurée pour recevoir un fil de laiton torsadé. Arc de jointure arrondi, formé en fait de trois branches accolées dont les latérales se détacheront pour border le plateau alors que l'arc proprement dit se poursuit pour aboutir au quillon enroulé. Les deux zones ainsi ménagées sont ornées de motifs végétaux et de rosaces découpés à jour (509a). Lame droite de 86 cm à 4 pans creux séparés deux à deux par une gouttière sur les officiers supérieurs. Lame de 77 cm à légère flêche avec pointe en lange de carpe sur les modèle d'officiers subalternes et d'adjudants. Fourreau cuir à trois garnitures laiton avec bracelets d'anneaux ciselés sur les sabres d'officiers supérieuers. L'illustration présente un sabre subalterne.pour les grades inférieurs. Les lames peuvent être signées : *M^{re} d'Armes de $Chât^{t}$ adj^{d} $d'Inf^{rie}$ Mle 1845.*

509a

- **a**, détail de l'ornementation de la coquille.

509

Evolution du modèle 1845

En 1855, le fourreau de cuir est remplacé par un fourreau en tôle d'acier (Modèle 1855), sauf pour les adjudants qui conserveront le fourreau de cuir jusqu'en 1870. En 1885, ce dernier fourreau ne conservera qu'un seul bracelet de bélière. En 1882, un nouveau modèle de sabre sera attribué aux officiers.

Sous une forme ou sous une autre, ces sabres seront en service de 1846 jusqu'après la Première Guerre Mondiale.

En 1846, la Manufacture de Châtellerault en fournit 600 exemplaires et entre le 1er août 1914 et la fin de 1916, 33 036 sabres seront issus de cette manufacture. La Manufacture de Klingenthal produira également de nombreux exemplaires signés *Coulaux Ainé et Fils.* On notera que les fourbisseurs Manceaux, Barré, Delacour… de Paris, Granger et la Société Balp de Saint-Etienne produiront également ce modèle et ses variantes.

510

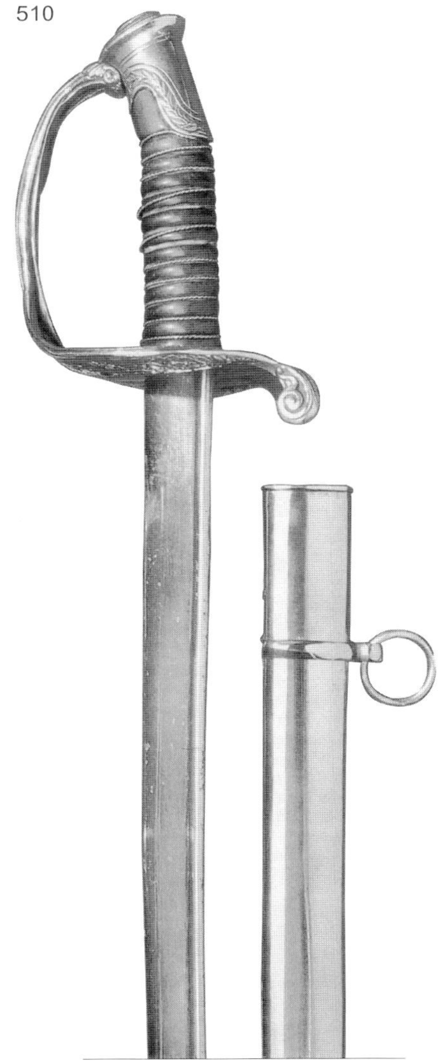

510 - Sabre d'adjudant.
Fabrication postérieure à 1885. Remarquer le fourreau de fer à un seul bracelet de bélière. (M.R.).

511

511 - Sabre d'officier général.
Second Empire.
Sabre conforme au modèle 1845, à monture de bronze doré. Remarquer la lame d'officier supérieur portant les inscriptions
dans la gouttière supérieure :
Mre Imp^{le} de Chât^{t} 7bre 1869
et dans l'inférieure :
O. Généraux (xx) *1869* (J.C. D.).

Famille V – Branche secondaire s'arrêtant à la bague médiane de l'arc de jointure

Les sabres de cette Famille sont d'une facture toute différente des sabres des Familles précédentes et annoncent une conception typiquement française et moderne. On a créé deux Sous-familles en se basant, entre autres, sur la présence ou l'absence de coquille.

Sous-famille V a – Petite branche secondaire placée dans le même plan que le plateau et à calotte à courte queue

Il s'agit du sabre des officiers d'Infanterie, modèle 1821. Ce sabre fut proposé le 15 janvier 1817 et ne fut adopté que le 31 juillet 1821, par le maréchal Victor. La position de la petite branche (voir p. 42, n° 9) fait qu'elle reste invisible lorsque le sabre est vu de profil, comme on le sait sur la majorité des photographies qui sont faites de ce modèle. Ce sabre d'Infanterie restera en service, moyennant quelques modifications de détail, jusque dans les premières années de la IIIème République.

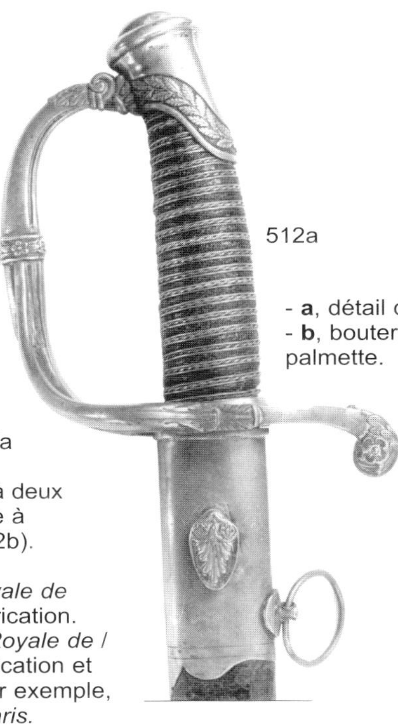

512a

512b

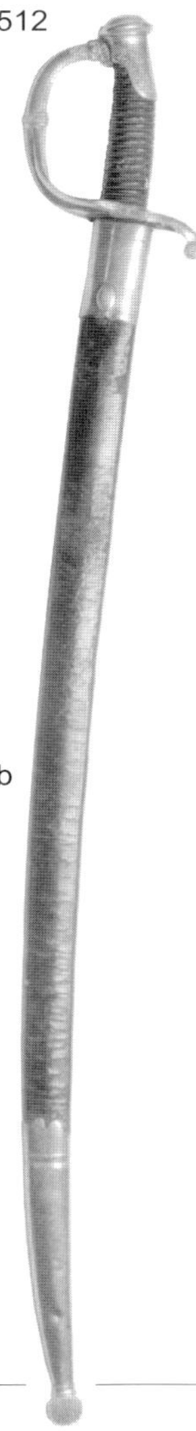

512

- **a**, détail de la monture ;
- **b**, bouterolle à dard en palmette.

512 - Sabre des officiers d'Infanterie.
Modèle 1821.
Monture en laiton doré. Calotte, à bourrelet et dôme-bouton de rivure, légèrement inclinée vers l'avant et courte queue à bordure de feuilles de laurier (512a). Poignée recouverte de cuir mince, filigrané de fils de laiton doré, faisant le plus souvent 18 spires. Arc de jointure arqué, à bague médiane ciselée d'une fleurette et de feuilles stylisées. De cette bague part une petite branche secondaire, nommée pontet ou étrier, longue de 4 cm, rejoignant le plateau dans le même plan. Plateau ovalaire, terminé par une fleurette et de petites feuilles, dans le même esprit que le motif figurant au début de l'arc. Quillon recourbé. Lame dite en demi-espadon, forme voisine de la lame à la Montmorency, longue de 75,8 cm, à pan creux et gouttière au fort, flèche de 2 cm. Fourreau de cuir, à deux garnitures : chape à crochet en écusson, bouterolle à bouton en guise de dard, plus ou moins ciselé (512b).

Marquages - Sur le dos de la lame : *Manf^{ture} Royale de Klingenthal* suivi du mois et du millésime de la fabrication. Ou bien sur le talon de la lame on lit : *Manuf^{re} / Royale de / Klingenthal /* puis le mois et le millésime de la fabrication et sur l'autre face, la signature du fourbisseur, soit par exemple, *Juste / Fourbisseur / rue St Honoré / N° 137 / à Paris.*

Ce sabre sera en usage dans l'armée jusqu'en 1845 et même jusqu'en 1870 pour les sous-officiers de la Garde de Napoléon III. Il est évident que pendant ce demi-siècle, il subit quelques modifications de détail et d'attribution.

Maryse Razo et M. de Fromont (1993) font les distinctions suivantes :

Sabre de luxe :
Les officiers achetant leur sabre, ont une certaine latitude pour l'enjoliver. La basane de la poignée peut être remplacée par du galuchat, voire de la corne ; le filigrane peut être très élaboré, constitué de plusieurs fils de laiton, torsadés ou non ; la lame peut être bleuie et gravée de motifs dorés et la bouterolle du fourreau est alors terminée par un bouton ciselé (512b).

514

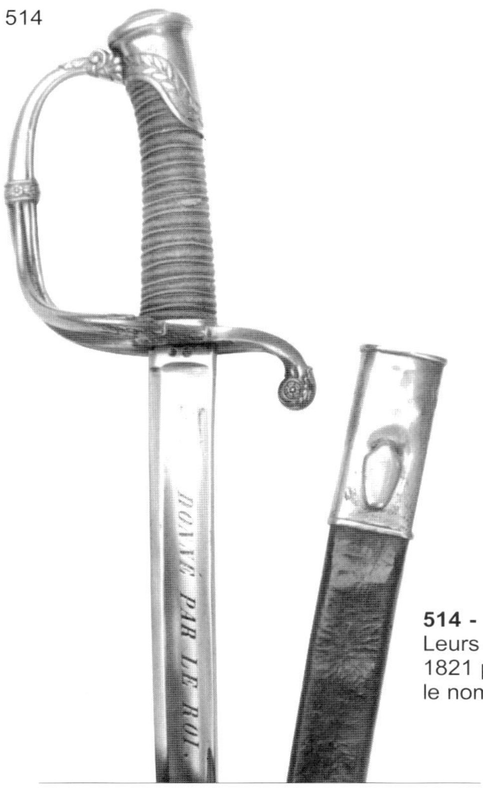

513 - Sabre d'officier de la Garde Royale.
Ce sabre d'officier d'Infanterie modèle 1821 attribué à la Garde Royale de la Restauration, présente les caractéristiques suivantes. Monture en laiton doré très comparable à celle des sabres d'officiers. En revanche, la lame bleuie parfois sur presque la moitié de sa longueur porte en or, sur le *recto*, la signature *MANUF^{RE} Royale de Klingenthal - Coulaux Frères*, au talon puis, les Armes de France, *GARDE ROYALE*, fleur de lys et chiffre de Louis XVIII. Sur le *verso* on trouve la signature *Duc Fourbisseur rue St Honoré N° 251*, puis, de nouveau, les Armes de France, 6e d'Infanterie (par exemple), fleur de lys et le chiffre de Louis XVIII . Fourreau de cuir noir vernissé, à deux garnitures de laiton doré, gravées de feuillages, chape à bouton ciselé de fleurette et bouterolle ornée d'une palmette (non figuré).

Nota - Certains exemplaires sont signés :
Juste / Fourbisseur / rue S^{t} honoré / N° 137 / à Paris.
Les lames portent généralement la signature de la *Manufacture de Klingenthal.*

514 - Sabre des élèves officiers des Écoles militaires.
Leurs études terminées, le roi offrait aux jeunes officiers un sabre du modèle 1821 portant gravée sur la lame, la mention *" DONNE PAR LE ROI "* et parfois le nom du récipiendaire. Ce sabre fut donné de 1822 à 1830.

515

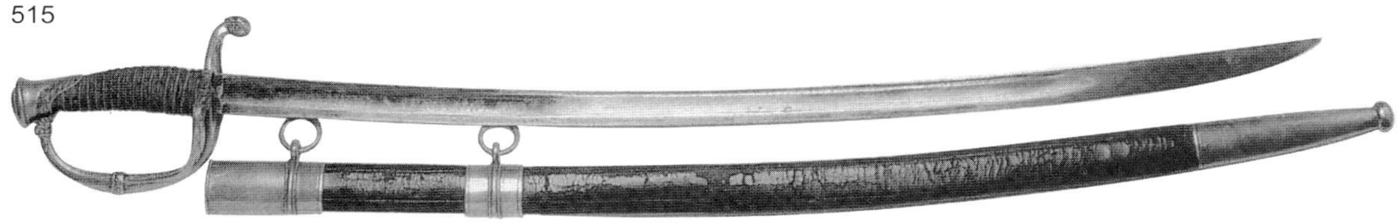

515 - Sabre de la Garde Nationale de Louis-Philippe.
Modèle 1821.
Sous Louis-Philippe, ce sabre reste doré pour les officiers mais non doré pour les sous-officiers qui en bénéficient. L'instruction de Mai 1845 intéresse une modification du fourreau. Celui-ci devient à trois garnitures : chape sans crochet mais une bague située à 5 cm de l'embouchure, à jonc de 0,8 cm d'épaisseur, portant l'anneau de bélière ainsi que la seconde garniture et courte bouterolle à bouton. La lame perd sa gouttière.
On rencontre des sabres du modèle 1821/1845 très ornés, notamment sur la bouterolle et sur la lame : celle-ci est bleuie et gravée de motifs dorés, avec l'inscription *GARDE NATIONALE.* En plus de ces sabres de luxe, il y a tous les sabres qui portent, gravés à l'acide, en caractères semi-gothiques : *Garde Nationale, Liberté, Ordre public Egalité* (515a)... souvent accompagnés du Coq de la Monarchie de Juillet.

Il faut aussi citer les sabres donnés aux Volontaires de la Charte
En récompense de leur fidélité, Louis-Philippe offrit ce sabre aux *150 Volontaires de la Charte* après la Révolution de Juillet 1830.

515a

- **a**, exemples de lames gravées à l'acide *Garde Nationale* sur une face et sur l'autre *Liberté - Ordre Public.*

516 - Modèle 1821 - 1854.
La Garde Impériale étant rétablie par Napoléon III, les sous-officiers se voient attribuer le 13 mai 1854, une modification du modèle 1821 que nous nommerons modèle 1821/1854.
Ce modèle n'est pas doré, le fourreau perd ses anneaux de bélière et un crochet est de nouveau brasé sur la chape. Puis, le 12 février 1855, la poignée devient en corne de buffle, le cuir du fourreau est estampé, le bouton de chape est maintenu par une goupille et le 30 janvier 1865, une vis maintient la chape sur le fourreau au lieu d'être uniquement collée.

Ce sabre est d'abord attribué aux adjudants, sous-chef de musique, musicien, sergent major, sergent, caporal tambour. Le 29 septembre 1855, reçoivent ce sabre, les maîtres ouvriers de l'Infanterie de la Garde, voltigeurs, chasseurs à pied, zouaves. Dans les premières années de la IIIème République, ce sabre des sous-officiers de la Garde Impériale seront encore portés par les sous-officiers du Service de la Justice Militaire.

Nota - Les premières fabrications sont signées de *Klingenthal* mais à partir du 27 mars 1856, les lames portent la signature de la *Manufacture de Châtellerault.*

Sous-famille V b – Branche secondaire limitant une coquille ciselée et ajourée, calotte à longue queue

Les sabres décrits forment un ensemble très homogène : la coquille est placée entre la bordure du plateau et la branche secondaire.

Les sabres d'officiers de Marine constituent la majorité de ces modèles à partir de 1837 jusqu'à nos jours. Contrairement aux sabres d'officiers d'Infanterie modèle 1821, la calotte est pourvue d'une longue queue généralement ciselée de coquilles Saint-Jacques pour la Marine. Le fourreau est souvent ciselé d'un décor constitué de plantes aquatiques : Sagittaire, Rubanier, Massette, Sparganium, Typha, toutes plantes vivant en eau douce, ce qui semble paradoxal mais qui sont le symbole de l'eau primordiale.

517 - Sabre d'officier de Marine.
Modèle du 20 juillet 1837.
Calotte légèrement convexe ciselée de feuilles radiantes. Longue queue décorée d'une coquille Saint-Jacques placée sur feuilles d'acanthe près de la calotte, puis de simples filets, alors que la bordure est à frise d'écailles (517a, 517b). Poignée en corne noire, à huit ou dix cannelures en général, sans filigrane. Arc de jointure à œillet supérieur permettant de passer la dragonne, rainuré jusqu'à la bague médiane ; de celle-ci part la branche secondaire, bordant la coquille.
Coquille ciselée à jour de feuilles d'acanthe ; dans la partie axiale, est placée une ancre étalinguée de 4,5 cm (517c). Contre-garde ornée d'une coquille Saint-Jacques flanquée de deux feuilles d'acanthe (517d). Quillon court, incliné et terminé par une tête de monstre marin. Lame plate de 70 à 80 cm, large au talon de 2,7 à 2,8 cm à légère flèche de 1,5 cm environ, dos à jonc, gouttière dont se détache à 31,5 cm de l'extrémité, une arête qui rejoint, en oblique, la pointe (518e). Fourreau à couture dorsale ; chape à bordure près de l'embouchure, puis trois bagues délimitant un carré à bouton et anneau de bélière et un rectangle à l'ancre étalinguée ciselée sur fond sablé (517f) ; bracelet et bouterolle unis, à bordure (517g) ; dard asymétrique décoré d'une coquille Saint-Jacques (517h).

517

517a

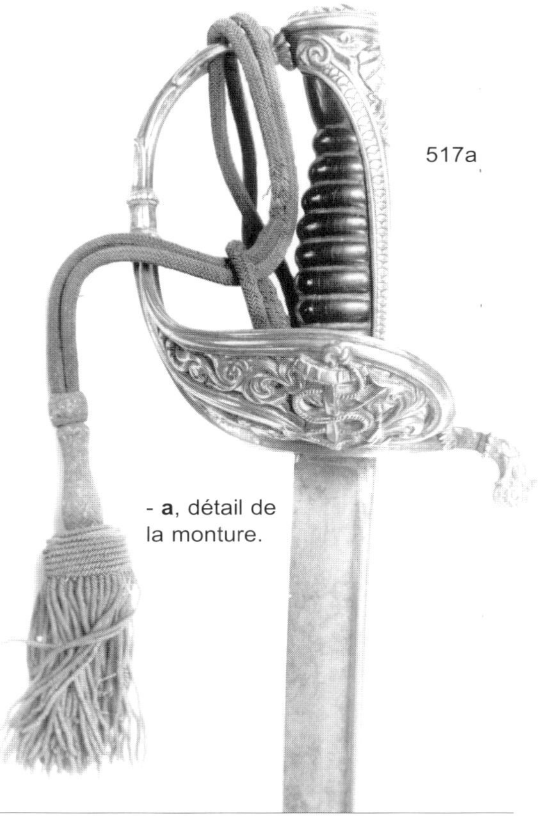

- **a**, détail de la monture.

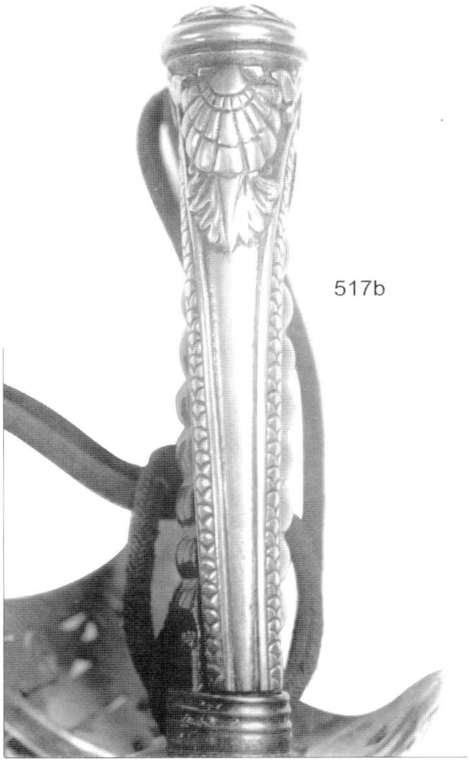

517b

517c

517d

517e

Marquage - Dos de la lame signé : Coulaux Frères à Klingenthal. Ce sabre sera en service jusque sous la Seconde République (1848-1852) selon le décret du 17 avril 1850.

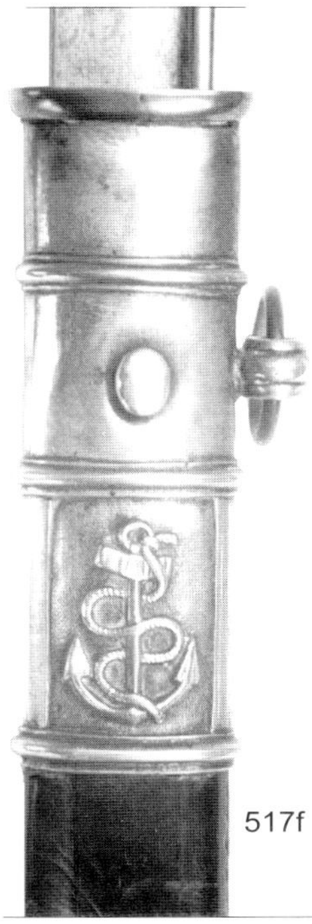

517f

517g

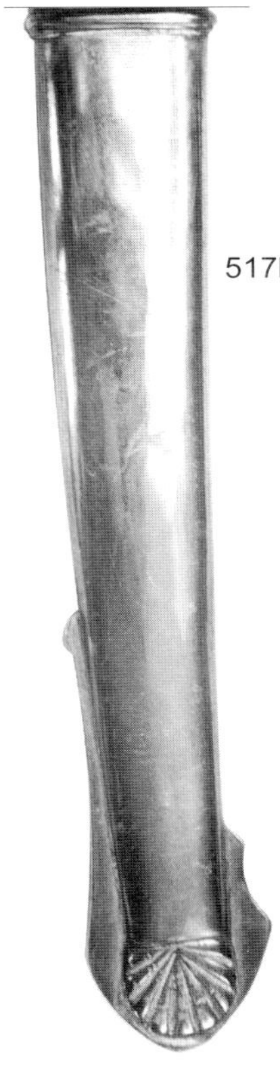

517h

- **b**, décoration de la queue de calotte ;
- **c**, détail de la coquille à l'ancre étalinguée ;
- **d**, contre-garde à la coquille Saint-Jacques ;
- **e**, extrémités de la lame à arête se séparant de la bordure pour rejoindre la pointe médiane ;
- **f**, détail de la chape ;
- **g**, bracelet médian uni :
- **h**, bouterolle avec son dard à la coquille Saint-Jacques.

518 - Variantes du Modèle 1837.
On remarquera sur la chape, la coquille Saint-Jacques surmontant l'ancre, elle-même flanquée de deux rameaux d'une plante aquatique que l'on peut assimiler à une sagittaire caractérisée par ses fruits groupés par trois (518a). (M.R.).

518

518a

- **a,** remarquer les branches de sagittaire flanquant l'ancre.

519

519 - Sabre d'officier de Marine.
Modèle 1837, variante.
Exemplaire très comparable au sabre précédent. Le décor de la chape est bien composé de l'ancre surmontée de la coquille, mais les végétaux la flanquant s'apparentent au laurier. Partie supérieure de l'arc de jointure encordée (M.R.).

520 - Sabre d'officier supérieur de Marine.
Modèle 1837, modifié sous le règne de Louis-Philippe.
Monture en laiton doré. Calotte à couvercle décoré d'un médaillon central. Poignée de bois cannelé en oblique, non filigrané. Longue queue de calotte ciselée de palmes nouées par un nœud, d'un médaillon ovale, de feuilles de chêne (520a,). Bague basale ornée de perles et pourvue d'un œillet pour passer la dragonne. Arc de jointure ciselé en suite. Coquille de garde ajourée, ornée en son milieu d'un écusson à l'ancre de marine étalinguée, surmontée de la couronne royale, flanqué de feuilles (520b,). Quillon à enroulement figuré. Fourreau de cuir noirci, à trois garnitures : chape à bouton de suspente en coquille, piton à méplats, encart dans lequel est inscrite une ancre surmontée de la couronne royale sur fond de deux pavillons dont on devine les trois couleurs nationales de la Monarchie de Juillet et se terminant par une bague à bordures striées stylisant des orins (520c). Bague médiane en suite. Bouterolle à bague, ciselée

520

520a

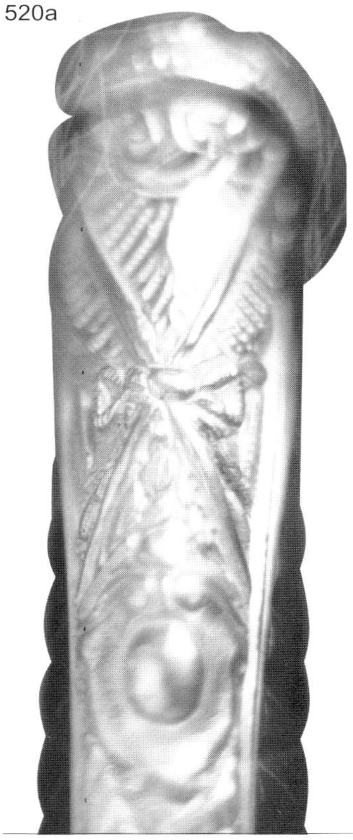

520b

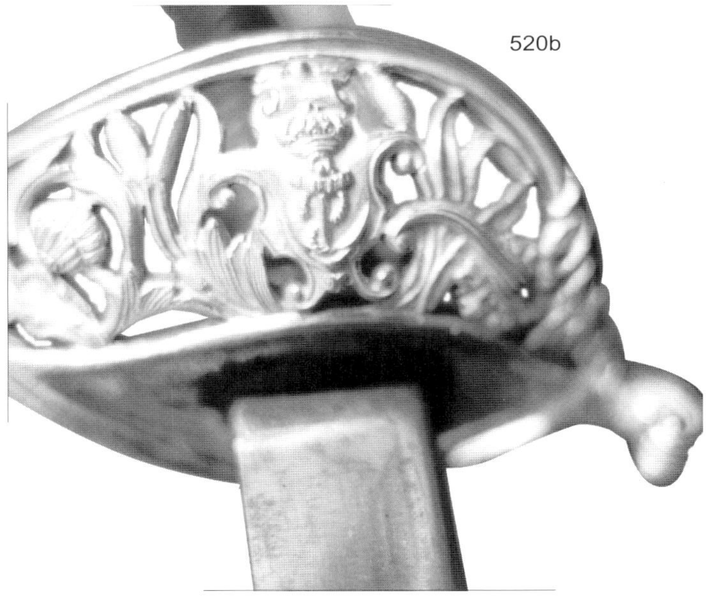

520c

520d

d'un typha et se terminant par des palmes. Dard asymétrique, à la coquille Saint-Jacques, à bordure ondulée (520d). Lame longue de 65,5 cm,
large au talon de 2,8 cm, à contre-tranchant et arrête décentrée. Le dos de la lame porte la signature
J.J. Bisch à Boersch et Klingenthal.

- **a**, queue de calotte ;
- **b**, garde ;
- **c**, chape ;
- **d**, bouterolle.

521 - Sabre d'officier de Marine.
Modèle 1837, modifié sous le règne de Louis-Philippe.
Ce sabre est évidemment très voisin du sabre précédent, mais est moins bien traité et la coquille de la garde n'a pas été modifiée, mais la couronne royale a disparu.

- la longue queue de calotte est toujours ornée d'une frise de coquilles Saint-Jacques,
- la bouterolle n'est pas décorée,
- l'encart de la chape chez le sabre d'officier supérieur de Marine est également présent, avec son ancre étalinguée et ses pavillons (521a).

521a

521

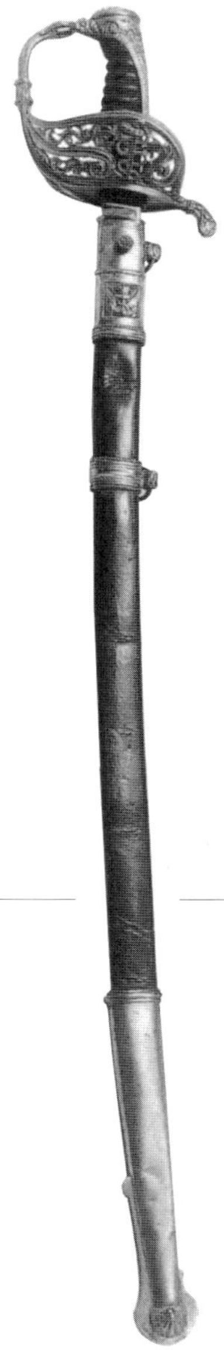

522 - Sabre d'officier de Marine.
Seconde République, 1848-1852.
Modèle 1837 avec quelques modifications dont la queue de calotte ornée de deux drapeaux croisés, placés sous une petite coquille Saint-Jacques. La caractéristique essentielle est le remplacement de la couronne royale par un enroulement de l'orin sur la coquille et la chape (522a, 522b).

522b

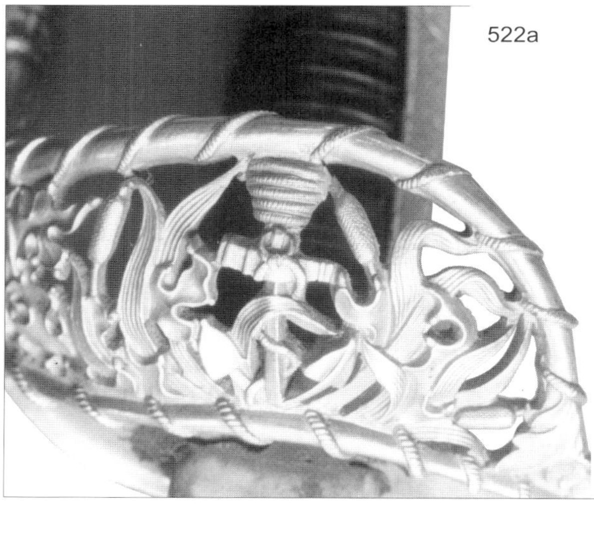

522a

523

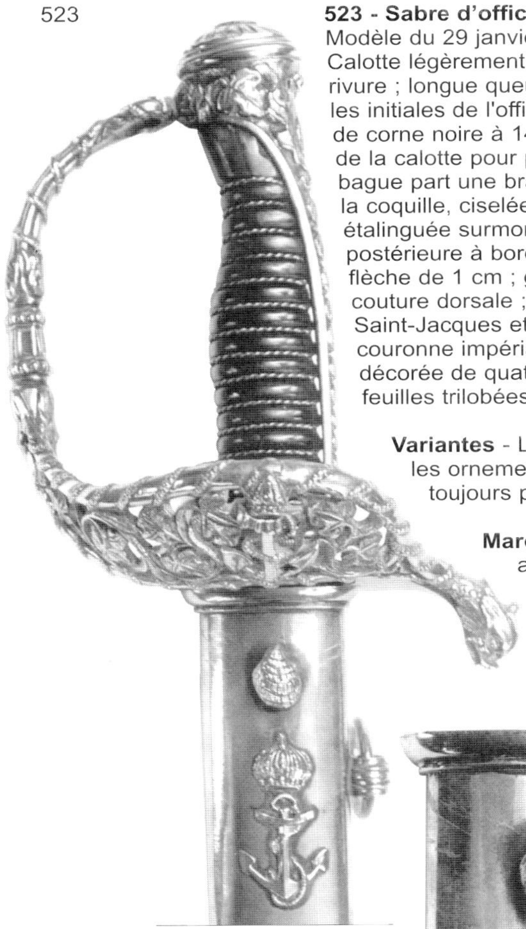

523 - Sabre d'officier de Marine.
Modèle du 29 janvier 1853.
Calotte légèrement convexe, ciselée de côtes radiantes aboutissant à un petit bouton de rivure ; longue queue décorée de feuilles d'eau, d'un petit médaillon destiné à recevoir les initiales de l'officier et au-dessous, une frise de 13 coquilles Saint-Jacques. Poignée de corne noire à 14 spires de fil torsadé. Arc de jointure encordé, muni d'un œillet près de la calotte pour passer la dragonne, régulièrement arqué à bague médiane ; de cette bague part une branche secondaire également encordée limitant la partie supérieure de la coquille, ciselée à jour de feuilles d'eau, avec dans l'axe du sabre une ancre étalinguée surmontée de la couronne impériale. Quillon à tête de dauphin. Coquille postérieure à bordure encordée. Lame de 67,5 cm, largeur au talon de 2,5 cm, légère flèche de 1 cm ; gouttière dorsale, pan creux, pointe médiane. Fourreau de cuir à couture dorsale ; trois garnitures : chape à embouchure bordée, à bouton en coquille Saint-Jacques et anneau de bélière, ornée de l'ancre étalinguée surmontée de la couronne impériale (523a) ; bracelet ciselé d'une feuille d'eau (523b) bouterolle décorée de quatre feuilles d'eau ; dard orné d'une coquille de vannet surmontant des feuilles trilobées (523c).

Variantes - Les variantes sont nombreuses et intéressent la longueur de la lame, les ornementations de l'arc de jointure, la coquille... Seule l'ancre couronnée est toujours présente.

Marquage - Le dos de la lame est signé : *Manufacture de Klingenthal* avec la date de fabrication.

523a

523b

523c

- **a**, détail de l'ornementation de la chape ;
- **b**, décor du bracelet médian à la feuille d'eau ;
- **c**, décor de la bouterolle aux feuilles d'eau et dard à la coquille de vannet. Le vannet est un bivalve aux deux coquilles convexes.

524

524 - Variante du Modèle 1853.

Queue de calotte à 13 coquilles Saint-Jacques. Arc de jointure et bord de la coquille de garde encordée (524a). Quillon en tête de dauphin (524b). Lame à flèche de 0,8 cm, longue de 68 cm, large au talon de 2,4 cm, à deux pans creux séparés par une arête qui se prolonge jusqu'à la partie médiane. Fourreau en tôle d'acier noirci. Chape et bracelet ciselés de feuilles d'acanthe et entre deux bourrelets simulants des cordages, l'ancre étalinguée surmontée de la couronne impériale (524c) ; bouterolle décorée de rameaux de sagittaire, avec à sa base un vannet ; dard asymétrique à coquille Saint-Jacques placée sur un motif trifolié (524d).

Marquage - Pas de marquage de manufacture mais poinçon de Joseph Antoine Bisch sur la talon de la lame.

524a

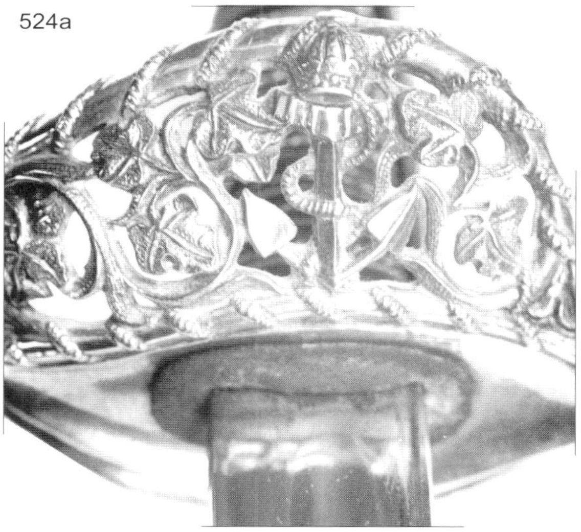

- **a**, sous le Second Empire l'ancre est couronnée ;
- **b**, détail du quillon ;
- **c**, bracelet médian au même décor que la chape ;
- **d**, bouterolle ornée de branches de sagittaire.

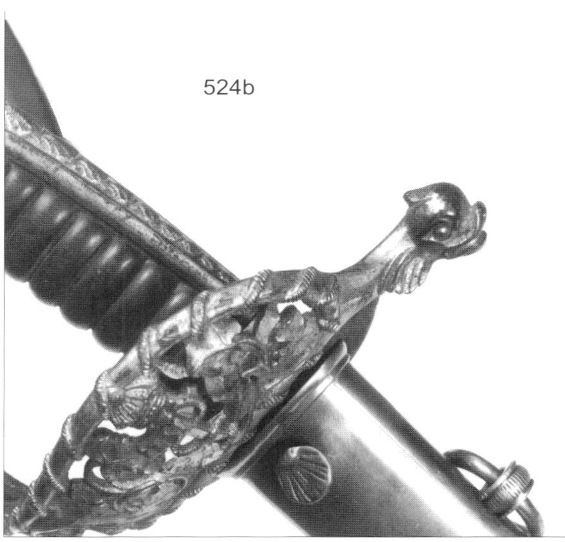

524b

524c

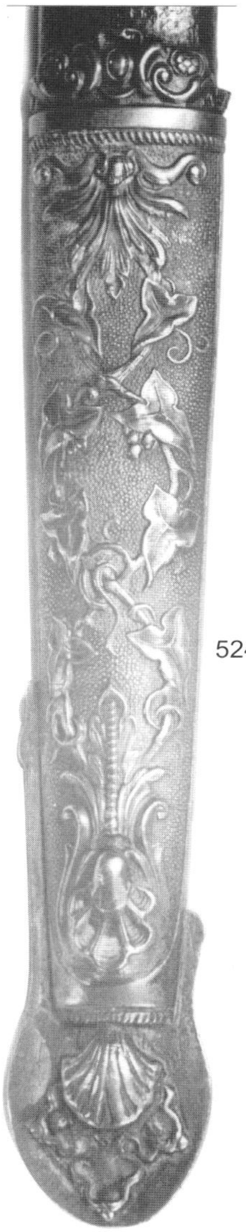

524d

525 - Modèle 1853, modifié par le décret du 10 octobre 1870.
Monture dorée, très comparable au modèle 1853, avec des différences mineures : suppression de la couronne impériale, calotte à décor radiant à éléments en creux, queue à frise de 14 coquilles Saint-Jacques. Poignée de corne noire à 12 cannelures, sans filigrane. Lame nickelée de 75 cm, large au talon de 2,5 cm à flèche de 1 cm ; dos plat à gouttière d'où part, à environ 23 cm de l'extrémité, une arête rejoignant la pointe médiane ; pan creux. Fourreau en cuir comprimé, à couture dorsale masquée ; chape à bordure près de l'embouchure, simplement décorée d'une ancre étalinguée, avec anneau de bélière ; bracelet intermédiaire à vis de fixation sur l'envers du fourreau ainsi que sur la bouterolle ; dard asymétrique orné d'une coquille Saint-Jacques placée sur trois feuilles d'acanthe (non figuré).

Marquages - Signé sur le dos de la lame : *Coulaux et Cie. Klingenthal*

526

526 - Sabre d'officier supérieur de la Marine.
Troisième République.
Ce modèle, tout en conservant l'esprit des modèles antérieurs est remarquable par les particularités qu'il présente : queue de calotte ornée d'une ove, puis d'un trophée d'armes et de Typhas. Arc de jointure non encordé, à longue bague médiane. Coquille ciselée à jour de l'ancre de 1 cm. Lame à deux pans creux séparés par une arête arrivant jusqu'à la pointe médiane. Fourreau de cuir noirci, à couture dorsale ; longue chape entièrement décorée avec, dans un médaillon, un trophée d'armes ; bouton et anneau de bélière (526a) ; bracelet à l'ancre étalinguée flanquée d'un rameau de chêne (à gauche) et d'un rameau de laurier (à droite) (526b) ; bouterolle ciselée de feuillages stylisés, avec en sa partie supérieure dans un cartouche, une boussole sur fond de pavillons, glaive, trompette, compas, gaffe... dard en lyre, ciselées de feuillages (526c).
On notera que ce sabre n'est pas orné de feuilles aquatiques lesquelles sont remplacées, des rameaux de chêne et de laurier, ces derniers étant généralement utilisés pour les armes des troupes terrestres.

Marquage - Lame signée sur le dos : *Coulaux et Cie Klingenthal.*

526a

526b

526c

- **a**, ornementation de la chape au trophée ;
- **b**, bracelet médian à l'ancre flanquée de rameaux de chêne (à gauche) et de laurier (à droite) ;
- **c**, bouterolle au trophée de pavillon, longue-vue, gaffe ... sous une boussole...

527

527a

- **a**, chape à la coquille Saint-Jacques et à l'ancre ;
- **b**, bracelet décoré en suite plus une feuille de vigne ;
- **c**, bouterolle décorée en suite avec quatre inflorescences et dard à la coquille Saint-Jacques.

527c

527 - Sabre d'officier supérieur de Marine.
Troisième République.
Monture en laiton doré. Calotte prolongée d'une longue queue à frise de coquilles Saint-Jacques. Poignée en corne non filigranée. Garde décorée du feuillage et de l'ancre classique, en découpe. Quillon en tête de dauphin. Lame droite. Fourreau à trois garnitures : chape à bouton en coquille Saint-Jacques et anneau de bélière, à l'ancre étalinguée, l'ensemble étant orné de feuillage stylisé (527a). Bracelet médian à la feuille d'eau (527b). Bouterolle décorée en suite avec de petites palmettes pouvant être assimilées à des inflorescences. Dard asymétrique à coquille (527c).

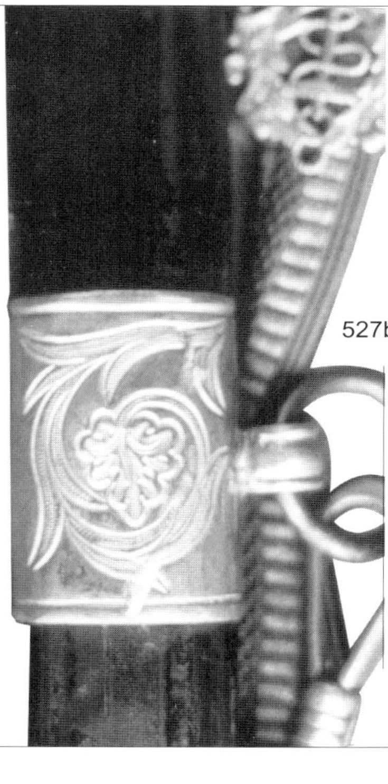

527b

528

528a

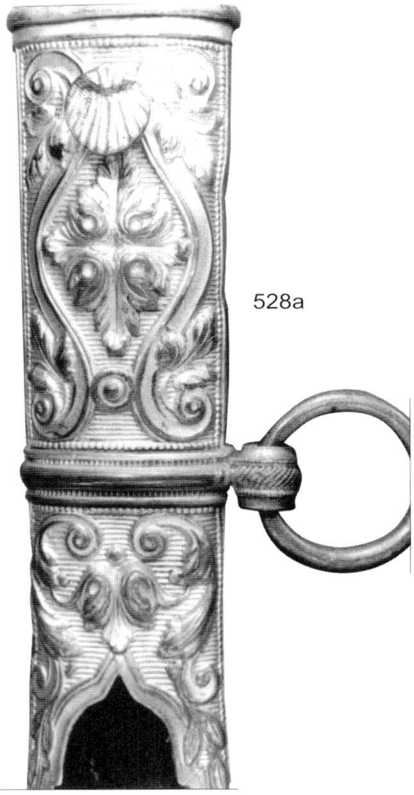

528 - Sabre d'officier supérieur de la Marine
Troisième République. Monture en laiton doré, de forme classique, seules les ornementations sont particulières : poignée en corne noire non filigranée, partie supérieure de l'arc de jointure décorée de feuilles de vigne, quillon en tête de monstre marin... Lame blanche. Fourreau à trois garnitures ciselées : chape à bouton en coquille Saint-Jacques, avec comme motif principal quatre feuilles d'acanthe stylisées, disposées en croix et portant le bracelet à anneau de bélière (528a) ; garniture médiane présentant parmi des arabesques foliacées, un médaillon dans lequel sont inscrits une étoile, vraisemblablement l'étoile polaire, une mapemonde sur son pied, surmontant une petite boussole, l'ensemble placé sur un trophée d'armes blanches et de fanions (528b) ; bouterolle à arabesques du modèle de celles du bracelet médian et on remarquera tout spécialement la feuille de vigne et les quatre grains de raisin, juste avant le dard. Notons encore que ce dard ne représente pas une coquille Saint-Jacques ni un vannet mais un motif en palmette foliacée (528c). (M.R.).

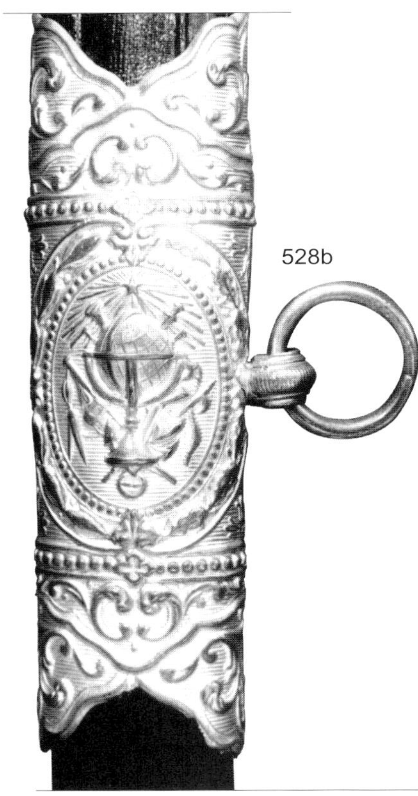

528b

528c

- **a**, détail du décor de la chape ;
- **b**, bracelet médian orné d'un médaillon aux accessoires marins ;
- **c**, bouterolle avec son dard à palmette.

529a

529b

529c

530

529 - Sabre d'amiral. Troisième République. Monture en bronze doré. La garde est très comparable aux sabres de Marine de la même période. La lame est signée *Coulaux et Cie à Klingenthal*. C'est le fourreau qui retiendra toute notre attention par son travail d'orfèvre : chape traitée en résille, à anneau de bélière en médaillon ovale à l'ancre de marine flanquée à gauche d'un rameau de chêne et à droite d'un rameau de laurier (529a). Le bracelet est également ciselé à jour (529b) ; la bouterolle reprend le motif du bracelet avec un médaillon très ovalaire décoré de feuilles d'acanthe disposées en croix ; la partie inférieure de cette garniture est ciselée de feuilles et se termine par un dard asymétrique orné d'une coquille de vannet sur les deux faces (529c) (M.R.)

- **a**, chape en résille à médaillon à l'ancre flanquée d'un rameau de chêne (à gauche, et d'un rameau de laurier (à droite) ;
- **b**, bracelet médian au même médaillon que la chape ;
- **c**, bouterolle à médaillon à feuilles d'acanthe et dard orné d'un vannet, (*verso*).

530 - Sabre d'officier supérieur de Marine. Milieu du XXème siècle. Modèle très ouvragé bien dans la tradition des fabrications antérieures. Dans ce modèle, la coquille est rès aplatie et la chape richement ciselée.

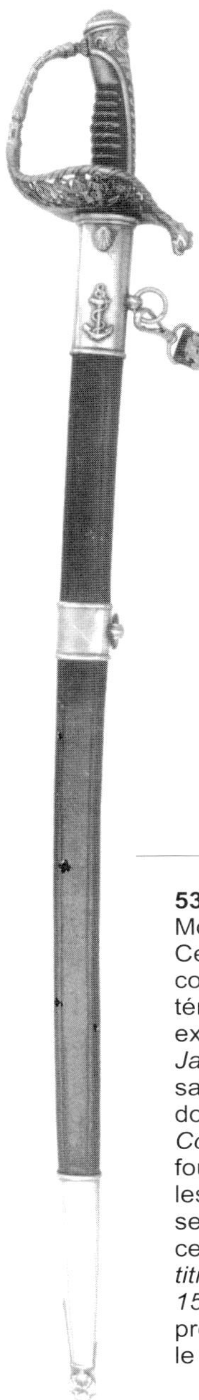

531 - Sabre d'officier de Marine.
Fabrication actuelle
Bouton de chape en coquille et maintien des ornementations traditionnelles (531a).

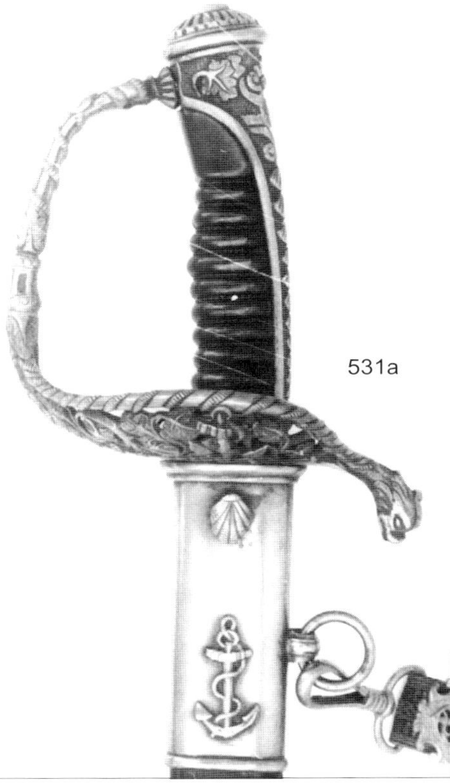

531a

- **a**, on remarquera la pérénité de l'ornementation de la monture, la queue de calotte à frise de coquilles Saint-Jacques.

532 - Sabres des élèves de l'Ecole Navale.
Modèle du 3 juin 1854.
Ce sabre, très voisin de celui des officiers a été conservé sous la IIIème République comme en témoigne l'inscription figurant sur la lame d'un exemplaire de 1901 : *Manuf^{re} d'Armes de Chat^{lt} Janvier 1901, Ecole M^{le} 1854.* Précédemment, le sabre de l'Ecole était fabriqué par la manufacture dont on trouve la signature : *Manufacture de Coulaux et Cie Klingenthal* et après 1901, par le fourbisseur *Carré à Paris*, jusqu'en 1914. Ensuite, les marquages deviennent de plus en plus rares et semblent avoir disparu à partir de 1920. Au sujet de ces sabres, Casanova (1996) écrit : *"On note à ce titre des commandes annuelles de l'ordre de 100 à 150 pièces passées par l'Ecole Navale... "* A chaque promotion, on peut déceler des variantes portant sur le quillon, le bouton ou le dard, (non figuré).

533

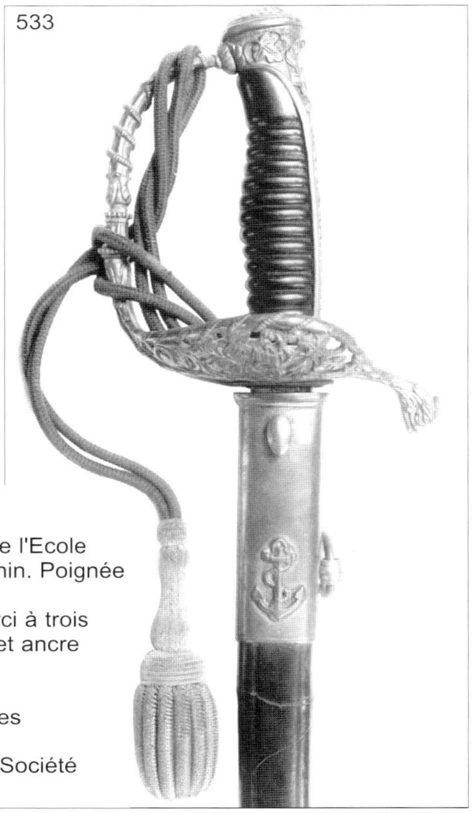

533 - Sabre d'officier de Marine.
Modèle du 26 octobre 1963.
Monture en laiton doré à ornementation comparable à celle des officiers de l'Ecole Navale du modèle 1854, avec ancre étalinguée et quillon en tête de dauphin. Poignée en acétate noir filigranée. Lame droite de 70 cm, large au talon de 2,4 cm, à flèche de 2,2 cm. Fourreau de cuir noirci à trois garnitures de laiton : chape à bouton en forme de coquille Saint-Jacques et ancre étalinguée.
Ce modèle, d'une facture inférieure à celle des modèles précédents, est réglementaire pour les officiers, les ingénieurs mécaniciens, les officiers des équipages et chefs de musique.
Ces sabres sont actuellement fabriqués par les Etablissements Balp et la Société Hostin.

534

534 - Sabre d'officier de l'Ecole Navale.
XXème siècle - Réglementaire.
Monture en bronze d'aluminium. Coquille ajourée ciselée de rinceaux et quillon en tête d'animal fabuleux. Poignée noire non filigranée. Lame de 70 cm. Fourreau de cuir noir avec renfort d'acier à l'intérieur. Chape à bouton en coquille et anneau ainsi que sur le bracelet médian.

535

535 - Sabre d'officier supérieur de la Marine.
Modèle actuel proposé par les Etablissement Balp de Saint-Etienne.

Les sabres d'officier d'Infanterie de Marine

Les sabres d'officier d'Infanterie de Marine ont des affinités certaines avec les sabres des officiers de Chasseurs de Vincennes et les sabres d'officiers supérieurs d'Infanterie du modèle 1845. Rappelons qu'en 1902, le corps de l'Infanterie de Marine fut rattaché sous le nom de *Troupes coloniales* au Département de la Guerre.

Nota - Les fourreaux métalliques sont souvent noircis, ainsi que les sabres de bord dans le seul but de les protéger des phénomènes de corrosion dus aux agressions du climat marin.

L'Infanterie de Marine, au cours des temps, porta des noms ou surnoms divers : Marsouins, fusiliers marins, voire troupes embarquées. Ce corps prit part à de nombreuses opérations militaires notamment dans les expéditions lointaines.

Les armes de la troupe étaient identiques à celles de l'Armée de Terre ; glaives ou briquets dont la croisière, en général, est frappée d'un poinçon à l'ancre non étalinguée (voir p. 235 - 237, n° 414, 416, 417.).

536

536 - Sabre d'officier d'Infanterie de Marine.

Modèle du 31 décembre 1845. Monture en laiton doré. Calotte bombée prolongée d'une longue queue ciselée en sa partie supérieure de feuille de laurier (536a) entourant un médaillon pour y inscrire les initiales de l'officier. Poignée en corne noire à 14 spires d'un fil torsadé. Arc de jointure à œillet supérieur formé par une feuille d'acanthe, à bague médiane où arrive la branche secondaire limitant la coquille. Celle-ci est ciselée à jour d'une ancre de 3 cm de haut, étalinguée. Cette ancre est placée sur un fond de quatre drapeaux à longues cravates (536b). A droite, un rameau de laurier avec ses fruits et à gauche un second rameau et des volutes issues du brin supplémentaire incorporé à la coquille. Intérieur de la coquille à décor allégé (536c). Quillon enroulé à feuilles d'acanthe et fleur latérale. Lame de 76 cm, large au talon de 3 cm avec une légère flèche d'environ un centimètre et arête médiane aboutissant à la pointe. Sur le dos on relève la signature *Manuf^{re} de Klingenthal Coulaux aîné et Cie*. Fourreau de cuir épais, à trois garnitures de laiton : chape à bague de 0,8 cm de largeur, bracelet médian à bague et bouterolle à bague supérieure et dard globuleux (536d).

536a

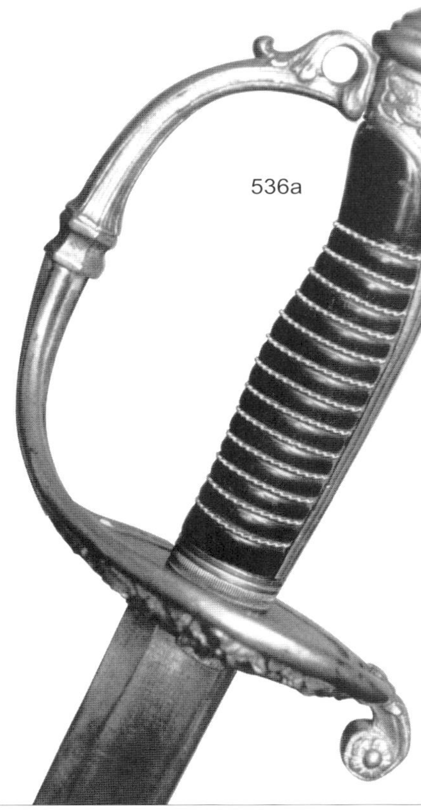

- **a**, vue da la calotte et de sa décoration. Remarquer le haut de l'arc de jointure avec une feuille d'acanthe recourbée faisant office d'anneau pour passer la dragonne.

536d

- **b**, coquille, face externe ;
- **c**, coquille, face interne ;
- **d**, bouterolle emboulée.

536b

536c

537 - Sabre d'officier supérieur d'Infanterie de Marine.
Modèle du 31 décembre 1845.
Ce modèle se différencie surtout de celui des officiers de rang inférieur par la lame qui peut atteindre 86 cm. Cette lame est à deux tranchants et à trois gouttières sur presque toute sa longueur. Fourreau à bagues ciselées et dard de fer en lyre (non figuré).

538 - Sabre d'officier d'Infanterie de Marine.
Modèle du 18 janvier 1856.
La principale différence de ce modèle par rapport aux antérieurs est la couronne impériale, surmontant l'ancre. Parfois, le fourreau au lieu d'être en cuir est en tôle d'acier noircie (non figuré).

539 - Sabre d'officier supérieur du Génie.
Règne de Louis-Philippe.
Monture en laiton doré. Calotte en tête de lion et longue queue ciselée. Branche principale arrondie, se divisant pour former les bordures du plateau et les coquilles, la coquille frontale étant la plus importante ; ces coquilles sont à décor ajouré représentant une cuirasse couronnée flanquée de rameaux de chêne. Quillon recourbé. Lame droite à deux pans creux, dos plat. Fourreau de cuir à trois garnitures de laiton doré, moulurées sur les bords.

Marquages - Sur le dos de la lame : *A. BISH à BOERSCH St ETIENNE* et sur la chape : *ROUART à PARIS*

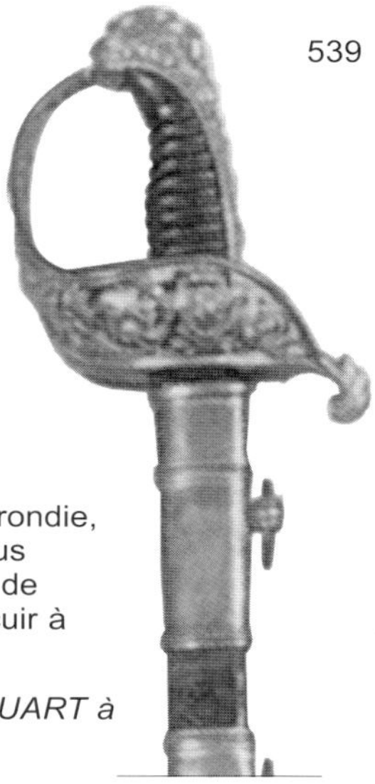
539

540 - Sabre d'officier d'Infanterie de Marine.
Modèle 1870 - 1882.
La monture de ce sabre est très comparable à celle du modèle 1845, mais avec les modifications suivantes : filigrane parfois de fils sinueux et faisant 13 spires, plus une virole striée verticalement. Coquille à une ancre étalinguée de 4 cm de haut, avec une coquille Saint-Jacques à son sommet ; le brin supplémentaire incorporé à la coquille se poursuit par des volutes réduites par rapport au modèle 1845 (540a).
- intérieur de la coquille ciselé du même décor que la face externe
- lame de 76 - 77,5 cm, large au talon de 3 cm et flèche identique.
- fourreau à trois garnitures : chape à bague ciselée de 0,7 cm de largeur, de laiton poli (540b), ainsi que sur le bracelet médian (540c) ; bouterolle simplement ornée de deux traits et terminée par un dard en olive de fer placée transversalement.

Ce sabre est également signé de la *Manufacture de Klingenthal* mais la ciselure est d'une qualité inférieure aux modèles antérieurs. Le fourreau peut être en tôle d'acier noircie.

Nota - Bien que réglementaire, ce modèle peut varier quelque peu dans son décor, étant souvent réalisé par des fourbisseurs privés, notamment par E. Bidal établi au 3, rue de Richelieu à Paris (540d).

540a

540b

540c

540d

- **a**, décor de la coquille ;
- **b**, décor de la chape ;
- **c**, bracelet au même décor que le bracelet de la chape ;
- **d**, signature de E. Bidal.

Sabres d'officiers des Chasseurs de Vincennes, modèle 1838

541

Ce sabre, proposé par Manceaux, fut attribué à une Compagnie expérimentale de 200 chasseurs, créée à Vincennes en 1838, sous le nom de Tirailleurs de Vincennes. Cette Compagnie fut également équipée de la carabine rayée du modèle 1838 particulièrement efficace. En 1840, le duc d'Orléans crée dix bataillons qui seront connus sous le nom de Chasseurs d'Orléans jusqu'en 1848. Puis il prendront le nom de Chasseurs à pied. Pendant toute cette période, les Chasseurs seront engagés dans la Campagne d'Algérie.

En 1854, dix autres bataillons seront formés, incluant celui de la Garde Impériale.

Le sabre des Chasseurs dits de Vincennes fait donc référence à la Première Compagnie levée en 1838.

541 - Sabre d'officier des Chasseurs de Vincennes.
La calotte présente beaucoup d'affinités avec le modèle 1821, (voir p. 282, n° 515), mais elle en diffère par une longue queue ; la bordure est ornée d'une frise de feuilles de laurier stylisées ou de feuilles de chêne ; le dos de la calotte porte un cor de chasse chez les modèles les plus anciens et ciselé ou non d'une guirlande, l'arc de jointure à son extrémité supérieure est en " S " enjolivé d'une étoile. Il existe une bague ornée d'une étoile ou d'une fleurette aux deux tiers de la longueur de l'arc. La branche secondaire part de la bague chez les modèles les plus anciens ou de la partie supérieure de la branche principale, donnant l'impression qu'elle passe sous la bague ; la branche secondaire se poursuit pour aboutir à la base du quillon, délimitant une coquille entre elle et la croisière. Dans cette coquille un motif découpé représente quatre drapeaux dont seulement deux sont déployés, entremêlés de deux rameaux de laurier ; dans les modèles les plus anciens, la branche de laurier de gauche, est remplacée par un rameau de chêne, (541a). La coquille postérieure est réduite. Le court quillon se termine en roulette ornée d'un motif végétal (gland ?) ou d'une étoile.
Les lames mesurent 76 cm, à dos en jonc, à rainure, à arête prolongée jusqu'à la pointe ménageant un important contre-tranchant. Elles sont souvent signées de la *Manufacture de Klingenthal*. Les fourreaux sont primitivement en cuir à trois garnitures, la chape portant un anneau de bélière, la bouterolle se terminant par un dard en fer. Plus tard, les fourreaux seront d'acier à deux bracelets munis chacun d'un anneau de bélière.

- **a**, détail de la coquille.

541a

Mf^{ture} N^{ale} du Klingenthal Coulaux Frères Entrep^{rs}. Poinçons SH. DD... Fourreau à alèses de bois de hêtre recouvertes d'une tôle de 0,95 mm d'épaisseur. Cuvette fixée par deux rivets. Bracelets plats portant les anneaux de bélière. Dard asymétrique. Poids total : 1,650 Kg.

Nota - Le modèle de ce sabre est conservé à la Manufacture Nationale de Saint-Etienne.

547 - Sabre d'officier de Gendarmerie du Grand-Duché de Berg
Modèle de l'An IX, de la Cavalerie légère
Napoléon Ier érigea le Duché de Berg, en Grand Duché (1806) et créa une Gendarmerie locale comme en témoigne l'inscription *GENDARMERIE* sur le *recto* de la lame et sur le *verso GRAND DUCHE DE BERG* (547a et 547b).
M. G. de Melun.

547

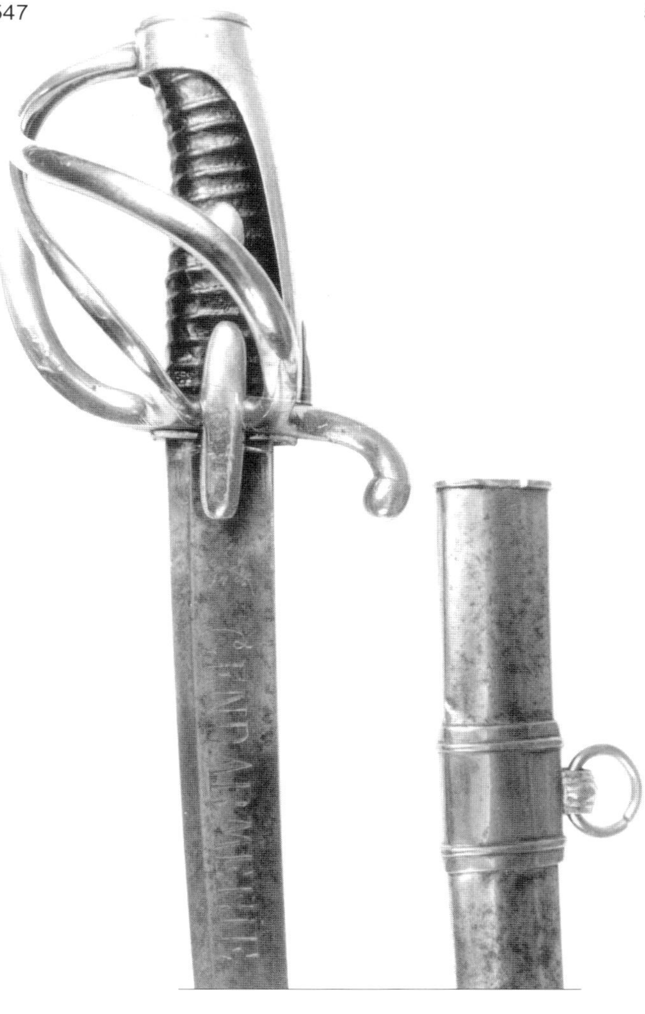

547a

547b

- **a**, *recto* de la lame du sabre de Gendarme du Grand-Duché de Berg ;
- **b**, *verso* de la lame figurée en a.

548a

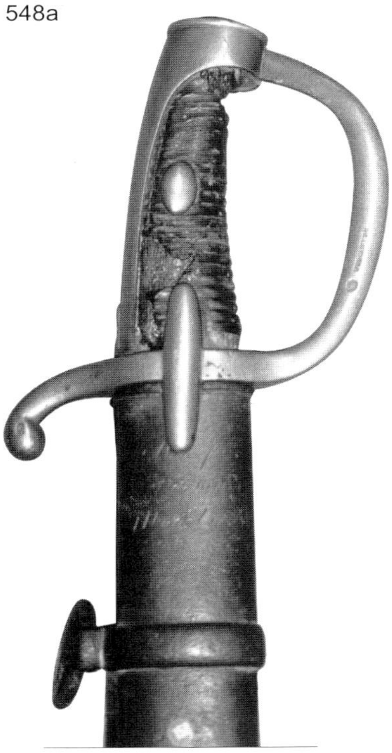

548 - Sabre de Cavalerie légère, troupe.

Modèle de l'An XI (1).

Monture très comparable à celle du modèle de l'An IX. Poignée ficelée, sans filigrane, à basane collée avec un clou de fixation. Lame de 86,6 - 89 cm, large au talon de 3,5 - 3,7 cm, à flèche de 5,18 - 5,3 cm. La grande différence entre les modèles de l'An IX et de l'An XI réside dans la structure du fourreau : le revêtement de tôle de celui-ci a une épaisseur de 2,5 mm et la cuvette est vissée. Cet alourdissement du fourreau fait que son poids est de 2,769 kg alors que celui de l'An IX ne dépassait pas 1,650 kg. D'autre part, les bracelets sont plats pour les exemplaires de l'An IX et bombés sur ceux de l'An XI.

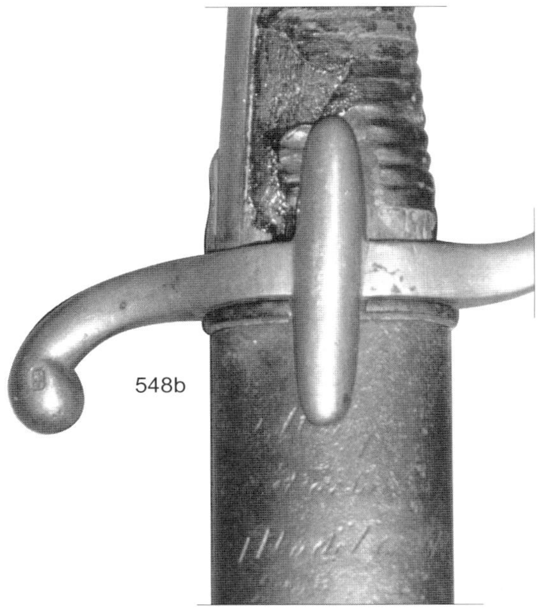

548b

548c

- **a**, monture (*verso*) ;
- **b**, haut du fourreau (*verso*) où l'on lit aisément sur la troisième ligne *Modèle* ;
- **c**, dard asymétrique.

Ce sabre a été mis en service en 1803 et porte alors la signature de la *Manufacture Impériale de Klingenthal* en général ou de la *Manufacture de Versailles* auxquelles viendra s'ajouter la signature de la *Manufacture de Turin* lors de l'occupation de l'Italie du Nord par les troupes françaises. Ce sabre perdurera sous la Restauration et portera alors la signature de la *Manufacture Royale de Klingenthal.* Parmi les poinçons on peut reconnaître ceux de Krantz, Bick, Lobstein...

(1) Le modèle est conservé à la Manufacture Nationale de Saint-Etienne.

549

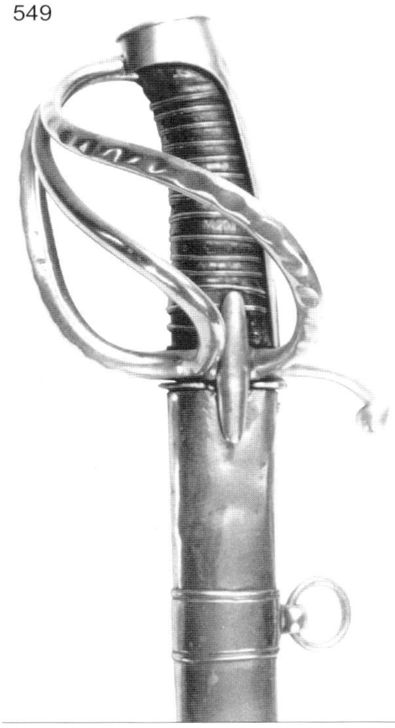

549 - Sabre d'officier de Cavalerie légère.

Modèle de l'An XI.

Monture de laiton doré. Calotte ovale à bouton de rivure plat à fine bordure en relief et longue queue guillochée. Poignée à filigrane de fils simples jumelés et faisant 13 spires. Arc de jointure guilloché sur la tranche ainsi que la branche secondaire supérieure sur la face frontale, branche secondaire inférieure unie. Les deux branches secondaires sont soudées de part et d'autre de l'oreillon frontal en navette très allongée de 5,5 sur 1,2 cm. Quillon infléchi, guilloché sur le dessus. Lame de 89 cm de longueur, large au talon de 3,2 cm ; contre-tranchant d'environ 26 cm. La lame est bleuie sur 25 cm et gravée de trophées et de motifs végétaux classiques dorés. Fourreau de laiton doré, à cuvette vissée ; deux bracelets de laiton plats, de 2,7 cm de hauteur, bordés d'un bourrelet de 0,3 cm ; anneaux de laiton ; dard asymétrique en fer.

<u>Nota</u> - Il est évident que ces guillochures faites à la meule et qui se rencontrent assez souvent ne sont qu'une fantaisie d'officier. Aucun marquage.

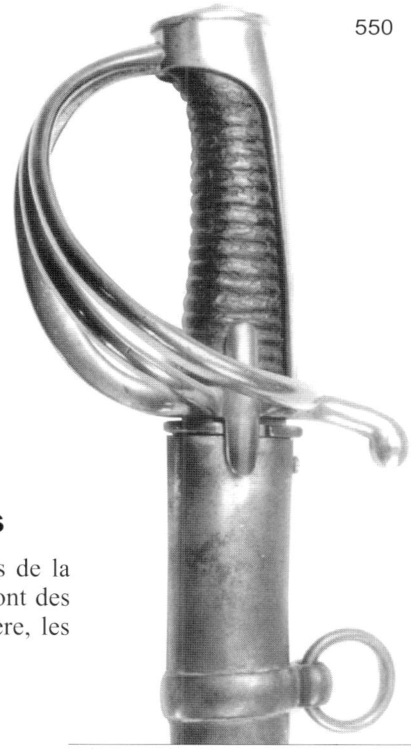
550

550 - Sabre de Cavalerie légère, troupe.
Modèle de l'An XI, Restauration.
Monture en laiton. Calotte ovale, débordant la largeur de la poignée. Longue queue unie ; bouton de rivure arrondi du côté dos et terminé en pointe à l'autre extrémité. Poignée ficelée, recouverte de basane, dépourvue du clou à tête ovale. Deux branches secondaires très voisines l'une de l'autre soudées à l'arc de jointure sur 5 cm, puis se séparant pour venir rejoindre la croisière, de part et d'autre des oreillons en batonnets à bouts arrondis très allongés, de 5 cm sur 1 cm. Lame à 86,3 cm, flèche de 4,5 cm, large au talon de 3,5 cm, contre-tranchant d'environ 26 cm. Fourreau en tôle d'acier pesante ; cuvette vissée, deux bracelets de bélière larges de 1,5 cm et légèrement convexes ; fort dard en fer, très élargi sur le devant.

Marquage - Pas de marquage sur la lame. Poinçon vraisemblablement de J.G. Bick sur le dard, à poinçon en " C " dans lequel se trouve une étoile, sur la calotte.

Famille III - Oreillons en navette ou ovoïdes

Ces sabres sont très voisins, quant à leur monture, des sabres de la Famille II, mais les oreillons sont en navette et parfois ovoïdes. Ce sont des sabres de l'An IX ou dérivés de ce modèle, armant la Cavalerie légère, les Chasseurs à cheval, les Chevaux-légers, les Lanciers.

551 - Sabre d'officier général d'Artillerie à cheval (1).
Premier Empire.
Monture à la chasseur, en laiton doré. Longue queue de calotte timbrée du " N " napoléonien. Poignée filigranée portant dans une rosette ovale *DROUOT GENERAL D'ARTILLERIE*. Branches secondaires ciselées de feuilles de chêne. Oreillons ovalaires ornés de deux canons brochant le feuillage. Lame de 88 cm, à faible courbure, gravée d'une tête d'oriental et, dans un cartouche, une calligraphie énigmatique. Fourreau en acier noirci avec cuvette et bracelets en laiton, ceux-ci ciselés de deux canons en sautoir. Très petit dard en fer (non figuré).

552 - Sabre d'officier d'Artillerie à cheval.
Fin du Premier Empire.
Monture à la chasseur, de laiton doré. Calotte à longue queue frappée d'un " N " lauré. Branches secondaires ciselées de canons en sautoir sur palme. Lame de 84 cm, à jonc, à faible courbure signée *MANUFACTURE DE KLINGENTHAL COULAUX FRERES*, gravée à l'acide de trophées, de feuillages et de l'aigle couronnée posée sur ses foudres, l'ensemble doré. Fourreau d'acier à double bracelet de bélière en laiton, ciselés de deux canons en sautoir, dard asymétrique en fer. Ce sabre peut être attribué à la Garde Impériale (non figuré).

553 - Sabre d'officier d'Artillerie.
Premier Empire.
Monture à deux branches secondaires rejoignant conjointement l'arc de jointure. Calotte à longue queue ciselée de feuilles d'acanthe, de palmette. Fusée recouverte de cuir, filigranée. Oreillons en navette, apparaissant entre les deux branches secondaires et ciselés de deux fûts de canon en gloire, flanqués de rameaux de laurier avec au-dessus la couronne impériale et au-dessous, des boulets. Lame à deux pans creux, légèrement courbée. Fourreau de laiton simplement orné de traits figurant les deux bracelets de bélière et signé, près de l'embouchure *SPOL / A METZ* (M.R.). Sabre à rapprocher des n° 551 et 552.

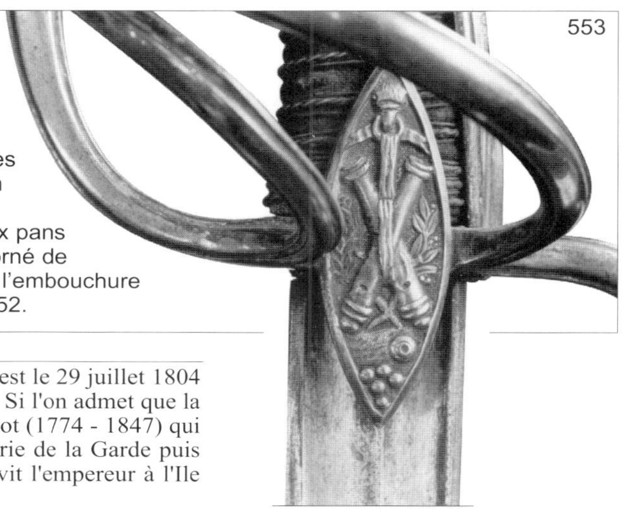
553

(1) Ce sabre peut être attribué à la Garde Impériale. En effet, c'est le 29 juillet 1804 que l'Artillerie à cheval entre dans la composition de la Garde. Si l'on admet que la rosette est d'époque, ce sabre aurait appartenu au général Drouot (1774 - 1847) qui participa à toutes les guerres de l'Empire. Colonel de l'Artillerie de la Garde puis comme général et aide de Camp de Napoléon en 1813, il suivit l'empereur à l'Ile d'Elbe et participa à la bataille de Waterloo.

554

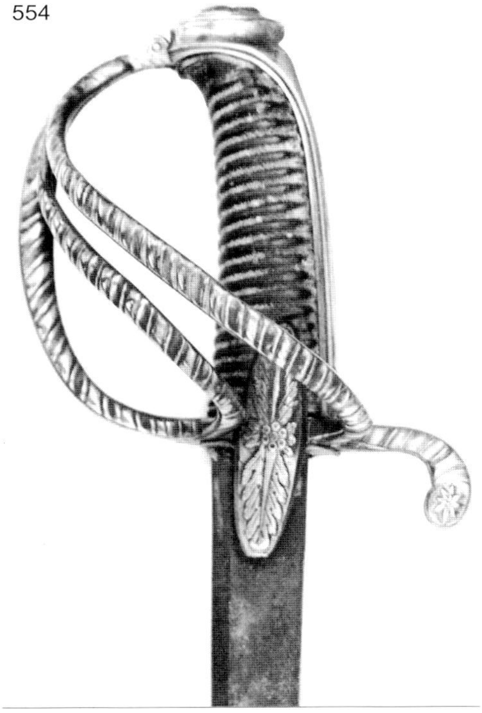

554 - Sabre d'officier de Cavalerie légère.
Type An IX. Premier Empire.
Monture en laiton. Calotte en cuillère, à longue queue. Poignée encordée, recouverte de basane,. Arc de jointure arrondi, recevant les deux branches secondaires en sa moitié supérieure. Toutes les branches sont décorées à la meule et les oreillons en navette sont ornés d'une composition végétale. Lame demi-courbe, anciennement bleuie et gravée de motifs dorés.

555 - Sabre d'officier de Cavalerie légère
Variante tardive du modèle de l'An XI.
Monture à la chasseur, en laiton à bouton de rivure en demi-olive. Calotte ovale, longue queue bordée d'un trait ainsi que toutes les branches, les oreillons et le quillon. Poignée recouverte de basane sans filigrane. Les deux branches secondaires, à tronc commun partent à 1 cm de la calotte et aboutissent sur la croisière, de part et d'autre des oreillons en navette (5,5 cm sur 1,5 cm). Quillon recourbé. Lame de 83,5 cm de longueur, large au talon de 3 cm, flèche de 5 cm, contre-tranchant d'environ 22 cm, à pan creux. Fourreau en tôle d'acier légère, cuvette rivée en acier ; deux bracelets de bélière en laiton, larges de 1,2 cm ; anneaux de laiton ; dard asymétrique en fer.

555

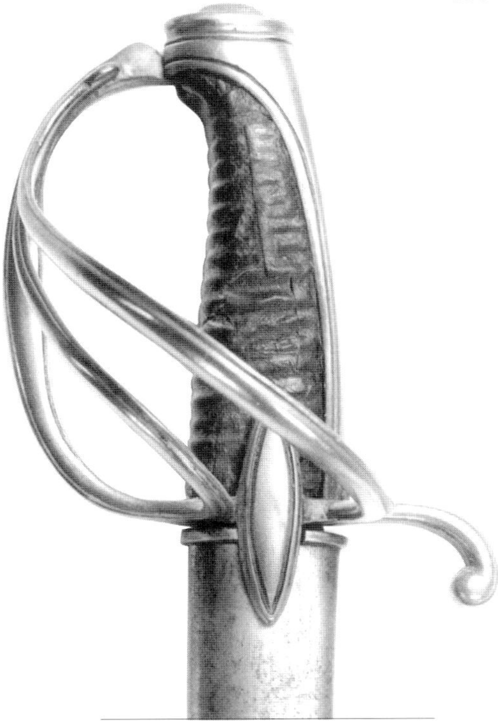

556 - Sabre d'officier de Cavalerie légère ou de chasseur.
Modèle de l'An IX. Monture en laiton. Calotte à longue queue bordée d'un filet. Arc de jointure, branches secondaires et oreillons en navettes également bordés d'un filet marginal. Poignée recouverte de basane avec ou sans filigrane de laiton plus ou moins élaboré. Quillon très recourbé terminé par un bouton. Lame d'environ 88 cm, à pan creux, souvent de Solingen et décorée à l'acide ou bleuie à motifs gravés et dorés ; sur ces lames on rencontre souvent des inscriptions cabalistiques et signée au talon " K " et " S " en tôle d'acier légère ; cuvettes et bracelets en laiton ; les bracelets sont de largeur et de décors variés.

556

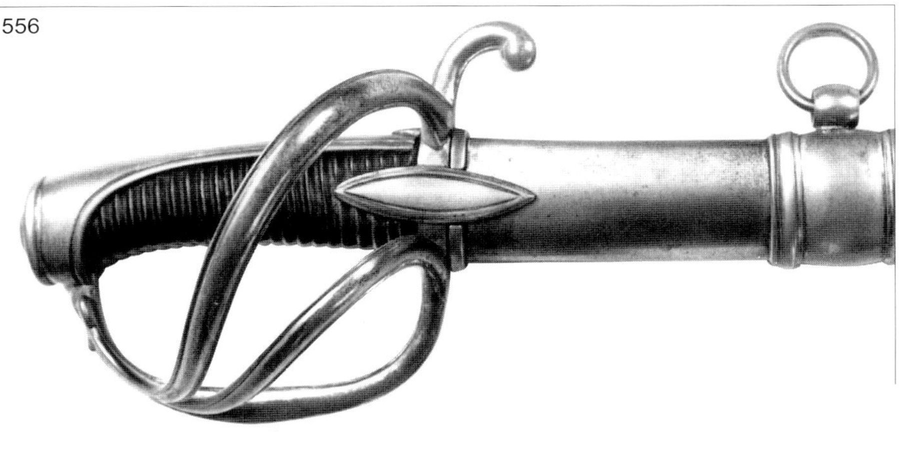

557

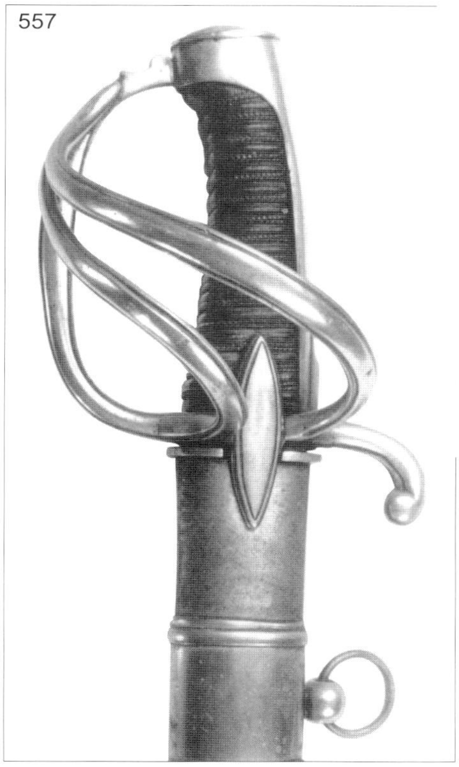

557 - Sabre d'officier de Cavalerie légère.
Premier Empire - Modèle de l'An XI.
Monture en laiton doré, décorée au trait. Calotte à bouton de rivure en navette bombée, débordant la poignée, à longue queue à fort tenon entrant dans la croisière. Poignée ficelée, recouverte de basane à seize spires d'un filigrane en chaînette. Arc de jointure arqué. Oreillons en navette (1,8 sur 6 cm), décorés d'un liseré. Deux branches secondaires à filet partant de l'arc de jointure à 2 cm de la calotte se déployant en " S " et soudées de part et d'autre de l'oreillon frontal. Quillon recourbé. Lame de 80 cm de long, large au talon de 3,8 cm, flèche de 4 cm et contre-tranchant bien marqué de 23,8 cm et se terminant en biseau. Fourreau en tôle d'acier légère, à cuvette en laiton doré ainsi que les bracelets de bélière de 5,5 cm de hauteur à deux bourrelets aux extrémités. Dard asymétrique en fer.

Marquage - Aucun marquage de manufacture ni de fourbisseur. Mais certaines lames sont bleuies et portent des inscriptions cabalistiques dorées (voir J. Lhoste et J.J. Buigné, 1999 p. 110).

558 - Sabre d'officier de Hussards.
Premier Empire.
Monture en laiton doré, décorée au trait. Calotte en cuillère, à bouton de rivure et longue queue. Poignée à fort décrochement supérieur, d'ébène finement quadrillé. Branches secondaires à filet. Arc de jointure et quillon sans décor. Oreillons en navette étroite. Lame à jonc, bleuie et gravée du Hussard chargeant et de motifs végétaux stylisés. Fourreau à deux bracelets de bélière et anneaux en fer.

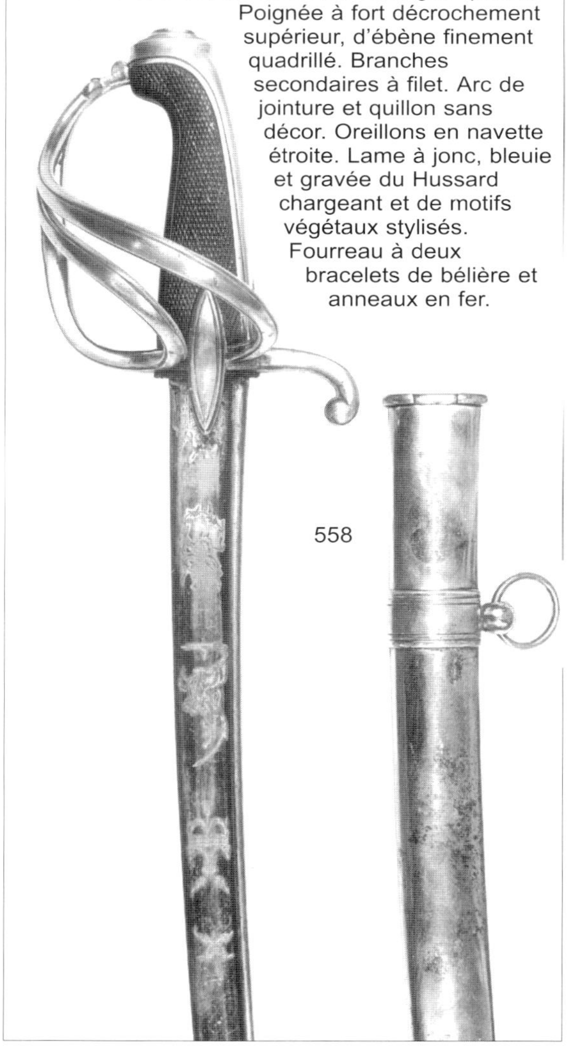

558

559 - Sabre du général d'Autancourt.
Premier Empire.
La description de ce sabre n'est donnée qu'à titre d'information car il est unique et bien répertorié.
Pierre d'Autancourt, est officier de police et de sûreté à l'Armée du Nord en 1793, capitaine aux Gendarmes d'Elite en 1801, major au 1er Chevaux-légers des Lanciers polonais en 1808 et à ce titre participe à la Campagne d'Allemagne en 1809 et à celle de Russie en 1812. D'Autancourt est promu général de brigade le 28 novembre 1814... En 1815 il reprend du service... et servira plus tard, Louis-Philippe !
Ce sabre lui fut remis en 1814. Il est d'une facture exceptionnelle :
Monture à la Chasseur en acier bruni. Calotte ornée d'un bourrelet et surmontée d'une fleur d'héliotrope dorée. Longue queue de calotte bordée de noir, ciselée sur le dos de feuilles de chêne à partir d'un gland. Arc de jointure arqué et deux branches secondaires décorées de chevrons. Poignée recouverte de galuchat apparaissant entre les spires d'un filigrane en chaînette bordé d'un fil simple. Oreillons en navette, le frontal rehaussé d'or et portant les initiales du général surmontées d'une couronne de baron, le postérieur à fleur stylisée. Lame de Solingen, demi-courbe à pan creux limité par une gouttière, dos à jonc ; le quart basal de la lame est gravé de trophée, carquois, flèche, volutes en or sur fond bleui. Fourreau en acier bruni à deux bracelets dorés, ciselés d'une frise de motifs végétaux. Dard en coquille Saint-Jacques dorée. Ce fourreau est signé, au-dessus du premier bracelet : *Manufacture / Imp^{le} à / Versailles* en lettres manuscrites.
Cette description doit pouvoir se faire une idée de la richesse de ce sabre dont on trouvera la figuration grâce à B. Bruel (1997). (non figuré).

560

560 - Sabre d'officier supérieur de Gendarmerie du Grand-Duché de Berg.
Vers 1806.
Monture à la chasseur en laiton, à longue queue de calotte et les deux branches secondaires ornées d'une frise de feuilles sinueuses. Poignée encordée, recouverte de basane filigranée de fils torsadés entourés de deux fils simples. Oreillons en navette, le frontal étant gravé d'une fleur centrale et de feuilles de laurier avec des fruits. Lame portant sur le *recto GENDARMERIE* (560a) et sur le *verso GRAND DUCHE DE BERG* (560b). (M. G. De Melun).

<u>Nota</u> - Rappelons que le Duché de Berg, situé sur la rive droite du Rhin avec comme capitale Düsseldorf fut conquit par Napoléon Ier qui l'agrandit et le transforma en Grand Duché. La souveraineté de cet état fut confiée à Joachim Murat, beau-frère de l'Empereur. (voir également le sabre p. 303).

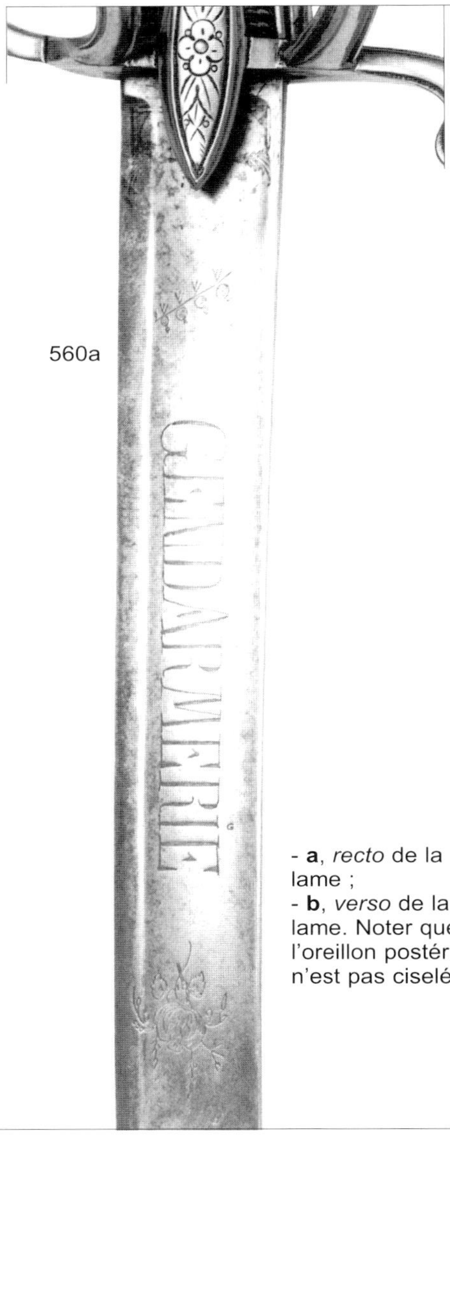

560a

560b

- **a**, *recto* de la lame ;
- **b**, *verso* de la lame. Noter que l'oreillon postérieur n'est pas ciselé

561 - Sabre d'officier supérieur de Cavalerie légère.
Modèle de l'An XI.
Monture en laiton. Calotte en cuillère à longue queue. Poignée recouverte de basane, à 22 spires d'un fil torsadé bordé de fil simple (561a). Arc de jointure arrondi et branches secondaires ciselées d'une frise constituée de disques se chevauchant. Oreillons en navette décorés d'une fleur à quatre pétales en leur centre, puis de feuilles en crosses se terminant par une pomme de pin (561a). Quillon recourbé. Lame à pan creux, dos à jonc, longue de 79 cm, large au talon de 3 cm, sans décor. Fourreau en tôle d'acier, à deux bracelets de bélière ornés, selon toute vraisemblance, de feuilles et de fruits de laurier hautement stylisés (M.R.).

561

561a

- **a**, oreillons en navette, ciselés d'une fleurette médiane, de volutes terminées par une pomme de pin.

562 - Sabre d'officier de Cavalerie de ligne.
Restauration.
B. Bruel décrit ce sabre en ces termes :
La monture, en bronze doré est du modèle à la chasseur (deux branches secondaires rejoignant l'arc de jointure à sa partie supérieure). *Les branches sont finement ciselées de feuilles de chêne et de laurier... la calotte à longue queue s'enrichit des Armes de France ciselées. Les oreillons* (en navette) *portent, au sommet une couronne princière, les Armes du Berry*, Armes de France à la bordure engrêlée de gueule qui surmonte la croix de l'Ordre Royal du Saint Esprit...
La lame... porte gravée à l'acide les grandes armes royales anglaises associées au monogramme G R (pour Georges Rex). Cette lame est signée *Prosser, maker to the prince Regent, Charing Cross, London*. Fourreau à trois garnitures dorées et percées à jour, le bracelet portant les Armes de France.

Nota - Ce sabre très représentatif d'un sabre d'émigré est figuré par B. Bruel (1997, p. 40 - 41)

563

563 - Sabre d'officier supérieur de la Garde Royale.
Restauration.
Monture en laiton doré. Calotte à longue queue, ciselée à son sommet d'un médaillon aux trois fleurs de lys, encadré de deux branches de laurier partant d'un rameau médian, puis d'une longue palmette (563a). Arc de jointure finement ciselé de feuillage stylisé (563b) ainsi que les deux branches secondaires et le quillon recourbé. Poignée encordée, recouverte de gros galuchat, non filigranée (563c). Oreillons en navettes à fleurette centrale, à six pétales, flanquée de deux palmettes (563d). Lame à pan creux, longue de 84 cm à flèche de 5,9 cm, gravée à l'acide de feuillage sur les 2/3 de sa longueur. Fourreau en acier bruni, à bracelets en laiton, légèrement bombés et anneaux de bélière (M.R.).

- **a**, vue générale de la queue de calotte avec le médaillon aux Armes de France ;
- **b**, ciselure de l'arc de jointure ;
- **c**, poignée recouverte de gros galuchat ;
- **d**, oreillon frontal, ciselé.

563a

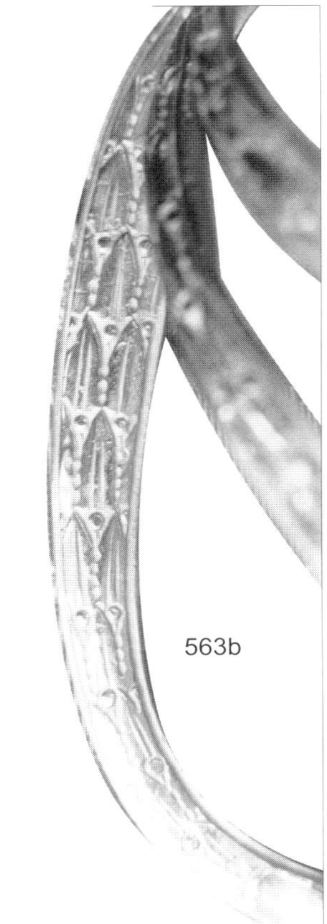

563b

563c

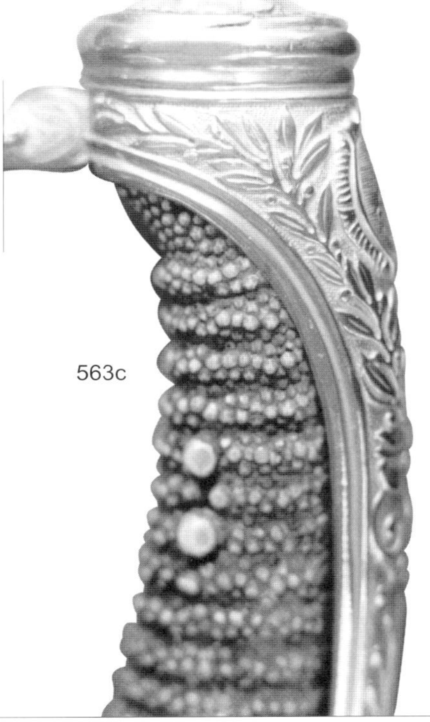

563d

564

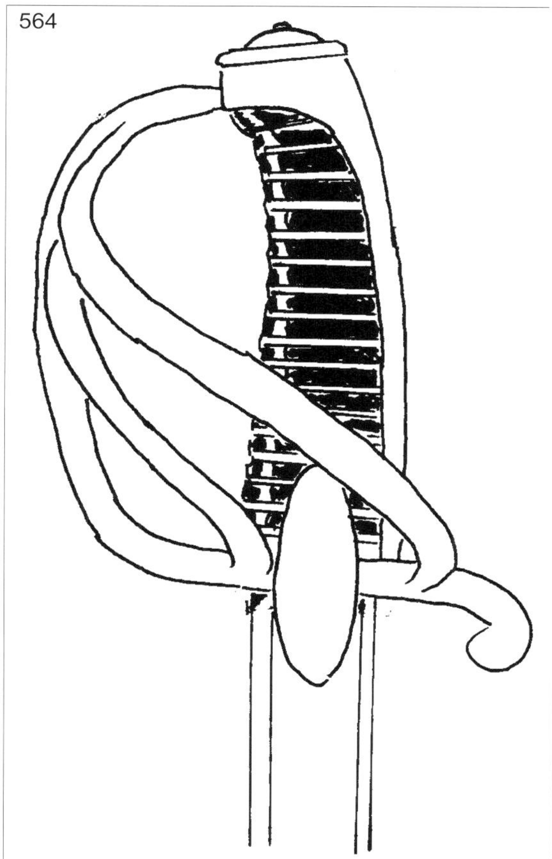

564 - Sabre d'officier de Lanciers de la Garde Royale.
Restauration.
Monture en fer poli. Calotte à bourrelet ; pièce de rivure en dôme, longue queue. Poignée recouverte de chagrin noir, filigranée d'un fil d'argent faisant environ 20 spires. Deux branches secondaires rejoignent l'arc de jointure en sa moitié supérieure. Oreillons en navette bombée. Quillon infléchi. Lame à pan creux et gouttière dorsale, bleuie à décoration dorée. Au *recto*, à partir du talon, on trouve la signature : *Manuf^{re} / Royale / du Klingenthal / Coulaux / Frères* puis les Armes de France, l'inscription, dans un cartouche *REGIT. DE. LANCIERS* accompagnée de petites fleurs de lys, d'une grande fleur de lys flanquée de rameaux de laurier, des " L " entrelacées de Louis XVIII avec fleur de lys, la couronne royale et une dernière fleur de lys.
Sur le *verso*, à partir du talon, on trouve *Duc / fourbisseur / Rue / S^{t} honoré / en face celle / de Richelieu / N° 251 / Paris*. Les autres décors sont semblables à ceux du *recto*, sauf *REGIT DE LANCIERS* qui est remplacé par *" GARDE ROYALE "*.
Fourreau en tôle d'acier, à deux bracelets de bélière, dard asymétrique.

565

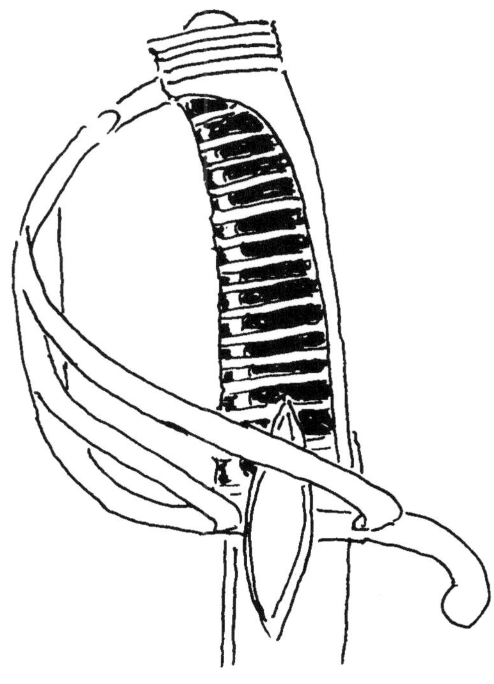

565 - Sabre d'officier de Lanciers.
1810 - 1830.
Monture en fer. Calotte rainurée, à bouton de rivure et longue queue. Poignée recouverte de basane et filigranée. Arc de jointure régulièrement arqué. Branches secondaires se réunissant sur la moitié supérieure de l'arc de jointure. Oreillons généralement en navette décorée au trait. Quillon recourbé. Lame demi-courbe, à pan creux. Fourreau en tôle d'acier, à deux bracelets, dard asymétrique.

<u>Nota</u> - La monture, toujours en fer et à trois branches, peut subir quelques variations : les oreillons peuvent être ovales ou disparaître, le quillon peut être très long.
La lame est généralement signée, sur le dos
Manuf^{re} R^{ale} du Klingenthal (suivi de la date de fabrication).

566

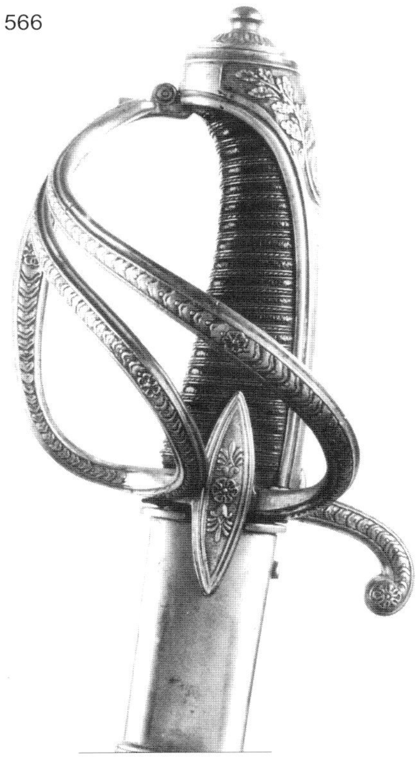

566a

566b

566 - Sabre d'officier supérieur de Cavalerie légère.
Restauration.
Ce sabre est du type de l'An IX, dit à la chasseur. Monture en laiton entièrement ciselée d'écailles. Calotte à gros bouton de rivure, prolongée d'une longue queue. Poignée recouverte de chagrin filigrané. Deux branches secondaires jointives rejoignent la moitié supérieure de l'arc. Quillon en crosse. Oreillons en navette avec une fleurette centrale flanquée de feuillage (563a). Lame de 88 cm, légèrement courbée, à dos plat signé *Manufacture Nationale de Klingenthal Coulaux Frères*, à contre-tranchant, gravé du faisceau de licteur, de grenades, de personnages, de poignard,... le tout doré (566a). Fourreau de fer à deux larges bracelets de laiton ciselés de motifs végétaux (566b).

- **a**, détail de la monture et de la décoration de la face frontale de la lame ;
- **b**, bracelet ciselé de palmettes et de feuilles d'acanthe.

567

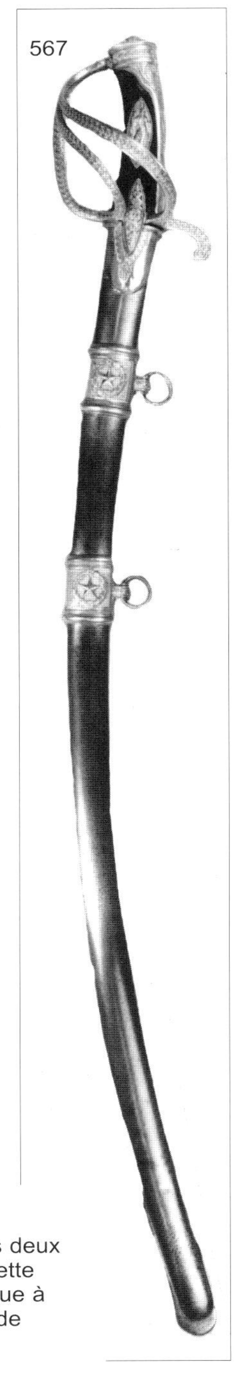

567 - Sabre d'officier d'Eclaireur.
Premier Empire.
Monture en laiton doré. Calotte à longue queue ciselée de feuillage ainsi que l'arc de jointure, les deux branches secondaires et le quillon. Poignée recouverte de chagrin, garnie d'un médaillon en navette décorée d'Athéna en arme. Oreillons en navettes ciselées. Lame demi-courbe. Fourreau métallique à deux bracelets en laiton ciselé, frappés d'une étoile à cinq branches inscrite dans une couronne de feuillage. Dard asymétrique (non figuré).

568 - Sabre d'officier d'Etat-Major de Cavalerie Légère.
Restauration.
Ce sabre est du type de l'An XI, dit à la Chasseur. Monture en laiton doré, entièrement ciselé d'écailles. Calotte à bouton de rivure très aplati et longue queue. Poignée recouverte de chagrin filigrané. Les deux branches secondaires rejoignent l'arc de jointure en sa moitié supérieure. Quillon recourbé. Oreillons en navettes décorés en leur milieu d'une fleurette flanquée de feuillage. Lame en Damas de 86,4 cm, courbée, à dos arrondi et contre-tranchant, dorée, avec un grand cartouche bleui orné d'un petit trophée de flèches d'un carquois et d'un soleil. Fourreau en tôle d'acier, à deux bracelets de laiton ciselé en ronde bosse, de casque, obusier, glaive... L'expert J.C. Dey ne garantit pas que les bracelets soient d'origine (568a).

568

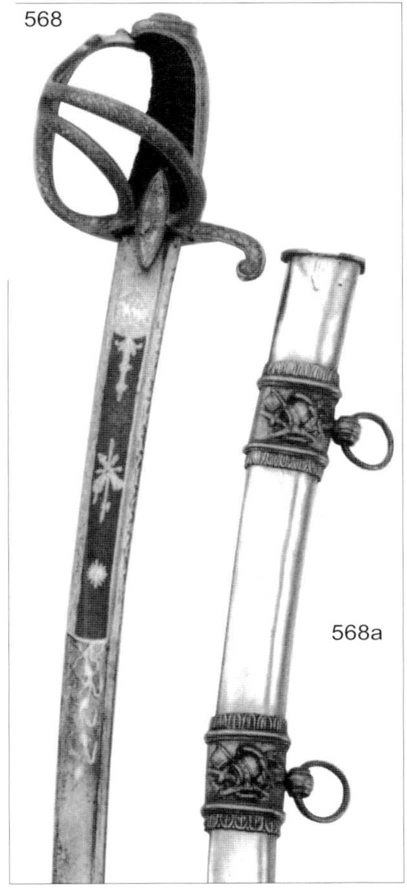

568a

569 - Sabre d'officier de Chevaux-légers.
Restauration.
Bronze doré. Calotte légèrement bombée décorée d'une rosace suivie d'une longue queue ciselée. Poignée recouverte de basane à filigrane de laiton torsadé faisant une quinzaine de spires. Arc de jointure arrondi et quillon recourbé. Oreillons en navette à fleurette médiane (569a). Deux branches secondaires partant de la partie supérieure de l'arc : toute la monture est décorée de frises de ronds enfilés. Lame de 84,3 cm de long, large au talon de 3 cm, à flèche de 5,5 cm, bleuie et gravée de motifs dorés classiques sur le 1/3 basal. Fourreau de fer, à deux bracelets de laiton à médaillon au profil de Henri IV sur la face frontale (569b) et d'une face léonine sur la face postérieure (M.R.).

Nota - Rappelons que les Chevaux-légers ont été introduits, en 1593, dans la Maison Rouge du Roi de France. Dissous en 1787, ils ont été rétablis en 1814 par Louis XVIII.

569

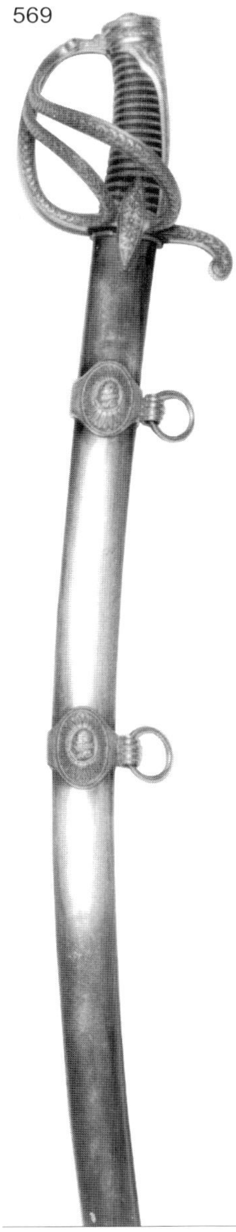

569a

- **a**, détail de la monture et notamment de l'oreillon frontal ;
- **b**, bracelet au profil de Henri IV.

569b

570

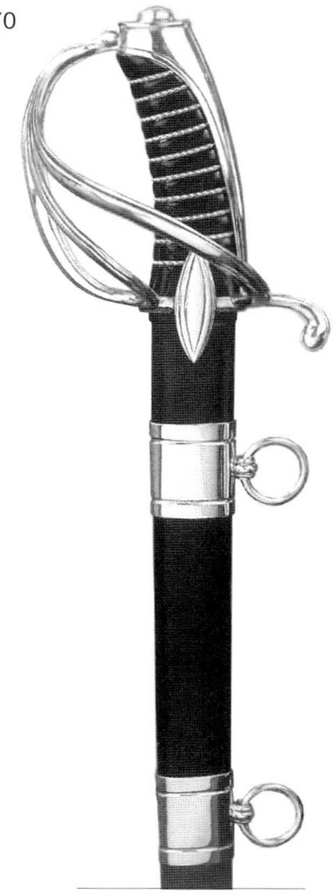

570 - Sabre d'officier (Modèle F1).
1974.
La création de ce sabre fut faite à la demande du chef d'Etat-Major, le général Alain de Boissieu, gendre du Général de Gaulle. En effet, si les officiers de Marine avaient conservé leur sabre et les Forces aériennes leur poignard, l'Infanterie était privée de sabre depuis 1934.
Le modèle choisi est inspiré du sabre d'officier de Cavalerie légère de l'An XI, à monture à la Chasseur. D. Venner (1986, p. 228) n'est pas tendre à son égard. Il écrit :
" On est évidemment très loin de l'An XI, pour l'esthétique et la finition. La lame est rachitique et les matières utilisées sentent un peu la " grande surface ". C'est un sabre à l'image de notre époque : toc et simulation. Il peut quand même sabrer une bouteille de champagne, j'ai essayé ". Jugement bien sévère... mais chacun sait que la copie ne vaut pas l'original !

Ce sabre est donc à calotte à longue queue, à trois branches de garde, les deux secondaires prenant naissance au sommet de l'arc de jointure, soudées sur la croisière, de part et d'autre de l'oreillon frontal en navette. Quillon long et recourbé. Poignée en acétate noir, filigrané. Lame de 80 cm, large de 2,7 cm, à pan creux et courbé. Fourreau en tôle d'acier revêtue d'une résine noire satinée, à deux larges bracelets en laiton poli et se terminant par un dard asymétrique.

571 - Sabre du marquis Jules de Mornay.
Premier Empire.
Sabre à calotte à longue queue simplement décorée au trait. Poignée recouverte de basane filigranée. Branche principale arrondie puis arrivant en angle droit sur la croisière avec deux branches secondaires rejoignant le haut de l'arc. Oreillons en navette très allongées. Ce sabre est surtout remarquable par ses inscriptions.

Marquages - Sur la queue de calotte est inscrit le nom de son propriétaire : *Le Mi*^{S} *Jules de Mornay* et sur la lame traitée dans le style des lames des sabres de la Garde Impériale on trouve dans un long cartouche terminé par l'aigle couronnée, sur le *verso* : *Au grand Napoléon et Empereur des Français et Roi d'Italie* et sur le *recto* : *Pour l'honneur de la Patrie, le 1er Rg*^{t} *de Chasseurs à cheval* le tout en caractères de grande taille et évidés. De plus, la lame est signée des Frères Weyersberg fabricants à Solingen. Fourreau en tôle d'acier à deux bracelets. (non figuré).

572

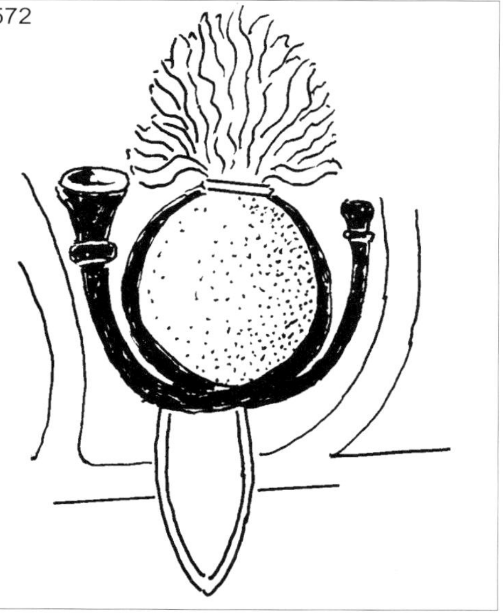

572 - Sabre d'officier de Compagnie d'Elite de Chasseur.
Restauration.
Ce sabre conservé au musée de l'Armée à Paris est du type de l'An XI mais entre les deux branches secondaire est incluse une grenade enflammée entourée d'un corps de chasse.

B - Croisière sans oreillon

Chapitre III

Monture à deux branches secondaires arrivant isolément sur l'arc de jointure, calotte à longue queue

Chez les sabres décrits dans ce chapitre et les suivants, les oreillons ont disparu sauf chez quelques armes de la Sous-famille VI a. Ces sabres ont été classés en fonction de la nature des matériaux utilisés et des caractères particuliers de la monture. Cinq Familles ont été ainsi définies.

Les familles V, VI et VII sont divisées en Sous-familles dont on trouvera l'énumération aux pages indiquées ci-dessus.

Famille IV - Poignée en laiton moulé

Cette famille est surtout composée de sabres de bord, obtenus par moulage, à la fin du XVIIIème siècle.

573

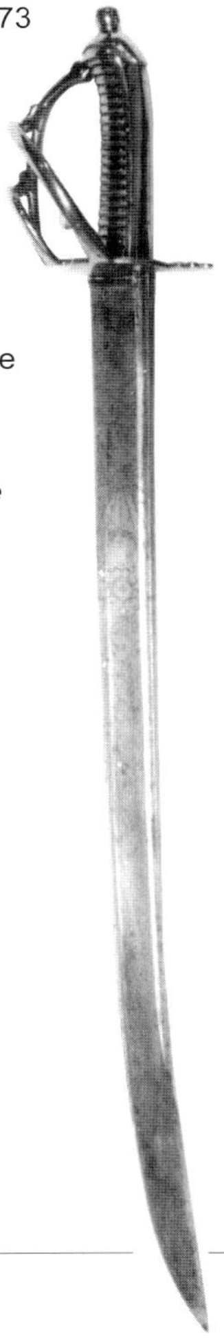

573a

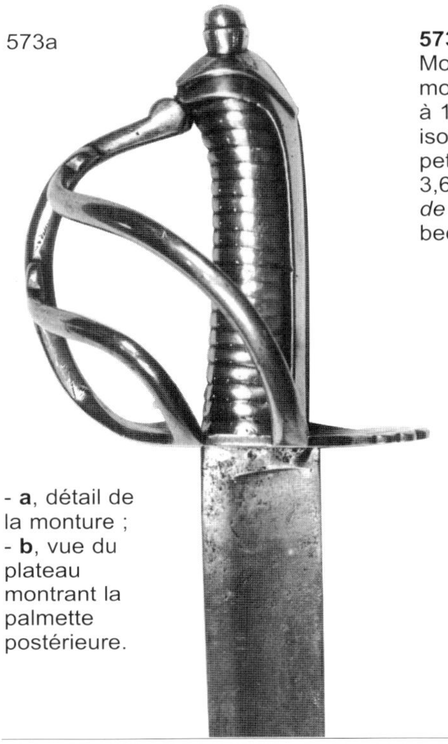

573 - Sabre de bord.
Modèle 1782.
monture en laiton moulé avec calotte à longue queue, poignée à 18 cannelures (570a). Branches secondaires rejoignant isolément l'arc de jointure. Plateau ovale terminé par une petite palmette (570b). Lame de 62,3 cm, large au talon de 3,6 cm et flèche de 1,12 cm à pan creux, signée *Manufacture de Klingenthal* et poinçons de Juchereau et de Bisch sur le bec du plateau. Chape du fourreau à fente pour la languette.

573b

- **a**, détail de la monture ;
- **b**, vue du plateau montrant la palmette postérieure.

574

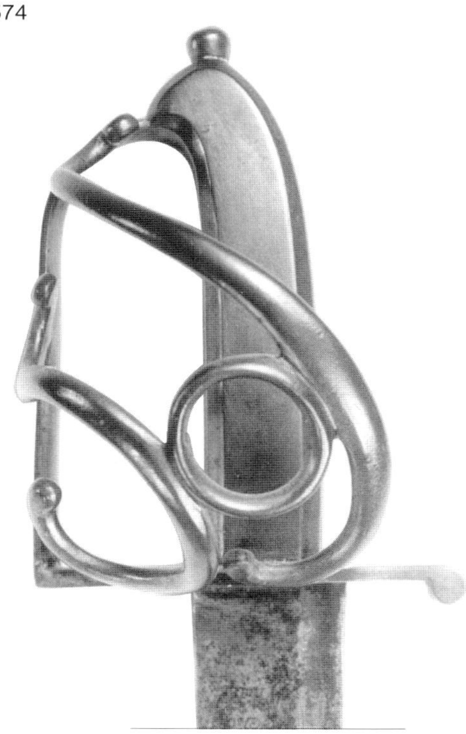

574 - Sabre de bord.
Fin du XVIIIème siècle.
Laiton moulé. Poignée à pans, unis, à bouton de rivure. Arc de jointure arrivant en angle droit sur la croisière rectiligne et terminée par un bouton pendant. Les deux branches secondaires sont réunies entre elles par un cercle. Lame à longue et large gouttière dorsale, signée au talon *Cassaignard à Nantes* et gravée d'un quartier de lune et du bras armé sur les deux faces.

Ce sabre peut être attribué à l'armement des navires marchands.

575 - Sabre de Grenadier.
Fin XVIIIème siècle.
Monture en bronze doré. Poignée en faisceau de licteur. Calotte en tête de lion et prolongée d'une longue queue. Arc de jointure légèrement arrondi. La branche secondaire supérieure, ciselée de feuillage, maintient un médaillon finement décoré de canon, obusiers, drapeaux, lances ... La seconde branche est transformée en un rameau feuillu. Plateau à, décor de faisceaux de licteur. Quillon en grenade enflammée. Lame à pan creux, bleuie et décorée de grottesques dorés (non figuré).

Famille V - Poignée en bois recouvert de basane filigranée, branches secondaires nues

Sont introduits dans cette famille des sabres dont la poignée est en bois recouvert de basane et filigranée. Le sabre de Mineur a été rattaché à cette Famille bien que possédant un brin unissant le haut de la branche secondaire supérieure à la base de la branche secondaire inférieure.

576 - Sabre d'officier de Volontaires Nationaux.
Après 1792.
Remarque préliminaire : ce sabre de grand luxe a sûrement perdu sa garde d'origine qui fut remplacée par une garde à trois branches simples, par son possesseur lui-même après un combat ou par un restaurateur visant à l'économie. Tout le reste de l'arme est conforme au type général de cette classe de sabres.

576

576a

576b

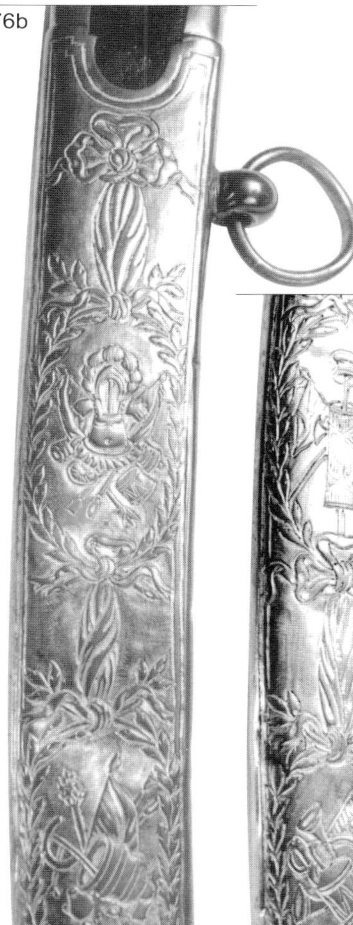

576c

Monture en laiton doré. Calotte en tête de lion et longue queue. Poignée recouverte de basane et filigranée. Lame de 80,8 cm, large au talon de 3,7 cm et flèche de 4,2 cm ; cette lame est bleuie et gravée sur les deux faces de la devise *VAINCRE . OU . MOURIR* en lettres d'or (576a). Fourreau de cuir à deux garnitures dorées réunies par deux cornières de laiton ; chape longue de 19,2 cm, à anneau soudé, décorée de rinceaux et dans un médaillon bordé de feuilles de laurier, la République couronnée de laurier et tenant dans sa main gauche une pique portant un bonnet phrygien, puis les lettres " F N " (vraisemblablement les initiales de l'officier) ; bouterolle de 51,5 cm, à anneau de bélière soudé à sa partie supérieure, à cinq médaillons également limités par des feuilles de laurier incluant de haut en bas, un casque empanaché sur fond de drapeaux (576b), de canons, de tambour, un trophée à l'obusier et drapeau, un faisceau de licteur sur fond de drapeaux et coiffé du bonnet phrygien et d'un trophée aux sabres, aux instruments de musique et d'un drapeau (576c). Dard en fer, asymétrique. Le *verso* du fourreau est uni.

- **a**, lame gravée de la devise bien connue : Vaincre ou mourir ;
- **b**, partie supérieure de la bouterolle ciselée de casque, d'instruments de musique ... l'ensemble enrubané ;
- **c**, partie inférieure de la bouterolle avec le faisceau de licteur, le bonnet phrygien sur deux drapeaux, instruments de musique ...

577

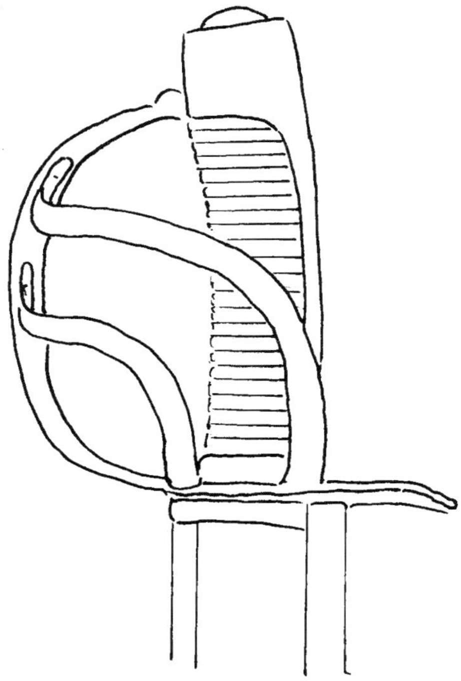

577 - Sabre de bord.
Vers 1790 - 1800.
Monture en fer. Calotte subcylindrique prolongée d'une longue queue. Arc de jointure régulièrement arrondie et deux branches secondaires en " S " venant isolément rejoindre l'arc. Poignée filigranée. Plateau de garde ovoïde finissant en bec ou se trouvent les poinçons de Mouton et de Levavasseur. Lame à dos plat et large à pan creux.

578

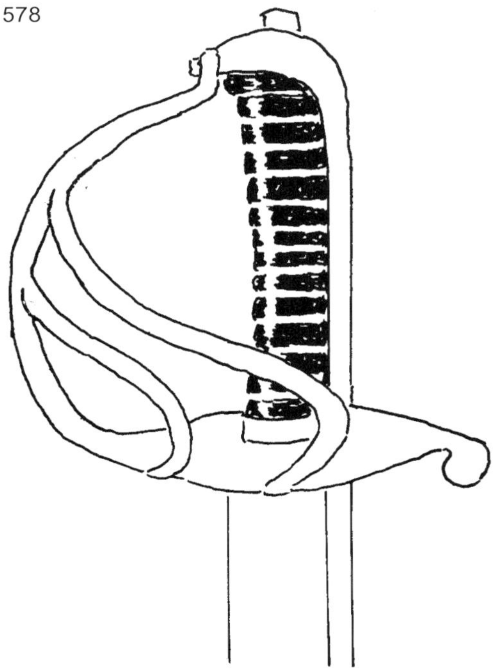

578 - Sabre de bord.
Modèle 1796.
Monture en laiton. Calotte simple, à gros bouton de rivure et longue queue. Poignée recouverte de basane maintenue par un filigrane de fils torsadés faisant une douzaine de spires. Branche principale régulièrement arquée, vissée sur la calotte. Plateau de garde ovoïde terminé par un court quillon en goutte d'eau. Deux branches secondaires en " S " rejoignant séparément la principale. Lame plate à gouttière dorsale (d'après M. Bottet).

579

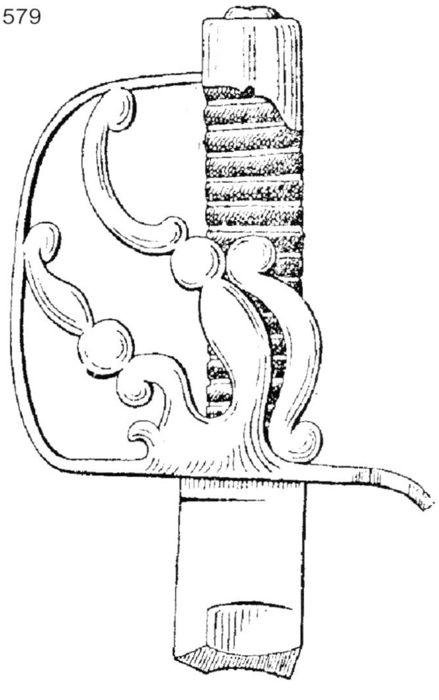

579 -Sabre d'officier de Chasseurs à cheval.
Vers 1790.
Monture en laiton. Calotte à courte queue. Poignée recouverte de basane filigranée. Arc de jointure courbé, recevant isolément deux branches secondaires constituées d'arcures et de disques. Lame à pan creux. Ce sabre est évidemment une fantaisie du fourbisseur et/ou de l'officier (d'après M. Bottet).

580

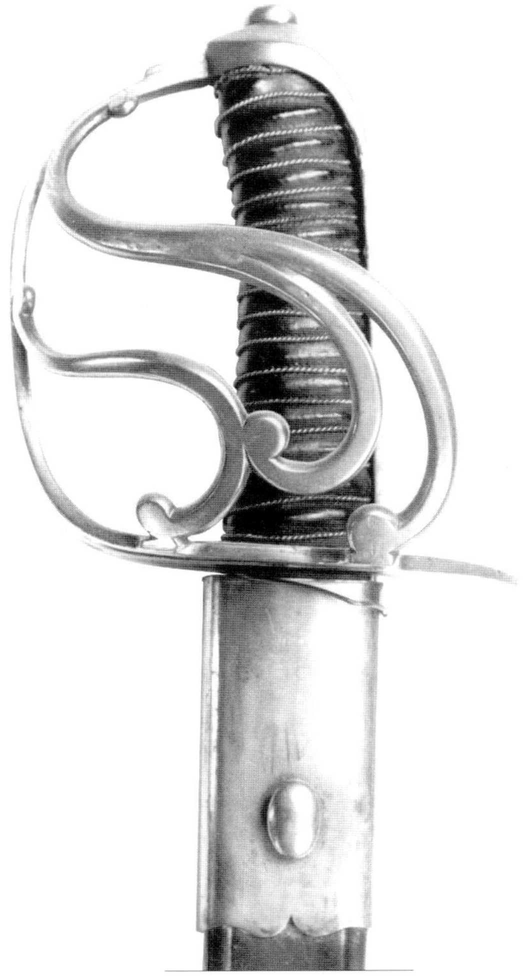

580 - Sabre dit de Mineur.
Fin de l'Ancien Régime, vers 1780.
Ce sabre n'étant pas réglementaire, on donnera la description d'un exemplaire type et de quelques variantes.
Monture en laiton. Calotte à bouton de rivure en cuillère prolongée par une longue queue. Poignée recouverte de basane, à filigrane plus ou moins élaboré faisant en général, entre 10 et 15 spires. Arc de jointure régulièrement courbé. Deux branches secondaires en " S " rejoignent séparément l'arc, l'externe donnant naissance à un brin en crosse fixé sur la petite branche basale. Plateau ovalaire, découpé de part et d'autre de la fusée de deux longues échancrures consolidées en leur milieu d'un anneau, type a, p. 21. Lame demi-courbe, de 60 à 75 cm de longueur, d'origines très diverses généralement à pan creux et gouttière dorsale au fort, pouvant être bleuie et gravée de motifs dorés pour les officiers. Fourreau de cuir noirci ou fauve ; chape à bouton avec parfois un anneau de bélière, bracelet éventuel à anneau, bouterolle courte à dard de forme variable mais asymétrique et ne protégeant que sa partie antérieure et basale.

Marquage - Certains exemples sont signés *Klingenthal.* On peut également trouver les poinçons de de Cabannes (vers 1791), de J.J. Mouton (vers 1793), de Levavasseur (vers 1798).

Nota - A l'origine, ce sabre fut attribué aux six Compagnies de Mineurs commandées par le comte de Gribeauval et rattachées au Corps Royal de l'Artillerie. Mais, selon toute évidence, du fait que ce sabre se rencontre assez souvent, d'autres Corps ont dû l'adopter et on en trouve la trace jusque sous la Restauration.

581

581 - Variante du sabre de Mineur.
Remarquer la lame à pan creux poinçonnée " MR ", signature de Levavasseur et l'arc de jointure droit en sa partie médiane.

582

582 - Sabre de Mineur.
Fin du XVIIIème siècle.
Monture en laiton doré. Ce sabre fantaisie a troqué sa simple calotte pour un casque à l'antique.

583 - Dague de Chasse.
Fin du XVIIIème siècle.
Monture d'un sabre de Mineur reconverti en dague de chasse comme permettent de le supposer les caractères suivants : lame de 50,8 cm, large au talon de 2,5 cm et la gravure d'un sanglier (flèche) entre deux entrelacs sur la lame droite. On lit sur le talon :
Bidault four / bisseur au / Dauphin Royal / Rüe St Honoré / attenant La / Ruë du champ / fleuri près La / barrière des / Sergents à Paris.

583

584

584 - Sabre d'officier de Chasseurs à pied.
Modèle 1882.
Monture en fer. Poignée renflée en son milieu, recouverte de basane et filigranée . Les deux branches secondaires partent d'un plateau terminé par un quillon recourbé. Lame droite marquée : *Coulaux à Klingenthal.* Fourreau de fer à un bracelet étroit et anneau de bélière en fer.

Famille VI - Poignée en bois recouvert de basane filigranée, branches secondaires partant d'une palmette

La palmette qui orne la garde, est un motif décoratif que l'on rencontre assez souvent sur les montures des sabres à partir du XVIIIème siècle. La Famille VI présente une palmette se prolongeant en deux branches secondaires qui rejoignent isolément l'arc de jointure. Il semble que les palmettes soient pleines chez les premiers modèles et que chez les modèles les plus tardifs, elles soient ajourées pour des raisons esthétiques ou pour alléger l'arme. Les sabres les plus anciens ont des oreillons, surtout basaux sur la croisière.

En conséquence, deux Sous-familles ont été créées :

Sous-famille VI a - Palmette pleine

La garde à palmette est plus ou moins ornée : palmes bien limitées par des sillons avec ornements divers dont les Armes de France ou une grenade... Sur certains exemplaires on peut découvrir de petits oreillons plus ou moins masqués par la monture.

585

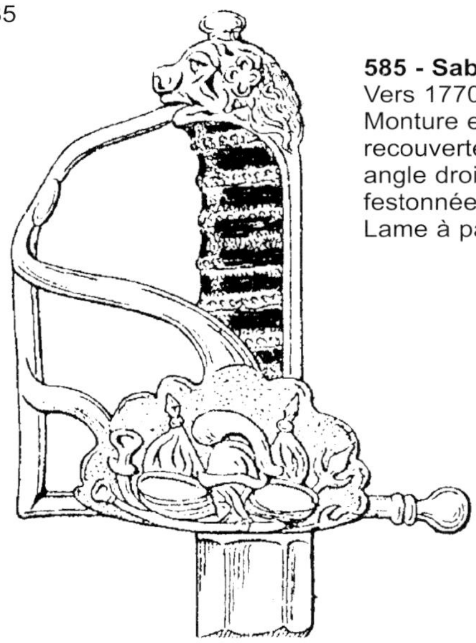

585 - Sabre d'officier de Dragons.
Vers 1770.
Monture en laiton. Calotte en tête de lion et longue queue. Poignée encordée , recouverte de basane, à 8/9 spires d'un filigrane torsadé. Arc de jointure arrivant en angle droit sur la croisière. Branches secondaires partant d'une coquille à bordure festonnée, ornée d'un casque empanaché, de deux drapeaux, de tambours ...
Lame à pan creux et gouttière dorsale (d'après M. Bottet).

586

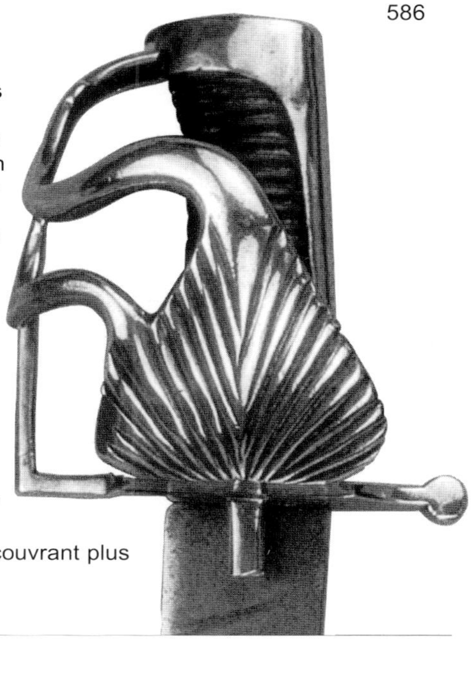

586 -Sabre des Volontaires de Clermont-Prince.
Corps fondé en 1758. Sabre du modèle 1762. Monture en laiton. Calotte plate, à rivure nettement déportée vers la droite, longue queue cloutée à sa base. Poignée massive, filigranée. Arc de jointure peu courbé, légèrement oblique et arrivant en angle presque droit sur la croisière rectiligne et terminé par un bouton. Deux branches secondaires en " S " partent de la coquille triangulaire à fines rainures en éventail. Oreillons basaux en baguettes, à extrémités carrées. Lame plate, de 92 cm, large au talon de 4,1 cm, et flèche de 5 cm. La lame décrite par C. Ariès (XIII,5, fig. 1 h) est gravée du haut vers le talon, d'un motif floral, d'un cavalier sur un cheval cabré et de l'inscription *Régiments / Volontaires / De / Clermont / Prince*.
Fourreau de cuir, à deux garnitures : chape à crevé et longue bouterolle recouvrant plus de la moitié de sa longueur et terminée par un dard asymétrique.
Les fourbisseurs sont PWF et Jean Knecht de Solingen.

587

587 - Sabres des Gardes Suisses (1) **de la Maison du Roi.**
Modèle 1763.
Monture en laiton. Calotte en cuillère surmontée d'un bouton de rivure et à longue queue. Poignée totalement filigranée de fils torsadés disposés en chaînette. Arc de jointure vissé sur la calotte, courbée en sa partie supérieure puis droit et arrivant à 90° sur la croisière. Coquille de garde en palmette, symétrique, pleine et composée d'une dizaine d'éléments. De cette coquille partent deux branches secondaires, courbée en " S " et rejoignant séparément l'arc. Oreillon inférieur. Court quillon terminé par un petit disque. Poucier à la base de la fusée sur la face postérieure. Lame plate ou à section hexagonale, longue d'environ 70,5 cm, large au talon de 3,4 cm, à très légère flèche de 0,17 - 0,19 cm, gravée *REGT DES GARDES SUISSES* sur une face et sur l'autre la désinence de la Compagnie. Fourreau à deux garnitures, chape à fente, bouterolle à bouton.

588 - Sabre des Grenadiers à cheval de la Maison du Roi.
Modèle de la réforme du duc de Choiseul de 1763. Monture en laiton. Calotte à gros bouton de rivure et longue queue unie. Poignée légèrement inclinée vers l'avant, recouverte de basane et filigranée de 17 spires. Arc de jointure droit et faisant un angle droit en rejoignant la croisière. Coquille en palme dont les 1er et le 3ème éléments s'allongent pour former les deux branches secondaires qui rejoignent l'arc de jointure séparément. Petits oreillons à extrémités arrondies, le postérieur portant un poucier de laiton. Lame blanche à pan creux, longue de 81,2 cm, large au talon de 3,6 cm, flèche de 4,1 cm. Fourreau inconnu.

588

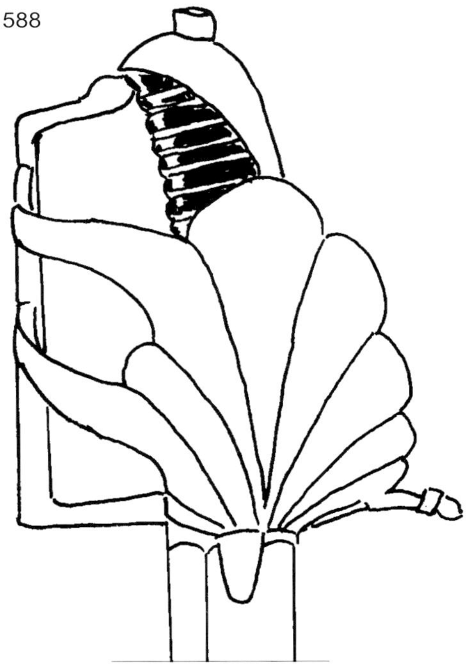

589 - Sabre d'officier de Carabiniers.
Vers 1770.
Calotte à longue queue, en laiton moulé, ainsi que la poignée à sept pans. Arc de jointure arrondi au sommet, puis droit en faisant un angle droit avec la croisière. Coquille de garde asymétrique d'où partent deux branches plates secondaires rejoignant isolément l'arc. Cette coquille dont la base est ornée des Armes de France couronnées et flanquées de quatre drapeaux vient s'appuyer sur l'oreillon frontal en losange curviligne. Lame presque droite, à pan creux et gouttière dorsale. Fourreau en cuir de vache noirci à trois garnitures, la bouterolle terminée par un dard ne protégeant que sa face antérieure et sa base (M.E. - S de P.) (non figuré).

Nota - On connaît d'autres modèles dont la fusée est filigranée et dont la lame est nettement courbée.

(1) La garde suisse a été créée en 1567, pour assurer la protection du roi. En 1616, un régiment permanent est formé et en 1671, Louis XIV incorpore les Gardes Suisses à la Maison du Roi.
Les Gardes sont massacrés le 10 août 1792, lors de la prise des Tuileries. Bonaparte fut témoin du carnage et devenu empereur puis déchu, il écrira dans ses mémoires : *Jamais, aucun de mes champs de bataille ne me donna l'idée d'autant de cadavres que m'en présentèrent les masses de Suisses.*

590

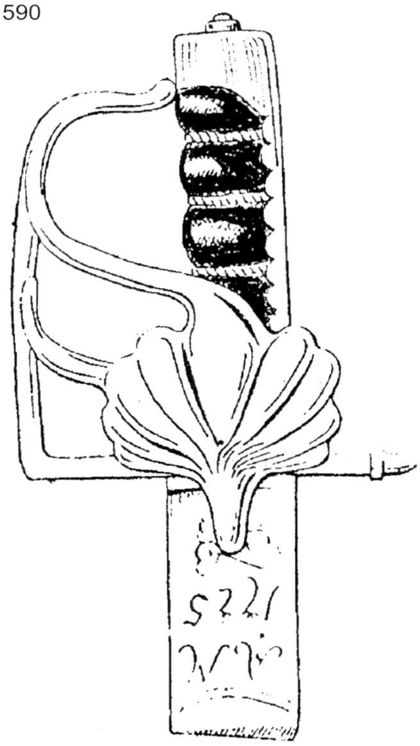

590 - Sabre d'officier de Cavalerie.
Lame marquée 1775.
Monture en laiton. Queue de calotte à bordure festonnée. Poignée encordée à seulement cinq spires d'un filigrane torsadé. Arc de jointure arrivant en angle droit sur la croisière. Les deux branches secondaires partent d'une coquille en palmette de neuf éléments et arrivant isolément sur l'arc. On notera la présence d'un petit oreillon basal formé par la queue de palmette. Lame plate (d'après M. Bottet).

591 - Sabre de Gardes Suisses de la Maison du Roi.
Modèle 1763, variante.
Monture en laiton. Calotte en cuillère à longue queue. Poignée recouverte de basane et finement filigranée. Arc de jointure courbé recevant isolément les deux branches secondaires partant d'une palmette à onze éléments. Lame plate de 70 cm (d'après M. Bottet).

591

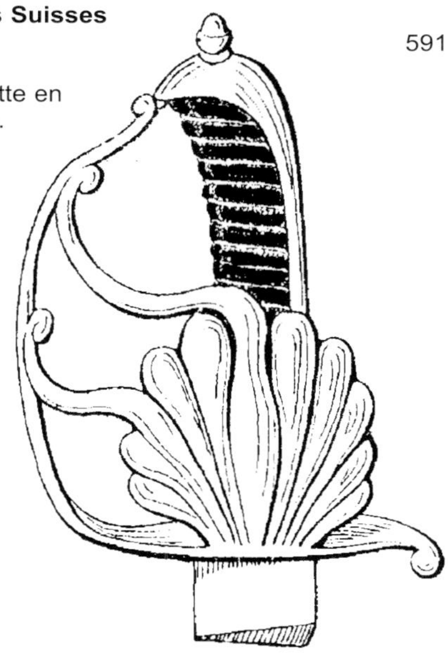

592

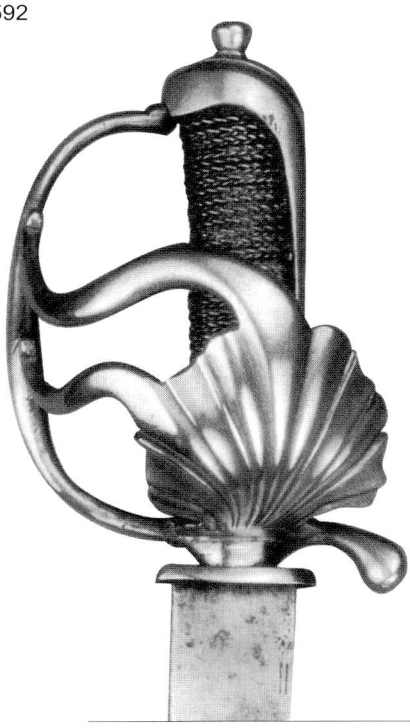

592 - Sabre des Gardes Suisses de la Maison du Roi.
Variante du modèle 1763.
Les principales différences entre ce modèle et celui de 1783 sont les suivantes : l'arc de jointure arrondi est fixé sur la calotte par un tenon, les oreillons ont disparu, le court quillon est légèrement courbé. Lame peu fléchée (1cm), de 80 à 83 cm de longueur, plate et gravée du Régiment ou de la Compagnie.

593

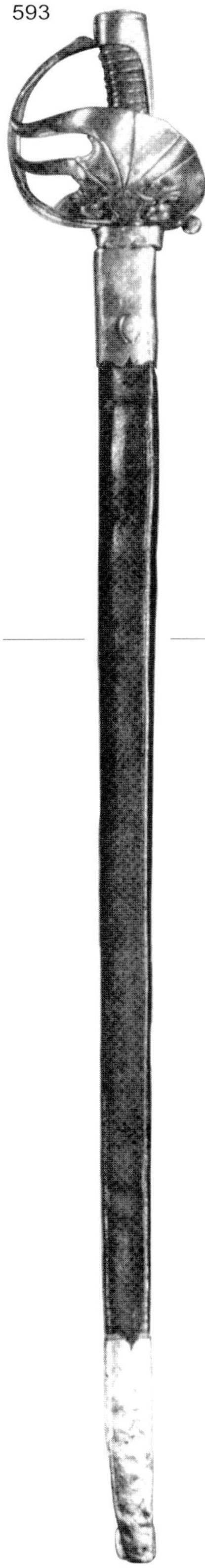

593 - Sabre de Carabiniers de Monsieur.
Modèle 1777.
Monture en bronze doré. Calotte plate à petit bouton de rivure et longue queue. Poignée recouverte de basane, à filigrane simple. Branche principale régulièrement arquée. Coquille asymétrique donnant naissance à deux branches secondaires plates rejoignant séparément l'arc de jointure et décorée à sa base, des Armes de France couronnées flanquées de deux grenades enflammées (593a). Contre-garde de taille réduite ciselée en palmette et lanière de cuir en guise de poucier, fixée à la base de la fusée. Couvre-chape. Lame longue de 70 à 82 cm, en général droite et gravée du talon à la pointe : *MANUFRE / DE / KLINGENTHAL*, d'une grenade, *REGIMENT / DES CARABINIERS / DE / MONSIEUR*, Armes de France et cordon du Saint-Esprit, *VIVAT / REGUM / CARISSIMUS*. Fourreau de cuir à bouton et bouterolle à dard protégeant sa face antérieure et sa base.

593a

Variante - Chez des exemplaires de facture comparable, la lame peut être gravée REGIMENT DES *CARABINIERS DE MONSIEUR* et signée au talon *DE / PRESOLLE / MARCHAND / FOURBISSEUR / DU / CORPS / DES / CARABINIERS / DE / MONSIEUR*. Poinçon de Bisch, contrôleur à Klingenthal à partir de 1783.

Nota - On trouve également des exemplaires de ce sabre dont les Armes de France ont été martelées, montrant ainsi que ce modèle a été porté après 1792.

- **a**, détail de l'ornement figurant à la base de la coquille.

594 - Sabre de Carabinier.
An IV.
Monture souvent en laiton riche en cuivre rouge. Calotte ovalaire à longue queue. Poignée recouverte de basane et filigranée de fils de laiton torsadés faisant douze spires. Arc de jointure régulièrement arrondi, recevant isolément les deux branches secondaires. Ces deux branches ne sont que des prolongements de la coquille rayonnante ornée de la grenade dite, à cause de sa figuration très stylisée, *pomme de pin* par dérision. Court quillon arrondi. Contre-garde en palmette, couvre-chape, poucier en languette de cuir. Lame plate, droite, longue de 97,5 cm, large au talon de 3,1 cm et gravée en capitales, entre une grenade et un bonnet phrygien, *CARABINIERS DE LA REPUBLIQUE FRANCAISE* sur les deux faces. On trouve également le poinçon au licteur et le " B " de Bisch. Fourreau de cuir noir à deux petites garnitures, chape à bouton, bouterolle à dard antérieur seulement.

594

595

595 - Variante.
Vue permettant de bien voir la *pomme de pin* et le demi-plateau postérieur. Pas de couvre-chape.

Sous-famille VI b - Palmette découpée

La structure de la monture et en particulier de la coquille est bien caractérisée : la palmette a toutes ses branches ajourées. Leur forme est parfois variable, possédant de 5 à 9 éléments et c'est toujours de cette coquille que partent les deux branches secondaires arrivant isolément sur l'arc de jointure.

596

596a

596b

596c

596 - Sabres d'Infanterie.
Vers 1780.
Monture en fer. Calotte à longue queue à double rainure latérale. Arc de jointure courbé en sa partie supérieure, puis droite et faisant un angle droit avec la croisière. Garde en palmette découpée, à sept éléments d'où partent deux branches secondaires rejoignant la principale, isolément. Poignée recouverte de basane filigranée. Petit plateau de garde découpé. Lame à pan creux, légèrement courbée. Fourreau de cuir noir à deux garnitures.

Nota - La palmette peut varier dans sa forme générale (596a, 596b, 596c).

597 - Sabre supposé d'officier de la Garde à cheval du Directoire.
Monture en général dorée. Calotte à bouton de rivure en demi-olive, à longue queue simplement décorée d'un filet marginal. Poignée recouverte de basane à seize spires de filigrane en chaînette. Arc de jointure aplati, se prolongeant par un plateau découpé de quatre vides ovalaires. Deux branches secondaires fixées séparément sur l'arc de jointure, la supérieure formant la bordure d'un médaillon à neuf découpes délimitant une palmette. La seconde branche secondaire, courte, rejoint le bas du médaillon. Lame longue de 87 cm, large au talon de 3cm, plate et pointe dans le prolongement du dos. Poinçon " MR ", de B.P.C. Levavasseur. Fourreau de cuir noir, à trois garnitures décorées au trait : chape à anneau de bélière en fer ; bracelet médian placé à seulement 8,4 cm de la chape ; bouterolle de 8 cm, avec dard seulement à son extrémité et se prolongeant sur la face antérieure.

Remarque - A la fin de l'Ancien Régime, il existait une monture, en fer, très comparable à ce sabre en laiton doré du Directoire.

597

598

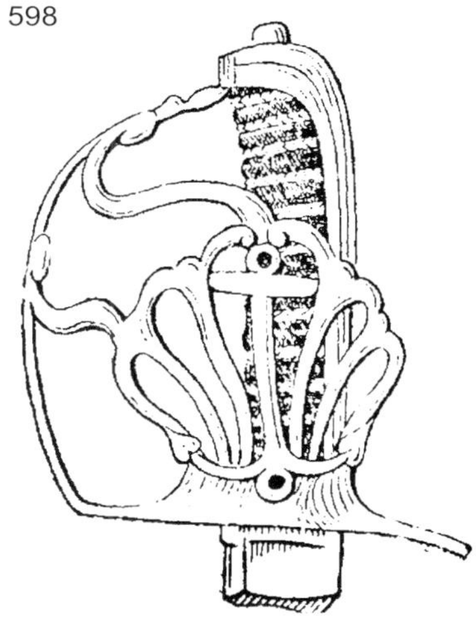

598 - Sabre de bord.
Vers 1788.
Monture en laiton. Calotte plate à bouton de rivure et longue queue à filet. Poignée recouverte de basane filigranée. Arc de jointure arqué. Deux branches secondaires rejoignant parallèlement l'arc de jointure. Cette coquille à palmette est construite autour d'une ancre de marine à un organeau aux deux extrémités de la verge.

Nota - L'organeau placé entre les bras de l'ancre est destiné à fixer un filin qui permet, en le tirant, de libérer l'ancre trop engagée. Cette figuration est exceptionnelle.

Famille VII - Poignée en bois recouvert de basane filigranée, branches secondaires portant l'emblème du corps

Ces sabres d'officiers sont généralement connus sous le terme générique de petits Montmorency. Ils furent créés sous le règne de Louis XVI. Ainsi que tous les sabres réunis dans le Chapitre III, les branches secondaires viennent rejoindre isolément l'arc de jointure et soutiennent un symbole relatif au régime ou le plus souvent au corps auquel appartient son possesseur. En conséquence, les sabres ont été regroupés en fonction de la nature de cet emblème, de la façon suivante :

Coquille :

Les autres parties de la monture subissent également des variations notables dont on n'a pas tenu compte dans la définition des Sous-familles, plusieurs coquilles identiques pouvant être assorties de calotte simple, de calotte figurative, tels tête de lion, casque à l'antique...

L'arc de jointure et le plateau de garde sont découpés dans la même plaque métallique. La poignée est recouverte de basane filigranée. La lame est à pan creux, avec ou sans gouttière dorsale au fort. Le fourreau est toujours de cuir, à deux ou trois garnitures.

Sous-famille VII a - Au soleil

Les sabres présentant cette caractéristique sont assez fréquemment rencontrés mais présentent des variations concernant surtout la forme de la calotte et la figuration du soleil.

La calotte - Celle-ci peut être simple du type de celle des sabres de Mineurs, ou en tête de félin aplatie, ou encore en casque à l'antique. La queue de calotte est très généralement longue mais il n'est pas impossible d'en rencontrer de courte.

Le soleil - Les soleils que nous avons pu examiner ont figure humaine ainsi que le présentent les astrologues et les alchimistes. Les rayons sont de types différents : les uns, bien détachés les uns des autres sont au nombre d'une douzaine. Souvent ils sont plus figuratifs, chaque rayon étant strié, séparé du rayon voisin. Plus rarement, le soleil est orné de rayons jointifs, abondamment striés ; dans ce cas, les rayons sont plus larges et peuvent être au nombre d'une quinzaine.

De plus, les plateaux peuvent être des types a, b, d, (voir p. 21) et les lames en général du type Montmorency peuvent également être de récupération et dans ce cas d'un modèle indéterminable le plus souvent.

599

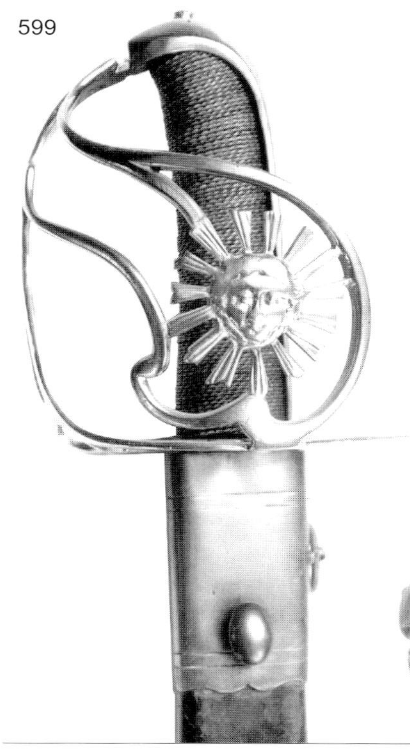

599 - Sabre d'officier d'Infanterie.
Règne de Louis XVI.
Monture en laiton doré. Calotte en cuillère se prolongeant par une longue queue unie. Poignée totalement filigranée d'un fil de laiton torsadé. Arc de jointure ajouré de trois découpes ovales plus une ronde à l'entrée du plateau du type e (p. 21). Motif de garde représentant un soleil à quinze rayons et fixé par une petite branche supplémentaire rejoignant la branche secondaire supérieure. Lame de 69 cm, large au talon de 3,3 cm avec une flèche de 1 cm, à la Montmorency et contre-tranchant d'environ 18 cm. Chape du fourreau décorée de deux doubles traits, à bouton et anneau de bélière ; bouterolle de 15 cm au maximum, à dard frontal et basal.

600 - Variante.
Calotte en cuillère et longue queue. Remarquer le brin qui relie les deux branches secondaires et le soleil placé de travers, preuve d'un travail artisanal. La prise de vue permet de découvrir le plateau au soleil. Fourreau à trois garnitures. (J.C. D.).

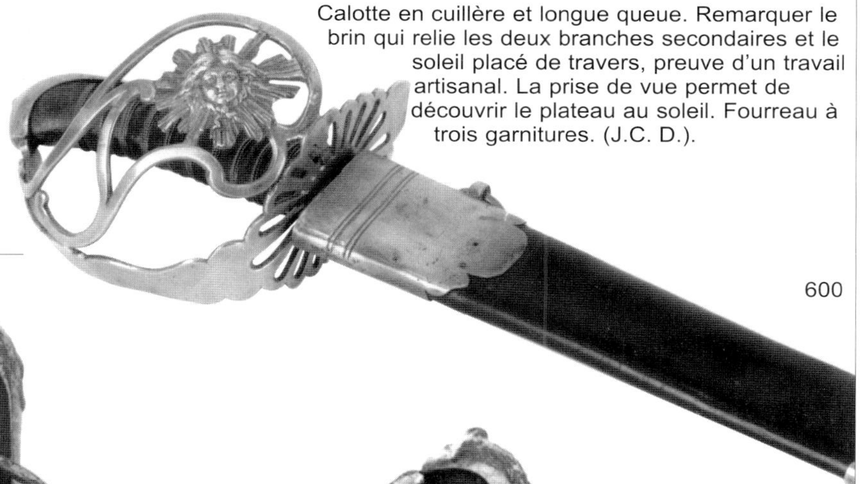

600

601

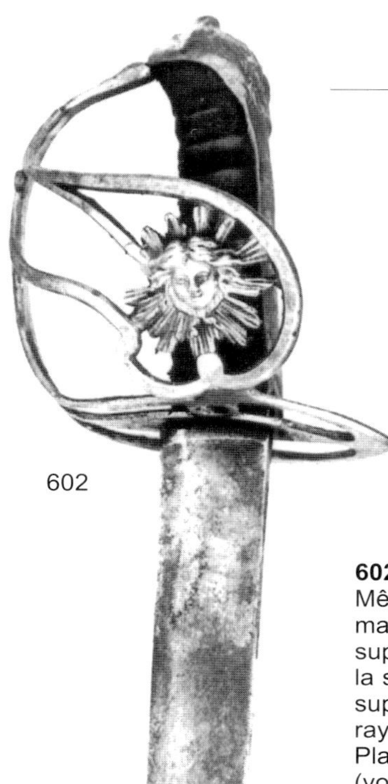

602

601 - Variante.
Même structure générale mais caractérisée par un soleil à douze rayons bien séparés les uns des autres et par une calotte en dépouille de lion. Plateau du type d (voir p. 21).

602 - Variante.
Même structure mais avec un brin supplémentaire sur la secondaire supérieure. Soleil à rayons striés. Plateau du type a (voir p. 21).

603 - Sabre pouvant être attribué à un officier des Gardes du Roi Louis XVI (*). Monture en bronze doré. Calotte à courte queue et fusée filigranée. Arc de jointure droit sur le cinquième de sa longueur. Garde fondue, finement reciselée, avec un soleil rayonnant maintenu par une branche ornée de feuillage et terminée par une tête d'oiseau ; la seconde branche est transformée en un rameau à feuilles lobées (chêne ?). Quillon feuillu, terminé par une tête d'animal mythique tenant une boule dans sa gueule (603a). Forte lame de 89 cm de long et de 5 cm de large au talon, à légère flèche et présentant des inscriptions en langue allemande, mais devenues illisibles par usure. Fourreau long de 96 cm, de cuir à trois garnitures : chape à trophée d'armes portant un médaillon avec l'inscription de la devise *PRO / DEO ET / PATRIA* ; large bracelet orné d'une chouette (603b) ; bouterolle décorée d'un trophée incluant un chapeau de Gueux (603c). (Dessin fournis par J.C. Dey et description à partir de documents signés Pétard).

603

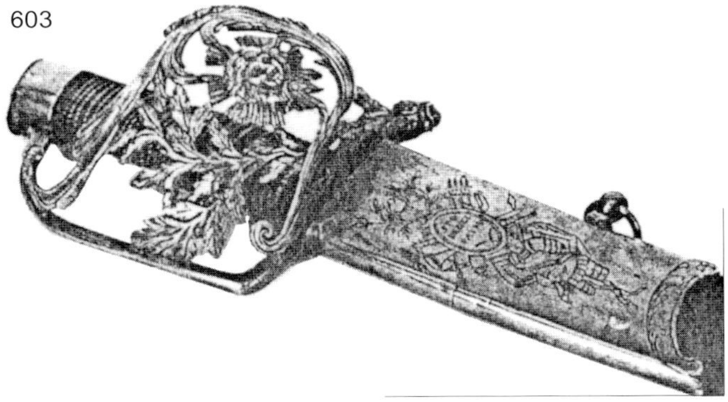

603a

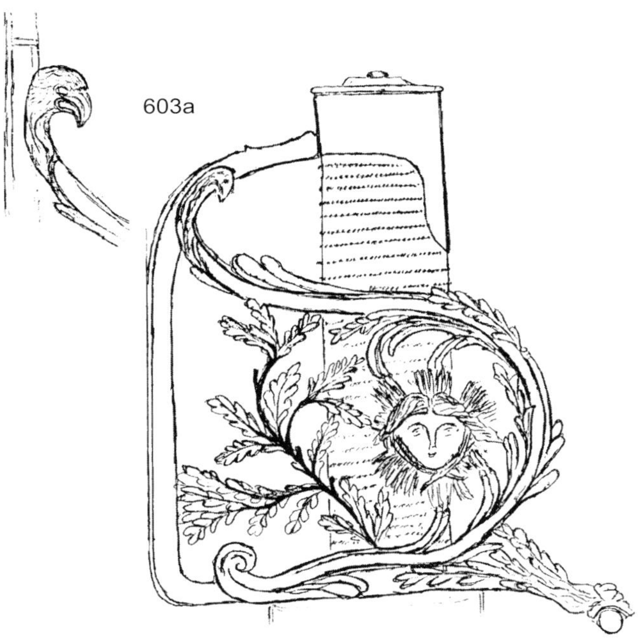

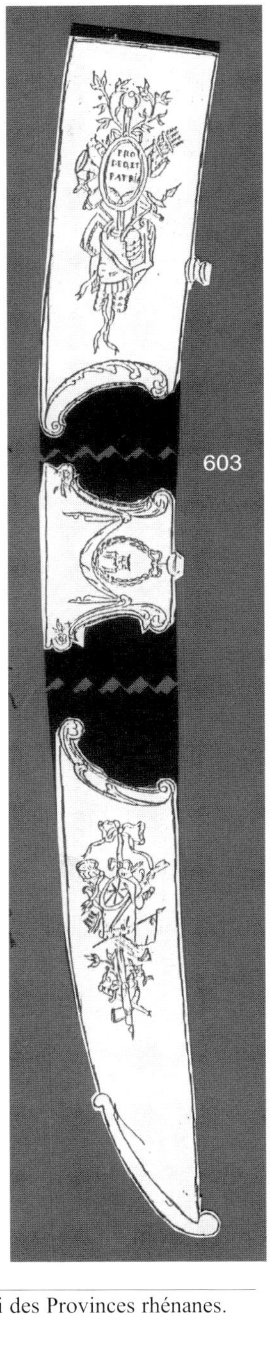

603

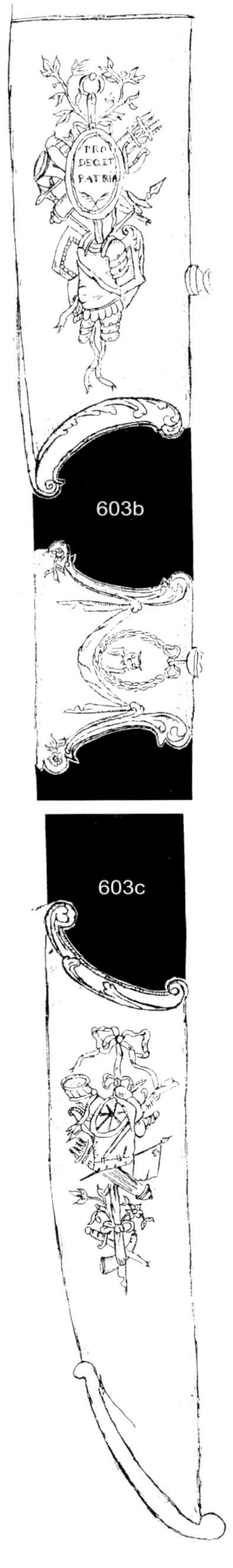

603b

603c

(*) ou de Régiment du Roi des Provinces rhénanes.

604 - Variante.
Soleil placé sur les branches secondaires d'un sabre de Mineur classique. Plateau découpé du type a (voir p. 21). (J.C. D.).

604

606 - Variante.
Vers 1792.
Calotte en casque empanaché, à longue queue. Soleil à une quinzaine de rayons striés, fixé sur les branches de garde de la disposition de celle des sabres de Mineurs. Plateau du type a (voir p. 21).
Lame de récupération à gouttière médiane.

606

607

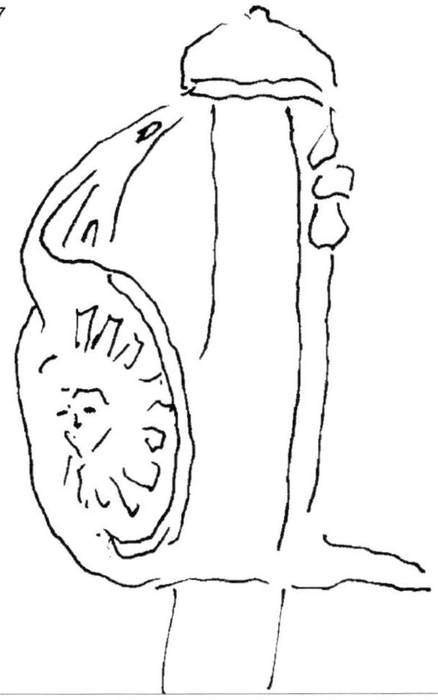

607 - Sabre d'officier à la coquille au soleil.
Ancien régime puis période révolutionnaire.
Petit Montmorency classique au soleil dont la calotte et la queue ont été remplacées : le casque a fait place au bonnet et la queue de calotte est ornée sur toute sa longueur d'une pique, la pointe et le gland rejoignant le bonnet. (G. Levalley, 1945, pl. XII, p. 4).

608 - Variante.
Règne de Louis XVI. Calotte au casque empanaché. On notera la structure de la branche secondaire supérieure ajourée et le soleil à figure très anthropomorphique. Lame plate.

608

609 - Sabre d'officier de Volontaires Nationaux.
Ce sabre est remarquable par sa calotte en tête d'aigle à longue queue et par le soleil qui, souvent, orne les sabres du règne de Louis XVI.

609

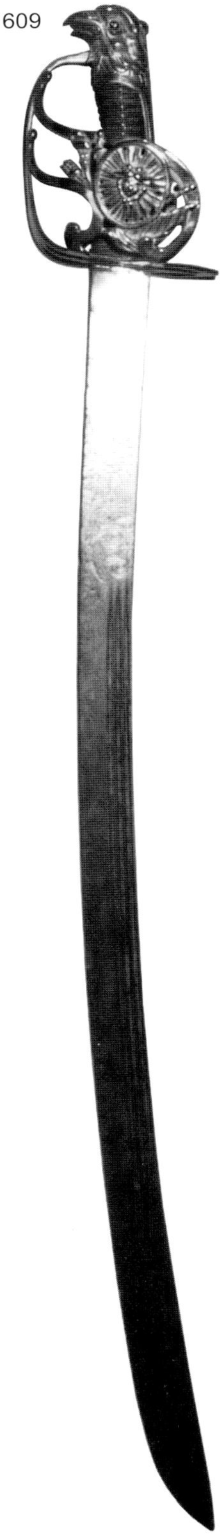

Sous-famille VII b - Au cornet ou au cor (1)

Les sabres au cornet ou au cor subissent la même évolution que les sabres à décor au soleil : les sabres de troupe ont une monture très comparable à celle des sabres de Mineurs alors que les sabres des officiers s'enjolivent. En effet, le cor est associé à des drapeaux, des têtes de lion… et les calotte deviennent au casque, au lion… A été inclus dans cette Sous-famille VII b, un type de sabre dont la branche supérieure est transformée en deux cornets. On connaît différents exemplaires possédant cette monture.

610

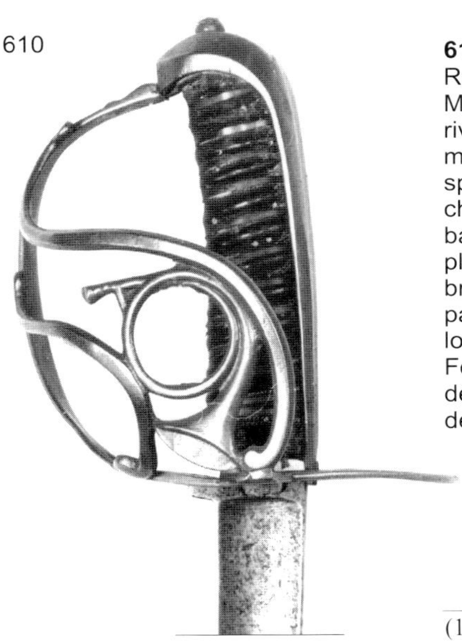

610 - Sabre de Chasseur à pied.
Règne de Louis XVI - Troupe.
Monture en laiton. Calotte arrondie, à bouton de rivure et longue queue du modèle des sabres des mineurs. Poignée recouverte de basane à onze spires formées de deux fils torsadés disposés en chaînette, plus trois spires serrées en guise de virole basale. Arc de jointure modérément arrondi, plein et plateau du type a (voir p. 21). Entre les deux branches secondaires est fixé un cor de Chasseur, le pavillon vers le bas. Lame blanche à la Montmorency, longue de 71,5 cm, large au talon de 3,4 cm. Fourreau de cuir noirci à trois garnitures à bordures découpées en accolade ; chape à bouton et anneau de bélière, bracelet quadrangulaire à anneau, bouterolle à dard frontal de laiton.

(1) Instrument à vent à corps en une spirale : cor, ou deux spirales : cornet.

611 - Sabre de Chasseur.
1791 - Troupe.
Monture en laiton. Calotte du type des sabres de Mineurs. Poignée filigranée. Arc de jointure droit sur les deux tiers de sa longueur, plein, suivi d'un plateau du type a (voir p. 21). Entre les deux branches secondaires est inscrit un cornet, le pavillon dirigé vers le haut et l'embouchure vers le bas. Lame à la Montmorency, au poinçon M.R. pour de Cabannes en 1791.

611

612

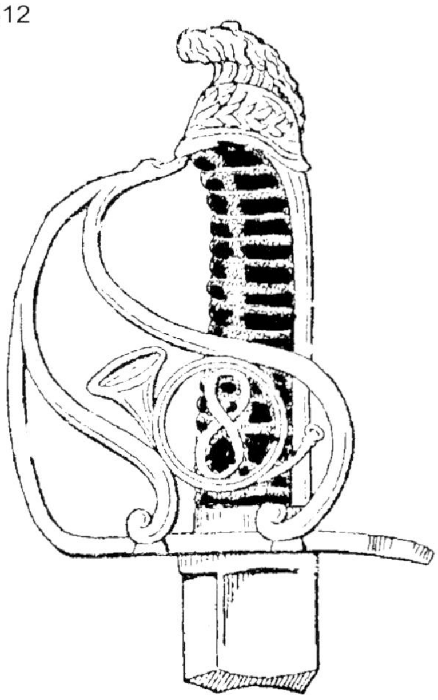

612 - Sabre d'officier de Chasseurs.
1785 - 1791.
Monture en laiton. Calotte en casque empanaché et longue queue. Poignée recouverte de basane et filigranée. Arc de jointure courbé, recevant isolément les deux branches secondaires entre lesquelles est inscrit un cornet portant le chiffre 8 désignant soit la Compagnie, soit le Régiment. Lame à pan creux.

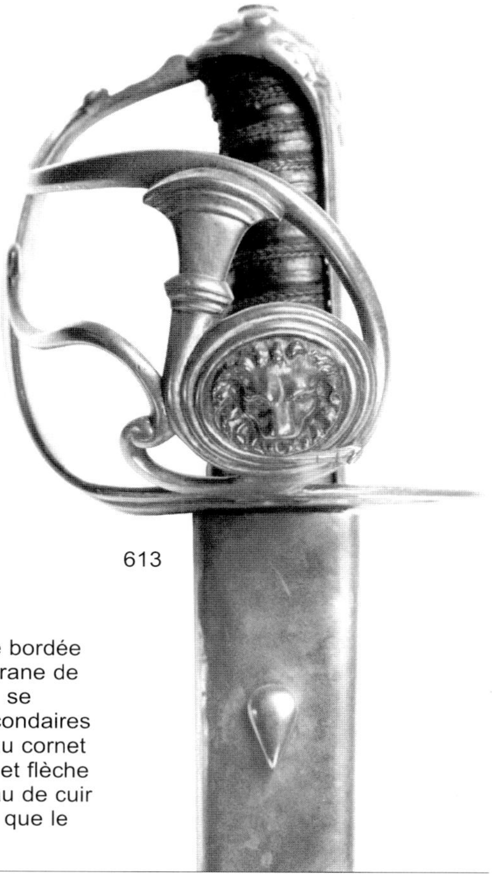

613

613 - Sabre d'officier de Chasseurs à pied.
Fin du XVIIIème siècle.
Monture en cuivre rouge. Calotte en tête de lion aplatie et longue queue bordée de filets. Poignée encordée, recouverte de basane, à 10 spires d'un filigrane de cuivre rouge torsadé et disposé en chaînette. Arc de jointure arqué, uni, se prolongeant en un plateau découpé du type a (voir p. 21). Branches secondaires arrivant isolément sur l'arc de jointure et soudées sur celui-ci. Coquille au cornet enserrant un masque au lion. Lame de 72 cm, large au talon de 3,7 cm et flèche de 1,5 cm, blanche, à pan creux et gouttière dorsale sur 37 cm. Fourreau de cuir à deux garnitures en cuivre rouge : chape à bouton, dard ne protégeant que le côté antérieur de la bouterolle.

614 - Variante.
Monture en laiton identique à l'exemplaire précédent, mais avec l'arc de jointure cannelé et un plateau du type d (voir p. 21). Lame à la Montmorency, au poinçon M.R. pour de Cabannes en 1791.

614

615

615 - Variante.
Le montage très spécial de ce sabre a été décrit p. 21.

616 - Sabre de Chasseur à pied.
Epoque révolutionnaire.
Monture en laiton. Calotte du type de celle des sabres de Mineurs. Poignée recouverte de basane et filigranée de deux fils torsadés disposés en chaînette. Arc de jointure arrondi, plein, prolongé par un plateau du type a (voir p. 21). Deux branches secondaires en " S " entre lesquelles est inscrit un cor de Chasseur, le pavillon tourné vers le bas soutenant un bonnet phrygien en son centre.
Lame blanche à la Montmorency de 71,8 cm, large au talon de 3,5 cm, flèche de 2 cm. Fourreau de cuir noirci à trois garnitures à double rainures, à bordures découpées en accolade : chape à bouton et anneau de bélière, bracelet à anneau, bouterolle à dard arrondi frontal.

616

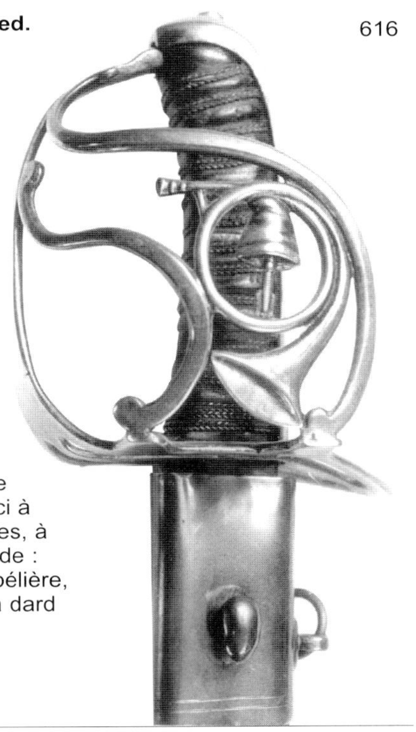

617

617a

- **a**, cuirasse inscrite dans le cornet. Remarquer le bonnet phrygien surmontant la cuirasse entre les six drapeaux.

614 - Sabre d'officier de Chasseur.
Période révolutionnaire.
Ce sabre présente toutes les caractéristiques d'un sabre d'Ancien Régime : calotte en casque à crête avec un masque léonin ébauché frontal et longue queue. Poignée de cuir encordé, à filigrane de fils torsadés faisant une douzaine de spires. Arc de jointure découpé avec le plateau ajouré du type a (voir p. 21). Deux branches secondaires en " S " arrivant séparément sur l'arc et soutenant un médaillon constitué d'une cuirasse surmontée d'un bonnet phrygien - seule allusion à la Révolution - sur fond de six drapeaux, l'ensemble étant inscrit dans un cor (617a). Lame à pan creux, à la Montmorency. Fourreau à deux garnitures de laiton, la chape étant à bouton et à anneau de bélière.

Variantes - Il existe, pour le moins, des modèles à calotte simple, unie, à pans. Un exemplaire est poinçonné sur la lame des lettres M.R.

618

618 - Variante.
Calotte en cuillère unie. Le médaillon à la cuirasse surmonté du bonnet phrygien flanqué de six drapeaux inscrits dans un cor a été soudé sur les branches secondaires.

619

619 - Sabre d'officier supérieur de Chasseurs.
Avant 1792.
Monture en laiton doré. Calotte en tête d'aigle au cou emplumé et prolongée, sans césure, en queue unie. Poignée recouverte de basane à filigrane de deux fils d'argent torsadés, faisant dix spires et virole basale. Arc de jointure uni. Motif de la coquille constitué d'un cor dans lequel est inscrit une cuirasse surmontée d'un bonnet phrygien (en place du casque classique des insignes d'Etat-Major) sur fond de six drapeaux. Lame de 72,5 cm, large au talon de 3,3 cm, contre-tranchant de 17 cm environ, décoré d'un aplat doré sur 12 cm de long et 2 cm de large se terminant par des entrelacs. Fourreau à chape munie d'un bouton et d'un anneau de bélière, bracelet à anneau, longue bouterolle de 19,5 cm au maximum et terminée par un dard asymétrique en fer.

620

620 - Sabre d'officier de la Garde Nationale de Paris.
Vers 1792.
Monture en laiton. Calotte en casque à cimier et longue queue. Poignée recouverte de basane filigranée. Arc de jointure se prolongeant en un plateau évidé du type a (voir p. 21). Les deux branches secondaires arrivant séparément sur l'arc supportent un cornet au pavillon brisé, dans lequel est inscrit la Nef parisienne, l'ensemble souligné par deux rameaux de laurier noués d'un ruban. Lame à la Montmorency gravée sur les deux faces *VIVE LA NATION*.

<u>Nota</u> - Il est évident que la décoration des branches a été rajoutée, probablement lors de la période révolutionnaire (J.C. D.).

621

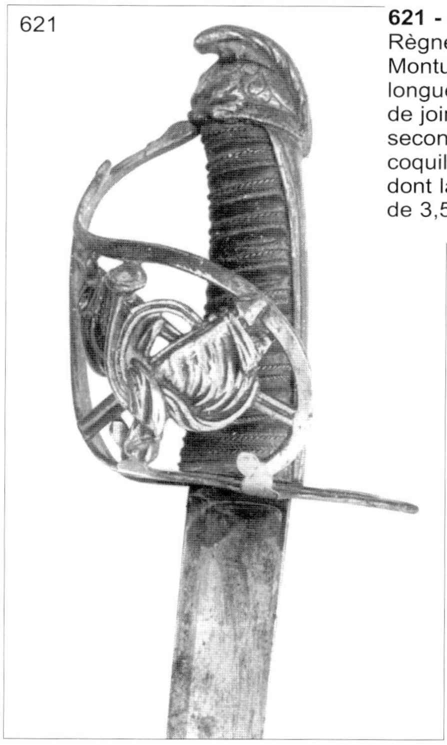

621 - Sabre d'officier de Chasseur à pied.
Règne de Louis XVI, vers 1789 - 1791.
Monture en laiton. Calotte en casque empanaché à masque léonin frontal à longue queue de calotte. Poignée recouverte de basane à filigrane torsadé. Arc de jointure peu arqué, à plateau de garde du type a (voir p. 21). Branches secondaires séparées, mais l'inférieure est partiellement cachée par le décor de la coquille. Ce décor est constitué d'un cornet de Chasseur traversé par un drapeau dont la hampe est coiffée d'un bonnet phrygien. Lame de 73,5 cm, large au talon de 3,5 cm, à la Montmorency.

622

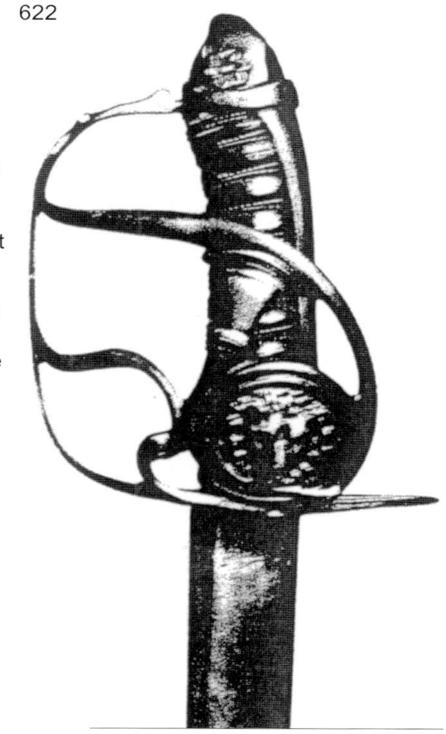

622 - Sabre d'officier de Chasseurs.
Période révolutionnaire.
Sabre à calotte en bonnet révolutionnaire à la cocarde, médaillon à la face léonine inscrite dans le cornet dont le pavillon est dirigé vers le haut. Ce modèle est figuré par G. Lavallay (op. cit. 1945, pl. XIII).

623 - Variante.
Autour de 1792. Calotte en bonnet phrygien, à longue queue ciselée d'un pique à glands. Poignée filigranée. Arc de jointure arrondi, plein, uni et plateau du type a (voir p. 21). Motif de la coquille à la tête de lion. Lame longue de 73,5 cm, large au talon de 3,5 cm (in Ph. Bordes, op. cit. p. 89, fig. 62) (non figuré).

624 - Sabre d'officier de Chasseurs à cheval.
Règne de Louis XVI.
Laiton. Calotte au casque empanaché, à grille et longue queue. Poignée ficelée, recouverte de basane, filigrane faisant une douzaine de spires. Plateau ajouré du type a (voir p. 21). Branche secondaire supérieure transformée en deux cornets liés par un ruban, la branche inférieure n'étant constituée que par un pan de celui-ci. Lame bleuie au tiers, gravée de trophées et de divers motifs dorés classiques. Fourreau de cuir noir, à trois garnitures : chape pourvue d'un bouton et d'un anneau de bélière et bracelet à anneau.

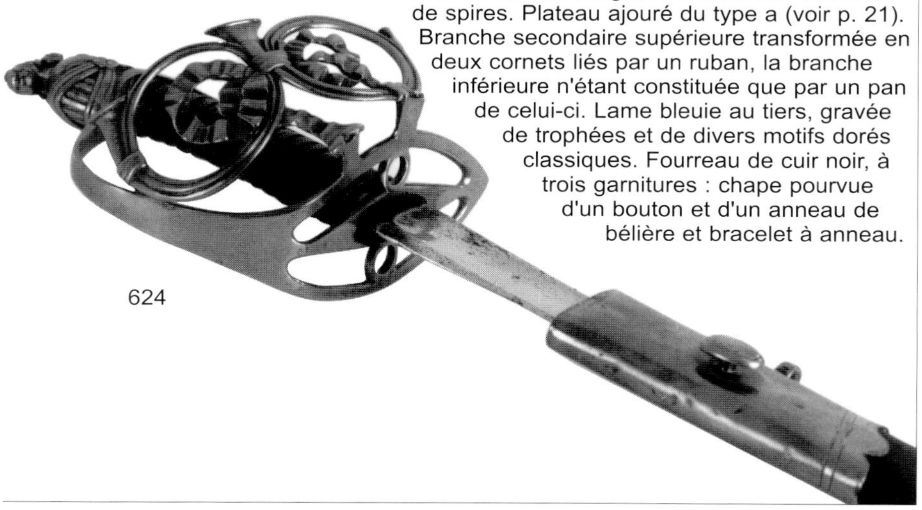

624

625 - Variantes.
Un sabre de ce modèle est figuré par Bordes (1987, p. 89). Il diffère du modèle décrit précédemment, sur les points suivants :
- calotte en casque empanaché, mais sans grille ;
- fusée dont le diamètre est inférieur à la largeur du casque ;
- filigrane de fils de laiton très serrés, torsadés ;
- lame bleuie, gravée au tiers de trophées où dominent les instruments de musique militaire.

Sabre fantaisie pour les chefs de musique ? (non figuré).

Sous-famille VII c - A la grenade

Les sabres à la grenade enflammée sont en général attribués aux Grenadiers. Cette grenade est insérée à la base de la monture, entre les deux branches secondaires. Ainsi que pour les sabres réunis dans le Chapitre III, la calotte est très variable de forme : simple, en bonnet, en casque à l'antique, en tête de lion, .. Parfois, la grenade est décorée de fleurs de lys ou de la nef symbolique de la Ville de Paris.

626

626 - Sabre d'officier de Grenadiers.
Règne de Louis XVI.
Monture en laiton. Calotte du type des sabres de Mineurs, prolongée par une longue queue unie. Poignée ficelée, recouverte de basane, à sept spires d'un filigrane constitué d'un ruban de laiton bordé d'un fil de laiton torsadé. Arc de jointure plein à trois rainures. Plateau du type a (voir p. 21). Motif au soleil dans la partie postérieure arrondie et, près de l'arc de jointure, entre les deux branches secondaires se trouve une grenade enflammée. Lame blanche, longue de 65,5 cm, large au talon de 3,5 cm, flèche de 1,3 cm, pan creux et contre-tranchant de 14 cm.

627 - Sabre d'officier de Grenadier de Picardie.
1779.
Monture en laiton. Calotte en tête de lion et longue queue. Poignée recouverte de basane filigranée. Arc de jointure arrivant en angle droit sur la croisière. Les deux branches secondaires en " S ", rejoignent isolément l'arc et entre elles est insérée une grenade enflammée. Lame à pan creux. Fourreau à deux garnitures à chape à bouton et anneau de bélière.

Nota - Selon M. Bottet, ce sabre appartint au capitaine M. de Parfontaine.

627

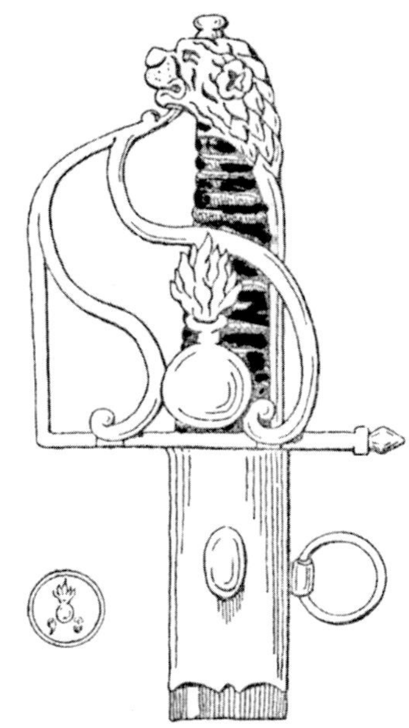

629 - Sabre d'officier de Grenadiers.
Règne de Louis XVI.
Calotte en bonnet de Grenadier, à longue queue. Poignée à sept spires constituées d'un fil en chaînette bordé de deux fils simples. Arc de jointure régulièrement arrondi. La grenade enflammée est maintenue par les deux branches secondaires. Lame à la Montmorency, blanche, longue de 69 cm,

628

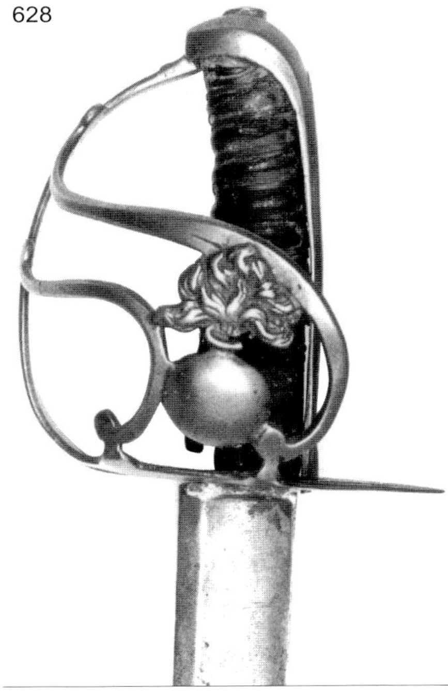

628 - Sabre de Grenadier pour la troupe.
Règne de Louis XVI.
Modèle de sabre de Mineur en laiton. Arc de jointure arrondi. Poignée filigranée d'une quinzaine de spires d'un fil torsadé. Grenade enflammée brasée sur la branche supplémentaire et sur les deux branches secondaires. Plateau du type a (voir p. 21). Lame à la Montmorency, blanche, longue d'environ 70 cm et large au talon de 3,3 cm. Fourreau à trois garnitures, chape à bouton et anneau de bélière, bracelet trapézoïdal également à anneau et dard frontal et basal en laiton.

large au talon de 3,3 cm. Fourreau à trois garnitures décorées au trait et bordures découpées en accolades ; chape à bouton et anneau de bélière en laiton, bracelet trapézoïdal également à anneau et dard en laiton frontal et basal.

629

630

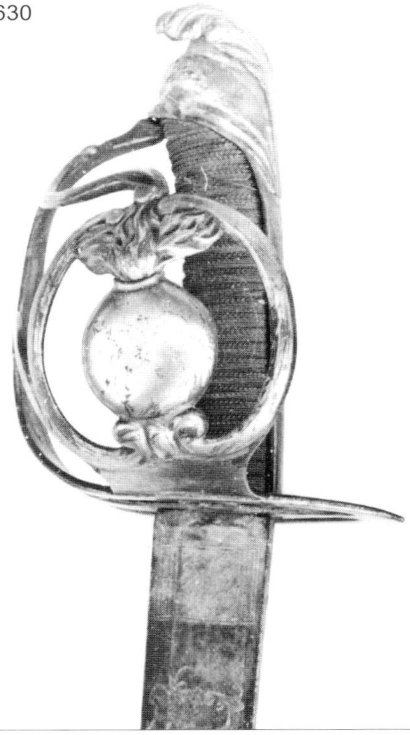

630 - Variante. Monture en laiton doré. Calotte au casque empanaché à longue queue. Poignée filigranée de nombreuses spires. Arc de jointure droite en sa partie médiane et prolongé par un plateau du type a (voir p. 21). Grand médaillon incluant partiellement les branches secondaires et fixé au plateau par une languette sur la face frontale ; dans le médaillon est inscrite une grosse grenade enflammée. Lame à la Montmorency, bleuie et gravée de rinceau dorés.

631 - Sabre d'officier de Grenadiers.
Règne de Louis XVI.
Monture en laiton doré. Calotte en casque à l'antique, décoré frontalement d'une face léonine. Longue queue de calotte unie. Poignée entièrement filigranée. Arc de jointure plein mais profondément rainuré. Coquille à la grenade enflammée sur fond de quatre drapeaux en sautoir. Plateau de garde du type d (voir p. 21). et quillon en palmette. Lame de 72 cm, large au talon de 3,4 cm, flèche de 2,5 cm, à pan creux, gouttière au fort et contre-tranchant de 22 cm environ. Cette lame est bleuie sur 24,4 cm et gravée de motifs dorés : végétaux entourant un trophée constitué d'un tambour sur fond de canon, d'étendard, de trompette...

631

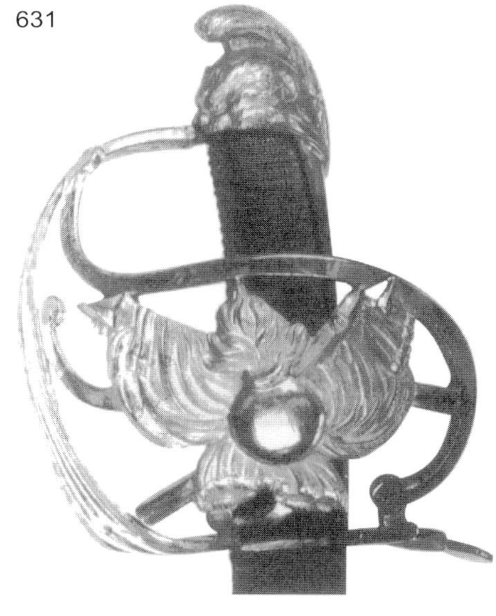

632

632 - Sabre d'officier supérieur des Volontaires Nationaux.
Après 1792.
Monture en laiton doré. Calotte à tête de lion et longue queue. Poignée encordée et filigranée de fils torsadés. Sur les branches secondaires, un trophée constitué d'une grenade placée sur deux drapeaux, avec à gauche un rameau de chêne et à droite un rameau de laurier. Arc de jointure traité façon ruban, ruban qui borde le plateau, lui-même constitué de trois soleils superposés (629a). Lame bleuie, gravée sur le *recto*, dans un cartouche
VIVRE LIBRE OU MOURIR / POUR LA NATION, LA LOI ET LE ROY
à noter que le mot ROY a été effacé après 1792 (629b). Sur le verso (629c), du talon vers la pointe, on rencontre la personnalisation de la Nation dont le nom figure sur le bouclier, un soleil, deux petits bandeaux portant : *JECLORIFIE* (sic) pour *je glorifie* sur le supérieur et sur l'inférieur *POUR LA NATION*, un trophée, un nœud de ruban, un coq surmontant une petite banderole où se trouve inscrite *LA VIGILANCE* (J.C. D.).

632a

- **a**, plateau aux soleils enrubannés ;
- **b**, *recto* de la lame ; remarquer *ET LE ROY* effacé ;
- **c**, *verso* de la lame avec deux petites banderolles, l'une portant je glorifie et l'autre pour la Nation.

632b

632c

633

633 - Sabre d'officier de Grenadiers.
Epoque révolutionnaire au Brabant.
Sabre de facture française. Monture en laiton. Calotte en casque à grille et empanaché. Poignée filigranée. Arc de jointure régulièrement arqué. Les deux branches secondaires sont travaillées, la supérieure en feuille d'acanthe, l'inférieure en rameau de laurier. Entre les deux secondaires est inscrite une grenade enflammée. Forte lame portant *VAINCRE OU MOURIR / POUR LA PATRIE* et les Armes du Brabant. Fourreau à trois garnitures.

634 - Sabre d'officier de Grenadier à cheval.
Règne de Louis XVI.
Monture en laiton doré. Calotte en dépouille de lion avec la crinière débordant sur la longue queue. Poignée ficelée à onze spires d'un filigrane formé d'un ruban de laiton bordé d'un mince fil torsadé. Arc de jointure uni. Plateau de garde du type a (voir p. 21). Motif de la coquille constitué par une grenade enflammée frappée de trois fleurs de lys. Lame à la Montmorency de 75,5 cm, large au talon de 3,8 cm, flèche de 2 cm et contre-tranchant d'environ 17 cm.

634

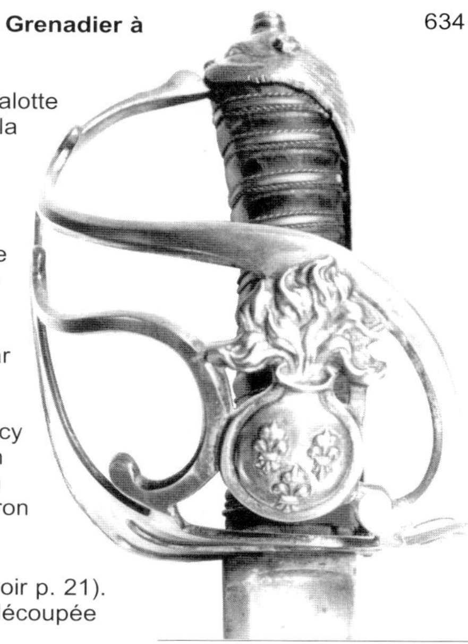

635 - Variante.
Monture comparable au sabre précédent, mais avec un plateau du type d (voir p. 21). Fourreau de cuir à deux garnitures : chape à anneau de bélière et bouton, découpée en accolade (non figuré).

636

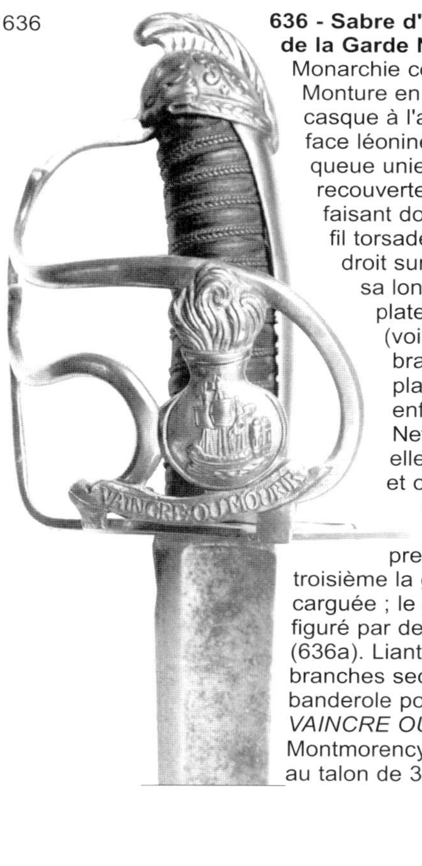

636 - Sabre d'officier de Grenadiers de la Garde Nationale de Paris.
Monarchie constitutionnelle.
Monture en laiton. Calotte au casque à l'antique, empanaché, à face léonine frontale. Longue queue unie. Poignée ficelée, recouverte de basane, à filigrane faisant douze spires d'un double fil torsadé. Arc de jointure plat, droit sur plus des deux tiers de sa longueur et suivi par un plateau de garde du type a (voir p. 21). Entre les deux branches secondaires est placée une grenade enflammée frappée de la Nef parisienne à trois mâts ; elle vogue vers la gauche et on reconnaît un hunier carré et une flamme sur chacun des deux premiers mâts et sur le troisième la grande voile est carguée ; le château arrière est figuré par deux tourelles jumelées (636a). Liant la base des deux branches secondaires, une banderole porte la devise : *VAINCRE OU MOURIR*. Lame à la Montmorency longue de 73 cm, large au talon de 3,7 cm et flèche de 1 cm.

636a

- **a**, nef parisienne sur la grenade d'un sabre du type *mineur* de la Garde Municipale de Paris, vers 1790 - 1792.

637

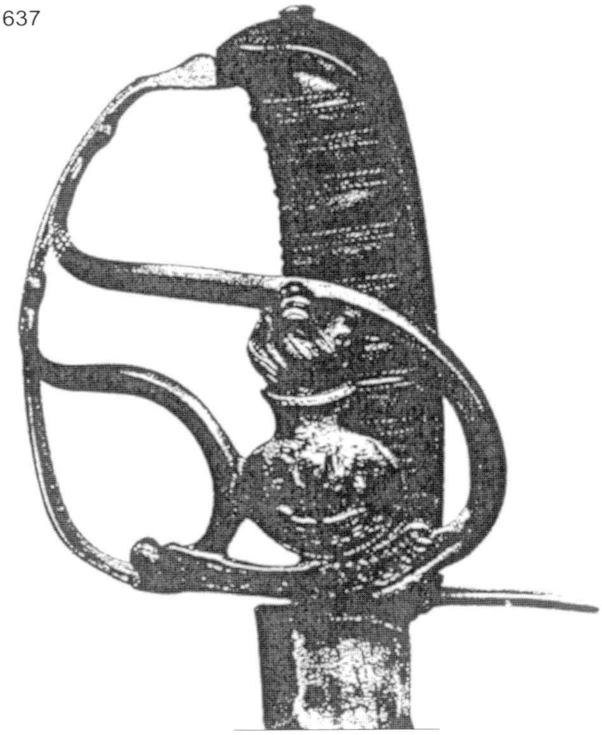

637 - Variante.
Troupe - Vers 1790.
Poignée ficelée recouverte de basane filigranée. Calotte en cuillére, à longue queue. Sous la grenade à la nef parisienne, dans une banderolle on lit : *VAINCRE OU / MOURIR* (Lavallay, 1945, op. cit., Pl. IV).

638

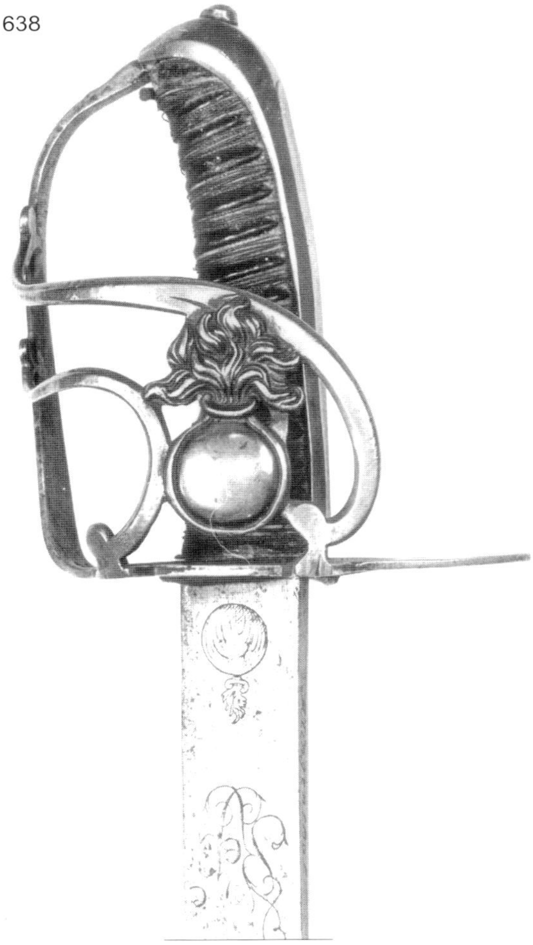

638 - Sabre de sous-officier ou d'officier de Grenadiers.
Après 1792.
Monture en laiton. Aspect général de la garde comparable à celle des sabres de Mineurs. Arc de jointure droit sur les deux tiers basaux de sa longueur. Grenade enflammée brasée sur le brin supplémentaire et sur les deux branches secondaires. Lame plate, gravée d'une grenade enflammée et de l'abréviation *REP. Fran*[se] pour Première République.

Sous-famille VII d - Au casque à l'antique

Les sabres dont la coquille s'orne d'un casque appartiennent à des officiers. Ce décor a été créé sous Louis XVI et se perpétue en période révolutionnaire. On trouve alors le casque associé à de nouveaux symboles, dont le serpent pouvant figurer le despotisme ou la médecine !. Ces sabres richement ornés resteront en service après 1792, après l'effacement des lys et du mot roy.

639

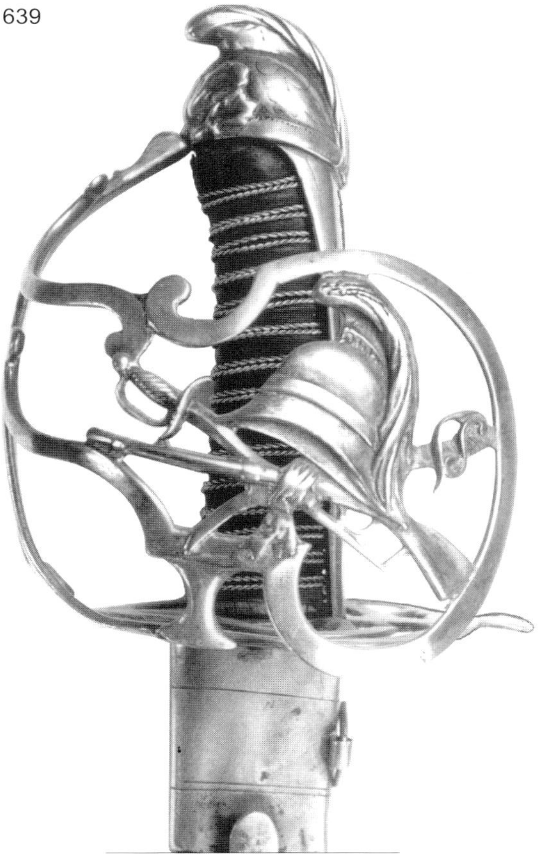

639 - Sabre d'officier monté.
Règne de Louis XVI.
Monture en laiton doré. Calotte au casque empanaché à masque léonin frontal. Poignée recouverte de basane à 15 spires d'un double fil d'argent torsadé et disposé en chaînette. Arc de jointure régulièrement arqué, découpé d'un triangle basal. Branches secondaires séparées en arrivant sur l'arc et, postérieurement, formant une sorte de médaillon dans lequel sont inscrits un casque empanaché, un sabre, une épée, un fusil, ces trois dernières armes étant nouées par un ruban. Plateau du type c (voir p. 21). Lame à la Montmorency de 75 cm, large au talon de 3,5 cm et flèche de 1,5 cm. Poinçon ovale (7,5 x 6,5 mm) bordé de perles mais illisible. Fourreau de cuir noirci à trois garnitures rainurées, à bordures découpées en accolade : chape à bouton et anneau de bélière, bracelet à anneau, petite bouterolle à dard frontal de laiton.

640

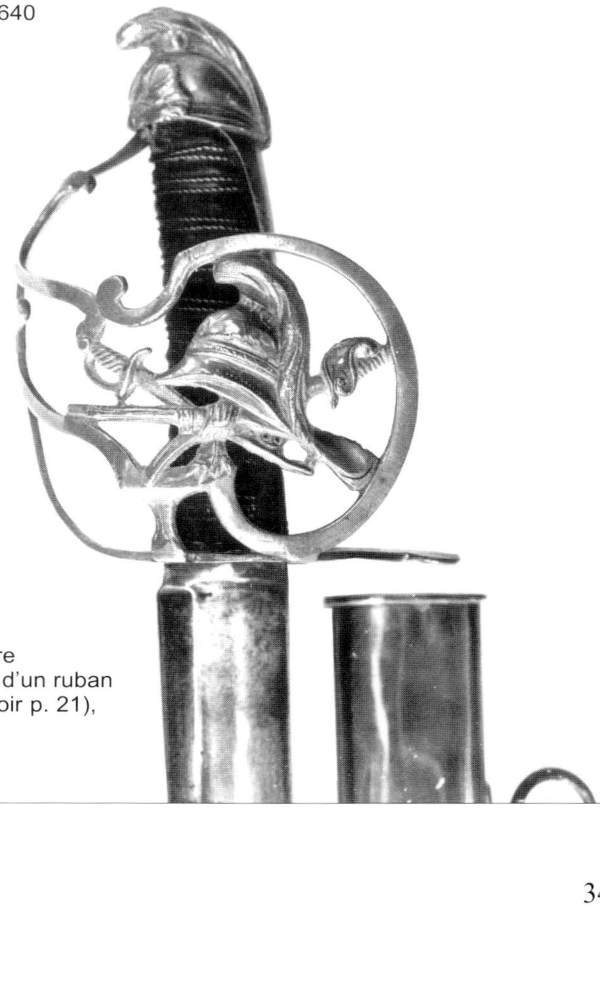

640 - Variante.
Ph. Bordes décrit (loc. cit. p. 91, fig. 69) un sabre à monture comparable mais avec des différences : filigrane composé d'un ruban de cuivre bordé de deux fils torsadés, plateau du type e (voir p. 21), lame de 78,8 cm, large au talon de 3,6 cm.

641

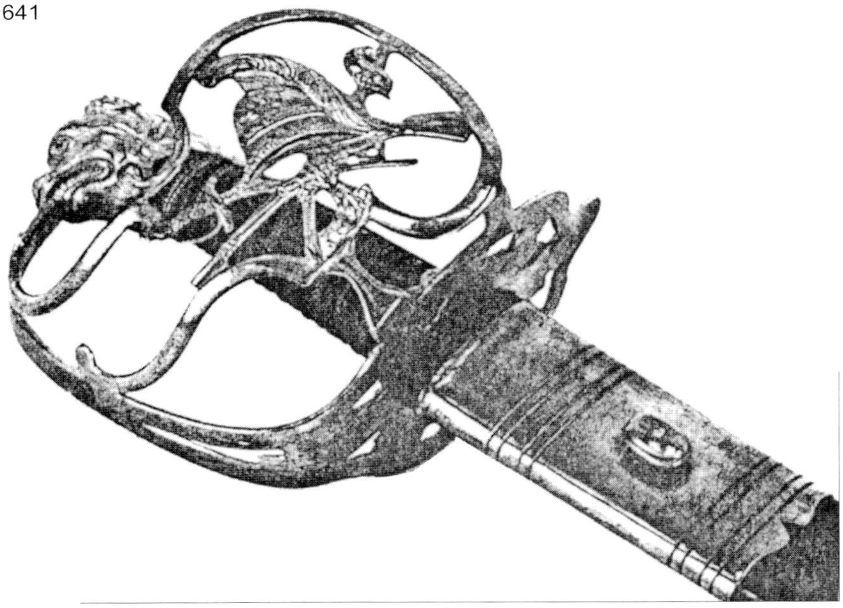

641 - Variante.
Vers 1792.
Monture en laiton doré. Calotte en tête de lion et longue queue. Arc de jointure arqué et ajouré d'un triangle basal. Poignée filigranée. Coquille au casque empanaché sur faisceau d'armes, l'ensemble lié par son milieu ainsi que chez les sabres précédents. Lame à la Montmorency gravée sur une face *O VINCERE* et sur l'autre *O MORIRE*. Fourreau de cuir à deux garnitures à bordure découpée en accolade et décorées au trait ; chape à bouton.

642

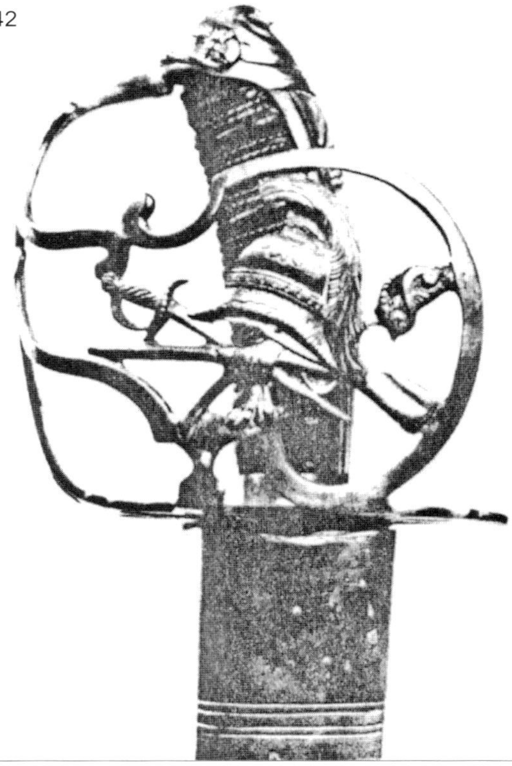

642 - Variante.
Vers 1792.
Même ornementation générale que les sabres précédents mais la calotte est en bonnet phrygien à cocarde. Lame bleuie avec l'inscription *VIVRE LIBRE OU MOURIR* en lettres d'or. Fourreau à trois garnitures et attelles.

643

643 - Sabre d'officier supérieur.
Monarchie Constitutionnelle et en usage après 1792.
Monture en laiton doré. Calotte en tête de lion et longue queue à pan. Poignée entièrement filigranée de fils en chaînette de deux grosseurs différentes. Arc de jointure régulièrement arrondi. Branches secondaires travaillées en feuilles d'acanthe pour la postérieure et en rameau de laurier pour l'antérieure, entourant un casque empanaché à jugulaire (643a). Plateau de garde constitué de deux rameaux de laurier liés par un ruban au niveau du bec, l'ensemble étant ciselé à jour. Lame de 67,3 cm, large au talon de 4,2 cm du type Montmorency, bleuie avec rinceaux, trophée... et une fleur de lys légèrement rayée près du talon. Dos de la lame signé *Klingenthal.* Fourreau à deux garnitures simplement ornées au trait, chape à bouton et anneau de bélière.

- **a**, détail de la garde. Noter sur le talon de la lame, la fleur de lys légèrement biffée.

643a

644

644 - Sabre d'officier général.
Monarchie constitutionnelle.
Monture en laiton doré. Calotte à la tête de lion et longue queue à filets. Filigrane constitué d'un ruban de laiton bordé d'un fil torsadé. Entre les deux branches secondaires est inséré un casque empanaché à jugulaire, sur fond de rameaux de laurier et d'une lance passant dans l'anneau formé par un serpent (*). Plateau ciselé à jour de deux rameaux de laurier à partir de la fusée et sur la partie postérieure se trouve une cuirasse, un casque à jugulaire et un médaillon à tête de gorgone grimaçante (644a). Lame à la Montmorency, bleuie décorée de motifs végétaux, de trophée et dans deux couronnes enrubannées, sur une face, sous une fleur de lys, : *l'Honneur / est Mon guide* (644b) et sur l'autre *Pour / La Nation / La Loy / et / Le Roy (*644c). Fourreau à trois garnitures simplement décorées au trait, chape à bouton et anneau de bélière.

- **a**, vue du plateau montrant son ornementation découpée, avec au premier plan, la tête de méduse ;
- **b**, *recto* de la lame richement décoré avec la devise de l'officier : *L'honneur / est Mon / Guide* ;
- **c**, *verso* de la lame avec la profession de foi : *Pour / la Nation / La loy / et / Le Roy.*

644a

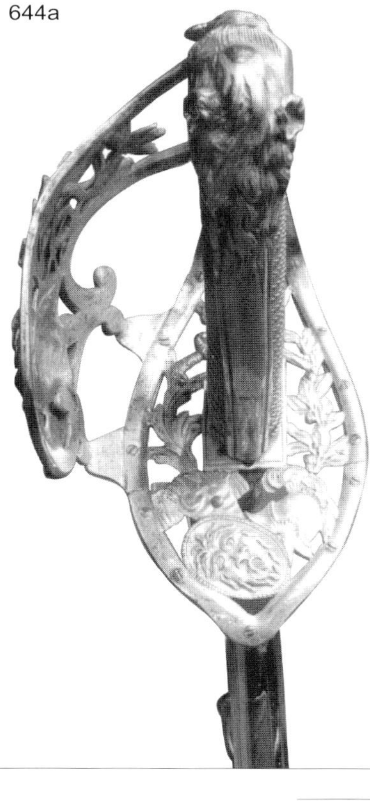

644b

644c

(*) Le serpent pourrait, selon G. Lavallay, symboliser la médecine. Il écrit : *Faut-il voir dans ce reptile un emblème de la médecine ? Peut-être ; car un sabre analogue, figurant dans le Catalogue du Musée de L'Artillerie, appartenait à Joseph Biston, Administrateur général des Hôpitaux.*

645

645 - Variante.
Monture simplifiée. Le plateau est au soleil sans autres fioritures (J.C. D).

646

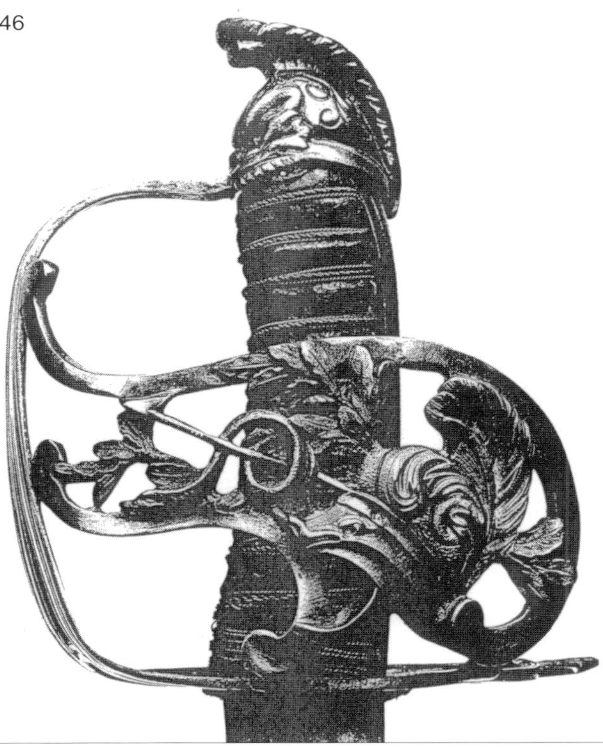

646 - Variante.
Ce sabre d'Etat-Major est vraisembablement d'avant 1792. Monture en laiton. Calotte en casque empanaché. Poignée recouverte de basane filigranee d'un fil torsadé plus gros que les deux autres fils torsadés l'accompagnant. Entre les deux branches secondaires même décor que pour les trois sabres précédent (Lavallay, op. cit., Pl. IX).

Sous-famille VII e - A la cuirasse à l'antique

Avant 1792, la cuirasse ou corselet de fer est surmontée d'un casque empanaché. Après cette date, le casque est remplacé par le bonnet phrygien. Ces sabres étant choisis par des officiers supérieurs, voire d'Etat-Major, sont finement ciselés et les calottes sont au casque, à la tête de lion, à la tête d'aigle…

647

647 - Sabre d'officier supérieur.
Règne de Louis XVI.
Monture en laiton doré. Calotte au casque empanaché et longue queue. Poignée filigranée de deux fils de laiton torsadés sur une basane encordée. Coquille à la cuirasse à l'antique et au casque empanaché, placée sur un faisceau de quatre drapeaux, d'une épée, d'un fusil et d'un tambour comme éléments décoratifs principaux. Lame à la Montmorency, blanche.

Variantes - Il existe des modèles dont la calotte est en tête de lion.

648

648 - Variante. Ce sabre du règne de Louis XVI, présente une monture comparable à celle du précédent mais il est plus finement ciselé et la calotte est en casque à cimier. La similitude des coquilles apporte la preuve que ce modèle fut réglementaire ou, pour le moins, traditionnel.

649

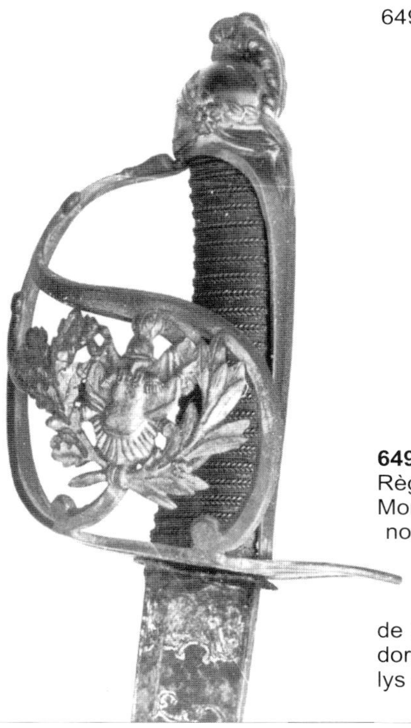

649 - Sabre d'officier supérieur.
Règne de Louis XVI, en usage avant et après 1792.
Monture en laiton doré. Calotte en casque à grille et longue queue. Poignée à nombreuses spires d'un filigrane en chaînette. Arc de jointure arrondi. Entre les deux branches secondaires, une cuirasse au casque empanaché reposant sur quatre drapeaux à pic en losange, l'ensemble flanqué d'un rameau de chêne (à gauche) et de laurier (à droite). Plateau du type a (voir p. 21). Lame de 79,3 cm, large au talon de 3,5 cm, à la Montmorency, bleuie gravée de motifs dorés et portant sur la face externe *POUR LA NATION* avec une petite fleur de lys et sur l'autre face *LA LOI et LE ROY*, le mot *ROY* ayant été gratté en 1792.

Sous-famille VIII a - Aux vainqueurs de la Bastille

On rencontre de rares sabres commémorant la prise de la Bastille, le 14 juillet 1789. Selon toute vraisemblance, ces sabres sont portés par des officiers de la Garde Nationale. Les sabres dont nous avons retrouvé l'existence sont les suivants :

664 - Sabre d'officier de la Garde Nationale ou des Volontaires Nationaux. Monture en laiton doré. Calotte à tête de lion et longue queue. Poignée filigranée. Arc de jointure légèrement arqué et suivi d'un plateau de garde du type d (voir p. 21). Les deux branches secondaires, en deux parties, arrivent sur l'arc en crosse et s'arrondissent sans discontinuité sur l'arrière. Le motif majeur ornant la coquille est constitué par la façade d'une forteresse dont une tour est surmontée d'un coq et l'autre du bonnet phrygien ; sur la gauche, on reconnaît une hache, une baïonnette, un canon, cinq boulets, un tonneau de poudre, un drapeau, des piques... Lame blanche, longue de 81 cm, large au talon de 4 cm, à pan creux du type des lames de Chasseurs à cheval. Ce sabre est conservé au Musée Carnavalet de Paris (inv. OA 94). Voir également Ph. Bordes, op. cit. p. 55, fig. 10 (non figuré).

665

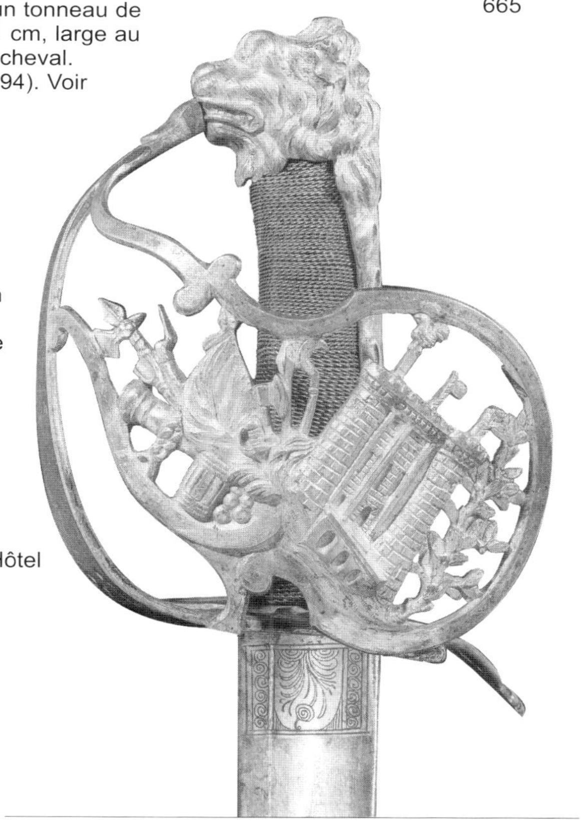

665 - Variante.
Même monture que l'exemplaire précédent, mais le bonnet phrygien surmontant la tour de droite a été brisé, volontairement ou non. Lame gravée et dorée sur la moitié des sa longueur et portant l'inscription *LA VIGILANCE / ET LA FORCE*, avec un coq symbolisant la vigilance et un lion, la force. Fourreau de cuir, à quatre garnitures.

Cet exemplaire a été présenté à l'Hôtel Drouot, le 25 janvier 1988.

666

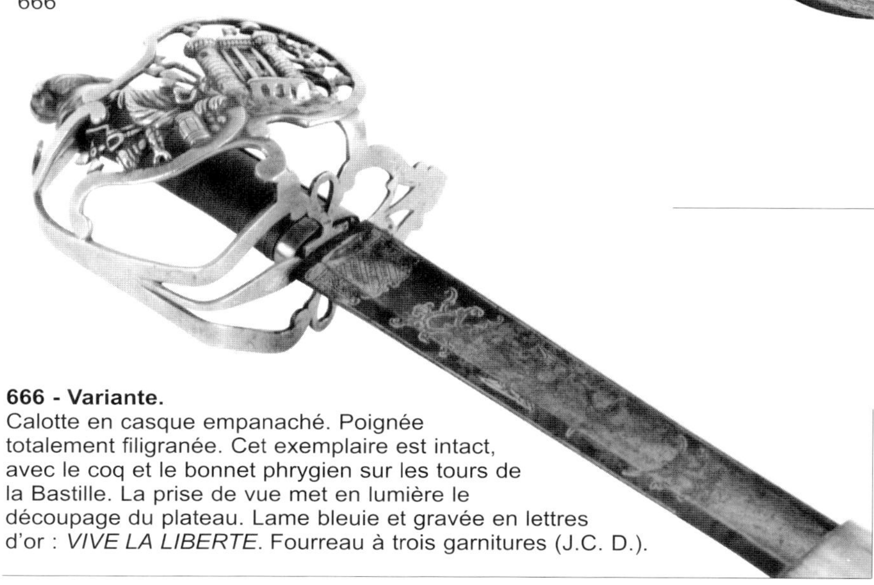

666 - Variante.
Calotte en casque empanaché. Poignée totalement filigranée. Cet exemplaire est intact, avec le coq et le bonnet phrygien sur les tours de la Bastille. La prise de vue met en lumière le découpage du plateau. Lame bleuie et gravée en lettres d'or : *VIVE LA LIBERTE*. Fourreau à trois garnitures (J.C. D.).

667 - Variante.
Coquille identique à l'exemplaire précédent mais avec une calotte en casque à grille. Plateau du type d (voir p. 21), terminé par un quillon recourbé. Lame de 76,2 cm, large au talon de 3,3 cm, gravée sur la face interne *BASTILLE 14 JUILLET 1789* (non figuré).

668 - Autre modèle.
Calotte à tête de lion. Plateau du type d. Entre les branches secondaires cannelées est placée une petite Bastille d'un modèle différent de celui des modèles précédents. De plus, une banderole part du sommet de la branche secondaire supérieure et s'étend sous la Bastille. Sur cette banderole, on lit : *VIVRE LIBRE OU MOURIR*. Lame de 68 cm, large au talon de 3,3 cm, à la Montmorency gravée sur la face externe d'un coq et d'un lion affrontés ; sur la face interne figure une femme représentant la Nation soulignée d'une banderole où on lit *LIBERTE* (voir Ph. Bordes, op. cit. p. 55, fig. 11).

668

On trouve encore un sabre dit aux trois médaillons : le supérieur est au portrait de La Fayette, les deux autres, placés au-dessous, figurent, l'un à droite, la Bastille et l'autre les Armes de Paris. (voir ci-après).

Sous-famille VIII b - A La Fayette

Un hommage est rendu à La Fayette en le faisant figurer sur la coquille des sabres de la Garde Nationale vers 1790. Le profil du commandant en chef de la Garde Nationale est inscrit dans un ovale flanqué de drapeaux. Les calottes des montures en laiton doré ou non, sont en casque avec ou sans grille, en tête de lion. La disgrâce de La Fayette en 1792, se traduira par l'effacement ou le découpage des médaillons sur lesquels il figurait…

669

669 - Sabre d'officier de la Garde Nationale.
Avant 1792.
Monture en laiton doré. Calotte en tête de lion, à longue queue cannelée longitudinalement. Poignée recouverte de basane filigranée de deux fils torsadés disposés en chaînette. Arc de jointure arqué, plateau découpé de deux faisceaux de licteur surmontés de haches inversées du type i (voir p. 21). Branche secondaire supérieure en cercle et rejoignant le plateau avant de continuer pour former la secondaire inférieure. La coquille est constituée du médaillon ovale au portrait de La Fayette, surmonté d'un nœud de ruban, placé au-dessus de deux médaillons légèrement plus petits figurant les Armes de Paris et la porte de la Bastille, l'ensemble sur fond de six drapeaux ; sur la gauche, le petit coq national ; sous les médaillons, deux feuilles d'acanthe. Lame de 74,2 cm, large au talon de 3,3 cm.

670 - Variante.
Il s'agit du même sabre que précédemment, mais le portrait de La Fayette a été remplacé par une plaque de métal unie. A noter que le plateau de garde est orné d'une hache, de deux grenades et de deux épées croisées en découpe (voir p. 21 g). La monture de ce sabre est figurée par Ph. Bordes (op. cit., p. 59, fig. 18a), (non figuré).

671

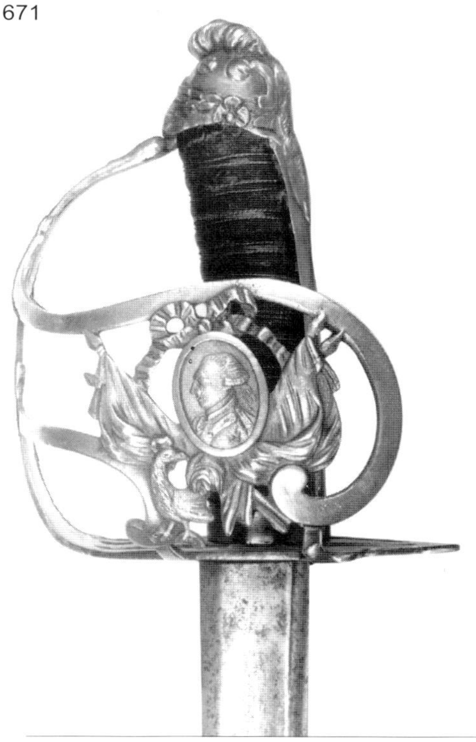

671 - Sabre d'officier de la Garde Nationale.
Avant 1792.
Monture en laiton doré. Calotte en casque empanaché et longue queue unie. Poignée filigranée d'un ruban de cuivre rouge bordé d'un fin fil torsadé et virole basale. Arc de jointure plein et uni suivi d'un plateau du type a (voir p. 21). Entre les deux branches, profil de La Fayette en habit de commandant en chef de la Garde Nationale (1) dans un médaillon surmonté d'un noeud de ruban. Ce médaillon est flanqué de quatre drapeaux avec au-dessous, du côté gauche, le petit coq national. Lame à la Montmorency, longue de 73 cm, large au talon de 3,4 cm, flèche de 3 cm.

672

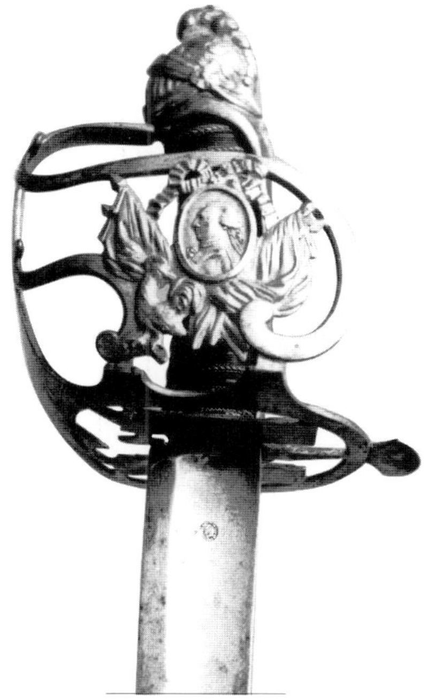

672 - Variante.
Avant 1792.
Monture identique au sabre précédent, mais avec une calotte en casque à grille, un plateau de garde garni de deux faisceaux de licteur avec haches affrontées et à l'arrière la massue d'Hercule. Quillon en palmette précise Ph. Bordes (op. cit., p. 59, fig. 17).

(1) Marie, Joseph, Paul, Yves, Roch, Gilbert MOTIER, marquis de La Fayette est né en Auvergne à Chaviniac en 1757 et mort à Paris en 1834. Les principaux évènements de sa longue carrière sont les suivants :
En 1773, il est nommé lieutenant et en 1777, ami de Franklin, il part pour l'Amérique aider les insurgés contre les Anglais. Revenu en France, en 1782, il devient maréchal de camp. En 1789, élu député de la Noblesse aux Etats Généraux, appartient à une loge franc-maçonne et le 15 juillet, il reçoit le commandement de la Garde Nationale. Partisan d'une monarchie parlementaire, le 17 juillet 1791, il fait tirer sur les manifestants du Champ de Mars qui demandent la destitution de Louis XVI. Il fonde le Club des Feuillants (ou Fayettistes). Nommé commandant de l'Armée du Centre puis du Nord, il se prononce contre la destitution du roi. Suspetct aux yeux des factions révolutionnaires, il doit fuir à l'étranger. Il sera interné cinq années dans les cachots autrichiens. La Fayette reste assez discret jusqu'aux Cent Jours (20 mars - 22 juin 1815) pendant lesquels il demande l'abdication de Napoléon 1er. Il revient à la vie publique sous la Restauration et en 1818, il se fait élire député de la Sarthe, puis en 1827, député de Meaux. Pendant cette période, il adhère à la Charbonnerie et à ce titre participe à plusieurs insurrections avortées. En 1830 il apporte son concours à l'élimination de Charles X du trône de France.

673

673 - Variante.
Avant 1792.
Lavallay (La Giberne, 1903-1904, IV, p. 146) décrit un plateau de garde en ces termes *" Plateau ajouré et découpé de manière à reproduire la devise si usitée alors : Vaincre ou Mourir "* (*in* Lavallay (op. cit. pl. VI).

674 - Variante.
Avant 1792.
Même décoration que les autres variantes, mais ici, la calotte est en tête de lion. De plus la forme de l'arc de jointure est différente : il arrive en angle droit sur le plateau de garde et la branche secondaire est masquée par le décor de la coquille.

674

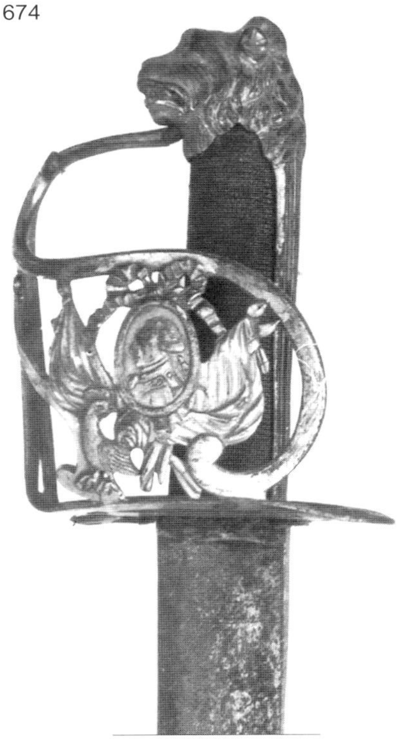

675 - Variante.
Après 1792.
En fait, il ne s'agit pas d'une *variante*, mais d'une amputation signalée par Ph. Bordes : le médaillon au portrait de La Fayette a été découpé et enlevé. En revanche, une grenade enflammée se trouve entre l'arc de jointure et la secondaire inférieure (Ph. Bordes, loc. cit. p. 59, fig. 17a). Lame de 71,4 cm, large au talon de 3,3 cm (non figuré).

676

676 - Variante.
Après 1792.
Modèle dit au portrait de La Fayette mais dont le visage a été martelé. Plateau de garde aux faisceaux de licteur. Sur la lame persiste l'inscription, sur une face *LA NATION. LA LOI . ET. LE ROI* et sur l'autre *VAINCRE OU MOURIR*. Fourreau de cuir à deux garnitures (J.C. D.).

677 - Variante.
Après 1792.
Exemplaire dont le médaillon au portrait de La Fayette a été remplacé par un autre en cuivre rouge au portrait de la République.
Monture en laiton, calotte en casque empanaché, plateau ajouré du type d (voir p. 21). Lame à la Montmorency, blanche. Fourreau à deux garnitures.

677

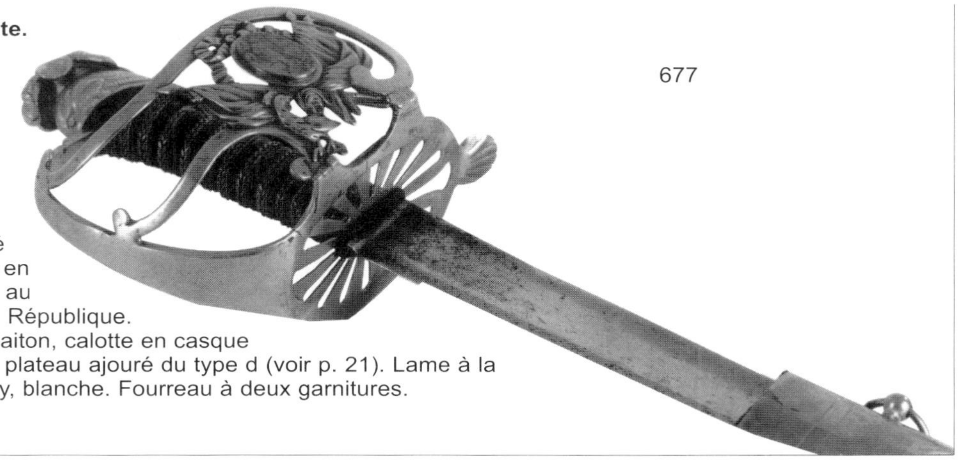

Sous-famille VIII c - Aux symboles maçonniques

Les symboles maçonniques, sous la forme du triangle ou des mains, paume contre paume, figurent sur quelques sabres de la Garde Nationale ou des Volontaires. Les calottes sont alors en tête de lion, de coq. Ces sabres sont particulièrement soignés.

678

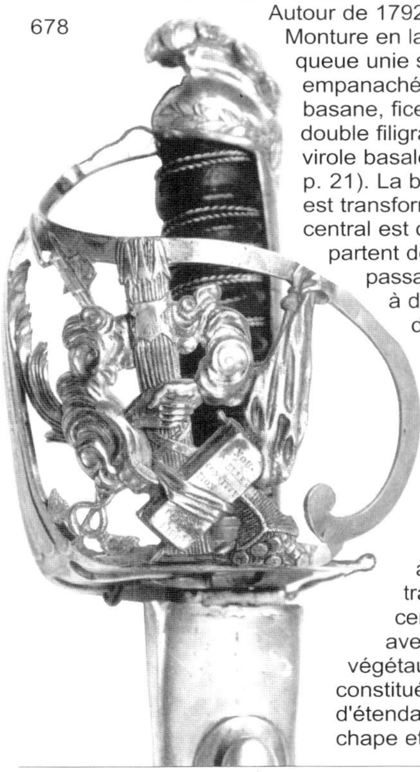

678 - Sabre d'officier de la Garde Nationale
Autour de 1792.
Monture en laiton doré. Calotte à longue queue unie suivant un casque empanaché. Poignée recouverte de basane, ficelée, à neuf spires d'un double filigrane de laiton torsadé et virole basale. Plateau du type a (voir p. 21). La branche secondaire basale est transformée en nuages. Le motif central est constitué de nuages d'où partent deux mains maçonniques passant sur un faisceau de licteur ; à droite, un étendard surmonté d'un bonnet phrygien ; au-dessous, un livre porte l'inscription *NOUVELLE CONSTITUTION D'UN PEUPLE LIBRE*, encadrée d'une corne d'abondance et du caducée ailé. Lame longue de 73 cm, large au talon de 3,3 cm, flèche de 1,5 cm, à pan creux et contre-tranchant d'une vingtaine de centimètres, bleuie sur 22,5 cm avec des gravures à motifs végétaux flanquant un trophée constitué de tambour, de trompettes, d'étendards,... Fourreau à bouton de chape et bouterolle sans dard.

679 - Sabre d'officier des Volontaires Nationaux.
Autour de 1790.
Monture en laiton doré. Calotte en tête de coq, à longue queue emplumé. Médaillon au triangle maçonnique coiffé du bonnet phrygien et flanqué de deux rameaux de laurier (d'après Ph. Bordes, op. cit.).

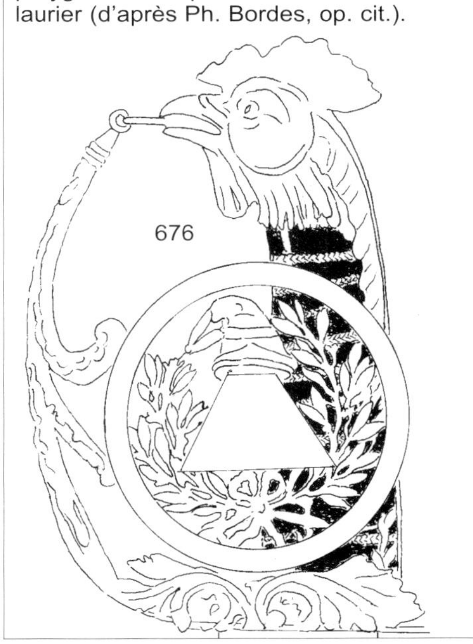
676

680 - Variante.
Ph. Bordes (op. cit. p. 73, fig. 41) figure un sabre à coquille voisin du n° 678 conservé au Musée Carnavalet de Paris (inv. OA 115). Ce sabre à monture en cuivre rouge est à calotte en tête de lion et à longue queue. La poignée est recouverte de basane, à seulement six spires d'un gros fil torsadé. Les principales différences par rapport à l'exemplaire précédemment décrit réside dans l'intitulé du livre qui ne porte que *La Loi* et dans la corne d'abondance d'où ne s'échappe qu'un fruit. Lame de 80,7 cm, large au talon de 4,3 cm, du type hussard. Sur la face interne, en plus de trophées et de rinceaux, on lit *POUR LA LOI* et sur la face externe *POUR LA PATRIE* (non figuré).

Sous-famille VIII d - Au bonnet phrygien

Le bonnet phrygien est rarement présenté seul comme symbole révolutionnaire et ne vient qu'en complément du décor principal. Aussi, le trouvera-t-on figuré dans d'autres Sous-familles, présentant des sabres après 1792.

681

681 - Sabre d'officier de la Garde Nationale.
Autour de 1792.
Monture en laiton. Calotte à longue queue, à tête de lion. Poignée recouverte de basane à nombreuses spires d'un double filigrane en chaînette. Arc de jointure plein, suivi d'un plateau ajouré. Médaillon à chaîne brisée, symbole de la liberté, dans lequel est inscrit la massue d'Hercule coiffée du bonnet phrygien et de fins rameaux, le tout en découpe. Sur la branche secondaire supérieure est inscrit *VIVRE LIBRE . OU MOURIR.* Lame à pans creux, bleuie, décorée de feuillages dorés et de la devise *VAINCRE OU MOURIR.*

682 - Sabre d'officier de la Garde Nationale ou de Volontaires.
Après 1792.
Monture en laiton doré. Calotte en casque à grille et empanaché, à longue queue. Poignée recouverte de basane à nombreuses spires, d'un double fil torsadé disposé en chaînette. Arc de jointure peu courbé se poursuivant par un plateau voisin du type d (voir p. 21). Les branches secondaires sont travaillées en rameaux à feuilles d'acanthe et maintiennent en leur centre un petit bonnet révolutionnaire (679a). Lame à la Montmorency de 71 cm, large au talon de 3,4 cm, bleuie et gravée de motifs dorés : du côté interne figure un médaillon ovale dans lequel se trouve un cœur posé sur un brasier avec au-dessus, des éclairs sortant des nuées, puis deux bandeaux dans lesquels on lit *JE SUIS FERME / COMME UNE ROCHE* et du côté externe, après un trophée composé de deux timbales, d'une trompette et de deux guidons, on lit *POUR LE SALUT / DE MA PATRIE* (Musée de l'Armée).

682a

Croquis du motif central qui peut être assimilé à un bonnet

683

683 - Sabre d'officier des Volontaires Nationaux.
Vers 1792.
Ce sabre est un montage d'époque révolutionnaire : sur une monture à arc de jointure arrivant en angle droit sur la croisière à oreillon, est soudé sur les deux branches secondaires un médaillon au trophée d'armes, de drapeaux, de feuillage surmonté d'un bonnet phrygien. Sur la banderolle bordant ce trophée, on lit : *VIVRE LIBRE OU MOURIR* (J.C. D.).

Sous-famille VIII e - Aux animaux

Le décor aux animaux est assez fréquent et intéresse le cheval, le lion, le coq, l'aigle… mais aussi certains animaux mythiques comme la chimère, l'Hydre de Lerne… Ainsi que dans les autres Sous-familles des sabres de cette époque, la calotte est soit en tête de lion, en casque à l'antique ou des plus simples. Parfois les branches secondaires sont ciselées d'ornements et quand il s'agit de feuillages, il n'est pas aisé d'identifier le nombre des branches secondaires.

Au cheval et au cavalier

684

684 - Sabre d'officier de Cavalerie des Volontaires Nationaux.
Règne de Louis XVI, vers 1790.
Monture en laiton dite au *cavalier sabrant*. Calotte en tête de lion, à longue queue. Poignée filigranée. Arc de jointure régulièrement arqué, suivi d'un plateau du type d (voir p. 21). Coquille à grand motif soutenu par les deux branches secondaires, représentant un cavalier chargeant, le sabre levé ; à ses pieds, un cheval et son cavalier sont abattus ; sur la gauche, un fort tronc de chêne avec ses ramures. Lame anciennement gravée de motifs aujourd'hui effacés. Fourreau de bois recouvert de cuir, à attelles et trois garnitures décorées au trait.

685

685 - Variante.
Monture voisine de celle du sabre précédent mais le tronc de chêne est remplacé par des rameaux feuillus.

686 - Variante.
Cette variante est décrite et figurée par Ph. Bordes (op. cit. p. 84, fig. 54). La monture est en laiton, la calotte est en casque à grille. Le motif de la coquille est un cavalier portant *habit, chapeau, culotte et bottes* écrit l'auteur. Le cavalier renversé est coiffé d'une sorte de schako. La lame à la Montmorency apporte les plus intéressantes précisions.
Ph. Bordes les décrit ainsi *" A l'intérieur, cartouche rocaille ; inscription LA LOI / LA PATRIE... A l'extérieur, trompette et fleurette ; " Américain " avec coiffure à plumes dressées, vêtements à franges... "*
La figuration d'un Indien rappelle la Guerre d'Indépendance. Ce sabre n'a-t-il pas été conçu en mémoire de la Campagne américaine engagée par Louis XVI ? Et pris comme symbole de la liberté ? (non figuré).

Au lion grimpant

687

687 - Sabre d'officier de la Garde Nationale ou des Volontaires Nationaux dit au lion grimpant.
Vers 1790 - 1792.
Monture en laiton. Calotte en cuillère simple à longue queue unie. Poignée encordée recouverte de basane et filigranée d'un fil de laiton torsadé. Arc de jointure se prolongeant par un plateau ajouré du type a (voir p. 21). Entre les deux branches secondaires, petit *lion grimpant* sur un tronc d'arbre. Entre l'arc et la branche secondaire inférieure, est inséré un petit coq. Lame à la Montmorency, blanche. Fourreau à deux garnitures fixées par martelage, chape à anneau de bélière.

688 - Variante.
Ph. Bordes (op. cit. p. 81, fig. 46) signale un sabre à décor identique mais dont la lame longue de 72,1 cm et large au talon de 3,2 cm est bleuie, gravée de rinceaux sur les deux faces et ajoute l'auteur *" au talon* (ornée) *d'un cartouche garni de canons, piques à large fer et drapeau "* (non figuré).

689

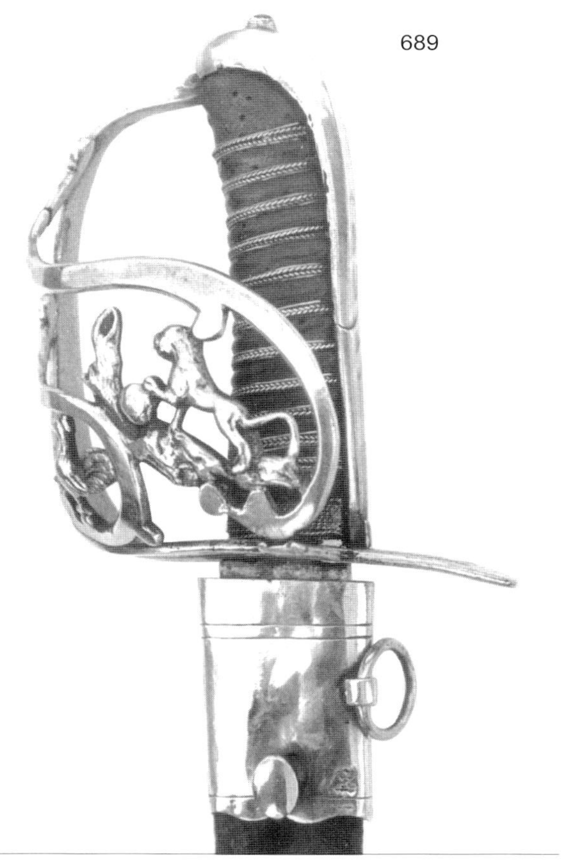

689 - Variante.
Monture en laiton. Il semble que ce sabre de l'ancien Régime ait été actualisé sous la Révolution en y ajoutant la coquille emblématique au lion grimpant et au coq, très à la mode après 1789. Chape à bouton et anneau.

690

690 - Variante.
Calotte en tête de lion, à longue queue rainurée. Même décor que les précédents. Lame blanche à la Montmorency, de 65,5 cm, large au talon de 3,5 cm et flèche de 2,8 cm, sur laquelle on lit *LA PATRIE.* Fourreau à deux garnitures unies, à bordures découpées en accolade : chape de 12,3 cm, à bouton en écusson, bouterolle à longueur maximum de 17 cm, dard frontal en laiton.

691 - Variante.
Vers 1790.
Ce modèle possède une calotte en casque empanaché et la lame porte l'inscription *Vaincre ou mourir* en petites capitales (G. Lavallay, op. cit. pl. XII).

691

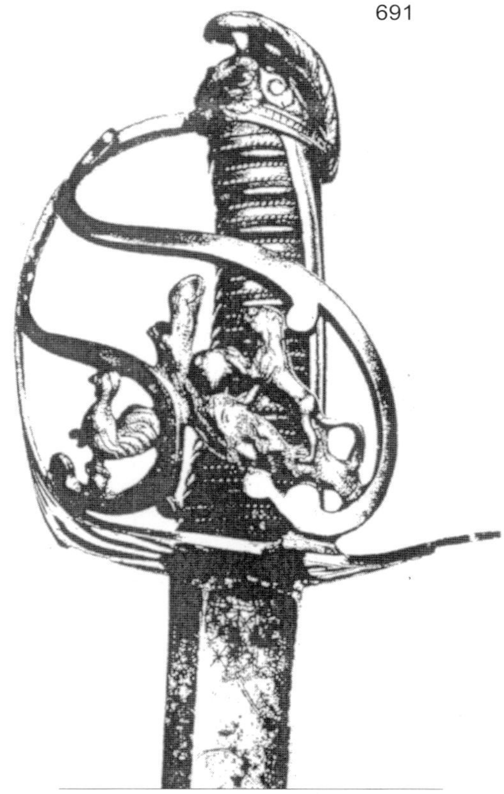

692

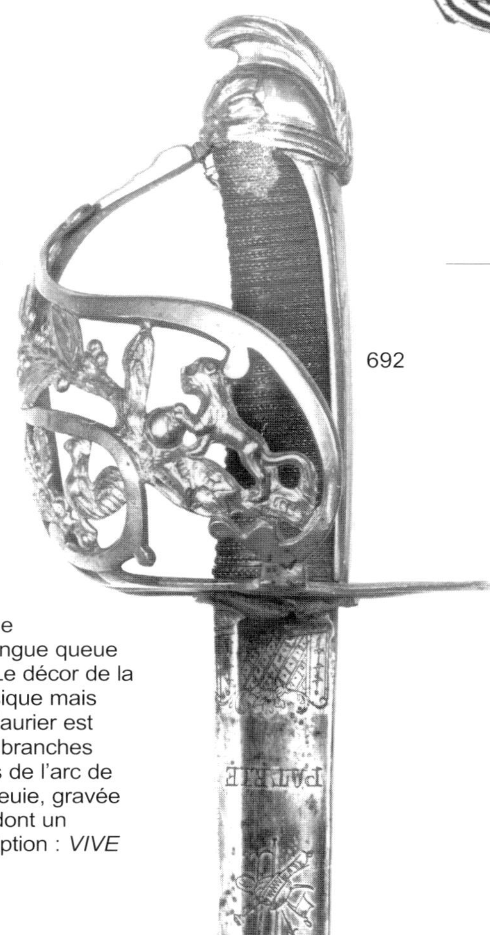

692 - Variante.
Vers 1790.
Calotte en casque empanaché, à longue queue ornée d'un filet. Le décor de la coquille est classique mais une branche de laurier est inscrite entre les branches secondaires près de l'arc de jointure. Lame bleuie, gravée de motifs dorés dont un trophée et l'inscription : *VIVE LA PATRIE..*

693

Au lion marchant

693 - Sabre d'officier de la Garde Nationale ou des Volontaires Nationaux.
Avant 1792.
Monture en laiton doré. Calotte au casque à grille à panache et longue queue. Poignée ficelée à neuf spires d'un filigrane torsadé. Arc de jointure peu arqué en sa partie médiane et plateau du type d (voir p. 21). Branches secondaires bien séparées supportant un lion placé horizontalement portant sur l'épaule droite la massue d'Hercule surmontée du bonnet phrygien tenue par la patte droite. Devant le lion, un arbre qui pourrait être un chêne. Lame à la Montmorency, blanche. Chape du fourreau à anneau et crochet, vraissemblalement rapporté.

694 - Variante.
Avant 1792.
Sabre à coquille au même décor que le précédent avec les différences suivantes : la calotte est en tête de lion, la poignée entièrement filigranée de fin fil de laiton et surtout les gravures de la lame qui permettent de préciser l'époque de sa mise en service. En effet, on lit l'inscription : *POUR / LA / NATION / LA LOI / ET LE ROY* sur une face et sur l'autre des gravures représentant le soleil, une allégorie représentant la nation, des fleurs de lys... (Ph. Bordes, op. cit. p. 79, fig. 44).

694

695

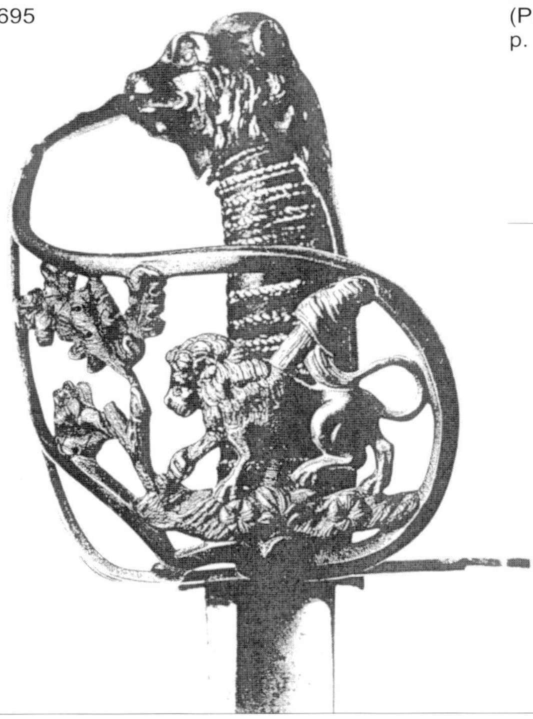

695 - Variante.
Même décor que le précédent. Poignée recouverte de basane filigranée d'un double fil de laiton torsadé. Lame à pan creux (Lavallay, op. cit. pl. X).

696

696 - Sabre d'officier de la Garde Nationale ou des Volontaires Nationaux.
Autour de 1792.
Monture en laiton. Calotte au casque à masque frontal et longue queue. Poignée recouverte de basane, à filigrane en ruban de cuivre bordé de deux fils torsadés. Arc de jointure droit en sa partie médiane, suivi d'un plateau découpé du type d (voir p. 21) et bec en palmette. Entre les deux branches secondaires est inscrit un lion passant sur un fond de feuilles de chêne, précédé d'un rameau (ou d'un arbre) de la même espèce. Lame blanche de 72 cm, large au talon de 3,2 cm.

Au coq

On a déjà vu le coq comme symbole secondaire. Chez les sabres regroupés ci-après, le coq national devient l'emblème principal.

697

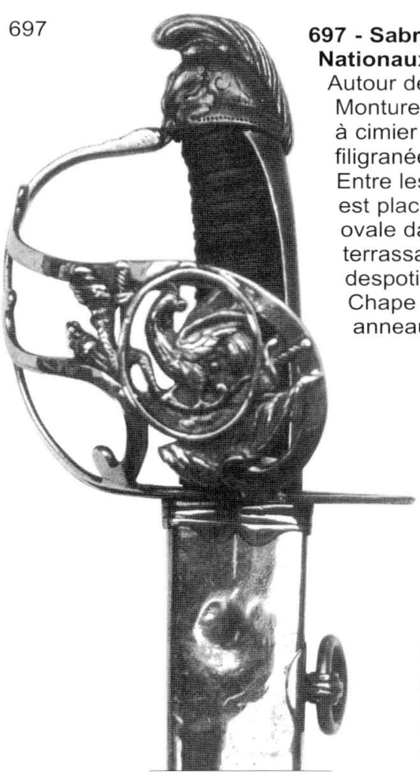

697 - Sabre d'officier des Volontaires Nationaux.
Autour de 1792.
Monture en laiton. Calotte en casque à cimier et longue queue. Poignée filigranée. Arc de jointure arrondi. Entre les deux branches secondaires est placé un médaillon légèrement ovale dans lequel est inscrit un coq terrassant le serpent symbolisant le despotisme. Lame à la Montmorency. Chape et fourreau à bouton et anneau de bélière.

698

698 - Variante.
Avant et après 1792.
Cet exemplaire possède une calotte en tête de lion. La coquille présente la même ornementation que le modèle précédent.

699 - Variante.
Avant et après 1792.
Ce sabre décrit par Ph. Bordes (op. cit. p. 82, fig. 52) présente les particularités suivantes : calotte en tête d'aigle, à longue queue, arc de jointure, à gorge et rainures latérales à sa base et suivi d'un plateau du type d (voir p. 21). On retrouve le coq du médaillon terrassant le serpent et les mêmes symboles latéraux des deux sabres précédents mais placés dans un ordre différent à gauche. Lame du type des Chasseurs à cheval, longue de 80 cm, large au talon de 3,7 cm, bleuie et gravée en lettres d'or *VAINCRE OU MOURIR* accompagné de trophée, d'un sabre à la turque... Cette description a été modifiée en 1792 : la devise a été effacée, ne laissant que le mot *VAINC*.

699

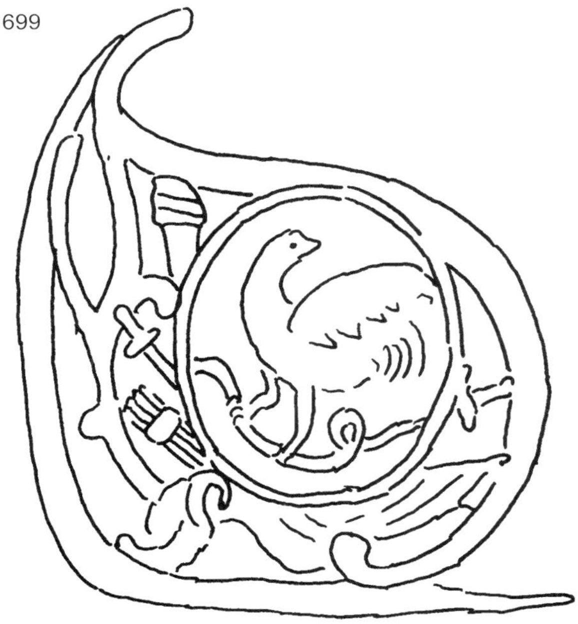

Dessin schématique de la garde au coq (d'après Ph. Bordes, 1987).

700

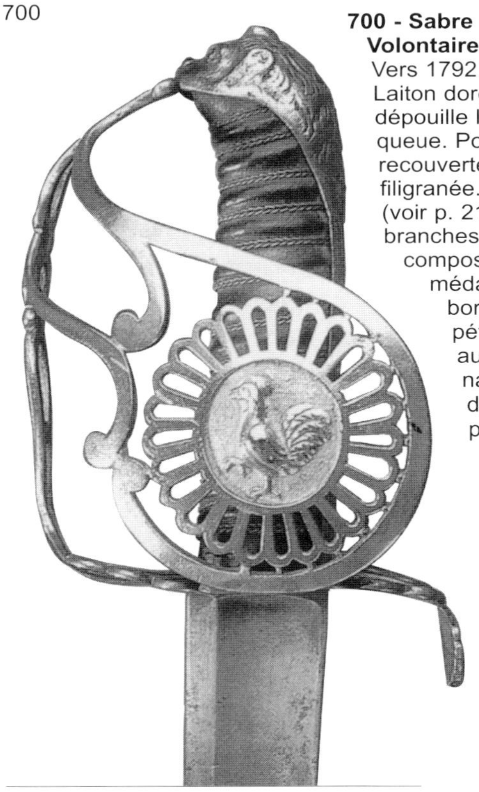

700 - Sabre d'officier des Volontaires Nationaux.
Vers 1792.
Laiton doré. Calotte en dépouille léonine et longue queue. Poignée ficelée, recouverte de basane et filigranée. Plateau du type d (voir p. 21). Entre les deux branches secondaires composites est inscrit un médaillon à large bordure à nombreux pétales ajourés avec au centre plein, le coq national, la patte droite levée. Lame à pan creux.

701

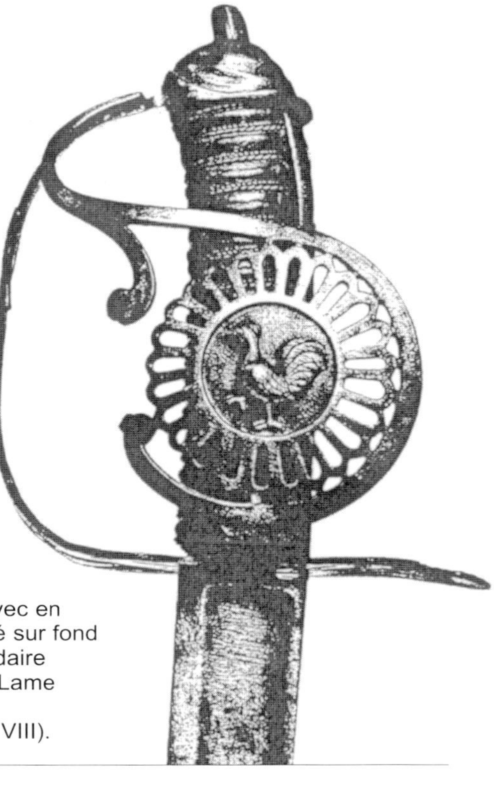

701 - Variante.
Vers 1792.
Laiton. Calotte en bonnet et longue queue. Poignée recouverte de basane ficelée et filigranée. La branche secondaire supérieure entoure un médaillon régulièrement ajouré, avec en son centre un coq placé sur fond d'or. La branche secondaire inférieure a été brisée. Lame blanche à pan creux (G. Lavallay, op. cit. pl. VIII).

A l'aigle

702

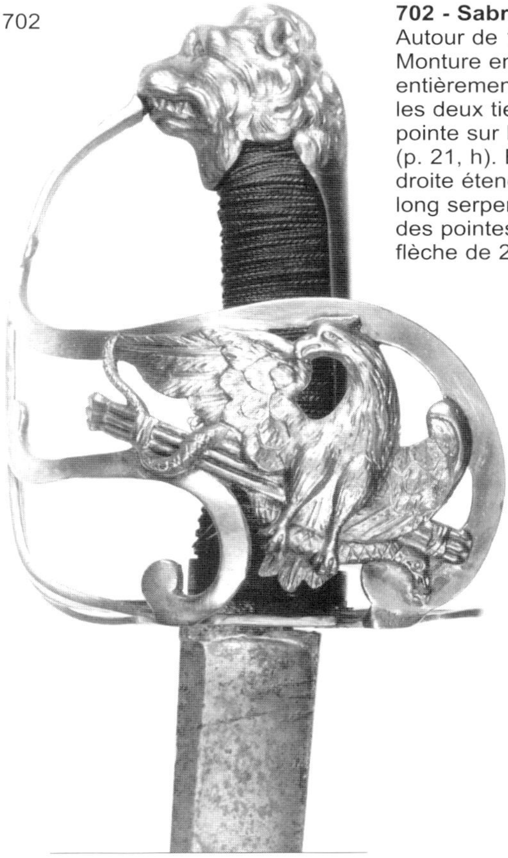

702 - Sabre d'officier des Volontaires Nationaux.
Autour de 1792.
Monture en laiton doré. Calotte en tête de lion, à longue queue unie. Poignée entièrement filigranée de fil très fin torsadé. Arc de jointure presque droit dans les deux tiers basaux de sa longueur. Plateau de garde à pique médiane, la pointe sur le bec et, de part et d'autre, deux ailes symbolisant la victoire (p. 21, h). Entre les deux branches secondaires bien séparées, une aigle à l'aile droite étendue et à l'aile gauche repliée. Les serres du rapace maintiennent un long serpent figurant l'ennemi vaincu (702a). En fond, trois piques liées près des pointes et près du talon. Lame blanche de 70 cm, large au talon de 3,3 cm, flèche de 2,5 cm.

702a

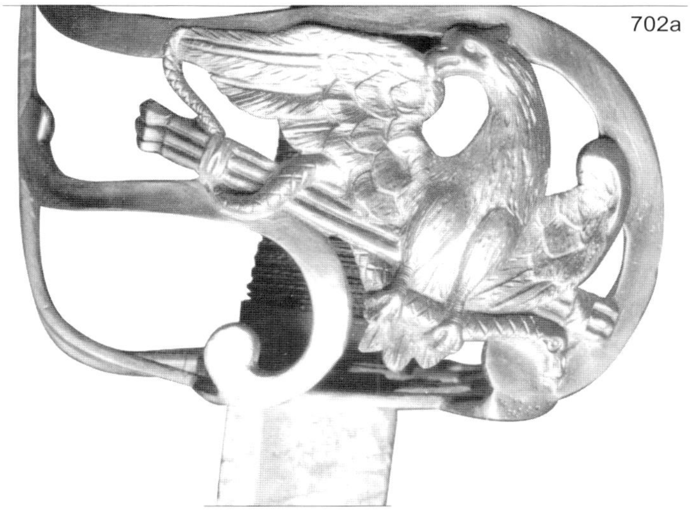

- **a**, détail de la coquille : l'aigle dominant le serpent.

703 - Variante.
Après 1792.
Même décor avec les différences suivantes : calotte plate à trois rainures plus une marginale qui se poursuit sur la longue queue ; arc de jointure arrondi ; lame du type Montmorency, marquée au talon … *MON* (pour Simon ?) arquebusier actif à Paris entre 1775 et 1800 ; lame longue de 76,2 cm, large au talon de 3,4 cm (*in* Ph. Bordes, op. cit. p. 83, fig. 52) (non figuré).

A la Chimère

704

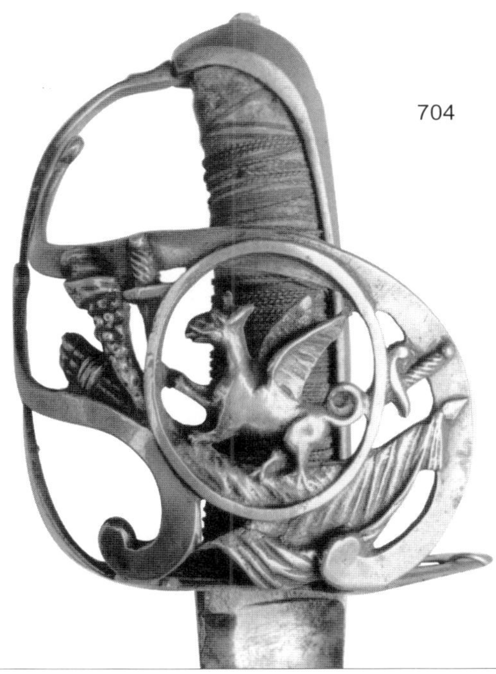

704 - Sabre d'officier de la Garde Nationale ou des Volontaires Nationaux.
1789 - 1792.
Monture en laiton du type de celle des sabres de mineurs. Poignée finement filigranée. Arc de jointure arqué. Entre les deux branches secondaires bien séparées est placé un médaillon à bordure unie dans lequel est inscrite une chimère ailée et à queue enroulée. Sur la gauche du médaillon sont insérés, de bas en haut, un faisceau, une massue d'Hercule coiffée d'un bonnet phrygien et un glaive. Du côté droit, un glaive est posé au-dessus d'un drapeau. Lame de 72,8 cm, large au talon de 3,5 cm (Musée de la Révolution française de Vizille).

A l'Hydre de Lerne

Sabre de la Garde Nationale ou des Volontaires Nationaux

Trois exemplaires ornés de l'Hydre de Lerne sont connus : un est au Musée de l'Armée (inv. J 738), un autre au Musée Carnavalet de Paris et le dernier dans une collection particulière (cf. Bordes, Op. cit. p. 75).

705

705 - Sabre d'officier supérieur de la Garde Nationale.
Monarchie constitutionnelle.
Monture en laiton doré. Calotte en tête de bélier, à longue queue. Poignée filigranée. Plateau de garde portant l'inscription *FETE DE LA FEDERATION LE 14 JUILLET 1790.* Coquille montrant Iolaos (705a) terrassant l'Hydre de Lerne (B.C.).

705a

- **a**, dessin semi-schématique montrant Iolaos (Hercule) brûlant les têtes de l'Hydre de Lerne figurant les forces anti-révolutionnaires.

706 - Variante, vers 1790 - 1791.
Monture en laiton doré. Calotte en tête de bélier. Poignée filigranée. Plateau ovale avec en sa partie antérieure un médaillon, en ce qui concerne l'exemplaire en collection privée, portant l'inscription en capitales *14 / JUILLET / FEDERATION / FRATERNITE / UNION.* Le sujet de la coquille est Hercule, brûlant les têtes de l'Hydre (voir J. Lhoste et J.J. Buigné, 1999). A noter que la fleur de lys qui était ciselée sur la poitrine d'Iolaos a été limée ainsi que le sexe d'Hercule !
Quant à l'exemplaire conservé au Musée Carnavalet, la calotte est en casque à grille et la lame est signée Bida de Bordeaux (non figuré).

707 - Variante.
Vers 1792.
Monture en laiton. Calotte en tête de lion à poignée entièrement filigranée. Coquille représentant Hercule brandissant sa massue pour abattre l'Hydre symbole de l'Ancien Régime. Lame légèrement courbée, gravée sur une face *VAINCRE OU MOURIR POUR LA LOY* et sur l'autre *UN HOMME EN ARME.* Fourreau de cuir à deux garnitures (non figuré).

Sous-famille VIII f - A l'arbre de la Liberté

L'arbre est rarement seul sur la garde des sabres. Un seul exemple sera figuré grâce à l'amabilité de Ph. Bordes.

708

708 - Sabre d'officier d'Infanterie.
1789 - 1792.
Monture en laiton du type de celle des sabres de Mineurs. Poignée ficelée, recouverte de basane et filigranée. Arc de jointure arqué et prolongé d'un plateau du type a (voir p. 21). Entre les deux branches secondaires bien séparées, un arbre du type héraldique terrassé (1) est inséré. Lame de 75 cm, large au talon de 4 cm. Le symbolisme de l'arbre, très ancien, a été exposé dans les " Armes Blanches " (J. Lhoste et J.J. Buigné, seconde édition, 1999).

(1) Signifie figuré avec la terre autour du pied de l'arbre !

Sous-famille VIII g - Aux figures allégoriques

Athéna, la déesse de la guerre, a été souvent transformée en personnalisation de la Nation : elle a gardé son casque ou fut coiffée d'un bonnet ; elle a gardé sa lance ou celle-ci est transformée en baguette, elle a perdu son bouclier. Après 1792 le profil d'une femme symbolisera la République.

709

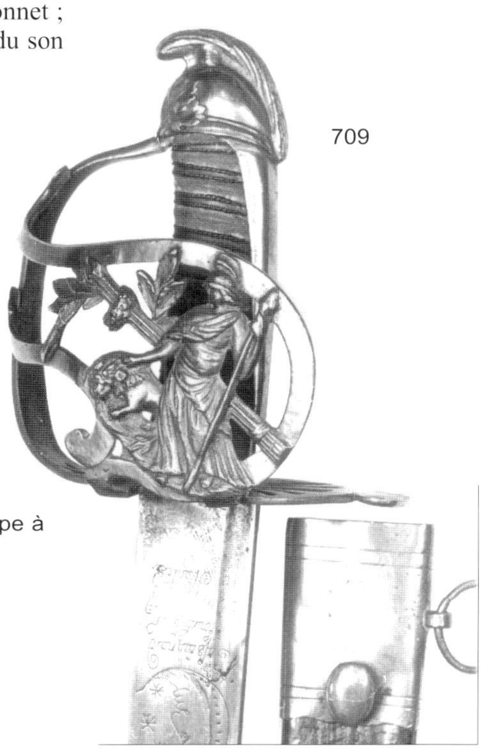

709 - Sabre d'officier des Volontaires Nationaux.
Après 1792.
Monture en laiton doré. Calotte en casque empanaché à face léonine frontale. Longue queue unie. Poignée recouverte de basane, à filigrane composé d'un ruban de cuivre rouge bordé d'un fil torsadé également de fil de cuivre rouge. Arc de jointure arrondi, uni, suivi d'un plateau de garde du type d (voir p. 21). Entre les deux branches secondaires bien séparées, est placée Athéna en pied, casquée mais sans cuirasse, tenant de sa main gauche une pique et reposant sa main droite sur un lion dominé, représentant l'ennemi. Derrière la déesse, un faisceau de quatre lances passant dans une couronne de laurier et liées près de leur talon. Sous les piques, passe un rameau de laurier. Lame signée Cassaignard et gravée de ses motifs favoris : croissant de lune, étoiles, bras armés. Fourreau à deux garnitures et chape à anneau de bélière et bouton.

710

710 - Variante, sabre d'officier de chasseur à cheval des Volontaires Nationaux.
Après 1792.
Monture très comparable au sabre précédent. Seule la lame est remarquable : longue de 78,5 cm, large au talon de 3,7 cm et flèche de 2 cm ; bleuie sur 23 cm, elle est gravée en letttres capitales dorées, sur le *recto L' / HONNEUR / ME FAIT / SERVIR* et sur *le verso JE / COMBAS / POUR LA / PATRIE* (710a et 710b).

710a

- **a**, *recto* de la lame.

710b

- **b**, *verso* de la lame.

711

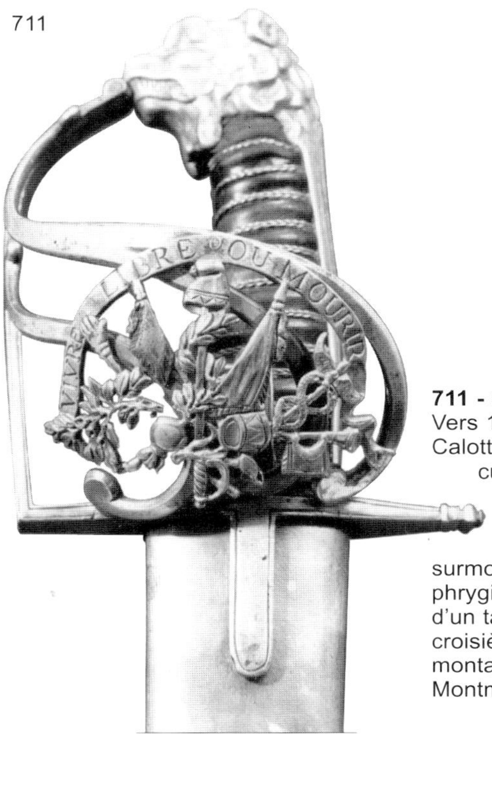

711 - Sabre d'officier de la Garde Nationale.
Vers 1792.
Calotte en tête de lion, prolongée d'une longue queue. Fusée recouverte de cuir, à une dizaine de spires d'un filigrane constitué de deux fils torsadés jumelés. Arc de jointure droit sur les deux tiers et recevant, séparément, les deux branches secondaires. La garde est ornée d'une banderolle portant la devise *VIVRE LIBRE OU MOURIR* et surmontant un trophée constitué d'une épée centrale portant un bonnet phrygien encadrée de deux drapeaux, d'une main de justice, d'une grenade, d'un tambour, d'un caducée, d'une trompette et son fanion ... A noter que la croisière porte deux oreillons en baguettes à extrémités arrondie, basaux, montage tout-à-fait exceptionnel sur ce type de sabre. Lame nue, à la Montmorency. Fourreau à deux garnitures de laiton uni.

712 - Variante.
Sabre d'officier de Chasseur à cheval des Volontaires Nationaux.
Après 1792.
Même décor à la déesse Athéna, mais portant cuirasse. Calotte en casque à grille. Arc de jointure ajouré à sa base d'une feuille d'acanthe. Plateau du type d. Lame de 81 cm, large au talon de 3,6 cm, à pan creux et sans gouttière (*in* Ph. Bordes, op. cit. p. 84, fig. 56), (non figuré).

713 - Variante.
Sabre d'officier des Volontaires Nationaux.
Après 1792.
Même décor à la déesse Athéna, sans cuirasse. Calotte en casque à grille. Arc de jointure ajouré longitudinalement sur la moitié de sa longueur. Plateau du type d. Lame de 69,2 cm, large au talon de 3,3 cm (*in* La Giberne, 1911 - 1912, XII, p. 18), (non figuré).

714 - Variante.
Sabre d'officier des Volontaires Nationaux.
Après 1792.
Exemplaire très voisin du modèle de référence avec l'arc de jointure décoré d'une feuille d'acanthe découpée et plateau de garde voisin du type d. Sur la branche secondaire supérieure, en relief, on lit . *L'UNION . FAIT . LA . FORCE.* (*in* Ph. Bordes, op. cit. p. 85, fig. 56 et 56b), (non figuré).

715

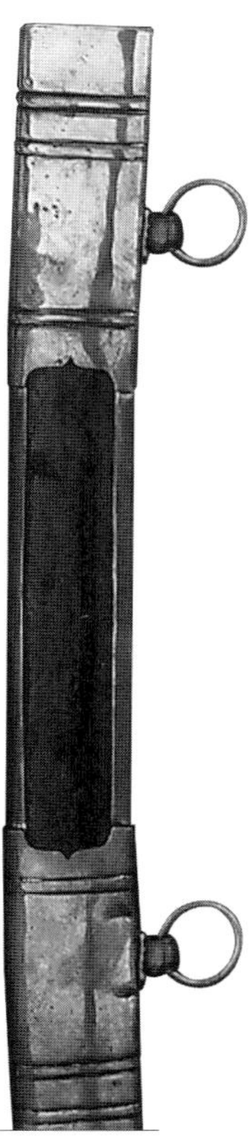

715 - Variante des n° 709 et 710.
Sabre d'officier des Volontaires Nationaux.
Monture en laiton doré. Ce sabre remarquable par sa lame de 90 cm bleuie au tiers et gravée de motifs dorés, à partir du talon, d'une étoile de grande taille, de deux petits disques représentant peut-être des planètes, d'un croissant de lune regardant une étoile, de deux petites étoiles, d'un soleil, d'autres petits disques et d'une tête de Turc. Fourreau à trois garnitures reliées par des attelles (Sotheby's, Sussex).

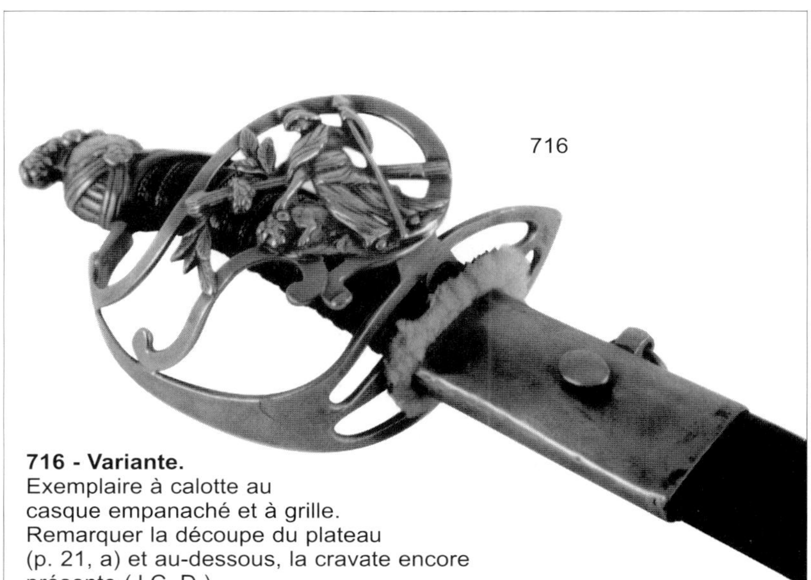

716

716 - Variante.
Exemplaire à calotte au casque empanaché et à grille. Remarquer la découpe du plateau (p. 21, a) et au-dessous, la cravate encore présente (J.C. D.).

717

717 - Sabre d'officier des Volontaires Nationaux.
Après 1792.
Monture en laiton. Calotte en bonnet phrygien à la cocarde. Poignée filigranée. Arc de jointure régulièrement arqué et plateau ajouré. Entre les deux branches secondaires, dans un médaillon ovale est inscrit le profil de la République. Le médaillon est surmonté d'un noeud de ruban et flanqué de quatre drapeaux. En bas, à gauche, le petit coq national.

Sous-famille VII h - Aux Armes de France

Louis XVIII étant installé sur le trône de France en 1814, on " oublie " la période impériale et l'on crée de nouveaux sabres à coquille et à deux branches arrivant isolément sur l'arc de jointure, selon la tradition établie sous le règne de Louis XVI. Cette conception sera limitée à quelques corps et sera de courte durée.

718

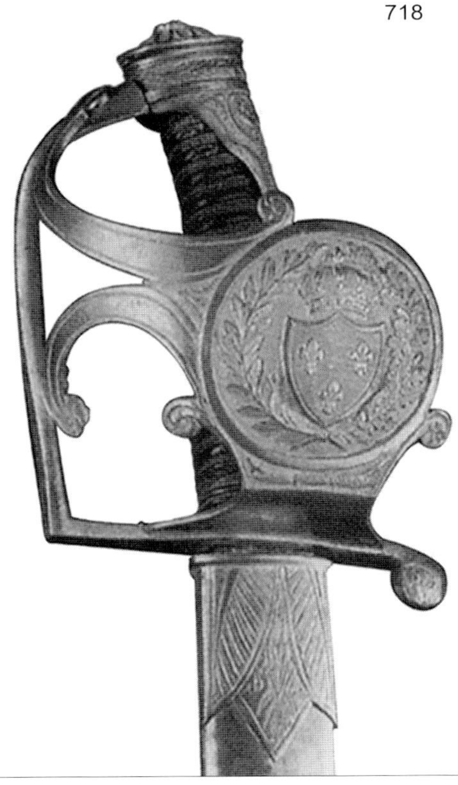

718 - Sabre d'officier des Gendarmes de la Garde de la Maison du Roi.
Modèle 1814.
Monture en bronze. Calotte en dôme orné de feuilles disposées en étoile et à bordure également foliacée. Longue calotte ciselée. Poignée richement filigranée. Branche principale unie arrivant en angle droit sur la croisière et recevant deux branches secondaires arquées en sens contraire. Ces branches soutiennent un médaillon aux Armes de France couronnées, flanquées d'un rameau de laurier (à gauche) et de chêne (à droite). Ce médaillon repose sur une embase ornée d'un bâton de foudre. Lame droite, longue de 92,5 à 97,5 cm, à deux pans creux, pouvant être signée " S " et " K " pour Schimmelbusch et Kirschbaum de Solingen et gravée de motifs variés dorés sur fond bleui. Fourreau en tôle de fer, sans bracelet, les deux anneaux de bélière étant soutenus par deux pitons soudés sur le côté. Embouchure en laiton, généralement ciselée de trois feuilles (ou de toiles d'araignée ?). Dard asymétrique.

Rappelons que les gendarmes de la Garde du Roi ont été créés en 1602 et dissous en 1787. Ce corps est reconstitué par Louis XVIII en 1814, au sein de la Maison du Roi.

Chapitre IV

Monture à deux branches secondaires convergeant sur l'arc de jointure, calotte à longue queue

Ce chapitre réunit les sabres à montures dont les deux branches secondaires arrivent conjointement sur l'arc de jointure.

Ces sabres se distinguent donc aisément de ceux qui sont groupés dans le Chapitre III. En effet, chez ceux-ci, rappelons-le, les branches secondaires arrivent isolément sur l'arc.

En ce qui concerne les Familles, elles ont été définies de la façon suivante :

La grande majorité de ces sabres a été en service depuis le règne de Louis XVI jusqu'à la période révolutionnaire.

Famille IX - Poignée en laiton moulé

Dans cette famille on trouvera des sabres de bord de la fin du XVIIIème siècle.

719

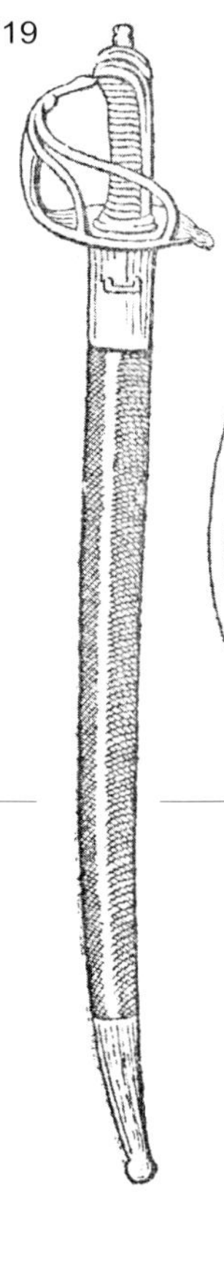

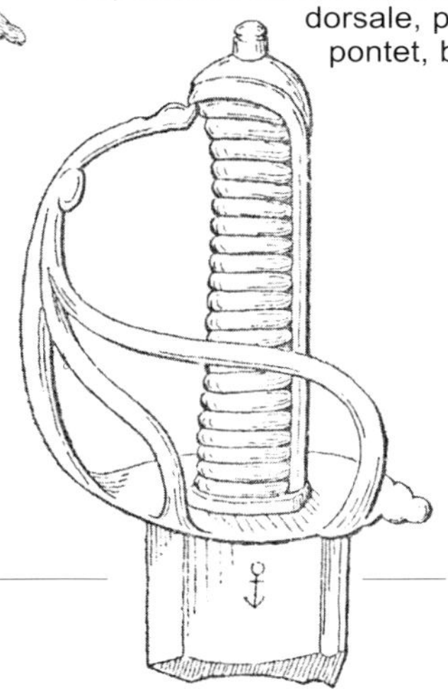

719 - Sabre de Bord.
Modèle 1783.
Monture et poignée en laiton doré. Calotte à gros bouton de rivure, longue queue et poignée à une vingtaine de cordons figurés. Branche principale régulièrement arquée. Deux branches secondaires en " S " partant du plateau de garde et rejoignant conjointement l'arc à son tiers supérieur. Plateau ovoïde terminé par une petite palmette. Lame large,à pan creux et gouttière dorsale, poinçonnée d'une petite ancre de marine. Fourreau de cuir noir, chape à pontet, bouterolle terminée par un gros bouton (d'après. Boutet).

720

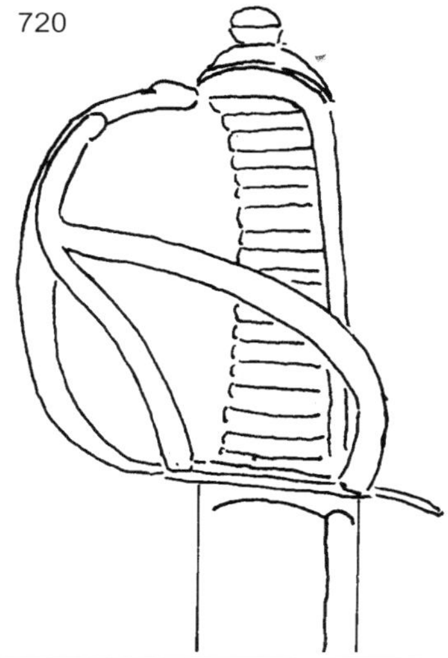

720 - Sabre de Bord.
Modèle 1792.
Monture et poignée en laiton doré. Calotte et fusée comparables à celle du modèle 1783. Arc arrondi. Branches secondaires se réunissant pour n'en former qu'une seule avant de rejoindre le haut de l'arc de jointure. Lame de 60 cm environ, large au talon de 5 cm, flèche de 1,12 cm, à pan creux et gouttière.

Marquages - Sur le dos de la lame *Mf^{ture} de Klingenthal - Couleaux frères entrep^{rs}* et dans un cartouche *Lame de / France*.

Famille X - Poignée en bois recouvert ou non de basane, à coquille en palmette ajourée ou à l'emblème du corps

La palmette ajourée a été associée dans la même famille aux sabres à garde au cornet, à la grenade, à l'ancre de Marine, au soleil, ce qui donne les cinq Sous-familles suivantes :

Coquille :

A noter que certains de ces décors ont déjà été rencontrés au Chapitre III : palmette (Famille VI) p. 321, cornet (Sous-famille VIb) p. 330, grenade (Sous-famille VIc) p. 336.

Sous-famille X a - A palmette ajourée

Les gardes de sabres à palmette découpée sont apparues sous le règne de Louis XVI.

721

721 - Sabre d'officier d'Infanterie.
Règne de Louis XVI.
Monture en laiton. Calotte légèrement bombée à longue queue simple. Poignée de bois noirci cannelé verticalement. Arc de jointure légèrement arrondi et rejoignant le plateau par un angle droit. Les deux branches secondaires jointives sur le sommet de l'arc, arrivant sur une palmette ajourée à neuf éléments. Lame à la Montmorency.

722

722 - Sabre d'officier d'Infanterie.
Fin du règne de Louis XVI. Monture en laiton. Calotte à bouton de rivure en losange, longue queue. Poignée ficelée, recouverte de basane à dix spires d'un fil de laiton torsadé bordées d'un fil simple. Arc de jointure arrondi, prolongé par un petit plateau ajouré. Coquille en palmette à six éléments d'où partent les branches secondaires ornées à leurs sommets d'arcures. Lame à la Montmorency, longue de 67 cm. Fourreau de cuir, chape à anneau de bélière.

723

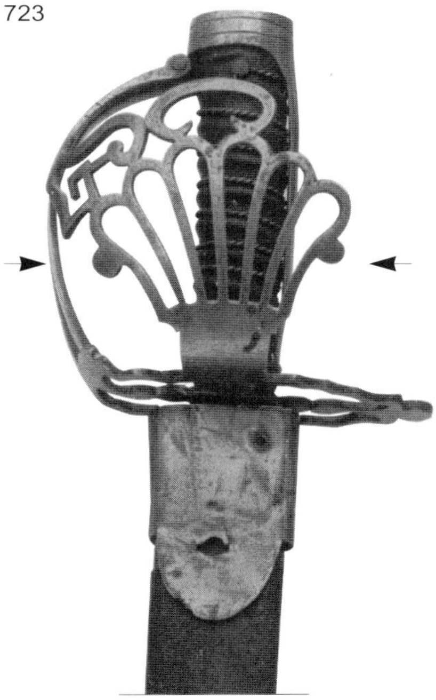

723 - Variante.
Officier de Volontaires.
Epoque révolutionnaire.
Remarquer que les deux décors arrondis, de part et d'autre de la palmette, ne sont pas ajourés (voir flèches). Fourreau à deux garnitures, courte chape à bouton.

724 - Variante
Officier de Volontaires.
Vers 1790.
Monture en laiton doré, très comparable à celle du sabre n° 725 avec les différences suivantes : coquille en palmette composée de sept éléments principaux, surmontée d'un petit lion découpé et situé entre les deux branches secondaires. Entre l'arc et la coquille remarquer la décoration en " S " évidé. Lame à la Montmorency, bleuie au tiers et décorée de motifs végétaux dorés.

724

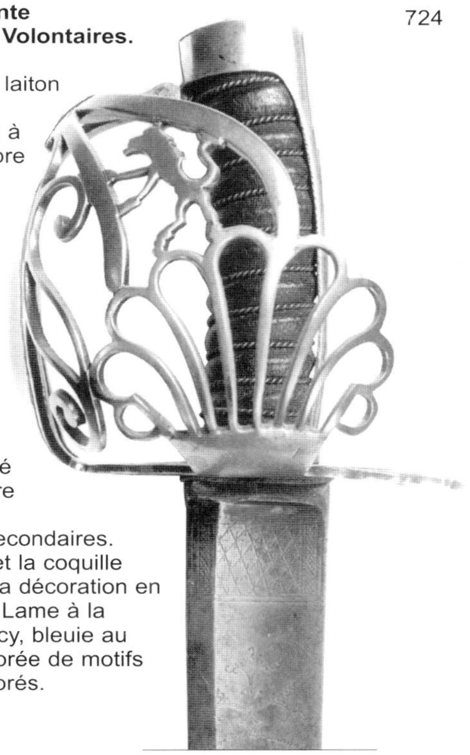

725

725 - Variante
Toute la monture est finement ciselée. De plus, sur cette photographie, on découvre la structure de l'arc de jointure prolongé du plateau découpé (voisin du type b, voir p. 21). Fourreau à deux garnitures, dont la chape est à bouton ovale (J.C. D.).

726 - Sabre de cavalerie de gaucher.
Monture de la fin du XVIIIème siècle. Calotte à pièce de rivure hexagonale, à longue queue. Fusée de cuir à sept spires d'un filigrane de fil de laiton torsadé en chaînette. Bague basale à deux fines rainures. Arc de jointure évidé en triangle, se poursuivant pour former le plateau ajouré selon la mode de l'époque. Coquille en palmette à sept palmes ajourées, liée à l'arc de jointure par deux branches secondaires fusionnées. Forte lame de 94 cm, à deux gouttières sur toute sa longueur, signée au talon " S " et " K " pour Schitzer et Kirchbaum, fourbisseurs à Solingen et travaillant au début du XIXème siècle. Il s'agit donc d'une lame de circonstance.

726

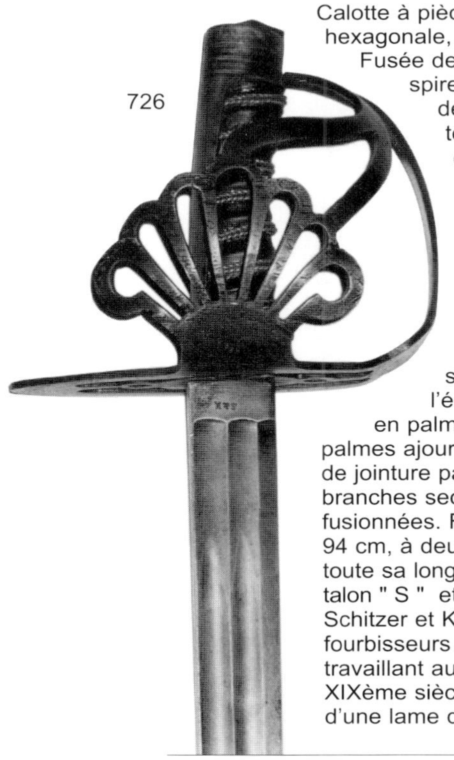

Sous-famille X b - Au cornet ou au cor

Les sabres de cette sous-famille, au cornet ou au cor, portent l'emblème des Chasseurs et en conséquence sont attribuables à ce corps ainsi qu'aux Carabiniers.

727

727 - Sabre d'officier de Carabinier d'Infanterie légère.
Règne de Louis XVI et Révolution.
Monture en laiton doré. Calotte plate à double rainure et longue queue unie. Poignée d'ébène cannelée verticalement. Arc de jointure droit sur les deux tiers de sa longueur. Dans un médaillon formé d'un cor de Chasseur, le pavillon vers le haut, est inscrite une grenade enflammée. Dans l'intervalle triangulaire des branches secondaires jointives se trouve une fleur de lys qui a été brisée sur l'exemplaire dont nous donnons la photographie, mais Ph. Bordes (op. cit. p. 53, fig. 7)) montre un exemplaire où cette fleur est bien présente. Entre l'arc et la secondaire inférieure est insérée une branche de renfort en " S " ajourée. Lame à la Montmorency gravée au talon d'un disque en pointillés (flèche).

728 - Variante.
Vers 1790 - 1792.
Même monture générale que celle du sabre précédent avec les modifications suivantes : calotte en casque empanaché, poignée recouverte de basane à filigrane d'un double fil de laiton torsadé faisant une dizaine de spires et surtout par la suppression de la fleur de lys, qui dans l'angle formé par les deux branches secondaires, est remplacée par le bonnet révolutionnaire.

728

729 - Variante.
Vers 1790 - 1792.
Même monture que celle du sabre précédent mais à calotte plate, à la grenade striée de cercles concentriques, au plateau du type b, (voir p. 21) mais à bordure festonnée. Lame plate de 72,3 cm et large au talon de 4 cm (cf. Ph. Bordes, p. 88, fig. 63) (non figuré).

730

730 - Sabre de Chasseur à pied.
Règne de Louis XVI.
Monture en laiton. Calotte en cuillère et longue queue. Grand cornet placé entre les deux branches secondaires, le pavillon en haut, avec une face léonine inscrite en son centre. Plateau du type a (voir p. 21). Forte lame à gouttière dorsale.

Sous-famille X c - A la grenade

Les sabres les plus anciens appartiennent à l'Infanterie légère. Le dernier sabre décrit date de la Restauration et a été attribué à la Gendarmerie à cheval.

731

731 - Sabre d'officier de Grenadiers.
Vers 1792.
Monture en laiton. Calotte en cuillère à longue queue du type des sabres de Mineurs. Poignée filigranée. Arc de jointure arrondi, plateau de garde du type a (voir p. 21). Entre les deux branches secondaires jointives est inscrite une grenade enflammée flanquée de deux rameaux de laurier, noués à leur base. Dans l'angle supérieur formé par les secondaires se trouve le bonnet phrygien surmontant la massue d'Hercule. Lame à la Montmorency.

732 - Sabre d'officier de Grenadiers.
Avant 1792.
Modèle plus luxueux que le précédent, à la même ornementation, sauf la calotte en tête de lion et la présence d'une petite branche supplémentaire partant de la jonction des secondaires et, après une courbe, rejoignant la base de l'arc. De plus, la lame de 76,5 cm, large au talon de 4 cm porte sur la face interne *POUR / LA/ NATION / LA LOI / ET / LE ROY.*
Sur la face externe des fleurs de lys et un personnage que Ph. Bordes qualifie de Turc. (op. cit. P. 90-91, fig. 66), (non figuré).

733 - Sabre de la Gendarmerie à cheval.
Restauration.
Monture de laiton. Calotte ovale surmontée d'un losange en léger relief servant de bouton de rivure, et prolongée, sans césure, par une longue queue unie. Poignée recouverte de basane, à filigrane constitué de deux fils torsadés faisant dix spires, sans virole, mais avec à sa place cinq spires serrées du filigrane. Arc de jointure fixé dans la calotte par un tenon, régulièrement arrondi et s'élargissant à sa base pour former un plateau ovalaire se terminant en bec. Branches secondaires arrivant en crosse sur le bord du plateau, se rejoignant avant d'être soudées sur l'arc à deux centimètres de la calotte. Entre ces branches est inscrite un grosse grenade enflammée. Lame droite de 83 cm, large au talon de 3,3 cm et poinçonnée de " C " et " F ", surmontés d'une étoile, signifiant Coulaux Frère vers 1816 et du " D " dans un cercle de perles. Le dos de la lame est signé *Manuf^{re} R^{le} du Klingenthal - Coulaux Frères.* Fourreau à trois garnitures.

733

Sous famille X d - A l'ancre de marine

Les sabres de la fin du XVIIIème siècle étant souvent fabriqués à l'entreprise,sont très variables dans leur structure. Ces sabres sont en principe réservés aux matelots, les officiers se voyant attribuer des sabres de même monture mais à calotte en tête de lion, sans queue de calotte (monture dite en couteau de chasse).

734

734 - Sabre de bord.
Vers 1770 - 1780.
Monture en laiton. Calotte à bouton de rivure et à longue queue. Poignée recouverte de basane à une dizaine de spires d'un filigrane disposé en chaînette. Arc de jointure finissant en angle droit sur la croisière. Deux branches secondaires : la branche supérieure arrondie à sa base en épousant la courbe des branches de l'ancre ; l'inférieure, tourmentée, est en forme de " L ", le jas de l'ancre réunissant les deux secondaires (734a). Lame à pan creux légèrement courbée à trois gouttières dorsales, longue de 75 à 81 cm. Fourreau à trois garnitures de laiton, dard apical et frontal.

Marquage - Souvent la lame est signée de *Cassagnard*

- **a**, détail de la monture. Large lame à trois rainures dorsales signées *Cassagnard / à Nantes* au talon, gravure au soleil, au bras armé ...

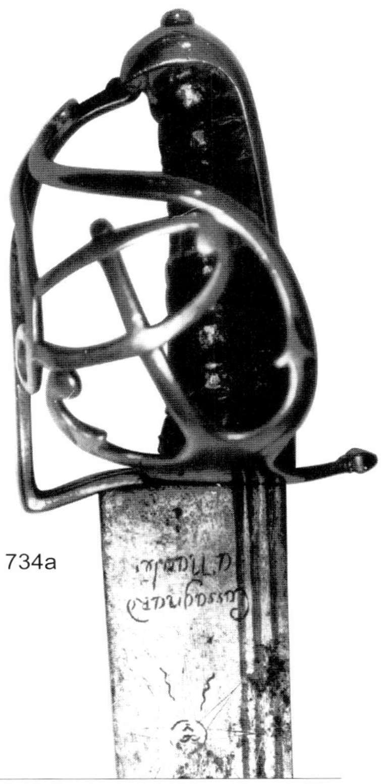

734a

735 - Sabre de bord de Grenadiers de la Marine.
Vers 1770 - 1780.
Monture en laiton. Ce modèle est conçu sur le modèle du sabre précédent. Il s'en distingue essentiellement par la présence d'une ancre de marine dont la verge est surchargée d'une grenade enflammée (M.E. - S. de P.).

<u>Nota</u> - Pour les Grenadiers de la Marine, une grenade enflammée est fixée sur la verge et pour les officiers, la calotte est en tête de lion (voir C. Aries, XVI, 1970).

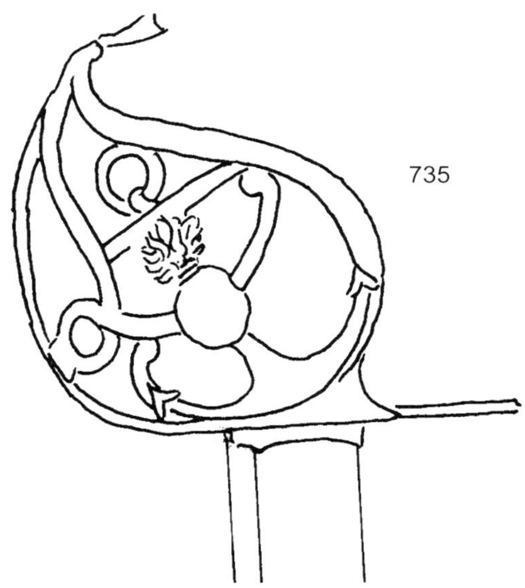

735

Sous-famille X e - Au soleil

Le sabre décrit est évidemment un remontage révolutionnaire.

736

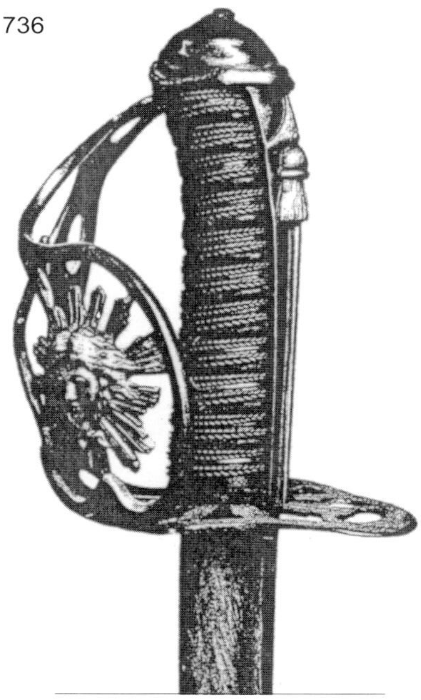

736 - Sabre d'officier d'Infanterie
Règne de Louis XVI actualisé probablement entre 1789 et 1792.
On remarquera que la coquille au soleil royal subsiste mais que la calotte est transformée en bonnet révolutionnaire et que sur la longue queue est placée une lance (Lavallay, 1945, pl. XIII).

Famille XI - Poignée en bois généralement recouvert de basane, à coquille décorée de sujets d'inspiration politique

Les sabres de la Famille XI, créés sous l'Ancien Régime, bénéficient d'un grand développement en période révolutionnaire. Après la fleur de lys royale, les porteurs de sabre affichèrent leurs opinions politiques sur la garde de leurs armes. Les ornementations suivantes ont été relevées, chacune donnant naissance à une Sous-famille :

Sous-famille XI a - Au lys et aux Armes de France

Il est évident que la présence d'une fleur de lys sur la garde d'un sabre indique que celui-ci fut conçu sous le règne de Louis XVI.

737

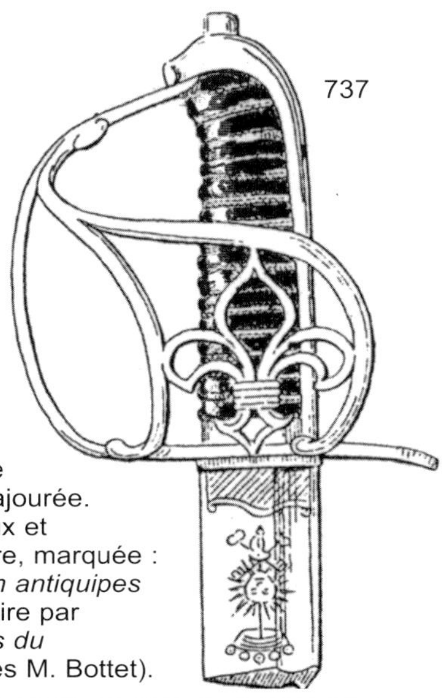

737 - Sabre d'officier de Dragons ou de Chasseurs à Cheval.
Vers 1780. Monture en laiton. Calotte simple à longue queue. Poignée filigranée. Entre les branches secondaires jointives est fixée une fleur de lys ajourée. Lame à pan creux et profonde gouttière, marquée : *Custodes Regum antiquipes* qui peut se traduire par *Anciens gardiens du Royaume* (d'après M. Bottet).

738

738 - Variante.
On remarquera que dans ce modèle, la garde, ans sa totalité est plus robuste que dans l'exemplaire n° 737.

739

739 - Sabre d'officier de Gendarmerie de France.
Vers 1766.
Monture en fer. Calotte plate à pièce de rivure en losange et longue queue. Poignée plus large à sa partie supérieure qu'à sa base, à filigrane en double fil torsadé de fer et disposé en chaînette formant 13 spires. Arc de jointure et les deux branches secondaires sont jointives sur les armes de France, et découpées en anneaux formant une chaînette. Plateau découpé, ménageant deux bordures courbes et couvre - chape. Longue lame droite et plane, gravée sous un buste de Gendarme, de cinq cartouches quadrangulaires en descendant vers le talon, avec *GENDAR/MERIE,* un motif foliacé, la devise *VAINCRE / OU / MOURIR*, un autre motif foliacé et enfin la signature *Lebeck / marchand / fourbisseur / à / Nancy* (M.E. - S. de P.).

740 - Sabre d'officier d'Infanterie.
Fin du XVIIIème siècle.
Monture à la Mineur, en laiton. Calotte en casque empanaché, à longue queue. Poignée entièrement filigranée. Entre les deux branches secondaires jointives, à partir du plateau, on trouve un faisceau ce flèches représentant les forces royales, puis les Armes de France couronnées, et dans l'angle supérieur, un cœur enflammé entre deux mains, vraisemblablement symbole catholique (M.E. - S. de P.).

Nota - On trouvera p. 385, n° 760, le même type de monture mais avec en plus, un dauphin entre l'arc et la secondaire antérieure. L'exemplaire présenté possède une calotte plate, non figurative.

740

Dessin schématique de l'emblème ornant la garde.

741 - Sabre de sous-officier de la Milice Royale à cheval.
Règne de Louis XVI.
Monture en laiton. Calotte plate à longue queue. Poignée filigranée. Arc de jointure longuement évidé d'un triangle et plateau du type a. La branche secondaire se déploie en multiples branches avant d'être rivée sur le haut de l'arc et maintient une fleur de lys surmontée d'une couronne de feuillage très stylisé. Lame à pan creux, longue de 81 cm, large au talon de 3,8 cm, à flèche de 4 cm, bleuie et gravée de rinceaux, de trophées où l'on distingue des tambours, des piques, des guidons...

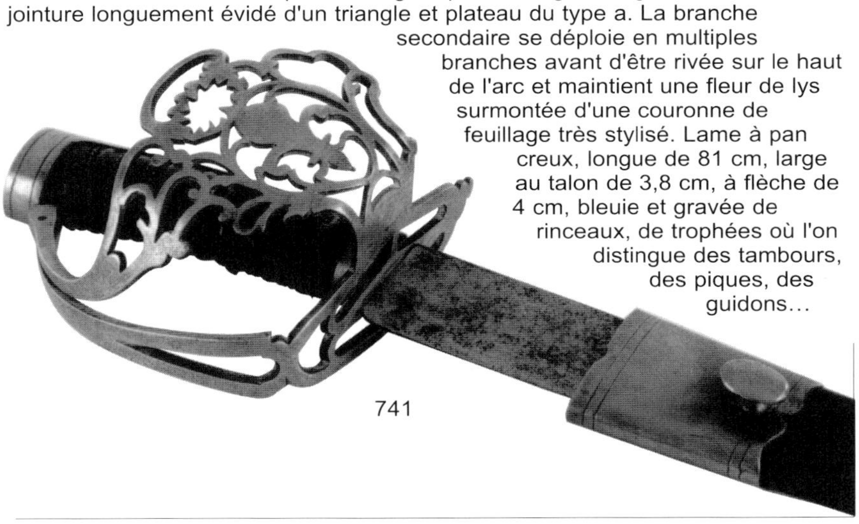

741

742

742 - Variante.
Après 1789.
Même monture que celle de la figure précédente mais la fleur de lys est gravée en majuscules *LIBERE / OU LA / MORT.*

Sous-famille XI b - Au profil de La Fayette

Les sabres au profil de La Fayette, profil parfois jumelé avec celui de Louis XVI, datent des années 1789-1792, lors de la période de la monarchie constitutionnelle.

743

743 - Sabre d'officier de la Garde Nationale.
Monarchie constitutionnelle.

Monture en laiton. Calotte en tête de lion, et longue queue. Poignée filigranée de deux fils torsadés disposés en chaînette. Entre les deux branches secondaires jointives, sur un fond de canon et de boulets, d'armes blanches, de drapeaux est placé un bouclier octogonal curviligne aux profils de La Fayette et de Louis XVI. P. Martin (1985, p. 128) précise que la *" lame* (est) *marquée Garde Nationale... soleil, des fleurs de lys et un personnage à l'orientale brandissant un sabre* ".

744

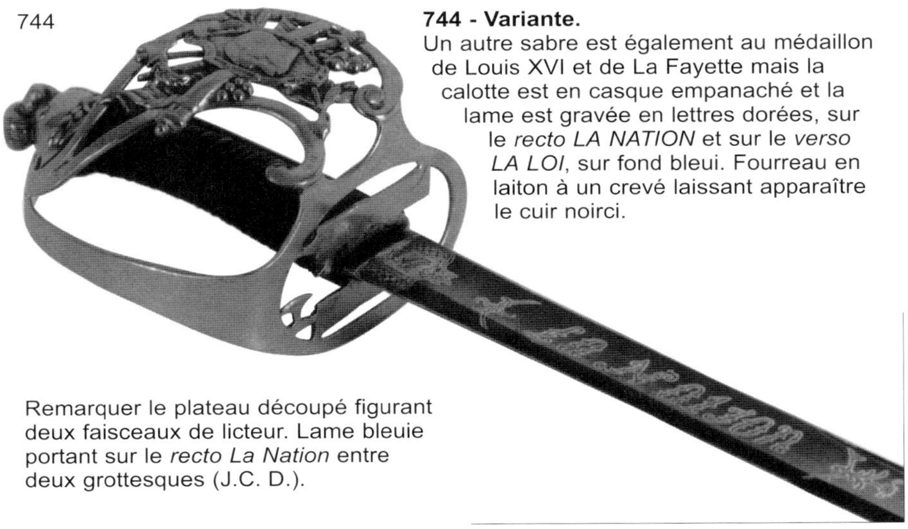

744 - Variante.
Un autre sabre est également au médaillon de Louis XVI et de La Fayette mais la calotte est en casque empanaché et la lame est gravée en lettres dorées, sur le *recto LA NATION* et sur le *verso LA LOI*, sur fond bleui. Fourreau en laiton à un crevé laissant apparaître le cuir noirci.

Remarquer le plateau découpé figurant deux faisceaux de licteur. Lame bleuie portant sur le *recto La Nation* entre deux grottesques (J.C. D.).

Sous-famille XI c - Aux Trois Ordres

Pour commémorer la réunion des Ordres, de la Noblesse, du Clergé et du Tiers-Etat aux Etats-Généraux des sabres ont été ornés du symbole de ces Ordres, la crosse, l'épée et la bêche, liées par un ruban.

Ce que nous voudrions montrer, c'est l'évolution de ce décor au cours des ans, évolution qui traduit la pluralité des opinions de leurs propriétaires entre 1789 et 1792.

En 1789, sous la monarchie constitutionnelle, les fleurs de lys subsistent. Aussi trouve-t-on sur des sabres aux Trois Ordres de cette époque des montures fleurdelisées. Cette catégorie de sabres semble, au moins partiellement, être des sabres anciens modifiés. Puis, les symboles sont placés en bonne place sur la garde, insérés entre les deux branches secondaires. L'évolution politique des porteurs de ces sabres est évidente lorsqu'on examine avec soin le manche de la bêche : d'abord terminé par une fleur de lys, cette ornementation sera remplacée par le bonnet révolutionnaire ou une sorte de trèfle. Les sabres décrits ci-après en donnent des exemples.

745

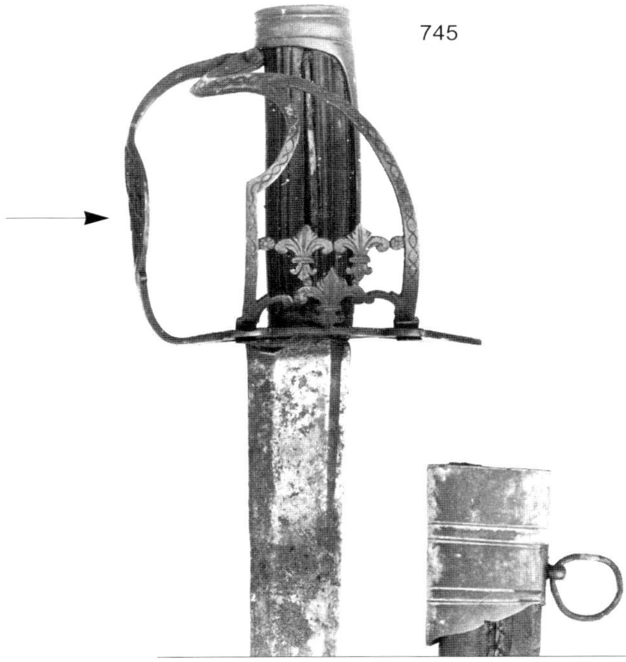

745 - Sabre aux Trois Ordres de Troupe à pied.
Monarchie Constitutionnelle. Monture en laiton. Calotte plate, rainurée à longue queue. Poignée d'ébène cannelée verticalement. Branche principale élargie en son milieu de manière à former un médaillon ajouré dans lequel est inscrit le symbole des Trois -Ordres (flèches). Deux branches secondaires jointives ornées de guirlandes et maintenant à leur base les trois fleurs royales. A noter que les deux branches secondaires sont pourvues d'un tenon qui entre dans un orifice ménagé sur la face frontale du plateau, lui-même du type d (voir p. 21). Lame de 75,5 cm, large au talon de 3,8 cm du type Montmorency. Fourreau à deux garnitures, chape à anneaux de bélière.

746

746 - Sabre d'officier aux Trois Ordres.
Monarchie Constitutionnelle.
Monture en laiton. Calotte plate, rainurée, à longue queue. Poignée recouverte de basane, à huit spires de fil de laiton torsadé. Arc de jointure élargi, en sa partie médiane en un médaillon dans lequel est inscrit en découpe le symbole des Trois Ordres (flèche). Entre les deux branches secondaires sont placées, également en découpe, trois fleurs de lys sous une couronne royale. Entre l'arc et la première secondaire, une branche supplémentaire de renfort en " S " ajouré est insérée. Plateau voisin du type b. Lame à la Montmorency de 76,5 cm, large au talon de 4,2 cm. Chape du fourreau à bouton et anneau de bélière.

747 - Sabre d'officier d'Infanterie aux Trois Ordres.
En service sous la Monarchie Constitutionnelle puis ultérieurement.
Monture en laiton. Calotte plate rainurée à longue queue.
Poignée filigranée d'un ruban de laiton souligné d'un fil torsadé.
Arc de jointure droit en sa partie médiane. Deux branches secondaires ornées d'un filet sinueux maintenant, en découpe, le symbole des Trois Ordres.
Remarquer que le manche de la bêche se termine en fleur de lys indiquant que ce sabre fut fait avant 1792, mais que vers cette date, la fleur fut modifiée en lui coupant les pétales latéraux (voir la flèche). Une branche supplémentaire en " S " ajouré jouxte l'arc.
Plateau voisin du type b. Lame à la Montmorency de 76 cm et large au talon de 3,7 cm. Fourreau de cuir noirci à trois garnitures réunies par des attelles en laiton dont la chape à anneau de bélière.

747

748

748 - Sabre (d'Infanterie) aux Trois Ordres.
Vers 1790, d'interprétation rustique.
Monture en laiton. Calotte ovale suivie d'une longue queue unie. Poignée ficelée, recouverte de basane à filigrane constitué de trois fils jointifs faisant 12 spires, et virole basale. Arc de jointure droit sur 10 cm évidé d'un long triangle. Plateau du type b. Deux branches secondaires partent d'un large talon lié à la partie antérieure du plateau et se réunissent avant de rejoindre l'arc sur lequel elles sont rivées. Entre les secondaires, le symbole des Trois Ordres apparaît en découpe assez grossière ; au sommet du manche de la bêche se trouve une sorte de trèfle. Lame plate, longue de 68 cm, large au talon de 3,9 cm, flèche de 2 cm. Fourreau de cuir noirci à deux garnitures, dont une chape à bouton.

749

Remarquer la calotte en casque et le manche de la bêche qui n'est plus en fleur de lys (J.C. D.)

749 - Variante.
Sabre très voisin du précédent, il possède une calotte en casque à l'antique et une lame réglementaire à la Montmorency.

750

750 - Sabre d'officier d'Infanterie aux Trois Ordres.
Vers 1792.
Même monture que le sabre de la n° 751, mais de meilleure finition. On remarquera surtout que le manche de la bêche se termine par un bonnet éliminant la fleur de lys primitive.

751

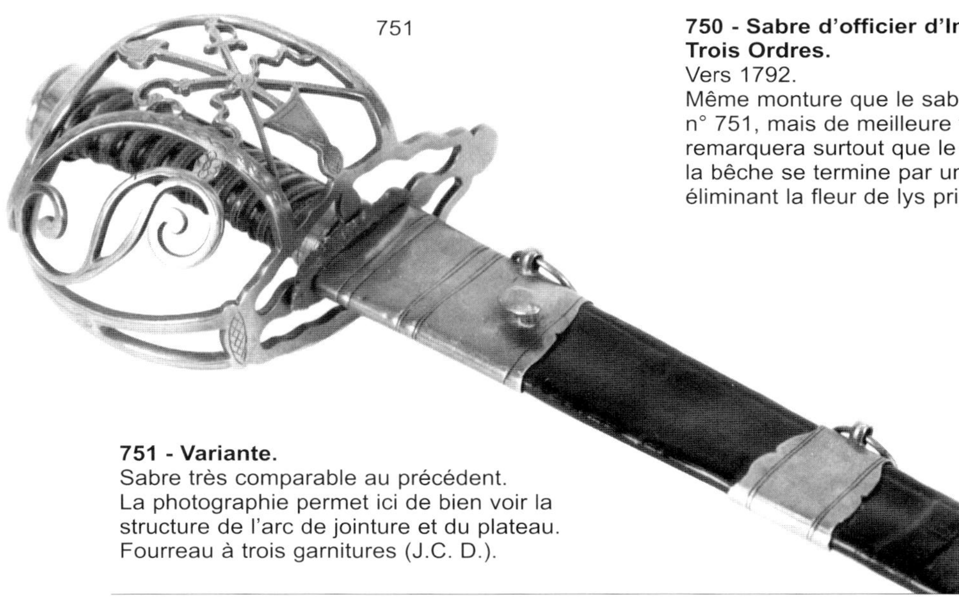

751 - Variante.
Sabre très comparable au précédent.
La photographie permet ici de bien voir la structure de l'arc de jointure et du plateau.
Fourreau à trois garnitures (J.C. D.).

Sous-famille XI d - Au lion

Sachant que le lion, depuis tout temps, est le symbole de la puissance, de la force, du courage, destructeur du mal et de l'ignorance, on ne sera pas surpris de le voir figurer sur les sabres révolutionnaires. En effet, on trouve sa tête ornant les calottes sur de nombreuses armes blanches, mais aussi sur la coquille accompagnée de divers emblèmes révolutionnaires : coq, bonnet phrygien…

752

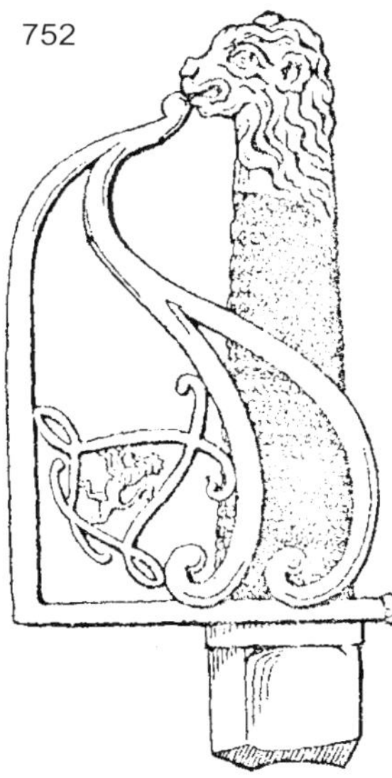

752 - Sabre d'officier de Régiment Nassau-Saarbruck.
Vers 1780.
Monture en fer. Calotte en tête de lion, à courte queue. Poignée filigranée. Arc de jointure arrivant en angle droit sur la croisière, celle-ci étant prolongée par un quillon en forme de poisson. Deux branches secondaires fusionnant avant d'atteindre le haut de l'arc. Entre celui-ci et la première secondaire sont inscrits les deux " L " avec entre eux le lion batave. Lame à pan creux (d'après M. Bottet).

754 - Variante.
Avant 1792.
Monture au lion grimpant. Calotte au casque empanaché à face frontale en tête de lion en relief. Fusée à filigrane double en chaînette, avec dans les intervalles des fils simples, (Ph. Bordes, op. cit. p. 76, fig. 43 ter). plateau du type d, *mais à rebord rubanné en relief à l'intérieur, avec retour en crosse du côté de la branche principale. Cette dernière est évidée et porte à l'intérieur un rameau de chêne de forme triangulaire* selon l'auteur. Lame de 75,2 cm, large au talon de 3,3 cm (non figuré).

755

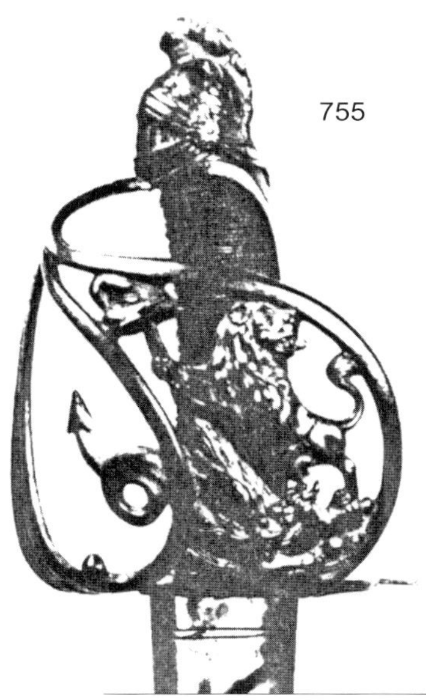

755 - Variante.
Vers 1792.
Monture au lion grimpant. Calotte en casque à l'antique, à grille frontale et grand panache.

753 - Sabre d'officier de la Garde Nationale ou des Volontaires Nationaux.
Avant et après 1792.
Monture en laiton dite au *lion grimpant*. Calotte au casque à l'antique orné d'un masque léonin frontal et longue queue unie. Poignée ficelée à onze spires d'un fil fin torsadé disposé en chaînette, plus deux spires serrées basales servant de virole. Arc de jointure arrondi, à bords cannelés sur toute sa longueur, plus trois courtes cannelures dans sa moitié inférieure. Plateau du type d. Entre les deux branches secondaires jointives est inséré un lion la tête tournée vers l'arrière et brandissant de ses deux pattes antérieures la massue d'Hercule surmontée d'un bonnet phrygien ; sa partie postérieure gauche repose sur un canon sous lequel on trouve deux boulets et sa patte postérieure droite écrase la tête d'un dragon ailé crachant le feu, et dont la queue enroulée est terminée par un dard en forme de fer de flèche apparaissant dans le vide situé entre l'arc de jointure et la deuxième branche. Lame à la Montmorency ornée de motifs au trait : feuillage, petit trophée…, longue de 71 cm, large au talon de 3,3 cm, à flèche de 2 cm. Cette lame est marquée sur le renfort du talon *HOWER*. Fourreau de cuir de vache fort et noirci, à trois garnitures décorées au trait, à bordures découpées en accolade : chape à bouton et anneau de bélière, bracelet trapézoïdal à anneau, bouterolle à dard mince, frontal et basal.

753

756

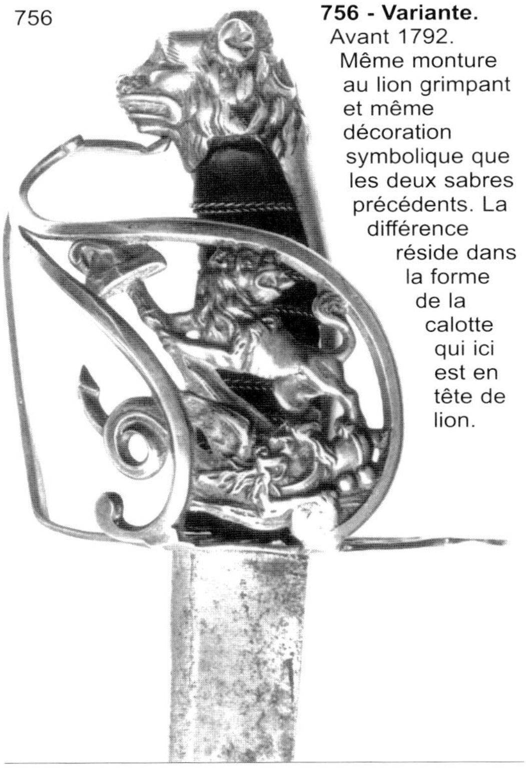

756 - Variante.
Avant 1792.
Même monture au lion grimpant et même décoration symbolique que les deux sabres précédents. La différence réside dans la forme de la calotte qui ici est en tête de lion.

757 - Variante.
Monture au lion grimpant. Calotte en tête de lion.
Lame marquée : *VIVRE LIBRE POUR LA PATRIE* en lettres d'or sur fond bleuï. (Lavallay, op. cit. pl. VIII, p. 92-93), (non figuré).

758 - Variante.
Avant 1792.
Monture au lion grimpant. Sabre à calotte en tête de lion, à longue queue, à plateau ajouré orné d'une grenade et de deux épées croisées et figuré par Ph. Bordes, (op. cit. p.77, fig. 43 bis). Ce qui fait l'originalité de ce sabre est l'inscription gravée sur la lame :

CONSACRE...DANS LE CHAMP DE LA CONFEDERATION FRANCAISE LE 14 JUILLET 1790 PAR J.P. BROCARD COLONEL DE LA GARDE NATIONALE DE BREUVANNE CHEF LIEU DE CANTON

sur une face, et sur l'autre

POUR / LA LOY / LE ROY / ET LA / NATION

le fourbisseur étant *Carrière / md fourbisseur / place du pont / St Michel à la / garde d'or / à Paris*
Ces inscriptions nous démontrent que ce sabre exceptionnel a été porté par un officier de la Garde Nationale du temps de la Monarchie Constitutionnelle (non figuré).

759 - Variante.
Après 1792.
Calotte en côtes de melon, à bouton de rivure et longue queue décorée de filets latéraux. Poignée à gros filigrane d'argent en chaînette. Ornements de la garde identique aux sabres précédents. Lame à la Montmorency, bleuie au tiers gravée de motifs dorés : rinceaux et inscription *LA NATION* sur le *recto* et sur le *verso* une femme représentant la Nation et l'inscription *LA LOY* (non figuré).

Les sabres qui viennent d'être décrits sont généralement désignés sous le nom de *sabres au lion grimpant*. Ils ont été en service à partir de la fin de 1789 jusque vers 1795.

Sous-famille XI e - Au dauphin (1)

Il faut rappeler que le Dauphiné, rattaché à la couronne de France en 1349, est la première province à se rebeller en 1788 contre le pouvoir central. En effet, le 18 juin 1788, c'est la manifestation révolutionnaire dite Journée des Tuiles et le mois suivant, c'est la réunion des Trois Ordres au Château de Vizille. Le 4 août 1789, le Dauphiné qui jusqu'à cette date bénéficiait d'un privilège lui laissant une certaine autonomie, est rattaché, sans restriction à la France. En conséquence, les Dauphinois participent aux guerres révolutionnaires et lèvent une légion dont les officiers et les sous-officiers, pour le moins, portent un sabre à l'emblème de leur province : le dauphin.

On remarquera que dans les armes du Dauphin de France, le dauphin est placé la tête en haut alors que sur les sabres, cet animal est toujours placé la tête en bas. Est-ce voulu ? On notera encore que sur certains sabres, le dauphin est couronné. Sur d'autres, la couronne a disparu. Peut-on en conclure qu'il s'agit de sabres fabriqués après 1792 ? On peut le penser...

Les armoiries du dauphin de France écartelé le premier & dernier azur à trois fleurs de lys d'or, les deux autres d'or à un dauphin d'azur (H. de Bara, 1581)

(1) Nos ancêtres ont toujours considéré le dauphin comme un poisson, d'où sa représentation le corps couvert d'écailles. Pour les naturalistes, rappelons que le dauphin est un mammifère marin voisin des cachalots, des baleines... de la famille des cétacés et en conséquence tout à fait dépourvus d'écailles.

760

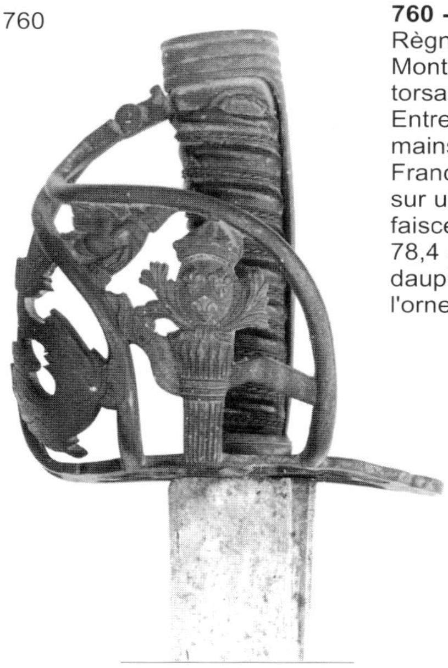

760 - Sabre d'officier monté de la Légion du Dauphiné.
Règne de Louis XVI.
Monture en laiton. Calotte plate et longue queue unie. Poignée filigranée d'un fil fin torsadé. Arc de jointure arrondi, ajouré à sa base d'un triangle et plateau du type a. Entre les branches secondaires jointives, se trouvent près de leur jonction, deux mains soutenant trois cœurs enflammés ; dans la partie centrale, les Armes de France couronnées entre deux palmes nouées à leur base, l'ensemble reposant sur un faisceau de flèches maintenu par une banderole. Il est évident que le faisceau fait allusion à toutes les forces humaines pour soutenir le roi. Lame de 78,4 cm, large au talon de 4,1 cm. Noter que l'exemplaire figuré est orné d'un dauphin entre l'arc et la première secondaire : il s'agit donc d'une modification de l'ornementation destinée à un officier de la Légion du Dauphiné.

761 - Sabre d'officier de la Légion du Dauphiné.
Règne de Louis XVI.
Monture en laiton. Calotte plate à bouton de rivure et bordure guillochée et longue queue. Poignée totalement filigranée de fils de laiton torsadés et bague basale. Petit dauphin ciselé et couronné placé entre les deux branches secondaires. Branche supplémentaire en " S " évidé entre l'arc de jointure et la secondaire antérieure. Lame à la Montmorency (d'après M. Bottet).

761

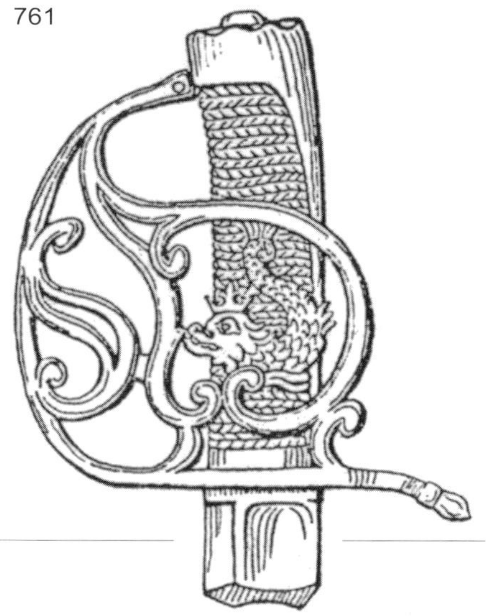

762

762 - Variante.
Monture au Dauphin couronné très comparable à celle du sabre présenté par M. Bottet. Noter que la poignée est encordée, recouverte de basane et filigranée.

763

763 - Sabre d'officier monté de la Légion du Dauphiné.
Règne de Louis XVI.
Monture en laiton. Calotte ovoïde à longue queue unie. Poignée recouverte de basane, à dix spires de ruban de laiton bordé d'un fil torsadé. Arc de jointure droit dans les deux tiers inférieurs, uni, découpé en triangle à sa base. Plateau de garde voisin du type b. Entre les deux branches secondaires se trouve la silhouette d'un dauphin couronné. Entre l'arc et la branche secondaire antérieure est insérée une branche supplémentaire en " S " évidé de renfort. Lame de 79 cm, large au talon de 4 cm, flèche de 2,5 cm et contre-tranchant d'environ 18 cm. Remarquer la rusticité de la monture.

764 - Variante.
Une variante possède le même type de garde que le n° 763, mais les branches sont rainurées et le dauphin est ciselé, notamment le corps est couvert d'écailles ; la couronne est également ciselée ainsi que la pièce qui rattache la queue du dauphin à la branche secondaire, cette ciselure figurant une fleur à trois pétales qui peut être interprétée comme une fleur de lys simplifiée (Ph. Bordes, 1987, p. 53), (non figuré).

765

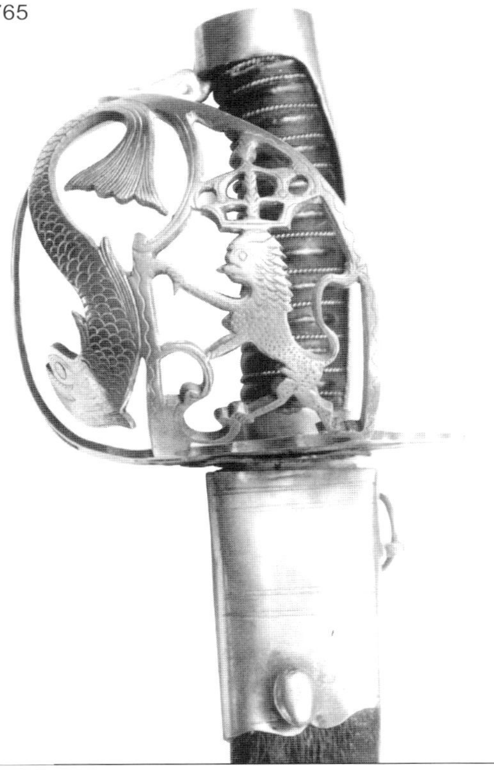

765 - Sabre d'officier monté de la Légion du Dauphiné.
Avant 1792.
Monture en laiton. Calotte plate, à petit bouton de rivure et prolongée d'une longue queue unie. Poignée ficelée, à douze spires d'un petit fil torsadé. Bague basale. Arc de jointure élargi en son milieu formant un médaillon dans lequel est inscrit un coq. Entre les deux branches secondaires se trouve un lion couronné en découpe et entre l'arc et la branche secondaire antérieure, un grand dauphin est inclus (765a). Plateau de garde voisin du type b. Lame de 80,5 cm, large au talon de 3,7 cm, flèche de 2,5 cm et contre-tranchant d'environ 19 cm. Fourreau à trois garnitures de laiton rainurées transversalement : chape à bouton et anneau de bélière ainsi que le bracelet en trapèze : bouterolle avec une longueur maximale de 25 cm y compris le dard de laiton asymétrique.

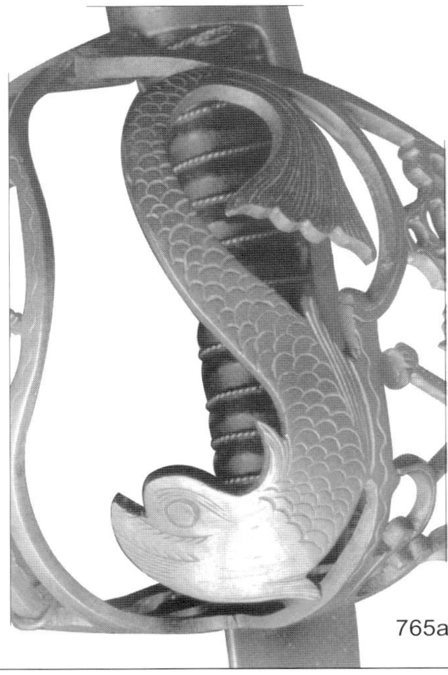

765a

766 - Variante.
Un modèle très voisin est figuré par Ph. Bordes (1987, p. 63, fig. 24). Ce modèle diffère du précédent par la fusée dont le filigrane est constitué d'un ruban de laiton souligné d'un fil torsadé et par le dauphin qui compte un nombre réduit d'écailles (non figuré).

767 - Sabre de sous-officier de la Légion du Dauphiné.
Avant 1792.
Monture en laiton. Calotte plate, à gros bouton de rivure et longue queue. Poignée d'ébène. Arc de jointure entier et plateau de garde du type a (voir p. 21). Entre les deux branches secondaires, petit dauphin en découpe, soigneusement ciselé et dépourvu de couronne. Lame à la Montmorency. Fourreau à deux garnitures de laiton, chape à bouton.

767

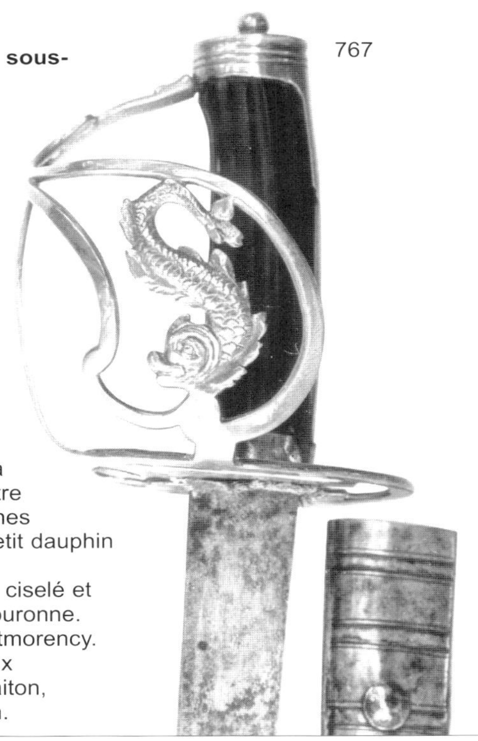

768 - Variante.
Un sabre présente le même dauphin et les mêmes branches de garde que le modèle précédent mais en diffère par la calotte convexe et par une poignée filigranée (Ph. Bordes, 1987, p. 53, fig. 8), (non figuré).

Sous-famille XI f - Aux oiseaux

Rappelons que les oiseaux pris comme emblème par les révolutionnaires de la fin du XVIIIème siècle sont :

- ***le coq***, associé depuis toujours au soleil qu'il salue à son lever
- ***l'aigle***, qui est l'oiseau souverain, aux vertus multiples et qui devait devenir l'emblème de l'Empire,
- ***le pélican***, oiseau fétiche, véritable porte-bonheur, symbole du sacrifice, les anciens croyant qu'il nourrissait ses petits de ses entrailles…

Pour un examen approfondi du symbolisme de ces oiseaux se reporter à l'ouvrage de J. Lhoste et J.J. Buigné (1999).

Il faut constater que bien souvent, les oiseaux symboles sont associés à d'autres attributs révolutionnaires.

769 - Sabre d'officier de la Garde Nationale.
Avant 1792.
Monture en laiton. Calotte en casque empanaché à grille et longue queue rainurée. Poignée filigranée. Arc de jointure arqué, suivi d'un plateau du type d. Entre les branches secondaires jointives, un médaillon à bordure cloutée porte l'inscription : *FIDELITE / A LA LOI / ET AU / ROY.* Ce médaillon est tenu par la gueule d'un lion, avec à gauche et en haut, un bonnet phrygien emmanché sur la massue d'Hercule, une hache et une branche de laurier ; en bas, le coq national, une giberne, une crosse de fusil et un canon ; sur la droite, un sabre… Lame à pan creux de 78,5 cm ; la longueur de la lame incite à penser que ce sabre appartenait à un officier monté.
Noter que sur l'exemplaire figuré, le mot *roy* a été biffé, vraisemblablement en 1792.

769

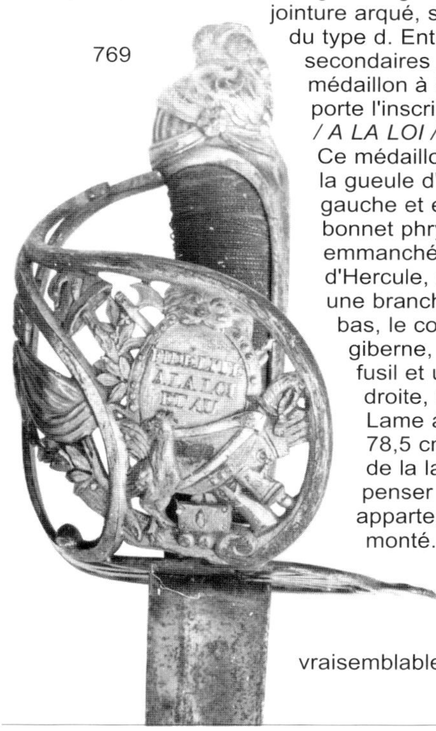

770

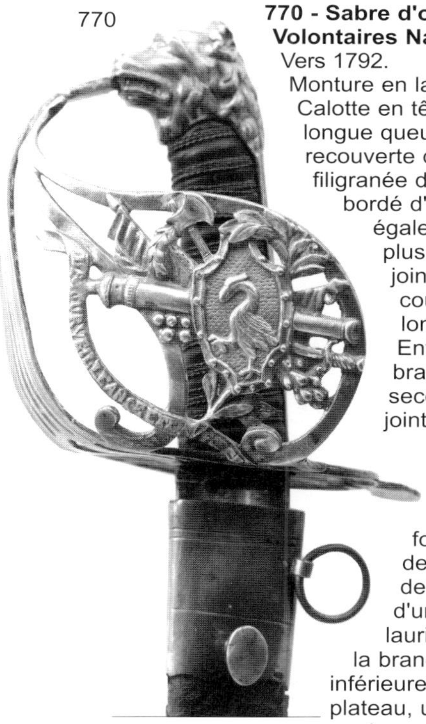

770 - Sabre d'officier des Volontaires Nationaux.
Vers 1792.
Monture en laiton doré. Calotte en tête de lion, et longue queue. Poignée recouverte de basane et filigranée d'un fil torsadé bordé d'un fil également torsadé plus fin. Arc de jointure peu courbé, strié longitudinalement. Entre les deux branches secondaires jointives est inscrit un bouclier octogonal curviligne au pélican sur fond de canon, de hache, d'épée, de cinq boulets, d'un rameau de laurier… Partant de la branche secondaire inférieure et rejoignant le plateau, une banderole sur laquelle ont lit :
LA SURVEILLANCE ET LA FORCE.
Lame à la Montmorency, longue de 71,5 cm et large au talon de 3,4 cm.

771 - Variante.
Après 1792.
Même monture mais avec sur le médaillon l'inscription *VIVRE / LIBRE / OU / MOURIRE* sic (voir La Giberne, 1910 - 1911, XI, P. 39), (non figuré).

772 - Variante.
Après 1792.
Calotte en tête de lion et dans le médaillon, l'inscription : *FIDELITE / A LA LOI* (*in* Ph. Bordes, op. cit. P. 81, fig. 49 a), (non figuré).

773 - Variante.
Après 1792.
Monture identique au sabre précédent. mais dont la calotte et en bonnet révolutionnaire.

773

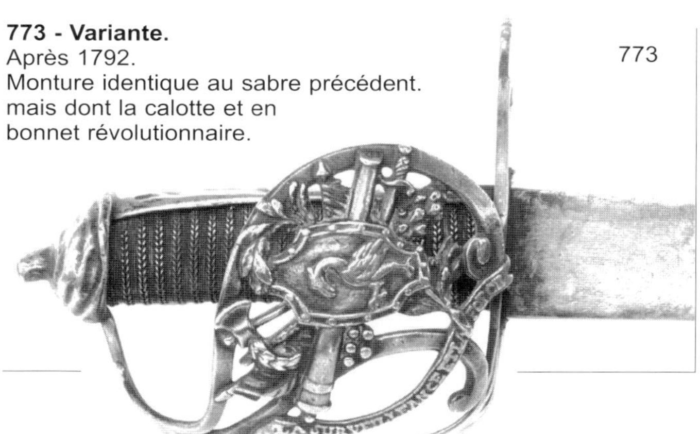

Sous-famille XI g - Aux figures allégoriques

Les figures allégoriques figurant sur les sabres dits *" à grosse coquille "* sont celles d'Athéna, de la Renommée exceptionnellement, de la Victoire, de la République… (voir J. Lhoste et J.J. Buigné, 1999).

774

774 - Sabre d'officier de la Garde Nationale.
Monarchie constitutionnelle.
Monture en laiton. Calotte en tête de lion. Poignée filigranée. Arc de jointure arqué. Deux branches secondaires jointives, la supérieure très ample. Le décor de la coquille, est formé : d'un autel d'où s'échappent des flammes, d'une femme, casquée et vêtue comme Athéna qui personnifie certainement la Nation prêtant serment sur l'autel de la liberté de sa main gauche et qui tient de sa main droite une pique coiffée du bonnet phrygien ; derrière elle, à ses pieds, les Armes de France couronnées ; quelques feuilles de laurier complètent le décor. Sur une banderole, parallèle à l'arc, on lit : *VIVE LA FRANCE ET LA LIBERTE.* Lame de 77 cm, large au talon de 3,6 cm, bleuie au tiers, gravée de motifs dorés. On reconnaît, sur la face extérieure, un bras armé, un chapeau de Gueux, trois fleurs de lys, un trophée de tambour, trompette, drapeaux et sur la face interne on lit : *POUR / LA / LIBERTE / ET LA / GLOIRE / DE LA / NATION.*

775 - Variante.
Avant et après 1792.
Motif de la coquille identique à celle du sabre précédent, mais avec une calotte en casque à grille et panache. Lame de 75,7 cm, large au talon de 3,4 cm, bleuie et gravée en lettres d'or sur la face extérieure l'*HONNEUR . EST . MON . GUIDE* et sur la face intérieure *POUR . LA . NATION . LA . LOI . ET . LE ROI.* Mais cet exemplaire a été mutilé en 1792 : le mot roi a été martelé, les Armes de France enlevées et la lance a été coupée au-dessous de la main de *" la Nation "*, la transformant en un court bâton coiffé du bonnet phrygien (*in* Ph. Bores, op. cit. p. 71, fig. 35 a), (non figuré).

776

776 - Variante.
1792.
Monture en laiton. Calotte en casque à très grand panache et à face léonine. Le motif de la garde représente Athéna prêtant serment sur un autel d'où jaillit une flamme ; de sa main droite elle tient un bâton coiffé du bonnet phrygien. Ce motif est souligné d'une banderolle sur laquelle est inscrit : *VIVE LA FRANCE ET LA LIBERTE.* Noter que les armes de France ont été enlevées.

777

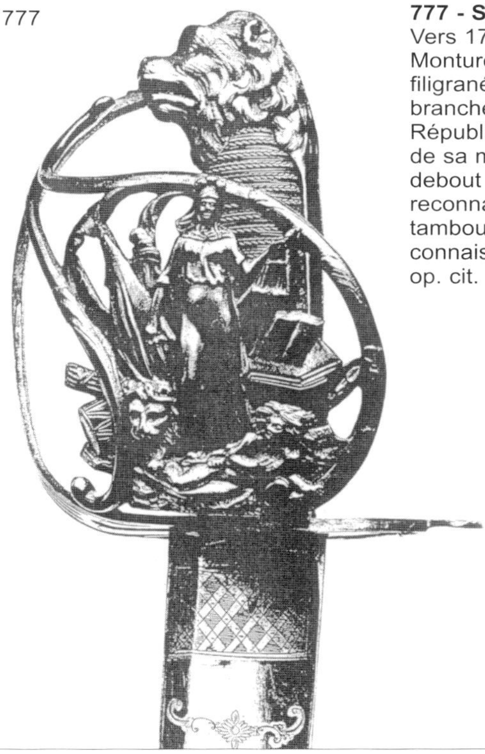

777 - Sabre d'officier des Volontaires Nationaux.
Vers 1790.
Monture en laiton. Calotte en tête de lion et longue queue. Poignée totalement filigranée de fils torsadés en alternance avec des fils simples. Entre les branches secondaires est inscrite une femme, selon toute vraisemblance, la République coiffée du bonnet phrygien, tenant de sa main droite un glaive et de sa main gauche la balance de la Justice. Ce personnage central est debout sur une femme terrassée, hirsute, représentant la tyrannie. On reconnaît encore, de gauche à droite, un drapeau, un faisceau de licteur, un tambour, une face léonine, un livre ouvert symbolisant la loi ou la connaissance. Lame bleuie gravée de motifs dorés classiques. (G. Lavallay, op. cit. Pl. XI).

778 - Variante.
Vers 1790.
Même monture que le sabre précédent. Cette photographie montre la totalité du mot *Constitution* en lettres dorées sur la lame à fond bleuï et le fourreau à trois garnitures (J.C. D.).

778

779

779 - Sabre d'officier des Volontaires Nationaux.
Monarchie constitutionnelle.
Monture en laiton. Calotte en casque à l'antique empanaché et à grille frontale. Les branches secondaires relativement ténues sont peu distinctes des ramifications d'enjolivure. Le motif symbolique est inscrit dans un cercle rejoignant deux arceaux soudés sur le plateau ajouré.
Ce motif représente vraisemblablement Athéna assise au centre d'un faisceau de drapeaux, d'armes... sur la gauche on décèle un blason couronné qui peut être d'un marquis conquis par des idées révolutionnaires modérées !

780

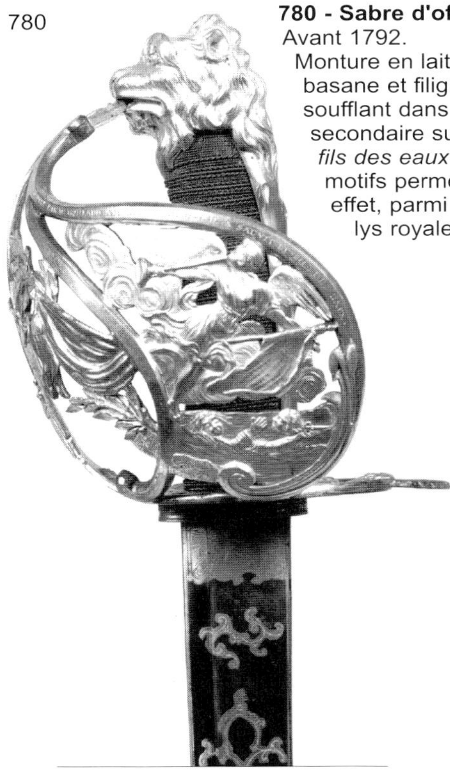

780 - Sabre d'officier supérieur de la Garde Nationale ou Volontaires Nationaux..
Avant 1792.
Monture en laiton doré. Calotte en tête de lion et longue queue. Poignée recouverte de basane et filigranée. Le motif principal décorant la coquille est la Renommée ailée soufflant dans sa trompette vers des nuages et soutenue par un drapeau. Sur la branche secondaire supérieure, on lit : *" offert à M. Guillaud par M. Delorme de qui il a sauvé le fils des eaux "* inscription de reconnaissance personnelle. Lame bleuie et décorée de motifs permettant de dater ce sabre de la période de la monarchie constitutionnelle. En effet, parmi les décors se trouve le chapeau des gueux surmontant les trois fleurs de lys royales...

781 - Sabre d'officier des Volontaires Nationaux.
Autour de 1792.
Monture en laiton. Calotte en casque empanaché, suivi d'une longue queue. Arc de jointure régulièrement arrondi et plateau du type d (voir p. 21). A la jonction des deux branches secondaires sont figurés des nuages masquant partiellement le soleil. Entre les secondaires, Athéna en pied, le bras gauche portant son bouclier et la main droite posée sur un faisceau de licteur ; à sa gauche un lion couché et abattu (?) ; entre l'arc et la secondaire inférieure deux drapeaux et un rameau de laurier. Aux pieds d'Athéna, un canon et deux boulets. Lame à la Montmorency.

781

782 - Variante.
Ph. Bordes, op. cit. p. 86, fig. 57, montre un sabre du même modèle mais à calotte en tête de lion. L'auteur précise que l'arc de jointure possède à sa base, un cœur ajouré ; la lame mesure 73 cm et sa largeur au talon est de 3,3 cm, (non figuré).

Sous-famille XI h - Aux symboles révolutionnaires

783

783 - Sabres d'officier des Volontaires Nationaux.
Epoque révolutionnaire.
Monture en laiton. Calotte en tête de lion, à longue queue. Poignée filigranée. Arc de jointure guilloché. Branche secondaire supérieure décorée de feuillage ; branche secondaire inférieure entièrement transformée en rameau de chêne celui-ci rejoignant le sommet de l'arc. D'autres feuillages de fantaisie viennent s'appuyer sur la base de l'arc. L'ornement symbolique principal est un bonnet phrygien inscrit dans un médaillon entouré de feuilles de chêne et de laurier, avec à sa gauche un autre bonnet placé sur une pique. Lame à pan creux peu marqué, gravée au talon d'un faisceau de licteur chapeauté d'un bonnet phrygien situé au centre d'un trophée de drapeaux (MAP 06383 - K 19202).

784 - Variante.
Un autre sabre à monture identique à celle du précédent possède un médaillon au faisceau de licteur en place du bonnet phrygien (MAP 05885 - K 19209).

784

variante aux faisceaux de licteur.

Chapitre V

Monture à deux branches secondaires arrivant isolément sur l'arc de jointure, calotte à courte queue

On trouvera dans ce Chapitre V, des sabres qui se différencient des sabres décrits dans le Chapitre IV par la présence d'une courte queue de calotte, voire totalement absente. Les branches secondaires arrivant isolément sur l'arc peuvent porter des ornements divers.

Famille XII - Ornementations diverses

Cette Famille XII étant assez polymorphe et comprenant des sabres de la fin du XVIIIème siècle, aussi bien que du Premier Empire et de la Restauration, des Sous-familles ont été créées en fonction de leurs caractéristiques les plus évidentes :

Sous-famille XII a - A décor symbolique

Cette Sous-famille regroupe quelques sabres dont la coquille est ornée d'un soleil, pour le plus ancien ou d'emblèmes maçonniques, de ruches et d'abeilles. On sait que cet insecte fut de tout temps le symbole, voire le modèle, d'une organisation sociale...

785 - Sabre d'officier d'Infanterie.
Règne de Louis XVI. Monture en laiton. Calotte plate à courte queue. Poignée recouverte de basane et filigranée. Branches secondaires arrivant isolément sur l'arc de jointure régulièrement arrondi et supportant un soleil. Plateau du type d (voir p. 21). Lame droite à gouttière. Fourreau à deux garnitures avec chape à bouton en écusson (785a).

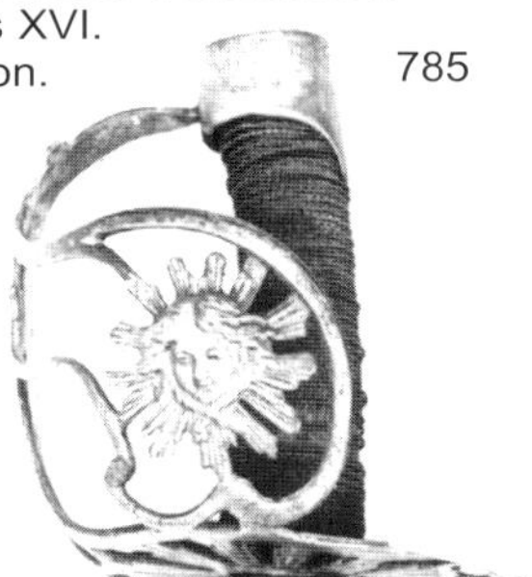
785

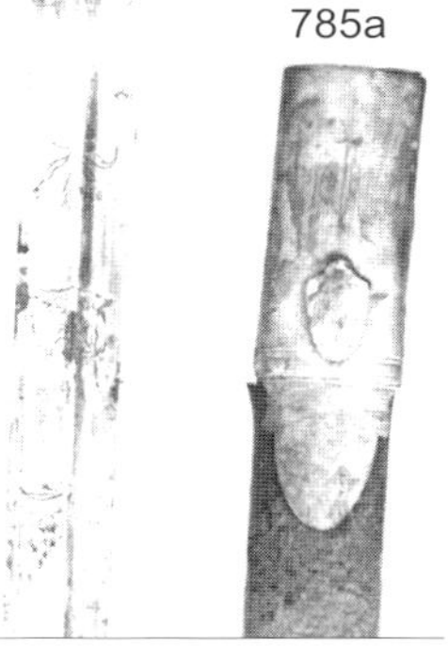
785a

786 - Sabre d'officier des Volontaires Nationaux.
Monarchie constitutionnelle. Monture en laiton doré. Calotte à tête de lion, sans queue. Poignée entièrement filigranée de fils torsadés. Arc de jointure ciselé à jour sur sa moitié basale, de rubans et de feuilles d'acanthe, précédant un plateau du type d. Branche secondaire supérieure ornée de feuilles d'acanthe, ainsi que la petite branche secondaire inférieure. Entre ces deux branches, grand décor foliacé avec une ruche en son milieu, posée sur un faisceau de licteur et un drapeau. Autour de la ruche volètent quatorze abeilles. Lame de 70,5 cm, large au talon de 4 cm, à la Montmorency, bleuie et gravée sur le *recto VAINCRE / OU / MOURIR* et de divers motifs dorés et sur le *verso*, une femme personnifiant la Nation avec au-dessous, *LA LOI / ET / LE ROI*, le mot roi ayant été biffé ultérieurement.

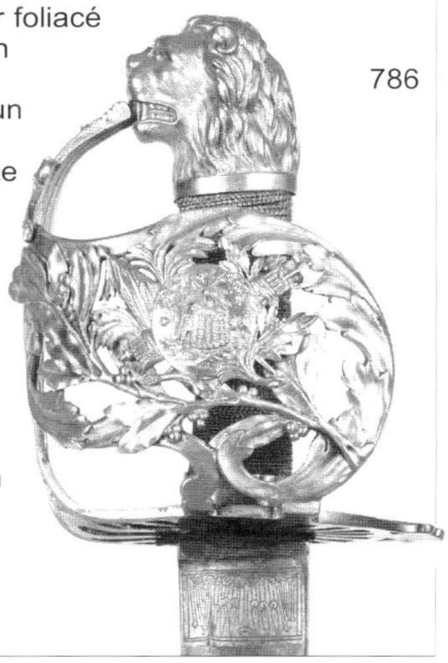
786

Sous-famille XII b - A palmette

Le sabre de la Maréchaussée du modèle 1783 est bien connu et se rencontre assez fréquemment. Le sabre d'honneur est beaucoup plus rare et sa monture est en argent.

787 - Sabre de troupe de la Maréchaussée.
Modèle 1783.
Monture en laiton. Calotte aplatie à bouton de rivure plus ou moins important et courte queue. Poignée totalement filigranée. Arc de jointure régulièrement arqué, recevant séparément les deux branches secondaires en " S " partant de la coquille. Cette coquille cannelée est prolongée par deux petites branches arquées. Plateau de garde se terminant par un quillon très court. Lame plate (787b) de 88 cm, large au talon de près de 3 cm et se terminant par un biseau de 17,5 cm. Cette lame peut être marquée *Mf*^{ture} *R*^{le} *du Klingenthal mai 1790* ou simplement *Manufacture De Klingenthal* ou encore, en période révolutionnaire ne porter que le poinçon au licteur... Fourreau de cuir noirci sans alèse, primitivement à deux garnitures : courte chape de 4 cm en arrière et de 6,4 cm en avant étant prolongée d'une partie arrondie et bouton ; bouterolle de 8,5 cm terminée par un bouton, puis à trois garnitures (787a). La lame peut être à pan (787c).

787

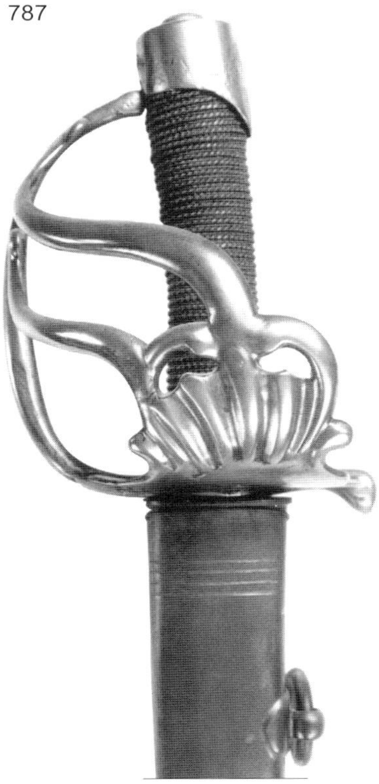

- **a**, détail de la monture.

787a

- **a**, ultérieurement le fourreau sera à trois garnitures comme le montre la figure ci-jointe.

787c

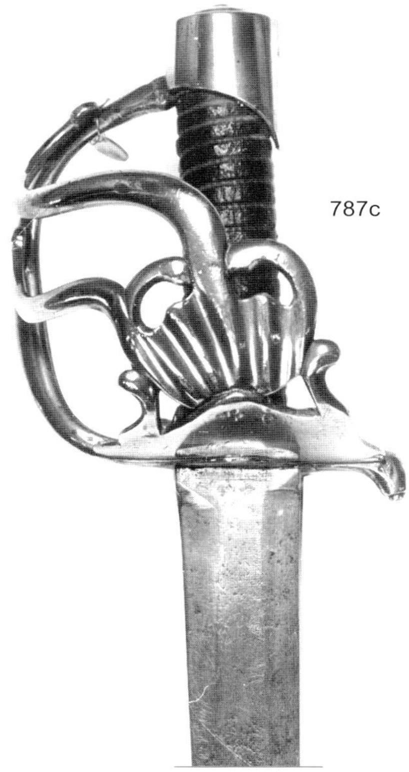

- **c**, monture légèrement différente et lame à méplat médian, dont la poignée est recouverte de cuir filigrané.

787b

- **b**, monture légèrement différente et montrant que la lame des premières fabrications était plate.

788 - Sabre d'Honneur de Grosse Cavalerie, Consulat.
Monture en argent poinçonné. Calotte ciselée de feuillage en rosace, cerclée d'une moulure striée ; queue courte, très échancrée latéralement. Poignée ficelée, recouverte de basane, à filigrane disposé en chaînette. Arc de jointure faisant un angle droit avec la croisière. Coquille formée d'une palmette à cinq feuilles bien détachées les unes des autres, reliées à leur partie supérieure par de petites branches avec ou sans fioritures et reposant sur un triangle à grosse rainure marginale. Quillon terminé par un aplat. Lame à pan creux. Fourreau en tôle d'acier à deux bracelets plats, argentés, bouterolle d'argent terminée par un bouton en guise de dard.

Marquage - Le dessous du plateau de Garde peut être signé *Entse Boutet MF^{re} à Versailles.*

788

789

789 - Variante pour officier
Les palmes sont réunies entre elles par de petites bandes ornées d'une demi-fleurette pouvant être assimilée à une demi-marguerite. Fourreau d'acier orné de garnitures en laiton doré, le bracelet de bélière étant décoré d'un trophée aux drapeaux, avec au-dessus la dédicace (illisible).

Autre variante - Certains modèles ont les palmes réunies par des anneaux (voir C. Ariès, XV, 7, 1970).

B. Malvaux (1993) décrit deux sabres de ce types :

790 - Sabre d'honneur de Casimir Girod de Novillard.
Lame signée sur le dos *Manufacture Nale du Kligenthal Coulaux Frères Ent^{rs}* avec les poinçons de Jean-Jacques Mouton et de Pierre Claude Levavasseur (vente J.C. D. Bayeux), (non figuré).

791 - Sabre d'Honneur du capitaine Deblon.
Dont la lame est signée *Mfture du Klingenthal Coulaux Frères Entrep^{rs}* avec les mêmes poinçons que le sabre précédent, mais le fourreau est signé de la *Mafre de Versailles*. Les dédicaces sont inscrites sur la face frontale du fourreau (voir J. Lhoste et J.J. Buigné, 1999, p. 130), (non figuré).

Sous-famille XII c - A la grenade

Cette Sous-famille réunit principalement les sabres des Grenadiers à cheval des Gardes Consulaire et Impériale mais aussi des sabres de la Garde Nationale parisienne.

Les sabres décrits ci-après ont tous une grenade inscrite dans un médaillon (a). L'arc de jointure s'élargit en un plateau terminé par un quillon infléchi ciselé d'une palmette. Deux branches secondaires : la supérieure part de la calotte et, après une courbe en " S " vient contourner, plus ou moins, un médaillon dans lequel une grenade découpée est inscrite ; l'inférieure part du milieu de l'arc et vient soutenir la base du médaillon. Lame plate ou à pan creux et à dos, de 96,5 à 97,5 cm, large au talon de 2,8 à 3 cm, flèche de 2,2 à 3 cm. On peut distinguer (cf. M. Pétard, 1990) trois types de ce modèle, surtout en considérant la structure du fourreau (b).

Type 1 - de l'An IX (1801-1802).
Fourreau de cuir à trois garnitures unies, découpées marginalement en accolade. (b1).

Type 2 - de l'An XI (1802-1803).
Fourreau de laiton à attelles ménageant un crevé de 18-20 cm, parfois renforcé par des bagues de laiton. (b2).

Type 3 - de 1810.
Fourreau à deux crevés de chacun 8,4 cm et bracelets de bélière de 2,2 cm de hauteur, (n°794 et b3). (M. G. de Melun).

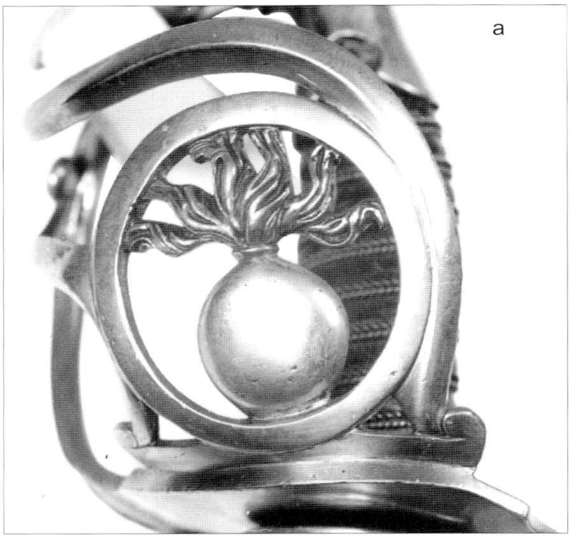

- **a**, médaillon à la grenade des montures des gardes consulaire et impériale.

Ces sabres ont été dessinés par M. Boutet et souvent portent sa signature. En 1806 ils armèrent les Dragons et la Gendarmerie d'Elite de la Garde Impériale.

Hélas les exemplaires répondant strictement aux descriptions originales sont rares. Beaucoup ont été réparés et le *Tarif des réparations* de 1806, *indique qu' avec des débris de vieux sabres* (on peut) *en former un nouveau complet avec des parties jugées bonnes, remonter le tout et nettoyer les pièces au sablon* ... Et puis, il y a les réparations ultérieures.

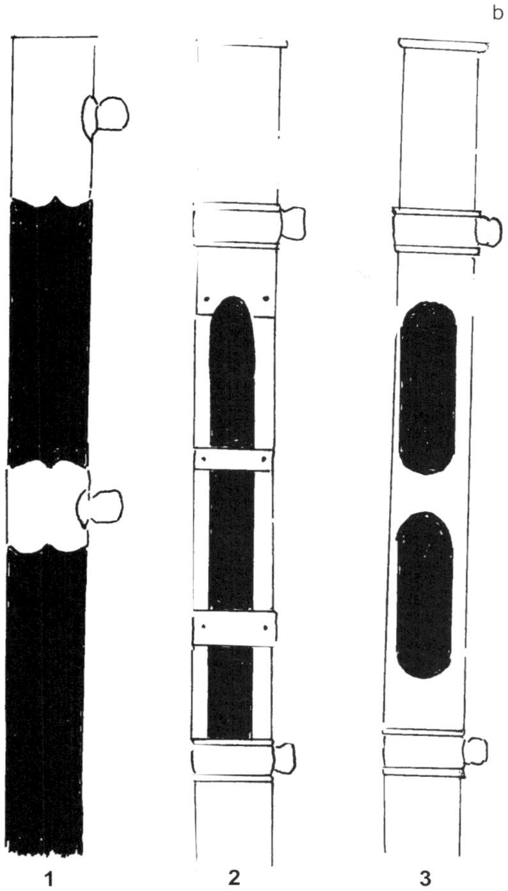

- **b**, schémas des fourreaux des sabres de la Garde consulaire et de la Garde Impériale :
  - 1, **type 1**, 1801-1802 ;
  - 2, **type 2**, 1802-1803 ;
  - 3, **type 3**, 1810.

Description du modèle pour la Troupe.
Monture en laiton, à deux branches secondaires. Calotte à bourrelets, à dôme-bouton de rivure et courte queue. Poignée encordée recouverte de basane, à une vingtaine de tours d'un filigrane constitué de deux fils de laiton torsadés. Arc de jointure régulièrement arrondi, les deux branches secondaires fusionnant dans la partie supérieure de l'arc. Plateau ovoïde terminé par un bec uni. Lame de 92 cm, large au talon de 3,1 cm avec une flèche d'environ 3,8 cm, à pan creux et gouttière au fort. A noter que les lames fabriquées avant 1860 ont le pan creux qui se termine en carré près du talon, alors que chez les lames postérieures à 1860, le pan creux se termine en fuseau (810a, b, et c.). Fourreau en tôle d'acier, à cuvette rivée et deux bracelets en jonc. Dard asymétrique. Poids total : 2,055 kg.

Les lames proviennent de la *Manufacture Royale de Klingenthal* et portent le mois et l'année de leur confection. Plus tard, la Manufacture de Châtellerault prendra la relève et on trouve des lames encore datées de janvier 1904.

808

808 - Modèle 1822, troupe.
Cet exemplaire est signé, sur le dos de la lame *M^{re} d'Armes de Chât. Avril 1875, Cavrie 1re M^{le} 1822.*

809

809 - Sabre de Cavalerie légère du modèle 1822, troupe.
Le second anneau de bélière du fourreau sera supprimé à partir de 1882.

810

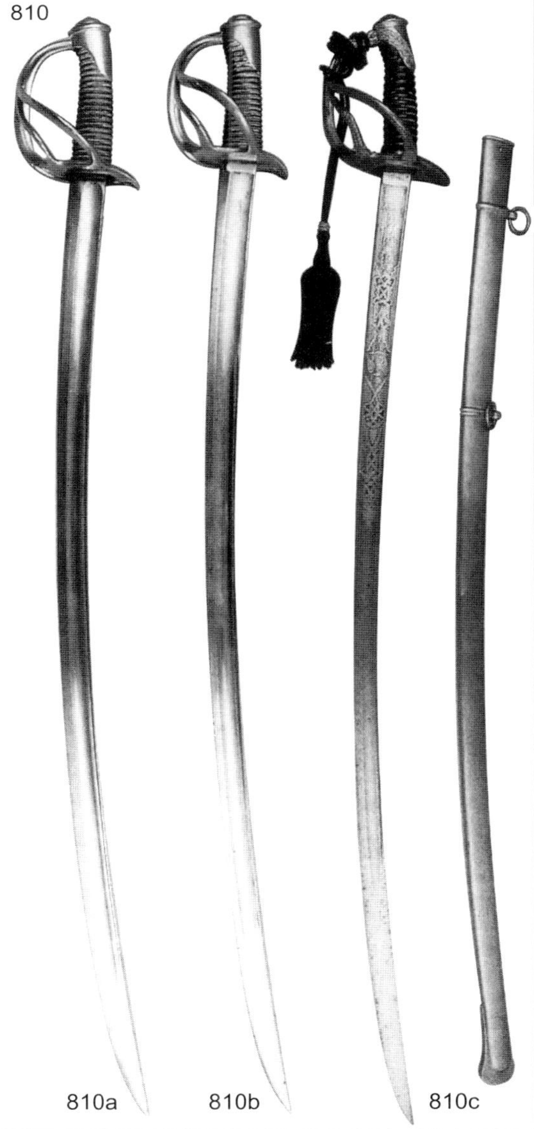

810a 810b 810c

810 - différents types de modèle 1822.
- **a**, vers 1850 ;
- **b**, avant 1860. Remarquer le talon de la lame ;
- **c**, après 1860. Sabre d'officier de Cavalerie légère avec lame gravée et dorée sur le tiers de sa longueur (voir n° 811).

811

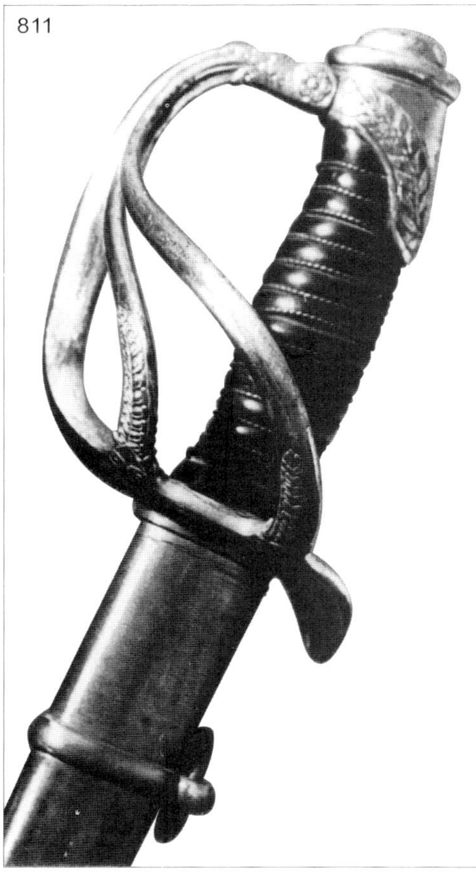

811 - Sabre d'officier de Cavalerie légère.
Modèle 1822.
Ce modèle est l'un des premiers qui soit devenu réglementaire pour les officiers. Il se distingue du modèle de troupe par les enjolivements suivants :
Monture dorée. Calotte dont le dôme-bouton de rivure est placé sur un support à deux niveaux vers 1840. Courte queue bordée d'une guirlande de feuilles de laurier. Arc de jointure ciselé d'une rosette et de petites feuilles d'acanthe à son extrémité supérieure. Branches secondaires décorées à leur base d'une palme partant d'une demi-fleur de lys en 1822 et à partir de 1830, de feuilles d'acanthe. Plateau de garde se terminant en " bec " celui-ci orné d'une palmette à sept branches surmontée en 1822 d'une demi-fleur de lys et à partir de 1831, de deux petites feuilles d'acanthe coincées entre deux rosettes (811a) et en 1840 de feuilles d'acanthe disposées en quatre lobes subégaux.

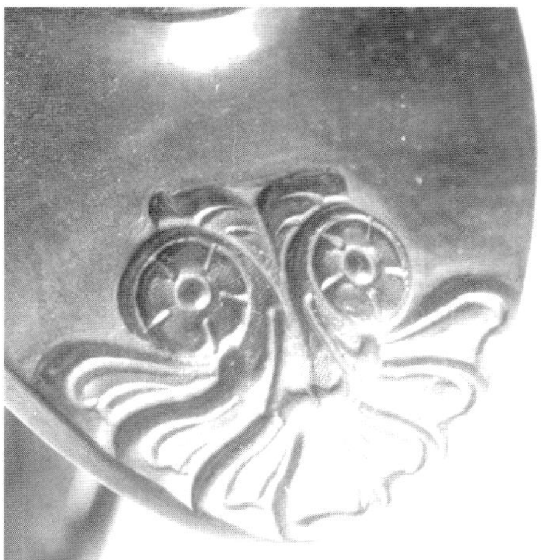

811a, décor du plateau en 1831

Selon la circulaire du 28 septembre 1820, toutes les lames des modèles 1822 devaient être fournies par la Manufacture de Klingenthal et signée *Coulaux et Cie*. Avec la création de la Manufacture de Châtellerault, cette décision ne fut plus respectée. On trouve donc des lames portant la signature de l'une ou de l'autre manufacture, signature suivie de la date de la mise en service. Parfois, à la suite de la signature, se trouvent les deux lettres *S.a.* qui signifie : *soie en acier.*

812

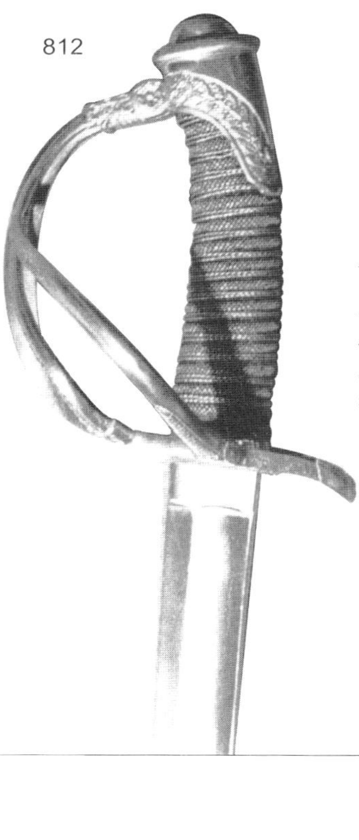

812 - Variante.
Ce sabre d'officier de Cavalerie légère du modèle 1822 est une fantaisie de son propriétaire, probablement datant de 1850-1860. Toute la monture est dorée à la feuille, la poignée est recouverte de galuchat filigrané et les ciselures sont très finement travaillées.

813 - Sabre des élèves des écoles militaires.
Modèle 1822.
Stiot et Pétard (1988) rappellent que le modèle 1822 d'officier de Cavalerie légère est remis aux élèves de l'Ecole Militaire de Saint-Cyr nommés sous-lieutenants dans le corps d'Etat-Major et la Cavalerie de 1822 à 1830 avec la mention *DONNE PAR LE ROI* gravé en majuscules et signé du fourbisseur *Manceaux / à / Paris* sur le plateau et sur le haut du fourreau. Quelques exemplaires de ce sabre furent également offerts aux *Volontaires de la Charte* (non figuré).

816 - Sabre d'officier d'Artillerie.
Modèle 1822 - 1899.
Monture identique au modèle 1822, à un arc de jointure et deux branches secondaires en laiton doré. Calotte à courte queue, ciselée d'une bordure de feuilles de laurier. Plateau à bec orné de feuilles d'acanthe disposées en croix, au départ d'une palmette (813a). Lame demi-courbe à dos plat, pan creux et gouttière, signée sur le dos, par exemple : *Manufacture Nationale d'Armes de Châtellerault 9bre 1911, OFF^{ier} d'ART^{rie} Mle 1822-99,* l'ensemble sur une seule ligne. Fourreau en tôle d'acier à un bracelet de bélière, dard asymétrique.
En 1914, ce sabre est attribué aux corps suivant : officiers de Chasseurs et de Hussards, Chasseurs d'Afrique, Spahis algériens (1), Médecins auxiliaires des troupes à cheval, Vétérinaires militaires, Officiers interprètes, Fonctionnaires de la Télégraphie militaire, Officier de Cavalerie instructeur d'Equitation, Officiers du cadre de Cavalerie de l'Ecole Spéciale militaire, de l'Ecole d'Application de Cavalerie, de l'Ecole militaire préparatoire de Cavalerie.

816a

- **a**, ornementation du bec.

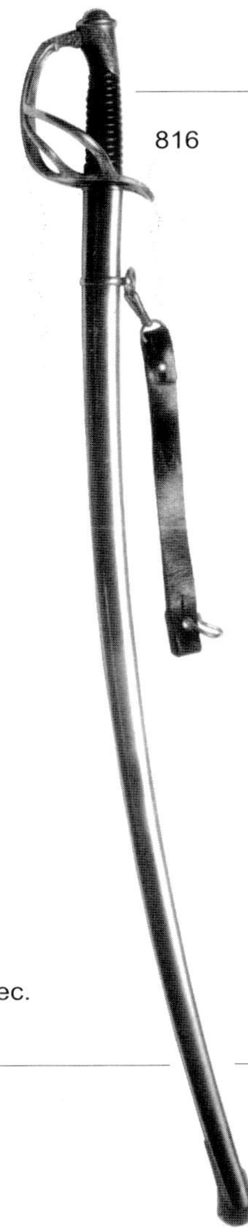
816

(1) Pour les Chasseurs d'Afrique et pour les Spahis les sabres du modèle 1822 qui leur sont attribués ont souvent une lame finement gravée à l'acide, voire bleuie et gravée de motifs dorés.

817

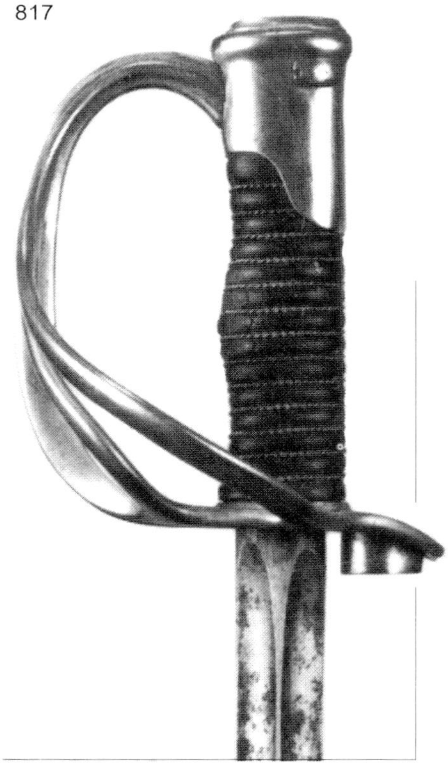

817 - Sabre lance Arcelin.
Modèle 1856.
Monture en laiton. Calotte à bourrelets et courte queue, avec son système de verrouillage pour la fixation sur le mousqueton. Poignée recouverte de cuir, renflée vers la garde et filigranée d'un fil de laiton torsadé. Arc de jointure arqué sur lequel fusionnent en sa moitié supérieure, deux branches secondaires. Plateau ovale. Bec muni d'une bague évidée pour recevoir le canon du mousqueton. Lame droite de 100 cm à section triangulaire du type Préval. Fourreau en tôle d'acier à deux bracelets de bélière et dard en lyre. Poids : environ 2 kg. Le mousqueton Arcellin et son sabre ont été mis à l'essai chez les Carabiniers, les Hussards et aux Dragons de la Garde.

818 - Sabre d'officier de Cavalerie légère de la Garde Impériale.
Second Empire.
Monture de laiton doré. Calotte à bourrelet ciselé en segments s'ajustant aux secteurs dont le dôme servant de bouton de rivure est décoré ; queue courte, bordée d'une frise de feuilles de laurier avec sur la partie dorsale, une aigle aux ailes déployées. Poignée recouverte de galuchat ou de roussette, à double filigrane formant chaînette et onze spires. Deux branches secondaires partent du sommet de la branche principale, ciselées sur les deux faces d'une frise de chevrons. Plateau décoré à sa partie postérieure d'une palmette. Lame et fourreau du modèle 1822. Modèle de fourbisseur (M.E. - S. de P.), (non figuré).

Avant d'adopter le sabre du modèle 1882, la Manufacture de Châtellerault étudia différentes montures dont celle signalée par E.B.Bredène et A. de Jenlis (Tradition, 146, p. 29). Ce sabre est du type de Cavalerie légère à deux branches secondaires, à calotte à courte queue. Il porte le N° 427 avec sur le dos de la lame de 87 cm : *Mre d'Armes de Chât[t] avril 1880 Cav[rie] légère*. On peut considérer ce sabre comme l'ancêtre du modèle 1882.

Sabre de Cavalerie Modèle 1822

Ce modèle fut créé selon la note ministérielle du 22 août 1882. Monture de laiton, à deux branches secondaires. Calotte convexe, à queue courte, recouvrant les quatre premiers tours du filigrane de laiton de la poignée recouverte de basane. Les deux branches secondaires sont amorcées dès le sommet de l'arc de jointure. Petit plateau de garde terminé en languette, sans véritable quillon. Lame droite, légère, à pan creux et dos plat. Fourreau en tôle d'acier à un seul bracelet pour le port à la selle, devenu obligatoire entre 1884 et 1887. La longueur de la lame est de de 95 cm pour le sabre de Cavalerie de réserve (Cuirassiers), 92,5 cm pour le qsabre de Dragons et de 87 cm pour le sabre de Cavalerie légère.

819 - Sabre de Cavalerie de réserve, pour les Cuirassiers.
Modèle 1882.
Poids du sabre : 1,120 kg, poids du fourreau : 0,730 kg. Lame de 95 cm (non figuré).

820 - Sabres des Dragons.
Modèle 1882.
Poids du sabre : 1,076 kg, poids du fourreau : 0,710 kg. Lame de 92,5 cm (non figuré).
Ces deux derniers sabres n'ont pas été mis en service, la lame étant jugée trop fragile.

821 - Sabres de Cavalerie.
Modèle 1854/1882.
Ce sabre est composé d'une lame de Cavalerie de Réserve du modèle 1854 à double pans creux et d'une monture 1882. Longueur de la lame : 95 cm pour les cuirassiers et 92,5 cm pour les Dragons, (non figuré).

822 - Sabre de Cavalerie légère.
Modèle 1822 - 1883.
La Cavalerie légère désirant un sabre à lame droite et les livraisons du modèle 1882 tardant, il fut décidé de transformer, en 1883, le sabre de cavalerie légère modèle 1822 en redressant la lame et le fourreau et en raccourcissant la lame de 50mm (87 cm au lieu de 92 cm). Ces lames ne supportant pas ce traitement, il ne fut réalisé que 3000 exemplaires environ de ce modèle, (non figuré).

823 - Sabre d'officiers de Cavalerie.
Modèle du 29 août 1883.
Même monture générale que pour la troupe, mais réalisation plus soignée : la calotte est à bordure ciselée de feuilles de laurier, la base des deux branches secondaires est décorée d'une palme, la fusée est en corne noire cannelée et filigranée, (non figuré).

824 - Sabre de Cavalerie.
Modèle 1896.
Pour mémoire ce sabre était monté avec la lame et le fourreau du modèle 1882, (non figuré).

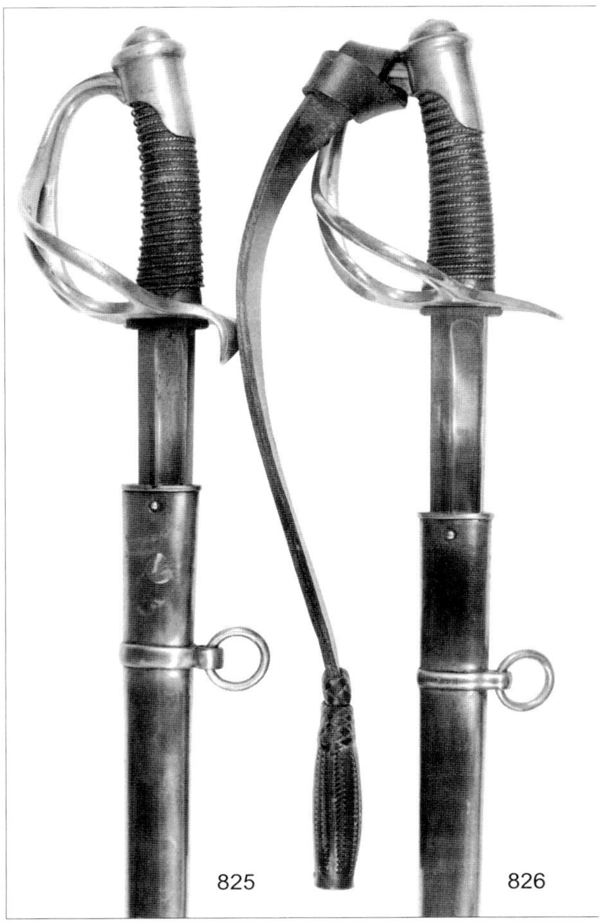

825 826

825 - Sabre de Cavalerie légère.
Modèle 1882.
Lame de 87 cm. Poids du sabre : 1,080 kg, poids du fourreau : 0,670 kg. L'exemplaire figuré est signé de la *Manufacture de Châtellerault* et porte la date de 1885.

826 - Sabre de Dragons.
Modèle 1882.
Monture en laiton. Calotte à bordure supérieure surmontée d'une pièce de rivure en dôme et courte queue de calotte unie à bordure découpée en " S ". Poignée recouverte de basane filigranée d'un fil de laiton torsadé faisant 19 spires. Arc de jointure modérément arrondi recevant dans sa moitié supérieure les deux branches secondaires partant du plateau, terminé par un bec uni. Lame droite, longue de 92,5 cm à pan creux marquée sur le dos : *Manufacture d'armes de Châtellerault août 1886 Dragon Mle 1882,* par exemple. Fourreau de tôle d'acier à un bracelet étroit portant le seul anneau de bélière.

<u>Nota</u> - L'exemplaire figuré porte sur la monture et le fourreau les initiales S.M. (1) et est pourvu de sa dragonne réglementaire en cuir terminée par un gland tressé.

(1) Les Notes ministérielles des 12 octobre 1887 et 9 mars 1906 créent une catégorie spéciale de 330 sabres de corvée par régiment marqués des lettres S.M.

828 - Sabre d'officier de Cavalerie.
Modèle 1883.
Ce sabre a des affinités évidentes avec le modèle 1882.
Monture en bronze demi-rosé. Calotte décentrée par rapport à la fusée et partie inférieure ainsi que le bas des branches secondaires ornées de frises de feuilles de laurier.
Poignée en corne à 14 - 15 torons filigranés de fil de laiton doré et torsadé.
Plateau à bec décoré d'un motif feuillu. Lame droite allégée par rapport à celle du modèle 1882, à pan creux portant sur le dos de la lame *Manufacture d'armes de Châtellerault* (mois, millésime par exemple : *septembre 1886*), *officier de Cavalerie M*^{le} *1883*.
La lame porte également la signature du fourbisseur : *Entreprise Henry*.
Fourreau en tôle d'acier nickelée, à un anneau de bélière et dard en lyre.

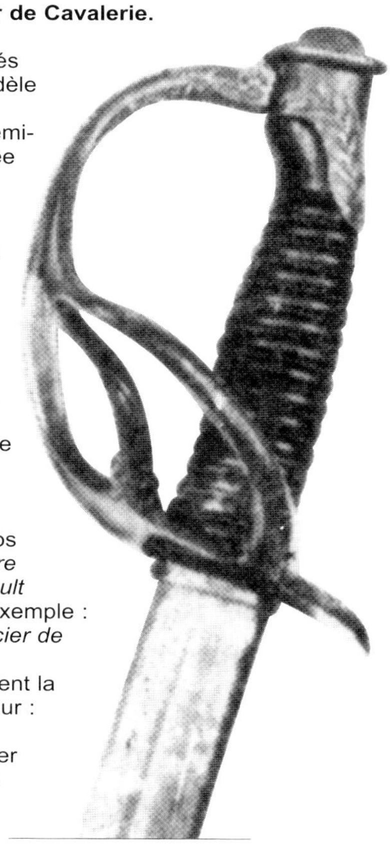

828

829

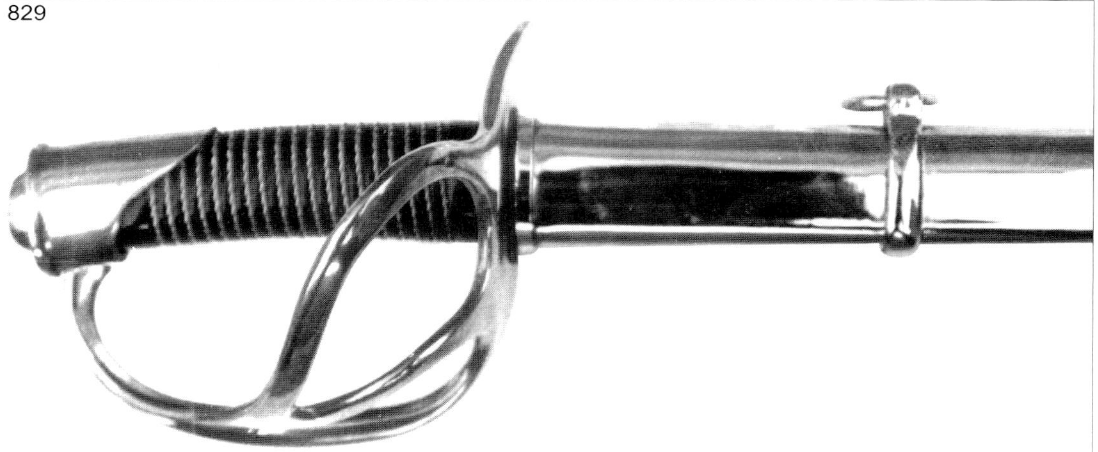

829 - Sabre d'officier de Cavalerie.
Modèle 1822 - 1961.
Monture en laiton. Calotte à bordure supérieure, pièce de rivure en dôme et courte queue à bordure presque droite. Poignée recouverte de basane, filigranée d'un fil de laiton torsadé faisant 19 spires. Arc de jointure modérément arrondi, recevant, vers la moitié de sa longueur deux branches secondaires partant du plateau terminé par un bec. Lame courbée portant la signature de Balp. Fourreau en tôle d'acier nickelée à un bracelet étroit portant l'anneau de bélière. Modèle réglementaire pour les sous-officiers de la Garde Républicaine.

Sous-famille XIV b - Branches secondaires portant un symbole de corps ou des armoiries

La monture des sabres de cette Sous-famille XIV b, possède en général, insérées entre les branches secondaires, une ancre, des armoiries royales de la Restauration ou celles du Second Empire.

831

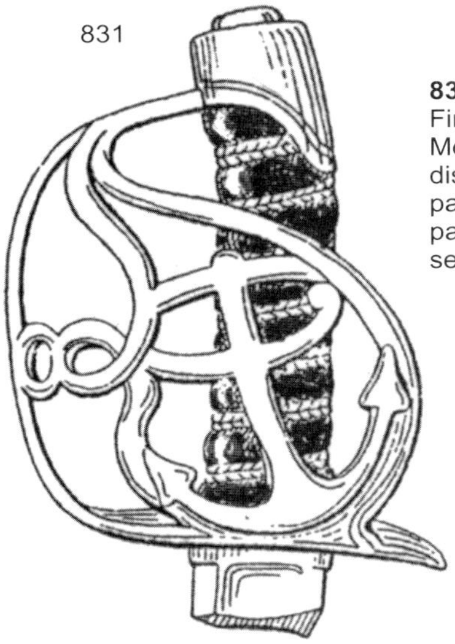

831 - Sabre de bord.
Fin du XVIIIème siècle.
Monture en laiton. Ce sabre se distingue du modèle p. 376, n° 735 par une queue de calotte courte et par un oeuillet réunissant la branches secondaire en " L " à l'arc de jointure.

833 - Sabre de la Garde Royale de Paris.
Première Restauration.
Monture de bronze. Calotte à queue courte recouverte de basane à fil torsadé faisant une quinzaine de spires. Arc de jointure peu arqué, droit sur les deux tiers et faisant un angle droit avec le plateau, celui-ci terminé par un quillon semi-circulaire ciselé d'une palmette à neuf feuilles. Deux branches secondaires partent du haut de la branche principale, à arête médiane, rejoignant le bord du plateau par deux petits enroulements symétriques. Entre les deux branches secondaires est fixée une grosse fleur de lys. Une banderole portant *GARDE ROYALE DE PARIS* part de la branche principale, passe sous la première branche secondaire et vient finir sur la deuxième branche secondaire. Lame longue et droite, à deux pans creux et pointe dans le prolongement du dos. Fourreau de cuir noir, à trois garnitures simples, longue bouterolle à dard asymétrique.

Marquages - Signature de Manceaux sur le bas de la branche principale. Sur le dos de la lame *Manuf^{re} R^{le} du Klingenthal, décembre 1814,* et au talon, *poinçons de Lobstein, de Joseph Bisch, de Borson.*
La Garde Royale de Paris est le nom donné au retour de Louis XVIII à la Gendarmerie Impériale de Paris créée sous le Premier Empire. En 1816 elle prendra le nom de Gendarmerie Royale de Paris. En 1830 Louis-Philippe mettra un terme à l'existence de cette formation.

833

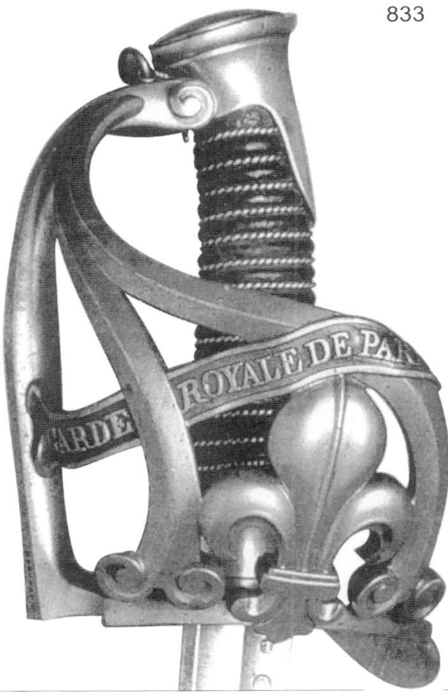

Remarquer la banderolle soudée à l'arc de jointure passant sous la première branche secondaire et s'enroulant sur la seconde. Le fourreau est à trois garnitures et anneaux de bélière (d'après A. de Jenlis et M. de Fromont, 1995.)

834

834 - Sabre d'officier des Cuirassiers de la Garde Royale.
1816-1830.
Monture en laiton doré. Calotte à bourrelet surmonté d'un dôme-bouton de rivure, à courte queue bordée d'une frise de feuilles de laurier. Poignée recouverte de chagrin ou de roussette, filigranée d'un fil torsadé de laiton doré. Arc de jointure régulièrement arqué. Entre les deux branches secondaires sont insérées les grandes Armes de France (831a) flanquées de deux branches de laurier et de six drapeaux. Plateau de garde à bec ciselé d'une palme. Lame à dos rond, droite, à deux pans creux et signée de la *Manufacture Royale du Klingenthal* avec la date de fabrication, par exemple *juin 1820*. Fourreau en tôle d'acier, à deux bracelets étroits et dard en lyre.

834a

- **a**, détail simplifié des grandes Armes de France.

Nota - Sous le Second Empire, les Cent Gardes seront équipés d'un sabre élaboré sur un modèle voisin, mais à quatre branches.

835 - Sabre d'officier de Cavalerie de la Garde Impériale.
Second Empire.
Monture en bronze doré. Calotte ciselée en quartiers et bordure à facettes. Courte queue à bordure de feuilles de laurier, avec sur le dos une aigle impériale posée sur un bâton de foudre surmontant une fleurette. Poignée en corne, à filigrane en chaînette. Branche principale régulièrement arquée ; deux branches secondaires ornées, sur les deux faces, d'une frise de feuilles disposées en chevron ; entre ces deux branches est inséré un médaillon ciselé à jour de l'aigle impériale couronnée. Lame droite, à deux pans creux, gravée à l'acide de feuillages,. Fourreau en tôle d'acier poli à deux bracelets étroits (M.E. - S. de P.), (non figuré).

Marquages - Sur la lame, gravée à l'acide *H. Rouher Major du 7e Lancier*, évidemment le nom du propriétaire de l'arme.

Chapitre VII

Monture à un arc de jointure principal et un arc secondaire réunis entre eux par une branche secondaire oblique, calotte à courte queue

Ces sabres sont facilement identifiables : ils ne possèdent pas de queue de calotte ou d'une queue très-courte et l'arc de jointure normal est accompagné d'un second arc frontal unissant le plateau de garde à la calotte. De plus, une branche secondaire part de la base de l'arc frontal pour rejoindre le haut de l'arc principal, l'ensemble dessinant la lettre " N ". On donne souvent le nom de *monture en demi-panier* à ce dispositif. Avec le temps, le plateau a évolué dans sa forme (fig 1, a, b, c,) ainsi que la forme des lames (fig. 2 a, à f) et des fourreaux (fig. 3 a et b).

Cette catégorie de sabres a été attribuée aux Dragons de Choiseul, ministre de la Guerre en 1761. On peut rencontrer des sabres d'Infanterie ayant le même type de monture mais il s'agit alors de sabres fantaisie armant une milice ou une garde locale. Bien qu'en service jusque sous la Restauration, les sabres de ce type sont très homogènes et tous ont été regroupés dans la Famille XV.

Nota - M. pétard (1992) a retracé l'évolution des sabres réglementaires dans une intéressante étude que le lecteur pourra consulter avec profit. Cet auteur cite également un sabre à garde en panier complet (fig. 4).

Présentement, seuls les sabres les plus fréquemment rencontrés seront décrits ci-après.

Famille XV - Monture en demi-panier

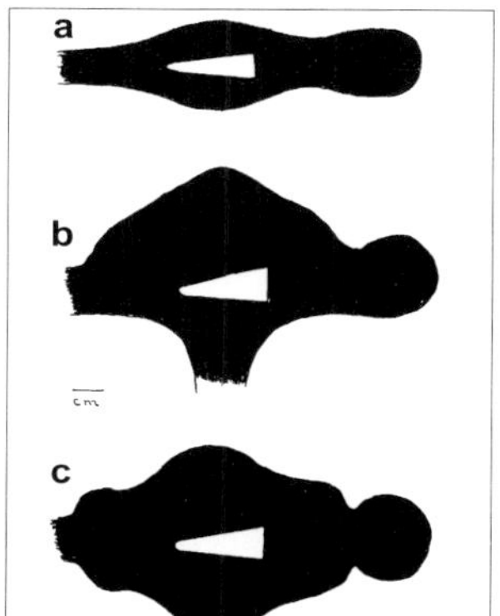

Figure 1

Figure 1 - Plateaux de garde de sabres à monture en demi-panier.

- **a**, modèle étroit et symétrique ;
- **b**, modèles de la fin du XVIII ème siècle, au moins depuis 1765 jusqu'après 1790. On remarquera l'asymétrie du plateau de garde beaucoup plus large dans sa partie opposée à l'arc de jointure secondaire ;
- **c**, ce plateau de la même époque a retrouvé sa symétrie, ce qui le rend plus efficace pour la protection de la main. Cette modification semble être une inovation du fourbisseur.

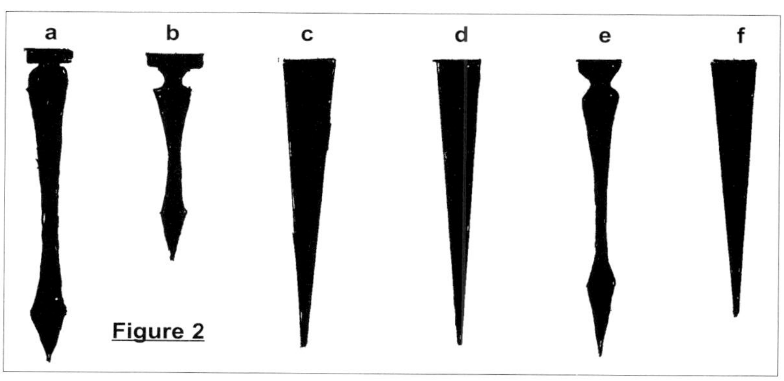

Figure 2

Figure 2 - Coupes des lames de sabres de Dragons à garde en demi-panier.

- **a**, modèle 1764 ;
- **b**, vers 1766 ;
- **c**, sabre des Dragons d'Aubigné, vers 1764 ;
- **d**, modèle 1768 ;
- **e**, entre 1784 et 1788 ;
- **f**, modèle 1790.

La forme des lames ne peut aider à préciser le modèle, celles-ci étant bien souvent récupérées sur des sabres plus anciens.

Figure 3

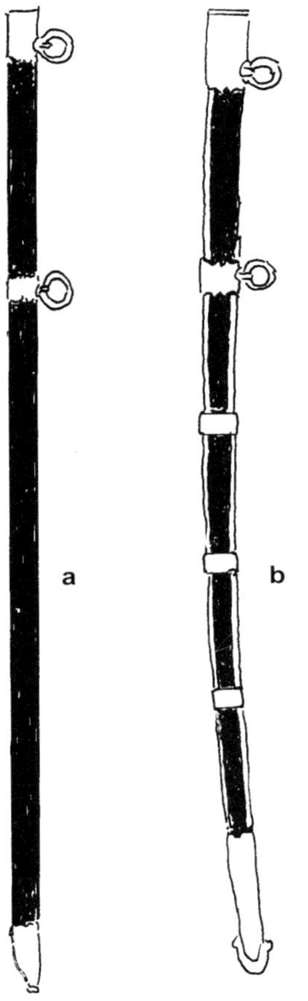

Figure 3 - Les deux principaux fourreaux des sabres de Dragons.

- **a**, à trois garnitures et dard à bouton. La longueur de la chape peut varier ainsi que l'emplacement du bracelet de bélière ;
- **b**, à trois garnitures principales plus trois bagues de renfort et des attelles latérales. Dard en lyre asymétrique. C'est le type qui équipera les sabres entre 1784 et 1788.

Figure 4

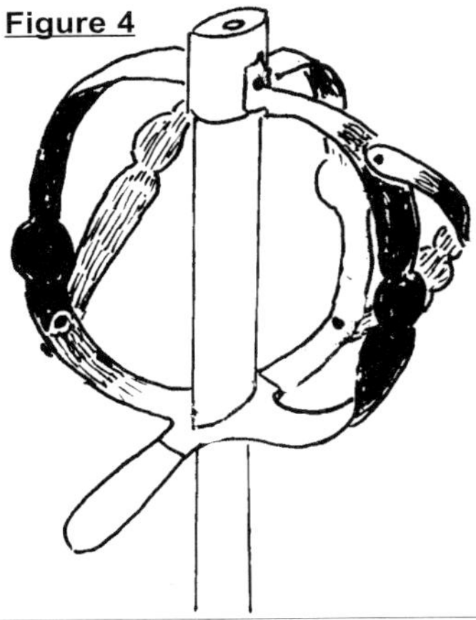

Figure 4 - Monture en panier intégral du sabre de Dragons, modèle 1764.

Ce sabre donnera naissance aux sabres à demi-panier qui survivront jusqu'à la fin du siècle. La garde se compose de trois arcs de jointure et de deux bandes obliques.

836

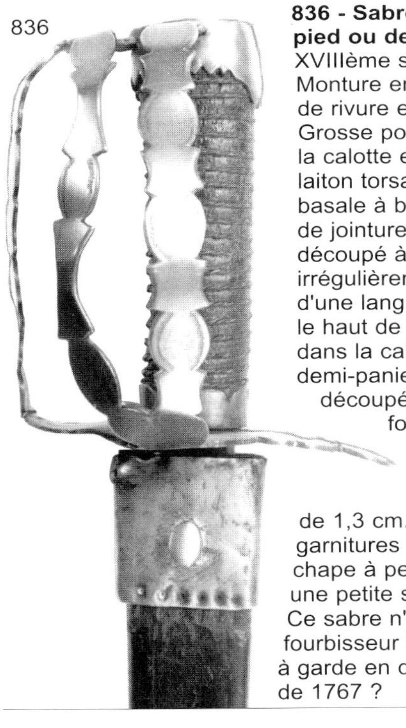

836 - Sabre d'Infanterie, de Gendarme à pied ou de milice.
XVIIIème siècle.
Monture en laiton. Calotte aplatie, à bouton de rivure en navette, à très courte queue. Grosse poignée (3,5 cm de diamètre près de la calotte et 3,3 cm à sa base). Filigrane de laiton torsadé, faisant 19 spires. Virole basale à bordure lobée irrégulièrement. Arc de jointure droit suivi d'un plateau non découpé à bordure également irrégulièrement lobée. Deux branches partent d'une languette liée au plateau : l'une rejoint le haut de l'arc, l'autre est fixée par un tenon dans la calotte. C'est de fait, une coquille en demi-panier. Les trois branches sont découpées latéralement de manière à former une succession d'ovales et de pièces concaves sur toute leur longueur. Lame plate, de 72 cm, large au talon de 3,1 cm, flèche de 1,3 cm. Fourreau de cuir noirci, à deux garnitures de laiton fixées par martelage : chape à petit bouton, bouterolle terminée par une petite sphère.
Ce sabre n'est-il qu'une fantaisie d'un fourbisseur inventif ou annonce-t-il les modèles à garde en demi-panier qui furent créés à partir de 1767 ?

837

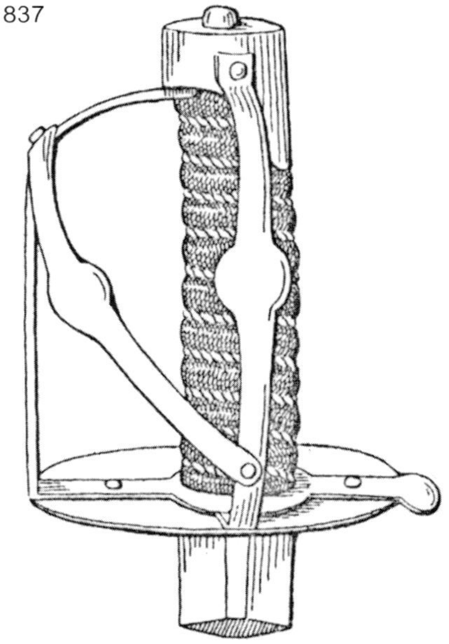

837 - Sabre de Choiseul-Dragon.
1761.
Monture en fer. Calotte à courte queue. Poignée recouverte de basane à dix spires d'un assez gros fil torsadé. Arc de jointure se prolongeant à travers un plateau ovale et fixé sur celui-ci par deux rivets. Branches secondaires également rivées : la grande sur la calotte, l'oblique sur l'arc et sur la base de la grande. Lame à section hexagonale (d'après M. Bottet).

Nota - Au sujet de ce sabre, D. Venner écrit (op. cit.) : *Monture en fer du type* Pallasch *primitif de Choiseul Dragons* (*successeur* de Bellisle Dragons, ex Dragons de Firmacon), acheté par cette unité *à Solingen en 1761, pendant la Guerre de Sept ans...* D'où sa facture germanique.

838 - Sabre de Dragons.
Modèle 1767.
On ne peut faire mieux, pour décrire ce sabre que de se reporter au texte officiel du 25 avril 1767 : " *Le sabre pour les Dragons sera à poignée de cuir, crénelé d'un fil de fer retors, longue de 5 pouces et demi (149 mm), la garde sera à bandelettes de fer plat, faite en demi-panier, à trois branches ; l'extrémité de la branche de côté, au travers de laquelle la soie de la lame doit passer, sera aplatie pour former une demi-coquille qui mettra le pouce à couvert* (838a)*; la garde sera montée sur une lame à pointe relevée, longue de 36 pouces (975 mm) pleine et à dos de largeur de 29 mm. Le fourreau sera de cuir de vache noirci, fort, sans éclisse, garni d'un bout de cuivre de la hauteur de 3 pouces (81 mm) et d'une chape de fer de 3 pouces et demi (95 mm) de long ; il y sera soudé en dedans une cuvette d'un pouce (27 mm) ou environ pour embrasser l'épaisseur du cuir, et faciliter l'entrée de la lame ; il sera garni de bélières et d'anneaux de fer soudé, pour y passer les courroies de support du ceinturon* ".

Nota - La lame peut être marquée, entre deux fleurs de lys : *Regt . De . Choiseul . Dragons* en grands caractères fleuris sur le *recto* et, également entre deux fleurs de lys, *Vive* (Armes de France) *Le Roy* sur le *verso*.

838

838a

839

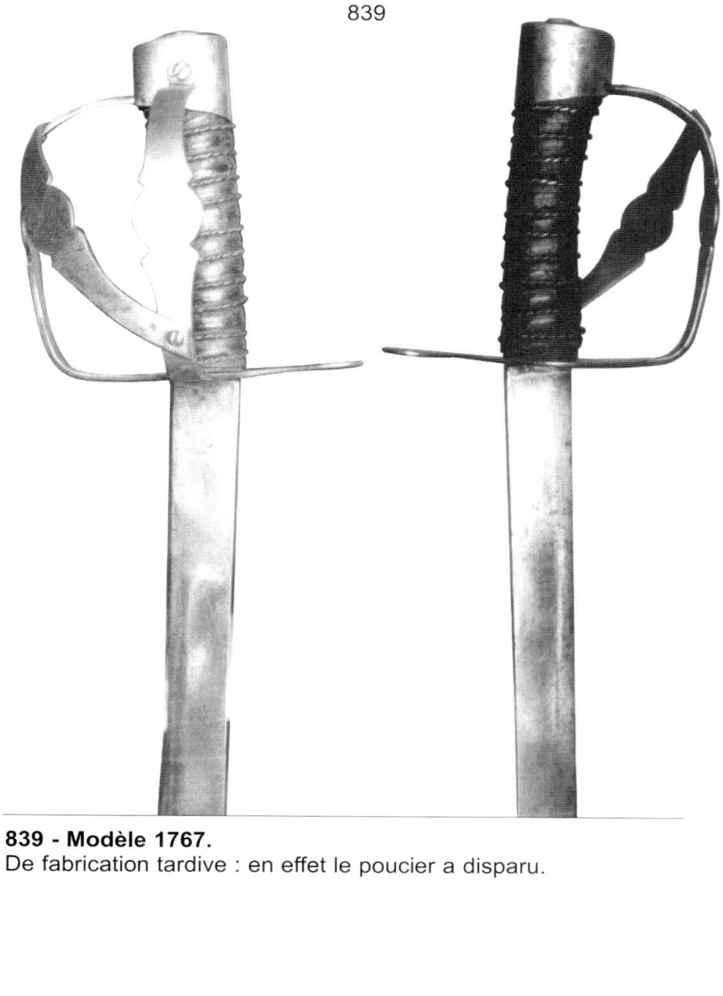

839 - Modèle 1767.
De fabrication tardive : en effet le poucier a disparu.

840

840 - Modèle 1767.
Monture du sabre d'officier. Quillon et lame pouvant être poinçonnés " M " pour S. F. Beaumaretz et " LW " pour B. C. Levasseur.

841 - Sabre de Dragon.
Vers 1767 - 1768.
Monture de fer en demi-panier. Arc de jointure principal entrant par un tenon dans la calotte à courte queue et prolongé par un plateau asymétrique du type b. Arc de jointure secondaire issu de la partie frontale du plateau et fixé sur la calotte par une vis. La branche secondaire oblique est vissée à ses deux extrémités sur les arcs de jointure. Quillon à poinçon indéfinissable sur le dessus et marqué, sur le dessous A 41 énigmatique. Poignée recouverte de cuir. Gros filigrane de fer torsadé faisant une dizaine de spires. Lame plate, à dos, longue de 97,5 cm, large au talon de 3 cm, et marquée au talon du poinçon au petit R et à la grosse couronne de Louis Antoine Gau. Fourreau de cuir noir à trois petites garnitures (non figuré).

842 - Sabre de Dragons.
Modèle 1770.
Monture en fer. Calotte courte, à bouton de rivure. Poignée recouverte de cuir et filigranée de fer. Trois branches à renflement médian. Arc de jointure régulièrement arqué à ses extrémités ; arc secondaire vertical vissé sur la calotte ; branche oblique brasée sur le bas de la verticale et sur l'arc. Plateau du type b (voir p. 409). Lame droite plate. Fourreau à trois petites garnitures de laiton.

842

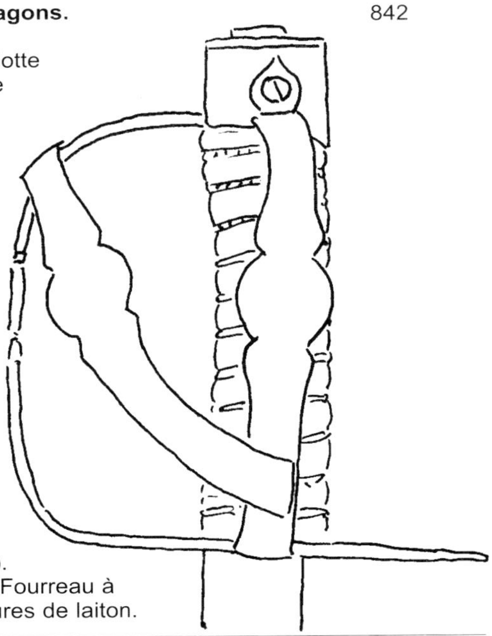

843

Sabre de Dragons, modèle 1776

843 - Sabre de Dragons de Montmorency.
1780 - 1786.
Monture en fer. Branche oblique frontale vissée sur les deux arcs. Poignée à filigrane constitué d'un fil non torsadé. Poinçon de Gau de Vaumarin sur la lame. Fourreau à chape et bracelets de fer, plus trois bagues de renfort en laiton avec attelles latérales allant de la chape à la bouterolle et attelles allant du bracelet à la bouterolle, celle-ci terminée par un petit dard asymétrique.

Remarque - Ce sabre est très semblable à celui du 2ème Chasseur (voir p. 415, n° 851).

844 - Sabre des Chasseurs des Evêchés.
1784 - 1787.
Monture en laiton. Branche oblique vissée sur la calotte. Poignée à filigrane non torsadé. Lame à pan creux (non figuré).

Marquage : Sur des exemplaires conservés au Musée de la Manufacture de Saint-Etienne, on trouve sur la lame, les poinçons de Gau de Vaumarin, de Bisch et du " K " couronné de Klingenthal.

Remarque : Ce modèle ne diffère de celui des Dragons de Montmorency que par sa monture en laiton mais sa structure générale est très voisine (voir p. 415, n° 851).

845

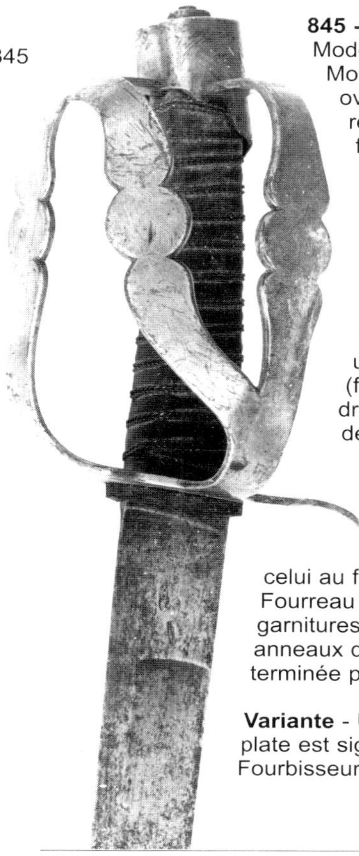

845 - Sabre de Dragons.
Modèle 1781.
Monture en laiton. Calotte ovoïde, très courte. Poignée recouverte de cuir, à filigrane de fil de laiton torsadé. Arc de jointure et branches secondaires verticales entrant dans la calotte par tenon : toutes les branches de garde sont découpées dans la même plaque de métal. Petit plateau prolongé par une sorte de quillon plat (fig. 1b, p. 409). Lame droite, à pan creux, longue de 97,5 cm, large au talon de 2,9 cm, marquée sur le dos *Manufacture de Klingenthal* avec les poinçons de Juchereau, de Bisch et celui au faisceau de licteur. Fourreau de cuir noir, à trois garnitures de laiton et deux anneaux de bélière, bouterolle terminée par un bouton de fer.

Variante - Un exemplaire à lame plate est signée GOZE marchand Fourbisseur à Metz.

846

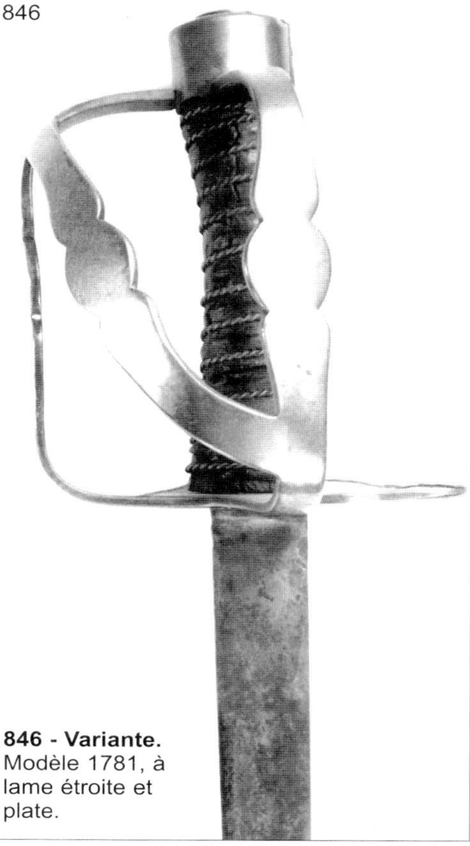

846 - Variante.
Modèle 1781, à lame étroite et plate.

847

847a

847 - Sabre de Chasseur à cheval.
Modèle 1790.
Monture en laiton. Calotte à section carrée, unie, à courte queue. Poignée recouverte de basane filigranée en chaînette faisant 13 spires. Arc de jointure fiché dans la calotte, à arrondi médian. Branche secondaire verticale également fichée dans la calotte, découpée seulement du côté antérieur et donnant naissance, à la branche secondaire oblique, à arrondi médian, fixée sur l'arc de jointure : Plateau du type b. Court quillon légèrement incliné, arrondi ou en palmette. Lame de 92 cm, large au talon de 4 cm, à flèche d'environ 2 cm, à dos plat et large pan creux commençant à 5,5 cm du talon ; contre-tranchant d'environ 17 cm. Fourreau à trois garnitures de laiton ou de fer : chape et bracelet unis, à bordure découpée en accolade, piton de laiton, anneaux de bélière en fer ; bouterolle de 27 cm, protégée en avant et à son extrémité par un dard de fer frontal.

847a, variante à garde plus étroite.

Marquages - Sur le quillon présence éventuelle d'un " M " pour Mouton. Sur le plat de la lame, poinçons de Jean Georges Bick, de de Cabannes possibles.

On consultera avec intérêt l'étude de G. Jaeger (1998) sur le sabre des chasseurs à cheval de 1790.

848

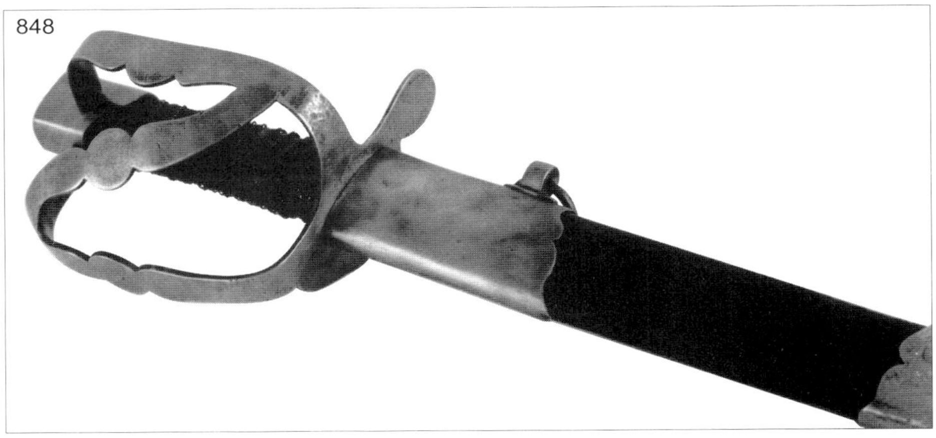

848 - Cette vue de sabre du modèle 1790 montre bien la forme de l'arc de jointure, l'asymétrie du plateau et la partie supérieure du fourreau (J.C. D.).

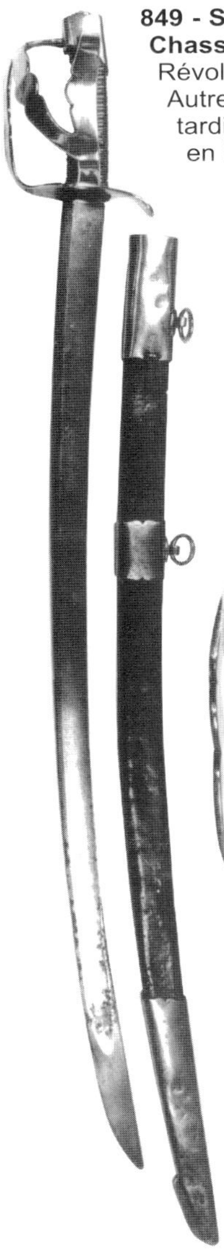

849 - Sabre de Chasseurs à cheval.
Révolution.
Autres fabrications tardives à monture en laiton.

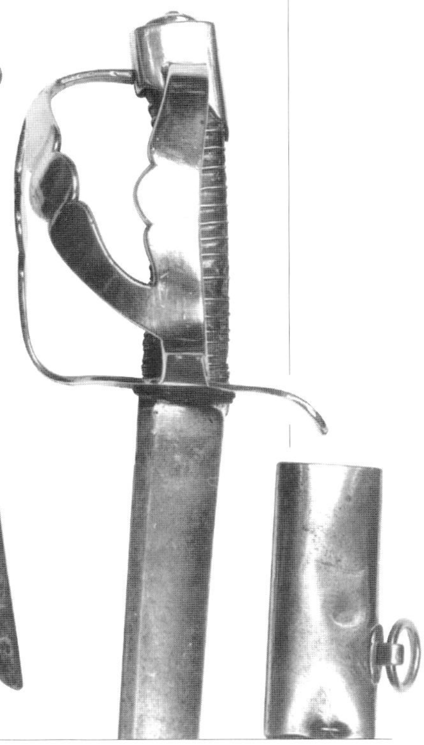

850

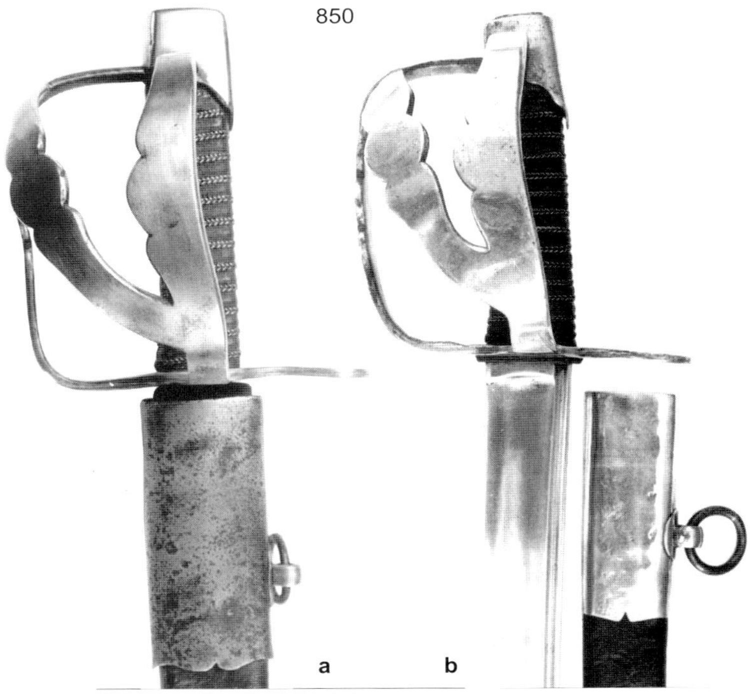

850 - Variantes.
Les fabrications tardives sont moins étriquées que les plus anciennes (850a). On remarquera la lame à forte gouttière de l'exemplaire de droite (850b).

851

851 - Sabre du 2ème Régiment de Chasseurs à cheval Troupe.
Consulat - Empire.
Poignée recouverte de basane avec filigrane en laiton torsadé. Monture en fer découpé. Garde en panier à trois branches à quillon poinçonné " M " (Mouton) et " L W " (Levavasseur). Lame droite, à dos plat, marquée : " *Manuf...* " à pan creux et gouttières, poinçonnés au talon " M " et " L W ". Fourreau en bois recouvert de cuir à deux garnitures, deux anneaux et dard en fer, attelles dissymétriques, trois anneaux de renfort et longue bouterolle en laiton.
Ce sabre est dérivé des sabres des Dragons de Montmorency qui devinrent Chasseurs des Evêchés en 1788. En 1799, le 2ème Chasseur est à l'Armée du Midi qui, avec Bonaparte se couvre de gloire au cours de la seconde Campagne d'Italie. Le régiment a pour chef de brigade Le Marois et fait partie du Corps Desaix. Le Marois est en outre Aide de Camp du 1er Consul depuis 1795. On sait le rôle décisif joué par Desaix à la bataille de Marengo le 18 juin 1800. Le 1er Consul n'avait pas grand chose à refuser à son Aide de Camp et au Régiment qui avait contribué à le tirer d'un pas fâcheux. Aussi, un décret du 30 thermidor AN X (18 août 1802) confirme le privilège du 2ème Chasseur qui conserve en principe son sabre.

852
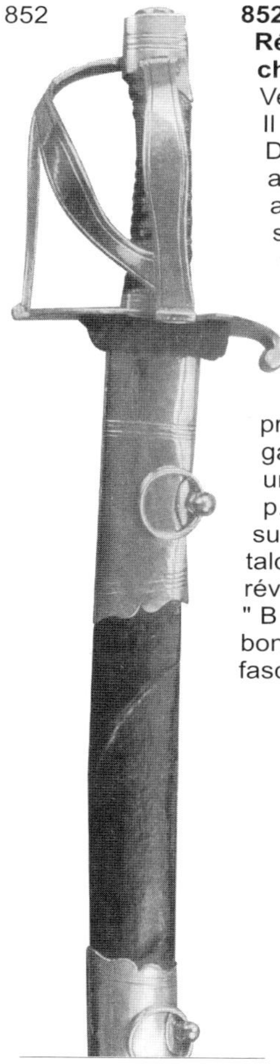

852 - Sabre d'officier du 2ème Régiment de Chasseurs à cheval.
Vers 1792.
Il existe de rares sabres de Dragons que l'on peut assimiler à des sabres attribués aux officiers subalternes. Ces sabres se distinguent des sabres de la troupe par une ornementation des branches de gardes ornées d'un liseré. Ce liseré est également présent sur la plateau de garde, ovalaire et terminé par un quillon arrondi (type c, p. 409). La lame est signée sur le dos *Klingenthal* et au talon des poinçons de l'époque révolutionnaire tel " MK " ou le " B " de Bisch surmonté d'un bonnet phrygien (voir l'Ariès, fasc. XI, pl. 8, fig. 4).

853 - Officier du 2ème Régiment de Chasseurs à cheval.
Vers 1803 - 1805.
Monture de fer très comparable à celle du modèle du Consulat : l'arc de jointure secondaire est vissé sur la calotte, la branche oblique est rivée sur l'arc principal à sa partie supérieure et sur la base de l'arc secondaire (853a). Lame de 92 cm, à flèche de 2,5 cm, blanche, à pan creux et gouttière au fort, signée *Coulaux Frères.*
Fourreau de cuir à chape et bracelet de bélière en fer ; trois anneaux de renfort en laiton placés à espaces réguliers ; attelle antérieure complète, la postérieure dépassant nettement le bracelet de bélière. (M.R.).

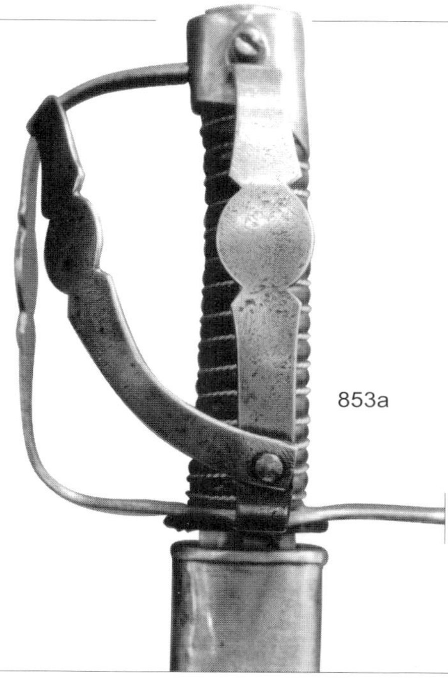
853a

853
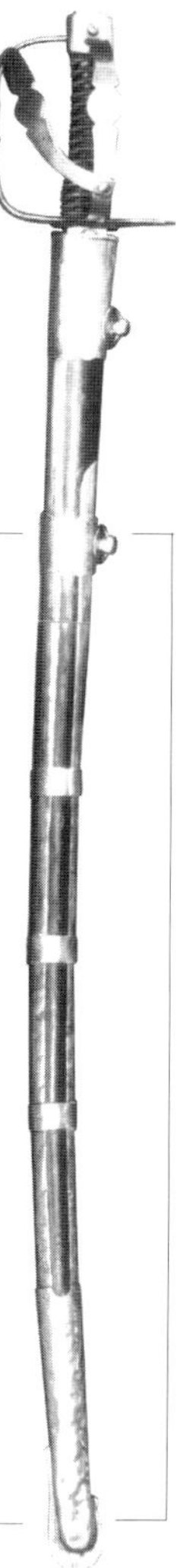

854 - Sabres d'officier du 2ème Régiment de Chasseurs à cheval.
Modèle du Consulat, début du Premier Empire.
Monture en fer. Calotte subquadrangulaire à courte queue. Poignée recouverte de basane, à 15 spires d'un fil de laiton torsadé. Arc de jointure secondaire vissé sur la calotte ; branche oblique soudée à sa partie supérieure sur l'arc de jointure principal et rivée à sa base sur l'arc secondaire. Lame presque droite : pour une longueur de 97,5 cm, la flèche est d'à peine 2,5 cm, à pan creux, gouttière au fort, marquée sur une face (854a) :

2ème Régiment de chasseurs / à cheval 1673

et sur l'autre (854b)

Bonaparte premier consul de la / République française . 30 thermidor an 10

et sur le dos

M[fre] *Imp*[le] *de Klingenthal Coulaux Frères.*

Fourreau de cuir à garnitures de laiton : chape et bracelet à anneau de bélière et 3 bagues de renfort précédent une longue bouterolle terminée par un dard asymétrique ; tout le fourreau est renforcé par des attelles de laiton : la frontale protège tout le côté antérieur, la postérieure s'arrête quelques centimètres au-dessus du bracelet de bélière (M.R.).

854

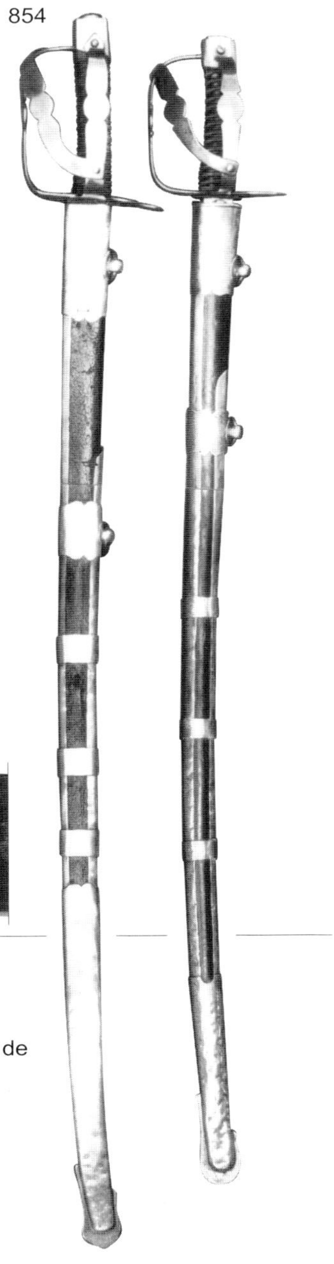

Gravures de la lame.

854a

854b

855 - Sabre du Régiment des Chasseurs à cheval de la Reine (2ème Rgt).
Restauration.
Ce sabre est très comparable aux sabres du modèle 1768. Lame gravée, sur une face de trophée, de motifs végétaux, des Armes de France surmontant *Vive / Le / Roi* et sur l'autre face : *Le Roi au Régiment de la Reine - 2e Chasseur 1673.*
Fourreau du même modèle que le sabre du Consulat (non figuré).

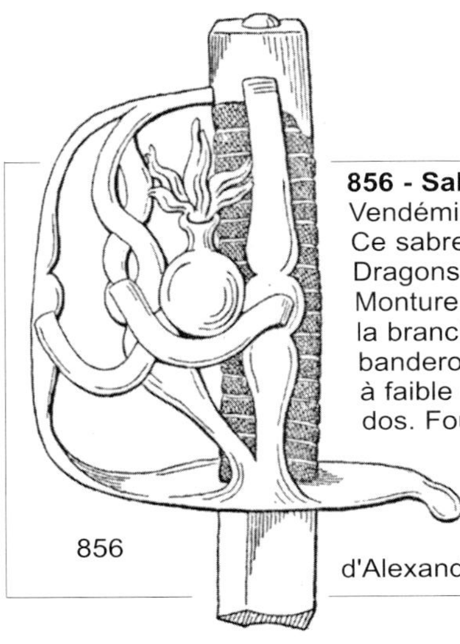

856

856 - Sabre de Gendarme de la Garde du Directoire.
Vendémiaire An III - An V.
Ce sabre est une modification enjolivée des sabres à monture en demi-panier des Dragons.
Monture en laiton. Calotte quadrangulaire, plate à bouton de rivure et courte queue. Entre la branche oblique et l'arc de jointure frontal est inscrite une grenade soulignée de deux banderoles sinueuses, la supérieure portant *GRENADIER DE LA GENDARMERIE*. Lame à faible courbure, à pan creux et gouttière dorsale et pointe dans le prolongement du dos. Fourreau de cuir noirci, chape à bouton en olive et anneau de bélière, bracelet à anneau, petite bouterolle avec un dard protégeant la base et le côté antérieur du fourreau (M.E. - S. de P.).

Nota - La lame peut être marquée du A couronné, vraisemblablement d'Alexandre qui fut contrôleur à la Manufacture de Maubeuge autour de 1792.

Sixième partie

Sabres à quatre branches : un arc de jointure plus trois branches secondaires

I - Branches secondaires arrivant isolément sur l'arc de jointure

II - Branches secondaires convergeant sur l'arc de jointure

La sixième partie est consacrée à l'étude des sabres à quatre branches : l'arc de jointure et les trois branches secondaires. Les branches secondaires partent du plateau de garde ou d'une coquille, d'un médaillon... et arrivent soit isolément, soit convergeant vers la partie supérieure de l'arc.

On retrouve, dans la monture, les détails caractéristiques qui ont été déjà exploités pour proposer un classement facilitant l'identification des sabres de la Cinquième Partie.

A savoir :

889

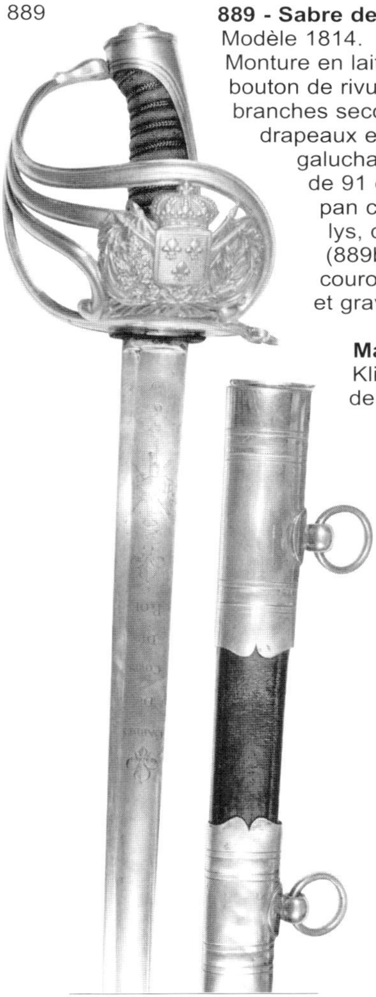

889 - Sabre de troupe.
Modèle 1814.
Monture en laiton doré. Longue queue de calotte, celle-ci ovoïde à petit dôme servant de bouton de rivure. Trois branches secondaires partent du haut de l'arc de jointure. Fixé sur les branches secondaires, un macaron découpé, aux Armes de France couronnées sur fond de drapeaux et de rameaux de chêne et de laurier (889a). Poignée de bois recouverte de galuchat gris foncé ou de roussette, filigranée. Quillon découpé en fleur de lys. Lame de 91 cm, en général, de longueur, mais variable en fonction de la taille du garde, à pan creux, à dos plat et marquée sur la face frontale, de haut en bas, d'une fleur de lys, de *GARDES / DU CORPS / DU / ROI*, une seconde fleur de lys et d'un trophée (889b) ; sur l'autre face, de haut en bas, se trouve un soleil, les Armes de France couronnées et un trophée (889c). Fourreau en cuir noirci à trois garnitures découpées et gravées de deux doubles traits ; longue bouterolle à dard en forme de boules.

Marquages - Lame signée, en abrégé, sur le dos, Manufacture Royale du Klingenthal (889d) et sur le talon ainsi que sur le quillon, les poinçons de Bick et de Borson (889b).

- **a**, détail de la garde des 1er et 2ème modèles ;
- **b**, *recto* de la lame. Noter les poinçons de Bick et de Borson ;
- **c**, *verso* des lames des 1er et 2ème modèles ;
- **d**, signature de la manufacture sur le dos de la lame.

889b

889c

889d

889a

890

890 - Sabre des Gardes du Corps du Roi.
Second modèle, 1815.
Monture identique à celle du modèle 1814, mais le fourreau de cuir est remplacé par un fourreau en tôle d'acier, long de 98 cm, à deux bracelets de bélière et chape de 3,8 cm de haut et ornée de feuille d'acanthe dorées.

891

891 - Sabre Gardes du Corps du Roi. (Réutilisation).
Restauration.
Sabre tout à fait comparable à celui des Gardes du Roi. Les particularités essentielles sont l'absence de bracelet sur le fourreau de fer et du décor de l'entrée du fourreau. Certains sabres de Garde du Corps furent réutilisés, sous la Restauration, en raison du prestige lié à cette arme.

892 - Sabre d'officier de la Garde Royale.
Restauration.
Ce sabre présente une monture identique à celle du sabre 1814 mais avec les différences suivantes : *VIVE LE ROI* écrit en caractères massifs, dorés et inscrits en un cartouche bleui (892a) sur le *recto*. Les fleurs de lys également dorées sont placées sur fond bleui sur le *verso* (892b). Les bracelets de bélières sont remarquables : ils sont limités par deux couronnes striées verticalement, de part et d'autre de trois frises de feuilles et de fruits de laurier (892c).

892a

892b

- **a**, lame à cartouche bleui, avec en lettres d'or sur le *recto VIVE LE ROI* ;
- **b**, sur le *verso* trois fleur de lis.

892c

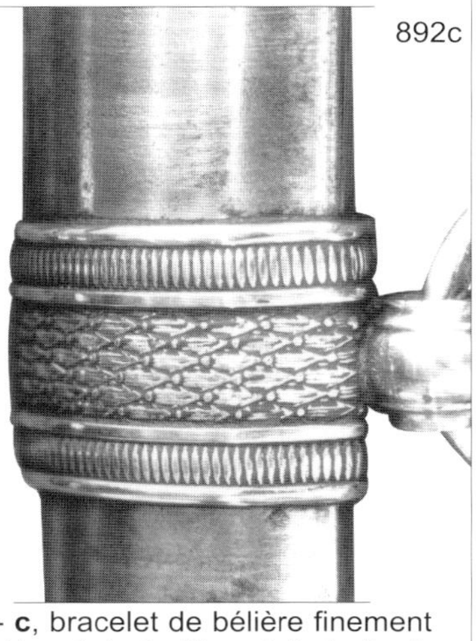

- **c**, bracelet de bélière finement décoré de feuilles et de fruits de laurier stylisés.

Famille IV - Calotte à courte queue

Rappelons que les sabres de cette Famille IV possèdent les branches secondaires convergeant sur l'arc de jointure. Mais dans un cas, ces branches partent du plateau de garde et dans l'autre cas, elles partent d'un médaillon.

D'où la création de deux Sous-familles :

Sous famille IV a - Branches secondaire nues

Plusieurs des modèles réunis dans cette Sous-famille IVa, sont des sabres attribués à la Cavalerie de la ligne. C'est dans cette Sous-famille que l'on trouvera également le fameux modèle 1822. Les schémas ci-contre, des calottes et de l'arrivée de l'arc de jointure donnent une idée des différents types de sabres de cette Sous-famille IVa.

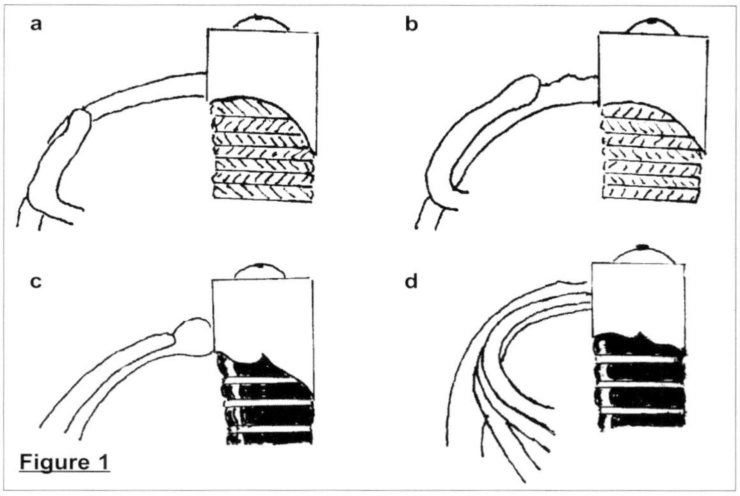

Figure 1 - Différentes connexions entre l'arc de jointure et les branches secondaires pour les sabres à courte queue de calotte de l'An III (1795) à l'An XIII (1805).

- **a**, branche secondaire supérieure arrivant à plusieurs centimètres de la calotte.
Exemple : sabre de la Maréchaussée Cavalerie de Ligne An IX ;
- **b** et **c**, branche secondaire arrivant à moins d'un centimètre de la calotte.
Exemples : sabre de Cavalerie légère, modèles de l'An IX, Dragon de l'An IX, et de l'An XI, sabres de Cuirassiers, de Carabiniers, modèles de l'An IX et de l'An XI ;
- **d**, branches secondaires arrivant conjointement sur l'arc au niveau de la calotte.
Exemples : Gendarmerie de l'An VII, sabres de Gendarmerie d'Armée de l'An IX, de Cavalerie de Ligne de l'An XIII.

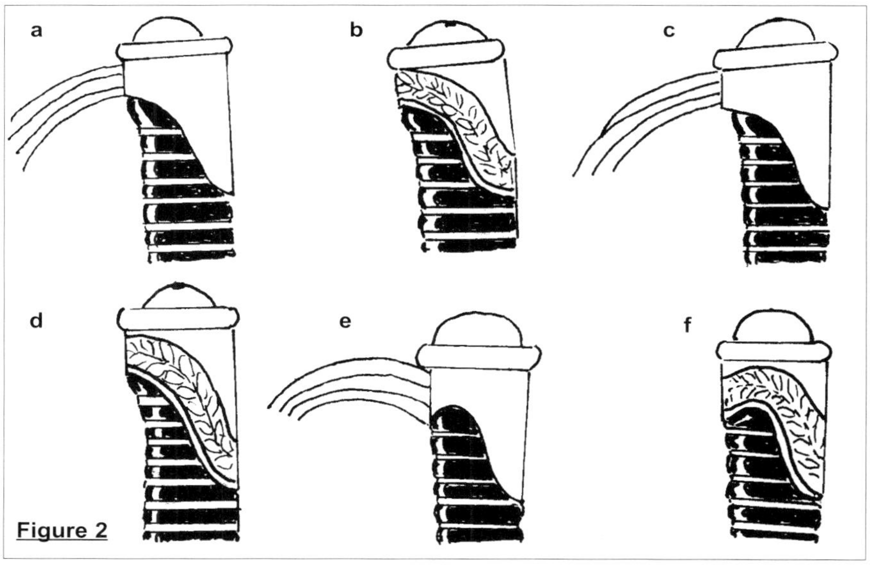

Figure 2 - Sabres à trois branches secondaires rejoignant conjointement la calotte à bourrelet et courte queue des modèles 1816, 1822, 1854.

- modèle 1816, troupe (**a**) et officier (**b**) ;
- modèle 1822, troupe (**c**) et officier (**d**) ;
- modèle 1854, troupe (**e**) et officier (**f**).

893 - Sabre des brigades à cheval de la Gendarmerie Nationale.
1796 - 1798.
Ce sabre donnera naissance à plusieurs modèles dont l'identification n'est pas toujours aisée.
Monture en laiton. Calotte très courte, à bouton de rivure ovalaire, à bordure découpée en accolade. Poignée finement filigranée. Branche principale régulièrement arquée, s'élargissant pour former le plateau légèrement infléchi en sa partie postérieure. Trois branches secondaires en " S " fusionnent pour rejoindre la partie supérieure de l'arc de jointure (fig. 1, d) et emboulées au départ du plateau. Lame droite, plate, longue de 97-97,3 cm, large au talon de 3,2 cm, à pointe dans le prolongement du dos. Fourreau de cuir avec alèses, à deux garnitures : chape à bouton, courte bouterolle protégée par un dard asymétrique.

Marquages - Sur le quillon, poinçon éventuel de Liorard, fourbisseur à Paris de la fin du XVIIIème siècle à 1800. Lame signée de la *Manufacture Nationale du Klingenthal - Coulaux Frères Entrepreneurs* (non figuré).

Les sabres de Cavalerie de Ligne des Modèles An IX, An XI et An XIII

Dans son rapport du 8 septembre 1800, la Commission des Armes à feu et Armes blanches décide de remplacer le sabre du modèle 1784 dit improprement arco. C'est le Gassendi, chef de la Division de l'Artillerie au Ministère de la Guerre qui est chargé de la refonte de l'armement. Un nouveau sabre est adopté pour la cavalerie et les Dragons. Mais ce sabre est très rapidement critiqué : sa lame est trop lourde, son fourreau trop fragile. Connu sous le nom de modèle de l'An IX, il sera modifié en l'An XI, puis en l'An XIII. Il sera mis en servivce au début de la Restauration

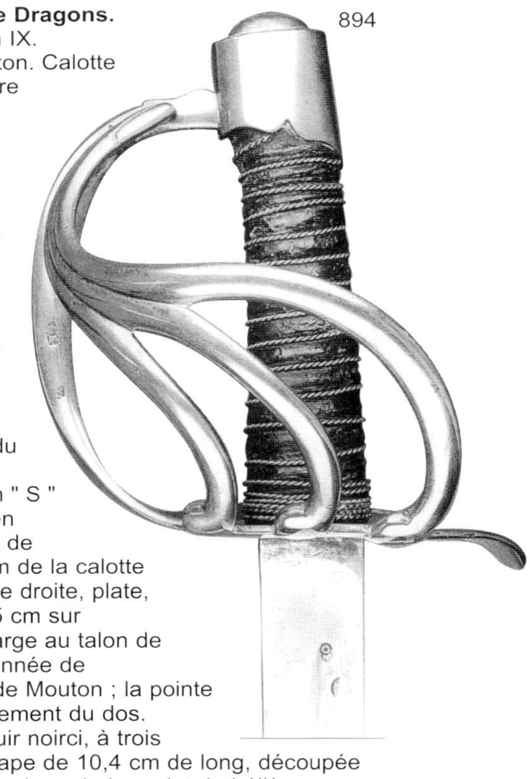
894

894 - Sabre de Dragons.
Modèle de l'An IX.
Monture en laiton. Calotte courte à bordure découpée en accolade. Poignée recouverte de basane liée d'un fil torsadé faisant 18 spires plus trois serrées faisant office de virole. Trois branches secondaires à rouleau à leur base, partant du plateau, se développant en " S " et fusionnant en rejoignant l'arc de jointure à 6 mm de la calotte (fig. 1, b). Lame droite, plate, longue de 97,5 cm sur l'exemplaire, large au talon de 3,3 cm, poinçonnée de Levavasseur, de Mouton ; la pointe est en prolongement du dos. Fourreau de cuir noirci, à trois garnitures : chape de 10,4 cm de long, découpée en accolade ainsi que le bracelet de bélière, bouterolle d'une quinzaine de centimètres terminée par un rectangle de fer, en guise de dard.

895 - Sabre de Cavalerie modèle An IX.
Monture en laiton. Calotte courte, subcylindrique, à bord inférieur découpé en accolade. Poignée recouverte de basane maintenue par un fil de laiton torsadé faisant une douzaine de spires, plus quatre ou cinq serrées tenant lieu de virole. Arc de jointure régulièrement arqué et trois branches secondaires arrivant conjointement sur lui, à 14 - 21 mm de la calotte (voir p. 435, fig. 1. a) et emboulées à leur jonction avec le plateau de garde. Lame plate de 97 - 97,3 cm et de 3,25 cm environ de largeur au talon. Fourreau à alèses, de tôle de 0,95 - 1,3 mm d'épaisseur ; dard asymétrique ou en lyre, à deux bracelets de 2,25 - 2,45 cm de hauteur (non figuré).

896 - Sabre de Gendarmerie d'Armée.
Modèle de l'An IX.
M. Pétard écrit (1995, Tradition, 98, 18-221 : *Durant la Révolution puis sous l'Empire, des compagnies à cheval dites de Gendarmerie d'Armée sont détachées de leur jurisprudence de Police Départementale pour le service au sein des Armées en Campagne. Cette gendarmerie de Guerre ne pouvait se satisfaire du sabre relativement léger de 1783 ou de son successeur de 1798 ...* D'où le sabre dit de l'An IX décrit ci-après.

896

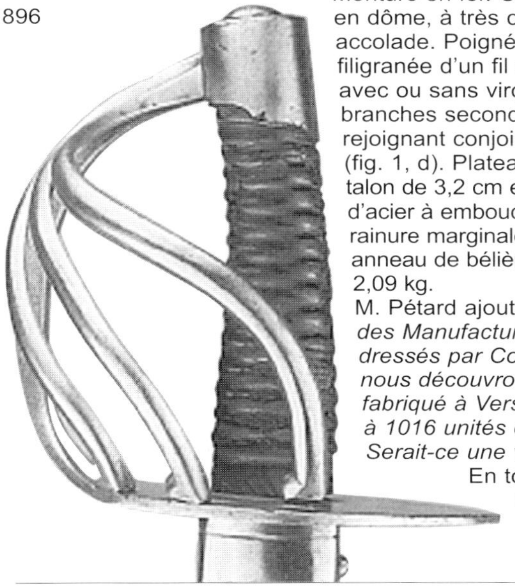

Monture en fer. Calotte surmontée d'une pièce de rivure en dôme, à très courte queue et bordure découpée en accolade. Poignée ficelée, recouverte de basane et filigranée d'un fil de laiton torsadé faisant 13 - 15 spires, avec ou sans virole. Arc de jointure arrondi avec les trois branches secondaires emboulées à leur base et rejoignant conjointement l'arc à son tiers supérieur (fig. 1, d). Plateau ovalaire. Lame plate, droite, large au talon de 3,2 cm et longue de 97 cm. Fourreau de tôle d'acier à embouchure renforcée d'une bague à double rainure marginale et deux bracelets en suite à piton et anneau de bélière. Dard en lyre asymétrique. Poids : 2,09 kg.
M. Pétard ajoute (op. cit.) : *Dans les bilans comptables des Manufactures de Versailles et de Klingenthal dressés par Cotty à partir des archives ministérielles, nous découvrons un sabre de Gendarmerie de l'An IX fabriqué à Versailles à 1016 unités en 1812 et 14 unités l'année suivante... Serait-ce une version de notre sabre de fer ... ?*
En tout état de cause, ce sabre est très voisin du sabre de Cavalerie de l'An IX.

897 - Sabre de Dragons.
Modèle de l'An XI.
Monture très comparable à celle du modèle de l'An IX, mais avec les modifications suivantes :
- filigrane ne faisant que 10 spires, la onzième disparaissant sous une virole.
- lame longue de 97,5 cm, large au talon de 3,8 cm, à deux pans creux, poinçonnée de Lobstein, de Bick, de Krantz ; sur le dos on trouve la signature de la *Manufacture de Klingenthal* suivie de l'année de fabrication.
- fourreau de cuir noirci, à trois garnitures de laiton : chape de 10 cm au maximum, bracelet de 4 cm, bouterolle de 15,6 cm de longueur terminée par un bouton en olive (non figuré).

898 - Sabre de Cavalerie de Ligne (Cuirassiers).
Modèle de l'An XI.
Sabre très comparable au modèle de l'An IX, mais avec les différences suivantes : branches secondaires arrivant à 0,6 cm de la calotte (voir p 435, fig. 1, b). Lame à deux pans creux, longue de 97,45 cm, large au talon de 3,82 cm. Fourreau de tôle d'acier de 2,5 cm d'épaisseur, à deux bracelets de 1,3 cm de hauteur (non figuré).

Marquages - Lame signée de la Manufacture de Klingenthal ou de Versailles.

899 - Sabre de Dragons.
Modèle de l'An XIII.
Même monture que pour le modèle de l'An XI, avec les modifications suivantes : branches secondaires fusionnées rejoignant la principale au niveau de la calotte (voir p. 435, fig. 1 d,). Poignée recouverte de basane à 12 spires d'un filigrane torsadé avec une virole de laiton basale. Lame droite à double pan creux, longue de 97,45 cm, large au talon de 3,83 cm. Forreau en cuir à trois arnitures laiton. A partir de 1816, les lames des sabres du modèle de l'An XIII (Dragons et Cavalerie) verront leur pointe ramenées dans l'axe de la lame. Le fourreau en cuir sera abandonné à la même époque chez les Dragons pour le fourreau en acier du modèle 1816.

900

900 - Sabre de Cavalerie (Cuirassiers).
Modèle de l'An XIII.
Sabre identique au modèle de Dragons An XIII. Fourreau en tôle d'acier de 2,4 mm d'épaisseur, bracelet à jonc de 1,35 cm de hauteur, dard en lyre.

Marquages - Dos de la lame signé *Mfture Imple du Klingenthal* (mois et date de fabrication), par exemple décembre 1812 ou pendant les Cent jours 1er juin 1815. Sur le plat de la lame, au talon, poinçons de de Beaumont, de Mouton. Sous la Restauration, sur le dos de la lame *Mfture Rle du Klingenthal* (mois et date) et poinçons de Borson, de Bick.

901 - Sabre d'essai de Dragons.
Troisième République.
Dans les années 1880, les autorités militaires ont expérimenté de nouveaux modèles pour remplacer les sabres 1812, 1822, 1854. E.B. Bredene et A de Jenlis signalent un sabre d'essai, portant le n° 514 et signé sur le dos de la lame de 92 cm *Mre d'armes de Châtt mai 1880 Dragon*. Ce sabre, à trois branches secondaires et courte queue en calotte fut abandonné, probablement à cause de sa lame trop longue, (non figuré).

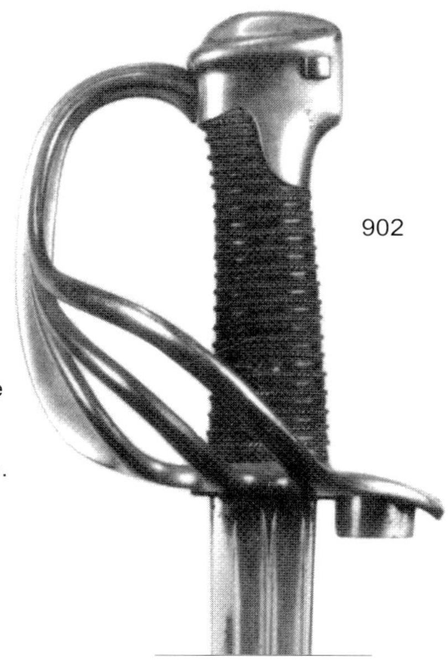

902

902 - Sabre-lance Chassepot.
1862.
Monture en laiton. Calotte globuleuse avec le système de fixation sur le fusil, et très courte queue. Poignée recouverte de cuir et filigranée de laiton torsadé. Arc de jointure arqué recevant en sa moitié supérieure trois: branches secondaires. Plateau muni d'une bague pour le passage du canon. Lame droite de 97,5 cm à deux pans creux. Fourreau en tôle d'acier à deux bracelets de bélière et dard en lyre. Poids : environ 2 kg. Fait pour le " fusil-lance " Chassepot mis à l'essai chez les Dragons de la Garde.

Les sabres des Gardes du Corps de Monsieur

Par ordonnance du 25 juillet 1814, les deux compagnies des Gardes du Corps de Monsieur sont rétablies *Sa Majesté ayant jugé à propos de rétablir sa Maison Militaire, et voulant en même temps, que les deux Compagnies des Gardes du Corps de Monsieur reçoivent l'organisation convenable et ordonne ce qui suit ...* (M. Malvaux, 1994). Au décès de Louis XVIII, Monsieur monte sur le trône et devient Charles X. Les Gardes du Corps de Monsieur sont incorporés, le 24 mai 1826, aux Gardes du Corps du Roi.

903 - Sabre d'officier de la Garde du Corps de Monsieur.
Modèle 1814.
Le modèle d'officier conserve la même ligne générale que celle du sabre de troupe, mais avec plus de raffinement. Ainsi, la courte calotte est frappée des initiales du comte d'Artois flanquées de rameaux de laurier et de petites quintefeuilles. Poignée ficelée, recouverte de veau chagriné, à quinze spires d'un filigrane doré en chaînette. Trois branches secondaires ornées de feuilles de laurier ciselées. Lame de 91 - 94 cm de longueur, large au talon de 3,1 cm bleuie et dorée au tiers, portant, sur une face *VIVE LE ROI* encadré de fleurs de lys et d'un trophée, les Armes de Monsieur et un soleil. Talon de la lame doré signé *Coulaux Frères* sur une face et sur l'autre *Manufacture Royale de Klingenthal.* Fourreau en tôle d'acier ; pitons anneaux de 1,5 cm de large et bracelets en laiton surdoré, ceux-ci décorés de branches de laurier et d'une quintefeuille, à bordure moulurée et dard emboulé (non figuré).

904 - Sabre de Garde du Corps de Monsieur, Troupe.
Modèle 1814.
Monture en laiton ciselé. Calotte légèrement inclinée vers l'avant, à sommet débordant la courte queue unie. Poignée recouverte de maroquin filigrané. Arc de jointure régulièrement courbé recevant en sa moitié supérieure les trois branches secondaires ; ces branches partant du plateau sont ornées à leur base, d'une rondelle prolongée d'une feuille.
Quillon plat découpé en fleur de lys et sur l'exemplaire présenté est inscrit le nombre 70.
Lame légèrement courbée, longue de 95 cm, à dos plat marqué :
MANUF^{re} R^{ale} du Klingenthal Mai 1816.
Dans le pan creux de la lame, sur le *verso,* on lit *GARDES / DU / CORPS / DE / MONSIEUR*
entre deux fleurs de lys avec un trophée d'armes près du talon (904a). Fourreau en tôle d'acier, à deux bracelets à anneaux de laiton, dard en olive de fer. En haut du fourreau, sur le *verso*, la signature de *MANCEAUX*.

904

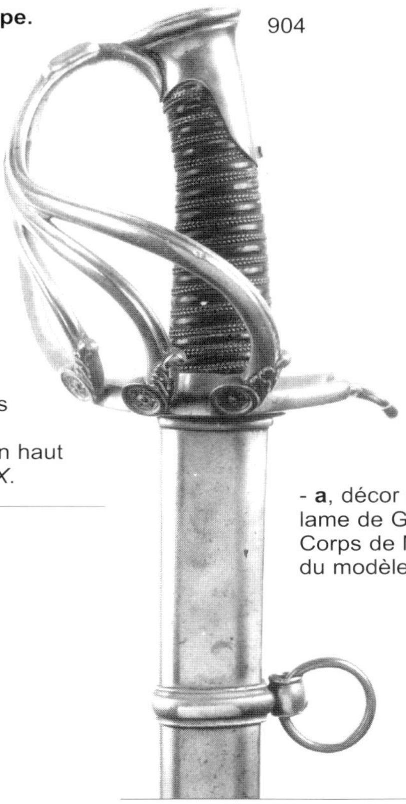

904a

- **a**, décor de la lame de Garde du Corps de Monsieur du modèle 1816.

905

905 - Sabre d'officier de Cavalerie légère.
Modèle 1817.
Monture en laiton doré. Calotte à fleur de lys. Lame de 90 cm à flèche de 6 cm, portant l'inscription en romaines dorées *VIVE LE ROY* dans un cartouche bleui sur une face, et sur l'autre, trois fleurs de lys dorées sur fond bleui.
On rencontre également des sabres dont la lame à deux pans creux est simplement gravée des motifs classiques.
Ce sabre n'eut pas un grand succès, les officiers de Cavalerie légère préfèrent conserver le sabre d'officier de l'An IX, avant d'adopter le sabre d'officier du modèle 1822. Il fut par contre en vogue dans la Gendarmerie.
Des sabres de ce modèle ont été produits par la Manufacture de Klingenthal.

906 - Sabre d'officier supérieur de Cavalerie légère.
Modèle 1817.
Monture en laiton doré. Calotte à fleur de lys, à courte queue mais avec un médaillon dorsal pour y graver les initiales ou les armoiries du propriétaire. Les trois branches secondaires partent de rosaces décorant la face frontale du demi-plateau. Poignée recouverte de chagrin, filigrane de fils torsadés limités par un fil de bordure simple. Quillon recourbé. Lame de 90 cm, à flèche de 6 cm, large au talon de 3,3 cm, gravée sur une face de *VIVE LE ROI* en romaines dorées dans un cartouche bleui et sur l'autre face de trois fleurs de lys d'or sur fond bleui et signée Duc. Fourreau d'acier, à deux bracelets de laiton également ciselés de feuillage (M.E. - S. de P.).

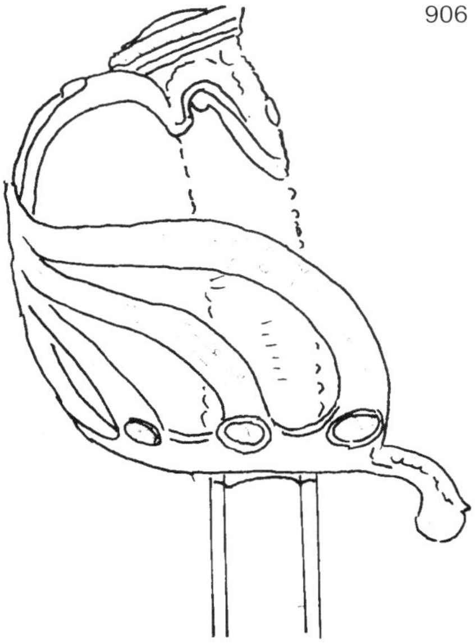
906

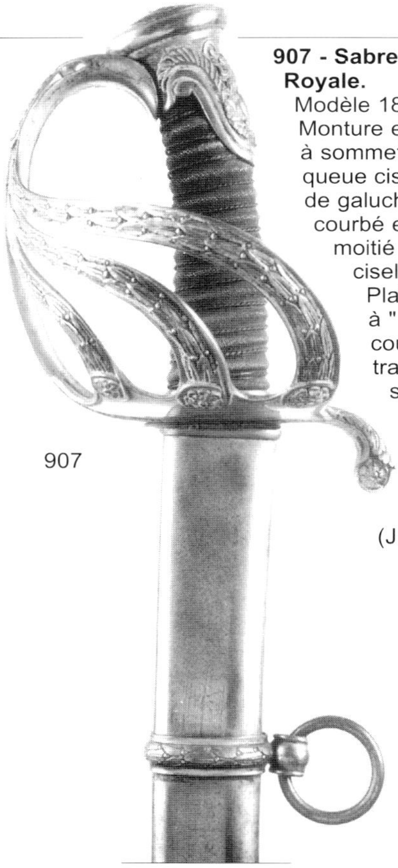
907

907 - Sabre d'officier supérieur de la Gendarmerie Royale.
Modèle 1817.
Monture en laiton doré. Calotte inclinée vers l'avant à sommet à la fleur de lys et débordant la courte queue ciselée d'une grenade. Poignée recouverte de galuchat filigrané. Arc de jointure régulièrement courbé et trois branches secondaires jointives à la moitié supérieure de l'arc, celles-ci entièrement ciselées de feuilles ou de fruits de laurier. Plateau prolongé par un quillon plat rarement à " boule ". Lame de 91,2 cm, légèrement courbée, à pan creux, gouttière et contre-tranchant. Cette lame, bleuie au tiers, porte sur les deux faces *GENDARMERIE ROYALE* dans un cartouche (907a) Le talon doré est signé *Coulaux / Frères / à / Klingenthal.* Fourreau de tôle d'acier à deux étroits bracelets en laiton ciselés de feuilles et de fruits de laurier (907b), (J.C. D.).

907b

907a

908 - Sabre d'officier de Gendarmerie.
Modèle 1817.
Ce sabre du type de Cavalerie légère est dit également de la Gendarmerie des Chasses.
Monture en laiton doré. Calotte frappée d'une fleur de lys (905a), à queue courte. Poignée recouverte de chagrin, à filigrane d'argent ou doré. Arc de jointure nettement recourbé sur la calotte et recevant les trois branches secondaires à environ un centimètre de cette calotte. Les branches secondaires, jointives, sont à deux gouges longitudinales. Petit plateau de garde uniquement frontal. Quillon recourbé. Lame de 90 cm de long, de 3,3 cm de large au talon, signée sur le dos de la *Manufacture de Klingenthal*, *Coulaux Frères* et, au talon, *Duc de Paris*. Flèche de 6 cm, gravée *GENDARMERIE ROYALE*. Fourreau de bois recouvert de tôle d'acier, cuvette maintenue par une vis, bracelets de laiton doré, à jonc, anneaux de bélière et dard asymétrique. Ce sabre ne fut en service que pendant cinq années.

908a

- **a**, dessus de la calotte

Variante - Sur les lames des Gendarmes de Province l'inscription *GENDARMERIE ROYALE* est remplacée par *GENDARMERIE DEPARTEMENTALE* ou par *GENDARMERIE RURALE*.

908

909 - Sabre d'officier supérieur des Haras.
Modèle 1817 de Cavalerie de Ligne modifié.
Monture en laiton. Calotte gravée de la fleur de lys et courte queue décor de feuilles (909a). Poignée recouverte de chagrin filigrané d'argent. Arc de jointure régulièrement courbé, recevant en sa moitié supérieure les trois branches secondaires jointives et ciselées d'une frise de feuilles de laurier sur toute leur longueur. Plateau décoré de tête de chevaux. Lame légèrement courbée, longue de 91,7 cm, finement gravée de motifs dorés parmi lesquels on reconnaît, sur le *recto* *Coulaux / Frères* et sur le *verso, à Klingenthal* (909b et 909c). Fourreau de tôle d'acier à deux bracelets et anneaux de bélière.

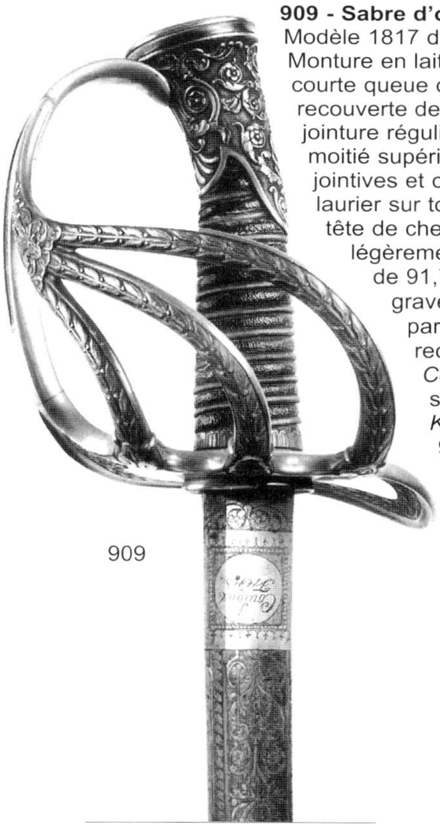

909

909a

909b

909c

- **a**, détail de la calotte vue de dos. Remarquer la structure du filigrane.
- **b**, décor du *recto* de la lame ;
- **c**, décor du *verso*.

910

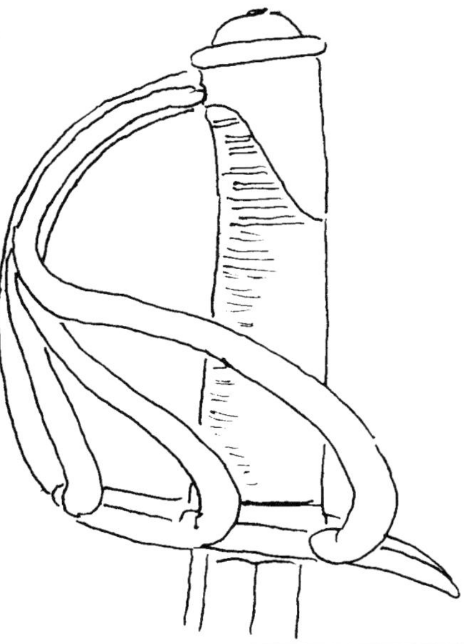

910 - Sabre de Cavalerie de Ligne.
Modèle 1816, troupe.
Monture en laiton. Calotte à bourrelet surmonté d'une pièce de rivure en dôme, à courte queue. Poignée recouverte de basane maintenue par un filigrane faisant 21 spires d'un fil torsadé. Arc de jointure régulièrement arqué à trois branches secondaires arrivant jointives sur la calotte. Plateau à bec. Lame droite de 97,5 cm, à deux pans creux se raccordant en arc de cercle sur l'arête médiane et se terminant brusquement près du talon. Cette lame est généralement signée sur le dos : *Manuf^{re} R^{ale} du Klingenthal* (mois et date). Fourreau en tôle d'acier à deux anneaux de bélière et dard en lyre.

911 - Sabre de Cavalerie de ligne.
Modèle 1816, officier.
Monture comparable à celle de la monture du sabre de troupe, mais avec les modifications suivantes : monture en laiton doré, branches secondaires partant du plateau, calotte et base des branches secondaires ornées de rameaux de laurier stylisés. Les premiers exemplaires fabriqués portaient une fleur de lys entre deux rameaux de laurier sur l'arc de jointure (non figuré).

912

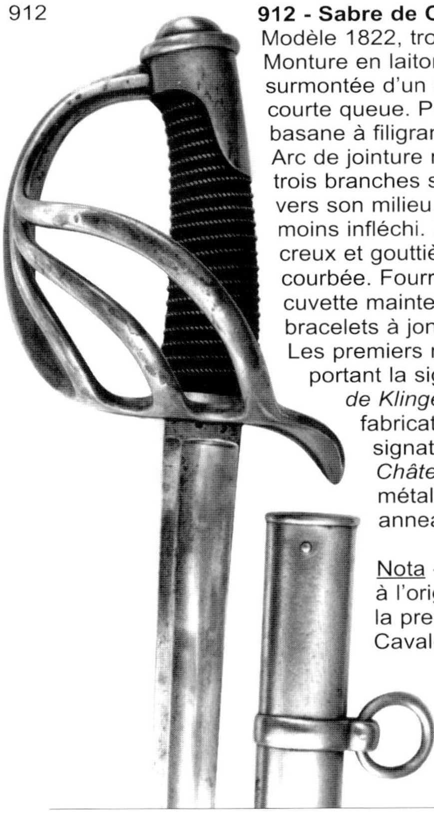

912 - Sabre de Cavalerie de Ligne.
Modèle 1822, troupe.
Monture en laiton. Calotte à bourrelet surmontée d'un dôme-bouton de rivure, à courte queue. Poignée recouverte de basane à filigrane de fil de laiton torsadé. Arc de jointure régulièrement courbé, les trois branches secondaires le rejoignant vers son milieu. Plateau à bec, plus ou moins infléchi. Lame de 97,46 cm, à pan creux et gouttière au fort, légèrement courbée. Fourreau en tôle d'acier, à cuvette maintenue par deux rivets, bracelets à jonc et dard asymétrique. Les premiers modèles ont une lame portant la signature : *Manuf^{re} R^{ale} de Klingenthal*, avec la date de fabrication et plus tard, la signature de la *Manufacture de Châtellerault* et les fourreaux métalliques sont à deux anneaux de bélière.

Nota - Ce modèle de sabre fut, à l'origine, très critiqué : c'était la première fois qu'un sabre de Cavalerie de Ligne était courbé, trop pour frapper d'estoc et pas assez pour frapper de taille, d'où son surnom de " *bancal* " qui voulait être péjoratif.

913

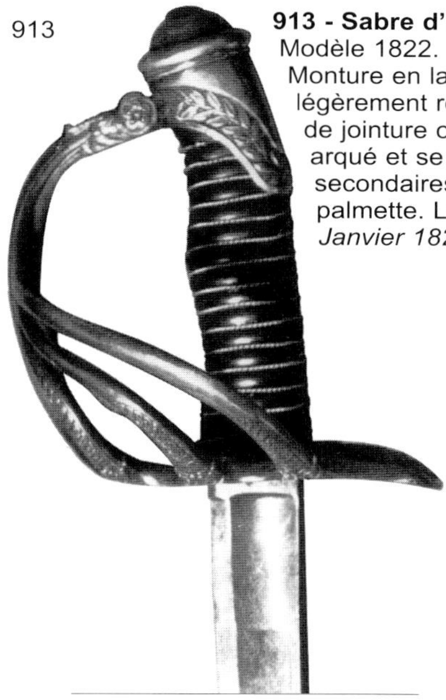

913 - Sabre d'officier de Cavalerie de Ligne.
Modèle 1822.
Monture en laiton. Calotte à courte queue à bordure de feuilles de laurier. Poignée légèrement renflée en sa partie médiane, à quatorze spires d'un fin fil de laiton torsadé. Arc de jointure orné d'une fleurette et de quelques feuilles à sa partie supérieure, régulièrement arqué et se prolongeant par un plateau terminé par un bec. Les deux ou trois branches secondaires rejoignent l'arc vers son milieu et sont décorées à leur base d'une longue palmette. Lame à pan creux, droite, pouvant être signée : *Manuf^{re} R^{ale} de Klingenthal, Janvier 1824.* Fourreau en tôle d'acier à deux bracelets de bélière.

914

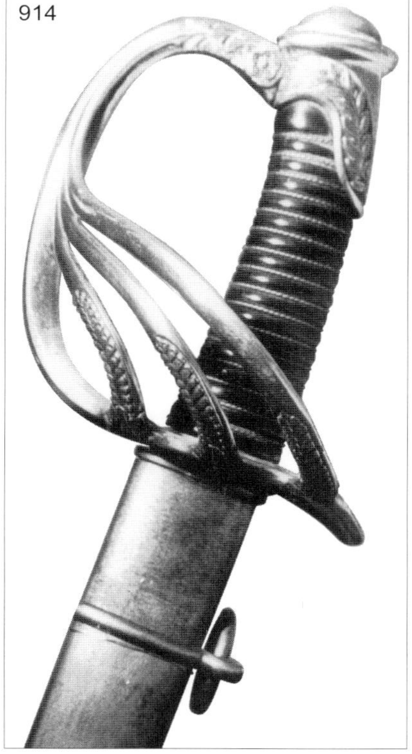

914 - Sabres d'officier de Cavalerie de Ligne.
Modèle 1822.
Facture générale comparable à celle du sabre de troupe, avec les modifications suivantes : monture en bronze doré ; calotte à courte queue bordée d'une frise de feuilles de laurier ; arc de jointure à rosace et motif foliacé à sa partie supérieure ; trois branches secondaires ornées de palmes à leur base.

915

915 - Sabre des Elèves des Ecoles Militaires.
Modèle 1822 d'officier de Cavalerie de Ligne.
Sabre offert par le Roi aux élèves sortant de l'Ecole Royale de Saint-Cyr entre 1823 et 1831.
Monture en laiton doré.
Lame de 97,5 cm et marquée *DONNE PAR LE ROI*, en capitales penchées.
L'exemplaire figuré est signé sur le dos : *MANUF^{re} R^{ale} de Klingenthal 8bre 1823.*
Le nom du fourbisseur *MANCEAUX à Paris* est gravé sur le bec du plateau et sur le haut du fourreau.

916 - Sabre d'officier supérieur de Cavalerie de Ligne.
Modèle 1822.
Ce sabre diffère du sabre d'officier par la richesse des gravures de sa lame signée *Coulaux Frères* à *Klingenthal*. En effet, sur une face sont gravés des attributs militaires, une charge de cavaliers dans un cartouche, un écu avec *Honneur et Patrie* sur un faisceau de drapeaux, une étoile et une Légion d'Honneur. Sur l'autre face les gravures représentent des attributs militaires, un combat dans un cartouche, les Armes de France avec la croix de Saint-Louis.
Bel exemple d'une fantaisie d'officier supérieur de la Restauration (non figuré).

917

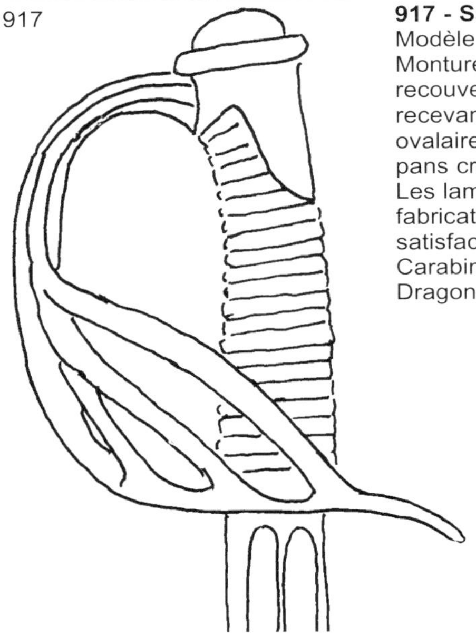

Sabre de Carabinier du modèle 1854, troupe.

917 - Sabre de Carabiniers et de Cavalerie de Réserve.
Modèle 1854, troupe.
Monture en laiton. Calotte à dessus bombé, à bourrelet et courte queue. Poignée recouverte de basane à fil de laiton torsadé faisant 21 spires. Arc de jointure recevant les trois branches secondaires vers son milieu, partant d'un plateau ovalaire terminé par un bec. Lame en acier fondu, longue de 100 cm, à deux pans creux. Fourreau en tôle d'acier, à deux bracelets à jonc et dard en lyre. Les lames sont marquées *Mre Imp^{ale} de Chat^{t}* avec le mois et le millésime de la fabrication. Le sabre de Cavalerie de ligne modèle 1822 ne donnant pas satisfaction, un modèle de sabre à lame droite fut adopté en 1854 pour les Carabiniers. En 1863 il fut attribué à la Cavalerie de Réserve (cuirassiers) puis au Dragons après raccourcissement de la lame.

918 - Sabre d'officier de Carabiniers.
Modèle 1854.
Monture en laiton doré et de même forme générale que pour le modèle de la troupe, mais la queue de calotte est ornée d'une frise de feuilles de laurier, la base des branches secondaires est ciselée, *recto - verso* d'une palme et le bec du plateau, de quatre feuilles d'acanthe d'où part une palmette à sept éléments.

918

Quillon de sabre d'officier de Carabiniers du modèle 1854.

919

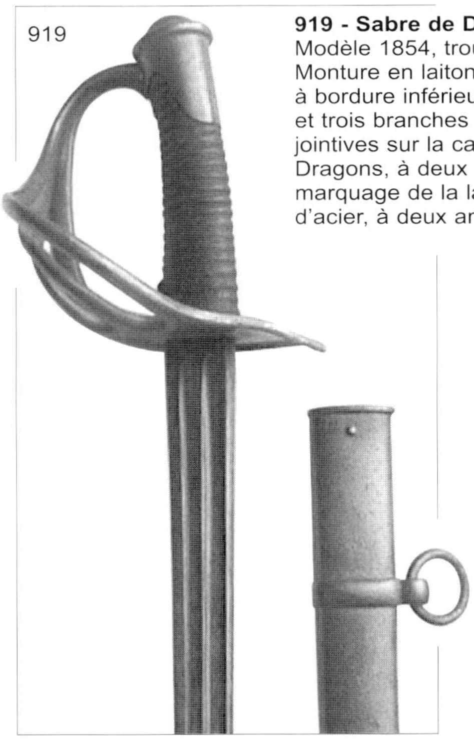

919 - Sabre de Dragons.
Modèle 1854, troupe.
Monture en laiton. Calotte à bourrelet surmontée d'un dôme-bouton de rivure. Courte queue à bordure inférieure échancrée en " S ". Poignée filigranée faisant 21 spires. Arc de jointure et trois branches secondaires partant, sans rouleaux, du plateau ovalaire à bec, et arrivant jointives sur la calotte. Lame primitivement de 100 cm et ramenée à 97,5 cm pour les Dragons, à deux pans creux et pointe en prolongement de l'arête médiane, droite. marquage de la lame : *Mre Imp^{ale} de Chat^{t} déc^{bre} 1863 Caval^{er} M^{le} 1854*. Fourreau en tôle d'acier, à deux anneaux de bélière et dard en lyre.

920 - Sabre d'officier de Dragons.
Modèle 1854.
Monture en laiton doré et de forme générale comparable au sabre de la troupe, avec les différences suivantes : calotte et base des branches secondaires ornées de frises de feuilles de laurier stylisées *recto* et *verso* ; haut de l'arc de jointure décoré d'une fleurette suivie de quelques feuilles : bec du plateau ciselé d'une palmette dont la base rejoint quatre feuilles d'acanthe stylisées. Fusée en corne de buffle. Lame portant l'inscription sur le dos *Mre Imp^{ale} de Chat^{t}* (mois, millésime). *O. de Dragon M^{le} 1854* (non figuré).

921 - Sabres réglementaires de Cavalerie.
Modèle 1854.
La fabrication de ces sabres fut confiée à la Manufacture de Châtellerault. En 1856, 1729 sabres de Carabiniers et 200 sabres de Dragons furent produits. La fabrication se poursuivit jusqu'à l'adoption des modèles 1882 rappelle Maryse Raso (1993), (non figuré).

922 - Sabre d'Etat-Major.
Modèle 1855.
Même monture que le sabre d'officier de Carabinier, modèle 1854, mais à lame de 84,5 cm de longueur. Fourreau à deux bracelets jusqu'en 1882, puis à un seul (non figuré).

923

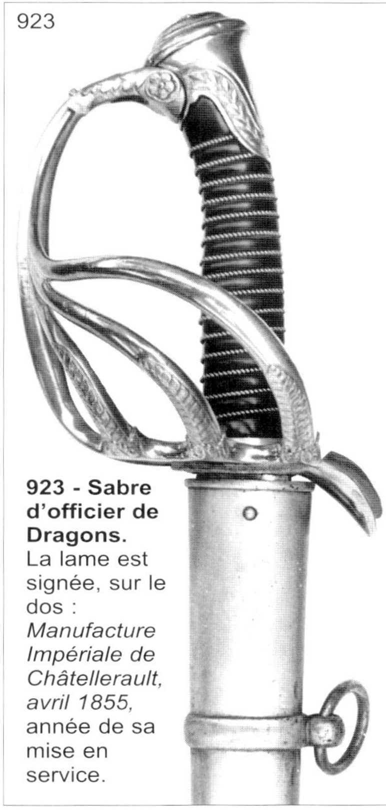

923 - Sabre d'officier de Dragons.
La lame est signée, sur le dos : *Manufacture Impériale de Châtellerault, avril 1855,* année de sa mise en service.

924 - Sabre d'officiers généraux et d'Etat-Major.
Modèle 1855.
Monture en laiton du type 1854. Calotte en dôme-bouton de rivure et courte queue bordée d'une frise de feuilles de laurier. Poignée à une quinzaine de spires d'un fil de laiton torsadé. Arc de jointure peu arqué en sa partie médiane, orné d'une fleurette et de quelques feuilles à son sommet et s'élargissant à sa base pour former un plateau terminé par un bec. Les trois branches secondaires sont décorées à leur base d'une longue palmette. Lame droite, de 94,5 cm, signée : *Manufacture Impériale de Chatellerault, avril 1855.* Ce sabre peut comporter des étoiles de grade sur le plateau.

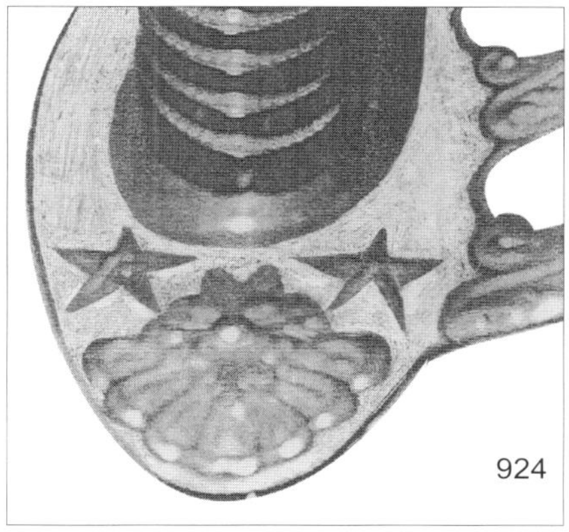

924

926

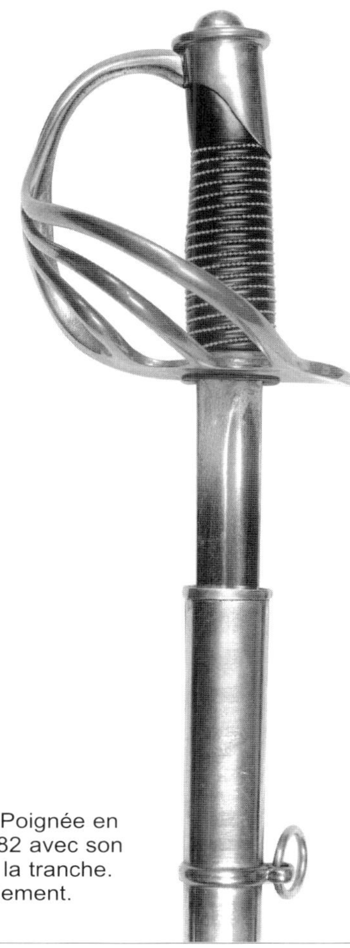

925 - Sabres d'officiers élèves des Ecoles militaires.
Modèle 1961.
Sabre à quatre branches de garde dont la monture en laiton est très comparable au sabre d'officier modèle 1961 (n° 926) du modèle 1961. Lame droite de 85 cm, à pan creux pour l'Infanterie, lame courbe de 90 cm pour la Cavalerie et l'Artillerie, signée *Etablissements Balp à Saint-Etienne.*

926 - Sabre du Modèle 1961, officier.
Monture bronze à quatre branches sans décor. Poignée en corne avec filigrane. Lame droite du modèle 1882 avec son marquage *" sabre droit d'officier Mle 1961 "* sur la tranche. Fourreau droit à un annneau. En service actuellement.

Sous-famille IV b - Branches secondaires partant d'une coquille ou d'un médaillon

Les sabres les plus remarquables de cette Sous-famille sont évidemment les sabres des Cent Gardes du Second Empire qui sont de fait, la réplique bonapartiste des sabres des Gardes du Corps de la Restauration.

927

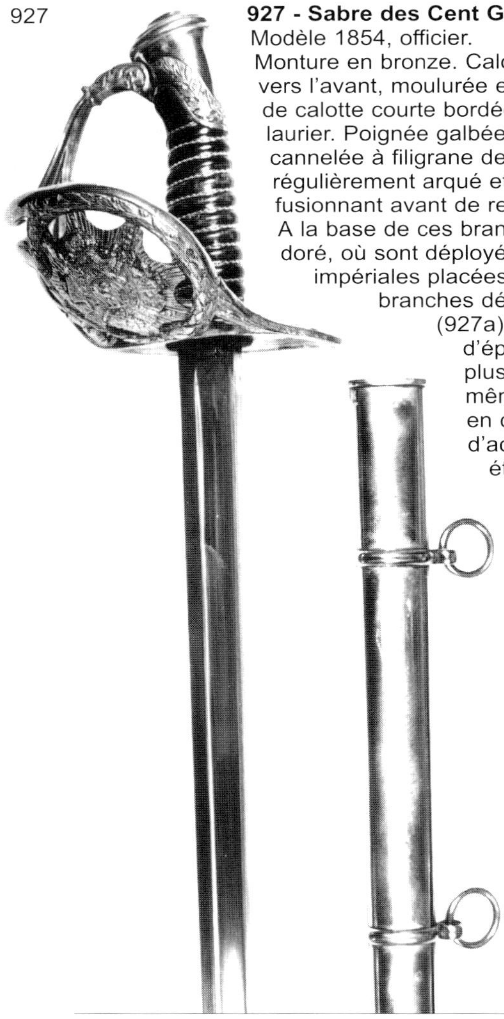

927 - Sabre des Cent Gardes.
Modèle 1854, officier.
Monture en bronze. Calotte légèrement infléchie vers l'avant, moulurée et bouton de rivure. Queue de calotte courte bordée d'une frise de feuilles de laurier. Poignée galbée en corne de buffle noire cannelée à filigrane de fil doré. Arc de jointure régulièrement arqué et trois branches secondaires fusionnant avant de rejoindre l'arc vers son milieu. A la base de ces branches est fixé un médaillon doré, où sont déployées les grandes armes impériales placées sur une sorte d'étoile à huit branches découpées sur fond rayonnant (927a). Lame de 100 cm, de 1 cm d'épaisseur, droite, à dos plat plus large que la lame elle-même, faisant ainsi un " T " en coupe. Fourreau en tôle d'acier à deux bracelets, en jonc étroit. Cuvette en contre-garde. Dard en lyre.

927a

- **a**, dessin simplifié du macaron aux grandes Armes Impériales.

928 - Variantes.
Ces sabres peuvent être issus des Manufactures d'Etat mais aussi de fourbisseurs de renom d'où des variations, notamment dans la monture, mais la forme générale reste la même. Ainsi, les branches secondaires, dorées comme toute la monture, peuvent être ciselées sur toute leur longueur de rameaux de chêne, la partie postérieure du plateau décorée de feuilles d'acanthe, les grandes Armes Impériales ne sont pas inscrites dans la couronne de laurier : elles sont découpées et fixées entre les deux branches secondaires supérieures. La lame, à pan creux, peut être bleuie, décorée de feuillages et du " N " napoléonien dorés. Le fourreau, en acier bien poli, à deux bracelets étroits de bélière, est pourvu d'une cuvette en laiton doré.

929 - Sabre de sous-officier des Cent Gardes.
Le sabre de sous-officier du modèle des Cent Gardes, modèle 1854 est très comparable à celui des officiers mais la monture n'est pas dorée (non figuré).

Marquages - Toutes les lames sont gravées sur le dos *Manu^{fre} Imp^{ale} de Châtellerault* suivi du mois et du millésime de la fabrication.

Ces sabres de bronze ciselé et de belle facture ont été attribués à l'Escadron des Cent Gardes créé le 24 mars 1854 qui fait partie de la Garde Impériale. Il compte 210 gardes, 19 officiers, 12 sous-officiers et 4 trompettes. Cet escadron sera dissous le 1er octobre 1870.

Septième partie

Sabres à cinq branches : un arc de jointure plus quatre branches secondaires

I - Branches secondaires arrivant isolément sur l'arc de jointure

II - Branches secondaires fusionnant en atteignant l'arc de jointure

III - Trois arcs de jointure et deux branches secondaires obliques

Les sabres entrant dans cette Septième Partie ont bien quatre branches secondaires frontales mais celles-ci sont disposées différemment : chez certaines montures, ainsi que dans la Sixième Partie, elles arrivent isolément sur l'arc de jointure alors que chez d'autres, elles arrivent conjointement sur cet arc. A ces deux types de montures nous avons ajouté celles à trois arcs de jointure reliées par des branches secondaires obliques. .

En conséquence trois chapitres ont été créés :

Chapitre I

Branches secondaires arrivant isolément sur l'arc de jointure

Ce type de monture est fréquent de la fin du règne de Louis XVI jusque sous la Restauration et facilement identifiable : sur un arc de jointure peu courbé, quatre branches secondaires, presque parallèles, arrivent isolément. Ce type de monture est associé soit à une longue queue de calotte, soit à une courte queue.

Cette caractéristique a conduit à former deux Familles :

Famille I - Calotte à longue queue

Les sabres de cette Famille I, bien que peu nombreux se rencontrent depuis la Révolution jusque sous la Restauration.

Deux Sous-familles ont été définies :

Sous-famille I a - Garde en coquille Saint-Jacques ajourée

Ces sabres sont évidemment des dérivés du modèle dit à Garde de Bataille de 1784 (voir p. 453, n° 945).

930

930 - Sabre d'officier de la Garde Nationale.
Vers 1791.
Monture en laiton. Calotte en tête de lion à longue queue. Poignée entièrement filigranée de fils de laiton torsadés. Lame à pan creux poinçonnée de de Cabanne.

931

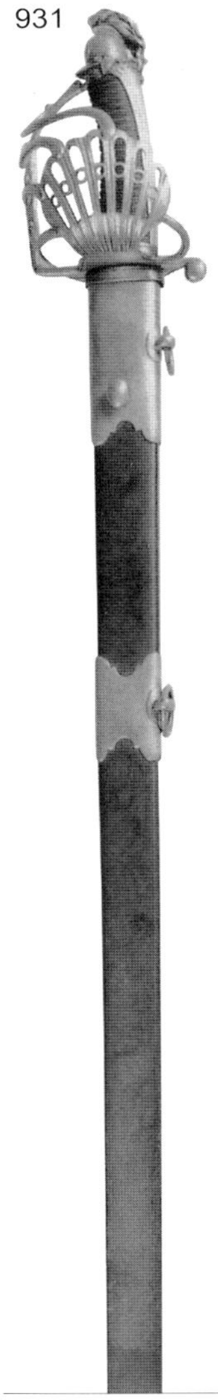

932

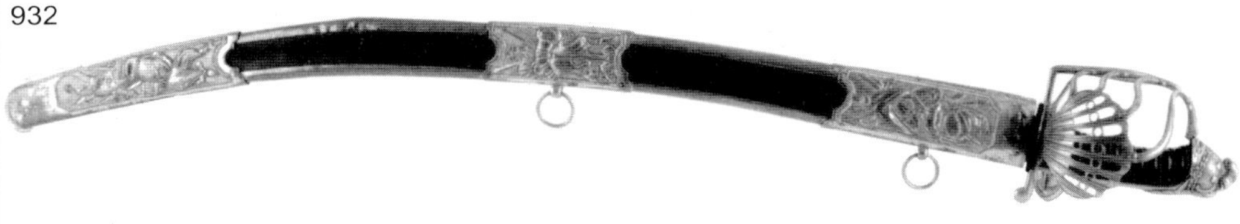

932 - Sabre d'officier supérieur de Cavalerie.
Fin XVIIIème début du XIXème siècle.
Monture en laiton doré. Calotte au casque à l'antique prolongée d'une longue queue, le reste de la monture étant conforme au modèle 1784. Large lame courbée. Fourreau à trois garnitures ciselées de trophées militaires, les garnitures étant réunies latéralement par des attelles de laiton. Dard asymétrique (J.C. D.).

931 - Sabre d'adjudant général du 20 thermidor An VI.
Ce modèle se distingue de la famille des sabres dits *de bataille* par sa calotte en casque empanaché et sa longue queue. Poignée recouverte de basane et filigranée. Lame droite à un seul tranchant, à double pan creux, gravée de motifs dorés classiques sur fond bleui sur le tiers environ de sa longueur. Fourreau à trois garnitures découpées, chape à bouton et anneau de bélière.
B. Croissy, en présentant cette arme fait le commentaire suivant : *Ce sabre peut être attribué aux adjudants généraux : dans le règlement concernant les uniformes des généraux et officiers d'Etat-Major, du 20 thermidor An VI, il est écrit : l'Adjudant général porte un sabre à garde de bataille à lame droite.* En effet, ce sabre présente quelques variations qui le différencient de tous les sabres d'officiers supérieurs.

933 - Sabre d'officier de Dragon.
Garde du modèle 1790 - 1800.
Monture à *garde de bataille mais à longue queue de* calotte. Dans le présent exemplaire cette garde a été montée sur une lame droite d'une longueur exceptionnelle, voisinant les 110 cm. Fourreau métallique à deux anneaux de bélière, une bague de renfort et un dard asymétrique.

933

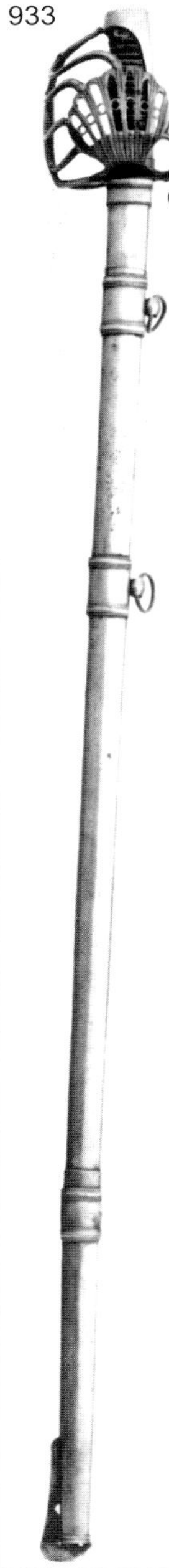

934

934 - Sabre d'officier.
Modèle 1814.
Ce sabre est très voisin du sabre modèle 1784, dit à garde de bataille : une seule différence, la calotte est à longue queue.

Sous-famille I b - Garde à motif révolutionnaire ou à médaillon armorié

A part le sabre de la Garde Nationale datant de la Monarchie constitutionnelle, les sabres ici regroupés sont essentiellement de la Restauration et ceux-ci portent dans un médaillon les armoiries du fondateur du Régiment.

935

935 - Sabre d'officier de la Garde Nationale ou de Volontaires.
1790.
Monture en laiton. Calotte en casque à l'antique à grille et longue queue. Sur les branches secondaires est placé un médaillon où deux représentants de la Convention prêtent serment sur l'autel de la Patrie marqué à sa base 1790.

936

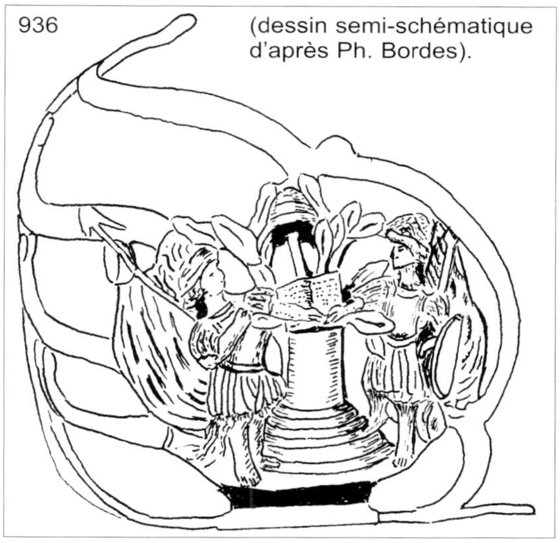

(dessin semi-schématique d'après Ph. Bordes).

936 - Variante.
Une garde est figurée par Ph. Bordes (1987), au même sujet (Musée de l'Armée, inv. J 477). L'auteur décrit la saynète de la façon suivante : *Sur un autel circulaire à cinq marches est posé, de face, un livre ouvert, appuyé sur un bâton qui porte un bonnet phrygien... De part et d'autre deux hommes prêtent serment. Ils sont costumés à l'antique... Celui de gauche a la main sur le cœur, le bras gauche tendu sur l'autel. Celui de droite a un glaive du côté gauche, un bouclier au bras gauche, le bras droit tendu sur l'autel...* Il est difficile de préciser quels sont ces personnages, mais on peut émettre l'hypothèse que celui de gauche personnifie les forces civiles et la loi et celui de droite, les forces armées.

937

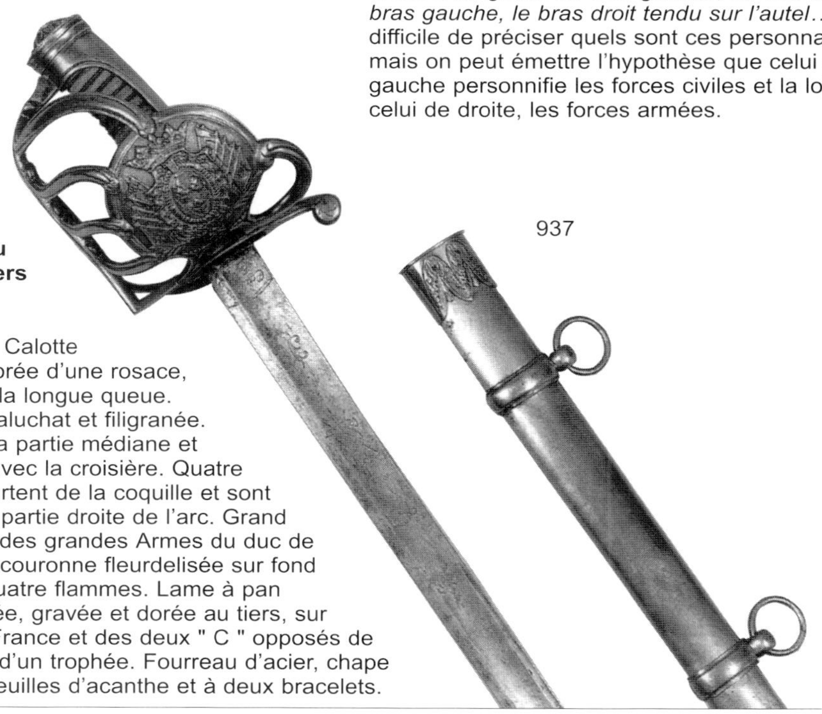

937 - Sabre d'officier du Régiment des Cuirassiers de Berry.
Restauration.
Monture en bronze doré. Calotte légèrement bombée décorée d'une rosace, gravée au trait ainsi que la longue queue. Poignée recouverte de galuchat et filigranée. Arc de jointure droit en sa partie médiane et faisant un angle de 90° avec la croisière. Quatre branches secondaires partent de la coquille et sont brasées isolément sur la partie droite de l'arc. Grand médaillon arrondi, ciselé des grandes Armes du duc de Berry, surmontées d'une couronne fleurdelisée sur fond de huit drapeaux et de quatre flammes. Lame à pan creux, légèrement courbée, gravée et dorée au tiers, sur une face des Armes de France et des deux " C " opposés de Charles X, et sur l'autre, d'un trophée. Fourreau d'acier, chape de laiton doré ciselé de feuilles d'acanthe et à deux bracelets.

938

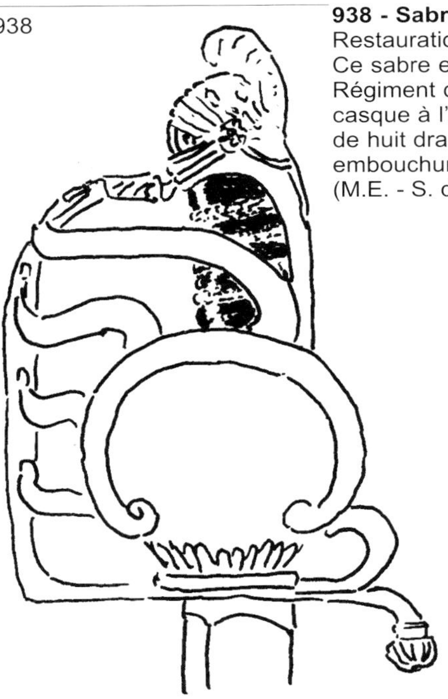

938 - Sabre d'officier de Cuirassiers du Régiment d'Artois.
Restauration.
Ce sabre est de même conception que le sabre d'officier des Cuirassiers du Régiment d'Angoulême (p. 462, n° 971). Il en diffère essentiellement par sa calotte en casque à l'antique, à longue queue. Le médaillon est aux Armes de France flanquées de huit drapeaux fleurdelisés. Petit contre-clavier en palmette. Fourreau en acier à embouchure ornée d'une bague aux Armes de France, ainsi que les deux bracelets (M.E. - S. de P.).

939 - Sabre d'officier du régiment des Cuirassiers de Condé.
Modèle 1817.
Monture en laiton doré. Calotte ornée d'une fleurette à huit pétales et à longue queue. Grand médaillon soutenu par quatre branches secondaires soudées séparément sur l'arc de jointure formant un angle droit avec la croisière. Le médaillon est aux armes des Condé ; ces armes sont entourées du collier de l'Ordre du Roi, de la Toison d'Or, de la Grand Croix de Saint-Louis et couronnées ; l'ensemble est sur fond de dix étendards. Lame droite à deux pans creux. Fourreau de tôle d'acier à deux bracelets en laiton doré, ou de fer, et anneaux de bélière en fer (M.E. - S. de P.), (non figuré).

940 - Sabre d'officier du Royal Carabinier.
1816 - 1817.
Monture de laiton doré. Calotte frappée de la fleur de lys, à longue queue. Poignée filigranée. Quatre branches secondaires soudées isolément sur l'arc de jointure et soutenant un médaillon à la grenade à trois fleurs de lys. Lame plate, gravée

Officier Royal Carabiniers

sur une face et sur l'autre :

Vivat Regnum carissimus,

chaque inscription étant entre les Armes de France. Fourreau de cuir noirci, à trois garnitures : chape portant le premier anneau de bélière, long bracelet portant le second et très longue bouterolle terminée par un dard asymétrique (non figuré).

Famille II - Calotte à courte queue

Cette Famille II est composée de nombreux sabres qui ont été classés en fonction de l'ornementation de la coquille de garde :

Sous-famille II a - Garde en coquille Saint-Jacques ajourée sans autre décor

Les sabres regroupés dans cette Sous-famille ont en commun l'arc de jointure où viennent se fixer, isolément quatre branches secondaires dans un plan sub-horizontal. Le sabre type est le modèle 1784 qui sera en service, moyennant quelques modifications, jusque vers 1830. La coquille Saint-Jacques est ajourée et sans autres ornementations.

Ce type de sabre, vraisemblablement inspiré d'un modèle germanique, a été attribué en 1787, à la Cavalerie et aux Dragons par le Maréchal de Ségur, ministre de la Guerre de Louis XVI.

951 - Sabre d'officier de Grenadiers à cheval de la Garde Impériale.
Premier Empire.
Ce sabre est caractérisé par la présence d'une petite grenade enflammée placée sur la coquille. La lame signée Duc porte l'inscription, sur une face *GARDE IMPERIALE* et sur l'autre *GRENADIERS A CHEVAL*. Les deux faces sont décorées près du talon, de feuillages stylisés et sur la partie distale d'un " N " dans une couronne de feuilles de laurier surmontée d'une aigle couronnée. Fourreau de cuir, à longues garnitures plus ou moins décorées, dard asymétrique.

951

952 - Sabre d'officier de Cavalerie de Ligne.
Premier Empire.
Monture en laiton. Calotte simple à courte queue. Poignée recouverte de basane filigranée. Grenade enflammée appliquée sur la coquille Saint-Jacques qui se trouve ainsi presque entièrement masquée (952a). Les quatre branches secondaires arrivent bien isolément sur l'arc de jointure. Couvre-chape sous le plateau. Lame de 89,2 cm, légèrement courbée, à deux pans creux, bleuie au tiers et gravée de motifs dorés traditionnels. Fourreau de laiton à deux crevés situés entre les deux bracelets bombés, dard en fer asymétrique.

952

952a

- **a**, vue permettant de voir toute la grenade. Lame à double pan creux, signée " S " et " K ", de Solingen.

953 - Variante officier supérieur.
Premier Empire.
Monture en laiton. Calotte à pans, légèrement bombée et courte queue. Poignée recouverte de basane filigranée. Garde en coquille Saint-Jacques ajourée frappée d'une grenade enflammée en argent et liée à l'arc de jointure par quatre branches secondaires isolées. Lame de 96,1 cm, droite à pan creux. Fourreau de cuir à trois longues garnitures de laiton décorées de doubles traits. Deux anneaux de bélière et dard en fer.

953
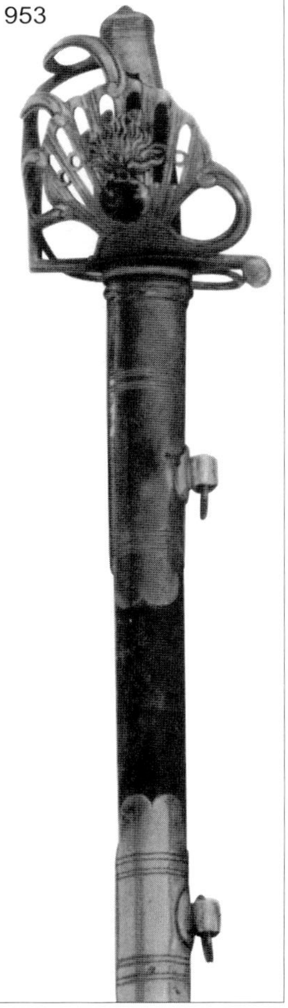

Sous-famille II c - Garde à médaillon orné d'un symbole de corps ou d'armoiries

Les symboles de corps sont des canons pour l'Artillerie, les grenades pour les Carabiniers, la Cavalerie de Ligne, la Gendarmerie, les Grenadiers sous le Premier Empire, les Dragons, les Carabiniers sous la Restauration, la Monarchie de Juillet... Les armoiries sont celles des régiments de duc de Berry, de Condé, d'Angoulême, d'Artois... et la croix enflammée des Mousquetaires de la Restauration.

954

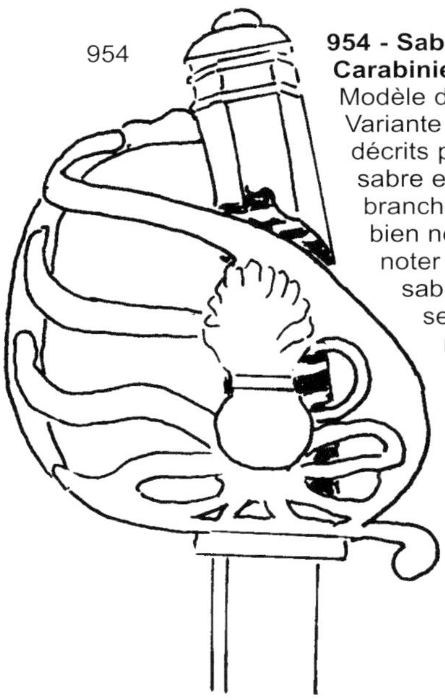

954 - Sabre d'officier de Carabiniers.
Modèle de 1809 - 1810. Variante des modèles décrits p. 456. Le présent sabre est à trois branches secondaires bien nettes. Il est à noter que ce type de sabre, ayant été en service de la fin du règne de Louis XVI jusqu'à la fin du Premier Empire, cela peut expliquer les variantes.

955

955 - Sabre d'officier supérieur de Carabiniers.
Vers 1805.
Monture en laiton doré, du type garde de bataille. Calotte carrée ou octogonale, recouverte d'une coiffe à la grenade cachant la rivure et courte queue. Médaillon en métal argenté, à bordure de feuilles de laurier, avec le " N " napoléonien surmonté de la couronne impériale, flanqué de deux grenades, le tout sur fond rayonnant. Lame à deux pans creux ornés de drapeaux et de motifs foliacés. Sur les deux faces, à partir du talon *Officier / de / Carabiniers*, branches de laurier, aigle impériale couronnée puis motifs foliacés. Fourreau de cuir, à chape et bracelet rainurés portant chacun un anneau de bélière et longue bouterolle à dard presque symétrique (M.E. - S. de P.).

956 - Sabre du Train d'Artillerie de la Garde Impériale.
Premier Empire.
Monture en laiton doré. Calotte à pans et courte queue. Poignée légèrement inclinée vers l'avant, recouverte de basane et filigranée. Les quatre branches secondaires supportent un médaillon dans lequel est inscrite une aigle couronnée posée sur deux canons croisés, l'ensemble entièrement découpé. Quillon rabattu en angle droit, à bouton ciselé. Lame entièrement décorée de gravure de végétaux stylisés dans sa partie basale puis, dans un cartouche, sur le *recto*, *TRAIN D'ARTILLERIE* qui se termine par le " N " napoléonien dans une couronne surmontée de l'aigle couronnée, le tout en or sur fond bleui. La lame est signée sur le talon *Duc / fourbisseur / de la / Garde / impériale / rue St / Honoré en / face celle / de la loi / N° 251 / à Paris.* Fourreau de cuir noir à trois longues garnitures rainurées.

Nota - Seuls trois exemplaires de ce sabre étaient connus en 1987 selon B. Croissy : *un est conservé au Musée de Nancy, un second fut vendu par Christies, à Londres, le 8 Mai 1985 en provenance de la collection Calvin Bullock et un troisième fut présenté à l'hôtel des Ventes de Saint-Germain- en-Laye le 25 janvier 1987.*

956

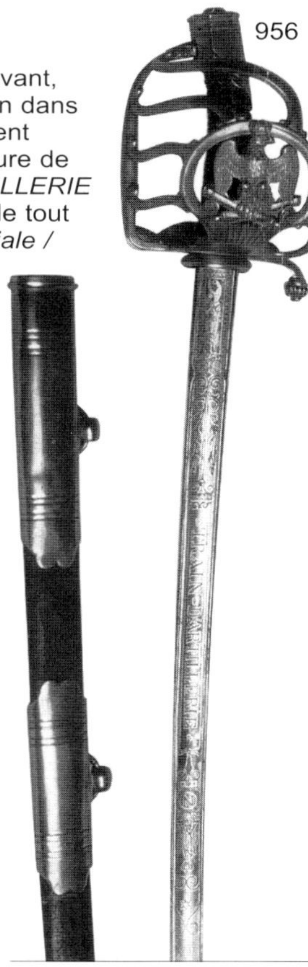

957 - Variantes.
Au lieu d'être à pans, la calotte peut être arrondie, surmontée d'une rosette à 18 pétales et à bordure ornée de feuilles en quinconce. Médaillon à deux canons en sautoir surmontés d'une aigle couronnée, l'ensemble ciselé à jour. Lame marquée *GARDE IMPERIALE* dans un cartouche, à décor plus allégé que pour la lame du sabre précédent. Fourreau métallique, à deux bracelets ciselés de deux canons en sautoir.

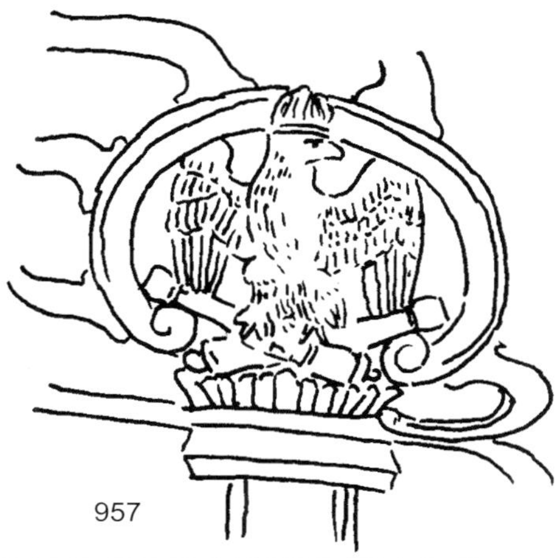

957

958

958 - Sabre d'officier d'Etat-Major.
Fin du Premier Empire.
Ca sabre à garde de bataille en cuivre est orné d'un foudre de Jupiter sur la coquille. La fine lame à jonc est décorée à l'acide de trophées d'armes. Le fourreau est en fer avec d'importantes garnitures de cuivre ciselées de motifs floraux.

959

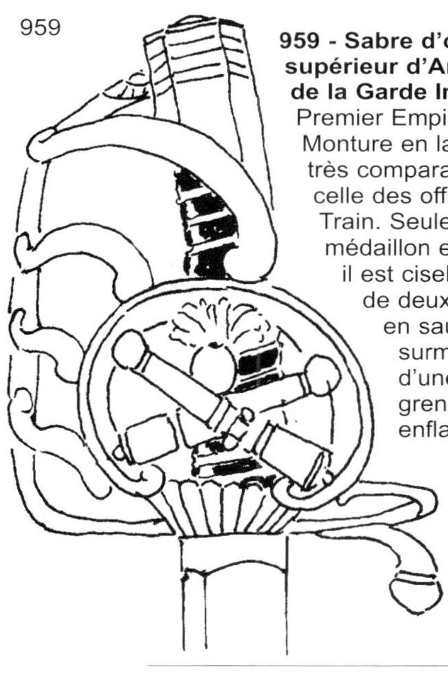

959 - Sabre d'officier supérieur d'Artillerie de la Garde Impériale.
Premier Empire.
Monture en laiton doré, très comparable à celle des officiers du Train. Seule le médaillon en diffère : il est ciselé à jour de deux canons en sautoir surmontés d'une petite grenade enflammée.

960

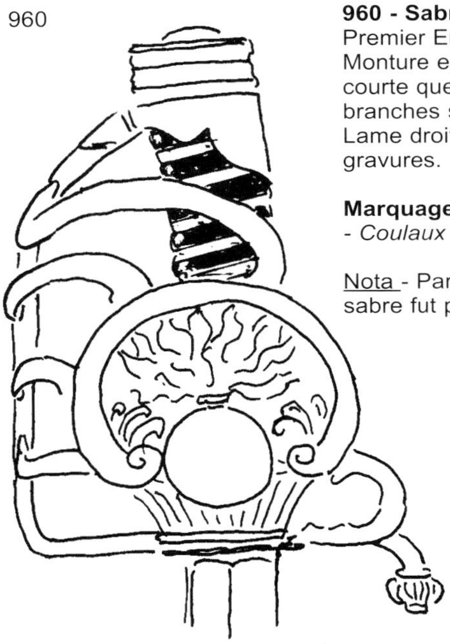

960 - Sabre d'officier de la Cavalerie de Ligne.
Premier Empire.
Monture en laiton doré. Calotte ovale ou à huit pans, moulurée, à bouton de rivure et courte queue découpée en accolade. Poignée filigranée de fils dorés. Les quatre branches secondaires soutiennent un médaillon à la grenade enflammée découpée. Lame droite à deux pans creux, longue de 89 - 97,5 cm, parfois surchargée de gravures. Fourreau de laiton à deux bracelets. Dard en lyre (M.E. - S. de P.).

Marquages - Souvent la lame est signée sur le dos de la *Manufacture de Klingenthal - Coulaux Frères*, avec le poinçon de F.X. Bisch.

Nota - Parfois la lame est demi-courbe. Dans ce cas le dard est asymétrique et le sabre fut porté par les Cuirassiers et les Dragons.

961

961 - Sabre d'officier de la Gendarmerie d'Elite.
Premier Empire.
Le sabre attribué par R. Duplan (Gaz. Armes, n°179, p. 60) à un officier de la Gendarmerie d'Elite du Premier Empire est conforme au modéle de celui des Grenadiers à cheval, mais particulièrement bien réalisé.

963

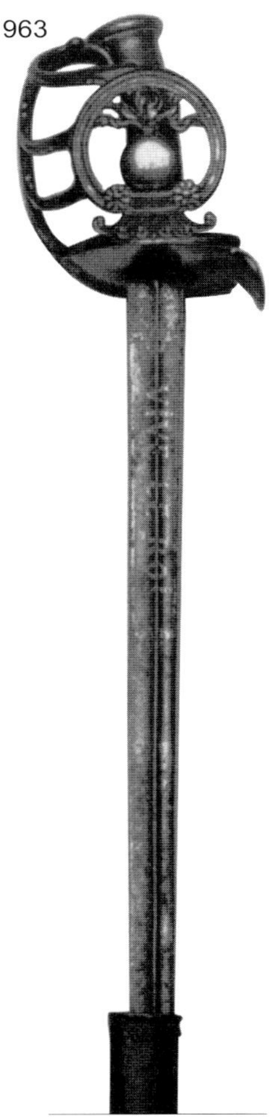

962 - Variantes.
M. Bottet signale une variante du modèle des carabiniers à monture de laiton argenté qui serait attribuée à la Gendarmerie d'Elite sous le Premier Empire (non figuré).

963 - Sabre d'officier de Grenadiers à cheval de la Garde Royale ou de Carabiniers.
1814.
Monture en laiton. Calotte ovale à courte queue. Poignée recouverte de basane, filigranée. Arc de jointure régulièrement arqué s'élargissant pour former le plateau de garde terminé par un bec incliné à 45°. Trois branches secondaires échelonnées tout au long de l'arc et rejoignant un médaillon dans lequel est inscrite une grenade enflammée découpée. Lame droite, à dos rond, contre-tranchant et deux pans creux, bleuie et dorée ; sur le *recto* est inscrit *VIVE LE ROI* et au talon *Coulaux Frères* ; sur le *verso MANUF*^{re} *ROYALE DU KLINGENTHAL* et trois fleurs de lys. Fourreau en fer à deux bracelets.

964

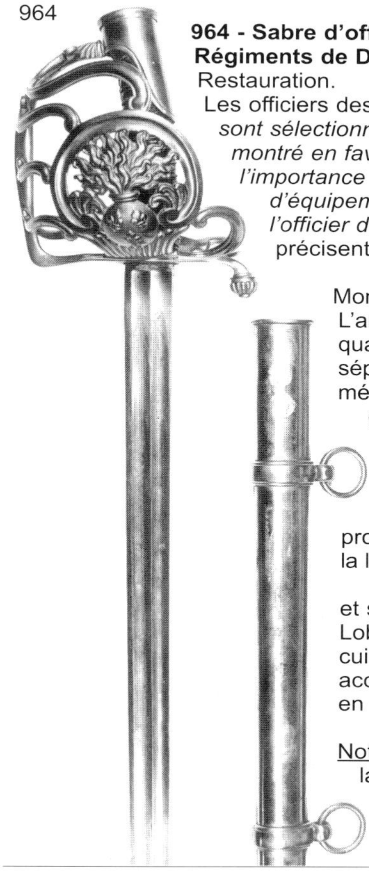

964 - Sabre d'officier de la Compagnie d'Elite des Régiments de Dragons.

Restauration.

Les officiers des Compagnies d'Elite, chez les Dragons, *sont sélectionnés en fonction de la fidélité qu'ils ont montré en faveur de Louis XVIII et aussi en fonction de l'importance de leurs revenus, une partie de leurs frais d'équipement étant à leur charge... ce qui distingue l'officier de la Compagnie d'Elite, c'est son sabre* précisent de Jenlis et de Fromont (1994).

Monture en laiton doré. Calotte à courte queue. L'arc de jointure est relié à un médaillon par quatre petites branches secondaires bien séparées les unes des autres. Dans le médaillon est inscrite, en découpe, une grenade enflammée frappée de trois fleurs de lys (964a). Contre-garde en petite coquille. Quillon recourbé, terminé par un bouton ciselé. Lame de 95 cm, à dos plat, deux pans creux et pointe dans le prolongement de l'arête médiane. Sur le dos de la lame, on lit :

$MANUF^{re}$ R^{le} du Klingenthal Juillet 1814

et sur le talon, on trouve les poinçons de Lobstein, Borson et J.G. Bick. Fourreau de cuir noirci, à trois garnitures à bordure en accolade ; bouterolle terminée par un bouton en guise de dard.

Nota - Chez les sabres de fabrication tardive, la calotte peut être à pans et le fourreau de fer, à un ou deux bracelets.

964a

- **a**, détail de la monture

965

965a

965 - Sabre d'officier de Cavalerie de Ligne.

Restauration.

Ce sabre à garde de bataille à coquille à la grenade enflammée qui présente la particularité d'être frappée d'une seule et grande fleur de lys (969a). L'expert J.C. Dey voit dans cet exemplaire une fabrication postérieure, un remontage ... il serait intéressant de connaître, s'il existe, d'autres exemplaires de ce type.

- **a**, détail de la coquille

966

966 - Sabre d'officier de la Garde Municipale de Paris.
Monture en laiton doré, inspirée de celle adoptée pour les mousquetaires et les Grenadiers à cheval sous la Restauration.
On lit sur la grenade la date du début de la Révolution de Juillet *29 juillet /1830* (966a), et sur la lame l'attribution du sabre : *GARDE MUNICIPALE DE PARIS / 29 JUILLET 1830* (966b).
(M. G. de Melun).

966a

- **a**, détail de la coquille avec la grenade portant la date du début de la Révolution soit : 29 juillet 1830 ;
- **b**, inscription figurant sur la lame.

966b

967

967a

967 - Sabre d'officier de Cuirassier du Régiment d'Angoulême
Restauration.
Monture en bronze doré. Arc de jointure venant rejoindre la croisière par un angle droit et quatre branches secondaires en " S " et partant d'un médaillon qui tient lieu de coquille dans laquelle sont inscrites les Armes de France couronnées. Ces armes sont placées sur fond de huit drapeaux fleurdelisés et de deux ancres en sautoir dont seuls les bras sont visibles (971a). Lame demi-courbe, à pan creux signé Coulaux Frère.
Fourreau en général d'acier à deux bracelets dorés simples et bouterolle globuleuse.

Nota - Ce modèle est assez variable dans les détails ; la calotte à pans peut être à queue longue ou courte, frappée ou non d'une fleur de lys ; le fourreau peut être de cuir à trois garnitures et dard plat, asymétrique. Les ancres peuvent avoir été supprimées (M.E. - S. de P.).

968

968 - Sabre de lieutenant (1) **de la Première Compagnie de mousquetaires.**
Modèle du 15 Juin 1814.
Monture en laiton doré. Calotte à courte queue. Poignée recouverte de chagrin filigrané. Arc de jointure droit sur les trois quarts inférieurs et faisant un angle à 90° avec la croisière. Le médaillon est doré, à la croix enflammée, dont les branches sont terminées par une fleur de lys (968a). De ce médaillon partent les quatre branches secondaires, courbées en " S ", rainurées marginalement et rejoignant l'arc, séparément. Quillon recourbé, à bouton pendant, ciselé en fleurette. Lame à deux pans creux, de 90 cm de longueur, signée *Frères WEYERSBERG à SOLINGEN.* Fourreau de cuir noirci à trois garnitures ; chape à bourrelets marginaux ainsi que le bracelet, bouterolle à gros bouton de fer.

968a

- **a**, détail de la croix enflammée

969a

969b

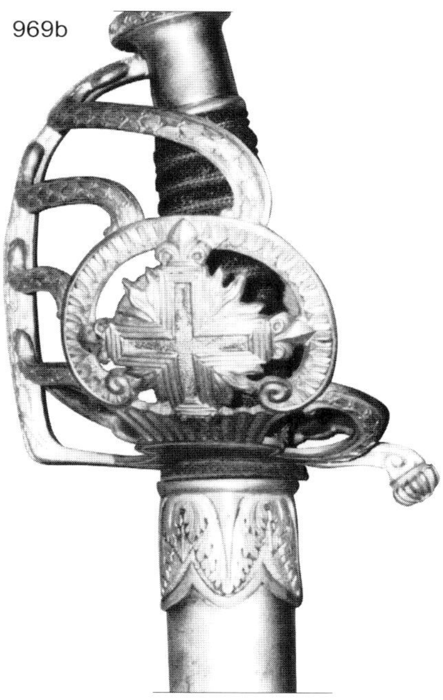

969 - Sabres d'officiers de la Première Compagnie des Mousquetaires.
Modèle du 15 juin 1814.
Même monture générale que pour les lieutenants, mais les sabres des officiers ont une croix enflammée dorée (969a). Les officiers de rang supérieur ont un sabre à garde plus décorée : le dôme de la calotte est orné de la croix enflammée ; l'arc de jointure et les quatre branches secondaires sont ciselées d'oves stylisées ; la bordure du médaillon est décorée d'une frise dorée, rappelant peut-être, la ligne sinueuse ornant les anciennes épées de la Garde du Roy (969b). La lame est signée sur le dos *Frères WEYERSBERG à SOLINGEN* et sur le talon Duc. Fourreaux identiques à celui des lieutenants mais à embouchure décorée de feuilles d'acanthe dorées (J.C. D.).

969a - Sabre d'officier : la croix enflammée est en laiton doré.

969b - Sabre d'officier supérieur : les fleurs de lis et la croix centrale sont argentées et les branches secondaires sont ciselées d'oves.

(1) Les lieutenants des Mousquetaires formaient *la troupe*. Les Mousquetaires de la Première Compagnie étaient également désignés sous le nom de Mousquetaires gris en accord avec le couleur de la robe de leurs chevaux. La monture du sabre peut être signée Duc.

970

970 - Sabre d'officier de la Deuxième Compagnie de Mousquetaire. (1)
Modèle 1814 - 1815.
Monture en laiton argenté. Calotte ovale, à pans, courte queue décorée d'un liseré. Le modèle présenté ci-contre possède une garde en coquille ajourée portant la croix enflammée.
Lame droite de 89,3 cm, à deux pans creux, gravée de motifs dorés sur fond bleui. Parfois, on rencontre des exemplaires dont la lame porte l'inscription, en capitales évidées *MOUSQUETAIRES DU ROI* avec deux croix enflammées l'encadrant. Fourreau de cuir noirci, à trois longues garnitures argentées : bouterolle à dard en fer ainsi que les anneaux de bélière.

Variantes - Certains sabres ont des fourreaux d'acier bleui, les lames peuvent être légèrement courbées et dans ce cas, le dard devient asymétrique.

(1) La Seconde Compagnie de Mousquetaires se nomme également les Mousquetaires Noirs, en accord avec la couleur de la robe de leurs chevaux.

Chapitre II

Branches secondaires fusionnant en atteignant l'arc de jointure

Les sabres de ce Chapitre II sont essentiellement des sabres qui ont été en service jusqu'après la fin de la Guerre de 1914-1918. L'arc de jointure et les branches secondaires sont découpés dans une plaque de métal avant d'être mis en forme pour obtenir la monture. Ces sabres sont peu réglementés et pouvant être commandés à des fourbisseurs privés, ceux-ci prenaient quelque liberté avec le modèle, les enjolivaient notamment en multipliant les branches secondaires. On parle alors de sabres à branches multiples.

Famille III - Calotte à longue queue

Les sabres regroupés ci-après ne sont que des variantes du modèle 1882 à quatre branches secondaires fusionnant avec l'arc de jointure.

971

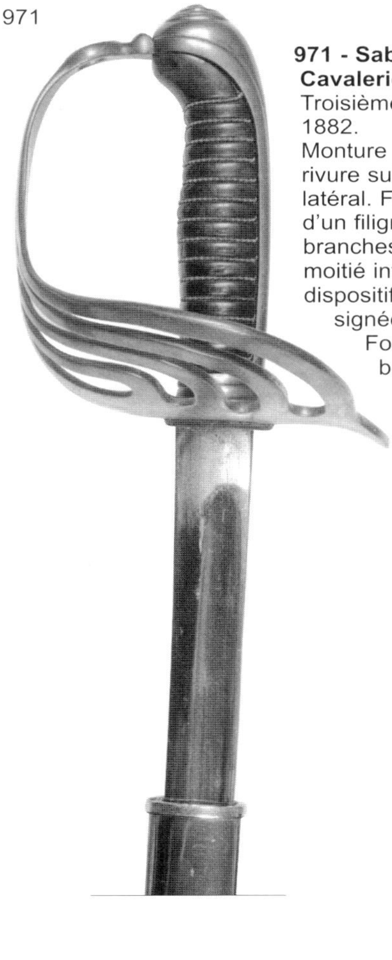

971 - Sabre de fantaisie d'officier de Cavalerie.
Troisième République, dérivé du modèle 1882.
Monture en bronze. Calotte à bouton de rivure suivie d'une longue queue à filet latéral. Fusée de corne noire à treize spires d'un filigrane de laiton torsadé. Quatre branches secondaires fusionnant dans la moitié inférieure de l'arc de jointure, selon le dispositif dit *"à la Saumur"*. Lame nickelée signée *Coulaux et Cie à Klingenthal.* Fourreau nickelé à un bracelet de bélière.

972

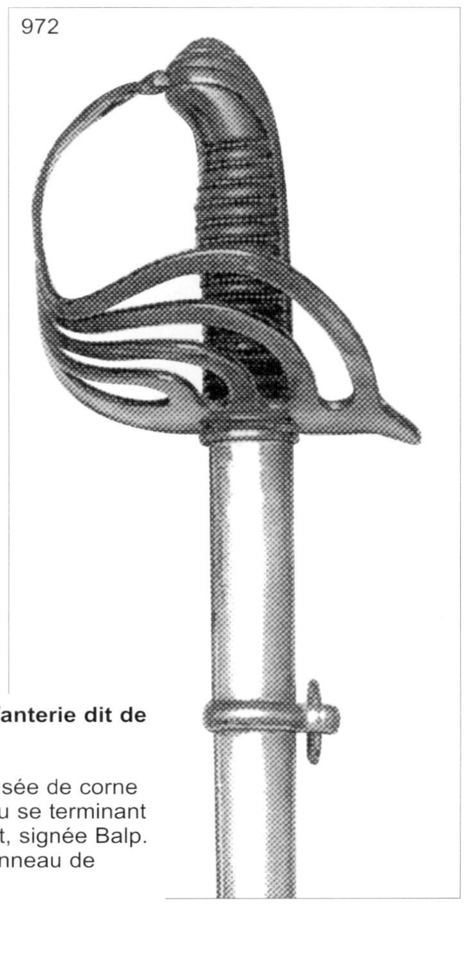

972 - Sabre d'officier d'Infanterie dit de Nancy.
Modèle 1882, variante.
Monture en laiton rouge. Fusée de corne noire filigranée. Petit plateau se terminant par un bec. Lame à dos plat, signée Balp. Fourreau de fer à un seul anneau de bélière.

Chapitre III
Trois arcs de jointure et deux branches secondaires obliques

La structure de ces sabres est directement inspirée de celle des wallonnes du début du XVIIIème siècle dont on trouvera la figuration dans " Les Epées " de J. Lhoste (op. cit. p. 120, figure 188, p. 121, figure 192, p. 122, figure 193, ...). La grande différence qui nous incite à placer ces armes dans les sabres est la suivante : les wallonnes ont un pommeau sphérique alors que les présentes armes ont une calotte plate à longue queue. Leur garde de fer s'inscrit nettement dans la tradition des sabres de la Gendarmerie.

Famille V - Monture en fer

Ces sabres sont à calotte à longue queue, à poignée filigranée et à cinq branches.

973

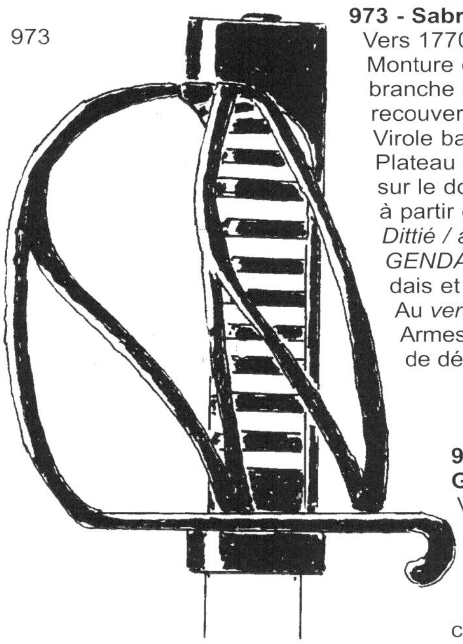

973 - Sabre d'officier de Gendarmerie.
Vers 1770 - 1785.
Monture d'acier bronzé, à trois arcs de jointure réunis, deux par deux, par une branche secondaire oblique. Calotte plate à longue queue. Poignée non ficelée, recouverte de veau et filigranée d'un fil d'argent faisant une dizaine de spires. Virole basale fixant une petite lanière de cuir faisant office de pièce de pouce. Plateau cordiforme et couvre-chape. Lame plate à dos, longue de 88 cm et gravée sur le dos *fait à Klingenthal*. Sur le *recto*, doré on trouve les inscriptions suivantes, à partir du talon :
Dittié / armurier / de la Gendarmerie / à Luneville, puis *OFFICIER / DE LA / GENDARM / ERIE*, les Armes royales, *VIVAT / REGUM / CARISIMUS*, le soleil, le dais et un motif floral.
Au *verso*, dans le même ordre *OFFICIER / DE LA / GENDARM / ERIE* , puis les Armes royales, *VAINCRE / OU / MOURIR*, le soleil, le dais, le motif floral (pour plus de détails, voir M. Pétard, 1994, Tradition, 84, 26 - 31).

974 - Sabre d'officier de Gendarmerie.
Vers 1780 - 1788.
Monture en acier bronzé, à trois arcs de jointure réunis entre eux par deux branches secondaires obliques ainsi que chez le modèle de 1770. Calotte à bouton de rivure très aplati et longue queue. Poignée non ficelée, recouverte de veau et filigranée d'un simple fil de fer faisant dix spires. Virole basale. Plateau cordiforme et couvre-chape. Lame à dos, à double pan creux, longue de 97 cm lorsqu'elle est en son intégrité, gravée au *recto*, à partir du talon, de trophée, *VAINCRE OU MOURIR* dans le sens de la longueur en cursives majuscules et sur le *verso*, à partir du talon d'un trophée, des Armes royales, *VIVE LE ROY*.

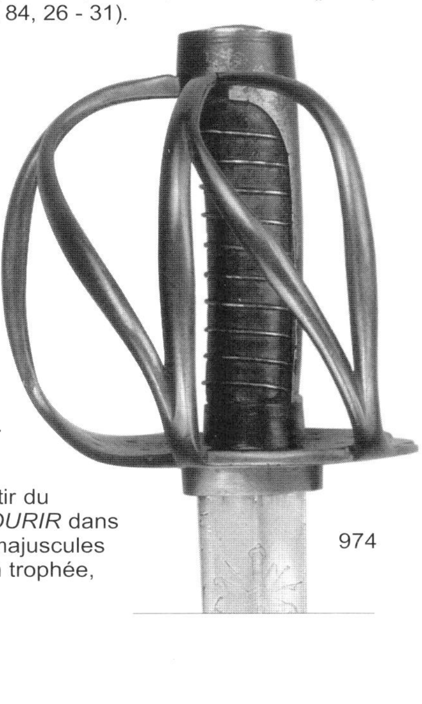

974

Huitième partie

Sabres à monture à contre-garde

Sabres à monture à contre-garde

R.F.

Depuis la création des sabres modernes, c'est-à-dire depuis le XVIIème siècle environ jusqu'en 1831, aucun sabre n'était pourvu d'une contre-garde vraiment fonctionnelle. Ce n'est qu'à cette date que le général Préval propose un sabre à garde et contre-garde symétriques comptant deux fois trois branches secondaires de part et d'autre d'une grosse coquille symétrique. Un modèle simple est prévu pour la troupe et un modèle enjolivé pour les officiers. Fabriqué vers 1833, ce sabre ne fut jamais adopté par le ministre de la Guerre de l'époque.

Peut-être inspiré par le modèle du général, à la fin du XIXème siècle et en particulier après la défaite de 1870, l'esprit de revanche qui anima tous les officiers se traduisit, entre autres, par une explosion des sabres de fantaisie à branches secondaires multiples. En effet, ces branches, y compris l'arc de jointure, peuvent atteindre le nombre de dix dont certaines en contre-garde. Il est parfois difficile de se retrouver devant une telle diversité, aussi en nous inspirant des travaux de R.D. Stiot (in Ariès, XXIX, 5 et 6, 1985) nous proposons le tableau d'identification ci-après. On remarquera que tantôt la garde est symétrique, tantôt asymétrique. Nous précisons, avec prudence, la période de mise en service, avant ou après 1882.

En effet, en 1882, pour mettre de l'ordre dans l'armement des officiers, un modèle, sans branches de contre-garde devient réglementaire (voir p. 404 - 406). Mais cette réglementation est peu observée et de nouveau, des sabres de fantaisie à contre-garde réapparaissent…

Cette multiplication des modèles est bien entretenue par les fourbisseurs privés qui rivalisent d'imagination pour attirer l'attention des officiers.

Chapitre Unique

Sabres à monture à contre-garde

Les sabres de cette Partie sont présentés en deux Familles ainsi définies :

Classification des sabres à contre-garde

	Nombre de branches secondaires frontales	Nombre de branches secondaires en contre-garde	Observations
Famille I Calotte à longue queue			
	3	1	avant et après 1882
	3	2	variante 1882
	4	4	avant 1882
	6	1	après 1882
Famille II Calotte à courte queue			
	2	1	modèle 1923
	2	2	modèle 1896
	3	1	modèle 1882, dit de Nancy
	3½	1	après 1882
	3	2	général de brigade 1882 (variante)
	3	3	modèle 1831
	3½	2½	modèle 1855
	4	1	modèle 1882 (variante)
	4	1½	modèle dit de Saumur
	4	2	modèle Dérué 1890
	4	2½	avant 1882
	4	4	modèle 1882 (variante)
	5	1	modèle 1882 (variante)
	6	1	avant 1882

Famille I - Calotte à longue queue

Les sabres regroupés dans cette famille à une exception près sont dérivés du modèle 1882 et parfois même en sont contemporains. Le modèle 1923 présente des variantes les rattachant au passé, telle celle à calotte en tête d'animal.

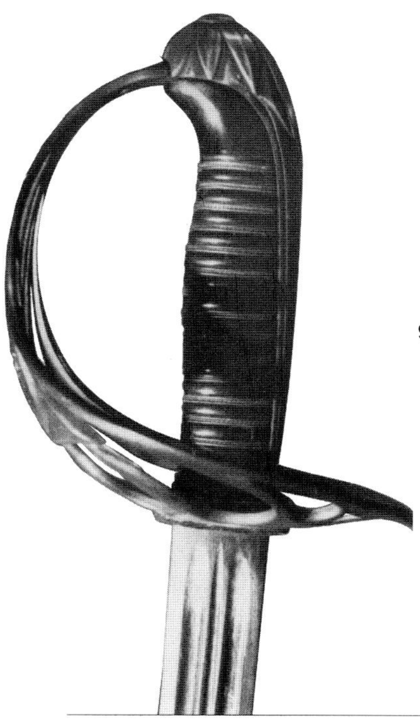

975 - Sabre d'officier supérieur de Chasseurs.
Troisième République.
Monture en métal blanc. Calotte ornée de feuilles triangulaires et prolongée d'une longue queue bordée d'un filet. Poignée à une quinzaine de spires d'un double fil torsadé. Arc de jointure divisé en deux petites branches à sa partie basale ; trois branches secondaires et demie en garde frontale portant un cor dans lequel est inscrit " L ", " C ", vraisemblablement les initiales de l'officier ; deux branches secondaires et demie en contre-garde. Lame droite, longue de 86 cm, à trois rainures et gravée *O. Supérieur M^le^ 1855*, signée *Manufacture d'Armes de Châtellerault, mars 1882.* Fourreau de tôle d'acier à un bracelet de bélière.

975

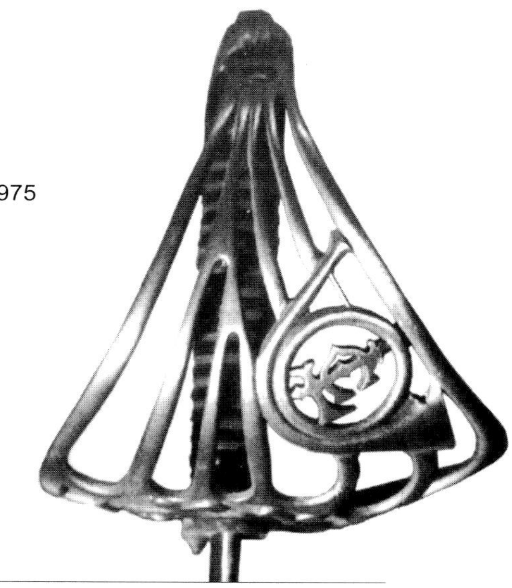

Sabres d'officiers d'Infanterie

Dérivés du modèle 1882. Ce modèle dans sa forme réglementaire est caractérisée par une monture à trois branches secondaires s'unissant vers la moitié de la longueur de l'arc de jointure. Mais la fantaisie des fourbisseurs encouragée par le désir des officiers de briller dans les salons de la Belle Epoque, a entraîné la création de nombreuses variantes . Le premier modèle (3/1) est dit sabre de Nancy. En comptant l'arc de jointure, le modèle 4/4 totalise donc neuf branches de garde.

Variation de la monture

Si la calotte, la fusée et le quillon ne varient guère, il n'en est pas de même pour la garde. En effet, si le modèle réglementaire ne possède que des branches secondaires frontales, de nombreuses variantes en possèdent en contre-garde ainsi que le montre le tableau ci-après :

Branches secondaires frontales	Branches secondaires en contre-garde	Métal utilisé
3	1	laiton ou fer
3	2	bronze de nickel
4	1	bronze de nickel
4	2	bronze de nickel
4	4	bronze de nickel
5	1	bronze de nickel

Variation de la lame

Les lames réglementaires sont à double tranchant mais les lames à dos deviennent de plus en plus fréquentes. Quant à leur longueur elle est fixée par décret et est en rapport avec la taille de l'officier, soit : 75, 80, 85 et 90 cm. De plus, on rencontre des lames enjolivées de gravures à l'acide, parfois dorées… et certains officiers y font graver leur nom.

Nota - Si le modèle réglementaire est produit par les Manufactures de Klingenthal et de Châtellerault, en revanche, les variantes relèvent de la fabrication privée notamment des fourbisseurs suivants à Saint-Etienne, Balp, Manufacture d'Armes et de Cycles ; à Paris, Gustave Barré, Gustave Barré et fils, Barat, Blackes et Berger, Delacour et Blakes, Lacroix, Lefèvre-Lacroix, Renoud ...

976

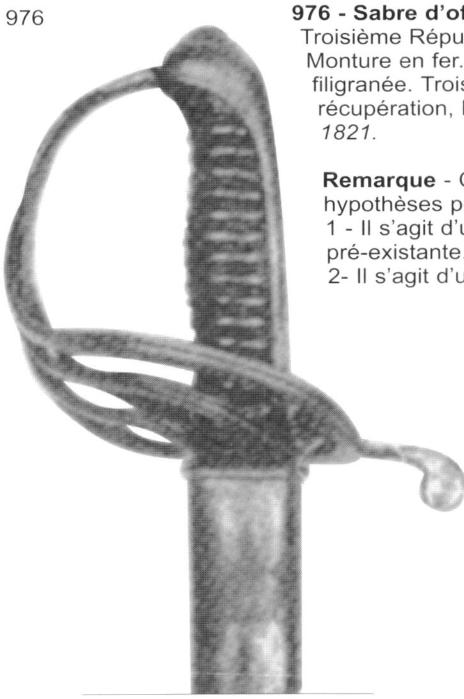

976 - Sabre d'officier d'Infanterie.
Troisième République.
Monture en fer. Calotte en cuillère à longue queue. Poignée en balustre de corne et filigranée. Trois branches secondaires frontales et une en contre-garde. Lame de récupération, longue de 87 cm, et marquée *Manufacture Royale de Klingenthal, avril 1821.*

Remarque - Ce sabre a été remonté ainsi que cela se rencontre souvent. Deux hypothèses peuvent être envisagées :
1 - Il s'agit d'un remontage d'époque pour réemploi d'une lame convenable, pré-existante.
2- Il s'agit d'un remontage de fantaisie…

977

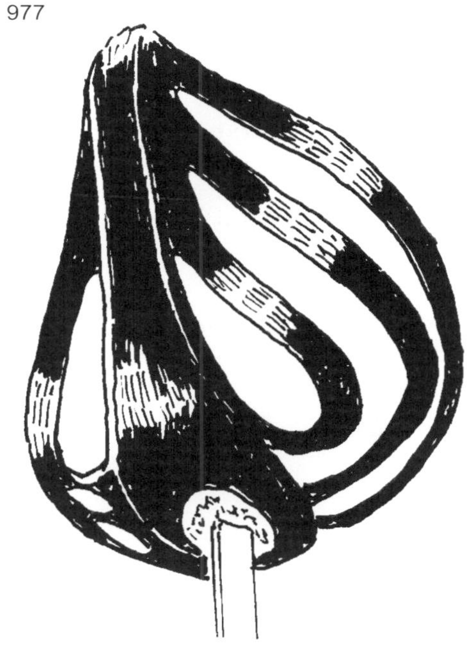

977 - Sabre d'officier d'Infanterie dit de Nancy.
Variante du modèle 1882.
Monture en laiton. Calotte surmontée d'un dôme-bouton de rivure, prolongée d'une longue queue à liseré marginal. Toute la monture est découpée dans la même plaque de métal ; arc de jointure plat, trois branches secondaires frontales et une branche en contre-garde dont l'espace ménagé le long du plateau se poursuit par deux découpes ovalaires. Fourreau en tôle d'acier à bracelet de bélière unique.

Nota - Ce modèle a été fabriqué également pour la Cavalerie mais dans ce cas, la lame est d'une plus grande longueur.

978

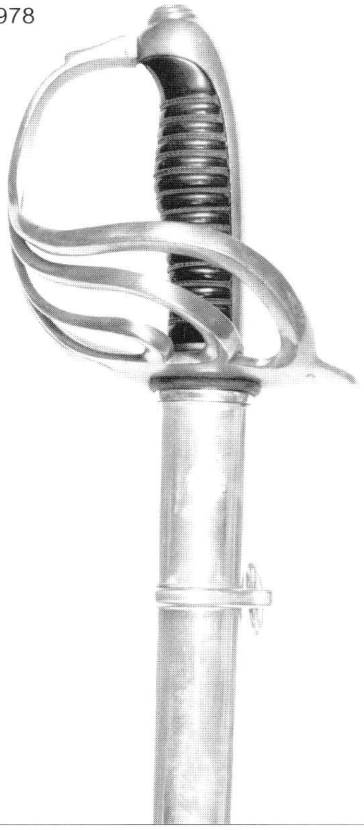

978 - Sabre de général de Brigade de Cavalerie.
Dérivé du modèle 1882. Monture de bronze. Calotte prolongée d'une longue queue unie. Poignée de corne à triple filigrane faisant 13 spires. Trois branches secondaires frontales et deux branches secondaires en contre-garde. Les deux étoiles du grade sont fixées sur le bec du plateau (979a). Lame à pan creux, marquée sur le dos *Officier de Cavalerie, Mle 1882, Manufacture Nationale de Châtellerault 1909*. Fourreau nickelé à un anneau de bélière.

978a

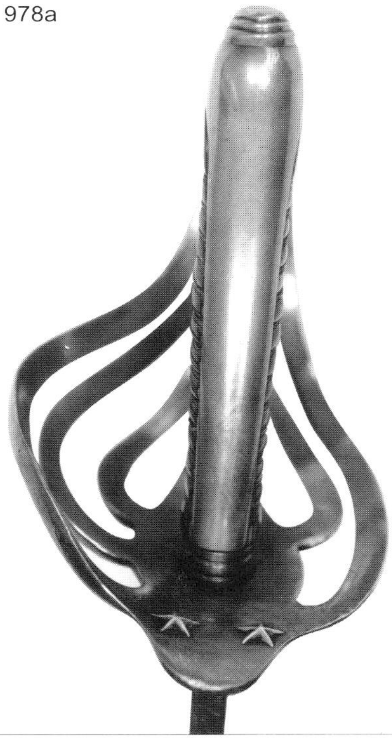

- **a**, remarquer les deux étoiles fixées sur le bec du plateau.

979 - Sabre d'officier d'Infanterie.
Modèle 1882. Variante.
Remarquer la calotte à trois torons (979a) et les quatre branches frontales et les deux branches en contre-garde.

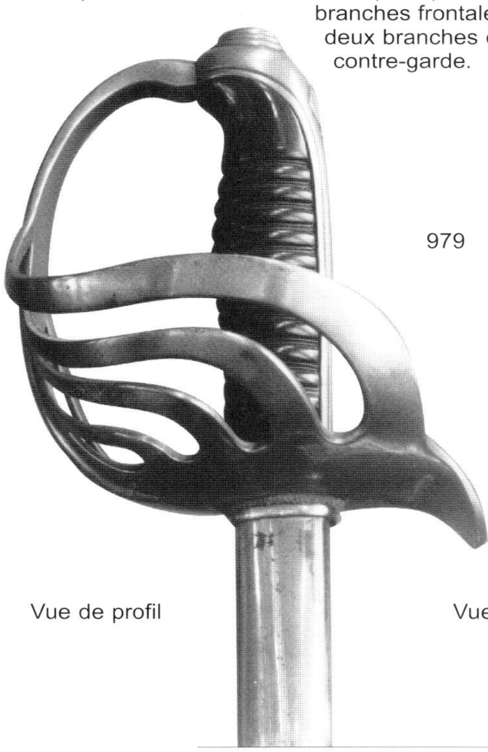

Vue de profil

979

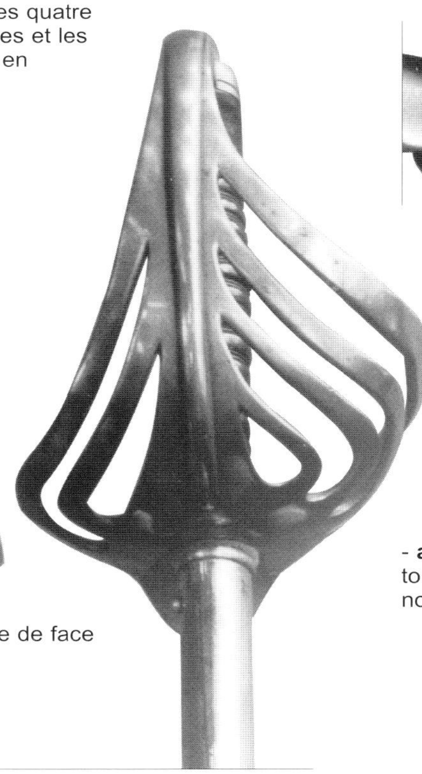

Vue de face

979a

- **a**, détail de la calotte à trois torons, de la fusée de corne noircie et du filigrane très fin.

980

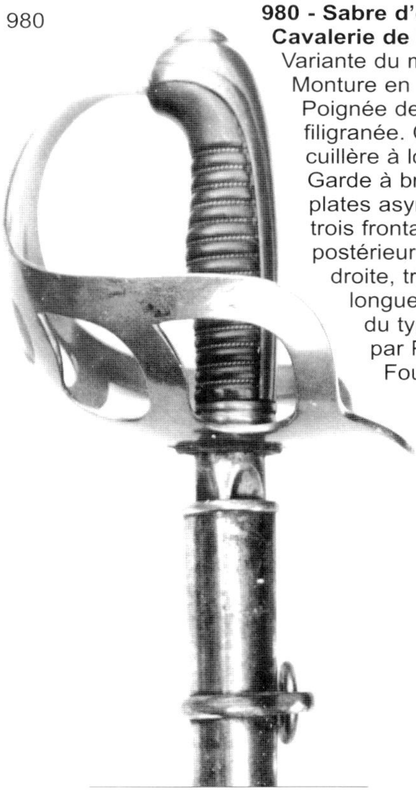

980 - Sabre d'officier de Cavalerie de Ligne.
Variante du modèle 1882. Monture en laiton. Poignée de corne filigranée. Calotte en cuillère à longue queue. Garde à branches plates asymétriques, trois frontales et une postérieure. Lame droite, triangulaire, longue de 97 cm, du type préconisé par Préval. Fourreau d'acier à un bracelet et anneau de bélière.

981

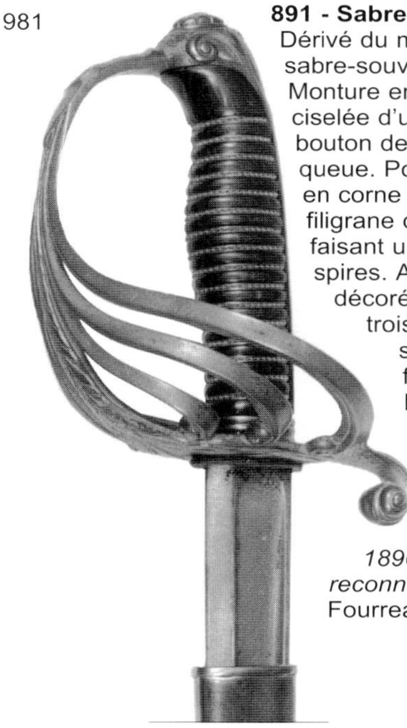

891 - Sabre d'officier.
Dérivé du modèle 1882, sabre-souvenir. Monture en bronze. Calotte ciselée d'une crosse à bouton de rivure et longue queue. Poignée en balustre en corne noire, à fin filigrane de fil torsadé faisant une quinzaine de spires. Arc de jointure décoré de feuillage avec trois branches secondaires frontales et deux branches en contre-garde. Quillon enroulé. Lame gravée à l'acide : *A.L. Moriot, Mai 1890. Souvenir et reconnaissance.* Fourreau d'acier poli.

Sabres d'officiers supérieurs

Modèle 1923.

R. Stiot (in Aries XXIX, 8) signale que *pour les officiers supérieurs ils* (les fourbisseurs) *ont en outre réalisé cinq types principaux de calotte à longue queue très ornementés : à tête de lion, à tête de loup,* Saumur, *à oreilles et le type Moisant.*

Précisons que le type Saumur possède une queue de calotte simplement ornée à sa partie supérieure de rameaux de laurier et de chêne traditionnels. Le type à oreilles a une calotte en dôme décoré d'un motif foliacé et un bourrelet de feuilles de laurier, toute la queue étant ornée de rinceaux avec deux débordements, les oreilles, qui embrassent partiellement la fusée. Le type dit prudemment Moisant possède également une calotte en dôme à motif foliacé mais à bourrelet lisse, à arête médiane et une queue ornée de palmettes ménageant un espace pour y placer le blason ou les initiales de son possesseur.

982

982 - Sabre d'officier supérieur.
Modèle 1923. Monture en bronze généralement doré. Calotte en tête de lion suivie d'une longue queue. Poignée en corne noircie et filigranée d'un fil torsadé. Arc de jointure et branches secondaires identiques au modèle des officiers de rang inférieur, deux frontales et une postérieure. Lame droite, longue de 89,5 cm, à pan creux nickelée et signée *Couleaux et Cie à Klingenthal.* Fourreau en acier également nickelé.

Nota - Il existe des sabres de même facture mais à courte queue de calotte.

983 - Sabre d'officier de la Garde Républicaine.
Modèle 1923.
Monture en bronze à trois branches disposées ainsi que chez le n° 982, mais à longue queue de calotte. Lame nickelée signée *Couleaux et Cie à Klingenthal* ornée sur une face de trophée, et sur l'autre de l'inscription *Garde Républicaine . Grandeur . Renommée* (non figuré).

Famille II - Courte queue de calotte

Les principaux représentants de cette famille de sabres sont les modèles 1831, 1896 et 1923 et leurs nombreuses variantes.

984

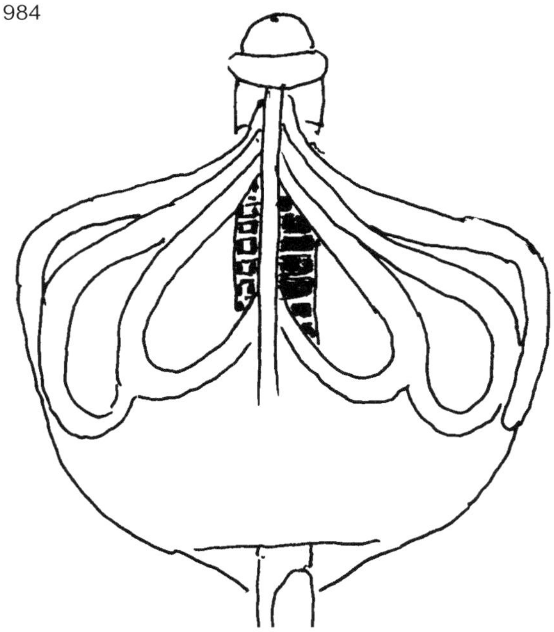

984 - Sabre d'essai de Cavalerie de Ligne.
Modèle 1831, troupe.
Monture volumineuse en laiton. Calotte à bourrelet surmontée d'un dôme-bouton de rivure et à courte queue. Poignée recouverte de basane filigranée. Plateau de garde transformé en une grosse coquille d'où partent trois branches secondaires frontales et trois branches postérieures. La coquille se termine par un bec. Lame droite, longue de 100 cm, à section triangulaire, conçue pour frapper d'estoc, ce qui est surprenant pour une arme de cavalier. Fourreau en tôle d'acier, à un ou deux bracelets à jonc et dard en lyre.

Nota - Ce sabre n'a pas été adopté par les autorités militaires de la Restauration. Sur la lame de l'exemplaire conservé au Musée de l'Empéri, la lame est signée de la *Manufacture de Châtellerault* et sur le fourreau, au-dessus du premier bracelet on trouve

Manceau / à Paris / 1833

griffe du fourbisseur.

985 - Sabre d'essai de Cavalerie de Ligne.
Modèle 1831, officier.
Ce sabre a la même monture que le sabre de troupe, avec les modifications suivantes :
- courte calotte à frise de feuilles de laurier,
- arc de jointure commençant par une rosette et deux petites feuilles,
- base externe des six branches secondaires ornées de palme,
- bec de la coquille orné d'un motif où domine une palmette (985a).

985

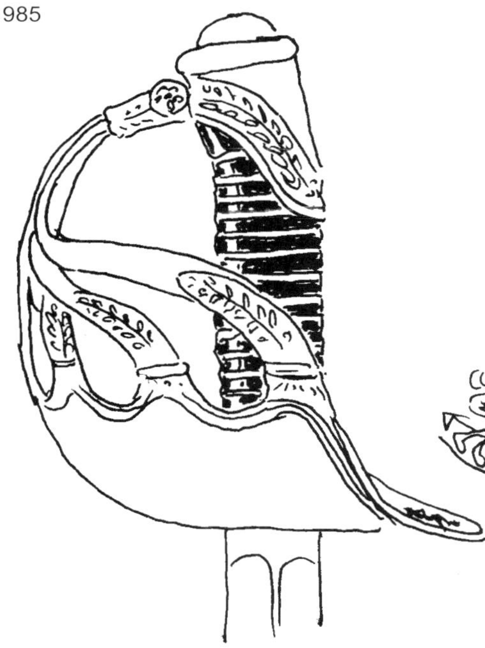

985a

La lame est identique à celle de la troupe mais marquée au talon *Manufacture / de / Klingenthal / Coulaux / aîné et Cie.*

986 - Sabre d'officier de Cavalerie de Ligne.
Troisième République. Sabre de fantaisie. Monture en bronze. Calotte en disque surmonté d'un dôme-bouton de rivure, courte queue à bordure inférieure ornée d'une frise de feuilles de laurier. Poignée en corne noire, lisse jusqu'à la fin de la queue de calotte puis quadrillée jusqu'à sa base. Trois branches secondaires ciselées à leur base de feuillage stylisé, fusionnant dans la moitié de l'arc de jointure, lui-même orné d'une fleurette et d'une petite feuille avant le tenon ; une branche en contre-garde. Lame à pan creux et dos plat. Fourreau de fer à un bracelet de bélière.

986

987 - Sabre d'officier de Cavalerie.
Dit " Type Dérué " 1890. Métal doré ou non. Calotte à bourrelet surmonté d'un dôme et d'un bouton de rivure, généralement à queue courte. Poignée en balustre recouverte de peau de cheval et filigranée d'un fil de laiton torsadé. Quatre branches secondaires frontales, arc de jointure plat et deux branches en contre-garde découpées dans la même plaque de métal ainsi que le plateau. Lame droite, à dos plus large que la lame ce qui donne " T " en coupe ; cette lame est prévue en trois longueurs différentes. Fourreau en tôle à un anneau de bélière et dard en lyre.

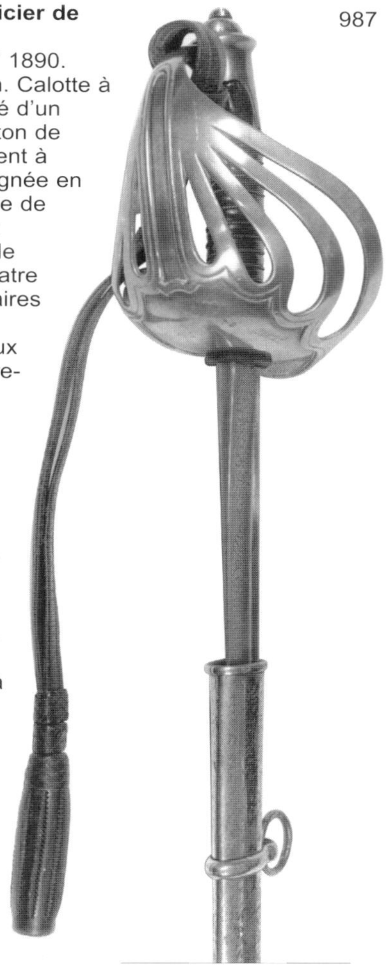

987

988 - Sabre d'officier d'Infanterie.
Dit " Type Dérué ", 1890.
Même monture que le sabre de Cavalerie mais avec une fusée en corne de buffle noircie et filigranée et une lame plus courte.

Nota - Ces sabres furent imaginés par le colonel Dérué, après la défaite de 1870. *Aucun texte ni aucune table n'ont jamais été rédigés* précise C. Ariès (XIII - 8) ce qui fait que ces sabres sont assez variables de structure. De plus, ils ont été fabriqués par des fourbisseurs privés et l'on rencontre des lames légèrement courbées et des calottes à longue queue. Certains exemplaires sont signés *F. Delacour fourbisseur à Paris.*

Un autre exemplaire fut offert comme récompense et porte l'inscription : *Courses de Bonneval, Prix de la 4ème Brigade de Cavalerie, 1er Sept. 1894* (non figuré).

989 - Variante.
Sabre du Type Dérué (989a) mais dont le plateau a été allégé grâce à une série de découpes (989b).

989a

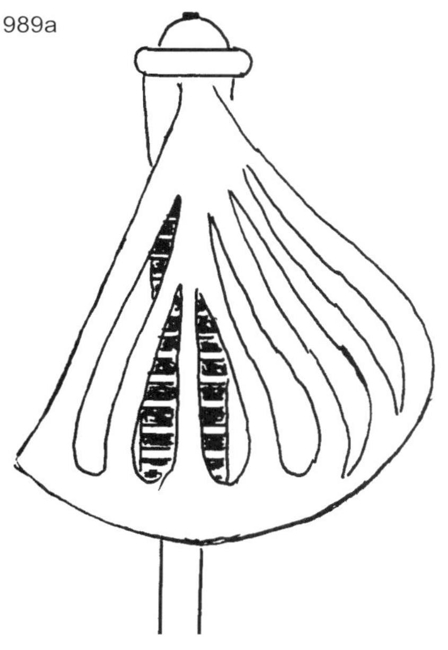

989b

- **a**, vue frontale de la monture montrant les deux branches de contre-garde, l'arc de jointure et les quatre branches de garde, l'ensemble découpé dans la même plaque métallique.
- **b**, dessin semi-schématique montrant les découpes d'allègement ménagées sur la partie marginale de la garde.

990

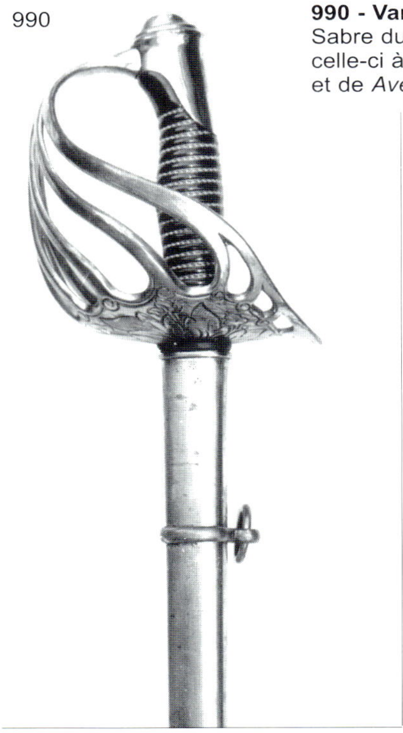

990 - Variante.
Sabre du Type Dérué à plateau à décor floral. Lame de 93 cm, à dos plat plus large que celle-ci à section en " T ", signée de la *Manufacture d'Armes de Châtellerault* mai *1891* et de *Ave Maria* entre deux décors de rinceaux.

Sabre de Cavalerie du modèle 1896

Le Comité de Cavalerie, devant l'échec du sabre du modèle 1882, demanda, à la fin du XIXème siècle, l'étude d'un nouveau modèle qui protégerait bien la main du combattant *en tierce* et *en quarte*. C'est ainsi que naquit le modèle 1896, suivi plus tard, du modèle 1923. Ce sabre devait dans l'esprit de ses promoteurs, remplacer tous les modèles de Cavalerie antérieurs. En fait, ce ne fut pas un succès et la Cavalerie de Ligne resta fidèle aux modèles 1854, 1854/1882 et même au modèle 1822 qui perdurera encore pendant longtemps dans la Garde Républicaine.

991 - Sabre d'officier de Cavalerie de Ligne.
Modèle 1896.
La fabrication de ce sabre suit les directives de la note ministérielle du 22 décembre 1896 confirmée par la Note du 20 août 1897.
Monture en laiton doré.
Calotte à bourrelet surmonté d'un dôme-bouton de rivure, à courte queue ornée d'un rameau de laurier sur la face frontale (991a) et d'un rameau de chêne sur la face postérieure (voir p. 477). Poignée en corne de buffle noircie, filigranée d'un fin fil de laiton torsadé bordé d'un fil simple. Monture d'une seule pièce incluant l'arc de jointure à brise lame médian et, de part et d'autre, deux entailles dégageant deux branches secondaires. A la base de chaque branche, sur la face externe, un rameau de laurier est ciselé. Au centre du plateau, sur la face supérieure, autour de l'orifice permettant de passer la lame, se trouve toute une frise de feuilles et de fruits stylisés (laurier ?) et sur le bec est ciselé (991b) une superbe tête de Gorgone (991d). La face inférieure du plateau présente presque la même décoration, la tête de Gorgone en moins. Lame de trois longueurs différentes : 95, 90 et 85 cm, droite, à pan creux et dos plat pouvant être signée :
M^le^ 1896 (1ère, 2ème ou 3ème) *Taille - Manuf^re^ d'armes de Chât^lt^* (millésime) ou plus tard, sur une seule ligne : *Sabre d'officier de Cavalerie Mle 1896* (...) *Taille Manufacture nationale d'armes de Châtellerault* (millésime).
Fourreau en tôle d'acier à un bracelet de bélière, dard en lyre.

Ce sabre a été attribué aux Cuirassiers et aux Dragons en remplacement du sabre modèle 1882.

Remarque - Un autre marquage est plus énigmatique : sur la lame d'un sabre daté de 1904, issu de la Manufacture d'Armes de Châtellerault, décorée de feuillage se trouve, sur une face le nombre de *6609 De Nathan*, sur l'autre face la date *10 février 1906* et au talon *Fauré Lepage*. Est-ce un sabre souvenir ? Mais on ne saurait dire pour quel événement...

991

Deux sabres d'officier du modèle 1896
On décèle quelques variantes entre ces deux sabres, notamment sur la lame de l'exemplaire de droite. La décoration, gravée à l'acide n'est qu'une fantaisie de l'officier.

En guise de conclusion

L'histoire des sabres, sous leur forme moderne, ne commence guère qu'avec le début du XVIIème siècle. En effet, avant cette époque, et même après, les armes blanches ont généralement une lame droite, descendantes des *longues épées gauloises* en ce qui concerne notre pays (voir J. Lhoste, 1997, p. 55). Cette catégorie d'épée évoluera et donnera naissance aux *Wallonnes* de fer qui seront en service dès la fin du XVIIème (op. cit. p. 118). Dans la même période, les montures en laiton apparaissent (modèle 1695) préludant ce qu'il est convenu de nommer les *fortes épées* que d'aucuns considèrent déjà comme des sabres (op. cit. p. 134). Les fortes épées disparaîtront de l'armement avec le XVIIIème siècle.

Nous savons que ce sont les cavaliers hongrois qui ont introduit le sabre moderne, à lame courbe, dans l'armée française, au tout début du XVIIème siècle. Ce sabre eut bientôt des modèles de fabrication française qui armèrent les Hussards, les Chasseurs à cheval, les officiers montés de l'Infanterie... Ce succès est dû à la plus grande efficacité de la lame courbe dans les charges de Cavalerie qui fauche l'adversaire en utilisant la vitesse du cheval.

Le sabre fut bientôt attribué aux différents corps de l'armée et il s'en suivit des règlements qui définissaient le type de la monture, la forme et la longueur de la lame. Il faut souligner que si ces règlements étaient assez bien respectés pour la troupe, il n'en était pas de même pour de nombreux officiers. En effet, ceux-ci préféraient une arme personnalisée qu'ils commandaient à une manufacture ou à un fourbisseur privé qui n'hésitait pas à enjoliver les modèles préconisés par les autorités militaires, d'où les nombreuses variantes dont certaines sont décrites dans les pages précédentes.

On aura noté que ces variantes intéressent surtout les sabres de l'époque révolutionnaire et du Premier Empire pour resurgir, modestement entre 1870 et 1913, lors de la Belle Epoque !

La Guerre de 1914 - 1918 met fin, presque totalement, à la fantaisie des officiers qui se voient attribuer des sabres réglementaires issus des modèles 1882 et 1896. Quant à la troupe, depuis le début du XIXème siècle, elle est armée de briquets ou de glaives dont l'efficacité comme arme de combat se révèle bien peu performante. L'arme d'assaut privilégiée devient la baïonnette qui, elle aussi, perdra de son intérêt contre les armées blindées et motorisées de la Guerre de 1939 - 1945. Pourtant, le sabre perdurera, mais deviendra l'apanage de certains corps, comme la Garde Républicaine, les officiers de la Marine... et en souvenir du passé, un sabre est offert aux jeunes officiers à leur sortie des Ecoles Militaires, sabre qui ne se montre guère que dans certaines cérémonies et lors des grandes parades.

Ainsi, l'histoire du sabre qui débute en France au tout début du XVIIème siècle, se termine comme arme de combat autour de 1939. En suivant son évolution on parcourt, non seulement l'Histoire de l'Armée française, celle des techniques guerrières, mais aussi une tranche de plus de trois cents ans de l'Histoire de France.

Annexe I
Glossaire

En l'absence d'une terminologie plus traditionnelle qu'officielle il nous a semblé utile de définir les termes qui ont été employés pour la description des sabres.

Acier : alliage de fer et de carbone.

Alaises : languettes de bois aminci glissées dans les fourreaux de métal ou cuir pour les renforcer, éviter les pliures et maintenir la lame dans les fourreaux métalliques.

Anneaux de suspente : ou de bélière. Anneaux fixés par l'intermédiaire de *pitons* sur les *garnitures du fourreau* ou sur celui-ci s'il est métallique.

Arc de jointure : branche simple qui relie la *calotte* à la *croisière* ou au *plateau*. L'arc peut recevoir des *branches secondaires*.

Arco : nom donné à un alliage d'un laiton riche en cuivre. Par extension, nom donné aux sabres fabriqués avec cet alliage.

Argent : du latin *argentum*, s'associe au cuivre pour lui donner plus de dureté.

Bague : pièce en anneau permettant de renforcer les fourreaux de cuir.

Basane : peau de mouton tannée.

Batte : partie de la cuvette logée à l'entrée du fourreau.

Baudrier : ceinturon auquel est suspendu le porte épée ou sabre.

Bavette : voir ***batte.***

Bec : terme utilisé pour désigner l'extrémité aplatie du *plateau* dépourvu de *quillon*

Bouterolle : Garniture protégeant la base du fourreau, généralement pourvue d'un *dard.*

Bouton de chape : de formes variées, permettant le port du sabre en le passant et le fixant dans le gousset du ceinturon ou du baudrier.

Bouton de rivure : partie supérieure de la calotte destinée à recevoir la rivure de la soie de la lame. Se dit également de la partie apparente de la soie matée sur la calotte, dite encore *goutte de suif.*

Bracelets : garnitures du fourreau, généralement pourvues d'un *anneau de suspente*. Ces bracelets sont souvent absents pour les sabres de l'Infanterie.

Branches secondaires : toutes branches partant du plateau ou d'une coquille et rejoignant l'*arc de jointure.*

Branche simple : branche(s) secondaire(s) arrivant sur l'arc de jointure, sans autres pièces de garde.

Brasures : assemblage de deux pièces par apport d'un autre métal en fusion.

Bronze : alliage de cuivre et d'étain. Aspect souvent très semblable à celui du laiton.

Bronze d'aluminium : alliage de cuivre et d'aluminium. cet alliage est d'emploi relativement récent.

Calotte : partie métallique coiffant le poignée ou la fusée et sur laquelle la lame est matée.

Calotte à courte queue : voir également ***jupe*** : calotte à prolongement dorsal n'atteignant pas la *croisière.*

Calotte à longue queue : prolongement de la calotte, en *gouttière* assurant la protection du dos de la poignée jusqu'à la *croisière.*

Calotte en cuillère : se dit d'une calotte à longue queue dont la calotte et la queue sont d'une seule pièce.

Cannelures : Ce dit des moulures horizontales, que l'on trouve notamment sur les poignées de glaives en laiton moulé.

Chagrin : cuir grenu fait de peau de mouton, chèvre, âne.

Chape : garniture supérieure des fourreaux de cuir assurant la protection de l'entrée du fourreau et portant un système de suspente : *anneau, crochet, bouton, pontet.*

Ciselure : travail d'un métal à l'aide d'un ciseau, d'un ciselet pour l'enjoliver de motifs variés, parfois d'une finesse remarquable.

Contre-tranchant : partie dorsale affûtée de la lame.

Coquille : terme général utilisé pour désigner la partie frontale pleine ou ajourée qui protège la main du combattant et d'où partent d'éventuelles branches secondaires.

Cravate : pièces de laine ou de drap enfilée sur la soie de la lame avant le montage pour assurer l'étanchéité à l'entrée du fourreau.

Crochets de suspente : crochets fixés de part et d'autre de la chape pour suspendre le sabre au ceinturon ou au baudrier.

Croisière : prolongement horizontal de l'arc de jointure aboutissant au *quillon.*

Cuvette : pièce métallique placée sur ou dans la *chape*, pouvant être munie de lames souples pour maintenir en place la lame de l'arme et éviter toute vibration (voir ***batte***).

Damas : acier très fin.

Damasquinage : incrustation de filets d'or, d'argent, de cuivre par martelage sur une surface préalablement ciselée.

Dard : voir ***bouterolle***. Le dard peut être globuleux, basal, basal et frontal, en lyre symétrique ou asymétrique. Il peut être en laiton ou en fer, pour assurer la protection de l'extrémité de la bouterolle.

Dorure : s'effectue au XVIIIème siècle par la technique au mercure ou au nitrate de mercure en bains. Peut se faire également par application de feuilles d'or dont l'épaisseur peut être réduite à 1/10000 de mm sur le cuivre, le laiton, le bronze.

Dos de la lame : partie plate ou bombée, dite alors à jonc, qui peut recevoir la signature de la Manufacture.

Ecailles : se dit d'une ornementation qui rappelle les écailles des poissons.

Eclisses : lattes de bois constituant l'armature des fourreaux de cuir pour leur donner une rigidité convenable.

Emboutissage : action de déformer une pièce de métal à l'aide d'une presse.

Estampage : pièce de métal réalisée par frappe à l'aide d'une masse.

Figuratif : signifie en particulier pour la *calotte*, que sa forme représente un motif particulier : tête humaine ou animale, casque à l'antique, bonnet...

Filigrane : fils de laiton, de fer ou d'argent, torsadé ou non, petit ruban métallique, l'ensemble disposé d'une façon plus ou moins décorative. Le filigrane permet une meilleure tenue en main.

Flèche : mesure indiquant la courbure : c'est la distance comprise entre le talon et la pointe de la lame (corde) ; la flèche est la distance qui existe entre cette ligne et la courbure maximale de la lame (arc).

Fleurette : figuration d'une petite fleur à quatre pétales ou plus. Peut se trouver sur le haut de l'arc de jointure, sur le quillon...

Garde : voir ***coquille*** et ***branches secondaires***

Garnitures : pièces métalliques enserrant les fourreaux de cuir : *chapes, bracelets,* éventuellement *bagues, bouterolles* ; à noter que sur les fourreaux métalliques et notamment de laiton, ces garnitures peuvent être figurées.

Goutte de suif : voir ***Bouton de rivure***

Gouttières : rainures plus ou moins profondes, plus ou moins longues, courant le long du dos de la lame

Gravure : décoration au trait fin obtenu avec un burin, un ciseau... Technique généralement utilisée pour inscrire les devises ou les affectations régimentaires des sabres.

Jonc : se dit des lames dont le dos est bombé. Cette forme renforce la lame ce qui permet de l'alléger.

Jupe : nom donné à des calottes totalement dépourvues de queue. On parle alors de calotte à jupe.

Lame : pièce principale de l'arme blanche, courbe ou droite, affûtée ou non ; on distingue trois parties : le talon faisant suite à la soie, le fort ou partie médiane et le faible au voisinage de la pointe. Voir également : ***dos, tranchant, contre-tranchant, pointe, gouttières, flèche, soie.***

Laiton : du latin *latun* qui signifie cuivre. En fait le laiton moderne est un alliage de cuivre (7/10ème) et de zinc (3/10ème) moins malléable que le cuivre seul.

Médaillon : se dit d'un ornement plus ou moins circulaire dans lequel sont inscrites : les Armes de France, de gentilshommes ayant la responsabilité d'un régiment, ou encore l'emblème du corps auquel appartient l'officier, ...

Monture : ensemble des pièces métalliques constituant la garde : calotte, arc de jointure, croisière, plateau, quillon, oreillons, coquille, branches secondaires.

Nœud de corps : partie médiane de la croisière, généralement placée dans l'axe de la poignée.

Or : du latin *aurum*, métal précieux très ductible et malléable. S'utilise généralement associé à l'argent, au zinc, au nickel ou au cuivre pour lui donner plus de dureté.

Oreillons : pièces verticales situées sur la croisière, servant de renfort à la poignée et assurant un meilleur maintien du sabre dans le fourreau.

Palmette : ornementation figurant des feuilles de palme.

Pan creux : dépression longitudinale de la lame entre le dos et le tranchant, sur chaque face, et obtenu par meulage.

Pan creux double : dit aussi ***à deux pans creux*** : dans ce cas le pan creux possède une arête médiane longitudinale sur chaque face de la lame.

Plateau : élargissement de la croisière pour assurer une meilleure protection de la main faisant suite en général, à l'*arc de jointure*, mais peut avoir été rapporté.

Poignée : partie de la monture en bois ou métallique, saisie par la main. Lorsqu'elle est en bois, elle peut être recouverte de *basane*, de *galuchat... filigranée* ou non. Le terme *fusée* a été réservé pour les épées.

Pointe : désigne l'extrémité de la lame. La pointe peut être médiane ou dans le prolongement du dos.

Pommeau : extrémité de la poignée sur une monture entièrement métallique.

Poucier : anneau de métal ou de cuir placé du côté contre-garde qui servait dans certains modèles (jusque vers 1750) à passer le pouce pour assurer un bon maintien de l'arme. Ce dispositif a été abandonné car le pouce, trop maintenu, pouvait être cassé.

Quillon (voir ***croisière***) : le quillon peut être droit, infléchi ou en " S " et plus ou moins orné.

Rouelle : ornementation circulaire terminant un quillon.

Soie : partie de la lame passant dans la poignée et matée sur la *calotte*.

Sustente : du latin *sustentare*, soutenir. Terme ancien datant des XII-XVIèmes siècles. Se dit des *crochets*, des *anneaux*, des *boutons* servant au port de l'épée ou du sabre.

Tranchant : partie affûtée de la lame opposée au dos.

Virole : bague métallique placée généralement au bas de la *poignée* pour maintenir le filigrane.

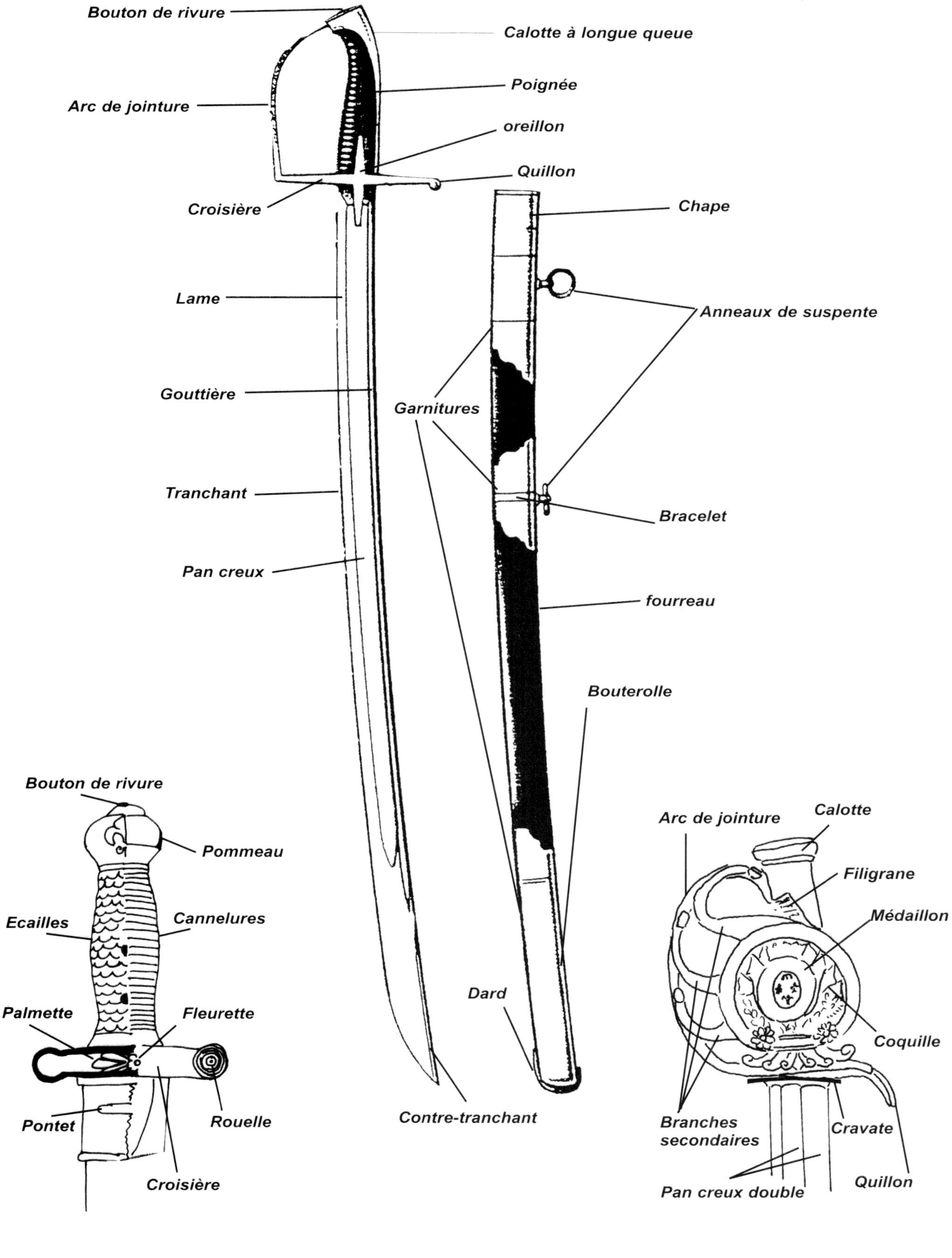
Bouton de rivure
Calotte à longue queue
Poignée
Arc de jointure
oreillon
Quillon
Croisière
Chape
Lame
Anneaux de suspente
Gouttière
Garnitures
Tranchant
Bracelet
Pan creux
fourreau
Bouterolle
Bouton de rivure
Pommeau
Ecailles
Cannelures
Palmette
Fleurette
Pontet
Rouelle
Croisière
Dard
Contre-tranchant
Arc de jointure
Calotte
Filigrane
Médaillon
Coquille
Branches secondaires
Cravate
Pan creux double
Quillon

Annexe II

Bibliographie

A

ANONYME (1794) - *Procédés de fabrication des Armes blanches* publié par ordre du Comité de Salut public.
A Paris, de l'Imprimerie du Département de la Guerre, rue Michodière, N° 3, l'An 2 Républicain

ARIES, C (1973 et suivant) - *Armes Blanches militaires françaises.* Librairie Petitot, Paris.

ARIES, C. et BOUDRIOT J. (1979) - *Les Glaives à la romaine de l'Artillerie, 1771-1772.*
Gaz. Armes, **76**, 32-36

B

BLONDIEAU C. (1989) - *Un pionnier et sa collection : Pierre Petitot,* Gaz. Armes, **195**, 17.

BOTTET M. (1968) - *L'Arme blanche des Armées françaises, 1789-1870,* F. de Nobèle Edit. Paris, 104 p.

BORDES Ph. (1987) - *Aux Armes Citoyens !* ,
Musée de la Révolution française
38220 Château de Vizille, Edit. 105 p.

BOSC E. (1984, réédition) - *Dictionnaire général d'Archéologie*
Hugues de Fleurville, Edit.

BOUDRIOT J. (1978) - *L'armement des Gardes du Corps sous la Première et la Seconde Restaurations*
Gaz. Armes, **66**, 28-35.

BRUEL, B (1996) - *Sabres d'Etat-Major, variantes du modèle Vendémiaire An XII* - Tradition, **107**, 45-47.

(1996) - *Un sabre des officiers de la Maréchaussée modèle 1787* - Tradition, **112-113**, 31-32.

(1996) - *Comme le roi Murat*, Tradition, **115**, 46-47

(1997) - *Un rare sabre d'officier de Cavalerie légère Aux Armes de France et de Berry* - Tradition, **119**, 40-41

(1997) - *Un sabre à la chasseur attribué au général d'Autancourt* - Tradition, **129**, 39-40

BUIGNE, J.J. - voir Lhoste et Buigné

C

CASANOVA D. (1991) - *Le sabre d'Arcole*
Gaz. Armes, **207**, 13-15

(1992) - *Un sabre du Directoire*
Gaz. Armes, **223**, 38-4

(1992) - *Le sabre de la Cavalerie légère, modèle 1822*
Gaz. Armes, **233**, 17-20

(1994) - *Le sabre des Marins de la Garde Impériale*
Gaz. Armes, **241**, 46-49

(1996) - *Le sabre d'officier de Marine, modèle 1837*
Tradition, **111**, 39-41.

Cottereau G. (1995) - *Elèves de l'Ecole de Mars*
Carnet Sabret, 414-419.

D

DELPERIER L. (1988) - *Le Sabre des troupes à pied, modèle 1831*
Tradition, **13**, 36-41.

F

FROMONT M. (de) voir de Jenlis et Maryse Raso.

J

JAEGER G. (1997) - *Sabre de Dragons, modèle An IV*
Tradition, **128**, 43-44.
(1997) - *Sabre de Carabiniers, modèle An IX*
Tradition, **130**, 42-44.
(1997) - *Sabre modèle 1790 pour Chasseurs à cheval*
Tradition, **132**, 5-9.
(1998) - *Les sabres de Grenadiers du Régiment de Nassau*
Tradition, **133**, 33-37.
(1998) - *Les Volontaires étrangers de Clermont-Prince Sabre d'officier d'Infanterie*
Tradition, **135**, 9-12.
(1998) - *Le Sabre d'Etat-Major, règlement du 1er Vendémiaire An XII -*
Tradition, **136**, 41-45.
(1999) - *Sabre attribué aux officiers des Régiments d'Eclaireurs*
Tradition, **141**, 11-15.
(1999) - *Le Briquet de la vieille Garde*
Tradition, **144**, 11-14.
(2000) - *Les sabres utilisés par la Marine sous le Premier Empire*
Tradition, **159**, 31-35.

JAEGER G. et de BRUEL B. (1997) - *Les sabres de Dragons, modèle An IX et modèle An XI*
Tradition, **124-125**, 28-29.

de JENLIS A. et de FROMONT M. (1994) - *Le sabre d'officier de Cavalerie, modèle 1817*
Gaz. Armes, **248**, 37-39.
(1994) - *Le Sabre des officiers de la Compagnie d'Elite des Régiments de Dragons*
Gaz. Armes, **250**, 33-35.
(1995) - *Sabre de la Garde Royale, 1814*
Gaz. Armes, **252**, 23-35.

K

KRÄHENBÜHL A. (1999) - *Ame du port de Toulon, un mystérieux marquage du sabre de bord, modèle 1782.*
Gaz. Armes, **305**.

L

LAMOUREUX P. et MALVAUX B. (1993) - *Un faussaire malchanceux*
Tradition, **77**, 30-33.

LAVALLEY G. (1945) - *L'Arme blanche sous la Révolution*
L. Jouan, Libraire Editeur, Caen.

LE DIBERDER G. (1989) - *Les Armées françaises à l'époque révolutionnaire 1789-1804*
Collection du Musée de l'Armée,
P.R.E.A.L. 36 Avenue Jean Jaurès, 94110 Arcueil

LHOSTE J. (1996) - *Le glaive de Mars, un emblème ou une arme ?*
Aux Armes d'Antan, cat. N° 10, 10-11.
(1997) - *Les épées portées en France des origines à nos jours*
Editions du Portail, 38353 La Tour du Pin, 479 p.
(2000) - *Un témoin des Guerres de l'Ouest*
Souvenir chouan de Bretagne, Bull. n°14, p. 8.
LHOSTE J. et BUIGNE J.J. (1994) - *Armes blanches françaises : symbolisme, inscriptions, marquages, fourbisseurs, manufactures.*
Edition du Portail, 38353 La Tour du Pin, 176 p.
LHOSTE J. et BUIGNE J.J. (1999) - *Armes blanches françaises : symbolisme, inscriptions, marquages, fourbisseurs, manufactures - Seconde édition, 256 p.*
Editions du Portail, 38353 La Tour du Pin.

M

MAINDRON M. (1890) - *Les Armes*
Librairie-Imprimeurs réunis,
7, rue Saint Benoît Paris 343p.

MALVAUX B. (1993) - *Bayeux : l'argent au prix de l'or*
Tradition, **74**, 32-34.
(1993) -*Les plaques de ceinturons des Marins de la Garde Impériale*
Tradition, **84**, 8-9.
(1994) - *Les sabres des Gardes de Monsieur*
Tradition, **85**, 11-16.

MARGERARD J.P. (1945) - *Armement et Equipement de l'Infanterie française*
Les Editions Militaires illustrés, 82, Rue de Laureston. Paris.

MARTIN J.P. (1985) - *Les Armes blanches de la Révolution*
Edition du Portail, 38353 La Tour du Pin, 155 p.

P

PETARD M. (1976) - *Le Dragon de Ligne*
Gaz. Uniformes, **29**, 19-28.
(1979) - *Le sabre d'Artillerie de ...1767*
Gaz. Armes, **76**, 12-19.
(1990) - *L'officier de Carabinier en 1809-1815*
Tradition, **38**, 4-9.
(1991) - *Le sabre de parade des officiers de Grenadiers à cheval de la Garde des Consuls*
Tradition, **50**, 4-9
(1991) - *Le sabre des Cuirassiers*
Tradition, **54-55**, 28-33.
(1991) - *Sabre de 1772 pour les troupes de Marine*
Tradition, **58**, 36-40.
(1992) - *Le sabre de 1767 pour les Dragons*
Tradition, **62**, 4-12.
(1993) - *Le sabre de Tambour-Major de 1822 à 1871*
Tradition, **80**, 12-18.
(1994) - *Le sabre des Marins de la Garde Impériale*
Tradition, **89**, 12-18.

(1995) - *Le Sabre de 1776 pour les Hussards*
Tradition, **96**, 34-39.
(1995) - *Un sabre de "Gendarmerie d'Armée" du modèle de l'An IX.*
Tradition, Mars, 18-22.

PIGEARD A. (1997) - *Sapeur du 14ème de Ligne, Premier Empire*
Tradition, **120**, 20-21.

POLLINO A. (1987) - *Les mérovingiens*
Musée d'Histoire et d'Archéologie,
Bastion Saint-André d'Antibes, Editeur, 211 p.

R

RASO Maryse (1993) - *Le sabre d'officier d'Infanterie 1882*
Gaz. Armes, **232**, 41-43.
(1993) - *Un sabre réglementaire français de Cavalerie , modèle 1854.*
Gaz. Armes, **234**, 33-35.
(1994) - *Les sabres de Garde du Corps du Roi*
Gaz. Armes, **240**, 50-53.

RASO Maryse. et de FROMONT M.
(1993) *Les sabres de Vincennes*
Gaz. Armes, **237**, 42-45.
(1993) - *Les sabres des Canonniers montés, modèle 1829*
Gaz. Armes, **238**, 43-46.
(1993) - *Sabre d'officier d'Infanterie, modèle 1821 et ses variantes*
Gaz. Armes, **239**, 38-41.
(1993) - *Le sabre des Cent-Gardes*, 1854-1870
Gaz. Armes, **235**, 46-49.

S

STIOT R. et PETARD M. (1988) - *Les sabres des troupes montées*
Gaz. Armes, **181**, 15-18.

V

VENNER D. (1986) - *Les Armes blanches*
Jacques Grancher, Edit. Paris, 319 p.

Remerciements

Un tel ouvrage sur les sabres français n'aurait pu être réalisé sans le concours de nombreux spécialistes qu'ils soient conservateurs de musées, militaires, antiquaires ou collectionneurs. A tous, nous adressons nos plus vifs remerciements pour l'aide apportée.

En premier lieu nous citerons trois personnalités qui, hélas, nous ont quittés :

Raoul Brunon, Conservateur du Musée de l'Empéri de Salon-de-Provence qui nous a toujours réservé le meilleur accueil,
le Docteur Pascal Villanova, Expert agréé par la Chambre Nationale des Experts spécialisés, qui exerçait son commerce sous l'enseigne de " *La Dragonne* ", à Cannes,
et **Guy Estarellas**, collectionneur averti.

Nous gardons un souvenir ému de ces trois personnages et nous leur rendons un respectueux hommage.

Parmi tous ceux qui, d'une façon ou d'une autre nous ont aidés, nous voulons citer :

- **Contre-amiral François Bellec**, Conservateur du Musée de la Marine, Palais de Chaillot à Paris.
- **Jean-Claude Dey** qui a aidé à la réalisation de cet ouvrage par ses recherches historiques et photographiques.
- **Christian Blondieau**, " *Au Képi Rouge* ", Village Suisse à Paris.
- **Charles Bouché**, " *Gloire du Passé* ", 41, rue Richelieu, Paris.
- **Gaëtan Brunel**, compétent et fidèle collaborateur de J.C. Dey.
- **Bernard Croissy**, Expert agréé en Armes anciennes et Souvenirs historiques
- **Général Desjardins**, Commandant de l'Ecole des officiers de Gendarmerie, à Melun.
- **Maréchal des Logis-chef B. Dupuis**, Musée de la Gendarmerie à Melun.
- **Roby Joffo** *Antiqu'hair*, rue du Laos à Paris.
- **Bernard Lemoine**, Centre hospitalier Emile Roux, Limiel-Brevanes, spécialiste des glaives du Second Empire.
- **Bernard Michaud**, *L'Echoppe Marine*, Village Suisse à Paris.
- **Maryse Raso,** Expert de la Compagnie d'Expertise en Antiquité et Objets d'Art, *Aux Armes d' Antan*, Village Suisse à Paris et qui nous a permis de photographier toutes les armes en sa possession.
- **G. Richard-Collin**, Collectionneur averti.
- **Maurice Ritter** Directeur Honoraire de l'Institut National de la Recherche Agronomique et numismate, à Antibes.
- **Jean François Teulière**, *Kanata-Ya*, Versailles.

Nos remerciements vont également à :

- **Marc de Fromont** qui connaît bien les armes et qui sait les photographier avec talent.
- **Christophe Georgioux** qui nous a signalé avec célérité toutes les armes intéressantes qui passaient entres ses mains au " Hussard ".
- **Patrick Smokvina** photographe spécialisé au " Hussard " et qui aime le travail bien fait.
- **Philippe Simon** du " Hussard " pour sa contribution à la réalisation de cet ouvrage.

La présentation de l'ouvrage a été réalisée par **Françoise Fusy** qui a su harmoniser les illustrations à partir de documents parfois très hétorogènes et bien entendu, **Paule Lhoste**, qui à son habitude, a relu les épreuves avec l'attention et le sens critique qui lui sont coutumiers.

Crédit photographique

Les photographies illustrant cet ouvrage ont été rassemblées grâce à la bienveillance de musées nationaux et régionaux, d'antiquaires spécialisés et de collectionneurs avertis. Nous donnons ci-après l'origine de ces illustrations.

B. C.	Bernard Croissy, *Courbevoie.*
B. L.	Bernard Lemoine, 1, avenue de Verdun, 94450 *Limeil-Brevannes*
B. M.	Bernard Michaud, *l'Echoppe Marine, Village Suisse, Paris.*
C. B.	Christian Blondieau, *Au Képi Rouge, Village Suisse, Paris*
Ch. B.	Charles Bouché, *Gloire du passé.*
J.C. D.	Jean-Claude Dey, *Marnes La Coquette.*
J.F. T.	Jean-François Teulière, *Versailles.*
M.E. - S de P.	Musée de l'Empéri, *Salon-de-Provence.*
M.G. de Melun	Musée de la Gendarmerie de *Melun.*
M.N.N. de Juan-les-Pins	Musée Naval et Napoléonien d'*Antibes / Juan les Pins.*
M. R.	Maryse Raso, *Aux Armes d'Antan, Village Suisse, Paris.*
Ph. B.	Philippe Bordes, *Musée de la Révolution française du Château de Vizille*
P. V.	Pascal Villanova, *La Dragonne, Cannes.*
R. J.	Roby Joffo, *Antiqu'hair, rue du Laos, Paris.*
S.P.R.N.M.	Service Photographique de la Réunion des Musées Nationaux.

Les photographies et les descriptions qui ne sont pas signalées par un sigle précisant leur origine, proviennent de collectionneurs qui ont préféré conserver leur anonymat ou des réserves du " *Hussard* ".

Cet ouvrage a été publié par les Editions du Portail - B. P. 69 - 38353 La Tour du Pin cedex

Conception et réalisation : Le Hussard
Photogravure, Studio Artis
Photo de couverture, Marc de Fromont
Achevé d'imprimer en France le 30 10 2001 aux Imprimeries de Champagne - 52, Langres
Dépôt légal 4e trimestre 2001

Les Editions du Portail sont un département de la S.A. Le Hussard

Diffusion :
- Histoire et Documents, B. P. 31, 94430 Chennevières s/Marne cedex - Tél. 01 45 76 77 41
- SA Le Hussard, B. P. 69, 38353 La Tour du Pin cedex - Tél. 04 74 83 20 75

ISBN 2-86551-048-4 / Prix : 52 €